GORDIAN®

Plumbing Costs with RSMeans data

Brian Adams, Senior Editor

2020

43rd annual edition

Chief Data Officer
Noam Reininger

Vice President, Data
Tim Duggan

Principal Engineer
Bob Mewis *(1, 4)*

Contributing Editors
Brian Adams *(21, 22)*
Paul Cowan
Christopher Babbitt
Sam Babbitt
Stephen Bell
Michelle Curran
Antonio D'Aulerio *(26, 27, 28, 48)*
Matthew Doheny *(8, 9, 10)*
John Gomes *(13, 41)*
Derrick Hale, PE *(2, 31, 32, 33, 34, 35, 44, 46)*
Barry Hutchinson
Joseph Kelble *(14, 23, 25)*
Scott Keller *(3, 5)*
Charles Kibbee
Gerard Lafond, PE
Thomas Lane *(6, 7)*
Thomas Lyons
Jake MacDonald *(11, 12)*
John Melin, P.E.
Elisa Mello
Matthew Sorrentino
Kevin Souza
David Yazbek

Production Manager
Debbie Panarelli

Production
Jonathan Forgit
Sharon Larsen
Sheryl Rose
Janice Thalin

Data Quality Manager
Joseph Ingargiola

Innovation
Ray Diwakar
Kedar Gaikwad
Srini Narla

Cover Design
Blaire Collins

Data Analytics
David Byars
Ellen D'amico
Thomas Hauger
Cameron Jagoe
Matthew Kelliher-Gibson
Renee Rudicil

**Numbers in italics are the divisional responsibilities for each editor. Please contact the designated editor directly with any questions.*

RSMeans data from Gordian
Construction Publishers & Consultants
1099 Hingham Street, Suite 201
Rockland, MA 02370
United States of America
1.800.448.8182
RSMeans.com

Cover photo © iStock/DenBoma

Printed in the United States of America
ISSN 1537-8411
ISBN 978-1-950656-15-8

0210 $319.99 per copy (in United States)
Price is subject to change without prior notice.

Related Data and Services

Our engineers recommend the following products and services to complement *Plumbing Costs with RSMeans data:*

Annual Cost Data Books

2020 Assemblies Costs with RSMeans data
2020 Square Foot Costs with RSMeans data

Reference Books

Estimating Building Costs
RSMeans Estimating Handbook
Green Building: Project Planning & Estimating
How to Estimate with RSMeans data
Plan Reading & Material Takeoff
Project Scheduling & Management for Construction
Universal Design Ideas for Style, Comfort & Safety

Seminars and In-House Training

Unit Price Estimating
Training for our online estimating solution
Practical Project Management for Construction Professionals
Scheduling with MSProject for Construction Professionals
Mechanical & Electrical Estimating

RSMeans data

For access to the latest cost data, an intuitive search, and an easy-to-use estimate builder, take advantage of the time savings available from our online application.

To learn more visit: **RSMeans.com/online**

Enterprise Solutions

Building owners, facility managers, building product manufacturers, and attorneys across the public and private sectors engage with RSMeans data Enterprise to solve unique challenges where trusted construction cost data is critical.

To learn more visit: **RSMeans.com/Enterprise**

Custom Built Data Sets

Building and Space Models: Quickly plan construction costs across multiple locations based on geography, project size, building system component, product options, and other variables for precise budgeting and cost control.

Predictive Analytics: Accurately plan future builds with custom graphical interactive dashboards, negotiate future costs of tenant build-outs, and identify and compare national account pricing.

Consulting

Building Product Manufacturing Analytics: Validate your claims and assist with new product launches.

Third-Party Legal Resources: Used in cases of construction cost or estimate disputes, construction product failure vs. installation failure, eminent domain, class action construction product liability, and more.

API

For resellers or internal application integration, RSMeans data is offered via API. Deliver Unit, Assembly, and Square Foot Model data within your interface. To learn more about how you can provide your customers with the latest in localized construction cost data visit: **RSMeans.com/API**

Table of Contents

Foreword

The Value of RSMeans data from Gordian

Since 1942, RSMeans data has been the industry-standard materials, labor, and equipment cost information database for contractors, facility owners and managers, architects, engineers, and anyone else that requires the latest localized construction cost information. More than 75 years later, the objective remains the same: to provide facility and construction professionals with the most current and comprehensive construction cost database possible.

With the constant influx of new construction methods and materials, in addition to ever-changing labor and material costs, last year's cost data is not reliable for today's designs, estimates, or budgets. Gordian's cost engineers apply real-world construction experience to identify and quantify new building products and methodologies, adjust productivity rates, and adjust costs to local market conditions across the nation. This adds up to more than 22,000 hours in cost research annually. This unparalleled construction cost expertise is why so many facility and construction professionals rely on RSMeans data year over year.

About Gordian

Gordian originated in the spirit of innovation and a strong commitment to helping clients reach and exceed their construction goals. In 1982, Gordian's chairman and founder, Harry H. Mellon, created Job Order Contracting while serving as chief engineer at the Supreme Headquarters Allied Powers Europe. Job Order Contracting is a unique indefinite delivery/indefinite quantity (IDIQ) process, which enables facility owners to complete a substantial number of repair, maintenance, and construction projects with a single, competitively awarded contract. Realizing facility and infrastructure owners across various industries could greatly benefit from the time and cost saving advantages of this innovative construction procurement solution, he established Gordian in 1990.

Continuing the commitment to providing the most relevant and accurate facility and construction data, software, and expertise in the industry, Gordian enhanced the fortitude of its data with the acquisition of RSMeans in 2014. And in an effort to expand its facility management capabilities, Gordian acquired Sightlines, the leading provider of facilities benchmarking data and analysis, in 2015.

Our Offerings

Gordian is the leader in facility and construction cost data, software, and expertise for all phases of the building life cycle. From planning to design, procurement, construction, and operations, Gordian's solutions help clients maximize efficiency, optimize cost savings, and increase building quality with its highly specialized data engineers, software, and unique proprietary data sets.

Our Commitment

At Gordian, we do more than talk about the quality of our data and the usefulness of its application. We stand behind all of our RSMeans data—from historical cost indexes to construction materials and techniques—to craft current costs and predict future trends. If you have any questions about our products or services, please call us toll-free at 800.448.8182 or visit our website at gordian.com.

MasterFormat® 2016/ MasterFormat® 2018 Comparison Table

This table compares the 2016 edition of the Construction Specifications Institute's MasterFormat® to the expanded 2018 edition. For your convenience, all revised 2016 numbers and titles are listed along with the corresponding 2018 numbers and titles.

CSI 2016 MF ID	CSI 2016 MF Description	CSI 2018 MF ID	CSI 2018 MF Description
015632	Temporary Security	015733	Temporary Security
019308	Facility Maintenance Equipment	019308	Facilities Maintenance, Equipment
024200	Removal and Salvage of Construction Materials	024200	Removal and Diversion of Construction Materials
026600	Landfill Construction and Storage	026600	Landfills
040130	Unit Masonry Cleaning	04012052	Cleaning Masonry
068010	Composite Decking	067300	Composite Decking
072127	Reflective Insulation	072153	Reflective Insulation
072610	Above-Grade Vapor Retarders	072613	Above-Grade Vapor Retarders
074473	Metal Faced Panels	074433	Metal Faced Panels
075430	Ketone Ethylene Ester Roofing	075416	Ketone Ethylene Ester Roofing
077280	Vents	077280	Vent Options
081410	Wood Doors	081410	Doors, Wood
083410	Special Function Doors	083410	Specialized Function Doors
087125	Weatherstripping	087125	Door Weatherstripping
087530	Weatherstripping	087530	Window Weatherstripping
096223	Bamboo Flooring	096436	Bamboo Flooring
096720	Epoxy-Marble Chip Flooring	096716	Epoxy-Marble Chip Flooring
099103	Paint Restoration	090190	Maintenance of Painting and Coating
102833	Laundry Accessories	102823	Laundry Accessories
114700	Ice Machines	114700	Ice Making Machines
117610	Operating Room Equipment	117610	Equipment for Operating Rooms
117710	Radiology Equipment	117710	Equipment for Radiology
122310	Wood Interior Shutters	122313	Wood Interior Shutters
123580	Commercial Kitchen Casework	123539	Commercial Kitchen Casework
124636	Desk Accessories	124113	Desk Accessories
125273	Multiple Seating	126000	Multiple Seating
141210	Dumbwaiters	141000	Dumbwaiters
211113	Facility Water Distribution Piping	211113	Facility Fire Suppression Piping
233715	Louvers	233715	Air Outlets and Inlets, HVAC Louvers
260580	Wiring Connections	260583	Wiring Connections
270110	Operation and Maintenance of Communications Systems	270110	Operation and Maintenance of Communication Systems
272123	Data Communications Switches and Hubs	272129	Data Communications Switches and Hubs
283149	Carbon-Monoxide Detection Sensors	284611	Carbon-Monoxide Detection Sensors
284621	Fire Alarm	284620	Fire Alarm
316233	Drilled Micropiles	316333	Drilled Micropiles
323420	Fabricated Pedestrian Bridges	323413	Fabricated Pedestrian Bridges
333633	Utility Septic Tank Drainage Field	333633	Ground-Level AWWA D110 Type III Prestressed Conc. Wastewater Storage Tanks
337543	Shunt Reactors	337253	Shunt Reactors
350100	Operation and Maint. of Waterway & Marine Construction	350100	Operation and Maintenance of Waterway and Marine Construction

How the Cost Data Is Built: An Overview

Unit Prices*

All cost data have been divided into 50 divisions according to the MasterFormat® system of classification and numbering.

Assemblies*

The cost data in this section have been organized in an "Assemblies" format. These assemblies are the functional elements of a building and are arranged according to the 7 elements of the UNIFORMAT II classification system. For a complete explanation of a typical "Assembly", see "RSMeans data: Assemblies—How They Work."

*Residential Models**

Model buildings for four classes of construction—economy, average, custom, and luxury—are developed and shown with complete costs per square foot.

*Commercial/Industrial/ Institutional Models**

This section contains complete costs for 77 typical model buildings expressed as costs per square foot.

*Green Commercial/Industrial/ Institutional Models**

This section contains complete costs for 25 green model buildings expressed as costs per square foot.

*References**

This section includes information on Equipment Rental Costs, Crew Listings, Historical Cost Indexes, City Cost Indexes, Location Factors, Reference Tables, and Change Orders, as well as a listing of abbreviations.

- **Equipment Rental Costs:** Included are the average costs to rent and operate hundreds of pieces of construction equipment.
- **Crew Listings**: This section lists all the crews referenced in the cost data. A crew is composed of more than one trade classification and/or the addition of power equipment to any trade classification. Power equipment is included in the cost of the crew. Costs are shown both with bare labor rates and with the installing contractor's overhead and profit added. For each, the total crew cost per eight-hour day and the composite cost per labor-hour are listed.

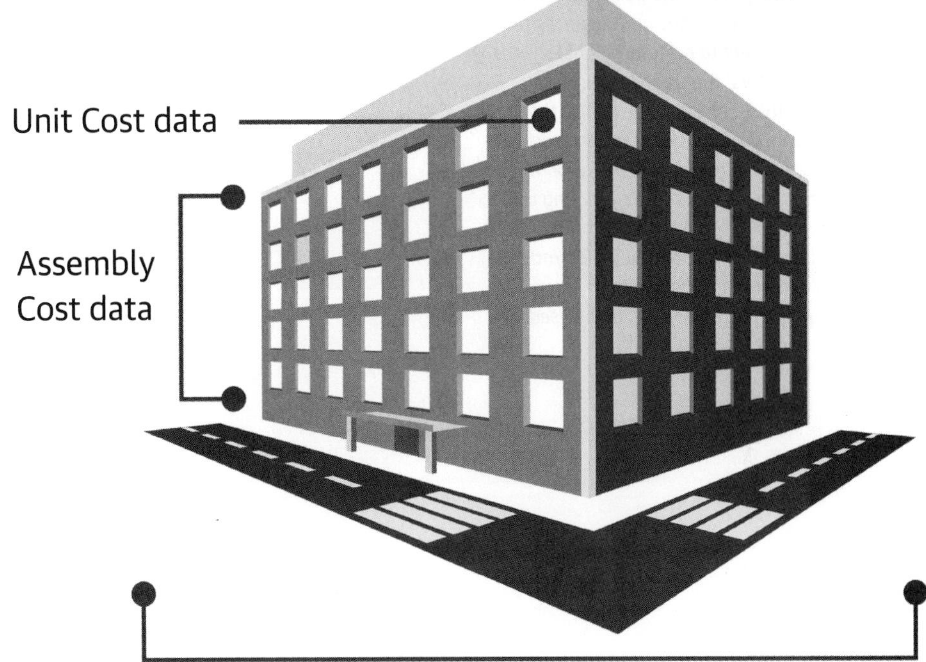

- **Historical Cost Indexes**: These indexes provide you with data to adjust construction costs over time.
- **City Cost Indexes**: All costs in this data set are U.S. national averages. Costs vary by region. You can adjust for this by CSI Division to over 730 cities in 900+ 3-digit zip codes throughout the U.S. and Canada by using this data.
- **Location Factors**: You can adjust total project costs to over 730 cities in 900+ 3-digit zip codes throughout the U.S. and Canada by using the weighted number, which applies across all divisions.
- **Reference Tables**: At the beginning of selected major classifications in the Unit Prices are reference numbers indicators. These numbers refer you to related information in the Reference Section. In this section, you'll find reference tables, explanations, and estimating information that support how we develop the unit price data, technical data, and estimating procedures.
- **Change Orders**: This section includes information on the factors that influence the pricing of change orders.
- **Abbreviations**: A listing of abbreviations used throughout this information, along with the terms they represent, is included.

Index (printed versions only)

A comprehensive listing of all terms and subjects will help you quickly find what you need when you are not sure where it occurs in MasterFormat®.

Conclusion

This information is designed to be as comprehensive and easy to use as possible.

The Construction Specifications Institute (CSI) and Construction Specifications Canada (CSC) have produced the 2018 edition of MasterFormat®, a system of titles and numbers used extensively to organize construction information.

All unit prices in the RSMeans cost data are now arranged in the 50-division MasterFormat® 2018 system.

* Not all information is available in all data sets

Note: The material prices in RSMeans cost data are "contractor's prices." They are the prices that contractors can expect to pay at the lumberyards, suppliers'/distributors' warehouses, etc. Small orders of specialty items would be higher than the costs shown, while very large orders, such as truckload lots, would be less. The variation would depend on the size, timing, and negotiating power of the contractor. The labor costs are primarily for new construction or major renovation rather than repairs or minor alterations. With reasonable exercise of judgment, the figures can be used for any building work.

Estimating with RSMeans data: Unit Prices

Following these steps will allow you to complete an accurate estimate using RSMeans data: Unit Prices.

1. Scope Out the Project

- Think through the project and identify the CSI divisions needed in your estimate.
- Identify the individual work tasks that will need to be covered in your estimate.
- The Unit Price data have been divided into 50 divisions according to CSI MasterFormat® 2018.
- In printed versions, the Unit Price Section Table of Contents on page 1 may also be helpful when scoping out your project.
- Experienced estimators find it helpful to begin with Division 2 and continue through completion. Division 1 can be estimated after the full project scope is known.

2. Quantify

- Determine the number of units required for each work task that you identified.
- Experienced estimators include an allowance for waste in their quantities. (Waste is not included in our Unit Price line items unless otherwise stated.)

3. Price the Quantities

- Use the search tools available to locate individual Unit Price line items for your estimate.
- Reference Numbers indicated within a Unit Price section refer to additional information that you may find useful.
- The crew indicates who is performing the work for that task. Crew codes are expanded in the Crew Listings in the Reference Section to include all trades and equipment that comprise the crew.
- The Daily Output is the amount of work the crew is expected to complete in one day.
- The Labor-Hours value is the amount of time it will take for the crew to install one unit of work.
- The abbreviated Unit designation indicates the unit of measure upon which the crew, productivity, and prices are based.
- Bare Costs are shown for materials, labor, and equipment needed to complete the Unit Price line item. Bare costs do not include waste, project overhead, payroll insurance, payroll taxes, main office overhead, or profit.
- The Total Incl O&P cost is the billing rate or invoice amount of the installing contractor or subcontractor who performs the work for the Unit Price line item.

4. Multiply

- Multiply the total number of units needed for your project by the Total Incl O&P cost for each Unit Price line item.
- Be careful that your take off unit of measure matches the unit of measure in the Unit column.
- The price you calculate is an estimate for a completed item of work.
- Keep scoping individual tasks, determining the number of units required for those tasks, matching each task with individual Unit Price line items, and multiplying quantities by Total Incl O&P costs.
- An estimate completed in this manner is priced as if a subcontractor, or set of subcontractors, is performing the work. The estimate does not yet include Project Overhead or Estimate Summary components such as general contractor markups on subcontracted work, general contractor office overhead and profit, contingency, and location factors.

5. Project Overhead

- Include project overhead items from Division 1-General Requirements.
- These items are needed to make the job run. They are typically, but not always, provided by the general contractor. Items include, but are not limited to, field personnel, insurance, performance bond, permits, testing, temporary utilities, field office and storage facilities, temporary scaffolding and platforms, equipment mobilization and demobilization, temporary roads and sidewalks, winter protection, temporary barricades and fencing, temporary security, temporary signs, field engineering and layout, final cleaning, and commissioning.
- Each item should be quantified and matched to individual Unit Price line items in Division 1, then priced and added to your estimate.
- An alternate method of estimating project overhead costs is to apply a percentage of the total project cost—usually 5% to 15% with an average of 10% (see General Conditions).
- Include other project related expenses in your estimate such as:
  - Rented equipment not itemized in the Crew Listings
  - Rubbish handling throughout the project (see section 02 41 19.19)

6. Estimate Summary

- Include sales tax as required by laws of your state or county.
- Include the general contractor's markup on self-performed work, usually 5% to 15% with an average of 10%.
- Include the general contractor's markup on subcontracted work, usually 5% to 15% with an average of 10%.
- Include the general contractor's main office overhead and profit:
  - RSMeans data provides general guidelines on the general contractor's main office overhead (see section 01 31 13.60 and Reference Number R013113-50).
  - Markups will depend on the size of the general contractor's operations, projected annual revenue, the level of risk, and the level of competition in the local area and for this project in particular.
- Include a contingency, usually 3% to 5%, if appropriate.
- Adjust your estimate to the project's location by using the City Cost Indexes or the Location Factors in the Reference Section:
  - Look at the rules in "How to Use the City Cost Indexes" to see how to apply the Indexes for your location.
  - When the proper Index or Factor has been identified for the project's location, convert it to a multiplier by dividing it by 100, then multiply that multiplier by your estimated total cost. The original estimated total cost will now be adjusted up or down from the national average to a total that is appropriate for your location.

Editors' Note:
We urge you to spend time reading and understanding the supporting material. An accurate estimate requires experience, knowledge, and careful calculation. The more you know about how we at RSMeans developed the data, the more accurate your estimate will be. In addition, it is important to take into consideration the reference material such as Equipment Listings, Crew Listings, City Cost Indexes, Location Factors, and Reference Tables.

How to Use the Cost Data: The Details

What's Behind the Numbers? The Development of Cost Data

RSMeans data engineers continually monitor developments in the construction industry in order to ensure reliable, thorough, and up-to-date cost information. While overall construction costs may vary relative to general economic conditions, price fluctuations within the industry are dependent upon many factors. Individual price variations may, in fact, be opposite to overall economic trends. Therefore, costs are constantly tracked and complete updates are performed yearly. Also, new items are frequently added in response to changes in materials and methods.

Costs in U.S. Dollars

All costs represent U.S. national averages and are given in U.S. dollars. The City Cost Index (CCI) with RSMeans data can be used to adjust costs to a particular location. The CCI for Canada can be used to adjust U.S. national averages to local costs in Canadian dollars. No exchange rate conversion is necessary because it has already been factored in.

G The processes or products identified by the green symbol in our publications have been determined to be environmentally responsible and/or resource-efficient solely by RSMeans data engineering staff. The inclusion of the green symbol does not represent compliance with any specific industry association or standard.

Material Costs

RSMeans data engineers contact manufacturers, dealers, distributors, and contractors all across the U.S. and Canada to determine national average material costs. If you have access to current material costs for your specific location, you may wish to make adjustments to reflect differences from the national average. Included within material costs are fasteners for a normal installation. RSMeans data engineers use manufacturers' recommendations, written specifications, and/or standard construction practices for the sizing and spacing of fasteners. Adjustments to material costs may be required for your specific application or location. The manufacturer's warranty is assumed. Extended warranties are not included in the material costs. **Material costs do not include sales tax.**

Labor Costs

Labor costs are based upon a mathematical average of trade-specific wages in 30 major U.S. cities. The type of wage (union, open shop, or residential) is identified on the inside back cover of printed publications or selected by the estimator when using the electronic products. Markups for the wages can also be found on the inside back cover of printed publications and/or under the labor references found in the electronic products.

- If wage rates in your area vary from those used, or if rate increases are expected within a given year, labor costs should be adjusted accordingly.

Labor costs reflect productivity based on actual working conditions. In addition to actual installation, these figures include time spent during a normal weekday on tasks, such as material receiving and handling, mobilization at the site, site movement, breaks, and cleanup.

Productivity data is developed over an extended period so as not to be influenced by abnormal variations and reflects a typical average.

Equipment Costs

Equipment costs include not only rental but also operating costs for equipment under normal use. The operating costs include parts and labor for routine servicing, such as the repair and replacement of pumps, filters, and worn lines. Normal operating expendables, such as fuel, lubricants, tires, and electricity (where applicable), are also included. Extraordinary operating expendables with highly variable wear patterns, such as diamond bits and blades, are excluded. These costs are included under materials. Equipment rental rates are obtained from industry sources throughout North America—contractors, suppliers, dealers, manufacturers, and distributors.

Rental rates can also be treated as reimbursement costs for contractor-owned equipment. Owned equipment costs include depreciation, loan payments, interest, taxes, insurance, storage, and major repairs.

Equipment costs do not include operators' wages.

Equipment Cost/Day—The cost of equipment required for each crew is included in the Crew Listings in the Reference Section (small tools that are considered essential everyday tools are not listed out separately). The Crew Listings itemize specialized tools and heavy equipment along with labor trades. The daily cost of itemized equipment included in a crew is based on dividing the weekly bare rental rate by 5 (number of working days per week), then adding the hourly operating cost times 8 (the number of hours per day). This Equipment Cost/Day is shown in the last column of the Equipment Rental Costs in the Reference Section.

Mobilization, Demobilization—The cost to move construction equipment from an equipment yard or rental company to the job site and back again is not included in equipment costs. Mobilization (to the site) and demobilization (from the site) costs can be found in the Unit Price Section. If a piece of equipment is already at the job site, it is not appropriate to utilize mobilization or demobilization costs again in an estimate.

Overhead and Profit

Total Cost including O&P for the installing contractor is shown in the last column of the Unit Price and/or Assemblies. This figure is the sum of the bare material cost plus 10% for profit, the bare labor cost plus total overhead and profit, and the bare equipment cost plus 10% for profit. Details for the calculation of overhead and profit on labor are shown on the inside back cover of the printed product and in the Reference Section of the electronic product.

General Conditions

Cost data in this data set are presented in two ways: Bare Costs and Total Cost including O&P (Overhead and Profit). General Conditions, or General Requirements, of the contract should also be added to the Total Cost including O&P when applicable. Costs for General Conditions are listed in Division 1 of the Unit Price Section and in the Reference Section.

General Conditions for the installing contractor may range from 0% to 10% of the Total Cost including O&P. For the general or prime contractor, costs for General Conditions may range from 5% to 15% of the Total Cost including O&P, with a figure of 10% as the most typical allowance. If applicable, the Assemblies and Models sections use costs that include the installing contractor's overhead and profit (O&P).

Factors Affecting Costs

Costs can vary depending upon a number of variables. Here's a listing of some factors that affect costs and points to consider.

Quality—The prices for materials and the workmanship upon which productivity is based represent sound construction work. They are also in line with industry standard and manufacturer specifications and are frequently used by federal, state, and local governments.

Overtime—We have made no allowance for overtime. If you anticipate premium time or work beyond normal working hours, be sure to make an appropriate adjustment to your labor costs.

Productivity—The productivity, daily output, and labor-hour figures for each line item are based on an eight-hour work day in daylight hours in moderate temperatures and up to a 14' working height unless otherwise indicated. For work that extends beyond normal work hours or is performed under adverse conditions, productivity may decrease.

Size of Project—The size, scope of work, and type of construction project will have a significant impact on cost. Economies of scale can reduce costs for large projects. Unit costs can often run higher for small projects.

Location—Material prices are for metropolitan areas. However, in dense urban areas, traffic and site storage limitations may increase costs. Beyond a 20-mile radius of metropolitan areas, extra trucking or transportation charges may also increase the material costs slightly. On the other hand, lower wage rates may be in effect. Be sure to consider both of these factors when preparing an estimate, particularly if the job site is located in a central city or remote rural location. In addition, highly specialized subcontract items may require travel and per-diem expenses for mechanics.

Other Factors—

- season of year
- contractor management
- weather conditions
- local union restrictions
- building code requirements
- availability of:
  - adequate energy
  - skilled labor
  - building materials
- owner's special requirements/restrictions
- safety requirements
- environmental considerations
- access

Unpredictable Factors—General business conditions influence "in-place" costs of all items. Substitute materials and construction methods may have to be employed. These may affect the installed cost and/or life cycle costs. Such factors may be difficult to evaluate and cannot necessarily be predicted on the basis of the job's location in a particular section of the country. Thus, where these factors apply, you may find significant but unavoidable cost variations for which you will have to apply a measure of judgment to your estimate.

Rounding of Costs

In printed publications only, all unit prices in excess of $5.00 have been rounded to make them easier to use and still maintain adequate precision of the results.

How Subcontracted Items Affect Costs

A considerable portion of all large construction jobs is usually subcontracted. In fact, the percentage done by subcontractors is constantly increasing and may run over 90%. Since the workers employed by these companies do nothing else but install their particular products, they soon become experts in that line. As a result, installation by these firms is accomplished so efficiently that the total in-place cost, even with the general contractor's overhead and profit, is no more, and often less, than if the principal contractor had handled the installation. Companies that deal with construction specialties are anxious to have their products perform well and, consequently, the installation will be the best possible.

Contingencies

The allowance for contingencies generally provides for unforeseen construction difficulties. On alterations or repair jobs, 20% is not too much. If drawings are final and only field contingencies are being considered, 2% or 3% is probably sufficient and often nothing needs to be added. Contractually, changes in plans will be covered by extras. The contractor should consider inflationary price trends and possible material shortages during the course of the job. These escalation factors are dependent upon both economic conditions and the anticipated time between the estimate and actual construction. If drawings are not complete or approved, or a budget cost is wanted, it is wise to add 5% to 10%. Contingencies, then, are a matter of judgment.

Important Estimating Considerations

The productivity, or daily output, of each craftsman or crew assumes a well-managed job where tradesmen with the proper tools and equipment, along with the appropriate construction materials, are present. Included are daily set-up and cleanup time, break time, and plan layout time. Unless otherwise indicated, time for material movement on site (for items that can be transported by hand) of up to 200' into the building and to the first or second floor is also included. If material has to be transported by other means, over greater distances, or to higher floors, an additional allowance should be considered by the estimator.

While horizontal movement is typically a sole function of distances, vertical transport introduces other variables that can significantly impact productivity. In an occupied building, the use of elevators (assuming access, size, and required protective measures are acceptable) must be understood at the time of the estimate. For new construction, hoist wait and cycle times can easily be 15 minutes and may result in scheduled access extending beyond the normal work day. Finally, all vertical transport will impose strict weight limits likely to preclude the use of any motorized material handling.

The productivity, or daily output, also assumes installation that meets manufacturer/designer/standard specifications. A time allowance for quality control checks, minor adjustments, and any task required to ensure proper function or operation is also included. For items that require connections to services, time is included for positioning, leveling, securing the unit, and making all the necessary connections (and start up where applicable) to ensure a complete installation. Estimating of the services themselves (electrical, plumbing, water, steam, hydraulics, dust collection, etc.) is separate.

In some cases, the estimator must consider the use of a crane and an appropriate crew for the installation of large or heavy items. For those situations where a crane is not included in the assigned crew and as part of the line item cost, then equipment rental costs, mobilization and demobilization costs, and operator and support personnel costs must be considered.

Labor-Hours

The labor-hours expressed in this publication are derived by dividing the total daily labor-hours for the crew by the daily output. Based on average installation time and the assumptions listed above, the labor-hours include: direct labor, indirect labor, and nonproductive time. A typical day for a craftsman might include but is not limited to:

- Direct Work
  - Measuring and layout
  - Preparing materials
  - Actual installation
  - Quality assurance/quality control
- Indirect Work
  - Reading plans or specifications
  - Preparing space
  - Receiving materials
  - Material movement
  - Giving or receiving instruction
  - Miscellaneous
- Non-Work
  - Chatting
  - Personal issues
  - Breaks
  - Interruptions (i.e., sickness, weather, material or equipment shortages, etc.)

If any of the items for a typical day do not apply to the particular work or project situation, the estimator should make any necessary adjustments.

Final Checklist

Estimating can be a straightforward process provided you remember the basics. Here's a checklist of some of the steps you should remember to complete before finalizing your estimate.

Did you remember to:

- factor in the City Cost Index for your locale?
- take into consideration which items have been marked up and by how much?
- mark up the entire estimate sufficiently for your purposes?
- read the background information on techniques and technical matters that could impact your project time span and cost?
- include all components of your project in the final estimate?
- double check your figures for accuracy?
- call RSMeans data engineers if you have any questions about your estimate or the data you've used? Remember, Gordian stands behind all of our products, including our extensive RSMeans data solutions. If you have any questions about your estimate, about the costs you've used from our data, or even about the technical aspects of the job that may affect your estimate, feel free to call the Gordian RSMeans editors at 1.800.448.8182.

Unit Price Section

Table of Contents

RSMeans data: Unit Prices—How They Work

All RSMeans data: Unit Prices are organized in the same way.

03 30 Cast-In-Place Concrete

03 30 53 – Miscellaneous Cast-In-Place Concrete

03 30 53.40 Concrete In Place		Crew	Daily Output	Labor-Hours	Unit	Material (2020 Bare Costs)	Labor (2020 Bare Costs)	Equipment (2020 Bare Costs)	Total (2020 Bare Costs)	Total Incl O&P
0010	CONCRETE IN PLACE									
0020	Including forms (4 uses), Grade 60 rebar, concrete (Portland cement									
0050	Type I), placement and finishing unless otherwise indicated									
3540	Equipment pad (3000 psi), 3′ x 3′ x 6″ thick	C-14H	45	1.067	Ea.	50.50	55	.60	106.10	139
3550	4′ x 4′ x 6″ thick		30	1.600		78	82.50	.90	161.40	210
3560	5′ x 5′ x 8″ thick		18	2.667		138	138	1.49	277.49	360
3570	6′ x 6′ x 8″ thick		14	3.429		190	177	1.92	368.92	475
3580	8′ x 8′ x 10″ thick		8	6		395	310	3.36	708.36	905
3590	10′ x 10′ x 12″ thick		5	9.600		695	495	5.40	1,195.40	1,500
3800	Footings (3000 psi), spread under 1 C.Y.	C-14C	28	4	C.Y.	203	201	.96	404.96	525
3825	1 C.Y. to 5 C.Y.		43	2.605		240	131	.63	371.63	465
3850	Over 5 C.Y. R033053-60		75	1.493		220	75	.36	295.36	355
3900	Footings, strip (3000 psi), 18″ x 9″, unreinforced	C-14L	40	2.400		154	118	.67	272.67	350
3920	18″ x 9″, reinforced	C-14C	35	3.200		181	161	.77	342.77	440

It is important to understand the structure of RSMeans data: Unit Prices so that you can find information easily and use it correctly.

1 Line Numbers

Line Numbers consist of 12 characters, which identify a unique location in the database for each task. The first 6 or 8 digits conform to the Construction Specifications Institute MasterFormat® 2018. The remainder of the digits are a further breakdown in order to arrange items in understandable groups of similar tasks. Line numbers are consistent across all of our publications, so a line number in any of our products will always refer to the same item of work.

2 Descriptions

Descriptions are shown in a hierarchical structure to make them readable. In order to read a complete description, read up through the indents to the top of the section. Include everything that is above and to the left that is not contradicted by information below. For instance, the complete description for line 03 30 53.40 3550 is "Concrete in place, including forms (4 uses), Grade 60 rebar, concrete (Portland cement Type 1), placement and finishing unless otherwise indicated; Equipment pad (3000 psi), 4' × 4' × 6" thick."

3 RSMeans data

When using **RSMeans data**, it is important to read through an entire section to ensure that you use the data that most closely matches your work. Note that sometimes there is additional information shown in the section that may improve your price. There are frequently lines that further describe, add to, or adjust data for specific situations.

4 Reference Information

Gordian's RSMeans engineers have created **reference** information to assist you in your estimate. **If** there is information that applies to a section, it will be indicated at the start of the section. The Reference Section is located in the back of the data set.

5 Crews

Crews include labor and/or equipment necessary to accomplish each task. In this case, Crew C-14H is used. Gordian's RSMeans staff selects a crew to represent the workers and equipment that are typically used for that task. In this case, Crew C-14H consists of one carpenter foreman (outside), two carpenters, one rodman, one laborer, one cement finisher, and one gas engine vibrator. Details of all crews can be found in the Reference Section.

Crews - Standard

Crew No.	Bare Costs		Incl. Subs O & P		Cost Per Labor-Hour	
Crew C-14H	**Hr.**	**Daily**	**Hr.**	**Daily**	**Bare Costs**	**Incl. O&P**
1 Carpenter Foreman (outside)	$55.15	$441.20	$82.85	$662.80	$51.65	$77.36
2 Carpenters	53.15	850.40	79.85	1277.60		
1 Rodman (reinf.)	56.40	451.20	84.90	679.20		
1 Laborer	42.10	336.80	63.25	506.00		
1 Cement Finisher	49.95	399.60	73.45	587.60		
1 Gas Engine Vibrator		26.85		29.54	.56	.62
48 L.H., Daily Totals		$2506.05		$3742.74	$52.21	$77.97

6 Daily Output

The **Daily Output** is the amount of work that the crew can do in a normal 8-hour workday, including mobilization, layout, movement of materials, and cleanup. In this case, crew C-14H can install thirty 4' × 4' × 6" thick concrete pads in a day. Daily output is variable and based on many factors, including the size of the job, location, and environmental conditions. RSMeans data represents work done in daylight (or adequate lighting) and temperate conditions.

7 Labor-Hours

The figure in the **Labor-Hours** column is the amount of labor required to perform one unit of work–in this case the amount of labor required to construct one 4' × 4' equipment pad. This figure is calculated by dividing the number of hours of labor in the crew by the daily output (48 labor-hours divided by 30 pads = 1.6 hours of labor per pad). Multiply 1.6 times 60 to see the value in minutes: 60 × 1.6 = 96 minutes. Note: the labor-hour figure is not dependent on the crew size. A change in crew size will result in a corresponding change in daily output, but the labor-hours per unit of work will not change.

8 Unit of Measure

All RSMeans data: Unit Prices include the typical **Unit of Measure** used for estimating that item. For concrete-in-place the typical unit is cubic yards (C.Y.) or each (Ea.). For installing broadloom carpet it is square yard and for gypsum board it is square foot. The estimator needs to take special care that the unit in the data matches the unit in the take-off. Unit conversions may be found in the Reference Section.

9 Bare Costs

Bare Costs are the costs of materials, labor, and equipment that the installing contractor pays. They represent the cost, in U.S. dollars, for one unit of work. They do not include any markups for profit or labor burden.

10 Bare Total

The **Total column** represents the total bare cost for the installing contractor in U.S. dollars. In this case, the sum of $78 for material + $82.50 for labor + $.90 for equipment is $161.40.

11 Total Incl O&P

The **Total Incl O&P column** is the total cost, including overhead and profit, that the installing contractor will charge the customer. This represents the cost of materials plus 10% profit, the cost of labor plus labor burden and 10% profit, and the cost of equipment plus 10% profit. It does not include the general contractor's overhead and profit. Note: See the inside back cover of the printed product or the Reference Section of the electronic product for details on how the labor burden is calculated.

National Average

The RSMeans data in our print publications represent a "national average" cost. This data should be modified to the project location using the ***City Cost Indexes*** *or* ***Location Factors*** *tables found in the Reference Section. Use the Location Factors to adjust estimate totals if the project covers multiple trades. Use the City Cost Indexes (CCI) for single trade projects or projects where a more detailed analysis is required. All figures in the two tables are derived from the same research. The last row of data in the CCI–the weighted average–is the same as the numbers reported for each location in the location factor table.*

RSMeans data: Unit Prices—How They Work (Continued)

Project Name: Pre-Engineered Steel Building		Architect: As Shown						
Location:	Anywhere, USA						01/01/20	STD
1 Line Number	Description	Qty	Unit	Material	Labor	Equipment	SubContract	Estimate Total
03 30 53.40 3940	Strip footing, 12" x 24", reinforced	15	C.Y.	$2,565.00	$1,770.00	$8.40	$0.00	
03 30 53.40 3950	Strip footing, 12" x 36", reinforced	34	C.Y.	$5,610.00	$3,196.00	$15.30	$0.00	
03 11 13.65 3000	Concrete slab edge forms	500	L.F.	$165.00	$1,345.00	$0.00	$0.00	
03 22 11.10 0200	Welded wire fabric reinforcing	150	C.S.F.	$2,872.50	$4,350.00	$0.00	$0.00	
03 31 13.35 0300	Ready mix concrete, 4000 psi for slab on grade	278	C.Y.	$35,306.00	$0.00	$0.00	$0.00	
03 31 13.70 4300	Place, strike off & consolidate concrete slab	278	C.Y.	$0.00	$5,309.80	$136.22	$0.00	
03 35 13.30 0250	Machine float & trowel concrete slab	15,000	S.F.	$0.00	$9,900.00	$750.00	$0.00	
03 15 16.20 0140	Cut control joints in concrete slab	950	L.F.	$47.50	$418.00	$57.00	$0.00	
03 39 23.13 0300	Sprayed concrete curing membrane	150	C.S.F.	$1,867.50	$1,065.00	$0.00	$0.00	
Division 03	**Subtotal**			**$48,433.50**	**$27,353.80**	**$966.92**	**$0.00**	**$76,754.22**
08 36 13.10 2650	Manual 10' x 10' steel sectional overhead door	8	Ea.	$11,000.00	$3,760.00	$0.00	$0.00	
08 36 13.10 2860	Insulation and steel back panel for OH door	800	S.F.	$4,000.00	$0.00	$0.00	$0.00	
Division 08	**Subtotal**			**$15,000.00**	**$3,760.00**	**$0.00**	**$0.00**	**$18,760.00**
13 34 19.50 1100	Pre-Engineered Steel Building, 100' x 150' x 24'	15,000	SF Flr.	$0.00	$0.00	$0.00	$367,500.00	
13 34 19.50 6050	Framing for PESB door opening, 3' x 7'	4	Opng.	$0.00	$0.00	$0.00	$2,320.00	
13 34 19.50 6100	Framing for PESB door opening, 10' x 10'	8	Opng.	$0.00	$0.00	$0.00	$9,600.00	
13 34 19.50 6200	Framing for PESB window opening, 4' x 3'	6	Opng.	$0.00	$0.00	$0.00	$3,450.00	
13 34 19.50 5750	PESB door, 3' x 7', single leaf	4	Opng.	$2,920.00	$736.00	$0.00	$0.00	
13 34 19.50 7750	PESB sliding window, 4' x 3' with screen	6	Opng.	$2,940.00	$630.00	$67.80	$0.00	
13 34 19.50 6550	PESB gutter, eave type, 26 ga., painted	300	L.F.	$2,415.00	$864.00	$0.00	$0.00	
13 34 19.50 8650	PESB roof vent, 12" wide x 10' long	15	Ea.	$570.00	$3,465.00	$0.00	$0.00	
13 34 19.50 6900	PESB insulation, vinyl faced, 4" thick	27,400	S.F.	$11,782.00	$10,138.00	$0.00	$0.00	
Division 13	**Subtotal**			**$20,627.00**	**$15,833.00**	**$67.80**	**$382,870.00**	**$419,397.80**
		Subtotal		**$84,060.50**	**$46,946.80**	**$1,034.72**	**$382,870.00**	**$514,912.02**
Division 01 2	**General Requirements @ 7%**			**5,884.24**	**3,286.28**	**72.43**	**26,800.90**	
		Estimate Subtotal		**$89,944.74**	**$50,233.08**	**$1,107.15**	**$409,670.90**	**$514,912.02**
	3	**Sales Tax @ 5%**		**4,497.24**		**55.36**	**10,241.77**	
		Subtotal A		**94,441.97**	**50,233.08**	**1,162.51**	**419,912.67**	
	4	**GC O & P**		**9,444.20**	**25,769.57**	**116.25**	**41,991.27**	
		Subtotal B		**103,886.17**	**76,002.64**	**1,278.76**	**461,903.94**	**$643,071.51**
	5	**Contingency @ 5%**						**32,153.58**
		Subtotal C						**$675,225.09**
	6	**Bond @ $12/1000 +10% O&P**						**8,912.97**
		Subtotal D						**$684,138.06**
	7	**Location Adjustment Factor**			**115.50**			**106,041.40**
		Grand Total						**$790,179.46**

This estimate is based on an interactive spreadsheet. You are free to download it and adjust it to your methodology. A copy of this spreadsheet is available at **RSMeans.com/2020books.**

Sample Estimate

This sample demonstrates the elements of an estimate, including a tally of the RSMeans data lines and a summary of the markups on a contractor's work to arrive at a total cost to the owner. The Location Factor with RSMeans data is added at the bottom of the estimate to adjust the cost of the work to a specific location.

1 Work Performed

The body of the estimate shows the RSMeans data selected, including the line number, a brief description of each item, its take-off unit and quantity, and the bare costs of materials, labor, and equipment. This estimate also includes a column titled "SubContract." This data is taken from the column "Total Incl O&P" and represents the total that a subcontractor would charge a general contractor for the work, including the sub's markup for overhead and profit.

2 Division 1, General Requirements

This is the first division numerically but the last division estimated. Division 1 includes project-wide needs provided by the general contractor. These requirements vary by project but may include temporary facilities and utilities, security, testing, project cleanup, etc. For small projects a percentage can be used—typically between 5% and 15% of project cost. For large projects the costs may be itemized and priced individually.

3 Sales Tax

If the work is subject to state or local sales taxes, the amount must be added to the estimate. Sales tax may be added to material costs, equipment costs, and subcontracted work. In this case, sales tax was added in all three categories. It was assumed that approximately half the subcontracted work would be material cost, so the tax was applied to 50% of the subcontract total.

4 GC O&P

This entry represents the general contractor's markup on material, labor, equipment, and subcontractor costs. Our standard markup on materials, equipment, and subcontracted work is 10%. In this estimate, the markup on the labor performed by the GC's workers uses "Skilled Workers Average" shown in Column F on the table "Installing Contractor's Overhead & Profit," which can be found on the inside back cover of the printed product or in the Reference Section of the electronic product.

5 Contingency

A factor for contingency may be added to any estimate to represent the cost of unknowns that may occur between the time that the estimate is performed and the time the project is constructed. The amount of the allowance will depend on the stage of design at which the estimate is done and the contractor's assessment of the risk involved. Refer to section 01 21 16.50 for contingency allowances.

6 Bonds

Bond costs should be added to the estimate. The figures here represent a typical performance bond, ensuring the owner that if the general contractor does not complete the obligations in the construction contract the bonding company will pay the cost for completion of the work.

7 Location Adjustment

Published prices are based on national average costs. If necessary, adjust the total cost of the project using a location factor from the "Location Factor" table or the "City Cost Index" table. Use location factors if the work is general, covering multiple trades. If the work is by a single trade (e.g., masonry) use the more specific data found in the "City Cost Indexes."

Estimating Tips

01 20 00 Price and Payment Procedures

- Allowances that should be added to estimates to cover contingencies and job conditions that are not included in the national average material and labor costs are shown in Section 01 21.
- When estimating historic preservation projects (depending on the condition of the existing structure and the owner's requirements), a 15–20% contingency or allowance is recommended, regardless of the stage of the drawings.

01 30 00 Administrative Requirements

- Before determining a final cost estimate, it is good practice to review all the items listed in Subdivisions 01 31 and 01 32 to make final adjustments for items that may need customizing to specific job conditions.
- Requirements for initial and periodic submittals can represent a significant cost to the General Requirements of a job. Thoroughly check the submittal specifications when estimating a project to determine any costs that should be included.

01 40 00 Quality Requirements

- All projects will require some degree of quality control. This cost is not included in the unit cost of construction listed in each division. Depending upon the terms of the contract, the various costs of inspection and testing can be the responsibility of either the owner or the contractor. Be sure to include the required costs in your estimate.

01 50 00 Temporary Facilities and Controls

- Barricades, access roads, safety nets, scaffolding, security, and many more requirements for the execution of a safe project are elements of direct cost. These costs can easily be overlooked when preparing an estimate. When looking through the major classifications of this subdivision, determine which items apply to each division in your estimate.
- Construction equipment rental costs can be found in the Reference Section in Section 01 54 33. Operators' wages are not included in equipment rental costs.
- Equipment mobilization and demobilization costs are not included in equipment rental costs and must be considered separately.
- The cost of small tools provided by the installing contractor for his workers is covered in the "Overhead" column on the "Installing Contractor's Overhead and Profit" table that lists labor trades, base rates, and markups. Therefore, it is included in the "Total Incl. O&P" cost of any unit price line item.

01 70 00 Execution and Closeout Requirements

- When preparing an estimate, thoroughly read the specifications to determine the requirements for Contract Closeout. Final cleaning, record documentation, operation and maintenance data, warranties and bonds, and spare parts and maintenance materials can all be elements of cost for the completion of a contract. Do not overlook these in your estimate.

Reference Numbers

Reference numbers are shown at the beginning of some major classifications. These numbers refer to related items in the Reference Section. The reference information may be an estimating procedure, an alternate pricing method, or technical information.

Note: Not all subdivisions listed here necessarily appear. ■

01 11 Summary of Work

01 11 31 – Professional Consultants

01 11 31.10 Architectural Fees		Crew	Daily Output	Labor-Hours	Unit	Material	2020 Bare Costs Labor	Equipment	Total	Total Incl O&P
0010	**ARCHITECTURAL FEES**									
0020	For new construction									
0060	Minimum				Project				4.90%	4.90%
0090	Maximum								16%	16%
0100	For alteration work, to $500,000, add to new construction fee								50%	50%
0150	Over $500,000, add to new construction fee								25%	25%

01 11 31.20 Construction Management Fees		Crew	Daily Output	Labor-Hours	Unit	Material	Labor	Equipment	Total	Total Incl O&P
0010	**CONSTRUCTION MANAGEMENT FEES**									
0020	$1,000,000 job, minimum				Project				4.50%	4.50%
0050	Maximum								7.50%	7.50%
0300	$50,000,000 job, minimum								2.50%	2.50%
0350	Maximum								4%	4%

01 11 31.30 Engineering Fees		Crew	Daily Output	Labor-Hours	Unit	Material	Labor	Equipment	Total	Total Incl O&P
0010	**ENGINEERING FEES** R011110-30									
0020	Educational planning consultant, minimum				Project				.50%	.50%
0100	Maximum				"				2.50%	2.50%
0200	Electrical, minimum				Contrct				4.10%	4.10%
0300	Maximum								10.10%	10.10%
0400	Elevator & conveying systems, minimum								2.50%	2.50%
0500	Maximum								5%	5%
0600	Food service & kitchen equipment, minimum								8%	8%
0700	Maximum								12%	12%
0800	Landscaping & site development, minimum								2.50%	2.50%
0900	Maximum								6%	6%
1000	Mechanical (plumbing & HVAC), minimum								4.10%	4.10%
1100	Maximum								10.10%	10.10%
1200	Structural, minimum				Project				1%	1%
1300	Maximum				"				2.50%	2.50%

01 21 Allowances

01 21 16 – Contingency Allowances

01 21 16.50 Contingencies		Crew	Daily Output	Labor-Hours	Unit	Material	Labor	Equipment	Total	Total Incl O&P
0010	**CONTINGENCIES**, Add to estimate									
0020	Conceptual stage				Project				20%	20%
0050	Schematic stage								15%	15%
0100	Preliminary working drawing stage (Design Dev.)								10%	10%
0150	Final working drawing stage								3%	3%

01 21 53 – Factors Allowance

01 21 53.50 Factors		Crew	Daily Output	Labor-Hours	Unit	Material	Labor	Equipment	Total	Total Incl O&P
0010	**FACTORS** Cost adjustments R012153-10									
0100	Add to construction costs for particular job requirements									
0500	Cut & patch to match existing construction, add, minimum				Costs	2%	3%			
0550	Maximum					5%	9%			
0800	Dust protection, add, minimum					1%	2%			
0850	Maximum					4%	11%			
1100	Equipment usage curtailment, add, minimum					1%	1%			
1150	Maximum					3%	10%			
1400	Material handling & storage limitation, add, minimum					1%	1%			
1450	Maximum					6%	7%			
1700	Protection of existing work, add, minimum					2%	2%			

01 21 Allowances

01 21 53 – Factors Allowance

01 21 53.50 Factors		Crew	Daily Output	Labor-Hours	Unit	Material	2020 Bare Costs Labor	Equipment	Total	Total Incl O&P
1750	Maximum				Costs	5%	7%			
2000	Shift work requirements, add, minimum						5%			
2050	Maximum						30%			
2300	Temporary shoring and bracing, add, minimum					2%	5%			
2350	Maximum				↓	5%	12%			
01 21 53.60 Security Factors										
0010	**SECURITY FACTORS** R012153-60									
0100	Additional costs due to security requirements									
0110	Daily search of personnel, supplies, equipment and vehicles									
0120	Physical search, inventory and doc of assets, at entry				Costs		30%			
0130	At entry and exit						50%			
0140	Physical search, at entry						6.25%			
0150	At entry and exit						12.50%			
0160	Electronic scan search, at entry						2%			
0170	At entry and exit						4%			
0180	Visual inspection only, at entry						.25%			
0190	At entry and exit						.50%			
0200	ID card or display sticker only, at entry						.12%			
0210	At entry and exit				↓		.25%			
0220	Day 1 as described below, then visual only for up to 5 day job duration									
0230	Physical search, inventory and doc of assets, at entry				Costs		5%			
0240	At entry and exit						10%			
0250	Physical search, at entry						1.25%			
0260	At entry and exit						2.50%			
0270	Electronic scan search, at entry						.42%			
0280	At entry and exit				↓		.83%			
0290	Day 1 as described below, then visual only for 6-10 day job duration									
0300	Physical search, inventory and doc of assets, at entry				Costs		2.50%			
0310	At entry and exit						5%			
0320	Physical search, at entry						.63%			
0330	At entry and exit						1.25%			
0340	Electronic scan search, at entry						.21%			
0350	At entry and exit				↓		.42%			
0360	Day 1 as described below, then visual only for 11-20 day job duration									
0370	Physical search, inventory and doc of assets, at entry				Costs		1.25%			
0380	At entry and exit						2.50%			
0390	Physical search, at entry						.31%			
0400	At entry and exit						.63%			
0410	Electronic scan search, at entry						.10%			
0420	At entry and exit				↓		.21%			
0430	Beyond 20 days, costs are negligible									
0440	Escort required to be with tradesperson during work effort				Costs		6.25%			

01 21 55 – Job Conditions Allowance

01 21 55.50 Job Conditions		Crew	Daily Output	Labor-Hours	Unit	Material	Labor	Equipment	Total	Total Incl O&P
0010	**JOB CONDITIONS** Modifications to applicable									
0020	cost summaries									
0100	Economic conditions, favorable, deduct				Project				2%	2%
0200	Unfavorable, add								5%	5%
0300	Hoisting conditions, favorable, deduct								2%	2%
0400	Unfavorable, add								5%	5%
0700	Labor availability, surplus, deduct								1%	1%
0800	Shortage, add				↓				10%	10%

01 21 Allowances

01 21 55 – Job Conditions Allowance

01 21 55.50 Job Conditions		Crew	Daily Output	Labor-Hours	Unit	Material	2020 Bare Costs Labor	Equipment	Total	Total Incl O&P
0900	Material storage area, available, deduct				Project				1%	1%
1000	Not available, add								2%	2%
1100	Subcontractor availability, surplus, deduct								5%	5%
1200	Shortage, add								12%	12%
1300	Work space, available, deduct								2%	2%
1400	Not available, add								5%	5%

01 21 57 – Overtime Allowance

01 21 57.50 Overtime		Crew	Daily Output	Labor-Hours	Unit	Material	Labor	Equipment	Total	Total Incl O&P
0010	**OVERTIME** for early completion of projects or where R012909-90									
0020	labor shortages exist, add to usual labor, up to				Costs		100%			

01 21 63 – Taxes

01 21 63.10 Taxes		Crew	Daily Output	Labor-Hours	Unit	Material	Labor	Equipment	Total	Total Incl O&P
0010	**TAXES** R012909-80									
0020	Sales tax, State, average				%	5.08%				
0050	Maximum R012909-85					7.50%				
0200	Social Security, on first $118,500 of wages						7.65%			
0300	Unemployment, combined Federal and State, minimum R012909-86						.60%			
0350	Average						9.60%			
0400	Maximum						12%			

01 31 Project Management and Coordination

01 31 13 – Project Coordination

01 31 13.20 Field Personnel		Crew	Daily Output	Labor-Hours	Unit	Material	Labor	Equipment	Total	Total Incl O&P
0010	**FIELD PERSONNEL**									
0020	Clerk, average				Week		495		495	750
0100	Field engineer, junior engineer						1,241		1,241	1,877
0120	Engineer						1,825		1,825	2,775
0140	Senior engineer						2,400		2,400	3,625
0160	General purpose laborer, average	1 Clab	.20	40			1,675		1,675	2,525
0180	Project manager, minimum						2,175		2,175	3,300
0200	Average						2,500		2,500	3,800
0220	Maximum						2,850		2,850	4,325
0240	Superintendent, minimum						2,125		2,125	3,225
0260	Average						2,325		2,325	3,525
0280	Maximum						2,650		2,650	4,025
0290	Timekeeper, average						1,350		1,350	2,050

01 31 13.30 Insurance		Crew	Daily Output	Labor-Hours	Unit	Material	Labor	Equipment	Total	Total Incl O&P
0010	**INSURANCE** R013113-40									
0020	Builders risk, standard, minimum				Job				.24%	.24%
0050	Maximum R013113-50								.64%	.64%
0200	All-risk type, minimum								.25%	.25%
0250	Maximum R013113-60								.62%	.62%
0400	Contractor's equipment floater, minimum				Value				.50%	.50%
0450	Maximum				"				1.50%	1.50%
0600	Public liability, average				Job				2.02%	2.02%
0800	Workers' compensation & employer's liability, average									
0850	by trade, carpentry, general				Payroll		11.97%			
1000	Electrical						4.91%			
1150	Insulation						10.07%			
1450	Plumbing						5.77%			

01 31 Project Management and Coordination

01 31 13 – Project Coordination

01 31 13.30 Insurance

Line	Description	Crew	Daily Output	Labor-Hours	Unit	Material	2020 Bare Costs Labor	Equipment	Total	Total Incl O&P
1550	Sheet metal work (HVAC)				Payroll		7.56%			

01 31 13.50 General Contractor's Mark-Up

Line	Description	Crew	Daily Output	Labor-Hours	Unit	Material	2020 Bare Costs Labor	Equipment	Total	Total Incl O&P
0010	**GENERAL CONTRACTOR'S MARK-UP** on Change Orders									
0200	Extra work, by subcontractors, add				%				10%	10%
0250	By General Contractor, add								15%	15%
0400	Omitted work, by subcontractors, deduct all but								5%	5%
0450	By General Contractor, deduct all but								7.50%	7.50%
0600	Overtime work, by subcontractors, add								15%	15%
0650	By General Contractor, add				↓				10%	10%

01 31 13.80 Overhead and Profit

Line	Description	Crew	Daily Output	Labor-Hours	Unit	Material	2020 Bare Costs Labor	Equipment	Total	Total Incl O&P
0010	**OVERHEAD & PROFIT** Allowance to add to items in this R013113-50									
0020	book that do not include Subs O&P, average				%				25%	
0100	Allowance to add to items in this book that R013113-55									
0110	do include Subs O&P, minimum				%				5%	5%
0150	Average								10%	10%
0200	Maximum								15%	15%
0300	Typical, by size of project, under $100,000								30%	
0350	$500,000 project								25%	
0400	$2,000,000 project								20%	
0450	Over $10,000,000 project				↓				15%	

01 31 13.90 Performance Bond

Line	Description	Crew	Daily Output	Labor-Hours	Unit	Material	2020 Bare Costs Labor	Equipment	Total	Total Incl O&P
0010	**PERFORMANCE BOND** R013113-80									
0020	For buildings, minimum				Job				.60%	.60%
0100	Maximum				"				2.50%	2.50%

01 31 14 – Facilities Services Coordination

01 31 14.20 Lock Out/Tag Out

Line	Description	Crew	Daily Output	Labor-Hours	Unit	Material	2020 Bare Costs Labor	Equipment	Total	Total Incl O&P
0010	**LOCK OUT / TAG OUT**									
0020	Miniature circuit breaker lock out device	1 Elec	220	.036	Ea.	24	2.23		26.23	30
0030	Miniature pin circuit breaker lock out device		220	.036		19.90	2.23		22.13	25.50
0040	Single circuit breaker lock out device		220	.036		21	2.23		23.23	26.50
0050	Multi-pole circuit breaker lock out device (15 to 225 Amp)		210	.038		19.90	2.34		22.24	25.50
0060	Large 3 pole circuit breaker lock out device (over 225 Amp)		210	.038		21	2.34		23.34	27
0080	Square D I-Line circuit breaker lock out device		210	.038		33	2.34		35.34	39.50
0090	Lock out disconnect switch, 30 to 100 Amp		330	.024		11	1.49		12.49	14.30
0100	100 to 400 Amp		330	.024		11	1.49		12.49	14.30
0110	Over 400 Amp		330	.024		11	1.49		12.49	14.30
0120	Lock out hasp for multiple lockout tags		200	.040		5.60	2.45		8.05	9.80
0130	Electrical cord plug lock out device		220	.036		10.50	2.23		12.73	14.85
0140	Electrical plug prong lock out device (3-wire grounding plug)		220	.036		6.95	2.23		9.18	10.95
0150	Wall switch lock out	↓	200	.040		25	2.45		27.45	31.50
0160	Fire alarm pull station lock out	1 Stpi	200	.040		18.75	2.62		21.37	24.50
0170	Sprinkler valve tamper and flow switch lock out device	1 Skwk	220	.036		16.60	1.99		18.59	21.50
0180	Lock out sign		330	.024		18.90	1.33		20.23	23
0190	Lock out tag	↓	440	.018	↓	5	1		6	7

01 32 Construction Progress Documentation

01 32 33 – Photographic Documentation

01 32 33.50 Photographs		Crew	Daily Output	Labor-Hours	Unit	2020 Bare Costs Material	Labor	Equipment	Total	Total Incl O&P
0010	**PHOTOGRAPHS**									
0020	8″ x 10″, 4 shots, 2 prints ea., std. mounting				Set	545			545	600
0100	Hinged linen mounts					550			550	605
0200	8″ x 10″, 4 shots, 2 prints each, in color					535			535	590
0300	For I.D. slugs, add to all above					5.10			5.10	5.60
1500	Time lapse equipment, camera and projector, buy				Ea.	2,800			2,800	3,100
1550	Rent per month				″	1,325			1,325	1,450
1700	Cameraman and processing, black & white				Day	1,300			1,300	1,425
1720	Color				″	1,500			1,500	1,650

01 41 Regulatory Requirements

01 41 26 – Permit Requirements

01 41 26.50 Permits		Crew	Daily Output	Labor-Hours	Unit	Material	Labor	Equipment	Total	Total Incl O&P
0010	**PERMITS**									
0020	Rule of thumb, most cities, minimum				Job				.50%	.50%
0100	Maximum				″				2%	2%

01 51 Temporary Utilities

01 51 13 – Temporary Electricity

01 51 13.80 Temporary Utilities		Crew	Daily Output	Labor-Hours	Unit	Material	Labor	Equipment	Total	Total Incl O&P
0010	**TEMPORARY UTILITIES**									
0350	Lighting, lamps, wiring, outlets, 40,000 S.F. building, 8 strings	1 Elec	34	.235	CSF Flr	5.85	14.45		20.30	28
0360	16 strings	″	17	.471		11.75	29		40.75	56
0400	Power for temp lighting only, 6.6 KWH, per month								.92	1.01
0430	11.8 KWH, per month								1.65	1.82
0450	23.6 KWH, per month								3.30	3.63
0600	Power for job duration incl. elevator, etc., minimum								53	58
0650	Maximum								110	121
0675	Temporary cooling				Ea.	1,025			1,025	1,125

01 52 Construction Facilities

01 52 13 – Field Offices and Sheds

01 52 13.20 Office and Storage Space		Crew	Daily Output	Labor-Hours	Unit	Material	Labor	Equipment	Total	Total Incl O&P
0010	**OFFICE AND STORAGE SPACE**									
0020	Office trailer, furnished, no hookups, 20′ x 8′, buy	2 Skwk	1	16	Ea.	9,525	880		10,405	11,800
0250	Rent per month					195			195	214
0300	32′ x 8′, buy	2 Skwk	.70	22.857		15,300	1,250		16,550	18,700
0350	Rent per month					245			245	269
0400	50′ x 10′, buy	2 Skwk	.60	26.667		30,300	1,475		31,775	35,500
0450	Rent per month					355			355	390
0500	50′ x 12′, buy	2 Skwk	.50	32		25,800	1,750		27,550	31,100
0550	Rent per month					460			460	505
0700	For air conditioning, rent per month, add					51.50			51.50	56.50
0800	For delivery, add per mile				Mile	12.20			12.20	13.40
0900	Bunk house trailer, 8′ x 40′ duplex dorm with kitchen, no hookups, buy	2 Carp	1	16	Ea.	89,000	850		89,850	99,000
0910	9 man with kitchen and bath, no hookups, buy		1	16		91,000	850		91,850	101,500
0920	18 man sleeper with bath, no hookups, buy		1	16		98,000	850		98,850	109,500

01 52 Construction Facilities

01 52 13 – Field Offices and Sheds

01 52 13.20 Office and Storage Space

	01 52 13.20 Office and Storage Space	Crew	Daily Output	Labor-Hours	Unit	2020 Bare Costs Material	Labor	Equipment	Total	Total Incl O&P
1000	Portable buildings, prefab, on skids, economy, 8′ x 8′	2 Carp	265	.060	S.F.	26.50	3.21		29.71	34
1100	Deluxe, 8′ x 12′		150	.107	″	29.50	5.65		35.15	40.50
1200	Storage boxes, 20′ x 8′, buy	2 Skwk	1.80	8.889	Ea.	3,150	490		3,640	4,200
1250	Rent per month					86			86	94.50
1300	40′ x 8′, buy	2 Skwk	1.40	11.429		3,800	625		4,425	5,125
1350	Rent per month					126			126	139

01 54 Construction Aids

01 54 16 – Temporary Hoists

01 54 16.50 Weekly Forklift Crew

	01 54 16.50 Weekly Forklift Crew	Crew	Daily Output	Labor-Hours	Unit	Material	Labor	Equipment	Total	Total Incl O&P
0010	**WEEKLY FORKLIFT CREW**									
0100	All-terrain forklift, 45′ lift, 35′ reach, 9000 lb. capacity	A-3P	.20	40	Week		2,125	1,875	4,000	5,225

01 54 19 – Temporary Cranes

01 54 19.50 Daily Crane Crews

	01 54 19.50 Daily Crane Crews	Crew	Daily Output	Labor-Hours	Unit	Material	Labor	Equipment	Total	Total Incl O&P
0010	**DAILY CRANE CREWS** for small jobs, portal to portal R015433-15									
0100	12-ton truck-mounted hydraulic crane	A-3H	1	8	Day		475	725	1,200	1,500
0200	25-ton	A-3I	1	8			475	800	1,275	1,600
0300	40-ton	A-3J	1	8			475	1,275	1,750	2,100
0400	55-ton	A-3K	1	16			880	1,500	2,380	2,950
0500	80-ton	A-3L	1	16			880	2,250	3,130	3,775
0900	If crane is needed on a Saturday, Sunday or Holiday									
0910	At time-and-a-half, add				Day		50%			
0920	At double time, add				″		100%			

01 54 19.60 Monthly Tower Crane Crew

	01 54 19.60 Monthly Tower Crane Crew	Crew	Daily Output	Labor-Hours	Unit	Material	Labor	Equipment	Total	Total Incl O&P
0010	**MONTHLY TOWER CRANE CREW**, excludes concrete footing									
0100	Static tower crane, 130′ high, 106′ jib, 6200 lb. capacity	A-3N	.05	176	Month		10,400	37,300	47,700	56,500

01 54 23 – Temporary Scaffolding and Platforms

01 54 23.70 Scaffolding

	01 54 23.70 Scaffolding	Crew	Daily Output	Labor-Hours	Unit	Material	Labor	Equipment	Total	Total Incl O&P
0010	**SCAFFOLDING** R015423-10									
0015	Steel tube, regular, no plank, labor only to erect & dismantle									
0090	Building exterior, wall face, 1 to 5 stories, 6′-4″ x 5′ frames	3 Carp	8	3	C.S.F.		159		159	240
0200	6 to 12 stories	4 Carp	8	4			213		213	320
0301	13 to 20 stories	5 Clab	8	5			211		211	315
0460	Building interior, wall face area, up to 16′ high	3 Carp	12	2			106		106	160
0560	16′ to 40′ high		10	2.400			128		128	192
0800	Building interior floor area, up to 30′ high		150	.160	C.C.F.		8.50		8.50	12.80
0900	Over 30′ high	4 Carp	160	.200	″		10.65		10.65	15.95
0906	Complete system for face of walls, no plank, material only rent/mo				C.S.F.	34			34	37.50
0908	Interior spaces, no plank, material only rent/mo				C.C.F.	3.88			3.88	4.27
0910	Steel tubular, heavy duty shoring, buy									
0920	Frames 5′ high 2′ wide				Ea.	89.50			89.50	98.50
0925	5′ high 4′ wide					106			106	116
0930	6′ high 2′ wide					107			107	118
0935	6′ high 4′ wide					115			115	126
0940	Accessories									
0945	Cross braces				Ea.	17.05			17.05	18.80
0950	U-head, 8″ x 8″					20.50			20.50	22.50
0955	J-head, 4″ x 8″					14.90			14.90	16.40
0960	Base plate, 8″ x 8″					16.05			16.05	17.65
0965	Leveling jack					35			35	38

01 54 Construction Aids

01 54 23 – Temporary Scaffolding and Platforms

01 54 23.70 Scaffolding		Crew	Daily Output	Labor-Hours	Unit	2020 Bare Costs Material	Labor	Equipment	Total	Total Incl O&P
1000	Steel tubular, regular, buy									
1100	Frames 3′ high 5′ wide				Ea.	92.50			92.50	102
1150	5′ high 5′ wide					108			108	119
1200	6′-4″ high 5′ wide					90.50			90.50	99.50
1350	7′-6″ high 6′ wide					151			151	166
1500	Accessories, cross braces					18.85			18.85	20.50
1550	Guardrail post					23.50			23.50	25.50
1600	Guardrail 7′ section					8.20			8.20	9.05
1650	Screw jacks & plates					26			26	28.50
1700	Sidearm brackets					23.50			23.50	26
1750	8″ casters					37.50			37.50	41
1800	Plank 2″ x 10″ x 16′-0″					64.50			64.50	71
1900	Stairway section					292			292	320
1910	Stairway starter bar					32.50			32.50	35.50
1920	Stairway inside handrail					54.50			54.50	60
1930	Stairway outside handrail					87.50			87.50	96
1940	Walk-thru frame guardrail				↓	42			42	46.50
2000	Steel tubular, regular, rent/mo.									
2100	Frames 3′ high 5′ wide				Ea.	4.50			4.50	4.95
2150	5′ high 5′ wide					4.50			4.50	4.95
2200	6′-4″ high 5′ wide					5.70			5.70	6.30
2250	7′-6″ high 6′ wide					9.90			9.90	10.90
2500	Accessories, cross braces					.90			.90	.99
2550	Guardrail post					.90			.90	.99
2600	Guardrail 7′ section					.90			.90	.99
2650	Screw jacks & plates					1.80			1.80	1.98
2700	Sidearm brackets					1.80			1.80	1.98
2750	8″ casters					7.20			7.20	7.90
2800	Outrigger for rolling tower					2.70			2.70	2.97
2850	Plank 2″ x 10″ x 16′-0″					9.95			9.95	10.90
2900	Stairway section					31.50			31.50	35
2940	Walk-thru frame guardrail				↓	2.25			2.25	2.48
3000	Steel tubular, heavy duty shoring, rent/mo.									
3250	5′ high 2′ & 4′ wide				Ea.	8.45			8.45	9.30
3300	6′ high 2′ & 4′ wide					8.45			8.45	9.30
3500	Accessories, cross braces					.90			.90	.99
3600	U-head, 8″ x 8″					2.50			2.50	2.75
3650	J-head, 4″ x 8″					2.50			2.50	2.75
3700	Base plate, 8″ x 8″					.90			.90	.99
3750	Leveling jack					2.47			2.47	2.72
5700	Planks, 2″ x 10″ x 16′-0″, labor only to erect & remove to 50′ H	3 Carp	72	.333			17.70		17.70	26.50
5800	Over 50′ high	4 Carp	80	.400	↓		21.50		21.50	32
6000	Heavy duty shoring for elevated slab forms to 8′-2″ high, floor area									
6100	Labor only to erect & dismantle	4 Carp	16	2	C.S.F.		106		106	160
6110	Materials only, rent/mo.				″	43.50			43.50	48
6500	To 14′-8″ high									
6600	Labor only to erect & dismantle	4 Carp	10	3.200	C.S.F.		170		170	256
6610	Materials only, rent/mo				″	63.50			63.50	70

01 54 Construction Aids

01 54 23 – Temporary Scaffolding and Platforms

01 54 23.75 Scaffolding Specialties		Crew	Daily Output	Labor-Hours	Unit	2020 Bare Costs Material	Labor	Equipment	Total	Total Incl O&P
0010	**SCAFFOLDING SPECIALTIES**									
1200	Sidewalk bridge, heavy duty steel posts & beams, including									
1210	parapet protection & waterproofing (material cost is rent/month)									
1220	8′ to 10′ wide, 2 posts	3 Carp	15	1.600	L.F.	46	85		131	179
1230	3 posts	"	10	2.400	"	71	128		199	270
1500	Sidewalk bridge using tubular steel scaffold frames including									
1510	planking (material cost is rent/month)	3 Carp	45	.533	L.F.	8.40	28.50		36.90	52
1600	For 2 uses per month, deduct from all above					50%				
1700	For 1 use every 2 months, add to all above					100%				
1900	Catwalks, 20″ wide, no guardrails, 7′ span, buy				Ea.	153			153	168
2000	10′ span, buy					213			213	234
3720	Putlog, standard, 8′ span, with hangers, buy					80			80	88
3730	Rent per month					16.30			16.30	17.95
3750	12′ span, buy					101			101	111
3755	Rent per month					20.50			20.50	22.50
3760	Trussed type, 16′ span, buy					262			262	288
3770	Rent per month					24.50			24.50	27
3790	22′ span, buy					284			284	315
3795	Rent per month					32.50			32.50	35.50
3800	Rolling ladders with handrails, 30″ wide, buy, 2 step					320			320	355
4000	7 step					1,125			1,125	1,250
4050	10 step					1,625			1,625	1,775
4100	Rolling towers, buy, 5′ wide, 7′ long, 10′ high					1,375			1,375	1,500
4200	For additional 5′ high sections, to buy					255			255	280
4300	Complete incl. wheels, railings, outriggers,									
4350	21′ high, to buy				Ea.	2,325			2,325	2,550
4400	Rent/month = 5% of purchase cost				"	116			116	128

01 54 36 – Equipment Mobilization

01 54 36.50 Mobilization

		Crew	Daily Output	Labor-Hours	Unit	Material	Labor	Equipment	Total	Total Incl O&P
0010	**MOBILIZATION** (Use line item again for demobilization) R015436-50									
0015	Up to 25 mi. haul dist. (50 mi. RT for mob/demob crew)									
1200	Small equipment, placed in rear of, or towed by pickup truck	A-3A	4	2	Ea.		106	44	150	207
1300	Equipment hauled on 3-ton capacity towed trailer	A-3Q	2.67	3			159	92.50	251.50	340
1400	20-ton capacity	B-34U	2	8			410	221	631	855
1500	40-ton capacity	B-34N	2	8			425	340	765	1,000
1600	50-ton capacity	B-34V	1	24			1,300	985	2,285	3,000
1700	Crane, truck-mounted, up to 75 ton (driver only)	1 Eqhv	4	2			118		118	177
1800	Over 75 ton (with chase vehicle)	A-3E	2.50	6.400			345	70.50	415.50	600
2400	Crane, large lattice boom, requiring assembly	B-34W	.50	144			7,425	7,025	14,450	18,800
2500	For each additional 5 miles haul distance, add						10%	10%		
3000	For large pieces of equipment, allow for assembly/knockdown									
3100	For mob/demob of micro-tunneling equip, see Section 33 05 07.36									

01 55 Vehicular Access and Parking

01 55 23 – Temporary Roads

01 55 23.50 Roads and Sidewalks		Crew	Daily Output	Labor-Hours	Unit	2020 Bare Costs Material	Labor	Equipment	Total	Total Incl O&P
0010	**ROADS AND SIDEWALKS** Temporary									
0050	Roads, gravel fill, no surfacing, 4" gravel depth	B-14	715	.067	S.Y.	3.38	2.97	.30	6.65	8.50
0100	8" gravel depth	"	615	.078	"	6.75	3.45	.35	10.55	13.05
1000	Ramp, 3/4" plywood on 2" x 6" joists, 16" OC	2 Carp	300	.053	S.F.	1.62	2.83		4.45	6.05
1100	On 2" x 10" joists, 16" OC	"	275	.058	"	2.25	3.09		5.34	7.10

01 56 Temporary Barriers and Enclosures

01 56 13 – Temporary Air Barriers

01 56 13.60 Tarpaulins

		Crew	Daily Output	Labor-Hours	Unit	Material	Labor	Equipment	Total	Total Incl O&P
0010	**TARPAULINS**									
0020	Cotton duck, 10-13.13 oz./S.Y., 6' x 8'				S.F.	.75			.75	.83
0050	30' x 30'					.63			.63	.69
0100	Polyvinyl coated nylon, 14-18 oz., minimum					1.49			1.49	1.64
0150	Maximum					1.49			1.49	1.64
0200	Reinforced polyethylene 3 mils thick, white					.05			.05	.06
0300	4 mils thick, white, clear or black					.16			.16	.18
0400	5.5 mils thick, clear					.19			.19	.21
0500	White, fire retardant					.60			.60	.66
0600	12 mils, oil resistant, fire retardant					.50			.50	.55
0700	8.5 mils, black					.26			.26	.29
0710	Woven polyethylene, 6 mils thick					.08			.08	.09
0730	Polyester reinforced w/integral fastening system, 11 mils thick					.20			.20	.22
0740	Polyethylene, reflective, 23 mils thick				↓	1.34			1.34	1.47

01 56 13.90 Winter Protection

		Crew	Daily Output	Labor-Hours	Unit	Material	Labor	Equipment	Total	Total Incl O&P
0010	**WINTER PROTECTION**									
0100	Framing to close openings	2 Clab	500	.032	S.F.	.51	1.35		1.86	2.58
0200	Tarpaulins hung over scaffolding, 8 uses, not incl. scaffolding		1500	.011		.25	.45		.70	.95
0250	Tarpaulin polyester reinf. w/integral fastening system, 11 mils thick		1600	.010		.22	.42		.64	.87
0300	Prefab fiberglass panels, steel frame, 8 uses	↓	1200	.013	↓	2.78	.56		3.34	3.90

01 56 16 – Temporary Dust Barriers

01 56 16.10 Dust Barriers, Temporary

		Crew	Daily Output	Labor-Hours	Unit	Material	Labor	Equipment	Total	Total Incl O&P
0010	**DUST BARRIERS, TEMPORARY**, erect and dismantle									
0020	Spring loaded telescoping pole & head, to 12', erect and dismantle	1 Clab	240	.033	Ea.		1.40		1.40	2.11
0025	Cost per day (based upon 250 days)				Day	.22			.22	.24
0030	To 21', erect and dismantle	1 Clab	240	.033	Ea.		1.40		1.40	2.11
0035	Cost per day (based upon 250 days)				Day	.72			.72	.79
0040	Accessories, caution tape reel, erect and dismantle	1 Clab	480	.017	Ea.		.70		.70	1.05
0045	Cost per day (based upon 250 days)				Day	.33			.33	.37
0060	Foam rail and connector, erect and dismantle	1 Clab	240	.033	Ea.		1.40		1.40	2.11
0065	Cost per day (based upon 250 days)				Day	.10			.10	.11
0070	Caution tape	1 Clab	384	.021	C.L.F.	2.49	.88		3.37	4.06
0080	Zipper, standard duty		60	.133	Ea.	10.55	5.60		16.15	20
0090	Heavy duty		48	.167	"	10.65	7		17.65	22.50
0100	Polyethylene sheet, 4 mil		37	.216	Sq.	2.62	9.10		11.72	16.60
0110	6 mil	↓	37	.216	"	4	9.10		13.10	18.10
1000	Dust partition, 6 mil polyethylene, 1" x 3" frame	2 Carp	2000	.008	S.F.	.33	.43		.76	1
1080	2" x 4" frame	"	2000	.008	"	.39	.43		.82	1.07
1085	Negative air machine, 1800 CFM				Ea.	890			890	975
1090	Adhesive strip application, 2" width	1 Clab	192	.042	C.L.F.	6.75	1.75		8.50	10.10
4000	Dust & infectious control partition, adj. to 10' high, obscured, 4' panel	2 Carp	90	.178	Ea.	580	9.45		589.45	655

01 56 Temporary Barriers and Enclosures

01 56 16 – Temporary Dust Barriers

01 56 16.10 Dust Barriers, Temporary		Crew	Daily Output	Labor-Hours	Unit	2020 Bare Costs Material	Labor	Equipment	Total	Total Incl O&P
4010	3′ panel	2 Carp	90	.178	Ea.	550	9.45		559.45	620
4020	2′ panel		90	.178		430	9.45		439.45	490
4030	1′ panel	1 Carp	90	.089		285	4.72		289.72	320
4040	6″ panel	″	90	.089		265	4.72		269.72	299
4050	2′ panel with HEPA filtered discharge port	2 Carp	90	.178		500	9.45		509.45	565
4060	3′ panel with 32″ door		90	.178		895	9.45		904.45	1,000
4070	4′ panel with 36″ door		90	.178		995	9.45		1,004.45	1,125
4080	4′-6″ panel with 44″ door		90	.178		1,200	9.45		1,209.45	1,350
4090	Hinged corner		80	.200		185	10.65		195.65	220
4100	Outside corner		80	.200		150	10.65		160.65	181
4110	T post		80	.200		150	10.65		160.65	181
4120	Accessories, ceiling grid clip		360	.044		7.45	2.36		9.81	11.75
4130	Panel locking clip		360	.044		5.15	2.36		7.51	9.20
4140	Panel joint closure strip		360	.044		8	2.36		10.36	12.35
4150	Screw jack		360	.044		6.65	2.36		9.01	10.85
4160	Digital pressure difference guage					275			275	305
4180	Combination lockset	1 Carp	13	.615		200	32.50		232.50	269
4185	Sealant tape, 2″ wide	1 Clab	192	.042	C.L.F.	6.75	1.75		8.50	10.10
4190	System in place, including door and accessories									
4200	Based upon 25 uses	2 Carp	51	.314	L.F.	9.85	16.65		26.50	36
4210	Based upon 50 uses		51	.314		4.94	16.65		21.59	30.50
4230	Based upon 100 uses		51	.314		2.47	16.65		19.12	27.50

01 56 23 – Temporary Barricades

01 56 23.10 Barricades		Crew	Daily Output	Labor-Hours	Unit	Material	Labor	Equipment	Total	Total Incl O&P
0010	**BARRICADES**									
0020	5′ high, 3 rail @ 2″ x 8″, fixed	2 Carp	20	.800	L.F.	6.75	42.50		49.25	71.50
0150	Movable		30	.533		5.60	28.50		34.10	48.50
1000	Guardrail, wooden, 3′ high, 1″ x 6″ on 2″ x 4″ posts		200	.080		1.44	4.25		5.69	8
1100	2″ x 6″ on 4″ x 4″ posts		165	.097		2.80	5.15		7.95	10.85
1200	Portable metal with base pads, buy					14.45			14.45	15.90
1250	Typical installation, assume 10 reuses	2 Carp	600	.027		2.31	1.42		3.73	4.67
1300	Barricade tape, polyethylene, 7 mil, 3″ wide x 300′ long roll	1 Clab	128	.063	Ea.	7.50	2.63		10.13	12.20
3000	Detour signs, set up and remove									
3010	Reflective aluminum, MUTCD, 24″ x 24″, post mounted	1 Clab	20	.400	Ea.	2.80	16.85		19.65	28.50
4000	Roof edge portable barrier stands and warning flags, 50 uses	1 Rohe	9100	.001	L.F.	.07	.03		.10	.12
4010	100 uses	″	9100	.001	″	.03	.03		.06	.09

01 56 26 – Temporary Fencing

01 56 26.50 Temporary Fencing		Crew	Daily Output	Labor-Hours	Unit	Material	Labor	Equipment	Total	Total Incl O&P
0010	**TEMPORARY FENCING**									
0020	Chain link, 11 ga., 4′ high	2 Clab	400	.040	L.F.	2	1.68		3.68	4.73
0100	6′ high		300	.053		4.96	2.25		7.21	8.80
0200	Rented chain link, 6′ high, to 1000′ (up to 12 mo.)		400	.040		3.37	1.68		5.05	6.25
0250	Over 1000′ (up to 12 mo.)		300	.053		4.14	2.25		6.39	7.90
0350	Plywood, painted, 2″ x 4″ frame, 4′ high	A-4	135	.178		6.90	8.95		15.85	21
0400	4″ x 4″ frame, 8′ high	″	110	.218		13.45	10.95		24.40	31
0500	Wire mesh on 4″ x 4″ posts, 4′ high	2 Carp	100	.160		11.15	8.50		19.65	25
0550	8′ high	″	80	.200		16.60	10.65		27.25	34

01 56 Temporary Barriers and Enclosures

01 56 29 – Temporary Protective Walkways

01 56 29.50 Protection		Crew	Daily Output	Labor-Hours	Unit	Material	Labor	Equipment	Total	Total Incl O&P
						2020 Bare Costs				
0010	**PROTECTION**									
0020	Stair tread, 2" x 12" planks, 1 use	1 Carp	75	.107	Tread	5.15	5.65		10.80	14.20
0100	Exterior plywood, 1/2" thick, 1 use		65	.123		2.05	6.55		8.60	12.10
0200	3/4" thick, 1 use		60	.133		2.84	7.10		9.94	13.75
2200	Sidewalks, 2" x 12" planks, 2 uses		350	.023	S.F.	.86	1.22		2.08	2.78
2300	Exterior plywood, 2 uses, 1/2" thick		750	.011		.34	.57		.91	1.23
2400	5/8" thick		650	.012		.38	.65		1.03	1.40
2500	3/4" thick		600	.013		.47	.71		1.18	1.58

01 57 Temporary Controls

01 57 33 – Temporary Security

01 57 33.50 Watchman

Line	Description	Crew	Daily Output	Labor-Hours	Unit	Material	Labor	Equipment	Total	Total Incl O&P
0010	**WATCHMAN**									
0020	Service, monthly basis, uniformed person, minimum				Hr.				27.65	30.40
0100	Maximum								56	61.62
0200	Person and command dog, minimum								31	34
0300	Maximum								60	65

01 58 Project Identification

01 58 13 – Temporary Project Signage

01 58 13.50 Signs

Line	Description	Crew	Daily Output	Labor-Hours	Unit	Material	Labor	Equipment	Total	Total Incl O&P
0010	**SIGNS**									
0020	High intensity reflectorized, no posts, buy				Ea.	27			27	29.50

01 66 Product Storage and Handling Requirements

01 66 19 – Material Handling

01 66 19.10 Material Handling

Line	Description	Crew	Daily Output	Labor-Hours	Unit	Material	Labor	Equipment	Total	Total Incl O&P
0010	**MATERIAL HANDLING**									
0020	Above 2nd story, via stairs, per C.Y. of material per floor	2 Clab	145	.110	C.Y.		4.65		4.65	7
0030	Via elevator, per C.Y. of material		240	.067			2.81		2.81	4.22
0050	Distances greater than 200', per C.Y. of material per each addl 200'		300	.053			2.25		2.25	3.37

01 74 Cleaning and Waste Management

01 74 13 – Progress Cleaning

01 74 13.20 Cleaning Up

Line	Description	Crew	Daily Output	Labor-Hours	Unit	Material	Labor	Equipment	Total	Total Incl O&P
0010	**CLEANING UP**									
0020	After job completion, allow, minimum				Job				.30%	.30%
0040	Maximum				"				1%	1%
0050	Cleanup of floor area, continuous, per day, during const.	A-5	24	.750	M.S.F.	2.47	32	2.04	36.51	53
0100	Final by GC at end of job	"	11.50	1.565	"	2.62	67	4.26	73.88	108

01 91 Commissioning

01 91 13 – General Commissioning Requirements

01 91 13.50 Building Commissioning		Crew	Daily Output	Labor-Hours	Unit	2020 Bare Costs Material	Labor	Equipment	Total	Total Incl O&P
0010	**BUILDING COMMISSIONING**									
0100	Systems operation and verification during turnover				%				.25%	.25%
0150	Including all systems subcontractors								.50%	.50%
0200	Systems design assistance, operation, verification and training								.50%	.50%
0250	Including all systems subcontractors								1%	1%

01 93 Facility Maintenance

01 93 13 – Facility Maintenance Procedures

01 93 13.15 Mechanical Facilities Maintenance

Line	Description	Crew	Daily Output	Labor-Hours	Unit	Material	Labor	Equipment	Total	Total Incl O&P
0010	**MECHANICAL FACILITIES MAINTENANCE**									
0100	Air conditioning system maintenance									
0130	Belt, replace	1 Stpi	15	.533	Ea.		35		35	52.50
0170	Fan, clean		16	.500			33		33	49
0180	Filter, remove, clean, replace		12	.667			43.50		43.50	65.50
0190	Flexible coupling alignment, inspect		40	.200			13.10		13.10	19.60
0200	Gas leak, locate and repair		4	2			131		131	196
0250	Pump packing gland, remove and replace		11	.727			47.50		47.50	71.50
0270	Tighten		32	.250			16.40		16.40	24.50
0290	Pump, disassemble and assemble		4	2			131		131	196
0300	Air pressure regulator, disassemble, clean, assemble	1 Skwk	4	2			110		110	166
0310	Repair or replace part	"	6	1.333			73		73	111
0320	Purging system	1 Stpi	16	.500			33		33	49
0400	Compressor, air, remove or install fan wheel	1 Skwk	20	.400			22		22	33
0410	Disassemble or assemble 2 cylinder, 2 stage		4	2			110		110	166
0420	4 cylinder, 4 stage		1	8			440		440	665
0430	Repair or replace part		2	4			219		219	330
0700	Demolition, for mech. demolition see Section 23 05 05.10 or 22 05 05.10									
0800	Ductwork, clean									
0810	Rectangular									
0820	6" G	1 Shee	187.50	.043	L.F.		2.66		2.66	4.03
0830	8" G		140.63	.057			3.54		3.54	5.40
0840	10" G		112.50	.071			4.43		4.43	6.70
0850	12" G		93.75	.085			5.30		5.30	8.05
0860	14" G		80.36	.100			6.20		6.20	9.40
0870	16" G		70.31	.114			7.10		7.10	10.75
0900	Round									
0910	4" G	1 Shee	358.10	.022	L.F.		1.39		1.39	2.11
0920	6" G		238.73	.034			2.09		2.09	3.17
0930	8" G		179.05	.045			2.78		2.78	4.22
0940	10" G		143.24	.056			3.48		3.48	5.30
0950	12" G		119.37	.067			4.18		4.18	6.35
0960	16" G		89.52	.089			5.55		5.55	8.45
1000	Expansion joint, not screwed, install or remove	1 Stpi	3	2.667	Ea.		175		175	262
1010	Repack	"	6	1.333	"		87.50		87.50	131
1200	Fire protection equipment									
1220	Fire hydrant, replace	Q-1	3	5.333	Ea.		310		310	465
1230	Service, lubricate, inspect, flush, clean	1 Plum	7	1.143			73.50		73.50	110
1240	Test	"	11	.727			47		47	70
1310	Inspect valves, pressure, nozzle	1 Spri	4	2	System		127		127	190
1800	Plumbing fixtures, for installation see Section 22 41 00									
1801	Plumbing fixtures									

01 93 Facility Maintenance

01 93 13 – Facility Maintenance Procedures

01 93 13.15 Mechanical Facilities Maintenance		Crew	Daily Output	Labor-Hours	Unit	2020 Bare Costs Material	Labor	Equipment	Total	Total Incl O&P
1850	Open drain with toilet auger	1 Plum	16	.500	Ea.		32		32	48
1870	Plaster trap, clean	"	6	1.333			86		86	129
1881	Clean commode	1 Clab	20	.400		.16	16.85		17.01	25.50
1882	Clean commode seat		48	.167		.08	7		7.08	10.65
1884	Clean double sink		20	.400		.16	16.85		17.01	25.50
1886	Clean bathtub		12	.667		.32	28		28.32	42.50
1888	Clean fiberglass tub/shower		8	1		.40	42		42.40	64
1889	Clean faucet set		32	.250		.06	10.55		10.61	15.85
1891	Clean shower head		80	.100		.01	4.21		4.22	6.35
1893	Clean water heater		16	.500		.57	21		21.57	32
1900	Relief valve, test and adjust	1 Stpi	20	.400			26		26	39
1910	Clean pump, heater, or motor for whirlpool	1 Clab	16	.500		.08	21		21.08	31.50
1920	Clean thermal cover for whirlpool	"	16	.500		.81	21		21.81	32.50
2000	Repair or replace, steam trap	1 Stpi	8	1			65.50		65.50	98
2020	Y-type or bell strainer		6	1.333			87.50		87.50	131
2040	Water trap or vacuum breaker, screwed joints		13	.615			40.50		40.50	60.50
2100	Steam specialties, clean									
2120	Air separator with automatic trap, 1" fittings	1 Stpi	12	.667	Ea.		43.50		43.50	65.50
2130	Bucket trap, 2" pipe		7	1.143			75		75	112
2150	Drip leg, 2" fitting		45	.178			11.65		11.65	17.45
2200	Thermodynamic trap, 1" fittings		50	.160			10.50		10.50	15.70
2210	Thermostatic		65	.123			8.05		8.05	12.05
2240	Screen and seat in Y-type strainer, plug type		25	.320			21		21	31.50
2242	Screen and seat in Y-type strainer, flange type		12	.667			43.50		43.50	65.50
2500	Valve, replace broken handwheel		24	.333			22		22	32.50
3000	Valve, overhaul, regulator, relief, flushometer, mixing									
3040	Cold water, gas	1 Stpi	5	1.600	Ea.		105		105	157
3050	Hot water, steam		3	2.667			175		175	262
3080	Globe, gate, check up to 4" cold water, gas		10	.800			52.50		52.50	78.50
3090	Hot water, steam		5	1.600			105		105	157
3100	Over 4" ID hot or cold line		1.40	5.714			375		375	560
3120	Remove and replace, gate, globe or check up to 4"	Q-5	6	2.667			157		157	235
3130	Over 4"	"	2	8			470		470	705
3150	Repack up to 4"	1 Stpi	13	.615			40.50		40.50	60.50
3160	Over 4"	"	4	2			131		131	196

Division 2 Existing Conditions

Estimating Tips

02 30 00 Subsurface Investigation

In preparing estimates on structures involving earthwork or foundations, all information concerning soil characteristics should be obtained. Look particularly for hazardous waste, evidence of prior dumping of debris, and previous stream beds.

02 40 00 Demolition and Structure Moving

The costs shown for selective demolition do not include rubbish handling or disposal. These items should be estimated separately using RSMeans data or other sources.

- Historic preservation often requires that the contractor remove materials from the existing structure, rehab them, and replace them. The estimator must be aware of any related measures and precautions that must be taken when doing selective demolition and cutting and patching. Requirements may include special handling and storage, as well as security.
- In addition to Subdivision 02 41 00, you can find selective demolition items in each division. Example: Roofing demolition is in Division 7.
- Absent of any other specific reference, an approximate demolish-in-place cost can be obtained by halving the new-install labor cost. To remove for reuse, allow the entire new-install labor figure.

02 40 00 Building Deconstruction

This section provides costs for the careful dismantling and recycling of most low-rise building materials.

02 50 00 Containment of Hazardous Waste

This section addresses on-site hazardous waste disposal costs.

02 80 00 Hazardous Material Disposal/Remediation

This subdivision includes information on hazardous waste handling, asbestos remediation, lead remediation, and mold remediation. See reference numbers R028213-20 and R028319-60 for further guidance in using these unit price lines.

02 90 00 Monitoring Chemical Sampling, Testing Analysis

This section provides costs for on-site sampling and testing hazardous waste.

Reference Numbers

Reference numbers are shown at the beginning of some major classifications. These numbers refer to related items in the Reference Section. The reference information may be an estimating procedure, an alternate pricing method, or technical information.

Note: Not all subdivisions listed here necessarily appear. ■

02 21 Surveys

02 21 13 – Site Surveys

02 21 13.09 Topographical Surveys

	02 21 13.09 Topographical Surveys	Crew	Daily Output	Labor-Hours	Unit	Material	2020 Bare Costs Labor	Equipment	Total	Total Incl O&P
0010	**TOPOGRAPHICAL SURVEYS**									
0020	Topographical surveying, conventional, minimum	A-7	3.30	7.273	Acre	23.50	415	9.30	447.80	660
0100	Maximum	A-8	.60	53.333	″	63	2,975	51	3,089	4,600

02 21 13.13 Boundary and Survey Markers

	02 21 13.13 Boundary and Survey Markers	Crew	Daily Output	Labor-Hours	Unit	Material	Labor	Equipment	Total	Total Incl O&P
0010	**BOUNDARY AND SURVEY MARKERS**									
0300	Lot location and lines, large quantities, minimum	A-7	2	12	Acre	36	690	15.35	741.35	1,075
0320	Average	″	1.25	19.200		63.50	1,100	24.50	1,188	1,750
0400	Small quantities, maximum	A-8	1	32		76.50	1,775	30.50	1,882	2,800
0600	Monuments, 3′ long	A-7	10	2.400	Ea.	35.50	138	3.07	176.57	249
0800	Property lines, perimeter, cleared land	″	1000	.024	L.F.	.08	1.38	.03	1.49	2.19
0900	Wooded land	A-8	875	.037	″	.10	2.04	.04	2.18	3.21

02 21 13.16 Aerial Surveys

	02 21 13.16 Aerial Surveys	Crew	Daily Output	Labor-Hours	Unit	Material	Labor	Equipment	Total	Total Incl O&P
0010	**AERIAL SURVEYS**									
1500	Aerial surveying, including ground control, minimum fee, 10 acres				Total				4,700	4,700
1510	100 acres								9,400	9,400
1550	From existing photography, deduct								1,625	1,625
1600	2′ contours, 10 acres				Acre				470	470
1850	100 acres								94	94
2000	1000 acres								90	90
2050	10,000 acres								85	85

02 41 Demolition

02 41 13 – Selective Site Demolition

02 41 13.17 Demolish, Remove Pavement and Curb

	02 41 13.17 Demolish, Remove Pavement and Curb	Crew	Daily Output	Labor-Hours	Unit	Material	Labor	Equipment	Total	Total Incl O&P
0010	**DEMOLISH, REMOVE PAVEMENT AND CURB** R024119-10									
5010	Pavement removal, bituminous roads, up to 3″ thick	B-38	690	.058	S.Y.		2.76	1.80	4.56	6.10
5050	4″-6″ thick	″	420	.095	″		4.53	2.96	7.49	10.05

02 41 13.23 Utility Line Removal

	02 41 13.23 Utility Line Removal	Crew	Daily Output	Labor-Hours	Unit	Material	Labor	Equipment	Total	Total Incl O&P
0010	**UTILITY LINE REMOVAL**									
0015	No hauling, abandon catch basin or manhole	B-6	7	3.429	Ea.		157	30.50	187.50	269
0020	Remove existing catch basin or manhole, masonry		4	6			274	53.50	327.50	470
0030	Catch basin or manhole frames and covers, stored		13	1.846			84.50	16.45	100.95	145
0040	Remove and reset		7	3.429			157	30.50	187.50	269
0900	Hydrants, fire, remove only	B-21A	5	8			420	85.50	505.50	720
0950	Remove and reset	″	2	20			1,050	214	1,264	1,800
2900	Pipe removal, sewer/water, no excavation, 12″ diameter	B-6	175	.137	L.F.		6.25	1.22	7.47	10.75
2930	15″-18″ diameter	B-12Z	150	.160			7.65	9.75	17.40	22
2960	21″-24″ diameter		120	.200			9.55	12.20	21.75	28
3000	27″-36″ diameter		90	.267			12.75	16.30	29.05	37
3200	Steel, welded connections, 4″ diameter	B-6	160	.150			6.85	1.34	8.19	11.75
3300	10″ diameter	″	80	.300			13.70	2.67	16.37	23.50

02 41 13.30 Minor Site Demolition

	02 41 13.30 Minor Site Demolition	Crew	Daily Output	Labor-Hours	Unit	Material	Labor	Equipment	Total	Total Incl O&P
0010	**MINOR SITE DEMOLITION** R024119-10									
4000	Sidewalk removal, bituminous, 2″ thick	B-6	350	.069	S.Y.		3.14	.61	3.75	5.35
4010	2-1/2″ thick		325	.074			3.38	.66	4.04	5.75
4100	Concrete, plain, 4″		160	.150			6.85	1.34	8.19	11.75
4110	Plain, 5″		140	.171			7.85	1.53	9.38	13.45
4120	Plain, 6″		120	.200			9.15	1.78	10.93	15.65

02 41 Demolition

02 41 19 – Selective Demolition

02 41 19.19 Selective Demolition		Crew	Daily Output	Labor-Hours	Unit	Material	Labor	Equipment	Total	Total Incl O&P
						2020 Bare Costs				
0010	**SELECTIVE DEMOLITION**, Rubbish Handling R024119-10									
0020	The following are to be added to the demolition prices									
0600	Dumpster, weekly rental, 1 dump/week, 6 C.Y. capacity (2 tons)				Week	415			415	455
0700	10 C.Y. capacity (3 tons)					480			480	530
0725	20 C.Y. capacity (5 tons) R024119-20					565			565	625
0800	30 C.Y. capacity (7 tons)					730			730	800
0840	40 C.Y. capacity (10 tons)					775			775	850
2000	Load, haul, dump and return, 0'-50' haul, hand carried	2 Clab	24	.667	C.Y.		28		28	42
2005	Wheeled		37	.432			18.20		18.20	27.50
2040	0'-100' haul, hand carried		16.50	.970			41		41	61.50
2045	Wheeled		25	.640			27		27	40.50
2050	Forklift	A-3R	25	.320			16.95	10.25	27.20	37
2080	Haul and return, add per each extra 100' haul, hand carried	2 Clab	35.50	.451			18.95		18.95	28.50
2085	Wheeled		54	.296			12.45		12.45	18.75
2120	For travel in elevators, up to 10 floors, add		140	.114			4.81		4.81	7.25
2130	0'-50' haul, incl. up to 5 riser stairs, hand carried		23	.696			29.50		29.50	44
2135	Wheeled		35	.457			19.25		19.25	29
2140	6-10 riser stairs, hand carried		22	.727			30.50		30.50	46
2145	Wheeled		34	.471			19.80		19.80	30
2150	11-20 riser stairs, hand carried		20	.800			33.50		33.50	50.50
2155	Wheeled		31	.516			21.50		21.50	32.50
2160	21-40 riser stairs, hand carried		16	1			42		42	63.50
2165	Wheeled		24	.667			28		28	42
2170	0'-100' haul, incl. 5 riser stairs, hand carried		15	1.067			45		45	67.50
2175	Wheeled		23	.696			29.50		29.50	44
2180	6-10 riser stairs, hand carried		14	1.143			48		48	72.50
2185	Wheeled		21	.762			32		32	48
2190	11-20 riser stairs, hand carried		12	1.333			56		56	84.50
2195	Wheeled		18	.889			37.50		37.50	56
2200	21-40 riser stairs, hand carried		8	2			84		84	127
2205	Wheeled		12	1.333			56		56	84.50
2210	Haul and return, add per each extra 100' haul, hand carried		35.50	.451			18.95		18.95	28.50
2215	Wheeled		54	.296			12.45		12.45	18.75
2220	For each additional flight of stairs, up to 5 risers, add		550	.029	Flight		1.22		1.22	1.84
2225	6-10 risers, add		275	.058			2.45		2.45	3.68
2230	11-20 risers, add		138	.116			4.88		4.88	7.35
2235	21-40 risers, add		69	.232			9.75		9.75	14.65
3000	Loading & trucking, including 2 mile haul, chute loaded	B-16	45	.711	C.Y.		31.50	12.75	44.25	61.50
3040	Hand loading truck, 50' haul	"	48	.667			29.50	11.95	41.45	57.50
3080	Machine loading truck	B-17	120	.267			12.40	5.35	17.75	24.50
5000	Haul, per mile, up to 8 C.Y. truck	B-34B	1165	.007			.34	.49	.83	1.04
5100	Over 8 C.Y. truck	"	1550	.005			.25	.37	.62	.79

02 41 19.20 Selective Demolition, Dump Charges

Code	Description	Crew	Daily Output	Labor-Hours	Unit	Material	Labor	Equipment	Total	Total Incl O&P
0010	**SELECTIVE DEMOLITION, DUMP CHARGES** R024119-10									
0100	Building construction materials				Ton	74			74	81
0300	Rubbish only					63			63	69.50
0500	Reclamation station, usual charge					74			74	81

02 41 19.27 Selective Demolition, Torch Cutting

Code	Description	Crew	Daily Output	Labor-Hours	Unit	Material	Labor	Equipment	Total	Total Incl O&P
0010	**SELECTIVE DEMOLITION, TORCH CUTTING** R024119-10									
0020	Steel, 1" thick plate	E-25	333	.024	L.F.	.90	1.43	.04	2.37	3.27
0040	1" diameter bar	"	600	.013	Ea.	.15	.79	.02	.96	1.43
1000	Oxygen lance cutting, reinforced concrete walls									

02 41 Demolition

02 41 19 – Selective Demolition

02 41 19.27 Selective Demolition, Torch Cutting		Crew	Daily Output	Labor-Hours	Unit	Material	2020 Bare Costs Labor	Equipment	Total	Total Incl O&P
1040	12"-16" thick walls	1 Clab	10	.800	L.F.		33.50		33.50	50.50
1080	24" thick walls	"	6	1.333	"		56		56	84.50

02 65 Underground Storage Tank Removal

02 65 10 – Underground Tank and Contaminated Soil Removal

02 65 10.30 Removal of Underground Storage Tanks

		Crew	Daily Output	Labor-Hours	Unit	Material	Labor	Equipment	Total	Total Incl O&P
0010	REMOVAL OF UNDERGROUND STORAGE TANKS R026510-20									
0011	Petroleum storage tanks, non-leaking									
0100	Excavate & load onto trailer									
0110	3,000 gal. to 5,000 gal. tank G	B-14	4	12	Ea.		530	53.50	583.50	855
0120	6,000 gal. to 8,000 gal. tank G	B-3A	3	13.333			600	229	829	1,150
0130	9,000 gal. to 12,000 gal. tank G	"	2	20			900	345	1,245	1,725
0190	Known leaking tank, add				%				100%	100%
0200	Remove sludge, water and remaining product from bottom									
0201	of tank with vacuum truck									
0300	3,000 gal. to 5,000 gal. tank G	A-13	5	1.600	Ea.		85	148	233	289
0310	6,000 gal. to 8,000 gal. tank G		4	2			106	184	290	360
0320	9,000 gal. to 12,000 gal. tank G		3	2.667			141	246	387	480
0390	Dispose of sludge off-site, average				Gal.				6.25	6.80
0400	Insert inert solid CO_2 "dry ice" into tank									
0401	For cleaning/transporting tanks (1.5 lb./100 gal. cap) G	1 Clab	500	.016	Lb.	1.22	.67		1.89	2.35
1020	Haul tank to certified salvage dump, 100 miles round trip									
1023	3,000 gal. to 5,000 gal. tank				Ea.				760	830
1026	6,000 gal. to 8,000 gal. tank								880	960
1029	9,000 gal. to 12,000 gal. tank								1,050	1,150
1100	Disposal of contaminated soil to landfill									
1110	Minimum				C.Y.				145	160
1111	Maximum				"				400	440
1120	Disposal of contaminated soil to									
1121	bituminous concrete batch plant									
1130	Minimum				C.Y.				80	88
1131	Maximum				"				115	125
2010	Decontamination of soil on site incl poly tarp on top/bottom									
2011	Soil containment berm and chemical treatment									
2020	Minimum G	B-11C	100	.160	C.Y.	7.65	7.90	2.14	17.69	22.50
2021	Maximum G	"	100	.160		9.90	7.90	2.14	19.94	25
2050	Disposal of decontaminated soil, minimum								135	150
2055	Maximum								400	440

02 81 Transportation and Disposal of Hazardous Materials

02 81 20 – Hazardous Waste Handling

02 81 20.10 Hazardous Waste Cleanup/Pickup/Disposal

		Crew	Daily Output	Labor-Hours	Unit	Material	Labor	Equipment	Total	Total Incl O&P
0010	HAZARDOUS WASTE CLEANUP/PICKUP/DISPOSAL									
0100	For contractor rental equipment, i.e., dozer,									
0110	Front end loader, dump truck, etc., see 01 54 33 Reference Section									
1000	Solid pickup									
1100	55 gal. drums				Ea.				240	265
1120	Bulk material, minimum				Ton				190	210
1130	Maximum				"				595	655

02 81 Transportation and Disposal of Hazardous Materials

02 81 20 – Hazardous Waste Handling

02 81 20.10 Hazardous Waste Cleanup/Pickup/Disposal		Crew	Daily Output	Labor-Hours	Unit	Material	Labor	Equipment	Total	Total Incl O&P
						2020 Bare Costs				
1200	Transportation to disposal site									
1220	Truckload = 80 drums or 25 C.Y. or 18 tons									
1260	Minimum				Mile				3.95	4.45
1270	Maximum				"				7.25	7.98
3000	Liquid pickup, vacuum truck, stainless steel tank									
3100	Minimum charge, 4 hours									
3110	1 compartment, 2200 gallon				Hr.				140	155
3120	2 compartment, 5000 gallon				"				200	225
3400	Transportation in 6900 gallon bulk truck				Mile				7.95	8.75
3410	In teflon lined truck				"				10.20	11.25
5000	Heavy sludge or dry vacuumable material				Hr.				140	155
6000	Dumpsite disposal charge, minimum				Ton				140	155
6020	Maximum				"				415	455

02 82 Asbestos Remediation

02 82 13 – Asbestos Abatement

02 82 13.39 Asbestos Remediation Plans and Methods

Line	Description	Crew	Daily Output	Labor-Hours	Unit	Material	Labor	Equipment	Total	Total Incl O&P
0010	**ASBESTOS REMEDIATION PLANS AND METHODS**									
0100	Building Survey-Commercial Building				Ea.				2,200	2,400
0200	Asbestos Abatement Remediation Plan				"				1,350	1,475

02 82 13.41 Asbestos Abatement Equipment

Line	Description	Crew	Daily Output	Labor-Hours	Unit	Material	Labor	Equipment	Total	Total Incl O&P
0010	**ASBESTOS ABATEMENT EQUIPMENT** R028213-20									
0011	Equipment and supplies, buy									
0200	Air filtration device, 2000 CFM				Ea.	950			950	1,050
0250	Large volume air sampling pump, minimum					315			315	350
0260	Maximum					345			345	380
0300	Airless sprayer unit, 2 gun					2,625			2,625	2,875
0350	Light stand, 500 watt					41			41	45
0400	Personal respirators									
0410	Negative pressure, 1/2 face, dual operation, minimum				Ea.	30			30	33
0420	Maximum					34			34	37.50
0450	P.A.P.R., full face, minimum					780			780	855
0460	Maximum					1,225			1,225	1,325
0470	Supplied air, full face, including air line, minimum					289			289	320
0480	Maximum					500			500	550
0500	Personnel sampling pump					237			237	261
1500	Power panel, 20 unit, including GFI					475			475	525
1600	Shower unit, including pump and filters					1,125			1,125	1,250
1700	Supplied air system (type C)					3,700			3,700	4,075
1750	Vacuum cleaner, HEPA, 16 gal., stainless steel, wet/dry					1,275			1,275	1,400
1760	55 gallon					1,475			1,475	1,600
1800	Vacuum loader, 9-18 ton/hr.					99,000			99,000	109,000
1900	Water atomizer unit, including 55 gal. drum					305			305	340
2000	Worker protection, whole body, foot, head cover & gloves, plastic					7.45			7.45	8.20
2500	Respirator, single use					30			30	33
2550	Cartridge for respirator					5.15			5.15	5.65
2570	Glove bag, 7 mil, 50" x 64"					8.75			8.75	9.60
2580	10 mil, 44" x 60"					7.90			7.90	8.65
2590	6 mil, 44" x 60"					4.83			4.83	5.30
6000	Disposable polyethylene bags, 6 mil, 3 C.F.					.77			.77	.85
6300	Disposable fiber drums, 3 C.F.					18.90			18.90	21

02 82 Asbestos Remediation

02 82 13 – Asbestos Abatement

02 82 13.41 Asbestos Abatement Equipment		Crew	Daily Output	Labor-Hours	Unit	Material	2020 Bare Costs Labor	Equipment	Total	Total Incl O&P
6400	Pressure sensitive caution labels, 3" x 5"				Ea.	3.08			3.08	3.39
6450	11" x 17"					7.75			7.75	8.55
6500	Negative air machine, 1800 CFM					890			890	975

02 82 13.42 Preparation of Asbestos Containment Area		Crew	Daily Output	Labor-Hours	Unit	Material	Labor	Equipment	Total	Total Incl O&P
0010	**PREPARATION OF ASBESTOS CONTAINMENT AREA** R028213-20									
0100	Pre-cleaning, HEPA vacuum and wet wipe, flat surfaces	A-9	12000	.005	S.F.	.01	.31		.32	.49
0200	Protect carpeted area, 2 layers 6 mil poly on 3/4" plywood	"	1000	.064		2.10	3.76		5.86	8.05
0300	Separation barrier, 2" x 4" @ 16", 1/2" plywood ea. side, 8' high	2 Carp	400	.040		3.08	2.13		5.21	6.60
0310	12' high		320	.050		3.02	2.66		5.68	7.30
0320	16' high		200	.080		2.99	4.25		7.24	9.70
0400	Personnel decontam. chamber, 2" x 4" @ 16", 3/4" ply ea. side		280	.057		3.55	3.04		6.59	8.45
0450	Waste decontam. chamber, 2" x 4" studs @ 16", 3/4" ply ea. side		360	.044		3.55	2.36		5.91	7.45
0500	Cover surfaces with polyethylene sheeting									
0501	Including glue and tape									
0550	Floors, each layer, 6 mil	A-9	8000	.008	S.F.	.04	.47		.51	.77
0551	4 mil		9000	.007		.03	.42		.45	.67
0560	Walls, each layer, 6 mil		6000	.011		.04	.63		.67	1.01
0561	4 mil		7000	.009		.03	.54		.57	.85
0570	For heights above 14', add						20%			
0575	For heights above 20', add						30%			
0580	For fire retardant poly, add					100%				
0590	For large open areas, deduct					10%	20%			
0600	Seal floor penetrations with foam firestop to 36 sq. in.	2 Carp	200	.080	Ea.	13.90	4.25		18.15	21.50
0610	36 sq. in. to 72 sq. in.		125	.128		28	6.80		34.80	40.50
0615	72 sq. in. to 144 sq. in.		80	.200		55.50	10.65		66.15	77
0620	Wall penetrations, to 36 sq. in.		180	.089		13.90	4.72		18.62	22.50
0630	36 sq. in. to 72 sq. in.		100	.160		28	8.50		36.50	43.50
0640	72 sq. in. to 144 sq. in.		60	.267		55.50	14.15		69.65	82.50
0800	Caulk seams with latex	1 Carp	230	.035	L.F.	.17	1.85		2.02	2.97
0900	Set up neg. air machine, 1-2k CFM/25 M.C.F. volume	1 Asbe	4.30	1.860	Ea.		109		109	168
0950	Set up and remove portable shower unit	2 Asbe	4	4	"		235		235	360

02 82 13.43 Bulk Asbestos Removal		Crew	Daily Output	Labor-Hours	Unit	Material	Labor	Equipment	Total	Total Incl O&P
0010	**BULK ASBESTOS REMOVAL**									
0020	Includes disposable tools and 2 suits and 1 respirator filter/day/worker									
0200	Boiler insulation	A-9	480	.133	S.F.	.40	7.85		8.25	12.45
0210	With metal lath, add				%				50%	50%
0300	Boiler breeching or flue insulation	A-9	520	.123	S.F.	.31	7.25		7.56	11.45
0310	For active boiler, add				%				100%	100%
0400	Duct or AHU insulation	A-10B	440	.073	S.F.	.18	4.28		4.46	6.75
0500	Duct vibration isolation joints, up to 24 sq. in. duct	A-9	56	1.143	Ea.	2.87	67		69.87	106
0520	25 sq. in. to 48 sq. in. duct		48	1.333		3.35	78.50		81.85	124
0530	49 sq. in. to 76 sq. in. duct		40	1.600		4.01	94		98.01	148
0600	Pipe insulation, air cell type, up to 4" diameter pipe		900	.071	L.F.	.18	4.17		4.35	6.60
0610	4" to 8" diameter pipe		800	.080		.20	4.70		4.90	7.40
0620	10" to 12" diameter pipe		700	.091		.23	5.35		5.58	8.50
0630	14" to 16" diameter pipe		550	.116		.29	6.85		7.14	10.80
0650	Over 16" diameter pipe		650	.098	S.F.	.25	5.80		6.05	9.15
0700	With glove bag up to 3" diameter pipe		200	.320	L.F.	9.40	18.80		28.20	39.50
1000	Pipe fitting insulation up to 4" diameter pipe		320	.200	Ea.	.50	11.75		12.25	18.60
1100	6" to 8" diameter pipe		304	.211		.53	12.35		12.88	19.60
1110	10" to 12" diameter pipe		192	.333		.84	19.55		20.39	31
1120	14" to 16" diameter pipe		128	.500		1.25	29.50		30.75	46.50

02 82 Asbestos Remediation

02 82 13 – Asbestos Abatement

02 82 13.43 Bulk Asbestos Removal

	02 82 13.43 Bulk Asbestos Removal	Crew	Daily Output	Labor-Hours	Unit	Material	Labor	Equipment	Total	Total Incl O&P
1130	Over 16" diameter pipe	A-9	176	.364	S.F.	.91	21.50		22.41	34
1200	With glove bag, up to 8" diameter pipe		75	.853	L.F.	6.55	50		56.55	84
2000	Scrape foam fireproofing from flat surface		2400	.027	S.F.	.07	1.57		1.64	2.47
2100	Irregular surfaces		1200	.053		.13	3.13		3.26	4.96
3000	Remove cementitious material from flat surface		1800	.036		.09	2.09		2.18	3.31
3100	Irregular surface		1000	.064		.11	3.76		3.87	5.90
6000	Remove contaminated soil from crawl space by hand		400	.160	C.F.	.40	9.40		9.80	14.85
6100	With large production vacuum loader	A-12	700	.091	"	.23	5.35	1.05	6.63	9.65
7000	Radiator backing, not including radiator removal	A-9	1200	.053	S.F.	.13	3.13		3.26	4.96
9000	For type B (supplied air) respirator equipment, add				%				10%	10%

Column headers: Crew, Daily Output, Labor-Hours, Unit; 2020 Bare Costs: Material, Labor, Equipment, Total; Total Incl O&P.

02 82 13.44 Demolition In Asbestos Contaminated Area

	02 82 13.44 Demolition In Asbestos Contaminated Area	Crew	Daily Output	Labor-Hours	Unit	Material	Labor	Equipment	Total	Total Incl O&P
0010	**DEMOLITION IN ASBESTOS CONTAMINATED AREA**									
0200	Ceiling, including suspension system, plaster and lath	A-9	2100	.030	S.F.	.08	1.79		1.87	2.83
0210	Finished plaster, leaving wire lath		585	.109		.27	6.40		6.67	10.15
0220	Suspended acoustical tile		3500	.018		.05	1.07		1.12	1.70
0230	Concealed tile grid system		3000	.021		.05	1.25		1.30	1.98
0240	Metal pan grid system		1500	.043		.11	2.51		2.62	3.97
0250	Gypsum board		2500	.026		.06	1.50		1.56	2.38
0260	Lighting fixtures up to 2' x 4'		72	.889	Ea.	2.23	52		54.23	82.50
0400	Partitions, non load bearing									
0410	Plaster, lath, and studs	A-9	690	.093	S.F.	.90	5.45		6.35	9.35
0450	Gypsum board and studs	"	1390	.046	"	.12	2.70		2.82	4.28
9000	For type B (supplied air) respirator equipment, add				%				10%	10%

02 82 13.45 OSHA Testing

	02 82 13.45 OSHA Testing	Crew	Daily Output	Labor-Hours	Unit	Material	Labor	Equipment	Total	Total Incl O&P
0010	**OSHA TESTING**									
0100	Certified technician, minimum				Day		340		340	340
0110	Maximum						670		670	670
0121	Industrial hygienist, minimum						385		385	385
0130	Maximum						755		755	755
0200	Asbestos sampling and PCM analysis, NIOSH 7400, minimum	1 Asbe	8	1	Ea.	15.90	58.50		74.40	108
0210	Maximum		4	2		27	117		144	210
1000	Cleaned area samples		8	1		203	58.50		261.50	315
1100	PCM air sample analysis, NIOSH 7400, minimum		8	1		17.70	58.50		76.20	110
1110	Maximum		4	2		2.30	117		119.30	183
1200	TEM air sample analysis, NIOSH 7402, minimum								80	106
1210	Maximum								360	450

02 82 13.46 Decontamination of Asbestos Containment Area

	02 82 13.46 Decontamination of Asbestos Containment Area	Crew	Daily Output	Labor-Hours	Unit	Material	Labor	Equipment	Total	Total Incl O&P
0010	**DECONTAMINATION OF ASBESTOS CONTAINMENT AREA**									
0100	Spray exposed substrate with surfactant (bridging)									
0200	Flat surfaces	A-9	6000	.011	S.F.	.41	.63		1.04	1.41
0250	Irregular surfaces		4000	.016	"	.45	.94		1.39	1.94
0300	Pipes, beams, and columns		2000	.032	L.F.	.60	1.88		2.48	3.54
1000	Spray encapsulate polyethylene sheeting		8000	.008	S.F.	.41	.47		.88	1.17
1100	Roll down polyethylene sheeting		8000	.008	"		.47		.47	.72
1500	Bag polyethylene sheeting		400	.160	Ea.	1.01	9.40		10.41	15.50
2000	Fine clean exposed substrate, with nylon brush		2400	.027	S.F.		1.57		1.57	2.40
2500	Wet wipe substrate		4800	.013			.78		.78	1.20
2600	Vacuum surfaces, fine brush		6400	.010			.59		.59	.90
3000	Structural demolition									
3100	Wood stud walls	A-9	2800	.023	S.F.		1.34		1.34	2.06
3500	Window manifolds, not incl. window replacement		4200	.015			.89		.89	1.37
3600	Plywood carpet protection		2000	.032			1.88		1.88	2.88

02 82 Asbestos Remediation

02 82 13 – Asbestos Abatement

02 82 13.46 Decontamination of Asbestos Containment Area		Crew	Daily Output	Labor-Hours	Unit	2020 Bare Costs Material	Labor	Equipment	Total	Total Incl O&P
4000	Remove custom decontamination facility	A-10A	8	3	Ea.	15.15	176		191.15	288
4100	Remove portable decontamination facility	3 Asbe	12	2	"	14.05	117		131.05	195
5000	HEPA vacuum, shampoo carpeting	A-9	4800	.013	S.F.	.11	.78		.89	1.32
9000	Final cleaning of protected surfaces	A-10A	8000	.003	"		.18		.18	.27

02 82 13.47 Asbestos Waste Pkg., Handling, and Disp.

		Crew	Daily Output	Labor-Hours	Unit	Material	Labor	Equipment	Total	Total Incl O&P
0010	**ASBESTOS WASTE PACKAGING, HANDLING, AND DISPOSAL**									
0100	Collect and bag bulk material, 3 C.F. bags, by hand	A-9	400	.160	Ea.	.77	9.40		10.17	15.25
0200	Large production vacuum loader	A-12	880	.073		.99	4.27	.84	6.10	8.55
1000	Double bag and decontaminate	A-9	960	.067		.77	3.91		4.68	6.85
2000	Containerize bagged material in drums, per 3 C.F. drum	"	800	.080		18.90	4.70		23.60	28
3000	Cart bags 50′ to dumpster	2 Asbe	400	.040	↓		2.35		2.35	3.60
5000	Disposal charges, not including haul, minimum				C.Y.				61	67
5020	Maximum				"				355	395
9000	For type B (supplied air) respirator equipment, add				%				10%	10%

02 82 13.48 Asbestos Encapsulation With Sealants

		Crew	Daily Output	Labor-Hours	Unit	Material	Labor	Equipment	Total	Total Incl O&P
0010	**ASBESTOS ENCAPSULATION WITH SEALANTS**									
0100	Ceilings and walls, minimum	A-9	21000	.003	S.F.	.42	.18		.60	.73
0110	Maximum		10600	.006	"	.52	.35		.87	1.11
0300	Pipes to 12″ diameter including minor repairs, minimum		800	.080	L.F.	.57	4.70		5.27	7.85
0310	Maximum	↓	400	.160	"	1.27	9.40		10.67	15.80

02 87 Biohazard Remediation

02 87 13 – Mold Remediation

02 87 13.16 Mold Remediation Preparation and Containment

		Crew	Daily Output	Labor-Hours	Unit	Material	Labor	Equipment	Total	Total Incl O&P
0010	**MOLD REMEDIATION PREPARATION AND CONTAINMENT**									
6010	Preparation of mold containment area									
6100	Pre-cleaning, HEPA vacuum and wet wipe, flat surfaces	A-9	12000	.005	S.F.	.01	.31		.32	.49
6300	Separation barrier, 2″ x 4″ @ 16″, 1/2″ plywood ea. side, 8′ high	2 Carp	400	.040		3.70	2.13		5.83	7.25
6310	12′ high		320	.050		3.52	2.66		6.18	7.85
6320	16′ high		200	.080		2.40	4.25		6.65	9.05
6400	Personnel decontam. chamber, 2″ x 4″ @ 16″, 3/4″ ply ea. side		280	.057		4.47	3.04		7.51	9.50
6450	Waste decontam. chamber, 2″ x 4″ studs @ 16″, 3/4″ ply each side	↓	360	.044	↓	4.54	2.36		6.90	8.55
6500	Cover surfaces with polyethylene sheeting									
6501	Including glue and tape									
6550	Floors, each layer, 6 mil	A-9	8000	.008	S.F.	.04	.47		.51	.77
6551	4 mil		9000	.007		.03	.42		.45	.67
6560	Walls, each layer, 6 mil		6000	.011		.04	.63		.67	1.01
6561	4 mil	↓	7000	.009	↓	.03	.54		.57	.85
6570	For heights above 14′, add						20%			
6575	For heights above 20′, add						30%			
6580	For fire retardant poly, add					100%				
6590	For large open areas, deduct					10%	20%			
6600	Seal floor penetrations with foam firestop to 36 sq. in.	2 Carp	200	.080	Ea.	13.90	4.25		18.15	21.50
6610	36 sq. in. to 72 sq. in.		125	.128		28	6.80		34.80	40.50
6615	72 sq. in. to 144 sq. in.		80	.200		55.50	10.65		66.15	77
6620	Wall penetrations, to 36 sq. in.		180	.089		13.90	4.72		18.62	22.50
6630	36 sq. in. to 72 sq. in.		100	.160		28	8.50		36.50	43.50
6640	72 sq. in. to 144 sq. in.	↓	60	.267	↓	55.50	14.15		69.65	82.50
6800	Caulk seams with latex caulk	1 Carp	230	.035	L.F.	.17	1.85		2.02	2.97
6900	Set up neg. air machine, 1-2k CFM/25 M.C.F. volume	1 Asbe	4.30	1.860	Ea.		109		109	168

02 87 Biohazard Remediation

02 87 13 – Mold Remediation

02 87 13.33 Removal and Disposal of Materials With Mold		Crew	Daily Output	Labor-Hours	Unit	2020 Bare Costs Material	Labor	Equipment	Total	Total Incl O&P
0010	**REMOVAL AND DISPOSAL OF MATERIALS WITH MOLD**									
0015	Demolition in mold contaminated area									
0200	Ceiling, including suspension system, plaster and lath	A-9	2100	.030	S.F.	.08	1.79		1.87	2.83
0210	Finished plaster, leaving wire lath		585	.109		.27	6.40		6.67	10.15
0220	Suspended acoustical tile		3500	.018		.05	1.07		1.12	1.70
0230	Concealed tile grid system		3000	.021		.05	1.25		1.30	1.98
0240	Metal pan grid system		1500	.043		.11	2.51		2.62	3.97
0250	Gypsum board		2500	.026		.06	1.50		1.56	2.38
0255	Plywood		2500	.026		.06	1.50		1.56	2.38
0260	Lighting fixtures up to 2′ x 4′		72	.889	Ea.	2.23	52		54.23	82.50
0400	Partitions, non load bearing									
0410	Plaster, lath, and studs	A-9	690	.093	S.F.	.90	5.45		6.35	9.35
0450	Gypsum board and studs		1390	.046		.12	2.70		2.82	4.28
0465	Carpet & pad		1390	.046		.12	2.70		2.82	4.28
0600	Pipe insulation, air cell type, up to 4″ diameter pipe		900	.071	L.F.	.18	4.17		4.35	6.60
0610	4″ to 8″ diameter pipe		800	.080		.20	4.70		4.90	7.40
0620	10″ to 12″ diameter pipe		700	.091		.23	5.35		5.58	8.50
0630	14″ to 16″ diameter pipe		550	.116		.29	6.85		7.14	10.80
0650	Over 16″ diameter pipe		650	.098	S.F.	.25	5.80		6.05	9.15
9000	For type B (supplied air) respirator equipment, add				%				10%	10%

Division Notes

	CREW	DAILY OUTPUT	LABOR-HOURS	UNIT	BARE COSTS MAT.	LABOR	EQUIP.	TOTAL	TOTAL INCL O&P

Estimating Tips

General

- Carefully check all the plans and specifications. Concrete often appears on drawings other than structural drawings, including mechanical and electrical drawings for equipment pads. The cost of cutting and patching is often difficult to estimate. See Subdivision 03 81 for Concrete Cutting, Subdivision 02 41 19.16 for Cutout Demolition, Subdivision 03 05 05.10 for Concrete Demolition, and Subdivision 02 41 19.19 for Rubbish Handling (handling, loading, and hauling of debris).
- Always obtain concrete prices from suppliers near the job site. A volume discount can often be negotiated, depending upon competition in the area. Remember to add for waste, particularly for slabs and footings on grade.

03 10 00 Concrete Forming and Accessories

- A primary cost for concrete construction is forming. Most jobs today are constructed with prefabricated forms. The selection of the forms best suited for the job and the total square feet of forms required for efficient concrete forming and placing are key elements in estimating concrete construction. Enough forms must be available for erection to make efficient use of the concrete placing equipment and crew.
- Concrete accessories for forming and placing depend upon the systems used. Study the plans and specifications to ensure that all special accessory requirements have been included in the cost estimate, such as anchor bolts, inserts, and hangers.
- Included within costs for forms-in-place are all necessary bracing and shoring.

03 20 00 Concrete Reinforcing

- Ascertain that the reinforcing steel supplier has included all accessories, cutting, bending, and an allowance for lapping, splicing, and waste. A good rule of thumb is 10% for lapping, splicing, and waste. Also, 10% waste should be allowed for welded wire fabric.
- The unit price items in the subdivisions for Reinforcing In Place, Glass Fiber Reinforcing, and Welded Wire Fabric include the labor to install accessories such as beam and slab bolsters, high chairs, and bar ties and tie wire. The material cost for these accessories is not included; they may be obtained from the Accessories Subdivisions.

03 30 00 Cast-In-Place Concrete

- When estimating structural concrete, pay particular attention to requirements for concrete additives, curing methods, and surface treatments. Special consideration for climate, hot or cold, must be included in your estimate. Be sure to include requirements for concrete placing equipment and concrete finishing.
- For accurate concrete estimating, the estimator must consider each of the following major components individually: forms, reinforcing steel, ready-mix concrete, placement of the concrete, and finishing of the top surface. For faster estimating, Subdivision 03 30 53.40 for Concrete-In-Place can be used; here, various items of concrete work are presented that include the costs of all five major components (unless specifically stated otherwise).

03 40 00 Precast Concrete
03 50 00 Cast Decks and Underlayment

- The cost of hauling precast concrete structural members is often an important factor. For this reason, it is important to get a quote from the nearest supplier. It may become economically feasible to set up precasting beds on the site if the hauling costs are prohibitive.

Reference Numbers

Reference numbers are shown at the beginning of some major classifications. These numbers refer to related items in the Reference Section. The reference information may be an estimating procedure, an alternate pricing method, or technical information.

Note: Not all subdivisions listed here necessarily appear. ■

03 11 Concrete Forming

03 11 13 – Structural Cast-In-Place Concrete Forming

03 11 13.40 Forms In Place, Equipment Foundations		Crew	Daily Output	Labor-Hours	Unit	Material	2020 Bare Costs Labor	Equipment	Total	Total Incl O&P
0010	**FORMS IN PLACE, EQUIPMENT FOUNDATIONS**									
0020	1 use	C-2	160	.300	SFCA	2.95	15.50		18.45	27
0050	2 use		190	.253		1.62	13.05		14.67	21.50
0100	3 use		200	.240		1.18	12.40		13.58	19.90
0150	4 use		205	.234		.96	12.10		13.06	19.20

03 11 13.45 Forms In Place, Footings

Line	Description	Crew	Daily Output	Labor-Hours	Unit	Material	Labor	Equipment	Total	Total Incl O&P
0010	**FORMS IN PLACE, FOOTINGS**									
0020	Continuous wall, plywood, 1 use	C-1	375	.085	SFCA	6.85	4.30		11.15	14
0050	2 use		440	.073		3.77	3.66		7.43	9.65
0100	3 use		470	.068		2.74	3.43		6.17	8.15
0150	4 use		485	.066		2.23	3.32		5.55	7.45
5000	Spread footings, job-built lumber, 1 use		305	.105		2.28	5.30		7.58	10.45
5050	2 use		371	.086		1.27	4.35		5.62	7.95
5100	3 use		401	.080		.91	4.02		4.93	7.05
5150	4 use		414	.077		.74	3.89		4.63	6.65

03 11 13.65 Forms In Place, Slab On Grade

Line	Description	Crew	Daily Output	Labor-Hours	Unit	Material	Labor	Equipment	Total	Total Incl O&P
0010	**FORMS IN PLACE, SLAB ON GRADE**									
3000	Edge forms, wood, 4 use, on grade, to 6" high	C-1	600	.053	L.F.	.33	2.69		3.02	4.40
6000	Trench forms in floor, wood, 1 use		160	.200	SFCA	2.05	10.10		12.15	17.40
6050	2 use		175	.183		1.13	9.20		10.33	15.10
6100	3 use		180	.178		.82	8.95		9.77	14.35
6150	4 use		185	.173		.67	8.70		9.37	13.85

03 15 Concrete Accessories

03 15 05 – Concrete Forming Accessories

03 15 05.75 Sleeves and Chases

Line	Description	Crew	Daily Output	Labor-Hours	Unit	Material	Labor	Equipment	Total	Total Incl O&P
0010	**SLEEVES AND CHASES**									
0100	Plastic, 1 use, 12" long, 2" diameter	1 Carp	100	.080	Ea.	2.32	4.25		6.57	8.95
0150	4" diameter		90	.089		4.97	4.72		9.69	12.55
0200	6" diameter		75	.107		10.85	5.65		16.50	20.50
0250	12" diameter		60	.133		28.50	7.10		35.60	41.50
5000	Sheet metal, 2" diameter G		100	.080		1.54	4.25		5.79	8.10
5100	4" diameter G		90	.089		1.93	4.72		6.65	9.20
5150	6" diameter G		75	.107		2.03	5.65		7.68	10.75
5200	12" diameter G		60	.133		4.24	7.10		11.34	15.30
6000	Steel pipe, 2" diameter G		100	.080		3.26	4.25		7.51	10
6100	4" diameter G		90	.089		17.20	4.72		21.92	26
6150	6" diameter G		75	.107		36.50	5.65		42.15	48.50
6200	12" diameter G		60	.133		85	7.10		92.10	104

03 15 16 – Concrete Construction Joints

03 15 16.20 Control Joints, Saw Cut

Line	Description	Crew	Daily Output	Labor-Hours	Unit	Material	Labor	Equipment	Total	Total Incl O&P
0010	**CONTROL JOINTS, SAW CUT**									
0100	Sawcut control joints in green concrete									
0120	1" depth	C-27	2000	.008	L.F.	.03	.40	.06	.49	.69
0140	1-1/2" depth		1800	.009		.05	.44	.06	.55	.78
0160	2" depth		1600	.010		.07	.50	.07	.64	.88
0180	Sawcut joint reservoir in cured concrete									
0182	3/8" wide x 3/4" deep, with single saw blade	C-27	1000	.016	L.F.	.05	.80	.11	.96	1.36
0184	1/2" wide x 1" deep, with double saw blades		900	.018		.10	.89	.12	1.11	1.56
0186	3/4" wide x 1-1/2" deep, with double saw blades		800	.020		.20	1	.14	1.34	1.84

03 15 Concrete Accessories

03 15 16 – Concrete Construction Joints

03 15 16.20 Control Joints, Saw Cut		Crew	Daily Output	Labor-Hours	Unit	2020 Bare Costs Material	Labor	Equipment	Total	Total Incl O&P
0190	Water blast joint to wash away laitance, 2 passes	C-29	2500	.003	L.F.		.13	.04	.17	.24
0200	Air blast joint to blow out debris and air dry, 2 passes	C-28	2000	.004			.20	.02	.22	.31
0300	For backer rod, see Section 07 91 23.10									
0342	For joint sealant, see Section 07 92 13.20									

03 15 19 – Cast-In Concrete Anchors

03 15 19.10 Anchor Bolts

			Crew	Daily Output	Labor-Hours	Unit	Material	Labor	Equipment	Total	Total Incl O&P
0010	**ANCHOR BOLTS**										
0015	Made from recycled materials										
0025	Single bolts installed in fresh concrete, no templates										
0030	Hooked w/nut and washer, 1/2" diameter, 8" long	G	1 Carp	132	.061	Ea.	1.62	3.22		4.84	6.60
0040	12" long	G		131	.061		1.80	3.25		5.05	6.85
0050	5/8" diameter, 8" long	G		129	.062		3.39	3.30		6.69	8.70
0060	12" long	G		127	.063		4.18	3.35		7.53	9.65
0070	3/4" diameter, 8" long	G		127	.063		4.18	3.35		7.53	9.65
0080	12" long	G		125	.064		5.20	3.40		8.60	10.85
0090	2-bolt pattern, including job-built 2-hole template, per set										
0100	J-type, incl. hex nut & washer, 1/2" diameter x 6" long	G	1 Carp	21	.381	Set	5.60	20.50		26.10	36.50
0110	12" long	G		21	.381		6.35	20.50		26.85	37.50
0120	18" long	G		21	.381		7.40	20.50		27.90	38.50
0130	3/4" diameter x 8" long	G		20	.400		11.10	21.50		32.60	44
0140	12" long	G		20	.400		13.15	21.50		34.65	46.50
0150	18" long	G		20	.400		16.30	21.50		37.80	50
0160	1" diameter x 12" long	G		19	.421		26.50	22.50		49	62.50
0170	18" long	G		19	.421		31	22.50		53.50	68
0180	24" long	G		19	.421		37.50	22.50		60	74.50
0190	36" long	G		18	.444		50.50	23.50		74	91
0200	1-1/2" diameter x 18" long	G		17	.471		47.50	25		72.50	89.50
0210	24" long	G		16	.500		56	26.50		82.50	102
0300	L-type, incl. hex nut & washer, 3/4" diameter x 12" long	G		20	.400		18.20	21.50		39.70	52
0310	18" long	G		20	.400		22.50	21.50		44	57
0320	24" long	G		20	.400		27	21.50		48.50	61.50
0330	30" long	G		20	.400		33.50	21.50		55	69
0340	36" long	G		20	.400		38	21.50		59.50	73.50
0350	1" diameter x 12" long	G		19	.421		24.50	22.50		47	60.50
0360	18" long	G		19	.421		29.50	22.50		52	66
0370	24" long	G		19	.421		36	22.50		58.50	73
0380	30" long	G		19	.421		42	22.50		64.50	80
0390	36" long	G		18	.444		47.50	23.50		71	88
0400	42" long	G		18	.444		57.50	23.50		81	98.50
0410	48" long	G		18	.444		64	23.50		87.50	106
0420	1-1/4" diameter x 18" long	G		18	.444		43.50	23.50		67	83.50
0430	24" long	G		18	.444		51.50	23.50		75	92
0440	30" long	G		17	.471		59	25		84	103
0450	36" long	G		17	.471		66.50	25		91.50	111
0460	42" long	G	2 Carp	32	.500		75	26.50		101.50	123
0470	48" long	G		32	.500		85	26.50		111.50	134
0480	54" long	G		31	.516		100	27.50		127.50	151
0490	60" long	G		31	.516		109	27.50		136.50	161
0500	1-1/2" diameter x 18" long	G		33	.485		47	26		73	90
0510	24" long	G		32	.500		54	26.50		80.50	99.50
0520	30" long	G		31	.516		61	27.50		88.50	108
0530	36" long	G		30	.533		69.50	28.50		98	119

03 15 Concrete Accessories

03 15 19 – Cast-In Concrete Anchors

03 15 19.10 Anchor Bolts			Crew	Daily Output	Labor-Hours	Unit	2020 Bare Costs Material	Labor	Equipment	Total	Total Incl O&P
0540	42″ long	G	2 Carp	30	.533	Set	79	28.50		107.50	129
0550	48″ long	G		29	.552		88	29.50		117.50	141
0560	54″ long	G		28	.571		107	30.50		137.50	163
0570	60″ long	G		28	.571		117	30.50		147.50	174
0580	1-3/4″ diameter x 18″ long	G		31	.516		71.50	27.50		99	120
0590	24″ long	G		30	.533		83.50	28.50		112	135
0600	30″ long	G		29	.552		96.50	29.50		126	150
0610	36″ long	G		28	.571		110	30.50		140.50	167
0620	42″ long	G		27	.593		123	31.50		154.50	184
0630	48″ long	G		26	.615		135	32.50		167.50	198
0640	54″ long	G		26	.615		167	32.50		199.50	233
0650	60″ long	G		25	.640		180	34		214	249
0660	2″ diameter x 24″ long	G		27	.593		135	31.50		166.50	197
0670	30″ long	G		27	.593		152	31.50		183.50	215
0680	36″ long	G		26	.615		167	32.50		199.50	232
0690	42″ long	G		25	.640		185	34		219	255
0700	48″ long	G		24	.667		213	35.50		248.50	287
0710	54″ long	G		23	.696		253	37		290	335
0720	60″ long	G		23	.696		272	37		309	355
0730	66″ long	G		22	.727		291	38.50		329.50	380
0740	72″ long	G	↓	21	.762	↓	320	40.50		360.50	410
1000	4-bolt pattern, including job-built 4-hole template, per set										
1100	J-type, incl. hex nut & washer, 1/2″ diameter x 6″ long	G	1 Carp	19	.421	Set	8.50	22.50		31	43
1110	12″ long	G		19	.421		9.95	22.50		32.45	44.50
1120	18″ long	G		18	.444		12.10	23.50		35.60	49
1130	3/4″ diameter x 8″ long	G		17	.471		19.45	25		44.45	59
1140	12″ long	G		17	.471		23.50	25		48.50	63.50
1150	18″ long	G		17	.471		30	25		55	70.50
1160	1″ diameter x 12″ long	G		16	.500		50	26.50		76.50	95
1170	18″ long	G		15	.533		59.50	28.50		88	108
1180	24″ long	G		15	.533		72	28.50		100.50	122
1190	36″ long	G		15	.533		98	28.50		126.50	151
1200	1-1/2″ diameter x 18″ long	G		13	.615		92	32.50		124.50	150
1210	24″ long	G		12	.667		109	35.50		144.50	173
1300	L-type, incl. hex nut & washer, 3/4″ diameter x 12″ long	G		17	.471		33.50	25		58.50	74.50
1310	18″ long	G		17	.471		42.50	25		67.50	84
1320	24″ long	G		17	.471		51	25		76	93.50
1330	30″ long	G		16	.500		64	26.50		90.50	111
1340	36″ long	G		16	.500		73	26.50		99.50	120
1350	1″ diameter x 12″ long	G		16	.500		46	26.50		72.50	90.50
1360	18″ long	G		15	.533		56.50	28.50		85	105
1370	24″ long	G		15	.533		69.50	28.50		98	119
1380	30″ long	G		15	.533		81.50	28.50		110	132
1390	36″ long	G		15	.533		92.50	28.50		121	145
1400	42″ long	G		14	.571		112	30.50		142.50	169
1410	48″ long	G		14	.571		125	30.50		155.50	184
1420	1-1/4″ diameter x 18″ long	G		14	.571		84.50	30.50		115	139
1430	24″ long	G		14	.571		100	30.50		130.50	156
1440	30″ long	G		13	.615		115	32.50		147.50	176
1450	36″ long	G	↓	13	.615		131	32.50		163.50	193
1460	42″ long	G	2 Carp	25	.640		147	34		181	213
1470	48″ long	G		24	.667		168	35.50		203.50	237
1480	54″ long	G	↓	23	.696	↓	197	37		234	273

03 15 Concrete Accessories

03 15 19 – Cast-In Concrete Anchors

03 15 19.10 Anchor Bolts			Crew	Daily Output	Labor-Hours	Unit	2020 Bare Costs Material	Labor	Equipment	Total	Total Incl O&P
1490	60" long	G	2 Carp	23	.696	Set	216	37		253	294
1500	1-1/2" diameter x 18" long	G		25	.640		91	34		125	151
1510	24" long	G		24	.667		105	35.50		140.50	169
1520	30" long	G		23	.696		119	37		156	187
1530	36" long	G		22	.727		136	38.50		174.50	208
1540	42" long	G		22	.727		155	38.50		193.50	228
1550	48" long	G		21	.762		173	40.50		213.50	252
1560	54" long	G		20	.800		211	42.50		253.50	296
1570	60" long	G		20	.800		231	42.50		273.50	320
1580	1-3/4" diameter x 18" long	G		22	.727		140	38.50		178.50	212
1590	24" long	G		21	.762		164	40.50		204.50	242
1600	30" long	G		21	.762		191	40.50		231.50	271
1610	36" long	G		20	.800		217	42.50		259.50	305
1620	42" long	G		19	.842		244	45		289	335
1630	48" long	G		18	.889		267	47		314	365
1640	54" long	G		18	.889		330	47		377	435
1650	60" long	G		17	.941		360	50		410	470
1660	2" diameter x 24" long	G		19	.842		267	45		312	360
1670	30" long	G		18	.889		300	47		347	400
1680	36" long	G		18	.889		330	47		377	435
1690	42" long	G		17	.941		370	50		420	480
1700	48" long	G		16	1		425	53		478	545
1710	54" long	G		15	1.067		505	56.50		561.50	640
1720	60" long	G		15	1.067		540	56.50		596.50	680
1730	66" long	G		14	1.143		580	60.50		640.50	725
1740	72" long	G		14	1.143		635	60.50		695.50	785
1990	For galvanized, add					Ea.	75%				

03 15 19.45 Machinery Anchors			Crew	Daily Output	Labor-Hours	Unit	Material	Labor	Equipment	Total	Total Incl O&P
0010	**MACHINERY ANCHORS**, heavy duty, incl. sleeve, floating base nut,										
0020	lower stud & coupling nut, fiber plug, connecting stud, washer & nut.										
0030	For flush mounted embedment in poured concrete heavy equip. pads.										
0200	Stud & bolt, 1/2" diameter	G	E-16	40	.400	Ea.	61.50	23.50	3.68	88.68	109
0300	5/8" diameter	G		35	.457		67	26.50	4.20	97.70	120
0500	3/4" diameter	G		30	.533		79.50	31	4.90	115.40	142
0600	7/8" diameter	G		25	.640		93.50	37.50	5.90	136.90	168
0800	1" diameter	G		20	.800		102	47	7.35	156.35	194
0900	1-1/4" diameter	G		15	1.067		133	62.50	9.80	205.30	256

03 21 Reinforcement Bars

03 21 11 – Plain Steel Reinforcement Bars

03 21 11.60 Reinforcing In Place			Crew	Daily Output	Labor-Hours	Unit	Material	Labor	Equipment	Total	Total Incl O&P
0010	**REINFORCING IN PLACE**, 50-60 ton lots, A615 Grade 60										
0020	Includes labor, but not material cost, to install accessories										
0030	Made from recycled materials										
0502	Footings, #4 to #7	G	4 Rodm	4200	.008	Lb.	.56	.43		.99	1.27
0552	#8 to #18	G		7200	.004		.56	.25		.81	1
0602	Slab on grade, #3 to #7	G		4200	.008		.56	.43		.99	1.27
0900	For other than 50-60 ton lots										
1000	Under 10 ton job, #3 to #7, add						25%	10%			
1010	#8 to #18, add						20%	10%			
1050	10-50 ton job, #3 to #7, add						10%				

03 21 Reinforcement Bars

03 21 11 – Plain Steel Reinforcement Bars

03 21 11.60 Reinforcing In Place		Crew	Daily Output	Labor-Hours	Unit	2020 Bare Costs Material	Labor	Equipment	Total	Total Incl O&P
1060	#8 to #18, add					5%				
1100	60-100 ton job, #3 to #7, deduct					5%				
1110	#8 to #18, deduct					10%				
1150	Over 100 ton job, #3 to #7, deduct					10%				
1160	#8 to #18, deduct					15%				

03 22 Fabric and Grid Reinforcing

03 22 11 – Plain Welded Wire Fabric Reinforcing

03 22 11.10 Plain Welded Wire Fabric

		Crew	Daily Output	Labor-Hours	Unit	Material	Labor	Equipment	Total	Total Incl O&P
0010	**PLAIN WELDED WIRE FABRIC** ASTM A185									
0020	Includes labor, but not material cost, to install accessories									
0030	Made from recycled materials									
0050	Sheets									
0100	6 x 6 - W1.4 x W1.4 (10 x 10) 21 lb./C.S.F. G	2 Rodm	35	.457	C.S.F.	16.25	26		42.25	57

03 22 13 – Galvanized Welded Wire Fabric Reinforcing

03 22 13.10 Galvanized Welded Wire Fabric

		Crew	Daily Output	Labor-Hours	Unit	Material	Labor	Equipment	Total	Total Incl O&P
0010	**GALVANIZED WELDED WIRE FABRIC**									
0100	Add to plain welded wire pricing for galvanized welded wire				Lb.	.26			.26	.28

03 22 16 – Epoxy-Coated Welded Wire Fabric Reinforcing

03 22 16.10 Epoxy-Coated Welded Wire Fabric

		Crew	Daily Output	Labor-Hours	Unit	Material	Labor	Equipment	Total	Total Incl O&P
0010	**EPOXY-COATED WELDED WIRE FABRIC**									
0100	Add to plain welded wire pricing for epoxy-coated welded wire				Lb.	.54			.54	.60

03 30 Cast-In-Place Concrete

03 30 53 – Miscellaneous Cast-In-Place Concrete

03 30 53.40 Concrete In Place

		Crew	Daily Output	Labor-Hours	Unit	Material	Labor	Equipment	Total	Total Incl O&P
0010	**CONCRETE IN PLACE**									
0020	Including forms (4 uses), Grade 60 rebar, concrete (Portland cement									
0050	Type I), placement and finishing unless otherwise indicated									
3540	Equipment pad (3000 psi), 3′ x 3′ x 6″ thick	C-14H	45	1.067	Ea.	50.50	55	.60	106.10	139
3550	4′ x 4′ x 6″ thick		30	1.600		78	82.50	.90	161.40	210
3560	5′ x 5′ x 8″ thick		18	2.667		138	138	1.49	277.49	360
3570	6′ x 6′ x 8″ thick		14	3.429		190	177	1.92	368.92	475
3580	8′ x 8′ x 10″ thick		8	6		395	310	3.36	708.36	905
3590	10′ x 10′ x 12″ thick	↓	5	9.600	↓	695	495	5.40	1,195.40	1,500
3800	Footings (3000 psi), spread under 1 C.Y.	C-14C	28	4	C.Y.	203	201	.96	404.96	525
3825	1 C.Y. to 5 C.Y.		43	2.605		240	131	.63	371.63	465
3850	Over 5 C.Y. R033053-60	↓	75	1.493		220	75	.36	295.36	355
3900	Footings, strip (3000 psi), 18″ x 9″, unreinforced	C-14L	40	2.400		154	118	.67	272.67	350
3920	18″ x 9″, reinforced	C-14C	35	3.200		181	161	.77	342.77	440
3925	20″ x 10″, unreinforced	C-14L	45	2.133		150	105	.60	255.60	325
3930	20″ x 10″, reinforced	C-14C	40	2.800		172	141	.67	313.67	400
3935	24″ x 12″, unreinforced	C-14L	55	1.745		148	86	.49	234.49	293
3940	24″ x 12″, reinforced	C-14C	48	2.333		171	118	.56	289.56	365
3945	36″ x 12″, unreinforced	C-14L	70	1.371		144	67.50	.38	211.88	261
3950	36″ x 12″, reinforced	C-14C	60	1.867		165	94	.45	259.45	320
4000	Foundation mat (3000 psi), under 10 C.Y.		38.67	2.896		246	146	.70	392.70	490
4050	Over 20 C.Y.	↓	56.40	1.986	↓	217	100	.48	317.48	390

03 30 Cast-In-Place Concrete

03 30 53 – Miscellaneous Cast-In-Place Concrete

03 30 53.40 Concrete In Place		Crew	Daily Output	Labor-Hours	Unit	2020 Bare Costs Material	Labor	Equipment	Total	Total Incl O&P
4650	Slab on grade (3500 psi), not including finish, 4" thick	C-14E	60.75	1.449	C.Y.	147	74	.45	221.45	273
4700	6" thick	"	92	.957	↓	141	49	.30	190.30	230
4701	Thickened slab edge (3500 psi), for slab on grade poured									
4702	monolithically with slab; depth is in addition to slab thickness;									
4703	formed vertical outside edge, earthen bottom and inside slope									
4705	8" deep x 8" wide bottom, unreinforced	C-14L	2190	.044	L.F.	4.07	2.16	.01	6.24	7.75
4710	8" x 8", reinforced	C-14C	1670	.067		6.70	3.38	.02	10.10	12.45
4715	12" deep x 12" wide bottom, unreinforced	C-14L	1800	.053		8.30	2.63	.01	10.94	13.10
4720	12" x 12", reinforced	C-14C	1310	.086		13.15	4.31	.02	17.48	21
4725	16" deep x 16" wide bottom, unreinforced	C-14L	1440	.067		14	3.29	.02	17.31	20.50
4730	16" x 16", reinforced	C-14C	1120	.100		19.75	5.05	.02	24.82	29.50
4735	20" deep x 20" wide bottom, unreinforced	C-14L	1150	.083		21	4.12	.02	25.14	29.50
4740	20" x 20", reinforced	C-14C	920	.122		28.50	6.15	.03	34.68	40.50
4745	24" deep x 24" wide bottom, unreinforced	C-14L	930	.103		30	5.10	.03	35.13	40.50
4750	24" x 24", reinforced	C-14C	740	.151	↓	39.50	7.60	.04	47.14	55

03 31 Structural Concrete

03 31 13 – Heavyweight Structural Concrete

03 31 13.35 Heavyweight Concrete, Ready Mix		Crew	Daily Output	Labor-Hours	Unit	Material	Labor	Equipment	Total	Total Incl O&P
0010	**HEAVYWEIGHT CONCRETE, READY MIX**, delivered									
0012	Includes local aggregate, sand, Portland cement (Type I) and water									
0015	Excludes all additives and treatments									
0020	2000 psi				C.Y.	107			107	117
0100	2500 psi					110			110	121
0150	3000 psi					129			129	142
0200	3500 psi					124			124	137
0300	4000 psi					127			127	140
1000	For high early strength (Portland cement Type III), add					10%				
1300	For winter concrete (hot water), add					5.35			5.35	5.90
1410	For mid-range water reducer, add					4.18			4.18	4.60
1420	For high-range water reducer/superplasticizer, add					6.40			6.40	7
1430	For retarder, add					3.45			3.45	3.80
1440	For non-Chloride accelerator, add					6.80			6.80	7.45
1450	For Chloride accelerator, per 1%, add					4.10			4.10	4.51
1460	For fiber reinforcing, synthetic (1 lb./C.Y.), add					8.10			8.10	8.90
1500	For Saturday delivery, add				↓	9.20			9.20	10.10
1510	For truck holding/waiting time past 1st hour per load, add				Hr.	109			109	120
1520	For short load (less than 4 C.Y.), add per load				Ea.	85			85	93.50
2000	For all lightweight aggregate, add				C.Y.	45%				

03 31 13.70 Placing Concrete		Crew	Daily Output	Labor-Hours	Unit	Material	Labor	Equipment	Total	Total Incl O&P
0010	**PLACING CONCRETE**									
0020	Includes labor and equipment to place, level (strike off) and consolidate									
1200	Duct bank, direct chute	C-6	155	.310	C.Y.		13.55	.35	13.90	21
1900	Footings, continuous, shallow, direct chute	"	120	.400			17.50	.45	17.95	26.50
1950	Pumped	C-20	150	.427			19.25	3.09	22.34	32.50
2000	With crane and bucket	C-7	90	.800			36.50	12	48.50	68
2100	Footings, continuous, deep, direct chute	C-6	140	.343			15	.38	15.38	23
2150	Pumped	C-20	160	.400			18.05	2.90	20.95	30
2200	With crane and bucket	C-7	110	.655			30	9.80	39.80	56
2400	Footings, spread, under 1 C.Y., direct chute	C-6	55	.873			38	.98	38.98	58
2450	Pumped	C-20	65	.985	↓		44.50	7.15	51.65	74.50

03 31 Structural Concrete

03 31 13 – Heavyweight Structural Concrete

03 31 13.70 Placing Concrete		Crew	Daily Output	Labor-Hours	Unit	Material	2020 Bare Costs Labor	Equipment	Total	Total Incl O&P
2500	With crane and bucket	C-7	45	1.600	C.Y.		73	24	97	137
2600	Over 5 C.Y., direct chute	C-6	120	.400			17.50	.45	17.95	26.50
2650	Pumped	C-20	150	.427			19.25	3.09	22.34	32.50
2700	With crane and bucket	C-7	100	.720			33	10.80	43.80	61.50
2900	Foundation mats, over 20 C.Y., direct chute	C-6	350	.137			6	.15	6.15	9.15
2950	Pumped	C-20	400	.160			7.25	1.16	8.41	12.10
3000	With crane and bucket	C-7	300	.240	↓		11	3.60	14.60	20.50

03 35 Concrete Finishing

03 35 13 – High-Tolerance Concrete Floor Finishing

03 35 13.30 Finishing Floors, High Tolerance		Crew	Daily Output	Labor-Hours	Unit	Material	Labor	Equipment	Total	Total Incl O&P
0010	**FINISHING FLOORS, HIGH TOLERANCE**									
0012	Finishing of fresh concrete flatwork requires that concrete									
0013	first be placed, struck off & consolidated									
0015	Basic finishing for various unspecified flatwork									
0100	Bull float only	C-10	4000	.006	S.F.		.28		.28	.42
0125	Bull float & manual float		2000	.012			.57		.57	.84
0150	Bull float, manual float & broom finish, w/edging & joints		1850	.013			.61		.61	.91
0200	Bull float, manual float & manual steel trowel	↓	1265	.019	↓		.90		.90	1.33
0210	For specified Random Access Floors in ACI Classes 1, 2, 3 and 4 to achieve									
0215	Composite Overall Floor Flatness and Levelness values up to FF35/FL25									
0250	Bull float, machine float & machine trowel (walk-behind)	C-10C	1715	.014	S.F.		.66	.05	.71	1.04
0300	Power screed, bull float, machine float & trowel (walk-behind)	C-10D	2400	.010			.47	.07	.54	.77
0350	Power screed, bull float, machine float & trowel (ride-on)	C-10E	4000	.006	↓		.28	.06	.34	.49
0352	For specified Random Access Floors in ACI Classes 5, 6, 7 and 8 to achieve									
0354	Composite Overall Floor Flatness and Levelness values up to FF50/FL50									
0356	Add for two-dimensional restraightening after power float	C-10	6000	.004	S.F.		.19		.19	.28
0358	For specified Random or Defined Access Floors in ACI Class 9 to achieve									
0360	Composite Overall Floor Flatness and Levelness values up to FF100/FL100									
0362	Add for two-dimensional restraightening after bull float & power float	C-10	3000	.008	S.F.		.38		.38	.56
0364	For specified Superflat Defined Access Floors in ACI Class 9 to achieve									
0366	Minimum Floor Flatness and Levelness values of FF100/FL100									
0368	Add for 2-dim'l restraightening after bull float, power float, power trowel	C-10	2000	.012	S.F.		.57		.57	.84

03 63 Epoxy Grouting

03 63 05 – Grouting of Dowels and Fasteners

03 63 05.10 Epoxy Only		Crew	Daily Output	Labor-Hours	Unit	Material	Labor	Equipment	Total	Total Incl O&P
0010	**EPOXY ONLY**									
1500	Chemical anchoring, epoxy cartridge, excludes layout, drilling, fastener									
1530	For fastener 3/4" diam. x 6" embedment	2 Skwk	72	.222	Ea.	4.78	12.20		16.98	23.50
1535	1" diam. x 8" embedment		66	.242		7.15	13.30		20.45	28
1540	1-1/4" diam. x 10" embedment		60	.267		14.35	14.65		29	38
1545	1-3/4" diam. x 12" embedment		54	.296		24	16.25		40.25	51
1550	14" embedment		48	.333		28.50	18.30		46.80	59
1555	2" diam. x 12" embedment		42	.381		38	21		59	73.50
1560	18" embedment	↓	32	.500	↓	48	27.50		75.50	94

03 82 Concrete Boring

03 82 13 – Concrete Core Drilling

03 82 13.10 Core Drilling		Crew	Daily Output	Labor-Hours	Unit	Material	2020 Bare Costs Labor	Equipment	Total	Total Incl O&P
0010	**CORE DRILLING**									
0015	Includes bit cost, layout and set-up time									
0020	Reinforced concrete slab, up to 6" thick									
0100	1" diameter core	B-89A	17	.941	Ea.	.20	45.50	6.70	52.40	76.50
0150	For each additional inch of slab thickness in same hole, add		1440	.011		.03	.54	.08	.65	.94
0200	2" diameter core		16.50	.970		.26	47	6.90	54.16	79
0250	For each additional inch of slab thickness in same hole, add		1080	.015		.04	.72	.11	.87	1.25
0300	3" diameter core		16	1		.40	48.50	7.15	56.05	81.50
0350	For each additional inch of slab thickness in same hole, add		720	.022		.07	1.08	.16	1.31	1.86
0500	4" diameter core		15	1.067		.45	51.50	7.60	59.55	87
0550	For each additional inch of slab thickness in same hole, add		480	.033		.07	1.62	.24	1.93	2.78
0700	6" diameter core		14	1.143		.65	55.50	8.15	64.30	93
0750	For each additional inch of slab thickness in same hole, add		360	.044		.11	2.15	.32	2.58	3.72
0900	8" diameter core		13	1.231		1.07	59.50	8.80	69.37	101
0950	For each additional inch of slab thickness in same hole, add		288	.056		.18	2.69	.40	3.27	4.70
1100	10" diameter core		12	1.333		1.20	64.50	9.50	75.20	109
1150	For each additional inch of slab thickness in same hole, add		240	.067		.20	3.23	.48	3.91	5.60
1300	12" diameter core		11	1.455		1.84	70.50	10.40	82.74	119
1350	For each additional inch of slab thickness in same hole, add		206	.078		.31	3.77	.55	4.63	6.65
1500	14" diameter core		10	1.600		1.84	77.50	11.40	90.74	132
1550	For each additional inch of slab thickness in same hole, add		180	.089		.31	4.31	.63	5.25	7.55
1700	18" diameter core		9	1.778		3.06	86	12.70	101.76	147
1750	For each additional inch of slab thickness in same hole, add		144	.111		.51	5.40	.79	6.70	9.55
1754	24" diameter core		8	2		4.15	97	14.30	115.45	166
1756	For each additional inch of slab thickness in same hole, add		120	.133		.69	6.45	.95	8.09	11.55
1760	For horizontal holes, add to above						20%	20%		
1770	Prestressed hollow core plank, 8" thick									
1780	1" diameter core	B-89A	17.50	.914	Ea.	.27	44.50	6.55	51.32	74.50
1790	For each additional inch of plank thickness in same hole, add		3840	.004		.03	.20	.03	.26	.37
1794	2" diameter core		17.25	.928		.35	45	6.60	51.95	75.50
1796	For each additional inch of plank thickness in same hole, add		2880	.006		.04	.27	.04	.35	.50
1800	3" diameter core		17	.941		.53	45.50	6.70	52.73	77
1810	For each additional inch of plank thickness in same hole, add		1920	.008		.07	.40	.06	.53	.75
1820	4" diameter core		16.50	.970		.60	47	6.90	54.50	79.50
1830	For each additional inch of plank thickness in same hole, add		1280	.013		.07	.61	.09	.77	1.09
1840	6" diameter core		15.50	1.032		.87	50	7.35	58.22	84.50
1850	For each additional inch of plank thickness in same hole, add		960	.017		.11	.81	.12	1.04	1.47
1860	8" diameter core		15	1.067		1.43	51.50	7.60	60.53	88
1870	For each additional inch of plank thickness in same hole, add		768	.021		.18	1.01	.15	1.34	1.88
1880	10" diameter core		14	1.143		1.60	55.50	8.15	65.25	94.50
1890	For each additional inch of plank thickness in same hole, add		640	.025		.20	1.21	.18	1.59	2.25
1900	12" diameter core		13.50	1.185		2.45	57.50	8.45	68.40	98.50
1910	For each additional inch of plank thickness in same hole, add		548	.029		.31	1.42	.21	1.94	2.70

03 82 16 – Concrete Drilling

03 82 16.10 Concrete Impact Drilling		Crew	Daily Output	Labor-Hours	Unit	Material	Labor	Equipment	Total	Total Incl O&P
0010	**CONCRETE IMPACT DRILLING**									
0020	Includes bit cost, layout and set-up time, no anchors									
0050	Up to 4" deep in concrete/brick floors/walls									
0100	Holes, 1/4" diameter	1 Carp	75	.107	Ea.	.07	5.65		5.72	8.60
0150	For each additional inch of depth in same hole, add		430	.019		.02	.99		1.01	1.51
0200	3/8" diameter		63	.127		.05	6.75		6.80	10.20
0250	For each additional inch of depth in same hole, add		340	.024		.01	1.25		1.26	1.89

03 82 Concrete Boring

03 82 16 – Concrete Drilling

03 82 16.10 Concrete Impact Drilling		Crew	Daily Output	Labor-Hours	Unit	Material	2020 Bare Costs Labor	Equipment	Total	Total Incl O&P
0300	1/2" diameter	1 Carp	50	.160	Ea.	.05	8.50		8.55	12.85
0350	For each additional inch of depth in same hole, add		250	.032		.01	1.70		1.71	2.57
0400	5/8" diameter		48	.167		.10	8.85		8.95	13.40
0450	For each additional inch of depth in same hole, add		240	.033		.02	1.77		1.79	2.69
0500	3/4" diameter		45	.178		.11	9.45		9.56	14.30
0550	For each additional inch of depth in same hole, add		220	.036		.03	1.93		1.96	2.93
0600	7/8" diameter		43	.186		.17	9.90		10.07	15.05
0650	For each additional inch of depth in same hole, add		210	.038		.04	2.03		2.07	3.09
0700	1" diameter		40	.200		.18	10.65		10.83	16.15
0750	For each additional inch of depth in same hole, add		190	.042		.04	2.24		2.28	3.41
0800	1-1/4" diameter		38	.211		.33	11.20		11.53	17.15
0850	For each additional inch of depth in same hole, add		180	.044		.08	2.36		2.44	3.64
0900	1-1/2" diameter		35	.229		.43	12.15		12.58	18.70
0950	For each additional inch of depth in same hole, add	↓	165	.048	↓	.11	2.58		2.69	3.99
1000	For ceiling installations, add						40%			

Estimating Tips

05 05 00 Common Work Results for Metals

- Nuts, bolts, washers, connection angles, and plates can add a significant amount to both the tonnage of a structural steel job and the estimated cost. As a rule of thumb, add 10% to the total weight to account for these accessories.
- Type 2 steel construction, commonly referred to as "simple construction," consists generally of field-bolted connections with lateral bracing supplied by other elements of the building, such as masonry walls or x-bracing. The estimator should be aware, however, that shop connections may be accomplished by welding or bolting. The method may be particular to the fabrication shop and may have an impact on the estimated cost.

05 10 00 Structural Steel

- Steel items can be obtained from two sources: a fabrication shop or a metals service center. Fabrication shops can fabricate items under more controlled conditions than crews in the field can. They are also more efficient and can produce items more economically. Metal service centers serve as a source of long mill shapes to both fabrication shops and contractors.
- Most line items in this structural steel subdivision, and most items in 05 50 00 Metal Fabrications, are indicated as being shop fabricated. The bare material cost for these shop fabricated items is the "Invoice Cost" from the shop and includes the mill base price of steel plus mill extras, transportation to the shop, shop drawings and detailing where warranted, shop fabrication and handling, sandblasting and a shop coat of primer paint, all necessary structural bolts, and delivery to the job site. The bare labor cost and bare equipment cost for these shop fabricated items are for field installation or erection.
- Line items in Subdivision 05 12 23.40 Lightweight Framing, and other items scattered in Division 5, are indicated as being field fabricated. The bare material cost for these field fabricated items is the "Invoice Cost" from the metals service center and includes the mill base price of steel plus mill extras, transportation to the metals service center, material handling, and delivery of long lengths of mill shapes to the job site. Material costs for structural bolts and welding rods should be added to the estimate. The bare labor cost and bare equipment cost for these items are for both field fabrication and field installation or erection, and include time for cutting, welding, and drilling in the fabricated metal items. Drilling into concrete and fasteners to fasten field fabricated items to other work is not included and should be added to the estimate.

05 20 00 Steel Joist Framing

- In any given project the total weight of open web steel joists is determined by the loads to be supported and the design. However, economies can be realized in minimizing the amount of labor used to place the joists. This is done by maximizing the joist spacing and therefore minimizing the number of joists required to be installed on the job. Certain spacings and locations may be required by the design, but in other cases maximizing the spacing and keeping it as uniform as possible will keep the costs down.

05 30 00 Steel Decking

- The takeoff and estimating of a metal deck involve more than the area of the floor or roof and the type of deck specified or shown on the drawings. Many different sizes and types of openings may exist. Small openings for individual pipes or conduits may be drilled after the floor/roof is installed, but larger openings may require special deck lengths as well as reinforcing or structural support. The estimator should determine who will be supplying this reinforcing. Additionally, some deck terminations are part of the deck package, such as screed angles and pour stops, and others will be part of the steel contract, such as angles attached to structural members and cast-in-place angles and plates. The estimator must ensure that all pieces are accounted for in the complete estimate.

05 50 00 Metal Fabrications

- The most economical steel stairs are those that use common materials, standard details, and most importantly, a uniform and relatively simple method of field assembly. Commonly available A36/A992 channels and plates are very good choices for the main stringers of the stairs, as are angles and tees for the carrier members. Risers and treads are usually made by specialty shops, and it is most economical to use a typical detail in as many places as possible. The stairs should be pre-assembled and shipped directly to the site. The field connections should be simple and straightforward enough to be accomplished efficiently, and with minimum equipment and labor.

Reference Numbers

Reference numbers are shown at the beginning of some major classifications. These numbers refer to related items in the Reference Section. The reference information may be an estimating procedure, an alternate pricing method, or technical information.

Note: Not all subdivisions listed here necessarily appear. ■

05 05 19 – Post-Installed Concrete Anchors

05 05 19.10 Chemical Anchors			Crew	Daily Output	Labor-Hours	Unit	Material	Labor	Equipment	Total	Total Incl O&P
							2020 Bare Costs				
0010	**CHEMICAL ANCHORS**										
0020	Includes layout & drilling										
1430	Chemical anchor, w/rod & epoxy cartridge, 3/4" diameter x 9-1/2" long		B-89A	27	.593	Ea.	10.15	28.50	4.23	42.88	59.50
1435	1" diameter x 11-3/4" long			24	.667		20.50	32.50	4.76	57.76	76.50
1440	1-1/4" diameter x 14" long			21	.762		39	37	5.45	81.45	105
1445	1-3/4" diameter x 15" long			20	.800		68.50	39	5.70	113.20	140
1450	18" long			17	.941		82.50	45.50	6.70	134.70	167
1455	2" diameter x 18" long			16	1		113	48.50	7.15	168.65	206
1460	24" long		↓	15	1.067	↓	148	51.50	7.60	207.10	249

05 05 19.20 Expansion Anchors			Crew	Daily Output	Labor-Hours	Unit	Material	Labor	Equipment	Total	Total Incl O&P
0010	**EXPANSION ANCHORS**										
0100	Anchors for concrete, brick or stone, no layout and drilling										
0200	Expansion shields, zinc, 1/4" diameter, 1-5/16" long, single	G	1 Carp	90	.089	Ea.	.46	4.72		5.18	7.60
0300	1-3/8" long, double	G		85	.094		.66	5		5.66	8.25
0400	3/8" diameter, 1-1/2" long, single	G		85	.094		.71	5		5.71	8.30
0500	2" long, double	G		80	.100		1.20	5.30		6.50	9.30
0600	1/2" diameter, 2-1/16" long, single	G		80	.100		1.28	5.30		6.58	9.40
0700	2-1/2" long, double	G		75	.107		2.17	5.65		7.82	10.90
0800	5/8" diameter, 2-5/8" long, single	G		75	.107		2.22	5.65		7.87	10.95
0900	2-3/4" long, double	G		70	.114		3.64	6.05		9.69	13.15
1000	3/4" diameter, 2-3/4" long, single	G		70	.114		3.66	6.05		9.71	13.20
1100	3-15/16" long, double	G	↓	65	.123	↓	5.65	6.55		12.20	16.10
2100	Hollow wall anchors for gypsum wall board, plaster or tile										
2300	1/8" diameter, short	G	1 Carp	160	.050	Ea.	.31	2.66		2.97	4.33
2400	Long	G		150	.053		.32	2.83		3.15	4.61
2500	3/16" diameter, short	G		150	.053		.58	2.83		3.41	4.90
2600	Long	G		140	.057		.75	3.04		3.79	5.40
2700	1/4" diameter, short	G		140	.057		.65	3.04		3.69	5.30
2800	Long	G		130	.062		1.03	3.27		4.30	6.05
3000	Toggle bolts, bright steel, 1/8" diameter, 2" long	G		85	.094		.21	5		5.21	7.75
3100	4" long	G		80	.100		.26	5.30		5.56	8.30
3200	3/16" diameter, 3" long	G		80	.100		.30	5.30		5.60	8.35
3300	6" long	G		75	.107		.49	5.65		6.14	9.05
3400	1/4" diameter, 3" long	G		75	.107		.44	5.65		6.09	9
3500	6" long	G		70	.114		.57	6.05		6.62	9.80
3600	3/8" diameter, 3" long	G		70	.114		.88	6.05		6.93	10.10
3700	6" long	G		60	.133		1.58	7.10		8.68	12.40
3800	1/2" diameter, 4" long	G		60	.133		1.86	7.10		8.96	12.70
3900	6" long	G	↓	50	.160	↓	2.37	8.50		10.87	15.40
4000	Nailing anchors										
4100	Nylon nailing anchor, 1/4" diameter, 1" long		1 Carp	3.20	2.500	C	19.75	133		152.75	222
4200	1-1/2" long			2.80	2.857		22	152		174	252
4300	2" long			2.40	3.333		24	177		201	293
4400	Metal nailing anchor, 1/4" diameter, 1" long	G		3.20	2.500		20	133		153	222
4500	1-1/2" long	G		2.80	2.857		25	152		177	256
4600	2" long	G	↓	2.40	3.333	↓	43	177		220	315
5000	Screw anchors for concrete, masonry,										
5100	stone & tile, no layout or drilling included										
5700	Lag screw shields, 1/4" diameter, short	G	1 Carp	90	.089	Ea.	.47	4.72		5.19	7.60
5800	Long	G		85	.094		.56	5		5.56	8.10
5900	3/8" diameter, short	G		85	.094		.74	5		5.74	8.30
6000	Long	G	↓	80	.100	↓	1.05	5.30		6.35	9.15

05 05 19 – Post-Installed Concrete Anchors

05 05 19.20 Expansion Anchors			Crew	Daily Output	Labor-Hours	Unit	Material	2020 Bare Costs Labor	Equipment	Total	Total Incl O&P
6100	1/2" diameter, short	G	1 Carp	80	.100	Ea.	.98	5.30		6.28	9.10
6200	Long	G		75	.107		1.52	5.65		7.17	10.15
6300	5/8" diameter, short	G		70	.114		1.54	6.05		7.59	10.85
6400	Long	G		65	.123		2.21	6.55		8.76	12.30
6600	Lead, #6 & #8, 3/4" long	G		260	.031		.18	1.64		1.82	2.66
6700	#10 - #14, 1-1/2" long	G		200	.040		.48	2.13		2.61	3.72
6800	#16 & #18, 1-1/2" long	G		160	.050		.42	2.66		3.08	4.45
6900	Plastic, #6 & #8, 3/4" long			260	.031		.05	1.64		1.69	2.52
7000	#8 & #10, 7/8" long			240	.033		.05	1.77		1.82	2.72
7100	#10 & #12, 1" long			220	.036		.07	1.93		2	2.98
7200	#14 & #16, 1-1/2" long		↓	160	.050	↓	.07	2.66		2.73	4.07
8000	Wedge anchors, not including layout or drilling										
8050	Carbon steel, 1/4" diameter, 1-3/4" long	G	1 Carp	150	.053	Ea.	.62	2.83		3.45	4.94
8100	3-1/4" long	G		140	.057		.82	3.04		3.86	5.45
8150	3/8" diameter, 2-1/4" long	G		145	.055		.44	2.93		3.37	4.90
8200	5" long	G		140	.057		.78	3.04		3.82	5.40
8250	1/2" diameter, 2-3/4" long	G		140	.057		.99	3.04		4.03	5.65
8300	7" long	G		125	.064		1.70	3.40		5.10	6.95
8350	5/8" diameter, 3-1/2" long	G		130	.062		2.05	3.27		5.32	7.15
8400	8-1/2" long	G		115	.070		4.36	3.70		8.06	10.35
8450	3/4" diameter, 4-1/4" long	G		115	.070		2.97	3.70		6.67	8.80
8500	10" long	G		95	.084		6.75	4.48		11.23	14.15
8550	1" diameter, 6" long	G		100	.080		8.80	4.25		13.05	16.05
8575	9" long	G		85	.094		11.40	5		16.40	20
8600	12" long	G		75	.107		12.30	5.65		17.95	22
8650	1-1/4" diameter, 9" long	G		70	.114		23.50	6.05		29.55	34.50
8700	12" long	G	↓	60	.133	↓	30	7.10		37.10	43.50
8750	For type 303 stainless steel, add						350%				
8800	For type 316 stainless steel, add						450%				
8950	Self-drilling concrete screw, hex washer head, 3/16" diam. x 1-3/4" long	G	1 Carp	300	.027	Ea.	.18	1.42		1.60	2.33
8960	2-1/4" long	G		250	.032		.24	1.70		1.94	2.82
8970	Phillips flat head, 3/16" diam. x 1-3/4" long	G		300	.027		.19	1.42		1.61	2.34
8980	2-1/4" long	G	↓	250	.032	↓	.23	1.70		1.93	2.81

05 05 21 – Fastening Methods for Metal

05 05 21.15 Drilling Steel		Crew	Daily Output	Labor-Hours	Unit	Material	Labor	Equipment	Total	Total Incl O&P
0010	**DRILLING STEEL**									
1910	Drilling & layout for steel, up to 1/4" deep, no anchor									
1920	Holes, 1/4" diameter	1 Sswk	112	.071	Ea.	.10	4.12		4.22	6.55
1925	For each additional 1/4" depth, add		336	.024		.10	1.37		1.47	2.26
1930	3/8" diameter		104	.077		.08	4.43		4.51	7.05
1935	For each additional 1/4" depth, add		312	.026		.08	1.48		1.56	2.41
1940	1/2" diameter		96	.083		.08	4.80		4.88	7.65
1945	For each additional 1/4" depth, add		288	.028		.08	1.60		1.68	2.60
1950	5/8" diameter		88	.091		.16	5.25		5.41	8.35
1955	For each additional 1/4" depth, add		264	.030		.16	1.75		1.91	2.91
1960	3/4" diameter		80	.100		.17	5.75		5.92	9.25
1965	For each additional 1/4" depth, add		240	.033		.17	1.92		2.09	3.19
1970	7/8" diameter		72	.111		.24	6.40		6.64	10.30
1975	For each additional 1/4" depth, add		216	.037		.24	2.14		2.38	3.62
1980	1" diameter		64	.125		.26	7.20		7.46	11.60
1985	For each additional 1/4" depth, add	↓	192	.042		.26	2.40		2.66	4.05
1990	For drilling up, add				↓		40%			

05 05 Common Work Results for Metals

05 05 21 – Fastening Methods for Metal

05 05 21.90 Welding Steel		Crew	Daily Output	Labor-Hours	Unit	2020 Bare Costs Material	Labor	Equipment	Total	Total Incl O&P
0010	**WELDING STEEL**, Structural R050521-20									
0020	Field welding, 1/8" E6011, cost per welder, no operating engineer	E-14	8	1	Hr.	5.55	59.50	18.40	83.45	120
0200	With 1/2 operating engineer	E-13	8	1.500		5.55	86	18.40	109.95	159
0300	With 1 operating engineer	E-12	8	2		5.55	112	18.40	135.95	198
0500	With no operating engineer, 2# weld rod per ton	E-14	8	1	Ton	5.55	59.50	18.40	83.45	120
0600	8# E6011 per ton	"	2	4		22	238	73.50	333.50	480
0800	With one operating engineer per welder, 2# E6011 per ton	E-12	8	2		5.55	112	18.40	135.95	198
0900	8# E6011 per ton	"	2	8		22	450	73.50	545.50	795
1200	Continuous fillet, down welding									
1300	Single pass, 1/8" thick, 0.1#/L.F.	E-14	150	.053	L.F.	.28	3.17	.98	4.43	6.35
1400	3/16" thick, 0.2#/L.F.		75	.107		.55	6.35	1.96	8.86	12.70
1500	1/4" thick, 0.3#/L.F.		50	.160		.83	9.50	2.94	13.27	19.05
1610	5/16" thick, 0.4#/L.F.		38	.211		1.11	12.50	3.87	17.48	25
1800	3 passes, 3/8" thick, 0.5#/L.F.		30	.267		1.38	15.85	4.90	22.13	32
2010	4 passes, 1/2" thick, 0.7#/L.F.		22	.364		1.93	21.50	6.70	30.13	43.50
2200	5 to 6 passes, 3/4" thick, 1.3#/L.F.		12	.667		3.59	39.50	12.25	55.34	79.50
2400	8 to 11 passes, 1" thick, 2.4#/L.F.		6	1.333		6.65	79.50	24.50	110.65	158
2600	For vertical joint welding, add						20%			
2700	Overhead joint welding, add						300%			
2900	For semi-automatic welding, obstructed joints, deduct						5%			
3000	Exposed joints, deduct						15%			
4000	Cleaning and welding plates, bars, or rods									
4010	to existing beams, columns, or trusses	E-14	12	.667	L.F.	1.38	39.50	12.25	53.13	77

05 05 23 – Metal Fastenings

05 05 23.30 Lag Screws

Line	Description	Crew	Daily Output	Labor-Hours	Unit	Material	Labor	Equipment	Total	Total Incl O&P
0010	**LAG SCREWS**									
0020	Steel, 1/4" diameter, 2" long G	1 Carp	200	.040	Ea.	.10	2.13		2.23	3.30
0100	3/8" diameter, 3" long G		150	.053		.31	2.83		3.14	4.60
0200	1/2" diameter, 3" long G		130	.062		.72	3.27		3.99	5.70
0300	5/8" diameter, 3" long G		120	.067		1.26	3.54		4.80	6.70

05 05 23.35 Machine Screws

Line	Description	Crew	Daily Output	Labor-Hours	Unit	Material	Labor	Equipment	Total	Total Incl O&P
0010	**MACHINE SCREWS**									
0020	Steel, round head, #8 x 1" long G	1 Carp	4.80	1.667	C	3.98	88.50		92.48	137
0110	#8 x 2" long G		2.40	3.333		8.80	177		185.80	276
0200	#10 x 1" long G		4	2		4.89	106		110.89	165
0300	#10 x 2" long G		2	4		9.80	213		222.80	330

05 05 23.50 Powder Actuated Tools and Fasteners

Line	Description	Crew	Daily Output	Labor-Hours	Unit	Material	Labor	Equipment	Total	Total Incl O&P
0010	**POWDER ACTUATED TOOLS & FASTENERS**									
0020	Stud driver, .22 caliber, single shot				Ea.	159			159	175
0100	.27 caliber, semi automatic, strip				"	460			460	505
0300	Powder load, single shot, .22 cal, power level 2, brown				C	6.25			6.25	6.85
0400	Strip, .27 cal, power level 4, red					10.30			10.30	11.35
0600	Drive pin, .300 x 3/4" long G	1 Carp	4.80	1.667		4.48	88.50		92.98	138
0700	.300 x 3" long with washer G	"	4	2		11.95	106		117.95	173

05 05 23.55 Rivets

Line	Description	Crew	Daily Output	Labor-Hours	Unit	Material	Labor	Equipment	Total	Total Incl O&P
0010	**RIVETS**									
0100	Aluminum rivet & mandrel, 1/2" grip length x 1/8" diameter G	1 Carp	4.80	1.667	C	8.15	88.50		96.65	142
0200	3/16" diameter G		4	2		11.65	106		117.65	173
0300	Aluminum rivet, steel mandrel, 1/8" diameter G		4.80	1.667		10.30	88.50		98.80	144
0400	3/16" diameter G		4	2		18.35	106		124.35	180
0500	Copper rivet, steel mandrel, 1/8" diameter G		4.80	1.667		10.30	88.50		98.80	144

05 05 Common Work Results for Metals

05 05 23 – Metal Fastenings

05 05 23.55 Rivets			Crew	Daily Output	Labor-Hours	Unit	Material	Labor	Equipment	Total	Total Incl O&P
							2020 Bare Costs				
0800	Stainless rivet & mandrel, 1/8" diameter	G	1 Carp	4.80	1.667	C	25	88.50		113.50	161
0900	3/16" diameter	G		4	2		41	106		147	205
1000	Stainless rivet, steel mandrel, 1/8" diameter	G		4.80	1.667		16.70	88.50		105.20	151
1100	3/16" diameter	G		4	2		26	106		132	189
1200	Steel rivet and mandrel, 1/8" diameter	G		4.80	1.667		8.15	88.50		96.65	142
1300	3/16" diameter	G		4	2		11	106		117	172
1400	Hand riveting tool, standard					Ea.	78			78	86
1500	Deluxe						415			415	460
1600	Power riveting tool, standard						545			545	600
1700	Deluxe						1,575			1,575	1,725
05 05 23.70 Structural Blind Bolts											
0010	**STRUCTURAL BLIND BOLTS**										
0100	1/4" diameter x 1/4" grip	G	1 Sswk	240	.033	Ea.	1.84	1.92		3.76	5.05
0150	1/2" grip	G		216	.037		1.89	2.14		4.03	5.45
0200	3/8" diameter x 1/2" grip	G		232	.034		3.50	1.99		5.49	6.95
0250	3/4" grip	G		208	.038		3.14	2.22		5.36	6.95
0300	1/2" diameter x 1/2" grip	G		224	.036		6.25	2.06		8.31	10.15
0350	3/4" grip	G		200	.040		9.10	2.31		11.41	13.60
0400	5/8" diameter x 3/4" grip	G		216	.037		11.25	2.14		13.39	15.75
0450	1" grip	G		192	.042		15.70	2.40		18.10	21
05 05 23.90 Welding Rod											
0010	**WELDING ROD**										
0020	Steel, type 6011, 1/8" diam., less than 500#					Lb.	2.76			2.76	3.04
0100	500# to 2,000#						2.49			2.49	2.74
0200	2,000# to 5,000#						2.34			2.34	2.57
0300	5/32" diam., less than 500#						2.46			2.46	2.71
0310	500# to 2,000#						2.22			2.22	2.44
0320	2,000# to 5,000#						2.09			2.09	2.30
0400	3/16" diam., less than 500#						2.75			2.75	3.03
0500	500# to 2,000#						2.48			2.48	2.73
0600	2,000# to 5,000#						2.33			2.33	2.56
0620	Steel, type 6010, 1/8" diam., less than 500#						2.49			2.49	2.74
0630	500# to 2,000#						2.24			2.24	2.46
0640	2,000# to 5,000#						2.11			2.11	2.32
0650	Steel, type 7018 Low Hydrogen, 1/8" diam., less than 500#						2.94			2.94	3.24
0660	500# to 2,000#						2.65			2.65	2.92
0670	2,000# to 5,000#						2.49			2.49	2.74
0700	Steel, type 7024 Jet Weld, 1/8" diam., less than 500#						2.55			2.55	2.81
0710	500# to 2,000#						2.30			2.30	2.53
0720	2,000# to 5,000#						2.16			2.16	2.38
1550	Aluminum, type 4043 TIG, 1/8" diam., less than 10#						5.20			5.20	5.75
1560	10# to 60#						4.70			4.70	5.15
1570	Over 60#						4.42			4.42	4.86
1600	Aluminum, type 5356 TIG, 1/8" diam., less than 10#						5.55			5.55	6.10
1610	10# to 60#						4.99			4.99	5.50
1620	Over 60#						4.69			4.69	5.15
1900	Cast iron, type 8 Nickel, 1/8" diam., less than 500#						22.50			22.50	24.50
1910	500# to 1,000#						20			20	22
1920	Over 1,000#						18.90			18.90	21
2000	Stainless steel, type 316/316L, 1/8" diam., less than 500#						7.15			7.15	7.85
2100	500# to 1,000#						6.45			6.45	7.10
2220	Over 1,000#						6.05			6.05	6.65

05 12 Structural Steel Framing

05 12 23 – Structural Steel for Buildings

05 12 23.40 Lightweight Framing

	05 12 23.40 Lightweight Framing		Crew	Daily Output	Labor-Hours	Unit	Material	2020 Bare Costs Labor	Equipment	Total	Total Incl O&P
0010	**LIGHTWEIGHT FRAMING**										
0015	Made from recycled materials										
0400	Angle framing, field fabricated, 4″ and larger	G	E-3	440	.055	Lb.	.83	3.18	.33	4.34	6.25
0450	Less than 4″ angles	G		265	.091		.86	5.30	.56	6.72	9.85
0600	Channel framing, field fabricated, 8″ and larger	G		500	.048		.86	2.80	.29	3.95	5.65
0650	Less than 8″ channels	G		335	.072		.86	4.17	.44	5.47	8
1000	Continuous slotted channel framing system, shop fab, simple framing	G	2 Sswk	2400	.007		4.45	.38		4.83	5.50
1200	Complex framing	G	″	1600	.010		5.05	.58		5.63	6.45
1250	Plate & bar stock for reinforcing beams and trusses	G					1.58			1.58	1.74
1300	Cross bracing, rods, shop fabricated, 3/4″ diameter	G	E-3	700	.034		1.72	2	.21	3.93	5.25
1310	7/8″ diameter	G		850	.028		1.72	1.65	.17	3.54	4.67
1320	1″ diameter	G		1000	.024		1.72	1.40	.15	3.27	4.25
1330	Angle, 5″ x 5″ x 3/8″	G		2800	.009		1.72	.50	.05	2.27	2.74
1350	Hanging lintels, shop fabricated	G		850	.028		1.72	1.65	.17	3.54	4.67
1380	Roof frames, shop fabricated, 3′-0″ square, 5′ span	G	E-2	4200	.013		1.72	.76	.40	2.88	3.52
1400	Tie rod, not upset, 1-1/2″ to 4″ diameter, with turnbuckle	G	2 Sswk	800	.020		1.87	1.15		3.02	3.86
1420	No turnbuckle	G		700	.023		1.80	1.32		3.12	4.05
1500	Upset, 1-3/4″ to 4″ diameter, with turnbuckle	G		800	.020		1.87	1.15		3.02	3.86
1520	No turnbuckle	G		700	.023		1.80	1.32		3.12	4.05

05 12 23.60 Pipe Support Framing

	05 12 23.60 Pipe Support Framing		Crew	Daily Output	Labor-Hours	Unit	Material	Labor	Equipment	Total	Total Incl O&P
0010	**PIPE SUPPORT FRAMING**										
0020	Under 10#/L.F., shop fabricated	G	E-4	3900	.008	Lb.	1.93	.48	.04	2.45	2.91
0200	10.1 to 15#/L.F.	G		4300	.007		1.90	.43	.03	2.36	2.81
0400	15.1 to 20#/L.F.	G		4800	.007		1.87	.39	.03	2.29	2.69
0600	Over 20#/L.F.	G		5400	.006		1.84	.34	.03	2.21	2.59

05 54 Metal Floor Plates

05 54 13 – Floor Plates

05 54 13.20 Checkered Plates

	05 54 13.20 Checkered Plates		Crew	Daily Output	Labor-Hours	Unit	Material	Labor	Equipment	Total	Total Incl O&P
0010	**CHECKERED PLATES**, steel, field fabricated										
0015	Made from recycled materials										
0020	1/4″ & 3/8″, 2000 to 5000 S.F., bolted	G	E-4	2900	.011	Lb.	.80	.64	.05	1.49	1.95
0100	Welded	G		4400	.007	″	.91	.42	.03	1.36	1.70
0300	Pit or trench cover and frame, 1/4″ plate, 2′ to 3′ wide	G		100	.320	S.F.	10.95	18.60	1.47	31.02	42.50
0400	For galvanizing, add	G				Lb.	.32			.32	.35
0500	Platforms, 1/4″ plate, no handrails included, rectangular	G	E-4	4200	.008		3.55	.44	.04	4.03	4.65
0600	Circular	G	″	2500	.013		4.44	.74	.06	5.24	6.10

Estimating Tips

06 05 00 Common Work Results for Wood, Plastics, and Composites

- Common to any wood-framed structure are the accessory connector items such as screws, nails, adhesives, hangers, connector plates, straps, angles, and hold-downs. For typical wood-framed buildings, such as residential projects, the aggregate total for these items can be significant, especially in areas where seismic loading is a concern. For floor and wall framing, the material cost is based on 10 to 25 lbs. of accessory connectors per MBF. Hold-downs, hangers, and other connectors should be taken off by the piece.

  Included with material costs are fasteners for a normal installation. Gordian's RSMeans engineers use manufacturers' recommendations, written specifications, and/or standard construction practice for the sizing and spacing of fasteners. Prices for various fasteners are shown for informational purposes only. Adjustments should be made if unusual fastening conditions exist.

06 10 00 Carpentry

- Lumber is a traded commodity and therefore sensitive to supply and demand in the marketplace. Even with "budgetary" estimating of wood-framed projects, it is advisable to call local suppliers for the latest market pricing.
- The common quantity unit for wood-framed projects is "thousand board feet" (MBF). A board foot is a volume of wood—1" x 1' x 1' or 144 cubic inches. Board-foot quantities are generally calculated using nominal material dimensions—dressed sizes are ignored. Board foot per lineal foot of any stick of lumber can be calculated by dividing the nominal cross-sectional area by 12. As an example, 2,000 lineal feet of 2 x 12 equates to 4 MBF by dividing the nominal area, 2 x 12, by 12, which equals 2, and multiplying that by 2,000 to give 4,000 board feet. This simple rule applies to all nominal dimensioned lumber.
- Waste is an issue of concern at the quantity takeoff for any area of construction. Framing lumber is sold in even foot lengths, i.e., 8', 10', 12', 14', 16', and depending on spans, wall heights, and the grade of lumber, waste is inevitable. A rule of thumb for lumber waste is 5–10% depending on material quality and the complexity of the framing.
- Wood in various forms and shapes is used in many projects, even where the main structural framing is steel, concrete, or masonry. Plywood as a back-up partition material and 2x boards used as blocking and cant strips around roof edges are two common examples. The estimator should ensure that the costs of all wood materials are included in the final estimate.

06 20 00 Finish Carpentry

- It is necessary to consider the grade of workmanship when estimating labor costs for erecting millwork and an interior finish. In practice, there are three grades: premium, custom, and economy. The RSMeans daily output for base and case moldings is in the range of 200 to 250 L.F. per carpenter per day. This is appropriate for most average custom-grade projects. For premium projects, an adjustment to productivity of 25–50% should be made, depending on the complexity of the job.

Reference Numbers

Reference numbers are shown at the beginning of some major classifications. These numbers refer to related items in the Reference Section. The reference information may be an estimating procedure, an alternate pricing method, or technical information.

Note: Not all subdivisions listed here necessarily appear. ■

06 11 Wood Framing

06 11 10 – Framing with Dimensional, Engineered or Composite Lumber

06 11 10.24 Miscellaneous Framing		Crew	Daily Output	Labor-Hours	Unit	Material	2020 Bare Costs Labor	Equipment	Total	Total Incl O&P
0010	MISCELLANEOUS FRAMING									
8500	Firestops, 2" x 4"	2 Carp	.51	31.373	M.B.F.	710	1,675		2,385	3,275
8505	Pneumatic nailed		.62	25.806		720	1,375		2,095	2,850
8520	2" x 6"		.60	26.667		685	1,425		2,110	2,875
8525	Pneumatic nailed		.73	21.858		695	1,150		1,845	2,525
8540	2" x 8"		.60	26.667		750	1,425		2,175	2,950
8560	2" x 12"		.70	22.857		870	1,225		2,095	2,775
8600	Nailers, treated, wood construction, 2" x 4"		.53	30.189		860	1,600		2,460	3,350
8605	Pneumatic nailed		.64	25.157		865	1,325		2,190	2,950
8620	2" x 6"		.75	21.333		760	1,125		1,885	2,550
8625	Pneumatic nailed	↓	.90	17.778	↓	770	945		1,715	2,275

06 16 Sheathing

06 16 36 – Wood Panel Product Sheathing

06 16 36.10 Sheathing		Crew	Daily Output	Labor-Hours	Unit	Material	Labor	Equipment	Total	Total Incl O&P
0010	SHEATHING									
0012	Plywood on roofs, CDX									
0050	3/8" thick	2 Carp	1525	.010	S.F.	.53	.56		1.09	1.42
0055	Pneumatic nailed		1860	.009		.53	.46		.99	1.27
0200	5/8" thick		1300	.012		.77	.65		1.42	1.83
0205	Pneumatic nailed		1586	.010		.77	.54		1.31	1.66
0300	3/4" thick		1200	.013		.95	.71		1.66	2.10
0305	Pneumatic nailed		1464	.011		.95	.58		1.53	1.91
0500	Plywood on walls, with exterior CDX, 3/8" thick		1200	.013		.53	.71		1.24	1.64
0505	Pneumatic nailed		1488	.011		.53	.57		1.10	1.44
0700	5/8" thick		1050	.015		.77	.81		1.58	2.07
0705	Pneumatic nailed		1302	.012		.77	.65		1.42	1.83
0800	3/4" thick		975	.016		.95	.87		1.82	2.35
0805	Pneumatic nailed	↓	1209	.013	↓	.95	.70		1.65	2.10

Estimating Tips

07 10 00 Dampproofing and Waterproofing

- Be sure of the job specifications before pricing this subdivision. The difference in cost between waterproofing and dampproofing can be great. Waterproofing will hold back standing water. Dampproofing prevents the transmission of water vapor. Also included in this section are vapor retarding membranes.

07 20 00 Thermal Protection

- Insulation and fireproofing products are measured by area, thickness, volume, or R-value. Specifications may give only what the specific R-value should be in a certain situation. The estimator may need to choose the type of insulation to meet that R-value.

07 30 00 Steep Slope Roofing
07 40 00 Roofing and Siding Panels

- Many roofing and siding products are bought and sold by the square. One square is equal to an area that measures 100 square feet.

  This simple change in unit of measure could create a large error if the estimator is not observant. Accessories necessary for a complete installation must be figured into any calculations for both material and labor.

07 50 00 Membrane Roofing
07 60 00 Flashing and Sheet Metal
07 70 00 Roofing and Wall Specialties and Accessories

- The items in these subdivisions compose a roofing system. No one component completes the installation, and all must be estimated. Built-up or single-ply membrane roofing systems are made up of many products and installation trades. Wood blocking at roof perimeters or penetrations, parapet coverings, reglets, roof drains, gutters, downspouts, sheet metal flashing, skylights, smoke vents, and roof hatches all need to be considered along with the roofing material. Several different installation trades will need to work together on the roofing system. Inherent difficulties in the scheduling and coordination of various trades must be accounted for when estimating labor costs.

07 90 00 Joint Protection

- To complete the weather-tight shell, the sealants and caulkings must be estimated. Where different materials meet—at expansion joints, at flashing penetrations, and at hundreds of other locations throughout a construction project—caulking and sealants provide another line of defense against water penetration. Often, an entire system is based on the proper location and placement of caulking or sealants. The detailed drawings that are included as part of a set of architectural plans show typical locations for these materials. When caulking or sealants are shown at typical locations, this means the estimator must include them for all the locations where this detail is applicable. Be careful to keep different types of sealants separate, and remember to consider backer rods and primers if necessary.

Reference Numbers

Reference numbers are shown at the beginning of some major classifications. These numbers refer to related items in the Reference Section. The reference information may be an estimating procedure, an alternate pricing method, or technical information.

Note: Not all subdivisions listed here necessarily appear. ■

07 65 Flexible Flashing

07 65 10 – Sheet Metal Flashing

07 65 10.10 Sheet Metal Flashing and Counter Flashing		Crew	Daily Output	Labor-Hours	Unit	Material	2020 Bare Costs Labor	Equipment	Total	Total Incl O&P
0010	**SHEET METAL FLASHING AND COUNTER FLASHING**									
0011	Including up to 4 bends									
0020	Aluminum, mill finish, .013" thick	1 Rofc	145	.055	S.F.	1.01	2.55		3.56	5.30
0030	.016" thick		145	.055		1.15	2.55		3.70	5.50
0060	.019" thick		145	.055		1.50	2.55		4.05	5.85
0100	.032" thick		145	.055		1.38	2.55		3.93	5.75
0200	.040" thick		145	.055		2.33	2.55		4.88	6.75
0300	.050" thick		145	.055		3.22	2.55		5.77	7.75
0325	Mill finish 5" x 7" step flashing, .016" thick		1920	.004	Ea.	.15	.19		.34	.49
0350	Mill finish 12" x 12" step flashing, .016" thick		1600	.005	"	.60	.23		.83	1.04
0400	Painted finish, add				S.F.	.34			.34	.37
1000	Mastic-coated 2 sides, .005" thick	1 Rofc	330	.024		1.89	1.12		3.01	3.93
1100	.016" thick		330	.024		2.09	1.12		3.21	4.15
1600	Copper, 16 oz. sheets, under 1000 lb.		115	.070		8.15	3.21		11.36	14.30
1700	Over 4000 lb.		155	.052		8.15	2.38		10.53	12.95
1900	20 oz. sheets, under 1000 lb.		110	.073		10.45	3.36		13.81	17.05
2000	Over 4000 lb.		145	.055		9.95	2.55		12.50	15.15
2200	24 oz. sheets, under 1000 lb.		105	.076		15.30	3.52		18.82	22.50
2300	Over 4000 lb.		135	.059		14.55	2.74		17.29	20.50
2500	32 oz. sheets, under 1000 lb.		100	.080		21	3.70		24.70	29
2600	Over 4000 lb.		130	.062		19.75	2.84		22.59	26
5800	Lead, 2.5 lb./S.F., up to 12" wide		135	.059		6.10	2.74		8.84	11.20
5900	Over 12" wide		135	.059		4.05	2.74		6.79	9
8900	Stainless steel sheets, 32 ga.		155	.052		3.51	2.38		5.89	7.80
9000	28 ga.		155	.052		5	2.38		7.38	9.45
9100	26 ga.		155	.052		5.10	2.38		7.48	9.55
9200	24 ga.		155	.052		5.85	2.38		8.23	10.35
9400	Terne coated stainless steel, .015" thick, 28 ga.		155	.052		8.20	2.38		10.58	12.95
9500	.018" thick, 26 ga.		155	.052		9.15	2.38		11.53	14
9600	Zinc and copper alloy (brass), .020" thick		155	.052		10.10	2.38		12.48	15.05
9700	.027" thick		155	.052		12.25	2.38		14.63	17.45
9800	.032" thick		155	.052		15.45	2.38		17.83	21
9900	.040" thick		155	.052		20.50	2.38		22.88	26.50

07 65 13 – Laminated Sheet Flashing

07 65 13.10 Laminated Sheet Flashing		Crew	Daily Output	Labor-Hours	Unit	Material	Labor	Equipment	Total	Total Incl O&P
0010	**LAMINATED SHEET FLASHING**, Including up to 4 bends									
0500	Aluminum, fabric-backed 2 sides, mill finish, .004" thick	1 Rofc	330	.024	S.F.	1.58	1.12		2.70	3.59
0700	.005" thick		330	.024		1.88	1.12		3	3.92
0750	Mastic-backed, self adhesive		460	.017		3.41	.80		4.21	5.10
0800	Mastic-coated 2 sides, .004" thick		330	.024		1.58	1.12		2.70	3.59
2800	Copper, paperbacked 1 side, 2 oz.		330	.024		2.16	1.12		3.28	4.23
2900	3 oz.		330	.024		3.15	1.12		4.27	5.30
3100	Paperbacked 2 sides, 2 oz.		330	.024		2.39	1.12		3.51	4.48
3150	3 oz.		330	.024		2.24	1.12		3.36	4.31
3200	5 oz.		330	.024		3.65	1.12		4.77	5.85
3250	7 oz.		330	.024		6.85	1.12		7.97	9.35
3400	Mastic-backed 2 sides, copper, 2 oz.		330	.024		2.15	1.12		3.27	4.22
3500	3 oz.		330	.024		2.57	1.12		3.69	4.68
3700	5 oz.		330	.024		3.90	1.12		5.02	6.15
3800	Fabric-backed 2 sides, copper, 2 oz.		330	.024		2.05	1.12		3.17	4.11
4000	3 oz.		330	.024		2.88	1.12		4	5
4100	5 oz.		330	.024		4.03	1.12		5.15	6.30

07 65 Flexible Flashing

07 65 13 – Laminated Sheet Flashing

07 65 13.10 Laminated Sheet Flashing		Crew	Daily Output	Labor-Hours	Unit	2020 Bare Costs Material	Labor	Equipment	Total	Total Incl O&P
4300	Copper-clad stainless steel, .015" thick, under 500 lb.	1 Rofc	115	.070	S.F.	7.05	3.21		10.26	13.05
4400	Over 2000 lb.		155	.052		7.05	2.38		9.43	11.70
4600	.018" thick, under 500 lb.		100	.080		7.95	3.70		11.65	14.85
4700	Over 2000 lb.		145	.055		8.10	2.55		10.65	13.10
8550	Shower pan, 3 ply copper and fabric, 3 oz.		155	.052		4.12	2.38		6.50	8.45
8600	7 oz.		155	.052		5.45	2.38		7.83	9.90
9300	Stainless steel, paperbacked 2 sides, .005" thick	↓	330	.024	↓	4.04	1.12		5.16	6.30

07 65 19 – Plastic Sheet Flashing

07 65 19.10 Plastic Sheet Flashing and Counter Flashing

		Crew	Daily Output	Labor-Hours	Unit	Material	Labor	Equipment	Total	Total Incl O&P
0010	PLASTIC SHEET FLASHING AND COUNTER FLASHING									
7300	Polyvinyl chloride, black, 10 mil	1 Rofc	285	.028	S.F.	.27	1.30		1.57	2.44
7400	20 mil		285	.028		.26	1.30		1.56	2.43
7600	30 mil		285	.028		.33	1.30		1.63	2.50
7700	60 mil		285	.028		.88	1.30		2.18	3.11
7900	Black or white for exposed roofs, 60 mil	↓	285	.028	↓	1.11	1.30		2.41	3.36
8060	PVC tape, 5" x 45 mils, for joint covers, 100 L.F./roll				Ea.	183			183	201
8850	Polyvinyl chloride, 30 mil	1 Rofc	160	.050	S.F.	1.47	2.31		3.78	5.45

07 71 Roof Specialties

07 71 16 – Manufactured Counterflashing Systems

07 71 16.20 Pitch Pockets, Variable Sizes

		Crew	Daily Output	Labor-Hours	Unit	Material	Labor	Equipment	Total	Total Incl O&P
0010	PITCH POCKETS, VARIABLE SIZES									
0100	Adjustable, 4" to 7", welded corners, 4" deep	1 Rofc	48	.167	Ea.	21.50	7.70		29.20	36.50
0200	Side extenders, 6"	"	240	.033	"	3.34	1.54		4.88	6.20

07 72 Roof Accessories

07 72 53 – Snow Guards

07 72 53.10 Snow Guard Options

		Crew	Daily Output	Labor-Hours	Unit	Material	Labor	Equipment	Total	Total Incl O&P
0010	SNOW GUARD OPTIONS									
0100	Slate & asphalt shingle roofs, fastened with nails	1 Rofc	160	.050	Ea.	12.45	2.31		14.76	17.50
0200	Standing seam metal roofs, fastened with set screws		48	.167		17.75	7.70		25.45	32.50
0300	Surface mount for metal roofs, fastened with solder		48	.167	↓	7.75	7.70		15.45	21.50
0400	Double rail pipe type, including pipe	↓	130	.062	L.F.	35	2.84		37.84	43

07 76 Roof Pavers

07 76 16 – Roof Decking Pavers

07 76 16.10 Roof Pavers and Supports

		Crew	Daily Output	Labor-Hours	Unit	Material	Labor	Equipment	Total	Total Incl O&P
0010	ROOF PAVERS AND SUPPORTS									
1000	Roof decking pavers, concrete blocks, 2" thick, natural	1 Clab	115	.070	S.F.	3.58	2.93		6.51	8.35
1100	Colors		115	.070	"	3.71	2.93		6.64	8.50
1200	Support pedestal, bottom cap		960	.008	Ea.	2.66	.35		3.01	3.46
1300	Top cap		960	.008		4.88	.35		5.23	5.90
1400	Leveling shims, 1/16"		1920	.004		1.22	.18		1.40	1.60
1500	1/8"		1920	.004		1.22	.18		1.40	1.60
1600	Buffer pad		960	.008	↓	2.54	.35		2.89	3.32
1700	PVC legs (4" SDR 35)		2880	.003	Inch	.14	.12		.26	.33
2000	Alternate pricing method, system in place	↓	101	.079	S.F.	7.20	3.33		10.53	12.90

07 84 Firestopping

07 84 13 – Penetration Firestopping

07 84 13.10 Firestopping		Crew	Daily Output	Labor-Hours	Unit	2020 Bare Costs Material	Labor	Equipment	Total	Total Incl O&P
0010	**FIRESTOPPING**									
0100	Metallic piping, non insulated									
0110	Through walls, 2" diameter	1 Carp	16	.500	Ea.	3.67	26.50		30.17	44
0120	4" diameter		14	.571		8.20	30.50		38.70	54.50
0130	6" diameter		12	.667		12.60	35.50		48.10	67
0140	12" diameter		10	.800		25	42.50		67.50	91.50
0150	Through floors, 2" diameter		32	.250		1.86	13.30		15.16	22
0160	4" diameter		28	.286		3.28	15.20		18.48	26.50
0170	6" diameter		24	.333		6.45	17.70		24.15	33.50
0180	12" diameter		20	.400		12.80	21.50		34.30	46
0190	Metallic piping, insulated									
0200	Through walls, 2" diameter	1 Carp	16	.500	Ea.	8.60	26.50		35.10	49.50
0210	4" diameter		14	.571		12.55	30.50		43.05	59.50
0220	6" diameter		12	.667		16.75	35.50		52.25	71.50
0230	12" diameter		10	.800		28.50	42.50		71	95.50
0240	Through floors, 2" diameter		32	.250		3.27	13.30		16.57	23.50
0250	4" diameter		28	.286		6.30	15.20		21.50	30
0260	6" diameter		24	.333		8.70	17.70		26.40	36
0270	12" diameter		20	.400		12.95	21.50		34.45	46
0280	Non metallic piping, non insulated									
0290	Through walls, 2" diameter	1 Carp	12	.667	Ea.	52.50	35.50		88	111
0300	4" diameter		10	.800		101	42.50		143.50	175
0310	6" diameter		8	1		186	53		239	285
0330	Through floors, 2" diameter		16	.500		36.50	26.50		63	80
0340	4" diameter		6	1.333		62	71		133	175
0350	6" diameter		6	1.333		135	71		206	255
0370	Ductwork, insulated & non insulated, round									
0380	Through walls, 6" diameter	1 Carp	12	.667	Ea.	16.55	35.50		52.05	71
0390	12" diameter		10	.800		28.50	42.50		71	95.50
0400	18" diameter		8	1		33.50	53		86.50	117
0410	Through floors, 6" diameter		16	.500		8.50	26.50		35	49.50
0420	12" diameter		14	.571		14.60	30.50		45.10	61.50
0430	18" diameter		12	.667		17.60	35.50		53.10	72.50
0440	Ductwork, insulated & non insulated, rectangular									
0450	With stiffener/closure angle, through walls, 6" x 12"	1 Carp	8	1	Ea.	21	53		74	103
0460	12" x 24"		6	1.333		36	71		107	146
0470	24" x 48"		4	2		84	106		190	252
0480	With stiffener/closure angle, through floors, 6" x 12"		10	.800		10.40	42.50		52.90	75.50
0490	12" x 24"		8	1		20	53		73	102
0500	24" x 48"		6	1.333		42	71		113	152
0510	Multi trade openings									
0520	Through walls, 6" x 12"	1 Carp	2	4	Ea.	52	213		265	380
0530	12" x 24"	"	1	8		189	425		614	850
0540	24" x 48"	2 Carp	1	16		670	850		1,520	2,025
0550	48" x 96"	"	.75	21.333		2,550	1,125		3,675	4,500
0560	Through floors, 6" x 12"	1 Carp	2	4		44.50	213		257.50	370
0570	12" x 24"	"	1	8		69.50	425		494.50	715
0580	24" x 48"	2 Carp	.75	21.333		154	1,125		1,279	1,875
0590	48" x 96"	"	.50	32		365	1,700		2,065	2,950
0600	Structural penetrations, through walls									
0610	Steel beams, W8 x 10	1 Carp	8	1	Ea.	75	53		128	163
0620	W12 x 14		6	1.333		109	71		180	225
0630	W21 x 44		5	1.600		154	85		239	297

07 84 Firestopping

07 84 13 – Penetration Firestopping

07 84 13.10 Firestopping		Crew	Daily Output	Labor-Hours	Unit	2020 Bare Costs Material	Labor	Equipment	Total	Total Incl O&P
0640	W36 x 135	1 Carp	3	2.667	Ea.	229	142		371	465
0650	Bar joists, 18" deep		6	1.333		42	71		113	153
0660	24" deep		6	1.333		59	71		130	171
0670	36" deep		5	1.600		88.50	85		173.50	226
0680	48" deep	↓	4	2	↓	120	106		226	292
0690	Construction joints, floor slab at exterior wall									
0700	Precast, brick, block or drywall exterior									
0710	2" wide joint	1 Carp	125	.064	L.F.	10.30	3.40		13.70	16.45
0720	4" wide joint	"	75	.107	"	16.60	5.65		22.25	27
0730	Metal panel, glass or curtain wall exterior									
0740	2" wide joint	1 Carp	40	.200	L.F.	12.45	10.65		23.10	29.50
0750	4" wide joint	"	25	.320	"	18.65	17		35.65	46
0760	Floor slab to drywall partition									
0770	Flat joint	1 Carp	100	.080	L.F.	14.55	4.25		18.80	22.50
0780	Fluted joint		50	.160		16.35	8.50		24.85	31
0790	Etched fluted joint	↓	75	.107	↓	20.50	5.65		26.15	31
0800	Floor slab to concrete/masonry partition									
0810	Flat joint	1 Carp	75	.107	L.F.	15.45	5.65		21.10	25.50
0820	Fluted joint	"	50	.160	"	17.65	8.50		26.15	32.50
0830	Concrete/CMU wall joints									
0840	1" wide	1 Carp	100	.080	L.F.	20.50	4.25		24.75	29
0850	2" wide		75	.107		30	5.65		35.65	41.50
0860	4" wide	↓	50	.160	↓	36	8.50		44.50	52.50
0870	Concrete/CMU floor joints									
0880	1" wide	1 Carp	200	.040	L.F.	20.50	2.13		22.63	26
0890	2" wide		150	.053		25	2.83		27.83	32
0900	4" wide	↓	100	.080	↓	36.50	4.25		40.75	46.50

07 91 Preformed Joint Seals

07 91 13 – Compression Seals

07 91 13.10 Compression Seals		Crew	Daily Output	Labor-Hours	Unit	Material	Labor	Equipment	Total	Total Incl O&P
0010	**COMPRESSION SEALS**									
4900	O-ring type cord, 1/4"	1 Bric	472	.017	L.F.	.40	.88		1.28	1.78
4910	1/2"		440	.018		1.01	.95		1.96	2.54
4920	3/4"		424	.019		2	.98		2.98	3.69
4930	1"		408	.020		3.93	1.02		4.95	5.85
4940	1-1/4"		384	.021		9.40	1.08		10.48	12
4950	1-1/2"		368	.022		11.50	1.13		12.63	14.35
4960	1-3/4"		352	.023		13.50	1.18		14.68	16.65
4970	2"	↓	344	.023	↓	22.50	1.21		23.71	27

07 91 16 – Joint Gaskets

07 91 16.10 Joint Gaskets		Crew	Daily Output	Labor-Hours	Unit	Material	Labor	Equipment	Total	Total Incl O&P
0010	**JOINT GASKETS**									
4400	Joint gaskets, neoprene, closed cell w/adh, 1/8" x 3/8"	1 Bric	240	.033	L.F.	.36	1.73		2.09	3.03
4500	1/4" x 3/4"		215	.037		.66	1.94		2.60	3.67
4700	1/2" x 1"		200	.040		1.71	2.08		3.79	5.05
4800	3/4" x 1-1/2"	↓	165	.048	↓	1.85	2.52		4.37	5.85

07 91 Preformed Joint Seals

07 91 23 – Backer Rods

07 91 23.10 Backer Rods		Crew	Daily Output	Labor-Hours	Unit	2020 Bare Costs Material	Labor	Equipment	Total	Total Incl O&P
0010	BACKER RODS									
0030	Backer rod, polyethylene, 1/4" diameter	1 Bric	4.60	1.739	C.L.F.	2.78	90.50		93.28	140
0050	1/2" diameter		4.60	1.739		4.10	90.50		94.60	142
0070	3/4" diameter		4.60	1.739		6.70	90.50		97.20	144
0090	1" diameter		4.60	1.739		14.80	90.50		105.30	153

07 91 26 – Joint Fillers

07 91 26.10 Joint Fillers

Line	Description	Crew	Daily Output	Labor-Hours	Unit	Material	Labor	Equipment	Total	Total Incl O&P
0010	JOINT FILLERS									
4360	Butyl rubber filler, 1/4" x 1/4"	1 Bric	290	.028	L.F.	.23	1.44		1.67	2.43
4365	1/2" x 1/2"		250	.032		.92	1.67		2.59	3.53
4370	1/2" x 3/4"		210	.038		1.38	1.98		3.36	4.52
4375	3/4" x 3/4"		230	.035		2.06	1.81		3.87	5
4380	1" x 1"		180	.044		2.75	2.31		5.06	6.55
4390	For coloring, add					12%				
4980	Polyethylene joint backing, 1/4" x 2"	1 Bric	2.08	3.846	C.L.F.	14.25	200		214.25	320
4990	1/4" x 6"		1.28	6.250	"	32.50	325		357.50	530
5600	Silicone, room temp vulcanizing foam seal, 1/4" x 1/2"		1312	.006	L.F.	.46	.32		.78	.98
5610	1/2" x 1/2"		656	.012		.92	.64		1.56	1.97
5620	1/2" x 3/4"		442	.018		1.37	.94		2.31	2.94
5630	3/4" x 3/4"		328	.024		2.06	1.27		3.33	4.19
5640	1/8" x 1"		1312	.006		.46	.32		.78	.98
5650	1/8" x 3"		442	.018		1.37	.94		2.31	2.94
5670	1/4" x 3"		295	.027		2.75	1.41		4.16	5.15
5680	1/4" x 6"		148	.054		5.50	2.81		8.31	10.30
5690	1/2" x 6"		82	.098		11	5.10		16.10	19.80
5700	1/2" x 9"		52.50	.152		16.50	7.95		24.45	30
5710	1/2" x 12"		33	.242		22	12.60		34.60	43

07 92 Joint Sealants

07 92 13 – Elastomeric Joint Sealants

07 92 13.20 Caulking and Sealant Options

Line	Description	Crew	Daily Output	Labor-Hours	Unit	Material	Labor	Equipment	Total	Total Incl O&P
0010	CAULKING AND SEALANT OPTIONS									
0050	Latex acrylic based, bulk				Gal.	31			31	34
0055	Bulk in place 1/4" x 1/4" bead	1 Bric	300	.027	L.F.	.10	1.39		1.49	2.21
0060	1/4" x 3/8"		294	.027		.16	1.42		1.58	2.33
0065	1/4" x 1/2"		288	.028		.22	1.45		1.67	2.43
0075	3/8" x 3/8"		284	.028		.24	1.47		1.71	2.49
0080	3/8" x 1/2"		280	.029		.32	1.49		1.81	2.61
0085	3/8" x 5/8"		276	.029		.40	1.51		1.91	2.74
0095	3/8" x 3/4"		272	.029		.48	1.53		2.01	2.85
0100	1/2" x 1/2"		275	.029		.43	1.51		1.94	2.78
0105	1/2" x 5/8"		269	.030		.54	1.55		2.09	2.94
0110	1/2" x 3/4"		263	.030		.65	1.58		2.23	3.11
0115	1/2" x 7/8"		256	.031		.76	1.63		2.39	3.30
0120	1/2" x 1"		250	.032		.87	1.67		2.54	3.47
0125	3/4" x 3/4"		244	.033		.97	1.71		2.68	3.66
0130	3/4" x 1"		225	.036		1.30	1.85		3.15	4.24
0135	1" x 1"		200	.040		1.73	2.08		3.81	5.05
0190	Cartridges				Gal.	33.50			33.50	37
0200	11 fl. oz. cartridge				Ea.	2.89			2.89	3.18

07 92 Joint Sealants

07 92 13 – Elastomeric Joint Sealants

07 92 13.20 Caulking and Sealant Options		Crew	Daily Output	Labor-Hours	Unit	Material	Labor	Equipment	Total	Total Incl O&P
						2020 Bare Costs				
0500	1/4" x 1/2"	1 Bric	288	.028	L.F.	.24	1.45		1.69	2.45
0600	1/2" x 1/2"		275	.029		.47	1.51		1.98	2.82
0800	3/4" x 3/4"		244	.033		1.06	1.71		2.77	3.76
0900	3/4" x 1"		225	.036		1.42	1.85		3.27	4.37
1000	1" x 1"		200	.040		1.77	2.08		3.85	5.10
1400	Butyl based, bulk				Gal.	43.50			43.50	48
1500	Cartridges				"	43.50			43.50	48
1700	1/4" x 1/2", 154 L.F./gal.	1 Bric	288	.028	L.F.	.28	1.45		1.73	2.50
1800	1/2" x 1/2", 77 L.F./gal.	"	275	.029	"	.57	1.51		2.08	2.92
2300	Polysulfide compounds, 1 component, bulk				Gal.	88.50			88.50	97.50
2600	1 or 2 component, in place, 1/4" x 1/4", 308 L.F./gal.	1 Bric	300	.027	L.F.	.29	1.39		1.68	2.42
2700	1/2" x 1/4", 154 L.F./gal.		288	.028		.57	1.45		2.02	2.82
2900	3/4" x 3/8", 68 L.F./gal.		272	.029		1.30	1.53		2.83	3.75
3000	1" x 1/2", 38 L.F./gal.		250	.032		2.33	1.67		4	5.10
3200	Polyurethane, 1 or 2 component				Gal.	54			54	59.50
3300	Cartridges				"	67.50			67.50	74.50
3500	Bulk, in place, 1/4" x 1/4"	1 Bric	300	.027	L.F.	.18	1.39		1.57	2.29
3655	1/2" x 1/4"		288	.028		.35	1.45		1.80	2.58
3800	3/4" x 3/8"		272	.029		.80	1.53		2.33	3.19
3900	1" x 1/2"		250	.032		1.40	1.67		3.07	4.06
4100	Silicone rubber, bulk				Gal.	68.50			68.50	75.50
4200	Cartridges				"	51.50			51.50	56.50

07 92 16 – Rigid Joint Sealants

07 92 16.10 Rigid Joint Sealants		Crew	Daily Output	Labor-Hours	Unit	Material	Labor	Equipment	Total	Total Incl O&P
0010	**RIGID JOINT SEALANTS**									
5800	Tapes, sealant, PVC foam adhesive, 1/16" x 1/4"				C.L.F.	9.05			9.05	10
5900	1/16" x 1/2"					9.30			9.30	10.25
5950	1/16" x 1"					16.20			16.20	17.85
6000	1/8" x 1/2"					9.20			9.20	10.10

07 92 19 – Acoustical Joint Sealants

07 92 19.10 Acoustical Sealant		Crew	Daily Output	Labor-Hours	Unit	Material	Labor	Equipment	Total	Total Incl O&P
0010	**ACOUSTICAL SEALANT**									
0020	Acoustical sealant, elastomeric, cartridges				Ea.	8.65			8.65	9.55
0025	In place, 1/4" x 1/4"	1 Bric	300	.027	L.F.	.35	1.39		1.74	2.49
0030	1/4" x 1/2"		288	.028		.71	1.45		2.16	2.97
0035	1/2" x 1/2"		275	.029		1.41	1.51		2.92	3.86
0040	1/2" x 3/4"		263	.030		2.12	1.58		3.70	4.73
0045	3/4" x 3/4"		244	.033		3.18	1.71		4.89	6.10
0050	1" x 1"		200	.040		5.65	2.08		7.73	9.40

Division Notes

	CREW	DAILY OUTPUT	LABOR-HOURS	UNIT	BARE COSTS MAT.	LABOR	EQUIP.	TOTAL	TOTAL INCL O&P

Estimating Tips

General

- Room Finish Schedule: A complete set of plans should contain a room finish schedule. If one is not available, it would be well worth the time and effort to obtain one.

09 20 00 Plaster and Gypsum Board

- Lath is estimated by the square yard plus a 5% allowance for waste. Furring, channels, and accessories are measured by the linear foot. An extra foot should be allowed for each accessory miter or stop.
- Plaster is also estimated by the square yard. Deductions for openings vary by preference, from zero deduction to 50% of all openings over 2 feet in width. The estimator should allow one extra square foot for each linear foot of horizontal interior or exterior angle located below the ceiling level. Also, double the areas of small radius work.
- Drywall accessories, studs, track, and acoustical caulking are all measured by the linear foot. Drywall taping is figured by the square foot. Gypsum wallboard is estimated by the square foot. No material deductions should be made for door or window openings under 32 S.F.

09 60 00 Flooring

- Tile and terrazzo areas are taken off on a square foot basis. Trim and base materials are measured by the linear foot. Accent tiles are listed per each. Two basic methods of installation are used. Mud set is approximately 30% more expensive than thin set. The cost of grout is included with tile unit price lines unless otherwise noted. In terrazzo work, be sure to include the linear footage of embedded decorative strips, grounds, machine rubbing, and power cleanup.
- Wood flooring is available in strip, parquet, or block configuration. The latter two types are set in adhesives with quantities estimated by the square foot. The laying pattern will influence labor costs and material waste. In addition to the material and labor for laying wood floors, the estimator must make allowances for sanding and finishing these areas, unless the flooring is prefinished.
- Sheet flooring is measured by the square yard. Roll widths vary, so consideration should be given to use the most economical width, as waste must be figured into the total quantity. Consider also the installation methods available—direct glue down or stretched. Direct glue-down installation is assumed with sheet carpet unit price lines unless otherwise noted.

09 70 00 Wall Finishes

- Wall coverings are estimated by the square foot. The area to be covered is measured—length by height of the wall above the baseboards—to calculate the square footage of each wall. This figure is divided by the number of square feet in the single roll which is being used. Deduct, in full, the areas of openings such as doors and windows. Where a pattern match is required allow 25–30% waste.

09 80 00 Acoustic Treatment

- Acoustical systems fall into several categories. The takeoff of these materials should be by the square foot of area with a 5% allowance for waste. Do not forget about scaffolding, if applicable, when estimating these systems.

09 90 00 Painting and Coating

- New line items created for cut-ins with reference diagram.
- A major portion of the work in painting involves surface preparation. Be sure to include cleaning, sanding, filling, and masking costs in the estimate.
- Protection of adjacent surfaces is not included in painting costs. When considering the method of paint application, an important factor is the amount of protection and masking required. These must be estimated separately and may be the determining factor in choosing the method of application.

Reference Numbers

Reference numbers are shown at the beginning of some major classifications. These numbers refer to related items in the Reference Section. The reference information may be an estimating procedure, an alternate pricing method, or technical information.

Note: Not all subdivisions listed here necessarily appear. ■

09 22 Supports for Plaster and Gypsum Board

09 22 03 – Fastening Methods for Finishes

09 22 03.20 Drilling Plaster/Drywall		Crew	Daily Output	Labor-Hours	Unit	2020 Bare Costs Material	Labor	Equipment	Total	Total Incl O&P
0010	**DRILLING PLASTER/DRYWALL**									
1100	Drilling & layout for drywall/plaster walls, up to 1" deep, no anchor									
1200	Holes, 1/4" diameter	1 Carp	150	.053	Ea.	.01	2.83		2.84	4.27
1300	3/8" diameter		140	.057		.01	3.04		3.05	4.57
1400	1/2" diameter		130	.062		.01	3.27		3.28	4.92
1500	3/4" diameter		120	.067		.01	3.54		3.55	5.30
1600	1" diameter		110	.073		.02	3.87		3.89	5.80
1700	1-1/4" diameter		100	.080		.04	4.25		4.29	6.45
1800	1-1/2" diameter		90	.089		.05	4.72		4.77	7.15
1900	For ceiling installations, add						40%			

09 91 Painting

09 91 23 – Interior Painting

09 91 23.52 Miscellaneous, Interior		Crew	Daily Output	Labor-Hours	Unit	Material	Labor	Equipment	Total	Total Incl O&P
0010	**MISCELLANEOUS, INTERIOR**									
3800	Grilles, per side, oil base, primer coat, brushwork	1 Pord	520	.015	S.F.	.14	.68		.82	1.17
3850	Spray		1140	.007		.15	.31		.46	.62
3880	Paint 1 coat, brushwork		520	.015		.25	.68		.93	1.30
3900	Spray		1140	.007		.28	.31		.59	.77
3920	Paint 2 coats, brushwork		325	.025		.49	1.09		1.58	2.17
3940	Spray		650	.012		.56	.55		1.11	1.43
3950	Prime & paint 1 coat		325	.025		.39	1.09		1.48	2.06
3960	Prime & paint 2 coats		270	.030		.39	1.32		1.71	2.38
4500	Louvers, 1 side, primer, brushwork		524	.015		.09	.68		.77	1.11
4520	Paint 1 coat, brushwork		520	.015		.11	.68		.79	1.14
4530	Spray		1140	.007		.12	.31		.43	.60
4540	Paint 2 coats, brushwork		325	.025		.22	1.09		1.31	1.87
4550	Spray		650	.012		.24	.55		.79	1.07
4560	Paint 3 coats, brushwork		270	.030		.32	1.32		1.64	2.31
4570	Spray		500	.016		.36	.71		1.07	1.45
5000	Pipe, 1"-4" diameter, primer or sealer coat, oil base, brushwork	2 Pord	1250	.013	L.F.	.09	.57		.66	.95
5100	Spray		2165	.007		.09	.33		.42	.59
5200	Paint 1 coat, brushwork		1250	.013		.13	.57		.70	.99
5300	Spray		2165	.007		.11	.33		.44	.62
5350	Paint 2 coats, brushwork		775	.021		.23	.92		1.15	1.61
5400	Spray		1240	.013		.25	.57		.82	1.13
5420	Paint 3 coats, brushwork		775	.021		.34	.92		1.26	1.73
5450	5"-8" diameter, primer or sealer coat, brushwork		620	.026		.18	1.15		1.33	1.90
5500	Spray		1085	.015		.30	.65		.95	1.30
5550	Paint 1 coat, brushwork		620	.026		.35	1.15		1.50	2.08
5600	Spray		1085	.015		.38	.65		1.03	1.39
5650	Paint 2 coats, brushwork		385	.042		.45	1.85		2.30	3.25
5700	Spray		620	.026		.51	1.15		1.66	2.26
5720	Paint 3 coats, brushwork		385	.042		.68	1.85		2.53	3.49
5750	9"-12" diameter, primer or sealer coat, brushwork		415	.039		.28	1.71		1.99	2.85
5800	Spray		725	.022		.41	.98		1.39	1.91
5850	Paint 1 coat, brushwork		415	.039		.35	1.71		2.06	2.94
6000	Spray		725	.022		.39	.98		1.37	1.89
6200	Paint 2 coats, brushwork		260	.062		.68	2.73		3.41	4.81
6250	Spray		415	.039		.76	1.71		2.47	3.38
6270	Paint 3 coats, brushwork		260	.062		1.01	2.73		3.74	5.15

09 91 Painting

09 91 23 – Interior Painting

09 91 23.52 Miscellaneous, Interior		Crew	Daily Output	Labor-Hours	Unit	2020 Bare Costs Material	Labor	Equipment	Total	Total Incl O&P
6300	13"-16" diameter, primer or sealer coat, brushwork	2 Pord	310	.052	L.F.	.37	2.29		2.66	3.81
6350	Spray		540	.030		.41	1.32		1.73	2.41
6400	Paint 1 coat, brushwork		310	.052		.47	2.29		2.76	3.93
6450	Spray		540	.030		.52	1.32		1.84	2.53
6500	Paint 2 coats, brushwork		195	.082		.91	3.64		4.55	6.40
6550	Spray		310	.052		1.01	2.29		3.30	4.52
6600	Radiators, per side, primer, brushwork	1 Pord	520	.015	S.F.	.09	.68		.77	1.12
6620	Paint, 1 coat		520	.015		.08	.68		.76	1.11
6640	2 coats		340	.024		.22	1.04		1.26	1.79
6660	3 coats		283	.028		.32	1.26		1.58	2.22

Division Notes

		CREW	DAILY OUTPUT	LABOR-HOURS	UNIT	BARE COSTS MAT.	BARE COSTS LABOR	BARE COSTS EQUIP.	BARE COSTS TOTAL	TOTAL INCL O&P

Division 10 Specialties

Estimating Tips

General

- The items in this division are usually priced per square foot or each.
- Many items in Division 10 require some type of support system or special anchors that are not usually furnished with the item. The required anchors must be added to the estimate in the appropriate division.
- Some items in Division 10, such as lockers, may require assembly before installation. Verify the amount of assembly required. Assembly can often exceed installation time.

10 20 00 Interior Specialties

- Support angles and blocking are not included in the installation of toilet compartments, shower/dressing compartments, or cubicles. Appropriate line items from Division 5 or 6 may need to be added to support the installations.
- Toilet partitions are priced by the stall. A stall consists of a side wall, pilaster, and door with hardware. Toilet tissue holders and grab bars are extra.
- The required acoustical rating of a folding partition can have a significant impact on costs. Verify the sound transmission coefficient rating of the panel priced against the specification requirements.
- Grab bar installation does not include supplemental blocking or backing to support the required load. When grab bars are installed at an existing facility, provisions must be made to attach the grab bars to a solid structure.

Reference Numbers

Reference numbers are shown at the beginning of some major classifications. These numbers refer to related items in the Reference Section. The reference information may be an estimating procedure, an alternate pricing method, or technical information.

Note: Not all subdivisions listed here necessarily appear. ■

10 21 Compartments and Cubicles

10 21 13 – Toilet Compartments

10 21 13.13 Metal Toilet Compartments		Crew	Daily Output	Labor-Hours	Unit	2020 Bare Costs Material	Labor	Equipment	Total	Total Incl O&P
0010	METAL TOILET COMPARTMENTS									
0110	Cubicles, ceiling hung									
0200	Powder coated steel	2 Carp	4	4	Ea.	575	213		788	955
0500	Stainless steel	"	4	4		1,100	213		1,313	1,525
0600	For handicap units, add					465			465	510
0900	Floor and ceiling anchored									
1000	Powder coated steel	2 Carp	5	3.200	Ea.	580	170		750	895
1300	Stainless steel	"	5	3.200		1,225	170		1,395	1,600
1400	For handicap units, add					370			370	405
1610	Floor anchored									
1700	Powder coated steel	2 Carp	7	2.286	Ea.	615	121		736	860
2000	Stainless steel	"	7	2.286		1,350	121		1,471	1,650
2100	For handicap units, add					360			360	395
2200	For juvenile units, deduct					44.50			44.50	49
2450	Floor anchored, headrail braced									
2500	Powder coated steel	2 Carp	6	2.667	Ea.	385	142		527	640
2804	Stainless steel	"	6	2.667		1,050	142		1,192	1,400
2900	For handicap units, add					325			325	360
3000	Wall hung partitions, powder coated steel	2 Carp	7	2.286		675	121		796	930
3300	Stainless steel	"	7	2.286		1,775	121		1,896	2,125
3400	For handicap units, add					325			325	360
4000	Screens, entrance, floor mounted, 58" high, 48" wide									
4200	Powder coated steel	2 Carp	15	1.067	Ea.	240	56.50		296.50	350
4500	Stainless steel	"	15	1.067	"	970	56.50		1,026.50	1,150
4650	Urinal screen, 18" wide									
4704	Powder coated steel	2 Carp	6.15	2.602	Ea.	226	138		364	455
5004	Stainless steel	"	6.15	2.602	"	535	138		673	795
5100	Floor mounted, headrail braced									
5300	Powder coated steel	2 Carp	8	2	Ea.	234	106		340	415
5600	Stainless steel	"	8	2	"	580	106		686	800
5750	Pilaster, flush									
5800	Powder coated steel	2 Carp	10	1.600	Ea.	275	85		360	430
6100	Stainless steel		10	1.600		610	85		695	805
6300	Post braced, powder coated steel		10	1.600		167	85		252	310
6800	Powder coated steel		10	1.600		176	85		261	320
7800	Wedge type, powder coated steel		10	1.600		138	85		223	280
8100	Stainless steel		10	1.600		615	85		700	805

10 21 13.16 Plastic-Laminate-Clad Toilet Compartments		Crew	Daily Output	Labor-Hours	Unit	Material	Labor	Equipment	Total	Total Incl O&P
0010	PLASTIC-LAMINATE-CLAD TOILET COMPARTMENTS									
0110	Cubicles, ceiling hung									
0300	Plastic laminate on particle board	2 Carp	4	4	Ea.	550	213		763	925
0600	For handicap units, add				"	465			465	510
0900	Floor and ceiling anchored									
1100	Plastic laminate on particle board	2 Carp	5	3.200	Ea.	800	170		970	1,125
1400	For handicap units, add				"	370			370	405
1610	Floor mounted									
1800	Plastic laminate on particle board	2 Carp	7	2.286	Ea.	540	121		661	780
2450	Floor mounted, headrail braced									
2600	Plastic laminate on particle board	2 Carp	6	2.667	Ea.	810	142		952	1,100
3400	For handicap units, add					325			325	360
4300	Entrance screen, floor mtd., plas. lam., 58" high, 48" wide	2 Carp	15	1.067		520	56.50		576.50	655
4800	Urinal screen, 18" wide, ceiling braced, plastic laminate		8	2		210	106		316	390

10 21 Compartments and Cubicles

10 21 13 – Toilet Compartments

10 21 13.16 Plastic-Laminate-Clad Toilet Compartments

	10 21 13.16 Plastic-Laminate-Clad Toilet Compartments	Crew	Daily Output	Labor-Hours	Unit	Material	Labor	Equipment	Total	Total Incl O&P
						2020 Bare Costs				
5400	Floor mounted, headrail braced	2 Carp	8	2	Ea.	208	106		314	390
5900	Pilaster, flush, plastic laminate		10	1.600		430	85		515	605
6400	Post braced, plastic laminate		10	1.600		237	85		322	390
6700	Wall hung, bracket supported									
6900	Plastic laminate on particle board	2 Carp	10	1.600	Ea.	97	85		182	234
7450	Flange supported									
7500	Plastic laminate on particle board	2 Carp	10	1.600	Ea.	251	85		336	405

10 21 13.19 Plastic Toilet Compartments

	10 21 13.19 Plastic Toilet Compartments	Crew	Daily Output	Labor-Hours	Unit	Material	Labor	Equipment	Total	Total Incl O&P
0010	**PLASTIC TOILET COMPARTMENTS**									
0110	Cubicles, ceiling hung									
0250	Phenolic	2 Carp	4	4	Ea.	885	213		1,098	1,300
0260	Polymer plastic	"	4	4		1,025	213		1,238	1,450
0600	For handicap units, add					465			465	510
0900	Floor and ceiling anchored									
1050	Phenolic	2 Carp	5	3.200	Ea.	940	170		1,110	1,275
1060	Polymer plastic	"	5	3.200		1,200	170		1,370	1,575
1400	For handicap units, add					370			370	405
1610	Floor mounted									
1750	Phenolic	2 Carp	7	2.286	Ea.	880	121		1,001	1,150
1760	Polymer plastic	"	7	2.286		855	121		976	1,125
2100	For handicap units, add					360			360	395
2200	For juvenile units, deduct					44.50			44.50	49
2450	Floor mounted, headrail braced									
2550	Phenolic	2 Carp	6	2.667	Ea.	740	142		882	1,025
3600	Polymer plastic		6	2.667		870	142		1,012	1,175
3810	Entrance screen, polymer plastic, flr. mtd., 48"x58"		6	2.667		495	142		637	760
3820	Entrance screen, polymer plastic, flr. to clg pilaster, 48"x58"		6	2.667		590	142		732	865
6110	Urinal screen, polymer plastic, pilaster flush, 18" w		6	2.667		430	142		572	690
7110	Wall hung		6	2.667		167	142		309	395
7710	Flange mounted		6	2.667		920	142		1,062	1,250

10 21 13.40 Stone Toilet Compartments

	10 21 13.40 Stone Toilet Compartments	Crew	Daily Output	Labor-Hours	Unit	Material	Labor	Equipment	Total	Total Incl O&P
0010	**STONE TOILET COMPARTMENTS**									
0100	Cubicles, ceiling hung, marble	2 Marb	2	8	Ea.	1,650	405		2,055	2,450
0600	For handicap units, add					465			465	510
0800	Floor & ceiling anchored, marble	2 Marb	2.50	6.400		1,775	325		2,100	2,450
1400	For handicap units, add					370			370	405
1600	Floor mounted, marble	2 Marb	3	5.333		1,050	271		1,321	1,550
2400	Floor mounted, headrail braced, marble	"	3	5.333		1,225	271		1,496	1,750
2900	For handicap units, add					325			325	360
4100	Entrance screen, floor mounted marble, 58" high, 48" wide	2 Marb	9	1.778		735	90.50		825.50	945
4600	Urinal screen, 18" wide, ceiling braced, marble	D-1	6	2.667		770	125		895	1,025
5100	Floor mounted, headrail braced									
5200	Marble	D-1	6	2.667	Ea.	660	125		785	915
5700	Pilaster, flush, marble		9	1.778		860	83.50		943.50	1,075
6200	Post braced, marble		9	1.778		845	83.50		928.50	1,050

10 28 Toilet, Bath, and Laundry Accessories

10 28 13 – Toilet Accessories

10 28 13.13	Commercial Toilet Accessories	Crew	Daily Output	Labor-Hours	Unit	2020 Bare Costs Material	Labor	Equipment	Total	Total Incl O&P
0010	**COMMERCIAL TOILET ACCESSORIES**									
0200	Curtain rod, stainless steel, 5' long, 1" diameter	1 Carp	13	.615	Ea.	26.50	32.50		59	78.50
0300	1-1/4" diameter	"	13	.615	"	29	32.50		61.50	81
0500	Dispenser units, combined soap & towel dispensers,									
0510	Mirror and shelf, flush mounted	1 Carp	10	.800	Ea.	345	42.50		387.50	445
0600	Towel dispenser and waste receptacle,									
0610	18 gallon capacity	1 Carp	10	.800	Ea.	365	42.50		407.50	470
0800	Grab bar, straight, 1-1/4" diameter, stainless steel, 18" long		24	.333		30	17.70		47.70	59
0900	24" long		23	.348		29	18.50		47.50	60
1000	30" long		22	.364		26.50	19.35		45.85	58
1100	36" long		20	.400		34.50	21.50		56	70
1105	42" long		20	.400		40	21.50		61.50	76
1120	Corner, 36" long		20	.400		95	21.50		116.50	136
1200	1-1/2" diameter, 24" long		23	.348		33.50	18.50		52	65
1300	36" long		20	.400		37	21.50		58.50	72.50
1310	42" long		18	.444		33.50	23.50		57	72.50
1500	Tub bar, 1-1/4" diameter, 24" x 36"		14	.571		90	30.50		120.50	145
1600	Plus vertical arm		12	.667		100	35.50		135.50	163
1900	End tub bar, 1" diameter, 90° angle, 16" x 32"		12	.667		111	35.50		146.50	175
2300	Hand dryer, surface mounted, electric, 115 volt, 20 amp		4	2		425	106		531	625
2400	230 volt, 10 amp		4	2		645	106		751	870
2450	Hand dryer, touch free, 1400 watt, 81,000 rpm		4	2		1,350	106		1,456	1,625
2600	Hat and coat strip, stainless steel, 4 hook, 36" long		24	.333		68	17.70		85.70	102
2700	6 hook, 60" long		20	.400		118	21.50		139.50	162
3000	Mirror, with stainless steel 3/4" square frame, 18" x 24"		20	.400		51.50	21.50		73	88.50
3100	36" x 24"		15	.533		104	28.50		132.50	157
3200	48" x 24"		10	.800		186	42.50		228.50	268
3300	72" x 24"		6	1.333		305	71		376	440
3500	With 5" stainless steel shelf, 18" x 24"		20	.400		185	21.50		206.50	235
3600	36" x 24"		15	.533		236	28.50		264.50	300
3700	48" x 24"		10	.800		246	42.50		288.50	335
3800	72" x 24"		6	1.333		277	71		348	410
4100	Mop holder strip, stainless steel, 5 holders, 48" long		20	.400		76.50	21.50		98	117
4200	Napkin/tampon dispenser, recessed		15	.533		560	28.50		588.50	660
4220	Semi-recessed		6.50	1.231		370	65.50		435.50	505
4250	Napkin receptacle, recessed		6.50	1.231		171	65.50		236.50	287
4300	Robe hook, single, regular		96	.083		20.50	4.43		24.93	29
4400	Heavy duty, concealed mounting		56	.143		25.50	7.60		33.10	39.50
4600	Soap dispenser, chrome, surface mounted, liquid		20	.400		58.50	21.50		80	96
5000	Recessed stainless steel, liquid		10	.800		163	42.50		205.50	243
5600	Shelf, stainless steel, 5" wide, 18 ga., 24" long		24	.333		82	17.70		99.70	117
5700	48" long		16	.500		168	26.50		194.50	225
5800	8" wide shelf, 18 ga., 24" long		22	.364		63	19.35		82.35	98.50
5900	48" long		14	.571		145	30.50		175.50	205
6000	Toilet seat cover dispenser, stainless steel, recessed		20	.400		181	21.50		202.50	231
6050	Surface mounted		15	.533		32.50	28.50		61	78.50
6200	Double roll		24	.333		23	17.70		40.70	51.50
6240	Plastic, twin/jumbo dbl. roll		24	.333		30.50	17.70		48.20	60
6400	Towel bar, stainless steel, 18" long		23	.348		40	18.50		58.50	72
6500	30" long		21	.381		48.50	20.50		69	83.50
6700	Towel dispenser, stainless steel, surface mounted		16	.500		41.50	26.50		68	86
6800	Flush mounted, recessed		10	.800		67.50	42.50		110	139
6900	Plastic, touchless, battery operated		16	.500		107	26.50		133.50	158

10 28 Toilet, Bath, and Laundry Accessories

10 28 13 – Toilet Accessories

10 28 13.13 Commercial Toilet Accessories

	10 28 13.13 Commercial Toilet Accessories	Crew	Daily Output	Labor-Hours	Unit	Material	2020 Bare Costs Labor	Equipment	Total	Total Incl O&P
7000	Towel holder, hotel type, 2 guest size	1 Carp	20	.400	Ea.	58	21.50		79.50	95.50
7200	Towel shelf, stainless steel, 24" long, 8" wide		20	.400		65	21.50		86.50	104
7400	Tumbler holder, for tumbler only		30	.267		37	14.15		51.15	62.50
7410	Tumbler holder, recessed		20	.400		9.40	21.50		30.90	42.50
7500	Soap, tumbler & toothbrush		30	.267		20	14.15		34.15	43.50
7510	Tumbler & toothbrush holder		20	.400		13.45	21.50		34.95	47
8000	Waste receptacles, stainless steel, with top, 13 gallon		10	.800		335	42.50		377.50	430
8100	36 gallon		8	1		460	53		513	585
9996	Bathroom access., grab bar, straight, 1-1/2" diam., SS, 42" L install only		18	.444			23.50		23.50	35.50

10 28 16 – Bath Accessories

10 28 16.20 Medicine Cabinets

	10 28 16.20 Medicine Cabinets	Crew	Daily Output	Labor-Hours	Unit	Material	Labor	Equipment	Total	Total Incl O&P
0010	**MEDICINE CABINETS**									
0020	With mirror, sst frame, 16" x 22", unlighted	1 Carp	14	.571	Ea.	98.50	30.50		129	154
0100	Wood frame		14	.571		128	30.50		158.50	187
0300	Sliding mirror doors, 20" x 16" x 4-3/4", unlighted		7	1.143		130	60.50		190.50	235
0400	24" x 19" x 8-1/2", lighted		5	1.600		219	85		304	370
0600	Triple door, 30" x 32", unlighted, plywood body		7	1.143		355	60.50		415.50	480
0700	Steel body		7	1.143		355	60.50		415.50	480
0900	Oak door, wood body, beveled mirror, single door		7	1.143		170	60.50		230.50	278
1000	Double door		6	1.333		370	71		441	510
1200	Hotel cabinets, stainless, with lower shelf, unlighted		10	.800		218	42.50		260.50	305
1300	Lighted		5	1.600		330	85		415	490

10 28 19 – Tub and Shower Enclosures

10 28 19.10 Partitions, Shower

	10 28 19.10 Partitions, Shower	Crew	Daily Output	Labor-Hours	Unit	Material	Labor	Equipment	Total	Total Incl O&P
0010	**PARTITIONS, SHOWER** floor mounted, no plumbing									
0400	Cabinet, one piece, fiberglass, 32" x 32"	2 Carp	5	3.200	Ea.	625	170		795	945
0420	36" x 36"		5	3.200		550	170		720	860
0440	36" x 48"		5	3.200		1,375	170		1,545	1,775
0460	Acrylic, 32" x 32"		5	3.200		335	170		505	625
0480	36" x 36"		5	3.200		1,050	170		1,220	1,400
0500	36" x 48"		5	3.200		1,300	170		1,470	1,675
0520	Shower door for above, clear plastic, 24" wide	1 Carp	8	1		185	53		238	284
0540	28" wide		8	1		241	53		294	345
0560	Tempered glass, 24" wide		8	1		260	53		313	365
0580	28" wide		8	1		293	53		346	400
2400	Glass stalls, with doors, no receptors, chrome on brass	2 Shee	3	5.333		1,625	330		1,955	2,300
2700	Anodized aluminum	"	4	4		1,300	249		1,549	1,825
2900	Marble shower stall, stock design, with shower door	2 Marb	1.20	13.333		2,275	680		2,955	3,525
3000	With curtain		1.30	12.308		2,000	625		2,625	3,150
3200	Receptors, precast terrazzo, 32" x 32"		14	1.143		300	58		358	420
3300	48" x 34"		9.50	1.684		440	85.50		525.50	610
3500	Plastic, simulated terrazzo receptor, 32" x 32"		14	1.143		172	58		230	277
3600	32" x 48"		12	1.333		315	68		383	450
3800	Precast concrete, colors, 32" x 32"		14	1.143		220	58		278	330
3900	48" x 48"		8	2		310	102		412	495
4100	Shower doors, economy plastic, 24" wide	1 Shee	9	.889		134	55.50		189.50	231
4200	Tempered glass door, economy		8	1		279	62.50		341.50	400
4400	Folding, tempered glass, aluminum frame		6	1.333		430	83		513	600
4500	Sliding, tempered glass, 48" opening		6	1.333		575	83		658	755
4700	Deluxe, tempered glass, chrome on brass frame, 42" to 44"		8	1		485	62.50		547.50	630
4800	39" to 48" wide		1	8		700	500		1,200	1,525
4850	On anodized aluminum frame, obscure glass		2	4		570	249		819	1,000

10 28 Toilet, Bath, and Laundry Accessories

10 28 19 – Tub and Shower Enclosures

10 28 19.10 Partitions, Shower		Crew	Daily Output	Labor-Hours	Unit	2020 Bare Costs Material	Labor	Equipment	Total	Total Incl O&P
4900	Clear glass	1 Shee	1	8	Ea.	690	500		1,190	1,525
5100	Shower enclosure, tempered glass, anodized alum. frame									
5120	2 panel & door, corner unit, 32″ x 32″	1 Shee	2	4	Ea.	965	249		1,214	1,425
5140	Neo-angle corner unit, 16″ x 24″ x 16″	″	2	4		1,025	249		1,274	1,500
5200	Shower surround, 3 wall, polypropylene, 32″ x 32″	1 Carp	4	2		635	106		741	860
5220	PVC, 32″ x 32″		4	2		405	106		511	605
5240	Fiberglass		4	2		390	106		496	590
5250	2 wall, polypropylene, 32″ x 32″		4	2		310	106		416	500
5270	PVC		4	2		370	106		476	565
5290	Fiberglass		4	2		375	106		481	570
5300	Tub doors, tempered glass & frame, obscure glass	1 Shee	8	1		219	62.50		281.50	335
5400	Clear glass		6	1.333		510	83		593	685
5600	Chrome plated, brass frame, obscure glass		8	1		289	62.50		351.50	415
5700	Clear glass		6	1.333		710	83		793	905
5900	Tub/shower enclosure, temp. glass, alum. frame, obscure glass		2	4		395	249		644	810
6200	Clear glass		1.50	5.333		820	330		1,150	1,400
6500	On chrome-plated brass frame, obscure glass		2	4		540	249		789	975
6600	Clear glass		1.50	5.333		1,150	330		1,480	1,775
6800	Tub surround, 3 wall, polypropylene	1 Carp	4	2		251	106		357	435
6900	PVC		4	2		365	106		471	565
7000	Fiberglass, obscure glass		4	2		385	106		491	585
7100	Clear glass		3	2.667		650	142		792	930

10 44 Fire Protection Specialties

10 44 13 – Fire Protection Cabinets

10 44 13.53 Fire Equipment Cabinets		Crew	Daily Output	Labor-Hours	Unit	Material	Labor	Equipment	Total	Total Incl O&P
0010	**FIRE EQUIPMENT CABINETS**, not equipped, 20 ga. steel box D4020–310									
0040	Recessed, D.S. glass in door, box size given									
1000	Portable extinguisher, single, 8″ x 12″ x 27″, alum. door & frame	Q-12	8	2	Ea.	142	114		256	325
1100	Steel door and frame		8	2		164	114		278	350
1200	Stainless steel door and frame		8	2		227	114		341	420
2000	Portable extinguisher, large, 8″ x 12″ x 36″, alum. door & frame		8	2		281	114		395	480
2100	Steel door and frame D4020–330		8	2		248	114		362	445
2200	Stainless steel door and frame		8	2		410	114		524	625
2500	8″ x 16″ x 38″, aluminum door & frame		8	2		251	114		365	445
2600	Steel door and frame		8	2		246	114		360	440
2700	Fire blanket & extinguisher cab, inc blanket, rec stl., 14″ x 40″ x 8″		7	2.286		219	130		349	435
2800	Fire blanket cab, inc blanket, surf mtd, stl, 15″x10″x5″, w/pwdr coat fin		8	2		103	114		217	284
3000	Hose rack assy., 1-1/2″ valve & 100′ hose, 24″ x 40″ x 5-1/2″									
3100	Aluminum door and frame	Q-12	6	2.667	Ea.	565	152		717	850
3200	Steel door and frame D4020–410		6	2.667		259	152		411	515
3300	Stainless steel door and frame		6	2.667		590	152		742	880
4000	Hose rack assy., 2-1/2″ x 1-1/2″ valve, 100′ hose, 24″ x 40″ x 8″									
4100	Aluminum door and frame	Q-12	6	2.667	Ea.	545	152		697	830
4200	Steel door and frame R211226-10		6	2.667		385	152		537	655
4300	Stainless steel door and frame		6	2.667		760	152		912	1,075
5000	Hose rack assy., 2-1/2″ x 1-1/2″ valve, 100′ hose									
5010	and extinguisher, 30″ x 40″ x 8″									
5100	Aluminum door and frame	Q-12	5	3.200	Ea.	725	182		907	1,075
5200	Steel door and frame		5	3.200		299	182		481	605
5300	Stainless steel door and frame		5	3.200		630	182		812	970

10 44 Fire Protection Specialties

10 44 13 – Fire Protection Cabinets

10 44 13.53 Fire Equipment Cabinets		Crew	Daily Output	Labor-Hours	Unit	2020 Bare Costs Material	Labor	Equipment	Total	Total Incl O&P
6000	Hose rack assy., 1-1/2" valve, 100' hose									
6010	and 2-1/2" FD valve, 24" x 44" x 8"									
6100	Aluminum door and frame	Q-12	5	3.200	Ea.	590	182		772	925
6200	Steel door and frame		5	3.200		375	182		557	690
6300	Stainless steel door and frame		5	3.200		855	182		1,037	1,225
7000	Hose rack assy., 1-1/2" valve & 100' hose, 2-1/2" FD valve R211226-20									
7010	and extinguisher, 30" x 44" x 8"									
7100	Aluminum door and frame	Q-12	5	3.200	Ea.	765	182		947	1,125
7200	Steel door and frame		5	3.200		465	182		647	785
7300	Stainless steel door and frame		5	3.200		720	182		902	1,075
8000	Valve cabinet for 2-1/2" FD angle valve, 18" x 18" x 8"									
8100	Aluminum door and frame	Q-12	12	1.333	Ea.	251	76		327	390
8200	Steel door and frame		12	1.333		207	76		283	340
8300	Stainless steel door and frame		12	1.333		335	76		411	485

10 44 16 – Fire Extinguishers

10 44 16.13 Portable Fire Extinguishers

Code	Description	Crew	Daily Output	Labor-Hours	Unit	Material	Labor	Equipment	Total	Total Incl O&P
0010	**PORTABLE FIRE EXTINGUISHERS**									
0140	CO_2, with hose and "H" horn, 10 lb.				Ea.	300			300	330
0160	15 lb.					385			385	425
0180	20 lb.					425			425	470
1000	Dry chemical, pressurized									
1040	Standard type, portable, painted, 2-1/2 lb.				Ea.	44			44	48.50
1060	5 lb.					62			62	68
1080	10 lb.					96			96	106
1100	20 lb.					138			138	152
1120	30 lb.					430			430	470
1300	Standard type, wheeled, 150 lb.					1,500			1,500	1,650
2000	ABC all purpose type, portable, 2-1/2 lb.					23.50			23.50	26
2060	5 lb.					28.50			28.50	31.50
2080	9-1/2 lb.					52			52	57
2100	20 lb.					91			91	100
5000	Pressurized water, 2-1/2 gallon, stainless steel					105			105	115
5060	With anti-freeze					118			118	130
9400	Installation of extinguishers, 12 or more, on nailable surface	1 Carp	30	.267			14.15		14.15	21.50
9420	On masonry or concrete	"	15	.533			28.50		28.50	42.50

10 44 16.16 Wheeled Fire Extinguisher Units

Code	Description	Crew	Daily Output	Labor-Hours	Unit	Material	Labor	Equipment	Total	Total Incl O&P
0010	**WHEELED FIRE EXTINGUISHER UNITS**									
0350	CO_2, portable, with swivel horn									
0360	Wheeled type, cart mounted, 50 lb.				Ea.	1,375			1,375	1,525
0400	100 lb.				"	4,250			4,250	4,675
2200	ABC all purpose type									
2300	Wheeled, 45 lb.				Ea.	775			775	850
2360	150 lb.				"	1,975			1,975	2,150

Division Notes

		CREW	DAILY OUTPUT	LABOR-HOURS	UNIT	BARE COSTS MAT.	LABOR	EQUIP.	TOTAL	TOTAL INCL O&P

Division 11 Equipment

Estimating Tips

General

- The items in this division are usually priced per square foot or each. Many of these items are purchased by the owner for installation by the contractor. Check the specifications for responsibilities and include time for receiving, storage, installation, and mechanical and electrical hookups in the appropriate divisions.
- Many items in Division 11 require some type of support system that is not usually furnished with the item. Examples of these systems include blocking for the attachment of casework and support angles for ceiling-hung projection screens. The required blocking or supports must be added to the estimate in the appropriate division.
- Some items in Division 11 may require assembly or electrical hookups. Verify the amount of assembly required or the need for a hard electrical connection and add the appropriate costs.

Reference Numbers

Reference numbers are shown at the beginning of some major classifications. These numbers refer to related items in the Reference Section. The reference information may be an estimating procedure, an alternate pricing method, or technical information.

11 11 Vehicle Service Equipment

11 11 13 – Compressed-Air Vehicle Service Equipment

11 11 13.10 Compressed Air Equipment		Crew	Daily Output	Labor-Hours	Unit	Material	2020 Bare Costs Labor	Equipment	Total	Total Incl O&P
0010	**COMPRESSED AIR EQUIPMENT**									
0030	Compressors, electric, 1-1/2 HP, standard controls	L-4	1.50	16	Ea.	505	800		1,305	1,750
0550	Dual controls		1.50	16		1,050	800		1,850	2,350
0600	5 HP, 115/230 volt, standard controls		1	24		1,725	1,200		2,925	3,725
0650	Dual controls	↓	1	24	↓	3,225	1,200		4,425	5,375

11 11 19 – Vehicle Lubrication Equipment

11 11 19.10 Lubrication Equipment		Crew	Daily Output	Labor-Hours	Unit	Material	Labor	Equipment	Total	Total Incl O&P
0010	**LUBRICATION EQUIPMENT**									
3000	Lube equipment, 3 reel type, with pumps, not including piping	L-4	.50	48	Set	10,400	2,400		12,800	15,100
3700	Pump lubrication, pneumatic, not incl. air compressor									
3710	Oil/gear lube	Q-1	9.60	1.667	Ea.	890	96.50		986.50	1,125
3720	Grease	"	9.60	1.667	"	1,200	96.50		1,296.50	1,475

11 11 33 – Vehicle Spray Painting Equipment

11 11 33.10 Spray Painting Equipment		Crew	Daily Output	Labor-Hours	Unit	Material	Labor	Equipment	Total	Total Incl O&P
0010	**SPRAY PAINTING EQUIPMENT**									
4000	Spray painting booth, 26' long, complete	L-4	.40	60	Ea.	11,600	3,000		14,600	17,200

11 21 Retail and Service Equipment

11 21 53 – Barber and Beauty Shop Equipment

11 21 53.10 Barber Equipment		Crew	Daily Output	Labor-Hours	Unit	Material	Labor	Equipment	Total	Total Incl O&P
0010	**BARBER EQUIPMENT**									
0020	Chair, hydraulic, movable, minimum	1 Carp	24	.333	Ea.	640	17.70		657.70	730
0050	Maximum	"	16	.500		4,150	26.50		4,176.50	4,625
0500	Sink, hair washing basin, rough plumbing not incl.	1 Plum	8	1		530	64.50		594.50	680
1000	Sterilizer, liquid solution for tools				↓	171			171	189

11 21 73 – Commercial Laundry and Dry Cleaning Equipment

11 21 73.16 Drying and Conditioning Equipment		Crew	Daily Output	Labor-Hours	Unit	Material	Labor	Equipment	Total	Total Incl O&P
0010	**DRYING AND CONDITIONING EQUIPMENT**									
0100	Dryers, not including rough-in									
1500	Industrial, 30 lb. capacity	1 Plum	2	4	Ea.	3,350	258		3,608	4,050
1600	50 lb. capacity	"	1.70	4.706	"	4,550	305		4,855	5,475

11 21 73.26 Commercial Washers and Extractors		Crew	Daily Output	Labor-Hours	Unit	Material	Labor	Equipment	Total	Total Incl O&P
0010	**COMMERCIAL WASHERS AND EXTRACTORS**, not including rough-in									
6000	Combination washer/extractor, 20 lb. capacity	L-6	1.50	8	Ea.	6,725	505		7,230	8,150
6100	30 lb. capacity		.80	15		10,300	950		11,250	12,700
6200	50 lb. capacity		.68	17.647		12,900	1,125		14,025	15,900
6300	75 lb. capacity		.30	40		19,200	2,525		21,725	24,900
6350	125 lb. capacity	↓	.16	75	↓	29,900	4,750		34,650	40,000

11 21 73.33 Coin-Operated Laundry Equipment		Crew	Daily Output	Labor-Hours	Unit	Material	Labor	Equipment	Total	Total Incl O&P
0010	**COIN-OPERATED LAUNDRY EQUIPMENT**									
0990	Dryer, gas fired									
1000	Commercial, 30 lb. capacity, coin operated, single	1 Plum	3	2.667	Ea.	3,475	172		3,647	4,075
1100	Double stacked	"	2	4	"	8,275	258		8,533	9,475
5290	Clothes washer									
5300	Commercial, coin operated, average	1 Plum	3	2.667	Ea.	1,350	172		1,522	1,750

11 21 Retail and Service Equipment

11 21 83 – Photo Processing Equipment

11 21 83.13 Darkroom Equipment		Crew	Daily Output	Labor-Hours	Unit	Material	2020 Bare Costs Labor	Equipment	Total	Total Incl O&P
0010	DARKROOM EQUIPMENT									
0020	Developing sink, 5" deep, 24" x 48"	Q-1	2	8	Ea.	500	465		965	1,250
0050	48" x 52"		1.70	9.412		1,250	545		1,795	2,200
0200	10" deep, 24" x 48"		1.70	9.412		1,625	545		2,170	2,625
0250	24" x 108"	↓	1.50	10.667	↓	3,875	620		4,495	5,175

11 30 Residential Equipment

11 30 13 – Residential Appliances

11 30 13.15 Cooking Equipment		Crew	Daily Output	Labor-Hours	Unit	Material	Labor	Equipment	Total	Total Incl O&P
0010	COOKING EQUIPMENT									
0020	Cooking range, 30" free standing, 1 oven, minimum	2 Clab	10	1.600	Ea.	465	67.50		532.50	610
0050	Maximum		4	4		2,200	168		2,368	2,675
0150	2 oven, minimum		10	1.600		1,050	67.50		1,117.50	1,275
0200	Maximum	↓	10	1.600	↓	3,425	67.50		3,492.50	3,850
11 30 13.16 Refrigeration Equipment										
0010	REFRIGERATION EQUIPMENT									
5200	Icemaker, automatic, 20 lbs./day	1 Plum	7	1.143	Ea.	1,375	73.50		1,448.50	1,600
5350	51 lbs./day	"	2	4	"	1,400	258		1,658	1,925
11 30 13.17 Kitchen Cleaning Equipment										
0010	KITCHEN CLEANING EQUIPMENT									
2750	Dishwasher, built-in, 2 cycles, minimum	L-1	4	4	Ea.	310	252		562	715
2800	Maximum		2	8		445	505		950	1,250
2950	4 or more cycles, minimum		4	4		400	252		652	815
2960	Average		4	4		530	252		782	960
3000	Maximum	↓	2	8	↓	1,850	505		2,355	2,775
11 30 13.18 Waste Disposal Equipment										
0010	WASTE DISPOSAL EQUIPMENT									
3300	Garbage disposal, sink type, minimum	L-1	10	1.600	Ea.	111	101		212	272
3350	Maximum	"	10	1.600	"	216	101		317	385
11 30 13.19 Kitchen Ventilation Equipment										
0010	KITCHEN VENTILATION EQUIPMENT									
4150	Hood for range, 2 speed, vented, 30" wide, minimum	L-3	5	3.200	Ea.	107	184		291	395
4200	Maximum		3	5.333		1,000	305		1,305	1,550
4300	42" wide, minimum		5	3.200		151	184		335	440
4330	Custom		5	3.200		1,800	184		1,984	2,250
4350	Maximum	↓	3	5.333		2,175	305		2,480	2,850
4500	For ventless hood, 2 speed, add					19.05			19.05	21
4650	For vented 1 speed, deduct from maximum				↓	74			74	81.50
11 30 13.24 Washers										
0010	WASHERS									
5000	Residential, 4 cycle, average	1 Plum	3	2.667	Ea.	990	172		1,162	1,350
6650	Washing machine, automatic, minimum		3	2.667		560	172		732	870
6700	Maximum	↓	1	8	↓	1,400	515		1,915	2,325
11 30 13.25 Dryers										
0010	DRYERS									
0500	Gas fired residential, 16 lb. capacity, average	1 Plum	3	2.667	Ea.	720	172		892	1,050
7450	Vent kits for dryers	1 Carp	10	.800	"	48	42.50		90.50	117

11 30 Residential Equipment

11 30 15 – Miscellaneous Residential Appliances

11 30 15.13 Sump Pumps		Crew	Daily Output	Labor-Hours	Unit	2020 Bare Costs Material	Labor	Equipment	Total	Total Incl O&P
0010	**SUMP PUMPS**									
6400	Cellar drainer, pedestal, 1/3 HP, molded PVC base	1 Plum	3	2.667	Ea.	141	172		313	410
6450	Solid brass	″	2	4	″	243	258		501	650
6460	Sump pump, see also Section 22 14 29.16									

11 30 15.23 Water Heaters		Crew	Daily Output	Labor-Hours	Unit	Material	Labor	Equipment	Total	Total Incl O&P
0010	**WATER HEATERS**									
6900	Electric, glass lined, 30 gallon, minimum	L-1	5	3.200	Ea.	970	201		1,171	1,375
6950	Maximum		3	5.333		1,350	335		1,685	1,975
7100	80 gallon, minimum		2	8		1,950	505		2,455	2,900
7150	Maximum		1	16		2,725	1,000		3,725	4,500
7180	Gas, glass lined, 30 gallon, minimum	2 Plum	5	3.200		1,900	206		2,106	2,375
7220	Maximum		3	5.333		2,625	345		2,970	3,425
7260	50 gallon, minimum		2.50	6.400		1,750	410		2,160	2,550
7300	Maximum		1.50	10.667		2,450	685		3,135	3,725
7310	Water heater, see also Section 22 33 30.13									

11 30 15.43 Air Quality		Crew	Daily Output	Labor-Hours	Unit	Material	Labor	Equipment	Total	Total Incl O&P
0010	**AIR QUALITY**									
2450	Dehumidifier, portable, automatic, 15 pint	1 Elec	4	2	Ea.	208	123		331	410
2550	40 pint		3.75	2.133		247	131		378	465
3550	Heater, electric, built-in, 1250 watt, ceiling type, minimum		4	2		125	123		248	320
3600	Maximum		3	2.667		184	164		348	445
3700	Wall type, minimum		4	2		221	123		344	425
3750	Maximum		3	2.667		198	164		362	460
3900	1500 watt wall type, with blower		4	2		186	123		309	385
3950	3000 watt		3	2.667		515	164		679	810
4850	Humidifier, portable, 8 gallons/day					158			158	174
5000	15 gallons/day					207			207	227

11 32 Unit Kitchens

11 32 13 – Metal Unit Kitchens

11 32 13.10 Commercial Unit Kitchens		Crew	Daily Output	Labor-Hours	Unit	Material	Labor	Equipment	Total	Total Incl O&P
0010	**COMMERCIAL UNIT KITCHENS**									
1500	Combination range, refrigerator and sink, 30″ wide, minimum	L-1	2	8	Ea.	710	505		1,215	1,525
1550	Maximum		1	16		1,100	1,000		2,100	2,700
1570	60″ wide, average		1.40	11.429		1,225	720		1,945	2,425
1590	72″ wide, average		1.20	13.333		2,000	840		2,840	3,450
1600	Office model, 48″ wide		2	8		1,425	505		1,930	2,325
1620	Refrigerator and sink only		2.40	6.667		2,625	420		3,045	3,525
1640	Combination range, refrigerator, sink, microwave									
1660	Oven and ice maker	L-1	.80	20	Ea.	4,900	1,250		6,150	7,250

11 41 Foodservice Storage Equipment

11 41 13 – Refrigerated Food Storage Cases

11 41 13.20 Refrigerated Food Storage Equipment		Crew	Daily Output	Labor-Hours	Unit	Material	Labor	Equipment	Total	Total Incl O&P
						2020 Bare Costs				
0010	REFRIGERATED FOOD STORAGE EQUIPMENT									
2350	Cooler, reach-in, beverage, 6' long	Q-1	6	2.667	Ea.	3,125	155		3,280	3,650
4300	Freezers, reach-in, 44 C.F.		4	4		4,425	232		4,657	5,225
4500	68 C.F.	↓	3	5.333	↓	5,575	310		5,885	6,600

11 44 Food Cooking Equipment

11 44 13 – Commercial Ranges

11 44 13.10 Cooking Equipment

		Crew	Daily Output	Labor-Hours	Unit	Material	Labor	Equipment	Total	Total Incl O&P
0010	COOKING EQUIPMENT									
0020	Bake oven, gas, one section	Q-1	8	2	Ea.	5,675	116		5,791	6,425
0300	Two sections		7	2.286		9,375	133		9,508	10,500
0600	Three sections	↓	6	2.667		12,400	155		12,555	13,800
0900	Electric convection, single deck	L-7	4	7		5,575	360		5,935	6,650
6350	Kettle, w/steam jacket, tilting, w/positive lock, SS, 20 gallons		7	4		8,650	205		8,855	9,825
6600	60 gallons	↓	6	4.667	↓	17,100	239		17,339	19,200

11 46 Food Dispensing Equipment

11 46 83 – Ice Machines

11 46 83.10 Commercial Ice Equipment

		Crew	Daily Output	Labor-Hours	Unit	Material	Labor	Equipment	Total	Total Incl O&P
0010	COMMERCIAL ICE EQUIPMENT									
5800	Ice cube maker, 50 lbs./day	Q-1	6	2.667	Ea.	1,800	155		1,955	2,200
6050	500 lbs./day	"	4	4	"	2,600	232		2,832	3,200

11 48 Foodservice Cleaning and Disposal Equipment

11 48 13 – Commercial Dishwashers

11 48 13.10 Dishwashers

		Crew	Daily Output	Labor-Hours	Unit	Material	Labor	Equipment	Total	Total Incl O&P
0010	DISHWASHERS									
2700	Dishwasher, commercial, rack type									
2720	10 to 12 racks/hour	Q-1	3.20	5	Ea.	3,450	290		3,740	4,225
2730	Energy star rated, 35 to 40 racks/hour G		1.30	12.308		5,500	715		6,215	7,125
2740	50 to 60 racks/hour G	↓	1.30	12.308		10,600	715		11,315	12,700
2800	Automatic, 190 to 230 racks/hour	L-6	.35	34.286		14,600	2,175		16,775	19,300
2820	235 to 275 racks/hour		.25	48		28,000	3,050		31,050	35,400
2840	8,750 to 12,500 dishes/hour	↓	.10	120	↓	47,500	7,600		55,100	64,000

11 53 Laboratory Equipment

11 53 13 – Laboratory Fume Hoods

11 53 13.13 Recirculating Laboratory Fume Hoods

		Crew	Daily Output	Labor-Hours	Unit	Material	Labor	Equipment	Total	Total Incl O&P
0010	RECIRCULATING LABORATORY FUME HOODS									
0670	Service fixtures, average				Ea.	365			365	405
0680	For sink assembly with hot and cold water, add	1 Plum	1.40	5.714	"	780	370		1,150	1,400

11 53 Laboratory Equipment

11 53 19 – Laboratory Sterilizers

11 53 19.13 Sterilizers		Crew	Daily Output	Labor-Hours	Unit	Material	2020 Bare Costs Labor	Equipment	Total	Total Incl O&P
0010	**STERILIZERS**									
0700	Glassware washer, undercounter, minimum	L-1	1.80	8.889	Ea.	6,850	560		7,410	8,350
0710	Maximum	"	1	16		15,200	1,000		16,200	18,200
1850	Utensil washer-sanitizer	1 Plum	2	4	↓	8,725	258		8,983	9,975

11 53 23 – Laboratory Refrigerators

11 53 23.13 Refrigerators		Crew	Daily Output	Labor-Hours	Unit	Material	Labor	Equipment	Total	Total Incl O&P
0010	**REFRIGERATORS**									
1200	Blood bank, 28.6 C.F. emergency signal				Ea.	14,400			14,400	15,900
1210	Reach-in, 16.9 C.F.				"	9,400			9,400	10,300

11 53 33 – Emergency Safety Appliances

11 53 33.13 Emergency Equipment		Crew	Daily Output	Labor-Hours	Unit	Material	Labor	Equipment	Total	Total Incl O&P
0010	**EMERGENCY EQUIPMENT**									
1400	Safety equipment, eye wash, hand held				Ea.	400			400	440
1450	Deluge shower				"	840			840	920

11 53 43 – Service Fittings and Accessories

11 53 43.13 Fittings		Crew	Daily Output	Labor-Hours	Unit	Material	Labor	Equipment	Total	Total Incl O&P
0010	**FITTINGS**									
1600	Sink, one piece plastic, flask wash, hose, free standing	1 Plum	1.60	5	Ea.	2,075	320		2,395	2,750
1610	Epoxy resin sink, 25" x 16" x 10"	"	2	4	"	227	258		485	635
8000	Alternate pricing method: as percent of lab furniture									
8050	Installation, not incl. plumbing & duct work				% Furn.				22%	22%
8100	Plumbing, final connections, simple system								10%	10%
8110	Moderately complex system								15%	15%
8120	Complex system								20%	20%
8150	Electrical, simple system								10%	10%
8160	Moderately complex system								20%	20%
8170	Complex system				↓				35%	35%

11 71 Medical Sterilizing Equipment

11 71 10 – Medical Sterilizers & Distillers

11 71 10.10 Sterilizers and Distillers		Crew	Daily Output	Labor-Hours	Unit	Material	Labor	Equipment	Total	Total Incl O&P
0010	**STERILIZERS AND DISTILLERS**									
0700	Distiller, water, steam heated, 50 gal. capacity	1 Plum	1.40	5.714	Ea.	25,500	370		25,870	28,700
5600	Sterilizers, floor loading, 26" x 62" x 42", single door, steam					155,000			155,000	170,500
5650	Double door, steam					239,000			239,000	262,500
5800	General purpose, 20" x 20" x 38", single door					12,500			12,500	13,700
6000	Portable, counter top, steam, minimum					2,900			2,900	3,175
6020	Maximum					5,075			5,075	5,575
6050	Portable, counter top, gas, 17" x 15" x 32-1/2"					45,400			45,400	49,900
6150	Manual washer/sterilizer, 16" x 16" x 26"	1 Plum	2	4	↓	62,000	258		62,258	69,000
6200	Steam generators, electric 10 kW to 180 kW, freestanding									
6250	Minimum	1 Elec	3	2.667	Ea.	12,200	164		12,364	13,600
6300	Maximum	"	.70	11.429		22,700	700		23,400	26,100
8200	Bed pan washer-sanitizer	1 Plum	2	4	↓	9,775	258		10,033	11,100

11 73 Patient Care Equipment

11 73 10 – Patient Treatment Equipment

11 73 10.10 Treatment Equipment		Crew	Daily Output	Labor-Hours	Unit	2020 Bare Costs Material	Labor	Equipment	Total	Total Incl O&P
0010	**TREATMENT EQUIPMENT**									
1800	Heat therapy unit, humidified, 26" x 78" x 28"				Ea.	4,250			4,250	4,675
2100	Hubbard tank with accessories, stainless steel,									
2110	125 GPM at 45 psi water pressure				Ea.	15,100			15,100	16,600
2150	For electric overhead hoist, add					3,275			3,275	3,600
3600	Paraffin bath, 126°F, auto controlled					280			280	305
4600	Station, dietary, medium, with ice					22,200			22,200	24,400
8400	Whirlpool bath, mobile, sst, 18" x 24" x 60"					6,550			6,550	7,200
8450	Fixed, incl. mixing valves	1 Plum	2	4	↓	4,825	258		5,083	5,700

11 74 Dental Equipment

11 74 10 – Dental Office Equipment

11 74 10.10 Diagnostic and Treatment Equipment		Crew	Daily Output	Labor-Hours	Unit	Material	Labor	Equipment	Total	Total Incl O&P
0010	**DIAGNOSTIC AND TREATMENT EQUIPMENT**									
0020	Central suction system, minimum	1 Plum	1.20	6.667	Ea.	1,175	430		1,605	1,925
0100	Maximum	"	.90	8.889		6,425	575		7,000	7,925
0300	Air compressor, minimum	1 Skwk	.80	10		2,850	550		3,400	3,950
0400	Maximum		.50	16		9,075	880		9,955	11,300
0600	Chair, electric or hydraulic, minimum		.50	16		2,625	880		3,505	4,200
0700	Maximum		.25	32		3,975	1,750		5,725	7,025
2000	Light, ceiling mounted, minimum		8	1		875	55		930	1,050
2100	Maximum	↓	8	1		2,050	55		2,105	2,325
2200	Unit light, minimum	2 Skwk	5.33	3.002		775	165		940	1,100
2210	Maximum		5.33	3.002		1,475	165		1,640	1,850
2220	Track light, minimum		3.20	5		1,725	274		1,999	2,325
2230	Maximum	↓	3.20	5		2,300	274		2,574	2,950
2300	Sterilizers, steam portable, minimum					1,875			1,875	2,050
2350	Maximum					6,350			6,350	7,000
2600	Steam, institutional					2,650			2,650	2,900
2650	Dry heat, electric, portable, 3 trays				↓	1,375			1,375	1,500

11 76 Operating Room Equipment

11 76 10 – Equipment for Operating Rooms

11 76 10.10 Surgical Equipment		Crew	Daily Output	Labor-Hours	Unit	Material	Labor	Equipment	Total	Total Incl O&P
0010	**SURGICAL EQUIPMENT**									
5000	Scrub, surgical, stainless steel, single station, minimum	1 Plum	3	2.667	Ea.	5,875	172		6,047	6,725
5100	Maximum				"	9,000			9,000	9,900

11 78 Mortuary Equipment

11 78 13 – Mortuary Refrigerators

11 78 13.10 Mortuary and Autopsy Equipment		Crew	Daily Output	Labor-Hours	Unit	Material	2020 Bare Costs Labor	Equipment	Total	Total Incl O&P
0010	**MORTUARY AND AUTOPSY EQUIPMENT**									
0015	Autopsy table, standard	1 Plum	1	8	Ea.	10,000	515		10,515	11,800
0020	Deluxe	"	.60	13.333	"	16,400	860		17,260	19,300

11 81 Facility Maintenance Equipment

11 81 19 – Vacuum Cleaning Systems

11 81 19.10 Vacuum Cleaning		Crew	Daily Output	Labor-Hours	Unit	Material	Labor	Equipment	Total	Total Incl O&P
0010	**VACUUM CLEANING**									
0020	Central, 3 inlet, residential	1 Skwk	.90	8.889	Total	1,325	490		1,815	2,200
0200	Commercial		.70	11.429		1,350	625		1,975	2,450
0400	5 inlet system, residential		.50	16		2,025	880		2,905	3,550
0600	7 inlet system, commercial		.40	20		2,600	1,100		3,700	4,500
0800	9 inlet system, residential	↓	.30	26.667	↓	4,225	1,475		5,700	6,850

11 91 Religious Equipment

11 91 13 – Baptisteries

11 91 13.10 Baptistry		Crew	Daily Output	Labor-Hours	Unit	Material	Labor	Equipment	Total	Total Incl O&P
0010	**BAPTISTRY**									
0150	Fiberglass, 3′-6″ deep, x 13′-7″ long,									
0160	steps at both ends, incl. plumbing, minimum	L-8	1	20	Ea.	6,300	1,100		7,400	8,625
0200	Maximum	"	.70	28.571		10,500	1,575		12,075	13,900
0250	Add for filter, heater and lights				↓	1,850			1,850	2,025

11 91 23 – Sanctuary Equipment

11 91 23.10 Sanctuary Furnishings		Crew	Daily Output	Labor-Hours	Unit	Material	Labor	Equipment	Total	Total Incl O&P
0010	**SANCTUARY FURNISHINGS**									
0020	Altar, wood, custom design, plain	1 Carp	1.40	5.714	Ea.	2,725	305		3,030	3,450
0050	Deluxe	"	.20	40	"	13,100	2,125		15,225	17,600

11 98 Detention Equipment

11 98 30 – Detention Cell Equipment

11 98 30.10 Cell Equipment		Crew	Daily Output	Labor-Hours	Unit	Material	Labor	Equipment	Total	Total Incl O&P
0010	**CELL EQUIPMENT**									
3000	Toilet apparatus including wash basin, average	L-8	1.50	13.333	Ea.	3,550	740		4,290	5,000

Estimating Tips

General

- The items in this division are usually priced per square foot or each. Most of these items are purchased by the owner and installed by the contractor. Do not assume the items in Division 12 will be purchased and installed by the contractor. Check the specifications for responsibilities and include receiving, storage, installation, and mechanical and electrical hookups in the appropriate divisions.
- Some items in this division require some type of support system that is not usually furnished with the item. Examples of these systems include blocking for the attachment of casework and heavy drapery rods. The required blocking must be added to the estimate in the appropriate division.

Reference Numbers

Reference numbers are shown at the beginning of some major classifications. These numbers refer to related items in the Reference Section. The reference information may be an estimating procedure, an alternate pricing method, or technical information.

12 32 Manufactured Wood Casework

12 32 23 – Hardwood Casework

12 32 23.10 Manufactured Wood Casework, Stock Units		Crew	Daily Output	Labor-Hours	Unit	Material	2020 Bare Costs Labor	Equipment	Total	Total Incl O&P
0010	**MANUFACTURED WOOD CASEWORK, STOCK UNITS**									
0700	Kitchen base cabinets, hardwood, not incl. counter tops,									
0710	24" deep, 35" high, prefinished									
0800	One top drawer, one door below, 12" wide	2 Carp	24.80	.645	Ea.	325	34.50		359.50	405
0840	18" wide		23.30	.687		299	36.50		335.50	385
0880	24" wide		22.30	.717		435	38		473	540
1000	Four drawers, 12" wide		24.80	.645		330	34.50		364.50	415
1040	18" wide		23.30	.687		370	36.50		406.50	460
1060	24" wide		22.30	.717		415	38		453	520
1200	Two top drawers, two doors below, 27" wide		22	.727		455	38.50		493.50	560
1260	36" wide		20.30	.788		570	42		612	695
1300	48" wide		18.90	.847		695	45		740	835
1500	Range or sink base, two doors below, 30" wide		21.40	.748		460	39.50		499.50	565
1540	36" wide		20.30	.788		495	42		537	610
1580	48" wide	↓	18.90	.847		545	45		590	670
1800	For sink front units, deduct				↓	188			188	207
9000	For deluxe models of all cabinets, add					40%				
9500	For custom built in place, add					25%	10%			
9558	Rule of thumb, kitchen cabinets not including									
9560	appliances & counter top, minimum	2 Carp	30	.533	L.F.	215	28.50		243.50	279
9600	Maximum	"	25	.640	"	460	34		494	555

12 32 23.30 Manufactured Wood Casework Vanities		Crew	Daily Output	Labor-Hours	Unit	Material	Labor	Equipment	Total	Total Incl O&P
0010	**MANUFACTURED WOOD CASEWORK VANITIES**									
8000	Vanity bases, 2 doors, 30" high, 21" deep, 24" wide	2 Carp	20	.800	Ea.	420	42.50		462.50	525
8050	30" wide		16	1		460	53		513	585
8100	36" wide		13.33	1.200		400	64		464	535
8150	48" wide	↓	11.43	1.400		610	74.50		684.50	780
9000	For deluxe models of all vanities, add to above					40%				
9500	For custom built in place, add to above				↓	25%	10%			

Estimating Tips

General

- The items and systems in this division are usually estimated, purchased, supplied, and installed as a unit by one or more subcontractors. The estimator must ensure that all parties are operating from the same set of specifications and assumptions, and that all necessary items are estimated and will be provided. Many times the complex items and systems are covered, but the more common ones, such as excavation or a crane, are overlooked for the very reason that everyone assumes nobody could miss them. The estimator should be the central focus and be able to ensure that all systems are complete.
- It is important to consider factors such as site conditions, weather, shape and size of building, as well as labor availability as they may impact the overall cost of erecting special structures and systems included in this division.
- Another area where problems can develop in this division is at the interface between systems. The estimator must ensure, for instance, that anchor bolts, nuts, and washers are estimated and included for the air-supported structures and pre-engineered buildings to be bolted to their foundations. Utility supply is a common area where essential items or pieces of equipment can be missed or overlooked because each subcontractor may feel it is another's responsibility. The estimator should also be aware of certain items which may be supplied as part of a package but installed by others, and ensure that the installing contractor's estimate includes the cost of installation. Conversely, the estimator must also ensure that items are not costed by two different subcontractors, resulting in an inflated overall estimate.

13 30 00 Special Structures

- The foundations and floor slab, as well as rough mechanical and electrical, should be estimated, as this work is required for the assembly and erection of the structure. Generally, as noted in the data set, the pre-engineered building comes as a shell. Pricing is based on the size and structural design parameters stated in the reference section. Additional features, such as windows and doors with their related structural framing, must also be included by the estimator. Here again, the estimator must have a clear understanding of the scope of each portion of the work and all the necessary interfaces.

Reference Numbers

Reference numbers are shown at the beginning of some major classifications. These numbers refer to related items in the Reference Section. The reference information may be an estimating procedure, an alternate pricing method, or technical information.

Note: Not all subdivisions listed here necessarily appear. ■

13 11 Swimming Pools

13 11 13 – Below-Grade Swimming Pools

13 11 13.50 Swimming Pools		Crew	Daily Output	Labor-Hours	Unit	2020 Bare Costs Material	Labor	Equipment	Total	Total Incl O&P
0010	**SWIMMING POOLS** Residential in-ground, vinyl lined									
0020	Concrete sides, w/equip, sand bottom	B-52	300	.187	SF Surf	28.50	9.10	1.95	39.55	47
0100	Metal or polystyrene sides R131113-20	B-14	410	.117		24	5.20	.52	29.72	35
0200	Add for vermiculite bottom					1.82			1.82	2
0500	Gunite bottom and sides, white plaster finish									
0600	12′ x 30′ pool	B-52	145	.386	SF Surf	53	18.80	4.03	75.83	91
0720	16′ x 32′ pool		155	.361		48	17.60	3.77	69.37	83
0750	20′ x 40′ pool		250	.224		42.50	10.90	2.34	55.74	66
0810	Concrete bottom and sides, tile finish									
0820	12′ x 30′ pool	B-52	80	.700	SF Surf	53.50	34	7.30	94.80	118
0830	16′ x 32′ pool		95	.589		44.50	28.50	6.15	79.15	99
0840	20′ x 40′ pool		130	.431		35.50	21	4.50	61	75.50
1100	Motel, gunite with plaster finish, incl. medium									
1150	capacity filtration & chlorination	B-52	115	.487	SF Surf	65.50	23.50	5.10	94.10	113
1200	Municipal, gunite with plaster finish, incl. high									
1250	capacity filtration & chlorination	B-52	100	.560	SF Surf	84.50	27.50	5.85	117.85	140
1350	Add for formed gutters				L.F.	124			124	137
1360	Add for stainless steel gutters				″	370			370	405
1600	For water heating system, see Section 23 52 28.10									
1700	Filtration and deck equipment only, as % of total				Total				20%	20%
1800	Automatic vacuum, hand tools, etc., 20′ x 40′ pool				SF Pool				.56	.62
1900	5,000 S.F. pool				″				.11	.12
3000	Painting pools, preparation + 3 coats, 20′ x 40′ pool, epoxy	2 Pord	.33	48.485	Total	1,450	2,150		3,600	4,800
3100	Rubber base paint, 18 gallons	″	.33	48.485		1,250	2,150		3,400	4,575
3500	42′ x 82′ pool, 75 gallons, epoxy paint	3 Pord	.14	171		6,150	7,600		13,750	18,100
3600	Rubber base paint	″	.14	171		5,150	7,600		12,750	17,000

13 11 23 – On-Grade Swimming Pools

13 11 23.50 Swimming Pools		Crew	Daily Output	Labor-Hours	Unit	Material	Labor	Equipment	Total	Total Incl O&P
0010	**SWIMMING POOLS** Residential above ground, steel construction									
0100	Round, 15′ diam.	B-80A	3	8	Ea.	895	335	273	1,503	1,800
0120	18′ diam.		2.50	9.600		1,025	405	330	1,760	2,100
0140	21′ diam.		2	12		1,175	505	410	2,090	2,500
0160	24′ diam.		1.80	13.333		1,350	560	455	2,365	2,850
0180	27′ diam.		1.50	16		1,550	675	545	2,770	3,325
0200	30′ diam.		1	24		1,575	1,000	820	3,395	4,175
0220	Oval, 12′ x 24′		2.30	10.435		1,700	440	355	2,495	2,900
0240	15′ x 30′		1.80	13.333		1,925	560	455	2,940	3,475
0260	18′ x 33′		1	24		2,150	1,000	820	3,970	4,775

13 11 46 – Swimming Pool Accessories

13 11 46.50 Swimming Pool Equipment		Crew	Daily Output	Labor-Hours	Unit	Material	Labor	Equipment	Total	Total Incl O&P
0010	**SWIMMING POOL EQUIPMENT**									
0020	Diving stand, stainless steel, 3 meter	2 Carp	.40	40	Ea.	18,000	2,125		20,125	23,000
0300	1 meter	″	2.70	5.926		10,900	315		11,215	12,500
2100	Lights, underwater, 12 volt, with transformer, 300 watt	1 Elec	1	8		365	490		855	1,125
2200	110 volt, 500 watt, standard		1	8		291	490		781	1,050
2400	Low water cutoff type		1	8		310	490		800	1,075
2800	Heaters, see Section 23 52 28.10									

13 12 Fountains

13 12 13 – Exterior Fountains

13 12 13.10 Outdoor Fountains		Crew	Daily Output	Labor-Hours	Unit	Material	Labor	Equipment	Total	Total Incl O&P
0010	OUTDOOR FOUNTAINS									
0100	Outdoor fountain, 48" high with bowl and figures	2 Clab	2	8	Ea.	435	335		770	985
0200	Commercial, concrete or cast stone, 40-60" H, simple		2	8		955	335		1,290	1,550
0220	Average		2	8		1,850	335		2,185	2,525
0240	Ornate		2	8		4,200	335		4,535	5,125
0260	Metal, 72" high		2	8		1,500	335		1,835	2,150
0280	90" high		2	8		2,250	335		2,585	2,975
0300	120" high		2	8		5,275	335		5,610	6,300
0320	Resin or fiberglass, 40-60" H, wall type		2	8		675	335		1,010	1,250
0340	Waterfall type		2	8		1,150	335		1,485	1,750

(Material, Labor, Equipment, Total: 2020 Bare Costs)

13 12 23 – Interior Fountains

13 12 23.10 Indoor Fountains		Crew	Daily Output	Labor-Hours	Unit	Material	Labor	Equipment	Total	Total Incl O&P
0010	INDOOR FOUNTAINS									
0100	Commercial, floor type, resin or fiberglass, lighted, cascade type	2 Clab	2	8	Ea.	425	335		760	970
0120	Tiered type		2	8		490	335		825	1,050
0140	Waterfall type		2	8		315	335		650	850

13 17 Tubs and Pools

13 17 13 – Hot Tubs

13 17 13.10 Redwood Hot Tub System		Crew	Daily Output	Labor-Hours	Unit	Material	Labor	Equipment	Total	Total Incl O&P
0010	REDWOOD HOT TUB SYSTEM									
7050	4′ diameter x 4′ deep	Q-1	1	16	Ea.	3,250	930		4,180	4,975
7100	5′ diameter x 4′ deep		1	16		4,125	930		5,055	5,925
7150	6′ diameter x 4′ deep		.80	20		4,975	1,150		6,125	7,200
7200	8′ diameter x 4′ deep		.80	20		7,300	1,150		8,450	9,750

13 17 33 – Whirlpool Tubs

13 17 33.10 Whirlpool Bath		Crew	Daily Output	Labor-Hours	Unit	Material	Labor	Equipment	Total	Total Incl O&P
0010	WHIRLPOOL BATH									
6000	Whirlpool, bath with vented overflow, molded fiberglass									
6100	66" x 36" x 24"	Q-1	1	16	Ea.	910	930		1,840	2,400
6400	72" x 36" x 21"		1	16		1,450	930		2,380	3,000
6500	60" x 34" x 21"		1	16		1,450	930		2,380	3,000
6600	72" x 42" x 23"		1	16		1,350	930		2,280	2,900
6710	For color, add					10%				
6711	For designer colors and trim, add					25%				

13 24 Special Activity Rooms

13 24 16 – Saunas

13 24 16.50 Saunas and Heaters		Crew	Daily Output	Labor-Hours	Unit	Material	Labor	Equipment	Total	Total Incl O&P
0010	SAUNAS AND HEATERS									
0020	Prefabricated, incl. heater & controls, 7′ high, 6′ x 4′, C/C	L-7	2.20	12.727	Ea.	5,250	650		5,900	6,750
0050	6′ x 4′, C/P		2	14		4,325	715		5,040	5,850
0400	6′ x 5′, C/C		2	14		5,300	715		6,015	6,900
0450	6′ x 5′, C/P		2	14		5,225	715		5,940	6,825
0600	6′ x 6′, C/C		1.80	15.556		6,950	795		7,745	8,850
0650	6′ x 6′, C/P		1.80	15.556		5,525	795		6,320	7,275
0800	6′ x 9′, C/C		1.60	17.500		7,325	895		8,220	9,400
0850	6′ x 9′, C/P		1.60	17.500		6,500	895		7,395	8,500

13 24 Special Activity Rooms

13 24 16 – Saunas

13 24 16.50 Saunas and Heaters		Crew	Daily Output	Labor-Hours	Unit	2020 Bare Costs Material	Labor	Equipment	Total	Total Incl O&P
1000	8′ x 12′, C/C	L-7	1.10	25.455	Ea.	11,500	1,300		12,800	14,700
1050	8′ x 12′, C/P		1.10	25.455		8,575	1,300		9,875	11,400
1200	8′ x 8′, C/C		1.40	20		9,600	1,025		10,625	12,100
1250	8′ x 8′, C/P		1.40	20		7,200	1,025		8,225	9,450
1400	8′ x 10′, C/C		1.20	23.333		8,925	1,200		10,125	11,600
1450	8′ x 10′, C/P		1.20	23.333		8,375	1,200		9,575	11,000
1600	10′ x 12′, C/C		1	28		13,500	1,425		14,925	17,000
1650	10′ x 12′, C/P		1	28		13,500	1,425		14,925	17,100
2500	Heaters only (incl. above), wall mounted, to 200 C.F.					1,025			1,025	1,125
2750	To 300 C.F.					1,150			1,150	1,250
3000	Floor standing, to 720 C.F., 10,000 watts, w/controls	1 Elec	3	2.667		2,575	164		2,739	3,100
3250	To 1,000 C.F., 16,000 watts	"	3	2.667		3,725	164		3,889	4,350

13 24 26 – Steam Baths

13 24 26.50 Steam Baths and Components		Crew	Daily Output	Labor-Hours	Unit	Material	Labor	Equipment	Total	Total Incl O&P
0010	**STEAM BATHS AND COMPONENTS**									
0020	Heater, timer & head, single, to 140 C.F.	1 Plum	1.20	6.667	Ea.	2,350	430		2,780	3,250
0500	To 300 C.F.		1.10	7.273		2,700	470		3,170	3,650
1000	Commercial size, with blow-down assembly, to 800 C.F.		.90	8.889		5,800	575		6,375	7,250
1500	To 2,500 C.F.		.80	10		8,200	645		8,845	10,000
2000	Multiple, motels, apts., 2 baths, w/blow-down assm., 500 C.F.	Q-1	1.30	12.308		6,650	715		7,365	8,400
2500	4 baths	"	.70	22.857		10,500	1,325		11,825	13,500
2700	Conversion unit for residential tub, including door					4,050			4,050	4,450

13 34 Fabricated Engineered Structures

13 34 23 – Fabricated Structures

13 34 23.10 Comfort Stations		Crew	Daily Output	Labor-Hours	Unit	Material	Labor	Equipment	Total	Total Incl O&P
0010	**COMFORT STATIONS** Prefab., stock, w/doors, windows & fixt.									
0100	Not incl. interior finish or electrical									
0300	Mobile, on steel frame, 2 unit				S.F.	190			190	209
0350	7 unit					320			320	350
0400	Permanent, including concrete slab, 2 unit	B-12J	50	.320		251	16.20	16.95	284.15	320
0500	6 unit	"	43	.372		201	18.85	19.70	239.55	271
0600	Alternate pricing method, mobile, 2 fixture				Fixture	6,675			6,675	7,350
0650	7 fixture					11,600			11,600	12,800
0700	Permanent, 2 unit	B-12J	.70	22.857		20,700	1,150	1,200	23,050	25,900
0750	6 unit	"	.50	32		17,800	1,625	1,700	21,125	23,900

13 42 Building Modules

13 42 63 – Detention Cell Modules

13 42 63.16 Steel Detention Cell Modules		Crew	Daily Output	Labor-Hours	Unit	Material	Labor	Equipment	Total	Total Incl O&P
0010	**STEEL DETENTION CELL MODULES**									
2000	Cells, prefab., 5′ to 6′ wide, 7′ to 8′ high, 7′ to 8′ deep,									
2010	bar front, cot, not incl. plumbing	E-4	1.50	21.333	Ea.	9,650	1,250	98	10,998	12,700

13 47 Facility Protection

13 47 13 – Cathodic Protection

13 47 13.16 Cathodic Prot. for Underground Storage Tanks		Crew	Daily Output	Labor-Hours	Unit	2020 Bare Costs Material	Labor	Equipment	Total	Total Incl O&P
0010	**CATHODIC PROTECTION FOR UNDERGROUND STORAGE TANKS**									
1000	Anodes, magnesium type, 9 #	R-15	18.50	2.595	Ea.	38.50	156	15.05	209.55	291
1010	17 #		13	3.692		81.50	222	21.50	325	445
1020	32 #		10	4.800		123	288	28	439	595
1030	48 #		7.20	6.667		160	400	38.50	598.50	815
1100	Graphite type w/epoxy cap, 3" x 60" (32 #)	R-22	8.40	4.438		148	249		397	535
1110	4" x 80" (68 #)		6	6.213		246	350		596	790
1120	6" x 72" (80 #)		5.20	7.169		1,525	405		1,930	2,275
1130	6" x 36" (45 #)		9.60	3.883		770	218		988	1,175
2000	Rectifiers, silicon type, air cooled, 28 V/10 A	R-19	3.50	5.714		2,325	350		2,675	3,075
2010	20 V/20 A		3.50	5.714		2,325	350		2,675	3,100
2100	Oil immersed, 28 V/10 A		3	6.667		2,875	410		3,285	3,775
2110	20 V/20 A		3	6.667		3,250	410		3,660	4,175
3000	Anode backfill, coke breeze	R-22	3850	.010	Lb.	.30	.54		.84	1.14
4000	Cable, HMWPE, No. 8		2.40	15.533	M.L.F.	605	870		1,475	1,975
4010	No. 6		2.40	15.533		820	870		1,690	2,200
4020	No. 4		2.40	15.533		1,225	870		2,095	2,650
4030	No. 2		2.40	15.533		1,900	870		2,770	3,400
4040	No. 1		2.20	16.945		2,575	950		3,525	4,275
4050	No. 1/0		2.20	16.945		3,350	950		4,300	5,100
4060	No. 2/0		2.20	16.945		5,250	950		6,200	7,200
4070	No. 4/0		2	18.640		7,225	1,050		8,275	9,500
5000	Test station, 7 terminal box, flush curb type w/lockable cover	R-19	12	1.667	Ea.	79.50	102		181.50	241
5010	Reference cell, 2" diam. PVC conduit, cplg., plug, set flush	"	4.80	4.167	"	158	256		414	555

Division Notes

	CREW	DAILY OUTPUT	LABOR-HOURS	UNIT	BARE COSTS MAT.	BARE COSTS LABOR	BARE COSTS EQUIP.	BARE COSTS TOTAL	TOTAL INCL O&P

Division 14 Conveying Equipment

Estimating Tips

General

- Many products in Division 14 will require some type of support or blocking for installation not included with the item itself. Examples are supports for conveyors or tube systems, attachment points for lifts, and footings for hoists or cranes. Add these supports in the appropriate division.

14 10 00 Dumbwaiters
14 20 00 Elevators

- Dumbwaiters and elevators are estimated and purchased in a method similar to buying a car. The manufacturer has a base unit with standard features. Added to this base unit price will be whatever options the owner or specifications require. Increased load capacity, additional vertical travel, additional stops, higher speed, and cab finish options are items to be considered. When developing an estimate for dumbwaiters and elevators, remember that some items needed by the installers may have to be included as part of the general contract.

  Examples are:
  - ☐ shaftway
  - ☐ rail support brackets
  - ☐ machine room
  - ☐ electrical supply
  - ☐ sill angles
  - ☐ electrical connections
  - ☐ pits
  - ☐ roof penthouses
  - ☐ pit ladders

  Check the job specifications and drawings before pricing.

- Installation of elevators and handicapped lifts in historic structures can require significant additional costs. The associated structural requirements may involve cutting into and repairing finishes, moldings, flooring, etc. The estimator must account for these special conditions.

14 30 00 Escalators and Moving Walks

- Escalators and moving walks are specialty items installed by specialty contractors. There are numerous options associated with these items. For specific options, contact a manufacturer or contractor. In a method similar to estimating dumbwaiters and elevators, you should verify the extent of general contract work and add items as necessary.

14 40 00 Lifts
14 90 00 Other Conveying Equipment

- Products such as correspondence lifts, chutes, and pneumatic tube systems, as well as other items specified in this subdivision, may require trained installers. The general contractor might not have any choice as to who will perform the installation or when it will be performed. Long lead times are often required for these products, making early decisions in scheduling necessary.

Reference Numbers

Reference numbers are shown at the beginning of some major classifications. These numbers refer to related items in the Reference Section. The reference information may be an estimating procedure, an alternate pricing method, or technical information.

Note: Not all subdivisions listed here necessarily appear. ■

14 45 Vehicle Lifts

14 45 10 – Hydraulic Vehicle Lifts

14 45 10.10 Hydraulic Lifts		Crew	Daily Output	Labor-Hours	Unit	2020 Bare Costs Material	Labor	Equipment	Total	Total Incl O&P
0010	**HYDRAULIC LIFTS**									
2200	Single post, 8,000 lb. capacity	L-4	.40	60	Ea.	7,100	3,000		10,100	12,400
2810	Double post, 6,000 lb. capacity		2.67	8.989		9,775	450		10,225	11,500
2815	9,000 lb. capacity		2.29	10.480		23,200	525		23,725	26,300
2820	15,000 lb. capacity		2	12		26,200	600		26,800	29,800
2822	Four post, 26,000 lb. capacity		1.80	13.333		15,100	665		15,765	17,600
2825	30,000 lb. capacity		1.60	15		58,000	750		58,750	65,000
2830	Ramp style, 4 post, 25,000 lb. capacity		2	12		22,700	600		23,300	25,900
2835	35,000 lb. capacity		1	24		106,000	1,200		107,200	118,500
2840	50,000 lb. capacity		1	24		121,000	1,200		122,200	135,500
2845	75,000 lb. capacity	▼	1	24		141,000	1,200		142,200	157,000
2850	For drive thru tracks, add, minimum					1,150			1,150	1,275
2855	Maximum					1,975			1,975	2,175
2860	Ramp extensions, 3′ (set of 2)					1,200			1,200	1,325
2865	Rolling jack platform					4,175			4,175	4,575
2870	Electric/hydraulic jacking beam					11,400			11,400	12,600
2880	Scissor lift, portable, 6,000 lb. capacity				▼	10,900			10,900	12,000

Estimating Tips

Pipe for fire protection and all uses is located in Subdivisions 21 11 13 and 22 11 13.

The labor adjustment factors listed in Subdivision 22 01 02.20 also apply to Division 21.

Many, but not all, areas in the U.S. require backflow protection in the fire system. Insurance underwriters may have specific requirements for the type of materials to be installed or design requirements based on the hazard to be protected. Local jurisdictions may have requirements not covered by code. It is advisable to be aware of any special conditions.

For your reference, the following is a list of the most applicable Fire Codes and Standards, which may be purchased from the NFPA, 1 Batterymarch Park, Quincy, MA 02169-7471.

- NFPA 1: Uniform Fire Code
- NFPA 10: Portable Fire Extinguishers
- NFPA 11: Low-, Medium-, and High-Expansion Foam
- NFPA 12: Carbon Dioxide Extinguishing Systems (Also companion 12A)
- NFPA 13: Installation of Sprinkler Systems (Also companion 13D, 13E, and 13R)
- NFPA 14: Installation of Standpipe and Hose Systems
- NFPA 15: Water Spray Fixed Systems for Fire Protection
- NFPA 16: Installation of Foam-Water Sprinkler and Foam-Water Spray Systems
- NFPA 17: Dry Chemical Extinguishing Systems (Also companion 17A)
- NFPA 18: Wetting Agents
- NFPA 20: Installation of Stationary Pumps for Fire Protection
- NFPA 22: Water Tanks for Private Fire Protection
- NFPA 24: Installation of Private Fire Service Mains and their Appurtenances
- NFPA 25: Inspection, Testing and Maintenance of Water-Based Fire Protection

Reference Numbers

Reference numbers are shown at the beginning of some major classifications. These numbers refer to related items in the Reference Section. The reference information may be an estimating procedure, an alternate pricing method, or technical information.

Note: "Powered in part by CINX™, based on liscensed proprietary information of Harrison Publishing House, Inc."

Note: Trade Service, in part, has been used as a reference source for some of the material prices used in Division 21.

21 05 23 – General-Duty Valves for Water-Based Fire-Suppression Piping

21 05 23.50 General-Duty Valves		Crew	Daily Output	Labor-Hours	Unit	2020 Bare Costs Material	Labor	Equipment	Total	Total Incl O&P
0010	**GENERAL-DUTY VALVES**, for water-based fire suppression									
6200	Valves and components									
6210	Wet alarm, includes									
6220	retard chamber, trim, gauges, alarm line strainer									
6260	3″ size	Q-12	3	5.333	Ea.	1,725	305		2,030	2,350
6280	4″ size	"	2	8		2,325	455		2,780	3,225
6300	6″ size	Q-13	4	8		2,000	480		2,480	2,925
6320	8″ size	"	3	10.667		2,425	640		3,065	3,650
6400	Dry alarm, includes									
6405	retard chamber, trim, gauges, alarm line strainer									
6410	1-1/2″ size	Q-12	3	5.333	Ea.	5,100	305		5,405	6,050
6420	2″ size		3	5.333		5,100	305		5,405	6,050
6430	3″ size		3	5.333		5,175	305		5,480	6,150
6440	4″ size		2	8		5,525	455		5,980	6,750
6450	6″ size	Q-13	3	10.667		6,375	640		7,015	7,975
6460	8″ size	"	3	10.667		9,350	640		9,990	11,300
6500	Check, swing, C.I. body, brass fittings, auto. ball drip									
6520	4″ size	Q-12	3	5.333	Ea.	420	305		725	915
6540	6″ size	Q-13	4	8		780	480		1,260	1,575
6580	8″ size	"	3	10.667		1,225	640		1,865	2,325
6800	Check, wafer, butterfly type, C.I. body, bronze fittings									
6820	4″ size	Q-12	4	4	Ea.	1,375	228		1,603	1,850
6840	6″ size	Q-13	5.50	5.818		1,975	350		2,325	2,700
6860	8″ size		5	6.400		3,425	385		3,810	4,350
6880	10″ size		4.50	7.111		2,650	430		3,080	3,575
8700	Floor control valve, includes trim and gauges, 2″ size	Q-12	6	2.667		1,000	152		1,152	1,325
8710	2-1/2″ size		6	2.667		1,150	152		1,302	1,475
8720	3″ size		6	2.667		1,150	152		1,302	1,475
8730	4″ size		6	2.667		1,150	152		1,302	1,475
8740	6″ size		5	3.200		1,150	182		1,332	1,525
8800	Flow control valve, includes trim and gauges, 2″ size		2	8		5,475	455		5,930	6,700
8820	3″ size		1.50	10.667		6,275	605		6,880	7,800
8840	4″ size	Q-13	2.80	11.429		6,650	690		7,340	8,350
8860	6″ size	"	2	16		8,325	965		9,290	10,600
9200	Pressure operated relief valve, brass body	1 Spri	18	.444		695	28		723	805
9600	Waterflow indicator, vane type, with recycling retard and									
9610	two single pole retard switches, 2″ thru 6″ pipe size	1 Spri	8	1	Ea.	171	63.50		234.50	284

21 05 53 – Identification For Fire-Suppression Piping and Equipment

21 05 53.50 Identification		Crew	Daily Output	Labor-Hours	Unit	Material	Labor	Equipment	Total	Total Incl O&P
0010	**IDENTIFICATION**, for fire suppression piping and equipment									
3010	Plates and escutcheons for identification of fire dept. service/connections									
3100	Wall mount, round, aluminum									
3110	4″	1 Plum	96	.083	Ea.	25.50	5.35		30.85	36.50
3120	6″	"	96	.083	"	67	5.35		72.35	81.50
3200	Wall mount, round, cast brass									
3210	2-1/2″	1 Plum	70	.114	Ea.	61.50	7.35		68.85	79
3220	3″		70	.114		72	7.35		79.35	90
3230	4″		70	.114		123	7.35		130.35	147
3240	6″		70	.114		144	7.35		151.35	169
3250	For polished brass, add					25%				
3260	For rough chrome, add					33%				
3270	For polished chrome, add					55%				

21 05 Common Work Results for Fire Suppression

21 05 53 – Identification For Fire-Suppression Piping and Equipment

21 05 53.50 Identification		Crew	Daily Output	Labor-Hours	Unit	Material	2020 Bare Costs Labor	Equipment	Total	Total Incl O&P
3300	Wall mount, square, cast brass									
3310	2-1/2"	1 Plum	70	.114	Ea.	164	7.35		171.35	192
3320	3"	"	70	.114	"	175	7.35		182.35	203
3330	For polished brass, add					15%				
3340	For rough chrome, add					25%				
3350	For polished chrome, add					33%				
3400	Wall mount, cast brass, multiple outlets									
3410	rect. 2 way	Q-1	5	3.200	Ea.	227	186		413	530
3420	rect. 3 way		4	4		535	232		767	935
3430	rect. 4 way		4	4		670	232		902	1,075
3440	square 4 way		4	4		665	232		897	1,075
3450	rect. 6 way		3	5.333		915	310		1,225	1,475
3460	For polished brass, add					10%				
3470	For rough chrome, add					20%				
3480	For polished chrome, add					25%				
3500	Base mount, free standing fdc, cast brass									
3510	4"	1 Plum	60	.133	Ea.	123	8.60		131.60	149
3520	6"	"	60	.133	"	164	8.60		172.60	194
3530	For polished brass, add					25%				
3540	For rough chrome, add					30%				
3550	For polished chrome, add					45%				

21 11 Facility Fire-Suppression Water-Service Piping

21 11 11 – Fire-Suppression, Pipe Fittings, Grooved Joint

21 11 11.05 Corrosion Monitors		Crew	Daily Output	Labor-Hours	Unit	Material	Labor	Equipment	Total	Total Incl O&P
0010	**CORROSION MONITORS**, pipe, in line spool									
1100	Powder coated, schedule 10									
1140	2"	1 Plum	17	.471	Ea.	180	30.50		210.50	244
1150	2-1/2"	Q-1	27	.593		188	34.50		222.50	259
1160	3"		22	.727		196	42		238	279
1170	4"		17	.941		216	54.50		270.50	320
1180	6"	Q-2	17	1.412		227	85		312	375
1190	8"	"	14	1.714		255	103		358	435
1200	Powder coated, schedule 40									
1240	2"	1 Plum	17	.471	Ea.	180	30.50		210.50	244
1250	2-1/2"	Q-1	27	.593		188	34.50		222.50	259
1260	3"		22	.727		196	42		238	279
1270	4"		17	.941		216	54.50		270.50	320
1280	6"	Q-2	17	1.412		227	85		312	375
1290	8"	"	14	1.714		255	103		358	435
1300	Galvanized, schedule 10									
1340	2"	1 Plum	17	.471	Ea.	188	30.50		218.50	253
1350	2-1/2"	Q-1	27	.593		200	34.50		234.50	272
1360	3"		22	.727		208	42		250	292
1370	4"		17	.941		227	54.50		281.50	330
1380	6"	Q-2	17	1.412		247	85		332	400
1390	8"	"	14	1.714		274	103		377	455
1400	Galvanized, schedule 40									
1440	2"	1 Plum	17	.471	Ea.	180	30.50		210.50	244
1450	2-1/2"	Q-1	27	.593		193	34.50		227.50	264
1460	3"		22	.727		196	42		238	279

21 11 11.05 Corrosion Monitors		Crew	Daily Output	Labor-Hours	Unit	Material	2020 Bare Costs Labor	Equipment	Total	Total Incl O&P
1470	4"	Q-1	17	.941	Ea.	216	54.50		270.50	320
1480	6"	Q-2	17	1.412		227	85		312	375
1490	8"	"	14	1.714	↓	255	103		358	435
1500	Mechanical tee, painted									
1540	2"	1 Plum	50	.160	Ea.	153	10.30		163.30	183
1550	2-1/2"	Q-1	80	.200		161	11.60		172.60	194
1560	3"		67	.239		165	13.85		178.85	202
1570	4"	↓	50	.320		169	18.55		187.55	213
1580	6"	Q-2	50	.480		172	29		201	233
1590	8"	"	42	.571	↓	196	34.50		230.50	268
1600	Mechanical tee, galvanized									
1640	2"	1 Plum	50	.160	Ea.	165	10.30		175.30	196
1650	2-1/2"	Q-1	80	.200		172	11.60		183.60	207
1660	3"		67	.239		192	13.85		205.85	232
1670	4"	↓	50	.320		180	18.55		198.55	226
1680	6"	Q-2	50	.480		192	29		221	254
1690	8"	"	42	.571	↓	216	34.50		250.50	289

21 11 11.16 Pipe Fittings, Grooved Joint		Crew	Daily Output	Labor-Hours	Unit	Material	Labor	Equipment	Total	Total Incl O&P
0010	**PIPE FITTINGS, GROOVED JOINT,** For fire-suppression									
0020	Fittings, ductile iron									
0030	Coupling required at joints not incl. in fitting price.									
0034	Add 1 coupling, material only, per joint for installed price.									
0038	For standard grooved joint materials see Div. 22 11 13.48									
0040	90° elbow									
0110	2"	1 Plum	25	.320	Ea.	22	20.50		42.50	55.50
0120	2-1/2"	Q-1	40	.400		29.50	23		52.50	67
0130	3"		33	.485		46.50	28		74.50	93.50
0140	4"		25	.640		54.50	37		91.50	116
0150	5"	↓	20	.800		128	46.50		174.50	210
0160	6"	Q-2	25	.960		150	57.50		207.50	252
0170	8"	"	21	1.143	↓	292	68.50		360.50	425
0200	45° elbow									
0210	2"	1 Plum	25	.320	Ea.	22	20.50		42.50	55.50
0220	2-1/2"	Q-1	40	.400		29.50	23		52.50	67
0230	3"		33	.485		38	28		66	84
0240	4"		25	.640		54.50	37		91.50	116
0250	5"	↓	20	.800		128	46.50		174.50	210
0260	6"	Q-2	25	.960		150	57.50		207.50	252
0270	8"	"	21	1.143	↓	296	68.50		364.50	430
0300	Tee									
0310	2"	1 Plum	17	.471	Ea.	34	30.50		64.50	82.50
0320	2-1/2"	Q-1	27	.593		44	34.50		78.50	100
0330	3"		22	.727		61	42		103	130
0340	4"		17	.941		92	54.50		146.50	183
0350	5"	↓	13	1.231		212	71.50		283.50	340
0360	6"	Q-2	17	1.412		244	85		329	395
0370	8"	"	14	1.714	↓	520	103		623	725
0400	Cap									
0410	1-1/4"	1 Plum	76	.105	Ea.	12.15	6.80		18.95	23.50
0420	1-1/2"		63	.127		12.80	8.20		21	26.50
0430	2"	↓	47	.170		15.20	10.95		26.15	33
0440	2-1/2"	Q-1	76	.211	↓	17.55	12.20		29.75	37.50

21 11 11.16	Pipe Fittings, Grooved Joint	Crew	Daily Output	Labor-Hours	Unit	2020 Bare Costs Material	Labor	Equipment	Total	Total Incl O&P
0450	3"	Q-1	63	.254	Ea.	21	14.75		35.75	45
0460	4"		47	.340		30.50	19.75		50.25	63
0470	5"		37	.432		61	25		86	105
0480	6"	Q-2	47	.511		68	30.50		98.50	121
0490	8"	"	39	.615		127	37		164	196
0500	Coupling, rigid									
0510	1-1/4"	1 Plum	100	.080	Ea.	27.50	5.15		32.65	37.50
0520	1-1/2"		67	.119		36.50	7.70		44.20	51.50
0530	2"		50	.160		39.50	10.30		49.80	58.50
0540	2-1/2"	Q-1	80	.200		46.50	11.60		58.10	69
0550	3"		67	.239		50.50	13.85		64.35	76
0560	4"		50	.320		70.50	18.55		89.05	106
0570	5"		40	.400		74.50	23		97.50	117
0580	6"	Q-2	50	.480		89.50	29		118.50	142
0590	8"	"	42	.571		150	34.50		184.50	217
0700	End of run fitting									
0710	1-1/4" x 1/2" NPT	1 Plum	76	.105	Ea.	41.50	6.80		48.30	55.50
0714	1-1/4" x 3/4" NPT		76	.105		41.50	6.80		48.30	55.50
0718	1-1/4" x 1" NPT		76	.105		41.50	6.80		48.30	55.50
0722	1-1/2" x 1/2" NPT		63	.127		41.50	8.20		49.70	58
0726	1-1/2" x 3/4" NPT		63	.127		41.50	8.20		49.70	58
0730	1-1/2" x 1" NPT		63	.127		41.50	8.20		49.70	58
0734	2" x 1/2" NPT		47	.170		44	10.95		54.95	65
0738	2" x 3/4" NPT		47	.170		44	10.95		54.95	65
0742	2" x 1" NPT		47	.170		44	10.95		54.95	65
0746	2-1/2" x 1/2" NPT		40	.200		52.50	12.90		65.40	77
0750	2-1/2" x 3/4" NPT		40	.200		52.50	12.90		65.40	77
0754	2-1/2" x 1" NPT		40	.200		52.50	12.90		65.40	77
0800	Drain elbow									
0820	2-1/2"	Q-1	40	.400	Ea.	103	23		126	149
0830	3"		33	.485		132	28		160	188
0840	4"		25	.640		134	37		171	204
0850	6"	Q-2	25	.960		235	57.50		292.50	345
1000	Valves, grooved joint									
1002	Coupling required at joints not incl. in fitting price.									
1004	Add 1 coupling, material only, per joint for installed price.									
1010	Ball valve with weatherproof actuator									
1020	1-1/4"	1 Plum	39	.205	Ea.	156	13.20		169.20	192
1030	1-1/2"		31	.258		185	16.65		201.65	228
1040	2"		24	.333		223	21.50		244.50	277
1100	Butterfly valve, high pressure, with actuator									
1110	Supervised open									
1120	2"	1 Plum	23	.348	Ea.	415	22.50		437.50	495
1130	2-1/2"	Q-1	38	.421		430	24.50		454.50	510
1140	3"		30	.533		465	31		496	555
1150	4"		22	.727		490	42		532	600
1160	5"		19	.842		730	49		779	880
1170	6"	Q-2	23	1.043		695	63		758	855
1180	8"		18	1.333		1,050	80		1,130	1,300
1190	10"		15	1.600		2,575	96		2,671	2,975
1200	12"		12	2		3,675	120		3,795	4,200
1300	Gate valve, OS&Y									
1310	2-1/2"	Q-1	35	.457	Ea.	470	26.50		496.50	555

21 11 11.16 Pipe Fittings, Grooved Joint		Crew	Daily Output	Labor-Hours	Unit	Material	2020 Bare Costs Labor	Equipment	Total	Total Incl O&P
1320	3"	Q-1	28	.571	Ea.	645	33		678	760
1330	4"		20	.800		760	46.50		806.50	905
1340	6"	Q-2	21	1.143		965	68.50		1,033.50	1,150
1350	8"		16	1.500		1,500	90		1,590	1,775
1360	10"		13	1.846		2,025	111		2,136	2,400
1370	12"		10	2.400		3,175	144		3,319	3,700
1400	Gate valve, non-rising stem									
1410	2-1/2"	Q-1	36	.444	Ea.	435	26		461	520
1420	3"		29	.552		585	32		617	695
1430	4"		21	.762		695	44		739	830
1440	6"	Q-2	22	1.091		895	65.50		960.50	1,075
1450	8"		17	1.412		1,325	85		1,410	1,575
1460	10"		14	1.714		1,925	103		2,028	2,250
1470	12"		11	2.182		2,675	131		2,806	3,150
2000	Alarm check valve, pre-trimmed									
2010	1-1/2"	Q-1	6	2.667	Ea.	1,925	155		2,080	2,350
2020	2"	"	5	3.200		1,925	186		2,111	2,400
2030	2-1/2"	Q-2	7	3.429		1,925	206		2,131	2,425
2040	3"		6	4		1,925	241		2,166	2,475
2050	4"		5	4.800		2,125	289		2,414	2,750
2060	6"		4	6		2,450	360		2,810	3,250
2070	8"		2	12		3,400	720		4,120	4,825
2200	Dry valve, pre-trimmed									
2210	1-1/2"	Q-1	5	3.200	Ea.	2,875	186		3,061	3,450
2220	2"	"	4	4		2,875	232		3,107	3,525
2230	2-1/2"	Q-2	6	4		2,925	241		3,166	3,575
2240	3"		5	4.800		2,925	289		3,214	3,650
2250	4"		4	6		3,125	360		3,485	4,000
2260	6"		2	12		3,600	720		4,320	5,050
2270	8"		1	24		5,300	1,450		6,750	7,975
2300	Deluge valve, pre-trimmed, with electric solenoid									
2310	1-1/2"	Q-1	5	3.200	Ea.	4,125	186		4,311	4,800
2320	2"	"	4	4		4,125	232		4,357	4,875
2330	2-1/2"	Q-2	6	4		4,175	241		4,416	4,950
2340	3"		5	4.800		4,175	289		4,464	5,025
2350	4"		4	6		4,975	360		5,335	6,025
2360	6"		2	12		5,950	720		6,670	7,625
2370	8"		1	24		6,725	1,450		8,175	9,550
2400	Preaction valve									
2410	Valve has double interlock,									
2420	pneumatic/electric actuation and trim.									
2430	1-1/2"	Q-1	5	3.200	Ea.	4,225	186		4,411	4,925
2440	2"	"	4	4		4,225	232		4,457	5,000
2450	2-1/2"	Q-2	6	4		4,225	241		4,466	5,000
2460	3"		5	4.800		4,225	289		4,514	5,075
2470	4"		4	6		4,650	360		5,010	5,650
2480	6"		2	12		4,650	720		5,370	6,175
2490	8"		1	24		5,100	1,450		6,550	7,775
3000	Fittings, ductile iron, ready to install									
3010	Includes bolts & grade "E" gaskets									
3012	Add 1 coupling, material only, per joint for installed price.									
3200	90° elbow									
3210	1-1/4"	1 Plum	88	.091	Ea.	113	5.85		118.85	133

21 11 11 – Fire-Suppression, Pipe Fittings, Grooved Joint

21 11 11.16 Pipe Fittings, Grooved Joint		Crew	Daily Output	Labor-Hours	Unit	2020 Bare Costs Material	Labor	Equipment	Total	Total Incl O&P
3220	1-1/2"	1 Plum	72	.111	Ea.	116	7.15		123.15	138
3230	2"		60	.133		117	8.60		125.60	141
3240	2-1/2"	Q-1	85	.188		141	10.90		151.90	171
3300	45° elbow									
3310	1-1/4"	1 Plum	88	.091	Ea.	113	5.85		118.85	133
3320	1-1/2"		72	.111		116	7.15		123.15	138
3330	2"		60	.133		117	8.60		125.60	141
3340	2-1/2"	Q-1	85	.188		141	10.90		151.90	171
3400	Tee									
3410	1-1/4"	1 Plum	60	.133	Ea.	174	8.60		182.60	204
3420	1-1/2"		48	.167		180	10.75		190.75	214
3430	2"		38	.211		182	13.55		195.55	221
3440	2-1/2"	Q-1	60	.267		210	15.45		225.45	254
3500	Coupling									
3510	1-1/4"	1 Plum	200	.040	Ea.	42.50	2.58		45.08	50.50
3520	1-1/2"		134	.060		43	3.85		46.85	53.50
3530	2"		100	.080		49	5.15		54.15	61
3540	2-1/2"	Q-1	160	.100		56	5.80		61.80	70
3550	3"		134	.119		63	6.95		69.95	80
3560	4"		100	.160		88	9.30		97.30	111
3570	5"		80	.200		117	11.60		128.60	146
3580	6"	Q-2	100	.240		140	14.45		154.45	176
3590	8"	"	84	.286		140	17.20		157.20	180
4000	For seismic bracing, see Section 22 05 48.40									
5000	For hangers and supports, see Section 22 05 29.10									

21 11 13 – Facility Fire Suppression Piping

21 11 13.16 Pipe, Plastic		Crew	Daily Output	Labor-Hours	Unit	Material	Labor	Equipment	Total	Total Incl O&P
0010	**PIPE, PLASTIC**									
0020	CPVC, fire suppression (C-UL-S, FM, NFPA 13, 13D & 13R)									
0030	Socket joint, no couplings or hangers									
0100	SDR 13.5 (ASTM F442)									
0120	3/4" diameter	Q-12	420	.038	L.F.	1.46	2.17		3.63	4.86
0130	1" diameter		340	.047		2.26	2.68		4.94	6.50
0140	1-1/4" diameter		260	.062		3.58	3.50		7.08	9.20
0150	1-1/2" diameter		190	.084		4.94	4.79		9.73	12.65
0160	2" diameter		140	.114		7.70	6.50		14.20	18.25
0170	2-1/2" diameter		130	.123		13.55	7		20.55	25.50
0180	3" diameter		120	.133		20.50	7.60		28.10	34

21 11 13.18 Pipe Fittings, Plastic		Crew	Daily Output	Labor-Hours	Unit	Material	Labor	Equipment	Total	Total Incl O&P
0010	**PIPE FITTINGS, PLASTIC**									
0020	CPVC, fire suppression (C-UL-S, FM, NFPA 13, 13D & 13R)									
0030	Socket joint									
0100	90° elbow									
0120	3/4"	1 Plum	26	.308	Ea.	1.88	19.85		21.73	31.50
0130	1"		22.70	.352		4.12	22.50		26.62	38.50
0140	1-1/4"		20.20	.396		5.20	25.50		30.70	44
0150	1-1/2"		18.20	.440		7.40	28.50		35.90	50.50
0160	2"	Q-1	33.10	.483		9.20	28		37.20	52
0170	2-1/2"		24.20	.661		17.70	38.50		56.20	77
0180	3"		20.80	.769		24	44.50		68.50	93.50
0200	45° elbow									
0210	3/4"	1 Plum	26	.308	Ea.	2.58	19.85		22.43	32.50

21 11 13.18 Pipe Fittings, Plastic		Crew	Daily Output	Labor-Hours	Unit	2020 Bare Costs Material	Labor	Equipment	Total	Total Incl O&P
0220	1"	1 Plum	22.70	.352	Ea.	3.03	22.50		25.53	37.50
0230	1-1/4"		20.20	.396		4.38	25.50		29.88	43
0240	1-1/2"		18.20	.440		6.10	28.50		34.60	49.50
0250	2"	Q-1	33.10	.483		7.60	28		35.60	50.50
0260	2-1/2"		24.20	.661		13.65	38.50		52.15	72.50
0270	3"		20.80	.769		19.55	44.50		64.05	88.50
0300	Tee									
0310	3/4"	1 Plum	17.30	.462	Ea.	2.58	30		32.58	47.50
0320	1"		15.20	.526		5.10	34		39.10	56.50
0330	1-1/4"		13.50	.593		7.65	38		45.65	65.50
0340	1-1/2"		12.10	.661		11.25	42.50		53.75	76.50
0350	2"	Q-1	20	.800		16.65	46.50		63.15	88
0360	2-1/2"		16.20	.988		27	57.50		84.50	116
0370	3"		13.90	1.151		42	67		109	146
0400	Tee, reducing x any size									
0420	1"	1 Plum	15.20	.526	Ea.	4.32	34		38.32	56
0430	1-1/4"		13.50	.593		7.90	38		45.90	65.50
0440	1-1/2"		12.10	.661		9.60	42.50		52.10	74.50
0450	2"	Q-1	20	.800		18.60	46.50		65.10	90
0460	2-1/2"		16.20	.988		21	57.50		78.50	109
0470	3"		13.90	1.151		24.50	67		91.50	127
0500	Coupling									
0510	3/4"	1 Plum	26	.308	Ea.	1.81	19.85		21.66	31.50
0520	1"		22.70	.352		2.39	22.50		24.89	36.50
0530	1-1/4"		20.20	.396		3.48	25.50		28.98	42
0540	1-1/2"		18.20	.440		4.96	28.50		33.46	48
0550	2"	Q-1	33.10	.483		6.70	28		34.70	49.50
0560	2-1/2"		24.20	.661		10.25	38.50		48.75	69
0570	3"		20.80	.769		13.30	44.50		57.80	81.50
0600	Coupling, reducing									
0610	1" x 3/4"	1 Plum	22.70	.352	Ea.	2.39	22.50		24.89	36.50
0620	1-1/4" x 1"		20.20	.396		3.61	25.50		29.11	42
0630	1-1/2" x 3/4"		18.20	.440		5.40	28.50		33.90	48.50
0640	1-1/2" x 1"		18.20	.440		5.20	28.50		33.70	48.50
0650	1-1/2" x 1-1/4"		18.20	.440		4.96	28.50		33.46	48
0660	2" x 1"	Q-1	33.10	.483		6.95	28		34.95	49.50
0670	2" x 1-1/2"	"	33.10	.483		6.70	28		34.70	49.50
0700	Cross									
0720	3/4"	1 Plum	13	.615	Ea.	4.06	39.50		43.56	64
0730	1"		11.30	.708		5.10	45.50		50.60	74
0740	1-1/4"		10.10	.792		7	51		58	84
0750	1-1/2"		9.10	.879		9.65	56.50		66.15	95.50
0760	2"	Q-1	16.60	.964		15.85	56		71.85	101
0770	2-1/2"	"	12.10	1.322		35	76.50		111.50	153
0800	Cap									
0820	3/4"	1 Plum	52	.154	Ea.	1.09	9.90		10.99	16.05
0830	1"		45	.178		1.55	11.45		13	18.85
0840	1-1/4"		40	.200		2.52	12.90		15.42	22
0850	1-1/2"		36.40	.220		3.48	14.15		17.63	25
0860	2"	Q-1	66	.242		5.20	14.05		19.25	27
0870	2-1/2"		48.40	.331		7.55	19.15		26.70	37
0880	3"		41.60	.385		12.15	22.50		34.65	47
0900	Adapter, sprinkler head, female w/metal thd. insert (s x FNPT)									

21 11 Facility Fire-Suppression Water-Service Piping

21 11 13 – Facility Fire Suppression Piping

21 11 13.18 Pipe Fittings, Plastic		Crew	Daily Output	Labor-Hours	Unit	2020 Bare Costs Material	Labor	Equipment	Total	Total Incl O&P
0920	3/4" x 1/2"	1 Plum	52	.154	Ea.	4.96	9.90		14.86	20.50
0930	1" x 1/2"		45	.178		5.25	11.45		16.70	23
0940	1" x 3/4"		45	.178		8.25	11.45		19.70	26.50

21 11 16 – Facility Fire Hydrants

21 11 16.50 Fire Hydrants for Buildings		Crew	Daily Output	Labor-Hours	Unit	Material	Labor	Equipment	Total	Total Incl O&P
0010	FIRE HYDRANTS FOR BUILDINGS									
3750	Hydrants, wall, w/caps, single, flush, polished brass									
3800	2-1/2" x 2-1/2"	Q-12	5	3.200	Ea.	273	182		455	575
3840	2-1/2" x 3"		5	3.200		495	182		677	820
3860	3" x 3"		4.80	3.333		400	190		590	725
3900	For polished chrome, add					20%				
3950	Double, flush, polished brass									
4000	2-1/2" x 2-1/2" x 4"	Q-12	5	3.200	Ea.	775	182		957	1,125
4040	2-1/2" x 2-1/2" x 6"		4.60	3.478		1,300	198		1,498	1,725
4080	3" x 3" x 4"		4.90	3.265		1,175	186		1,361	1,550
4120	3" x 3" x 6"		4.50	3.556		1,575	202		1,777	2,025
4200	For polished chrome, add					10%				
4350	Double, projecting, polished brass									
4400	2-1/2" x 2-1/2" x 4"	Q-12	5	3.200	Ea.	290	182		472	595
4450	2-1/2" x 2-1/2" x 6"	"	4.60	3.478	"	595	198		793	945
4460	Valve control, dbl. flush/projecting hydrant, cap &									
4470	chain, extension rod & cplg., escutcheon, polished brass	Q-12	8	2	Ea.	268	114		382	465
4480	Four-way square, flush, polished brass									
4540	2-1/2" (4) x 6"	Q-12	3.60	4.444	Ea.	3,900	253		4,153	4,650

21 11 19 – Fire-Department Connections

21 11 19.50 Connections for the Fire-Department		Crew	Daily Output	Labor-Hours	Unit	Material	Labor	Equipment	Total	Total Incl O&P
0010	CONNECTIONS FOR THE FIRE-DEPARTMENT									
0020	For fire pro. cabinets, see Section 10 44 13.53									
4000	Storz type, with cap and chain									
4100	2-1/2" Storz x 2-1/2" F NPT, silver powder coat	Q-12	4.80	3.333	Ea.	69	190		259	360
4200	4" Storz x 1-1/2" F NPT, silver powder coat		4.80	3.333		133	190		323	430
4300	4" Storz x 2-1/2" F NPT, silver powder coat		4.80	3.333		138	190		328	435
4400	4" Storz x 3" F NPT, silver powder coat		4.80	3.333		148	190		338	445
4500	4" Storz x 4" F NPT, red powder coat		4.80	3.333		166	190		356	465
4600	4" Storz x 4" F NPT, silver powder coat		4.80	3.333		162	190		352	460
4700	4" Storz x 6" F NPT, red powder coat		4.80	3.333		228	190		418	535
4800	4" Storz x 6" F NPT, silver powder coat		4.80	3.333		226	190		416	535
4900	5" Storz x 2-1/2" F NPT, silver powder coat		4.80	3.333		198	190		388	500
5000	5" Storz x 3" F NPT, silver powder coat		4.80	3.333		208	190		398	515
5100	5" Storz x 4" F NPT, red powder coat		4.80	3.333		182	190		372	485
5200	5" Storz x 4" F NPT, silver powder coat		4.80	3.333		183	190		373	485
5300	5" Storz x 6" F NPT, red powder coat		4.80	3.333		273	190		463	585
5400	5" Storz x 6" F NPT, silver powder coat		4.80	3.333		274	190		464	585
6000	Roof manifold, horiz., brass, without valves & caps									
6040	2-1/2" x 2-1/2" x 4"	Q-12	4.80	3.333	Ea.	228	190		418	535
6060	2-1/2" x 2-1/2" x 6"		4.60	3.478		236	198		434	555
6080	2-1/2" x 2-1/2" x 2-1/2" x 4"		4.60	3.478		380	198		578	715
6090	2-1/2" x 2-1/2" x 2-1/2" x 6"		4.60	3.478		405	198		603	740
7000	Sprinkler line tester, cast brass					38.50			38.50	42.50
7140	Standpipe connections, wall, w/plugs & chains									
7160	Single, flush, brass, 2-1/2" x 2-1/2", Fire Dept Conn.	Q-12	5	3.200	Ea.	186	182		368	480
7180	2-1/2" x 3"	"	5	3.200	"	191	182		373	485

21 11 Facility Fire-Suppression Water-Service Piping

21 11 19 – Fire-Department Connections

21 11 19.50	Connections for the Fire-Department	Crew	Daily Output	Labor-Hours	Unit	Material	2020 Bare Costs Labor	Equipment	Total	Total Incl O&P
7240	For polished chrome, add					15%				
7280	Double, flush, polished brass									
7300	2-1/2" x 2-1/2" x 4"	Q-12	5	3.200	Ea.	775	182		957	1,125
7330	2-1/2" x 2-1/2" x 6"		4.60	3.478		855	198		1,053	1,225
7340	3" x 3" x 4"		4.90	3.265		1,150	186		1,336	1,550
7370	3" x 3" x 6"		4.50	3.556		1,300	202		1,502	1,750
7400	For polished chrome, add					15%				
7440	For sill cock combination, add				Ea.	101			101	112
7580	Double projecting, polished brass									
7600	2-1/2" x 2-1/2" x 4"	Q-12	5	3.200	Ea.	610	182		792	945
7630	2-1/2" x 2-1/2" x 6"	"	4.60	3.478	"	1,025	198		1,223	1,425
7680	For polished chrome, add					15%				
7900	Three way, flush, polished brass									
7920	2-1/2" (3) x 4"	Q-12	4.80	3.333	Ea.	2,000	190		2,190	2,475
7930	2-1/2" (3) x 6"	"	4.60	3.478		2,150	198		2,348	2,675
8000	For polished chrome, add					9%				
8020	Three way, projecting, polished brass									
8040	2-1/2" (3) x 4"	Q-12	4.80	3.333	Ea.	910	190		1,100	1,275
8070	2-1/2" (3) x 6"	"	4.60	3.478		1,800	198		1,998	2,275
8100	For polished chrome, add					12%				
8200	Four way, square, flush, polished brass,									
8240	2-1/2" (4) x 6"	Q-12	3.60	4.444	Ea.	1,750	253		2,003	2,300
8300	For polished chrome, add				"	10%				
8550	Wall, vertical, flush, cast brass									
8600	Two way, 2-1/2" x 2-1/2" x 4"	Q-12	5	3.200	Ea.	400	182		582	715
8660	Four way, 2-1/2" (4) x 6"		3.80	4.211		1,350	240		1,590	1,825
8680	Six way, 2-1/2" (6) x 6"		3.40	4.706		1,600	268		1,868	2,150
8700	For polished chrome, add					10%				
8800	Free standing siamese unit, polished brass, two way									
8820	2-1/2" x 2-1/2" x 4"	Q-12	2.50	6.400	Ea.	750	365		1,115	1,375
8850	2-1/2" x 2-1/2" x 6"		2	8		830	455		1,285	1,600
8860	3" x 3" x 4"		2.50	6.400		570	365		935	1,175
8890	3" x 3" x 6"		2	8		1,475	455		1,930	2,300
8940	For polished chrome, add					12%				
9100	Free standing siamese unit, polished brass, three way									
9120	2-1/2" x 2-1/2" x 2-1/2" x 6"	Q-12	2	8	Ea.	980	455		1,435	1,750
9160	For polished chrome, add				"	15%				

21 12 Fire-Suppression Standpipes

21 12 13 – Fire-Suppression Hoses and Nozzles

21 12 13.50	Fire Hoses and Nozzles	Crew	Daily Output	Labor-Hours	Unit	Material	Labor	Equipment	Total	Total Incl O&P
0010	**FIRE HOSES AND NOZZLES**									
0200	Adapters, rough brass, straight hose threads R211226-10									
0220	One piece, female to male, rocker lugs									
0240	1" x 1" R211226-20				Ea.	56.50			56.50	62
0260	1-1/2" x 1"					45.50			45.50	50
0280	1-1/2" x 1-1/2"					10.15			10.15	11.20
0300	2" x 1-1/2"					78.50			78.50	86.50
0320	2" x 2"					42.50			42.50	46.50
0340	2-1/2" x 1-1/2"					32			32	35
0360	3" x 1-1/2"					56.50			56.50	62.50

21 12 Fire-Suppression Standpipes

21 12 13 – Fire-Suppression Hoses and Nozzles

21 12 13.50 Fire Hoses and Nozzles		Crew	Daily Output	Labor-Hours	Unit	2020 Bare Costs Material	Labor	Equipment	Total	Total Incl O&P
0380	2-1/2" x 2-1/2"				Ea.	19.55			19.55	21.50
0400	3" x 2-1/2"					141			141	156
0420	3" x 3"					85			85	93.50
0500	For polished brass, add					50%				
0520	For polished chrome, add				▼	75%				
0700	One piece, female to male, hexagon									
0740	1-1/2" x 3/4"				Ea.	50			50	55
0760	2" x 1-1/2"					114			114	125
0780	2-1/2" x 1"					196			196	215
0800	2-1/2" x 1-1/2"					64.50			64.50	71
0820	2-1/2" x 2"					45.50			45.50	50
0840	3" x 2-1/2"					85.50			85.50	94
0900	For polished chrome, add				▼	75%				
1100	Swivel, female to female, pin lugs									
1120	1-1/2" x 1-1/2"				Ea.	74.50			74.50	82
1200	2-1/2" x 2-1/2"					147			147	162
1260	For polished brass, add					50%				
1280	For polished chrome, add				▼	75%				
1400	Couplings, sngl. & dbl. jacket, pin lug or rocker lug, cast brass									
1410	1-1/2"				Ea.	63			63	69.50
1420	2-1/2"				"	51.50			51.50	57
1500	For polished brass, add					20%				
1520	For polished chrome, add					40%				
1580	Reducing, F x M, interior installation, cast brass									
1590	2" x 1-1/2"				Ea.	83			83	91.50
1600	2-1/2" x 1-1/2"					19.50			19.50	21.50
1680	For polished brass, add					50%				
1720	For polished chrome, add				▼	75%				
2200	Hose, less couplings									
2260	Synthetic jacket, lined, 300 lb. test, 1-1/2" diameter	Q-12	2600	.006	L.F.	3.31	.35		3.66	4.16
2270	2" diameter		2200	.007		2.59	.41		3	3.47
2280	2-1/2" diameter		2200	.007		5.90	.41		6.31	7.10
2290	3" diameter		2200	.007		3.27	.41		3.68	4.22
2360	High strength, 500 lb. test, 1-1/2" diameter		2600	.006		2.59	.35		2.94	3.37
2380	2-1/2" diameter	▼	2200	.007	▼	6.05	.41		6.46	7.25
5000	Nipples, straight hose to tapered iron pipe, brass									
5060	Female to female, 1-1/2" x 1-1/2"				Ea.	26			26	28.50
5100	2-1/2" x 2-1/2"					45			45	49.50
5190	For polished chrome, add					75%				
5200	Double male or male to female, 1" x 1"					56.50			56.50	62
5220	1-1/2" x 1"					72			72	79.50
5230	1-1/2" x 1-1/2"					21			21	23
5260	2" x 1-1/2"					82			82	90
5270	2" x 2"					115			115	127
5280	2-1/2" x 1-1/2"					70.50			70.50	77.50
5300	2-1/2" x 2"					56			56	62
5310	2-1/2" x 2-1/2"					32			32	35
5340	For polished chrome, add				▼	75%				
5600	Nozzles, brass									
5620	Adjustable fog, 3/4" booster line				Ea.	150			150	165
5630	1" booster line					145			145	159
5640	1-1/2" leader line					117			117	128
5660	2-1/2" direct connection				▼	199			199	219

21 12 Fire-Suppression Standpipes

21 12 13 – Fire-Suppression Hoses and Nozzles

21 12 13.50 Fire Hoses and Nozzles		Crew	Daily Output	Labor-Hours	Unit	Material	2020 Bare Costs Labor	Equipment	Total	Total Incl O&P
5680	2-1/2" playpipe nozzle				Ea.	238			238	262
5780	For chrome plated, add					8%				
5850	Electrical fire, adjustable fog, no shock									
5900	1-1/2"				Ea.	455			455	500
5920	2-1/2"					900			900	990
5980	For polished chrome, add					6%				
6200	Heavy duty, comb. adj. fog and str. stream, with handle									
6210	1" booster line				Ea.	208			208	229
6240	1-1/2"					440			440	485
6260	2-1/2", for playpipe					675			675	745
6280	2-1/2" direct connection					470			470	520
6300	2-1/2" playpipe combination					605			605	665
6480	For polished chrome, add					7%				
6500	Plain fog, polished brass, 1-1/2"					154			154	170
6540	Chrome plated, 1-1/2"					128			128	141
6700	Plain stream, polished brass, 1-1/2" x 10"					58			58	63.50
6760	2-1/2" x 15" x 7/8" or 1-1/2"					109			109	119
6860	For polished chrome, add					20%				
7000	Underwriters playpipe, 2-1/2" x 30" with 1-1/8" tip				Ea.	640			640	705
9200	Storage house, hose only, primed steel					1,175			1,175	1,275
9220	Aluminum					2,550			2,550	2,800
9280	Hose and hydrant house, primed steel					1,350			1,350	1,475
9300	Aluminum					1,625			1,625	1,800
9340	Tools, crowbar and brackets	1 Carp	12	.667		97.50	35.50		133	160
9360	Combination hydrant wrench and spanner					39.50			39.50	43.50
9380	Fire axe and brackets									
9400	6 lb.	1 Carp	12	.667	Ea.	137	35.50		172.50	204
9500	For fire equipment cabinets, Section 10 44 13.53									

21 12 16 – Fire-Suppression Hose Reels

21 12 16.50 Fire-Suppression Hose Reels		Crew	Daily Output	Labor-Hours	Unit	Material	Labor	Equipment	Total	Total Incl O&P
0010	**FIRE-SUPPRESSION HOSE REELS**									
2990	Hose reel, swinging, for 1-1/2" polyester neoprene lined hose									
3000	50' long	Q-12	14	1.143	Ea.	158	65		223	272
3020	100' long		14	1.143		248	65		313	370
3060	For 2-1/2" cotton rubber hose, 75' long		14	1.143		296	65		361	425
3100	150' long		14	1.143		296	65		361	425

21 12 19 – Fire-Suppression Hose Racks

21 12 19.50 Fire Hose Racks		Crew	Daily Output	Labor-Hours	Unit	Material	Labor	Equipment	Total	Total Incl O&P
0010	**FIRE HOSE RACKS**									
2600	Hose rack, swinging, for 1-1/2" diameter hose,									
2620	Enameled steel, 50' and 75' lengths of hose	Q-12	20	.800	Ea.	72	45.50		117.50	148
2640	100' and 125' lengths of hose		20	.800		89	45.50		134.50	166
2680	Chrome plated, 50' and 75' lengths of hose		20	.800		73.50	45.50		119	149
2700	100' and 125' lengths of hose		20	.800		153	45.50		198.50	237
2750	2-1/2" diameter, 100' hose		20	.800		104	45.50		149.50	183
2780	For hose rack nipple, 1-1/2" polished brass, add					32.50			32.50	35.50
2820	2-1/2" polished brass, add					55.50			55.50	61
2840	1-1/2" polished chrome, add					38.50			38.50	42
2860	2-1/2" polished chrome, add					75.50			75.50	83

21 12 Fire-Suppression Standpipes

21 12 23 – Fire-Suppression Hose Valves

21 12 23.70 Fire Hose Valves		Crew	Daily Output	Labor-Hours	Unit	2020 Bare Costs Material	Labor	Equipment	Total	Total Incl O&P
0010	**FIRE HOSE VALVES**									
0020	Angle, combination pressure adjust/restricting, rough brass R211226-10									
0030	1-1/2"	1 Spri	12	.667	Ea.	110	42		152	184
0040	2-1/2" R211226-20	"	7	1.143	"	202	72.50		274.50	330
0042	Nonpressure adjustable/restricting, rough brass									
0044	1-1/2"	1 Spri	12	.667	Ea.	48	42		90	116
0046	2-1/2"	"	7	1.143	"	143	72.50		215.50	265
0050	For polished brass, add					30%				
0060	For polished chrome, add					40%				
0080	Wheel handle, 300 lb., 1-1/2"	1 Spri	12	.667	Ea.	105	42		147	179
0090	2-1/2"	"	7	1.143	"	198	72.50		270.50	325
0100	For polished brass, add					35%				
0110	For polished chrome, add					50%				
1000	Ball drip, automatic, rough brass, 1/2"	1 Spri	20	.400	Ea.	20.50	25.50		46	60.50
1010	3/4"	"	20	.400	"	26	25.50		51.50	67
1100	Ball, 175 lb., sprinkler system, FM/UL, threaded, bronze									
1120	Slow close									
1150	1" size	1 Spri	19	.421	Ea.	305	26.50		331.50	375
1160	1-1/4" size		15	.533		330	33.50		363.50	410
1170	1-1/2" size		13	.615		425	39		464	530
1180	2" size		11	.727		525	46		571	650
1190	2-1/2" size	Q-12	15	1.067		710	60.50		770.50	870
1230	For supervisory switch kit, all sizes									
1240	One circuit, add	1 Spri	48	.167	Ea.	157	10.55		167.55	189
1280	Quarter turn for trim									
1300	1/2" size	1 Spri	22	.364	Ea.	42	23		65	80.50
1310	3/4" size		20	.400		45	25.50		70.50	87.50
1320	1" size		19	.421		50	26.50		76.50	95
1330	1-1/4" size		15	.533		81.50	33.50		115	141
1340	1-1/2" size		13	.615		102	39		141	172
1350	2" size		11	.727		122	46		168	203
1400	Caps, polished brass with chain, 3/4"					68			68	74.50
1420	1"					85			85	93.50
1440	1-1/2"					20.50			20.50	22.50
1460	2-1/2"					30			30	33
1480	3"					40			40	44
1900	Escutcheon plate, for angle valves, polished brass, 1-1/2"					16.60			16.60	18.25
1920	2-1/2"					24			24	26.50
1940	3"					31.50			31.50	34.50
1980	For polished chrome, add					15%				
2000	Foam, control valve, 3"	1 Spri	6	1.333		2,250	84.50		2,334.50	2,600
2020	Supply valve, 2-1/2"		7	1.143		158	72.50		230.50	281
2040	Proportioner, 8"		2	4		3,600	253		3,853	4,350
2060	Oscillating foam monitor with electric remote control	Q-12	5.33	3.002		17,900	171		18,071	20,000
3000	Gate, hose, wheel handle, N.R.S., rough brass, 1-1/2"	1 Spri	12	.667		166	42		208	245
3040	2-1/2", 300 lb.	"	7	1.143		217	72.50		289.50	345
3080	For polished brass, add					40%				
3090	For polished chrome, add					50%				
3800	Hydrant, screw type, crank handle, brass									
3840	2-1/2" size	Q-12	11	1.455	Ea.	375	83		458	535
3880	For chrome, same price									
4200	Hydrolator, vent and draining, rough brass, 1-1/2"	1 Spri	12	.667	Ea.	115	42		157	189

21 12 Fire-Suppression Standpipes

21 12 23 – Fire-Suppression Hose Valves

21 12 23.70 Fire Hose Valves		Crew	Daily Output	Labor-Hours	Unit	2020 Bare Costs Material	Labor	Equipment	Total	Total Incl O&P
4280	For polished brass, add				Ea.	50%				
4290	For polished chrome, add					90%				
5000	Pressure reducing rough brass, 1-1/2"	1 Spri	12	.667		325	42		367	420
5020	2-1/2"	"	7	1.143		445	72.50		517.50	600
5080	For polished brass, add					105%				
5090	For polished chrome, add					140%				
8000	Wye, leader line, ball type, swivel female x male x male									
8040	2-1/2" x 1-1/2" x 1-1/2" polished brass				Ea.	360			360	395
8060	2-1/2" x 1-1/2" x 1-1/2" polished chrome				"	310			310	340

21 13 Fire-Suppression Sprinkler Systems

21 13 13 – Wet-Pipe Sprinkler Systems

21 13 13.50 Wet-Pipe Sprinkler System Components		Crew	Daily Output	Labor-Hours	Unit	Material	Labor	Equipment	Total	Total Incl O&P
0010	**WET-PIPE SPRINKLER SYSTEM COMPONENTS**									
1100	Alarm, electric pressure switch (circuit closer) R211313-10	1 Spri	26	.308	Ea.	112	19.45		131.45	152
1140	For explosion proof, max 20 psi, contacts close or open		26	.308		740	19.45		759.45	845
1220	Water motor gong		4	2		480	127		607	720
1900	Flexible sprinkler head connectors									
1910	Braided stainless steel hose with mounting bracket									
1920	1/2" and 3/4" outlet size									
1940	40" length	1 Spri	30	.267	Ea.	73	16.85		89.85	106
1960	60" length	"	22	.364	"	85	23		108	128
1982	May replace hard-pipe armovers									
1984	for wet and pre-action systems.									
2000	Release, emergency, manual, for hydraulic or pneumatic system	1 Spri	12	.667	Ea.	206	42		248	290
2060	Release, thermostatic, for hydraulic or pneumatic release line		20	.400		800	25.50		825.50	920
2200	Sprinkler cabinets, 6 head capacity		16	.500		77.50	31.50		109	133
2260	12 head capacity		16	.500		88	31.50		119.50	145
2340	Sprinkler head escutcheons, standard, brass tone, 1" size		40	.200		3.56	12.65		16.21	23
2360	Chrome, 1" size		40	.200		3.50	12.65		16.15	23
2400	Recessed type, bright brass		40	.200		11	12.65		23.65	31
2440	Chrome or white enamel		40	.200		4.22	12.65		16.87	23.50
2600	Sprinkler heads, not including supply piping									
3700	Standard spray, pendent or upright, brass, 135°F to 286°F									
3720	1/2" NPT, K5.6	1 Spri	16	.500	Ea.	16.75	31.50		48.25	66
3730	1/2" NPT, 7/16" orifice		16	.500		16.80	31.50		48.30	66
3740	1/2" NPT,K5.6		16	.500		11.20	31.50		42.70	60
3760	1/2" NPT, 17/32" orifice		16	.500		13.75	31.50		45.25	62.50
3780	3/4" NPT, 17/32" orifice		16	.500		13.45	31.50		44.95	62.50
3800	For open sprinklers, deduct					15%				
3840	For chrome, add				Ea.	4.06			4.06	4.47
3920	For 360°F, same cost									
3930	For 400°F	1 Spri	16	.500	Ea.	110	31.50		141.50	169
3940	For 500°F	"	16	.500	"	110	31.50		141.50	169
4200	Sidewall, vertical brass, 135°F to 286°F									
4240	1/2" NPT, 1/2" orifice	1 Spri	16	.500	Ea.	29.50	31.50		61	80
4280	3/4" NPT, 17/32" orifice	"	16	.500		81.50	31.50		113	137
4360	For satin chrome, add					4.66			4.66	5.15
4400	For 360°F, same cost									
4500	Sidewall, horizontal, brass, 135°F to 286°F									
4520	1/2" NPT, 1/2" orifice	1 Spri	16	.500	Ea.	29.50	31.50		61	80

21 13 Fire-Suppression Sprinkler Systems

21 13 13 – Wet-Pipe Sprinkler Systems

21 13 13.50 Wet-Pipe Sprinkler System Components		Crew	Daily Output	Labor-Hours	Unit	2020 Bare Costs Material	Labor	Equipment	Total	Total Incl O&P
4540	For 360°F, same cost									
4800	Recessed pendent, brass, 135°F to 286°F									
4820	1/2" NPT, K5.6	1 Spri	10	.800	Ea.	49.50	50.50		100	131
4830	1/2" NPT, 7/16" orifice		10	.800		22	50.50		72.50	100
4840	1/2" NPT, K5.6		10	.800		17.10	50.50		67.60	95
4860	1/2" NPT, 17/32" orifice		10	.800		50	50.50		100.50	131
4900	For satin chrome, add					6.30			6.30	6.90
5000	Recessed-vertical sidewall, brass, 135°F to 286°F									
5020	1/2" NPT, K5.6	1 Spri	10	.800	Ea.	36	50.50		86.50	116
5030	1/2" NPT, 7/16" orifice		10	.800		36	50.50		86.50	116
5040	1/2" NPT, K5.6		10	.800		36	50.50		86.50	116
5100	For bright nickel, same cost									
5600	Concealed, complete with cover plate									
5620	1/2" NPT, 1/2" orifice, 135°F to 212°F	1 Spri	9	.889	Ea.	27.50	56		83.50	115
5800	Window, brass, 1/2" NPT, 1/4" orifice		16	.500		40	31.50		71.50	91.50
5810	1/2" NPT, 5/16" orifice		16	.500		40	31.50		71.50	91.50
5820	1/2" NPT, 3/8" orifice		16	.500		40	31.50		71.50	91.50
5830	1/2" NPT, 7/16" orifice		16	.500		40	31.50		71.50	91.50
5840	1/2" NPT, 1/2" orifice		16	.500		42	31.50		73.50	94
5860	For polished chrome, add					5.10			5.10	5.60
5880	3/4" NPT, 5/8" orifice	1 Spri	16	.500		44	31.50		75.50	95.50
5890	3/4" NPT, 3/4" orifice	"	16	.500		44	31.50		75.50	95.50
6000	Sprinkler head guards, bright zinc, 1/2" NPT					5.70			5.70	6.30
6020	Bright zinc, 3/4" NPT					5.75			5.75	6.35
6025	Residential sprinkler components (one and two family)									
6026	Water motor alarm with strainer	1 Spri	4	2	Ea.	480	127		607	720
6027	Fast response, glass bulb, 135°F to 155°F									
6028	1/2" NPT, pendent, brass	1 Spri	16	.500	Ea.	34	31.50		65.50	85
6029	1/2" NPT, sidewall, brass		16	.500		36	31.50		67.50	87
6030	1/2" NPT, pendent, brass, extended coverage		16	.500		28.50	31.50		60	78.50
6031	1/2" NPT, sidewall, brass, extended coverage		16	.500		26	31.50		57.50	76.50
6032	3/4" NPT sidewall, brass, extended coverage		16	.500		29	31.50		60.50	79
6033	For chrome, add					15%				
6034	For polyester/teflon coating, add					20%				
6100	Sprinkler head wrenches, standard head				Ea.	28			28	30.50
6120	Recessed head					42			42	46
6160	Tamper switch (valve supervisory switch)	1 Spri	16	.500		265	31.50		296.50	340
6165	Flow switch (valve supervisory switch)	"	16	.500		265	31.50		296.50	340

21 13 16 – Dry-Pipe Sprinkler Systems

21 13 16.50 Dry-Pipe Sprinkler System Components		Crew	Daily Output	Labor-Hours	Unit	Material	Labor	Equipment	Total	Total Incl O&P
0010	**DRY-PIPE SPRINKLER SYSTEM COMPONENTS**									
0600	Accelerator	1 Spri	8	1	Ea.	900	63.50		963.50	1,075
0800	Air compressor for dry pipe system, automatic, complete									
0820	30 gal. system capacity, 3/4 HP	1 Spri	1.30	6.154	Ea.	1,200	390		1,590	1,900
0860	30 gal. system capacity, 1 HP		1.30	6.154		1,575	390		1,965	2,325
0910	30 gal. system capacity, 1-1/2 HP		1.30	6.154		1,000	390		1,390	1,675
0920	30 gal. system capacity, 2 HP		1.30	6.154		905	390		1,295	1,575
0960	Air pressure maintenance control		24	.333		375	21		396	440
1600	Dehydrator package, incl. valves and nipples R211313-20		12	.667		845	42		887	995
2600	Sprinkler heads, not including supply piping									
2640	Dry, pendent, 1/2" orifice, 3/4" or 1" NPT									
2660	3" to 6" length	1 Spri	14	.571	Ea.	145	36		181	214

21 13 16 – Dry-Pipe Sprinkler Systems

21 13 16.50 Dry-Pipe Sprinkler System Components		Crew	Daily Output	Labor-Hours	Unit	2020 Bare Costs Material	Labor	Equipment	Total	Total Incl O&P
2670	6-1/4" to 8" length	1 Spri	14	.571	Ea.	147	36		183	216
2680	8-1/4" to 12" length		14	.571		158	36		194	228
2690	12-1/4" to 15" length		14	.571		164	36		200	234
2700	15-1/4" to 18" length		14	.571		170	36		206	241
2710	18-1/4" to 21" length		13	.615		177	39		216	253
2720	21-1/4" to 24" length		13	.615		183	39		222	260
2730	24-1/4" to 27" length		13	.615		189	39		228	267
2740	27-1/4" to 30" length		13	.615		196	39		235	275
2750	30-1/4" to 33" length		13	.615		203	39		242	282
2760	33-1/4" to 36" length		13	.615		209	39		248	289
2780	36-1/4" to 39" length		12	.667		216	42		258	300
2790	39-1/4" to 42" length		12	.667		222	42		264	305
2800	For each inch or fraction, add					4.07			4.07	4.48
6330	Valves and components									
6340	Alarm test/shut off valve, 1/2"	1 Spri	20	.400	Ea.	26	25.50		51.50	66.50
8000	Dry pipe air check valve, 3" size	Q-12	2	8		2,100	455		2,555	3,000
8200	Dry pipe valve, incl. trim and gauges, 3" size		2	8		3,075	455		3,530	4,050
8220	4" size		1	16		3,300	910		4,210	5,000
8240	6" size	Q-13	2	16		4,000	965		4,965	5,850
8280	For accelerator trim with gauges, add	1 Spri	8	1		284	63.50		347.50	405

21 13 19 – Preaction Sprinkler Systems

21 13 19.50 Preaction Sprinkler System Components		Crew	Daily Output	Labor-Hours	Unit	Material	Labor	Equipment	Total	Total Incl O&P
0010	PREACTION SPRINKLER SYSTEM COMPONENTS									
3000	Preaction valve cabinet									
3100	Single interlock, pneum. release, panel, 1/2 HP comp. regul. air trim									
3110	1-1/2"	Q-12	3	5.333	Ea.	49,600	305		49,905	55,000
3120	2"		3	5.333		49,600	305		49,905	55,000
3130	2-1/2"		3	5.333		49,600	305		49,905	55,000
3140	3"		3	5.333		49,700	305		50,005	55,000
3150	4"		2	8		51,500	455		51,955	57,000
3160	6"	Q-13	4	8		54,000	480		54,480	60,000
3200	Double interlock, pneum. release, panel, 1/2 HP comp. regul. air trim									
3210	1-1/2"	Q-12	3	5.333	Ea.	48,200	305		48,505	53,500
3220	2"		3	5.333		48,200	305		48,505	53,500
3230	2-1/2"		3	5.333		48,200	305		48,505	53,500
3240	3"		3	5.333		48,300	305		48,605	53,500
3250	4"		2	8		50,500	455		50,955	56,500
3260	6"	Q-13	4	8		53,000	480		53,480	58,500

21 13 20 – On-Off Multicycle Sprinkler System

21 13 20.50 On-Off Multicycle Fire-Suppression Sprinkler Systems		Crew	Daily Output	Labor-Hours	Unit	Material	Labor	Equipment	Total	Total Incl O&P
0010	ON-OFF MULTICYCLE FIRE-SUPPRESSION SPRINKLER SYSTEMS									
8400	On-off multicycle package, includes swing check									
8420	and flow control valves with required trim									
8440	2" size	Q-12	2	8	Ea.	5,500	455		5,955	6,725
8460	3" size		1.50	10.667		6,025	605		6,630	7,525
8480	4" size		1	16		6,725	910		7,635	8,775
8500	6" size	Q-13	1.40	22.857		7,750	1,375		9,125	10,600

21 13 Fire-Suppression Sprinkler Systems

21 13 26 – Deluge Fire-Suppression Sprinkler Systems

21 13 26.50 Deluge Fire-Suppression Sprinkler Sys. Comp.		Crew	Daily Output	Labor-Hours	Unit	Material	Labor	Equipment	Total	Total Incl O&P
						2020 Bare Costs				
0010	**DELUGE FIRE-SUPPRESSION SPRINKLER SYSTEM COMPONENTS**									
1400	Deluge system, monitoring panel w/deluge valve & trim	1 Spri	18	.444	Ea.	7,475	28		7,503	8,275
6200	Valves and components									
7000	Deluge, assembly, incl. trim, pressure									
7020	operated relief, emergency release, gauges									
7040	2″ size	Q-12	2	8	Ea.	4,400	455		4,855	5,500
7060	3″ size		1.50	10.667		4,925	605		5,530	6,325
7080	4″ size		1	16		5,625	910		6,535	7,575
7100	6″ size	Q-13	1.80	17.778		6,650	1,075		7,725	8,925
7800	Pneumatic actuator, bronze, required on all									
7820	pneumatic release systems, any size deluge	1 Spri	18	.444	Ea.	480	28		508	570

21 13 39 – Foam-Water Systems

21 13 39.50 Foam-Water System Components		Crew	Daily Output	Labor-Hours	Unit	Material	Labor	Equipment	Total	Total Incl O&P
0010	**FOAM-WATER SYSTEM COMPONENTS**									
2600	Sprinkler heads, not including supply piping									
3600	Foam-water, pendent or upright, 1/2″ NPT	1 Spri	12	.667	Ea.	237	42		279	325

21 21 Carbon-Dioxide Fire-Extinguishing Systems

21 21 16 – Carbon-Dioxide Fire-Extinguishing Equipment

21 21 16.50 CO2 Fire Extinguishing System		Crew	Daily Output	Labor-Hours	Unit	Material	Labor	Equipment	Total	Total Incl O&P
0010	**CO_2 FIRE EXTINGUISHING SYSTEM**									
0042	For detectors and control stations, see Section 28 31 23.50									
0100	Control panel, single zone with batteries (2 zones det., 1 suppr.)	1 Elec	1	8	Ea.	1,050	490		1,540	1,875
0150	Multizone (4) with batteries (8 zones det., 4 suppr.)	"	.50	16		3,075	980		4,055	4,825
1000	Dispersion nozzle, CO_2, 3″ x 5″	1 Plum	18	.444		166	28.50		194.50	225
2000	Extinguisher, CO_2 system, high pressure, 75 lb. cylinder	Q-1	6	2.667		1,475	155		1,630	1,850
2100	100 lb. cylinder	"	5	3.200		1,950	186		2,136	2,425
3000	Electro/mechanical release	L-1	4	4		1,050	252		1,302	1,525
3400	Manual pull station	1 Plum	6	1.333		92.50	86		178.50	231
4000	Pneumatic damper release	"	8	1		182	64.50		246.50	297

21 22 Clean-Agent Fire-Extinguishing Systems

21 22 16 – Clean-Agent Fire-Extinguishing Equipment

21 22 16.50 Clean-Agent Extinguishing Systems		Crew	Daily Output	Labor-Hours	Unit	Material	Labor	Equipment	Total	Total Incl O&P
0010	**CLEAN-AGENT EXTINGUISHING SYSTEMS**									
0020	FM200 fire extinguishing system									
1100	Dispersion nozzle FM200, 1-1/2″	1 Plum	14	.571	Ea.	236	37		273	315
2400	Extinguisher, FM200 system, filled, with mounting bracket									
2460	26 lb. container	Q-1	8	2	Ea.	2,075	116		2,191	2,450
2480	44 lb. container		7	2.286		2,825	133		2,958	3,300
2500	63 lb. container		6	2.667		3,675	155		3,830	4,250
2520	101 lb. container		5	3.200		5,375	186		5,561	6,175
2540	196 lb. container		4	4		7,000	232		7,232	8,050
6000	FM200 system, simple nozzle layout, with broad dispersion				C.F.	1.90			1.90	2.09
6010	Extinguisher, FM200 system, filled, with mounting bracket									
6020	Complex nozzle layout and/or including underfloor dispersion				C.F.	3.78			3.78	4.16
6100	20,000 C.F. 2 exits, 8′ clng					2.08			2.08	2.29
6200	100,000 C.F. 4 exits, 8′ clng					1.88			1.88	2.07

21 22 Clean-Agent Fire-Extinguishing Systems

21 22 16 – Clean-Agent Fire-Extinguishing Equipment

21 22 16.50 Clean-Agent Extinguishing Systems		Crew	Daily Output	Labor-Hours	Unit	2020 Bare Costs Material	Labor	Equipment	Total	Total Incl O&P
6300	250,000 C.F. 6 exits, 8′ clng				C.F.	1.59			1.59	1.75
7010	HFC-227ea fire extinguishing system									
7100	Cylinders with clean-agent									
7110	Does not include pallete jack/fork lift rental fees									
7120	70 lb. cyl, w/35 lb. agent, no solenoid	Q-12	14	1.143	Ea.	2,500	65		2,565	2,850
7130	70 lb. cyl w/70 lb. agent, no solenoid		10	1.600		3,275	91		3,366	3,725
7140	70 lb. cyl w/35 lb. agent, w/solenoid		14	1.143		3,475	65		3,540	3,925
7150	70 lb. cyl w/70 lb. agent, w/solenoid		10	1.600		4,325	91		4,416	4,900
7220	250 lb. cyl, w/125 lb. agent, no solenoid		8	2		5,750	114		5,864	6,500
7230	250 lb. cyl, w/250 lb. agent, no solenoid		5	3.200		5,750	182		5,932	6,600
7240	250 lb. cyl, w/125 lb. agent, w/solenoid		8	2		7,425	114		7,539	8,350
7250	250 lb. cyl, w/250 lb. agent, w/solenoid		5	3.200		10,700	182		10,882	12,000
7320	560 lb. cyl, w/300 lb. agent, no solenoid		4	4		10,800	228		11,028	12,200
7330	560 lb. cyl, w/560 lb. agent, no solenoid		2.50	6.400		16,400	365		16,765	18,500
7340	560 lb. cyl, w/300 lb. agent, w/solenoid		4	4		13,700	228		13,928	15,400
7350	560 lb. cyl, w/560 lb. agent, w/solenoid		2.50	6.400		20,700	365		21,065	23,200
7420	1,200 lb. cyl, w/600 lb. agent, no solenoid	Q-13	4	8		20,500	480		20,980	23,200
7430	1,200 lb. cyl, w/1,200 lb. agent, no solenoid		3	10.667		33,600	640		34,240	38,000
7440	1,200 lb. cyl, w/600 lb. agent, w/solenoid		4	8		48,300	480		48,780	53,500
7450	1,200 lb. cyl, w/1,200 lb. agent, w/solenoid		3	10.667		75,000	640		75,640	83,500
7500	Accessories									
7510	Dispersion nozzle	1 Spri	16	.500	Ea.	98.50	31.50		130	156
7520	Agent release panel	1 Elec	4	2		61	123		184	250
7530	Maintenance switch	"	6	1.333		39	82		121	165
7540	Solenoid valve, 12v dc	1 Spri	8	1		233	63.50		296.50	350
7550	12v ac		8	1		500	63.50		563.50	645
7560	12v dc, explosion proof		8	1		395	63.50		458.50	530

21 31 Centrifugal Fire Pumps

21 31 13 – Electric-Drive, Centrifugal Fire Pumps

21 31 13.50 Electric-Drive Fire Pumps		Crew	Daily Output	Labor-Hours	Unit	Material	Labor	Equipment	Total	Total Incl O&P
0010	**ELECTRIC-DRIVE FIRE PUMPS** Including controller, fittings and relief valve									
3100	250 GPM, 55 psi, 15 HP, 3550 RPM, 2″ pump	Q-13	.70	45.714	Ea.	15,500	2,750		18,250	21,100
3200	500 GPM, 50 psi, 27 HP, 1770 RPM, 4″ pump		.68	47.059		15,400	2,825		18,225	21,200
3250	500 GPM, 100 psi, 47 HP, 3550 RPM, 3″ pump		.66	48.485		16,300	2,925		19,225	22,300
3300	500 GPM, 125 psi, 64 HP, 3550 RPM, 3″ pump		.62	51.613		18,300	3,100		21,400	24,900
3350	750 GPM, 50 psi, 44 HP, 1770 RPM, 5″ pump		.64	50		19,000	3,000		22,000	25,400
3400	750 GPM, 100 psi, 66 HP, 3550 RPM, 4″ pump		.58	55.172		18,500	3,325		21,825	25,400
3450	750 GPM, 165 psi, 120 HP, 3550 RPM, 4″ pump		.56	57.143		25,000	3,450		28,450	32,700
3500	1000 GPM, 50 psi, 48 HP, 1770 RPM, 5″ pump		.60	53.333		20,300	3,200		23,500	27,100
3550	1000 GPM, 100 psi, 86 HP, 3550 RPM, 5″ pump		.54	59.259		25,000	3,575		28,575	32,900
3600	1000 GPM, 150 psi, 142 HP, 3550 RPM, 5″ pump		.50	64		29,200	3,850		33,050	37,900
3650	1000 GPM, 200 psi, 245 HP, 1770 RPM, 6″ pump		.36	88.889		47,900	5,350		53,250	60,500
3660	1250 GPM, 75 psi, 75 HP, 1770 RPM, 5″ pump		.55	58.182		23,600	3,500		27,100	31,300
3700	1500 GPM, 50 psi, 66 HP, 1770 RPM, 6″ pump		.50	64		22,600	3,850		26,450	30,600
3750	1500 GPM, 100 psi, 139 HP, 1770 RPM, 6″ pump		.46	69.565		28,500	4,200		32,700	37,600
3800	1500 GPM, 150 psi, 200 HP, 1770 RPM, 6″ pump		.36	88.889		48,800	5,350		54,150	61,500
3850	1500 GPM, 200 psi, 279 HP, 1770 RPM, 6″ pump		.32	100		52,500	6,025		58,525	66,500
3900	2000 GPM, 100 psi, 167 HP, 1770 RPM, 6″ pump		.34	94.118		34,400	5,675		40,075	46,300
3950	2000 GPM, 150 psi, 292 HP, 1770 RPM, 6″ pump		.28	114		47,000	6,875		53,875	62,000
4000	2500 GPM, 100 psi, 213 HP, 1770 RPM, 8″ pump		.30	107		40,600	6,425		47,025	54,000

21 31 Centrifugal Fire Pumps

21 31 13 – Electric-Drive, Centrifugal Fire Pumps

21 31 13.50 Electric-Drive Fire Pumps		Crew	Daily Output	Labor-Hours	Unit	Material	2020 Bare Costs Labor	Equipment	Total	Total Incl O&P
4040	2500 GPM, 135 psi, 339 HP, 1770 RPM, 8" pump	Q-13	.26	123	Ea.	61,000	7,400		68,400	78,000
4100	3000 GPM, 100 psi, 250 HP, 1770 RPM, 8" pump		.28	114		62,000	6,875		68,875	78,500
4150	3000 GPM, 140 psi, 428 HP, 1770 RPM, 10" pump		.24	133		74,000	8,025		82,025	93,500
4200	3500 GPM, 100 psi, 300 HP, 1770 RPM, 10" pump		.26	123		66,500	7,400		73,900	84,000
4250	3500 GPM, 140 psi, 450 HP, 1770 RPM, 10" pump		.24	133		85,000	8,025		93,025	105,500
5000	For jockey pump 1", 3 HP, with control, add	Q-12	2	8		3,025	455		3,480	4,025

21 31 16 – Diesel-Drive, Centrifugal Fire Pumps

21 31 16.50 Diesel-Drive Fire Pumps		Crew	Daily Output	Labor-Hours	Unit	Material	Labor	Equipment	Total	Total Incl O&P
0010	**DIESEL-DRIVE FIRE PUMPS** Including controller, fittings and relief valve									
0050	500 GPM, 50 psi, 27 HP, 4" pump	Q-13	.64	50	Ea.	43,600	3,000		46,600	52,500
0100	500 GPM, 100 psi, 62 HP, 4" pump		.60	53.333		53,500	3,200		56,700	64,000
0150	500 GPM, 125 psi, 78 HP, 4" pump		.56	57.143		57,000	3,450		60,450	67,500
0200	750 GPM, 50 psi, 44 HP, 5" pump		.60	53.333		45,000	3,200		48,200	54,500
0250	750 GPM, 100 psi, 80 HP, 4" pump		.56	57.143		49,700	3,450		53,150	59,500
0300	750 GPM, 165 psi, 203 HP, 5" pump		.52	61.538		56,500	3,700		60,200	67,500
0350	1000 GPM, 50 psi, 48 HP, 5" pump		.58	55.172		48,600	3,325		51,925	58,500
0400	1000 GPM, 100 psi, 89 HP, 4" pump		.56	57.143		50,000	3,450		53,450	60,000
0450	1000 GPM, 150 psi, 148 HP, 4" pump		.48	66.667		54,500	4,025		58,525	66,000
0470	1000 GPM, 200 psi, 280 HP, 5" pump		.40	80		71,500	4,825		76,325	85,500
0480	1250 GPM, 75 psi, 75 HP, 5" pump		.54	59.259		52,500	3,575		56,075	63,000
0500	1500 GPM, 50 psi, 66 HP, 6" pump		.50	64		50,500	3,850		54,350	61,500
0550	1500 GPM, 100 psi, 140 HP, 6" pump		.46	69.565		54,000	4,200		58,200	66,000
0600	1500 GPM, 150 psi, 228 HP, 6" pump		.42	76.190		69,500	4,575		74,075	83,500
0650	1500 GPM, 200 psi, 279 HP, 6" pump		.38	84.211		97,000	5,075		102,075	114,000
0700	2000 GPM, 100 psi, 167 HP, 6" pump		.34	94.118		65,000	5,675		70,675	80,000
0750	2000 GPM, 150 psi, 284 HP, 6" pump		.30	107		82,500	6,425		88,925	100,500
0800	2500 GPM, 100 psi, 213 HP, 8" pump		.32	100		67,500	6,025		73,525	83,000
0820	2500 GPM, 150 psi, 365 HP, 8" pump		.26	123		88,500	7,400		95,900	108,000
0850	3000 GPM, 100 psi, 250 HP, 8" pump		.28	114		96,000	6,875		102,875	116,000
0900	3000 GPM, 150 psi, 384 HP, 10" pump		.20	160		118,000	9,625		127,625	144,000
0950	3500 GPM, 100 psi, 300 HP, 10" pump		.24	133		89,000	8,025		97,025	110,000
1000	3500 GPM, 150 psi, 518 HP, 10" pump		.20	160		119,000	9,625		128,625	145,500

Division Notes

	CREW	DAILY OUTPUT	LABOR-HOURS	UNIT	BARE COSTS MAT.	LABOR	EQUIP.	TOTAL	TOTAL INCL O&P

Estimating Tips

22 10 00 Plumbing Piping and Pumps

This subdivision is primarily basic pipe and related materials. The pipe may be used by any of the mechanical disciplines, i.e., plumbing, fire protection, heating, and air conditioning.

Note: CPVC plastic piping approved for fire protection is located in 21 11 13.

- The labor adjustment factors listed in Subdivision 22 01 02.20 apply throughout Divisions 21, 22, and 23. CAUTION: the correct percentage may vary for the same items. For example, the percentage add for the basic pipe installation should be based on the maximum height that the installer must install for that particular section. If the pipe is to be located 14' above the floor but it is suspended on threaded rod from beams, the bottom flange of which is 18' high (4' rods), then the height is actually 18' and the add is 20%. The pipe cover, however, does not have to go above the 14' and so the add should be 10%.
- Most pipe is priced first as straight pipe with a joint (coupling, weld, etc.) every 10' and a hanger usually every 10'. There are exceptions with hanger spacing such as for cast iron pipe (5') and plastic pipe (3 per 10'). Following each type of pipe there are several lines listing sizes and the amount to be subtracted to delete couplings and hangers. This is for pipe that is to be buried or supported together on trapeze hangers. The reason that the couplings are deleted is that these runs are usually long, and frequently longer lengths of pipe are used. By deleting the couplings, the estimator is expected to look up and add back the correct reduced number of couplings.
- When preparing an estimate, it may be necessary to approximate the fittings. Fittings usually run between 25% and 50% of the cost of the pipe. The lower percentage is for simpler runs, and the higher number is for complex areas, such as mechanical rooms.
- For historic restoration projects, the systems must be as invisible as possible, and pathways must be sought for pipes, conduit, and ductwork. While installations in accessible spaces (such as basements and attics) are relatively straightforward to estimate, labor costs may be more difficult to determine when delivery systems must be concealed.

22 40 00 Plumbing Fixtures

- Plumbing fixture costs usually require two lines: the fixture itself and its "rough-in, supply, and waste."
- In the Assemblies Section (Plumbing D2010) for the desired fixture, the System Components Group at the center of the page shows the fixture on the first line. The rest of the list (fittings, pipe, tubing, etc.) will total up to what we refer to in the Unit Price section as "Rough-in, supply, waste, and vent." Note that for most fixtures we allow a nominal 5' of tubing to reach from the fixture to a main or riser.
- Remember that gas- and oil-fired units need venting.

Reference Numbers

Reference numbers are shown at the beginning of some major classifications. These numbers refer to related items in the Reference Section. The reference information may be an estimating procedure, an alternate pricing method, or technical information.

Note: Not all subdivisions listed here necessarily appear. ■

Note: "Powered in part by CINX™, based on licensed proprietary information of Harrison Publishing House, Inc."

Note: Trade Service, in part, has been used as a reference source for some of the material prices used in Division 22.

22 01 Operation and Maintenance of Plumbing

22 01 02 – Labor Adjustments

22 01 02.10 Boilers, General		Crew	Daily Output	Labor-Hours	Unit	2020 Bare Costs Material	Labor	Equipment	Total	Total Incl O&P
0010	**BOILERS, GENERAL**, Prices do not include flue piping, elec. wiring,									
0020	gas or oil piping, boiler base, pad, or tankless unless noted									
0100	Boiler H.P.: 10 KW = 34 lb./steam/hr. = 33,475 BTU/hr.									
0150	To convert SFR to BTU rating: Hot water, 150 x SFR;									
0160	Forced hot water, 180 x SFR; steam, 240 x SFR									

22 01 02.20 Labor Adjustment Factors

		Crew	Daily Output	Labor-Hours	Unit	Material	Labor	Equipment	Total	Total Incl O&P
0010	**LABOR ADJUSTMENT FACTORS** (For Div. 21, 22 and 23) R220102-20									
0100	Labor factors: The below are reasonable suggestions, but									
0110	each project must be evaluated for its own peculiarities, and									
0120	the adjustments be increased or decreased depending on the									
0130	severity of the special conditions.									
1000	Add to labor for elevated installation (Above floor level)									
1080	10′ to 14.5′ high R221113-70						10%			
1100	15′ to 19.5′ high						20%			
1120	20′ to 24.5′ high						25%			
1140	25′ to 29.5′ high						35%			
1160	30′ to 34.5′ high						40%			
1180	35′ to 39.5′ high						50%			
1200	40′ and higher						55%			
2000	Add to labor for crawl space									
2100	3′ high						40%			
2140	4′ high						30%			
3000	Add to labor for multi-story building									
3010	For new construction (No elevator available)									
3100	Add for floors 3 thru 10						5%			
3110	Add for floors 11 thru 15						10%			
3120	Add for floors 16 thru 20						15%			
3130	Add for floors 21 thru 30						20%			
3140	Add for floors 31 and up						30%			
3170	For existing structure (Elevator available)									
3180	Add for work on floor 3 and above						2%			
4000	Add to labor for working in existing occupied buildings									
4100	Hospital						35%			
4140	Office building						25%			
4180	School						20%			
4220	Factory or warehouse						15%			
4260	Multi dwelling						15%			
5000	Add to labor, miscellaneous									
5100	Cramped shaft						35%			
5140	Congested area						15%			
5180	Excessive heat or cold						30%			
9000	Labor factors: The above are reasonable suggestions, but									
9010	each project should be evaluated for its own peculiarities.									
9100	Other factors to be considered are:									
9140	Movement of material and equipment through finished areas									
9180	Equipment room									
9220	Attic space									
9260	No service road									
9300	Poor unloading/storage area									
9340	Congested site area/heavy traffic									

22 05 Common Work Results for Plumbing

22 05 05 – Selective Demolition for Plumbing

22 05 05.10 Plumbing Demolition		Crew	Daily Output	Labor-Hours	Unit	Material	2020 Bare Costs Labor	Equipment	Total	Total Incl O&P
0010	**PLUMBING DEMOLITION** R220105-10									
0400	Air compressor, up thru 2 HP	Q-1	10	1.600	Ea.		93		93	139
0410	3 HP thru 7-1/2 HP R024119-10		5.60	2.857			166		166	248
0420	10 HP thru 15 HP		1.40	11.429			665		665	990
0430	20 HP thru 30 HP	Q-2	1.30	18.462			1,100		1,100	1,650
0500	Backflow preventer, up thru 2" diameter	1 Plum	17	.471			30.50		30.50	45.50
0510	2-1/2" thru 3" diameter	Q-1	10	1.600			93		93	139
0520	4" thru 6" diameter	"	5	3.200			186		186	278
0530	8" thru 10" diameter	Q-2	3	8			480		480	720
0700	Carriers and supports									
0710	Fountains, sinks, lavatories and urinals	1 Plum	14	.571	Ea.		37		37	55
0720	Water closets	"	12	.667	"		43		43	64.50
0730	Grinder pump or sewage ejector system									
0732	Simplex	Q-1	7	2.286	Ea.		133		133	198
0734	Duplex	"	2.80	5.714			330		330	495
0738	Hot water dispenser	1 Plum	36	.222			14.30		14.30	21.50
0740	Hydrant, wall		26	.308			19.85		19.85	29.50
0744	Ground		12	.667			43		43	64.50
0760	Cleanouts and drains, up thru 4" pipe diameter		10	.800			51.50		51.50	77
0764	5" thru 8" pipe diameter	Q-1	10	1.600			93		93	139
0780	Industrial safety fixtures	1 Plum	8	1			64.50		64.50	96.50
1020	Fixtures, including 10' piping									
1100	Bathtubs, cast iron	1 Plum	4	2	Ea.		129		129	193
1120	Fiberglass		6	1.333			86		86	129
1140	Steel		5	1.600			103		103	154
1150	Bidet	Q-1	7	2.286			133		133	198
1200	Lavatory, wall hung	1 Plum	10	.800			51.50		51.50	77
1220	Counter top		8	1			64.50		64.50	96.50
1300	Sink, single compartment		8	1			64.50		64.50	96.50
1320	Double compartment		7	1.143			73.50		73.50	110
1340	Shower, stall and receptor	Q-1	6	2.667			155		155	231
1350	Group	"	7	2.286			133		133	198
1400	Water closet, floor mounted	1 Plum	8	1			64.50		64.50	96.50
1420	Wall mounted	"	7	1.143			73.50		73.50	110
1440	Wash fountain, 36" diameter	Q-2	8	3			180		180	270
1442	54" diameter	"	7	3.429			206		206	310
1500	Urinal, floor mounted	1 Plum	4	2			129		129	193
1520	Wall mounted	"	7	1.143			73.50		73.50	110
1590	Whirl pool or hot tub	Q-1	2.60	6.154			355		355	535
1600	Water fountains, free standing	1 Plum	8	1			64.50		64.50	96.50
1620	Wall or deck mounted		6	1.333			86		86	129
1800	Medical gas specialties		8	1			64.50		64.50	96.50
1810	Plumbing demo, floor drain, remove		12	.667			43		43	64.50
1820	Plumbing demo, roof drain, remove		9	.889			57.50		57.50	85.50
1900	Piping fittings, single connection, up thru 1-1/2" diameter		30	.267			17.20		17.20	25.50
1910	2" thru 4" diameter		14	.571			37		37	55
1980	Pipe hanger/support removal		80	.100			6.45		6.45	9.65
1990	Glass pipe with fittings, 1" thru 3" diameter		200	.040	L.F.		2.58		2.58	3.86
1992	4" thru 6" diameter		150	.053			3.44		3.44	5.15
2000	Piping, metal, up thru 1-1/2" diameter		200	.040			2.58		2.58	3.86
2050	2" thru 3-1/2" diameter		150	.053			3.44		3.44	5.15
2100	4" thru 6" diameter	2 Plum	100	.160			10.30		10.30	15.45
2150	8" thru 14" diameter	"	60	.267			17.20		17.20	25.50

22 05 05.10 Plumbing Demolition		Crew	Daily Output	Labor-Hours	Unit	Material	2020 Bare Costs Labor	Equipment	Total	Total Incl O&P
2153	16" thru 20" diameter	Q-18	70	.343	L.F.		21	1.52	22.52	33
2155	24" thru 26" diameter		55	.436			26.50	1.93	28.43	42
2156	30" thru 36" diameter		40	.600			36.50	2.66	39.16	58
2160	Plastic pipe with fittings, up thru 1-1/2" diameter	1 Plum	250	.032			2.06		2.06	3.09
2162	2" thru 3" diameter	"	200	.040			2.58		2.58	3.86
2164	4" thru 6" diameter	Q-1	200	.080			4.64		4.64	6.95
2166	8" thru 14" diameter		150	.107			6.20		6.20	9.25
2168	16" diameter		100	.160			9.30		9.30	13.90
2170	Prison fixtures, lavatory or sink		18	.889	Ea.		51.50		51.50	77
2172	Shower		5.60	2.857			166		166	248
2174	Urinal or water closet		13	1.231			71.50		71.50	107
2180	Pumps, all fractional horse-power		12	1.333			77.50		77.50	116
2184	1 HP thru 5 HP		6	2.667			155		155	231
2186	7-1/2 HP thru 15 HP		2.50	6.400			370		370	555
2188	20 HP thru 25 HP	Q-2	4	6			360		360	540
2190	30 HP thru 60 HP		.80	30			1,800		1,800	2,700
2192	75 HP thru 100 HP		.60	40			2,400		2,400	3,600
2194	150 HP		.50	48			2,875		2,875	4,325
2198	Pump, sump or submersible	1 Plum	12	.667			43		43	64.50
2200	Receptors and interceptors, up thru 20 GPM	"	8	1			64.50		64.50	96.50
2204	25 thru 100 GPM	Q-1	6	2.667			155		155	231
2208	125 thru 300 GPM	"	2.40	6.667			385		385	580
2211	325 thru 500 GPM	Q-2	2.60	9.231			555		555	830
2212	Deduct for salvage, aluminum scrap				Ton				635	695
2214	Brass scrap								2,700	2,975
2216	Copper scrap								4,475	4,925
2218	Lead scrap								850	935
2220	Steel scrap								204	224
2230	Temperature maintenance cable	1 Plum	1200	.007	L.F.		.43		.43	.64
2240	Toilet partitions, see Section 10 21 13									
2250	Water heater, 40 gal.	1 Plum	6	1.333	Ea.		86		86	129
3100	Tanks, water heaters and liquid containers									
3110	Up thru 45 gallons	Q-1	22	.727	Ea.		42		42	63
3120	50 thru 120 gallons		14	1.143			66.50		66.50	99
3130	130 thru 240 gallons		7.60	2.105			122		122	183
3140	250 thru 500 gallons		5.40	2.963			172		172	257
3150	600 thru 1,000 gallons	Q-2	1.60	15			900		900	1,350
3160	1,100 thru 2,000 gallons		.70	34.286			2,050		2,050	3,075
3170	2,100 thru 4,000 gallons		.50	48			2,875		2,875	4,325
6000	Remove and reset fixtures, easy access	1 Plum	6	1.333			86		86	129
6100	Difficult access	"	4	2			129		129	193
6500	Solar heating system									
6510	Solar panel, hot water system	1 Plum	304	.026	S.F.		1.70		1.70	2.54
7910	Solar panel for pool	"	300	.027	"		1.72		1.72	2.57
8000	Sprinkler system									
8100	Exposed wet/dry	1 Plum	2500	.003	SF Flr.		.21		.21	.31
8200	Concealed wet/dry	"	1250	.006	"		.41		.41	.62
9100	Valve, metal valves or strainers and similar, up thru 1-1/2" diameter	1 Stpi	28	.286	Ea.		18.75		18.75	28
9110	2" thru 3" diameter	Q-1	11	1.455			84.50		84.50	126
9120	4" thru 6" diameter	"	8	2			116		116	174
9130	8" thru 14" diameter	Q-2	8	3			180		180	270
9140	16" thru 20" diameter		2	12			720		720	1,075
9150	24" diameter		1.20	20			1,200		1,200	1,800

22 05 Common Work Results for Plumbing

22 05 05 – Selective Demolition for Plumbing

22 05 05.10 Plumbing Demolition		Crew	Daily Output	Labor-Hours	Unit	Material	2020 Bare Costs Labor	Equipment	Total	Total Incl O&P
9200	Valve, plastic, up thru 1-1/2" diameter	1 Plum	42	.190	Ea.		12.30		12.30	18.35
9210	2" thru 3" diameter		15	.533			34.50		34.50	51.50
9220	4" thru 6" diameter		12	.667			43		43	64.50
9300	Vent flashing and caps		55	.145			9.35		9.35	14.05
9350	Water filter, commercial, 1" thru 1-1/2"	Q-1	2	8			465		465	695
9360	2" thru 2-1/2"	"	1.60	10			580		580	870
9400	Water heaters									
9410	Up thru 245 GPH	Q-1	2.40	6.667	Ea.		385		385	580
9420	250 thru 756 GPH		1.60	10			580		580	870
9430	775 thru 1,640 GPH		.80	20			1,150		1,150	1,725
9440	1,650 thru 4,000 GPH	Q-2	.50	48			2,875		2,875	4,325
9470	Water softener	Q-1	2	8			465		465	695

22 05 23 – General-Duty Valves for Plumbing Piping

22 05 23.10 Valves, Brass		Crew	Daily Output	Labor-Hours	Unit	Material	Labor	Equipment	Total	Total Incl O&P
0010	**VALVES, BRASS**									
0032	For motorized valves, see Section 23 09 53.10									
0500	Gas cocks, threaded									
0510	1/4"	1 Plum	26	.308	Ea.	15.80	19.85		35.65	47
0520	3/8"		24	.333		15.80	21.50		37.30	49.50
0530	1/2"		24	.333		13.40	21.50		34.90	47
0540	3/4"		22	.364		18.10	23.50		41.60	55
0550	1"		19	.421		34	27		61	78
0560	1-1/4"		15	.533		57	34.50		91.50	114
0570	1-1/2"		13	.615		85.50	39.50		125	154
0580	2"		11	.727		111	47		158	192
0672	For larger sizes use lubricated plug valve, Section 23 05 23.70									

22 05 23.20 Valves, Bronze		Crew	Daily Output	Labor-Hours	Unit	Material	Labor	Equipment	Total	Total Incl O&P
0010	**VALVES, BRONZE** R220523-80									
1020	Angle, 150 lb., rising stem, threaded									
1030	1/8" R220523-90	1 Plum	24	.333	Ea.	142	21.50		163.50	188
1040	1/4"		24	.333		162	21.50		183.50	210
1050	3/8"		24	.333		165	21.50		186.50	213
1060	1/2"		22	.364		169	23.50		192.50	221
1070	3/4"		20	.400		230	26		256	291
1080	1"		19	.421		315	27		342	385
1090	1-1/4"		15	.533		355	34.50		389.50	445
1100	1-1/2"		13	.615		525	39.50		564.50	635
1102	Soldered same price as threaded									
1110	2"	1 Plum	11	.727	Ea.	845	47		892	1,000
1300	Ball									
1304	Soldered									
1312	3/8"	1 Plum	21	.381	Ea.	19.70	24.50		44.20	58
1316	1/2"		18	.444		18.70	28.50		47.20	63.50
1320	3/4"		17	.471		33	30.50		63.50	82
1324	1"		15	.533		44	34.50		78.50	100
1328	1-1/4"		13	.615		46	39.50		85.50	110
1332	1-1/2"		11	.727		86.50	47		133.50	166
1336	2"		9	.889		92	57.50		149.50	187
1340	2-1/2"		7	1.143		485	73.50		558.50	645
1344	3"		5	1.600		570	103		673	780
1350	Single union end									
1358	3/8"	1 Plum	21	.381	Ea.	26.50	24.50		51	65.50

22 05 23.20 Valves, Bronze		Crew	Daily Output	Labor-Hours	Unit	Material	Labor	Equipment	Total	Total Incl O&P
						2020 Bare Costs				
1362	1/2"	1 Plum	18	.444	Ea.	29.50	28.50		58	75.50
1366	3/4"		17	.471		56	30.50		86.50	108
1370	1"		15	.533		65.50	34.50		100	124
1374	1-1/4"		13	.615		102	39.50		141.50	172
1378	1-1/2"		11	.727		145	47		192	230
1382	2"		9	.889		194	57.50		251.50	299
1398	Threaded, 150 psi									
1400	1/4"	1 Plum	24	.333	Ea.	19.55	21.50		41.05	53.50
1430	3/8"		24	.333		18.25	21.50		39.75	52
1450	1/2"		22	.364		18	23.50		41.50	55
1460	3/4"		20	.400		35.50	26		61.50	77.50
1470	1"		19	.421		32	27		59	75.50
1480	1-1/4"		15	.533		50	34.50		84.50	107
1490	1-1/2"		13	.615		75.50	39.50		115	143
1500	2"		11	.727		83.50	47		130.50	162
1510	2-1/2"		9	.889		264	57.50		321.50	375
1520	3"		8	1		425	64.50		489.50	560
1522	Solder the same price as threaded									
1600	Butterfly, 175 psi, full port, solder or threaded ends									
1610	Stainless steel disc and stem									
1620	1/4"	1 Plum	24	.333	Ea.	24	21.50		45.50	58.50
1630	3/8"		24	.333		16.85	21.50		38.35	50.50
1640	1/2"		22	.364		21	23.50		44.50	58.50
1650	3/4"		20	.400		29	26		55	70.50
1660	1"		19	.421		36	27		63	80
1670	1-1/4"		15	.533		69	34.50		103.50	128
1680	1-1/2"		13	.615		73.50	39.50		113	141
1690	2"		11	.727		93	47		140	173
1750	Check, swing, class 150, regrinding disc, threaded									
1800	1/8"	1 Plum	24	.333	Ea.	82.50	21.50		104	123
1830	1/4"		24	.333		76.50	21.50		98	117
1840	3/8"		24	.333		71	21.50		92.50	111
1850	1/2"		24	.333		77	21.50		98.50	117
1860	3/4"		20	.400		94	26		120	142
1870	1"		19	.421		178	27		205	237
1880	1-1/4"		15	.533		210	34.50		244.50	283
1890	1-1/2"		13	.615		300	39.50		339.50	390
1900	2"		11	.727		380	47		427	490
1910	2-1/2"	Q-1	15	1.067		925	62		987	1,125
1920	3"	"	13	1.231		1,275	71.50		1,346.50	1,500
2000	For 200 lb., add					5%	10%			
2040	For 300 lb., add					15%	15%			
2060	Check swing, 300 lb., lead free unless noted, sweat, 3/8" size	1 Plum	24	.333	Ea.	96	21.50		117.50	138
2070	1/2"		24	.333		96	21.50		117.50	138
2080	3/4"		20	.400		130	26		156	182
2090	1"		19	.421		191	27		218	251
2100	1-1/4"		15	.533		269	34.50		303.50	350
2110	1-1/2"		13	.615		315	39.50		354.50	410
2120	2"		11	.727		465	47		512	580
2130	2-1/2", not lead free	Q-1	15	1.067		750	62		812	920
2140	3", not lead free	"	13	1.231		950	71.50		1,021.50	1,150
2850	Gate, N.R.S., soldered, 125 psi									
2900	3/8"	1 Plum	24	.333	Ea.	75.50	21.50		97	115

22 05 23.20 Valves, Bronze		Crew	Daily Output	Labor-Hours	Unit	Material	Labor	Equipment	Total	Total Incl O&P
						2020 Bare Costs				
2920	1/2"	1 Plum	24	.333	Ea.	60.50	21.50		82	98.50
2940	3/4"		20	.400		74	26		100	120
2950	1"		19	.421		78.50	27		105.50	127
2960	1-1/4"		15	.533		152	34.50		186.50	219
2970	1-1/2"		13	.615		184	39.50		223.50	263
2980	2"		11	.727		213	47		260	305
2990	2-1/2"	Q-1	15	1.067		515	62		577	660
3000	3"	"	13	1.231		570	71.50		641.50	735
3350	Threaded, class 150									
3410	1/4"	1 Plum	24	.333	Ea.	81	21.50		102.50	121
3420	3/8"		24	.333		98.50	21.50		120	140
3430	1/2"		24	.333		78.50	21.50		100	119
3440	3/4"		20	.400		104	26		130	153
3450	1"		19	.421		134	27		161	189
3460	1-1/4"		15	.533		137	34.50		171.50	203
3470	1-1/2"		13	.615		269	39.50		308.50	355
3480	2"		11	.727		234	47		281	325
3490	2-1/2"	Q-1	15	1.067		700	62		762	865
3500	3"	"	13	1.231		865	71.50		936.50	1,050
3600	Gate, flanged, 150 lb.									
3610	1"	1 Plum	7	1.143	Ea.	1,175	73.50		1,248.50	1,375
3620	1-1/2"		6	1.333		1,450	86		1,536	1,725
3630	2"		5	1.600		2,500	103		2,603	2,900
3634	2-1/2"	Q-1	5	3.200		1,975	186		2,161	2,450
3640	3"	"	4.50	3.556		3,850	206		4,056	4,550
3850	Rising stem, soldered, 300 psi									
3900	3/8"	1 Plum	24	.333	Ea.	155	21.50		176.50	203
3920	1/2"		24	.333		157	21.50		178.50	205
3940	3/4"		20	.400		169	26		195	225
3950	1"		19	.421		229	27		256	293
3960	1-1/4"		15	.533		315	34.50		349.50	400
3970	1-1/2"		13	.615		385	39.50		424.50	485
3980	2"		11	.727		615	47		662	750
3990	2-1/2"	Q-1	15	1.067		1,325	62		1,387	1,550
4000	3"	"	13	1.231		2,025	71.50		2,096.50	2,350
4250	Threaded, class 150									
4310	1/4"	1 Plum	24	.333	Ea.	90	21.50		111.50	131
4320	3/8"		24	.333		90	21.50		111.50	131
4330	1/2"		24	.333		69.50	21.50		91	109
4340	3/4"		20	.400		81.50	26		107.50	128
4350	1"		19	.421		125	27		152	179
4360	1-1/4"		15	.533		132	34.50		166.50	197
4370	1-1/2"		13	.615		167	39.50		206.50	243
4380	2"		11	.727		224	47		271	315
4390	2-1/2"	Q-1	15	1.067		670	62		732	835
4400	3"	"	13	1.231		935	71.50		1,006.50	1,125
4500	For 300 psi, threaded, add					100%	15%			
4540	For chain operated type, add					15%				
4850	Globe, class 150, rising stem, threaded									
4920	1/4"	1 Plum	24	.333	Ea.	123	21.50		144.50	168
4940	3/8"		24	.333		121	21.50		142.50	166
4950	1/2"		24	.333		106	21.50		127.50	149
4960	3/4"		20	.400		172	26		198	228

22 05 23.20 Valves, Bronze		Crew	Daily Output	Labor-Hours	Unit	Material	Labor	Equipment	Total	Total Incl O&P
						2020 Bare Costs				
4970	1"	1 Plum	19	.421	Ea.	239	27		266	305
4980	1-1/4"		15	.533		292	34.50		326.50	370
4990	1-1/2"		13	.615		460	39.50		499.50	565
5000	2"		11	.727		690	47		737	830
5010	2-1/2"	Q-1	15	1.067		1,475	62		1,537	1,725
5020	3"	"	13	1.231		2,125	71.50		2,196.50	2,425
5120	For 300 lb. threaded, add					50%	15%			
5600	Relief, pressure & temperature, self-closing, ASME, threaded									
5640	3/4"	1 Plum	28	.286	Ea.	257	18.40		275.40	310
5650	1"		24	.333		425	21.50		446.50	495
5660	1-1/4"		20	.400		805	26		831	925
5670	1-1/2"		18	.444		1,550	28.50		1,578.50	1,750
5680	2"		16	.500		1,675	32		1,707	1,900
5950	Pressure, poppet type, threaded									
6000	1/2"	1 Plum	30	.267	Ea.	90	17.20		107.20	125
6040	3/4"	"	28	.286	"	104	18.40		122.40	143
6400	Pressure, water, ASME, threaded									
6440	3/4"	1 Plum	28	.286	Ea.	119	18.40		137.40	159
6450	1"		24	.333		340	21.50		361.50	405
6460	1-1/4"		20	.400		485	26		511	575
6470	1-1/2"		18	.444		750	28.50		778.50	865
6480	2"		16	.500		1,075	32		1,107	1,250
6490	2-1/2"		15	.533		4,400	34.50		4,434.50	4,900
6900	Reducing, water pressure									
6920	300 psi to 25-75 psi, threaded or sweat									
6940	1/2"	1 Plum	24	.333	Ea.	470	21.50		491.50	550
6950	3/4"		20	.400		550	26		576	645
6960	1"		19	.421		850	27		877	975
6970	1-1/4"		15	.533		1,475	34.50		1,509.50	1,675
6980	1-1/2"		13	.615		2,225	39.50		2,264.50	2,500
6990	2"		11	.727		3,350	47		3,397	3,750
7100	For built-in by-pass or 10-35 psi, add					53			53	58.50
7700	High capacity, 250 psi to 25-75 psi, threaded									
7740	1/2"	1 Plum	24	.333	Ea.	890	21.50		911.50	1,000
7780	3/4"		20	.400		780	26		806	895
7790	1"		19	.421		1,050	27		1,077	1,225
7800	1-1/4"		15	.533		1,825	34.50		1,859.50	2,075
7810	1-1/2"		13	.615		2,675	39.50		2,714.50	3,000
7820	2"		11	.727		3,900	47		3,947	4,375
7830	2-1/2"		9	.889		5,575	57.50		5,632.50	6,200
7840	3"		8	1		6,575	64.50		6,639.50	7,350
7850	3" flanged (iron body)	Q-1	10	1.600		6,575	93		6,668	7,375
7860	4" flanged (iron body)	"	8	2		8,450	116		8,566	9,475
7920	For higher pressure, add					25%				
8000	Silent check, bronze trim									
8010	Compact wafer type, for 125 or 150 lb. flanges									
8020	1-1/2"	1 Plum	11	.727	Ea.	490	47		537	610
8021	2"	"	9	.889		520	57.50		577.50	655
8022	2-1/2"	Q-1	9	1.778		545	103		648	755
8023	3"		8	2		605	116		721	845
8024	4"		5	3.200		1,050	186		1,236	1,425
8025	5"	Q-2	6	4		1,000	241		1,241	1,450
8026	6"	"	5	4.800		1,200	289		1,489	1,750

22 05 23.20 Valves, Bronze		Crew	Daily Output	Labor-Hours	Unit	Material	2020 Bare Costs Labor	Equipment	Total	Total Incl O&P
8050	For 250 or 300 lb. flanges, thru 6" no change									
8060	Full flange wafer type, 150 lb.									
8063	1-1/2"	1 Plum	11	.727	Ea.	395	47		442	505
8064	2"	"	9	.889		535	57.50		592.50	670
8065	2-1/2"	Q-1	9	1.778		600	103		703	815
8066	3"		8	2		665	116		781	905
8067	4"		5	3.200		980	186		1,166	1,350
8068	5"	Q-2	6	4		1,275	241		1,516	1,750
8069	6"	"	5	4.800		1,675	289		1,964	2,275
8080	For 300 lb., add					40%	10%			
8100	Globe type, 150 lb.									
8110	2"	1 Plum	9	.889	Ea.	620	57.50		677.50	765
8111	2-1/2"	Q-1	9	1.778		765	103		868	1,000
8112	3"		8	2		915	116		1,031	1,175
8113	4"		5	3.200		1,275	186		1,461	1,675
8114	5"	Q-2	6	4		1,550	241		1,791	2,050
8115	6"	"	5	4.800		2,150	289		2,439	2,800
8130	For 300 lb., add					20%	10%			
8140	Screwed end type, 250 lb.									
8141	1/2"	1 Plum	24	.333	Ea.	56	21.50		77.50	93.50
8142	3/4"		20	.400		56	26		82	100
8143	1"		19	.421		63.50	27		90.50	111
8144	1-1/4"		15	.533		88	34.50		122.50	148
8145	1-1/2"		13	.615		98.50	39.50		138	169
8146	2"		11	.727		133	47		180	216
8350	Tempering, water, sweat connections									
8400	1/2"	1 Plum	24	.333	Ea.	118	21.50		139.50	162
8440	3/4"	"	20	.400	"	167	26		193	222
8650	Threaded connections									
8700	1/2"	1 Plum	24	.333	Ea.	162	21.50		183.50	210
8740	3/4"		20	.400		1,050	26		1,076	1,225
8750	1"		19	.421		1,200	27		1,227	1,350
8760	1-1/4"		15	.533		1,975	34.50		2,009.50	2,225
8770	1-1/2"		13	.615		2,025	39.50		2,064.50	2,275
8780	2"		11	.727		3,000	47		3,047	3,375
8800	Water heater water & gas safety shut off									
8810	Protection against a leaking water heater									
8814	Shut off valve	1 Plum	16	.500	Ea.	196	32		228	264
8818	Water heater dam		32	.250		32.50	16.10		48.60	59.50
8822	Gas control wiring harness		32	.250		24.50	16.10		40.60	51
8830	Whole house flood safety shut off									
8834	Connections									
8838	3/4" NPT	1 Plum	12	.667	Ea.	1,025	43		1,068	1,200
8842	1" NPT		11	.727		1,100	47		1,147	1,275
8846	1-1/4" NPT		10	.800		1,100	51.50		1,151.50	1,275

22 05 23.40 Valves, Lined, Corrosion Resistant/High Purity

		Crew	Daily Output	Labor-Hours	Unit	Material	Labor	Equipment	Total	Total Incl O&P
0010	VALVES, LINED, CORROSION RESISTANT/HIGH PURITY									
3500	Check lift, 125 lb., cast iron flanged									
3510	Horizontal PPL or SL lined									
3530	1"	1 Plum	14	.571	Ea.	660	37		697	780
3540	1-1/2"		11	.727		795	47		842	945
3550	2"		8	1		930	64.50		994.50	1,125

22 05 23 – General-Duty Valves for Plumbing Piping

22 05 23.40 Valves, Lined, Corrosion Resistant/High Purity		Crew	Daily Output	Labor-Hours	Unit	Material	2020 Bare Costs Labor	Equipment	Total	Total Incl O&P
3560	2-1/2"	Q-1	5	3.200	Ea.	1,200	186		1,386	1,600
3570	3"		4.50	3.556		1,525	206		1,731	1,975
3590	4"		3	5.333		2,000	310		2,310	2,675
3610	6"	Q-2	3	8		3,400	480		3,880	4,450
3620	8"	"	2.50	9.600		7,475	575		8,050	9,100
4250	Vertical PPL or SL lined									
4270	1"	1 Plum	14	.571	Ea.	665	37		702	785
4290	1-1/2"		11	.727		770	47		817	920
4300	2"		8	1		910	64.50		974.50	1,100
4310	2-1/2"	Q-1	5	3.200		1,325	186		1,511	1,725
4320	3"		4.50	3.556		1,375	206		1,581	1,800
4340	4"		3	5.333		1,875	310		2,185	2,550
4360	6"	Q-2	3	8		2,825	480		3,305	3,850
4370	8"	"	2.50	9.600		5,625	575		6,200	7,075
5000	Clamp type, ductile iron, 150 lb. flanged									
5010	TFE lined									
5030	1" size, lever handle	1 Plum	9	.889	Ea.	1,375	57.50		1,432.50	1,575
5050	1-1/2" size, lever handle		6	1.333		1,800	86		1,886	2,100
5060	2" size, lever handle		5	1.600		2,175	103		2,278	2,550
5080	3" size, lever handle	Q-1	4.50	3.556		2,750	206		2,956	3,325
5100	4" size, gear operated	"	3	5.333		3,950	310		4,260	4,825
5120	6" size, gear operated	Q-2	3	8		13,700	480		14,180	15,800
5130	8" size, gear operated	"	2.50	9.600		14,900	575		15,475	17,200
6000	Diaphragm type, cast iron, 125 lb. flanged									
6010	PTFE or VITON, lined									
6030	1" size, handwheel operated	1 Plum	9	.889	Ea.	345	57.50		402.50	465
6050	1-1/2" size, handwheel operated		6	1.333		410	86		496	580
6060	2" size, handwheel operated		5	1.600		465	103		568	670
6080	3" size, handwheel operated	Q-1	4.50	3.556		775	206		981	1,150
6100	4" size, handwheel operated	"	3	5.333		1,400	310		1,710	2,025
6120	6" size, handwheel operated	Q-2	3	8		2,400	480		2,880	3,350
6130	8" size, handwheel operated	"	2.50	9.600		4,825	575		5,400	6,175

22 05 23.60 Valves, Plastic

		Crew	Daily Output	Labor-Hours	Unit	Material	Labor	Equipment	Total	Total Incl O&P
0010	**VALVES, PLASTIC** R220523-90									
1100	Angle, PVC, threaded									
1110	1/4"	1 Plum	26	.308	Ea.	60	19.85		79.85	95.50
1120	1/2"		26	.308		81.50	19.85		101.35	119
1130	3/4"		25	.320		96.50	20.50		117	137
1140	1"		23	.348		117	22.50		139.50	163
1150	Ball, PVC, socket or threaded, true union									
1230	1/2"	1 Plum	26	.308	Ea.	39	19.85		58.85	72.50
1240	3/4"		25	.320		50	20.50		70.50	86
1250	1"		23	.348		55.50	22.50		78	94.50
1260	1-1/4"		21	.381		94	24.50		118.50	140
1270	1-1/2"		20	.400		86	26		112	134
1280	2"		17	.471		136	30.50		166.50	195
1290	2-1/2"	Q-1	26	.615		236	35.50		271.50	315
1300	3"		24	.667		272	38.50		310.50	355
1310	4"		20	.800		455	46.50		501.50	575
1360	For PVC, flanged, add					100%	15%			
1450	Double union 1/2"	1 Plum	26	.308		37.50	19.85		57.35	70.50
1460	3/4"		25	.320		43.50	20.50		64	79

22 05 23.60 Valves, Plastic		Crew	Daily Output	Labor-Hours	Unit	Material	Labor	Equipment	Total	Total Incl O&P
						2020 Bare Costs				
1470	1"	1 Plum	23	.348	Ea.	57	22.50		79.50	96.50
1480	1-1/4"		21	.381		72.50	24.50		97	116
1490	1-1/2"		20	.400		95	26		121	143
1500	2"	↓	17	.471	↓	109	30.50		139.50	166
1650	CPVC, socket or threaded, single union									
1700	1/2"	1 Plum	26	.308	Ea.	61.50	19.85		81.35	97
1720	3/4"		25	.320		81	20.50		101.50	120
1730	1"		23	.348		92.50	22.50		115	136
1750	1-1/4"		21	.381		144	24.50		168.50	196
1760	1-1/2"		20	.400		146	26		172	200
1770	2"	↓	17	.471		202	30.50		232.50	268
1780	3"	Q-1	24	.667		755	38.50		793.50	890
1840	For CPVC, flanged, add					65%	15%			
1880	For true union, socket or threaded, add				↓	50%	5%			
2050	Polypropylene, threaded									
2100	1/4"	1 Plum	26	.308	Ea.	44	19.85		63.85	78
2120	3/8"		26	.308		45.50	19.85		65.35	79.50
2130	1/2"		26	.308		39.50	19.85		59.35	73
2140	3/4"		25	.320		49	20.50		69.50	85
2150	1"		23	.348		55.50	22.50		78	94.50
2160	1-1/4"		21	.381		74	24.50		98.50	118
2170	1-1/2"		20	.400		91.50	26		117.50	140
2180	2"	↓	17	.471		123	30.50		153.50	182
2190	3"	Q-1	24	.667		300	38.50		338.50	390
2200	4"	"	20	.800	↓	500	46.50		546.50	620
2550	PVC, three way, socket or threaded									
2600	1/2"	1 Plum	26	.308	Ea.	57	19.85		76.85	92
2640	3/4"		25	.320		67.50	20.50		88	106
2650	1"		23	.348		81.50	22.50		104	123
2660	1-1/2"		20	.400		117	26		143	168
2670	2"	↓	17	.471		155	30.50		185.50	217
2680	3"	Q-1	24	.667		465	38.50		503.50	570
2740	For flanged, add				↓	60%	15%			
3150	Ball check, PVC, socket or threaded									
3200	1/4"	1 Plum	26	.308	Ea.	48	19.85		67.85	82.50
3220	3/8"		26	.308		48	19.85		67.85	82.50
3240	1/2"		26	.308		46.50	19.85		66.35	80.50
3250	3/4"		25	.320		52	20.50		72.50	88
3260	1"		23	.348		65	22.50		87.50	105
3270	1-1/4"		21	.381		124	24.50		148.50	173
3280	1-1/2"		20	.400		109	26		135	159
3290	2"	↓	17	.471		149	30.50		179.50	209
3310	3"	Q-1	24	.667		415	38.50		453.50	515
3320	4"	"	20	.800		585	46.50		631.50	710
3360	For PVC, flanged, add				↓	50%	15%			
3750	CPVC, socket or threaded									
3800	1/2"	1 Plum	26	.308	Ea.	66.50	19.85		86.35	103
3840	3/4"		25	.320		85.50	20.50		106	126
3850	1"		23	.348		104	22.50		126.50	148
3860	1-1/2"		20	.400		175	26		201	231
3870	2"	↓	17	.471		227	30.50		257.50	295
3880	3"	Q-1	24	.667		700	38.50		738.50	830
3920	4"	"	20	.800	↓	1,000	46.50		1,046.50	1,175

22 05 23.60 Valves, Plastic		Crew	Daily Output	Labor-Hours	Unit	2020 Bare Costs Material	Labor	Equipment	Total	Total Incl O&P
3930	For CPVC, flanged, add				Ea.	40%	15%			
4340	Polypropylene, threaded									
4360	1/2"	1 Plum	26	.308	Ea.	49.50	19.85		69.35	84
4400	3/4"		25	.320		69	20.50		89.50	107
4440	1"		23	.348		73	22.50		95.50	114
4450	1-1/2"		20	.400		141	26		167	194
4460	2"		17	.471		179	30.50		209.50	243
4500	For polypropylene flanged, add					200%	15%			
4850	Foot valve, PVC, socket or threaded									
4900	1/2"	1 Plum	34	.235	Ea.	62	15.15		77.15	91
4930	3/4"		32	.250		70	16.10		86.10	101
4940	1"		28	.286		91.50	18.40		109.90	128
4950	1-1/4"		27	.296		173	19.10		192.10	219
4960	1-1/2"		26	.308		175	19.85		194.85	223
4970	2"		24	.333		202	21.50		223.50	254
4980	3"		20	.400		480	26		506	570
4990	4"		18	.444		845	28.50		873.50	975
5000	For flanged, add					25%	10%			
5050	CPVC, socket or threaded									
5060	1/2"	1 Plum	34	.235	Ea.	77.50	15.15		92.65	108
5070	3/4"		32	.250		99.50	16.10		115.60	134
5080	1"		28	.286		122	18.40		140.40	162
5090	1-1/4"		27	.296		194	19.10		213.10	242
5100	1-1/2"		26	.308		194	19.85		213.85	243
5110	2"		24	.333		243	21.50		264.50	299
5120	3"		20	.400		495	26		521	585
5130	4"		18	.444		895	28.50		923.50	1,025
5140	For flanged, add					25%	10%			
5280	Needle valve, PVC, threaded									
5300	1/4"	1 Plum	26	.308	Ea.	66	19.85		85.85	102
5340	3/8"		26	.308		76.50	19.85		96.35	114
5360	1/2"		26	.308		76	19.85		95.85	113
5380	For polypropylene, add					10%				
5800	Y check, PVC, socket or threaded									
5820	1/2"	1 Plum	26	.308	Ea.	68	19.85		87.85	105
5840	3/4"		25	.320		74	20.50		94.50	112
5850	1"		23	.348		82	22.50		104.50	124
5860	1-1/4"		21	.381		129	24.50		153.50	179
5870	1-1/2"		20	.400		141	26		167	194
5880	2"		17	.471		175	30.50		205.50	238
5890	2-1/2"		15	.533		370	34.50		404.50	460
5900	3"	Q-1	24	.667		365	38.50		403.50	460
5910	4"	"	20	.800		610	46.50		656.50	740
5960	For PVC flanged, add					45%	15%			
6350	Y sediment strainer, PVC, socket or threaded									
6400	1/2"	1 Plum	26	.308	Ea.	56.50	19.85		76.35	91.50
6440	3/4"		24	.333		65.50	21.50		87	105
6450	1"		23	.348		67.50	22.50		90	108
6460	1-1/4"		21	.381		113	24.50		137.50	162
6470	1-1/2"		20	.400		133	26		159	185
6480	2"		17	.471		142	30.50		172.50	202
6490	2-1/2"		15	.533		345	34.50		379.50	430
6500	3"	Q-1	24	.667		340	38.50		378.50	435

22 05 23 – General-Duty Valves for Plumbing Piping

22 05 23.60 Valves, Plastic		Crew	Daily Output	Labor-Hours	Unit	Material	Labor	Equipment	Total	Total Incl O&P
						2020 Bare Costs				
6510	4"	Q-1	20	.800	Ea.	580	46.50		626.50	705
6560	For PVC, flanged, add					55%	15%			

22 05 29 – Hangers and Supports for Plumbing Piping and Equipment

22 05 29.10 Hangers & Supp. for Plumb'g/HVAC Pipe/Equip.

Line	Description	Crew	Daily Output	Labor-Hours	Unit	Material	Labor	Equipment	Total	Total Incl O&P
0010	**HANGERS AND SUPPORTS FOR PLUMB'G/HVAC PIPE/EQUIP.**									
0011	TYPE numbers per MSS-SP58									
0050	Brackets									
0060	Beam side or wall, malleable iron, TYPE 34									
0070	3/8" threaded rod size	1 Plum	48	.167	Ea.	5.10	10.75		15.85	21.50
0080	1/2" threaded rod size		48	.167		4	10.75		14.75	20.50
0090	5/8" threaded rod size		48	.167		11.70	10.75		22.45	29
0100	3/4" threaded rod size		48	.167		21.50	10.75		32.25	39.50
0110	7/8" threaded rod size		48	.167		13.95	10.75		24.70	31.50
0120	For concrete installation, add						30%			
0150	Wall, welded steel, medium, TYPE 32									
0160	0 size, 12" wide, 18" deep	1 Plum	34	.235	Ea.	183	15.15		198.15	225
0170	1 size, 18" wide, 24" deep		34	.235		218	15.15		233.15	263
0180	2 size, 24" wide, 30" deep		34	.235		289	15.15		304.15	340
0200	Beam attachment, welded, TYPE 22									
0202	3/8"	Q-15	80	.200	Ea.	12	11.60	1.33	24.93	32
0203	1/2"		76	.211		13.15	12.20	1.40	26.75	34
0204	5/8"		72	.222		13.55	12.90	1.48	27.93	36
0205	3/4"		68	.235		14.05	13.65	1.56	29.26	37.50
0206	7/8"		64	.250		19.30	14.50	1.66	35.46	44.50
0207	1"		56	.286		41.50	16.55	1.90	59.95	72.50
0300	Clamps									
0310	C-clamp, for mounting on steel beam flange, w/locknut, TYPE 23									
0320	3/8" threaded rod size	1 Plum	160	.050	Ea.	3.38	3.22		6.60	8.55
0330	1/2" threaded rod size		160	.050		4.76	3.22		7.98	10.05
0340	5/8" threaded rod size		160	.050		5.30	3.22		8.52	10.60
0350	3/4" threaded rod size		160	.050		6.85	3.22		10.07	12.30
0352	7/8" threaded rod size		140	.057		17.30	3.68		20.98	24.50
0400	High temperature to 1050°F, alloy steel									
0410	4" pipe size	Q-1	106	.151	Ea.	28	8.75		36.75	43.50
0420	6" pipe size		106	.151		43	8.75		51.75	60
0430	8" pipe size		97	.165		48	9.55		57.55	67.50
0440	10" pipe size		84	.190		81.50	11.05		92.55	106
0450	12" pipe size		72	.222		82	12.90		94.90	110
0460	14" pipe size		64	.250		253	14.50		267.50	300
0470	16" pipe size		56	.286		269	16.55		285.55	320
0480	Beam clamp, flange type, TYPE 25									
0482	For 3/8" bolt	1 Plum	48	.167	Ea.	4.85	10.75		15.60	21.50
0483	For 1/2" bolt		44	.182		6.15	11.70		17.85	24.50
0484	For 5/8" bolt		40	.200		6.10	12.90		19	26
0485	For 3/4" bolt		36	.222		6.10	14.30		20.40	28
0486	For 1" bolt		32	.250		6.15	16.10		22.25	31
0500	I-beam, for mounting on bottom flange, strap iron, TYPE 21									
0530	4" flange size	1 Plum	93	.086	Ea.	6.75	5.55		12.30	15.70
0540	5" flange size		92	.087		7.40	5.60		13	16.55
0550	6" flange size		90	.089		8.40	5.75		14.15	17.80
0560	7" flange size		88	.091		9.35	5.85		15.20	19.05
0570	8" flange size		86	.093		10.05	6		16.05	20

22 05 29 – Hangers and Supports for Plumbing Piping and Equipment

22 05 29.10 Hangers & Supp. for Plumb'g/HVAC Pipe/Equip.		Crew	Daily Output	Labor-Hours	Unit	Material	2020 Bare Costs Labor	Equipment	Total	Total Incl O&P
0600	One hole, vertical mounting, malleable iron									
0610	1/2" pipe size	1 Plum	160	.050	Ea.	1.27	3.22		4.49	6.20
0620	3/4" pipe size		145	.055		1.35	3.56		4.91	6.80
0630	1" pipe size		136	.059		1.46	3.79		5.25	7.25
0640	1-1/4" pipe size		128	.063		2.56	4.03		6.59	8.85
0650	1-1/2" pipe size		120	.067		2.96	4.30		7.26	9.70
0660	2" pipe size		112	.071		4.27	4.60		8.87	11.60
0670	2-1/2" pipe size		104	.077		8.45	4.96		13.41	16.65
0680	3" pipe size		96	.083		9.85	5.35		15.20	18.85
0690	3-1/2" pipe size		90	.089		9.55	5.75		15.30	19.05
0700	4" pipe size		84	.095		13.50	6.15		19.65	24
0750	Riser or extension pipe, carbon steel, TYPE 8									
0756	1/2" pipe size	1 Plum	52	.154	Ea.	2.51	9.90		12.41	17.60
0760	3/4" pipe size		48	.167		3.48	10.75		14.23	19.95
0770	1" pipe size		47	.170		4.04	10.95		14.99	21
0780	1-1/4" pipe size		46	.174		5.45	11.20		16.65	23
0790	1-1/2" pipe size		45	.178		5.65	11.45		17.10	23.50
0800	2" pipe size		43	.186		5.65	12		17.65	24
0810	2-1/2" pipe size		41	.195		6.65	12.60		19.25	26
0820	3" pipe size		40	.200		6.70	12.90		19.60	26.50
0830	3-1/2" pipe size		39	.205		6.05	13.20		19.25	26.50
0840	4" pipe size		38	.211		9.20	13.55		22.75	30.50
0850	5" pipe size		37	.216		14.30	13.95		28.25	37
0860	6" pipe size		36	.222		16.30	14.30		30.60	39.50
0870	8" pipe size		34	.235		30	15.15		45.15	55.50
0880	10" pipe size		32	.250		30.50	16.10		46.60	57.50
0890	12" pipe size		28	.286		43.50	18.40		61.90	75.50
0900	For plastic coating 3/4" to 4", add					190%				
0910	For copper plating 3/4" to 4", add					58%				
0950	Two piece, complete, carbon steel, medium weight, TYPE 4									
0960	1/2" pipe size	Q-1	137	.117	Ea.	2.67	6.75		9.42	13.10
0970	3/4" pipe size		134	.119		2.73	6.95		9.68	13.35
0980	1" pipe size		132	.121		2.51	7.05		9.56	13.25
0990	1-1/4" pipe size		130	.123		3.30	7.15		10.45	14.35
1000	1-1/2" pipe size		126	.127		4.40	7.35		11.75	15.85
1010	2" pipe size		124	.129		5.20	7.50		12.70	16.90
1020	2-1/2" pipe size		120	.133		5.60	7.75		13.35	17.70
1030	3" pipe size		117	.137		5.25	7.95		13.20	17.65
1040	3-1/2" pipe size		114	.140		11.80	8.15		19.95	25
1050	4" pipe size		110	.145		9	8.45		17.45	22.50
1060	5" pipe size		106	.151		17	8.75		25.75	32
1070	6" pipe size		104	.154		22	8.90		30.90	37.50
1080	8" pipe size		100	.160		21.50	9.30		30.80	38
1090	10" pipe size		96	.167		51.50	9.65		61.15	71
1100	12" pipe size		89	.180		77.50	10.45		87.95	101
1110	14" pipe size		82	.195		87	11.30		98.30	112
1120	16" pipe size		68	.235		136	13.65		149.65	171
1130	For galvanized, add					45%				
1150	Insert, concrete									
1160	Wedge type, carbon steel body, malleable iron nut, galvanized									
1170	1/4" threaded rod size	1 Plum	96	.083	Ea.	8.40	5.35		13.75	17.30
1180	3/8" threaded rod size		96	.083		20.50	5.35		25.85	30.50
1190	1/2" threaded rod size		96	.083		39	5.35		44.35	51

22 05 Common Work Results for Plumbing

22 05 29 – Hangers and Supports for Plumbing Piping and Equipment

22 05 29.10 Hangers & Supp. for Plumb'g/HVAC Pipe/Equip.		Crew	Daily Output	Labor-Hours	Unit	2020 Bare Costs Material	Labor	Equipment	Total	Total Incl O&P
1200	5/8" threaded rod size	1 Plum	96	.083	Ea.	6.55	5.35		11.90	15.30
1210	3/4" threaded rod size		96	.083		12.30	5.35		17.65	21.50
1220	7/8" threaded rod size		96	.083		12.30	5.35		17.65	21.50
1250	Pipe guide sized for insulation									
1260	No. 1, 1" pipe size, 1" thick insulation	1 Stpi	26	.308	Ea.	143	20		163	187
1270	No. 2, 1-1/4"-2" pipe size, 1" thick insulation		23	.348		169	23		192	220
1280	No. 3, 1-1/4"-2" pipe size, 1-1/2" thick insulation		21	.381		169	25		194	224
1290	No. 4, 2-1/2"-3-1/2" pipe size, 1-1/2" thick insulation		18	.444		183	29		212	245
1300	No. 5, 4"-5" pipe size, 1-1/2" thick insulation		16	.500		197	33		230	265
1310	No. 6, 5"-6" pipe size, 2" thick insulation	Q-5	21	.762		226	45		271	315
1320	No. 7, 8" pipe size, 2" thick insulation		16	1		292	59		351	410
1330	No. 8, 10" pipe size, 2" thick insulation		12	1.333		435	78.50		513.50	600
1340	No. 9, 12" pipe size, 2" thick insulation	Q-6	17	1.412		435	86.50		521.50	610
1350	No. 10, 12"-14" pipe size, 2-1/2" thick insulation		16	1.500		520	92		612	705
1360	No. 11, 16" pipe size, 2-1/2" thick insulation		10.50	2.286		520	140		660	780
1370	No. 12, 16"-18" pipe size, 3" thick insulation		9	2.667		740	163		903	1,050
1380	No. 13, 20" pipe size, 3" thick insulation		7.50	3.200		740	196		936	1,100
1390	No. 14, 24" pipe size, 3" thick insulation		7	3.429		1,025	210		1,235	1,475
1400	Bands									
1410	Adjustable band, carbon steel, for non-insulated pipe, TYPE 7									
1420	1/2" pipe size	Q-1	142	.113	Ea.	.38	6.55		6.93	10.20
1430	3/4" pipe size		140	.114		.38	6.65		7.03	10.30
1440	1" pipe size		137	.117		.38	6.75		7.13	10.55
1450	1-1/4" pipe size		134	.119		.46	6.95		7.41	10.85
1460	1-1/2" pipe size		131	.122		.41	7.10		7.51	11.05
1470	2" pipe size		129	.124		.41	7.20		7.61	11.20
1480	2-1/2" pipe size		125	.128		.74	7.40		8.14	11.90
1490	3" pipe size		122	.131		.79	7.60		8.39	12.25
1500	3-1/2" pipe size		119	.134		.95	7.80		8.75	12.70
1510	4" pipe size		114	.140		1.29	8.15		9.44	13.60
1520	5" pipe size		110	.145		2.23	8.45		10.68	15.10
1530	6" pipe size		108	.148		2.68	8.60		11.28	15.80
1540	8" pipe size		104	.154		4.08	8.90		12.98	17.85
1550	For copper plated, add					50%				
1560	For galvanized, add					30%				
1570	For plastic coating, add					30%				
1600	Adjusting nut malleable iron, steel band, TYPE 9									
1610	1/2" pipe size, galvanized band	Q-1	137	.117	Ea.	5.75	6.75		12.50	16.50
1620	3/4" pipe size, galvanized band		135	.119		5.70	6.85		12.55	16.60
1630	1" pipe size, galvanized band		132	.121		5.40	7.05		12.45	16.45
1640	1-1/4" pipe size, galvanized band		129	.124		5.85	7.20		13.05	17.15
1650	1-1/2" pipe size, galvanized band		126	.127		5.60	7.35		12.95	17.15
1660	2" pipe size, galvanized band		124	.129		6.35	7.50		13.85	18.15
1670	2-1/2" pipe size, galvanized band		120	.133		9.95	7.75		17.70	22.50
1680	3" pipe size, galvanized band		117	.137		10.40	7.95		18.35	23.50
1690	3-1/2" pipe size, galvanized band		114	.140		32	8.15		40.15	47
1700	4" pipe size, cadmium plated band		110	.145		28	8.45		36.45	43.50
1740	For plastic coated band, add					35%				
1750	For completely copper coated, add					45%				
1800	Clevis, adjustable, carbon steel, for non-insulated pipe, TYPE 1									
1810	1/2" pipe size	Q-1	137	.117	Ea.	1.12	6.75		7.87	11.40
1820	3/4" pipe size		135	.119		1.19	6.85		8.04	11.60
1830	1" pipe size		132	.121		1.51	7.05		8.56	12.15

22 05 29 – Hangers and Supports for Plumbing Piping and Equipment

22 05 29.10 Hangers & Supp. for Plumb'g/HVAC Pipe/Equip.		Crew	Daily Output	Labor-Hours	Unit	2020 Bare Costs Material	Labor	Equipment	Total	Total Incl O&P
1840	1-1/4" pipe size	Q-1	129	.124	Ea.	1.68	7.20		8.88	12.60
1850	1-1/2" pipe size		126	.127		1.15	7.35		8.50	12.25
1860	2" pipe size		124	.129		2.15	7.50		9.65	13.55
1870	2-1/2" pipe size		120	.133		3.55	7.75		11.30	15.45
1880	3" pipe size		117	.137		4.22	7.95		12.17	16.50
1890	3-1/2" pipe size		114	.140		4.50	8.15		12.65	17.15
1900	4" pipe size		110	.145		3.15	8.45		11.60	16.10
1910	5" pipe size		106	.151		4.27	8.75		13.02	17.80
1920	6" pipe size		104	.154		5.40	8.90		14.30	19.30
1930	8" pipe size		100	.160		10.60	9.30		19.90	25.50
1940	10" pipe size		96	.167		17.05	9.65		26.70	33
1950	12" pipe size		89	.180		22	10.45		32.45	40
1960	14" pipe size		82	.195		36	11.30		47.30	57
1970	16" pipe size		68	.235		70.50	13.65		84.15	98
1971	18" pipe size		54	.296		90	17.20		107.20	125
1972	20" pipe size		38	.421		139	24.50		163.50	190
1980	For galvanized, add					66%				
1990	For copper plated 1/2" to 4", add					77%				
2000	For light weight 1/2" to 4", deduct					13%				
2010	Insulated pipe type, 3/4" to 12" pipe, add					180%				
2020	Insulated pipe type, chrome-moly U-strap, add					530%				
2250	Split ring, malleable iron, for non-insulated pipe, TYPE 11									
2260	1/2" pipe size	Q-1	137	.117	Ea.	5.05	6.75		11.80	15.70
2270	3/4" pipe size		135	.119		5.15	6.85		12	16
2280	1" pipe size		132	.121		5.45	7.05		12.50	16.50
2290	1-1/4" pipe size		129	.124		6.80	7.20		14	18.20
2300	1-1/2" pipe size		126	.127		8.25	7.35		15.60	20
2310	2" pipe size		124	.129		9.35	7.50		16.85	21.50
2320	2-1/2" pipe size		120	.133		13.45	7.75		21.20	26.50
2330	3" pipe size		117	.137		16.85	7.95		24.80	30.50
2340	3-1/2" pipe size		114	.140		17.25	8.15		25.40	31
2350	4" pipe size		110	.145		17.55	8.45		26	32
2360	5" pipe size		106	.151		23.50	8.75		32.25	38.50
2370	6" pipe size		104	.154		52	8.90		60.90	71
2380	8" pipe size		100	.160		78	9.30		87.30	99.50
2390	For copper plated, add					8%				
2400	Channels, steel, 3/4" x 1-1/2"	1 Plum	80	.100	L.F.	2.98	6.45		9.43	12.95
2404	1-1/2" x 1-1/2"		70	.114		3.68	7.35		11.03	15.05
2408	1-7/8" x 1-1/2"		60	.133		21	8.60		29.60	36.50
2412	3" x 1-1/2"		50	.160		19.85	10.30		30.15	37.50
2416	Hangers, trapeze channel support, 12 ga. 1-1/2" x 1-1/2", 12" wide, steel		8.80	.909	Ea.	26	58.50		84.50	116
2418	18" wide, steel		8.30	.964		27.50	62		89.50	124
2430	Spring nuts, long, 1/4"		120	.067		.65	4.30		4.95	7.15
2432	3/8"		100	.080		.75	5.15		5.90	8.55
2434	1/2"		80	.100		.83	6.45		7.28	10.55
2436	5/8"		80	.100		3.71	6.45		10.16	13.75
2438	3/4"		75	.107		6.20	6.85		13.05	17.15
2440	Spring nuts, short, 1/4"		120	.067		1.36	4.30		5.66	7.95
2442	3/8"		100	.080		1.70	5.15		6.85	9.55
2444	1/2"		80	.100		1.81	6.45		8.26	11.65
2500	Washer, flat steel									
2502	3/8"	1 Plum	240	.033	Ea.	.08	2.15		2.23	3.30
2503	1/2"		220	.036		.21	2.34		2.55	3.74

22 05 29 – Hangers and Supports for Plumbing Piping and Equipment

22 05 29.10 Hangers & Supp. for Plumb'g/HVAC Pipe/Equip.		Crew	Daily Output	Labor-Hours	Unit	Material	2020 Bare Costs Labor	Equipment	Total	Total Incl O&P
2504	5/8"	1 Plum	200	.040	Ea.	.35	2.58		2.93	4.25
2505	3/4"		180	.044		.61	2.86		3.47	4.96
2506	7/8"		160	.050		.43	3.22		3.65	5.30
2507	1"		140	.057		1.38	3.68		5.06	7
2508	1-1/4"		120	.067		1.52	4.30		5.82	8.10
2520	Nut, steel, hex									
2522	3/8"	1 Plum	200	.040	Ea.	.22	2.58		2.80	4.10
2523	1/2"		180	.044		.53	2.86		3.39	4.87
2524	5/8"		160	.050		.91	3.22		4.13	5.80
2525	3/4"		140	.057		1.42	3.68		5.10	7.05
2526	7/8"		120	.067		2.09	4.30		6.39	8.75
2527	1"		100	.080		3.15	5.15		8.30	11.15
2528	1-1/4"		80	.100		8.10	6.45		14.55	18.60
2532	Turnbuckle, TYPE 13									
2534	3/8"	1 Plum	80	.100	Ea.	5.10	6.45		11.55	15.25
2535	1/2"		72	.111		5.65	7.15		12.80	16.95
2536	5/8"		64	.125		10.05	8.05		18.10	23
2537	3/4"		56	.143		12.80	9.20		22	28
2538	7/8"		48	.167		33.50	10.75		44.25	53
2539	1"		40	.200		45.50	12.90		58.40	69.50
2540	1-1/4"		32	.250		84	16.10		100.10	117
2650	Rods, carbon steel									
2660	Continuous thread									
2670	1/4" thread size	1 Plum	144	.056	L.F.	2.43	3.58		6.01	8
2680	3/8" thread size		144	.056		2.59	3.58		6.17	8.20
2690	1/2" thread size		144	.056		4.07	3.58		7.65	9.85
2700	5/8" thread size		144	.056		5.75	3.58		9.33	11.70
2710	3/4" thread size		144	.056		10.60	3.58		14.18	17.05
2720	7/8" thread size		144	.056		13.35	3.58		16.93	20
2721	1" thread size	Q-1	160	.100		22.50	5.80		28.30	33.50
2722	1-1/8" thread size	"	120	.133		24.50	7.75		32.25	38.50
2725	1/4" thread size, bright finish	1 Plum	144	.056		1.57	3.58		5.15	7.10
2726	1/2" thread size, bright finish	"	144	.056		5.10	3.58		8.68	10.95
2730	For galvanized, add					40%				
2820	Rod couplings									
2821	3/8"	1 Plum	60	.133	Ea.	1.74	8.60		10.34	14.75
2822	1/2"		54	.148		2.43	9.55		11.98	16.95
2823	5/8"		48	.167		3.09	10.75		13.84	19.50
2824	3/4"		44	.182		4.34	11.70		16.04	22.50
2825	7/8"		40	.200		7.35	12.90		20.25	27.50
2826	1"		34	.235		9.70	15.15		24.85	33
2827	1-1/8"		30	.267		16.40	17.20		33.60	43.50
2860	Pipe hanger assy, adj. clevis, saddle, rod, clamp, insul. allowance									
2864	1/2" pipe size	Q-5	35	.457	Ea.	26	27		53	69
2866	3/4" pipe size		34.80	.460		27.50	27		54.50	70.50
2868	1" pipe size		34.60	.462		27	27.50		54.50	70.50
2869	1-1/4" pipe size		34.30	.466		27	27.50		54.50	71
2870	1-1/2" pipe size		33.90	.472		27.50	28		55.50	71.50
2872	2" pipe size		33.30	.480		30	28.50		58.50	75.50
2874	2-1/2" pipe size		32.30	.495		33.50	29		62.50	80
2876	3" pipe size		31.20	.513		36	30.50		66.50	85
2880	4" pipe size		30.70	.521		38.50	31		69.50	88.50
2884	6" pipe size		29.80	.537		49.50	31.50		81	102

22 05 29.10 Hangers & Supp. for Plumb'g/HVAC Pipe/Equip.		Crew	Daily Output	Labor-Hours	Unit	Material	2020 Bare Costs Labor	Equipment	Total	Total Incl O&P
2888	8" pipe size	Q-5	28	.571	Ea.	67.50	33.50		101	125
2892	10" pipe size		25.20	.635		74.50	37.50		112	138
2896	12" pipe size		23.20	.690		117	40.50		157.50	189
2900	Rolls									
2910	Adjustable yoke, carbon steel with CI roll, TYPE 43									
2918	2" pipe size	Q-1	140	.114	Ea.	11.70	6.65		18.35	23
2920	2-1/2" pipe size		137	.117		12.40	6.75		19.15	24
2930	3" pipe size		131	.122		13.25	7.10		20.35	25
2940	3-1/2" pipe size		124	.129		18.45	7.50		25.95	31.50
2950	4" pipe size		117	.137		18.15	7.95		26.10	32
2960	5" pipe size		110	.145		30	8.45		38.45	45.50
2970	6" pipe size		104	.154		40.50	8.90		49.40	58
2980	8" pipe size		96	.167		58	9.65		67.65	78.50
2990	10" pipe size		80	.200		74.50	11.60		86.10	99.50
3000	12" pipe size		68	.235		108	13.65		121.65	140
3010	14" pipe size		56	.286		185	16.55		201.55	229
3020	16" pipe size		48	.333		224	19.35		243.35	275
3050	Chair, carbon steel with CI roll									
3060	2" pipe size	1 Plum	68	.118	Ea.	14.65	7.60		22.25	27.50
3070	2-1/2" pipe size		65	.123		15.75	7.95		23.70	29
3080	3" pipe size		62	.129		16.80	8.30		25.10	31
3090	3-1/2" pipe size		60	.133		20.50	8.60		29.10	35.50
3100	4" pipe size		58	.138		19.90	8.90		28.80	35.50
3110	5" pipe size		56	.143		23	9.20		32.20	39.50
3120	6" pipe size		53	.151		31	9.75		40.75	48.50
3130	8" pipe size		50	.160		42	10.30		52.30	61.50
3140	10" pipe size		48	.167		56.50	10.75		67.25	78
3150	12" pipe size		46	.174		86	11.20		97.20	111
3170	Single pipe roll (see line 2650 for rods), TYPE 41, 1" pipe size	Q-1	137	.117		9.85	6.75		16.60	21
3180	1-1/4" pipe size		131	.122		10.45	7.10		17.55	22
3190	1-1/2" pipe size		129	.124		10.45	7.20		17.65	22
3200	2" pipe size		124	.129		10.45	7.50		17.95	22.50
3210	2-1/2" pipe size		118	.136		11.15	7.85		19	24
3220	3" pipe size		115	.139		11.75	8.05		19.80	25
3230	3-1/2" pipe size		113	.142		12.80	8.20		21	26.50
3240	4" pipe size		112	.143		13	8.30		21.30	26.50
3250	5" pipe size		110	.145		15.10	8.45		23.55	29.50
3260	6" pipe size		101	.158		30	9.20		39.20	47
3270	8" pipe size		90	.178		32.50	10.30		42.80	51
3280	10" pipe size		80	.200		42	11.60		53.60	63.50
3290	12" pipe size		68	.235		68.50	13.65		82.15	95.50
3300	Saddles (add vertical pipe riser, usually 3" diameter)									
3310	Pipe support, complete, adjust., CI saddle, TYPE 36									
3320	2-1/2" pipe size	1 Plum	96	.083	Ea.	111	5.35		116.35	130
3330	3" pipe size		88	.091		113	5.85		118.85	133
3340	3-1/2" pipe size		79	.101		111	6.55		117.55	132
3350	4" pipe size		68	.118		160	7.60		167.60	187
3360	5" pipe size		64	.125		169	8.05		177.05	198
3370	6" pipe size		59	.136		165	8.75		173.75	194
3380	8" pipe size		53	.151		170	9.75		179.75	202
3390	10" pipe size		50	.160		198	10.30		208.30	232
3400	12" pipe size		48	.167		209	10.75		219.75	246
3450	For standard pipe support, one piece, CI, deduct					34%				

22 05 29 – Hangers and Supports for Plumbing Piping and Equipment

22 05 29.10 Hangers & Supp. for Plumb'g/HVAC Pipe/Equip.		Crew	Daily Output	Labor-Hours	Unit	2020 Bare Costs Material	Labor	Equipment	Total	Total Incl O&P
3460	For stanchion support, CI with steel yoke, deduct					60%				
3550	Insulation shield 1" thick, 1/2" pipe size, TYPE 40	1 Asbe	100	.080	Ea.	3.37	4.69		8.06	10.90
3560	3/4" pipe size		100	.080		4.06	4.69		8.75	11.65
3570	1" pipe size		98	.082		4.39	4.79		9.18	12.20
3580	1-1/4" pipe size		98	.082		4.68	4.79		9.47	12.50
3590	1-1/2" pipe size		96	.083		4.35	4.89		9.24	12.30
3600	2" pipe size		96	.083		4.53	4.89		9.42	12.50
3610	2-1/2" pipe size		94	.085		4.67	4.99		9.66	12.80
3620	3" pipe size		94	.085		5.55	4.99		10.54	13.75
3630	2" thick, 3-1/2" pipe size		92	.087		5.95	5.10		11.05	14.35
3640	4" pipe size		92	.087		9.85	5.10		14.95	18.70
3650	5" pipe size		90	.089		12.05	5.20		17.25	21.50
3660	6" pipe size		90	.089		13.85	5.20		19.05	23.50
3670	8" pipe size		88	.091		19.65	5.35		25	29.50
3680	10" pipe size		88	.091		27.50	5.35		32.85	38.50
3690	12" pipe size		86	.093		29.50	5.45		34.95	41
3700	14" pipe size		86	.093		57.50	5.45		62.95	71.50
3710	16" pipe size		84	.095		56.50	5.60		62.10	71
3720	18" pipe size		84	.095		59	5.60		64.60	73
3730	20" pipe size		82	.098		64.50	5.70		70.20	80
3732	24" pipe size		80	.100		76	5.85		81.85	93
3750	Covering protection saddle, TYPE 39									
3760	1" covering size									
3770	3/4" pipe size	1 Plum	68	.118	Ea.	5.60	7.60		13.20	17.50
3780	1" pipe size		68	.118		5.60	7.60		13.20	17.50
3790	1-1/4" pipe size		68	.118		5.60	7.60		13.20	17.50
3800	1-1/2" pipe size		66	.121		6	7.80		13.80	18.30
3810	2" pipe size		66	.121		6	7.80		13.80	18.30
3820	2-1/2" pipe size		64	.125		6.05	8.05		14.10	18.75
3830	3" pipe size		64	.125		8	8.05		16.05	21
3840	3-1/2" pipe size		62	.129		9.55	8.30		17.85	23
3850	4" pipe size		62	.129		8.85	8.30		17.15	22
3860	5" pipe size		60	.133		9.65	8.60		18.25	23.50
3870	6" pipe size		60	.133		11.55	8.60		20.15	25.50
3900	1-1/2" covering size									
3910	3/4" pipe size	1 Plum	68	.118	Ea.	7.20	7.60		14.80	19.25
3920	1" pipe size		68	.118		7.15	7.60		14.75	19.25
3930	1-1/4" pipe size		68	.118		7.20	7.60		14.80	19.25
3940	1-1/2" pipe size		66	.121		6.70	7.80		14.50	19.10
3950	2" pipe size		66	.121		7.25	7.80		15.05	19.65
3960	2-1/2" pipe size		64	.125		8.70	8.05		16.75	21.50
3970	3" pipe size		64	.125		8.70	8.05		16.75	21.50
3980	3-1/2" pipe size		62	.129		9	8.30		17.30	22.50
3990	4" pipe size		62	.129		9.15	8.30		17.45	22.50
4000	5" pipe size		60	.133		9.90	8.60		18.50	24
4010	6" pipe size		60	.133		12.45	8.60		21.05	26.50
4020	8" pipe size		58	.138		15.60	8.90		24.50	30.50
4022	10" pipe size		56	.143		15.60	9.20		24.80	31
4024	12" pipe size		54	.148		37.50	9.55		47.05	56
4028	2" covering size									
4029	2-1/2" pipe size	1 Plum	62	.129	Ea.	9.50	8.30		17.80	23
4032	3" pipe size		60	.133		10.40	8.60		19	24.50
4033	4" pipe size		58	.138		10.40	8.90		19.30	25

22 05 29.10	Hangers & Supp. for Plumb'g/HVAC Pipe/Equip.	Crew	Daily Output	Labor-Hours	Unit	2020 Bare Costs Material	Labor	Equipment	Total	Total Incl O&P
4034	6" pipe size	1 Plum	56	.143	Ea.	14.90	9.20		24.10	30
4035	8" pipe size		54	.148		17.70	9.55		27.25	34
4080	10" pipe size		58	.138		20	8.90		28.90	35.50
4090	12" pipe size		56	.143		42	9.20		51.20	60.50
4100	14" pipe size		56	.143		43	9.20		52.20	61.50
4110	16" pipe size		54	.148		57	9.55		66.55	77.50
4120	18" pipe size		54	.148		61	9.55		70.55	82
4130	20" pipe size		52	.154		68	9.90		77.90	90
4150	24" pipe size		50	.160		79.50	10.30		89.80	103
4160	30" pipe size		48	.167		87.50	10.75		98.25	112
4180	36" pipe size		45	.178		101	11.45		112.45	128
4186	2-1/2" covering size									
4187	3" pipe size	1 Plum	58	.138	Ea.	10.60	8.90		19.50	25
4188	4" pipe size		56	.143		11.55	9.20		20.75	26.50
4189	6" pipe size		52	.154		17.65	9.90		27.55	34.50
4190	8" pipe size		48	.167		19.70	10.75		30.45	37.50
4191	10" pipe size		44	.182		21.50	11.70		33.20	41.50
4192	12" pipe size		40	.200		50	12.90		62.90	74.50
4193	14" pipe size		36	.222		50	14.30		64.30	76.50
4194	16" pipe size		32	.250		60	16.10		76.10	90
4195	18" pipe size		28	.286		67	18.40		85.40	101
4200	Sockets									
4210	Rod end, malleable iron, TYPE 16									
4220	1/4" thread size	1 Plum	240	.033	Ea.	1.64	2.15		3.79	5
4230	3/8" thread size		240	.033		1.67	2.15		3.82	5.05
4240	1/2" thread size		230	.035		2.20	2.24		4.44	5.75
4250	5/8" thread size		225	.036		4.51	2.29		6.80	8.40
4260	3/4" thread size		220	.036		5.90	2.34		8.24	10
4270	7/8" thread size		210	.038		8.65	2.46		11.11	13.20
4290	Strap, 1/2" pipe size, TYPE 26	Q-1	142	.113		1.94	6.55		8.49	11.95
4300	3/4" pipe size		140	.114		1.97	6.65		8.62	12.05
4310	1" pipe size		137	.117		2.77	6.75		9.52	13.20
4320	1-1/4" pipe size		134	.119		2.84	6.95		9.79	13.45
4330	1-1/2" pipe size		131	.122		3.08	7.10		10.18	14
4340	2" pipe size		129	.124		3.13	7.20		10.33	14.20
4350	2-1/2" pipe size		125	.128		4.91	7.40		12.31	16.50
4360	3" pipe size		122	.131		6.25	7.60		13.85	18.25
4370	3-1/2" pipe size		119	.134		7	7.80		14.80	19.35
4380	4" pipe size		114	.140		7.35	8.15		15.50	20.50
4400	U-bolt, carbon steel									
4410	Standard, with nuts, TYPE 42									
4420	1/2" pipe size	1 Plum	160	.050	Ea.	1.41	3.22		4.63	6.35
4430	3/4" pipe size		158	.051		1.22	3.26		4.48	6.20
4450	1" pipe size		152	.053		1.25	3.39		4.64	6.50
4460	1-1/4" pipe size		148	.054		1.57	3.48		5.05	6.95
4470	1-1/2" pipe size		143	.056		1.67	3.61		5.28	7.25
4480	2" pipe size		139	.058		1.81	3.71		5.52	7.55
4490	2-1/2" pipe size		134	.060		2.96	3.85		6.81	9
4500	3" pipe size		128	.063		3.31	4.03		7.34	9.70
4510	3-1/2" pipe size		122	.066		4.29	4.23		8.52	11
4520	4" pipe size		117	.068		4.42	4.41		8.83	11.45
4530	5" pipe size		114	.070		4.39	4.52		8.91	11.60
4540	6" pipe size		111	.072		7.35	4.64		11.99	15.05

22 05 29.10 Hangers & Supp. for Plumb'g/HVAC Pipe/Equip.		Crew	Daily Output	Labor-Hours	Unit	2020 Bare Costs Material	Labor	Equipment	Total	Total Incl O&P
4550	8" pipe size	1 Plum	109	.073	Ea.	8.85	4.73		13.58	16.80
4560	10" pipe size		107	.075		15.15	4.82		19.97	24
4570	12" pipe size		104	.077		21	4.96		25.96	30.50
4580	For plastic coating on 1/2" thru 6" size, add					150%				
4700	U-hook, carbon steel, requires mounting screws or bolts									
4710	3/4" thru 2" pipe size									
4720	6" long	1 Plum	96	.083	Ea.	1.16	5.35		6.51	9.35
4730	8" long		96	.083		1.46	5.35		6.81	9.65
4740	10" long		96	.083		1.59	5.35		6.94	9.80
4750	12" long		96	.083		1.75	5.35		7.10	10
4760	For copper plated, add					50%				
7000	Roof supports									
7006	Duct									
7010	Rectangular, open, 12" off roof									
7020	To 18" wide	Q-9	26	.615	Ea.	163	34.50		197.50	232
7030	To 24" wide	"	22	.727		190	41		231	271
7040	To 36" wide	Q-10	30	.800		217	46.50		263.50	310
7050	To 48" wide		28	.857		244	50		294	345
7060	To 60" wide		24	1		271	58		329	385
7100	Equipment									
7120	Equipment support	Q-5	20	.800	Ea.	71	47		118	149
7300	Pipe									
7310	Roller type									
7320	Up to 2-1/2" diam. pipe									
7324	3-1/2" off roof	Q-5	24	.667	Ea.	22.50	39.50		62	83.50
7326	Up to 10" off roof	"	20	.800	"	28	47		75	101
7340	2-1/2" to 3-1/2" diam. pipe									
7342	Up to 16" off roof	Q-5	18	.889	Ea.	49.50	52.50		102	133
7360	4" to 5" diam. pipe									
7362	Up to 12" off roof	Q-5	16	1	Ea.	70.50	59		129.50	166
7400	Strut/channel type									
7410	Up to 2-1/2" diam. pipe									
7424	3-1/2" off roof	Q-5	24	.667	Ea.	20	39.50		59.50	81
7426	Up to 10" off roof	"	20	.800	"	25.50	47		72.50	98.50
7440	Strut and roller type									
7452	2-1/2" to 3-1/2" diam. pipe									
7454	Up to 16" off roof	Q-5	18	.889	Ea.	71	52.50		123.50	157
7460	Strut and hanger type									
7470	Up to 3" diam. pipe									
7474	Up to 8" off roof	Q-5	19	.842	Ea.	45	49.50		94.50	124
8000	Pipe clamp, plastic, 1/2" CTS	1 Plum	80	.100		.20	6.45		6.65	9.85
8010	3/4" CTS		73	.110		.24	7.05		7.29	10.80
8020	1" CTS		68	.118		.54	7.60		8.14	11.95
8080	Economy clamp, 1/4" CTS		175	.046		.05	2.95		3	4.47
8090	3/8" CTS		168	.048		.05	3.07		3.12	4.65
8100	1/2" CTS		160	.050		.05	3.22		3.27	4.88
8110	3/4" CTS		145	.055		.05	3.56		3.61	5.35
8200	Half clamp, 1/2" CTS		80	.100		.07	6.45		6.52	9.75
8210	3/4" CTS		73	.110		.10	7.05		7.15	10.65
8300	Suspension clamp, 1/2" CTS		80	.100		.24	6.45		6.69	9.90
8310	3/4" CTS		73	.110		.22	7.05		7.27	10.80
8320	1" CTS		68	.118		.53	7.60		8.13	11.95
8400	Insulator, 1/2" CTS		80	.100		.36	6.45		6.81	10.05

22 05 Common Work Results for Plumbing

22 05 29 – Hangers and Supports for Plumbing Piping and Equipment

22 05 29.10 Hangers & Supp. for Plumb'g/HVAC Pipe/Equip.		Crew	Daily Output	Labor-Hours	Unit	Material	2020 Bare Costs Labor	Equipment	Total	Total Incl O&P
8410	3/4" CTS	1 Plum	73	.110	Ea.	.37	7.05		7.42	10.95
8420	1" CTS		68	.118		.39	7.60		7.99	11.80
8500	J hook clamp with nail, 1/2" CTS		240	.033		.09	2.15		2.24	3.31
8501	3/4" CTS		240	.033		.11	2.15		2.26	3.33
8800	Wire cable support system									
8810	Cable with hook terminal and locking device									
8830	2 mm (.079") diam. cable (100 lb. cap.)									
8840	1 m (3.3') length, with hook	1 Shee	96	.083	Ea.	3.90	5.20		9.10	12.15
8850	2 m (6.6') length, with hook		84	.095		4.53	5.95		10.48	14
8860	3 m (9.9') length, with hook		72	.111		4.47	6.90		11.37	15.40
8870	5 m (16.4') length, with hook	Q-9	60	.267		5.55	14.95		20.50	28.50
8880	10 m (32.8') length, with hook	"	30	.533		8.05	30		38.05	54.50
8900	3 mm (.118") diam. cable (200 lb. cap.)									
8910	1 m (3.3') length, with hook	1 Shee	96	.083	Ea.	9.80	5.20		15	18.65
8920	2 m (6.6') length, with hook		84	.095		10.85	5.95		16.80	21
8930	3 m (9.9') length, with hook		72	.111		11.80	6.90		18.70	23.50
8940	5 m (16.4') length, with hook	Q-9	60	.267		10.40	14.95		25.35	34
8950	10 m (32.8') length, with hook	"	30	.533		13.90	30		43.90	61
9000	Cable system accessories									
9010	Anchor bolt, 3/8", with nut	1 Shee	140	.057	Ea.	1.13	3.56		4.69	6.65
9020	Air duct corner protector		160	.050		1.51	3.12		4.63	6.40
9030	Air duct support attachment		140	.057		1.82	3.56		5.38	7.40
9040	Flange clip, hammer-on style									
9044	For flange thickness 3/32"-9/64", 160 lb. cap.	1 Shee	180	.044	Ea.	.44	2.77		3.21	4.68
9048	For flange thickness 1/8"-1/4", 200 lb. cap.		160	.050		.32	3.12		3.44	5.10
9052	For flange thickness 5/16"-1/2", 200 lb. cap.		150	.053		.65	3.32		3.97	5.75
9056	For flange thickness 9/16"-3/4", 200 lb. cap.		140	.057		.89	3.56		4.45	6.40
9060	Wire insulation protection tube		180	.044	L.F.	.54	2.77		3.31	4.79
9070	Wire cutter				Ea.	44			44	48.50

22 05 33 – Heat Tracing for Plumbing Piping

22 05 33.20 Temperature Maintenance Cable		Crew	Daily Output	Labor-Hours	Unit	Material	Labor	Equipment	Total	Total Incl O&P
0010	TEMPERATURE MAINTENANCE CABLE									
0040	Components									
0080	Heating cable									
0100	208 V									
0150	140°F	Q-1	1060.80	.015	L.F.	10.15	.87		11.02	12.50
0200	120 V									
0220	125°F	Q-1	1060.80	.015	L.F.	9.50	.87		10.37	11.75
0300	Power kit w/1 end seal	1 Elec	48.80	.164	Ea.	136	10.05		146.05	165
0310	Splice kit		35.50	.225		161	13.85		174.85	198
0320	End seal		160	.050		12.75	3.07		15.82	18.55
0330	Tee kit w/1 end seal		26.80	.299		166	18.30		184.30	210
0340	Powered splice w/2 end seals		20	.400		118	24.50		142.50	167
0350	Powered tee kit w/3 end seals		18	.444		152	27.50		179.50	208
0360	Cross kit w/2 end seals		18.60	.430		156	26.50		182.50	212
0500	Recommended thickness of fiberglass insulation									
0510	Pipe size									
0520	1/2" to 1" use 1" insulation									
0530	1-1/4" to 2" use 1-1/2" insulation									
0540	2-1/2" to 6" use 2" insulation									
0560	NOTE: For pipe sizes 1-1/4" and smaller use 1/4" larger diameter									
0570	insulation to allow room for installation over cable.									

22 05 48 – Vibration and Seismic Controls for Plumbing Piping and Equipment

22 05 48.40 Vibration Absorbers		Crew	Daily Output	Labor-Hours	Unit	2020 Bare Costs Material	Labor	Equipment	Total	Total Incl O&P
0010	**VIBRATION ABSORBERS**									
0100	Hangers, neoprene flex									
0200	10-120 lb. capacity				Ea.	24			24	26.50
0220	75-550 lb. capacity					42			42	46
0240	250-1,100 lb. capacity					78			78	86
0260	1,000-4,000 lb. capacity					144			144	158
0500	Spring flex, 60 lb. capacity					35.50			35.50	39.50
0520	450 lb. capacity					64.50			64.50	71
0540	900 lb. capacity					81			81	89
0560	1,100-1,300 lb. capacity					78			78	86
0600	Rubber in shear									
0610	45-340 lb., up to 1/2" rod size	1 Stpi	22	.364	Ea.	23	24		47	61
0620	130-700 lb., up to 3/4" rod size		20	.400		50	26		76	94
0630	50-1,000 lb., up to 3/4" rod size		18	.444		66	29		95	116
1000	Mounts, neoprene, 45-380 lb. capacity					17.35			17.35	19.10
1020	250-1,100 lb. capacity					63			63	69
1040	1,000-4,000 lb. capacity					116			116	127
1100	Spring flex, 60 lb. capacity					79			79	86.50
1120	165 lb. capacity					82			82	90
1140	260 lb. capacity					80.50			80.50	89
1160	450 lb. capacity					115			115	127
1180	600 lb. capacity					100			100	110
1200	750 lb. capacity					157			157	173
1220	900 lb. capacity					152			152	167
1240	1,100 lb. capacity					175			175	192
1260	1,300 lb. capacity					168			168	184
1280	1,500 lb. capacity					178			178	195
1300	1,800 lb. capacity					238			238	262
1320	2,200 lb. capacity					299			299	330
1340	2,600 lb. capacity					330			330	365
1399	Spring type									
1400	2 piece									
1410	50-1,000 lb.	1 Stpi	12	.667	Ea.	210	43.50		253.50	297
1420	1,100-1,600 lb.	"	12	.667	"	226	43.50		269.50	315
1500	Double spring open									
1510	150-450 lb.	1 Stpi	24	.333	Ea.	144	22		166	191
1520	500-1,000 lb.		24	.333		144	22		166	191
1530	1,100-1,600 lb.		24	.333		144	22		166	191
1540	1,700-2,400 lb.		24	.333		156	22		178	205
1550	2,500-3,400 lb.		24	.333		266	22		288	325
2000	Pads, cork rib, 18" x 18" x 1", 10-50 psi					196			196	215
2020	18" x 36" x 1", 10-50 psi					390			390	430
2100	Shear flexible pads, 18" x 18" x 3/8", 20-70 psi					78.50			78.50	86.50
2120	18" x 36" x 3/8", 20-70 psi					195			195	215
2150	Laminated neoprene and cork									
2160	1" thick	1 Stpi	16	.500	S.F.	87.50	33		120.50	145
2200	Neoprene elastomer isolation bearing pad									
2230	464 psi, 5/8" x 19-11/16" x 39-3/8"	1 Stpi	16	.500	Ea.	345	33		378	430

22 05 48 – Vibration and Seismic Controls for Plumbing Piping and Equipment

22 05 48.40 Vibration Absorbers		Crew	Daily Output	Labor-Hours	Unit	Material	2020 Bare Costs Labor	Equipment	Total	Total Incl O&P
3000	Note overlap in capacities due to deflections									

22 05 53 – Identification for Plumbing Piping and Equipment

22 05 53.10 Piping System Identification Labels

		Crew	Daily Output	Labor-Hours	Unit	Material	Labor	Equipment	Total	Total Incl O&P
0010	**PIPING SYSTEM IDENTIFICATION LABELS**									
0100	Indicate contents and flow direction									
0106	Pipe markers									
0110	Plastic snap around									
0114	1/2" pipe	1 Plum	80	.100	Ea.	6.90	6.45		13.35	17.25
0116	3/4" pipe		80	.100		7.25	6.45		13.70	17.60
0118	1" pipe		80	.100		7.25	6.45		13.70	17.60
0120	2" pipe		75	.107		8.60	6.85		15.45	19.75
0122	3" pipe		70	.114		11.25	7.35		18.60	23.50
0124	4" pipe		60	.133		11.55	8.60		20.15	25.50
0126	6" pipe		60	.133		12.15	8.60		20.75	26.50
0128	8" pipe		56	.143		17.10	9.20		26.30	32.50
0130	10" pipe		56	.143		18.80	9.20		28	34.50
0200	Over 10" pipe size		50	.160		22	10.30		32.30	39.50
1110	Self adhesive									
1114	1" pipe	1 Plum	80	.100	Ea.	3.29	6.45		9.74	13.25
1116	2" pipe		75	.107		3.46	6.85		10.31	14.10
1118	3" pipe		70	.114		5.30	7.35		12.65	16.85
1120	4" pipe		60	.133		5.30	8.60		13.90	18.65
1122	6" pipe		60	.133		5.30	8.60		13.90	18.70
1124	8" pipe		56	.143		7.60	9.20		16.80	22
1126	10" pipe		56	.143		7.60	9.20		16.80	22
1200	Over 10" pipe size		50	.160		10.65	10.30		20.95	27
2000	Valve tags									
2010	Numbered plus identifying legend									
2100	Brass, 2" diameter	1 Plum	40	.200	Ea.	3.16	12.90		16.06	23
2200	Plastic, 1-1/2" diameter	"	40	.200	"	2.99	12.90		15.89	22.50

22 05 76 – Facility Drainage Piping Cleanouts

22 05 76.10 Cleanouts

		Crew	Daily Output	Labor-Hours	Unit	Material	Labor	Equipment	Total	Total Incl O&P
0010	**CLEANOUTS**									
0060	Floor type									
0080	Round or square, scoriated nickel bronze top									
0100	2" pipe size	1 Plum	10	.800	Ea.	183	51.50		234.50	279
0120	3" pipe size		8	1		264	64.50		328.50	385
0140	4" pipe size		6	1.333		284	86		370	440
0160	5" pipe size		4	2		485	129		614	730
0180	6" pipe size	Q-1	6	2.667		530	155		685	810
0200	8" pipe size	"	4	4		845	232		1,077	1,275
0340	Recessed for tile, same price									
0980	Round top, recessed for terrazzo									
1000	2" pipe size	1 Plum	9	.889	Ea.	545	57.50		602.50	685
1080	3" pipe size		6	1.333		620	86		706	810
1100	4" pipe size		4	2		710	129		839	975
1120	5" pipe size	Q-1	6	2.667		1,175	155		1,330	1,525
1140	6" pipe size		5	3.200		1,175	186		1,361	1,575
1160	8" pipe size		4	4		1,100	232		1,332	1,575
2000	Round scoriated nickel bronze top, extra heavy duty									
2060	2" pipe size	1 Plum	9	.889	Ea.	335	57.50		392.50	455
2080	3" pipe size		6	1.333		355	86		441	520

22 05 76.10 Cleanouts		Crew	Daily Output	Labor-Hours	Unit	2020 Bare Costs Material	Labor	Equipment	Total	Total Incl O&P
2100	4" pipe size	1 Plum	4	2	Ea.	500	129		629	745
2120	5" pipe size	Q-1	6	2.667		730	155		885	1,025
2140	6" pipe size		5	3.200		730	186		916	1,075
2160	8" pipe size		4	4		905	232		1,137	1,350
4000	Wall type, square smooth cover, over wall frame									
4060	2" pipe size	1 Plum	14	.571	Ea.	355	37		392	445
4080	3" pipe size		12	.667		345	43		388	445
4100	4" pipe size		10	.800		360	51.50		411.50	475
4120	5" pipe size		9	.889		585	57.50		642.50	725
4140	6" pipe size		8	1		655	64.50		719.50	820
4160	8" pipe size	Q-1	11	1.455		885	84.50		969.50	1,100
5000	Extension, CI; bronze countersunk plug, 8" long									
5040	2" pipe size	1 Plum	16	.500	Ea.	200	32		232	268
5060	3" pipe size		14	.571		273	37		310	355
5080	4" pipe size		13	.615		234	39.50		273.50	320
5100	5" pipe size		12	.667		380	43		423	480
5120	6" pipe size		11	.727		710	47		757	850

22 05 76.20 Cleanout Tees		Crew	Daily Output	Labor-Hours	Unit	Material	Labor	Equipment	Total	Total Incl O&P
0010	**CLEANOUT TEES**									
0100	Cast iron, B&S, with countersunk plug									
0200	2" pipe size	1 Plum	4	2	Ea.	143	129		272	350
0220	3" pipe size		3.60	2.222		181	143		324	415
0240	4" pipe size		3.30	2.424		275	156		431	535
0260	5" pipe size	Q-1	5.50	2.909		545	169		714	855
0280	6" pipe size	"	5	3.200		790	186		976	1,150
0300	8" pipe size	Q-3	5	6.400		1,100	395		1,495	1,825
0500	For round smooth access cover, same price									
0600	For round scoriated access cover, same price									
0700	For square smooth access cover, add				Ea.	60%				
2000	Cast iron, no hub									
2010	Cleanout tee, with 2 couplings									
2012	2"	Q-1	22	.727	Ea.	48.50	42		90.50	117
2014	3"		19	.842		61	49		110	140
2016	4"		16.50	.970		83.50	56		139.50	176
2018	6"	Q-2	20	1.200		201	72		273	330
2040	Cleanout plug, no hub, with 1 coupling									
2042	2"	Q-1	44	.364	Ea.	11.85	21		32.85	44.50
2046	3"		38	.421		22	24.50		46.50	61
2048	4"		33	.485		36	28		64	82
2050	6"	Q-2	40	.600		130	36		166	197
2052	8"	"	33	.727		182	44		226	266
4000	Plastic, tees and adapters. Add plugs									
4010	ABS, DWV									
4020	Cleanout tee, 1-1/2" pipe size	1 Plum	15	.533	Ea.	14.15	34.50		48.65	67
4030	2" pipe size	Q-1	27	.593		12.65	34.50		47.15	65.50
4040	3" pipe size		21	.762		25.50	44		69.50	94
4050	4" pipe size		16	1		64.50	58		122.50	158
4100	Cleanout plug, 1-1/2" pipe size	1 Plum	32	.250		2.72	16.10		18.82	27
4110	2" pipe size	Q-1	56	.286		3.28	16.55		19.83	28.50
4120	3" pipe size		36	.444		5.35	26		31.35	44.50
4130	4" pipe size		30	.533		9.15	31		40.15	56.50
4180	Cleanout adapter fitting, 1-1/2" pipe size	1 Plum	32	.250		4.18	16.10		20.28	28.50

22 05 Common Work Results for Plumbing

22 05 76 – Facility Drainage Piping Cleanouts

22 05 76.20 Cleanout Tees		Crew	Daily Output	Labor-Hours	Unit	Material	Labor	Equipment	Total	Total Incl O&P
						2020 Bare Costs				
4190	2" pipe size	Q-1	56	.286	Ea.	6.10	16.55		22.65	31.50
4200	3" pipe size		36	.444		15.60	26		41.60	55.50
4210	4" pipe size		30	.533		29	31		60	78.50
5000	PVC, DWV									
5010	Cleanout tee, 1-1/2" pipe size	1 Plum	15	.533	Ea.	11.90	34.50		46.40	64.50
5020	2" pipe size	Q-1	27	.593		9.15	34.50		43.65	61.50
5030	3" pipe size		21	.762		27	44		71	96
5040	4" pipe size		16	1		36.50	58		94.50	128
5090	Cleanout plug, 1-1/2" pipe size	1 Plum	32	.250		1.87	16.10		17.97	26
5100	2" pipe size	Q-1	56	.286		2.09	16.55		18.64	27.50
5110	3" pipe size		36	.444		3.73	26		29.73	42.50
5120	4" pipe size		30	.533		5.50	31		36.50	52.50
5130	6" pipe size		24	.667		17.75	38.50		56.25	77.50
5170	Cleanout adapter fitting, 1-1/2" pipe size	1 Plum	32	.250		2.51	16.10		18.61	27
5180	2" pipe size	Q-1	56	.286		3.08	16.55		19.63	28.50
5190	3" pipe size		36	.444		8.45	26		34.45	48
5200	4" pipe size		30	.533		13.90	31		44.90	62
5210	6" pipe size		24	.667		40	38.50		78.50	102
5300	Cleanout tee, with plug									
5310	4"	Q-1	14	1.143	Ea.	36.50	66.50		103	139
5320	6"	"	8	2		85	116		201	268
5330	8"	Q-2	11	2.182		129	131		260	340
5340	12"	"	10	2.400		650	144		794	930
5400	Polypropylene, Schedule 40									
5410	Cleanout tee with plug									
5420	1-1/2"	1 Plum	10	.800	Ea.	35.50	51.50		87	116
5430	2"	Q-1	17	.941		36.50	54.50		91	122
5440	3"		11	1.455		78.50	84.50		163	213
5450	4"		9	1.778		107	103		210	271

22 07 Plumbing Insulation

22 07 16 – Plumbing Equipment Insulation

22 07 16.10 Insulation for Plumbing Equipment		Crew	Daily Output	Labor-Hours	Unit	Material	Labor	Equipment	Total	Total Incl O&P
0010	**INSULATION FOR PLUMBING EQUIPMENT**									
2900	Domestic water heater wrap kit									
2920	1-1/2" with vinyl jacket, 20 to 60 gal. G	1 Plum	8	1	Ea.	16.70	64.50		81.20	115
2925	50 to 80 gal. G	"	8	1	"	28	64.50		92.50	128

22 07 19 – Plumbing Piping Insulation

22 07 19.10 Piping Insulation		Crew	Daily Output	Labor-Hours	Unit	Material	Labor	Equipment	Total	Total Incl O&P
0010	**PIPING INSULATION**									
0110	Insulation req'd. is based on the surface size/area to be covered									
0230	Insulated protectors (ADA)									
0235	For exposed piping under sinks or lavatories									
0240	Vinyl coated foam, velcro tabs									
0245	P Trap, 1-1/4" or 1-1/2"	1 Plum	32	.250	Ea.	16.95	16.10		33.05	42.50
0260	Valve and supply cover									
0265	1/2", 3/8", and 7/16" pipe size	1 Plum	32	.250	Ea.	16.40	16.10		32.50	42
0280	Tailpiece offset (wheelchair)									
0285	1-1/4" pipe size	1 Plum	32	.250	Ea.	13.15	16.10		29.25	38.50
0600	Pipe covering (price copper tube one size less than IPS)									

22 07 19 – Plumbing Piping Insulation

22 07 19.10 Piping Insulation			Crew	Daily Output	Labor-Hours	Unit	2020 Bare Costs Material	Labor	Equipment	Total	Total Incl O&P
1000	Mineral wool										
1010	Preformed, 1200°F, plain										
1014	1" wall										
1016	1/2" iron pipe size	G	Q-14	230	.070	L.F.	1.92	3.67		5.59	7.75
1018	3/4" iron pipe size	G		220	.073		1.95	3.84		5.79	8.05
1022	1" iron pipe size	G		210	.076		2.05	4.02		6.07	8.40
1024	1-1/4" iron pipe size	G		205	.078		2.11	4.12		6.23	8.60
1026	1-1/2" iron pipe size	G		205	.078		1.99	4.12		6.11	8.50
1028	2" iron pipe size	G		200	.080		2.69	4.22		6.91	9.45
1030	2-1/2" iron pipe size	G		190	.084		2.76	4.44		7.20	9.85
1032	3" iron pipe size	G		180	.089		2.96	4.69		7.65	10.45
1034	4" iron pipe size	G		150	.107		3.64	5.65		9.29	12.65
1036	5" iron pipe size	G		140	.114		4.12	6.05		10.17	13.80
1038	6" iron pipe size	G		120	.133		4.40	7.05		11.45	15.65
1040	7" iron pipe size	G		110	.145		5.15	7.70		12.85	17.45
1042	8" iron pipe size	G		100	.160		7.10	8.45		15.55	21
1044	9" iron pipe size	G		90	.178		7.45	9.40		16.85	22.50
1046	10" iron pipe size	G	↓	90	.178	↓	8.45	9.40		17.85	23.50
1050	1-1/2" wall										
1052	1/2" iron pipe size	G	Q-14	225	.071	L.F.	3.27	3.75		7.02	9.35
1054	3/4" iron pipe size	G		215	.074		3.36	3.93		7.29	9.75
1056	1" iron pipe size	G		205	.078		3.40	4.12		7.52	10.05
1058	1-1/4" iron pipe size	G		200	.080		3.47	4.22		7.69	10.30
1060	1-1/2" iron pipe size	G		200	.080		3.64	4.22		7.86	10.50
1062	2" iron pipe size	G		190	.084		3.86	4.44		8.30	11.05
1064	2-1/2" iron pipe size	G		180	.089		4.20	4.69		8.89	11.80
1066	3" iron pipe size	G		165	.097		4.38	5.10		9.48	12.65
1068	4" iron pipe size	G		140	.114		5.25	6.05		11.30	15.05
1070	5" iron pipe size	G		130	.123		5.60	6.50		12.10	16.10
1072	6" iron pipe size	G		110	.145		5.80	7.70		13.50	18.15
1074	7" iron pipe size	G		100	.160		6.85	8.45		15.30	20.50
1076	8" iron pipe size	G		90	.178		7.60	9.40		17	23
1078	9" iron pipe size	G		85	.188		8.75	9.95		18.70	25
1080	10" iron pipe size	G		80	.200		9.60	10.55		20.15	27
1082	12" iron pipe size	G		75	.213		10.95	11.25		22.20	29.50
1084	14" iron pipe size	G		70	.229		12.55	12.05		24.60	32.50
1086	16" iron pipe size	G		65	.246		13.80	13		26.80	35
1088	18" iron pipe size	G		60	.267		16.05	14.05		30.10	39
1090	20" iron pipe size	G		55	.291		17.15	15.35		32.50	42.50
1092	22" iron pipe size	G		50	.320		19.60	16.90		36.50	47.50
1094	24" iron pipe size	G	↓	45	.356	↓	21	18.75		39.75	52
1100	2" wall										
1102	1/2" iron pipe size	G	Q-14	220	.073	L.F.	4.23	3.84		8.07	10.55
1104	3/4" iron pipe size	G		210	.076		4.39	4.02		8.41	11
1106	1" iron pipe size	G		200	.080		4.65	4.22		8.87	11.60
1108	1-1/4" iron pipe size	G		190	.084		4.99	4.44		9.43	12.30
1110	1-1/2" iron pipe size	G		190	.084		5.20	4.44		9.64	12.50
1112	2" iron pipe size	G		180	.089		5.45	4.69		10.14	13.20
1114	2-1/2" iron pipe size	G		170	.094		6.15	4.97		11.12	14.45
1116	3" iron pipe size	G		160	.100		6.60	5.30		11.90	15.35
1118	4" iron pipe size	G		130	.123		7.70	6.50		14.20	18.40
1120	5" iron pipe size	G		120	.133		8.55	7.05		15.60	20
1122	6" iron pipe size	G	↓	100	.160	↓	8.85	8.45		17.30	22.50

22 07 19.10 Piping Insulation			Crew	Daily Output	Labor-Hours	Unit	Material	2020 Bare Costs Labor	Equipment	Total	Total Incl O&P
1124	8" iron pipe size	G	Q-14	80	.200	L.F.	10.75	10.55		21.30	28
1126	10" iron pipe size	G		70	.229		13.15	12.05		25.20	33
1128	12" iron pipe size	G		65	.246		14.60	13		27.60	36
1130	14" iron pipe size	G		60	.267		16.30	14.05		30.35	39.50
1132	16" iron pipe size	G		55	.291		18.45	15.35		33.80	44
1134	18" iron pipe size	G		50	.320		19.95	16.90		36.85	48
1136	20" iron pipe size	G		45	.356		23	18.75		41.75	54.50
1138	22" iron pipe size	G		45	.356		25	18.75		43.75	56.50
1140	24" iron pipe size	G	↓	40	.400	↓	26.50	21		47.50	61.50
1150	4" wall										
1152	1-1/4" iron pipe size	G	Q-14	170	.094	L.F.	13.45	4.97		18.42	22.50
1154	1-1/2" iron pipe size	G		165	.097		14	5.10		19.10	23.50
1156	2" iron pipe size	G		155	.103		14.25	5.45		19.70	24
1158	4" iron pipe size	G		105	.152		18	8.05		26.05	32
1160	6" iron pipe size	G		75	.213		21	11.25		32.25	41
1162	8" iron pipe size	G		60	.267		24	14.05		38.05	48
1164	10" iron pipe size	G		50	.320		29	16.90		45.90	58
1166	12" iron pipe size	G		45	.356		32	18.75		50.75	64
1168	14" iron pipe size	G		40	.400		34	21		55	70
1170	16" iron pipe size	G		35	.457		38.50	24		62.50	79.50
1172	18" iron pipe size	G		32	.500		41.50	26.50		68	86
1174	20" iron pipe size	G		30	.533		44.50	28		72.50	92
1176	22" iron pipe size	G		28	.571		50	30		80	102
1178	24" iron pipe size	G	↓	26	.615	↓	53	32.50		85.50	109
4280	Cellular glass, closed cell foam, all service jacket, sealant,										
4281	working temp. (-450°F to +900°F), 0 water vapor transmission										
4284	1" wall										
4286	1/2" iron pipe size	G	Q-14	120	.133	L.F.	9.40	7.05		16.45	21
4300	1-1/2" wall										
4301	1" iron pipe size	G	Q-14	105	.152	L.F.	10.40	8.05		18.45	24
4304	2-1/2" iron pipe size	G		90	.178		13.70	9.40		23.10	29.50
4306	3" iron pipe size	G		85	.188		17.50	9.95		27.45	34.50
4308	4" iron pipe size	G		70	.229		24.50	12.05		36.55	45.50
4310	5" iron pipe size	G	↓	65	.246	↓	25.50	13		38.50	48
4320	2" wall										
4322	1" iron pipe size	G	Q-14	100	.160	L.F.	16.45	8.45		24.90	31
4324	2-1/2" iron pipe size	G		85	.188		21	9.95		30.95	39
4326	3" iron pipe size	G		80	.200		21	10.55		31.55	39.50
4328	4" iron pipe size	G		65	.246		29	13		42	52
4330	5" iron pipe size	G		60	.267		31.50	14.05		45.55	56
4332	6" iron pipe size	G		50	.320		36.50	16.90		53.40	66
4336	8" iron pipe size	G		40	.400		42	21		63	78.50
4338	10" iron pipe size	G	↓	35	.457	↓	45	24		69	86.50
4350	2-1/2" wall										
4360	12" iron pipe size	G	Q-14	32	.500	L.F.	75	26.50		101.50	123
4362	14" iron pipe size	G	"	28	.571	"	82.50	30		112.50	138
4370	3" wall										
4378	6" iron pipe size	G	Q-14	48	.333	L.F.	55	17.60		72.60	87.50
4380	8" iron pipe size	G		38	.421		60	22		82	100
4382	10" iron pipe size	G		33	.485		69	25.50		94.50	115
4384	16" iron pipe size	G		25	.640		98.50	34		132.50	161
4386	18" iron pipe size	G		22	.727		127	38.50		165.50	199
4388	20" iron pipe size	G	↓	20	.800	↓	135	42		177	214

22 07 19.10 Piping Insulation			Crew	Daily Output	Labor-Hours	Unit	2020 Bare Costs Material	Labor	Equipment	Total	Total Incl O&P
4400	3-1/2" wall										
4412	12" iron pipe size	G	Q-14	27	.593	L.F.	79.50	31.50		111	136
4414	14" iron pipe size	G	"	25	.640	"	96.50	34		130.50	158
4430	4" wall										
4446	16" iron pipe size	G	Q-14	22	.727	L.F.	92.50	38.50		131	161
4448	18" iron pipe size	G		20	.800		118	42		160	195
4450	20" iron pipe size	G		18	.889		135	47		182	220
4480	Fittings, average with fabric and mastic										
4484	1" wall										
4486	1/2" iron pipe size	G	1 Asbe	40	.200	Ea.	7.40	11.75		19.15	26
4500	1-1/2" wall										
4502	1" iron pipe size	G	1 Asbe	38	.211	Ea.	8.15	12.35		20.50	28
4504	2-1/2" iron pipe size	G		32	.250		18.40	14.65		33.05	42.50
4506	3" iron pipe size	G		30	.267		18.20	15.65		33.85	44
4508	4" iron pipe size	G		28	.286		19.30	16.75		36.05	46.50
4510	5" iron pipe size	G		24	.333		25	19.55		44.55	57.50
4520	2" wall										
4522	1" iron pipe size	G	1 Asbe	36	.222	Ea.	13.10	13.05		26.15	34.50
4524	2-1/2" iron pipe size	G		30	.267		20	15.65		35.65	46
4526	3" iron pipe size	G		28	.286		22.50	16.75		39.25	50.50
4528	4" iron pipe size	G		24	.333		25.50	19.55		45.05	58
4530	5" iron pipe size	G		22	.364		42.50	21.50		64	80
4532	6" iron pipe size	G		20	.400		51.50	23.50		75	93
4536	8" iron pipe size	G		12	.667		89	39		128	158
4538	10" iron pipe size	G		8	1		102	58.50		160.50	202
4550	2-1/2" wall										
4560	12" iron pipe size	G	1 Asbe	6	1.333	Ea.	189	78		267	330
4562	14" iron pipe size	G	"	4	2	"	207	117		324	410
4570	3" wall										
4578	6" iron pipe size	G	1 Asbe	16	.500	Ea.	63.50	29.50		93	115
4580	8" iron pipe size	G		10	.800		94.50	47		141.50	176
4582	10" iron pipe size	G		6	1.333		128	78		206	261
4900	Calcium silicate, with 8 oz. canvas cover										
5100	1" wall, 1/2" iron pipe size	G	Q-14	170	.094	L.F.	4.29	4.97		9.26	12.35
5130	3/4" iron pipe size	G		170	.094		4.32	4.97		9.29	12.40
5140	1" iron pipe size	G		170	.094		4.23	4.97		9.20	12.30
5150	1-1/4" iron pipe size	G		165	.097		4.29	5.10		9.39	12.55
5160	1-1/2" iron pipe size	G		165	.097		4.34	5.10		9.44	12.60
5170	2" iron pipe size	G		160	.100		5	5.30		10.30	13.60
5180	2-1/2" iron pipe size	G		160	.100		5.30	5.30		10.60	13.95
5190	3" iron pipe size	G		150	.107		5.80	5.65		11.45	15
5200	4" iron pipe size	G		140	.114		7.15	6.05		13.20	17.10
5210	5" iron pipe size	G		135	.119		7.55	6.25		13.80	17.90
5220	6" iron pipe size	G		130	.123		8	6.50		14.50	18.75
5280	1-1/2" wall, 1/2" iron pipe size	G		150	.107		4.65	5.65		10.30	13.75
5310	3/4" iron pipe size	G		150	.107		4.74	5.65		10.39	13.85
5320	1" iron pipe size	G		150	.107		5.15	5.65		10.80	14.30
5330	1-1/4" iron pipe size	G		145	.110		5.55	5.80		11.35	15.05
5340	1-1/2" iron pipe size	G		145	.110		5.95	5.80		11.75	15.50
5350	2" iron pipe size	G		140	.114		6.55	6.05		12.60	16.45
5360	2-1/2" iron pipe size	G		140	.114		7.15	6.05		13.20	17.10
5370	3" iron pipe size	G		135	.119		7.50	6.25		13.75	17.85
5380	4" iron pipe size	G		125	.128		8.70	6.75		15.45	19.90

22 07 19.10 Piping Insulation			Crew	Daily Output	Labor-Hours	Unit	2020 Bare Costs Material	Labor	Equipment	Total	Total Incl O&P
5390	5" iron pipe size	G	Q-14	120	.133	L.F.	9.80	7.05		16.85	21.50
5400	6" iron pipe size	G		110	.145		10.10	7.70		17.80	23
5460	2" wall, 1/2" iron pipe size	G		135	.119		7.20	6.25		13.45	17.50
5490	3/4" iron pipe size	G		135	.119		7.55	6.25		13.80	17.90
5500	1" iron pipe size	G		135	.119		7.95	6.25		14.20	18.35
5510	1-1/4" iron pipe size	G		130	.123		8.50	6.50		15	19.30
5520	1-1/2" iron pipe size	G		130	.123		8.90	6.50		15.40	19.70
5530	2" iron pipe size	G		125	.128		9.40	6.75		16.15	20.50
5540	2-1/2" iron pipe size	G		125	.128		11.10	6.75		17.85	22.50
5550	3" iron pipe size	G		120	.133		11.20	7.05		18.25	23
5560	4" iron pipe size	G		115	.139		12.95	7.35		20.30	25.50
5570	5" iron pipe size	G		110	.145		14.75	7.70		22.45	28
5580	6" iron pipe size	G	↓	105	.152	↓	16.15	8.05		24.20	30
5600	Calcium silicate, no cover										
5720	1" wall, 1/2" iron pipe size	G	Q-14	180	.089	L.F.	3.91	4.69		8.60	11.50
5740	3/4" iron pipe size	G		180	.089		3.91	4.69		8.60	11.50
5750	1" iron pipe size	G		180	.089		3.74	4.69		8.43	11.30
5760	1-1/4" iron pipe size	G		175	.091		3.82	4.83		8.65	11.60
5770	1-1/2" iron pipe size	G		175	.091		3.85	4.83		8.68	11.65
5780	2" iron pipe size	G		170	.094		4.42	4.97		9.39	12.50
5790	2-1/2" iron pipe size	G		170	.094		4.69	4.97		9.66	12.80
5800	3" iron pipe size	G		160	.100		5.05	5.30		10.35	13.65
5810	4" iron pipe size	G		150	.107		6.30	5.65		11.95	15.55
5820	5" iron pipe size	G		145	.110		6.55	5.80		12.35	16.15
5830	6" iron pipe size	G		140	.114		6.90	6.05		12.95	16.80
5900	1-1/2" wall, 1/2" iron pipe size	G		160	.100		4.14	5.30		9.44	12.65
5920	3/4" iron pipe size	G		160	.100		4.20	5.30		9.50	12.70
5930	1" iron pipe size	G		160	.100		4.57	5.30		9.87	13.15
5940	1-1/4" iron pipe size	G		155	.103		4.94	5.45		10.39	13.80
5950	1-1/2" iron pipe size	G		155	.103		5.30	5.45		10.75	14.20
5960	2" iron pipe size	G		150	.107		5.85	5.65		11.50	15.10
5970	2-1/2" iron pipe size	G		150	.107		6.40	5.65		12.05	15.65
5980	3" iron pipe size	G		145	.110		6.65	5.80		12.45	16.30
5990	4" iron pipe size	G		135	.119		7.70	6.25		13.95	18.10
6000	5" iron pipe size	G		130	.123		8.70	6.50		15.20	19.50
6010	6" iron pipe size	G		120	.133		8.90	7.05		15.95	20.50
6020	7" iron pipe size	G		115	.139		10.45	7.35		17.80	22.50
6030	8" iron pipe size	G		105	.152		11.75	8.05		19.80	25.50
6040	9" iron pipe size	G		100	.160		14.10	8.45		22.55	28.50
6050	10" iron pipe size	G		95	.168		15.55	8.90		24.45	31
6060	12" iron pipe size	G		90	.178		18.50	9.40		27.90	35
6070	14" iron pipe size	G		85	.188		21	9.95		30.95	38.50
6080	16" iron pipe size	G		80	.200		23.50	10.55		34.05	41.50
6090	18" iron pipe size	G		75	.213		26	11.25		37.25	46
6120	2" wall, 1/2" iron pipe size	G		145	.110		6.55	5.80		12.35	16.15
6140	3/4" iron pipe size	G		145	.110		6.90	5.80		12.70	16.50
6150	1" iron pipe size	G		145	.110		7.25	5.80		13.05	16.95
6160	1-1/4" iron pipe size	G		140	.114		7.75	6.05		13.80	17.80
6170	1-1/2" iron pipe size	G		140	.114		8.10	6.05		14.15	18.15
6180	2" iron pipe size	G		135	.119		8.50	6.25		14.75	18.95
6190	2-1/2" iron pipe size	G		135	.119		10.25	6.25		16.50	21
6200	3" iron pipe size	G		130	.123		10.30	6.50		16.80	21.50
6210	4" iron pipe size	G	↓	125	.128	↓	11.90	6.75		18.65	23.50

22 07 19.10 Piping Insulation			Crew	Daily Output	Labor-Hours	Unit	2020 Bare Costs Material	Labor	Equipment	Total	Total Incl O&P
6220	5" iron pipe size	G	Q-14	120	.133	L.F.	13.55	7.05		20.60	26
6230	6" iron pipe size	G		115	.139		14.80	7.35		22.15	27.50
6240	7" iron pipe size	G		110	.145		16.05	7.70		23.75	29.50
6250	8" iron pipe size	G		105	.152		18.05	8.05		26.10	32.50
6260	9" iron pipe size	G		100	.160		20	8.45		28.45	35
6270	10" iron pipe size	G		95	.168		22	8.90		30.90	38
6280	12" iron pipe size	G		90	.178		24.50	9.40		33.90	41.50
6290	14" iron pipe size	G		85	.188		27	9.95		36.95	45.50
6300	16" iron pipe size	G		80	.200		30	10.55		40.55	49
6310	18" iron pipe size	G		75	.213		32.50	11.25		43.75	53.50
6320	20" iron pipe size	G		65	.246		40	13		53	64
6330	22" iron pipe size	G		60	.267		44.50	14.05		58.55	70.50
6340	24" iron pipe size	G		55	.291		45.50	15.35		60.85	73.50
6360	3" wall, 1/2" iron pipe size	G		115	.139		11.85	7.35		19.20	24.50
6380	3/4" iron pipe size	G		115	.139		11.90	7.35		19.25	24.50
6390	1" iron pipe size	G		115	.139		12	7.35		19.35	24.50
6400	1-1/4" iron pipe size	G		110	.145		12.20	7.70		19.90	25
6410	1-1/2" iron pipe size	G		110	.145		12.20	7.70		19.90	25
6420	2" iron pipe size	G		105	.152		12.65	8.05		20.70	26.50
6430	2-1/2" iron pipe size	G		105	.152		14.90	8.05		22.95	28.50
6440	3" iron pipe size	G		100	.160		15	8.45		23.45	29.50
6450	4" iron pipe size	G		95	.168		19.15	8.90		28.05	34.50
6460	5" iron pipe size	G		90	.178		21	9.40		30.40	37.50
6470	6" iron pipe size	G		90	.178		23.50	9.40		32.90	40
6480	7" iron pipe size	G		85	.188		26	9.95		35.95	44
6490	8" iron pipe size	G		85	.188		28	9.95		37.95	46
6500	9" iron pipe size	G		80	.200		31.50	10.55		42.05	50.50
6510	10" iron pipe size	G		75	.213		33.50	11.25		44.75	54.50
6520	12" iron pipe size	G		70	.229		37	12.05		49.05	59.50
6530	14" iron pipe size	G		65	.246		41.50	13		54.50	66
6540	16" iron pipe size	G		60	.267		46	14.05		60.05	72
6550	18" iron pipe size	G		55	.291		50.50	15.35		65.85	79
6560	20" iron pipe size	G		50	.320		60	16.90		76.90	92
6570	22" iron pipe size	G		45	.356		65	18.75		83.75	101
6580	24" iron pipe size	G	▼	40	.400	▼	70	21		91	110
6600	Fiberglass, with all service jacket										
6640	1/2" wall, 1/2" iron pipe size	G	Q-14	250	.064	L.F.	.75	3.38		4.13	6.05
6660	3/4" iron pipe size	G		240	.067		.86	3.52		4.38	6.35
6670	1" iron pipe size	G		230	.070		.89	3.67		4.56	6.65
6680	1-1/4" iron pipe size	G		220	.073		.95	3.84		4.79	6.95
6690	1-1/2" iron pipe size	G		220	.073		1.08	3.84		4.92	7.10
6700	2" iron pipe size	G		210	.076		1.16	4.02		5.18	7.45
6710	2-1/2" iron pipe size	G		200	.080		1.21	4.22		5.43	7.85
6840	1" wall, 1/2" iron pipe size	G		240	.067		.92	3.52		4.44	6.40
6860	3/4" iron pipe size	G		230	.070		.98	3.67		4.65	6.75
6870	1" iron pipe size	G		220	.073		1.08	3.84		4.92	7.10
6880	1-1/4" iron pipe size	G		210	.076		1.16	4.02		5.18	7.45
6890	1-1/2" iron pipe size	G		210	.076		1.25	4.02		5.27	7.55
6900	2" iron pipe size	G		200	.080		1.78	4.22		6	8.45
6910	2-1/2" iron pipe size	G		190	.084		1.79	4.44		6.23	8.75
6920	3" iron pipe size	G		180	.089		1.94	4.69		6.63	9.35
6930	3-1/2" iron pipe size	G		170	.094		2.32	4.97		7.29	10.20
6940	4" iron pipe size	G	▼	150	.107	▼	2.58	5.65		8.23	11.50

22 07 19.10 Piping Insulation			Crew	Daily Output	Labor-Hours	Unit	2020 Bare Costs Material	Labor	Equipment	Total	Total Incl O&P
6950	5" iron pipe size	G	Q-14	140	.114	L.F.	2.84	6.05		8.89	12.35
6960	6" iron pipe size	G		120	.133		3.07	7.05		10.12	14.20
6970	7" iron pipe size	G		110	.145		3.65	7.70		11.35	15.80
6980	8" iron pipe size	G		100	.160		5	8.45		13.45	18.45
6990	9" iron pipe size	G		90	.178		5.30	9.40		14.70	20
7000	10" iron pipe size	G		90	.178		5.35	9.40		14.75	20.50
7010	12" iron pipe size	G		80	.200		5.85	10.55		16.40	22.50
7020	14" iron pipe size	G		80	.200		7.05	10.55		17.60	24
7030	16" iron pipe size	G		70	.229		9	12.05		21.05	28.50
7040	18" iron pipe size	G		70	.229		10	12.05		22.05	29.50
7050	20" iron pipe size	G		60	.267		11.15	14.05		25.20	34
7060	24" iron pipe size	G		60	.267		13.60	14.05		27.65	36.50
7080	1-1/2" wall, 1/2" iron pipe size	G		230	.070		2.06	3.67		5.73	7.90
7100	3/4" iron pipe size	G		220	.073		2.06	3.84		5.90	8.15
7110	1" iron pipe size	G		210	.076		2.21	4.02		6.23	8.60
7120	1-1/4" iron pipe size	G		200	.080		2.40	4.22		6.62	9.15
7130	1-1/2" iron pipe size	G		200	.080		2.81	4.22		7.03	9.60
7140	2" iron pipe size	G		190	.084		2.79	4.44		7.23	9.85
7150	2-1/2" iron pipe size	G		180	.089		2.98	4.69		7.67	10.50
7160	3" iron pipe size	G		170	.094		3.12	4.97		8.09	11.10
7170	3-1/2" iron pipe size	G		160	.100		3.42	5.30		8.72	11.85
7180	4" iron pipe size	G		140	.114		3.95	6.05		10	13.60
7190	5" iron pipe size	G		130	.123		3.97	6.50		10.47	14.30
7200	6" iron pipe size	G		110	.145		4.20	7.70		11.90	16.40
7210	7" iron pipe size	G		100	.160		4.66	8.45		13.11	18.10
7220	8" iron pipe size	G		90	.178		5.90	9.40		15.30	21
7230	9" iron pipe size	G		85	.188		6.10	9.95		16.05	22
7240	10" iron pipe size	G		80	.200		6.35	10.55		16.90	23
7250	12" iron pipe size	G		75	.213		7.20	11.25		18.45	25
7260	14" iron pipe size	G		70	.229		8.65	12.05		20.70	28
7270	16" iron pipe size	G		65	.246		11.30	13		24.30	32.50
7280	18" iron pipe size	G		60	.267		12.65	14.05		26.70	35.50
7290	20" iron pipe size	G		55	.291		12.95	15.35		28.30	37.50
7300	24" iron pipe size	G		50	.320		15.85	16.90		32.75	43.50
7320	2" wall, 1/2" iron pipe size	G		220	.073		3.20	3.84		7.04	9.40
7340	3/4" iron pipe size	G		210	.076		3.30	4.02		7.32	9.80
7350	1" iron pipe size	G		200	.080		3.52	4.22		7.74	10.35
7360	1-1/4" iron pipe size	G		190	.084		3.71	4.44		8.15	10.90
7370	1-1/2" iron pipe size	G		190	.084		4.47	4.44		8.91	11.70
7380	2" iron pipe size	G		180	.089		4.08	4.69		8.77	11.70
7390	2-1/2" iron pipe size	G		170	.094		4.39	4.97		9.36	12.50
7400	3" iron pipe size	G		160	.100		4.67	5.30		9.97	13.25
7410	3-1/2" iron pipe size	G		150	.107		5.05	5.65		10.70	14.20
7420	4" iron pipe size	G		130	.123		6.25	6.50		12.75	16.85
7430	5" iron pipe size	G		120	.133		6.20	7.05		13.25	17.65
7440	6" iron pipe size	G		100	.160		6.90	8.45		15.35	20.50
7450	7" iron pipe size	G		90	.178		7.30	9.40		16.70	22.50
7460	8" iron pipe size	G		80	.200		7.80	10.55		18.35	25
7470	9" iron pipe size	G		75	.213		8.55	11.25		19.80	26.50
7480	10" iron pipe size	G		70	.229		9.30	12.05		21.35	29
7490	12" iron pipe size	G		65	.246		10.45	13		23.45	31.50
7500	14" iron pipe size	G		60	.267		13.60	14.05		27.65	36.50
7510	16" iron pipe size	G		55	.291		14.90	15.35		30.25	40

22 07 19.10 Piping Insulation			Crew	Daily Output	Labor-Hours	Unit	Material	2020 Bare Costs Labor	Equipment	Total	Total Incl O&P
7520	18" iron pipe size	G	Q-14	50	.320	L.F.	16.60	16.90		33.50	44.50
7530	20" iron pipe size	G		45	.356		18.90	18.75		37.65	50
7540	24" iron pipe size	G		40	.400		20	21		41	54.50
7560	2-1/2" wall, 1/2" iron pipe size	G		210	.076		3.79	4.02		7.81	10.30
7562	3/4" iron pipe size	G		200	.080		3.96	4.22		8.18	10.85
7564	1" iron pipe size	G		190	.084		4.13	4.44		8.57	11.35
7566	1-1/4" iron pipe size	G		185	.086		4.28	4.56		8.84	11.70
7568	1-1/2" iron pipe size	G		180	.089		4.50	4.69		9.19	12.15
7570	2" iron pipe size	G		170	.094		4.72	4.97		9.69	12.85
7572	2-1/2" iron pipe size	G		160	.100		5.45	5.30		10.75	14.10
7574	3" iron pipe size	G		150	.107		5.70	5.65		11.35	14.95
7576	3-1/2" iron pipe size	G		140	.114		6.25	6.05		12.30	16.10
7578	4" iron pipe size	G		120	.133		6.55	7.05		13.60	18.05
7580	5" iron pipe size	G		110	.145		5.65	7.70		13.35	18.05
7582	6" iron pipe size	G		90	.178		9.45	9.40		18.85	25
7584	7" iron pipe size	G		80	.200		9.50	10.55		20.05	26.50
7586	8" iron pipe size	G		70	.229		10.05	12.05		22.10	29.50
7588	9" iron pipe size	G		65	.246		11	13		24	32
7590	10" iron pipe size	G		60	.267		12	14.05		26.05	34.50
7592	12" iron pipe size	G		55	.291		14.55	15.35		29.90	39.50
7594	14" iron pipe size	G		50	.320		17.15	16.90		34.05	45
7596	16" iron pipe size	G		45	.356		19.70	18.75		38.45	50.50
7598	18" iron pipe size	G		40	.400		21.50	21		42.50	56
7602	24" iron pipe size	G		30	.533		28	28		56	74
7620	3" wall, 1/2" iron pipe size	G		200	.080		4.85	4.22		9.07	11.85
7622	3/4" iron pipe size	G		190	.084		5.15	4.44		9.59	12.45
7624	1" iron pipe size	G		180	.089		5.45	4.69		10.14	13.20
7626	1-1/4" iron pipe size	G		175	.091		5.55	4.83		10.38	13.50
7628	1-1/2" iron pipe size	G		170	.094		5.80	4.97		10.77	14.05
7630	2" iron pipe size	G		160	.100		6.25	5.30		11.55	15
7632	2-1/2" iron pipe size	G		150	.107		6.50	5.65		12.15	15.80
7634	3" iron pipe size	G		140	.114		6.95	6.05		13	16.85
7636	3-1/2" iron pipe size	G		130	.123		7.65	6.50		14.15	18.40
7638	4" iron pipe size	G		110	.145		8.20	7.70		15.90	21
7640	5" iron pipe size	G		100	.160		9.30	8.45		17.75	23
7642	6" iron pipe size	G		80	.200		9.95	10.55		20.50	27
7644	7" iron pipe size	G		70	.229		11.40	12.05		23.45	31
7646	8" iron pipe size	G		60	.267		12.40	14.05		26.45	35
7648	9" iron pipe size	G		55	.291		13.35	15.35		28.70	38
7650	10" iron pipe size	G		50	.320		14.35	16.90		31.25	42
7652	12" iron pipe size	G		45	.356		17.90	18.75		36.65	48.50
7654	14" iron pipe size	G		40	.400		21	21		42	55.50
7656	16" iron pipe size	G		35	.457		23.50	24		47.50	63
7658	18" iron pipe size	G		32	.500		25	26.50		51.50	68
7660	20" iron pipe size	G		30	.533		27.50	28		55.50	73
7662	24" iron pipe size	G		28	.571		34.50	30		64.50	84.50
7664	26" iron pipe size	G		26	.615		38.50	32.50		71	92
7666	30" iron pipe size	G	↓	24	.667	↓	48	35		83	107
7800	For fiberglass with standard canvas jacket, deduct						5%				
7802	For fittings, add 3 L.F. for each fitting										
7804	plus 4 L.F. for each flange of the fitting										
7820	Polyethylene tubing flexible closed cell foam, UV resistant										
7828	Standard temperature (-90°F to +212°F)										

22 07 19.10 Piping Insulation			Crew	Daily Output	Labor-Hours	Unit	Material	2020 Bare Costs Labor	Equipment	Total	Total Incl O&P
7830	3/8" wall, 1/8" iron pipe size	G	1 Asbe	130	.062	L.F.	.33	3.61		3.94	5.90
7831	1/4" iron pipe size	G		130	.062		.34	3.61		3.95	5.90
7832	3/8" iron pipe size	G		130	.062		.33	3.61		3.94	5.90
7833	1/2" iron pipe size	G		126	.063		.43	3.72		4.15	6.15
7834	3/4" iron pipe size	G		122	.066		.45	3.85		4.30	6.40
7835	1" iron pipe size	G		120	.067		.55	3.91		4.46	6.60
7836	1-1/4" iron pipe size	G		118	.068		.70	3.98		4.68	6.85
7837	1-1/2" iron pipe size	G		118	.068		.76	3.98		4.74	6.95
7838	2" iron pipe size	G		116	.069		1.12	4.05		5.17	7.45
7839	2-1/2" iron pipe size	G		114	.070		1.39	4.12		5.51	7.85
7840	3" iron pipe size	G		112	.071		2.26	4.19		6.45	8.95
7842	1/2" wall, 1/8" iron pipe size	G		120	.067		.42	3.91		4.33	6.45
7843	1/4" iron pipe size	G		120	.067		.52	3.91		4.43	6.55
7844	3/8" iron pipe size	G		120	.067		.55	3.91		4.46	6.60
7845	1/2" iron pipe size	G		118	.068		.63	3.98		4.61	6.80
7846	3/4" iron pipe size	G		116	.069		.64	4.05		4.69	6.90
7847	1" iron pipe size	G		114	.070		.71	4.12		4.83	7.10
7848	1-1/4" iron pipe size	G		112	.071		.82	4.19		5.01	7.35
7849	1-1/2" iron pipe size	G		110	.073		.95	4.27		5.22	7.60
7850	2" iron pipe size	G		108	.074		1.61	4.34		5.95	8.40
7851	2-1/2" iron pipe size	G		106	.075		2.06	4.43		6.49	9.05
7852	3" iron pipe size	G		104	.077		2.68	4.51		7.19	9.90
7853	3-1/2" iron pipe size	G		102	.078		3.01	4.60		7.61	10.35
7854	4" iron pipe size	G		100	.080		3.59	4.69		8.28	11.15
7855	3/4" wall, 1/8" iron pipe size	G		110	.073		.63	4.27		4.90	7.25
7856	1/4" iron pipe size	G		110	.073		.68	4.27		4.95	7.30
7857	3/8" iron pipe size	G		108	.074		.90	4.34		5.24	7.65
7858	1/2" iron pipe size	G		106	.075		1.07	4.43		5.50	8
7859	3/4" iron pipe size	G		104	.077		1.27	4.51		5.78	8.35
7860	1" iron pipe size	G		102	.078		1.49	4.60		6.09	8.70
7861	1-1/4" iron pipe size	G		100	.080		1.76	4.69		6.45	9.15
7862	1-1/2" iron pipe size	G		100	.080		1.89	4.69		6.58	9.30
7863	2" iron pipe size	G		98	.082		2.87	4.79		7.66	10.50
7864	2-1/2" iron pipe size	G		96	.083		3.56	4.89		8.45	11.40
7865	3" iron pipe size	G		94	.085		3.81	4.99		8.80	11.85
7866	3-1/2" iron pipe size	G		92	.087		5.10	5.10		10.20	13.45
7867	4" iron pipe size	G		90	.089		5.70	5.20		10.90	14.25
7868	1" wall, 1/4" iron pipe size	G		100	.080		1.56	4.69		6.25	8.90
7869	3/8" iron pipe size	G		98	.082		1.51	4.79		6.30	9
7870	1/2" iron pipe size	G		96	.083		1.97	4.89		6.86	9.65
7871	3/4" iron pipe size	G		94	.085		2.26	4.99		7.25	10.15
7872	1" iron pipe size	G		92	.087		2.76	5.10		7.86	10.90
7873	1-1/4" iron pipe size	G		90	.089		3.13	5.20		8.33	11.45
7874	1-1/2" iron pipe size	G		90	.089		3.46	5.20		8.66	11.80
7875	2" iron pipe size	G		88	.091		4.64	5.35		9.99	13.30
7876	2-1/2" iron pipe size	G		86	.093		6.30	5.45		11.75	15.30
7877	3" iron pipe size	G	↓	84	.095	↓	6.90	5.60		12.50	16.20
7878	Contact cement, quart can	G				Ea.	13.20			13.20	14.55
7879	Rubber tubing, flexible closed cell foam										
7880	3/8" wall, 1/4" iron pipe size	G	1 Asbe	120	.067	L.F.	.33	3.91		4.24	6.35
7900	3/8" iron pipe size	G		120	.067		.46	3.91		4.37	6.50
7910	1/2" iron pipe size	G		115	.070		.50	4.08		4.58	6.80
7920	3/4" iron pipe size	G	↓	115	.070	↓	.57	4.08		4.65	6.90

22 07 Plumbing Insulation

22 07 19 – Plumbing Piping Insulation

22 07 19.10 Piping Insulation			Crew	Daily Output	Labor-Hours	Unit	Material	2020 Bare Costs Labor	Equipment	Total	Total Incl O&P
7930	1" iron pipe size	G	1 Asbe	110	.073	L.F.	.60	4.27		4.87	7.20
7940	1-1/4" iron pipe size	G		110	.073		.68	4.27		4.95	7.30
7950	1-1/2" iron pipe size	G		110	.073		.79	4.27		5.06	7.40
8100	1/2" wall, 1/4" iron pipe size	G		90	.089		.99	5.20		6.19	9.10
8120	3/8" iron pipe size	G		90	.089		1.05	5.20		6.25	9.15
8130	1/2" iron pipe size	G		89	.090		1.06	5.25		6.31	9.25
8140	3/4" iron pipe size	G		89	.090		1.22	5.25		6.47	9.45
8150	1" iron pipe size	G		88	.091		.86	5.35		6.21	9.15
8160	1-1/4" iron pipe size	G		87	.092		1.21	5.40		6.61	9.65
8170	1-1/2" iron pipe size	G		87	.092		1.89	5.40		7.29	10.40
8180	2" iron pipe size	G		86	.093		2.38	5.45		7.83	11
8190	2-1/2" iron pipe size	G		86	.093		2.60	5.45		8.05	11.25
8200	3" iron pipe size	G		85	.094		2.78	5.50		8.28	11.55
8210	3-1/2" iron pipe size	G		85	.094		3.81	5.50		9.31	12.70
8220	4" iron pipe size	G		80	.100		4.12	5.85		9.97	13.55
8230	5" iron pipe size	G		80	.100		5.60	5.85		11.45	15.15
8240	6" iron pipe size	G		75	.107		5.60	6.25		11.85	15.75
8300	3/4" wall, 1/4" iron pipe size	G		90	.089		.94	5.20		6.14	9.05
8320	3/8" iron pipe size	G		90	.089		1.10	5.20		6.30	9.20
8330	1/2" iron pipe size	G		89	.090		1.13	5.25		6.38	9.35
8340	3/4" iron pipe size	G		89	.090		2.13	5.25		7.38	10.45
8350	1" iron pipe size	G		88	.091		2.10	5.35		7.45	10.50
8360	1-1/4" iron pipe size	G		87	.092		2.46	5.40		7.86	11
8370	1-1/2" iron pipe size	G		87	.092		3.07	5.40		8.47	11.70
8380	2" iron pipe size	G		86	.093		4.27	5.45		9.72	13.10
8390	2-1/2" iron pipe size	G		86	.093		4.89	5.45		10.34	13.80
8400	3" iron pipe size	G		85	.094		5.50	5.50		11	14.55
8410	3-1/2" iron pipe size	G		85	.094		5.20	5.50		10.70	14.25
8420	4" iron pipe size	G		80	.100		6.80	5.85		12.65	16.50
8430	5" iron pipe size	G		80	.100		8.15	5.85		14	17.95
8440	6" iron pipe size	G		80	.100		9.90	5.85		15.75	19.90
8444	1" wall, 1/2" iron pipe size	G		86	.093		3.18	5.45		8.63	11.90
8445	3/4" iron pipe size	G		84	.095		3.90	5.60		9.50	12.90
8446	1" iron pipe size	G		84	.095		3.49	5.60		9.09	12.45
8447	1-1/4" iron pipe size	G		82	.098		3.83	5.70		9.53	13
8448	1-1/2" iron pipe size	G		82	.098		6.15	5.70		11.85	15.55
8449	2" iron pipe size	G		80	.100		7.55	5.85		13.40	17.30
8450	2-1/2" iron pipe size	G		80	.100		8.45	5.85		14.30	18.30
8456	Rubber insulation tape, 1/8" x 2" x 30'	G				Ea.	23			23	25
8460	Polyolefin tubing, flexible closed cell foam, UV stabilized, work										
8462	temp. -165°F to +210°F, 0 water vapor transmission										
8464	3/8" wall, 1/8" iron pipe size	G	1 Asbe	140	.057	L.F.	.49	3.35		3.84	5.70
8466	1/4" iron pipe size	G		140	.057		.51	3.35		3.86	5.70
8468	3/8" iron pipe size	G		140	.057		.57	3.35		3.92	5.80
8470	1/2" iron pipe size	G		136	.059		.63	3.45		4.08	6
8472	3/4" iron pipe size	G		132	.061		.63	3.55		4.18	6.15
8474	1" iron pipe size	G		130	.062		.84	3.61		4.45	6.45
8476	1-1/4" iron pipe size	G		128	.063		1.03	3.67		4.70	6.80
8478	1-1/2" iron pipe size	G		128	.063		1.26	3.67		4.93	7.05
8480	2" iron pipe size	G		126	.063		1.52	3.72		5.24	7.35
8482	2-1/2" iron pipe size	G		123	.065		2.21	3.81		6.02	8.30
8484	3" iron pipe size	G		121	.066		2.46	3.88		6.34	8.65
8486	4" iron pipe size	G		118	.068		5.05	3.98		9.03	11.65

22 07 19.10 Piping Insulation			Crew	Daily Output	Labor-Hours	Unit	Material	2020 Bare Costs Labor	Equipment	Total	Total Incl O&P
8500	1/2" wall, 1/8" iron pipe size	G	1 Asbe	130	.062	L.F.	.72	3.61		4.33	6.35
8502	1/4" iron pipe size	G		130	.062		.77	3.61		4.38	6.40
8504	3/8" iron pipe size	G		130	.062		.83	3.61		4.44	6.45
8506	1/2" iron pipe size	G		128	.063		.89	3.67		4.56	6.65
8508	3/4" iron pipe size	G		126	.063		1.05	3.72		4.77	6.85
8510	1" iron pipe size	G		123	.065		1.19	3.81		5	7.15
8512	1-1/4" iron pipe size	G		121	.066		1.41	3.88		5.29	7.50
8514	1-1/2" iron pipe size	G		119	.067		1.69	3.94		5.63	7.90
8516	2" iron pipe size	G		117	.068		2.14	4.01		6.15	8.50
8518	2-1/2" iron pipe size	G		114	.070		2.90	4.12		7.02	9.50
8520	3" iron pipe size	G		112	.071		3.77	4.19		7.96	10.60
8522	4" iron pipe size	G		110	.073		5.20	4.27		9.47	12.30
8534	3/4" wall, 1/8" iron pipe size	G		120	.067		1.10	3.91		5.01	7.20
8536	1/4" iron pipe size	G		120	.067		1.16	3.91		5.07	7.30
8538	3/8" iron pipe size	G		117	.068		1.33	4.01		5.34	7.60
8540	1/2" iron pipe size	G		114	.070		1.47	4.12		5.59	7.90
8542	3/4" iron pipe size	G		112	.071		1.59	4.19		5.78	8.20
8544	1" iron pipe size	G		110	.073		2.24	4.27		6.51	9
8546	1-1/4" iron pipe size	G		108	.074		2.95	4.34		7.29	9.90
8548	1-1/2" iron pipe size	G		108	.074		3.50	4.34		7.84	10.50
8550	2" iron pipe size	G		106	.075		4.42	4.43		8.85	11.65
8552	2-1/2" iron pipe size	G		104	.077		5.55	4.51		10.06	13.10
8554	3" iron pipe size	G		102	.078		7.20	4.60		11.80	15
8556	4" iron pipe size	G		100	.080		9.15	4.69		13.84	17.25
8570	1" wall, 1/8" iron pipe size	G		110	.073		1.98	4.27		6.25	8.75
8572	1/4" iron pipe size	G		108	.074		2.37	4.34		6.71	9.25
8574	3/8" iron pipe size	G		106	.075		2.08	4.43		6.51	9.10
8576	1/2" iron pipe size	G		104	.077		2.12	4.51		6.63	9.30
8578	3/4" iron pipe size	G		102	.078		2.82	4.60		7.42	10.15
8580	1" iron pipe size	G		100	.080		3.16	4.69		7.85	10.70
8582	1-1/4" iron pipe size	G		97	.082		3.79	4.84		8.63	11.60
8584	1-1/2" iron pipe size	G		97	.082		4.32	4.84		9.16	12.20
8586	2" iron pipe size	G		95	.084		5.60	4.94		10.54	13.75
8588	2-1/2" iron pipe size	G		93	.086		7.95	5.05		13	16.50
8590	3" iron pipe size	G	↓	91	.088	↓	9.70	5.15		14.85	18.60
8606	Contact adhesive (R-320)	G				Qt.	19.40			19.40	21.50
8608	Contact adhesive (R-320)	G				Gal.	77.50			77.50	85.50
8610	NOTE: Preslit/preglued vs. unslit, same price										

22 07 19.30 Piping Insulation Protective Jacketing, PVC		Crew	Daily Output	Labor-Hours	Unit	Material	Labor	Equipment	Total	Total Incl O&P
0010	PIPING INSULATION PROTECTIVE JACKETING, PVC									
0100	PVC, white, 48" lengths cut from roll goods									
0120	20 mil thick									
0140	Size based on OD of insulation									
0150	1-1/2" ID	Q-14	270	.059	L.F.	.29	3.13		3.42	5.10
0152	2" ID		260	.062		.39	3.25		3.64	5.40
0154	2-1/2" ID		250	.064		.45	3.38		3.83	5.70
0156	3" ID		240	.067		.53	3.52		4.05	6
0158	3-1/2" ID		230	.070		.61	3.67		4.28	6.30
0160	4" ID		220	.073		.68	3.84		4.52	6.65
0162	4-1/2" ID		210	.076		.77	4.02		4.79	7
0164	5" ID		200	.080		.84	4.22		5.06	7.40
0166	5-1/2" ID	↓	190	.084	↓	.92	4.44		5.36	7.80

22 07 19 – Plumbing Piping Insulation

22 07 19.30 Piping Insulation Protective Jacketing, PVC		Crew	Daily Output	Labor-Hours	Unit	Material	Labor	Equipment	Total	Total Incl O&P
						2020 Bare Costs				
0168	6" ID	Q-14	180	.089	L.F.	1	4.69		5.69	8.30
0170	6-1/2" ID		175	.091		1.08	4.83		5.91	8.60
0172	7" ID		170	.094		1.15	4.97		6.12	8.90
0174	7-1/2" ID		164	.098		1.25	5.15		6.40	9.30
0176	8" ID		161	.099		1.33	5.25		6.58	9.50
0178	8-1/2" ID		158	.101		1.40	5.35		6.75	9.75
0180	9" ID		155	.103		1.48	5.45		6.93	10
0182	9-1/2" ID		152	.105		1.56	5.55		7.11	10.25
0184	10" ID		149	.107		1.63	5.65		7.28	10.50
0186	10-1/2" ID		146	.110		1.72	5.80		7.52	10.80
0188	11" ID		143	.112		1.80	5.90		7.70	11.05
0190	11-1/2" ID		140	.114		1.87	6.05		7.92	11.30
0192	12" ID		137	.117		1.95	6.15		8.10	11.60
0194	12-1/2" ID		134	.119		2.03	6.30		8.33	11.95
0195	13" ID		132	.121		2.11	6.40		8.51	12.10
0196	13-1/2" ID		132	.121		2.19	6.40		8.59	12.20
0198	14" ID		130	.123		2.27	6.50		8.77	12.45
0200	15" ID		128	.125		2.42	6.60		9.02	12.80
0202	16" ID		126	.127		2.58	6.70		9.28	13.15
0204	17" ID		124	.129		2.74	6.80		9.54	13.45
0206	18" ID		122	.131		2.90	6.90		9.80	13.85
0208	19" ID		120	.133		3.05	7.05		10.10	14.15
0210	20" ID		118	.136		3.24	7.15		10.39	14.55
0212	21" ID		116	.138		3.37	7.30		10.67	14.90
0214	22" ID		114	.140		3.54	7.40		10.94	15.25
0216	23" ID		112	.143		3.70	7.55		11.25	15.65
0218	24" ID		110	.145		3.86	7.70		11.56	16.05
0220	25" ID		108	.148		4.02	7.80		11.82	16.40
0222	26" ID		106	.151		4.18	7.95		12.13	16.85
0224	27" ID		104	.154		4.34	8.10		12.44	17.20
0226	28" ID		102	.157		4.49	8.30		12.79	17.65
0228	29" ID		100	.160		4.66	8.45		13.11	18.10
0230	30" ID		98	.163		4.82	8.60		13.42	18.55
0300	For colors, add				Ea.	10%				
1000	30 mil thick									
1010	Size based on OD of insulation									
1020	2" ID	Q-14	260	.062	L.F.	.55	3.25		3.80	5.60
1022	2-1/2" ID		250	.064		.67	3.38		4.05	5.95
1024	3" ID		240	.067		.80	3.52		4.32	6.30
1026	3-1/2" ID		230	.070		.91	3.67		4.58	6.65
1028	4" ID		220	.073		1.02	3.84		4.86	7
1030	4-1/2" ID		210	.076		1.14	4.02		5.16	7.40
1032	5" ID		200	.080		1.26	4.22		5.48	7.90
1034	5-1/2" ID		190	.084		1.38	4.44		5.82	8.30
1036	6" ID		180	.089		1.49	4.69		6.18	8.85
1038	6-1/2" ID		175	.091		1.61	4.83		6.44	9.15
1040	7" ID		170	.094		1.73	4.97		6.70	9.55
1042	7-1/2" ID		164	.098		1.86	5.15		7.01	9.95
1044	8" ID		161	.099		1.97	5.25		7.22	10.20
1046	8-1/2" ID		158	.101		2.09	5.35		7.44	10.50
1048	9" ID		155	.103		2.21	5.45		7.66	10.80
1050	9-1/2" ID		152	.105		2.33	5.55		7.88	11.10
1052	10" ID		149	.107		2.44	5.65		8.09	11.40

22 07 19.30	Piping Insulation Protective Jacketing, PVC	Crew	Daily Output	Labor-Hours	Unit	2020 Bare Costs Material	Labor	Equipment	Total	Total Incl O&P
1054	10-1/2" ID	Q-14	146	.110	L.F.	2.56	5.80		8.36	11.70
1056	11" ID		143	.112		2.69	5.90		8.59	12
1058	11-1/2" ID		140	.114		2.80	6.05		8.85	12.35
1060	12" ID		137	.117		2.91	6.15		9.06	12.65
1062	12-1/2" ID		134	.119		3.04	6.30		9.34	13.05
1063	13" ID		132	.121		3.16	6.40		9.56	13.30
1064	13-1/2" ID		132	.121		3.28	6.40		9.68	13.40
1066	14" ID		130	.123		3.39	6.50		9.89	13.70
1068	15" ID		128	.125		3.63	6.60		10.23	14.15
1070	16" ID		126	.127		3.87	6.70		10.57	14.55
1072	17" ID		124	.129		4.11	6.80		10.91	14.95
1074	18" ID		122	.131		4.34	6.90		11.24	15.40
1076	19" ID		120	.133		4.58	7.05		11.63	15.85
1078	20" ID		118	.136		4.80	7.15		11.95	16.30
1080	21" ID		116	.138		5.05	7.30		12.35	16.75
1082	22" ID		114	.140		5.25	7.40		12.65	17.15
1084	23" ID		112	.143		5.55	7.55		13.10	17.70
1086	24" ID		110	.145		5.75	7.70		13.45	18.15
1088	25" ID		108	.148		6	7.80		13.80	18.60
1090	26" ID		106	.151		6.25	7.95		14.20	19.10
1092	27" ID		104	.154		6.50	8.10		14.60	19.60
1094	28" ID		102	.157		6.70	8.30		15	20
1096	29" ID		100	.160		6.95	8.45		15.40	20.50
1098	30" ID		98	.163		7.20	8.60		15.80	21
1300	For colors, add				Ea.	10%				
2000	PVC, white, fitting covers									
2020	Fiberglass insulation inserts included with sizes 1-3/4" thru 9-3/4"									
2030	Size is based on OD of insulation									
2040	90° elbow fitting									
2060	1-3/4"	Q-14	135	.119	Ea.	.55	6.25		6.80	10.20
2062	2"		130	.123		.68	6.50		7.18	10.70
2064	2-1/4"		128	.125		.79	6.60		7.39	11
2068	2-1/2"		126	.127		.84	6.70		7.54	11.20
2070	2-3/4"		123	.130		.99	6.85		7.84	11.65
2072	3"		120	.133		.96	7.05		8.01	11.85
2074	3-3/8"		116	.138		1.13	7.30		8.43	12.45
2076	3-3/4"		113	.142		1.21	7.45		8.66	12.80
2078	4-1/8"		110	.145		1.60	7.70		9.30	13.55
2080	4-3/4"		105	.152		1.93	8.05		9.98	14.45
2082	5-1/4"		100	.160		2.25	8.45		10.70	15.45
2084	5-3/4"		95	.168		2.84	8.90		11.74	16.75
2086	6-1/4"		90	.178		4.73	9.40		14.13	19.60
2088	6-3/4"		87	.184		4.99	9.70		14.69	20.50
2090	7-1/4"		85	.188		6.25	9.95		16.20	22
2092	7-3/4"		83	.193		6.60	10.15		16.75	23
2094	8-3/4"		80	.200		8.45	10.55		19	25.50
2096	9-3/4"		77	.208		11.30	10.95		22.25	29.50
2098	10-7/8"		74	.216		12.60	11.40		24	31.50
2100	11-7/8"		71	.225		14.45	11.90		26.35	34
2102	12-7/8"		68	.235		19.95	12.40		32.35	41
2104	14-1/8"		66	.242		21	12.80		33.80	42.50
2106	15-1/8"		64	.250		22.50	13.20		35.70	45
2108	16-1/8"		63	.254		24.50	13.40		37.90	47.50

22 07 19.30 Piping Insulation Protective Jacketing, PVC		Crew	Daily Output	Labor-Hours	Unit	Material	Labor	Equipment	Total	Total Incl O&P
						2020 Bare Costs				
2110	17-1/8"	Q-14	62	.258	Ea.	27	13.60		40.60	51
2112	18-1/8"		61	.262		36.50	13.85		50.35	61.50
2114	19-1/8"		60	.267		47.50	14.05		61.55	73.50
2116	20-1/8"	↓	59	.271	↓	61	14.30		75.30	89
2200	45° elbow fitting									
2220	1-3/4" thru 9-3/4" same price as 90° elbow fitting									
2320	10-7/8"	Q-14	74	.216	Ea.	12.60	11.40		24	31.50
2322	11-7/8"		71	.225		13.85	11.90		25.75	33.50
2324	12-7/8"		68	.235		15.50	12.40		27.90	36
2326	14-1/8"		66	.242		17.65	12.80		30.45	39
2328	15-1/8"		64	.250		19	13.20		32.20	41.50
2330	16-1/8"		63	.254		21.50	13.40		34.90	44.50
2332	17-1/8"		62	.258		24.50	13.60		38.10	48
2334	18-1/8"		61	.262		30	13.85		43.85	54.50
2336	19-1/8"		60	.267		40.50	14.05		54.55	66.50
2338	20-1/8"	↓	59	.271	↓	46	14.30		60.30	72.50
2400	Tee fitting									
2410	1-3/4"	Q-14	96	.167	Ea.	1.04	8.80		9.84	14.65
2412	2"		94	.170		1.18	9		10.18	15.10
2414	2-1/4"		91	.176		1.27	9.30		10.57	15.65
2416	2-1/2"		88	.182		1.39	9.60		10.99	16.30
2418	2-3/4"		85	.188		1.53	9.95		11.48	16.95
2420	3"		82	.195		1.66	10.30		11.96	17.65
2422	3-3/8"		79	.203		1.91	10.70		12.61	18.50
2424	3-3/4"		76	.211		2.37	11.10		13.47	19.65
2426	4-1/8"		73	.219		2.57	11.55		14.12	20.50
2428	4-3/4"		70	.229		3.19	12.05		15.24	22
2430	5-1/4"		67	.239		3.84	12.60		16.44	23.50
2432	5-3/4"		63	.254		5.10	13.40		18.50	26
2434	6-1/4"		60	.267		6.70	14.05		20.75	29
2436	6-3/4"		59	.271		8.30	14.30		22.60	31
2438	7-1/4"		57	.281		13.40	14.80		28.20	38
2440	7-3/4"		54	.296		14.70	15.65		30.35	40
2442	8-3/4"		52	.308		17.85	16.25		34.10	44.50
2444	9-3/4"		50	.320		21	16.90		37.90	49
2446	10-7/8"		48	.333		21.50	17.60		39.10	50.50
2448	11-7/8"		47	.340		23.50	17.95		41.45	53.50
2450	12-7/8"		46	.348		26	18.35		44.35	56.50
2452	14-1/8"		45	.356		28.50	18.75		47.25	60.50
2454	15-1/8"		44	.364		31	19.20		50.20	63.50
2456	16-1/8"		43	.372		33	19.65		52.65	66.50
2458	17-1/8"		42	.381		35.50	20		55.50	70
2460	18-1/8"		41	.390		39	20.50		59.50	74.50
2462	19-1/8"		40	.400		42.50	21		63.50	79.50
2464	20-1/8"	↓	39	.410	↓	47	21.50		68.50	85
4000	Mechanical grooved fitting cover, including insert									
4020	90° elbow fitting									
4030	3/4" & 1"	Q-14	140	.114	Ea.	5.05	6.05		11.10	14.80
4040	1-1/4" & 1-1/2"		135	.119		6.20	6.25		12.45	16.40
4042	2"		130	.123		8.60	6.50		15.10	19.45
4044	2-1/2"		125	.128		9.60	6.75		16.35	21
4046	3"		120	.133		10.75	7.05		17.80	22.50
4048	3-1/2"	↓	115	.139	↓	12.45	7.35		19.80	25

22 07 19.30 Piping Insulation Protective Jacketing, PVC

	22 07 19.30 Piping Insulation Protective Jacketing, PVC	Crew	Daily Output	Labor-Hours	Unit	2020 Bare Costs Material	Labor	Equipment	Total	Total Incl O&P
4050	4"	Q-14	110	.145	Ea.	13.85	7.70		21.55	27
4052	5"		100	.160		17.25	8.45		25.70	32
4054	6"		90	.178		25.50	9.40		34.90	42.50
4056	8"		80	.200		27.50	10.55		38.05	46.50
4058	10"		75	.213		35.50	11.25		46.75	56.50
4060	12"		68	.235		51	12.40		63.40	75
4062	14"		65	.246		61	13		74	87.50
4064	16"		63	.254		83	13.40		96.40	112
4066	18"		61	.262		114	13.85		127.85	148
4100	45° elbow fitting									
4120	3/4" & 1"	Q-14	140	.114	Ea.	4.53	6.05		10.58	14.25
4130	1-1/4" & 1-1/2"		135	.119		5.60	6.25		11.85	15.75
4140	2"		130	.123		8.40	6.50		14.90	19.20
4142	2-1/2"		125	.128		9.35	6.75		16.10	20.50
4144	3"		120	.133		9.60	7.05		16.65	21.50
4146	3-1/2"		115	.139		12.15	7.35		19.50	24.50
4148	4"		110	.145		15.45	7.70		23.15	29
4150	5"		100	.160		16.75	8.45		25.20	31.50
4152	6"		90	.178		23.50	9.40		32.90	40.50
4154	8"		80	.200		24.50	10.55		35.05	43
4156	10"		75	.213		31.50	11.25		42.75	52.50
4158	12"		68	.235		47.50	12.40		59.90	71
4160	14"		65	.246		64	13		77	90.50
4162	16"		63	.254		78.50	13.40		91.90	107
4164	18"		61	.262		101	13.85		114.85	133
4200	Tee fitting									
4220	3/4" & 1"	Q-14	93	.172	Ea.	6.60	9.10		15.70	21
4230	1-1/4" & 1-1/2"		90	.178		8.10	9.40		17.50	23.50
4240	2"		87	.184		13.55	9.70		23.25	30
4242	2-1/2"		84	.190		15.10	10.05		25.15	32
4244	3"		80	.200		16.20	10.55		26.75	34
4246	3-1/2"		77	.208		17.35	10.95		28.30	36
4248	4"		73	.219		21.50	11.55		33.05	42
4250	5"		67	.239		23.50	12.60		36.10	45.50
4252	6"		60	.267		37	14.05		51.05	62
4254	8"		54	.296		38.50	15.65		54.15	66.50
4256	10"		50	.320		46.50	16.90		63.40	77
4258	12"		46	.348		72	18.35		90.35	108
4260	14"		43	.372		101	19.65		120.65	141
4262	16"		42	.381		105	20		125	147
4264	18"		41	.390		127	20.50		147.50	171

22 07 19.40 Pipe Insulation Protective Jacketing, Aluminum

	22 07 19.40 Pipe Insulation Protective Jacketing, Aluminum	Crew	Daily Output	Labor-Hours	Unit	Material	Labor	Equipment	Total	Total Incl O&P
0010	**PIPE INSULATION PROTECTIVE JACKETING, ALUMINUM**									
0100	Metal roll jacketing									
0120	Aluminum with polykraft moisture barrier									
0140	Smooth, based on OD of insulation, .016" thick									
0180	1/2" ID	Q-14	220	.073	L.F.	.29	3.84		4.13	6.20
0190	3/4" ID		215	.074		.37	3.93		4.30	6.45
0200	1" ID		210	.076		.45	4.02		4.47	6.65
0210	1-1/4" ID		205	.078		.54	4.12		4.66	6.90
0220	1-1/2" ID		202	.079		.62	4.18		4.80	7.10
0230	1-3/4" ID		199	.080		.79	4.24		5.03	7.35

22 07 19.40 Pipe Insulation Protective Jacketing, Aluminum		Crew	Daily Output	Labor-Hours	Unit	2020 Bare Costs Material	Labor	Equipment	Total	Total Incl O&P
0240	2″ ID	Q-14	195	.082	L.F.	.88	4.33		5.21	7.60
0250	2-1/4″ ID		191	.084		.98	4.42		5.40	7.90
0260	2-1/2″ ID		187	.086		.97	4.52		5.49	8
0270	2-3/4″ ID		184	.087		1.06	4.59		5.65	8.20
0280	3″ ID		180	.089		1.27	4.69		5.96	8.60
0290	3-1/4″ ID		176	.091		1.37	4.80		6.17	8.85
0300	3-1/2″ ID		172	.093		1.33	4.91		6.24	9
0310	3-3/4″ ID		169	.095		1.41	5		6.41	9.20
0320	4″ ID		165	.097		1.50	5.10		6.60	9.50
0330	4-1/4″ ID		161	.099		1.74	5.25		6.99	9.95
0340	4-1/2″ ID		157	.102		1.68	5.40		7.08	10.10
0350	4-3/4″ ID		154	.104		1.93	5.50		7.43	10.50
0360	5″ ID		150	.107		2.03	5.65		7.68	10.90
0370	5-1/4″ ID		146	.110		2.12	5.80		7.92	11.25
0380	5-1/2″ ID		143	.112		2.02	5.90		7.92	11.25
0390	5-3/4″ ID		139	.115		2.33	6.10		8.43	11.90
0400	6″ ID		135	.119		2.42	6.25		8.67	12.25
0410	6-1/4″ ID		133	.120		2.51	6.35		8.86	12.50
0420	6-1/2″ ID		131	.122		2.37	6.45		8.82	12.50
0430	7″ ID		128	.125		2.52	6.60		9.12	12.90
0440	7-1/4″ ID		125	.128		2.61	6.75		9.36	13.20
0450	7-1/2″ ID		123	.130		2.69	6.85		9.54	13.50
0460	8″ ID		121	.132		2.86	7		9.86	13.85
0470	8-1/2″ ID		119	.134		3.03	7.10		10.13	14.25
0480	9″ ID		116	.138		3.20	7.30		10.50	14.70
0490	9-1/2″ ID		114	.140		3.38	7.40		10.78	15.05
0500	10″ ID		112	.143		3.55	7.55		11.10	15.50
0510	10-1/2″ ID		110	.145		4.09	7.70		11.79	16.30
0520	11″ ID		107	.150		3.88	7.90		11.78	16.35
0530	11-1/2″ ID		105	.152		4.06	8.05		12.11	16.80
0540	12″ ID		103	.155		4.24	8.20		12.44	17.25
0550	12-1/2″ ID		100	.160		4.41	8.45		12.86	17.80
0560	13″ ID		99	.162		4.58	8.55		13.13	18.15
0570	14″ ID		98	.163		4.93	8.60		13.53	18.65
0580	15″ ID		96	.167		5.25	8.80		14.05	19.30
0590	16″ ID		95	.168		5.60	8.90		14.50	19.80
0600	17″ ID		93	.172		5.95	9.10		15.05	20.50
0610	18″ ID		92	.174		6.20	9.20		15.40	21
0620	19″ ID		90	.178		6.65	9.40		16.05	21.50
0630	20″ ID		89	.180		7	9.50		16.50	22.50
0640	21″ ID		87	.184		7.35	9.70		17.05	23
0650	22″ ID		86	.186		7.65	9.80		17.45	23.50
0660	23″ ID		84	.190		8	10.05		18.05	24.50
0670	24″ ID		83	.193		8.35	10.15		18.50	25
0710	For smooth .020″ thick, add					27%	10%			
0720	For smooth .024″ thick, add					52%	20%			
0730	For smooth .032″ thick, add					104%	33%			
0800	For stucco embossed, add					1%				
0820	For corrugated, add					2.50%				
0900	White aluminum with polysurlyn moisture barrier									
0910	Smooth, % is an add to polykraft lines of same thickness									
0940	For smooth .016″ thick, add				L.F.	35%				
0960	For smooth .024″ thick, add				″	22%				

22 07 19 – Plumbing Piping Insulation

22 07 19.40 Pipe Insulation Protective Jacketing, Aluminum		Crew	Daily Output	Labor-Hours	Unit	Material	2020 Bare Costs Labor	Equipment	Total	Total Incl O&P
1000	Aluminum fitting covers									
1010	Size is based on OD of insulation									
1020	90° LR elbow, 2 piece									
1100	1-1/2"	Q-14	140	.114	Ea.	4.37	6.05		10.42	14.05
1110	1-3/4"		135	.119		4.37	6.25		10.62	14.40
1120	2"		130	.123		5.50	6.50		12	16
1130	2-1/4"		128	.125		5.50	6.60		12.10	16.20
1140	2-1/2"		126	.127		5.50	6.70		12.20	16.35
1150	2-3/4"		123	.130		5.50	6.85		12.35	16.60
1160	3"		120	.133		5.95	7.05		13	17.35
1170	3-1/4"		117	.137		5.95	7.20		13.15	17.65
1180	3-1/2"		115	.139		7.05	7.35		14.40	19
1190	3-3/4"		113	.142		7.25	7.45		14.70	19.40
1200	4"		110	.145		7.55	7.70		15.25	20
1210	4-1/4"		108	.148		7.80	7.80		15.60	20.50
1220	4-1/2"		106	.151		7.80	7.95		15.75	21
1230	4-3/4"		104	.154		7.80	8.10		15.90	21
1240	5"		102	.157		8.75	8.30		17.05	22.50
1250	5-1/4"		100	.160		10.10	8.45		18.55	24
1260	5-1/2"		97	.165		12.60	8.70		21.30	27
1270	5-3/4"		95	.168		12.60	8.90		21.50	27.50
1280	6"		92	.174		10.75	9.20		19.95	26
1290	6-1/4"		90	.178		10.75	9.40		20.15	26
1300	6-1/2"		87	.184		12.95	9.70		22.65	29
1310	7"		85	.188		15.70	9.95		25.65	32.50
1320	7-1/4"		84	.190		15.70	10.05		25.75	33
1330	7-1/2"		83	.193		23	10.15		33.15	41
1340	8"		82	.195		17.20	10.30		27.50	35
1350	8-1/2"		80	.200		32	10.55		42.55	51
1360	9"		78	.205		32	10.85		42.85	51.50
1370	9-1/2"		77	.208		23.50	10.95		34.45	43
1380	10"		76	.211		23.50	11.10		34.60	43
1390	10-1/2"		75	.213		24	11.25		35.25	44
1400	11"		74	.216		24	11.40		35.40	44
1410	11-1/2"		72	.222		27	11.75		38.75	48
1420	12"		71	.225		27	11.90		38.90	48.50
1430	12-1/2"		69	.232		48.50	12.25		60.75	72
1440	13"		68	.235		48.50	12.40		60.90	72
1450	14"		66	.242		65	12.80		77.80	91
1460	15"		64	.250		68	13.20		81.20	95
1470	16"		63	.254		74	13.40		87.40	102
2000	45° elbow, 2 piece									
2010	2-1/2"	Q-14	126	.127	Ea.	4.54	6.70		11.24	15.30
2020	2-3/4"		123	.130		4.54	6.85		11.39	15.55
2030	3"		120	.133		5.15	7.05		12.20	16.45
2040	3-1/4"		117	.137		5.15	7.20		12.35	16.75
2050	3-1/2"		115	.139		5.85	7.35		13.20	17.70
2060	3-3/4"		113	.142		5.85	7.45		13.30	17.90
2070	4"		110	.145		5.95	7.70		13.65	18.35
2080	4-1/4"		108	.148		6.80	7.80		14.60	19.45
2090	4-1/2"		106	.151		6.80	7.95		14.75	19.70
2100	4-3/4"		104	.154		6.80	8.10		14.90	19.90
2110	5"		102	.157		7.95	8.30		16.25	21.50

22 07 19.40 Pipe Insulation Protective Jacketing, Aluminum		Crew	Daily Output	Labor-Hours	Unit	Material	2020 Bare Costs Labor	Equipment	Total	Total Incl O&P
2120	5-1/4"	Q-14	100	.160	Ea.	7.95	8.45		16.40	21.50
2130	5-1/2"		97	.165		8.35	8.70		17.05	22.50
2140	6"		92	.174		8.35	9.20		17.55	23.50
2150	6-1/2"		87	.184		11.55	9.70		21.25	27.50
2160	7"		85	.188		11.55	9.95		21.50	28
2170	7-1/2"		83	.193		11.65	10.15		21.80	28.50
2180	8"		82	.195		11.65	10.30		21.95	28.50
2190	8-1/2"		80	.200		14.50	10.55		25.05	32
2200	9"		78	.205		14.50	10.85		25.35	32.50
2210	9-1/2"		77	.208		20	10.95		30.95	39
2220	10"		76	.211		20	11.10		31.10	39
2230	10-1/2"		75	.213		18.80	11.25		30.05	38
2240	11"		74	.216		18.80	11.40		30.20	38
2250	11-1/2"		72	.222		22.50	11.75		34.25	43
2260	12"		71	.225		22.50	11.90		34.40	43.50
2270	13"		68	.235		26.50	12.40		38.90	48
2280	14"		66	.242		32.50	12.80		45.30	55.50
2290	15"		64	.250		50	13.20		63.20	75.50
2300	16"		63	.254		54.50	13.40		67.90	80.50
2310	17"		62	.258		53	13.60		66.60	79.50
2320	18"		61	.262		60.50	13.85		74.35	88
2330	19"		60	.267		74	14.05		88.05	103
2340	20"		59	.271		71.50	14.30		85.80	101
2350	21"		58	.276		77	14.55		91.55	108
3000	Tee, 4 piece									
3010	2-1/2"	Q-14	88	.182	Ea.	29.50	9.60		39.10	47.50
3020	2-3/4"		86	.186		29.50	9.80		39.30	47.50
3030	3"		84	.190		33.50	10.05		43.55	52.50
3040	3-1/4"		82	.195		33.50	10.30		43.80	53
3050	3-1/2"		80	.200		35	10.55		45.55	54.50
3060	4"		78	.205		35.50	10.85		46.35	55.50
3070	4-1/4"		76	.211		37	11.10		48.10	58
3080	4-1/2"		74	.216		37	11.40		48.40	58.50
3090	4-3/4"		72	.222		37	11.75		48.75	59
3100	5"		70	.229		38.50	12.05		50.55	61
3110	5-1/4"		68	.235		38.50	12.40		50.90	61.50
3120	5-1/2"		66	.242		40.50	12.80		53.30	64
3130	6"		64	.250		40.50	13.20		53.70	65
3140	6-1/2"		60	.267		44.50	14.05		58.55	70
3150	7"		58	.276		44.50	14.55		59.05	71
3160	7-1/2"		56	.286		49	15.10		64.10	77
3170	8"		54	.296		49	15.65		64.65	78
3180	8-1/2"		52	.308		50.50	16.25		66.75	80.50
3190	9"		50	.320		50.50	16.90		67.40	81.50
3200	9-1/2"		49	.327		37.50	17.25		54.75	67.50
3210	10"		48	.333		37.50	17.60		55.10	68
3220	10-1/2"		47	.340		40	17.95		57.95	71.50
3230	11"		46	.348		40	18.35		58.35	72
3240	11-1/2"		45	.356		42	18.75		60.75	75.50
3250	12"		44	.364		42	19.20		61.20	76
3260	13"		43	.372		44	19.65		63.65	78.50
3270	14"		42	.381		46.50	20		66.50	82
3280	15"		41	.390		50	20.50		70.50	86.50

22 07 19.40 Pipe Insulation Protective Jacketing, Aluminum		Crew	Daily Output	Labor-Hours	Unit	2020 Bare Costs Material	Labor	Equipment	Total	Total Incl O&P
3290	16"	Q-14	40	.400	Ea.	51.50	21		72.50	89
3300	17"		39	.410		59	21.50		80.50	98
3310	18"		38	.421		61.50	22		83.50	102
3320	19"		37	.432		68	23		91	110
3330	20"		36	.444		69.50	23.50		93	113
3340	22"		35	.457		88.50	24		112.50	134
3350	23"		34	.471		90.50	25		115.50	138
3360	24"		31	.516		93	27		120	144

22 07 19.50 Pipe Insulation Protective Jacketing, St. Stl.

Line	Description	Crew	Daily Output	Labor-Hours	Unit	Material	Labor	Equipment	Total	Total Incl O&P
0010	PIPE INSULATION PROTECTIVE JACKETING, STAINLESS STEEL									
0100	Metal roll jacketing									
0120	Type 304 with moisture barrier									
0140	Smooth, based on OD of insulation, .010" thick									
0260	2-1/2" ID	Q-14	250	.064	L.F.	3.10	3.38		6.48	8.60
0270	2-3/4" ID		245	.065		3.37	3.45		6.82	9
0280	3" ID		240	.067		3.64	3.52		7.16	9.40
0290	3-1/4" ID		235	.068		3.95	3.59		7.54	9.85
0300	3-1/2" ID		230	.070		4.19	3.67		7.86	10.25
0310	3-3/4" ID		225	.071		4.50	3.75		8.25	10.70
0320	4" ID		220	.073		4.75	3.84		8.59	11.15
0330	4-1/4" ID		215	.074		5	3.93		8.93	11.55
0340	4-1/2" ID		210	.076		5.30	4.02		9.32	12
0350	5" ID		200	.080		5.85	4.22		10.07	12.95
0360	5-1/2" ID		190	.084		6.40	4.44		10.84	13.85
0370	6" ID		180	.089		6.95	4.69		11.64	14.85
0380	6-1/2" ID		175	.091		7.50	4.83		12.33	15.65
0390	7" ID		170	.094		8.05	4.97		13.02	16.50
0400	7-1/2" ID		164	.098		8.60	5.15		13.75	17.35
0410	8" ID		161	.099		9.15	5.25		14.40	18.10
0420	8-1/2" ID		158	.101		9.70	5.35		15.05	18.85
0430	9" ID		155	.103		10.25	5.45		15.70	19.65
0440	9-1/2" ID		152	.105		10.80	5.55		16.35	20.50
0450	10" ID		149	.107		11.35	5.65		17	21
0460	10-1/2" ID		146	.110		11.90	5.80		17.70	22
0470	11" ID		143	.112		12.45	5.90		18.35	23
0480	12" ID		137	.117		13.55	6.15		19.70	24.50
0490	13" ID		132	.121		14.65	6.40		21.05	26
0500	14" ID		130	.123		15.75	6.50		22.25	27.50
0700	For smooth .016" thick, add					45%	33%			
1000	Stainless steel, Type 316, fitting covers									
1010	Size is based on OD of insulation									
1020	90° LR elbow, 2 piece									
1100	1-1/2"	Q-14	126	.127	Ea.	12.90	6.70		19.60	24.50
1110	2-3/4"		123	.130		13.20	6.85		20.05	25
1120	3"		120	.133		13.80	7.05		20.85	26
1130	3-1/4"		117	.137		13.80	7.20		21	26.50
1140	3-1/2"		115	.139		14.50	7.35		21.85	27
1150	3-3/4"		113	.142		15.35	7.45		22.80	28.50
1160	4"		110	.145		16.85	7.70		24.55	30.50
1170	4-1/4"		108	.148		22.50	7.80		30.30	36.50
1180	4-1/2"		106	.151		22.50	7.95		30.45	37
1190	5"		102	.157		23	8.30		31.30	37.50

22 07 Plumbing Insulation

22 07 19 – Plumbing Piping Insulation

22 07 19.50 Pipe Insulation Protective Jacketing, St. Stl.		Crew	Daily Output	Labor-Hours	Unit	Material	Labor	Equipment	Total	Total Incl O&P
						2020 Bare Costs				
1200	5-1/2"	Q-14	97	.165	Ea.	34	8.70		42.70	51
1210	6"		92	.174		37.50	9.20		46.70	55
1220	6-1/2"		87	.184		51.50	9.70		61.20	72
1230	7"		85	.188		51.50	9.95		61.45	72.50
1240	7-1/2"		83	.193		60	10.15		70.15	82
1250	8"		80	.200		60	10.55		70.55	82.50
1260	8-1/2"		80	.200		62.50	10.55		73.05	84.50
1270	9"		78	.205		95	10.85		105.85	121
1280	9-1/2"		77	.208		93	10.95		103.95	119
1290	10"		76	.211		93	11.10		104.10	119
1300	10-1/2"		75	.213		111	11.25		122.25	140
1310	11"		74	.216		106	11.40		117.40	135
1320	12"		71	.225		121	11.90		132.90	151
1330	13"		68	.235		167	12.40		179.40	203
1340	14"	▼	66	.242	▼	168	12.80		180.80	205
2000	45° elbow, 2 piece									
2010	2-1/2"	Q-14	126	.127	Ea.	11	6.70		17.70	22.50
2020	2-3/4"		123	.130		11	6.85		17.85	22.50
2030	3"		120	.133		11.85	7.05		18.90	24
2040	3-1/4"		117	.137		11.85	7.20		19.05	24
2050	3-1/2"		115	.139		12	7.35		19.35	24.50
2060	3-3/4"		113	.142		12	7.45		19.45	24.50
2070	4"		110	.145		15.40	7.70		23.10	29
2080	4-1/4"		108	.148		21.50	7.80		29.30	36
2090	4-1/2"		106	.151		21.50	7.95		29.45	36.50
2100	4-3/4"		104	.154		21.50	8.10		29.60	36.50
2110	5"		102	.157		22	8.30		30.30	36.50
2120	5-1/2"		97	.165		22	8.70		30.70	37.50
2130	6"		92	.174		26	9.20		35.20	42.50
2140	6-1/2"		87	.184		26	9.70		35.70	43.50
2150	7"		85	.188		44	9.95		53.95	64
2160	7-1/2"		83	.193		45	10.15		55.15	65
2170	8"		82	.195		45	10.30		55.30	65.50
2180	8-1/2"		80	.200		52.50	10.55		63.05	74
2190	9"		78	.205		52.50	10.85		63.35	74.50
2200	9-1/2"		77	.208		62	10.95		72.95	85
2210	10"		76	.211		62	11.10		73.10	85
2220	10-1/2"		75	.213		74.50	11.25		85.75	99.50
2230	11"		74	.216		74.50	11.40		85.90	99.50
2240	12"		71	.225		81	11.90		92.90	107
2250	13"	▼	68	.235	▼	95	12.40		107.40	124

22 11 13.14 Pipe, Brass

22 11 13.14 Pipe, Brass		Crew	Daily Output	Labor-Hours	Unit	2020 Bare Costs Material	Labor	Equipment	Total	Total Incl O&P
0010	**PIPE, BRASS**, Plain end									
0900	Field threaded, coupling & clevis hanger assembly 10′ OC									
0920	Regular weight									
1120	1/2″ diameter	1 Plum	48	.167	L.F.	8.30	10.75		19.05	25
1140	3/4″ diameter		46	.174		10.95	11.20		22.15	29
1160	1″ diameter		43	.186		7.15	12		19.15	26
1180	1-1/4″ diameter	Q-1	72	.222		23.50	12.90		36.40	45.50
1200	1-1/2″ diameter		65	.246		28	14.30		42.30	52
1220	2″ diameter		53	.302		18.85	17.50		36.35	47
1240	2-1/2″ diameter		41	.390		32.50	22.50		55	70
1260	3″ diameter		31	.516		43.50	30		73.50	93
1300	4″ diameter	Q-2	37	.649		151	39		190	225
1930	To delete coupling & hanger, subtract									
1940	1/2″ diam.					40%	46%			
1950	3/4″ diam. to 1-1/2″ diam.					39%	39%			
1960	2″ diam. to 4″ diam.					48%	35%			

22 11 13.16 Pipe Fittings, Brass

22 11 13.16 Pipe Fittings, Brass		Crew	Daily Output	Labor-Hours	Unit	Material	Labor	Equipment	Total	Total Incl O&P
0010	**PIPE FITTINGS, BRASS**, Rough bronze, threaded, lead free.									
1000	Standard wt., 90° elbow									
1040	1/8″	1 Plum	13	.615	Ea.	22.50	39.50		62	84
1060	1/4″		13	.615		22.50	39.50		62	84
1080	3/8″		13	.615		22.50	39.50		62	84
1100	1/2″		12	.667		22.50	43		65.50	89
1120	3/4″		11	.727		30	47		77	103
1140	1″		10	.800		48.50	51.50		100	131
1160	1-1/4″	Q-1	17	.941		78.50	54.50		133	168
1180	1-1/2″		16	1		97.50	58		155.50	194
1200	2″		14	1.143		157	66.50		223.50	272
1220	2-1/2″		11	1.455		380	84.50		464.50	545
1240	3″		8	2		580	116		696	815
1260	4″	Q-2	11	2.182		1,175	131		1,306	1,500
1280	5″		8	3		2,800	180		2,980	3,350
1300	6″		7	3.429		4,800	206		5,006	5,575
1500	45° elbow, 1/8″	1 Plum	13	.615		27.50	39.50		67	90
1540	1/4″		13	.615		27.50	39.50		67	90
1560	3/8″		13	.615		27.50	39.50		67	90
1580	1/2″		12	.667		27.50	43		70.50	95
1600	3/4″		11	.727		39	47		86	113
1620	1″		10	.800		66.50	51.50		118	151
1640	1-1/4″	Q-1	17	.941		106	54.50		160.50	198
1660	1-1/2″		16	1		133	58		191	234
1680	2″		14	1.143		215	66.50		281.50	335
1700	2-1/2″		11	1.455		410	84.50		494.50	580
1720	3″		8	2		545	116		661	775
1740	4″	Q-2	11	2.182		1,225	131		1,356	1,550
1760	5″		8	3		2,275	180		2,455	2,775
1780	6″		7	3.429		3,150	206		3,356	3,775
2000	Tee, 1/8″	1 Plum	9	.889		26.50	57.50		84	115
2040	1/4″		9	.889		26.50	57.50		84	115
2060	3/8″		9	.889		26.50	57.50		84	115
2080	1/2″		8	1		26.50	64.50		91	126
2100	3/4″		7	1.143		37.50	73.50		111	151

22 11 13.16 Pipe Fittings, Brass		Crew	Daily Output	Labor-Hours	Unit	Material	Labor	Equipment	Total	Total Incl O&P
						2020 Bare Costs				
2120	1"	1 Plum	6	1.333	Ea.	67.50	86		153.50	203
2140	1-1/4"	Q-1	10	1.600		116	93		209	267
2160	1-1/2"		9	1.778		131	103		234	298
2180	2"		8	2		217	116		333	415
2200	2-1/2"		7	2.286		515	133		648	770
2220	3"		5	3.200		790	186		976	1,150
2240	4"	Q-2	7	3.429		1,950	206		2,156	2,450
2260	5"		5	4.800		3,350	289		3,639	4,100
2280	6"		4	6		5,525	360		5,885	6,625
2500	Coupling, 1/8"	1 Plum	26	.308		18.70	19.85		38.55	50
2540	1/4"		22	.364		18.70	23.50		42.20	55.50
2560	3/8"		18	.444		18.70	28.50		47.20	63.50
2580	1/2"		15	.533		18.80	34.50		53.30	72
2600	3/4"		14	.571		26.50	37		63.50	84
2620	1"		13	.615		45	39.50		84.50	109
2640	1-1/4"	Q-1	22	.727		75	42		117	146
2660	1-1/2"		20	.800		97.50	46.50		144	177
2680	2"		18	.889		161	51.50		212.50	254
2700	2-1/2"		14	1.143		275	66.50		341.50	405
2720	3"		10	1.600		380	93		473	560
2740	4"	Q-2	12	2		780	120		900	1,025
2760	5"		10	2.400		1,625	144		1,769	2,000
2780	6"		9	2.667		2,325	160		2,485	2,825
3000	Union, 125 lb.									
3020	1/8"	1 Plum	12	.667	Ea.	41	43		84	110
3040	1/4"		12	.667		43	43		86	112
3060	3/8"		12	.667		43	43		86	112
3080	1/2"		11	.727		43	47		90	118
3100	3/4"		10	.800		59.50	51.50		111	143
3120	1"		9	.889		91.50	57.50		149	187
3140	1-1/4"	Q-1	16	1		132	58		190	233
3160	1-1/2"		15	1.067		158	62		220	266
3180	2"		13	1.231		208	71.50		279.50	335
3200	2-1/2"		10	1.600		725	93		818	940
3220	3"		7	2.286		970	133		1,103	1,275
3240	4"	Q-2	10	2.400		2,800	144		2,944	3,325

22 11 13.23 Pipe/Tube, Copper		Crew	Daily Output	Labor-Hours	Unit	Material	Labor	Equipment	Total	Total Incl O&P
0010	**PIPE/TUBE, COPPER**, Solder joints R221113-50									
0100	Solder									
0120	Solder, lead free, roll				Lb.	14.25			14.25	15.70
1000	Type K tubing, couplings & clevis hanger assemblies 10' OC									
1100	1/4" diameter R221113-70	1 Plum	84	.095	L.F.	4.25	6.15		10.40	13.85
1120	3/8" diameter		82	.098		4.76	6.30		11.06	14.65
1140	1/2" diameter		78	.103		5.30	6.60		11.90	15.70
1160	5/8" diameter		77	.104		6.75	6.70		13.45	17.40
1180	3/4" diameter		74	.108		8.65	6.95		15.60	19.95
1200	1" diameter		66	.121		12.80	7.80		20.60	26
1220	1-1/4" diameter		56	.143		15.30	9.20		24.50	30.50
1240	1-1/2" diameter		50	.160		18.60	10.30		28.90	36
1260	2" diameter		40	.200		27	12.90		39.90	49.50
1280	2-1/2" diameter	Q-1	60	.267		44	15.45		59.45	71.50
1300	3" diameter		54	.296		58	17.20		75.20	89

22 11 Facility Water Distribution

22 11 13 – Facility Water Distribution Piping

22 11 13.23 Pipe/Tube, Copper		Crew	Daily Output	Labor-Hours	Unit	2020 Bare Costs Material	Labor	Equipment	Total	Total Incl O&P
1320	3-1/2" diameter	Q-1	42	.381	L.F.	80	22		102	121
1330	4" diameter		38	.421		97.50	24.50		122	144
1340	5" diameter		32	.500		140	29		169	198
1360	6" diameter	Q-2	38	.632		207	38		245	284
1380	8" diameter	"	34	.706		360	42.50		402.50	465
1390	For other than full hard temper, add					13%				
1440	For silver solder, add						15%			
1800	For medical clean (oxygen class), add					12%				
1950	To delete cplgs. & hngrs., 1/4"-1" pipe, subtract					27%	60%			
1960	1-1/4" -3" pipe, subtract					14%	52%			
1970	3-1/2"-5" pipe, subtract					10%	60%			
1980	6"-8" pipe, subtract					19%	53%			
2000	Type L tubing, couplings & clevis hanger assemblies 10' OC									
2100	1/4" diameter	1 Plum	88	.091	L.F.	2.67	5.85		8.52	11.70
2120	3/8" diameter		84	.095		3.34	6.15		9.49	12.85
2140	1/2" diameter		81	.099		3.68	6.35		10.03	13.60
2160	5/8" diameter		79	.101		5.75	6.55		12.30	16.05
2180	3/4" diameter		76	.105		4.71	6.80		11.51	15.35
2200	1" diameter		68	.118		7.60	7.60		15.20	19.70
2220	1-1/4" diameter		58	.138		12	8.90		20.90	26.50
2240	1-1/2" diameter		52	.154		11.60	9.90		21.50	27.50
2260	2" diameter		42	.190		18.85	12.30		31.15	39.50
2280	2-1/2" diameter	Q-1	62	.258		30	14.95		44.95	55.50
2300	3" diameter		56	.286		46.50	16.55		63.05	76
2320	3-1/2" diameter		43	.372		60	21.50		81.50	98.50
2340	4" diameter		39	.410		65.50	24		89.50	108
2360	5" diameter		34	.471		148	27.50		175.50	203
2380	6" diameter	Q-2	40	.600		151	36		187	220
2400	8" diameter	"	36	.667		250	40		290	335
2410	For other than full hard temper, add					21%				
2590	For silver solder, add						15%			
2900	For medical clean (oxygen class), add					12%				
2940	To delete cplgs. & hngrs., 1/4"-1" pipe, subtract					37%	63%			
2960	1-1/4"-3" pipe, subtract					12%	53%			
2970	3-1/2"-5" pipe, subtract					12%	63%			
2980	6"-8" pipe, subtract					24%	55%			
3000	Type M tubing, couplings & clevis hanger assemblies 10' OC									
3100	1/4" diameter	1 Plum	90	.089	L.F.	3.72	5.75		9.47	12.65
3120	3/8" diameter		87	.092		4.09	5.95		10.04	13.35
3140	1/2" diameter		84	.095		3.56	6.15		9.71	13.10
3160	5/8" diameter		81	.099		4.94	6.35		11.29	15
3180	3/4" diameter		78	.103		5.05	6.60		11.65	15.45
3200	1" diameter		70	.114		8.50	7.35		15.85	20.50
3220	1-1/4" diameter		60	.133		12.30	8.60		20.90	26.50
3240	1-1/2" diameter		54	.148		14.45	9.55		24	30
3260	2" diameter		44	.182		21.50	11.70		33.20	41
3280	2-1/2" diameter	Q-1	64	.250		31.50	14.50		46	56.50
3300	3" diameter		58	.276		39.50	16		55.50	67.50
3320	3-1/2" diameter		45	.356		58.50	20.50		79	95
3340	4" diameter		40	.400		75	23		98	117
3360	5" diameter		36	.444		138	26		164	191
3370	6" diameter	Q-2	42	.571		197	34.50		231.50	269
3380	8" diameter	"	38	.632		325	38		363	415

22 11 Facility Water Distribution

22 11 13 – Facility Water Distribution Piping

22 11 13.23 Pipe/Tube, Copper		Crew	Daily Output	Labor-Hours	Unit	2020 Bare Costs Material	Labor	Equipment	Total	Total Incl O&P
3440	For silver solder, add						15%			
3960	To delete cplgs. & hngrs., 1/4"-1" pipe, subtract					35%	65%			
3970	1-1/4"-3" pipe, subtract					19%	56%			
3980	3-1/2"-5" pipe, subtract					13%	65%			
3990	6"-8" pipe, subtract					28%	58%			
4000	Type DWV tubing, couplings & clevis hanger assemblies 10' OC									
4100	1-1/4" diameter	1 Plum	60	.133	L.F.	12.45	8.60		21.05	26.50
4120	1-1/2" diameter		54	.148		12.60	9.55		22.15	28
4140	2" diameter		44	.182		18.30	11.70		30	37.50
4160	3" diameter	Q-1	58	.276		29	16		45	56
4180	4" diameter		40	.400		60.50	23		83.50	101
4200	5" diameter		36	.444		110	26		136	160
4220	6" diameter	Q-2	42	.571		161	34.50		195.50	229
4240	8" diameter	"	38	.632		510	38		548	620
4730	To delete cplgs. & hngrs., 1-1/4"-2" pipe, subtract					16%	53%			
4740	3"-4" pipe, subtract					13%	60%			
4750	5"-8" pipe, subtract					23%	58%			
5200	ACR tubing, type L, hard temper, cleaned and									
5220	capped, no couplings or hangers									
5240	3/8" OD				L.F.	1.76			1.76	1.94
5250	1/2" OD					2.64			2.64	2.90
5260	5/8" OD					3.30			3.30	3.63
5270	3/4" OD					4.48			4.48	4.93
5280	7/8" OD					5.05			5.05	5.55
5290	1-1/8" OD					7.20			7.20	7.90
5300	1-3/8" OD					9.65			9.65	10.60
5310	1-5/8" OD					12.60			12.60	13.85
5320	2-1/8" OD					20.50			20.50	22.50
5330	2-5/8" OD					27.50			27.50	30.50
5340	3-1/8" OD					34			34	37.50
5350	3-5/8" OD					63.50			63.50	70
5360	4-1/8" OD					65			65	72
5380	ACR tubing, type L, hard, cleaned and capped									
5381	No couplings or hangers									
5384	3/8"	1 Stpi	160	.050	L.F.	1.76	3.28		5.04	6.85
5385	1/2"		160	.050		2.64	3.28		5.92	7.80
5386	5/8"		160	.050		3.30	3.28		6.58	8.55
5387	3/4"		130	.062		4.48	4.03		8.51	11
5388	7/8"		130	.062		5.05	4.03		9.08	11.60
5389	1-1/8"		115	.070		7.20	4.56		11.76	14.70
5390	1-3/8"		100	.080		9.65	5.25		14.90	18.45
5391	1-5/8"		90	.089		12.60	5.85		18.45	22.50
5392	2-1/8"		80	.100		20.50	6.55		27.05	32.50
5393	2-5/8"	Q-5	125	.128		27.50	7.55		35.05	42
5394	3-1/8"		105	.152		34	9		43	51
5395	4-1/8"		95	.168		65	9.95		74.95	87
5800	Refrigeration tubing, dryseal, 50' coils									
5840	1/8" OD				Coil	34			34	37.50
5850	3/16" OD					46			46	50.50
5860	1/4" OD					46.50			46.50	51
5870	5/16" OD					71			71	78
5880	3/8" OD					64.50			64.50	71
5890	1/2" OD					93			93	102

22 11 13.23 Pipe/Tube, Copper

	22 11 13.23 Pipe/Tube, Copper	Crew	Daily Output	Labor-Hours	Unit	Material	Labor	Equipment	Total	Total Incl O&P
5900	5/8" OD				Coil	124			124	136
5910	3/4" OD					145			145	159
5920	7/8" OD					147			147	162
5930	1-1/8" OD					330			330	365
5940	1-3/8" OD					555			555	615
5950	1-5/8" OD					710			710	780

22 11 13.25 Pipe/Tube Fittings, Copper

	22 11 13.25 Pipe/Tube Fittings, Copper	Crew	Daily Output	Labor-Hours	Unit	Material (2020 Bare Costs)	Labor (2020 Bare Costs)	Equipment (2020 Bare Costs)	Total (2020 Bare Costs)	Total Incl O&P
0010	**PIPE/TUBE FITTINGS, COPPER**, Wrought unless otherwise noted									
0020	For silver solder, add						15%			
0040	Solder joints, copper x copper									
0070	90° elbow, 1/4"	1 Plum	22	.364	Ea.	3.88	23.50		27.38	39.50
0090	3/8"		22	.364		4.15	23.50		27.65	39.50
0100	1/2"		20	.400		1.26	26		27.26	40
0110	5/8"		19	.421		2.77	27		29.77	43.50
0120	3/4"		19	.421		2.68	27		29.68	43.50
0130	1"		16	.500		7.05	32		39.05	56
0140	1-1/4"		15	.533		11.65	34.50		46.15	64.50
0150	1-1/2"		13	.615		16.10	39.50		55.60	77
0160	2"		11	.727		29.50	47		76.50	103
0170	2-1/2"	Q-1	13	1.231		51	71.50		122.50	163
0180	3"		11	1.455		86.50	84.50		171	221
0190	3-1/2"		10	1.600		281	93		374	450
0200	4"		9	1.778		236	103		339	415
0210	5"		6	2.667		775	155		930	1,075
0220	6"	Q-2	9	2.667		1,025	160		1,185	1,375
0230	8"	"	8	3		3,800	180		3,980	4,475
0250	45° elbow, 1/4"	1 Plum	22	.364		7.75	23.50		31.25	43.50
0270	3/8"		22	.364		6.80	23.50		30.30	42.50
0280	1/2"		20	.400		2.75	26		28.75	41.50
0290	5/8"		19	.421		11.45	27		38.45	53
0300	3/4"		19	.421		4.46	27		31.46	45.50
0310	1"		16	.500		11.20	32		43.20	60.50
0320	1-1/4"		15	.533		16	34.50		50.50	69
0330	1-1/2"		13	.615		18.20	39.50		57.70	79.50
0340	2"		11	.727		30.50	47		77.50	104
0350	2-1/2"	Q-1	13	1.231		49.50	71.50		121	162
0360	3"		13	1.231		88	71.50		159.50	204
0370	3-1/2"		10	1.600		122	93		215	273
0380	4"		9	1.778		173	103		276	345
0390	5"		6	2.667		635	155		790	930
0400	6"	Q-2	9	2.667		995	160		1,155	1,350
0410	8"	"	8	3		4,300	180		4,480	5,000
0450	Tee, 1/4"	1 Plum	14	.571		8.45	37		45.45	64.50
0470	3/8"		14	.571		6.85	37		43.85	62.50
0480	1/2"		13	.615		2.44	39.50		41.94	62
0490	5/8"		12	.667		16.25	43		59.25	82.50
0500	3/4"		12	.667		6.05	43		49.05	71
0510	1"		10	.800		17.05	51.50		68.55	96
0520	1-1/4"		9	.889		24.50	57.50		82	113
0530	1-1/2"		8	1		36.50	64.50		101	137
0540	2"		7	1.143		58	73.50		131.50	174
0550	2-1/2"	Q-1	8	2		111	116		227	296

22 11 13.25 Pipe/Tube Fittings, Copper		Crew	Daily Output	Labor-Hours	Unit	2020 Bare Costs Material	Labor	Equipment	Total	Total Incl O&P
0560	3"	Q-1	7	2.286	Ea.	158	133		291	370
0570	3-1/2"		6	2.667		475	155		630	750
0580	4"		5	3.200		365	186		551	680
0590	5"		4	4		1,075	232		1,307	1,550
0600	6"	Q-2	6	4		1,475	241		1,716	1,975
0610	8"	"	5	4.800		5,875	289		6,164	6,900
0612	Tee, reducing on the outlet, 1/4"	1 Plum	15	.533		17.05	34.50		51.55	70.50
0613	3/8"		15	.533		15.55	34.50		50.05	68.50
0614	1/2"		14	.571		15.60	37		52.60	72
0615	5/8"		13	.615		30.50	39.50		70	93
0616	3/4"		12	.667		8	43		51	73.50
0617	1"		11	.727		29	47		76	102
0618	1-1/4"		10	.800		32.50	51.50		84	113
0619	1-1/2"		9	.889		32	57.50		89.50	121
0620	2"		8	1		64.50	64.50		129	168
0621	2-1/2"	Q-1	9	1.778		137	103		240	305
0622	3"		8	2		163	116		279	355
0623	4"		6	2.667		299	155		454	560
0624	5"		5	3.200		1,725	186		1,911	2,175
0625	6"	Q-2	7	3.429		2,150	206		2,356	2,675
0626	8"	"	6	4		10,800	241		11,041	12,300
0630	Tee, reducing on the run, 1/4"	1 Plum	15	.533		23.50	34.50		58	77.50
0631	3/8"		15	.533		31	34.50		65.50	85.50
0632	1/2"		14	.571		20.50	37		57.50	77.50
0633	5/8"		13	.615		29	39.50		68.50	91.50
0634	3/4"		12	.667		21	43		64	87.50
0635	1"		11	.727		26	47		73	98.50
0636	1-1/4"		10	.800		41	51.50		92.50	122
0637	1-1/2"		9	.889		72	57.50		129.50	165
0638	2"		8	1		82.50	64.50		147	188
0639	2-1/2"	Q-1	9	1.778		177	103		280	350
0640	3"		8	2		320	116		436	530
0641	4"		6	2.667		585	155		740	875
0642	5"		5	3.200		1,950	186		2,136	2,425
0643	6"	Q-2	7	3.429		2,950	206		3,156	3,550
0644	8"	"	6	4		10,200	241		10,441	11,600
0650	Coupling, 1/4"	1 Plum	24	.333		1.07	21.50		22.57	33
0670	3/8"		24	.333		1.53	21.50		23.03	33.50
0680	1/2"		22	.364		.96	23.50		24.46	36
0690	5/8"		21	.381		4.23	24.50		28.73	41
0700	3/4"		21	.381		2.71	24.50		27.21	39.50
0710	1"		18	.444		5.30	28.50		33.80	49
0715	1-1/4"		17	.471		7.80	30.50		38.30	54
0716	1-1/2"		15	.533		10.80	34.50		45.30	63.50
0718	2"		13	.615		15.30	39.50		54.80	76.50
0721	2-1/2"	Q-1	15	1.067		40.50	62		102.50	137
0722	3"		13	1.231		47	71.50		118.50	159
0724	3-1/2"		8	2		124	116		240	310
0726	4"		7	2.286		154	133		287	370
0728	5"		6	2.667		228	155		383	480
0731	6"	Q-2	8	3		355	180		535	660
0732	8"	"	7	3.429		1,275	206		1,481	1,700
0741	Coupling, reducing, concentric									

22 11 13.25 Pipe/Tube Fittings, Copper		Crew	Daily Output	Labor-Hours	Unit	Material	2020 Bare Costs Labor	Equipment	Total	Total Incl O&P
0743	1/2"	1 Plum	23	.348	Ea.	2.85	22.50		25.35	36.50
0745	3/4"		21.50	.372		5.65	24		29.65	42
0747	1"		19.50	.410		8.80	26.50		35.30	49
0748	1-1/4"		18	.444		12.15	28.50		40.65	56.50
0749	1-1/2"		16	.500		17.05	32		49.05	67
0751	2"		14	.571		25.50	37		62.50	83
0752	2-1/2"		13	.615		56	39.50		95.50	121
0753	3"	Q-1	14	1.143		44.50	66.50		111	148
0755	4"	"	8	2		133	116		249	320
0757	5"	Q-2	7.50	3.200		755	192		947	1,125
0759	6"		7	3.429		1,200	206		1,406	1,625
0761	8"		6.50	3.692		3,825	222		4,047	4,525
0771	Cap, sweat									
0773	1/2"	1 Plum	40	.200	Ea.	1.27	12.90		14.17	20.50
0775	3/4"		38	.211		2.36	13.55		15.91	23
0777	1"		32	.250		5.50	16.10		21.60	30
0778	1-1/4"		29	.276		6.95	17.80		24.75	34
0779	1-1/2"		26	.308		12.80	19.85		32.65	43.50
0781	2"		22	.364		19.70	23.50		43.20	56.50
0791	Flange, sweat									
0793	3"	Q-1	22	.727	Ea.	275	42		317	370
0795	4"		18	.889		415	51.50		466.50	530
0797	5"		12	1.333		720	77.50		797.50	905
0799	6"	Q-2	18	1.333		750	80		830	945
0801	8"	"	16	1.500		1,275	90		1,365	1,525
0850	Unions, 1/4"	1 Plum	21	.381		41	24.50		65.50	82
0870	3/8"		21	.381		41.50	24.50		66	82
0880	1/2"		19	.421		22.50	27		49.50	65
0890	5/8"		18	.444		89	28.50		117.50	141
0900	3/4"		18	.444		23	28.50		51.50	68
0910	1"		15	.533		47.50	34.50		82	104
0920	1-1/4"		14	.571		78.50	37		115.50	142
0930	1-1/2"		12	.667		118	43		161	195
0940	2"		10	.800		152	51.50		203.50	245
0950	2-1/2"	Q-1	12	1.333		385	77.50		462.50	540
0960	3"	"	10	1.600		1,000	93		1,093	1,250
0980	Adapter, copper x male IPS, 1/4"	1 Plum	20	.400		13.55	26		39.55	53.50
0990	3/8"		20	.400		8.50	26		34.50	48
1000	1/2"		18	.444		3.64	28.50		32.14	47
1010	3/4"		17	.471		6.10	30.50		36.60	52
1020	1"		15	.533		15.55	34.50		50.05	68.50
1030	1-1/4"		13	.615		26.50	39.50		66	88.50
1040	1-1/2"		12	.667		26.50	43		69.50	93.50
1050	2"		11	.727		44.50	47		91.50	119
1060	2-1/2"	Q-1	10.50	1.524		182	88.50		270.50	330
1070	3"		10	1.600		226	93		319	390
1080	3-1/2"		9	1.778		219	103		322	395
1090	4"		8	2		247	116		363	445
1200	5", cast		6	2.667		2,200	155		2,355	2,650
1210	6", cast	Q-2	8.50	2.824		2,475	170		2,645	2,975
1214	Adapter, copper x female IPS									
1216	1/2"	1 Plum	18	.444	Ea.	5.75	28.50		34.25	49.50
1218	3/4"		17	.471		7.90	30.50		38.40	54

22 11 13.25 Pipe/Tube Fittings, Copper		Crew	Daily Output	Labor-Hours	Unit	2020 Bare Costs Material	Labor	Equipment	Total	Total Incl O&P
1220	1"	1 Plum	15	.533	Ea.	18.15	34.50		52.65	71.50
1221	1-1/4"		13	.615		26	39.50		65.50	88
1222	1-1/2"		12	.667		41.50	43		84.50	110
1224	2"		11	.727		56	47		103	132
1250	Cross, 1/2"		10	.800		28.50	51.50		80	109
1260	3/4"		9.50	.842		55	54.50		109.50	142
1270	1"		8	1		93.50	64.50		158	200
1280	1-1/4"		7.50	1.067		135	69		204	251
1290	1-1/2"		6.50	1.231		192	79.50		271.50	330
1300	2"		5.50	1.455		365	94		459	540
1310	2-1/2"	Q-1	6.50	2.462		835	143		978	1,125
1320	3"	"	5.50	2.909		710	169		879	1,025
1500	Tee fitting, mechanically formed (Type 1, 'branch sizes up to 2 in.')									
1520	1/2" run size, 3/8" to 1/2" branch size	1 Plum	80	.100	Ea.		6.45		6.45	9.65
1530	3/4" run size, 3/8" to 3/4" branch size		60	.133			8.60		8.60	12.85
1540	1" run size, 3/8" to 1" branch size		54	.148			9.55		9.55	14.30
1550	1-1/4" run size, 3/8" to 1-1/4" branch size		48	.167			10.75		10.75	16.10
1560	1-1/2" run size, 3/8" to 1-1/2" branch size		40	.200			12.90		12.90	19.30
1570	2" run size, 3/8" to 2" branch size		35	.229			14.75		14.75	22
1580	2-1/2" run size, 1/2" to 2" branch size		32	.250			16.10		16.10	24
1590	3" run size, 1" to 2" branch size		26	.308			19.85		19.85	29.50
1600	4" run size, 1" to 2" branch size		24	.333			21.50		21.50	32
1640	Tee fitting, mechanically formed (Type 2, branches 2-1/2" thru 4")									
1650	2-1/2" run size, 2-1/2" branch size	1 Plum	12.50	.640	Ea.		41.50		41.50	61.50
1660	3" run size, 2-1/2" to 3" branch size		12	.667			43		43	64.50
1670	3-1/2" run size, 2-1/2" to 3-1/2" branch size		11	.727			47		47	70
1680	4" run size, 2-1/2" to 4" branch size		10.50	.762			49		49	73.50
1698	5" run size, 2" to 4" branch size		9.50	.842			54.50		54.50	81
1700	6" run size, 2" to 4" branch size		8.50	.941			60.50		60.50	91
1710	8" run size, 2" to 4" branch size		7	1.143			73.50		73.50	110
1800	ACR fittings, OD size									
1802	Tee, straight									
1808	5/8"	1 Stpi	12	.667	Ea.	2.96	43.50		46.46	69
1810	3/4"		12	.667		22	43.50		65.50	89.50
1812	7/8"		10	.800		7.15	52.50		59.65	86.50
1813	1"		10	.800		60	52.50		112.50	145
1814	1-1/8"		10	.800		25.50	52.50		78	107
1816	1-3/8"		9	.889		29.50	58.50		88	120
1818	1-5/8"		8	1		53	65.50		118.50	156
1820	2-1/8"		7	1.143		71	75		146	191
1822	2-5/8"	Q-5	8	2		153	118		271	345
1824	3-1/8"		7	2.286		220	135		355	445
1826	4-1/8"		5	3.200		415	189		604	745
1830	90° elbow									
1836	5/8"	1 Stpi	19	.421	Ea.	6.10	27.50		33.60	48
1838	3/4"		19	.421		11.05	27.50		38.55	53.50
1840	7/8"		16	.500		11	33		44	61
1842	1-1/8"		16	.500		14.70	33		47.70	65
1844	1-3/8"		15	.533		14.20	35		49.20	68
1846	1-5/8"		13	.615		22	40.50		62.50	85
1848	2-1/8"		11	.727		40.50	47.50		88	116
1850	2-5/8"	Q-5	13	1.231		86	72.50		158.50	204
1852	3-1/8"		11	1.455		104	86		190	243

22 11 13.25 Pipe/Tube Fittings, Copper		Crew	Daily Output	Labor-Hours	Unit	2020 Bare Costs Material	Labor	Equipment	Total	Total Incl O&P
1854	4-1/8"	Q-5	9	1.778	Ea.	228	105		333	410
1860	Coupling									
1866	5/8"	1 Stpi	21	.381	Ea.	1.27	25		26.27	39
1868	3/4"		21	.381		3.46	25		28.46	41.50
1870	7/8"		18	.444		2.49	29		31.49	46
1871	1"		18	.444		8.55	29		37.55	53
1872	1-1/8"		18	.444		5.25	29		34.25	49.50
1874	1-3/8"		17	.471		9.20	31		40.20	56
1876	1-5/8"		15	.533		12.15	35		47.15	66
1878	2-1/8"		13	.615		20.50	40.50		61	83
1880	2-5/8"	Q-5	15	1.067		39.50	63		102.50	138
1882	3-1/8"		13	1.231		53.50	72.50		126	168
1884	4-1/8"		7	2.286		118	135		253	330
2000	DWV, solder joints, copper x copper									
2030	90° elbow, 1-1/4"	1 Plum	13	.615	Ea.	18.50	39.50		58	80
2050	1-1/2"		12	.667		24.50	43		67.50	91.50
2070	2"		10	.800		46	51.50		97.50	128
2090	3"	Q-1	10	1.600		85.50	93		178.50	233
2100	4"	"	9	1.778		545	103		648	755
2150	45° elbow, 1-1/4"	1 Plum	13	.615		15.15	39.50		54.65	76
2170	1-1/2"		12	.667		14.05	43		57.05	80
2180	2"		10	.800		29	51.50		80.50	109
2190	3"	Q-1	10	1.600		59	93		152	204
2200	4"	"	9	1.778		94	103		197	258
2250	Tee, sanitary, 1-1/4"	1 Plum	9	.889		28.50	57.50		86	117
2270	1-1/2"		8	1		35.50	64.50		100	136
2290	2"		7	1.143		55	73.50		128.50	171
2310	3"	Q-1	7	2.286		213	133		346	435
2330	4"	"	6	2.667		520	155		675	805
2400	Coupling, 1-1/4"	1 Plum	14	.571		7.75	37		44.75	63.50
2420	1-1/2"		13	.615		9.60	39.50		49.10	70
2440	2"		11	.727		13.35	47		60.35	84.50
2460	3"	Q-1	11	1.455		30.50	84.50		115	160
2480	4"	"	10	1.600		68	93		161	214
2602	Traps, see Section 22 13 16.60									
3500	Compression joint fittings									
3510	As used for plumbing and oil burner work									
3520	Fitting price includes nuts and sleeves									
3540	Sleeve, 1/8"				Ea.	.15			.15	.17
3550	3/16"					.15			.15	.17
3560	1/4"					.05			.05	.06
3570	5/16"					.15			.15	.17
3580	3/8"					.30			.30	.33
3600	1/2"					.41			.41	.45
3620	Nut, 1/8"					.20			.20	.22
3630	3/16"					.23			.23	.25
3640	1/4"					.21			.21	.23
3650	5/16"					.34			.34	.37
3660	3/8"					.36			.36	.40
3670	1/2"					.64			.64	.70
3710	Union, 1/8"	1 Plum	26	.308		1.40	19.85		21.25	31
3720	3/16"		24	.333		1.33	21.50		22.83	33.50
3730	1/4"		24	.333		1.80	21.50		23.30	34

22 11 Facility Water Distribution

22 11 13 – Facility Water Distribution Piping

22 11 13.25 Pipe/Tube Fittings, Copper		Crew	Daily Output	Labor-Hours	Unit	Material	Labor	Equipment	Total	Total Incl O&P
						2020 Bare Costs				
3740	5/16"	1 Plum	23	.348	Ea.	2.30	22.50		24.80	36
3750	3/8"		22	.364		2.04	23.50		25.54	37
3760	1/2"		22	.364		3.21	23.50		26.71	38.50
3780	5/8"		21	.381		2.94	24.50		27.44	39.50
3820	Union tee, 1/8"		17	.471		5.30	30.50		35.80	51.50
3830	3/16"		16	.500		3.06	32		35.06	51.50
3840	1/4"		15	.533		3.49	34.50		37.99	55.50
3850	5/16"		15	.533		3.89	34.50		38.39	56
3860	3/8"		15	.533		3.58	34.50		38.08	55.50
3870	1/2"		15	.533		7.85	34.50		42.35	60
3910	Union elbow, 1/4"		24	.333		1.94	21.50		23.44	34
3920	5/16"		23	.348		2.73	22.50		25.23	36.50
3930	3/8"		22	.364		3.70	23.50		27.20	39
3940	1/2"		22	.364		5.55	23.50		29.05	41
3980	Female connector, 1/8"		26	.308		.97	19.85		20.82	30.50
4000	3/16" x 1/8"		24	.333		1.37	21.50		22.87	33.50
4010	1/4" x 1/8"		24	.333		1.53	21.50		23.03	33.50
4020	1/4"		24	.333		2.06	21.50		23.56	34.50
4030	3/8" x 1/4"		22	.364		2.06	23.50		25.56	37.50
4040	1/2" x 3/8"		22	.364		3.27	23.50		26.77	38.50
4050	5/8" x 1/2"		21	.381		3.85	24.50		28.35	40.50
4090	Male connector, 1/8"		26	.308		1.41	19.85		21.26	31
4100	3/16" x 1/8"		24	.333		.90	21.50		22.40	33
4110	1/4" x 1/8"		24	.333		1.19	21.50		22.69	33.50
4120	1/4"		24	.333		1.19	21.50		22.69	33.50
4130	5/16" x 1/8"		23	.348		1.41	22.50		23.91	35
4140	5/16" x 1/4"		23	.348		1.24	22.50		23.74	35
4150	3/8" x 1/8"		22	.364		1.67	23.50		25.17	37
4160	3/8" x 1/4"		22	.364		1.91	23.50		25.41	37
4170	3/8"		22	.364		1.92	23.50		25.42	37
4180	3/8" x 1/2"		22	.364		2.61	23.50		26.11	38
4190	1/2" x 3/8"		22	.364		2.55	23.50		26.05	38
4200	1/2"		22	.364		2.85	23.50		26.35	38
4210	5/8" x 1/2"		21	.381		3.30	24.50		27.80	40
4240	Male elbow, 1/8"		26	.308		2.30	19.85		22.15	32
4250	3/16" x 1/8"		24	.333		1.97	21.50		23.47	34
4260	1/4" x 1/8"		24	.333		1.91	21.50		23.41	34
4270	1/4"		24	.333		2.17	21.50		23.67	34.50
4280	3/8" x 1/4"		22	.364		2.76	23.50		26.26	38
4290	3/8"		22	.364		4.73	23.50		28.23	40
4300	1/2" x 1/4"		22	.364		4.58	23.50		28.08	40
4310	1/2" x 3/8"		22	.364		7.20	23.50		30.70	43
4340	Female elbow, 1/8"		26	.308		3.86	19.85		23.71	34
4350	1/4" x 1/8"		24	.333		2.81	21.50		24.31	35
4360	1/4"		24	.333		3.68	21.50		25.18	36
4370	3/8" x 1/4"		22	.364		3.62	23.50		27.12	39
4380	1/2" x 3/8"		22	.364		4.84	23.50		28.34	40.50
4390	1/2"		22	.364		6.50	23.50		30	42
4420	Male run tee, 1/4" x 1/8"		15	.533		3.43	34.50		37.93	55.50
4430	5/16" x 1/8"		15	.533		5.15	34.50		39.65	57
4440	3/8" x 1/4"		15	.533		5.30	34.50		39.80	57.50
4480	Male branch tee, 1/4" x 1/8"		15	.533		3.98	34.50		38.48	56
4490	1/4"		15	.533		4.28	34.50		38.78	56

22 11 Facility Water Distribution

22 11 13 – Facility Water Distribution Piping

22 11 13.25	Pipe/Tube Fittings, Copper	Crew	Daily Output	Labor-Hours	Unit	2020 Bare Costs Material	Labor	Equipment	Total	Total Incl O&P
4500	3/8" x 1/4"	1 Plum	15	.533	Ea.	3.90	34.50		38.40	56
4510	1/2" x 3/8"		15	.533		6.05	34.50		40.55	58
4520	1/2"		15	.533		9.70	34.50		44.20	62
4800	Flare joint fittings									
4810	Refrigeration fittings									
4820	Flare joint nuts and labor not incl. in price. Add 1 nut per jnt.									
4830	90° elbow, 1/4"				Ea.	2.51			2.51	2.76
4840	3/8"					2.07			2.07	2.28
4850	1/2"					2.66			2.66	2.93
4860	5/8"					3.66			3.66	4.03
4870	3/4"					11.90			11.90	13.10
5030	Tee, 1/4"					2.20			2.20	2.42
5040	5/16"					2.58			2.58	2.84
5050	3/8"					2.43			2.43	2.67
5060	1/2"					3.67			3.67	4.04
5070	5/8"					5.30			5.30	5.80
5080	3/4"					6.75			6.75	7.45
5140	Union, 3/16"					.91			.91	1
5150	1/4"					.69			.69	.76
5160	5/16"					1.27			1.27	1.40
5170	3/8"					.96			.96	1.06
5180	1/2"					1.43			1.43	1.57
5190	5/8"					2.29			2.29	2.52
5200	3/4"					7.80			7.80	8.60
5260	Long flare nut, 3/16"	1 Stpi	42	.190		.40	12.50		12.90	19.15
5270	1/4"		41	.195		.72	12.80		13.52	19.95
5280	5/16"		40	.200		1.30	13.10		14.40	21
5290	3/8"		39	.205		2.06	13.45		15.51	22.50
5300	1/2"		38	.211		2.39	13.80		16.19	23
5310	5/8"		37	.216		4.09	14.15		18.24	25.50
5320	3/4"		34	.235		16.20	15.40		31.60	41
5380	Short flare nut, 3/16"		42	.190		4.11	12.50		16.61	23
5390	1/4"		41	.195		.50	12.80		13.30	19.70
5400	5/16"		40	.200		.60	13.10		13.70	20.50
5410	3/8"		39	.205		.69	13.45		14.14	21
5420	1/2"		38	.211		1.18	13.80		14.98	22
5430	5/8"		36	.222		1.65	14.55		16.20	24
5440	3/4"		34	.235		5.75	15.40		21.15	29.50
5500	90° elbow flare by MIPS, 1/4"					2.10			2.10	2.31
5510	3/8"					2.04			2.04	2.24
5520	1/2"					3.68			3.68	4.05
5530	5/8"					6.95			6.95	7.65
5540	3/4"					12.30			12.30	13.55
5600	Flare by FIPS, 1/4"					3.78			3.78	4.16
5610	3/8"					3.41			3.41	3.75
5620	1/2"					4.83			4.83	5.30
5670	Flare by sweat, 1/4"					2.29			2.29	2.52
5680	3/8"					5.05			5.05	5.60
5690	1/2"					6.25			6.25	6.85
5700	5/8"					16.50			16.50	18.15
5760	Tee flare by IPS, 1/4"					3.14			3.14	3.45
5770	3/8"					4.92			4.92	5.40
5780	1/2"					8.30			8.30	9.10

22 11 13.25 Pipe/Tube Fittings, Copper		Crew	Daily Output	Labor-Hours	Unit	2020 Bare Costs Material	Labor	Equipment	Total	Total Incl O&P
5790	5/8"				Ea.	6.45			6.45	7.10
5850	Connector, 1/4"					.74			.74	.81
5860	3/8"					1.10			1.10	1.21
5870	1/2"					1.85			1.85	2.04
5880	5/8"					3.77			3.77	4.15
5890	3/4"					5.65			5.65	6.20
5950	Seal cap, 1/4"					.48			.48	.53
5960	3/8"					.95			.95	1.05
5970	1/2"					.97			.97	1.07
5980	5/8"					2.38			2.38	2.62
5990	3/4"					6.05			6.05	6.65
6000	Water service fittings									
6010	Flare joints nut and labor are included in the fitting price.									
6020	90° elbow, C x C, 3/8"	1 Plum	19	.421	Ea.	57	27		84	103
6030	1/2"		18	.444		70.50	28.50		99	121
6040	3/4"		16	.500		88	32		120	145
6050	1"		15	.533		148	34.50		182.50	215
6080	2"		10	.800		615	51.50		666.50	755
6090	90° elbow, C x MPT, 3/8"		19	.421		43	27		70	88
6100	1/2"		18	.444		53.50	28.50		82	102
6110	3/4"		16	.500		62	32		94	116
6120	1"		15	.533		170	34.50		204.50	239
6130	1-1/4"		13	.615		375	39.50		414.50	475
6140	1-1/2"		12	.667		335	43		378	430
6150	2"		10	.800		305	51.50		356.50	410
6160	90° elbow, C x FPT, 3/8"		19	.421		46.50	27		73.50	91.50
6170	1/2"		18	.444		57.50	28.50		86	106
6180	3/4"		16	.500		71	32		103	126
6190	1"		15	.533		164	34.50		198.50	232
6200	1-1/4"		13	.615		124	39.50		163.50	196
6210	1-1/2"		12	.667		475	43		518	590
6220	2"		10	.800		685	51.50		736.50	830
6230	Tee, C x C x C, 3/8"		13	.615		90	39.50		129.50	159
6240	1/2"		12	.667		103	43		146	178
6250	3/4"		11	.727		125	47		172	207
6260	1"		10	.800		187	51.50		238.50	282
6330	Tube nut, C x nut seat, 3/8"		40	.200		22	12.90		34.90	43.50
6340	1/2"		38	.211		22	13.55		35.55	44.50
6350	3/4"		34	.235		18.65	15.15		33.80	43
6360	1"		32	.250		38	16.10		54.10	65.50
6380	Coupling, C x C, 3/8"		19	.421		55.50	27		82.50	102
6390	1/2"		18	.444		63	28.50		91.50	113
6400	3/4"		16	.500		80	32		112	136
6410	1"		15	.533		148	34.50		182.50	214
6420	1-1/4"		12	.667		239	43		282	330
6430	1-1/2"		12	.667		360	43		403	465
6440	2"		10	.800		545	51.50		596.50	675
6450	Adapter, C x FPT, 3/8"		19	.421		42	27		69	86.50
6460	1/2"		18	.444		47.50	28.50		76	95.50
6470	3/4"		16	.500		57.50	32		89.50	112
6480	1"		15	.533		123	34.50		157.50	187
6490	1-1/4"		13	.615		108	39.50		147.50	179
6500	1-1/2"		12	.667		261	43		304	350

22 11 13.25 Pipe/Tube Fittings, Copper

	22 11 13.25 Pipe/Tube Fittings, Copper	Crew	Daily Output	Labor-Hours	Unit	2020 Bare Costs Material	Labor	Equipment	Total	Total Incl O&P
6510	2"	1 Plum	10	.800	Ea.	297	51.50		348.50	400
6520	Adapter, C x MPT, 3/8"		19	.421		34.50	27		61.50	78.50
6530	1/2"		18	.444		42.50	28.50		71	90
6540	3/4"		16	.500		59	32		91	113
6550	1"		15	.533		104	34.50		138.50	166
6560	1-1/4"		13	.615		227	39.50		266.50	310
6570	1-1/2"		12	.667		261	43		304	355
6580	2"		10	.800		315	51.50		366.50	420
6992	Tube connector fittings, See Section 22 11 13.76 for plastic ftng.									
7000	Insert type brass/copper, 100 psi @ 180°F, CTS									
7010	Adapter MPT 3/8" x 1/2" CTS	1 Plum	29	.276	Ea.	2.90	17.80		20.70	29.50
7020	1/2" x 1/2"		26	.308		3	19.85		22.85	33
7030	3/4" x 1/2"		26	.308		3.82	19.85		23.67	33.50
7040	3/4" x 3/4"		25	.320		4.52	20.50		25.02	36
7050	Adapter CTS 1/2" x 1/2" sweat		24	.333		3.82	21.50		25.32	36
7060	3/4" x 3/4" sweat		22	.364		1.43	23.50		24.93	36.50
7070	Coupler center set 3/8" CTS		25	.320		1.43	20.50		21.93	32.50
7080	1/2" CTS		23	.348		3.67	22.50		26.17	37.50
7090	3/4" CTS		22	.364		1.47	23.50		24.97	36.50
7100	Elbow 90°, copper 3/8"		25	.320		2.98	20.50		23.48	34.50
7110	1/2" CTS		23	.348		2	22.50		24.50	35.50
7120	3/4" CTS		22	.364		2.63	23.50		26.13	38
7130	Tee copper 3/8" CTS		17	.471		3.79	30.50		34.29	49.50
7140	1/2" CTS		15	.533		2.67	34.50		37.17	54.50
7150	3/4" CTS		14	.571		4.16	37		41.16	59.50
7160	3/8" x 3/8" x 1/2"		16	.500		4.21	32		36.21	52.50
7170	1/2" x 3/8" x 1/2"		15	.533		2.84	34.50		37.34	54.50
7180	3/4" x 1/2" x 3/4"		14	.571		3.91	37		40.91	59.50

22 11 13.27 Pipe/Tube, Grooved Joint for Copper

	22 11 13.27 Pipe/Tube, Grooved Joint for Copper	Crew	Daily Output	Labor-Hours	Unit	Material	Labor	Equipment	Total	Total Incl O&P
0010	**PIPE/TUBE, GROOVED JOINT FOR COPPER**									
4000	Fittings: coupling material required at joints not incl. in fitting price.									
4001	Add 1 selected coupling, material only, per joint for installed price.									
4010	Coupling, rigid style									
4018	2" diameter	1 Plum	50	.160	Ea.	30.50	10.30		40.80	49
4020	2-1/2" diameter	Q-1	80	.200		34	11.60		45.60	55
4022	3" diameter		67	.239		38.50	13.85		52.35	62.50
4024	4" diameter		50	.320		57.50	18.55		76.05	91.50
4026	5" diameter		40	.400		96	23		119	141
4028	6" diameter	Q-2	50	.480		127	29		156	183
4100	Elbow, 90° or 45°									
4108	2" diameter	1 Plum	25	.320	Ea.	55.50	20.50		76	92
4110	2-1/2" diameter	Q-1	40	.400		60	23		83	101
4112	3" diameter		33	.485		84	28		112	135
4114	4" diameter		25	.640		193	37		230	268
4116	5" diameter		20	.800		545	46.50		591.50	670
4118	6" diameter	Q-2	25	.960		870	57.50		927.50	1,050
4200	Tee									
4208	2" diameter	1 Plum	17	.471	Ea.	91	30.50		121.50	146
4210	2-1/2" diameter	Q-1	27	.593		96.50	34.50		131	158
4212	3" diameter		22	.727		144	42		186	221
4214	4" diameter		17	.941		315	54.50		369.50	425
4216	5" diameter		13	1.231		880	71.50		951.50	1,075

22 11 13.27 Pipe/Tube, Grooved Joint for Copper

	22 11 13.27 Pipe/Tube, Grooved Joint for Copper	Crew	Daily Output	Labor-Hours	Unit	Material (2020 Bare Costs)	Labor (2020 Bare Costs)	Equipment (2020 Bare Costs)	Total (2020 Bare Costs)	Total Incl O&P
4218	6" diameter	Q-2	17	1.412	Ea.	1,075	85		1,160	1,325
4300	Reducer, concentric									
4310	3" x 2-1/2" diameter	Q-1	35	.457	Ea.	79.50	26.50		106	127
4312	4" x 2-1/2" diameter		32	.500		163	29		192	223
4314	4" x 3" diameter		29	.552		163	32		195	227
4316	5" x 3" diameter		25	.640		455	37		492	555
4318	5" x 4" diameter		22	.727		455	42		497	565
4320	6" x 3" diameter	Q-2	28	.857		495	51.50		546.50	620
4322	6" x 4" diameter		26	.923		495	55.50		550.50	630
4324	6" x 5" diameter		24	1		495	60		555	635
4350	Flange, w/groove gasket									
4351	ANSI class 125 and 150									
4355	2" diameter	1 Plum	23	.348	Ea.	229	22.50		251.50	286
4356	2-1/2" diameter	Q-1	37	.432		239	25		264	300
4358	3" diameter		31	.516		250	30		280	320
4360	4" diameter		23	.696		274	40.50		314.50	360
4362	5" diameter		19	.842		370	49		419	480
4364	6" diameter	Q-2	23	1.043		400	63		463	535

22 11 13.29 Pipe, Fittings and Valves, Copper, Pressed-Joint

	22 11 13.29 Pipe, Fittings and Valves, Copper, Pressed-Joint	Crew	Daily Output	Labor-Hours	Unit	Material	Labor	Equipment	Total	Total Incl O&P
0010	**PIPE, FITTINGS AND VALVES, COPPER, PRESSED-JOINT**									
0040	Pipe/tube includes coupling & clevis type hanger assy's, 10' OC									
0120	Type K									
0130	1/2" diameter	1 Plum	78	.103	L.F.	5.50	6.60		12.10	15.95
0134	3/4" diameter		74	.108		8.85	6.95		15.80	20
0138	1" diameter		66	.121		13.25	7.80		21.05	26.50
0142	1-1/4" diameter		56	.143		15.75	9.20		24.95	31
0146	1-1/2" diameter		50	.160		19.75	10.30		30.05	37.50
0150	2" diameter		40	.200		28.50	12.90		41.40	51
0154	2-1/2" diameter	Q-1	60	.267		49	15.45		64.45	76.50
0158	3" diameter		54	.296		64.50	17.20		81.70	96.50
0162	4" diameter		38	.421		98	24.50		122.50	145
0180	To delete cplgs. & hngrs., 1/2" pipe, subtract					19%	48%			
0184	3/4" -2" pipe, subtract					14%	46%			
0186	2-1/2"-4" pipe, subtract					24%	34%			
0220	Type L									
0230	1/2" diameter	1 Plum	81	.099	L.F.	3.90	6.35		10.25	13.85
0234	3/4" diameter		76	.105		4.93	6.80		11.73	15.55
0238	1" diameter		68	.118		8.05	7.60		15.65	20
0242	1-1/4" diameter		58	.138		12.45	8.90		21.35	27
0246	1-1/2" diameter		52	.154		12.80	9.90		22.70	29
0250	2" diameter		42	.190		20	12.30		32.30	40.50
0254	2-1/2" diameter	Q-1	62	.258		35	14.95		49.95	61
0258	3" diameter		56	.286		53	16.55		69.55	83.50
0262	4" diameter		39	.410		66	24		90	108
0280	To delete cplgs. & hngrs., 1/2" pipe, subtract					21%	52%			
0284	3/4"-2" pipe, subtract					17%	46%			
0286	2-1/2"-4" pipe, subtract					23%	35%			
0320	Type M									
0330	1/2" diameter	1 Plum	84	.095	L.F.	3.78	6.15		9.93	13.35
0334	3/4" diameter		78	.103		5.25	6.60		11.85	15.70
0338	1" diameter		70	.114		8.95	7.35		16.30	21
0342	1-1/4" diameter		60	.133		12.75	8.60		21.35	27

22 11 13.29 Pipe, Fittings and Valves, Copper, Pressed-Joint		Crew	Daily Output	Labor-Hours	Unit	2020 Bare Costs Material	Labor	Equipment	Total	Total Incl O&P
0346	1-1/2" diameter	1 Plum	54	.148	L.F.	15.65	9.55		25.20	31.50
0350	2" diameter		44	.182		22.50	11.70		34.20	42.50
0354	2-1/2" diameter	Q-1	64	.250		36.50	14.50		51	61.50
0358	3" diameter		58	.276		46.50	16		62.50	75
0362	4" diameter		40	.400		75.50	23		98.50	118
0380	To delete cplgs. & hngrs., 1/2" pipe, subtract					32%	49%			
0384	3/4"-2" pipe, subtract					21%	46%			
0386	2-1/2"-4" pipe, subtract					25%	36%			
1600	Fittings									
1610	Press joints, copper x copper									
1620	Note: Reducing fittings show most expensive size combination.									
1800	90° elbow, 1/2"	1 Plum	36.60	.219	Ea.	3.58	14.10		17.68	25
1810	3/4"		27.50	.291		5.80	18.75		24.55	34.50
1820	1"		25.90	.309		11.60	19.90		31.50	43
1830	1-1/4"		20.90	.383		23	24.50		47.50	62.50
1840	1-1/2"		18.30	.437		44	28		72	90.50
1850	2"		15.70	.510		62	33		95	117
1860	2-1/2"	Q-1	25.90	.618		179	36		215	250
1870	3"		22	.727		225	42		267	310
1880	4"		16.30	.982		278	57		335	390
2000	45° elbow, 1/2"	1 Plum	36.60	.219		4.20	14.10		18.30	25.50
2010	3/4"		27.50	.291		4.95	18.75		23.70	33.50
2020	1"		25.90	.309		15.80	19.90		35.70	47.50
2030	1-1/4"		20.90	.383		23	24.50		47.50	62
2040	1-1/2"		18.30	.437		36.50	28		64.50	82.50
2050	2"		15.70	.510		51	33		84	105
2060	2-1/2"	Q-1	25.90	.618		120	36		156	186
2070	3"		22	.727		169	42		211	248
2080	4"		16.30	.982		238	57		295	345
2200	Tee, 1/2"	1 Plum	27.50	.291		5.45	18.75		24.20	34
2210	3/4"		20.70	.386		9.40	25		34.40	48
2220	1"		19.40	.412		17.15	26.50		43.65	59
2230	1-1/4"		15.70	.510		29.50	33		62.50	81.50
2240	1-1/2"		13.80	.580		56.50	37.50		94	118
2250	2"		11.80	.678		69.50	43.50		113	142
2260	2-1/2"	Q-1	19.40	.825		224	48		272	320
2270	3"		16.50	.970		276	56		332	390
2280	4"		12.20	1.311		395	76		471	545
2400	Tee, reducing on the outlet									
2410	3/4"	1 Plum	20.70	.386	Ea.	8.15	25		33.15	46.50
2420	1"		19.40	.412		19.95	26.50		46.45	62
2430	1-1/4"		15.70	.510		29	33		62	81
2440	1-1/2"		13.80	.580		61	37.50		98.50	123
2450	2"		11.80	.678		96	43.50		139.50	172
2460	2-1/2"	Q-1	19.40	.825		300	48		348	400
2470	3"		16.50	.970		360	56		416	480
2480	4"		12.20	1.311		450	76		526	610
2600	Tee, reducing on the run									
2610	3/4"	1 Plum	20.70	.386	Ea.	16	25		41	55
2620	1"		19.40	.412		30.50	26.50		57	73.50
2630	1-1/4"		15.70	.510		61	33		94	116
2640	1-1/2"		13.80	.580		93.50	37.50		131	159
2650	2"		11.80	.678		102	43.50		145.50	179

22 11 13 – Facility Water Distribution Piping

22 11 13.29 Pipe, Fittings and Valves, Copper, Pressed-Joint		Crew	Daily Output	Labor-Hours	Unit	2020 Bare Costs Material	Labor	Equipment	Total	Total Incl O&P
2660	2-1/2"	Q-1	19.40	.825	Ea.	335	48		383	440
2670	3"		16.50	.970		400	56		456	525
2680	4"		12.20	1.311		525	76		601	690
2800	Coupling, 1/2"	1 Plum	36.60	.219		3.16	14.10		17.26	24.50
2810	3/4"		27.50	.291		4.88	18.75		23.63	33.50
2820	1"		25.90	.309		9.85	19.90		29.75	41
2830	1-1/4"		20.90	.383		12.30	24.50		36.80	50.50
2840	1-1/2"		18.30	.437		22.50	28		50.50	67
2850	2"		15.70	.510		29	33		62	80.50
2860	2-1/2"	Q-1	25.90	.618		89	36		125	152
2870	3"		22	.727		113	42		155	188
2880	4"		16.30	.982		160	57		217	261
3000	Union, 1/2"	1 Plum	36.60	.219		28	14.10		42.10	51.50
3010	3/4"		27.50	.291		35	18.75		53.75	66.50
3020	1"		25.90	.309		56.50	19.90		76.40	92
3030	1-1/4"		20.90	.383		82.50	24.50		107	128
3040	1-1/2"		18.30	.437		108	28		136	161
3050	2"		15.70	.510		174	33		207	240
3200	Adapter, tube to MPT									
3210	1/2"	1 Plum	15.10	.530	Ea.	4.10	34		38.10	55.50
3220	3/4"		13.60	.588		7.60	38		45.60	65
3230	1"		11.60	.690		13.70	44.50		58.20	81.50
3240	1-1/4"		9.90	.808		29.50	52		81.50	111
3250	1-1/2"		9	.889		41.50	57.50		99	131
3260	2"		7.90	1.013		80	65.50		145.50	186
3270	2-1/2"	Q-1	13.40	1.194		181	69.50		250.50	305
3280	3"		10.60	1.509		228	87.50		315.50	380
3290	4"		7.70	2.078		277	121		398	485
3400	Adapter, tube to FPT									
3410	1/2"	1 Plum	15.10	.530	Ea.	5.05	34		39.05	56.50
3420	3/4"		13.60	.588		8	38		46	65.50
3430	1"		11.60	.690		15.15	44.50		59.65	83
3440	1-1/4"		9.90	.808		34.50	52		86.50	116
3450	1-1/2"		9	.889		49	57.50		106.50	140
3460	2"		7.90	1.013		82.50	65.50		148	189
3470	2-1/2"	Q-1	13.40	1.194		203	69.50		272.50	325
3480	3"		10.60	1.509		335	87.50		422.50	495
3490	4"		7.70	2.078		430	121		551	650
3600	Flange									
3620	1"	1 Plum	36.20	.221	Ea.	147	14.25		161.25	183
3630	1-1/4"		29.30	.273		210	17.60		227.60	258
3640	1-1/2"		25.60	.313		232	20		252	285
3650	2"		22	.364		261	23.50		284.50	320
3660	2-1/2"	Q-1	36.20	.442		273	25.50		298.50	340
3670	3"		30.80	.519		330	30		360	410
3680	4"		22.80	.702		370	40.50		410.50	465
3800	Cap, 1/2"	1 Plum	53.10	.151		6.85	9.70		16.55	22
3810	3/4"		39.80	.201		11.50	12.95		24.45	32
3820	1"		37.50	.213		17.75	13.75		31.50	40
3830	1-1/4"		30.40	.263		21	16.95		37.95	48.50
3840	1-1/2"		26.60	.301		32.50	19.40		51.90	65
3850	2"		22.80	.351		39.50	22.50		62	77.50
3860	2-1/2"	Q-1	37.50	.427		124	25		149	174

22 11 13.29 Pipe, Fittings and Valves, Copper, Pressed-Joint		Crew	Daily Output	Labor-Hours	Unit	Material	2020 Bare Costs Labor	Equipment	Total	Total Incl O&P
3870	3"	Q-1	31.90	.502	Ea.	157	29		186	217
3880	4"		23.60	.678		192	39.50		231.50	270
4000	Reducer									
4010	3/4"	1 Plum	27.50	.291	Ea.	14.45	18.75		33.20	44
4020	1"		25.90	.309		26.50	19.90		46.40	59
4030	1-1/4"		20.90	.383		39	24.50		63.50	80
4040	1-1/2"		18.30	.437		46.50	28		74.50	93.50
4050	2"		15.70	.510		64.50	33		97.50	120
4060	2-1/2"	Q-1	25.90	.618		179	36		215	251
4070	3"		22	.727		242	42		284	330
4080	4"		16.30	.982		310	57		367	430
4100	Stub out, 1/2"	1 Plum	50	.160		9.65	10.30		19.95	26
4110	3/4"		35	.229		14.90	14.75		29.65	38.50
4120	1"		30	.267		20.50	17.20		37.70	48
6000	Valves									
6200	Ball valve									
6210	1/2"	1 Plum	25.60	.313	Ea.	40	20		60	73.50
6220	3/4"		19.20	.417		53	27		80	98
6230	1"		18.10	.442		63.50	28.50		92	113
6240	1-1/4"		14.70	.544		103	35		138	167
6250	1-1/2"		12.80	.625		254	40.50		294.50	340
6260	2"		11	.727		365	47		412	475
6400	Check valve									
6410	1/2"	1 Plum	30	.267	Ea.	34.50	17.20		51.70	63.50
6420	3/4"		22.50	.356		44.50	23		67.50	83.50
6430	1"		21.20	.377		51	24.50		75.50	93
6440	1-1/4"		17.20	.465		74.50	30		104.50	127
6450	1-1/2"		15	.533		105	34.50		139.50	168
6460	2"		12.90	.620		191	40		231	270
6600	Butterfly valve, lug type									
6660	2-1/2"	Q-1	9	1.778	Ea.	163	103		266	335
6670	3"		8	2		200	116		316	395
6680	4"		5	3.200		250	186		436	555

22 11 13.44 Pipe, Steel		Crew	Daily Output	Labor-Hours	Unit	Material	Labor	Equipment	Total	Total Incl O&P
0010	**PIPE, STEEL** R221113-50									
0020	All pipe sizes are to Spec. A-53 unless noted otherwise R221113-70									
0032	Schedule 10, see Line 22 11 13.48 0500									
0050	Schedule 40, threaded, with couplings, and clevis hanger									
0060	assemblies sized for covering, 10' OC									
0540	Black, 1/4" diameter	1 Plum	66	.121	L.F.	6.15	7.80		13.95	18.50
0550	3/8" diameter		65	.123		6.85	7.95		14.80	19.40
0560	1/2" diameter		63	.127		4.03	8.20		12.23	16.70
0570	3/4" diameter		61	.131		4.40	8.45		12.85	17.50
0580	1" diameter		53	.151		7.80	9.75		17.55	23
0590	1-1/4" diameter	Q-1	89	.180		8.20	10.45		18.65	24.50
0600	1-1/2" diameter		80	.200		8.75	11.60		20.35	27
0610	2" diameter		64	.250		6.90	14.50		21.40	29
0620	2-1/2" diameter		50	.320		11.40	18.55		29.95	40.50
0630	3" diameter		43	.372		14.35	21.50		35.85	48.50
0640	3-1/2" diameter		40	.400		21	23		44	58
0650	4" diameter		36	.444		23	26		49	64
0809	A-106, gr. A/B, seamless w/cplgs. & clevis hanger assemblies									

22 11 13.44 Pipe, Steel		Crew	Daily Output	Labor-Hours	Unit	2020 Bare Costs Material	Labor	Equipment	Total	Total Incl O&P
0811	1/4" diameter	1 Plum	66	.121	L.F.	9.15	7.80		16.95	22
0812	3/8" diameter		65	.123		9.20	7.95		17.15	22
0813	1/2" diameter		63	.127		9.40	8.20		17.60	22.50
0814	3/4" diameter		61	.131		12.30	8.45		20.75	26
0815	1" diameter		53	.151		11.30	9.75		21.05	27
0816	1-1/4" diameter	Q-1	89	.180		13.20	10.45		23.65	30
0817	1-1/2" diameter		80	.200		19.55	11.60		31.15	39
0819	2" diameter		64	.250		15.70	14.50		30.20	39
0821	2-1/2" diameter		50	.320		21	18.55		39.55	51
0822	3" diameter		43	.372		26.50	21.50		48	61.50
0823	4" diameter		36	.444		41	26		67	83.50
1220	To delete coupling & hanger, subtract									
1230	1/4" diam. to 3/4" diam.					31%	56%			
1240	1" diam. to 1-1/2" diam.					23%	51%			
1250	2" diam. to 4" diam.					23%	41%			
1280	All pipe sizes are to Spec. A-53 unless noted otherwise									
1281	Schedule 40, threaded, with couplings and clevis hanger									
1282	assemblies sized for covering, 10' OC									
1290	Galvanized, 1/4" diameter	1 Plum	66	.121	L.F.	8.10	7.80		15.90	20.50
1300	3/8" diameter		65	.123		8.35	7.95		16.30	21
1310	1/2" diameter		63	.127		4.10	8.20		12.30	16.75
1320	3/4" diameter		61	.131		5	8.45		13.45	18.15
1330	1" diameter		53	.151		7.90	9.75		17.65	23.50
1340	1-1/4" diameter	Q-1	89	.180		8.85	10.45		19.30	25.50
1350	1-1/2" diameter		80	.200		9.45	11.60		21.05	28
1360	2" diameter		64	.250		7.90	14.50		22.40	30
1370	2-1/2" diameter		50	.320		13.15	18.55		31.70	42.50
1380	3" diameter		43	.372		16	21.50		37.50	50
1390	3-1/2" diameter		40	.400		23	23		46	59.50
1400	4" diameter		36	.444		25.50	26		51.50	66.50
1750	To delete coupling & hanger, subtract									
1760	1/4" diam. to 3/4" diam.					31%	56%			
1770	1" diam. to 1-1/2" diam.					23%	51%			
1780	2" diam. to 4" diam.					23%	41%			
1900	Pipe nipple std black 2" long, 1/2" diameter	1 Plum	19	.421	Ea.	1.23	27		28.23	42
1910	Pipe nipple std black 2" long, 1" diameter	"	15	.533		2.42	34.50		36.92	54
1920	3" long, 2-1/2" diameter	Q-1	16	1		12.80	58		70.80	101
2000	Welded, sch. 40, on yoke & roll hanger assy's, sized for covering, 10' OC									
2040	Black, 1" diameter	Q-15	93	.172	L.F.	7.15	10	1.14	18.29	24
2050	1-1/4" diameter		84	.190		8.20	11.05	1.27	20.52	27
2060	1-1/2" diameter		76	.211		8.65	12.20	1.40	22.25	29.50
2070	2" diameter		61	.262		6.60	15.20	1.74	23.54	32
2080	2-1/2" diameter		47	.340		11.80	19.75	2.26	33.81	45
2090	3" diameter		43	.372		16.65	21.50	2.47	40.62	53.50
2100	3-1/2" diameter		39	.410		17.45	24	2.73	44.18	57.50
2110	4" diameter		37	.432		24	25	2.88	51.88	67
2120	5" diameter		32	.500		37	29	3.33	69.33	87.50
2130	6" diameter	Q-16	36	.667		45	40	2.95	87.95	113
2140	8" diameter		29	.828		71.50	50	3.67	125.17	157
2150	10" diameter		24	1		90	60	4.43	154.43	194
2160	12" diameter		19	1.263		107	76	5.60	188.60	237
2170	14" diameter (two rod roll type hanger for 14" diam. and up)		15	1.600		104	96	7.10	207.10	267
2180	16" diameter (two rod roll type hanger)		13	1.846		172	111	8.20	291.20	365

22 11 13.44 Pipe, Steel		Crew	Daily Output	Labor-Hours	Unit	Material	2020 Bare Costs Labor	Equipment	Total	Total Incl O&P
2190	18" diameter (two rod roll type hanger)	Q-16	11	2.182	L.F.	147	131	9.65	287.65	370
2200	20" diameter (two rod roll type hanger)		9	2.667		156	160	11.80	327.80	425
2220	24" diameter (two rod roll type hanger)		8	3		203	180	13.30	396.30	510
2560	To delete hanger, subtract									
2570	1" diam. to 1-1/2" diam.					15%	34%			
2580	2" diam. to 3-1/2" diam.					9%	21%			
2590	4" diam. to 12" diam.					5%	12%			
2596	14" diam. to 24" diam.					3%	10%			
3250	Flanged, 150 lb. weld neck, on yoke & roll hangers									
3260	sized for covering, 10' OC									
3290	Black, 1" diameter	Q-15	70	.229	L.F.	10.60	13.25	1.52	25.37	33
3300	1-1/4" diameter		64	.250		11.65	14.50	1.66	27.81	36
3310	1-1/2" diameter		58	.276		12.10	16	1.83	29.93	39.50
3320	2" diameter		45	.356		15.25	20.50	2.36	38.11	50.50
3330	2-1/2" diameter		36	.444		23	26	2.96	51.96	67.50
3340	3" diameter		32	.500		28.50	29	3.33	60.83	78.50
3350	3-1/2" diameter		29	.552		31	32	3.67	66.67	86
3360	4" diameter		26	.615		33	35.50	4.09	72.59	94.50
3370	5" diameter		21	.762		51.50	44	5.05	100.55	129
3380	6" diameter	Q-16	25	.960		62.50	57.50	4.25	124.25	160
3390	8" diameter		19	1.263		101	76	5.60	182.60	231
3400	10" diameter		16	1.500		141	90	6.65	237.65	297
3410	12" diameter		14	1.714		175	103	7.60	285.60	355
3470	For 300 lb. flanges, add					63%				
3480	For 600 lb. flanges, add					310%				
3960	To delete flanges & hanger, subtract									
3970	1" diam. to 2" diam.					76%	65%			
3980	2-1/2" diam. to 4" diam.					62%	59%			
3990	5" diam. to 12" diam.					60%	46%			
4750	Schedule 80, threaded, with couplings, and clevis hanger assemblies									
4760	sized for covering, 10' OC									
4790	Black, 1/4" diameter	1 Plum	54	.148	L.F.	7.60	9.55		17.15	22.50
4800	3/8" diameter		53	.151		9.70	9.75		19.45	25
4810	1/2" diameter		52	.154		4.87	9.90		14.77	20
4820	3/4" diameter		50	.160		6.45	10.30		16.75	22.50
4830	1" diameter		45	.178		7.35	11.45		18.80	25.50
4840	1-1/4" diameter	Q-1	75	.213		9.40	12.35		21.75	29
4850	1-1/2" diameter		69	.232		10.35	13.45		23.80	31.50
4860	2" diameter		56	.286		13.20	16.55		29.75	39.50
4870	2-1/2" diameter		44	.364		22	21		43	55.50
4880	3" diameter		38	.421		27	24.50		51.50	66.50
4890	3-1/2" diameter		35	.457		32.50	26.50		59	75.50
4900	4" diameter		32	.500		36	29		65	83.50
5430	To delete coupling & hanger, subtract									
5440	1/4" diam. to 1/2" diam.					31%	54%			
5450	3/4" diam. to 1-1/2" diam.					28%	49%			
5460	2" diam. to 4" diam.					21%	40%			
5510	Galvanized, 1/4" diameter	1 Plum	54	.148	L.F.	9.90	9.55		19.45	25
5520	3/8" diameter		53	.151		10.70	9.75		20.45	26.50
5530	1/2" diameter		52	.154		5.65	9.90		15.55	21
5540	3/4" diameter		50	.160		7.10	10.30		17.40	23.50
5550	1" diameter		45	.178		8.25	11.45		19.70	26.50
5560	1-1/4" diameter	Q-1	75	.213		10.70	12.35		23.05	30.50

22 11 13 – Facility Water Distribution Piping

22 11 13.44 Pipe, Steel		Crew	Daily Output	Labor-Hours	Unit	Material	2020 Bare Costs Labor	Equipment	Total	Total Incl O&P
5570	1-1/2" diameter	Q-1	69	.232	L.F.	11.90	13.45		25.35	33
5580	2" diameter		56	.286		15.45	16.55		32	42
5590	2-1/2" diameter		44	.364		25	21		46	59
5600	3" diameter		38	.421		31	24.50		55.50	70.50
5610	3-1/2" diameter		35	.457		38	26.50		64.50	81.50
5620	4" diameter		32	.500		39.50	29		68.50	87
5930	To delete coupling & hanger, subtract									
5940	1/4" diam. to 1/2" diam.					31%	54%			
5950	3/4" diam. to 1-1/2" diam.					28%	49%			
5960	2" diam. to 4" diam.					21%	40%			
6000	Welded, on yoke & roller hangers									
6010	sized for covering, 10' OC									
6040	Black, 1" diameter	Q-15	85	.188	L.F.	7.15	10.90	1.25	19.30	25.50
6050	1-1/4" diameter		79	.203		9	11.75	1.35	22.10	29
6060	1-1/2" diameter		72	.222		9.95	12.90	1.48	24.33	32
6070	2" diameter		57	.281		12	16.30	1.87	30.17	40
6080	2-1/2" diameter		44	.364		20.50	21	2.42	43.92	56.50
6090	3" diameter		40	.400		25	23	2.66	50.66	65
6100	3-1/2" diameter		34	.471		30	27.50	3.13	60.63	77.50
6110	4" diameter		33	.485		32	28	3.22	63.22	81
6120	5" diameter, A-106B		26	.615		57.50	35.50	4.09	97.09	122
6130	6" diameter, A-106B	Q-16	30	.800		67.50	48	3.54	119.04	150
6140	8" diameter, A-106B		25	.960		103	57.50	4.25	164.75	205
6150	10" diameter, A-106B		20	1.200		153	72	5.30	230.30	282
6160	12" diameter, A-106B		15	1.600		345	96	7.10	448.10	530
6540	To delete hanger, subtract									
6550	1" diam. to 1-1/2" diam.					30%	14%			
6560	2" diam. to 3" diam.					23%	9%			
6570	3-1/2" diam. to 5" diam.					12%	6%			
6580	6" diam. to 12" diam.					10%	4%			
7250	Flanged, 300 lb. weld neck, on yoke & roll hangers									
7260	sized for covering, 10' OC									
7290	Black, 1" diameter	Q-15	66	.242	L.F.	14.40	14.05	1.61	30.06	38.50
7300	1-1/4" diameter		61	.262		16.25	15.20	1.74	33.19	43
7310	1-1/2" diameter		54	.296		17.20	17.20	1.97	36.37	46.50
7320	2" diameter		42	.381		22	22	2.53	46.53	60
7330	2-1/2" diameter		33	.485		31.50	28	3.22	62.72	80
7340	3" diameter		29	.552		34.50	32	3.67	70.17	90
7350	3-1/2" diameter		24	.667		47	38.50	4.43	89.93	114
7360	4" diameter		23	.696		49	40.50	4.63	94.13	120
7370	5" diameter		19	.842		82.50	49	5.60	137.10	170
7380	6" diameter	Q-16	23	1.043		92.50	63	4.62	160.12	201
7390	8" diameter		17	1.412		149	85	6.25	240.25	298
7400	10" diameter		14	1.714		241	103	7.60	351.60	425
7410	12" diameter		12	2		440	120	8.85	568.85	675
7470	For 600 lb. flanges, add					100%				
7940	To delete flanges & hanger, subtract									
7950	1" diam. to 1-1/2" diam.					75%	66%			
7960	2" diam. to 3" diam.					62%	60%			
7970	3-1/2" diam. to 5" diam.					54%	66%			
7980	6" diam. to 12" diam.					55%	62%			
9000	Threading pipe labor, one end, all schedules through 80									
9010	1/4" through 3/4" pipe size	1 Plum	80	.100	Ea.		6.45		6.45	9.65

22 11 13.44 Pipe, Steel		Crew	Daily Output	Labor-Hours	Unit	Material	2020 Bare Costs Labor	Equipment	Total	Total Incl O&P
9020	1" through 2" pipe size	1 Plum	73	.110	Ea.		7.05		7.05	10.55
9030	2-1/2" pipe size		53	.151			9.75		9.75	14.55
9040	3" pipe size		50	.160			10.30		10.30	15.45
9050	3-1/2" pipe size	Q-1	89	.180			10.45		10.45	15.60
9060	4" pipe size		73	.219			12.70		12.70	19
9070	5" pipe size		53	.302			17.50		17.50	26
9080	6" pipe size		46	.348			20		20	30
9090	8" pipe size		29	.552			32		32	48
9100	10" pipe size		21	.762			44		44	66
9110	12" pipe size		13	1.231			71.50		71.50	107
9120	Cutting pipe labor, one cut									
9124	Shop fabrication, machine cut									
9126	Schedule 40, straight pipe									
9128	2" pipe size or less	1 Stpi	62	.129	Ea.		8.45		8.45	12.65
9130	2-1/2" pipe size		56	.143			9.35		9.35	14
9132	3" pipe size		42	.190			12.50		12.50	18.70
9134	4" pipe size		31	.258			16.90		16.90	25.50
9136	5" pipe size		26	.308			20		20	30
9138	6" pipe size		19	.421			27.50		27.50	41.50
9140	8" pipe size		14	.571			37.50		37.50	56
9142	10" pipe size		10	.800			52.50		52.50	78.50
9144	12" pipe size		7	1.143			75		75	112
9146	14" pipe size	Q-5	10.50	1.524			90		90	135
9148	16" pipe size		8.60	1.860			110		110	164
9150	18" pipe size		7	2.286			135		135	202
9152	20" pipe size		5.80	2.759			163		163	244
9154	24" pipe size		4	4			236		236	355
9160	Schedule 80, straight pipe									
9164	2" pipe size or less	1 Stpi	42	.190	Ea.		12.50		12.50	18.70
9166	2-1/2" pipe size		37	.216			14.15		14.15	21
9168	3" pipe size		31	.258			16.90		16.90	25.50
9170	4" pipe size		23	.348			23		23	34
9172	5" pipe size		18	.444			29		29	43.50
9174	6" pipe size		14.60	.548			36		36	54
9176	8" pipe size		10	.800			52.50		52.50	78.50
9178	10" pipe size	Q-5	14	1.143			67.50		67.50	101
9180	12" pipe size	"	10	1.600			94.50		94.50	141
9200	Welding labor per joint									
9210	Schedule 40									
9230	1/2" pipe size	Q-15	32	.500	Ea.		29	3.33	32.33	47
9240	3/4" pipe size		27	.593			34.50	3.94	38.44	56
9250	1" pipe size		23	.696			40.50	4.63	45.13	65.50
9260	1-1/4" pipe size		20	.800			46.50	5.30	51.80	75.50
9270	1-1/2" pipe size		19	.842			49	5.60	54.60	79
9280	2" pipe size		16	1			58	6.65	64.65	94.50
9290	2-1/2" pipe size		13	1.231			71.50	8.20	79.70	116
9300	3" pipe size		12	1.333			77.50	8.85	86.35	126
9310	4" pipe size		10	1.600			93	10.65	103.65	151
9320	5" pipe size		9	1.778			103	11.80	114.80	167
9330	6" pipe size		8	2			116	13.30	129.30	189
9340	8" pipe size		5	3.200			186	21.50	207.50	300
9350	10" pipe size		4	4			232	26.50	258.50	375
9360	12" pipe size		3	5.333			310	35.50	345.50	505

22 11 13.44 Pipe, Steel

22 11 13.44 Pipe, Steel		Crew	Daily Output	Labor-Hours	Unit	Material	2020 Bare Costs Labor	Equipment	Total	Total Incl O&P
9370	14" pipe size	Q-15	2.60	6.154	Ea.		355	41	396	580
9380	16" pipe size		2.20	7.273			420	48.50	468.50	685
9390	18" pipe size		2	8			465	53	518	755
9400	20" pipe size		1.80	8.889			515	59	574	835
9410	22" pipe size		1.70	9.412			545	62.50	607.50	885
9420	24" pipe size		1.50	10.667			620	71	691	1,000
9450	Schedule 80									
9460	1/2" pipe size	Q-15	27	.593	Ea.		34.50	3.94	38.44	56
9470	3/4" pipe size		23	.696			40.50	4.63	45.13	65.50
9480	1" pipe size		20	.800			46.50	5.30	51.80	75.50
9490	1-1/4" pipe size		19	.842			49	5.60	54.60	79
9500	1-1/2" pipe size		18	.889			51.50	5.90	57.40	83.50
9510	2" pipe size		15	1.067			62	7.10	69.10	100
9520	2-1/2" pipe size		12	1.333			77.50	8.85	86.35	126
9530	3" pipe size		11	1.455			84.50	9.65	94.15	137
9540	4" pipe size		8	2			116	13.30	129.30	189
9550	5" pipe size		6	2.667			155	17.75	172.75	251
9560	6" pipe size		5	3.200			186	21.50	207.50	300
9570	8" pipe size		4	4			232	26.50	258.50	375
9580	10" pipe size		3	5.333			310	35.50	345.50	505
9590	12" pipe size		2	8			465	53	518	755
9600	14" pipe size	Q-16	2.60	9.231			555	41	596	875
9610	16" pipe size		2.30	10.435			630	46	676	990
9620	18" pipe size		2	12			720	53	773	1,125
9630	20" pipe size		1.80	13.333			800	59	859	1,275
9640	22" pipe size		1.60	15			900	66.50	966.50	1,425
9650	24" pipe size		1.50	16			960	71	1,031	1,525

22 11 13.45 Pipe Fittings, Steel, Threaded

22 11 13.45 Pipe Fittings, Steel, Threaded		Crew	Daily Output	Labor-Hours	Unit	Material	Labor	Equipment	Total	Total Incl O&P
0010	**PIPE FITTINGS, STEEL, THREADED**									
0020	Cast iron									
0040	Standard weight, black									
0060	90° elbow, straight									
0070	1/4"	1 Plum	16	.500	Ea.	11.95	32		43.95	61
0080	3/8"		16	.500		17.30	32		49.30	67
0090	1/2"		15	.533		7.60	34.50		42.10	60
0100	3/4"		14	.571		7.90	37		44.90	63.50
0110	1"		13	.615		9.35	39.50		48.85	70
0120	1-1/4"	Q-1	22	.727		13.25	42		55.25	77.50
0130	1-1/2"		20	.800		18.35	46.50		64.85	89.50
0140	2"		18	.889		28.50	51.50		80	109
0150	2-1/2"		14	1.143		68.50	66.50		135	175
0160	3"		10	1.600		113	93		206	263
0170	3-1/2"		8	2		305	116		421	510
0180	4"		6	2.667		209	155		364	460
0250	45° elbow, straight									
0260	1/4"	1 Plum	16	.500	Ea.	14.75	32		46.75	64.50
0270	3/8"		16	.500		15.85	32		47.85	65.50
0280	1/2"		15	.533		11.60	34.50		46.10	64.50
0300	3/4"		14	.571		11.65	37		48.65	68
0320	1"		13	.615		13.70	39.50		53.20	74.50
0330	1-1/4"	Q-1	22	.727		18.40	42		60.40	83
0340	1-1/2"		20	.800		30.50	46.50		77	103

22 11 13.45 Pipe Fittings, Steel, Threaded		Crew	Daily Output	Labor-Hours	Unit	2020 Bare Costs Material	Labor	Equipment	Total	Total Incl O&P
0350	2″	Q-1	18	.889	Ea.	35	51.50		86.50	116
0360	2-1/2″		14	1.143		91.50	66.50		158	200
0370	3″		10	1.600		145	93		238	298
0380	3-1/2″		8	2		340	116		456	550
0400	4″		6	2.667		300	155		455	560
0500	Tee, straight									
0510	1/4″	1 Plum	10	.800	Ea.	18.70	51.50		70.20	97.50
0520	3/8″		10	.800		18.20	51.50		69.70	97
0530	1/2″		9	.889		11.80	57.50		69.30	98.50
0540	3/4″		9	.889		13.75	57.50		71.25	101
0550	1″		8	1		12.25	64.50		76.75	110
0560	1-1/4″	Q-1	14	1.143		22.50	66.50		89	124
0570	1-1/2″		13	1.231		29	71.50		100.50	139
0580	2″		11	1.455		40.50	84.50		125	171
0590	2-1/2″		9	1.778		105	103		208	270
0600	3″		6	2.667		161	155		316	410
0610	3-1/2″		5	3.200		325	186		511	640
0620	4″		4	4		315	232		547	690
0660	Tee, reducing, run or outlet									
0661	1/2″	1 Plum	9	.889	Ea.	29	57.50		86.50	118
0662	3/4″		9	.889		21	57.50		78.50	109
0663	1″		8	1		28.50	64.50		93	128
0664	1-1/4″	Q-1	14	1.143		40	66.50		106.50	143
0665	1-1/2″		13	1.231		41	71.50		112.50	152
0666	2″		11	1.455		51.50	84.50		136	183
0667	2-1/2″		9	1.778		119	103		222	284
0668	3″		6	2.667		208	155		363	460
0669	3-1/2″		5	3.200		515	186		701	845
0670	4″		4	4		455	232		687	845
0674	Reducer, concentric									
0675	3/4″	1 Plum	18	.444	Ea.	21	28.50		49.50	66.50
0676	1″	″	15	.533		14.35	34.50		48.85	67.50
0677	1-1/4″	Q-1	26	.615		37	35.50		72.50	94
0678	1-1/2″		24	.667		57.50	38.50		96	121
0679	2″		21	.762		71.50	44		115.50	145
0680	2-1/2″		18	.889		101	51.50		152.50	189
0681	3″		14	1.143		162	66.50		228.50	277
0682	3-1/2″		12	1.333		320	77.50		397.50	465
0683	4″		10	1.600		320	93		413	490
0687	Reducer, eccentric									
0688	3/4″	1 Plum	16	.500	Ea.	44.50	32		76.50	97
0689	1″	″	14	.571		48	37		85	108
0690	1-1/4″	Q-1	25	.640		73.50	37		110.50	137
0691	1-1/2″		22	.727		97	42		139	169
0692	2″		20	.800		135	46.50		181.50	218
0693	2-1/2″		16	1		198	58		256	305
0694	3″		12	1.333		300	77.50		377.50	445
0695	3-1/2″		10	1.600		455	93		548	640
0696	4″		9	1.778		585	103		688	800
0700	Standard weight, galvanized cast iron									
0720	90° elbow, straight									
0730	1/4″	1 Plum	16	.500	Ea.	21.50	32		53.50	71.50
0740	3/8″		16	.500		21.50	32		53.50	71.50

22 11 13.45 Pipe Fittings, Steel, Threaded		Crew	Daily Output	Labor-Hours	Unit	Material	2020 Bare Costs Labor	Equipment	Total	Total Incl O&P
0750	1/2"	1 Plum	15	.533	Ea.	22	34.50		56.50	75.50
0760	3/4"		14	.571		21	37		58	78
0770	1"	↓	13	.615		28	39.50		67.50	90
0780	1-1/4"	Q-1	22	.727		37.50	42		79.50	104
0790	1-1/2"		20	.800		59.50	46.50		106	135
0800	2"		18	.889		76	51.50		127.50	161
0810	2-1/2"		14	1.143		156	66.50		222.50	270
0820	3"		10	1.600		237	93		330	400
0830	3-1/2"		8	2		380	116		496	590
0840	4"	↓	6	2.667	↓	435	155		590	710
0900	45° elbow, straight									
0910	1/4"	1 Plum	16	.500	Ea.	23	32		55	73.50
0920	3/8"		16	.500		20	32		52	70
0930	1/2"		15	.533		21	34.50		55.50	75
0940	3/4"		14	.571		24.50	37		61.50	81.50
0950	1"	↓	13	.615		34	39.50		73.50	97
0960	1-1/4"	Q-1	22	.727		52	42		94	120
0970	1-1/2"		20	.800		69	46.50		115.50	146
0980	2"		18	.889		96.50	51.50		148	183
0990	2-1/2"		14	1.143		136	66.50		202.50	249
1000	3"		10	1.600		305	93		398	475
1010	3-1/2"		8	2		525	116		641	755
1020	4"	↓	6	2.667	↓	475	155		630	755
1100	Tee, straight									
1110	1/4"	1 Plum	10	.800	Ea.	20	51.50		71.50	99
1120	3/8"		10	.800		22.50	51.50		74	102
1130	1/2"		9	.889		26.50	57.50		84	115
1140	3/4"		9	.889		34.50	57.50		92	124
1150	1"	↓	8	1		32.50	64.50		97	133
1160	1-1/4"	Q-1	14	1.143		57	66.50		123.50	162
1170	1-1/2"		13	1.231		75.50	71.50		147	190
1180	2"		11	1.455		94	84.50		178.50	229
1190	2-1/2"		9	1.778		191	103		294	365
1200	3"		6	2.667		455	155		610	730
1210	3-1/2"		5	3.200		535	186		721	870
1220	4"	↓	4	4	↓	590	232		822	995
1300	Extra heavy weight, black									
1310	Couplings, steel straight									
1320	1/4"	1 Plum	19	.421	Ea.	5.55	27		32.55	46.50
1330	3/8"		19	.421		6.05	27		33.05	47
1340	1/2"		19	.421		8.20	27		35.20	49.50
1350	3/4"		18	.444		8.75	28.50		37.25	52.50
1360	1"	↓	15	.533		11.15	34.50		45.65	64
1370	1-1/4"	Q-1	26	.615		17.85	35.50		53.35	73
1380	1-1/2"		24	.667		17.85	38.50		56.35	77.50
1390	2"		21	.762		27	44		71	96
1400	2-1/2"		18	.889		40.50	51.50		92	122
1410	3"		14	1.143		48	66.50		114.50	152
1420	3-1/2"		12	1.333		64.50	77.50		142	187
1430	4"	↓	10	1.600	↓	76	93		169	223
1510	90° elbow, straight									
1520	1/2"	1 Plum	15	.533	Ea.	39	34.50		73.50	94.50
1530	3/4"	↓	14	.571	↓	39.50	37		76.50	98.50

22 11 13.45 Pipe Fittings, Steel, Threaded		Crew	Daily Output	Labor-Hours	Unit	2020 Bare Costs Material	Labor	Equipment	Total	Total Incl O&P
1540	1"	1 Plum	13	.615	Ea.	48	39.50		87.50	113
1550	1-1/4"	Q-1	22	.727		71.50	42		113.50	142
1560	1-1/2"		20	.800		88.50	46.50		135	167
1580	2"		18	.889		109	51.50		160.50	197
1590	2-1/2"		14	1.143		266	66.50		332.50	390
1600	3"		10	1.600		360	93		453	535
1610	4"		6	2.667		785	155		940	1,100
1650	45° elbow, straight									
1660	1/2"	1 Plum	15	.533	Ea.	55.50	34.50		90	113
1670	3/4"		14	.571		53.50	37		90.50	114
1680	1"		13	.615		64.50	39.50		104	131
1690	1-1/4"	Q-1	22	.727		106	42		148	180
1700	1-1/2"		20	.800		117	46.50		163.50	199
1710	2"		18	.889		166	51.50		217.50	260
1720	2-1/2"		14	1.143		287	66.50		353.50	415
1800	Tee, straight									
1810	1/2"	1 Plum	9	.889	Ea.	61	57.50		118.50	153
1820	3/4"		9	.889		61	57.50		118.50	153
1830	1"		8	1		73.50	64.50		138	178
1840	1-1/4"	Q-1	14	1.143		110	66.50		176.50	220
1850	1-1/2"		13	1.231		141	71.50		212.50	262
1860	2"		11	1.455		175	84.50		259.50	320
1870	2-1/2"		9	1.778		370	103		473	560
1880	3"		6	2.667		525	155		680	805
1890	4"		4	4		1,025	232		1,257	1,475
4000	Standard weight, black									
4010	Couplings, steel straight, merchants									
4030	1/4"	1 Plum	19	.421	Ea.	1.48	27		28.48	42
4040	3/8"		19	.421		1.79	27		28.79	42.50
4050	1/2"		19	.421		1.91	27		28.91	42.50
4060	3/4"		18	.444		2.42	28.50		30.92	45.50
4070	1"		15	.533		3.39	34.50		37.89	55
4080	1-1/4"	Q-1	26	.615		4.32	35.50		39.82	58.50
4090	1-1/2"		24	.667		5.50	38.50		44	64
4100	2"		21	.762		7.85	44		51.85	74.50
4110	2-1/2"		18	.889		25.50	51.50		77	105
4120	3"		14	1.143		36	66.50		102.50	139
4130	3-1/2"		12	1.333		64	77.50		141.50	186
4140	4"		10	1.600		64	93		157	209
4166	Plug, 1/4"	1 Plum	38	.211		3.37	13.55		16.92	24
4167	3/8"		38	.211		3.03	13.55		16.58	24
4168	1/2"		38	.211		3.37	13.55		16.92	24
4169	3/4"		32	.250		8.90	16.10		25	34
4170	1"		30	.267		9.50	17.20		26.70	36
4171	1-1/4"	Q-1	52	.308		10.90	17.85		28.75	38.50
4172	1-1/2"		48	.333		15.50	19.35		34.85	46
4173	2"		42	.381		20	22		42	55
4176	2-1/2"		36	.444		29.50	26		55.50	71
4180	4"		20	.800		62.50	46.50		109	138
4200	Standard weight, galvanized									
4210	Couplings, steel straight, merchants									
4230	1/4"	1 Plum	19	.421	Ea.	1.72	27		28.72	42.50
4240	3/8"		19	.421		2.20	27		29.20	43

22 11 13.45 Pipe Fittings, Steel, Threaded		Crew	Daily Output	Labor-Hours	Unit	Material	2020 Bare Costs Labor	Equipment	Total	Total Incl O&P
4250	1/2"	1 Plum	19	.421	Ea.	2.33	27		29.33	43
4260	3/4"		18	.444		2.91	28.50		31.41	46
4270	1"		15	.533		4.07	34.50		38.57	56
4280	1-1/4"	Q-1	26	.615		5.20	35.50		40.70	59.50
4290	1-1/2"		24	.667		6.45	38.50		44.95	65
4300	2"		21	.762		9.65	44		53.65	76.50
4310	2-1/2"		18	.889		27	51.50		78.50	107
4320	3"		14	1.143		36	66.50		102.50	139
4330	3-1/2"		12	1.333		73.50	77.50		151	197
4340	4"		10	1.600		73.50	93		166.50	220
4370	Plug, galvanized, square head									
4374	1/2"	1 Plum	38	.211	Ea.	8.75	13.55		22.30	30
4375	3/4"		32	.250		8.15	16.10		24.25	33
4376	1"		30	.267		8.15	17.20		25.35	34.50
4377	1-1/4"	Q-1	52	.308		13.60	17.85		31.45	41.50
4378	1-1/2"		48	.333		18.35	19.35		37.70	49
4379	2"		42	.381		23	22		45	58
4380	2-1/2"		36	.444		48	26		74	91.50
4381	3"		28	.571		61.50	33		94.50	118
4382	4"		20	.800		169	46.50		215.50	256
4385	8"		20	.800		127	46.50		173.50	210
4700	Nipple, black									
4710	1/2" x 4" long	1 Plum	19	.421	Ea.	3.40	27		30.40	44
4712	3/4" x 4" long		18	.444		4.11	28.50		32.61	47.50
4714	1" x 4" long		15	.533		5.70	34.50		40.20	58
4716	1-1/4" x 4" long	Q-1	26	.615		7.15	35.50		42.65	61.50
4718	1-1/2" x 4" long		24	.667		8.40	38.50		46.90	67.50
4720	2" x 4" long		21	.762		11.75	44		55.75	79
4722	2-1/2" x 4" long		18	.889		32	51.50		83.50	112
4724	3" x 4" long		14	1.143		40.50	66.50		107	144
4726	4" x 4" long		10	1.600		48.50	93		141.50	193
4800	Nipple, galvanized									
4810	1/2" x 4" long	1 Plum	19	.421	Ea.	4.17	27		31.17	45
4812	3/4" x 4" long		18	.444		5.15	28.50		33.65	48.50
4814	1" x 4" long		15	.533		6.90	34.50		41.40	59
4816	1-1/4" x 4" long	Q-1	26	.615		8.50	35.50		44	63
4818	1-1/2" x 4" long		24	.667		10.75	38.50		49.25	70
4820	2" x 4" long		21	.762		13.65	44		57.65	81
4822	2-1/2" x 4" long		18	.889		36	51.50		87.50	117
4824	3" x 4" long		14	1.143		47.50	66.50		114	152
4826	4" x 4" long		10	1.600		64	93		157	210
5000	Malleable iron, 150 lb.									
5020	Black									
5040	90° elbow, straight									
5060	1/4"	1 Plum	16	.500	Ea.	5.90	32		37.90	54.50
5070	3/8"		16	.500		5.90	32		37.90	54.50
5080	1/2"		15	.533		4	34.50		38.50	56
5090	3/4"		14	.571		4.12	37		41.12	59.50
5100	1"		13	.615		8.35	39.50		47.85	68.50
5110	1-1/4"	Q-1	22	.727		13.85	42		55.85	78.50
5120	1-1/2"		20	.800		14.60	46.50		61.10	85.50
5130	2"		18	.889		25.50	51.50		77	105
5140	2-1/2"		14	1.143		71	66.50		137.50	178

22 11 13 – Facility Water Distribution Piping

22 11 13.45 Pipe Fittings, Steel, Threaded		Crew	Daily Output	Labor-Hours	Unit	2020 Bare Costs Material	Labor	Equipment	Total	Total Incl O&P
5150	3"	Q-1	10	1.600	Ea.	104	93		197	254
5160	3-1/2"		8	2		287	116		403	490
5170	4"		6	2.667		196	155		351	445
5250	45° elbow, straight									
5270	1/4"	1 Plum	16	.500	Ea.	8.85	32		40.85	58
5280	3/8"		16	.500		8.85	32		40.85	58
5290	1/2"		15	.533		6.75	34.50		41.25	59
5300	3/4"		14	.571		8.35	37		45.35	64
5310	1"		13	.615		10.50	39.50		50	71
5320	1-1/4"	Q-1	22	.727		18.55	42		60.55	83.50
5330	1-1/2"		20	.800		23	46.50		69.50	95
5340	2"		18	.889		34.50	51.50		86	115
5350	2-1/2"		14	1.143		100	66.50		166.50	209
5360	3"		10	1.600		131	93		224	283
5370	3-1/2"		8	2		261	116		377	460
5380	4"		6	2.667		256	155		411	510
5450	Tee, straight									
5470	1/4"	1 Plum	10	.800	Ea.	8.55	51.50		60.05	86.50
5480	3/8"		10	.800		8.55	51.50		60.05	86.50
5490	1/2"		9	.889		5.45	57.50		62.95	91.50
5500	3/4"		9	.889		7.85	57.50		65.35	94
5510	1"		8	1		13.40	64.50		77.90	111
5520	1-1/4"	Q-1	14	1.143		21.50	66.50		88	123
5530	1-1/2"		13	1.231		27	71.50		98.50	137
5540	2"		11	1.455		46	84.50		130.50	177
5550	2-1/2"		9	1.778		99	103		202	263
5560	3"		6	2.667		146	155		301	390
5570	3-1/2"		5	3.200		340	186		526	655
5580	4"		4	4		355	232		587	735
5601	Tee, reducing, on outlet									
5602	1/2"	1 Plum	9	.889	Ea.	12.75	57.50		70.25	99.50
5603	3/4"		9	.889		13.15	57.50		70.65	100
5604	1"		8	1		17.30	64.50		81.80	116
5605	1-1/4"	Q-1	14	1.143		37	66.50		103.50	140
5606	1-1/2"		13	1.231		38	71.50		109.50	149
5607	2"		11	1.455		54	84.50		138.50	186
5608	2-1/2"		9	1.778		136	103		239	305
5609	3"		6	2.667		205	155		360	455
5610	3-1/2"		5	3.200		490	186		676	815
5611	4"		4	4		415	232		647	800
5650	Coupling									
5670	1/4"	1 Plum	19	.421	Ea.	7.30	27		34.30	48.50
5680	3/8"		19	.421		7.30	27		34.30	48.50
5690	1/2"		19	.421		5.65	27		32.65	46.50
5700	3/4"		18	.444		6.60	28.50		35.10	50.50
5710	1"		15	.533		9.90	34.50		44.40	62.50
5720	1-1/4"	Q-1	26	.615		11.95	35.50		47.45	66.50
5730	1-1/2"		24	.667		15.80	38.50		54.30	75.50
5740	2"		21	.762		23.50	44		67.50	91.50
5750	2-1/2"		18	.889		64.50	51.50		116	148
5760	3"		14	1.143		87.50	66.50		154	195
5770	3-1/2"		12	1.333		161	77.50		238.50	293
5780	4"		10	1.600		176	93		269	330

22 11 13 – Facility Water Distribution Piping

22 11 13.45 Pipe Fittings, Steel, Threaded		Crew	Daily Output	Labor-Hours	Unit	2020 Bare Costs Material	Labor	Equipment	Total	Total Incl O&P
5840	Reducer, concentric, 1/4"	1 Plum	19	.421	Ea.	7.70	27		34.70	49
5850	3/8"		19	.421		8.40	27		35.40	50
5860	1/2"		19	.421		7	27		34	48
5870	3/4"		16	.500		8	32		40	57
5880	1"		15	.533		12.65	34.50		47.15	65.50
5890	1-1/4"	Q-1	26	.615		15.65	35.50		51.15	71
5900	1-1/2"		24	.667		21	38.50		59.50	81
5910	2"		21	.762		30	44		74	99
5911	2-1/2"		18	.889		68.50	51.50		120	152
5912	3"		14	1.143		95.50	66.50		162	204
5913	3-1/2"		12	1.333		310	77.50		387.50	455
5914	4"		10	1.600		230	93		323	390
5981	Bushing, 1/4"	1 Plum	19	.421		1.09	27		28.09	41.50
5982	3/8"		19	.421		5.70	27		32.70	47
5983	1/2"		19	.421		6	27		33	47
5984	3/4"		16	.500		6.95	32		38.95	55.50
5985	1"		15	.533		10.70	34.50		45.20	63.50
5986	1-1/4"	Q-1	26	.615		13.25	35.50		48.75	68
5987	1-1/2"		24	.667		11.45	38.50		49.95	70.50
5988	2"		21	.762		14.35	44		58.35	82
5989	Cap, 1/4"	1 Plum	38	.211		6.85	13.55		20.40	28
5991	3/8"		38	.211		6.45	13.55		20	27.50
5992	1/2"		38	.211		4.13	13.55		17.68	25
5993	3/4"		32	.250		5.60	16.10		21.70	30
5994	1"		30	.267		6.75	17.20		23.95	33
5995	1-1/4"	Q-1	52	.308		8.90	17.85		26.75	36.50
5996	1-1/2"		48	.333		12.25	19.35		31.60	42.50
5997	2"		42	.381		17.85	22		39.85	52.50
6000	For galvanized elbows, tees, and couplings, add					20%				
6058	For galvanized reducers, caps and bushings, add					20%				
6100	90° elbow, galvanized, 150 lb., reducing									
6110	3/4" x 1/2"	1 Plum	15.40	.519	Ea.	11.70	33.50		45.20	63
6112	1" x 3/4"		14	.571		15.05	37		52.05	71.50
6114	1" x 1/2"		14.50	.552		15.95	35.50		51.45	70.50
6116	1-1/4" x 1"	Q-1	24.20	.661		25.50	38.50		64	85.50
6118	1-1/4" x 3/4"		25.40	.630		30.50	36.50		67	88
6120	1-1/4" x 1/2"		26.20	.611		32	35.50		67.50	88.50
6122	1-1/2" x 1-1/4"		21.60	.741		40.50	43		83.50	109
6124	1-1/2" x 1"		23.50	.681		40.50	39.50		80	104
6126	1-1/2" x 3/4"		24.60	.650		40.50	37.50		78	101
6128	2" x 1-1/2"		20.50	.780		47	45.50		92.50	120
6130	2" x 1-1/4"		21	.762		54	44		98	126
6132	2" x 1"		22.80	.702		56	40.50		96.50	123
6134	2" x 3/4"		23.90	.669		57	39		96	121
6136	2-1/2" x 2"		12.30	1.301		160	75.50		235.50	289
6138	2-1/2" x 1-1/2"		12.50	1.280		178	74		252	305
6140	3" x 2-1/2"		8.60	1.860		290	108		398	480
6142	3" x 2"		11.80	1.356		252	78.50		330.50	395
6144	4" x 3"		8.20	1.951		655	113		768	895
6160	90° elbow, black, 150 lb., reducing									
6170	1" x 3/4"	1 Plum	14	.571	Ea.	10.25	37		47.25	66.50
6174	1-1/2" x 1"	Q-1	23.50	.681		24.50	39.50		64	86
6178	1-1/2" x 3/4"		24.60	.650		28	37.50		65.50	87

22 11 13.45 Pipe Fittings, Steel, Threaded		Crew	Daily Output	Labor-Hours	Unit	2020 Bare Costs Material	Labor	Equipment	Total	Total Incl O&P
6182	2" x 1-1/2"	Q-1	20.50	.780	Ea.	35.50	45.50		81	107
6186	2" x 1"		22.80	.702		40	40.50		80.50	105
6190	2" x 3/4"		23.90	.669		42.50	39		81.50	105
6194	2-1/2" x 2"		12.30	1.301		97	75.50		172.50	219
7000	Union, with brass seat									
7010	1/4"	1 Plum	15	.533	Ea.	26.50	34.50		61	80.50
7020	3/8"		15	.533		21	34.50		55.50	75
7030	1/2"		14	.571		19.15	37		56.15	76
7040	3/4"		13	.615		18.95	39.50		58.45	80.50
7050	1"		12	.667		24.50	43		67.50	91.50
7060	1-1/4"	Q-1	21	.762		41	44		85	112
7070	1-1/2"		19	.842		51.50	49		100.50	130
7080	2"		17	.941		59.50	54.50		114	147
7090	2-1/2"		13	1.231		176	71.50		247.50	300
7100	3"		9	1.778		184	103		287	355
7120	Union, galvanized									
7124	1/2"	1 Plum	14	.571	Ea.	25.50	37		62.50	83
7125	3/4"		13	.615		29	39.50		68.50	91.50
7126	1"		12	.667		38.50	43		81.50	107
7127	1-1/4"	Q-1	21	.762		55.50	44		99.50	127
7128	1-1/2"		19	.842		67	49		116	147
7129	2"		17	.941		77	54.50		131.50	167
7130	2-1/2"		13	1.231		268	71.50		339.50	400
7131	3"		9	1.778		375	103		478	565
7500	Malleable iron, 300 lb.									
7520	Black									
7540	90° elbow, straight, 1/4"	1 Plum	16	.500	Ea.	20.50	32		52.50	70.50
7560	3/8"		16	.500		18.35	32		50.35	68
7570	1/2"		15	.533		24	34.50		58.50	77.50
7580	3/4"		14	.571		27	37		64	84.50
7590	1"		13	.615		34.50	39.50		74	97.50
7600	1-1/4"	Q-1	22	.727		50	42		92	118
7610	1-1/2"		20	.800		59	46.50		105.50	135
7620	2"		18	.889		85	51.50		136.50	171
7630	2-1/2"		14	1.143		220	66.50		286.50	340
7640	3"		10	1.600		253	93		346	415
7650	4"		6	2.667		595	155		750	885
7700	45° elbow, straight, 1/4"	1 Plum	16	.500		30.50	32		62.50	82
7720	3/8"		16	.500		30.50	32		62.50	81.50
7730	1/2"		15	.533		34	34.50		68.50	89
7740	3/4"		14	.571		38	37		75	97
7750	1"		13	.615		42	39.50		81.50	106
7760	1-1/4"	Q-1	22	.727		67.50	42		109.50	137
7770	1-1/2"		20	.800		88	46.50		134.50	166
7780	2"		18	.889		133	51.50		184.50	223
7790	2-1/2"		14	1.143		297	66.50		363.50	425
7800	3"		10	1.600		390	93		483	570
7810	4"		6	2.667		890	155		1,045	1,200
7850	Tee, straight, 1/4"	1 Plum	10	.800		26	51.50		77.50	106
7870	3/8"		10	.800		27.50	51.50		79	108
7880	1/2"		9	.889		35	57.50		92.50	124
7890	3/4"		9	.889		38	57.50		95.50	127
7900	1"		8	1		45.50	64.50		110	147

22 11 Facility Water Distribution

22 11 13 – Facility Water Distribution Piping

22 11 13.45 Pipe Fittings, Steel, Threaded		Crew	Daily Output	Labor-Hours	Unit	2020 Bare Costs Material	Labor	Equipment	Total	Total Incl O&P
7910	1-1/4"	Q-1	14	1.143	Ea.	65	66.50		131.50	171
7920	1-1/2"		13	1.231		76	71.50		147.50	191
7930	2"		11	1.455		112	84.50		196.50	250
7940	2-1/2"		9	1.778		273	103		376	455
7950	3"		6	2.667		390	155		545	660
7960	4"		4	4		1,100	232		1,332	1,575
8050	Couplings, straight, 1/4"	1 Plum	19	.421		18.60	27		45.60	61
8070	3/8"		19	.421		18.60	27		45.60	61
8080	1/2"		19	.421		21	27		48	63.50
8090	3/4"		18	.444		24	28.50		52.50	69.50
8100	1"		15	.533		27.50	34.50		62	81.50
8110	1-1/4"	Q-1	26	.615		32.50	35.50		68	89.50
8120	1-1/2"		24	.667		48.50	38.50		87	112
8130	2"		21	.762		70	44		114	143
8140	2-1/2"		18	.889		129	51.50		180.50	219
8150	3"		14	1.143		186	66.50		252.50	305
8160	4"		10	1.600		355	93		448	530
8162	6"		10	1.600		355	93		448	530
8200	Galvanized									
8220	90° elbow, straight, 1/4"	1 Plum	16	.500	Ea.	37	32		69	88.50
8222	3/8"		16	.500		37.50	32		69.50	89.50
8224	1/2"		15	.533		44.50	34.50		79	101
8226	3/4"		14	.571		50	37		87	110
8228	1"		13	.615		64	39.50		103.50	130
8230	1-1/4"	Q-1	22	.727		101	42		143	174
8232	1-1/2"		20	.800		109	46.50		155.50	190
8234	2"		18	.889		184	51.50		235.50	279
8236	2-1/2"		14	1.143		410	66.50		476.50	555
8238	3"		10	1.600		485	93		578	675
8240	4"		6	2.667		1,450	155		1,605	1,825
8280	45° elbow, straight									
8282	1/2"	1 Plum	15	.533	Ea.	66.50	34.50		101	125
8284	3/4"		14	.571		77	37		114	140
8286	1"		13	.615		84.50	39.50		124	152
8288	1-1/4"	Q-1	22	.727		131	42		173	207
8290	1-1/2"		20	.800		169	46.50		215.50	256
8292	2"		18	.889		234	51.50		285.50	335
8310	Tee, straight, 1/4"	1 Plum	10	.800		52.50	51.50		104	135
8312	3/8"		10	.800		53.50	51.50		105	136
8314	1/2"		9	.889		67	57.50		124.50	160
8316	3/4"		9	.889		74	57.50		131.50	167
8318	1"		8	1		91	64.50		155.50	197
8320	1-1/4"	Q-1	14	1.143		130	66.50		196.50	242
8322	1-1/2"		13	1.231		135	71.50		206.50	256
8324	2"		11	1.455		216	84.50		300.50	365
8326	2-1/2"		9	1.778		685	103		788	905
8328	3"		6	2.667		775	155		930	1,075
8330	4"		4	4		2,000	232		2,232	2,550
8380	Couplings, straight, 1/4"	1 Plum	19	.421		24.50	27		51.50	67
8382	3/8"		19	.421		35	27		62	79
8384	1/2"		19	.421		36	27		63	80
8386	3/4"		18	.444		40.50	28.50		69	87.50
8388	1"		15	.533		53.50	34.50		88	111

22 11 13.45 Pipe Fittings, Steel, Threaded		Crew	Daily Output	Labor-Hours	Unit	2020 Bare Costs Material	Labor	Equipment	Total	Total Incl O&P
8390	1-1/4"	Q-1	26	.615	Ea.	69.50	35.50		105	130
8392	1-1/2"		24	.667		98.50	38.50		137	166
8394	2"		21	.762		119	44		163	197
8396	2-1/2"		18	.889		315	51.50		366.50	420
8398	3"		14	1.143		375	66.50		441.50	510
8399	4"		10	1.600		570	93		663	765
8529	Black									
8530	Reducer, concentric, 1/4"	1 Plum	19	.421	Ea.	25	27		52	68
8531	3/8"		19	.421		24.50	27		51.50	67.50
8532	1/2"		17	.471		29.50	30.50		60	78
8533	3/4"		16	.500		34.50	32		66.50	85.50
8534	1"		15	.533		54.50	34.50		89	112
8535	1-1/4"	Q-1	26	.615		74.50	35.50		110	136
8536	1-1/2"		24	.667		76	38.50		114.50	142
8537	2"		21	.762		100	44		144	177
8550	Cap, 1/4"	1 Plum	38	.211		19.15	13.55		32.70	41.50
8551	3/8"		38	.211		19.15	13.55		32.70	41.50
8552	1/2"		34	.235		18.95	15.15		34.10	43.50
8553	3/4"		32	.250		24	16.10		40.10	50.50
8554	1"		30	.267		31.50	17.20		48.70	60
8555	1-1/4"	Q-1	52	.308		35	17.85		52.85	65
8556	1-1/2"		48	.333		51	19.35		70.35	85
8557	2"		42	.381		70.50	22		92.50	111
8570	Plug, 1/4"	1 Plum	38	.211		3.68	13.55		17.23	24.50
8571	3/8"		38	.211		3.50	13.55		17.05	24.50
8572	1/2"		34	.235		3.59	15.15		18.74	26.50
8573	3/4"		32	.250		4.48	16.10		20.58	29
8574	1"		30	.267		6.85	17.20		24.05	33
8575	1-1/4"	Q-1	52	.308		14.05	17.85		31.90	42
8576	1-1/2"		48	.333		16.15	19.35		35.50	47
8577	2"		42	.381		25.50	22		47.50	61
9500	Union with brass seat, 1/4"	1 Plum	15	.533		45	34.50		79.50	101
9530	3/8"		15	.533		45.50	34.50		80	102
9540	1/2"		14	.571		36.50	37		73.50	95.50
9550	3/4"		13	.615		40.50	39.50		80	104
9560	1"		12	.667		53	43		96	123
9570	1-1/4"	Q-1	21	.762		86.50	44		130.50	161
9580	1-1/2"		19	.842		90.50	49		139.50	173
9590	2"		17	.941		112	54.50		166.50	205
9600	2-1/2"		13	1.231		360	71.50		431.50	505
9610	3"		9	1.778		470	103		573	670
9620	4"		5	3.200		1,475	186		1,661	1,900
9630	Union, all iron, 1/4"	1 Plum	15	.533		69.50	34.50		104	128
9650	3/8"		15	.533		69.50	34.50		104	128
9660	1/2"		14	.571		73	37		110	136
9670	3/4"		13	.615		78.50	39.50		118	146
9680	1"		12	.667		96	43		139	171
9690	1-1/4"	Q-1	21	.762		148	44		192	229
9700	1-1/2"		19	.842		180	49		229	271
9710	2"		17	.941		228	54.50		282.50	335
9720	2-1/2"		13	1.231		405	71.50		476.50	550
9730	3"		9	1.778		770	103		873	1,000
9750	For galvanized unions, add					15%				

22 11 13 – Facility Water Distribution Piping

22 11 13.45 Pipe Fittings, Steel, Threaded		Crew	Daily Output	Labor-Hours	Unit	2020 Bare Costs Material	Labor	Equipment	Total	Total Incl O&P
9757	Forged steel, 3000 lb.									
9758	Black									
9760	90° elbow, 1/4"	1 Plum	16	.500	Ea.	19.25	32		51.25	69
9761	3/8"		16	.500		19.25	32		51.25	69
9762	1/2"		15	.533		15	34.50		49.50	68
9763	3/4"		14	.571		18.45	37		55.45	75.50
9764	1"		13	.615		29.50	39.50		69	92
9765	1-1/4"	Q-1	22	.727		52.50	42		94.50	121
9766	1-1/2"		20	.800		67.50	46.50		114	144
9767	2"		18	.889		82	51.50		133.50	167
9780	45° elbow, 1/4"	1 Plum	16	.500		24.50	32		56.50	75
9781	3/8"		16	.500		24.50	32		56.50	75
9782	1/2"		15	.533		24	34.50		58.50	78
9783	3/4"		14	.571		28	37		65	85.50
9784	1"		13	.615		38	39.50		77.50	102
9785	1-1/4"	Q-1	22	.727		52.50	42		94.50	121
9786	1-1/2"		20	.800		75.50	46.50		122	153
9787	2"		18	.889		104	51.50		155.50	191
9800	Tee, 1/4"	1 Plum	10	.800		23.50	51.50		75	103
9801	3/8"		10	.800		23.50	51.50		75	103
9802	1/2"		9	.889		21	57.50		78.50	109
9803	3/4"		9	.889		28.50	57.50		86	117
9804	1"		8	1		37	64.50		101.50	138
9805	1-1/4"	Q-1	14	1.143		73	66.50		139.50	179
9806	1-1/2"		13	1.231		85	71.50		156.50	201
9807	2"		11	1.455		99.50	84.50		184	235
9820	Reducer, concentric, 1/4"	1 Plum	19	.421		12.85	27		39.85	54.50
9821	3/8"		19	.421		13.40	27		40.40	55.50
9822	1/2"		17	.471		13.40	30.50		43.90	60.50
9823	3/4"		16	.500		15.90	32		47.90	65.50
9824	1"		15	.533		20.50	34.50		55	74.50
9825	1-1/4"	Q-1	26	.615		36	35.50		71.50	93
9826	1-1/2"		24	.667		39	38.50		77.50	101
9827	2"		21	.762		57	44		101	129
9840	Cap, 1/4"	1 Plum	38	.211		8.10	13.55		21.65	29.50
9841	3/8"		38	.211		7.75	13.55		21.30	29
9842	1/2"		34	.235		7.55	15.15		22.70	31
9843	3/4"		32	.250		10.50	16.10		26.60	35.50
9844	1"		30	.267		16.10	17.20		33.30	43
9845	1-1/4"	Q-1	52	.308		24.50	17.85		42.35	53
9846	1-1/2"		48	.333		29	19.35		48.35	60.50
9847	2"		42	.381		42	22		64	79
9860	Plug, 1/4"	1 Plum	38	.211		4.09	13.55		17.64	25
9861	3/8"		38	.211		3.72	13.55		17.27	24.50
9862	1/2"		34	.235		3.58	15.15		18.73	26.50
9863	3/4"		32	.250		5.55	16.10		21.65	30
9864	1"		30	.267		8.50	17.20		25.70	35
9865	1-1/4"	Q-1	52	.308		16.40	17.85		34.25	44.50
9866	1-1/2"		48	.333		18.65	19.35		38	49.50
9867	2"		42	.381		29.50	22		51.50	65.50
9880	Union, bronze seat, 1/4"	1 Plum	15	.533		65	34.50		99.50	123
9881	3/8"		15	.533		65	34.50		99.50	123
9882	1/2"		14	.571		62.50	37		99.50	124

22 11 13.45 Pipe Fittings, Steel, Threaded		Crew	Daily Output	Labor-Hours	Unit	2020 Bare Costs Material	Labor	Equipment	Total	Total Incl O&P
9883	3/4"	1 Plum	13	.615	Ea.	84.50	39.50		124	153
9884	1"		12	.667		94.50	43		137.50	169
9885	1-1/4"	Q-1	21	.762		173	44		217	256
9886	1-1/2"		19	.842		189	49		238	281
9887	2"		17	.941		224	54.50		278.50	330
9900	Coupling, 1/4"	1 Plum	19	.421		7.60	27		34.60	49
9901	3/8"		19	.421		7.60	27		34.60	49
9902	1/2"		17	.471		6.20	30.50		36.70	52.50
9903	3/4"		16	.500		8	32		40	57
9904	1"		15	.533		13.90	34.50		48.40	67
9905	1-1/4"	Q-1	26	.615		23.50	35.50		59	79
9906	1-1/2"		24	.667		30	38.50		68.50	91
9907	2"		21	.762		40	44		84	111

22 11 13.47 Pipe Fittings, Steel		Crew	Daily Output	Labor-Hours	Unit	Material	Labor	Equipment	Total	Total Incl O&P
0010	**PIPE FITTINGS, STEEL**, flanged, welded & special									
0020	Flanged joints, CI, standard weight, black. One gasket & bolt									
0040	set, mat'l only, required at each joint, not included (see line 0620)									
0060	90° elbow, straight, 1-1/2" pipe size	Q-1	14	1.143	Ea.	690	66.50		756.50	860
0080	2" pipe size		13	1.231		400	71.50		471.50	545
0090	2-1/2" pipe size		12	1.333		430	77.50		507.50	590
0100	3" pipe size		11	1.455		360	84.50		444.50	520
0110	4" pipe size		8	2		445	116		561	665
0120	5" pipe size		7	2.286		1,050	133		1,183	1,350
0130	6" pipe size	Q-2	9	2.667		695	160		855	1,000
0140	8" pipe size		8	3		1,200	180		1,380	1,600
0150	10" pipe size		7	3.429		2,625	206		2,831	3,200
0160	12" pipe size		6	4		5,300	241		5,541	6,175
0171	90° elbow, reducing									
0172	2-1/2" by 2" pipe size	Q-1	12	1.333	Ea.	1,350	77.50		1,427.50	1,600
0173	3" by 2-1/2" pipe size		11	1.455		1,325	84.50		1,409.50	1,575
0174	4" by 3" pipe size		8	2		985	116		1,101	1,250
0175	5" by 3" pipe size		7	2.286		2,175	133		2,308	2,575
0176	6" by 4" pipe size	Q-2	9	2.667		1,225	160		1,385	1,600
0177	8" by 6" pipe size		8	3		1,775	180		1,955	2,250
0178	10" by 8" pipe size		7	3.429		3,400	206		3,606	4,050
0179	12" by 10" pipe size		6	4		6,525	241		6,766	7,525
0200	45° elbow, straight, 1-1/2" pipe size	Q-1	14	1.143		840	66.50		906.50	1,025
0220	2" pipe size		13	1.231		575	71.50		646.50	740
0230	2-1/2" pipe size		12	1.333		615	77.50		692.50	790
0240	3" pipe size		11	1.455		600	84.50		684.50	785
0250	4" pipe size		8	2		675	116		791	915
0260	5" pipe size		7	2.286		1,625	133		1,758	1,975
0270	6" pipe size	Q-2	9	2.667		1,100	160		1,260	1,450
0280	8" pipe size		8	3		1,600	180		1,780	2,050
0290	10" pipe size		7	3.429		3,400	206		3,606	4,050
0300	12" pipe size		6	4		5,200	241		5,441	6,075
0310	Cross, straight									
0311	2-1/2" pipe size	Q-1	6	2.667	Ea.	1,275	155		1,430	1,625
0312	3" pipe size		5	3.200		1,325	186		1,511	1,725
0313	4" pipe size		4	4		1,725	232		1,957	2,250
0314	5" pipe size		3	5.333		3,950	310		4,260	4,800
0315	6" pipe size	Q-2	5	4.800		3,625	289		3,914	4,425

22 11 Facility Water Distribution

22 11 13 – Facility Water Distribution Piping

22 11 13.47 Pipe Fittings, Steel		Crew	Daily Output	Labor-Hours	Unit	2020 Bare Costs Material	Labor	Equipment	Total	Total Incl O&P
0316	8" pipe size	Q-2	4	6	Ea.	5,875	360		6,235	7,000
0317	10" pipe size		3	8		6,975	480		7,455	8,400
0318	12" pipe size		2	12		11,500	720		12,220	13,700
0350	Tee, straight, 1-1/2" pipe size	Q-1	10	1.600		795	93		888	1,025
0370	2" pipe size		9	1.778		435	103		538	635
0380	2-1/2" pipe size		8	2		635	116		751	875
0390	3" pipe size		7	2.286		445	133		578	690
0400	4" pipe size		5	3.200		675	186		861	1,025
0410	5" pipe size		4	4		1,825	232		2,057	2,350
0420	6" pipe size	Q-2	6	4		985	241		1,226	1,425
0430	8" pipe size		5	4.800		1,675	289		1,964	2,275
0440	10" pipe size		4	6		4,525	360		4,885	5,525
0450	12" pipe size		3	8		7,100	480		7,580	8,525
0459	Tee, reducing on outlet									
0460	2-1/2" by 2" pipe size	Q-1	8	2	Ea.	1,375	116		1,491	1,700
0461	3" by 2-1/2" pipe size		7	2.286		1,325	133		1,458	1,650
0462	4" by 3" pipe size		5	3.200		1,400	186		1,586	1,800
0463	5" by 4" pipe size		4	4		3,100	232		3,332	3,775
0464	6" by 4" pipe size	Q-2	6	4		1,350	241		1,591	1,825
0465	8" by 6" pipe size		5	4.800		1,125	289		1,414	1,675
0466	10" by 8" pipe size		4	6		4,775	360		5,135	5,800
0467	12" by 10" pipe size		3	8		8,325	480		8,805	9,875
0476	Reducer, concentric									
0477	3" by 2-1/2"	Q-1	12	1.333	Ea.	800	77.50		877.50	995
0478	4" by 3"		9	1.778		895	103		998	1,150
0479	5" by 4"		8	2		1,375	116		1,491	1,700
0480	6" by 4"	Q-2	10	2.400		1,200	144		1,344	1,550
0481	8" by 6"		9	2.667		1,475	160		1,635	1,875
0482	10" by 8"		8	3		3,100	180		3,280	3,700
0483	12" by 10"		7	3.429		5,375	206		5,581	6,225
0492	Reducer, eccentric									
0493	4" by 3"	Q-1	8	2	Ea.	1,450	116		1,566	1,775
0494	5" by 4"	"	7	2.286		2,275	133		2,408	2,700
0495	6" by 4"	Q-2	9	2.667		1,375	160		1,535	1,775
0496	8" by 6"		8	3		1,750	180		1,930	2,200
0497	10" by 8"		7	3.429		4,425	206		4,631	5,175
0498	12" by 10"		6	4		6,050	241		6,291	7,000
0500	For galvanized elbows and tees, add					100%				
0520	For extra heavy weight elbows and tees, add					140%				
0620	Gasket and bolt set, 150 lb., 1/2" pipe size	1 Plum	20	.400		3.83	26		29.83	42.50
0622	3/4" pipe size		19	.421		4.13	27		31.13	45
0624	1" pipe size		18	.444		4.62	28.50		33.12	48
0626	1-1/4" pipe size		17	.471		4.36	30.50		34.86	50.50
0628	1-1/2" pipe size		15	.533		4.38	34.50		38.88	56.50
0630	2" pipe size		13	.615		8.90	39.50		48.40	69.50
0640	2-1/2" pipe size		12	.667		8.45	43		51.45	74
0650	3" pipe size		11	.727		9.10	47		56.10	80
0660	3-1/2" pipe size		9	.889		20	57.50		77.50	108
0670	4" pipe size		8	1		17.90	64.50		82.40	116
0680	5" pipe size		7	1.143		27	73.50		100.50	140
0690	6" pipe size		6	1.333		28.50	86		114.50	160
0700	8" pipe size		5	1.600		31	103		134	188
0710	10" pipe size		4.50	1.778		57	115		172	234

22 11 Facility Water Distribution

22 11 13 – Facility Water Distribution Piping

22 11 13.47 Pipe Fittings, Steel		Crew	Daily Output	Labor-Hours	Unit	Material	Labor	Equipment	Total	Total Incl O&P
						2020 Bare Costs				
0720	12″ pipe size	1 Plum	4.20	1.905	Ea.	59	123		182	249
0730	14″ pipe size		4	2		54.50	129		183.50	253
0740	16″ pipe size		3	2.667		65.50	172		237.50	330
0750	18″ pipe size		2.70	2.963		127	191		318	425
0760	20″ pipe size		2.30	3.478		206	224		430	560
0780	24″ pipe size		1.90	4.211		257	271		528	690
0790	26″ pipe size		1.60	5		350	320		670	865
0810	30″ pipe size		1.40	5.714		680	370		1,050	1,300
0830	36″ pipe size		1.10	7.273		1,275	470		1,745	2,100
0850	For 300 lb. gasket set, add					40%				
2000	Flanged unions, 125 lb., black, 1/2″ pipe size	1 Plum	17	.471	Ea.	103	30.50		133.50	159
2040	3/4″ pipe size		17	.471		143	30.50		173.50	203
2050	1″ pipe size		16	.500		138	32		170	200
2060	1-1/4″ pipe size	Q-1	28	.571		164	33		197	230
2070	1-1/2″ pipe size		27	.593		151	34.50		185.50	218
2080	2″ pipe size		26	.615		178	35.50		213.50	250
2090	2-1/2″ pipe size		24	.667		241	38.50		279.50	325
2100	3″ pipe size		22	.727		273	42		315	365
2110	3-1/2″ pipe size		18	.889		460	51.50		511.50	585
2120	4″ pipe size		16	1		370	58		428	495
2130	5″ pipe size		14	1.143		865	66.50		931.50	1,050
2140	6″ pipe size	Q-2	19	1.263		810	76		886	1,000
2150	8″ pipe size	″	16	1.500		1,875	90		1,965	2,175
2200	For galvanized unions, add					150%				
2290	Threaded flange									
2300	Cast iron									
2310	Black, 125 lb., per flange									
2320	1″ pipe size	1 Plum	27	.296	Ea.	53	19.10		72.10	87
2330	1-1/4″ pipe size	Q-1	44	.364		69.50	21		90.50	108
2340	1-1/2″ pipe size		40	.400		59	23		82	99
2350	2″ pipe size		36	.444		64.50	26		90.50	109
2360	2-1/2″ pipe size		28	.571		75	33		108	132
2370	3″ pipe size		20	.800		97	46.50		143.50	177
2380	3-1/2″ pipe size		16	1		138	58		196	239
2390	4″ pipe size		12	1.333		119	77.50		196.50	247
2400	5″ pipe size		10	1.600		184	93		277	340
2410	6″ pipe size	Q-2	14	1.714		209	103		312	385
2420	8″ pipe size		12	2		330	120		450	540
2430	10″ pipe size		10	2.400		585	144		729	855
2440	12″ pipe size		8	3		1,325	180		1,505	1,725
2460	For galvanized flanges, add					95%				
2490	Blind flange									
2492	Cast iron									
2494	Black, 125 lb., per flange									
2496	1″ pipe size	1 Plum	27	.296	Ea.	92.50	19.10		111.60	131
2500	1-1/2″ pipe size	Q-1	40	.400		100	23		123	145
2502	2″ pipe size		36	.444		117	26		143	168
2504	2-1/2″ pipe size		28	.571		125	33		158	187
2506	3″ pipe size		20	.800		155	46.50		201.50	240
2508	4″ pipe size		12	1.333		198	77.50		275.50	335
2510	5″ pipe size		10	1.600		320	93		413	495
2512	6″ pipe size	Q-2	14	1.714		350	103		453	540
2514	8″ pipe size		12	2		550	120		670	785

22 11 13.47 Pipe Fittings, Steel		Crew	Daily Output	Labor-Hours	Unit	2020 Bare Costs Material	Labor	Equipment	Total	Total Incl O&P
2516	10″ pipe size	Q-2	10	2.400	Ea.	815	144		959	1,100
2518	12″ pipe size		8	3		1,575	180		1,755	2,000
2520	For galvanized flanges, add					80%				
2570	Threaded flange									
2580	Forged steel									
2590	Black, 150 lb., per flange									
2600	1/2″ pipe size	1 Plum	30	.267	Ea.	27.50	17.20		44.70	56
2610	3/4″ pipe size		28	.286		27.50	18.40		45.90	58
2620	1″ pipe size		27	.296		27.50	19.10		46.60	59
2630	1-1/4″ pipe size	Q-1	44	.364		27.50	21		48.50	62
2640	1-1/2″ pipe size		40	.400		27.50	23		50.50	65
2650	2″ pipe size		36	.444		31	26		57	72.50
2660	2-1/2″ pipe size		28	.571		42.50	33		75.50	96.50
2670	3″ pipe size		20	.800		43	46.50		89.50	117
2690	4″ pipe size		12	1.333		48.50	77.50		126	169
2700	5″ pipe size		10	1.600		79.50	93		172.50	227
2710	6″ pipe size	Q-2	14	1.714		85	103		188	248
2720	8″ pipe size		12	2		149	120		269	345
2730	10″ pipe size		10	2.400		266	144		410	510
2860	Black, 300 lb., per flange									
2870	1/2″ pipe size	1 Plum	30	.267	Ea.	30.50	17.20		47.70	59
2880	3/4″ pipe size		28	.286		30.50	18.40		48.90	61
2890	1″ pipe size		27	.296		30.50	19.10		49.60	62
2900	1-1/4″ pipe size	Q-1	44	.364		30.50	21		51.50	65
2910	1-1/2″ pipe size		40	.400		30.50	23		53.50	68
2920	2″ pipe size		36	.444		40	26		66	82.50
2930	2-1/2″ pipe size		28	.571		49.50	33		82.50	104
2940	3″ pipe size		20	.800		50.50	46.50		97	125
2960	4″ pipe size		12	1.333		72.50	77.50		150	196
2970	6″ pipe size	Q-2	14	1.714		157	103		260	325
3000	Weld joint, butt, carbon steel, standard weight									
3040	90° elbow, long radius									
3050	1/2″ pipe size	Q-15	16	1	Ea.	44	58	6.65	108.65	143
3060	3/4″ pipe size		16	1		44	58	6.65	108.65	143
3070	1″ pipe size		16	1		26.50	58	6.65	91.15	123
3080	1-1/4″ pipe size		14	1.143		26.50	66.50	7.60	100.60	136
3090	1-1/2″ pipe size		13	1.231		26.50	71.50	8.20	106.20	145
3100	2″ pipe size		10	1.600		22.50	93	10.65	126.15	175
3110	2-1/2″ pipe size		8	2		35	116	13.30	164.30	227
3120	3″ pipe size		7	2.286		30	133	15.20	178.20	248
3130	4″ pipe size		5	3.200		56.50	186	21.50	264	365
3136	5″ pipe size		4	4		112	232	26.50	370.50	500
3140	6″ pipe size	Q-16	5	4.800		126	289	21.50	436.50	590
3150	8″ pipe size		3.75	6.400		237	385	28.50	650.50	865
3160	10″ pipe size		3	8		470	480	35.50	985.50	1,275
3170	12″ pipe size		2.50	9.600		665	575	42.50	1,282.50	1,650
3180	14″ pipe size		2	12		1,025	720	53	1,798	2,250
3190	16″ pipe size		1.50	16		1,400	960	71	2,431	3,075
3191	18″ pipe size		1.25	19.200		1,750	1,150	85	2,985	3,750
3192	20″ pipe size		1.15	20.870		2,425	1,250	92.50	3,767.50	4,625
3194	24″ pipe size		1.02	23.529		3,625	1,425	104	5,154	6,250
3200	45° elbow, long									
3210	1/2″ pipe size	Q-15	16	1	Ea.	65.50	58	6.65	130.15	166

	22 11 13.47 Pipe Fittings, Steel	Crew	Daily Output	Labor-Hours	Unit	2020 Bare Costs Material	Labor	Equipment	Total	Total Incl O&P
3220	3/4" pipe size	Q-15	16	1	Ea.	65.50	58	6.65	130.15	166
3230	1" pipe size		16	1		23	58	6.65	87.65	120
3240	1-1/4" pipe size		14	1.143		23	66.50	7.60	97.10	133
3250	1-1/2" pipe size		13	1.231		23	71.50	8.20	102.70	142
3260	2" pipe size		10	1.600		23	93	10.65	126.65	176
3270	2-1/2" pipe size		8	2		28	116	13.30	157.30	220
3280	3" pipe size		7	2.286		29	133	15.20	177.20	247
3290	4" pipe size		5	3.200		52	186	21.50	259.50	360
3296	5" pipe size		4	4		79	232	26.50	337.50	460
3300	6" pipe size	Q-16	5	4.800		103	289	21.50	413.50	565
3310	8" pipe size		3.75	6.400		171	385	28.50	584.50	795
3320	10" pipe size		3	8		340	480	35.50	855.50	1,125
3330	12" pipe size		2.50	9.600		480	575	42.50	1,097.50	1,450
3340	14" pipe size		2	12		650	720	53	1,423	1,850
3341	16" pipe size		1.50	16		1,150	960	71	2,181	2,800
3342	18" pipe size		1.25	19.200		1,650	1,150	85	2,885	3,625
3343	20" pipe size		1.15	20.870		1,700	1,250	92.50	3,042.50	3,850
3345	24" pipe size		1.05	22.857		2,525	1,375	101	4,001	4,925
3346	26" pipe size		.85	28.235		2,825	1,700	125	4,650	5,825
3347	30" pipe size		.45	53.333		3,125	3,200	236	6,561	8,475
3349	36" pipe size		.38	63.158		3,425	3,800	280	7,505	9,750
3350	Tee, straight									
3352	For reducing tees and concentrics see starting line 4600									
3360	1/2" pipe size	Q-15	10	1.600	Ea.	116	93	10.65	219.65	279
3370	3/4" pipe size		10	1.600		116	93	10.65	219.65	279
3380	1" pipe size		10	1.600		57.50	93	10.65	161.15	214
3390	1-1/4" pipe size		9	1.778		72	103	11.80	186.80	246
3400	1-1/2" pipe size		8	2		72	116	13.30	201.30	268
3410	2" pipe size		6	2.667		57.50	155	17.75	230.25	315
3420	2-1/2" pipe size		5	3.200		79.50	186	21.50	287	390
3430	3" pipe size		4	4		88.50	232	26.50	347	470
3440	4" pipe size		3	5.333		124	310	35.50	469.50	640
3446	5" pipe size		2.50	6.400		205	370	42.50	617.50	825
3450	6" pipe size	Q-16	3	8		213	480	35.50	728.50	995
3460	8" pipe size		2.50	9.600		370	575	42.50	987.50	1,325
3470	10" pipe size		2	12		730	720	53	1,503	1,950
3480	12" pipe size		1.60	15		1,025	900	66.50	1,991.50	2,550
3481	14" pipe size		1.30	18.462		1,800	1,100	82	2,982	3,725
3482	16" pipe size		1	24		2,000	1,450	106	3,556	4,475
3483	18" pipe size		.80	30		3,175	1,800	133	5,108	6,350
3484	20" pipe size		.75	32		5,000	1,925	142	7,067	8,525
3486	24" pipe size		.70	34.286		6,450	2,050	152	8,652	10,300
3487	26" pipe size		.55	43.636		7,075	2,625	193	9,893	11,900
3488	30" pipe size		.30	80		7,750	4,800	355	12,905	16,100
3490	36" pipe size		.25	96		8,550	5,775	425	14,750	18,500
3491	Eccentric reducer, 1-1/2" pipe size	Q-15	14	1.143		44.50	66.50	7.60	118.60	156
3492	2" pipe size		11	1.455		58	84.50	9.65	152.15	200
3493	2-1/2" pipe size		9	1.778		63.50	103	11.80	178.30	237
3494	3" pipe size		8	2		83	116	13.30	212.30	280
3495	4" pipe size		6	2.667		166	155	17.75	338.75	435
3496	6" pipe size	Q-16	5	4.800		223	289	21.50	533.50	700
3497	8" pipe size		4	6		258	360	26.50	644.50	855
3498	10" pipe size		3	8		440	480	35.50	955.50	1,250

22 11 13 – Facility Water Distribution Piping

22 11 13.47 Pipe Fittings, Steel		Crew	Daily Output	Labor-Hours	Unit	Material	2020 Bare Costs Labor	Equipment	Total	Total Incl O&P
3499	12" pipe size	Q-16	2.50	9.600	Ea.	680	575	42.50	1,297.50	1,650
3501	Cap, 1-1/2" pipe size	Q-15	28	.571		20.50	33	3.80	57.30	76
3502	2" pipe size		22	.727		22.50	42	4.84	69.34	93
3503	2-1/2" pipe size		18	.889		23.50	51.50	5.90	80.90	109
3504	3" pipe size		16	1		23.50	58	6.65	88.15	120
3505	4" pipe size		12	1.333		34	77.50	8.85	120.35	163
3506	6" pipe size	Q-16	10	2.400		58.50	144	10.65	213.15	292
3507	8" pipe size		8	3		89	180	13.30	282.30	380
3508	10" pipe size		6	4		160	241	17.70	418.70	555
3509	12" pipe size		5	4.800		247	289	21.50	557.50	725
3511	14" pipe size		4	6		480	360	26.50	866.50	1,100
3512	16" pipe size		4	6		535	360	26.50	921.50	1,150
3513	18" pipe size		3	8		515	480	35.50	1,030.50	1,325
3517	Weld joint, butt, carbon steel, extra strong									
3519	90° elbow, long									
3520	1/2" pipe size	Q-15	13	1.231	Ea.	58.50	71.50	8.20	138.20	181
3530	3/4" pipe size		12	1.333		58.50	77.50	8.85	144.85	190
3540	1" pipe size		11	1.455		28.50	84.50	9.65	122.65	168
3550	1-1/4" pipe size		10	1.600		28.50	93	10.65	132.15	182
3560	1-1/2" pipe size		9	1.778		28.50	103	11.80	143.30	199
3570	2" pipe size		8	2		29	116	13.30	158.30	221
3580	2-1/2" pipe size		7	2.286		40.50	133	15.20	188.70	260
3590	3" pipe size		6	2.667		52	155	17.75	224.75	310
3600	4" pipe size		4	4		86	232	26.50	344.50	470
3606	5" pipe size		3.50	4.571		206	265	30.50	501.50	655
3610	6" pipe size	Q-16	4.50	5.333		217	320	23.50	560.50	745
3620	8" pipe size		3.50	6.857		410	410	30.50	850.50	1,100
3630	10" pipe size		2.50	9.600		870	575	42.50	1,487.50	1,875
3640	12" pipe size		2.25	10.667		1,075	640	47.50	1,762.50	2,175
3650	45° elbow, long									
3660	1/2" pipe size	Q-15	13	1.231	Ea.	65	71.50	8.20	144.70	188
3670	3/4" pipe size		12	1.333		65	77.50	8.85	151.35	197
3680	1" pipe size		11	1.455		30.50	84.50	9.65	124.65	170
3690	1-1/4" pipe size		10	1.600		30.50	93	10.65	134.15	184
3700	1-1/2" pipe size		9	1.778		30.50	103	11.80	145.30	201
3710	2" pipe size		8	2		30.50	116	13.30	159.80	222
3720	2-1/2" pipe size		7	2.286		67.50	133	15.20	215.70	289
3730	3" pipe size		6	2.667		39	155	17.75	211.75	294
3740	4" pipe size		4	4		61.50	232	26.50	320	440
3746	5" pipe size		3.50	4.571		146	265	30.50	441.50	590
3750	6" pipe size	Q-16	4.50	5.333		169	320	23.50	512.50	690
3760	8" pipe size		3.50	6.857		291	410	30.50	731.50	970
3770	10" pipe size		2.50	9.600		560	575	42.50	1,177.50	1,525
3780	12" pipe size		2.25	10.667		830	640	47.50	1,517.50	1,925
3800	Tee, straight									
3810	1/2" pipe size	Q-15	9	1.778	Ea.	143	103	11.80	257.80	325
3820	3/4" pipe size		8.50	1.882		137	109	12.50	258.50	330
3830	1" pipe size		8	2		54.50	116	13.30	183.80	248
3840	1-1/4" pipe size		7	2.286		54	133	15.20	202.20	274
3850	1-1/2" pipe size		6	2.667		54.50	155	17.75	227.25	310
3860	2" pipe size		5	3.200		61	186	21.50	268.50	370
3870	2-1/2" pipe size		4	4		102	232	26.50	360.50	485
3880	3" pipe size		3.50	4.571		128	265	30.50	423.50	570

22 11 13.47 Pipe Fittings, Steel		Crew	Daily Output	Labor-Hours	Unit	Material	2020 Bare Costs Labor	Equipment	Total	Total Incl O&P
3890	4″ pipe size	Q-15	2.50	6.400	Ea.	153	370	42.50	565.50	770
3896	5″ pipe size		2.25	7.111		390	410	47.50	847.50	1,100
3900	6″ pipe size	Q-16	2.25	10.667		300	640	47.50	987.50	1,350
3910	8″ pipe size		2	12		575	720	53	1,348	1,775
3920	10″ pipe size		1.75	13.714		865	825	61	1,751	2,250
3930	12″ pipe size		1.50	16		1,275	960	71	2,306	2,925
4000	Eccentric reducer, 1-1/2″ pipe size	Q-15	10	1.600		20.50	93	10.65	124.15	174
4010	2″ pipe size		9	1.778		56.50	103	11.80	171.30	230
4020	2-1/2″ pipe size		8	2		82.50	116	13.30	211.80	279
4030	3″ pipe size		7	2.286		69	133	15.20	217.20	291
4040	4″ pipe size		5	3.200		111	186	21.50	318.50	425
4046	5″ pipe size		4.70	3.404		300	197	22.50	519.50	650
4050	6″ pipe size	Q-16	4.50	5.333		305	320	23.50	648.50	840
4060	8″ pipe size		3.50	6.857		405	410	30.50	845.50	1,100
4070	10″ pipe size		2.50	9.600		690	575	42.50	1,307.50	1,675
4080	12″ pipe size		2.25	10.667		880	640	47.50	1,567.50	1,975
4090	14″ pipe size		2.10	11.429		1,650	685	50.50	2,385.50	2,875
4100	16″ pipe size		1.90	12.632		2,125	760	56	2,941	3,500
4151	Cap, 1-1/2″ pipe size	Q-15	24	.667		25.50	38.50	4.43	68.43	91
4152	2″ pipe size		18	.889		22.50	51.50	5.90	79.90	108
4153	2-1/2″ pipe size		16	1		30.50	58	6.65	95.15	128
4154	3″ pipe size		14	1.143		34	66.50	7.60	108.10	145
4155	4″ pipe size		10	1.600		45.50	93	10.65	149.15	201
4156	6″ pipe size	Q-16	9	2.667		95.50	160	11.80	267.30	360
4157	8″ pipe size		7	3.429		143	206	15.20	364.20	485
4158	10″ pipe size		5	4.800		217	289	21.50	527.50	695
4159	12″ pipe size		4	6		294	360	26.50	680.50	895
4190	Weld fittings, reducing, standard weight									
4200	Welding ring w/spacer pins, 2″ pipe size				Ea.	2.23			2.23	2.45
4210	2-1/2″ pipe size					2.16			2.16	2.38
4220	3″ pipe size					2.25			2.25	2.48
4230	4″ pipe size					2.81			2.81	3.09
4236	5″ pipe size					3.46			3.46	3.81
4240	6″ pipe size					3.29			3.29	3.62
4250	8″ pipe size					3.65			3.65	4.02
4260	10″ pipe size					4.22			4.22	4.64
4270	12″ pipe size					4.93			4.93	5.40
4280	14″ pipe size					5.65			5.65	6.25
4290	16″ pipe size					6.65			6.65	7.30
4300	18″ pipe size					7.35			7.35	8.05
4310	20″ pipe size					9.25			9.25	10.20
4330	24″ pipe size					12.85			12.85	14.10
4340	26″ pipe size					15.85			15.85	17.40
4350	30″ pipe size					18.40			18.40	20.50
4370	36″ pipe size					23			23	25.50
4600	Tee, reducing on outlet									
4601	2-1/2″ x 2″ pipe size	Q-15	5	3.200	Ea.	106	186	21.50	313.50	420
4602	3″ x 2-1/2″ pipe size		4	4		151	232	26.50	409.50	540
4604	4″ x 3″ pipe size		3	5.333		122	310	35.50	467.50	640
4605	5″ x 4″ pipe size		2.50	6.400		340	370	42.50	752.50	975
4606	6″ x 5″ pipe size	Q-16	3	8		475	480	35.50	990.50	1,275
4607	8″ x 6″ pipe size		2.50	9.600		585	575	42.50	1,202.50	1,550
4608	10″ x 8″ pipe size		2	12		840	720	53	1,613	2,050

22 11 13.47 Pipe Fittings, Steel		Crew	Daily Output	Labor-Hours	Unit	Material	Labor	Equipment	Total	Total Incl O&P
						2020 Bare Costs				
4609	12" x 10" pipe size	Q-16	1.60	15	Ea.	1,325	900	66.50	2,291.50	2,875
4610	16" x 12" pipe size		1.50	16		2,475	960	71	3,506	4,250
4611	14" x 12" pipe size		1.52	15.789		2,400	950	70	3,420	4,125
4618	Reducer, concentric									
4619	2-1/2" by 2" pipe size	Q-15	10	1.600	Ea.	42	93	10.65	145.65	197
4620	3" by 2-1/2" pipe size		9	1.778		34.50	103	11.80	149.30	205
4621	3-1/2" by 3" pipe size		8	2		112	116	13.30	241.30	310
4622	4" by 2-1/2" pipe size		7	2.286		57.50	133	15.20	205.70	278
4623	5" by 3" pipe size		7	2.286		144	133	15.20	292.20	375
4624	6" by 4" pipe size	Q-16	6	4		120	241	17.70	378.70	510
4625	8" by 6" pipe size		5	4.800		122	289	21.50	432.50	590
4626	10" by 8" pipe size		4	6		254	360	26.50	640.50	850
4627	12" by 10" pipe size		3	8		292	480	35.50	807.50	1,075
4660	Reducer, eccentric									
4662	3" x 2" pipe size	Q-15	8	2	Ea.	75	116	13.30	204.30	271
4664	4" x 3" pipe size		6	2.667		91	155	17.75	263.75	350
4666	4" x 2" pipe size		6	2.667		111	155	17.75	283.75	375
4670	6" x 4" pipe size	Q-16	5	4.800		188	289	21.50	498.50	660
4672	6" x 3" pipe size		5	4.800		295	289	21.50	605.50	780
4676	8" x 6" pipe size		4	6		240	360	26.50	626.50	835
4678	8" x 4" pipe size		4	6		440	360	26.50	826.50	1,050
4682	10" x 8" pipe size		3	8		315	480	35.50	830.50	1,100
4684	10" x 6" pipe size		3	8		430	480	35.50	945.50	1,225
4688	12" x 10" pipe size		2.50	9.600		490	575	42.50	1,107.50	1,450
4690	12" x 8" pipe size		2.50	9.600		770	575	42.50	1,387.50	1,750
4691	14" x 12" pipe size		2.20	10.909		860	655	48.50	1,563.50	1,975
4693	16" x 14" pipe size		1.80	13.333		1,025	800	59	1,884	2,400
4694	16" x 12" pipe size		2	12		1,750	720	53	2,523	3,050
4696	18" x 16" pipe size		1.60	15		1,075	900	66.50	2,041.50	2,625
5000	Weld joint, socket, forged steel, 3000 lb., schedule 40 pipe									
5010	90° elbow, straight									
5020	1/4" pipe size	Q-15	22	.727	Ea.	29.50	42	4.84	76.34	101
5030	3/8" pipe size		22	.727		29.50	42	4.84	76.34	101
5040	1/2" pipe size		20	.800		16.45	46.50	5.30	68.25	93.50
5050	3/4" pipe size		20	.800		13.65	46.50	5.30	65.45	90.50
5060	1" pipe size		20	.800		22	46.50	5.30	73.80	99.50
5070	1-1/4" pipe size		18	.889		42.50	51.50	5.90	99.90	130
5080	1-1/2" pipe size		16	1		41.50	58	6.65	106.15	140
5090	2" pipe size		12	1.333		62	77.50	8.85	148.35	194
5100	2-1/2" pipe size		10	1.600		206	93	10.65	309.65	380
5110	3" pipe size		8	2		355	116	13.30	484.30	580
5120	4" pipe size		6	2.667		925	155	17.75	1,097.75	1,275
5130	45° elbow, straight									
5134	1/4" pipe size	Q-15	22	.727	Ea.	29.50	42	4.84	76.34	101
5135	3/8" pipe size		22	.727		29.50	42	4.84	76.34	101
5136	1/2" pipe size		20	.800		22	46.50	5.30	73.80	99.50
5137	3/4" pipe size		20	.800		25.50	46.50	5.30	77.30	103
5140	1" pipe size		20	.800		33.50	46.50	5.30	85.30	112
5150	1-1/4" pipe size		18	.889		46	51.50	5.90	103.40	134
5160	1-1/2" pipe size		16	1		56	58	6.65	120.65	156
5170	2" pipe size		12	1.333		90.50	77.50	8.85	176.85	225
5180	2-1/2" pipe size		10	1.600		238	93	10.65	341.65	410
5190	3" pipe size		8	2		395	116	13.30	524.30	625

22 11 Facility Water Distribution

22 11 13 – Facility Water Distribution Piping

22 11 13.47 Pipe Fittings, Steel		Crew	Daily Output	Labor-Hours	Unit	Material	Labor	Equipment	Total	Total Incl O&P
						2020 Bare Costs				
5200	4" pipe size	Q-15	6	2.667	Ea.	775	155	17.75	947.75	1,100
5250	Tee, straight									
5254	1/4" pipe size	Q-15	15	1.067	Ea.	32.50	62	7.10	101.60	136
5255	3/8" pipe size		15	1.067		32.50	62	7.10	101.60	136
5256	1/2" pipe size		13	1.231		20.50	71.50	8.20	100.20	139
5257	3/4" pipe size		13	1.231		25	71.50	8.20	104.70	144
5260	1" pipe size		13	1.231		34	71.50	8.20	113.70	154
5270	1-1/4" pipe size		12	1.333		52.50	77.50	8.85	138.85	184
5280	1-1/2" pipe size		11	1.455		68.50	84.50	9.65	162.65	212
5290	2" pipe size		8	2		100	116	13.30	229.30	299
5300	2-1/2" pipe size		6	2.667		246	155	17.75	418.75	520
5310	3" pipe size		5	3.200		590	186	21.50	797.50	945
5320	4" pipe size		4	4		1,100	232	26.50	1,358.50	1,600
5350	For reducing sizes, add					60%				
5450	Couplings									
5451	1/4" pipe size	Q-15	23	.696	Ea.	19.25	40.50	4.63	64.38	86.50
5452	3/8" pipe size		23	.696		19.25	40.50	4.63	64.38	86.50
5453	1/2" pipe size		21	.762		8.75	44	5.05	57.80	81
5454	3/4" pipe size		21	.762		11.30	44	5.05	60.35	84
5460	1" pipe size		20	.800		12.45	46.50	5.30	64.25	89
5470	1-1/4" pipe size		20	.800		22	46.50	5.30	73.80	100
5480	1-1/2" pipe size		18	.889		24.50	51.50	5.90	81.90	111
5490	2" pipe size		14	1.143		39	66.50	7.60	113.10	150
5500	2-1/2" pipe size		12	1.333		87.50	77.50	8.85	173.85	222
5510	3" pipe size		9	1.778		196	103	11.80	310.80	385
5520	4" pipe size		7	2.286		295	133	15.20	443.20	540
5570	Union, 1/4" pipe size		21	.762		40.50	44	5.05	89.55	116
5571	3/8" pipe size		21	.762		40.50	44	5.05	89.55	116
5572	1/2" pipe size		19	.842		34	49	5.60	88.60	117
5573	3/4" pipe size		19	.842		39.50	49	5.60	94.10	123
5574	1" pipe size		19	.842		51	49	5.60	105.60	136
5575	1-1/4" pipe size		17	.941		82	54.50	6.25	142.75	179
5576	1-1/2" pipe size		15	1.067		89.50	62	7.10	158.60	199
5577	2" pipe size		11	1.455		109	84.50	9.65	203.15	257
5600	Reducer, 1/4" pipe size		23	.696		45.50	40.50	4.63	90.63	116
5601	3/8" pipe size		23	.696		48.50	40.50	4.63	93.63	119
5602	1/2" pipe size		21	.762		31	44	5.05	80.05	106
5603	3/4" pipe size		21	.762		31	44	5.05	80.05	106
5604	1" pipe size		21	.762		36.50	44	5.05	85.55	112
5605	1-1/4" pipe size		19	.842		55	49	5.60	109.60	140
5607	1-1/2" pipe size		17	.941		53	54.50	6.25	113.75	146
5608	2" pipe size		13	1.231		58.50	71.50	8.20	138.20	181
5612	Cap, 1/4" pipe size		46	.348		18.10	20	2.31	40.41	52.50
5613	3/8" pipe size		46	.348		18.10	20	2.31	40.41	52.50
5614	1/2" pipe size		42	.381		10.95	22	2.53	35.48	48
5615	3/4" pipe size		42	.381		12.75	22	2.53	37.28	50
5616	1" pipe size		42	.381		19.35	22	2.53	43.88	57.50
5617	1-1/4" pipe size		38	.421		23	24.50	2.80	50.30	64.50
5618	1-1/2" pipe size		34	.471		32	27.50	3.13	62.63	79.50
5619	2" pipe size		26	.615		49	35.50	4.09	88.59	112
5630	T-O-L, 1/4" pipe size, nozzle		23	.696		8.25	40.50	4.63	53.38	74.50
5631	3/8" pipe size, nozzle		23	.696		8.40	40.50	4.63	53.53	75
5632	1/2" pipe size, nozzle		22	.727		8.25	42	4.84	55.09	77.50

22 11 13.47 Pipe Fittings, Steel		Crew	Daily Output	Labor-Hours	Unit	Material	Labor (2020 Bare Costs)	Equipment	Total	Total Incl O&P
5633	3/4" pipe size, nozzle	Q-15	21	.762	Ea.	9.50	44	5.05	58.55	82
5634	1" pipe size, nozzle		20	.800		11.10	46.50	5.30	62.90	87.50
5635	1-1/4" pipe size, nozzle		18	.889		15.80	51.50	5.90	73.20	101
5636	1-1/2" pipe size, nozzle		16	1		17.55	58	6.65	82.20	114
5637	2" pipe size, nozzle		12	1.333		19.15	77.50	8.85	105.50	147
5638	2-1/2" pipe size, nozzle		10	1.600		67	93	10.65	170.65	225
5639	4" pipe size, nozzle		6	2.667		151	155	17.75	323.75	415
5640	W-O-L, 1/4" pipe size, nozzle		23	.696		17.60	40.50	4.63	62.73	85
5641	3/8" pipe size, nozzle		23	.696		16.65	40.50	4.63	61.78	84
5642	1/2" pipe size, nozzle		22	.727		17.10	42	4.84	63.94	87
5643	3/4" pipe size, nozzle		21	.762		18.05	44	5.05	67.10	91.50
5644	1" pipe size, nozzle		20	.800		18.85	46.50	5.30	70.65	96
5645	1-1/4" pipe size, nozzle		18	.889		21.50	51.50	5.90	78.90	108
5646	1-1/2" pipe size, nozzle		16	1		22.50	58	6.65	87.15	119
5647	2" pipe size, nozzle		12	1.333		22.50	77.50	8.85	108.85	151
5648	2-1/2" pipe size, nozzle		10	1.600		51.50	93	10.65	155.15	208
5649	3" pipe size, nozzle		8	2		56.50	116	13.30	185.80	251
5650	4" pipe size, nozzle		6	2.667		71.50	155	17.75	244.25	330
5651	5" pipe size, nozzle		5	3.200		176	186	21.50	383.50	495
5652	6" pipe size, nozzle		4	4		199	232	26.50	457.50	595
5653	8" pipe size, nozzle		3	5.333		380	310	35.50	725.50	920
5654	10" pipe size, nozzle		2.60	6.154		525	355	41	921	1,150
5655	12" pipe size, nozzle		2.20	7.273		1,000	420	48.50	1,468.50	1,775
5674	S-O-L, 1/4" pipe size, outlet		23	.696		9.80	40.50	4.63	54.93	76.50
5675	3/8" pipe size, outlet		23	.696		9.80	40.50	4.63	54.93	76.50
5676	1/2" pipe size, outlet		22	.727		10.25	42	4.84	57.09	79.50
5677	3/4" pipe size, outlet		21	.762		10.40	44	5.05	59.45	83
5678	1" pipe size, outlet		20	.800		11.50	46.50	5.30	63.30	88
5679	1-1/4" pipe size, outlet		18	.889		19.25	51.50	5.90	76.65	105
5680	1-1/2" pipe size, outlet		16	1		19.25	58	6.65	83.90	115
5681	2" pipe size, outlet	↓	12	1.333	↓	19.75	77.50	8.85	106.10	147
6000	Weld-on flange, forged steel									
6020	Slip-on, 150 lb. flange (welded front and back)									
6050	1/2" pipe size	Q-15	18	.889	Ea.	19.85	51.50	5.90	77.25	106
6060	3/4" pipe size		18	.889		19.85	51.50	5.90	77.25	106
6070	1" pipe size		17	.941		19.85	54.50	6.25	80.60	110
6080	1-1/4" pipe size		16	1		19.85	58	6.65	84.50	116
6090	1-1/2" pipe size		15	1.067		19.85	62	7.10	88.95	122
6100	2" pipe size		12	1.333		26	77.50	8.85	112.35	155
6110	2-1/2" pipe size		10	1.600		38.50	93	10.65	142.15	193
6120	3" pipe size		9	1.778		32	103	11.80	146.80	203
6130	3-1/2" pipe size		7	2.286		34	133	15.20	182.20	252
6140	4" pipe size		6	2.667		33	155	17.75	205.75	287
6150	5" pipe size	↓	5	3.200		67.50	186	21.50	275	375
6160	6" pipe size	Q-16	6	4		54	241	17.70	312.70	440
6170	8" pipe size		5	4.800		94.50	289	21.50	405	560
6180	10" pipe size		4	6		166	360	26.50	552.50	755
6190	12" pipe size		3	8		249	480	35.50	764.50	1,025
6191	14" pipe size		2.50	9.600		280	575	42.50	897.50	1,225
6192	16" pipe size	↓	1.80	13.333	↓	495	800	59	1,354	1,800
6200	300 lb. flange									
6210	1/2" pipe size	Q-15	17	.941	Ea.	24.50	54.50	6.25	85.25	115
6220	3/4" pipe size	↓	17	.941	↓	24.50	54.50	6.25	85.25	115

22 11 13 – Facility Water Distribution Piping

22 11 13.47 Pipe Fittings, Steel		Crew	Daily Output	Labor-Hours	Unit	Material	Labor	Equipment	Total	Total Incl O&P
						2020 Bare Costs				
6230	1" pipe size	Q-15	16	1	Ea.	24.50	58	6.65	89.15	121
6240	1-1/4" pipe size		13	1.231		24.50	71.50	8.20	104.20	143
6250	1-1/2" pipe size		12	1.333		24.50	77.50	8.85	110.85	153
6260	2" pipe size		11	1.455		36.50	84.50	9.65	130.65	177
6270	2-1/2" pipe size		9	1.778		36.50	103	11.80	151.30	207
6280	3" pipe size		7	2.286		38	133	15.20	186.20	257
6290	4" pipe size		6	2.667		57.50	155	17.75	230.25	315
6300	5" pipe size		4	4		111	232	26.50	369.50	495
6310	6" pipe size	Q-16	5	4.800		108	289	21.50	418.50	575
6320	8" pipe size		4	6		184	360	26.50	570.50	770
6330	10" pipe size		3.40	7.059		325	425	31.50	781.50	1,025
6340	12" pipe size		2.80	8.571		340	515	38	893	1,175
6400	Welding neck, 150 lb. flange									
6410	1/2" pipe size	Q-15	40	.400	Ea.	25.50	23	2.66	51.16	65.50
6420	3/4" pipe size		36	.444		25.50	26	2.96	54.46	70
6430	1" pipe size		32	.500		25.50	29	3.33	57.83	75
6440	1-1/4" pipe size		29	.552		25.50	32	3.67	61.17	80
6450	1-1/2" pipe size		26	.615		25.50	35.50	4.09	65.09	86
6460	2" pipe size		20	.800		34	46.50	5.30	85.80	113
6470	2-1/2" pipe size		16	1		37	58	6.65	101.65	135
6480	3" pipe size		14	1.143		41	66.50	7.60	115.10	152
6500	4" pipe size		10	1.600		42	93	10.65	145.65	197
6510	5" pipe size		8	2		66.50	116	13.30	195.80	262
6520	6" pipe size	Q-16	10	2.400		63.50	144	10.65	218.15	298
6530	8" pipe size		7	3.429		118	206	15.20	339.20	455
6540	10" pipe size		6	4		188	241	17.70	446.70	585
6550	12" pipe size		5	4.800		296	289	21.50	606.50	780
6551	14" pipe size		4.50	5.333		440	320	23.50	783.50	990
6552	16" pipe size		3	8		595	480	35.50	1,110.50	1,425
6553	18" pipe size		2.50	9.600		800	575	42.50	1,417.50	1,800
6554	20" pipe size		2.30	10.435		970	630	46	1,646	2,075
6556	24" pipe size		2	12		1,300	720	53	2,073	2,575
6557	26" pipe size		1.70	14.118		1,425	850	62.50	2,337.50	2,925
6558	30" pipe size		.90	26.667		1,625	1,600	118	3,343	4,325
6559	36" pipe size		.75	32		1,875	1,925	142	3,942	5,100
6560	300 lb. flange									
6570	1/2" pipe size	Q-15	36	.444	Ea.	34.50	26	2.96	63.46	80
6580	3/4" pipe size		34	.471		34.50	27.50	3.13	65.13	82.50
6590	1" pipe size		30	.533		34.50	31	3.55	69.05	88.50
6600	1-1/4" pipe size		28	.571		34.50	33	3.80	71.30	91.50
6610	1-1/2" pipe size		24	.667		34.50	38.50	4.43	77.43	101
6620	2" pipe size		18	.889		45.50	51.50	5.90	102.90	134
6630	2-1/2" pipe size		14	1.143		52	66.50	7.60	126.10	164
6640	3" pipe size		12	1.333		44.50	77.50	8.85	130.85	175
6650	4" pipe size		8	2		76	116	13.30	205.30	272
6660	5" pipe size		7	2.286		111	133	15.20	259.20	335
6670	6" pipe size	Q-16	9	2.667		113	160	11.80	284.80	380
6680	8" pipe size		6	4		216	241	17.70	474.70	620
6690	10" pipe size		5	4.800		415	289	21.50	725.50	910
6700	12" pipe size		4	6		450	360	26.50	836.50	1,075
7740	Plain ends for plain end pipe, mechanically coupled									
7750	Cplg. & labor required at joints not included, add 1 per									
7760	joint for installed price, see line 9180									

22 11 13 – Facility Water Distribution Piping

22 11 13.47 Pipe Fittings, Steel		Crew	Daily Output	Labor-Hours	Unit	Material	Labor	Equipment	Total	Total Incl O&P
						2020 Bare Costs				
7770	Malleable iron, painted, unless noted otherwise									
7800	90° elbow, 1"				Ea.	154			154	170
7810	1-1/2"					182			182	201
7820	2"					271			271	299
7830	2-1/2"					320			320	350
7840	3"					330			330	365
7860	4"					375			375	410
7870	5" welded steel					430			430	470
7880	6"					540			540	595
7890	8" welded steel					1,025			1,025	1,125
7900	10" welded steel					1,300			1,300	1,425
7910	12" welded steel					1,450			1,450	1,600
7970	45° elbow, 1"					92.50			92.50	102
7980	1-1/2"					137			137	151
7990	2"					288			288	315
8000	2-1/2"					288			288	315
8010	3"					330			330	365
8030	4"					345			345	380
8040	5" welded steel					430			430	470
8050	6"					490			490	540
8060	8"					575			575	635
8070	10" welded steel					595			595	655
8080	12" welded steel					1,075			1,075	1,175
8140	Tee, straight 1"					172			172	189
8150	1-1/2"					222			222	244
8160	2"					222			222	244
8170	2-1/2"					289			289	320
8180	3"					445			445	490
8200	4"					635			635	700
8210	5" welded steel					870			870	955
8220	6"					755			755	830
8230	8" welded steel					1,100			1,100	1,200
8240	10" welded steel					1,725			1,725	1,900
8250	12" welded steel					2,000			2,000	2,200
8340	Segmentally welded steel, painted									
8390	Wye 2"				Ea.	251			251	276
8400	2-1/2"					251			251	276
8410	3"					300			300	330
8430	4"					390			390	430
8440	5"					570			570	625
8450	6"					695			695	760
8460	8"					1,025			1,025	1,125
8470	10"					1,675			1,675	1,850
8480	12"					1,725			1,725	1,900
8540	Wye, lateral 2"					299			299	330
8550	2-1/2"					345			345	380
8560	3"					410			410	450
8580	4"					560			560	620
8590	5"					955			955	1,050
8600	6"					985			985	1,075
8610	8"					1,650			1,650	1,825
8620	10"					2,425			2,425	2,675
8630	12"					3,025			3,025	3,325

22 11 13 – Facility Water Distribution Piping

22 11 13.47 Pipe Fittings, Steel		Crew	Daily Output	Labor-Hours	Unit	2020 Bare Costs Material	Labor	Equipment	Total	Total Incl O&P
8690	Cross, 2"				Ea.	181			181	199
8700	2-1/2"					181			181	199
8710	3"					216			216	237
8730	4"					490			490	540
8740	5"					560			560	615
8750	6"					710			710	780
8760	8"					1,100			1,100	1,225
8770	10"					1,425			1,425	1,575
8780	12"					1,825			1,825	2,025
8800	Tees, reducing 2" x 1"					235			235	259
8810	2" x 1-1/2"					235			235	259
8820	3" x 1"					217			217	238
8830	3" x 1-1/2"					298			298	330
8840	3" x 2"					216			216	237
8850	4" x 1"					315			315	350
8860	4" x 1-1/2"					315			315	350
8870	4" x 2"					470			470	515
8880	4" x 2-1/2"					485			485	530
8890	4" x 3"					485			485	530
8900	6" x 2"					400			400	440
8910	6" x 3"					298			298	330
8920	6" x 4"					380			380	420
8930	8" x 2"					575			575	635
8940	8" x 3"					625			625	690
8950	8" x 4"					670			670	735
8960	8" x 5"					470			470	515
8970	8" x 6"					640			640	705
8980	10" x 4"					530			530	585
8990	10" x 6"					545			545	600
9000	10" x 8"					850			850	940
9010	12" x 6"					1,325			1,325	1,450
9020	12" x 8"					1,050			1,050	1,150
9030	12" x 10"					1,025			1,025	1,150
9080	Adapter nipples 3" long									
9090	1"				Ea.	27			27	29.50
9100	1-1/2"					27			27	29.50
9110	2"					27			27	29.50
9120	2-1/2"					31			31	34.50
9130	3"					38			38	41.50
9140	4"					61.50			61.50	67.50
9150	6"					157			157	172
9180	Coupling, mechanical, plain end pipe to plain end pipe or fitting									
9190	1"	Q-1	29	.552	Ea.	93.50	32		125.50	151
9200	1-1/2"		28	.571		93.50	33		126.50	153
9210	2"		27	.593		93.50	34.50		128	155
9220	2-1/2"		26	.615		93.50	35.50		129	157
9230	3"		25	.640		136	37		173	206
9240	3-1/2"		24	.667		157	38.50		195.50	231
9250	4"		22	.727		157	42		199	236
9260	5"	Q-2	28	.857		223	51.50		274.50	325
9270	6"		24	1		273	60		333	390
9280	8"		19	1.263		480	76		556	645
9290	10"		16	1.500		625	90		715	820

22 11 13.47 Pipe Fittings, Steel		Crew	Daily Output	Labor-Hours	Unit	2020 Bare Costs Material	Labor	Equipment	Total	Total Incl O&P
9300	12"	Q-2	12	2	Ea.	785	120		905	1,050
9310	Outlets for precut holes through pipe wall									
9331	Strapless type, with gasket									
9332	4" to 8" pipe x 1/2"	1 Plum	13	.615	Ea.	102	39.50		141.50	172
9333	4" to 8" pipe x 3/4"		13	.615		109	39.50		148.50	180
9334	10" pipe and larger x 1/2"		11	.727		102	47		149	182
9335	10" pipe and larger x 3/4"		11	.727		109	47		156	190
9341	Thermometer wells with gasket									
9342	4" to 8" pipe, 6" stem	1 Plum	14	.571	Ea.	153	37		190	223
9343	8" pipe and larger, 6" stem	"	13	.615	"	153	39.50		192.50	228
9800	Mech. tee, cast iron, grooved or threaded, w/nuts and bolts									
9805	2" x 1"	1 Plum	50	.160	Ea.	101	10.30		111.30	126
9810	2" x 1-1/4"		50	.160		117	10.30		127.30	143
9815	2" x 1-1/2"		50	.160		125	10.30		135.30	153
9820	2-1/2" x 1"	Q-1	80	.200		106	11.60		117.60	133
9825	2-1/2" x 1-1/4"		80	.200		156	11.60		167.60	189
9830	2-1/2" x 1-1/2"		80	.200		156	11.60		167.60	189
9835	2-1/2" x 2"		80	.200		106	11.60		117.60	133
9840	3" x 1"		80	.200		115	11.60		126.60	143
9845	3" x 1-1/4"		80	.200		153	11.60		164.60	186
9850	3" x 1-1/2"		80	.200		153	11.60		164.60	186
9852	3" x 2"		80	.200		184	11.60		195.60	219
9856	4" x 1"		67	.239		142	13.85		155.85	178
9858	4" x 1-1/4"		67	.239		187	13.85		200.85	227
9860	4" x 1-1/2"		67	.239		191	13.85		204.85	231
9862	4" x 2"		67	.239		196	13.85		209.85	236
9864	4" x 2-1/2"		67	.239		203	13.85		216.85	244
9866	4" x 3"		67	.239		207	13.85		220.85	249
9870	6" x 1-1/2"	Q-2	50	.480		239	29		268	305
9872	6" x 2"		50	.480		242	29		271	310
9874	6" x 2-1/2"		50	.480		246	29		275	315
9876	6" x 3"		50	.480		274	29		303	345
9878	6" x 4"		50	.480		305	29		334	380
9880	8" x 2"		42	.571		415	34.50		449.50	505
9882	8" x 2-1/2"		42	.571		415	34.50		449.50	505
9884	8" x 3"		42	.571		435	34.50		469.50	530
9886	8" x 4"		42	.571		890	34.50		924.50	1,025
9940	For galvanized fittings for plain end pipe, add					20%				

22 11 13.48 Pipe, Fittings and Valves, Steel, Grooved-Joint		Crew	Daily Output	Labor-Hours	Unit	Material	Labor	Equipment	Total	Total Incl O&P
0010	**PIPE, FITTINGS AND VALVES, STEEL, GROOVED-JOINT** R221113-70									
0012	Fittings are ductile iron. Steel fittings noted.									
0020	Pipe includes coupling & clevis type hanger assemblies, 10' OC									
0500	Schedule 10, black									
0550	2" diameter	1 Plum	43	.186	L.F.	9.40	12		21.40	28.50
0560	2-1/2" diameter	Q-1	61	.262		13.25	15.20		28.45	37.50
0570	3" diameter		55	.291		14.65	16.85		31.50	41.50
0580	3-1/2" diameter		53	.302		18.40	17.50		35.90	46.50
0590	4" diameter		49	.327		15.15	18.95		34.10	45
0600	5" diameter		40	.400		18.75	23		41.75	55
0610	6" diameter	Q-2	46	.522		24.50	31.50		56	74
0620	8" diameter	"	41	.585		36.50	35		71.50	92.50
0700	To delete couplings & hangers, subtract									

22 11 13.48	Pipe, Fittings and Valves, Steel, Grooved-Joint	Crew	Daily Output	Labor-Hours	Unit	Material	2020 Bare Costs Labor	Equipment	Total	Total Incl O&P
0710	2" diam. to 5" diam.					25%	20%			
0720	6" diam. to 8" diam.					27%	15%			
1000	Schedule 40, black									
1040	3/4" diameter	1 Plum	71	.113	L.F.	6.55	7.25		13.80	18.05
1050	1" diameter		63	.127		6.45	8.20		14.65	19.35
1060	1-1/4" diameter		58	.138		7.70	8.90		16.60	22
1070	1-1/2" diameter		51	.157		8.40	10.10		18.50	24.50
1080	2" diameter		40	.200		9.70	12.90		22.60	30
1090	2-1/2" diameter	Q-1	57	.281		16.50	16.30		32.80	42.50
1100	3" diameter		50	.320		22	18.55		40.55	52
1110	4" diameter		45	.356		26	20.50		46.50	59.50
1120	5" diameter		37	.432		41	25		66	82.50
1130	6" diameter	Q-2	42	.571		52.50	34.50		87	109
1140	8" diameter		37	.649		84.50	39		123.50	152
1150	10" diameter		31	.774		114	46.50		160.50	195
1160	12" diameter		27	.889		128	53.50		181.50	220
1170	14" diameter		20	1.200		138	72		210	260
1180	16" diameter		17	1.412		215	85		300	365
1190	18" diameter		14	1.714		207	103		310	380
1200	20" diameter		12	2		252	120		372	455
1210	24" diameter		10	2.400		284	144		428	525
1740	To delete coupling & hanger, subtract									
1750	3/4" diam. to 2" diam.					65%	27%			
1760	2-1/2" diam. to 5" diam.					41%	18%			
1770	6" diam. to 12" diam.					31%	13%			
1780	14" diam. to 24" diam.					35%	10%			
1800	Galvanized									
1840	3/4" diameter	1 Plum	71	.113	L.F.	7.15	7.25		14.40	18.70
1850	1" diameter		63	.127		7.10	8.20		15.30	20
1860	1-1/4" diameter		58	.138		8.50	8.90		17.40	22.50
1870	1-1/2" diameter		51	.157		9.25	10.10		19.35	25.50
1880	2" diameter		40	.200		10.80	12.90		23.70	31
1890	2-1/2" diameter	Q-1	57	.281		16.70	16.30		33	43
1900	3" diameter		50	.320		19.65	18.55		38.20	49.50
1910	4" diameter		45	.356		23	20.50		43.50	56.50
1920	5" diameter		37	.432		30.50	25		55.50	71
1930	6" diameter	Q-2	42	.571		39.50	34.50		74	95
1940	8" diameter		37	.649		58.50	39		97.50	123
1950	10" diameter		31	.774		114	46.50		160.50	195
1960	12" diameter		27	.889		137	53.50		190.50	231
2540	To delete coupling & hanger, subtract									
2550	3/4" diam. to 2" diam.					36%	27%			
2560	2-1/2" diam. to 5" diam.					19%	18%			
2570	6" diam. to 12" diam.					14%	13%			
2600	Schedule 80, black									
2610	3/4" diameter	1 Plum	65	.123	L.F.	8	7.95		15.95	20.50
2650	1" diameter		61	.131		8.50	8.45		16.95	22
2660	1-1/4" diameter		55	.145		10.55	9.35		19.90	25.50
2670	1-1/2" diameter		49	.163		11.75	10.50		22.25	28.50
2680	2" diameter		38	.211		14.20	13.55		27.75	36
2690	2-1/2" diameter	Q-1	54	.296		22	17.20		39.20	49.50
2700	3" diameter		48	.333		27	19.35		46.35	59
2710	4" diameter		44	.364		35.50	21		56.50	70.50

22 11 Facility Water Distribution

22 11 13 – Facility Water Distribution Piping

22 11 13.48 Pipe, Fittings and Valves, Steel, Grooved-Joint		Crew	Daily Output	Labor-Hours	Unit	2020 Bare Costs Material	Labor	Equipment	Total	Total Incl O&P
2720	5″ diameter	Q-1	35	.457	L.F.	63	26.50		89.50	109
2730	6″ diameter	Q-2	40	.600		72.50	36		108.50	134
2740	8″ diameter		35	.686		113	41.50		154.50	187
2750	10″ diameter		29	.828		169	50		219	260
2760	12″ diameter		24	1		365	60		425	490
3240	To delete coupling & hanger, subtract									
3250	3/4″ diam. to 2″ diam.					30%	25%			
3260	2-1/2″ diam. to 5″ diam.					14%	17%			
3270	6″ diam. to 12″ diam.					12%	12%			
3300	Galvanized									
3310	3/4″ diameter	1 Plum	65	.123	L.F.	8.60	7.95		16.55	21.50
3350	1″ diameter		61	.131		9.40	8.45		17.85	23
3360	1-1/4″ diameter		55	.145		11.85	9.35		21.20	27
3370	1-1/2″ diameter		46	.174		13.30	11.20		24.50	31.50
3380	2″ diameter		38	.211		17	13.55		30.55	39
3390	2-1/2″ diameter	Q-1	54	.296		26	17.20		43.20	54
3400	3″ diameter		48	.333		32	19.35		51.35	64
3410	4″ diameter		44	.364		39.50	21		60.50	75
3420	5″ diameter		35	.457		47	26.50		73.50	91
3430	6″ diameter	Q-2	40	.600		65	36		101	126
3440	8″ diameter		35	.686		120	41.50		161.50	194
3450	10″ diameter		29	.828		273	50		323	375
3460	12″ diameter		24	1		590	60		650	740
3920	To delete coupling & hanger, subtract									
3930	3/4″ diam. to 2″ diam.					30%	25%			
3940	2-1/2″ diam. to 5″ diam.					15%	17%			
3950	6″ diam. to 12″ diam.					11%	12%			
3990	Fittings: coupling material required at joints not incl. in fitting price.									
3994	Add 1 selected coupling, material only, per joint for installed price.									
4000	Elbow, 90° or 45°, painted									
4030	3/4″ diameter	1 Plum	50	.160	Ea.	82.50	10.30		92.80	106
4040	1″ diameter		50	.160		44.50	10.30		54.80	64.50
4050	1-1/4″ diameter		40	.200		44.50	12.90		57.40	68.50
4060	1-1/2″ diameter		33	.242		44.50	15.60		60.10	72.50
4070	2″ diameter		25	.320		44.50	20.50		65	80
4080	2-1/2″ diameter	Q-1	40	.400		44.50	23		67.50	83.50
4090	3″ diameter		33	.485		78	28		106	128
4100	4″ diameter		25	.640		84.50	37		121.50	149
4110	5″ diameter		20	.800		201	46.50		247.50	291
4120	6″ diameter	Q-2	25	.960		236	57.50		293.50	345
4130	8″ diameter		21	1.143		490	68.50		558.50	645
4140	10″ diameter		18	1.333		900	80		980	1,100
4150	12″ diameter		15	1.600		975	96		1,071	1,225
4170	14″ diameter		12	2		1,025	120		1,145	1,300
4180	16″ diameter		11	2.182		1,125	131		1,256	1,425
4190	18″ diameter	Q-3	14	2.286		1,700	140		1,840	2,050
4200	20″ diameter		12	2.667		2,225	164		2,389	2,700
4210	24″ diameter		10	3.200		3,225	196		3,421	3,850
4250	For galvanized elbows, add					26%				
4690	Tee, painted									
4700	3/4″ diameter	1 Plum	38	.211	Ea.	88.50	13.55		102.05	118
4740	1″ diameter		33	.242		68.50	15.60		84.10	98.50
4750	1-1/4″ diameter		27	.296		68.50	19.10		87.60	104

22 11 13.48 Pipe, Fittings and Valves, Steel, Grooved-Joint		Crew	Daily Output	Labor-Hours	Unit	2020 Bare Costs Material	Labor	Equipment	Total	Total Incl O&P
4760	1-1/2" diameter	1 Plum	22	.364	Ea.	68.50	23.50		92	110
4770	2" diameter		17	.471		68.50	30.50		99	121
4780	2-1/2" diameter	Q-1	27	.593		68.50	34.50		103	127
4790	3" diameter		22	.727		93.50	42		135.50	166
4800	4" diameter		17	.941		142	54.50		196.50	238
4810	5" diameter		13	1.231		330	71.50		401.50	470
4820	6" diameter	Q-2	17	1.412		380	85		465	545
4830	8" diameter		14	1.714		835	103		938	1,075
4840	10" diameter		12	2		1,050	120		1,170	1,350
4850	12" diameter		10	2.400		1,375	144		1,519	1,750
4851	14" diameter		9	2.667		1,375	160		1,535	1,750
4852	16" diameter		8	3		1,550	180		1,730	1,975
4853	18" diameter	Q-3	10	3.200		1,925	196		2,121	2,425
4854	20" diameter		9	3.556		2,775	218		2,993	3,375
4855	24" diameter		8	4		4,225	245		4,470	5,025
4900	For galvanized tees, add					24%				
4906	Couplings, rigid style, painted									
4908	1" diameter	1 Plum	100	.080	Ea.	34.50	5.15		39.65	45.50
4909	1-1/4" diameter		100	.080		32	5.15		37.15	42.50
4910	1-1/2" diameter		67	.119		34.50	7.70		42.20	49.50
4912	2" diameter		50	.160		43.50	10.30		53.80	63.50
4914	2-1/2" diameter	Q-1	80	.200		49.50	11.60		61.10	72
4916	3" diameter		67	.239		57.50	13.85		71.35	83.50
4918	4" diameter		50	.320		79.50	18.55		98.05	116
4920	5" diameter		40	.400		102	23		125	148
4922	6" diameter	Q-2	50	.480		135	29		164	192
4924	8" diameter		42	.571		213	34.50		247.50	286
4926	10" diameter		35	.686		410	41.50		451.50	510
4928	12" diameter		32	.750		460	45		505	575
4930	14" diameter		24	1		495	60		555	635
4931	16" diameter		20	1.200		450	72		522	605
4932	18" diameter		18	1.333		530	80		610	705
4933	20" diameter		16	1.500		725	90		815	930
4934	24" diameter		13	1.846		930	111		1,041	1,200
4940	Flexible, standard, painted									
4950	3/4" diameter	1 Plum	100	.080	Ea.	25	5.15		30.15	35
4960	1" diameter		100	.080		25	5.15		30.15	35
4970	1-1/4" diameter		80	.100		32.50	6.45		38.95	45
4980	1-1/2" diameter		67	.119		35	7.70		42.70	50
4990	2" diameter		50	.160		38	10.30		48.30	57
5000	2-1/2" diameter	Q-1	80	.200		44	11.60		55.60	66
5010	3" diameter		67	.239		48.50	13.85		62.35	74
5020	3-1/2" diameter		57	.281		69.50	16.30		85.80	101
5030	4" diameter		50	.320		70	18.55		88.55	105
5040	5" diameter		40	.400		105	23		128	151
5050	6" diameter	Q-2	50	.480		125	29		154	180
5070	8" diameter		42	.571		202	34.50		236.50	274
5090	10" diameter		35	.686		330	41.50		371.50	420
5110	12" diameter		32	.750		375	45		420	480
5120	14" diameter		24	1		530	60		590	670
5130	16" diameter		20	1.200		695	72		767	870
5140	18" diameter		18	1.333		810	80		890	1,000
5150	20" diameter		16	1.500		1,275	90		1,365	1,525

22 11 13.48 Pipe, Fittings and Valves, Steel, Grooved-Joint		Crew	Daily Output	Labor-Hours	Unit	2020 Bare Costs Material	Labor	Equipment	Total	Total Incl O&P
5160	24″ diameter	Q-2	13	1.846	Ea.	1,400	111		1,511	1,700
5176	Lightweight style, painted									
5178	1-1/2″ diameter	1 Plum	67	.119	Ea.	31	7.70		38.70	46
5180	2″ diameter	″	50	.160		37.50	10.30		47.80	57
5182	2-1/2″ diameter	Q-1	80	.200		44	11.60		55.60	65.50
5184	3″ diameter		67	.239		48.50	13.85		62.35	74
5186	3-1/2″ diameter		57	.281		59.50	16.30		75.80	89.50
5188	4″ diameter		50	.320		70	18.55		88.55	105
5190	5″ diameter		40	.400		104	23		127	150
5192	6″ diameter	Q-2	50	.480		124	29		153	179
5194	8″ diameter		42	.571		200	34.50		234.50	272
5196	10″ diameter		35	.686		410	41.50		451.50	510
5198	12″ diameter		32	.750		455	45		500	570
5200	For galvanized couplings, add					33%				
5220	Tee, reducing, painted									
5225	2″ x 1-1/2″ diameter	Q-1	38	.421	Ea.	144	24.50		168.50	195
5226	2-1/2″ x 2″ diameter		28	.571		144	33		177	208
5227	3″ x 2-1/2″ diameter		23	.696		127	40.50		167.50	201
5228	4″ x 3″ diameter		18	.889		171	51.50		222.50	265
5229	5″ x 4″ diameter		15	1.067		365	62		427	495
5230	6″ x 4″ diameter	Q-2	18	1.333		400	80		480	560
5231	8″ x 6″ diameter		15	1.600		835	96		931	1,075
5232	10″ x 8″ diameter		13	1.846		920	111		1,031	1,200
5233	12″ x 10″ diameter		11	2.182		1,125	131		1,256	1,450
5234	14″ x 12″ diameter		10	2.400		1,475	144		1,619	1,825
5235	16″ x 12″ diameter		9	2.667		1,225	160		1,385	1,600
5236	18″ x 12″ diameter	Q-3	12	2.667		1,475	164		1,639	1,875
5237	18″ x 16″ diameter		11	2.909		1,875	178		2,053	2,325
5238	20″ x 16″ diameter		10	3.200		2,450	196		2,646	3,000
5239	24″ x 20″ diameter		9	3.556		3,875	218		4,093	4,575
5240	Reducer, concentric, painted									
5241	2-1/2″ x 2″ diameter	Q-1	43	.372	Ea.	52	21.50		73.50	89.50
5242	3″ x 2-1/2″ diameter		35	.457		62.50	26.50		89	108
5243	4″ x 3″ diameter		29	.552		75	32		107	131
5244	5″ x 4″ diameter		22	.727		103	42		145	177
5245	6″ x 4″ diameter	Q-2	26	.923		120	55.50		175.50	215
5246	8″ x 6″ diameter		23	1.043		310	63		373	435
5247	10″ x 8″ diameter		20	1.200		605	72		677	775
5248	12″ x 10″ diameter		16	1.500		1,075	90		1,165	1,325
5255	Eccentric, painted									
5256	2-1/2″ x 2″ diameter	Q-1	42	.381	Ea.	107	22		129	151
5257	3″ x 2-1/2″ diameter		34	.471		123	27.50		150.50	176
5258	4″ x 3″ diameter		28	.571		149	33		182	214
5259	5″ x 4″ diameter		21	.762		202	44		246	288
5260	6″ x 4″ diameter	Q-2	25	.960		235	57.50		292.50	345
5261	8″ x 6″ diameter		22	1.091		475	65.50		540.50	620
5262	10″ x 8″ diameter		19	1.263		1,250	76		1,326	1,500
5263	12″ x 10″ diameter		15	1.600		1,725	96		1,821	2,050
5270	Coupling, reducing, painted									
5272	2″ x 1-1/2″ diameter	1 Plum	52	.154	Ea.	50	9.90		59.90	70
5274	2-1/2″ x 2″ diameter	Q-1	82	.195		64.50	11.30		75.80	87.50
5276	3″ x 2″ diameter		69	.232		73	13.45		86.45	100
5278	4″ x 2″ diameter		52	.308		115	17.85		132.85	154

22 11 13 – Facility Water Distribution Piping

22 11 13.48 Pipe, Fittings and Valves, Steel, Grooved-Joint		Crew	Daily Output	Labor-Hours	Unit	Material	Labor	Equipment	Total	Total Incl O&P
						2020 Bare Costs				
5280	5″ x 4″ diameter	Q-1	42	.381	Ea.	130	22		152	176
5282	6″ x 4″ diameter	Q-2	52	.462		195	28		223	257
5284	8″ x 6″ diameter	″	44	.545		291	33		324	370
5290	Outlet coupling, painted									
5294	1-1/2″ x 1″ pipe size	1 Plum	65	.123	Ea.	66	7.95		73.95	84.50
5296	2″ x 1″ pipe size	″	48	.167		67	10.75		77.75	90
5298	2-1/2″ x 1″ pipe size	Q-1	78	.205		103	11.90		114.90	131
5300	2-1/2″ x 1-1/4″ pipe size		70	.229		116	13.25		129.25	147
5302	3″ x 1″ pipe size		65	.246		131	14.30		145.30	166
5304	4″ x 3/4″ pipe size		48	.333		145	19.35		164.35	188
5306	4″ x 1-1/2″ pipe size		46	.348		205	20		225	256
5308	6″ x 1-1/2″ pipe size	Q-2	44	.545		290	33		323	370
5320	Outlet, strap-on T, painted									
5324	Threaded female branch									
5330	2″ pipe x 1-1/2″ branch size	1 Plum	50	.160	Ea.	47	10.30		57.30	67.50
5334	2-1/2″ pipe x 1-1/2″ branch size	Q-1	80	.200		57.50	11.60		69.10	81
5338	3″ pipe x 1-1/2″ branch size		67	.239		59.50	13.85		73.35	86
5342	3″ pipe x 2″ branch size		59	.271		68.50	15.75		84.25	99
5346	4″ pipe x 2″ branch size		50	.320		72	18.55		90.55	107
5350	4″ pipe x 2-1/2″ branch size		47	.340		75	19.75		94.75	112
5354	4″ pipe x 3″ branch size		43	.372		80	21.50		101.50	121
5358	5″ pipe x 3″ branch size		40	.400		98.50	23		121.50	143
5362	6″ pipe x 2″ branch size	Q-2	50	.480		90	29		119	142
5366	6″ pipe x 3″ branch size		47	.511		103	30.50		133.50	159
5370	6″ pipe x 4″ branch size		44	.545		114	33		147	175
5374	8″ pipe x 2-1/2″ branch size		42	.571		157	34.50		191.50	225
5378	8″ pipe x 4″ branch size		38	.632		170	38		208	244
5390	Grooved branch									
5394	2″ pipe x 1-1/4″ branch size	1 Plum	78	.103	Ea.	47	6.60		53.60	62
5398	2″ pipe x 1-1/2″ branch size	″	50	.160		47	10.30		57.30	67.50
5402	2-1/2″ pipe x 1-1/2″ branch size	Q-1	80	.200		57.50	11.60		69.10	81
5406	3″ pipe x 1-1/2″ branch size		67	.239		59.50	13.85		73.35	86
5410	3″ pipe x 2″ branch size		59	.271		68.50	15.75		84.25	99
5414	4″ pipe x 2″ branch size		50	.320		72	18.55		90.55	107
5418	4″ pipe x 2-1/2″ branch size		47	.340		75	19.75		94.75	112
5422	4″ pipe x 3″ branch size		43	.372		80	21.50		101.50	121
5426	5″ pipe x 3″ branch size		40	.400		98.50	23		121.50	143
5430	6″ pipe x 2″ branch size	Q-2	50	.480		90	29		119	142
5434	6″ pipe x 3″ branch size		47	.511		103	30.50		133.50	159
5438	6″ pipe x 4″ branch size		44	.545		114	33		147	175
5442	8″ pipe x 2-1/2″ branch size		42	.571		157	34.50		191.50	225
5446	8″ pipe x 4″ branch size		38	.632		170	38		208	244
5750	Flange, w/groove gasket, black steel									
5754	See Line 22 11 13.47 0620 for gasket & bolt set									
5760	ANSI class 125 and 150, painted									
5780	2″ pipe size	1 Plum	23	.348	Ea.	145	22.50		167.50	194
5790	2-1/2″ pipe size	Q-1	37	.432		181	25		206	237
5800	3″ pipe size		31	.516		195	30		225	259
5820	4″ pipe size		23	.696		260	40.50		300.50	345
5830	5″ pipe size		19	.842		300	49		349	405
5840	6″ pipe size	Q-2	23	1.043		330	63		393	455
5850	8″ pipe size		17	1.412		370	85		455	535
5860	10″ pipe size		14	1.714		585	103		688	800

22 11 13.48 Pipe, Fittings and Valves, Steel, Grooved-Joint		Crew	Daily Output	Labor-Hours	Unit	2020 Bare Costs Material	Labor	Equipment	Total	Total Incl O&P
5870	12" pipe size	Q-2	12	2	Ea.	765	120		885	1,025
5880	14" pipe size		10	2.400		1,400	144		1,544	1,775
5890	16" pipe size		9	2.667		1,625	160		1,785	2,050
5900	18" pipe size		6	4		2,000	241		2,241	2,550
5910	20" pipe size		5	4.800		2,425	289		2,714	3,075
5920	24" pipe size		4.50	5.333		3,100	320		3,420	3,875
5940	ANSI class 350, painted									
5946	2" pipe size	1 Plum	23	.348	Ea.	182	22.50		204.50	234
5948	2-1/2" pipe size	Q-1	37	.432		210	25		235	269
5950	3" pipe size		31	.516		287	30		317	360
5952	4" pipe size		23	.696		380	40.50		420.50	480
5954	5" pipe size		19	.842		435	49		484	555
5956	6" pipe size	Q-2	23	1.043		505	63		568	650
5958	8" pipe size		17	1.412		580	85		665	760
5960	10" pipe size		14	1.714		920	103		1,023	1,175
5962	12" pipe size		12	2		985	120		1,105	1,250
6100	Cross, painted									
6110	2" diameter	1 Plum	12.50	.640	Ea.	122	41.50		163.50	196
6112	2-1/2" diameter	Q-1	20	.800		122	46.50		168.50	204
6114	3" diameter		15.50	1.032		216	60		276	330
6116	4" diameter		12.50	1.280		360	74		434	505
6118	6" diameter	Q-2	12.50	1.920		935	115		1,050	1,200
6120	8" diameter		10.50	2.286		1,225	137		1,362	1,550
6122	10" diameter		9	2.667		2,025	160		2,185	2,475
6124	12" diameter		7.50	3.200		2,925	192		3,117	3,525
7400	Suction diffuser									
7402	Grooved end inlet x flanged outlet									
7410	3" x 3"	Q-1	27	.593	Ea.	1,175	34.50		1,209.50	1,350
7412	4" x 4"		19	.842		1,600	49		1,649	1,825
7414	5" x 5"		14	1.143		1,850	66.50		1,916.50	2,150
7416	6" x 6"	Q-2	20	1.200		2,350	72		2,422	2,675
7418	8" x 8"		15	1.600		4,375	96		4,471	4,950
7420	10" x 10"		12	2		5,925	120		6,045	6,700
7422	12" x 12"		9	2.667		9,775	160		9,935	10,900
7424	14" x 14"		7	3.429		9,525	206		9,731	10,800
7426	16" x 14"		6	4		9,950	241		10,191	11,300
7500	Strainer, tee type, painted									
7506	2" pipe size	1 Plum	21	.381	Ea.	785	24.50		809.50	895
7508	2-1/2" pipe size	Q-1	30	.533		825	31		856	950
7510	3" pipe size		28	.571		925	33		958	1,075
7512	4" pipe size		20	.800		1,050	46.50		1,096.50	1,225
7514	5" pipe size		15	1.067		1,525	62		1,587	1,775
7516	6" pipe size	Q-2	23	1.043		1,650	63		1,713	1,900
7518	8" pipe size		16	1.500		2,525	90		2,615	2,900
7520	10" pipe size		13	1.846		3,700	111		3,811	4,250
7522	12" pipe size		10	2.400		4,750	144		4,894	5,450
7524	14" pipe size		8	3		12,900	180		13,080	14,500
7526	16" pipe size		7	3.429		16,200	206		16,406	18,100
7570	Expansion joint, max. 3" travel									
7572	2" diameter	1 Plum	24	.333	Ea.	990	21.50		1,011.50	1,125
7574	3" diameter	Q-1	31	.516		1,125	30		1,155	1,300
7576	4" diameter	"	23	.696		1,500	40.50		1,540.50	1,700
7578	6" diameter	Q-2	38	.632		2,325	38		2,363	2,600

22 11 13 – Facility Water Distribution Piping

22 11 13.48 Pipe, Fittings and Valves, Steel, Grooved-Joint		Crew	Daily Output	Labor-Hours	Unit	2020 Bare Costs Material	Labor	Equipment	Total	Total Incl O&P
7790	Valves: coupling material required at joints not incl. in valve price.									
7794	Add 1 selected coupling, material only, per joint for installed price.									
7800	Ball valve w/handle, carbon steel trim									
7810	1-1/2" pipe size	1 Plum	31	.258	Ea.	187	16.65		203.65	231
7812	2" pipe size	"	24	.333		178	21.50		199.50	228
7814	2-1/2" pipe size	Q-1	38	.421		450	24.50		474.50	530
7816	3" pipe size		31	.516		730	30		760	845
7818	4" pipe size		23	.696		1,125	40.50		1,165.50	1,300
7820	6" pipe size	Q-2	24	1		3,475	60		3,535	3,925
7830	With gear operator									
7834	2-1/2" pipe size	Q-1	38	.421	Ea.	910	24.50		934.50	1,025
7836	3" pipe size		31	.516		1,300	30		1,330	1,475
7838	4" pipe size		23	.696		1,650	40.50		1,690.50	1,850
7840	6" pipe size	Q-2	24	1		3,875	60		3,935	4,375
7870	Check valve									
7874	2-1/2" pipe size	Q-1	38	.421	Ea.	375	24.50		399.50	450
7876	3" pipe size		31	.516		440	30		470	530
7878	4" pipe size		23	.696		465	40.50		505.50	575
7880	5" pipe size		18	.889		775	51.50		826.50	930
7882	6" pipe size	Q-2	24	1		920	60		980	1,100
7884	8" pipe size		19	1.263		1,250	76		1,326	1,500
7886	10" pipe size		16	1.500		3,375	90		3,465	3,850
7888	12" pipe size		13	1.846		4,000	111		4,111	4,575
7900	Plug valve, balancing, w/lever operator									
7906	3" pipe size	Q-1	31	.516	Ea.	925	30		955	1,075
7908	4" pipe size	"	23	.696		1,150	40.50		1,190.50	1,300
7909	6" pipe size	Q-2	24	1		1,750	60		1,810	2,025
7916	With gear operator									
7920	3" pipe size	Q-1	31	.516	Ea.	1,750	30		1,780	1,975
7922	4" pipe size	"	23	.696		1,825	40.50		1,865.50	2,050
7924	6" pipe size	Q-2	24	1		2,450	60		2,510	2,800
7926	8" pipe size		19	1.263		3,775	76		3,851	4,275
7928	10" pipe size		16	1.500		4,775	90		4,865	5,375
7930	12" pipe size		13	1.846		7,275	111		7,386	8,175
8000	Butterfly valve, 2 position handle, with standard trim									
8010	1-1/2" pipe size	1 Plum	30	.267	Ea.	350	17.20		367.20	410
8020	2" pipe size	"	23	.348		350	22.50		372.50	420
8030	3" pipe size	Q-1	30	.533		505	31		536	600
8050	4" pipe size	"	22	.727		555	42		597	675
8070	6" pipe size	Q-2	23	1.043		1,125	63		1,188	1,325
8080	8" pipe size		18	1.333		1,450	80		1,530	1,725
8090	10" pipe size		15	1.600		2,000	96		2,096	2,350
8200	With stainless steel trim									
8240	1-1/2" pipe size	1 Plum	30	.267	Ea.	445	17.20		462.20	515
8250	2" pipe size	"	23	.348		445	22.50		467.50	525
8270	3" pipe size	Q-1	30	.533		600	31		631	705
8280	4" pipe size	"	22	.727		650	42		692	780
8300	6" pipe size	Q-2	23	1.043		1,200	63		1,263	1,425
8310	8" pipe size		18	1.333		3,000	80		3,080	3,425
8320	10" pipe size		15	1.600		4,800	96		4,896	5,425
8322	12" pipe size		12	2		5,750	120		5,870	6,500
8324	14" pipe size		10	2.400		7,375	144		7,519	8,325
8326	16" pipe size		9	2.667		12,500	160		12,660	13,900

22 11 13.48 Pipe, Fittings and Valves, Steel, Grooved-Joint		Crew	Daily Output	Labor-Hours	Unit	Material	2020 Bare Costs Labor	Equipment	Total	Total Incl O&P
8328	18" pipe size	Q-3	12	2.667	Ea.	13,100	164		13,264	14,600
8330	20" pipe size		10	3.200		16,100	196		16,296	18,000
8332	24" pipe size		9	3.556		21,600	218		21,818	24,000
8336	Note: sizes 8" up w/manual gear operator									
9000	Cut one groove, labor									
9010	3/4" pipe size	Q-1	152	.105	Ea.		6.10		6.10	9.15
9020	1" pipe size		140	.114			6.65		6.65	9.90
9030	1-1/4" pipe size		124	.129			7.50		7.50	11.20
9040	1-1/2" pipe size		114	.140			8.15		8.15	12.20
9050	2" pipe size		104	.154			8.90		8.90	13.35
9060	2-1/2" pipe size		96	.167			9.65		9.65	14.45
9070	3" pipe size		88	.182			10.55		10.55	15.80
9080	3-1/2" pipe size		83	.193			11.20		11.20	16.75
9090	4" pipe size		78	.205			11.90		11.90	17.80
9100	5" pipe size		72	.222			12.90		12.90	19.30
9110	6" pipe size		70	.229			13.25		13.25	19.85
9120	8" pipe size		54	.296			17.20		17.20	25.50
9130	10" pipe size		38	.421			24.50		24.50	36.50
9140	12" pipe size		30	.533			31		31	46.50
9150	14" pipe size		20	.800			46.50		46.50	69.50
9160	16" pipe size		19	.842			49		49	73
9170	18" pipe size		18	.889			51.50		51.50	77
9180	20" pipe size		17	.941			54.50		54.50	81.50
9190	24" pipe size		15	1.067			62		62	92.50
9210	Roll one groove									
9220	3/4" pipe size	Q-1	266	.060	Ea.		3.49		3.49	5.20
9230	1" pipe size		228	.070			4.07		4.07	6.10
9240	1-1/4" pipe size		200	.080			4.64		4.64	6.95
9250	1-1/2" pipe size		178	.090			5.20		5.20	7.80
9260	2" pipe size		116	.138			8		8	11.95
9270	2-1/2" pipe size		110	.145			8.45		8.45	12.65
9280	3" pipe size		100	.160			9.30		9.30	13.90
9290	3-1/2" pipe size		94	.170			9.85		9.85	14.75
9300	4" pipe size		86	.186			10.80		10.80	16.15
9310	5" pipe size		84	.190			11.05		11.05	16.55
9320	6" pipe size		80	.200			11.60		11.60	17.35
9330	8" pipe size		66	.242			14.05		14.05	21
9340	10" pipe size		58	.276			16		16	24
9350	12" pipe size		46	.348			20		20	30
9360	14" pipe size		30	.533			31		31	46.50
9370	16" pipe size		28	.571			33		33	49.50
9380	18" pipe size		27	.593			34.50		34.50	51.50
9390	20" pipe size		25	.640			37		37	55.50
9400	24" pipe size		23	.696			40.50		40.50	60.50

22 11 13.60 Tubing, Stainless Steel		Crew	Daily Output	Labor-Hours	Unit	Material	Labor	Equipment	Total	Total Incl O&P
0010	**TUBING, STAINLESS STEEL**									
5000	Tubing									
5010	Type 304, no joints, no hangers									
5020	.035 wall									
5021	1/4"	1 Plum	160	.050	L.F.	6.70	3.22		9.92	12.20
5022	3/8"		160	.050		6.25	3.22		9.47	11.70
5023	1/2"		160	.050		10.60	3.22		13.82	16.45

22 11 13.60 Tubing, Stainless Steel		Crew	Daily Output	Labor-Hours	Unit	Material	Labor	Equipment	Total	Total Incl O&P
						2020 Bare Costs				
5024	5/8"	1 Plum	160	.050	L.F.	10.25	3.22		13.47	16.05
5025	3/4"		133	.060		10.80	3.88		14.68	17.65
5026	7/8"		133	.060		11.05	3.88		14.93	17.95
5027	1"	↓	114	.070	↓	14.65	4.52		19.17	23
5040	.049 wall									
5041	1/4"	1 Plum	160	.050	L.F.	7.05	3.22		10.27	12.60
5042	3/8"		160	.050		10.60	3.22		13.82	16.45
5043	1/2"		160	.050		10.45	3.22		13.67	16.25
5044	5/8"		160	.050		11.70	3.22		14.92	17.65
5045	3/4"		133	.060		12.40	3.88		16.28	19.45
5046	7/8"		133	.060		10.80	3.88		14.68	17.65
5047	1"	↓	114	.070	↓	12	4.52		16.52	19.95
5060	.065 wall									
5061	1/4"	1 Plum	160	.050	L.F.	7.60	3.22		10.82	13.20
5062	3/8"		160	.050		10.60	3.22		13.82	16.45
5063	1/2"		160	.050		10.15	3.22		13.37	15.95
5064	5/8"		160	.050		13	3.22		16.22	19.10
5065	3/4"		133	.060		14.55	3.88		18.43	22
5066	7/8"		133	.060		13.20	3.88		17.08	20.50
5067	1"	↓	114	.070	↓	12.55	4.52		17.07	20.50
5210	Type 316									
5220	.035 wall									
5221	1/4"	1 Plum	160	.050	L.F.	6.10	3.22		9.32	11.50
5222	3/8"		160	.050		6.30	3.22		9.52	11.75
5223	1/2"		160	.050		9.45	3.22		12.67	15.20
5224	5/8"		160	.050		9.80	3.22		13.02	15.60
5225	3/4"		133	.060		12.15	3.88		16.03	19.15
5226	7/8"		133	.060		18.10	3.88		21.98	25.50
5227	1"	↓	114	.070	↓	16.30	4.52		20.82	24.50
5240	.049 wall									
5241	1/4"	1 Plum	160	.050	L.F.	5.30	3.22		8.52	10.65
5242	3/8"		160	.050		9.80	3.22		13.02	15.55
5243	1/2"		160	.050		10.35	3.22		13.57	16.15
5244	5/8"		160	.050		11.25	3.22		14.47	17.15
5245	3/4"		133	.060		12.20	3.88		16.08	19.20
5246	7/8"		133	.060		20.50	3.88		24.38	28.50
5247	1"	↓	114	.070	↓	14.55	4.52		19.07	23
5260	.065 wall									
5261	1/4"	1 Plum	160	.050	L.F.	7.20	3.22		10.42	12.75
5262	3/8"		160	.050		10.35	3.22		13.57	16.20
5263	1/2"		160	.050		10.05	3.22		13.27	15.85
5264	5/8"		160	.050		12.10	3.22		15.32	18.10
5265	3/4"		133	.060		14.05	3.88		17.93	21.50
5266	7/8"		133	.060		21	3.88		24.88	29
5267	1"	↓	114	.070	↓	18.25	4.52		22.77	27

22 11 13.61 Tubing Fittings, Stainless Steel		Crew	Daily Output	Labor-Hours	Unit	Material	Labor	Equipment	Total	Total Incl O&P
0010	**TUBING FITTINGS, STAINLESS STEEL**									
8200	Tube fittings, compression type									
8202	Type 316									
8204	90° elbow									
8206	1/4"	1 Plum	24	.333	Ea.	17.30	21.50		38.80	51
8207	3/8"	↓	22	.364	↓	21.50	23.50		45	58.50

22 11 13.61 Tubing Fittings, Stainless Steel

Line	22 11 13.61 Tubing Fittings, Stainless Steel	Crew	Daily Output	Labor-Hours	Unit	Material (2020 Bare Costs)	Labor (2020 Bare Costs)	Equipment (2020 Bare Costs)	Total (2020 Bare Costs)	Total Incl O&P
8208	1/2″	1 Plum	22	.364	Ea.	35	23.50		58.50	73.50
8209	5/8″		21	.381		39.50	24.50		64	80
8210	3/4″		21	.381		62	24.50		86.50	105
8211	7/8″		20	.400		95	26		121	144
8212	1″		20	.400		118	26		144	169
8220	Union tee									
8222	1/4″	1 Plum	15	.533	Ea.	24.50	34.50		59	78.50
8224	3/8″		15	.533		31.50	34.50		66	86
8225	1/2″		15	.533		48.50	34.50		83	105
8226	5/8″		14	.571		54.50	37		91.50	115
8227	3/4″		14	.571		73	37		110	135
8228	7/8″		13	.615		141	39.50		180.50	215
8229	1″		13	.615		156	39.50		195.50	232
8234	Union									
8236	1/4″	1 Plum	24	.333	Ea.	12	21.50		33.50	45
8237	3/8″		22	.364		17.10	23.50		40.60	54
8238	1/2″		22	.364		25.50	23.50		49	63
8239	5/8″		21	.381		33	24.50		57.50	73
8240	3/4″		21	.381		41.50	24.50		66	82
8241	7/8″		20	.400		67.50	26		93.50	113
8242	1″		20	.400		71.50	26		97.50	117
8250	Male connector									
8252	1/4″ x 1/4″	1 Plum	24	.333	Ea.	7.70	21.50		29.20	40.50
8253	3/8″ x 3/8″		22	.364		12	23.50		35.50	48
8254	1/2″ x 1/2″		22	.364		17.75	23.50		41.25	54.50
8256	3/4″ x 3/4″		21	.381		27	24.50		51.50	66.50
8258	1″ x 1″		20	.400		47.50	26		73.50	90.50

22 11 13.64 Pipe, Stainless Steel

Line	22 11 13.64 Pipe, Stainless Steel	Crew	Daily Output	Labor-Hours	Unit	Material	Labor	Equipment	Total	Total Incl O&P
0010	**PIPE, STAINLESS STEEL** R221113-70									
0020	Welded, with clevis type hanger assemblies, 10′ OC									
0500	Schedule 5, type 304									
0540	1/2″ diameter	Q-15	128	.125	L.F.	13.70	7.25	.83	21.78	27
0550	3/4″ diameter		116	.138		18.70	8	.92	27.62	33.50
0560	1″ diameter		103	.155		17.05	9	1.03	27.08	33.50
0570	1-1/4″ diameter		93	.172		20.50	10	1.14	31.64	39
0580	1-1/2″ diameter		85	.188		23	10.90	1.25	35.15	43
0590	2″ diameter		69	.232		33.50	13.45	1.54	48.49	58.50
0600	2-1/2″ diameter		53	.302		47.50	17.50	2.01	67.01	80.50
0610	3″ diameter		48	.333		73.50	19.35	2.22	95.07	112
0620	4″ diameter		44	.364		127	21	2.42	150.42	173
0630	5″ diameter		36	.444		154	26	2.96	182.96	211
0640	6″ diameter	Q-16	42	.571		320	34.50	2.53	357.03	405
0650	8″ diameter		34	.706		490	42.50	3.13	535.63	605
0660	10″ diameter		26	.923		300	55.50	4.09	359.59	420
0670	12″ diameter		21	1.143		395	68.50	5.05	468.55	545
0700	To delete hangers, subtract									
0710	1/2″ diam. to 1-1/2″ diam.					8%	19%			
0720	2″ diam. to 5″ diam.					4%	9%			
0730	6″ diam. to 12″ diam.					3%	4%			
0750	For small quantities, add				L.F.	10%				
1250	Schedule 5, type 316									
1290	1/2″ diameter	Q-15	128	.125	L.F.	13.05	7.25	.83	21.13	26

22 11 13.64 Pipe, Stainless Steel		Crew	Daily Output	Labor-Hours	Unit	Material	Labor	Equipment	Total	Total Incl O&P
						2020 Bare Costs				
1300	3/4" diameter	Q-15	116	.138	L.F.	21	8	.92	29.92	36.50
1310	1" diameter		103	.155		21.50	9	1.03	31.53	38.50
1320	1-1/4" diameter		93	.172		25	10	1.14	36.14	43.50
1330	1-1/2" diameter		85	.188		44	10.90	1.25	56.15	66
1340	2" diameter		69	.232		59	13.45	1.54	73.99	86.50
1350	2-1/2" diameter		53	.302		91.50	17.50	2.01	111.01	129
1360	3" diameter		48	.333		115	19.35	2.22	136.57	158
1370	4" diameter		44	.364		139	21	2.42	162.42	187
1380	5" diameter		36	.444		159	26	2.96	187.96	217
1390	6" diameter	Q-16	42	.571		190	34.50	2.53	227.03	263
1400	8" diameter		34	.706		286	42.50	3.13	331.63	380
1410	10" diameter		26	.923		335	55.50	4.09	394.59	455
1420	12" diameter		21	1.143		430	68.50	5.05	503.55	585
1490	For small quantities, add					10%				
1940	To delete hanger, subtract									
1950	1/2" diam. to 1-1/2" diam.					5%	19%			
1960	2" diam. to 5" diam.					3%	9%			
1970	6" diam. to 12" diam.					2%	4%			
2000	Schedule 10, type 304									
2040	1/4" diameter	Q-15	131	.122	L.F.	9.10	7.10	.81	17.01	21.50
2050	3/8" diameter		128	.125		8.15	7.25	.83	16.23	20.50
2060	1/2" diameter		125	.128		13.95	7.40	.85	22.20	27.50
2070	3/4" diameter		113	.142		16.40	8.20	.94	25.54	31.50
2080	1" diameter		100	.160		15.30	9.30	1.06	25.66	32
2090	1-1/4" diameter		91	.176		25.50	10.20	1.17	36.87	45
2100	1-1/2" diameter		83	.193		22	11.20	1.28	34.48	42
2110	2" diameter		67	.239		23.50	13.85	1.59	38.94	48.50
2120	2-1/2" diameter		51	.314		42	18.20	2.09	62.29	75.50
2130	3" diameter		46	.348		50.50	20	2.31	72.81	88
2140	4" diameter		42	.381		44	22	2.53	68.53	84.50
2150	5" diameter		35	.457		66.50	26.50	3.04	96.04	116
2160	6" diameter	Q-16	40	.600		60	36	2.66	98.66	123
2170	8" diameter		33	.727		102	44	3.22	149.22	181
2180	10" diameter		25	.960		140	57.50	4.25	201.75	245
2190	12" diameter		21	1.143		174	68.50	5.05	247.55	300
2250	For small quantities, add					10%				
2650	To delete hanger, subtract									
2660	1/4" diam. to 3/4" diam.					9%	22%			
2670	1" diam. to 2" diam.					4%	15%			
2680	2-1/2" diam. to 5" diam.					3%	8%			
2690	6" diam. to 12" diam.					3%	4%			
2750	Schedule 10, type 316									
2790	1/4" diameter	Q-15	131	.122	L.F.	7.15	7.10	.81	15.06	19.35
2800	3/8" diameter		128	.125		8.70	7.25	.83	16.78	21.50
2810	1/2" diameter		125	.128		11.10	7.40	.85	19.35	24
2820	3/4" diameter		113	.142		20.50	8.20	.94	29.64	36
2830	1" diameter		100	.160		23.50	9.30	1.06	33.86	41
2840	1-1/4" diameter		91	.176		22.50	10.20	1.17	33.87	41.50
2850	1-1/2" diameter		83	.193		30.50	11.20	1.28	42.98	51.50
2860	2" diameter		67	.239		32	13.85	1.59	47.44	57.50
2870	2-1/2" diameter		51	.314		39	18.20	2.09	59.29	72.50
2880	3" diameter		46	.348		59	20	2.31	81.31	97.50
2890	4" diameter		42	.381		59	22	2.53	83.53	101

22 11 13.64 Pipe, Stainless Steel		Crew	Daily Output	Labor-Hours	Unit	2020 Bare Costs Material	Labor	Equipment	Total	Total Incl O&P
2900	5″ diameter	Q-15	35	.457	L.F.	56.50	26.50	3.04	86.04	105
2910	6″ diameter	Q-16	40	.600		91	36	2.66	129.66	157
2920	8″ diameter		33	.727		133	44	3.22	180.22	215
2930	10″ diameter		25	.960		174	57.50	4.25	235.75	283
2940	12″ diameter		21	1.143		218	68.50	5.05	291.55	350
2990	For small quantities, add					10%				
3430	To delete hanger, subtract									
3440	1/4″ diam. to 3/4″ diam.					6%	22%			
3450	1″ diam. to 2″ diam.					3%	15%			
3460	2-1/2″ diam. to 5″ diam.					2%	8%			
3470	6″ diam. to 12″ diam.					2%	4%			
3500	Threaded, couplings and clevis hanger assemblies, 10′ OC									
3520	Schedule 40, type 304									
3540	1/4″ diameter	1 Plum	54	.148	L.F.	10.20	9.55		19.75	25.50
3550	3/8″ diameter		53	.151		11.15	9.75		20.90	27
3560	1/2″ diameter		52	.154		12.40	9.90		22.30	28.50
3570	3/4″ diameter		51	.157		20.50	10.10		30.60	37.50
3580	1″ diameter		45	.178		24.50	11.45		35.95	44
3590	1-1/4″ diameter	Q-1	76	.211		39	12.20		51.20	61.50
3600	1-1/2″ diameter		69	.232		40	13.45		53.45	64.50
3610	2″ diameter		57	.281		52	16.30		68.30	82
3620	2-1/2″ diameter		44	.364		98	21		119	140
3630	3″ diameter		38	.421		113	24.50		137.50	162
3640	4″ diameter	Q-2	51	.471		131	28.50		159.50	187
3740	For small quantities, add					10%				
4200	To delete couplings & hangers, subtract									
4210	1/4″ diam. to 3/4″ diam.					15%	56%			
4220	1″ diam. to 2″ diam.					18%	49%			
4230	2-1/2″ diam. to 4″ diam.					34%	40%			
4250	Schedule 40, type 316									
4290	1/4″ diameter	1 Plum	54	.148	L.F.	23.50	9.55		33.05	40
4300	3/8″ diameter		53	.151		25.50	9.75		35.25	42.50
4310	1/2″ diameter		52	.154		28.50	9.90		38.40	46.50
4320	3/4″ diameter		51	.157		32.50	10.10		42.60	50.50
4330	1″ diameter		45	.178		45	11.45		56.45	66.50
4340	1-1/4″ diameter	Q-1	76	.211		74	12.20		86.20	99.50
4350	1-1/2″ diameter		69	.232		67.50	13.45		80.95	94
4360	2″ diameter		57	.281		90.50	16.30		106.80	124
4370	2-1/2″ diameter		44	.364		130	21		151	175
4380	3″ diameter		38	.421		152	24.50		176.50	204
4390	4″ diameter	Q-2	51	.471		187	28.50		215.50	249
4490	For small quantities, add					10%				
4900	To delete couplings & hangers, subtract									
4910	1/4″ diam. to 3/4″ diam.					12%	56%			
4920	1″ diam. to 2″ diam.					14%	49%			
4930	2-1/2″ diam. to 4″ diam.					27%	40%			
5000	Schedule 80, type 304									
5040	1/4″ diameter	1 Plum	53	.151	L.F.	21	9.75		30.75	37.50
5050	3/8″ diameter		52	.154		37	9.90		46.90	56
5060	1/2″ diameter		51	.157		31.50	10.10		41.60	49.50
5070	3/4″ diameter		48	.167		38	10.75		48.75	57.50
5080	1″ diameter		43	.186		55	12		67	78.50
5090	1-1/4″ diameter	Q-1	73	.219		59	12.70		71.70	83.50

22 11 13.64 Pipe, Stainless Steel		Crew	Daily Output	Labor-Hours	Unit	2020 Bare Costs Material	Labor	Equipment	Total	Total Incl O&P
5100	1-1/2" diameter	Q-1	67	.239	L.F.	67.50	13.85		81.35	95
5110	2" diameter	↓	54	.296		108	17.20		125.20	145
5190	For small quantities, add				↓	10%				
5700	To delete couplings & hangers, subtract									
5710	1/4" diam. to 3/4" diam.					10%	53%			
5720	1" diam. to 2" diam.					14%	47%			
5750	Schedule 80, type 316									
5790	1/4" diameter	1 Plum	53	.151	L.F.	42.50	9.75		52.25	61.50
5800	3/8" diameter		52	.154		42.50	9.90		52.40	61.50
5810	1/2" diameter		51	.157		54.50	10.10		64.60	75
5820	3/4" diameter		48	.167		54	10.75		64.75	75.50
5830	1" diameter	↓	43	.186		51.50	12		63.50	75
5840	1-1/4" diameter	Q-1	73	.219		77.50	12.70		90.20	104
5850	1-1/2" diameter		67	.239		104	13.85		117.85	135
5860	2" diameter	↓	54	.296		117	17.20		134.20	154
5950	For small quantities, add				↓	10%				
7000	To delete couplings & hangers, subtract									
7010	1/4" diam. to 3/4" diam.					9%	53%			
7020	1" diam. to 2" diam.					14%	47%			
8000	Weld joints with clevis type hanger assemblies, 10' OC									
8010	Schedule 40, type 304									
8050	1/8" pipe size	Q-15	126	.127	L.F.	7.55	7.35	.84	15.74	20
8060	1/4" pipe size		125	.128		8.85	7.40	.85	17.10	22
8070	3/8" pipe size		122	.131		9.55	7.60	.87	18.02	23
8080	1/2" pipe size		118	.136		10.25	7.85	.90	19	24
8090	3/4" pipe size		109	.147		17.70	8.50	.98	27.18	33.50
8100	1" pipe size		95	.168		19.80	9.75	1.12	30.67	38
8110	1-1/4" pipe size		86	.186		30.50	10.80	1.24	42.54	51
8120	1-1/2" pipe size		78	.205		32	11.90	1.36	45.26	54.50
8130	2" pipe size		62	.258		38.50	14.95	1.72	55.17	67
8140	2-1/2" pipe size		49	.327		66	18.95	2.17	87.12	103
8150	3" pipe size		44	.364		69.50	21	2.42	92.92	111
8160	3-1/2" pipe size		44	.364		88	21	2.42	111.42	131
8170	4" pipe size		39	.410		71	24	2.73	97.73	117
8180	5" pipe size	↓	32	.500		121	29	3.33	153.33	180
8190	6" pipe size	Q-16	37	.649		160	39	2.87	201.87	238
8191	6" pipe size		37	.649		157	39	2.87	198.87	234
8200	8" pipe size		29	.828		164	50	3.67	217.67	260
8210	10" pipe size		24	1		273	60	4.43	337.43	395
8220	12" pipe size	↓	20	1.200	↓	515	72	5.30	592.30	680
8300	Schedule 40, type 316									
8310	1/8" pipe size	Q-15	126	.127	L.F.	20	7.35	.84	28.19	34
8320	1/4" pipe size		125	.128		22	7.40	.85	30.25	36
8330	3/8" pipe size		122	.131		23.50	7.60	.87	31.97	38.50
8340	1/2" pipe size		118	.136		26	7.85	.90	34.75	41
8350	3/4" pipe size		109	.147		29	8.50	.98	38.48	46
8360	1" pipe size		95	.168		39.50	9.75	1.12	50.37	59.50
8370	1-1/4" pipe size		86	.186		63	10.80	1.24	75.04	87
8380	1-1/2" pipe size		78	.205		57.50	11.90	1.36	70.76	83
8390	2" pipe size		62	.258		74	14.95	1.72	90.67	106
8400	2-1/2" pipe size		49	.327		91.50	18.95	2.17	112.62	132
8410	3" pipe size		44	.364		99.50	21	2.42	122.92	143
8420	3-1/2" pipe size	↓	44	.364	↓	109	21	2.42	132.42	154

22 11 13.64 Pipe, Stainless Steel		Crew	Daily Output	Labor-Hours	Unit	Material	2020 Bare Costs Labor	Equipment	Total	Total Incl O&P
8430	4″ pipe size	Q-15	39	.410	L.F.	114	24	2.73	140.73	165
8440	5″ pipe size	↓	32	.500		180	29	3.33	212.33	245
8450	6″ pipe size	Q-16	37	.649		263	39	2.87	304.87	350
8460	8″ pipe size		29	.828		271	50	3.67	324.67	375
8470	10″ pipe size		24	1		335	60	4.43	399.43	465
8480	12″ pipe size	↓	20	1.200	↓	455	72	5.30	532.30	615
8500	Schedule 80, type 304									
8510	1/4″ pipe size	Q-15	110	.145	L.F.	19.40	8.45	.97	28.82	35
8520	3/8″ pipe size		109	.147		35	8.50	.98	44.48	53
8530	1/2″ pipe size		106	.151		29.50	8.75	1	39.25	46.50
8540	3/4″ pipe size		96	.167		35	9.65	1.11	45.76	54
8550	1″ pipe size		87	.184		50.50	10.65	1.22	62.37	73
8560	1-1/4″ pipe size		81	.198		49.50	11.45	1.31	62.26	73
8570	1-1/2″ pipe size		74	.216		55	12.55	1.44	68.99	80.50
8580	2″ pipe size		58	.276		93.50	16	1.83	111.33	129
8590	2-1/2″ pipe size		46	.348		104	20	2.31	126.31	147
8600	3″ pipe size		41	.390		123	22.50	2.60	148.10	172
8610	4″ pipe size	↓	33	.485		154	28	3.22	185.22	215
8630	6″ pipe size	Q-16	30	.800	↓	279	48	3.54	330.54	380
8640	Schedule 80, type 316									
8650	1/4″ pipe size	Q-15	110	.145	L.F.	40.50	8.45	.97	49.92	58
8660	3/8″ pipe size		109	.147		40.50	8.50	.98	49.98	58.50
8670	1/2″ pipe size		106	.151		52	8.75	1	61.75	71.50
8680	3/4″ pipe size		96	.167		51	9.65	1.11	61.76	71.50
8690	1″ pipe size		87	.184		45.50	10.65	1.22	57.37	67.50
8700	1-1/4″ pipe size		81	.198		63.50	11.45	1.31	76.26	88
8710	1-1/2″ pipe size		74	.216		87	12.55	1.44	100.99	116
8720	2″ pipe size		58	.276		94	16	1.83	111.83	129
8730	2-1/2″ pipe size		46	.348		95.50	20	2.31	117.81	138
8740	3″ pipe size		41	.390		148	22.50	2.60	173.10	200
8760	4″ pipe size	↓	33	.485		195	28	3.22	226.22	260
8770	6″ pipe size	Q-16	30	.800	↓	340	48	3.54	391.54	450
9100	Threading pipe labor, sst, one end, schedules 40 & 80									
9110	1/4″ through 3/4″ pipe size	1 Plum	61.50	.130	Ea.		8.40		8.40	12.55
9120	1″ through 2″ pipe size		55.90	.143			9.20		9.20	13.80
9130	2-1/2″ pipe size		41.50	.193			12.40		12.40	18.60
9140	3″ pipe size	↓	38.50	.208			13.40		13.40	20
9150	3-1/2″ pipe size	Q-1	68.40	.234			13.55		13.55	20.50
9160	4″ pipe size		73	.219			12.70		12.70	19
9170	5″ pipe size		40.70	.393			23		23	34
9180	6″ pipe size		35.40	.452			26		26	39
9190	8″ pipe size		22.30	.717			41.50		41.50	62.50
9200	10″ pipe size		16.10	.994			57.50		57.50	86.50
9210	12″ pipe size	↓	12.30	1.301	↓		75.50		75.50	113
9250	Welding labor per joint for stainless steel									
9260	Schedule 5 and 10									
9270	1/4″ pipe size	Q-15	36	.444	Ea.		26	2.96	28.96	42
9280	3/8″ pipe size		35	.457			26.50	3.04	29.54	43
9290	1/2″ pipe size		35	.457			26.50	3.04	29.54	43
9300	3/4″ pipe size		28	.571			33	3.80	36.80	53.50
9310	1″ pipe size		25	.640			37	4.26	41.26	60
9320	1-1/4″ pipe size		22	.727			42	4.84	46.84	68.50
9330	1-1/2″ pipe size	↓	21	.762	↓		44	5.05	49.05	71.50

22 11 13.64 Pipe, Stainless Steel		Crew	Daily Output	Labor-Hours	Unit	Material	2020 Bare Costs Labor	Equipment	Total	Total Incl O&P
9340	2" pipe size	Q-15	18	.889	Ea.		51.50	5.90	57.40	83.50
9350	2-1/2" pipe size		12	1.333			77.50	8.85	86.35	126
9360	3" pipe size		9.73	1.644			95.50	10.95	106.45	155
9370	4" pipe size		7.37	2.171			126	14.45	140.45	204
9380	5" pipe size		6.15	2.602			151	17.30	168.30	245
9390	6" pipe size		5.71	2.802			163	18.65	181.65	264
9400	8" pipe size		3.69	4.336			251	29	280	405
9410	10" pipe size		2.91	5.498			320	36.50	356.50	515
9420	12" pipe size		2.31	6.926			400	46	446	650
9500	Schedule 40									
9510	1/4" pipe size	Q-15	28	.571	Ea.		33	3.80	36.80	53.50
9520	3/8" pipe size		27	.593			34.50	3.94	38.44	56
9530	1/2" pipe size		25.40	.630			36.50	4.19	40.69	59
9540	3/4" pipe size		22.22	.720			42	4.79	46.79	68
9550	1" pipe size		20.25	.790			46	5.25	51.25	74.50
9560	1-1/4" pipe size		18.82	.850			49.50	5.65	55.15	80
9570	1-1/2" pipe size		17.78	.900			52	6	58	84.50
9580	2" pipe size		15.09	1.060			61.50	7.05	68.55	100
9590	2-1/2" pipe size		7.96	2.010			117	13.35	130.35	189
9600	3" pipe size		6.43	2.488			144	16.55	160.55	234
9610	4" pipe size		4.88	3.279			190	22	212	310
9620	5" pipe size		4.26	3.756			218	25	243	355
9630	6" pipe size		3.77	4.244			246	28	274	400
9640	8" pipe size		2.44	6.557			380	43.50	423.50	620
9650	10" pipe size		1.92	8.333			485	55.50	540.50	785
9660	12" pipe size		1.52	10.526			610	70	680	990
9750	Schedule 80									
9760	1/4" pipe size	Q-15	21.55	.742	Ea.		43	4.94	47.94	70
9770	3/8" pipe size		20.75	.771			44.50	5.15	49.65	72.50
9780	1/2" pipe size		19.54	.819			47.50	5.45	52.95	77
9790	3/4" pipe size		17.09	.936			54.50	6.25	60.75	88.50
9800	1" pipe size		15.58	1.027			59.50	6.85	66.35	96.50
9810	1-1/4" pipe size		14.48	1.105			64	7.35	71.35	104
9820	1-1/2" pipe size		13.68	1.170			68	7.80	75.80	111
9830	2" pipe size		11.61	1.378			80	9.15	89.15	130
9840	2-1/2" pipe size		6.12	2.614			152	17.40	169.40	246
9850	3" pipe size		4.94	3.239			188	21.50	209.50	305
9860	4" pipe size		3.75	4.267			247	28.50	275.50	400
9870	5" pipe size		3.27	4.893			284	32.50	316.50	460
9880	6" pipe size		2.90	5.517			320	36.50	356.50	520
9890	8" pipe size		1.87	8.556			495	57	552	810
9900	10" pipe size		1.48	10.811			625	72	697	1,025
9910	12" pipe size		1.17	13.675			795	91	886	1,275
9920	Schedule 160, 1/2" pipe size		17	.941			54.50	6.25	60.75	88.50
9930	3/4" pipe size		14.81	1.080			62.50	7.20	69.70	102
9940	1" pipe size		13.50	1.185			68.50	7.90	76.40	112
9950	1-1/4" pipe size		12.55	1.275			74	8.50	82.50	120
9960	1-1/2" pipe size		11.85	1.350			78.50	9	87.50	127
9970	2" pipe size		10	1.600			93	10.65	103.65	151
9980	3" pipe size		4.28	3.738			217	25	242	355
9990	4" pipe size		3.25	4.923			286	32.50	318.50	460

22 11 13.66 Pipe Fittings, Stainless Steel		Crew	Daily Output	Labor-Hours	Unit	Material	Labor	Equipment	Total	Total Incl O&P
						2020 Bare Costs				
0010	**PIPE FITTINGS, STAINLESS STEEL**									
0100	Butt weld joint, schedule 5, type 304									
0120	90° elbow, long									
0140	1/2"	Q-15	17.50	.914	Ea.	18.05	53	6.10	77.15	106
0150	3/4"		14	1.143		18.05	66.50	7.60	92.15	127
0160	1"		12.50	1.280		19.05	74	8.50	101.55	141
0170	1-1/4"		11	1.455		26	84.50	9.65	120.15	165
0180	1-1/2"		10.50	1.524		22	88.50	10.15	120.65	168
0190	2"		9	1.778		26	103	11.80	140.80	196
0200	2-1/2"		6	2.667		60	155	17.75	232.75	315
0210	3"		4.86	3.292		56	191	22	269	370
0220	3-1/2"		4.27	3.747		165	217	25	407	535
0230	4"		3.69	4.336		94	251	29	374	510
0240	5"		3.08	5.195		350	300	34.50	684.50	875
0250	6"	Q-16	4.29	5.594		271	335	25	631	830
0260	8"		2.76	8.696		570	525	38.50	1,133.50	1,450
0270	10"		2.18	11.009		880	660	49	1,589	2,025
0280	12"		1.73	13.873		1,250	835	61.50	2,146.50	2,700
0320	For schedule 5, type 316, add					30%				
0600	45° elbow, long									
0620	1/2"	Q-15	17.50	.914	Ea.	18.05	53	6.10	77.15	106
0630	3/4"		14	1.143		18.05	66.50	7.60	92.15	127
0640	1"		12.50	1.280		19.05	74	8.50	101.55	141
0650	1-1/4"		11	1.455		26	84.50	9.65	120.15	165
0660	1-1/2"		10.50	1.524		22	88.50	10.15	120.65	168
0670	2"		9	1.778		26	103	11.80	140.80	196
0680	2-1/2"		6	2.667		60	155	17.75	232.75	315
0690	3"		4.86	3.292		45	191	22	258	360
0700	3-1/2"		4.27	3.747		165	217	25	407	535
0710	4"		3.69	4.336		76	251	29	356	490
0720	5"		3.08	5.195		281	300	34.50	615.50	800
0730	6"	Q-16	4.29	5.594		190	335	25	550	745
0740	8"		2.76	8.696		400	525	38.50	963.50	1,275
0750	10"		2.18	11.009		700	660	49	1,409	1,825
0760	12"		1.73	13.873		875	835	61.50	1,771.50	2,275
0800	For schedule 5, type 316, add					25%				
1100	Tee, straight									
1130	1/2"	Q-15	11.66	1.372	Ea.	54	79.50	9.15	142.65	189
1140	3/4"		9.33	1.715		54	99.50	11.40	164.90	221
1150	1"		8.33	1.921		57	111	12.75	180.75	244
1160	1-1/4"		7.33	2.183		46	127	14.50	187.50	256
1170	1-1/2"		7	2.286		45	133	15.20	193.20	264
1180	2"		6	2.667		47	155	17.75	219.75	305
1190	2-1/2"		4	4		113	232	26.50	371.50	500
1200	3"		3.24	4.938		86	286	33	405	560
1210	3-1/2"		2.85	5.614		237	325	37.50	599.50	785
1220	4"		2.46	6.504		128	375	43.50	546.50	755
1230	5"		2	8		405	465	53	923	1,200
1240	6"	Q-16	2.85	8.421		320	505	37.50	862.50	1,150
1250	8"		1.84	13.043		680	785	58	1,523	2,000
1260	10"		1.45	16.552		1,100	995	73.50	2,168.50	2,800
1270	12"		1.15	20.870		1,525	1,250	92.50	2,867.50	3,650

22 11 13.66 Pipe Fittings, Stainless Steel		Crew	Daily Output	Labor-Hours	Unit	2020 Bare Costs Material	Labor	Equipment	Total	Total Incl O&P
1320	For schedule 5, type 316, add				Ea.	25%				
2000	Butt weld joint, schedule 10, type 304									
2020	90° elbow, long									
2040	1/2"	Q-15	17	.941	Ea.	16.55	54.50	6.25	77.30	107
2050	3/4"		14	1.143		12.75	66.50	7.60	86.85	121
2060	1"		12.50	1.280		14.65	74	8.50	97.15	137
2070	1-1/4"		11	1.455		24	84.50	9.65	118.15	163
2080	1-1/2"		10.50	1.524		17	88.50	10.15	115.65	162
2090	2"		9	1.778		20	103	11.80	134.80	189
2100	2-1/2"		6	2.667		55	155	17.75	227.75	310
2110	3"		4.86	3.292		51.50	191	22	264.50	365
2120	3-1/2"		4.27	3.747		152	217	25	394	520
2130	4"		3.69	4.336		72.50	251	29	352.50	485
2140	5"		3.08	5.195		320	300	34.50	654.50	845
2150	6"	Q-16	4.29	5.594		208	335	25	568	760
2160	8"		2.76	8.696		440	525	38.50	1,003.50	1,325
2170	10"		2.18	11.009		680	660	49	1,389	1,800
2180	12"		1.73	13.873		965	835	61.50	1,861.50	2,375
2500	45° elbow, long									
2520	1/2"	Q-15	17.50	.914	Ea.	16.55	53	6.10	75.65	104
2530	3/4"		14	1.143		16.55	66.50	7.60	90.65	126
2540	1"		12.50	1.280		14.65	74	8.50	97.15	137
2550	1-1/4"		11	1.455		24	84.50	9.65	118.15	163
2560	1-1/2"		10.50	1.524		17	88.50	10.15	115.65	162
2570	2"		9	1.778		20	103	11.80	134.80	189
2580	2-1/2"		6	2.667		55	155	17.75	227.75	310
2590	3"		4.86	3.292		41.50	191	22	254.50	355
2600	3-1/2"		4.27	3.747		152	217	25	394	520
2610	4"		3.69	4.336		58.50	251	29	338.50	470
2620	5"		3.08	5.195		257	300	34.50	591.50	770
2630	6"	Q-16	4.29	5.594		147	335	25	507	695
2640	8"		2.76	8.696		310	525	38.50	873.50	1,175
2650	10"		2.18	11.009		540	660	49	1,249	1,650
2660	12"		1.73	13.873		675	835	61.50	1,571.50	2,075
2670	Reducer, concentric									
2674	1" x 3/4"	Q-15	13.25	1.208	Ea.	31	70	8.05	109.05	148
2676	2" x 1-1/2"	"	9.75	1.641		25.50	95	10.90	131.40	182
2678	6" x 4"	Q-16	4.91	4.888		76.50	294	21.50	392	550
2680	Caps									
2682	1"	Q-15	25	.640	Ea.	32	37	4.26	73.26	95
2684	1-1/2"		21	.762		38.50	44	5.05	87.55	114
2685	2"		18	.889		36	51.50	5.90	93.40	123
2686	4"		7.38	2.168		57.50	126	14.40	197.90	267
2687	6"	Q-16	8.58	2.797		92	168	12.40	272.40	365
3000	Tee, straight									
3030	1/2"	Q-15	11.66	1.372	Ea.	49.50	79.50	9.15	138.15	184
3040	3/4"		9.33	1.715		49.50	99.50	11.40	160.40	216
3050	1"		8.33	1.921		44	111	12.75	167.75	230
3060	1-1/4"		7.33	2.183		55.50	127	14.50	197	266
3070	1-1/2"		7	2.286		47	133	15.20	195.20	267
3080	2"		6	2.667		49.50	155	17.75	222.25	305
3090	2-1/2"		4	4		63	232	26.50	321.50	445
3100	3"		3.24	4.938		82	286	33	401	555

22 11 Facility Water Distribution

22 11 13 – Facility Water Distribution Piping

22 11 13.66 Pipe Fittings, Stainless Steel		Crew	Daily Output	Labor-Hours	Unit	Material	Labor	Equipment	Total	Total Incl O&P
						2020 Bare Costs				
3110	3-1/2"	Q-15	2.85	5.614	Ea.	121	325	37.50	483.50	660
3120	4"		2.46	6.504		99	375	43.50	517.50	720
3130	5"		2	8		375	465	53	893	1,175
3140	6"	Q-16	2.85	8.421		247	505	37.50	789.50	1,075
3150	8"		1.84	13.043		525	785	58	1,368	1,825
3151	10"		1.45	16.552		850	995	73.50	1,918.50	2,525
3152	12"		1.15	20.870		1,175	1,250	92.50	2,517.50	3,275
3154	For schedule 10, type 316, add					25%				
3281	Butt weld joint, schedule 40, type 304									
3284	90° elbow, long, 1/2"	Q-15	12.70	1.260	Ea.	19.30	73	8.40	100.70	139
3288	3/4"		11.10	1.441		19.30	83.50	9.60	112.40	157
3289	1"		10.13	1.579		20	91.50	10.50	122	171
3290	1-1/4"		9.40	1.702		26.50	98.50	11.30	136.30	190
3300	1-1/2"		8.89	1.800		21	104	11.95	136.95	192
3310	2"		7.55	2.119		30.50	123	14.10	167.60	233
3320	2-1/2"		3.98	4.020		58	233	26.50	317.50	445
3330	3"		3.21	4.984		74.50	289	33	396.50	555
3340	3-1/2"		2.83	5.654		276	330	37.50	643.50	835
3350	4"		2.44	6.557		129	380	43.50	552.50	760
3360	5"		2.13	7.512		430	435	50	915	1,175
3370	6"	Q-16	2.83	8.481		375	510	37.50	922.50	1,225
3380	8"		1.83	13.115		635	790	58	1,483	1,925
3390	10"		1.44	16.667		1,325	1,000	74	2,399	3,050
3400	12"		1.14	21.053		1,700	1,275	93.50	3,068.50	3,875
3410	For schedule 40, type 316, add					25%				
3460	45° elbow, long, 1/2"	Q-15	12.70	1.260	Ea.	19.30	73	8.40	100.70	139
3470	3/4"		11.10	1.441		19.30	83.50	9.60	112.40	157
3480	1"		10.13	1.579		20	91.50	10.50	122	171
3490	1-1/4"		9.40	1.702		26.50	98.50	11.30	136.30	190
3500	1-1/2"		8.89	1.800		21	104	11.95	136.95	192
3510	2"		7.55	2.119		30.50	123	14.10	167.60	233
3520	2-1/2"		3.98	4.020		58	233	26.50	317.50	445
3530	3"		3.21	4.984		59	289	33	381	535
3540	3-1/2"		2.83	5.654		276	330	37.50	643.50	835
3550	4"		2.44	6.557		92	380	43.50	515.50	720
3560	5"		2.13	7.512		305	435	50	790	1,050
3570	6"	Q-16	2.83	8.481		266	510	37.50	813.50	1,100
3580	8"		1.83	13.115		445	790	58	1,293	1,725
3590	10"		1.44	16.667		935	1,000	74	2,009	2,600
3600	12"		1.14	21.053		1,200	1,275	93.50	2,568.50	3,300
3610	For schedule 40, type 316, add					25%				
3660	Tee, straight 1/2"	Q-15	8.46	1.891	Ea.	49.50	110	12.60	172.10	232
3670	3/4"		7.40	2.162		49.50	125	14.40	188.90	258
3680	1"		6.74	2.374		52.50	138	15.80	206.30	281
3690	1-1/4"		6.27	2.552		122	148	16.95	286.95	375
3700	1-1/2"		5.92	2.703		56	157	17.95	230.95	315
3710	2"		5.03	3.181		83.50	184	21	288.50	390
3720	2-1/2"		2.65	6.038		106	350	40	496	685
3730	3"		2.14	7.477		107	435	49.50	591.50	820
3740	3-1/2"		1.88	8.511		201	495	56.50	752.50	1,025
3750	4"		1.62	9.877		201	575	65.50	841.50	1,150
3760	5"		1.42	11.268		465	655	75	1,195	1,575
3770	6"	Q-16	1.88	12.766		375	770	56.50	1,201.50	1,625

22 11 13.66 Pipe Fittings, Stainless Steel		Crew	Daily Output	Labor-Hours	Unit	Material	Labor	Equipment	Total	Total Incl O&P
						2020 Bare Costs				
3780	8"	Q-16	1.22	19.672	Ea.	755	1,175	87	2,017	2,700
3790	10"		.96	25		1,475	1,500	111	3,086	4,000
3800	12"		.76	31.579		1,950	1,900	140	3,990	5,150
3810	For schedule 40, type 316, add					25%				
3820	Tee, reducing on outlet, 3/4" x 1/2"	Q-15	7.73	2.070	Ea.	57	120	13.75	190.75	258
3822	1" x 1/2"		7.24	2.210		71.50	128	14.70	214.20	287
3824	1" x 3/4"		6.96	2.299		66	133	15.30	214.30	289
3826	1-1/4" x 1"		6.43	2.488		162	144	16.55	322.55	410
3828	1-1/2" x 1/2"		6.58	2.432		97.50	141	16.15	254.65	335
3830	1-1/2" x 3/4"		6.35	2.520		91	146	16.75	253.75	335
3832	1-1/2" x 1"		6.18	2.589		70	150	17.20	237.20	320
3834	2" x 1"		5.50	2.909		125	169	19.35	313.35	410
3836	2" x 1-1/2"		5.30	3.019		105	175	20	300	400
3838	2-1/2" x 2"		3.15	5.079		158	295	34	487	650
3840	3" x 1-1/2"		2.72	5.882		160	340	39	539	730
3842	3" x 2"		2.65	6.038		133	350	40	523	715
3844	4" x 2"		2.10	7.619		335	440	50.50	825.50	1,075
3846	4" x 3"		1.77	9.040		242	525	60	827	1,125
3848	5" x 4"		1.48	10.811		560	625	72	1,257	1,625
3850	6" x 3"	Q-16	2.19	10.959		615	660	48.50	1,323.50	1,725
3852	6" x 4"		2.04	11.765		535	710	52	1,297	1,700
3854	8" x 4"		1.46	16.438		1,250	990	73	2,313	2,925
3856	10" x 8"		.69	34.783		2,100	2,100	154	4,354	5,600
3858	12" x 10"		.55	43.636		2,800	2,625	193	5,618	7,225
3950	Reducer, concentric, 3/4" x 1/2"	Q-15	11.85	1.350		36.50	78.50	9	124	167
3952	1" x 3/4"		10.60	1.509		38.50	87.50	10.05	136.05	185
3954	1-1/4" x 3/4"		10.19	1.570		94.50	91	10.45	195.95	252
3956	1-1/4" x 1"		9.76	1.639		48	95	10.90	153.90	207
3958	1-1/2" x 3/4"		9.88	1.619		80	94	10.75	184.75	241
3960	1-1/2" x 1"		9.47	1.690		64.50	98	11.25	173.75	230
3962	2" x 1"		8.65	1.850		32.50	107	12.30	151.80	211
3964	2" x 1-1/2"		8.16	1.961		32	114	13.05	159.05	220
3966	2-1/2" x 1"		5.71	2.802		139	163	18.65	320.65	415
3968	2-1/2" x 2"		5.21	3.071		70.50	178	20.50	269	370
3970	3" x 1"		4.88	3.279		95.50	190	22	307.50	415
3972	3" x 1-1/2"		4.72	3.390		48.50	197	22.50	268	375
3974	3" x 2"		4.51	3.548		46	206	23.50	275.50	385
3976	4" x 2"		3.69	4.336		65	251	29	345	480
3978	4" x 3"		2.77	5.776		48.50	335	38.50	422	595
3980	5" x 3"		2.56	6.250		290	365	41.50	696.50	910
3982	5" x 4"		2.27	7.048		234	410	47	691	920
3984	6" x 3"	Q-16	3.57	6.723		150	405	30	585	805
3986	6" x 4"		3.19	7.524		123	455	33.50	611.50	845
3988	8" x 4"		2.44	9.836		325	590	43.50	958.50	1,300
3990	8" x 6"		2.22	10.811		242	650	48	940	1,300
3992	10" x 6"		1.91	12.565		450	755	55.50	1,260.50	1,675
3994	10" x 8"		1.61	14.907		375	895	66	1,336	1,850
3995	12" x 6"		1.63	14.724		825	885	65	1,775	2,300
3996	12" x 8"		1.41	17.021		640	1,025	75.50	1,740.50	2,300
3997	12" x 10"		1.27	18.898		450	1,125	83.50	1,658.50	2,275
4000	Socket weld joint, 3,000 lb., type 304									
4100	90° elbow									
4140	1/4"	Q-15	13.47	1.188	Ea.	45	69	7.90	121.90	161

22 11 13 – Facility Water Distribution Piping

22 11 13.66 Pipe Fittings, Stainless Steel		Crew	Daily Output	Labor-Hours	Unit	Material	Labor	Equipment	Total	Total Incl O&P
						2020 Bare Costs				
4150	3/8"	Q-15	12.97	1.234	Ea.	59	71.50	8.20	138.70	181
4160	1/2"		12.21	1.310		63	76	8.70	147.70	193
4170	3/4"		10.68	1.498		74	87	9.95	170.95	222
4180	1"		9.74	1.643		111	95.50	10.90	217.40	278
4190	1-1/4"		9.05	1.768		194	103	11.75	308.75	380
4200	1-1/2"		8.55	1.871		236	109	12.45	357.45	435
4210	2"		7.26	2.204		380	128	14.65	522.65	625
4300	45° elbow									
4340	1/4"	Q-15	13.47	1.188	Ea.	85	69	7.90	161.90	205
4350	3/8"		12.97	1.234		85	71.50	8.20	164.70	210
4360	1/2"		12.21	1.310		85	76	8.70	169.70	217
4370	3/4"		10.68	1.498		96.50	87	9.95	193.45	247
4380	1"		9.74	1.643		140	95.50	10.90	246.40	310
4390	1-1/4"		9.05	1.768		230	103	11.75	344.75	420
4400	1-1/2"		8.55	1.871		231	109	12.45	352.45	430
4410	2"		7.26	2.204		420	128	14.65	562.65	665
4500	Tee									
4540	1/4"	Q-15	8.97	1.784	Ea.	60	103	11.85	174.85	234
4550	3/8"		8.64	1.852		72	107	12.30	191.30	254
4560	1/2"		8.13	1.968		88	114	13.10	215.10	282
4570	3/4"		7.12	2.247		102	130	14.95	246.95	325
4580	1"		6.48	2.469		137	143	16.40	296.40	385
4590	1-1/4"		6.03	2.653		243	154	17.65	414.65	515
4600	1-1/2"		5.69	2.812		350	163	18.70	531.70	650
4610	2"		4.83	3.313		530	192	22	744	895
5000	Socket weld joint, 3,000 lb., type 316									
5100	90° elbow									
5140	1/4"	Q-15	13.47	1.188	Ea.	55.50	69	7.90	132.40	173
5150	3/8"		12.97	1.234		64.50	71.50	8.20	144.20	187
5160	1/2"		12.21	1.310		78	76	8.70	162.70	209
5170	3/4"		10.68	1.498		103	87	9.95	199.95	254
5180	1"		9.74	1.643		146	95.50	10.90	252.40	315
5190	1-1/4"		9.05	1.768		260	103	11.75	374.75	450
5200	1-1/2"		8.55	1.871		293	109	12.45	414.45	500
5210	2"		7.26	2.204		500	128	14.65	642.65	755
5300	45° elbow									
5340	1/4"	Q-15	13.47	1.188	Ea.	110	69	7.90	186.90	233
5350	3/8"		12.97	1.234		110	71.50	8.20	189.70	237
5360	1/2"		12.21	1.310		112	76	8.70	196.70	247
5370	3/4"		10.68	1.498		124	87	9.95	220.95	277
5380	1"		9.74	1.643		187	95.50	10.90	293.40	360
5390	1-1/4"		9.05	1.768		267	103	11.75	381.75	460
5400	1-1/2"		8.55	1.871		300	109	12.45	421.45	505
5410	2"		7.26	2.204		450	128	14.65	592.65	695
5500	Tee									
5540	1/4"	Q-15	8.97	1.784	Ea.	75	103	11.85	189.85	251
5550	3/8"		8.64	1.852		91.50	107	12.30	210.80	276
5560	1/2"		8.13	1.968		102	114	13.10	229.10	297
5570	3/4"		7.12	2.247		126	130	14.95	270.95	350
5580	1"		6.48	2.469		193	143	16.40	352.40	445
5590	1-1/4"		6.03	2.653		305	154	17.65	476.65	590
5600	1-1/2"		5.69	2.812		430	163	18.70	611.70	740
5610	2"		4.83	3.313		690	192	22	904	1,075

22 11 13.66 Pipe Fittings, Stainless Steel		Crew	Daily Output	Labor-Hours	Unit	Material	2020 Bare Costs Labor	Equipment	Total	Total Incl O&P
5700	For socket weld joint, 6,000 lb., type 304 and 316, add				Ea.	100%				
6000	Threaded companion flange									
6010	Stainless steel, 150 lb., type 304									
6020	1/2″ diam.	1 Plum	30	.267	Ea.	46	17.20		63.20	76
6030	3/4″ diam.		28	.286		51	18.40		69.40	83.50
6040	1″ diam.		27	.296		56	19.10		75.10	90
6050	1-1/4″ diam.	Q-1	44	.364		72.50	21		93.50	111
6060	1-1/2″ diam.		40	.400		72.50	23		95.50	114
6070	2″ diam.		36	.444		94.50	26		120.50	143
6080	2-1/2″ diam.		28	.571		132	33		165	196
6090	3″ diam.		20	.800		139	46.50		185.50	222
6110	4″ diam.		12	1.333		189	77.50		266.50	325
6130	6″ diam.	Q-2	14	1.714		340	103		443	530
6140	8″ diam.	″	12	2		655	120		775	900
6150	For type 316, add					40%				
6260	Weld flanges, stainless steel, type 304									
6270	Slip on, 150 lb. (welded, front and back)									
6280	1/2″ diam.	Q-15	12.70	1.260	Ea.	41	73	8.40	122.40	163
6290	3/4″ diam.		11.11	1.440		42	83.50	9.60	135.10	182
6300	1″ diam.		10.13	1.579		46.50	91.50	10.50	148.50	200
6310	1-1/4″ diam.		9.41	1.700		62.50	98.50	11.30	172.30	229
6320	1-1/2″ diam.		8.89	1.800		62.50	104	11.95	178.45	238
6330	2″ diam.		7.55	2.119		81	123	14.10	218.10	289
6340	2-1/2″ diam.		3.98	4.020		113	233	26.50	372.50	505
6350	3″ diam.		3.21	4.984		122	289	33	444	605
6370	4″ diam.		2.44	6.557		166	380	43.50	589.50	800
6390	6″ diam.	Q-16	1.89	12.698		252	765	56.50	1,073.50	1,500
6400	8″ diam.	″	1.22	19.672		475	1,175	87	1,737	2,400
6410	For type 316, add					40%				
6530	Weld neck 150 lb.									
6540	1/2″ diam.	Q-15	25.40	.630	Ea.	21.50	36.50	4.19	62.19	83
6550	3/4″ diam.		22.22	.720		26	42	4.79	72.79	96.50
6560	1″ diam.		20.25	.790		29.50	46	5.25	80.75	106
6570	1-1/4″ diam.		18.82	.850		41.50	49.50	5.65	96.65	126
6580	1-1/2″ diam.		17.78	.900		41.50	52	6	99.50	130
6590	2″ diam.		15.09	1.060		46.50	61.50	7.05	115.05	151
6600	2-1/2″ diam.		7.96	2.010		75.50	117	13.35	205.85	272
6610	3″ diam.		6.43	2.488		76	144	16.55	236.55	320
6630	4″ diam.		4.88	3.279		110	190	22	322	430
6640	5″ diam.		4.26	3.756		139	218	25	382	505
6650	6″ diam.	Q-16	5.66	4.240		163	255	18.80	436.80	580
6652	8″ diam.		3.65	6.575		286	395	29	710	935
6654	10″ diam.		2.88	8.333		400	500	37	937	1,225
6656	12″ diam.		2.28	10.526		740	635	46.50	1,421.50	1,825
6670	For type 316, add					23%				
7000	Threaded joint, 150 lb., type 304									
7030	90° elbow									
7040	1/8″	1 Plum	13	.615	Ea.	29.50	39.50		69	92
7050	1/4″		13	.615		29.50	39.50		69	92
7070	3/8″		13	.615		34.50	39.50		74	97.50
7080	1/2″		12	.667		30.50	43		73.50	98
7090	3/4″		11	.727		36.50	47		83.50	111
7100	1″		10	.800		50.50	51.50		102	133

22 11 13 – Facility Water Distribution Piping

22 11 13.66 Pipe Fittings, Stainless Steel		Crew	Daily Output	Labor-Hours	Unit	Material	Labor	Equipment	Total	Total Incl O&P
						2020 Bare Costs				
7110	1-1/4"	Q-1	17	.941	Ea.	78.50	54.50		133	168
7120	1-1/2"		16	1		90	58		148	186
7130	2"		14	1.143		130	66.50		196.50	242
7140	2-1/2"		11	1.455		315	84.50		399.50	475
7150	3"		8	2		460	116		576	680
7160	4"	Q-2	11	2.182		775	131		906	1,050
7180	45° elbow									
7190	1/8"	1 Plum	13	.615	Ea.	43	39.50		82.50	107
7200	1/4"		13	.615		43	39.50		82.50	107
7210	3/8"		13	.615		43.50	39.50		83	108
7220	1/2"		12	.667		44	43		87	113
7230	3/4"		11	.727		48	47		95	123
7240	1"		10	.800		55.50	51.50		107	138
7250	1-1/4"	Q-1	17	.941		77.50	54.50		132	167
7260	1-1/2"		16	1		100	58		158	197
7270	2"		14	1.143		143	66.50		209.50	257
7280	2-1/2"		11	1.455		445	84.50		529.50	615
7290	3"		8	2		655	116		771	895
7300	4"	Q-2	11	2.182		1,175	131		1,306	1,500
7320	Tee, straight									
7330	1/8"	1 Plum	9	.889	Ea.	45.50	57.50		103	136
7340	1/4"		9	.889		45.50	57.50		103	136
7350	3/8"		9	.889		48.50	57.50		106	139
7360	1/2"		8	1		46	64.50		110.50	148
7370	3/4"		7	1.143		49.50	73.50		123	165
7380	1"		6.50	1.231		64	79.50		143.50	190
7390	1-1/4"	Q-1	11	1.455		107	84.50		191.50	244
7400	1-1/2"		10	1.600		138	93		231	291
7410	2"		9	1.778		171	103		274	340
7420	2-1/2"		7	2.286		455	133		588	700
7430	3"		5	3.200		690	186		876	1,050
7440	4"	Q-2	7	3.429		1,700	206		1,906	2,175
7460	Coupling, straight									
7470	1/8"	1 Plum	19	.421	Ea.	10.85	27		37.85	52.50
7480	1/4"		19	.421		13.35	27		40.35	55
7490	3/8"		19	.421		16	27		43	58
7500	1/2"		19	.421		21.50	27		48.50	64
7510	3/4"		18	.444		28.50	28.50		57	74
7520	1"		15	.533		45.50	34.50		80	102
7530	1-1/4"	Q-1	26	.615		88.50	35.50		124	151
7540	1-1/2"		24	.667		82	38.50		120.50	149
7550	2"		21	.762		137	44		181	217
7560	2-1/2"		18	.889		320	51.50		371.50	425
7570	3"		14	1.143		435	66.50		501.50	580
7580	4"	Q-2	16	1.500		605	90		695	800
7600	Reducer, concentric, 1/2"	1 Plum	12	.667		23	43		66	89.50
7610	3/4"		11	.727		29.50	47		76.50	103
7612	1"		10	.800		49	51.50		100.50	131
7614	1-1/4"	Q-1	17	.941		104	54.50		158.50	196
7616	1-1/2"		16	1		114	58		172	212
7618	2"		14	1.143		177	66.50		243.50	294
7620	2-1/2"		11	1.455		470	84.50		554.50	640
7622	3"		8	2		520	116		636	750

22 11 13.66 Pipe Fittings, Stainless Steel		Crew	Daily Output	Labor-Hours	Unit	Material	2020 Bare Costs Labor	Equipment	Total	Total Incl O&P
7624	4"	Q-2	11	2.182	Ea.	865	131		996	1,150
7710	Union									
7720	1/8"	1 Plum	12	.667	Ea.	53.50	43		96.50	124
7730	1/4"		12	.667		53.50	43		96.50	124
7740	3/8"		12	.667		63	43		106	134
7750	1/2"		11	.727		78	47		125	156
7760	3/4"		10	.800		107	51.50		158.50	194
7770	1"		9	.889		155	57.50		212.50	256
7780	1-1/4"	Q-1	16	1		385	58		443	510
7790	1-1/2"		15	1.067		415	62		477	550
7800	2"		13	1.231		525	71.50		596.50	680
7810	2-1/2"		10	1.600		1,000	93		1,093	1,250
7820	3"		7	2.286		1,325	133		1,458	1,675
7830	4"	Q-2	10	2.400		1,825	144		1,969	2,250
7838	Caps									
7840	1/2"	1 Plum	24	.333	Ea.	14.25	21.50		35.75	47.50
7841	3/4"		22	.364		20.50	23.50		44	57.50
7842	1"		20	.400		33	26		59	75
7843	1-1/2"	Q-1	32	.500		83.50	29		112.50	136
7844	2"	"	28	.571		106	33		139	166
7845	4"	Q-2	22	1.091		415	65.50		480.50	555
7850	For 150 lb., type 316, add					25%				

22 11 13.74 Pipe, Plastic		Crew	Daily Output	Labor-Hours	Unit	Material	Labor	Equipment	Total	Total Incl O&P
0010	**PIPE, PLASTIC** R221113-70									
0020	Fiberglass reinforced, couplings 10' OC, clevis hanger assy's, 3 per 10'									
0080	General service									
0120	2" diameter	Q-1	59	.271	L.F.	17.50	15.75		33.25	43
0140	3" diameter		52	.308		27.50	17.85		45.35	57
0150	4" diameter		48	.333		23	19.35		42.35	54.50
0160	6" diameter		39	.410		43	24		67	83
0170	8" diameter	Q-2	49	.490		67.50	29.50		97	118
0180	10" diameter		41	.585		95.50	35		130.50	158
0190	12" diameter		36	.667		125	40		165	198
0600	PVC, high impact/pressure, cplgs. 10' OC, clevis hanger assy's, 3 per 10'									
1020	Schedule 80									
1070	1/2" diameter	1 Plum	50	.160	L.F.	6.85	10.30		17.15	23
1080	3/4" diameter		47	.170		7.95	10.95		18.90	25
1090	1" diameter		43	.186		13.25	12		25.25	32.50
1100	1-1/4" diameter		39	.205		15.40	13.20		28.60	37
1110	1-1/2" diameter		34	.235		16.75	15.15		31.90	41
1120	2" diameter	Q-1	55	.291		20.50	16.85		37.35	48
1140	3" diameter		50	.320		40.50	18.55		59.05	73
1150	4" diameter		46	.348		43	20		63	77
1170	6" diameter		38	.421		66.50	24.50		91	110
1730	To delete coupling & hangers, subtract									
1740	1/2" diam.					62%	80%			
1750	3/4" diam. to 1-1/4" diam.					58%	73%			
1760	1-1/2" diam. to 6" diam.					40%	57%			
1800	PVC, couplings 10' OC, clevis hanger assemblies, 3 per 10'									
1820	Schedule 40									
1860	1/2" diameter	1 Plum	54	.148	L.F.	4.90	9.55		14.45	19.70
1870	3/4" diameter		51	.157		5.40	10.10		15.50	21

22 11 13.74 Pipe, Plastic		Crew	Daily Output	Labor-Hours	Unit	Material	Labor (2020 Bare Costs)	Equipment	Total	Total Incl O&P
1880	1" diameter	1 Plum	46	.174	L.F.	9.45	11.20		20.65	27
1890	1-1/4" diameter		42	.190		9.95	12.30		22.25	29.50
1900	1-1/2" diameter		36	.222		10.25	14.30		24.55	33
1910	2" diameter	Q-1	59	.271		12.15	15.75		27.90	37
1920	2-1/2" diameter		56	.286		20	16.55		36.55	47.50
1930	3" diameter		53	.302		22	17.50		39.50	50.50
1940	4" diameter		48	.333		14.80	19.35		34.15	45.50
1950	5" diameter		43	.372		23.50	21.50		45	58.50
1960	6" diameter		39	.410		30.50	24		54.50	69
1970	8" diameter	Q-2	48	.500		37	30		67	86
1980	10" diameter		43	.558		77	33.50		110.50	135
1990	12" diameter		42	.571		93	34.50		127.50	154
2000	14" diameter		31	.774		151	46.50		197.50	236
2010	16" diameter		23	1.043		220	63		283	335
2340	To delete coupling & hangers, subtract									
2360	1/2" diam. to 1-1/4" diam.					65%	74%			
2370	1-1/2" diam. to 6" diam.					44%	57%			
2380	8" diam. to 12" diam.					41%	53%			
2390	14" diam. to 16" diam.					48%	45%			
2420	Schedule 80									
2440	1/4" diameter	1 Plum	58	.138	L.F.	4.46	8.90		13.36	18.20
2450	3/8" diameter		55	.145		4.46	9.35		13.81	18.95
2460	1/2" diameter		50	.160		5.15	10.30		15.45	21
2470	3/4" diameter		47	.170		5.55	10.95		16.50	22.50
2480	1" diameter		43	.186		9.90	12		21.90	29
2490	1-1/4" diameter		39	.205		10.45	13.20		23.65	31.50
2500	1-1/2" diameter		34	.235		10.65	15.15		25.80	34.50
2510	2" diameter	Q-1	55	.291		11.55	16.85		28.40	38
2520	2-1/2" diameter		52	.308		19.55	17.85		37.40	48
2530	3" diameter		50	.320		24	18.55		42.55	54.50
2540	4" diameter		46	.348		19.30	20		39.30	51
2550	5" diameter		42	.381		29.50	22		51.50	65.50
2560	6" diameter		38	.421		37	24.50		61.50	77
2570	8" diameter	Q-2	47	.511		47	30.50		77.50	97.50
2580	10" diameter		42	.571		110	34.50		144.50	173
2590	12" diameter		38	.632		134	38		172	204
2830	To delete coupling & hangers, subtract									
2840	1/4" diam. to 1/2" diam.					66%	80%			
2850	3/4" diam. to 1-1/4" diam.					61%	73%			
2860	1-1/2" diam. to 6" diam.					41%	57%			
2870	8" diam. to 12" diam.					31%	50%			
2900	Schedule 120									
2910	1/2" diameter	1 Plum	50	.160	L.F.	5.60	10.30		15.90	21.50
2950	3/4" diameter		47	.170		6.20	10.95		17.15	23
2960	1" diameter		43	.186		10.60	12		22.60	29.50
2970	1-1/4" diameter		39	.205		11.90	13.20		25.10	33
2980	1-1/2" diameter		33	.242		12.25	15.60		27.85	37
2990	2" diameter	Q-1	54	.296		14.35	17.20		31.55	41.50
3000	2-1/2" diameter		52	.308		25	17.85		42.85	54.50
3010	3" diameter		49	.327		36	18.95		54.95	68
3020	4" diameter		45	.356		25.50	20.50		46	59
3030	6" diameter		37	.432		63	25		88	107
3240	To delete coupling & hangers, subtract									

22 11 13.74 Pipe, Plastic		Crew	Daily Output	Labor-Hours	Unit	Material	Labor	Equipment	Total	Total Incl O&P
						2020 Bare Costs				
3250	1/2" diam. to 1-1/4" diam.					52%	74%			
3260	1-1/2" diam. to 4" diam.					30%	57%			
3270	6" diam.					17%	50%			
3300	PVC, pressure, couplings 10' OC, clevis hanger assy's, 3 per 10'									
3310	SDR 26, 160 psi									
3350	1-1/4" diameter	1 Plum	42	.190	L.F.	9.85	12.30		22.15	29
3360	1-1/2" diameter	"	36	.222		10.60	14.30		24.90	33
3370	2" diameter	Q-1	59	.271		11.05	15.75		26.80	35.50
3380	2-1/2" diameter		56	.286		19.35	16.55		35.90	46.50
3390	3" diameter		53	.302		22	17.50		39.50	50.50
3400	4" diameter		48	.333		16.35	19.35		35.70	47
3420	6" diameter	↓	39	.410		33.50	24		57.50	72.50
3430	8" diameter	Q-2	48	.500	↓	49	30		79	98.50
3660	To delete coupling & clevis hanger assy's, subtract									
3670	1-1/4" diam.					63%	68%			
3680	1-1/2" diam. to 4" diam.					48%	57%			
3690	6" diam. to 8" diam.					60%	54%			
3720	SDR 21, 200 psi, 1/2" diameter	1 Plum	54	.148	L.F.	4.80	9.55		14.35	19.60
3740	3/4" diameter		51	.157		5.05	10.10		15.15	20.50
3750	1" diameter		46	.174		8.90	11.20		20.10	26.50
3760	1-1/4" diameter		42	.190		10	12.30		22.30	29.50
3770	1-1/2" diameter	↓	36	.222		10.55	14.30		24.85	33
3780	2" diameter	Q-1	59	.271		10.45	15.75		26.20	35
3790	2-1/2" diameter		56	.286		19.85	16.55		36.40	47
3800	3" diameter		53	.302		21	17.50		38.50	49
3810	4" diameter		48	.333		16.95	19.35		36.30	47.50
3830	6" diameter	↓	39	.410		29.50	24		53.50	68
3840	8" diameter	Q-2	48	.500	↓	48	30		78	97.50
4000	To delete coupling & hangers, subtract									
4010	1/2" diam. to 3/4" diam.					71%	77%			
4020	1" diam. to 1-1/4" diam.					63%	70%			
4030	1-1/2" diam. to 6" diam.					44%	57%			
4040	8" diam.					46%	54%			
4100	DWV type, schedule 40, couplings 10' OC, clevis hanger assy's, 3 per 10'									
4210	ABS, schedule 40, foam core type									
4212	Plain end black									
4214	1-1/2" diameter	1 Plum	39	.205	L.F.	8.60	13.20		21.80	29.50
4216	2" diameter	Q-1	62	.258		9.25	14.95		24.20	32.50
4218	3" diameter		56	.286		17.60	16.55		34.15	44.50
4220	4" diameter		51	.314		9.80	18.20		28	38
4222	6" diameter	↓	42	.381	↓	21.50	22		43.50	56.50
4240	To delete coupling & hangers, subtract									
4244	1-1/2" diam. to 6" diam.					43%	48%			
4400	PVC									
4410	1-1/4" diameter	1 Plum	42	.190	L.F.	9.10	12.30		21.40	28.50
4420	1-1/2" diameter	"	36	.222		8.25	14.30		22.55	30.50
4460	2" diameter	Q-1	59	.271		9.15	15.75		24.90	33.50
4470	3" diameter		53	.302		17.45	17.50		34.95	45
4480	4" diameter		48	.333		19.20	19.35		38.55	50
4490	6" diameter	↓	39	.410		18.60	24		42.60	56
4500	8" diameter	Q-2	48	.500	↓	25.50	30		55.50	73
4510	To delete coupling & hangers, subtract									
4520	1-1/4" diam. to 1-1/2" diam.					48%	60%			

22 11 Facility Water Distribution

22 11 13 – Facility Water Distribution Piping

22 11 13.74 Pipe, Plastic		Crew	Daily Output	Labor-Hours	Unit	2020 Bare Costs Material	Labor	Equipment	Total	Total Incl O&P
4530	2" diam. to 8" diam.					42%	54%			
4532	to delete hangers, 2" diam. to 8" diam.	Q-1	50	.320	L.F.	5.50	18.55		24.05	34
4550	PVC, schedule 40, foam core type									
4552	Plain end, white									
4554	1-1/2" diameter	1 Plum	39	.205	L.F.	8.40	13.20		21.60	29
4556	2" diameter	Q-1	62	.258		8.95	14.95		23.90	32.50
4558	3" diameter		56	.286		17.05	16.55		33.60	44
4560	4" diameter		51	.314		18.55	18.20		36.75	47.50
4562	6" diameter		42	.381		17.25	22		39.25	52
4564	8" diameter	Q-2	51	.471		23.50	28.50		52	68
4568	10" diameter		48	.500		27.50	30		57.50	75
4570	12" diameter		46	.522		30.50	31.50		62	80.50
4580	To delete coupling & hangers, subtract									
4582	1-1/2" diam. to 2" diam.					58%	54%			
4584	3" diam. to 12" diam.					46%	42%			
4800	PVC, clear pipe, cplgs. 10' OC, clevis hanger assy's 3 per 10', Sched. 40									
4840	1/4" diameter	1 Plum	59	.136	L.F.	4.93	8.75		13.68	18.50
4850	3/8" diameter		56	.143		5.05	9.20		14.25	19.35
4860	1/2" diameter		54	.148		5.90	9.55		15.45	21
4870	3/4" diameter		51	.157		6.60	10.10		16.70	22.50
4880	1" diameter		46	.174		11.60	11.20		22.80	29.50
4890	1-1/4" diameter		42	.190		13.10	12.30		25.40	33
4900	1-1/2" diameter		36	.222		13.95	14.30		28.25	37
4910	2" diameter	Q-1	59	.271		16.45	15.75		32.20	41.50
4920	2-1/2" diameter		56	.286		28	16.55		44.55	56
4930	3" diameter		53	.302		32.50	17.50		50	62
4940	3-1/2" diameter		50	.320		40	18.55		58.55	72
4950	4" diameter		48	.333		28.50	19.35		47.85	60
5250	To delete coupling & hangers, subtract									
5260	1/4" diam. to 3/8" diam.					60%	81%			
5270	1/2" diam. to 3/4" diam.					41%	77%			
5280	1" diam. to 1-1/2" diam.					26%	67%			
5290	2" diam. to 4" diam.					16%	58%			
5300	CPVC, socket joint, couplings 10' OC, clevis hanger assemblies, 3 per 10'									
5302	Schedule 40									
5304	1/2" diameter	1 Plum	54	.148	L.F.	5.80	9.55		15.35	20.50
5305	3/4" diameter		51	.157		6.95	10.10		17.05	23
5306	1" diameter		46	.174		11.35	11.20		22.55	29.50
5307	1-1/4" diameter		42	.190		12.85	12.30		25.15	32.50
5308	1-1/2" diameter		36	.222		12.45	14.30		26.75	35
5309	2" diameter	Q-1	59	.271		15.70	15.75		31.45	41
5310	2-1/2" diameter		56	.286		28	16.55		44.55	56
5311	3" diameter		53	.302		32	17.50		49.50	61
5312	4" diameter		48	.333		32.50	19.35		51.85	65
5314	6" diameter		43	.372		55.50	21.50		77	93.50
5318	To delete coupling & hangers, subtract									
5319	1/2" diam. to 3/4" diam.					37%	77%			
5320	1" diam. to 1-1/4" diam.					27%	70%			
5321	1-1/2" diam. to 3" diam.					21%	57%			
5322	4" diam. to 6" diam.					16%	57%			
5324	Schedule 80									
5325	1/2" diameter	1 Plum	50	.160	L.F.	6.10	10.30		16.40	22
5326	3/4" diameter		47	.170		6.95	10.95		17.90	24

22 11 13.74 Pipe, Plastic		Crew	Daily Output	Labor-Hours	Unit	2020 Bare Costs Material	Labor	Equipment	Total	Total Incl O&P
5327	1" diameter	1 Plum	43	.186	L.F.	12	12		24	31
5328	1-1/4" diameter		39	.205		13.55	13.20		26.75	34.50
5329	1-1/2" diameter		34	.235		14.70	15.15		29.85	38.50
5330	2" diameter	Q-1	55	.291		17.30	16.85		34.15	44.50
5331	2-1/2" diameter		52	.308		30.50	17.85		48.35	60
5332	3" diameter		50	.320		32.50	18.55		51.05	63.50
5333	4" diameter		46	.348		31.50	20		51.50	64.50
5334	6" diameter		38	.421		61.50	24.50		86	104
5335	8" diameter	Q-2	47	.511		170	30.50		200.50	233
5339	To delete couplings & hangers, subtract									
5340	1/2" diam. to 3/4" diam.					44%	77%			
5341	1" diam. to 1-1/4" diam.					32%	71%			
5342	1-1/2" diam. to 4" diam.					25%	58%			
5343	6" diam. to 8" diam.					20%	53%			
5360	CPVC, threaded, couplings 10' OC, clevis hanger assemblies, 3 per 10'									
5380	Schedule 40									
5460	1/2" diameter	1 Plum	54	.148	L.F.	6.65	9.55		16.20	21.50
5470	3/4" diameter		51	.157		8.45	10.10		18.55	24.50
5480	1" diameter		46	.174		12.90	11.20		24.10	31
5490	1-1/4" diameter		42	.190		14.05	12.30		26.35	34
5500	1-1/2" diameter		36	.222		13.45	14.30		27.75	36.50
5510	2" diameter	Q-1	59	.271		16.90	15.75		32.65	42
5520	2-1/2" diameter		56	.286		29.50	16.55		46.05	57.50
5530	3" diameter		53	.302		33.50	17.50		51	63
5540	4" diameter		48	.333		40	19.35		59.35	73.50
5550	6" diameter		43	.372		59	21.50		80.50	97.50
5730	To delete coupling & hangers, subtract									
5740	1/2" diam. to 3/4" diam.					37%	77%			
5750	1" diam. to 1-1/4" diam.					27%	70%			
5760	1-1/2" diam. to 3" diam.					21%	57%			
5770	4" diam. to 6" diam.					16%	57%			
5800	Schedule 80									
5860	1/2" diameter	1 Plum	50	.160	L.F.	6.95	10.30		17.25	23
5870	3/4" diameter		47	.170		8.45	10.95		19.40	25.50
5880	1" diameter		43	.186		13.55	12		25.55	33
5890	1-1/4" diameter		39	.205		14.75	13.20		27.95	36
5900	1-1/2" diameter		34	.235		15.70	15.15		30.85	40
5910	2" diameter	Q-1	55	.291		18.55	16.85		35.40	46
5920	2-1/2" diameter		52	.308		31.50	17.85		49.35	61.50
5930	3" diameter		50	.320		34.50	18.55		53.05	66
5940	4" diameter		46	.348		39	20		59	73
5950	6" diameter		38	.421		65	24.50		89.50	108
5960	8" diameter	Q-2	47	.511		168	30.50		198.50	231
6060	To delete couplings & hangers, subtract									
6070	1/2" diam. to 3/4" diam.					44%	77%			
6080	1" diam. to 1-1/4" diam.					32%	71%			
6090	1-1/2" diam. to 4" diam.					25%	58%			
6100	6" diam. to 8" diam.					20%	53%			
6240	CTS, 1/2" diameter	1 Plum	54	.148	L.F.	4.80	9.55		14.35	19.60
6250	3/4" diameter		51	.157		10.05	10.10		20.15	26
6260	1" diameter		46	.174		13.85	11.20		25.05	32
6270	1-1/4" diameter		42	.190		18.65	12.30		30.95	39
6280	1-1/2" diameter		36	.222		22	14.30		36.30	46

22 11 13 – Facility Water Distribution Piping

22 11 13.74 Pipe, Plastic		Crew	Daily Output	Labor-Hours	Unit	2020 Bare Costs Material	Labor	Equipment	Total	Total Incl O&P
6290	2" diameter	Q-1	59	.271	L.F.	33	15.75		48.75	59.50
6370	To delete coupling & hangers, subtract									
6380	1/2" diam.					51%	79%			
6390	3/4" diam.					40%	76%			
6392	1" thru 2" diam.					72%	68%			
6500	Residential installation, plastic pipe									
6510	Couplings 10' OC, strap hangers 3 per 10'									
6520	PVC, Schedule 40									
6530	1/2" diameter	1 Plum	138	.058	L.F.	1.06	3.74		4.80	6.75
6540	3/4" diameter		128	.063		1.30	4.03		5.33	7.50
6550	1" diameter		119	.067		1.87	4.33		6.20	8.55
6560	1-1/4" diameter		111	.072		2.13	4.64		6.77	9.30
6570	1-1/2" diameter		104	.077		2.53	4.96		7.49	10.20
6580	2" diameter	Q-1	197	.081		3.45	4.71		8.16	10.85
6590	2-1/2" diameter		162	.099		5.70	5.75		11.45	14.85
6600	4" diameter		123	.130		8.15	7.55		15.70	20.50
6700	PVC, DWV, Schedule 40									
6720	1-1/4" diameter	1 Plum	100	.080	L.F.	2.14	5.15		7.29	10.05
6730	1-1/2" diameter	"	94	.085		1.54	5.50		7.04	9.90
6740	2" diameter	Q-1	178	.090		2.16	5.20		7.36	10.20
6760	4" diameter	"	110	.145		6.20	8.45		14.65	19.45
7280	PEX, flexible, no couplings or hangers									
7282	Note: For labor costs add 25% to the couplings and fittings labor total.									
7285	For fittings see section 23 83 16.10 7000									
7300	Non-barrier type, hot/cold tubing rolls									
7310	1/4" diameter x 100'				L.F.	.56			.56	.62
7350	3/8" diameter x 100'					.55			.55	.61
7360	1/2" diameter x 100'					.76			.76	.84
7370	1/2" diameter x 500'					.74			.74	.81
7380	1/2" diameter x 1000'					.72			.72	.79
7400	3/4" diameter x 100'					1.04			1.04	1.14
7410	3/4" diameter x 500'					1.17			1.17	1.29
7420	3/4" diameter x 1000'					1.17			1.17	1.29
7460	1" diameter x 100'					2.02			2.02	2.22
7470	1" diameter x 300'					2.02			2.02	2.22
7480	1" diameter x 500'					2.03			2.03	2.23
7500	1-1/4" diameter x 100'					3.44			3.44	3.78
7510	1-1/4" diameter x 300'					3.44			3.44	3.78
7540	1-1/2" diameter x 100'					4.71			4.71	5.20
7550	1-1/2" diameter x 300'					4.72			4.72	5.20
7596	Most sizes available in red or blue									
7700	Non-barrier type, hot/cold tubing straight lengths									
7710	1/2" diameter x 20'				L.F.	.68			.68	.75
7750	3/4" diameter x 20'					1.20			1.20	1.32
7760	1" diameter x 20'					2.04			2.04	2.24
7770	1-1/4" diameter x 20'					3.65			3.65	4.02
7780	1-1/2" diameter x 20'					4.73			4.73	5.20
7790	2" diameter					9.30			9.30	10.25
7796	Most sizes available in red or blue									
9000	Polypropylene pipe									
9002	For fusion weld fittings and accessories see line 22 11 13.76 9400									
9004	Note: sizes 1/2" thru 4" use socket fusion									
9005	Sizes 6" thru 10" use butt fusion									

22 11 13 – Facility Water Distribution Piping

22 11 13.74 Pipe, Plastic

	22 11 13.74 Pipe, Plastic	Crew	Daily Output	Labor-Hours	Unit	2020 Bare Costs Material	Labor	Equipment	Total	Total Incl O&P
9010	SDR 7.4 (domestic hot water piping)									
9011	Enhanced to minimize thermal expansion and high temperature life									
9016	13′ lengths, size is ID, includes joints 13′ OC and hangers 3 per 10′									
9020	3/8″ diameter	1 Plum	53	.151	L.F.	3.20	9.75		12.95	18.05
9022	1/2″ diameter		52	.154		3.67	9.90		13.57	18.90
9024	3/4″ diameter		50	.160		4.25	10.30		14.55	20
9026	1″ diameter		45	.178		5.40	11.45		16.85	23
9028	1-1/4″ diameter		40	.200		7.90	12.90		20.80	28
9030	1-1/2″ diameter		35	.229		10.45	14.75		25.20	33.50
9032	2″ diameter	Q-1	58	.276		14.50	16		30.50	40
9034	2-1/2″ diameter		55	.291		19.55	16.85		36.40	47
9036	3″ diameter		52	.308		25.50	17.85		43.35	54.50
9038	3-1/2″ diameter		49	.327		35	18.95		53.95	66.50
9040	4″ diameter		46	.348		37	20		57	71
9042	6″ diameter		39	.410		50	24		74	90.50
9044	8″ diameter	Q-2	48	.500		75	30		105	128
9046	10″ diameter	″	43	.558		118	33.50		151.50	180
9050	To delete joint & hangers, subtract									
9052	3/8″ diam. to 1″ diam.					45%	65%			
9054	1-1/4″ diam. to 4″ diam.					15%	45%			
9056	6″ diam. to 10″ diam.					5%	24%			
9060	SDR 11 (domestic cold water piping)									
9062	13′ lengths, size is ID, includes joints 13′ OC and hangers 3 per 10′									
9064	1/2″ diameter	1 Plum	57	.140	L.F.	3.33	9.05		12.38	17.20
9066	3/4″ diameter		54	.148		3.70	9.55		13.25	18.35
9068	1″ diameter		49	.163		4.43	10.50		14.93	20.50
9070	1-1/4″ diameter		45	.178		6.45	11.45		17.90	24.50
9072	1-1/2″ diameter		40	.200		7.95	12.90		20.85	28
9074	2″ diameter	Q-1	62	.258		10.95	14.95		25.90	34.50
9076	2-1/2″ diameter		59	.271		14.05	15.75		29.80	39
9078	3″ diameter		56	.286		18.65	16.55		35.20	45.50
9080	3-1/2″ diameter		53	.302		26.50	17.50		44	55
9082	4″ diameter		50	.320		29.50	18.55		48.05	60.50
9084	6″ diameter		47	.340		34.50	19.75		54.25	67
9086	8″ diameter	Q-2	51	.471		52.50	28.50		81	100
9088	10″ diameter	″	46	.522		78.50	31.50		110	134
9090	To delete joint & hangers, subtract									
9092	1/2″ diam. to 1″ diam.					45%	65%			
9094	1-1/4″ diam. to 4″ diam.					15%	45%			
9096	6″ diam. to 10″ diam.					5%	24%			

22 11 13.76 Pipe Fittings, Plastic

	22 11 13.76 Pipe Fittings, Plastic	Crew	Daily Output	Labor-Hours	Unit	Material	Labor	Equipment	Total	Total Incl O&P
0010	**PIPE FITTINGS, PLASTIC**									
0030	Epoxy resin, fiberglass reinforced, general service									
0100	3″	Q-1	20.80	.769	Ea.	118	44.50		162.50	196
0110	4″		16.50	.970		121	56		177	217
0120	6″		10.10	1.584		234	92		326	395
0130	8″	Q-2	9.30	2.581		430	155		585	705
0140	10″		8.50	2.824		540	170		710	850
0150	12″		7.60	3.158		775	190		965	1,150
0170	Elbow, 90°, flanged									
0172	2″	Q-1	23	.696	Ea.	253	40.50		293.50	340
0173	3″		16	1		290	58		348	405

22 11 13 – Facility Water Distribution Piping

22 11 13.76 Pipe Fittings, Plastic		Crew	Daily Output	Labor-Hours	Unit	Material	Labor	Equipment	Total	Total Incl O&P
						2020 Bare Costs				
0174	4"	Q-1	13	1.231	Ea.	380	71.50		451.50	520
0176	6"		8	2		685	116		801	930
0177	8"	Q-2	9	2.667		1,250	160		1,410	1,625
0178	10"		7	3.429		1,700	206		1,906	2,175
0179	12"		5	4.800		2,275	289		2,564	2,925
0186	Elbow, 45°, flanged									
0188	2"	Q-1	23	.696	Ea.	253	40.50		293.50	340
0189	3"		16	1		290	58		348	405
0190	4"		13	1.231		380	71.50		451.50	520
0192	6"		8	2		685	116		801	930
0193	8"	Q-2	9	2.667		1,250	160		1,410	1,625
0194	10"		7	3.429		1,750	206		1,956	2,225
0195	12"		5	4.800		2,375	289		2,664	3,050
0352	Tee, flanged									
0354	2"	Q-1	17	.941	Ea.	345	54.50		399.50	455
0355	3"		10	1.600		455	93		548	640
0356	4"		8	2		510	116		626	735
0358	6"		5	3.200		875	186		1,061	1,250
0359	8"	Q-2	6	4		1,650	241		1,891	2,175
0360	10"		5	4.800		2,400	289		2,689	3,075
0361	12"		4	6		3,325	360		3,685	4,200
0365	Wye, flanged									
0367	2"	Q-1	17	.941	Ea.	680	54.50		734.50	825
0368	3"		10	1.600		935	93		1,028	1,175
0369	4"		8	2		1,250	116		1,366	1,550
0371	6"		5	3.200		1,775	186		1,961	2,225
0372	8"	Q-2	6	4		2,875	241		3,116	3,525
0373	10"		5	4.800		4,475	289		4,764	5,350
0374	12"		4	6		6,500	360		6,860	7,700
0380	Couplings									
0410	2"	Q-1	33.10	.483	Ea.	27	28		55	71.50
0420	3"		20.80	.769		31.50	44.50		76	102
0430	4"		16.50	.970		43.50	56		99.50	132
0440	6"		10.10	1.584		101	92		193	249
0450	8"	Q-2	9.30	2.581		171	155		326	420
0460	10"		8.50	2.824		233	170		403	510
0470	12"		7.60	3.158		335	190		525	655
0473	High corrosion resistant couplings, add					30%				
0474	Reducer, concentric, flanged									
0475	2" x 1-1/2"	Q-1	30	.533	Ea.	660	31		691	770
0476	3" x 2"		24	.667		715	38.50		753.50	850
0477	4" x 3"		19	.842		805	49		854	960
0479	6" x 4"		15	1.067		805	62		867	980
0480	8" x 6"	Q-2	16	1.500		1,325	90		1,415	1,575
0481	10" x 8"		13	1.846		1,225	111		1,336	1,525
0482	12" x 10"		11	2.182		1,900	131		2,031	2,300
0486	Adapter, bell x male or female									
0488	2"	Q-1	28	.571	Ea.	33.50	33		66.50	86
0489	3"		20	.800		50.50	46.50		97	125
0491	4"		17	.941		59	54.50		113.50	147
0492	6"		12	1.333		138	77.50		215.50	268
0493	8"	Q-2	15	1.600		198	96		294	360
0494	10"	"	11	2.182		285	131		416	510

22 11 13.76 Pipe Fittings, Plastic		Crew	Daily Output	Labor-Hours	Unit	2020 Bare Costs Material	Labor	Equipment	Total	Total Incl O&P
0528	Flange									
0532	2"	Q-1	46	.348	Ea.	43.50	20		63.50	78
0533	3"		32	.500		61	29		90	111
0534	4"		26	.615		77.50	35.50		113	139
0536	6"		16	1		147	58		205	249
0537	8"	Q-2	18	1.333		225	80		305	370
0538	10"		14	1.714		320	103		423	505
0539	12"		10	2.400		400	144		544	655
2100	PVC schedule 80, socket joint									
2110	90° elbow, 1/2"	1 Plum	30.30	.264	Ea.	2.67	17		19.67	28.50
2130	3/4"		26	.308		3.42	19.85		23.27	33.50
2140	1"		22.70	.352		5.55	22.50		28.05	40
2150	1-1/4"		20.20	.396		7.35	25.50		32.85	46
2160	1-1/2"		18.20	.440		7.90	28.50		36.40	51
2170	2"	Q-1	33.10	.483		9.55	28		37.55	52.50
2180	3"		20.80	.769		25	44.50		69.50	94.50
2190	4"		16.50	.970		38	56		94	126
2200	6"		10.10	1.584		109	92		201	257
2210	8"	Q-2	9.30	2.581		299	155		454	560
2250	45° elbow, 1/2"	1 Plum	30.30	.264		5.05	17		22.05	31
2270	3/4"		26	.308		7.70	19.85		27.55	38
2280	1"		22.70	.352		11.55	22.50		34.05	46.50
2290	1-1/4"		20.20	.396		14.60	25.50		40.10	54
2300	1-1/2"		18.20	.440		17.35	28.50		45.85	61.50
2310	2"	Q-1	33.10	.483		22.50	28		50.50	67
2320	3"		20.80	.769		57	44.50		101.50	130
2330	4"		16.50	.970		104	56		160	198
2340	6"		10.10	1.584		130	92		222	282
2350	8"	Q-2	9.30	2.581		283	155		438	540
2400	Tee, 1/2"	1 Plum	20.20	.396		7.55	25.50		33.05	46.50
2420	3/4"		17.30	.462		7.90	30		37.90	53
2430	1"		15.20	.526		9.90	34		43.90	62
2440	1-1/4"		13.50	.593		27	38		65	86.50
2450	1-1/2"		12.10	.661		27	42.50		69.50	94
2460	2"	Q-1	20	.800		34	46.50		80.50	107
2470	3"		13.90	1.151		46	67		113	151
2480	4"		11	1.455		53.50	84.50		138	185
2490	6"		6.70	2.388		183	139		322	410
2500	8"	Q-2	6.20	3.871		420	233		653	810
2510	Flange, socket, 150 lb., 1/2"	1 Plum	55.60	.144		14.55	9.25		23.80	30
2514	3/4"		47.60	.168		15.55	10.85		26.40	33.50
2518	1"		41.70	.192		17.35	12.35		29.70	37.50
2522	1-1/2"		33.30	.240		18.30	15.50		33.80	43
2526	2"	Q-1	60.60	.264		24.50	15.30		39.80	49.50
2530	4"		30.30	.528		52.50	30.50		83	104
2534	6"		18.50	.865		82.50	50		132.50	166
2538	8"	Q-2	17.10	1.404		147	84.50		231.50	288
2550	Coupling, 1/2"	1 Plum	30.30	.264		5.10	17		22.10	31
2570	3/4"		26	.308		7.55	19.85		27.40	38
2580	1"		22.70	.352		7.45	22.50		29.95	42
2590	1-1/4"		20.20	.396		10.15	25.50		35.65	49
2600	1-1/2"		18.20	.440		11.40	28.50		39.90	55
2610	2"	Q-1	33.10	.483		12.25	28		40.25	55.50

22 11 Facility Water Distribution

22 11 13 – Facility Water Distribution Piping

22 11 13.76 Pipe Fittings, Plastic		Crew	Daily Output	Labor-Hours	Unit	2020 Bare Costs Material	Labor	Equipment	Total	Total Incl O&P
2620	3″	Q-1	20.80	.769	Ea.	33.50	44.50		78	104
2630	4″		16.50	.970		43.50	56		99.50	132
2640	6″		10.10	1.584		93	92		185	241
2650	8″	Q-2	9.30	2.581		122	155		277	365
2660	10″		8.50	2.824		420	170		590	715
2670	12″		7.60	3.158		485	190		675	815
2700	PVC (white), schedule 40, socket joints									
2760	90° elbow, 1/2″	1 Plum	33.30	.240	Ea.	.52	15.50		16.02	23.50
2770	3/4″		28.60	.280		.59	18.05		18.64	27.50
2780	1″		25	.320		1.05	20.50		21.55	32
2790	1-1/4″		22.20	.360		1.83	23		24.83	37
2800	1-1/2″		20	.400		1.99	26		27.99	40.50
2810	2″	Q-1	36.40	.440		3.11	25.50		28.61	41.50
2820	2-1/2″		26.70	.599		9.60	35		44.60	62.50
2830	3″		22.90	.699		11.35	40.50		51.85	73
2840	4″		18.20	.879		20.50	51		71.50	99
2850	5″		12.10	1.322		52	76.50		128.50	173
2860	6″		11.10	1.441		64.50	83.50		148	196
2870	8″	Q-2	10.30	2.330		166	140		306	395
2980	45° elbow, 1/2″	1 Plum	33.30	.240		.86	15.50		16.36	24
2990	3/4″		28.60	.280		1.34	18.05		19.39	28.50
3000	1″		25	.320		1.60	20.50		22.10	33
3010	1-1/4″		22.20	.360		2.23	23		25.23	37.50
3020	1-1/2″		20	.400		2.80	26		28.80	41.50
3030	2″	Q-1	36.40	.440		3.65	25.50		29.15	42
3040	2-1/2″		26.70	.599		9.50	35		44.50	62.50
3050	3″		22.90	.699		14.65	40.50		55.15	76.50
3060	4″		18.20	.879		26.50	51		77.50	106
3070	5″		12.10	1.322		52	76.50		128.50	173
3080	6″		11.10	1.441		65.50	83.50		149	197
3090	8″	Q-2	10.30	2.330		157	140		297	385
3180	Tee, 1/2″	1 Plum	22.20	.360		.65	23		23.65	35.50
3190	3/4″		19	.421		.75	27		27.75	41.50
3200	1″		16.70	.479		1.39	31		32.39	47.50
3210	1-1/4″		14.80	.541		2.16	35		37.16	54.50
3220	1-1/2″		13.30	.602		2.64	39		41.64	61
3230	2″	Q-1	24.20	.661		3.84	38.50		42.34	61.50
3240	2-1/2″		17.80	.899		12.60	52		64.60	92
3250	3″		15.20	1.053		17.15	61		78.15	110
3260	4″		12.10	1.322		30	76.50		106.50	148
3270	5″		8.10	1.975		72.50	115		187.50	251
3280	6″		7.40	2.162		101	125		226	300
3290	8″	Q-2	6.80	3.529		235	212		447	580
3380	Coupling, 1/2″	1 Plum	33.30	.240		.34	15.50		15.84	23.50
3390	3/4″		28.60	.280		.47	18.05		18.52	27.50
3400	1″		25	.320		.83	20.50		21.33	32
3410	1-1/4″		22.20	.360		1.13	23		24.13	36
3420	1-1/2″		20	.400		1.22	26		27.22	40
3430	2″	Q-1	36.40	.440		1.86	25.50		27.36	40
3440	2-1/2″		26.70	.599		4.11	35		39.11	56.50
3450	3″		22.90	.699		7.70	40.50		48.20	69
3460	4″		18.20	.879		9.60	51		60.60	87
3470	5″		12.10	1.322		16.95	76.50		93.45	134

22 11 13.76 Pipe Fittings, Plastic		Crew	Daily Output	Labor-Hours	Unit	Material	2020 Bare Costs Labor	Equipment	Total	Total Incl O&P
3480	6″	Q-1	11.10	1.441	Ea.	30.50	83.50		114	159
3490	8″	Q-2	10.30	2.330		54.50	140		194.50	270
3600	Cap, schedule 40, PVC socket 1/2″	1 Plum	60.60	.132		.47	8.50		8.97	13.25
3610	3/4″		51.90	.154		.55	9.95		10.50	15.45
3620	1″		45.50	.176		.86	11.35		12.21	17.90
3630	1-1/4″		40.40	.198		1.21	12.75		13.96	20.50
3640	1-1/2″		36.40	.220		1.34	14.15		15.49	22.50
3650	2″	Q-1	66.10	.242		1.60	14.05		15.65	23
3660	2-1/2″		48.50	.330		5.05	19.15		24.20	34
3670	3″		41.60	.385		5.55	22.50		28.05	39.50
3680	4″		33.10	.483		12.65	28		40.65	56
3690	6″		20.20	.792		30.50	46		76.50	103
3700	8″	Q-2	18.60	1.290		76.50	77.50		154	200
3710	Reducing insert, schedule 40, socket weld									
3712	3/4″	1 Plum	31.50	.254	Ea.	.53	16.35		16.88	25
3713	1″		27.50	.291		.97	18.75		19.72	29
3715	1-1/2″		22	.364		1.37	23.50		24.87	36.50
3716	2″	Q-1	40	.400		2.27	23		25.27	37
3717	4″		20	.800		12	46.50		58.50	82.50
3718	6″		12.20	1.311		30	76		106	147
3719	8″	Q-2	11.30	2.124		110	128		238	310
3730	Reducing insert, socket weld x female/male thread									
3732	1/2″	1 Plum	38.30	.209	Ea.	2.37	13.45		15.82	22.50
3733	3/4″		32.90	.243		1.47	15.65		17.12	25
3734	1″		28.80	.278		2.06	17.90		19.96	29.50
3736	1-1/2″		23	.348		3.70	22.50		26.20	37.50
3737	2″	Q-1	41.90	.382		3.96	22		25.96	37.50
3738	4″	″	20.90	.766		39	44.50		83.50	110
3742	Male adapter, socket weld x male thread									
3744	1/2″	1 Plum	38.30	.209	Ea.	.45	13.45		13.90	20.50
3745	3/4″		32.90	.243		.51	15.65		16.16	24
3746	1″		28.80	.278		.91	17.90		18.81	28
3748	1-1/2″		23	.348		1.58	22.50		24.08	35
3749	2″	Q-1	41.90	.382		1.94	22		23.94	35
3750	4″	″	20.90	.766		10.75	44.50		55.25	78.50
3754	Female adapter, socket weld x female thread									
3756	1/2″	1 Plum	38.30	.209	Ea.	.57	13.45		14.02	20.50
3757	3/4″		32.90	.243		.73	15.65		16.38	24.50
3758	1″		28.80	.278		.84	17.90		18.74	28
3760	1-1/2″		23	.348		1.49	22.50		23.99	35
3761	2″	Q-1	41.90	.382		2	22		24	35
3762	4″	″	20.90	.766		11.25	44.50		55.75	79
3800	PVC, schedule 80, socket joints									
3810	Reducing insert									
3812	3/4″	1 Plum	28.60	.280	Ea.	1.47	18.05		19.52	28.50
3813	1″		25	.320		4.21	20.50		24.71	35.50
3815	1-1/2″		20	.400		9.85	26		35.85	49.50
3816	2″	Q-1	36.40	.440		14	25.50		39.50	53.50
3817	4″		18.20	.879		53.50	51		104.50	135
3818	6″		11.10	1.441		64	83.50		147.50	196
3819	8″	Q-2	10.20	2.353		430	142		572	680
3830	Reducing insert, socket weld x female/male thread									
3832	1/2″	1 Plum	34.80	.230	Ea.	10.35	14.80		25.15	33.50

22 11 13 – Facility Water Distribution Piping

22 11 13.76 Pipe Fittings, Plastic		Crew	Daily Output	Labor-Hours	Unit	Material	Labor	Equipment	Total	Total Incl O&P
						2020 Bare Costs				
3833	3/4"	1 Plum	29.90	.268	Ea.	6.15	17.25		23.40	33
3834	1"		26.10	.307		9.65	19.75		29.40	40
3836	1-1/2"		20.90	.383		12.15	24.50		36.65	50.50
3837	2"	Q-1	38	.421		17.75	24.50		42.25	56
3838	4"	"	19	.842		86	49		135	168
3844	Adapter, male socket x male thread									
3846	1/2"	1 Plum	34.80	.230	Ea.	3.89	14.80		18.69	26.50
3847	3/4"		29.90	.268		4.29	17.25		21.54	30.50
3848	1"		26.10	.307		7.40	19.75		27.15	37.50
3850	1-1/2"		20.90	.383		12.45	24.50		36.95	50.50
3851	2"	Q-1	38	.421		18.05	24.50		42.55	56.50
3852	4"	"	19	.842		40.50	49		89.50	118
3860	Adapter, female socket x female thread									
3862	1/2"	1 Plum	34.80	.230	Ea.	4.66	14.80		19.46	27
3863	3/4"		29.90	.268		6.95	17.25		24.20	33.50
3864	1"		26.10	.307		10.20	19.75		29.95	41
3866	1-1/2"		20.90	.383		20.50	24.50		45	59.50
3867	2"	Q-1	38	.421		35.50	24.50		60	75.50
3868	4"	"	19	.842		108	49		157	192
3872	Union, socket joints									
3874	1/2"	1 Plum	25.80	.310	Ea.	8.80	20		28.80	39.50
3875	3/4"		22.10	.362		11.15	23.50		34.65	47.50
3876	1"		19.30	.415		12.70	26.50		39.20	54
3878	1-1/2"		15.50	.516		28.50	33.50		62	81.50
3879	2"	Q-1	28.10	.569		39	33		72	92
3888	Cap									
3890	1/2"	1 Plum	54.50	.147	Ea.	5.40	9.45		14.85	20
3891	3/4"		46.70	.171		4.34	11.05		15.39	21.50
3892	1"		41	.195		8.40	12.60		21	28
3894	1-1/2"		32.80	.244		12.45	15.70		28.15	37
3895	2"	Q-1	59.50	.269		29	15.60		44.60	55.50
3896	4"		30	.533		75.50	31		106.50	130
3897	6"		18.20	.879		218	51		269	315
3898	8"	Q-2	16.70	1.437		280	86.50		366.50	440
4500	DWV, ABS, non pressure, socket joints									
4540	1/4 bend, 1-1/4"	1 Plum	20.20	.396	Ea.	4.48	25.50		29.98	43
4560	1-1/2"	"	18.20	.440		3.44	28.50		31.94	46.50
4570	2"	Q-1	33.10	.483		5.30	28		33.30	48
4580	3"		20.80	.769		13.45	44.50		57.95	82
4590	4"		16.50	.970		27.50	56		83.50	115
4600	6"		10.10	1.584		118	92		210	267
4650	1/8 bend, same as 1/4 bend									
4800	Tee, sanitary									
4820	1-1/4"	1 Plum	13.50	.593	Ea.	5.95	38		43.95	63.50
4830	1-1/2"	"	12.10	.661		5.15	42.50		47.65	69.50
4840	2"	Q-1	20	.800		7.90	46.50		54.40	78
4850	3"		13.90	1.151		21.50	67		88.50	124
4860	4"		11	1.455		38.50	84.50		123	169
4862	Tee, sanitary, reducing, 2" x 1-1/2"		22	.727		7	42		49	70.50
4864	3" x 2"		15.30	1.046		13.40	60.50		73.90	106
4868	4" x 3"		12.10	1.322		37.50	76.50		114	157
4870	Combination Y and 1/8 bend									
4872	1-1/2"	1 Plum	12.10	.661	Ea.	12.25	42.50		54.75	77.50

22 11 13.76 Pipe Fittings, Plastic		Crew	Daily Output	Labor-Hours	Unit	2020 Bare Costs Material	Labor	Equipment	Total	Total Incl O&P
4874	2"	Q-1	20	.800	Ea.	13.70	46.50		60.20	84.50
4876	3"		13.90	1.151		32	67		99	136
4878	4"		11	1.455		62	84.50		146.50	194
4880	3" x 1-1/2"		15.50	1.032		32.50	60		92.50	126
4882	4" x 3"		12.10	1.322		49.50	76.50		126	170
4900	Wye, 1-1/4"	1 Plum	13.50	.593		6.85	38		44.85	64.50
4902	1-1/2"	"	12.10	.661		7.85	42.50		50.35	72.50
4904	2"	Q-1	20	.800		10.20	46.50		56.70	80.50
4906	3"		13.90	1.151		24	67		91	127
4908	4"		11	1.455		49	84.50		133.50	180
4910	6"		6.70	2.388		146	139		285	370
4918	3" x 1-1/2"		15.50	1.032		19.40	60		79.40	111
4920	4" x 3"		12.10	1.322		38.50	76.50		115	158
4922	6" x 4"		6.90	2.319		116	134		250	330
4930	Double wye, 1-1/2"	1 Plum	9.10	.879		25	56.50		81.50	113
4932	2"	Q-1	16.60	.964		30	56		86	117
4934	3"		10.40	1.538		70.50	89		159.50	212
4936	4"		8.25	1.939		138	112		250	320
4940	2" x 1-1/2"		16.80	.952		27.50	55		82.50	113
4942	3" x 2"		10.60	1.509		49	87.50		136.50	185
4944	4" x 3"		8.45	1.893		109	110		219	284
4946	6" x 4"		7.25	2.207		156	128		284	365
4950	Reducer bushing, 2" x 1-1/2"		36.40	.440		2.72	25.50		28.22	41
4952	3" x 1-1/2"		27.30	.586		11.95	34		45.95	64
4954	4" x 2"		18.20	.879		22.50	51		73.50	101
4956	6" x 4"		11.10	1.441		62.50	83.50		146	194
4960	Couplings, 1-1/2"	1 Plum	18.20	.440		1.66	28.50		30.16	44.50
4962	2"	Q-1	33.10	.483		2.22	28		30.22	44.50
4963	3"		20.80	.769		6.30	44.50		50.80	74
4964	4"		16.50	.970		11.35	56		67.35	96.50
4966	6"		10.10	1.584		47.50	92		139.50	190
4970	2" x 1-1/2"		33.30	.480		4.80	28		32.80	47
4972	3" x 1-1/2"		21	.762		13.85	44		57.85	81.50
4974	4" x 3"		16.70	.958		21	55.50		76.50	107
4978	Closet flange, 4"	1 Plum	32	.250		11.60	16.10		27.70	37
4980	4" x 3"	"	34	.235		14.10	15.15		29.25	38
5000	DWV, PVC, schedule 40, socket joints									
5040	1/4 bend, 1-1/4"	1 Plum	20.20	.396	Ea.	8.80	25.50		34.30	47.50
5060	1-1/2"	"	18.20	.440		2.48	28.50		30.98	45
5070	2"	Q-1	33.10	.483		3.91	28		31.91	46.50
5080	3"		20.80	.769		11.65	44.50		56.15	80
5090	4"		16.50	.970		23	56		79	109
5100	6"		10.10	1.584		79.50	92		171.50	226
5105	8"	Q-2	9.30	2.581		122	155		277	365
5106	10"	"	8.50	2.824		365	170		535	655
5110	1/4 bend, long sweep, 1-1/2"	1 Plum	18.20	.440		5.85	28.50		34.35	49
5112	2"	Q-1	33.10	.483		6.50	28		34.50	49
5114	3"		20.80	.769		15.05	44.50		59.55	83.50
5116	4"		16.50	.970		28.50	56		84.50	116
5150	1/8 bend, 1-1/4"	1 Plum	20.20	.396		6.05	25.50		31.55	44.50
5170	1-1/2"	"	18.20	.440		2.46	28.50		30.96	45
5180	2"	Q-1	33.10	.483		3.64	28		31.64	46
5190	3"		20.80	.769		10.45	44.50		54.95	78.50

22 11 13.76	Pipe Fittings, Plastic	Crew	Daily Output	Labor-Hours	Unit	Material	2020 Bare Costs Labor	Equipment	Total	Total Incl O&P
5200	4"	Q-1	16.50	.970	Ea.	18.95	56		74.95	105
5210	6"		10.10	1.584		71.50	92		163.50	217
5215	8"	Q-2	9.30	2.581		117	155		272	360
5216	10"		8.50	2.824		238	170		408	515
5217	12"		7.60	3.158		294	190		484	610
5250	Tee, sanitary 1-1/4"	1 Plum	13.50	.593		9.45	38		47.45	67.50
5254	1-1/2"	"	12.10	.661		4.41	42.50		46.91	69
5255	2"	Q-1	20	.800		6.50	46.50		53	76.50
5256	3"		13.90	1.151		17.05	67		84.05	119
5257	4"		11	1.455		31	84.50		115.50	161
5259	6"		6.70	2.388		126	139		265	345
5261	8"	Q-2	6.20	3.871		293	233		526	675
5276	Tee, sanitary, reducing									
5281	2" x 1-1/2" x 1-1/2"	Q-1	23	.696	Ea.	5.70	40.50		46.20	67
5282	2" x 1-1/2" x 2"		22	.727		6.95	42		48.95	70.50
5283	2" x 2" x 1-1/2"		22	.727		5.75	42		47.75	69.50
5284	3" x 3" x 1-1/2"		15.50	1.032		12.45	60		72.45	103
5285	3" x 3" x 2"		15.30	1.046		12.85	60.50		73.35	105
5286	4" x 4" x 1-1/2"		12.30	1.301		33	75.50		108.50	149
5287	4" x 4" x 2"		12.20	1.311		27	76		103	144
5288	4" x 4" x 3"		12.10	1.322		36.50	76.50		113	156
5291	6" x 6" x 4"		6.90	2.319		122	134		256	335
5294	Tee, double sanitary									
5295	1-1/2"	1 Plum	9.10	.879	Ea.	9.80	56.50		66.30	96
5296	2"	Q-1	16.60	.964		13.20	56		69.20	98
5297	3"		10.40	1.538		37	89		126	175
5298	4"		8.25	1.939		59.50	112		171.50	234
5303	Wye, reducing									
5304	2" x 1-1/2" x 1-1/2"	Q-1	23	.696	Ea.	11.15	40.50		51.65	73
5305	2" x 2" x 1-1/2"		22	.727		9.75	42		51.75	73.50
5306	3" x 3" x 2"		15.30	1.046		31.50	60.50		92	126
5307	4" x 4" x 2"		12.20	1.311		23.50	76		99.50	140
5309	4" x 4" x 3"		12.10	1.322		31.50	76.50		108	150
5314	Combination Y & 1/8 bend, 1-1/2"	1 Plum	12.10	.661		10.75	42.50		53.25	76
5315	2"	Q-1	20	.800		12.90	46.50		59.40	83.50
5317	3"		13.90	1.151		29	67		96	132
5318	4"		11	1.455		58	84.50		142.50	190
5319	6"		6.70	2.388		189	139		328	415
5320	8"	Q-2	6.20	3.871		287	233		520	665
5321	10"		5.70	4.211		975	253		1,228	1,450
5322	12"		5.10	4.706		1,225	283		1,508	1,775
5324	Combination Y & 1/8 bend, reducing									
5325	2" x 2" x 1-1/2"	Q-1	22	.727	Ea.	15.15	42		57.15	79.50
5327	3" x 3" x 1-1/2"		15.50	1.032		27	60		87	119
5328	3" x 3" x 2"		15.30	1.046		19.95	60.50		80.45	113
5329	4" x 4" x 2"		12.20	1.311		30.50	76		106.50	148
5331	Wye, 1-1/4"	1 Plum	13.50	.593		12.10	38		50.10	70.50
5332	1-1/2"	"	12.10	.661		8.20	42.50		50.70	73
5333	2"	Q-1	20	.800		7.90	46.50		54.40	78
5334	3"		13.90	1.151		21.50	67		88.50	124
5335	4"		11	1.455		39	84.50		123.50	169
5336	6"		6.70	2.388		114	139		253	335
5337	8"	Q-2	6.20	3.871		217	233		450	590

	22 11 13.76 Pipe Fittings, Plastic	Crew	Daily Output	Labor-Hours	Unit	Material	2020 Bare Costs Labor	Equipment	Total	Total Incl O&P
5338	10"	Q-2	5.70	4.211	Ea.	475	253		728	905
5339	12"		5.10	4.706		770	283		1,053	1,275
5341	2" x 1-1/2"	Q-1	22	.727		9.85	42		51.85	74
5342	3" x 1-1/2"		15.50	1.032		14.60	60		74.60	106
5343	4" x 3"		12.10	1.322		32	76.50		108.50	150
5344	6" x 4"		6.90	2.319		87	134		221	297
5345	8" x 6"	Q-2	6.40	3.750		188	226		414	545
5347	Double wye, 1-1/2"	1 Plum	9.10	.879		18.25	56.50		74.75	105
5348	2"	Q-1	16.60	.964		20.50	56		76.50	106
5349	3"		10.40	1.538		42.50	89		131.50	181
5350	4"		8.25	1.939		85.50	112		197.50	263
5353	Double wye, reducing									
5354	2" x 2" x 1-1/2" x 1-1/2"	Q-1	16.80	.952	Ea.	18.60	55		73.60	103
5355	3" x 3" x 2" x 2"		10.60	1.509		31.50	87.50		119	166
5356	4" x 4" x 3" x 3"		8.45	1.893		68	110		178	239
5357	6" x 6" x 4" x 4"		7.25	2.207		239	128		367	455
5374	Coupling, 1-1/4"	1 Plum	20.20	.396		5.70	25.50		31.20	44.50
5376	1-1/2"	"	18.20	.440		1.19	28.50		29.69	44
5378	2"	Q-1	33.10	.483		1.62	28		29.62	44
5380	3"		20.80	.769		5.65	44.50		50.15	73
5390	4"		16.50	.970		9.65	56		65.65	94.50
5400	6"		10.10	1.584		31.50	92		123.50	173
5402	8"	Q-2	9.30	2.581		57	155		212	295
5404	2" x 1-1/2"	Q-1	33.30	.480		3.63	28		31.63	45.50
5406	3" x 1-1/2"		21	.762		10.65	44		54.65	78
5408	4" x 3"		16.70	.958		17.35	55.50		72.85	102
5410	Reducer bushing, 2" x 1-1/4"		36.50	.438		3.38	25.50		28.88	41.50
5411	2" x 1-1/2"		36.40	.440		2.07	25.50		27.57	40.50
5412	3" x 1-1/2"		27.30	.586		10.10	34		44.10	62
5413	3" x 2"		27.10	.590		5.25	34		39.25	57.50
5414	4" x 2"		18.20	.879		17.35	51		68.35	95.50
5415	4" x 3"		16.70	.958		8.90	55.50		64.40	93
5416	6" x 4"		11.10	1.441		45	83.50		128.50	175
5418	8" x 6"	Q-2	10.20	2.353		89	142		231	310
5425	Closet flange 4"	Q-1	32	.500		13.70	29		42.70	58.50
5426	4" x 3"	"	34	.471		13.50	27.50		41	56
5450	Solvent cement for PVC, industrial grade, per quart				Qt.	31			31	34
5500	CPVC, Schedule 80, threaded joints									
5540	90° elbow, 1/4"	1 Plum	32	.250	Ea.	13	16.10		29.10	38.50
5560	1/2"		30.30	.264		7.55	17		24.55	34
5570	3/4"		26	.308		11.30	19.85		31.15	42
5580	1"		22.70	.352		15.85	22.50		38.35	51.50
5590	1-1/4"		20.20	.396		30.50	25.50		56	71.50
5600	1-1/2"		18.20	.440		33	28.50		61.50	78.50
5610	2"	Q-1	33.10	.483		44	28		72	90.50
5620	2-1/2"		24.20	.661		137	38.50		175.50	208
5630	3"		20.80	.769		147	44.50		191.50	228
5640	4"		16.50	.970		230	56		286	335
5650	6"		10.10	1.584		267	92		359	430
5660	45° elbow same as 90° elbow									
5700	Tee, 1/4"	1 Plum	22	.364	Ea.	25.50	23.50		49	63
5702	1/2"		20.20	.396		25.50	25.50		51	66
5704	3/4"		17.30	.462		36.50	30		66.50	84.50

22 11 13.76 Pipe Fittings, Plastic		Crew	Daily Output	Labor-Hours	Unit	2020 Bare Costs Material	Labor	Equipment	Total	Total Incl O&P
5706	1"	1 Plum	15.20	.526	Ea.	39.50	34		73.50	94
5708	1-1/4"		13.50	.593		39.50	38		77.50	101
5710	1-1/2"		12.10	.661		41.50	42.50		84	110
5712	2"	Q-1	20	.800		46	46.50		92.50	120
5714	2-1/2"		16.20	.988		225	57.50		282.50	335
5716	3"		13.90	1.151		263	67		330	390
5718	4"		11	1.455		625	84.50		709.50	810
5720	6"		6.70	2.388		710	139		849	985
5730	Coupling, 1/4"	1 Plum	32	.250		16.65	16.10		32.75	42.50
5732	1/2"		30.30	.264		13.75	17		30.75	40.50
5734	3/4"		26	.308		22	19.85		41.85	54
5736	1"		22.70	.352		25	22.50		47.50	61.50
5738	1-1/4"		20.20	.396		26.50	25.50		52	67
5740	1-1/2"		18.20	.440		28.50	28.50		57	74
5742	2"	Q-1	33.10	.483		33.50	28		61.50	79
5744	2-1/2"		24.20	.661		60.50	38.50		99	124
5746	3"		20.80	.769		70	44.50		114.50	145
5748	4"		16.50	.970		143	56		199	241
5750	6"		10.10	1.584		196	92		288	355
5752	8"	Q-2	9.30	2.581		410	155		565	680
5900	CPVC, Schedule 80, socket joints									
5904	90° elbow, 1/4"	1 Plum	32	.250	Ea.	12.45	16.10		28.55	37.50
5906	1/2"		30.30	.264		4.88	17		21.88	31
5908	3/4"		26	.308		6.25	19.85		26.10	36.50
5910	1"		22.70	.352		9.90	22.50		32.40	45
5912	1-1/4"		20.20	.396		21.50	25.50		47	61.50
5914	1-1/2"		18.20	.440		24	28.50		52.50	68.50
5916	2"	Q-1	33.10	.483		29	28		57	73.50
5918	2-1/2"		24.20	.661		66.50	38.50		105	131
5920	3"		20.80	.769		75	44.50		119.50	150
5922	4"		16.50	.970		135	56		191	233
5924	6"		10.10	1.584		272	92		364	435
5926	8"		9.30	1.720		665	100		765	880
5930	45° elbow, 1/4"	1 Plum	32	.250		18.50	16.10		34.60	44.50
5932	1/2"		30.30	.264		5.95	17		22.95	32
5934	3/4"		26	.308		7.80	19.85		27.65	38
5936	1"		22.70	.352		13.70	22.50		36.20	49
5938	1-1/4"		20.20	.396		27	25.50		52.50	67.50
5940	1-1/2"		18.20	.440		27.50	28.50		56	73
5942	2"	Q-1	33.10	.483		27.50	28		55.50	72
5944	2-1/2"		24.20	.661		63.50	38.50		102	127
5946	3"		20.80	.769		81.50	44.50		126	157
5948	4"		16.50	.970		98.50	56		154.50	192
5950	6"		10.10	1.584		310	92		402	480
5952	8"		9.30	1.720		715	100		815	935
5960	Tee, 1/4"	1 Plum	22	.364		11.45	23.50		34.95	47.50
5962	1/2"		20.20	.396		11.45	25.50		36.95	50.50
5964	3/4"		17.30	.462		11.65	30		41.65	57.50
5966	1"		15.20	.526		14.25	34		48.25	66.50
5968	1-1/4"		13.50	.593		30	38		68	90
5970	1-1/2"		12.10	.661		34.50	42.50		77	102
5972	2"	Q-1	20	.800		34	46.50		80.50	107
5974	2-1/2"		16.20	.988		97.50	57.50		155	193

22 11 13.76 Pipe Fittings, Plastic		Crew	Daily Output	Labor-Hours	Unit	2020 Bare Costs				Total Incl O&P
						Material	Labor	Equipment	Total	
5976	3"	Q-1	13.90	1.151	Ea.	97.50	67		164.50	207
5978	4"		11	1.455		130	84.50		214.50	269
5980	6"		6.70	2.388		340	139		479	575
5982	8"	Q-2	6.20	3.871		965	233		1,198	1,400
5990	Coupling, 1/4"	1 Plum	32	.250		13.25	16.10		29.35	38.50
5992	1/2"		30.30	.264		5.15	17		22.15	31
5994	3/4"		26	.308		7.20	19.85		27.05	37.50
5996	1"		22.70	.352		9.70	22.50		32.20	44.50
5998	1-1/4"		20.20	.396		14.55	25.50		40.05	54
6000	1-1/2"		18.20	.440		18.25	28.50		46.75	62.50
6002	2"	Q-1	33.10	.483		21.50	28		49.50	65.50
6004	2-1/2"		24.20	.661		47.50	38.50		86	110
6006	3"		20.80	.769		51.50	44.50		96	124
6008	4"		16.50	.970		67.50	56		123.50	158
6010	6"		10.10	1.584		159	92		251	310
6012	8"	Q-2	9.30	2.581		430	155		585	700
6200	CTS, 100 psi at 180°F, hot and cold water									
6230	90° elbow, 1/2"	1 Plum	20	.400	Ea.	.32	26		26.32	39
6250	3/4"		19	.421		.52	27		27.52	41
6251	1"		16	.500		1.65	32		33.65	50
6252	1-1/4"		15	.533		3.04	34.50		37.54	55
6253	1-1/2"		14	.571		5.50	37		42.50	61
6254	2"	Q-1	23	.696		10.55	40.50		51.05	72
6260	45° elbow, 1/2"	1 Plum	20	.400		.40	26		26.40	39
6280	3/4"		19	.421		.69	27		27.69	41.50
6281	1"		16	.500		1.87	32		33.87	50
6282	1-1/4"		15	.533		3.75	34.50		38.25	55.50
6283	1-1/2"		14	.571		5.45	37		42.45	61
6284	2"	Q-1	23	.696		12.40	40.50		52.90	74
6290	Tee, 1/2"	1 Plum	13	.615		.40	39.50		39.90	60
6310	3/4"		12	.667		.75	43		43.75	65.50
6311	1"		11	.727		3.80	47		50.80	74
6312	1-1/4"		10	.800		5.85	51.50		57.35	83.50
6313	1-1/2"		10	.800		7.65	51.50		59.15	85.50
6314	2"	Q-1	17	.941		12.35	54.50		66.85	95
6320	Coupling, 1/2"	1 Plum	22	.364		.26	23.50		23.76	35.50
6340	3/4"		21	.381		.34	24.50		24.84	37
6341	1"		18	.444		1.55	28.50		30.05	44.50
6342	1-1/4"		17	.471		2.01	30.50		32.51	47.50
6343	1-1/2"		16	.500		3.07	32		35.07	51.50
6344	2"	Q-1	28	.571		5.40	33		38.40	55.50
6360	Solvent cement for CPVC, commercial grade, per quart				Qt.	51.50			51.50	56.50
7340	PVC flange, slip-on, Sch 80 std., 1/2"	1 Plum	22	.364	Ea.	14.65	23.50		38.15	51
7350	3/4"		21	.381		15.65	24.50		40.15	54
7360	1"		18	.444		17.45	28.50		45.95	62
7370	1-1/4"		17	.471		17.95	30.50		48.45	65.50
7380	1-1/2"		16	.500		18.35	32		50.35	68
7390	2"	Q-1	26	.615		24.50	35.50		60	80.50
7400	2-1/2"		24	.667		37.50	38.50		76	99.50
7410	3"		18	.889		41.50	51.50		93	123
7420	4"		15	1.067		52.50	62		114.50	151
7430	6"		10	1.600		83	93		176	230
7440	8"	Q-2	11	2.182		148	131		279	360

22 11 13.76 Pipe Fittings, Plastic		Crew	Daily Output	Labor-Hours	Unit	2020 Bare Costs Material	Labor	Equipment	Total	Total Incl O&P
7550	Union, schedule 40, socket joints, 1/2"	1 Plum	19	.421	Ea.	5.20	27		32.20	46.50
7560	3/4"		18	.444		5.80	28.50		34.30	49.50
7570	1"		15	.533		6	34.50		40.50	58
7580	1-1/4"		14	.571		17.85	37		54.85	74.50
7590	1-1/2"		13	.615		20	39.50		59.50	81.50
7600	2"	Q-1	20	.800		27	46.50		73.50	99.50
7992	Polybutyl/polyethyl pipe, for copper fittings see Line 22 11 13.25 7000									
8000	Compression type, PVC, 160 psi cold water									
8010	Coupling, 3/4" CTS	1 Plum	21	.381	Ea.	4.27	24.50		28.77	41
8020	1" CTS		18	.444		6	28.50		34.50	49.50
8030	1-1/4" CTS		17	.471		8.35	30.50		38.85	54.50
8040	1-1/2" CTS		16	.500		10	32		42	59
8050	2" CTS		15	.533		15.85	34.50		50.35	69
8060	Female adapter, 3/4" FPT x 3/4" CTS		23	.348		6.35	22.50		28.85	40.50
8070	3/4" FPT x 1" CTS		21	.381		7.35	24.50		31.85	44.50
8080	1" FPT x 1" CTS		20	.400		7.35	26		33.35	46.50
8090	1-1/4" FPT x 1-1/4" CTS		18	.444		9.60	28.50		38.10	53.50
8100	1-1/2" FPT x 1-1/2" CTS		16	.500		10.95	32		42.95	60
8110	2" FPT x 2" CTS		13	.615		16.05	39.50		55.55	77
8130	Male adapter, 3/4" MPT x 3/4" CTS		23	.348		5.30	22.50		27.80	39.50
8140	3/4" MPT x 1" CTS		21	.381		6.25	24.50		30.75	43.50
8150	1" MPT x 1" CTS		20	.400		6.25	26		32.25	45.50
8160	1-1/4" MPT x 1-1/4" CTS		18	.444		8.45	28.50		36.95	52.50
8170	1-1/2" MPT x 1-1/2" CTS		16	.500		10.15	32		42.15	59
8180	2" MPT x 2" CTS		13	.615		13.15	39.50		52.65	74
8200	Spigot adapter, 3/4" IPS x 3/4" CTS		23	.348		2.22	22.50		24.72	36
8210	3/4" IPS x 1" CTS		21	.381		2.71	24.50		27.21	39.50
8220	1" IPS x 1" CTS		20	.400		2.70	26		28.70	41.50
8230	1-1/4" IPS x 1-1/4" CTS		18	.444		4.07	28.50		32.57	47.50
8240	1-1/2" IPS x 1-1/2" CTS		16	.500		4.26	32		36.26	52.50
8250	2" IPS x 2" CTS		13	.615		5.30	39.50		44.80	65.50
8270	Price includes insert stiffeners									
8280	250 psi is same price as 160 psi									
8300	Insert type, nylon, 160 & 250 psi, cold water									
8310	Clamp ring stainless steel, 3/4" IPS	1 Plum	115	.070	Ea.	2.78	4.48		7.26	9.75
8320	1" IPS		107	.075		3.07	4.82		7.89	10.60
8330	1-1/4" IPS		101	.079		2.85	5.10		7.95	10.80
8340	1-1/2" IPS		95	.084		3.95	5.45		9.40	12.45
8350	2" IPS		85	.094		4.74	6.05		10.79	14.30
8370	Coupling, 3/4" IPS		22	.364		.95	23.50		24.45	36
8380	1" IPS		19	.421		.99	27		27.99	41.50
8390	1-1/4" IPS		18	.444		1.47	28.50		29.97	44.50
8400	1-1/2" IPS		17	.471		1.73	30.50		32.23	47.50
8410	2" IPS		16	.500		3.36	32		35.36	51.50
8430	Elbow, 90°, 3/4" IPS		22	.364		1.89	23.50		25.39	37
8440	1" IPS		19	.421		2.10	27		29.10	43
8450	1-1/4" IPS		18	.444		2.35	28.50		30.85	45.50
8460	1-1/2" IPS		17	.471		2.77	30.50		33.27	48.50
8470	2" IPS		16	.500		3.86	32		35.86	52.50
8490	Male adapter, 3/4" IPS x 3/4" MPT		25	.320		.95	20.50		21.45	32
8500	1" IPS x 1" MPT		21	.381		.98	24.50		25.48	37.50
8510	1-1/4" IPS x 1-1/4" MPT		20	.400		1.55	26		27.55	40
8520	1-1/2" IPS x 1-1/2" MPT		18	.444		1.73	28.50		30.23	45

22 11 13.76 Pipe Fittings, Plastic		Crew	Daily Output	Labor-Hours	Unit	Material	Labor	Equipment	Total	Total Incl O&P
						2020 Bare Costs				
8530	2" IPS x 2" MPT	1 Plum	15	.533	Ea.	3.33	34.50		37.83	55
8550	Tee, 3/4" IPS		14	.571		1.83	37		38.83	57
8560	1" IPS		13	.615		2.40	39.50		41.90	62
8570	1-1/4" IPS		12	.667		3.73	43		46.73	68.50
8580	1-1/2" IPS		11	.727		4.24	47		51.24	74.50
8590	2" IPS		10	.800		8.35	51.50		59.85	86
8610	Insert type, PVC, 100 psi @ 180°F, hot & cold water									
8620	Coupler, male, 3/8" CTS x 3/8" MPT	1 Plum	29	.276	Ea.	.76	17.80		18.56	27.50
8630	3/8" CTS x 1/2" MPT		28	.286		.76	18.40		19.16	28.50
8640	1/2" CTS x 1/2" MPT		27	.296		.78	19.10		19.88	29.50
8650	1/2" CTS x 3/4" MPT		26	.308		2.46	19.85		22.31	32
8660	3/4" CTS x 1/2" MPT		25	.320		2.22	20.50		22.72	33.50
8670	3/4" CTS x 3/4" MPT		25	.320		.93	20.50		21.43	32
8700	Coupling, 3/8" CTS x 1/2" CTS		25	.320		4.44	20.50		24.94	36
8710	1/2" CTS		23	.348		5.65	22.50		28.15	40
8730	3/4" CTS		22	.364		12.35	23.50		35.85	48.50
8750	Elbow 90°, 3/8" CTS		25	.320		4.15	20.50		24.65	35.50
8760	1/2" CTS		23	.348		5.10	22.50		27.60	39
8770	3/4" CTS		22	.364		7.65	23.50		31.15	43.50
8800	Rings, crimp, copper, 3/8" CTS		120	.067		.18	4.30		4.48	6.65
8810	1/2" CTS		117	.068		.20	4.41		4.61	6.80
8820	3/4" CTS		115	.070		.26	4.48		4.74	7
8850	Reducer tee, bronze, 3/8" x 3/8" x 1/2" CTS		17	.471		5.50	30.50		36	51.50
8860	1/2" x 1/2" x 3/4" CTS		15	.533		7.40	34.50		41.90	59.50
8870	3/4" x 1/2" x 1/2" CTS		14	.571		7.40	37		44.40	63
8890	3/4" x 3/4" x 1/2" CTS		14	.571		7.50	37		44.50	63.50
8900	1" x 1/2" x 1/2" CTS		14	.571		12.40	37		49.40	68.50
8930	Tee, 3/8" CTS		17	.471		3.24	30.50		33.74	49
8940	1/2" CTS		15	.533		2.92	34.50		37.42	54.50
8950	3/4" CTS		14	.571		3.61	37		40.61	59
8960	Copper rings included in fitting price									
9000	Flare type, assembled, acetal, hot & cold water									
9010	Coupling, 1/4" & 3/8" CTS	1 Plum	24	.333	Ea.	3.55	21.50		25.05	36
9020	1/2" CTS		22	.364		4.07	23.50		27.57	39.50
9030	3/4" CTS		21	.381		6	24.50		30.50	43
9040	1" CTS		18	.444		7.65	28.50		36.15	51.50
9050	Elbow 90°, 1/4" CTS		26	.308		3.94	19.85		23.79	34
9060	3/8" CTS		24	.333		4.23	21.50		25.73	36.50
9070	1/2" CTS		22	.364		5	23.50		28.50	40.50
9080	3/4" CTS		21	.381		7.65	24.50		32.15	45
9090	1" CTS		18	.444		9.90	28.50		38.40	54
9110	Tee, 1/4" CTS		16	.500		4.33	32		36.33	53
9114	3/8" CTS		15	.533		4.42	34.50		38.92	56.50
9120	1/2" CTS		14	.571		5.55	37		42.55	61
9130	3/4" CTS		13	.615		8.70	39.50		48.20	69
9140	1" CTS		12	.667		11.70	43		54.70	77.50
9400	Polypropylene, fittings and accessories									
9404	Fittings fusion welded, sizes are ID									
9408	Note: sizes 1/2" thru 4" use socket fusion									
9410	Sizes 6" thru 10" use butt fusion									
9416	Coupling									
9420	3/8"	1 Plum	39	.205	Ea.	.92	13.20		14.12	21
9422	1/2"		37.40	.214		1.23	13.80		15.03	22

22 11 13.76 Pipe Fittings, Plastic		Crew	Daily Output	Labor-Hours	Unit	2020 Bare Costs Material	Labor	Equipment	Total	Total Incl O&P
9424	3/4"	1 Plum	35.40	.226	Ea.	1.36	14.55		15.91	23.50
9426	1"		29.70	.269		1.79	17.35		19.14	28
9428	1-1/4"		27.60	.290		2.14	18.70		20.84	30.50
9430	1-1/2"		24.80	.323		4.51	21		25.51	36
9432	2"	Q-1	43	.372		9	21.50		30.50	42.50
9434	2-1/2"		35.60	.449		10.10	26		36.10	50
9436	3"		30.90	.518		22	30		52	69.50
9438	3-1/2"		27.80	.576		36	33.50		69.50	89.50
9440	4"		25	.640		47.50	37		84.50	108
9442	Reducing coupling, female to female									
9446	2" to 1-1/2"	Q-1	49	.327	Ea.	13	18.95		31.95	43
9448	2-1/2" to 2"		41.20	.388		14.25	22.50		36.75	49
9450	3" to 2-1/2"		33.10	.483		19.25	28		47.25	63
9470	Reducing bushing, female to female									
9472	1/2" to 3/8"	1 Plum	38.20	.209	Ea.	1.23	13.50		14.73	21.50
9474	3/4" to 3/8" or 1/2"		36.50	.219		1.36	14.15		15.51	22.50
9476	1" to 3/4" or 1/2"		33.10	.242		1.81	15.60		17.41	25.50
9478	1-1/4" to 3/4" or 1"		28.70	.279		2.79	17.95		20.74	30
9480	1-1/2" to 1/2" thru 1-1/4"		26.20	.305		4.62	19.70		24.32	34.50
9482	2" to 1/2" thru 1-1/2"	Q-1	43	.372		9.25	21.50		30.75	42.50
9484	2-1/2" to 1/2" thru 2"		41.20	.388		10.35	22.50		32.85	45
9486	3" to 1-1/2" thru 2-1/2"		33.10	.483		23	28		51	67.50
9488	3-1/2" to 2" thru 3"		29.20	.548		37	32		69	88
9490	4" to 2-1/2" thru 3-1/2"		26.20	.611		57.50	35.50		93	117
9491	6" to 4" SDR 7.4		16.50	.970		69	56		125	160
9492	6" to 4" SDR 11		16.50	.970		69	56		125	160
9493	8" to 6" SDR 7.4		10.10	1.584		102	92		194	250
9494	8" to 6" SDR 11		10.10	1.584		69.50	92		161.50	215
9495	10" to 8" SDR 7.4		7.80	2.051		139	119		258	330
9496	10" to 8" SDR 11		7.80	2.051		100	119		219	288
9500	90° elbow									
9504	3/8"	1 Plum	39	.205	Ea.	1.26	13.20		14.46	21
9506	1/2"		37.40	.214		1.31	13.80		15.11	22
9508	3/4"		35.40	.226		1.68	14.55		16.23	24
9510	1"		29.70	.269		2.42	17.35		19.77	28.50
9512	1-1/4"		27.60	.290		3.73	18.70		22.43	32
9514	1-1/2"		24.80	.323		8	21		29	40
9516	2"	Q-1	43	.372		12.30	21.50		33.80	46
9518	2-1/2"		35.60	.449		27.50	26		53.50	69
9520	3"		30.90	.518		45.50	30		75.50	95
9522	3-1/2"		27.80	.576		64.50	33.50		98	121
9524	4"		25	.640		99.50	37		136.50	165
9526	6" SDR 7.4		5.55	2.883		113	167		280	375
9528	6" SDR 11		5.55	2.883		89	167		256	350
9530	8" SDR 7.4	Q-2	8.10	2.963		295	178		473	590
9532	8" SDR 11		8.10	2.963		229	178		407	520
9534	10" SDR 7.4		7.50	3.200		390	192		582	720
9536	10" SDR 11		7.50	3.200		365	192		557	695
9551	45° elbow									
9554	3/8"	1 Plum	39	.205	Ea.	1.26	13.20		14.46	21
9556	1/2"		37.40	.214		1.31	13.80		15.11	22
9558	3/4"		35.40	.226		1.68	14.55		16.23	24
9564	1"		29.70	.269		2.42	17.35		19.77	28.50

22 11 13.76 Pipe Fittings, Plastic		Crew	Daily Output	Labor-Hours	Unit	Material	Labor	Equipment	Total	Total Incl O&P
						2020 Bare Costs				
9566	1-1/4"	1 Plum	27.60	.290	Ea.	3.73	18.70		22.43	32
9568	1-1/2"		24.80	.323		8	21		29	40
9570	2"	Q-1	43	.372		12.15	21.50		33.65	46
9572	2-1/2"		35.60	.449		27	26		53	68.50
9574	3"		30.90	.518		50	30		80	100
9576	3-1/2"		27.80	.576		71	33.50		104.50	128
9578	4"		25	.640		109	37		146	176
9580	6" SDR 7.4		5.55	2.883		126	167		293	390
9582	6" SDR 11		5.55	2.883		104	167		271	365
9584	8" SDR 7.4	Q-2	8.10	2.963		274	178		452	565
9586	8" SDR 11		8.10	2.963		239	178		417	530
9588	10" SDR 7.4		7.50	3.200		440	192		632	775
9590	10" SDR 11		7.50	3.200		365	192		557	695
9600	Tee									
9604	3/8"	1 Plum	26	.308	Ea.	1.68	19.85		21.53	31.50
9606	1/2"		24.90	.321		2.31	20.50		22.81	33.50
9608	3/4"		23.70	.338		2.42	22		24.42	35
9610	1"		19.90	.402		3.05	26		29.05	42.50
9612	1-1/4"		18.50	.432		4.69	28		32.69	46.50
9614	1-1/2"		16.60	.482		13.40	31		44.40	61.50
9616	2"	Q-1	26.80	.597		18.25	34.50		52.75	72
9618	2-1/2"		23.80	.672		30	39		69	91.50
9620	3"		20.60	.777		60	45		105	134
9622	3-1/2"		18.40	.870		94	50.50		144.50	179
9624	4"		16.70	.958		110	55.50		165.50	204
9626	6" SDR 7.4		3.70	4.324		137	251		388	525
9628	6" SDR 11		3.70	4.324		167	251		418	560
9630	8" SDR 7.4	Q-2	5.40	4.444		375	267		642	815
9632	8" SDR 11		5.40	4.444		330	267		597	765
9634	10" SDR 7.4		5	4.800		645	289		934	1,150
9636	10" SDR 11		5	4.800		560	289		849	1,050
9638	For reducing tee use same tee price									
9660	End cap									
9662	3/8"	1 Plum	78	.103	Ea.	1.80	6.60		8.40	11.90
9664	1/2"		74.60	.107		1.90	6.90		8.80	12.45
9666	3/4"		71.40	.112		2.42	7.20		9.62	13.45
9668	1"		59.50	.134		2.92	8.65		11.57	16.15
9670	1-1/4"		55.60	.144		4.62	9.25		13.87	19
9672	1-1/2"		49.50	.162		6.35	10.40		16.75	22.50
9674	2"	Q-1	80	.200		10.70	11.60		22.30	29
9676	2-1/2"		71.40	.224		15.45	13		28.45	36.50
9678	3"		61.70	.259		35	15.05		50.05	61
9680	3-1/2"		55.20	.290		42	16.80		58.80	71
9682	4"		50	.320		64	18.55		82.55	98.50
9684	6" SDR 7.4		26.50	.604		88	35		123	149
9686	6" SDR 11		26.50	.604		71	35		106	131
9688	8" SDR 7.4	Q-2	16.30	1.472		88	88.50		176.50	230
9690	8" SDR 11		16.30	1.472		77.50	88.50		166	218
9692	10" SDR 7.4		14.90	1.611		132	97		229	290
9694	10" SDR 11		14.90	1.611		94	97		191	248
9800	Accessories and tools									
9802	Pipe clamps for suspension, not including rod or beam clamp									
9804	3/8"	1 Plum	74	.108	Ea.	2.25	6.95		9.20	12.95

22 11 Facility Water Distribution

22 11 13 – Facility Water Distribution Piping

22 11 13.76 Pipe Fittings, Plastic		Crew	Daily Output	Labor-Hours	Unit	Material	2020 Bare Costs Labor	Equipment	Total	Total Incl O&P
9805	1/2″	1 Plum	70	.114	Ea.	2.83	7.35		10.18	14.10
9806	3/4″		68	.118		2.69	7.60		10.29	14.30
9807	1″		66	.121		3.55	7.80		11.35	15.60
9808	1-1/4″		64	.125		3.64	8.05		11.69	16.05
9809	1-1/2″		62	.129		3.98	8.30		12.28	16.85
9810	2″	Q-1	110	.145		4.94	8.45		13.39	18.10
9811	2-1/2″		104	.154		6.35	8.90		15.25	20.50
9812	3″		98	.163		6.85	9.45		16.30	21.50
9813	3-1/2″		92	.174		7.50	10.10		17.60	23.50
9814	4″		86	.186		8.40	10.80		19.20	25.50
9815	6″		70	.229		10.20	13.25		23.45	31
9816	8″	Q-2	100	.240		36.50	14.45		50.95	62
9817	10″	″	94	.255		42	15.35		57.35	69
9820	Pipe cutter									
9822	For 3/8″ thru 1-1/4″				Ea.	113			113	124
9824	For 1-1/2″ thru 4″				″	296			296	325
9826	Note: Pipes may be cut with standard									
9827	iron saw with blades for plastic.									
9982	For plastic hangers see Line 22 05 29.10 8000									
9986	For copper/brass fittings see Line 22 11 13.25 7000									

22 11 13.78 Pipe, High Density Polyethylene Plastic (HDPE)		Crew	Daily Output	Labor-Hours	Unit	Material	Labor	Equipment	Total	Total Incl O&P
0010	**PIPE, HIGH DENSITY POLYETHYLENE PLASTIC (HDPE)**									
0020	Not incl. hangers, trenching, backfill, hoisting or digging equipment.									
0030	Standard length is 40′, add a weld for each joint									
0035	For HDPE weld machine see 015433401685 in equipment rental									
0040	Single wall									
0050	Straight									
0054	1″ diameter DR 11				L.F.	.89			.89	.98
0058	1-1/2″ diameter DR 11					1.14			1.14	1.25
0062	2″ diameter DR 11					1.89			1.89	2.08
0066	3″ diameter DR 11					2.29			2.29	2.52
0070	3″ diameter DR 17					1.83			1.83	2.01
0074	4″ diameter DR 11					3.83			3.83	4.21
0078	4″ diameter DR 17					3.83			3.83	4.21
0082	6″ diameter DR 11					9.50			9.50	10.45
0086	6″ diameter DR 17					6.30			6.30	6.90
0090	8″ diameter DR 11					15.80			15.80	17.40
0094	8″ diameter DR 26					7.40			7.40	8.15
0098	10″ diameter DR 11					25			25	27
0102	10″ diameter DR 26					11.45			11.45	12.60
0106	12″ diameter DR 11					36			36	40
0110	12″ diameter DR 26					17.15			17.15	18.85
0114	16″ diameter DR 11					55			55	61
0118	16″ diameter DR 26					25			25	27
0122	18″ diameter DR 11					70.50			70.50	77.50
0126	18″ diameter DR 26					32.50			32.50	35.50
0130	20″ diameter DR 11					85.50			85.50	94.50
0134	20″ diameter DR 26					38			38	42
0138	22″ diameter DR 11					105			105	115
0142	22″ diameter DR 26					47.50			47.50	52.50
0146	24″ diameter DR 11					124			124	136
0150	24″ diameter DR 26					55			55	61

22 11 13.78 Pipe, High Density Polyethylene Plastic (HDPE)	Crew	Daily Output	Labor-Hours	Unit	Material	2020 Bare Costs Labor	Equipment	Total	Total Incl O&P
0154 28" diameter DR 17				L.F.	114			114	126
0158 28" diameter DR 26					76			76	84
0162 30" diameter DR 21					107			107	117
0166 30" diameter DR 26					87.50			87.50	96.50
0170 36" diameter DR 26					87.50			87.50	96.50
0174 42" diameter DR 26					170			170	186
0178 48" diameter DR 26					223			223	245
0182 54" diameter DR 26				▼	280			280	310
0300 90° elbow									
0304 1" diameter DR 11				Ea.	5.50			5.50	6.05
0308 1-1/2" diameter DR 11					7.30			7.30	8.05
0312 2" diameter DR 11					7.30			7.30	8.05
0316 3" diameter DR 11					14.60			14.60	16.05
0320 3" diameter DR 17					14.60			14.60	16.05
0324 4" diameter DR 11					20.50			20.50	22.50
0328 4" diameter DR 17					20.50			20.50	22.50
0332 6" diameter DR 11					46.50			46.50	51.50
0336 6" diameter DR 17					46.50			46.50	51.50
0340 8" diameter DR 11					115			115	127
0344 8" diameter DR 26					102			102	112
0348 10" diameter DR 11					430			430	475
0352 10" diameter DR 26					400			400	440
0356 12" diameter DR 11					455			455	500
0360 12" diameter DR 26					410			410	450
0364 16" diameter DR 11					540			540	590
0368 16" diameter DR 26					530			530	585
0372 18" diameter DR 11					675			675	745
0376 18" diameter DR 26					635			635	700
0380 20" diameter DR 11					795			795	875
0384 20" diameter DR 26					770			770	845
0388 22" diameter DR 11					820			820	905
0392 22" diameter DR 26					795			795	875
0396 24" diameter DR 11					930			930	1,025
0400 24" diameter DR 26					900			900	990
0404 28" diameter DR 17					1,125			1,125	1,250
0408 28" diameter DR 26					1,050			1,050	1,175
0412 30" diameter DR 17					1,600			1,600	1,750
0416 30" diameter DR 26					1,450			1,450	1,600
0420 36" diameter DR 26					1,850			1,850	2,050
0424 42" diameter DR 26					2,400			2,400	2,625
0428 48" diameter DR 26					2,775			2,775	3,075
0432 54" diameter DR 26				▼	6,625			6,625	7,300
0500 45° elbow									
0512 2" diameter DR 11				Ea.	5.85			5.85	6.45
0516 3" diameter DR 11					14.60			14.60	16.05
0520 3" diameter DR 17					14.60			14.60	16.05
0524 4" diameter DR 11					20.50			20.50	22.50
0528 4" diameter DR 17					20.50			20.50	22.50
0532 6" diameter DR 11					46.50			46.50	51.50
0536 6" diameter DR 17					46.50			46.50	51.50
0540 8" diameter DR 11					115			115	127
0544 8" diameter DR 26					67			67	74
0548 10" diameter DR 11				▼	430			430	475

22 11 13.78 Pipe, High Density Polyethylene Plastic (HDPE)		Crew	Daily Output	Labor-Hours	Unit	2020 Bare Costs Material	Labor	Equipment	Total	Total Incl O&P
0552	10″ diameter DR 26				Ea.	400			400	440
0556	12″ diameter DR 11					455			455	500
0560	12″ diameter DR 26					410			410	450
0564	16″ diameter DR 11					239			239	263
0568	16″ diameter DR 26					226			226	248
0572	18″ diameter DR 11					253			253	279
0576	18″ diameter DR 26					232			232	255
0580	20″ diameter DR 11					385			385	425
0584	20″ diameter DR 26					365			365	400
0588	22″ diameter DR 11					530			530	585
0592	22″ diameter DR 26					505			505	555
0596	24″ diameter DR 11					650			650	715
0600	24″ diameter DR 26					625			625	685
0604	28″ diameter DR 17					740			740	810
0608	28″ diameter DR 26					715			715	790
0612	30″ diameter DR 17					915			915	1,000
0616	30″ diameter DR 26					890			890	980
0620	36″ diameter DR 26					1,125			1,125	1,250
0624	42″ diameter DR 26					1,450			1,450	1,600
0628	48″ diameter DR 26					1,575			1,575	1,725
0632	54″ diameter DR 26				▼	2,100			2,100	2,325
0700	Tee									
0704	1″ diameter DR 11				Ea.	7.60			7.60	8.40
0708	1-1/2″ diameter DR 11					10.70			10.70	11.75
0712	2″ diameter DR 11					9.15			9.15	10.10
0716	3″ diameter DR 11					16.75			16.75	18.45
0720	3″ diameter DR 17					16.75			16.75	18.45
0724	4″ diameter DR 11					24.50			24.50	27
0728	4″ diameter DR 17					24.50			24.50	27
0732	6″ diameter DR 11					61			61	67
0736	6″ diameter DR 17					61			61	67
0740	8″ diameter DR 11					151			151	166
0744	8″ diameter DR 17					151			151	166
0748	10″ diameter DR 11					450			450	495
0752	10″ diameter DR 17					450			450	495
0756	12″ diameter DR 11					600			600	660
0760	12″ diameter DR 17					600			600	660
0764	16″ diameter DR 11					315			315	350
0768	16″ diameter DR 17					261			261	287
0772	18″ diameter DR 11					445			445	490
0776	18″ diameter DR 17					365			365	405
0780	20″ diameter DR 11					545			545	600
0784	20″ diameter DR 17					445			445	490
0788	22″ diameter DR 11					700			700	770
0792	22″ diameter DR 17					555			555	610
0796	24″ diameter DR 11					870			870	955
0800	24″ diameter DR 17					730			730	805
0804	28″ diameter DR 17					1,425			1,425	1,550
0812	30″ diameter DR 17					1,625			1,625	1,775
0820	36″ diameter DR 17					2,675			2,675	2,950
0824	42″ diameter DR 26					2,975			2,975	3,275
0828	48″ diameter DR 26				▼	3,200			3,200	3,525
1000	Flange adptr, w/back-up ring and 1/2 cost of plated bolt set									

22 11 Facility Water Distribution

22 11 13 – Facility Water Distribution Piping

22 11 13.78 Pipe, High Density Polyethylene Plastic (HDPE)		Crew	Daily Output	Labor-Hours	Unit	Material	Labor	Equipment	Total	Total Incl O&P
						2020 Bare Costs				
1004	1" diameter DR 11				Ea.	29			29	32
1008	1-1/2" diameter DR 11					29			29	32
1012	2" diameter DR 11					18.30			18.30	20
1016	3" diameter DR 11					21.50			21.50	23.50
1020	3" diameter DR 17					21.50			21.50	23.50
1024	4" diameter DR 11					29			29	32
1028	4" diameter DR 17					29			29	32
1032	6" diameter DR 11					41			41	45.50
1036	6" diameter DR 17					41			41	45.50
1040	8" diameter DR 11					59.50			59.50	65.50
1044	8" diameter DR 26					59.50			59.50	65.50
1048	10" diameter DR 11					94.50			94.50	104
1052	10" diameter DR 26					94.50			94.50	104
1056	12" diameter DR 11					139			139	153
1060	12" diameter DR 26					139			139	153
1064	16" diameter DR 11					297			297	325
1068	16" diameter DR 26					297			297	325
1072	18" diameter DR 11					385			385	425
1076	18" diameter DR 26					385			385	425
1080	20" diameter DR 11					535			535	590
1084	20" diameter DR 26					535			535	590
1088	22" diameter DR 11					580			580	640
1092	22" diameter DR 26					580			580	640
1096	24" diameter DR 17					630			630	695
1100	24" diameter DR 32.5					630			630	695
1104	28" diameter DR 15.5					855			855	940
1108	28" diameter DR 32.5					855			855	940
1112	30" diameter DR 11					990			990	1,100
1116	30" diameter DR 21					990			990	1,100
1120	36" diameter DR 26					1,075			1,075	1,200
1124	42" diameter DR 26					1,225			1,225	1,350
1128	48" diameter DR 26					1,500			1,500	1,650
1132	54" diameter DR 26					1,825			1,825	2,025
1200	Reducer									
1208	2" x 1-1/2" diameter DR 11				Ea.	8.75			8.75	9.65
1212	3" x 2" diameter DR 11					8.75			8.75	9.65
1216	4" x 2" diameter DR 11					10.20			10.20	11.25
1220	4" x 3" diameter DR 11					13.15			13.15	14.45
1224	6" x 4" diameter DR 11					30.50			30.50	33.50
1228	8" x 6" diameter DR 11					46.50			46.50	51.50
1232	10" x 8" diameter DR 11					80.50			80.50	88.50
1236	12" x 8" diameter DR 11					131			131	144
1240	12" x 10" diameter DR 11					105			105	116
1244	14" x 12" diameter DR 11					117			117	128
1248	16" x 14" diameter DR 11					149			149	164
1252	18" x 16" diameter DR 11					181			181	199
1256	20" x 18" diameter DR 11					360			360	395
1260	22" x 20" diameter DR 11					440			440	485
1264	24" x 22" diameter DR 11					495			495	545
1268	26" x 24" diameter DR 11					585			585	640
1272	28" x 24" diameter DR 11					745			745	820
1276	32" x 28" diameter DR 17					965			965	1,050
1280	36" x 32" diameter DR 17					1,325			1,325	1,450

22 11 13.78 Pipe, High Density Polyethylene Plastic (HDPE)		Crew	Daily Output	Labor-Hours	Unit	Material	2020 Bare Costs Labor	Equipment	Total	Total Incl O&P
4000	Welding labor per joint, not including welding machine									
4010	Pipe joint size (cost based on thickest wall for each diam.)									
4030	1" pipe size	4 Skwk	273	.117	Ea.		6.45		6.45	9.70
4040	1-1/2" pipe size		175	.183			10.05		10.05	15.15
4050	2" pipe size		128	.250			13.70		13.70	20.50
4060	3" pipe size		100	.320			17.55		17.55	26.50
4070	4" pipe size		77	.416			23		23	34.50
4080	6" pipe size	5 Skwk	63	.635			35		35	52.50
4090	8" pipe size		48	.833			45.50		45.50	69
4100	10" pipe size		40	1			55		55	83
4110	12" pipe size	6 Skwk	41	1.171			64		64	97
4120	16" pipe size		34	1.412			77.50		77.50	117
4130	18" pipe size		32	1.500			82.50		82.50	124
4140	20" pipe size	8 Skwk	37	1.730			95		95	143
4150	22" pipe size		35	1.829			100		100	152
4160	24" pipe size		34	1.882			103		103	156
4170	28" pipe size		33	1.939			106		106	161
4180	30" pipe size		32	2			110		110	166
4190	36" pipe size		31	2.065			113		113	171
4200	42" pipe size		30	2.133			117		117	177
4210	48" pipe size	9 Skwk	33	2.182			120		120	181
4220	54" pipe size	"	31	2.323			127		127	193
5000	Dual wall contained pipe									
5040	Straight									
5054	1" DR 11 x 3" DR 11				L.F.	5.35			5.35	5.90
5058	1" DR 11 x 4" DR 11					5.65			5.65	6.25
5062	1-1/2" DR 11 x 4" DR 17					6.25			6.25	6.90
5066	2" DR 11 x 4" DR 17					6.60			6.60	7.25
5070	2" DR 11 x 6" DR 17					10			10	11
5074	3" DR 11 x 6" DR 17					11.20			11.20	12.30
5078	3" DR 11 x 6" DR 26					9.55			9.55	10.50
5086	3" DR 17 x 8" DR 17					12.15			12.15	13.40
5090	4" DR 11 x 8" DR 17					16.85			16.85	18.55
5094	4" DR 17 x 8" DR 26					13.20			13.20	14.50
5098	6" DR 11 x 10" DR 17					26			26	29
5102	6" DR 17 x 10" DR 26					17.80			17.80	19.60
5106	6" DR 26 x 10" DR 26					16.70			16.70	18.35
5110	8" DR 17 x 12" DR 26					25.50			25.50	28.50
5114	8" DR 26 x 12" DR 32.5					23.50			23.50	26
5118	10" DR 17 x 14" DR 26					38.50			38.50	42
5122	10" DR 17 x 16" DR 26					27.50			27.50	30.50
5126	10" DR 26 x 16" DR 26					36.50			36.50	40
5130	12" DR 26 x 16" DR 26					38.50			38.50	42.50
5134	12" DR 17 x 18" DR 26					53.50			53.50	58.50
5138	12" DR 26 x 18" DR 26					50.50			50.50	56
5142	14" DR 26 x 20" DR 32.5					51.50			51.50	57
5146	16" DR 26 x 22" DR 32.5					57			57	63
5150	18" DR 26 x 24" DR 32.5					71			71	78.50
5154	20" DR 32.5 x 28" DR 32.5					80.50			80.50	88.50
5158	22" DR 32.5 x 30" DR 32.5					80.50			80.50	88.50
5162	24" DR 32.5 x 32" DR 32.5					83.50			83.50	92
5166	36" DR 32.5 x 42" DR 32.5					97.50			97.50	107
5300	Force transfer coupling									

22 11 13.78 Pipe, High Density Polyethylene Plastic (HDPE)		Crew	Daily Output	Labor-Hours	Unit	Material	2020 Bare Costs Labor	Equipment	Total	Total Incl O&P
5354	1" DR 11 x 3" DR 11				Ea.	188			188	207
5358	1" DR 11 x 4" DR 17					221			221	243
5362	1-1/2" DR 11 x 4" DR 17					212			212	233
5366	2" DR 11 x 4" DR 17					212			212	233
5370	2" DR 11 x 6" DR 17					365			365	400
5374	3" DR 11 x 6" DR 17					390			390	425
5378	3" DR 11 x 6" DR 26					390			390	425
5382	3" DR 11 x 8" DR 11					390			390	430
5386	3" DR 11 x 8" DR 17					425			425	465
5390	4" DR 11 x 8" DR 17					435			435	475
5394	4" DR 17 x 8" DR 26					440			440	480
5398	6" DR 11 x 10" DR 17					520			520	570
5402	6" DR 17 x 10" DR 26					495			495	545
5406	6" DR 26 x 10" DR 26					675			675	745
5410	8" DR 17 x 12" DR 26					720			720	790
5414	8" DR 26 x 12" DR 32.5					830			830	910
5418	10" DR 17 x 14" DR 26					980			980	1,075
5422	10" DR 17 x 16" DR 26					990			990	1,100
5426	10" DR 26 x 16" DR 26					1,100			1,100	1,200
5430	12" DR 26 x 16" DR 26					1,350			1,350	1,475
5434	12" DR 17 x 18" DR 26					1,350			1,350	1,475
5438	12" DR 26 x 18" DR 26					1,475			1,475	1,625
5442	14" DR 26 x 20" DR 32.5					1,650			1,650	1,800
5446	16" DR 26 x 22" DR 32.5					1,825			1,825	2,000
5450	18" DR 26 x 24" DR 32.5					2,000			2,000	2,200
5454	20" DR 32.5 x 28" DR 32.5					2,200			2,200	2,425
5458	22" DR 32.5 x 30" DR 32.5					2,425			2,425	2,675
5462	24" DR 32.5 x 32" DR 32.5					2,675			2,675	2,950
5466	36" DR 32.5 x 42" DR 32.5				▼	2,875			2,875	3,175
5600	90° elbow									
5654	1" DR 11 x 3" DR 11				Ea.	206			206	226
5658	1" DR 11 x 4" DR 17					243			243	267
5662	1-1/2" DR 11 x 4" DR 17					250			250	275
5666	2" DR 11 x 4" DR 17					244			244	269
5670	2" DR 11 x 6" DR 17					355			355	390
5674	3" DR 11 x 6" DR 17					380			380	420
5678	3" DR 17 x 6" DR 26					390			390	425
5682	3" DR 17 x 8" DR 11					520			520	575
5686	3" DR 17 x 8" DR 17					555			555	610
5690	4" DR 11 x 8" DR 17					470			470	520
5694	4" DR 17 x 8" DR 26					450			450	495
5698	6" DR 11 x 10" DR 17					820			820	905
5702	6" DR 17 x 10" DR 26					835			835	920
5706	6" DR 26 x 10" DR 26					815			815	900
5710	8" DR 17 x 12" DR 26					1,100			1,100	1,200
5714	8" DR 26 x 12" DR 32.5					905			905	995
5718	10" DR 17 x 14" DR 26					1,075			1,075	1,175
5722	10" DR 17 x 16" DR 26					1,050			1,050	1,150
5726	10" DR 26 x 16" DR 26					1,025			1,025	1,150
5730	12" DR 26 x 16" DR 26					1,050			1,050	1,175
5734	12" DR 17 x 18" DR 26					1,150			1,150	1,275
5738	12" DR 26 x 18" DR 26					1,175			1,175	1,300
5742	14" DR 26 x 20" DR 32.5				▼	1,275			1,275	1,400

22 11 13.78 Pipe, High Density Polyethylene Plastic (HDPE)		Crew	Daily Output	Labor-Hours	Unit	2020 Bare Costs Material	Labor	Equipment	Total	Total Incl O&P
5746	16" DR 26 x 22" DR 32.5				Ea.	1,325			1,325	1,450
5750	18" DR 26 x 24" DR 32.5					1,375			1,375	1,500
5754	20" DR 32.5 x 28" DR 32.5					1,425			1,425	1,550
5758	22" DR 32.5 x 30" DR 32.5					1,475			1,475	1,600
5762	24" DR 32.5 x 32" DR 32.5					1,525			1,525	1,675
5766	36" DR 32.5 x 42" DR 32.5				↓	1,575			1,575	1,725
5800	45° elbow									
5804	1" DR 11 x 3" DR 11				Ea.	222			222	244
5808	1" DR 11 x 4" DR 17					195			195	215
5812	1-1/2" DR 11 x 4" DR 17					199			199	219
5816	2" DR 11 x 4" DR 17					190			190	209
5820	2" DR 11 x 6" DR 17					272			272	299
5824	3" DR 11 x 6" DR 17					292			292	320
5828	3" DR 17 x 6" DR 26					295			295	325
5832	3" DR 17 x 8" DR 11					375			375	410
5836	3" DR 17 x 8" DR 17					420			420	465
5840	4" DR 11 x 8" DR 17					345			345	380
5844	4" DR 17 x 8" DR 26					350			350	385
5848	6" DR 11 x 10" DR 17					545			545	600
5852	6" DR 17 x 10" DR 26					555			555	610
5856	6" DR 26 x 10" DR 26					625			625	685
5860	8" DR 17 x 12" DR 26					740			740	815
5864	8" DR 26 x 12" DR 32.5					695			695	765
5868	10" DR 17 x 14" DR 26					785			785	860
5872	10" DR 17 x 16" DR 26					755			755	830
5876	10" DR 26 x 16" DR 26					800			800	880
5880	12" DR 26 x 16" DR 26					720			720	795
5884	12" DR 17 x 18" DR 26					1,025			1,025	1,125
5888	12" DR 26 x 18" DR 26					910			910	1,000
5892	14" DR 26 x 20" DR 32.5					990			990	1,100
5896	16" DR 26 x 22" DR 32.5					1,025			1,025	1,125
5900	18" DR 26 x 24" DR 32.5					1,100			1,100	1,200
5904	20" DR 32.5 x 28" DR 32.5					1,075			1,075	1,200
5908	22" DR 32.5 x 30" DR 32.5					1,125			1,125	1,225
5912	24" DR 32.5 x 32" DR 32.5					1,150			1,150	1,275
5916	36" DR 32.5 x 42" DR 32.5				↓	1,200			1,200	1,325
6000	Access port with 4" riser									
6050	1" DR 11 x 4" DR 17				Ea.	284			284	310
6054	1-1/2" DR 11 x 4" DR 17					286			286	315
6058	2" DR 11 x 6" DR 17					350			350	385
6062	3" DR 11 x 6" DR 17					355			355	390
6066	3" DR 17 x 6" DR 26					345			345	380
6070	3" DR 17 x 8" DR 11					355			355	390
6074	3" DR 17 x 8" DR 17					370			370	405
6078	4" DR 11 x 8" DR 17					380			380	415
6082	4" DR 17 x 8" DR 26					360			360	400
6086	6" DR 11 x 10" DR 17					440			440	485
6090	6" DR 17 x 10" DR 26					405			405	445
6094	6" DR 26 x 10" DR 26					425			425	465
6098	8" DR 17 x 12" DR 26					455			455	500
6102	8" DR 26 x 12" DR 32.5				↓	445			445	490
6200	End termination with vent plug									
6204	1" DR 11 x 3" DR 11				Ea.	116			116	128

22 11 13.78 Pipe, High Density Polyethylene Plastic (HDPE)		Crew	Daily Output	Labor-Hours	Unit	Material	2020 Bare Costs Labor	Equipment	Total	Total Incl O&P
6208	1" DR 11 x 4" DR 17				Ea.	141			141	155
6212	1-1/2" DR 11 x 4" DR 17					141			141	155
6216	2" DR 11 x 4" DR 17					141			141	155
6220	2" DR 11 x 6" DR 17					286			286	315
6224	3" DR 11 x 6" DR 17					286			286	315
6228	3" DR 17 x 6" DR 26					286			286	315
6232	3" DR 17 x 8" DR 11					276			276	305
6236	3" DR 17 x 8" DR 17					310			310	340
6240	4" DR 11 x 8" DR 17					370			370	405
6244	4" DR 17 x 8" DR 26					345			345	380
6248	6" DR 11 x 10" DR 17					400			400	440
6252	6" DR 17 x 10" DR 26					370			370	410
6256	6" DR 26 x 10" DR 26					540			540	595
6260	8" DR 17 x 12" DR 26					565			565	625
6264	8" DR 26 x 12" DR 32.5					680			680	745
6268	10" DR 17 x 14" DR 26					900			900	990
6272	10" DR 17 x 16" DR 26					850			850	935
6276	10" DR 26 x 16" DR 26					955			955	1,050
6280	12" DR 26 x 16" DR 26					1,050			1,050	1,175
6284	12" DR 17 x 18" DR 26					1,225			1,225	1,350
6288	12" DR 26 x 18" DR 26					1,350			1,350	1,475
6292	14" DR 26 x 20" DR 32.5					1,525			1,525	1,675
6296	16" DR 26 x 22" DR 32.5					1,700			1,700	1,850
6300	18" DR 26 x 24" DR 32.5					1,900			1,900	2,100
6304	20" DR 32.5 x 28" DR 32.5					2,125			2,125	2,325
6308	22" DR 32.5 x 30" DR 32.5					2,375			2,375	2,600
6312	24" DR 32.5 x 32" DR 32.5					2,650			2,650	2,925
6316	36" DR 32.5 x 42" DR 32.5					2,925			2,925	3,225
6600	Tee									
6604	1" DR 11 x 3" DR 11				Ea.	239			239	263
6608	1" DR 11 x 4" DR 17					260			260	286
6612	1-1/2" DR 11 x 4" DR 17					269			269	296
6616	2" DR 11 x 4" DR 17					244			244	268
6620	2" DR 11 x 6" DR 17					390			390	430
6624	3" DR 11 x 6" DR 17					460			460	505
6628	3" DR 17 x 6" DR 26					435			435	480
6632	3" DR 17 x 8" DR 11					535			535	590
6636	3" DR 17 x 8" DR 17					580			580	640
6640	4" DR 11 x 8" DR 17					700			700	770
6644	4" DR 17 x 8" DR 26					700			700	770
6648	6" DR 11 x 10" DR 17					910			910	1,000
6652	6" DR 17 x 10" DR 26					915			915	1,000
6656	6" DR 26 x 10" DR 26					1,125			1,125	1,225
6660	8" DR 17 x 12" DR 26					1,125			1,125	1,225
6664	8" DR 26 x 12" DR 32.5					1,450			1,450	1,600
6668	10" DR 17 x 14" DR 26					1,675			1,675	1,850
6672	10" DR 17 x 16" DR 26					1,875			1,875	2,050
6676	10" DR 26 x 16" DR 26					1,875			1,875	2,050
6680	12" DR 26 x 16" DR 26					2,125			2,125	2,350
6684	12" DR 17 x 18" DR 26					2,125			2,125	2,350
6688	12" DR 26 x 18" DR 26					2,425			2,425	2,650
6692	14" DR 26 x 20" DR 32.5					2,725			2,725	3,000
6696	16" DR 26 x 22" DR 32.5					3,075			3,075	3,400

22 11 13.78 Pipe, High Density Polyethylene Plastic (HDPE)		Crew	Daily Output	Labor-Hours	Unit	2020 Bare Costs Material	Labor	Equipment	Total	Total Incl O&P
6700	18" DR 26 x 24" DR 32.5				Ea.	3,500			3,500	3,850
6704	20" DR 32.5 x 28" DR 32.5					3,950			3,950	4,350
6708	22" DR 32.5 x 30" DR 32.5					4,475			4,475	4,925
6712	24" DR 32.5 x 32" DR 32.5					5,075			5,075	5,575
6716	36" DR 32.5 x 42" DR 32.5				↓	5,825			5,825	6,400
6800	Wye									
6816	2" DR 11 x 4" DR 17				Ea.	595			595	655
6820	2" DR 11 x 6" DR 17					735			735	810
6824	3" DR 11 x 6" DR 17					780			780	855
6828	3" DR 17 x 6" DR 26					780			780	855
6832	3" DR 17 x 8" DR 11					830			830	910
6836	3" DR 17 x 8" DR 17					900			900	990
6840	4" DR 11 x 8" DR 17					990			990	1,100
6844	4" DR 17 x 8" DR 26					625			625	685
6848	6" DR 11 x 10" DR 17					1,525			1,525	1,675
6852	6" DR 17 x 10" DR 26					950			950	1,050
6856	6" DR 26 x 10" DR 26					1,325			1,325	1,450
6860	8" DR 17 x 12" DR 26					1,850			1,850	2,025
6864	8" DR 26 x 12" DR 32.5					1,525			1,525	1,675
6868	10" DR 17 x 14" DR 26					1,625			1,625	1,800
6872	10" DR 17 x 16" DR 26					1,750			1,750	1,925
6876	10" DR 26 x 16" DR 26					1,900			1,900	2,100
6880	12" DR 26 x 16" DR 26					2,050			2,050	2,250
6884	12" DR 17 x 18" DR 26					2,200			2,200	2,425
6888	12" DR 26 x 18" DR 26					2,400			2,400	2,625
6892	14" DR 26 x 20" DR 32.5					2,550			2,550	2,825
6896	16" DR 26 x 22" DR 32.5					2,700			2,700	2,975
6900	18" DR 26 x 24" DR 32.5					3,100			3,100	3,400
6904	20" DR 32.5 x 28" DR 32.5					3,225			3,225	3,550
6908	22" DR 32.5 x 30" DR 32.5					3,500			3,500	3,850
6912	24" DR 32.5 x 32" DR 32.5				↓	3,775			3,775	4,150
9000	Welding labor per joint, not including welding machine									
9010	Pipe joint size, outer pipe (cost based on the thickest walls)									
9020	Straight pipe									
9050	3" pipe size	4 Skwk	96	.333	Ea.		18.30		18.30	27.50
9060	4" pipe size	"	77	.416			23		23	34.50
9070	6" pipe size	5 Skwk	60	.667			36.50		36.50	55.50
9080	8" pipe size	"	40	1			55		55	83
9090	10" pipe size	6 Skwk	41	1.171			64		64	97
9100	12" pipe size		39	1.231			67.50		67.50	102
9110	14" pipe size		38	1.263			69.50		69.50	105
9120	16" pipe size	↓	35	1.371			75		75	114
9130	18" pipe size	8 Skwk	45	1.422			78		78	118
9140	20" pipe size		42	1.524			83.50		83.50	126
9150	22" pipe size		40	1.600			88		88	133
9160	24" pipe size		38	1.684			92.50		92.50	140
9170	28" pipe size		37	1.730			95		95	143
9180	30" pipe size		36	1.778			97.50		97.50	147
9190	32" pipe size		35	1.829			100		100	152
9200	42" pipe size	↓	32	2	↓		110		110	166

22 11 19 – Domestic Water Piping Specialties

22 11 19.10 Flexible Connectors

	22 11 19.10 Flexible Connectors	Crew	Daily Output	Labor-Hours	Unit	2020 Bare Costs Material	Labor	Equipment	Total	Total Incl O&P
0010	**FLEXIBLE CONNECTORS**, Corrugated, 5/8" OD, 3/4" ID									
0050	Gas, seamless brass, steel fittings									
0200	12" long	1 Plum	36	.222	Ea.	6.80	14.30		21.10	29
0220	18" long		36	.222		8.55	14.30		22.85	31
0240	24" long		34	.235		10	15.15		25.15	33.50
0260	30" long		34	.235		12.45	15.15		27.60	36
0280	36" long		32	.250		11.90	16.10		28	37
0320	48" long		30	.267		15.10	17.20		32.30	42
0340	60" long		30	.267		17.95	17.20		35.15	45.50
0360	72" long		30	.267		21	17.20		38.20	48.50
2000	Water, copper tubing, dielectric separators									
2100	12" long	1 Plum	36	.222	Ea.	8.20	14.30		22.50	30.50
2220	15" long		36	.222		9.15	14.30		23.45	31.50
2240	18" long		36	.222		9.10	14.30		23.40	31.50
2260	24" long		34	.235		12.20	15.15		27.35	36

22 11 19.14 Flexible Metal Hose

	22 11 19.14 Flexible Metal Hose	Crew	Daily Output	Labor-Hours	Unit	Material	Labor	Equipment	Total	Total Incl O&P
0010	**FLEXIBLE METAL HOSE**, Connectors, standard lengths									
0100	Bronze braided, bronze ends									
0120	3/8" diameter x 12"	1 Stpi	26	.308	Ea.	28.50	20		48.50	61
0140	1/2" diameter x 12"		24	.333		29	22		51	64.50
0160	3/4" diameter x 12"		20	.400		39.50	26		65.50	82.50
0180	1" diameter x 18"		19	.421		51	27.50		78.50	97.50
0200	1-1/2" diameter x 18"		13	.615		57	40.50		97.50	123
0220	2" diameter x 18"		11	.727		93.50	47.50		141	175
1000	Carbon steel ends									
1020	1/4" diameter x 12"	1 Stpi	28	.286	Ea.	36	18.75		54.75	67.50
1040	3/8" diameter x 12"		26	.308		37	20		57	70.50
1060	1/2" diameter x 12"		24	.333		32	22		54	68
1080	1/2" diameter x 24"		24	.333		61.50	22		83.50	101
1120	3/4" diameter x 12"		20	.400		47	26		73	90.50
1140	3/4" diameter x 24"		20	.400		75	26		101	122
1160	3/4" diameter x 36"		20	.400		82	26		108	130
1180	1" diameter x 18"		19	.421		65.50	27.50		93	114
1200	1" diameter x 30"		19	.421		69.50	27.50		97	118
1220	1" diameter x 36"		19	.421		104	27.50		131.50	156
1240	1-1/4" diameter x 18"		15	.533		88.50	35		123.50	150
1260	1-1/4" diameter x 36"		15	.533		141	35		176	208
1280	1-1/2" diameter x 18"		13	.615		121	40.50		161.50	194
1300	1-1/2" diameter x 36"		13	.615		145	40.50		185.50	221
1320	2" diameter x 24"		11	.727		182	47.50		229.50	272
1340	2" diameter x 36"		11	.727		183	47.50		230.50	274
1360	2-1/2" diameter x 24"		9	.889		350	58.50		408.50	470
1380	2-1/2" diameter x 36"		9	.889		126	58.50		184.50	226
1400	3" diameter x 24"		7	1.143		465	75		540	620
1420	3" diameter x 36"		7	1.143		1,025	75		1,100	1,250
2000	Carbon steel braid, carbon steel solid ends									
2100	1/2" diameter x 12"	1 Stpi	24	.333	Ea.	50.50	22		72.50	88.50
2120	3/4" diameter x 12"		20	.400		78.50	26		104.50	126
2140	1" diameter x 12"		19	.421		112	27.50		139.50	165
2160	1-1/4" diameter x 12"		15	.533		55	35		90	113
2180	1-1/2" diameter x 12"		13	.615		60.50	40.50		101	127
3000	Stainless steel braid, welded on carbon steel ends									

22 11 19 – Domestic Water Piping Specialties

22 11 19.14 Flexible Metal Hose		Crew	Daily Output	Labor-Hours	Unit	Material	2020 Bare Costs Labor	Equipment	Total	Total Incl O&P
3100	1/2″ diameter x 12″	1 Stpi	24	.333	Ea.	74.50	22		96.50	115
3120	3/4″ diameter x 12″		20	.400		90	26		116	138
3140	3/4″ diameter x 24″		20	.400		103	26		129	152
3160	3/4″ diameter x 36″		20	.400		119	26		145	170
3180	1″ diameter x 12″		19	.421		108	27.50		135.50	161
3200	1″ diameter x 24″		19	.421		115	27.50		142.50	168
3220	1″ diameter x 36″		19	.421		147	27.50		174.50	204
3240	1-1/4″ diameter x 12″		15	.533		151	35		186	219
3260	1-1/4″ diameter x 24″		15	.533		178	35		213	249
3280	1-1/4″ diameter x 36″		15	.533		200	35		235	273
3300	1-1/2″ diameter x 12″		13	.615		62.50	40.50		103	130
3320	1-1/2″ diameter x 24″		13	.615		186	40.50		226.50	266
3340	1-1/2″ diameter x 36″		13	.615		226	40.50		266.50	310
3400	Metal stainless steel braid, over corrugated stainless steel, flanged ends									
3410	150 psi									
3420	1/2″ diameter x 12″	1 Stpi	24	.333	Ea.	91	22		113	133
3430	1″ diameter x 12″		20	.400		136	26		162	189
3440	1-1/2″ diameter x 12″		15	.533		146	35		181	213
3450	2-1/2″ diameter x 9″		12	.667		64.50	43.50		108	137
3460	3″ diameter x 9″		9	.889		32	58.50		90.50	122
3470	4″ diameter x 9″		7	1.143		41	75		116	157
3480	4″ diameter x 30″		5	1.600		410	105		515	605
3490	4″ diameter x 36″		4.80	1.667		430	109		539	635
3500	6″ diameter x 11″		5	1.600		75	105		180	239
3510	6″ diameter x 36″		3.80	2.105		525	138		663	780
3520	8″ diameter x 12″		4	2		296	131		427	520
3530	10″ diameter x 13″		3	2.667		213	175		388	495
3540	12″ diameter x 14″	Q-5	4	4		335	236		571	725
6000	Molded rubber with helical wire reinforcement									
6010	150 psi									
6020	1-1/2″ diameter x 12″	1 Stpi	15	.533	Ea.	31	35		66	86.50
6030	2″ diameter x 12″		12	.667		78	43.50		121.50	152
6040	3″ diameter x 12″		8	1		81.50	65.50		147	188
6050	4″ diameter x 12″		6	1.333		103	87.50		190.50	245
6060	6″ diameter x 18″		4	2		155	131		286	365
6070	8″ diameter x 24″		3	2.667		217	175		392	500
6080	10″ diameter x 24″		2	4		263	262		525	680
6090	12″ diameter x 24″	Q-5	3	5.333		300	315		615	800
7000	Molded teflon with stainless steel flanges									
7010	150 psi									
7020	2-1/2″ diameter x 3-3/16″	Q-1	7.80	2.051	Ea.	725	119		844	975
7030	3″ diameter x 3-5/8″		6.50	2.462		550	143		693	820
7040	4″ diameter x 3-5/8″		5	3.200		700	186		886	1,050
7050	6″ diameter x 4″		4.30	3.721		975	216		1,191	1,400
7060	8″ diameter x 6″		3.80	4.211		1,550	244		1,794	2,075

22 11 19.18 Mixing Valve		Crew	Daily Output	Labor-Hours	Unit	Material	Labor	Equipment	Total	Total Incl O&P
0010	**MIXING VALVE**, Automatic, water tempering.									
0040	1/2″ size	1 Stpi	19	.421	Ea.	625	27.50		652.50	730
0050	3/4″ size		18	.444		660	29		689	770
0100	1″ size		16	.500		930	33		963	1,075
0120	1-1/4″ size		13	.615		1,250	40.50		1,290.50	1,425
0140	1-1/2″ size		10	.800		1,525	52.50		1,577.50	1,750

22 11 19.18 Mixing Valve

22 11 19.18 Mixing Valve		Crew	Daily Output	Labor-Hours	Unit	2020 Bare Costs Material	Labor	Equipment	Total	Total Incl O&P
0160	2″ size	1 Stpi	8	1	Ea.	1,900	65.50		1,965.50	2,200
0170	2-1/2″ size		6	1.333		1,900	87.50		1,987.50	2,225
0180	3″ size		4	2		4,550	131		4,681	5,200
0190	4″ size		3	2.667		4,550	175		4,725	5,250

22 11 19.22 Pressure Reducing Valve

22 11 19.22 Pressure Reducing Valve		Crew	Daily Output	Labor-Hours	Unit	Material	Labor	Equipment	Total	Total Incl O&P
0010	**PRESSURE REDUCING VALVE**, Steam, pilot operated.									
0100	Threaded, iron body									
0200	1-1/2″ size	1 Stpi	8	1	Ea.	2,125	65.50		2,190.50	2,425
0220	2″ size	″	5	1.600	″	2,450	105		2,555	2,850
1000	Flanged, iron body, 125 lb. flanges									
1020	2″ size	1 Stpi	8	1	Ea.	3,400	65.50		3,465.50	3,850
1040	2-1/2″ size	″	4	2		2,525	131		2,656	2,975
1060	3″ size	Q-5	4.50	3.556		3,225	210		3,435	3,875
1080	4″ size	″	3	5.333		4,050	315		4,365	4,925
1500	For 250 lb. flanges, add					5%				

22 11 19.26 Pressure Regulators

22 11 19.26 Pressure Regulators		Crew	Daily Output	Labor-Hours	Unit	Material	Labor	Equipment	Total	Total Incl O&P
0010	**PRESSURE REGULATORS**									
0100	Gas appliance regulators									
0106	Main burner and pilot applications									
0108	Rubber seat poppet type									
0109	1/8″ pipe size	1 Stpi	24	.333	Ea.	19.65	22		41.65	54
0110	1/4″ pipe size		24	.333		23.50	22		45.50	58
0112	3/8″ pipe size		24	.333		24	22		46	58.50
0113	1/2″ pipe size		24	.333		26.50	22		48.50	61.50
0114	3/4″ pipe size		20	.400		31	26		57	73
0122	Lever action type									
0123	3/8″ pipe size	1 Stpi	24	.333	Ea.	30	22		52	65.50
0124	1/2″ pipe size		24	.333		30	22		52	65.50
0125	3/4″ pipe size		20	.400		58	26		84	103
0126	1″ pipe size		19	.421		58	27.50		85.50	106
0132	Double diaphragm type									
0133	3/8″ pipe size	1 Stpi	24	.333	Ea.	46	22		68	83.50
0134	1/2″ pipe size		24	.333		59.50	22		81.50	98
0135	3/4″ pipe size		20	.400		96	26		122	144
0136	1″ pipe size		19	.421		110	27.50		137.50	163
0137	1-1/4″ pipe size		15	.533		395	35		430	490
0138	1-1/2″ pipe size		13	.615		725	40.50		765.50	860
0139	2″ pipe size		11	.727		725	47.50		772.50	870
0140	2-1/2″ pipe size	Q-5	15	1.067		1,400	63		1,463	1,650
0141	3″ pipe size		13	1.231		1,400	72.50		1,472.50	1,650
0142	4″ pipe size (flanged)		8	2		2,500	118		2,618	2,925
0160	Main burner only									
0162	Straight-thru-flow design									
0163	1/2″ pipe size	1 Stpi	24	.333	Ea.	48.50	22		70.50	86
0164	3/4″ pipe size		20	.400		66.50	26		92.50	112
0165	1″ pipe size		19	.421		88	27.50		115.50	138
0166	1-1/4″ pipe size		15	.533		88	35		123	149
0200	Oil, light, hot water, ordinary steam, threaded									
0220	Bronze body, 1/4″ size	1 Stpi	24	.333	Ea.	197	22		219	250
0230	3/8″ size		24	.333		190	22		212	242
0240	1/2″ size		24	.333		233	22		255	290
0250	3/4″ size		20	.400		278	26		304	345

22 11 Facility Water Distribution

22 11 19 – Domestic Water Piping Specialties

22 11 19.26 Pressure Regulators		Crew	Daily Output	Labor-Hours	Unit	2020 Bare Costs Material	Labor	Equipment	Total	Total Incl O&P
0260	1″ size	1 Stpi	19	.421	Ea.	430	27.50		457.50	510
0270	1-1/4″ size		15	.533		575	35		610	685
0320	Iron body, 1/4″ size		24	.333		128	22		150	173
0330	3/8″ size		24	.333		172	22		194	223
0340	1/2″ size		24	.333		157	22		179	206
0350	3/4″ size		20	.400		191	26		217	249
0360	1″ size		19	.421		247	27.50		274.50	315
0370	1-1/4″ size		15	.533		345	35		380	435
9002	For water pressure regulators, see Section 22 05 23.20									

22 11 19.30 Pressure and Temperature Safety Plug		Crew	Daily Output	Labor-Hours	Unit	Material	Labor	Equipment	Total	Total Incl O&P
0010	PRESSURE & TEMPERATURE SAFETY PLUG									
1000	3/4″ external thread, 3/8″ diam. element									
1020	Carbon steel									
1050	7-1/2″ insertion	1 Stpi	32	.250	Ea.	50.50	16.40		66.90	80
1120	304 stainless steel									
1150	7-1/2″ insertion	1 Stpi	32	.250	Ea.	48	16.40		64.40	77
1220	316 stainless steel									
1250	7-1/2″ insertion	1 Stpi	32	.250	Ea.	58.50	16.40		74.90	88.50

22 11 19.32 Pressure and Temperature Measurement Plug		Crew	Daily Output	Labor-Hours	Unit	Material	Labor	Equipment	Total	Total Incl O&P
0010	PRESSURE & TEMPERATURE MEASUREMENT PLUG									
0020	A permanent access port for insertion of									
0030	a pressure or temperature measuring probe									
0100	Plug, brass									
0110	1/4″ MNPT, 1-1/2″ long	1 Stpi	32	.250	Ea.	6.55	16.40		22.95	31.50
0120	3″ long		31	.258		12.20	16.90		29.10	39
0140	1/2″ MNPT, 1-1/2″ long		30	.267		9	17.50		26.50	36
0150	3″ long		29	.276		14.80	18.10		32.90	43.50
0200	Pressure gauge probe adapter									
0210	1/8″ diameter, 1-1/2″ probe				Ea.	21			21	23.50
0220	3″ probe				"	35			35	38.50
0300	Temperature gauge, 5″ stem									
0310	Analog				Ea.	10.90			10.90	12
0330	Digital				"	29.50			29.50	32.50
0400	Pressure gauge, compound									
0410	1/4″ MNPT				Ea.	32.50			32.50	35.50
0500	Pressure and temperature test kit									
0510	Contains 2 thermometers and 2 pressure gauges									
0520	Kit				Ea.	355			355	390

22 11 19.34 Sleeves and Escutcheons		Crew	Daily Output	Labor-Hours	Unit	Material	Labor	Equipment	Total	Total Incl O&P
0010	SLEEVES & ESCUTCHEONS									
0100	Pipe sleeve									
0110	Steel, w/water stop, 12″ long, with link seal									
0120	2″ diam. for 1/2″ carrier pipe	1 Plum	8.40	.952	Ea.	64.50	61.50		126	163
0130	2-1/2″ diam. for 3/4″ carrier pipe		8	1		73	64.50		137.50	177
0140	2-1/2″ diam. for 1″ carrier pipe		8	1		69	64.50		133.50	173
0150	3″ diam. for 1-1/4″ carrier pipe		7.20	1.111		89	71.50		160.50	205
0160	3-1/2″ diam. for 1-1/2″ carrier pipe		6.80	1.176		97	76		173	219
0170	4″ diam. for 2″ carrier pipe		6	1.333		98	86		184	237
0180	4″ diam. for 2-1/2″ carrier pipe		6	1.333		98	86		184	237
0190	5″ diam. for 3″ carrier pipe		5.40	1.481		115	95.50		210.50	270
0200	6″ diam. for 4″ carrier pipe		4.80	1.667		138	107		245	310
0210	10″ diam. for 6″ carrier pipe	Q-1	8	2		271	116		387	470

22 11 19.34 Sleeves and Escutcheons		Crew	Daily Output	Labor-Hours	Unit	2020 Bare Costs Material	Labor	Equipment	Total	Total Incl O&P
0220	12″ diam. for 8″ carrier pipe	Q-1	7.20	2.222	Ea.	370	129		499	605
0230	14″ diam. for 10″ carrier pipe		6.40	2.500		385	145		530	635
0240	16″ diam. for 12″ carrier pipe		5.80	2.759		425	160		585	710
0250	18″ diam. for 14″ carrier pipe		5.20	3.077		780	178		958	1,125
0260	24″ diam. for 18″ carrier pipe		4	4		1,125	232		1,357	1,600
0270	24″ diam. for 20″ carrier pipe		4	4		975	232		1,207	1,425
0280	30″ diam. for 24″ carrier pipe		3.20	5		1,400	290		1,690	1,950
0500	Wall sleeve									
0510	Ductile iron with rubber gasket seal									
0520	3″	1 Plum	8.40	.952	Ea.	213	61.50		274.50	325
0530	4″		7.20	1.111		227	71.50		298.50	355
0540	6″		6	1.333		276	86		362	435
0550	8″		4	2		1,225	129		1,354	1,550
0560	10″		3	2.667		1,500	172		1,672	1,900
0570	12″		2.40	3.333		1,750	215		1,965	2,250
5000	Escutcheon									
5100	Split ring, pipe									
5110	Chrome plated									
5120	1/2″	1 Plum	160	.050	Ea.	1.10	3.22		4.32	6.05
5130	3/4″		160	.050		1.29	3.22		4.51	6.25
5140	1″		135	.059		1.43	3.82		5.25	7.25
5150	1-1/2″		115	.070		1.86	4.48		6.34	8.75
5160	2″		100	.080		2.16	5.15		7.31	10.10
5170	4″		80	.100		5	6.45		11.45	15.15
5180	6″		68	.118		7.90	7.60		15.50	20
5400	Shallow flange type									
5410	Chrome plated steel									
5420	1/2″ CTS	1 Plum	180	.044	Ea.	.33	2.86		3.19	4.65
5430	3/4″ CTS		180	.044		.37	2.86		3.23	4.70
5440	1/2″ IPS		180	.044		.40	2.86		3.26	4.73
5450	3/4″ IPS		180	.044		.44	2.86		3.30	4.77
5460	1″ IPS		175	.046		.48	2.95		3.43	4.94
5470	1-1/2″ IPS		170	.047		.81	3.03		3.84	5.45
5480	2″ IPS		160	.050		.99	3.22		4.21	5.90

22 11 19.38 Water Supply Meters

Line	Description	Crew	Daily Output	Labor-Hours	Unit	Material	Labor	Equipment	Total	Total Incl O&P
0010	**WATER SUPPLY METERS**									
1000	Detector, serves dual systems such as fire and domestic or									
1020	process water, wide range cap., UL and FM approved									
1100	3″ mainline x 2″ by-pass, 400 GPM	Q-1	3.60	4.444	Ea.	8,025	258		8,283	9,200
1140	4″ mainline x 2″ by-pass, 700 GPM	″	2.50	6.400		8,025	370		8,395	9,375
1180	6″ mainline x 3″ by-pass, 1,600 GPM	Q-2	2.60	9.231		12,300	555		12,855	14,300
1220	8″ mainline x 4″ by-pass, 2,800 GPM		2.10	11.429		18,200	685		18,885	21,000
1260	10″ mainline x 6″ by-pass, 4,400 GPM		2	12		26,000	720		26,720	29,700
1300	10″ x 12″ mainlines x 6″ by-pass, 5,400 GPM		1.70	14.118		35,300	850		36,150	40,100
2000	Domestic/commercial, bronze									
2020	Threaded									
2060	5/8″ diameter, to 20 GPM	1 Plum	16	.500	Ea.	54	32		86	108
2080	3/4″ diameter, to 30 GPM		14	.571		98.50	37		135.50	163
2100	1″ diameter, to 50 GPM		12	.667		149	43		192	229
2300	Threaded/flanged									
2340	1-1/2″ diameter, to 100 GPM	1 Plum	8	1	Ea.	365	64.50		429.50	495
2360	2″ diameter, to 160 GPM	″	6	1.333	″	495	86		581	675

22 11 Facility Water Distribution

22 11 19 – Domestic Water Piping Specialties

22 11 19.38 Water Supply Meters

Line	Description	Crew	Daily Output	Labor-Hours	Unit	Material (2020 Bare Costs)	Labor (2020 Bare Costs)	Equipment (2020 Bare Costs)	Total (2020 Bare Costs)	Total Incl O&P
2600	Flanged, compound									
2640	3" diameter, 320 GPM	Q-1	3	5.333	Ea.	2,650	310		2,960	3,400
2660	4" diameter, to 500 GPM		1.50	10.667		4,250	620		4,870	5,600
2680	6" diameter, to 1,000 GPM		1	16		6,850	930		7,780	8,950
2700	8" diameter, to 1,800 GPM		.80	20		10,700	1,150		11,850	13,500
7000	Turbine									
7260	Flanged									
7300	2" diameter, to 160 GPM	1 Plum	7	1.143	Ea.	615	73.50		688.50	785
7320	3" diameter, to 450 GPM	Q-1	3.60	4.444		1,100	258		1,358	1,575
7340	4" diameter, to 650 GPM	"	2.50	6.400		1,825	370		2,195	2,550
7360	6" diameter, to 1,800 GPM	Q-2	2.60	9.231		3,000	555		3,555	4,125
7380	8" diameter, to 2,500 GPM		2.10	11.429		5,150	685		5,835	6,700
7400	10" diameter, to 5,500 GPM		1.70	14.118		6,925	850		7,775	8,900

22 11 19.42 Backflow Preventers

Line	Description	Crew	Daily Output	Labor-Hours	Unit	Material	Labor	Equipment	Total	Total Incl O&P
0010	**BACKFLOW PREVENTERS**, Includes valves									
0020	and four test cocks, corrosion resistant, automatic operation									
1000	Double check principle									
1010	Threaded, with ball valves									
1020	3/4" pipe size	1 Plum	16	.500	Ea.	241	32		273	315
1030	1" pipe size		14	.571		253	37		290	335
1040	1-1/2" pipe size		10	.800		620	51.50		671.50	760
1050	2" pipe size		7	1.143		665	73.50		738.50	845
1080	Threaded, with gate valves									
1100	3/4" pipe size	1 Plum	16	.500	Ea.	1,150	32		1,182	1,300
1120	1" pipe size		14	.571		1,150	37		1,187	1,325
1140	1-1/2" pipe size		10	.800		1,225	51.50		1,276.50	1,425
1160	2" pipe size		7	1.143		1,825	73.50		1,898.50	2,100
1200	Flanged, valves are gate									
1210	3" pipe size	Q-1	4.50	3.556	Ea.	2,975	206		3,181	3,575
1220	4" pipe size	"	3	5.333		3,000	310		3,310	3,775
1230	6" pipe size	Q-2	3	8		4,675	480		5,155	5,850
1240	8" pipe size		2	12		7,975	720		8,695	9,850
1250	10" pipe size		1	24		12,000	1,450		13,450	15,400
1300	Flanged, valves are OS&Y									
1370	1" pipe size	1 Plum	5	1.600	Ea.	1,275	103		1,378	1,550
1374	1-1/2" pipe size		5	1.600		1,350	103		1,453	1,625
1378	2" pipe size		4.80	1.667		1,950	107		2,057	2,300
1380	3" pipe size	Q-1	4.50	3.556		2,850	206		3,056	3,450
1400	4" pipe size	"	3	5.333		4,675	310		4,985	5,600
1420	6" pipe size	Q-2	3	8		6,700	480		7,180	8,100
1430	8" pipe size	"	2	12		14,000	720		14,720	16,500
4000	Reduced pressure principle									
4100	Threaded, bronze, valves are ball									
4120	3/4" pipe size	1 Plum	16	.500	Ea.	530	32		562	635
4140	1" pipe size		14	.571		560	37		597	670
4150	1-1/4" pipe size		12	.667		1,125	43		1,168	1,300
4160	1-1/2" pipe size		10	.800		1,150	51.50		1,201.50	1,350
4180	2" pipe size		7	1.143		1,325	73.50		1,398.50	1,575
5000	Flanged, bronze, valves are OS&Y									
5060	2-1/2" pipe size	Q-1	5	3.200	Ea.	5,475	186		5,661	6,300
5080	3" pipe size		4.50	3.556		6,600	206		6,806	7,550
5100	4" pipe size		3	5.333		7,725	310		8,035	8,975

22 11 19 – Domestic Water Piping Specialties

22 11 19.42 Backflow Preventers		Crew	Daily Output	Labor-Hours	Unit	Material	Labor	Equipment	Total	Total Incl O&P
						2020 Bare Costs				
5120	6" pipe size	Q-2	3	8	Ea.	12,000	480		12,480	13,900
5200	Flanged, iron, valves are gate									
5210	2-1/2" pipe size	Q-1	5	3.200	Ea.	3,200	186		3,386	3,775
5220	3" pipe size		4.50	3.556		3,300	206		3,506	3,950
5230	4" pipe size		3	5.333		3,775	310		4,085	4,650
5240	6" pipe size	Q-2	3	8		5,675	480		6,155	6,975
5250	8" pipe size		2	12		9,575	720		10,295	11,600
5260	10" pipe size		1	24		16,200	1,450		17,650	20,100
5600	Flanged, iron, valves are OS&Y									
5660	2-1/2" pipe size	Q-1	5	3.200	Ea.	3,300	186		3,486	3,900
5680	3" pipe size		4.50	3.556		3,275	206		3,481	3,900
5700	4" pipe size		3	5.333		4,725	310		5,035	5,675
5720	6" pipe size	Q-2	3	8		5,825	480		6,305	7,150
5740	8" pipe size		2	12		10,300	720		11,020	12,400
5760	10" pipe size		1	24		19,800	1,450		21,250	24,000

22 11 19.50 Vacuum Breakers

Line	Description	Crew	Daily Output	Labor-Hours	Unit	Material	Labor	Equipment	Total	Total Incl O&P
0010	**VACUUM BREAKERS** R221113-40									
0013	See also backflow preventers Section 22 11 19.42									
1000	Anti-siphon continuous pressure type									
1010	Max. 150 psi - 210°F									
1020	Bronze body									
1030	1/2" size	1 Stpi	24	.333	Ea.	192	22		214	244
1040	3/4" size		20	.400		192	26		218	250
1050	1" size		19	.421		197	27.50		224.50	258
1060	1-1/4" size		15	.533		390	35		425	485
1070	1-1/2" size		13	.615		475	40.50		515.50	585
1080	2" size		11	.727		490	47.50		537.50	610
1200	Max. 125 psi with atmospheric vent									
1210	Brass, in-line construction									
1220	1/4" size	1 Stpi	24	.333	Ea.	143	22		165	190
1230	3/8" size	"	24	.333		143	22		165	190
1260	For polished chrome finish, add					13%				
2000	Anti-siphon, non-continuous pressure type									
2010	Hot or cold water 125 psi - 210°F									
2020	Bronze body									
2030	1/4" size	1 Stpi	24	.333	Ea.	88.50	22		110.50	130
2040	3/8" size		24	.333		88.50	22		110.50	130
2050	1/2" size		24	.333		99	22		121	142
2060	3/4" size		20	.400		119	26		145	169
2070	1" size		19	.421		183	27.50		210.50	244
2080	1-1/4" size		15	.533		320	35		355	410
2090	1-1/2" size		13	.615		350	40.50		390.50	445
2100	2" size		11	.727		585	47.50		632.50	715
2110	2-1/2" size		8	1		1,675	65.50		1,740.50	1,950
2120	3" size		6	1.333		2,225	87.50		2,312.50	2,575
2150	For polished chrome finish, add					50%				
3000	Air gap fitting									
3020	1/2" NPT size	1 Plum	19	.421	Ea.	61.50	27		88.50	108
3030	1" NPT size	"	15	.533		70.50	34.50		105	129
3040	2" NPT size	Q-1	21	.762		138	44		182	218
3050	3" NPT size		14	1.143		273	66.50		339.50	400
3060	4" NPT size		10	1.600		273	93		366	440

22 11 19.54 Water Hammer Arresters/Shock Absorbers		Crew	Daily Output	Labor-Hours	Unit	2020 Bare Costs Material	Labor	Equipment	Total	Total Incl O&P
0010	**WATER HAMMER ARRESTERS/SHOCK ABSORBERS**									
0490	Copper									
0500	3/4" male IPS for 1 to 11 fixtures	1 Plum	12	.667	Ea.	31	43		74	98.50
0600	1" male IPS for 12 to 32 fixtures		8	1		51.50	64.50		116	153
0700	1-1/4" male IPS for 33 to 60 fixtures		8	1		50.50	64.50		115	153
0800	1-1/2" male IPS for 61 to 113 fixtures		8	1		73	64.50		137.50	177
0900	2" male IPS for 114 to 154 fixtures		8	1		108	64.50		172.50	215
1000	2-1/2" male IPS for 155 to 330 fixtures		4	2		330	129		459	560
4000	Bellows type									
4010	3/4" FNPT, to 11 fixture units	1 Plum	10	.800	Ea.	355	51.50		406.50	465
4020	1" FNPT, to 32 fixture units		8.80	.909		725	58.50		783.50	885
4030	1" FNPT, to 60 fixture units		8.80	.909		1,075	58.50		1,133.50	1,300
4040	1" FNPT, to 113 fixture units		8.80	.909		2,725	58.50		2,783.50	3,075
4050	1" FNPT, to 154 fixture units		6.60	1.212		3,075	78		3,153	3,500
4060	1-1/2" FNPT, to 300 fixture units		5	1.600		3,775	103		3,878	4,300
22 11 19.64 Hydrants										
0010	**HYDRANTS**									
0050	Wall type, moderate climate, bronze, encased									
0200	3/4" IPS connection	1 Plum	16	.500	Ea.	1,025	32		1,057	1,175
0300	1" IPS connection		14	.571		1,125	37		1,162	1,300
0500	Anti-siphon type, 3/4" connection		16	.500		815	32		847	945
1000	Non-freeze, bronze, exposed									
1100	3/4" IPS connection, 4" to 9" thick wall	1 Plum	14	.571	Ea.	470	37		507	570
1120	10" to 14" thick wall		12	.667		445	43		488	555
1140	15" to 19" thick wall		12	.667		550	43		593	670
1160	20" to 24" thick wall		10	.800		830	51.50		881.50	985
1200	For 1" IPS connection, add					15%	10%			
1240	For 3/4" adapter type vacuum breaker, add				Ea.	83			83	91.50
1280	For anti-siphon type, add				"	176			176	194
2000	Non-freeze bronze, encased, anti-siphon type									
2100	3/4" IPS connection, 5" to 9" thick wall	1 Plum	14	.571	Ea.	1,275	37		1,312	1,450
2120	10" to 14" thick wall		12	.667		1,625	43		1,668	1,875
2140	15" to 19" thick wall		12	.667		1,725	43		1,768	1,975
2160	20" to 24" thick wall		10	.800		1,800	51.50		1,851.50	2,050
2200	For 1" IPS connection, add					10%	10%			
3000	Ground box type, bronze frame, 3/4" IPS connection									
3080	Non-freeze, all bronze, polished face, set flush									
3100	2' depth of bury	1 Plum	8	1	Ea.	1,325	64.50		1,389.50	1,550
3120	3' depth of bury		8	1		1,425	64.50		1,489.50	1,675
3140	4' depth of bury		8	1		1,525	64.50		1,589.50	1,775
3160	5' depth of bury		7	1.143		1,625	73.50		1,698.50	1,875
3180	6' depth of bury		7	1.143		1,700	73.50		1,773.50	1,950
3200	7' depth of bury		6	1.333		1,675	86		1,761	1,975
3220	8' depth of bury		5	1.600		1,775	103		1,878	2,100
3240	9' depth of bury		4	2		1,875	129		2,004	2,275
3260	10' depth of bury		4	2		1,975	129		2,104	2,350
3400	For 1" IPS connection, add					15%	10%			
3450	For 1-1/4" IPS connection, add					325%	14%			
3500	For 1-1/2" IPS connection, add					370%	18%			
3550	For 2" IPS connection, add					445%	24%			
3600	For tapped drain port in box, add					107			107	118
4000	Non-freeze, CI body, bronze frame & scoriated cover									

22 11 Facility Water Distribution

22 11 19 – Domestic Water Piping Specialties

22 11 19.64 Hydrants		Crew	Daily Output	Labor-Hours	Unit	2020 Bare Costs Material	Labor	Equipment	Total	Total Incl O&P
4010	with hose storage									
4100	2′ depth of bury	1 Plum	7	1.143	Ea.	2,200	73.50		2,273.50	2,525
4120	3′ depth of bury		7	1.143		2,300	73.50		2,373.50	2,625
4140	4′ depth of bury		7	1.143		2,375	73.50		2,448.50	2,700
4160	5′ depth of bury		6.50	1.231		2,425	79.50		2,504.50	2,775
4180	6′ depth of bury		6	1.333		2,450	86		2,536	2,825
4200	7′ depth of bury		5.50	1.455		2,550	94		2,644	2,950
4220	8′ depth of bury		5	1.600		2,625	103		2,728	3,025
4240	9′ depth of bury		4.50	1.778		2,825	115		2,940	3,300
4260	10′ depth of bury		4	2		2,825	129		2,954	3,300
4280	For 1″ IPS connection, add					505			505	555
4300	For tapped drain port in box, add					113			113	124
5000	Moderate climate, all bronze, polished face									
5020	and scoriated cover, set flush									
5100	3/4″ IPS connection	1 Plum	16	.500	Ea.	830	32		862	965
5120	1″ IPS connection	″	14	.571		2,000	37		2,037	2,275
5200	For tapped drain port in box, add					113			113	124
6000	Ground post type, all non-freeze, all bronze, aluminum casing									
6010	guard, exposed head, 3/4″ IPS connection									
6100	2′ depth of bury	1 Plum	8	1	Ea.	1,275	64.50		1,339.50	1,500
6120	3′ depth of bury		8	1		1,300	64.50		1,364.50	1,550
6140	4′ depth of bury		8	1		1,475	64.50		1,539.50	1,725
6160	5′ depth of bury		7	1.143		1,450	73.50		1,523.50	1,700
6180	6′ depth of bury		7	1.143		1,775	73.50		1,848.50	2,050
6200	7′ depth of bury		6	1.333		1,675	86		1,761	1,950
6220	8′ depth of bury		5	1.600		1,775	103		1,878	2,100
6240	9′ depth of bury		4	2		1,875	129		2,004	2,275
6260	10′ depth of bury		4	2		2,000	129		2,129	2,375
6300	For 1″ IPS connection, add					40%	10%			
6350	For 1-1/4″ IPS connection, add					140%	14%			
6400	For 1-1/2″ IPS connection, add					225%	18%			
6450	For 2″ IPS connection, add					315%	24%			

22 11 23 – Domestic Water Pumps

22 11 23.10 General Utility Pumps		Crew	Daily Output	Labor-Hours	Unit	Material	Labor	Equipment	Total	Total Incl O&P
0010	**GENERAL UTILITY PUMPS**									
2000	Single stage									
3000	Double suction,									
3140	50 HP, 5″ D x 6″ S	Q-2	.33	72.727	Ea.	9,900	4,375		14,275	17,500
3180	60 HP, 6″ D x 8″ S	Q-3	.30	107		17,500	6,550		24,050	29,100
3190	75 HP, to 2,500 GPM		.28	114		16,500	7,000		23,500	28,600
3220	100 HP, to 3,000 GPM		.26	123		20,300	7,550		27,850	33,600
3240	150 HP, to 4,000 GPM		.24	133		23,600	8,175		31,775	38,200
4000	Centrifugal, end suction, mounted on base									
4010	Horizontal mounted, with drip proof motor, rated @ 100′ head									
4020	Vertical split case, single stage									
4040	100 GPM, 5 HP, 1-1/2″ discharge	Q-1	1.70	9.412	Ea.	4,725	545		5,270	6,000
4050	200 GPM, 10 HP, 2″ discharge		1.30	12.308		4,925	715		5,640	6,500
4060	250 GPM, 10 HP, 3″ discharge		1.28	12.500		5,500	725		6,225	7,125
4070	300 GPM, 15 HP, 2″ discharge	Q-2	1.56	15.385		6,475	925		7,400	8,475
4080	500 GPM, 20 HP, 4″ discharge		1.44	16.667		7,725	1,000		8,725	10,000
4090	750 GPM, 30 HP, 4″ discharge		1.20	20		8,325	1,200		9,525	11,000
4100	1,050 GPM, 40 HP, 5″ discharge		1	24		9,900	1,450		11,350	13,100

22 11 23.10 General Utility Pumps

	22 11 23.10 General Utility Pumps	Crew	Daily Output	Labor-Hours	Unit	2020 Bare Costs Material	Labor	Equipment	Total	Total Incl O&P
4110	1,500 GPM, 60 HP, 6" discharge	Q-2	.60	40	Ea.	17,500	2,400		19,900	22,900
4120	2,000 GPM, 75 HP, 6" discharge		.50	48		18,900	2,875		21,775	25,100
4130	3,000 GPM, 100 HP, 8" discharge		.40	60		22,500	3,600		26,100	30,200
4200	Horizontal split case, single stage									
4210	100 GPM, 7.5 HP, 1-1/2" discharge	Q-1	1.70	9.412	Ea.	5,250	545		5,795	6,600
4220	250 GPM, 15 HP, 2-1/2" discharge	"	1.30	12.308		7,800	715		8,515	9,650
4230	500 GPM, 20 HP, 4" discharge	Q-2	1.60	15		10,600	900		11,500	13,100
4240	750 GPM, 25 HP, 5" discharge		1.54	15.584		10,700	935		11,635	13,100
4250	1,000 GPM, 40 HP, 5" discharge		1.20	20		13,400	1,200		14,600	16,500
4260	1,500 GPM, 50 HP, 6" discharge	Q-3	1.42	22.535		17,000	1,375		18,375	20,800
4270	2,000 GPM, 75 HP, 8" discharge		1.14	28.070		21,900	1,725		23,625	26,700
4280	3,000 GPM, 100 HP, 10" discharge		.96	33.333		22,800	2,050		24,850	28,100
4290	3,500 GPM, 150 HP, 10" discharge		.86	37.209		37,700	2,275		39,975	44,900
4300	4,000 GPM, 200 HP, 10" discharge		.66	48.485		27,800	2,975		30,775	35,100
4330	Horizontal split case, two stage, 500' head									
4340	100 GPM, 40 HP, 1-1/2" discharge	Q-2	1.70	14.118	Ea.	17,800	850		18,650	20,900
4350	200 GPM, 50 HP, 1-1/2" discharge	"	1.44	16.667		19,600	1,000		20,600	23,000
4360	300 GPM, 75 HP, 2" discharge	Q-3	1.57	20.382		25,600	1,250		26,850	30,000
4370	400 GPM, 100 HP, 3" discharge		1.14	28.070		31,100	1,725		32,825	36,800
4380	800 GPM, 200 HP, 4" discharge		.86	37.209		41,800	2,275		44,075	49,400
5000	Centrifugal, in-line									
5006	Vertical mount, iron body, 125 lb. flgd, 1,800 RPM TEFC mtr									
5010	Single stage									
5012	.5 HP, 1-1/2" suction & discharge	Q-1	3.20	5	Ea.	1,525	290		1,815	2,100
5014	.75 HP, 2" suction & discharge		2.80	5.714		2,950	330		3,280	3,750
5015	1 HP, 3" suction & discharge		2.50	6.400		3,250	370		3,620	4,125
5016	1.5 HP, 3" suction & discharge		2.30	6.957		3,200	405		3,605	4,125
5018	2 HP, 4" suction & discharge		2.10	7.619		4,450	440		4,890	5,550
5020	3 HP, 5" suction & discharge		1.90	8.421		4,900	490		5,390	6,100
5030	5 HP, 6" suction & discharge		1.60	10		5,550	580		6,130	6,975
5040	7.5 HP, 6" suction & discharge		1.30	12.308		6,400	715		7,115	8,125
5050	10 HP, 8" suction & discharge	Q-2	1.80	13.333		8,500	800		9,300	10,600
5060	15 HP, 8" suction & discharge		1.70	14.118		10,000	850		10,850	12,300
5064	20 HP, 8" suction & discharge		1.60	15		12,300	900		13,200	14,900
5070	30 HP, 8" suction & discharge		1.50	16		14,700	960		15,660	17,700
5080	40 HP, 8" suction & discharge		1.40	17.143		17,700	1,025		18,725	21,100
5090	50 HP, 8" suction & discharge		1.30	18.462		18,000	1,100		19,100	21,500
5094	60 HP, 8" suction & discharge		.90	26.667		20,100	1,600		21,700	24,500
5100	75 HP, 8" suction & discharge	Q-3	.60	53.333		21,600	3,275		24,875	28,700
5110	100 HP, 8" suction & discharge	"	.50	64		24,600	3,925		28,525	32,900

22 11 23.11 Miscellaneous Pumps

	22 11 23.11 Miscellaneous Pumps	Crew	Daily Output	Labor-Hours	Unit	Material	Labor	Equipment	Total	Total Incl O&P
0010	**MISCELLANEOUS PUMPS**									
0020	Water pump, portable, gasoline powered									
0100	170 GPH, 2" discharge	Q-1	11	1.455	Ea.	780	84.50		864.50	980
0110	343 GPH, 3" discharge		10.50	1.524		955	88.50		1,043.50	1,175
0120	608 GPH, 4" discharge		10	1.600		1,550	93		1,643	1,875
0500	Pump, propylene body, housing and impeller									
0510	22 GPM, 1/3 HP, 40' HD	1 Plum	5	1.600	Ea.	920	103		1,023	1,150
0520	33 GPM, 1/2 HP, 40' HD	Q-1	5	3.200		840	186		1,026	1,200
0530	53 GPM, 3/4 HP, 40' HD	"	4	4		1,025	232		1,257	1,475
0600	Rotary pump, CI									
0614	282 GPH, 3/4 HP, 1" discharge	Q-1	4.50	3.556	Ea.	1,200	206		1,406	1,625

22 11 23.11 Miscellaneous Pumps		Crew	Daily Output	Labor-Hours	Unit	Material	2020 Bare Costs Labor	Equipment	Total	Total Incl O&P
0618	277 GPH, 1 HP, 1″ discharge	Q-1	4	4	Ea.	1,200	232		1,432	1,675
0624	1,100 GPH, 1.5 HP, 1-1/4″ discharge		3.60	4.444		2,975	258		3,233	3,650
0628	1,900 GPH, 2 HP, 1-1/4″ discharge		3.20	5		4,200	290		4,490	5,050
1000	Turbine pump, CI									
1010	50 GPM, 2 HP, 3″ discharge	Q-1	.80	20	Ea.	4,075	1,150		5,225	6,200
1020	100 GPM, 3 HP, 4″ discharge	Q-2	.96	25		6,750	1,500		8,250	9,675
1030	250 GPM, 15 HP, 6″ discharge	″	.94	25.532		8,350	1,525		9,875	11,500
1040	500 GPM, 25 HP, 6″ discharge	Q-3	1.22	26.230		8,450	1,600		10,050	11,700
1050	1,000 GPM, 50 HP, 8″ discharge		1.14	28.070		11,600	1,725		13,325	15,400
1060	2,000 GPM, 100 HP, 10″ discharge		1	32		13,800	1,975		15,775	18,200
1070	3,000 GPM, 150 HP, 10″ discharge		.80	40		22,200	2,450		24,650	28,100
1080	4,000 GPM, 200 HP, 12″ discharge		.70	45.714		24,300	2,800		27,100	30,900
1090	6,000 GPM, 300 HP, 14″ discharge		.60	53.333		37,000	3,275		40,275	45,600
1100	10,000 GPM, 300 HP, 18″ discharge		.58	55.172		42,800	3,375		46,175	52,000
2000	Centrifugal stainless steel pumps									
2100	100 GPM, 100′ TDH, 3 HP	Q-1	1.80	8.889	Ea.	12,100	515		12,615	14,100
2130	250 GPM, 100′ TDH, 10 HP	″	1.28	12.500		17,000	725		17,725	19,700
2160	500 GPM, 100′ TDH, 20 HP	Q-2	1.44	16.667		19,400	1,000		20,400	22,800
2200	Vertical turbine stainless steel pumps									
2220	100 GPM, 100′ TDH, 7.5 HP	Q-2	.95	25.263	Ea.	68,000	1,525		69,525	77,000
2240	250 GPM, 100′ TDH, 15 HP		.94	25.532		75,000	1,525		76,525	85,000
2260	500 GPM, 100′ TDH, 20 HP		.93	25.806		82,500	1,550		84,050	93,000
2280	750 GPM, 100′ TDH, 30 HP	Q-3	1.19	26.891		89,500	1,650		91,150	101,000
2300	1,000 GPM, 100′ TDH, 40 HP	″	1.16	27.586		97,000	1,700		98,700	109,000
2320	100 GPM, 200′ TDH, 15 HP	Q-2	.94	25.532		104,000	1,525		105,525	117,000
2340	250 GPM, 200′ TDH, 25 HP	Q-3	1.22	26.230		111,500	1,600		113,100	125,000
2360	500 GPM, 200′ TDH, 40 HP		1.16	27.586		118,500	1,700		120,200	133,000
2380	750 GPM, 200′ TDH, 60 HP		1.13	28.319		120,000	1,725		121,725	134,500
2400	1,000 GPM, 200′ TDH, 75 HP		1.12	28.571		123,500	1,750		125,250	138,000

22 11 23.13 Domestic-Water Packaged Booster Pumps		Crew	Daily Output	Labor-Hours	Unit	Material	Labor	Equipment	Total	Total Incl O&P
0010	**DOMESTIC-WATER PACKAGED BOOSTER PUMPS**									
0200	Pump system, with diaphragm tank, control, press. switch									
0300	1 HP pump	Q-1	1.30	12.308	Ea.	7,500	715		8,215	9,325
0400	1-1/2 HP pump		1.25	12.800		7,575	740		8,315	9,450
0420	2 HP pump		1.20	13.333		7,300	775		8,075	9,175
0440	3 HP pump		1.10	14.545		7,900	845		8,745	9,950
0460	5 HP pump	Q-2	1.50	16		8,200	960		9,160	10,500
0480	7-1/2 HP pump		1.42	16.901		9,725	1,025		10,750	12,200
0500	10 HP pump		1.34	17.910		10,100	1,075		11,175	12,800
2000	Pump system, variable speed, base, controls, starter									
2010	Duplex, 100′ head									
2020	400 GPM, 7-1/2 HP, 4″ discharge	Q-2	.70	34.286	Ea.	44,200	2,050		46,250	52,000
2025	Triplex, 100′ head									
2030	1,000 GPM, 15 HP, 6″ discharge	Q-2	.50	48	Ea.	49,200	2,875		52,075	58,500
2040	1,700 GPM, 30 HP, 6″ discharge	″	.30	80	″	72,000	4,800		76,800	86,000

22 12 Facility Potable-Water Storage Tanks

22 12 21 – Facility Underground Potable-Water Storage Tanks

22 12 21.13 Fiberglass, Undrgrnd Pot.-Water Storage Tanks		Crew	Daily Output	Labor-Hours	Unit	2020 Bare Costs Material	Labor	Equipment	Total	Total Incl O&P
0010	**FIBERGLASS, UNDERGROUND POTABLE-WATER STORAGE TANKS**									
0020	Excludes excavation, backfill & piping									
0030	Single wall									
2000	600 gallon capacity	B-21B	3.75	10.667	Ea.	4,125	490	126	4,741	5,425
2010	1,000 gallon capacity		3.50	11.429		5,275	525	135	5,935	6,725
2020	2,000 gallon capacity		3.25	12.308		7,500	565	145	8,210	9,250
2030	4,000 gallon capacity		3	13.333		10,100	610	157	10,867	12,200
2040	6,000 gallon capacity		2.65	15.094		11,400	695	178	12,273	13,800
2050	8,000 gallon capacity		2.30	17.391		13,900	800	205	14,905	16,700
2060	10,000 gallon capacity		2	20		15,400	920	236	16,556	18,600
2070	12,000 gallon capacity		1.50	26.667		21,600	1,225	315	23,140	25,900
2080	15,000 gallon capacity		1	40		24,500	1,825	470	26,795	30,300
2090	20,000 gallon capacity		.75	53.333		31,800	2,450	630	34,880	39,400
2100	25,000 gallon capacity		.50	80		47,400	3,675	940	52,015	58,500
2110	30,000 gallon capacity		.35	114		57,000	5,250	1,350	63,600	72,000
2120	40,000 gallon capacity		.30	133		80,000	6,125	1,575	87,700	99,000

22 12 23 – Facility Indoor Potable-Water Storage Tanks

22 12 23.13 Facility Steel, Indoor Pot.-Water Storage Tanks		Crew	Daily Output	Labor-Hours	Unit	Material	Labor	Equipment	Total	Total Incl O&P
0010	**FACILITY STEEL, INDOOR POT.-WATER STORAGE TANKS**									
2000	Galvanized steel, 15 gal., 14" diam. x 26" LOA	1 Plum	12	.667	Ea.	1,500	43		1,543	1,725
2060	30 gal., 14" diam. x 49" LOA		11	.727		1,775	47		1,822	2,025
2080	80 gal., 20" diam. x 64" LOA		9	.889		2,800	57.50		2,857.50	3,150
2100	135 gal., 24" diam. x 75" LOA		6	1.333		4,025	86		4,111	4,550
2120	240 gal., 30" diam. x 86" LOA		4	2		7,850	129		7,979	8,825
2140	300 gal., 36" diam. x 76" LOA		3	2.667		11,200	172		11,372	12,600
2160	400 gal., 36" diam. x 100" LOA	Q-1	4	4		13,700	232		13,932	15,400
2180	500 gal., 36" diam. x 126" LOA	"	3	5.333		17,000	310		17,310	19,200
3000	Glass lined, P.E., 80 gal., 20" diam. x 60" LOA	1 Plum	9	.889		4,050	57.50		4,107.50	4,525
3060	140 gal., 24" diam. x 75" LOA		6	1.333		5,575	86		5,661	6,250
3080	200 gal., 30" diam. x 71" LOA		4	2		8,050	129		8,179	9,050
3100	350 gal., 36" diam. x 86" LOA		3	2.667		9,250	172		9,422	10,500
3120	450 gal., 42" diam. x 79" LOA	Q-1	4	4		13,300	232		13,532	15,000
3140	600 gal., 48" diam. x 81" LOA		3	5.333		15,900	310		16,210	18,000
3160	750 gal., 48" diam. x 105" LOA		3	5.333		17,800	310		18,110	20,100
3180	900 gal., 54" diam. x 95" LOA		2.50	6.400		25,500	370		25,870	28,700
3200	1,500 gal., 54" diam. x 153" LOA		2	8		32,200	465		32,665	36,100
3220	1,800 gal., 54" diam. x 181" LOA		1.50	10.667		35,700	620		36,320	40,200
3240	2,000 gal., 60" diam. x 165" LOA		1	16		41,000	930		41,930	46,500
3260	3,500 gal., 72" diam. x 201" LOA	Q-2	1.50	16		51,000	960		51,960	57,500

22 13 Facility Sanitary Sewerage

22 13 16 – Sanitary Waste and Vent Piping

22 13 16.20 Pipe, Cast Iron		Crew	Daily Output	Labor-Hours	Unit	Material	Labor	Equipment	Total	Total Incl O&P
0010	**PIPE, CAST IRON**, Soil, on clevis hanger assemblies, 5' OC R221113-70									
0020	Single hub, service wt., lead & oakum joints 10' OC									
2120	2" diameter R221316-10	Q-1	63	.254	L.F.	17.10	14.75		31.85	41
2140	3" diameter		60	.267		26.50	15.45		41.95	52
2160	4" diameter R221316-20		55	.291		24.50	16.85		41.35	52.50
2180	5" diameter	Q-2	76	.316		31	19		50	62.50
2200	6" diameter	"	73	.329		42.50	19.80		62.30	76

22 13 16.20 Pipe, Cast Iron		Crew	Daily Output	Labor-Hours	Unit	Material	2020 Bare Costs Labor	Equipment	Total	Total Incl O&P
2220	8″ diameter	Q-3	59	.542	L.F.	64.50	33.50		98	121
2240	10″ diameter		54	.593		102	36.50		138.50	168
2260	12″ diameter		48	.667		145	41		186	221
2261	15″ diameter		40	.800		173	49		222	264
2320	For service weight, double hub, add					10%				
2340	For extra heavy, single hub, add					48%	4%			
2360	For extra heavy, double hub, add					71%	4%			
2400	Lead for caulking (1#/diam. in.)	Q-1	160	.100	Lb.	1.01	5.80		6.81	9.80
2420	Oakum for caulking (1/8#/diam. in.)	"	40	.400	"	4.37	23		27.37	39.50
2960	To delete hangers, subtract									
2970	2″ diam. to 4″ diam.					16%	19%			
2980	5″ diam. to 8″ diam.					14%	14%			
2990	10″ diam. to 15″ diam.					13%	19%			
3000	Single hub, service wt., push-on gasket joints 10′ OC									
3010	2″ diameter	Q-1	66	.242	L.F.	18.30	14.05		32.35	41
3020	3″ diameter		63	.254		28	14.75		42.75	53
3030	4″ diameter		57	.281		26.50	16.30		42.80	53.50
3040	5″ diameter	Q-2	79	.304		34	18.25		52.25	65
3050	6″ diameter	"	75	.320		45.50	19.25		64.75	79
3060	8″ diameter	Q-3	62	.516		71	31.50		102.50	126
3070	10″ diameter		56	.571		114	35		149	178
3080	12″ diameter		49	.653		160	40		200	236
3082	15″ diameter		40	.800		191	49		240	284
3100	For service weight, double hub, add					65%				
3110	For extra heavy, single hub, add					48%	4%			
3120	For extra heavy, double hub, add					29%	4%			
3130	To delete hangers, subtract									
3140	2″ diam. to 4″ diam.					12%	21%			
3150	5″ diam. to 8″ diam.					10%	16%			
3160	10″ diam. to 15″ diam.					9%	21%			
4000	No hub, couplings 10′ OC									
4100	1-1/2″ diameter	Q-1	71	.225	L.F.	16.85	13.05		29.90	38
4120	2″ diameter		67	.239		20	13.85		33.85	42.50
4140	3″ diameter		64	.250		27	14.50		41.50	51
4160	4″ diameter		58	.276		25	16		41	51.50
4180	5″ diameter	Q-2	83	.289		36.50	17.40		53.90	66.50
4200	6″ diameter	"	79	.304		43.50	18.25		61.75	75.50
4220	8″ diameter	Q-3	69	.464		71	28.50		99.50	121
4240	10″ diameter		61	.525		117	32		149	177
4244	12″ diameter		58	.552		135	34		169	200
4248	15″ diameter		52	.615		207	38		245	285
4280	To delete hangers, subtract									
4290	1-1/2″ diam. to 6″ diam.					22%	47%			
4300	8″ diam. to 10″ diam.					21%	44%			
4310	12″ diam. to 15″ diam.					19%	40%			

22 13 16.30 Pipe Fittings, Cast Iron		Crew	Daily Output	Labor-Hours	Unit	Material	Labor	Equipment	Total	Total Incl O&P
0010	**PIPE FITTINGS, CAST IRON**, Soil									
0040	Hub and spigot, service weight, lead & oakum joints									
0080	1/4 bend, 2″	Q-1	16	1	Ea.	23.50	58		81.50	113
0120	3″		14	1.143		31	66.50		97.50	133
0140	4″		13	1.231		49	71.50		120.50	161
0160	5″	Q-2	18	1.333		68	80		148	195

22 13 16.30	Pipe Fittings, Cast Iron	Crew	Daily Output	Labor-Hours	Unit	2020 Bare Costs Material	Labor	Equipment	Total	Total Incl O&P
0180	6″	Q-2	17	1.412	Ea.	85	85		170	221
0200	8″	Q-3	11	2.909		256	178		434	550
0220	10″		10	3.200		370	196		566	700
0224	12″		9	3.556		510	218		728	885
0226	15″		7	4.571		1,750	280		2,030	2,350
0242	Short sweep, CI, 90°, 2″	Q-1	16	1		23	58		81	112
0243	3″		14	1.143		41	66.50		107.50	144
0244	4″		13	1.231		62.50	71.50		134	176
0245	6″	Q-2	17	1.412		127	85		212	267
0246	8″	Q-3	11	2.909		275	178		453	570
0247	10″		10	3.200		565	196		761	915
0248	12″		9	3.556		1,325	218		1,543	1,775
0251	Long sweep elbow									
0252	2″	Q-1	16	1	Ea.	35.50	58		93.50	126
0253	3″		14	1.143		50.50	66.50		117	155
0254	4″		13	1.231		74.50	71.50		146	189
0255	6″	Q-2	17	1.412		150	85		235	292
0256	8″	Q-3	11	2.909		340	178		518	635
0257	10″		10	3.200		570	196		766	925
0258	12″		9	3.556		925	218		1,143	1,350
0259	15″		7	4.571		2,425	280		2,705	3,075
0266	Closet bend, 3″ diameter with flange 10″ x 16″	Q-1	14	1.143		128	66.50		194.50	240
0268	16″ x 16″		12	1.333		139	77.50		216.50	269
0270	Closet bend, 4″ diameter, 2-1/2″ x 4″ ring, 6″ x 16″		13	1.231		110	71.50		181.50	228
0280	8″ x 16″		13	1.231		97.50	71.50		169	214
0290	10″ x 12″		12	1.333		129	77.50		206.50	258
0300	10″ x 18″		11	1.455		130	84.50		214.50	269
0310	12″ x 16″		11	1.455		110	84.50		194.50	247
0330	16″ x 16″		10	1.600		143	93		236	297
0340	1/8 bend, 2″		16	1		16.65	58		74.65	105
0350	3″		14	1.143		26	66.50		92.50	128
0360	4″		13	1.231		38	71.50		109.50	149
0380	5″	Q-2	18	1.333		54	80		134	180
0400	6″	″	17	1.412		64.50	85		149.50	198
0420	8″	Q-3	11	2.909		193	178		371	480
0440	10″		10	3.200		277	196		473	600
0460	12″		9	3.556		525	218		743	905
0461	15″		7	4.571		1,200	280		1,480	1,725
0500	Sanitary tee, 2″	Q-1	10	1.600		32.50	93		125.50	175
0540	3″		9	1.778		53	103		156	212
0620	4″		8	2		65.50	116		181.50	246
0700	5″	Q-2	12	2		129	120		249	320
0800	6″	″	11	2.182		146	131		277	355
0880	8″	Q-3	7	4.571		385	280		665	845
0881	10″		7	4.571		710	280		990	1,200
0882	12″		6	5.333		1,250	325		1,575	1,875
0883	15″		4	8		2,500	490		2,990	3,475
0900	Sanitary tee, tapped									
0901	2″ x 2″	Q-1	9	1.778	Ea.	43	103		146	201
0902	3″ x 2″		8	2		48	116		164	227
0903	4″ x 2″		7	2.286		79.50	133		212.50	286
0910	Sanitary cross, tapped [double tapped sanitary tee]									
0911	2″ x 2″	Q-1	7	2.286	Ea.	67	133		200	272

22 13 Facility Sanitary Sewerage

22 13 16 – Sanitary Waste and Vent Piping

22 13 16.30 Pipe Fittings, Cast Iron		Crew	Daily Output	Labor-Hours	Unit	Material	2020 Bare Costs Labor	Equipment	Total	Total Incl O&P
0912	3″ x 2″	Q-1	6	2.667	Ea.	86	155		241	325
0913	4″ x 2″		5	3.200		84	186		270	370
0940	Sanitary tee, reducing									
0942	3″ x 2″	Q-1	10	1.600	Ea.	45.50	93		138.50	189
0943	4″ x 3″		9	1.778		59	103		162	219
0944	4″ x 2″		9	1.778		54.50	103		157.50	214
0945	5″ x 3″	Q-2	12.50	1.920		117	115		232	300
0946	6″ x 4″		10	2.400		165	144		309	400
0947	6″ x 3″		11	2.182		118	131		249	325
0948	6″ x 2″		12.50	1.920		114	115		229	299
0949	8″ x 6″	Q-3	9	3.556		325	218		543	685
0950	8″ x 5″		9	3.556		325	218		543	685
0951	8″ x 4″		10	3.200		247	196		443	565
0954	10″ x 6″		8	4		550	245		795	970
0958	12″ x 8″		7	4.571		985	280		1,265	1,500
0962	15″ x 10″		5	6.400		2,100	395		2,495	2,900
1000	Tee, 2″	Q-1	10	1.600		47	93		140	191
1060	3″		9	1.778		70.50	103		173.50	232
1120	4″		8	2		88.50	116		204.50	271
1200	5″	Q-2	12	2		187	120		307	385
1300	6″	″	11	2.182		186	131		317	400
1380	8″	Q-3	7	4.571		370	280		650	825
1400	Combination Y and 1/8 bend									
1420	2″	Q-1	10	1.600	Ea.	41	93		134	184
1460	3″		9	1.778		62	103		165	222
1520	4″		8	2		85.50	116		201.50	268
1540	5″	Q-2	12	2		162	120		282	360
1560	6″		11	2.182		205	131		336	420
1580	8″		7	3.429		505	206		711	870
1582	12″	Q-3	6	5.333		1,025	325		1,350	1,625
1584	Combination Y & 1/8 bend, reducing									
1586	3″ x 2″	Q-1	10	1.600	Ea.	46	93		139	190
1587	4″ x 2″		9.50	1.684		63	97.50		160.50	216
1588	4″ x 3″		9	1.778		74	103		177	236
1589	6″ x 2″	Q-2	12.50	1.920		133	115		248	320
1590	6″ x 3″		12	2		141	120		261	335
1591	6″ x 4″		11	2.182		148	131		279	360
1592	8″ x 2″	Q-3	11	2.909		279	178		457	570
1593	8″ x 4″		10	3.200		246	196		442	565
1594	8″ x 6″		9	3.556		340	218		558	695
1600	Double Y, 2″	Q-1	8	2		72	116		188	253
1610	3″		7	2.286		90	133		223	297
1620	4″		6.50	2.462		118	143		261	345
1630	5″	Q-2	9	2.667		210	160		370	470
1640	6″	″	8	3		310	180		490	610
1650	8″	Q-3	5.50	5.818		745	355		1,100	1,350
1660	10″		5	6.400		1,675	395		2,070	2,450
1670	12″		4.50	7.111		1,925	435		2,360	2,775
1676	Combination double Y & 1/8 bend									
1678	2″	Q-1	8	2	Ea.	83.50	116		199.50	266
1680	3″		7	2.286		107	133		240	315
1682	4″		6.50	2.462		172	143		315	405
1684	6″	Q-2	8	3		610	180		790	940

22 13 16.30 Pipe Fittings, Cast Iron		Crew	Daily Output	Labor-Hours	Unit	Material	Labor	Equipment	Total	Total Incl O&P
						2020 Bare Costs				
1690	Combination double Y & 1/8 bend, CI, reducing									
1692	3″ x 2″	Q-1	8	2	Ea.	93.50	116		209.50	277
1694	4″ x 2″		7	2.286		109	133		242	320
1696	4″ x 3″		7	2.286		124	133		257	335
1698	6″ x 4″	Q-2	9	2.667		390	160		550	670
1700	Double Y, CI, reducing									
1702	3″ x 2″	Q-1	8	2	Ea.	76	116		192	258
1703	4″ x 2″		7	2.286		92.50	133		225.50	300
1704	4″ x 3″		7	2.286		98	133		231	305
1706	6″ x 3″	Q-2	9.50	2.526		236	152		388	485
1707	6″ x 4″	″	9	2.667		219	160		379	480
1708	8″ x 4″	Q-3	8.50	3.765		515	231		746	915
1709	8″ x 6″		8	4		530	245		775	950
1711	10″ x 6″		8	4		1,150	245		1,395	1,625
1712	10″ x 8″		6.50	4.923		1,600	300		1,900	2,200
1713	12″ x 6″		7.50	4.267		1,975	262		2,237	2,550
1714	12″ x 8″		6	5.333		1,950	325		2,275	2,650
1740	Reducer, 3″ x 2″	Q-1	15	1.067		23	62		85	118
1750	4″ x 2″		14.50	1.103		26	64		90	125
1760	4″ x 3″		14	1.143		29.50	66.50		96	132
1770	5″ x 2″		14	1.143		61.50	66.50		128	167
1780	5″ x 3″		13.50	1.185		66	68.50		134.50	176
1790	5″ x 4″		13	1.231		38.50	71.50		110	149
1800	6″ x 2″		13.50	1.185		59	68.50		127.50	168
1810	6″ x 3″		13	1.231		61	71.50		132.50	174
1830	6″ x 4″		12.50	1.280		60	74		134	177
1840	6″ x 5″		11	1.455		64.50	84.50		149	197
1880	8″ x 3″	Q-2	13.50	1.778		116	107		223	288
1900	8″ x 4″		13	1.846		100	111		211	276
1920	8″ x 5″		12	2		106	120		226	297
1940	8″ x 6″		12	2		100	120		220	290
1942	10″ x 4″		12.50	1.920		151	115		266	340
1943	10″ x 6″		11.50	2.087		169	126		295	375
1944	10″ x 8″		9.50	2.526		169	152		321	415
1945	12″ x 4″		11.50	2.087		246	126		372	460
1946	12″ x 6″		11	2.182		263	131		394	485
1947	12″ x 8″		9	2.667		271	160		431	540
1948	12″ x 10″		8.50	2.824		275	170		445	560
1949	15″ x 6″	Q-3	12	2.667		510	164		674	805
1950	15″ x 8″		10	3.200		560	196		756	910
1951	15″ x 10″		9.50	3.368		580	207		787	945
1952	15″ x 12″		9	3.556		585	218		803	970
1960	Increaser, 2″ x 3″	Q-1	15	1.067		55	62		117	153
1980	2″ x 4″		14	1.143		55	66.50		121.50	159
2000	2″ x 5″		13	1.231		62.50	71.50		134	176
2020	3″ x 4″		13	1.231		60	71.50		131.50	173
2040	3″ x 5″		13	1.231		62.50	71.50		134	176
2060	3″ x 6″		12	1.333		80.50	77.50		158	205
2070	4″ x 5″		13	1.231		71	71.50		142.50	185
2080	4″ x 6″		12	1.333		81.50	77.50		159	206
2090	4″ x 8″	Q-2	13	1.846		170	111		281	355
2100	5″ x 6″	Q-1	11	1.455		118	84.50		202.50	255
2110	5″ x 8″	Q-2	12	2		183	120		303	380

22 13 16.30 Pipe Fittings, Cast Iron		Crew	Daily Output	Labor-Hours	Unit	2020 Bare Costs Material	Labor	Equipment	Total	Total Incl O&P
2120	6″ x 8″	Q-2	12	2	Ea.	195	120		315	395
2130	6″ x 10″		8	3		350	180		530	655
2140	8″ x 10″		6.50	3.692		360	222		582	730
2150	10″ x 12″		5.50	4.364		640	262		902	1,100
2500	Y, 2″	Q-1	10	1.600		29.50	93		122.50	172
2510	3″		9	1.778		55	103		158	215
2520	4″		8	2		72	116		188	253
2530	5″	Q-2	12	2		128	120		248	320
2540	6″	"	11	2.182		170	131		301	385
2550	8″	Q-3	7	4.571		415	280		695	875
2560	10″		6	5.333		670	325		995	1,225
2570	12″		5	6.400		1,550	395		1,945	2,325
2580	15″		4	8		3,350	490		3,840	4,425
2581	Y, reducing									
2582	3″ x 2″	Q-1	10	1.600	Ea.	42.50	93		135.50	186
2584	4″ x 2″		9.50	1.684		57.50	97.50		155	209
2586	4″ x 3″		9	1.778		63	103		166	224
2588	6″ x 2″	Q-2	12.50	1.920		113	115		228	298
2590	6″ x 3″		12.25	1.959		116	118		234	305
2592	6″ x 4″		12	2		115	120		235	305
2594	8″ x 2″	Q-3	11	2.909		255	178		433	545
2596	8″ x 3″		10.50	3.048		253	187		440	560
2598	8″ x 4″		10	3.200		219	196		415	535
2600	8″ x 6″		9	3.556		271	218		489	625
2602	10″ x 3″		10	3.200		380	196		576	715
2604	10″ x 4″		9.50	3.368		375	207		582	720
2606	10″ x 6″		8	4		410	245		655	815
2608	10″ x 8″		8	4		560	245		805	980
2610	12″ x 4″		9	3.556		635	218		853	1,025
2612	12″ x 6″		8.50	3.765		650	231		881	1,050
2614	12″ x 8″		8	4		780	245		1,025	1,225
2616	12″ x 10″		7.50	4.267		1,375	262		1,637	1,925
2618	15″ x 4″		7	4.571		2,025	280		2,305	2,675
2620	15″ x 6″		6.50	4.923		2,125	300		2,425	2,800
2622	15″ x 8″		6.50	4.923		2,175	300		2,475	2,825
2624	15″ x 10″		5	6.400		2,200	395		2,595	3,025
2626	15″ x 12″		4.50	7.111		2,300	435		2,735	3,175
3000	For extra heavy, add					44%	4%			
3600	Hub and spigot, service weight gasket joint									
3605	Note: gaskets and joint labor have									
3606	been included with all listed fittings.									
3610	1/4 bend, 2″	Q-1	20	.800	Ea.	35	46.50		81.50	108
3620	3″		17	.941		46.50	54.50		101	133
3630	4″		15	1.067		68	62		130	168
3640	5″	Q-2	21	1.143		97.50	68.50		166	210
3650	6″	"	19	1.263		116	76		192	241
3660	8″	Q-3	12	2.667		325	164		489	600
3670	10″		11	2.909		485	178		663	795
3680	12″		10	3.200		655	196		851	1,025
3690	15″		8	4		1,925	245		2,170	2,500
3692	Short sweep, CI, 90°, 2″	Q-1	20	.800		34.50	46.50		81	108
3693	3″		17	.941		56	54.50		110.50	143
3694	4″		15	1.067		81.50	62		143.50	183

22 13 16.30 Pipe Fittings, Cast Iron		Crew	Daily Output	Labor-Hours	Unit	2020 Bare Costs Material	Labor	Equipment	Total	Total Incl O&P
3695	6″	Q-2	19	1.263	Ea.	158	76		234	287
3696	8″	Q-3	12	2.667		345	164		509	620
3697	10″		11	2.909		680	178		858	1,000
3698	12″		10	3.200		1,450	196		1,646	1,900
3700	Closet bend, 3″ diameter with ring 10″ x 16″	Q-1	17	.941		143	54.50		197.50	239
3710	16″ x 16″		15	1.067		155	62		217	263
3730	Closet bend, 4″ diameter, 1″ x 4″ ring, 6″ x 16″		15	1.067		129	62		191	235
3740	8″ x 16″		15	1.067		117	62		179	221
3750	10″ x 12″		14	1.143		148	66.50		214.50	262
3760	10″ x 18″		13	1.231		149	71.50		220.50	271
3770	12″ x 16″		13	1.231		129	71.50		200.50	249
3780	16″ x 16″		12	1.333		162	77.50		239.50	295
3786	Long sweep elbow									
3787	2″	Q-1	20	.800	Ea.	47.50	46.50		94	122
3788	3″		17	.941		65.50	54.50		120	154
3789	4″		15	1.067		93.50	62		155.50	196
3790	6″	Q-2	19	1.263		180	76		256	315
3791	8″	Q-3	12	2.667		405	164		569	690
3792	10″		11	2.909		685	178		863	1,025
3793	12″		10	3.200		1,075	196		1,271	1,475
3794	15″		8	4		2,600	245		2,845	3,225
3800	1/8 bend, 2″	Q-1	20	.800		28.50	46.50		75	101
3810	3″		17	.941		41	54.50		95.50	127
3820	4″		15	1.067		57	62		119	156
3830	5″	Q-2	21	1.143		84	68.50		152.50	196
3840	6″	"	19	1.263		95.50	76		171.50	219
3850	8″	Q-3	12	2.667		261	164		425	530
3860	10″		11	2.909		390	178		568	695
3870	12″		10	3.200		670	196		866	1,025
3880	15″		8	4		1,375	245		1,620	1,875
3882	Sanitary tee, tapped									
3884	2″ x 2″	Q-1	11	1.455	Ea.	66.50	84.50		151	199
3886	3″ x 2″		10	1.600		75	93		168	221
3888	4″ x 2″		9	1.778		110	103		213	275
3890	Sanitary cross, tapped [double tapped sanitary tee]									
3892	2″ x 2″	Q-1	9	1.778	Ea.	90.50	103		193.50	254
3894	3″ x 2″		8	2		113	116		229	298
3896	4″ x 2″		7	2.286		115	133		248	325
3900	Sanitary tee, 2″		12	1.333		56	77.50		133.50	178
3910	3″		10	1.600		83	93		176	231
3920	4″		9	1.778		103	103		206	268
3930	5″	Q-2	13	1.846		188	111		299	375
3940	6″	"	11	2.182		208	131		339	425
3950	8″	Q-3	8.50	3.765		520	231		751	920
3952	10″		8	4		940	245		1,185	1,400
3954	12″		7	4.571		1,550	280		1,830	2,125
3956	15″		6	5.333		2,850	325		3,175	3,650
3960	Sanitary tee, reducing									
3961	3″ x 2″	Q-1	10.50	1.524	Ea.	72	88.50		160.50	212
3962	4″ x 2″		10	1.600		85.50	93		178.50	233
3963	4″ x 3″		9.50	1.684		93	97.50		190.50	248
3964	5″ x 3″	Q-2	13.50	1.778		162	107		269	340
3965	6″ x 2″		13	1.846		157	111		268	340

22 13 16.30	Pipe Fittings, Cast Iron	Crew	Daily Output	Labor-Hours	Unit	2020 Bare Costs Material	Labor	Equipment	Total	Total Incl O&P
3966	6″ x 3″	Q-2	12.50	1.920	Ea.	164	115		279	355
3967	6″ x 4″		12	2		215	120		335	415
3968	8″ x 4″	Q-3	10.50	3.048		335	187		522	645
3969	8″ x 5″		10	3.200		425	196		621	760
3970	8″ x 6″		9.50	3.368		425	207		632	775
3971	10″ x 6″		9	3.556		695	218		913	1,100
3972	12″ x 8″		8.50	3.765		1,200	231		1,431	1,675
3973	15″ x 10″		6	5.333		2,375	325		2,700	3,125
3980	Tee, 2″	Q-1	12	1.333		70.50	77.50		148	194
3990	3″		10	1.600		101	93		194	250
4000	4″		9	1.778		126	103		229	293
4010	5″	Q-2	13	1.846		246	111		357	435
4020	6″	″	11	2.182		248	131		379	470
4030	8″	Q-3	8	4		505	245		750	920
4060	Combination Y and 1/8 bend									
4070	2″	Q-1	12	1.333	Ea.	64.50	77.50		142	187
4080	3″		10	1.600		92.50	93		185.50	241
4090	4″		9	1.778		124	103		227	290
4100	5″	Q-2	13	1.846		222	111		333	410
4110	6″	″	11	2.182		267	131		398	490
4120	8″	Q-3	8	4		640	245		885	1,075
4121	12″	″	7	4.571		1,325	280		1,605	1,875
4130	Combination Y & 1/8 bend, reducing									
4132	3″ x 2″	Q-1	10.50	1.524	Ea.	73	88.50		161.50	212
4134	4″ x 2″		10	1.600		94	93		187	242
4136	4″ x 3″		9.50	1.684		108	97.50		205.50	265
4138	6″ x 2″	Q-2	13	1.846		175	111		286	360
4140	6″ x 3″		12.50	1.920		187	115		302	380
4142	6″ x 4″		12	2		198	120		318	400
4144	8″ x 2″	Q-3	11	2.909		380	178		558	680
4146	8″ x 4″		10.50	3.048		335	187		522	645
4148	8″ x 6″		9.50	3.368		435	207		642	790
4160	Double Y, 2″	Q-1	10	1.600		107	93		200	257
4170	3″		8	2		135	116		251	325
4180	4″		7	2.286		175	133		308	390
4190	5″	Q-2	10	2.400		300	144		444	545
4200	6″	″	9	2.667		405	160		565	685
4210	8″	Q-3	6	5.333		950	325		1,275	1,550
4220	10″		5	6.400		2,025	395		2,420	2,825
4230	12″		4.50	7.111		2,375	435		2,810	3,250
4234	Combination double Y & 1/8 bend									
4235	2″	Q-1	10	1.600	Ea.	119	93		212	270
4236	3″		8	2		153	116		269	340
4237	4″		7	2.286		229	133		362	450
4238	6″	Q-2	9	2.667		700	160		860	1,000
4242	Combination double Y & 1/8 bend, CI, reducing									
4243	3″ x 2″	Q-1	9	1.778	Ea.	132	103		235	300
4244	4″ x 2″		8	2		151	116		267	340
4245	4″ x 3″		7.50	2.133		173	124		297	375
4246	6″ x 4″	Q-2	10	2.400		460	144		604	720
4248	Double Y, CI, reducing									
4249	3″ x 2″	Q-1	9	1.778	Ea.	115	103		218	280
4250	4″ x 2″		8	2		135	116		251	325

22 13 16.30 Pipe Fittings, Cast Iron

		Crew	Daily Output	Labor-Hours	Unit	Material	Labor	Equipment	Total	Total Incl O&P
						2020 Bare Costs				
4251	4″ x 3″	Q-1	7.50	2.133	Ea.	132	124		256	330
4252	6″ x 3″	Q-2	10.50	2.286		298	137		435	530
4253	6″ x 4″	″	10	2.400		288	144		432	530
4254	8″ x 4″	Q-3	9.50	3.368		620	207		827	995
4255	8″ x 6″		9	3.556		660	218		878	1,050
4256	10″ x 6″		8.50	3.765		1,325	231		1,556	1,800
4257	10″ x 8″		7.50	4.267		1,850	262		2,112	2,425
4258	12″ x 6″		8	4		2,175	245		2,420	2,775
4259	12″ x 8″		7	4.571		2,225	280		2,505	2,875
4260	Reducer, 3″ x 2″	Q-1	17	.941		50	54.50		104.50	137
4270	4″ x 2″		16.50	.970		56.50	56		112.50	147
4280	4″ x 3″		16	1		64	58		122	158
4290	5″ x 2″		16	1		103	58		161	201
4300	5″ x 3″		15.50	1.032		111	60		171	212
4310	5″ x 4″		15	1.067		87	62		149	189
4320	6″ x 2″		15.50	1.032		102	60		162	202
4330	6″ x 3″		15	1.067		107	62		169	211
4336	6″ x 4″		14	1.143		110	66.50		176.50	220
4340	6″ x 5″		13	1.231		125	71.50		196.50	245
4360	8″ x 3″	Q-2	15	1.600		199	96		295	365
4370	8″ x 4″		15	1.600		187	96		283	350
4380	8″ x 5″		14	1.714		204	103		307	380
4390	8″ x 6″		14	1.714		199	103		302	375
4394	10″ x 4″		13.50	1.778		284	107		391	475
4395	10″ x 6″		13	1.846		315	111		426	510
4396	10″ x 8″		12.50	1.920		350	115		465	560
4397	12″ x 4″		12	2		410	120		530	630
4398	12″ x 6″		11.50	2.087		440	126		566	675
4399	12″ x 8″		11	2.182		485	131		616	730
4400	12″ x 10″		10.50	2.286		535	137		672	795
4401	15″ x 6″	Q-3	13	2.462		715	151		866	1,025
4402	15″ x 8″		12	2.667		800	164		964	1,125
4403	15″ x 10″		11	2.909		870	178		1,048	1,225
4404	15″ x 12″		10	3.200		905	196		1,101	1,300
4430	Increaser, 2″ x 3″	Q-1	17	.941		70	54.50		124.50	159
4440	2″ x 4″		16	1		74	58		132	168
4450	2″ x 5″		15	1.067		92.50	62		154.50	195
4460	3″ x 4″		15	1.067		79	62		141	180
4470	3″ x 5″		15	1.067		92.50	62		154.50	195
4480	3″ x 6″		14	1.143		111	66.50		177.50	222
4490	4″ x 5″		15	1.067		101	62		163	204
4500	4″ x 6″		14	1.143		112	66.50		178.50	223
4510	4″ x 8″	Q-2	15	1.600		237	96		333	405
4520	5″ x 6″	Q-1	13	1.231		149	71.50		220.50	270
4530	5″ x 8″	Q-2	14	1.714		250	103		353	430
4540	6″ x 8″		14	1.714		263	103		366	445
4550	6″ x 10″		10	2.400		465	144		609	725
4560	8″ x 10″		8.50	2.824		475	170		645	780
4570	10″ x 12″		7.50	3.200		785	192		977	1,150
4600	Y, 2″	Q-1	12	1.333		53.50	77.50		131	175
4610	3″		10	1.600		85.50	93		178.50	233
4620	4″		9	1.778		110	103		213	275
4630	5″	Q-2	13	1.846		188	111		299	375

22 13 16.30 Pipe Fittings, Cast Iron		Crew	Daily Output	Labor-Hours	Unit	2020 Bare Costs Material	Labor	Equipment	Total	Total Incl O&P
4640	6″	Q-2	11	2.182	Ea.	232	131		363	450
4650	8″	Q-3	8	4		550	245		795	970
4660	10″		7	4.571		900	280		1,180	1,400
4670	12″		6	5.333		1,850	325		2,175	2,525
4672	15″		5	6.400		3,700	395		4,095	4,675
4680	Y, reducing									
4681	3″ x 2″	Q-1	10.50	1.524	Ea.	69.50	88.50		158	209
4682	4″ x 2″		10	1.600		88	93		181	236
4683	4″ x 3″		9.50	1.684		97.50	97.50		195	253
4684	6″ x 2″	Q-2	13	1.846		156	111		267	335
4685	6″ x 3″		12.50	1.920		162	115		277	350
4686	6″ x 4″		12	2		164	120		284	360
4687	8″ x 2″	Q-3	11	2.909		335	178		513	630
4688	8″ x 3″		10.75	2.977		335	183		518	645
4689	8″ x 4″		10.50	3.048		305	187		492	615
4690	8″ x 6″		9.50	3.368		370	207		577	715
4691	10″ x 3″		10	3.200		510	196		706	855
4692	10″ x 4″		9.50	3.368		510	207		717	870
4693	10″ x 6″		9	3.556		555	218		773	935
4694	10″ x 8″		8.80	3.636		740	223		963	1,150
4695	12″ x 4″		9.50	3.368		800	207		1,007	1,200
4696	12″ x 6″		9	3.556		830	218		1,048	1,225
4697	12″ x 8″		8.50	3.765		995	231		1,226	1,450
4698	12″ x 10″		8	4		1,625	245		1,870	2,175
4699	15″ x 4″		8	4		2,225	245		2,470	2,825
4700	15″ x 6″		7.50	4.267		2,325	262		2,587	2,975
4701	15″ x 8″		7	4.571		2,400	280		2,680	3,075
4702	15″ x 10″		6	5.333		2,475	325		2,800	3,225
4703	15″ x 12″		5	6.400		2,625	395		3,020	3,475
4900	For extra heavy, add					44%	4%			
4940	Gasket and making push-on joint									
4950	2″	Q-1	40	.400	Ea.	11.75	23		34.75	47.50
4960	3″		35	.457		15.15	26.50		41.65	56
4970	4″		32	.500		19.05	29		48.05	64.50
4980	5″	Q-2	43	.558		30	33.50		63.50	83
4990	6″	″	40	.600		31	36		67	88
5000	8″	Q-3	32	1		67.50	61.50		129	167
5010	10″		29	1.103		114	67.50		181.50	227
5020	12″		25	1.280		146	78.50		224.50	279
5022	15″		21	1.524		175	93.50		268.50	330
5030	Note: gaskets and joint labor have									
5040	been included with all listed fittings.									
5990	No hub									
6000	Cplg. & labor required at joints not incl. in fitting									
6010	price. Add 1 coupling per joint for installed price									
6020	1/4 bend, 1-1/2″				Ea.	11.65			11.65	12.80
6060	2″					12.60			12.60	13.85
6080	3″					17.65			17.65	19.40
6120	4″					26			26	28.50
6140	5″					65			65	71.50
6160	6″					63			63	69.50
6180	8″					178			178	195
6181	10″					385			385	420

22 13 16.30 Pipe Fittings, Cast Iron		Crew	Daily Output	Labor-Hours	Unit	Material	Labor	Equipment	Total	Total Incl O&P
						2020 Bare Costs				
6182	12"				Ea.	1,150			1,150	1,275
6183	15"					1,475			1,475	1,625
6184	1/4 bend, long sweep, 1-1/2"					29.50			29.50	32.50
6186	2"					28			28	30.50
6188	3"					33.50			33.50	37
6189	4"					53			53	58.50
6190	5"					103			103	114
6191	6"					118			118	129
6192	8"					310			310	345
6193	10"					655			655	725
6200	1/8 bend, 1-1/2"					9.70			9.70	10.70
6210	2"					10.90			10.90	12
6212	3"					14.55			14.55	16
6214	4"					19.15			19.15	21
6216	5"					40.50			40.50	44.50
6218	6"					42.50			42.50	46.50
6220	8"					122			122	134
6222	10"					233			233	256
6364	Closet flange									
6366	4"				Ea.	17.90			17.90	19.70
6370	Closet bend, no hub									
6376	4" x 16"				Ea.	95			95	105
6380	Sanitary tee, tapped, 1-1/2"					23.50			23.50	26
6382	2" x 1-1/2"					21			21	23
6384	2"					22			22	24
6386	3" x 2"					33.50			33.50	36.50
6388	3"					56			56	62
6390	4" x 1-1/2"					30			30	32.50
6392	4" x 2"					34			34	37
6393	4"					34			34	37
6394	6" x 1-1/2"					76.50			76.50	84
6396	6" x 2"					77.50			77.50	85
6459	Sanitary tee, 1-1/2"					16.25			16.25	17.85
6460	2"					17.40			17.40	19.15
6470	3"					21.50			21.50	23.50
6472	4"					41			41	45
6474	5"					97			97	107
6476	6"					97			97	106
6478	8"					395			395	435
6480	10"					465			465	515
6724	Sanitary tee, reducing									
6725	3" x 2"				Ea.	19			19	21
6726	4" x 3"					31.50			31.50	34.50
6727	5" x 4"					74.50			74.50	82
6728	6" x 3"					72.50			72.50	80
6729	8" x 4"					210			210	231
6730	Y, 1-1/2"					16.45			16.45	18.10
6740	2"					16.10			16.10	17.70
6750	3"					23.50			23.50	26
6760	4"					37.50			37.50	41.50
6762	5"					90.50			90.50	99.50
6764	6"					100			100	110
6768	8"					239			239	263

22 13 16.30 Pipe Fittings, Cast Iron		Crew	Daily Output	Labor-Hours	Unit	Material	2020 Bare Costs Labor	Equipment	Total	Total Incl O&P
6769	10"				Ea.	540			540	595
6770	12"					1,025			1,025	1,150
6771	15"					2,325			2,325	2,550
6791	Y, reducing, 3" x 2"					17.40			17.40	19.15
6792	4" x 2"					25			25	27.50
6793	5" x 2"					56			56	61.50
6794	6" x 2"					62.50			62.50	68.50
6795	6" x 4"					79.50			79.50	87.50
6796	8" x 4"					139			139	153
6797	8" x 6"					168			168	185
6798	10" x 6"					385			385	425
6799	10" x 8"					455			455	500
6800	Double Y, 2"					26			26	28.50
6920	3"					48			48	52.50
7000	4"					96			96	106
7100	6"					173			173	190
7120	8"					490			490	540
7200	Combination Y and 1/8 bend									
7220	1-1/2"				Ea.	17.80			17.80	19.60
7260	2"					18.40			18.40	20
7320	3"					29.50			29.50	32.50
7400	4"					56.50			56.50	62.50
7480	5"					116			116	127
7500	6"					153			153	169
7520	8"					360			360	395
7800	Reducer, 3" x 2"					9.05			9.05	9.95
7820	4" x 2"					13.70			13.70	15.05
7840	4" x 3"					13.70			13.70	15.05
7842	6" x 3"					37			37	41
7844	6" x 4"					37.50			37.50	41
7846	6" x 5"					38			38	41.50
7848	8" x 2"					58.50			58.50	64.50
7850	8" x 3"					54.50			54.50	60
7852	8" x 4"					57			57	63
7854	8" x 5"					64.50			64.50	71
7856	8" x 6"					63.50			63.50	70
7858	10" x 4"					112			112	123
7860	10" x 6"					119			119	131
7862	10" x 8"					139			139	153
7864	12" x 4"					235			235	259
7866	12" x 6"					250			250	275
7868	12" x 8"					257			257	283
7870	12" x 10"					262			262	289
7872	15" x 4"					490			490	540
7874	15" x 6"					465			465	510
7876	15" x 8"					550			550	605
7878	15" x 10"					550			550	605
7880	15" x 12"					555			555	610
8000	Coupling, standard (by CISPI Mfrs.)									
8020	1-1/2"	Q-1	48	.333	Ea.	16.70	19.35		36.05	47.50
8040	2"		44	.364		17.80	21		38.80	51
8080	3"		38	.421		20	24.50		44.50	58.50
8120	4"		33	.485		23	28		51	67.50

22 13 16 – Sanitary Waste and Vent Piping

22 13 16.30 Pipe Fittings, Cast Iron

	22 13 16.30 Pipe Fittings, Cast Iron	Crew	Daily Output	Labor-Hours	Unit	Material (2020 Bare Costs)	Labor (2020 Bare Costs)	Equipment (2020 Bare Costs)	Total (2020 Bare Costs)	Total Incl O&P
8160	5″	Q-2	44	.545	Ea.	58	33		91	113
8180	6″	″	40	.600		60	36		96	120
8200	8″	Q-3	33	.970		111	59.50		170.50	211
8220	10″	″	26	1.231		149	75.50		224.50	277
8300	Coupling, cast iron clamp & neoprene gasket (by MG)									
8310	1-1/2″	Q-1	48	.333	Ea.	9.30	19.35		28.65	39.50
8320	2″		44	.364		12.90	21		33.90	45.50
8330	3″		38	.421		11.55	24.50		36.05	49
8340	4″		33	.485		18.85	28		46.85	62.50
8350	5″	Q-2	44	.545		33	33		66	85
8360	6″	″	40	.600		36	36		72	93.50
8380	8″	Q-3	33	.970		110	59.50		169.50	210
8400	10″	″	26	1.231		181	75.50		256.50	315
8410	Reducing, no hub									
8416	2″ x 1-1/2″	Q-1	44	.364	Ea.	10.50	21		31.50	43
8600	Coupling, stainless steel, heavy duty									
8620	1-1/2″	Q-1	48	.333	Ea.	5.90	19.35		25.25	35.50
8630	2″		44	.364		6.40	21		27.40	38.50
8640	2″ x 1-1/2″		44	.364		11.45	21		32.45	44
8650	3″		38	.421		6.70	24.50		31.20	44
8660	4″		33	.485		7.55	28		35.55	50.50
8670	4″ x 3″		33	.485		18.85	28		46.85	62.50
8680	5″	Q-2	44	.545		16.20	33		49.20	67
8690	6″	″	40	.600		18.15	36		54.15	74
8700	8″	Q-3	33	.970		30.50	59.50		90	123
8710	10″		26	1.231		39	75.50		114.50	156
8712	12″		22	1.455		69.50	89		158.50	211
8715	15″		18	1.778		111	109		220	285

22 13 16.40 Pipe Fittings, Cast Iron for Drainage

	22 13 16.40 Pipe Fittings, Cast Iron for Drainage	Crew	Daily Output	Labor-Hours	Unit	Material	Labor	Equipment	Total	Total Incl O&P
0010	**PIPE FITTINGS, CAST IRON FOR DRAINAGE**, Special									
0020	Cast iron, drainage, threaded, black									
0030	90° elbow, straight									
0031	1-1/2″ pipe size	Q-1	20	.800	Ea.	45.50	46.50		92	120
0032	2″ pipe size		18	.889		68.50	51.50		120	152
0033	3″ pipe size		10	1.600		256	93		349	420
0034	4″ pipe size		6	2.667		405	155		560	675
0040	90° long turn elbow, straight									
0041	1-1/2″ pipe size	Q-1	20	.800	Ea.	63.50	46.50		110	139
0042	2″ pipe size		18	.889		94.50	51.50		146	181
0043	3″ pipe size		10	1.600		345	93		438	520
0044	4″ pipe size		6	2.667		605	155		760	895
0050	90° street elbow, straight									
0051	1-1/2″ pipe size	Q-1	20	.800	Ea.	64.50	46.50		111	141
0052	2″ pipe size	″	18	.889	″	85.50	51.50		137	171
0060	45° elbow									
0061	1-1/2″ pipe size	Q-1	20	.800	Ea.	44.50	46.50		91	118
0062	2″ pipe size		18	.889		64	51.50		115.50	148
0063	3″ pipe size		10	1.600		250	93		343	415
0064	4″ pipe size		6	2.667		390	155		545	660
0070	45° street elbow									
0071	1-1/2″ pipe size	Q-1	20	.800	Ea.	60.50	46.50		107	136
0072	2″ pipe size	″	18	.889	″	97.50	51.50		149	184

22 13 16.40 Pipe Fittings, Cast Iron for Drainage		Crew	Daily Output	Labor-Hours	Unit	2020 Bare Costs Material	Labor	Equipment	Total	Total Incl O&P
0092	Tees, straight									
0093	1-1/2" pipe size	Q-1	13	1.231	Ea.	74.50	71.50		146	189
0094	2" pipe size	"	11	1.455	"	123	84.50		207.50	261
0100	TY's, straight									
0101	1-1/2" pipe size	Q-1	13	1.231	Ea.	73	71.50		144.50	188
0102	2" pipe size		11	1.455		120	84.50		204.50	259
0103	3" pipe size		6	2.667		455	155		610	730
0104	4" pipe size		4	4		655	232		887	1,075
0120	45° Y branch, straight									
0121	1-1/2" pipe size	Q-1	13	1.231	Ea.	89.50	71.50		161	206
0122	2" pipe size		11	1.455		160	84.50		244.50	300
0123	3" pipe size		6	2.667		575	155		730	865
0124	4" pipe size		4	4		860	232		1,092	1,300
0147	Double Y branch, straight									
0148	1-1/2" pipe size	Q-1	10	1.600	Ea.	228	93		321	390
0149	2" pipe size	"	7	2.286	"	244	133		377	465
0160	P trap									
0161	1-1/2" pipe size	Q-1	15	1.067	Ea.	133	62		195	239
0162	2" pipe size		13	1.231		228	71.50		299.50	360
0163	3" pipe size		7	2.286		710	133		843	980
0164	4" pipe size		5	3.200		1,700	186		1,886	2,150
0180	Tucker connection									
0181	1-1/2" pipe size	Q-1	24	.667	Ea.	132	38.50		170.50	203
0182	2" pipe size	"	21	.762	"	163	44		207	245
0205	Cast iron, drainage, threaded, galvanized									
0206	90° elbow, straight									
0207	1-1/2" pipe size	Q-1	20	.800	Ea.	69.50	46.50		116	146
0208	2" pipe size		18	.889		97	51.50		148.50	184
0209	3" pipe size		10	1.600		385	93		478	560
0210	4" pipe size		6	2.667		715	155		870	1,025
0216	90° long turn elbow, straight									
0217	1-1/2" pipe size	Q-1	20	.800	Ea.	76.50	46.50		123	154
0218	2" pipe size		18	.889		131	51.50		182.50	221
0219	3" pipe size		10	1.600		350	93		443	530
0220	4" pipe size		6	2.667		565	155		720	855
0226	90° street elbow, straight									
0227	1-1/2" pipe size	Q-1	20	.800	Ea.	85	46.50		131.50	163
0228	2" pipe size	"	18	.889	"	133	51.50		184.50	224
0236	45° elbow									
0237	1-1/2" pipe size	Q-1	20	.800	Ea.	66.50	46.50		113	143
0238	2" pipe size		18	.889		95	51.50		146.50	181
0239	3" pipe size		10	1.600		330	93		423	505
0240	4" pipe size		6	2.667		540	155		695	825
0246	45° street elbow									
0247	1-1/2" pipe size	Q-1	20	.800	Ea.	88	46.50		134.50	167
0248	2" pipe size	"	18	.889	"	136	51.50		187.50	226
0268	Tees, straight									
0269	1-1/2" pipe size	Q-1	13	1.231	Ea.	101	71.50		172.50	218
0270	2" pipe size	"	11	1.455	"	169	84.50		253.50	310
0276	TY's, straight									
0277	1-1/2" pipe size	Q-1	13	1.231	Ea.	101	71.50		172.50	218
0278	2" pipe size		11	1.455		167	84.50		251.50	310
0279	3" pipe size		6	2.667		685	155		840	985

22 13 Facility Sanitary Sewerage

22 13 16 – Sanitary Waste and Vent Piping

22 13 16.40	Pipe Fittings, Cast Iron for Drainage	Crew	Daily Output	Labor-Hours	Unit	2020 Bare Costs Material	Labor	Equipment	Total	Total Incl O&P
0280	4" pipe size	Q-1	4	4	Ea.	805	232		1,037	1,225
0296	45° Y branch, straight									
0297	1-1/2" pipe size	Q-1	13	1.231	Ea.	113	71.50		184.50	232
0298	2" pipe size		11	1.455		227	84.50		311.50	375
0299	3" pipe size		6	2.667		805	155		960	1,125
0300	4" pipe size		4	4		1,275	232		1,507	1,750
0323	Double Y branch, straight									
0324	1-1/2" pipe size	Q-1	10	1.600	Ea.	153	93		246	310
0325	2" pipe size	"	7	2.286	"	380	133		513	615
0336	P trap									
0337	1-1/2" pipe size	Q-1	15	1.067	Ea.	186	62		248	298
0338	2" pipe size		13	1.231		225	71.50		296.50	355
0339	3" pipe size		7	2.286		1,075	133		1,208	1,375
0340	4" pipe size		5	3.200		1,650	186		1,836	2,100
0356	Tucker connection									
0357	1-1/2" pipe size	Q-1	24	.667	Ea.	206	38.50		244.50	285
0358	2" pipe size	"	21	.762	"	310	44		354	405
1000	Drip pan elbow (safety valve discharge elbow)									
1010	Cast iron, threaded inlet									
1014	2-1/2"	Q-1	8	2	Ea.	1,600	116		1,716	1,950
1015	3"		6.40	2.500		1,725	145		1,870	2,125
1017	4"		4.80	3.333		2,325	193		2,518	2,850
1018	Cast iron, flanged inlet									
1019	6"	Q-2	3.60	6.667	Ea.	2,525	400		2,925	3,375
1020	8"	"	2.60	9.231	"	2,675	555		3,230	3,775

22 13 16.50 Shower Drains

Line	Description	Crew	Daily Output	Labor-Hours	Unit	Material	Labor	Equipment	Total	Total Incl O&P
0010	**SHOWER DRAINS**									
2780	Shower, with strainer, uniform diam. trap, bronze top									
2800	2" and 3" pipe size	Q-1	8	2	Ea.	360	116		476	570
2820	4" pipe size	"	7	2.286		415	133		548	655
2840	For galvanized body, add					239			239	263
2860	With strainer, backwater valve, drum trap									
2880	1-1/2", 2" & 3" pipe size	Q-1	8	2	Ea.	405	116		521	620
2890	4" pipe size	"	7	2.286		565	133		698	820
2900	For galvanized body, add					189			189	208

22 13 16.60 Traps

Line	Description	Crew	Daily Output	Labor-Hours	Unit	Material	Labor	Equipment	Total	Total Incl O&P
0010	**TRAPS**									
0030	Cast iron, service weight									
0050	Running P trap, without vent									
1100	2"	Q-1	16	1	Ea.	167	58		225	271
1140	3"		14	1.143		167	66.50		233.50	283
1150	4"		13	1.231		167	71.50		238.50	291
1160	6"	Q-2	17	1.412		770	85		855	975
1180	Running trap, single hub, with vent									
2080	3" pipe size, 3" vent	Q-1	14	1.143	Ea.	133	66.50		199.50	245
2120	4" pipe size, 4" vent	"	13	1.231		187	71.50		258.50	315
2140	5" pipe size, 4" vent	Q-2	11	2.182		299	131		430	525
2160	6" pipe size, 4" vent		10	2.400		805	144		949	1,100
2180	6" pipe size, 6" vent		8	3		865	180		1,045	1,225
2200	8" pipe size, 4" vent	Q-3	10	3.200		3,500	196		3,696	4,150
2220	8" pipe size, 6" vent	"	8	4		2,725	245		2,970	3,375
2300	For double hub, vent, add					10%	20%			

22 13 16.60 Traps		Crew	Daily Output	Labor-Hours	Unit	Material	2020 Bare Costs Labor	Equipment	Total	Total Incl O&P
2800	S trap,									
2850	4" pipe size	Q-1	13	1.231	Ea.	89	71.50		160.50	205
3000	P trap, B&S, 2" pipe size		16	1		41.50	58		99.50	133
3040	3" pipe size		14	1.143		61.50	66.50		128	167
3060	4" pipe size		13	1.231		88.50	71.50		160	205
3080	5" pipe size	Q-2	18	1.333		197	80		277	335
3100	6" pipe size	"	17	1.412		274	85		359	425
3120	8" pipe size	Q-3	11	2.909		825	178		1,003	1,175
3130	10" pipe size	"	10	3.200		1,625	196		1,821	2,100
3150	P trap, no hub, 1-1/2" pipe size	Q-1	17	.941		21.50	54.50		76	105
3160	2" pipe size		16	1		20	58		78	109
3170	3" pipe size		14	1.143		39.50	66.50		106	143
3180	4" pipe size		13	1.231		82	71.50		153.50	197
3190	6" pipe size	Q-2	17	1.412		199	85		284	345
3350	Deep seal trap, B&S									
3400	1-1/4" pipe size	Q-1	14	1.143	Ea.	69	66.50		135.50	175
3410	1-1/2" pipe size		14	1.143		69	66.50		135.50	175
3420	2" pipe size		14	1.143		59.50	66.50		126	164
3440	3" pipe size		12	1.333		76	77.50		153.50	200
3460	4" pipe size		11	1.455		120	84.50		204.50	258
3500	For trap primer connection, add					170			170	187
3540	For trap with floor cleanout, add					70%	5%			
3580	For trap with adjustable cleanout, add	Q-1	10	1.600	Ea.	197	93		290	355
4700	Copper, drainage, drum trap									
4800	3" x 5" solid, 1-1/2" pipe size	1 Plum	16	.500	Ea.	156	32		188	220
4840	3" x 6" swivel, 1-1/2" pipe size	"	16	.500	"	315	32		347	395
5100	P trap, standard pattern									
5200	1-1/4" pipe size	1 Plum	18	.444	Ea.	102	28.50		130.50	156
5240	1-1/2" pipe size		17	.471		111	30.50		141.50	169
5260	2" pipe size		15	.533		184	34.50		218.50	254
5280	3" pipe size		11	.727		525	47		572	650
5340	With cleanout, swivel joint and slip joint									
5360	1-1/4" pipe size	1 Plum	18	.444	Ea.	124	28.50		152.50	179
5400	1-1/2" pipe size		17	.471		199	30.50		229.50	265
5420	2" pipe size		15	.533		176	34.50		210.50	246
5750	Chromed brass, tubular, P trap, without cleanout, 20 ga.									
5800	1-1/4" pipe size	1 Plum	18	.444	Ea.	15.45	28.50		43.95	60
5840	1-1/2" pipe size	"	17	.471	"	17.60	30.50		48.10	65
5900	With cleanout, 20 ga.									
5940	1-1/4" pipe size	1 Plum	18	.444	Ea.	22.50	28.50		51	68
6000	1-1/2" pipe size	"	17	.471	"	27	30.50		57.50	75
6350	S trap, without cleanout, 20 ga.									
6400	1-1/4" pipe size	1 Plum	18	.444	Ea.	70	28.50		98.50	120
6440	1-1/2" pipe size	"	17	.471	"	43	30.50		73.50	92.50
6550	With cleanout, 20 ga.									
6600	1-1/4" pipe size	1 Plum	18	.444	Ea.	60.50	28.50		89	110
6640	1-1/2" pipe size	"	17	.471		48	30.50		78.50	98
6660	Corrosion resistant, glass, P trap, 1-1/2" pipe size	Q-1	17	.941		82.50	54.50		137	173
6670	2" pipe size		16	1		108	58		166	206
6680	3" pipe size		14	1.143		222	66.50		288.50	345
6690	4" pipe size		13	1.231		330	71.50		401.50	470
6700	6" pipe size	Q-2	17	1.412		1,275	85		1,360	1,525
6710	ABS DWV P trap, solvent weld joint									

22 13 16.60 Traps		Crew	Daily Output	Labor-Hours	Unit	Material	Labor	Equipment	Total	Total Incl O&P
						2020 Bare Costs				
6720	1-1/2" pipe size	1 Plum	18	.444	Ea.	10.50	28.50		39	54.50
6722	2" pipe size		17	.471		14.05	30.50		44.55	61
6724	3" pipe size		15	.533		55.50	34.50		90	113
6726	4" pipe size		14	.571		114	37		151	181
6732	PVC DWV P trap, solvent weld joint									
6733	1-1/2" pipe size	1 Plum	18	.444	Ea.	8.95	28.50		37.45	53
6734	2" pipe size		17	.471		11.15	30.50		41.65	58
6735	3" pipe size		15	.533		38	34.50		72.50	93
6736	4" pipe size		14	.571		85	37		122	149
6760	PP DWV, dilution trap, 1-1/2" pipe size		16	.500		274	32		306	350
6770	P trap, 1-1/2" pipe size		17	.471		68	30.50		98.50	121
6780	2" pipe size		16	.500		101	32		133	159
6790	3" pipe size		14	.571		186	37		223	259
6800	4" pipe size		13	.615		310	39.50		349.50	405
6830	S trap, 1-1/2" pipe size		16	.500		55.50	32		87.50	109
6840	2" pipe size		15	.533		83	34.50		117.50	143
6850	Universal trap, 1-1/2" pipe size		14	.571		112	37		149	178
6860	PVC DWV hub x hub, basin trap, 1-1/4" pipe size		18	.444		54	28.50		82.50	102
6870	Sink P trap, 1-1/2" pipe size		18	.444		14.45	28.50		42.95	59
6880	Tubular S trap, 1-1/2" pipe size		17	.471		26.50	30.50		57	74.50
6890	PVC sch. 40 DWV, drum trap									
6900	1-1/2" pipe size	1 Plum	16	.500	Ea.	37	32		69	88.50
6910	P trap, 1-1/2" pipe size		18	.444		8.80	28.50		37.30	52.50
6920	2" pipe size		17	.471		11.85	30.50		42.35	58.50
6930	3" pipe size		15	.533		40.50	34.50		75	96.50
6940	4" pipe size		14	.571		91.50	37		128.50	156
6950	P trap w/clean out, 1-1/2" pipe size		18	.444		14.90	28.50		43.40	59.50
6960	2" pipe size		17	.471		24.50	30.50		55	72.50
6970	P trap adjustable, 1-1/2" pipe size		17	.471		11.60	30.50		42.10	58.50
6980	P trap adj. w/union & cleanout, 1-1/2" pipe size		16	.500		47.50	32		79.50	101
7000	Trap primer, flow through type, 1/2" diameter		24	.333		47	21.50		68.50	84
7100	With sediment strainer		22	.364		51	23.50		74.50	91
7450	Trap primer distribution unit									
7500	2 openings	1 Plum	18	.444	Ea.	29.50	28.50		58	75.50
7540	3 openings		17	.471		32.50	30.50		63	81.50
7560	4 openings		16	.500		37.50	32		69.50	89
7850	Trap primer manifold									
7900	2 outlet	1 Plum	18	.444	Ea.	62.50	28.50		91	112
7940	4 outlet		16	.500		100	32		132	158
7960	6 outlet		15	.533		139	34.50		173.50	205
7980	8 outlet		13	.615		178	39.50		217.50	255

22 13 16.80 Vent Flashing and Caps		Crew	Daily Output	Labor-Hours	Unit	Material	Labor	Equipment	Total	Total Incl O&P
0010	**VENT FLASHING AND CAPS**									
0120	Vent caps									
0140	Cast iron									
0160	1-1/4" to 1-1/2" pipe	1 Plum	23	.348	Ea.	41.50	22.50		64	79
0170	2" to 2-1/8" pipe		22	.364		48	23.50		71.50	88
0180	2-1/2" to 3-5/8" pipe		21	.381		52	24.50		76.50	94
0190	4" to 4-1/8" pipe		19	.421		75	27		102	124
0200	5" to 6" pipe		17	.471		104	30.50		134.50	160
0300	PVC									
0320	1-1/4" to 1-1/2" pipe	1 Plum	24	.333	Ea.	13.70	21.50		35.20	47

22 13 16 – Sanitary Waste and Vent Piping

22 13 16.80 Vent Flashing and Caps		Crew	Daily Output	Labor-Hours	Unit	Material	Labor	Equipment	Total	Total Incl O&P
						2020 Bare Costs				
0330	2" to 2-1/8" pipe	1 Plum	23	.348	Ea.	15.50	22.50		38	50.50
0900	Vent flashing									
1000	Aluminum with lead ring									
1020	1-1/4" pipe	1 Plum	20	.400	Ea.	5.95	26		31.95	45
1030	1-1/2" pipe		20	.400		5.75	26		31.75	45
1040	2" pipe		18	.444		5.65	28.50		34.15	49
1050	3" pipe		17	.471		6.25	30.50		36.75	52.50
1060	4" pipe		16	.500		7.55	32		39.55	56.50
1350	Copper with neoprene ring									
1400	1-1/4" pipe	1 Plum	20	.400	Ea.	74	26		100	120
1430	1-1/2" pipe		20	.400		74	26		100	120
1440	2" pipe		18	.444		74	28.50		102.50	125
1450	3" pipe		17	.471		89.50	30.50		120	144
1460	4" pipe		16	.500		89.50	32		121.50	147
2000	Galvanized with neoprene ring									
2020	1-1/4" pipe	1 Plum	20	.400	Ea.	16.40	26		42.40	56.50
2030	1-1/2" pipe		20	.400		17.45	26		43.45	57.50
2040	2" pipe		18	.444		16	28.50		44.50	60.50
2050	3" pipe		17	.471		19.30	30.50		49.80	66.50
2060	4" pipe		16	.500		17.65	32		49.65	67.50
2980	Neoprene, one piece									
3000	1-1/4" pipe	1 Plum	24	.333	Ea.	3.29	21.50		24.79	35.50
3030	1-1/2" pipe		24	.333		3.35	21.50		24.85	35.50
3040	2" pipe		23	.348		4.79	22.50		27.29	39
3050	3" pipe		21	.381		7	24.50		31.50	44
3060	4" pipe		20	.400		10.45	26		36.45	50
4000	Lead, 4#, 8" skirt, vent through roof									
4100	2" pipe	1 Plum	18	.444	Ea.	43	28.50		71.50	90.50
4110	3" pipe		17	.471		48.50	30.50		79	99
4120	4" pipe		16	.500		56	32		88	110
4130	6" pipe		14	.571		84	37		121	148

22 13 19 – Sanitary Waste Piping Specialties

22 13 19.13 Sanitary Drains		Crew	Daily Output	Labor-Hours	Unit	Material	Labor	Equipment	Total	Total Incl O&P
0010	**SANITARY DRAINS**									
0400	Deck, auto park, CI, 13" top									
0440	3", 4", 5", and 6" pipe size	Q-1	8	2	Ea.	1,875	116		1,991	2,225
0480	For galvanized body, add				"	1,125			1,125	1,250
0800	Promenade, heelproof grate, CI, 14" top									
0840	2", 3", and 4" pipe size	Q-1	10	1.600	Ea.	750	93		843	965
0860	5" and 6" pipe size		9	1.778		935	103		1,038	1,175
0880	8" pipe size		8	2		1,050	116		1,166	1,325
0940	For galvanized body, add					590			590	645
0960	With polished bronze top, 2"-3"-4" diam.					1,325			1,325	1,450
1200	Promenade, heelproof grate, CI, lateral, 14" top									
1240	2", 3" and 4" pipe size	Q-1	10	1.600	Ea.	985	93		1,078	1,225
1260	5" and 6" pipe size		9	1.778		1,100	103		1,203	1,375
1280	8" pipe size		8	2		1,275	116		1,391	1,575
1340	For galvanized body, add					645			645	710
1360	For polished bronze top, add					1,000			1,000	1,100
1500	Promenade, slotted grate, CI, 11" top									
1540	2", 3", 4", 5", and 6" pipe size	Q-1	12	1.333	Ea.	555	77.50		632.50	725
1600	For galvanized body, add					335			335	365

22 13 Facility Sanitary Sewerage

22 13 19 – Sanitary Waste Piping Specialties

22 13 19.13 Sanitary Drains		Crew	Daily Output	Labor-Hours	Unit	2020 Bare Costs Material	2020 Bare Costs Labor	2020 Bare Costs Equipment	Total	Total Incl O&P
1640	With polished bronze top				Ea.	920			920	1,000
2000	Floor, medium duty, CI, deep flange, 7" diam. top									
2040	2" and 3" pipe size	Q-1	12	1.333	Ea.	227	77.50		304.50	365
2080	For galvanized body, add					140			140	154
2120	With polished bronze top					375			375	410
2160	Heavy duty, CI, 12" diam. anti-tilt grate									
2180	2", 3", 4", 5" and 6" pipe size	Q-1	10	1.600	Ea.	665	93		758	870
2220	For galvanized body, add					425			425	470
2240	With polished bronze top					1,000			1,000	1,100
2300	Extra-heavy duty, CI, 15" anti-tilt grate									
2320	4", 5", 6", and 8" pipe size	Q-1	8	2	Ea.	1,625	116		1,741	1,950
2360	For galvanized body, add					700			700	770
2380	With polished bronze top					1,975			1,975	2,175
2400	Heavy duty, with sediment bucket, CI, 12" diam. loose grate									
2420	2", 3", 4", 5", and 6" pipe size	Q-1	9	1.778	Ea.	935	103		1,038	1,175
2440	For galvanized body, add					635			635	700
2460	With polished bronze top					1,150			1,150	1,250
2500	Heavy duty, cleanout & trap w/bucket, CI, 15" top									
2540	2", 3", and 4" pipe size	Q-1	6	2.667	Ea.	8,125	155		8,280	9,175
2560	For galvanized body, add					2,425			2,425	2,675
2580	With polished bronze top					9,700			9,700	10,700
2600	Medium duty, with perforated SS basket, CI, body,									
2610	18" top for refuse container washing area									
2620	2" thru 6" pipe size	Q-1	4	4	Ea.	4,525	232		4,757	5,325
2630	Acid resistant									
2638	PVC									
2640	2", 3" and 4" pipe size	Q-1	16	1	Ea.	385	58		443	510
2644	Cast iron, epoxy coated									
2646	2", 3" and 4" pipe size	Q-1	14	1.143	Ea.	1,375	66.50		1,441.50	1,600
2650	PVC or ABS thermoplastic									
2660	3" and 4" pipe size	Q-1	16	1	Ea.	415	58		473	540
2680	Extra heavy duty, oil intercepting, gas seal cone,									
2690	with cleanout, loose grate, CI, body 16" top									
2700	3" and 4" diameter outlet, 4" slab depth	Q-1	4	4	Ea.	9,225	232		9,457	10,400
2720	4" diameter outlet, 8" slab depth		3	5.333		8,550	310		8,860	9,875
2740	4" diam. outlet, 10"-12" slab depth, 16" top		2	8		11,600	465		12,065	13,400
2910	Prison cell, vandal-proof, 1-1/2", and 2" diam. pipe		12	1.333		575	77.50		652.50	750
2920	3" pipe size		10	1.600		570	93		663	765
2930	Trap drain, light duty, backwater valve CI top									
2950	8" diameter top, 2" pipe size	Q-1	12	1.333	Ea.	480	77.50		557.50	640
2960	10" diameter top, 3" pipe size		10	1.600		655	93		748	865
2970	12" diameter top, 4" pipe size		8	2		920	116		1,036	1,175

22 13 19.14 Floor Receptors		Crew	Daily Output	Labor-Hours	Unit	Material	Labor	Equipment	Total	Total Incl O&P
0010	**FLOOR RECEPTORS**, For connection to 2", 3" & 4" diameter pipe									
0200	12-1/2" square top, 25 sq. in. open area	Q-1	10	1.600	Ea.	980	93		1,073	1,225
0300	For grate with 4" diameter x 3-3/4" high funnel, add					330			330	360
0400	For grate with 6" diameter x 6" high funnel, add					264			264	290
0500	For full hinged grate with open center, add					103			103	114
0600	For aluminum bucket, add					189			189	208
0700	For acid-resisting bucket, add					325			325	355
0900	For stainless steel mesh bucket liner, add					282			282	310
1000	For bronze antisplash dome strainer, add					135			135	148

22 13 Facility Sanitary Sewerage

22 13 19 – Sanitary Waste Piping Specialties

22 13 19.14 Floor Receptors

22 13 19.14 Floor Receptors		Crew	Daily Output	Labor-Hours	Unit	2020 Bare Costs Material	Labor	Equipment	Total	Total Incl O&P
1100	For partial solid cover, add				Ea.	68.50			68.50	75.50
1200	For trap primer connection, add					112			112	123
2000	12-5/8" diameter top, 40 sq. in. open area	Q-1	10	1.600		960	93		1,053	1,200
2100	For options, add same prices as square top									
3000	8" x 4" rectangular top, 7.5 sq. in. open area	Q-1	14	1.143	Ea.	1,050	66.50		1,116.50	1,250
3100	For trap primer connections, add					110			110	121
4000	24" x 16" rectangular top, 70 sq. in. open area	Q-1	4	4		5,375	232		5,607	6,250
4100	For trap primer connection, add					226			226	248

22 13 19.15 Sink Waste Treatment

22 13 19.15 Sink Waste Treatment		Crew	Daily Output	Labor-Hours	Unit	Material	Labor	Equipment	Total	Total Incl O&P
0010	**SINK WASTE TREATMENT**, System for commercial kitchens									
0100	includes clock timer & fittings									
0200	System less chemical, wall mounted cabinet	1 Plum	16	.500	Ea.	505	32		537	605
2000	Chemical, 1 gallon, add					41			41	45
2100	6 gallons, add					185			185	203
2200	15 gallons, add					500			500	550
2300	30 gallons, add					940			940	1,025
2400	55 gallons, add					1,600			1,600	1,750

22 13 19.39 Floor Drain Trap Seal

22 13 19.39 Floor Drain Trap Seal		Crew	Daily Output	Labor-Hours	Unit	Material	Labor	Equipment	Total	Total Incl O&P
0010	**FLOOR DRAIN TRAP SEAL**									
0100	Inline									
0110	2"	1 Plum	20	.400	Ea.	35	26		61	77
0120	3"		16	.500		43	32		75	95
0130	3.5"		14	.571		42.50	37		79.50	102
0140	4"		10	.800		53	51.50		104.50	136

22 13 23 – Sanitary Waste Interceptors

22 13 23.10 Interceptors

22 13 23.10 Interceptors		Crew	Daily Output	Labor-Hours	Unit	Material	Labor	Equipment	Total	Total Incl O&P
0010	**INTERCEPTORS**									
0150	Grease, fabricated steel, 4 GPM, 8 lb. fat capacity	1 Plum	4	2	Ea.	1,575	129		1,704	1,925
0200	7 GPM, 14 lb. fat capacity		4	2		2,050	129		2,179	2,475
1000	10 GPM, 20 lb. fat capacity		4	2		2,575	129		2,704	3,025
1040	15 GPM, 30 lb. fat capacity		4	2		3,575	129		3,704	4,150
1060	20 GPM, 40 lb. fat capacity		3	2.667		4,650	172		4,822	5,375
1080	25 GPM, 50 lb. fat capacity	Q-1	3.50	4.571		4,925	265		5,190	5,825
1100	35 GPM, 70 lb. fat capacity		3	5.333		4,900	310		5,210	5,875
1120	50 GPM, 100 lb. fat capacity		2	8		8,075	465		8,540	9,575
1140	75 GPM, 150 lb. fat capacity		2	8		14,500	465		14,965	16,600
1160	100 GPM, 200 lb. fat capacity		2	8		20,100	465		20,565	22,900
1180	150 GPM, 300 lb. fat capacity		2	8		21,700	465		22,165	24,600
1200	200 GPM, 400 lb. fat capacity		1.50	10.667		31,200	620		31,820	35,300
1220	250 GPM, 500 lb. fat capacity		1.30	12.308		34,300	715		35,015	38,800
1240	300 GPM, 600 lb. fat capacity		1	16		40,400	930		41,330	45,900
1260	400 GPM, 800 lb. fat capacity	Q-2	1.20	20		49,000	1,200		50,200	56,000
1280	500 GPM, 1,000 lb. fat capacity	"	1	24		58,500	1,450		59,950	66,500
1580	For seepage pan, add					7%				
3000	Hair, cast iron, 1-1/4" and 1-1/2" pipe connection	1 Plum	8	1	Ea.	560	64.50		624.50	710
3100	For chrome-plated cast iron, add					375			375	410
3200	For polished bronze, add					1,025			1,025	1,125
3400	Lint interceptor, fabricated steel									
3410	Size based on 10 GPM per machine									
3420	30 GPM, 2" pipe size	Q-1	3	5.333	Ea.	6,500	310		6,810	7,650
3430	70 GPM, 3" pipe size		2.50	6.400		7,575	370		7,945	8,875
3440	100 GPM, 4" pipe size		2	8		8,925	465		9,390	10,500

22 13 Facility Sanitary Sewerage

22 13 23 – Sanitary Waste Interceptors

22 13 23.10 Interceptors		Crew	Daily Output	Labor-Hours	Unit	Material	2020 Bare Costs Labor	Equipment	Total	Total Incl O&P
3450	200 GPM, 4″ pipe size	Q-1	1.50	10.667	Ea.	9,950	620		10,570	11,800
3460	300 GPM, 6″ pipe size		1	16		12,500	930		13,430	15,200
3470	400 GPM, 6″ pipe size	Q-2	1.20	20		14,300	1,200		15,500	17,500
3480	500 GPM, 6″ pipe size	″	1	24		14,800	1,450		16,250	18,500
4000	Oil, fabricated steel, 10 GPM, 2″ pipe size	1 Plum	4	2		3,300	129		3,429	3,850
4100	15 GPM, 2″ or 3″ pipe size		4	2		4,525	129		4,654	5,200
4120	20 GPM, 2″ or 3″ pipe size		3	2.667		6,150	172		6,322	7,000
4140	25 GPM, 2″ or 3″ pipe size	Q-1	3.50	4.571		5,950	265		6,215	6,950
4160	35 GPM, 2″, 3″, or 4″ pipe size		3	5.333		7,250	310		7,560	8,450
4180	50 GPM, 2″, 3″, or 4″ pipe size		2	8		9,750	465		10,215	11,400
4200	75 GPM, 3″ pipe size		2	8		18,600	465		19,065	21,200
4220	100 GPM, 3″ pipe size		2	8		18,300	465		18,765	20,800
4240	150 GPM, 4″ pipe size		2	8		22,700	465		23,165	25,600
4260	200 GPM, 4″ pipe size		1.50	10.667		32,000	620		32,620	36,100
4280	250 GPM, 5″ pipe size		1.30	12.308		37,100	715		37,815	41,900
4300	300 GPM, 5″ pipe size		1	16		42,000	930		42,930	47,600
4320	400 GPM, 6″ pipe size	Q-2	1.20	20		54,000	1,200		55,200	61,500
4340	500 GPM, 6″ pipe size	″	1	24		68,000	1,450		69,450	77,000
5000	Sand interceptor, fabricated steel									
5020	20 GPM, 4″ pipe size	Q-1	3	5.333	Ea.	8,975	310		9,285	10,300
5030	50 GPM, 4″ pipe size		2.50	6.400		12,500	370		12,870	14,300
5040	150 GPM, 4″ pipe size		2	8		37,700	465		38,165	42,200
5050	250 GPM, 6″ pipe size		1.30	12.308		39,200	715		39,915	44,200
5060	500 GPM, 6″ pipe size	Q-2	1	24		50,500	1,450		51,950	57,500
6000	Solids, precious metals recovery, CI, 1-1/4″ to 2″ pipe	1 Plum	4	2		755	129		884	1,025
6100	Dental lab., large, CI, 1-1/2″ to 2″ pipe	″	3	2.667		2,650	172		2,822	3,150

22 13 26 – Sanitary Waste Separators

22 13 26.10 Separators		Crew	Daily Output	Labor-Hours	Unit	Material	Labor	Equipment	Total	Total Incl O&P
0010	**SEPARATORS**, Entrainment eliminator, steel body, 150 PSIG									
0100	1/4″ size	1 Stpi	24	.333	Ea.	276	22		298	340
0120	1/2″ size		24	.333		284	22		306	350
0140	3/4″ size		20	.400		284	26		310	350
0160	1″ size		19	.421		310	27.50		337.50	380
0180	1-1/4″ size		15	.533		325	35		360	415
0200	1-1/2″ size		13	.615		360	40.50		400.50	455
0220	2″ size		11	.727		400	47.50		447.50	510

22 13 29 – Sanitary Sewerage Pumps

22 13 29.13 Wet-Pit-Mounted, Vertical Sewerage Pumps		Crew	Daily Output	Labor-Hours	Unit	Material	Labor	Equipment	Total	Total Incl O&P
0010	**WET-PIT-MOUNTED, VERTICAL SEWERAGE PUMPS**									
0020	Controls incl. alarm/disconnect panel w/wire. Excavation not included									
0260	Simplex, 9 GPM at 60 PSIG, 91 gal. tank				Ea.	3,525			3,525	3,900
0300	Unit with manway, 26″ ID, 18″ high					4,150			4,150	4,575
0340	26″ ID, 36″ high					4,075			4,075	4,500
0380	43″ ID, 4′ high					4,250			4,250	4,675
0600	Simplex, 9 GPM at 60 PSIG, 150 gal. tank, indoor					4,275			4,275	4,700
0700	Unit with manway, 26″ ID, 36″ high					4,975			4,975	5,475
0740	26″ ID, 4′ high					5,025			5,025	5,525
2000	Duplex, 18 GPM at 60 PSIG, 150 gal. tank, indoor					8,175			8,175	9,000
2060	Unit with manway, 43″ ID, 4′ high					8,975			8,975	9,875
2400	For core only					2,050			2,050	2,250
3000	Indoor residential type installation									
3020	Simplex, 9 GPM at 60 PSIG, 91 gal. HDPE tank				Ea.	3,625			3,625	3,975

22 13 29 – Sanitary Sewerage Pumps

22 13 29.14 Sewage Ejector Pumps		Crew	Daily Output	Labor-Hours	Unit	Material	Labor	Equipment	Total	Total Incl O&P
						2020 Bare Costs				
0010	**SEWAGE EJECTOR PUMPS**, With operating and level controls									
0100	Simplex system incl. tank, cover, pump 15′ head									
0500	37 gal. PE tank, 12 GPM, 1/2 HP, 2″ discharge	Q-1	3.20	5	Ea.	505	290		795	990
0510	3″ discharge		3.10	5.161		550	299		849	1,050
0530	87 GPM, .7 HP, 2″ discharge		3.20	5		770	290		1,060	1,275
0540	3″ discharge		3.10	5.161		835	299		1,134	1,375
0600	45 gal. coated stl. tank, 12 GPM, 1/2 HP, 2″ discharge		3	5.333		900	310		1,210	1,450
0610	3″ discharge		2.90	5.517		935	320		1,255	1,500
0630	87 GPM, .7 HP, 2″ discharge		3	5.333		1,150	310		1,460	1,750
0640	3″ discharge		2.90	5.517		1,225	320		1,545	1,825
0660	134 GPM, 1 HP, 2″ discharge		2.80	5.714		1,250	330		1,580	1,875
0680	3″ discharge		2.70	5.926		1,325	345		1,670	1,975
0700	70 gal. PE tank, 12 GPM, 1/2 HP, 2″ discharge		2.60	6.154		970	355		1,325	1,600
0710	3″ discharge		2.40	6.667		1,050	385		1,435	1,725
0730	87 GPM, .7 HP, 2″ discharge		2.50	6.400		1,275	370		1,645	1,975
0740	3″ discharge		2.30	6.957		1,325	405		1,730	2,075
0760	134 GPM, 1 HP, 2″ discharge		2.20	7.273		1,375	420		1,795	2,125
0770	3″ discharge		2	8		1,450	465		1,915	2,300
0800	75 gal. coated stl. tank, 12 GPM, 1/2 HP, 2″ discharge		2.40	6.667		1,075	385		1,460	1,750
0810	3″ discharge		2.20	7.273		1,125	420		1,545	1,850
0830	87 GPM, .7 HP, 2″ discharge		2.30	6.957		1,350	405		1,755	2,100
0840	3″ discharge		2.10	7.619		1,425	440		1,865	2,225
0860	134 GPM, 1 HP, 2″ discharge		2	8		1,450	465		1,915	2,300
0880	3″ discharge	↓	1.80	8.889	↓	1,525	515		2,040	2,450
1040	Duplex system incl. tank, covers, pumps									
1060	110 gal. fiberglass tank, 24 GPM, 1/2 HP, 2″ discharge	Q-1	1.60	10	Ea.	1,950	580		2,530	3,025
1080	3″ discharge		1.40	11.429		2,050	665		2,715	3,250
1100	174 GPM, .7 HP, 2″ discharge		1.50	10.667		2,525	620		3,145	3,700
1120	3″ discharge		1.30	12.308		2,625	715		3,340	3,950
1140	268 GPM, 1 HP, 2″ discharge		1.20	13.333		2,725	775		3,500	4,150
1160	3″ discharge	↓	1	16		2,825	930		3,755	4,500
1260	135 gal. coated stl. tank, 24 GPM, 1/2 HP, 2″ discharge	Q-2	1.70	14.118		2,000	850		2,850	3,475
2000	3″ discharge		1.60	15		2,125	900		3,025	3,700
2640	174 GPM, .7 HP, 2″ discharge		1.60	15		2,625	900		3,525	4,225
2660	3″ discharge		1.50	16		2,775	960		3,735	4,500
2700	268 GPM, 1 HP, 2″ discharge		1.30	18.462		2,850	1,100		3,950	4,775
3040	3″ discharge		1.10	21.818		3,025	1,300		4,325	5,300
3060	275 gal. coated stl. tank, 24 GPM, 1/2 HP, 2″ discharge		1.50	16		2,500	960		3,460	4,200
3080	3″ discharge		1.40	17.143		2,550	1,025		3,575	4,350
3100	174 GPM, .7 HP, 2″ discharge		1.40	17.143		3,225	1,025		4,250	5,100
3120	3″ discharge		1.30	18.462		3,475	1,100		4,575	5,475
3140	268 GPM, 1 HP, 2″ discharge		1.10	21.818		3,550	1,300		4,850	5,875
3160	3″ discharge	↓	.90	26.667	↓	3,750	1,600		5,350	6,500
3260	Pump system accessories, add									
3300	Alarm horn and lights, 115 V mercury switch	Q-1	8	2	Ea.	102	116		218	286
3340	Switch, mag. contactor, alarm bell, light, 3 level control		5	3.200		495	186		681	825
3380	Alternator, mercury switch activated	↓	4	4	↓	895	232		1,127	1,325

22 14 Facility Storm Drainage

22 14 23 – Storm Drainage Piping Specialties

22 14 23.33 Backwater Valves		Crew	Daily Output	Labor-Hours	Unit	2020 Bare Costs Material	Labor	Equipment	Total	Total Incl O&P
0010	**BACKWATER VALVES**, CI Body									
6980	Bronze gate and automatic flapper valves									
7000	3″ and 4″ pipe size	Q-1	13	1.231	Ea.	2,525	71.50		2,596.50	2,875
7100	5″ and 6″ pipe size	″	13	1.231	″	3,850	71.50		3,921.50	4,350
7240	Bronze flapper valve, bolted cover									
7260	2″ pipe size	Q-1	16	1	Ea.	735	58		793	895
7280	3″ pipe size		14.50	1.103		1,150	64		1,214	1,375
7300	4″ pipe size		13	1.231		1,300	71.50		1,371.50	1,525
7320	5″ pipe size	Q-2	18	1.333		1,600	80		1,680	1,875
7340	6″ pipe size	″	17	1.412		2,275	85		2,360	2,625
7360	8″ pipe size	Q-3	10	3.200		3,225	196		3,421	3,825
7380	10″ pipe size	″	9	3.556		5,250	218		5,468	6,100
7500	For threaded cover, same cost									
7540	Revolving disk type, same cost as flapper type									

22 14 26 – Facility Storm Drains

22 14 26.13 Roof Drains		Crew	Daily Output	Labor-Hours	Unit	Material	Labor	Equipment	Total	Total Incl O&P
0010	**ROOF DRAINS**									
0140	Cornice, CI, 45° or 90° outlet									
0200	3″ and 4″ pipe size	Q-1	12	1.333	Ea.	385	77.50		462.50	540
0260	For galvanized body, add					111			111	122
0280	For polished bronze dome, add					114			114	126
3860	Roof, flat metal deck, CI body, 12″ CI dome									
3880	2″ pipe size	Q-1	15	1.067	Ea.	340	62		402	470
3890	3″ pipe size		14	1.143		410	66.50		476.50	555
3900	4″ pipe size		13	1.231		510	71.50		581.50	665
3910	5″ pipe size		12	1.333		660	77.50		737.50	845
3920	6″ pipe size		10	1.600		865	93		958	1,100
4280	Integral expansion joint, CI body, 12″ CI dome									
4300	2″ pipe size	Q-1	8	2	Ea.	765	116		881	1,025
4320	3″ pipe size		7	2.286		790	133		923	1,075
4340	4″ pipe size		6	2.667		845	155		1,000	1,150
4360	5″ pipe size		4	4		970	232		1,202	1,425
4380	6″ pipe size		3	5.333		1,050	310		1,360	1,650
4400	8″ pipe size		3	5.333		1,600	310		1,910	2,250
4440	For galvanized body, add					435			435	475
4620	Main, all aluminum, 12″ low profile dome									
4640	2″, 3″ and 4″ pipe size	Q-1	14	1.143	Ea.	590	66.50		656.50	745
4660	5″ and 6″ pipe size		13	1.231		780	71.50		851.50	960
4680	8″ pipe size		10	1.600		785	93		878	1,000
4690	Main, CI body, 12″ poly. dome, 2″, 3″, & 4″ pipe		8	2		435	116		551	655
4710	5″ and 6″ pipe size		6	2.667		625	155		780	920
4720	8″ pipe size		4	4		815	232		1,047	1,250
4730	For underdeck clamp, add		22	.727		300	42		342	395
4740	For vandalproof dome, add					75.50			75.50	83
4750	For galvanized body, add					615			615	680
4760	Main, ABS body and dome, 2″ pipe size	Q-1	14	1.143		126	66.50		192.50	237
4780	3″ pipe size		14	1.143		126	66.50		192.50	238
4800	4″ pipe size		14	1.143		126	66.50		192.50	237
4820	For underdeck clamp, add		24	.667		28	38.50		66.50	88.50
4900	Terrace planting area, with perforated overflow, CI									
4920	2″, 3″ and 4″ pipe size	Q-1	8	2	Ea.	765	116		881	1,025

22 14 Facility Storm Drainage

22 14 26 – Facility Storm Drains

22 14 26.16 Facility Area Drains		Crew	Daily Output	Labor-Hours	Unit	2020 Bare Costs Material	Labor	Equipment	Total	Total Incl O&P
0010	**FACILITY AREA DRAINS**									
4980	Scupper floor, oblique strainer, CI									
5000	6" x 7" top, 2", 3" and 4" pipe size	Q-1	16	1	Ea.	405	58		463	530
5100	8" x 12" top, 5" and 6" pipe size	"	14	1.143		790	66.50		856.50	965
5160	For galvanized body, add					40%				
5200	For polished bronze strainer, add					85%				
22 14 26.19 Facility Trench Drains										
0010	**FACILITY TRENCH DRAINS**									
5980	Trench, floor, heavy duty, modular, CI, 12" x 12" top									
6000	2", 3", 4", 5" & 6" pipe size	Q-1	8	2	Ea.	1,100	116		1,216	1,400
6100	For unit with polished bronze top		8	2		1,625	116		1,741	1,975
6200	For 12" extension section, CI top		8	2		1,100	116		1,216	1,400
6240	For 12" extension section, polished bronze top		8	2		1,825	116		1,941	2,175
6600	Trench, floor, for cement concrete encasement									
6610	Not including trenching or concrete									
6640	Polyester polymer concrete									
6650	4" internal width, with grate									
6660	Light duty steel grate	Q-1	120	.133	L.F.	68.50	7.75		76.25	86.50
6670	Medium duty steel grate		115	.139		129	8.05		137.05	154
6680	Heavy duty iron grate		110	.145		122	8.45		130.45	147
6700	12" internal width, with grate									
6770	Heavy duty galvanized grate	Q-1	80	.200	L.F.	171	11.60		182.60	205
6800	Fiberglass									
6810	8" internal width, with grate									
6820	Medium duty galvanized grate	Q-1	115	.139	L.F.	122	8.05		130.05	146
6830	Heavy duty iron grate	"	110	.145	"	201	8.45		209.45	234

22 14 29 – Sump Pumps

22 14 29.13 Wet-Pit-Mounted, Vertical Sump Pumps		Crew	Daily Output	Labor-Hours	Unit	Material	Labor	Equipment	Total	Total Incl O&P
0010	**WET-PIT-MOUNTED, VERTICAL SUMP PUMPS**									
0400	Molded PVC base, 21 GPM at 15' head, 1/3 HP	1 Plum	5	1.600	Ea.	141	103		244	310
0800	Iron base, 21 GPM at 15' head, 1/3 HP		5	1.600		155	103		258	325
1200	Solid brass, 21 GPM at 15' head, 1/3 HP		5	1.600		243	103		346	420
2000	Sump pump, single stage									
2010	25 GPM, 1 HP, 1-1/2" discharge	Q-1	1.80	8.889	Ea.	3,825	515		4,340	5,000
2020	75 GPM, 1-1/2 HP, 2" discharge		1.50	10.667		4,050	620		4,670	5,375
2030	100 GPM, 2 HP, 2-1/2" discharge		1.30	12.308		4,125	715		4,840	5,625
2040	150 GPM, 3 HP, 3" discharge		1.10	14.545		4,125	845		4,970	5,825
2050	200 GPM, 3 HP, 3" discharge		1	16		4,375	930		5,305	6,225
2060	300 GPM, 10 HP, 4" discharge	Q-2	1.20	20		4,725	1,200		5,925	7,000
2070	500 GPM, 15 HP, 5" discharge		1.10	21.818		5,375	1,300		6,675	7,875
2080	800 GPM, 20 HP, 6" discharge		1	24		6,350	1,450		7,800	9,125
2090	1,000 GPM, 30 HP, 6" discharge		.85	28.235		6,975	1,700		8,675	10,200
2100	1,600 GPM, 50 HP, 8" discharge		.72	33.333		10,900	2,000		12,900	15,000
2110	2,000 GPM, 60 HP, 8" discharge	Q-3	.85	37.647		11,100	2,300		13,400	15,700
2202	For general purpose float switch, copper coated float, add	Q-1	5	3.200		107	186		293	395

22 14 Facility Storm Drainage

22 14 29 – Sump Pumps

22 14 29.16 Submersible Sump Pumps		Crew	Daily Output	Labor-Hours	Unit	Material	Labor	Equipment	Total	Total Incl O&P
0010	**SUBMERSIBLE SUMP PUMPS**									
1000	Elevator sump pumps, automatic									
1010	Complete systems, pump, oil detector, controls and alarm									
1020	1-1/2" discharge, does not include the sump pit/tank									
1040	1/3 HP, 115 V	1 Plum	4.40	1.818	Ea.	1,925	117		2,042	2,275
1050	1/2 HP, 115 V		4	2		2,025	129		2,154	2,425
1060	1/2 HP, 230 V		4	2		2,000	129		2,129	2,400
1070	3/4 HP, 115 V		3.60	2.222		2,100	143		2,243	2,550
1080	3/4 HP, 230 V		3.60	2.222		2,075	143		2,218	2,500
1100	Sump pump only									
1110	1/3 HP, 115 V	1 Plum	6.40	1.250	Ea.	179	80.50		259.50	320
1120	1/2 HP, 115 V		5.80	1.379		254	89		343	415
1130	1/2 HP, 230 V		5.80	1.379		294	89		383	460
1140	3/4 HP, 115 V		5.40	1.481		340	95.50		435.50	520
1150	3/4 HP, 230 V		5.40	1.481		380	95.50		475.50	565
1200	Oil detector, control and alarm only									
1210	115 V	1 Plum	8	1	Ea.	1,775	64.50		1,839.50	2,050
1220	230 V	"	8	1	"	1,775	64.50		1,839.50	2,050
7000	Sump pump, automatic									
7100	Plastic, 1-1/4" discharge, 1/4 HP	1 Plum	6.40	1.250	Ea.	168	80.50		248.50	305
7140	1/3 HP		6	1.333		224	86		310	375
7160	1/2 HP		5.40	1.481		245	95.50		340.50	415
7180	1-1/2" discharge, 1/2 HP		5.20	1.538		325	99		424	505
7500	Cast iron, 1-1/4" discharge, 1/4 HP		6	1.333		225	86		311	375
7540	1/3 HP		6	1.333		264	86		350	420
7560	1/2 HP		5	1.600		320	103		423	505

22 14 53 – Rainwater Storage Tanks

22 14 53.13 Fiberglass, Rainwater Storage Tank		Crew	Daily Output	Labor-Hours	Unit	Material	Labor	Equipment	Total	Total Incl O&P
0010	**FIBERGLASS, RAINWATER STORAGE TANK**									
2000	600 gallon	B-21B	3.75	10.667	Ea.	4,125	490	126	4,741	5,425
2010	1,000 gallon		3.50	11.429		5,275	525	135	5,935	6,725
2020	2,000 gallon		3.25	12.308		7,500	565	145	8,210	9,250
2030	4,000 gallon		3	13.333		10,100	610	157	10,867	12,200
2040	6,000 gallon		2.65	15.094		11,400	695	178	12,273	13,800
2050	8,000 gallon		2.30	17.391		13,900	800	205	14,905	16,700
2060	10,000 gallon		2	20		15,400	920	236	16,556	18,600
2070	12,000 gallon		1.50	26.667		21,600	1,225	315	23,140	25,900
2080	15,000 gallon		1	40		24,500	1,825	470	26,795	30,300
2090	20,000 gallon		.75	53.333		31,800	2,450	630	34,880	39,400
2100	25,000 gallon		.50	80		47,400	3,675	940	52,015	58,500
2110	30,000 gallon		.35	114		57,000	5,250	1,350	63,600	72,000
2120	40,000 gallon		.30	133		80,000	6,125	1,575	87,700	99,000

22 15 General Service Compressed-Air Systems

22 15 13 – General Service Compressed-Air Piping

22 15 13.10 Compressor Accessories		Crew	Daily Output	Labor-Hours	Unit	Material	2020 Bare Costs Labor	Equipment	Total	Total Incl O&P
0010	**COMPRESSOR ACCESSORIES**									
3460	Air filter, regulator, lubricator combination									
3470	Flush mount									
3480	Adjustable range 0-140 psi									
3500	1/8" NPT, 34 SCFM	1 Stpi	17	.471	Ea.	107	31		138	163
3510	1/4" NPT, 61 SCFM		17	.471		192	31		223	257
3520	3/8" NPT, 85 SCFM		16	.500		196	33		229	265
3530	1/2" NPT, 150 SCFM		15	.533		210	35		245	284
3540	3/4" NPT, 171 SCFM		14	.571		260	37.50		297.50	340
3550	1" NPT, 150 SCFM		13	.615		415	40.50		455.50	515
4000	Couplers, air line, sleeve type									
4010	Female, connection size NPT									
4020	1/4"	1 Stpi	38	.211	Ea.	7.80	13.80		21.60	29
4030	3/8"		36	.222		12.60	14.55		27.15	36
4040	1/2"		35	.229		20.50	15		35.50	45
4050	3/4"		34	.235		22.50	15.40		37.90	48
4100	Male									
4110	1/4"	1 Stpi	38	.211	Ea.	7.95	13.80		21.75	29.50
4120	3/8"		36	.222		11.25	14.55		25.80	34.50
4130	1/2"		35	.229		19.90	15		34.90	44.50
4140	3/4"		34	.235		22.50	15.40		37.90	47.50
4150	Coupler, combined male and female halves									
4160	1/2"	1 Stpi	17	.471	Ea.	40.50	31		71.50	90.50
4170	3/4"	"	15	.533	"	45	35		80	102

22 15 19 – General Service Packaged Air Compressors and Receivers

22 15 19.10 Air Compressors		Crew	Daily Output	Labor-Hours	Unit	Material	2020 Bare Costs Labor	Equipment	Total	Total Incl O&P
0010	**AIR COMPRESSORS**									
5250	Air, reciprocating air cooled, splash lubricated, tank mounted									
5300	Single stage, 1 phase, 140 psi									
5303	1/2 HP, 17 gal. tank	1 Stpi	3	2.667	Ea.	2,000	175		2,175	2,450
5305	3/4 HP, 30 gal. tank		2.60	3.077		1,700	202		1,902	2,175
5307	1 HP, 30 gal. tank		2.20	3.636		2,075	238		2,313	2,650
5309	2 HP, 30 gal. tank	Q-5	4	4		2,575	236		2,811	3,200
5310	3 HP, 30 gal. tank		3.60	4.444		2,700	262		2,962	3,375
5314	3 HP, 60 gal. tank		3.50	4.571		3,175	270		3,445	3,875
5320	5 HP, 60 gal. tank		3.20	5		3,375	295		3,670	4,175
5330	5 HP, 80 gal. tank		3	5.333		3,325	315		3,640	4,125
5340	7.5 HP, 80 gal. tank		2.60	6.154		4,550	365		4,915	5,550
5600	2 stage pkg., 3 phase									
5650	6 CFM at 125 psi, 1-1/2 HP, 60 gal. tank	Q-5	3	5.333	Ea.	2,575	315		2,890	3,325
5670	10.9 CFM at 125 psi, 3 HP, 80 gal. tank		1.50	10.667		3,300	630		3,930	4,600
5680	38.7 CFM at 125 psi, 10 HP, 120 gal. tank		.60	26.667		6,275	1,575		7,850	9,250
5690	105 CFM at 125 psi, 25 HP, 250 gal. tank	Q-6	.60	40		16,300	2,450		18,750	21,600
5800	With single stage pump									
5850	8.3 CFM at 125 psi, 2 HP, 80 gal. tank	Q-6	3.50	6.857	Ea.	3,175	420		3,595	4,125
5860	38.7 CFM at 125 psi, 10 HP, 120 gal. tank	"	.90	26.667	"	5,800	1,625		7,425	8,825
6000	Reciprocating, 2 stage, tank mtd, 3 ph, cap rated @ 175 PSIG									
6050	Pressure lubricated, hvy. duty, 9.7 CFM, 3 HP, 120 gal. tank	Q-5	1.30	12.308	Ea.	4,975	725		5,700	6,550
6054	5 CFM, 1-1/2 HP, 80 gal. tank		2.80	5.714		3,100	335		3,435	3,900
6056	6.4 CFM, 2 HP, 80 gal. tank		2	8		3,175	470		3,645	4,200
6058	8.1 CFM, 3 HP, 80 gal. tank		1.70	9.412		3,400	555		3,955	4,550
6059	14.8 CFM, 5 HP, 80 gal. tank		1	16		3,425	945		4,370	5,175

22 15 General Service Compressed-Air Systems

22 15 19 – General Service Packaged Air Compressors and Receivers

22 15 19.10 Air Compressors		Crew	Daily Output	Labor-Hours	Unit	2020 Bare Costs Material	Labor	Equipment	Total	Total Incl O&P
6060	16.5 CFM, 5 HP, 120 gal. tank	Q-5	1	16	Ea.	5,625	945		6,570	7,600
6063	13 CFM, 6 HP, 80 gal. tank		.90	17.778		5,225	1,050		6,275	7,325
6066	19.8 CFM, 7.5 HP, 80 gal. tank		.80	20		5,150	1,175		6,325	7,425
6070	25.8 CFM, 7-1/2 HP, 120 gal. tank		.80	20		5,400	1,175		6,575	7,700
6078	34.8 CFM, 10 HP, 80 gal. tank		.70	22.857		7,800	1,350		9,150	10,600
6080	34.8 CFM, 10 HP, 120 gal. tank		.60	26.667		8,325	1,575		9,900	11,500
6090	53.7 CFM, 15 HP, 120 gal. tank	Q-6	.80	30		8,750	1,825		10,575	12,400
6100	76.7 CFM, 20 HP, 120 gal. tank		.70	34.286		12,600	2,100		14,700	17,000
6104	76.7 CFM, 20 HP, 240 gal. tank		.68	35.294		12,200	2,150		14,350	16,600
6110	90.1 CFM, 25 HP, 120 gal. tank		.63	38.095		12,700	2,325		15,025	17,500
6120	101 CFM, 30 HP, 120 gal. tank		.57	42.105		11,300	2,575		13,875	16,300
6130	101 CFM, 30 HP, 250 gal. tank		.52	46.154		12,800	2,825		15,625	18,300
6200	Oil-less, 13.6 CFM, 5 HP, 120 gal. tank	Q-5	.88	18.182		13,700	1,075		14,775	16,600
6210	13.6 CFM, 5 HP, 250 gal. tank		.80	20		14,600	1,175		15,775	17,900
6220	18.2 CFM, 7.5 HP, 120 gal. tank		.73	21.918		13,600	1,300		14,900	16,900
6230	18.2 CFM, 7.5 HP, 250 gal. tank		.67	23.881		14,600	1,400		16,000	18,200
6250	30.5 CFM, 10 HP, 120 gal. tank		.57	28.070		16,100	1,650		17,750	20,200
6260	30.5 CFM, 10 HP, 250 gal. tank		.53	30.189		17,000	1,775		18,775	21,500
6270	41.3 CFM, 15 HP, 120 gal. tank	Q-6	.70	34.286		17,400	2,100		19,500	22,400
6280	41.3 CFM, 15 HP, 250 gal. tank	"	.67	35.821		18,500	2,200		20,700	23,600

22 31 Domestic Water Softeners

22 31 13 – Residential Domestic Water Softeners

22 31 13.10 Residential Water Softeners		Crew	Daily Output	Labor-Hours	Unit	Material	Labor	Equipment	Total	Total Incl O&P
0010	**RESIDENTIAL WATER SOFTENERS**									
7350	Water softener, automatic, to 30 grains per gallon	2 Plum	5	3.200	Ea.	395	206		601	745
7400	To 100 grains per gallon	"	4	4	"	940	258		1,198	1,400

22 31 16 – Commercial Domestic Water Softeners

22 31 16.10 Water Softeners		Crew	Daily Output	Labor-Hours	Unit	Material	Labor	Equipment	Total	Total Incl O&P
0010	**WATER SOFTENERS**									
5800	Softener systems, automatic, intermediate sizes									
5820	available, may be used in multiples.									
6000	Hardness capacity between regenerations and flow									
6060	40,000 grains, 14 GPM	Q-1	4	4	Ea.	2,525	232		2,757	3,125
6070	50,000 grains, 17 GPM		3.60	4.444		2,775	258		3,033	3,425
6080	90,000 grains, 25 GPM		2.80	5.714		5,950	330		6,280	7,025
6100	150,000 grains, 37 GPM cont., 51 GPM peak		1.20	13.333		4,925	775		5,700	6,575
6200	300,000 grains, 81 GPM cont., 113 GPM peak		1	16		8,525	930		9,455	10,800
6300	750,000 grains, 160 GPM cont., 230 GPM peak		.80	20		12,600	1,150		13,750	15,600
6400	900,000 grains, 185 GPM cont., 270 GPM peak		.70	22.857		20,300	1,325		21,625	24,300
8000	Water treatment, salts, 50 lb. bag									
8020	Salt, water softener, bag, pelletized				Lb.	3.68			3.68	4.05
8030	Salt, water softener, bag, crystal rock salt				"	.23			.23	.25

22 32 Domestic Water Filtration Equipment

22 32 19 – Domestic-Water Off-Floor Cartridge Filters

22 32 19.10 Water Filters		Crew	Daily Output	Labor-Hours	Unit	2020 Bare Costs Material	Labor	Equipment	Total	Total Incl O&P
0010	**WATER FILTERS**, Purification and treatment									
1000	Cartridge style, dirt and rust type	1 Plum	12	.667	Ea.	80.50	43		123.50	154
1200	Replacement cartridge		32	.250		16.05	16.10		32.15	41.50
1600	Taste and odor type		12	.667		115	43		158	191
1700	Replacement cartridge		32	.250		42	16.10		58.10	70
3000	Central unit, dirt/rust/odor/taste/scale		4	2		271	129		400	490
3100	Replacement cartridge, standard		20	.400		92.50	26		118.50	141
3600	Replacement cartridge, heavy duty		20	.400		84	26		110	131
8000	Commercial, fully automatic or push button automatic									
8200	Iron removal, 660 GPH, 1″ pipe size	Q-1	1.50	10.667	Ea.	1,750	620		2,370	2,850
8240	1,500 GPH, 1-1/4″ pipe size		1	16		2,950	930		3,880	4,650
8280	2,340 GPH, 1-1/2″ pipe size		.80	20		3,250	1,150		4,400	5,300
8320	3,420 GPH, 2″ pipe size		.60	26.667		5,975	1,550		7,525	8,900
8360	4,620 GPH, 2-1/2″ pipe size		.50	32		9,500	1,850		11,350	13,300
8500	Neutralizer for acid water, 780 GPH, 1″ pipe size		1.50	10.667		1,700	620		2,320	2,775
8540	1,140 GPH, 1-1/4″ pipe size		1	16		1,900	930		2,830	3,500
8580	1,740 GPH, 1-1/2″ pipe size		.80	20		2,775	1,150		3,925	4,775
8620	2,520 GPH, 2″ pipe size		.60	26.667		3,600	1,550		5,150	6,275
8660	3,480 GPH, 2-1/2″ pipe size		.50	32		5,975	1,850		7,825	9,350
8800	Sediment removal, 780 GPH, 1″ pipe size		1.50	10.667		1,625	620		2,245	2,700
8840	1,140 GPH, 1-1/4″ pipe size		1	16		1,925	930		2,855	3,525
8880	1,740 GPH, 1-1/2″ pipe size		.80	20		2,550	1,150		3,700	4,550
8920	2,520 GPH, 2″ pipe size		.60	26.667		3,700	1,550		5,250	6,400
8960	3,480 GPH, 2-1/2″ pipe size		.50	32		5,825	1,850		7,675	9,175
9200	Taste and odor removal, 660 GPH, 1″ pipe size		1.50	10.667		2,250	620		2,870	3,400
9240	1,500 GPH, 1-1/4″ pipe size		1	16		3,875	930		4,805	5,650
9280	2,340 GPH, 1-1/2″ pipe size		.80	20		4,425	1,150		5,575	6,575
9320	3,420 GPH, 2″ pipe size		.60	26.667		6,775	1,550		8,325	9,775
9360	4,620 GPH, 2-1/2″ pipe size		.50	32		9,050	1,850		10,900	12,700

22 33 Electric Domestic Water Heaters

22 33 13 – Instantaneous Electric Domestic Water Heaters

22 33 13.10 Hot Water Dispensers

		Crew	Daily Output	Labor-Hours	Unit	Material	Labor	Equipment	Total	Total Incl O&P
0010	**HOT WATER DISPENSERS**									
0160	Commercial, 100 cup, 11.3 amp	1 Plum	14	.571	Ea.	460	37		497	560
3180	Household, 60 cup	"	14	.571	"	244	37		281	325

22 33 13.20 Instantaneous Elec. Point-Of-Use Water Heaters

		Crew	Daily Output	Labor-Hours	Unit	Material	Labor	Equipment	Total	Total Incl O&P
0010	**INSTANTANEOUS ELECTRIC POINT-OF-USE WATER HEATERS**									
8965	Point of use, electric, glass lined									
8969	Energy saver									
8970	2.5 gal. single element G	1 Plum	2.80	2.857	Ea.	370	184		554	680
8971	4 gal. single element G		2.80	2.857		375	184		559	685
8974	6 gal. single element G		2.50	3.200		415	206		621	765
8975	10 gal. single element G		2.50	3.200		470	206		676	825
8976	15 gal. single element G		2.40	3.333		545	215		760	920
8977	20 gal. single element G		2.40	3.333		675	215		890	1,050
8978	30 gal. single element G		2.30	3.478		700	224		924	1,100
8979	40 gal. single element G		2.20	3.636		1,250	234		1,484	1,725
8988	Commercial (ASHRAE energy std. 90)									
8989	6 gallon G	1 Plum	2.50	3.200	Ea.	1,000	206		1,206	1,400
8990	10 gallon G		2.50	3.200		1,175	206		1,381	1,575

22 33 Electric Domestic Water Heaters

22 33 13 – Instantaneous Electric Domestic Water Heaters

22 33 13.20 Instantaneous Elec. Point-Of-Use Water Heaters		Crew	Daily Output	Labor-Hours	Unit	2020 Bare Costs Material	Labor	Equipment	Total	Total Incl O&P
8991	15 gallon G	1 Plum	2.40	3.333	Ea.	1,150	215		1,365	1,600
8992	20 gallon G		2.40	3.333		1,350	215		1,565	1,800
8993	30 gallon G		2.30	3.478		2,825	224		3,049	3,425
8995	Under the sink, copper, w/bracket									
8996	2.5 gallon G	1 Plum	4	2	Ea.	405	129		534	640

22 33 30 – Residential, Electric Domestic Water Heaters

22 33 30.13 Residential, Small-Capacity Elec. Water Heaters		Crew	Daily Output	Labor-Hours	Unit	Material	Labor	Equipment	Total	Total Incl O&P
0010	**RESIDENTIAL, SMALL-CAPACITY ELECTRIC DOMESTIC WATER HEATERS**									
1000	Residential, electric, glass lined tank, 5 yr., 10 gal., single element D2020–210	1 Plum	2.30	3.478	Ea.	470	224		694	850
1040	20 gallon, single element		2.20	3.636		675	234		909	1,100
1060	30 gallon, double element		2.20	3.636		1,075	234		1,309	1,525
1080	40 gallon, double element		2	4		1,250	258		1,508	1,750
1100	52 gallon, double element		2	4		1,425	258		1,683	1,950
1120	66 gallon, double element		1.80	4.444		1,925	286		2,211	2,550
1140	80 gallon, double element		1.60	5		2,175	320		2,495	2,875
1180	120 gallon, double element		1.40	5.714		3,000	370		3,370	3,850

22 33 33 – Light-Commercial Electric Domestic Water Heaters

22 33 33.10 Commercial Electric Water Heaters		Crew	Daily Output	Labor-Hours	Unit	Material	Labor	Equipment	Total	Total Incl O&P
0010	**COMMERCIAL ELECTRIC WATER HEATERS**									
4000	Commercial, 100° rise. NOTE: for each size tank, a range of									
4010	heaters between the ones shown is available									
4020	Electric									
4100	5 gal., 3 kW, 12 GPH, 208 volt	1 Plum	2	4	Ea.	4,600	258		4,858	5,425
4120	10 gal., 6 kW, 25 GPH, 208 volt		2	4		5,025	258		5,283	5,925
4130	30 gal., 24 kW, 98 GPH, 208 volt		1.92	4.167		8,100	269		8,369	9,300
4136	40 gal., 36 kW, 148 GPH, 208 volt		1.88	4.255		11,300	274		11,574	12,800
4140	50 gal., 9 kW, 37 GPH, 208 volt		1.80	4.444		6,900	286		7,186	8,025
4160	50 gal., 36 kW, 148 GPH, 208 volt		1.80	4.444		11,100	286		11,386	12,600
4180	80 gal., 12 kW, 49 GPH, 208 volt		1.50	5.333		8,500	345		8,845	9,875
4200	80 gal., 36 kW, 148 GPH, 208 volt		1.50	5.333		12,100	345		12,445	13,800
4220	100 gal., 36 kW, 148 GPH, 208 volt		1.20	6.667		13,200	430		13,630	15,100
4240	120 gal., 36 kW, 148 GPH, 208 volt		1.20	6.667		13,700	430		14,130	15,700
4260	150 gal., 15 kW, 61 GPH, 480 volt		1	8		34,500	515		35,015	38,700
4280	150 gal., 120 kW, 490 GPH, 480 volt		1	8		47,300	515		47,815	53,000
4300	200 gal., 15 kW, 61 GPH, 480 volt	Q-1	1.70	9.412		36,700	545		37,245	41,200
4320	200 gal., 120 kW, 490 GPH, 480 volt		1.70	9.412		48,400	545		48,945	54,000
4340	250 gal., 15 kW, 61 GPH, 480 volt		1.50	10.667		38,000	620		38,620	42,700
4360	250 gal., 150 kW, 615 GPH, 480 volt		1.50	10.667		48,100	620		48,720	54,000
4380	300 gal., 30 kW, 123 GPH, 480 volt		1.30	12.308		41,900	715		42,615	47,100
4400	300 gal., 180 kW, 738 GPH, 480 volt		1.30	12.308		67,500	715		68,215	75,500
4420	350 gal., 30 kW, 123 GPH, 480 volt		1.10	14.545		39,600	845		40,445	44,800
4440	350 gal., 180 kW, 738 GPH, 480 volt		1.10	14.545		56,000	845		56,845	63,000
4460	400 gal., 30 kW, 123 GPH, 480 volt		1	16		50,500	930		51,430	57,000
4480	400 gal., 210 kW, 860 GPH, 480 volt		1	16		80,500	930		81,430	90,000
4500	500 gal., 30 kW, 123 GPH, 480 volt		.80	20		57,500	1,150		58,650	64,500
4520	500 gal., 240 kW, 984 GPH, 480 volt		.80	20		97,500	1,150		98,650	109,000
4540	600 gal., 30 kW, 123 GPH, 480 volt	Q-2	1.20	20		66,500	1,200		67,700	75,000
4560	600 gal., 300 kW, 1,230 GPH, 480 volt		1.20	20		112,500	1,200		113,700	125,500
4580	700 gal., 30 kW, 123 GPH, 480 volt		1	24		63,500	1,450		64,950	71,500
4600	700 gal., 300 kW, 1,230 GPH, 480 volt		1	24		95,000	1,450		96,450	106,500
4620	800 gal., 60 kW, 245 GPH, 480 volt		.90	26.667		79,000	1,600		80,600	89,000
4640	800 gal., 300 kW, 1,230 GPH, 480 volt		.90	26.667		97,000	1,600		98,600	109,500

22 33 Electric Domestic Water Heaters

22 33 33 – Light-Commercial Electric Domestic Water Heaters

22 33 33.10 Commercial Electric Water Heaters		Crew	Daily Output	Labor-Hours	Unit	2020 Bare Costs Material	Labor	Equipment	Total	Total Incl O&P
4660	1,000 gal., 60 kW, 245 GPH, 480 volt	Q-2	.70	34.286	Ea.	75,000	2,050		77,050	85,500
4680	1,000 gal., 480 kW, 1,970 GPH, 480 volt		.70	34.286		124,500	2,050		126,550	140,000
4700	1,200 gal., 60 kW, 245 GPH, 480 volt		.60	40		81,500	2,400		83,900	93,500
4720	1,200 gal., 480 kW, 1,970 GPH, 480 volt		.60	40		128,000	2,400		130,400	144,000
4740	1,500 gal., 60 kW, 245 GPH, 480 volt		.50	48		107,500	2,875		110,375	122,500
4760	1,500 gal., 480 kW, 1,970 GPH, 480 volt		.50	48		154,500	2,875		157,375	174,500
5400	Modulating step control for under 90 kW, 2-5 steps	1 Elec	5.30	1.509		880	92.50		972.50	1,100
5440	For above 90 kW, 1 through 5 steps beyond standard, add		3.20	2.500		272	153		425	525
5460	For above 90 kW, 6 through 10 steps beyond standard, add		2.70	2.963		550	182		732	875
5480	For above 90 kW, 11 through 18 steps beyond standard, add		1.60	5		820	305		1,125	1,350

22 34 Fuel-Fired Domestic Water Heaters

22 34 13 – Instantaneous, Tankless, Gas Domestic Water Heaters

22 34 13.10 Instantaneous, Tankless, Gas Water Heaters		Crew	Daily Output	Labor-Hours	Unit	Material	Labor	Equipment	Total	Total Incl O&P
0010	**INSTANTANEOUS, TANKLESS, GAS WATER HEATERS**									
9410	Natural gas/propane, 3.2 GPM G	1 Plum	2	4	Ea.	575	258		833	1,025
9420	6.4 GPM G		1.90	4.211		780	271		1,051	1,275
9430	8.4 GPM G		1.80	4.444		880	286		1,166	1,400
9440	9.5 GPM G		1.60	5		1,050	320		1,370	1,625

22 34 30 – Residential Gas Domestic Water Heaters

22 34 30.13 Residential, Atmos, Gas Domestic Wtr Heaters		Crew	Daily Output	Labor-Hours	Unit	Material	Labor	Equipment	Total	Total Incl O&P
0010	**RESIDENTIAL, ATMOSPHERIC, GAS DOMESTIC WATER HEATERS**									
2000	Gas fired, foam lined tank, 10 yr., vent not incl.									
2040	30 gallon	1 Plum	2	4	Ea.	2,100	258		2,358	2,700
2060	40 gallon		1.90	4.211		1,825	271		2,096	2,400
2080	50 gallon		1.80	4.444		1,950	286		2,236	2,575
2090	60 gallon		1.70	4.706		1,900	305		2,205	2,550
2100	75 gallon		1.50	5.333		2,650	345		2,995	3,450
2120	100 gallon		1.30	6.154		3,150	395		3,545	4,075
2900	Water heater, safety-drain pan, 26" round		20	.400		21.50	26		47.50	62
3000	Tank leak safety, water & gas shut off see 22 05 23.20 8800									

22 34 36 – Commercial Gas Domestic Water Heaters

22 34 36.13 Commercial, Atmos., Gas Domestic Water Htrs.		Crew	Daily Output	Labor-Hours	Unit	Material	Labor	Equipment	Total	Total Incl O&P
0010	**COMMERCIAL, ATMOSPHERIC, GAS DOMESTIC WATER HEATERS**									
6000	Gas fired, flush jacket, std. controls, vent not incl.									
6040	75 MBH input, 73 GPH	1 Plum	1.40	5.714	Ea.	3,725	370		4,095	4,650
6060	98 MBH input, 95 GPH		1.40	5.714		8,800	370		9,170	10,200
6080	120 MBH input, 110 GPH		1.20	6.667		9,000	430		9,430	10,500
6100	120 MBH input, 115 GPH		1.10	7.273		8,725	470		9,195	10,300
6120	140 MBH input, 130 GPH		1	8		10,800	515		11,315	12,600
6140	155 MBH input, 150 GPH		.80	10		12,100	645		12,745	14,300
6160	180 MBH input, 170 GPH		.70	11.429		11,700	735		12,435	14,000
6180	200 MBH input, 192 GPH		.60	13.333		12,100	860		12,960	14,700
6200	250 MBH input, 245 GPH		.50	16		12,700	1,025		13,725	15,600
6220	260 MBH input, 250 GPH	Q-1	.80	20		13,100	1,150		14,250	16,100
6240	360 MBH input, 360 GPH		.80	20		15,400	1,150		16,550	18,600
6260	500 MBH input, 480 GPH		.70	22.857		21,600	1,325		22,925	25,800
6280	725 MBH input, 690 GPH		.60	26.667		26,900	1,550		28,450	31,900
6900	For low water cutoff, add	1 Plum	8	1		385	64.50		449.50	520
6960	For bronze body hot water circulator, add	"	4	2		2,250	129		2,379	2,675

22 34 Fuel-Fired Domestic Water Heaters

22 34 46 – Oil-Fired Domestic Water Heaters

22 34 46.10 Residential Oil-Fired Water Heaters		Crew	Daily Output	Labor-Hours	Unit	Material	2020 Bare Costs Labor	Equipment	Total	Total Incl O&P
0010	**RESIDENTIAL OIL-FIRED WATER HEATERS**									
3000	Oil fired, glass lined tank, 5 yr., vent not included, 30 gallon D2020–260	1 Plum	2	4	Ea.	1,400	258		1,658	1,925
3040	50 gallon		1.80	4.444		1,375	286		1,661	1,950
3060	70 gallon	↓	1.50	5.333	↓	2,375	345		2,720	3,150

22 34 46.20 Commercial Oil-Fired Water Heaters

Code	Description	Crew	Daily Output	Labor-Hours	Unit	Material	Labor	Equipment	Total	Total Incl O&P
0010	**COMMERCIAL OIL-FIRED WATER HEATERS**									
8000	Oil fired, glass lined, UL listed, std. controls, vent not incl.									
8060	140 gal., 140 MBH input, 134 GPH	Q-1	2.13	7.512	Ea.	25,200	435		25,635	28,400
8080	140 gal., 199 MBH input, 191 GPH		2	8		26,100	465		26,565	29,400
8100	140 gal., 255 MBH input, 247 GPH		1.60	10		26,800	580		27,380	30,400
8120	140 gal., 270 MBH input, 259 GPH		1.20	13.333		33,100	775		33,875	37,700
8140	140 gal., 400 MBH input, 384 GPH		1	16		34,000	930		34,930	38,800
8160	140 gal., 540 MBH input, 519 GPH		.96	16.667		35,600	965		36,565	40,600
8180	140 gal., 720 MBH input, 691 GPH		.92	17.391		36,200	1,000		37,200	41,300
8200	221 gal., 300 MBH input, 288 GPH		.88	18.182		47,900	1,050		48,950	54,000
8220	221 gal., 600 MBH input, 576 GPH		.86	18.605		61,500	1,075		62,575	69,000
8240	221 gal., 800 MBH input, 768 GPH	↓	.82	19.512		54,000	1,125		55,125	60,500
8260	201 gal., 1,000 MBH input, 960 GPH	Q-2	1.26	19.048		55,000	1,150		56,150	62,000
8280	201 gal., 1,250 MBH input, 1,200 GPH		1.22	19.672		56,000	1,175		57,175	63,500
8300	201 gal., 1,500 MBH input, 1,441 GPH		1.16	20.690		60,500	1,250		61,750	68,500
8320	411 gal., 600 MBH input, 576 GPH		1.12	21.429		61,000	1,300		62,300	69,000
8340	411 gal., 800 MBH input, 768 GPH		1.08	22.222		63,500	1,325		64,825	71,500
8360	411 gal., 1,000 MBH input, 960 GPH		1.04	23.077		64,500	1,400		65,900	73,000
8380	411 gal., 1,250 MBH input, 1,200 GPH		.98	24.490		66,000	1,475		67,475	75,000
8400	397 gal., 1,500 MBH input, 1,441 GPH		.92	26.087		70,000	1,575		71,575	79,500
8420	397 gal., 1,750 MBH input, 1,681 GPH		.86	27.907		72,000	1,675		73,675	82,000
8430	397 gal., 2,000 MBH input, 1,921 GPH		.82	29.268		78,000	1,750		79,750	88,500
8440	375 gal., 2,250 MBH input, 2,161 GPH		.76	31.579		80,500	1,900		82,400	91,500
8450	375 gal., 2,500 MBH input, 2,401 GPH	↓	.82	29.268	↓	83,500	1,750		85,250	94,000
8500	Oil fired, polymer lined									
8510	400 MBH, 125 gallon	Q-2	1	24	Ea.	37,300	1,450		38,750	43,200
8520	400 MBH, 600 gallon		.80	30		68,500	1,800		70,300	78,000
8530	800 MBH, 400 gallon		.67	35.821		62,500	2,150		64,650	71,500
8540	800 MBH, 600 gallon		.60	40		81,500	2,400		83,900	93,000
8550	1,000 MBH, 600 gallon		.50	48		83,500	2,875		86,375	96,500
8560	1,200 MBH, 900 gallon	↓	.40	60		92,500	3,600		96,100	107,000
8900	For low water cutoff, add	1 Plum	8	1		405	64.50		469.50	545
8960	For bronze body hot water circulator, add	"	4	2	↓	1,050	129		1,179	1,350

22 35 Domestic Water Heat Exchangers

22 35 30 – Water Heating by Steam

22 35 30.10 Water Heating Transfer Package

Code	Description	Crew	Daily Output	Labor-Hours	Unit	Material	Labor	Equipment	Total	Total Incl O&P
0010	**WATER HEATING TRANSFER PACKAGE**, Complete controls,									
0020	expansion tank, converter, air separator									
1000	Hot water, 180°F enter, 200°F leaving, 15# steam									
1010	One pump system, 28 GPM	Q-6	.75	32	Ea.	21,800	1,950		23,750	26,900
1020	35 GPM		.70	34.286		24,700	2,100		26,800	30,400
1040	55 GPM		.65	36.923		28,000	2,250		30,250	34,200
1060	130 GPM		.55	43.636		35,500	2,675		38,175	43,100
1080	255 GPM	↓	.40	60	↓	46,600	3,675		50,275	56,500

22 35 Domestic Water Heat Exchangers

22 35 30 – Water Heating by Steam

22 35 30.10 Water Heating Transfer Package		Crew	Daily Output	Labor-Hours	Unit	2020 Bare Costs Material	Labor	Equipment	Total	Total Incl O&P
1100	550 GPM	Q-6	.30	80	Ea.	67,000	4,900		71,900	81,500
1120	800 GPM		.25	96		74,000	5,875		79,875	90,500
1220	Two pump system, 28 GPM		.70	34.286		29,700	2,100		31,800	35,800
1240	35 GPM		.65	36.923		36,100	2,250		38,350	43,100
1260	55 GPM		.60	40		35,900	2,450		38,350	43,100
1280	130 GPM		.50	48		49,400	2,925		52,325	59,000
1300	255 GPM		.35	68.571		65,000	4,200		69,200	78,000
1320	550 GPM		.25	96		79,500	5,875		85,375	96,000
1340	800 GPM		.20	120		102,500	7,350		109,850	123,500

22 35 43 – Domestic Water Heat Reclaimers

22 35 43.10 Drainwater Heat Recovery

Line	Description	Crew	Daily Output	Labor-Hours	Unit	Material	Labor	Equipment	Total	Total Incl O&P
0010	**DRAINWATER HEAT RECOVERY**									
9005	Drainwater heat recov unit, copp coil type, for 1/2" supply, 3" waste	1 Plum	3	2.667	Ea.	505	172		677	810
9010	For 1/2" supply, 4" waste		3	2.667		665	172		837	985
9020	For 3/4" supply, 3" waste		3	2.667		705	172		877	1,025
9030	For 3/4" supply, 4" waste		3	2.667		745	172		917	1,075
9040	For 1" supply, 4" waste, double manifold		3	2.667		1,375	172		1,547	1,750

22 41 Residential Plumbing Fixtures

22 41 06 – Plumbing Fixtures General

22 41 06.10 Plumbing Fixture Notes

Line	Description	Crew	Daily Output	Labor-Hours	Unit	Material	Labor	Equipment	Total	Total Incl O&P
0010	**PLUMBING FIXTURE NOTES**, Incl. trim fittings unless otherwise noted R224000-30									
0080	For rough-in, supply, waste, and vent, see add for each type									
0122	For electric water coolers, see Section 22 47 16.10									
0160	For color, unless otherwise noted, add				Ea.	20%				

22 41 13 – Residential Water Closets, Urinals, and Bidets

22 41 13.13 Water Closets

Line	Description	Crew	Daily Output	Labor-Hours	Unit	Material	Labor	Equipment	Total	Total Incl O&P
0010	**WATER CLOSETS** D2010–110									
0022	For seats, see Section 22 41 13.44									
0032	For automatic flush, see Line 22 42 39.10 0972									
0150	Tank type, vitreous china, incl. seat, supply pipe w/stop, 1.6 gpf or noted									
0200	Wall hung R224000-30									
0400	Two piece, close coupled	Q-1	5.30	3.019	Ea.	420	175		595	720
0960	For rough-in, supply, waste, vent and carrier	"	2.73	5.861	"	1,275	340		1,615	1,900
0999	Floor mounted									
1020	One piece, low profile	Q-1	5.30	3.019	Ea.	900	175		1,075	1,250
1050	One piece		5.30	3.019		945	175		1,120	1,300
1100	Two piece, close coupled		5.30	3.019		219	175		394	505
1102	Economy		5.30	3.019		121	175		296	395
1110	Two piece, close coupled, dual flush		5.30	3.019		280	175		455	570
1140	Two piece, close coupled, 1.28 gpf, ADA G		5.30	3.019		320	175		495	610
1960	For color, add					30%				
1961	For designer colors and trim, add					55%				
1980	For rough-in, supply, waste and vent	Q-1	3.05	5.246	Ea.	340	305		645	830

22 41 13.19 Bidets

Line	Description	Crew	Daily Output	Labor-Hours	Unit	Material	Labor	Equipment	Total	Total Incl O&P
0010	**BIDETS**									
0180	Vitreous china, with trim on fixture	Q-1	5	3.200	Ea.	685	186		871	1,025
0200	With trim for wall mounting	"	5	3.200		805	186		991	1,175
9590	For color, add					40%				
9591	For designer colors and trim, add					50%				

22 41 Residential Plumbing Fixtures

22 41 13 – Residential Water Closets, Urinals, and Bidets

22 41 13.19 Bidets		Crew	Daily Output	Labor-Hours	Unit	2020 Bare Costs Material	Labor	Equipment	Total	Total Incl O&P
9600	For rough-in, supply, waste and vent, add	Q-1	1.78	8.989	Ea.	430	520		950	1,250

22 41 13.44 Toilet Seats

		Crew	Daily Output	Labor-Hours	Unit	Material	Labor	Equipment	Total	Total Incl O&P
0010	**TOILET SEATS**									
0100	Molded composition, white									
0150	Industrial, w/o cover, open front, regular bowl	1 Plum	24	.333	Ea.	22	21.50		43.50	56.50
0200	With self-sustaining hinge		24	.333		23.50	21.50		45	58
0220	With self-sustaining check hinge		24	.333		23	21.50		44.50	57
0240	Extra heavy, with check hinge		24	.333		22.50	21.50		44	57
0260	Elongated bowl, same price									
0300	Junior size, w/o cover, open front	1 Plum	24	.333	Ea.	43.50	21.50		65	80
0320	Regular primary bowl, open front		24	.333		43.50	21.50		65	80
0340	Regular baby bowl, open front, check hinge		24	.333		37.50	21.50		59	73.50
0380	Open back & front, w/o cover, reg. or elongated bowl		24	.333		21	21.50		42.50	55
0400	Residential									
0420	Regular bowl, w/cover, closed front	1 Plum	24	.333	Ea.	29.50	21.50		51	64
0440	Open front	"	24	.333	"	26.50	21.50		48	61
0460	Elongated bowl, add					25%				
0500	Self-raising hinge, w/o cover, open front									
0520	Regular bowl	1 Plum	24	.333	Ea.	103	21.50		124.50	145
0540	Elongated bowl	"	24	.333	"	25	21.50		46.50	59.50
0700	Molded wood, white, with cover									
0720	Closed front, regular bowl, square back	1 Plum	24	.333	Ea.	10.15	21.50		31.65	43
0740	Extended back		24	.333		14.10	21.50		35.60	47.50
0780	Elongated bowl, square back		24	.333		13.50	21.50		35	47
0800	Open front		24	.333		14.20	21.50		35.70	47.50
0850	Decorator styles									
0890	Vinyl top, patterned	1 Plum	24	.333	Ea.	21.50	21.50		43	56
0900	Vinyl padded, plain colors, regular bowl		24	.333		21.50	21.50		43	56
0930	Elongated bowl		24	.333		25.50	21.50		47	60
1000	Solid plastic, white									
1030	Industrial, w/o cover, open front, regular bowl	1 Plum	24	.333	Ea.	26.50	21.50		48	61.50
1080	Extra heavy, concealed check hinge		24	.333		21	21.50		42.50	55.50
1100	Self-sustaining hinge		24	.333		25	21.50		46.50	59.50
1150	Elongated bowl		24	.333		26	21.50		47.50	60.50
1170	Concealed check		24	.333		17.60	21.50		39.10	51.50
1190	Self-sustaining hinge, concealed check		24	.333		55.50	21.50		77	93
1220	Residential, with cover, closed front, regular bowl		24	.333		41.50	21.50		63	77.50
1240	Elongated bowl		24	.333		50	21.50		71.50	87
1260	Open front, regular bowl		24	.333		43	21.50		64.50	79
1280	Elongated bowl		24	.333		52	21.50		73.50	89.50

22 41 16 – Residential Lavatories and Sinks

22 41 16.13 Lavatories

		Crew	Daily Output	Labor-Hours	Unit	Material	Labor	Equipment	Total	Total Incl O&P
0010	**LAVATORIES**, With trim, white unless noted otherwise D2010–310									
0500	Vanity top, porcelain enamel on cast iron									
0600	20" x 18" R224000-30	Q-1	6.40	2.500	Ea.	305	145		450	555
0640	33" x 19" oval		6.40	2.500		520	145		665	790
0680	20" x 17" oval		6.40	2.500		117	145		262	345
0720	19" round		6.40	2.500		455	145		600	715
0760	20" x 12" triangular bowl		6.40	2.500		239	145		384	480
0860	For color, add					25%				
0861	For designer colors and trim, add					70%				
1000	Cultured marble, 19" x 17", single bowl	Q-1	6.40	2.500	Ea.	124	145		269	355

22 41 16.13 Lavatories		Crew	Daily Output	Labor-Hours	Unit	2020 Bare Costs Material	Labor	Equipment	Total	Total Incl O&P
1040	25" x 19", single bowl	Q-1	6.40	2.500	Ea.	150	145		295	380
1080	31" x 19", single bowl		6.40	2.500		166	145		311	400
1120	25" x 22", single bowl		6.40	2.500		164	145		309	395
1160	37" x 22", single bowl		6.40	2.500		203	145		348	440
1200	49" x 22", single bowl		6.40	2.500		242	145		387	485
1580	For color, same price									
1900	Stainless steel, self-rimming, 25" x 22", single bowl, ledge	Q-1	6.40	2.500	Ea.	315	145		460	565
1960	17" x 22", single bowl		6.40	2.500		305	145		450	550
2040	18-3/4" round		6.40	2.500		820	145		965	1,125
2600	Steel, enameled, 20" x 17", single bowl		5.80	2.759		126	160		286	380
2660	19" round		5.80	2.759		162	160		322	415
2720	18" round		5.80	2.759		90	160		250	340
2860	For color, add					10%				
2861	For designer colors and trim, add					20%				
2900	Vitreous china, 20" x 16", single bowl	Q-1	5.40	2.963	Ea.	211	172		383	490
2960	20" x 17", single bowl		5.40	2.963		121	172		293	390
3020	19" round, single bowl		5.40	2.963		117	172		289	385
3080	19" x 16", single bowl		5.40	2.963		220	172		392	500
3140	17" x 14", single bowl		5.40	2.963		154	172		326	425
3200	22" x 13", single bowl		5.40	2.963		216	172		388	495
3560	For color, add					50%				
3561	For designer colors and trim, add					100%				
3580	Rough-in, supply, waste and vent for all above lavatories	Q-1	2.30	6.957	Ea.	278	405		683	910
4000	Wall hung									
4040	Porcelain enamel on cast iron, 16" x 14", single bowl	Q-1	8	2	Ea.	435	116		551	650
4060	18" x 15", single bowl		8	2		350	116		466	560
4120	19" x 17", single bowl		8	2		405	116		521	620
4180	20" x 18", single bowl		8	2		246	116		362	445
4240	22" x 19", single bowl		8	2		690	116		806	935
4580	For color, add					30%				
4581	For designer colors and trim, add					75%				
6000	Vitreous china, 18" x 15", single bowl with backsplash	Q-1	7	2.286	Ea.	170	133		303	385
6060	19" x 17", single bowl		7	2.286		122	133		255	330
6120	20" x 18", single bowl		7	2.286		247	133		380	470
6210	27" x 20", ADA compliant		7	2.286		930	133		1,063	1,225
6500	For color, add					30%				
6501	For designer colors and trim, add					50%				
6960	Rough-in, supply, waste and vent for above lavatories	Q-1	1.66	9.639	Ea.	475	560		1,035	1,350
7000	Pedestal type									
7600	Vitreous china, 27" x 21", white	Q-1	6.60	2.424	Ea.	685	141		826	965
7610	27" x 21", colored		6.60	2.424		875	141		1,016	1,175
7620	27" x 21", premium color		6.60	2.424		995	141		1,136	1,300
7660	26" x 20", white		6.60	2.424		665	141		806	940
7670	26" x 20", colored		6.60	2.424		855	141		996	1,150
7680	26" x 20", premium color		6.60	2.424		1,025	141		1,166	1,325
7700	24" x 20", white		6.60	2.424		465	141		606	720
7710	24" x 20", colored		6.60	2.424		580	141		721	845
7720	24" x 20", premium color		6.60	2.424		640	141		781	915
7760	21" x 18", white		6.60	2.424		258	141		399	495
7770	21" x 18", colored		6.60	2.424		289	141		430	525
7990	Rough-in, supply, waste and vent for pedestal lavatories		1.66	9.639		475	560		1,035	1,350

22 41 Residential Plumbing Fixtures

22 41 16 – Residential Lavatories and Sinks

22 41 16.16 Sinks		Crew	Daily Output	Labor-Hours	Unit	2020 Bare Costs Material	Labor	Equipment	Total	Total Incl O&P
0010	**SINKS**, With faucets and drain D2010–410									
2000	Kitchen, counter top style, PE on CI, 24" x 21" single bowl	Q-1	5.60	2.857	Ea.	310	166		476	590
2100	31" x 22" single bowl		5.60	2.857		860	166		1,026	1,200
2200	32" x 21" double bowl		4.80	3.333		390	193		583	720
2310	For color, add					20%				
2311	For designer colors and trim, add					50%				
3000	Stainless steel, self rimming, 19" x 18" single bowl	Q-1	5.60	2.857	Ea.	625	166		791	935
3100	25" x 22" single bowl		5.60	2.857		690	166		856	1,000
3200	33" x 22" double bowl		4.80	3.333		1,000	193		1,193	1,400
3300	43" x 22" double bowl		4.80	3.333		1,150	193		1,343	1,575
3400	22" x 43" triple bowl		4.40	3.636		1,225	211		1,436	1,675
3500	Corner double bowl each 14" x 16"		4.80	3.333		805	193		998	1,175
4000	Steel, enameled, with ledge, 24" x 21" single bowl		5.60	2.857		535	166		701	840
4100	32" x 21" double bowl		4.80	3.333		520	193		713	860
4960	For color sinks except stainless steel, add					10%				
4961	For designer colors and trim, add					20%				
4980	For rough-in, supply, waste and vent, counter top sinks	Q-1	2.14	7.477		320	435		755	1,000
5000	Kitchen, raised deck, PE on CI									
5100	32" x 21", dual level, double bowl	Q-1	2.60	6.154	Ea.	485	355		840	1,075
5200	42" x 21", double bowl & disposer well	"	2.20	7.273		1,000	420		1,420	1,725
5700	For color, add					20%				
5701	For designer colors and trim, add					50%				
5790	For rough-in, supply, waste & vent, sinks	Q-1	1.85	8.649		320	500		820	1,100

22 41 19 – Residential Bathtubs

22 41 19.10 Baths		Crew	Daily Output	Labor-Hours	Unit	Material	Labor	Equipment	Total	Total Incl O&P
0010	**BATHS** D2010–510									
0100	Tubs, recessed porcelain enamel on cast iron, with trim									
0180	48" x 42"	Q-1	4	4	Ea.	3,150	232		3,382	3,800
0220	72" x 36"	"	3	5.333	"	2,975	310		3,285	3,750
0300	Mat bottom									
0340	4'-6" long	Q-1	5	3.200	Ea.	1,500	186		1,686	1,925
0380	5' long		4.40	3.636		1,275	211		1,486	1,725
0420	5'-6" long		4	4		1,975	232		2,207	2,500
0480	Above floor drain, 5' long		4	4		825	232		1,057	1,250
0560	Corner 48" x 44"		4.40	3.636		2,925	211		3,136	3,525
0750	For color, add					30%				
0760	For designer colors & trim, add					60%				
2000	Enameled formed steel, 4'-6" long	Q-1	5.80	2.759	Ea.	515	160		675	805
2300	Above floor drain, 5' long	"	5.50	2.909	"	590	169		759	905
2350	For color, add					10%				
4000	Soaking, acrylic, w/pop-up drain 66" x 36" x 20" deep	Q-1	5.50	2.909	Ea.	1,725	169		1,894	2,150
4100	60" x 42" x 20" deep		5	3.200		1,325	186		1,511	1,725
4200	72" x 42" x 23" deep		4.80	3.333		2,200	193		2,393	2,725
4310	For color, add					5%				
4311	For designer colors & trim, add					20%				
4600	Module tub & showerwall surround, molded fiberglass									
4610	5' long x 34" wide x 76" high	Q-1	4	4	Ea.	790	232		1,022	1,225
4620	For color, add					10%				
4621	For designer colors and trim, add					25%				
4750	ADA compliant with 1-1/2" OD grab bar, antiskid bottom									
4760	60" x 32-3/4" x 72" high	Q-1	4	4	Ea.	595	232		827	1,000
4770	60" x 30" x 71" high with molded seat		3.50	4.571		755	265		1,020	1,225

22 41 Residential Plumbing Fixtures

22 41 19 – Residential Bathtubs

22 41 19.10 Baths		Crew	Daily Output	Labor-Hours	Unit	Material	Labor	Equipment	Total	Total Incl O&P
						2020 Bare Costs				
9600	Rough-in, supply, waste and vent, for all above tubs, add	Q-1	2.07	7.729	Ea.	445	450		895	1,150

22 41 23 – Residential Showers

22 41 23.20 Showers

		Crew	Daily Output	Labor-Hours	Unit	Material	Labor	Equipment	Total	Total Incl O&P
0010	**SHOWERS** D2010–710									
1500	Stall, with drain only. Add for valve and door/curtain									
1510	Baked enamel, molded stone receptor, 30" square	Q-1	5.20	3.077	Ea.	1,300	178		1,478	1,700
1520	32" square		5	3.200		1,175	186		1,361	1,550
1530	36" square		4.80	3.333		2,950	193		3,143	3,525
1540	Terrazzo receptor, 32" square		5	3.200		1,375	186		1,561	1,800
1560	36" square		4.80	3.333		1,650	193		1,843	2,125
1580	36" corner angle		4.80	3.333		2,050	193		2,243	2,550
1600	For color, add					10%				
1601	For designer colors and trim, add					15%				
1604	For thermostatic valve, add				Ea.	1,225			1,225	1,350
3000	Fiberglass, one piece, with 3 walls, 32" x 32" square	Q-1	5.50	2.909		355	169		524	645
3100	36" x 36" square	"	5.50	2.909		465	169		634	765
3200	ADA compliant, 1-1/2" OD grab bars, nonskid floor									
3210	48" x 34-1/2" x 72" corner seat	Q-1	5	3.200	Ea.	665	186		851	1,025
3220	60" x 34-1/2" x 72" corner seat		4	4		765	232		997	1,200
3230	48" x 34-1/2" x 72" fold up seat		5	3.200		1,150	186		1,336	1,550
3250	64" x 65-3/4" x 81-1/2" fold. seat, ADA		3.80	4.211		1,475	244		1,719	2,000
3260	For thermostatic valve, add					1,225			1,225	1,350
4000	Polypropylene, stall only, w/molded-stone floor, 30" x 30"	Q-1	2	8		720	465		1,185	1,500
4100	32" x 32"	"	2	8		735	465		1,200	1,500
4110	For thermostatic valve, add					1,225			1,225	1,350
4200	Rough-in, supply, waste and vent for above showers	Q-1	2.05	7.805		410	455		865	1,125

22 41 23.40 Shower System Components

		Crew	Daily Output	Labor-Hours	Unit	Material	Labor	Equipment	Total	Total Incl O&P
0010	**SHOWER SYSTEM COMPONENTS**									
4500	Receptor only									
4510	For tile, 36" x 36"	1 Plum	4	2	Ea.	390	129		519	625
4520	Fiberglass receptor only, 32" x 32"		8	1		107	64.50		171.50	215
4530	34" x 34"		7.80	1.026		128	66		194	240
4540	36" x 36"		7.60	1.053		125	68		193	240
4600	Rectangular									
4620	32" x 48"	1 Plum	7.40	1.081	Ea.	154	69.50		223.50	273
4630	34" x 54"		7.20	1.111		176	71.50		247.50	300
4640	34" x 60"		7	1.143		187	73.50		260.50	315
5000	Built-in, head, arm, 2.5 GPM valve		4	2		104	129		233	305
5200	Head, arm, by-pass, integral stops, handles		3.60	2.222		296	143		439	540
5500	Head, water economizer, 1.6 GPM G		24	.333		53	21.50		74.50	90.50
5800	Mixing valve, built-in		6	1.333		144	86		230	288
5900	Exposed		6	1.333		710	86		796	915

22 41 36 – Residential Laundry Trays

22 41 36.10 Laundry Sinks

		Crew	Daily Output	Labor-Hours	Unit	Material	Labor	Equipment	Total	Total Incl O&P
0010	**LAUNDRY SINKS**, With trim D2010–420									
0020	Porcelain enamel on cast iron, black iron frame									
0050	24" x 21", single compartment	Q-1	6	2.667	Ea.	615	155		770	905
0100	26" x 21", single compartment		6	2.667		630	155		785	925
0200	48" x 20", double compartment		5	3.200		800	186		986	1,150
2000	Molded stone, on wall hanger or legs									
2020	22" x 23", single compartment	Q-1	6	2.667	Ea.	176	155		331	425

22 41 Residential Plumbing Fixtures

22 41 36 – Residential Laundry Trays

22 41 36.10 Laundry Sinks		Crew	Daily Output	Labor-Hours	Unit	2020 Bare Costs Material	Labor	Equipment	Total	Total Incl O&P
2100	45" x 21", double compartment	Q-1	5	3.200	Ea.	360	186		546	675
3000	Plastic, on wall hanger or legs									
3020	18" x 23", single compartment	Q-1	6.50	2.462	Ea.	145	143		288	375
3100	20" x 24", single compartment		6.50	2.462		165	143		308	395
3200	36" x 23", double compartment		5.50	2.909		219	169		388	495
3300	40" x 24", double compartment		5.50	2.909		287	169		456	570
5000	Stainless steel, counter top, 22" x 17" single compartment		6	2.667		77.50	155		232.50	315
5200	33" x 22", double compartment		5	3.200		93	186		279	380
9600	Rough-in, supply, waste and vent, for all laundry sinks		2.14	7.477		320	435		755	1,000

22 41 39 – Residential Faucets, Supplies and Trim

22 41 39.10 Faucets and Fittings		Crew	Daily Output	Labor-Hours	Unit	Material	Labor	Equipment	Total	Total Incl O&P
0010	**FAUCETS AND FITTINGS**									
0150	Bath, faucets, diverter spout combination, sweat	1 Plum	8	1	Ea.	87	64.50		151.50	192
0200	For integral stops, IPS unions, add					111			111	123
0300	Three valve combinations, spout, head, arm, flange, sweat	1 Plum	6	1.333		98.50	86		184.50	238
0400	For integral stops, IPS unions, add				Pr.	63.50			63.50	70
0420	Bath, press-bal mix valve w/diverter, spout, shower head, arm/flange	1 Plum	8	1	Ea.	185	64.50		249.50	300
0500	Drain, central lift, 1-1/2" IPS male		20	.400		50.50	26		76.50	94
0600	Trip lever, 1-1/2" IPS male		20	.400		60.50	26		86.50	105
0700	Pop up, 1-1/2" IPS male		18	.444		69	28.50		97.50	119
0800	Chain and stopper, 1-1/2" IPS male		24	.333		33	21.50		54.50	68.50
0810	Bidet									
0812	Fitting, over the rim, swivel spray/pop-up drain	1 Plum	8	1	Ea.	279	64.50		343.50	400
1000	Kitchen sink faucets, top mount, cast spout		10	.800		84	51.50		135.50	170
1100	For spray, add		24	.333		17.25	21.50		38.75	51
1110	For basket strainer w/tail piece, add		24	.333		15.50	21.50		37	49
1200	Wall type, swing tube spout		10	.800		74.50	51.50		126	159
1240	For soap dish, add					3.67			3.67	4.04
1250	For basket strainer w/tail piece, add					45.50			45.50	50
1300	Single control lever handle									
1310	With pull out spray									
1320	Polished chrome	1 Plum	10	.800	Ea.	196	51.50		247.50	293
2000	Laundry faucets, shelf type, IPS or copper unions		12	.667		62	43		105	133
2100	Lavatory faucet, centerset, without drain		10	.800		67.50	51.50		119	152
2120	With pop-up drain		6.66	1.201		54.50	77.50		132	176
2130	For acrylic handles, add					5.25			5.25	5.80
2150	Concealed, 12" centers	1 Plum	10	.800		90	51.50		141.50	176
2160	With pop-up drain	"	6.66	1.201		108	77.50		185.50	235
2210	Porcelain cross handles and pop-up drain									
2220	Polished chrome	1 Plum	6.66	1.201	Ea.	222	77.50		299.50	360
2230	Polished brass	"	6.66	1.201	"	293	77.50		370.50	440
2260	Single lever handle and pop-up drain									
2280	Satin nickel	1 Plum	6.66	1.201	Ea.	280	77.50		357.50	425
2290	Polished chrome		6.66	1.201		200	77.50		277.50	335
2600	Shelfback, 4" to 6" centers, 17 ga. tailpiece		10	.800		81	51.50		132.50	166
2650	With pop-up drain		6.66	1.201		101	77.50		178.50	227
2700	Shampoo faucet with supply tube		24	.333		47.50	21.50		69	84.50
2800	Self-closing, center set		10	.800		151	51.50		202.50	243
2810	Automatic sensor and operator, with faucet head		6.15	1.301		495	84		579	665
4000	Shower by-pass valve with union		18	.444		57.50	28.50		86	107
4100	Shower arm with flange and head		22	.364		21	23.50		44.50	58.50
4140	Shower, hand held, pin mount, massage action, chrome		22	.364		82	23.50		105.50	126

22 41 Residential Plumbing Fixtures

22 41 39 – Residential Faucets, Supplies and Trim

22 41 39.10 Faucets and Fittings		Crew	Daily Output	Labor-Hours	Unit	2020 Bare Costs Material	2020 Bare Costs Labor	2020 Bare Costs Equipment	2020 Bare Costs Total	Total Incl O&P
4142	Polished brass	1 Plum	22	.364	Ea.	162	23.50		185.50	213
4144	Shower, hand held, wall mtd, adj. spray, 2 wall mounts, chrome		20	.400		116	26		142	167
4146	Polished brass		20	.400		237	26		263	299
4148	Shower, hand held head, bar mounted 24", adj. spray, chrome		20	.400		180	26		206	237
4150	Polished brass		20	.400		370	26		396	445
4200	Shower thermostatic mixing valve, concealed, with shower head trim kit		8	1		385	64.50		449.50	515
4220	Shower pressure balancing mixing valve									
4230	With shower head, arm, flange and diverter tub spout									
4240	Chrome	1 Plum	6.14	1.303	Ea.	415	84		499	580
4250	Satin nickel		6.14	1.303		560	84		644	740
4260	Polished graphite		6.14	1.303		545	84		629	725
5000	Sillcock, compact, brass, IPS or copper to hose		24	.333		12.05	21.50		33.55	45.50
6000	Stop and waste valves, bronze									
6100	Angle, solder end 1/2"	1 Plum	24	.333	Ea.	28	21.50		49.50	63
6110	3/4"		20	.400		36	26		62	78
6300	Straightway, solder end 3/8"		24	.333		20.50	21.50		42	54.50
6310	1/2"		24	.333		20.50	21.50		42	54.50
6320	3/4"		20	.400		21	26		47	61.50
6410	Straightway, threaded 1/2"		24	.333		22	21.50		43.50	56
6420	3/4"		20	.400		24.50	26		50.50	65.50
6430	1"		19	.421		27	27		54	70
7800	Water closet, wax gasket		96	.083		1.67	5.35		7.02	9.90
7820	Gasket toilet tank to bowl		32	.250		3.15	16.10		19.25	27.50
7830	Replacement diaphragm washer assy for ballcock valve		12	.667		3.15	43		46.15	68
7850	Dual flush valve		12	.667		133	43		176	212
8000	Water supply stops, polished chrome plate									
8200	Angle, 3/8"	1 Plum	24	.333	Ea.	9.30	21.50		30.80	42
8300	1/2"		22	.364		10.05	23.50		33.55	46
8400	Straight, 3/8"		26	.308		9.40	19.85		29.25	40
8500	1/2"		24	.333		9.75	21.50		31.25	42.50
8600	Water closet, angle, w/flex riser, 3/8"		24	.333		34	21.50		55.50	69.50
9100	Miscellaneous									
9720	Teflon tape, 1/2" x 520" roll				Ea.	1			1	1.10

22 41 39.70 Washer/Dryer Accessories		Crew	Daily Output	Labor-Hours	Unit	Material	Labor	Equipment	Total	Total Incl O&P
0010	**WASHER/DRYER ACCESSORIES**									
1020	Valves ball type single lever									
1030	1/2" diam., IPS	1 Plum	21	.381	Ea.	65	24.50		89.50	108
1040	1/2" diam., solder	"	21	.381	"	65	24.50		89.50	108
1050	Recessed box, 16 ga., two hose valves and drain									
1060	1/2" size, 1-1/2" drain	1 Plum	18	.444	Ea.	152	28.50		180.50	210
1070	1/2" size, 2" drain	"	17	.471	"	135	30.50		165.50	195
1080	With grounding electric receptacle									
1090	1/2" size, 1-1/2" drain	1 Plum	18	.444	Ea.	167	28.50		195.50	226
1100	1/2" size, 2" drain	"	17	.471	"	179	30.50		209.50	243
1110	With grounding and dryer receptacle									
1120	1/2" size, 1-1/2" drain	1 Plum	18	.444	Ea.	205	28.50		233.50	268
1130	1/2" size, 2" drain	"	17	.471	"	207	30.50		237.50	274
1140	Recessed box, 16 ga., ball valves with single lever and drain									
1150	1/2" size, 1-1/2" drain	1 Plum	19	.421	Ea.	286	27		313	355
1160	1/2" size, 2" drain	"	18	.444	"	247	28.50		275.50	315
1170	With grounding electric receptacle									
1180	1/2" size, 1-1/2" drain	1 Plum	19	.421	Ea.	305	27		332	375

22 41 Residential Plumbing Fixtures

22 41 39 – Residential Faucets, Supplies and Trim

22 41 39.70 Washer/Dryer Accessories		Crew	Daily Output	Labor-Hours	Unit	2020 Bare Costs Material	Labor	Equipment	Total	Total Incl O&P
1190	1/2" size, 2" drain	1 Plum	18	.444	Ea.	275	28.50		303.50	350
1200	With grounding and dryer receptacles									
1210	1/2" size, 1-1/2" drain	1 Plum	19	.421	Ea.	272	27		299	340
1220	1/2" size, 2" drain	"	18	.444	"	300	28.50		328.50	375
1300	Recessed box, 20 ga., two hose valves and drain (economy type)									
1310	1/2" size, 1-1/2" drain	1 Plum	19	.421	Ea.	118	27		145	171
1320	1/2" size, 2" drain		18	.444		110	28.50		138.50	164
1330	Box with drain only		24	.333		68.50	21.50		90	107
1340	1/2" size, 1-1/2" ABS/PVC drain		19	.421		126	27		153	179
1350	1/2" size, 2" ABS/PVC drain		18	.444		134	28.50		162.50	191
1352	Box with drain and 15 A receptacle		24	.333		72.50	21.50		94	112
1360	1/2" size, 2" drain ABS/PVC, 15 A receptacle		24	.333		134	21.50		155.50	179
1400	Wall mounted									
1410	1/2" size, 1-1/2" plastic drain	1 Plum	19	.421	Ea.	34.50	27		61.50	78
1420	1/2" size, 2" plastic drain	"	18	.444	"	20	28.50		48.50	65
1500	Dryer vent kit									
1510	8' flex duct, clamps and outside hood	1 Plum	20	.400	Ea.	13.95	26		39.95	54
1980	Rough-in, supply, waste, and vent for washer boxes	"	3.46	2.310	"	355	149		504	615

22 42 Commercial Plumbing Fixtures

22 42 13 – Commercial Water Closets, Urinals, and Bidets

22 42 13.13 Water Closets		Crew	Daily Output	Labor-Hours	Unit	Material	Labor	Equipment	Total	Total Incl O&P
0010	**WATER CLOSETS**									
3000	Bowl only, with flush valve, seat, 1.6 gpf unless noted									
3100	Wall hung	Q-1	5.80	2.759	Ea.	1,100	160		1,260	1,450
3200	For rough-in, supply, waste and vent, single WC		2.56	6.250		1,300	365		1,665	2,000
3300	Floor mounted		5.80	2.759		360	160		520	635
3350	With wall outlet		5.80	2.759		570	160		730	865
3360	With floor outlet, 1.28 gpf G		5.80	2.759		550	160		710	845
3362	With floor outlet, 1.28 gpf, ADA G		5.80	2.759		570	160		730	870
3370	For rough-in, supply, waste and vent, single WC		2.84	5.634		385	325		710	910
3390	Floor mounted children's size, 10-3/4" high									
3392	With automatic flush sensor, 1.6 gpf	Q-1	6.20	2.581	Ea.	660	150		810	950
3396	With automatic flush sensor, 1.28 gpf		6.20	2.581		610	150		760	895
3400	For rough-in, supply, waste and vent, single WC		2.84	5.634		385	325		710	910
3500	Gang side by side carrier system, rough-in, supply, waste & vent									
3510	For single hook-up	Q-1	1.97	8.122	Ea.	1,550	470		2,020	2,400
3520	For each additional hook-up, add	"	2.14	7.477	"	1,475	435		1,910	2,250
3550	Gang back to back carrier system, rough-in, supply, waste & vent									
3560	For pair hook-up	Q-1	1.76	9.091	Pr.	2,075	525		2,600	3,075
3570	For each additional pair hook-up, add	"	1.81	8.840	"	2,000	515		2,515	2,975

22 42 13.16 Urinals		Crew	Daily Output	Labor-Hours	Unit	Material	Labor	Equipment	Total	Total Incl O&P
0010	**URINALS** D2010–210									
0102	For automatic flush see Line 22 42 39.10 0972									
3000	Wall hung, vitreous china, with self-closing valve R224000-30									
3100	Siphon jet type	Q-1	3	5.333	Ea.	315	310		625	810
3120	Blowout type		3	5.333		480	310		790	995
3140	Water saving .5 gpf G		3	5.333		595	310		905	1,125
3300	Rough-in, supply, waste & vent		2.83	5.654		755	330		1,085	1,325
5000	Stall type, vitreous china, includes valve		2.50	6.400		830	370		1,200	1,475
6980	Rough-in, supply, waste and vent		1.99	8.040		510	465		975	1,275

22 42 Commercial Plumbing Fixtures

22 42 13 – Commercial Water Closets, Urinals, and Bidets

22 42 13.16 Urinals		Crew	Daily Output	Labor-Hours	Unit	Material	Labor	Equipment	Total	Total Incl O&P
8000	Waterless (no flush) urinal									
8010	Wall hung									
8014	Fiberglass reinforced polyester									
8020	Standard unit G	Q-1	21.30	.751	Ea.	450	43.50		493.50	560
8030	ADA compliant unit G	"	21.30	.751		420	43.50		463.50	525
8070	For solid color, add G					64			64	70
8080	For 2" brass flange (new const.), add G	Q-1	96	.167	↓	20.50	9.65		30.15	37
8200	Vitreous china									
8220	ADA compliant unit, 14" G	Q-1	21.30	.751	Ea.	211	43.50		254.50	297
8250	ADA compliant unit, 15.5"	"	21.30	.751		272	43.50		315.50	365
8270	For solid color, add G					64			64	70
8290	Rough-in, supply, waste & vent G	Q-1	2.92	5.479	↓	720	320		1,040	1,275
8400	Trap liquid									
8410	1 quart G				Ea.	19.20			19.20	21
8420	1 gallon G				"	63			63	69.50

22 42 16 – Commercial Lavatories and Sinks

22 42 16.13 Lavatories		Crew	Daily Output	Labor-Hours	Unit	Material	Labor	Equipment	Total	Total Incl O&P
0010	**LAVATORIES**, With trim, white unless noted otherwise									
0020	Commercial lavatories same as residential. See Section 22 41 16									

22 42 16.16 Commercial Sinks		Crew	Daily Output	Labor-Hours	Unit	Material	Labor	Equipment	Total	Total Incl O&P
0010	**COMMERCIAL SINKS**									
5900	Scullery sink, stainless steel									
5910	1 bowl and drain board, 43" x 22" OD	Q-1	5.40	2.963	Ea.	2,600	172		2,772	3,125
5920	2 bowls and drain board, 49" x 22" OD		4.60	3.478		3,975	202		4,177	4,675
5930	3 bowls and drain board, 43" x 22" OD		4.20	3.810		4,425	221		4,646	5,200
5940	1 bowl and drain board, 50" x 28" OD, with legs	↓	5.40	2.963	↓	1,800	172		1,972	2,225

22 42 16.30 Classroom Sinks		Crew	Daily Output	Labor-Hours	Unit	Material	Labor	Equipment	Total	Total Incl O&P
0010	**CLASSROOM SINKS**									
6020	Countertop, stainless steel									
6024	with faucet, bubbler and strainer, ADA compliant									
6036	25" x 17" single bowl	Q-1	5.20	3.077	Ea.	1,175	178		1,353	1,575
6040	28" x 22" single bowl		5.20	3.077		1,275	178		1,453	1,700
6044	31" x 19" single bowl		5.20	3.077		1,325	178		1,503	1,750
6070	37" x 17" double bowl		4.40	3.636		2,000	211		2,211	2,525
6100	For rough-in, supply, waste and vent, counter top classroom sinks	↓	2.14	7.477	↓	320	435		755	1,000

22 42 16.34 Laboratory Countertops and Sinks		Crew	Daily Output	Labor-Hours	Unit	Material	Labor	Equipment	Total	Total Incl O&P
0010	**LABORATORY COUNTERTOPS AND SINKS**									
0050	Laboratory sinks, corrosion resistant									
1000	Stainless steel sink, bench mounted, with									
1020	plug & waste fitting with 1-1/2" straight threads									
1030	Single bowl, 2 drainboards, backnut & strainer									
1050	18-1/2" x 15-1/2" x 12-1/2" sink, 54" x 24" OD	Q-1	3	5.333	Ea.	1,150	310		1,460	1,725
1100	Single bowl, single drainboard, backnut & strainer									
1130	18-1/2" x 15-1/2" x 12-1/2" sink, 47" x 24" OD	Q-1	3	5.333	Ea.	920	310		1,230	1,475
1146	Double bowl, single drainboard, backnut & strainer									
1150	18-1/2" x 15-1/2" x 12-1/2" sink, 70" x 24" OD	Q-1	3	5.333	Ea.	1,250	310		1,560	1,850
1280	Polypropylene									
1290	Flanged 1-1/4" wide, rectangular with strainer									
1300	plug & waste fitting, 1-1/2" straight threads									
1320	12" x 12" x 8" sink, 14-1/2" x 14-1/2" OD	Q-1	4	4	Ea.	267	232		499	640
1340	16" x 16" x 8" sink, 18-1/2" x 18-1/2" OD	↓	4	4	↓	375	232		607	760

22 42 Commercial Plumbing Fixtures

22 42 16 – Commercial Lavatories and Sinks

22 42 16.34 Laboratory Countertops and Sinks		Crew	Daily Output	Labor-Hours	Unit	Material	2020 Bare Costs Labor	Equipment	Total	Total Incl O&P
1360	21" x 18" x 10" sink, 23-1/2" x 20-1/2" OD	Q-1	4	4	Ea.	395	232		627	780
1490	For rough-in, supply, waste & vent, add		2.02	7.921		229	460		689	940
1600	Polypropylene									
1620	Cup sink, oval, integral strainers									
1640	6" x 3" I.D., 7" x 4" OD	Q-1	6	2.667	Ea.	149	155		304	395
1660	9" x 3" I.D., 10" x 4-1/2" OD	"	6	2.667		176	155		331	425
1740	1-1/2" diam. x 11" long					44.50			44.50	48.50
1980	For rough-in, supply, waste & vent, add	Q-1	1.70	9.412		256	545		801	1,100

22 42 16.40 Service Sinks		Crew	Daily Output	Labor-Hours	Unit	Material	Labor	Equipment	Total	Total Incl O&P
0010	**SERVICE SINKS**									
6650	Service, floor, corner, PE on CI, 28" x 28"	Q-1	4.40	3.636	Ea.	1,125	211		1,336	1,575
6750	Vinyl coated rim guard, add					65.50			65.50	72
6755	Mop sink, molded stone, 22" x 18"	1 Plum	3.33	2.402		545	155		700	830
6760	Mop sink, molded stone, 24" x 36"		3.33	2.402		279	155		434	535
6770	Mop sink, molded stone, 24" x 36", w/rim 3 sides		3.33	2.402		264	155		419	520
6790	For rough-in, supply, waste & vent, floor service sinks	Q-1	1.64	9.756		1,075	565		1,640	2,025
7000	Service, wall, PE on CI, roll rim, 22" x 18"		4	4		930	232		1,162	1,375
7100	24" x 20"		4	4		950	232		1,182	1,400
7600	For stainless steel rim guard, two sides only, add					49			49	53.50
7800	For stainless steel rim guard, front only, add					57			57	62.50
8600	Vitreous china, 22" x 20"	Q-1	4	4		875	232		1,107	1,300
8960	For stainless steel rim guard, front or one side, add					65.50			65.50	72
8980	For rough-in, supply, waste & vent, wall service sinks	Q-1	1.30	12.308		1,425	715		2,140	2,650

22 42 23 – Commercial Showers

22 42 23.30 Group Showers		Crew	Daily Output	Labor-Hours	Unit	Material	Labor	Equipment	Total	Total Incl O&P
0010	**GROUP SHOWERS**									
6000	Group, w/pressure balancing valve, rough-in and rigging not included									
6800	Column, 6 heads, no receptors, less partitions	Q-1	3	5.333	Ea.	9,675	310		9,985	11,100
6900	With stainless steel partitions		1	16		12,600	930		13,530	15,300
7600	5 heads, no receptors, less partitions		3	5.333		6,650	310		6,960	7,775
7620	4 heads (1 ADA compliant) no receptors, less partitions ♿		3	5.333		6,050	310		6,360	7,125
7700	With stainless steel partitions		1	16		6,600	930		7,530	8,675
8000	Wall, 2 heads, no receptors, less partitions		4	4		2,825	232		3,057	3,450
8100	With stainless steel partitions		2	8		6,200	465		6,665	7,500

22 42 33 – Wash Fountains

22 42 33.20 Commercial Wash Fountains		Crew	Daily Output	Labor-Hours	Unit	Material	Labor	Equipment	Total	Total Incl O&P
0010	**COMMERCIAL WASH FOUNTAINS** D2010–610									
1900	Group, foot control									
2000	Precast terrazzo, circular, 36" diam., 5 or 6 persons	Q-2	3	8	Ea.	7,675	480		8,155	9,150
2100	54" diam. for 8 or 10 persons		2.50	9.600		10,600	575		11,175	12,500
2400	Semi-circular, 36" diam. for 3 persons		3	8		6,250	480		6,730	7,600
2500	54" diam. for 4 or 5 persons		2.50	9.600		9,700	575		10,275	11,600
2700	Quarter circle (corner), 54" diam. for 3 persons		3.50	6.857		7,675	410		8,085	9,075
3000	Stainless steel, circular, 36" diameter		3.50	6.857		6,775	410		7,185	8,075
3100	54" diameter		2.80	8.571		8,575	515		9,090	10,200
3400	Semi-circular, 36" diameter		3.50	6.857		5,625	410		6,035	6,825
3500	54" diameter		2.80	8.571		7,150	515		7,665	8,650
5000	Thermoplastic, pre-assembled, circular, 36" diameter		6	4		4,700	241		4,941	5,525
5100	54" diameter		4	6		5,450	360		5,810	6,550
5400	Semi-circular, 36" diameter		6	4		4,750	241		4,991	5,575
5600	54" diameter		4	6		6,250	360		6,610	7,425

22 42 Commercial Plumbing Fixtures

22 42 33 – Wash Fountains

22 42 33.20 Commercial Wash Fountains		Crew	Daily Output	Labor-Hours	Unit	2020 Bare Costs Material	Labor	Equipment	Total	Total Incl O&P
5610	Group, infrared control, barrier free									
5614	Precast terrazzo									
5620	Semi-circular 36″ diam. for 3 persons	Q-2	3	8	Ea.	8,475	480		8,955	10,000
5630	46″ diam. for 4 persons		2.80	8.571		9,100	515		9,615	10,800
5640	Circular, 54″ diam. for 8 persons, button control		2.50	9.600		11,200	575		11,775	13,200
5700	Rough-in, supply, waste and vent for above wash fountains	Q-1	1.82	8.791		505	510		1,015	1,325
6200	Duo for small washrooms, stainless steel		2	8		3,300	465		3,765	4,325
6400	Bowl with backsplash		2	8		2,400	465		2,865	3,325
6500	Rough-in, supply, waste & vent for duo fountains		2.02	7.921		268	460		728	985

22 42 39 – Commercial Faucets, Supplies, and Trim

22 42 39.10 Faucets and Fittings

Line	Description	Crew	Daily Output	Labor-Hours	Unit	Material	Labor	Equipment	Total	Total Incl O&P
0010	**FAUCETS AND FITTINGS**									
0840	Flush valves, with vacuum breaker									
0850	Water closet									
0860	Exposed, rear spud	1 Plum	8	1	Ea.	146	64.50		210.50	258
0870	Top spud		8	1		197	64.50		261.50	315
0880	Concealed, rear spud		8	1		213	64.50		277.50	330
0890	Top spud		8	1		173	64.50		237.50	287
0900	Wall hung		8	1		199	64.50		263.50	315
0910	Dual flush flushometer		12	.667		245	43		288	335
0912	Flushometer retrofit kit		18	.444		18.15	28.50		46.65	63
0920	Urinal									
0930	Exposed, stall	1 Plum	8	1	Ea.	197	64.50		261.50	315
0940	Wall (washout)		8	1		156	64.50		220.50	269
0950	Pedestal, top spud		8	1		137	64.50		201.50	247
0960	Concealed, stall		8	1		170	64.50		234.50	284
0970	Wall (washout)		8	1		183	64.50		247.50	298
0971	Automatic flush sensor and operator for									
0972	urinals or water closets, standard	1 Plum	8	1	Ea.	485	64.50		549.50	630
0980	High efficiency water saving									
0984	Water closets, 1.28 gpf	1 Plum	8	1	Ea.	425	64.50		489.50	565
0988	Urinals, .5 gpf	″	8	1	″	425	64.50		489.50	565
2790	Faucets for lavatories									
2800	Self-closing, center set	1 Plum	10	.800	Ea.	151	51.50		202.50	243
2810	Automatic sensor and operator, with faucet head		6.15	1.301		495	84		579	665
3000	Service sink faucet, cast spout, pail hook, hose end		14	.571		76.50	37		113.50	139

22 42 39.30 Carriers and Supports

Line	Description	Crew	Daily Output	Labor-Hours	Unit	Material	Labor	Equipment	Total	Total Incl O&P
0010	**CARRIERS AND SUPPORTS**, For plumbing fixtures									
0500	Drinking fountain, wall mounted									
0600	Plate type with studs, top back plate	1 Plum	7	1.143	Ea.	61.50	73.50		135	178
0700	Top front and back plate		7	1.143		153	73.50		226.50	278
0800	Top & bottom, front & back plates, w/bearing jacks		7	1.143		181	73.50		254.50	310
3000	Lavatory, concealed arm									
3050	Floor mounted, single									
3100	High back fixture	1 Plum	6	1.333	Ea.	655	86		741	850
3200	Flat slab fixture		6	1.333		575	86		661	765
3220	ADA compliant		6	1.333		665	86		751	865
3250	Floor mounted, back to back									
3300	High back fixtures	1 Plum	5	1.600	Ea.	1,075	103		1,178	1,325
3400	Flat slab fixtures		5	1.600		1,325	103		1,428	1,600
3430	ADA compliant		5	1.600		810	103		913	1,050
3500	Wall mounted, in stud or masonry									

22 42 39.30 Carriers and Supports		Crew	Daily Output	Labor-Hours	Unit	2020 Bare Costs Material	Labor	Equipment	Total	Total Incl O&P
3600	High back fixture	1 Plum	6	1.333	Ea.	345	86		431	510
3700	Flat slab fixture	"	6	1.333	"	275	86		361	435
4000	Exposed arm type, floor mounted									
4100	Single high back or flat slab fixture	1 Plum	6	1.333	Ea.	950	86		1,036	1,175
4200	Back to back, high back or flat slab fixtures		5	1.600		1,700	103		1,803	2,000
4300	Wall mounted, high back or flat slab lavatory		6	1.333		750	86		836	955
4600	Sink, floor mounted									
4650	Exposed arm system									
4700	Single heavy fixture	1 Plum	5	1.600	Ea.	645	103		748	865
4750	Single heavy sink with slab		5	1.600		1,600	103		1,703	1,900
4800	Back to back, standard fixtures		5	1.600		845	103		948	1,075
4850	Back to back, heavy fixtures		5	1.600		1,300	103		1,403	1,575
4900	Back to back, heavy sink with slab		5	1.600		1,825	103		1,928	2,175
4950	Exposed offset arm system									
5000	Single heavy deep fixture	1 Plum	5	1.600	Ea.	1,200	103		1,303	1,475
5100	Plate type system									
5200	With bearing jacks, single fixture	1 Plum	5	1.600	Ea.	1,600	103		1,703	1,900
5300	With exposed arms, single heavy fixture		5	1.600		1,250	103		1,353	1,550
5400	Wall mounted, exposed arms, single heavy fixture		5	1.600		450	103		553	650
6000	Urinal, floor mounted, 2" or 3" coupling, blowout type		6	1.333		805	86		891	1,025
6100	With fixture or hanger bolts, blowout or washout		6	1.333		575	86		661	760
6200	With bearing plate		6	1.333		640	86		726	835
6300	Wall mounted, plate type system		6	1.333		430	86		516	600
6980	Water closet, siphon jet									
7000	Horizontal, adjustable, caulk									
7040	Single, 4" pipe size	1 Plum	5.33	1.501	Ea.	1,125	96.50		1,221.50	1,375
7050	4" pipe size, ADA compliant		5.33	1.501		825	96.50		921.50	1,050
7060	5" pipe size		5.33	1.501		1,125	96.50		1,221.50	1,375
7100	Double, 4" pipe size		5	1.600		1,625	103		1,728	1,925
7110	4" pipe size, ADA compliant		5	1.600		1,625	103		1,728	1,950
7120	5" pipe size		5	1.600		1,925	103		2,028	2,275
7160	Horizontal, adjustable, extended, caulk									
7180	Single, 4" pipe size	1 Plum	5.33	1.501	Ea.	1,500	96.50		1,596.50	1,800
7200	5" pipe size		5.33	1.501		1,875	96.50		1,971.50	2,225
7240	Double, 4" pipe size		5	1.600		2,150	103		2,253	2,500
7260	5" pipe size		5	1.600		2,850	103		2,953	3,275
7400	Vertical, adjustable, caulk or thread									
7440	Single, 4" pipe size	1 Plum	5.33	1.501	Ea.	1,250	96.50		1,346.50	1,525
7460	5" pipe size		5.33	1.501		1,575	96.50		1,671.50	1,875
7480	6" pipe size		5	1.600		1,625	103		1,728	1,950
7520	Double, 4" pipe size		5	1.600		2,300	103		2,403	2,675
7540	5" pipe size		5	1.600		2,325	103		2,428	2,700
7560	6" pipe size		4	2		2,575	129		2,704	3,025
7600	Vertical, adjustable, extended, caulk									
7620	Single, 4" pipe size	1 Plum	5.33	1.501	Ea.	1,125	96.50		1,221.50	1,400
7640	5" pipe size		5.33	1.501		890	96.50		986.50	1,125
7680	6" pipe size		5	1.600		1,725	103		1,828	2,050
7720	Double, 4" pipe size		5	1.600		2,300	103		2,403	2,700
7740	5" pipe size		5	1.600		1,250	103		1,353	1,525
7760	6" pipe size		4	2		1,350	129		1,479	1,700
7780	Water closet, blow out									
7800	Vertical offset, caulk or thread									
7820	Single, 4" pipe size	1 Plum	5.33	1.501	Ea.	1,225	96.50		1,321.50	1,500

22 42 Commercial Plumbing Fixtures

22 42 39 – Commercial Faucets, Supplies, and Trim

22 42 39.30 Carriers and Supports		Crew	Daily Output	Labor-Hours	Unit	Material	2020 Bare Costs Labor	Equipment	Total	Total Incl O&P
7840	Double, 4" pipe size	1 Plum	5	1.600	Ea.	2,075	103		2,178	2,450
7880	Vertical offset, extended, caulk									
7900	Single, 4" pipe size	1 Plum	5.33	1.501	Ea.	1,525	96.50		1,621.50	1,825
7920	Double, 4" pipe size	"	5	1.600	"	2,400	103		2,503	2,775
7960	Vertical, for floor mounted back-outlet									
7980	Single, 4" thread, 2" vent	1 Plum	5.33	1.501	Ea.	790	96.50		886.50	1,025
8000	Double, 4" thread, 2" vent	"	6	1.333	"	2,325	86		2,411	2,675
8040	Vertical, for floor mounted back-outlet, extended									
8060	Single, 4" caulk, 2" vent	1 Plum	6	1.333	Ea.	790	86		876	1,000
8080	Double, 4" caulk, 2" vent	"	6	1.333	"	2,325	86		2,411	2,675
8200	Water closet, residential									
8220	Vertical centerline, floor mount									
8240	Single, 3" caulk, 2" or 3" vent	1 Plum	6	1.333	Ea.	875	86		961	1,100
8260	4" caulk, 2" or 4" vent		6	1.333		1,125	86		1,211	1,375
8280	3" copper sweat, 3" vent		6	1.333		785	86		871	995
8300	4" copper sweat, 4" vent		6	1.333		950	86		1,036	1,175
8400	Vertical offset, floor mount									
8420	Single, 3" or 4" caulk, vent	1 Plum	4	2	Ea.	1,100	129		1,229	1,400
8440	3" or 4" copper sweat, vent		5	1.600		1,100	103		1,203	1,350
8460	Double, 3" or 4" caulk, vent		4	2		1,875	129		2,004	2,250
8480	3" or 4" copper sweat, vent		5	1.600		1,875	103		1,978	2,200
9000	Water cooler (electric), floor mounted									
9100	Plate type with bearing plate, single	1 Plum	6	1.333	Ea.	480	86		566	660
9140	Plate type with bearing plate, back to back	"	4	2	"	635	129		764	890

22 43 Healthcare Plumbing Fixtures

22 43 13 – Healthcare Water Closets

22 43 13.40 Water Closets		Crew	Daily Output	Labor-Hours	Unit	Material	Labor	Equipment	Total	Total Incl O&P
0010	**WATER CLOSETS**									
1000	Bowl only, 1 piece, w/seat and flush valve, ADA compliant, 18" high									
1030	Floor mounted									
1150	With wall outlet	Q-1	5.30	3.019	Ea.	320	175		495	610
1180	For rough-in, supply, waste and vent		2.84	5.634		385	325		710	910
1200	With floor outlet		5.30	3.019		370	175		545	665
1800	For rough-in, supply, waste and vent		3.05	5.246		340	305		645	830
3100	Wall hung		5.80	2.759		1,100	160		1,260	1,450
3150	Hospital type, slotted rim for bed pan									
3156	Elongated bowl, top spud	Q-1	5.80	2.759	Ea.	650	160		810	955
3160	Elongated bowl, rear spud		5.80	2.759		395	160		555	670
3200	For rough-in, supply, waste and vent, single WC		2.56	6.250		1,300	365		1,665	2,000
3300	Floor mounted									
3320	Bariatric (1,200 lb. capacity), elongated bowl, ADA compliant	Q-1	4.60	3.478	Ea.	2,875	202		3,077	3,450
3360	Hospital type, slotted rim for bed pan									
3370	Elongated bowl, top spud	Q-1	5	3.200	Ea.	330	186		516	640
3380	Elongated bowl, rear spud		5	3.200		405	186		591	725
3500	For rough-in, supply, waste and vent		3.05	5.246		340	305		645	830

22 43 Healthcare Plumbing Fixtures

22 43 16 – Healthcare Sinks

22 43 16.10 Sinks		Crew	Daily Output	Labor-Hours	Unit	Material (2020 Bare Costs)	Labor (2020 Bare Costs)	Equipment (2020 Bare Costs)	Total (2020 Bare Costs)	Total Incl O&P
0010	**SINKS**									
0020	Vitreous china									
6702	Hospital type, without trim (see Section 22 41 39.10)									
6710	20" x 18", contoured splash shield	Q-1	8	2	Ea.	96.50	116		212.50	280
6730	28" x 20", surgeon, side decks		8	2		560	116		676	795
6740	28" x 22", surgeon scrub-up, deep bowl		8	2		855	116		971	1,125
6750	20" x 27", patient, ADA compliant		7	2.286		490	133		623	740
6760	30" x 22", all purpose		7	2.286		780	133		913	1,050
6770	30" x 22", plaster work		7	2.286		710	133		843	980
6820	20" x 24" clinic service, liquid/solid waste		6	2.667		985	155		1,140	1,300

22 43 19 – Healthcare Bathtubs

22 43 19.10 Bathtubs		Crew	Daily Output	Labor-Hours	Unit	Material	Labor	Equipment	Total	Total Incl O&P
0010	**BATHTUBS**									
5002	Hospital type, with trim, see Section 22 41 39.10									
5050	Bathing pool, porcelain enamel on cast iron, grab bars									
5060	pop-up drain, 72" x 36"	Q-1	3	5.333	Ea.	3,850	310		4,160	4,725
5100	Perineal (sitz), vitreous china		3	5.333		1,300	310		1,610	1,900
5120	For pedestal, vitreous china, add		8	2		261	116		377	460
5300	Whirlpool, porcelain enamel on cast iron, 72" x 36"		1	16		4,775	930		5,705	6,650
5310	For color, add					5%				
5311	For designer colors and trim, add					15%				

22 43 23 – Healthcare Showers

22 43 23.10 Showers		Crew	Daily Output	Labor-Hours	Unit	Material	Labor	Equipment	Total	Total Incl O&P
0010	**SHOWERS**									
5950	Module, ADA compl, SS panel, fixed & hand held head, control									
5960	valves, grab bar, curtain & rod, folding seat	1 Plum	4	2	Ea.	1,625	129		1,754	1,975

22 43 39 – Healthcare Faucets

22 43 39.10 Faucets and Fittings		Crew	Daily Output	Labor-Hours	Unit	Material	Labor	Equipment	Total	Total Incl O&P
0010	**FAUCETS AND FITTINGS**									
2850	Medical, bedpan cleanser, with pedal valve,	1 Plum	12	.667	Ea.	815	43		858	960
2860	With screwdriver stop valve		12	.667		420	43		463	530
2870	With self-closing spray valve		12	.667		260	43		303	350
2900	Faucet, gooseneck spout, wrist handles, grid drain		10	.800		202	51.50		253.50	299
2940	Mixing valve, knee action, screwdriver stops		4	2		450	129		579	690

22 45 Emergency Plumbing Fixtures

22 45 13 – Emergency Showers

22 45 13.10 Emergency Showers		Crew	Daily Output	Labor-Hours	Unit	Material	Labor	Equipment	Total	Total Incl O&P
0010	**EMERGENCY SHOWERS**, Rough-in not included									
5000	Shower, single head, drench, ball valve, pull, freestanding	Q-1	4	4	Ea.	390	232		622	775
5200	Horizontal or vertical supply		4	4		615	232		847	1,025
6000	Multi-nozzle, eye/face wash combination		4	4		760	232		992	1,175
6400	Multi-nozzle, 12 spray, shower only		4	4		2,100	232		2,332	2,675
6600	For freeze-proof, add		6	2.667		505	155		660	785
8000	Walk-thru decontamination with eye-face wash		2	8		4,150	465		4,615	5,275
8200	For freeze proof, add		4	4		600	232		832	1,000

22 45 Emergency Plumbing Fixtures

22 45 16 – Eyewash Equipment

	22 45 16.10 Eyewash Safety Equipment	Crew	Daily Output	Labor-Hours	Unit	2020 Bare Costs Material	Labor	Equipment	Total	Total Incl O&P
0010	**EYEWASH SAFETY EQUIPMENT**, Rough-in not included									
1000	Eye wash fountain									
1400	Plastic bowl, pedestal mounted	Q-1	4	4	Ea.	335	232		567	710
1600	Unmounted		4	4		259	232		491	630
1800	Wall mounted		4	4		490	232		722	885
2000	Stainless steel, pedestal mounted		4	4		360	232		592	740
2200	Unmounted		4	4		298	232		530	675
2400	Wall mounted		4	4		325	232		557	700

22 45 19 – Self-Contained Eyewash Equipment

	22 45 19.10 Self-Contained Eyewash Safety Equipment	Crew	Daily Output	Labor-Hours	Unit	Material	Labor	Equipment	Total	Total Incl O&P
0010	**SELF-CONTAINED EYEWASH SAFETY EQUIPMENT**									
3000	Eye wash, portable, self-contained				Ea.	1,950			1,950	2,150

22 45 26 – Eye/Face Wash Equipment

	22 45 26.10 Eye/Face Wash Safety Equipment	Crew	Daily Output	Labor-Hours	Unit	Material	Labor	Equipment	Total	Total Incl O&P
0010	**EYE/FACE WASH SAFETY EQUIPMENT**, Rough-in not included									
4000	Eye and face wash, combination fountain									
4200	Stainless steel, pedestal mounted	Q-1	4	4	Ea.	1,150	232		1,382	1,600
4400	Unmounted		4	4		297	232		529	670
4600	Wall mounted		4	4		284	232		516	660

22 46 Security Plumbing Fixtures

22 46 13 – Security Water Closets and Urinals

	22 46 13.10 Security Water Closets and Urinals	Crew	Daily Output	Labor-Hours	Unit	Material	Labor	Equipment	Total	Total Incl O&P
0010	**SECURITY WATER CLOSETS AND URINALS**, Stainless steel									
2000	Urinal, back supply and flush									
2200	Wall hung	Q-1	4	4	Ea.	3,250	232		3,482	3,925
2240	Stall		2.50	6.400		4,300	370		4,670	5,275
2300	For urinal rough-in, supply, waste and vent		1.49	10.738		395	625		1,020	1,375
3000	Water closet, integral seat, back supply and flush									
3300	Wall hung, wall outlet	Q-1	5.80	2.759	Ea.	1,700	160		1,860	2,125
3400	Floor mount, wall outlet		5.80	2.759		2,150	160		2,310	2,600
3440	Floor mount, floor outlet		5.80	2.759		2,500	160		2,660	3,000
3480	For recessed tissue holder, add					172			172	189
3500	For water closet rough-in, supply, waste and vent	Q-1	1.19	13.445		380	780		1,160	1,600
5000	Water closet and lavatory units, push button filler valves,									
5010	soap & paper holders, seat									
5300	Wall hung	Q-1	5	3.200	Ea.	3,775	186		3,961	4,425
5400	Floor mount		5	3.200		3,775	186		3,961	4,425
6300	For unit rough-in, supply, waste and vent		1	16		445	930		1,375	1,875

22 46 16 – Security Lavatories and Sinks

	22 46 16.13 Security Lavatories	Crew	Daily Output	Labor-Hours	Unit	Material	Labor	Equipment	Total	Total Incl O&P
0010	**SECURITY LAVATORIES**, Stainless steel									
1000	Lavatory, wall hung, push button filler valve									
1100	Rectangular bowl	Q-1	8	2	Ea.	1,350	116		1,466	1,650
1200	Oval bowl		8	2		1,325	116		1,441	1,625
1240	Oval bowl, corner mount		8	2		1,650	116		1,766	2,000
1300	For lavatory rough-in, supply, waste and vent		1.50	10.667		375	620		995	1,325

22 46 Security Plumbing Fixtures

22 46 63 – Security Service Sink

22 46 63.10 Security Service Sink		Crew	Daily Output	Labor-Hours	Unit	Material	2020 Bare Costs Labor	Equipment	Total	Total Incl O&P
0010	**SECURITY SERVICE SINK**, Stainless steel									
1700	Service sink, with soap dish									
1740	24" x 19" size	Q-1	3	5.333	Ea.	2,650	310		2,960	3,400
1790	For sink rough-in, supply, waste and vent	"	.89	17.978	"	760	1,050		1,810	2,375

22 46 73 – Security Shower

22 46 73.10 Security Shower		Crew	Daily Output	Labor-Hours	Unit	Material	Labor	Equipment	Total	Total Incl O&P
0010	**SECURITY SHOWER**, Stainless steel									
1800	Shower cabinet, unitized									
1840	36" x 36" x 88"	Q-1	2.20	7.273	Ea.	10,100	420		10,520	11,700
1900	Shower package for built-in									
1940	Hot & cold valves, recessed soap dish	Q-1	6	2.667	Ea.	440	155		595	715

22 47 Drinking Fountains and Water Coolers

22 47 13 – Drinking Fountains

22 47 13.10 Drinking Water Fountains		Crew	Daily Output	Labor-Hours	Unit	Material	Labor	Equipment	Total	Total Incl O&P
0010	**DRINKING WATER FOUNTAINS**, For connection to cold water supply									
0802	For remote water chiller, see Section 22 47 23.10									
1000	Wall mounted, non-recessed R224000-30									
1200	Aluminum,									
1280	Dual bubbler type	1 Plum	3.20	2.500	Ea.	2,625	161		2,786	3,125
1400	Bronze, with no back		4	2		1,125	129		1,254	1,425
1600	Cast iron, enameled, low back, single bubbler D2010–810		4	2		1,225	129		1,354	1,550
1640	Dual bubbler type		3.20	2.500		1,725	161		1,886	2,150
1680	Triple bubbler type		3.20	2.500		2,150	161		2,311	2,625
1800	Cast aluminum, enameled, for correctional institutions		4	2		1,400	129		1,529	1,750
2000	Fiberglass, 12" back, single bubbler unit		4	2		2,000	129		2,129	2,425
2040	Dual bubbler		3.20	2.500		2,575	161		2,736	3,075
2080	Triple bubbler		3.20	2.500		2,825	161		2,986	3,350
2200	Polymarble, no back, single bubbler		4	2		935	129		1,064	1,225
2240	Dual bubbler		3.20	2.500		2,250	161		2,411	2,725
2280	Triple bubbler		3.20	2.500		2,175	161		2,336	2,650
2400	Precast stone, no back		4	2		1,050	129		1,179	1,350
2700	Stainless steel, single bubbler, no back		4	2		915	129		1,044	1,200
2740	With back		4	2		1,100	129		1,229	1,400
2780	Dual handle, ADA compliant		4	2		760	129		889	1,025
2820	Dual level, ADA compliant		3.20	2.500		1,600	161		1,761	2,000
2840	Vandal resistant type		4	2		680	129		809	940
3300	Vitreous china									
3340	7" back	1 Plum	4	2	Ea.	575	129		704	830
3940	For vandal-resistant bottom plate, add					75			75	82.50
3960	For freeze-proof valve system, add	1 Plum	2	4		770	258		1,028	1,225
3980	For rough-in, supply and waste, add	"	2.21	3.620		229	233		462	600
4000	Wall mounted, semi-recessed									
4200	Poly-marble, single bubbler	1 Plum	4	2	Ea.	960	129		1,089	1,250
4600	Stainless steel, satin finish, single bubbler		4	2		1,400	129		1,529	1,750
4900	Vitreous china, single bubbler		4	2		935	129		1,064	1,225
5980	For rough-in, supply and waste, add		1.83	4.372		229	282		511	670
6000	Wall mounted, fully recessed									
6400	Poly-marble, single bubbler	1 Plum	4	2	Ea.	1,775	129		1,904	2,150
6440	For water glass filler, add					95.50			95.50	105

22 47 Drinking Fountains and Water Coolers

22 47 13 – Drinking Fountains

22 47 13.10 Drinking Water Fountains		Crew	Daily Output	Labor-Hours	Unit	Material	2020 Bare Costs Labor	Equipment	Total	Total Incl O&P
6800	Stainless steel, single bubbler	1 Plum	4	2	Ea.	1,750	129		1,879	2,125
6900	Fountain and cuspidor combination		2	4		3,150	258		3,408	3,850
7560	For freeze-proof valve system, add		2	4		985	258		1,243	1,450
7580	For rough-in, supply and waste, add	↓	1.83	4.372	↓	229	282		511	670
7600	Floor mounted, pedestal type									
7700	Aluminum, architectural style, CI base	1 Plum	2	4	Ea.	2,650	258		2,908	3,300
7780	ADA compliant unit		2	4		1,675	258		1,933	2,225
8000	Bronze, architectural style		2	4		2,625	258		2,883	3,250
8040	Enameled steel cylindrical column style		2	4		2,325	258		2,583	2,950
8200	Precast stone/concrete, cylindrical column		1	8		1,625	515		2,140	2,550
8240	ADA compliant unit		1	8		3,250	515		3,765	4,350
8400	Stainless steel, architectural style		2	4		2,150	258		2,408	2,750
8600	Enameled iron, heavy duty service, 2 bubblers		2	4		3,675	258		3,933	4,425
8660	4 bubblers		2	4		4,650	258		4,908	5,475
8880	For freeze-proof valve system, add		2	4		700	258		958	1,150
8900	For rough-in, supply and waste, add	↓	1.83	4.372	↓	229	282		511	670
9100	Deck mounted									
9500	Stainless steel, circular receptor	1 Plum	4	2	Ea.	455	129		584	700
9540	14" x 9" receptor		4	2		380	129		509	610
9580	25" x 17" deep receptor, with water glass filler		3	2.667		300	172		472	585
9760	White enameled steel, 14" x 9" receptor		4	2		400	129		529	635
9860	White enameled cast iron, 24" x 16" receptor		3	2.667		525	172		697	835
9980	For rough-in, supply and waste, add	↓	1.83	4.372	↓	229	282		511	670

22 47 16 – Pressure Water Coolers

22 47 16.10 Electric Water Coolers

		Crew	Daily Output	Labor-Hours	Unit	Material	Labor	Equipment	Total	Total Incl O&P
0010	**ELECTRIC WATER COOLERS** D2010–820									
0100	Wall mounted, non-recessed									
0140	4 GPH	Q-1	4	4	Ea.	700	232		932	1,125
0160	8 GPH, barrier free, sensor operated		4	4		1,125	232		1,357	1,600
0180	8.2 GPH		4	4		1,025	232		1,257	1,475
0220	14.3 GPH		4	4		1,075	232		1,307	1,525
0600	8 GPH hot and cold water	↓	4	4		1,050	232		1,282	1,500
0640	For stainless steel cabinet, add					93			93	102
1000	Dual height, 8.2 GPH	Q-1	3.80	4.211		2,025	244		2,269	2,600
1040	14.3 GPH	"	3.80	4.211		1,750	244		1,994	2,300
1240	For stainless steel cabinet, add					213			213	235
2600	ADA compliant, 8 GPH	Q-1	4	4		1,100	232		1,332	1,550
3000	Simulated recessed, 8 GPH		4	4		795	232		1,027	1,225
3040	11.5 GPH	↓	4	4		1,025	232		1,257	1,475
3200	For glass filler, add					102			102	112
3240	For stainless steel cabinet, add					86.50			86.50	95.50
3300	Semi-recessed, 8.1 GPH	Q-1	4	4		915	232		1,147	1,350
3320	12 GPH	"	4	4		1,025	232		1,257	1,475
3340	For glass filler, add					172			172	189
3360	For stainless steel cabinet, add					161			161	177
3400	Full recessed, stainless steel, 8 GPH	Q-1	3.50	4.571		2,325	265		2,590	2,975
3420	11.5 GPH	"	3.50	4.571		1,775	265		2,040	2,350
3460	For glass filler, add					197			197	216
3600	For mounting can only				↓	218			218	240
4600	Floor mounted, flush-to-wall									
4640	4 GPH	1 Plum	3	2.667	Ea.	840	172		1,012	1,175
4680	8.2 GPH	↓	3	2.667	↓	935	172		1,107	1,275

22 47 Drinking Fountains and Water Coolers

22 47 16 – Pressure Water Coolers

22 47 16.10 Electric Water Coolers		Crew	Daily Output	Labor-Hours	Unit	2020 Bare Costs Material	Labor	Equipment	Total	Total Incl O&P
4720	14.3 GPH	1 Plum	3	2.667	Ea.	1,075	172		1,247	1,425
4960	14 GPH hot and cold water		3	2.667		1,200	172		1,372	1,575
4980	For stainless steel cabinet, add					141			141	155
5000	Dual height, 8.2 GPH	1 Plum	2	4		1,225	258		1,483	1,725
5040	14.3 GPH	"	2	4		1,275	258		1,533	1,775
5120	For stainless steel cabinet, add					209			209	229
5600	Explosion proof, 16 GPH	1 Plum	3	2.667		2,525	172		2,697	3,025
6000	Refrigerator compartment type, 4.5 GPH		3	2.667		1,550	172		1,722	1,975
6600	Bottle supply type, 1.0 GPH		4	2		415	129		544	650
6640	Hot and cold, 1.0 GPH		4	2		570	129		699	825
9800	For supply, waste & vent, all coolers		2.21	3.620		229	233		462	600

22 47 23 – Remote Water Coolers

22 47 23.10 Remote Water Coolers		Crew	Daily Output	Labor-Hours	Unit	Material	Labor	Equipment	Total	Total Incl O&P
0010	**REMOTE WATER COOLERS**, 80°F inlet									
0100	Air cooled, 50°F outlet, 115 V, 4.1 GPH	1 Plum	6	1.333	Ea.	580	86		666	770
0200	5.7 GPH		5.50	1.455		1,075	94		1,169	1,325
0300	8.0 GPH		5	1.600		825	103		928	1,050
0400	10.0 GPH		4.50	1.778		1,075	115		1,190	1,350
0500	13.4 GPH		4	2		1,800	129		1,929	2,175
0700	29 GPH	Q-1	5	3.200		1,925	186		2,111	2,400
1000	230 V, 32 GPH	"	5	3.200		2,000	186		2,186	2,475

22 51 Swimming Pool Plumbing Systems

22 51 19 – Swimming Pool Water Treatment Equipment

22 51 19.50 Swimming Pool Filtration Equipment		Crew	Daily Output	Labor-Hours	Unit	Material	Labor	Equipment	Total	Total Incl O&P
0010	**SWIMMING POOL FILTRATION EQUIPMENT**									
0900	Filter system, sand or diatomite type, incl. pump, 6,000 gal./hr.	2 Plum	1.80	8.889	Total	2,375	575		2,950	3,475
1020	Add for chlorination system, 800 S.F. pool		3	5.333	Ea.	235	345		580	775
1040	5,000 S.F. pool		3	5.333	"	1,950	345		2,295	2,650

22 52 Fountain Plumbing Systems

22 52 16 – Fountain Pumps

22 52 16.10 Fountain Water Pumps		Crew	Daily Output	Labor-Hours	Unit	Material	Labor	Equipment	Total	Total Incl O&P
0010	**FOUNTAIN WATER PUMPS**									
0100	Pump w/controls									
0200	Single phase, 100′ cord, 1/2 HP pump	2 Skwk	4.40	3.636	Ea.	1,325	199		1,524	1,750
0300	3/4 HP pump		4.30	3.721		1,375	204		1,579	1,825
0400	1 HP pump		4.20	3.810		1,925	209		2,134	2,450
0500	1-1/2 HP pump		4.10	3.902		2,575	214		2,789	3,175
0600	2 HP pump		4	4		4,350	219		4,569	5,100
0700	Three phase, 200′ cord, 5 HP pump		3.90	4.103		6,250	225		6,475	7,225
0800	7-1/2 HP pump		3.80	4.211		12,600	231		12,831	14,200
0900	10 HP pump		3.70	4.324		15,200	237		15,437	17,200
1000	15 HP pump		3.60	4.444		21,500	244		21,744	24,000
2000	DESIGN NOTE: Use two horsepower per surface acre.									

22 52 Fountain Plumbing Systems

22 52 33 – Fountain Ancillary

22 52 33.10 Fountain Miscellaneous		Crew	Daily Output	Labor-Hours	Unit	Material	2020 Bare Costs Labor	Equipment	Total	Total Incl O&P
0010	**FOUNTAIN MISCELLANEOUS**									
1300	Lights w/mounting kits, 200 watt	2 Skwk	18	.889	Ea.	1,200	49		1,249	1,400
1400	300 watt		18	.889		1,350	49		1,399	1,575
1500	500 watt		18	.889		1,525	49		1,574	1,750
1600	Color blender	↓	12	1.333	↓	600	73		673	770

22 62 Vacuum Systems for Laboratory and Healthcare Facilities

22 62 19 – Vacuum Equipment for Laboratory and Healthcare Facilities

22 62 19.70 Healthcare Vacuum Equipment		Crew	Daily Output	Labor-Hours	Unit	Material	Labor	Equipment	Total	Total Incl O&P
0010	**HEALTHCARE VACUUM EQUIPMENT**									
0300	Dental oral									
0310	Duplex									
0330	165 SCFM with 77 gal. separator	Q-2	1.30	18.462	Ea.	52,500	1,100		53,600	59,000
1100	Vacuum system									
1110	Vacuum outlet alarm panel	1 Plum	3.20	2.500	Ea.	1,100	161		1,261	1,450
2000	Medical, with receiver									
2100	Rotary vane type, lubricated, with controls									
2110	Simplex									
2120	1.5 HP, 80 gal. tank	Q-1	8	2	Ea.	6,650	116		6,766	7,500
2130	2 HP, 80 gal. tank		7	2.286		6,900	133		7,033	7,775
2140	3 HP, 80 gal. tank		6.60	2.424		7,275	141		7,416	8,200
2150	5 HP, 80 gal. tank	↓	6	2.667	↓	7,925	155		8,080	8,950
2200	Duplex									
2210	1 HP, 80 gal. tank	Q-1	8.40	1.905	Ea.	10,600	110		10,710	11,900
2220	1.5 HP, 80 gal. tank		7.80	2.051		10,900	119		11,019	12,200
2230	2 HP, 80 gal. tank		6.80	2.353		11,800	136		11,936	13,200
2240	3 HP, 120 gal. tank		6	2.667		12,800	155		12,955	14,300
2250	5 HP, 120 gal. tank	↓	5.40	2.963		14,600	172		14,772	16,300
2260	7.5 HP, 200 gal. tank	Q-2	7	3.429		20,900	206		21,106	23,200
2270	10 HP, 200 gal. tank		6.40	3.750		24,100	226		24,326	26,800
2280	15 HP, 200 gal. tank		5.80	4.138		43,100	249		43,349	47,800
2290	20 HP, 200 gal. tank		5	4.800		49,100	289		49,389	54,500
2300	25 HP, 200 gal. tank	↓	4	6	↓	57,000	360		57,360	63,500
2400	Triplex									
2410	7.5 HP, 200 gal. tank	Q-2	6.40	3.750	Ea.	33,100	226		33,326	36,700
2420	10 HP, 200 gal. tank		5.70	4.211		38,500	253		38,753	42,800
2430	15 HP, 200 gal. tank		4.90	4.898		65,000	295		65,295	72,000
2440	20 HP, 200 gal. tank		4	6		74,500	360		74,860	82,000
2450	25 HP, 200 gal. tank	↓	3.70	6.486	↓	86,000	390		86,390	95,000
2500	Quadruplex									
2510	7.5 HP, 200 gal. tank	Q-2	5.30	4.528	Ea.	41,200	272		41,472	45,700
2520	10 HP, 200 gal. tank		4.70	5.106		47,500	305		47,805	53,000
2530	15 HP, 200 gal. tank		4	6		86,000	360		86,360	95,000
2540	20 HP, 200 gal. tank		3.60	6.667		98,000	400		98,400	108,000
2550	25 HP, 200 gal. tank	↓	3.20	7.500	↓	113,500	450		113,950	125,500
4000	Liquid ring type, water sealed, with controls									
4200	Duplex									
4210	1.5 HP, 120 gal. tank	Q-1	6	2.667	Ea.	21,000	155		21,155	23,300
4220	3 HP, 120 gal. tank	"	5.50	2.909		21,900	169		22,069	24,400
4230	4 HP, 120 gal. tank	Q-2	6.80	3.529	↓	23,400	212		23,612	26,000

22 62 Vacuum Systems for Laboratory and Healthcare Facilities

22 62 19 – Vacuum Equipment for Laboratory and Healthcare Facilities

22 62 19.70 Healthcare Vacuum Equipment		Crew	Daily Output	Labor-Hours	Unit	2020 Bare Costs Material	Labor	Equipment	Total	Total Incl O&P
4240	5 HP, 120 gal. tank	Q-2	6.50	3.692	Ea.	24,700	222		24,922	27,500
4250	7.5 HP, 200 gal. tank		6.20	3.871		30,000	233		30,233	33,400
4260	10 HP, 200 gal. tank		5.80	4.138		37,500	249		37,749	41,600
4270	15 HP, 200 gal. tank		5.10	4.706		45,400	283		45,683	50,500
4280	20 HP, 200 gal. tank		4.60	5.217		58,500	315		58,815	65,000
4290	30 HP, 200 gal. tank		4	6		70,000	360		70,360	77,500

22 63 Gas Systems for Laboratory and Healthcare Facilities

22 63 13 – Gas Piping for Laboratory and Healthcare Facilities

22 63 13.70 Healthcare Gas Piping		Crew	Daily Output	Labor-Hours	Unit	Material	Labor	Equipment	Total	Total Incl O&P
0010	**HEALTHCARE GAS PIPING**									
0030	Air compressor intake filter									
0034	Rooftop									
0036	Filter/silencer									
0040	1″	1 Stpi	10	.800	Ea.	254	52.50		306.50	360
0044	1-1/4″		9.60	.833		261	54.50		315.50	370
0048	1-1/2″		9.20	.870		280	57		337	395
0052	2″		8.80	.909		280	59.50		339.50	400
0056	2-1/2″		8.40	.952		375	62.50		437.50	505
0060	3″		8	1		450	65.50		515.50	595
0064	4″		7.60	1.053		465	69		534	620
0068	5″		7.40	1.081		655	71		726	825
0076	Inline									
0080	Filter/silencer									
0084	2″	1 Stpi	8.20	.976	Ea.	850	64		914	1,025
0088	2-1/2″		8	1		870	65.50		935.50	1,050
0090	3″		7.60	1.053		965	69		1,034	1,150
0092	4″		7.20	1.111		1,125	73		1,198	1,350
0094	5″		6.80	1.176		1,225	77		1,302	1,475
0096	6″		6.40	1.250		1,375	82		1,457	1,625
1000	Nitrogen or oxygen system									
1010	Cylinder manifold									
1020	5 cylinder	1 Plum	.80	10	Ea.	5,900	645		6,545	7,450
1026	10 cylinder	″	.40	20		7,100	1,300		8,400	9,725
1050	Nitrogen generator, 30 LPM	Q-5	.40	40		28,900	2,350		31,250	35,300
1900	Vaporizers									
1910	LOX vaporizers									
1920	Nominal capacity									
1930	1410 SCFM	Q-1	2	8	Ea.	1,900	465		2,365	2,800
1940	5650 SCFM		1.40	11.429		4,400	665		5,065	5,850
1950	12,703 SCFM		.80	20		6,750	1,150		7,900	9,150
1980	Removal of LOX vaporizers									
1982	Nominal capacity									
1986	1410 SCFM	Q-1	4	4	Ea.		232		232	345
1990	5650 SCFM		2.80	5.714			330		330	495
1994	12,703 SCFM		1.60	10			580		580	870
3000	Outlets and valves									
3010	Recessed, wall mounted									
3012	Single outlet	1 Plum	3.20	2.500	Ea.	70.50	161		231.50	320
3100	Ceiling outlet									
3190	Zone valve with box									

	22 63 13.70 Healthcare Gas Piping	Crew	Daily Output	Labor-Hours	Unit	2020 Bare Costs Material	Labor	Equipment	Total	Total Incl O&P
3192	Cleaned for oxygen service, not including gauges									
3194	1/2" valve size	1 Plum	4.60	1.739	Ea.	222	112		334	410
3196	3/4" valve size		4.30	1.860		245	120		365	450
3198	1" valve size		4	2		271	129		400	490
3202	1-1/4" valve size		3.80	2.105		305	136		441	540
3206	1-1/2" valve size		3.60	2.222		340	143		483	585
3210	2" valve size		3.20	2.500		415	161		576	695
3214	2-1/2" valve size		3.10	2.581		1,000	166		1,166	1,350
3218	3" valve size		3	2.667		1,400	172		1,572	1,800
3224	Gauges for zone valve box									
3226	0-100 psi (O2, air, N2, CO_2)	1 Plum	16	.500	Ea.	17.10	32		49.10	67
3228	Vacuum, WAGD (Waste Anesthesia Gas)		16	.500		17.10	32		49.10	67
3230	0-300 psi (Nitrogen)		16	.500		17.10	32		49.10	67
4000	Alarm panel, medical gases and vacuum									
4010	Alarm panel	1 Plum	3.20	2.500	Ea.	1,100	161		1,261	1,450
4030	Master alarm panel									
4034	Can also monitor area alarms									
4038	and communicate with PC-based alarm monitor.									
4040	10 signal	1 Elec	3.20	2.500	Ea.	1,025	153		1,178	1,350
4044	20 signal		3	2.667		1,175	164		1,339	1,550
4048	30 signal		2.80	2.857		1,575	175		1,750	1,975
4052	40 signal		2.60	3.077		1,775	189		1,964	2,225
4056	50 signal		2.40	3.333		1,900	205		2,105	2,400
4060	60 signal		2.20	3.636		2,225	223		2,448	2,775
4100	Area alarm panel									
4104	Does not include specific gas transducers.									
4108	3 module alarm panel									
4112	P-P-P	1 Elec	2.80	2.857	Ea.	1,375	175		1,550	1,750
4116	P-P-V		2.80	2.857		930	175		1,105	1,275
4120	D-D-D		2.80	2.857		1,025	175		1,200	1,375
4130	6 module alarm panel									
4132	4-P, 3-V	1 Elec	2.20	3.636	Ea.	1,450	223		1,673	1,925
4136	3-P, 2-V, B		2.20	3.636		1,075	223		1,298	1,525
4140	5-D, B		2.20	3.636		1,425	223		1,648	1,900
4144	6-D		2.20	3.636		1,900	223		2,123	2,425
4148	3-P, 3-V		2.20	3.636		1,775	223		1,998	2,275
4170	Note: P=pressure, V=vacuum, B=blank, D=dual display									
4180	Alarm transducers, gas specific									
4182	Oxygen	1 Elec	24	.333	Ea.	157	20.50		177.50	204
4184	Vacuum		24	.333		153	20.50		173.50	200
4186	Nitrous oxide		24	.333		153	20.50		173.50	200
4188	Medical air		24	.333		157	20.50		177.50	204
4190	Carbon dioxide		24	.333		157	20.50		177.50	204
4192	Nitrogen		24	.333		157	20.50		177.50	204
4194	WAGD		24	.333		167	20.50		187.50	215
4300	Ball valves cleaned for oxygen service									
4310	with copper extensions and gauge port									
4320	1/4" diam.	1 Plum	24	.333	Ea.	63.50	21.50		85	102
4330	1/2" diam.		22	.364		60	23.50		83.50	101
4334	3/4" diam.		20	.400		77.50	26		103.50	124
4338	1" diam.		19	.421		112	27		139	164
4342	1-1/4" diam.		15	.533		135	34.50		169.50	201
4346	1-1/2" diam.		13	.615		179	39.50		218.50	256

22 63 13.70 Healthcare Gas Piping		Crew	Daily Output	Labor-Hours	Unit	2020 Bare Costs Material	Labor	Equipment	Total	Total Incl O&P
4350	2" diam.	1 Plum	11	.727	Ea.	305	47		352	405
4354	2-1/2" diam.	Q-1	15	1.067		845	62		907	1,025
4358	3" diam.		13	1.231		1,250	71.50		1,321.50	1,475
4362	4" diam.		10	1.600		2,400	93		2,493	2,775
5000	Manifold									
5010	Automatic switchover type									
5020	Note: Both a control panel and header assembly are required									
5030	Control panel									
5040	Oxygen	Q-1	4	4	Ea.	4,825	232		5,057	5,675
5060	Header assembly									
5066	Oxygen									
5070	2 x 2	Q-1	6	2.667	Ea.	790	155		945	1,100
5074	3 x 3		5.50	2.909		930	169		1,099	1,275
5078	4 x 4		5	3.200		1,225	186		1,411	1,625
5082	5 x 5		4.50	3.556		1,350	206		1,556	1,800
5086	6 x 6		4	4		1,700	232		1,932	2,225
5090	7 x 7		3.50	4.571		1,850	265		2,115	2,425
7000	Medical air compressors									
7020	Oil-less with inlet filter, duplexed dryers and aftercoolers									
7030	Duplex systems, horizontal tank, 208/230/460/575 V, 3 Ph.									
7040	1 HP, 80 gal. tank, 3.8 ACFM @50PSIG, 2.7 ACFM @100PSIG	Q-5	4	4	Ea.	29,900	236		30,136	33,300
7050	1.5 HP, 80 gal. tank, 6.8 ACFM @50PSIG, 4.5 ACFM @100PSIG		3.80	4.211		35,700	248		35,948	39,600
7060	2 HP, 80 gal. tank, 8.2 ACFM @50PSIG, 6.3 ACFM @100PSIG		3.60	4.444		35,800	262		36,062	39,800
7070	3 HP, 120 gal. tank, 11.2 ACFM @50PSIG, 9.4 ACFM @100PSIG		3.20	5		27,700	295		27,995	30,900
7080	5 HP, 120 gal. tank, 17.4 ACFM @50PSIG, 15.6 ACFM @100PSIG		2.80	5.714		46,200	335		46,535	51,500
7090	7.5 HP, 250 gal. tank, 31.6 ACFM @50PSIG, 25.9 ACFM @100PSIG		2.40	6.667		61,500	395		61,895	68,000
7100	10 HP, 250 gal. tank, 43 ACFM @50PSIG, 35.2 ACFM @100PSIG		2	8		66,500	470		66,970	73,500
7110	15 HP, 250 gal. tank, 69 ACFM @50PSIG, 56.5 ACFM @100PSIG		1.80	8.889		80,000	525		80,525	89,000
7200	Aftercooler, air-cooled									
7210	Steel manifold, copper tube, aluminum fins									
7220	35 SCFM @ 100PSIG	Q-5	4	4	Ea.	1,025	236		1,261	1,475
7300	Air dryer system									
7310	Refrigerated type									
7320	Flow @ 125 psi									
7330	20 SCFM	Q-5	6	2.667	Ea.	1,525	157		1,682	1,900
7332	25 SCFM		5.80	2.759		1,600	163		1,763	2,000
7334	35 SCFM		5.40	2.963		1,950	175		2,125	2,400
7336	50 SCFM		5	3.200		2,425	189		2,614	2,950
7338	75 SCFM		4.60	3.478		3,025	205		3,230	3,625
7340	100 SCFM		4.30	3.721		3,625	220		3,845	4,325
7342	125 SCFM		4	4		4,675	236		4,911	5,500
7360	Desiccant type									
7362	Flow with energy saving controls									
7366	40 SCFM	Q-5	3.60	4.444	Ea.	6,200	262		6,462	7,225
7368	60 SCFM		3.40	4.706		6,750	278		7,028	7,875
7370	90 SCFM		3.20	5		8,025	295		8,320	9,275
7372	115 SCFM		3	5.333		8,625	315		8,940	9,950
7374	165 SCFM		2.90	5.517		9,275	325		9,600	10,700
7378	260 SCFM	Q-6	3.60	6.667		10,900	410		11,310	12,600
7382	370 SCFM		3.40	7.059		12,700	430		13,130	14,600
7386	590 SCFM		3.20	7.500		15,800	460		16,260	18,000
7390	1130 SCFM		3	8		24,400	490		24,890	27,500
7460	Dew point monitor									

22 63 Gas Systems for Laboratory and Healthcare Facilities

22 63 13 – Gas Piping for Laboratory and Healthcare Facilities

22 63 13.70 Healthcare Gas Piping		Crew	Daily Output	Labor-Hours	Unit	Material	Labor	Equipment	Total	Total Incl O&P
						2020 Bare Costs				
7466	LCD readout, high dew point alarm, probe included									
7470	Monitor	1 Stpi	2	4	Ea.	3,250	262		3,512	3,975

22 66 Chemical-Waste Systems for Lab. and Healthcare Facilities

22 66 53 – Laboratory Chemical-Waste and Vent Piping

22 66 53.30 Glass Pipe

Line	Description	Crew	Daily Output	Labor-Hours	Unit	Material	Labor	Equipment	Total	Total Incl O&P
0010	**GLASS PIPE**, Borosilicate, couplings & clevis hanger assemblies, 10' OC R221113-70									
0020	Drainage									
1100	1-1/2" diameter	Q-1	52	.308	L.F.	15.30	17.85		33.15	43.50
1120	2" diameter		44	.364		19.50	21		40.50	53
1140	3" diameter		39	.410		26	24		50	64
1160	4" diameter		30	.533		42.50	31		73.50	93.50
1180	6" diameter		26	.615		75	35.50		110.50	136
1870	To delete coupling & hanger, subtract									
1880	1-1/2" diam. to 2" diam.					19%	22%			
1890	3" diam. to 6" diam.					20%	17%			
2000	Process supply (pressure), beaded joints									
2040	1/2" diameter	1 Plum	36	.222	L.F.	9.55	14.30		23.85	32
2060	3/4" diameter		31	.258		10.55	16.65		27.20	36.50
2080	1" diameter		27	.296		24.50	19.10		43.60	55.50
2100	1-1/2" diameter	Q-1	47	.340		17	19.75		36.75	48
2120	2" diameter		39	.410		21	24		45	58.50
2140	3" diameter		34	.471		29.50	27.50		57	73.50
2160	4" diameter		25	.640		42.50	37		79.50	103
2180	6" diameter		21	.762		128	44		172	207
2860	To delete coupling & hanger, subtract									
2870	1/2" diam. to 1" diam.					25%	33%			
2880	1-1/2" diam. to 3" diam.					22%	21%			
2890	4" diam. to 6" diam.					23%	15%			
3800	Conical joint, transparent									
3980	6" diameter	Q-1	21	.762	L.F.	168	44		212	251
4500	To delete couplings & hangers, subtract									
4530	6" diam.					22%	26%			

22 66 53.40 Pipe Fittings, Glass

Line	Description	Crew	Daily Output	Labor-Hours	Unit	Material	Labor	Equipment	Total	Total Incl O&P
0010	**PIPE FITTINGS, GLASS**									
0020	Drainage, beaded ends									
0040	Coupling & labor required at joints not incl. in fitting									
0050	price. Add 1 per joint for installed price									
0070	90° bend or sweep, 1-1/2"				Ea.	36.50			36.50	40
0090	2"					46.50			46.50	51.50
0100	3"					76			76	83.50
0110	4"					121			121	134
0120	6" (sweep only)					360			360	395
0200	45° bend or sweep same as 90°									
0350	Tee, single sanitary, 1-1/2"				Ea.	59			59	65
0370	2"					59			59	65
0380	3"					86			86	94.50
0390	4"					157			157	173
0400	6"					420			420	465
0410	Tee, straight, 1-1/2"					73			73	80
0430	2"					73			73	80

22 66 53 – Laboratory Chemical-Waste and Vent Piping

22 66 53.40 Pipe Fittings, Glass

	22 66 53.40 Pipe Fittings, Glass	Crew	Daily Output	Labor-Hours	Unit	Material	Labor	Equipment	Total	Total Incl O&P
						2020 Bare Costs				
0440	3"				Ea.	106			106	116
0450	4"					139			139	153
0460	6"					455			455	500
0500	Coupling, stainless steel, TFE seal ring									
0520	1-1/2"	Q-1	32	.500	Ea.	27.50	29		56.50	74
0530	2"		30	.533		35	31		66	85
0540	3"		25	.640		33.50	37		70.50	92
0550	4"		23	.696		80.50	40.50		121	149
0560	6"		20	.800		181	46.50		227.50	269
0600	Coupling, stainless steel, bead to plain end									
0610	1-1/2"	Q-1	36	.444	Ea.	36.50	26		62.50	79
0620	2"		34	.471		44.50	27.50		72	89.50
0630	3"		29	.552		75.50	32		107.50	131
0640	4"		27	.593		112	34.50		146.50	175
0650	6"		24	.667		380	38.50		418.50	475
2350	Coupling, Viton liner, for temperatures to 400°F									
2370	1/2"	Q-1	40	.400	Ea.	71	23		94	113
2380	3/4"		37	.432		81	25		106	127
2390	1"		35	.457		135	26.50		161.50	188
2400	1-1/2"		32	.500		27.50	29		56.50	74
2410	2"		30	.533		34.50	31		65.50	84.50
2420	3"		25	.640		46	37		83	106
2430	4"		23	.696		79.50	40.50		120	148
2440	6"		20	.800		201	46.50		247.50	291
2550	For beaded joint armored fittings, add					200%				
2600	Conical ends. Flange set, gasket & labor not incl. in fitting									
2620	price. Add 1 per joint for installed price.									
2650	90° sweep elbow, 1"				Ea.	118			118	130
2670	1-1/2"					275			275	305
2680	2"					246			246	270
2690	3"					445			445	485
2700	4"					795			795	875
2710	6"					1,175			1,175	1,300
2750	Cross (straight), add					55%				
2850	Tee, add					20%				

22 66 53.60 Corrosion Resistant Pipe

	22 66 53.60 Corrosion Resistant Pipe	Crew	Daily Output	Labor-Hours	Unit	Material	Labor	Equipment	Total	Total Incl O&P
0010	**CORROSION RESISTANT PIPE**, No couplings or hangers R221113-70									
0020	Iron alloy, drain, mechanical joint									
1000	1-1/2" diameter	Q-1	70	.229	L.F.	84.50	13.25		97.75	113
1100	2" diameter		66	.242		87	14.05		101.05	117
1120	3" diameter		60	.267		95	15.45		110.45	128
1140	4" diameter		52	.308		117	17.85		134.85	156
1980	Iron alloy, drain, B&S joint									
2000	2" diameter	Q-1	54	.296	L.F.	85.50	17.20		102.70	120
2100	3" diameter		52	.308		82.50	17.85		100.35	117
2120	4" diameter		48	.333		101	19.35		120.35	140
2140	6" diameter	Q-2	59	.407		152	24.50		176.50	204
2160	8" diameter	"	54	.444		315	26.50		341.50	385
2980	Plastic, epoxy, fiberglass filament wound, B&S joint									
3000	2" diameter	Q-1	62	.258	L.F.	12.35	14.95		27.30	36
3100	3" diameter		51	.314		14.40	18.20		32.60	43
3120	4" diameter		45	.356		20.50	20.50		41	53.50

22 66 53 – Laboratory Chemical-Waste and Vent Piping

22 66 53.60 Corrosion Resistant Pipe		Crew	Daily Output	Labor-Hours	Unit	Material	Labor	Equipment	Total	Total Incl O&P
						2020 Bare Costs				
3140	6" diameter	Q-1	32	.500	L.F.	29	29		58	75
3160	8" diameter	Q-2	38	.632		45	38		83	107
3180	10" diameter		32	.750		62	45		107	136
3200	12" diameter		28	.857		74.50	51.50		126	159
3980	Polyester, fiberglass filament wound, B&S joint									
4000	2" diameter	Q-1	62	.258	L.F.	13.40	14.95		28.35	37.50
4100	3" diameter		51	.314		17.55	18.20		35.75	46.50
4120	4" diameter		45	.356		25.50	20.50		46	59.50
4140	6" diameter		32	.500		37	29		66	84.50
4160	8" diameter	Q-2	38	.632		88	38		126	154
4180	10" diameter		32	.750		118	45		163	198
4200	12" diameter		28	.857		130	51.50		181.50	220
4980	Polypropylene, acid resistant, fire retardant, Schedule 40									
5000	1-1/2" diameter	Q-1	68	.235	L.F.	9.05	13.65		22.70	30.50
5100	2" diameter		62	.258		14.55	14.95		29.50	38.50
5120	3" diameter		51	.314		25	18.20		43.20	54.50
5140	4" diameter		45	.356		32	20.50		52.50	66
5160	6" diameter		32	.500		63.50	29		92.50	114
5980	Proxylene, fire retardant, Schedule 40									
6000	1-1/2" diameter	Q-1	68	.235	L.F.	14	13.65		27.65	36
6100	2" diameter		62	.258		19.20	14.95		34.15	43.50
6120	3" diameter		51	.314		34.50	18.20		52.70	65
6140	4" diameter		45	.356		49	20.50		69.50	85
6160	6" diameter		32	.500		83	29		112	135
6820	For Schedule 80, add					35%	2%			

22 66 53.70 Pipe Fittings, Corrosion Resistant		Crew	Daily Output	Labor-Hours	Unit	Material	Labor	Equipment	Total	Total Incl O&P
0010	**PIPE FITTINGS, CORROSION RESISTANT**									
0030	Iron alloy									
0050	Mechanical joint									
0060	1/4 bend, 1-1/2"	Q-1	12	1.333	Ea.	125	77.50		202.50	253
0080	2"		10	1.600		204	93		297	365
0090	3"		9	1.778		245	103		348	425
0100	4"		8	2		282	116		398	485
0110	1/8 bend, 1-1/2"		12	1.333		79.50	77.50		157	204
0130	2"		10	1.600		136	93		229	289
0140	3"		9	1.778		182	103		285	355
0150	4"		8	2		243	116		359	440
0160	Tee and Y, sanitary, straight									
0170	1-1/2"	Q-1	8	2	Ea.	136	116		252	325
0180	2"		7	2.286		182	133		315	400
0190	3"		6	2.667		282	155		437	540
0200	4"		5	3.200		520	186		706	850
0360	Coupling, 1-1/2"		14	1.143		75	66.50		141.50	182
0380	2"		12	1.333		85	77.50		162.50	210
0390	3"		11	1.455		89.50	84.50		174	225
0400	4"		10	1.600		101	93		194	250
0500	Bell & Spigot									
0510	1/4 and 1/16 bend, 2"	Q-1	16	1	Ea.	119	58		177	218
0520	3"		14	1.143		278	66.50		344.50	405
0530	4"		13	1.231		285	71.50		356.50	420
0540	6"	Q-2	17	1.412		565	85		650	745
0550	8"	"	12	2		2,325	120		2,445	2,725

22 66 53.70 Pipe Fittings, Corrosion Resistant		Crew	Daily Output	Labor-Hours	Unit	Material	2020 Bare Costs Labor	Equipment	Total	Total Incl O&P
0620	1/8 bend, 2"	Q-1	16	1	Ea.	134	58		192	234
0640	3"		14	1.143		248	66.50		314.50	370
0650	4"		13	1.231		246	71.50		317.50	375
0660	6"	Q-2	17	1.412		470	85		555	645
0680	8"	"	12	2		1,850	120		1,970	2,200
0700	Tee, sanitary, 2"	Q-1	10	1.600		254	93		347	420
0710	3"		9	1.778		825	103		928	1,050
0720	4"		8	2		685	116		801	930
0730	6"	Q-2	11	2.182		855	131		986	1,150
0740	8"	"	8	3		2,350	180		2,530	2,850
1800	Y, sanitary, 2"	Q-1	10	1.600		268	93		361	435
1820	3"		9	1.778		480	103		583	680
1830	4"		8	2		430	116		546	650
1840	6"	Q-2	11	2.182		1,425	131		1,556	1,775
1850	8"	"	8	3		3,950	180		4,130	4,625
3000	Epoxy, filament wound									
3030	Quick-lock joint									
3040	90° elbow, 2"	Q-1	28	.571	Ea.	100	33		133	161
3060	3"		16	1		116	58		174	214
3070	4"		13	1.231		158	71.50		229.50	280
3080	6"		8	2		230	116		346	425
3090	8"	Q-2	9	2.667		425	160		585	705
3100	10"		7	3.429		535	206		741	895
3110	12"		6	4		765	241		1,006	1,200
3120	45° elbow, 2"	Q-1	28	.571		77.50	33		110.50	135
3130	3"		16	1		119	58		177	218
3140	4"		13	1.231		122	71.50		193.50	241
3150	6"		8	2		230	116		346	425
3160	8"	Q-2	9	2.667		425	160		585	705
3170	10"		7	3.429		535	206		741	895
3180	12"		6	4		765	241		1,006	1,200
3190	Tee, 2"	Q-1	19	.842		239	49		288	335
3200	3"		11	1.455		289	84.50		373.50	445
3210	4"		9	1.778		350	103		453	540
3220	6"		5	3.200		585	186		771	925
3230	8"	Q-2	6	4		665	241		906	1,100
3240	10"		5	4.800		925	289		1,214	1,450
3250	12"		4	6		1,425	360		1,785	2,125
4000	Polypropylene, acid resistant									
4020	Non-pressure, electrofusion joints									
4050	1/4 bend, 1-1/2"	1 Plum	16	.500	Ea.	16.55	32		48.55	66.50
4060	2"	Q-1	28	.571		33	33		66	85.50
4080	3"		17	.941		35.50	54.50		90	121
4090	4"		14	1.143		58	66.50		124.50	163
4110	6"		8	2		138	116		254	325
4150	1/4 bend, long sweep									
4170	1-1/2"	1 Plum	16	.500	Ea.	18.70	32		50.70	68.50
4180	2"	Q-1	28	.571		33	33		66	85.50
4200	3"		17	.941		41	54.50		95.50	127
4210	4"		14	1.143		59.50	66.50		126	165
4250	1/8 bend, 1-1/2"	1 Plum	16	.500		17.90	32		49.90	67.50
4260	2"	Q-1	28	.571		20	33		53	71.50
4280	3"		17	.941		37.50	54.50		92	123

22 66 53.70 Pipe Fittings, Corrosion Resistant		Crew	Daily Output	Labor-Hours	Unit	Material	Labor	Equipment	Total	Total Incl O&P
						2020 Bare Costs				
4290	4"	Q-1	14	1.143	Ea.	42	66.50		108.50	145
4310	6"		8	2		116	116		232	300
4400	Tee, sanitary									
4420	1-1/2"	1 Plum	10	.800	Ea.	20.50	51.50		72	99.50
4430	2"	Q-1	17	.941		25	54.50		79.50	109
4450	3"		11	1.455		50	84.50		134.50	181
4460	4"		9	1.778		75	103		178	237
4480	6"		5	3.200		495	186		681	825
4490	Tee, sanitary reducing, 2" x 2" x 1-1/2"		17	.941		25	54.50		79.50	109
4491	3" x 3" x 1-1/2"		11	1.455		44.50	84.50		129	175
4492	3" x 3" x 2"		11	1.455		50	84.50		134.50	181
4493	4" x 4" x 2"		10	1.600		67	93		160	213
4494	4" x 4" x 3"		9	1.778		73	103		176	235
4496	6" x 6" x 4"		5	3.200		240	186		426	540
4650	Wye 45°, 1-1/2"	1 Plum	10	.800		23	51.50		74.50	103
4652	2"	Q-1	17	.941		31.50	54.50		86	116
4653	3"		11	1.455		53.50	84.50		138	185
4654	4"		9	1.778		77.50	103		180.50	239
4656	6"		5	3.200		200	186		386	500
4660	Wye, reducing									
4662	2" x 2" x 1-1/2"	Q-1	17	.941	Ea.	30	54.50		84.50	115
4666	3" x 3" x 2"		11	1.455		60.50	84.50		145	193
4668	4" x 4" x 2"		10	1.600		71.50	93		164.50	218
4669	4" x 4" x 3"		9	1.778		74.50	103		177.50	236
4671	6" x 6" x 2"		6	2.667		121	155		276	365
4673	6" x 6" x 3"		5.50	2.909		134	169		303	400
4675	6" x 6" x 4"		5	3.200		139	186		325	430
4678	Combination Y & 1/8 bend									
4681	1-1/2"	1 Plum	10	.800	Ea.	28	51.50		79.50	108
4683	2"	Q-1	17	.941		35.50	54.50		90	121
4684	3"		11	1.455		60.50	84.50		145	193
4685	4"		9	1.778		83	103		186	246
4689	Combination Y & 1/8 bend, reducing									
4692	2" x 2" x 1-1/2"	Q-1	17	.941	Ea.	33	54.50		87.50	118
4694	3" x 3" x 1-1/2"		12	1.333		51	77.50		128.50	172
4695	3" x 3" x 2"		11	1.455		53.50	84.50		138	185
4697	4" x 4" x 2"		10	1.600		75.50	93		168.50	222
4699	4" x 4" x 3"		9	1.778		77.50	103		180.50	239
4710	Hub adapter									
4712	1-1/2"	1 Plum	16	.500	Ea.	38.50	32		70.50	90.50
4713	2"	Q-1	28	.571		44.50	33		77.50	98.50
4714	3"		17	.941		55	54.50		109.50	142
4715	4"		14	1.143		73	66.50		139.50	180
4719	Mechanical joint adapter									
4721	1-1/2"	1 Plum	16	.500	Ea.	24	32		56	74.50
4722	2"	Q-1	28	.571		25	33		58	77
4723	3"		17	.941		35.50	54.50		90	121
4724	4"		14	1.143		53.50	66.50		120	158
4728	Couplings									
4731	1-1/2"	1 Plum	16	.500	Ea.	13.40	32		45.40	62.50
4732	2"	Q-1	28	.571		16.60	33		49.60	68
4733	3"		17	.941		21.50	54.50		76	105
4734	4"		14	1.143		30	66.50		96.50	133

22 66 Chemical-Waste Systems for Lab. and Healthcare Facilities

22 66 53 – Laboratory Chemical-Waste and Vent Piping

22 66 53.70 Pipe Fittings, Corrosion Resistant		Crew	Daily Output	Labor-Hours	Unit	2020 Bare Costs Material	Labor	Equipment	Total	Total Incl O&P
4736	6″	Q-1	8	2	Ea.	48	116		164	227

22 66 83 – Chemical-Waste Tanks

22 66 83.13 Chemical-Waste Dilution Tanks

Line	Description	Crew	Daily Output	Labor-Hours	Unit	Material	Labor	Equipment	Total	Total Incl O&P
0010	**CHEMICAL-WASTE DILUTION TANKS**									
7000	Tanks, covers included									
7800	Polypropylene									
7810	Continuous service to 200°F									
7830	2 gallon, 8″ x 8″ x 8″	Q-1	20	.800	Ea.	121	46.50		167.50	204
7850	7 gallon, 12″ x 12″ x 12″		20	.800		189	46.50		235.50	278
7870	16 gallon, 18″ x 12″ x 18″		17	.941		247	54.50		301.50	355
8010	33 gallon, 24″ x 18″ x 18″		12	1.333		320	77.50		397.50	465
8070	44 gallon, 24″ x 18″ x 24″		10	1.600		385	93		478	565
8080	89 gallon, 36″ x 24″ x 24″		8	2		540	116		656	770
8150	Polyethylene, heavy duty walls									
8160	Continuous service to 180°F									
8180	5 gallon, 12″ x 6″ x 18″	Q-1	20	.800	Ea.	53	46.50		99.50	128
8210	15 gallon, 14″ I.D. x 27″ deep		17	.941		78.50	54.50		133	168
8230	55 gallon, 22″ I.D. x 36″ deep		10	1.600		180	93		273	335
8250	100 gallon, 28″ I.D. x 42″ deep		8	2		390	116		506	605
8270	200 gallon, 36″ I.D. x 48″ deep		6	2.667		610	155		765	900
8290	360 gallon, 48″ I.D. x 48″ deep		5	3.200		750	186		936	1,100

Division Notes

I	CREW	DAILY OUTPUT	LABOR-HOURS	UNIT	BARE COSTS MAT.	LABOR	EQUIP.	TOTAL	TOTAL INCL O&P

Division 23

Heating, Ventilating & Air Conditioning

Estimating Tips

The labor adjustment factors listed in Subdivision 22 01 02.20 also apply to Division 23.

23 10 00 Facility Fuel Systems

- The prices in this subdivision for above- and below-ground storage tanks do not include foundations or hold-down slabs, unless noted. The estimator should refer to Divisions 3 and 31 for foundation system pricing. In addition to the foundations, required tank accessories, such as tank gauges, leak detection devices, and additional manholes and piping, must be added to the tank prices.

23 50 00 Central Heating Equipment

- When estimating the cost of an HVAC system, check to see who is responsible for providing and installing the temperature control system. It is possible to overlook controls, assuming that they would be included in the electrical estimate.
- When looking up a boiler, be careful on specified capacity. Some manufacturers rate their products on output while others use input.
- Include HVAC insulation for pipe, boiler, and duct (wrap and liner).
- Be careful when looking up mechanical items to get the correct pressure rating and connection type (thread, weld, flange).

23 70 00 Central HVAC Equipment

- Combination heating and cooling units are sized by the air conditioning requirements. (See Reference No. R236000-20 for the preliminary sizing guide.)
- A ton of air conditioning is nominally 400 CFM.
- Rectangular duct is taken off by the linear foot for each size, but its cost is usually estimated by the pound. Remember that SMACNA standards now base duct on internal pressure.
- Prefabricated duct is estimated and purchased like pipe: straight sections and fittings.
- Note that cranes or other lifting equipment are not included on any lines in Division 23. For example, if a crane is required to lift a heavy piece of pipe into place high above a gym floor, or to put a rooftop unit on the roof of a four-story building, etc., it must be added. Due to the potential for extreme variation—from nothing additional required to a major crane or helicopter—we feel that including a nominal amount for "lifting contingency" would be useless and detract from the accuracy of the estimate. When using equipment rental cost data from RSMeans, do not forget to include the cost of the operator(s).

Reference Numbers

Reference numbers are shown at the beginning of some major classifications. These numbers refer to related items in the Reference Section. The reference information may be an estimating procedure, an alternate pricing method, or technical information.

Note: Not all subdivisions listed here necessarily appear. ■

Note: "Powered in part by CINX™, based on licensed proprietary information of Harrison Publishing House, Inc."

Note: Trade Service, in part, has been used as a reference source for some of the material prices used in Division 23.

23 05 Common Work Results for HVAC

23 05 02 – HVAC General

23 05 02.10 Air Conditioning, General		Crew	Daily Output	Labor-Hours	Unit	Material	Labor	Equipment	Total	Total Incl O&P
						2020 Bare Costs				
0010	**AIR CONDITIONING, GENERAL** Prices are for standard efficiencies (SEER 13)									
0020	for upgrade to SEER 14 add					10%				

23 05 05 – Selective Demolition for HVAC

23 05 05.10 HVAC Demolition

23 05 05.10 HVAC Demolition		Crew	Daily Output	Labor-Hours	Unit	Material	Labor	Equipment	Total	Total Incl O&P
0010	**HVAC DEMOLITION** R220105-10									
0100	Air conditioner, split unit, 3 ton	Q-5	2	8	Ea.		470		470	705
0150	Package unit, 3 ton R024119-10	Q-6	3	8			490		490	735
0190	Rooftop, self contained, up to 5 ton	1 Plum	1.20	6.667			430		430	645
0250	Air curtain	Q-9	20	.800	L.F.		45		45	68
0254	Air filters, up thru 16,000 CFM		20	.800	Ea.		45		45	68
0256	20,000 thru 60,000 CFM		16	1			56		56	85
0297	Boiler blowdown	Q-5	8	2			118		118	177
0298	Boilers									
0300	Electric, up thru 148 kW	Q-19	2	12	Ea.		715		715	1,075
0310	150 thru 518 kW	"	1	24			1,425		1,425	2,150
0320	550 thru 2,000 kW	Q-21	.40	80			4,900		4,900	7,325
0330	2,070 kW and up	"	.30	107			6,525		6,525	9,750
0340	Gas and/or oil, up thru 150 MBH	Q-7	2.20	14.545			910		910	1,350
0350	160 thru 2,000 MBH		.80	40			2,500		2,500	3,725
0360	2,100 thru 4,500 MBH		.50	64			4,000		4,000	5,975
0370	4,600 thru 7,000 MBH		.30	107			6,650		6,650	9,950
0380	7,100 thru 12,000 MBH		.16	200			12,500		12,500	18,700
0390	12,200 thru 25,000 MBH		.12	267			16,600		16,600	24,900
0400	Central station air handler unit, up thru 15 ton	Q-5	1.60	10			590		590	885
0410	17.5 thru 30 ton	"	.80	20			1,175		1,175	1,775
0430	Computer room unit									
0434	Air cooled split, up thru 10 ton	Q-5	.67	23.881	Ea.		1,400		1,400	2,100
0436	12 thru 23 ton		.53	30.189			1,775		1,775	2,675
0440	Chilled water, up thru 10 ton		1.30	12.308			725		725	1,075
0444	12 thru 23 ton		1	16			945		945	1,425
0450	Glycol system, up thru 10 ton		.53	30.189			1,775		1,775	2,675
0454	12 thru 23 ton		.40	40			2,350		2,350	3,525
0460	Water cooled, not including condenser, up thru 10 ton		.80	20			1,175		1,175	1,775
0464	12 thru 23 ton		.60	26.667			1,575		1,575	2,350
0600	Condenser, up thru 50 ton		1	16			945		945	1,425
0610	51 thru 100 ton	Q-6	.90	26.667			1,625		1,625	2,450
0620	101 thru 1,000 ton	"	.70	34.286			2,100		2,100	3,150
0660	Condensing unit, up thru 10 ton	Q-5	1.25	12.800			755		755	1,125
0670	11 thru 50 ton	"	.40	40			2,350		2,350	3,525
0680	60 thru 100 ton	Q-6	.30	80			4,900		4,900	7,325
0700	Cooling tower, up thru 400 ton		.80	30			1,825		1,825	2,750
0710	450 thru 600 ton		.53	45.283			2,775		2,775	4,150
0720	700 thru 1,300 ton		.40	60			3,675		3,675	5,500
0780	Dehumidifier, up thru 155 lb./hr.	Q-1	8	2			116		116	174
0790	240 lb./hr. and up	"	2	8			465		465	695
1560	Ductwork									
1570	Metal, steel, sst, fabricated	Q-9	1000	.016	Lb.		.90		.90	1.36
1580	Aluminum, fabricated		485	.033	"		1.85		1.85	2.81
1590	Spiral, prefabricated		400	.040	L.F.		2.24		2.24	3.40
1600	Fiberglass, prefabricated		400	.040			2.24		2.24	3.40
1610	Flex, prefabricated		500	.032			1.79		1.79	2.72
1620	Glass fiber reinforced plastic, prefabricated		280	.057			3.20		3.20	4.86

23 05 05.10 HVAC Demolition

23 05 05.10 HVAC Demolition		Crew	Daily Output	Labor-Hours	Unit	Material	2020 Bare Costs Labor	Equipment	Total	Total Incl O&P
1630	Diffusers, registers or grills, up thru 20" max dimension	1 Shee	50	.160	Ea.		9.95		9.95	15.10
1640	21 thru 36" max dimension		36	.222			13.85		13.85	21
1650	Above 36" max dimension		30	.267			16.60		16.60	25
1700	Evaporator, up thru 12,000 BTUH	Q-5	5.30	3.019			178		178	267
1710	12,500 thru 30,000 BTUH	"	2.70	5.926			350		350	525
1720	31,000 BTUH and up	Q-6	1.50	16			980		980	1,475
1730	Evaporative cooler, up thru 5 HP	Q-9	2.70	5.926			330		330	505
1740	10 thru 30 HP	"	.67	23.881			1,350		1,350	2,025
1750	Exhaust systems									
1760	Exhaust components	1 Shee	8	1	System		62.50		62.50	94.50
1770	Weld fume hoods	"	20	.400	Ea.		25		25	38
2120	Fans, up thru 1 HP or 2,000 CFM	Q-9	8	2			112		112	170
2124	1-1/2 thru 10 HP or 20,000 CFM		5.30	3.019			169		169	257
2128	15 thru 30 HP or above 20,000 CFM		4	4			224		224	340
2150	Fan coil air conditioner, chilled water, up thru 7.5 ton	Q-5	14	1.143			67.50		67.50	101
2154	Direct expansion, up thru 10 ton		8	2			118		118	177
2158	11 thru 30 ton		2	8			470		470	705
2170	Flue shutter damper	Q-9	8	2			112		112	170
2200	Furnace, electric	Q-20	2	10			570		570	865
2300	Gas or oil, under 120 MBH	Q-9	4	4			224		224	340
2340	Over 120 MBH	"	3	5.333			299		299	455
2730	Heating and ventilating unit	Q-5	2.70	5.926			350		350	525
2740	Heater, electric, wall, baseboard and quartz	1 Elec	10	.800			49		49	73
2750	Heater, electric, unit, cabinet, fan and convector	"	8	1			61.50		61.50	91.50
2760	Heat exchanger, shell and tube type	Q-5	1.60	10			590		590	885
2770	Plate type	Q-6	.60	40			2,450		2,450	3,675
2810	Heat pump									
2820	Air source, split, 4 thru 10 ton	Q-5	.90	17.778	Ea.		1,050		1,050	1,575
2830	15 thru 25 ton	Q-6	.80	30			1,825		1,825	2,750
2850	Single package, up thru 12 ton	Q-5	1	16			945		945	1,425
2860	Water source, up thru 15 ton	"	.90	17.778			1,050		1,050	1,575
2870	20 thru 50 ton	Q-6	.80	30			1,825		1,825	2,750
2910	Heat recovery package, up thru 20,000 CFM	Q-5	2	8			470		470	705
2920	25,000 CFM and up		1.20	13.333			785		785	1,175
2930	Heat transfer package, up thru 130 GPM		.80	20			1,175		1,175	1,775
2934	255 thru 800 GPM		.42	38.095			2,250		2,250	3,375
2940	Humidifier		10.60	1.509			89		89	133
2961	Hydronic unit heaters, up thru 200 MBH		14	1.143			67.50		67.50	101
2962	Above 200 MBH		8	2			118		118	177
2964	Valance units		32	.500			29.50		29.50	44
2966	Radiant floor heating									
2967	System valves, controls, manifolds	Q-5	16	1	Ea.		59		59	88.50
2968	Per room distribution		8	2			118		118	177
2970	Hydronic heating, baseboard radiation		16	1			59		59	88.50
2976	Convectors and free standing radiators		18	.889			52.50		52.50	78.50
2980	Induced draft fan, up thru 1 HP	Q-9	4.60	3.478			195		195	296
2984	1-1/2 HP thru 7-1/2 HP	"	2.20	7.273			410		410	620
2988	Infrared unit	Q-5	16	1			59		59	88.50
2992	Louvers	1 Shee	46	.174	S.F.		10.85		10.85	16.45
3000	Mechanical equipment, light items. Unit is weight, not cooling.	Q-5	.90	17.778	Ton		1,050		1,050	1,575
3600	Heavy items		1.10	14.545	"		860		860	1,275
3720	Make-up air unit, up thru 6,000 CFM		3	5.333	Ea.		315		315	470
3730	6,500 thru 30,000 CFM		1.60	10			590		590	885

23 05 05 – Selective Demolition for HVAC

23 05 05.10 HVAC Demolition		Crew	Daily Output	Labor-Hours	Unit	Material	2020 Bare Costs Labor	Equipment	Total	Total Incl O&P
3740	35,000 thru 75,000 CFM	Q-6	1	24	Ea.		1,475		1,475	2,200
3800	Mixing boxes, constant and VAV	Q-9	18	.889			50		50	75.50
4000	Packaged terminal air conditioner, up thru 18,000 BTUH	Q-5	8	2			118		118	177
4010	24,000 thru 48,000 BTUH	"	2.80	5.714			335		335	505
5000	Refrigerant compressor, reciprocating or scroll									
5010	Up thru 5 ton	1 Stpi	6	1.333	Ea.		87.50		87.50	131
5020	5.08 thru 10 ton	Q-5	6	2.667			157		157	235
5030	15 thru 50 ton	"	3	5.333			315		315	470
5040	60 thru 130 ton	Q-6	2.80	8.571			525		525	785
5090	Remove refrigerant from system	1 Stpi	40	.200	Lb.		13.10		13.10	19.60
5100	Rooftop air conditioner, up thru 10 ton	Q-5	1.40	11.429	Ea.		675		675	1,000
5110	12 thru 40 ton	Q-6	1	24			1,475		1,475	2,200
5120	50 thru 140 ton		.50	48			2,925		2,925	4,400
5130	150 thru 300 ton		.30	80			4,900		4,900	7,325
6000	Self contained single package air conditioner, up thru 10 ton	Q-5	1.60	10			590		590	885
6010	15 thru 60 ton	Q-6	1.20	20			1,225		1,225	1,825
6100	Space heaters, up thru 200 MBH	Q-5	10	1.600			94.50		94.50	141
6110	Over 200 MBH		5	3.200			189		189	283
6200	Split ductless, both sections		8	2			118		118	177
6300	Steam condensate meter	1 Stpi	11	.727			47.50		47.50	71.50
6600	Thru-the-wall air conditioner	L-2	8	2			93		93	141
7000	Vent chimney, prefabricated, up thru 12″ diameter	Q-9	94	.170	V.L.F.		9.55		9.55	14.50
7010	14″ thru 36″ diameter		40	.400			22.50		22.50	34
7020	38″ thru 48″ diameter		32	.500			28		28	42.50
7030	54″ thru 60″ diameter	Q-10	14	1.714			99.50		99.50	151
7400	Ventilators, up thru 14″ neck diameter	Q-9	58	.276	Ea.		15.45		15.45	23.50
7410	16″ thru 50″ neck diameter		40	.400			22.50		22.50	34
7450	Relief vent, up thru 24″ x 96″		22	.727			41		41	62
7460	48″ x 60″ thru 96″ x 144″		10	1.600			89.50		89.50	136
8000	Water chiller up thru 10 ton	Q-5	2.50	6.400			380		380	565
8010	15 thru 100 ton	Q-6	.48	50			3,050		3,050	4,575
8020	110 thru 500 ton	Q-7	.29	110			6,875		6,875	10,300
8030	600 thru 1000 ton		.23	139			8,675		8,675	13,000
8040	1100 ton and up		.20	160			9,975		9,975	14,900
8400	Window air conditioner	1 Carp	16	.500			26.50		26.50	40
8401	HVAC demo, water chiller, 1100 ton and up	"	12	.667	Ton		35.50		35.50	53

23 05 23 – General-Duty Valves for HVAC Piping

23 05 23.20 Valves, Bronze/Brass		Crew	Daily Output	Labor-Hours	Unit	Material	Labor	Equipment	Total	Total Incl O&P
0010	**VALVES, BRONZE/BRASS**									
0020	Brass									
1300	Ball combination valves, shut-off and union									
1310	Solder, with strainer, drain and PT ports									
1320	1/2″	1 Stpi	17	.471	Ea.	70	31		101	123
1330	3/4″		16	.500		76.50	33		109.50	134
1340	1″		14	.571		104	37.50		141.50	171
1350	1-1/4″		12	.667		133	43.50		176.50	212
1360	1-1/2″		10	.800		195	52.50		247.50	294
1370	2″		8	1		245	65.50		310.50	370
1410	Threaded, with strainer, drain and PT ports									
1420	1/2″	1 Stpi	20	.400	Ea.	70	26		96	116
1430	3/4″		18	.444		76.50	29		105.50	128
1440	1″		17	.471		104	31		135	161

23 05 23.20 Valves, Bronze/Brass		Crew	Daily Output	Labor-Hours	Unit	2020 Bare Costs Material	Labor	Equipment	Total	Total Incl O&P
1450	1-1/4"	1 Stpi	12	.667	Ea.	133	43.50		176.50	212
1460	1-1/2"		11	.727		195	47.50		242.50	287
1470	2"		9	.889		245	58.50		303.50	355

23 05 23.30 Valves, Iron Body		Crew	Daily Output	Labor-Hours	Unit	Material	Labor	Equipment	Total	Total Incl O&P
0010	**VALVES, IRON BODY** R220523-90									
0022	For grooved joint, see Section 22 11 13.48									
0100	Angle, 125 lb.									
0110	Flanged									
0116	2"	1 Plum	5	1.600	Ea.	1,525	103		1,628	1,825
0118	4"	Q-1	3	5.333		2,600	310		2,910	3,350
0120	6"	Q-2	3	8		5,700	480		6,180	7,000
0122	8"	"	2.50	9.600		8,150	575		8,725	9,825
1020	Butterfly, wafer type, gear actuator, 200 lb.									
1030	2"	1 Plum	14	.571	Ea.	118	37		155	185
1040	2-1/2"	Q-1	9	1.778		125	103		228	291
1050	3"		8	2		124	116		240	310
1060	4"		5	3.200		139	186		325	430
1070	5"	Q-2	5	4.800		157	289		446	600
1080	6"		5	4.800		177	289		466	625
1090	8"		4.50	5.333		220	320		540	720
1100	10"		4	6		295	360		655	865
1110	12"		3	8		685	480		1,165	1,475
1200	Wafer type, lever actuator, 200 lb.									
1220	2"	1 Plum	14	.571	Ea.	203	37		240	278
1230	2-1/2"	Q-1	9	1.778		206	103		309	380
1240	3"		8	2		258	116		374	455
1250	4"		5	3.200		265	186		451	570
1260	5"	Q-2	5	4.800		395	289		684	865
1270	6"		5	4.800		445	289		734	920
1280	8"		4.50	5.333		700	320		1,020	1,250
1290	10"		4	6		740	360		1,100	1,350
1300	12"		3	8		1,250	480		1,730	2,100
1650	Gate, 125 lb., N.R.S.									
2150	Flanged									
2200	2"	1 Plum	5	1.600	Ea.	610	103		713	830
2240	2-1/2"	Q-1	5	3.200		630	186		816	970
2260	3"		4.50	3.556		710	206		916	1,100
2280	4"		3	5.333		1,000	310		1,310	1,575
2290	5"	Q-2	3.40	7.059		1,725	425		2,150	2,525
2300	6"		3	8		1,725	480		2,205	2,625
2320	8"		2.50	9.600		2,950	575		3,525	4,125
2340	10"		2.20	10.909		5,175	655		5,830	6,675
2360	12"		1.70	14.118		7,125	850		7,975	9,100
2420	For 250 lb. flanged, add					200%	10%			
3550	OS&Y, 125 lb., flanged									
3600	2"	1 Plum	5	1.600	Ea.	400	103		503	595
3640	2-1/2"	Q-1	5	3.200		430	186		616	750
3660	3"		4.50	3.556		460	206		666	820
3670	3-1/2"		3	5.333		770	310		1,080	1,300
3680	4"		3	5.333		685	310		995	1,225
3690	5"	Q-2	3.40	7.059		1,075	425		1,500	1,825
3700	6"		3	8		1,075	480		1,555	1,925

23 05 23 – General-Duty Valves for HVAC Piping

23 05 23.30 Valves, Iron Body		Crew	Daily Output	Labor-Hours	Unit	Material	Labor	Equipment	Total	Total Incl O&P
						2020 Bare Costs				
3720	8"	Q-2	2.50	9.600	Ea.	1,925	575		2,500	3,000
3740	10"		2.20	10.909		3,525	655		4,180	4,850
3760	12"		1.70	14.118		4,800	850		5,650	6,550
3900	For 175 lb., flanged, add					200%	10%			
4350	Globe, OS&Y									
4540	Class 125, flanged									
4550	2"	1 Plum	5	1.600	Ea.	935	103		1,038	1,175
4560	2-1/2"	Q-1	5	3.200		940	186		1,126	1,300
4570	3"		4.50	3.556		1,150	206		1,356	1,550
4580	4"		3	5.333		1,625	310		1,935	2,275
4590	5"	Q-2	3.40	7.059		2,975	425		3,400	3,900
4600	6"		3	8		2,975	480		3,455	4,000
4610	8"		2.50	9.600		5,850	575		6,425	7,300
4612	10"		2.20	10.909		11,500	655		12,155	13,700
4614	12"		1.70	14.118		11,000	850		11,850	13,400
5040	Class 250, flanged									
5050	2"	1 Plum	4.50	1.778	Ea.	1,500	115		1,615	1,825
5060	2-1/2"	Q-1	4.50	3.556		1,950	206		2,156	2,450
5070	3"		4	4		2,025	232		2,257	2,575
5080	4"		2.70	5.926		2,950	345		3,295	3,775
5090	5"	Q-2	3	8		6,225	480		6,705	7,575
5100	6"		2.70	8.889		6,475	535		7,010	7,925
5110	8"		2.20	10.909		8,975	655		9,630	10,900
5120	10"		2	12		12,900	720		13,620	15,300
5130	12"		1.60	15		24,200	900		25,100	28,100
5240	Valve sprocket rim w/chain, for 2" valve	1 Stpi	30	.267		185	17.50		202.50	229
5250	2-1/2" valve		27	.296		180	19.40		199.40	227
5260	3-1/2" valve		25	.320		194	21		215	245
5270	6" valve		20	.400		192	26		218	251
5280	8" valve		18	.444		225	29		254	291
5290	12" valve		16	.500		237	33		270	310
5300	16" valve		12	.667		258	43.50		301.50	350
5310	20" valve		10	.800		258	52.50		310.50	365
5320	36" valve		8	1		405	65.50		470.50	545
5450	Swing check, 125 lb., threaded									
5500	2"	1 Plum	11	.727	Ea.	575	47		622	700
5540	2-1/2"	Q-1	15	1.067		710	62		772	875
5550	3"		13	1.231		770	71.50		841.50	955
5560	4"		10	1.600		1,225	93		1,318	1,500
5950	Flanged									
5994	1"	1 Plum	7	1.143	Ea.	289	73.50		362.50	430
5998	1-1/2"		6	1.333		475	86		561	655
6000	2"		5	1.600		510	103		613	720
6040	2-1/2"	Q-1	5	3.200		365	186		551	680
6050	3"		4.50	3.556		490	206		696	850
6060	4"		3	5.333		785	310		1,095	1,325
6070	6"	Q-2	3	8		1,325	480		1,805	2,175
6080	8"		2.50	9.600		2,525	575		3,100	3,650
6090	10"		2.20	10.909		4,275	655		4,930	5,675
6100	12"		1.70	14.118		7,000	850		7,850	8,975
6102	14"		1.55	15.484		13,400	930		14,330	16,200
6104	16"		1.40	17.143		17,000	1,025		18,025	20,300
6110	18"		1.30	18.462		25,900	1,100		27,000	30,200

23 05 23 – General-Duty Valves for HVAC Piping

23 05 23.30 Valves, Iron Body		Crew	Daily Output	Labor-Hours	Unit	2020 Bare Costs Material	Labor	Equipment	Total	Total Incl O&P
6112	20"	Q-2	1	24	Ea.	49,000	1,450		50,450	56,000
6114	24"		.75	32		66,000	1,925		67,925	75,500
6160	For 250 lb. flanged, add					200%	20%			
6600	Silent check, bronze trim									
6610	Compact wafer type, for 125 or 150 lb. flanges									
6630	1-1/2"	1 Plum	11	.727	Ea.	162	47		209	248
6640	2"	"	9	.889		198	57.50		255.50	305
6650	2-1/2"	Q-1	9	1.778		215	103		318	390
6660	3"		8	2		236	116		352	435
6670	4"		5	3.200		290	186		476	600
6680	5"	Q-2	6	4		390	241		631	785
6690	6"		6	4		530	241		771	945
6700	8"		4.50	5.333		885	320		1,205	1,450
6710	10"		4	6		1,550	360		1,910	2,250
6720	12"		3	8		2,875	480		3,355	3,875
6740	For 250 or 300 lb. flanges, thru 6" no change									
6741	For 8" and 10", add				Ea.	11%	10%			
6750	Twin disc									
6752	2"	1 Plum	9	.889	Ea.	425	57.50		482.50	550
6754	4"	Q-1	5	3.200		705	186		891	1,050
6756	6"	Q-2	5	4.800		1,050	289		1,339	1,575
6758	8"		4.50	5.333		1,600	320		1,920	2,250
6760	10"		4	6		2,575	360		2,935	3,375
6762	12"		3	8		3,325	480		3,805	4,400
6764	18"		1.50	16		14,400	960		15,360	17,300
6766	24"		.75	32		20,300	1,925		22,225	25,200
6800	Full flange type, 150 lb.									
6900	Globe type, 125 lb.									
6911	2-1/2"	Q-1	9	1.778	Ea.	490	103		593	695
6912	3"		8	2		440	116		556	655
6913	4"		5	3.200		770	186		956	1,125
6914	5"	Q-2	6	4		1,100	241		1,341	1,550
6915	6"		5	4.800		1,350	289		1,639	1,900
6916	8"		4.50	5.333		2,450	320		2,770	3,175
6917	10"		4	6		3,100	360		3,460	3,950
6918	12"		3	8		5,325	480		5,805	6,600
6940	For 250 lb., add					40%	10%			
6980	Screwed end type, 125 lb.									
6981	1"	1 Plum	19	.421	Ea.	97.50	27		124.50	148
6982	1-1/4"		15	.533		114	34.50		148.50	177
6983	1-1/2"		13	.615		139	39.50		178.50	213
6984	2"		11	.727		190	47		237	279

23 05 23.70 Valves, Semi-Steel		Crew	Daily Output	Labor-Hours	Unit	Material	Labor	Equipment	Total	Total Incl O&P
0010	**VALVES, SEMI-STEEL** R220523-90									
1020	Lubricated plug valve, threaded, 200 psi									
1030	1/2"	1 Plum	18	.444	Ea.	119	28.50		147.50	174
1040	3/4"		16	.500		125	32		157	186
1050	1"		14	.571		135	37		172	203
1060	1-1/4"		12	.667		159	43		202	240
1070	1-1/2"		11	.727		154	47		201	240
1080	2"		8	1		208	64.50		272.50	325
1090	2-1/2"	Q-1	5	3.200		299	186		485	610

23 05 23.70 Valves, Semi-Steel		Crew	Daily Output	Labor-Hours	Unit	Material	2020 Bare Costs Labor	Equipment	Total	Total Incl O&P
1100	3"	Q-1	4.50	3.556	Ea.	425	206		631	780
6990	Flanged, 200 psi									
7000	2"	1 Plum	8	1	Ea.	277	64.50		341.50	400
7010	2-1/2"	Q-1	5	3.200		390	186		576	710
7020	3"		4.50	3.556		470	206		676	830
7030	4"		3	5.333		585	310		895	1,100
7036	5"	↓	2.50	6.400		1,325	370		1,695	2,000
7040	6"	Q-2	3	8		1,675	480		2,155	2,550
7050	8"		2.50	9.600		2,225	575		2,800	3,325
7060	10"		2.20	10.909		3,650	655		4,305	5,000
7070	12"	↓	1.70	14.118	↓	5,625	850		6,475	7,475

23 05 23.90 Valves, Stainless Steel		Crew	Daily Output	Labor-Hours	Unit	Material	2020 Bare Costs Labor	Equipment	Total	Total Incl O&P
0010	**VALVES, STAINLESS STEEL** R220523-90									
1700	Check, 200 lb., threaded									
1710	1/4"	1 Plum	24	.333	Ea.	145	21.50		166.50	191
1720	1/2"		22	.364		145	23.50		168.50	194
1730	3/4"		20	.400		149	26		175	203
1750	1"		19	.421		194	27		221	255
1760	1-1/2"		13	.615		365	39.50		404.50	460
1770	2"	↓	11	.727	↓	620	47		667	750
1800	150 lb., flanged									
1810	2-1/2"	Q-1	5	3.200	Ea.	1,925	186		2,111	2,400
1820	3"		4.50	3.556		1,975	206		2,181	2,475
1830	4"	↓	3	5.333		2,950	310		3,260	3,700
1840	6"	Q-2	3	8		5,200	480		5,680	6,450
1850	8"	"	2.50	9.600	↓	10,600	575		11,175	12,600
2100	Gate, OS&Y, 150 lb., flanged									
2120	1/2"	1 Plum	18	.444	Ea.	515	28.50		543.50	610
2140	3/4"		16	.500		495	32		527	595
2150	1"		14	.571		625	37		662	745
2160	1-1/2"		11	.727		1,200	47		1,247	1,400
2170	2"	↓	8	1		1,425	64.50		1,489.50	1,675
2180	2-1/2"	Q-1	5	3.200		1,800	186		1,986	2,275
2190	3"		4.50	3.556		1,950	206		2,156	2,450
2200	4"		3	5.333		2,675	310		2,985	3,425
2205	5"	↓	2.80	5.714		5,125	330		5,455	6,150
2210	6"	Q-2	3	8		5,000	480		5,480	6,225
2220	8"		2.50	9.600		8,675	575		9,250	10,400
2230	10"		2.30	10.435		15,200	630		15,830	17,600
2240	12"	↓	1.90	12.632		20,200	760		20,960	23,400
2260	For 300 lb., flanged, add				↓	120%	15%			
2600	600 lb., flanged									
2620	1/2"	1 Plum	16	.500	Ea.	184	32		216	250
2640	3/4"		14	.571		199	37		236	274
2650	1"		12	.667		239	43		282	330
2660	1-1/2"		10	.800		380	51.50		431.50	495
2670	2"	↓	7	1.143		525	73.50		598.50	690
2680	2-1/2"	Q-1	4	4		6,625	232		6,857	7,650
2690	3"	"	3.60	4.444	↓	6,625	258		6,883	7,675
3100	Globe, OS&Y, 150 lb., flanged									
3120	1/2"	1 Plum	18	.444	Ea.	495	28.50		523.50	585
3140	3/4"	↓	16	.500	↓	535	32		567	640

23 05 23 – General-Duty Valves for HVAC Piping

23 05 23.90 Valves, Stainless Steel		Crew	Daily Output	Labor-Hours	Unit	Material	2020 Bare Costs Labor	Equipment	Total	Total Incl O&P
3150	1"	1 Plum	14	.571	Ea.	700	37		737	825
3160	1-1/2"		11	.727		1,100	47		1,147	1,275
3170	2"		8	1		1,425	64.50		1,489.50	1,675
3180	2-1/2"	Q-1	5	3.200		3,075	186		3,261	3,650
3190	3"		4.50	3.556		3,075	206		3,281	3,675
3200	4"		3	5.333		4,925	310		5,235	5,875
3210	6"	Q-2	3	8		8,275	480		8,755	9,825

23 05 23.94 Hospital Type Valves		Crew	Daily Output	Labor-Hours	Unit	Material	Labor	Equipment	Total	Total Incl O&P
0010	**HOSPITAL TYPE VALVES**									
0300	Chiller valves									
0330	Manual operation balancing valve									
0340	2-1/2" line size	Q-1	9	1.778	Ea.	620	103		723	835
0350	3" line size		8	2		665	116		781	905
0360	4" line size		5	3.200		920	186		1,106	1,275
0370	5" line size	Q-2	5	4.800		1,150	289		1,439	1,700
0380	6" line size		5	4.800		1,425	289		1,714	2,000
0390	8" line size		4.50	5.333		3,100	320		3,420	3,900
0400	10" line size		4	6		5,075	360		5,435	6,125
0410	12" line size		3	8		6,725	480		7,205	8,125
0600	Automatic flow limiting valve									
0620	2-1/2" line size	Q-1	8	2	Ea.	620	116		736	860
0630	3" line size		7	2.286		1,000	133		1,133	1,300
0640	4" line size		6	2.667		1,425	155		1,580	1,800
0645	5" line size	Q-2	5	4.800		1,800	289		2,089	2,400
0650	6" line size		5	4.800		2,400	289		2,689	3,050
0660	8" line size		4.50	5.333		3,600	320		3,920	4,425
0670	10" line size		4	6		5,625	360		5,985	6,750
0680	12" line size		3	8		9,550	480		10,030	11,200

23 05 93 – Testing, Adjusting, and Balancing for HVAC

23 05 93.50 Piping, Testing		Crew	Daily Output	Labor-Hours	Unit	Material	Labor	Equipment	Total	Total Incl O&P
0010	**PIPING, TESTING**									
0100	Nondestructive testing									
0120	1" - 4" pipe									
0140	0-250 L.F.	1 Stpi	1.33	6.015	Ea.		395		395	590
0160	250-500 L.F.	"	.80	10			655		655	980
0180	500-1000 L.F.	Q-5	1.14	14.035			830		830	1,250
0200	1000-2000 L.F.		.80	20			1,175		1,175	1,775
0320	0-250 L.F.		1	16			945		945	1,425
0340	250-500 L.F.		.73	21.918			1,300		1,300	1,925
0360	500-1000 L.F.		.53	30.189			1,775		1,775	2,675
0380	1000-2000 L.F.		.38	42.105			2,475		2,475	3,725
1000	Pneumatic pressure test, includes soaping joints									
1120	1" - 4" pipe									
1140	0-250 L.F.	Q-5	2.67	5.993	Ea.	12.30	355		367.30	545
1160	250-500 L.F.		1.33	12.030		24.50	710		734.50	1,075
1180	500-1000 L.F.		.80	20		37	1,175		1,212	1,825
1200	1000-2000 L.F.		.50	32		49	1,900		1,949	2,875
1300	6" - 10" pipe									
1320	0-250 L.F.	Q-5	1.33	12.030	Ea.	12.30	710		722.30	1,075
1340	250-500 L.F.		.67	23.881		24.50	1,400		1,424.50	2,125
1360	500-1000 L.F.		.40	40		49	2,350		2,399	3,575
1380	1000-2000 L.F.		.25	64		61.50	3,775		3,836.50	5,725

23 05 Common Work Results for HVAC

23 05 93 – Testing, Adjusting, and Balancing for HVAC

23 05 93.50	Piping, Testing	Crew	Daily Output	Labor-Hours	Unit	2020 Bare Costs Material	Labor	Equipment	Total	Total Incl O&P
2110	2" diam.	1 Stpi	8	1	Ea.	17.25	65.50		82.75	117
2120	3" diam.		8	1		17.25	65.50		82.75	117
2130	4" diam.		8	1		26	65.50		91.50	127
2140	6" diam.		8	1		26	65.50		91.50	127
2150	8" diam.		6.60	1.212		26	79.50		105.50	148
2160	10" diam.		6	1.333		34.50	87.50		122	169
3000	Liquid penetration of welds									
3110	2" diam.	1 Stpi	14	.571	Ea.	13.05	37.50		50.55	70.50
3120	3" diam.		13.60	.588		13.05	38.50		51.55	72
3130	4" diam.		13.40	.597		13.05	39		52.05	73
3140	6" diam.		13.20	.606		13.05	39.50		52.55	74
3150	8" diam.		13	.615		19.60	40.50		60.10	82
3160	10" diam.		12.80	.625		19.60	41		60.60	83

23 07 HVAC Insulation

23 07 13 – Duct Insulation

23 07 13.10	Duct Thermal Insulation		Crew	Daily Output	Labor-Hours	Unit	Material	Labor	Equipment	Total	Total Incl O&P
0010	**DUCT THERMAL INSULATION**										
0110	Insulation req'd. is based on the surface size/area to be covered										
3730	Sheet insulation										
3760	Polyethylene foam, closed cell, UV resistant										
3770	Standard temperature (-90°F to +212°F)										
3771	1/4" thick	G	Q-14	450	.036	S.F.	1.83	1.88		3.71	4.89
3772	3/8" thick	G		440	.036		2.61	1.92		4.53	5.80
3773	1/2" thick	G		420	.038		3.21	2.01		5.22	6.60
3774	3/4" thick	G		400	.040		4.58	2.11		6.69	8.30
3775	1" thick	G		380	.042		6.20	2.22		8.42	10.20
3776	1-1/2" thick	G		360	.044		9.80	2.35		12.15	14.35
3777	2" thick	G		340	.047		12.95	2.48		15.43	18.05
3778	2-1/2" thick	G		320	.050		16.60	2.64		19.24	22.50
3779	Adhesive (see line 7878)										
3780	Foam, rubber										
3782	1" thick	G	1 Stpi	50	.160	S.F.	3.05	10.50		13.55	19.05
7000	Board insulation										
7020	Mineral wool, 1200° F										
7022	6 lb. density, plain										
7024	1" thick	G	Q-14	370	.043	S.F.	.34	2.28		2.62	3.87
7026	1-1/2" thick	G		350	.046		.51	2.41		2.92	4.26
7028	2" thick	G		330	.048		.68	2.56		3.24	4.68
7030	3" thick	G		300	.053		.84	2.81		3.65	5.25
7032	4" thick	G		280	.057		1.37	3.02		4.39	6.15
7038	8 lb. density, plain										
7040	1" thick	G	Q-14	360	.044	S.F.	.42	2.35		2.77	4.06
7042	1-1/2" thick	G		340	.047		.61	2.48		3.09	4.48
7044	2" thick	G		320	.050		.83	2.64		3.47	4.96
7046	3" thick	G		290	.055		1.25	2.91		4.16	5.85
7048	4" thick	G		270	.059		1.62	3.13		4.75	6.60
7060	10 lb. density, plain										
7062	1" thick	G	Q-14	350	.046	S.F.	.58	2.41		2.99	4.34
7064	1-1/2" thick	G		330	.048		.87	2.56		3.43	4.89
7066	2" thick	G		310	.052		1.16	2.72		3.88	5.45

23 07 HVAC Insulation

23 07 13 – Duct Insulation

23 07 13.10 Duct Thermal Insulation		Crew	Daily Output	Labor-Hours	Unit	Material	2020 Bare Costs Labor	Equipment	Total	Total Incl O&P
7068	3″ thick G	Q-14	280	.057	S.F.	1.75	3.02		4.77	6.55
7070	4″ thick G		260	.062		2.22	3.25		5.47	7.45
7878	Contact cement, quart can				Ea.	13.20			13.20	14.55

23 07 16 – HVAC Equipment Insulation

23 07 16.10 HVAC Equipment Thermal Insulation		Crew	Daily Output	Labor-Hours	Unit	Material	Labor	Equipment	Total	Total Incl O&P
0010	**HVAC EQUIPMENT THERMAL INSULATION**									
0110	Insulation req'd. is based on the surface size/area to be covered									
1000	Boiler, 1-1/2″ calcium silicate only G	Q-14	110	.145	S.F.	4.47	7.70		12.17	16.70
1020	Plus 2″ fiberglass G	″	80	.200	″	6.05	10.55		16.60	23
2000	Breeching, 2″ calcium silicate									
2020	Rectangular G	Q-14	42	.381	S.F.	8.65	20		28.65	40.50
2040	Round G	″	38.70	.413	″	9	22		31	43.50
2300	Calcium silicate block, +200°F to +1,200°F									
2310	On irregular surfaces, valves and fittings									
2340	1″ thick G	Q-14	30	.533	S.F.	3.90	28		31.90	47.50
2360	1-1/2″ thick G		25	.640		4.40	34		38.40	57
2380	2″ thick G		22	.727		4.19	38.50		42.69	63.50
2400	3″ thick G		18	.889		6.45	47		53.45	79
2410	On plane surfaces									
2420	1″ thick G	Q-14	126	.127	S.F.	3.90	6.70		10.60	14.60
2430	1-1/2″ thick G		120	.133		4.40	7.05		11.45	15.65
2440	2″ thick G		100	.160		4.19	8.45		12.64	17.55
2450	3″ thick G		70	.229		6.45	12.05		18.50	25.50

23 09 Instrumentation and Control for HVAC

23 09 13 – Instrumentation and Control Devices for HVAC

23 09 13.60 Water Level Controls		Crew	Daily Output	Labor-Hours	Unit	Material	Labor	Equipment	Total	Total Incl O&P
0010	**WATER LEVEL CONTROLS**									
3000	Low water cut-off for hot water boiler, 50 psi maximum									
3100	1″ top & bottom equalizing pipes, manual reset	1 Stpi	14	.571	Ea.	460	37.50		497.50	560
3200	1″ top & bottom equalizing pipes		14	.571		420	37.50		457.50	515
3300	2-1/2″ side connection for nipple-to-boiler		14	.571		415	37.50		452.50	510

23 09 53 – Pneumatic and Electric Control System for HVAC

23 09 53.10 Control Components		Crew	Daily Output	Labor-Hours	Unit	Material	Labor	Equipment	Total	Total Incl O&P
0010	**CONTROL COMPONENTS** R230500-10									
0680	Controller for VAV box, includes actuator	1 Stpi	7.30	1.096	Ea.	281	72		353	420
2000	Gauges, pressure or vacuum									
2100	2″ diameter dial	1 Stpi	32	.250	Ea.	10.55	16.40		26.95	36
2200	2-1/2″ diameter dial		32	.250		12.80	16.40		29.20	38.50
2300	3-1/2″ diameter dial		32	.250		18.15	16.40		34.55	44.50
2400	4-1/2″ diameter dial		32	.250		20	16.40		36.40	46.50
2700	Flanged iron case, black ring									
2800	3-1/2″ diameter dial	1 Stpi	32	.250	Ea.	101	16.40		117.40	137
2900	4-1/2″ diameter dial		32	.250		105	16.40		121.40	140
3000	6″ diameter dial		32	.250		167	16.40		183.40	209
3010	Steel case, 0-300 psi									
3012	2″ diameter dial	1 Stpi	16	.500	Ea.	8.60	33		41.60	58.50
3014	4″ diameter dial	″	16	.500	″	23	33		56	74.50
3020	Aluminum case, 0-300 psi									
3022	3-1/2″ diameter dial	1 Stpi	16	.500	Ea.	12.60	33		45.60	63

23 09 53.10 Control Components		Crew	Daily Output	Labor-Hours	Unit	2020 Bare Costs Material	Labor	Equipment	Total	Total Incl O&P
3024	4-1/2" diameter dial	1 Stpi	16	.500	Ea.	122	33		155	183
3026	6" diameter dial		16	.500		193	33		226	262
3028	8-1/2" diameter dial		16	.500		335	33		368	415
3030	Brass case, 0-300 psi									
3032	2" diameter dial	1 Stpi	16	.500	Ea.	49	33		82	103
3034	4-1/2" diameter dial	"	16	.500	"	104	33		137	163
3040	Steel case, high pressure, 0-10,000 psi									
3042	4-1/2" diameter dial	1 Stpi	16	.500	Ea.	315	33		348	395
3044	6-1/2" diameter dial		16	.500		325	33		358	405
3046	8-1/2" diameter dial		16	.500		430	33		463	525
3080	Pressure gauge, differential, magnehelic									
3084	0-2" W.C., with air filter kit	1 Stpi	6	1.333	Ea.	84.50	87.50		172	224
3300	For compound pressure-vacuum, add					18%				
4000	Thermometers									
4100	Dial type, 3-1/2" diameter, vapor type, union connection	1 Stpi	32	.250	Ea.	226	16.40		242.40	274
4120	Liquid type, union connection		32	.250		435	16.40		451.40	505
4500	Stem type, 6-1/2" case, 2" stem, 1/2" NPT		32	.250		60	16.40		76.40	90.50
4520	4" stem, 1/2" NPT		32	.250		73.50	16.40		89.90	106
4600	9" case, 3-1/2" stem, 3/4" NPT		28	.286		93.50	18.75		112.25	131
4620	6" stem, 3/4" NPT		28	.286		111	18.75		129.75	151
4640	8" stem, 3/4" NPT		28	.286		187	18.75		205.75	234
4660	12" stem, 1" NPT		26	.308		164	20		184	211
4670	Bi-metal, dial type, steel case brass stem									
4672	2" dial, 4" - 9" stem	1 Stpi	16	.500	Ea.	40	33		73	93
4673	2-1/2" dial, 4" - 9" stem		16	.500		40	33		73	93
4674	3-1/2" dial, 4" - 9" stem		16	.500		49	33		82	103
4680	Mercury filled, industrial, union connection type									
4682	Angle stem, 7" scale	1 Stpi	16	.500	Ea.	61	33		94	116
4683	9" scale		16	.500		167	33		200	233
4684	12" scale		16	.500		214	33		247	285
4686	Straight stem, 7" scale		16	.500		163	33		196	228
4687	9" scale		16	.500		145	33		178	209
4688	12" scale		16	.500		171	33		204	237
4690	Mercury filled, industrial, separable socket type, with well									
4692	Angle stem, with socket, 7" scale	1 Stpi	16	.500	Ea.	198	33		231	267
4693	9" scale		16	.500		197	33		230	266
4694	12" scale		16	.500		234	33		267	305
4696	Straight stem, with socket, 7" scale		16	.500		154	33		187	218
4697	9" scale		16	.500		158	33		191	223
4698	12" scale		16	.500		189	33		222	257
6000	Valves, motorized zone									
6100	Sweat connections, 1/2" C x C	1 Stpi	20	.400	Ea.	136	26		162	188
6110	3/4" C x C		20	.400		138	26		164	191
6120	1" C x C		19	.421		185	27.50		212.50	245
6140	1/2" C x C, with end switch, 2 wire		20	.400		137	26		163	190
6150	3/4" C x C, with end switch, 2 wire		20	.400		118	26		144	169
6160	1" C x C, with end switch, 2 wire		19	.421		179	27.50		206.50	238

23 11 Facility Fuel Piping

23 11 13 – Facility Fuel-Oil Piping

23 11 13.10 Fuel Oil Specialties		Crew	Daily Output	Labor-Hours	Unit	2020 Bare Costs Material	Labor	Equipment	Total	Total Incl O&P
0010	**FUEL OIL SPECIALTIES**									
0020	Foot valve, single poppet, metal to metal construction									
0040	Bevel seat, 1/2" diameter	1 Stpi	20	.400	Ea.	80.50	26		106.50	128
0060	3/4" diameter		18	.444		113	29		142	168
1000	Oil filters, 3/8" IPT, 20 gal. per hour		20	.400		43	26		69	86.50
2000	Remote tank gauging system, self contained									
3000	Valve, ball check, globe type, 3/8" diameter	1 Stpi	24	.333	Ea.	13.05	22		35.05	47
3500	Fusible, 3/8" diameter		24	.333		14.70	22		36.70	48.50
3600	1/2" diameter		24	.333		40	22		62	76.50
3610	3/4" diameter		20	.400		79	26		105	126
3620	1" diameter		19	.421		229	27.50		256.50	294
4000	Nonfusible, 3/8" diameter		24	.333		27	22		49	62.50
4500	Shutoff, gate type, lever handle, spring-fusible kit									
4520	1/4" diameter	1 Stpi	14	.571	Ea.	44	37.50		81.50	104
4540	3/8" diameter		12	.667		42.50	43.50		86	113
4560	1/2" diameter		10	.800		60.50	52.50		113	145
4570	3/4" diameter		8	1		113	65.50		178.50	222
5000	Vent alarm, whistling signal					38			38	42
5500	Vent protector/breather, 1-1/4" diameter	1 Stpi	32	.250		22.50	16.40		38.90	49

23 11 23 – Facility Natural-Gas Piping

23 11 23.10 Gas Meters		Crew	Daily Output	Labor-Hours	Unit	Material	Labor	Equipment	Total	Total Incl O&P
0010	**GAS METERS**									
4000	Residential									
4010	Gas meter, residential, 3/4" pipe size	1 Plum	14	.571	Ea.	305	37		342	390
4020	Gas meter, residential, 1" pipe size		12	.667		290	43		333	385
4030	Gas meter, residential, 1-1/4" pipe size		10	.800		254	51.50		305.50	355

23 11 23.20 Gas Piping, Flexible (Csst)		Crew	Daily Output	Labor-Hours	Unit	Material	Labor	Equipment	Total	Total Incl O&P
0010	**GAS PIPING, FLEXIBLE (CSST)**									
0100	Tubing with lightning protection									
0110	3/8"	1 Stpi	65	.123	L.F.	3.22	8.05		11.27	15.60
0120	1/2"		62	.129		3.62	8.45		12.07	16.65
0130	3/4"		60	.133		4.71	8.75		13.46	18.30
0140	1"		55	.145		7.25	9.55		16.80	22
0150	1-1/4"		50	.160		8.40	10.50		18.90	25
0160	1-1/2"		45	.178		14.75	11.65		26.40	33.50
0170	2"		40	.200		21	13.10		34.10	43
0200	Tubing for underground/underslab burial									
0210	3/8"	1 Stpi	65	.123	L.F.	4.48	8.05		12.53	17
0220	1/2"		62	.129		5.10	8.45		13.55	18.30
0230	3/4"		60	.133		6.50	8.75		15.25	20.50
0240	1"		55	.145		8.75	9.55		18.30	24
0250	1-1/4"		50	.160		11.75	10.50		22.25	28.50
0260	1-1/2"		45	.178		22	11.65		33.65	41.50
0270	2"		40	.200		26	13.10		39.10	48.50
3000	Fittings									
3010	Straight									
3100	Tube to NPT									
3110	3/8"	1 Stpi	29	.276	Ea.	13.35	18.10		31.45	41.50
3120	1/2"		27	.296		14.35	19.40		33.75	45
3130	3/4"		25	.320		19.60	21		40.60	53
3140	1"		23	.348		30.50	23		53.50	67.50
3150	1-1/4"		20	.400		68	26		94	114

23 11 Facility Fuel Piping

23 11 23 – Facility Natural-Gas Piping

23 11 23.20 Gas Piping, Flexible (Csst)		Crew	Daily Output	Labor-Hours	Unit	Material	Labor	Equipment	Total	Total Incl O&P
						2020 Bare Costs				
3160	1-1/2"	1 Stpi	17	.471	Ea.	138	31		169	198
3170	2"	↓	15	.533	↓	235	35		270	310
3200	Coupling									
3210	3/8"	1 Stpi	29	.276	Ea.	23	18.10		41.10	52.50
3220	1/2"		27	.296		26.50	19.40		45.90	58
3230	3/4"		25	.320		36.50	21		57.50	71.50
3240	1"		23	.348		56.50	23		79.50	96
3250	1-1/4"		20	.400		125	26		151	177
3260	1-1/2"		17	.471		262	31		293	335
3270	2"	↓	15	.533	↓	445	35		480	545
3300	Flange fitting									
3310	3/8"	1 Stpi	25	.320	Ea.	20.50	21		41.50	54
3320	1/2"		22	.364		19.50	24		43.50	57
3330	3/4"		19	.421		25.50	27.50		53	69.50
3340	1"		16	.500		35	33		68	87.50
3350	1-1/4"	↓	12	.667	↓	81	43.50		124.50	155
3400	90° flange valve									
3410	3/8"	1 Stpi	25	.320	Ea.	39	21		60	74.50
3420	1/2"		22	.364		39.50	24		63.50	79
3430	3/4"	↓	19	.421	↓	48.50	27.50		76	95
4000	Tee									
4120	1/2"	1 Stpi	20.50	.390	Ea.	41.50	25.50		67	84.50
4130	3/4"		19	.421		55.50	27.50		83	103
4140	1"	↓	17.50	.457	↓	99.50	30		129.50	155
5000	Reducing									
5110	Tube to NPT									
5120	3/4" to 1/2" NPT	1 Stpi	26	.308	Ea.	21	20		41	53
5130	1" to 3/4" NPT	"	24	.333	"	32.50	22		54.50	68.50
5200	Reducing tee									
5210	1/2" x 3/8" x 3/8"	1 Stpi	21	.381	Ea.	55	25		80	98
5220	3/4" x 1/2" x 1/2"		20	.400		51.50	26		77.50	96
5230	1" x 3/4" x 1/2"		18	.444		85.50	29		114.50	138
5240	1-1/4" x 1-1/4" x 1"		15.60	.513		228	33.50		261.50	300
5250	1-1/2" x 1-1/2" x 1-1/4"		13.30	.602		350	39.50		389.50	445
5260	2" x 2" x 1-1/2"	↓	11.80	.678	↓	525	44.50		569.50	640
5300	Manifold with four ports and mounting bracket									
5302	Labor to mount manifold does not include making pipe									
5304	connections which are included in fitting labor.									
5310	3/4" x 1/2" x 1/2" (4)	1 Stpi	76	.105	Ea.	42	6.90		48.90	57
5330	1-1/4" x 1" x 3/4" (4)		72	.111		60.50	7.30		67.80	77.50
5350	2" x 1-1/2" x 1" (4)	↓	68	.118	↓	76	7.70		83.70	95
5600	Protective striker plate									
5610	Quarter plate, 3" x 2"	1 Stpi	88	.091	Ea.	.93	5.95		6.88	9.90
5620	Half plate, 3" x 7"		82	.098		2.30	6.40		8.70	12.10
5630	Full plate, 3" x 12"	↓	78	.103	↓	3.92	6.70		10.62	14.35

23 12 Facility Fuel Pumps

23 12 13 – Facility Fuel-Oil Pumps

	23 12 13.10 Pump and Motor Sets	Crew	Daily Output	Labor-Hours	Unit	Material	2020 Bare Costs Labor	Equipment	Total	Total Incl O&P
0010	**PUMP AND MOTOR SETS**									
1810	Light fuel and diesel oils									
1820	20 GPH, 1/3 HP	Q-5	6	2.667	Ea.	1,600	157		1,757	2,000
1850	145 GPH, 1/2 HP	″	4	4	″	1,700	236		1,936	2,225

23 13 Facility Fuel-Storage Tanks

23 13 13 – Facility Underground Fuel-Oil, Storage Tanks

	23 13 13.09 Single-Wall Steel Fuel-Oil Tanks	Crew	Daily Output	Labor-Hours	Unit	Material	Labor	Equipment	Total	Total Incl O&P
0010	**SINGLE-WALL STEEL FUEL-OIL TANKS**									
5000	Tanks, steel ugnd., sti-p3, not incl. hold-down bars									
5500	Excavation, pad, pumps and piping not included									
5510	Single wall, 500 gallon capacity, 7 ga. shell	Q-5	2.70	5.926	Ea.	1,875	350		2,225	2,600
5520	1,000 gallon capacity, 7 ga. shell	″	2.50	6.400		2,925	380		3,305	3,800
5530	2,000 gallon capacity, 1/4″ thick shell	Q-7	4.60	6.957		3,725	435		4,160	4,750
5535	2,500 gallon capacity, 7 ga. shell	Q-5	3	5.333		5,675	315		5,990	6,700
5540	5,000 gallon capacity, 1/4″ thick shell	Q-7	3.20	10		13,400	625		14,025	15,600
5560	10,000 gallon capacity, 1/4″ thick shell		2	16		10,800	1,000		11,800	13,400
5580	15,000 gallon capacity, 5/16″ thick shell		1.70	18.824		11,200	1,175		12,375	14,100
5600	20,000 gallon capacity, 5/16″ thick shell		1.50	21.333		22,600	1,325		23,925	26,900
5610	25,000 gallon capacity, 3/8″ thick shell		1.30	24.615		29,600	1,525		31,125	34,900
5620	30,000 gallon capacity, 3/8″ thick shell		1.10	29.091		35,000	1,825		36,825	41,200
5630	40,000 gallon capacity, 3/8″ thick shell		.90	35.556		44,500	2,225		46,725	52,500
5640	50,000 gallon capacity, 3/8″ thick shell		.80	40		49,800	2,500		52,300	58,000

	23 13 13.13 Dbl-Wall Steel, Undrgrnd Fuel-Oil, Stor. Tanks	Crew	Daily Output	Labor-Hours	Unit	Material	Labor	Equipment	Total	Total Incl O&P
0010	**DOUBLE-WALL STEEL, UNDERGROUND FUEL-OIL, STORAGE TANKS**									
6200	Steel, underground, 360°, double wall, UL listed,									
6210	with sti-P3 corrosion protection,									
6220	(dielectric coating, cathodic protection, electrical									
6230	isolation) 30 year warranty,									
6240	not incl. manholes or hold-downs.									
6250	500 gallon capacity	Q-5	2.40	6.667	Ea.	3,325	395		3,720	4,250
6260	1,000 gallon capacity	″	2.25	7.111		3,675	420		4,095	4,650
6270	2,000 gallon capacity	Q-7	4.16	7.692		4,850	480		5,330	6,050
6280	3,000 gallon capacity		3.90	8.205		6,950	510		7,460	8,425
6290	4,000 gallon capacity		3.64	8.791		8,200	550		8,750	9,850
6300	5,000 gallon capacity		2.91	10.997		8,800	685		9,485	10,700
6310	6,000 gallon capacity		2.42	13.223		11,600	825		12,425	14,000
6320	8,000 gallon capacity		2.08	15.385		11,200	960		12,160	13,700
6330	10,000 gallon capacity		1.82	17.582		13,900	1,100		15,000	17,000
6340	12,000 gallon capacity		1.70	18.824		16,600	1,175		17,775	20,100
6350	15,000 gallon capacity		1.33	24.060		20,700	1,500		22,200	25,100
6360	20,000 gallon capacity		1.33	24.060		34,200	1,500		35,700	40,000
6370	25,000 gallon capacity		1.16	27.586		38,100	1,725		39,825	44,500
6380	30,000 gallon capacity		1.03	31.068		37,800	1,950		39,750	44,500
6390	40,000 gallon capacity		.80	40		115,000	2,500		117,500	129,500
6395	50,000 gallon capacity		.73	43.836		139,500	2,725		142,225	157,500
6400	For hold-downs 500-2,000 gal., add		16	2	Set	226	125		351	435
6410	For hold-downs 3,000-6,000 gal., add		12	2.667		385	166		551	670
6420	For hold-downs 8,000-12,000 gal., add		11	2.909		455	182		637	770
6430	For hold-downs 15,000 gal., add		9	3.556		675	222		897	1,075
6440	For hold-downs 20,000 gal., add		8	4		780	250		1,030	1,225

23 13 13.13 Dbl-Wall Steel, Undrgrnd Fuel-Oil, Stor. Tanks		Crew	Daily Output	Labor-Hours	Unit	Material	2020 Bare Costs Labor	Equipment	Total	Total Incl O&P
6450	For hold-downs 20,000 gal. plus, add	Q-7	6	5.333	Set	1,000	335		1,335	1,600
6500	For manways, add				Ea.	1,050			1,050	1,175
6600	In place with hold-downs									
6652	550 gallon capacity	Q-5	1.84	8.696	Ea.	3,550	515		4,065	4,675

23 13 13.23 Glass-Fiber-Reinfcd-Plastic, Fuel-Oil, Storage		Crew	Daily Output	Labor-Hours	Unit	Material	Labor	Equipment	Total	Total Incl O&P
0010	**GLASS-FIBER-REINFCD-PLASTIC, UNDERGRND. FUEL-OIL, STORAGE**									
0210	Fiberglass, underground, single wall, UL listed, not including									
0220	manway or hold-down strap									
0225	550 gallon capacity	Q-5	2.67	5.993	Ea.	4,075	355		4,430	5,025
0230	1,000 gallon capacity	"	2.46	6.504		5,275	385		5,660	6,375
0240	2,000 gallon capacity	Q-7	4.57	7.002		8,275	435		8,710	9,750
0245	3,000 gallon capacity		3.90	8.205		8,400	510		8,910	10,000
0250	4,000 gallon capacity		3.55	9.014		11,100	560		11,660	13,000
0255	5,000 gallon capacity		3.20	10		10,400	625		11,025	12,300
0260	6,000 gallon capacity		2.67	11.985		11,800	750		12,550	14,100
0270	8,000 gallon capacity		2.29	13.974		14,900	870		15,770	17,700
0280	10,000 gallon capacity		2	16		15,600	1,000		16,600	18,700
0282	12,000 gallon capacity		1.88	17.021		20,900	1,050		21,950	24,600
0284	15,000 gallon capacity		1.68	19.048		27,900	1,200		29,100	32,500
0290	20,000 gallon capacity		1.45	22.069		33,200	1,375		34,575	38,600
0300	25,000 gallon capacity		1.28	25		58,500	1,550		60,050	67,000
0320	30,000 gallon capacity		1.14	28.070		104,500	1,750		106,250	117,500
0340	40,000 gallon capacity		.89	35.955		114,000	2,250		116,250	129,000
0360	48,000 gallon capacity		.81	39.506		177,000	2,475		179,475	198,000
0500	For manway, fittings and hold-downs, add					20%	15%			
0600	For manways, add					2,525			2,525	2,775
1000	For helical heating coil, add	Q-5	2.50	6.400		5,450	380		5,830	6,575
1020	Fiberglass, underground, double wall, UL listed									
1030	includes manways, not incl. hold-down straps									
1040	600 gallon capacity	Q-5	2.42	6.612	Ea.	9,100	390		9,490	10,600
1050	1,000 gallon capacity	"	2.25	7.111		12,400	420		12,820	14,300
1060	2,500 gallon capacity	Q-7	4.16	7.692		17,300	480		17,780	19,700
1070	3,000 gallon capacity		3.90	8.205		19,400	510		19,910	22,100
1080	4,000 gallon capacity		3.64	8.791		19,600	550		20,150	22,300
1090	6,000 gallon capacity		2.42	13.223		26,000	825		26,825	29,800
1100	8,000 gallon capacity		2.08	15.385		33,000	960		33,960	37,700
1110	10,000 gallon capacity		1.82	17.582		38,800	1,100		39,900	44,300
1120	12,000 gallon capacity		1.70	18.824		48,300	1,175		49,475	55,000
1122	15,000 gallon capacity		1.52	21.053		67,000	1,325		68,325	76,000
1124	20,000 gallon capacity		1.33	24.060		79,500	1,500		81,000	90,000
1126	25,000 gallon capacity		1.16	27.586		92,000	1,725		93,725	103,500
1128	30,000 gallon capacity		1.03	31.068		110,500	1,950		112,450	124,500
1140	For hold-down straps, add					2%	10%			
1150	For hold-downs 500-4,000 gal., add	Q-7	16	2	Set	505	125		630	740
1160	For hold-downs 5,000-15,000 gal., add		8	4		1,000	250		1,250	1,475
1170	For hold-downs 20,000 gal., add		5.33	6.004		1,500	375		1,875	2,200
1180	For hold-downs 25,000 gal., add		4	8		2,025	500		2,525	2,975
1190	For hold-downs 30,000 gal., add		2.60	12.308		3,025	770		3,795	4,475
2210	Fiberglass, underground, single wall, UL listed, including									
2220	hold-down straps, no manways									
2225	550 gallon capacity	Q-5	2	8	Ea.	4,600	470		5,070	5,750
2230	1,000 gallon capacity	"	1.88	8.511		5,775	500		6,275	7,100

23 13 Facility Fuel-Storage Tanks

23 13 13 – Facility Underground Fuel-Oil, Storage Tanks

23 13 13.23 Glass-Fiber-Reinfcd-Plastic, Fuel-Oil, Storage		Crew	Daily Output	Labor-Hours	Unit	2020 Bare Costs Material	Labor	Equipment	Total	Total Incl O&P
2240	2,000 gallon capacity	Q-7	3.55	9.014	Ea.	8,775	560		9,335	10,500
2250	4,000 gallon capacity		2.90	11.034		11,600	690		12,290	13,800
2260	6,000 gallon capacity		2	16		12,800	1,000		13,800	15,600
2270	8,000 gallon capacity		1.78	17.978		15,900	1,125		17,025	19,200
2280	10,000 gallon capacity		1.60	20		16,600	1,250		17,850	20,200
2282	12,000 gallon capacity		1.52	21.053		21,900	1,325		23,225	26,100
2284	15,000 gallon capacity		1.39	23.022		28,900	1,425		30,325	34,000
2290	20,000 gallon capacity		1.14	28.070		34,700	1,750		36,450	40,800
2300	25,000 gallon capacity		.96	33.333		60,500	2,075		62,575	69,500
2320	30,000 gallon capacity		.80	40		107,500	2,500		110,000	121,500
3020	Fiberglass, underground, double wall, UL listed									
3030	includes manways and hold-down straps									
3040	600 gallon capacity	Q-5	1.86	8.602	Ea.	9,600	510		10,110	11,400
3050	1,000 gallon capacity	"	1.70	9.412		12,900	555		13,455	15,000
3060	2,500 gallon capacity	Q-7	3.29	9.726		17,800	605		18,405	20,500
3070	3,000 gallon capacity		3.13	10.224		19,900	640		20,540	22,800
3080	4,000 gallon capacity		2.93	10.922		20,100	680		20,780	23,100
3090	6,000 gallon capacity		1.86	17.204		27,000	1,075		28,075	31,300
3100	8,000 gallon capacity		1.65	19.394		34,000	1,200		35,200	39,200
3110	10,000 gallon capacity		1.48	21.622		39,800	1,350		41,150	45,800
3120	12,000 gallon capacity		1.40	22.857		49,300	1,425		50,725	56,000
3122	15,000 gallon capacity		1.28	25		68,000	1,550		69,550	77,500
3124	20,000 gallon capacity		1.06	30.189		81,000	1,875		82,875	92,000
3126	25,000 gallon capacity		.90	35.556		94,000	2,225		96,225	107,000
3128	30,000 gallon capacity		.74	43.243		113,500	2,700		116,200	129,000

23 13 23 – Facility Aboveground Fuel-Oil, Storage Tanks

23 13 23.16 Horizontal, Stl, Abvgrd Fuel-Oil, Storage Tanks		Crew	Daily Output	Labor-Hours	Unit	Material	Labor	Equipment	Total	Total Incl O&P
0010	**HORIZONTAL, STEEL, ABOVEGROUND FUEL-OIL, STORAGE TANKS**									
3000	Steel, storage, aboveground, including cradles, coating,									
3020	fittings, not including foundation, pumps or piping									
3040	Single wall, 275 gallon	Q-5	5	3.200	Ea.	510	189		699	845
3060	550 gallon	"	2.70	5.926		4,400	350		4,750	5,375
3080	1,000 gallon	Q-7	5	6.400		7,175	400		7,575	8,475
3100	1,500 gallon		4.75	6.737		10,100	420		10,520	11,700
3120	2,000 gallon		4.60	6.957		13,500	435		13,935	15,600
3320	Double wall, 500 gallon capacity	Q-5	2.40	6.667		2,000	395		2,395	2,800
3330	2,000 gallon capacity	Q-7	4.15	7.711		6,650	480		7,130	8,050
3340	4,000 gallon capacity		3.60	8.889		14,500	555		15,055	16,700
3350	6,000 gallon capacity		2.40	13.333		16,400	830		17,230	19,400
3360	8,000 gallon capacity		2	16		19,500	1,000		20,500	23,000
3370	10,000 gallon capacity		1.80	17.778		30,500	1,100		31,600	35,200
3380	15,000 gallon capacity		1.50	21.333		41,100	1,325		42,425	47,200
3390	20,000 gallon capacity		1.30	24.615		48,400	1,525		49,925	55,500
3400	25,000 gallon capacity		1.15	27.826		60,000	1,725		61,725	68,500
3410	30,000 gallon capacity		1	32		67,000	2,000		69,000	76,500

23 13 23.26 Horizontal, Conc., Abvgrd Fuel-Oil, Stor. Tanks		Crew	Daily Output	Labor-Hours	Unit	Material	Labor	Equipment	Total	Total Incl O&P
0010	**HORIZONTAL, CONCRETE, ABOVEGROUND FUEL-OIL, STORAGE TANKS**									
0050	Concrete, storage, aboveground, including pad & pump									
0100	500 gallon	F-3	2	20	Ea.	10,500	1,075	236	11,811	13,400
0200	1,000 gallon	"	2	20		14,700	1,075	236	16,011	18,000
0300	2,000 gallon	F-4	2	24		18,900	1,300	490	20,690	23,300

23 21 20.10 Air Control

	23 21 20.10 Air Control	Crew	Daily Output	Labor-Hours	Unit	Material	Labor	Equipment	Total	Total Incl O&P
						2020 Bare Costs				
0010	**AIR CONTROL**									
0030	Air separator, with strainer									
0040	2" diameter	Q-5	6	2.667	Ea.	1,575	157		1,732	1,950
0080	2-1/2" diameter		5	3.200		1,750	189		1,939	2,200
0100	3" diameter		4	4		2,700	236		2,936	3,325
1000	Micro-bubble separator for total air removal, closed loop system									
1010	Requires bladder type tank in system.									
1020	Water (hot or chilled) or glycol system									
1030	Threaded									
1040	3/4" diameter	1 Stpi	20	.400	Ea.	89.50	26		115.50	138
1050	1" diameter		19	.421		104	27.50		131.50	156
1060	1-1/4" diameter		16	.500		139	33		172	202
1070	1-1/2" diameter		13	.615		182	40.50		222.50	261
1080	2" diameter		11	.727		1,225	47.50		1,272.50	1,425
1090	2-1/2" diameter	Q-5	15	1.067		1,475	63		1,538	1,725
1100	3" diameter		13	1.231		1,975	72.50		2,047.50	2,275
1110	4" diameter		10	1.600		2,175	94.50		2,269.50	2,550
1800	With drain/dismantle/cleaning access flange									
1810	Flanged									
1820	2" diameter	Q-5	4.60	3.478	Ea.	3,125	205		3,330	3,750
1830	2-1/2" diameter		4	4		3,375	236		3,611	4,075
1840	3" diameter		3.60	4.444		4,550	262		4,812	5,425
1850	4" diameter		2.50	6.400		5,225	380		5,605	6,300
1860	5" diameter	Q-6	2.80	8.571		8,700	525		9,225	10,400
1870	6" diameter	"	2.60	9.231		11,200	565		11,765	13,100

23 21 20.14 Air Purging Scoop

	23 21 20.14 Air Purging Scoop	Crew	Daily Output	Labor-Hours	Unit	Material	Labor	Equipment	Total	Total Incl O&P
0010	**AIR PURGING SCOOP**, With tappings.									
0020	For air vent and expansion tank connection									
0100	1" pipe size, threaded	1 Stpi	19	.421	Ea.	29.50	27.50		57	73.50
0110	1-1/4" pipe size, threaded		15	.533		29.50	35		64.50	84.50
0120	1-1/2" pipe size, threaded		13	.615		64.50	40.50		105	132
0130	2" pipe size, threaded		11	.727		74	47.50		121.50	153

23 21 20.18 Automatic Air Vent

	23 21 20.18 Automatic Air Vent	Crew	Daily Output	Labor-Hours	Unit	Material	Labor	Equipment	Total	Total Incl O&P
0010	**AUTOMATIC AIR VENT**									
0020	Cast iron body, stainless steel internals, float type									
0180	1/2" NPT inlet, 250 psi	1 Stpi	10	.800	Ea.	375	52.50		427.50	495
0220	3/4" NPT inlet, 250 psi	"	10	.800	"	375	52.50		427.50	495
0600	Forged steel body, stainless steel internals, float type									
0640	1/2" NPT inlet, 750 psi	1 Stpi	12	.667	Ea.	1,250	43.50		1,293.50	1,450
0680	3/4" NPT inlet, 750 psi	"	12	.667	"	1,250	43.50		1,293.50	1,450
1100	Formed steel body, noncorrosive									
1110	1/8" NPT inlet, 150 psi	1 Stpi	32	.250	Ea.	15.70	16.40		32.10	42
1120	1/4" NPT inlet, 150 psi		32	.250		53.50	16.40		69.90	83.50
1130	3/4" NPT inlet, 150 psi		32	.250		53.50	16.40		69.90	83.50
1300	Chrome plated brass, automatic/manual, for radiators									
1310	1/8" NPT inlet, nickel plated brass	1 Stpi	32	.250	Ea.	9.25	16.40		25.65	34.50

23 21 20.26 Circuit Setter

	23 21 20.26 Circuit Setter	Crew	Daily Output	Labor-Hours	Unit	Material	Labor	Equipment	Total	Total Incl O&P
0010	**CIRCUIT SETTER**, Balance valve									
0018	Threaded									
0019	1/2" pipe size	1 Stpi	22	.364	Ea.	84.50	24		108.50	129
0020	3/4" pipe size	"	20	.400	"	90	26		116	138

23 21 20 – Hydronic HVAC Piping Specialties

23 21 20.30 Cocks, Drains and Specialties		Crew	Daily Output	Labor-Hours	Unit	Material	Labor	Equipment	Total	Total Incl O&P
						2020 Bare Costs				
0010	**COCKS, DRAINS AND SPECIALTIES**									
1000	Boiler drain									
1010	Pipe thread to hose									
1020	Bronze									
1030	1/2" size	1 Stpi	36	.222	Ea.	15.25	14.55		29.80	39
1040	3/4" size	"	34	.235	"	16.50	15.40		31.90	41
1100	Solder to hose									
1110	Bronze									
1120	1/2" size	1 Stpi	46	.174	Ea.	15.25	11.40		26.65	34
1130	3/4" size	"	44	.182	"	16.60	11.90		28.50	36
1600	With built-in vacuum breaker									
1610	1/2" I.P. or solder	1 Stpi	36	.222	Ea.	38	14.55		52.55	63.50
1630	With tamper proof vacuum breaker									
1640	1/2" I.P. or solder	1 Stpi	36	.222	Ea.	51.50	14.55		66.05	79
1650	3/4" I.P. or solder	"	34	.235	"	56.50	15.40		71.90	85
3000	Cocks									
3010	Air, lever or tee handle									
3020	Bronze, single thread									
3030	1/8" size	1 Stpi	52	.154	Ea.	12.90	10.10		23	29.50
3040	1/4" size		46	.174		13.40	11.40		24.80	32
3050	3/8" size		40	.200		14	13.10		27.10	35
3060	1/2" size	↓	36	.222	↓	16.50	14.55		31.05	40
3100	Bronze, double thread									
3110	1/8" size	1 Stpi	26	.308	Ea.	16.20	20		36.20	48
3120	1/4" size		22	.364		16.85	24		40.85	54
3130	3/8" size		18	.444		17.85	29		46.85	63
3140	1/2" size	↓	15	.533	↓	22	35		57	76.50
4500	Gauge cock, brass									
4510	1/4" FPT	1 Stpi	24	.333	Ea.	13.60	22		35.60	47.50
4512	1/4" MPT	"	24	.333	"	17.10	22		39.10	51.50
4600	Pigtail, steam syphon									
4604	1/4"	1 Stpi	24	.333	Ea.	24.50	22		46.50	59.50
4650	Snubber valve									
4654	1/4"	1 Stpi	22	.364	Ea.	10.25	24		34.25	47
4660	Nipple, black steel									
4664	1/4" x 3"	1 Stpi	37	.216	Ea.	3.51	14.15		17.66	25

23 21 20.34 Dielectric Unions

		Crew	Daily Output	Labor-Hours	Unit	Material	Labor	Equipment	Total	Total Incl O&P
0010	**DIELECTRIC UNIONS**, Standard gaskets for water and air									
0020	250 psi maximum pressure									
0280	Female IPT to sweat, straight									
0300	1/2" pipe size	1 Plum	24	.333	Ea.	8.40	21.50		29.90	41.50
0340	3/4" pipe size		20	.400		9.40	26		35.40	49
0360	1" pipe size		19	.421		12.65	27		39.65	54.50
0380	1-1/4" pipe size		15	.533		17.35	34.50		51.85	70.50
0400	1-1/2" pipe size		13	.615		29.50	39.50		69	92
0420	2" pipe size	↓	11	.727	↓	40	47		87	114
0580	Female IPT to brass pipe thread, straight									
0600	1/2" pipe size	1 Plum	24	.333	Ea.	18.10	21.50		39.60	52
0640	3/4" pipe size		20	.400		19.85	26		45.85	60.50
0660	1" pipe size		19	.421		37	27		64	81
0680	1-1/4" pipe size		15	.533		37.50	34.50		72	92.50
0700	1-1/2" pipe size	↓	13	.615	↓	54.50	39.50		94	120

23 21 20 – Hydronic HVAC Piping Specialties

23 21 20.34 Dielectric Unions

23 21 20.34 Dielectric Unions		Crew	Daily Output	Labor-Hours	Unit	2020 Bare Costs Material	Labor	Equipment	Total	Total Incl O&P
0720	2" pipe size	1 Plum	11	.727	Ea.	110	47		157	191
0780	Female IPT to female IPT, straight									
0800	1/2" pipe size	1 Plum	24	.333	Ea.	15.90	21.50		37.40	49.50
0840	3/4" pipe size		20	.400		18.25	26		44.25	58.50
0860	1" pipe size		19	.421		24	27		51	66.50
0880	1-1/4" pipe size		15	.533		31.50	34.50		66	86.50
0900	1-1/2" pipe size		13	.615		63.50	39.50		103	129
0920	2" pipe size	↓	11	.727	↓	94	47		141	173
2000	175 psi maximum pressure									
2180	Female IPT to sweat									
2240	2" pipe size	1 Plum	9	.889	Ea.	188	57.50		245.50	292
2260	2-1/2" pipe size	Q-1	15	1.067		222	62		284	335
2280	3" pipe size		14	1.143		305	66.50		371.50	435
2300	4" pipe size	↓	11	1.455	↓	805	84.50		889.50	1,000
2480	Female IPT to brass pipe									
2500	1-1/2" pipe size	1 Plum	11	.727	Ea.	218	47		265	310
2540	2" pipe size	"	9	.889		229	57.50		286.50	340
2560	2-1/2" pipe size	Q-1	15	1.067		320	62		382	445
2580	3" pipe size		14	1.143		330	66.50		396.50	460
2600	4" pipe size	↓	11	1.455	↓	470	84.50		554.50	645

23 21 20.42 Expansion Joints

23 21 20.42 Expansion Joints		Crew	Daily Output	Labor-Hours	Unit	Material	Labor	Equipment	Total	Total Incl O&P
0010	**EXPANSION JOINTS**									
0100	Bellows type, neoprene cover, flanged spool									
0140	6" face to face, 1-1/4" diameter	1 Stpi	11	.727	Ea.	264	47.50		311.50	360
0160	1-1/2" diameter	"	10.60	.755		264	49.50		313.50	365
0180	2" diameter	Q-5	13.30	1.203		267	71		338	400
0190	2-1/2" diameter		12.40	1.290		276	76		352	420
0200	3" diameter		11.40	1.404		310	83		393	465
0210	4" diameter		8.40	1.905		335	112		447	540
0220	5" diameter		7.60	2.105		410	124		534	635
0230	6" diameter		6.80	2.353		420	139		559	675
0240	8" diameter		5.40	2.963		490	175		665	800
0250	10" diameter		5	3.200		675	189		864	1,025
0260	12" diameter		4.60	3.478		775	205		980	1,150
0480	10" face to face, 2" diameter		13	1.231		385	72.50		457.50	535
0500	2-1/2" diameter		12	1.333		405	78.50		483.50	565
0520	3" diameter		11	1.455		410	86		496	585
0540	4" diameter		8	2		470	118		588	690
0560	5" diameter		7	2.286		560	135		695	815
0580	6" diameter		6	2.667		580	157		737	870
0600	8" diameter		5	3.200		690	189		879	1,050
0620	10" diameter		4.60	3.478		760	205		965	1,150
0640	12" diameter		4	4		945	236		1,181	1,400
0660	14" diameter		3.80	4.211		1,175	248		1,423	1,650
0680	16" diameter		2.90	5.517		1,350	325		1,675	1,950
0700	18" diameter		2.50	6.400		1,525	380		1,905	2,250
0720	20" diameter		2.10	7.619		1,600	450		2,050	2,425
0740	24" diameter		1.80	8.889		1,875	525		2,400	2,825
0760	26" diameter		1.40	11.429		2,075	675		2,750	3,300
0780	30" diameter		1.20	13.333		2,325	785		3,110	3,750
0800	36" diameter	↓	1	16	↓	2,875	945		3,820	4,600

23 21 Hydronic Piping and Pumps

23 21 20 – Hydronic HVAC Piping Specialties

23 21 20.46 Expansion Tanks		Crew	Daily Output	Labor-Hours	Unit	2020 Bare Costs Material	Labor	Equipment	Total	Total Incl O&P
0010	EXPANSION TANKS									
1502	Plastic, corrosion resistant, see Section 22 66 83.13									
1505	Aboveground fuel-oil, storage tanks, see Section 23 13 23									
1507	Underground fuel-oil storage tanks, see Section 23 13 13									
1512	Tank leak detection systems, see Section 28 33 33.50									
2000	Steel, liquid expansion, ASME, painted, 15 gallon capacity	Q-5	17	.941	Ea.	855	55.50		910.50	1,025
2020	24 gallon capacity		14	1.143		760	67.50		827.50	935
2040	30 gallon capacity		12	1.333		985	78.50		1,063.50	1,200
2060	40 gallon capacity		10	1.600		1,150	94.50		1,244.50	1,425
2360	Galvanized									
2370	15 gallon capacity	Q-5	17	.941	Ea.	1,275	55.50		1,330.50	1,475
2380	24 gallon capacity		14	1.143		1,550	67.50		1,617.50	1,825
2390	30 gallon capacity		12	1.333		1,600	78.50		1,678.50	1,900
3000	Steel ASME expansion, rubber diaphragm, 19 gal. cap. accept.		12	1.333		3,200	78.50		3,278.50	3,650
3020	31 gallon capacity		8	2		3,575	118		3,693	4,125
3040	61 gallon capacity		6	2.667		5,225	157		5,382	5,975

23 21 20.50 Float Valves		Crew	Daily Output	Labor-Hours	Unit	Material	Labor	Equipment	Total	Total Incl O&P
0010	FLOAT VALVES									
0020	With ball and bracket									
0030	Single seat, threaded									
0040	Brass body									
0050	1/2"	1 Stpi	11	.727	Ea.	108	47.50		155.50	191
0060	3/4"		9	.889		119	58.50		177.50	218
0070	1"		7	1.143		166	75		241	295
0080	1-1/2"		4.50	1.778		238	117		355	435
0090	2"		3.60	2.222		235	146		381	475
0300	For condensate receivers, CI, in-line mount									
0320	1" inlet	1 Stpi	7	1.143	Ea.	154	75		229	281
0360	For condensate receiver, CI, external float, flanged tank mount									
0370	3/4" inlet	1 Stpi	5	1.600	Ea.	117	105		222	285

23 21 20.54 Flow Check Control		Crew	Daily Output	Labor-Hours	Unit	Material	Labor	Equipment	Total	Total Incl O&P
0010	FLOW CHECK CONTROL									
0100	Bronze body, soldered									
0110	3/4" size	1 Stpi	20	.400	Ea.	77	26		103	124
0120	1" size	"	19	.421	"	93	27.50		120.50	144
0200	Cast iron body, threaded									
0210	3/4" size	1 Stpi	20	.400	Ea.	59.50	26		85.50	105
0220	1" size		19	.421		67	27.50		94.50	116
0230	1-1/4" size		15	.533		81.50	35		116.50	142
0240	1-1/2" size		13	.615		125	40.50		165.50	198
0250	2" size		11	.727		179	47.50		226.50	269

23 21 20.58 Hydronic Heating Control Valves		Crew	Daily Output	Labor-Hours	Unit	Material	Labor	Equipment	Total	Total Incl O&P
0010	HYDRONIC HEATING CONTROL VALVES									
0050	Hot water, nonelectric, thermostatic									
0100	Radiator supply, 1/2" diameter	1 Stpi	24	.333	Ea.	78.50	22		100.50	119
0120	3/4" diameter	"	20	.400	"	74.50	26		100.50	121
1000	Manual, radiator supply									
1010	1/2" pipe size, angle union	1 Stpi	24	.333	Ea.	71	22		93	111
1020	3/4" pipe size, angle union	"	20	.400	"	89.50	26		115.50	138
1100	Radiator, balancing, straight, sweat connections									
1110	1/2" pipe size	1 Stpi	24	.333	Ea.	25	22		47	60

23 21 Hydronic Piping and Pumps

23 21 20 – Hydronic HVAC Piping Specialties

23 21 20.58 Hydronic Heating Control Valves

	23 21 20.58 Hydronic Heating Control Valves	Crew	Daily Output	Labor-Hours	Unit	2020 Bare Costs Material	Labor	Equipment	Total	Total Incl O&P
1120	3/4″ pipe size	1 Stpi	20	.400	Ea.	34.50	26		60.50	77
1200	Steam, radiator, supply									
1210	1/2″ pipe size, angle union	1 Stpi	24	.333	Ea.	67	22		89	107
1220	3/4″ pipe size, angle union	″	20	.400	″	73.50	26		99.50	120
8000	System balancing and shut-off									
8020	Butterfly, quarter turn, calibrated, threaded or solder									
8040	Bronze, -30°F to +350°F, pressure to 175 psi									
8060	1/2″ size	1 Stpi	22	.364	Ea.	21	24		45	59
8070	3/4″ size	″	20	.400	″	29	26		55	71

23 21 20.66 Monoflow Tee Fitting

	23 21 20.66 Monoflow Tee Fitting	Crew	Daily Output	Labor-Hours	Unit	Material	Labor	Equipment	Total	Total Incl O&P
0010	**MONOFLOW TEE FITTING**									
1100	For one pipe hydronic, supply and return									
1110	Copper, soldered									
1120	3/4″ x 1/2″ size	1 Stpi	13	.615	Ea.	21	40.50		61.50	83.50

23 21 20.74 Strainers, Basket Type

	23 21 20.74 Strainers, Basket Type	Crew	Daily Output	Labor-Hours	Unit	Material	Labor	Equipment	Total	Total Incl O&P
0010	**STRAINERS, BASKET TYPE**, Perforated stainless steel basket									
0100	Brass or monel available									
2000	Simplex style									
2300	Bronze body									
2320	Screwed, 3/8″ pipe size	1 Stpi	22	.364	Ea.	191	24		215	246
2340	1/2″ pipe size		20	.400		199	26		225	258
2360	3/4″ pipe size		17	.471		305	31		336	380
2380	1″ pipe size		15	.533		380	35		415	475
2400	1-1/4″ pipe size		13	.615		415	40.50		455.50	520
2420	1-1/2″ pipe size		12	.667		425	43.50		468.50	530
2440	2″ pipe size		10	.800		625	52.50		677.50	765
2460	2-1/2″ pipe size	Q-5	15	1.067		1,050	63		1,113	1,250
2480	3″ pipe size	″	14	1.143		1,550	67.50		1,617.50	1,800
2600	Flanged, 2″ pipe size	1 Stpi	6	1.333		860	87.50		947.50	1,075
2620	2-1/2″ pipe size	Q-5	4.50	3.556		1,500	210		1,710	1,975
2640	3″ pipe size		3.50	4.571		1,700	270		1,970	2,250
2660	4″ pipe size		3	5.333		2,575	315		2,890	3,300
2680	5″ pipe size	Q-6	3.40	7.059		4,100	430		4,530	5,175
2700	6″ pipe size		3	8		5,000	490		5,490	6,225
2710	8″ pipe size		2.50	9.600		8,150	585		8,735	9,850
3600	Iron body									
3700	Screwed, 3/8″ pipe size	1 Stpi	22	.364	Ea.	165	24		189	217
3720	1/2″ pipe size		20	.400		170	26		196	226
3740	3/4″ pipe size		17	.471		217	31		248	284
3760	1″ pipe size		15	.533		221	35		256	296
3780	1-1/4″ pipe size		13	.615		289	40.50		329.50	380
3800	1-1/2″ pipe size		12	.667		320	43.50		363.50	415
3820	2″ pipe size		10	.800		380	52.50		432.50	500
3840	2-1/2″ pipe size	Q-5	15	1.067		510	63		573	655
3860	3″ pipe size	″	14	1.143		615	67.50		682.50	780
4000	Flanged, 2″ pipe size	1 Stpi	6	1.333		560	87.50		647.50	745
4020	2-1/2″ pipe size	Q-5	4.50	3.556		755	210		965	1,150
4040	3″ pipe size		3.50	4.571		790	270		1,060	1,275
4060	4″ pipe size		3	5.333		1,225	315		1,540	1,825
4080	5″ pipe size	Q-6	3.40	7.059		1,825	430		2,255	2,675
4100	6″ pipe size		3	8		2,350	490		2,840	3,300
4120	8″ pipe size		2.50	9.600		4,225	585		4,810	5,525

23 21 20.74	Strainers, Basket Type	Crew	Daily Output	Labor-Hours	Unit	2020 Bare Costs Material	Labor	Equipment	Total	Total Incl O&P
4140	10″ pipe size	Q-6	2.20	10.909	Ea.	9,300	665		9,965	11,200
7000	Stainless steel body									
7200	Screwed, 1″ pipe size	1 Stpi	15	.533	Ea.	465	35		500	570
7210	1-1/4″ pipe size		13	.615		725	40.50		765.50	860
7220	1-1/2″ pipe size		12	.667		725	43.50		768.50	865
7240	2″ pipe size		10	.800		1,075	52.50		1,127.50	1,275
7260	2-1/2″ pipe size	Q-5	15	1.067		1,525	63		1,588	1,775
7280	3″ pipe size	″	14	1.143		2,100	67.50		2,167.50	2,425
7400	Flanged, 2″ pipe size	1 Stpi	6	1.333		1,650	87.50		1,737.50	1,950
7420	2-1/2″ pipe size	Q-5	4.50	3.556		3,225	210		3,435	3,850
7440	3″ pipe size		3.50	4.571		3,300	270		3,570	4,025
7460	4″ pipe size		3	5.333		5,175	315		5,490	6,175
7480	6″ pipe size	Q-6	3	8		10,900	490		11,390	12,700
7500	8″ pipe size	″	2.50	9.600		15,300	585		15,885	17,800
8100	Duplex style									
8200	Bronze body									
8240	Screwed, 3/4″ pipe size	1 Stpi	16	.500	Ea.	1,325	33		1,358	1,525
8260	1″ pipe size		14	.571		1,325	37.50		1,362.50	1,525
8280	1-1/4″ pipe size		12	.667		2,700	43.50		2,743.50	3,050
8300	1-1/2″ pipe size		11	.727		2,700	47.50		2,747.50	3,050
8320	2″ pipe size		9	.889		4,225	58.50		4,283.50	4,725
8340	2-1/2″ pipe size	Q-5	14	1.143		5,450	67.50		5,517.50	6,100
8420	Flanged, 2″ pipe size	1 Stpi	6	1.333		5,225	87.50		5,312.50	5,875
8440	2-1/2″ pipe size	Q-5	4.50	3.556		7,225	210		7,435	8,275
8460	3″ pipe size		3.50	4.571		7,875	270		8,145	9,050
8480	4″ pipe size		3	5.333		11,500	315		11,815	13,200
8500	5″ pipe size	Q-6	3.40	7.059		25,900	430		26,330	29,000
8520	6″ pipe size	″	3	8		25,100	490		25,590	28,400
8700	Iron body									
8740	Screwed, 3/4″ pipe size	1 Stpi	16	.500	Ea.	1,675	33		1,708	1,900
8760	1″ pipe size		14	.571		1,675	37.50		1,712.50	1,900
8780	1-1/4″ pipe size		12	.667		1,850	43.50		1,893.50	2,100
8800	1-1/2″ pipe size		11	.727		1,850	47.50		1,897.50	2,100
8820	2″ pipe size		9	.889		3,100	58.50		3,158.50	3,500
8840	2-1/2″ pipe size	Q-5	14	1.143		3,425	67.50		3,492.50	3,875
9000	Flanged, 2″ pipe size	1 Stpi	6	1.333		2,825	87.50		2,912.50	3,225
9020	2-1/2″ pipe size	Q-5	4.50	3.556		3,550	210		3,760	4,225
9040	3″ pipe size		3.50	4.571		3,875	270		4,145	4,650
9060	4″ pipe size		3	5.333		6,525	315		6,840	7,650
9080	5″ pipe size	Q-6	3.40	7.059		14,300	430		14,730	16,300
9100	6″ pipe size		3	8		14,300	490		14,790	16,400
9120	8″ pipe size		2.50	9.600		25,100	585		25,685	28,500
9140	10″ pipe size		2.20	10.909		40,700	665		41,365	45,800
9160	12″ pipe size		1.70	14.118		44,800	865		45,665	50,500
9170	14″ pipe size		1.40	17.143		52,500	1,050		53,550	59,500
9180	16″ pipe size		1	24		64,500	1,475		65,975	73,000
9700	Stainless steel body									
9740	Screwed, 1″ pipe size	1 Stpi	14	.571	Ea.	3,250	37.50		3,287.50	3,625
9760	1-1/2″ pipe size		11	.727		4,875	47.50		4,922.50	5,425
9780	2″ pipe size		9	.889		7,175	58.50		7,233.50	7,975
9860	Flanged, 2″ pipe size		6	1.333		7,400	87.50		7,487.50	8,275
9880	2-1/2″ pipe size	Q-5	4.50	3.556		12,400	210		12,610	14,000
9900	3″ pipe size		3.50	4.571		13,500	270		13,770	15,300

23 21 20.74 Strainers, Basket Type		Crew	Daily Output	Labor-Hours	Unit	Material	Labor	Equipment	Total	Total Incl O&P
						2020 Bare Costs				
9920	4″ pipe size	Q-5	3	5.333	Ea.	17,600	315		17,915	19,900
9940	6″ pipe size	Q-6	3	8		26,900	490		27,390	30,300
9960	8″ pipe size	″	2.50	9.600		69,000	585		69,585	77,000

23 21 20.76 Strainers, Y Type, Bronze Body		Crew	Daily Output	Labor-Hours	Unit	Material	Labor	Equipment	Total	Total Incl O&P
0010	**STRAINERS, Y TYPE, BRONZE BODY**									
0050	Screwed, 125 lb., 1/4″ pipe size	1 Stpi	24	.333	Ea.	44	22		66	81
0070	3/8″ pipe size		24	.333		45.50	22		67.50	83
0100	1/2″ pipe size		20	.400		45.50	26		71.50	89
0120	3/4″ pipe size		19	.421		57	27.50		84.50	105
0140	1″ pipe size		17	.471		66.50	31		97.50	120
0150	1-1/4″ pipe size		15	.533		134	35		169	201
0160	1-1/2″ pipe size		14	.571		144	37.50		181.50	214
0180	2″ pipe size		13	.615		190	40.50		230.50	270
0182	3″ pipe size		12	.667		1,200	43.50		1,243.50	1,400
0200	300 lb., 2-1/2″ pipe size	Q-5	17	.941		865	55.50		920.50	1,025
0220	3″ pipe size		16	1		1,325	59		1,384	1,575
0240	4″ pipe size		15	1.067		2,425	63		2,488	2,775
0500	For 300 lb. rating 1/4″ thru 2″, add					15%				
1000	Flanged, 150 lb., 1-1/2″ pipe size	1 Stpi	11	.727	Ea.	545	47.50		592.50	670
1020	2″ pipe size	″	8	1		630	65.50		695.50	790
1030	2-1/2″ pipe size	Q-5	5	3.200		1,125	189		1,314	1,525
1040	3″ pipe size		4.50	3.556		1,375	210		1,585	1,850
1060	4″ pipe size		3	5.333		2,125	315		2,440	2,825
1080	5″ pipe size	Q-6	3.40	7.059		3,025	430		3,455	3,975
1100	6″ pipe size		3	8		4,100	490		4,590	5,250
1106	8″ pipe size		2.60	9.231		4,450	565		5,015	5,750
1500	For 300 lb. rating, add					40%				

23 21 20.78 Strainers, Y Type, Iron Body		Crew	Daily Output	Labor-Hours	Unit	Material	Labor	Equipment	Total	Total Incl O&P
0010	**STRAINERS, Y TYPE, IRON BODY**									
0050	Screwed, 250 lb., 1/4″ pipe size	1 Stpi	20	.400	Ea.	20	26		46	61
0070	3/8″ pipe size		20	.400		20	26		46	61
0100	1/2″ pipe size		20	.400		20	26		46	61
0120	3/4″ pipe size		18	.444		23.50	29		52.50	69
0140	1″ pipe size		16	.500		27.50	33		60.50	79
0150	1-1/4″ pipe size		15	.533		43.50	35		78.50	101
0160	1-1/2″ pipe size		12	.667		55.50	43.50		99	127
0180	2″ pipe size		8	1		83	65.50		148.50	189
0200	2-1/2″ pipe size	Q-5	12	1.333		288	78.50		366.50	435
0220	3″ pipe size		11	1.455		310	86		396	475
0240	4″ pipe size		5	3.200		530	189		719	865
0500	For galvanized body, add					50%				
1000	Flanged, 125 lb., 1-1/2″ pipe size	1 Stpi	11	.727	Ea.	167	47.50		214.50	256
1020	2″ pipe size	″	8	1		177	65.50		242.50	293
1030	2-1/2″ pipe size	Q-5	5	3.200		161	189		350	460
1040	3″ pipe size		4.50	3.556		188	210		398	520
1060	4″ pipe size		3	5.333		375	315		690	885
1080	5″ pipe size	Q-6	3.40	7.059		405	430		835	1,100
1100	6″ pipe size		3	8		675	490		1,165	1,475
1120	8″ pipe size		2.50	9.600		1,300	585		1,885	2,300
1140	10″ pipe size		2	12		2,300	735		3,035	3,625
1160	12″ pipe size		1.70	14.118		3,525	865		4,390	5,175
1170	14″ pipe size		1.30	18.462		6,500	1,125		7,625	8,850

23 21 20 – Hydronic HVAC Piping Specialties

23 21 20.78 Strainers, Y Type, Iron Body		Crew	Daily Output	Labor-Hours	Unit	Material (2020 Bare Costs)	Labor (2020 Bare Costs)	Equipment (2020 Bare Costs)	Total (2020 Bare Costs)	Total Incl O&P
1180	16″ pipe size	Q-6	1	24	Ea.	9,225	1,475		10,700	12,400
1500	For 250 lb. rating, add					20%				
2000	For galvanized body, add					50%				
2500	For steel body, add					40%				

23 21 20.80 Suction Diffusers		Crew	Daily Output	Labor-Hours	Unit	Material	Labor	Equipment	Total	Total Incl O&P
0010	**SUCTION DIFFUSERS**									
0100	Cast iron body with integral straightening vanes, strainer									
1000	Flanged									
1010	2″ inlet, 1-1/2″ pump side	1 Stpi	6	1.333	Ea.	415	87.50		502.50	585
1020	2″ pump side	″	5	1.600		490	105		595	695
1030	3″ inlet, 2″ pump side	Q-5	6.50	2.462	↓	520	145		665	790

23 21 20.84 Thermoflo Indicator		Crew	Daily Output	Labor-Hours	Unit	Material	Labor	Equipment	Total	Total Incl O&P
0010	**THERMOFLO INDICATOR**, For balancing									
1000	Sweat connections, 1-1/4″ pipe size	1 Stpi	12	.667	Ea.	790	43.50		833.50	935
1020	1-1/2″ pipe size	″	10	.800	″	805	52.50		857.50	965

23 21 20.88 Venturi Flow		Crew	Daily Output	Labor-Hours	Unit	Material	Labor	Equipment	Total	Total Incl O&P
0010	**VENTURI FLOW**, Measuring device									
0050	1/2″ diameter	1 Stpi	24	.333	Ea.	310	22		332	375
0100	3/4″ diameter		20	.400		284	26		310	355
0120	1″ diameter		19	.421		305	27.50		332.50	375
0140	1-1/4″ diameter		15	.533		380	35		415	470
0160	1-1/2″ diameter		13	.615		380	40.50		420.50	475
0180	2″ diameter	↓	11	.727		405	47.50		452.50	515
0200	2-1/2″ diameter	Q-5	16	1		555	59		614	700
0220	3″ diameter		14	1.143		570	67.50		637.50	730
0240	4″ diameter	↓	11	1.455		855	86		941	1,075
0260	5″ diameter	Q-6	4	6		1,125	365		1,490	1,775
0280	6″ diameter		3.50	6.857		1,250	420		1,670	2,000
0300	8″ diameter		3	8		1,600	490		2,090	2,475
0320	10″ diameter		2	12		3,775	735		4,510	5,250
0330	12″ diameter		1.80	13.333		5,225	815		6,040	6,975
0340	14″ diameter		1.60	15		5,925	920		6,845	7,900
0350	16″ diameter	↓	1.40	17.143		6,775	1,050		7,825	9,025
0500	For meter, add				↓	2,425			2,425	2,650

23 21 23 – Hydronic Pumps

23 21 23.13 In-Line Centrifugal Hydronic Pumps		Crew	Daily Output	Labor-Hours	Unit	Material	Labor	Equipment	Total	Total Incl O&P
0010	**IN-LINE CENTRIFUGAL HYDRONIC PUMPS**									
0600	Bronze, sweat connections, 1/40 HP, in line									
0640	3/4″ size	Q-1	16	1	Ea.	275	58		333	385
1000	Flange connection, 3/4″ to 1-1/2″ size									
1040	1/12 HP	Q-1	6	2.667	Ea.	710	155		865	1,025
1060	1/8 HP		6	2.667		1,250	155		1,405	1,600
1100	1/3 HP		6	2.667		1,400	155		1,555	1,750
1140	2″ size, 1/6 HP		5	3.200		1,750	186		1,936	2,200
1180	2-1/2″ size, 1/4 HP		5	3.200		2,200	186		2,386	2,700
1220	3″ size, 1/4 HP		4	4		2,400	232		2,632	2,975
1260	1/3 HP		4	4		2,825	232		3,057	3,475
1300	1/2 HP		4	4		2,950	232		3,182	3,575
1340	3/4 HP		4	4		3,000	232		3,232	3,650
1380	1 HP	↓	4	4	↓	4,775	232		5,007	5,600
2000	Cast iron, flange connection									

23 21 23 – Hydronic Pumps

23 21 23.13 In-Line Centrifugal Hydronic Pumps		Crew	Daily Output	Labor-Hours	Unit	2020 Bare Costs Material	Labor	Equipment	Total	Total Incl O&P
2040	3/4" to 1-1/2" size, in line, 1/12 HP	Q-1	6	2.667	Ea.	470	155		625	750
2060	1/8 HP		6	2.667		780	155		935	1,100
2100	1/3 HP		6	2.667		870	155		1,025	1,175
2140	2" size, 1/6 HP		5	3.200		955	186		1,141	1,325
2180	2-1/2" size, 1/4 HP		5	3.200		1,150	186		1,336	1,525
2220	3" size, 1/4 HP		4	4		1,150	232		1,382	1,625
2260	1/3 HP		4	4		1,525	232		1,757	2,025
2300	1/2 HP		4	4		1,575	232		1,807	2,075
2340	3/4 HP		4	4		1,825	232		2,057	2,350
2380	1 HP	↓	4	4	↓	2,625	232		2,857	3,225
2600	For nonferrous impeller, add					3%				
3000	High head, bronze impeller									
3030	1-1/2" size, 1/2 HP	Q-1	5	3.200	Ea.	1,400	186		1,586	1,825
3040	1-1/2" size, 3/4 HP		5	3.200		1,625	186		1,811	2,075
3050	2" size, 1 HP		4	4		1,975	232		2,207	2,525
3090	2" size, 1-1/2 HP	↓	4	4	↓	2,400	232		2,632	2,975
4000	Close coupled, end suction, bronze impeller									
4040	1-1/2" size, 1-1/2 HP, to 40 GPM	Q-1	3	5.333	Ea.	2,500	310		2,810	3,225
4090	2" size, 2 HP, to 50 GPM		3	5.333		3,025	310		3,335	3,800
4100	2" size, 3 HP, to 90 GPM		2.30	6.957		3,075	405		3,480	3,975
4190	2-1/2" size, 3 HP, to 150 GPM		2	8		3,350	465		3,815	4,400
4300	3" size, 5 HP, to 225 GPM		1.80	8.889		3,900	515		4,415	5,075
4410	3" size, 10 HP, to 350 GPM		1.60	10		6,100	580		6,680	7,575
4420	4" size, 7-1/2 HP, to 350 GPM	↓	1.60	10		6,250	580		6,830	7,750
4520	4" size, 10 HP, to 600 GPM	Q-2	1.70	14.118		6,475	850		7,325	8,400
4530	5" size, 15 HP, to 1,000 GPM		1.70	14.118		6,500	850		7,350	8,425
4610	5" size, 20 HP, to 1,350 GPM		1.50	16		6,925	960		7,885	9,050
4620	5" size, 25 HP, to 1,550 GPM	↓	1.50	16	↓	8,325	960		9,285	10,600
5000	Base mounted, bronze impeller, coupling guard									
5040	1-1/2" size, 1-1/2 HP, to 40 GPM	Q-1	2.30	6.957	Ea.	7,300	405		7,705	8,625
5090	2" size, 2 HP, to 50 GPM		2.30	6.957		8,175	405		8,580	9,600
5100	2" size, 3 HP, to 90 GPM		2	8		9,000	465		9,465	10,600
5190	2-1/2" size, 3 HP, to 150 GPM		1.80	8.889		10,400	515		10,915	12,200
5300	3" size, 5 HP, to 225 GPM		1.60	10		11,300	580		11,880	13,300
5410	4" size, 5 HP, to 350 GPM		1.50	10.667		13,100	620		13,720	15,300
5420	4" size, 7-1/2 HP, to 350 GPM	↓	1.50	10.667		14,100	620		14,720	16,400
5520	5" size, 10 HP, to 600 GPM	Q-2	1.60	15		19,400	900		20,300	22,700
5530	5" size, 15 HP, to 1,000 GPM		1.60	15		20,900	900		21,800	24,400
5610	6" size, 20 HP, to 1,350 GPM		1.40	17.143		22,600	1,025		23,625	26,500
5620	6" size, 25 HP, to 1,550 GPM	↓	1.40	17.143	↓	26,600	1,025		27,625	30,900
5800	The above pump capacities are based on 1,800 RPM,									
5810	at a 60' head. Increasing the RPM									
5820	or decreasing the head will increase the GPM.									

23 21 29 – Automatic Condensate Pump Units

23 21 29.10 Condensate Removal Pump System		Crew	Daily Output	Labor-Hours	Unit	Material	Labor	Equipment	Total	Total Incl O&P
0010	**CONDENSATE REMOVAL PUMP SYSTEM**									
0020	Pump with 1 gal. ABS tank									
0100	115 V									
0120	1/50 HP, 200 GPH G	1 Stpi	12	.667	Ea.	188	43.50		231.50	272
0140	1/18 HP, 270 GPH G		10	.800		186	52.50		238.50	284
0160	1/5 HP, 450 GPH G	↓	8	1	↓	425	65.50		490.50	570
0200	230 V									

23 21 Hydronic Piping and Pumps

23 21 29 – Automatic Condensate Pump Units

23 21 29.10 Condensate Removal Pump System		Crew	Daily Output	Labor-Hours	Unit	2020 Bare Costs Material	Labor	Equipment	Total	Total Incl O&P
0240	1/18 HP, 270 GPH	1 Stpi	10	.800	Ea.	208	52.50		260.50	310
0260	1/5 HP, 450 GPH G	"	8	1	"	495	65.50		560.50	645

23 23 Refrigerant Piping

23 23 23 – Refrigerants

23 23 23.10 Anti-Freeze		Crew	Daily Output	Labor-Hours	Unit	Material	Labor	Equipment	Total	Total Incl O&P
0010	**ANTI-FREEZE**, Inhibited									
0900	Ethylene glycol concentrated									
1000	55 gallon drums, small quantities				Gal.	12.90			12.90	14.20
1200	Large quantities					11.80			11.80	12.95
2000	Propylene glycol, for solar heat, small quantities					23			23	25.50
2100	Large quantities					9.90			9.90	10.90

23 34 HVAC Fans

23 34 14 – Blower HVAC Fans

23 34 14.10 Blower Type HVAC Fans		Crew	Daily Output	Labor-Hours	Unit	Material	Labor	Equipment	Total	Total Incl O&P
0010	**BLOWER TYPE HVAC FANS**									
2500	Ceiling fan, right angle, extra quiet, 0.10" S.P.									
2520	95 CFM	Q-20	20	1	Ea.	305	57		362	420
2540	210 CFM		19	1.053		360	60		420	485
2560	385 CFM		18	1.111		455	63.50		518.50	600
2640	For wall or roof cap, add	1 Shee	16	.500		305	31		336	385
2660	For straight thru fan, add					10%				
2680	For speed control switch, add	1 Elec	16	.500		166	30.50		196.50	229

23 34 23 – HVAC Power Ventilators

23 34 23.10 HVAC Power Circulators and Ventilators		Crew	Daily Output	Labor-Hours	Unit	Material	Labor	Equipment	Total	Total Incl O&P
0010	**HVAC POWER CIRCULATORS AND VENTILATORS**									
6650	Residential, bath exhaust, grille, back draft damper									
6660	50 CFM	Q-20	24	.833	Ea.	61	47.50		108.50	139
6670	110 CFM		22	.909		116	52		168	207
6680	Light combination, squirrel cage, 100 watt, 70 CFM		24	.833		123	47.50		170.50	207
6700	Light/heater combination, ceiling mounted									
6710	70 CFM, 1,450 watt	Q-20	24	.833	Ea.	139	47.50		186.50	225
6800	Heater combination, recessed, 70 CFM		24	.833		83.50	47.50		131	164
6820	With 2 infrared bulbs		23	.870		100	49.50		149.50	185
6940	Residential roof jacks and wall caps									
6944	Wall cap with back draft damper									
6946	3" & 4" diam. round duct	1 Shee	11	.727	Ea.	26.50	45.50		72	97.50
6948	6" diam. round duct	"	11	.727	"	76	45.50		121.50	153
6958	Roof jack with bird screen and back draft damper									
6960	3" & 4" diam. round duct	1 Shee	11	.727	Ea.	18.85	45.50		64.35	89.50
6962	3-1/4" x 10" rectangular duct	"	10	.800	"	35	50		85	114
6980	Transition									
6982	3-1/4" x 10" to 6" diam. round	1 Shee	20	.400	Ea.	36	25		61	77.50

23 35 Special Exhaust Systems

23 35 16 – Engine Exhaust Systems

23 35 16.10 Engine Exhaust Removal Systems		Crew	Daily Output	Labor-Hours	Unit	Material	Labor (2020 Bare Costs)	Equipment (2020 Bare Costs)	Total (2020 Bare Costs)	Total Incl O&P
0010	**ENGINE EXHAUST REMOVAL SYSTEMS** D3090–320									
0500	Engine exhaust, garage, in-floor system									
0510	Single tube outlet assemblies									
0520	For transite pipe ducting, self-storing tube									
0530	3″ tubing adapter plate	1 Shee	16	.500	Ea.	262	31		293	335
0540	4″ tubing adapter plate		16	.500		266	31		297	340
0550	5″ tubing adapter plate		16	.500		264	31		295	340
0600	For vitrified tile ducting									
0610	3″ tubing adapter plate, self-storing tube	1 Shee	16	.500	Ea.	260	31		291	335
0620	4″ tubing adapter plate, self-storing tube		16	.500		265	31		296	340
0660	5″ tubing adapter plate, self-storing tube		16	.500		265	31		296	340
0800	Two tube outlet assemblies									
0810	For transite pipe ducting, self-storing tube									
0820	3″ tubing, dual exhaust adapter plate	1 Shee	16	.500	Ea.	267	31		298	340
0850	For vitrified tile ducting									
0860	3″ tubing, dual exhaust, self-storing tube	1 Shee	16	.500	Ea.	300	31		331	380
0870	3″ tubing, double outlet, non-storing tubes	″	16	.500	″	300	31		331	380
0900	Accessories for metal tubing (overhead systems also)									
0910	Adapters, for metal tubing end									
0920	3″ tail pipe type				Ea.	51			51	56.50
0930	4″ tail pipe type					56			56	62
0940	5″ tail pipe type					57			57	63
0990	5″ diesel stack type					283			283	310
1000	6″ diesel stack type					335			335	370
1100	Bullnose (guide) required for in-floor assemblies									
1110	3″ tubing size				Ea.	30.50			30.50	33.50
1120	4″ tubing size					31.50			31.50	35
1130	5″ tubing size					33.50			33.50	36.50
1150	Plain rings, for tubing end									
1160	3″ tubing size				Ea.	23			23	25
1170	4″ tubing size				″	38.50			38.50	42.50
1200	Tubing, galvanized, flexible (for overhead systems also)									
1210	3″ ID				L.F.	10.80			10.80	11.85
1220	4″ ID					13.20			13.20	14.55
1230	5″ ID					15.55			15.55	17.15
1240	6″ ID					18			18	19.80
1250	Stainless steel, flexible (for overhead system also)									
1260	3″ ID				L.F.	24			24	26.50
1270	4″ ID					33			33	36
1280	5″ ID					37.50			37.50	41
1290	6″ ID					43.50			43.50	47.50
1500	Engine exhaust, garage, overhead components, for neoprene tubing									
1510	Alternate metal tubing & accessories see above									
1550	Adapters, for neoprene tubing end									
1560	3″ tail pipe, adjustable, neoprene				Ea.	58.50			58.50	64
1570	3″ tail pipe, heavy wall neoprene					66			66	72.50
1580	4″ tail pipe, heavy wall neoprene					103			103	113
1590	5″ tail pipe, heavy wall neoprene					117			117	129
1650	Connectors, tubing									
1660	3″ interior, aluminum				Ea.	23			23	25.50
1670	4″ interior, aluminum					38.50			38.50	42.50
1710	5″ interior, neoprene					59			59	64.50
1750	3″ spiralock, neoprene					23			23	25

23 35 Special Exhaust Systems

23 35 16 – Engine Exhaust Systems

23 35 16.10 Engine Exhaust Removal Systems		Crew	Daily Output	Labor-Hours	Unit	Material	2020 Bare Costs Labor	Equipment	Total	Total Incl O&P
1760	4″ spiralock, neoprene				Ea.	38.50			38.50	42.50
1780	Y for 3″ ID tubing, neoprene, dual exhaust					194			194	213
1790	Y for 4″ ID tubing, aluminum, dual exhaust				↓	191			191	210
1850	Elbows, aluminum, splice into tubing for strap									
1860	3″ neoprene tubing size				Ea.	54			54	59.50
1870	4″ neoprene tubing size					36			36	40
1900	Flange assemblies, connect tubing to overhead duct				↓	69.50			69.50	76.50
2000	Hardware and accessories									
2020	Cable, galvanized, 1/8″ diameter				L.F.	.47			.47	.52
2040	Cleat, tie down cable or rope				Ea.	5.90			5.90	6.50
2060	Pulley					8.15			8.15	9
2080	Pulley hook, universal				↓	5.90			5.90	6.50
2100	Rope, nylon, 1/4″ diameter				L.F.	.41			.41	.45
2120	Winch, 1″ diameter				Ea.	119			119	131
2150	Lifting strap, mounts on neoprene									
2160	3″ tubing size				Ea.	29.50			29.50	32.50
2170	4″ tubing size					29.50			29.50	32.50
2180	5″ tubing size					29.50			29.50	32.50
2190	6″ tubing size				↓	29.50			29.50	32.50
2200	Tubing, neoprene, 11′ lengths									
2210	3″ ID				L.F.	12.25			12.25	13.50
2220	4″ ID					16.90			16.90	18.55
2230	5″ ID				↓	29			29	32
2500	Engine exhaust, thru-door outlet									
2510	3″ tube size	1 Carp	16	.500	Ea.	64	26.50		90.50	111
2530	4″ tube size	"	16	.500	"	69.50	26.50		96	117
3000	Tubing, exhaust, flex hose, with									
3010	coupler, damper and tail pipe adapter									
3020	Neoprene									
3040	3″ x 20′	1 Shee	6	1.333	Ea.	360	83		443	525
3050	4″ x 15′		5.40	1.481		420	92.50		512.50	605
3060	4″ x 20′		5	1.600		505	99.50		604.50	705
3070	5″ x 15′	↓	4.40	1.818	↓	645	113		758	880
3100	Galvanized									
3110	3″ x 20′	1 Shee	6	1.333	Ea.	315	83		398	475
3120	4″ x 17′		5.60	1.429		350	89		439	515
3130	4″ x 20′		5	1.600		385	99.50		484.50	575
3140	5″ x 17′	↓	4.60	1.739	↓	415	108		523	620

23 35 43 – Welding Fume Elimination Systems

23 35 43.10 Welding Fume Elimination System Components		Crew	Daily Output	Labor-Hours	Unit	Material	Labor	Equipment	Total	Total Incl O&P
0010	**WELDING FUME ELIMINATION SYSTEM COMPONENTS**									
7500	Welding fume elimination accessories for garage exhaust systems									
7600	Cut off (blast gate)									
7610	3″ tubing size, 3″ x 6″ opening	1 Shee	24	.333	Ea.	27.50	21		48.50	62
7620	4″ tubing size, 4″ x 8″ opening		24	.333		28	21		49	62.50
7630	5″ tubing size, 5″ x 10″ opening		24	.333		32	21		53	67
7640	6″ tubing size		24	.333		33.50	21		54.50	68.50
7650	8″ tubing size	↓	24	.333	↓	43.50	21		64.50	79
7700	Hoods, magnetic, with handle & screen									
7710	3″ tubing size, 3″ x 6″ opening	1 Shee	24	.333	Ea.	94.50	21		115.50	136
7720	4″ tubing size, 4″ x 8″ opening		24	.333		102	21		123	145
7730	5″ tubing size, 5″ x 10″ opening	↓	24	.333	↓	102	21		123	145

23 38 Ventilation Hoods

23 38 13 – Commercial-Kitchen Hoods

23 38 13.10 Hood and Ventilation Equipment		Crew	Daily Output	Labor-Hours	Unit	Material (2020 Bare Costs)	Labor (2020 Bare Costs)	Equipment (2020 Bare Costs)	Total (2020 Bare Costs)	Total Incl O&P
0010	**HOOD AND VENTILATION EQUIPMENT**									
2970	Exhaust hood, sst, gutter on all sides, 4′ x 4′ x 2′	1 Carp	1.80	4.444	Ea.	4,075	236		4,311	4,825
2980	4′ x 4′ x 7′	"	1.60	5		6,750	266		7,016	7,825
7800	Vent hood, wall canopy with fire protection, 30″	L-3A	9	1.333		390	76.50		466.50	545
7810	Without fire protection, 36″		10	1.200		430	69		499	575
7820	Island canopy with fire protection, 30″		7	1.714		760	98.50		858.50	985
7830	Without fire protection, 36″		8	1.500		785	86.50		871.50	990
7840	Back shelf with fire protection, 30″		11	1.091		410	63		473	545
7850	Without fire protection, black, 36″		12	1		450	57.50		507.50	585
7852	Without fire protection, stainless steel, 36″		12	1		470	57.50		527.50	600
7860	Range hood & CO_2 system, 30″	1 Carp	2.50	3.200		4,325	170		4,495	5,000
7950	Hood fire protection system, electric stove	Q-1	3	5.333		2,025	310		2,335	2,700
7952	Hood fire protection system, gas stove	"	3	5.333		2,350	310		2,660	3,050

23 51 Breechings, Chimneys, and Stacks

23 51 13 – Draft Control Devices

23 51 13.16 Vent Dampers

Line	Description	Crew	Daily Output	Labor-Hours	Unit	Material	Labor	Equipment	Total	Total Incl O&P
0010	**VENT DAMPERS**									
5000	Vent damper, bi-metal, gas, 3″ diameter	Q-9	24	.667	Ea.	71	37.50		108.50	135
5010	4″ diameter	"	24	.667	"	70	37.50		107.50	134

23 51 13.19 Barometric Dampers

Line	Description	Crew	Daily Output	Labor-Hours	Unit	Material	Labor	Equipment	Total	Total Incl O&P
0010	**BAROMETRIC DAMPERS**									
1000	Barometric, gas fired system only, 6″ size for 5″ and 6″ pipes	1 Shee	20	.400	Ea.	105	25		130	154
1020	7″ size, for 6″ and 7″ pipes		19	.421		115	26		141	166
1040	8″ size, for 7″ and 8″ pipes		18	.444		148	27.50		175.50	205
1060	9″ size, for 8″ and 9″ pipes		16	.500		164	31		195	228
2000	All fuel, oil, oil/gas, coal									
2020	10″ for 9″ and 10″ pipes	1 Shee	15	.533	Ea.	255	33		288	330
2040	12″ for 11″ and 12″ pipes		15	.533		335	33		368	415
2060	14″ for 13″ and 14″ pipes		14	.571		435	35.50		470.50	530
2080	16″ for 15″ and 16″ pipes		13	.615		605	38.50		643.50	725
2100	18″ for 17″ and 18″ pipes		12	.667		805	41.50		846.50	950
2120	20″ for 19″ and 21″ pipes		10	.800		965	50		1,015	1,125
2140	24″ for 22″ and 25″ pipes	Q-9	12	1.333		1,175	75		1,250	1,425
2160	28″ for 26″ and 30″ pipes		10	1.600		1,475	89.50		1,564.50	1,725
2180	32″ for 31″ and 34″ pipes		8	2		1,875	112		1,987	2,225
3260	For thermal switch for above, add	1 Shee	24	.333		105	21		126	148

23 51 23 – Gas Vents

23 51 23.10 Gas Chimney Vents

Line	Description	Crew	Daily Output	Labor-Hours	Unit	Material	Labor	Equipment	Total	Total Incl O&P
0010	**GAS CHIMNEY VENTS**, Prefab metal, UL listed									
0020	Gas, double wall, galvanized steel									
0080	3″ diameter	Q-9	72	.222	V.L.F.	7.60	12.45		20.05	27.50
0100	4″ diameter		68	.235		9.35	13.20		22.55	30.50
0120	5″ diameter		64	.250		10.60	14		24.60	33
0140	6″ diameter		60	.267		12.65	14.95		27.60	36.50
0160	7″ diameter		56	.286		24.50	16		40.50	51.50
0180	8″ diameter		52	.308		26	17.25		43.25	54.50
0200	10″ diameter		48	.333		49.50	18.70		68.20	83
0220	12″ diameter		44	.364		58.50	20.50		79	95.50
0240	14″ diameter		42	.381		97.50	21.50		119	140

23 51 23.10 Gas Chimney Vents		Crew	Daily Output	Labor-Hours	Unit	Material	2020 Bare Costs Labor	Equipment	Total	Total Incl O&P
0260	16″ diameter	Q-9	40	.400	V.L.F.	140	22.50		162.50	188
0280	18″ diameter		38	.421		173	23.50		196.50	226
0300	20″ diameter	Q-10	36	.667		205	39		244	285
0320	22″ diameter		34	.706		261	41		302	350
0340	24″ diameter		32	.750		320	43.50		363.50	415
0600	For 4″, 5″ and 6″ oval, add					50%				
0650	Gas, double wall, galvanized steel, fittings									
0660	Elbow 45°, 3″ diameter	Q-9	36	.444	Ea.	13.45	25		38.45	53
0670	4″ diameter		34	.471		16.45	26.50		42.95	58
0680	5″ diameter		32	.500		19.05	28		47.05	63.50
0690	6″ diameter		30	.533		23.50	30		53.50	71.50
0700	7″ diameter		28	.571		38	32		70	90.50
0710	8″ diameter		26	.615		50	34.50		84.50	108
0720	10″ diameter		24	.667		106	37.50		143.50	174
0730	12″ diameter		22	.727		113	41		154	186
0740	14″ diameter		21	.762		178	42.50		220.50	261
0750	16″ diameter		20	.800		231	45		276	320
0760	18″ diameter		19	.842		305	47		352	405
0770	20″ diameter	Q-10	18	1.333		340	77.50		417.50	490
0780	22″ diameter		17	1.412		545	82		627	725
0790	24″ diameter		16	1.500		695	87		782	895
0916	Adjustable length									
0918	3″ diameter, to 12″	Q-9	36	.444	Ea.	16.95	25		41.95	56.50
0920	4″ diameter, to 12″		34	.471		19.70	26.50		46.20	61.50
0924	6″ diameter, to 12″		30	.533		25.50	30		55.50	73.50
0928	8″ diameter, to 12″		26	.615		47.50	34.50		82	105
0930	10″ diameter, to 18″		24	.667		148	37.50		185.50	219
0932	12″ diameter, to 18″		22	.727		150	41		191	227
0936	16″ diameter, to 18″		20	.800		300	45		345	400
0938	18″ diameter, to 18″		19	.842		340	47		387	445
0944	24″ diameter, to 18″	Q-10	16	1.500		655	87		742	850
0950	Elbow 90°, adjustable, 3″ diameter	Q-9	36	.444		23	25		48	63.50
0960	4″ diameter		34	.471		27	26.50		53.50	69.50
0970	5″ diameter		32	.500		32.50	28		60.50	78.50
0980	6″ diameter		30	.533		39	30		69	88.50
0990	7″ diameter		28	.571		67	32		99	123
1010	8″ diameter		26	.615		68.50	34.50		103	128
1020	Wall thimble, 4 to 7″ adjustable, 3″ diameter		36	.444		14.40	25		39.40	54
1022	4″ diameter		34	.471		16.20	26.50		42.70	58
1024	5″ diameter		32	.500		19.40	28		47.40	64
1026	6″ diameter		30	.533		20	30		50	67.50
1028	7″ diameter		28	.571		41.50	32		73.50	94.50
1030	8″ diameter		26	.615		51	34.50		85.50	109
1040	Roof flashing, 3″ diameter		36	.444		7.85	25		32.85	46.50
1050	4″ diameter		34	.471		9.15	26.50		35.65	50
1060	5″ diameter		32	.500		27.50	28		55.50	73
1070	6″ diameter		30	.533		22	30		52	69.50
1080	7″ diameter		28	.571		29	32		61	80
1090	8″ diameter		26	.615		31	34.50		65.50	86.50
1100	10″ diameter		24	.667		41.50	37.50		79	102
1110	12″ diameter		22	.727		58	41		99	126
1120	14″ diameter		20	.800		132	45		177	213
1130	16″ diameter		18	.889		158	50		208	250

23 51 23.10 Gas Chimney Vents		Crew	Daily Output	Labor-Hours	Unit	Material	2020 Bare Costs Labor	Equipment	Total	Total Incl O&P
1140	18" diameter	Q-9	16	1	Ea.	222	56		278	330
1150	20" diameter	Q-10	18	1.333		288	77.50		365.50	435
1160	22" diameter		14	1.714		360	99.50		459.50	545
1170	24" diameter		12	2		415	116		531	630
1200	Tee, 3" diameter	Q-9	27	.593		35.50	33		68.50	89.50
1210	4" diameter		26	.615		38	34.50		72.50	94
1220	5" diameter		25	.640		40	36		76	98.50
1230	6" diameter		24	.667		45.50	37.50		83	107
1240	7" diameter		23	.696		65	39		104	131
1250	8" diameter		22	.727		72.50	41		113.50	142
1260	10" diameter		21	.762		191	42.50		233.50	275
1270	12" diameter		20	.800		195	45		240	283
1280	14" diameter		18	.889		350	50		400	460
1290	16" diameter		16	1		520	56		576	655
1300	18" diameter		14	1.143		625	64		689	780
1310	20" diameter	Q-10	17	1.412		855	82		937	1,075
1320	22" diameter		13	1.846		1,100	107		1,207	1,375
1330	24" diameter		12	2		1,275	116		1,391	1,575
1460	Tee cap, 3" diameter	Q-9	45	.356		2.38	19.95		22.33	32.50
1470	4" diameter		42	.381		2.57	21.50		24.07	35.50
1480	5" diameter		40	.400		3.31	22.50		25.81	37.50
1490	6" diameter		37	.432		4.71	24.50		29.21	42
1500	7" diameter		35	.457		8.65	25.50		34.15	48.50
1510	8" diameter		34	.471		8.65	26.50		35.15	49.50
1520	10" diameter		32	.500		69.50	28		97.50	119
1530	12" diameter		30	.533		69	30		99	122
1540	14" diameter		28	.571		72.50	32		104.50	128
1550	16" diameter		25	.640		74	36		110	136
1560	18" diameter		24	.667		87.50	37.50		125	153
1570	20" diameter	Q-10	27	.889		97.50	51.50		149	186
1580	22" diameter		22	1.091		171	63.50		234.50	284
1590	24" diameter		21	1.143		200	66.50		266.50	320
1750	Top, 3" diameter	Q-9	46	.348		17.75	19.50		37.25	49
1760	4" diameter		44	.364		18.65	20.50		39.15	51.50
1770	5" diameter		42	.381		17.95	21.50		39.45	52.50
1780	6" diameter		40	.400		23.50	22.50		46	60
1790	7" diameter		38	.421		51.50	23.50		75	92.50
1800	8" diameter		36	.444		63	25		88	108
1810	10" diameter		34	.471		103	26.50		129.50	154
1820	12" diameter		32	.500		129	28		157	185
1830	14" diameter		30	.533		205	30		235	271
1840	16" diameter		28	.571		260	32		292	335
1850	18" diameter		26	.615		360	34.50		394.50	450
1860	20" diameter	Q-10	28	.857		555	50		605	685
1870	22" diameter		22	1.091		830	63.50		893.50	1,000
1880	24" diameter		20	1.200		1,050	70		1,120	1,250
1900	Gas, double wall, galvanized steel, oval									
1904	4" x 1'	Q-9	68	.235	V.L.F.	18.60	13.20		31.80	40.50
1906	5" x 1'		64	.250		92.50	14		106.50	124
1908	5"/6" x 1'		60	.267		46.50	14.95		61.45	73.50
1910	Oval fittings									
1912	Adjustable length									
1914	4" diameter to 12" long	Q-9	34	.471	Ea.	19.35	26.50		45.85	61.50

23 51 23 – Gas Vents

23 51 23.10 Gas Chimney Vents		Crew	Daily Output	Labor-Hours	Unit	Material	Labor	Equipment	Total	Total Incl O&P
						2020 Bare Costs				
1916	5″ diameter to 12″ long	Q-9	32	.500	Ea.	21.50	28		49.50	66
1918	5″/6″ diameter to 12″ long		30	.533		26	30		56	74
1920	Elbow 45°									
1922	4″	Q-9	34	.471	Ea.	31.50	26.50		58	74.50
1924	5″		32	.500		62.50	28		90.50	111
1926	5″/6″		30	.533		64	30		94	116
1930	Elbow 45°, flat									
1932	4″	Q-9	34	.471	Ea.	31	26.50		57.50	74
1934	5″		32	.500		62.50	28		90.50	111
1936	5″/6″		30	.533		65	30		95	117
1940	Top									
1942	4″	Q-9	44	.364	Ea.	37	20.50		57.50	71.50
1944	5″		42	.381		35.50	21.50		57	71.50
1946	5″/6″		40	.400		46	22.50		68.50	85
1950	Adjustable flashing									
1952	4″	Q-9	34	.471	Ea.	13.25	26.50		39.75	54.50
1954	5″		32	.500		39	28		67	85
1956	5″/6″		30	.533		40	30		70	89.50
1960	Tee									
1962	4″	Q-9	26	.615	Ea.	46	34.50		80.50	103
1964	5″		25	.640		95.50	36		131.50	160
1966	5″/6″		24	.667		97	37.50		134.50	163
1970	Tee with short snout									
1972	4″	Q-9	26	.615	Ea.	49	34.50		83.50	107

23 51 26 – All-Fuel Vent Chimneys

23 51 26.10 All-Fuel Vent Chimneys, Press. Tight, Dbl. Wall		Crew	Daily Output	Labor-Hours	Unit	Material	Labor	Equipment	Total	Total Incl O&P
0010	**ALL-FUEL VENT CHIMNEYS, PRESSURE TIGHT, DOUBLE WALL**									
3200	All fuel, pressure tight, double wall, 1″ insulation, UL listed, 1,400°F.									
3210	304 stainless steel liner, aluminized steel outer jacket									
3220	6″ diameter	Q-9	60	.267	L.F.	60.50	14.95		75.45	89
3221	8″ diameter		52	.308		65	17.25		82.25	97.50
3222	10″ diameter		48	.333		72.50	18.70		91.20	109
3223	12″ diameter		44	.364		70.50	20.50		91	109
3224	14″ diameter		42	.381		90.50	21.50		112	132
3225	16″ diameter		40	.400		106	22.50		128.50	151
3226	18″ diameter		38	.421		114	23.50		137.50	162
3227	20″ diameter	Q-10	36	.667		136	39		175	209
3228	24″ diameter	″	32	.750		174	43.50		217.50	257
3260	For 316 stainless steel liner, add					30%				
3280	All fuel, pressure tight, double wall fittings									
3284	304 stainless steel inner, aluminized steel jacket									
3288	Adjustable 20″/29″ section									
3292	6″ diameter	Q-9	30	.533	Ea.	204	30		234	270
3293	8″ diameter		26	.615		218	34.50		252.50	293
3294	10″ diameter		24	.667		245	37.50		282.50	325
3295	12″ diameter		22	.727		279	41		320	365
3296	14″ diameter		21	.762		295	42.50		337.50	390
3297	16″ diameter		20	.800		345	45		390	450
3298	18″ diameter		19	.842		395	47		442	505
3299	20″ diameter	Q-10	18	1.333		430	77.50		507.50	595
3300	24″ diameter	″	16	1.500		555	87		642	740
3350	Elbow 90° fixed									

23 51 26.10 All-Fuel Vent Chimneys, Press. Tight, Dbl. Wall		Crew	Daily Output	Labor-Hours	Unit	Material	2020 Bare Costs Labor	Equipment	Total	Total Incl O&P
3354	6" diameter	Q-9	30	.533	Ea.	420	30		450	510
3355	8" diameter		26	.615		475	34.50		509.50	575
3356	10" diameter		24	.667		540	37.50		577.50	645
3357	12" diameter		22	.727		610	41		651	730
3358	14" diameter		21	.762		690	42.50		732.50	825
3359	16" diameter		20	.800		780	45		825	930
3360	18" diameter		19	.842		880	47		927	1,050
3361	20" diameter	Q-10	18	1.333		995	77.50		1,072.50	1,225
3362	24" diameter	"	16	1.500		1,275	87		1,362	1,525
3380	For 316 stainless steel liner, add					30%				
3400	Elbow 45°									
3404	6" diameter	Q-9	30	.533	Ea.	212	30		242	279
3405	8" diameter		26	.615		238	34.50		272.50	315
3406	10" diameter		24	.667		271	37.50		308.50	355
3407	12" diameter		22	.727		310	41		351	400
3408	14" diameter		21	.762		350	42.50		392.50	450
3409	16" diameter		20	.800		395	45		440	500
3410	18" diameter		19	.842		445	47		492	560
3411	20" diameter	Q-10	18	1.333		515	77.50		592.50	690
3412	24" diameter	"	16	1.500		660	87		747	855
3430	For 316 stainless steel liner, add					30%				
3450	Tee 90°									
3454	6" diameter	Q-9	24	.667	Ea.	252	37.50		289.50	335
3455	8" diameter		22	.727		275	41		316	360
3456	10" diameter		21	.762		310	42.50		352.50	405
3457	12" diameter		20	.800		360	45		405	465
3458	14" diameter		18	.889		405	50		455	525
3459	16" diameter		16	1		455	56		511	590
3460	18" diameter		14	1.143		535	64		599	685
3461	20" diameter	Q-10	17	1.412		610	82		692	795
3462	24" diameter	"	12	2		760	116		876	1,000
3480	For tee cap, add					35%	20%			
3500	For 316 stainless steel liner, add					30%				
3520	Plate support, galvanized									
3524	6" diameter	Q-9	26	.615	Ea.	125	34.50		159.50	191
3525	8" diameter		22	.727		147	41		188	223
3526	10" diameter		20	.800		160	45		205	244
3527	12" diameter		18	.889		168	50		218	261
3528	14" diameter		17	.941		198	53		251	298
3529	16" diameter		16	1		209	56		265	315
3530	18" diameter		15	1.067		220	60		280	335
3531	20" diameter	Q-10	16	1.500		231	87		318	385
3532	24" diameter	"	14	1.714		240	99.50		339.50	415
3570	Bellows, lined									
3574	6" diameter	Q-9	30	.533	Ea.	1,350	30		1,380	1,525
3575	8" diameter		26	.615		1,400	34.50		1,434.50	1,600
3576	10" diameter		24	.667		1,425	37.50		1,462.50	1,625
3577	12" diameter		22	.727		1,475	41		1,516	1,675
3578	14" diameter		21	.762		1,525	42.50		1,567.50	1,750
3579	16" diameter		20	.800		1,575	45		1,620	1,825
3580	18" diameter		19	.842		1,650	47		1,697	1,875
3581	20" diameter	Q-10	18	1.333		1,675	77.50		1,752.50	1,975
3590	For all 316 stainless steel construction, add					55%				

23 51 Breechings, Chimneys, and Stacks

23 51 26 – All-Fuel Vent Chimneys

23 51 26.10 All-Fuel Vent Chimneys, Press. Tight, Dbl. Wall		Crew	Daily Output	Labor-Hours	Unit	2020 Bare Costs Material	Labor	Equipment	Total	Total Incl O&P
3600	Ventilated roof thimble, 304 stainless steel									
3620	6" diameter	Q-9	26	.615	Ea.	264	34.50		298.50	345
3624	8" diameter		22	.727		259	41		300	345
3625	10" diameter		20	.800		272	45		317	365
3626	12" diameter		18	.889		282	50		332	385
3627	14" diameter		17	.941		305	53		358	415
3628	16" diameter		16	1		325	56		381	445
3629	18" diameter		15	1.067		350	60		410	475
3630	20" diameter	Q-10	16	1.500		370	87		457	540
3631	24" diameter	"	14	1.714		410	99.50		509.50	605
3650	For 316 stainless steel, add					30%				
3670	Exit cone, 316 stainless steel only									
3674	6" diameter	Q-9	46	.348	Ea.	199	19.50		218.50	249
3675	8" diameter		42	.381		204	21.50		225.50	258
3676	10" diameter		40	.400		210	22.50		232.50	265
3677	12" diameter		38	.421		230	23.50		253.50	289
3678	14" diameter		37	.432		239	24.50		263.50	300
3679	16" diameter		36	.444		294	25		319	365
3680	18" diameter		35	.457		320	25.50		345.50	395
3681	20" diameter	Q-10	28	.857		385	50		435	500
3682	24" diameter	"	26	.923		510	53.50		563.50	640
3720	Roof guide, 304 stainless steel									
3724	6" diameter	Q-9	25	.640	Ea.	93	36		129	157
3725	8" diameter		21	.762		108	42.50		150.50	184
3726	10" diameter		19	.842		119	47		166	202
3727	12" diameter		17	.941		123	53		176	215
3728	14" diameter		16	1		143	56		199	242
3729	16" diameter		15	1.067		151	60		211	257
3730	18" diameter		14	1.143		160	64		224	274
3731	20" diameter	Q-10	15	1.600		169	93		262	325
3732	24" diameter	"	13	1.846		176	107		283	355
3750	For 316 stainless steel, add					30%				
3770	Rain cap with bird screen									
3774	6" diameter	Q-9	46	.348	Ea.	305	19.50		324.50	365
3775	8" diameter		42	.381		350	21.50		371.50	420
3776	10" diameter		40	.400		410	22.50		432.50	485
3777	12" diameter		38	.421		475	23.50		498.50	555
3778	14" diameter		37	.432		545	24.50		569.50	635
3779	16" diameter		36	.444		620	25		645	720
3780	18" diameter		35	.457		705	25.50		730.50	815
3781	20" diameter	Q-10	28	.857		800	50		850	955
3782	24" diameter	"	26	.923		965	53.50		1,018.50	1,125

23 51 26.30 All-Fuel Vent Chimneys, Double Wall, St. Stl.		Crew	Daily Output	Labor-Hours	Unit	Material	Labor	Equipment	Total	Total Incl O&P
0010	**ALL-FUEL VENT CHIMNEYS, DOUBLE WALL, STAINLESS STEEL**									
7780	All fuel, pressure tight, double wall, 4" insulation, UL listed, 1,400°F.									
7790	304 stainless steel liner, aluminized steel outer jacket									
7800	6" diameter	Q-9	60	.267	V.L.F.	64.50	14.95		79.45	93.50
7804	8" diameter		52	.308		77	17.25		94.25	111
7806	10" diameter		48	.333		87.50	18.70		106.20	125
7808	12" diameter		44	.364		101	20.50		121.50	142
7810	14" diameter		42	.381		113	21.50		134.50	157
7880	For 316 stainless steel liner, add				L.F.	30%				

23 51 Breechings, Chimneys, and Stacks

23 51 26 – All-Fuel Vent Chimneys

23 51 26.30 All-Fuel Vent Chimneys, Double Wall, St. Stl.		Crew	Daily Output	Labor-Hours	Unit	Material	2020 Bare Costs Labor	Equipment	Total	Total Incl O&P
8000	All fuel, double wall, stainless steel fittings									
8010	Roof support, 6" diameter	Q-9	30	.533	Ea.	120	30		150	178
8030	8" diameter		26	.615		136	34.50		170.50	202
8040	10" diameter		24	.667		151	37.50		188.50	223
8050	12" diameter		22	.727		157	41		198	234
8060	14" diameter		21	.762		167	42.50		209.50	248
8100	Elbow 45°, 6" diameter		30	.533		252	30		282	325
8140	8" diameter		26	.615		286	34.50		320.50	370
8160	10" diameter		24	.667		330	37.50		367.50	420
8180	12" diameter		22	.727		375	41		416	470
8200	14" diameter		21	.762		410	42.50		452.50	515
8300	Insulated tee, 6" diameter		30	.533		296	30		326	370
8360	8" diameter		26	.615		330	34.50		364.50	420
8380	10" diameter		24	.667		350	37.50		387.50	440
8400	12" diameter		22	.727		420	41		461	520
8420	14" diameter		20	.800		490	45		535	610
8500	Boot tee, 6" diameter		28	.571		610	32		642	720
8520	8" diameter		24	.667		670	37.50		707.50	790
8530	10" diameter		22	.727		770	41		811	905
8540	12" diameter		20	.800		905	45		950	1,075
8550	14" diameter		18	.889		945	50		995	1,125
8600	Rain cap with bird screen, 6" diameter		30	.533		297	30		327	370
8640	8" diameter		26	.615		350	34.50		384.50	440
8660	10" diameter		24	.667		410	37.50		447.50	505
8680	12" diameter		22	.727		470	41		511	580
8700	14" diameter		21	.762		535	42.50		577.50	655
8800	Flat roof flashing, 6" diameter		30	.533		111	30		141	168
8840	8" diameter		26	.615		121	34.50		155.50	186
8860	10" diameter		24	.667		132	37.50		169.50	202
8880	12" diameter		22	.727		145	41		186	222
8900	14" diameter		21	.762		147	42.50		189.50	227

23 52 Heating Boilers

23 52 13 – Electric Boilers

23 52 13.10 Electric Boilers, ASME		Crew	Daily Output	Labor-Hours	Unit	Material	Labor	Equipment	Total	Total Incl O&P
0010	**ELECTRIC BOILERS, ASME**, Standard controls and trim D3020–102									
1000	Steam, 6 KW, 20.5 MBH	Q-19	1.20	20	Ea.	4,450	1,200		5,650	6,650
1040	9 KW, 30.7 MBH		1.20	20		4,400	1,200		5,600	6,625
1060	18 KW, 61.4 MBH		1.20	20		4,525	1,200		5,725	6,750
1080	24 KW, 81.8 MBH		1.10	21.818		5,250	1,300		6,550	7,725
1120	36 KW, 123 MBH		1.10	21.818		5,800	1,300		7,100	8,350
2000	Hot water, 7.5 KW, 25.6 MBH		1.30	18.462		5,425	1,100		6,525	7,600
2020	15 KW, 51.2 MBH		1.30	18.462		5,450	1,100		6,550	7,650
2040	30 KW, 102 MBH		1.20	20		5,850	1,200		7,050	8,200
2060	45 KW, 164 MBH		1.20	20		5,700	1,200		6,900	8,025
2070	60 KW, 205 MBH		1.20	20		5,800	1,200		7,000	8,175
2080	75 KW, 256 MBH		1.10	21.818		6,150	1,300		7,450	8,700

23 52 16 – Condensing Boilers

23 52 16.24 Condensing Boilers			Crew	Daily Output	Labor-Hours	Unit	Material	Labor (2020 Bare Costs)	Equipment	Total	Total Incl O&P
0010	**CONDENSING BOILERS**, Cast iron, high efficiency										
0020	Packaged with standard controls, circulator and trim										
0030	Intermittent (spark) pilot, natural or LP gas										
0040	Hot water, DOE MBH output (AFUE %)										
0100	42 MBH (84.0%)	G	Q-5	1.80	8.889	Ea.	1,625	525		2,150	2,575
0120	57 MBH (84.3%)	G		1.60	10		1,775	590		2,365	2,825
0140	85 MBH (84.0%)	G		1.40	11.429		1,975	675		2,650	3,175
0160	112 MBH (83.7%)	G		1.20	13.333		2,225	785		3,010	3,625
0180	140 MBH (83.3%)	G	Q-6	1.60	15		2,500	920		3,420	4,125
0200	167 MBH (83.0%)	G		1.40	17.143		2,825	1,050		3,875	4,700
0220	194 MBH (82.7%)	G		1.20	20		3,125	1,225		4,350	5,250

23 52 19 – Pulse Combustion Boilers

23 52 19.20 Pulse Type Combustion Boilers			Crew	Daily Output	Labor-Hours	Unit	Material	Labor	Equipment	Total	Total Incl O&P
0010	**PULSE TYPE COMBUSTION BOILERS**, High efficiency										
7990	Special feature gas fired boilers										
8000	Pulse combustion, standard controls/trim										
8010	Hot water, DOE MBH output (AFUE %)										
8030	71 MBH (95.2%)		Q-5	1.60	10	Ea.	3,975	590		4,565	5,225
8050	94 MBH (95.3%)	G		1.40	11.429		4,250	675		4,925	5,675
8080	139 MBH (95.6%)	G		1.20	13.333		4,950	785		5,735	6,625
8090	207 MBH (95.4%)			1.16	13.793		5,600	815		6,415	7,400
8120	270 MBH (96.4%)			1.12	14.286		7,750	845		8,595	9,775
8130	365 MBH (91.7%)			1.07	14.953		8,750	880		9,630	11,000

23 52 23 – Cast-Iron Boilers

23 52 23.20 Gas-Fired Boilers		Crew	Daily Output	Labor-Hours	Unit	Material	Labor	Equipment	Total	Total Incl O&P
0010	**GAS-FIRED BOILERS**, Natural or propane, standard controls, packaged									
1000	Cast iron, with insulated jacket									
2000	Steam, gross output, 81 MBH	Q-7	1.40	22.857	Ea.	2,750	1,425		4,175	5,150
2020	102 MBH		1.30	24.615		2,625	1,525		4,150	5,200
2040	122 MBH		1	32		3,250	2,000		5,250	6,575
2060	163 MBH		.90	35.556		3,400	2,225		5,625	7,075
2080	203 MBH		.90	35.556		3,950	2,225		6,175	7,675
2100	240 MBH		.85	37.647		4,025	2,350		6,375	7,950
2120	280 MBH		.80	40		4,525	2,500		7,025	8,700
2140	320 MBH		.70	45.714		5,200	2,850		8,050	9,975
3000	Hot water, gross output, 80 MBH		1.46	21.918		1,975	1,375		3,350	4,225
3020	100 MBH		1.35	23.704		2,575	1,475		4,050	5,050
3040	122 MBH		1.10	29.091		2,825	1,825		4,650	5,825
3060	163 MBH		1	32		3,275	2,000		5,275	6,600
3080	203 MBH		1	32		3,950	2,000		5,950	7,350
3100	240 MBH		.95	33.684		3,825	2,100		5,925	7,375
3120	280 MBH		.90	35.556		4,375	2,225		6,600	8,150
3140	320 MBH		.80	40		4,275	2,500		6,775	8,425
7000	For tankless water heater, add					10%				
7050	For additional zone valves up to 312 MBH, add					199			199	219

23 52 23.40 Oil-Fired Boilers		Crew	Daily Output	Labor-Hours	Unit	Material	Labor	Equipment	Total	Total Incl O&P
0010	**OIL-FIRED BOILERS**, Standard controls, flame retention burner, packaged									
1000	Cast iron, with insulated flush jacket									
2000	Steam, gross output, 109 MBH	Q-7	1.20	26.667	Ea.	2,375	1,675		4,050	5,100
2020	144 MBH		1.10	29.091		2,650	1,825		4,475	5,650
2040	173 MBH		1	32		3,000	2,000		5,000	6,300

23 52 23 – Cast-Iron Boilers

23 52 23.40 Oil-Fired Boilers		Crew	Daily Output	Labor-Hours	Unit	Material	Labor	Equipment	Total	Total Incl O&P
						2020 Bare Costs				
2060	207 MBH	Q-7	.90	35.556	Ea.	3,225	2,225		5,450	6,850
3000	Hot water, same price as steam									
4000	For tankless coil in smaller sizes, add				Ea.	15%				

23 52 26 – Steel Boilers

23 52 26.40 Oil-Fired Boilers

		Crew	Daily Output	Labor-Hours	Unit	Material	Labor	Equipment	Total	Total Incl O&P
0010	**OIL-FIRED BOILERS**, Standard controls, flame retention burner									
5000	Steel, with insulated flush jacket									
7000	Hot water, gross output, 103 MBH	Q-6	1.60	15	Ea.	2,000	920		2,920	3,575
7020	122 MBH		1.45	16.506		2,150	1,000		3,150	3,875
7040	137 MBH		1.36	17.595		2,325	1,075		3,400	4,150
7060	168 MBH		1.30	18.405		2,275	1,125		3,400	4,200
7080	225 MBH	↓	1.22	19.704		4,000	1,200		5,200	6,200
7340	For tankless coil in steam or hot water, add				↓	7%				

23 52 28 – Swimming Pool Boilers

23 52 28.10 Swimming Pool Heaters

		Crew	Daily Output	Labor-Hours	Unit	Material	Labor	Equipment	Total	Total Incl O&P
0010	**SWIMMING POOL HEATERS**, Not including wiring, external									
0020	piping, base or pad									
0160	Gas fired, input, 155 MBH	Q-6	1.50	16	Ea.	1,850	980		2,830	3,500
0200	199 MBH		1	24		1,975	1,475		3,450	4,375
0220	250 MBH		.70	34.286		2,175	2,100		4,275	5,525
0240	300 MBH		.60	40		2,275	2,450		4,725	6,175
0260	399 MBH		.50	48		2,550	2,925		5,475	7,200
0280	500 MBH		.40	60		8,500	3,675		12,175	14,900
0300	650 MBH		.35	68.571		9,050	4,200		13,250	16,200
0320	750 MBH		.33	72.727		9,875	4,450		14,325	17,600
0360	990 MBH		.22	109		13,300	6,675		19,975	24,600
0370	1,260 MBH		.21	114		17,700	7,000		24,700	30,000
0380	1,440 MBH		.19	126		19,000	7,725		26,725	32,500
0400	1,800 MBH		.14	171		21,000	10,500		31,500	38,800
0410	2,070 MBH	↓	.13	185		24,800	11,300		36,100	44,200
2000	Electric, 12 KW, 4,800 gallon pool	Q-19	3	8		2,200	480		2,680	3,150
2020	15 KW, 7,200 gallon pool		2.80	8.571		2,225	510		2,735	3,225
2040	24 KW, 9,600 gallon pool		2.40	10		2,575	600		3,175	3,750
2060	30 KW, 12,000 gallon pool		2	12		2,650	715		3,365	4,000
2080	36 KW, 14,400 gallon pool		1.60	15		3,075	895		3,970	4,725
2100	57 KW, 24,000 gallon pool	↓	1.20	20	↓	3,875	1,200		5,075	6,025
9000	To select pool heater: 12 BTUH x S.F. pool area									
9010	X temperature differential = required output									
9050	For electric, KW = gallons x 2.5 divided by 1,000									
9100	For family home type pool, double the									
9110	Rated gallon capacity = 1/2°F rise per hour									

23 52 88 – Burners

23 52 88.10 Replacement Type Burners

		Crew	Daily Output	Labor-Hours	Unit	Material	Labor	Equipment	Total	Total Incl O&P
0010	**REPLACEMENT TYPE BURNERS**									
0990	Residential, conversion, gas fired, LP or natural									
1000	Gun type, atmospheric input 50 to 225 MBH	Q-1	2.50	6.400	Ea.	1,025	370		1,395	1,675
1020	100 to 400 MBH	"	2	8	"	1,725	465		2,190	2,600
3000	Flame retention oil fired assembly, input									
3020	.50 to 2.25 GPH	Q-1	2.40	6.667	Ea.	360	385		745	975
3040	2.0 to 5.0 GPH	"	2	8		335	465		800	1,075
4600	Gas safety, shut off valve, 3/4" threaded	1 Stpi	20	.400	↓	190	26		216	247

23 52 Heating Boilers

23 52 88 – Burners

23 52 88.10 Replacement Type Burners		Crew	Daily Output	Labor-Hours	Unit	2020 Bare Costs Material	Labor	Equipment	Total	Total Incl O&P
4610	1" threaded	1 Stpi	19	.421	Ea.	183	27.50		210.50	243
4620	1-1/4" threaded		15	.533		206	35		241	280
4630	1-1/2" threaded		13	.615		223	40.50		263.50	305
4640	2" threaded		11	.727		250	47.50		297.50	345
4650	2-1/2" threaded	Q-1	15	1.067		286	62		348	410
4660	3" threaded		13	1.231		395	71.50		466.50	540
4670	4" flanged		3	5.333		2,800	310		3,110	3,550
4680	6" flanged	Q-2	3	8		6,250	480		6,730	7,600

23 54 Furnaces

23 54 16 – Fuel-Fired Furnaces

23 54 16.14 Condensing Furnaces		Crew	Daily Output	Labor-Hours	Unit	Material	Labor	Equipment	Total	Total Incl O&P
0010	**CONDENSING FURNACES**, High efficiency									
0020	Oil fired, packaged, complete									
0030	Upflow									
0040	Output @ 95% A.F.U.E.									
0100	49 MBH @ 1,000 CFM	Q-9	3.70	4.324	Ea.	7,375	243		7,618	8,475
0110	73.5 MBH @ 2,000 CFM		3.60	4.444		7,825	249		8,074	9,000
0120	96 MBH @ 2,000 CFM		3.40	4.706		7,825	264		8,089	9,025
0130	115.6 MBH @ 2,000 CFM		3.40	4.706		7,825	264		8,089	9,025
0140	147 MBH @ 2,000 CFM		3.30	4.848		18,400	272		18,672	20,700
0150	192 MBH @ 4,000 CFM		2.60	6.154		19,000	345		19,345	21,400
0170	231.5 MBH @ 4,000 CFM		2.30	6.957		19,000	390		19,390	21,500
0260	For variable speed motor, add					785			785	865
0270	Note: Also available in horizontal, counterflow and lowboy configurations.									

23 55 Fuel-Fired Heaters

23 55 23 – Gas-Fired Radiant Heaters

23 55 23.10 Infrared Type Heating Units		Crew	Daily Output	Labor-Hours	Unit	Material	Labor	Equipment	Total	Total Incl O&P
0010	**INFRARED TYPE HEATING UNITS**									
0020	Gas fired, unvented, electric ignition, 100% shutoff.									
0030	Piping and wiring not included									
0100	Input, 30 MBH	Q-5	6	2.667	Ea.	1,050	157		1,207	1,400
0120	45 MBH		5	3.200		995	189		1,184	1,375
0140	50 MBH		4.50	3.556		1,000	210		1,210	1,425
0160	60 MBH		4	4		1,025	236		1,261	1,475
0180	75 MBH		3	5.333		1,025	315		1,340	1,600
0200	90 MBH		2.50	6.400		1,050	380		1,430	1,725
0220	105 MBH		2	8		1,100	470		1,570	1,925
0240	120 MBH		2	8		1,625	470		2,095	2,500
1000	Gas fired, vented, electric ignition, tubular									
1020	Piping and wiring not included, 20' to 80' lengths									
1030	Single stage, input, 60 MBH	Q-6	4.50	5.333	Ea.	1,550	325		1,875	2,200
1040	80 MBH		3.90	6.154		1,550	375		1,925	2,275
1050	100 MBH		3.40	7.059		1,550	430		1,980	2,350
1060	125 MBH		2.90	8.276		1,525	505		2,030	2,425
1070	150 MBH		2.70	8.889		1,525	545		2,070	2,500
1080	170 MBH		2.50	9.600		1,525	585		2,110	2,550
1090	200 MBH		2.20	10.909		1,725	665		2,390	2,900

23 55 Fuel-Fired Heaters

23 55 23 – Gas-Fired Radiant Heaters

23 55 23.10 Infrared Type Heating Units		Crew	Daily Output	Labor-Hours	Unit	2020 Bare Costs Material	Labor	Equipment	Total	Total Incl O&P
1100	Note: Final pricing may vary due to									
1110	tube length and configuration package selected									
1130	Two stage, input, 60 MBH high, 45 MBH low	Q-6	4.50	5.333	Ea.	1,900	325		2,225	2,575
1140	80 MBH high, 60 MBH low		3.90	6.154		1,900	375		2,275	2,650
1150	100 MBH high, 65 MBH low		3.40	7.059		1,900	430		2,330	2,725
1160	125 MBH high, 95 MBH low		2.90	8.276		1,925	505		2,430	2,850
1170	150 MBH high, 100 MBH low		2.70	8.889		1,925	545		2,470	2,925
1180	170 MBH high, 125 MBH low		2.50	9.600		2,125	585		2,710	3,225
1190	200 MBH high, 150 MBH low		2.20	10.909		2,125	665		2,790	3,350
1220	Note: Final pricing may vary due to									
1230	tube length and configuration package selected									

23 56 Solar Energy Heating Equipment

23 56 16 – Packaged Solar Heating Equipment

23 56 16.40 Solar Heating Systems		Crew	Daily Output	Labor-Hours	Unit	Material	Labor	Equipment	Total	Total Incl O&P
0010	SOLAR HEATING SYSTEMS D2020–265									
0020	System/package prices, not including connecting									
0030	pipe, insulation, or special heating/plumbing fixtures D2020–270									
0152	For solar ultraviolet pipe insulation see Section 22 07 19.10									
0500	Hot water, standard package, low temperature D2020–275									
0540	1 collector, circulator, fittings, 65 gal. tank G	Q-1	.50	32	Ea.	4,400	1,850		6,250	7,600
0580	2 collectors, circulator, fittings, 120 gal. tank D2020–280 G		.40	40		5,750	2,325		8,075	9,800
0620	3 collectors, circulator, fittings, 120 gal. tank G		.34	47.059		7,850	2,725		10,575	12,700
0700	Medium temperature package D2020–285									
0720	1 collector, circulator, fittings, 80 gal. tank G	Q-1	.50	32	Ea.	5,450	1,850		7,300	8,750
0740	2 collectors, circulator, fittings, 120 gal. tank D2020–290 G		.40	40		7,400	2,325		9,725	11,600
0780	3 collectors, circulator, fittings, 120 gal. tank G		.30	53.333		8,250	3,100		11,350	13,700
0980	For each additional 120 gal. tank, add D2020–295 G					1,775			1,775	1,975

23 56 19 – Solar Heating Components

23 56 19.50 Solar Heating Ancillary		Crew	Daily Output	Labor-Hours	Unit	Material	Labor	Equipment	Total	Total Incl O&P
0010	SOLAR HEATING ANCILLARY									
2300	Circulators, air D3010–650									
2310	Blowers									
2330	100-300 S.F. system, 1/10 HP D3010–660 G	Q-9	16	1	Ea.	266	56		322	380
2340	300-500 S.F. system, 1/5 HP G		15	1.067		355	60		415	480
2350	Two speed, 100-300 S.F., 1/10 HP D3010–675 G		14	1.143		147	64		211	259
2400	Reversible fan, 20" diameter, 2 speed G		18	.889		116	50		166	204
2550	Booster fan 6" diameter, 120 CFM G		16	1		38	56		94	127
2570	6" diameter, 225 CFM G		16	1		46.50	56		102.50	136
2580	8" diameter, 150 CFM G		16	1		42.50	56		98.50	132
2590	8" diameter, 310 CFM G		14	1.143		65.50	64		129.50	169
2600	8" diameter, 425 CFM G		14	1.143		73.50	64		137.50	178
2650	Rheostat G		32	.500		15.95	28		43.95	60
2660	Shutter/damper G		12	1.333		58.50	75		133.50	178
2670	Shutter motor G		16	1		146	56		202	246
2800	Circulators, liquid, 1/25 HP, 5.3 GPM G	Q-1	14	1.143		205	66.50		271.50	325
2820	1/20 HP, 17 GPM G		12	1.333		284	77.50		361.50	425
2850	1/20 HP, 17 GPM, stainless steel G		12	1.333		262	77.50		339.50	405
2870	1/12 HP, 30 GPM G		10	1.600		360	93		453	535
3000	Collector panels, air with aluminum absorber plate									

23 56 Solar Energy Heating Equipment

23 56 19 – Solar Heating Components

23 56 19.50 Solar Heating Ancillary			Crew	Daily Output	Labor-Hours	Unit	2020 Bare Costs Material	Labor	Equipment	Total	Total Incl O&P
3010	Wall or roof mount										
3040	Flat black, plastic glazing										
3080	4′ x 8′	G	Q-9	6	2.667	Ea.	685	150		835	980
3100	4′ x 10′	G		5	3.200	"	840	179		1,019	1,200
3200	Flush roof mount, 10′ to 16′ x 22″ wide	G		96	.167	L.F.	165	9.35		174.35	195
3210	Manifold, by L.F. width of collectors	G		160	.100	"	161	5.60		166.60	187
3300	Collector panels, liquid with copper absorber plate										
3320	Black chrome, tempered glass glazing										
3330	Alum. frame, 4′ x 8′, 5/32″ single glazing	G	Q-1	9.50	1.684	Ea.	1,075	97.50		1,172.50	1,325
3390	Alum. frame, 4′ x 10′, 5/32″ single glazing	G		6	2.667		1,225	155		1,380	1,575
3450	Flat black, alum. frame, 3.5′ x 7.5′	G		9	1.778		815	103		918	1,050
3500	4′ x 8′	G		5.50	2.909		1,025	169		1,194	1,375
3520	4′ x 10′	G		10	1.600		1,225	93		1,318	1,500
3540	4′ x 12.5′	G		5	3.200		1,300	186		1,486	1,700
3550	Liquid with fin tube absorber plate										
3560	Alum. frame 4′ x 8′ tempered glass	G	Q-1	10	1.600	Ea.	605	93		698	805
3580	Liquid with vacuum tubes, 4′ x 6′-10″	G		9	1.778		915	103		1,018	1,150
3600	Liquid, full wetted, plastic, alum. frame, 4′ x 10′	G		5	3.200		335	186		521	650
3650	Collector panel mounting, flat roof or ground rack	G		7	2.286		253	133		386	475
3670	Roof clamps	G		70	.229	Set	3.27	13.25		16.52	23.50
3700	Roof strap, teflon	G	1 Plum	205	.039	L.F.	25	2.51		27.51	31.50
3900	Differential controller with two sensors										
3930	Thermostat, hard wired	G	1 Plum	8	1	Ea.	112	64.50		176.50	220
3950	Line cord and receptacle	G		12	.667		154	43		197	234
4050	Pool valve system	G		2.50	3.200		182	206		388	510
4070	With 12 VAC actuator	G		2	4		335	258		593	750
4080	Pool pump system, 2″ pipe size	G		6	1.333		205	86		291	355
4100	Five station with digital read-out	G		3	2.667		271	172		443	555
4150	Sensors										
4200	Brass plug, 1/2″ MPT	G	1 Plum	32	.250	Ea.	16.70	16.10		32.80	42.50
4210	Brass plug, reversed	G		32	.250		26.50	16.10		42.60	53
4220	Freeze prevention	G		32	.250		25	16.10		41.10	51.50
4240	Screw attached	G		32	.250		10.05	16.10		26.15	35
4250	Brass, immersion	G		32	.250		15.50	16.10		31.60	41
4300	Heat exchanger										
4315	includes coil, blower, circulator										
4316	and controller for DHW and space hot air										
4330	Fluid to air coil, up flow, 45 MBH	G	Q-1	4	4	Ea.	330	232		562	710
4380	70 MBH	G		3.50	4.571		375	265		640	805
4400	80 MBH	G		3	5.333		495	310		805	1,000
4580	Fluid to fluid package includes two circulating pumps										
4590	expansion tank, check valve, relief valve										
4600	controller, high temperature cutoff and sensors	G	Q-1	2.50	6.400	Ea.	810	370		1,180	1,450
4650	Heat transfer fluid										
4700	Propylene glycol, inhibited anti-freeze	G	1 Plum	28	.286	Gal.	23	18.40		41.40	53
4800	Solar storage tanks, knocked down										
4810	Air, galvanized steel clad, double wall, 4″ fiberglass insulation										
5120	45 mil reinforced polypropylene lining,										
5140	4′ high, 4′ x 4′ = 64 C.F./450 gallons	G	Q-9	2	8	Ea.	3,700	450		4,150	4,750
5150	4′ x 8′ = 128 C.F./900 gallons	G		1.50	10.667		5,875	600		6,475	7,375
5160	4′ x 12′ = 190 C.F./1,300 gallons	G		1.30	12.308		7,425	690		8,115	9,225
5170	8′ x 8′ = 250 C.F./1,700 gallons	G		1	16		7,700	895		8,595	9,825
5190	6′-3″ high, 7′ x 7′ = 306 C.F./2,000 gallons	G	Q-10	1.20	20		13,900	1,175		15,075	17,100

23 56 Solar Energy Heating Equipment

23 56 19 – Solar Heating Components

23 56 19.50 Solar Heating Ancillary			Crew	Daily Output	Labor-Hours	Unit	2020 Bare Costs Material	Labor	Equipment	Total	Total Incl O&P
5200	7′ x 10′-6″ = 459 C.F./3,000 gallons	G	Q-10	.80	30	Ea.	17,300	1,750		19,050	21,700
5210	7′ x 14′ = 613 C.F./4,000 gallons	G		.60	40		22,400	2,325		24,725	28,100
5220	10′-6″ x 10′-6″ = 689 C.F./4,500 gallons	G		.50	48		20,800	2,800		23,600	27,000
5230	10′-6″ x 14′ = 919 C.F./6,000 gallons	G		.40	60		24,200	3,500		27,700	32,000
5240	14′ x 14′ = 1,225 C.F./8,000 gallons	G	Q-11	.40	80		29,000	4,750		33,750	39,100
5250	14′ x 17′-6″ = 1,531 C.F./10,000 gallons	G		.30	107		31,400	6,325		37,725	44,200
5260	17′-6″ x 17′-6″ = 1,914 C.F./12,500 gallons	G		.25	128		35,500	7,600		43,100	50,500
5270	17′-6″ x 21′ = 2,297 C.F./15,000 gallons	G		.20	160		40,300	9,500		49,800	59,000
5280	21′ x 21′ = 2,756 C.F./18,000 gallons	G		.18	178		43,100	10,500		53,600	63,500
5290	30 mil reinforced Hypalon lining, add						.02%				
7000	Solar control valves and vents										
7050	Air purger, 1″ pipe size	G	1 Plum	12	.667	Ea.	46.50	43		89.50	116
7070	Air eliminator, automatic 3/4″ size	G		32	.250		31	16.10		47.10	58
7090	Air vent, automatic, 1/8″ fitting	G		32	.250		17.90	16.10		34	43.50
7100	Manual, 1/8″ NPT	G		32	.250		2.69	16.10		18.79	27
7120	Backflow preventer, 1/2″ pipe size	G		16	.500		90.50	32		122.50	148
7130	3/4″ pipe size	G		16	.500		89	32		121	146
7150	Balancing valve, 3/4″ pipe size	G		20	.400		59.50	26		85.50	104
7180	Draindown valve, 1/2″ copper tube	G		9	.889		219	57.50		276.50	325
7200	Flow control valve, 1/2″ pipe size	G		22	.364		140	23.50		163.50	189
7220	Expansion tank, up to 5 gal.	G		32	.250		70.50	16.10		86.60	102
7250	Hydronic controller (aquastat)	G		8	1		217	64.50		281.50	335
7400	Pressure gauge, 2″ dial	G		32	.250		25	16.10		41.10	51.50
7450	Relief valve, temp. and pressure 3/4″ pipe size	G		30	.267		25.50	17.20		42.70	53.50
7500	Solenoid valve, normally closed										
7520	Brass, 3/4″ NPT, 24V	G	1 Plum	9	.889	Ea.	137	57.50		194.50	236
7530	1″ NPT, 24V	G		9	.889		224	57.50		281.50	335
7750	Vacuum relief valve, 3/4″ pipe size	G		32	.250		31	16.10		47.10	58
7800	Thermometers										
7820	Digital temperature monitoring, 4 locations	G	1 Plum	2.50	3.200	Ea.	138	206		344	460
7900	Upright, 1/2″ NPT	G		8	1		39.50	64.50		104	140
7970	Remote probe, 2″ dial	G		8	1		34.50	64.50		99	135
7990	Stem, 2″ dial, 9″ stem	G		16	.500		22.50	32		54.50	73
8250	Water storage tank with heat exchanger and electric element										
8270	66 gal. with 2″ x 2 lb. density insulation	G	1 Plum	1.60	5	Ea.	2,150	320		2,470	2,850
8300	80 gal. with 2″ x 2 lb. density insulation	G		1.60	5		1,950	320		2,270	2,625
8380	120 gal. with 2″ x 2 lb. density insulation	G		1.40	5.714		1,900	370		2,270	2,650
8400	120 gal. with 2″ x 2 lb. density insul., 40 S.F. heat coil	G		1.40	5.714		2,825	370		3,195	3,675
8500	Water storage module, plastic										
8600	Tubular, 12″ diameter, 4′ high	G	1 Carp	48	.167	Ea.	99	8.85		107.85	122
8610	12″ diameter, 8′ high	G		40	.200		157	10.65		167.65	188
8620	18″ diameter, 5′ high	G		38	.211		176	11.20		187.20	210
8630	18″ diameter, 10′ high	G		32	.250		263	13.30		276.30	310
8640	58″ diameter, 5′ high	G	2 Carp	32	.500		815	26.50		841.50	935
8650	Cap, 12″ diameter	G					20			20	22
8660	18″ diameter	G					25.50			25.50	28

23 57 Heat Exchangers for HVAC

23 57 16 – Steam-to-Water Heat Exchangers

23 57 16.10 Shell/Tube Type Steam-to-Water Heat Exch.		Crew	Daily Output	Labor-Hours	Unit	2020 Bare Costs Material	Labor	Equipment	Total	Total Incl O&P
0010	**SHELL AND TUBE TYPE STEAM-TO-WATER HEAT EXCHANGERS**									
0016	Shell & tube type, 2 or 4 pass, 3/4" OD copper tubes,									
0020	C.I. heads, C.I. tube sheet, steel shell									
0100	Hot water 40°F to 180°F, by steam at 10 psi									
0120	8 GPM	Q-5	6	2.667	Ea.	2,750	157		2,907	3,250
0140	10 GPM		5	3.200		4,125	189		4,314	4,825
0160	40 GPM		4	4		6,400	236		6,636	7,375
0500	For bronze head and tube sheet, add					50%				

23 57 19 – Liquid-to-Liquid Heat Exchangers

23 57 19.13 Plate-Type, Liquid-to-Liquid Heat Exchangers		Crew	Daily Output	Labor-Hours	Unit	Material	Labor	Equipment	Total	Total Incl O&P
0010	**PLATE-TYPE, LIQUID-TO-LIQUID HEAT EXCHANGERS**									
3000	Plate type,									
3100	400 GPM	Q-6	.80	30	Ea.	45,700	1,825		47,525	53,500
3120	800 GPM	"	.50	48		79,000	2,925		81,925	91,000
3140	1,200 GPM	Q-7	.34	94.118		117,000	5,875		122,875	138,000
3160	1,800 GPM	"	.24	133		155,500	8,325		163,825	183,500

23 57 19.16 Shell-Type, Liquid-to-Liquid Heat Exchangers		Crew	Daily Output	Labor-Hours	Unit	Material	Labor	Equipment	Total	Total Incl O&P
0010	**SHELL-TYPE, LIQUID-TO-LIQUID HEAT EXCHANGERS**									
1000	Hot water 40°F to 140°F, by water at 200°F									
1020	7 GPM	Q-5	6	2.667	Ea.	3,225	157		3,382	3,750
1040	16 GPM		5	3.200		4,550	189		4,739	5,275
1060	34 GPM		4	4		6,875	236		7,111	7,925
1080	55 GPM		3	5.333		9,750	315		10,065	11,200
1100	74 GPM		1.50	10.667		12,500	630		13,130	14,600
1120	86 GPM		1.40	11.429		16,700	675		17,375	19,400
1140	112 GPM	Q-6	2	12		20,800	735		21,535	24,000
1160	126 GPM		1.80	13.333		25,900	815		26,715	29,700
1180	152 GPM		1	24		33,100	1,475		34,575	38,600

23 72 Air-to-Air Energy Recovery Equipment

23 72 16 – Heat-Pipe Air-To-Air Energy-Recovery Equipment

23 72 16.10 Heat Pipes		Crew	Daily Output	Labor-Hours	Unit	Material	Labor	Equipment	Total	Total Incl O&P
0010	**HEAT PIPES**									
8000	Heat pipe type, glycol, 50% efficient									
8010	100 MBH, 1,700 CFM	1 Stpi	.80	10	Ea.	4,625	655		5,280	6,075
8020	160 MBH, 2,700 CFM		.60	13.333		6,225	875		7,100	8,150
8030	620 MBH, 4,000 CFM		.40	20		9,400	1,300		10,700	12,300

23 81 Decentralized Unitary HVAC Equipment

23 81 13 – Packaged Terminal Air-Conditioners

23 81 13.10 Packaged Cabinet Type Air-Conditioners		Crew	Daily Output	Labor-Hours	Unit	Material	Labor	Equipment	Total	Total Incl O&P
0010	**PACKAGED CABINET TYPE AIR-CONDITIONERS**, Cabinet, wall sleeve,									
0100	louver, electric heat, thermostat, manual changeover, 208 V									
0200	6,000 BTUH cooling, 8,800 BTU heat	Q-5	6	2.667	Ea.	725	157		882	1,025
0220	9,000 BTUH cooling, 13,900 BTU heat		5	3.200		975	189		1,164	1,350
0240	12,000 BTUH cooling, 13,900 BTU heat		4	4		1,550	236		1,786	2,050
0260	15,000 BTUH cooling, 13,900 BTU heat		3	5.333		1,525	315		1,840	2,175

23 81 Decentralized Unitary HVAC Equipment

23 81 19 – Self-Contained Air-Conditioners

23 81 19.10 Window Unit Air Conditioners		Crew	Daily Output	Labor-Hours	Unit	2020 Bare Costs Material	Labor	Equipment	Total	Total Incl O&P
0010	**WINDOW UNIT AIR CONDITIONERS**									
4000	Portable/window, 15 amp, 125 V grounded receptacle required									
4060	5,000 BTUH	1 Carp	8	1	Ea.	305	53		358	420
4340	6,000 BTUH		8	1		239	53		292	345
4480	8,000 BTUH		6	1.333		385	71		456	530
4500	10,000 BTUH		6	1.333		650	71		721	820
4520	12,000 BTUH	L-2	8	2		2,000	93		2,093	2,350
4600	Window/thru-the-wall, 15 amp, 230 V grounded receptacle required									
4780	18,000 BTUH	L-2	6	2.667	Ea.	715	124		839	975
4940	25,000 BTUH		4	4		1,075	186		1,261	1,475
4960	29,000 BTUH		4	4		1,050	186		1,236	1,425

23 81 43 – Air-Source Unitary Heat Pumps

23 81 43.10 Air-Source Heat Pumps		Crew	Daily Output	Labor-Hours	Unit	2020 Bare Costs Material	Labor	Equipment	Total	Total Incl O&P
0010	**AIR-SOURCE HEAT PUMPS**, Not including interconnecting tubing									
1000	Air to air, split system, not including curbs, pads, fan coil and ductwork									
1010	For curbs/pads see Section 23 91 10									
1012	Outside condensing unit only, for fan coil see Section 23 82 19.10									
1015	1.5 ton cooling, 7 MBH heat @ 0°F	Q-5	2.40	6.667	Ea.	1,625	395		2,020	2,400
1020	2 ton cooling, 8.5 MBH heat @ 0°F		2	8		1,775	470		2,245	2,650
1030	2.5 ton cooling, 10 MBH heat @ 0°F		1.60	10		1,950	590		2,540	3,025
1040	3 ton cooling, 13 MBH heat @ 0°F		1.20	13.333		2,150	785		2,935	3,525
1050	3.5 ton cooling, 18 MBH heat @ 0°F		1	16		2,275	945		3,220	3,925
1054	4 ton cooling, 24 MBH heat @ 0°F		.80	20		2,475	1,175		3,650	4,500
1060	5 ton cooling, 27 MBH heat @ 0°F		.50	32		2,700	1,900		4,600	5,800
1500	Single package, not including curbs, pads, or plenums									
1502	0.5 ton cooling, supplementary heat included	Q-5	8	2	Ea.	3,550	118		3,668	4,075
1504	0.75 ton cooling, supplementary heat included		6	2.667		4,150	157		4,307	4,800
1506	1 ton cooling, supplementary heat included		4	4		3,425	236		3,661	4,125
1510	1.5 ton cooling, 5 MBH heat @ 0°F		1.55	10.323		3,375	610		3,985	4,625
1520	2 ton cooling, 6.5 MBH heat @ 0°F		1.50	10.667		3,325	630		3,955	4,625
1540	2.5 ton cooling, 8 MBH heat @ 0°F		1.40	11.429		3,300	675		3,975	4,625
1560	3 ton cooling, 10 MBH heat @ 0°F		1.20	13.333		3,775	785		4,560	5,325
1570	3.5 ton cooling, 11 MBH heat @ 0°F		1	16		4,125	945		5,070	5,975
1580	4 ton cooling, 13 MBH heat @ 0°F		.96	16.667		4,425	985		5,410	6,350
1620	5 ton cooling, 27 MBH heat @ 0°F		.65	24.615		5,125	1,450		6,575	7,825
1640	7.5 ton cooling, 35 MBH heat @ 0°F		.40	40		7,900	2,350		10,250	12,200
6000	Air to water, single package, excluding storage tank and ductwork									
6010	Includes circulating water pump, air duct connections, digital temperature									
6020	controller with remote tank temp. probe and sensor for storage tank.									
6040	Water heating - air cooling capacity									
6110	35.5 MBH heat water, 2.3 ton cool air	Q-5	1.60	10	Ea.	17,300	590		17,890	20,000
6120	58 MBH heat water, 3.8 ton cool air		1.10	14.545		19,800	860		20,660	23,000
6130	76 MBH heat water, 4.9 ton cool air		.87	18.391		23,600	1,075		24,675	27,500
6140	98 MBH heat water, 6.5 ton cool air		.62	25.806		30,500	1,525		32,025	35,800
6150	113 MBH heat water, 7.4 ton cool air		.59	27.119		33,300	1,600		34,900	39,000
6160	142 MBH heat water, 9.2 ton cool air		.52	30.769		40,500	1,825		42,325	47,300
6170	171 MBH heat water, 11.1 ton cool air		.49	32.653		47,500	1,925		49,425	55,000

23 81 Decentralized Unitary HVAC Equipment

23 81 46 – Water-Source Unitary Heat Pumps

23 81 46.10 Water Source Heat Pumps		Crew	Daily Output	Labor-Hours	Unit	2020 Bare Costs Material	Labor	Equipment	Total	Total Incl O&P
0010	**WATER SOURCE HEAT PUMPS**, Not incl. connecting tubing or water source									
2000	Water source to air, single package									
2100	1 ton cooling, 13 MBH heat @ 75°F	Q-5	2	8	Ea.	2,125	470		2,595	3,050
2120	1.5 ton cooling, 17 MBH heat @ 75°F		1.80	8.889		2,075	525		2,600	3,050
2140	2 ton cooling, 19 MBH heat @ 75°F		1.70	9.412		2,475	555		3,030	3,550
2160	2.5 ton cooling, 25 MBH heat @ 75°F		1.60	10		2,700	590		3,290	3,850
2180	3 ton cooling, 27 MBH heat @ 75°F		1.40	11.429		2,750	675		3,425	4,025
2190	3.5 ton cooling, 29 MBH heat @ 75°F		1.30	12.308		2,975	725		3,700	4,350
2200	4 ton cooling, 31 MBH heat @ 75°F		1.20	13.333		3,125	785		3,910	4,600
2220	5 ton cooling, 29 MBH heat @ 75°F		.90	17.778		3,400	1,050		4,450	5,300
3960	For supplementary heat coil, add					10%				

23 82 Convection Heating and Cooling Units

23 82 19 – Fan Coil Units

23 82 19.10 Fan Coil Air Conditioning		Crew	Daily Output	Labor-Hours	Unit	Material	Labor	Equipment	Total	Total Incl O&P
0010	**FAN COIL AIR CONDITIONING**									
0030	Fan coil AC, cabinet mounted, filters and controls									
0320	1 ton cooling	Q-5	6	2.667	Ea.	1,750	157		1,907	2,150
0940	Direct expansion, for use w/air cooled condensing unit, 1-1/2 ton cooling		5	3.200		685	189		874	1,050
0950	2 ton cooling		4.80	3.333		670	197		867	1,025
0960	2-1/2 ton cooling		4.40	3.636		785	215		1,000	1,175
0970	3 ton cooling		3.80	4.211		960	248		1,208	1,425
0980	3-1/2 ton cooling		3.60	4.444		990	262		1,252	1,500
0990	4 ton cooling		3.40	4.706		995	278		1,273	1,525
1000	5 ton cooling		3	5.333		1,250	315		1,565	1,850
1500	For hot water coil, add					40%	10%			

23 82 29 – Radiators

23 82 29.10 Hydronic Heating		Crew	Daily Output	Labor-Hours	Unit	Material	Labor	Equipment	Total	Total Incl O&P
0010	**HYDRONIC HEATING**, Terminal units, not incl. main supply pipe									
1000	Radiation									
1100	Panel, baseboard, C.I., including supports, no covers	Q-5	46	.348	L.F.	49	20.50		69.50	84.50
3000	Radiators, cast iron									
3100	Free standing or wall hung, 6 tube, 25" high	Q-5	96	.167	Section	61	9.85		70.85	81.50
3150	4 tube, 25" high		96	.167		47.50	9.85		57.35	66.50
3200	4 tube, 19" high		96	.167		42	9.85		51.85	60.50
3250	Adj. brackets, 2 per wall radiator up to 30 sections	1 Stpi	32	.250	Ea.	66	16.40		82.40	97
3500	Recessed, 20" high x 5" deep, without grille	Q-5	60	.267	Section	51.50	15.75		67.25	80.50
3525	Free standing or wall hung, 30" high	"	60	.267		46.50	15.75		62.25	74.50
3600	For inlet grille, add					6.40			6.40	7.05
9500	To convert SFR to BTU rating: Hot water, 150 x SFR									
9510	Forced hot water, 180 x SFR; steam, 240 x SFR									

23 82 33 – Convectors

23 82 33.10 Convector Units		Crew	Daily Output	Labor-Hours	Unit	Material	Labor	Equipment	Total	Total Incl O&P
0010	**CONVECTOR UNITS**, Terminal units, not incl. main supply pipe									
2204	Convector, multifin, 2 pipe w/cabinet									
2210	17" H x 24" L	Q-5	10	1.600	Ea.	110	94.50		204.50	262
2214	17" H x 36" L		8.60	1.860		165	110		275	345
2218	17" H x 48" L		7.40	2.162		220	128		348	435
2222	21" H x 24" L		9	1.778		124	105		229	293

23 82 Convection Heating and Cooling Units

23 82 33 – Convectors

23 82 33.10 Convector Units		Crew	Daily Output	Labor-Hours	Unit	2020 Bare Costs Material	Labor	Equipment	Total	Total Incl O&P
2228	21" H x 48" L	Q-5	6.80	2.353	Ea.	247	139		386	480
2240	For knob operated damper, add					140%				
2241	For metal trim strips, add	Q-5	64	.250	Ea.	14.25	14.75		29	37.50
2243	For snap-on inlet grille, add					10%	10%			
2245	For hinged access door, add	Q-5	64	.250	Ea.	40.50	14.75		55.25	66.50
2246	For air chamber, auto-venting, add	"	58	.276	"	8.75	16.30		25.05	34

23 82 36 – Finned-Tube Radiation Heaters

23 82 36.10 Finned Tube Radiation		Crew	Daily Output	Labor-Hours	Unit	Material	Labor	Equipment	Total	Total Incl O&P
0010	**FINNED TUBE RADIATION**, Terminal units, not incl. main supply pipe									
1310	Baseboard, pkgd, 1/2" copper tube, alum. fin, 7" high	Q-5	60	.267	L.F.	11.95	15.75		27.70	36.50
1320	3/4" copper tube, alum. fin, 7" high	"	58	.276	"	8.05	16.30		24.35	33.50
1381	Rough in baseboard panel & fin tube, supply & balance valves	1 Stpi	1.06	7.547	Ea.	279	495		774	1,050
1500	Note: fin tube may also require corners, caps, etc.									

23 83 Radiant Heating Units

23 83 16 – Radiant-Heating Hydronic Piping

23 83 16.10 Radiant Floor Heating		Crew	Daily Output	Labor-Hours	Unit	Material	Labor	Equipment	Total	Total Incl O&P
0010	**RADIANT FLOOR HEATING**									
0100	Tubing, PEX (cross-linked polyethylene)									
0110	Oxygen barrier type for systems with ferrous materials									
0120	1/2"	Q-5	800	.020	L.F.	1.05	1.18		2.23	2.93
0130	3/4"		535	.030		1.55	1.76		3.31	4.35
0140	1"		400	.040		2.32	2.36		4.68	6.10
0200	Non barrier type for ferrous free systems									
0210	1/2"	Q-5	800	.020	L.F.	.57	1.18		1.75	2.40
0220	3/4"		535	.030		1	1.76		2.76	3.74
0230	1"		400	.040		1.79	2.36		4.15	5.50
1000	Manifolds									
1110	Brass									
1120	With supply and return valves, flow meter, thermometer,									
1122	auto air vent and drain/fill valve.									
1130	1", 2 circuit	Q-5	14	1.143	Ea.	355	67.50		422.50	495
1140	1", 3 circuit		13.50	1.185		405	70		475	550
1150	1", 4 circuit		13	1.231		435	72.50		507.50	590
1154	1", 5 circuit		12.50	1.280		525	75.50		600.50	690
1158	1", 6 circuit		12	1.333		590	78.50		668.50	765
1162	1", 7 circuit		11.50	1.391		640	82		722	825
1166	1", 8 circuit		11	1.455		685	86		771	885
1172	1", 9 circuit		10.50	1.524		760	90		850	970
1174	1", 10 circuit		10	1.600		795	94.50		889.50	1,025
1178	1", 11 circuit		9.50	1.684		820	99.50		919.50	1,050
1182	1", 12 circuit		9	1.778		930	105		1,035	1,175
1610	Copper manifold header (cut to size)									
1620	1" header, 12 circuit 1/2" sweat outlets	Q-5	3.33	4.805	Ea.	117	283		400	555
1630	1-1/4" header, 12 circuit 1/2" sweat outlets		3.20	5		136	295		431	590
1640	1-1/4" header, 12 circuit 3/4" sweat outlets		3	5.333		143	315		458	625
1650	1-1/2" header, 12 circuit 3/4" sweat outlets		3.10	5.161		173	305		478	645
1660	2" header, 12 circuit 3/4" sweat outlets		2.90	5.517		257	325		582	770
3000	Valves									
3110	Thermostatic zone valve actuator with end switch	Q-5	40	.400	Ea.	51.50	23.50		75	92

23 83 Radiant Heating Units

23 83 16 – Radiant-Heating Hydronic Piping

23 83 16.10 Radiant Floor Heating		Crew	Daily Output	Labor-Hours	Unit	2020 Bare Costs Material	Labor	Equipment	Total	Total Incl O&P
3114	Thermostatic zone valve actuator	Q-5	36	.444	Ea.	98	26		124	147
3120	Motorized straight zone valve with operator complete									
3130	3/4"	Q-5	35	.457	Ea.	160	27		187	217
3140	1"		32	.500		174	29.50		203.50	235
3150	1-1/4"	↓	29.60	.541	↓	219	32		251	289
3500	4 way mixing valve, manual, brass									
3530	1"	Q-5	13.30	1.203	Ea.	230	71		301	360
3540	1-1/4"		11.40	1.404		249	83		332	395
3550	1-1/2"		11	1.455		315	86		401	475
3560	2"		10.60	1.509		455	89		544	635
3800	Mixing valve motor, 4 way for valves, 1" and 1-1/4"		34	.471		390	28		418	465
3810	Mixing valve motor, 4 way for valves, 1-1/2" and 2"	↓	30	.533	↓	435	31.50		466.50	525
5000	Radiant floor heating, zone control panel									
5120	4 zone actuator valve control, expandable	Q-5	20	.800	Ea.	163	47		210	251
5130	6 zone actuator valve control, expandable		18	.889		271	52.50		323.50	375
6070	Thermal track, straight panel for long continuous runs, 5.333 S.F.		40	.400		36	23.50		59.50	75.50
6080	Thermal track, utility panel, for direction reverse at run end, 5.333 S.F.		40	.400		36	23.50		59.50	75.50
6090	Combination panel, for direction reverse plus straight run, 5.333 S.F.	↓	40	.400	↓	36	23.50		59.50	75.50
7000	PEX tubing fittings									
7100	Compression type									
7116	Coupling									
7120	1/2" x 1/2"	1 Stpi	27	.296	Ea.	6.95	19.40		26.35	36.50
7124	3/4" x 3/4"	"	23	.348	"	16.35	23		39.35	52
7130	Adapter									
7132	1/2" x female sweat 1/2"	1 Stpi	27	.296	Ea.	4.51	19.40		23.91	34
7134	1/2" x female sweat 3/4"		26	.308		5.05	20		25.05	35.50
7136	5/8" x female sweat 3/4"	↓	24	.333	↓	7.25	22		29.25	40.50
7140	Elbow									
7142	1/2" x female sweat 1/2"	1 Stpi	27	.296	Ea.	7.30	19.40		26.70	37
7144	1/2" x female sweat 3/4"		26	.308		8.55	20		28.55	39.50
7146	5/8" x female sweat 3/4"	↓	24	.333	↓	9.60	22		31.60	43
7200	Insert type									
7206	PEX x male NPT									
7210	1/2" x 1/2"	1 Stpi	29	.276	Ea.	2.94	18.10		21.04	30
7220	3/4" x 3/4"		27	.296		4.35	19.40		23.75	34
7230	1" x 1"	↓	26	.308	↓	7.30	20		27.30	38
7300	PEX coupling									
7310	1/2" x 1/2"	1 Stpi	30	.267	Ea.	.64	17.50		18.14	26.50
7320	3/4" x 3/4"		29	.276		.81	18.10		18.91	28
7330	1" x 1"	↓	28	.286	↓	1.40	18.75		20.15	29.50
7400	PEX stainless crimp ring									
7410	1/2" x 1/2"	1 Stpi	86	.093	Ea.	.56	6.10		6.66	9.75
7420	3/4" x 3/4"		84	.095		.77	6.25		7.02	10.20
7430	1" x 1"	↓	82	.098	↓	1.11	6.40		7.51	10.75

23 83 33 – Electric Radiant Heaters

23 83 33.10 Electric Heating		Crew	Daily Output	Labor-Hours	Unit	Material	Labor	Equipment	Total	Total Incl O&P
0010	**ELECTRIC HEATING**, not incl. conduit or feed wiring									
1100	Rule of thumb: Baseboard units, including control	1 Elec	4.40	1.818	kW	116	112		228	294
1300	Baseboard heaters, 2' long, 350 watt		8	1	Ea.	29	61.50		90.50	123
1400	3' long, 750 watt		8	1		35	61.50		96.50	130
1600	4' long, 1,000 watt		6.70	1.194		40.50	73.50		114	154
1800	5' long, 935 watt	↓	5.70	1.404	↓	44	86		130	177

23 83 Radiant Heating Units

23 83 33 – Electric Radiant Heaters

23 83 33.10 Electric Heating		Crew	Daily Output	Labor-Hours	Unit	2020 Bare Costs Material	Labor	Equipment	Total	Total Incl O&P
2000	6′ long, 1,500 watt	1 Elec	5	1.600	Ea.	57	98		155	209
2200	7′ long, 1,310 watt		4.40	1.818		61	112		173	234
2400	8′ long, 2,000 watt		4	2		70	123		193	260
2600	9′ long, 1,680 watt		3.60	2.222		86.50	136		222.50	298
2800	10′ long, 1,875 watt		3.30	2.424		162	149		311	400
2950	Wall heaters with fan, 120 to 277 volt									
3160	Recessed, residential, 750 watt	1 Elec	6	1.333	Ea.	94.50	82		176.50	226
3170	1,000 watt		6	1.333		94.50	82		176.50	226
3180	1,250 watt		5	1.600		107	98		205	264
3190	1,500 watt		4	2		107	123		230	300
3600	Thermostats, integral		16	.500		29	30.50		59.50	77.50
3800	Line voltage, 1 pole		8	1		16.30	61.50		77.80	109
3810	2 pole		8	1		23.50	61.50		85	118
4000	Heat trace system, 400 degree									
4020	115 V, 2.5 watts/L.F.	1 Elec	530	.015	L.F.	11	.93		11.93	13.50
4030	5 watts/L.F.		530	.015		11	.93		11.93	13.50
4050	10 watts/L.F.		530	.015		9.20	.93		10.13	11.50
4060	208 V, 5 watts/L.F.		530	.015		11	.93		11.93	13.50
4080	480 V, 8 watts/L.F.		530	.015		11	.93		11.93	13.50
4200	Heater raceway									
4260	Heat transfer cement									
4280	1 gallon				Ea.	64.50			64.50	71
4300	5 gallon				"	252			252	277
4320	Cable tie									
4340	3/4″ pipe size	1 Elec	470	.017	Ea.	.02	1.04		1.06	1.57
4360	1″ pipe size		444	.018		.02	1.11		1.13	1.67
4380	1-1/4″ pipe size		400	.020		.02	1.23		1.25	1.85
4400	1-1/2″ pipe size		355	.023		.02	1.38		1.40	2.08
4420	2″ pipe size		320	.025		.05	1.53		1.58	2.34
4440	3″ pipe size		160	.050		.05	3.07		3.12	4.63
4460	4″ pipe size		100	.080		.06	4.91		4.97	7.35
4480	Thermostat NEMA 3R, 22 amp, 0-150 degree, 10′ cap.		8	1		193	61.50		254.50	305
4500	Thermostat NEMA 4X, 25 amp, 40 degree, 5-1/2′ cap.		7	1.143		248	70		318	375
4520	Thermostat NEMA 4X, 22 amp, 25-325 degree, 10′ cap.		7	1.143		660	70		730	830
4540	Thermostat NEMA 4X, 22 amp, 15-140 degree		6	1.333		560	82		642	735
4580	Thermostat NEMA 4, 7, 9, 22 amp, 25-325 degree, 10′ cap.		3.60	2.222		810	136		946	1,100
4600	Thermostat NEMA 4, 7, 9, 22 amp, 15-140 degree		3	2.667		815	164		979	1,150
4720	Fiberglass application tape, 36 yard roll		11	.727		75	44.50		119.50	149
5000	Radiant heating ceiling panels, 2′ x 4′, 500 watt		16	.500		405	30.50		435.50	490
5050	750 watt		16	.500		430	30.50		460.50	520
5200	For recessed plaster frame, add		32	.250		138	15.35		153.35	175
5300	Infrared quartz heaters, 120 volts, 1,000 watt		6.70	1.194		350	73.50		423.50	495
5350	1,500 watt		5	1.600		335	98		433	515
5400	240 volts, 1,500 watt		5	1.600		405	98		503	590
5450	2,000 watt		4	2		335	123		458	555
5500	3,000 watt		3	2.667		330	164		494	610
5550	4,000 watt		2.60	3.077		380	189		569	700
5570	Modulating control		.80	10		141	615		756	1,075
5600	Unit heaters, heavy duty, with fan & mounting bracket									
5650	Single phase, 208-240-277 volt, 3 kW	1 Elec	6	1.333	Ea.	705	82		787	895
5750	5 kW		5.50	1.455		800	89		889	1,025
5800	7 kW		5	1.600		1,375	98		1,473	1,650
5850	10 kW		4	2		1,100	123		1,223	1,375

23 83 33.10 Electric Heating		Crew	Daily Output	Labor-Hours	Unit	Material	2020 Bare Costs Labor	Equipment	Total	Total Incl O&P
5950	15 kW	1 Elec	3.80	2.105	Ea.	1,850	129		1,979	2,225
6000	480 volt, 3 kW		6	1.333		470	82		552	635
6020	4 kW		5.80	1.379		700	84.50		784.50	895
6040	5 kW		5.50	1.455		710	89		799	915
6060	7 kW		5	1.600		1,150	98		1,248	1,425
6080	10 kW		4	2		1,100	123		1,223	1,400
6100	13 kW		3.80	2.105		1,950	129		2,079	2,350
6120	15 kW		3.70	2.162		2,150	133		2,283	2,575
6140	20 kW		3.50	2.286		3,075	140		3,215	3,600
6300	3 phase, 208-240 volt, 5 kW		5.50	1.455		575	89		664	770
6320	7 kW		5	1.600		940	98		1,038	1,175
6340	10 kW		4	2		805	123		928	1,075
6360	15 kW		3.70	2.162		1,600	133		1,733	1,950
6380	20 kW		3.50	2.286		2,500	140		2,640	2,950
6400	25 kW		3.30	2.424		3,725	149		3,874	4,325
6500	480 volt, 5 kW		5.50	1.455		650	89		739	850
6520	7 kW		5	1.600		930	98		1,028	1,175
6540	10 kW		4	2		860	123		983	1,125
6560	13 kW		3.80	2.105		1,650	129		1,779	2,025
6580	15 kW		3.70	2.162		1,475	133		1,608	1,825
6600	20 kW	↓	3.50	2.286	↓	2,325	140		2,465	2,750
6800	Vertical discharge heaters, with fan									
6820	Single phase, 208-240-277 volt, 10 kW	1 Elec	4	2	Ea.	860	123		983	1,125
6840	15 kW		3.70	2.162		1,650	133		1,783	2,000
6900	3 phase, 208-240 volt, 10 kW		4	2		1,050	123		1,173	1,350
6920	15 kW		3.70	2.162		1,600	133		1,733	1,975
6940	20 kW		3.50	2.286		2,925	140		3,065	3,425
7100	480 volt, 10 kW		4	2		1,200	123		1,323	1,500
7120	15 kW		3.70	2.162		1,950	133		2,083	2,350
7140	20 kW		3.50	2.286		2,650	140		2,790	3,125
7160	25 kW	↓	3.30	2.424	↓	3,450	149		3,599	4,025
7900	Cabinet convector heaters, 240 volt, three phase,									
7920	3′ long, 2,000 watt	1 Elec	5.30	1.509	Ea.	2,300	92.50		2,392.50	2,675
7940	3,000 watt		5.30	1.509		2,400	92.50		2,492.50	2,800
7960	4,000 watt		5.30	1.509		2,475	92.50		2,567.50	2,850
7980	6,000 watt		4.60	1.739		2,675	107		2,782	3,100
8000	8,000 watt		4.60	1.739		2,750	107		2,857	3,175
8020	4′ long, 4,000 watt		4.60	1.739		2,500	107		2,607	2,900
8040	6,000 watt		4	2		2,600	123		2,723	3,025
8060	8,000 watt		4	2		2,675	123		2,798	3,100
8080	10,000 watt	↓	4	2	↓	2,700	123		2,823	3,150
8100	Available also in 208 or 277 volt									
8200	Cabinet unit heaters, 120 to 277 volt, 1 pole,									
8220	wall mounted, 2 kW	1 Elec	4.60	1.739	Ea.	2,075	107		2,182	2,450
8230	3 kW		4.60	1.739		2,075	107		2,182	2,425
8240	4 kW		4.40	1.818		2,150	112		2,262	2,550
8250	5 kW		4.40	1.818		2,275	112		2,387	2,675
8260	6 kW		4.20	1.905		2,250	117		2,367	2,650
8270	8 kW		4	2		2,300	123		2,423	2,700
8280	10 kW		3.80	2.105		2,350	129		2,479	2,775
8290	12 kW		3.50	2.286		2,575	140		2,715	3,050
8300	13.5 kW		2.90	2.759		3,650	169		3,819	4,275
8310	16 kW	↓	2.70	2.963	↓	3,025	182		3,207	3,625

23 83 Radiant Heating Units

23 83 33 – Electric Radiant Heaters

23 83 33.10 Electric Heating		Crew	Daily Output	Labor-Hours	Unit	2020 Bare Costs Material	Labor	Equipment	Total	Total Incl O&P
8320	20 kW	1 Elec	2.30	3.478	Ea.	3,825	213		4,038	4,525
8330	24 kW		1.90	4.211		4,000	258		4,258	4,800
8350	Recessed, 2 kW		4.40	1.818		2,275	112		2,387	2,675
8370	3 kW		4.40	1.818		2,325	112		2,437	2,725
8380	4 kW		4.20	1.905		2,200	117		2,317	2,600
8390	5 kW		4.20	1.905		2,300	117		2,417	2,700
8400	6 kW		4	2		2,350	123		2,473	2,750
8410	8 kW		3.80	2.105		2,525	129		2,654	3,000
8420	10 kW		3.50	2.286		3,150	140		3,290	3,675
8430	12 kW		2.90	2.759		3,175	169		3,344	3,750
8440	13.5 kW		2.70	2.963		3,050	182		3,232	3,625
8450	16 kW		2.30	3.478		3,250	213		3,463	3,900
8460	20 kW		1.90	4.211		3,775	258		4,033	4,550
8470	24 kW		1.60	5		3,725	305		4,030	4,550
8490	Ceiling mounted, 2 kW		3.20	2.500		2,100	153		2,253	2,525
8510	3 kW		3.20	2.500		2,200	153		2,353	2,650
8520	4 kW		3	2.667		2,250	164		2,414	2,725
8530	5 kW		3	2.667		2,325	164		2,489	2,800
8540	6 kW		2.80	2.857		2,325	175		2,500	2,825
8550	8 kW		2.40	3.333		2,450	205		2,655	3,000
8560	10 kW		2.20	3.636		3,050	223		3,273	3,675
8570	12 kW		2	4		2,625	245		2,870	3,275
8580	13.5 kW		1.50	5.333		2,550	325		2,875	3,275
8590	16 kW		1.30	6.154		2,675	380		3,055	3,475
8600	20 kW		.90	8.889		3,700	545		4,245	4,875
8610	24 kW		.60	13.333		3,725	820		4,545	5,325
8630	208 to 480 V, 3 pole									
8650	Wall mounted, 2 kW	1 Elec	4.60	1.739	Ea.	2,025	107		2,132	2,400
8670	3 kW		4.60	1.739		2,375	107		2,482	2,775
8680	4 kW		4.40	1.818		2,425	112		2,537	2,850
8690	5 kW		4.40	1.818		2,500	112		2,612	2,925
8700	6 kW		4.20	1.905		2,525	117		2,642	2,950
8710	8 kW		4	2		2,625	123		2,748	3,050
8720	10 kW		3.80	2.105		2,775	129		2,904	3,250
8730	12 kW		3.50	2.286		2,825	140		2,965	3,300
8740	13.5 kW		2.90	2.759		2,825	169		2,994	3,350
8750	16 kW		2.70	2.963		2,925	182		3,107	3,500
8760	20 kW		2.30	3.478		4,475	213		4,688	5,250
8770	24 kW		1.90	4.211		4,200	258		4,458	4,975
8790	Recessed, 2 kW		4.40	1.818		2,250	112		2,362	2,650
8810	3 kW		4.40	1.818		2,300	112		2,412	2,725
8820	4 kW		4.20	1.905		2,400	117		2,517	2,825
8830	5 kW		4.20	1.905		2,425	117		2,542	2,850
8840	6 kW		4	2		2,350	123		2,473	2,750
8850	8 kW		3.80	2.105		2,425	129		2,554	2,875
8860	10 kW		3.50	2.286		2,450	140		2,590	2,900
8870	12 kW		2.90	2.759		2,750	169		2,919	3,250
8880	13.5 kW		2.70	2.963		2,375	182		2,557	2,900
8890	16 kW		2.30	3.478		5,175	213		5,388	6,025
8900	20 kW		1.90	4.211		4,300	258		4,558	5,100
8920	24 kW		1.60	5		4,350	305		4,655	5,250
8940	Ceiling mount, 2 kW		3.20	2.500		2,325	153		2,478	2,775
8950	3 kW		3.20	2.500		2,150	153		2,303	2,575

23 83 Radiant Heating Units

23 83 33 – Electric Radiant Heaters

23 83 33.10 Electric Heating		Crew	Daily Output	Labor-Hours	Unit	Material	Labor	Equipment	Total	Total Incl O&P
						2020 Bare Costs				
8960	4 kW	1 Elec	3	2.667	Ea.	2,475	164		2,639	2,975
8970	5 kW		3	2.667		2,175	164		2,339	2,625
8980	6 kW		2.80	2.857		2,475	175		2,650	2,975
8990	8 kW		2.40	3.333		2,550	205		2,755	3,125
9000	10 kW		2.20	3.636		2,350	223		2,573	2,925
9020	13.5 kW		1.50	5.333		3,150	325		3,475	3,950
9030	16 kW		1.30	6.154		4,300	380		4,680	5,300
9040	20 kW		.90	8.889		4,300	545		4,845	5,525
9060	24 kW		.60	13.333		5,325	820		6,145	7,075
9230	13.5 kW, 40,956 BTU		2.20	3.636		2,075	223		2,298	2,600
9250	24 kW, 81,912 BTU	▼	2	4	▼	2,025	245		2,270	2,600

23 91 Prefabricated Equipment Supports

23 91 10 – Prefabricated Curbs, Pads and Stands

23 91 10.10 Prefabricated Pads and Stands		Crew	Daily Output	Labor-Hours	Unit	Material	Labor	Equipment	Total	Total Incl O&P
0010	**PREFABRICATED PADS AND STANDS**									
6000	Pad, fiberglass reinforced concrete with polystyrene foam core									
6220	30" x 36"	1 Shee	8	1	Ea.	51	62.50		113.50	151
6340	36" x 54"	Q-9	6	2.667	"	96	150		246	330

Division Notes

	CREW	DAILY OUTPUT	LABOR-HOURS	UNIT	BARE COSTS MAT.	BARE COSTS LABOR	BARE COSTS EQUIP.	BARE COSTS TOTAL	TOTAL INCL O&P

Estimating Tips

26 05 00 Common Work Results for Electrical

- Conduit should be taken off in three main categories—power distribution, branch power, and branch lighting—so the estimator can concentrate on systems and components, therefore making it easier to ensure all items have been accounted for.
- For cost modifications for elevated conduit installation, add the percentages to labor according to the height of installation and only to the quantities exceeding the different height levels, not to the total conduit quantities. Refer to subdivision 26 01 02.20 for labor adjustment factors.
- Remember that aluminum wiring of equal ampacity is larger in diameter than copper and may require larger conduit.
- If more than three wires at a time are being pulled, deduct percentages from the labor hours of that grouping of wires.
- When taking off grounding systems, identify separately the type and size of wire, and list each unique type of ground connection.
- The estimator should take the weights of materials into consideration when completing a takeoff. Topics to consider include: How will the materials be supported? What methods of support are available? How high will the support structure have to reach? Will the final support structure be able to withstand the total burden? Is the support material included or separate from the fixture, equipment, and material specified?
- Do not overlook the costs for equipment used in the installation. If scaffolding or highlifts are available in the field, contractors may use them in lieu of the proposed ladders and rolling staging.

26 20 00 Low-Voltage Electrical Transmission

- Supports and concrete pads may be shown on drawings for the larger equipment, or the support system may be only a piece of plywood for the back of a panelboard. In either case, they must be included in the costs.

26 40 00 Electrical and Cathodic Protection

- When taking off cathodic protection systems, identify the type and size of cable, and list each unique type of anode connection.

26 50 00 Lighting

- Fixtures should be taken off room by room using the fixture schedule, specifications, and the ceiling plan. For large concentrations of lighting fixtures in the same area, deduct the percentages from labor hours.

Reference Numbers

Reference numbers are shown at the beginning of some major classifications. These numbers refer to related items in the Reference Section. The reference information may be an estimating procedure, an alternate pricing method, or technical information.

Note: Not all subdivisions listed here necessarily appear. ■

Note: "Powered in part by CINX™, based on licensed proprietary information of Harrison Publishing House, Inc."

Note: The following companies, in part, have been used as a reference source for some of the material prices used in Division 26:

Electriflex

Trade Service

26 05 33.95 Cutting and Drilling		Crew	Daily Output	Labor-Hours	Unit	Material	2020 Bare Costs Labor	Equipment	Total	Total Incl O&P
0010	**CUTTING AND DRILLING**									
0100	Hole drilling to 10′ high, concrete wall									
0110	8″ thick, 1/2″ pipe size	R-31	12	.667	Ea.	.27	41	5.20	46.47	67
0120	3/4″ pipe size		12	.667		.27	41	5.20	46.47	67
0130	1″ pipe size		9.50	.842		.35	51.50	6.60	58.45	84.50
0140	1-1/4″ pipe size		9.50	.842		.35	51.50	6.60	58.45	84.50
0150	1-1/2″ pipe size		9.50	.842		.35	51.50	6.60	58.45	84.50
0160	2″ pipe size		4.40	1.818		.53	112	14.20	126.73	182
0170	2-1/2″ pipe size		4.40	1.818		.53	112	14.20	126.73	182
0180	3″ pipe size		4.40	1.818		.53	112	14.20	126.73	182
0190	3-1/2″ pipe size		3.30	2.424		.60	149	18.95	168.55	243
0200	4″ pipe size		3.30	2.424		.60	149	18.95	168.55	243
0500	12″ thick, 1/2″ pipe size		9.40	.851		.40	52	6.65	59.05	85
0520	3/4″ pipe size		9.40	.851		.40	52	6.65	59.05	85
0540	1″ pipe size		7.30	1.096		.52	67	8.55	76.07	110
0560	1-1/4″ pipe size		7.30	1.096		.52	67	8.55	76.07	110
0570	1-1/2″ pipe size		7.30	1.096		.52	67	8.55	76.07	110
0580	2″ pipe size		3.60	2.222		.80	136	17.35	154.15	223
0590	2-1/2″ pipe size		3.60	2.222		.80	136	17.35	154.15	223
0600	3″ pipe size		3.60	2.222		.80	136	17.35	154.15	223
0610	3-1/2″ pipe size		2.80	2.857		.90	175	22.50	198.40	286
0630	4″ pipe size		2.50	3.200		.90	196	25	221.90	320
0650	16″ thick, 1/2″ pipe size		7.60	1.053		.54	64.50	8.20	73.24	106
0670	3/4″ pipe size		7	1.143		.54	70	8.95	79.49	114
0690	1″ pipe size		6	1.333		.69	82	10.40	93.09	134
0710	1-1/4″ pipe size		5.50	1.455		.69	89	11.35	101.04	146
0730	1-1/2″ pipe size		5.50	1.455		.69	89	11.35	101.04	146
0750	2″ pipe size		3	2.667		1.06	164	21	186.06	268
0770	2-1/2″ pipe size		2.70	2.963		1.06	182	23	206.06	298
0790	3″ pipe size		2.50	3.200		1.06	196	25	222.06	320
0810	3-1/2″ pipe size		2.30	3.478		1.20	213	27	241.20	350
0830	4″ pipe size		2	4		1.20	245	31	277.20	400
0850	20″ thick, 1/2″ pipe size		6.40	1.250		.67	76.50	9.75	86.92	125
0870	3/4″ pipe size		6	1.333		.67	82	10.40	93.07	134
0890	1″ pipe size		5	1.600		.86	98	12.50	111.36	161
0910	1-1/4″ pipe size		4.80	1.667		.86	102	13	115.86	167
0930	1-1/2″ pipe size		4.60	1.739		.86	107	13.60	121.46	175
0950	2″ pipe size		2.70	2.963		1.33	182	23	206.33	298
0970	2-1/2″ pipe size		2.40	3.333		1.33	205	26	232.33	335
0990	3″ pipe size		2.20	3.636		1.33	223	28.50	252.83	360
1010	3-1/2″ pipe size		2	4		1.50	245	31	277.50	400
1030	4″ pipe size		1.70	4.706		1.50	289	37	327.50	470
1050	24″ thick, 1/2″ pipe size		5.50	1.455		.81	89	11.35	101.16	146
1070	3/4″ pipe size		5.10	1.569		.81	96	12.25	109.06	157
1090	1″ pipe size		4.30	1.860		1.04	114	14.55	129.59	187
1110	1-1/4″ pipe size		4	2		1.04	123	15.60	139.64	201
1130	1-1/2″ pipe size		4	2		1.04	123	15.60	139.64	201
1150	2″ pipe size		2.40	3.333		1.59	205	26	232.59	335
1170	2-1/2″ pipe size		2.20	3.636		1.59	223	28.50	253.09	365
1190	3″ pipe size		2	4		1.59	245	31	277.59	400
1210	3-1/2″ pipe size		1.80	4.444		1.80	273	34.50	309.30	445
1230	4″ pipe size		1.50	5.333		1.80	325	41.50	368.30	535
1500	Brick wall, 8″ thick, 1/2″ pipe size	↓	18	.444	↓	.27	27.50	3.47	31.24	44.50

26 05 33 – Raceway and Boxes for Electrical Systems

26 05 33.95 Cutting and Drilling		Crew	Daily Output	Labor-Hours	Unit	2020 Bare Costs Material	Labor	Equipment	Total	Total Incl O&P
1520	3/4" pipe size	R-31	18	.444	Ea.	.27	27.50	3.47	31.24	44.50
1540	1" pipe size		13.30	.602		.35	37	4.70	42.05	60.50
1560	1-1/4" pipe size		13.30	.602		.35	37	4.70	42.05	60.50
1580	1-1/2" pipe size		13.30	.602		.35	37	4.70	42.05	60.50
1600	2" pipe size		5.70	1.404		.53	86	10.95	97.48	141
1620	2-1/2" pipe size		5.70	1.404		.53	86	10.95	97.48	141
1640	3" pipe size		5.70	1.404		.53	86	10.95	97.48	141
1660	3-1/2" pipe size		4.40	1.818		.60	112	14.20	126.80	182
1680	4" pipe size		4	2		.60	123	15.60	139.20	201
1700	12" thick, 1/2" pipe size		14.50	.552		.40	34	4.31	38.71	55.50
1720	3/4" pipe size		14.50	.552		.40	34	4.31	38.71	55.50
1740	1" pipe size		11	.727		.52	44.50	5.70	50.72	73.50
1760	1-1/4" pipe size		11	.727		.52	44.50	5.70	50.72	73.50
1780	1-1/2" pipe size		11	.727		.52	44.50	5.70	50.72	73.50
1800	2" pipe size		5	1.600		.80	98	12.50	111.30	161
1820	2-1/2" pipe size		5	1.600		.80	98	12.50	111.30	161
1840	3" pipe size		5	1.600		.80	98	12.50	111.30	161
1860	3-1/2" pipe size		3.80	2.105		.90	129	16.45	146.35	211
1880	4" pipe size		3.30	2.424		.90	149	18.95	168.85	243
1900	16" thick, 1/2" pipe size		12.30	.650		.54	40	5.10	45.64	65.50
1920	3/4" pipe size		12.30	.650		.54	40	5.10	45.64	65.50
1940	1" pipe size		9.30	.860		.69	53	6.70	60.39	86.50
1960	1-1/4" pipe size		9.30	.860		.69	53	6.70	60.39	86.50
1980	1-1/2" pipe size		9.30	.860		.69	53	6.70	60.39	86.50
2000	2" pipe size		4.40	1.818		1.06	112	14.20	127.26	183
2010	2-1/2" pipe size		4.40	1.818		1.06	112	14.20	127.26	183
2030	3" pipe size		4.40	1.818		1.06	112	14.20	127.26	183
2050	3-1/2" pipe size		3.30	2.424		1.20	149	18.95	169.15	243
2070	4" pipe size		3	2.667		1.20	164	21	186.20	268
2090	20" thick, 1/2" pipe size		10.70	.748		.67	46	5.85	52.52	75.50
2110	3/4" pipe size		10.70	.748		.67	46	5.85	52.52	75.50
2130	1" pipe size		8	1		.86	61.50	7.80	70.16	101
2150	1-1/4" pipe size		8	1		.86	61.50	7.80	70.16	101
2170	1-1/2" pipe size		8	1		.86	61.50	7.80	70.16	101
2190	2" pipe size		4	2		1.33	123	15.60	139.93	202
2210	2-1/2" pipe size		4	2		1.33	123	15.60	139.93	202
2230	3" pipe size		4	2		1.33	123	15.60	139.93	202
2250	3-1/2" pipe size		3	2.667		1.50	164	21	186.50	269
2270	4" pipe size		2.70	2.963		1.50	182	23	206.50	298
2290	24" thick, 1/2" pipe size		9.40	.851		.81	52	6.65	59.46	85.50
2310	3/4" pipe size		9.40	.851		.81	52	6.65	59.46	85.50
2330	1" pipe size		7.10	1.127		1.04	69	8.80	78.84	114
2350	1-1/4" pipe size		7.10	1.127		1.04	69	8.80	78.84	114
2370	1-1/2" pipe size		7.10	1.127		1.04	69	8.80	78.84	114
2390	2" pipe size		3.60	2.222		1.59	136	17.35	154.94	224
2410	2-1/2" pipe size		3.60	2.222		1.59	136	17.35	154.94	224
2430	3" pipe size		3.60	2.222		1.59	136	17.35	154.94	224
2450	3-1/2" pipe size		2.80	2.857		1.80	175	22.50	199.30	287
2470	4" pipe size		2.50	3.200		1.80	196	25	222.80	320
3000	Knockouts to 8' high, metal boxes & enclosures									
3020	With hole saw, 1/2" pipe size	1 Elec	53	.151	Ea.		9.25		9.25	13.80
3040	3/4" pipe size		47	.170			10.45		10.45	15.55
3050	1" pipe size		40	.200			12.25		12.25	18.25

26 05 33 – Raceway and Boxes for Electrical Systems

26 05 33.95 Cutting and Drilling		Crew	Daily Output	Labor-Hours	Unit	2020 Bare Costs Material	Labor	Equipment	Total	Total Incl O&P
3060	1-1/4" pipe size	1 Elec	36	.222	Ea.		13.65		13.65	20.50
3070	1-1/2" pipe size		32	.250			15.35		15.35	23
3080	2" pipe size		27	.296			18.20		18.20	27
3090	2-1/2" pipe size		20	.400			24.50		24.50	36.50
4010	3" pipe size		16	.500			30.50		30.50	45.50
4030	3-1/2" pipe size		13	.615			38		38	56
4050	4" pipe size		11	.727			44.50		44.50	66.50

26 05 83 – Wiring Connections

26 05 83.10 Motor Connections

		Crew	Daily Output	Labor-Hours	Unit	Material	Labor	Equipment	Total	Total Incl O&P
0010	**MOTOR CONNECTIONS**									
0020	Flexible conduit and fittings, 115 volt, 1 phase, up to 1 HP motor	1 Elec	8	1	Ea.	5.90	61.50		67.40	98
0050	2 HP motor		6.50	1.231		10.40	75.50		85.90	123
0100	3 HP motor		5.50	1.455		9.20	89		98.20	143
0110	230 volt, 3 phase, 3 HP motor		6.78	1.180		7.15	72.50		79.65	116
0112	5 HP motor		5.47	1.463		6.05	89.50		95.55	141
0114	7-1/2 HP motor		4.61	1.735		9.35	106		115.35	169
0120	10 HP motor		4.20	1.905		17.40	117		134.40	193
0150	15 HP motor		3.30	2.424		17.40	149		166.40	240
0200	25 HP motor		2.70	2.963		29.50	182		211.50	305
0400	50 HP motor		2.20	3.636		54.50	223		277.50	390
0600	100 HP motor		1.50	5.333		121	325		446	620

26 05 90 – Residential Applications

26 05 90.10 Residential Wiring

		Crew	Daily Output	Labor-Hours	Unit	Material	Labor	Equipment	Total	Total Incl O&P
0010	**RESIDENTIAL WIRING**									
0020	20' avg. runs and #14/2 wiring incl. unless otherwise noted									
1000	Service & panel, includes 24' SE-AL cable, service eye, meter,									
1010	Socket, panel board, main bkr., ground rod, 15 or 20 amp									
1020	1-pole circuit breakers, and misc. hardware									
1100	100 amp, with 10 branch breakers	1 Elec	1.19	6.723	Ea.	335	410		745	980
1110	With PVC conduit and wire		.92	8.696		370	535		905	1,200
1120	With RGS conduit and wire		.73	10.959		570	670		1,240	1,625
1150	150 amp, with 14 branch breakers		1.03	7.767		805	475		1,280	1,600
1170	With PVC conduit and wire		.82	9.756		875	600		1,475	1,850
1180	With RGS conduit and wire		.67	11.940		1,200	735		1,935	2,425
1200	200 amp, with 18 branch breakers	2 Elec	1.80	8.889		1,025	545		1,570	1,925
1220	With PVC conduit and wire		1.46	10.959		1,100	670		1,770	2,200
1230	With RGS conduit and wire		1.24	12.903		1,500	790		2,290	2,825
1800	Lightning surge suppressor	1 Elec	32	.250		85	15.35		100.35	117
2000	Switch devices									
2100	Single pole, 15 amp, ivory, with a 1-gang box, cover plate,									
2110	Type NM (Romex) cable	1 Elec	17.10	.468	Ea.	15.55	28.50		44.05	59.50
2120	Type MC cable		14.30	.559		26	34.50		60.50	80
2130	EMT & wire		5.71	1.401		36	86		122	168
2150	3-way, #14/3, type NM cable		14.55	.550		10.10	33.50		43.60	61
2170	Type MC cable		12.31	.650		24.50	40		64.50	86.50
2180	EMT & wire		5	1.600		30.50	98		128.50	180
2200	4-way, #14/3, type NM cable		14.55	.550		18.20	33.50		51.70	70
2220	Type MC cable		12.31	.650		32.50	40		72.50	95.50
2230	EMT & wire		5	1.600		38.50	98		136.50	189
2250	S.P., 20 amp, #12/2, type NM cable		13.33	.600		12.30	37		49.30	68.50
2270	Type MC cable		11.43	.700		22	43		65	88
2280	EMT & wire		4.85	1.649		35	101		136	190

26 05 90.10 Residential Wiring		Crew	Daily Output	Labor-Hours	Unit	Material	2020 Bare Costs Labor	Equipment	Total	Total Incl O&P
2290	S.P. rotary dimmer, 600 W, no wiring	1 Elec	17	.471	Ea.	32	29		61	78
2300	S.P. rotary dimmer, 600 W, type NM cable		14.55	.550		35.50	33.50		69	89
2320	Type MC cable		12.31	.650		46	40		86	111
2330	EMT & wire		5	1.600		57.50	98		155.50	209
2350	3-way rotary dimmer, type NM cable		13.33	.600		24	37		61	81.50
2370	Type MC cable		11.43	.700		34.50	43		77.50	102
2380	EMT & wire		4.85	1.649		46	101		147	202
2400	Interval timer wall switch, 20 amp, 1-30 min., #12/2									
2410	Type NM cable	1 Elec	14.55	.550	Ea.	59	33.50		92.50	115
2420	Type MC cable		12.31	.650		64.50	40		104.50	131
2430	EMT & wire		5	1.600		81.50	98		179.50	236
2500	Decorator style									
2510	S.P., 15 amp, type NM cable	1 Elec	17.10	.468	Ea.	21	28.50		49.50	65.50
2520	Type MC cable		14.30	.559		32	34.50		66.50	86
2530	EMT & wire		5.71	1.401		41.50	86		127.50	174
2550	3-way, #14/3, type NM cable		14.55	.550		15.65	33.50		49.15	67.50
2570	Type MC cable		12.31	.650		30	40		70	92.50
2580	EMT & wire		5	1.600		36	98		134	186
2600	4-way, #14/3, type NM cable		14.55	.550		24	33.50		57.50	76
2620	Type MC cable		12.31	.650		38	40		78	102
2630	EMT & wire		5	1.600		44	98		142	195
2650	S.P., 20 amp, #12/2, type NM cable		13.33	.600		17.85	37		54.85	74.50
2670	Type MC cable		11.43	.700		27.50	43		70.50	94
2680	EMT & wire		4.85	1.649		40.50	101		141.50	196
2700	S.P., slide dimmer, type NM cable		17.10	.468		37.50	28.50		66	84
2720	Type MC cable		14.30	.559		48	34.50		82.50	104
2730	EMT & wire		5.71	1.401		59.50	86		145.50	194
2750	S.P., touch dimmer, type NM cable		17.10	.468		53	28.50		81.50	101
2770	Type MC cable		14.30	.559		64	34.50		98.50	121
2780	EMT & wire		5.71	1.401		75	86		161	211
2800	3-way touch dimmer, type NM cable		13.33	.600		49.50	37		86.50	110
2820	Type MC cable		11.43	.700		60.50	43		103.50	131
2830	EMT & wire		4.85	1.649		71.50	101		172.50	230
3000	Combination devices									
3100	S.P. switch/15 amp recpt., ivory, 1-gang box, plate									
3110	Type NM cable	1 Elec	11.43	.700	Ea.	22	43		65	88
3120	Type MC cable		10	.800		32.50	49		81.50	109
3130	EMT & wire		4.40	1.818		44	112		156	214
3150	S.P. switch/pilot light, type NM cable		11.43	.700		23.50	43		66.50	89.50
3170	Type MC cable		10	.800		34	49		83	111
3180	EMT & wire		4.43	1.806		45	111		156	215
3190	2-S.P. switches, 2-#14/2, no wiring		14	.571		13	35		48	66.50
3200	2-S.P. switches, 2-#14/2, type NM cables		10	.800		24	49		73	99.50
3220	Type MC cable		8.89	.900		38.50	55		93.50	125
3230	EMT & wire		4.10	1.951		46	120		166	229
3250	3-way switch/15 amp recpt., #14/3, type NM cable		10	.800		30	49		79	106
3270	Type MC cable		8.89	.900		44	55		99	131
3280	EMT & wire		4.10	1.951		50	120		170	233
3300	2-3 way switches, 2-#14/3, type NM cables		8.89	.900		38	55		93	124
3320	Type MC cable		8	1		60	61.50		121.50	158
3330	EMT & wire		4	2		56.50	123		179.50	246
3350	S.P. switch/20 amp recpt., #12/2, type NM cable		10	.800		40	49		89	117
3370	Type MC cable		8.89	.900		45.50	55		100.50	132

26 05 90.10 Residential Wiring		Crew	Daily Output	Labor-Hours	Unit	2020 Bare Costs Material	Labor	Equipment	Total	Total Incl O&P
3380	EMT & wire	1 Elec	4.10	1.951	Ea.	62.50	120		182.50	247
3400	Decorator style									
3410	S.P. switch/15 amp recpt., type NM cable	1 Elec	11.43	.700	Ea.	27.50	43		70.50	94
3420	Type MC cable		10	.800		38	49		87	115
3430	EMT & wire		4.40	1.818		49.50	112		161.50	221
3450	S.P. switch/pilot light, type NM cable		11.43	.700		29	43		72	95.50
3470	Type MC cable		10	.800		39.50	49		88.50	117
3480	EMT & wire		4.40	1.818		50.50	112		162.50	222
3500	2-S.P. switches, 2-#14/2, type NM cables		10	.800		29.50	49		78.50	106
3520	Type MC cable		8.89	.900		44.50	55		99.50	131
3530	EMT & wire		4.10	1.951		51.50	120		171.50	235
3550	3-way/15 amp recpt., #14/3, type NM cable		10	.800		35.50	49		84.50	112
3570	Type MC cable		8.89	.900		49.50	55		104.50	137
3580	EMT & wire		4.10	1.951		55.50	120		175.50	239
3650	2-3 way switches, 2-#14/3, type NM cables		8.89	.900		43.50	55		98.50	130
3670	Type MC cable		8	1		65.50	61.50		127	164
3680	EMT & wire		4	2		62	123		185	252
3700	S.P. switch/20 amp recpt., #12/2, type NM cable		10	.800		45.50	49		94.50	123
3720	Type MC cable		8.89	.900		51	55		106	138
3730	EMT & wire		4.10	1.951		68	120		188	253
4000	Receptacle devices									
4010	Duplex outlet, 15 amp recpt., ivory, 1-gang box, plate									
4015	Type NM cable	1 Elec	14.55	.550	Ea.	8.90	33.50		42.40	60
4020	Type MC cable		12.31	.650		19.55	40		59.55	81
4030	EMT & wire		5.33	1.501		29.50	92		121.50	170
4050	With #12/2, type NM cable		12.31	.650		10.50	40		50.50	71
4070	Type MC cable		10.67	.750		19.95	46		65.95	90.50
4080	EMT & wire		4.71	1.699		33	104		137	191
4100	20 amp recpt., #12/2, type NM cable		12.31	.650		19.50	40		59.50	81
4120	Type MC cable		10.67	.750		29	46		75	101
4130	EMT & wire		4.71	1.699		42	104		146	201
4140	For GFI see Section 26 05 90.10 line 4300 below									
4150	Decorator style, 15 amp recpt., type NM cable	1 Elec	14.55	.550	Ea.	14.45	33.50		47.95	66
4170	Type MC cable		12.31	.650		25	40		65	87
4180	EMT & wire		5.33	1.501		35	92		127	176
4200	With #12/2, type NM cable		12.31	.650		16.05	40		56.05	77
4220	Type MC cable		10.67	.750		25.50	46		71.50	96.50
4230	EMT & wire		4.71	1.699		38.50	104		142.50	198
4250	20 amp recpt., #12/2, type NM cable		12.31	.650		25	40		65	87
4270	Type MC cable		10.67	.750		34.50	46		80.50	107
4280	EMT & wire		4.71	1.699		47.50	104		151.50	208
4300	GFI, 15 amp recpt., type NM cable		12.31	.650		21	40		61	82.50
4320	Type MC cable		10.67	.750		31.50	46		77.50	103
4330	EMT & wire		4.71	1.699		41.50	104		145.50	201
4350	GFI with #12/2, type NM cable		10.67	.750		22.50	46		68.50	93
4370	Type MC cable		9.20	.870		32	53.50		85.50	115
4380	EMT & wire		4.21	1.900		45	117		162	224
4400	20 amp recpt., #12/2, type NM cable		10.67	.750		54	46		100	128
4420	Type MC cable		9.20	.870		63.50	53.50		117	150
4430	EMT & wire		4.21	1.900		76.50	117		193.50	258
4500	Weather-proof cover for above receptacles, add		32	.250		2.02	15.35		17.37	25
4550	Air conditioner outlet, 20 amp-240 volt recpt.									
4560	30' of #12/2, 2 pole circuit breaker									

26 05 90.10 Residential Wiring		Crew	Daily Output	Labor-Hours	Unit	Material	Labor	Equipment	Total	Total Incl O&P
						2020 Bare Costs				
4570	Type NM cable	1 Elec	10	.800	Ea.	61	49		110	140
4580	Type MC cable		9	.889		72	54.50		126.50	160
4590	EMT & wire		4	2		84	123		207	275
4600	Decorator style, type NM cable		10	.800		66	49		115	146
4620	Type MC cable		9	.889		77	54.50		131.50	166
4630	EMT & wire		4	2		88.50	123		211.50	281
4650	Dryer outlet, 30 amp-240 volt recpt., 20′ of #10/3									
4660	2 pole circuit breaker									
4670	Type NM cable	1 Elec	6.41	1.248	Ea.	54.50	76.50		131	174
4680	Type MC cable		5.71	1.401		62.50	86		148.50	197
4690	EMT & wire		3.48	2.299		73	141		214	291
4700	Range outlet, 50 amp-240 volt recpt., 30′ of #8/3									
4710	Type NM cable	1 Elec	4.21	1.900	Ea.	82.50	117		199.50	265
4720	Type MC cable		4	2		133	123		256	330
4730	EMT & wire		2.96	2.703		105	166		271	360
4750	Central vacuum outlet, type NM cable		6.40	1.250		58	76.50		134.50	178
4770	Type MC cable		5.71	1.401		70.50	86		156.50	206
4780	EMT & wire		3.48	2.299		88	141		229	305
4800	30 amp-110 volt locking recpt., #10/2 circ. bkr.									
4810	Type NM cable	1 Elec	6.20	1.290	Ea.	67	79		146	192
4830	EMT & wire	"	3.20	2.500	"	98	153		251	335
4900	Low voltage outlets									
4910	Telephone recpt., 20′ of 4/C phone wire	1 Elec	26	.308	Ea.	8.80	18.90		27.70	37.50
4920	TV recpt., 20′ of RG59U coax wire, F type connector	"	16	.500	"	17.60	30.50		48.10	65
4950	Door bell chime, transformer, 2 buttons, 60′ of bellwire									
4970	Economy model	1 Elec	11.50	.696	Ea.	57.50	42.50		100	127
4980	Custom model		11.50	.696		111	42.50		153.50	186
4990	Luxury model, 3 buttons		9.50	.842		188	51.50		239.50	284
6000	Lighting outlets									
6050	Wire only (for fixture), type NM cable	1 Elec	32	.250	Ea.	5.95	15.35		21.30	29.50
6070	Type MC cable		24	.333		11.30	20.50		31.80	43
6080	EMT & wire		10	.800		20.50	49		69.50	95.50
6100	Box (4″), and wire (for fixture), type NM cable		25	.320		15.05	19.65		34.70	45.50
6120	Type MC cable		20	.400		20.50	24.50		45	59
6130	EMT & wire		11	.727		29.50	44.50		74	99
6200	Fixtures (use with line 6050 or 6100 above)									
6210	Canopy style, economy grade	1 Elec	40	.200	Ea.	22	12.25		34.25	43
6220	Custom grade		40	.200		53	12.25		65.25	76.50
6250	Dining room chandelier, economy grade		19	.421		82	26		108	129
6260	Custom grade		19	.421		320	26		346	395
6270	Luxury grade		15	.533		1,300	32.50		1,332.50	1,475
6310	Kitchen fixture (fluorescent), economy grade		30	.267		72.50	16.35		88.85	104
6320	Custom grade		25	.320		147	19.65		166.65	191
6350	Outdoor, wall mounted, economy grade		30	.267		30.50	16.35		46.85	58
6360	Custom grade		30	.267		119	16.35		135.35	155
6370	Luxury grade		25	.320		247	19.65		266.65	300
6410	Outdoor PAR floodlights, 1 lamp, 150 watt		20	.400		28	24.50		52.50	67
6420	2 lamp, 150 watt each		20	.400		45	24.50		69.50	86
6430	For infrared security sensor, add		32	.250		95	15.35		110.35	127
6450	Outdoor, quartz-halogen, 300 watt flood		20	.400		40	24.50		64.50	81
6600	Recessed downlight, round, pre-wired, 50 or 75 watt trim		30	.267		68	16.35		84.35	99.50
6610	With shower light trim		30	.267		93.50	16.35		109.85	128
6620	With wall washer trim		28	.286		94	17.55		111.55	129

26 05 90.10 Residential Wiring		Crew	Daily Output	Labor-Hours	Unit	Material	2020 Bare Costs Labor	Equipment	Total	Total Incl O&P
6630	With eye-ball trim	1 Elec	28	.286	Ea.	85	17.55		102.55	120
6700	Porcelain lamp holder		40	.200		2.76	12.25		15.01	21.50
6710	With pull switch		40	.200		10.70	12.25		22.95	30
6750	Fluorescent strip, 2-20 watt tube, wrap around diffuser, 24″		24	.333		45.50	20.50		66	80.50
6770	2-34 watt tubes, 48″		20	.400		160	24.50		184.50	213
6800	Bathroom heat lamp, 1-250 watt		28	.286		33.50	17.55		51.05	63
6810	2-250 watt lamps		28	.286		64.50	17.55		82.05	96.50
6820	For timer switch, see Section 26 05 90.10 line 2400									
6900	Outdoor post lamp, incl. post, fixture, 35′ of #14/2									
6910	Type NM cable	1 Elec	3.50	2.286	Ea.	325	140		465	565
6920	Photo-eye, add		27	.296		27	18.20		45.20	57
6950	Clock dial time switch, 24 hr., w/enclosure, type NM cable		11.43	.700		72.50	43		115.50	144
6970	Type MC cable		11	.727		83	44.50		127.50	158
6980	EMT & wire		4.85	1.649		93	101		194	253
7000	Alarm systems									
7050	Smoke detectors, box, #14/3, type NM cable	1 Elec	14.55	.550	Ea.	33.50	33.50		67	87
7070	Type MC cable		12.31	.650		44	40		84	108
7080	EMT & wire		5	1.600		50	98		148	201
7090	For relay output to security system, add					10.25			10.25	11.25
8000	Residential equipment									
8050	Disposal hook-up, incl. switch, outlet box, 3′ of flex									
8060	20 amp-1 pole circ. bkr., and 25′ of #12/2									
8070	Type NM cable	1 Elec	10	.800	Ea.	29	49		78	105
8080	Type MC cable		8	1		39	61.50		100.50	135
8090	EMT & wire		5	1.600		55	98		153	206
8100	Trash compactor or dishwasher hook-up, incl. outlet box,									
8110	3′ of flex, 15 amp-1 pole circ. bkr., and 25′ of #14/2									
8120	Type NM cable	1 Elec	10	.800	Ea.	16.05	49		65.05	90.50
8130	Type MC cable		8	1		28	61.50		89.50	123
8140	EMT & wire		5	1.600		41.50	98		139.50	192
8150	Hot water sink dispenser hook-up, use line 8100									
8200	Vent/exhaust fan hook-up, type NM cable	1 Elec	32	.250	Ea.	5.95	15.35		21.30	29.50
8220	Type MC cable		24	.333		11.30	20.50		31.80	43
8230	EMT & wire		10	.800		20.50	49		69.50	95.50
8250	Bathroom vent fan, 50 CFM (use with above hook-up)									
8260	Economy model	1 Elec	15	.533	Ea.	18.80	32.50		51.30	69
8270	Low noise model		15	.533		47.50	32.50		80	101
8280	Custom model		12	.667		117	41		158	190
8300	Bathroom or kitchen vent fan, 110 CFM									
8310	Economy model	1 Elec	15	.533	Ea.	66.50	32.50		99	122
8320	Low noise model	″	15	.533	″	96.50	32.50		129	155
8350	Paddle fan, variable speed (w/o lights)									
8360	Economy model (AC motor)	1 Elec	10	.800	Ea.	136	49		185	222
8362	With light kit		10	.800		176	49		225	266
8370	Custom model (AC motor)		10	.800		345	49		394	455
8372	With light kit		10	.800		385	49		434	500
8380	Luxury model (DC motor)		8	1		315	61.50		376.50	435
8382	With light kit		8	1		355	61.50		416.50	480
8390	Remote speed switch for above, add		12	.667		40.50	41		81.50	106
8500	Whole house exhaust fan, ceiling mount, 36″, variable speed									
8510	Remote switch, incl. shutters, 20 amp-1 pole circ. bkr.									
8520	30′ of #12/2, type NM cable	1 Elec	4	2	Ea.	1,375	123		1,498	1,700
8530	Type MC cable		3.50	2.286		1,400	140		1,540	1,750

26 05 Common Work Results for Electrical

26 05 90 – Residential Applications

26 05 90.10 Residential Wiring		Crew	Daily Output	Labor-Hours	Unit	Material	Labor	Equipment	Total	Total Incl O&P
						2020 Bare Costs				
8540	EMT & wire	1 Elec	3	2.667	Ea.	1,425	164		1,589	1,800
8600	Whirlpool tub hook-up, incl. timer switch, outlet box									
8610	3′ of flex, 20 amp-1 pole GFI circ. bkr.									
8620	30′ of #12/2, type NM cable	1 Elec	5	1.600	Ea.	129	98		227	288
8630	Type MC cable		4.20	1.905		136	117		253	325
8640	EMT & wire		3.40	2.353		149	144		293	380
8650	Hot water heater hook-up, incl. 1-2 pole circ. bkr., box;									
8660	3′ of flex, 20′ of #10/2, type NM cable	1 Elec	5	1.600	Ea.	29	98		127	178
8670	Type MC cable		4.20	1.905		42	117		159	220
8680	EMT & wire		3.40	2.353		48	144		192	268
9000	Heating/air conditioning									
9050	Furnace/boiler hook-up, incl. firestat, local on-off switch									
9060	Emergency switch, and 40′ of type NM cable	1 Elec	4	2	Ea.	57	123		180	246
9070	Type MC cable		3.50	2.286		72	140		212	288
9080	EMT & wire		1.50	5.333		93.50	325		418.50	590
9100	Air conditioner hook-up, incl. local 60 amp disc. switch									
9110	3′ sealtite, 40 amp, 2 pole circuit breaker									
9130	40′ of #8/2, type NM cable	1 Elec	3.50	2.286	Ea.	144	140		284	365
9140	Type MC cable		3	2.667		212	164		376	475
9150	EMT & wire		1.30	6.154		185	380		565	765
9200	Heat pump hook-up, 1-40 & 1-100 amp 2 pole circ. bkr.									
9210	Local disconnect switch, 3′ sealtite									
9220	40′ of #8/2 & 30′ of #3/2									
9230	Type NM cable	1 Elec	1.30	6.154	Ea.	520	380		900	1,125
9240	Type MC cable		1.08	7.407		550	455		1,005	1,275
9250	EMT & wire		.94	8.511		535	520		1,055	1,350
9500	Thermostat hook-up, using low voltage wire									
9520	Heating only, 25′ of #18-3	1 Elec	24	.333	Ea.	6.75	20.50		27.25	38
9530	Heating/cooling, 25′ of #18-4	″	20	.400	″	8.60	24.50		33.10	46

26 24 Switchboards and Panelboards

26 24 19 – Motor-Control Centers

26 24 19.40 Motor Starters and Controls		Crew	Daily Output	Labor-Hours	Unit	Material	Labor	Equipment	Total	Total Incl O&P
0010	**MOTOR STARTERS AND CONTROLS**									
0050	Magnetic, FVNR, with enclosure and heaters, 480 volt									
0080	2 HP, size 00	1 Elec	3.50	2.286	Ea.	196	140		336	425
0100	5 HP, size 0		2.30	3.478		345	213		558	700
0200	10 HP, size 1		1.60	5		262	305		567	745
0300	25 HP, size 2	2 Elec	2.20	7.273		490	445		935	1,200
0400	50 HP, size 3		1.80	8.889		800	545		1,345	1,700
0500	100 HP, size 4		1.20	13.333		1,775	820		2,595	3,175
0600	200 HP, size 5		.90	17.778		4,150	1,100		5,250	6,200
0610	400 HP, size 6		.80	20		18,300	1,225		19,525	21,900
0620	NEMA 7, 5 HP, size 0	1 Elec	1.60	5		1,375	305		1,680	1,950
0630	10 HP, size 1	″	1.10	7.273		1,425	445		1,870	2,250
0640	25 HP, size 2	2 Elec	1.80	8.889		2,300	545		2,845	3,325
0650	50 HP, size 3		1.20	13.333		3,450	820		4,270	5,025
0660	100 HP, size 4		.90	17.778		5,575	1,100		6,675	7,750
0670	200 HP, size 5		.50	32		13,300	1,975		15,275	17,500
0700	Combination, with motor circuit protectors, 5 HP, size 0	1 Elec	1.80	4.444		985	273		1,258	1,475
0800	10 HP, size 1	″	1.30	6.154		1,025	380		1,405	1,675

26 24 Switchboards and Panelboards

26 24 19 – Motor-Control Centers

26 24 19.40 Motor Starters and Controls		Crew	Daily Output	Labor-Hours	Unit	Material	Labor	Equipment	Total	Total Incl O&P
						2020 Bare Costs				
0900	25 HP, size 2	2 Elec	2	8	Ea.	1,425	490		1,915	2,300
1000	50 HP, size 3		1.32	12.121		2,075	745		2,820	3,375
1200	100 HP, size 4	↓	.80	20		4,450	1,225		5,675	6,725
1220	NEMA 7, 5 HP, size 0	1 Elec	1.30	6.154		3,550	380		3,930	4,475
1230	10 HP, size 1	"	1	8		3,650	490		4,140	4,725
1240	25 HP, size 2	2 Elec	1.32	12.121		4,875	745		5,620	6,450
1250	50 HP, size 3		.80	20		8,050	1,225		9,275	10,700
1260	100 HP, size 4		.60	26.667		12,500	1,625		14,125	16,200
1270	200 HP, size 5	↓	.40	40		27,200	2,450		29,650	33,700
1400	Combination, with fused switch, 5 HP, size 0	1 Elec	1.80	4.444		590	273		863	1,050
1600	10 HP, size 1	"	1.30	6.154		630	380		1,010	1,250
1800	25 HP, size 2	2 Elec	2	8		1,025	490		1,515	1,850
2000	50 HP, size 3		1.32	12.121		1,725	745		2,470	3,000
2200	100 HP, size 4	↓	.80	20	↓	3,025	1,225		4,250	5,150
3500	Magnetic FVNR with NEMA 12, enclosure & heaters, 480 volt									
3600	5 HP, size 0	1 Elec	2.20	3.636	Ea.	237	223		460	590
3700	10 HP, size 1	"	1.50	5.333		355	325		680	875
3800	25 HP, size 2	2 Elec	2	8		665	490		1,155	1,475
3900	50 HP, size 3		1.60	10		1,025	615		1,640	2,050
4000	100 HP, size 4		1	16		2,450	980		3,430	4,150
4100	200 HP, size 5	↓	.80	20		5,900	1,225		7,125	8,300
4200	Combination, with motor circuit protectors, 5 HP, size 0	1 Elec	1.70	4.706		785	289		1,074	1,300
4300	10 HP, size 1	"	1.20	6.667		815	410		1,225	1,500
4400	25 HP, size 2	2 Elec	1.80	8.889		1,225	545		1,770	2,150
4500	50 HP, size 3		1.20	13.333		2,000	820		2,820	3,400
4600	100 HP, size 4	↓	.74	21.622		4,475	1,325		5,800	6,900
4700	Combination, with fused switch, 5 HP, size 0	1 Elec	1.70	4.706		745	289		1,034	1,250
4800	10 HP, size 1	"	1.20	6.667		775	410		1,185	1,450
4900	25 HP, size 2	2 Elec	1.80	8.889		1,175	545		1,720	2,100
5000	50 HP, size 3		1.20	13.333		1,900	820		2,720	3,300
5100	100 HP, size 4	↓	.74	21.622	↓	3,825	1,325		5,150	6,200
5200	Factory installed controls, adders to size 0 thru 5									
5300	Start-stop push button	1 Elec	32	.250	Ea.	49.50	15.35		64.85	77.50
5400	Hand-off-auto-selector switch		32	.250		49.50	15.35		64.85	77.50
5500	Pilot light		32	.250		93	15.35		108.35	125
5600	Start-stop-pilot		32	.250		143	15.35		158.35	180
5700	Auxiliary contact, NO or NC		32	.250		68	15.35		83.35	98
5800	NO-NC	↓	32	.250	↓	136	15.35		151.35	173

26 29 Low-Voltage Controllers

26 29 13 – Enclosed Controllers

26 29 13.20 Control Stations		Crew	Daily Output	Labor-Hours	Unit	Material	Labor	Equipment	Total	Total Incl O&P
0010	**CONTROL STATIONS**									
0050	NEMA 1, heavy duty, stop/start	1 Elec	8	1	Ea.	153	61.50		214.50	260
0100	Stop/start, pilot light		6.20	1.290		208	79		287	345
0200	Hand/off/automatic		6.20	1.290		113	79		192	242
0400	Stop/start/reverse		5.30	1.509		206	92.50		298.50	365
0500	NEMA 7, heavy duty, stop/start		6	1.333		555	82		637	730
0600	Stop/start, pilot light	↓	4	2	↓	680	123		803	930

Division 28

Electronic Safety & Security

Estimating Tips

- When estimating material costs for electronic safety and security systems, it is always prudent to obtain manufacturers' quotations for equipment prices and special installation requirements that may affect the total cost.
- Fire alarm systems consist of control panels, annunciator panels, batteries with rack, charger, and fire alarm actuating and indicating devices. Some fire alarm systems include speakers, telephone lines, door closer controls, and other components. Be careful not to overlook the costs related to installation for these items. Also be aware of costs for integrated automation instrumentation and terminal devices, control equipment, control wiring, and programming. Insurance underwriters may have specific requirements for the type of materials to be installed or design requirements based on the hazard to be protected. Local jurisdictions may have requirements not covered by code. It is advisable to be aware of any special conditions.
- Security equipment includes items such as CCTV, access control, and other detection and identification systems to perform alert and alarm functions. Be sure to consider the costs related to installation for this security equipment, such as for integrated automation instrumentation and terminal devices, control equipment, control wiring, and programming.

Reference Numbers

Reference numbers are shown at the beginning of some major classifications. These numbers refer to related items in the Reference Section. The reference information may be an estimating procedure, an alternate pricing method, or technical information.

Note: Trade Service, in part, has been used as a reference source for some of the material prices used in Division 28.

28 31 Intrusion Detection

28 31 16 – Intrusion Detection Systems Infrastructure

28 31 16.50 Intrusion Detection		Crew	Daily Output	Labor-Hours	Unit	2020 Bare Costs Material	Labor	Equipment	Total	Total Incl O&P
0010	**INTRUSION DETECTION**, not including wires & conduits									
0100	Burglar alarm, battery operated, mechanical trigger	1 Elec	4	2	Ea.	289	123		412	505
0200	Electrical trigger		4	2		345	123		468	565
0400	For outside key control, add		8	1		89.50	61.50		151	190
0600	For remote signaling circuitry, add		8	1		143	61.50		204.50	249
0800	Card reader, flush type, standard		2.70	2.963		725	182		907	1,075
1000	Multi-code		2.70	2.963		1,250	182		1,432	1,650

28 42 Gas Detection and Alarm

28 42 15 – Gas Detection Sensors

28 42 15.50 Tank Leak Detection Systems		Crew	Daily Output	Labor-Hours	Unit	Material	Labor	Equipment	Total	Total Incl O&P
0010	**TANK LEAK DETECTION SYSTEMS** Liquid and vapor									
0100	For hydrocarbons and hazardous liquids/vapors									
0120	Controller, data acquisition, incl. printer, modem, RS232 port									
0140	24 channel, for use with all probes				Ea.	3,125			3,125	3,450
0160	9 channel, for external monitoring				"	1,225			1,225	1,350
0200	Probes									
0210	Well monitoring									
0220	Liquid phase detection				Ea.	725			725	800
0230	Hydrocarbon vapor, fixed position					870			870	960
0240	Hydrocarbon vapor, float mounted					615			615	675
0250	Both liquid and vapor hydrocarbon					645			645	710
0300	Secondary containment, liquid phase									
0310	Pipe trench/manway sump				Ea.	765			765	840
0320	Double wall pipe and manual sump					640			640	705
0330	Double wall fiberglass annular space					430			430	470
0340	Double wall steel tank annular space					282			282	310
0500	Accessories									
0510	Modem, non-dedicated phone line				Ea.	281			281	310
0600	Monitoring, internal									
0610	Automatic tank gauge, incl. overfill				Ea.	1,325			1,325	1,450
0620	Product line				"	1,375			1,375	1,525
0700	Monitoring, special									
0710	Cathodic protection				Ea.	695			695	765
0720	Annular space chemical monitor				"	945			945	1,050

28 46 Fire Detection and Alarm

28 46 11 – Fire Sensors and Detectors

28 46 11.27 Other Sensors		Crew	Daily Output	Labor-Hours	Unit	Material	Labor	Equipment	Total	Total Incl O&P
0010	**OTHER SENSORS**									
5200	Smoke detector, ceiling type	1 Elec	6.20	1.290	Ea.	120	79		199	251
5240	Smoke detector, addressable type		6	1.333		224	82		306	370
5400	Duct type		3.20	2.500		295	153		448	555
5420	Duct addressable type		3.20	2.500		520	153		673	805

28 46 11.50 Fire and Heat Detectors		Crew	Daily Output	Labor-Hours	Unit	Material	Labor	Equipment	Total	Total Incl O&P
0010	**FIRE & HEAT DETECTORS**									
5000	Detector, rate of rise	1 Elec	8	1	Ea.	52	61.50		113.50	149
5100	Fixed temp fire alarm	"	7	1.143	"	42	70		112	151

28 46 20 – Fire Alarm

28 46 20.50 Alarm Panels and Devices		Crew	Daily Output	Labor-Hours	Unit	Material	Labor	Equipment	Total	Total Incl O&P
						2020 Bare Costs				
0010	**ALARM PANELS AND DEVICES**, not including wires & conduits									
3600	4 zone	2 Elec	2	8	Ea.	410	490		900	1,175
3800	8 zone		1	16		875	980		1,855	2,425
3810	5 zone		1.50	10.667		700	655		1,355	1,750
3900	10 zone		1.25	12.800		1,025	785		1,810	2,325
4000	12 zone		.67	23.988		2,350	1,475		3,825	4,775
4020	Alarm device, tamper, flow	1 Elec	8	1		230	61.50		291.50	345
4025	Fire alarm, loop expander card		16	.500		680	30.50		710.50	795
4050	Actuating device		8	1		340	61.50		401.50	460
4200	Battery and rack		4	2		460	123		583	695
4400	Automatic charger		8	1		525	61.50		586.50	670
4600	Signal bell		8	1		66.50	61.50		128	165
4610	Fire alarm signal bell 10" red 20-24 V P		8	1		154	61.50		215.50	261
4800	Trouble buzzer or manual station		8	1		88.50	61.50		150	189
5425	Duct smoke and heat detector 2 wire		8	1		125	61.50		186.50	230
5430	Fire alarm duct detector controller		3	2.667		254	164		418	525
5435	Fire alarm duct detector sensor kit		8	1		75	61.50		136.50	174
5440	Remote test station for smoke detector duct type		5.30	1.509		54.50	92.50		147	198
5460	Remote fire alarm indicator light		5.30	1.509		25	92.50		117.50	166
5600	Strobe and horn		5.30	1.509		137	92.50		229.50	289
5610	Strobe and horn (ADA type)		5.30	1.509		165	92.50		257.50	320
5620	Visual alarm (ADA type)		6.70	1.194		119	73.50		192.50	240
5800	electric bell		6.70	1.194		54	73.50		127.50	169
6000	Door holder, electro-magnetic		4	2		92	123		215	284
6200	Combination holder and closer		3.20	2.500		146	153		299	390
6600	Drill switch		8	1		400	61.50		461.50	530
6800	Master box		2.70	2.963		6,675	182		6,857	7,600
7000	Break glass station		8	1		58	61.50		119.50	155
7800	Remote annunciator, 8 zone lamp		1.80	4.444		212	273		485	640
8000	12 zone lamp	2 Elec	2.60	6.154		415	380		795	1,025
8200	16 zone lamp	"	2.20	7.273		370	445		815	1,075

Division Notes

I	CREW	DAILY OUTPUT	LABOR-HOURS	UNIT	BARE COSTS MAT.	BARE COSTS LABOR	BARE COSTS EQUIP.	BARE COSTS TOTAL	TOTAL INCL O&P

Division 31 Earthwork

Estimating Tips

31 05 00 Common Work Results for Earthwork

- Estimating the actual cost of performing earthwork requires careful consideration of the variables involved. This includes items such as type of soil, whether water will be encountered, dewatering, whether banks need bracing, disposal of excavated earth, and length of haul to fill or spoil sites, etc. If the project has large quantities of cut or fill, consider raising or lowering the site to reduce costs, while paying close attention to the effect on site drainage and utilities.
- If the project has large quantities of fill, creating a borrow pit on the site can significantly lower the costs.
- It is very important to consider what time of year the project is scheduled for completion. Bad weather can create large cost overruns from dewatering, site repair, and lost productivity from cold weather.

Reference Numbers

Reference numbers are shown at the beginning of some major classifications. These numbers refer to related items in the Reference Section. The reference information may be an estimating procedure, an alternate pricing method, or technical information.

Note: Not all subdivisions listed here necessarily appear. ■

31 23 16.13 Excavating, Trench		Crew	Daily Output	Labor-Hours	Unit	Material	2020 Bare Costs Labor	Equipment	Total	Total Incl O&P
0010	**EXCAVATING, TRENCH** G1030-805									
0011	Or continuous footing									
0020	Common earth with no sheeting or dewatering included									
0050	1′ to 4′ deep, 3/8 C.Y. excavator	B-11C	150	.107	B.C.Y.		5.25	1.43	6.68	9.45
0060	1/2 C.Y. excavator	B-11M	200	.080			3.95	1.16	5.11	7.20
0090	4′ to 6′ deep, 1/2 C.Y. excavator	"	200	.080			3.95	1.16	5.11	7.20
0100	5/8 C.Y. excavator	B-12Q	250	.064			3.24	2.39	5.63	7.50
0300	1/2 C.Y. excavator, truck mounted	B-12J	200	.080			4.05	4.23	8.28	10.70
0500	6′ to 10′ deep, 3/4 C.Y. excavator	B-12F	225	.071			3.60	3.09	6.69	8.80
0600	1 C.Y. excavator, truck mounted	B-12K	400	.040			2.03	2.43	4.46	5.70
0900	10′ to 14′ deep, 3/4 C.Y. excavator	B-12F	200	.080			4.05	3.47	7.52	9.85
1000	1-1/2 C.Y. excavator	B-12B	540	.030			1.50	1.27	2.77	3.65
1300	14′ to 20′ deep, 1 C.Y. excavator	B-12A	320	.050			2.53	2.49	5.02	6.50
1340	20′ to 24′ deep, 1 C.Y. excavator	"	288	.056			2.81	2.76	5.57	7.25
1352	4′ to 6′ deep, 1/2 C.Y. excavator w/trench box	B-13H	188	.085			4.31	5.15	9.46	12.10
1354	5/8 C.Y. excavator	"	235	.068			3.45	4.10	7.55	9.65
1362	6′ to 10′ deep, 3/4 C.Y. excavator w/trench box	B-13G	212	.075			3.82	3.83	7.65	9.95
1374	10′ to 14′ deep, 3/4 C.Y. excavator w/trench box	"	188	.085			4.31	4.32	8.63	11.20
1376	1-1/2 C.Y. excavator	B-13E	508	.032			1.60	1.59	3.19	4.13
1381	14′ to 20′ deep, 1 C.Y. excavator w/trench box	B-13D	301	.053			2.69	3.03	5.72	7.35
1386	20′ to 24′ deep, 1 C.Y. excavator w/trench box	"	271	.059			2.99	3.37	6.36	8.20
1400	By hand with pick and shovel 2′ to 6′ deep, light soil	1 Clab	8	1			42		42	63.50
1500	Heavy soil	"	4	2			84		84	127
1700	For tamping backfilled trenches, air tamp, add	A-1G	100	.080	E.C.Y.		3.37	.54	3.91	5.65
1900	Vibrating plate, add	B-18	180	.133	"		5.70	.92	6.62	9.55
2100	Trim sides and bottom for concrete pours, common earth		1500	.016	S.F.		.68	.11	.79	1.15
2300	Hardpan		600	.040	"		1.71	.28	1.99	2.87
5020	Loam & sandy clay with no sheeting or dewatering included									
5050	1′ to 4′ deep, 3/8 C.Y. tractor loader/backhoe	B-11C	162	.099	B.C.Y.		4.88	1.32	6.20	8.75
5060	1/2 C.Y. excavator	B-11M	216	.074			3.66	1.08	4.74	6.70
5080	4′ to 6′ deep, 1/2 C.Y. excavator	"	216	.074			3.66	1.08	4.74	6.70
5090	5/8 C.Y. excavator	B-12Q	276	.058			2.94	2.17	5.11	6.80
5130	1/2 C.Y. excavator, truck mounted	B-12J	216	.074			3.75	3.92	7.67	9.90
5140	6′ to 10′ deep, 3/4 C.Y. excavator	B-12F	243	.066			3.33	2.86	6.19	8.15
5160	1 C.Y. excavator, truck mounted	B-12K	432	.037			1.88	2.25	4.13	5.30
5190	10′ to 14′ deep, 3/4 C.Y. excavator	B-12F	216	.074			3.75	3.21	6.96	9.15
5210	1-1/2 C.Y. excavator	B-12B	583	.027			1.39	1.18	2.57	3.38
5250	14′ to 20′ deep, 1 C.Y. excavator	B-12A	346	.046			2.34	2.30	4.64	6.05
5300	20′ to 24′ deep, 1 C.Y. excavator	"	311	.051			2.61	2.56	5.17	6.70
5352	4′ to 6′ deep, 1/2 C.Y. excavator w/trench box	B-13H	205	.078			3.95	4.70	8.65	11.05
5354	5/8 C.Y. excavator	"	257	.062			3.15	3.75	6.90	8.85
5362	6′ to 10′ deep, 3/4 C.Y. excavator w/trench box	B-13G	231	.069			3.51	3.51	7.02	9.10
5370	10′ to 14′ deep, 3/4 C.Y. excavator w/trench box	"	205	.078			3.95	3.96	7.91	10.25
5374	1-1/2 C.Y. excavator	B-13E	554	.029			1.46	1.45	2.91	3.79
5382	14′ to 20′ deep, 1 C.Y. excavator w/trench box	B-13D	329	.049			2.46	2.78	5.24	6.75
5392	20′ to 24′ deep, 1 C.Y. excavator w/trench box	"	295	.054			2.75	3.10	5.85	7.50
6020	Sand & gravel with no sheeting or dewatering included									
6050	1′ to 4′ deep, 3/8 C.Y. excavator	B-11C	165	.097	B.C.Y.		4.79	1.30	6.09	8.65
6060	1/2 C.Y. excavator	B-11M	220	.073			3.60	1.06	4.66	6.55
6080	4′ to 6′ deep, 1/2 C.Y. excavator	"	220	.073			3.60	1.06	4.66	6.55
6090	5/8 C.Y. excavator	B-12Q	275	.058			2.95	2.18	5.13	6.80
6130	1/2 C.Y. excavator, truck mounted	B-12J	220	.073			3.68	3.85	7.53	9.75
6140	6′ to 10′ deep, 3/4 C.Y. excavator	B-12F	248	.065			3.27	2.80	6.07	8

31 23 16.13 Excavating, Trench		Crew	Daily Output	Labor-Hours	Unit	Material	2020 Bare Costs Labor	Equipment	Total	Total Incl O&P
6160	1 C.Y. excavator, truck mounted	B-12K	440	.036	B.C.Y.		1.84	2.21	4.05	5.20
6190	10′ to 14′ deep, 3/4 C.Y. excavator	B-12F	220	.073			3.68	3.16	6.84	8.95
6210	1-1/2 C.Y. excavator	B-12B	594	.027			1.36	1.16	2.52	3.31
6250	14′ to 20′ deep, 1 C.Y. excavator	B-12A	352	.045			2.30	2.26	4.56	5.95
6300	20′ to 24′ deep, 1 C.Y. excavator	"	317	.050			2.56	2.51	5.07	6.60
6352	4′ to 6′ deep, 1/2 C.Y. excavator w/trench box	B-13H	209	.077			3.88	4.61	8.49	10.90
6354	5/8 C.Y. excavator	"	261	.061			3.10	3.69	6.79	8.70
6362	6′ to 10′ deep, 3/4 C.Y. excavator w/trench box	B-13G	236	.068			3.43	3.44	6.87	8.95
6370	10′ to 14′ deep, 3/4 C.Y. excavator w/trench box	"	209	.077			3.88	3.89	7.77	10.05
6374	1-1/2 C.Y. excavator	B-13E	564	.028			1.44	1.43	2.87	3.72
6382	14′ to 20′ deep, 1 C.Y. excavator w/trench box	B-13D	334	.048			2.43	2.73	5.16	6.65
6392	20′ to 24′ deep, 1 C.Y. excavator w/trench box	"	301	.053	↓		2.69	3.03	5.72	7.35
7020	Dense hard clay with no sheeting or dewatering included									
7050	1′ to 4′ deep, 3/8 C.Y. excavator	B-11C	132	.121	B.C.Y.		6	1.62	7.62	10.80
7060	1/2 C.Y. excavator	B-11M	176	.091			4.49	1.32	5.81	8.20
7080	4′ to 6′ deep, 1/2 C.Y. excavator	"	176	.091			4.49	1.32	5.81	8.20
7090	5/8 C.Y. excavator	B-12Q	220	.073			3.68	2.72	6.40	8.50
7130	1/2 C.Y. excavator, truck mounted	B-12J	176	.091			4.60	4.81	9.41	12.20
7140	6′ to 10′ deep, 3/4 C.Y. excavator	B-12F	198	.081			4.09	3.51	7.60	10
7160	1 C.Y. excavator, truck mounted	B-12K	352	.045			2.30	2.77	5.07	6.50
7190	10′ to 14′ deep, 3/4 C.Y. excavator	B-12F	176	.091			4.60	3.94	8.54	11.25
7210	1-1/2 C.Y. excavator	B-12B	475	.034			1.71	1.45	3.16	4.15
7250	14′ to 20′ deep, 1 C.Y. excavator	B-12A	282	.057			2.87	2.82	5.69	7.40
7300	20′ to 24′ deep, 1 C.Y. excavator	"	254	.063	↓		3.19	3.13	6.32	8.20

31 23 16.14 Excavating, Utility Trench		Crew	Daily Output	Labor-Hours	Unit	Material	Labor	Equipment	Total	Total Incl O&P
0010	**EXCAVATING, UTILITY TRENCH** G1030–805									
0011	Common earth									
0050	Trenching with chain trencher, 12 HP, operator walking									
0100	4″ wide trench, 12″ deep	B-53	800	.010	L.F.		.53	.20	.73	1.01
0150	18″ deep		750	.011			.57	.21	.78	1.08
0200	24″ deep		700	.011			.61	.22	.83	1.16
0300	6″ wide trench, 12″ deep		650	.012			.65	.24	.89	1.25
0350	18″ deep		600	.013			.71	.26	.97	1.35
0400	24″ deep		550	.015			.77	.29	1.06	1.46
0450	36″ deep		450	.018			.94	.35	1.29	1.79
0600	8″ wide trench, 12″ deep		475	.017			.89	.33	1.22	1.69
0650	18″ deep		400	.020			1.06	.39	1.45	2.01
0700	24″ deep		350	.023			1.21	.45	1.66	2.30
0750	36″ deep	↓	300	.027	↓		1.41	.52	1.93	2.69
1000	Backfill by hand including compaction, add									
1050	4″ wide trench, 12″ deep	A-1G	800	.010	L.F.		.42	.07	.49	.70
1100	18″ deep		530	.015			.64	.10	.74	1.06
1150	24″ deep		400	.020			.84	.14	.98	1.42
1300	6″ wide trench, 12″ deep		540	.015			.62	.10	.72	1.05
1350	18″ deep		405	.020			.83	.13	.96	1.40
1400	24″ deep		270	.030			1.25	.20	1.45	2.09
1450	36″ deep		180	.044			1.87	.30	2.17	3.14
1600	8″ wide trench, 12″ deep		400	.020			.84	.14	.98	1.42
1650	18″ deep		265	.030			1.27	.20	1.47	2.13
1700	24″ deep		200	.040			1.68	.27	1.95	2.83
1750	36″ deep	↓	135	.059	↓		2.49	.40	2.89	4.19
2000	Chain trencher, 40 HP operator riding									

31 23 Excavation and Fill

31 23 16 – Excavation

31 23 16.14 Excavating, Utility Trench		Crew	Daily Output	Labor-Hours	Unit	Material	2020 Bare Costs Labor	Equipment	Total	Total Incl O&P
2050	6″ wide trench and backfill, 12″ deep	B-54	1200	.007	L.F.		.35	.34	.69	.90
2100	18″ deep		1000	.008			.42	.40	.82	1.07
2150	24″ deep		975	.008			.44	.41	.85	1.10
2200	36″ deep		900	.009			.47	.45	.92	1.19
2250	48″ deep		750	.011			.57	.54	1.11	1.44
2300	60″ deep		650	.012			.65	.62	1.27	1.66
2400	8″ wide trench and backfill, 12″ deep		1000	.008			.42	.40	.82	1.07
2450	18″ deep		950	.008			.45	.42	.87	1.14
2500	24″ deep		900	.009			.47	.45	.92	1.19
2550	36″ deep		800	.010			.53	.50	1.03	1.34
2600	48″ deep		650	.012			.65	.62	1.27	1.66
2700	12″ wide trench and backfill, 12″ deep		975	.008			.44	.41	.85	1.10
2750	18″ deep		860	.009			.49	.47	.96	1.25
2800	24″ deep		800	.010			.53	.50	1.03	1.34
2850	36″ deep		725	.011			.58	.55	1.13	1.48
3000	16″ wide trench and backfill, 12″ deep		835	.010			.51	.48	.99	1.29
3050	18″ deep		750	.011			.57	.54	1.11	1.44
3100	24″ deep		700	.011			.61	.57	1.18	1.54
3200	Compaction with vibratory plate, add								35%	35%
5100	Hand excavate and trim for pipe bells after trench excavation									
5200	8″ pipe	1 Clab	155	.052	L.F.		2.17		2.17	3.26
5300	18″ pipe	″	130	.062	″		2.59		2.59	3.89

31 23 19 – Dewatering

31 23 19.20 Dewatering Systems		Crew	Daily Output	Labor-Hours	Unit	Material	Labor	Equipment	Total	Total Incl O&P
0010	**DEWATERING SYSTEMS**									
0020	Excavate drainage trench, 2′ wide, 2′ deep	B-11C	90	.178	C.Y.		8.80	2.38	11.18	15.75
0100	2′ wide, 3′ deep, with backhoe loader	″	135	.119			5.85	1.58	7.43	10.55
0200	Excavate sump pits by hand, light soil	1 Clab	7.10	1.127			47.50		47.50	71.50
0300	Heavy soil	″	3.50	2.286			96		96	145
0500	Pumping 8 hrs., attended 2 hrs./day, incl. 20 L.F.									
0550	of suction hose & 100 L.F. discharge hose									
0600	2″ diaphragm pump used for 8 hrs.	B-10H	4	3	Day		156	24.50	180.50	260
0650	4″ diaphragm pump used for 8 hrs.	B-10I	4	3			156	37	193	274
0800	8 hrs. attended, 2″ diaphragm pump	B-10H	1	12			620	98.50	718.50	1,050
0900	3″ centrifugal pump	B-10J	1	12			620	92	712	1,025
1000	4″ diaphragm pump	B-10I	1	12			620	148	768	1,100
1100	6″ centrifugal pump	B-10K	1	12			620	294	914	1,250
1300	CMP, incl. excavation 3′ deep, 12″ diameter	B-6	115	.209	L.F.	12.40	9.55	1.86	23.81	30
1400	18″ diameter		100	.240	″	18.80	11	2.14	31.94	39.50
1600	Sump hole construction, incl. excavation and gravel, pit		1250	.019	C.F.	1.10	.88	.17	2.15	2.72
1700	With 12″ gravel collar, 12″ pipe, corrugated, 16 ga.		70	.343	L.F.	23	15.70	3.05	41.75	52.50
1800	15″ pipe, corrugated, 16 ga.		55	.436		30	19.95	3.89	53.84	67.50
1900	18″ pipe, corrugated, 16 ga.		50	.480		35	22	4.28	61.28	76
2000	24″ pipe, corrugated, 14 ga.		40	.600		42	27.50	5.35	74.85	93
2200	Wood lining, up to 4′ x 4′, add		300	.080	SFCA	16.40	3.66	.71	20.77	24.50
9950	See Section 31 23 19.40 for wellpoints									
9960	See Section 31 23 19.30 for deep well systems									

31 23 19.30 Wells		Crew	Daily Output	Labor-Hours	Unit	Material	Labor	Equipment	Total	Total Incl O&P
0010	**WELLS**									
0011	For dewatering 10′ to 20′ deep, 2′ diameter									
0020	with steel casing, minimum	B-6	165	.145	V.L.F.	37.50	6.65	1.30	45.45	53
0050	Average		98	.245		44.50	11.20	2.18	57.88	68

31 23 19.30 Wells		Crew	Daily Output	Labor-Hours	Unit	2020 Bare Costs Material	Labor	Equipment	Total	Total Incl O&P
0100	Maximum	B-6	49	.490	V.L.F.	48	22.50	4.36	74.86	91
0300	For dewatering pumps see 01 54 33 in Reference Section									
0500	For domestic water wells, see Section 33 21 13.10									

31 23 19.40 Wellpoints

		Crew	Daily Output	Labor-Hours	Unit	Material	Labor	Equipment	Total	Total Incl O&P
0010	**WELLPOINTS** R312319-90									
0011	For equipment rental, see 01 54 33 in Reference Section									
0100	Installation and removal of single stage system									
0110	Labor only, 0.75 labor-hours per L.F.	1 Clab	10.70	.748	LF Hdr		31.50		31.50	47.50
0200	2.0 labor-hours per L.F.	"	4	2	"		84		84	127
0400	Pump operation, 4 @ 6 hr. shifts									
0410	Per 24 hr. day	4 Eqlt	1.27	25.197	Day		1,325		1,325	2,000
0500	Per 168 hr. week, 160 hr. straight, 8 hr. double time		.18	178	Week		9,425		9,425	14,100
0550	Per 4.3 week month		.04	800	Month		42,400		42,400	63,500
0600	Complete installation, operation, equipment rental, fuel &									
0610	removal of system with 2" wellpoints 5' OC									
0700	100' long header, 6" diameter, first month	4 Eqlt	3.23	9.907	LF Hdr	160	525		685	960
0800	Thereafter, per month		4.13	7.748		128	410		538	755
1000	200' long header, 8" diameter, first month		6	5.333		154	283		437	590
1100	Thereafter, per month		8.39	3.814		72	202		274	380
1300	500' long header, 8" diameter, first month		10.63	3.010		56	160		216	300
1400	Thereafter, per month		20.91	1.530		40	81		121	165
1600	1,000' long header, 10" diameter, first month		11.62	2.754		48	146		194	271
1700	Thereafter, per month		41.81	.765		24	40.50		64.50	87
1900	Note: above figures include pumping 168 hrs. per week,									
1910	the pump operator, and one stand-by pump.									

31 23 23 – Fill

31 23 23.15 Borrow, Loading and/or Spreading

		Crew	Daily Output	Labor-Hours	Unit	Material	Labor	Equipment	Total	Total Incl O&P
0010	**BORROW, LOADING AND/OR SPREADING**									
0020	Material only, bank run gravel				Ton	13.30			13.30	14.60
0500	Haul 2 mi. spread, 200 HP dozer, bank run gravel	B-15	1100	.025			1.28	2.41	3.69	4.56
1000	Hand spread, bank run gravel	A-5	33	.545			23.50	1.48	24.98	36.50
1800	Delivery charge, minimum 20 tons, 1 hr. round trip, add	B-34B	130	.062			3.01	4.41	7.42	9.35
1820	1-1/2 hr. round trip, add		93	.086			4.21	6.15	10.36	13.10
1840	2 hr. round trip, add		65	.123			6	8.80	14.80	18.75

31 23 23.16 Fill By Borrow and Utility Bedding

		Crew	Daily Output	Labor-Hours	Unit	Material	Labor	Equipment	Total	Total Incl O&P
0010	**FILL BY BORROW AND UTILITY BEDDING**									
0049	Utility bedding, for pipe & conduit, not incl. compaction G1030-805									
0050	Crushed or screened bank run gravel	B-6	150	.160	L.C.Y.	21	7.30	1.43	29.73	35.50
0100	Crushed stone 3/4" to 1/2"		150	.160		27.50	7.30	1.43	36.23	43
0200	Sand, dead or bank		150	.160		18.25	7.30	1.43	26.98	32.50
0500	Compacting bedding in trench	A-1D	90	.089	E.C.Y.		3.74	.35	4.09	6
0600	If material source exceeds 2 miles, add for extra mileage.									
0610	See Section 31 23 23.20 for hauling mileage add.									

31 23 23.17 General Fill

		Crew	Daily Output	Labor-Hours	Unit	Material	Labor	Equipment	Total	Total Incl O&P
0010	**GENERAL FILL**									
0011	Spread dumped material, no compaction									
0020	By dozer	B-10B	1000	.012	L.C.Y.		.62	1.50	2.12	2.58
0100	By hand	1 Clab	12	.667	"		28		28	42
0500	Gravel fill, compacted, under floor slabs, 4" deep	B-37	10000	.005	S.F.	.49	.21	.03	.73	.89
0600	6" deep		8600	.006		.74	.25	.03	1.02	1.21
0700	9" deep		7200	.007		1.23	.30	.04	1.57	1.83

31 23 Excavation and Fill

31 23 23 – Fill

31 23 23.17 General Fill		Crew	Daily Output	Labor-Hours	Unit	Material	2020 Bare Costs Labor	Equipment	Total	Total Incl O&P
0800	12" deep	B-37	6000	.008	S.F.	1.72	.35	.04	2.11	2.47
1000	Alternate pricing method, 4" deep		120	.400	E.C.Y.	37	17.70	2.13	56.83	69.50
1100	6" deep		160	.300		37	13.30	1.60	51.90	62
1200	9" deep		200	.240		37	10.60	1.28	48.88	58
1300	12" deep		220	.218		37	9.65	1.16	47.81	56.50
1400	Granular fill				L.C.Y.	24.50			24.50	27

31 23 23.20 Hauling		Crew	Daily Output	Labor-Hours	Unit	Material	Labor	Equipment	Total	Total Incl O&P
0010	**HAULING**									
0011	Excavated or borrow, loose cubic yards									
0012	no loading equipment, including hauling, waiting, loading/dumping									
0013	time per cycle (wait, load, travel, unload or dump & return)									
0014	8 C.Y. truck, 15 MPH avg., cycle 0.5 miles, 10 min. wait/ld./uld.	B-34A	320	.025	L.C.Y.		1.22	1.34	2.56	3.30
0016	cycle 1 mile		272	.029			1.44	1.57	3.01	3.89
0018	cycle 2 miles		208	.038			1.88	2.06	3.94	5.10
0020	cycle 4 miles		144	.056			2.72	2.97	5.69	7.35
0022	cycle 6 miles		112	.071			3.50	3.82	7.32	9.45
0024	cycle 8 miles		88	.091			4.45	4.86	9.31	12
0026	20 MPH avg., cycle 0.5 mile		336	.024			1.17	1.27	2.44	3.15
0028	cycle 1 mile		296	.027			1.32	1.45	2.77	3.57
0030	cycle 2 miles		240	.033			1.63	1.78	3.41	4.40
0032	cycle 4 miles		176	.045			2.22	2.43	4.65	6
0034	cycle 6 miles		136	.059			2.88	3.15	6.03	7.75
0036	cycle 8 miles		112	.071			3.50	3.82	7.32	9.45
0044	25 MPH avg., cycle 4 miles		192	.042			2.04	2.23	4.27	5.50
0046	cycle 6 miles		160	.050			2.45	2.68	5.13	6.60
0048	cycle 8 miles		128	.063			3.06	3.34	6.40	8.25
0050	30 MPH avg., cycle 4 miles		216	.037			1.81	1.98	3.79	4.90
0052	cycle 6 miles		176	.045			2.22	2.43	4.65	6
0054	cycle 8 miles		144	.056			2.72	2.97	5.69	7.35
0114	15 MPH avg., cycle 0.5 mile, 15 min. wait/ld./uld.		224	.036			1.75	1.91	3.66	4.72
0116	cycle 1 mile		200	.040			1.96	2.14	4.10	5.30
0118	cycle 2 miles		168	.048			2.33	2.55	4.88	6.30
0120	cycle 4 miles		120	.067			3.26	3.57	6.83	8.80
0122	cycle 6 miles		96	.083			4.08	4.46	8.54	11
0124	cycle 8 miles		80	.100			4.90	5.35	10.25	13.25
0126	20 MPH avg., cycle 0.5 mile		232	.034			1.69	1.85	3.54	4.56
0128	cycle 1 mile		208	.038			1.88	2.06	3.94	5.10
0130	cycle 2 miles		184	.043			2.13	2.33	4.46	5.75
0132	cycle 4 miles		144	.056			2.72	2.97	5.69	7.35
0134	cycle 6 miles		112	.071			3.50	3.82	7.32	9.45
0136	cycle 8 miles		96	.083			4.08	4.46	8.54	11
0144	25 MPH avg., cycle 4 miles		152	.053			2.58	2.82	5.40	6.95
0146	cycle 6 miles		128	.063			3.06	3.34	6.40	8.25
0148	cycle 8 miles		112	.071			3.50	3.82	7.32	9.45
0150	30 MPH avg., cycle 4 miles		168	.048			2.33	2.55	4.88	6.30
0152	cycle 6 miles		144	.056			2.72	2.97	5.69	7.35
0154	cycle 8 miles		120	.067			3.26	3.57	6.83	8.80
0214	15 MPH avg., cycle 0.5 mile, 20 min. wait/ld./uld.		176	.045			2.22	2.43	4.65	6
0216	cycle 1 mile		160	.050			2.45	2.68	5.13	6.60
0218	cycle 2 miles		136	.059			2.88	3.15	6.03	7.75
0220	cycle 4 miles		104	.077			3.77	4.12	7.89	10.20
0222	cycle 6 miles		88	.091			4.45	4.86	9.31	12

31 23 23.20 Hauling		Crew	Daily Output	Labor-Hours	Unit	Material	2020 Bare Costs Labor	Equipment	Total	Total Incl O&P
0224	cycle 8 miles	B-34A	72	.111	L.C.Y.		5.45	5.95	11.40	14.70
0226	20 MPH avg., cycle 0.5 mile		176	.045			2.22	2.43	4.65	6
0228	cycle 1 mile		168	.048			2.33	2.55	4.88	6.30
0230	cycle 2 miles		144	.056			2.72	2.97	5.69	7.35
0232	cycle 4 miles		120	.067			3.26	3.57	6.83	8.80
0234	cycle 6 miles		96	.083			4.08	4.46	8.54	11
0236	cycle 8 miles		88	.091			4.45	4.86	9.31	12
0244	25 MPH avg., cycle 4 miles		128	.063			3.06	3.34	6.40	8.25
0246	cycle 6 miles		112	.071			3.50	3.82	7.32	9.45
0248	cycle 8 miles		96	.083			4.08	4.46	8.54	11
0250	30 MPH avg., cycle 4 miles		136	.059			2.88	3.15	6.03	7.75
0252	cycle 6 miles		120	.067			3.26	3.57	6.83	8.80
0254	cycle 8 miles		104	.077			3.77	4.12	7.89	10.20
0314	15 MPH avg., cycle 0.5 mile, 25 min. wait/ld./uld.		144	.056			2.72	2.97	5.69	7.35
0316	cycle 1 mile		128	.063			3.06	3.34	6.40	8.25
0318	cycle 2 miles		112	.071			3.50	3.82	7.32	9.45
0320	cycle 4 miles		96	.083			4.08	4.46	8.54	11
0322	cycle 6 miles		80	.100			4.90	5.35	10.25	13.25
0324	cycle 8 miles		64	.125			6.10	6.70	12.80	16.50
0326	20 MPH avg., cycle 0.5 mile		144	.056			2.72	2.97	5.69	7.35
0328	cycle 1 mile		136	.059			2.88	3.15	6.03	7.75
0330	cycle 2 miles		120	.067			3.26	3.57	6.83	8.80
0332	cycle 4 miles		104	.077			3.77	4.12	7.89	10.20
0334	cycle 6 miles		88	.091			4.45	4.86	9.31	12
0336	cycle 8 miles		80	.100			4.90	5.35	10.25	13.25
0344	25 MPH avg., cycle 4 miles		112	.071			3.50	3.82	7.32	9.45
0346	cycle 6 miles		96	.083			4.08	4.46	8.54	11
0348	cycle 8 miles		88	.091			4.45	4.86	9.31	12
0350	30 MPH avg., cycle 4 miles		112	.071			3.50	3.82	7.32	9.45
0352	cycle 6 miles		104	.077			3.77	4.12	7.89	10.20
0354	cycle 8 miles		96	.083			4.08	4.46	8.54	11
0414	15 MPH avg., cycle 0.5 mile, 30 min. wait/ld./uld.		120	.067			3.26	3.57	6.83	8.80
0416	cycle 1 mile		112	.071			3.50	3.82	7.32	9.45
0418	cycle 2 miles		96	.083			4.08	4.46	8.54	11
0420	cycle 4 miles		80	.100			4.90	5.35	10.25	13.25
0422	cycle 6 miles		72	.111			5.45	5.95	11.40	14.70
0424	cycle 8 miles		64	.125			6.10	6.70	12.80	16.50
0426	20 MPH avg., cycle 0.5 mile		120	.067			3.26	3.57	6.83	8.80
0428	cycle 1 mile		112	.071			3.50	3.82	7.32	9.45
0430	cycle 2 miles		104	.077			3.77	4.12	7.89	10.20
0432	cycle 4 miles		88	.091			4.45	4.86	9.31	12
0434	cycle 6 miles		80	.100			4.90	5.35	10.25	13.25
0436	cycle 8 miles		72	.111			5.45	5.95	11.40	14.70
0444	25 MPH avg., cycle 4 miles		96	.083			4.08	4.46	8.54	11
0446	cycle 6 miles		88	.091			4.45	4.86	9.31	12
0448	cycle 8 miles		80	.100			4.90	5.35	10.25	13.25
0450	30 MPH avg., cycle 4 miles		96	.083			4.08	4.46	8.54	11
0452	cycle 6 miles		88	.091			4.45	4.86	9.31	12
0454	cycle 8 miles		80	.100			4.90	5.35	10.25	13.25
0514	15 MPH avg., cycle 0.5 mile, 35 min. wait/ld./uld.		104	.077			3.77	4.12	7.89	10.20
0516	cycle 1 mile		96	.083			4.08	4.46	8.54	11
0518	cycle 2 miles		88	.091			4.45	4.86	9.31	12
0520	cycle 4 miles		72	.111			5.45	5.95	11.40	14.70

31 23 23.20 Hauling		Crew	Daily Output	Labor-Hours	Unit	Material	2020 Bare Costs Labor	2020 Bare Costs Equipment	Total	Total Incl O&P
0522	cycle 6 miles	B-34A	64	.125	L.C.Y.		6.10	6.70	12.80	16.50
0524	cycle 8 miles		56	.143			7	7.65	14.65	18.90
0526	20 MPH avg., cycle 0.5 mile		104	.077			3.77	4.12	7.89	10.20
0528	cycle 1 mile		96	.083			4.08	4.46	8.54	11
0530	cycle 2 miles		96	.083			4.08	4.46	8.54	11
0532	cycle 4 miles		80	.100			4.90	5.35	10.25	13.25
0534	cycle 6 miles		72	.111			5.45	5.95	11.40	14.70
0536	cycle 8 miles		64	.125			6.10	6.70	12.80	16.50
0544	25 MPH avg., cycle 4 miles		88	.091			4.45	4.86	9.31	12
0546	cycle 6 miles		80	.100			4.90	5.35	10.25	13.25
0548	cycle 8 miles		72	.111			5.45	5.95	11.40	14.70
0550	30 MPH avg., cycle 4 miles		88	.091			4.45	4.86	9.31	12
0552	cycle 6 miles		80	.100			4.90	5.35	10.25	13.25
0554	cycle 8 miles		72	.111			5.45	5.95	11.40	14.70
1014	12 C.Y. truck, cycle 0.5 mile, 15 MPH avg., 15 min. wait/ld./uld.	B-34B	336	.024			1.17	1.71	2.88	3.63
1016	cycle 1 mile		300	.027			1.31	1.91	3.22	4.06
1018	cycle 2 miles		252	.032			1.55	2.27	3.82	4.83
1020	cycle 4 miles		180	.044			2.18	3.18	5.36	6.75
1022	cycle 6 miles		144	.056			2.72	3.98	6.70	8.45
1024	cycle 8 miles		120	.067			3.26	4.77	8.03	10.15
1025	cycle 10 miles		96	.083			4.08	5.95	10.03	12.65
1026	20 MPH avg., cycle 0.5 mile		348	.023			1.13	1.65	2.78	3.50
1028	cycle 1 mile		312	.026			1.26	1.84	3.10	3.90
1030	cycle 2 miles		276	.029			1.42	2.08	3.50	4.41
1032	cycle 4 miles		216	.037			1.81	2.65	4.46	5.65
1034	cycle 6 miles		168	.048			2.33	3.41	5.74	7.25
1036	cycle 8 miles		144	.056			2.72	3.98	6.70	8.45
1038	cycle 10 miles		120	.067			3.26	4.77	8.03	10.15
1040	25 MPH avg., cycle 4 miles		228	.035			1.72	2.51	4.23	5.35
1042	cycle 6 miles		192	.042			2.04	2.98	5.02	6.35
1044	cycle 8 miles		168	.048			2.33	3.41	5.74	7.25
1046	cycle 10 miles		144	.056			2.72	3.98	6.70	8.45
1050	30 MPH avg., cycle 4 miles		252	.032			1.55	2.27	3.82	4.83
1052	cycle 6 miles		216	.037			1.81	2.65	4.46	5.65
1054	cycle 8 miles		180	.044			2.18	3.18	5.36	6.75
1056	cycle 10 miles		156	.051			2.51	3.67	6.18	7.80
1060	35 MPH avg., cycle 4 miles		264	.030			1.48	2.17	3.65	4.61
1062	cycle 6 miles		228	.035			1.72	2.51	4.23	5.35
1064	cycle 8 miles		204	.039			1.92	2.81	4.73	5.95
1066	cycle 10 miles		180	.044			2.18	3.18	5.36	6.75
1068	cycle 20 miles		120	.067			3.26	4.77	8.03	10.15
1069	cycle 30 miles		84	.095			4.66	6.80	11.46	14.50
1070	cycle 40 miles		72	.111			5.45	7.95	13.40	16.90
1072	40 MPH avg., cycle 6 miles		240	.033			1.63	2.39	4.02	5.05
1074	cycle 8 miles		216	.037			1.81	2.65	4.46	5.65
1076	cycle 10 miles		192	.042			2.04	2.98	5.02	6.35
1078	cycle 20 miles		120	.067			3.26	4.77	8.03	10.15
1080	cycle 30 miles		96	.083			4.08	5.95	10.03	12.65
1082	cycle 40 miles		72	.111			5.45	7.95	13.40	16.90
1084	cycle 50 miles		60	.133			6.55	9.55	16.10	20.50
1094	45 MPH avg., cycle 8 miles		216	.037			1.81	2.65	4.46	5.65
1096	cycle 10 miles		204	.039			1.92	2.81	4.73	5.95
1098	cycle 20 miles		132	.061			2.97	4.34	7.31	9.20

31 23 23.20 Hauling		Crew	Daily Output	Labor-Hours	Unit	Material	2020 Bare Costs Labor	Equipment	Total	Total Incl O&P
1100	cycle 30 miles	B-34B	108	.074	L.C.Y.		3.63	5.30	8.93	11.30
1102	cycle 40 miles		84	.095			4.66	6.80	11.46	14.50
1104	cycle 50 miles		72	.111			5.45	7.95	13.40	16.90
1106	50 MPH avg., cycle 10 miles		216	.037			1.81	2.65	4.46	5.65
1108	cycle 20 miles		144	.056			2.72	3.98	6.70	8.45
1110	cycle 30 miles		108	.074			3.63	5.30	8.93	11.30
1112	cycle 40 miles		84	.095			4.66	6.80	11.46	14.50
1114	cycle 50 miles		72	.111			5.45	7.95	13.40	16.90
1214	15 MPH avg., cycle 0.5 mile, 20 min. wait/ld./uld.		264	.030			1.48	2.17	3.65	4.61
1216	cycle 1 mile		240	.033			1.63	2.39	4.02	5.05
1218	cycle 2 miles		204	.039			1.92	2.81	4.73	5.95
1220	cycle 4 miles		156	.051			2.51	3.67	6.18	7.80
1222	cycle 6 miles		132	.061			2.97	4.34	7.31	9.20
1224	cycle 8 miles		108	.074			3.63	5.30	8.93	11.30
1225	cycle 10 miles		96	.083			4.08	5.95	10.03	12.65
1226	20 MPH avg., cycle 0.5 mile		264	.030			1.48	2.17	3.65	4.61
1228	cycle 1 mile		252	.032			1.55	2.27	3.82	4.83
1230	cycle 2 miles		216	.037			1.81	2.65	4.46	5.65
1232	cycle 4 miles		180	.044			2.18	3.18	5.36	6.75
1234	cycle 6 miles		144	.056			2.72	3.98	6.70	8.45
1236	cycle 8 miles		132	.061			2.97	4.34	7.31	9.20
1238	cycle 10 miles		108	.074			3.63	5.30	8.93	11.30
1240	25 MPH avg., cycle 4 miles		192	.042			2.04	2.98	5.02	6.35
1242	cycle 6 miles		168	.048			2.33	3.41	5.74	7.25
1244	cycle 8 miles		144	.056			2.72	3.98	6.70	8.45
1246	cycle 10 miles		132	.061			2.97	4.34	7.31	9.20
1250	30 MPH avg., cycle 4 miles		204	.039			1.92	2.81	4.73	5.95
1252	cycle 6 miles		180	.044			2.18	3.18	5.36	6.75
1254	cycle 8 miles		156	.051			2.51	3.67	6.18	7.80
1256	cycle 10 miles		144	.056			2.72	3.98	6.70	8.45
1260	35 MPH avg., cycle 4 miles		216	.037			1.81	2.65	4.46	5.65
1262	cycle 6 miles		192	.042			2.04	2.98	5.02	6.35
1264	cycle 8 miles		168	.048			2.33	3.41	5.74	7.25
1266	cycle 10 miles		156	.051			2.51	3.67	6.18	7.80
1268	cycle 20 miles		108	.074			3.63	5.30	8.93	11.30
1269	cycle 30 miles		72	.111			5.45	7.95	13.40	16.90
1270	cycle 40 miles		60	.133			6.55	9.55	16.10	20.50
1272	40 MPH avg., cycle 6 miles		192	.042			2.04	2.98	5.02	6.35
1274	cycle 8 miles		180	.044			2.18	3.18	5.36	6.75
1276	cycle 10 miles		156	.051			2.51	3.67	6.18	7.80
1278	cycle 20 miles		108	.074			3.63	5.30	8.93	11.30
1280	cycle 30 miles		84	.095			4.66	6.80	11.46	14.50
1282	cycle 40 miles		72	.111			5.45	7.95	13.40	16.90
1284	cycle 50 miles		60	.133			6.55	9.55	16.10	20.50
1294	45 MPH avg., cycle 8 miles		180	.044			2.18	3.18	5.36	6.75
1296	cycle 10 miles		168	.048			2.33	3.41	5.74	7.25
1298	cycle 20 miles		120	.067			3.26	4.77	8.03	10.15
1300	cycle 30 miles		96	.083			4.08	5.95	10.03	12.65
1302	cycle 40 miles		72	.111			5.45	7.95	13.40	16.90
1304	cycle 50 miles		60	.133			6.55	9.55	16.10	20.50
1306	50 MPH avg., cycle 10 miles		180	.044			2.18	3.18	5.36	6.75
1308	cycle 20 miles		132	.061			2.97	4.34	7.31	9.20
1310	cycle 30 miles		96	.083			4.08	5.95	10.03	12.65

31 23 23.20 Hauling		Crew	Daily Output	Labor-Hours	Unit	Material	2020 Bare Costs Labor	Equipment	Total	Total Incl O&P
1312	cycle 40 miles	B-34B	84	.095	L.C.Y.		4.66	6.80	11.46	14.50
1314	cycle 50 miles		72	.111			5.45	7.95	13.40	16.90
1414	15 MPH avg., cycle 0.5 mile, 25 min. wait/ld./uld.		204	.039			1.92	2.81	4.73	5.95
1416	cycle 1 mile		192	.042			2.04	2.98	5.02	6.35
1418	cycle 2 miles		168	.048			2.33	3.41	5.74	7.25
1420	cycle 4 miles		132	.061			2.97	4.34	7.31	9.20
1422	cycle 6 miles		120	.067			3.26	4.77	8.03	10.15
1424	cycle 8 miles		96	.083			4.08	5.95	10.03	12.65
1425	cycle 10 miles		84	.095			4.66	6.80	11.46	14.50
1426	20 MPH avg., cycle 0.5 mile		216	.037			1.81	2.65	4.46	5.65
1428	cycle 1 mile		204	.039			1.92	2.81	4.73	5.95
1430	cycle 2 miles		180	.044			2.18	3.18	5.36	6.75
1432	cycle 4 miles		156	.051			2.51	3.67	6.18	7.80
1434	cycle 6 miles		132	.061			2.97	4.34	7.31	9.20
1436	cycle 8 miles		120	.067			3.26	4.77	8.03	10.15
1438	cycle 10 miles		96	.083			4.08	5.95	10.03	12.65
1440	25 MPH avg., cycle 4 miles		168	.048			2.33	3.41	5.74	7.25
1442	cycle 6 miles		144	.056			2.72	3.98	6.70	8.45
1444	cycle 8 miles		132	.061			2.97	4.34	7.31	9.20
1446	cycle 10 miles		108	.074			3.63	5.30	8.93	11.30
1450	30 MPH avg., cycle 4 miles		168	.048			2.33	3.41	5.74	7.25
1452	cycle 6 miles		156	.051			2.51	3.67	6.18	7.80
1454	cycle 8 miles		132	.061			2.97	4.34	7.31	9.20
1456	cycle 10 miles		120	.067			3.26	4.77	8.03	10.15
1460	35 MPH avg., cycle 4 miles		180	.044			2.18	3.18	5.36	6.75
1462	cycle 6 miles		156	.051			2.51	3.67	6.18	7.80
1464	cycle 8 miles		144	.056			2.72	3.98	6.70	8.45
1466	cycle 10 miles		132	.061			2.97	4.34	7.31	9.20
1468	cycle 20 miles		96	.083			4.08	5.95	10.03	12.65
1469	cycle 30 miles		72	.111			5.45	7.95	13.40	16.90
1470	cycle 40 miles		60	.133			6.55	9.55	16.10	20.50
1472	40 MPH avg., cycle 6 miles		168	.048			2.33	3.41	5.74	7.25
1474	cycle 8 miles		156	.051			2.51	3.67	6.18	7.80
1476	cycle 10 miles		144	.056			2.72	3.98	6.70	8.45
1478	cycle 20 miles		96	.083			4.08	5.95	10.03	12.65
1480	cycle 30 miles		84	.095			4.66	6.80	11.46	14.50
1482	cycle 40 miles		60	.133			6.55	9.55	16.10	20.50
1484	cycle 50 miles		60	.133			6.55	9.55	16.10	20.50
1494	45 MPH avg., cycle 8 miles		156	.051			2.51	3.67	6.18	7.80
1496	cycle 10 miles		144	.056			2.72	3.98	6.70	8.45
1498	cycle 20 miles		108	.074			3.63	5.30	8.93	11.30
1500	cycle 30 miles		84	.095			4.66	6.80	11.46	14.50
1502	cycle 40 miles		72	.111			5.45	7.95	13.40	16.90
1504	cycle 50 miles		60	.133			6.55	9.55	16.10	20.50
1506	50 MPH avg., cycle 10 miles		156	.051			2.51	3.67	6.18	7.80
1508	cycle 20 miles		120	.067			3.26	4.77	8.03	10.15
1510	cycle 30 miles		96	.083			4.08	5.95	10.03	12.65
1512	cycle 40 miles		72	.111			5.45	7.95	13.40	16.90
1514	cycle 50 miles		60	.133			6.55	9.55	16.10	20.50
1614	15 MPH avg., cycle 0.5 mile, 30 min. wait/ld./uld.		180	.044			2.18	3.18	5.36	6.75
1616	cycle 1 mile		168	.048			2.33	3.41	5.74	7.25
1618	cycle 2 miles		144	.056			2.72	3.98	6.70	8.45
1620	cycle 4 miles		120	.067			3.26	4.77	8.03	10.15

31 23 23.20 Hauling		Crew	Daily Output	Labor-Hours	Unit	Material	2020 Bare Costs Labor	Equipment	Total	Total Incl O&P
1622	cycle 6 miles	B-34B	108	.074	L.C.Y.		3.63	5.30	8.93	11.30
1624	cycle 8 miles		84	.095			4.66	6.80	11.46	14.50
1625	cycle 10 miles		84	.095			4.66	6.80	11.46	14.50
1626	20 MPH avg., cycle 0.5 mile		180	.044			2.18	3.18	5.36	6.75
1628	cycle 1 mile		168	.048			2.33	3.41	5.74	7.25
1630	cycle 2 miles		156	.051			2.51	3.67	6.18	7.80
1632	cycle 4 miles		132	.061			2.97	4.34	7.31	9.20
1634	cycle 6 miles		120	.067			3.26	4.77	8.03	10.15
1636	cycle 8 miles		108	.074			3.63	5.30	8.93	11.30
1638	cycle 10 miles		96	.083			4.08	5.95	10.03	12.65
1640	25 MPH avg., cycle 4 miles		144	.056			2.72	3.98	6.70	8.45
1642	cycle 6 miles		132	.061			2.97	4.34	7.31	9.20
1644	cycle 8 miles		108	.074			3.63	5.30	8.93	11.30
1646	cycle 10 miles		108	.074			3.63	5.30	8.93	11.30
1650	30 MPH avg., cycle 4 miles		144	.056			2.72	3.98	6.70	8.45
1652	cycle 6 miles		132	.061			2.97	4.34	7.31	9.20
1654	cycle 8 miles		120	.067			3.26	4.77	8.03	10.15
1656	cycle 10 miles		108	.074			3.63	5.30	8.93	11.30
1660	35 MPH avg., cycle 4 miles		156	.051			2.51	3.67	6.18	7.80
1662	cycle 6 miles		144	.056			2.72	3.98	6.70	8.45
1664	cycle 8 miles		132	.061			2.97	4.34	7.31	9.20
1666	cycle 10 miles		120	.067			3.26	4.77	8.03	10.15
1668	cycle 20 miles		84	.095			4.66	6.80	11.46	14.50
1669	cycle 30 miles		72	.111			5.45	7.95	13.40	16.90
1670	cycle 40 miles		60	.133			6.55	9.55	16.10	20.50
1672	40 MPH avg., cycle 6 miles		144	.056			2.72	3.98	6.70	8.45
1674	cycle 8 miles		132	.061			2.97	4.34	7.31	9.20
1676	cycle 10 miles		120	.067			3.26	4.77	8.03	10.15
1678	cycle 20 miles		96	.083			4.08	5.95	10.03	12.65
1680	cycle 30 miles		72	.111			5.45	7.95	13.40	16.90
1682	cycle 40 miles		60	.133			6.55	9.55	16.10	20.50
1684	cycle 50 miles		48	.167			8.15	11.95	20.10	25.50
1694	45 MPH avg., cycle 8 miles		144	.056			2.72	3.98	6.70	8.45
1696	cycle 10 miles		132	.061			2.97	4.34	7.31	9.20
1698	cycle 20 miles		96	.083			4.08	5.95	10.03	12.65
1700	cycle 30 miles		84	.095			4.66	6.80	11.46	14.50
1702	cycle 40 miles		60	.133			6.55	9.55	16.10	20.50
1704	cycle 50 miles		60	.133			6.55	9.55	16.10	20.50
1706	50 MPH avg., cycle 10 miles		132	.061			2.97	4.34	7.31	9.20
1708	cycle 20 miles		108	.074			3.63	5.30	8.93	11.30
1710	cycle 30 miles		84	.095			4.66	6.80	11.46	14.50
1712	cycle 40 miles		72	.111			5.45	7.95	13.40	16.90
1714	cycle 50 miles		60	.133			6.55	9.55	16.10	20.50
2000	Hauling, 8 C.Y. truck, small project cost per hour	B-34A	8	1	Hr.		49	53.50	102.50	133
2100	12 C.Y. truck	B-34B	8	1			49	71.50	120.50	153
2150	16.5 C.Y. truck	B-34C	8	1			49	78.50	127.50	160
2175	18 C.Y. 8 wheel truck	B-34I	8	1			49	89	138	172
2200	20 C.Y. truck	B-34D	8	1			49	80.50	129.50	162
2300	Grading at dump, or embankment if required, by dozer	B-10B	1000	.012	L.C.Y.		.62	1.50	2.12	2.58
2310	Spotter at fill or cut, if required	1 Clab	8	1	Hr.		42		42	63.50
9014	18 C.Y. truck, 8 wheels,15 min. wait/ld./uld.,15 MPH, cycle 0.5 mi.	B-34I	504	.016	L.C.Y.		.78	1.42	2.20	2.72
9016	cycle 1 mile		450	.018			.87	1.59	2.46	3.05
9018	cycle 2 miles		378	.021			1.04	1.89	2.93	3.63

31 23 23.20 Hauling		Crew	Daily Output	Labor-Hours	Unit	Material	2020 Bare Costs Labor	Equipment	Total	Total Incl O&P
9020	cycle 4 miles	B-34I	270	.030	L.C.Y.		1.45	2.64	4.09	5.10
9022	cycle 6 miles		216	.037			1.81	3.31	5.12	6.35
9024	cycle 8 miles		180	.044			2.18	3.97	6.15	7.60
9025	cycle 10 miles		144	.056			2.72	4.96	7.68	9.55
9026	20 MPH avg., cycle 0.5 mile		522	.015			.75	1.37	2.12	2.62
9028	cycle 1 mile		468	.017			.84	1.52	2.36	2.93
9030	cycle 2 miles		414	.019			.95	1.72	2.67	3.32
9032	cycle 4 miles		324	.025			1.21	2.20	3.41	4.23
9034	cycle 6 miles		252	.032			1.55	2.83	4.38	5.45
9036	cycle 8 miles		216	.037			1.81	3.31	5.12	6.35
9038	cycle 10 miles		180	.044			2.18	3.97	6.15	7.60
9040	25 MPH avg., cycle 4 miles		342	.023			1.14	2.09	3.23	4.02
9042	cycle 6 miles		288	.028			1.36	2.48	3.84	4.77
9044	cycle 8 miles		252	.032			1.55	2.83	4.38	5.45
9046	cycle 10 miles		216	.037			1.81	3.31	5.12	6.35
9050	30 MPH avg., cycle 4 miles		378	.021			1.04	1.89	2.93	3.63
9052	cycle 6 miles		324	.025			1.21	2.20	3.41	4.23
9054	cycle 8 miles		270	.030			1.45	2.64	4.09	5.10
9056	cycle 10 miles		234	.034			1.67	3.05	4.72	5.85
9060	35 MPH avg., cycle 4 miles		396	.020			.99	1.80	2.79	3.46
9062	cycle 6 miles		342	.023			1.14	2.09	3.23	4.02
9064	cycle 8 miles		288	.028			1.36	2.48	3.84	4.77
9066	cycle 10 miles		270	.030			1.45	2.64	4.09	5.10
9068	cycle 20 miles		162	.049			2.42	4.41	6.83	8.45
9070	cycle 30 miles		126	.063			3.11	5.65	8.76	10.90
9072	cycle 40 miles		90	.089			4.35	7.95	12.30	15.20
9074	40 MPH avg., cycle 6 miles		360	.022			1.09	1.98	3.07	3.81
9076	cycle 8 miles		324	.025			1.21	2.20	3.41	4.23
9078	cycle 10 miles		288	.028			1.36	2.48	3.84	4.77
9080	cycle 20 miles		180	.044			2.18	3.97	6.15	7.60
9082	cycle 30 miles		144	.056			2.72	4.96	7.68	9.55
9084	cycle 40 miles		108	.074			3.63	6.60	10.23	12.70
9086	cycle 50 miles		90	.089			4.35	7.95	12.30	15.20
9094	45 MPH avg., cycle 8 miles		324	.025			1.21	2.20	3.41	4.23
9096	cycle 10 miles		306	.026			1.28	2.33	3.61	4.49
9098	cycle 20 miles		198	.040			1.98	3.60	5.58	6.95
9100	cycle 30 miles		144	.056			2.72	4.96	7.68	9.55
9102	cycle 40 miles		126	.063			3.11	5.65	8.76	10.90
9104	cycle 50 miles		108	.074			3.63	6.60	10.23	12.70
9106	50 MPH avg., cycle 10 miles		324	.025			1.21	2.20	3.41	4.23
9108	cycle 20 miles		216	.037			1.81	3.31	5.12	6.35
9110	cycle 30 miles		162	.049			2.42	4.41	6.83	8.45
9112	cycle 40 miles		126	.063			3.11	5.65	8.76	10.90
9114	cycle 50 miles		108	.074			3.63	6.60	10.23	12.70
9214	20 min. wait/ld./uld.,15 MPH, cycle 0.5 mi.		396	.020			.99	1.80	2.79	3.46
9216	cycle 1 mile		360	.022			1.09	1.98	3.07	3.81
9218	cycle 2 miles		306	.026			1.28	2.33	3.61	4.49
9220	cycle 4 miles		234	.034			1.67	3.05	4.72	5.85
9222	cycle 6 miles		198	.040			1.98	3.60	5.58	6.95
9224	cycle 8 miles		162	.049			2.42	4.41	6.83	8.45
9225	cycle 10 miles		144	.056			2.72	4.96	7.68	9.55
9226	20 MPH avg., cycle 0.5 mile		396	.020			.99	1.80	2.79	3.46
9228	cycle 1 mile		378	.021			1.04	1.89	2.93	3.63

31 23 23.20 Hauling		Crew	Daily Output	Labor-Hours	Unit	Material	2020 Bare Costs Labor	Equipment	Total	Total Incl O&P
9230	cycle 2 miles	B-34I	324	.025	L.C.Y.		1.21	2.20	3.41	4.23
9232	cycle 4 miles		270	.030			1.45	2.64	4.09	5.10
9234	cycle 6 miles		216	.037			1.81	3.31	5.12	6.35
9236	cycle 8 miles		198	.040			1.98	3.60	5.58	6.95
9238	cycle 10 miles		162	.049			2.42	4.41	6.83	8.45
9240	25 MPH avg., cycle 4 miles		288	.028			1.36	2.48	3.84	4.77
9242	cycle 6 miles		252	.032			1.55	2.83	4.38	5.45
9244	cycle 8 miles		216	.037			1.81	3.31	5.12	6.35
9246	cycle 10 miles		198	.040			1.98	3.60	5.58	6.95
9250	30 MPH avg., cycle 4 miles		306	.026			1.28	2.33	3.61	4.49
9252	cycle 6 miles		270	.030			1.45	2.64	4.09	5.10
9254	cycle 8 miles		234	.034			1.67	3.05	4.72	5.85
9256	cycle 10 miles		216	.037			1.81	3.31	5.12	6.35
9260	35 MPH avg., cycle 4 miles		324	.025			1.21	2.20	3.41	4.23
9262	cycle 6 miles		288	.028			1.36	2.48	3.84	4.77
9264	cycle 8 miles		252	.032			1.55	2.83	4.38	5.45
9266	cycle 10 miles		234	.034			1.67	3.05	4.72	5.85
9268	cycle 20 miles		162	.049			2.42	4.41	6.83	8.45
9270	cycle 30 miles		108	.074			3.63	6.60	10.23	12.70
9272	cycle 40 miles		90	.089			4.35	7.95	12.30	15.20
9274	40 MPH avg., cycle 6 miles		288	.028			1.36	2.48	3.84	4.77
9276	cycle 8 miles		270	.030			1.45	2.64	4.09	5.10
9278	cycle 10 miles		234	.034			1.67	3.05	4.72	5.85
9280	cycle 20 miles		162	.049			2.42	4.41	6.83	8.45
9282	cycle 30 miles		126	.063			3.11	5.65	8.76	10.90
9284	cycle 40 miles		108	.074			3.63	6.60	10.23	12.70
9286	cycle 50 miles		90	.089			4.35	7.95	12.30	15.20
9294	45 MPH avg., cycle 8 miles		270	.030			1.45	2.64	4.09	5.10
9296	cycle 10 miles		252	.032			1.55	2.83	4.38	5.45
9298	cycle 20 miles		180	.044			2.18	3.97	6.15	7.60
9300	cycle 30 miles		144	.056			2.72	4.96	7.68	9.55
9302	cycle 40 miles		108	.074			3.63	6.60	10.23	12.70
9304	cycle 50 miles		90	.089			4.35	7.95	12.30	15.20
9306	50 MPH avg., cycle 10 miles		270	.030			1.45	2.64	4.09	5.10
9308	cycle 20 miles		198	.040			1.98	3.60	5.58	6.95
9310	cycle 30 miles		144	.056			2.72	4.96	7.68	9.55
9312	cycle 40 miles		126	.063			3.11	5.65	8.76	10.90
9314	cycle 50 miles		108	.074			3.63	6.60	10.23	12.70
9414	25 min. wait/ld./uld.,15 MPH, cycle 0.5 mi.		306	.026			1.28	2.33	3.61	4.49
9416	cycle 1 mile		288	.028			1.36	2.48	3.84	4.77
9418	cycle 2 miles		252	.032			1.55	2.83	4.38	5.45
9420	cycle 4 miles		198	.040			1.98	3.60	5.58	6.95
9422	cycle 6 miles		180	.044			2.18	3.97	6.15	7.60
9424	cycle 8 miles		144	.056			2.72	4.96	7.68	9.55
9425	cycle 10 miles		126	.063			3.11	5.65	8.76	10.90
9426	20 MPH avg., cycle 0.5 mile		324	.025			1.21	2.20	3.41	4.23
9428	cycle 1 mile		306	.026			1.28	2.33	3.61	4.49
9430	cycle 2 miles		270	.030			1.45	2.64	4.09	5.10
9432	cycle 4 miles		234	.034			1.67	3.05	4.72	5.85
9434	cycle 6 miles		198	.040			1.98	3.60	5.58	6.95
9436	cycle 8 miles		180	.044			2.18	3.97	6.15	7.60
9438	cycle 10 miles		144	.056			2.72	4.96	7.68	9.55
9440	25 MPH avg., cycle 4 miles		252	.032			1.55	2.83	4.38	5.45

31 23 23.20 Hauling		Crew	Daily Output	Labor-Hours	Unit	Material	2020 Bare Costs Labor	Equipment	Total	Total Incl O&P
9442	cycle 6 miles	B-34I	216	.037	L.C.Y.		1.81	3.31	5.12	6.35
9444	cycle 8 miles		198	.040			1.98	3.60	5.58	6.95
9446	cycle 10 miles		180	.044			2.18	3.97	6.15	7.60
9450	30 MPH avg., cycle 4 miles		252	.032			1.55	2.83	4.38	5.45
9452	cycle 6 miles		234	.034			1.67	3.05	4.72	5.85
9454	cycle 8 miles		198	.040			1.98	3.60	5.58	6.95
9456	cycle 10 miles		180	.044			2.18	3.97	6.15	7.60
9460	35 MPH avg., cycle 4 miles		270	.030			1.45	2.64	4.09	5.10
9462	cycle 6 miles		234	.034			1.67	3.05	4.72	5.85
9464	cycle 8 miles		216	.037			1.81	3.31	5.12	6.35
9466	cycle 10 miles		198	.040			1.98	3.60	5.58	6.95
9468	cycle 20 miles		144	.056			2.72	4.96	7.68	9.55
9470	cycle 30 miles		108	.074			3.63	6.60	10.23	12.70
9472	cycle 40 miles		90	.089			4.35	7.95	12.30	15.20
9474	40 MPH avg., cycle 6 miles		252	.032			1.55	2.83	4.38	5.45
9476	cycle 8 miles		234	.034			1.67	3.05	4.72	5.85
9478	cycle 10 miles		216	.037			1.81	3.31	5.12	6.35
9480	cycle 20 miles		144	.056			2.72	4.96	7.68	9.55
9482	cycle 30 miles		126	.063			3.11	5.65	8.76	10.90
9484	cycle 40 miles		90	.089			4.35	7.95	12.30	15.20
9486	cycle 50 miles		90	.089			4.35	7.95	12.30	15.20
9494	45 MPH avg., cycle 8 miles		234	.034			1.67	3.05	4.72	5.85
9496	cycle 10 miles		216	.037			1.81	3.31	5.12	6.35
9498	cycle 20 miles		162	.049			2.42	4.41	6.83	8.45
9500	cycle 30 miles		126	.063			3.11	5.65	8.76	10.90
9502	cycle 40 miles		108	.074			3.63	6.60	10.23	12.70
9504	cycle 50 miles		90	.089			4.35	7.95	12.30	15.20
9506	50 MPH avg., cycle 10 miles		234	.034			1.67	3.05	4.72	5.85
9508	cycle 20 miles		180	.044			2.18	3.97	6.15	7.60
9510	cycle 30 miles		144	.056			2.72	4.96	7.68	9.55
9512	cycle 40 miles		108	.074			3.63	6.60	10.23	12.70
9514	cycle 50 miles		90	.089			4.35	7.95	12.30	15.20
9614	30 min. wait/ld./uld.,15 MPH, cycle 0.5 mi.		270	.030			1.45	2.64	4.09	5.10
9616	cycle 1 mile		252	.032			1.55	2.83	4.38	5.45
9618	cycle 2 miles		216	.037			1.81	3.31	5.12	6.35
9620	cycle 4 miles		180	.044			2.18	3.97	6.15	7.60
9622	cycle 6 miles		162	.049			2.42	4.41	6.83	8.45
9624	cycle 8 miles		126	.063			3.11	5.65	8.76	10.90
9625	cycle 10 miles		126	.063			3.11	5.65	8.76	10.90
9626	20 MPH avg., cycle 0.5 mile		270	.030			1.45	2.64	4.09	5.10
9628	cycle 1 mile		252	.032			1.55	2.83	4.38	5.45
9630	cycle 2 miles		234	.034			1.67	3.05	4.72	5.85
9632	cycle 4 miles		198	.040			1.98	3.60	5.58	6.95
9634	cycle 6 miles		180	.044			2.18	3.97	6.15	7.60
9636	cycle 8 miles		162	.049			2.42	4.41	6.83	8.45
9638	cycle 10 miles		144	.056			2.72	4.96	7.68	9.55
9640	25 MPH avg., cycle 4 miles		216	.037			1.81	3.31	5.12	6.35
9642	cycle 6 miles		198	.040			1.98	3.60	5.58	6.95
9644	cycle 8 miles		180	.044			2.18	3.97	6.15	7.60
9646	cycle 10 miles		162	.049			2.42	4.41	6.83	8.45
9650	30 MPH avg., cycle 4 miles		216	.037			1.81	3.31	5.12	6.35
9652	cycle 6 miles		198	.040			1.98	3.60	5.58	6.95
9654	cycle 8 miles		180	.044			2.18	3.97	6.15	7.60

31 23 23.20 Hauling		Crew	Daily Output	Labor-Hours	Unit	Material	2020 Bare Costs Labor	Equipment	Total	Total Incl O&P
9656	cycle 10 miles	B-34I	162	.049	L.C.Y.		2.42	4.41	6.83	8.45
9660	35 MPH avg., cycle 4 miles		234	.034			1.67	3.05	4.72	5.85
9662	cycle 6 miles		216	.037			1.81	3.31	5.12	6.35
9664	cycle 8 miles		198	.040			1.98	3.60	5.58	6.95
9666	cycle 10 miles		180	.044			2.18	3.97	6.15	7.60
9668	cycle 20 miles		126	.063			3.11	5.65	8.76	10.90
9670	cycle 30 miles		108	.074			3.63	6.60	10.23	12.70
9672	cycle 40 miles		90	.089			4.35	7.95	12.30	15.20
9674	40 MPH avg., cycle 6 miles		216	.037			1.81	3.31	5.12	6.35
9676	cycle 8 miles		198	.040			1.98	3.60	5.58	6.95
9678	cycle 10 miles		180	.044			2.18	3.97	6.15	7.60
9680	cycle 20 miles		144	.056			2.72	4.96	7.68	9.55
9682	cycle 30 miles		108	.074			3.63	6.60	10.23	12.70
9684	cycle 40 miles		90	.089			4.35	7.95	12.30	15.20
9686	cycle 50 miles		72	.111			5.45	9.90	15.35	19.05
9694	45 MPH avg., cycle 8 miles		216	.037			1.81	3.31	5.12	6.35
9696	cycle 10 miles		198	.040			1.98	3.60	5.58	6.95
9698	cycle 20 miles		144	.056			2.72	4.96	7.68	9.55
9700	cycle 30 miles		126	.063			3.11	5.65	8.76	10.90
9702	cycle 40 miles		108	.074			3.63	6.60	10.23	12.70
9704	cycle 50 miles		90	.089			4.35	7.95	12.30	15.20
9706	50 MPH avg., cycle 10 miles		198	.040			1.98	3.60	5.58	6.95
9708	cycle 20 miles		162	.049			2.42	4.41	6.83	8.45
9710	cycle 30 miles		126	.063			3.11	5.65	8.76	10.90
9712	cycle 40 miles		108	.074			3.63	6.60	10.23	12.70
9714	cycle 50 miles	▼	90	.089	▼		4.35	7.95	12.30	15.20

Division Notes

	CREW	DAILY OUTPUT	LABOR-HOURS	UNIT	BARE COSTS MAT.	LABOR	EQUIP.	TOTAL	TOTAL INCL O&P

Division 32 Exterior Improvements

Estimating Tips

32 01 00 Operations and Maintenance of Exterior Improvements

- Recycling of asphalt pavement is becoming very popular and is an alternative to removal and replacement. It can be a good value engineering proposal if removed pavement can be recycled, either at the project site or at another site that is reasonably close to the project site. Sections on repair of flexible and rigid pavement are included.

32 10 00 Bases, Ballasts, and Paving

- When estimating paving, keep in mind the project schedule. Also note that prices for asphalt and concrete are generally higher in the cold seasons. Lines for pavement markings, including tactile warning systems and fence lines, are included.

32 90 00 Planting

- The timing of planting and guarantee specifications often dictate the costs for establishing tree and shrub growth and a stand of grass or ground cover. Establish the work performance schedule to coincide with the local planting season. Maintenance and growth guarantees can add 20–100% to the total landscaping cost and can be contractually cumbersome. The cost to replace trees and shrubs can be as high as 5% of the total cost, depending on the planting zone, soil conditions, and time of year.

Reference Numbers

Reference numbers are shown at the beginning of some major classifications. These numbers refer to related items in the Reference Section. The reference information may be an estimating procedure, an alternate pricing method, or technical information.

Note: Not all subdivisions listed here necessarily appear. ■

32 12 16.13 Plant-Mix Asphalt Paving

	32 12 16.13 Plant-Mix Asphalt Paving	Crew	Daily Output	Labor-Hours	Unit	2020 Bare Costs Material	Labor	Equipment	Total	Total Incl O&P
0010	**PLANT-MIX ASPHALT PAVING**									
0020	For highways and large paved areas, exludes hauling									
0025	See Section 31 23 23.20 for hauling costs									
0080	Binder course, 1-1/2" thick	B-25	7725	.011	S.Y.	5.55	.53	.35	6.43	7.25
0120	2" thick		6345	.014		7.40	.64	.43	8.47	9.55
0130	2-1/2" thick		5620	.016		9.25	.72	.48	10.45	11.75
0160	3" thick		4905	.018		11.05	.83	.55	12.43	14.05
0170	3-1/2" thick		4520	.019		12.90	.90	.60	14.40	16.20
0200	4" thick		4140	.021		14.75	.98	.65	16.38	18.45
0300	Wearing course, 1" thick	B-25B	10575	.009		3.66	.43	.28	4.37	4.98
0340	1-1/2" thick		7725	.012		6.15	.59	.38	7.12	8.05
0380	2" thick		6345	.015		8.25	.71	.46	9.42	10.70
0420	2-1/2" thick		5480	.018		10.20	.83	.54	11.57	13.05
0460	3" thick		4900	.020		12.15	.92	.60	13.67	15.40
0470	3-1/2" thick		4520	.021		14.25	1	.65	15.90	17.90
0480	4" thick		4140	.023		16.30	1.09	.71	18.10	20.50
0500	Open graded friction course	B-25C	5000	.010		2.55	.45	.47	3.47	4
0800	Alternate method of figuring paving costs									
0810	Binder course, 1-1/2" thick	B-25	630	.140	Ton	68	6.45	4.29	78.74	89
0811	2" thick		690	.128		68	5.90	3.91	77.81	87.50
0812	3" thick		800	.110		68	5.10	3.37	76.47	86
0813	4" thick		900	.098		68	4.53	3	75.53	84.50
0850	Wearing course, 1" thick	B-25B	575	.167		73.50	7.85	5.10	86.45	98.50
0851	1-1/2" thick		630	.152		73.50	7.20	4.66	85.36	97
0852	2" thick		690	.139		73.50	6.55	4.26	84.31	95.50
0853	2-1/2" thick		765	.125		73.50	5.90	3.84	83.24	94
0854	3" thick		800	.120		73.50	5.65	3.67	82.82	93.50
1000	Pavement replacement over trench, 2" thick	B-37	90	.533	S.Y.	7.60	23.50	2.84	33.94	47
1050	4" thick		70	.686		15.10	30.50	3.65	49.25	66
1080	6" thick		55	.873		24	38.50	4.65	67.15	89.50

32 12 16.14 Asphaltic Concrete Paving

	32 12 16.14 Asphaltic Concrete Paving	Crew	Daily Output	Labor-Hours	Unit	2020 Bare Costs Material	Labor	Equipment	Total	Total Incl O&P
0011	**ASPHALTIC CONCRETE PAVING**, parking lots & driveways									
0015	No asphalt hauling included									
0018	Use 6.05 C.Y. per inch per M.S.F. for hauling									
0020	6" stone base, 2" binder course, 1" topping	B-25C	9000	.005	S.F.	1.92	.25	.26	2.43	2.78
0025	2" binder course, 2" topping		9000	.005		2.36	.25	.26	2.87	3.27
0030	3" binder course, 2" topping		9000	.005		2.78	.25	.26	3.29	3.73
0035	4" binder course, 2" topping		9000	.005		3.19	.25	.26	3.70	4.17
0040	1-1/2" binder course, 1" topping		9000	.005		1.72	.25	.26	2.23	2.56
0042	3" binder course, 1" topping		9000	.005		2.33	.25	.26	2.84	3.24
0045	3" binder course, 3" topping		9000	.005		3.23	.25	.26	3.74	4.22
0050	4" binder course, 3" topping		9000	.005		3.63	.25	.26	4.14	4.67
0055	4" binder course, 4" topping		9000	.005		4.08	.25	.26	4.59	5.15
0300	Binder course, 1-1/2" thick		35000	.001		.62	.06	.07	.75	.85
0400	2" thick		25000	.002		.80	.09	.09	.98	1.12
0500	3" thick		15000	.003		1.23	.15	.16	1.54	1.76
0600	4" thick		10800	.004		1.62	.21	.22	2.05	2.34
0800	Sand finish course, 3/4" thick		41000	.001		.31	.06	.06	.43	.48
0900	1" thick		34000	.001		.38	.07	.07	.52	.60
1000	Fill pot holes, hot mix, 2" thick	B-16	4200	.008		.82	.34	.14	1.30	1.56
1100	4" thick		3500	.009		1.20	.41	.16	1.77	2.11
1120	6" thick		3100	.010		1.61	.46	.18	2.25	2.66

32 12 Flexible Paving

32 12 16 – Asphalt Paving

32 12 16.14 Asphaltic Concrete Paving		Crew	Daily Output	Labor-Hours	Unit	2020 Bare Costs Material	Labor	Equipment	Total	Total Incl O&P
1140	Cold patch, 2″ thick	B-51	3000	.016	S.F.	.87	.69	.07	1.63	2.07
1160	4″ thick		2700	.018		1.66	.77	.07	2.50	3.07
1180	6″ thick		1900	.025		2.58	1.09	.10	3.77	4.59

32 84 Planting Irrigation

32 84 13 – Drip Irrigation

32 84 13.10 Subsurface Drip Irrigation			Crew	Daily Output	Labor-Hours	Unit	Material	Labor	Equipment	Total	Total Incl O&P
0010	**SUBSURFACE DRIP IRRIGATION**										
0011	Looped grid, pressure compensating										
0100	Preinserted PE emitter, line, hand bury, irregular area, small	G	3 Skwk	1200	.020	L.F.	.26	1.10		1.36	1.95
0150	Medium	G		1800	.013		.26	.73		.99	1.40
0200	Large	G		2520	.010		.26	.52		.78	1.08
0250	Rectangular area, small	G		2040	.012		.26	.65		.91	1.27
0300	Medium	G		2640	.009		.26	.50		.76	1.04
0350	Large	G		3600	.007		.26	.37		.63	.84
0400	Install in trench, irregular area, small	G		4050	.006		.26	.33		.59	.78
0450	Medium	G		7488	.003		.26	.18		.44	.56
0500	Large	G		16560	.001		.26	.08		.34	.41
0550	Rectangular area, small	G		8100	.003		.26	.16		.42	.54
0600	Medium			21960	.001		.26	.06		.32	.38
0650	Large	G		33264	.001		.26	.04		.30	.35
0700	Trenching and backfill	G	B-53	500	.016			.85	.31	1.16	1.62
0800	Vinyl tubing, 1/4″, material only	G					.08			.08	.09
0850	Supply tubing, 1/2″, material only, 100′ coil	G					.13			.13	.14
0900	500′ coil	G					.12			.12	.13
0950	Compression fittings	G	1 Skwk	90	.089	Ea.	1.71	4.88		6.59	9.25
1000	Barbed fittings, 1/4″	G		360	.022		.12	1.22		1.34	1.97
1100	Flush risers	G		60	.133		3.76	7.30		11.06	15.20
1150	Flush ends, figure eight	G		180	.044		.55	2.44		2.99	4.30
1300	Auto flush, spring loaded	G		90	.089		2.30	4.88		7.18	9.90
1350	Volumetric	G		90	.089		8	4.88		12.88	16.15
1400	Air relief valve, inline with compensation tee, 1/2″	G		45	.178		18.70	9.75		28.45	35.50
1450	1″	G		30	.267		17.45	14.65		32.10	41
1500	Round box for flush ends, 6″	G		30	.267		8.10	14.65		22.75	31
1600	Screen filter, 3/4″ screen	G		12	.667		10.20	36.50		46.70	66.50
1650	1″ disk	G		8	1		28	55		83	114
1700	1-1/2″ disk	G		4	2		105	110		215	281
1750	2″ disk	G		3	2.667		147	146		293	385
1800	Typical installation 18″ OC, small, minimum					S.F.				1.53	2.35
1850	Maximum									1.78	2.75
1900	Large, minimum									1.48	2.28
2000	Maximum									1.71	2.64
2100	For non-pressure compensating systems, deduct									10%	10%

32 84 23 – Underground Sprinklers

32 84 23.10 Sprinkler Irrigation System		Crew	Daily Output	Labor-Hours	Unit	Material	Labor	Equipment	Total	Total Incl O&P
0010	**SPRINKLER IRRIGATION SYSTEM** G2050–710									
0011	For lawns									
0800	Residential system, custom, 1″ supply	B-20	2000	.012	S.F.	.27	.56		.83	1.15
0900	1-1/2″ supply	″	1800	.013	″	.42	.63		1.05	1.40
1020	Pop up spray head w/risers, hi-pop, full circle pattern, 4″	2 Skwk	76	.211	Ea.	4.85	11.55		16.40	23

32 84 23.10 Sprinkler Irrigation System		Crew	Daily Output	Labor-Hours	Unit	Material	2020 Bare Costs Labor	Equipment	Total	Total Incl O&P
1030	1/2 circle pattern, 4″	2 Skwk	76	.211	Ea.	7	11.55		18.55	25
1040	6″, full circle pattern		76	.211		12.40	11.55		23.95	31
1050	1/2 circle pattern, 6″		76	.211		12.95	11.55		24.50	31.50
1060	12″, full circle pattern		76	.211		15.75	11.55		27.30	35
1070	1/2 circle pattern, 12″		76	.211		15.95	11.55		27.50	35
1080	Pop up bubbler head w/risers, hi-pop bubbler head, 4″		76	.211		5.60	11.55		17.15	23.50
1110	Impact full/part circle sprinklers, 28′-54′ @ 25-60 psi		37	.432		21.50	23.50		45	59.50
1120	Spaced 37′-49′ @ 25-50 psi		37	.432		22	23.50		45.50	60
1130	Spaced 43′-61′ @ 30-60 psi		37	.432		67.50	23.50		91	110
1140	Spaced 54′-78′ @ 40-80 psi		37	.432		124	23.50		147.50	172
1145	Impact rotor pop-up full/part commercial circle sprinklers									
1150	Spaced 42′-65′ @ 35-80 psi	2 Skwk	25	.640	Ea.	17.70	35		52.70	72.50
1160	Spaced 48′-76′ @ 45-85 psi	″	25	.640	″	19.45	35		54.45	74.50
1165	Impact rotor pop-up part. circle comm., 53′-75′, 55-100 psi, w/accessories									
1180	Sprinkler, premium, pop-up rotator, 50′-100′	2 Skwk	25	.640	Ea.	96.50	35		131.50	159
1250	Plastic case, 2 nozzle, metal cover		25	.640		105	35		140	169
1260	Rubber cover		25	.640		104	35		139	167
1270	Iron case, 2 nozzle, metal cover		22	.727		149	40		189	225
1280	Rubber cover		22	.727		148	40		188	224
1282	Impact rotor pop-up full circle commercial, 39′-99′, 30-100 psi									
1284	Plastic case, metal cover	2 Skwk	25	.640	Ea.	74	35		109	135
1286	Rubber cover		25	.640		108	35		143	172
1288	Iron case, metal cover		22	.727		143	40		183	218
1290	Rubber cover		22	.727		168	40		208	246
1292	Plastic case, 2 nozzle, metal cover		22	.727		107	40		147	178
1294	Rubber cover		22	.727		109	40		149	181
1296	Iron case, 2 nozzle, metal cover		20	.800		142	44		186	224
1298	Rubber cover		20	.800		148	44		192	230
1305	Electric remote control valve, plastic, 3/4″		18	.889		20.50	49		69.50	96
1310	1″		18	.889		25	49		74	101
1320	1-1/2″		18	.889		99	49		148	183
1330	2″		18	.889		126	49		175	212
1335	Quick coupling valves, brass, locking cover									
1340	Inlet coupling valve, 3/4″	2 Skwk	18.75	.853	Ea.	24.50	47		71.50	98
1350	1″		18.75	.853		33.50	47		80.50	108
1360	Controller valve boxes, 6″ round boxes		18.75	.853		7.55	47		54.55	79.50
1370	10″ round boxes		14.25	1.123		17.95	61.50		79.45	113
1380	12″ square box		9.75	1.641		19.10	90		109.10	157
1388	Electromech. control, 14 day 3-60 min., auto start to 23/day									
1390	4 station	2 Skwk	1.04	15.385	Ea.	82	845		927	1,375
1400	7 station		.64	25		156	1,375		1,531	2,250
1410	12 station		.40	40		184	2,200		2,384	3,525
1420	Dual programs, 18 station		.24	66.667		217	3,650		3,867	5,775
1430	23 station		.16	100		267	5,475		5,742	8,600
1435	Backflow preventer, bronze, 0-175 psi, w/valves, test cocks									
1440	3/4″	2 Skwk	6	2.667	Ea.	93.50	146		239.50	325
1450	1″		6	2.667		106	146		252	340
1460	1-1/2″		6	2.667		269	146		415	515
1470	2″		6	2.667		395	146		541	655
1475	Pressure vacuum breaker, brass, 15-150 psi									
1480	3/4″	2 Skwk	6	2.667	Ea.	24	146		170	248
1490	1″		6	2.667		69	146		215	297
1500	1-1/2″		6	2.667		85	146		231	315

32 84 23 – Underground Sprinklers

32 84 23.10 Sprinkler Irrigation System		Crew	Daily Output	Labor-Hours	Unit	Material	Labor	Equipment	Total	Total Incl O&P
						2020 Bare Costs				
1510	2"	2 Skwk	6	2.667	Ea.	157	146		303	395
6000	Riser mounted gear drive sprinkler, nozzle and case									
6200	Plastic, medium volume	1 Skwk	30	.267	Ea.	16.90	14.65		31.55	40.50
7000	Riser mounted impact sprinkler, body, part or full circle									
7200	Full circle plastic, low/medium volume	1 Skwk	25	.320	Ea.	16.75	17.55		34.30	45
7400	Brass, low/medium volume		25	.320		44.50	17.55		62.05	75
7600	Medium volume		25	.320		59	17.55		76.55	91
7800	Female thread		25	.320		105	17.55		122.55	143
8000	Riser mounted impact sprinkler, body, full circle only									
8200	Plastic, low/medium volume	1 Skwk	25	.320	Ea.	10	17.55		27.55	37.50
8400	Brass, low/medium volume		25	.320		42.50	17.55		60.05	73
8600	High volume		25	.320		88	17.55		105.55	123
8800	Very high volume		25	.320		168	17.55		185.55	212

Division Notes

		CREW	DAILY OUTPUT	LABOR-HOURS	UNIT	BARE COSTS MAT.	BARE COSTS LABOR	BARE COSTS EQUIP.	BARE COSTS TOTAL	TOTAL INCL O&P

Estimating Tips

33 10 00 Water Utilities
33 30 00 Sanitary Sewerage Utilities
33 40 00 Storm Drainage Utilities

- Never assume that the water, sewer, and drainage lines will go in at the early stages of the project. Consider the site access needs before dividing the site in half with open trenches, loose pipe, and machinery obstructions. Always inspect the site to establish that the site drawings are complete. Check off all existing utilities on your drawings as you locate them. Be especially careful with underground utilities because appurtenances are sometimes buried during regrading or repaving operations. If you find any discrepancies, mark up the site plan for further research. Differing site conditions can be very costly if discovered later in the project.
- See also Section 33 01 00 for restoration of pipe where removal/replacement may be undesirable. Use of new types of piping materials can reduce the overall project cost. Owners/design engineers should consider the installing contractor as a valuable source of current information on utility products and local conditions that could lead to significant cost savings.

Reference Numbers

Reference numbers are shown at the beginning of some major classifications. These numbers refer to related items in the Reference Section. The reference information may be an estimating procedure, an alternate pricing method, or technical information.

Note: Not all subdivisions listed here necessarily appear. ■

Note: "Powered in part by CINX™, based on licensed proprietary information of Harrison Publishing House, Inc."

Note: Trade Service, in part, has been used as a reference source for some of the material prices used in Division 33

33 01 10.10 Corrosion Resistance

	33 01 10.10 Corrosion Resistance	Crew	Daily Output	Labor-Hours	Unit	Material (2020 Bare Costs)	Labor (2020 Bare Costs)	Equipment (2020 Bare Costs)	Total (2020 Bare Costs)	Total Incl O&P
0010	**CORROSION RESISTANCE**									
0012	Wrap & coat, add to pipe, 4″ diameter				L.F.	2.38			2.38	2.62
0020	5″ diameter					3.04			3.04	3.34
0040	6″ diameter					3.67			3.67	4.04
0060	8″ diameter					5			5	5.50
0080	10″ diameter					6.30			6.30	6.95
0100	12″ diameter					7.30			7.30	8
0120	14″ diameter					8.35			8.35	9.20
0140	16″ diameter					10.55			10.55	11.60
0160	18″ diameter					11			11	12.10
0180	20″ diameter					12.25			12.25	13.50
0200	24″ diameter					14.75			14.75	16.20
0220	Small diameter pipe, 1″ diameter, add					1.01			1.01	1.11
0240	2″ diameter					1.28			1.28	1.41
0260	2-1/2″ diameter					1.61			1.61	1.77
0280	3″ diameter					1.93			1.93	2.12
0300	Fittings, field covered, add				S.F.	9.40			9.40	10.35
0500	Coating, bituminous, per diameter inch, 1 coat, add				L.F.	.21			.21	.23
0540	3 coat					.57			.57	.63
0560	Coal tar epoxy, per diameter inch, 1 coat, add					.21			.21	.23
0600	3 coat					.66			.66	.73
1000	Polyethylene H.D. extruded, 0.025″ thk., 1/2″ diameter, add					.10			.10	.11
1020	3/4″ diameter					.15			.15	.17
1040	1″ diameter					.19			.19	.21
1060	1-1/4″ diameter					.25			.25	.28
1080	1-1/2″ diameter					.28			.28	.31
1100	0.030″ thk., 2″ diameter					.39			.39	.43
1120	2-1/2″ diameter					.50			.50	.55
1140	0.035″ thk., 3″ diameter					.56			.56	.62
1160	3-1/2″ diameter					.65			.65	.72
1180	4″ diameter					.81			.81	.89
1200	0.040″ thk., 5″ diameter					.93			.93	1.02
1220	6″ diameter					1.14			1.14	1.25
1240	8″ diameter					1.46			1.46	1.61
1260	10″ diameter					1.79			1.79	1.97
1280	12″ diameter					2.15			2.15	2.37
1300	0.060″ thk., 14″ diameter					2.76			2.76	3.04
1320	16″ diameter					3.16			3.16	3.48
1340	18″ diameter					3.30			3.30	3.63
1360	20″ diameter					3.67			3.67	4.04
1380	Fittings, field wrapped, add				S.F.	4.30			4.30	4.73

33 01 10.20 Pipe Repair

	33 01 10.20 Pipe Repair	Crew	Daily Output	Labor-Hours	Unit	Material	Labor	Equipment	Total	Total Incl O&P
0010	**PIPE REPAIR**									
0020	Not including excavation or backfill									
0100	Clamp, stainless steel, lightweight, for steel pipe									
0110	3″ long, 1/2″ diameter pipe	1 Plum	34	.235	Ea.	11.15	15.15		26.30	35
0120	3/4″ diameter pipe		32	.250		14.10	16.10		30.20	39.50
0130	1″ diameter pipe		30	.267		12.50	17.20		29.70	39.50
0140	1-1/4″ diameter pipe		28	.286		13.15	18.40		31.55	42
0150	1-1/2″ diameter pipe		26	.308		16.85	19.85		36.70	48
0160	2″ diameter pipe		24	.333		18.15	21.50		39.65	52
0170	2-1/2″ diameter pipe		23	.348		17.40	22.50		39.90	52.50

33 01 10 – Operation and Maintenance of Water Utilities

33 01 10.20 Pipe Repair		Crew	Daily Output	Labor-Hours	Unit	Material	2020 Bare Costs Labor	Equipment	Total	Total Incl O&P
0180	3" diameter pipe	1 Plum	22	.364	Ea.	24.50	23.50		48	62
0190	3-1/2" diameter pipe		21	.381		25	24.50		49.50	64
0200	4" diameter pipe	B-20	44	.545		23	25.50		48.50	63.50
0210	5" diameter pipe		42	.571		26.50	27		53.50	69.50
0220	6" diameter pipe		38	.632		29.50	29.50		59	77
0230	8" diameter pipe		30	.800		35.50	37.50		73	96
0240	10" diameter pipe		28	.857		120	40.50		160.50	193
0250	12" diameter pipe		24	1		131	47		178	216
0260	14" diameter pipe		22	1.091		166	51.50		217.50	261
0270	16" diameter pipe		20	1.200		179	56.50		235.50	282
0280	18" diameter pipe		18	1.333		206	62.50		268.50	320
0290	20" diameter pipe		16	1.500		234	70.50		304.50	365
0300	24" diameter pipe		14	1.714		238	80.50		318.50	385
0360	For 6" long, add					100%	40%			
0370	For 9" long, add					200%	100%			
0380	For 12" long, add					300%	150%			
0390	For 18" long, add					500%	200%			
0400	Pipe freezing for live repairs of systems 3/8" to 6"									
0410	Note: Pipe freezing can also be used to install a valve into a live system									
0420	Pipe freezing each side 3/8"	2 Skwk	8	2	Ea.	550	110		660	770
0425	Pipe freezing each side 3/8", second location same kit		8	2		32	110		142	202
0430	Pipe freezing each side 3/4"		8	2		515	110		625	735
0435	Pipe freezing each side 3/4", second location same kit		8	2		32	110		142	202
0440	Pipe freezing each side 1-1/2"		6	2.667		515	146		661	790
0445	Pipe freezing each side 1-1/2" , second location same kit		6	2.667		32	146		178	257
0450	Pipe freezing each side 2"		6	2.667		945	146		1,091	1,275
0455	Pipe freezing each side 2", second location same kit		6	2.667		32	146		178	257
0460	Pipe freezing each side 2-1/2" to 3"		6	2.667		955	146		1,101	1,275
0465	Pipe freezing each side 2-1/2" to 3", second location same kit		6	2.667		64	146		210	292
0470	Pipe freezing each side 4"		4	4		1,575	219		1,794	2,050
0475	Pipe freezing each side 4", second location same kit		4	4		73	219		292	410
0480	Pipe freezing each side 5" to 6"		4	4		4,400	219		4,619	5,175
0485	Pipe freezing each side 5" to 6", second location same kit		4	4		218	219		437	570
0490	Pipe freezing extra 20 lb. CO_2 cylinders (3/8" to 2" - 1 ea, 3" - 2 ea)					231			231	254
0500	Pipe freezing extra 50 lb. CO_2 cylinders (4" - 2 ea, 5"-6" - 6 ea)					530			530	585
1000	Clamp, stainless steel, with threaded service tap									
1040	Full seal for iron, steel, PVC pipe									
1100	6" long, 2" diameter pipe	1 Plum	17	.471	Ea.	115	30.50		145.50	173
1110	2-1/2" diameter pipe		16	.500		114	32		146	174
1120	3" diameter pipe		15.60	.513		112	33		145	173
1130	3-1/2" diameter pipe		15	.533		135	34.50		169.50	200
1140	4" diameter pipe	B-20	32	.750		125	35.50		160.50	190
1150	6" diameter pipe		28	.857		164	40.50		204.50	241
1160	8" diameter pipe		21	1.143		176	53.50		229.50	275
1170	10" diameter pipe		20	1.200		232	56.50		288.50	340
1180	12" diameter pipe		17	1.412		266	66.50		332.50	390
1200	8" long, 2" diameter pipe	1 Plum	11.72	.683		117	44		161	195
1210	2-1/2" diameter pipe		11	.727		127	47		174	210
1220	3" diameter pipe		10.75	.744		125	48		173	210
1230	3-1/2" diameter pipe		10.34	.774		132	50		182	220
1240	4" diameter pipe	B-20	22	1.091		158	51.50		209.50	252
1250	6" diameter pipe		19.31	1.243		162	58.50		220.50	267
1260	8" diameter pipe		14.48	1.657		211	78		289	350

33 01 10.20 Pipe Repair		Crew	Daily Output	Labor-Hours	Unit	Material	Labor	Equipment	Total	Total Incl O&P
						2020 Bare Costs				
1270	10″ diameter pipe	B-20	13.80	1.739	Ea.	243	82		325	390
1280	12″ diameter pipe		11.72	2.048		305	96.50		401.50	485
1300	12″ long, 2″ diameter pipe	1 Plum	9.44	.847		248	54.50		302.50	355
1310	2-1/2″ diameter pipe		8.89	.900		204	58		262	310
1320	3″ diameter pipe		8.67	.923		191	59.50		250.50	299
1330	3-1/2″ diameter pipe		8.33	.960		228	62		290	345
1340	4″ diameter pipe	B-20	17.78	1.350		237	63.50		300.50	355
1350	6″ diameter pipe		15.56	1.542		243	72.50		315.50	375
1360	8″ diameter pipe		11.67	2.057		288	96.50		384.50	460
1370	10″ diameter pipe		11.11	2.160		470	102		572	675
1380	12″ diameter pipe		9.44	2.542		475	120		595	705
1400	20″ long, 2″ diameter pipe	1 Plum	8.10	.988		231	63.50		294.50	350
1410	2-1/2″ diameter pipe		7.62	1.050		272	67.50		339.50	400
1420	3″ diameter pipe		7.43	1.077		345	69.50		414.50	485
1430	3-1/2″ diameter pipe		7.14	1.120		310	72		382	450
1440	4″ diameter pipe	B-20	15.24	1.575		340	74		414	485
1450	6″ diameter pipe		13.33	1.800		405	84.50		489.50	575
1460	8″ diameter pipe		10	2.400		495	113		608	715
1470	10″ diameter pipe		9.52	2.521		555	119		674	795
1480	12″ diameter pipe		8.10	2.963		660	139		799	935
1600	Clamp, stainless steel, single section									
1640	Full seal for iron, steel, PVC pipe									
1700	6″ long, 2″ diameter pipe	1 Plum	17	.471	Ea.	91	30.50		121.50	146
1710	2-1/2″ diameter pipe		16	.500		94	32		126	151
1720	3″ diameter pipe		15.60	.513		86.50	33		119.50	145
1730	3-1/2″ diameter pipe		15	.533		113	34.50		147.50	176
1740	4″ diameter pipe	B-20	32	.750		123	35.50		158.50	188
1750	6″ diameter pipe		27	.889		152	42		194	231
1760	8″ diameter pipe		21	1.143		161	53.50		214.50	258
1770	10″ diameter pipe		20	1.200		227	56.50		283.50	335
1780	12″ diameter pipe		17	1.412		218	66.50		284.50	340
1800	8″ long, 2″ diameter pipe	1 Plum	11.72	.683		126	44		170	205
1805	2-1/2″ diameter pipe		11.03	.725		129	46.50		175.50	212
1810	3″ diameter pipe		10.76	.743		136	48		184	222
1815	3-1/2″ diameter pipe		10.34	.774		124	50		174	212
1820	4″ diameter pipe	B-20	22.07	1.087		147	51		198	239
1825	6″ diameter pipe		19.31	1.243		187	58.50		245.50	294
1830	8″ diameter pipe		14.48	1.657		223	78		301	365
1835	10″ diameter pipe		13.79	1.740		300	82		382	460
1840	12″ diameter pipe		11.72	2.048		335	96.50		431.50	515
1850	12″ long, 2″ diameter pipe	1 Plum	9.44	.847		190	54.50		244.50	290
1855	2-1/2″ diameter pipe		8.89	.900		196	58		254	305
1860	3″ diameter pipe		8.67	.923		215	59.50		274.50	325
1865	3-1/2″ diameter pipe		8.33	.960		197	62		259	310
1870	4″ diameter pipe	B-20	17.78	1.350		224	63.50		287.50	345
1875	6″ diameter pipe		15.56	1.542		305	72.50		377.50	445
1880	8″ diameter pipe		11.67	2.057		330	96.50		426.50	510
1885	10″ diameter pipe		11.11	2.160		450	102		552	645
1890	12″ diameter pipe		9.44	2.542		535	120		655	770
1900	20″ long, 2″ diameter pipe	1 Plum	8.10	.988		197	63.50		260.50	315
1905	2-1/2″ diameter pipe		7.62	1.050		218	67.50		285.50	340
1910	3″ diameter pipe		7.43	1.077		250	69.50		319.50	380
1915	3-1/2″ diameter pipe		7.14	1.120		320	72		392	465

33 01 10 – Operation and Maintenance of Water Utilities

33 01 10.20 Pipe Repair		Crew	Daily Output	Labor-Hours	Unit	2020 Bare Costs Material	Labor	Equipment	Total	Total Incl O&P
1920	4″ diameter pipe	B-20	15.24	1.575	Ea.	425	74		499	575
1925	6″ diameter pipe		13.33	1.800		465	84.50		549.50	640
1930	8″ diameter pipe		10	2.400		555	113		668	780
1935	10″ diameter pipe		9.52	2.521		700	119		819	950
1940	12″ diameter pipe		8.10	2.963		830	139		969	1,125
2000	Clamp, stainless steel, two section									
2040	Full seal, for iron, steel, PVC pipe									
2100	8″ long, 4″ diameter pipe	B-20	24	1	Ea.	230	47		277	325
2110	6″ diameter pipe		20	1.200		264	56.50		320.50	375
2120	8″ diameter pipe		13	1.846		305	87		392	470
2130	10″ diameter pipe		12	2		293	94		387	460
2140	12″ diameter pipe		10	2.400		395	113		508	605
2200	10″ long, 4″ diameter pipe		16	1.500		300	70.50		370.50	435
2210	6″ diameter pipe		13	1.846		340	87		427	505
2220	8″ diameter pipe		9	2.667		380	125		505	605
2230	10″ diameter pipe		8	3		490	141		631	750
2240	12″ diameter pipe		7	3.429		565	161		726	870
2242	Clamp, stainless steel, three section									
2244	Full seal, for iron, steel, PVC pipe									
2250	10″ long, 14″ diameter pipe	B-20	6.40	3.750	Ea.	655	176		831	985
2260	16″ diameter pipe		6	4		695	188		883	1,050
2270	18″ diameter pipe		5	4.800		795	226		1,021	1,225
2280	20″ diameter pipe		4.60	5.217		905	245		1,150	1,375
2290	24″ diameter pipe		4	6		1,250	282		1,532	1,800
2320	For 12″ long, add to 10″					15%	25%			
2330	For 20″ long, add to 10″					70%	55%			
8000	For internal cleaning and inspection, see Section 33 01 30.11									
8100	For pipe testing, see Section 23 05 93.50									

33 01 30 – Operation and Maintenance of Sewer Utilities

33 01 30.11 Television Inspection of Sewers		Crew	Daily Output	Labor-Hours	Unit	Material	Labor	Equipment	Total	Total Incl O&P
0010	TELEVISION INSPECTION OF SEWERS									
0100	Pipe internal cleaning & inspection, cleaning, pressure pipe systems									
0120	Pig method, lengths 1000′ to 10,000′									
0140	4″ diameter thru 24″ diameter, minimum				L.F.				3.60	4.14
0160	Maximum				″				18	21
6000	Sewage/sanitary systems									
6100	Power rodder with header & cutters									
6110	Mobilization charge, minimum				Total				695	800
6120	Mobilization charge, maximum				″				9,125	10,600
6140	Cleaning 4″-12″ diameter				L.F.				6	6.60
6190	14″-24″ diameter								8	8.80
6240	30″ diameter								9.60	10.50
6250	36″ diameter								12	13.20
6260	48″ diameter								16	17.60
6270	60″ diameter								24	26.50
6280	72″ diameter								48	53
9000	Inspection, television camera with video									
9060	up to 500 linear feet				Total				715	820

33 01 Operation and Maintenance of Utilities

33 01 30 – Operation and Maintenance of Sewer Utilities

33 01 30.23 Pipe Bursting		Crew	Daily Output	Labor-Hours	Unit	2020 Bare Costs Material	Labor	Equipment	Total	Total Incl O&P
0010	**PIPE BURSTING**									
0011	300′ runs, replace with HDPE pipe									
0020	Not including excavation, backfill, shoring, or dewatering									
0100	6″ to 15″ diameter, minimum				L.F.				200	220
0200	Maximum								550	605
0300	18″ to 36″ diameter, minimum								650	715
0400	Maximum								1,050	1,150
0500	Mobilize and demobilize, minimum				Job				3,000	3,300
0600	Maximum				″				32,100	35,400

33 01 30.72 Cured-In-Place Pipe Lining		Crew	Daily Output	Labor-Hours	Unit	Material	Labor	Equipment	Total	Total Incl O&P
0010	**CURED-IN-PLACE PIPE LINING**									
0011	Not incl. bypass or cleaning									
0020	Less than 10,000 L.F., urban, 6″ to 10″	C-17E	130	.615	L.F.	9.40	34	.28	43.68	62
0050	10″ to 12″		125	.640		11.60	35.50	.29	47.39	66.50
0070	12″ to 16″		115	.696		11.85	38.50	.31	50.66	71.50
0100	16″ to 20″		95	.842		13.95	46.50	.38	60.83	86
0200	24″ to 36″		90	.889		15.30	49	.40	64.70	91.50
0300	48″ to 72″		80	1		24	55.50	.45	79.95	110

33 01 30.74 Sliplining, Excludes Cleaning		Crew	Daily Output	Labor-Hours	Unit	Material	Labor	Equipment	Total	Total Incl O&P
0010	**SLIPLINING, excludes cleaning** and video inspection									
0020	Pipe relined with one pipe size smaller than original (4″ for 6″)									
0100	6″ diameter, original size	B-6B	600	.080	L.F.	5.60	3.42	1.67	10.69	13.20
0150	8″ diameter, original size		600	.080		9.60	3.42	1.67	14.69	17.55
0200	10″ diameter, original size		600	.080		11.25	3.42	1.67	16.34	19.40
0250	12″ diameter, original size		400	.120		16	5.15	2.50	23.65	28
0300	14″ diameter, original size		400	.120		15.60	5.15	2.50	23.25	27.50
0350	16″ diameter, original size	B-6C	300	.160		18.30	6.85	7.75	32.90	39
0400	18″ diameter, original size	″	300	.160		28	6.85	7.75	42.60	50
1000	Pipe HDPE lining, make service line taps	B-6	4	6	Ea.	87	274	53.50	414.50	565

33 05 Common Work Results for Utilities

33 05 07 – Trenchless Installation of Utility Piping

33 05 07.23 Utility Boring and Jacking		Crew	Daily Output	Labor-Hours	Unit	Material	Labor	Equipment	Total	Total Incl O&P
0010	**UTILITY BORING AND JACKING**									
0011	Casing only, 100′ minimum,									
0020	not incl. jacking pits or dewatering									
0100	Roadwork, 1/2″ thick wall, 24″ diameter casing	B-42	20	3.200	L.F.	122	152	52	326	420
0200	36″ diameter		16	4		215	190	65	470	595
0300	48″ diameter		15	4.267		300	203	69.50	572.50	710
0500	Railroad work, 24″ diameter		15	4.267		122	203	69.50	394.50	515
0600	36″ diameter		14	4.571		215	217	74.50	506.50	650
0700	48″ diameter		12	5.333		300	253	87	640	805
0900	For ledge, add								20%	20%
1000	Small diameter boring, 3″, sandy soil	B-82	900	.018		22	.85	.20	23.05	26
1040	Rocky soil	″	500	.032		22	1.52	.36	23.88	27
1100	Prepare jacking pits, incl. mobilization & demobilization, minimum				Ea.				3,225	3,700
1101	Maximum				″				22,000	25,500

33 05 Common Work Results for Utilities

33 05 07 – Trenchless Installation of Utility Piping

33 05 07.36 Microtunneling		Crew	Daily Output	Labor-Hours	Unit	2020 Bare Costs Material	Labor	Equipment	Total	Total Incl O&P
0010	**MICROTUNNELING**									
0011	Not including excavation, backfill, shoring,									
0020	or dewatering, average 50′/day, slurry method									
0100	24″ to 48″ outside diameter, minimum				L.F.				965	965
0110	Adverse conditions, add				"				500	500
1000	Rent microtunneling machine, average monthly lease				Month				97,500	107,000
1010	Operating technician				Day				630	705
1100	Mobilization and demobilization, minimum				Job				41,200	45,900
1110	Maximum				"				445,500	490,500

33 05 61 – Concrete Manholes

33 05 61.10 Storm Drainage Manholes, Frames and Covers		Crew	Daily Output	Labor-Hours	Unit	2020 Bare Costs Material	Labor	Equipment	Total	Total Incl O&P
0010	**STORM DRAINAGE MANHOLES, FRAMES & COVERS**									
0020	Excludes footing, excavation, backfill (See line items for frame & cover)									
0050	Brick, 4′ inside diameter, 4′ deep	D-1	1	16	Ea.	615	750		1,365	1,825
0100	6′ deep		.70	22.857		870	1,075		1,945	2,575
0150	8′ deep		.50	32		1,125	1,500		2,625	3,500
0200	For depths over 8′, add		4	4	V.L.F.	95.50	188		283.50	390
0400	Concrete blocks (radial), 4′ ID, 4′ deep		1.50	10.667	Ea.	430	500		930	1,225
0500	6′ deep		1	16		565	750		1,315	1,775
0600	8′ deep		.70	22.857		705	1,075		1,780	2,400
0700	For depths over 8′, add		5.50	2.909	V.L.F.	71.50	136		207.50	286
0800	Concrete, cast in place, 4′ x 4′, 8″ thick, 4′ deep	C-14H	2	24	Ea.	580	1,250	13.45	1,843.45	2,500
0900	6′ deep		1.50	32		830	1,650	17.90	2,497.90	3,400
1000	8′ deep		1	48		1,200	2,475	27	3,702	5,075
1100	For depths over 8′, add		8	6	V.L.F.	133	310	3.36	446.36	615
1110	Precast, 4′ ID, 4′ deep	B-22	4.10	7.317	Ea.	880	360	69	1,309	1,600
1120	6′ deep		3	10		1,025	495	94.50	1,614.50	1,975
1130	8′ deep		2	15		1,200	740	142	2,082	2,600
1140	For depths over 8′, add		16	1.875	V.L.F.	134	92.50	17.70	244.20	305
1150	5′ ID, 4′ deep	B-6	3	8	Ea.	1,725	365	71.50	2,161.50	2,525
1160	6′ deep		2	12		2,075	550	107	2,732	3,225
1170	8′ deep		1.50	16		2,625	730	143	3,498	4,125
1180	For depths over 8′, add		12	2	V.L.F.	325	91.50	17.80	434.30	515
1190	6′ ID, 4′ deep		2	12	Ea.	2,600	550	107	3,257	3,800
1200	6′ deep		1.50	16		2,975	730	143	3,848	4,525
1210	8′ deep		1	24		3,675	1,100	214	4,989	5,925
1220	For depths over 8′, add		8	3	V.L.F.	410	137	26.50	573.50	685
1250	Slab tops, precast, 8″ thick									
1300	4′ diameter manhole	B-6	8	3	Ea.	282	137	26.50	445.50	545
1400	5′ diameter manhole		7.50	3.200		440	146	28.50	614.50	735
1500	6′ diameter manhole		7	3.429		705	157	30.50	892.50	1,050
3800	Steps, heavyweight cast iron, 7″ x 9″	1 Bric	40	.200		18.05	10.40		28.45	35.50
3900	8″ x 9″		40	.200		21.50	10.40		31.90	40
3928	12″ x 10-1/2″		40	.200		29.50	10.40		39.90	48.50
4000	Standard sizes, galvanized steel		40	.200		26.50	10.40		36.90	45
4100	Aluminum		40	.200		30	10.40		40.40	49
4150	Polyethylene		40	.200		32.50	10.40		42.90	52

33 05 Common Work Results for Utilities

33 05 63 – Concrete Vaults and Chambers

33 05 63.13 Precast Concrete Utility Structures		Crew	Daily Output	Labor-Hours	Unit	Material	Labor	Equipment	Total	Total Incl O&P
						2020 Bare Costs				
0010	**PRECAST CONCRETE UTILITY STRUCTURES**, 6″ thick									
0050	5′ x 10′ x 6′ high, ID	B-13	2	28	Ea.	1,850	1,300	290	3,440	4,300
0100	6′ x 10′ x 6′ high, ID		2	28		1,925	1,300	290	3,515	4,375
0150	5′ x 12′ x 6′ high, ID		2	28		2,050	1,300	290	3,640	4,500
0200	6′ x 12′ x 6′ high, ID		1.80	31.111		2,275	1,425	325	4,025	5,025
0250	6′ x 13′ x 6′ high, ID		1.50	37.333		3,000	1,725	385	5,110	6,300
0300	8′ x 14′ x 7′ high, ID		1	56		3,250	2,575	580	6,405	8,100
0350	Hand hole, precast concrete, 1-1/2″ thick									
0400	1′-0″ x 2′-0″ x 1′-9″, ID, light duty	B-1	4	6	Ea.	540	257		797	980
0450	4′-6″ x 3′-2″ x 2′-0″, OD, heavy duty	B-6	3	8	″	1,625	365	71.50	2,061.50	2,400

33 05 97 – Identification and Signage for Utilities

33 05 97.05 Utility Connection

Line	Description	Crew	Daily Output	Labor-Hours	Unit	Material	Labor	Equipment	Total	Total Incl O&P
0010	**UTILITY CONNECTION**									
0020	Water, sanitary, stormwater, gas, single connection	B-14	1	48	Ea.	3,050	2,125	214	5,389	6,775
0030	Telecommunication	″	3	16	″	400	710	71	1,181	1,600

33 05 97.10 Utility Accessories

Line	Description	Crew	Daily Output	Labor-Hours	Unit	Material	Labor	Equipment	Total	Total Incl O&P
0010	**UTILITY ACCESSORIES**									
0400	Underground tape, detectable, reinforced, alum. foil core, 2″	1 Clab	150	.053	C.L.F.	9	2.25		11.25	13.25
0500	6″	″	140	.057	″	38	2.41		40.41	45

33 11 Groundwater Sources

33 11 13 – Potable Water Supply Wells

33 11 13.10 Wells and Accessories

Line	Description	Crew	Daily Output	Labor-Hours	Unit	Material	Labor	Equipment	Total	Total Incl O&P
0010	**WELLS & ACCESSORIES**									
0011	Domestic									
0100	Drilled, 4″ to 6″ diameter	B-23	120	.333	L.F.		14.15	13.25	27.40	36
0200	8″ diameter	″	95.20	.420	″		17.85	16.70	34.55	45.50
0400	Gravel pack well, 40′ deep, incl. gravel & casing, complete									
0500	24″ diameter casing x 18″ diameter screen	B-23	.13	308	Total	38,500	13,100	12,200	63,800	75,500
0600	36″ diameter casing x 18″ diameter screen		.12	333	″	40,800	14,200	13,300	68,300	81,000
0800	Observation wells, 1-1/4″ riser pipe		163	.245	V.L.F.	19.90	10.45	9.75	40.10	48.50
0900	For flush Buffalo roadway box, add	1 Skwk	16.60	.482	Ea.	53	26.50		79.50	98
1200	Test well, 2-1/2″ diameter, up to 50′ deep (15 to 50 GPM)	B-23	1.51	26.490	″	830	1,125	1,050	3,005	3,775
1300	Over 50′ deep, add	″	121.80	.328	L.F.	22	13.95	13.05	49	60
1500	Pumps, installed in wells to 100′ deep, 4″ submersible									
1510	1/2 HP	Q-1	3.22	4.969	Ea.	660	288		948	1,150
1520	3/4 HP		2.66	6.015		795	350		1,145	1,400
1600	1 HP		2.29	6.987		975	405		1,380	1,675
1700	1-1/2 HP	Q-22	1.60	10		2,150	580	294	3,024	3,575
1800	2 HP		1.33	12.030		1,800	700	355	2,855	3,450
1900	3 HP		1.14	14.035		2,425	815	415	3,655	4,350
2000	5 HP		1.14	14.035		2,850	815	415	4,080	4,800
3000	Pump, 6″ submersible, 25′ to 150′ deep, 25 HP, 103 to 400 GPM		.89	17.978		9,325	1,050	530	10,905	12,400
3100	25′ to 500′ deep, 30 HP, 104 to 400 GPM		.73	21.918		11,000	1,275	645	12,920	14,700
8000	Steel well casing	B-23A	3020	.008	Lb.	1.29	.38	.29	1.96	2.31
8110	Well screen assembly, stainless steel, 2″ diameter		273	.088	L.F.	81.50	4.19	3.23	88.92	100
8120	3″ diameter		253	.095		143	4.52	3.49	151.01	168
8130	4″ diameter		200	.120		178	5.70	4.41	188.11	209
8140	5″ diameter		168	.143		184	6.80	5.25	196.05	218
8150	6″ diameter		126	.190		204	9.10	7	220.10	246

33 11 Groundwater Sources

33 11 13 – Potable Water Supply Wells

33 11 13.10 Wells and Accessories		Crew	Daily Output	Labor-Hours	Unit	2020 Bare Costs Material	Labor	Equipment	Total	Total Incl O&P
8160	8″ diameter	B-23A	98.50	.244	L.F.	274	11.60	8.95	294.55	325
8170	10″ diameter		73	.329		340	15.65	12.10	367.75	410
8180	12″ diameter		62.50	.384		395	18.30	14.10	427.40	480
8190	14″ diameter		54.30	.442		445	21	16.25	482.25	540
8200	16″ diameter		48.30	.497		490	23.50	18.25	531.75	595
8210	18″ diameter		39.20	.612		620	29	22.50	671.50	755
8220	20″ diameter		31.20	.769		700	36.50	28.50	765	855
8230	24″ diameter		23.80	1.008		855	48	37	940	1,050
8240	26″ diameter		21	1.143		960	54.50	42	1,056.50	1,175
8244	Well casing or drop pipe, PVC, 1/2″ diameter		550	.044		1.31	2.08	1.60	4.99	6.30
8245	3/4″ diameter		550	.044		1.32	2.08	1.60	5	6.35
8246	1″ diameter		550	.044		1.36	2.08	1.60	5.04	6.40
8247	1-1/4″ diameter		520	.046		1.71	2.20	1.70	5.61	7.05
8248	1-1/2″ diameter		490	.049		1.81	2.33	1.80	5.94	7.45
8249	1-3/4″ diameter		380	.063		1.85	3.01	2.32	7.18	9.10
8250	2″ diameter		280	.086		2.12	4.08	3.15	9.35	11.90
8252	3″ diameter		260	.092		4.16	4.40	3.39	11.95	14.90
8254	4″ diameter		205	.117		4.46	5.60	4.30	14.36	18
8255	5″ diameter		170	.141		4.48	6.75	5.20	16.43	20.50
8256	6″ diameter		130	.185		8.60	8.80	6.80	24.20	30
8258	8″ diameter		100	.240		13	11.45	8.80	33.25	41
8260	10″ diameter		73	.329		19.90	15.65	12.10	47.65	59
8262	12″ diameter		62	.387		24	18.45	14.25	56.70	69
8300	Slotted PVC, 1-1/4″ diameter		521	.046		2.89	2.20	1.69	6.78	8.35
8310	1-1/2″ diameter		488	.049		3.27	2.34	1.81	7.42	9.10
8320	2″ diameter		273	.088		4.09	4.19	3.23	11.51	14.35
8330	3″ diameter		253	.095		6.75	4.52	3.49	14.76	18.05
8340	4″ diameter		200	.120		8.35	5.70	4.41	18.46	22.50
8350	5″ diameter		168	.143		15.05	6.80	5.25	27.10	32.50
8360	6″ diameter		126	.190		16.30	9.10	7	32.40	39
8370	8″ diameter		98.50	.244		24	11.60	8.95	44.55	53.50
8400	Artificial gravel pack, 2″ screen, 6″ casing	B-23B	174	.138		4.68	6.55	6.40	17.63	22
8405	8″ casing		111	.216		6.30	10.30	10.05	26.65	33.50
8410	10″ casing		74.50	.322		8.80	15.35	14.95	39.10	49
8415	12″ casing		60	.400		14.05	19.05	18.60	51.70	64.50
8420	14″ casing		50.20	.478		14.65	23	22	59.65	74.50
8425	16″ casing		40.70	.590		18.55	28	27.50	74.05	92.50
8430	18″ casing		36	.667		22.50	32	31	85.50	106
8435	20″ casing		29.50	.814		26	39	38	103	128
8440	24″ casing		25.70	.934		30	44.50	43.50	118	147
8445	26″ casing		24.60	.976		33	46.50	45.50	125	156
8450	30″ casing		20	1.200		38	57	55.50	150.50	189
8455	36″ casing		16.40	1.463		40.50	69.50	68	178	225
8500	Develop well		8	3	Hr.	300	143	139	582	695
8550	Pump test well		8	3		96.50	143	139	378.50	475
8560	Standby well	B-23A	8	3		105	143	110	358	450
8570	Standby, drill rig		8	3			143	110	253	335
8580	Surface seal well, concrete filled		1	24	Ea.	1,025	1,150	880	3,055	3,850
8590	Well test pump, install & remove	B-23	1	40			1,700	1,600	3,300	4,300
8600	Well sterilization, chlorine	2 Clab	1	16		141	675		816	1,150
8610	Well water pressure switch	1 Clab	12	.667		98	28		126	150
8630	Well water pressure switch with manual reset	″	12	.667		130	28		158	185
9950	See Section 31 23 19.40 for wellpoints									

33 11 Groundwater Sources

33 11 13 – Potable Water Supply Wells

33 11 13.10 Wells and Accessories		Crew	Daily Output	Labor-Hours	Unit	Material	2020 Bare Costs Labor	Equipment	Total	Total Incl O&P
9960	See Section 31 23 19.30 for drainage wells									

33 11 13.20 Water Supply Wells, Pumps		Crew	Daily Output	Labor-Hours	Unit	Material	Labor	Equipment	Total	Total Incl O&P
0010	**WATER SUPPLY WELLS, PUMPS**									
0011	With pressure control									
1000	Deep well, jet, 42 gal. galvanized tank									
1040	3/4 HP	1 Plum	.80	10	Ea.	1,125	645		1,770	2,225
3000	Shallow well, jet, 30 gal. galvanized tank									
3040	1/2 HP	1 Plum	2	4	Ea.	915	258		1,173	1,375

33 14 Water Utility Transmission and Distribution

33 14 13 – Public Water Utility Distribution Piping

33 14 13.15 Water Supply, Ductile Iron Pipe		Crew	Daily Output	Labor-Hours	Unit	Material	2020 Bare Costs Labor	Equipment	Total	Total Incl O&P
0010	**WATER SUPPLY, DUCTILE IRON PIPE** R331113-80									
0020	Not including excavation or backfill									
2000	Pipe, class 50 water piping, 18′ lengths									
2020	Mechanical joint, 4″ diameter	B-21A	200	.200	L.F.	42.50	10.45	2.14	55.09	64.50
2040	6″ diameter		160	.250		51	13.05	2.68	66.73	78.50
2060	8″ diameter		133.33	.300		53.50	15.70	3.21	72.41	86
2080	10″ diameter		114.29	.350		71	18.30	3.75	93.05	110
2100	12″ diameter		105.26	.380		90.50	19.85	4.07	114.42	134
2120	14″ diameter		100	.400		105	21	4.28	130.28	152
2140	16″ diameter		72.73	.550		112	29	5.90	146.90	173
2160	18″ diameter		68.97	.580		129	30.50	6.20	165.70	194
2170	20″ diameter		57.14	.700		132	36.50	7.50	176	208
2180	24″ diameter		47.06	.850		168	44.50	9.10	221.60	262
3000	Push-on joint, 4″ diameter		400	.100		23	5.25	1.07	29.32	34.50
3020	6″ diameter		333.33	.120		23.50	6.25	1.29	31.04	37
3040	8″ diameter		200	.200		32.50	10.45	2.14	45.09	54
3060	10″ diameter		181.82	.220		50	11.50	2.36	63.86	75
3080	12″ diameter		160	.250		53	13.05	2.68	68.73	80.50
3100	14″ diameter		133.33	.300		53	15.70	3.21	71.91	85
3120	16″ diameter		114.29	.350		55	18.30	3.75	77.05	92
3140	18″ diameter		100	.400		61	21	4.28	86.28	103
3160	20″ diameter		88.89	.450		63.50	23.50	4.82	91.82	110
3180	24″ diameter	↓	76.92	.520	↓	86.50	27	5.55	119.05	142
8000	Piping, fittings, mechanical joint, AWWA C110									
8006	90° bend, 4″ diameter	B-20A	16	2	Ea.	176	101		277	345
8020	6″ diameter		12.80	2.500		256	126		382	470
8040	8″ diameter	↓	10.67	2.999		500	152		652	775
8060	10″ diameter	B-21A	11.43	3.500		730	183	37.50	950.50	1,125
8080	12″ diameter		10.53	3.799		960	199	40.50	1,199.50	1,400
8100	14″ diameter		10	4		1,375	209	43	1,627	1,875
8120	16″ diameter		7.27	5.502		1,750	288	59	2,097	2,425
8140	18″ diameter		6.90	5.797		2,400	305	62	2,767	3,175
8160	20″ diameter		5.71	7.005		2,875	365	75	3,315	3,775
8180	24″ diameter	↓	4.70	8.511		4,975	445	91	5,511	6,250
8200	Wye or tee, 4″ diameter	B-20A	10.67	2.999		390	152		542	650
8220	6″ diameter		8.53	3.751		585	190		775	925
8240	8″ diameter	↓	7.11	4.501		890	228		1,118	1,325
8260	10″ diameter	B-21A	7.62	5.249		1,175	274	56	1,505	1,775
8280	12″ diameter	↓	7.02	5.698	↓	1,975	298	61	2,334	2,675

33 14 13.15 Water Supply, Ductile Iron Pipe		Crew	Daily Output	Labor-Hours	Unit	Material	Labor (2020 Bare Costs)	Equipment	Total	Total Incl O&P
8300	14″ diameter	B-21A	6.67	5.997	Ea.	2,425	315	64	2,804	3,225
8320	16″ diameter		4.85	8.247		2,750	430	88.50	3,268.50	3,775
8340	18″ diameter		4.60	8.696		4,500	455	93	5,048	5,725
8360	20″ diameter		3.81	10.499		6,525	550	112	7,187	8,125
8380	24″ diameter		3.14	12.739		8,800	665	136	9,601	10,800
8398	45° bend, 4″ diameter	B-20A	16	2		206	101		307	380
8400	6″ diameter		12.80	2.500		295	126		421	515
8405	8″ diameter		10.67	2.999		430	152		582	700
8410	12″ diameter	B-21A	10.53	3.799		875	199	40.50	1,114.50	1,300
8420	16″ diameter		7.27	5.502		1,725	288	59	2,072	2,400
8430	20″ diameter		5.71	7.005		2,450	365	75	2,890	3,325
8440	24″ diameter		4.70	8.511		3,450	445	91	3,986	4,575
8450	Decreaser, 6″ x 4″ diameter	B-20A	14.22	2.250		262	114		376	460
8460	8″ x 6″ diameter	″	11.64	2.749		350	139		489	595
8470	10″ x 6″ diameter	B-21A	13.33	3.001		425	157	32	614	740
8480	12″ x 6″ diameter		12.70	3.150		620	165	33.50	818.50	965
8490	16″ x 6″ diameter		10	4		1,075	209	43	1,327	1,525
8500	20″ x 6″ diameter		8.42	4.751		1,800	248	51	2,099	2,425
8552	For water utility valves see Section 33 14 19									
8700	Joint restraint, ductile iron mechanical joints									
8710	4″ diameter	B-20A	32	1	Ea.	32.50	50.50		83	112
8720	6″ diameter		25.60	1.250		41	63		104	140
8730	8″ diameter		21.33	1.500		59	76		135	179
8740	10″ diameter		18.28	1.751		102	88.50		190.50	246
8750	12″ diameter		16.84	1.900		117	96		213	273
8760	14″ diameter		16	2		177	101		278	345
8770	16″ diameter		11.64	2.749		190	139		329	415
8780	18″ diameter		11.03	2.901		267	147		414	515
8785	20″ diameter		9.14	3.501		330	177		507	630
8790	24″ diameter		7.53	4.250		450	215		665	815
9600	Steel sleeve with tap, 4″ diameter	B-20	3	8		505	375		880	1,125
9620	6″ diameter		2	12		555	565		1,120	1,475
9630	8″ diameter		2	12		720	565		1,285	1,650

33 14 13.20 Water Supply, Polyethylene Pipe, C901		Crew	Daily Output	Labor-Hours	Unit	Material	Labor	Equipment	Total	Total Incl O&P
0010	**WATER SUPPLY, POLYETHYLENE PIPE, C901**									
0020	Not including excavation or backfill									
1000	Piping, 160 psi, 3/4″ diameter	Q-1A	525	.019	L.F.	.40	1.24		1.64	2.29
1120	1″ diameter		485	.021		.58	1.34		1.92	2.64
1140	1-1/2″ diameter		450	.022		.96	1.44		2.40	3.22
1160	2″ diameter		365	.027		1.44	1.78		3.22	4.24
2000	Fittings, insert type, nylon, 160 & 250 psi, cold water									
2220	Clamp ring, stainless steel, 3/4″ diameter	Q-1A	345	.029	Ea.	1.23	1.88		3.11	4.16
2240	1″ diameter		321	.031		1.44	2.02		3.46	4.60
2260	1-1/2″ diameter		285	.035		1.91	2.28		4.19	5.50
2280	2″ diameter		255	.039		2.63	2.54		5.17	6.70
2300	Coupling, 3/4″ diameter		66	.152		1	9.85		10.85	15.80
2320	1″ diameter		57	.175		1.41	11.40		12.81	18.60
2340	1-1/2″ diameter		51	.196		3.18	12.70		15.88	22.50
2360	2″ diameter		48	.208		4.38	13.50		17.88	25
2400	Elbow, 90°, 3/4″ diameter		66	.152		1.76	9.85		11.61	16.65
2420	1″ diameter		57	.175		2.09	11.40		13.49	19.35
2440	1-1/2″ diameter		51	.196		4.41	12.70		17.11	24

33 14 13.20 Water Supply, Polyethylene Pipe, C901		Crew	Daily Output	Labor-Hours	Unit	Material	Labor	Equipment	Total	Total Incl O&P
						2020 Bare Costs				
2460	2″ diameter	Q-1A	48	.208	Ea.	9.55	13.50		23.05	30.50

33 14 13.25 Water Supply, Polyvinyl Chloride Pipe		Crew	Daily Output	Labor-Hours	Unit	Material	Labor	Equipment	Total	Total Incl O&P
0010	WATER SUPPLY, POLYVINYL CHLORIDE PIPE R331113-80									
0020	Not including excavation or backfill, unless specified									
2100	PVC pipe, Class 150, 1-1/2″ diameter	Q-1A	750	.013	L.F.	.61	.86		1.47	1.96
2120	2″ diameter		686	.015		.88	.95		1.83	2.39
2140	2-1/2″ diameter		500	.020		1.18	1.30		2.48	3.24
2160	3″ diameter	B-20	430	.056		1.59	2.62		4.21	5.70
3010	AWWA C905, PR 100, DR 25									
3030	14″ diameter	B-20A	213	.150	L.F.	14.20	7.60		21.80	27
3040	16″ diameter		200	.160		17.30	8.10		25.40	31
3050	18″ diameter		160	.200		27.50	10.10		37.60	45.50
3060	20″ diameter		133	.241		30.50	12.15		42.65	52
3070	24″ diameter		107	.299		45	15.10		60.10	72
3080	30″ diameter		80	.400		74.50	20		94.50	113
3090	36″ diameter		80	.400		115	20		135	158
3100	42″ diameter		60	.533		153	27		180	209
3200	48″ diameter		60	.533		200	27		227	261
3960	Pressure pipe, class 200, ASTM 2241, SDR 21, 3/4″ diameter	Q-1A	1000	.010		.35	.65		1	1.36
3980	1″ diameter		900	.011		.35	.72		1.07	1.47
4000	1-1/2″ diameter		750	.013		1.12	.86		1.98	2.52
4010	2″ diameter		686	.015		1.24	.95		2.19	2.78
4020	2-1/2″ diameter		500	.020		2.32	1.30		3.62	4.49
4030	3″ diameter	B-20A	430	.074		2.40	3.76		6.16	8.30
4040	4″ diameter		375	.085		3.48	4.31		7.79	10.30
4050	6″ diameter		316	.101		7.25	5.10		12.35	15.60
4060	8″ diameter		260	.123		8.80	6.20		15	19
4090	Including trenching to 3′ deep, 3/4″ diameter	Q-1C	300	.080		.35	4.61	6.25	11.21	14.15
4100	1″ diameter		280	.086		.35	4.94	6.70	11.99	15.15
4110	1-1/2″ diameter		260	.092		1.12	5.30	7.20	13.62	17.10
4120	2″ diameter		220	.109		1.24	6.30	8.50	16.04	20
4130	2-1/2″ diameter		200	.120		2.32	6.90	9.35	18.57	23
4140	3″ diameter		175	.137		2.40	7.90	10.70	21	26
4150	4″ diameter		150	.160		3.48	9.20	12.50	25.18	31.50
4160	6″ diameter		125	.192		7.25	11.05	15	33.30	41
4165	Fittings									
4170	Elbow, 90°, 3/4″	Q-1A	114	.088	Ea.	1.22	5.70		6.92	9.85
4180	1″		100	.100		3.58	6.50		10.08	13.65
4190	1-1/2″		80	.125		15.40	8.10		23.50	29
4200	2″		72	.139		19.25	9		28.25	34.50
4210	3″	B-20A	46	.696		25.50	35		60.50	80.50
4220	4″		36	.889		40.50	45		85.50	112
4230	6″		24	1.333		71	67.50		138.50	179
4240	8″		14	2.286		153	116		269	340
4250	Elbow, 45°, 3/4″	Q-1A	114	.088		2.49	5.70		8.19	11.25
4260	1″		100	.100		5.70	6.50		12.20	16
4270	1-1/2″		80	.125		15.15	8.10		23.25	29
4280	2″		72	.139		17.25	9		26.25	32.50
4290	2-1/2″		54	.185		19.55	12		31.55	39.50
4300	3″	B-20A	46	.696		22	35		57	77
4310	4″		36	.889		36	45		81	107
4320	6″		24	1.333		76	67.50		143.50	185

33 14 Water Utility Transmission and Distribution

33 14 13 – Public Water Utility Distribution Piping

33 14 13.25 Water Supply, Polyvinyl Chloride Pipe		Crew	Daily Output	Labor-Hours	Unit	2020 Bare Costs Material	Labor	Equipment	Total	Total Incl O&P
4330	8"	B-20A	14	2.286	Ea.	161	116		277	350
4340	Tee, 3/4"	Q-1A	76	.132		3.08	8.55		11.63	16.15
4350	1"		66	.152		11.90	9.85		21.75	28
4360	1-1/2"		54	.185		25	12		37	45.50
4370	2"		48	.208		24	13.50		37.50	46.50
4380	2-1/2"		36	.278		29	18		47	59
4390	3"	B-20A	30	1.067		34	54		88	118
4400	4"		24	1.333		47	67.50		114.50	153
4410	6"		14.80	2.162		109	109		218	284
4420	8"		9	3.556		250	180		430	545
4430	Coupling, 3/4"	Q-1A	114	.088		1.79	5.70		7.49	10.45
4440	1"		100	.100		7.05	6.50		13.55	17.45
4450	1-1/2"		80	.125		9.80	8.10		17.90	23
4460	2"		72	.139		11.05	9		20.05	25.50
4470	2-1/2"		54	.185		11.40	12		23.40	30.50
4480	3"	B-20A	46	.696		17.40	35		52.40	71.50
4490	4"		36	.889		29.50	45		74.50	100
4500	6"		24	1.333		44.50	67.50		112	150
4510	8"		14	2.286		94	116		210	276
4520	Pressure pipe Class 150, SDR 18, AWWA C900, 4" diameter		380	.084	L.F.	3.03	4.26		7.29	9.75
4530	6" diameter		316	.101		5	5.10		10.10	13.15
4540	8" diameter		264	.121		8.60	6.15		14.75	18.65
4550	10" diameter		220	.145		12.65	7.35		20	25
4560	12" diameter		186	.172		17.75	8.70		26.45	32.50
8000	Fittings with rubber gasket									
8003	Class 150, DR 18									
8006	90° bend , 4" diameter	B-20	100	.240	Ea.	42.50	11.30		53.80	63.50
8020	6" diameter		90	.267		74	12.55		86.55	100
8040	8" diameter		80	.300		139	14.10		153.10	175
8060	10" diameter		50	.480		365	22.50		387.50	435
8080	12" diameter		30	.800		445	37.50		482.50	545
8100	Tee, 4" diameter		90	.267		63	12.55		75.55	88
8120	6" diameter		80	.300		139	14.10		153.10	175
8140	8" diameter		70	.343		199	16.10		215.10	243
8160	10" diameter		40	.600		660	28		688	770
8180	12" diameter		20	1.200		905	56.50		961.50	1,075
8200	45° bend, 4" diameter		100	.240		41.50	11.30		52.80	62.50
8220	6" diameter		90	.267		77	12.55		89.55	103
8240	8" diameter		50	.480		137	22.50		159.50	184
8260	10" diameter		50	.480		305	22.50		327.50	370
8280	12" diameter		30	.800		400	37.50		437.50	495
8300	Reducing tee 6" x 4"		100	.240		126	11.30		137.30	156
8320	8" x 6"		90	.267		202	12.55		214.55	241
8330	10" x 6"		90	.267		340	12.55		352.55	395
8340	10" x 8"		90	.267		365	12.55		377.55	420
8350	12" x 6"		90	.267		420	12.55		432.55	480
8360	12" x 8"		90	.267		440	12.55		452.55	505
8400	Tapped service tee (threaded type) 6" x 6" x 3/4"		100	.240		96	11.30		107.30	123
8430	6" x 6" x 1"		90	.267		96	12.55		108.55	125
8440	6" x 6" x 1-1/2"		90	.267		90.50	12.55		103.05	118
8450	6" x 6" x 2"		90	.267		96	12.55		108.55	125
8460	8" x 8" x 3/4"		90	.267		126	12.55		138.55	158
8470	8" x 8" x 1"		90	.267		126	12.55		138.55	158

33 14 13.25 Water Supply, Polyvinyl Chloride Pipe

	33 14 13.25 Water Supply, Polyvinyl Chloride Pipe	Crew	Daily Output	Labor-Hours	Unit	2020 Bare Costs Material	Labor	Equipment	Total	Total Incl O&P
8480	8" x 8" x 1-1/2"	B-20	90	.267	Ea.	126	12.55		138.55	158
8490	8" x 8" x 2"		90	.267		126	12.55		138.55	158
8500	Repair coupling 4"		100	.240		23.50	11.30		34.80	43
8520	6" diameter		90	.267		36.50	12.55		49.05	59.50
8540	8" diameter		50	.480		90	22.50		112.50	133
8560	10" diameter		50	.480		175	22.50		197.50	227
8580	12" diameter		50	.480		265	22.50		287.50	325
8600	Plug end 4"		100	.240		21	11.30		32.30	40
8620	6" diameter		90	.267		43	12.55		55.55	66
8640	8" diameter		50	.480		63.50	22.50		86	104
8660	10" diameter		50	.480		97.50	22.50		120	141
8680	12" diameter		50	.480		120	22.50		142.50	166
8700	PVC pipe, joint restraint									
8710	4" diameter	B-20A	32	1	Ea.	46.50	50.50		97	128
8720	6" diameter		25.60	1.250		58	63		121	159
8730	8" diameter		21.33	1.500		84.50	76		160.50	207
8740	10" diameter		18.28	1.751		139	88.50		227.50	286
8750	12" diameter		16.84	1.900		146	96		242	305
8760	14" diameter		16	2		201	101		302	375
8770	16" diameter		11.64	2.749		270	139		409	505
8780	18" diameter		11.03	2.901		335	147		482	585
8785	20" diameter		9.14	3.501		445	177		622	755
8790	24" diameter		7.53	4.250		515	215		730	885

33 14 13.35 Water Supply, HDPE

	33 14 13.35 Water Supply, HDPE	Crew	Daily Output	Labor-Hours	Unit	Material	Labor	Equipment	Total	Total Incl O&P
0010	**WATER SUPPLY, HDPE**									
0011	Butt fusion joints, SDR 21 40' lengths not including excavation or backfill									
0100	4" diameter	B-22A	400	.100	L.F.	2.59	4.85	2.01	9.45	12.35
0200	6" diameter		380	.105		5.60	5.10	2.12	12.82	16.20
0300	8" diameter		320	.125		9.60	6.05	2.52	18.17	22.50
0400	10" diameter		300	.133		11.25	6.45	2.68	20.38	25
0500	12" diameter		260	.154		16	7.45	3.10	26.55	32
0600	14" diameter	B-22B	220	.182		15.60	8.80	6.60	31	37.50
0700	16" diameter		180	.222		18.30	10.75	8.10	37.15	45
0800	18" diameter		140	.286		28	13.85	10.40	52.25	63.50
0850	20" diameter		130	.308		34	14.90	11.20	60.10	72.50
0900	24" diameter		100	.400		55.50	19.40	14.55	89.45	106
1000	Fittings									
1100	Elbows, 90 degrees									
1200	4" diameter	B-22A	32	1.250	Ea.	17.25	60.50	25	102.75	138
1300	6" diameter		28	1.429		47.50	69	28.50	145	188
1400	8" diameter		24	1.667		113	81	33.50	227.50	282
1500	10" diameter		18	2.222		254	108	44.50	406.50	490
1600	12" diameter		12	3.333		297	162	67	526	640
1700	14" diameter	B-22B	9	4.444		630	215	162	1,007	1,200
1800	16" diameter		6	6.667		935	325	242	1,502	1,775
1900	18" diameter		4	10		1,050	485	365	1,900	2,275
2000	24" diameter		3	13.333		1,825	645	485	2,955	3,500
2100	Tees									
2200	4" diameter	B-22A	30	1.333	Ea.	22.50	64.50	27	114	152
2300	6" diameter		26	1.538		74	74.50	31	179.50	227
2400	8" diameter		22	1.818		130	88	36.50	254.50	315
2500	10" diameter		15	2.667		172	129	53.50	354.50	440

33 14 Water Utility Transmission and Distribution

33 14 13 – Public Water Utility Distribution Piping

33 14 13.35 Water Supply, HDPE		Crew	Daily Output	Labor-Hours	Unit	Material	2020 Bare Costs Labor	Equipment	Total	Total Incl O&P
2600	12″ diameter	B-22A	10	4	Ea.	400	194	80.50	674.50	820
2700	14″ diameter	B-22B	8	5		475	242	182	899	1,075
2800	16″ diameter		6	6.667		555	325	242	1,122	1,350
2900	18″ diameter		4	10		670	485	365	1,520	1,875
3000	24″ diameter		2	20		895	970	725	2,590	3,225
4100	Caps									
4110	4″ diameter	B-22A	34	1.176	Ea.	15.40	57	23.50	95.90	128
4120	6″ diameter		30	1.333		33.50	64.50	27	125	164
4130	8″ diameter		26	1.538		57	74.50	31	162.50	209
4150	10″ diameter		20	2		139	97	40	276	345
4160	12″ diameter		14	2.857		168	138	57.50	363.50	455

33 14 13.45 Water Supply, Copper Pipe

Line	Description	Crew	Daily Output	Labor-Hours	Unit	Material	Labor	Equipment	Total	Total Incl O&P
0010	**WATER SUPPLY, COPPER PIPE**									
0020	Not including excavation or backfill									
2000	Tubing, type K, 20′ joints, 3/4″ diameter	Q-1	400	.040	L.F.	7.05	2.32		9.37	11.20
2200	1″ diameter		320	.050		9.70	2.90		12.60	15
3000	1-1/2″ diameter		265	.060		14.95	3.50		18.45	21.50
3020	2″ diameter		230	.070		23	4.04		27.04	31.50
3040	2-1/2″ diameter		146	.110		35	6.35		41.35	48
3060	3″ diameter		134	.119		48.50	6.95		55.45	63.50
4012	4″ diameter		95	.168		80	9.75		89.75	103
4016	6″ diameter	Q-2	80	.300		134	18.05		152.05	175
4018	8″ diameter	″	80	.300		325	18.05		343.05	380
5000	Tubing, type L									
5108	2″ diameter	Q-1	230	.070	L.F.	14.70	4.04		18.74	22
6010	3″ diameter		134	.119		36.50	6.95		43.45	51
6012	4″ diameter		95	.168		48	9.75		57.75	67.50
6016	6″ diameter	Q-2	80	.300		112	18.05		130.05	150
7165	Fittings, brass, corporation stops, no lead, 3/4″ diameter	1 Plum	19	.421	Ea.	76.50	27		103.50	125
7166	1″ diameter		16	.500		98	32		130	156
7167	1-1/2″ diameter		13	.615		208	39.50		247.50	289
7168	2″ diameter		11	.727		330	47		377	430
7170	Curb stops, no lead, 3/4″ diameter		19	.421		96	27		123	146
7171	1″ diameter		16	.500		142	32		174	204
7172	1-1/2″ diameter		13	.615		269	39.50		308.50	355
7173	2″ diameter		11	.727		325	47		372	425

33 14 17 – Site Water Utility Service Laterals

33 14 17.15 Tapping, Crosses and Sleeves

Line	Description	Crew	Daily Output	Labor-Hours	Unit	Material	Labor	Equipment	Total	Total Incl O&P
0010	**TAPPING, CROSSES AND SLEEVES**									
4000	Drill and tap pressurized main (labor only)									
4100	6″ main, 1″ to 2″ service	Q-1	3	5.333	Ea.		310		310	465
4150	8″ main, 1″ to 2″ service	″	2.75	5.818	″		335		335	505
4500	Tap and insert gate valve									
4600	8″ main, 4″ branch	B-21	3.20	8.750	Ea.		425	59	484	705
4650	6″ branch		2.70	10.370			505	70	575	835
4700	10″ main, 4″ branch		2.70	10.370			505	70	575	835
4750	6″ branch		2.35	11.915			580	80.50	660.50	965
4800	12″ main, 6″ branch		2.35	11.915			580	80.50	660.50	965
4850	8″ branch		2.35	11.915			580	80.50	660.50	965
7020	Crosses, 4″ x 4″		37	.757		1,450	37	5.10	1,492.10	1,625
7030	6″ x 4″		25	1.120		1,650	54.50	7.55	1,712.05	1,925
7040	6″ x 6″		25	1.120		1,725	54.50	7.55	1,787.05	2,000

33 14 17.15 Tapping, Crosses and Sleeves		Crew	Daily Output	Labor-Hours	Unit	2020 Bare Costs Material	Labor	Equipment	Total	Total Incl O&P
7060	8″ x 6″	B-21	21	1.333	Ea.	2,050	65	9	2,124	2,350
7080	8″ x 8″		21	1.333		2,175	65	9	2,249	2,500
7100	10″ x 6″		21	1.333		4,150	65	9	4,224	4,675
7120	10″ x 10″		21	1.333		4,475	65	9	4,549	5,025
7140	12″ x 6″		18	1.556		4,125	76	10.50	4,211.50	4,675
7160	12″ x 12″		18	1.556		5,300	76	10.50	5,386.50	5,950
7180	14″ x 6″		16	1.750		10,200	85.50	11.80	10,297.30	11,300
7200	14″ x 14″		16	1.750		10,800	85.50	11.80	10,897.30	12,000
7220	16″ x 6″		14	2		10,900	97.50	13.50	11,011	12,200
7240	16″ x 10″		14	2		11,100	97.50	13.50	11,211	12,400
7260	16″ x 16″		14	2		11,700	97.50	13.50	11,811	13,000
7280	18″ x 6″		10	2.800		16,400	137	18.90	16,555.90	18,200
7300	18″ x 12″		10	2.800		16,500	137	18.90	16,655.90	18,400
7320	18″ x 18″		10	2.800		17,100	137	18.90	17,255.90	19,000
7340	20″ x 6″		8	3.500		13,200	171	23.50	13,394.50	14,900
7360	20″ x 12″		8	3.500		14,200	171	23.50	14,394.50	15,900
7380	20″ x 20″		8	3.500		21,000	171	23.50	21,194.50	23,400
7400	24″ x 6″		6	4.667		17,100	228	31.50	17,359.50	19,200
7420	24″ x 12″		6	4.667		17,400	228	31.50	17,659.50	19,600
7440	24″ x 18″		6	4.667		27,100	228	31.50	27,359.50	30,200
7460	24″ x 24″		6	4.667		27,300	228	31.50	27,559.50	30,500
7600	Cut-in sleeves with rubber gaskets, 4″		18	1.556		560	76	10.50	646.50	740
7620	6″		12	2.333		705	114	15.75	834.75	965
7640	8″		10	2.800		895	137	18.90	1,050.90	1,200
7660	10″		10	2.800		1,250	137	18.90	1,405.90	1,600
7680	12″		9	3.111		1,800	152	21	1,973	2,225
7800	Cut-in valves with rubber gaskets, 4″		18	1.556		560	76	10.50	646.50	740
7820	6″		12	2.333		715	114	15.75	844.75	975
7840	8″		10	2.800		1,000	137	18.90	1,155.90	1,325
7860	10″		10	2.800		1,775	137	18.90	1,930.90	2,175
7880	12″		9	3.111		2,350	152	21	2,523	2,825
7900	Tapping valve 4″, MJ, ductile iron		18	1.556		455	76	10.50	541.50	625
7920	6″, MJ, ductile iron		12	2.333		620	114	15.75	749.75	870
7940	Tapping valve 8″, MJ, ductile iron		10	2.800		840	137	18.90	995.90	1,150
7960	Tapping valve 10″, MJ, ductile iron		10	2.800		1,175	137	18.90	1,330.90	1,525
7980	Tapping valve 12″, MJ, ductile iron		8	3.500		1,875	171	23.50	2,069.50	2,325
8000	Sleeves with rubber gaskets, 4″ x 4″		37	.757		1,050	37	5.10	1,092.10	1,200
8010	6″ x 4″		25	1.120		985	54.50	7.55	1,047.05	1,175
8020	6″ x 6″		25	1.120		1,300	54.50	7.55	1,362.05	1,525
8030	8″ x 4″		21	1.333		1,025	65	9	1,099	1,225
8040	8″ x 6″		21	1.333		910	65	9	984	1,100
8060	8″ x 8″		21	1.333		1,350	65	9	1,424	1,575
8070	10″ x 4″		21	1.333		830	65	9	904	1,025
8080	10″ x 6″		21	1.333		1,025	65	9	1,099	1,225
8090	10″ x 8″		21	1.333		1,525	65	9	1,599	1,800
8100	10″ x 10″		21	1.333		2,225	65	9	2,299	2,550
8110	12″ x 4″		18	1.556		1,375	76	10.50	1,461.50	1,625
8120	12″ x 6″		18	1.556		1,425	76	10.50	1,511.50	1,675
8130	12″ x 8″		18	1.556		1,500	76	10.50	1,586.50	1,775
8135	12″ x 10″		18	1.556		2,600	76	10.50	2,686.50	2,975
8140	12″ x 12″		18	1.556		2,725	76	10.50	2,811.50	3,125
8160	14″ x 6″		16	1.750		1,600	85.50	11.80	1,697.30	1,900
8180	14″ x 14″		16	1.750		2,575	85.50	11.80	2,672.30	3,000

33 14 Water Utility Transmission and Distribution

33 14 17 – Site Water Utility Service Laterals

33 14 17.15 Tapping, Crosses and Sleeves		Crew	Daily Output	Labor-Hours	Unit	2020 Bare Costs Material	Labor	Equipment	Total	Total Incl O&P
8200	16″ x 6″	B-21	14	2	Ea.	1,875	97.50	13.50	1,986	2,225
8220	16″ x 10″		14	2		3,750	97.50	13.50	3,861	4,275
8240	16″ x 16″		14	2		9,700	97.50	13.50	9,811	10,900
8260	18″ x 6″		10	2.800		1,150	137	18.90	1,305.90	1,500
8280	18″ x 12″		10	2.800		2,925	137	18.90	3,080.90	3,425
8300	18″ x 18″		10	2.800		8,650	137	18.90	8,805.90	9,750
8320	20″ x 6″		8	3.500		2,425	171	23.50	2,619.50	2,950
8340	20″ x 12″		8	3.500		4,750	171	23.50	4,944.50	5,500
8360	20″ x 20″		8	3.500		10,600	171	23.50	10,794.50	11,900
8380	24″ x 6″		6	4.667		2,325	228	31.50	2,584.50	2,925
8400	24″ x 12″		6	4.667		6,350	228	31.50	6,609.50	7,350
8420	24″ x 18″		6	4.667		11,300	228	31.50	11,559.50	12,800
8440	24″ x 24″		6	4.667		13,300	228	31.50	13,559.50	15,000
8800	Hydrant valve box, 6′ long	B-20	20	1.200		230	56.50		286.50	340
8820	8′ long		18	1.333		305	62.50		367.50	430
8830	Valve box w/lid 4′ deep		14	1.714		96.50	80.50		177	227
8840	Valve box and large base w/lid		14	1.714		325	80.50		405.50	480

33 14 19 – Valves and Hydrants for Water Utility Service

33 14 19.10 Valves		Crew	Daily Output	Labor-Hours	Unit	Material	Labor	Equipment	Total	Total Incl O&P
0010	**VALVES**, water distribution									
0011	See Sections 22 05 23.20 and 22 05 23.60									
3000	Butterfly valves with boxes, cast iron, mechanical joint									
3100	4″ diameter	B-6	6	4	Ea.	550	183	35.50	768.50	925
3140	6″ diameter		6	4		725	183	35.50	943.50	1,100
3180	8″ diameter		6	4		910	183	35.50	1,128.50	1,325
3300	10″ diameter		6	4		1,475	183	35.50	1,693.50	1,925
3340	12″ diameter		6	4		1,650	183	35.50	1,868.50	2,150
3400	14″ diameter		4	6		2,825	274	53.50	3,152.50	3,575
3440	16″ diameter		4	6		4,150	274	53.50	4,477.50	5,050
3460	18″ diameter		4	6		4,950	274	53.50	5,277.50	5,925
3480	20″ diameter		4	6		6,525	274	53.50	6,852.50	7,650
3500	24″ diameter		4	6		10,800	274	53.50	11,127.50	12,400
3510	30″ diameter		4	6		11,600	274	53.50	11,927.50	13,300
3520	36″ diameter		4	6		14,400	274	53.50	14,727.50	16,300
3530	42″ diameter		4	6		18,900	274	53.50	19,227.50	21,300
3540	48″ diameter		4	6		24,300	274	53.50	24,627.50	27,200
3600	With lever operator									
3610	4″ diameter	B-6	6	4	Ea.	455	183	35.50	673.50	815
3614	6″ diameter		6	4		625	183	35.50	843.50	1,000
3616	8″ diameter		6	4		815	183	35.50	1,033.50	1,200
3618	10″ diameter		6	4		1,375	183	35.50	1,593.50	1,825
3620	12″ diameter		6	4		1,550	183	35.50	1,768.50	2,050
3622	14″ diameter		4	6		2,500	274	53.50	2,827.50	3,225
3624	16″ diameter		4	6		3,825	274	53.50	4,152.50	4,675
3626	18″ diameter		4	6		4,625	274	53.50	4,952.50	5,575
3628	20″ diameter		4	6		6,200	274	53.50	6,527.50	7,300
3630	24″ diameter		4	6		10,500	274	53.50	10,827.50	12,100
3700	Check valves, flanged									
3710	4″ diameter	B-6	6	4	Ea.	785	183	35.50	1,003.50	1,175
3714	6″ diameter		6	4		1,325	183	35.50	1,543.50	1,775
3716	8″ diameter		6	4		2,525	183	35.50	2,743.50	3,100
3718	10″ diameter		6	4		4,275	183	35.50	4,493.50	5,025

33 14 Water Utility Transmission and Distribution

33 14 19 – Valves and Hydrants for Water Utility Service

33 14 19.10 Valves		Crew	Daily Output	Labor-Hours	Unit	Material	Labor	Equipment	Total	Total Incl O&P
						2020 Bare Costs				
3720	12" diameter	B-6	6	4	Ea.	7,000	183	35.50	7,218.50	8,025
3726	18" diameter		4	6		25,900	274	53.50	26,227.50	29,000
3730	24" diameter		4	6		66,000	274	53.50	66,327.50	73,000
3800	Gate valves, C.I., 125 psi, mechanical joint, w/boxes									
3810	4" diameter	B-6	6	4	Ea.	780	183	35.50	998.50	1,175
3814	6" diameter		6	4		1,175	183	35.50	1,393.50	1,625
3816	8" diameter		6	4		2,025	183	35.50	2,243.50	2,550
3818	10" diameter		6	4		3,625	183	35.50	3,843.50	4,300
3820	12" diameter		6	4		4,900	183	35.50	5,118.50	5,700
3822	14" diameter		4	6		10,700	274	53.50	11,027.50	12,300
3824	16" diameter		4	6		15,100	274	53.50	15,427.50	17,100
3826	18" diameter		4	6		24,500	274	53.50	24,827.50	27,500
3828	20" diameter		4	6		25,600	274	53.50	25,927.50	28,700
3830	24" diameter		4	6		38,300	274	53.50	38,627.50	42,700
3831	30" diameter		4	6		59,000	274	53.50	59,327.50	65,500
3832	36" diameter		4	6		86,000	274	53.50	86,327.50	95,000
3880	Sleeve, for tapping mains, 8" x 4", add					1,025			1,025	1,125
3884	10" x 6", add					1,025			1,025	1,125
3888	12" x 6", add					1,425			1,425	1,550
3892	12" x 8", add					1,500			1,500	1,650

33 14 19.20 Valves		Crew	Daily Output	Labor-Hours	Unit	Material	Labor	Equipment	Total	Total Incl O&P
0010	**VALVES**									
0011	Special trim or use									
9000	Valves, gate valve, N.R.S. PIV with post, 4" diameter	B-6	6	4	Ea.	2,000	183	35.50	2,218.50	2,525
9020	6" diameter		6	4		2,400	183	35.50	2,618.50	2,975
9040	8" diameter		6	4		3,250	183	35.50	3,468.50	3,875
9060	10" diameter		6	4		4,850	183	35.50	5,068.50	5,650
9080	12" diameter		6	4		6,125	183	35.50	6,343.50	7,050
9100	14" diameter		6	4		11,700	183	35.50	11,918.50	13,200
9120	OS&Y, 4" diameter		6	4		920	183	35.50	1,138.50	1,325
9140	6" diameter		6	4		1,275	183	35.50	1,493.50	1,725
9160	8" diameter		6	4		1,650	183	35.50	1,868.50	2,150
9180	10" diameter		6	4		2,575	183	35.50	2,793.50	3,175
9200	12" diameter		6	4		4,025	183	35.50	4,243.50	4,750
9220	14" diameter		4	6		6,150	274	53.50	6,477.50	7,250
9400	Check valves, rubber disc, 2-1/2" diameter		6	4		475	183	35.50	693.50	840
9420	3" diameter		6	4		595	183	35.50	813.50	970
9440	4" diameter		6	4		785	183	35.50	1,003.50	1,175
9480	6" diameter		6	4		1,325	183	35.50	1,543.50	1,775
9500	8" diameter		6	4		2,525	183	35.50	2,743.50	3,100
9520	10" diameter		6	4		4,275	183	35.50	4,493.50	5,025
9540	12" diameter		6	4		7,000	183	35.50	7,218.50	8,025
9542	14" diameter		4	6		6,750	274	53.50	7,077.50	7,900
9700	Detector check valves, reducing, 4" diameter		6	4		1,125	183	35.50	1,343.50	1,575
9720	6" diameter		6	4		1,925	183	35.50	2,143.50	2,450
9740	8" diameter		6	4		2,875	183	35.50	3,093.50	3,500
9760	10" diameter		6	4		5,025	183	35.50	5,243.50	5,850
9800	Galvanized, 4" diameter		6	4		2,025	183	35.50	2,243.50	2,550
9820	6" diameter		6	4		2,700	183	35.50	2,918.50	3,275
9840	8" diameter		6	4		4,025	183	35.50	4,243.50	4,750
9860	10" diameter		6	4		5,425	183	35.50	5,643.50	6,275

33 14 Water Utility Transmission and Distribution

33 14 19 – Valves and Hydrants for Water Utility Service

	33 14 19.30 Fire Hydrants	Crew	Daily Output	Labor-Hours	Unit	Material (2020 Bare Costs)	Labor (2020 Bare Costs)	Equipment (2020 Bare Costs)	Total (2020 Bare Costs)	Total Incl O&P
0010	**FIRE HYDRANTS** G3010–410									
0020	Mechanical joints unless otherwise noted									
1000	Fire hydrants, two way; excavation and backfill not incl.									
1100	4-1/2" valve size, depth 2'-0"	B-21	10	2.800	Ea.	2,900	137	18.90	3,055.90	3,400
1200	4'-6"		9	3.111		3,000	152	21	3,173	3,550
1260	6'-0"		7	4		3,400	195	27	3,622	4,075
1340	8'-0"		6	4.667		2,775	228	31.50	3,034.50	3,425
1420	10'-0"		5	5.600		4,000	273	38	4,311	4,875
2000	5-1/4" valve size, depth 2'-0"		10	2.800		2,375	137	18.90	2,530.90	2,850
2080	4'-0"		9	3.111		2,700	152	21	2,873	3,225
2160	6'-0"		7	4		2,900	195	27	3,122	3,500
2240	8'-0"		6	4.667		3,325	228	31.50	3,584.50	4,050
2320	10'-0"		5	5.600		3,700	273	38	4,011	4,525
2350	For threeway valves, add					7%				
2400	Lower barrel extensions with stems, 1'-0"	B-20	14	1.714		360	80.50		440.50	515
2440	2'-0"		13	1.846		440	87		527	615
2480	3'-0"		12	2		915	94		1,009	1,150
2520	4'-0"		10	2.400		835	113		948	1,075
5000	Indicator post									
5020	Adjustable, valve size 4" to 14", 4' bury	B-21	10	2.800	Ea.	1,475	137	18.90	1,630.90	1,850
5060	8' bury		7	4		1,325	195	27	1,547	1,775
5080	10' bury		6	4.667		1,775	228	31.50	2,034.50	2,325
5100	12' bury		5	5.600		1,900	273	38	2,211	2,550
5120	14' bury		4	7		1,825	340	47.50	2,212.50	2,575
5500	Non-adjustable, valve size 4" to 14", 3' bury		10	2.800		925	137	18.90	1,080.90	1,250
5520	3'-6" bury		10	2.800		925	137	18.90	1,080.90	1,250
5540	4' bury		9	3.111		925	152	21	1,098	1,275

33 16 Water Utility Storage Tanks

33 16 36 – Ground-Level Reinforced Concrete Water Storage Tanks

	33 16 36.16 Prestressed Conc. Water Storage Tanks	Crew	Daily Output	Labor-Hours	Unit	Material	Labor	Equipment	Total	Total Incl O&P
0010	**PRESTRESSED CONC. WATER STORAGE TANKS**									
0020	Not including fdn., pipe or pumps, 250,000 gallons				Ea.				299,000	329,500

33 31 Sanitary Sewerage Piping

33 31 11 – Public Sanitary Sewerage Gravity Piping

	33 31 11.10 Sewage Collection, Valves	Crew	Daily Output	Labor-Hours	Unit	Material	Labor	Equipment	Total	Total Incl O&P
0010	**SEWAGE COLLECTION, VALVES**									
1000	Backwater sewer line valve									
1010	Offset type, bronze swing check assy and bronze cover									
1020	B&S connections									
1040	2" size	2 Skwk	14	1.143	Ea.	770	62.50		832.50	940
1050	3" size		12.50	1.280		1,050	70		1,120	1,250
1060	4" size		11	1.455		1,150	80		1,230	1,375
1070	6" size	3 Skwk	15	1.600		1,825	88		1,913	2,125
1080	8" size	"	7.50	3.200		2,350	176		2,526	2,850
1110	Backwater drainage control									
1120	Offset type, w/bronze manually operated shear gate									
1130	B&S connections									

33 31 Sanitary Sewerage Piping

33 31 11 – Public Sanitary Sewerage Gravity Piping

33 31 11.10 Sewage Collection, Valves		Crew	Daily Output	Labor-Hours	Unit	2020 Bare Costs Material	Labor	Equipment	Total	Total Incl O&P
1160	4″ size	2 Skwk	11	1.455	Ea.	2,150	80		2,230	2,475
1170	6″ size	3 Skwk	15	1.600	″	3,250	88		3,338	3,700
33 31 11.15 Sewage Collection, Concrete Pipe										
0010	**SEWAGE COLLECTION, CONCRETE PIPE**									
0020	See Section 33 42 11.60 for sewage/drainage collection, concrete pipe									
33 31 11.25 Sewage Collection, Polyvinyl Chloride Pipe										
0010	**SEWAGE COLLECTION, POLYVINYL CHLORIDE PIPE**									
0020	Not including excavation or backfill									
2000	20′ lengths, SDR 35, B&S, 4″ diameter	B-20	375	.064	L.F.	1.67	3.01		4.68	6.35
2040	6″ diameter	″	350	.069		4.05	3.22		7.27	9.30
2120	10″ diameter	B-21	330	.085		11.30	4.14	.57	16.01	19.30
2160	12″ diameter		320	.088		15.90	4.27	.59	20.76	24.50
2170	14″ diameter		280	.100		19.45	4.88	.68	25.01	29.50
2200	15″ diameter		240	.117		22	5.70	.79	28.49	33.50
2250	16″ diameter		220	.127		29	6.20	.86	36.06	42
4000	Piping, DWV PVC, no exc./bkfill., 10′ L, Sch 40, 4″ diameter	B-20	375	.064		3.83	3.01		6.84	8.75
4010	6″ diameter		350	.069		8.25	3.22		11.47	13.90
4020	8″ diameter		335	.072		13.40	3.37		16.77	19.80

33 34 Onsite Wastewater Disposal

33 34 13 – Septic Tanks

33 34 13.13 Concrete Septic Tanks		Crew	Daily Output	Labor-Hours	Unit	Material	Labor	Equipment	Total	Total Incl O&P
0010	**CONCRETE SEPTIC TANKS** G3020–300									
0011	Not including excavation or piping									
0015	Septic tanks, precast, 1,000 gallon	B-21	8	3.500	Ea.	1,050	171	23.50	1,244.50	1,450
0060	1,500 gallon		7	4		1,625	195	27	1,847	2,125
0100	2,000 gallon		5	5.600		2,500	273	38	2,811	3,200
0200	5,000 gallon	B-13	3.50	16		6,325	735	166	7,226	8,225
0300	15,000 gallon, 4 piece	B-13B	1.70	32.941		23,700	1,525	575	25,800	29,000
0400	25,000 gallon, 4 piece		1.10	50.909		47,900	2,350	890	51,140	57,000
0500	40,000 gallon, 4 piece		.80	70		53,500	3,225	1,225	57,950	64,500
0520	50,000 gallon, 5 piece	B-13C	.60	93.333		61,500	4,300	3,800	69,600	78,000
0640	75,000 gallon, cast in place	C-14C	.25	448		74,500	22,600	108	97,208	116,000
0660	100,000 gallon	″	.15	747		92,500	37,600	179	130,279	158,000
1150	Leaching field chambers, 13′ x 3′-7″ x 1′-4″, standard	B-13	16	3.500		500	161	36.50	697.50	830
1200	Heavy duty, 8′ x 4′ x 1′-6″		14	4		310	184	41.50	535.50	660
1300	13′ x 3′-9″ x 1′-6″		12	4.667		1,350	215	48.50	1,613.50	1,850
1350	20′ x 4′ x 1′-6″		5	11.200		1,225	515	116	1,856	2,250
1400	Leaching pit, precast concrete, 3′ diameter, 3′ deep	B-21	8	3.500		735	171	23.50	929.50	1,100
1500	6′ diameter, 3′ section		4.70	5.957		940	290	40	1,270	1,500
2000	Velocity reducing pit, precast conc., 6′ diameter, 3′ deep		4.70	5.957		1,825	290	40	2,155	2,500
33 34 13.33 Polyethylene Septic Tanks										
0010	**POLYETHYLENE SEPTIC TANKS**									
0015	High density polyethylene, 1,000 gallon	B-21	8	3.500	Ea.	1,475	171	23.50	1,669.50	1,900
0020	1,250 gallon		8	3.500		1,325	171	23.50	1,519.50	1,725
0025	1,500 gallon		7	4		1,375	195	27	1,597	1,850

33 34 Onsite Wastewater Disposal

33 34 16 – Septic Tank Effluent Filters

33 34 16.13 Septic Tank Gravity Effluent Filters		Crew	Daily Output	Labor-Hours	Unit	Material	Labor	Equipment	Total	Total Incl O&P
						2020 Bare Costs				
0010	SEPTIC TANK GRAVITY EFFLUENT FILTERS									
3000	Effluent filter, 4" diameter	1 Skwk	8	1	Ea.	44	55		99	131
3020	6" diameter		7	1.143		42.50	62.50		105	142
3030	8" diameter		7	1.143		252	62.50		314.50	370
3040	8" diameter, very fine		7	1.143		500	62.50		562.50	645
3050	10" diameter, very fine		6	1.333		237	73		310	370
3060	10" diameter		6	1.333		276	73		349	415
3080	12" diameter		6	1.333		655	73		728	830
3090	15" diameter		5	1.600		1,075	88		1,163	1,325

33 34 51 – Drainage Field Systems

33 34 51.10 Drainage Field Excavation and Fill		Crew	Daily Output	Labor-Hours	Unit	Material	Labor	Equipment	Total	Total Incl O&P
0010	DRAINAGE FIELD EXCAVATION AND FILL									
2200	Septic tank & drainage field excavation with 3/4 C.Y. backhoe	B-12F	145	.110	C.Y.		5.60	4.79	10.39	13.60
2400	4' trench for disposal field, 3/4 C.Y. backhoe	"	335	.048	L.F.		2.42	2.07	4.49	5.90
2600	Gravel fill, run of bank	B-6	150	.160	C.Y.	16.90	7.30	1.43	25.63	31
2800	Crushed stone, 3/4"	"	150	.160	"	42	7.30	1.43	50.73	58.50

33 34 51.13 Utility Septic Tank Tile Drainage Field		Crew	Daily Output	Labor-Hours	Unit	Material	Labor	Equipment	Total	Total Incl O&P
0010	UTILITY SEPTIC TANK TILE DRAINAGE FIELD									
0015	Distribution box, concrete, 5 outlets	2 Clab	20	.800	Ea.	93.50	33.50		127	154
0020	7 outlets		16	1		103	42		145	178
0025	9 outlets		8	2		545	84		629	725
0115	Distribution boxes, HDPE, 5 outlets		20	.800		76.50	33.50		110	135
0117	6 outlets		15	1.067		81	45		126	157
0118	7 outlets		15	1.067		75	45		120	151
0120	8 outlets		10	1.600		79	67.50		146.50	188
0240	Distribution boxes, outlet flow leveler	1 Clab	50	.160		2.53	6.75		9.28	12.90
0300	Precast concrete, galley, 4' x 4' x 4'	B-21	16	1.750		246	85.50	11.80	343.30	410
0350	HDPE infiltration chamber 12" H x 15" W	2 Clab	300	.053	L.F.	7	2.25		9.25	11.05
0351	12" H x 15" W end cap	1 Clab	32	.250	Ea.	18.90	10.55		29.45	37
0355	chamber 12" H x 22" W	2 Clab	300	.053	L.F.	6.65	2.25		8.90	10.70
0356	12" H x 22" W end cap	1 Clab	32	.250	Ea.	16.60	10.55		27.15	34
0360	chamber 13" H x 34" W	2 Clab	300	.053	L.F.	14.30	2.25		16.55	19.10
0361	13" H x 34" W end cap	1 Clab	32	.250	Ea.	50.50	10.55		61.05	72
0365	chamber 16" H x 34" W	2 Clab	300	.053	L.F.	19.25	2.25		21.50	24.50
0366	16" H x 34" W end cap	1 Clab	32	.250	Ea.	16.55	10.55		27.10	34
0370	chamber 8" H x 16" W	2 Clab	300	.053	L.F.	9.90	2.25		12.15	14.25
0371	8" H x 16" W end cap	1 Clab	32	.250	Ea.	11.10	10.55		21.65	28

33 41 Subdrainage

33 41 16 – Subdrainage Piping

33 41 16.25 Piping, Subdrainage, Corrugated Metal		Crew	Daily Output	Labor-Hours	Unit	Material	Labor	Equipment	Total	Total Incl O&P
0010	PIPING, SUBDRAINAGE, CORRUGATED METAL G1030–805									
0021	Not including excavation and backfill									
2010	Aluminum, perforated									
2020	6" diameter, 18 ga.	B-20	380	.063	L.F.	6.40	2.97		9.37	11.50
2200	8" diameter, 16 ga.	"	370	.065		8.80	3.05		11.85	14.30
2220	10" diameter, 16 ga.	B-21	360	.078		11	3.79	.53	15.32	18.40
2240	12" diameter, 16 ga.		285	.098		12.30	4.79	.66	17.75	21.50
2260	18" diameter, 16 ga.		205	.137		18.50	6.65	.92	26.07	31.50
3000	Uncoated galvanized, perforated									

33 41 Subdrainage

33 41 16 – Subdrainage Piping

33 41 16.25 Piping, Subdrainage, Corrugated Metal		Crew	Daily Output	Labor-Hours	Unit	Material	2020 Bare Costs Labor	Equipment	Total	Total Incl O&P
3020	6″ diameter, 18 ga.	B-20	380	.063	L.F.	6.65	2.97		9.62	11.75
3200	8″ diameter, 16 ga.	″	370	.065		8.10	3.05		11.15	13.50
3220	10″ diameter, 16 ga.	B-21	360	.078		8.60	3.79	.53	12.92	15.75
3240	12″ diameter, 16 ga.		285	.098		9.55	4.79	.66	15	18.50
3260	18″ diameter, 16 ga.		205	.137		14.65	6.65	.92	22.22	27
4000	Steel, perforated, asphalt coated									
4020	6″ diameter, 18 ga.	B-20	380	.063	L.F.	6.35	2.97		9.32	11.45
4030	8″ diameter, 18 ga.	″	370	.065		9	3.05		12.05	14.50
4040	10″ diameter, 16 ga.	B-21	360	.078		10.05	3.79	.53	14.37	17.35
4050	12″ diameter, 16 ga.		285	.098		11.30	4.79	.66	16.75	20.50
4060	18″ diameter, 16 ga.		205	.137		17.80	6.65	.92	25.37	30.50

33 41 16.30 Piping, Subdrainage, Plastic		Crew	Daily Output	Labor-Hours	Unit	Material	Labor	Equipment	Total	Total Incl O&P
0010	**PIPING, SUBDRAINAGE, PLASTIC**									
0020	Not including excavation and backfill									
2100	Perforated PVC, 4″ diameter	B-14	314	.153	L.F.	1.67	6.75	.68	9.10	12.75
2110	6″ diameter		300	.160		4.05	7.10	.71	11.86	15.90
2120	8″ diameter		290	.166		6.10	7.30	.74	14.14	18.50
2130	10″ diameter		280	.171		10.45	7.60	.76	18.81	23.50
2140	12″ diameter		270	.178		14.75	7.85	.79	23.39	29

33 42 Stormwater Conveyance

33 42 11 – Stormwater Gravity Piping

33 42 11.40 Piping, Storm Drainage, Corrugated Metal		Crew	Daily Output	Labor-Hours	Unit	Material	Labor	Equipment	Total	Total Incl O&P
0010	**PIPING, STORM DRAINAGE, CORRUGATED METAL**									
0020	Not including excavation or backfill									
2000	Corrugated metal pipe, galvanized									
2020	Bituminous coated with paved invert, 20′ lengths									
2040	8″ diameter, 16 ga.	B-14	330	.145	L.F.	8.35	6.45	.65	15.45	19.50
2060	10″ diameter, 16 ga.		260	.185		8.65	8.15	.82	17.62	22.50
2080	12″ diameter, 16 ga.		210	.229		12.40	10.10	1.02	23.52	30
2100	15″ diameter, 16 ga.		200	.240		14.60	10.60	1.07	26.27	33
2120	18″ diameter, 16 ga.		190	.253		18.80	11.20	1.12	31.12	38.50
2140	24″ diameter, 14 ga.		160	.300		24.50	13.30	1.34	39.14	48.50
2160	30″ diameter, 14 ga.	B-13	120	.467		29	21.50	4.84	55.34	69.50
2180	36″ diameter, 12 ga.		120	.467		35	21.50	4.84	61.34	76
2200	48″ diameter, 12 ga.		100	.560		46.50	26	5.80	78.30	96
2220	60″ diameter, 10 ga.	B-13B	75	.747		80	34.50	13.05	127.55	153
2240	72″ diameter, 8 ga.	″	45	1.244		84.50	57.50	22	164	203
2250	End sections, 8″ diameter, 16 ga.	B-14	20	2.400	Ea.	42	106	10.70	158.70	217
2255	10″ diameter, 16 ga.		20	2.400		52	106	10.70	168.70	228
2260	12″ diameter, 16 ga.		18	2.667		107	118	11.85	236.85	305
2265	15″ diameter, 16 ga.		18	2.667		206	118	11.85	335.85	415
2270	18″ diameter, 16 ga.		16	3		240	133	13.35	386.35	480
2275	24″ diameter, 16 ga.	B-13	16	3.500		315	161	36.50	512.50	630
2280	30″ diameter, 16 ga.		14	4		520	184	41.50	745.50	890
2285	36″ diameter, 14 ga.		14	4		665	184	41.50	890.50	1,050
2290	48″ diameter, 14 ga.		10	5.600		1,850	258	58	2,166	2,500
2292	60″ diameter, 14 ga.		6	9.333		1,875	430	97	2,402	2,825
2294	72″ diameter, 14 ga.	B-13B	5	11.200		2,925	515	196	3,636	4,225
2300	Bends or elbows, 8″ diameter	B-14	28	1.714		110	76	7.65	193.65	243
2320	10″ diameter		25	1.920		137	85	8.55	230.55	287

33 42 Stormwater Conveyance

33 42 11 – Stormwater Gravity Piping

33 42 11.40 Piping, Storm Drainage, Corrugated Metal		Crew	Daily Output	Labor-Hours	Unit	2020 Bare Costs Material	Labor	Equipment	Total	Total Incl O&P
2340	12″ diameter, 16 ga.	B-14	23	2.087	Ea.	162	92.50	9.30	263.80	325
2342	18″ diameter, 16 ga.		20	2.400		231	106	10.70	347.70	425
2344	24″ diameter, 14 ga.		16	3		360	133	13.35	506.35	610
2346	30″ diameter, 14 ga.		15	3.200		430	142	14.25	586.25	705
2348	36″ diameter, 14 ga.	B-13	15	3.733		590	172	38.50	800.50	950
2350	48″ diameter, 12 ga.	"	12	4.667		760	215	48.50	1,023.50	1,200
2352	60″ diameter, 10 ga.	B-13B	10	5.600		1,025	258	98	1,381	1,625
2354	72″ diameter, 10 ga.	"	6	9.333		1,275	430	163	1,868	2,225
2360	Wyes or tees, 8″ diameter	B-14	25	1.920		156	85	8.55	249.55	310
2380	10″ diameter		21	2.286		194	101	10.15	305.15	375
2400	12″ diameter, 16 ga.		19	2.526		222	112	11.25	345.25	425
2410	18″ diameter, 16 ga.		16	3		305	133	13.35	451.35	555
2412	24″ diameter, 14 ga.		16	3		525	133	13.35	671.35	790
2414	30″ diameter, 14 ga.	B-13	12	4.667		665	215	48.50	928.50	1,100
2416	36″ diameter, 14 ga.		11	5.091		815	234	53	1,102	1,300
2418	48″ diameter, 12 ga.		10	5.600		1,100	258	58	1,416	1,650
2420	60″ diameter, 10 ga.	B-13B	8	7		1,600	320	123	2,043	2,400
2422	72″ diameter, 10 ga.	"	5	11.200		1,925	515	196	2,636	3,125
2500	Galvanized, uncoated, 20′ lengths									
2520	8″ diameter, 16 ga.	B-14	355	.135	L.F.	7.85	6	.60	14.45	18.25
2540	10″ diameter, 16 ga.		280	.171		8.40	7.60	.76	16.76	21.50
2560	12″ diameter, 16 ga.		220	.218		9.75	9.65	.97	20.37	26.50
2580	15″ diameter, 16 ga.		220	.218		11.05	9.65	.97	21.67	28
2600	18″ diameter, 16 ga.		205	.234		13.50	10.35	1.04	24.89	31.50
2620	24″ diameter, 14 ga.		175	.274		23	12.15	1.22	36.37	45
2640	30″ diameter, 14 ga.	B-13	130	.431		32	19.85	4.47	56.32	70
2660	36″ diameter, 12 ga.		130	.431		39	19.85	4.47	63.32	77
2680	48″ diameter, 12 ga.		110	.509		54	23.50	5.30	82.80	100
2690	60″ diameter, 10 ga.	B-13B	78	.718		84	33	12.55	129.55	155
2695	72″ diameter, 10 ga.	"	60	.933		109	43	16.35	168.35	202
2711	Bends or elbows, 12″ diameter, 16 ga.	B-14	30	1.600	Ea.	130	71	7.10	208.10	257
2712	15″ diameter, 16 ga.		25.04	1.917		167	85	8.55	260.55	320
2714	18″ diameter, 16 ga.		20	2.400		187	106	10.70	303.70	375
2716	24″ diameter, 14 ga.		16	3		281	133	13.35	427.35	525
2718	30″ diameter, 14 ga.		15	3.200		350	142	14.25	506.25	615
2720	36″ diameter, 14 ga.	B-13	15	3.733		485	172	38.50	695.50	835
2722	48″ diameter, 12 ga.		12	4.667		655	215	48.50	918.50	1,100
2724	60″ diameter, 10 ga.		10	5.600		1,025	258	58	1,341	1,575
2726	72″ diameter, 10 ga.		6	9.333		1,325	430	97	1,852	2,200
2728	Wyes or tees, 12″ diameter, 16 ga.	B-14	22.48	2.135		168	94.50	9.50	272	335
2730	18″ diameter, 16 ga.		15	3.200		261	142	14.25	417.25	515
2732	24″ diameter, 14 ga.		15	3.200		435	142	14.25	591.25	710
2734	30″ diameter, 14 ga.		14	3.429		565	152	15.25	732.25	865
2736	36″ diameter, 14 ga.	B-13	14	4		715	184	41.50	940.50	1,100
2738	48″ diameter, 12 ga.		12	4.667		1,050	215	48.50	1,313.50	1,525
2740	60″ diameter, 10 ga.		10	5.600		1,475	258	58	1,791	2,075
2742	72″ diameter, 10 ga.		6	9.333		1,800	430	97	2,327	2,725
2780	End sections, 8″ diameter	B-14	35	1.371		78.50	60.50	6.10	145.10	184
2785	10″ diameter		35	1.371		87	60.50	6.10	153.60	193
2790	12″ diameter		35	1.371		100	60.50	6.10	166.60	208
2800	18″ diameter		30	1.600		117	71	7.10	195.10	243
2810	24″ diameter	B-13	25	2.240		197	103	23	323	395
2820	30″ diameter		25	2.240		350	103	23	476	565

33 42 Stormwater Conveyance

33 42 11 – Stormwater Gravity Piping

33 42 11.40 Piping, Storm Drainage, Corrugated Metal

	33 42 11.40 Piping, Storm Drainage, Corrugated Metal	Crew	Daily Output	Labor-Hours	Unit	Material	Labor	Equipment	Total	Total Incl O&P
						2020 Bare Costs				
2825	36″ diameter	B-13	20	2.800	Ea.	470	129	29	628	740
2830	48″ diameter		10	5.600		940	258	58	1,256	1,475
2835	60″ diameter	B-13B	5	11.200		1,725	515	196	2,436	2,900
2840	72″ diameter	″	4	14		2,175	645	245	3,065	3,625
2850	Couplings, 12″ diameter					16.50			16.50	18.15
2855	18″ diameter					21.50			21.50	24
2860	24″ diameter					28			28	30.50
2865	30″ diameter					29			29	32
2870	36″ diameter					38.50			38.50	42
2875	48″ diameter					56.50			56.50	62
2880	60″ diameter					62.50			62.50	69
2885	72″ diameter					71			71	78

33 42 11.60 Sewage/Drainage Collection, Concrete Pipe

	33 42 11.60 Sewage/Drainage Collection, Concrete Pipe	Crew	Daily Output	Labor-Hours	Unit	Material	Labor	Equipment	Total	Total Incl O&P
0010	**SEWAGE/DRAINAGE COLLECTION, CONCRETE PIPE**									
0020	Not including excavation or backfill									
1000	Non-reinforced pipe, extra strength, B&S or T&G joints									
1010	6″ diameter	B-14	265.04	.181	L.F.	7.80	8	.81	16.61	21.50
1020	8″ diameter		224	.214		8.55	9.50	.95	19	24.50
1030	10″ diameter		216	.222		9.50	9.85	.99	20.34	26.50
1040	12″ diameter		200	.240		10.05	10.60	1.07	21.72	28
1050	15″ diameter		180	.267		14.40	11.80	1.19	27.39	35
1060	18″ diameter		144	.333		18.05	14.75	1.48	34.28	43.50
1070	21″ diameter		112	.429		18.95	18.95	1.91	39.81	51.50
1080	24″ diameter		100	.480		22	21	2.14	45.14	58.50
2000	Reinforced culvert, class 3, no gaskets									
2010	12″ diameter	B-14	150	.320	L.F.	16.15	14.15	1.42	31.72	41
2020	15″ diameter		150	.320		22	14.15	1.42	37.57	47
2030	18″ diameter		132	.364		24.50	16.10	1.62	42.22	53
2035	21″ diameter		120	.400		28.50	17.70	1.78	47.98	60
2040	24″ diameter		100	.480		28.50	21	2.14	51.64	66
2045	27″ diameter	B-13	92	.609		47.50	28	6.30	81.80	101
2050	30″ diameter		88	.636		52	29.50	6.60	88.10	108
2060	36″ diameter		72	.778		83	36	8.05	127.05	153
2070	42″ diameter	B-13B	68	.824		99.50	38	14.40	151.90	182
2080	48″ diameter		64	.875		100	40.50	15.30	155.80	187
2085	54″ diameter		56	1		147	46	17.50	210.50	249
2090	60″ diameter		48	1.167		181	53.50	20.50	255	300
2100	72″ diameter		40	1.400		273	64.50	24.50	362	425
2120	84″ diameter		32	1.750		405	80.50	30.50	516	600
2140	96″ diameter		24	2.333		485	107	41	633	740
2200	With gaskets, class 3, 12″ diameter	B-21	168	.167		17.80	8.15	1.13	27.08	33
2220	15″ diameter		160	.175		24	8.55	1.18	33.73	40.50
2230	18″ diameter		152	.184		27	9	1.24	37.24	45
2235	21″ diameter		152	.184		31.50	9	1.24	41.74	49.50
2240	24″ diameter		136	.206		35.50	10.05	1.39	46.94	55.50
2260	30″ diameter	B-13	88	.636		60	29.50	6.60	96.10	117
2270	36″ diameter	″	72	.778		92.50	36	8.05	136.55	164
2290	48″ diameter	B-13B	64	.875		114	40.50	15.30	169.80	202
2310	72″ diameter	″	40	1.400		294	64.50	24.50	383	450
2330	Flared ends, 12″ diameter	B-21	31	.903	Ea.	305	44	6.10	355.10	410
2340	15″ diameter		25	1.120		350	54.50	7.55	412.05	475
2400	18″ diameter		20	1.400		400	68.50	9.45	477.95	555

33 42 Stormwater Conveyance

33 42 11 – Stormwater Gravity Piping

33 42 11.60 Sewage/Drainage Collection, Concrete Pipe		Crew	Daily Output	Labor-Hours	Unit	Material	2020 Bare Costs Labor	Equipment	Total	Total Incl O&P
2420	24″ diameter	B-21	14	2	Ea.	455	97.50	13.50	566	665
2440	36″ diameter	B-13	10	5.600	↓	990	258	58	1,306	1,550
3080	Radius pipe, add to pipe prices, 12″ to 60″ diameter				L.F.	50%				
3090	Over 60″ diameter, add				″	20%				
3500	Reinforced elliptical, 8′ lengths, C507 class 3									
3520	14″ x 23″ inside, round equivalent 18″ diameter	B-21	82	.341	L.F.	40.50	16.65	2.30	59.45	72
3530	24″ x 38″ inside, round equivalent 30″ diameter	B-13	58	.966		69.50	44.50	10	124	154
3540	29″ x 45″ inside, round equivalent 36″ diameter		52	1.077		106	49.50	11.15	166.65	204
3550	38″ x 60″ inside, round equivalent 48″ diameter		38	1.474		171	68	15.30	254.30	305
3560	48″ x 76″ inside, round equivalent 60″ diameter		26	2.154		215	99	22.50	336.50	410
3570	58″ x 91″ inside, round equivalent 72″ diameter	↓	22	2.545	↓	330	117	26.50	473.50	565
3780	Concrete slotted pipe, class 4 mortar joint									
3800	12″ diameter	B-21	168	.167	L.F.	33	8.15	1.13	42.28	49.50
3840	18″ diameter	″	152	.184	″	37	9	1.24	47.24	55.50
3900	Concrete slotted pipe, Class 4 O-ring joint									
3940	12″ diameter	B-21	168	.167	L.F.	30.50	8.15	1.13	39.78	47
3960	18″ diameter	″	152	.184	″	32	9	1.24	42.24	50

33 42 13 – Stormwater Culverts

33 42 13.15 Oval Arch Culverts		Crew	Daily Output	Labor-Hours	Unit	Material	Labor	Equipment	Total	Total Incl O&P
0010	**OVAL ARCH CULVERTS**									
3000	Corrugated galvanized or aluminum, coated & paved									
3020	17″ x 13″, 16 ga., 15″ equivalent	B-14	200	.240	L.F.	14.80	10.60	1.07	26.47	33.50
3040	21″ x 15″, 16 ga., 18″ equivalent		150	.320		19.25	14.15	1.42	34.82	44
3060	28″ x 20″, 14 ga., 24″ equivalent		125	.384		23	17	1.71	41.71	52.50
3080	35″ x 24″, 14 ga., 30″ equivalent	↓	100	.480		30.50	21	2.14	53.64	68
3100	42″ x 29″, 12 ga., 36″ equivalent	B-13	100	.560		36	26	5.80	67.80	84.50
3120	49″ x 33″, 12 ga., 42″ equivalent		90	.622		43.50	28.50	6.45	78.45	98
3140	57″ x 38″, 12 ga., 48″ equivalent	↓	75	.747	↓	56	34.50	7.75	98.25	122
3160	Steel, plain oval arch culverts, plain									
3180	17″ x 13″, 16 ga., 15″ equivalent	B-14	225	.213	L.F.	13.25	9.45	.95	23.65	30
3200	21″ x 15″, 16 ga., 18″ equivalent		175	.274		17.25	12.15	1.22	30.62	38.50
3220	28″ x 20″, 14 ga., 24″ equivalent	↓	150	.320		27	14.15	1.42	42.57	52.50
3240	35″ x 24″, 14 ga., 30″ equivalent	B-13	108	.519		32	24	5.40	61.40	77
3260	42″ x 29″, 12 ga., 36″ equivalent		108	.519		40	24	5.40	69.40	86
3280	49″ x 33″, 12 ga., 42″ equivalent		92	.609		59.50	28	6.30	93.80	114
3300	57″ x 38″, 12 ga., 48″ equivalent		75	.747	↓	69	34.50	7.75	111.25	136
3320	End sections, 17″ x 13″		22	2.545	Ea.	139	117	26.50	282.50	360
3340	42″ x 29″	↓	17	3.294	″	425	152	34	611	735
3360	Multi-plate arch, steel	B-20	1690	.014	Lb.	1.39	.67		2.06	2.54

33 42 33 – Stormwater Curbside Drains and Inlets

33 42 33.13 Catch Basins		Crew	Daily Output	Labor-Hours	Unit	Material	Labor	Equipment	Total	Total Incl O&P
0010	**CATCH BASINS**									
0011	Not including footing & excavation									
1600	Frames & grates, C.I., 24″ square, 500 lb.	B-6	7.80	3.077	Ea.	375	141	27.50	543.50	650
1700	26″ D shape, 600 lb.		7	3.429		720	157	30.50	907.50	1,050
1800	Light traffic, 18″ diameter, 100 lb.		10	2.400		215	110	21.50	346.50	425
1900	24″ diameter, 300 lb.		8.70	2.759		199	126	24.50	349.50	435
2000	36″ diameter, 900 lb.		5.80	4.138		765	189	37	991	1,175
2100	Heavy traffic, 24″ diameter, 400 lb.		7.80	3.077		295	141	27.50	463.50	565
2200	36″ diameter, 1,150 lb.		3	8		935	365	71.50	1,371.50	1,650
2300	Mass. State standard, 26″ diameter, 475 lb.		7	3.429		750	157	30.50	937.50	1,100
2400	30″ diameter, 620 lb.	↓	7	3.429	↓	340	157	30.50	527.50	645

33 42 Stormwater Conveyance

33 42 33 – Stormwater Curbside Drains and Inlets

33 42 33.13 Catch Basins		Crew	Daily Output	Labor-Hours	Unit	2020 Bare Costs Material	Labor	Equipment	Total	Total Incl O&P
2500	Watertight, 24" diameter, 350 lb.	B-6	7.80	3.077	Ea.	415	141	27.50	583.50	695
2600	26" diameter, 500 lb.		7	3.429		450	157	30.50	637.50	765
2700	32" diameter, 575 lb.		6	4		960	183	35.50	1,178.50	1,375
2800	3 piece cover & frame, 10" deep,									
2900	1,200 lb., for heavy equipment	B-6	3	8	Ea.	1,200	365	71.50	1,636.50	1,950
3000	Raised for paving 1-1/4" to 2" high									
3100	4 piece expansion ring									
3200	20" to 26" diameter	1 Clab	3	2.667	Ea.	213	112		325	405
3300	30" to 36" diameter	"	3	2.667	"	295	112		407	495
3320	Frames and covers, existing, raised for paving, 2", including									
3340	row of brick, concrete collar, up to 12" wide frame	B-6	18	1.333	Ea.	52	61	11.90	124.90	162
3360	20" to 26" wide frame		11	2.182		78.50	100	19.45	197.95	258
3380	30" to 36" wide frame		9	2.667		97	122	24	243	315
3400	Inverts, single channel brick	D-1	3	5.333		112	250		362	505
3500	Concrete		5	3.200		124	150		274	365
3600	Triple channel, brick		2	8		182	375		557	770
3700	Concrete		3	5.333		156	250		406	550

33 52 Hydrocarbon Transmission and Distribution

33 52 13 – Liquid Hydrocarbon Piping

33 52 13.16 Gasoline Piping		Crew	Daily Output	Labor-Hours	Unit	Material	Labor	Equipment	Total	Total Incl O&P
0010	**GASOLINE PIPING**									
0020	Primary containment pipe, fiberglass-reinforced									
0030	Plastic pipe 15' & 30' lengths									
0040	2" diameter	Q-6	425	.056	L.F.	7	3.45		10.45	12.85
0050	3" diameter		400	.060		10.65	3.67		14.32	17.20
0060	4" diameter		375	.064		13.85	3.92		17.77	21
0100	Fittings									
0110	Elbows, 90° & 45°, bell ends, 2"	Q-6	24	1	Ea.	40.50	61		101.50	137
0120	3" diameter		22	1.091		56	66.50		122.50	162
0130	4" diameter		20	1.200		70	73.50		143.50	188
0200	Tees, bell ends, 2"		21	1.143		57	70		127	168
0210	3" diameter		18	1.333		68	81.50		149.50	197
0230	Flanges bell ends, 2"		24	1		31.50	61		92.50	126
0240	3" diameter		22	1.091		38.50	66.50		105	142
0250	4" diameter		20	1.200		43.50	73.50		117	158
0260	Sleeve couplings, 2"		21	1.143		12.05	70		82.05	118
0270	3" diameter		18	1.333		18.05	81.50		99.55	142
0280	4" diameter		15	1.600		22.50	98		120.50	172
0290	Threaded adapters, 2"		21	1.143		19.25	70		89.25	126
0300	3" diameter		18	1.333		32	81.50		113.50	158
0310	4" diameter		15	1.600		35.50	98		133.50	186
0320	Reducers, 2"		27	.889		27	54.50		81.50	111
0330	3" diameter		22	1.091		27	66.50		93.50	130
0340	4" diameter		20	1.200		36	73.50		109.50	150
1010	Gas station product line for secondary containment (double wall)									
1100	Fiberglass reinforced plastic pipe 25' lengths									
1120	Pipe, plain end, 3" diameter	Q-6	375	.064	L.F.	29.50	3.92		33.42	38.50
1130	4" diameter		350	.069		33.50	4.20		37.70	43.50
1140	5" diameter		325	.074		35.50	4.52		40.02	46.50
1150	6" diameter		300	.080		39.50	4.89		44.39	51

33 52 Hydrocarbon Transmission and Distribution

33 52 13 – Liquid Hydrocarbon Piping

33 52 13.16 Gasoline Piping		Crew	Daily Output	Labor-Hours	Unit	2020 Bare Costs Material	Labor	Equipment	Total	Total Incl O&P
1200	Fittings									
1230	Elbows, 90° & 45°, 3" diameter	Q-6	18	1.333	Ea.	131	81.50		212.50	266
1240	4" diameter		16	1.500		163	92		255	315
1250	5" diameter		14	1.714		170	105		275	345
1260	6" diameter		12	2		211	122		333	415
1270	Tees, 3" diameter		15	1.600		157	98		255	320
1280	4" diameter		12	2		184	122		306	385
1290	5" diameter		9	2.667		299	163		462	575
1300	6" diameter		6	4		340	245		585	740
1310	Couplings, 3" diameter		18	1.333		54.50	81.50		136	182
1320	4" diameter		16	1.500		117	92		209	266
1330	5" diameter		14	1.714		218	105		323	395
1340	6" diameter		12	2		320	122		442	535
1350	Cross-over nipples, 3" diameter		18	1.333		10.30	81.50		91.80	133
1360	4" diameter		16	1.500		12.55	92		104.55	151
1370	5" diameter		14	1.714		16.30	105		121.30	175
1380	6" diameter		12	2		17.55	122		139.55	202
1400	Telescoping, reducers, concentric 4" x 3"		18	1.333		44.50	81.50		126	171
1410	5" x 4"		17	1.412		94.50	86.50		181	233
1420	6" x 5"		16	1.500		237	92		329	400

33 52 16 – Gas Hydrocarbon Piping

33 52 16.20 Piping, Gas Service and Distribution, P.E.

		Crew	Daily Output	Labor-Hours	Unit	Material	Labor	Equipment	Total	Total Incl O&P
0010	**PIPING, GAS SERVICE AND DISTRIBUTION, POLYETHYLENE**									
0020	Not including excavation or backfill									
1000	60 psi coils, compression coupling @ 100', 1/2" diameter, SDR 11	B-20A	608	.053	L.F.	.49	2.66		3.15	4.53
1010	1" diameter, SDR 11		544	.059		1.14	2.97		4.11	5.70
1040	1-1/4" diameter, SDR 11		544	.059		1.64	2.97		4.61	6.25
1100	2" diameter, SDR 11		488	.066		2.48	3.31		5.79	7.70
1160	3" diameter, SDR 11		408	.078		6.20	3.96		10.16	12.75
1500	60 psi 40' joints with coupling, 3" diameter, SDR 11	B-21A	408	.098		7.35	5.15	1.05	13.55	16.95
1540	4" diameter, SDR 11		352	.114		13.35	5.95	1.22	20.52	25
1600	6" diameter, SDR 11		328	.122		35.50	6.40	1.31	43.21	50
1640	8" diameter, SDR 11		272	.147		54	7.70	1.58	63.28	72.50

33 52 16.23 Medium Density Polyethylene Piping

		Crew	Daily Output	Labor-Hours	Unit	Material	Labor	Equipment	Total	Total Incl O&P
2010	**MEDIUM DENSITY POLYETHYLENE PIPING**									
2020	ASTM D2513, not including excavation or backfill									
2200	Butt fused pipe									
2205	80psi coils, butt fusion joint @ 100', 1/2" CTS diameter, SDR 7	B-22C	2050	.008	L.F.	.18	.38	.08	.64	.85
2210	1" CTS diameter, SDR 11.5		1900	.008		.25	.41	.08	.74	.99
2215	80psi coils, butt fusion joint @ 100', IPS 1/2" diameter, SDR 9.3		1950	.008		.26	.40	.08	.74	.98
2220	3/4" diameter, SDR 11		1950	.008		.33	.40	.08	.81	1.05
2225	1" diameter, SDR 11		1850	.009		.80	.42	.08	1.30	1.60
2230	1-1/4" diameter, SDR 11		1850	.009		.80	.42	.08	1.30	1.60
2235	1-1/2" diameter, SDR 11		1750	.009		.98	.44	.09	1.51	1.85
2240	2" diameter, SDR 11		1750	.009		1.33	.44	.09	1.86	2.23
2245	80psi 40' lengths, butt fusion joint, IPS 3" diameter, SDR 11.5	B-22A	660	.061		2.55	2.94	1.22	6.71	8.55
2250	4" diameter, SDR 11.5	"	500	.080		4.02	3.88	1.61	9.51	12.05
2400	Socket fused pipe									
2405	80psi coils, socket fusion coupling @ 100', 1/2" CTS diameter, SDR 7	B-20	1050	.023	L.F.	.20	1.07		1.27	1.84
2410	1" CTS diameter, SDR 11.5		1000	.024		.28	1.13		1.41	2.01
2415	80psi coils, socket fusion coupling @ 100', IPS 1/2" diameter, SDR 9.3		1000	.024		.28	1.13		1.41	2.01
2420	3/4" diameter, SDR 11		1000	.024		.35	1.13		1.48	2.09

33 52 16 – Gas Hydrocarbon Piping

33 52 16.23	Medium Density Polyethylene Piping	Crew	Daily Output	Labor-Hours	Unit	Material	Labor	Equipment	Total	Total Incl O&P
						2020 Bare Costs				
2425	1" diameter, SDR 11	B-20	950	.025	L.F.	.82	1.19		2.01	2.70
2430	1-1/4" diameter, SDR 11		950	.025		.82	1.19		2.01	2.70
2435	1-1/2" diameter, SDR 11		900	.027		1.01	1.25		2.26	3
2440	2" diameter, SDR 11		850	.028		1.36	1.33		2.69	3.49
2445	80psi 40′ lengths, socket fusion coupling, IPS 3" diameter, SDR 11.5	B-21A	340	.118		3.03	6.15	1.26	10.44	13.90
2450	4" diameter, SDR 11.5	"	260	.154		4.91	8.05	1.65	14.61	19.25
2600	Compression coupled pipe									
2605	80psi coils, compression coupling @ 100′, 1/2" CTS diameter, SDR 7	B-20	2250	.011	L.F.	.31	.50		.81	1.10
2610	1" CTS diameter, SDR 11.5		2175	.011		.50	.52		1.02	1.33
2615	80psi coils, compression coupling @ 100′, IPS 1/2" diameter, SDR 9.3		2175	.011		.50	.52		1.02	1.33
2620	3/4" diameter, SDR 11		2100	.011		.57	.54		1.11	1.43
2625	1" diameter, SDR 11		2025	.012		1.13	.56		1.69	2.08
3000	Fittings, butt fusion									
3010	SDR 11, IPS unless noted CTS									
3100	Caps, 3/4" diameter	B-22C	28.50	.561	Ea.	3.36	27	5.45	35.81	50.50
3105	1" diameter		27	.593		3.06	28.50	5.80	37.36	53
3110	1-1/4" diameter		27	.593		4.71	28.50	5.80	39.01	55
3115	1-1/2" diameter		25.50	.627		3.67	30.50	6.10	40.27	57
3120	2" diameter		24	.667		6.05	32.50	6.50	45.05	62.50
3125	3" diameter		24	.667		11.50	32.50	6.50	50.50	68.50
3130	4" diameter		18	.889		24	43	8.65	75.65	101
3200	Reducers, 1" x 3/4" diameters		13.50	1.185		6.85	57.50	11.55	75.90	107
3205	1-1/4" x 1" diameters		13.50	1.185		7	57.50	11.55	76.05	107
3210	1-1/2" x 3/4" diameters		12.75	1.255		7.15	61	12.25	80.40	113
3215	1-1/2" x 1" diameters		12.75	1.255		7.15	61	12.25	80.40	113
3220	1-1/2" x 1-1/4" diameters		12.75	1.255		7.15	61	12.25	80.40	113
3225	2" x 1" diameters		12	1.333		7.70	64.50	13	85.20	120
3230	2" x 1-1/4" diameters		12	1.333		10.85	64.50	13	88.35	124
3235	2" x 1-1/2" diameters		12	1.333		11.95	64.50	13	89.45	125
3240	3" x 2" diameters		12	1.333		15.10	64.50	13	92.60	128
3245	4" x 2" diameters		9	1.778		22.50	86	17.35	125.85	174
3250	4" x 3" diameters		9	1.778		32	86	17.35	135.35	184
3300	Elbows, 90°, 3/4" diameter		14.25	1.123		4.92	54.50	10.95	70.37	99.50
3302	1" diameter		13.50	1.185		5.10	57.50	11.55	74.15	105
3304	1-1/4" diameter		13.50	1.185		9.75	57.50	11.55	78.80	110
3306	1-1/2" diameter		12.75	1.255		8.60	61	12.25	81.85	114
3308	2" diameter		12	1.333		11.40	64.50	13	88.90	124
3310	3" diameter		12	1.333		25.50	64.50	13	103	140
3312	4" diameter		9	1.778		34.50	86	17.35	137.85	187
3350	45°, 3" diameter		12	1.333		25.50	64.50	13	103	140
3352	4" diameter		9	1.778		34.50	86	17.35	137.85	187
3400	Tees, 3/4" diameter		9.50	1.684		16.95	81.50	16.40	114.85	160
3405	1" diameter		9	1.778		5.10	86	17.35	108.45	155
3410	1-1/4" diameter		9	1.778		10.25	86	17.35	113.60	160
3415	1-1/2" diameter		8.50	1.882		8.70	91.50	18.35	118.55	168
3420	2" diameter		8	2		12.70	97	19.50	129.20	182
3425	3" diameter		8	2		25	97	19.50	141.50	195
3430	4" diameter		6	2.667		34.50	129	26	189.50	261
3500	Tapping tees, high volume, butt fusion outlets									
3505	1-1/2" punch, 2" x 2" outlet	B-22C	11	1.455	Ea.	92.50	70.50	14.20	177.20	224
3510	1-7/8" punch, 3" x 2" outlet		11	1.455		93	70.50	14.20	177.70	224
3515	4" x 2" outlet		9	1.778		94.50	86	17.35	197.85	253
3550	For protective sleeves, add					2.92			2.92	3.21

33 52 16.23 Medium Density Polyethylene Piping		Crew	Daily Output	Labor-Hours	Unit	Material	Labor	Equipment	Total	Total Incl O&P
						2020 Bare Costs				
3600	Service saddles, saddle contour x outlet, butt fusion outlets									
3602	1-1/4" x 3/4" outlet	B-22C	17	.941	Ea.	6.65	45.50	9.20	61.35	86.50
3604	1-1/4" x 1" outlet		17	.941		6.65	45.50	9.20	61.35	86.50
3606	1-1/4" x 1-1/4" outlet		16	1		6.65	48.50	9.75	64.90	91
3608	1-1/2" x 3/4" outlet		16	1		6.65	48.50	9.75	64.90	91
3610	1-1/2" x 1-1/4" outlet		15	1.067		6.65	51.50	10.40	68.55	97
3612	2" x 3/4" outlet		12	1.333		6.65	64.50	13	84.15	119
3614	2" x 1" outlet		12	1.333		6.65	64.50	13	84.15	119
3616	2" x 1-1/4" outlet		11	1.455		6.65	70.50	14.20	91.35	129
3618	3" x 3/4" outlet		12	1.333		6.65	64.50	13	84.15	119
3620	3" x 1" outlet		12	1.333		6.65	64.50	13	84.15	119
3622	3" x 1-1/4" outlet		11	1.455		6.65	70.50	14.20	91.35	129
3624	4" x 3/4" outlet		10	1.600		6.65	77.50	15.60	99.75	141
3626	4" x 1" outlet		10	1.600		6.65	77.50	15.60	99.75	141
3628	4" x 1-1/4" outlet		9	1.778		6.65	86	17.35	110	156
3685	For protective sleeves, 3/4" diameter outlets, add					.86			.86	.95
3690	1" diameter outlets, add					1.45			1.45	1.60
3695	1-1/4" diameter outlets, add					2.49			2.49	2.74
3700	Branch saddles, contour x outlet, butt outlets, round base									
3702	2" x 2" outlet	B-22C	11	1.455	Ea.	17.30	70.50	14.20	102	141
3704	3" x 2" outlet		11	1.455		18.80	70.50	14.20	103.50	142
3706	3" x 3" outlet		11	1.455		27	70.50	14.20	111.70	151
3708	4" x 2" outlet		9	1.778		21.50	86	17.35	124.85	173
3710	4" x 3" outlet		9	1.778		27	86	17.35	130.35	179
3712	4" x 4" outlet		9	1.778		36	86	17.35	139.35	189
3750	Rectangular base, 2" x 2" outlet		11	1.455		16.90	70.50	14.20	101.60	140
3752	3" x 2" outlet		11	1.455		18.35	70.50	14.20	103.05	142
3754	3" x 3" outlet		11	1.455		26.50	70.50	14.20	111.20	151
3756	4" x 2" outlet		9	1.778		21	86	17.35	124.35	172
3758	4" x 3" outlet		9	1.778		25.50	86	17.35	128.85	177
3760	4" x 4" outlet		9	1.778		31	86	17.35	134.35	184
4000	Fittings, socket fusion									
4010	SDR 11, IPS unless noted CTS									
4100	Caps, 1/2" CTS diameter	B-20	30	.800	Ea.	2.57	37.50		40.07	59.50
4105	1/2" diameter		28.50	.842		2.43	39.50		41.93	62
4110	3/4" diameter		28.50	.842		3.05	39.50		42.55	63
4115	1" CTS diameter		28	.857		6.90	40.50		47.40	68
4120	1" diameter		27	.889		3.65	42		45.65	67
4125	1-1/4" diameter		27	.889		4.02	42		46.02	67.50
4130	1-1/2" diameter		25.50	.941		5.05	44.50		49.55	72
4135	2" diameter		24	1		4.84	47		51.84	76.50
4140	3" diameter		24	1		19.30	47		66.30	92
4145	4" diameter		18	1.333		33	62.50		95.50	131
4200	Reducers, 1/2" x 1/2" CTS diameters		14.25	1.684		6.40	79		85.40	126
4202	3/4" x 1/2" CTS diameters		14.25	1.684		5.70	79		84.70	125
4204	3/4" x 1/2" diameters		14.25	1.684		6.30	79		85.30	126
4206	1" CTS x 1/2" diameters		14	1.714		5.80	80.50		86.30	127
4208	1" CTS x 3/4" diameters		13.50	1.778		5.55	83.50		89.05	132
4210	1" x 1/2" CTS diameters		13.50	1.778		7.45	83.50		90.95	134
4212	1" x 1/2" diameters		13.50	1.778		6.15	83.50		89.65	133
4214	1" x 3/4" diameters		13.50	1.778		9	83.50		92.50	136
4216	1" x 1" CTS diameters		13.50	1.778		7.85	83.50		91.35	135
4218	1-1/4" x 1/2" CTS diameters		13.50	1.778		16.35	83.50		99.85	144

33 52 16.23 Medium Density Polyethylene Piping		Crew	Daily Output	Labor-Hours	Unit	Material	Labor	Equipment	Total	Total Incl O&P
						2020 Bare Costs				
4220	1-1/4" x 1/2" diameters	B-20	13.50	1.778	Ea.	8.65	83.50		92.15	136
4222	1-1/4" x 3/4" diameters		13.50	1.778		12.80	83.50		96.30	140
4224	1-1/4" x 1" CTS diameters		13.50	1.778		9.20	83.50		92.70	136
4226	1-1/4" x 1" diameters		13.50	1.778		10.90	83.50		94.40	138
4228	1-1/2" x 3/4" diameters		12.75	1.882		9.50	88.50		98	143
4230	1-1/2" x 1" diameters		12.75	1.882		8.25	88.50		96.75	142
4232	1-1/2" x 1-1/4" diameters		12.75	1.882		9.35	88.50		97.85	143
4234	2" x 3/4" diameters		12	2		10.30	94		104.30	153
4236	2" x 1" CTS diameters		12	2		9.75	94		103.75	153
4238	2" x 1" diameters		12	2		11.10	94		105.10	154
4240	2" x 1-1/4" diameters		12	2		10.10	94		104.10	153
4242	2" x 1-1/2" diameters		12	2		9.80	94		103.80	153
4244	3" x 2" diameters		12	2		11.75	94		105.75	155
4246	4" x 2" diameters		9	2.667		26.50	125		151.50	218
4248	4" x 3" diameters		9	2.667		30	125		155	222
4300	Couplings, 1/2" CTS diameter		15	1.600		1.86	75		76.86	115
4305	1/2" diameter		14.25	1.684		1.79	79		80.79	121
4310	3/4" diameter		14.25	1.684		2.25	79		81.25	121
4315	1" CTS diameter		14	1.714		3.08	80.50		83.58	124
4320	1" diameter		13.50	1.778		2.28	83.50		85.78	129
4325	1-1/4" diameter		13.50	1.778		2.34	83.50		85.84	129
4330	1-1/2" diameter		12.75	1.882		2.65	88.50		91.15	136
4335	2" diameter		12	2		2.86	94		96.86	145
4340	3" diameter		12	2		19.25	94		113.25	163
4345	4" diameter		9	2.667		35.50	125		160.50	228
4400	Elbows, 90°, 1/2" CTS diameter		15	1.600		3.41	75		78.41	117
4405	1/2" diameter		14.25	1.684		3.42	79		82.42	123
4410	3/4" diameter		14.25	1.684		4.07	79		83.07	123
4415	1" CTS diameter		14	1.714		3.37	80.50		83.87	125
4420	1" diameter		13.50	1.778		3.57	83.50		87.07	130
4425	1-1/4" diameter		13.50	1.778		4.04	83.50		87.54	130
4430	1-1/2" diameter		12.75	1.882		6.80	88.50		95.30	141
4435	2" diameter		12	2		7.40	94		101.40	150
4440	3" diameter		12	2		25	94		119	170
4445	4" diameter		9	2.667		71.50	125		196.50	268
4460	45°, 1" diameter		13.50	1.778		7	83.50		90.50	134
4465	2" diameter		12	2		9.45	94		103.45	152
4500	Tees, 1/2" CTS diameter		10	2.400		3.26	113		116.26	174
4505	1/2" diameter		9.50	2.526		3.07	119		122.07	182
4510	3/4" diameter		9.50	2.526		3.89	119		122.89	183
4515	1" CTS diameter		9.33	2.571		3.81	121		124.81	186
4520	1" diameter		9	2.667		4.55	125		129.55	194
4525	1-1/4" diameter		9	2.667		5.20	125		130.20	195
4530	1-1/2" diameter		8.50	2.824		8.95	133		141.95	210
4535	2" diameter		8	3		10.45	141		151.45	223
4540	3" diameter		8	3		38	141		179	254
4545	4" diameter		6	4		74.50	188		262.50	365
4600	Tapping tees, type II, 3/4" punch, socket fusion outlets									
4601	Saddle contour x outlet diameter, IPS unless noted CTS									
4602	1-1/4" x 1/2" CTS outlet	B-20	17	1.412	Ea.	14.95	66.50		81.45	116
4604	1-1/4" x 1/2" outlet		17	1.412		14.95	66.50		81.45	116
4606	1-1/4" x 3/4" outlet		17	1.412		14.95	66.50		81.45	116
4608	1-1/4" x 1" CTS outlet		17	1.412		15.55	66.50		82.05	117

33 52 16.23 Medium Density Polyethylene Piping		Crew	Daily Output	Labor-Hours	Unit	Material	2020 Bare Costs Labor	Equipment	Total	Total Incl O&P
4610	1-1/4" x 1" outlet	B-20	17	1.412	Ea.	15.55	66.50		82.05	117
4612	1-1/4" x 1-1/4" outlet		16	1.500		22	70.50		92.50	130
4614	1-1/2" x 1/2" CTS outlet		16	1.500		11.95	70.50		82.45	119
4616	1-1/2" x 1/2" outlet		16	1.500		11.95	70.50		82.45	119
4618	1-1/2" x 3/4" outlet		16	1.500		11.95	70.50		82.45	119
4620	1-1/2" x 1" CTS outlet		16	1.500		13.05	70.50		83.55	120
4622	1-1/2" x 1" outlet		16	1.500		13.05	70.50		83.55	120
4624	1-1/2" x 1-1/4" outlet		15	1.600		22	75		97	137
4626	2" x 1/2" CTS outlet		12	2		14	94		108	157
4628	2" x 1/2" outlet		12	2		14	94		108	157
4630	2" x 3/4" outlet		12	2		14	94		108	157
4632	2" x 1" CTS outlet		12	2		14.60	94		108.60	158
4634	2" x 1" outlet		12	2		14.60	94		108.60	158
4636	2" x 1-1/4" outlet		11	2.182		22.50	103		125.50	180
4638	3" x 1/2" CTS outlet		12	2		14.50	94		108.50	158
4640	3" x 1/2" outlet		12	2		14.50	94		108.50	158
4642	3" x 3/4" outlet		12	2		14.50	94		108.50	158
4644	3" x 1" CTS outlet		12	2		15.55	94		109.55	159
4646	3" x 1" outlet		12	2		15.10	94		109.10	159
4648	3" x 1-1/4" outlet		11	2.182		23.50	103		126.50	181
4650	4" x 1/2" CTS outlet		10	2.400		12.05	113		125.05	183
4652	4" x 1/2" outlet		10	2.400		12.05	113		125.05	183
4654	4" x 3/4" outlet		10	2.400		12.05	113		125.05	183
4656	4" x 1" CTS outlet		10	2.400		13.15	113		126.15	184
4658	4" x 1" outlet		10	2.400		17.50	113		130.50	189
4660	4" x 1-1/4" outlet		9	2.667		25.50	125		150.50	218
4685	For protective sleeves, 1/2" CTS to 3/4" diameter outlets, add					.86			.86	.95
4690	1" CTS & IPS diameter outlets, add					1.45			1.45	1.60
4695	1-1/4" diameter outlets, add					2.49			2.49	2.74
4700	Tapping tees, high volume, socket fusion outlets									
4705	1-1/2" punch, 2" x 1-1/4" outlet	B-20	11	2.182	Ea.	111	103		214	277
4710	1-7/8" punch, 3" x 1-1/4" outlet		11	2.182		111	103		214	277
4715	4" x 1-1/4" outlet		9	2.667		111	125		236	310
4750	For protective sleeves, add					2.92			2.92	3.21
4800	Service saddles, saddle contour x outlet, socket fusion outlets									
4801	IPS unless noted CTS									
4802	1-1/4" x 1/2" CTS outlet	B-20	17	1.412	Ea.	3.73	66.50		70.23	104
4804	1-1/4" x 1/2" outlet		17	1.412		3.73	66.50		70.23	104
4806	1-1/4" x 3/4" outlet		17	1.412		3.73	66.50		70.23	104
4808	1-1/4" x 1" CTS outlet		17	1.412		6.15	66.50		72.65	107
4810	1-1/4" x 1" outlet		17	1.412		6.15	66.50		72.65	107
4812	1-1/4" x 1-1/4" outlet		16	1.500		9.80	70.50		80.30	117
4814	1-1/2" x 1/2" CTS outlet		16	1.500		3.70	70.50		74.20	110
4816	1-1/2" x 1/2" outlet		16	1.500		3.70	70.50		74.20	110
4818	1-1/2" x 3/4" outlet		16	1.500		3.70	70.50		74.20	110
4820	1-1/2" x 1-1/4" outlet		15	1.600		9.80	75		84.80	124
4822	2" x 1/2" CTS outlet		12	2		3.73	94		97.73	146
4824	2" x 1/2" outlet		12	2		3.73	94		97.73	146
4826	2" x 3/4" outlet		12	2		3.73	94		97.73	146
4828	2" x 1" CTS outlet		12	2		6.15	94		100.15	149
4830	2" x 1" outlet		12	2		6.15	94		100.15	149
4832	2" x 1-1/4" outlet		11	2.182		9.80	103		112.80	166
4834	3" x 1/2" CTS outlet		12	2		3.70	94		97.70	146

33 52 16.23 Medium Density Polyethylene Piping		Crew	Daily Output	Labor-Hours	Unit	Material	Labor	Equipment	Total	Total Incl O&P
						2020 Bare Costs				
4836	3" x 1/2" outlet	B-20	12	2	Ea.	3.70	94		97.70	146
4838	3" x 3/4" outlet		12	2		3.73	94		97.73	146
4840	3" x 1" CTS outlet		12	2		6.10	94		100.10	149
4842	3" x 1" outlet		12	2		6.15	94		100.15	149
4844	3" x 1-1/4" outlet		11	2.182		9.90	103		112.90	166
4846	4" x 1/2" CTS outlet		10	2.400		3.70	113		116.70	174
4848	4" x 1/2" outlet		10	2.400		3.70	113		116.70	174
4850	4" x 3/4" outlet		10	2.400		3.73	113		116.73	174
4852	4" x 1" CTS outlet		10	2.400		6.10	113		119.10	177
4854	4" x 1" outlet		10	2.400		6.15	113		119.15	177
4856	4" x 1-1/4" outlet		9	2.667		9.90	125		134.90	200
4885	For protective sleeves, 1/2" CTS to 3/4" diameter outlets, add					.86			.86	.95
4890	1" CTS & IPS diameter outlets, add					1.45			1.45	1.60
4895	1-1/4" diameter outlets, add					2.49			2.49	2.74
4900	Spigot fittings, tees, SDR 7, 1/2" CTS diameter	B-20	10	2.400		25.50	113		138.50	198
4901	SDR 9.3, 1/2" diameter		9.50	2.526		11.30	119		130.30	191
4902	SDR 10, 1-1/4" diameter		9	2.667		11.30	125		136.30	201
4903	SDR 11, 3/4" diameter		9.50	2.526		23.50	119		142.50	205
4904	2" diameter		8	3		21.50	141		162.50	236
4906	SDR 11.5, 1" CTS diameter		9.33	2.571		60	121		181	248
4907	3" diameter		8	3		61	141		202	279
4908	4" diameter		6	4		157	188		345	455
4921	90° elbows, SDR 7, 1/2" CTS diameter		15	1.600		13.45	75		88.45	128
4922	SDR 9.3, 1/2" diameter		14.25	1.684		5.80	79		84.80	125
4923	SDR 10, 1-1/4" diameter		13.50	1.778		5.90	83.50		89.40	133
4924	SDR 11, 3/4" diameter		14.25	1.684		6.60	79		85.60	126
4925	2" diameter		12	2		25	94		119	170
4927	SDR 11.5, 1" CTS diameter		14	1.714		33	80.50		113.50	158
4928	3" diameter		12	2		56.50	94		150.50	204
4929	4" diameter		9	2.667		121	125		246	320
4935	Caps, SDR 7, 1/2" CTS diameter		30	.800		4.46	37.50		41.96	61.50
4936	SDR 9.3, 1/2" diameter		28.50	.842		4.46	39.50		43.96	64.50
4937	SDR 10, 1-1/4" diameter		27	.889		4.65	42		46.65	68
4938	SDR 11, 3/4" diameter		28.50	.842		4.77	39.50		44.27	65
4939	1" diameter		27	.889		5.30	42		47.30	69
4940	2" diameter		24	1		5.70	47		52.70	77.50
4942	SDR 11.5, 1" CTS diameter		28	.857		18.70	40.50		59.20	81
4943	3" diameter		24	1		30.50	47		77.50	105
4944	4" diameter		18	1.333		29	62.50		91.50	127
4946	SDR 13.5, 4" diameter		18	1.333		73.50	62.50		136	175
4950	Reducers, SDR 10 x SDR 11, 1-1/4" x 3/4" diameters		13.50	1.778		15.25	83.50		98.75	143
4951	1-1/4" x 1" diameters		13.50	1.778		15.25	83.50		98.75	143
4952	SDR 10 x SDR 11.5, 1-1/4" x 1" CTS diameters		13.50	1.778		15.25	83.50		98.75	143
4953	SDR 11 x SDR 7, 3/4" x 1/2" CTS diameters		14.25	1.684		6.80	79		85.80	127
4954	SDR 11 x SDR 9.3, 3/4" x 1/2" diameters		14.25	1.684		6.80	79		85.80	127
4955	SDR 11 x SDR 10, 2" x 1-1/4" diameters		12	2		15.80	94		109.80	159
4956	SDR 11 x SDR 11, 1" x 3/4" diameters		13.50	1.778		7.85	83.50		91.35	135
4957	2" x 3/4" diameters		12	2		15.60	94		109.60	159
4958	2" x 1" diameters		12	2		15.70	94		109.70	159
4959	SDR 11 x SDR 11.5, 1" x 1" CTS diameters		13.50	1.778		8	83.50		91.50	135
4960	2" x 1" CTS diameters		12	2		15.60	94		109.60	159
4962	SDR 11.5 x SDR 11, 3" x 2" diameters		12	2		21	94		115	165
4963	4" x 2" diameters		9	2.667		32	125		157	224

33 52 16.23 Medium Density Polyethylene Piping		Crew	Daily Output	Labor-Hours	Unit	2020 Bare Costs Material	Labor	Equipment	Total	Total Incl O&P
4964	SDR 11.5 x SDR 11.5, 4" x 3" diameters	B-20	9	2.667	Ea.	34.50	125		159.50	227
6100	Fittings, compression									
6101	MDPE gas pipe, ASTM D2513/ASTM F1924-98									
6102	Caps, SDR 7, 1/2" CTS diameter	B-20	60	.400	Ea.	12.85	18.80		31.65	42.50
6104	SDR 9.3, 1/2" IPS diameter		58	.414		23	19.45		42.45	55
6106	SDR 10, 1/2" CTS diameter		60	.400		11.35	18.80		30.15	41
6108	1-1/4" IPS diameter		54	.444		74.50	21		95.50	114
6110	SDR 11, 3/4" IPS diameter		58	.414		20	19.45		39.45	51.50
6112	1" IPS diameter		54	.444		32.50	21		53.50	67
6114	1-1/4" IPS diameter		54	.444		78	21		99	118
6116	1-1/2" IPS diameter		52	.462		98.50	21.50		120	141
6118	2" IPS diameter		50	.480		86	22.50		108.50	129
6120	SDR 11.5, 1" CTS diameter		56	.429		21	20		41	53.50
6122	SDR 12.5, 1" CTS diameter		56	.429		24	20		44	57
6202	Reducers, SDR 7 x SDR 10, 1/2" CTS x 1/2" CTS diameters		30	.800		17.35	37.50		54.85	75.50
6204	SDR 9.3 x SDR 7, 1/2" IPS x 1/2" CTS diameters		29	.828		51.50	39		90.50	115
6206	SDR 11 x SDR 7, 3/4" IPS x 1/2" CTS diameters		29	.828		51.50	39		90.50	115
6208	1" IPS x 1/2" CTS diameters		27	.889		60	42		102	129
6210	SDR 11 x SDR 9.3, 3/4" IPS x 1/2" IPS diameters		29	.828		50	39		89	114
6212	1" IPS x 1/2" IPS diameters		27	.889		64.50	42		106.50	134
6214	SDR 11 x SDR 10, 2" IPS x 1-1/4" IPS diameters		25	.960		95.50	45		140.50	173
6216	SDR 11 x SDR 11, 1" IPS x 3/4" IPS diameters		27	.889		60	42		102	129
6218	1-1/4" IPS x 1" IPS diameters		27	.889		58.50	42		100.50	128
6220	2" IPS x 1-1/4" IPS diameters		25	.960		95.50	45		140.50	173
6222	SDR 11 x SDR 11.5, 1" IPS x 1" CTS diameters		27	.889		46	42		88	114
6224	SDR 11 x SDR 12.5, 1" IPS x 1" CTS diameters		27	.889		56	42		98	125
6226	SDR 11.5 x SDR 7, 1" CTS x 1/2" CTS diameters		28	.857		34	40.50		74.50	98
6228	SDR 11.5 x SDR 9.3, 1" CTS x 1/2" IPS diameters		28	.857		39.50	40.50		80	104
6230	SDR 11.5 x SDR 11, 1" CTS x 3/4" IPS diameters		28	.857		55	40.50		95.50	121
6232	SDR 12.5 x SDR 7, 1" CTS x 1/2" CTS diameters		28	.857		36	40.50		76.50	100
6234	SDR 12.5 x SDR 9.3, 1" CTS x 1/2" IPS diameters		28	.857		40.50	40.50		81	105
6236	SDR 12.5 x SDR 11, 1" CTS x 3/4" IPS diameters		28	.857		34.50	40.50		75	98.50
6302	Couplings, SDR 7, 1/2" CTS diameter		30	.800		13.10	37.50		50.60	71
6304	SDR 9.3, 1/2" IPS diameter		29	.828		23.50	39		62.50	84.50
6306	SDR 10, 1/2" CTS diameter		30	.800		12.40	37.50		49.90	70
6308	SDR 11, 3/4" IPS diameter		29	.828		24	39		63	84.50
6310	1" IPS diameter		27	.889		32.50	42		74.50	99
6312	SDR 11.5, 1" CTS diameter		28	.857		25	40.50		65.50	88.50
6314	SDR 12.5, 1" CTS diameter		28	.857		28	40.50		68.50	91
6402	Repair couplings, SDR 10, 1-1/4" IPS diameter		27	.889		71.50	42		113.50	142
6404	SDR 11, 1-1/4" IPS diameter		27	.889		76	42		118	147
6406	1-1/2" IPS diameter		26	.923		81.50	43.50		125	155
6408	2" IPS diameter		25	.960		95	45		140	172
6502	Elbows, 90°, SDR 7, 1/2" CTS diameter		30	.800		22.50	37.50		60	81.50
6504	SDR 9.3, 1/2" IPS diameter		29	.828		41.50	39		80.50	104
6506	SDR 10, 1-1/4" IPS diameter		27	.889		102	42		144	175
6508	SDR 11, 3/4" IPS diameter		29	.828		37.50	39		76.50	99.50
6510	1" IPS diameter		27	.889		35	42		77	102
6512	1-1/4" IPS diameter		27	.889		101	42		143	174
6514	1-1/2" IPS diameter		26	.923		151	43.50		194.50	232
6516	2" IPS diameter		25	.960		124	45		169	205
6518	SDR 11.5, 1" CTS diameter		28	.857		37	40.50		77.50	101
6520	SDR 12.5, 1" CTS diameter		28	.857		53.50	40.50		94	120

33 52 Hydrocarbon Transmission and Distribution

33 52 16 – Gas Hydrocarbon Piping

33 52 16.23 Medium Density Polyethylene Piping		Crew	Daily Output	Labor-Hours	Unit	2020 Bare Costs Material	Labor	Equipment	Total	Total Incl O&P
6602	Tees, SDR 7, 1/2" CTS diameter	B-20	20	1.200	Ea.	42	56.50		98.50	131
6604	SDR 9.3, 1/2" IPS diameter		19.33	1.241		46	58.50		104.50	139
6606	SDR 10, 1-1/4" IPS diameter		18	1.333		127	62.50		189.50	235
6608	SDR 11, 3/4" IPS diameter		19.33	1.241		58.50	58.50		117	153
6610	1" IPS diameter		18	1.333		64.50	62.50		127	165
6612	1-1/4" IPS diameter		18	1.333		107	62.50		169.50	213
6614	1-1/2" IPS diameter		17.33	1.385		224	65		289	345
6616	2" IPS diameter		16.67	1.440		143	67.50		210.50	260
6618	SDR 11.5, 1" CTS diameter		18.67	1.286		59	60.50		119.50	156
6620	SDR 12.5, 1" CTS diameter		18.67	1.286		59	60.50		119.50	156
8100	Fittings, accessories									
8105	SDR 11, IPS unless noted CTS									
8110	Protective sleeves, for high volume tapping tees, butt fusion outlets	1 Skwk	16	.500	Ea.	5.45	27.50		32.95	47.50
8120	For tapping tees, socket outlets, CTS, 1/2" diameter		18	.444		1.42	24.50		25.92	38.50
8122	1" diameter		18	.444		2.51	24.50		27.01	40
8124	IPS, 1/2" diameter		18	.444		1.96	24.50		26.46	39
8126	3/4" diameter		18	.444		2.51	24.50		27.01	40
8128	1" diameter		18	.444		3.42	24.50		27.92	41
8130	1-1/4" diameter		18	.444		4.31	24.50		28.81	41.50
8205	Tapping tee test caps, type I, yellow PE cap, "aldyl style"		45	.178		44	9.75		53.75	63.50
8210	Type II, yellow polyethylene cap		45	.178		44	9.75		53.75	63.50
8215	High volume, yellow polyethylene cap		45	.178		56.50	9.75		66.25	77.50
8250	Quick connector, female x female inlets, 1/4" N.P.T.		50	.160		18.45	8.80		27.25	34
8255	Test hose, 24" length, 3/8" ID, male outlets, 1/4" N.P.T.		50	.160		25	8.80		33.80	41
8260	Quick connector and test hose assembly		50	.160		43.50	8.80		52.30	61.50
8305	Purge point caps, butt fusion, SDR 10, 1-1/4" diameter	B-22C	27	.593		32	28.50	5.80	66.30	85
8310	SDR 11, 1-1/4" diameter		27	.593		32	28.50	5.80	66.30	85
8315	2" diameter		24	.667		32.50	32.50	6.50	71.50	91
8320	3" diameter		24	.667		46	32.50	6.50	85	106
8325	4" diameter		18	.889		45	43	8.65	96.65	124
8340	Socket fusion, SDR 11, 1-1/4" diameter		27	.593		8.20	28.50	5.80	42.50	59
8345	2" diameter		24	.667		9.85	32.50	6.50	48.85	66.50
8350	3" diameter		24	.667		60.50	32.50	6.50	99.50	122
8355	4" diameter		18	.889		44.50	43	8.65	96.15	124
8360	Purge test quick connector, female x female inlets, 1/4" N.P.T.	1 Skwk	50	.160		18.45	8.80		27.25	34
8365	Purge test hose, 24" length, 3/8" ID, male outlets, 1/4" N.P.T.		50	.160		25	8.80		33.80	41
8370	Purge test quick connector and test hose assembly		50	.160		43.50	8.80		52.30	61.50
8405	Transition fittings, MDPE x zinc plated steel, SDR 7, 1/2" CTS x 1/2" MPT	B-22C	30	.533		25	26	5.20	56.20	72
8410	SDR 9.3, 1/2" IPS x 3/4" MPT		28.50	.561		39.50	27	5.45	71.95	90.50
8415	SDR 10, 1-1/4" IPS x 1-1/4" MPT		27	.593		37.50	28.50	5.80	71.80	91
8420	SDR 11, 3/4" IPS x 3/4" MPT		28.50	.561		20	27	5.45	52.45	69
8425	1" IPS x 1" MPT		27	.593		24.50	28.50	5.80	58.80	77
8430	1-1/4" IPS x 1-1/4" MPT		17	.941		46	45.50	9.20	100.70	130
8435	1-1/2" IPS x 1-1/2" MPT		25.50	.627		47	30.50	6.10	83.60	104
8440	2" IPS x 2" MPT		24	.667		53.50	32.50	6.50	92.50	115
8445	SDR 11.5, 1" CTS x 1" MPT		28	.571		33	27.50	5.55	66.05	84.50

33 52 16.26 High Density Polyethylene Piping		Crew	Daily Output	Labor-Hours	Unit	Material	Labor	Equipment	Total	Total Incl O&P
2010	**HIGH DENSITY POLYETHYLENE PIPING**									
2020	ASTM D2513, not including excavation or backfill									
2200	Butt fused pipe									
2205	125psi coils, butt fusion joint @ 100', 1/2" CTS diameter, SDR 7	B-22C	2050	.008	L.F.	.13	.38	.08	.59	.79
2210	160psi coils, butt fusion joint @ 100', IPS 1/2" diameter, SDR 9		1950	.008		.18	.40	.08	.66	.89

33 52 Hydrocarbon Transmission and Distribution

33 52 16 – Gas Hydrocarbon Piping

	33 52 16.26 High Density Polyethylene Piping	Crew	Daily Output	Labor-Hours	Unit	2020 Bare Costs Material	Labor	Equipment	Total	Total Incl O&P
2215	3/4″ diameter, SDR 11	B-22C	1950	.008	L.F.	.26	.40	.08	.74	.98
2220	1″ diameter, SDR 11		1850	.009		.42	.42	.08	.92	1.18
2225	1-1/4″ diameter, SDR 11		1850	.009		.62	.42	.08	1.12	1.40
2230	1-1/2″ diameter, SDR 11		1750	.009		1.17	.44	.09	1.70	2.06
2235	2″ diameter, SDR 11		1750	.009		.94	.44	.09	1.47	1.80
2240	3″ diameter, SDR 11		1650	.010		1.55	.47	.09	2.11	2.52
2245	160psi 40′ lengths, butt fusion joint, IPS 3″ diameter, SDR 11	B-22A	660	.061		1.69	2.94	1.22	5.85	7.60
2250	4″ diameter, SDR 11		500	.080		2.78	3.88	1.61	8.27	10.70
2255	6″ diameter, SDR 11		500	.080		6.05	3.88	1.61	11.54	14.25
2260	8″ diameter, SDR 11		420	.095		10.20	4.62	1.92	16.74	20.50
2265	10″ diameter, SDR 11		340	.118		24	5.70	2.37	32.07	37
2270	12″ diameter, SDR 11		300	.133		33	6.45	2.68	42.13	49
2400	Socket fused pipe									
2405	125psi coils, socket fusion coupling @ 100′, 1/2″ CTS diameter, SDR 7	B-20	1050	.023	L.F.	.15	1.07		1.22	1.79
2410	160psi coils, socket fusion coupling @ 100′, IPS 1/2″ diameter, SDR 9		1000	.024		.20	1.13		1.33	1.92
2415	3/4″ diameter, SDR 11		1000	.024		.28	1.13		1.41	2
2420	1″ diameter, SDR 11		950	.025		.44	1.19		1.63	2.27
2425	1-1/4″ diameter, SDR 11		950	.025		.64	1.19		1.83	2.50
2430	1-1/2″ diameter, SDR 11		900	.027		1.19	1.25		2.44	3.20
2435	2″ diameter, SDR 11		850	.028		.97	1.33		2.30	3.06
2440	3″ diameter, SDR 11		850	.028		1.66	1.33		2.99	3.83
2445	160psi 40′ lengths, socket fusion coupling, IPS 3″ diameter, SDR 11	B-21A	340	.118		1.98	6.15	1.26	9.39	12.75
2450	4″ diameter, SDR 11	″	260	.154		3.34	8.05	1.65	13.04	17.55
2600	Compression coupled pipe									
2610	160psi coils, compression coupling @ 100′, IPS 1/2″ diameter, SDR 9	B-20	1000	.024	L.F.	.42	1.13		1.55	2.16
2615	3/4″ diameter, SDR 11		1000	.024		.50	1.13		1.63	2.25
2620	1″ diameter, SDR 11		950	.025		.75	1.19		1.94	2.61
2625	1-1/4″ diameter, SDR 11		950	.025		1.38	1.19		2.57	3.31
2630	1-1/2″ diameter, SDR 11		900	.027		1.99	1.25		3.24	4.07
2635	2″ diameter, SDR 11		850	.028		1.89	1.33		3.22	4.08
3000	Fittings, butt fusion									
3010	SDR 11, IPS unless noted CTS									
3105	Caps, 1/2″ diameter	B-22C	28.50	.561	Ea.	6.60	27	5.45	39.05	54.50
3110	3/4″ diameter		28.50	.561		3.53	27	5.45	35.98	51
3115	1″ diameter		27	.593		3.62	28.50	5.80	37.92	54
3120	1-1/4″ diameter		27	.593		4.79	28.50	5.80	39.09	55
3125	1-1/2″ diameter		25.50	.627		4.42	30.50	6.10	41.02	57.50
3130	2″ diameter		24	.667		4.73	32.50	6.50	43.73	61
3135	3″ diameter		24	.667		7.15	32.50	6.50	46.15	63.50
3140	4″ diameter		18	.889		15.05	43	8.65	66.70	91
3145	6″ diameter	B-22A	18	2.222		38	108	44.50	190.50	253
3150	8″ diameter		15	2.667		62.50	129	53.50	245	320
3155	10″ diameter		12	3.333		340	162	67	569	690
3160	12″ diameter		10.50	3.810		365	185	76.50	626.50	760
3205	Reducers, 1/2″ x 1/2″ CTS diameters	B-22C	14.25	1.123		12.25	54.50	10.95	77.70	108
3210	3/4″ x 1/2″ CTS diameters		14.25	1.123		8.80	54.50	10.95	74.25	104
3215	1″ x 1/2″ CTS diameters		13.50	1.185		13.65	57.50	11.55	82.70	114
3220	1″ x 1/2″ diameters		13.50	1.185		12.25	57.50	11.55	81.30	113
3225	1″ x 3/4″ diameters		13.50	1.185		10.65	57.50	11.55	79.70	111
3230	1-1/4″ x 1″ diameters		13.50	1.185		10.80	57.50	11.55	79.85	111
3232	1-1/2″ x 3/4″ diameters		12.75	1.255		7.80	61	12.25	81.05	114
3233	1-1/2″ x 1″ diameters		12.75	1.255		7.80	61	12.25	81.05	114
3234	1-1/2″ x 1-1/4″ diameters		12.75	1.255		7.80	61	12.25	81.05	114

33 52 16.26 High Density Polyethylene Piping		Crew	Daily Output	Labor-Hours	Unit	2020 Bare Costs Material	Labor	Equipment	Total	Total Incl O&P
3235	2" x 1" diameters	B-22C	12	1.333	Ea.	8.60	64.50	13	86.10	121
3240	2" x 1-1/4" diameters		12	1.333		8.20	64.50	13	85.70	121
3245	2" x 1-1/2" diameters		12	1.333		9.05	64.50	13	86.55	122
3250	3" x 2" diameters		12	1.333		10.40	64.50	13	87.90	123
3255	4" x 2" diameters		9	1.778		14.40	86	17.35	117.75	165
3260	4" x 3" diameters		9	1.778		15.40	86	17.35	118.75	166
3262	6" x 3" diameters	B-22A	9	4.444		35.50	215	89.50	340	465
3265	6" x 4" diameters		9	4.444		48.50	215	89.50	353	475
3270	8" x 6" diameters		7.50	5.333		81	259	107	447	595
3275	10" x 8" diameters		6	6.667		174	325	134	633	825
3280	12" x 8" diameters		5.25	7.619		257	370	153	780	1,000
3285	12" x 10" diameters		5.25	7.619		120	370	153	643	855
3305	Elbows, 90°, 3/4" diameter	B-22C	14.25	1.123		5.85	54.50	10.95	71.30	101
3310	1" diameter		13.50	1.185		5.95	57.50	11.55	75	106
3315	1-1/4" diameter		13.50	1.185		6.60	57.50	11.55	75.65	106
3320	1-1/2" diameter		12.75	1.255		7.90	61	12.25	81.15	114
3325	2" diameter		12	1.333		8.25	64.50	13	85.75	121
3330	3" diameter		12	1.333		16.25	64.50	13	93.75	130
3335	4" diameter		9	1.778		20.50	86	17.35	123.85	172
3340	6" diameter	B-22A	9	4.444		55	215	89.50	359.50	485
3345	8" diameter		7.50	5.333		164	259	107	530	690
3350	10" diameter		6	6.667		410	325	134	869	1,075
3355	12" diameter		5.25	7.619		380	370	153	903	1,150
3380	45°, 3/4" diameter	B-22C	14.25	1.123		6.70	54.50	10.95	72.15	101
3385	1" diameter		13.50	1.185		6.70	57.50	11.55	75.75	107
3390	1-1/4" diameter		13.50	1.185		6.90	57.50	11.55	75.95	107
3395	1-1/2" diameter		12.75	1.255		9	61	12.25	82.25	115
3400	2" diameter		12	1.333		9.95	64.50	13	87.45	123
3405	3" diameter		12	1.333		16.45	64.50	13	93.95	130
3410	4" diameter		9	1.778		20.50	86	17.35	123.85	172
3415	6" diameter	B-22A	9	4.444		55	215	89.50	359.50	485
3420	8" diameter		7.50	5.333		166	259	107	532	690
3425	10" diameter		6	6.667		390	325	134	849	1,075
3430	12" diameter		5.25	7.619		760	370	153	1,283	1,550
3505	Tees, 1/2" CTS diameter	B-22C	10	1.600		8.05	77.50	15.60	101.15	143
3510	1/2" diameter		9.50	1.684		8.90	81.50	16.40	106.80	151
3515	3/4" diameter		9.50	1.684		6.70	81.50	16.40	104.60	148
3520	1" diameter		9	1.778		7.25	86	17.35	110.60	157
3525	1-1/4" diameter		9	1.778		7.35	86	17.35	110.70	157
3530	1-1/2" diameter		8.50	1.882		12.65	91.50	18.35	122.50	172
3535	2" diameter		8	2		10	97	19.50	126.50	179
3540	3" diameter		8	2		17.85	97	19.50	134.35	187
3545	4" diameter		6	2.667		27	129	26	182	253
3550	6" diameter	B-22A	6	6.667		66	325	134	525	705
3555	8" diameter		5	8		264	390	161	815	1,050
3560	10" diameter		4	10		505	485	201	1,191	1,500
3565	12" diameter		3.50	11.429		670	555	230	1,455	1,825
3600	Tapping tees, type II, 3/4" punch, butt fusion outlets									
3601	Saddle contour x outlet diameter, IPS unless noted CTS									
3602	1-1/4" x 1/2" CTS outlet	B-22C	17	.941	Ea.	11.20	45.50	9.20	65.90	91.50
3604	1-1/4" x 1/2" outlet		17	.941		11.20	45.50	9.20	65.90	91.50
3606	1-1/4" x 3/4" outlet		17	.941		11.20	45.50	9.20	65.90	91.50
3608	1-1/4" x 1" outlet		17	.941		13.05	45.50	9.20	67.75	93.50

33 52 16.26 High Density Polyethylene Piping		Crew	Daily Output	Labor-Hours	Unit	Material	2020 Bare Costs Labor	Equipment	Total	Total Incl O&P
3610	1-1/4" x 1-1/4" outlet	B-22C	16	1	Ea.	23.50	48.50	9.75	81.75	109
3612	1-1/2" x 1/2" CTS outlet		16	1		11.20	48.50	9.75	69.45	96
3614	1-1/2" x 1/2" outlet		16	1		11.20	48.50	9.75	69.45	96
3616	1-1/2" x 3/4" outlet		16	1		11.20	48.50	9.75	69.45	96
3618	1-1/2" x 1" outlet		16	1		13.05	48.50	9.75	71.30	98
3620	1-1/2" x 1-1/4" outlet		15	1.067		23.50	51.50	10.40	85.40	115
3622	2" x 1/2" CTS outlet		12	1.333		11.20	64.50	13	88.70	124
3624	2" x 1/2" outlet		12	1.333		11.20	64.50	13	88.70	124
3626	2" x 3/4" outlet		12	1.333		11.20	64.50	13	88.70	124
3628	2" x 1" outlet		12	1.333		13.05	64.50	13	90.55	126
3630	2" x 1-1/4" outlet		11	1.455		23.50	70.50	14.20	108.20	147
3632	3" x 1/2" CTS outlet		12	1.333		11.20	64.50	13	88.70	124
3634	3" x 1/2" outlet		12	1.333		11.20	64.50	13	88.70	124
3636	3" x 3/4" outlet		12	1.333		11.20	64.50	13	88.70	124
3638	3" x 1" outlet		12	1.333		13.15	64.50	13	90.65	126
3640	3" x 1-1/4" outlet		11	1.455		23.50	70.50	14.20	108.20	148
3642	4" x 1/2" CTS outlet		10	1.600		11.20	77.50	15.60	104.30	146
3644	4" x 1/2" outlet		10	1.600		11.20	77.50	15.60	104.30	146
3646	4" x 3/4" outlet		10	1.600		11.20	77.50	15.60	104.30	146
3648	4" x 1" outlet		10	1.600		13.05	77.50	15.60	106.15	149
3650	4" x 1-1/4" outlet		9	1.778		23.50	86	17.35	126.85	175
3652	6" x 1/2" CTS outlet	B-22A	10	4		11.20	194	80.50	285.70	390
3654	6" x 1/2" outlet		10	4		11.20	194	80.50	285.70	390
3656	6" x 3/4" outlet		10	4		11.20	194	80.50	285.70	390
3658	6" x 1" outlet		10	4		13.05	194	80.50	287.55	395
3660	6" x 1-1/4" outlet		9	4.444		23.50	215	89.50	328	450
3662	8" x 1/2" CTS outlet		8	5		11.20	242	101	354.20	490
3664	8" x 1/2" outlet		8	5		11.20	242	101	354.20	490
3666	8" x 3/4" outlet		8	5		11.20	242	101	354.20	490
3668	8" x 1" outlet		8	5		13.05	242	101	356.05	490
3670	8" x 1-1/4" outlet		7	5.714		23.50	277	115	415.50	570
3675	For protective sleeves, 1/2" to 3/4" diameter outlets, add					.86			.86	.95
3680	1" diameter outlets, add					1.45			1.45	1.60
3685	1-1/4" diameter outlets, add					2.49			2.49	2.74
3700	Tapping tees, high volume, butt fusion outlets									
3705	1-1/2" punch, 2" x 2" outlet	B-22C	11	1.455	Ea.	110	70.50	14.20	194.70	243
3710	1-7/8" punch, 3" x 2" outlet		11	1.455		110	70.50	14.20	194.70	243
3715	4" x 2" outlet		9	1.778		110	86	17.35	213.35	270
3720	6" x 2" outlet	B-22A	9	4.444		110	215	89.50	414.50	545
3725	8" x 2" outlet		7	5.714		110	277	115	502	665
3730	10" x 2" outlet		6	6.667		110	325	134	569	755
3735	12" x 2" outlet		5	8		110	390	161	661	885
3750	For protective sleeves, add					2.92			2.92	3.21
3800	Service saddles, saddle contour x outlet, butt fusion outlets									
3801	IPS unless noted CTS									
3802	1-1/4" x 3/4" outlet	B-22C	17	.941	Ea.	6.65	45.50	9.20	61.35	86.50
3804	1-1/4" x 1" outlet		16	1		6.65	48.50	9.75	64.90	91
3806	1-1/4" x 1-1/4" outlet		16	1		6.95	48.50	9.75	65.20	91.50
3808	1-1/2" x 3/4" outlet		16	1		6.65	48.50	9.75	64.90	91
3810	1-1/2" x 1-1/4" outlet		15	1.067		6.95	51.50	10.40	68.85	97
3812	2" x 3/4" outlet		12	1.333		6.65	64.50	13	84.15	119
3814	2" x 1" outlet		11	1.455		6.65	70.50	14.20	91.35	129
3816	2" x 1-1/4" outlet		11	1.455		6.95	70.50	14.20	91.65	129

	33 52 16.26 High Density Polyethylene Piping	Crew	Daily Output	Labor-Hours	Unit	Material	2020 Bare Costs Labor	Equipment	Total	Total Incl O&P
3818	3" x 3/4" outlet	B-22C	12	1.333	Ea.	6.65	64.50	13	84.15	119
3820	3" x 1" outlet		11	1.455		6.65	70.50	14.20	91.35	129
3822	3" x 1-1/4" outlet		11	1.455		6.95	70.50	14.20	91.65	129
3824	4" x 3/4" outlet		10	1.600		6.65	77.50	15.60	99.75	141
3826	4" x 1" outlet		9	1.778		6.65	86	17.35	110	156
3828	4" x 1-1/4" outlet		9	1.778		6.95	86	17.35	110.30	157
3830	6" x 3/4" outlet	B-22A	10	4		6.65	194	80.50	281.15	385
3832	6" x 1" outlet		9	4.444		6.65	215	89.50	311.15	430
3834	6" x 1-1/4" outlet		9	4.444		6.95	215	89.50	311.45	430
3836	8" x 3/4" outlet		8	5		6.65	242	101	349.65	485
3838	8" x 1" outlet		7	5.714		6.65	277	115	398.65	550
3840	8" x 1-1/4" outlet		7	5.714		6.95	277	115	398.95	550
3880	For protective sleeves, 3/4" diameter outlets, add					.86			.86	.95
3885	1" diameter outlets, add					1.45			1.45	1.60
3890	1-1/4" diameter outlets, add					2.49			2.49	2.74
3905	Branch saddles, contour x outlet, 2" x 2" outlet	B-22C	11	1.455		18.75	70.50	14.20	103.45	142
3910	3" x 2" outlet		11	1.455		18.75	70.50	14.20	103.45	142
3915	3" x 3" outlet		11	1.455		25	70.50	14.20	109.70	149
3920	4" x 2" outlet		9	1.778		18.75	86	17.35	122.10	170
3925	4" x 3" outlet		9	1.778		25	86	17.35	128.35	177
3930	4" x 4" outlet		9	1.778		34.50	86	17.35	137.85	187
3935	6" x 2" outlet	B-22A	9	4.444		18.75	215	89.50	323.25	445
3940	6" x 3" outlet		9	4.444		25	215	89.50	329.50	450
3945	6" x 4" outlet		9	4.444		34.50	215	89.50	339	460
3950	6" x 6" outlet		9	4.444		82	215	89.50	386.50	515
3955	8" x 2" outlet		7	5.714		18.75	277	115	410.75	565
3960	8" x 3" outlet		7	5.714		25	277	115	417	570
3965	8" x 4" outlet		7	5.714		34.50	277	115	426.50	580
3970	8" x 6" outlet		7	5.714		82	277	115	474	635
3980	Ball valves, full port, 3/4" diameter	B-22C	14.25	1.123		59	54.50	10.95	124.45	159
3982	1" diameter		13.50	1.185		59	57.50	11.55	128.05	164
3984	1-1/4" diameter		13.50	1.185		59.50	57.50	11.55	128.55	165
3986	1-1/2" diameter		12.75	1.255		86.50	61	12.25	159.75	200
3988	2" diameter		12	1.333		135	64.50	13	212.50	261
3990	3" diameter		12	1.333		320	64.50	13	397.50	460
3992	4" diameter		9	1.778		420	86	17.35	523.35	610
3994	6" diameter	B-22A	9	4.444		1,125	215	89.50	1,429.50	1,650
3996	8" diameter	"	7.50	5.333		1,800	259	107	2,166	2,500
4000	Fittings, socket fusion									
4010	SDR 11, IPS unless noted CTS									
4105	Caps, 1/2" CTS diameter	B-20	30	.800	Ea.	2.39	37.50		39.89	59
4110	1/2" diameter		28.50	.842		2.44	39.50		41.94	62
4115	3/4" diameter		28.50	.842		2.60	39.50		42.10	62.50
4120	1" diameter		27	.889		3.38	42		45.38	66.50
4125	1-1/4" diameter		27	.889		3.53	42		45.53	67
4130	1-1/2" diameter		25.50	.941		3.91	44.50		48.41	71
4135	2" diameter		24	1		4.01	47		51.01	75.50
4140	3" diameter		24	1		24.50	47		71.50	98
4145	4" diameter		18	1.333		41	62.50		103.50	140
4205	Reducers, 1/2" x 1/2" CTS diameters		14.25	1.684		6.05	79		85.05	126
4210	3/4" x 1/2" CTS diameters		14.25	1.684		6.15	79		85.15	126
4215	3/4" x 1/2" diameters		14.25	1.684		6.15	79		85.15	126
4220	1" x 1/2" CTS diameters		13.50	1.778		6	83.50		89.50	133

33 52 16.26 High Density Polyethylene Piping		Crew	Daily Output	Labor-Hours	Unit	2020 Bare Costs Material	Labor	Equipment	Total	Total Incl O&P
4225	1" x 1/2" diameters	B-20	13.50	1.778	Ea.	6.35	83.50		89.85	133
4230	1" x 3/4" diameters		13.50	1.778		3.99	83.50		87.49	130
4235	1-1/4" x 1/2" CTS diameters		13.50	1.778		6	83.50		89.50	133
4240	1-1/4" x 3/4" diameters		13.50	1.778		6.60	83.50		90.10	133
4245	1-1/4" x 1" diameters		13.50	1.778		5.80	83.50		89.30	132
4250	1-1/2" x 3/4" diameters		12.75	1.882		8.30	88.50		96.80	142
4252	1-1/2" x 1" diameters		12.75	1.882		8.25	88.50		96.75	142
4255	1-1/2" x 1-1/4" diameters		12.75	1.882		8.45	88.50		96.95	142
4260	2" x 3/4" diameters		12	2		8.85	94		102.85	152
4265	2" x 1" diameters		12	2		9	94		103	152
4270	2" x 1-1/4" diameters		12	2		9.10	94		103.10	152
4275	3" x 2" diameters		12	2		12.15	94		106.15	155
4280	4" x 2" diameters		9	2.667		28	125		153	220
4285	4" x 3" diameters		9	2.667		26	125		151	218
4305	Couplings, 1/2" CTS diameter		15	1.600		2.02	75		77.02	115
4310	1/2" diameter		14.25	1.684		2.02	79		81.02	121
4315	3/4" diameter		14.25	1.684		1.54	79		80.54	121
4320	1" diameter		13.50	1.778		1.54	83.50		85.04	128
4325	1-1/4" diameter		13.50	1.778		2.22	83.50		85.72	128
4330	1-1/2" diameter		12.75	1.882		2.49	88.50		90.99	136
4335	2" diameter		12	2		2.59	94		96.59	145
4340	3" diameter		12	2		11.45	94		105.45	155
4345	4" diameter		9	2.667		22.50	125		147.50	214
4405	Elbows, 90°, 1/2" CTS diameter		15	1.600		3.46	75		78.46	117
4410	1/2" diameter		14.25	1.684		3.48	79		82.48	123
4415	3/4" diameter		14.25	1.684		2.73	79		81.73	122
4420	1" diameter		13.50	1.778		3.36	83.50		86.86	130
4425	1-1/4" diameter		13.50	1.778		4.12	83.50		87.62	131
4430	1-1/2" diameter		12.75	1.882		6.85	88.50		95.35	141
4435	2" diameter		12	2		7	94		101	150
4440	3" diameter		12	2		25	94		119	170
4445	4" diameter		9	2.667		102	125		227	300
4450	45°, 3/4" diameter		14.25	1.684		7.05	79		86.05	127
4455	1" diameter		13.50	1.778		7.85	83.50		91.35	135
4460	1-1/4" diameter		13.50	1.778		6.60	83.50		90.10	133
4465	1-1/2" diameter		12.75	1.882		10.35	88.50		98.85	144
4470	2" diameter		12	2		7.70	94		101.70	150
4475	3" diameter		12	2		42.50	94		136.50	189
4480	4" diameter		9	2.667		55	125		180	250
4505	Tees, 1/2" CTS diameter		10	2.400		3.14	113		116.14	173
4510	1/2" diameter		9.50	2.526		2.95	119		121.95	182
4515	3/4" diameter		9.50	2.526		2.94	119		121.94	182
4520	1" diameter		9	2.667		3.62	125		128.62	193
4525	1-1/4" diameter		9	2.667		5.05	125		130.05	195
4530	1-1/2" diameter		8.50	2.824		8.45	133		141.45	209
4535	2" diameter		8	3		9.20	141		150.20	222
4540	3" diameter		8	3		37.50	141		178.50	253
4545	4" diameter		6	4		72.50	188		260.50	365
4600	Tapping tees, type II, 3/4" punch, socket fusion outlets									
4601	Saddle countour x outlet diameter, IPS unless noted CTS									
4602	1-1/4" x 1/2" CTS outlet	B-20	17	1.412	Ea.	11.25	66.50		77.75	112
4604	1-1/4" x 1/2" outlet		17	1.412		11.25	66.50		77.75	112
4606	1-1/4" x 3/4" outlet		17	1.412		11.25	66.50		77.75	112

33 52 16.26 High Density Polyethylene Piping		Crew	Daily Output	Labor-Hours	Unit	2020 Bare Costs Material	Labor	Equipment	Total	Total Incl O&P
4608	1-1/4" x 1" outlet	B-20	17	1.412	Ea.	13.15	66.50		79.65	114
4610	1-1/4" x 1-1/4" outlet		16	1.500		23.50	70.50		94	132
4612	1-1/2" x 1/2" CTS outlet		16	1.500		11.25	70.50		81.75	118
4614	1-1/2" x 1/2" outlet		16	1.500		11.25	70.50		81.75	118
4616	1-1/2" x 3/4" outlet		16	1.500		11.25	70.50		81.75	118
4618	1-1/2" x 1" outlet		16	1.500		13.15	70.50		83.65	120
4620	1-1/2" x 1-1/4" outlet		15	1.600		23.50	75		98.50	139
4622	2" x 1/2" CTS outlet		12	2		11.25	94		105.25	154
4624	2" x 1/2" outlet		12	2		11.25	94		105.25	154
4626	2" x 3/4" outlet		12	2		11.25	94		105.25	154
4628	2" x 1" outlet		12	2		13.15	94		107.15	156
4630	2" x 1-1/4" outlet		11	2.182		23.50	103		126.50	181
4632	3" x 1/2" CTS outlet		12	2		11.25	94		105.25	154
4634	3" x 1/2" outlet		12	2		11.25	94		105.25	154
4636	3" x 3/4" outlet		12	2		11.25	94		105.25	154
4638	3" x 1" outlet		12	2		13.15	94		107.15	156
4640	3" x 1-1/4" outlet		11	2.182		23.50	103		126.50	181
4642	4" x 1/2" CTS outlet		10	2.400		11.25	113		124.25	182
4644	4" x 1/2" outlet		10	2.400		11.25	113		124.25	182
4646	4" x 3/4" outlet		10	2.400		11.25	113		124.25	182
4648	4" x 1" outlet		10	2.400		13.15	113		126.15	184
4650	4" x 1-1/4" outlet		9	2.667		23.50	125		148.50	215
4652	6" x 1/2" CTS outlet	B-21A	10	4		11.25	209	43	263.25	375
4654	6" x 1/2" outlet	B-20	10	2.400		11.25	113		124.25	182
4656	6" x 3/4" outlet	B-21A	10	4		11.25	209	43	263.25	375
4658	6" x 1" outlet		10	4		13.15	209	43	265.15	375
4660	6" x 1-1/4" outlet		9	4.444		23.50	232	47.50	303	430
4662	8" x 1/2" CTS outlet		8	5		11.25	261	53.50	325.75	460
4664	8" x 1/2" outlet		8	5		11.25	261	53.50	325.75	460
4666	8" x 3/4" outlet		8	5		11.25	261	53.50	325.75	460
4668	8" x 1" outlet		8	5		13.15	261	53.50	327.65	465
4670	8" x 1-1/4" outlet		7	5.714		23.50	299	61	383.50	545
4675	For protective sleeves, 1/2" to 3/4" diameter outlets, add					.86			.86	.95
4680	1" diameter outlets, add					1.45			1.45	1.60
4685	1-1/4" diameter outlets, add					2.49			2.49	2.74
4700	Tapping tees, high volume, socket fusion outlets									
4705	1-1/2" punch, 2" x 1-1/4" outlet	B-20	11	2.182	Ea.	111	103		214	277
4710	1-7/8" punch, 3" x 1-1/4" outlet		11	2.182		111	103		214	277
4715	4" x 1-1/4" outlet		9	2.667		111	125		236	310
4720	6" x 1-1/4" outlet	B-21A	9	4.444		111	232	47.50	390.50	525
4725	8" x 1-1/4" outlet		7	5.714		111	299	61	471	640
4730	10" x 1-1/4" outlet		6	6.667		111	350	71.50	532.50	720
4735	12" x 1-1/4" outlet		5	8		111	420	85.50	616.50	840
4750	For protective sleeves, add					2.92			2.92	3.21
4800	Service saddles, saddle contour x outlet, socket fusion outlets									
4801	IPS unless noted CTS									
4802	1-1/4" x 1/2" CTS outlet	B-20	17	1.412	Ea.	5.95	66.50		72.45	107
4804	1-1/4" x 1/2" outlet		17	1.412		5.95	66.50		72.45	107
4806	1-1/4" x 3/4" outlet		17	1.412		5.95	66.50		72.45	107
4808	1-1/4" x 1" outlet		16	1.500		7	70.50		77.50	114
4810	1-1/4" x 1-1/4" outlet		16	1.500		9.90	70.50		80.40	117
4812	1-1/2" x 1/2" CTS outlet		16	1.500		5.95	70.50		76.45	113
4814	1-1/2" x 1/2" outlet		16	1.500		5.95	70.50		76.45	113

33 52 16.26 High Density Polyethylene Piping		Crew	Daily Output	Labor-Hours	Unit	2020 Bare Costs Material	Labor	Equipment	Total	Total Incl O&P
4816	1-1/2" x 3/4" outlet	B-20	16	1.500	Ea.	5.95	70.50		76.45	113
4818	1-1/2" x 1-1/4" outlet		15	1.600		9.90	75		84.90	124
4820	2" x 1/2" CTS outlet		12	2		5.95	94		99.95	149
4822	2" x 1/2" outlet		12	2		5.95	94		99.95	149
4824	2" x 3/4" outlet		12	2		5.95	94		99.95	149
4826	2" x 1" outlet		11	2.182		7	103		110	163
4828	2" x 1-1/4" outlet		11	2.182		9.90	103		112.90	166
4830	3" x 1/2" CTS outlet		12	2		5.95	94		99.95	149
4832	3" x 1/2" outlet		12	2		5.95	94		99.95	149
4834	3" x 3/4" outlet		12	2		5.95	94		99.95	149
4836	3" x 1" outlet		11	2.182		7	103		110	163
4838	3" x 1-1/4" outlet		11	2.182		9.90	103		112.90	166
4840	4" x 1/2" CTS outlet		10	2.400		5.95	113		118.95	177
4842	4" x 1/2" outlet		10	2.400		5.95	113		118.95	177
4844	4" x 3/4" outlet		10	2.400		5.95	113		118.95	177
4846	4" x 1" outlet		9	2.667		7	125		132	197
4848	4" x 1-1/4" outlet		9	2.667		9.90	125		134.90	200
4850	6" x 1/2" CTS outlet	B-21A	10	4		5.95	209	43	257.95	370
4852	6" x 1/2" outlet		10	4		5.95	209	43	257.95	370
4854	6" x 3/4" outlet		10	4		5.95	209	43	257.95	370
4856	6" x 1" outlet		9	4.444		7	232	47.50	286.50	410
4858	6" x 1-1/4" outlet		9	4.444		9.90	232	47.50	289.40	415
4860	8" x 1/2" CTS outlet		8	5		5.95	261	53.50	320.45	455
4862	8" x 1/2" outlet		8	5		5.95	261	53.50	320.45	455
4864	8" x 3/4" outlet		8	5		5.95	261	53.50	320.45	455
4866	8" x 1" outlet		7	5.714		7	299	61	367	525
4868	8" x 1-1/4" outlet		7	5.714		9.90	299	61	369.90	530
4880	For protective sleeves, 1/2" to 3/4" diameter outlets, add					.86			.86	.95
4885	1" diameter outlets, add					1.45			1.45	1.60
4890	1-1/4" diameter outlets, add					2.49			2.49	2.74
4900	Reducer tees, 1-1/4" x 3/4" x 3/4" diameters	B-20	9	2.667		7.35	125		132.35	197
4902	1-1/4" x 3/4" x 1" diameters		9	2.667		9.55	125		134.55	200
4904	1-1/4" x 3/4" x 1-1/4" diameters		9	2.667		9.55	125		134.55	200
4906	1-1/4" x 1" x 3/4" diameters		9	2.667		9.55	125		134.55	200
4908	1-1/4" x 1" x 1" diameters		9	2.667		7.60	125		132.60	197
4910	1-1/4" x 1" x 1-1/4" diameters		9	2.667		9.55	125		134.55	200
4912	1-1/4" x 1-1/4" x 3/4" diameters		9	2.667		7.20	125		132.20	197
4914	1-1/4" x 1-1/4" x 1" diameters		9	2.667		7.30	125		132.30	197
4916	1-1/2" x 3/4" x 3/4" diameters		8.50	2.824		12.15	133		145.15	213
4918	1-1/2" x 3/4" x 1" diameters		8.50	2.824		12.15	133		145.15	213
4920	1-1/2" x 3/4" x 1-1/4" diameters		8.50	2.824		12.15	133		145.15	213
4922	1-1/2" x 3/4" x 1-1/2" diameters		8.50	2.824		12.15	133		145.15	213
4924	1-1/2" x 1" x 3/4" diameters		8.50	2.824		12.15	133		145.15	213
4926	1-1/2" x 1" x 1" diameters		8.50	2.824		12.15	133		145.15	213
4928	1-1/2" x 1" x 1-1/4" diameters		8.50	2.824		12.15	133		145.15	213
4930	1-1/2" x 1" x 1-1/2" diameters		8.50	2.824		12.15	133		145.15	213
4932	1-1/2" x 1-1/4" x 3/4" diameters		8.50	2.824		10.15	133		143.15	211
4934	1-1/2" x 1-1/4" x 1" diameters		8.50	2.824		10.15	133		143.15	211
4936	1-1/2" x 1-1/4" x 1-1/4" diameters		8.50	2.824		10.15	133		143.15	211
4938	1-1/2" x 1-1/4" x 1-1/2" diameters		8.50	2.824		12.15	133		145.15	213
4940	1-1/2" x 1-1/2" x 3/4" diameters		8.50	2.824		9.75	133		142.75	211
4942	1-1/2" x 1-1/2" x 1" diameters		8.50	2.824		9.75	133		142.75	211
4944	1-1/2" x 1-1/2" x 1-1/4" diameters		8.50	2.824		9.75	133		142.75	211

33 52 16.26 High Density Polyethylene Piping		Crew	Daily Output	Labor-Hours	Unit	2020 Bare Costs Material	Labor	Equipment	Total	Total Incl O&P
4946	2" x 1-1/4" x 3/4" diameters	B-20	8	3	Ea.	11	141		152	224
4948	2" x 1-1/4" x 1" diameters		8	3		11	141		152	224
4950	2" x 1-1/4" x 1-1/4" diameters		8	3		11.45	141		152.45	225
4952	2" x 1-1/2" x 3/4" diameters		8	3		11	141		152	224
4954	2" x 1-1/2" x 1" diameters		8	3		11	141		152	224
4956	2" x 1-1/2" x 1-1/4" diameters		8	3		11	141		152	224
4958	2" x 2" x 3/4" diameters		8	3		10.95	141		151.95	224
4960	2" x 2" x 1" diameters		8	3		11.10	141		152.10	224
4962	2" x 2" x 1-1/4" diameters		8	3		11.45	141		152.45	225
8100	Fittings, accessories									
8105	SDR 11, IPS unless noted CTS									
8110	Flange adapters, 2" x 6" long	1 Skwk	32	.250	Ea.	14.55	13.70		28.25	36.50
8115	3" x 6" long		32	.250		17.65	13.70		31.35	40
8120	4" x 6" long		24	.333		22	18.30		40.30	51.50
8125	6" x 8" long	2 Skwk	24	.667		33.50	36.50		70	92
8130	8" x 9" long	"	20	.800		48	44		92	120
8135	Backup flanges, 2" diameter	1 Skwk	32	.250		19.55	13.70		33.25	42
8140	3" diameter		32	.250		26.50	13.70		40.20	49.50
8145	4" diameter		24	.333		51.50	18.30		69.80	84.50
8150	6" diameter	2 Skwk	24	.667		53	36.50		89.50	114
8155	8" diameter	"	20	.800		91	44		135	167
8200	Tapping tees, test caps									
8205	Type II, yellow polyethylene cap	1 Skwk	45	.178	Ea.	44	9.75		53.75	63.50
8210	High volume, black polyethylene cap		45	.178		56.50	9.75		66.25	77.50
8215	Quick connector, female x female inlets, 1/4" N.P.T.		50	.160		18.45	8.80		27.25	34
8220	Test hose, 24" length, 3/8" ID, male outlets, 1/4" N.P.T.		50	.160		25	8.80		33.80	41
8225	Quick connector and test hose assembly		50	.160		43.50	8.80		52.30	61.50
8300	Threaded transition fittings									
8302	HDPE x MPT zinc plated steel, SDR 7, 1/2" CTS x 1/2" MPT	B-22C	30	.533	Ea.	31.50	26	5.20	62.70	79
8304	SDR 9.3, 1/2" IPS x 3/4" MPT		28.50	.561		41.50	27	5.45	73.95	93
8306	SDR 11, 3/4" IPS x 3/4" MPT		28.50	.561		20.50	27	5.45	52.95	69.50
8308	1" IPS x 1" MPT		27	.593		26.50	28.50	5.80	60.80	79
8310	1-1/4" IPS x 1-1/4" MPT		27	.593		42.50	28.50	5.80	76.80	97
8312	1-1/2" IPS x 1-1/2" MPT		25.50	.627		49.50	30.50	6.10	86.10	107
8314	2" IPS x 2" MPT		24	.667		56.50	32.50	6.50	95.50	118
8322	HDPE x MPT 316 stainless steel, SDR 11, 3/4" IPS x 3/4" MPT		28.50	.561		27	27	5.45	59.45	77
8324	1" IPS x 1" MPT		27	.593		28	28.50	5.80	62.30	81
8326	1-1/4" IPS x 1-1/4" MPT		27	.593		34.50	28.50	5.80	68.80	88
8328	2" IPS x 2" MPT		24	.667		40.50	32.50	6.50	79.50	100
8330	3" IPS x 3" MPT		24	.667		82.50	32.50	6.50	121.50	147
8332	4" IPS x 4" MPT		18	.889		113	43	8.65	164.65	200
8334	6" IPS x 6" MPT	B-22A	18	2.222		208	108	44.50	360.50	440
8342	HDPE x FPT 316 stainless steel, SDR 11, 3/4" IPS x 3/4" FPT	B-22C	28.50	.561		35.50	27	5.45	67.95	86
8344	1" IPS x 1" FPT		27	.593		47.50	28.50	5.80	81.80	102
8346	1-1/4" IPS x 1-1/4" FPT		27	.593		76.50	28.50	5.80	110.80	134
8348	1-1/2" IPS x 1-1/2" FPT		25.50	.627		72.50	30.50	6.10	109.10	132
8350	2" IPS x 2" FPT		24	.667		96.50	32.50	6.50	135.50	162
8352	3" IPS x 3" FPT		24	.667		187	32.50	6.50	226	262
8354	4" IPS x 4" FPT		18	.889		269	43	8.65	320.65	370
8362	HDPE x MPT epoxy carbon steel, SDR 11, 3/4" IPS x 3/4" MPT		28.50	.561		20.50	27	5.45	52.95	69.50
8364	1" IPS x 1" MPT		27	.593		22.50	28.50	5.80	56.80	75
8366	1-1/4" IPS x 1-1/4" MPT		27	.593		24	28.50	5.80	58.30	76.50
8368	1-1/2" IPS x 1-1/2" MPT		25.50	.627		26	30.50	6.10	62.60	82

33 52 Hydrocarbon Transmission and Distribution

33 52 16 – Gas Hydrocarbon Piping

33 52 16.26 High Density Polyethylene Piping		Crew	Daily Output	Labor-Hours	Unit	2020 Bare Costs Material	Labor	Equipment	Total	Total Incl O&P
8370	2" IPS x 2" MPT	B-22C	24	.667	Ea.	28.50	32.50	6.50	67.50	86.50
8372	3" IPS x 3" MPT		24	.667		43.50	32.50	6.50	82.50	104
8374	4" IPS x 4" MPT		18	.889		59	43	8.65	110.65	140
8382	HDPE x FPT epoxy carbon steel, SDR 11, 3/4" IPS x 3/4" FPT		28.50	.561		25	27	5.45	57.45	74.50
8384	1" IPS x 1" FPT		27	.593		27	28.50	5.80	61.30	80
8386	1-1/4" IPS x 1-1/4" FPT		27	.593		58.50	28.50	5.80	92.80	114
8388	1-1/2" IPS x 1-1/2" FPT		25.50	.627		58.50	30.50	6.10	95.10	117
8390	2" IPS x 2" FPT		24	.667		63	32.50	6.50	102	125
8392	3" IPS x 3" FPT		24	.667		107	32.50	6.50	146	174
8394	4" IPS x 4" FPT		18	.889		134	43	8.65	185.65	222
8402	HDPE x MPT poly coated carbon steel, SDR 11, 1" IPS x 1" MPT		27	.593		22	28.50	5.80	56.30	74
8404	1-1/4" IPS x 1-1/4" MPT		27	.593		28.50	28.50	5.80	62.80	81.50
8406	1-1/2" IPS x 1-1/2" MPT		25.50	.627		34	30.50	6.10	70.60	90.50
8408	2" IPS x 2" MPT		24	.667		38.50	32.50	6.50	77.50	98
8410	3" IPS x 3" MPT		24	.667		75	32.50	6.50	114	138
8412	4" IPS x 4" MPT		18	.889		116	43	8.65	167.65	202
8414	6" IPS x 6" MPT	B-22A	18	2.222		270	108	44.50	422.50	510
8602	Socket fused HDPE x MPT brass, SDR 11, 3/4" IPS x 3/4" MPT	B-20	28.50	.842		12	39.50		51.50	72.50
8604	1" IPS x 3/4" MPT		27	.889		14.50	42		56.50	79
8606	1" IPS x 1" MPT		27	.889		14	42		56	78.50
8608	1-1/4" IPS x 3/4" MPT		27	.889		15	42		57	79.50
8610	1-1/4" IPS x 1" MPT		27	.889		15	42		57	79.50
8612	1-1/4" IPS x 1-1/4" MPT		27	.889		21.50	42		63.50	86.50
8614	1-1/2" IPS x 1-1/2" MPT		25.50	.941		29.50	44.50		74	99
8616	2" IPS x 2" MPT		24	1		30	47		77	104
8618	Socket fused HDPE x FPT brass, SDR 11, 3/4" IPS x 3/4" FPT		28.50	.842		12	39.50		51.50	72.50
8620	1" IPS x 1/2" FPT		27	.889		12	42		54	76
8622	1" IPS x 3/4" FPT		27	.889		14.50	42		56.50	79
8624	1" IPS x 1" FPT		27	.889		14	42		56	78.50
8626	1-1/4" IPS x 3/4" FPT		27	.889		15	42		57	79.50
8628	1-1/4" IPS x 1" FPT		27	.889		14.50	42		56.50	79
8630	1-1/4" IPS x 1-1/4" FPT		27	.889		21.50	42		63.50	86.50
8632	1-1/2" IPS x 1-1/2" FPT		25.50	.941		30	44.50		74.50	99.50
8634	2" IPS x 1-1/2" FPT		24	1		30.50	47		77.50	105
8636	2" IPS x 2" FPT		24	1		32	47		79	106

33 61 Hydronic Energy Distribution

33 61 13 – Underground Hydronic Energy Distribution

33 61 13.20 Pipe Conduit, Prefabricated/Preinsulated		Crew	Daily Output	Labor-Hours	Unit	Material	Labor	Equipment	Total	Total Incl O&P
0010	**PIPE CONDUIT, PREFABRICATED/PREINSULATED** R221113-70									
0020	Does not include trenching, fittings or crane.									
0300	For cathodic protection, add 12 to 14%									
0310	of total built-up price (casing plus service pipe)									
0580	Polyurethane insulated system, 250°F max. temp.									
0620	Black steel service pipe, standard wt., 1/2" insulation									
0660	3/4" diam. pipe size	Q-17	54	.296	L.F.	88.50	17.50	1.97	107.97	126
0670	1" diam. pipe size		50	.320		97.50	18.90	2.13	118.53	138
0680	1-1/4" diam. pipe size		47	.340		109	20	2.26	131.26	152
0690	1-1/2" diam. pipe size		45	.356		118	21	2.36	141.36	164
0700	2" diam. pipe size		42	.381		122	22.50	2.53	147.03	171
0710	2-1/2" diam. pipe size		34	.471		124	28	3.13	155.13	182

33 61 13.20 Pipe Conduit, Prefabricated/Preinsulated		Crew	Daily Output	Labor-Hours	Unit	Material	2020 Bare Costs Labor	Equipment	Total	Total Incl O&P
0720	3" diam. pipe size	Q-17	28	.571	L.F.	144	33.50	3.80	181.30	213
0730	4" diam. pipe size		22	.727		182	43	4.84	229.84	269
0740	5" diam. pipe size		18	.889		232	52.50	5.90	290.40	340
0750	6" diam. pipe size	Q-18	23	1.043		269	64	4.62	337.62	395
0760	8" diam. pipe size		19	1.263		395	77.50	5.60	478.10	555
0770	10" diam. pipe size		16	1.500		500	92	6.65	598.65	695
0780	12" diam. pipe size		13	1.846		625	113	8.20	746.20	865
0790	14" diam. pipe size		11	2.182		695	133	9.65	837.65	975
0800	16" diam. pipe size		10	2.400		800	147	10.65	957.65	1,100
0810	18" diam. pipe size		8	3		920	184	13.30	1,117.30	1,300
0820	20" diam. pipe size		7	3.429		1,025	210	15.20	1,250.20	1,450
0830	24" diam. pipe size		6	4		1,250	245	17.70	1,512.70	1,750
0900	For 1" thick insulation, add					10%				
0940	For 1-1/2" thick insulation, add					13%				
0980	For 2" thick insulation, add					20%				
1500	Gland seal for system, 3/4" diam. pipe size	Q-17	32	.500	Ea.	1,025	29.50	3.33	1,057.83	1,175
1510	1" diam. pipe size		32	.500		1,025	29.50	3.33	1,057.83	1,175
1540	1-1/4" diam. pipe size		30	.533		1,100	31.50	3.55	1,135.05	1,250
1550	1-1/2" diam. pipe size		30	.533		1,100	31.50	3.55	1,135.05	1,250
1560	2" diam. pipe size		28	.571		1,300	33.50	3.80	1,337.30	1,475
1570	2-1/2" diam. pipe size		26	.615		1,400	36.50	4.09	1,440.59	1,600
1580	3" diam. pipe size		24	.667		1,500	39.50	4.43	1,543.93	1,725
1590	4" diam. pipe size		22	.727		1,775	43	4.84	1,822.84	2,025
1600	5" diam. pipe size		19	.842		2,200	49.50	5.60	2,255.10	2,475
1610	6" diam. pipe size	Q-18	26	.923		2,325	56.50	4.09	2,385.59	2,650
1620	8" diam. pipe size		25	.960		2,700	58.50	4.25	2,762.75	3,075
1630	10" diam. pipe size		23	1.043		3,175	64	4.62	3,243.62	3,600
1640	12" diam. pipe size		21	1.143		3,500	70	5.05	3,575.05	3,950
1650	14" diam. pipe size		19	1.263		3,925	77.50	5.60	4,008.10	4,450
1660	16" diam. pipe size		18	1.333		4,650	81.50	5.90	4,737.40	5,225
1670	18" diam. pipe size		16	1.500		4,925	92	6.65	5,023.65	5,575
1680	20" diam. pipe size		14	1.714		5,600	105	7.60	5,712.60	6,325
1690	24" diam. pipe size		12	2		6,200	122	8.85	6,330.85	7,025
2000	Elbow, 45° for system									
2020	3/4" diam. pipe size	Q-17	14	1.143	Ea.	675	67.50	7.60	750.10	850
2040	1" diam. pipe size		13	1.231		690	72.50	8.20	770.70	880
2050	1-1/4" diam. pipe size		11	1.455		785	86	9.65	880.65	1,000
2060	1-1/2" diam. pipe size		9	1.778		790	105	11.80	906.80	1,050
2070	2" diam. pipe size		6	2.667		845	157	17.75	1,019.75	1,175
2080	2-1/2" diam. pipe size		4	4		930	236	26.50	1,192.50	1,400
2090	3" diam. pipe size		3.50	4.571		1,075	270	30.50	1,375.50	1,625
2100	4" diam. pipe size		3	5.333		1,250	315	35.50	1,600.50	1,875
2110	5" diam. pipe size		2.80	5.714		1,600	335	38	1,973	2,325
2120	6" diam. pipe size	Q-18	4	6		1,825	365	26.50	2,216.50	2,575
2130	8" diam. pipe size		3	8		2,650	490	35.50	3,175.50	3,675
2140	10" diam. pipe size		2.40	10		3,250	610	44.50	3,904.50	4,550
2150	12" diam. pipe size		2	12		4,275	735	53	5,063	5,850
2160	14" diam. pipe size		1.80	13.333		5,325	815	59	6,199	7,150
2170	16" diam. pipe size		1.60	15		6,325	920	66.50	7,311.50	8,400
2180	18" diam. pipe size		1.30	18.462		7,925	1,125	82	9,132	10,500
2190	20" diam. pipe size		1	24		10,000	1,475	106	11,581	13,300
2200	24" diam. pipe size		.70	34.286		12,600	2,100	152	14,852	17,100
2260	For elbow, 90°, add					25%				

33 61 Hydronic Energy Distribution

33 61 13 – Underground Hydronic Energy Distribution

33 61 13.20 Pipe Conduit, Prefabricated/Preinsulated		Crew	Daily Output	Labor-Hours	Unit	2020 Bare Costs Material	Labor	Equipment	Total	Total Incl O&P
2300	For tee, straight, add				Ea.	85%	30%			
2340	For tee, reducing, add					170%	30%			
2380	For weldolet, straight, add				↓	50%				
2400	Polyurethane insulation, 1"									
2410	FRP carrier and casing									
2420	4"	Q-5	18	.889	L.F.	56.50	52.50		109	141
2422	6"	Q-6	23	1.043		98.50	64		162.50	204
2424	8"		19	1.263		160	77.50		237.50	293
2426	10"		16	1.500		216	92		308	375
2428	12"	↓	13	1.846	↓	282	113		395	480
2430	FRP carrier and PVC casing									
2440	4"	Q-5	18	.889	L.F.	23.50	52.50		76	105
2444	8"	Q-6	19	1.263		48.50	77.50		126	170
2446	10"		16	1.500		71	92		163	215
2448	12"	↓	13	1.846	↓	97	113		210	275
2450	PVC carrier and casing									
2460	4"	Q-1	36	.444	L.F.	12.25	26		38.25	52
2462	6"	"	29	.552		17.05	32		49.05	67
2464	8"	Q-2	36	.667		23.50	40		63.50	86
2466	10"		32	.750		32.50	45		77.50	103
2468	12"	↓	31	.774	↓	34.50	46.50		81	108

33 63 Steam Energy Distribution

33 63 13 – Underground Steam and Condensate Distribution Piping

33 63 13.10 Calcium Silicate Insulated System		Crew	Daily Output	Labor-Hours	Unit	Material	Labor	Equipment	Total	Total Incl O&P
0010	**CALCIUM SILICATE INSULATED SYSTEM**									
0011	High temp. (1200 degrees F)									
2840	Steel casing with protective exterior coating									
2850	6-5/8" diameter	Q-18	52	.462	L.F.	130	28	2.04	160.04	188
2860	8-5/8" diameter		50	.480		141	29.50	2.13	172.63	201
2870	10-3/4" diameter		47	.511		165	31	2.26	198.26	231
2880	12-3/4" diameter		44	.545		180	33.50	2.42	215.92	251
2890	14" diameter		41	.585		203	36	2.59	241.59	280
2900	16" diameter		39	.615		218	37.50	2.73	258.23	300
2910	18" diameter		36	.667		240	41	2.95	283.95	330
2920	20" diameter		34	.706		271	43	3.13	317.13	365
2930	22" diameter		32	.750		375	46	3.32	424.32	480
2940	24" diameter		29	.828		430	50.50	3.67	484.17	550
2950	26" diameter		26	.923		490	56.50	4.09	550.59	630
2960	28" diameter		23	1.043		605	64	4.62	673.62	765
2970	30" diameter		21	1.143		650	70	5.05	725.05	825
2980	32" diameter		19	1.263		730	77.50	5.60	813.10	920
2990	34" diameter		18	1.333		735	81.50	5.90	822.40	940
3000	36" diameter	↓	16	1.500		790	92	6.65	888.65	1,025
3040	For multi-pipe casings, add					10%				
3060	For oversize casings, add				↓	2%				
3400	Steel casing gland seal, single pipe									
3420	6-5/8" diameter	Q-18	25	.960	Ea.	1,975	58.50	4.25	2,037.75	2,275
3440	8-5/8" diameter		23	1.043		2,325	64	4.62	2,393.62	2,650
3450	10-3/4" diameter		21	1.143		2,575	70	5.05	2,650.05	2,950
3460	12-3/4" diameter	↓	19	1.263	↓	3,050	77.50	5.60	3,133.10	3,475

33 63 Steam Energy Distribution

33 63 13 – Underground Steam and Condensate Distribution Piping

33 63 13.10 Calcium Silicate Insulated System		Crew	Daily Output	Labor-Hours	Unit	2020 Bare Costs Material	Labor	Equipment	Total	Total Incl O&P
3470	14" diameter	Q-18	17	1.412	Ea.	3,375	86.50	6.25	3,467.75	3,850
3480	16" diameter		16	1.500		3,950	92	6.65	4,048.65	4,500
3490	18" diameter		15	1.600		4,400	98	7.10	4,505.10	4,975
3500	20" diameter		13	1.846		4,850	113	8.20	4,971.20	5,525
3510	22" diameter		12	2		5,450	122	8.85	5,580.85	6,200
3520	24" diameter		11	2.182		6,125	133	9.65	6,267.65	6,950
3530	26" diameter		10	2.400		7,000	147	10.65	7,157.65	7,925
3540	28" diameter		9.50	2.526		7,900	155	11.20	8,066.20	8,950
3550	30" diameter		9	2.667		8,050	163	11.80	8,224.80	9,100
3560	32" diameter		8.50	2.824		9,050	173	12.50	9,235.50	10,200
3570	34" diameter		8	3		9,850	184	13.30	10,047.30	11,100
3580	36" diameter		7	3.429		10,500	210	15.20	10,725.20	11,800
3620	For multi-pipe casings, add					5%				
4000	Steel casing anchors, single pipe									
4020	6-5/8" diameter	Q-18	8	3	Ea.	1,775	184	13.30	1,972.30	2,250
4040	8-5/8" diameter		7.50	3.200		1,850	196	14.20	2,060.20	2,325
4050	10-3/4" diameter		7	3.429		2,400	210	15.20	2,625.20	2,975
4060	12-3/4" diameter		6.50	3.692		2,575	226	16.35	2,817.35	3,200
4070	14" diameter		6	4		3,050	245	17.70	3,312.70	3,725
4080	16" diameter		5.50	4.364		3,525	267	19.35	3,811.35	4,300
4090	18" diameter		5	4.800		3,975	294	21.50	4,290.50	4,850
4100	20" diameter		4.50	5.333		4,400	325	23.50	4,748.50	5,350
4110	22" diameter		4	6		4,900	365	26.50	5,291.50	5,950
4120	24" diameter		3.50	6.857		5,325	420	30.50	5,775.50	6,525
4130	26" diameter		3	8		6,025	490	35.50	6,550.50	7,400
4140	28" diameter		2.50	9.600		6,625	585	42.50	7,252.50	8,225
4150	30" diameter		2	12		7,050	735	53	7,838	8,900
4160	32" diameter		1.50	16		8,400	980	71	9,451	10,800
4170	34" diameter		1	24		9,450	1,475	106	11,031	12,700
4180	36" diameter		1	24		10,300	1,475	106	11,881	13,600
4220	For multi-pipe, add					5%	20%			
4800	Steel casing elbow									
4820	6-5/8" diameter	Q-18	15	1.600	Ea.	2,525	98	7.10	2,630.10	2,925
4830	8-5/8" diameter		15	1.600		2,700	98	7.10	2,805.10	3,100
4850	10-3/4" diameter		14	1.714		3,250	105	7.60	3,362.60	3,750
4860	12-3/4" diameter		13	1.846		3,850	113	8.20	3,971.20	4,400
4870	14" diameter		12	2		4,125	122	8.85	4,255.85	4,750
4880	16" diameter		11	2.182		4,500	133	9.65	4,642.65	5,150
4890	18" diameter		10	2.400		5,150	147	10.65	5,307.65	5,875
4900	20" diameter		9	2.667		5,500	163	11.80	5,674.80	6,300
4910	22" diameter		8	3		5,900	184	13.30	6,097.30	6,800
4920	24" diameter		7	3.429		6,500	210	15.20	6,725.20	7,475
4930	26" diameter		6	4		7,125	245	17.70	7,387.70	8,225
4940	28" diameter		5	4.800		7,700	294	21.50	8,015.50	8,925
4950	30" diameter		4	6		7,725	365	26.50	8,116.50	9,075
4960	32" diameter		3	8		8,775	490	35.50	9,300.50	10,400
4970	34" diameter		2	12		9,650	735	53	10,438	11,800
4980	36" diameter		2	12		10,300	735	53	11,088	12,500
5500	Black steel service pipe, std. wt., 1" thick insulation									
5510	3/4" diameter pipe size	Q-17	54	.296	L.F.	79.50	17.50	1.97	98.97	115
5540	1" diameter pipe size		50	.320		81.50	18.90	2.13	102.53	121
5550	1-1/4" diameter pipe size		47	.340		104	20	2.26	126.26	146
5560	1-1/2" diameter pipe size		45	.356		109	21	2.36	132.36	153

33 63 Steam Energy Distribution

33 63 13 – Underground Steam and Condensate Distribution Piping

33 63 13.10 Calcium Silicate Insulated System		Crew	Daily Output	Labor-Hours	Unit	2020 Bare Costs Material	Labor	Equipment	Total	Total Incl O&P
5570	2" diameter pipe size	Q-17	42	.381	L.F.	118	22.50	2.53	143.03	166
5580	2-1/2" diameter pipe size		34	.471		124	28	3.13	155.13	181
5590	3" diameter pipe size		28	.571		108	33.50	3.80	145.30	174
5600	4" diameter pipe size		22	.727		121	43	4.84	168.84	202
5610	5" diameter pipe size		18	.889		136	52.50	5.90	194.40	235
5620	6" diameter pipe size	Q-18	23	1.043		147	64	4.62	215.62	262
6000	Black steel service pipe, std. wt., 1-1/2" thick insul.									
6010	3/4" diameter pipe size	Q-17	54	.296	L.F.	94	17.50	1.97	113.47	131
6040	1" diameter pipe size		50	.320		99	18.90	2.13	120.03	140
6050	1-1/4" diameter pipe size		47	.340		107	20	2.26	129.26	149
6060	1-1/2" diameter pipe size		45	.356		106	21	2.36	129.36	150
6070	2" diameter pipe size		42	.381		113	22.50	2.53	138.03	161
6080	2-1/2" diameter pipe size		34	.471		104	28	3.13	135.13	159
6090	3" diameter pipe size		28	.571		115	33.50	3.80	152.30	182
6100	4" diameter pipe size		22	.727		132	43	4.84	179.84	214
6110	5" diameter pipe size		18	.889		157	52.50	5.90	215.40	258
6120	6" diameter pipe size	Q-18	23	1.043		164	64	4.62	232.62	281
6130	8" diameter pipe size		19	1.263		208	77.50	5.60	291.10	350
6140	10" diameter pipe size		16	1.500		278	92	6.65	376.65	450
6150	12" diameter pipe size		13	1.846		370	113	8.20	491.20	585
6190	For 2" thick insulation, add					15%				
6220	For 2-1/2" thick insulation, add					25%				
6260	For 3" thick insulation, add					30%				
6800	Black steel service pipe, ex. hvy. wt., 1" thick insul.									
6820	3/4" diameter pipe size	Q-17	50	.320	L.F.	95	18.90	2.13	116.03	136
6840	1" diameter pipe size		47	.340		101	20	2.26	123.26	143
6850	1-1/4" diameter pipe size		44	.364		112	21.50	2.42	135.92	158
6860	1-1/2" diameter pipe size		42	.381		114	22.50	2.53	139.03	161
6870	2" diameter pipe size		40	.400		125	23.50	2.66	151.16	175
6880	2-1/2" diameter pipe size		31	.516		102	30.50	3.43	135.93	162
6890	3" diameter pipe size		27	.593		118	35	3.94	156.94	187
6900	4" diameter pipe size		21	.762		129	45	5.05	179.05	215
6910	5" diameter pipe size		17	.941		156	55.50	6.25	217.75	262
6920	6" diameter pipe size	Q-18	22	1.091		166	66.50	4.83	237.33	288
7400	Black steel service pipe, ex. hvy. wt., 1-1/2" thick insul.									
7420	3/4" diameter pipe size	Q-17	50	.320	L.F.	70	18.90	2.13	91.03	108
7440	1" diameter pipe size		47	.340		88.50	20	2.26	110.76	129
7450	1-1/4" diameter pipe size		44	.364		97.50	21.50	2.42	121.42	142
7460	1-1/2" diameter pipe size		42	.381		96	22.50	2.53	121.03	142
7470	2" diameter pipe size		40	.400		93	23.50	2.66	119.16	140
7480	2-1/2" diameter pipe size		31	.516		95.50	30.50	3.43	129.43	154
7490	3" diameter pipe size		27	.593		105	35	3.94	143.94	173
7500	4" diameter pipe size		21	.762		134	45	5.05	184.05	220
7510	5" diameter pipe size		17	.941		185	55.50	6.25	246.75	293
7520	6" diameter pipe size	Q-18	22	1.091		197	66.50	4.83	268.33	320
7530	8" diameter pipe size		18	1.333		270	81.50	5.90	357.40	425
7540	10" diameter pipe size		15	1.600		300	98	7.10	405.10	490
7550	12" diameter pipe size		13	1.846		360	113	8.20	481.20	580
7590	For 2" thick insulation, add					13%				
7640	For 2-1/2" thick insulation, add					18%				
7680	For 3" thick insulation, add					24%				

33 63 13 – Underground Steam and Condensate Distribution Piping

33 63 13.20 Combined Steam Pipe With Condensate Return		Crew	Daily Output	Labor-Hours	Unit	Material	2020 Bare Costs Labor	Equipment	Total	Total Incl O&P
0010	**COMBINED STEAM PIPE WITH CONDENSATE RETURN**									
9010	8" and 4" in 24" case	Q-18	100	.240	L.F.	415	14.70	1.06	430.76	480
9020	6" and 3" in 20" case		104	.231		405	14.10	1.02	420.12	465
9030	3" and 1-1/2" in 16" case		110	.218		305	13.35	.97	319.32	355
9040	2" and 1-1/4" in 12-3/4" case		114	.211		255	12.90	.93	268.83	300
9050	1-1/2" and 1-1/4" in 10-3/4" case	↓	116	.207	↓	266	12.65	.92	279.57	315
9100	Steam pipe only (no return)									
9110	6" in 18" case	Q-18	108	.222	L.F.	330	13.60	.98	344.58	385
9120	4" in 14" case		112	.214		255	13.10	.95	269.05	300
9130	3" in 14" case		114	.211		242	12.90	.93	255.83	286
9140	2-1/2" in 12-3/4" case		118	.203		225	12.45	.90	238.35	268
9150	2" in 12-3/4" case	↓	122	.197	↓	213	12.05	.87	225.92	253

Estimating Tips

Products such as conveyors, material handling cranes and hoists, and other items specified in this division require trained installers. The general contractor may not have any choice as to who will perform the installation or when it will be performed. Long lead times are often required for these products, making early decisions in purchasing and scheduling necessary. The installation of this type of equipment may require the embedment of mounting hardware during the construction of floors, structural walls, or interior walls/partitions. Electrical connections will require coordination with the electrical contractor.

Reference Numbers

Reference numbers are shown at the beginning of some major classifications. These numbers refer to related items in the Reference Section. The reference information may be an estimating procedure, an alternate pricing method, or technical information.

41 22 Cranes and Hoists

41 22 23 – Hoists

41 22 23.10 Material Handling		Crew	Daily Output	Labor-Hours	Unit	Material	Labor	Equipment	Total	Total Incl O&P
						2020 Bare Costs				
0010	**MATERIAL HANDLING**, cranes, hoists and lifts									
1500	Cranes, portable hydraulic, floor type, 2,000 lb. capacity				Ea.	3,700			3,700	4,075
1600	4,000 lb. capacity					5,075			5,075	5,575
1800	Movable gantry type, 12′ to 15′ range, 2,000 lb. capacity					4,775			4,775	5,250
1900	6,000 lb. capacity					7,450			7,450	8,200
2100	Hoists, electric overhead, chain, hook hung, 15′ lift, 1 ton cap.					2,825			2,825	3,100
2200	3 ton capacity					3,800			3,800	4,200
2500	5 ton capacity				▼	8,350			8,350	9,200
2600	For hand-pushed trolley, add					15%				
2700	For geared trolley, add					30%				
2800	For motor trolley, add					75%				
3000	For lifts over 15′, 1 ton, add				L.F.	25.50			25.50	28
3100	5 ton, add				″	78.50			78.50	86
3300	Lifts, scissor type, portable, electric, 36″ high, 2,000 lb.				Ea.	3,875			3,875	4,250
3400	48″ high, 4,000 lb.				″	4,700			4,700	5,175

Estimating Tips

This section involves equipment and construction costs for air, noise, and odor pollution control systems. These systems may be interrelated and care must be taken that the complete systems are estimated. For example, air pollution equipment may include dust and air-entrained particles that have to be collected. The vacuum systems could be noisy, requiring silencers to reduce noise pollution, and the collected solids have to be disposed of to prevent solid pollution.

Reference Numbers

Reference numbers are shown at the beginning of some major classifications. These numbers refer to related items in the Reference Section. The reference information may be an estimating procedure, an alternate pricing method, or technical information.

44 11 Particulate Control Equipment

44 11 16 – Fugitive Dust Barrier Systems

44 11 16.10 Dust Collection Systems		Crew	Daily Output	Labor-Hours	Unit	Material	2020 Bare Costs Labor	Equipment	Total	Total Incl O&P
0010	**DUST COLLECTION SYSTEMS** Commercial/Industrial									
0120	Central vacuum units									
0130	Includes stand, filters and motorized shaker									
0200	500 CFM, 10" inlet, 2 HP	Q-20	2.40	8.333	Ea.	4,850	475		5,325	6,050
0220	1,000 CFM, 10" inlet, 3 HP		2.20	9.091		5,100	520		5,620	6,375
0240	1,500 CFM, 10" inlet, 5 HP		2	10		5,775	570		6,345	7,250
0260	3,000 CFM, 13" inlet, 10 HP		1.50	13.333		16,100	760		16,860	18,900
0280	5,000 CFM, 16" inlet, 2 @ 10 HP		1	20		17,100	1,150		18,250	20,500
1000	Vacuum tubing, galvanized									
1100	2-1/8" OD, 16 ga.	Q-9	440	.036	L.F.	3.60	2.04		5.64	7.05
1110	2-1/2" OD, 16 ga.		420	.038		3.98	2.14		6.12	7.60
1120	3" OD, 16 ga.		400	.040		4.80	2.24		7.04	8.70
1130	3-1/2" OD, 16 ga.		380	.042		6.80	2.36		9.16	11.10
1140	4" OD, 16 ga.		360	.044		7.40	2.49		9.89	11.90
1150	5" OD, 14 ga.		320	.050		13.30	2.80		16.10	18.85
1160	6" OD, 14 ga.		280	.057		14.85	3.20		18.05	21
1170	8" OD, 14 ga.		200	.080		21.50	4.49		25.99	30.50
1180	10" OD, 12 ga.		160	.100		48.50	5.60		54.10	62
1190	12" OD, 12 ga.		120	.133		51.50	7.50		59	68
1200	14" OD, 12 ga.		80	.200		65	11.20		76.20	88.50
1940	Hose, flexible wire reinforced rubber									
1956	3" diam.	Q-9	400	.040	L.F.	6.90	2.24		9.14	10.95
1960	4" diam.		360	.044		8.10	2.49		10.59	12.70
1970	5" diam.		320	.050		10.05	2.80		12.85	15.30
1980	6" diam.		280	.057		10.70	3.20		13.90	16.60
2000	90° elbow, slip fit									
2110	2-1/8" diam.	Q-9	70	.229	Ea.	13.05	12.80		25.85	34
2120	2-1/2" diam.		65	.246		18.25	13.80		32.05	41
2130	3" diam.		60	.267		24.50	14.95		39.45	49.50
2140	3-1/2" diam.		55	.291		30.50	16.30		46.80	58
2150	4" diam.		50	.320		38	17.95		55.95	69
2160	5" diam.		45	.356		72	19.95		91.95	109
2170	6" diam.		40	.400		99	22.50		121.50	143
2180	8" diam.		30	.533		189	30		219	253
2400	45° elbow, slip fit									
2410	2-1/8" diam.	Q-9	70	.229	Ea.	11.45	12.80		24.25	32
2420	2-1/2" diam.		65	.246		16.85	13.80		30.65	39.50
2430	3" diam.		60	.267		20.50	14.95		35.45	45
2440	3-1/2" diam.		55	.291		25.50	16.30		41.80	52.50
2450	4" diam.		50	.320		33	17.95		50.95	63.50
2460	5" diam.		45	.356		56	19.95		75.95	91.50
2470	6" diam.		40	.400		75.50	22.50		98	117
2480	8" diam.		35	.457		148	25.50		173.50	202
2800	90° TY, slip fit thru 6" diam.									
2810	2-1/8" diam.	Q-9	42	.381	Ea.	21	21.50		42.50	56
2820	2-1/2" diam.		39	.410		27.50	23		50.50	65
2830	3" diam.		36	.444		36.50	25		61.50	78.50
2840	3-1/2" diam.		33	.485		46.50	27		73.50	92.50
2850	4" diam.		30	.533		45.50	30		75.50	95.50
2860	5" diam.		27	.593		74.50	33		107.50	133
2870	6" diam.		24	.667		111	37.50		148.50	179
2880	8" diam., butt end					180			180	198
2890	10" diam., butt end					565			565	625

44 11 Particulate Control Equipment

44 11 16 – Fugitive Dust Barrier Systems

44 11 16.10 Dust Collection Systems		Crew	Daily Output	Labor-Hours	Unit	2020 Bare Costs Material	Labor	Equipment	Total	Total Incl O&P
2900	12″ diam., butt end				Ea.	565			565	625
2910	14″ diam., butt end					1,025			1,025	1,125
2920	6″ x 4″ diam., butt end					127			127	140
2930	8″ x 4″ diam., butt end					241			241	265
2940	10″ x 4″ diam., butt end					315			315	350
2950	12″ x 4″ diam., butt end					360			360	395
3100	90° elbow, butt end, segmented									
3110	8″ diam., butt end, segmented				Ea.	158			158	174
3120	10″ diam., butt end, segmented					470			470	515
3130	12″ diam., butt end, segmented					595			595	655
3140	14″ diam., butt end, segmented					745			745	820
3200	45° elbow, butt end, segmented									
3210	8″ diam., butt end, segmented				Ea.	148			148	163
3220	10″ diam., butt end, segmented					340			340	375
3230	12″ diam., butt end, segmented					345			345	380
3240	14″ diam., butt end, segmented					640			640	705
3400	All butt end fittings require one coupling per joint.									
3410	Labor for fitting included with couplings.									
3460	Compression coupling, galvanized, neoprene gasket									
3470	2-1/8″ diam.	Q-9	44	.364	Ea.	13.55	20.50		34.05	46
3480	2-1/2″ diam.		44	.364		13.55	20.50		34.05	46
3490	3″ diam.		38	.421		19.45	23.50		42.95	57.50
3500	3-1/2″ diam.		35	.457		22	25.50		47.50	63
3510	4″ diam.		33	.485		24	27		51	67
3520	5″ diam.		29	.552		27	31		58	77
3530	6″ diam.		26	.615		33	34.50		67.50	89
3540	8″ diam.		22	.727		57.50	41		98.50	125
3550	10″ diam.		20	.800		93.50	45		138.50	171
3560	12″ diam.		18	.889		115	50		165	202
3570	14″ diam.		16	1		150	56		206	250
3800	Air gate valves, galvanized									
3810	2-1/8″ diam.	Q-9	30	.533	Ea.	122	30		152	180
3820	2-1/2″ diam.		28	.571		133	32		165	195
3830	3″ diam.		26	.615		141	34.50		175.50	209
3840	4″ diam.		23	.696		164	39		203	239
3850	6″ diam.		18	.889		228	50		278	325

Division Notes

	CREW	DAILY OUTPUT	LABOR-HOURS	UNIT	BARE COSTS MAT.	BARE COSTS LABOR	BARE COSTS EQUIP.	BARE COSTS TOTAL	TOTAL INCL O&P

Estimating Tips

This division contains information about water and wastewater equipment and systems, which was formerly located in Division 44. The main areas of focus are total wastewater treatment plants and components of wastewater treatment plants. Also included in this section are oil/water separators for wastewater treatment.

Reference Numbers

Reference numbers are shown at the beginning of some major classifications. These numbers refer to related items in the Reference Section. The reference information may be an estimating procedure, an alternate pricing method, or technical information.

46 07 Packaged Water and Wastewater Treatment Equipment

46 07 53 – Packaged Wastewater Treatment Equipment

46 07 53.10 Biological Pkg. Wastewater Treatment Plants

		Crew	Daily Output	Labor-Hours	Unit	2020 Bare Costs Material	Labor	Equipment	Total	Total Incl O&P
0010	**BIOLOGICAL PACKAGED WASTEWATER TREATMENT PLANTS**									
0011	Not including fencing or external piping									
0020	Steel packaged, blown air aeration plants									
0100	1,000 GPD				Gal.				55	60.50
0200	5,000 GPD								22	24
0300	15,000 GPD								22	24
0400	30,000 GPD								15.40	16.95
0500	50,000 GPD								11	12.10
0600	100,000 GPD								9.90	10.90
0700	200,000 GPD								8.80	9.70
0800	500,000 GPD								7.70	8.45
1000	Concrete, extended aeration, primary and secondary treatment									
1010	10,000 GPD				Gal.				22	24
1100	30,000 GPD								15.40	16.95
1200	50,000 GPD								11	12.10
1400	100,000 GPD								9.90	10.90
1500	500,000 GPD								7.70	8.45
1700	Municipal wastewater treatment facility									
1720	1.0 MGD				Gal.				11	12.10
1740	1.5 MGD								10.60	11.65
1760	2.0 MGD								10	11
1780	3.0 MGD								7.80	8.60
1800	5.0 MGD								5.80	6.70
2000	Holding tank system, not incl. excavation or backfill									
2010	Recirculating chemical water closet	2 Plum	4	4	Ea.	550	258		808	990
2100	For voltage converter, add	"	16	1		320	64.50		384.50	445
2200	For high level alarm, add	1 Plum	7.80	1.026		123	66		189	235

46 07 53.20 Wastewater Treatment System

		Crew	Daily Output	Labor-Hours	Unit	Material	Labor	Equipment	Total	Total Incl O&P
0010	**WASTEWATER TREATMENT SYSTEM**									
0020	Fiberglass, 1,000 gallon	B-21	1.29	21.705	Ea.	4,450	1,050	147	5,647	6,650
0100	1,500 gallon	"	1.03	27.184	"	9,700	1,325	184	11,209	12,900

46 23 Grit Removal And Handling Equipment

46 23 23 – Vortex Grit Removal Equipment

46 23 23.10 Rainwater Filters

		Crew	Daily Output	Labor-Hours	Unit	Material	Labor	Equipment	Total	Total Incl O&P
0010	**RAINWATER FILTERS**									
0100	42 gal./min	B-21	3.50	8	Ea.	40,600	390	54	41,044	45,300
0200	65 gal./min		3.50	8		40,700	390	54	41,144	45,300
0300	208 gal./min		3.50	8		40,700	390	54	41,144	45,400

46 25 Oil and Grease Separation and Removal Equipment

46 25 13 – Coalescing Oil-Water Separators

46 25 13.20 Oil/Water Separators		Crew	Daily Output	Labor-Hours	Unit	2020 Bare Costs Material	Labor	Equipment	Total	Total Incl O&P
0010	**OIL/WATER SEPARATORS**									
0020	Underground, tank only									
0030	Excludes excavation, backfill & piping									
0100	200 GPM	B-21	3.50	8	Ea.	40,400	390	54	40,844	45,000
0110	400 GPM		3.25	8.615		85,500	420	58	85,978	94,500
0120	600 GPM		2.75	10.182		90,000	495	68.50	90,563.50	100,500
0130	800 GPM		2.50	11.200		108,000	545	75.50	108,620.50	119,500
0140	1,000 GPM		2	14		122,000	685	94.50	122,779.50	135,500
0150	1,200 GPM		1.50	18.667		134,000	910	126	135,036	148,500
0160	1,500 GPM	B-21	1	28	Ea.	154,000	1,375	189	155,564	171,500

Division Notes

		CREW	DAILY OUTPUT	LABOR-HOURS	UNIT	BARE COSTS MAT.	LABOR	EQUIP.	TOTAL	TOTAL INCL O&P

Assemblies Section

Table of Contents

RSMeans data: Assemblies—How They Work

Assemblies estimating provides a fast and reasonably accurate way to develop construction costs. An assembly is the grouping of individual work items—with appropriate quantities—to provide a cost for a major construction component in a convenient unit of measure.

An assemblies estimate is often used during early stages of design development to compare the cost impact of various design alternatives on total building cost.

Assemblies estimates are also used as an efficient tool to verify construction estimates.

Assemblies estimates do not require a completed design or detailed drawings. Instead, they are based on the general size of the structure and other known parameters of the project. The degree of accuracy of an assemblies estimate is generally within +/- 15%.

Most assemblies consist of three major elements: a graphic, the system components, and the cost data itself. The **Graphic** is a visual representation showing the typical appearance of the assembly

1 Unique 12-character Identifier

Our assemblies are identified by a **unique 12-character identifier**. The assemblies are numbered using UNIFORMAT II, ASTM Standard E1557. The first 5 characters represent this system to Level 3. The last 7 characters represent further breakdown in order to arrange items in understandable groups of similar tasks. Line numbers are consistent across all of our publications, so a line number in any assemblies data set will always refer to the same item.

2 Reference Box

Information is available in the Reference Section to assist the estimator with estimating procedures, alternate pricing methods, and additional technical information.

The **Reference Box** indicates the exact location of this information in the Reference Section. The "R" stands for "reference," and the remaining characters are the line numbers.

3 Narrative Descriptions

Our assemblies descriptions appear in two formats: narrative and table. **Narrative descriptions** are shown in a hierarchical structure to make them readable. In order to read a complete description, read up through the indents to the top of the section. Include everything that is above and to the left that is not contradicted by information below.

4 System Components

System components are listed separately to detail what is included in the development of the total system price.

Narrative Format

D40 Fire Protection

D4020 Standpipes

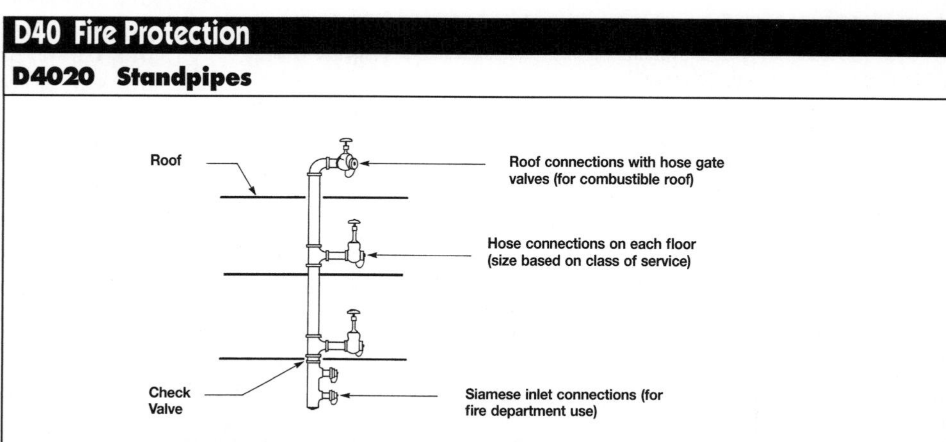

System Components	QUANTITY	UNIT	COST PER FLOOR MAT.	COST PER FLOOR INST.	COST PER FLOOR TOTAL
SYSTEM D4020 310 0560					
WET STANDPIPE RISER, CLASS I, STEEL, BLACK, SCH. 40 PIPE, 10' HEIGHT					
4" DIAMETER PIPE, ONE FLOOR					
Pipe, steel, black, schedule 40, threaded, 4" diam.	20.000	L.F.	510	770	1,280
Pipe, Tee, malleable iron, black, 150 lb. threaded, 4" pipe size	2.000	Ea.	780	690	1,470
Pipe, 90° elbow, malleable iron, black, 150 lb threaded, 4" pipe size	1.000	Ea.	216	231	447
Pipe, nipple, steel, black, schedule 40, 2-1/2" pipe size x 3" long	2.000	Ea.	28.20	174	202.20
Fire valve, gate, 300 lb., brass w/handwheel, 2-1/2" pipe size	1.000	Ea.	239	108	347
Fire valve, pressure reducing rgh brs, 2-1/2" pipe size	1.000	Ea.	980	216	1,196
Valve, swing check, w/ball drip, CI w/brs. ftngs., 4" pipe size	1.000	Ea.	460	455	915
Standpipe conn wall dble. flush brs. w/plugs & chains 2-1/2"x2-1/2"x4"	1.000	Ea.	850	273	1,123
Valve, swing check, bronze, 125 lb, regrinding disc, 2-1/2" pipe size	1.000	Ea.	1,025	92.50	1,117.50
Roof manifold, fire, w/valves & caps, horiz/vert brs 2-1/2"x2-1/2"x4"	1.000	Ea.	250	284	534
Fire, hydrolator, vent & drain, 2-1/2" pipe size	1.000	Ea.	126	63	189
Valve, gate, iron body 125 lb., OS&Y, threaded, 4" pipe size	1.000	Ea.	755	465	1,220
Tamper switch (valve supervisory switch)	1.000	Ea.	291	47.50	338.50
TOTAL			6,510.20	3,869	10,379.20

D4020 310	Wet Standpipe Risers, Class I		COST PER FLOOR MAT.	COST PER FLOOR INST.	COST PER FLOOR TOTAL
0550	Wet standpipe risers, Class I, steel, black, sch. 40, 10' height				
0560	4" diameter pipe, one floor		6,500	3,875	10,375
0580	Additional floors		1,525	1,175	2,700
0600	6" diameter pipe, one floor		10,200	6,750	16,950
0620	Additional floors		2,700	1,900	4,600
0640	8" diameter pipe, one floor	R211226-10	15,200	8,150	23,350
0660	Additional floors		3,900	2,300	6,200
0680		R211226-20			

For supplemental customizable square foot estimating forms, visit: **RSMeans.com/2020books**

in question. It is frequently accompanied by additional explanatory technical information describing the class of items. The **System Components** is a listing of the individual tasks that make up the assembly, including the quantity and unit of measure for each item, along with the cost of material and installation. The **Assemblies data** below lists prices for other similar systems with dimensional and/or size variations.

All of our assemblies costs represent the cost for the installing contractor. An allowance for profit has been added to all material, labor, and equipment rental costs. A markup for labor burdens, including workers' compensation, fixed overhead, and business overhead, is included with installation costs.

The information in RSMeans cost data represents a "national average" cost. This data should be modified to the project location using the **City Cost Indexes** or **Location Factors** tables found in the Reference Section.

Table Format

D20 Plumbing

D2010 Plumbing Fixtures

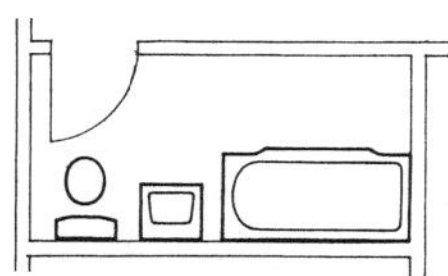

Example of Plumbing Cost Calculations: The bathroom system includes the individual fixtures such as bathtub, lavatory, shower and water closet. These fixtures are listed below as separate items merely as a checklist.

D2010 951 Plumbing Systems 20 Unit, 2 Story Apartment Building

	FIXTURE	SYSTEM	LINE	QUANTITY	UNIT	COST EACH MAT.	COST EACH INST.	COST EACH TOTAL
0440	Bathroom	D2010 926	3640	20	Ea.	66,500	55,000	121,500
0480	Bathtub							
0520	Booster pump[1]	not req'd.						
0560	Drinking fountain							
0600	Garbage disposal[1]	not incl.						
0660								
0680	Grease interceptor							
0720	Water heater	D2020 250	2140	1	Ea.	16,900	3,725	20,625
0760	Kitchen sink	D2010 410	1960	20	Ea.	29,600	19,400	49,000
0800	Laundry sink	D2010 420	1840	4	Ea.	5,050	3,825	8,875
0840	Lavatory							
0900								
0920	Roof drain, 1 floor	D2040 210	4200	2	Ea.	2,575	2,000	4,575
0960	Roof drain, add'l floor	D2040 210	4240	20	L.F.	540	510	1,050
1000	Service sink	D2010 440	4300	1	Ea.	2,675	1,350	4,025
1040	Sewage ejector[1]	not req'd.						
1080	Shower							
1100								
1160	Sump pump							
1200	Urinal							
1240	Water closet							
1320								
1360	**SUB TOTAL**					123,500	86,000	209,500
1481	Water controls	R221113-40		10%[2]		12,400	8,600	21,000
1521	Pipe & fittings[3]	R221113-40		30%[2]		37,100	25,800	62,900
1560	Other							
1601	Quality/complexity	R221113-40		15%[2]		18,600	12,900	31,500
1680								
1720	**TOTAL**					192,000	133,500	325,500
1741								

[1]**Note:** Cost for items such as booster pumps, backflow preventers, sewage ejectors, water meters, etc., may be obtained from the unit price section in the front of this data set.

Water controls, pipe and fittings, and the Quality/Complexity factors come from Table R221113-40.

[2]Percentage of subtotal.

[3]Long, easily discernable runs of pipe would be more accurately priced from Unit Price Section 22 11 13. If this is done, reduce the miscellaneous percentage in proportion.

5 Unit of Measure

All RSMeans data: Assemblies include a typical **Unit of Measure** used for estimating that item. For instance, for continuous footings or foundation walls the unit is linear feet (L.F.). For spread footings the unit is each (Ea.). The estimator needs to take special care that the unit in the data matches the unit in the takeoff. Abbreviations and unit conversions can be found in the Reference Section.

6 Table Descriptions

Table descriptions work similar to Narrative Descriptions, except that if there is a blank in the column at a particular line number, read up to the description above in the same column.

RSMeans data: Assemblies—How They Work (Continued)

Sample Estimate

This sample demonstrates the elements of an estimate, including a tally of the RSMeans data lines. Published assemblies costs include all markups for labor burden and profit for the installing contractor. This estimate adds a summary of the markups applied by a general contractor on the installing contractor's work. These figures represent the total cost to the owner. The location factor with RSMeans data is applied at the bottom of the estimate to adjust the cost of the work to a specific location.

	Project Name:	Interior Fit-out, ABC Office			
	Location:	**Anywhere, USA**	**Date: 1/1/2020**		**STD**
	Assembly Number	**Description**	**Qty.**	**Unit**	**Subtotal**
1	C1010 124 1200	Wood partition, 2 x 4 @ 16" OC w/5/8" FR gypsum board	560	S.F.	$2,996.00
	C1020 114 1800	Metal door & frame, flush hollow core, 3'-0" x 7'-0"	2	Ea.	$2,620.00
	C3010 230 0080	Painting, brushwork, primer & 2 coats	1,120	S.F.	$1,467.20
	C3020 410 0140	Carpet, tufted, nylon, roll goods, 12' wide, 26 oz	240	S.F.	$928.80
	C3030 210 6000	Acoustic ceilings, 24" x 48" tile, tee grid suspension	200	S.F.	$1,326.00
	D5020 125 0560	Receptacles incl plate, box, conduit, wire, 20 A duplex	8	Ea.	$2,428.00
	D5020 125 0720	Light switch incl plate, box, conduit, wire, 20 A single pole	2	Ea.	$591.00
	D5020 210 0560	Fluorescent fixtures, recess mounted, 20 per 1000 SF	200	S.F.	$2,320.00
		Assembly Subtotal			**$14,677.00**
		Sales Tax @ 2	**5 %**		**$ 366.93**
		General Requirements @ 3	**7 %**		**$ 1,027.39**
		Subtotal A			**$16,071.32**
		GC Overhead @ 4	**5 %**		**$ 803.57**
		Subtotal B			**$16,874.88**
		GC Profit @ 5	**5 %**		**$ 843.74**
		Subtotal C			**$17,718.62**
		Adjusted by Location Factor 6	**115.5**		**$ 20,465.01**
		Architects Fee @ 7	**8 %**		**$ 1,637.20**
		Contingency @ 8	**15 %**		**$ 3,069.75**
		Project Total Cost			**$ 25,171.96**

This estimate is based on an interactive spreadsheet. You are free to download it and adjust it to your methodology.
A copy of this spreadsheet is available at **RSMeans.com/2020books.**

1 Work Performed

The body of the estimate shows the RSMeans data selected, including line numbers, a brief description of each item, its takeoff quantity and unit, and the total installed cost, including the installing contractor's overhead and profit.

2 Sales Tax

If the work is subject to state or local sales taxes, the amount must be added to the estimate. In a conceptual estimate it can be assumed that one half of the total represents material costs. Therefore, apply the sales tax rate to 50% of the assembly subtotal.

3 General Requirements

This item covers project-wide needs provided by the general contractor. These items vary by project but may include temporary facilities and utilities, security, testing, project cleanup, etc. In assemblies estimates a percentage is used—typically between 5% and 15% of project cost.

4 General Contractor Overhead

This entry represents the general contractor's markup on all work to cover project administration costs.

5 General Contractor Profit

This entry represents the GC's profit on all work performed. The value included here can vary widely by project and is influenced by the GC's perception of the project's financial risk and market conditions.

6 Location Factor

RSMeans published data are based on national average costs. If necessary, adjust the total cost of the project using a location factor from the "Location Factor" table or the "City Cost Indexes" table found in the Reference Section. Use location factors if the work is general, covering the work of multiple trades. If the work is by a single trade (e.g., masonry) use the more specific data found in the City Cost Indexes.

To adjust costs by location factors, multiply the base cost by the factor and divide by 100.

7 Architect's Fee

If appropriate, add the design cost to the project estimate. These fees vary based on project complexity and size. Typical design and engineering fees can be found in the Reference Section.

8 Contingency

A factor for contingency may be added to any estimate to represent the cost of unknowns that may occur between the time that the estimate is performed and the time the project is constructed. The amount of the allowance will depend on the stage of design at which the estimate is done, as well as the contractor's assessment of the risk involved.

D2010 Plumbing Fixtures

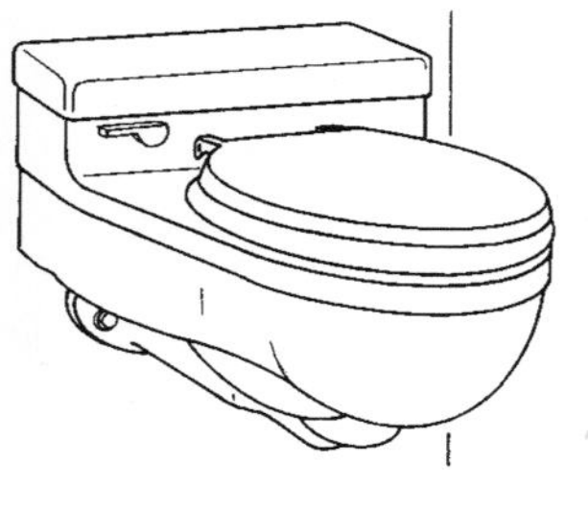

One Piece Wall Hung

Systems are complete with trim seat and rough-in (supply, waste and vent) for connection to supply branches and waste mains.

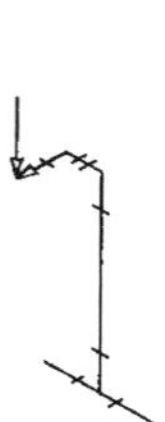

Supply

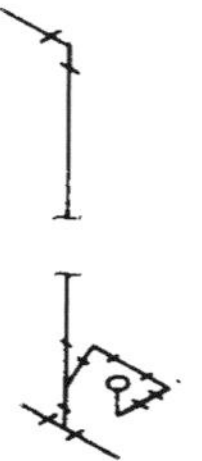

Waste/Vent

Floor Mount

System Components	QUANTITY	UNIT	COST EACH MAT.	COST EACH INST.	COST EACH TOTAL
SYSTEM D2010 110 1880					
WATER CLOSET, VITREOUS CHINA					
TANK TYPE, WALL HUNG, TWO PIECE					
Water closet, tank type vit china wall hung 2 pc. w/seat supply & stop	1.000	Ea.	460	262	722
Pipe Steel galvanized, schedule 40, threaded, 2″ diam.	4.000	L.F.	34.80	86	120.80
Pipe, CI soil, no hub, cplg 10′ OC, hanger 5′ OC, 4″ diam.	2.000	L.F.	55	48	103
Pipe, coupling, standard coupling, CI soil, no hub, 4″ diam.	2.000	Ea.	51	84	135
Copper tubing type L solder joint, hangar 10′ O.C., 1/2″ diam.	6.000	L.F.	24.30	57.30	81.60
Wrought copper 90° elbow for solder joints 1/2″ diam.	2.000	Ea.	2.78	77	79.78
Wrought copper Tee for solder joints 1/2″ diam.	1.000	Ea.	2.68	59.50	62.18
Supports/carrier, water closet, siphon jet, horiz, single, 4″ waste	1.000	Ea.	1,225	145	1,370
TOTAL			1,855.56	818.80	2,674.36

D2010 110	Water Closet Systems		COST EACH MAT.	COST EACH INST.	COST EACH TOTAL
1800	Water closet, vitreous china				
1840	Tank type, wall hung				
1880	Close coupled two piece	R221113 -40	1,850	820	2,670
1920	Floor mount, one piece		1,475	870	2,345
1960	One piece low profile	R224000 -30	1,425	870	2,295
2000	Two piece close coupled		675	870	1,545
2040	Bowl only with flush valve				
2080	Wall hung		2,775	930	3,705
2120	Floor mount		875	885	1,760
2160	Floor mount, ADA compliant with 18″ high bowl		885	905	1,790

D2010 Plumbing Fixtures

Systems are complete with trim, seat, flush valve and rough-in (supply, waste and vent) for connection to supply branches and waste mains.

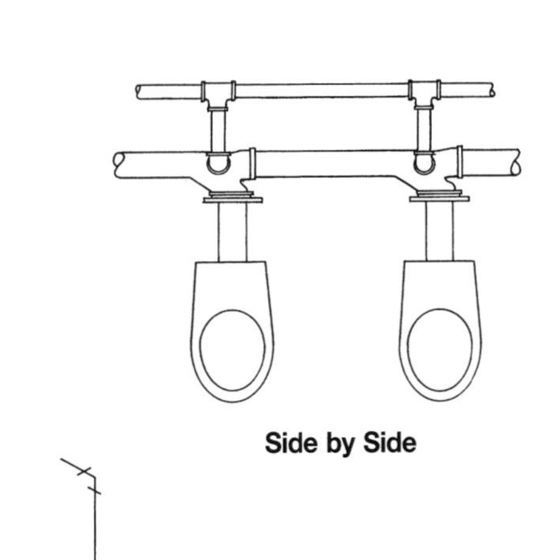
Side by Side

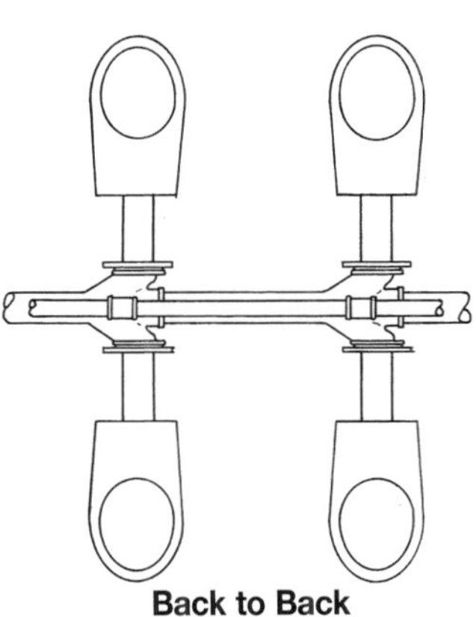
Back to Back

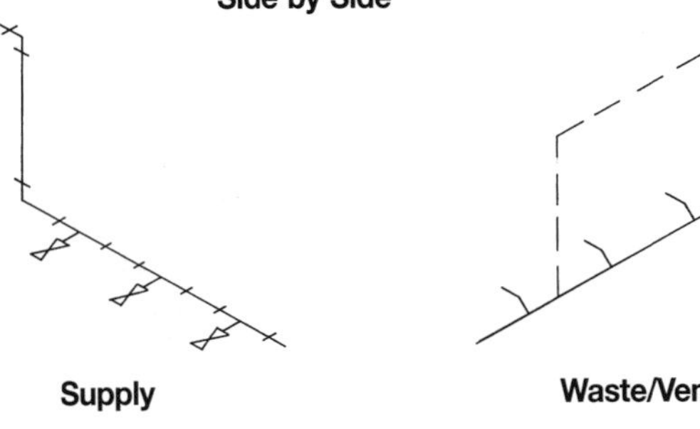

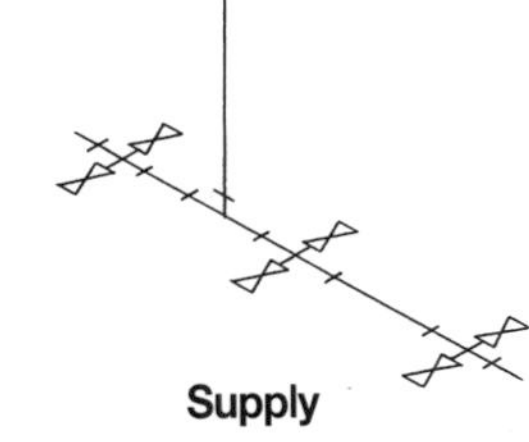

Supply

Waste/Vent

System Components	QUANTITY	UNIT	COST EACH MAT.	INST.	TOTAL
SYSTEM D2010 120 1760					
WATER CLOSETS, BATTERY MOUNT, WALL HUNG, SIDE BY SIDE, FIRST CLOSET					
Water closet, bowl only w/flush valve, seat, wall hung	1.000	Ea.	1,200	239	1,439
Pipe, CI soil, no hub, cplg 10′ OC, hanger 5′ OC, 4″ diam.	3.000	L.F.	82.50	72	154.50
Coupling, standard, CI, soil, no hub, 4″ diam.	2.000	Ea.	51	84	135
Copper tubing, type L, solder joints, hangers 10′ OC, 1″ diam.	6.000	L.F.	50.10	68.10	118.20
Copper tubing, type DWV, solder joints, hangers 10′OC, 2″ diam.	6.000	L.F.	120	105.30	225.30
Wrought copper 90° elbow for solder joints 1″ diam.	1.000	Ea.	7.75	48	55.75
Wrought copper Tee for solder joints 1″ diam.	1.000	Ea.	18.75	77	95.75
Support/carrier, siphon jet, horiz, adjustable single, 4″ pipe	1.000	Ea.	1,225	145	1,370
Valve, gate, bronze, 125 lb, NRS, soldered 1″ diam.	1.000	Ea.	86.50	40.50	127
Wrought copper, DWV, 90° elbow, 2″ diam.	1.000	Ea.	50.50	77	127.50
TOTAL			2,892.10	955.90	3,848

D2010 120	Water Closets, Group		COST EACH MAT.	INST.	TOTAL
1760	Water closets, battery mount, wall hung, side by side, first closet	R221113 -40	2,900	955	3,855
1800	Each additional water closet, add		2,775	905	3,680
3000	Back to back, first pair of closets	R224000 -30	4,675	1,275	5,950
3100	Each additional pair of closets, back to back		4,600	1,250	5,850
9000	Back to back, first pair of closets, auto sensor flush valve, 1.28 gpf		5,150	1,375	6,525
9100	Ea additional pair of cls, back to back, auto sensor flush valve, 1.28 gpf		4,975	1,275	6,250

D2010 Plumbing Fixtures

Systems are complete with trim, flush valve and rough-in (supply, waste and vent) for connection to supply branches and waste mains.

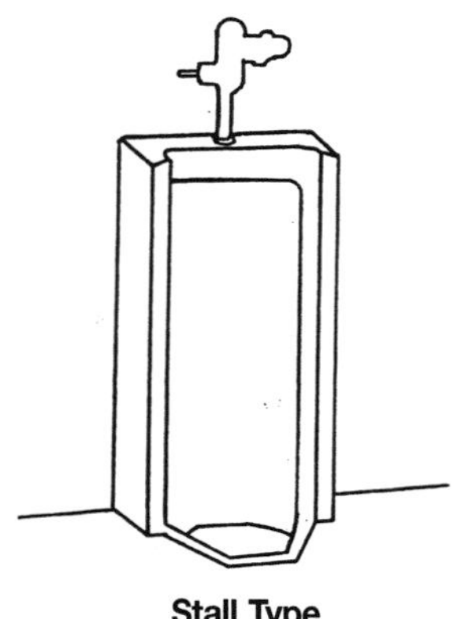

Stall Type

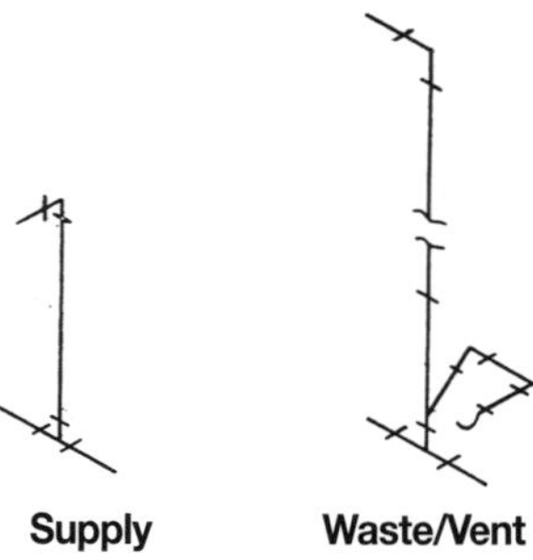

Supply Waste/Vent

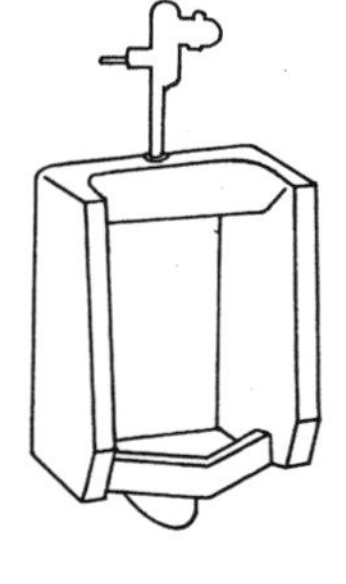

Wall Hung

System Components	QUANTITY	UNIT	COST EACH MAT.	COST EACH INST.	COST EACH TOTAL
SYSTEM D2010 210 2000					
URINAL, VITREOUS CHINA, WALL HUNG					
Urinal, wall hung, vitreous china, incl. hanger	1.000	Ea.	345	465	810
Pipe, steel, galvanized, schedule 40, threaded, 1-1/2″ diam.	5.000	L.F.	52	86.75	138.75
Copper tubing type DWV, solder joint, hangers 10′ OC, 2″ diam.	3.000	L.F.	60	52.65	112.65
Combination Y & 1/8 bend for CI soil pipe, no hub, 3″ diam.	1.000	Ea.	23.50		23.50
Pipe, CI, no hub, cplg. 10′ OC, hanger 5′ OC, 3″ diam.	4.000	L.F.	118	86	204
Pipe coupling standard, CI soil, no hub, 3″ diam.	2.000	Ea.	44	73	117
Copper tubing type L, solder joint, hanger 10′ OC 3/4″ diam.	5.000	L.F.	26	50.75	76.75
Wrought copper 90° elbow for solder joints 3/4″ diam.	1.000	Ea.	2.95	40.50	43.45
Wrought copper Tee for solder joints, 3/4″ diam.	1.000	Ea.	6.65	64.50	71.15
TOTAL			678.10	919.15	1,597.25

D2010 210	Urinal Systems		COST EACH MAT.	COST EACH INST.	COST EACH TOTAL
2000	Urinal, vitreous china, wall hung	R224000-30	680	920	1,600
2040	Stall type		1,475	1,100	2,575

D2010 Plumbing Fixtures

Systems are complete with trim, flush valve and rough-in (supply, waste and vent) for connection to supply branches and waste mains.

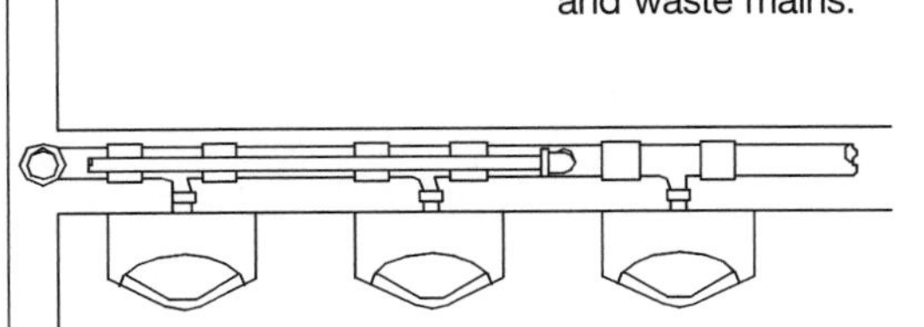

Side by Side

Back to Back

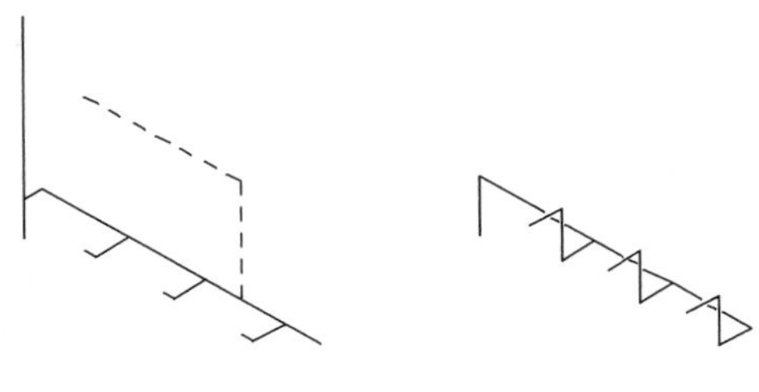

Waste/Vent

Supply

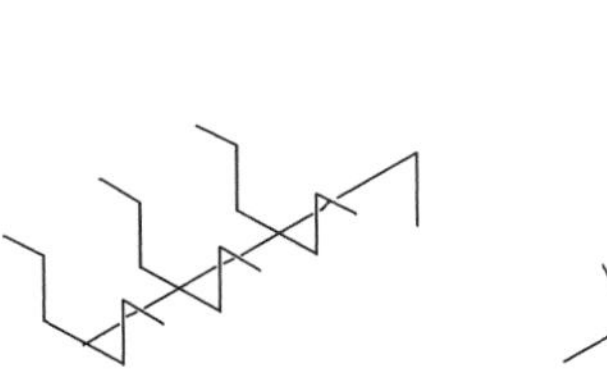

Supply

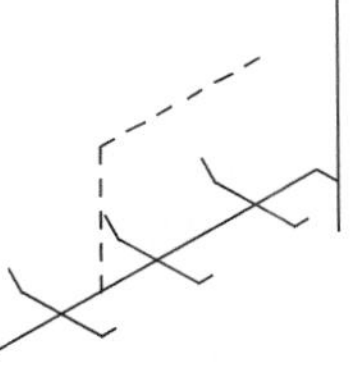

Waste/Vent

System Components	QUANTITY	UNIT	COST EACH MAT.	COST EACH INST.	COST EACH TOTAL
SYSTEM D2010 220 1760					
URINALS, BATTERY MOUNT, WALL HUNG, SIDE BY SIDE, FIRST URINAL					
Urinal, wall hung, vitreous china, with hanger & trim	1.000	Ea.	345	465	810
No hub cast iron soil pipe, 3″ diameter	4.000	L.F.	118	86	204
No hub cast iron sanitary tee, 3″ diameter	1.000	Ea.	23.50		23.50
No hub coupling, 3″ diameter	2.000	Ea.	44	73	117
Copper tubing, type L, 3/4″ diameter	5.000	L.F.	26	50.75	76.75
Copper tubing, type DWV, 2″ diameter	2.000	L.F.	40	35.10	75.10
Copper 90° elbow, 3/4″ diameter	2.000	Ea.	5.90	81	86.90
Copper 90° elbow, type DWV, 2″ diameter	2.000	Ea.	101	154	255
Galvanized steel pipe, 1-1/2″ diameter	5.000	L.F.	52	86.75	138.75
Cast iron drainage elbow, 90°, 1-1/2″ diameter	1.000	Ea.	6.65	64.50	71.15
TOTAL			762.05	1,096.10	1,858.15

D2010 220	Urinal Systems, Battery Mount		COST EACH MAT.	COST EACH INST.	COST EACH TOTAL
1760	Urinals, battery mount, side by side, first urinal	R221113 -40	760	1,100	1,860
1800	Each additional urinal, add		680	995	1,675
2000	Back to back, first pair of urinals	R224000 -30	1,300	1,675	2,975
2100	Each additional pair of urinals, back to back		985	1,325	2,310

D2010 Plumbing Fixtures

Systems are complete with trim and rough-in (supply, waste and vent) to connect to supply branches and waste mains.

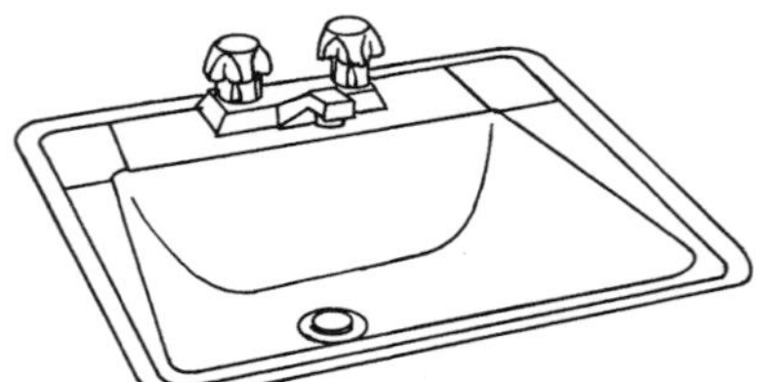

Vanity Top

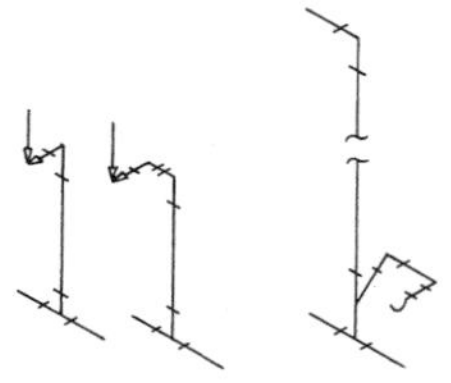

Supply   Waste/Vent

Wall Hung

System Components	QUANTITY	UNIT	COST EACH		
			MAT.	INST.	TOTAL
SYSTEM D2010 310 1560					
LAVATORY W/TRIM, VANITY TOP, P.E. ON C.I., 20″ X 18″					
Lavatory w/trim, PE on CI, white, vanity top, 20″ x 18″ oval	1.000	Ea.	340	217	557
Pipe, steel, galvanized, schedule 40, threaded, 1-1/4″ diam.	4.000	L.F.	38.80	62.40	101.20
Copper tubing type DWV, solder joint, hanger 10′ OC 1-1/4″ diam.	4.000	L.F.	54.80	51.40	106.20
Wrought copper DWV, Tee, sanitary, 1-1/4″ diam.	1.000	Ea.	31.50	85.50	117
P trap w/cleanout, 20 ga., 1-1/4″ diam.	1.000	Ea.	113	43	156
Copper tubing type L, solder joint, hanger 10′ OC 1/2″ diam.	10.000	L.F.	40.50	95.50	136
Wrought copper 90° elbow for solder joints 1/2″ diam.	2.000	Ea.	2.78	77	79.78
Wrought copper Tee for solder joints, 1/2″ diam.	2.000	Ea.	5.36	119	124.36
Stop, chrome, angle supply, 1/2″ diam.	2.000	Ea.	22.20	70	92.20
TOTAL			648.94	820.80	1,469.74

D2010 310	Lavatory Systems		COST EACH		
			MAT.	INST.	TOTAL
1560	Lavatory w/trim, vanity top, PE on CI, 20″ x 18″, Vanity top by others.	R221113 -40	650	820	1,470
1600	19″ x 16″ oval		440	820	1,260
1640	18″ round		810	820	1,630
1680	Cultured marble, 19″ x 17″		445	820	1,265
1720	25″ x 19″		475	820	1,295
1760	Stainless, self-rimming, 25″ x 22″		660	820	1,480
1800	17″ x 22″	R224000 -30	645	820	1,465
1840	Steel enameled, 20″ x 17″		450	845	1,295
1880	19″ round		485	845	1,330
1920	Vitreous china, 20″ x 16″		540	860	1,400
1960	19″ x 16″		550	860	1,410
2000	22″ x 13″		545	860	1,405
2040	Wall hung, PE on CI, 18″ x 15″		910	905	1,815
2080	19″ x 17″		970	905	1,875
2120	20″ x 18″		795	905	1,700
2160	Vitreous china, 18″ x 15″		715	930	1,645
2200	19″ x 17″		660	930	1,590
2240	24″ x 20″		800	930	1,730
2300	20″ x 27″, handicap		1,600	1,000	2,600

D2010 Plumbing Fixtures

Systems are complete with trim, flush valve and rough-in (supply, waste and vent) for connection to supply branches and waste mains.

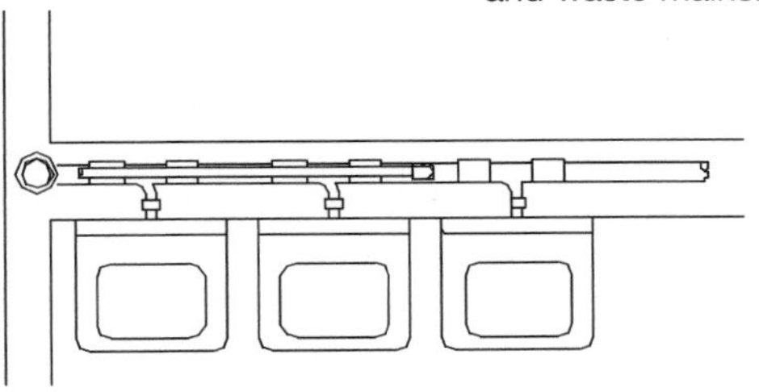

Side by Side

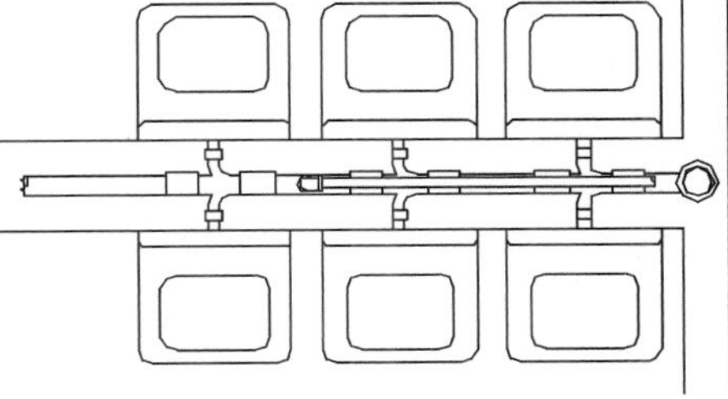

Back to Back

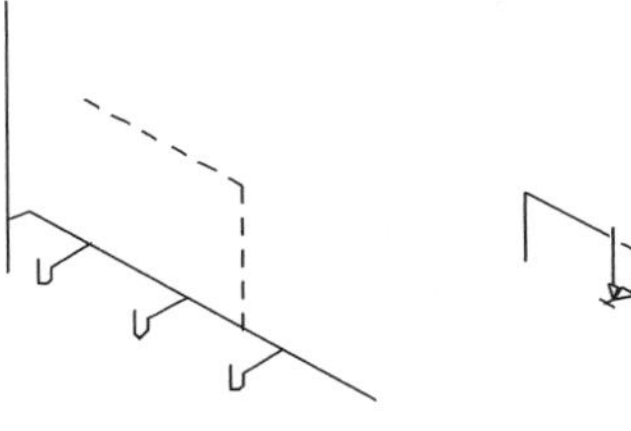

Waste/Vent

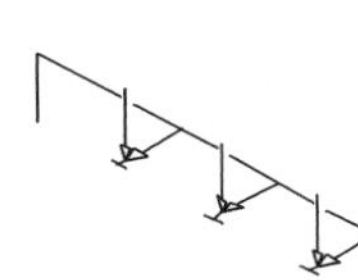

Supply
(Two supply systems required)

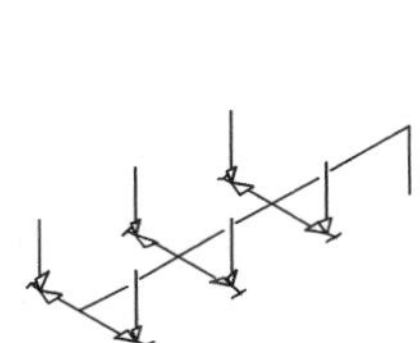

Supply
(Two supply systems required)

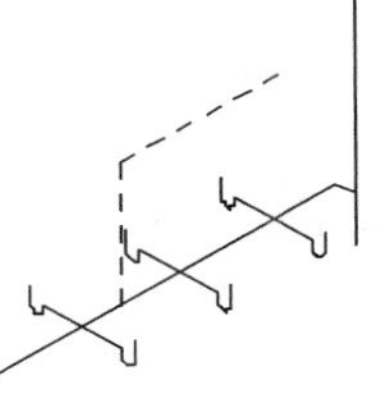

Waste/Vent

System Components	QUANTITY	UNIT	COST EACH MAT.	COST EACH INST.	COST EACH TOTAL
SYSTEM D2010 320 1760					
LAVATORIES, BATTERY MOUNT, WALL HUNG, SIDE BY SIDE, FIRST LAVATORY					
Lavatory w/trim wall hung PE on CI 20″ x 18″	1.000	Ea.	271	174	445
Stop, chrome, angle supply, 3/8″ diameter	2.000	Ea.	20.40	64	84.40
Concealed arm support	1.000	Ea.	720	129	849
P trap w/cleanout, 20 ga. C.P., 1-1/4″ diameter	1.000	Ea.	25	43	68
Copper tubing, type L, 1/2″ diameter	10.000	L.F.	40.50	95.50	136
Copper tubing, type DWV, 1-1/4″ diameter	4.000	L.F.	54.80	51.40	106.20
Copper 90° elbow, 1/2″ diameter	2.000	Ea.	2.78	77	79.78
Copper tee, 1/2″ diameter	2.000	Ea.	5.36	119	124.36
DWV copper sanitary tee, 1-1/4″ diameter	2.000	Ea.	63	171	234
Galvanized steel pipe, 1-1/4″ diameter	4.000	L.F.	38.80	62.40	101.20
Black cast iron 90° elbow, 1-1/4″ diameter	1.000	Ea.	14.60	63	77.60
TOTAL			1,256.24	1,049.30	2,305.54

D2010 320	Lavatory Systems, Battery Mount		COST EACH MAT.	COST EACH INST.	COST EACH TOTAL
1760	Lavatories, battery mount, side by side, first lavatory	R221113 -40	1,250	1,050	2,300
1800	Each additional lavatory, add		1,150	730	1,880
2000	Back to back, first pair of lavatories		2,100	1,675	3,775
2100	Each additional pair of lavatories, back to back		2,000	1,375	3,375

D2010 Plumbing Fixtures

Systems are complete with trim and rough-in (supply, waste and vent) to connect to supply branches and waste mains.

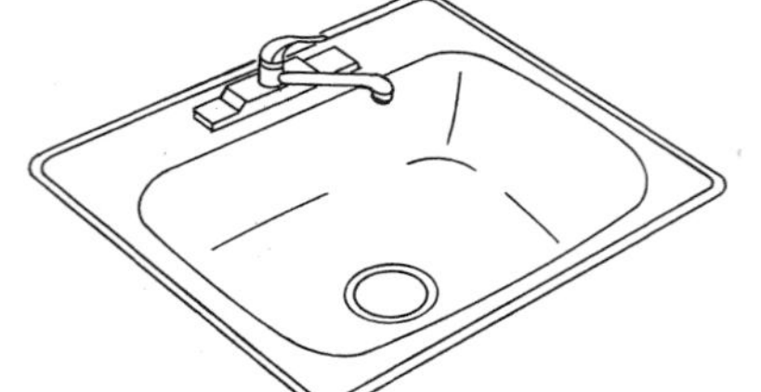

Countertop Single Bowl

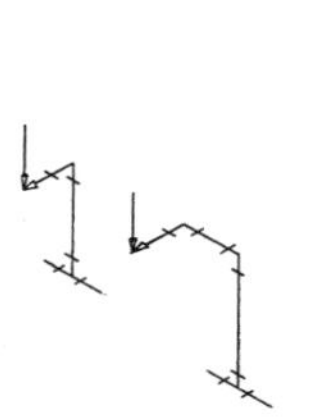

Supply

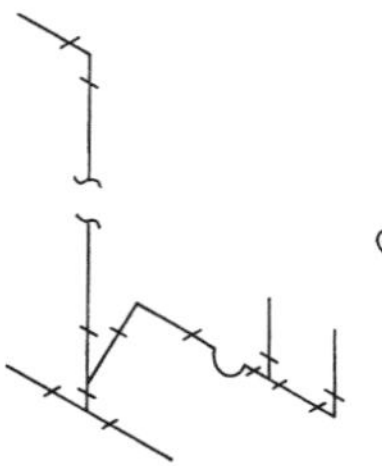

Waste/Vent

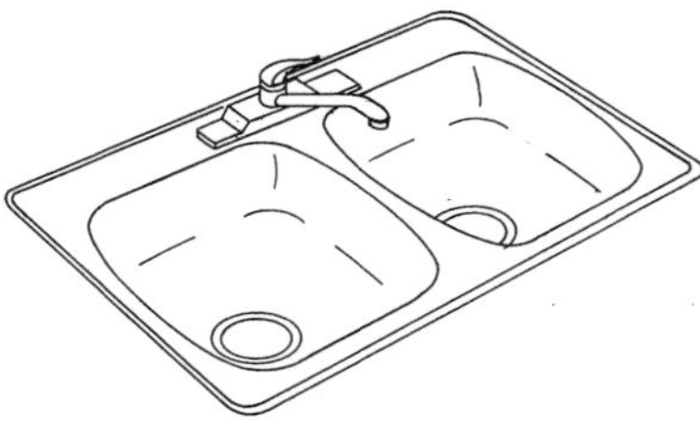

Countertop Double Bowl

System Components	QUANTITY	UNIT	COST EACH		
			MAT.	INST.	TOTAL
SYSTEM D2010 410 1720					
KITCHEN SINK W/TRIM, COUNTERTOP, P.E. ON C.I., 24″ X 21″, SINGLE BOWL					
Kitchen sink, counter top style, PE on CI, 24″ x 21″ single bowl	1.000	Ea.	340	248	588
Pipe, steel, galvanized, schedule 40, threaded, 1-1/4″ diam.	4.000	L.F.	38.80	62.40	101.20
Copper tubing, type DWV, solder, hangers 10′ OC 1-1/2″ diam.	6.000	L.F.	83.10	85.80	168.90
Wrought copper, DWV, Tee, sanitary, 1-1/2″ diam.	1.000	Ea.	39	96.50	135.50
P trap, standard, copper, 1-1/2″ diam.	1.000	Ea.	123	45.50	168.50
Copper tubing, type L, solder joints, hangers 10′ OC 1/2″ diam.	10.000	L.F.	40.50	95.50	136
Wrought copper 90° elbow for solder joints 1/2″ diam.	2.000	Ea.	2.78	77	79.78
Wrought copper Tee for solder joints, 1/2″ diam.	2.000	Ea.	5.36	119	124.36
Stop, angle supply, chrome, 1/2″ CTS	2.000	Ea.	22.20	70	92.20
TOTAL			694.74	899.70	1,594.44

D2010 410	Kitchen Sink Systems		COST EACH		
			MAT.	INST.	TOTAL
1720	Kitchen sink w/trim, countertop, PE on CI, 24″x21″, single bowl	R221113 -40	695	900	1,595
1760	30″ x 21″ single bowl		1,300	900	2,200
1800	32″ x 21″ double bowl		810	970	1,780
1880	Stainless steel, 19″ x 18″ single bowl		1,050	900	1,950
1920	25″ x 22″ single bowl		1,125	900	2,025
1960	33″ x 22″ double bowl		1,475	970	2,445
2000	43″ x 22″ double bowl		1,675	985	2,660
2040	44″ x 22″ triple bowl		1,750	1,025	2,775
2080	44″ x 24″ corner double bowl		1,275	985	2,260
2120	Steel, enameled, 24″ x 21″ single bowl		945	900	1,845
2160	32″ x 21″ double bowl		950	970	1,920
2240	Raised deck, PE on CI, 32″ x 21″, dual level, double bowl		925	1,225	2,150
2280	42″ x 21″ dual level, triple bowl		1,500	1,350	2,850

D2010 Plumbing Fixtures

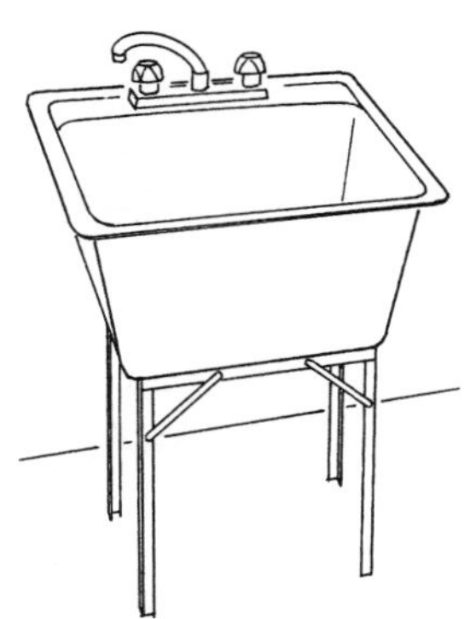

Single Compartment Sink

Systems are complete with trim and rough-in (supply, waste and vent) to connect to supply branches and waste mains.

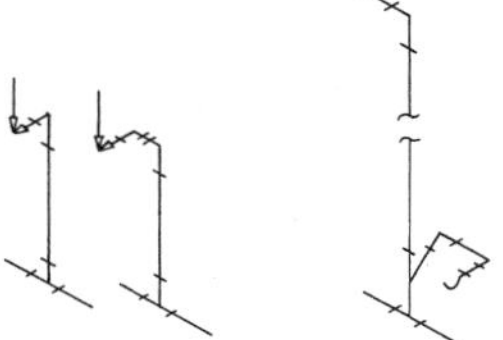

Supply   Waste/Vent

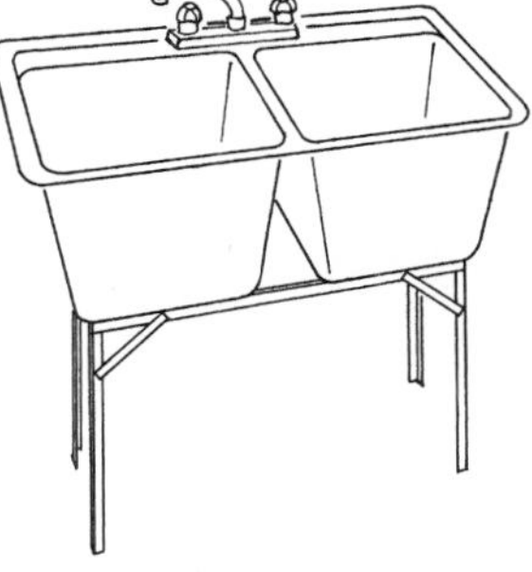

Double Compartment Sink

System Components	QUANTITY	UNIT	COST EACH MAT.	COST EACH INST.	COST EACH TOTAL
SYSTEM D2010 420 1760					
LAUNDRY SINK W/TRIM, PE ON CI, BLACK IRON FRAME					
24″ X 20″ OD, SINGLE COMPARTMENT					
Laundry sink PE on CI w/trim & frame, 24″ x 21″ OD, 1 compartment	1.000	Ea.	675	231	906
Pipe, steel, galvanized, schedule 40, threaded, 1-1/4″ diam	4.000	L.F.	38.80	62.40	101.20
Copper tubing, type DWV, solder joint, hanger 10′ OC 1-1/2″diam	6.000	L.F.	83.10	85.80	168.90
Wrought copper, DWV, Tee, sanitary, 1-1/2″ diam	1.000	Ea.	39	96.50	135.50
P trap, standard, copper, 1-1/2″ diam	1.000	Ea.	123	45.50	168.50
Copper tubing type L, solder joints, hangers 10′ OC, 1/2″ diam	10.000	L.F.	40.50	95.50	136
Wrought copper 90° elbow for solder joints 1/2″ diam	2.000	Ea.	2.78	77	79.78
Wrought copper Tee for solder joints, 1/2″ diam	2.000	Ea.	5.36	119	124.36
Stop, angle supply, 1/2″ diam	2.000	Ea.	22.20	70	92.20
TOTAL			1,029.74	882.70	1,912.44

D2010 420	Laundry Sink Systems		COST EACH MAT.	COST EACH INST.	COST EACH TOTAL
1740	Laundry sink w/trim, PE on CI, black iron frame	R221113 -40			
1760	24″ x 20″, single compartment		1,025	885	1,910
1800	24″ x 23″ single compartment	R224000 -30	1,050	885	1,935
1840	48″ x 21″ double compartment		1,250	960	2,210
1920	Molded stone, on wall, 22″ x 21″ single compartment		550	885	1,435
1960	45″x 21″ double compartment		775	960	1,735
2040	Plastic, on wall or legs, 18″ x 23″ single compartment		515	865	1,380
2080	20″ x 24″ single compartment		535	865	1,400
2120	36″ x 23″ double compartment		625	935	1,560
2160	40″ x 24″ double compartment		695	935	1,630

D2010 Plumbing Fixtures

Corrosion resistant laboratory sink systems are complete with trim and rough-in (supply, waste and vent) to connect to supply branches and waste mains.

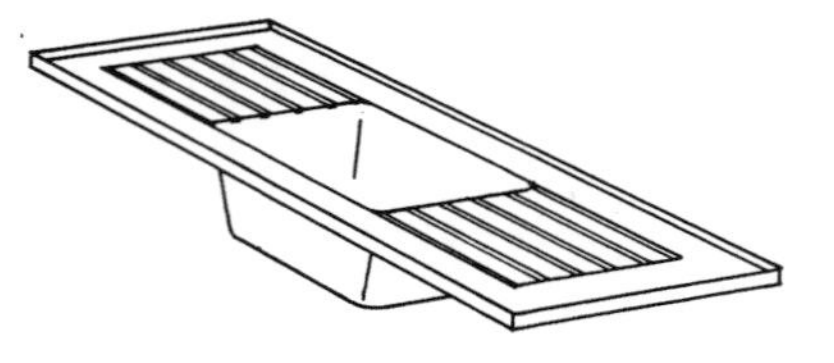

Laboratory Sink

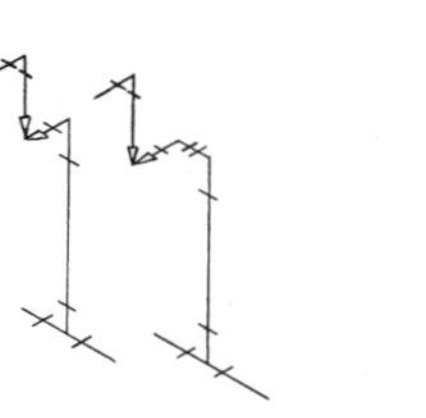

Supply Waste/Vent

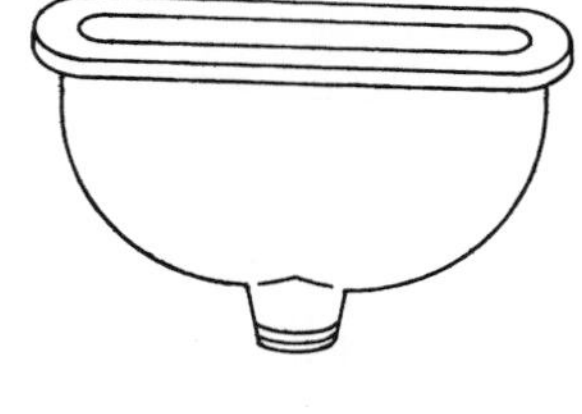

Polypropylene Cup Sink

System Components	QUANTITY	UNIT	COST EACH MAT.	COST EACH INST.	COST EACH TOTAL
SYSTEM D2010 430 1600					
LABORATORY SINK W/TRIM, STAINLESS STEEL, SINGLE BOWL					
DOUBLE DRAINBOARD, 54" X 24" O.D.					
Sink w/trim, stainless steel, 1 bowl, 2 drainboards 54" x 24" OD	1.000	Ea.	1,250	465	1,715
Pipe, polypropylene, schedule 40, acid resistant 1-1/2" diam.	10.000	L.F.	99.50	205	304.50
Tee, sanitary, polypropylene, acid resistant, 1-1/2" diam.	1.000	Ea.	22.50	77	99.50
P trap, polypropylene, acid resistant, 1-1/2" diam.	1.000	Ea.	75	45.50	120.50
Copper tubing type L, solder joint, hanger 10' O.C. 1/2" diam.	10.000	L.F.	40.50	95.50	136
Wrought copper 90° elbow for solder joints 1/2" diam.	2.000	Ea.	2.78	77	79.78
Wrought copper Tee for solder joints, 1/2" diam.	2.000	Ea.	5.36	119	124.36
Stop, angle supply, chrome, 1/2" diam.	2.000	Ea.	22.20	70	92.20
TOTAL			1,517.84	1,154	2,671.84

D2010 430	Laboratory Sink Systems	COST EACH MAT.	COST EACH INST.	COST EACH TOTAL
1580	Laboratory sink w/trim,			
1590	Stainless steel, single bowl, R221113-40			
1600	Double drainboard, 54" x 24" O.D.	1,525	1,150	2,675
1640	Single drainboard, 47" x 24"O.D.	1,275	1,150	2,425
1670	Stainless steel, double bowl,			
1680	70" x 24" O.D.	1,650	1,150	2,800
1750	Polyethylene, single bowl,			
1760	Flanged, 14-1/2" x 14-1/2" O.D.	560	1,025	1,585
1800	18-1/2" x 18-1/2" O.D.	685	1,025	1,710
1840	23-1/2" x 20-1/2" O.D.	705	1,025	1,730
1920	Polypropylene, cup sink, oval, 7" x 4" O.D.	455	915	1,370
1960	10" x 4-1/2" O.D.	485	915	1,400

D2010 Plumbing Fixtures

Corrosion resistant laboratory sink systems are complete with trim and rough–in (supply, waste and vent) to connect to supply branches and waste mains.

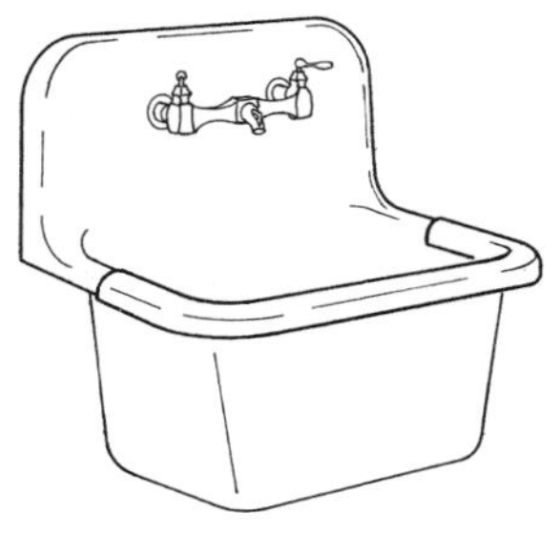

Wall Hung

Supply

Waste/Vent

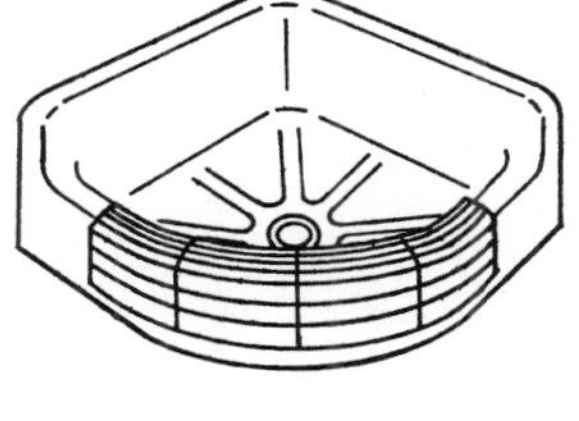

Corner, Floor

System Components

System Components	QUANTITY	UNIT	COST EACH MAT.	COST EACH INST.	COST EACH TOTAL
SYSTEM D2010 440 4260					
SERVICE SINK, PE ON CI, CORNER FLOOR, 28″X28″, W/RIM GUARD & TRIM					
Service sink, corner floor, PE on CI, 28″ x 28″, w/rim guard & trim	1.000	Ea.	1,250	315	1,565
Copper tubing type DWV, solder joint, hanger 10′OC 3″ diam.	6.000	L.F.	192	144	336
Copper tubing type DWV, solder joint, hanger 10′OC 2″ diam	4.000	L.F.	80	70.20	150.20
Wrought copper DWV, Tee, sanitary, 3″ diam.	1.000	Ea.	235	198	433
P trap with cleanout & slip joint, copper 3″ diam	1.000	Ea.	580	70	650
Copper tubing, type L, solder joints, hangers 10′ OC, 1/2″ diam	10.000	L.F.	40.50	95.50	136
Wrought copper 90° elbow for solder joints 1/2″ diam	2.000	Ea.	2.78	77	79.78
Wrought copper Tee for solder joints, 1/2″ diam	2.000	Ea.	5.36	119	124.36
Stop, angle supply, chrome, 1/2″ diam	2.000	Ea.	22.20	70	92.20
TOTAL			2,407.84	1,158.70	3,566.54

D2010 440 Service Sink Systems

D2010 440	Service Sink Systems		COST EACH MAT.	COST EACH INST.	COST EACH TOTAL
4260	Service sink w/trim, PE on CI, corner floor, 28″ x 28″, w/rim guard	R221113 -40	2,400	1,150	3,550
4300	Wall hung w/rim guard, 22″ x 18″		2,675	1,350	4,025
4340	24″ x 20″	R224000 -30	2,700	1,350	4,050
4380	Vitreous china, wall hung 22″ x 20″		2,625	1,350	3,975

D2010 Plumbing Fixtures

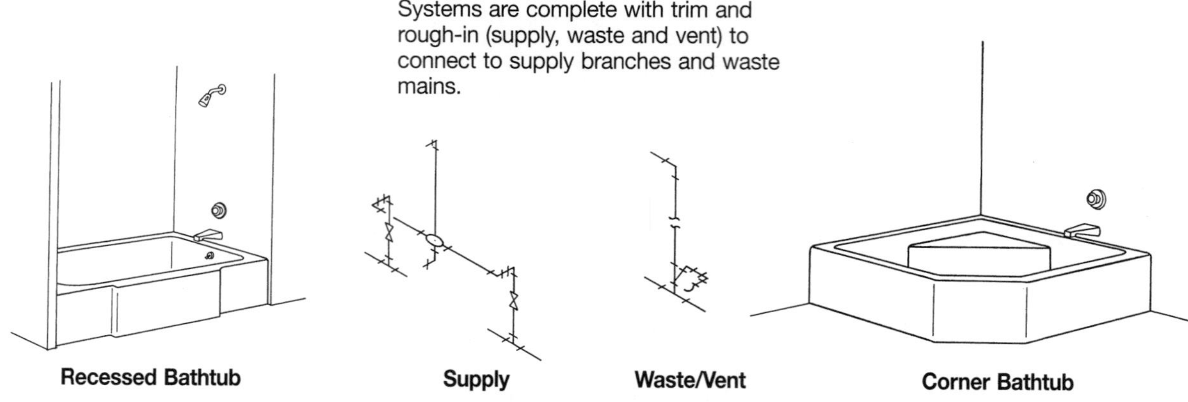

System Components	QUANTITY	UNIT	COST EACH MAT.	COST EACH INST.	COST EACH TOTAL
SYSTEM D2010 510 2000					
BATHTUB, RECESSED, PORCELAIN ENAMEL ON CAST IRON,, 48″ x 42″					
Bath tub, porcelain enamel on cast iron, w/fittings, 48″ x 42″	1.000	Ea.	3,450	345	3,795
Pipe, steel, galvanized, schedule 40, threaded, 1-1/4″ diam.	4.000	L.F.	38.80	62.40	101.20
Pipe, CI no hub soil w/couplings 10′ OC, hangers 5′ OC, 4″ diam.	3.000	L.F.	82.50	72	154.50
Combination Y and 1/8 bend for C.I. soil pipe, no hub, 4″ pipe size	1.000	Ea.	62.50		62.50
Drum trap, 3″ x 5″, copper, 1-1/2″ diam.	1.000	Ea.	172	48	220
Copper tubing type L, solder joints, hangers 10′ OC 1/2″ diam.	10.000	L.F.	40.50	95.50	136
Wrought copper 90° elbow, solder joints, 1/2″ diam.	2.000	Ea.	2.78	77	79.78
Wrought copper Tee, solder joints, 1/2″ diam.	2.000	Ea.	5.36	119	124.36
Stop, angle supply, 1/2″ diameter	2.000	Ea.	22.20	70	92.20
Copper tubing type DWV, solder joints, hanger 10′ OC 1-1/2″ diam.	3.000	L.F.	41.55	42.90	84.45
Pipe coupling, standard, C.I. soil no hub, 4″ pipe size	2.000	Ea.	51	84	135
TOTAL			3,969.19	1,015.80	4,984.99

D2010 510	Bathtub Systems		COST EACH MAT.	COST EACH INST.	COST EACH TOTAL
2000	Bathtub, recessed, P.E. on CI., 48″ x 42″	R224000-30	3,975	1,025	5,000
2040	72″ x 36″		3,800	1,125	4,925
2080	Mat bottom, 5′ long	R221113-40	1,925	985	2,910
2120	5′-6″ long		2,675	1,025	3,700
2160	Corner, 48″ x 42″		3,725	985	4,710
2200	Formed steel, enameled, 4′-6″ long		1,075	910	1,985

D2010 Plumbing Fixtures

Systems are complete with trim, flush valve and rough-in (supply, waste and vent) for connection to supply branches and waste mains.

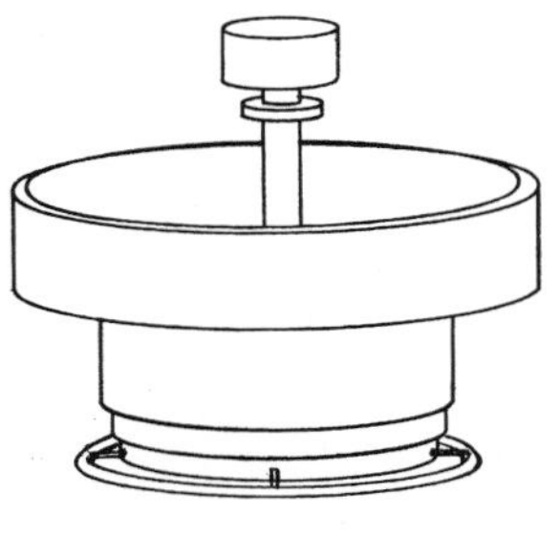

Circular Fountain

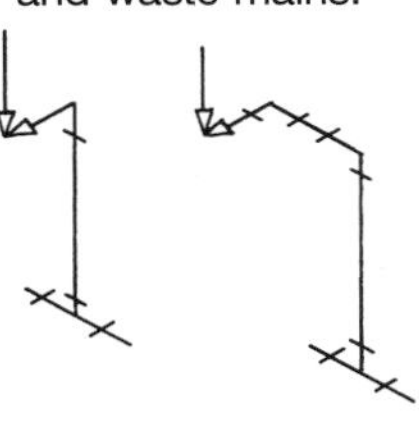

Supply

Waste/Vent

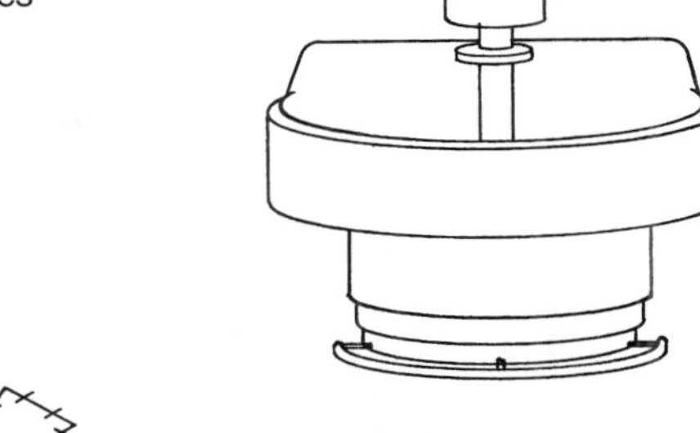

Semi-Circular Fountain

System Components	QUANTITY	UNIT	COST EACH MAT.	COST EACH INST.	COST EACH TOTAL
SYSTEM D2010 610 1760					
GROUP WASH FOUNTAIN, PRECAST TERRAZZO					
CIRCULAR, 36″ DIAMETER					
Wash fountain, group, precast terrazzo, foot control 36″ diam.	1.000	Ea.	8,425	720	9,145
Copper tubing type DWV, solder joint, hanger 10′ OC, 2″ diam.	10.000	L.F.	200	175.50	375.50
P trap, standard, copper, 2″ diam.	1.000	Ea.	202	51.50	253.50
Wrought copper, Tee, sanitary, 2″ diam.	1.000	Ea.	60.50	110	170.50
Copper tubing type L, solder joint, hanger 10′ OC 1/2″ diam.	20.000	L.F.	81	191	272
Wrought copper 90° elbow for solder joints 1/2″ diam.	3.000	Ea.	4.17	115.50	119.67
Wrought copper Tee for solder joints, 1/2″ diam.	2.000	Ea.	5.36	119	124.36
TOTAL			8,978.03	1,482.50	10,460.53

D2010 610	Group Wash Fountain Systems	COST EACH MAT.	COST EACH INST.	COST EACH TOTAL
1740	Group wash fountain, precast terrazzo (R221113-40)			
1760	Circular, 36″ diameter	8,975	1,475	10,450
1800	54″ diameter (R224000-30)	12,200	1,625	13,825
1840	Semi-circular, 36″ diameter	7,425	1,475	8,900
1880	54″ diameter	11,300	1,625	12,925
1960	Stainless steel, circular, 36″ diameter	8,000	1,375	9,375
2000	54″ diameter	9,975	1,525	11,500
2040	Semi-circular, 36″ diameter	6,750	1,375	8,125
2080	54″ diameter	8,425	1,525	9,950
2160	Thermoplastic, circular, 36″ diameter	5,725	1,125	6,850
2200	54″ diameter	6,550	1,300	7,850
2240	Semi-circular, 36″ diameter	5,775	1,125	6,900
2280	54″ diameter	7,425	1,300	8,725

D2010 Plumbing Fixtures

Systems are complete with trim and rough-in (supply, waste and vent) for connection to supply branches and waste mains.

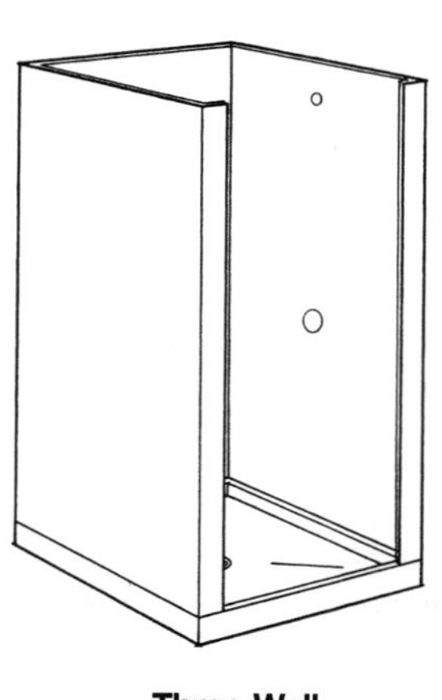
Three Wall

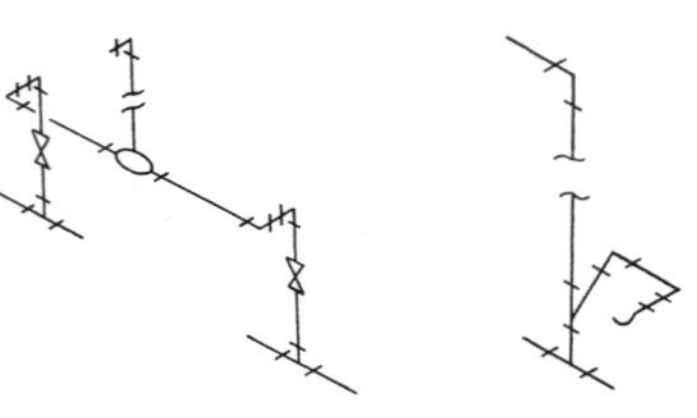
Supply

Waste/Vent

Corner Angle

System Components	QUANTITY	UNIT	COST EACH MAT.	COST EACH INST.	COST EACH TOTAL
SYSTEM D2010 710 1560					
SHOWER, STALL, BAKED ENAMEL, MOLDED STONE RECEPTOR, 30″ SQUARE					
Shower stall, enameled steel, molded stone receptor, 30″ square	1.000	Ea.	1,425	267	1,692
Copper tubing type DWV, solder joints, hangers 10′ OC, 2″ diam.	6.000	L.F.	83.10	85.80	168.90
Wrought copper DWV, Tee, sanitary, 2″ diam.	1.000	Ea.	39	96.50	135.50
Trap, standard, copper, 2″ diam.	1.000	Ea.	123	45.50	168.50
Copper tubing type L, solder joint, hanger 10′ OC 1/2″ diam.	16.000	L.F.	64.80	152.80	217.60
Wrought copper 90° elbow for solder joints 1/2″ diam.	3.000	Ea.	4.17	115.50	119.67
Wrought copper Tee for solder joints, 1/2″ diam.	2.000	Ea.	5.36	119	124.36
Stop and waste, straightway, bronze, solder joint 1/2″ diam.	2.000	Ea.	45	64	109
TOTAL			1,789.43	946.10	2,735.53

D2010 710	Shower Systems		COST EACH MAT.	COST EACH INST.	COST EACH TOTAL
1560	Shower, stall, baked enamel, molded stone receptor, 30″ square	R221113-40	1,800	945	2,745
1600	32″ square		1,650	955	2,605
1640	Terrazzo receptor, 32″ square	R224000-30	1,900	955	2,855
1680	36″ square		2,200	970	3,170
1720	36″ corner angle		2,425	400	2,825
1800	Fiberglass one piece, three walls, 32″ square		755	930	1,685
1840	36″ square		875	930	1,805
1880	Polypropylene, molded stone receptor, 30″ square		1,150	1,375	2,525
1920	32″ square		1,175	1,375	2,550
1960	Built-in head, arm, bypass, stops and handles		138	355	493
2050	Shower, stainless steel panels, handicap				
2100	w/fixed and handheld head, control valves, grab bar, and seat		4,475	4,250	8,725
2500	Shower, group with six heads, thermostatic mix valves & balancing valve		13,100	1,050	14,150
2520	Five heads		9,400	950	10,350

D2010 Plumbing Fixtures

Systems are complete with trim and rough-in (supply, waste and vent) to connect to supply branches and waste mains.

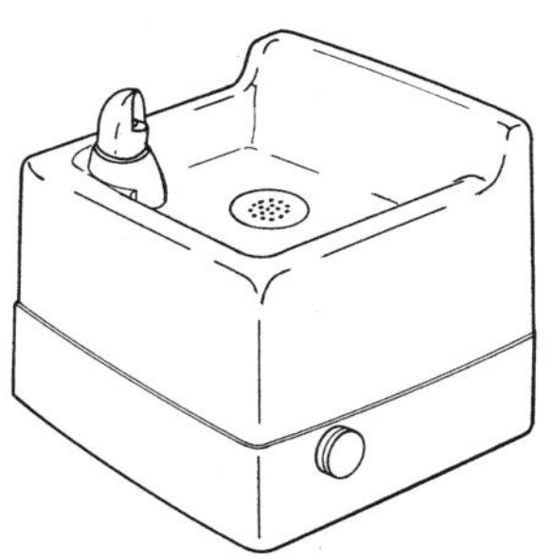

Wall Mounted, No Back

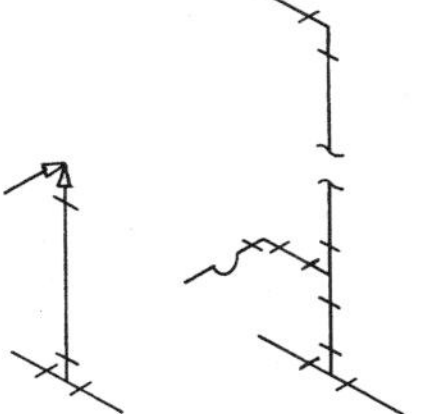

Supply Waste/Vent

Wall Mounted, Low Back

System Components	QUANTITY	UNIT	COST EACH MAT.	COST EACH INST.	COST EACH TOTAL
SYSTEM D2010 810 1800					
DRINKING FOUNTAIN, ONE BUBBLER, WALL MOUNTED					
NON RECESSED, BRONZE, NO BACK					
Drinking fountain, wall mount, bronze, 1 bubbler	1.000	Ea.	1,225	193	1,418
Copper tubing, type L, solder joint, hanger 10′ OC 3/8″ diam.	5.000	L.F.	18.35	46	64.35
Stop, supply, straight, chrome, 3/8″ diam.	1.000	Ea.	22.50	32	54.50
Wrought copper 90° elbow for solder joints 3/8″ diam.	1.000	Ea.	4.57	35	39.57
Wrought copper Tee for solder joints, 3/8″ diam.	1.000	Ea.	7.55	55	62.55
Copper tubing, type DWV, solder joint, hanger 10′ OC 1-1/4″ diam.	4.000	L.F.	54.80	51.40	106.20
P trap, standard, copper drainage, 1-1/4″ diam.	1.000	Ea.	113	43	156
Wrought copper, DWV, Tee, sanitary, 1-1/4″ diam.	1.000	Ea.	31.50	85.50	117
TOTAL			1,477.27	540.90	2,018.17

D2010 810	Drinking Fountain Systems	COST EACH MAT.	COST EACH INST.	COST EACH TOTAL
1740	Drinking fountain, one bubbler, wall mounted (R221113-40)			
1760	Non recessed			
1800	Bronze, no back (R224000-30)	1,475	540	2,015
1840	Cast iron, enameled, low back	1,600	540	2,140
1880	Fiberglass, 12″ back	2,475	540	3,015
1920	Stainless steel, no back	1,250	540	1,790
1960	Semi-recessed, poly marble	1,300	540	1,840
2040	Stainless steel	1,800	540	2,340
2080	Vitreous china	1,275	540	1,815
2120	Full recessed, poly marble	2,200	540	2,740
2200	Stainless steel	2,175	540	2,715
2240	Floor mounted, pedestal type, aluminum	3,175	735	3,910
2320	Bronze	3,125	735	3,860
2360	Stainless steel	2,625	735	3,360

D2010 Plumbing Fixtures

Systems are complete with trim and rough-in (supply, waste and vent) for connection to supply branches and waste mains.

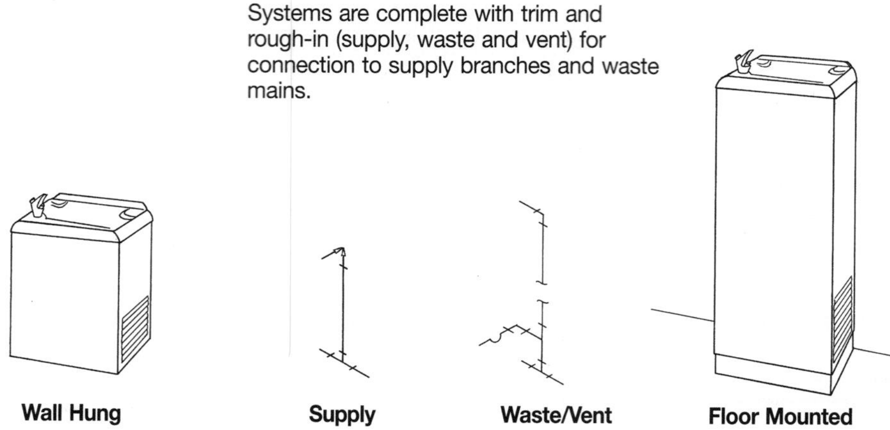

System Components	QUANTITY	UNIT	COST EACH MAT.	COST EACH INST.	COST EACH TOTAL
SYSTEM D2010 820 1840					
WATER COOLER, ELECTRIC, SELF CONTAINED, WALL HUNG, 8.2 G.P.H.					
Water cooler, wall mounted, 8.2 GPH	1.000	Ea.	1,125	345	1,470
Copper tubing type DWV, solder joint, hanger 10′ OC 1-1/4″ diam.	4.000	L.F.	54.80	51.40	106.20
Wrought copper DWV, Tee, sanitary 1-1/4″ diam.	1.000	Ea.	31.50	85.50	117
P trap, copper drainage, 1-1/4″ diam.	1.000	Ea.	113	43	156
Copper tubing type L, solder joint, hanger 10′ OC 3/8″ diam.	5.000	L.F.	18.35	46	64.35
Wrought copper 90° elbow for solder joints 3/8″ diam.	1.000	Ea.	4.57	35	39.57
Wrought copper Tee for solder joints, 3/8″ diam.	1.000	Ea.	7.55	55	62.55
Stop and waste, straightway, bronze, solder, 3/8″ diam.	1.000	Ea.	22.50	32	54.50
TOTAL			1,377.27	692.90	2,070.17

D2010 820	Water Cooler Systems		COST EACH MAT.	COST EACH INST.	COST EACH TOTAL
1840	Water cooler, electric, wall hung, 8.2 G.P.H.	R221113 -40	1,375	695	2,070
1880	Dual height, 14.3 G.P.H.		2,175	715	2,890
1920	Wheelchair type, 7.5 G.P.H.	R224000 -30	1,450	695	2,145
1960	Semi recessed, 8.1 G.P.H.		1,250	695	1,945
2000	Full recessed, 8 G.P.H.		2,825	745	3,570
2040	Floor mounted, 14.3 G.P.H.		1,425	605	2,030
2080	Dual height, 14.3 G.P.H.		1,650	735	2,385
2120	Refrigerated compartment type, 1.5 G.P.H.		1,975	605	2,580

D2010 Plumbing Fixtures

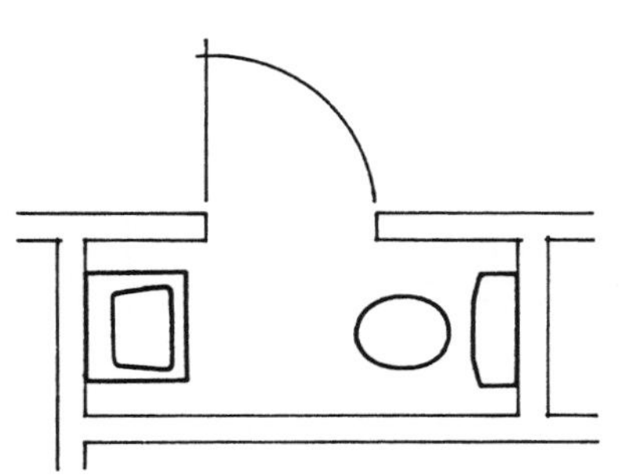

*Common wall is with an adjacent bathroom.

Two Fixture Bathroom Systems consisting of a lavatory, water closet, and rough-in service piping.

- Prices for plumbing and fixtures only.

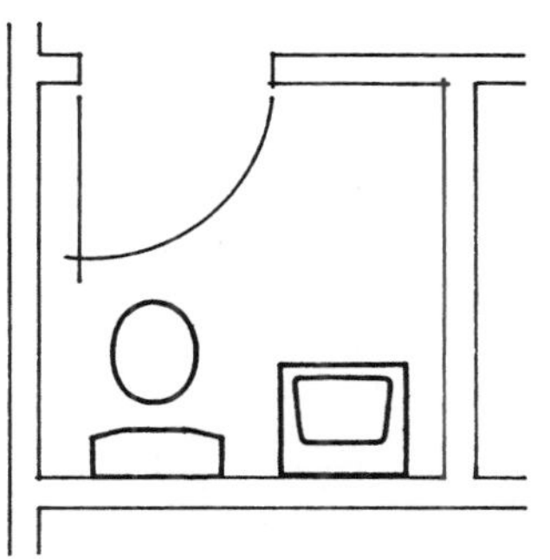

System Components	QUANTITY	UNIT	COST EACH MAT.	COST EACH INST.	COST EACH TOTAL
SYSTEM D2010 920 1180					
BATHROOM, LAVATORY & WATER CLOSET, 2 WALL PLUMBING, STAND ALONE					
Water closet, two piece, close coupled	1.000	Ea.	241	262	503
Water closet, rough-in waste & vent	1.000	Set	375	455	830
Lavatory w/ftngs., wall hung, white, PE on CI, 20″ x 18″	1.000	Ea.	271	174	445
Lavatory, rough-in waste & vent	1.000	Set	520	835	1,355
Copper tubing type L, solder joint, hanger 10′ OC 1/2″ diam.	10.000	L.F.	40.50	95.50	136
Pipe, steel, galvanized, schedule 40, threaded, 2″ diam.	12.000	L.F.	104.40	258	362.40
Pipe, CI soil, no hub, coupling 10′ OC, hanger 5′ OC, 4″ diam.	7.000	L.F.	189	178.50	367.50
TOTAL			1,740.90	2,258	3,998.90

D2010 920	Two Fixture Bathroom, Two Wall Plumbing		COST EACH MAT.	COST EACH INST.	COST EACH TOTAL
1180	Bathroom, lavatory & water closet, 2 wall plumbing, stand alone	R221113 -40	1,750	2,250	4,000
1200	Share common plumbing wall*		1,625	1,950	3,575

D2010 922	Two Fixture Bathroom, One Wall Plumbing		COST EACH MAT.	COST EACH INST.	COST EACH TOTAL
2220	Bathroom, lavatory & water closet, one wall plumbing, stand alone	R221113 -40	1,650	2,025	3,675
2240	Share common plumbing wall*		1,400	1,725	3,125
2260					
2280					

D2010 Plumbing Fixtures

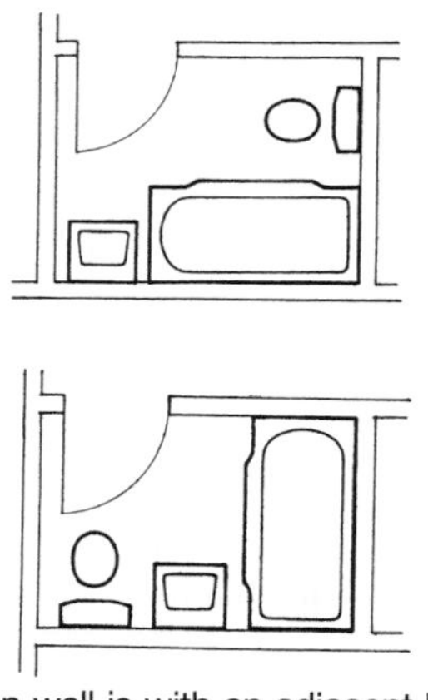

Three Fixture Bathroom Systems consisting of a lavatory, water closet, bathtub or shower and rough-in service piping.

- Prices for plumbing and fixtures only.

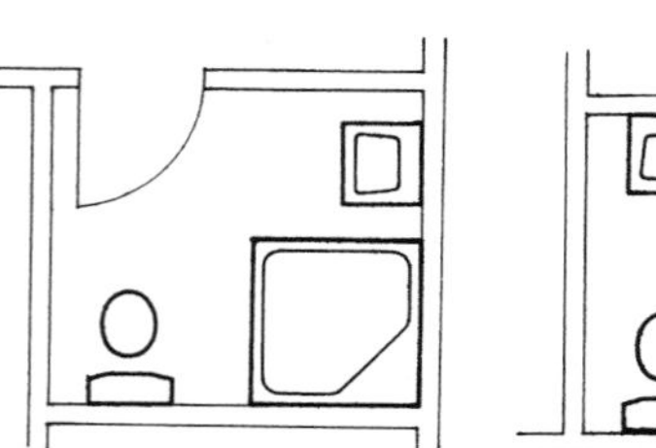

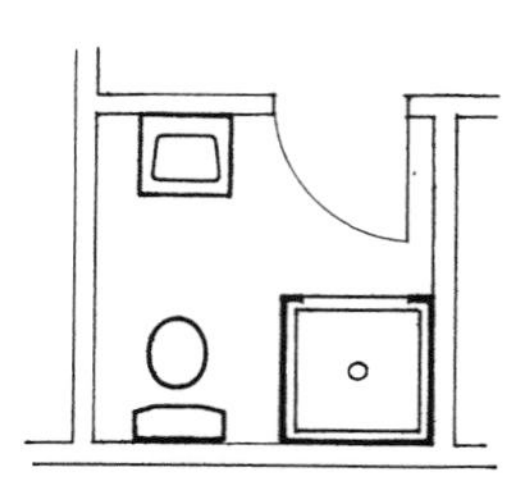

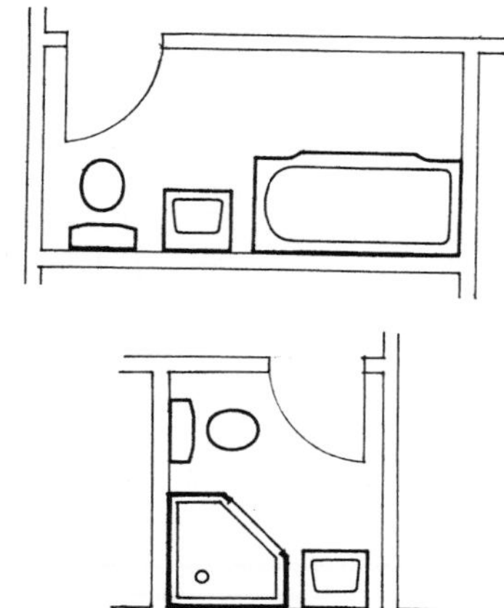

*Common wall is with an adjacent bathroom.

System Components	QUANTITY	UNIT	COST EACH		
			MAT.	INST.	TOTAL
SYSTEM D2010 924 1170					
BATHROOM, LAVATORY, WATER CLOSET & BATHTUB					
ONE WALL PLUMBING, STAND ALONE					
Wtr closet, rough-in, supply, waste and vent	1.000	Set	375	455	830
Wtr closet, 2 pc close cpld vit china flr mntd w/seat supply & stop	1.000	Ea.	241	262	503
Lavatory w/ftngs, wall hung, white, PE on CI, 20″ x 18″	1.000	Ea.	271	174	445
Lavatory, rough-in waste & vent	1.000	Set	520	835	1,355
Bathtub, white PE on CI, w/ftgs, mat bottom, recessed, 5′ long	1.000	Ea.	1,400	315	1,715
Baths, rough-in waste and vent	1.000	Set	490	670	1,160
TOTAL			3,297	2,711	6,008

D2010 924	Three Fixture Bathroom, One Wall Plumbing	COST EACH		
		MAT.	INST.	TOTAL
1150	Bathroom, three fixture, one wall plumbing			
1160	Lavatory, water closet & bathtub (R221113-40)			
1170	Stand alone	3,300	2,700	6,000
1180	Share common plumbing wall * (R224000-30)	2,800	1,900	4,700

D2010 926	Three Fixture Bathroom, Two Wall Plumbing	COST EACH		
		MAT.	INST.	TOTAL
2130	Bathroom, three fixture, two wall plumbing			
2140	Lavatory, water closet & bathtub (R221113-40)			
2160	Stand alone	3,275	2,675	5,950
2180	Long plumbing wall common *	2,925	2,125	5,050
3610	Lavatory, bathtub & water closet			
3620	Stand alone	3,475	3,050	6,525
3640	Long plumbing wall common *	3,325	2,750	6,075
4660	Water closet, corner bathtub & lavatory			
4680	Stand alone	5,100	2,700	7,800
4700	Long plumbing wall common *	4,600	2,050	6,650
6100	Water closet, stall shower & lavatory			
6120	Stand alone	3,425	3,050	6,475
6140	Long plumbing wall common *	3,250	2,800	6,050
7060	Lavatory, corner stall shower & water closet			
7080	Stand alone	4,100	2,700	6,800
7100	Short plumbing wall common *	3,525	1,800	5,325

D2010 Plumbing Fixtures

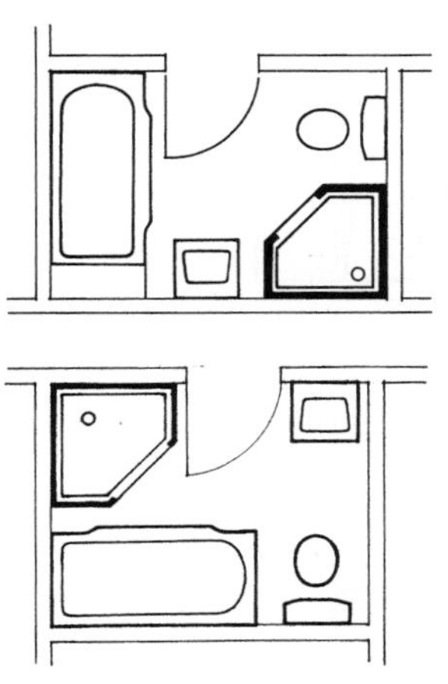

Four Fixture Bathroom Systems consisting of a lavatory, water closet, bathtub, shower and rough-in service piping.

- Prices for plumbing and fixtures only.

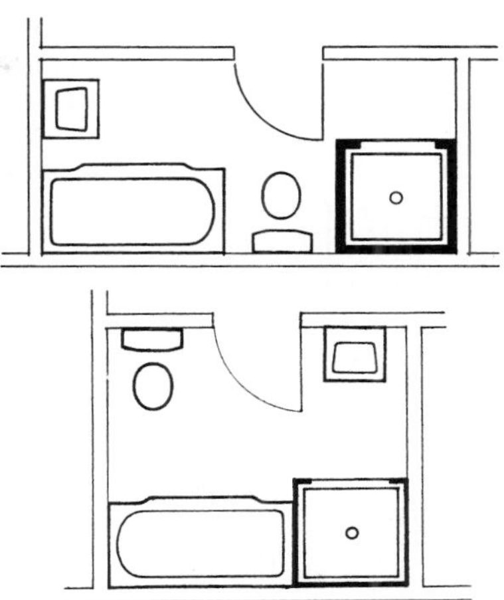

*Common wall is with an adjacent bathroom.

System Components	QUANTITY	UNIT	COST EACH MAT.	COST EACH INST.	COST EACH TOTAL
SYSTEM D2010 928 1160					
BATHROOM, BATHTUB, WATER CLOSET, STALL SHOWER & LAVATORY					
TWO WALL PLUMBING, STAND ALONE					
Wtr closet, 2 pc close cpld vit china flr mntd w/seat supply & stop	1.000	Ea.	241	262	503
Water closet, rough-in waste & vent	1.000	Set	375	455	830
Lavatory w/ftngs, wall hung, white PE on CI, 20″ x 18″	1.000	Ea.	271	174	445
Lavatory, rough-in waste & vent	1.000	Set	520	835	1,355
Bathtub, white PE on CI, w/ftgs, mat bottom, recessed, 5′ long	1.000	Ea.	1,400	315	1,715
Baths, rough-in waste and vent	1.000	Set	490	670	1,160
Shower stall, bkd enam, molded stone receptor, door & trim 32″ sq.	1.000	Ea.	1,275	278	1,553
Shower stall, rough-in supply, waste & vent	1.000	Set	450	675	1,125
TOTAL			5,022	3,664	8,686

D2010 928	Four Fixture Bathroom, Two Wall Plumbing		COST EACH MAT.	COST EACH INST.	COST EACH TOTAL
1140	Bathroom, four fixture, two wall plumbing	R221113 -40			
1150	Bathtub, water closet, stall shower & lavatory				
1160	Stand alone	R224000 -30	5,025	3,675	8,700
1180	Long plumbing wall common *		4,075	2,250	6,325
2260	Bathtub, lavatory, corner stall shower & water closet				
2280	Stand alone		5,525	2,925	8,450
2320	Long plumbing wall common *		5,050	2,250	7,300
3620	Bathtub, stall shower, lavatory & water closet				
3640	Stand alone		5,025	3,675	8,700
3660	Long plumbing wall (opp. door) common *		4,525	3,000	7,525

D2010 930	Four Fixture Bathroom, Three Wall Plumbing		COST EACH MAT.	COST EACH INST.	COST EACH TOTAL
4680	Bathroom, four fixture, three wall plumbing	R221113 -40			
4700	Bathtub, stall shower, lavatory & water closet				
4720	Stand alone		6,150	4,025	10,175
4760	Long plumbing wall (opposite door) common *		6,025	3,725	9,750

D2010 Plumbing Fixtures

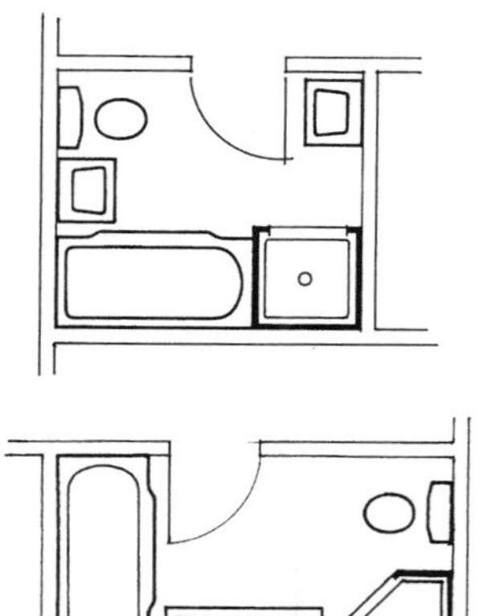

Five Fixture Bathroom Systems consisting of two lavatories, a water closet, bathtub, shower and rough-in service piping.

- Prices for plumbing and fixtures only.

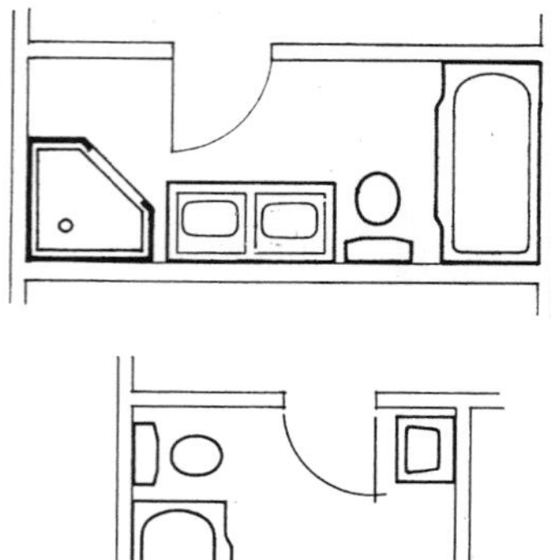

*Common wall is with an adjacent bathroom.

System Components	QUANTITY	UNIT	COST EACH MAT.	COST EACH INST.	COST EACH TOTAL
SYSTEM D2010 932 1360					
BATHROOM, BATHTUB, WATER CLOSET, STALL SHOWER & TWO LAVATORIES					
TWO WALL PLUMBING, STAND ALONE					
Wtr closet, 2 pc close cpld vit china flr mntd incl seat,supply & stop	1.000	Ea.	241	262	503
Water closet, rough-in waste & vent	1.000	Set	375	455	830
Lavatory w/ftngs, wall hung, white PE on CI, 20″ x 18″	2.000	Ea.	542	348	890
Lavatory, rough-in waste & vent	2.000	Set	1,040	1,670	2,710
Bathtub, white PE on CI, w/ftgs, mat bottom, recessed, 5′ long	1.000	Ea.	1,400	315	1,715
Baths, rough-in waste and vent	1.000	Set	490	670	1,160
Shower stall, bkd enam molded stone receptor, door & ftng, 32″ sq.	1.000	Ea.	1,275	278	1,553
Shower stall, rough-in supply, waste & vent	1.000	Set	450	675	1,125
TOTAL			5,813	4,673	10,486

D2010 932	Five Fixture Bathroom, Two Wall Plumbing		COST EACH MAT.	COST EACH INST.	COST EACH TOTAL
1320	Bathroom, five fixture, two wall plumbing				
1340	Bathtub, water closet, stall shower & two lavatories	R221113 -40			
1360	Stand alone		5,825	4,675	10,500
1400	One short plumbing wall common *	R224000 -30	5,325	4,000	9,325
1500	Bathtub, two lavatories, corner stall shower & water closet				
1520	Stand alone		6,800	4,675	11,475
1540	Long plumbing wall common*		6,150	3,650	9,800

D2010 934	Five Fixture Bathroom, Three Wall Plumbing		COST EACH MAT.	COST EACH INST.	COST EACH TOTAL
2360	Bathroom, five fixture, three wall plumbing				
2380	Water closet, bathtub, two lavatories & stall shower	R221113 -40			
2400	Stand alone		6,800	4,675	11,475
2440	One short plumbing wall common *		6,300	4,025	10,325

D2010 936	Five Fixture Bathroom, One Wall Plumbing		COST EACH MAT.	COST EACH INST.	COST EACH TOTAL
4080	Bathroom, five fixture, one wall plumbing				
4100	Bathtub, two lavatories, corner stall shower & water closet	R221113 -40			
4120	Stand alone		6,475	4,175	10,650
4160	Share common wall *		5,500	2,700	8,200

Example of Plumbing Cost Calculations: The bathroom system includes the individual fixtures such as bathtub, lavatory, shower and water closet. These fixtures are listed below as separate items merely as a checklist.

D2010 951 Plumbing Systems 20 Unit, 2 Story Apartment Building

	FIXTURE	SYSTEM	LINE	QUANTITY	UNIT	COST EACH		
						MAT.	INST.	TOTAL
0440	Bathroom	D2010 926	3640	20	Ea.	66,500	55,000	121,500
0480	Bathtub							
0520	Booster pump[1]	not req'd.						
0560	Drinking fountain							
0600	Garbage disposal[1]	not incl.						
0660								
0680	Grease interceptor							
0720	Water heater	D2020 250	2140	1	Ea.	16,900	3,725	20,625
0760	Kitchen sink	D2010 410	1960	20	Ea.	29,600	19,400	49,000
0800	Laundry sink	D2010 420	1840	4	Ea.	5,050	3,825	8,875
0840	Lavatory							
0900								
0920	Roof drain, 1 floor	D2040 210	4200	2	Ea.	2,575	2,000	4,575
0960	Roof drain, add'l floor	D2040 210	4240	20	L.F.	540	510	1,050
1000	Service sink	D2010 440	4300	1	Ea.	2,675	1,350	4,025
1040	Sewage ejector[1]	not req'd.						
1080	Shower							
1100								
1160	Sump pump							
1200	Urinal							
1240	Water closet							
1320								
1360	**SUB TOTAL**					123,500	86,000	209,500
1481	Water controls	R221113-40		10%[2]		12,400	8,600	21,000
1521	Pipe & fittings[3]	R221113-40		30%[2]		37,100	25,800	62,900
1560	Other							
1601	Quality/complexity	R221113-40		15%[2]		18,600	12,900	31,500
1680								
1720	**TOTAL**					192,000	133,500	325,500
1741								

[1]**Note:** Cost for items such as booster pumps, backflow preventers, sewage ejectors, water meters, etc., may be obtained from the unit price section in the front of this data set.

Water controls, pipe and fittings, and the Quality/Complexity factors come from Table R221113-40.

[2]Percentage of subtotal.

[3]Long, easily discernable runs of pipe would be more accurately priced from Unit Price Section 22 11 13. If this is done, reduce the miscellaneous percentage in proportion.

D2020 Domestic Water Distribution

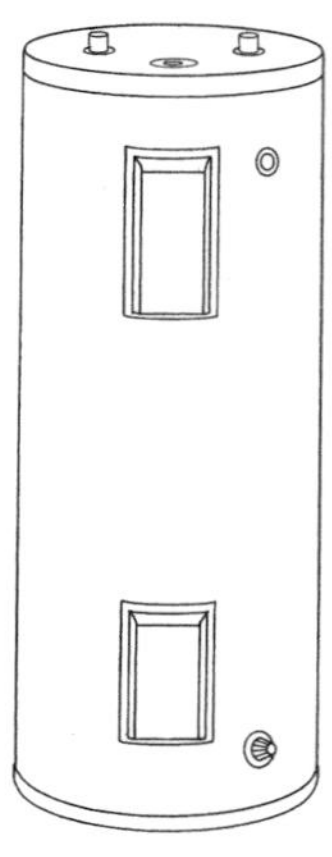

Installation includes piping and fittings within 10′ of heater. Electric water heaters do not require venting.

1 Kilowatt hour will raise:			
Gallons of Water	Degrees F	Gallons of Water	Degrees F
4.1	100°	6.8	60°
4.5	90°	8.2	50°
5.1	80°	10.0	40°
5.9	70°		

System Components	QUANTITY	UNIT	COST EACH MAT.	COST EACH INST.	COST EACH TOTAL
SYSTEM D2020 210 1780					
ELECTRIC WATER HEATER, RESIDENTIAL, 100°F RISE					
10 GALLON TANK, 7 GPH					
Water heater, residential electric, glass lined tank, 10 gal.	1.000	Ea.	515	335	850
Copper tubing, type L, solder joint, hanger 10′ OC 1/2″ diam.	30.000	L.F.	121.50	286.50	408
Wrought copper 90° elbow for solder joints 1/2″ diam.	4.000	Ea.	5.56	154	159.56
Wrought copper Tee for solder joints, 1/2″ diam.	2.000	Ea.	5.36	119	124.36
Union, wrought copper, 1/2″ diam.	2.000	Ea.	49	81	130
Valve, gate, bronze, 125 lb, NRS, soldered 1/2″ diam.	2.000	Ea.	133	64	197
Relief valve, bronze, press & temp, self-close, 3/4″ IPS	1.000	Ea.	283	27.50	310.50
Wrought copper adapter, CTS to MPT 3/4″ IPS	1.000	Ea.	6.70	45.50	52.20
Copper tubing, type L, solder joints, 3/4″ diam.	1.000	L.F.	5.20	10.15	15.35
Wrought copper 90° elbow for solder joints 3/4″ diam.	1.000	Ea.	2.95	40.50	43.45
TOTAL			1,127.27	1,163.15	2,290.42

D2020 210	Electric Water Heaters - Residential Systems		COST EACH MAT.	COST EACH INST.	COST EACH TOTAL
1760	Electric water heater, residential, 100°F rise				
1780	10 gallon tank, 7 GPH	R224000 -10	1,125	1,175	2,300
1820	20 gallon tank, 7 GPH	R224000 -20	1,425	1,250	2,675
1860	30 gallon tank, 7 GPH		2,100	1,300	3,400
1900	40 gallon tank, 8 GPH		2,500	1,425	3,925
1940	52 gallon tank, 10 GPH		2,700	1,450	4,150
1980	66 gallon tank, 13 GPH		4,150	1,650	5,800
2020	80 gallon tank, 16 GPH		4,450	1,725	6,175
2060	120 gallon tank, 23 GPH		6,400	2,025	8,425

D2020 Domestic Water Distribution

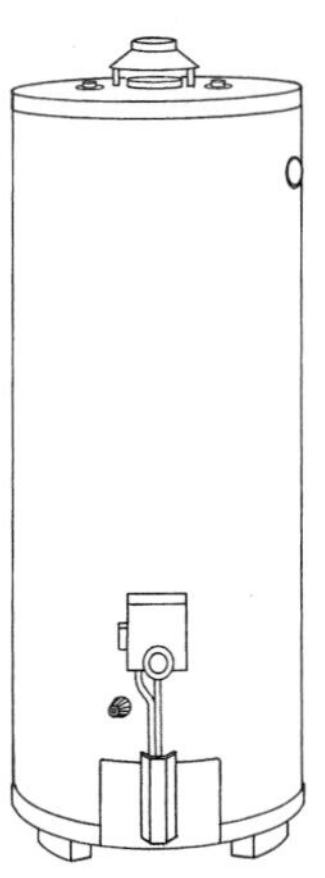

Installation includes piping and fittings within 10′ of heater. Gas heaters require vent piping (not included with these units).

System Components	QUANTITY	UNIT	COST EACH MAT.	COST EACH INST.	COST EACH TOTAL
SYSTEM D2020 220 2260					
GAS FIRED WATER HEATER, RESIDENTIAL, 100°F RISE					
30 GALLON TANK, 32 GPH					
Water heater, residential, gas, glass lined tank, 30 gallon	1.000	Ea.	2,325	385	2,710
Copper tubing, type L, solder joint, hanger 10′ OC, 3/4″ diam	33.000	L.F.	171.60	334.95	506.55
Wrought copper 90° elbow for solder joints, 3/4″ diam	5.000	Ea.	14.75	202.50	217.25
Wrought copper Tee for solder joints, 3/4″ diam	2.000	Ea.	13.30	129	142.30
Wrought copper union for soldered joints, 3/4″ diam.	2.000	Ea.	50	86	136
Valve bronze, 125 lb., NRS, soldered 3/4″ diam	2.000	Ea.	163	77	240
Relief valve, press & temp, bronze, self-close, 3/4″ diam	1.000	Ea.	283	27.50	310.50
Wrought copper, adapter, CTS to MPT 3/4″ IPS	1.000	Ea.	6.70	45.50	52.20
Pipe steel black, schedule 40, threaded, 1/2″ diam	10.000	L.F.	44.30	122.50	166.80
Pipe, 90° elbow, malleable iron black, 150 lb., threaded, 1/2″ diam	2.000	Ea.	8.80	103	111.80
Pipe, union with brass seat, malleable iron black, 1/2″ diam	1.000	Ea.	21	55	76
Valve, gas stop w/o check, brass, 1/2″ IPS	1.000	Ea.	14.75	32	46.75
TOTAL			3,116.20	1,599.95	4,716.15

D2020 220	Gas Fired Water Heaters - Residential Systems		COST EACH MAT.	COST EACH INST.	COST EACH TOTAL
2200	Gas fired water heater, residential, 100°F rise	R224000 -10			
2260	30 gallon tank, 32 GPH		3,125	1,600	4,725
2300	40 gallon tank, 32 GPH	R224000 -20	3,000	1,800	4,800
2340	50 gallon tank, 63 GPH		3,175	1,800	4,975
2380	75 gallon tank, 63 GPH		4,350	2,000	6,350
2420	100 gallon tank, 63 GPH		4,925	2,100	7,025

D2020 Domestic Water Distribution

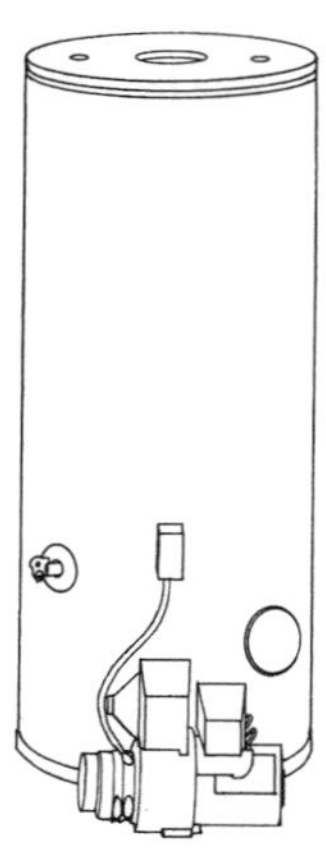

Installation includes piping and fittings within 10′ of heater. Oil fired heaters require vent piping (not included in these prices).

System Components	QUANTITY	UNIT	COST EACH MAT.	COST EACH INST.	COST EACH TOTAL
SYSTEM D2020 230 2220					
OIL FIRED WATER HEATER, RESIDENTIAL, 100°F RISE					
30 GALLON TANK, 103 GPH					
Water heater, residential, oil glass lined tank, 30 Gal	1.000	Ea.	1,550	385	1,935
Copper tubing, type L, solder joint, hanger 10′ O.C. 3/4″ diam.	33.000	L.F.	171.60	334.95	506.55
Wrought copper 90° elbow for solder joints 3/4″ diam.	5.000	Ea.	14.75	202.50	217.25
Wrought copper Tee for solder joints, 3/4″ diam.	2.000	Ea.	13.30	129	142.30
Wrought copper union for soldered joints, 3/4″ diam.	2.000	Ea.	50	86	136
Valve, gate, bronze, 125 lb, NRS, soldered 3/4″ diam.	2.000	Ea.	163	77	240
Relief valve, bronze, press & temp, self-close, 3/4″ IPS	1.000	Ea.	283	27.50	310.50
Wrought copper adapter, CTS to MPT, 3/4″ IPS	1.000	Ea.	6.70	45.50	52.20
Copper tubing, type L, solder joint, hanger 10′ OC 3/8″ diam.	10.000	L.F.	36.70	92	128.70
Wrought copper 90° elbow for solder joints 3/8″ diam.	2.000	Ea.	9.14	70	79.14
Valve, globe, fusible, 3/8″ diam.	1.000	Ea.	16.20	32.50	48.70
TOTAL			2,314.39	1,481.95	3,796.34

D2020 230	Oil Fired Water Heaters - Residential Systems	COST EACH MAT.	COST EACH INST.	COST EACH TOTAL
2200	Oil fired water heater, residential, 100°F rise			
2220	30 gallon tank, 103 GPH	2,325	1,475	3,800
2260	50 gallon tank, 145 GPH (R224000-20)	2,500	1,650	4,150
2300	70 gallon tank, 164 GPH	4,025	1,875	5,900
2340	85 gallon tank, 181 GPH	12,300	1,925	14,225

D2020 Domestic Water Distribution

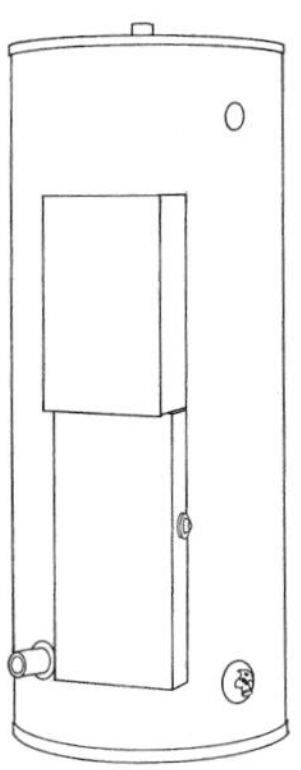

Systems below include piping and fittings within 10′ of heater. Electric water heaters do not require venting.

System Components	QUANTITY	UNIT	COST EACH MAT.	COST EACH INST.	COST EACH TOTAL
SYSTEM D2020 240 1820					
ELECTRIC WATER HEATER, COMMERCIAL, 100°F RISE					
50 GALLON TANK, 9 KW, 37 GPH					
Water heater, commercial, electric, 50 Gal, 9 KW, 37 GPH	1.000	Ea.	7,600	430	8,030
Copper tubing, type L, solder joint, hanger 10′ OC, 3/4″ diam	34.000	L.F.	176.80	345.10	521.90
Wrought copper 90° elbow for solder joints 3/4″ diam	5.000	Ea.	14.75	202.50	217.25
Wrought copper Tee for solder joints, 3/4″ diam	2.000	Ea.	13.30	129	142.30
Wrought copper union for soldered joints, 3/4″ diam.	2.000	Ea.	50	86	136
Valve, gate, bronze, 125 lb, NRS, soldered 3/4″ diam	2.000	Ea.	163	77	240
Relief valve, bronze, press & temp, self-close, 3/4″ IPS	1.000	Ea.	283	27.50	310.50
Wrought copper adapter, copper tubing to male, 3/4″ IPS	1.000	Ea.	6.70	45.50	52.20
TOTAL			8,307.55	1,342.60	9,650.15

D2020 240	Electric Water Heaters - Commercial Systems		COST EACH MAT.	COST EACH INST.	COST EACH TOTAL
1800	Electric water heater, commercial, 100°F rise				
1820	50 gallon tank, 9 KW 37 GPH		8,300	1,350	9,650
1860	80 gal, 12 KW 49 GPH	R224000-10	10,700	1,650	12,350
1900	36 KW 147 GPH		14,800	1,800	16,600
1940	120 gal, 36 KW 147 GPH	R224000-20	16,600	1,925	18,525
1980	150 gal, 120 KW 490 GPH		53,500	2,075	55,575
2020	200 gal, 120 KW 490 GPH		54,500	2,125	56,625
2060	250 gal, 150 KW 615 GPH		55,000	2,475	57,475
2100	300 gal, 180 KW 738 GPH		76,500	2,625	79,125
2140	350 gal, 30 KW 123 GPH		45,700	2,825	48,525
2180	180 KW 738 GPH		63,500	2,825	66,325
2220	500 gal, 30 KW 123 GPH		65,500	3,300	68,800
2260	240 KW 984 GPH		110,000	3,300	113,300
2300	700 gal, 30 KW 123 GPH		72,000	3,775	75,775
2340	300 KW 1230 GPH		107,000	3,775	110,775
2380	1000 gal, 60 KW 245 GPH		88,000	5,275	93,275
2420	480 KW 1970 GPH		142,500	5,300	147,800
2460	1500 gal, 60 KW 245 GPH		123,500	6,550	130,050
2500	480 KW 1970 GPH		175,500	6,550	182,050

D2020 Domestic Water Distribution

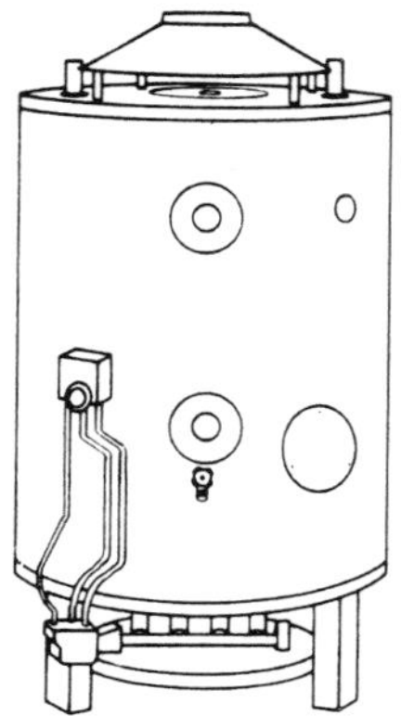

Units may be installed in multiples for increased capacity.

Included below is the heater with self-energizing gas controls, safety pilots, insulated jacket, hi-limit aquastat and pressure relief valve.

Installation includes piping and fittings within 10′ of heater. Gas heaters require vent piping (not included in these prices).

System Components	QUANTITY	UNIT	COST EACH MAT.	COST EACH INST.	COST EACH TOTAL
SYSTEM D2020 250 1780					
GAS FIRED WATER HEATER, COMMERCIAL, 100°F RISE					
75.5 MBH INPUT, 63 GPH					
Water heater, commercial, gas, 75.5 MBH, 63 GPH	1.000	Ea.	4,100	550	4,650
Copper tubing, type L, solder joint, hanger 10′ OC, 1-1/4″ diam	30.000	L.F.	396	399	795
Wrought copper 90° elbow for solder joints 1-1/4″ diam	4.000	Ea.	51.40	206	257.40
Wrought copper tee for solder joints, 1-1/4″ diam	2.000	Ea.	54	171	225
Wrought copper union for soldered joints, 1-1/4″ diam	2.000	Ea.	173	110	283
Valve, gate, bronze, 125 lb, NRS, soldered 1-1/4″ diam	2.000	Ea.	334	103	437
Relief valve, bronze, press & temp, self-close, 3/4″ IPS	1.000	Ea.	283	27.50	310.50
Copper tubing, type L, solder joints, 3/4″ diam	8.000	L.F.	41.60	81.20	122.80
Wrought copper 90° elbow for solder joints 3/4″ diam	1.000	Ea.	2.95	40.50	43.45
Wrought copper, adapter, CTS to MPT, 3/4″ IPS	1.000	Ea.	6.70	45.50	52.20
Pipe steel black, schedule 40, threaded, 3/4″ diam	10.000	L.F.	48.40	126.50	174.90
Pipe, 90° elbow, malleable iron black, 150 lb threaded, 3/4″ diam	2.000	Ea.	9.06	110	119.06
Pipe, union with brass seat, malleable iron black, 3/4″ diam	1.000	Ea.	21	59.50	80.50
Valve, gas stop w/o check, brass, 3/4″ IPS	1.000	Ea.	19.90	35	54.90
TOTAL			5,541.01	2,064.70	7,605.71

D2020 250	Gas Fired Water Heaters - Commercial Systems		COST EACH MAT.	COST EACH INST.	COST EACH TOTAL
1760	Gas fired water heater, commercial, 100°F rise				
1780	75.5 MBH input, 63 GPH		5,550	2,075	7,625
1820	95 MBH input, 86 GPH	R224000 -10	11,100	2,075	13,175
1860	100 MBH input, 91 GPH		11,300	2,150	13,450
1900	115 MBH input, 110 GPH		11,000	2,225	13,225
1980	155 MBH input, 150 GPH		14,800	2,500	17,300
2020	175 MBH input, 168 GPH	R224000 -20	14,600	2,675	17,275
2060	200 MBH input, 192 GPH		15,300	3,025	18,325
2100	240 MBH input, 230 GPH		15,900	3,300	19,200
2140	300 MBH input, 278 GPH		16,900	3,725	20,625
2180	390 MBH input, 374 GPH		19,900	3,775	23,675
2220	500 MBH input, 480 GPH		26,900	4,075	30,975
2260	600 MBH input, 576 GPH		32,700	4,425	37,125

D2020 Domestic Water Distribution

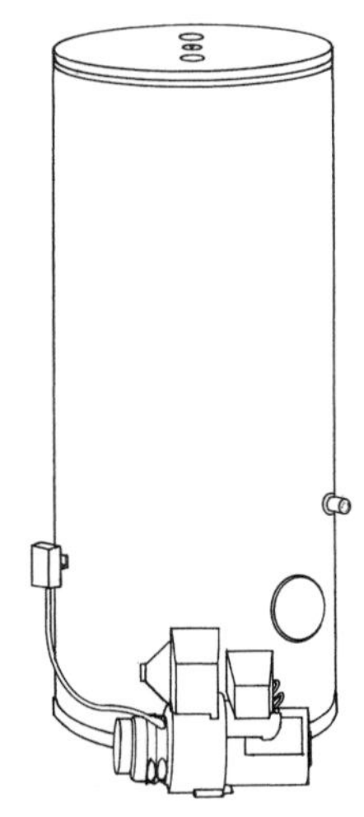

Units may be installed in multiples for increased capacity.

Included below is the heater, wired-in flame retention burners, cadmium cell primary controls, hi-limit controls, ASME pressure relief valves, draft controls, and insulated jacket.

Oil fired water heater systems include piping and fittings within 10′ of heater. Oil fired heaters require vent piping (not included in these systems).

System Components	QUANTITY	UNIT	COST EACH MAT.	COST EACH INST.	COST EACH TOTAL
SYSTEM D2020 260 1820					
OIL FIRED WATER HEATER, COMMERCIAL, 100°F RISE					
140 GAL., 140 MBH INPUT, 134 GPH					
Water heater, commercial, oil, 140 gal., 140 MBH input, 134 GPH	1.000	Ea.	27,700	650	28,350
Copper tubing, type L, solder joint, hanger 10′ OC, 3/4″ diam.	34.000	L.F.	176.80	345.10	521.90
Wrought copper 90° elbow for solder joints 3/4″ diam.	5.000	Ea.	14.75	202.50	217.25
Wrought copper Tee for solder joints, 3/4″ diam.	2.000	Ea.	13.30	129	142.30
Wrought copper union for soldered joints, 3/4″ diam.	2.000	Ea.	50	86	136
Valve, bronze, 125 lb, NRS, soldered 3/4″ diam.	2.000	Ea.	163	77	240
Relief valve, bronze, press & temp, self-close, 3/4″ IPS	1.000	Ea.	283	27.50	310.50
Wrought copper adapter, copper tubing to male, 3/4″ IPS	1.000	Ea.	6.70	45.50	52.20
Copper tubing, type L, solder joint, hanger 10′ OC, 3/8″ diam.	10.000	L.F.	36.70	92	128.70
Wrought copper 90° elbow for solder joints 3/8″ diam.	2.000	Ea.	9.14	70	79.14
Valve, globe, fusible, 3/8″ IPS	1.000	Ea.	16.20	32.50	48.70
TOTAL			28,469.59	1,757.10	30,226.69

D2020 260	Oil Fired Water Heaters - Commercial Systems	COST EACH MAT.	COST EACH INST.	COST EACH TOTAL
1800	Oil fired water heater, commercial, 100°F rise			
1820	140 gal., 140 MBH input, 134 GPH	28,500	1,750	30,250
1900	140 gal., 255 MBH input, 247 GPH (R224000-10)	30,900	2,200	33,100
1940	140 gal., 270 MBH input, 259 GPH	37,900	2,500	40,400
1980	140 gal., 400 MBH input, 384 GPH (R224000-20)	39,200	2,925	42,125
2060	140 gal., 720 MBH input, 691 GPH	41,600	3,025	44,625
2100	221 gal., 300 MBH input, 288 GPH	55,000	3,325	58,325
2140	221 gal., 600 MBH input, 576 GPH	70,000	3,375	73,375
2180	221 gal., 800 MBH input, 768 GPH	62,000	3,500	65,500
2220	201 gal., 1000 MBH input, 960 GPH	63,500	3,525	67,025
2260	201 gal., 1250 MBH input, 1200 GPH	65,000	3,600	68,600
2300	201 gal., 1500 MBH input, 1441 GPH	70,000	3,675	73,675
2340	411 gal., 600 MBH input, 576 GPH	70,500	3,775	74,275
2380	411 gal., 800 MBH input, 768 GPH	73,500	3,875	77,375
2420	411 gal., 1000 MBH input, 960 GPH	76,500	4,475	80,975
2460	411 gal., 1250 MBH input, 1200 GPH	78,500	4,600	83,100
2500	397 gal., 1500 MBH input, 1441 GPH	82,500	4,750	87,250

D2020 Domestic Water Distribution

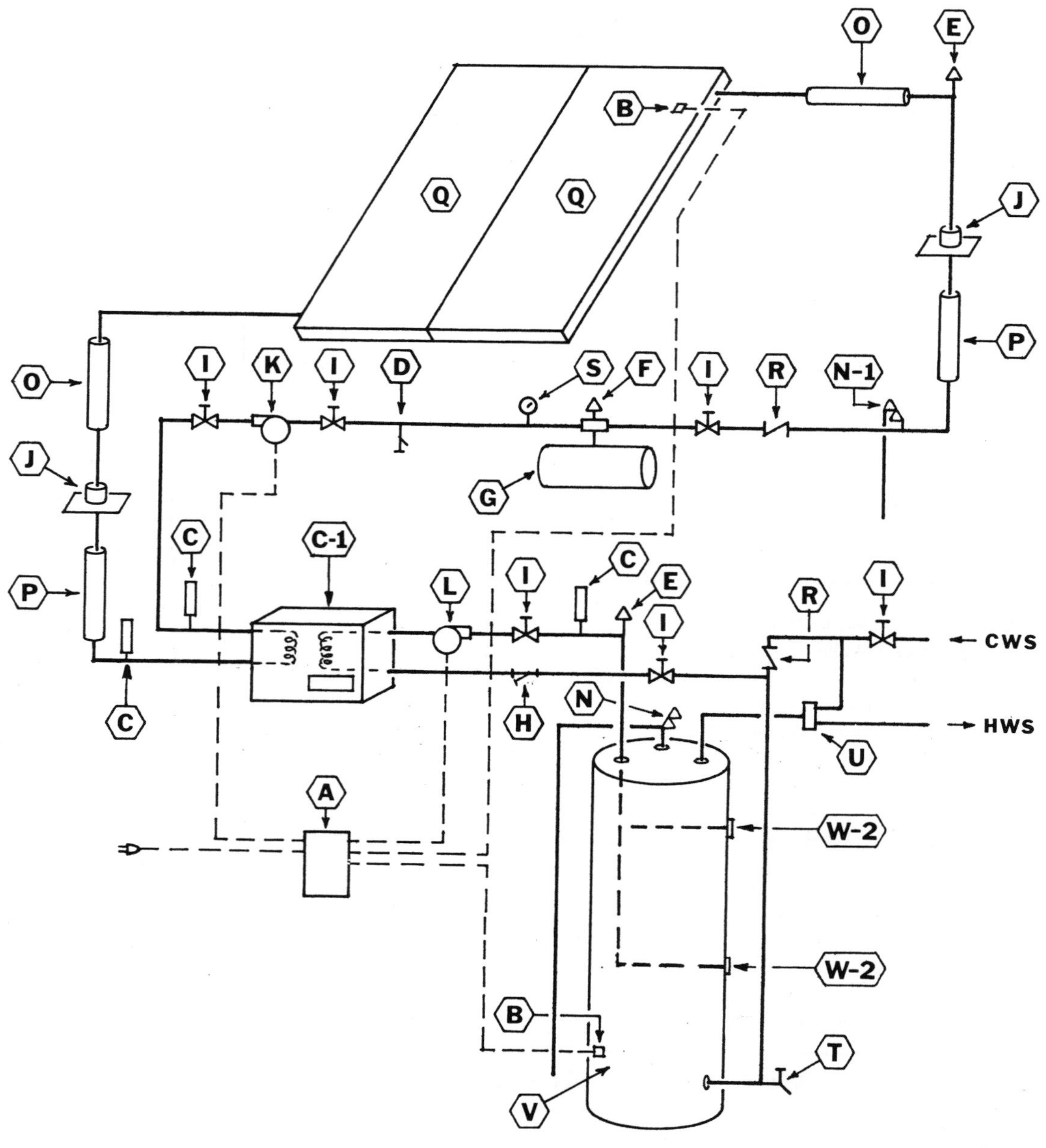

In this closed-loop indirect collection system, fluid with a low freezing temperature, such as propylene glycol, transports heat from the collectors to water storage. The transfer fluid is contained in a closed-loop consisting of collectors, supply and return piping, and a remote heat exchanger. The heat exchanger transfers heat energy from the fluid in the collector loop to potable water circulated in a storage loop. A typical two-or-three panel system contains 5 to 6 gallons of heat transfer fluid.

When the collectors become approximately 20°F warmer than the storage temperature, a controller activates the circulator on the collector and storage loops. The circulators will move the fluid and potable water through the heat exchanger until heat collection no longer occurs. At that point, the system shuts down. Since the heat transfer medium is a fluid with a very low freezing temperature, there is no need for it to be drained from the system between periods of collection.

D2020 Domestic Water Distribution

System Components		QUANTITY	UNIT	COST EACH MAT.	COST EACH INST.	COST EACH TOTAL
	SYSTEM D2020 265 2760					
	SOLAR, CLOSED LOOP, ADD-ON HOT WATER SYS., EXTERNAL HEAT EXCHANGER					
	3/4″ TUBING, TWO 3′X7′ BLACK CHROME COLLECTORS					
A,B,G,L,K,M	Heat exchanger fluid-fluid pkg incl 2 circulators, expansion tank, Check valve, relief valve, controller, hi temp cutoff, & 2 sensors	1.000	Ea.	890	555	1,445
C	Thermometer, 2″ dial	3.000	Ea.	75	144	219
D, T	Fill & drain valve, brass, 3/4″ connection	1.000	Ea.	13.25	32	45.25
E	Air vent, manual, 1/8″ fitting	2.000	Ea.	5.92	48	53.92
F	Air purger	1.000	Ea.	51	64.50	115.50
H	Strainer, Y type, bronze body, 3/4″ IPS	1.000	Ea.	63	41.50	104.50
I	Valve, gate, bronze, NRS, soldered 3/4″ diam	6.000	Ea.	489	231	720
J	Neoprene vent flashing	2.000	Ea.	23	77	100
N-1, N	Relief valve temp & press, 150 psi 210°F self-closing 3/4″ IPS	1.000	Ea.	28	25.50	53.50
O	Pipe covering, urethane, ultraviolet cover, 1″ wall 3/4″ diam	20.000	L.F.	62	141	203
P	Pipe covering, fiberglass, all service jacket, 1″ wall, 3/4″ diam	50.000	L.F.	54	282.50	336.50
Q	Collector panel solar energy blk chrome on copper, 1/8″ temp glass 3′x7′	2.000	Ea.	2,350	292	2,642
	Roof clamps for solar energy collector panels	2.000	Set	7.20	39.70	46.90
R	Valve, swing check, bronze, regrinding disc, 3/4″ diam	2.000	Ea.	206	77	283
S	Pressure gauge, 60 psi, 2″ dial	1.000	Ea.	27.50	24	51.50
U	Valve, water tempering, bronze, sweat connections, 3/4″ diam	1.000	Ea.	183	38.50	221.50
W-2, V	Tank water storage w/heating element, drain, relief valve, existing	1.000	Ea.			
	Copper tubing type L, solder joint, hanger 10′ OC 3/4″ diam	20.000	L.F.	104	203	307
	Copper tubing, type M, solder joint, hanger 10′ OC 3/4″ diam	70.000	L.F.	388.50	693	1,081.50
	Sensor wire, #22-2 conductor multistranded	.500	C.L.F.	6.50	36.50	43
	Solar energy heat transfer fluid, propylene glycol anti-freeze	6.000	Gal.	153	165	318
	Wrought copper fittings & solder, 3/4″ diam	76.000	Ea.	224.20	3,078	3,302.20
	TOTAL			5,404.07	6,288.70	11,692.77

D2020 265	Solar, Closed Loop, Add-On Hot Water Systems		COST EACH MAT.	COST EACH INST.	COST EACH TOTAL
2550	Solar, closed loop, add-on hot water system, external heat exchanger				
2570	3/8″ tubing, 3 ea. 4′ x 4′-4″ vacuum tube collectors		5,950	5,875	11,825
2580	1/2″ tubing, 4 ea. 4 x 4′-4″ vacuum tube collectors	R235616 -60	6,600	6,375	12,975
2600	2 ea. 3′x7′ black chrome collectors		4,950	6,000	10,950
2620	3 ea. 3′x7′ black chrome collectors		6,125	6,175	12,300
2640	2 ea. 3′x7′ flat black collectors		4,375	6,025	10,400
2660	3 ea. 3′x7′ flat black collectors		5,275	6,200	11,475
2700	3/4″ tubing, 3 ea. 3′x7′ black chrome collectors		6,575	6,450	13,025
2720	3 ea. 3′x7′ flat black absorber plate collectors		5,750	6,475	12,225
2740	2 ea. 4′x9′ flat black w/plastic glazing collectors		5,300	6,500	11,800
2760	2 ea. 3′x7′ black chrome collectors		5,400	6,300	11,700
2780	1″ tubing, 4 ea 2′x9′ plastic absorber & glazing collectors		6,650	7,300	13,950
2800	4 ea. 3′x7′ black chrome absorber collectors		8,700	7,325	16,025
2820	4 ea. 3′x7′ flat black absorber collectors		7,575	7,350	14,925

D2020 Domestic Water Distribution

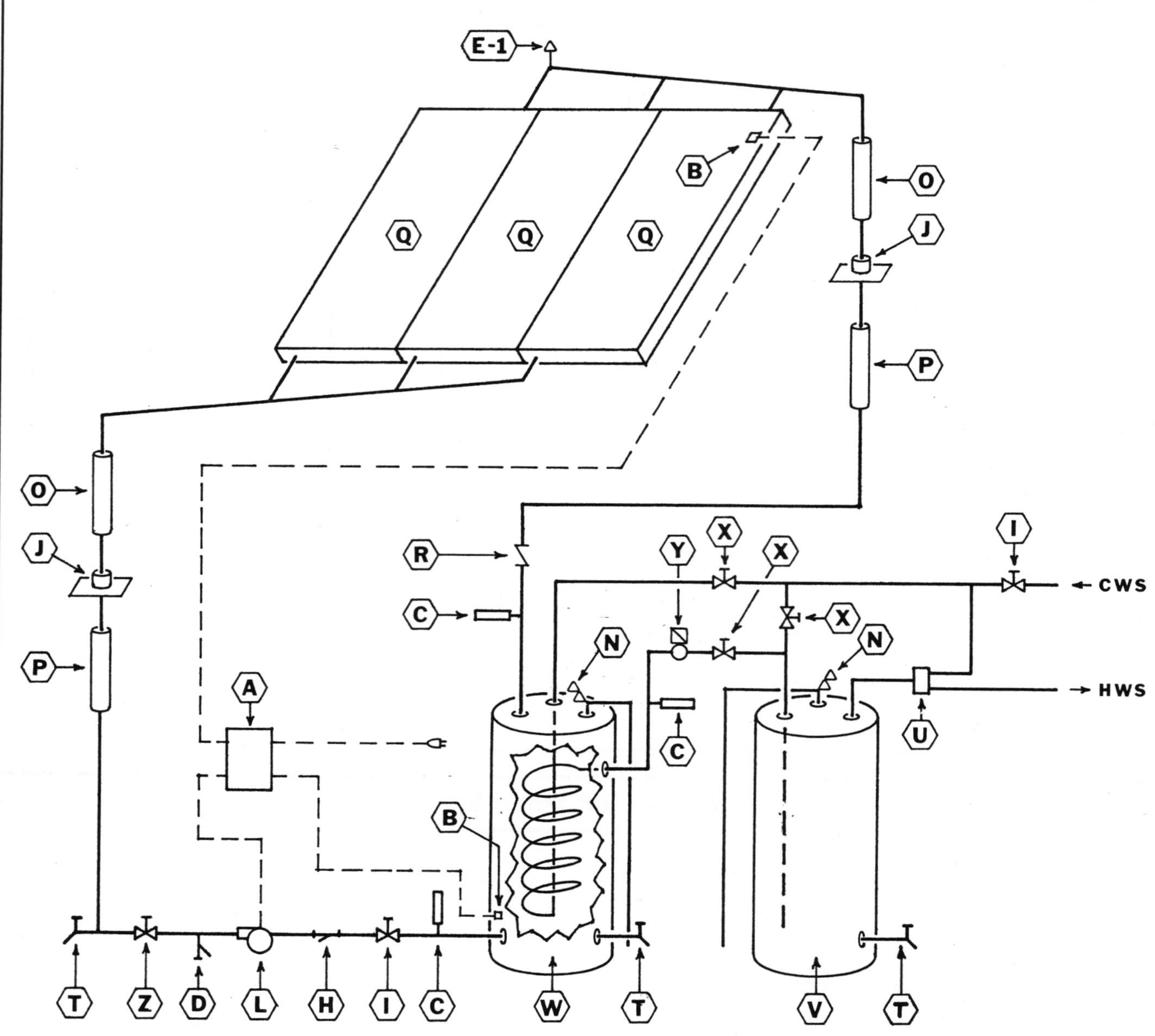

In the drainback indirect-collection system, the heat transfer fluid is distilled water contained in a loop consisting of collectors, supply and return piping, and an unpressurized holding tank. A large heat exchanger containing incoming potable water is immersed in the holding tank. When a controller activates solar collection, the distilled water is pumped through the collectors and heated and pumped back down to the holding tank. When the temperature differential between the water in the collectors and water in storage is such that collection no longer occurs, the pump turns off and gravity causes the distilled water in the collector loop to drain back to the holding tank. All the loop piping is pitched so that the water can drain out of the collectors and piping and not freeze there. As hot water is needed in the home, incoming water first flows through the holding tank with the immersed heat exchanger and is warmed and then flows through a conventional heater for any supplemental heating that is necessary.

D2020 Domestic Water Distribution

System Components

System Components	QUANTITY	UNIT	COST EACH MAT.	COST EACH INST.	COST EACH TOTAL
SYSTEM D2020 270 2760					
SOLAR, DRAINBACK, ADD ON, HOT WATER, IMMERSED HEAT EXCHANGER					
3/4″ TUBING, THREE EA 3′X7′ BLACK CHROME COLLECTOR					
A, B Differential controller 2 sensors, thermostat, solar energy system	1.000	Ea.	169	64.50	233.50
C Thermometer 2″ dial	3.000	Ea.	75	144	219
D, T Fill & drain valve, brass, 3/4″ connection	1.000	Ea.	13.25	32	45.25
E-1 Automatic air vent 1/8″ fitting	1.000	Ea.	19.70	24	43.70
H Strainer, Y type, bronze body, 3/4″ IPS	1.000	Ea.	63	41.50	104.50
I Valve, gate, bronze, NRS, soldered 3/4″ diam	2.000	Ea.	163	77	240
J Neoprene vent flashing	2.000	Ea.	23	77	100
L Circulator, solar heated liquid, 1/20 HP	1.000	Ea.	310	116	426
N Relief valve temp. & press. 150 psi 210°F self-closing 3/4″ IPS	1.000	Ea.	28	25.50	53.50
O Pipe covering, urethane, ultraviolet cover, 1″ wall, 3/4″ diam	20.000	L.F.	62	141	203
P Pipe covering, fiberglass, all service jacket, 1″ wall, 3/4″ diam	50.000	L.F.	54	282.50	336.50
Q Collector panel solar energy blk chrome on copper, 1/8″ temp glas 3′x7′	3.000	Ea.	3,525	438	3,963
Roof clamps for solar energy collector panels	3.000	Set	10.80	59.55	70.35
R Valve, swing check, bronze, regrinding disc, 3/4″ diam	1.000	Ea.	103	38.50	141.50
U Valve, water tempering, bronze sweat connections, 3/4″ diam	1.000	Ea.	183	38.50	221.50
V Tank, water storage w/heating element, drain, relief valve, existing	1.000	Ea.			
W Tank, water storage immersed heat exchr elec 2″x1/2# insul 120 gal	1.000	Ea.	2,100	550	2,650
X Valve, globe, bronze, rising stem, 3/4″ diam, soldered	3.000	Ea.	567	115.50	682.50
Y Flow control valve	1.000	Ea.	154	35	189
Z Valve, ball, bronze, solder 3/4″ diam, solar loop flow control	1.000	Ea.	39	38.50	77.50
Copper tubing, type L, solder joint, hanger 10′ OC 3/4″ diam	20.000	L.F.	104	203	307
Copper tubing, type M, solder joint, hanger 10′ OC 3/4″ diam	70.000	L.F.	388.50	693	1,081.50
Sensor wire, #22-2 conductor, multistranded	.500	C.L.F.	6.50	36.50	43
Wrought copper fittings & solder, 3/4″ diam	76.000	Ea.	224.20	3,078	3,302.20
TOTAL			8,384.95	6,349.05	14,734

D2020 270 Solar, Drainback, Hot Water Systems

D2020 270	Solar, Drainback, Hot Water Systems	COST EACH MAT.	COST EACH INST.	COST EACH TOTAL
2550	Solar, drainback, hot water, immersed heat exchanger			
2560	3/8″ tubing, 3 ea. 4′ x 4′-4″ vacuum tube collectors	7,650	5,725	13,375
2580	1/2″ tubing, 4 ea 4′ x 4′-4″ vacuum tube collectors, 80 gal tank (R235616 -60)	8,300	6,200	14,500
2600	120 gal tank	8,250	6,275	14,525
2640	2 ea. 3′x7′ blk chrome collectors, 80 gal tank	6,625	5,825	12,450
2660	3 ea. 3′x7′ blk chrome collectors, 120 gal tank	7,750	6,075	13,825
2700	2 ea. 3′x7′ flat blk collectors, 120 gal tank	6,025	5,925	11,950
2720	3 ea. 3′x7′ flat blk collectors, 120 gal tank	6,925	6,100	13,025
2760	3/4″ tubing, 3 ea 3′x7′ black chrome collectors, 120 gal tank	8,375	6,350	14,725
2780	3 ea. 3′x7′ flat black absorber collectors, 120 gal tank	7,550	6,375	13,925
2800	2 ea. 4′x9′ flat blk w/plastic glazing collectors 120 gal tank	7,100	6,400	13,500
2840	1″ tubing, 4 ea. 2′x9′ plastic absorber & glazing collectors, 120 gal tank	8,550	7,200	15,750
2860	4 ea. 3′x7′ black chrome absorber collectors, 120 gal tank	10,600	7,225	17,825
2880	4 ea. 3′x7′ flat black absorber collectors, 120 gal tank	9,475	7,250	16,725

D2020 Domestic Water Distribution

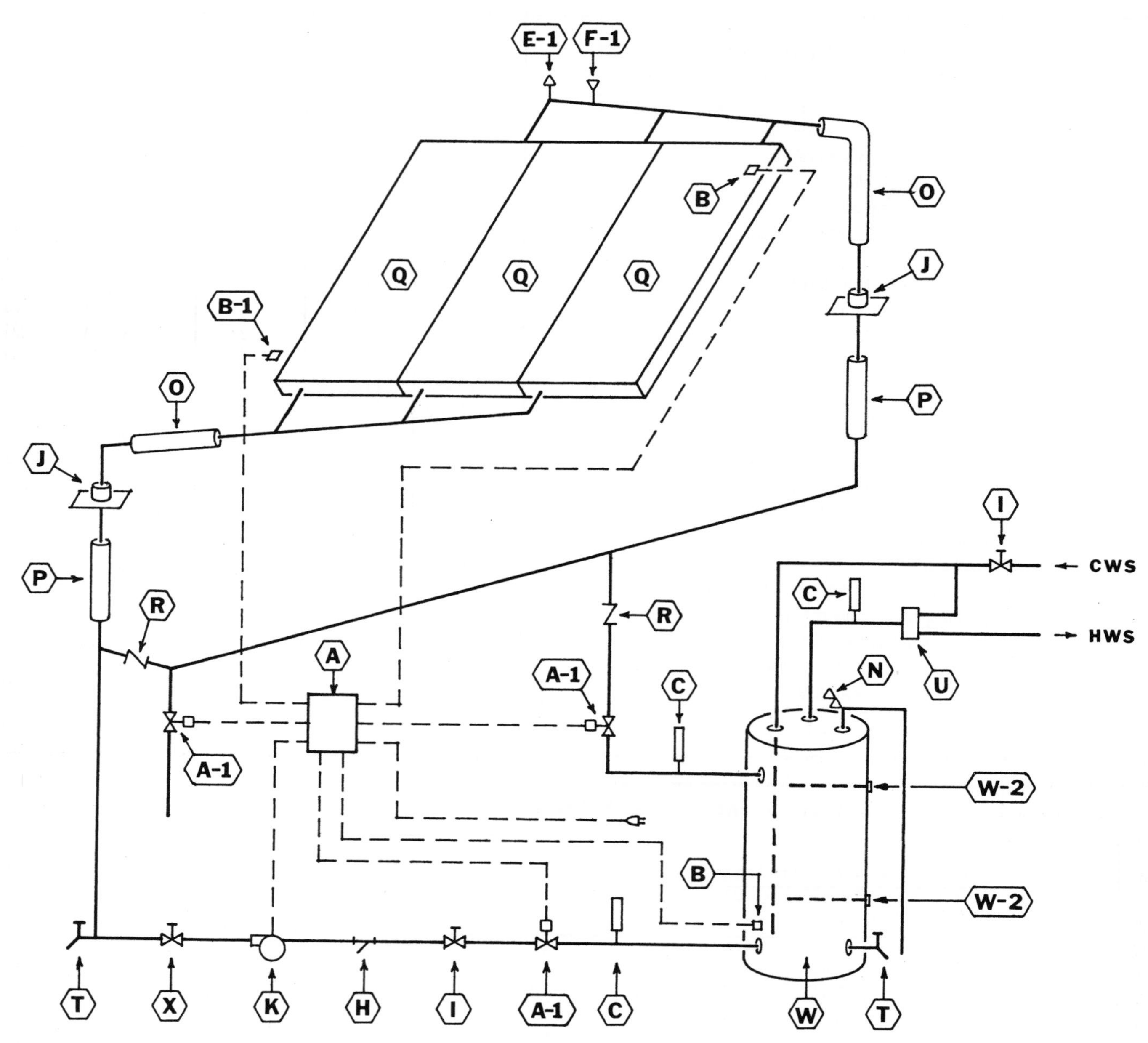

In the draindown direct-collection system, incoming domestic water is heated in the collectors. When the controller activates solar collection, domestic water is first heated as it flows through the collectors and is then pumped to storage. When conditions are no longer suitable for heat collection, the pump shuts off and the water in the loop drains down and out of the system by means of solenoid valves and properly pitched piping.

D2020 Domestic Water Distribution

System Components	QUANTITY	UNIT	COST EACH MAT.	COST EACH INST.	COST EACH TOTAL
SYSTEM D2020 275 2760					
SOLAR, DRAINDOWN, HOT WATER, DIRECT COLLECTION					
3/4″ TUBING, THREE 3′X7′ BLACK CHROME COLLECTORS					
A, B Differential controller, 2 sensors, thermostat, solar energy system	1.000	Ea.	169	64.50	233.50
A-1 Solenoid valve, solar heating loop, brass, 3/4″ diam, 24 volts	3.000	Ea.	450	256.50	706.50
B-1 Solar energy sensor, freeze prevention	1.000	Ea.	27.50	24	51.50
C Thermometer, 2″ dial	3.000	Ea.	75	144	219
E-1 Vacuum relief valve, 3/4″ diam	1.000	Ea.	34	24	58
F-1 Air vent, automatic, 1/8″ fitting	1.000	Ea.	19.70	24	43.70
H Strainer, Y type, bronze body, 3/4″ IPS	1.000	Ea.	63	41.50	104.50
I Valve, gate, bronze, NRS, soldered, 3/4″ diam	2.000	Ea.	163	77	240
J Vent flashing neoprene	2.000	Ea.	23	77	100
K Circulator, solar heated liquid, 1/25 HP	1.000	Ea.	225	99	324
N Relief valve temp & press 150 psi 210°F self-closing 3/4″ IPS	1.000	Ea.	28	25.50	53.50
O Pipe covering, urethane, ultraviolet cover, 1″ wall, 3/4″ diam	20.000	L.F.	62	141	203
P Pipe covering, fiberglass, all service jacket, 1″ wall, 3/4″ diam	50.000	L.F.	54	282.50	336.50
Roof clamps for solar energy collector panels	3.000	Set	10.80	59.55	70.35
Q Collector panel solar energy blk chrome on copper, 1/8″ temp glass 3′x7′	3.000	Ea.	3,525	438	3,963
R Valve, swing check, bronze, regrinding disc, 3/4″ diam, soldered	2.000	Ea.	206	77	283
T Drain valve, brass, 3/4″ connection	2.000	Ea.	26.50	64	90.50
U Valve, water tempering, bronze, sweat connections, 3/4″ diam	1.000	Ea.	183	38.50	221.50
W-2, W Tank, water storage elec elem 2″x1/2# insul 120 gal	1.000	Ea.	2,100	550	2,650
X Valve, globe, bronze, rising stem, 3/4″ diam, soldered	1.000	Ea.	189	38.50	227.50
Copper tubing, type L, solder joints, hangers 10′ OC 3/4″ diam	20.000	L.F.	104	203	307
Copper tubing, type M, solder joints, hangers 10′ OC 3/4″ diam	70.000	L.F.	388.50	693	1,081.50
Sensor wire, #22-2 conductor, multistranded	.500	C.L.F.	6.50	36.50	43
Wrought copper fittings & solder, 3/4″ diam	76.000	Ea.	224.20	3,078	3,302.20
TOTAL			8,356.70	6,556.55	14,913.25

D2020 275	Solar, Draindown, Hot Water Systems	COST EACH MAT.	COST EACH INST.	COST EACH TOTAL
2550	Solar, draindown, hot water			
2560	3/8″ tubing, 3 ea. 4′ x 4′-4″ vacuum tube collectors, 80 gal tank	7,400	6,250	13,650
2580	1/2″ tubing, 4 ea. 4′ x 4′-4″ vacuum tube collectors, 80 gal tank R235616 -60	8,400	6,425	14,825
2600	120 gal tank	8,350	6,475	14,825
2640	2 ea. 3′x7′ black chrome collectors, 80 gal tank	6,750	6,050	12,800
2660	3 ea. 3′x7′ black chrome collectors, 120 gal tank	7,875	6,300	14,175
2700	2 ea. 3′x7′ flat black collectors, 120 gal tank	6,150	6,150	12,300
2720	3 ea. 3′x7′ flat black collectors, 120 gal tank	6,950	6,275	13,225
2760	3/4″ tubing, 3 ea. 3′x7′ black chrome collectors, 120 gal tank	8,350	6,550	14,900
2780	3 ea. 3′x7′ flat collectors, 120 gal tank	7,525	6,575	14,100
2800	2 ea. 4′x9′ flat black & plastic glazing collectors, 120 gal tank	7,075	6,600	13,675
2840	1″ tubing, 4 ea. 2′x9′ plastic absorber & glazing collectors, 120 gal tank	12,300	8,700	21,000
2860	4 ea. 3′x7′ black chrome absorber collectors, 120 gal tank	11,900	8,225	20,125
2880	4 ea. 3′x7′ flat black absorber collectors, 120 gal tank	9,700	7,450	17,150

D2020 Domestic Water Distribution

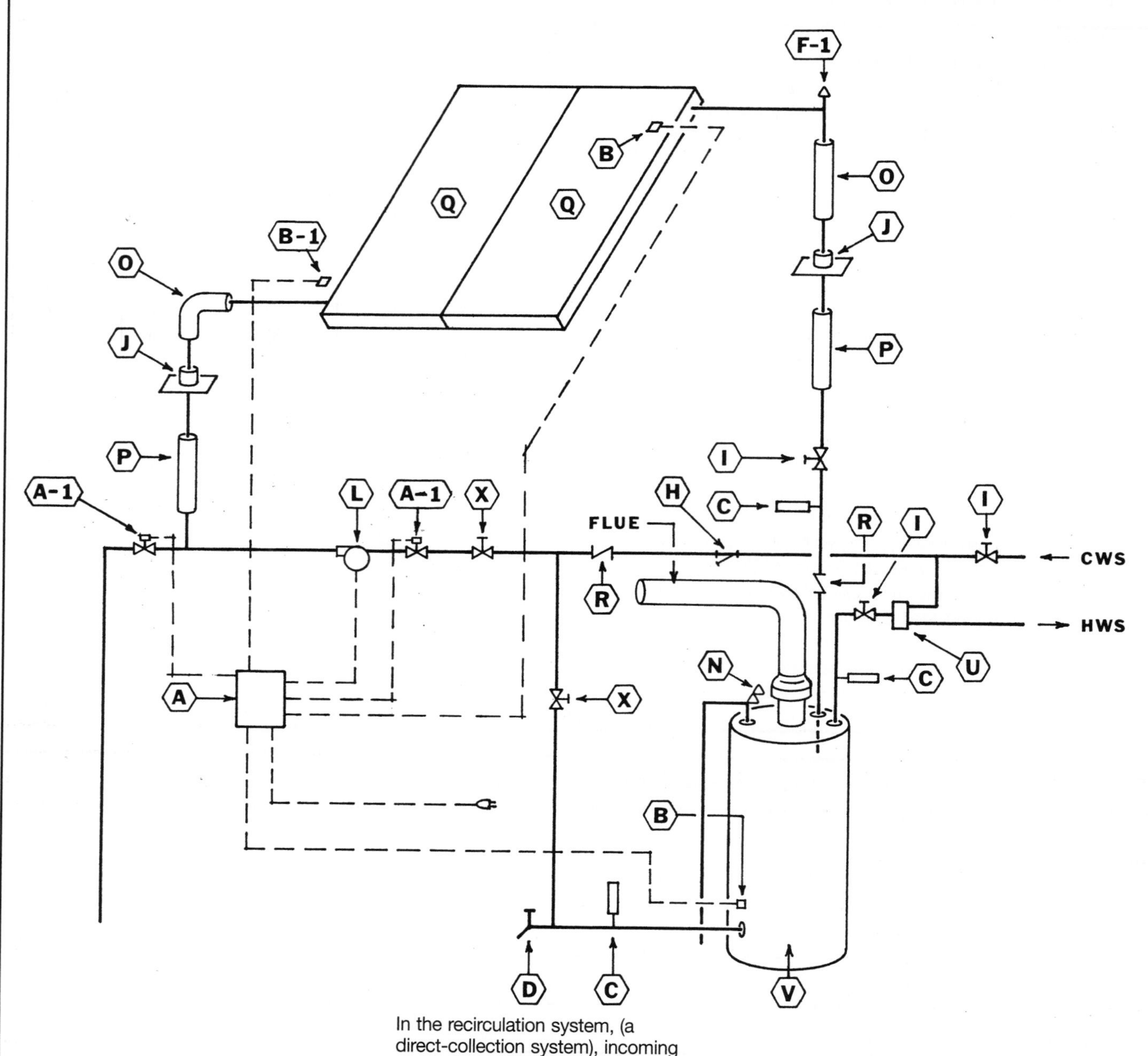

In the recirculation system, (a direct-collection system), incoming domestic water is heated in the collectors. When the controller activates solar collection, domestic water is heated as it flows through the collectors and then it flows back to storage. When conditions are not suitable for heat collection, the pump shuts off and the flow of the water stops. In this type of system, water remains in the collector loop at all times. A "frost sensor" at the collector activates the circulation of warm water from storage through the collectors when protection from freezing is required.

D2020 Domestic Water Distribution

System Components		QUANTITY	UNIT	COST EACH MAT.	COST EACH INST.	COST EACH TOTAL
	SYSTEM D2020 280 2820 **SOLAR, RECIRCULATION, HOT WATER** **3/4″ TUBING, TWO 3′X7′ BLACK CHROME COLLECTORS**					
A, B	Differential controller 2 sensors, thermostat, for solar energy system	1.000	Ea.	169	64.50	233.50
A-1	Solenoid valve, solar heating loop, brass, 3/4″ IPS, 24 volts	2.000	Ea.	300	171	471
B-1	Solar energy sensor freeze prevention	1.000	Ea.	27.50	24	51.50
C	Thermometer, 2″ dial	3.000	Ea.	75	144	219
D	Drain valve, brass, 3/4″ connection	1.000	Ea.	13.25	32	45.25
F-1	Air vent, automatic, 1/8″ fitting	1.000	Ea.	19.70	24	43.70
H	Strainer, Y type, bronze body, 3/4″ IPS	1.000	Ea.	63	41.50	104.50
I	Valve, gate, bronze, 125 lb, soldered 3/4″ diam	3.000	Ea.	244.50	115.50	360
J	Vent flashing, neoprene	2.000	Ea.	23	77	100
L	Circulator, solar heated liquid, 1/20 HP	1.000	Ea.	310	116	426
N	Relief valve, temp & press 150 psi 210° F self-closing 3/4″ IPS	2.000	Ea.	56	51	107
O	Pipe covering, urethane, ultraviolet cover, 1″ wall, 3/4″ diam	20.000	L.F.	62	141	203
P	Pipe covering, fiberglass, all service jacket, 1″ wall 3/4″ diam	50.000	L.F.	54	282.50	336.50
Q	Collector panel solar energy blk chrome on copper, 1/8″ temp glass 3′x7′	2.000	Ea.	2,350	292	2,642
	Roof clamps for solar energy collector panels	2.000	Set	7.20	39.70	46.90
R	Valve, swing check, bronze, 125 lb, regrinding disc, soldered 3/4″ diam	2.000	Ea.	206	77	283
U	Valve, water tempering, bronze, sweat connections, 3/4″ diam	1.000	Ea.	183	38.50	221.50
V	Tank, water storage, w/heating element, drain, relief valve, existing	1.000	Ea.			
X	Valve, globe, bronze, 125 lb, 3/4″ diam	2.000	Ea.	378	77	455
	Copper tubing, type L, solder joints, hangers 10′ OC 3/4″ diam	20.000	L.F.	104	203	307
	Copper tubing, type M, solder joints, hangers 10′ OC 3/4″ diam	70.000	L.F.	388.50	693	1,081.50
	Wrought copper fittings & solder, 3/4″ diam	76.000	Ea.	224.20	3,078	3,302.20
	Sensor wire, #22-2 conductor, multistranded	.500	C.L.F.	6.50	36.50	43
	TOTAL			5,264.35	5,818.70	11,083.05

D2020 280	Solar, Recirculation, Domestic Hot Water Systems	COST EACH MAT.	COST EACH INST.	COST EACH TOTAL
2550	Solar, recirculation, hot water			
2560	3/8″ tubing, 3 ea. 4′ x 4′-4″ vacuum tube collectors	5,700	5,425	11,125
2580	1/2″ tubing, 4 ea. 4′ x 4′-4″ vacuum tube collectors R235616-60	6,350	5,900	12,250
2640	2 ea. 3′x7′ black chrome collectors	4,700	5,525	10,225
2660	3 ea. 3′x7′ black chrome collectors	5,875	5,700	11,575
2700	2 ea. 3′x7′ flat black collectors	4,150	5,550	9,700
2720	3 ea. 3′x7′ flat black collectors	5,025	5,725	10,750
2760	3/4″ tubing, 3 ea. 3′x7′ black chrome collectors	6,450	5,975	12,425
2780	3 ea. 3′x7′ flat black absorber plate collectors	5,600	6,000	11,600
2800	2 ea. 4′x9′ flat black w/plastic glazing collectors	5,175	6,025	11,200
2820	2 ea. 3′x7′ black chrome collectors	5,275	5,825	11,100
2840	1″ tubing, 4 ea. 2′x9′ black plastic absorber & glazing collectors	6,850	6,825	13,675
2860	4 ea. 3′x7′ black chrome absorber collectors	8,875	6,850	15,725
2880	4 ea. 3′x7′ flat black absorber collectors	7,750	6,875	14,625

D2020 Domestic Water Distribution

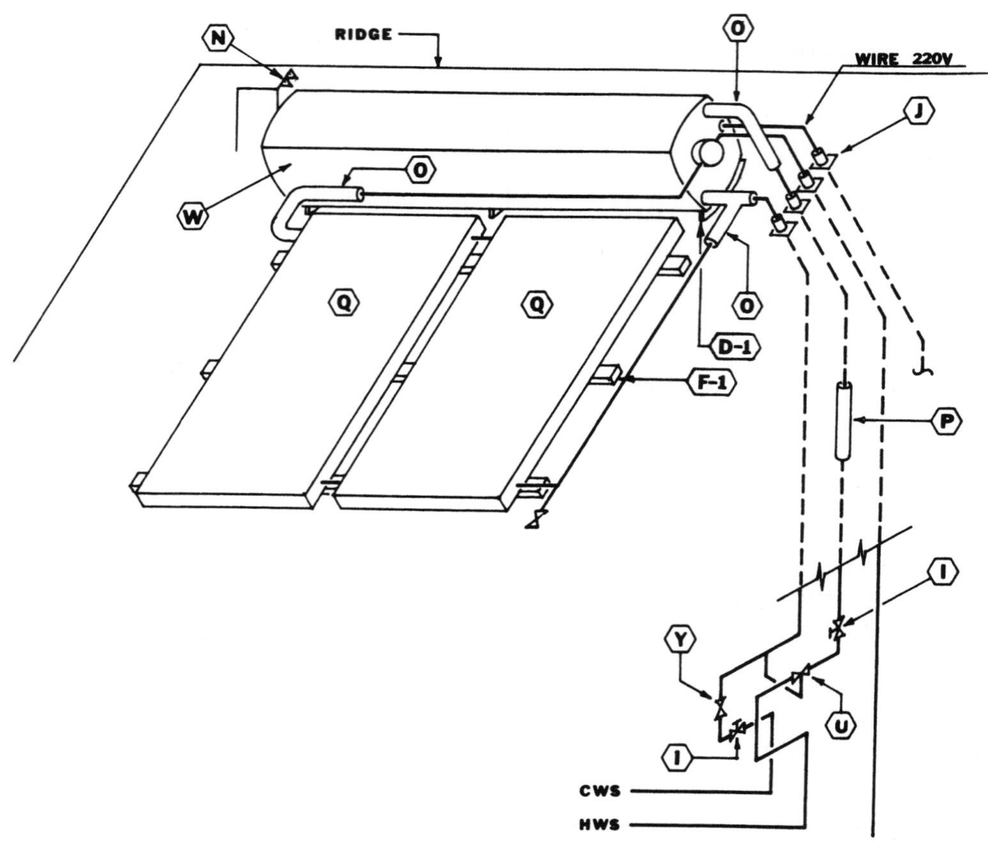

The thermosyphon domestic hot water system, a direct collection system, operates under city water pressure and does not require pumps for system operation. An insulated water storage tank is located above the collectors. As the sun heats the collectors, warm water in them rises by means of natural convection; the colder water in the storage tank flows into the collectors by means of gravity. As long as the sun is shining the water continues to flow through the collectors and to become warmer.

To prevent freezing, the system must be drained or the collectors covered with an insulated lid when the temperature drops below 32°F.

D2020 Domestic Water Distribution

System Components		QUANTITY	UNIT	COST EACH MAT.	COST EACH INST.	COST EACH TOTAL
	SYSTEM D2020 285 0960 SOLAR, THERMOSYPHON, WATER HEATER 3/4" TUBING, TWO 3'X7' BLACK CHROME COLLECTORS					
D-1	Framing lumber, fir, 2" x 6" x 8', tank cradle	.008	M.B.F.	6.04	8.20	14.24
F-1	Framing lumber, fir, 2" x 4" x 24', sleepers	.016	M.B.F.	12.48	24.80	37.28
I	Valve, gate, bronze, 125 lb, soldered 1/2" diam	2.000	Ea.	133	64	197
J	Vent flashing, neoprene	4.000	Ea.	46	154	200
O	Pipe covering, urethane, ultraviolet cover, 1" wall, 1/2"diam	40.000	L.F.	93.20	278	371.20
P	Pipe covering fiberglass all service jacket 1" wall 1/2" diam	160.000	L.F.	161.60	864	1,025.60
Q	Collector panel solar, blk chrome on copper, 3/16" temp glass 3'-6" x 7.5'	2.000	Ea.	1,790	308	2,098
Y	Flow control valve, globe, bronze, 125#, soldered, 1/2" diam	1.000	Ea.	117	32	149
U	Valve, water tempering, bronze,sweat connections, 1/2" diam	1.000	Ea.	130	32	162
W	Tank, water storage, solar energy system, 80 Gal, 2" x 1/2 lb insul	1.000	Ea.	2,150	480	2,630
	Copper tubing type L, solder joints, hangers 10' OC 1/2" diam	150.000	L.F.	607.50	1,432.50	2,040
	Copper tubing type M, solder joints, hangers 10' OC 1/2" diam	50.000	L.F.	196	460	656
	Sensor wire, #22-2 gauge multistranded	.500	C.L.F.	6.50	36.50	43
	Wrought copper fittings & solder, 1/2" diam	75.000	Ea.	104.25	2,887.50	2,991.75
	TOTAL			5,553.57	7,061.50	12,615.07

D2020 285	Thermosyphon, Hot Water		COST EACH MAT.	COST EACH INST.	COST EACH TOTAL
0960	Solar, thermosyphon, hot water, two collector system	R235616-60	5,550	7,050	12,600
0970					

D2020 Domestic Water Distribution

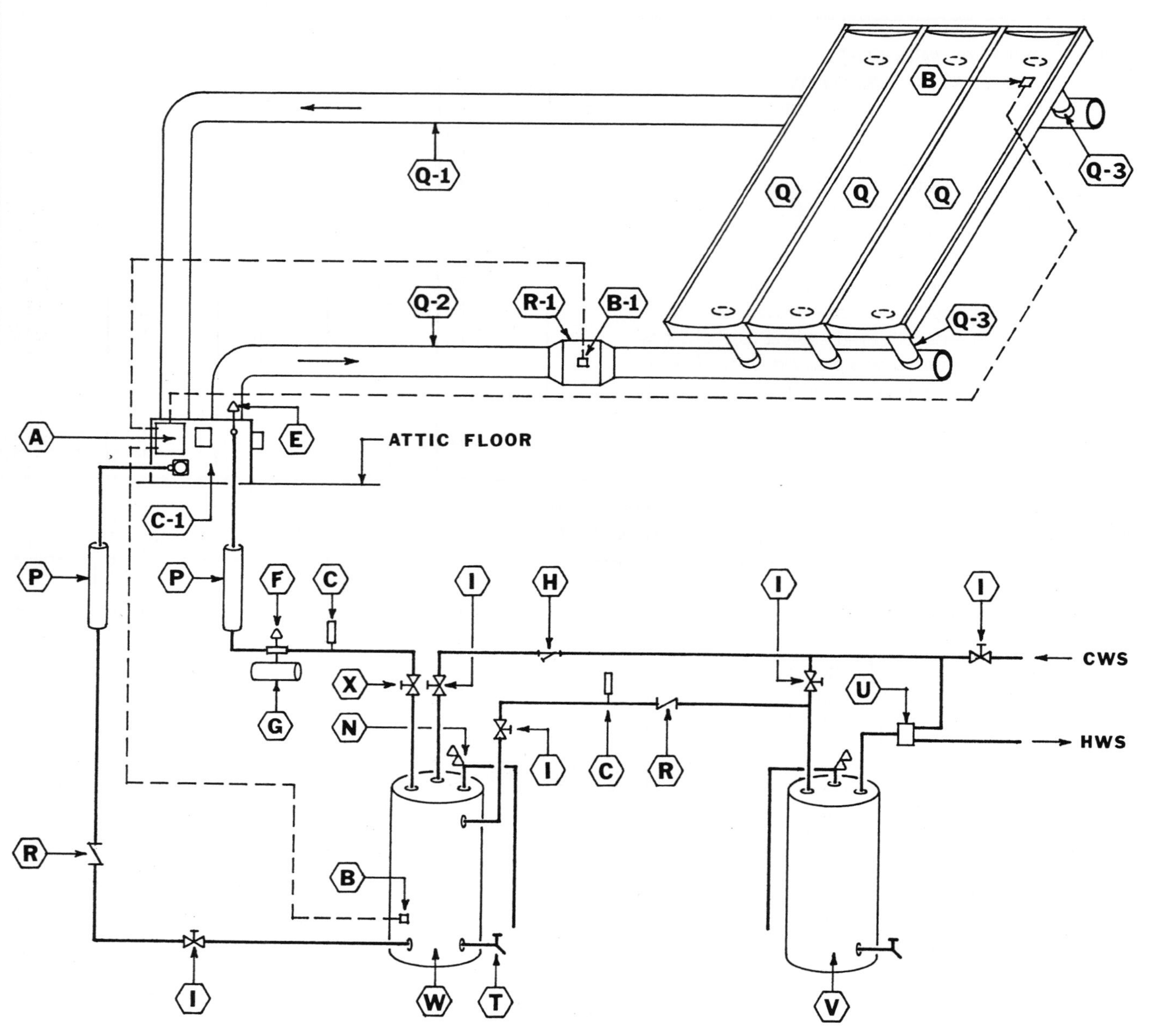

This domestic hot water pre-heat system includes heat exchanger with a circulating pump, blower, air-to-water, coil and controls, mounted in the upper collector manifold. Heat from the hot air coming out of the collectors is transferred through the heat exchanger. For each degree of DHW preheating gained, one degree less heating is needed from the fuel fired water heater. The system is simple, inexpensive to operate and can provide a substantial portion of DHW requirements for modest additional cost.

D2020 Domestic Water Distribution

System Components	QUANTITY	UNIT	COST EACH MAT.	COST EACH INST.	COST EACH TOTAL
SYSTEM D2020 290 2560					
SOLAR, HOT WATER, AIR TO WATER HEAT EXCHANGE					
THREE COLLECTORS, OPTICAL BLACK ON ALUM., 7.5′ x 3.5′,80 GAL TANK					
A, B Differential controller, 2 sensors, thermostat, solar energy system	1.000	Ea.	169	64.50	233.50
C Thermometer, 2″ dial	2.000	Ea.	50	96	146
C-1 Heat exchanger, air to fluid, up flow 70 MBH	1.000	Ea.	410	395	805
E Air vent, manual, for solar energy system 1/8″ fitting	1.000	Ea.	2.96	24	26.96
F Air purger	1.000	Ea.	51	64.50	115.50
G Expansion tank	1.000	Ea.	77.50	24	101.50
H Strainer, Y type, bronze body, 1/2″ IPS	1.000	Ea.	50	39	89
I Valve, gate, bronze, 125 lb, NRS, soldered 1/2″ diam	5.000	Ea.	332.50	160	492.50
N Relief valve, temp & pressure solar 150 psi 210°F self-closing	1.000	Ea.	28	25.50	53.50
P Pipe covering, fiberglass, all service jacket, 1″ wall, 1/2″ diam	60.000	L.F.	60.60	324	384.60
Q Collector panel solar energy, air, black on alum plate, 7.5′ x 3.5′	3.000	Ea.	2,685	462	3,147
B-1, R-1 Shutter damper for solar heater circulator	1.000	Ea.	64.50	113	177.50
B-1, R-1 Shutter motor for solar heater circulator	1.000	Ea.	161	85	246
R Backflow preventer, 1/2″ pipe size	2.000	Ea.	199	96	295
T Drain valve, brass, 3/4″ connection	1.000	Ea.	13.25	32	45.25
U Valve, water tempering, bronze, sweat connections, 1/2 ″ diam	1.000	Ea.	130	32	162
W Tank, water storage, solar energy system, 80 Gal, 2″ x 2 lb insul	1.000	Ea.	2,150	480	2,630
V Tank, water storage, w/heating element, drain, relief valve, existing	1.000	System			
X Valve, globe, bronze, 125 lb, rising stem, 1/2″ diam	1.000	Ea.	117	32	149
Copper tubing type M, solder joints, hangers 10′ OC, 1/2″ diam	50.000	L.F.	196	460	656
Copper tubing type L, solder joints, hangers 10′ OC 1/2″ diam	10.000	L.F.	40.50	95.50	136
Wrought copper fittings & solder, 1/2″ diam	10.000	Ea.	13.90	385	398.90
Sensor wire, #22-2 conductor multistranded	.500	C.L.F.	6.50	36.50	43
Q-1, Q-2 Ductwork, fiberglass, aluminized jacket, 1-1/2″ thick, 8″ diam	32.000	S.F.	228.80	241.60	470.40
Q-3 Manifold for flush mount solar energy collector panels, air	6.000	L.F.	1,068	51	1,119
TOTAL			8,305.01	3,818.10	12,123.11

D2020 290	Solar, Hot Water, Air To Water Heat Exchange		COST EACH MAT.	COST EACH INST.	COST EACH TOTAL
2550	Solar, hot water, air to water heat exchange				
2560	Three collectors, optical black on aluminum, 7.5′ x 3.5′, 80 Gal tank		8,300	3,825	12,125
2580	Four collectors, optical black on aluminum, 7.5′ x 3.5′, 80 Gal tank	R235616 -60	9,600	4,050	13,650
2600	Four collectors, optical black on aluminum, 7.5′ x 3.5′, 120 Gal tank		9,550	4,125	13,675

D2020 Domestic Water Distribution

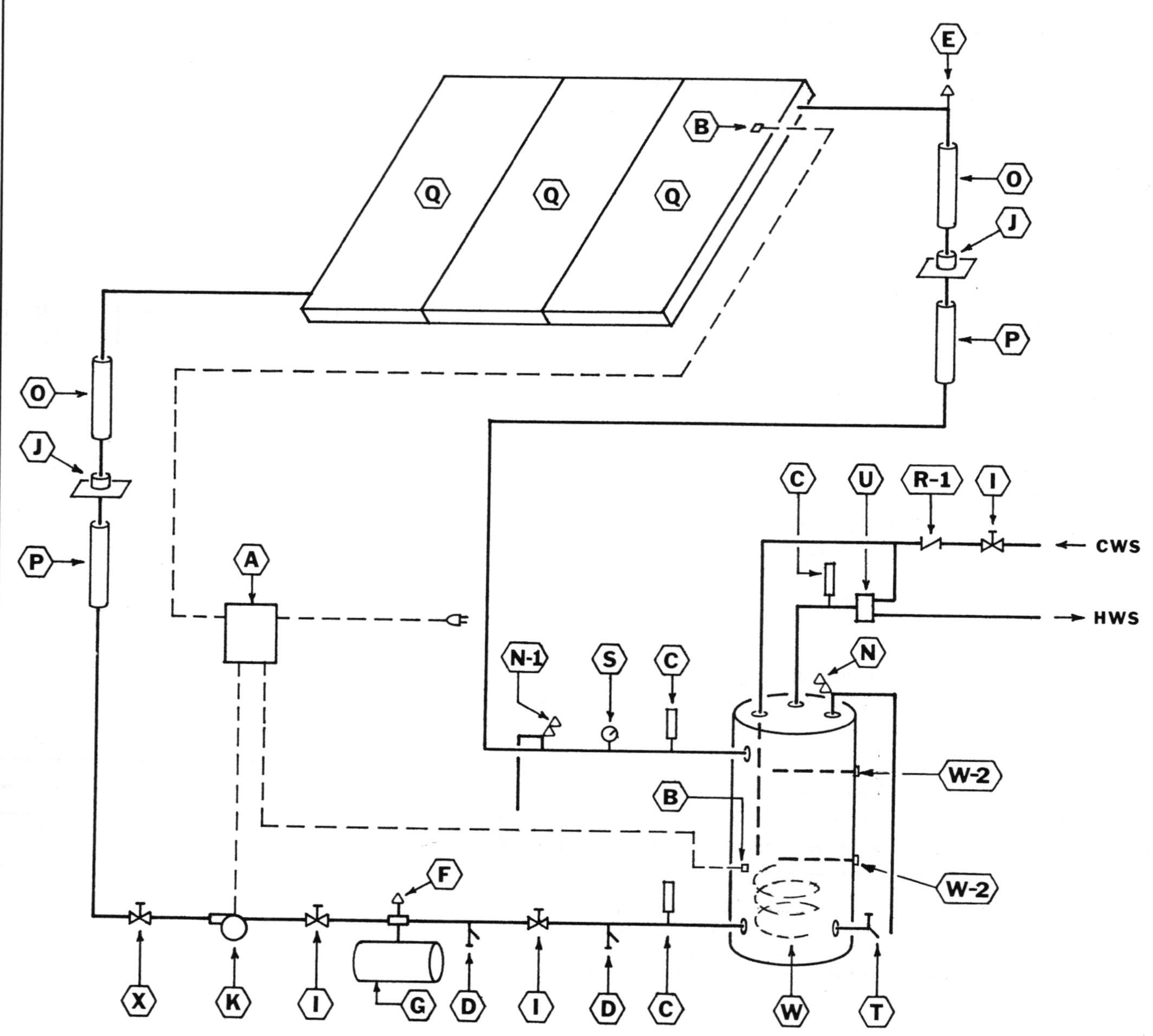

In this closed-loop indirect collection system, fluid with a low freezing temperature, such as propylene glycol, transports heat from the collectors to water storage. The transfer fluid is contained in a closed-loop consisting of collectors, supply and return piping, and a heat exchanger immersed in the storage tank. A typical two-or-three panel system contains 5 to 6 gallons of heat transfer fluid.

When the collectors become approximately 20°F warmer than the storage temperature, a controller activates the circulator. The circulator moves the fluid continuously through the collectors until the temperature difference between the collectors and storage is such that heat collection no longer occurs; at that point, the circulator shuts off. Since the heat transfer fluid has a very low freezing temperature, there is no need for it to be drained from the collectors between periods of collection.

D2020 Domestic Water Distribution

System Components		QUANTITY	UNIT	COST EACH MAT.	COST EACH INST.	COST EACH TOTAL
	SYSTEM D2020 295 2760 SOLAR, CLOSED LOOP, HOT WATER SYSTEM, IMMERSED HEAT EXCHANGER 3/4″ TUBING, THREE 3′ X 7′ BLACK CHROME COLLECTORS					
A, B	Differential controller, 2 sensors, thermostat, solar energy system	1.000	Ea.	169	64.50	233.50
C	Thermometer 2″ dial	3.000	Ea.	75	144	219
D, T	Fill & drain valves, brass, 3/4″ connection	3.000	Ea.	39.75	96	135.75
E	Air vent, manual, 1/8″ fitting	1.000	Ea.	2.96	24	26.96
F	Air purger	1.000	Ea.	51	64.50	115.50
G	Expansion tank	1.000	Ea.	77.50	24	101.50
I	Valve, gate, bronze, NRS, soldered 3/4″ diam	3.000	Ea.	244.50	115.50	360
J	Neoprene vent flashing	2.000	Ea.	23	77	100
K	Circulator, solar heated liquid, 1/25 HP	1.000	Ea.	225	99	324
N-1, N	Relief valve, temp & press 150 psi 210°F self-closing 3/4″ IPS	2.000	Ea.	56	51	107
O	Pipe covering, urethane ultraviolet cover, 1″ wall, 3/4″ diam	20.000	L.F.	62	141	203
P	Pipe covering, fiberglass, all service jacket, 1″ wall, 3/4″ diam	50.000	L.F.	54	282.50	336.50
	Roof clamps for solar energy collector panel	3.000	Set	10.80	59.55	70.35
Q	Collector panel solar blk chrome on copper, 1/8″ temp glass, 3′x7′	3.000	Ea.	3,525	438	3,963
R-1	Valve, swing check, bronze, regrinding disc, 3/4″ diam, soldered	1.000	Ea.	103	38.50	141.50
S	Pressure gauge, 60 psi, 2-1/2″ dial	1.000	Ea.	27.50	24	51.50
U	Valve, water tempering, bronze, sweat connections, 3/4″ diam	1.000	Ea.	183	38.50	221.50
W-2, W	Tank, water storage immersed heat exchr elec elem 2″x2# insul 120 Gal	1.000	Ea.	2,100	550	2,650
X	Valve, globe, bronze, rising stem, 3/4″ diam, soldered	1.000	Ea.	189	38.50	227.50
	Copper tubing type L, solder joint, hanger 10′ OC 3/4″ diam	20.000	L.F.	104	203	307
	Copper tubing, type M, solder joint, hanger 10′ OC 3/4″ diam	70.000	L.F.	388.50	693	1,081.50
	Sensor wire, #22-2 conductor multistranded	.500	C.L.F.	6.50	36.50	43
	Solar energy heat transfer fluid, propylene glycol, anti-freeze	6.000	Gal.	153	165	318
	Wrought copper fittings & solder, 3/4″ diam	76.000	Ea.	224.20	3,078	3,302.20
	TOTAL			8,094.21	6,545.55	14,639.76

D2020 295	Solar, Closed Loop, Hot Water Systems	COST EACH MAT.	COST EACH INST.	COST EACH TOTAL
2550	Solar, closed loop, hot water system, immersed heat exchanger			
2560	3/8″ tubing, 3 ea. 4′ x 4′-4″ vacuum tube collectors, 80 gal. tank	7,500	5,925	13,425
2580	1/2″ tubing, 4 ea. 4′ x 4′-4″ vacuum tube collectors, 80 gal. tank	8,150	6,400	14,550
2600	120 gal. tank	8,100	6,475	14,575
2640	2 ea. 3′x7′ black chrome collectors, 80 gal. tank (R235616-60)	6,575	6,125	12,700
2660	120 gal. tank	6,450	6,125	12,575
2700	2 ea. 3′x7′ flat black collectors, 120 gal. tank	5,900	6,125	12,025
2720	3 ea. 3′x7′ flat black collectors, 120 gal. tank	6,800	6,300	13,100
2760	3/4″ tubing, 3 ea. 3′x7′ black chrome collectors, 120 gal. tank	8,100	6,550	14,650
2780	3 ea. 3′x7′ flat black collectors, 120 gal. tank	7,250	6,575	13,825
2800	2 ea. 4′x9′ flat black w/plastic glazing collectors 120 gal. tank	6,825	6,600	13,425
2840	1″ tubing, 4 ea. 2′x9′ plastic absorber & glazing collectors 120 gal. tank	8,050	7,350	15,400
2860	4 ea. 3′x7′ black chrome collectors, 120 gal. tank	10,100	7,375	17,475
2880	4 ea. 3′x7′ flat black absorber collectors, 120 gal. tank	8,975	7,425	16,400

D2040 Rain Water Drainage

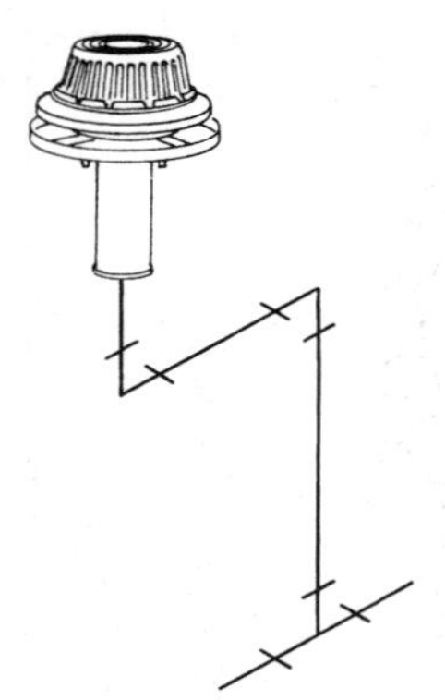

Design Assumptions: Vertical conductor size is based on a maximum rate of rainfall of 4″ per hour. To convert roof area to other rates multiply "Max. S.F. Roof Area" shown by four and divide the result by desired local rate. The answer is the local roof area that may be handled by the indicated pipe diameter.

Basic cost is for roof drain, 10′ of vertical leader and 10′ of horizontal, plus connection to the main.

Pipe Dia.	Max. S.F. Roof Area	Gallons per Min.
2″	544	23
3″	1610	67
4″	3460	144
5″	6280	261
6″	10,200	424
8″	22,000	913

System Components	QUANTITY	UNIT	COST EACH MAT.	COST EACH INST.	COST EACH TOTAL
SYSTEM D2040 210 1880					
ROOF DRAIN, DWV PVC PIPE, 2″ DIAM., 10′ HIGH					
Drain, roof, main, ABS, dome type 2″ pipe size	1.000	Ea.	138	99	237
Clamp, roof drain, underdeck	1.000	Ea.	30.50	58	88.50
Pipe, Tee, PVC DWV, schedule 40, 2″ pipe size	1.000	Ea.	7.15	69.50	76.65
Pipe, PVC, DWV, schedule 40, 2″ diam.	20.000	L.F.	201	470	671
Pipe, elbow, PVC schedule 40, 2″ diam.	2.000	Ea.	6.84	76	82.84
TOTAL			383.49	772.50	1,155.99

D2040 210	Roof Drain Systems	COST EACH MAT.	COST EACH INST.	COST EACH TOTAL
1880	Roof drain, DWV PVC, 2″ diam., piping, 10′ high	385	775	1,160
1920	For each additional foot add	10.05	23.50	33.55
1960	3″ diam., 10′ high	595	900	1,495
2000	For each additional foot add	19.20	26	45.20
2040	4″ diam., 10′ high	670	1,025	1,695
2080	For each additional foot add	21	29	50
2120	5″ diam., 10′ high	1,500	1,175	2,675
2160	For each additional foot add	26	32.50	58.50
2200	6″ diam., 10′ high	1,675	1,300	2,975
2240	For each additional foot add	20.50	35.50	56
2280	8″ diam., 10′ high	3,375	2,200	5,575
2320	For each additional foot add	41	45	86
3940	C.I., soil, single hub, service wt., 2″ diam. piping, 10′ high	835	845	1,680
3980	For each additional foot add	18.80	22	40.80
4120	3″ diam., 10′ high	1,175	910	2,085
4160	For each additional foot add	29	23	52
4200	4″ diam., 10′ high	1,275	1,000	2,275
4240	For each additional foot add	27	25.50	52.50
4280	5″ diam., 10′ high	1,700	1,100	2,800
4320	For each additional foot add	34	28.50	62.50
4360	6″ diam., 10′ high	2,250	1,175	3,425
4400	For each additional foot add	46.50	29.50	76
4440	8″ diam., 10′ high	4,225	2,425	6,650
4480	For each additional foot add	71	50	121
6040	Steel galv. sch 40 threaded, 2″ diam. piping, 10′ high	750	815	1,565
6080	For each additional foot add	8.70	21.50	30.20
6120	3″ diam., 10′ high	1,400	1,175	2,575
6160	For each additional foot add	17.60	32.50	50.10

D2040 Rain Water Drainage

D2040 210	Roof Drain Systems	COST EACH		
		MAT.	INST.	TOTAL
6200	4″ diam., 10′ high	2,150	1,525	3,675
6240	For each additional foot add	28	38.50	66.50
6280	5″ diam., 10′ high	2,200	1,250	3,450
6320	For each additional foot add	33.50	37.50	71
6360	6″ diam, 10′ high	2,825	1,625	4,450
6400	For each additional foot add	43.50	51.50	95
6440	8″ diam., 10′ high	5,050	2,375	7,425
6480	For each additional foot add	64	58.50	122.50

D3010 Energy Supply

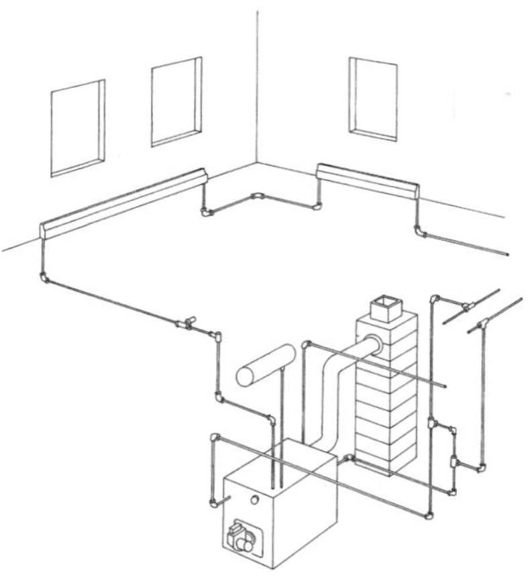

Basis for Heat Loss Estimate, Apartment Type Structures:

1. Masonry walls and flat roof are insulated. U factor is assumed at .08.
2. Window glass area taken as BOCA minimum, 1/10th of floor area. Double insulating glass with 1/4″ air space, U = .65.
3. Infiltration = 0.3 C.F. per hour per S.F. of net wall.
4. Concrete floor loss is 2 BTUH per S.F.
5. Temperature difference taken as 70°F.
6. Ventilating or makeup air has not been included and must be added if desired. Air shafts are not used.

System Components	QUANTITY	UNIT	COST EACH MAT.	COST EACH INST.	COST EACH TOTAL
SYSTEM D3010 510 1760					
HEATING SYSTEM, FIN TUBE RADIATION, FORCED HOT WATER					
1,000 S.F. AREA, 10,000 C.F. VOLUME					
Boiler, oil fired, CI, burner, ctrls/insul/breech/pipe/ftng/valves, 109 MBH	1.000	Ea.	4,550	4,375	8,925
Circulating pump, CI flange connection, 1/12 HP	1.000	Ea.	520	231	751
Expansion tank, painted steel, ASME 18 Gal capacity	1.000	Ea.	835	101	936
Storage tank, steel, above ground, 275 Gal capacity w/supports	1.000	Ea.	560	283	843
Copper tubing type L, solder joint, hanger 10′ OC, 3/4″ diam	100.000	L.F.	520	1,015	1,535
Radiation, 3/4″ copper tube w/alum fin baseboard pkg, 7″ high	30.000	L.F.	265.50	735	1,000.50
Pipe covering, calcium silicate w/cover, 1′ wall, 3/4′ diam	100.000	L.F.	472	765	1,237
TOTAL			7,722.50	7,505	15,227.50
COST PER S.F.			7.72	7.51	15.23

D3010 510	Apartment Building Heating - Fin Tube Radiation	COST PER S.F. MAT.	COST PER S.F. INST.	COST PER S.F. TOTAL
1740	Heating systems, fin tube radiation, forced hot water			
1760	1,000 S.F. area, 10,000 C.F. volume	7.73	7.50	15.23
1800	10,000 S.F. area, 100,000 C.F. volume	3.50	4.52	8.02
1840	20,000 S.F. area, 200,000 C.F. volume	3.81	5.05	8.86
1880	30,000 S.F. area, 300,000 C.F. volume	3.69	4.91	8.60

D3010 Energy Supply

Many styles of active solar energy systems exist. Those shown on the following page represent the majority of systems now being installed in different regions of the country.

The five active domestic hot water (DHW) systems which follow are typically specified as two or three panel systems with additional variations being the type of glazing and size of the storage tanks. Various combinations have been costed for the user's evaluation and comparison. The basic specifications from which the following systems were developed satisfy the construction detail requirements specified in the HUD Intermediate Minimum Property Standards (IMPS). If these standards are not complied with, the renewable energy system's costs could be significantly lower than shown.

To develop the system's specifications and costs it was necessary to make a number of assumptions about the systems. Certain systems are more appropriate to one climatic region than another or the systems may require modifications to be usable in particular locations. Specific instances in which the systems are not appropriate throughout the country as specified, or in which modification will be needed include the following:

- The freeze protection mechanisms provided in the DHW systems vary greatly. In harsh climates, where freeze is a major concern, a closed-loop indirect collection system may be more appropriate than a direct collection system.
- The thermosyphon water heater system described cannot be used when temperatures drop below 32°F.
- In warm climates it may be necessary to modify the systems installed to prevent overheating.

For each renewable resource (solar) system a schematic diagram and descriptive summary of the system is provided along with a list of all the components priced as part of the system. The costs were developed based on these specifications.

Considerations affecting costs which may increase or decrease beyond the estimates presented here include the following:

- Special structural qualities (allowance for earthquake, future expansion, high winds, and unusual spans or shapes);
- Isolated building site or rough terrain that would affect the transportation of personnel, material, or equipment;
- Unusual climatic conditions during the construction process;
- Substitution of other materials or system components for those used in the system specifications.

D3010 Energy Supply

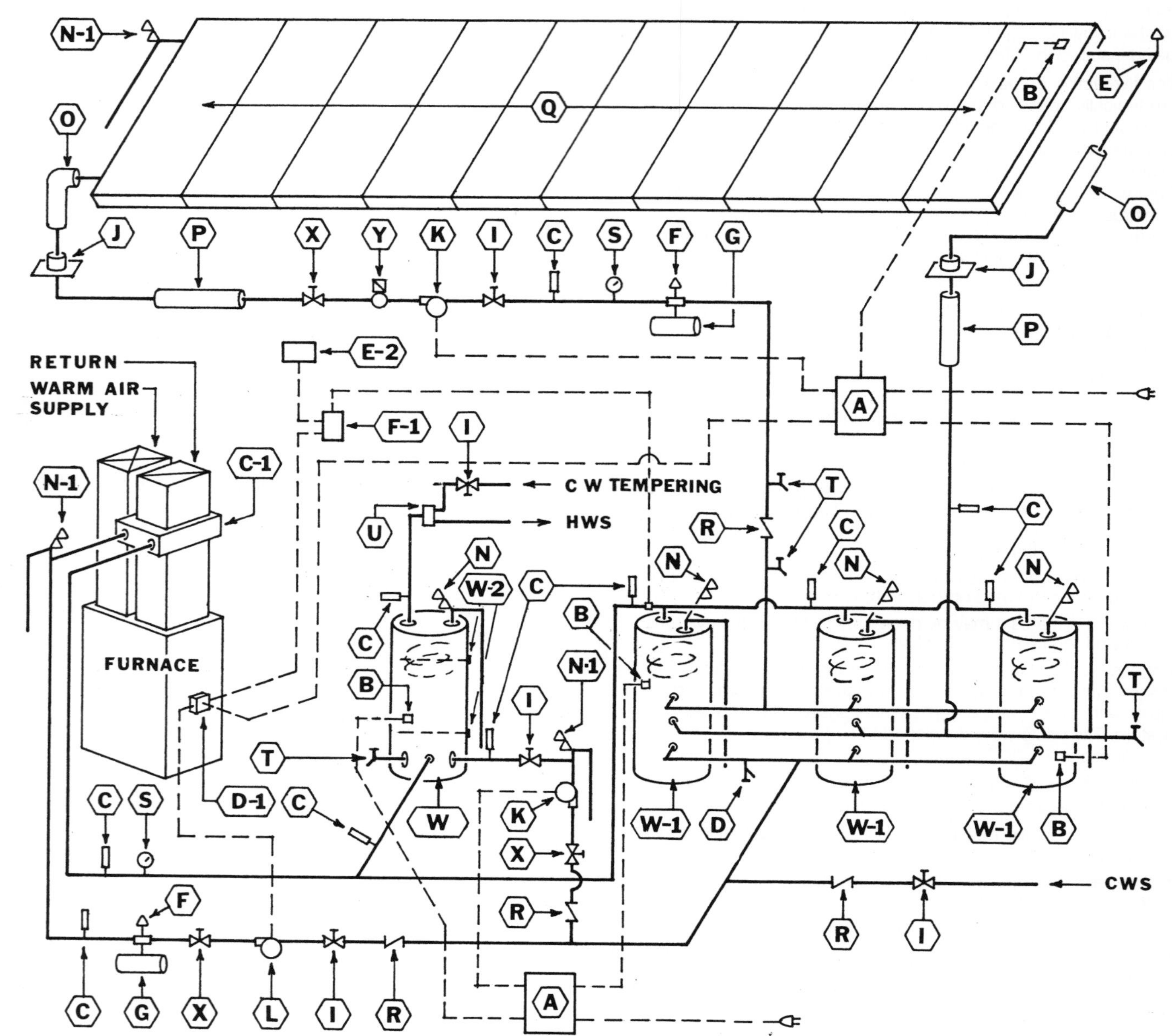

In this closed-loop indirect collection system, fluid with a low freezing temperature, propylene glycol, transports heat from the collectors to water storage. The transfer fluid is contained in a closed loop consisting of collectors, supply and return piping, and a heat exchanger immersed in the storage tank.

When the collectors become approximately 20°F warmer than the storage temperature, the controller activates the circulator. The circulator moves the fluid continuously until the temperature difference between fluid in the collectors and storage is such that the collection will no longer occur and then the circulator turns off. Since the heat transfer fluid has a very low freezing temperature, there is no need for it to be drained from the collectors between periods of collection.

D3010 Energy Supply

System Components

System Components	QUANTITY	UNIT	COST EACH MAT.	COST EACH INST.	COST EACH TOTAL
SYSTEM D3010 650 2750					
SOLAR, CLOSED LOOP, SPACE/HOT WATER					
1" TUBING, TEN 3'X7' BLK CHROME ON COPPER ABSORBER COLLECTORS					
A, B Differential controller 2 sensors, thermostat, solar energy system	2.000	Ea.	338	129	467
C Thermometer, 2" dial	10.000	Ea.	250	480	730
C-1 Heat exchanger, solar energy system, fluid to air, up flow, 80 MBH	1.000	Ea.	545	465	1,010
D, T Fill & drain valves, brass, 3/4" connection	5.000	Ea.	66.25	160	226.25
D-1 Fan center	1.000	Ea.	131	162	293
E Air vent, manual, 1/8" fitting	1.000	Ea.	2.96	24	26.96
E-2 Thermostat, 2 stage for sensing room temperature	1.000	Ea.	283	98	381
F Air purger	2.000	Ea.	102	129	231
F-1 Controller, liquid temperature, solar energy system	1.000	Ea.	129	154	283
G Expansion tank	2.000	Ea.	155	48	203
I Valve, gate, bronze, 125 lb, soldered, 1" diam	5.000	Ea.	432.50	202.50	635
J Vent flashing, neoprene	2.000	Ea.	23	77	100
K Circulator, solar heated liquid, 1/25 HP	2.000	Ea.	450	198	648
L Circulator, solar heated liquid, 1/20 HP	1.000	Ea.	310	116	426
N Relief valve, temp & pressure 150 psi 210°F self-closing	4.000	Ea.	112	102	214
N-1 Relief valve, pressure poppet, bronze, 30 psi, 3/4" IPS	3.000	Ea.	345	82.50	427.50
O Pipe covering, urethane, ultraviolet cover, 1" wall, 1" diam	50.000	L.F.	155	352.50	507.50
P Pipe covering, fiberglass, all service jacket, 1" wall, 1" diam	60.000	L.F.	64.80	339	403.80
Q Collector panel solar energy blk chrome on copper 1/8" temp glass 3'x7'	10.000	Ea.	11,750	1,460	13,210
Roof clamps for solar energy collector panel	10.000	Set	36	198.50	234.50
R Valve, swing check, bronze, 125 lb, regrinding disc, 3/4" & 1" diam	4.000	Ea.	784	162	946
S Pressure gage, 0-60 psi, for solar energy system	2.000	Ea.	55	48	103
U Valve, water tempering, bronze, sweat connections, 3/4" diam	1.000	Ea.	183	38.50	221.50
W-2, W-1, W Tank, water storage immersed heat xchr elec elem 2"x1/2# ins 120 gal	4.000	Ea.	8,400	2,200	10,600
X Valve, globe, bronze, 125 lb, rising stem, 1" diam	3.000	Ea.	789	121.50	910.50
Y Valve, flow control	1.000	Ea.	154	35	189
Copper tubing, type M, solder joint, hanger 10' OC 1" diam	110.000	L.F.	1,028.50	1,210	2,238.50
Copper tubing, type L, solder joint, hanger 10' OC 3/4" diam	20.000	L.F.	104	203	307
Wrought copper fittings & solder, 3/4" & 1" diam	121.000	Ea.	937.75	5,808	6,745.75
Sensor, wire, #22-2 conductor, multistranded	.700	C.L.F.	9.10	51.10	60.20
Ductwork, galvanized steel, for heat exchanger	8.000	Lb.	5.28	72	77.28
Solar energy heat transfer fluid propylene glycol, anti-freeze	25.000	Gal.	637.50	687.50	1,325
TOTAL			28,767.64	15,613.60	44,381.24

D3010 650	Solar, Closed Loop, Space/Hot Water Systems	COST EACH MAT.	COST EACH INST.	COST EACH TOTAL
2540	Solar, closed loop, space/hot water			
2550	1/2" tubing, 12 ea. 4'x4'4" vacuum tube collectors	26,600	14,600	41,200
2600	3/4" tubing, 12 ea. 4'x4'4" vacuum tube collectors R235616-60	28,000	15,000	43,000
2650	10 ea. 3' x 7' black chrome absorber collectors	27,700	14,600	42,300
2700	10 ea. 3' x 7' flat black absorber collectors	24,900	14,700	39,600
2750	1" tubing, 10 ea. 3' x 7' black chrome absorber collectors	28,800	15,600	44,400
2800	10 ea. 3' x 7' flat black absorber collectors	26,100	15,700	41,800
2850	6 ea. 4' x 9' flat black w/plastic glazing collectors	23,900	15,600	39,500
2900	12 ea. 2' x 9' plastic absorber and glazing collectors	25,200	15,900	41,100

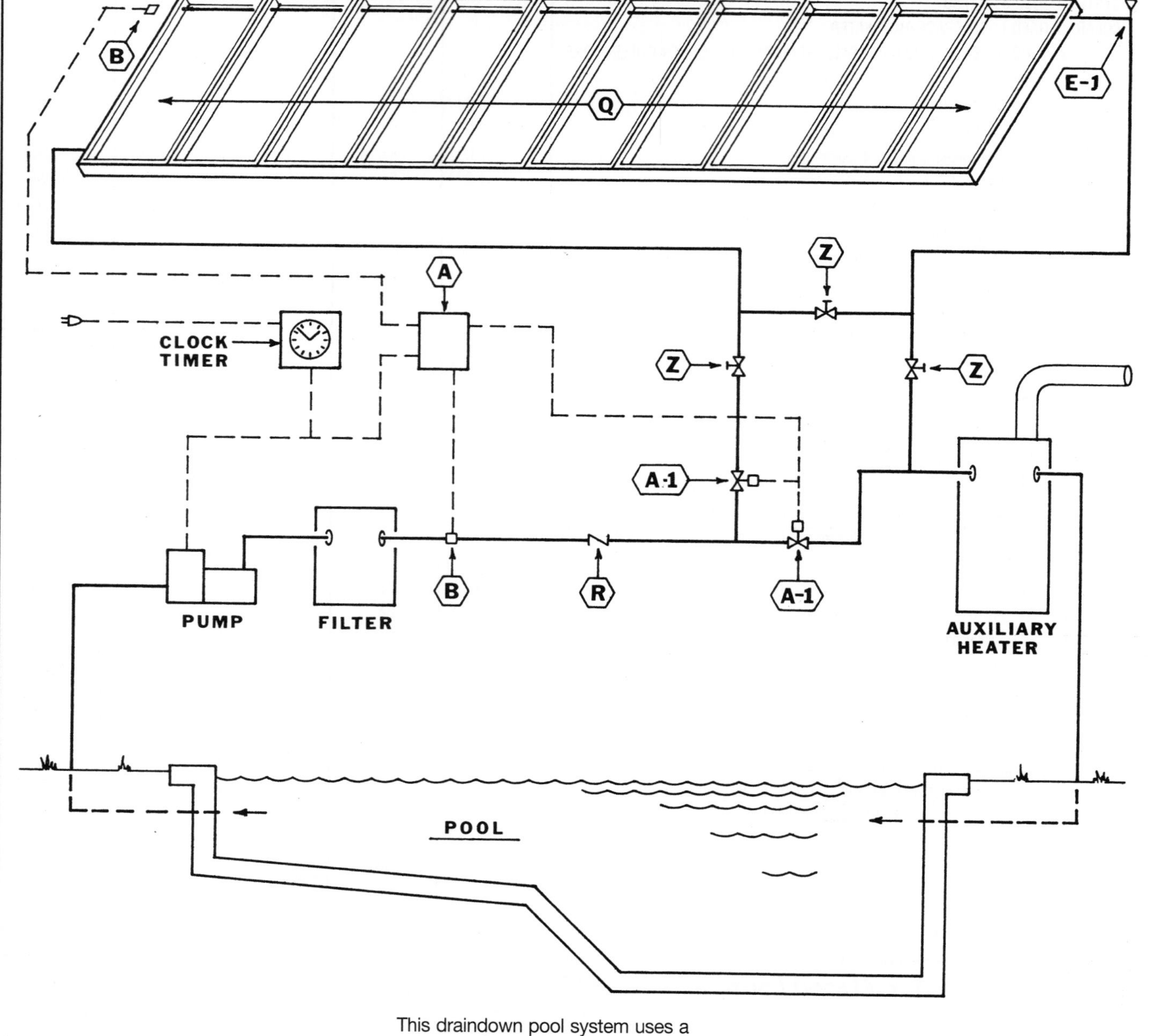

This draindown pool system uses a differential thermostat similar to those used in solar domestic hot water and space heating applications. To heat the pool, the pool water passes through the conventional pump-filter loop and then flows through the collectors. When collection is not possible, or when the pool temperature is reached, all water drains from the solar loop back to the pool through the existing piping. The modes are controlled by solenoid valves or other automatic valves in conjunction with a vacuum breaker relief valve, which facilitates draindown.

D3010 Energy Supply

System Components	QUANTITY	UNIT	COST EACH MAT.	COST EACH INST.	COST EACH TOTAL
SYSTEM D3010 660 2640					
SOLAR SWIMMING POOL HEATER, ROOF MOUNTED COLLECTORS					
TEN 4′ X 10′ FULLY WETTED UNGLAZED PLASTIC ABSORBERS					
A Differential thermostat/controller, 110V, adj pool pump system	1.000	Ea.	365	385	750
A-1 Solenoid valve, PVC, normally 1 open 1 closed (included)	2.000	Ea.			
B Sensor, thermistor type (included)	2.000	Ea.			
E-1 Valve, vacuum relief	1.000	Ea.	34	24	58
Q Collector panel, solar energy, plastic, liquid full wetted, 4′ x 10′	10.000	Ea.	3,700	2,780	6,480
R Valve, ball check, PVC, socket, 1-1/2″ diam	1.000	Ea.	120	38.50	158.50
Z Valve, ball, PVC, socket, 1-1/2″ diam	3.000	Ea.	285	115.50	400.50
Pipe, PVC, sch 40, 1-1/2″ diam	80.000	L.F.	904	1,720	2,624
Pipe fittings, PVC sch 40, socket joint, 1-1/2″ diam	10.000	Ea.	21.90	385	406.90
Sensor wire, #22-2 conductor, multistranded	.500	C.L.F.	6.50	36.50	43
Roof clamps for solar energy collector panels	10.000	Set	36	198.50	234.50
Roof strap, teflon for solar energy collector panels	26.000	L.F.	715	97.76	812.76
TOTAL			6,187.40	5,780.76	11,968.16

D3010 660	Solar Swimming Pool Heater Systems	COST EACH MAT.	COST EACH INST.	COST EACH TOTAL
2530	Solar swimming pool heater systems, roof mounted collectors			
2540	10 ea. 3′x7′ black chrome absorber, 1/8″ temp. glass	14,200	4,450	18,650
2560	10 ea. 4′x8′ black chrome absorber, 3/16″ temp. glass (R235616-60)	16,000	5,300	21,300
2580	10 ea. 3′8″x6′ flat black absorber, 3/16″ temp. glass	11,400	4,550	15,950
2600	10 ea. 4′x9′ flat black absorber, plastic glazing	13,700	5,525	19,225
2620	10 ea. 2′x9′ rubber absorber, plastic glazing	15,800	5,975	21,775
2640	10 ea. 4′x10′ fully wetted unglazed plastic absorber	6,175	5,775	11,950
2660	Ground mounted collectors			
2680	10 ea. 3′x7′ black chrome absorber, 1/8″ temp. glass	14,300	4,975	19,275
2700	10 ea. 4′x8′ black chrome absorber, 3/16″ temp. glass	16,100	5,825	21,925
2720	10 ea. 3′8″x6′ flat blk absorber, 3/16″ temp. glass	11,500	5,050	16,550
2740	10 ea. 4′x9′ flat blk absorber, plastic glazing	13,800	6,050	19,850
2760	10 ea. 2′x9′ rubber absorber, plastic glazing	15,900	6,300	22,200
2780	10 ea. 4′x10′ fully wetted unglazed plastic absorber	6,250	6,300	12,550

D3010 Energy Supply

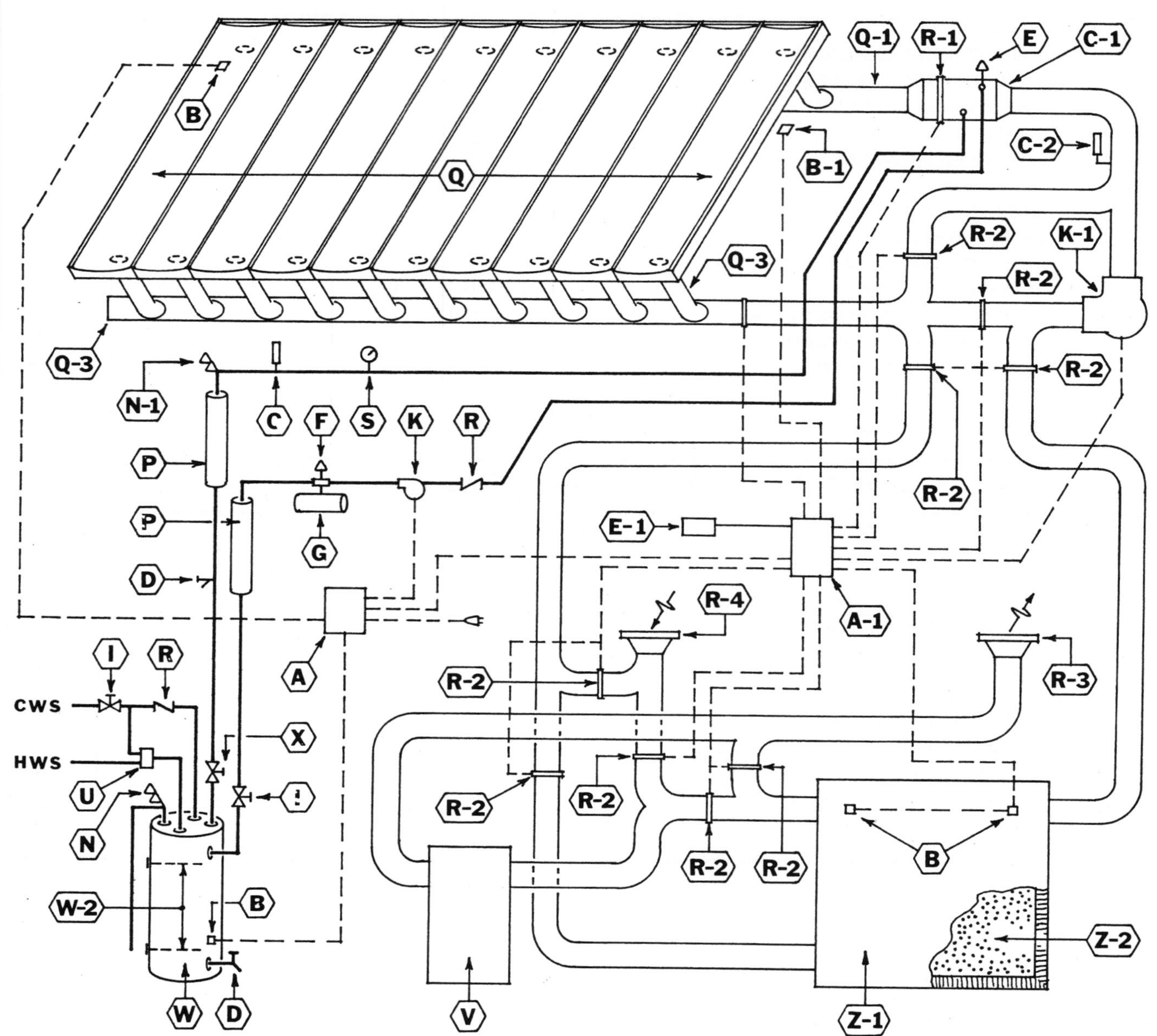

The complete Solar Air Heating System provides maximum savings of conventional fuel with both space heating and year-round domestic hot water heating. It allows for the home air conditioning to operate simultaneously and independently from the solar domestic water heating in summer. The system's modes of operation are:

Mode 1: The building is heated directly from the collectors with air circulated by the Solar Air Mover.

Mode 2: When heat is not needed in the building, the dampers change within the air mover to circulate the air from the collectors to the rock storage bin.

Mode 3: When heat is not available from the collector array and is available in rock storage, the air mover draws heated air from rock storage and directs it into the building. When heat is not available from the collectors or the rock storage bin, the auxiliary heating unit will provide heat for the building. The size of the collector array is typically 25% the size of the main floor area.

D3010 Energy Supply

System Components		QUANTITY	UNIT	COST EACH MAT.	COST EACH INST.	COST EACH TOTAL
	SYSTEM D3010 675 1210					
	SOLAR, SPACE/HOT WATER, AIR TO WATER HEAT EXCHANGE					
A, B	Differential controller 2 sensors thermos., solar energy sys liquid loop	1.000	Ea.	169	64.50	233.50
A-1, B	Differential controller 2 sensors 6 station solar energy sys air loop	1.000	Ea.	299	257	556
B-1	Solar energy sensor, freeze prevention	1.000	Ea.	27.50	24	51.50
C	Thermometer for solar energy system, 2″ dial	2.000	Ea.	50	96	146
C-1	Heat exchanger, solar energy system, air to fluid, up flow, 70 MBH	1.000	Ea.	410	395	805
D	Drain valve, brass, 3/4″ connection	2.000	Ea.	26.50	64	90.50
E	Air vent, manual, for solar energy system 1/8″ fitting	1.000	Ea.	2.96	24	26.96
E-1	Thermostat, 2 stage for sensing room temperature	1.000	Ea.	283	98	381
F	Air purger	1.000	Ea.	51	64.50	115.50
G	Expansion tank, for solar energy system	1.000	Ea.	77.50	24	101.50
I	Valve, gate, bronze, 125 lb, NRS, soldered 3/4″ diam	2.000	Ea.	163	77	240
K	Circulator, solar heated liquid, 1/25 HP	1.000	Ea.	225	99	324
N	Relief valve temp & press 150 psi 210°F self-closing, 3/4″ IPS	1.000	Ea.	28	25.50	53.50
N-1	Relief valve, pressure, poppet, bronze, 30 psi, 3/4″ IPS	1.000	Ea.	115	27.50	142.50
P	Pipe covering, fiberglass, all service jacket, 1″ wall, 3/4″ diam	60.000	L.F.	64.80	339	403.80
Q-3	Manifold for flush mount solar energy collector panels	20.000	L.F.	3,560	170	3,730
Q	Collector panel solar energy, air, black on alum. plate, 7.5′ x 3.5′	10.000	Ea.	8,950	1,540	10,490
R	Valve, swing check, bronze, 125 lb, regrinding disc, 3/4″ diam	2.000	Ea.	206	77	283
S	Pressure gage, 2″ dial, for solar energy system	1.000	Ea.	27.50	24	51.50
U	Valve, water tempering, bronze, sweat connections, 3/4″ diam	1.000	Ea.	183	38.50	221.50
W-2, W	Tank, water storage, solar, elec element 2″x1/2# insul, 80 Gal	1.000	Ea.	2,150	480	2,630
X	Valve, globe, bronze, 125 lb, soldered, 3/4″ diam	1.000	Ea.	189	38.50	227.50
	Copper tubing type L, solder joints, hangers 10′ OC 3/4″ diam	10.000	L.F.	52	101.50	153.50
	Copper tubing type M, solder joints, hangers 10′ OC 3/4″ diam	60.000	L.F.	333	594	927
	Wrought copper fittings & solder, 3/4″ diam	26.000	Ea.	76.70	1,053	1,129.70
	Sensor wire, #22-2 conductor multistranded	1.200	C.L.F.	15.60	87.60	103.20
Q-1	Duct work, rigid fiberglass, rectangular	400.000	S.F.	392	2,420	2,812
	Duct work, spiral preformed, steel, PVC coated both sides, 12″ x 10″	8.000	Ea.	204	452	656
R-2	Shutter/damper for solar heater circulator	9.000	Ea.	580.50	1,017	1,597.50
K-1	Shutter motor for solar heater circulator blower	9.000	Ea.	1,449	765	2,214
K-1	Fan, solar energy heated air circulator, space & DHW system	1.000	Ea.	1,775	2,725	4,500
Z-1	Tank, solar energy air storage, 6′-3″H 7′x7′ = 306 CF/2000 Gal	1.000	Ea.	15,300	1,775	17,075
Z-2	Crushed stone 1-1/2″	11.000	C.Y.	335.50	92.07	427.57
C-2	Themometer, remote probe, 2″ dial	1.000	Ea.	38	96.50	134.50
R-1	Solenoid valve	1.000	Ea.	150	85.50	235.50
V, R-4, R-3	Furnace, supply diffusers, return grilles, existing	1.000	Ea.	15,300	1,775	17,075
	TOTAL			53,259.06	17,086.17	70,345.23

D3010 675	Air To Water Heat Exchange		COST EACH MAT.	COST EACH INST.	COST EACH TOTAL
1210	Solar, air to water heat exchange, for space/hot water heating	R235616 -60	53,500	17,100	70,600
1220					

D3020 Heat Generating Systems

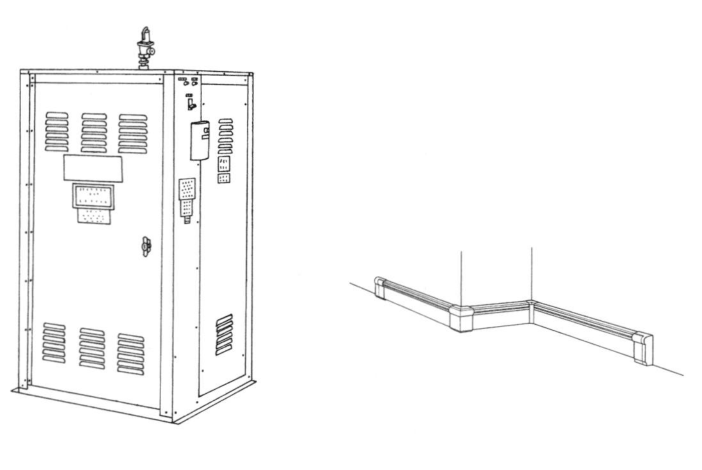

Small Electric Boiler System Considerations:

1. Terminal units are fin tube baseboard radiation rated at 720 BTU/hr with 200° water temperature or 820 BTU/hr steam.
2. Primary use being for residential or smaller supplementary areas, the floor levels are based on 7-1/2′ ceiling heights.
3. All distribution piping is copper for boilers through 205 MBH. All piping for larger systems is steel pipe.

System Components	QUANTITY	UNIT	COST EACH MAT.	COST EACH INST.	COST EACH TOTAL
SYSTEM D3020 102 1120					
SMALL HEATING SYSTEM, HYDRONIC, ELECTRIC BOILER					
1,480 S.F., 61 MBH, STEAM, 1 FLOOR					
Boiler, electric steam, std cntrls, trim, ftngs and valves, 18 KW, 61.4 MBH	1.000	Ea.	5,472.50	1,952.50	7,425
Copper tubing type L, solder joint, hanger 10′OC, 1-1/4″ diam	160.000	L.F.	2,112	2,128	4,240
Radiation, 3/4″ copper tube w/alum fin baseboard pkg 7″ high	60.000	L.F.	531	1,470	2,001
Rough in baseboard panel or fin tube with valves & traps	10.000	Set	3,050	7,400	10,450
Pipe covering, calcium silicate w/cover, 1″ wall 1-1/4″ diam	160.000	L.F.	755.20	1,256	2,011.20
Low water cut-off, quick hookup, in gage glass tappings	1.000	Ea.	375	49	424
TOTAL			12,295.70	14,255.50	26,551.20
COST PER S.F.			8.31	9.63	17.94

D3020 102	Small Heating Systems, Hydronic, Electric Boilers	COST PER S.F. MAT.	COST PER S.F. INST.	COST PER S.F. TOTAL
1100	Small heating systems, hydronic, electric boilers			
1120	Steam, 1 floor, 1480 S.F., 61 M.B.H.	8.31	9.66	17.97
1160	3,000 S.F., 123 M.B.H.	6.40	8.45	14.85
1200	5,000 S.F., 205 M.B.H.	5.30	7.80	13.10
1240	2 floors, 12,400 S.F., 512 M.B.H.	4.42	7.75	12.17
1280	3 floors, 24,800 S.F., 1023 M.B.H.	4.62	7.65	12.27
1360	Hot water, 1 floor, 1,000 S.F., 41 M.B.H.	14.25	5.35	19.60
1400	2,500 S.F., 103 M.B.H.	9.65	9.65	19.30
1440	2 floors, 4,850 S.F., 205 M.B.H.	8.90	11.55	20.45
1480	3 floors, 9,700 S.F., 410 M.B.H.	9.50	11.95	21.45

D3090 Other HVAC Systems/Equip

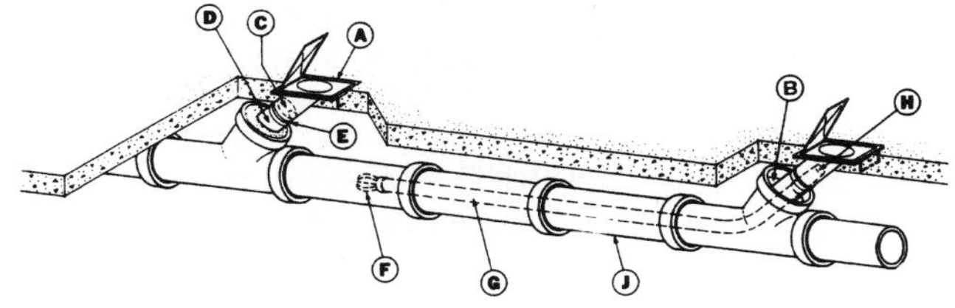

Cast Iron Garage Exhaust System

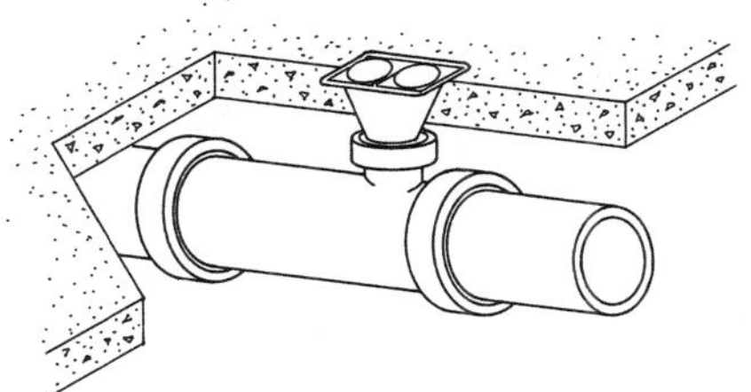

Dual Exhaust System

System Components		QUANTITY	UNIT	COST EACH MAT.	COST EACH INST.	COST EACH TOTAL
SYSTEM D3090 320 1040						
GARAGE, EXHAUST, SINGLE 3″ EXHAUST OUTLET, CARS & LIGHT TRUCKS						
A	Outlet top assy, for engine exhaust system with adapters and ftngs, 3″ diam	1.000	Ea.	286	47.50	333.50
F	Bullnose (guide) for engine exhaust system, 3″ diam	1.000	Ea.	33.50		33.50
G	Galvanized flexible tubing for engine exhaust system, 3″ diam	8.000	L.F.	94.80		94.80
H	Adapter for metal tubing end of engine exhaust system, 3″ tail pipe	1.000	Ea.	56.50		56.50
J	Pipe, sewer, cast iron, push-on joint, 8″ diam.	18.000	L.F.	1,404	855	2,259
	Excavating utility trench, chain trencher, 8″ wide, 24″ deep	18.000	L.F.		41.40	41.40
	Backfill utility trench by hand, incl. compaction, 8″ wide 24″ deep	18.000	L.F.		50.94	50.94
	Stand for blower, concrete over polystyrene core, 6″ high	1.000	Ea.	5.90	23.50	29.40
	AC&V duct spiral reducer 10″x8″	1.000	Ea.	24	42.50	66.50
	AC&V duct spiral reducer 12″x10″	1.000	Ea.	24.50	56.50	81
	AC&V duct spiral preformed 45° elbow, 8″ diam	1.000	Ea.	8.95	48.50	57.45
	AC&V utility fan, belt drive, 3 phase, 2000 CFM, 1 HP	2.000	Ea.	2,900	750	3,650
	Safety switch, heavy duty fused, 240V, 3 pole, 30 amp	1.000	Ea.	113	228	341
	TOTAL			4,951.15	2,143.84	7,094.99

D3090 320	Garage Exhaust Systems	COST PER BAY MAT.	COST PER BAY INST.	COST PER BAY TOTAL
1040	Garage, single 3″ exhaust outlet, cars & light trucks, one bay	4,950	2,150	7,100
1060	Additional bays up to seven bays	1,250	575	1,825
1500	4″ outlet, trucks, one bay	4,975	2,150	7,125
1520	Additional bays up to six bays	1,275	575	1,850
1600	5″ outlet, diesel trucks, one bay	5,250	2,150	7,400
1650	Additional single bays up to six bays	1,700	680	2,380
1700	Two adjoining bays	5,250	2,150	7,400
2000	Dual exhaust, 3″ outlets, pair of adjoining bays	6,250	2,675	8,925
2100	Additional pairs of adjoining bays	1,925	680	2,605

D4010 Sprinklers

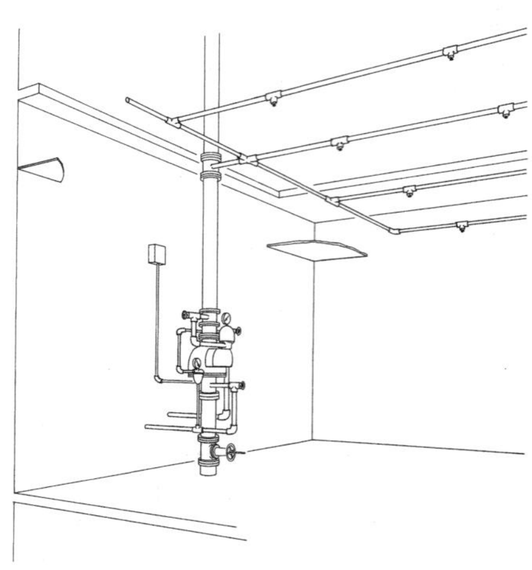

Dry Pipe System: A system employing automatic sprinklers attached to a piping system containing air under pressure, the release of which from the opening of sprinklers permits the water pressure to open a valve known as a "dry pipe valve". The water then flows into the piping system and out the opened sprinklers.

All areas are assumed to be open.

System Components	QUANTITY	UNIT	COST EACH MAT.	COST EACH INST.	COST EACH TOTAL
SYSTEM D4010 310 0580					
DRY PIPE SPRINKLER, STEEL, BLACK, SCH. 40 PIPE					
LIGHT HAZARD, ONE FLOOR, 2000 S.F.					
Valve, gate, iron body 125 lb., OS&Y, flanged, 4″ pipe size	1.000	Ea.	755	465	1,220
Backflow pvntr, auto, 2 OS&Y v, tst cocks, dbl chk, flgd, 4″	1.000	Ea.	5,125	465	5,590
Tamper switch (valve supervisory switch)	3.000	Ea.	873	142.50	1,015.50
Valve, swing check, bronze, 125 lb, regrinding disc, 2-1/2″ pipe size	1.000	Ea.	1,025	92.50	1,117.50
Valve, angle, bronze, 150 lb., rising stem, threaded, 2″ pipe size	1.000	Ea.	930	70	1,000
*Alarm valve, 2-1/2″ pipe size	1.000	Ea.	1,900	455	2,355
Alarm, water motor, complete with gong	1.000	Ea.	530	190	720
Fire alarm horn, electric	1.000	Ea.	59.50	109	168.50
Valve swing check w/balldrip CI with brass trim, 4″ pipe size	1.000	Ea.	460	455	915
Pipe, steel, black, schedule 40, 4″ diam.	8.000	L.F.	212	325.36	537.36
Dry pipe valve, trim & gauges, 4″ pipe size	1.000	Ea.	3,625	1,375	5,000
Pipe, steel, black, schedule 40, threaded, cplg & hngr 10′OC 2-1/2″ diam.	15.000	L.F.	188.25	420	608.25
Pipe, steel, black, schedule 40, threaded, cplg & hngr 10′OC 2″ diam.	9.375	L.F.	71.25	201.56	272.81
Pipe, steel, black, schedule 40, threaded, cplg & hngr 10′OC 1-1/4″ diam.	28.125	L.F.	253.13	438.75	691.88
Pipe, steel, black, schedule 40, threaded, cplg & hngr 10′OC 1″ diam.	84.000	L.F.	718.20	1,222.20	1,940.40
Pipe Tee, malleable iron black, 150 lb. threaded, 4″ pipe size	2.000	Ea.	780	690	1,470
Pipe Tee, malleable iron black, 150 lb. threaded, 2-1/2″ pipe size	2.000	Ea.	218	308	526
Pipe Tee, malleable iron black, 150 lb. threaded, 2″ pipe size	1.000	Ea.	50.50	126	176.50
Pipe Tee, malleable iron black, 150 lb. threaded, 1-1/4″ pipe size	4.000	Ea.	96	396	492
Pipe Tee, malleable iron black, 150 lb. threaded, 1″ pipe size	3.000	Ea.	44.25	289.50	333.75
Pipe 90° elbow malleable iron black, 150 lb. threaded, 1″ pipe size	5.000	Ea.	46	297.50	343.50
Sprinkler head dry K5.6, 1″ NPT, 3″ to 6″ length	12.000	Ea.	1,920	648	2,568
Air compressor, 200 Gal sprinkler system capacity, 1/3 HP	1.000	Ea.	1,325	585	1,910
*Standpipe connection, wall, flush, brs. w/plug & chain 2-1/2″x2-1/2″	1.000	Ea.	205	273	478
Valve gate bronze, 300 psi, NRS, class 150, threaded, 1″ pipe size	1.000	Ea.	148	40.50	188.50
TOTAL			21,558.08	10,080.37	31,638.45
COST PER S.F.			8.08	3.78	11.86

*Not included in systems under 2000 S.F.

D4010 310	Dry Pipe Sprinkler Systems		COST PER S.F. MAT.	COST PER S.F. INST.	COST PER S.F. TOTAL
0520	Dry pipe sprinkler systems, steel, black, sch. 40 pipe				
0530	Light hazard, one floor, 500 S.F.		14.80	7.20	22
0560	1000 S.F.	R211313 -10	8.95	4.22	13.17
0580	2000 S.F.		8.10	3.79	11.89

D4010 Sprinklers

D4010 310	Dry Pipe Sprinkler Systems	MAT.	INST.	TOTAL
		COST PER S.F.		
0600	5000 S.F. (R211313-20)	4.25	2.86	7.11
0620	10,000 S.F.	2.76	2.29	5.05
0640	50,000 S.F.	1.88	1.98	3.86
0660	Each additional floor, 500 S.F.	2.77	3.42	6.19
0680	1000 S.F.	2.38	2.78	5.16
0700	2000 S.F.	2.26	2.56	4.82
0720	5000 S.F.	1.85	2.20	4.05
0740	10,000 S.F.	1.59	2	3.59
0760	50,000 S.F.	1.45	1.76	3.21
1000	Ordinary hazard, one floor, 500 S.F.	15.05	7.25	22.30
1020	1000 S.F.	9.10	4.25	13.35
1040	2000 S.F.	8.65	4.44	13.09
1060	5000 S.F.	4.77	3.06	7.83
1080	10,000 S.F.	3.47	2.96	6.43
1100	50,000 S.F.	3.09	2.84	5.93
1140	Each additional floor, 500 S.F.	3.01	3.50	6.51
1160	1000 S.F.	2.60	3.10	5.70
1180	2000 S.F.	2.65	2.82	5.47
1200	5000 S.F.	2.44	2.44	4.88
1220	10,000 S.F.	2.18	2.31	4.49
1240	50,000 S.F.	2.13	2.05	4.18
1500	Extra hazard, one floor, 500 S.F.	18.45	9	27.45
1520	1000 S.F.	15.10	6.55	21.65
1540	2000 S.F.	9.45	5.65	15.10
1560	5000 S.F.	5.35	4.21	9.56
1580	10,000 S.F.	5.40	3.98	9.38
1600	50,000 S.F.	5.35	3.86	9.21
1660	Each additional floor, 500 S.F.	4.14	4.29	8.43
1680	1000 S.F.	3.64	4.02	7.66
1700	2000 S.F.	3.50	4.04	7.54
1720	5000 S.F.	2.97	3.54	6.51
1740	10,000 S.F.	3.64	3.23	6.87
1760	50,000 S.F.	3.67	3.13	6.80
2020	Grooved steel, black, sch. 40 pipe, light hazard, one floor, 2000 S.F.	8.35	3.69	12.04
2060	10,000 S.F.	2.81	1.98	4.79
2100	Each additional floor, 2000 S.F.	2.34	2.07	4.41
2150	10,000 S.F.	1.64	1.69	3.33
2200	Ordinary hazard, one floor, 2000 S.F.	8.65	3.90	12.55
2250	10,000 S.F.	3.41	2.56	5.97
2300	Each additional floor, 2000 S.F.	2.63	2.28	4.91
2350	10,000 S.F.	2.47	2.30	4.77
2400	Extra hazard, one floor, 2000 S.F.	9.50	4.86	14.36
2450	10,000 S.F.	4.82	3.30	8.12
2500	Each additional floor, 2000 S.F.	3.67	3.35	7.02
2550	10,000 S.F.	3.28	2.84	6.12
3050	Grooved steel, black, sch. 10 pipe, light hazard, one floor, 2000 S.F.	8.25	3.66	11.91
3100	10,000 S.F.	2.77	1.96	4.73
3150	Each additional floor, 2000 S.F.	2.26	2.03	4.29
3200	10,000 S.F.	1.60	1.67	3.27
3250	Ordinary hazard, one floor, 2000 S.F.	8.55	3.88	12.43
3300	10,000 S.F.	3.26	2.49	5.75
3350	Each additional floor, 2000 S.F.	2.56	2.26	4.82
3400	10,000 S.F.	2.17	2.22	4.39
3450	Extra hazard, one floor, 2000 S.F.	9.45	4.84	14.29
3500	10,000 S.F.	4.61	3.24	7.85
3550	Each additional floor, 2000 S.F.	3.61	3.33	6.94
3600	10,000 S.F.	3.18	2.80	5.98
4050	Copper tubing, type L, light hazard, one floor, 2000 S.F.	8.55	3.69	12.24

D4010 Sprinklers

D4010 310	Dry Pipe Sprinkler Systems	COST PER S.F.		
		MAT.	INST.	TOTAL
4100	10,000 S.F.	3.20	2.04	5.24
4150	Each additional floor, 2000 S.F.	2.56	2.11	4.67
4200	10,000 S.F.	2.03	1.76	3.79
4250	Ordinary hazard, one floor, 2000 S.F.	9	4.07	13.07
4300	10,000 S.F.	3.98	2.32	6.30
4350	Each additional floor, 2000 S.F.	3.47	2.39	5.86
4400	10,000 S.F.	2.76	2	4.76
4450	Extra hazard, one floor, 2000 S.F.	10	4.93	14.93
4500	10,000 S.F.	6.75	3.54	10.29
4550	Each additional floor, 2000 S.F.	4.33	3.37	7.70
4600	10,000 S.F.	4.53	3.05	7.58
5050	Copper tubing, type L, T-drill system, light hazard, one floor			
5060	2000 S.F.	8.60	3.46	12.06
5100	10,000 S.F.	3.08	1.71	4.79
5150	Each additional floor, 2000 S.F.	2.62	1.88	4.50
5200	10,000 S.F.	1.91	1.43	3.34
5250	Ordinary hazard, one floor, 2000 S.F.	8.80	3.53	12.33
5300	10,000 S.F.	3.75	2.15	5.90
5350	Each additional floor, 2000 S.F.	2.79	1.91	4.70
5400	10,000 S.F.	2.48	1.77	4.25
5450	Extra hazard, one floor, 2000 S.F.	9.45	4.16	13.61
5500	10,000 S.F.	5.45	2.69	8.14
5550	Each additional floor, 2000 S.F.	3.61	2.65	6.26
5600	10,000 S.F.	3.42	2.19	5.61

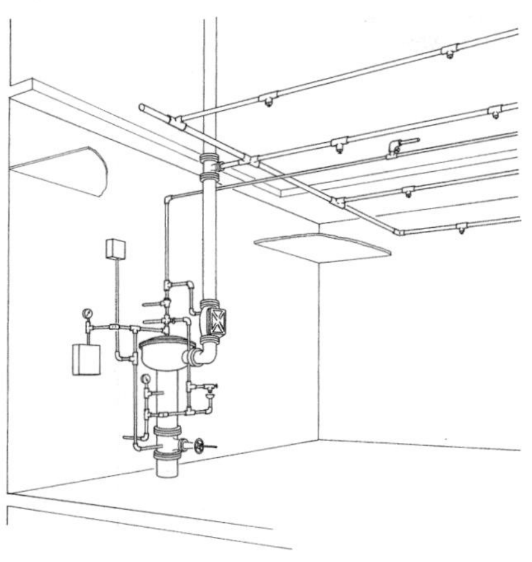

Pre-Action System: A system employing automatic sprinklers attached to a piping system containing air that may or may not be under pressure, with a supplemental heat responsive system of generally more sensitive characteristics than the automatic sprinklers themselves, installed in the same areas as the sprinklers. Actuation of the heat responsive system, as from a fire, opens a valve which permits water to flow into the sprinkler piping system and to be discharged from those sprinklers which were opened by heat from the fire.

All areas are assumed to be open.

System Components	QUANTITY	UNIT	COST EACH MAT.	COST EACH INST.	COST EACH TOTAL
SYSTEM D4010 350 0580					
PREACTION SPRINKLER SYSTEM, STEEL, BLACK, SCH. 40 PIPE					
LIGHT HAZARD, 1 FLOOR, 2000 S.F.					
Valve, gate, iron body 125 lb., OS&Y, flanged, 4″ pipe size	1.000	Ea.	755	465	1,220
Backflow pvntr, auto, 2 OS&Y v, tst cocks, dbl chk, flgd, 4″	1.000	Ea.	5,125	465	5,590
Tamper switch (valve supervisory switch)	3.000	Ea.	873	142.50	1,015.50
*Valve, swing check w/ball drip CI with brass trim 4″ pipe size	1.000	Ea.	460	455	915
Valve, swing check, bronze, 125 lb, regrinding disc, 2-1/2″ pipe size	1.000	Ea.	1,025	92.50	1,117.50
Valve, angle, bronze, 150 lb., rising stem, threaded, 2″ pipe size	1.000	Ea.	930	70	1,000
*Alarm valve, 2-1/2″ pipe size	1.000	Ea.	1,900	455	2,355
Alarm, water motor, complete with gong	1.000	Ea.	530	190	720
Fire alarm horn, electric	1.000	Ea.	59.50	109	168.50
Thermostatic release for release line	2.000	Ea.	1,760	76	1,836
Pipe, steel, black, schedule 40, 4″ diam.	8.000	L.F.	212	325.36	537.36
Dry pipe valve, trim & gauges, 4″ pipe size	1.000	Ea.	3,625	1,375	5,000
Pipe, steel, black, schedule 40, threaded, cplg. & hngr. 10′OC 2-1/2″ diam.	15.000	L.F.	188.25	420	608.25
Pipe steel black, schedule 40, threaded, cplg. & hngr. 10′OC 2″ diam.	9.375	L.F.	71.25	201.56	272.81
Pipe, steel, black, schedule 40, threaded, cplg. & hngr. 10′OC 1-1/4″ diam.	28.125	L.F.	253.13	438.75	691.88
Pipe, steel, black, schedule 40, threaded, cplg. & hngr. 10′OC 1″ diam.	84.000	L.F.	718.20	1,222.20	1,940.40
Pipe, Tee, malleable iron, black, 150 lb. threaded, 4″ diam.	2.000	Ea.	780	690	1,470
Pipe, Tee, malleable iron, black, 150 lb. threaded, 2-1/2″ pipe size	2.000	Ea.	218	308	526
Pipe, Tee, malleable iron, black, 150 lb. threaded, 2″ pipe size	1.000	Ea.	50.50	126	176.50
Pipe, Tee, malleable iron, black, 150 lb. threaded, 1-1/4″ pipe size	4.000	Ea.	96	396	492
Pipe, Tee, malleable iron, black, 150 lb. threaded, 1″ pipe size	3.000	Ea.	44.25	289.50	333.75
Pipe, 90° elbow, malleable iron, blk., 150 lb. threaded, 1″ pipe size	5.000	Ea.	46	297.50	343.50
Sprinkler head, std. spray, brass 135°-286°F 1/2″ NPT, 3/8″ orifice	12.000	Ea.	221.40	570	791.40
Air compressor auto complete 200 Gal sprinkler sys. cap., 1/3 HP	1.000	Ea.	1,325	585	1,910
*Standpipe conn.,wall, flush, brass w/plug & chain 2-1/2″ x 2-1/2″	1.000	Ea.	205	273	478
Valve, gate, bronze, 300 psi, NRS, class 150, threaded, 1″ pipe size	1.000	Ea.	148	40.50	188.50
TOTAL			21,619.48	10,078.37	31,697.85
COST PER S.F.			8.11	3.78	11.89

*Not included in systems under 2000 S.F.

D4010 350	Preaction Sprinkler Systems	COST PER S.F. MAT.	COST PER S.F. INST.	COST PER S.F. TOTAL
0520	Preaction sprinkler systems, steel, black, sch. 40 pipe			
0530	Light hazard, one floor, 500 S.F.	15	5.80	20.80

D4010 Sprinklers

D4010 350	Preaction Sprinkler Systems		COST PER S.F.		
			MAT.	INST.	TOTAL
0560	1000 S.F.	R211313 -10	7.95	4.27	12.22
0580	2000 S.F.		8.10	3.79	11.89
0600	5000 S.F.	R211313 -20	4.10	2.85	6.95
0620	10,000 S.F.		2.63	2.28	4.91
0640	50,000 S.F.		1.74	1.98	3.72
0660	Each additional floor, 500 S.F.		4.16	3.05	7.21
0680	1000 S.F.		2.40	2.78	5.18
0700	2000 S.F.		2.28	2.56	4.84
0720	5000 S.F.		2.33	2.24	4.57
0740	10,000 S.F.		1.98	2.02	4
0760	50,000 S.F.		1.43	1.82	3.25
1000	Ordinary hazard, one floor, 500 S.F.		15.15	6.25	21.40
1020	1000 S.F.		8.90	4.25	13.15
1040	2000 S.F.		8.80	4.44	13.24
1060	5000 S.F.		4.45	3.04	7.49
1080	10,000 S.F.		3.08	2.95	6.03
1100	50,000 S.F.		2.71	2.82	5.53
1140	Each additional floor, 500 S.F.		3.48	3.52	7
1160	1000 S.F.		2.31	2.81	5.12
1180	2000 S.F.		2.13	2.79	4.92
1200	5000 S.F.		2.26	2.62	4.88
1220	10,000 S.F.		1.92	2.67	4.59
1240	50,000 S.F.		1.93	2.42	4.35
1500	Extra hazard, one floor, 500 S.F.		20	8.05	28.05
1520	1000 S.F.		14.30	6	20.30
1540	2000 S.F.		8.85	5.60	14.45
1560	5000 S.F.		4.97	4.52	9.49
1580	10,000 S.F.		4.82	4.40	9.22
1600	50,000 S.F.		4.69	4.27	8.96
1660	Each additional floor, 500 S.F.		4.19	4.29	8.48
1680	1000 S.F.		3.25	4.03	7.28
1700	2000 S.F.		3	4.03	7.03
1720	5000 S.F.		2.44	3.56	6
1740	10,000 S.F.		2.88	3.27	6.15
1760	50,000 S.F.		2.75	3.09	5.84
2020	Grooved steel, black, sch. 40 pipe, light hazard, one floor, 2000 S.F.		8.35	3.69	12.04
2060	10,000 S.F.		2.68	1.97	4.65
2100	Each additional floor of 2000 S.F.		2.36	2.07	4.43
2150	10,000 S.F.		1.51	1.68	3.19
2200	Ordinary hazard, one floor, 2000 S.F.		8.45	3.89	12.34
2250	10,000 S.F.		3.02	2.55	5.57
2300	Each additional floor, 2000 S.F.		2.44	2.27	4.71
2350	10,000 S.F.		1.86	2.27	4.13
2400	Extra hazard, one floor, 2000 S.F.		8.90	4.83	13.73
2450	10,000 S.F.		4.06	3.27	7.33
2500	Each additional floor, 2000 S.F.		3.06	3.32	6.38
2550	10,000 S.F.		2.46	2.80	5.26
3050	Grooved steel, black, sch. 10 pipe light hazard, one floor, 2000 S.F.		8.30	3.66	11.96
3100	10,000 S.F.		2.64	1.95	4.59
3150	Each additional floor, 2000 S.F.		2.28	2.03	4.31
3200	10,000 S.F.		1.47	1.66	3.13
3250	Ordinary hazard, one floor, 2000 S.F.		8.30	3.65	11.95
3300	10,000 S.F.		2.61	2.48	5.09
3350	Each additional floor, 2000 S.F.		2.37	2.25	4.62
3400	10,000 S.F.		1.78	2.21	3.99
3450	Extra hazard, one floor, 2000 S.F.		8.85	4.81	13.66
3500	10,000 S.F.		3.79	3.20	6.99
3550	Each additional floor, 2000 S.F.		3	3.30	6.30

D4010 Sprinklers

D4010 350	Preaction Sprinkler Systems	COST PER S.F.		
		MAT.	INST.	TOTAL
3600	10,000 S.F.	2.36	2.76	5.12
4050	Copper tubing, type L, light hazard, one floor, 2000 S.F.	8.60	3.69	12.29
4100	10,000 S.F.	3.07	2.03	5.10
4150	Each additional floor, 2000 S.F.	2.61	2.11	4.72
4200	10,000 S.F.	1.57	1.74	3.31
4250	Ordinary hazard, one floor, 2000 S.F.	8.90	4.04	12.94
4300	10,000 S.F.	3.59	2.31	5.90
4350	Each additional floor, 2000 S.F.	2.66	2.12	4.78
4400	10,000 S.F.	2.15	1.83	3.98
4450	Extra hazard, one floor, 2000 S.F.	9.55	4.85	14.40
4500	10,000 S.F.	5.85	3.49	9.34
4550	Each additional floor, 2000 S.F.	3.72	3.34	7.06
4600	10,000 S.F.	3.71	3.01	6.72
5050	Copper tubing, type L, T-drill system, light hazard, one floor			
5060	2000 S.F.	8.65	3.46	12.11
5100	10,000 S.F.	2.95	1.70	4.65
5150	Each additional floor, 2000 S.F.	2.64	1.88	4.52
5200	10,000 S.F.	1.78	1.42	3.20
5250	Ordinary hazard, one floor, 2000 S.F.	8.60	3.52	12.12
5300	10,000 S.F.	3.36	2.14	5.50
5350	Each additional floor, 2000 S.F.	2.60	1.91	4.51
5400	10,000 S.F.	2.20	1.86	4.06
5450	Extra hazard, one floor, 2000 S.F.	8.85	4.13	12.98
5500	10,000 S.F.	4.56	2.64	7.20
5550	Each additional floor, 2000 S.F.	3	2.62	5.62
5600	10,000 S.F.	2.60	2.15	4.75

D4010 Sprinklers

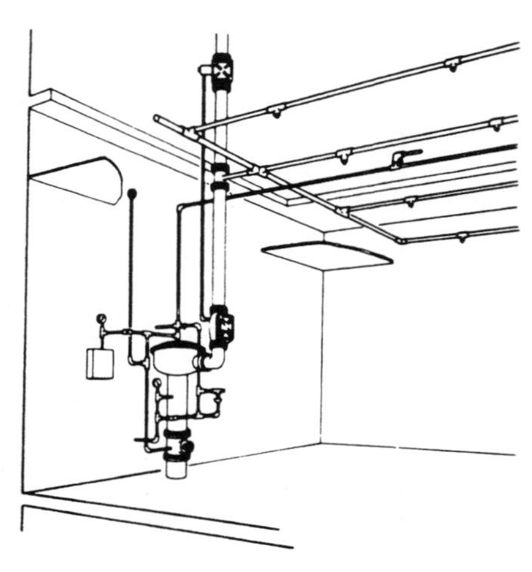

Deluge System: A system employing open sprinklers attached to a piping system connected to a water supply through a valve which is opened by the operation of a heat responsive system installed in the same areas as the sprinklers. When this valve opens, water flows into the piping system and discharges from all sprinklers attached thereto.

All areas are assumed to be open.

System Components	QUANTITY	UNIT	COST EACH MAT.	COST EACH INST.	COST EACH TOTAL
SYSTEM D4010 370 0580					
DELUGE SPRINKLER SYSTEM, STEEL BLACK SCH. 40 PIPE					
LIGHT HAZARD, 1 FLOOR, 2000 S.F.					
Valve, gate, iron body 125 lb., OS&Y, flanged, 4″ pipe size	1.000	Ea.	755	465	1,220
Backflow pvntr, auto, 2 OS&Y v, tst cocks, dbl chk, flgd, 4″	1.000	Ea.	5,125	465	5,590
Tamper switch (valve supervisory switch)	3.000	Ea.	873	142.50	1,015.50
Valve, swing check w/ball drip, CI w/brass ftngs., 4″ pipe size	1.000	Ea.	460	455	915
Valve, swing check, bronze, 125 lb, regrinding disc, 2-1/2″ pipe size	1.000	Ea.	1,025	92.50	1,117.50
Valve, angle, bronze, 150 lb., rising stem, threaded, 2″ pipe size	1.000	Ea.	930	70	1,000
*Alarm valve, 2-1/2″ pipe size	1.000	Ea.	1,900	455	2,355
Alarm, water motor, complete with gong	1.000	Ea.	530	190	720
Fire alarm horn, electric	1.000	Ea.	59.50	109	168.50
Thermostatic release for release line	2.000	Ea.	1,760	76	1,836
Pipe, steel, black, schedule 40, 4″ diam.	8.000	L.F.	212	325.36	537.36
Deluge valve trim, pressure relief, emergency release, gauge, 4″ pipe size	1.000	Ea.	6,200	1,375	7,575
Deluge system, monitoring panel w/deluge valve & trim	1.000	Ea.	8,225	42	8,267
Pipe, steel, black, schedule 40, threaded, cplg & hngr 10′ OC 2-1/2″ diam.	15.000	L.F.	188.25	420	608.25
Pipe, steel, black, schedule 40, threaded, cplg & hngr 10′ OC 2″ diam.	9.375	L.F.	71.25	201.56	272.81
Pipe, steel, black, schedule 40, threaded, cplg & hngr 10′ OC 1-1/4″ diam.	28.125	L.F.	253.13	438.75	691.88
Pipe, steel, black, schedule 40, threaded, cplg & hngr 10′ OC 1″ diam.	84.000	L.F.	718.20	1,222.20	1,940.40
Pipe, Tee, malleable iron, black, 150 lb. threaded, 4″ pipe size	2.000	Ea.	780	690	1,470
Pipe, Tee, malleable iron, black, 150 lb. threaded, 2-1/2″ pipe size	2.000	Ea.	218	308	526
Pipe, Tee, malleable iron, black, 150 lb. threaded, 2″ pipe size	1.000	Ea.	50.50	126	176.50
Pipe, Tee, malleable iron, black, 150 lb. threaded, 1-1/4″ pipe size	4.000	Ea.	96	396	492
Pipe, Tee, malleable iron, black, 150 lb. threaded, 1″ pipe size	3.000	Ea.	44.25	289.50	333.75
Pipe, 90° elbow, malleable iron, black, 150 lb. threaded 1″ pipe size	5.000	Ea.	46	297.50	343.50
Sprinkler head, std spray, brass 135°-286°F 1/2″ NPT, 3/8″ orifice	12.000	Ea.	221.40	570	791.40
Air compressor, auto, complete, 200 Gal sprinkler sys. cap., 1/3 HP	1.000	Ea.	1,325	585	1,910
*Standpipe connection, wall, flush w/plug & chain 2-1/2″ x 2-1/2″	1.000	Ea.	205	273	478
Valve, gate, bronze, 300 psi, NRS, class 150, threaded, 1″ pipe size	1.000	Ea.	148	40.50	188.50
TOTAL			32,419.48	10,120.37	42,539.85
COST PER S.F.			12.16	3.80	15.96
*Not included in systems under 2000 S.F.					

D4010 370	Deluge Sprinkler Systems	COST PER S.F. MAT.	COST PER S.F. INST.	COST PER S.F. TOTAL
0520	Deluge sprinkler systems, steel, black, sch. 40 pipe			
0530	Light hazard, one floor, 500 S.F.	29.50	5.85	35.35

D4010 Sprinklers

D4010 370	Deluge Sprinkler Systems	MAT.	INST.	TOTAL
		COST PER S.F.		
0560	1000 S.F. R211313-10	16.20	4.15	20.35
0580	2000 S.F.	12.15	3.79	15.94
0600	5000 S.F. R211313-20	5.70	2.86	8.56
0620	10,000 S.F.	3.45	2.28	5.73
0640	50,000 S.F.	1.90	1.98	3.88
0660	Each additional floor, 500 S.F.	3.34	3.05	6.39
0680	1000 S.F.	2.40	2.78	5.18
0700	2000 S.F.	2.28	2.56	4.84
0720	5000 S.F.	1.70	2.19	3.89
0740	10,000 S.F.	1.46	1.99	3.45
0760	50,000 S.F.	1.40	1.82	3.22
1000	Ordinary hazard, one floor, 500 S.F.	30.50	6.65	37.15
1020	1000 S.F.	16.15	4.28	20.43
1040	2000 S.F.	12.85	4.46	17.31
1060	5000 S.F.	6.05	3.05	9.10
1080	10,000 S.F.	3.90	2.95	6.85
1100	50,000 S.F.	2.92	2.86	5.78
1140	Each additional floor, 500 S.F.	3.48	3.52	7
1160	1000 S.F.	2.31	2.81	5.12
1180	2000 S.F.	2.13	2.79	4.92
1200	5000 S.F.	2.12	2.42	4.54
1220	10,000 S.F.	1.95	2.36	4.31
1240	50,000 S.F.	1.81	2.22	4.03
1500	Extra hazard, one floor, 500 S.F.	34.50	8.10	42.60
1520	1000 S.F.	22	6.20	28.20
1540	2000 S.F.	12.90	5.60	18.50
1560	5000 S.F.	6.35	4.18	10.53
1580	10,000 S.F.	5.45	4.02	9.47
1600	50,000 S.F.	5.15	3.93	9.08
1660	Each additional floor, 500 S.F.	4.19	4.29	8.48
1680	1000 S.F.	3.25	4.03	7.28
1700	2000 S.F.	3	4.03	7.03
1720	5000 S.F.	2.44	3.56	6
1740	10,000 S.F.	2.95	3.39	6.34
1760	50,000 S.F.	2.92	3.30	6.22
2000	Grooved steel, black, sch. 40 pipe, light hazard, one floor			
2020	2000 S.F.	12.40	3.71	16.11
2060	10,000 S.F.	3.53	1.99	5.52
2100	Each additional floor, 2,000 S.F.	2.36	2.07	4.43
2150	10,000 S.F.	1.51	1.68	3.19
2200	Ordinary hazard, one floor, 2000 S.F.	8.45	3.89	12.34
2250	10,000 S.F.	3.84	2.55	6.39
2300	Each additional floor, 2000 S.F.	2.44	2.27	4.71
2350	10,000 S.F.	1.86	2.27	4.13
2400	Extra hazard, one floor, 2000 S.F.	12.95	4.85	17.80
2450	10,000 S.F.	4.90	3.28	8.18
2500	Each additional floor, 2000 S.F.	3.06	3.32	6.38
2550	10,000 S.F.	2.46	2.80	5.26
3000	Grooved steel, black, sch. 10 pipe, light hazard, one floor			
3050	2000 S.F.	11.70	3.53	15.23
3100	10,000 S.F.	3.46	1.95	5.41
3150	Each additional floor, 2000 S.F.	2.28	2.03	4.31
3200	10,000 S.F.	1.47	1.66	3.13
3250	Ordinary hazard, one floor, 2000 S.F.	12.40	3.89	16.29
3300	10,000 S.F.	3.43	2.48	5.91
3350	Each additional floor, 2000 S.F.	2.37	2.25	4.62
3400	10,000 S.F.	1.78	2.21	3.99
3450	Extra hazard, one floor, 2000 S.F.	12.90	4.83	17.73

D4010 Sprinklers

D4010 370	Deluge Sprinkler Systems	COST PER S.F.		
		MAT.	INST.	TOTAL
3500	10,000 S.F.	4.61	3.20	7.81
3550	Each additional floor, 2000 S.F.	3	3.30	6.30
3600	10,000 S.F.	2.36	2.76	5.12
4000	Copper tubing, type L, light hazard, one floor			
4050	2000 S.F.	12.60	3.71	16.31
4100	10,000 S.F.	3.76	2.03	5.79
4150	Each additional floor, 2000 S.F.	2.58	2.11	4.69
4200	10,000 S.F.	1.57	1.74	3.31
4250	Ordinary hazard, one floor, 2000 S.F.	12.85	4.08	16.93
4300	10,000 S.F.	4.13	2.37	6.50
4350	Each additional floor, 2000 S.F.	2.56	2.13	4.69
4400	10,000 S.F.	2.03	1.88	3.91
4450	Extra hazard, one floor, 2000 S.F.	13.40	4.92	18.32
4500	10,000 S.F.	6.20	3.57	9.77
4550	Each additional floor, 2000 S.F.	3.53	3.39	6.92
4600	10,000 S.F.	3.51	3.06	6.57
5000	Copper tubing, type L, T-drill system, light hazard, one floor			
5050	2000 S.F.	12.65	3.48	16.13
5100	10,000 S.F.	3.77	1.70	5.47
5150	Each additional floor, 2000 S.F.	2.60	1.91	4.51
5200	10,000 S.F.	1.78	1.42	3.20
5250	Ordinary hazard, one floor, 2000 S.F.	10.40	3.32	13.72
5300	10,000 S.F.	4.18	2.14	6.32
5350	Each additional floor, 2000 S.F.	2.64	1.90	4.54
5400	10,000 S.F.	2.20	1.86	4.06
5450	Extra hazard, one floor, 2000 S.F.	10.65	3.93	14.58
5500	10,000 S.F.	5.40	2.64	8.04
5550	Each additional floor, 2000 S.F.	3	2.62	5.62
5600	10,000 S.F.	2.60	2.15	4.75

D4010 Sprinklers

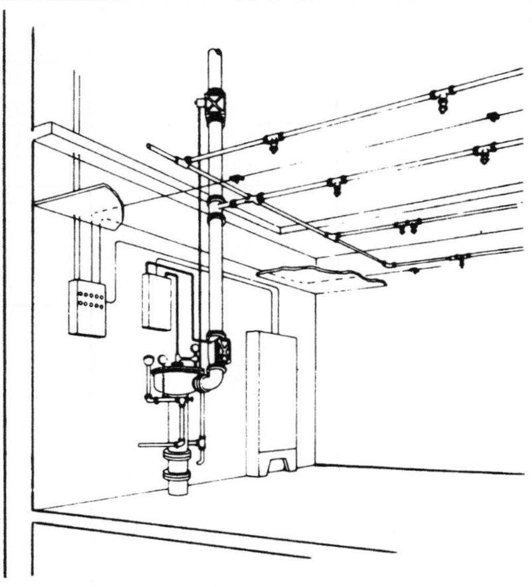

On-off multicycle sprinkler system is a fixed fire protection system utilizing water as its extinguishing agent. It is a time delayed, recycling, preaction type which automatically shuts the water off when heat is reduced below the detector operating temperature and turns the water back on when that temperature is exceeded.

The system senses a fire condition through a closed circuit electrical detector system which controls water flow to the fire automatically. Batteries supply up to 90 hour emergency power supply for system operation. The piping system is dry (until water is required) and is monitored with pressurized air. Should any leak in the system piping occur, an alarm will sound, but water will not enter the system until heat is sensed by a Firecycle detector.

All areas are assumed to be open.

System Components	QUANTITY	UNIT	COST EACH MAT.	COST EACH INST.	COST EACH TOTAL
SYSTEM D4010 390 0580					
ON-OFF MULTICYCLE SPRINKLER SYSTEM, STEEL, BLACK, SCH. 40 PIPE					
LIGHT HAZARD, ONE FLOOR, 2000 S.F.					
Valve, gate, iron body 125 lb., OS&Y, flanged, 4″ pipe size	1.000	Ea.	755	465	1,220
Backflow pvntr, auto, 2 OS&Y v, tst cocks, dbl chk, flgd, 4″	1.000	Ea.	5,125	465	5,590
Tamper switch (valve supervisory switch)	3.000	Ea.	873	142.50	1,015.50
Valve, angle, bronze, 150 lb., rising stem, threaded, 2″ pipe size	1.000	Ea.	930	70	1,000
Valve, swing check, bronze, 125 lb, regrinding disc, 2-1/2″ pipe size	1.000	Ea.	1,025	92.50	1,117.50
*Alarm valve, 2-1/2″ pipe size	1.000	Ea.	1,900	455	2,355
Alarm, water motor, complete with gong	1.000	Ea.	530	190	720
Pipe, steel, black, schedule 40, 4″ diam.	8.000	L.F.	212	325.36	537.36
Fire alarm, horn, electric	1.000	Ea.	59.50	109	168.50
Pipe, steel, black, schedule 40, threaded, cplg & hngr 10′ OC 2-1/2″ diam.	15.000	L.F.	188.25	420	608.25
Pipe, steel, black, schedule 40, threaded, cplg & hngr 10′ OC 2″ diam.	9.375	L.F.	71.25	201.56	272.81
Pipe, steel, black, schedule 40, threaded, cplg & hngr 10′ OC 1-1/4″ diam.	28.125	L.F.	253.13	438.75	691.88
Pipe, steel, black, schedule 40, threaded, cplg & hngr 10′ OC 1″ diam.	84.000	L.F.	718.20	1,222.20	1,940.40
Pipe, Tee, malleable iron, black, 150 lb. threaded, 4″ pipe size	2.000	Ea.	780	690	1,470
Pipe, Tee, malleable iron, black, 150 lb. threaded, 2-1/2″ pipe size	2.000	Ea.	218	308	526
Pipe, Tee, malleable iron, black, 150 lb. threaded, 2″ pipe size	1.000	Ea.	50.50	126	176.50
Pipe, Tee, malleable iron, black, 150 lb. threaded, 1-1/4″ pipe size	4.000	Ea.	96	396	492
Pipe, Tee, malleable iron, black, 150 lb. threaded, 1″ pipe size	3.000	Ea.	44.25	289.50	333.75
Pipe, 90° elbow, malleable iron, black, 150 lb. threaded, 1″ pipe size	5.000	Ea.	46	297.50	343.50
Sprinkler head std spray, brass 135°-286°F 1/2″ NPT, 3/8″ orifice	12.000	Ea.	221.40	570	791.40
Firecycle controls, incls panel, battery, solenoid valves, press switches	1.000	Ea.	24,600	2,900	27,500
Detector, firecycle system	2.000	Ea.	1,760	95	1,855
Firecycle pkg, swing check & flow control valves w/trim 4″ pipe size	1.000	Ea.	7,400	1,375	8,775
Air compressor, auto, complete, 200 Gal sprinkler sys. cap., 1/3 HP	1.000	Ea.	1,325	585	1,910
*Standpipe connection, wall, flush, brass w/plug & chain 2-1/2″x2-1/2″	1.000	Ea.	205	273	478
Valve, gate, bronze 300 psi, NRS, class 150, threaded, 1″ diam.	1.000	Ea.	148	40.50	188.50
TOTAL			49,534.48	12,542.37	62,076.85
COST PER S.F.			18.58	4.70	23.28

*Not included in systems under 2000 S.F.

D4010 390	On-off multicycle Sprinkler Systems	COST PER S.F. MAT.	COST PER S.F. INST.	COST PER S.F. TOTAL
0520	On-off multicycle sprinkler systems, steel, black, sch. 40 pipe			
0530	Light hazard, one floor, 500 S.F.	57.50	11.10	68.60

D4010 Sprinklers

D4010 390	On-off multicycle Sprinkler Systems		COST PER S.F.		
			MAT.	INST.	TOTAL
0560	1000 S.F.	R211313 -10	30	6.95	36.95
0580	2000 S.F.		18.56	4.69	23.25
0600	5000 S.F.	R211313 -20	8.30	3.22	11.52
0620	10,000 S.F.		4.81	2.48	7.29
0640	50,000 S.F.		2.22	2.02	4.24
0660	Each additional floor of 500 S.F.		3.34	3.06	6.40
0680	1000 S.F.		2.40	2.79	5.19
0700	2000 S.F.		1.95	2.55	4.50
0720	5000 S.F.		1.70	2.20	3.90
0740	10,000 S.F.		1.53	2	3.53
0760	50,000 S.F.		1.46	1.82	3.28
1000	Ordinary hazard, one floor, 500 S.F.		54.50	11.25	65.75
1020	1000 S.F.		30	6.90	36.90
1040	2000 S.F.		18.95	5.35	24.30
1060	5000 S.F.		8.65	3.41	12.06
1080	10,000 S.F.		5.25	3.15	8.40
1100	50,000 S.F.		3.45	3.21	6.66
1140	Each additional floor, 500 S.F.		3.48	3.53	7.01
1160	1000 S.F.		2.31	2.82	5.13
1180	2000 S.F.		2.31	2.57	4.88
1200	5000 S.F.		2.12	2.43	4.55
1220	10,000 S.F.		1.86	2.31	4.17
1240	50,000 S.F.		1.80	2.11	3.91
1500	Extra hazard, one floor, 500 S.F.		62.50	13.35	75.85
1520	1000 S.F.		35	8.65	43.65
1540	2000 S.F.		19.30	6.55	25.85
1560	5000 S.F.		8.90	4.54	13.44
1580	10,000 S.F.		6.90	4.57	11.47
1600	50,000 S.F.		5.65	5.05	10.70
1660	Each additional floor, 500 S.F.		4.19	4.30	8.49
1680	1000 S.F.		3.25	4.04	7.29
1700	2000 S.F.		3	4.04	7.04
1720	5000 S.F.		2.44	3.57	6.01
1740	10,000 S.F.		2.95	3.28	6.23
1760	50,000 S.F.		2.91	3.19	6.10
2020	Grooved steel, black, sch. 40 pipe, light hazard, one floor				
2030	2000 S.F.		16.60	4.40	21
2060	10,000 S.F.		5.20	3	8.20
2100	Each additional floor, 2000 S.F.		2.36	2.08	4.44
2150	10,000 S.F.		1.58	1.69	3.27
2200	Ordinary hazard, one floor, 2000 S.F.		18.90	4.82	23.72
2250	10,000 S.F.		5.60	2.90	8.50
2300	Each additional floor, 2000 S.F.		2.44	2.28	4.72
2350	10,000 S.F.		1.93	2.28	4.21
2400	Extra hazard, one floor, 2000 S.F.		17.10	5.55	22.65
2450	10,000 S.F.		6.15	3.44	9.59
2500	Each additional floor, 2000 S.F.		3.06	3.33	6.39
2550	10,000 S.F.		2.53	2.81	5.34
3050	Grooved steel, black, sch. 10 pipe, light hazard, one floor,				
3060	2000 S.F.		18.75	4.59	23.34
3100	10,000 S.F.		4.82	2.15	6.97
3150	Each additional floor, 2000 S.F.		2.28	2.04	4.32
3200	10,000 S.F.		1.54	1.67	3.21
3250	Ordinary hazard, one floor, 2000 S.F.		18.85	4.80	23.65
3300	10,000 S.F.		5.10	2.69	7.79
3350	Each additional floor, 2000 S.F.		2.37	2.26	4.63
3400	10,000 S.F.		1.85	2.22	4.07
3450	Extra hazard, one floor, 2000 S.F.		19.30	5.75	25.05

D4010 Sprinklers

D4010 390	On-off multicycle Sprinkler Systems	COST PER S.F. MAT.	INST.	TOTAL
3500	10,000 S.F.	5.95	3.38	9.33
3550	Each additional floor, 2000 S.F.	3	3.31	6.31
3600	10,000 S.F.	2.43	2.77	5.20
4060	Copper tubing, type L, light hazard, one floor, 2000 S.F.	19.05	4.62	23.67
4100	10,000 S.F.	5.25	2.23	7.48
4150	Each additional floor, 2000 S.F.	2.58	2.12	4.70
4200	10,000 S.F.	1.97	1.76	3.73
4250	Ordinary hazard, one floor, 2000 S.F.	19.40	4.97	24.37
4300	10,000 S.F.	5.75	2.51	8.26
4350	Each additional floor, 2000 S.F.	2.66	2.13	4.79
4400	10,000 S.F.	2.19	1.81	4
4450	Extra hazard, one floor, 2000 S.F.	20	5.80	25.80
4500	10,000 S.F.	8.10	3.73	11.83
4550	Each additional floor, 2000 S.F.	3.72	3.35	7.07
4600	10,000 S.F.	3.78	3.02	6.80
5060	Copper tubing, type L, T-drill system, light hazard, one floor 2000 S.F.	19.10	4.39	23.49
5100	10,000 S.F.	5.15	1.89	7.04
5150	Each additional floor, 2000 S.F.	2.83	1.97	4.80
5200	10,000 S.F.	1.85	1.43	3.28
5250	Ordinary hazard, one floor, 2000 S.F.	19.20	4.43	23.63
5300	10,000 S.F.	5.65	2.28	7.93
5350	Each additional floor, 2000 S.F.	2.72	1.89	4.61
5400	10,000 S.F.	2.40	1.81	4.21
5450	Extra hazard, one floor, 2000 S.F.	19.45	5	24.45
5500	10,000 S.F.	7.10	2.76	9.86
5550	Each additional floor, 2000 S.F.	3.15	2.58	5.73
5600	10,000 S.F.	2.86	2.11	4.97

D4010 Sprinklers

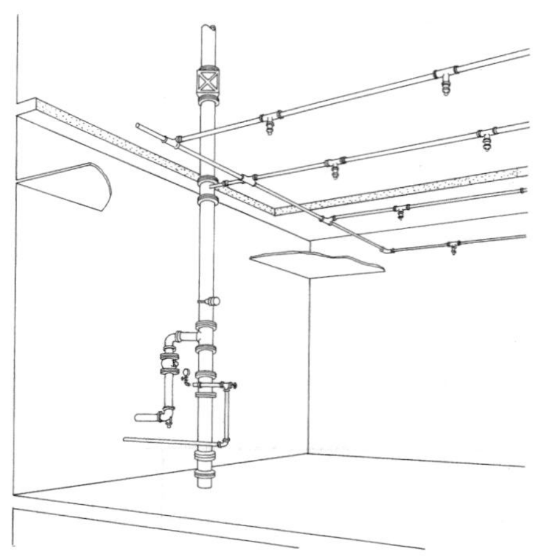

Wet Pipe System. A system employing automatic sprinklers attached to a piping system containing water and connected to a water supply so that water discharges immediately from sprinklers opened by heat from a fire.

All areas are assumed to be open.

System Components	QUANTITY	UNIT	COST EACH MAT.	COST EACH INST.	COST EACH TOTAL
SYSTEM D4010 410 0580					
WET PIPE SPRINKLER, STEEL, BLACK, SCH. 40 PIPE					
LIGHT HAZARD, ONE FLOOR, 2000 S.F.					
Valve, gate, iron body, 125 lb., OS&Y, flanged, 4″ diam.	1.000	Ea.	755	465	1,220
Backflow pvntr, auto, 2 OS&Y v, tst cocks, dbl chk, flgd, 4″	1.000	Ea.	5,125	465	5,590
Tamper switch (valve supervisory switch)	3.000	Ea.	873	142.50	1,015.50
Valve, swing check, bronze, 125 lb, regrinding disc, 2-1/2″ pipe size	1.000	Ea.	1,025	92.50	1,117.50
Valve, angle, bronze, 150 lb., rising stem, threaded, 2″ diam.	1.000	Ea.	930	70	1,000
*Alarm valve, 2-1/2″ pipe size	1.000	Ea.	1,900	455	2,355
Alarm, water motor, complete with gong	1.000	Ea.	530	190	720
Valve, swing check, w/balldrip CI with brass trim 4″ pipe size	1.000	Ea.	460	455	915
Pipe, steel, black, schedule 40, 4″ diam.	8.000	L.F.	212	325.36	537.36
Fire alarm horn, electric	1.000	Ea.	59.50	109	168.50
Pipe, steel, black, schedule 40, threaded, cplg & hngr 10′ OC, 2-1/2″ diam.	15.000	L.F.	188.25	420	608.25
Pipe, steel, black, schedule 40, threaded, cplg & hngr 10′ OC, 2″ diam.	9.375	L.F.	71.25	201.56	272.81
Pipe, steel, black, schedule 40, threaded, cplg & hngr 10′ OC, 1-1/4″ diam.	28.125	L.F.	253.13	438.75	691.88
Pipe, steel, black, schedule 40, threaded cplg & hngr 10′ OC, 1″ diam.	84.000	L.F.	718.20	1,222.20	1,940.40
Pipe Tee, malleable iron black, 150 lb. threaded, 4″ pipe size	2.000	Ea.	780	690	1,470
Pipe Tee, malleable iron black, 150 lb. threaded, 2-1/2″ pipe size	2.000	Ea.	218	308	526
Pipe Tee, malleable iron black, 150 lb. threaded, 2″ pipe size	1.000	Ea.	50.50	126	176.50
Pipe Tee, malleable iron black, 150 lb. threaded, 1-1/4″ pipe size	4.000	Ea.	96	396	492
Pipe Tee, malleable iron black, 150 lb. threaded, 1″ pipe size	3.000	Ea.	44.25	289.50	333.75
Pipe 90° elbow, malleable iron black, 150 lb. threaded, 1″ pipe size	5.000	Ea.	46	297.50	343.50
Sprinkler head, standard spray, brass 135°-286°F 1/2″ NPT, 3/8″ orifice	12.000	Ea.	221.40	570	791.40
Valve, gate, bronze, NRS, class 150, threaded, 1″ pipe size	1.000	Ea.	148	40.50	188.50
*Standpipe connection, wall, single, flush w/plug & chain 2-1/2″x2-1/2″	1.000	Ea.	205	273	478
TOTAL			14,909.48	8,042.37	22,951.85
COST PER S.F.			5.59	3.02	8.61
*Not included in systems under 2000 S.F.					

D4010 410	Wet Pipe Sprinkler Systems		COST PER S.F. MAT.	COST PER S.F. INST.	COST PER S.F. TOTAL
0520	Wet pipe sprinkler systems, steel, black, sch. 40 pipe				
0530	Light hazard, one floor, 500 S.F.		6.60	3.82	10.42
0560	1000 S.F.	R211313 -10	4.83	3.33	8.16
0580	2000 S.F.		5.59	3.02	8.61
0600	5000 S.F.	R211313 -20	2.83	2.53	5.36
0620	10,000 S.F.		1.80	2.12	3.92

D4010 Sprinklers

D4010 410	Wet Pipe Sprinkler Systems	COST PER S.F.		
		MAT.	INST.	TOTAL
0640	50,000 S.F.	1.25	1.93	3.18
0660	Each additional floor, 500 S.F.	2.05	3.06	5.11
0680	1000 S.F. R211313-40	1.77	2.83	4.60
0700	2000 S.F.	1.62	2.53	4.15
0720	5000 S.F.	1.17	2.17	3.34
0740	10,000 S.F.	1	1.97	2.97
0760	50,000 S.F.	.83	1.53	2.36
1000	Ordinary hazard, one floor, 500 S.F.	6.70	4.06	10.76
1020	1000 S.F.	4.71	3.27	7.98
1040	2000 S.F.	5.95	3.66	9.61
1060	5000 S.F.	3.18	2.72	5.90
1080	10,000 S.F.	2.25	2.79	5.04
1100	50,000 S.F.	2.17	2.74	4.91
1140	Each additional floor, 500 S.F.	2.16	3.46	5.62
1160	1000 S.F.	1.65	2.78	4.43
1180	2000 S.F.	1.80	2.78	4.58
1200	5000 S.F.	1.75	2.65	4.40
1220	10,000 S.F.	1.46	2.65	4.11
1240	50,000 S.F.	1.51	2.40	3.91
1500	Extra hazard, one floor, 500 S.F.	9	5.20	14.20
1520	1000 S.F.	8.70	4.68	13.38
1540	2000 S.F.	6.35	4.99	11.34
1560	5000 S.F.	3.87	4.45	8.32
1580	10,000 S.F.	3.92	4.27	8.19
1600	50,000 S.F.	3.95	4.12	8.07
1660	Each additional floor, 500 S.F.	2.87	4.23	7.10
1680	1000 S.F.	2.59	4	6.59
1700	2000 S.F.	2.34	4	6.34
1720	5000 S.F.	1.91	3.54	5.45
1740	10,000 S.F.	2.42	3.25	5.67
1760	50,000 S.F.	2.39	3.12	5.51
2020	Grooved steel, black sch. 40 pipe, light hazard, one floor, 2000 S.F.	5.85	2.92	8.77
2060	10,000 S.F.	1.85	1.81	3.66
2100	Each additional floor, 2000 S.F.	1.70	2.04	3.74
2150	10,000 S.F.	1.05	1.66	2.71
2200	Ordinary hazard, one floor, 2000 S.F.	5.95	3.12	9.07
2250	10,000 S.F.	2.19	2.39	4.58
2300	Each additional floor, 2000 S.F.	1.78	2.24	4.02
2350	10,000 S.F.	1.40	2.25	3.65
2400	Extra hazard, one floor, 2000 S.F.	4.13	3.84	7.97
2450	10,000 S.F.	2.57	3.04	5.61
2500	Each additional floor, 2000 S.F.	2.40	3.29	5.69
2550	10,000 S.F.	2	2.78	4.78
3050	Grooved steel, black sch. 10 pipe, light hazard, one floor, 2000 S.F.	3.52	2.67	6.19
3100	10,000 S.F.	1.81	1.79	3.60
3150	Each additional floor, 2000 S.F.	1.62	2	3.62
3200	10,000 S.F.	1.01	1.64	2.65
3250	Ordinary hazard, one floor, 2000 S.F.	5.85	3.10	8.95
3300	10,000 S.F.	2.04	2.32	4.36
3350	Each additional floor, 2000 S.F.	1.71	2.22	3.93
3400	10,000 S.F.	1.32	2.19	3.51
3450	Extra hazard, one floor, 2000 S.F.	6.30	4.04	10.34
3500	10,000 S.F.	2.96	3.04	6
3550	Each additional floor, 2000 S.F.	2.34	3.27	5.61
3600	10,000 S.F.	1.90	2.74	4.64
4050	Copper tubing, type L, light hazard, one floor, 2000 S.F.	6.05	2.92	8.97
4100	10,000 S.F.	2.24	1.87	4.11
4150	Each additional floor, 2000 S.F.	1.92	2.08	4

D4010 Sprinklers

D4010 410	Wet Pipe Sprinkler Systems	COST PER S.F.		
		MAT.	INST.	TOTAL
4200	10,000 S.F.	1.44	1.73	3.17
4250	Ordinary hazard, one floor, 2000 S.F.	6.25	3.29	9.54
4300	10,000 S.F.	2.61	2.21	4.82
4350	Each additional floor, 2000 S.F.	2.17	2.29	4.46
4400	10,000 S.F.	1.76	2.03	3.79
4450	Extra hazard, one floor, 2000 S.F.	6.85	4.13	10.98
4500	10,000 S.F.	4.64	3.39	8.03
4550	Each additional floor, 2000 S.F.	2.87	3.36	6.23
4600	10,000 S.F.	3.05	3.04	6.09
5050	Copper tubing, type L, T-drill system, light hazard, one floor			
5060	2000 S.F.	6.10	2.69	8.79
5100	10,000 S.F.	2.12	1.54	3.66
5150	Each additional floor, 2000 S.F.	1.98	1.85	3.83
5200	10,000 S.F.	1.32	1.40	2.72
5250	Ordinary hazard, one floor, 2000 S.F.	3.84	2.53	6.37
5300	10,000 S.F.	2.53	1.98	4.51
5350	Each additional floor, 2000 S.F.	1.94	1.87	3.81
5400	10,000 S.F.	1.74	1.84	3.58
5450	Extra hazard, one floor, 2000 S.F.	6.30	3.36	9.66
5500	10,000 S.F.	3.73	2.48	6.21
5550	Each additional floor, 2000 S.F.	5.10	2.89	7.99
5600	10,000 S.F.	2.14	2.13	4.27

D4010 Sprinklers

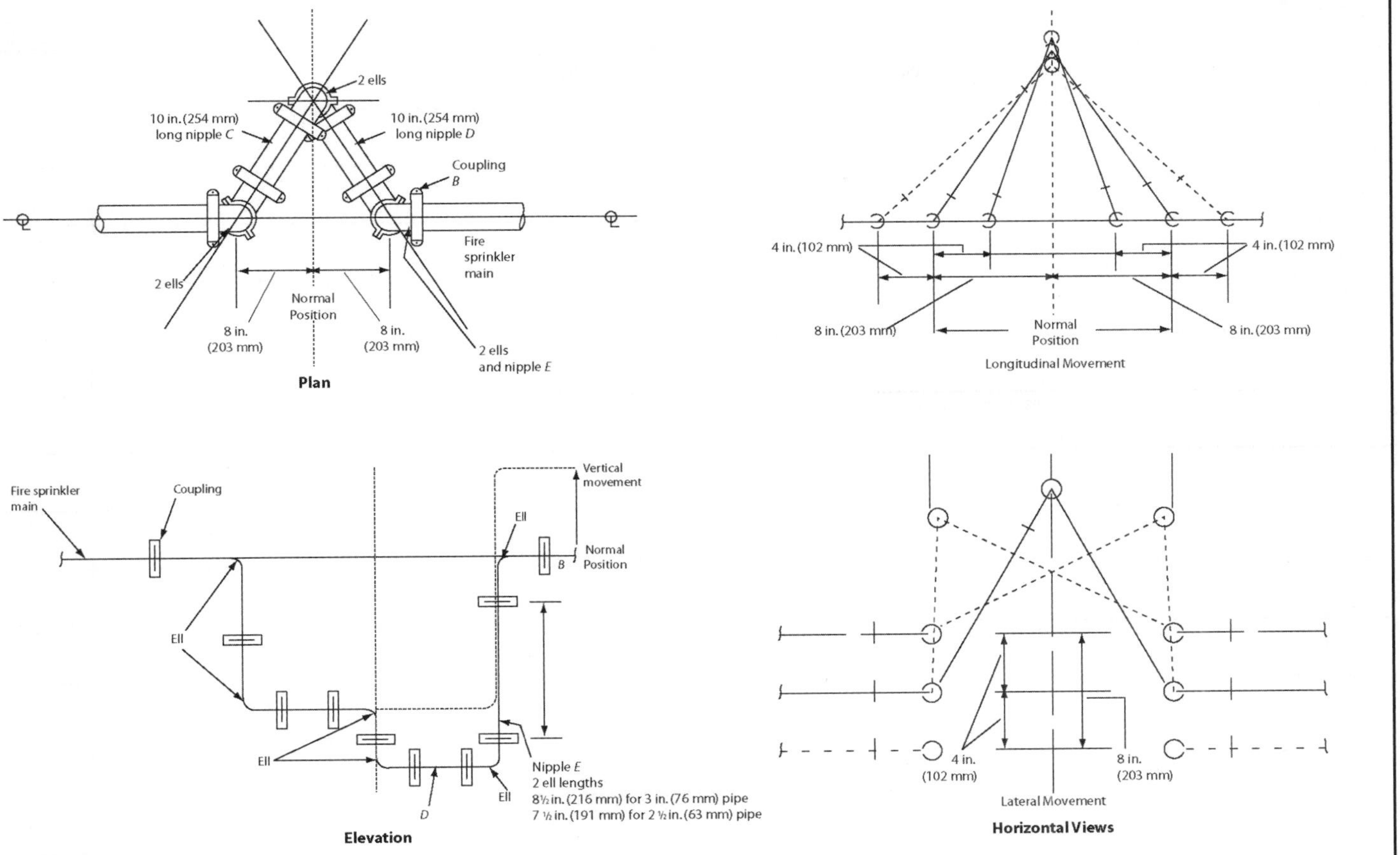

Plan

Longitudinal Movement

Elevation

Lateral Movement

Horizontal Views

System Components	QUANTITY	UNIT	COST PER EACH MAT.	COST PER EACH INST.	COST PER EACH TOTAL
SYSTEM D4010 412 1000					
WET SPRINKLER SYSTEM, SCHEDULE 10 BLACK STEEL GROOVED PIPE					
SEPARATION ASSEMBLY, 2 INCH DIAMETER					
Hanger support for 2 inch pipe	2.000	Ea.	66	85	151
Pipe roller support yoke	2.000	Ea.	25.70	19.80	45.50
Spool pieces 2 inch	3.000	Ea.	31.05	53.85	84.90
2 inch grooved elbows	6.000	Ea.	294	186	480
Coupling grooved joint 2 inch pipe	10.000	Ea.	480	154.50	634.50
Roll grooved joint labor only	8.000	Ea.		95.60	95.60
TOTAL			896.75	594.75	1,491.50

D4010 412	Wet Sprinkler Seismic Components	COST PER EACH MAT.	COST PER EACH INST.	COST PER EACH TOTAL
1000	Wet sprinkler sys, Sch. 10, blk steel grooved, separation assembly, 2 inch	895	595	1,490
1100	2-1/2 inch	985	660	1,645
1200	3 inch	1,300	755	2,055
1300	4 inch	1,600	945	2,545
1400	6 inch	3,325	1,350	4,675
1500	8 inch	5,975	1,600	7,575
2000	Wet sprinkler sys, Sch. 10, blk steel grooved, flexible coupling,2-1/2 inch	595	330	925
2100	3 inch	610	385	995
2200	4 inch	830	515	1,345
2300	6 inch	1,150	725	1,875

D4010 Sprinklers

D4010 412	Wet Sprinkler Seismic Components	COST PER EACH		
		MAT.	INST.	TOTAL
2400	8 inch	1,675	905	2,580
2990	Allowance for seismic movement by providing annular space in concrete wall			
2995	If additional cores are performed for the same size and at the same time reduce			
2996	price of additional cores by 50 %. If different size cores reduce by 30 %.			
3000	Wet pipe sprinkler systems, 2 inch black steel, annular space 6 inch wall	4.04	690	694.04
3010	8 inch wall	4.25	695	699.25
3020	10 inch wall	4.46	705	709.46
3030	12 inch wall	4.66	710	714.66
3100	Wet pipe sprinkler systems, 2-1/2/3 inch black steel, annular space 6″ wall	4.04	710	714.04
3110	8 inch wall	4.25	715	719.25
3120	10 inch wall	4.46	720	724.46
3130	12 inch wall	4.66	725	729.66
3200	Wet pipe sprinkler systems, 4 inch black steel, annular space 6″ wall	4.04	790	794.04
3210	8 inch wall	4.25	795	799.25
3220	10 inch wall	4.46	800	804.46
3230	12 inch wall	4.66	805	809.66
3300	Wet pipe sprinkler systems, 6 inch black steel, annular space 6″ wall	4.04	815	819.04
3310	8 inch wall	4.25	825	829.25
3320	10 inch wall	4.46	830	834.46
3330	12 inch wall	4.66	840	844.66
3400	Wet pipe sprinkler systems, 8 inch black steel, annular space 6″ wall	4.04	850	854.04
3410	8 inch wall	4.25	860	864.25
3420	10 inch wall	4.46	865	869.46
3430	12 inch wall	4.66	870	874.66
3996	Lateral, longitudinal and 4-way seismic strut braces			
4000	Wet pipe sprinkler systems, 2 inch pipe double lateral strut brace	130	325	455
4010	2 -1/2 inch	135	325	460
4020	3 inch	138	325	463
4030	4 inch	169	330	499
4040	6 inch	207	330	537
4050	8 inch	235	330	565
4100	Wet pipe sprinkler systems, 2 inch pipe longitudinal strut brace	130	325	455
4110	2 -1/2 inch	135	325	460
4120	3 inch	138	325	463
4130	4 inch	169	330	499
4140	6 inch	207	330	537
4150	8 inch	235	330	565
4200	Wet pipe sprinkler systems, 2 inch pipe single lateral strut brace	95.50	213	308.50
4210	2 -1/2 inch	100	214	314
4220	3 inch	104	214	318
4230	4 inch	132	218	350
4240	6 inch	170	218	388
4250	8 inch	198	219	417
4300	Wet pipe sprinkler systems, 2 inch pipe longitudinal wire brace	105	224	329
4310	2 -1/2 inch	110	224	334
4320	3 inch	114	225	339
4330	4 inch	146	232	378
4340	6 inch	182	230	412
4350	8 inch	210	230	440
4400	Wet pipe sprinkler systems, 2 inch pipe 4-way wire brace	150	345	495
4410	2 -1/2 inch	154	345	499
4420	3 inch	158	345	503
4430	4 inch	197	360	557
4440	6 inch	231	350	581
4450	8 inch	259	355	614
4500	Wet pipe sprinkler systems, 2 inch pipe 4-way strut brace	199	545	744
4510	2 -1/2 inch	204	545	749
4520	3 inch	207	545	752

D4010 Sprinklers

D4010 412	Wet Sprinkler Seismic Components	COST PER EACH		
		MAT.	INST.	TOTAL
4530	4 inch	243	550	793
4540	6 inch	281	550	831
4550	8 inch	310	550	860

D4010 Sprinklers

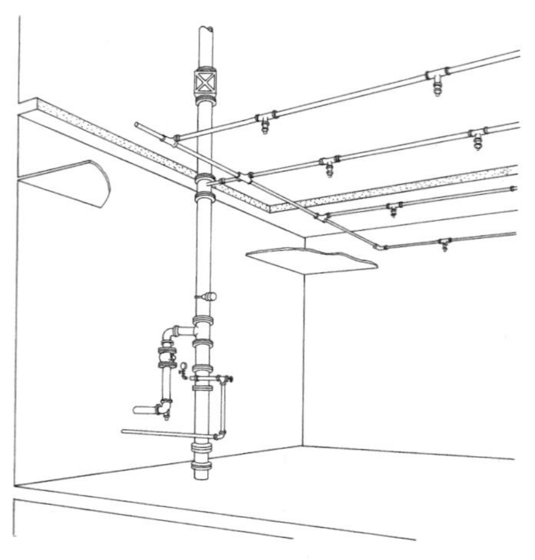

Wet Pipe System. A system employing automatic sprinklers attached to a piping system containing water and connected to a water supply so that water discharges immediately from sprinklers opened by heat from a fire.

All areas are assumed to be open.

System Components	QUANTITY	UNIT	COST PER S.F. MAT.	COST PER S.F. INST.	COST PER S.F. TOTAL
SYSTEM D4010 413 0580					
SEISMIC WET PIPE SPRINKLER, STEEL, BLACK, SCH. 40 PIPE					
LIGHT HAZARD, ONE FLOOR, 3′ FLEXIBLE FEED, 2000 S.F.					
Valve, gate, iron body, 125 lb., OS&Y, flanged, 4″ diam.	1.000	Ea.	.28	.17	.45
Backflow pvntr, auto, 2 OS&Y v, tst cocks, dbl chk, flgd, 4″	1.000	Ea.	1.92	.17	2.09
Tamper switch (valve supervisory switch)	3.000	Ea.	.33	.05	.38
Valve, swing check, bronze, 125 lb, regrinding disc, 2-1/2″ pipe size	1.000	Ea.	.38	.03	.41
Valve, angle, bronze, 150 lb., rising stem, threaded, 2″ diam.	1.000	Ea.	.35	.03	.38
*Alarm valve, 2-1/2″ pipe size	1.000	Ea.	.71	.17	.88
Alarm, water motor, complete with gong	1.000	Ea.	.20	.07	.27
Valve, swing check, w/balldrip CI with brass trim 4″ pipe size	1.000	Ea.	.17	.17	.34
Pipe, steel, black, schedule 40, 4″ diam.	8.000	L.F.	.08	.12	.20
Fire alarm horn, electric	1.000	Ea.	.02	.04	.06
Pipe, steel, black, schedule 40, threaded, cplg & hngr 10′ OC, 2-1/2″ diam.	15.000	L.F.	.07	.16	.23
Pipe, steel, black, schedule 40, threaded, cplg & hngr 10′ OC, 2″ diam.	9.375	L.F.	.03	.08	.11
Pipe, steel, black, schedule 40, threaded, cplg & hngr 10′ OC, 1-1/4″ diam.	28.125	L.F.	.09	.16	.25
Pipe Tee, malleable iron black, 150 lb. threaded, 4″ pipe size	2.000	Ea.	.29	.26	.55
Pipe Tee, malleable iron black, 150 lb. threaded, 2-1/2″ pipe size	2.000	Ea.	.08	.12	.20
Pipe Tee, malleable iron black, 150 lb. threaded, 2″ pipe size	1.000	Ea.	.02	.05	.07
Pipe Tee, malleable iron black, 150 lb. threaded, 1-1/4″ pipe size	4.000	Ea.	.04	.15	.19
Pipe Tee, malleable iron black, 150 lb. threaded, 1″ pipe size	3.000	Ea.	.02	.11	.13
Pipe 90° elbow, malleable iron black, 150 lb. threaded, 1″ pipe size	5.000	Ea.	.02	.11	.13
Sprinkler head, standard spray, brass 135°-286°F 1/2″ NPT, 3/8″ orifice	12.000	Ea.	.08	.21	.29
Valve, gate, bronze, NRS, class 150, threaded, 1″ pipe size	1.000	Ea.	.06	.02	.08
*Standpipe connection, wall, single, flush w/plug & chain 2-1/2″x2-1/2″	1.000	Ea.	.08	.10	.18
Flexible connector 36 inches long	12.000	Ea.	.36	.11	.47
TOTAL			5.68	2.66	8.34

*Not included in systems under 2000 S.F.

D4010 413	Seismic Wet Pipe Flexible Feed Sprinkler Systems	COST PER S.F. MAT.	COST PER S.F. INST.	COST PER S.F. TOTAL
0520	Seismic wet pipe 3′ flexible feed sprinkler systems, steel, black, sch. 40 pipe			
0525	Note: See D4010-412 for strut braces, flexible joints and separation assemblies			
0530	Light hazard, one floor, 3′ flexible feed, 500 S.F.	6.50	3.14	9.64
0560	1000 S.F.	4.87	2.89	7.76
0580	2000 S.F.	5.70	2.66	8.36
0600	5000 S.F.	2.95	2.21	5.16

D4010 Sprinklers

D4010 413	Seismic Wet Pipe Flexible Feed Sprinkler Systems	COST PER S.F.		
		MAT.	INST.	TOTAL
0620	10,000 S.F.	1.99	1.98	3.97
0640	50,000 S.F.	1.35	1.68	3.03
0660	Each additional floor, 500 S.F.	2.07	2.42	4.49
0680	1000 S.F.	1.93	2.43	4.36
0700	2000 S.F.	1.62	2.03	3.65
0720	5000 S.F.	1.29	1.85	3.14
0740	10,000 S.F.	1.21	1.84	3.05
0760	50,000 S.F.	1.03	1.43	2.46
1000	Ordinary hazard, one floor, 500 S.F.	6.75	3.42	10.17
1020	1000 S.F.	4.82	2.80	7.62
1040	2000 S.F.	6.15	3.37	9.52
1060	5000 S.F.	3.45	2.52	5.97
1080	10,000 S.F.	2.56	2.66	5.22
1100	50,000 S.F.	2.40	2.48	4.88
1140	Each additional floor, 500 S.F.	4.89	3.06	7.95
1160	1000 S.F.	1.76	2.31	4.07
1180	2000 S.F.	2.02	2.49	4.51
1200	5000 S.F.	2.07	2.45	4.52
1220	10,000 S.F.	1.79	2.53	4.32
1240	50,000 S.F.	1.76	2.14	3.90
1500	Extra hazard, one floor, 500 S.F.	9	4.23	13.23
1520	1000 S.F.	8.75	3.72	12.47
1540	2000 S.F.	6.70	4.54	11.24
1560	5000 S.F.	4.24	4.16	8.40
1580	10,000 S.F.	4.39	4.05	8.44
1600	50,000 S.F.	4.45	3.88	8.33
1660	Each additional floor, 500 S.F.	2.45	3.18	5.63
1680	1000 S.F.	2.84	3.07	5.91
1700	2000 S.F.	2.78	3.57	6.35
1720	5000 S.F.	2.33	3.25	5.58
1740	10,000 S.F.	2.91	3.04	5.95
1760	50,000 S.F.	2.88	2.87	5.75
2020	Grooved steel, black sch. 40 pipe, light hazard, one floor, 2000 S.F.	3.66	2.30	5.96
2060	10,000 S.F.	1.83	1.63	3.46
2100	Each additional floor, 2000 S.F.	1.76	1.64	3.40
2150	10,000 S.F.	1.05	1.49	2.54
2200	Ordinary hazard, one floor, 2000 S.F.	6.20	2.90	9.10
2250	10,000 S.F.	2.53	2.30	4.83
2300	Each additional floor, 2000 S.F.	2.05	2.02	4.07
2350	10,000 S.F.	1.76	2.17	3.93
2400	Extra hazard, one floor, 2000 S.F.	6.75	3.71	10.46
2450	10,000 S.F.	3.66	2.95	6.61
2500	Each additional floor, 2000 S.F.	2.88	2.96	5.84
2550	10,000 S.F.	2.51	2.64	5.15
3050	Grooved steel black sch. 10 pipe, light hazard, one floor, 2000 S.F.	5.85	2.49	8.34
3100	10,000 S.F.	2.03	1.69	3.72
3150	Each additional floor, 2000 S.F.	1.68	1.60	3.28
3200	10,000 S.F.	1.01	1.47	2.48
3250	Ordinary hazard, one floor, 2000 S.F.	6.15	2.88	9.03
3300	10,000 S.F.	2.45	2.24	4.69
3350	Each additional floor, 2000 S.F.	1.87	1.98	3.85
3400	10,000 S.F.	1.64	2.10	3.74
3450	Extra hazard, one floor, 2000 S.F.	6.70	3.69	10.39
3500	10,000 S.F.	3.45	2.89	6.34
3550	Each additional floor, 2000 S.F.	2.82	2.94	5.76
3600	10,000 S.F.	2.41	2.60	5.01

D4010 Sprinklers

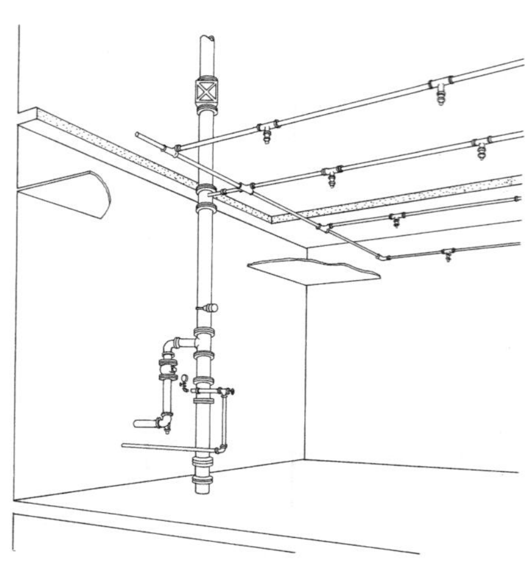

Wet Pipe System. A system employing automatic sprinklers attached to a piping system containing water and connected to a water supply so that water discharges immediately from sprinklers opened by heat from a fire.

All areas are assumed to be open.

System Components	QUANTITY	UNIT	COST PER S.F. MAT.	COST PER S.F. INST.	COST PER S.F. TOTAL
SYSTEM D4010 414 0580					
SEISMIC WET PIPE SPRINKLER, STEEL, BLACK, SCH. 40 PIPE					
LIGHT HAZARD, ONE FLOOR, 4′ FLEXIBLE FEED, 2000 S.F.					
Valve, gate, iron body, 125 lb., OS&Y, flanged, 4″ diam.	1.000	Ea.	.28	.17	.45
Backflow pvntr, auto, 2 OS&Y v, tst cocks, dbl chk, flgd, 4″	1.000	Ea.	1.92	.17	2.09
Tamper switch (valve supervisory switch)	3.000	Ea.	.33	.05	.38
Valve, swing check, bronze, 125 lb, regrinding disc, 2-1/2″ pipe size	1.000	Ea.	.38	.03	.41
Valve, angle, bronze, 150 lb., rising stem, threaded, 2″ diam.	1.000	Ea.	.35	.03	.38
*Alarm valve, 2-1/2″ pipe size	1.000	Ea.	.71	.17	.88
Alarm, water motor, complete with gong	1.000	Ea.	.20	.07	.27
Valve, swing check, w/balldrip CI with brass trim 4″ pipe size	1.000	Ea.	.17	.17	.34
Pipe, steel, black, schedule 40, 4″ diam.	8.000	L.F.	.08	.12	.20
Fire alarm horn, electric	1.000	Ea.	.02	.04	.06
Pipe, steel, black, schedule 40, threaded, cplg & hngr 10′ OC, 2-1/2″ diam.	15.000	L.F.	.07	.16	.23
Pipe, steel, black, schedule 40, threaded, cplg & hngr 10′ OC, 2″ diam.	9.375	L.F.	.03	.08	.11
Pipe, steel, black, schedule 40, threaded, cplg & hngr 10′ OC, 1-1/4″ diam.	28.125	L.F.	.09	.16	.25
Pipe Tee, malleable iron black, 150 lb. threaded, 4″ pipe size	2.000	Ea.	.29	.26	.55
Pipe Tee, malleable iron black, 150 lb. threaded, 2-1/2″ pipe size	2.000	Ea.	.08	.12	.20
Pipe Tee, malleable iron black, 150 lb. threaded, 2″ pipe size	1.000	Ea.	.02	.05	.07
Pipe Tee, malleable iron black, 150 lb. threaded, 1-1/4″ pipe size	4.000	Ea.	.04	.15	.19
Pipe Tee, malleable iron black, 150 lb. threaded, 1″ pipe size	3.000	Ea.	.02	.11	.13
Pipe 90° elbow, malleable iron black, 150 lb. threaded, 1″ pipe size	5.000	Ea.	.02	.11	.13
Sprinkler head, standard spray, brass 135°-286°F 1/2″ NPT, 3/8″ orifice	12.000	Ea.	.08	.21	.29
Valve, gate, bronze, NRS, class 150, threaded, 1″ pipe size	1.000	Ea.	.06	.02	.08
*Standpipe connection, wall, single, flush w/plug & chain 2-1/2″x2-1/2″	1.000	Ea.	.08	.10	.18
Flexible connector 48 inches long	12.000	Ea.	.42	.16	.58
TOTAL			5.74	2.71	8.45

*Not included in systems under 2000 S.F.

D4010 414	Seismic Wet Pipe Flexible Feed Sprinkler Systems	COST PER S.F. MAT.	COST PER S.F. INST.	COST PER S.F. TOTAL
0520	Seismic wet pipe 4′ flexible feed sprinkler systems, steel, black, sch. 40 pipe			
0525	Note: See D4010-412 for strut braces, flexible joints and separation assemblies			
0530	Light hazard, one floor, 4′ flexible feed, 500 S.F.	6.60	3.19	9.79
0560	1000 S.F.	4.93	2.94	7.87
0580	2000 S.F.	5.75	2.71	8.46
0600	5000 S.F.	3.02	2.26	5.28
0620	10,000 S.F.	2.04	2.01	4.05
0640	50,000 S.F.	1.41	1.72	3.13

D4010 Sprinklers

D4010 414	Seismic Wet Pipe Flexible Feed Sprinkler Systems	COST PER S.F.		
		MAT.	INST.	TOTAL
0660	Each additional floor, 500 S.F.	2.15	2.48	4.63
0680	1000 S.F.	2.01	2.49	4.50
0700	2000 S.F.	1.68	2.08	3.76
0720	5000 S.F.	1.36	1.90	3.26
0740	10,000 S.F.	1.26	1.87	3.13
0760	50,000 S.F.	1.09	1.47	2.56
1000	Ordinary hazard, one floor, 500 S.F.	6.80	3.48	10.28
1020	1000 S.F.	4.90	2.86	7.76
1040	2000 S.F.	6.25	3.43	9.68
1060	5000 S.F.	3.53	2.58	6.11
1080	10,000 S.F.	2.64	2.72	5.36
1100	50,000 S.F.	2.48	2.53	5.01
1140	Each additional floor, 500 S.F.	2.26	2.88	5.14
1160	1000 S.F.	1.84	2.37	4.21
1180	2000 S.F.	2.10	2.55	4.65
1200	5000 S.F.	2.15	2.51	4.66
1220	10,000 S.F.	1.87	2.59	4.46
1240	50,000 S.F.	1.84	2.19	4.03
1500	Extra hazard, one floor, 500 S.F.	9.15	4.31	13.46
1520	1000 S.F.	8.85	3.80	12.65
1540	2000 S.F.	6.80	4.62	11.42
1560	5000 S.F.	4.35	4.23	8.58
1580	10,000 S.F.	4.51	4.13	8.64
1600	50,000 S.F.	4.57	3.97	8.54
1660	Each additional floor, 500 S.F.	2.57	3.26	5.83
1680	1000 S.F.	2.96	3.15	6.11
1700	2000 S.F.	2.90	3.65	6.55
1720	5000 S.F.	2.44	3.32	5.76
1740	10,000 S.F.	3.03	3.12	6.15
1760	50,000 S.F.	3	2.96	5.96
2020	Grooved steel, black sch. 40 pipe, light hazard, one floor, 2000 S.F.	5.95	2.57	8.52
2060	10,000 S.F.	1.84	1.64	3.48
2100	Each additional floor, 2000 S.F.	1.82	1.69	3.51
2150	10,000 S.F.	1.06	1.50	2.56
2200	Ordinary hazard, one floor, 2000 S.F.	6.30	2.96	9.26
2250	10,000 S.F.	2.61	2.36	4.97
2300	Each additional floor, 2000 S.F.	2.13	2.08	4.21
2350	10,000 S.F.	1.84	2.23	4.07
2400	Extra hazard, one floor, 2000 S.F.	6.85	3.79	10.64
2450	10,000 S.F.	3.78	3.03	6.81
2500	Each additional floor, 2000 S.F.	3	3.04	6.04
2550	10,000 S.F.	2.63	2.72	5.35
3050	Grooved steel black sch. 10 pipe, light hazard, one floor, 2000 S.F.	5.90	2.54	8.44
3100	10,000 S.F.	2.08	1.72	3.80
3150	Each additional floor, 2000 S.F.	1.74	1.65	3.39
3200	10,000 S.F.	1.02	1.48	2.50
3250	Ordinary hazard, one floor, 2000 S.F.	6.20	2.94	9.14
3300	10,000 S.F.	2.53	2.30	4.83
3350	Each additional floor, 2000 S.F.	2.06	2.06	4.12
3400	10,000 S.F.	1.76	2.17	3.93
3450	Extra hazard, one floor, 2000 S.F.	6.80	3.77	10.57
3500	10,000 S.F.	3.57	2.97	6.54
3550	Each additional floor, 2000 S.F.	2.94	3.02	5.96
3600	10,000 S.F.	2.53	2.68	5.21

D4020 Standpipes

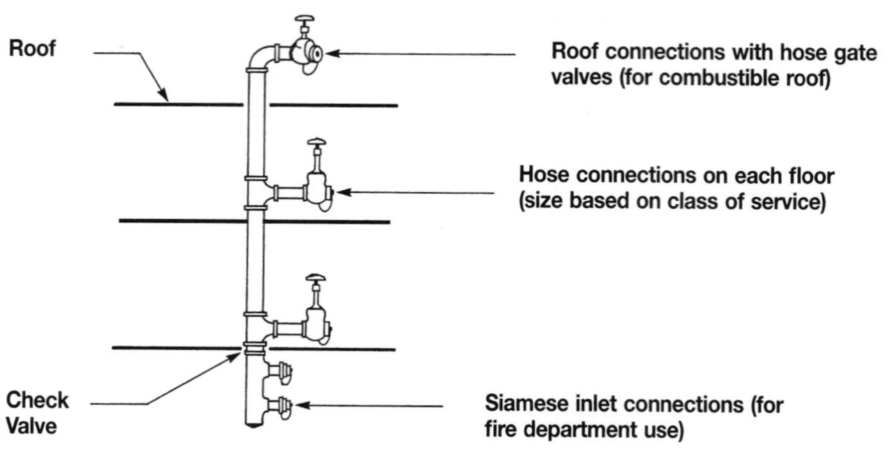

System Components	QUANTITY	UNIT	COST PER FLOOR MAT.	INST.	TOTAL
SYSTEM D4020 310 0560					
WET STANDPIPE RISER, CLASS I, STEEL, BLACK, SCH. 40 PIPE, 10' HEIGHT					
4" DIAMETER PIPE, ONE FLOOR					
Pipe, steel, black, schedule 40, threaded, 4" diam.	20.000	L.F.	510	770	1,280
Pipe, Tee, malleable iron, black, 150 lb. threaded, 4" pipe size	2.000	Ea.	780	690	1,470
Pipe, 90° elbow, malleable iron, black, 150 lb threaded, 4" pipe size	1.000	Ea.	216	231	447
Pipe, nipple, steel, black, schedule 40, 2-1/2" pipe size x 3" long	2.000	Ea.	28.20	174	202.20
Fire valve, gate, 300 lb., brass w/handwheel, 2-1/2" pipe size	1.000	Ea.	239	108	347
Fire valve, pressure reducing rgh brs, 2-1/2" pipe size	1.000	Ea.	980	216	1,196
Valve, swing check, w/ball drip, CI w/brs. ftngs., 4" pipe size	1.000	Ea.	460	455	915
Standpipe conn wall dble. flush brs. w/plugs & chains 2-1/2"x2-1/2"x4"	1.000	Ea.	850	273	1,123
Valve, swing check, bronze, 125 lb, regrinding disc, 2-1/2" pipe size	1.000	Ea.	1,025	92.50	1,117.50
Roof manifold, fire, w/valves & caps, horiz/vert brs 2-1/2"x2-1/2"x4"	1.000	Ea.	250	284	534
Fire, hydrolator, vent & drain, 2-1/2" pipe size	1.000	Ea.	126	63	189
Valve, gate, iron body 125 lb., OS&Y, threaded, 4" pipe size	1.000	Ea.	755	465	1,220
Tamper switch (valve supervisory switch)	1.000	Ea.	291	47.50	338.50
TOTAL			6,510.20	3,869	10,379.20

D4020 310	Wet Standpipe Risers, Class I		COST PER FLOOR MAT.	INST.	TOTAL
0550	Wet standpipe risers, Class I, steel, black, sch. 40, 10' height				
0560	4" diameter pipe, one floor		6,500	3,875	10,375
0580	Additional floors		1,525	1,175	2,700
0600	6" diameter pipe, one floor		10,200	6,750	16,950
0620	Additional floors	R211226 -10	2,700	1,900	4,600
0640	8" diameter pipe, one floor		15,200	8,150	23,350
0660	Additional floors	R211226 -20	3,900	2,300	6,200
0680					

D4020 310	Wet Standpipe Risers, Class II	COST PER FLOOR MAT.	INST.	TOTAL
1030	Wet standpipe risers, Class II, steel, black sch. 40, 10' height			
1040	2" diameter pipe, one floor	2,750	1,425	4,175
1060	Additional floors	880	530	1,410
1080	2-1/2" diameter pipe, one floor	3,600	2,025	5,625
1100	Additional floors	975	625	1,600
1120				

D4020 Standpipes

D4020 310	Wet Standpipe Risers, Class III	COST PER FLOOR		
		MAT.	INST.	TOTAL
1530	Wet standpipe risers, Class III, steel, black, sch. 40, 10′ height			
1540	4″ diameter pipe, one floor	6,675	3,875	10,550
1560	Additional floors	1,350	990	2,340
1580	6″ diameter pipe, one floor	10,400	6,750	17,150
1600	Additional floors	2,775	1,900	4,675
1620	8″ diameter pipe, one floor	15,300	8,150	23,450
1640	Additional floors	3,975	2,300	6,275

D4020 Standpipes

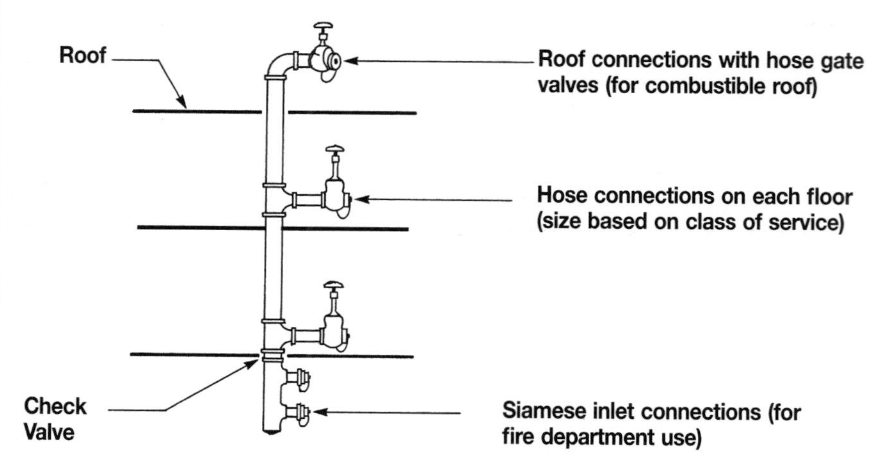

System Components	QUANTITY	UNIT	COST PER FLOOR		
			MAT.	INST.	TOTAL
SYSTEM D4020 330 0540					
DRY STANDPIPE RISER, CLASS I, PIPE, STEEL, BLACK, SCH 40, 10′ HEIGHT					
4″ DIAMETER PIPE, ONE FLOOR					
Pipe, steel, black, schedule 40, threaded, 4″ diam.	20.000	L.F.	510	770	1,280
Pipe, Tee, malleable iron, black, 150 lb. threaded, 4″ pipe size	2.000	Ea.	780	690	1,470
Pipe, 90° elbow, malleable iron, black, 150 lb threaded, 4″ pipe size	1.000	Ea.	216	231	447
Pipe, nipple, steel, black, schedule 40, 2-1/2″ pipe size x 3″ long	2.000	Ea.	28.20	174	202.20
Fire valve gate NRS 300 lb., brass w/handwheel, 2-1/2″ pipe size	1.000	Ea.	239	108	347
Tamper switch (valve supervisory switch)	1.000	Ea.	291	47.50	338.50
Fire valve, pressure reducing rgh brs, 2-1/2″ pipe size	1.000	Ea.	490	108	598
Standpipe conn wall dble. flush brs. w/plugs & chains 2-1/2″x2-1/2″x4″	1.000	Ea.	850	273	1,123
Valve swing check w/ball drip CI w/brs. ftngs., 4″pipe size	1.000	Ea.	460	455	915
Roof manifold, fire, w/valves & caps, horiz/vert brs 2-1/2″x2-1/2″x4″	1.000	Ea.	250	284	534
TOTAL			4,114.20	3,140.50	7,254.70

D4020 330	Dry Standpipe Risers, Class I		COST PER FLOOR		
			MAT.	INST.	TOTAL
0530	Dry standpipe riser, Class I, steel, black, sch. 40, 10′ height				
0540	4″ diameter pipe, one floor		4,125	3,150	7,275
0560	Additional floors		1,400	1,125	2,525
0580	6″ diameter pipe, one floor		7,775	5,375	13,150
0600	Additional floors	R211226 -10	2,575	1,825	4,400
0620	8″ diameter pipe, one floor		11,700	6,500	18,200
0640	Additional floors	R211226 -20	3,775	2,225	6,000
0660					

D4020 330	Dry Standpipe Risers, Class II	COST PER FLOOR		
		MAT.	INST.	TOTAL
1030	Dry standpipe risers, Class II, steel, black, sch. 40, 10′ height			
1040	2″ diameter pipe, one floor	2,200	1,425	3,625
1060	Additional floors	755	465	1,220
1080	2-1/2″ diameter pipe, one floor	3,000	1,675	4,675
1100	Additional floors	850	560	1,410
1120				

D4020 Standpipes

D4020 330	Dry Standpipe Risers, Class III	COST PER FLOOR		
		MAT.	INST.	TOTAL
1530	Dry standpipe risers, Class III, steel, black, sch. 40, 10′ height			
1540	4″ diameter pipe, one floor	3,875	3,025	6,900
1560	Additional floors	1,250	1,000	2,250
1580	6″ diameter pipe, one floor	7,850	5,375	13,225
1600	Additional floors	2,650	1,825	4,475
1620	8″ diameter pipe, one floor	11,800	6,500	18,300
1640	Additional floors	3,850	2,225	6,075

D4020 Standpipes

D4020 410	Fire Hose Equipment	COST EACH		
		MAT.	INST.	TOTAL
0100	Adapters, reducing, 1 piece, FxM, hexagon, cast brass, 2-1/2" x 1-1/2"	71		71
0200	Pin lug, 1-1/2" x 1"	50		50
0250	3" x 2-1/2"	156		156
0300	For polished chrome, add 75% mat.			
0400	Cabinets, D.S. glass in door, recessed, steel box, not equipped			
0500	Single extinguisher, steel door & frame	180	171	351
0550	Stainless steel door & frame	249	171	420
0600	Valve, 2-1/2" angle, steel door & frame	227	114	341
0650	Aluminum door & frame	276	114	390
0700	Stainless steel door & frame	370	114	484
0750	Hose rack assy, 2-1/2" x 1-1/2" valve & 100' hose, steel door & frame	425	228	653
0800	Aluminum door & frame	600	228	828
0850	Stainless steel door & frame	835	228	1,063
0900	Hose rack assy & extinguisher,2-1/2"x1-1/2" valve & hose,steel door & frame	330	273	603
0950	Aluminum	800	273	1,073
1000	Stainless steel	695	273	968
1550	Compressor, air, dry pipe system, automatic, 200 gal., 3/4 H.P.	1,325	585	1,910
1600	520 gal., 1 H.P.	1,750	585	2,335
1650	Alarm, electric pressure switch (circuit closer)	123	29	152
2500	Couplings, hose, rocker lug, cast brass, 1-1/2"	69.50		69.50
2550	2-1/2"	57		57
3000	Escutcheon plate, for angle valves, polished brass, 1-1/2"	18.25		18.25
3050	2-1/2"	26.50		26.50
3500	Fire pump, electric, w/controller, fittings, relief valve			
3550	4" pump, 30 HP, 500 GPM	16,900	4,250	21,150
3600	5" pump, 40 H.P., 1000 G.P.M.	22,300	4,825	27,125
3650	5" pump, 100 H.P., 1000 G.P.M.	27,500	5,350	32,850
3700	For jockey pump system, add	3,350	685	4,035
5000	Hose, per linear foot, synthetic jacket, lined,			
5100	300 lb. test, 1-1/2" diameter	3.64	.52	4.16
5150	2-1/2" diameter	6.50	.62	7.12
5200	500 lb. test, 1-1/2" diameter	2.85	.52	3.37
5250	2-1/2" diameter	6.65	.62	7.27
5500	Nozzle, plain stream, polished brass, 1-1/2" x 10"	63.50		63.50
5550	2-1/2" x 15" x 13/16" or 1-1/2"	119		119
5600	Heavy duty combination adjustable fog and straight stream w/handle 1-1/2"	485		485
5650	2-1/2" direct connection	520		520
6000	Rack, for 1-1/2" diameter hose 100 ft. long, steel	97.50	68.50	166
6050	Brass	168	68.50	236.50
6500	Reel, steel, for 50 ft. long 1-1/2" diameter hose	174	97.50	271.50
6550	For 75 ft. long 2-1/2" diameter hose	325	97.50	422.50
7050	Siamese, w/plugs & chains, polished brass, sidewalk, 4" x 2-1/2" x 2-1/2"	825	545	1,370
7100	6" x 2-1/2" x 2-1/2"	910	685	1,595
7200	Wall type, flush, 4" x 2-1/2" x 2-1/2"	850	273	1,123
7250	6" x 2-1/2" x 2-1/2"	940	297	1,237
7300	Projecting, 4" x 2-1/2" x 2-1/2"	670	273	943
7350	6" x 2-1/2" x 2-1/2"	1,125	297	1,422
7400	For chrome plate, add 15% mat.			
8000	Valves, angle, wheel handle, 300 Lb., rough brass, 1-1/2"	116	63	179
8050	2-1/2"	217	108	325
8100	Combination pressure restricting, 1-1/2"	121	63	184
8150	2-1/2"	222	108	330
8200	Pressure restricting, adjustable, satin brass, 1-1/2"	355	63	418
8250	2-1/2"	490	108	598
8300	Hydrolator, vent and drain, rough brass, 1-1/2"	126	63	189
8350	2-1/2"	126	63	189
8400	Cabinet assy, incls. adapter, rack, hose, and nozzle	1,025	410	1,435

D4090 Other Fire Protection Systems

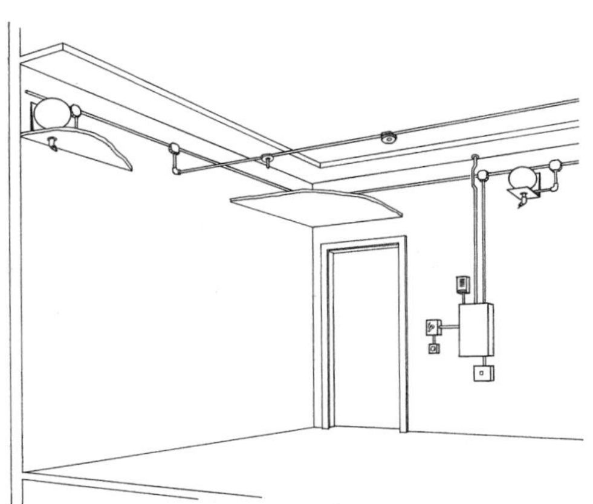

General: Automatic fire protection (suppression) systems other than water sprinklers may be desired for special environments, high risk areas, isolated locations or unusual hazards. Some typical applications would include:

Paint dip tanks
Securities vaults
Electronic data processing
Tape and data storage
Transformer rooms
Spray booths
Petroleum storage
High rack storage

Piping and wiring costs are dependent on the individual application and must be added to the component costs shown below.

All areas are assumed to be open.

D4090 910	Fire Suppression Unit Components	COST EACH		
		MAT.	INST.	TOTAL
0020	Detectors with brackets			
0040	Fixed temperature heat detector	46.50	104	150.50
0060	Rate of temperature rise detector	57	91.50	148.50
0080	Ion detector (smoke) detector	133	118	251
0200	Extinguisher agent			
0240	200 lb FM200, container	7,700	345	8,045
0280	75 lb carbon dioxide cylinder	1,625	231	1,856
0320	Dispersion nozzle			
0340	FM200 1-1/2" dispersion nozzle	260	55	315
0380	Carbon dioxide 3" x 5" dispersion nozzle	182	43	225
0420	Control station			
0440	Single zone control station with batteries	1,150	730	1,880
0470	Multizone (4) control station with batteries	3,375	1,450	4,825
0490				
0500	Electric mechanical release	1,150	375	1,525
0520				
0550	Manual pull station	102	129	231
0570				
0640	Battery standby power 10" x 10" x 17"	510	183	693
0700				
0740	Bell signalling device	73	91.50	164.50

D4090 920	FM200 Systems	COST PER C.F.		
		MAT.	INST.	TOTAL
0820	Average FM200 system, minimum			2.09
0840	Maximum			4.16

G1030 Site Earthwork

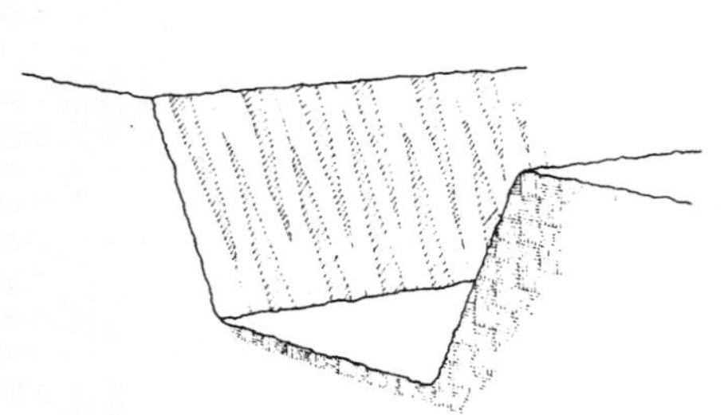

Trenching Systems are shown on a cost per linear foot basis. The systems include: excavation; backfill and removal of spoil; and compaction for various depths and trench bottom widths. The backfill has been reduced to accommodate a pipe of suitable diameter and bedding.

The slope for trench sides varies from none to 1:1.

The Expanded System Listing shows Trenching Systems that range from 2′ to 12′ in width. Depths range from 2′ to 25′.

System Components	QUANTITY	UNIT	COST PER L.F. EQUIP.	COST PER L.F. LABOR	COST PER L.F. TOTAL
SYSTEM G1030 805 1310					
TRENCHING, COMMON EARTH, NO SLOPE, 2′ WIDE, 2′ DP, 3/8 C.Y. BUCKET					
Excavation, trench, hyd. backhoe, track mtd., 3/8 C.Y. bucket	.148	B.C.Y.	.23	1.17	1.40
Backfill and load spoil, from stockpile	.153	L.C.Y.	.13	.36	.49
Compaction by vibrating plate, 6″ lifts, 4 passes	.118	E.C.Y.	.03	.43	.46
Remove excess spoil, 8 C.Y. dump truck, 2 mile roundtrip	.040	L.C.Y.	.16	.20	.36
TOTAL			.55	2.16	2.71

G1030 805	Trenching Common Earth	COST PER L.F. EQUIP.	COST PER L.F. LABOR	COST PER L.F. TOTAL
1310	Trenching, common earth, no slope, 2′ wide, 2′ deep, 3/8 C.Y. bucket	.55	2.16	2.71
1320	3′ deep, 3/8 C.Y. bucket	.77	3.22	3.99
1330	4′ deep, 3/8 C.Y. bucket	.98	4.31	5.29
1340	6′ deep, 3/8 C.Y. bucket	1.28	5.55	6.83
1350	8′ deep, 1/2 C.Y. bucket	1.65	7.35	9
1360	10′ deep, 1 C.Y. bucket	3.57	8.80	12.37
1400	4′ wide, 2′ deep, 3/8 C.Y. bucket	1.29	4.28	5.57
1410	3′ deep, 3/8 C.Y. bucket	1.72	6.40	8.12
1420	4′ deep, 1/2 C.Y. bucket	1.92	7.15	9.07
1430	6′ deep, 1/2 C.Y. bucket	3.18	11.50	14.68
1440	8′ deep, 1/2 C.Y. bucket	6.65	14.95	21.60
1450	10′ deep, 1 C.Y. bucket	8.10	18.55	26.65
1460	12′ deep, 1 C.Y. bucket	10.45	23.50	33.95
1470	15′ deep, 1-1/2 C.Y. bucket	7.65	21	28.65
1480	18′ deep, 2-1/2 C.Y. bucket	12.65	29.50	42.15
1520	6′ wide, 6′ deep, 5/8 C.Y. bucket w/trench box	9	17.30	26.30
1530	8′ deep, 3/4 C.Y. bucket	11.65	22.50	34.15
1540	10′ deep, 1 C.Y. bucket	11.35	23.50	34.85
1550	12′ deep, 1-1/2 C.Y. bucket	10.30	25	35.30
1560	16′ deep, 2-1/2 C.Y. bucket	16	31.50	47.50
1570	20′ deep, 3-1/2 C.Y. bucket	20.50	37.50	58
1580	24′ deep, 3-1/2 C.Y. bucket	24	45	69
1640	8′ wide, 12′ deep, 1-1/2 C.Y. bucket w/trench box	14.35	32	46.35
1650	15′ deep, 1-1/2 C.Y. bucket	18.60	42	60.60
1660	18′ deep, 2-1/2 C.Y. bucket	23	42	65
1680	24′ deep, 3-1/2 C.Y. bucket	32.50	58.50	91
1730	10′ wide, 20′ deep, 3-1/2 C.Y. bucket w/trench box	26.50	55.50	82
1740	24′ deep, 3-1/2 C.Y. bucket	39	66.50	105.50
1780	12′ wide, 20′ deep, 3-1/2 C.Y. bucket w/trench box	41.50	70.50	112
1790	25′ deep, bucket	51.50	89.50	141
1800	1/2 to 1 slope, 2′ wide, 2′ deep, 3/8 C.Y. bucket	.77	3.22	3.99
1810	3′ deep, 3/8 C.Y. bucket	1.25	5.65	6.90

G1030 Site Earthwork

G1030 805	Trenching Common Earth	COST PER L.F.		
		EQUIP.	LABOR	TOTAL
1820	4′ deep, 3/8 C.Y. bucket	1.84	8.60	10.44
1840	6′ deep, 3/8 C.Y. bucket	3.02	13.95	16.97
1860	8′ deep, 1/2 C.Y. bucket	4.75	22	26.75
1880	10′ deep, 1 C.Y. bucket	12.30	31	43.30
2300	4′ wide, 2′ deep, 3/8 C.Y. bucket	1.50	5.35	6.85
2310	3′ deep, 3/8 C.Y. bucket	2.20	8.85	11.05
2320	4′ deep, 1/2 C.Y. bucket	2.70	10.90	13.60
2340	6′ deep, 1/2 C.Y. bucket	5.30	20.50	25.80
2360	8′ deep, 1/2 C.Y. bucket	12.95	30	42.95
2380	10′ deep, 1 C.Y. bucket	18	42.50	60.50
2400	12′ deep, 1 C.Y. bucket	23.50	57.50	81
2430	15′ deep, 1-1/2 C.Y. bucket	21.50	61.50	83
2460	18′ deep, 2-1/2 C.Y. bucket	43.50	96	139.50
2840	6′ wide, 6′ deep, 5/8 C.Y. bucket w/trench box	13	25.50	38.50
2860	8′ deep, 3/4 C.Y. bucket	18.90	38.50	57.40
2880	10′ deep, 1 C.Y. bucket	15.45	39	54.45
2900	12′ deep, 1-1/2 C.Y. bucket	19.60	51	70.60
2940	16′ deep, 2-1/2 C.Y. bucket	36.50	75	111.50
2980	20′ deep, 3-1/2 C.Y. bucket	51	101	152
3020	24′ deep, 3-1/2 C.Y. bucket	72	138	210
3100	8′ wide, 12′ deep, 1-1/2 C.Y. bucket w/trench box	24	58	82
3120	15′ deep, 1-1/2 C.Y. bucket	35	84.50	119.50
3140	18′ deep, 2-1/2 C.Y. bucket	50.50	100	150.50
3180	24′ deep, 3-1/2 C.Y. bucket	80.50	151	231.50
3270	10′ wide, 20′ deep, 3-1/2 C.Y. bucket w/trench box	52	116	168
3280	24′ deep, 3-1/2 C.Y. bucket	89	165	254
3370	12′ wide, 20′ deep, 3-1/2 C.Y. bucket w/trench box	75	135	210
3380	25′ deep, 3-1/2 C.Y. bucket	104	193	297
3500	1 to 1 slope, 2′ wide, 2′ deep, 3/8 C.Y. bucket	.99	4.31	5.30
3520	3′ deep, 3/8 C.Y. bucket	3.14	9.85	12.99
3540	4′ deep, 3/8 C.Y. bucket	2.70	12.95	15.65
3560	6′ deep, 1/2 C.Y. bucket	3.02	13.90	16.92
3580	8′ deep, 1/2 C.Y. bucket	5.90	28	33.90
3600	10′ deep, 1 C.Y. bucket	21	53.50	74.50
3800	4′ wide, 2′ deep, 3/8 C.Y. bucket	1.72	6.45	8.17
3820	3′ deep, 3/8 C.Y. bucket	2.69	11.25	13.94
3840	4′ deep, 1/2 C.Y. bucket	3.47	14.60	18.07
3860	6′ deep, 1/2 C.Y. bucket	7.50	29.50	37
3880	8′ deep, 1/2 C.Y. bucket	19.30	45.50	64.80
3900	10′ deep, 1 C.Y. bucket	28	66	94
3920	12′ deep, 1 C.Y. bucket	41	95.50	136.50
3940	15′ deep, 1-1/2 C.Y. bucket	35.50	102	137.50
3960	18′ deep, 2-1/2 C.Y. bucket	59.50	132	191.50
4030	6′ wide, 6′ deep, 5/8 C.Y. bucket w/trench box	17.15	34.50	51.65
4040	8′ deep, 3/4 C.Y. bucket	25	47.50	72.50
4050	10′ deep, 1 C.Y. bucket	22.50	57.50	80
4060	12′ deep, 1-1/2 C.Y. bucket	30	77.50	107.50
4070	16′ deep, 2-1/2 C.Y. bucket	57.50	118	175.50
4080	20′ deep, 3-1/2 C.Y. bucket	82.50	164	246.50
4090	24′ deep, 3-1/2 C.Y. bucket	120	231	351
4500	8′ wide, 12′ deep, 1-1/2 C.Y. bucket w/trench box	34	84.50	118.50
4550	15′ deep, 1-1/2 C.Y. bucket	51	127	178
4600	18′ deep, 2-1/2 C.Y. bucket	76.50	155	231.50
4650	24′ deep, 3-1/2 C.Y. bucket	129	244	373
4800	10′ wide, 20′ deep, 3-1/2 C.Y. bucket w/trench box	77.50	177	254.50
4850	24′ deep, 3-1/2 C.Y. bucket	137	258	395
4950	12′ wide, 20′ deep, 3-1/2 C.Y. bucket w/trench box	108	199	307
4980	25′ deep, 3-1/2 C.Y. bucket	155	291	446

G1030 Site Earthwork

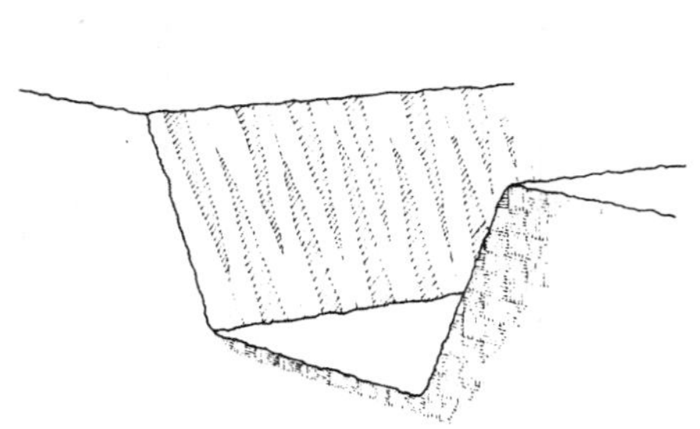

Trenching Systems are shown on a cost per linear foot basis. The systems include: excavation; backfill and removal of spoil; and compaction for various depths and trench bottom widths. The backfill has been reduced to accommodate a pipe of suitable diameter and bedding.

The slope for trench sides varies from none to 1:1.

The Expanded System Listing shows Trenching Systems that range from 2′ to 12′ in width. Depths range from 2′ to 25′.

System Components	QUANTITY	UNIT	COST PER L.F. EQUIP.	COST PER L.F. LABOR	COST PER L.F. TOTAL
SYSTEM G1030 806 1310					
TRENCHING, LOAM & SANDY CLAY, NO SLOPE, 2′ WIDE, 2′ DP, 3/8 C.Y. BUCKET					
Excavation, trench, hyd. backhoe, track mtd., 3/8 C.Y. bucket	.148	B.C.Y.	.21	1.08	1.29
Backfill and load spoil, from stockpile	.165	L.C.Y.	.14	.39	.53
Compaction by vibrating plate 18″ wide, 6″ lifts, 4 passes	.118	E.C.Y.	.03	.43	.46
Remove excess spoil, 8 C.Y. dump truck, 2 mile roundtrip	.042	L.C.Y.	.16	.21	.37
TOTAL			.54	2.11	2.65

G1030 806	Trenching Loam & Sandy Clay	COST PER L.F. EQUIP.	COST PER L.F. LABOR	COST PER L.F. TOTAL
1310	Trenching, loam & sandy clay, no slope, 2′ wide, 2′ deep, 3/8 C.Y. bucket	.54	2.11	2.65
1320	3′ deep, 3/8 C.Y. bucket	.81	3.44	4.25
1330	4′ deep, 3/8 C.Y. bucket	.97	4.20	5.15
1340	6′ deep, 3/8 C.Y. bucket	1.80	5	6.80
1350	8′ deep, 1/2 C.Y. bucket	2.35	6.60	8.95
1360	10′ deep, 1 C.Y. bucket	2.63	7.05	9.70
1400	4′ wide, 2′ deep, 3/8 C.Y. bucket	1.33	4.19	5.50
1410	3′ deep, 3/8 C.Y. bucket	1.74	6.30	8.05
1420	4′ deep, 1/2 C.Y. bucket	1.94	7.05	9
1430	6′ deep, 1/2 C.Y. bucket	4.26	10.40	14.65
1440	8′ deep, 1/2 C.Y. bucket	6.50	14.75	21.50
1450	10′ deep, 1 C.Y. bucket	6.25	15.05	21.50
1460	12′ deep, 1 C.Y. bucket	7.85	18.75	26.50
1470	15′ deep, 1-1/2 C.Y. bucket	7.95	22	30
1480	18′ deep, 2-1/2 C.Y. bucket	10.75	24	35
1520	6′ wide, 6′ deep, 5/8 C.Y. bucket w/trench box	8.75	16.95	25.50
1530	8′ deep, 3/4 C.Y. bucket	11.20	22	33
1540	10′ deep, 1 C.Y. bucket	10.40	22.50	33
1550	12′ deep, 1-1/2 C.Y. bucket	10.15	25	35
1560	16′ deep, 2-1/2 C.Y. bucket	15.65	32	47.50
1570	20′ deep, 3-1/2 C.Y. bucket	18.75	37.50	56.50
1580	24′ deep, 3-1/2 C.Y. bucket	23.50	46	69.50
1640	8′ wide, 12′ deep, 1-1/4 C.Y. bucket w/trench box	14.30	32	46.50
1650	15′ deep, 1-1/2 C.Y. bucket	17.50	41	58.50
1660	18′ deep, 2-1/2 C.Y. bucket	23.50	46	69.50
1680	24′ deep, 3-1/2 C.Y. bucket	32	59	91
1730	10′ wide, 20′ deep, 3-1/2 C.Y. bucket w/trench box	32	59.50	91.50
1740	24′ deep, 3-1/2 C.Y. bucket	40	73.50	114
1780	12′ wide, 20′ deep, 3-1/2 C.Y. bucket w/trench box	39	70.50	110
1790	25′ deep, 3-1/2 C.Y. bucket	50.50	91	142
1800	1/2:1 slope, 2′ wide, 2′ deep, 3/8 C.Y. bucket	.75	3.15	3.90
1810	3′ deep, 3/8 C.Y. bucket	1.23	5.50	6.75
1820	4′ deep, 3/8 C.Y. bucket	1.81	8.40	10.20
1840	6′ deep, 3/8 C.Y. bucket	4.33	12.50	16.85

G1030 Site Earthwork

G1030 806	Trenching Loam & Sandy Clay	COST PER L.F. EQUIP.	LABOR	TOTAL
1860	8' deep, 1/2 C.Y. bucket	6.85	20	27
1880	10' deep, 1 C.Y. bucket	9	25	34
2300	4' wide, 2' deep, 3/8 C.Y. bucket	1.54	5.25	6.80
2310	3' deep, 3/8 C.Y. bucket	2.23	8.65	10.90
2320	4' deep, 1/2 C.Y. bucket	2.71	10.75	13.45
2340	6' deep, 1/2 C.Y. bucket	7.20	18.45	25.50
2360	8' deep, 1/2 C.Y. bucket	12.60	30	42.50
2380	10' deep, 1 C.Y. bucket	13.75	34.50	48.50
2400	12' deep, 1 C.Y. bucket	23	57	80
2430	15' deep, 1-1/2 C.Y. bucket	22.50	63.50	86
2460	18' deep, 2-1/2 C.Y. bucket	42.50	97.50	140
2840	6' wide, 6' deep, 5/8 C.Y. bucket w/trench box	12.65	25.50	38
2860	8' deep, 3/4 C.Y. bucket	18.20	37.50	55.50
2880	10' deep, 1 C.Y. bucket	18.70	42.50	61
2900	12' deep, 1-1/2 C.Y. bucket	19.80	51.50	71.50
2940	16' deep, 2-1/2 C.Y. bucket	36	75.50	112
2980	20' deep, 3-1/2 C.Y. bucket	49.50	102	152
3020	24' deep, 3-1/2 C.Y. bucket	70	140	210
3100	8' wide, 12' deep, 1-1/2 C.Y. bucket w/trench box	24	58.50	82.50
3120	15' deep, 1-1/2 C.Y. bucket	32.50	82	115
3140	18' deep, 2-1/2 C.Y. bucket	49	101	150
3180	24' deep, 3-1/2 C.Y. bucket	78.50	153	232
3270	10' wide, 20' deep, 3-1/2 C.Y. bucket w/trench box	63	124	187
3280	24' deep, 3-1/2 C.Y. bucket	86.50	168	255
3320	12' wide, 20' deep, 3-1/2 C.Y. bucket w/trench box	69.50	135	205
3380	25' deep, 3-1/2 C.Y. bucket w/trench box	95	181	276
3500	1:1 slope, 2' wide, 2' deep, 3/8 C.Y. bucket	.97	4.20	5.15
3520	3' deep, 3/8 C.Y. bucket	1.71	7.90	9.60
3540	4' deep, 3/8 C.Y. bucket	2.66	12.60	15.25
3560	6' deep, 1/2 C.Y. bucket	4.33	12.50	16.85
3580	8' deep, 1/2 C.Y. bucket	11.35	33.50	45
3600	10' deep, 1 C.Y. bucket	15.35	43	58.50
3800	4' wide, 2' deep, 3/8 C.Y. bucket	1.74	6.30	8.05
3820	3' deep, 1/2 C.Y. bucket	2.71	11	13.70
3840	4' deep, 1/2 C.Y. bucket	3.48	14.40	17.90
3860	6' deep, 1/2 C.Y. bucket	10.15	26.50	36.50
3880	8' deep, 1/2 C.Y. bucket	18.75	45	64
3900	10' deep, 1 C.Y. bucket	21.50	54	75.50
3920	12' deep, 1 C.Y. bucket	30.50	76.50	107
3940	15' deep, 1-1/2 C.Y. bucket	37	105	142
3960	18' deep, 2-1/2 C.Y. bucket	58.50	134	193
4030	6' wide, 6' deep, 5/8 C.Y. bucket w/trench box	16.60	34.50	51
4040	8' deep, 3/4 C.Y. bucket	25	53	78
4050	10' deep, 1 C.Y. bucket	27	62	89
4060	12' deep, 1-1/2 C.Y. bucket	29.50	78	108
4070	16' deep, 2-1/2 C.Y. bucket	56	120	176
4080	20' deep, 3-1/2 C.Y. bucket	80.50	167	248
4090	24' deep, 3-1/2 C.Y. bucket	117	234	350
4500	8' wide, 12' deep, 1-1/4 C.Y. bucket w/trench box	33.50	85	119
4550	15' deep, 1-1/2 C.Y. bucket	48	123	171
4600	18' deep, 2-1/2 C.Y. bucket	75	157	232
4650	24' deep, 3-1/2 C.Y. bucket	125	248	375
4800	10' wide, 20' deep, 3-1/2 C.Y. bucket w/trench box	93.50	189	283
4850	24' deep, 3-1/2 C.Y. bucket	133	262	395
4950	12' wide, 20' deep, 3-1/2 C.Y. bucket w/trench box	100	200	300
4980	25' deep, 3-1/2 C.Y. bucket	151	295	445

G1030 Site Earthwork

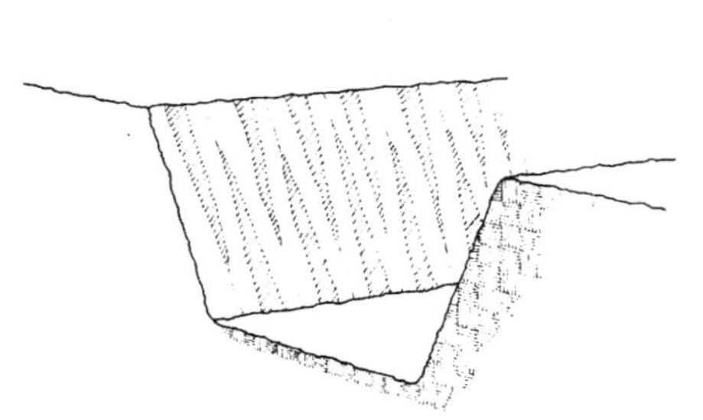

Trenching Systems are shown on a cost per linear foot basis. The systems include: excavation; backfill and removal of spoil; and compaction for various depths and trench bottom widths. The backfill has been reduced to accommodate a pipe of suitable diameter and bedding.

The slope for trench sides varies from none to 1:1.

The Expanded System Listing shows Trenching Systems that range from 2′ to 12′ in width. Depths range from 2′ to 25′.

System Components	QUANTITY	UNIT	COST PER L.F. EQUIP.	LABOR	TOTAL
SYSTEM G1030 807 1310					
TRENCHING, SAND & GRAVEL, NO SLOPE, 2′ WIDE, 2′ DEEP, 3/8 C.Y. BUCKET					
Excavation, trench, hyd. backhoe, track mtd., 3/8 C.Y. bucket	.148	B.C.Y.	.21	1.07	1.28
Backfill and load spoil, from stockpile	.140	L.C.Y.	.12	.33	.45
Compaction by vibrating plate 18″ wide, 6″ lifts, 4 passes	.118	E.C.Y.	.03	.43	.46
Remove excess spoil, 8 C.Y. dump truck, 2 mile roundtrip	.035	L.C.Y.	.14	.17	.31
TOTAL			.50	2	2.50

G1030 807	Trenching Sand & Gravel	COST PER L.F. EQUIP.	LABOR	TOTAL
1310	Trenching, sand & gravel, no slope, 2′ wide, 2′ deep, 3/8 C.Y. bucket	.50	2	2.50
1320	3′ deep, 3/8 C.Y. bucket	.75	3.32	4.07
1330	4′ deep, 3/8 C.Y. bucket	.89	3.99	4.88
1340	6′ deep, 3/8 C.Y. bucket	1.71	4.76	6.45
1350	8′ deep, 1/2 C.Y. bucket	2.23	6.30	8.55
1360	10′ deep, 1 C.Y. bucket	2.45	6.65	9.10
1400	4′ wide, 2′ deep, 3/8 C.Y. bucket	1.19	3.95	5.15
1410	3′ deep, 3/8 C.Y. bucket	1.57	5.95	7.50
1420	4′ deep, 1/2 C.Y. bucket	1.76	6.70	8.45
1430	6′ deep, 1/2 C.Y. bucket	4.03	9.90	13.95
1440	8′ deep, 1/2 C.Y. bucket	6.15	14	20
1450	10′ deep, 1 C.Y. bucket	5.85	14.20	20
1460	12′ deep, 1 C.Y. bucket	7.40	17.75	25
1470	15′ deep, 1-1/2 C.Y. bucket	7.45	20.50	28
1480	18′ deep, 2-1/2 C.Y. bucket	10.15	22.50	32.50
1520	6′ wide, 6′ deep, 5/8 C.Y. bucket w/trench box	8.25	16.05	24.50
1530	8′ deep, 3/4 C.Y. bucket	10.65	21	31.50
1540	10′ deep, 1 C.Y. bucket	9.80	21.50	31.50
1550	12′ deep, 1-1/2 C.Y. bucket	9.50	24	33.50
1560	16′ deep, 2 C.Y. bucket	14.90	30	45
1570	20′ deep, 3-1/2 C.Y. bucket	17.75	35	53
1580	24′ deep, 3-1/2 C.Y. bucket	22.50	43	65.50
1640	8′ wide, 12′ deep, 1-1/2 C.Y. bucket w/trench box	13.30	30	43.50
1650	15′ deep, 1-1/2 C.Y. bucket	16.25	38.50	55
1660	18′ deep, 2-1/2 C.Y. bucket	22.50	43	65.50
1680	24′ deep, 3-1/2 C.Y. bucket	30	55.50	85.50
1730	10′ wide, 20′ deep, 3-1/2 C.Y. bucket w/trench box	30	55.50	85.50
1740	24′ deep, 3-1/2 C.Y. bucket	37.50	68.50	106
1780	12′ wide, 20′ deep, 3-1/2 C.Y. bucket w/trench box	36.50	66	103
1790	25′ deep, 3-1/2 C.Y. bucket	47	85	132
1800	1/2:1 slope, 2′ wide, 2′ deep, 3/8 C.Y. bucket	.70	2.99	3.69
1810	3′ deep, 3/8 C.Y. bucket	1.14	5.25	6.40
1820	4′ deep, 3/8 C.Y. bucket	1.68	8	9.70
1840	6′ deep, 3/8 C.Y. bucket	4.12	11.95	16.05

G1030 Site Earthwork

G1030 807	Trenching Sand & Gravel	COST PER L.F. EQUIP.	COST PER L.F. LABOR	COST PER L.F. TOTAL
1860	8' deep, 1/2 C.Y. bucket	6.50	19.10	25.50
1880	10' deep, 1 C.Y. bucket	8.40	23.50	32
2300	4' wide, 2' deep, 3/8 C.Y. bucket	1.38	4.96	6.35
2310	3' deep, 3/8 C.Y. bucket	2.01	8.20	10.20
2320	4' deep, 1/2 C.Y. bucket	2.46	10.15	12.60
2340	6' deep, 1/2 C.Y. bucket	6.85	17.65	24.50
2360	8' deep, 1/2 C.Y. bucket	11.95	28.50	40.50
2380	10' deep, 1 C.Y. bucket	12.95	32.50	45.50
2400	12' deep, 1 C.Y. bucket	22	54	76
2430	15' deep, 1-1/2 C.Y. bucket	21	60	81
2460	18' deep, 2-1/2 C.Y. bucket	40.50	91.50	132
2840	6' wide, 6' deep, 5/8 C.Y. bucket w/trench box	12	24.50	36.50
2860	8' deep, 3/4 C.Y. bucket	17.30	35.50	53
2880	10' deep, 1 C.Y. bucket	17.70	40	57.50
2900	12' deep, 1-1/2 C.Y. bucket	18.60	48.50	67
2940	16' deep, 2 C.Y. bucket	34	71.50	106
2980	20' deep, 3-1/2 C.Y. bucket	47	96	143
3020	24' deep, 3-1/2 C.Y. bucket	66.50	132	199
3100	8' wide, 12' deep, 1-1/4 C.Y. bucket w/trench box	22.50	55	77.50
3120	15' deep, 1-1/2 C.Y. bucket	30.50	77	108
3140	18' deep, 2-1/2 C.Y. bucket	46.50	95.50	142
3180	24' deep, 3-1/2 C.Y. bucket	74	144	218
3270	10' wide, 20' deep, 3-1/2 C.Y. bucket w/trench box	59	116	175
3280	24' deep, 3-1/2 C.Y. bucket	82	157	239
3370	12' wide, 20' deep, 3-1/2 C.Y. bucket w/trench box	66	129	195
3380	25' deep, 3-1/2 C.Y. bucket	95	181	276
3500	1:1 slope, 2' wide, 2' deep, 3/8 C.Y. bucket	1.81	5.05	6.85
3520	3' deep, 3/8 C.Y. bucket	1.57	7.50	9.05
3540	4' deep, 3/8 C.Y. bucket	2.46	12.05	14.50
3560	6' deep, 3/8 C.Y. bucket	4.13	11.95	16.10
3580	8' deep, 1/2 C.Y. bucket	10.80	32	43
3600	10' deep, 1 C.Y. bucket	14.35	40.50	55
3800	4' wide, 2' deep, 3/8 C.Y. bucket	1.58	5.95	7.55
3820	3' deep, 3/8 C.Y. bucket	2.47	10.50	12.95
3840	4' deep, 1/2 C.Y. bucket	3.17	13.65	16.80
3860	6' deep, 1/2 C.Y. bucket	9.70	25.50	35
3880	8' deep, 1/2 C.Y. bucket	17.80	42.50	60.50
3900	10' deep, 1 C.Y. bucket	20	51	71
3920	12' deep, 1 C.Y. bucket	29	72.50	102
3940	15' deep, 1-1/2 C.Y. bucket	35	99.50	135
3960	18' deep, 2-1/2 C.Y. bucket	55.50	126	182
4030	6' wide, 6' deep, 5/8 C.Y. bucket w/trench box	15.75	32.50	48.50
4040	8' deep, 3/4 C.Y. bucket	24	50.50	74.50
4050	10' deep, 1 C.Y. bucket	25.50	58.50	84
4060	12' deep, 1-1/2 C.Y. bucket	27.50	73.50	101
4070	16' deep, 2 C.Y. bucket	53.50	113	167
4080	20' deep, 3-1/2 C.Y. bucket	76	157	233
4090	24' deep, 3-1/2 C.Y. bucket	110	220	330
4500	8' wide, 12' deep, 1-1/2 C.Y. bucket w/trench box	31.50	80	112
4550	15' deep, 1-1/2 C.Y. bucket	44.50	116	161
4600	18' deep, 2-1/2 C.Y. bucket	71	148	219
4650	24' deep, 3-1/2 C.Y. bucket	118	233	350
4800	10' wide, 20' deep, 3-1/2 C.Y. bucket w/trench box	88.50	177	266
4850	24' deep, 3-1/2 C.Y. bucket	126	246	370
4950	12' wide, 20' deep, 3-1/2 C.Y. bucket w/trench box	94.50	187	282
4980	25' deep, 3-1/2 C.Y. bucket	143	277	420

G1030 Site Earthwork

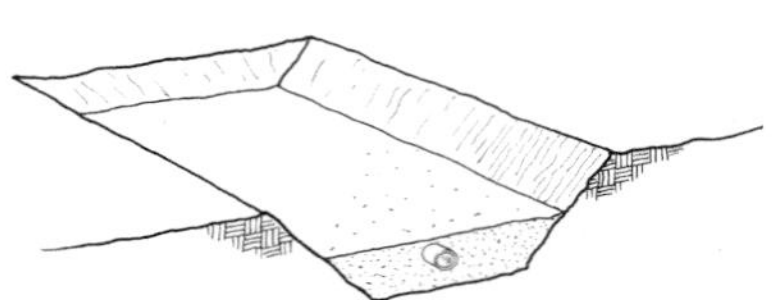

The Pipe Bedding System is shown for various pipe diameters. Compacted bank sand is used for pipe bedding and to fill 12″ over the pipe. No backfill is included. Various side slopes are shown to accommodate different soil conditions. Pipe sizes vary from 6″ to 84″ diameter.

System Components	QUANTITY	UNIT	COST PER L.F. MAT.	INST.	TOTAL
SYSTEM G1030 815 1440					
PIPE BEDDING, SIDE SLOPE 0 TO 1, 1′ WIDE, PIPE SIZE 6″ DIAMETER					
Borrow, bank sand, 2 mile haul, machine spread	.086	C.Y.	1.71	.72	2.43
Compaction, vibrating plate	.086	C.Y.		.23	.23
TOTAL			1.71	.95	2.66

G1030 815	Pipe Bedding	COST PER L.F. MAT.	INST.	TOTAL
1440	Pipe bedding, side slope 0 to 1, 1′ wide, pipe size 6″ diameter	1.71	.95	2.66
1460	2′ wide, pipe size 8″ diameter	3.70	2.05	5.75
1480	Pipe size 10″ diameter	3.78	2.09	5.87
1500	Pipe size 12″ diameter	3.86	2.14	6
1520	3′ wide, pipe size 14″ diameter	6.25	3.46	9.71
1540	Pipe size 15″ diameter	6.30	3.50	9.80
1560	Pipe size 16″ diameter	6.40	3.53	9.93
1580	Pipe size 18″ diameter	6.50	3.60	10.10
1600	4′ wide, pipe size 20″ diameter	9.25	5.10	14.35
1620	Pipe size 21″ diameter	9.30	5.15	14.45
1640	Pipe size 24″ diameter	9.50	5.25	14.75
1660	Pipe size 30″ diameter	9.70	5.35	15.05
1680	6′ wide, pipe size 32″ diameter	16.60	9.15	25.75
1700	Pipe size 36″ diameter	17	9.40	26.40
1720	7′ wide, pipe size 48″ diameter	27	14.95	41.95
1740	8′ wide, pipe size 60″ diameter	33	18.20	51.20
1760	10′ wide, pipe size 72″ diameter	45.50	25.50	71
1780	12′ wide, pipe size 84″ diameter	60.50	33.50	94
2140	Side slope 1/2 to 1, 1′ wide, pipe size 6″ diameter	3.19	1.77	4.96
2160	2′ wide, pipe size 8″ diameter	5.40	3.01	8.41
2180	Pipe size 10″ diameter	5.80	3.22	9.02
2200	Pipe size 12″ diameter	6.15	3.41	9.56
2220	3′ wide, pipe size 14″ diameter	8.90	4.93	13.83
2240	Pipe size 15″ diameter	9.10	5.05	14.15
2260	Pipe size 16″ diameter	9.35	5.15	14.50
2280	Pipe size 18″ diameter	9.80	5.45	15.25
2300	4′ wide, pipe size 20″ diameter	12.95	7.15	20.10
2320	Pipe size 21″ diameter	13.25	7.35	20.60
2340	Pipe size 24″ diameter	14.10	7.80	21.90
2360	Pipe size 30″ diameter	15.60	8.65	24.25
2380	6′ wide, pipe size 32″ diameter	23	12.75	35.75
2400	Pipe size 36″ diameter	24.50	13.55	38.05
2420	7′ wide, pipe size 48″ diameter	38	21	59
2440	8′ wide, pipe size 60″ diameter	48.50	27	75.50
2460	10′ wide, pipe size 72″ diameter	66.50	37	103.50
2480	12′ wide, pipe size 84″ diameter	87	48.50	135.50
2620	Side slope 1 to 1, 1′ wide, pipe size 6″ diameter	4.67	2.58	7.25
2640	2′ wide, pipe size 8″ diameter	7.20	3.98	11.18

G1030 Site Earthwork

G1030 815	Pipe Bedding	COST PER L.F. MAT.	COST PER L.F. INST.	COST PER L.F. TOTAL
2660	Pipe size 10" diameter	7.80	4.33	12.13
2680	Pipe size 12" diameter	8.50	4.72	13.22
2700	3' wide, pipe size 14" diameter	11.55	6.40	17.95
2720	Pipe size 15" diameter	11.90	6.60	18.50
2740	Pipe size 16" diameter	12.30	6.80	19.10
2760	Pipe size 18" diameter	13.15	7.25	20.40
2780	4' wide, pipe size 20" diameter	16.70	9.25	25.95
2800	Pipe size 21" diameter	17.15	9.50	26.65
2820	Pipe size 24" diameter	18.60	10.30	28.90
2840	Pipe size 30" diameter	21.50	11.90	33.40
2860	6' wide, pipe size 32" diameter	29.50	16.30	45.80
2880	Pipe size 36" diameter	32	17.70	49.70
2900	7' wide, pipe size 48" diameter	49.50	27.50	77
2920	8' wide, pipe size 60" diameter	64	35.50	99.50
2940	10' wide, pipe size 72" diameter	87.50	48.50	136
2960	12' wide, pipe size 84" diameter	114	63	177

G2040 Site Development

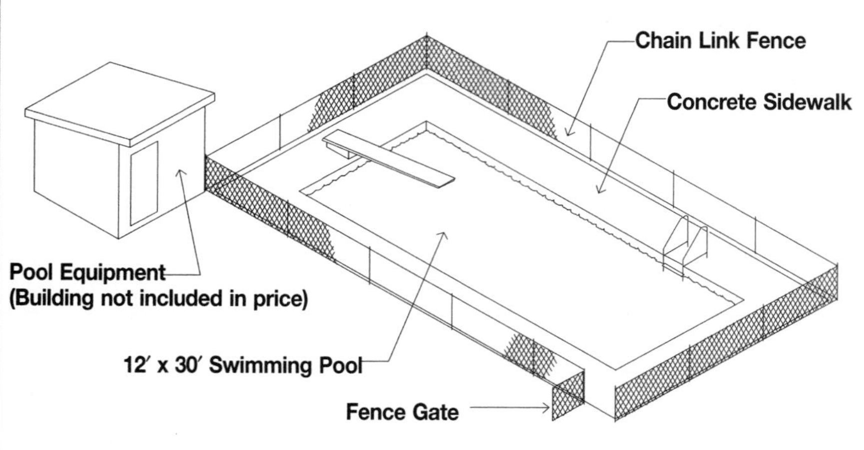

The Swimming Pool System is a complete package. Everything from excavation to deck hardware is included in system costs. Below are three basic types of pool systems: residential, motel, and municipal. Systems elements include: excavation, pool materials, installation, deck hardware, pumps and filters, sidewalk, and fencing.
The Expanded System Listing shows three basic types of pools with a variety of finishes and basic materials. These systems are either vinyl lined with metal sides; gunite shell with a cement plaster finish or tile finish; or concrete sided with vinyl lining. Pool sizes listed here vary from 12′ x 30′ to 60′ x 82.5′. All costs are on a per unit basis.

System Components	QUANTITY	UNIT	COST EACH MAT.	COST EACH INST.	COST EACH TOTAL
SYSTEM G2040 920 1000 SWIMMING POOL, RESIDENTIAL, CONC. SIDES, VINYL LINED, 12′ X 30′					
Swimming pool, residential, in-ground including equipment	360.000	S.F.	11,340	5,666.40	17,006.40
4″ thick reinforced concrete sidewalk, broom finish, no base	400.000	S.F.	1,016	1,156	2,172
Chain link fence, residential, 4′ high	124.000	L.F.	762.60	575.36	1,337.96
Fence gate, chain link, 4′ high	1.000	Ea.	101	185.50	286.50
TOTAL			13,219.60	7,583.26	20,802.86

G2040 920	Swimming Pools	COST EACH MAT.	COST EACH INST.	COST EACH TOTAL
1000	Swimming pool, residential class, concrete sides, vinyl lined, 12′ x 30′	13,200	7,575	20,775
1100	16′ x 32′	15,000	8,575	23,575
1200	20′ x 40′	19,300	10,900	30,200
1500	Tile finish, 12′ x 30′	28,800	31,900	60,700
1600	16′ x 32′	29,300	33,600	62,900
1700	20′ x 40′	43,800	47,300	91,100
2000	Metal sides, vinyl lined, 12′ x 30′	11,400	4,925	16,325
2100	16′ x 32′	12,900	5,525	18,425
2200	20′ x 40′	16,700	6,950	23,650
3000	Gunite shell, cement plaster finish, 12′ x 30′	22,900	13,600	36,500
3100	16′ x 32′	29,000	17,900	46,900
3200	20′ x 40′	40,100	17,600	57,700
4000	Motel class, concrete sides, vinyl lined, 20′ x 40′	27,700	15,100	42,800
4100	28′ x 60′	38,800	21,100	59,900
4500	Tile finish, 20′ x 40′	51,500	54,500	106,000
4600	28′ x 60′	81,500	88,500	170,000
5000	Metal sides, vinyl lined, 20′ x 40′	29,000	10,800	39,800
5100	28′ x 60′	38,900	14,500	53,400
6000	Gunite shell, cement plaster finish, 20′ x 40′	74,500	43,700	118,200
6100	28′ x 60′	111,500	65,000	176,500
7000	Municipal class, gunite shell, cement plaster finish, 42′ x 75′	297,500	154,000	451,500
7100	60′ x 82.5′	382,000	197,500	579,500
7500	Concrete walls, tile finish, 42′ x 75′	332,000	207,000	539,000
7600	60′ x 82.5′	432,000	275,000	707,000
7700	Tile finish and concrete gutter, 42′ x 75′	364,000	207,000	571,000
7800	60′ x 82.5′	471,000	275,000	746,000
7900	Tile finish and stainless gutter, 42′ x 75′	427,000	207,000	634,000
8000	60′ x 82.5′	631,500	317,500	949,000

G2050 Landscaping

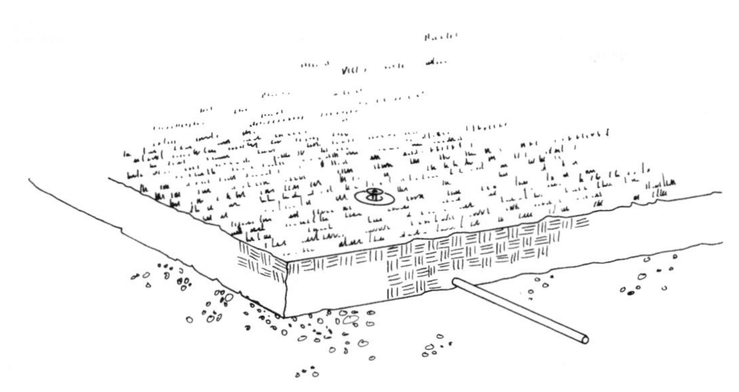

There are three basic types of Site Irrigation Systems: pop-up, riser mounted and quick coupling. Sprinkler heads are spray, impact or gear driven. Each system includes: the hardware for spraying the water; the pipe and fittings needed to deliver the water; and all other accessory equipment such as valves, couplings, nipples, and nozzles. Excavation heads and backfill costs are also included in the system.
The Expanded System Listing shows a wide variety of Site Irrigation Systems.

System Components	QUANTITY	UNIT	COST PER S.F. MAT.	INST.	TOTAL
SYSTEM G2050 710 1000					
SITE IRRIGATION, POP UP SPRAY, PLASTIC, 10′ RADIUS, 1000 S.F., PVC PIPE					
Excavation, chain trencher	54.000	L.F.		54.54	54.54
Pipe, fittings & nipples, PVC Schedule 40, 1″ diameter	5.000	Ea.	4.55	122.50	127.05
Fittings, bends or elbows, 4″ diameter	5.000	Ea.	62.25	277.50	339.75
Couplings PVC plastic, high pressure, 1″ diameter	5.000	Ea.	4.55	155	159.55
Valves, bronze, globe, 125 lb. rising stem, threaded, 1″ diameter	1.000	Ea.	263	40.50	303.50
Head & nozzle, pop-up spray, PVC plastic	5.000	Ea.	21.30	110	131.30
Backfill by hand with compaction	54.000	L.F.		37.80	37.80
Total cost per 1,000 S.F.			355.65	797.84	1,153.49
Total cost per S.F.			.36	.80	1.15

G2050 710	Site Irrigation	COST PER S.F. MAT.	INST.	TOTAL
1000	Site irrigation, pop up spray, 10′ radius, 1000 S.F., PVC pipe	.36	.80	1.16
1100	Polyethylene pipe	.38	.69	1.07
1200	14′ radius, 8000 S.F., PVC pipe	.13	.34	.47
1300	Polyethylene pipe	.12	.31	.43
1400	18′ square, 1000 S.F. PVC pipe	.35	.53	.88
1500	Polyethylene pipe	.35	.46	.81
1600	24′ square, 8000 S.F., PVC pipe	.11	.22	.33
1700	Polyethylene pipe	.10	.21	.31
1800	4′ x 30′ strip, 200 S.F., PVC pipe	1.69	2.25	3.94
1900	Polyethylene pipe	1.68	2.09	3.77
2000	6′ x 40′ strip, 800 S.F., PVC pipe	.45	.75	1.20
2200	Economy brass, 11′ radius, 1000 S.F., PVC pipe	.45	.76	1.21
2300	Polyethylene pipe	.45	.69	1.14
2400	14′ radius, 8000 S.F., PVC pipe	.16	.34	.50
2500	Polyethylene pipe	.15	.31	.46
2600	3′ x 28′ strip, 200 S.F., PVC pipe	1.90	2.25	4.15
2700	Polyethylene pipe	1.89	2.09	3.98
2800	7′ x 36′ strip, 1000 S.F., PVC pipe	.41	.59	1
2900	Polyethylene pipe	.41	.54	.95
3000	Hd brass, 11′ radius, 1000 S.F., PVC pipe	.49	.76	1.25
3100	Polyethylene pipe	.49	.69	1.18
3200	14′ radius, 8000 S.F., PVC pipe	.18	.34	.52
3300	Polyethylene pipe	.17	.31	.48
3400	Riser mounted spray, plastic, 10′ radius, 1000 S.F., PVC pipe	.38	.76	1.14
3500	Polyethylene pipe	.38	.69	1.07
3600	12′ radius, 5000 S.F., PVC pipe	.19	.46	.65

G2050 Landscaping

G2050 710	Site Irrigation	COST PER S.F.		
		MAT.	INST.	TOTAL
3700	Polyethylene pipe	.19	.43	.62
3800	19′ square, 2000 S.F., PVC pipe	.18	.24	.42
3900	Polyethylene pipe	.17	.22	.39
4000	24′ square, 8000 S.F., PVC pipe	.11	.22	.33
4100	Polyethylene pipe	.10	.21	.31
4200	5′ x 32′ strip, 300 S.F., PVC pipe	1.13	1.52	2.65
4300	Polyethylene pipe	1.13	1.41	2.54
4400	6′ x 40′ strip, 800 S.F., PVC pipe	.45	.75	1.20
4500	Polyethylene pipe	.45	.70	1.15
4600	Brass, 11′ radius, 1000 S.F., PVC pipe	.43	.76	1.19
4700	Polyethylene pipe	.43	.69	1.12
4800	14′ radius, 8000 S.F., PVC pipe	.15	.34	.49
4900	Polyethylene pipe	.14	.31	.45
5000	Pop up gear drive stream type, plastic, 30′ radius, 10,000 S.F., PVC pipe	.40	.66	1.06
5020	Polyethylene pipe	.15	.20	.35
5040	40,000 S.F., PVC pipe	.36	.66	1.02
5060	Polyethylene pipe	.13	.18	.31
5080	Riser mounted gear drive stream type, plas., 30′ rad., 10,000 S.F.,PVC pipe	.39	.66	1.05
5100	Polyethylene pipe	.14	.20	.34
5120	40,000 S.F., PVC pipe	.35	.66	1.01
5140	Polyethylene pipe	.12	.18	.30
5200	Q.C. valve thread type w/impact head, brass, 75′ rad, 20,000 S.F., PVC pipe	.17	.25	.42
5250	Polyethylene pipe	.07	.06	.13
5300	100,000 S.F., PVC pipe	.25	.41	.66
5350	Polyethylene pipe	.10	.09	.19
5400	Q.C. valve lug type w/impact head, brass, 75′ rad, 20,000 S.F., PVC pipe	.17	.25	.42
5450	Polyethylene pipe	.07	.06	.13
5500	100,000 S.F. PVC pipe	.25	.41	.66
5550	Polyethylene pipe	.10	.09	.19
6000	Site irrigation, pop up impact type, plastic, 40′ rad.,10000 S.F., PVC pipe	.25	.43	.68
6100	Polyethylene pipe	.08	.12	.20
6200	45,000 S.F., PVC pipe	.24	.47	.71
6300	Polyethylene pipe	.07	.12	.19
6400	High medium volume brass, 40′ radius, 10000 S.F., PVC pipe	.26	.43	.69
6500	Polyethylene pipe	.09	.12	.21
6600	45,000 S.F. PVC pipe	.25	.47	.72
6700	Polyethylene pipe	.08	.12	.20
6800	High volume brass, 60′ radius, 25,000 S.F., PVC pipe	.21	.33	.54
6900	Polyethylene pipe	.07	.08	.15
7000	100,000 S.F., PVC pipe	.20	.33	.53
7100	Polyethylene pipe	.08	.08	.16
7200	Riser mounted part/full impact type, plas., 40′ rad., 10,000 S.F., PVC pipe	.25	.43	.68
7300	Polyethylene pipe	.09	.26	.35
7400	45,000 S.F., PVC pipe	.24	.47	.71
7500	Polyethylene pipe	.07	.12	.19
7600	Low medium volume brass, 40′ radius, 10,000 S.F., PVC pipe	.25	.43	.68
7700	Polyethylene pipe	.08	.12	.20
7800	45,000 S.F., PVC pipe	.24	.47	.71
7900	Polyethylene pipe	.07	.12	.19
8000	Medium volume brass, 50′ radius, 30,000 S.F., PVC pipe	.21	.35	.56
8100	Polyethylene pipe	.07	.09	.16
8200	70,000 S.F., PVC pipe	.31	.57	.88
8300	Polyethylene pipe	.11	.14	.25
8400	Riser mounted full only impact type, plas., 40′ rad., 10,000 S.F., PVC pipe	.24	.43	.67
8500	Polyethylene pipe	.07	.12	.19
8600	45,000 S.F., PVC pipe	.23	.47	.70
8700	Polyethylene pipe	.06	.12	.18
8800	Low medium volume brass, 40′ radius, 10,000 S.F., PVC pipe	.25	.43	.68

G2050 Landscaping

G2050 710	Site Irrigation	COST PER S.F. MAT.	COST PER S.F. INST.	COST PER S.F. TOTAL
8900	Polyethylene pipe	.08	.12	.20
9000	45,000 S.F., PVC pipe	.24	.47	.71
9100	Polyethylene pipe	.07	.12	.19
9200	High volume brass, 80' radius, 75,000 S.F., PVC pipe	.15	.24	.39
9300	Polyethylene pipe	.01	.02	.03
9400	150,000 S.F., PVC pipe	.15	.22	.37
9500	Polyethylene pipe	.01	.02	.03
9600	Very high volume brass, 100' radius, 100,000 S.F., PVC pipe	.13	.18	.31
9700	Polyethylene pipe	.01	.02	.03
9800	200,000 S.F., PVC pipe	.13	.18	.31
9900	Polyethylene pipe	.01	.02	.03

G2050 Landscaping

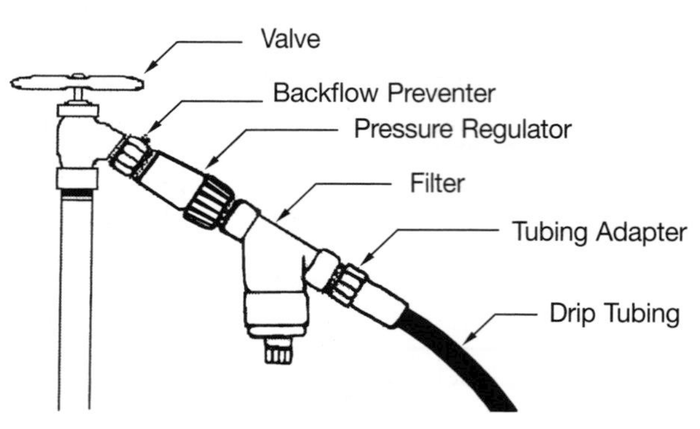

Drip irrigation systems can be for either lawns or individual plants. Shown below are drip irrigation systems for lawns. These drip type systems are most cost effective where there are water restrictions on use or the cost of water is very high. The supply uses PVC piping with the laterals polyethylene drip lines.

System Components	QUANTITY	UNIT	COST PER S.F. MAT.	INST.	TOTAL
SYSTEM G2050 720 1000					
SITE IRRIGATION, DRIP SYSTEM, 800 S.F. (20′ X 40′)					
Excavate supply and header lines 4″ x 18″ deep	50.000	L.F.		54	54
Excavate drip lines 4″ x 4″ deep	560.000	L.F.		907.20	907.20
Install supply manifold (stop valve, strainer, PRV, adapters)	1.000	Ea.	26.50	166	192.50
Install 3/4″ PVC supply line and risers to 4 ″ drip level	75.000	L.F.	24	165.75	189.75
Install 3/4″ PVC fittings for supply line	33.000	Ea.	60.39	547.80	608.19
Install PVC to drip line adapters	14.000	Ea.	26.18	232.40	258.58
Install 1/4 inch drip lines with hose clamps	560.000	L.F.	61.60	308	369.60
Backfill trenches	3.180	C.Y.		146.28	146.28
Cleanup area when completed	1.000	Ea.		127	127
Total cost per 800 S.F.			198.67	2,654.43	2,853.10
Total cost per S.F.			.25	3.32	3.57

G2050 720	Site Irrigation	COST PER S.F. MAT.	INST.	TOTAL
1000	Site irrigation, drip lawn watering system, 800 S.F. (20′ x 40′ area)	.25	3.31	3.56
1100	1000 S.F. (20′ x 50′ area)	.22	3	3.22
1200	1600 S.F. (40′ x 40′ area), 2 zone	.27	3.19	3.46
1300	2000 S.F. (40′ x 50′ area), 2 zone	.24	2.88	3.12
1400	2400 S.F. (60′ x 40′ area), 3 zone	.26	3.16	3.42
1500	3000 S.F. (60′ x 50′ area), 3 zone	.23	2.85	3.08
1600	3200 S.F. (2 x 40′ x 40′), 4 zone	.26	3.13	3.39
1700	4000 S.F. (2 x 40′ x 50′), 4 zone	.24	2.84	3.08
1800	4800 S.F. (2 x 60′ x 40′), 6 zone with control	.28	3.10	3.38
1900	6000 S.F. (2 x 60′ x 50′), 6 zone with control	.24	2.82	3.06
2000	6400 S.F. (4 x 40′ x 40′), 8 zone with control	.27	3.16	3.43
2100	8000 S.F. (4 x 40′ x 50′), 8 zone with control	.24	2.86	3.10
2200	3200 S.F. (2 x 40′ x 40′), 4 zone with control	.28	3.20	3.48
2300	4000 S.F. (2 x 40′ x 50′), 4 zone with control	.25	2.89	3.14
2400	2400 S.F. (60′ x 40′), 3 zone with control	.28	3.26	3.54
2500	3000 S.F. (60′ x 50′), 3 zone with control	.25	2.94	3.19
2600	4800 S.F. (2 x 60′ x 40′), 6 zone, manual	.26	3.07	3.33
2700	6000 S.F. (2 x 60′ x 50′), 6 zone, manual	.22	2.79	3.01
2800	1000 S.F. (10′ x 100′ area)	.13	1.87	2
2900	2000 S.F. (2 x 10′ x 100′ area)	.14	1.79	1.93

G3010 Water Supply

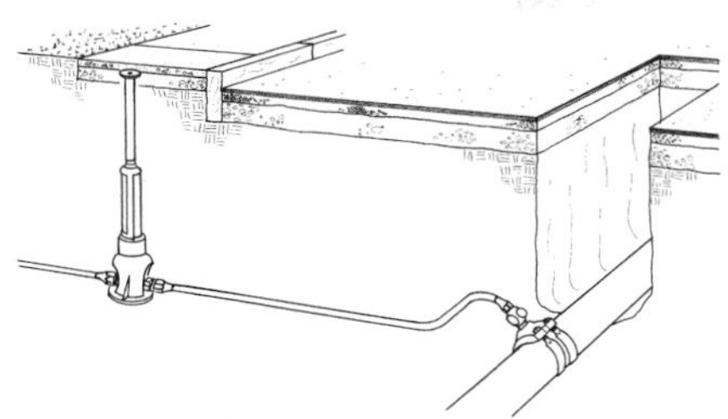

The Water Service Systems are for copper service taps from 1″ to 2″ diameter into pressurized mains from 6″ to 8″ diameter. Costs are given for two offsets with depths varying from 2′ to 10′. Included in system components are excavation and backfill and required curb stops with boxes.

System Components	QUANTITY	UNIT	COST EACH		
			MAT.	INST.	TOTAL
SYSTEM G3010 121 1000					
WATER SERVICE, LEAD FREE, 6″ MAIN, 1″ COPPER SERVICE, 10′ OFFSET, 2′ DEEP					
Trench excavation, 1/2 C.Y. backhoe, 1 laborer	2.220	C.Y.		15.94	15.94
Drill & tap pressurized main, 6″, 1″ to 2″ service	1.000	Ea.		465	465
Corporation stop, 1″ diameter	1.000	Ea.	108	48	156
Saddles, 3/4″ & 1″ diameter, add	1.000	Ea.	84		84
Copper tubing, type K, 1″ diameter	12.000	L.F.	127.80	52.08	179.88
Piping, curb stops, no lead, 1″ diameter	1.000	Ea.	156	48	204
Curb box, cast iron, 1/2″ to 1″ curb stops	1.000	Ea.	68.50	64.50	133
Backfill by hand, no compaction, heavy soil	2.890	C.Y.		8.33	8.33
Compaction, vibrating plate	2.220	C.Y.		21.22	21.22
TOTAL			544.30	723.07	1,267.37

G3010 121	Water Service, Lead Free	MAT.	INST.	TOTAL
1000	Water service, Lead Free, 6″ main, 1″ copper service, 10′ offset, 2′ deep	545	725	1,270
1040	4′ deep	545	770	1,315
1060	6′ deep	545	985	1,530
1080	8′ deep	545	1,100	1,645
1100	10′ deep	545	1,200	1,745
1200	20′ offset, 2′ deep	545	770	1,315
1240	4′ deep	545	860	1,405
1260	6′ deep	545	1,300	1,845
1280	8′ deep	545	1,500	2,045
1300	10′ deep	545	1,725	2,270
1400	1-1/2″ copper service, 10′ offset, 2′ deep	950	790	1,740
1440	4′ deep	950	835	1,785
1460	6′ deep	950	1,050	2,000
1480	8′ deep	950	1,150	2,100
1500	10′ deep	950	1,275	2,225
1520	20′ offset, 2′ deep	950	835	1,785
1530	4′ deep	950	925	1,875
1540	6′ deep	950	1,350	2,300
1560	8′ deep	950	1,575	2,525
1580	10′ deep	950	1,800	2,750
1600	2″ copper service, 10′ offset, 2′ deep	1,250	820	2,070
1610	4′ deep	1,250	865	2,115
1620	6′ deep	1,250	1,075	2,325
1630	8′ deep	1,250	1,175	2,425
1650	10′ deep	1,250	1,300	2,550
1660	20′ offset, 2′ deep	1,250	865	2,115
1670	4′ deep	1,250	955	2,205
1680	6′ deep	1,250	1,400	2,650

G3010 Water Supply

G3010 121	Water Service, Lead Free	COST EACH MAT.	INST.	TOTAL
1690	8' deep	1,250	1,600	2,850
1695	10' deep	1,250	1,825	3,075
1700	8" main, 1" copper service, 10' offset, 2' deep	545	765	1,310
1710	4' deep	545	810	1,355
1720	6' deep	545	1,025	1,570
1730	8' deep	545	1,125	1,670
1740	10' deep	545	1,250	1,795
1750	20' offset, 2' deep	545	810	1,355
1760	4' deep	545	900	1,445
1770	6' deep	545	1,325	1,870
1780	8' deep	545	1,550	2,095
1790	10' deep	545	1,775	2,320
1800	1-1/2" copper service, 10' offset, 2' deep	950	830	1,780
1810	4' deep	950	875	1,825
1820	6' deep	950	1,100	2,050
1830	8' deep	950	1,600	2,550
1840	10' deep	950	1,300	2,250
1850	20' offset, 2' deep	950	875	1,825
1860	4' deep	950	875	1,825
1870	6' deep	950	1,400	2,350
1880	8' deep	950	1,600	2,550
1890	10' deep	950	1,825	2,775
1900	2" copper service, 10' offset, 2' deep	1,250	860	2,110
1910	4' deep	1,250	905	2,155
1920	6' deep	1,250	1,125	2,375
1930	8' deep	1,250	1,225	2,475
1940	10' deep	1,250	1,325	2,575
1950	20' offset, 2' deep	1,250	905	2,155
1960	4' deep	1,250	995	2,245
1970	6' deep	1,250	1,425	2,675
1980	8' deep	1,250	1,650	2,900
1990	10' deep	1,250	1,850	3,100

G3010 Water Supply

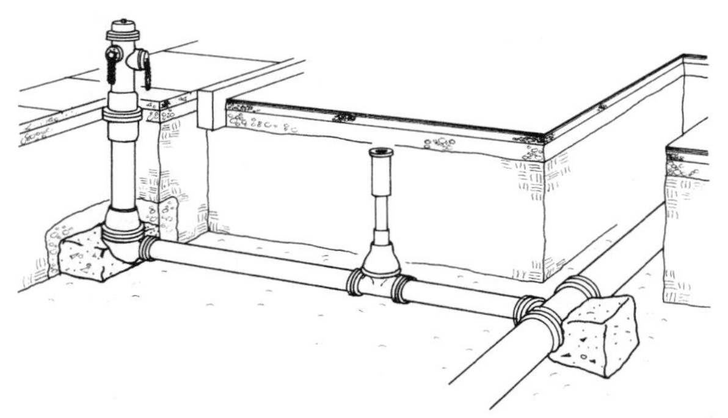

The Fire Hydrant Systems include: four different hydrants with three different lengths of offsets, at several depths of trenching. Excavation and backfill is included with each system as well as thrust blocks as necessary and pipe bedding. Finally spreading of excess material and fine grading complete the components.

System Components	QUANTITY	UNIT	COST PER EACH		
			MAT.	INST.	TOTAL
SYSTEM G3010 410 1500					
HYDRANT, 4-1/2" VALVE SIZE, TWO WAY, 10' OFFSET, 2' DEEP					
Excavation, trench, 1 C.Y. hydraulic backhoe	3.556	C.Y.		31.25	31.25
Sleeve, 12" x 6"	1.000	Ea.	1,550	125.55	1,675.55
Ductile iron pipe, 6" diameter	8.000	L.F.	448	180.40	628.40
Gate valve, 6"	1.000	Ea.	2,650	313	2,963
Hydrant valve box, 6' long	1.000	Ea.	253	85	338
4-1/2" valve size, depth 2'-0"	1.000	Ea.	3,175	226	3,401
Thrust blocks, at valve, shoe and sleeve	1.600	C.Y.	358.40	481.66	840.06
Borrow, crushed stone, 3/8"	4.089	C.Y.	145.16	34.22	179.38
Backfill and compact, dozer, air tamped	4.089	C.Y.		80.56	80.56
Spread excess excavated material	.204	C.Y.		9.38	9.38
Fine grade, hand	400.000	S.F.		294	294
TOTAL			8,579.56	1,861.02	10,440.58

G3010 410	Fire Hydrants	COST PER EACH		
		MAT.	INST.	TOTAL
1000	Hydrant, 4-1/2" valve size, two way, 0' offset, 2' deep	8,125	1,275	9,400
1100	4' deep	8,150	1,300	9,450
1200	6' deep	8,700	1,350	10,050
1300	8' deep	8,075	1,425	9,500
1400	10' deep	9,450	1,500	10,950
1500	10' offset, 2' deep	8,575	1,850	10,425
1600	4' deep	8,600	2,200	10,800
1700	6' deep	9,150	2,800	11,950
1800	8' deep	8,525	3,725	12,250
1900	10' deep	9,900	5,625	15,525
2000	20' offset, 2' deep	9,125	2,475	11,600
2100	4' deep	9,150	3,000	12,150
2200	6' deep	9,700	3,925	13,625
2300	8' deep	9,100	5,225	14,325
2400	10' deep	10,500	7,025	17,525
2500	Hydrant, 4-1/2" valve size, three way, 0' offset, 2' deep	8,350	1,275	9,625
2600	4' deep	8,375	1,300	9,675
2700	6' deep	8,975	1,375	10,350
2800	8' deep	8,300	1,450	9,750
2900	10' deep	9,775	1,550	11,325
3000	10' offset, 2' deep	8,800	1,875	10,675
3100	4' deep	8,825	2,225	11,050
3200	6' deep	9,425	2,825	12,250
3300	8' deep	8,750	3,750	12,500

G3010 Water Supply

G3010 410	Fire Hydrants	COST PER EACH		
		MAT.	INST.	TOTAL
3400	10′ deep	10,200	5,650	15,850
3500	20′ offset, 2′ deep	9,350	2,475	11,825
3600	4′ deep	9,375	3,025	12,400
3700	6′ deep	9,975	3,950	13,925
3800	8′ deep	9,300	5,275	14,575
3900	10′ deep	10,800	7,075	17,875
5000	5-1/4″ valve size, two way, 0′ offset, 2′ deep	7,575	1,275	8,850
5100	4′ deep	7,925	1,300	9,225
5200	6′ deep	8,125	1,350	9,475
5300	8′ deep	8,700	1,425	10,125
5400	10′ deep	9,100	1,500	10,600
5500	10′ offset, 2′ deep	8,025	1,850	9,875
5600	4′ deep	8,375	2,200	10,575
5700	6′ deep	8,575	2,800	11,375
5800	8′ deep	9,150	3,725	12,875
5900	10′ deep	9,550	5,625	15,175
6000	20′ offset, 2′ deep	8,575	2,475	11,050
6100	4′ deep	8,925	3,000	11,925
6200	6′ deep	9,125	3,925	13,050
6300	8′ deep	9,725	5,225	14,950
6400	10′ deep	10,100	7,025	17,125
6500	Three way, 0′ offset, 2′ deep	7,750	1,275	9,025
6600	4′ deep	8,125	1,300	9,425
6700	6′ deep	8,350	1,375	9,725
6800	8′ deep	8,975	1,450	10,425
6900	10′ deep	9,400	1,550	10,950
7000	10′ offset, 2′ deep	8,200	1,875	10,075
7100	4′ deep	8,575	2,225	10,800
7200	6′ deep	8,800	2,825	11,625
7300	8′ deep	9,425	3,750	13,175
7400	10′ deep	9,850	5,650	15,500
7500	20′ offset, 2′ deep	8,775	2,475	11,250
7600	4′ deep	9,150	2,725	11,875
7700	6′ deep	9,350	3,950	13,300

G3020 Sanitary Sewer

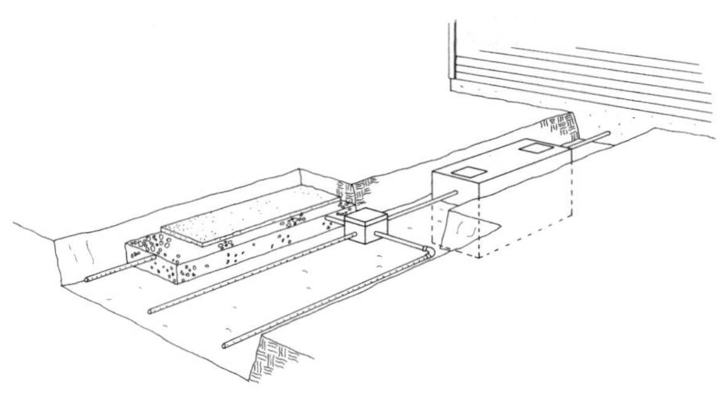

The Septic System includes: a septic tank; leaching field; concrete distribution boxes; plus excavation and gravel backfill.

The Expanded System Listing shows systems with tanks ranging from 1000 gallons to 2000 gallons. Tanks are either concrete or fiberglass. Cost is on a complete unit basis.

System Components	QUANTITY	UNIT	COST EACH MAT.	COST EACH INST.	COST EACH TOTAL
SYSTEM G3020 302 0300					
1000 GALLON TANK, LEACHING FIELD, 600 S.F., 50′ FROM BLDG.					
Mobilization	2.000	Ea.		1,706	1,706
Septic tank, precast, 1000 gallon	1.000	Ea.	1,175	283	1,458
Effluent filter	1.000	Ea.	48	83	131
Distribution box, precast, 5 outlets	1.000	Ea.	114	63.50	177.50
Flow leveler	3.000	Ea.	8.34	30.30	38.64
Sewer pipe, PVC, SDR 35, 4″ diameter	160.000	L.F.	294.40	724.80	1,019.20
Tee	1.000	Ea.	20.50	83	103.50
Elbow	2.000	Ea.	24.90	111	135.90
Viewport cap	1.000	Ea.	7.55	27.50	35.05
Filter fabric	67.000	S.Y.	115.91	28.14	144.05
Detectable marking tape	1.600	C.L.F.	15.84	5.39	21.23
Excavation	160.000	C.Y.		2,176	2,176
Backfill	133.000	L.C.Y.		420.28	420.28
Spoil	55.000	L.C.Y.		404.25	404.25
Compaction	113.000	E.C.Y.		1,020.39	1,020.39
Stone fill, 3/4″ to 1-1/2″	39.000	C.Y.	1,794	488.28	2,282.28
TOTAL			3,618.44	7,654.83	11,273.27

G3020 302	Septic Systems	COST EACH MAT.	COST EACH INST.	COST EACH TOTAL
0300	1000 gal. concrete septic tank, 600 S.F. leaching field, 50 feet from bldg.	3,625	7,650	11,275
0301	200 feet from building	3,900	9,500	13,400
0305	750 S.F. leaching field, 50 feet from building	4,125	8,325	12,450
0306	200 feet from building	4,400	10,200	14,600
0400	1250 gal. concrete septic tank, 600 S.F. leaching field, 50 feet from bldg.	4,025	7,950	11,975
0401	200 feet from building	4,300	9,800	14,100
0405	750 S.F. leaching field, 50 feet from building	4,525	8,625	13,150
0406	200 feet from building	4,800	10,500	15,300
0410	1000 S.F. leaching field, 50 feet from building	5,425	9,900	15,325
0411	200 feet from building	5,725	11,800	17,525
0500	1500 gal. concrete septic tank, 600 S.F. leaching field, 50 feet from bldg.	4,275	8,225	12,500
0501	200 feet from building	4,525	10,100	14,625
0505	750 S.F. leaching field, 50 feet from building	4,775	8,950	13,725
0506	200 feet from building	5,050	11,800	16,850
0510	1000 S.F. leaching field, 50 feet from building	5,650	10,200	15,850
0511	200 feet from building	5,950	12,100	18,050
0515	1100 S.F. leaching field, 50 feet from building	6,000	10,700	16,700
0516	200 feet from building	6,275	12,600	18,875
0600	2000 gal concrete tank, 1200 S.F. leaching field, 50 feet from building	7,325	11,300	18,625
0601	200 feet from building	7,600	13,100	20,700

G3020 Sanitary Sewer

G3020 302	Septic Systems	COST EACH		
		MAT.	INST.	TOTAL
0605	1500 S.F. leaching field, 50 feet from building	8,350	12,800	21,150
0606	200 feet from building	8,650	14,600	23,250
0710	2500 gal concrete tank, 1500 S.F. leaching field, 50 feet from building	8,200	13,500	21,700
0711	200 feet from building	8,500	15,400	23,900
0715	1600 S.F. leaching field, 50 feet from building	8,575	14,000	22,575
0716	200 feet from building	8,875	15,800	24,675
0720	1800 S.F. leaching field, 50 feet from building	9,275	15,000	24,275
0721	200 feet from building	9,575	16,900	26,475
1300	1000 gal. 2 compartment tank, 600 S.F. leaching field, 50 feet from bldg.	3,200	7,650	10,850
1301	200 feet from building	3,475	9,500	12,975
1305	750 S.F. leaching field, 50 feet from building	3,700	8,325	12,025
1306	200 feet from building	4,000	10,200	14,200
1400	1250 gal. 2 compartment tank, 600 S.F. leaching field, 50 feet from bldg.	3,600	7,950	11,550
1401	200 feet from building	3,875	9,800	13,675
1405	750 S.F. leaching field, 50 feet from building	4,100	8,625	12,725
1406	200 feet from building	4,375	10,500	14,875
1410	1000 S.F. leaching field, 50 feet from building	5,000	9,900	14,900
1411	200 feet from building	5,300	11,800	17,100
1500	1500 gal. 2 compartment tank, 600 S.F. leaching field, 50 feet from bldg.	4,200	8,225	12,425
1501	200 feet from building	4,450	10,100	14,550
1505	750 S.F. leaching field, 50 feet from building	4,700	8,950	13,650
1506	200 feet from building	4,975	11,800	16,775
1510	1000 S.F. leaching field, 50 feet from building	5,575	10,200	15,775
1511	200 feet from building	5,875	12,100	17,975
1515	1100 S.F. leaching field, 50 feet from building	5,925	10,700	16,625
1516	200 feet from building	6,200	12,600	18,800
1600	2000 gal 2 compartment tank, 1200 S.F. leaching field, 50 feet from bldg.	7,350	11,300	18,650
1601	200 feet from building	7,625	13,100	20,725
1605	1500 S.F. leaching field, 50 feet from building	8,375	12,800	21,175
1606	200 feet from building	8,675	14,600	23,275
1710	2500 gal 2 compartment tank, 1500 S.F. leaching field, 50 feet from bldg.	9,150	13,500	22,650
1711	200 feet from building	9,450	15,400	24,850
1715	1600 S.F. leaching field, 50 feet from building	9,525	14,000	23,525
1716	200 feet from building	9,825	15,800	25,625
1720	1800 S.F. leaching field, 50 feet from building	10,200	15,000	25,200
1721	200 feet from building	10,500	16,900	27,400

G3030 Storm Sewer

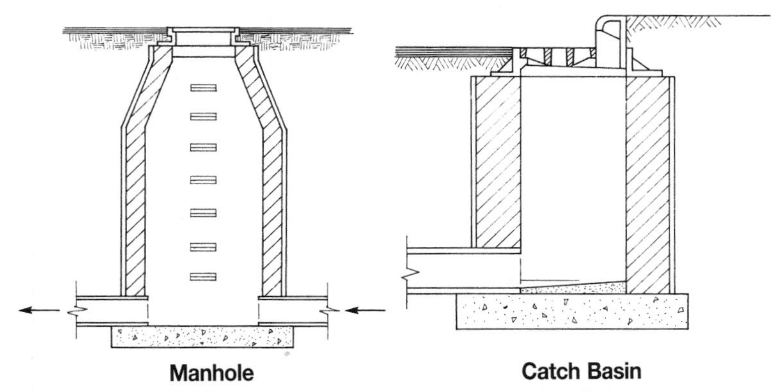

The Manhole and Catch Basin System includes: excavation with a backhoe; a formed concrete footing; frame and cover; cast iron steps and compacted backfill.

The Expanded System Listing shows manholes that have a 4′, 5′ and 6′ inside diameter riser. Depths range from 4′ to 14′. Construction material shown is either concrete, concrete block, precast concrete, or brick.

System Components	QUANTITY	UNIT	COST PER EACH MAT.	COST PER EACH INST.	COST PER EACH TOTAL
SYSTEM G3030 210 1920					
MANHOLE/CATCH BASIN, BRICK, 4′ I.D. RISER, 4′ DEEP					
Excavation, hydraulic backhoe, 3/8 C.Y. bucket	14.815	B.C.Y.		130.22	130.22
Trim sides and bottom of excavation	64.000	S.F.		73.60	73.60
Forms in place, manhole base, 4 uses	20.000	SFCA	16.40	117	133.40
Reinforcing in place footings, #4 to #7	.019	Ton	23.75	24.70	48.45
Concrete, 3000 psi	.925	C.Y.	131.35		131.35
Place and vibrate concrete, footing, direct chute	.925	C.Y.		53.72	53.72
Catch basin or MH, brick, 4′ ID, 4′ deep	1.000	Ea.	675	1,150	1,825
Catch basin or MH steps; heavy galvanized cast iron	1.000	Ea.	19.85	15.80	35.65
Catch basin or MH frame and cover	1.000	Ea.	325	241	566
Fill, granular	12.954	L.C.Y.	349.76		349.76
Backfill, spread with wheeled front end loader	12.954	L.C.Y.		37.31	37.31
Backfill compaction, 12″ lifts, air tamp	12.954	E.C.Y.		133.82	133.82
TOTAL			1,541.11	1,977.17	3,518.28

G3030 210	Manholes & Catch Basins	COST PER EACH MAT.	COST PER EACH INST.	COST PER EACH TOTAL
1920	Manhole/catch basin, brick, 4′ I.D. riser, 4′ deep	1,550	1,975	3,525
1940	6′ deep	2,150	2,725	4,875
1960	8′ deep	2,875	3,750	6,625
1980	10′ deep	3,375	4,650	8,025
3000	12′ deep	4,200	5,075	9,275
3020	14′ deep	5,175	7,100	12,275
3200	Block, 4′ I.D. riser, 4′ deep	1,325	1,575	2,900
3220	6′ deep	1,825	2,250	4,075
3240	8′ deep	2,425	3,100	5,525
3260	10′ deep	2,875	3,850	6,725
3280	12′ deep	3,650	4,900	8,550
3300	14′ deep	4,550	5,975	10,525
4620	Concrete, cast-in-place, 4′ I.D. riser, 4′ deep	1,500	2,700	4,200
4640	6′ deep	2,125	3,600	5,725
4660	8′ deep	2,975	5,250	8,225
4680	10′ deep	3,550	6,500	10,050
4700	12′ deep	4,475	8,075	12,550
4720	14′ deep	5,525	9,675	15,200
5820	Concrete, precast, 4′ I.D. riser, 4′ deep	1,825	1,450	3,275
5840	6′ deep	2,325	1,950	4,275

G3030 Storm Sewer

G3030 210	Manholes & Catch Basins	COST PER EACH		
		MAT.	INST.	TOTAL
5860	8′ deep	2,975	2,775	5,750
5880	10′ deep	3,550	3,400	6,950
5900	12′ deep	4,475	4,100	8,575
5920	14′ deep	5,525	5,350	10,875
6000	5′ I.D. riser, 4′ deep	2,850	1,575	4,425
6020	6′ deep	3,575	2,200	5,775
6040	8′ deep	4,650	2,900	7,550
6060	10′ deep	5,950	3,700	9,650
6080	12′ deep	7,400	4,725	12,125
6100	14′ deep	8,975	5,800	14,775
6200	6′ I.D. riser, 4′ deep	3,975	2,050	6,025
6220	6′ deep	4,825	2,700	7,525
6240	8′ deep	6,150	3,800	9,950
6260	10′ deep	7,700	4,825	12,525
6280	12′ deep	9,400	6,100	15,500
6300	14′ deep	11,300	7,425	18,725

Reference Section

All the reference information is in one section, making it easy to find what you need to know and easy to use the data set on a daily basis. This section is visually identified by a vertical black bar on the page edges.

In this Reference Section, we've included Equipment Rental Costs, a listing of rental and operating costs; Crew Listings, a full listing of all crews and equipment, and their costs; Historical Cost Indexes for cost comparisons over time; City Cost Indexes and Location Factors for adjusting costs to the region you are in; Reference Tables, where you will find explanations, estimating information and procedures, and technical data; Change Orders, information on pricing changes to contract documents; and an explanation of all the Abbreviations in the data set.

Table of Contents

Estimating Tips

- This section contains the average costs to rent and operate hundreds of pieces of construction equipment. This is useful information when one is estimating the time and material requirements of any particular operation in order to establish a unit or total cost. Bare equipment costs shown on a unit cost line include, not only rental, but also operating costs for equipment under normal use.

Rental Costs

- Equipment rental rates are obtained from the following industry sources throughout North America: contractors, suppliers, dealers, manufacturers, and distributors.
- Rental rates vary throughout the country, with larger cities generally having lower rates. Lease plans for new equipment are available for periods in excess of six months, with a percentage of payments applying toward purchase.
- Monthly rental rates vary from 2% to 5% of the purchase price of the equipment depending on the anticipated life of the equipment and its wearing parts.
- Weekly rental rates are about 1/3 of the monthly rates, and daily rental rates are about 1/3 of the weekly rate.
- Rental rates can also be treated as reimbursement costs for contractor-owned equipment. Owned equipment costs include depreciation, loan payments, interest, taxes, insurance, storage, and major repairs.

Operating Costs

- The operating costs include parts and labor for routine servicing, such as the repair and replacement of pumps, filters, and worn lines. Normal operating expendables, such as fuel, lubricants, tires, and electricity (where applicable), are also included.
- Extraordinary operating expendables with highly variable wear patterns, such as diamond bits and blades, are excluded. These costs can be found as material costs in the Unit Price section.
- The hourly operating costs listed do not include the operator's wages.

Equipment Cost/Day

- Any power equipment required by a crew is shown in the Crew Listings with a daily cost.
- This daily cost of equipment needed by a crew includes both the rental cost and the operating cost and is based on dividing the weekly rental rate by 5 (the number of working days in the week), then adding the hourly operating cost multiplied by 8 (the number of hours in a day). This "Equipment Cost/Day" is shown in the far right column of the Equipment Rental section.
- If equipment is needed for only one or two days, it is best to develop your own cost by including components for daily rent and hourly operating costs. This is important when the listed Crew for a task does not contain the equipment needed, such as a crane for lifting mechanical heating/cooling equipment up onto a roof.
- If the quantity of work is less than the crew's Daily Output shown for a Unit Price line item that includes a bare unit equipment cost, the recommendation is to estimate one day's rental cost and operating cost for equipment shown in the Crew Listing for that line item.
- Please note, in some cases the equipment description in the crew is followed by a time period in parenthesis. For example: (daily) or (monthly). In these cases the equipment cost/day is calculated by adding the rental cost per time period to the hourly operating cost multiplied by 8.

Mobilization, Demobilization Costs

- The cost to move construction equipment from an equipment yard or rental company to the job site and back again is not included in equipment rental costs listed in the Reference Section. It is also not included in the bare equipment cost of any Unit Price line item or in any equipment costs shown in the Crew Listings.
- Mobilization (to the site) and demobilization (from the site) costs can be found in the Unit Price section.
- If a piece of equipment is already at the job site, it is not appropriate to utilize mobilization or demobilization costs again in an estimate. ■

01 54 33 | Equipment Rental

		01 54 33 | Equipment Rental	UNIT	HOURLY OPER. COST	RENT PER DAY	RENT PER WEEK	RENT PER MONTH	EQUIPMENT COST/DAY	
10	0010	**CONCRETE EQUIPMENT RENTAL** without operators R015433-10							10
	0200	Bucket, concrete lightweight, 1/2 C.Y.	Ea.	.87	38.50	115	345	30	
	0300	1 C.Y.		.98	63.50	190	570	45.80	
	0400	1-1/2 C.Y.		1.23	60	180	540	45.85	
	0500	2 C.Y.		1.34	73.50	220	660	54.70	
	0580	8 C.Y.		6.47	93.50	280	840	107.75	
	0600	Cart, concrete, self-propelled, operator walking, 10 C.F.		2.85	175	525	1,575	127.85	
	0700	Operator riding, 18 C.F.		4.81	192	575	1,725	153.45	
	0800	Conveyer for concrete, portable, gas, 16" wide, 26' long		10.61	160	480	1,450	180.85	
	0900	46' long		10.98	175	525	1,575	192.85	
	1000	56' long		11.15	192	575	1,725	204.15	
	1100	Core drill, electric, 2-1/2 H.P., 1" to 8" bit diameter		1.56	83.50	250	750	62.50	
	1150	11 H.P., 8" to 18" cores		5.38	119	356.32	1,075	114.30	
	1200	Finisher, concrete floor, gas, riding trowel, 96" wide		9.64	153	459.60	1,375	169	
	1300	Gas, walk-behind, 3 blade, 36" trowel		2.03	96	287.50	865	73.75	
	1400	4 blade, 48" trowel		3.06	104	312.50	940	87	
	1500	Float, hand-operated (Bull float), 48" wide		.08	12.35	37	111	8.05	
	1570	Curb builder, 14 H.P., gas, single screw		14.00	253	760	2,275	263.95	
	1590	Double screw		15.00	253	760	2,275	272	
	1600	Floor grinder, concrete and terrazzo, electric, 22" path		3.03	134	401.75	1,200	104.60	
	1700	Edger, concrete, electric, 7" path		1.18	57.50	172.50	520	43.95	
	1750	Vacuum pick-up system for floor grinders, wet/dry		1.61	102	305.71	915	74.05	
	1800	Mixer, powered, mortar and concrete, gas, 6 C.F., 18 H.P.		7.39	97	291	875	117.35	
	1900	10 C.F., 25 H.P.		8.97	114	342.50	1,025	140.30	
	2000	16 C.F.		9.33	144	432.50	1,300	161.15	
	2100	Concrete, stationary, tilt drum, 2 C.Y.		7.21	80	240	720	105.70	
	2120	Pump, concrete, truck mounted, 4" line, 80' boom		29.79	287	860	2,575	410.35	
	2140	5" line, 110' boom		37.34	287	860	2,575	470.75	
	2160	Mud jack, 50 C.F. per hr.		6.43	228	685	2,050	188.45	
	2180	225 C.F. per hr.		8.52	293	880	2,650	244.15	
	2190	Shotcrete pump rig, 12 C.Y./hr.		13.93	223	670	2,000	245.40	
	2200	35 C.Y./hr.		15.75	287	860	2,575	298	
	2600	Saw, concrete, manual, gas, 18 H.P.		5.52	112	337	1,000	111.55	
	2650	Self-propelled, gas, 30 H.P.		7.87	81	242.71	730	111.50	
	2675	V-groove crack chaser, manual, gas, 6 H.P.		1.64	100	300	900	73.10	
	2700	Vibrators, concrete, electric, 60 cycle, 2 H.P.		.47	73	218.50	655	47.45	
	2800	3 H.P.		.56	73.50	221	665	48.70	
	2900	Gas engine, 5 H.P.		1.54	16.85	50.61	152	22.45	
	3000	8 H.P.		2.08	17.05	51.12	153	26.85	
	3050	Vibrating screed, gas engine, 8 H.P.		2.80	88	263.50	790	75.10	
	3120	Concrete transit mixer, 6 x 4, 250 H.P., 8 C.Y., rear discharge		50.57	70	210	630	446.55	
	3200	Front discharge		58.71	135	405	1,225	550.65	
	3300	6 x 6, 285 H.P., 12 C.Y., rear discharge		57.97	150	450	1,350	553.80	
	3400	Front discharge	↓	60.41	170	510	1,525	585.25	
20	0010	**EARTHWORK EQUIPMENT RENTAL** without operators R015433-10							20
	0040	Aggregate spreader, push type, 8' to 12' wide	Ea.	2.59	75	225	675	65.75	
	0045	Tailgate type, 8' wide		2.54	63.50	190	570	58.30	
	0055	Earth auger, truck mounted, for fence & sign posts, utility poles		13.81	150	450	1,350	200.50	
	0060	For borings and monitoring wells		42.52	83.50	250	750	390.20	
	0070	Portable, trailer mounted		2.29	100	300	900	78.35	
	0075	Truck mounted, for caissons, water wells		85.14	150	450	1,350	771.10	
	0080	Horizontal boring machine, 12" to 36" diameter, 45 H.P.		22.70	104	312	935	244	
	0090	12" to 48" diameter, 65 H.P.		31.16	108	325	975	314.25	
	0095	Auger, for fence posts, gas engine, hand held		.45	84	251.50	755	53.85	
	0100	Excavator, diesel hydraulic, crawler mounted, 1/2 C.Y. cap.		21.66	465	1,394.28	4,175	452.10	
	0120	5/8 C.Y. capacity		28.95	610	1,833.22	5,500	598.25	
	0140	3/4 C.Y. capacity		32.56	725	2,168.88	6,500	694.25	
	0150	1 C.Y. capacity	↓	41.09	780	2,333	7,000	795.30	

01 54 33 | Equipment Rental

20

	01 54 33 Equipment Rental	UNIT	HOURLY OPER. COST	RENT PER DAY	RENT PER WEEK	RENT PER MONTH	EQUIPMENT COST/DAY
0200	1-1/2 C.Y. capacity	Ea.	48.44	500	1,500	4,500	687.55
0300	2 C.Y. capacity		56.41	835	2,500	7,500	951.30
0320	2-1/2 C.Y. capacity		82.39	1,350	4,027.92	12,100	1,465
0325	3-1/2 C.Y. capacity		119.76	2,000	6,000	18,000	2,158
0330	4-1/2 C.Y. capacity		151.17	3,675	11,000	33,000	3,409
0335	6 C.Y. capacity		191.81	3,225	9,650	29,000	3,464
0340	7 C.Y. capacity		174.67	3,400	10,200	30,600	3,437
0342	Excavator attachments, bucket thumbs		3.39	258	774.60	2,325	182.05
0345	Grapples		3.13	222	666.16	2,000	158.30
0346	Hydraulic hammer for boom mounting, 4000 ft lb.		13.44	890	2,670	8,000	641.50
0347	5000 ft lb.		15.90	950	2,850	8,550	697.25
0348	8000 ft lb.		23.47	1,200	3,600	10,800	907.75
0349	12,000 ft lb.		25.64	1,100	3,333	10,000	871.75
0350	Gradall type, truck mounted, 3 ton @ 15′ radius, 5/8 C.Y.		43.31	835	2,500	7,500	846.50
0370	1 C.Y. capacity		59.22	835	2,500	7,500	973.80
0400	Backhoe-loader, 40 to 45 H.P., 5/8 C.Y. capacity		11.86	244	732.50	2,200	241.40
0450	45 H.P. to 60 H.P., 3/4 C.Y. capacity		17.97	117	350	1,050	213.75
0460	80 H.P., 1-1/4 C.Y. capacity		20.30	117	350	1,050	232.40
0470	112 H.P., 1-1/2 C.Y. capacity		32.89	610	1,833.22	5,500	629.75
0482	Backhoe-loader attachment, compactor, 20,000 lb.		6.42	155	464.76	1,400	144.30
0485	Hydraulic hammer, 750 ft lb.		3.67	107	320.17	960	93.40
0486	Hydraulic hammer, 1200 ft lb.		6.53	205	614.52	1,850	175.15
0500	Brush chipper, gas engine, 6″ cutter head, 35 H.P.		9.14	247	740	2,225	221.10
0550	Diesel engine, 12″ cutter head, 130 H.P.		23.60	340	1,020	3,050	392.80
0600	15″ cutter head, 165 H.P.		26.51	415	1,239.36	3,725	459.90
0750	Bucket, clamshell, general purpose, 3/8 C.Y.		1.40	91.50	275	825	66.15
0800	1/2 C.Y.		1.51	91.50	275	825	67.10
0850	3/4 C.Y.		1.64	91.50	275	825	68.10
0900	1 C.Y.		1.70	91.50	275	825	68.55
0950	1-1/2 C.Y.		2.78	91.50	275	825	77.25
1000	2 C.Y.		2.91	91.50	275	825	78.30
1010	Bucket, dragline, medium duty, 1/2 C.Y.		.82	91.50	275	825	61.55
1020	3/4 C.Y.		.78	91.50	275	825	61.25
1030	1 C.Y.		.80	91.50	275	825	61.35
1040	1-1/2 C.Y.		1.26	91.50	275	825	65.05
1050	2 C.Y.		1.29	91.50	275	825	65.30
1070	3 C.Y.		2.07	91.50	275	825	71.60
1200	Compactor, manually guided 2-drum vibratory smooth roller, 7.5 H.P.		7.20	181	542.50	1,625	166.10
1250	Rammer/tamper, gas, 8″		2.20	48	144.59	435	46.50
1260	15″		2.62	55	165.25	495	54
1300	Vibratory plate, gas, 18″ plate, 3000 lb. blow		2.12	24.50	72.81	218	31.55
1350	21″ plate, 5000 lb. blow		2.61	241	722.50	2,175	165.35
1370	Curb builder/extruder, 14 H.P., gas, single screw		13.99	253	760	2,275	263.95
1390	Double screw		14.99	253	760	2,275	271.95
1500	Disc harrow attachment, for tractor		.47	82.50	246.84	740	53.15
1810	Feller buncher, shearing & accumulating trees, 100 H.P.		39.08	460	1,380	4,150	588.60
1860	Grader, self-propelled, 25,000 lb.		33.25	1,100	3,333	10,000	932.60
1910	30,000 lb.		32.76	1,325	4,000	12,000	1,062
1920	40,000 lb.		51.73	1,550	4,667	14,000	1,347
1930	55,000 lb.		66.73	1,775	5,333	16,000	1,600
1950	Hammer, pavement breaker, self-propelled, diesel, 1000 to 1250 lb.		28.31	600	1,800	5,400	586.50
2000	1300 to 1500 lb.		42.67	1,000	3,020.94	9,075	945.55
2050	Pile driving hammer, steam or air, 4150 ft lb. @ 225 bpm		12.11	500	1,500	4,500	396.90
2100	8750 ft lb. @ 145 bpm		14.30	700	2,100	6,300	534.45
2150	15,000 ft lb. @ 60 bpm		14.63	835	2,500	7,500	617.05
2200	24,450 ft lb. @ 111 bpm		15.64	965	2,900	8,700	705.15
2250	Leads, 60′ high for pile driving hammers up to 20,000 ft lb.		3.66	300	900	2,700	209.25
2300	90′ high for hammers over 20,000 ft lb.	↓	5.43	540	1,620	4,850	367.45

01 54 33 | Equipment Rental

			UNIT	HOURLY OPER. COST	RENT PER DAY	RENT PER WEEK	RENT PER MONTH	EQUIPMENT COST/DAY	
20	2350	Diesel type hammer, 22,400 ft lb.	Ea.	17.76	490	1,471.74	4,425	436.45	20
	2400	41,300 ft lb.		25.61	620	1,859.04	5,575	576.65	
	2450	141,000 ft lb.		41.20	980	2,943.48	8,825	918.35	
	2500	Vib. elec. hammer/extractor, 200 kW diesel generator, 34 H.P.		41.25	715	2,143.06	6,425	758.60	
	2550	80 H.P.		72.81	1,025	3,098.40	9,300	1,202	
	2600	150 H.P.		134.74	2,000	5,964.42	17,900	2,271	
	2700	Hydro Excavator w/EXT boom 12 C.Y., 1200 gallons		37.70	1,600	4,800	14,400	1,262	
	2800	Log chipper, up to 22" diameter, 600 H.P.		46.13	305	915	2,750	552	
	2850	Logger, for skidding & stacking logs, 150 H.P.		43.40	930	2,785	8,350	904.20	
	2860	Mulcher, diesel powered, trailer mounted		17.99	305	915	2,750	326.90	
	2900	Rake, spring tooth, with tractor		14.67	370	1,110.26	3,325	339.45	
	3000	Roller, vibratory, tandem, smooth drum, 20 H.P.		7.78	320	967	2,900	255.65	
	3050	35 H.P.		10.10	260	779.76	2,350	236.75	
	3100	Towed type vibratory compactor, smooth drum, 50 H.P.		25.20	520	1,566	4,700	514.75	
	3150	Sheepsfoot, 50 H.P.		25.56	385	1,161.90	3,475	436.90	
	3170	Landfill compactor, 220 H.P.		69.80	1,650	4,985	15,000	1,555	
	3200	Pneumatic tire roller, 80 H.P.		12.88	405	1,213.54	3,650	345.75	
	3250	120 H.P.		19.33	665	1,988.14	5,975	552.25	
	3300	Sheepsfoot vibratory roller, 240 H.P.		62.02	1,425	4,260.30	12,800	1,348	
	3320	340 H.P.		83.57	2,175	6,500	19,500	1,969	
	3350	Smooth drum vibratory roller, 75 H.P.		23.27	655	1,962.32	5,875	578.60	
	3400	125 H.P.		27.53	740	2,220.52	6,650	664.35	
	3410	Rotary mower, brush, 60", with tractor		18.73	360	1,084.44	3,250	366.75	
	3420	Rototiller, walk-behind, gas, 5 H.P.		2.13	60	180	540	53.05	
	3422	8 H.P.		2.80	132	395	1,175	101.40	
	3440	Scrapers, towed type, 7 C.Y. capacity		6.42	127	382.14	1,150	127.80	
	3450	10 C.Y. capacity		7.18	170	511.24	1,525	159.70	
	3500	15 C.Y. capacity		7.38	196	588.70	1,775	176.75	
	3525	Self-propelled, single engine, 14 C.Y. capacity		132.89	2,225	6,660	20,000	2,395	
	3550	Dual engine, 21 C.Y. capacity		140.95	2,500	7,500	22,500	2,628	
	3600	31 C.Y. capacity		187.28	3,625	10,844.40	32,500	3,667	
	3640	44 C.Y. capacity		231.98	4,650	13,942.80	41,800	4,644	
	3650	Elevating type, single engine, 11 C.Y. capacity		61.68	1,075	3,200	9,600	1,133	
	3700	22 C.Y. capacity		114.28	1,625	4,850	14,600	1,884	
	3710	Screening plant, 110 H.P. w/5' x 10' screen		21.07	645	1,933	5,800	555.15	
	3720	5' x 16' screen		26.60	1,325	4,000	12,000	1,013	
	3850	Shovel, crawler-mounted, front-loading, 7 C.Y. capacity		218.00	3,925	11,773.92	35,300	4,099	
	3855	12 C.Y. capacity		335.89	5,450	16,318.24	49,000	5,951	
	3860	Shovel/backhoe bucket, 1/2 C.Y.		2.68	73	218.95	655	65.25	
	3870	3/4 C.Y.		2.66	82	245.81	735	70.40	
	3880	1 C.Y.		2.75	91	272.66	820	76.50	
	3890	1-1/2 C.Y.		2.94	107	320.17	960	87.60	
	3910	3 C.Y.		3.43	145	433.78	1,300	114.15	
	3950	Stump chipper, 18" deep, 30 H.P.		6.88	232	697	2,100	194.45	
	4110	Dozer, crawler, torque converter, diesel 80 H.P.		25.18	335	1,000	3,000	401.40	
	4150	105 H.P.		34.23	600	1,800	5,400	633.85	
	4200	140 H.P.		41.15	720	2,166	6,500	762.40	
	4260	200 H.P.		62.97	1,675	5,000	15,000	1,504	
	4310	300 H.P.		80.49	1,875	5,600	16,800	1,764	
	4360	410 H.P.		106.42	3,200	9,630	28,900	2,777	
	4370	500 H.P.		132.98	3,900	11,670	35,000	3,398	
	4380	700 H.P.		229.47	5,475	16,421.52	49,300	5,120	
	4400	Loader, crawler, torque conv., diesel, 1-1/2 C.Y., 80 H.P.		29.45	550	1,651	4,950	565.80	
	4450	1-1/2 to 1-3/4 C.Y., 95 H.P.		30.18	695	2,091.42	6,275	659.75	
	4510	1-3/4 to 2-1/4 C.Y., 130 H.P.		47.61	965	2,900	8,700	960.90	
	4530	2-1/2 to 3-1/4 C.Y., 190 H.P.		57.61	1,175	3,540	10,600	1,169	
	4560	3-1/2 to 5 C.Y., 275 H.P.		71.19	1,525	4,595.96	13,800	1,489	
	4610	Front end loader, 4WD, articulated frame, diesel, 1 to 1-1/4 C.Y., 70 H.P.	↓	16.58	282	846.90	2,550	302	

01 54 33 | Equipment Rental

		01 54 33 Equipment Rental	UNIT	HOURLY OPER. COST	RENT PER DAY	RENT PER WEEK	RENT PER MONTH	EQUIPMENT COST/DAY
20	4620	1-1/2 to 1-3/4 C.Y., 95 H.P.	Ea.	19.94	440	1,320	3,950	423.50
	4650	1-3/4 to 2 C.Y., 130 H.P.		21.00	395	1,187.72	3,575	405.55
	4710	2-1/2 to 3-1/2 C.Y., 145 H.P.		29.44	780	2,333	7,000	702.10
	4730	3 to 4-1/2 C.Y., 185 H.P.		31.99	890	2,667	8,000	789.35
	4760	5-1/4 to 5-3/4 C.Y., 270 H.P.		53.03	890	2,666	8,000	957.50
	4810	7 to 9 C.Y., 475 H.P.		90.90	2,550	7,667	23,000	2,261
	4870	9 to 11 C.Y., 620 H.P.		131.52	2,700	8,107.48	24,300	2,674
	4880	Skid-steer loader, wheeled, 10 C.F., 30 H.P. gas		9.54	169	506.07	1,525	177.55
	4890	1 C.Y., 78 H.P., diesel		18.38	420	1,265.18	3,800	400.10
	4892	Skid-steer attachment, auger		.74	145	433.50	1,300	92.65
	4893	Backhoe		.74	122	366.64	1,100	79.25
	4894	Broom		.70	140	420.25	1,250	89.70
	4895	Forks		.15	32	96	288	20.45
	4896	Grapple		.72	88.50	265.25	795	58.80
	4897	Concrete hammer		1.05	183	550	1,650	118.40
	4898	Tree spade		.60	103	309.84	930	66.75
	4899	Trencher		.65	102	305	915	66.20
	4900	Trencher, chain, boom type, gas, operator walking, 12 H.P.		4.16	206	618.25	1,850	156.95
	4910	Operator riding, 40 H.P.		16.64	450	1,343.75	4,025	401.85
	5000	Wheel type, diesel, 4′ deep, 12″ wide		68.50	965	2,891.84	8,675	1,126
	5100	6′ deep, 20″ wide		87.32	1,050	3,127.50	9,375	1,324
	5150	Chain type, diesel, 5′ deep, 8″ wide		16.25	360	1,084.44	3,250	346.90
	5200	Diesel, 8′ deep, 16″ wide		89.39	1,925	5,783.68	17,400	1,872
	5202	Rock trencher, wheel type, 6″ wide x 18″ deep		46.98	90	270	810	429.85
	5206	Chain type, 18″ wide x 7′ deep		104.38	283	850	2,550	1,005
	5210	Tree spade, self-propelled		13.65	230	690	2,075	247.20
	5250	Truck, dump, 2-axle, 12 ton, 8 C.Y. payload, 220 H.P.		23.88	395	1,185	3,550	428.05
	5300	Three axle dump, 16 ton, 12 C.Y. payload, 400 H.P.		44.50	360	1,084.44	3,250	572.90
	5310	Four axle dump, 25 ton, 18 C.Y. payload, 450 H.P.		49.85	525	1,575.02	4,725	713.80
	5350	Dump trailer only, rear dump, 16-1/2 C.Y.		5.73	151	454.43	1,375	136.70
	5400	20 C.Y.		6.18	170	511.24	1,525	151.70
	5450	Flatbed, single axle, 1-1/2 ton rating		19.00	73.50	221.02	665	196.15
	5500	3 ton rating		23.05	1,050	3,180	9,550	820.40
	5550	Off highway rear dump, 25 ton capacity		62.67	1,475	4,389.40	13,200	1,379
	5600	35 ton capacity		66.90	665	2,000	6,000	935.20
	5610	50 ton capacity		83.87	1,825	5,499.66	16,500	1,771
	5620	65 ton capacity		89.56	2,000	5,990.24	18,000	1,915
	5630	100 ton capacity		121.24	2,950	8,830.44	26,500	2,736
	6000	Vibratory plow, 25 H.P., walking	↓	6.77	300	900	2,700	234.20
40	0010	**GENERAL EQUIPMENT RENTAL** without operators R015433-10						
	0020	Aerial lift, scissor type, to 20′ high, 1200 lb. capacity, electric	Ea.	3.48	129	385.75	1,150	105
	0030	To 30′ high, 1200 lb. capacity		3.77	203	607.67	1,825	151.70
	0040	Over 30′ high, 1500 lb. capacity		5.13	243	727.50	2,175	186.60
	0070	Articulating boom, to 45′ high, 500 lb. capacity, diesel R015433-15		9.92	250	750	2,250	229.35
	0075	To 60′ high, 500 lb. capacity		13.66	300	900	2,700	289.25
	0080	To 80′ high, 500 lb. capacity		16.05	900	2,702.25	8,100	668.85
	0085	To 125′ high, 500 lb. capacity		18.34	1,525	4,603.50	13,800	1,067
	0100	Telescoping boom to 40′ high, 500 lb. capacity, diesel		11.24	315	945	2,825	278.90
	0105	To 45′ high, 500 lb. capacity		12.51	320	965	2,900	293.05
	0110	To 60′ high, 500 lb. capacity		16.36	300	900	2,700	310.90
	0115	To 80′ high, 500 lb. capacity		21.27	355	1,067	3,200	383.55
	0120	To 100′ high, 500 lb. capacity		28.71	865	2,587.75	7,775	747.25
	0125	To 120′ high, 500 lb. capacity		29.16	1,450	4,348.50	13,000	1,103
	0195	Air compressor, portable, 6.5 CFM, electric		.90	44.50	133	400	33.80
	0196	Gasoline		.65	56	167.50	505	38.70
	0200	Towed type, gas engine, 60 CFM		9.43	129	387.50	1,175	152.95
	0300	160 CFM	↓	10.47	198	595	1,775	202.80

01 54 33 | Equipment Rental

		01 54 33 | Equipment Rental	UNIT	HOURLY OPER. COST	RENT PER DAY	RENT PER WEEK	RENT PER MONTH	EQUIPMENT COST/DAY	
40	0400	Diesel engine, rotary screw, 250 CFM	Ea.	12.08	175	524	1,575	201.50	40
	0500	365 CFM		16.00	310	937	2,800	315.40	
	0550	450 CFM		19.95	277	832.25	2,500	326.05	
	0600	600 CFM		34.10	248	743.62	2,225	421.50	
	0700	750 CFM		34.62	435	1,306.50	3,925	538.25	
	0930	Air tools, breaker, pavement, 60 lb.		.57	81.50	245	735	53.50	
	0940	80 lb.		.56	81	242.50	730	53	
	0950	Drills, hand (jackhammer), 65 lb.		.67	68	203.50	610	46.05	
	0960	Track or wagon, swing boom, 4″ drifter		54.66	1,025	3,104	9,300	1,058	
	0970	5″ drifter		63.30	1,025	3,104	9,300	1,127	
	0975	Track mounted quarry drill, 6″ diameter drill		101.91	1,900	5,665	17,000	1,948	
	0980	Dust control per drill		1.04	25	75.50	227	23.40	
	0990	Hammer, chipping, 12 lb.		.60	46	138	415	32.40	
	1000	Hose, air with couplings, 50′ long, 3/4″ diameter		.07	12	36	108	7.75	
	1100	1″ diameter		.08	12.35	37	111	8	
	1200	1-1/2″ diameter		.22	37.50	112.50	340	24.25	
	1300	2″ diameter		.24	45	135	405	28.90	
	1400	2-1/2″ diameter		.36	57.50	172.50	520	37.35	
	1410	3″ diameter		.42	58.50	175	525	38.35	
	1450	Drill, steel, 7/8″ x 2′		.08	12.90	38.73	116	8.40	
	1460	7/8″ x 6′		.12	19.60	58.87	177	12.70	
	1520	Moil points		.03	7	21	63	4.40	
	1525	Pneumatic nailer w/accessories		.48	39.50	118	355	27.40	
	1530	Sheeting driver for 60 lb. breaker		.04	7.75	23.24	69.50	5	
	1540	For 90 lb. breaker		.13	10.50	31.50	94.50	7.35	
	1550	Spade, 25 lb.		.50	7.40	22.21	66.50	8.45	
	1560	Tamper, single, 35 lb.		.59	48.50	145.75	435	33.85	
	1570	Triple, 140 lb.		.89	61.50	184.87	555	44.05	
	1580	Wrenches, impact, air powered, up to 3/4″ bolt		.43	49.50	148.25	445	33.05	
	1590	Up to 1-1/4″ bolt		.58	79.50	238.50	715	52.30	
	1600	Barricades, barrels, reflectorized, 1 to 99 barrels		.03	4	12	36	2.65	
	1610	100 to 200 barrels		.02	4.41	13.22	39.50	2.85	
	1620	Barrels with flashers, 1 to 99 barrels		.03	6.40	19.16	57.50	4.10	
	1630	100 to 200 barrels		.03	5.10	15.34	46	3.30	
	1640	Barrels with steady burn type C lights		.05	8.45	25.30	76	5.45	
	1650	Illuminated board, trailer mounted, with generator		3.28	139	418.28	1,250	109.85	
	1670	Portable barricade, stock, with flashers, 1 to 6 units		.03	6.35	19.11	57.50	4.10	
	1680	25 to 50 units		.03	5.95	17.82	53.50	3.85	
	1685	Butt fusion machine, wheeled, 1.5 HP electric, 2″ - 8″ diameter pipe		2.63	225	675	2,025	156.05	
	1690	Tracked, 20 HP diesel, 4″-12″ diameter pipe		11.23	560	1,685	5,050	426.85	
	1695	83 HP diesel, 8″ - 24″ diameter pipe		51.32	1,100	3,325	9,975	1,076	
	1700	Carts, brick, gas engine, 1000 lb. capacity		2.94	65	195	585	62.55	
	1800	1500 lb., 7-1/2′ lift		2.92	69.50	208	625	64.95	
	1822	Dehumidifier, medium, 6 lb./hr., 150 CFM		1.19	76.50	229.28	690	55.35	
	1824	Large, 18 lb./hr., 600 CFM		2.19	585	1,750	5,250	367.55	
	1830	Distributor, asphalt, trailer mounted, 2000 gal., 38 H.P. diesel		10.99	355	1,058.62	3,175	299.65	
	1840	3000 gal., 38 H.P. diesel		12.87	380	1,136.08	3,400	330.15	
	1850	Drill, rotary hammer, electric		1.11	71.50	214	640	51.70	
	1860	Carbide bit, 1-1/2″ diameter, add to electric rotary hammer		.03	41.50	125	375	25.25	
	1865	Rotary, crawler, 250 H.P.		135.77	2,300	6,868.12	20,600	2,460	
	1870	Emulsion sprayer, 65 gal., 5 H.P. gas engine		2.77	107	320.17	960	86.15	
	1880	200 gal., 5 H.P. engine		7.22	179	537.06	1,600	165.20	
	1900	Floor auto-scrubbing machine, walk-behind, 28″ path		5.62	222	667	2,000	178.40	
	1930	Floodlight, mercury vapor, or quartz, on tripod, 1000 watt		.46	36.50	110	330	25.65	
	1940	2000 watt		.59	28	84.69	254	21.65	
	1950	Floodlights, trailer mounted with generator, 1 - 300 watt light		3.54	78.50	235.48	705	75.45	
	1960	2 - 1000 watt lights		4.49	87.50	262.33	785	88.35	
	2000	4 - 300 watt lights	↓	4.24	100	299.51	900	93.85	

01 54 33 | Equipment Rental

40

		UNIT	HOURLY OPER. COST	RENT PER DAY	RENT PER WEEK	RENT PER MONTH	EQUIPMENT COST/DAY
2005	Foam spray rig, incl. box trailer, compressor, generator, proportioner	Ea.	25.46	535	1,600.84	4,800	523.85
2015	Forklift, pneumatic tire, rough terr, straight mast, 5000 lb, 12′ lift, gas		18.59	219	655.83	1,975	279.90
2025	8000 lb, 12′ lift		22.68	360	1,084.44	3,250	398.35
2030	5000 lb, 12′ lift, diesel		15.41	244	733.29	2,200	269.90
2035	8000 lb, 12′ lift, diesel		16.70	277	831.40	2,500	299.90
2045	All terrain, telescoping boom, diesel, 5000 lb, 10′ reach, 19′ lift		17.20	233	700	2,100	277.60
2055	6600 lb, 29′ reach, 42′ lift		21.04	233	700	2,100	308.30
2065	10,000 lb, 31′ reach, 45′ lift		23.03	315	950	2,850	374.20
2070	Cushion tire, smooth floor, gas, 5000 lb capacity		8.23	247	741.50	2,225	214.10
2075	8000 lb capacity		11.33	275	826.25	2,475	255.90
2085	Diesel, 5000 lb capacity		7.73	210	629.25	1,900	187.65
2090	12,000 lb capacity		12.01	400	1,194.50	3,575	335
2095	20,000 lb capacity		17.20	660	1,980	5,950	533.60
2100	Generator, electric, gas engine, 1.5 kW to 3 kW		2.57	46.50	140	420	48.55
2200	5 kW		3.21	91	272.50	820	80.15
2300	10 kW		5.91	108	322.50	970	111.80
2400	25 kW		7.38	405	1,210	3,625	301.10
2500	Diesel engine, 20 kW		9.18	229	687.50	2,075	210.95
2600	50 kW		15.90	370	1,110	3,325	349.20
2700	100 kW		28.51	445	1,340.50	4,025	496.20
2800	250 kW		54.19	750	2,249.33	6,750	883.40
2850	Hammer, hydraulic, for mounting on boom, to 500 ft lb.		2.89	93.50	279.89	840	79.10
2860	1000 ft lb.		4.59	139	418.28	1,250	120.40
2900	Heaters, space, oil or electric, 50 MBH		1.46	46.50	140	420	39.65
3000	100 MBH		2.71	46.50	140	420	49.70
3100	300 MBH		7.90	135	405	1,225	144.20
3150	500 MBH		13.12	200	600	1,800	224.95
3200	Hose, water, suction with coupling, 20′ long, 2″ diameter		.02	5.65	17	51	3.55
3210	3″ diameter		.03	14.15	42.50	128	8.75
3220	4″ diameter		.03	28	84	252	17.05
3230	6″ diameter		.11	40.50	121.50	365	25.20
3240	8″ diameter		.27	53.50	160	480	34.15
3250	Discharge hose with coupling, 50′ long, 2″ diameter		.01	6.50	19.50	58.50	4
3260	3″ diameter		.01	7.35	22	66	4.50
3270	4″ diameter		.02	21	62.50	188	12.65
3280	6″ diameter		.06	29	87	261	17.90
3290	8″ diameter		.24	37.50	112.50	340	24.40
3295	Insulation blower		.83	117	350	1,050	76.65
3300	Ladders, extension type, 16′ to 36′ long		.18	41.50	125	375	26.45
3400	40′ to 60′ long		.64	120	360.50	1,075	77.20
3405	Lance for cutting concrete		2.20	65	195	585	56.60
3407	Lawn mower, rotary, 22″, 5 H.P.		1.05	38.50	115	345	31.40
3408	48″ self-propelled		2.89	138	415	1,250	106.10
3410	Level, electronic, automatic, with tripod and leveling rod		1.05	37.50	112	335	30.80
3430	Laser type, for pipe and sewer line and grade		2.17	117	350	1,050	87.35
3440	Rotating beam for interior control		.90	64	192.50	580	45.70
3460	Builder's optical transit, with tripod and rod		.10	37.50	112	335	23.20
3500	Light towers, towable, with diesel generator, 2000 watt		4.25	101	303.64	910	94.75
3600	4000 watt		4.50	165	495	1,475	135
3700	Mixer, powered, plaster and mortar, 6 C.F., 7 H.P.		2.05	83.50	250	750	66.40
3800	10 C.F., 9 H.P.		2.24	124	372.50	1,125	92.40
3850	Nailer, pneumatic		.48	33.50	100.18	300	23.85
3900	Paint sprayers complete, 8 CFM		.85	61.50	184.87	555	43.75
4000	17 CFM		1.60	110	330.50	990	78.85
4020	Pavers, bituminous, rubber tires, 8′ wide, 50 H.P., diesel		31.93	570	1,704.12	5,100	596.25
4030	10′ wide, 150 H.P.		95.62	1,950	5,835.32	17,500	1,932
4050	Crawler, 8′ wide, 100 H.P., diesel		87.59	2,050	6,170.98	18,500	1,935
4060	10′ wide, 150 H.P.		103.97	2,350	7,048.86	21,100	2,241

01 54 33 | Equipment Rental

			UNIT	HOURLY OPER. COST	RENT PER DAY	RENT PER WEEK	RENT PER MONTH	EQUIPMENT COST/DAY	
40	4070	Concrete paver, 12′ to 24′ wide, 250 H.P.	Ea.	87.62	1,675	5,060.72	15,200	1,713	40
	4080	Placer-spreader-trimmer, 24′ wide, 300 H.P.		117.51	2,550	7,668.54	23,000	2,474	
	4100	Pump, centrifugal gas pump, 1-1/2″ diam., 65 GPM		3.92	54	162.15	485	63.80	
	4200	2″ diameter, 130 GPM		4.98	44	132.50	400	66.35	
	4300	3″ diameter, 250 GPM		5.12	56	167.50	505	74.45	
	4400	6″ diameter, 1500 GPM		22.24	91.50	275	825	232.90	
	4500	Submersible electric pump, 1-1/4″ diameter, 55 GPM		.40	36	107.50	325	24.70	
	4600	1-1/2″ diameter, 83 GPM		.44	43.50	130	390	29.55	
	4700	2″ diameter, 120 GPM		1.64	61.50	185	555	50.15	
	4800	3″ diameter, 300 GPM		3.03	109	327.50	985	89.75	
	4900	4″ diameter, 560 GPM		14.75	61.50	185	555	155	
	5000	6″ diameter, 1590 GPM		22.08	65	195	585	215.60	
	5100	Diaphragm pump, gas, single, 1-1/2″ diameter		1.13	38.50	115	345	32	
	5200	2″ diameter		3.98	91.50	275	825	86.80	
	5300	3″ diameter		4.05	95	285	855	89.40	
	5400	Double, 4″ diameter		6.03	95	285	855	105.25	
	5450	Pressure washer 5 GPM, 3000 psi		3.87	110	330	990	96.95	
	5460	7 GPM, 3000 psi		4.94	85	255	765	90.50	
	5470	High pressure water jet 10 ksi		39.55	720	2,160	6,475	748.40	
	5480	40 ksi		27.88	980	2,940	8,825	811.05	
	5500	Trash pump, self-priming, gas, 2″ diameter		3.82	108	325	975	95.55	
	5600	Diesel, 4″ diameter		6.68	162	485	1,450	150.40	
	5650	Diesel, 6″ diameter		16.85	162	485	1,450	231.80	
	5655	Grout Pump		18.70	281	841.73	2,525	317.90	
	5700	Salamanders, L.P. gas fired, 100,000 BTU		2.88	57.50	172.50	520	57.55	
	5705	50,000 BTU		1.66	23.50	71	213	27.50	
	5720	Sandblaster, portable, open top, 3 C.F. capacity		.60	132	395	1,175	83.80	
	5730	6 C.F. capacity		1.00	132	395	1,175	87	
	5740	Accessories for above		.14	24	71.26	214	15.35	
	5750	Sander, floor		.77	73.50	220	660	50.15	
	5760	Edger		.52	35	105	315	25.15	
	5800	Saw, chain, gas engine, 18″ long		1.75	63.50	190	570	52	
	5900	Hydraulic powered, 36″ long		.78	58.50	175	525	41.25	
	5950	60″ long		.78	65	195	585	45.25	
	6000	Masonry, table mounted, 14″ diameter, 5 H.P.		1.32	76.50	230	690	56.55	
	6050	Portable cut-off, 8 H.P.		1.81	77.50	232.50	700	61	
	6100	Circular, hand held, electric, 7-1/4″ diameter		.23	13.85	41.50	125	10.10	
	6200	12″ diameter		.24	41	122.50	370	26.40	
	6250	Wall saw, w/hydraulic power, 10 H.P.		3.29	98.50	296	890	85.50	
	6275	Shot blaster, walk-behind, 20″ wide		4.73	281	841.73	2,525	206.20	
	6280	Sidewalk broom, walk-behind		2.24	82.50	247.87	745	67.50	
	6300	Steam cleaner, 100 gallons per hour		3.34	82.50	247.87	745	76.30	
	6310	200 gallons per hour		4.33	100	299.51	900	94.55	
	6340	Tar Kettle/Pot, 400 gallons		16.48	127	380	1,150	207.85	
	6350	Torch, cutting, acetylene-oxygen, 150′ hose, excludes gases		.45	15.30	45.96	138	12.80	
	6360	Hourly operating cost includes tips and gas		20.92	7	20.98	63	171.55	
	6410	Toilet, portable chemical		.13	23	69.20	208	14.90	
	6420	Recycle flush type		.16	28.50	85.72	257	18.45	
	6430	Toilet, fresh water flush, garden hose,		.19	34	102.25	305	22	
	6440	Hoisted, non-flush, for high rise		.16	28	83.66	251	18	
	6465	Tractor, farm with attachment		17.37	390	1,172.50	3,525	373.40	
	6480	Trailers, platform, flush deck, 2 axle, 3 ton capacity		1.69	94.50	284	850	70.30	
	6500	25 ton capacity		6.23	143	428.61	1,275	135.55	
	6600	40 ton capacity		8.04	203	609.35	1,825	186.20	
	6700	3 axle, 50 ton capacity		8.72	225	676.48	2,025	205.05	
	6800	75 ton capacity		11.08	300	898.54	2,700	268.30	
	6810	Trailer mounted cable reel for high voltage line work		5.89	28.50	85	255	64.10	
	6820	Trailer mounted cable tensioning rig	↓	11.67	28.50	85	255	110.40	

01 54 33 | Equipment Rental

		01 54 33 Equipment Rental	UNIT	HOURLY OPER. COST	RENT PER DAY	RENT PER WEEK	RENT PER MONTH	EQUIPMENT COST/DAY
40	6830	Cable pulling rig	Ea.	73.77	28.50	85	255	607.15
	6850	Portable cable/wire puller, 8000 lb max pulling capacity		3.70	120	360	1,075	101.60
	6900	Water tank trailer, engine driven discharge, 5000 gallons		7.16	158	475.09	1,425	152.25
	6925	10,000 gallons		9.75	215	645.50	1,925	207.10
	6950	Water truck, off highway, 6000 gallons		71.75	835	2,504.54	7,525	1,075
	7010	Tram car for high voltage line work, powered, 2 conductor		6.88	28.50	85	255	72.05
	7020	Transit (builder's level) with tripod		.10	17.55	52.67	158	11.30
	7030	Trench box, 3000 lb., 6′ x 8′		.56	96.50	290.22	870	62.50
	7040	7200 lb., 6′ x 20′		.72	187	560	1,675	117.75
	7050	8000 lb., 8′ x 16′		1.08	186	557.71	1,675	120.15
	7060	9500 lb., 8′ x 20′		1.20	232	697.14	2,100	149.05
	7065	11,000 lb., 8′ x 24′		1.26	219	655.83	1,975	141.25
	7070	12,000 lb., 10′ x 20′		1.49	263	790.09	2,375	169.95
	7100	Truck, pickup, 3/4 ton, 2 wheel drive		9.24	61.50	184.87	555	110.85
	7200	4 wheel drive		9.48	167	500	1,500	175.85
	7250	Crew carrier, 9 passenger		12.66	108	325	975	166.25
	7290	Flat bed truck, 20,000 lb. GVW		15.26	133	397.63	1,200	201.60
	7300	Tractor, 4 x 2, 220 H.P.		22.25	215	645.50	1,925	307.10
	7410	330 H.P.		32.33	294	883.04	2,650	435.25
	7500	6 x 4, 380 H.P.		36.09	340	1,022.47	3,075	493.25
	7600	450 H.P.		44.23	415	1,239.36	3,725	601.75
	7610	Tractor, with A frame, boom and winch, 225 H.P.		24.74	293	877.88	2,625	373.50
	7620	Vacuum truck, hazardous material, 2500 gallons		12.79	310	929.52	2,800	288.25
	7625	5,000 gallons		13.02	440	1,316.82	3,950	367.55
	7650	Vacuum, HEPA, 16 gallon, wet/dry		.85	122	365	1,100	79.80
	7655	55 gallon, wet/dry		.78	25.50	76.50	230	21.50
	7660	Water tank, portable		.73	160	480.25	1,450	101.90
	7690	Sewer/catch basin vacuum, 14 C.Y., 1500 gallons		17.31	665	1,988.14	5,975	536.15
	7700	Welder, electric, 200 amp		3.81	33.50	100	300	50.50
	7800	300 amp		5.55	103	310	930	106.40
	7900	Gas engine, 200 amp		8.95	58.50	175	525	106.55
	8000	300 amp		10.13	110	330	990	147
	8100	Wheelbarrow, any size		.06	11.15	33.50	101	7.20
	8200	Wrecking ball, 4000 lb.	↓	2.50	60	180	540	56
50	0010	**HIGHWAY EQUIPMENT RENTAL** without operators R015433-10						
	0050	Asphalt batch plant, portable drum mixer, 100 ton/hr.	Ea.	88.41	1,550	4,621.78	13,900	1,632
	0060	200 ton/hr.		101.99	1,650	4,931.62	14,800	1,802
	0070	300 ton/hr.		119.86	1,925	5,783.68	17,400	2,116
	0100	Backhoe attachment, long stick, up to 185 H.P., 10.5′ long		.37	25.50	76.43	229	18.25
	0140	Up to 250 H.P., 12′ long		.41	28.50	85.72	257	20.45
	0180	Over 250 H.P., 15′ long		.56	39	116.71	350	27.85
	0200	Special dipper arm, up to 100 H.P., 32′ long		1.16	79.50	238.58	715	56.95
	0240	Over 100 H.P., 33′ long		1.44	100	299.51	900	71.45
	0280	Catch basin/sewer cleaning truck, 3 ton, 9 C.Y., 1000 gal.		35.39	420	1,265.18	3,800	536.15
	0300	Concrete batch plant, portable, electric, 200 C.Y./hr.		24.18	560	1,678.30	5,025	529.15
	0520	Grader/dozer attachment, ripper/scarifier, rear mounted, up to 135 H.P.		3.15	63.50	190.04	570	63.20
	0540	Up to 180 H.P.		4.13	95.50	287.12	860	90.50
	0580	Up to 250 H.P.		5.85	153	459.60	1,375	138.75
	0700	Pvmt. removal bucket, for hyd. excavator, up to 90 H.P.		2.16	58	174.54	525	52.20
	0740	Up to 200 H.P.		2.31	74.50	223.08	670	63.05
	0780	Over 200 H.P.		2.52	91	273.69	820	74.90
	0900	Aggregate spreader, self-propelled, 187 H.P.		50.60	740	2,220.52	6,650	848.90
	1000	Chemical spreader, 3 C.Y.		3.17	96.50	290	870	83.35
	1900	Hammermill, traveling, 250 H.P.		67.35	515	1,550	4,650	848.80
	2000	Horizontal borer, 3″ diameter, 13 H.P. gas driven		5.42	232	695	2,075	182.35
	2150	Horizontal directional drill, 20,000 lb. thrust, 78 H.P. diesel		27.58	530	1,590	4,775	538.65
	2160	30,000 lb. thrust, 115 H.P.		33.90	615	1,850	5,550	641.20
	2170	50,000 lb. thrust, 170 H.P.	↓	48.60	710	2,135	6,400	815.80

01 54 33 | Equipment Rental

			UNIT	HOURLY OPER. COST	RENT PER DAY	RENT PER WEEK	RENT PER MONTH	EQUIPMENT COST/DAY
50	2190	Mud trailer for HDD, 1500 gallons, 175 H.P., gas	Ea.	25.50	175	525	1,575	309
	2200	Hydromulcher, diesel, 3000 gallon, for truck mounting		17.43	227	680	2,050	275.45
	2300	Gas, 600 gallon		7.49	95	285	855	116.95
	2400	Joint & crack cleaner, walk behind, 25 H.P.		3.16	45.50	136	410	52.45
	2500	Filler, trailer mounted, 400 gallons, 20 H.P.		8.35	147	440	1,325	154.75
	3000	Paint striper, self-propelled, 40 gallon, 22 H.P.		6.76	122	365	1,100	127.10
	3100	120 gallon, 120 H.P.		19.23	380	1,140	3,425	381.80
	3200	Post drivers, 6" I-Beam frame, for truck mounting		12.41	320	960	2,875	291.30
	3400	Road sweeper, self-propelled, 8' wide, 90 H.P.		35.91	715	2,143.06	6,425	715.85
	3450	Road sweeper, vacuum assisted, 4 C.Y., 220 gallons		58.28	670	2,013.96	6,050	869
	4000	Road mixer, self-propelled, 130 H.P.		46.23	825	2,478.72	7,425	865.60
	4100	310 H.P.		75.01	2,150	6,480.82	19,400	1,896
	4220	Cold mix paver, incl. pug mill and bitumen tank, 165 H.P.		94.97	2,325	6,945.58	20,800	2,149
	4240	Pavement brush, towed		3.43	100	299.51	900	87.30
	4250	Paver, asphalt, wheel or crawler, 130 H.P., diesel		94.23	2,275	6,816.48	20,400	2,117
	4300	Paver, road widener, gas, 1' to 6', 67 H.P.		46.66	975	2,917.66	8,750	956.80
	4400	Diesel, 2' to 14', 88 H.P.		56.38	1,150	3,459.88	10,400	1,143
	4600	Slipform pavers, curb and gutter, 2 track, 75 H.P.		57.83	1,250	3,769.72	11,300	1,217
	4700	4 track, 165 H.P.		35.69	845	2,530.36	7,600	791.55
	4800	Median barrier, 215 H.P.		58.43	1,350	4,027.92	12,100	1,273
	4901	Trailer, low bed, 75 ton capacity		10.71	282	846.90	2,550	255.05
	5000	Road planer, walk behind, 10" cutting width, 10 H.P.		2.45	243	730	2,200	165.65
	5100	Self-propelled, 12" cutting width, 64 H.P.		8.26	190	570	1,700	180.05
	5120	Traffic line remover, metal ball blaster, truck mounted, 115 H.P.		46.56	905	2,720	8,150	916.50
	5140	Grinder, truck mounted, 115 H.P.		50.89	905	2,720	8,150	951.15
	5160	Walk-behind, 11 H.P.		3.56	142	425	1,275	113.45
	5200	Pavement profiler, 4' to 6' wide, 450 H.P.		216.58	1,275	3,800	11,400	2,493
	5300	8' to 10' wide, 750 H.P.		331.58	1,325	3,975	11,900	3,448
	5400	Roadway plate, steel, 1" x 8' x 20'		.09	61	182.50	550	37.20
	5600	Stabilizer, self-propelled, 150 H.P.		41.14	1,025	3,100	9,300	949.10
	5700	310 H.P.		76.18	1,300	3,900	11,700	1,389
	5800	Striper, truck mounted, 120 gallon paint, 460 H.P.		48.74	340	1,015	3,050	592.95
	5900	Thermal paint heating kettle, 115 gallons		7.71	61.50	185	555	98.65
	6000	Tar kettle, 330 gallon, trailer mounted		12.27	96.50	290	870	156.20
	7000	Tunnel locomotive, diesel, 8 to 12 ton		29.76	620	1,859.04	5,575	609.85
	7005	Electric, 10 ton		29.25	705	2,117.24	6,350	657.40
	7010	Muck cars, 1/2 C.Y. capacity		2.30	26.50	80.04	240	34.40
	7020	1 C.Y. capacity		2.51	35	104.31	315	40.95
	7030	2 C.Y. capacity		2.66	39	116.71	350	44.60
	7040	Side dump, 2 C.Y. capacity		2.87	48	144.59	435	51.90
	7050	3 C.Y. capacity		3.85	53	159.05	475	62.65
	7060	5 C.Y. capacity		5.62	68.50	205.53	615	86.10
	7100	Ventilating blower for tunnel, 7-1/2 H.P.		2.14	52.50	158.02	475	48.70
	7110	10 H.P.		2.42	55	165.25	495	52.40
	7120	20 H.P.		3.54	71.50	214.82	645	71.30
	7140	40 H.P.		6.14	94.50	284.02	850	105.90
	7160	60 H.P.		8.69	102	304.68	915	130.45
	7175	75 H.P.		10.37	158	475.09	1,425	177.95
	7180	200 H.P.		20.78	310	934.68	2,800	353.20
	7800	Windrow loader, elevating	↓	53.94	1,650	4,975	14,900	1,427
60	0010	**LIFTING AND HOISTING EQUIPMENT RENTAL** without operators R015433-10						
	0150	Crane, flatbed mounted, 3 ton capacity	Ea.	14.41	201	604.19	1,825	236.10
	0200	Crane, climbing, 106' jib, 6000 lb. capacity, 410 fpm R312316-45		39.72	2,600	7,800	23,400	1,878
	0300	101' jib, 10,250 lb. capacity, 270 fpm		46.43	2,275	6,800	20,400	1,731
	0500	Tower, static, 130' high, 106' jib, 6200 lb. capacity at 400 fpm		45.16	2,250	6,715	20,100	1,704
	0520	Mini crawler spider crane, up to 24" wide, 1990 lb. lifting capacity		12.50	550	1,652.48	4,950	430.50
	0525	Up to 30" wide, 6450 lb. lifting capacity		14.52	655	1,962.32	5,875	508.65
	0530	Up to 52" wide, 6680 lb. lifting capacity	↓	23.10	800	2,401.26	7,200	665.05

01 54 33 | Equipment Rental

60

		UNIT	HOURLY OPER. COST	RENT PER DAY	RENT PER WEEK	RENT PER MONTH	EQUIPMENT COST/DAY
0535	Up to 55" wide, 8920 lb. lifting capacity	Ea.	25.79	885	2,659.46	7,975	738.25
0540	Up to 66" wide, 13,350 lb. lifting capacity		34.92	1,375	4,131.20	12,400	1,106
0600	Crawler mounted, lattice boom, 1/2 C.Y., 15 tons at 12' radius		36.96	830	2,483	7,450	792.30
0700	3/4 C.Y., 20 tons at 12' radius		50.42	930	2,790	8,375	961.35
0800	1 C.Y., 25 tons at 12' radius		67.42	985	2,950	8,850	1,129
0900	1-1/2 C.Y., 40 tons at 12' radius		66.31	1,125	3,375	10,100	1,206
1000	2 C.Y., 50 tons at 12' radius		88.77	1,325	4,000	12,000	1,510
1100	3 C.Y., 75 tons at 12' radius		75.26	2,325	7,000	21,000	2,002
1200	100 ton capacity, 60' boom		85.91	2,675	8,000	24,000	2,287
1300	165 ton capacity, 60' boom		106.10	3,000	9,000	27,000	2,649
1400	200 ton capacity, 70' boom		138.21	3,825	11,500	34,500	3,406
1500	350 ton capacity, 80' boom		182.20	4,175	12,500	37,500	3,958
1600	Truck mounted, lattice boom, 6 x 4, 20 tons at 10' radius		39.76	1,950	5,850	17,600	1,488
1700	25 tons at 10' radius		42.73	2,325	7,000	21,000	1,742
1800	8 x 4, 30 tons at 10' radius		45.54	2,500	7,500	22,500	1,864
1900	40 tons at 12' radius		48.55	2,725	8,200	24,600	2,028
2000	60 tons at 15' radius		53.69	1,650	4,950	14,900	1,419
2050	82 tons at 15' radius		59.43	1,775	5,350	16,100	1,545
2100	90 tons at 15' radius		66.39	1,950	5,825	17,500	1,696
2200	115 tons at 15' radius		74.90	2,175	6,525	19,600	1,904
2300	150 tons at 18' radius		81.09	2,700	8,100	24,300	2,269
2350	165 tons at 18' radius		87.03	2,425	7,275	21,800	2,151
2400	Truck mounted, hydraulic, 12 ton capacity		29.50	390	1,175	3,525	471
2500	25 ton capacity		36.36	485	1,450	4,350	580.85
2550	33 ton capacity		50.67	900	2,700	8,100	945.35
2560	40 ton capacity		49.47	900	2,700	8,100	935.80
2600	55 ton capacity		53.78	915	2,750	8,250	980.20
2700	80 ton capacity		75.71	1,475	4,400	13,200	1,486
2720	100 ton capacity		74.96	1,550	4,675	14,000	1,535
2740	120 ton capacity		102.81	1,825	5,500	16,500	1,922
2760	150 ton capacity		109.92	2,050	6,125	18,400	2,104
2800	Self-propelled, 4 x 4, with telescoping boom, 5 ton		15.14	430	1,285	3,850	378.10
2900	12-1/2 ton capacity		21.42	430	1,285	3,850	428.30
3000	15 ton capacity		34.42	450	1,350	4,050	545.35
3050	20 ton capacity		24.02	650	1,950	5,850	582.20
3100	25 ton capacity		36.69	1,425	4,250	12,800	1,144
3150	40 ton capacity		44.90	660	1,975	5,925	754.20
3200	Derricks, guy, 20 ton capacity, 60' boom, 75' mast		22.74	1,425	4,250	12,800	1,032
3300	100' boom, 115' mast		36.04	2,000	6,000	18,000	1,488
3400	Stiffleg, 20 ton capacity, 70' boom, 37' mast		25.41	615	1,850	5,550	573.25
3500	100' boom, 47' mast		39.32	665	2,000	6,000	714.55
3550	Helicopter, small, lift to 1250 lb. maximum, w/pilot		99.14	2,150	6,435	19,300	2,080
3600	Hoists, chain type, overhead, manual, 3/4 ton		.14	10.25	30.70	92	7.30
3900	10 ton		.79	6.20	18.59	56	10
4000	Hoist and tower, 5000 lb. cap., portable electric, 40' high		5.12	142	426	1,275	126.20
4100	For each added 10' section, add		.12	31.50	95	285	19.95
4200	Hoist and single tubular tower, 5000 lb. electric, 100' high		6.96	105	315	945	118.65
4300	For each added 6'-6" section, add		.21	38.50	115	345	24.65
4400	Hoist and double tubular tower, 5000 lb., 100' high		7.57	105	315	945	123.60
4500	For each added 6'-6" section, add		.23	41.50	125	375	26.80
4550	Hoist and tower, mast type, 6000 lb., 100' high		8.24	94.50	284	850	122.70
4570	For each added 10' section, add		.13	31.50	95	285	20.05
4600	Hoist and tower, personnel, electric, 2000 lb., 100' @ 125 fpm		17.50	25	75	225	155
4700	3000 lb., 100' @ 200 fpm		20.02	25	75	225	175.15
4800	3000 lb., 150' @ 300 fpm		22.22	25	75	225	192.75
4900	4000 lb., 100' @ 300 fpm		22.98	25	75	225	198.85
5000	6000 lb., 100' @ 275 fpm	▼	24.70	25	75	225	212.60
5100	For added heights up to 500', add	L.F.	.01	3.33	10	30	2.10

01 54 33 | Equipment Rental

			UNIT	HOURLY OPER. COST	RENT PER DAY	RENT PER WEEK	RENT PER MONTH	EQUIPMENT COST/DAY	
60	5200	Jacks, hydraulic, 20 ton	Ea.	.05	19.65	59	177	12.20	60
	5500	100 ton		.40	26	78.50	236	18.90	
	6100	Jacks, hydraulic, climbing w/50′ jackrods, control console, 30 ton cap.		2.17	31	93	279	35.90	
	6150	For each added 10′ jackrod section, add		.05	5	15	45	3.40	
	6300	50 ton capacity		3.48	33.50	100	300	47.85	
	6350	For each added 10′ jackrod section, add		.06	5	15	45	3.50	
	6500	125 ton capacity		9.10	51.50	155	465	103.85	
	6550	For each added 10′ jackrod section, add		.61	5	15	45	7.90	
	6600	Cable jack, 10 ton capacity with 200′ cable		1.82	35.50	107	320	35.95	
	6650	For each added 50′ of cable, add	▼	.22	15	45	135	10.75	
70	0010	**WELLPOINT EQUIPMENT RENTAL** without operators (R015433 -10)							70
	0020	Based on 2 months rental							
	0100	Combination jetting & wellpoint pump, 60 H.P. diesel	Ea.	15.67	298	895	2,675	304.35	
	0200	High pressure gas jet pump, 200 H.P., 300 psi	"	33.83	275	825	2,475	435.65	
	0300	Discharge pipe, 8″ diameter	L.F.	.01	1.40	4.20	12.60	.90	
	0350	12″ diameter		.01	2.07	6.20	18.60	1.35	
	0400	Header pipe, flows up to 150 GPM, 4″ diameter		.01	.73	2.20	6.60	.50	
	0500	400 GPM, 6″ diameter		.01	1.07	3.20	9.60	.70	
	0600	800 GPM, 8″ diameter		.01	1.40	4.20	12.60	.95	
	0700	1500 GPM, 10″ diameter		.01	1.73	5.20	15.60	1.15	
	0800	2500 GPM, 12″ diameter		.03	2.07	6.20	18.60	1.45	
	0900	4500 GPM, 16″ diameter		.03	2.40	7.20	21.50	1.70	
	0950	For quick coupling aluminum and plastic pipe, add	▼	.03	9.35	28	84	5.85	
	1100	Wellpoint, 25′ long, with fittings & riser pipe, 1-1/2″ or 2″ diameter	Ea.	.07	132	395	1,175	79.55	
	1200	Wellpoint pump, diesel powered, 4″ suction, 20 H.P.		7.00	150	450	1,350	146	
	1300	6″ suction, 30 H.P.		9.39	167	500	1,500	175.15	
	1400	8″ suction, 40 H.P.		12.73	250	750	2,250	251.80	
	1500	10″ suction, 75 H.P.		18.77	265	795	2,375	309.20	
	1600	12″ suction, 100 H.P.		27.24	298	895	2,675	396.90	
	1700	12″ suction, 175 H.P.	▼	38.98	315	950	2,850	501.80	
80	0010	**MARINE EQUIPMENT RENTAL** without operators (R015433 -10)							80
	0200	Barge, 400 Ton, 30′ wide x 90′ long	Ea.	17.63	1,200	3,588.98	10,800	858.85	
	0240	800 Ton, 45′ wide x 90′ long		22.14	1,475	4,415.22	13,200	1,060	
	2000	Tugboat, diesel, 100 H.P.		29.57	238	712.63	2,150	379.10	
	2040	250 H.P.		57.41	430	1,291	3,875	717.50	
	2080	380 H.P.		124.99	1,300	3,873	11,600	1,774	
	3000	Small work boat, gas, 16-foot, 50 H.P.		11.35	48	143.56	430	119.50	
	4000	Large, diesel, 48-foot, 200 H.P.	▼	74.68	1,375	4,105.38	12,300	1,418	

Crew No.	Bare Costs		Incl. Subs O&P		Cost Per Labor-Hour	
Crew A-1	**Hr.**	**Daily**	**Hr.**	**Daily**	**Bare Costs**	**Incl. O&P**
1 Building Laborer	$42.10	$336.80	$63.25	$506.00	$42.10	$63.25
1 Concrete Saw, Gas Manual		111.55		122.71	13.94	15.34
8 L.H., Daily Totals		$448.35		$628.71	$56.04	$78.59
Crew A-1A	**Hr.**	**Daily**	**Hr.**	**Daily**	**Bare Costs**	**Incl. O&P**
1 Skilled Worker	$54.85	$438.80	$82.95	$663.60	$54.85	$82.95
1 Shot Blaster, 20"		206.20		226.82	25.77	28.35
8 L.H., Daily Totals		$645.00		$890.42	$80.63	$111.30
Crew A-1B	**Hr.**	**Daily**	**Hr.**	**Daily**	**Bare Costs**	**Incl. O&P**
1 Building Laborer	$42.10	$336.80	$63.25	$506.00	$42.10	$63.25
1 Concrete Saw		111.50		122.65	13.94	15.33
8 L.H., Daily Totals		$448.30		$628.65	$56.04	$78.58
Crew A-1C	**Hr.**	**Daily**	**Hr.**	**Daily**	**Bare Costs**	**Incl. O&P**
1 Building Laborer	$42.10	$336.80	$63.25	$506.00	$42.10	$63.25
1 Chain Saw, Gas, 18"		52.00		57.20	6.50	7.15
8 L.H., Daily Totals		$388.80		$563.20	$48.60	$70.40
Crew A-1D	**Hr.**	**Daily**	**Hr.**	**Daily**	**Bare Costs**	**Incl. O&P**
1 Building Laborer	$42.10	$336.80	$63.25	$506.00	$42.10	$63.25
1 Vibrating Plate, Gas, 18"		31.55		34.70	3.94	4.34
8 L.H., Daily Totals		$368.35		$540.71	$46.04	$67.59
Crew A-1E	**Hr.**	**Daily**	**Hr.**	**Daily**	**Bare Costs**	**Incl. O&P**
1 Building Laborer	$42.10	$336.80	$63.25	$506.00	$42.10	$63.25
1 Vibrating Plate, Gas, 21"		165.35		181.88	20.67	22.74
8 L.H., Daily Totals		$502.15		$687.88	$62.77	$85.99
Crew A-1F	**Hr.**	**Daily**	**Hr.**	**Daily**	**Bare Costs**	**Incl. O&P**
1 Building Laborer	$42.10	$336.80	$63.25	$506.00	$42.10	$63.25
1 Rammer/Tamper, Gas, 8"		46.50		51.15	5.81	6.39
8 L.H., Daily Totals		$383.30		$557.15	$47.91	$69.64
Crew A-1G	**Hr.**	**Daily**	**Hr.**	**Daily**	**Bare Costs**	**Incl. O&P**
1 Building Laborer	$42.10	$336.80	$63.25	$506.00	$42.10	$63.25
1 Rammer/Tamper, Gas, 15"		54.00		59.40	6.75	7.42
8 L.H., Daily Totals		$390.80		$565.40	$48.85	$70.67
Crew A-1H	**Hr.**	**Daily**	**Hr.**	**Daily**	**Bare Costs**	**Incl. O&P**
1 Building Laborer	$42.10	$336.80	$63.25	$506.00	$42.10	$63.25
1 Exterior Steam Cleaner		76.30		83.93	9.54	10.49
8 L.H., Daily Totals		$413.10		$589.93	$51.64	$73.74
Crew A-1J	**Hr.**	**Daily**	**Hr.**	**Daily**	**Bare Costs**	**Incl. O&P**
1 Building Laborer	$42.10	$336.80	$63.25	$506.00	$42.10	$63.25
1 Cultivator, Walk-Behind, 5 H.P.		53.05		58.35	6.63	7.29
8 L.H., Daily Totals		$389.85		$564.36	$48.73	$70.54
Crew A-1K	**Hr.**	**Daily**	**Hr.**	**Daily**	**Bare Costs**	**Incl. O&P**
1 Building Laborer	$42.10	$336.80	$63.25	$506.00	$42.10	$63.25
1 Cultivator, Walk-Behind, 8 H.P.		101.40		111.54	12.68	13.94
8 L.H., Daily Totals		$438.20		$617.54	$54.77	$77.19
Crew A-1M	**Hr.**	**Daily**	**Hr.**	**Daily**	**Bare Costs**	**Incl. O&P**
1 Building Laborer	$42.10	$336.80	$63.25	$506.00	$42.10	$63.25
1 Snow Blower, Walk-Behind		67.50		74.25	8.44	9.28
8 L.H., Daily Totals		$404.30		$580.25	$50.54	$72.53

Crew No.	Bare Costs		Incl. Subs O&P		Cost Per Labor-Hour	
Crew A-2	**Hr.**	**Daily**	**Hr.**	**Daily**	**Bare Costs**	**Incl. O&P**
2 Laborers	$42.10	$673.60	$63.25	$1012.00	$43.82	$65.77
1 Truck Driver (light)	47.25	378.00	70.80	566.40		
1 Flatbed Truck, Gas, 1.5 Ton		196.15		215.76	8.17	8.99
24 L.H., Daily Totals		$1247.75		$1794.17	$51.99	$74.76
Crew A-2A	**Hr.**	**Daily**	**Hr.**	**Daily**	**Bare Costs**	**Incl. O&P**
2 Laborers	$42.10	$673.60	$63.25	$1012.00	$43.82	$65.77
1 Truck Driver (light)	47.25	378.00	70.80	566.40		
1 Flatbed Truck, Gas, 1.5 Ton		196.15		215.76		
1 Concrete Saw		111.50		122.65	12.82	14.10
24 L.H., Daily Totals		$1359.25		$1916.82	$56.64	$79.87
Crew A-2B	**Hr.**	**Daily**	**Hr.**	**Daily**	**Bare Costs**	**Incl. O&P**
1 Truck Driver (light)	$47.25	$378.00	$70.80	$566.40	$47.25	$70.80
1 Flatbed Truck, Gas, 1.5 Ton		196.15		215.76	24.52	26.97
8 L.H., Daily Totals		$574.15		$782.16	$71.77	$97.77
Crew A-3A	**Hr.**	**Daily**	**Hr.**	**Daily**	**Bare Costs**	**Incl. O&P**
1 Equip. Oper. (light)	$53.00	$424.00	$79.20	$633.60	$53.00	$79.20
1 Pickup Truck, 4x4, 3/4 Ton		175.85		193.44	21.98	24.18
8 L.H., Daily Totals		$599.85		$827.03	$74.98	$103.38
Crew A-3B	**Hr.**	**Daily**	**Hr.**	**Daily**	**Bare Costs**	**Incl. O&P**
1 Equip. Oper. (medium)	$56.75	$454.00	$84.85	$678.80	$52.85	$79.10
1 Truck Driver (heavy)	48.95	391.60	73.35	586.80		
1 Dump Truck, 12 C.Y., 400 H.P.		572.90		630.19		
1 F.E. Loader, W.M., 2.5 C.Y.		702.10		772.31	79.69	87.66
16 L.H., Daily Totals		$2120.60		$2668.10	$132.54	$166.76
Crew A-3C	**Hr.**	**Daily**	**Hr.**	**Daily**	**Bare Costs**	**Incl. O&P**
1 Equip. Oper. (light)	$53.00	$424.00	$79.20	$633.60	$53.00	$79.20
1 Loader, Skid Steer, 78 H.P.		400.10		440.11	50.01	55.01
8 L.H., Daily Totals		$824.10		$1073.71	$103.01	$134.21
Crew A-3D	**Hr.**	**Daily**	**Hr.**	**Daily**	**Bare Costs**	**Incl. O&P**
1 Truck Driver (light)	$47.25	$378.00	$70.80	$566.40	$47.25	$70.80
1 Pickup Truck, 4x4, 3/4 Ton		175.85		193.44		
1 Flatbed Trailer, 25 Ton		135.55		149.10	38.92	42.82
8 L.H., Daily Totals		$689.40		$908.94	$86.17	$113.62
Crew A-3E	**Hr.**	**Daily**	**Hr.**	**Daily**	**Bare Costs**	**Incl. O&P**
1 Equip. Oper. (crane)	$59.20	$473.60	$88.50	$708.00	$54.08	$80.92
1 Truck Driver (heavy)	48.95	391.60	73.35	586.80		
1 Pickup Truck, 4x4, 3/4 Ton		175.85		193.44	10.99	12.09
16 L.H., Daily Totals		$1041.05		$1488.23	$65.07	$93.01
Crew A-3F	**Hr.**	**Daily**	**Hr.**	**Daily**	**Bare Costs**	**Incl. O&P**
1 Equip. Oper. (crane)	$59.20	$473.60	$88.50	$708.00	$54.08	$80.92
1 Truck Driver (heavy)	48.95	391.60	73.35	586.80		
1 Pickup Truck, 4x4, 3/4 Ton		175.85		193.44		
1 Truck Tractor, 6x4, 380 H.P.		493.25		542.58		
1 Lowbed Trailer, 75 Ton		255.05		280.56	57.76	63.54
16 L.H., Daily Totals		$1789.35		$2311.36	$111.83	$144.46

Crew No.	Bare Costs		Incl. Subs O&P		Cost Per Labor-Hour	
Crew A-3G	**Hr.**	**Daily**	**Hr.**	**Daily**	**Bare Costs**	**Incl. O&P**
1 Equip. Oper. (crane)	$59.20	$473.60	$88.50	$708.00	$54.08	$80.92
1 Truck Driver (heavy)	48.95	391.60	73.35	586.80		
1 Pickup Truck, 4x4, 3/4 Ton		175.85		193.44		
1 Truck Tractor, 6x4, 450 H.P.		601.75		661.92		
1 Lowbed Trailer, 75 Ton		255.05		280.56	64.54	70.99
16 L.H., Daily Totals		$1897.85		$2430.72	$118.62	$151.92
Crew A-3H	**Hr.**	**Daily**	**Hr.**	**Daily**	**Bare Costs**	**Incl. O&P**
1 Equip. Oper. (crane)	$59.20	$473.60	$88.50	$708.00	$59.20	$88.50
1 Hyd. Crane, 12 Ton (Daily)		724.85		797.34	90.61	99.67
8 L.H., Daily Totals		$1198.45		$1505.34	$149.81	$188.17
Crew A-3I	**Hr.**	**Daily**	**Hr.**	**Daily**	**Bare Costs**	**Incl. O&P**
1 Equip. Oper. (crane)	$59.20	$473.60	$88.50	$708.00	$59.20	$88.50
1 Hyd. Crane, 25 Ton (Daily)		801.40		881.54	100.18	110.19
8 L.H., Daily Totals		$1275.00		$1589.54	$159.38	$198.69
Crew A-3J	**Hr.**	**Daily**	**Hr.**	**Daily**	**Bare Costs**	**Incl. O&P**
1 Equip. Oper. (crane)	$59.20	$473.60	$88.50	$708.00	$59.20	$88.50
1 Hyd. Crane, 40 Ton (Daily)		1272.00		1399.20	159.00	174.90
8 L.H., Daily Totals		$1745.60		$2107.20	$218.20	$263.40
Crew A-3K	**Hr.**	**Daily**	**Hr.**	**Daily**	**Bare Costs**	**Incl. O&P**
1 Equip. Oper. (crane)	$59.20	$473.60	$88.50	$708.00	$54.88	$82.03
1 Equip. Oper. (oiler)	50.55	404.40	75.55	604.40		
1 Hyd. Crane, 55 Ton (Daily)		1362.00		1498.20		
1 P/U Truck, 3/4 Ton (Daily)		142.15		156.37	94.01	103.41
16 L.H., Daily Totals		$2382.15		$2966.97	$148.88	$185.44
Crew A-3L	**Hr.**	**Daily**	**Hr.**	**Daily**	**Bare Costs**	**Incl. O&P**
1 Equip. Oper. (crane)	$59.20	$473.60	$88.50	$708.00	$54.88	$82.03
1 Equip. Oper. (oiler)	50.55	404.40	75.55	604.40		
1 Hyd. Crane, 80 Ton (Daily)		2101.00		2311.10		
1 P/U Truck, 3/4 Ton (Daily)		142.15		156.37	140.20	154.22
16 L.H., Daily Totals		$3121.15		$3779.86	$195.07	$236.24
Crew A-3M	**Hr.**	**Daily**	**Hr.**	**Daily**	**Bare Costs**	**Incl. O&P**
1 Equip. Oper. (crane)	$59.20	$473.60	$88.50	$708.00	$54.88	$82.03
1 Equip. Oper. (oiler)	50.55	404.40	75.55	604.40		
1 Hyd. Crane, 100 Ton (Daily)		2227.00		2449.70		
1 P/U Truck, 3/4 Ton (Daily)		142.15		156.37	148.07	162.88
16 L.H., Daily Totals		$3247.15		$3918.47	$202.95	$244.90
Crew A-3N	**Hr.**	**Daily**	**Hr.**	**Daily**	**Bare Costs**	**Incl. O&P**
1 Equip. Oper. (crane)	$59.20	$473.60	$88.50	$708.00	$59.20	$88.50
1 Tower Crane (monthly)		1693.00		1862.30	211.63	232.79
8 L.H., Daily Totals		$2166.60		$2570.30	$270.82	$321.29
Crew A-3P	**Hr.**	**Daily**	**Hr.**	**Daily**	**Bare Costs**	**Incl. O&P**
1 Equip. Oper. (light)	$53.00	$424.00	$79.20	$633.60	$53.00	$79.20
1 A.T. Forklift, 31' reach, 45' lift		374.20		411.62	46.77	51.45
8 L.H., Daily Totals		$798.20		$1045.22	$99.78	$130.65
Crew A-3Q	**Hr.**	**Daily**	**Hr.**	**Daily**	**Bare Costs**	**Incl. O&P**
1 Equip. Oper. (light)	$53.00	$424.00	$79.20	$633.60	$53.00	$79.20
1 Pickup Truck, 4x4, 3/4 Ton		175.85		193.44		
1 Flatbed Trailer, 3 Ton		70.30		77.33	30.77	33.85
8 L.H., Daily Totals		$670.15		$904.37	$83.77	$113.05

Crew No.	Bare Costs		Incl. Subs O&P		Cost Per Labor-Hour	
Crew A-3R	**Hr.**	**Daily**	**Hr.**	**Daily**	**Bare Costs**	**Incl. O&P**
1 Equip. Oper. (light)	$53.00	$424.00	$79.20	$633.60	$53.00	$79.20
1 Forklift, Smooth Floor, 8,000 Lb.		255.90		281.49	31.99	35.19
8 L.H., Daily Totals		$679.90		$915.09	$84.99	$114.39
Crew A-4	**Hr.**	**Daily**	**Hr.**	**Daily**	**Bare Costs**	**Incl. O&P**
2 Carpenters	$53.15	$850.40	$79.85	$1277.60	$50.23	$75.25
1 Painter, Ordinary	44.40	355.20	66.05	528.40		
24 L.H., Daily Totals		$1205.60		$1806.00	$50.23	$75.25
Crew A-5	**Hr.**	**Daily**	**Hr.**	**Daily**	**Bare Costs**	**Incl. O&P**
2 Laborers	$42.10	$673.60	$63.25	$1012.00	$42.67	$64.09
.25 Truck Driver (light)	47.25	94.50	70.80	141.60		
.25 Flatbed Truck, Gas, 1.5 Ton		49.04		53.94	2.72	3.00
18 L.H., Daily Totals		$817.14		$1207.54	$45.40	$67.09
Crew A-6	**Hr.**	**Daily**	**Hr.**	**Daily**	**Bare Costs**	**Incl. O&P**
1 Instrument Man	$54.85	$438.80	$82.95	$663.60	$52.92	$79.67
1 Rodman/Chainman	51.00	408.00	76.40	611.20		
1 Level, Electronic		30.80		33.88	1.93	2.12
16 L.H., Daily Totals		$877.60		$1308.68	$54.85	$81.79
Crew A-7	**Hr.**	**Daily**	**Hr.**	**Daily**	**Bare Costs**	**Incl. O&P**
1 Chief of Party	$66.05	$528.40	$99.00	$792.00	$57.30	$86.12
1 Instrument Man	54.85	438.80	82.95	663.60		
1 Rodman/Chainman	51.00	408.00	76.40	611.20		
1 Level, Electronic		30.80		33.88	1.28	1.41
24 L.H., Daily Totals		$1406.00		$2100.68	$58.58	$87.53
Crew A-8	**Hr.**	**Daily**	**Hr.**	**Daily**	**Bare Costs**	**Incl. O&P**
1 Chief of Party	$66.05	$528.40	$99.00	$792.00	$55.73	$83.69
1 Instrument Man	54.85	438.80	82.95	663.60		
2 Rodmen/Chainmen	51.00	816.00	76.40	1222.40		
1 Level, Electronic		30.80		33.88	.96	1.06
32 L.H., Daily Totals		$1814.00		$2711.88	$56.69	$84.75
Crew A-9	**Hr.**	**Daily**	**Hr.**	**Daily**	**Bare Costs**	**Incl. O&P**
1 Asbestos Foreman	$59.15	$473.20	$90.80	$726.40	$58.71	$90.14
7 Asbestos Workers	58.65	3284.40	90.05	5042.80		
64 L.H., Daily Totals		$3757.60		$5769.20	$58.71	$90.14
Crew A-10A	**Hr.**	**Daily**	**Hr.**	**Daily**	**Bare Costs**	**Incl. O&P**
1 Asbestos Foreman	$59.15	$473.20	$90.80	$726.40	$58.82	$90.30
2 Asbestos Workers	58.65	938.40	90.05	1440.80		
24 L.H., Daily Totals		$1411.60		$2167.20	$58.82	$90.30
Crew A-10B	**Hr.**	**Daily**	**Hr.**	**Daily**	**Bare Costs**	**Incl. O&P**
1 Asbestos Foreman	$59.15	$473.20	$90.80	$726.40	$58.77	$90.24
3 Asbestos Workers	58.65	1407.60	90.05	2161.20		
32 L.H., Daily Totals		$1880.80		$2887.60	$58.77	$90.24
Crew A-10C	**Hr.**	**Daily**	**Hr.**	**Daily**	**Bare Costs**	**Incl. O&P**
3 Asbestos Workers	$58.65	$1407.60	$90.05	$2161.20	$58.65	$90.05
1 Flatbed Truck, Gas, 1.5 Ton		196.15		215.76	8.17	8.99
24 L.H., Daily Totals		$1603.75		$2376.97	$66.82	$99.04

Crew No.	Bare Costs		Incl. Subs O&P		Cost Per Labor-Hour	

Crew A-10D	Hr.	Daily	Hr.	Daily	Bare Costs	Incl. O&P
2 Asbestos Workers	$58.65	$938.40	$90.05	$1440.80	$56.76	$86.04
1 Equip. Oper. (crane)	59.20	473.60	88.50	708.00		
1 Equip. Oper. (oiler)	50.55	404.40	75.55	604.40		
1 Hydraulic Crane, 33 Ton		945.35		1039.89	29.54	32.50
32 L.H., Daily Totals		$2761.75		$3793.09	$86.30	$118.53

Crew A-11	Hr.	Daily	Hr.	Daily	Bare Costs	Incl. O&P
1 Asbestos Foreman	$59.15	$473.20	$90.80	$726.40	$58.71	$90.14
7 Asbestos Workers	58.65	3284.40	90.05	5042.80		
2 Chip. Hammers, 12 Lb., Elec.		64.80		71.28	1.01	1.11
64 L.H., Daily Totals		$3822.40		$5840.48	$59.73	$91.26

Crew A-12	Hr.	Daily	Hr.	Daily	Bare Costs	Incl. O&P
1 Asbestos Foreman	$59.15	$473.20	$90.80	$726.40	$58.71	$90.14
7 Asbestos Workers	58.65	3284.40	90.05	5042.80		
1 Trk-Mtd Vac, 14 CY, 1500 Gal.		536.15		589.76		
1 Flatbed Truck, 20,000 GVW		201.60		221.76	11.53	12.68
64 L.H., Daily Totals		$4495.35		$6580.73	$70.24	$102.82

Crew A-13	Hr.	Daily	Hr.	Daily	Bare Costs	Incl. O&P
1 Equip. Oper. (light)	$53.00	$424.00	$79.20	$633.60	$53.00	$79.20
1 Trk-Mtd Vac, 14 CY, 1500 Gal.		536.15		589.76		
1 Flatbed Truck, 20,000 GVW		201.60		221.76	92.22	101.44
8 L.H., Daily Totals		$1161.75		$1445.13	$145.22	$180.64

Crew B-1	Hr.	Daily	Hr.	Daily	Bare Costs	Incl. O&P
1 Labor Foreman (outside)	$44.10	$352.80	$66.25	$530.00	$42.77	$64.25
2 Laborers	42.10	673.60	63.25	1012.00		
24 L.H., Daily Totals		$1026.40		$1542.00	$42.77	$64.25

Crew B-1A	Hr.	Daily	Hr.	Daily	Bare Costs	Incl. O&P
1 Labor Foreman (outside)	$44.10	$352.80	$66.25	$530.00	$42.77	$64.25
2 Laborers	42.10	673.60	63.25	1012.00		
2 Cutting Torches		25.60		28.16		
2 Sets of Gases		343.10		377.41	15.36	16.90
24 L.H., Daily Totals		$1395.10		$1947.57	$58.13	$81.15

Crew B-1B	Hr.	Daily	Hr.	Daily	Bare Costs	Incl. O&P
1 Labor Foreman (outside)	$44.10	$352.80	$66.25	$530.00	$46.88	$70.31
2 Laborers	42.10	673.60	63.25	1012.00		
1 Equip. Oper. (crane)	59.20	473.60	88.50	708.00		
2 Cutting Torches		25.60		28.16		
2 Sets of Gases		343.10		377.41		
1 Hyd. Crane, 12 Ton		471.00		518.10	26.24	28.86
32 L.H., Daily Totals		$2339.70		$3173.67	$73.12	$99.18

Crew B-1C	Hr.	Daily	Hr.	Daily	Bare Costs	Incl. O&P
1 Labor Foreman (outside)	$44.10	$352.80	$66.25	$530.00	$42.77	$64.25
2 Laborers	42.10	673.60	63.25	1012.00		
1 Telescoping Boom Lift, to 60'		310.90		341.99	12.95	14.25
24 L.H., Daily Totals		$1337.30		$1883.99	$55.72	$78.50

Crew B-1D	Hr.	Daily	Hr.	Daily	Bare Costs	Incl. O&P
2 Laborers	$42.10	$673.60	$63.25	$1012.00	$42.10	$63.25
1 Small Work Boat, Gas, 50 H.P.		119.50		131.45		
1 Pressure Washer, 7 GPM		90.50		99.55	13.13	14.44
16 L.H., Daily Totals		$883.60		$1243.00	$55.23	$77.69

Crew B-1E	Hr.	Daily	Hr.	Daily	Bare Costs	Incl. O&P
1 Labor Foreman (outside)	$44.10	$352.80	$66.25	$530.00	$42.60	$64.00
3 Laborers	42.10	1010.40	63.25	1518.00		
1 Work Boat, Diesel, 200 H.P.		1418.00		1559.80		
2 Pressure Washers, 7 GPM		181.00		199.10	49.97	54.97
32 L.H., Daily Totals		$2962.20		$3806.90	$92.57	$118.97

Crew B-1F	Hr.	Daily	Hr.	Daily	Bare Costs	Incl. O&P
2 Skilled Workers	$54.85	$877.60	$82.95	$1327.20	$50.60	$76.38
1 Laborer	42.10	336.80	63.25	506.00		
1 Small Work Boat, Gas, 50 H.P.		119.50		131.45		
1 Pressure Washer, 7 GPM		90.50		99.55	8.75	9.63
24 L.H., Daily Totals		$1424.40		$2064.20	$59.35	$86.01

Crew B-1G	Hr.	Daily	Hr.	Daily	Bare Costs	Incl. O&P
2 Laborers	$42.10	$673.60	$63.25	$1012.00	$42.10	$63.25
1 Small Work Boat, Gas, 50 H.P.		119.50		131.45	7.47	8.22
16 L.H., Daily Totals		$793.10		$1143.45	$49.57	$71.47

Crew B-1H	Hr.	Daily	Hr.	Daily	Bare Costs	Incl. O&P
2 Skilled Workers	$54.85	$877.60	$82.95	$1327.20	$50.60	$76.38
1 Laborer	42.10	336.80	63.25	506.00		
1 Small Work Boat, Gas, 50 H.P.		119.50		131.45	4.98	5.48
24 L.H., Daily Totals		$1333.90		$1964.65	$55.58	$81.86

Crew B-1J	Hr.	Daily	Hr.	Daily	Bare Costs	Incl. O&P
1 Labor Foreman (inside)	$42.60	$340.80	$64.00	$512.00	$42.35	$63.63
1 Laborer	42.10	336.80	63.25	506.00		
16 L.H., Daily Totals		$677.60		$1018.00	$42.35	$63.63

Crew B-1K	Hr.	Daily	Hr.	Daily	Bare Costs	Incl. O&P
1 Carpenter Foreman (inside)	$53.65	$429.20	$80.60	$644.80	$53.40	$80.22
1 Carpenter	53.15	425.20	79.85	638.80		
16 L.H., Daily Totals		$854.40		$1283.60	$53.40	$80.22

Crew B-2	Hr.	Daily	Hr.	Daily	Bare Costs	Incl. O&P
1 Labor Foreman (outside)	$44.10	$352.80	$66.25	$530.00	$42.50	$63.85
4 Laborers	42.10	1347.20	63.25	2024.00		
40 L.H., Daily Totals		$1700.00		$2554.00	$42.50	$63.85

Crew B-2A	Hr.	Daily	Hr.	Daily	Bare Costs	Incl. O&P
1 Labor Foreman (outside)	$44.10	$352.80	$66.25	$530.00	$42.77	$64.25
2 Laborers	42.10	673.60	63.25	1012.00		
1 Telescoping Boom Lift, to 60'		310.90		341.99	12.95	14.25
24 L.H., Daily Totals		$1337.30		$1883.99	$55.72	$78.50

Crew B-3	Hr.	Daily	Hr.	Daily	Bare Costs	Incl. O&P
1 Labor Foreman (outside)	$44.10	$352.80	$66.25	$530.00	$47.16	$70.72
2 Laborers	42.10	673.60	63.25	1012.00		
1 Equip. Oper. (medium)	56.75	454.00	84.85	678.80		
2 Truck Drivers (heavy)	48.95	783.20	73.35	1173.60		
1 Crawler Loader, 3 C.Y.		1169.00		1285.90		
2 Dump Trucks, 12 C.Y., 400 H.P.		1145.80		1260.38	48.23	53.05
48 L.H., Daily Totals		$4578.40		$5940.68	$95.38	$123.76

Crew B-3A	Hr.	Daily	Hr.	Daily	Bare Costs	Incl. O&P
4 Laborers	$42.10	$1347.20	$63.25	$2024.00	$45.03	$67.57
1 Equip. Oper. (medium)	56.75	454.00	84.85	678.80		
1 Hyd. Excavator, 1.5 C.Y.		687.55		756.30	17.19	18.91
40 L.H., Daily Totals		$2488.75		$3459.11	$62.22	$86.48

Crews - Standard

Crew No.	Bare Costs		Incl. Subs O&P		Cost Per Labor-Hour	

Crew B-3B	Hr.	Daily	Hr.	Daily	Bare Costs	Incl. O&P
2 Laborers	$42.10	$673.60	$63.25	$1012.00	$47.48	$71.17
1 Equip. Oper. (medium)	56.75	454.00	84.85	678.80		
1 Truck Driver (heavy)	48.95	391.60	73.35	586.80		
1 Backhoe Loader, 80 H.P.		232.40		255.64		
1 Dump Truck, 12 C.Y., 400 H.P.		572.90		630.19	25.17	27.68
32 L.H., Daily Totals		$2324.50		$3163.43	$72.64	$98.86

Crew B-3C	Hr.	Daily	Hr.	Daily	Bare Costs	Incl. O&P
3 Laborers	$42.10	$1010.40	$63.25	$1518.00	$45.76	$68.65
1 Equip. Oper. (medium)	56.75	454.00	84.85	678.80		
1 Crawler Loader, 4 C.Y.		1489.00		1637.90	46.53	51.18
32 L.H., Daily Totals		$2953.40		$3834.70	$92.29	$119.83

Crew B-4	Hr.	Daily	Hr.	Daily	Bare Costs	Incl. O&P
1 Labor Foreman (outside)	$44.10	$352.80	$66.25	$530.00	$43.58	$65.43
4 Laborers	42.10	1347.20	63.25	2024.00		
1 Truck Driver (heavy)	48.95	391.60	73.35	586.80		
1 Truck Tractor, 220 H.P.		307.10		337.81		
1 Flatbed Trailer, 40 Ton		186.20		204.82	10.28	11.30
48 L.H., Daily Totals		$2584.90		$3683.43	$53.85	$76.74

Crew B-5	Hr.	Daily	Hr.	Daily	Bare Costs	Incl. O&P
1 Labor Foreman (outside)	$44.10	$352.80	$66.25	$530.00	$46.57	$69.85
4 Laborers	42.10	1347.20	63.25	2024.00		
2 Equip. Oper. (medium)	56.75	908.00	84.85	1357.60		
1 Air Compressor, 250 cfm		201.50		221.65		
2 Breakers, Pavement, 60 lb.		107.00		117.70		
2 -50' Air Hoses, 1.5"		48.50		53.35		
1 Crawler Loader, 3 C.Y.		1169.00		1285.90	27.25	29.98
56 L.H., Daily Totals		$4134.00		$5590.20	$73.82	$99.83

Crew B-5A	Hr.	Daily	Hr.	Daily	Bare Costs	Incl. O&P
1 Labor Foreman (outside)	$44.10	$352.80	$66.25	$530.00	$46.76	$70.11
6 Laborers	42.10	2020.80	63.25	3036.00		
2 Equip. Oper. (medium)	56.75	908.00	84.85	1357.60		
1 Equip. Oper. (light)	53.00	424.00	79.20	633.60		
2 Truck Drivers (heavy)	48.95	783.20	73.35	1173.60		
1 Air Compressor, 365 cfm		315.40		346.94		
2 Breakers, Pavement, 60 lb.		107.00		117.70		
8 -50' Air Hoses, 1"		64.00		70.40		
2 Dump Trucks, 8 C.Y., 220 H.P.		856.10		941.71	13.98	15.38
96 L.H., Daily Totals		$5831.30		$8207.55	$60.74	$85.50

Crew B-5B	Hr.	Daily	Hr.	Daily	Bare Costs	Incl. O&P
1 Powderman	$54.85	$438.80	$82.95	$663.60	$52.53	$78.78
2 Equip. Oper. (medium)	56.75	908.00	84.85	1357.60		
3 Truck Drivers (heavy)	48.95	1174.80	73.35	1760.40		
1 F.E. Loader, W.M., 2.5 C.Y.		702.10		772.31		
3 Dump Trucks, 12 C.Y., 400 H.P.		1718.70		1890.57		
1 Air Compressor, 365 cfm		315.40		346.94	57.00	62.70
		$5257.80		$6791.42	$109.54	$141.49

Crew B-5C	Hr.	Daily	Hr.	Daily	Bare Costs	Incl. O&P
3 Laborers	$42.10	$1010.40	$63.25	$1518.00	$48.84	$73.17
1 Equip. Oper. (medium)	56.75	454.00	84.85	678.80		
2 Truck Drivers (heavy)	48.95	783.20	73.35	1173.60		
1 Equip. Oper. (crane)	59.20	473.60	88.50	708.00		
1 Equip. Oper. (oiler)	50.55	404.40	75.55	604.40		
2 Dump Trucks, 12 C.Y., 400 H.P.		1145.80		1260.38		
1 Crawler Loader, 4 C.Y.		1489.00		1637.90		
1 S.P. Crane, 4x4, 25 Ton		1144.00		1258.40	59.04	64.95
64 L.H., Daily Totals		$6904.40		$8839.48	$107.88	$138.12

Crew B-5D	Hr.	Daily	Hr.	Daily	Bare Costs	Incl. O&P
1 Labor Foreman (outside)	$44.10	$352.80	$66.25	$530.00	$46.87	$70.29
4 Laborers	42.10	1347.20	63.25	2024.00		
2 Equip. Oper. (medium)	56.75	908.00	84.85	1357.60		
1 Truck Driver (heavy)	48.95	391.60	73.35	586.80		
1 Air Compressor, 250 cfm		201.50		221.65		
2 Breakers, Pavement, 60 lb.		107.00		117.70		
2 -50' Air Hoses, 1.5"		48.50		53.35		
1 Crawler Loader, 3 C.Y.		1169.00		1285.90		
1 Dump Truck, 12 C.Y., 400 H.P.		572.90		630.19	32.80	36.07
64 L.H., Daily Totals		$5098.50		$6807.19	$79.66	$106.36

Crew B-5E	Hr.	Daily	Hr.	Daily	Bare Costs	Incl. O&P
1 Labor Foreman (outside)	$44.10	$352.80	$66.25	$530.00	$46.87	$70.29
4 Laborers	42.10	1347.20	63.25	2024.00		
2 Equip. Oper. (medium)	56.75	908.00	84.85	1357.60		
1 Truck Driver (heavy)	48.95	391.60	73.35	586.80		
1 Water Tank Trailer, 5000 Gal.		152.25		167.47		
1 High Press. Water Jet, 40 KSI		811.05		892.16		
2 -50' Air Hoses, 1.5"		48.50		53.35		
1 Crawler Loader, 3 C.Y.		1169.00		1285.90		
1 Dump Truck, 12 C.Y., 400 H.P.		572.90		630.19	43.03	47.33
64 L.H., Daily Totals		$5753.30		$7527.47	$89.90	$117.62

Crew B-6	Hr.	Daily	Hr.	Daily	Bare Costs	Incl. O&P
2 Laborers	$42.10	$673.60	$63.25	$1012.00	$45.73	$68.57
1 Equip. Oper. (light)	53.00	424.00	79.20	633.60		
1 Backhoe Loader, 48 H.P.		213.75		235.13	8.91	9.80
24 L.H., Daily Totals		$1311.35		$1880.72	$54.64	$78.36

Crew B-6A	Hr.	Daily	Hr.	Daily	Bare Costs	Incl. O&P
.5 Labor Foreman (outside)	$44.10	$176.40	$66.25	$265.00	$48.36	$72.49
1 Laborer	42.10	336.80	63.25	506.00		
1 Equip. Oper. (medium)	56.75	454.00	84.85	678.80		
1 Vacuum Truck, 5000 Gal.		367.55		404.31	18.38	20.22
20 L.H., Daily Totals		$1334.75		$1854.11	$66.74	$92.71

Crew B-6B	Hr.	Daily	Hr.	Daily	Bare Costs	Incl. O&P
2 Labor Foremen (outside)	$44.10	$705.60	$66.25	$1060.00	$42.77	$64.25
4 Laborers	42.10	1347.20	63.25	2024.00		
1 S.P. Crane, 4x4, 5 Ton		378.10		415.91		
1 Flatbed Truck, Gas, 1.5 Ton		196.15		215.76		
1 Butt Fusion Mach., 4"-12" diam.		426.85		469.54	20.86	22.94
48 L.H., Daily Totals		$3053.90		$4185.21	$63.62	$87.19

Crew No.	Bare Costs		Incl. Subs O&P		Cost Per Labor-Hour	
Crew B-6C	**Hr.**	**Daily**	**Hr.**	**Daily**	**Bare Costs**	**Incl. O&P**
2 Labor Foremen (outside)	$44.10	$705.60	$66.25	$1060.00	$42.77	$64.25
4 Laborers	42.10	1347.20	63.25	2024.00		
1 S.P. Crane, 4x4, 12 Ton		428.30		471.13		
1 Flatbed Truck, Gas, 3 Ton		820.40		902.44		
1 Butt Fusion Mach., 8"-24" diam.		1076.00		1183.60	48.43	53.27
48 L.H., Daily Totals		$4377.50		$5641.17	$91.20	$117.52
Crew B-6D	**Hr.**	**Daily**	**Hr.**	**Daily**	**Bare Costs**	**Incl. O&P**
0.5 Labor Foreman (outside)	$44.10	$176.40	$66.25	$265.00	$48.36	$72.49
1 Laborer	42.10	336.80	63.25	506.00		
1 Equip. Oper. (medium)	56.75	454.00	84.85	678.80		
1 Hydro Excavator, 12 C.Y.		1262.00		1388.20	63.10	69.41
20 L.H., Daily Totals		$2229.20		$2838.00	$111.46	$141.90
Crew B-7	**Hr.**	**Daily**	**Hr.**	**Daily**	**Bare Costs**	**Incl. O&P**
1 Labor Foreman (outside)	$44.10	$352.80	$66.25	$530.00	$44.88	$67.35
4 Laborers	42.10	1347.20	63.25	2024.00		
1 Equip. Oper. (medium)	56.75	454.00	84.85	678.80		
1 Brush Chipper, 12", 130 H.P.		392.80		432.08		
1 Crawler Loader, 3 C.Y.		1169.00		1285.90		
2 Chain Saws, Gas, 36" Long		82.50		90.75	34.26	37.68
48 L.H., Daily Totals		$3798.30		$5041.53	$79.13	$105.03
Crew B-7A	**Hr.**	**Daily**	**Hr.**	**Daily**	**Bare Costs**	**Incl. O&P**
2 Laborers	$42.10	$673.60	$63.25	$1012.00	$45.73	$68.57
1 Equip. Oper. (light)	53.00	424.00	79.20	633.60		
1 Rake w/Tractor		339.45		373.39		
2 Chain Saws, Gas, 18"		104.00		114.40	18.48	20.32
24 L.H., Daily Totals		$1541.05		$2133.40	$64.21	$88.89
Crew B-7B	**Hr.**	**Daily**	**Hr.**	**Daily**	**Bare Costs**	**Incl. O&P**
1 Labor Foreman (outside)	$44.10	$352.80	$66.25	$530.00	$45.46	$68.21
4 Laborers	42.10	1347.20	63.25	2024.00		
1 Equip. Oper. (medium)	56.75	454.00	84.85	678.80		
1 Truck Driver (heavy)	48.95	391.60	73.35	586.80		
1 Brush Chipper, 12", 130 H.P.		392.80		432.08		
1 Crawler Loader, 3 C.Y.		1169.00		1285.90		
2 Chain Saws, Gas, 36" Long		82.50		90.75		
1 Dump Truck, 8 C.Y., 220 H.P.		428.05		470.86	37.01	40.71
56 L.H., Daily Totals		$4617.95		$6099.19	$82.46	$108.91
Crew B-7C	**Hr.**	**Daily**	**Hr.**	**Daily**	**Bare Costs**	**Incl. O&P**
1 Labor Foreman (outside)	$44.10	$352.80	$66.25	$530.00	$45.46	$68.21
4 Laborers	42.10	1347.20	63.25	2024.00		
1 Equip. Oper. (medium)	56.75	454.00	84.85	678.80		
1 Truck Driver (heavy)	48.95	391.60	73.35	586.80		
1 Brush Chipper, 12", 130 H.P.		392.80		432.08		
1 Crawler Loader, 3 C.Y.		1169.00		1285.90		
2 Chain Saws, Gas, 36" Long		82.50		90.75		
1 Dump Truck, 12 C.Y., 400 H.P.		572.90		630.19	39.59	43.55
56 L.H., Daily Totals		$4762.80		$6258.52	$85.05	$111.76

Crew No.	Bare Costs		Incl. Subs O&P		Cost Per Labor-Hour	
Crew B-8	**Hr.**	**Daily**	**Hr.**	**Daily**	**Bare Costs**	**Incl. O&P**
1 Labor Foreman (outside)	$44.10	$352.80	$66.25	$530.00	$48.78	$73.09
2 Laborers	42.10	673.60	63.25	1012.00		
2 Equip. Oper. (medium)	56.75	908.00	84.85	1357.60		
1 Equip. Oper. (oiler)	50.55	404.40	75.55	604.40		
2 Truck Drivers (heavy)	48.95	783.20	73.35	1173.60		
1 Hyd. Crane, 25 Ton		580.85		638.93		
1 Crawler Loader, 3 C.Y.		1169.00		1285.90		
2 Dump Trucks, 12 C.Y., 400 H.P.		1145.80		1260.38	45.24	49.77
64 L.H., Daily Totals		$6017.65		$7862.81	$94.03	$122.86
Crew B-9	**Hr.**	**Daily**	**Hr.**	**Daily**	**Bare Costs**	**Incl. O&P**
1 Labor Foreman (outside)	$44.10	$352.80	$66.25	$530.00	$42.50	$63.85
4 Laborers	42.10	1347.20	63.25	2024.00		
1 Air Compressor, 250 cfm		201.50		221.65		
2 Breakers, Pavement, 60 lb.		107.00		117.70		
2 -50' Air Hoses, 1.5"		48.50		53.35	8.93	9.82
40 L.H., Daily Totals		$2057.00		$2946.70	$51.42	$73.67
Crew B-9A	**Hr.**	**Daily**	**Hr.**	**Daily**	**Bare Costs**	**Incl. O&P**
2 Laborers	$42.10	$673.60	$63.25	$1012.00	$44.38	$66.62
1 Truck Driver (heavy)	48.95	391.60	73.35	586.80		
1 Water Tank Trailer, 5000 Gal.		152.25		167.47		
1 Truck Tractor, 220 H.P.		307.10		337.81		
2 -50' Discharge Hoses, 3"		9.00		9.90	19.51	21.47
24 L.H., Daily Totals		$1533.55		$2113.99	$63.90	$88.08
Crew B-9B	**Hr.**	**Daily**	**Hr.**	**Daily**	**Bare Costs**	**Incl. O&P**
2 Laborers	$42.10	$673.60	$63.25	$1012.00	$44.38	$66.62
1 Truck Driver (heavy)	48.95	391.60	73.35	586.80		
2 -50' Discharge Hoses, 3"		9.00		9.90		
1 Water Tank Trailer, 5000 Gal.		152.25		167.47		
1 Truck Tractor, 220 H.P.		307.10		337.81		
1 Pressure Washer		96.95		106.65	23.55	25.91
24 L.H., Daily Totals		$1630.50		$2220.63	$67.94	$92.53
Crew B-9D	**Hr.**	**Daily**	**Hr.**	**Daily**	**Bare Costs**	**Incl. O&P**
1 Labor Foreman (outside)	$44.10	$352.80	$66.25	$530.00	$42.50	$63.85
4 Common Laborers	42.10	1347.20	63.25	2024.00		
1 Air Compressor, 250 cfm		201.50		221.65		
2 -50' Air Hoses, 1.5"		48.50		53.35		
2 Air Powered Tampers		67.70		74.47	7.94	8.74
40 L.H., Daily Totals		$2017.70		$2903.47	$50.44	$72.59
Crew B-9E	**Hr.**	**Daily**	**Hr.**	**Daily**	**Bare Costs**	**Incl. O&P**
1 Cement Finisher	$49.95	$399.60	$73.45	$587.60	$46.02	$68.35
1 Laborer	42.10	336.80	63.25	506.00		
1 Chip. Hammers, 12 Lb., Elec.		32.40		35.64	2.02	2.23
16 L.H., Daily Totals		$768.80		$1129.24	$48.05	$70.58
Crew B-10	**Hr.**	**Daily**	**Hr.**	**Daily**	**Bare Costs**	**Incl. O&P**
1 Equip. Oper. (medium)	$56.75	$454.00	$84.85	$678.80	$51.87	$77.65
.5 Laborer	42.10	168.40	63.25	253.00		
12 L.H., Daily Totals		$622.40		$931.80	$51.87	$77.65
Crew B-10A	**Hr.**	**Daily**	**Hr.**	**Daily**	**Bare Costs**	**Incl. O&P**
1 Equip. Oper. (medium)	$56.75	$454.00	$84.85	$678.80	$51.87	$77.65
.5 Laborer	42.10	168.40	63.25	253.00		
1 Roller, 2-Drum, W.B., 7.5 H.P.		166.10		182.71	13.84	15.23
12 L.H., Daily Totals		$788.50		$1114.51	$65.71	$92.88

Crew No.	Bare Costs		Incl. Subs O&P		Cost Per Labor-Hour	
Crew B-10B	**Hr.**	**Daily**	**Hr.**	**Daily**	**Bare Costs**	**Incl. O&P**
1 Equip. Oper. (medium)	$56.75	$454.00	$84.85	$678.80	$51.87	$77.65
.5 Laborer	42.10	168.40	63.25	253.00		
1 Dozer, 200 H.P.		1504.00		1654.40	125.33	137.87
12 L.H., Daily Totals		$2126.40		$2586.20	$177.20	$215.52

Crew B-10C	Hr.	Daily	Hr.	Daily	Bare Costs	Incl. O&P
1 Equip. Oper. (medium)	$56.75	$454.00	$84.85	$678.80	$51.87	$77.65
.5 Laborer	42.10	168.40	63.25	253.00		
1 Dozer, 200 H.P.		1504.00		1654.40		
1 Vibratory Roller, Towed, 23 Ton		514.75		566.23	168.23	185.05
12 L.H., Daily Totals		$2641.15		$3152.43	$220.10	$262.70

Crew B-10D	Hr.	Daily	Hr.	Daily	Bare Costs	Incl. O&P
1 Equip. Oper. (medium)	$56.75	$454.00	$84.85	$678.80	$51.87	$77.65
.5 Laborer	42.10	168.40	63.25	253.00		
1 Dozer, 200 H.P.		1504.00		1654.40		
1 Sheepsft. Roller, Towed		436.90		480.59	161.74	177.92
12 L.H., Daily Totals		$2563.30		$3066.79	$213.61	$255.57

Crew B-10E	Hr.	Daily	Hr.	Daily	Bare Costs	Incl. O&P
1 Equip. Oper. (medium)	$56.75	$454.00	$84.85	$678.80	$51.87	$77.65
.5 Laborer	42.10	168.40	63.25	253.00		
1 Tandem Roller, 5 Ton		255.65		281.21	21.30	23.43
12 L.H., Daily Totals		$878.05		$1213.02	$73.17	$101.08

Crew B-10F	Hr.	Daily	Hr.	Daily	Bare Costs	Incl. O&P
1 Equip. Oper. (medium)	$56.75	$454.00	$84.85	$678.80	$51.87	$77.65
.5 Laborer	42.10	168.40	63.25	253.00		
1 Tandem Roller, 10 Ton		236.75		260.43	19.73	21.70
12 L.H., Daily Totals		$859.15		$1192.22	$71.60	$99.35

Crew B-10G	Hr.	Daily	Hr.	Daily	Bare Costs	Incl. O&P
1 Equip. Oper. (medium)	$56.75	$454.00	$84.85	$678.80	$51.87	$77.65
.5 Laborer	42.10	168.40	63.25	253.00		
1 Sheepsfoot Roller, 240 H.P.		1348.00		1482.80	112.33	123.57
12 L.H., Daily Totals		$1970.40		$2414.60	$164.20	$201.22

Crew B-10H	Hr.	Daily	Hr.	Daily	Bare Costs	Incl. O&P
1 Equip. Oper. (medium)	$56.75	$454.00	$84.85	$678.80	$51.87	$77.65
.5 Laborer	42.10	168.40	63.25	253.00		
1 Diaphragm Water Pump, 2"		86.80		95.48		
1 -20' Suction Hose, 2"		3.55		3.90		
2 -50' Discharge Hoses, 2"		8.00		8.80	8.20	9.02
12 L.H., Daily Totals		$720.75		$1039.98	$60.06	$86.67

Crew B-10I	Hr.	Daily	Hr.	Daily	Bare Costs	Incl. O&P
1 Equip. Oper. (medium)	$56.75	$454.00	$84.85	$678.80	$51.87	$77.65
.5 Laborer	42.10	168.40	63.25	253.00		
1 Diaphragm Water Pump, 4"		105.25		115.78		
1 -20' Suction Hose, 4"		17.05		18.75		
2 -50' Discharge Hoses, 4"		25.30		27.83	12.30	13.53
12 L.H., Daily Totals		$770.00		$1094.16	$64.17	$91.18

Crew B-10J	Hr.	Daily	Hr.	Daily	Bare Costs	Incl. O&P
1 Equip. Oper. (medium)	$56.75	$454.00	$84.85	$678.80	$51.87	$77.65
.5 Laborer	42.10	168.40	63.25	253.00		
1 Centrifugal Water Pump, 3"		74.45		81.89		
1 -20' Suction Hose, 3"		8.75		9.63		
2 -50' Discharge Hoses, 3"		9.00		9.90	7.68	8.45
12 L.H., Daily Totals		$714.60		$1033.22	$59.55	$86.10

Crew B-10K	Hr.	Daily	Hr.	Daily	Bare Costs	Incl. O&P
1 Equip. Oper. (medium)	$56.75	$454.00	$84.85	$678.80	$51.87	$77.65
.5 Laborer	42.10	168.40	63.25	253.00		
1 Centr. Water Pump, 6"		232.90		256.19		
1 -20' Suction Hose, 6"		25.20		27.72		
2 -50' Discharge Hoses, 6"		35.80		39.38	24.49	26.94
12 L.H., Daily Totals		$916.30		$1255.09	$76.36	$104.59

Crew B-10L	Hr.	Daily	Hr.	Daily	Bare Costs	Incl. O&P
1 Equip. Oper. (medium)	$56.75	$454.00	$84.85	$678.80	$51.87	$77.65
.5 Laborer	42.10	168.40	63.25	253.00		
1 Dozer, 80 H.P.		401.40		441.54	33.45	36.80
12 L.H., Daily Totals		$1023.80		$1373.34	$85.32	$114.44

Crew B-10M	Hr.	Daily	Hr.	Daily	Bare Costs	Incl. O&P
1 Equip. Oper. (medium)	$56.75	$454.00	$84.85	$678.80	$51.87	$77.65
.5 Laborer	42.10	168.40	63.25	253.00		
1 Dozer, 300 H.P.		1764.00		1940.40	147.00	161.70
12 L.H., Daily Totals		$2386.40		$2872.20	$198.87	$239.35

Crew B-10N	Hr.	Daily	Hr.	Daily	Bare Costs	Incl. O&P
1 Equip. Oper. (medium)	$56.75	$454.00	$84.85	$678.80	$51.87	$77.65
.5 Laborer	42.10	168.40	63.25	253.00		
1 F.E. Loader, T.M., 1.5 C.Y.		565.80		622.38	47.15	51.87
12 L.H., Daily Totals		$1188.20		$1554.18	$99.02	$129.51

Crew B-10O	Hr.	Daily	Hr.	Daily	Bare Costs	Incl. O&P
1 Equip. Oper. (medium)	$56.75	$454.00	$84.85	$678.80	$51.87	$77.65
.5 Laborer	42.10	168.40	63.25	253.00		
1 F.E. Loader, T.M., 2.25 C.Y.		960.90		1056.99	80.08	88.08
12 L.H., Daily Totals		$1583.30		$1988.79	$131.94	$165.73

Crew B-10P	Hr.	Daily	Hr.	Daily	Bare Costs	Incl. O&P
1 Equip. Oper. (medium)	$56.75	$454.00	$84.85	$678.80	$51.87	$77.65
.5 Laborer	42.10	168.40	63.25	253.00		
1 Crawler Loader, 3 C.Y.		1169.00		1285.90	97.42	107.16
12 L.H., Daily Totals		$1791.40		$2217.70	$149.28	$184.81

Crew B-10Q	Hr.	Daily	Hr.	Daily	Bare Costs	Incl. O&P
1 Equip. Oper. (medium)	$56.75	$454.00	$84.85	$678.80	$51.87	$77.65
.5 Laborer	42.10	168.40	63.25	253.00		
1 Crawler Loader, 4 C.Y.		1489.00		1637.90	124.08	136.49
12 L.H., Daily Totals		$2111.40		$2569.70	$175.95	$214.14

Crew B-10R	Hr.	Daily	Hr.	Daily	Bare Costs	Incl. O&P
1 Equip. Oper. (medium)	$56.75	$454.00	$84.85	$678.80	$51.87	$77.65
.5 Laborer	42.10	168.40	63.25	253.00		
1 F.E. Loader, W.M., 1 C.Y.		302.00		332.20	25.17	27.68
12 L.H., Daily Totals		$924.40		$1264.00	$77.03	$105.33

Crew B-10S	Hr.	Daily	Hr.	Daily	Bare Costs	Incl. O&P
1 Equip. Oper. (medium)	$56.75	$454.00	$84.85	$678.80	$51.87	$77.65
.5 Laborer	42.10	168.40	63.25	253.00		
1 F.E. Loader, W.M., 1.5 C.Y.		423.50		465.85	35.29	38.82
12 L.H., Daily Totals		$1045.90		$1397.65	$87.16	$116.47

Crew B-10T	Hr.	Daily	Hr.	Daily	Bare Costs	Incl. O&P
1 Equip. Oper. (medium)	$56.75	$454.00	$84.85	$678.80	$51.87	$77.65
.5 Laborer	42.10	168.40	63.25	253.00		
1 F.E. Loader, W.M., 2.5 C.Y.		702.10		772.31	58.51	64.36
12 L.H., Daily Totals		$1324.50		$1704.11	$110.38	$142.01

Crew No.	Bare Costs		Incl. Subs O&P		Cost Per Labor-Hour	
Crew B-10U	**Hr.**	**Daily**	**Hr.**	**Daily**	**Bare Costs**	**Incl. O&P**
1 Equip. Oper. (medium)	$56.75	$454.00	$84.85	$678.80	$51.87	$77.65
.5 Laborer	42.10	168.40	63.25	253.00		
1 F.E. Loader, W.M., 5.5 C.Y.		957.50		1053.25	79.79	87.77
12 L.H., Daily Totals		$1579.90		$1985.05	$131.66	$165.42
Crew B-10V	**Hr.**	**Daily**	**Hr.**	**Daily**	**Bare Costs**	**Incl. O&P**
1 Equip. Oper. (medium)	$56.75	$454.00	$84.85	$678.80	$51.87	$77.65
.5 Laborer	42.10	168.40	63.25	253.00		
1 Dozer, 700 H.P.		5120.00		5632.00	426.67	469.33
12 L.H., Daily Totals		$5742.40		$6563.80	$478.53	$546.98
Crew B-10W	**Hr.**	**Daily**	**Hr.**	**Daily**	**Bare Costs**	**Incl. O&P**
1 Equip. Oper. (medium)	$56.75	$454.00	$84.85	$678.80	$51.87	$77.65
.5 Laborer	42.10	168.40	63.25	253.00		
1 Dozer, 105 H.P.		633.85		697.24	52.82	58.10
12 L.H., Daily Totals		$1256.25		$1629.04	$104.69	$135.75
Crew B-10X	**Hr.**	**Daily**	**Hr.**	**Daily**	**Bare Costs**	**Incl. O&P**
1 Equip. Oper. (medium)	$56.75	$454.00	$84.85	$678.80	$51.87	$77.65
.5 Laborer	42.10	168.40	63.25	253.00		
1 Dozer, 410 H.P.		2777.00		3054.70	231.42	254.56
12 L.H., Daily Totals		$3399.40		$3986.50	$283.28	$332.21
Crew B-10Y	**Hr.**	**Daily**	**Hr.**	**Daily**	**Bare Costs**	**Incl. O&P**
1 Equip. Oper. (medium)	$56.75	$454.00	$84.85	$678.80	$51.87	$77.65
.5 Laborer	42.10	168.40	63.25	253.00		
1 Vibr. Roller, Towed, 12 Ton		578.60		636.46	48.22	53.04
12 L.H., Daily Totals		$1201.00		$1568.26	$100.08	$130.69
Crew B-11A	**Hr.**	**Daily**	**Hr.**	**Daily**	**Bare Costs**	**Incl. O&P**
1 Equipment Oper. (med.)	$56.75	$454.00	$84.85	$678.80	$49.42	$74.05
1 Laborer	42.10	336.80	63.25	506.00		
1 Dozer, 200 H.P.		1504.00		1654.40	94.00	103.40
16 L.H., Daily Totals		$2294.80		$2839.20	$143.43	$177.45
Crew B-11B	**Hr.**	**Daily**	**Hr.**	**Daily**	**Bare Costs**	**Incl. O&P**
1 Equipment Oper. (light)	$53.00	$424.00	$79.20	$633.60	$47.55	$71.22
1 Laborer	42.10	336.80	63.25	506.00		
1 Air Powered Tamper		33.85		37.23		
1 Air Compressor, 365 cfm		315.40		346.94		
2 -50' Air Hoses, 1.5"		48.50		53.35	24.86	27.35
16 L.H., Daily Totals		$1158.55		$1577.13	$72.41	$98.57
Crew B-11C	**Hr.**	**Daily**	**Hr.**	**Daily**	**Bare Costs**	**Incl. O&P**
1 Equipment Oper. (med.)	$56.75	$454.00	$84.85	$678.80	$49.42	$74.05
1 Laborer	42.10	336.80	63.25	506.00		
1 Backhoe Loader, 48 H.P.		213.75		235.13	13.36	14.70
16 L.H., Daily Totals		$1004.55		$1419.93	$62.78	$88.75
Crew B-11J	**Hr.**	**Daily**	**Hr.**	**Daily**	**Bare Costs**	**Incl. O&P**
1 Equipment Oper. (med.)	$56.75	$454.00	$84.85	$678.80	$49.42	$74.05
1 Laborer	42.10	336.80	63.25	506.00		
1 Grader, 30,000 Lbs.		1062.00		1168.20		
1 Ripper, Beam & 1 Shank		90.50		99.55	72.03	79.23
16 L.H., Daily Totals		$1943.30		$2452.55	$121.46	$153.28

Crew No.	Bare Costs		Incl. Subs O&P		Cost Per Labor-Hour	
Crew B-11K	**Hr.**	**Daily**	**Hr.**	**Daily**	**Bare Costs**	**Incl. O&P**
1 Equipment Oper. (med.)	$56.75	$454.00	$84.85	$678.80	$49.42	$74.05
1 Laborer	42.10	336.80	63.25	506.00		
1 Trencher, Chain Type, 8' D		1872.00		2059.20	117.00	128.70
16 L.H., Daily Totals		$2662.80		$3244.00	$166.43	$202.75
Crew B-11L	**Hr.**	**Daily**	**Hr.**	**Daily**	**Bare Costs**	**Incl. O&P**
1 Equipment Oper. (med.)	$56.75	$454.00	$84.85	$678.80	$49.42	$74.05
1 Laborer	42.10	336.80	63.25	506.00		
1 Grader, 30,000 Lbs.		1062.00		1168.20	66.38	73.01
16 L.H., Daily Totals		$1852.80		$2353.00	$115.80	$147.06
Crew B-11M	**Hr.**	**Daily**	**Hr.**	**Daily**	**Bare Costs**	**Incl. O&P**
1 Equipment Oper. (med.)	$56.75	$454.00	$84.85	$678.80	$49.42	$74.05
1 Laborer	42.10	336.80	63.25	506.00		
1 Backhoe Loader, 80 H.P.		232.40		255.64	14.53	15.98
16 L.H., Daily Totals		$1023.20		$1440.44	$63.95	$90.03
Crew B-11N	**Hr.**	**Daily**	**Hr.**	**Daily**	**Bare Costs**	**Incl. O&P**
1 Labor Foreman (outside)	$44.10	$352.80	$66.25	$530.00	$50.14	$75.12
2 Equipment Operators (med.)	56.75	908.00	84.85	1357.60		
6 Truck Drivers (heavy)	48.95	2349.60	73.35	3520.80		
1 F.E. Loader, W.M., 5.5 C.Y.		957.50		1053.25		
1 Dozer, 410 H.P.		2777.00		3054.70		
6 Dump Trucks, Off Hwy., 50 Ton		10626.00		11688.60	199.45	219.40
72 L.H., Daily Totals		$17970.90		$21204.95	$249.60	$294.51
Crew B-11Q	**Hr.**	**Daily**	**Hr.**	**Daily**	**Bare Costs**	**Incl. O&P**
1 Equipment Operator (med.)	$56.75	$454.00	$84.85	$678.80	$51.87	$77.65
.5 Laborer	42.10	168.40	63.25	253.00		
1 Dozer, 140 H.P.		762.40		838.64	63.53	69.89
12 L.H., Daily Totals		$1384.80		$1770.44	$115.40	$147.54
Crew B-11R	**Hr.**	**Daily**	**Hr.**	**Daily**	**Bare Costs**	**Incl. O&P**
1 Equipment Operator (med.)	$56.75	$454.00	$84.85	$678.80	$51.87	$77.65
.5 Laborer	42.10	168.40	63.25	253.00		
1 Dozer, 200 H.P.		1504.00		1654.40	125.33	137.87
12 L.H., Daily Totals		$2126.40		$2586.20	$177.20	$215.52
Crew B-11S	**Hr.**	**Daily**	**Hr.**	**Daily**	**Bare Costs**	**Incl. O&P**
1 Equipment Operator (med.)	$56.75	$454.00	$84.85	$678.80	$51.87	$77.65
.5 Laborer	42.10	168.40	63.25	253.00		
1 Dozer, 300 H.P.		1764.00		1940.40		
1 Ripper, Beam & 1 Shank		90.50		99.55	154.54	170.00
12 L.H., Daily Totals		$2476.90		$2971.75	$206.41	$247.65
Crew B-11T	**Hr.**	**Daily**	**Hr.**	**Daily**	**Bare Costs**	**Incl. O&P**
1 Equipment Operator (med.)	$56.75	$454.00	$84.85	$678.80	$51.87	$77.65
.5 Laborer	42.10	168.40	63.25	253.00		
1 Dozer, 410 H.P.		2777.00		3054.70		
1 Ripper, Beam & 2 Shanks		138.75		152.63	242.98	267.28
12 L.H., Daily Totals		$3538.15		$4139.13	$294.85	$344.93
Crew B-11U	**Hr.**	**Daily**	**Hr.**	**Daily**	**Bare Costs**	**Incl. O&P**
1 Equipment Operator (med.)	$56.75	$454.00	$84.85	$678.80	$51.87	$77.65
.5 Laborer	42.10	168.40	63.25	253.00		
1 Dozer, 520 H.P.		3398.00		3737.80	283.17	311.48
12 L.H., Daily Totals		$4020.40		$4669.60	$335.03	$389.13

Crew No.	Bare Costs		Incl. Subs O&P		Cost Per Labor-Hour	
Crew B-11V	**Hr.**	**Daily**	**Hr.**	**Daily**	**Bare Costs**	**Incl. O&P**
3 Laborers	$42.10	$1010.40	$63.25	$1518.00	$42.10	$63.25
1 Roller, 2-Drum, W.B., 7.5 H.P.		166.10		182.71	6.92	7.61
24 L.H., Daily Totals		$1176.50		$1700.71	$49.02	$70.86
Crew B-11W	**Hr.**	**Daily**	**Hr.**	**Daily**	**Bare Costs**	**Incl. O&P**
1 Equipment Operator (med.)	$56.75	$454.00	$84.85	$678.80	$49.03	$73.47
1 Common Laborer	42.10	336.80	63.25	506.00		
10 Truck Drivers (heavy)	48.95	3916.00	73.35	5868.00		
1 Dozer, 200 H.P.		1504.00		1654.40		
1 Vibratory Roller, Towed, 23 Ton		514.75		566.23		
10 Dump Trucks, 8 C.Y., 220 H.P.		4280.50		4708.55	65.62	72.18
96 L.H., Daily Totals		$11006.05		$13981.98	$114.65	$145.65
Crew B-11Y	**Hr.**	**Daily**	**Hr.**	**Daily**	**Bare Costs**	**Incl. O&P**
1 Labor Foreman (outside)	$44.10	$352.80	$66.25	$530.00	$47.21	$70.78
5 Common Laborers	42.10	1684.00	63.25	2530.00		
3 Equipment Operators (med.)	56.75	1362.00	84.85	2036.40		
1 Dozer, 80 H.P.		401.40		441.54		
2 Rollers, 2-Drum, W.B., 7.5 H.P.		332.20		365.42		
4 Vibrating Plates, Gas, 21"		661.40		727.54	19.38	21.31
72 L.H., Daily Totals		$4793.80		$6630.90	$66.58	$92.10
Crew B-12A	**Hr.**	**Daily**	**Hr.**	**Daily**	**Bare Costs**	**Incl. O&P**
1 Equip. Oper. (crane)	$59.20	$473.60	$88.50	$708.00	$50.65	$75.88
1 Laborer	42.10	336.80	63.25	506.00		
1 Hyd. Excavator, 1 C.Y.		795.30		874.83	49.71	54.68
16 L.H., Daily Totals		$1605.70		$2088.83	$100.36	$130.55
Crew B-12B	**Hr.**	**Daily**	**Hr.**	**Daily**	**Bare Costs**	**Incl. O&P**
1 Equip. Oper. (crane)	$59.20	$473.60	$88.50	$708.00	$50.65	$75.88
1 Laborer	42.10	336.80	63.25	506.00		
1 Hyd. Excavator, 1.5 C.Y.		687.55		756.30	42.97	47.27
16 L.H., Daily Totals		$1497.95		$1970.31	$93.62	$123.14
Crew B-12C	**Hr.**	**Daily**	**Hr.**	**Daily**	**Bare Costs**	**Incl. O&P**
1 Equip. Oper. (crane)	$59.20	$473.60	$88.50	$708.00	$50.65	$75.88
1 Laborer	42.10	336.80	63.25	506.00		
1 Hyd. Excavator, 2 C.Y.		951.30		1046.43	59.46	65.40
16 L.H., Daily Totals		$1761.70		$2260.43	$110.11	$141.28
Crew B-12D	**Hr.**	**Daily**	**Hr.**	**Daily**	**Bare Costs**	**Incl. O&P**
1 Equip. Oper. (crane)	$59.20	$473.60	$88.50	$708.00	$50.65	$75.88
1 Laborer	42.10	336.80	63.25	506.00		
1 Hyd. Excavator, 3.5 C.Y.		2158.00		2373.80	134.88	148.36
16 L.H., Daily Totals		$2968.40		$3587.80	$185.53	$224.24
Crew B-12E	**Hr.**	**Daily**	**Hr.**	**Daily**	**Bare Costs**	**Incl. O&P**
1 Equip. Oper. (crane)	$59.20	$473.60	$88.50	$708.00	$50.65	$75.88
1 Laborer	42.10	336.80	63.25	506.00		
1 Hyd. Excavator, .5 C.Y.		452.10		497.31	28.26	31.08
16 L.H., Daily Totals		$1262.50		$1711.31	$78.91	$106.96
Crew B-12F	**Hr.**	**Daily**	**Hr.**	**Daily**	**Bare Costs**	**Incl. O&P**
1 Equip. Oper. (crane)	$59.20	$473.60	$88.50	$708.00	$50.65	$75.88
1 Laborer	42.10	336.80	63.25	506.00		
1 Hyd. Excavator, .75 C.Y.		694.25		763.67	43.39	47.73
16 L.H., Daily Totals		$1504.65		$1977.68	$94.04	$123.60

Crew No.	Bare Costs		Incl. Subs O&P		Cost Per Labor-Hour	
Crew B-12G	**Hr.**	**Daily**	**Hr.**	**Daily**	**Bare Costs**	**Incl. O&P**
1 Equip. Oper. (crane)	$59.20	$473.60	$88.50	$708.00	$50.65	$75.88
1 Laborer	42.10	336.80	63.25	506.00		
1 Crawler Crane, 15 Ton		792.30		871.53		
1 Clamshell Bucket, .5 C.Y.		67.10		73.81	53.71	59.08
16 L.H., Daily Totals		$1669.80		$2159.34	$104.36	$134.96
Crew B-12H	**Hr.**	**Daily**	**Hr.**	**Daily**	**Bare Costs**	**Incl. O&P**
1 Equip. Oper. (crane)	$59.20	$473.60	$88.50	$708.00	$50.65	$75.88
1 Laborer	42.10	336.80	63.25	506.00		
1 Crawler Crane, 25 Ton		1129.00		1241.90		
1 Clamshell Bucket, 1 C.Y.		68.55		75.41	74.85	82.33
16 L.H., Daily Totals		$2007.95		$2531.30	$125.50	$158.21
Crew B-12I	**Hr.**	**Daily**	**Hr.**	**Daily**	**Bare Costs**	**Incl. O&P**
1 Equip. Oper. (crane)	$59.20	$473.60	$88.50	$708.00	$50.65	$75.88
1 Laborer	42.10	336.80	63.25	506.00		
1 Crawler Crane, 20 Ton		961.35		1057.48		
1 Dragline Bucket, .75 C.Y.		61.25		67.38	63.91	70.30
16 L.H., Daily Totals		$1833.00		$2338.86	$114.56	$146.18
Crew B-12J	**Hr.**	**Daily**	**Hr.**	**Daily**	**Bare Costs**	**Incl. O&P**
1 Equip. Oper. (crane)	$59.20	$473.60	$88.50	$708.00	$50.65	$75.88
1 Laborer	42.10	336.80	63.25	506.00		
1 Gradall, 5/8 C.Y.		846.50		931.15	52.91	58.20
16 L.H., Daily Totals		$1656.90		$2145.15	$103.56	$134.07
Crew B-12K	**Hr.**	**Daily**	**Hr.**	**Daily**	**Bare Costs**	**Incl. O&P**
1 Equip. Oper. (crane)	$59.20	$473.60	$88.50	$708.00	$50.65	$75.88
1 Laborer	42.10	336.80	63.25	506.00		
1 Gradall, 3 Ton, 1 C.Y.		973.80		1071.18	60.86	66.95
16 L.H., Daily Totals		$1784.20		$2285.18	$111.51	$142.82
Crew B-12L	**Hr.**	**Daily**	**Hr.**	**Daily**	**Bare Costs**	**Incl. O&P**
1 Equip. Oper. (crane)	$59.20	$473.60	$88.50	$708.00	$50.65	$75.88
1 Laborer	42.10	336.80	63.25	506.00		
1 Crawler Crane, 15 Ton		792.30		871.53		
1 F.E. Attachment, .5 C.Y.		65.25		71.78	53.60	58.96
16 L.H., Daily Totals		$1667.95		$2157.30	$104.25	$134.83
Crew B-12M	**Hr.**	**Daily**	**Hr.**	**Daily**	**Bare Costs**	**Incl. O&P**
1 Equip. Oper. (crane)	$59.20	$473.60	$88.50	$708.00	$50.65	$75.88
1 Laborer	42.10	336.80	63.25	506.00		
1 Crawler Crane, 20 Ton		961.35		1057.48		
1 F.E. Attachment, .75 C.Y.		70.40		77.44	64.48	70.93
16 L.H., Daily Totals		$1842.15		$2348.93	$115.13	$146.81
Crew B-12N	**Hr.**	**Daily**	**Hr.**	**Daily**	**Bare Costs**	**Incl. O&P**
1 Equip. Oper. (crane)	$59.20	$473.60	$88.50	$708.00	$50.65	$75.88
1 Laborer	42.10	336.80	63.25	506.00		
1 Crawler Crane, 25 Ton		1129.00		1241.90		
1 F.E. Attachment, 1 C.Y.		76.50		84.15	75.34	82.88
16 L.H., Daily Totals		$2015.90		$2540.05	$125.99	$158.75
Crew B-12O	**Hr.**	**Daily**	**Hr.**	**Daily**	**Bare Costs**	**Incl. O&P**
1 Equip. Oper. (crane)	$59.20	$473.60	$88.50	$708.00	$50.65	$75.88
1 Laborer	42.10	336.80	63.25	506.00		
1 Crawler Crane, 40 Ton		1206.00		1326.60		
1 F.E. Attachment, 1.5 C.Y.		87.60		96.36	80.85	88.94
16 L.H., Daily Totals		$2104.00		$2636.96	$131.50	$164.81

Crew No.	Bare Costs		Incl. Subs O&P		Cost Per Labor-Hour	
Crew B-12P	**Hr.**	**Daily**	**Hr.**	**Daily**	**Bare Costs**	**Incl. O&P**
1 Equip. Oper. (crane)	$59.20	$473.60	$88.50	$708.00	$50.65	$75.88
1 Laborer	42.10	336.80	63.25	506.00		
1 Crawler Crane, 40 Ton		1206.00		1326.60		
1 Dragline Bucket, 1.5 C.Y.		65.05		71.56	79.44	87.38
16 L.H., Daily Totals		$2081.45		$2612.16	$130.09	$163.26
Crew B-12Q	**Hr.**	**Daily**	**Hr.**	**Daily**	**Bare Costs**	**Incl. O&P**
1 Equip. Oper. (crane)	$59.20	$473.60	$88.50	$708.00	$50.65	$75.88
1 Laborer	42.10	336.80	63.25	506.00		
1 Hyd. Excavator, 5/8 C.Y.		598.25		658.08	37.39	41.13
16 L.H., Daily Totals		$1408.65		$1872.08	$88.04	$117.00
Crew B-12S	**Hr.**	**Daily**	**Hr.**	**Daily**	**Bare Costs**	**Incl. O&P**
1 Equip. Oper. (crane)	$59.20	$473.60	$88.50	$708.00	$50.65	$75.88
1 Laborer	42.10	336.80	63.25	506.00		
1 Hyd. Excavator, 2.5 C.Y.		1465.00		1611.50	91.56	100.72
16 L.H., Daily Totals		$2275.40		$2825.50	$142.21	$176.59
Crew B-12T	**Hr.**	**Daily**	**Hr.**	**Daily**	**Bare Costs**	**Incl. O&P**
1 Equip. Oper. (crane)	$59.20	$473.60	$88.50	$708.00	$50.65	$75.88
1 Laborer	42.10	336.80	63.25	506.00		
1 Crawler Crane, 75 Ton		2002.00		2202.20		
1 F.E. Attachment, 3 C.Y.		114.15		125.57	132.26	145.49
16 L.H., Daily Totals		$2926.55		$3541.76	$182.91	$221.36
Crew B-12V	**Hr.**	**Daily**	**Hr.**	**Daily**	**Bare Costs**	**Incl. O&P**
1 Equip. Oper. (crane)	$59.20	$473.60	$88.50	$708.00	$50.65	$75.88
1 Laborer	42.10	336.80	63.25	506.00		
1 Crawler Crane, 75 Ton		2002.00		2202.20		
1 Dragline Bucket, 3 C.Y.		71.60		78.76	129.60	142.56
16 L.H., Daily Totals		$2884.00		$3494.96	$180.25	$218.44
Crew B-12Y	**Hr.**	**Daily**	**Hr.**	**Daily**	**Bare Costs**	**Incl. O&P**
1 Equip. Oper. (crane)	$59.20	$473.60	$88.50	$708.00	$47.80	$71.67
2 Laborers	42.10	673.60	63.25	1012.00		
1 Hyd. Excavator, 3.5 C.Y.		2158.00		2373.80	89.92	98.91
24 L.H., Daily Totals		$3305.20		$4093.80	$137.72	$170.57
Crew B-12Z	**Hr.**	**Daily**	**Hr.**	**Daily**	**Bare Costs**	**Incl. O&P**
1 Equip. Oper. (crane)	$59.20	$473.60	$88.50	$708.00	$47.80	$71.67
2 Laborers	42.10	673.60	63.25	1012.00		
1 Hyd. Excavator, 2.5 C.Y.		1465.00		1611.50	61.04	67.15
24 L.H., Daily Totals		$2612.20		$3331.50	$108.84	$138.81
Crew B-13	**Hr.**	**Daily**	**Hr.**	**Daily**	**Bare Costs**	**Incl. O&P**
1 Labor Foreman (outside)	$44.10	$352.80	$66.25	$530.00	$46.04	$69.04
4 Laborers	42.10	1347.20	63.25	2024.00		
1 Equip. Oper. (crane)	59.20	473.60	88.50	708.00		
1 Equip. Oper. (oiler)	50.55	404.40	75.55	604.40		
1 Hyd. Crane, 25 Ton		580.85		638.93	10.37	11.41
56 L.H., Daily Totals		$3158.85		$4505.34	$56.41	$80.45

Crew No.	Bare Costs		Incl. Subs O&P		Cost Per Labor-Hour	
Crew B-13A	**Hr.**	**Daily**	**Hr.**	**Daily**	**Bare Costs**	**Incl. O&P**
1 Labor Foreman (outside)	$44.10	$352.80	$66.25	$530.00	$48.53	$72.74
2 Laborers	42.10	673.60	63.25	1012.00		
2 Equipment Operators (med.)	56.75	908.00	84.85	1357.60		
2 Truck Drivers (heavy)	48.95	783.20	73.35	1173.60		
1 Crawler Crane, 75 Ton		2002.00		2202.20		
1 Crawler Loader, 4 C.Y.		1489.00		1637.90		
2 Dump Trucks, 8 C.Y., 220 H.P.		856.10		941.71	77.63	85.39
56 L.H., Daily Totals		$7064.70		$8855.01	$126.16	$158.13
Crew B-13B	**Hr.**	**Daily**	**Hr.**	**Daily**	**Bare Costs**	**Incl. O&P**
1 Labor Foreman (outside)	$44.10	$352.80	$66.25	$530.00	$46.04	$69.04
4 Laborers	42.10	1347.20	63.25	2024.00		
1 Equip. Oper. (crane)	59.20	473.60	88.50	708.00		
1 Equip. Oper. (oiler)	50.55	404.40	75.55	604.40		
1 Hyd. Crane, 55 Ton		980.20		1078.22	17.50	19.25
56 L.H., Daily Totals		$3558.20		$4944.62	$63.54	$88.30
Crew B-13C	**Hr.**	**Daily**	**Hr.**	**Daily**	**Bare Costs**	**Incl. O&P**
1 Labor Foreman (outside)	$44.10	$352.80	$66.25	$530.00	$46.04	$69.04
4 Laborers	42.10	1347.20	63.25	2024.00		
1 Equip. Oper. (crane)	59.20	473.60	88.50	708.00		
1 Equip. Oper. (oiler)	50.55	404.40	75.55	604.40		
1 Crawler Crane, 100 Ton		2287.00		2515.70	40.84	44.92
56 L.H., Daily Totals		$4865.00		$6382.10	$86.88	$113.97
Crew B-13D	**Hr.**	**Daily**	**Hr.**	**Daily**	**Bare Costs**	**Incl. O&P**
1 Laborer	$42.10	$336.80	$63.25	$506.00	$50.65	$75.88
1 Equip. Oper. (crane)	59.20	473.60	88.50	708.00		
1 Hyd. Excavator, 1 C.Y.		795.30		874.83		
1 Trench Box		117.75		129.53	57.07	62.77
16 L.H., Daily Totals		$1723.45		$2218.36	$107.72	$138.65
Crew B-13E	**Hr.**	**Daily**	**Hr.**	**Daily**	**Bare Costs**	**Incl. O&P**
1 Laborer	$42.10	$336.80	$63.25	$506.00	$50.65	$75.88
1 Equip. Oper. (crane)	59.20	473.60	88.50	708.00		
1 Hyd. Excavator, 1.5 C.Y.		687.55		756.30		
1 Trench Box		117.75		129.53	50.33	55.36
16 L.H., Daily Totals		$1615.70		$2099.83	$100.98	$131.24
Crew B-13F	**Hr.**	**Daily**	**Hr.**	**Daily**	**Bare Costs**	**Incl. O&P**
1 Laborer	$42.10	$336.80	$63.25	$506.00	$50.65	$75.88
1 Equip. Oper. (crane)	59.20	473.60	88.50	708.00		
1 Hyd. Excavator, 3.5 C.Y.		2158.00		2373.80		
1 Trench Box		117.75		129.53	142.23	156.46
16 L.H., Daily Totals		$3086.15		$3717.32	$192.88	$232.33
Crew B-13G	**Hr.**	**Daily**	**Hr.**	**Daily**	**Bare Costs**	**Incl. O&P**
1 Laborer	$42.10	$336.80	$63.25	$506.00	$50.65	$75.88
1 Equip. Oper. (crane)	59.20	473.60	88.50	708.00		
1 Hyd. Excavator, .75 C.Y.		694.25		763.67		
1 Trench Box		117.75		129.53	50.75	55.83
16 L.H., Daily Totals		$1622.40		$2107.20	$101.40	$131.70
Crew B-13H	**Hr.**	**Daily**	**Hr.**	**Daily**	**Bare Costs**	**Incl. O&P**
1 Laborer	$42.10	$336.80	$63.25	$506.00	$50.65	$75.88
1 Equip. Oper. (crane)	59.20	473.60	88.50	708.00		
1 Gradall, 5/8 C.Y.		846.50		931.15		
1 Trench Box		117.75		129.53	60.27	66.29
16 L.H., Daily Totals		$1774.65		$2274.68	$110.92	$142.17

Crew No.	Bare Costs		Incl. Subs O&P		Cost Per Labor-Hour	
Crew B-13I	**Hr.**	**Daily**	**Hr.**	**Daily**	**Bare Costs**	**Incl. O&P**
1 Laborer	$42.10	$336.80	$63.25	$506.00	$50.65	$75.88
1 Equip. Oper. (crane)	59.20	473.60	88.50	708.00		
1 Gradall, 3 Ton, 1 C.Y.		973.80		1071.18		
1 Trench Box		117.75		129.53	68.22	75.04
16 L.H., Daily Totals		$1901.95		$2414.70	$118.87	$150.92
Crew B-13J	**Hr.**	**Daily**	**Hr.**	**Daily**	**Bare Costs**	**Incl. O&P**
1 Laborer	$42.10	$336.80	$63.25	$506.00	$50.65	$75.88
1 Equip. Oper. (crane)	59.20	473.60	88.50	708.00		
1 Hyd. Excavator, 2.5 C.Y.		1465.00		1611.50		
1 Trench Box		117.75		129.53	98.92	108.81
16 L.H., Daily Totals		$2393.15		$2955.03	$149.57	$184.69
Crew B-13K	**Hr.**	**Daily**	**Hr.**	**Daily**	**Bare Costs**	**Incl. O&P**
2 Equip. Opers. (crane)	$59.20	$947.20	$88.50	$1416.00	$59.20	$88.50
1 Hyd. Excavator, .75 C.Y.		694.25		763.67		
1 Hyd. Hammer, 4000 ft-lb		641.50		705.65		
1 Hyd. Excavator, .75 C.Y.		694.25		763.67	126.88	139.56
16 L.H., Daily Totals		$2977.20		$3649.00	$186.07	$228.06
Crew B-13L	**Hr.**	**Daily**	**Hr.**	**Daily**	**Bare Costs**	**Incl. O&P**
2 Equip. Opers. (crane)	$59.20	$947.20	$88.50	$1416.00	$59.20	$88.50
1 Hyd. Excavator, 1.5 C.Y.		687.55		756.30		
1 Hyd. Hammer, 5000 ft-lb		697.25		766.98		
1 Hyd. Excavator, .75 C.Y.		694.25		763.67	129.94	142.93
16 L.H., Daily Totals		$3026.25		$3702.95	$189.14	$231.43
Crew B-13M	**Hr.**	**Daily**	**Hr.**	**Daily**	**Bare Costs**	**Incl. O&P**
2 Equip. Opers. (crane)	$59.20	$947.20	$88.50	$1416.00	$59.20	$88.50
1 Hyd. Excavator, 2.5 C.Y.		1465.00		1611.50		
1 Hyd. Hammer, 8000 ft-lb		907.75		998.52		
1 Hyd. Excavator, 1.5 C.Y.		687.55		756.30	191.27	210.40
16 L.H., Daily Totals		$4007.50		$4782.33	$250.47	$298.90
Crew B-13N	**Hr.**	**Daily**	**Hr.**	**Daily**	**Bare Costs**	**Incl. O&P**
2 Equip. Opers. (crane)	$59.20	$947.20	$88.50	$1416.00	$59.20	$88.50
1 Hyd. Excavator, 3.5 C.Y.		2158.00		2373.80		
1 Hyd. Hammer, 12,000 ft-lb		871.75		958.92		
1 Hyd. Excavator, 1.5 C.Y.		687.55		756.30	232.33	255.56
16 L.H., Daily Totals		$4664.50		$5505.03	$291.53	$344.06
Crew B-14	**Hr.**	**Daily**	**Hr.**	**Daily**	**Bare Costs**	**Incl. O&P**
1 Labor Foreman (outside)	$44.10	$352.80	$66.25	$530.00	$44.25	$66.41
4 Laborers	42.10	1347.20	63.25	2024.00		
1 Equip. Oper. (light)	53.00	424.00	79.20	633.60		
1 Backhoe Loader, 48 H.P.		213.75		235.13	4.45	4.90
48 L.H., Daily Totals		$2337.75		$3422.72	$48.70	$71.31
Crew B-14A	**Hr.**	**Daily**	**Hr.**	**Daily**	**Bare Costs**	**Incl. O&P**
1 Equip. Oper. (crane)	$59.20	$473.60	$88.50	$708.00	$53.50	$80.08
.5 Laborer	42.10	168.40	63.25	253.00		
1 Hyd. Excavator, 4.5 C.Y.		3409.00		3749.90	284.08	312.49
12 L.H., Daily Totals		$4051.00		$4710.90	$337.58	$392.57
Crew B-14B	**Hr.**	**Daily**	**Hr.**	**Daily**	**Bare Costs**	**Incl. O&P**
1 Equip. Oper. (crane)	$59.20	$473.60	$88.50	$708.00	$53.50	$80.08
.5 Laborer	42.10	168.40	63.25	253.00		
1 Hyd. Excavator, 6 C.Y.		3464.00		3810.40	288.67	317.53
12 L.H., Daily Totals		$4106.00		$4771.40	$342.17	$397.62

Crew No.	Bare Costs		Incl. Subs O&P		Cost Per Labor-Hour	
Crew B-14C	**Hr.**	**Daily**	**Hr.**	**Daily**	**Bare Costs**	**Incl. O&P**
1 Equip. Oper. (crane)	$59.20	$473.60	$88.50	$708.00	$53.50	$80.08
.5 Laborer	42.10	168.40	63.25	253.00		
1 Hyd. Excavator, 7 C.Y.		3437.00		3780.70	286.42	315.06
12 L.H., Daily Totals		$4079.00		$4741.70	$339.92	$395.14
Crew B-14F	**Hr.**	**Daily**	**Hr.**	**Daily**	**Bare Costs**	**Incl. O&P**
1 Equip. Oper. (crane)	$59.20	$473.60	$88.50	$708.00	$53.50	$80.08
.5 Laborer	42.10	168.40	63.25	253.00		
1 Hyd. Shovel, 7 C.Y.		4099.00		4508.90	341.58	375.74
12 L.H., Daily Totals		$4741.00		$5469.90	$395.08	$455.82
Crew B-14G	**Hr.**	**Daily**	**Hr.**	**Daily**	**Bare Costs**	**Incl. O&P**
1 Equip. Oper. (crane)	$59.20	$473.60	$88.50	$708.00	$53.50	$80.08
.5 Laborer	42.10	168.40	63.25	253.00		
1 Hyd. Shovel, 12 C.Y.		5951.00		6546.10	495.92	545.51
12 L.H., Daily Totals		$6593.00		$7507.10	$549.42	$625.59
Crew B-14J	**Hr.**	**Daily**	**Hr.**	**Daily**	**Bare Costs**	**Incl. O&P**
1 Equip. Oper. (medium)	$56.75	$454.00	$84.85	$678.80	$51.87	$77.65
.5 Laborer	42.10	168.40	63.25	253.00		
1 F.E. Loader, 8 C.Y.		2261.00		2487.10	188.42	207.26
12 L.H., Daily Totals		$2883.40		$3418.90	$240.28	$284.91
Crew B-14K	**Hr.**	**Daily**	**Hr.**	**Daily**	**Bare Costs**	**Incl. O&P**
1 Equip. Oper. (medium)	$56.75	$454.00	$84.85	$678.80	$51.87	$77.65
.5 Laborer	42.10	168.40	63.25	253.00		
1 F.E. Loader, 10 C.Y.		2674.00		2941.40	222.83	245.12
12 L.H., Daily Totals		$3296.40		$3873.20	$274.70	$322.77
Crew B-15	**Hr.**	**Daily**	**Hr.**	**Daily**	**Bare Costs**	**Incl. O&P**
1 Equipment Oper. (med.)	$56.75	$454.00	$84.85	$678.80	$50.20	$75.19
.5 Laborer	42.10	168.40	63.25	253.00		
2 Truck Drivers (heavy)	48.95	783.20	73.35	1173.60		
2 Dump Trucks, 12 C.Y., 400 H.P.		1145.80		1260.38		
1 Dozer, 200 H.P.		1504.00		1654.40	94.64	104.10
28 L.H., Daily Totals		$4055.40		$5020.18	$144.84	$179.29
Crew B-16	**Hr.**	**Daily**	**Hr.**	**Daily**	**Bare Costs**	**Incl. O&P**
1 Labor Foreman (outside)	$44.10	$352.80	$66.25	$530.00	$44.31	$66.53
2 Laborers	42.10	673.60	63.25	1012.00		
1 Truck Driver (heavy)	48.95	391.60	73.35	586.80		
1 Dump Truck, 12 C.Y., 400 H.P.		572.90		630.19	17.90	19.69
32 L.H., Daily Totals		$1990.90		$2758.99	$62.22	$86.22
Crew B-17	**Hr.**	**Daily**	**Hr.**	**Daily**	**Bare Costs**	**Incl. O&P**
2 Laborers	$42.10	$673.60	$63.25	$1012.00	$46.54	$69.76
1 Equip. Oper. (light)	53.00	424.00	79.20	633.60		
1 Truck Driver (heavy)	48.95	391.60	73.35	586.80		
1 Backhoe Loader, 48 H.P.		213.75		235.13		
1 Dump Truck, 8 C.Y., 220 H.P.		428.05		470.86	20.06	22.06
32 L.H., Daily Totals		$2131.00		$2938.38	$66.59	$91.82
Crew B-17A	**Hr.**	**Daily**	**Hr.**	**Daily**	**Bare Costs**	**Incl. O&P**
2 Labor Foremen (outside)	$44.10	$705.60	$66.25	$1060.00	$45.25	$68.09
6 Laborers	42.10	2020.80	63.25	3036.00		
1 Skilled Worker Foreman (out)	56.85	454.80	86.00	688.00		
1 Skilled Worker	54.85	438.80	82.95	663.60		
80 L.H., Daily Totals		$3620.00		$5447.60	$45.25	$68.09

Crew No.	Bare Costs		Incl. Subs O&P		Cost Per Labor-Hour	

Crew B-17B	Hr.	Daily	Hr.	Daily	Bare Costs	Incl. O&P
2 Laborers	$42.10	$673.60	$63.25	$1012.00	$46.54	$69.76
1 Equip. Oper. (light)	53.00	424.00	79.20	633.60		
1 Truck Driver (heavy)	48.95	391.60	73.35	586.80		
1 Backhoe Loader, 48 H.P.		213.75		235.13		
1 Dump Truck, 12 C.Y., 400 H.P.		572.90		630.19	24.58	27.04
32 L.H., Daily Totals		$2275.85		$3097.72	$71.12	$96.80

Crew B-18	Hr.	Daily	Hr.	Daily	Bare Costs	Incl. O&P
1 Labor Foreman (outside)	$44.10	$352.80	$66.25	$530.00	$42.77	$64.25
2 Laborers	42.10	673.60	63.25	1012.00		
1 Vibrating Plate, Gas, 21"		165.35		181.88	6.89	7.58
24 L.H., Daily Totals		$1191.75		$1723.89	$49.66	$71.83

Crew B-19	Hr.	Daily	Hr.	Daily	Bare Costs	Incl. O&P
1 Pile Driver Foreman (outside)	$56.20	$449.60	$87.30	$698.40	$55.24	$84.58
4 Pile Drivers	54.20	1734.40	84.20	2694.40		
2 Equip. Oper. (crane)	59.20	947.20	88.50	1416.00		
1 Equip. Oper. (oiler)	50.55	404.40	75.55	604.40		
1 Crawler Crane, 40 Ton		1206.00		1326.60		
1 Lead, 90' High		367.45		404.19		
1 Hammer, Diesel, 22k ft-lb		436.45		480.10	31.40	34.55
64 L.H., Daily Totals		$5545.50		$7624.09	$86.65	$119.13

Crew B-19A	Hr.	Daily	Hr.	Daily	Bare Costs	Incl. O&P
1 Pile Driver Foreman (outside)	$56.20	$449.60	$87.30	$698.40	$55.24	$84.58
4 Pile Drivers	54.20	1734.40	84.20	2694.40		
2 Equip. Oper. (crane)	59.20	947.20	88.50	1416.00		
1 Equip. Oper. (oiler)	50.55	404.40	75.55	604.40		
1 Crawler Crane, 75 Ton		2002.00		2202.20		
1 Lead, 90' High		367.45		404.19		
1 Hammer, Diesel, 41k ft-lb		576.65		634.32	46.03	50.64
64 L.H., Daily Totals		$6481.70		$8653.91	$101.28	$135.22

Crew B-19B	Hr.	Daily	Hr.	Daily	Bare Costs	Incl. O&P
1 Pile Driver Foreman (outside)	$56.20	$449.60	$87.30	$698.40	$55.24	$84.58
4 Pile Drivers	54.20	1734.40	84.20	2694.40		
2 Equip. Oper. (crane)	59.20	947.20	88.50	1416.00		
1 Equip. Oper. (oiler)	50.55	404.40	75.55	604.40		
1 Crawler Crane, 40 Ton		1206.00		1326.60		
1 Lead, 90' High		367.45		404.19		
1 Hammer, Diesel, 22k ft-lb		436.45		480.10		
1 Barge, 400 Ton		858.85		944.74	44.82	49.31
64 L.H., Daily Totals		$6404.35		$8568.83	$100.07	$133.89

Crew B-19C	Hr.	Daily	Hr.	Daily	Bare Costs	Incl. O&P
1 Pile Driver Foreman (outside)	$56.20	$449.60	$87.30	$698.40	$55.24	$84.58
4 Pile Drivers	54.20	1734.40	84.20	2694.40		
2 Equip. Oper. (crane)	59.20	947.20	88.50	1416.00		
1 Equip. Oper. (oiler)	50.55	404.40	75.55	604.40		
1 Crawler Crane, 75 Ton		2002.00		2202.20		
1 Lead, 90' High		367.45		404.19		
1 Hammer, Diesel, 41k ft-lb		576.65		634.32		
1 Barge, 400 Ton		858.85		944.74	59.45	65.40
64 L.H., Daily Totals		$7340.55		$9598.65	$114.70	$149.98

Crew B-20	Hr.	Daily	Hr.	Daily	Bare Costs	Incl. O&P
1 Labor Foreman (outside)	$44.10	$352.80	$66.25	$530.00	$47.02	$70.82
1 Skilled Worker	54.85	438.80	82.95	663.60		
1 Laborer	42.10	336.80	63.25	506.00		
24 L.H., Daily Totals		$1128.40		$1699.60	$47.02	$70.82

Crew No.	Bare Costs		Incl. Subs O&P		Cost Per Labor-Hour	

Crew B-20A	Hr.	Daily	Hr.	Daily	Bare Costs	Incl. O&P
1 Labor Foreman (outside)	$44.10	$352.80	$66.25	$530.00	$50.55	$75.78
1 Laborer	42.10	336.80	63.25	506.00		
1 Plumber	64.45	515.60	96.45	771.60		
1 Plumber Apprentice	51.55	412.40	77.15	617.20		
32 L.H., Daily Totals		$1617.60		$2424.80	$50.55	$75.78

Crew B-21	Hr.	Daily	Hr.	Daily	Bare Costs	Incl. O&P
1 Labor Foreman (outside)	$44.10	$352.80	$66.25	$530.00	$48.76	$73.34
1 Skilled Worker	54.85	438.80	82.95	663.60		
1 Laborer	42.10	336.80	63.25	506.00		
.5 Equip. Oper. (crane)	59.20	236.80	88.50	354.00		
.5 S.P. Crane, 4x4, 5 Ton		189.05		207.96	6.75	7.43
28 L.H., Daily Totals		$1554.25		$2261.55	$55.51	$80.77

Crew B-21A	Hr.	Daily	Hr.	Daily	Bare Costs	Incl. O&P
1 Labor Foreman (outside)	$44.10	$352.80	$66.25	$530.00	$52.28	$78.32
1 Laborer	42.10	336.80	63.25	506.00		
1 Plumber	64.45	515.60	96.45	771.60		
1 Plumber Apprentice	51.55	412.40	77.15	617.20		
1 Equip. Oper. (crane)	59.20	473.60	88.50	708.00		
1 S.P. Crane, 4x4, 12 Ton		428.30		471.13	10.71	11.78
40 L.H., Daily Totals		$2519.50		$3603.93	$62.99	$90.10

Crew B-21B	Hr.	Daily	Hr.	Daily	Bare Costs	Incl. O&P
1 Labor Foreman (outside)	$44.10	$352.80	$66.25	$530.00	$45.92	$68.90
3 Laborers	42.10	1010.40	63.25	1518.00		
1 Equip. Oper. (crane)	59.20	473.60	88.50	708.00		
1 Hyd. Crane, 12 Ton		471.00		518.10	11.78	12.95
40 L.H., Daily Totals		$2307.80		$3274.10	$57.70	$81.85

Crew B-21C	Hr.	Daily	Hr.	Daily	Bare Costs	Incl. O&P
1 Labor Foreman (outside)	$44.10	$352.80	$66.25	$530.00	$46.04	$69.04
4 Laborers	42.10	1347.20	63.25	2024.00		
1 Equip. Oper. (crane)	59.20	473.60	88.50	708.00		
1 Equip. Oper. (oiler)	50.55	404.40	75.55	604.40		
2 Cutting Torches		25.60		28.16		
2 Sets of Gases		343.10		377.41		
1 Lattice Boom Crane, 90 Ton		1696.00		1865.60	36.87	40.56
56 L.H., Daily Totals		$4642.70		$6137.57	$82.91	$109.60

Crew B-22	Hr.	Daily	Hr.	Daily	Bare Costs	Incl. O&P
1 Labor Foreman (outside)	$44.10	$352.80	$66.25	$530.00	$49.45	$74.35
1 Skilled Worker	54.85	438.80	82.95	663.60		
1 Laborer	42.10	336.80	63.25	506.00		
.75 Equip. Oper. (crane)	59.20	355.20	88.50	531.00		
.75 S.P. Crane, 4x4, 5 Ton		283.57		311.93	9.45	10.40
30 L.H., Daily Totals		$1767.18		$2542.53	$58.91	$84.75

Crew B-22A	Hr.	Daily	Hr.	Daily	Bare Costs	Incl. O&P
1 Labor Foreman (outside)	$44.10	$352.80	$66.25	$530.00	$48.47	$72.84
1 Skilled Worker	54.85	438.80	82.95	663.60		
2 Laborers	42.10	673.60	63.25	1012.00		
1 Equipment Operator, Crane	59.20	473.60	88.50	708.00		
1 S.P. Crane, 4x4, 5 Ton		378.10		415.91		
1 Butt Fusion Mach., 4"-12" diam.		426.85		469.54	20.12	22.14
40 L.H., Daily Totals		$2743.75		$3799.05	$68.59	$94.98

Crew No.	Bare Costs		Incl. Subs O&P		Cost Per Labor-Hour	
Crew B-22B	**Hr.**	**Daily**	**Hr.**	**Daily**	**Bare Costs**	**Incl. O&P**
1 Labor Foreman (outside)	$44.10	$352.80	$66.25	$530.00	$48.47	$72.84
1 Skilled Worker	54.85	438.80	82.95	663.60		
2 Laborers	42.10	673.60	63.25	1012.00		
1 Equip. Oper. (crane)	59.20	473.60	88.50	708.00		
1 S.P. Crane, 4x4, 5 Ton		378.10		415.91		
1 Butt Fusion Mach., 8"-24" diam.		1076.00		1183.60	36.35	39.99
40 L.H., Daily Totals		$3392.90		$4513.11	$84.82	$112.83
Crew B-22C	**Hr.**	**Daily**	**Hr.**	**Daily**	**Bare Costs**	**Incl. O&P**
1 Skilled Worker	$54.85	$438.80	$82.95	$663.60	$48.48	$73.10
1 Laborer	42.10	336.80	63.25	506.00		
1 Butt Fusion Mach., 2"-8" diam.		156.05		171.66	9.75	10.73
16 L.H., Daily Totals		$931.65		$1341.26	$58.23	$83.83
Crew B-23	**Hr.**	**Daily**	**Hr.**	**Daily**	**Bare Costs**	**Incl. O&P**
1 Labor Foreman (outside)	$44.10	$352.80	$66.25	$530.00	$42.50	$63.85
4 Laborers	42.10	1347.20	63.25	2024.00		
1 Drill Rig, Truck-Mounted		771.10		848.21		
1 Flatbed Truck, Gas, 3 Ton		820.40		902.44	39.79	43.77
40 L.H., Daily Totals		$3291.50		$4304.65	$82.29	$107.62
Crew B-23A	**Hr.**	**Daily**	**Hr.**	**Daily**	**Bare Costs**	**Incl. O&P**
1 Labor Foreman (outside)	$44.10	$352.80	$66.25	$530.00	$47.65	$71.45
1 Laborer	42.10	336.80	63.25	506.00		
1 Equip. Oper. (medium)	56.75	454.00	84.85	678.80		
1 Drill Rig, Truck-Mounted		771.10		848.21		
1 Pickup Truck, 3/4 Ton		110.85		121.94	36.75	40.42
24 L.H., Daily Totals		$2025.55		$2684.95	$84.40	$111.87
Crew B-23B	**Hr.**	**Daily**	**Hr.**	**Daily**	**Bare Costs**	**Incl. O&P**
1 Labor Foreman (outside)	$44.10	$352.80	$66.25	$530.00	$47.65	$71.45
1 Laborer	42.10	336.80	63.25	506.00		
1 Equip. Oper. (medium)	56.75	454.00	84.85	678.80		
1 Drill Rig, Truck-Mounted		771.10		848.21		
1 Pickup Truck, 3/4 Ton		110.85		121.94		
1 Centr. Water Pump, 6"		232.90		256.19	46.45	51.10
24 L.H., Daily Totals		$2258.45		$2941.14	$94.10	$122.55
Crew B-24	**Hr.**	**Daily**	**Hr.**	**Daily**	**Bare Costs**	**Incl. O&P**
1 Cement Finisher	$49.95	$399.60	$73.45	$587.60	$48.40	$72.18
1 Laborer	42.10	336.80	63.25	506.00		
1 Carpenter	53.15	425.20	79.85	638.80		
24 L.H., Daily Totals		$1161.60		$1732.40	$48.40	$72.18
Crew B-25	**Hr.**	**Daily**	**Hr.**	**Daily**	**Bare Costs**	**Incl. O&P**
1 Labor Foreman (outside)	$44.10	$352.80	$66.25	$530.00	$46.28	$69.41
7 Laborers	42.10	2357.60	63.25	3542.00		
3 Equip. Oper. (medium)	56.75	1362.00	84.85	2036.40		
1 Asphalt Paver, 130 H.P.		2117.00		2328.70		
1 Tandem Roller, 10 Ton		236.75		260.43		
1 Roller, Pneum. Whl., 12 Ton		345.75		380.32	30.68	33.74
88 L.H., Daily Totals		$6771.90		$9077.85	$76.95	$103.16
Crew B-25B	**Hr.**	**Daily**	**Hr.**	**Daily**	**Bare Costs**	**Incl. O&P**
1 Labor Foreman (outside)	$44.10	$352.80	$66.25	$530.00	$47.15	$70.70
7 Laborers	42.10	2357.60	63.25	3542.00		
4 Equip. Oper. (medium)	56.75	1816.00	84.85	2715.20		
1 Asphalt Paver, 130 H.P.		2117.00		2328.70		
2 Tandem Rollers, 10 Ton		473.50		520.85		
1 Roller, Pneum. Whl., 12 Ton		345.75		380.32	30.59	33.64
96 L.H., Daily Totals		$7462.65		$10017.08	$77.74	$104.34

Crew No.	Bare Costs		Incl. Subs O&P		Cost Per Labor-Hour	
Crew B-25C	**Hr.**	**Daily**	**Hr.**	**Daily**	**Bare Costs**	**Incl. O&P**
1 Labor Foreman (outside)	$44.10	$352.80	$66.25	$530.00	$47.32	$70.95
3 Laborers	42.10	1010.40	63.25	1518.00		
2 Equip. Oper. (medium)	56.75	908.00	84.85	1357.60		
1 Asphalt Paver, 130 H.P.		2117.00		2328.70		
1 Tandem Roller, 10 Ton		236.75		260.43	49.04	53.94
48 L.H., Daily Totals		$4624.95		$5994.73	$96.35	$124.89
Crew B-25D	**Hr.**	**Daily**	**Hr.**	**Daily**	**Bare Costs**	**Incl. O&P**
1 Labor Foreman (outside)	$44.10	$352.80	$66.25	$530.00	$47.54	$71.28
3 Laborers	42.10	1010.40	63.25	1518.00		
2.125 Equip. Oper. (medium)	56.75	964.75	84.85	1442.45		
.125 Truck Driver (heavy)	48.95	48.95	73.35	73.35		
.125 Truck Tractor, 6x4, 380 H.P.		61.66		67.82		
.125 Dist. Tanker, 3000 Gallon		41.27		45.40		
1 Asphalt Paver, 130 H.P.		2117.00		2328.70		
1 Tandem Roller, 10 Ton		236.75		260.43	49.13	54.05
50 L.H., Daily Totals		$4833.58		$6266.14	$96.67	$125.32
Crew B-25E	**Hr.**	**Daily**	**Hr.**	**Daily**	**Bare Costs**	**Incl. O&P**
1 Labor Foreman (outside)	$44.10	$352.80	$66.25	$530.00	$47.74	$71.58
3 Laborers	42.10	1010.40	63.25	1518.00		
2.250 Equip. Oper. (medium)	56.75	1021.50	84.85	1527.30		
.25 Truck Driver (heavy)	48.95	97.90	73.35	146.70		
.25 Truck Tractor, 6x4, 380 H.P.		123.31		135.64		
.25 Dist. Tanker, 3000 Gallon		82.54		90.79		
1 Asphalt Paver, 130 H.P.		2117.00		2328.70		
1 Tandem Roller, 10 Ton		236.75		260.43	49.22	54.15
52 L.H., Daily Totals		$5042.20		$6537.56	$96.97	$125.72
Crew B-26	**Hr.**	**Daily**	**Hr.**	**Daily**	**Bare Costs**	**Incl. O&P**
1 Labor Foreman (outside)	$44.10	$352.80	$66.25	$530.00	$46.96	$70.35
6 Laborers	42.10	2020.80	63.25	3036.00		
2 Equip. Oper. (medium)	56.75	908.00	84.85	1357.60		
1 Rodman (reinf.)	56.40	451.20	84.90	679.20		
1 Cement Finisher	49.95	399.60	73.45	587.60		
1 Grader, 30,000 Lbs.		1062.00		1168.20		
1 Paving Mach. & Equip.		2474.00		2721.40	40.18	44.20
88 L.H., Daily Totals		$7668.40		$10080.00	$87.14	$114.55
Crew B-26A	**Hr.**	**Daily**	**Hr.**	**Daily**	**Bare Costs**	**Incl. O&P**
1 Labor Foreman (outside)	$44.10	$352.80	$66.25	$530.00	$46.96	$70.35
6 Laborers	42.10	2020.80	63.25	3036.00		
2 Equip. Oper. (medium)	56.75	908.00	84.85	1357.60		
1 Rodman (reinf.)	56.40	451.20	84.90	679.20		
1 Cement Finisher	49.95	399.60	73.45	587.60		
1 Grader, 30,000 Lbs.		1062.00		1168.20		
1 Paving Mach. & Equip.		2474.00		2721.40		
1 Concrete Saw		111.50		122.65	41.45	45.59
88 L.H., Daily Totals		$7779.90		$10202.65	$88.41	$115.94
Crew B-26B	**Hr.**	**Daily**	**Hr.**	**Daily**	**Bare Costs**	**Incl. O&P**
1 Labor Foreman (outside)	$44.10	$352.80	$66.25	$530.00	$47.77	$71.55
6 Laborers	42.10	2020.80	63.25	3036.00		
3 Equip. Oper. (medium)	56.75	1362.00	84.85	2036.40		
1 Rodman (reinf.)	56.40	451.20	84.90	679.20		
1 Cement Finisher	49.95	399.60	73.45	587.60		
1 Grader, 30,000 Lbs.		1062.00		1168.20		
1 Paving Mach. & Equip.		2474.00		2721.40		
1 Concrete Pump, 110' Boom		470.75		517.83	41.74	45.91
96 L.H., Daily Totals		$8593.15		$11276.63	$89.51	$117.46

Crews - Standard

Crew No.	Bare Costs		Incl. Subs O&P		Cost Per Labor-Hour	
Crew B-26C	**Hr.**	**Daily**	**Hr.**	**Daily**	**Bare Costs**	**Incl. O&P**
1 Labor Foreman (outside)	$44.10	$352.80	$66.25	$530.00	$45.98	$68.89
6 Laborers	42.10	2020.80	63.25	3036.00		
1 Equip. Oper. (medium)	56.75	454.00	84.85	678.80		
1 Rodman (reinf.)	56.40	451.20	84.90	679.20		
1 Cement Finisher	49.95	399.60	73.45	587.60		
1 Paving Mach. & Equip.		2474.00		2721.40		
1 Concrete Saw		111.50		122.65	32.32	35.55
80 L.H., Daily Totals		$6263.90		$8355.65	$78.30	$104.45

Crew B-27	Hr.	Daily	Hr.	Daily	Bare Costs	Incl. O&P
1 Labor Foreman (outside)	$44.10	$352.80	$66.25	$530.00	$42.60	$64.00
3 Laborers	42.10	1010.40	63.25	1518.00		
1 Berm Machine		271.95		299.14	8.50	9.35
32 L.H., Daily Totals		$1635.15		$2347.15	$51.10	$73.35

Crew B-28	Hr.	Daily	Hr.	Daily	Bare Costs	Incl. O&P
2 Carpenters	$53.15	$850.40	$79.85	$1277.60	$49.47	$74.32
1 Laborer	42.10	336.80	63.25	506.00		
24 L.H., Daily Totals		$1187.20		$1783.60	$49.47	$74.32

Crew B-29	Hr.	Daily	Hr.	Daily	Bare Costs	Incl. O&P
1 Labor Foreman (outside)	$44.10	$352.80	$66.25	$530.00	$46.04	$69.04
4 Laborers	42.10	1347.20	63.25	2024.00		
1 Equip. Oper. (crane)	59.20	473.60	88.50	708.00		
1 Equip. Oper. (oiler)	50.55	404.40	75.55	604.40		
1 Gradall, 5/8 C.Y.		846.50		931.15	15.12	16.63
56 L.H., Daily Totals		$3424.50		$4797.55	$61.15	$85.67

Crew B-30	Hr.	Daily	Hr.	Daily	Bare Costs	Incl. O&P
1 Equip. Oper. (medium)	$56.75	$454.00	$84.85	$678.80	$51.55	$77.18
2 Truck Drivers (heavy)	48.95	783.20	73.35	1173.60		
1 Hyd. Excavator, 1.5 C.Y.		687.55		756.30		
2 Dump Trucks, 12 C.Y., 400 H.P.		1145.80		1260.38	76.39	84.03
24 L.H., Daily Totals		$3070.55		$3869.09	$127.94	$161.21

Crew B-31	Hr.	Daily	Hr.	Daily	Bare Costs	Incl. O&P
1 Labor Foreman (outside)	$44.10	$352.80	$66.25	$530.00	$44.71	$67.17
3 Laborers	42.10	1010.40	63.25	1518.00		
1 Carpenter	53.15	425.20	79.85	638.80		
1 Air Compressor, 250 cfm		201.50		221.65		
1 Sheeting Driver		7.35		8.09		
2 -50' Air Hoses, 1.5"		48.50		53.35	6.43	7.08
40 L.H., Daily Totals		$2045.75		$2969.89	$51.14	$74.25

Crew B-32	Hr.	Daily	Hr.	Daily	Bare Costs	Incl. O&P
1 Laborer	$42.10	$336.80	$63.25	$506.00	$53.09	$79.45
3 Equip. Oper. (medium)	56.75	1362.00	84.85	2036.40		
1 Grader, 30,000 Lbs.		1062.00		1168.20		
1 Tandem Roller, 10 Ton		236.75		260.43		
1 Dozer, 200 H.P.		1504.00		1654.40	87.59	96.34
32 L.H., Daily Totals		$4501.55		$5625.43	$140.67	$175.79

Crew B-32A	Hr.	Daily	Hr.	Daily	Bare Costs	Incl. O&P
1 Laborer	$42.10	$336.80	$63.25	$506.00	$51.87	$77.65
2 Equip. Oper. (medium)	56.75	908.00	84.85	1357.60		
1 Grader, 30,000 Lbs.		1062.00		1168.20		
1 Roller, Vibratory, 25 Ton		664.35		730.78	71.93	79.12
24 L.H., Daily Totals		$2971.15		$3762.59	$123.80	$156.77

Crew No.	Bare Costs		Incl. Subs O&P		Cost Per Labor-Hour	
Crew B-32B	**Hr.**	**Daily**	**Hr.**	**Daily**	**Bare Costs**	**Incl. O&P**
1 Laborer	$42.10	$336.80	$63.25	$506.00	$51.87	$77.65
2 Equip. Oper. (medium)	56.75	908.00	84.85	1357.60		
1 Dozer, 200 H.P.		1504.00		1654.40		
1 Roller, Vibratory, 25 Ton		664.35		730.78	90.35	99.38
24 L.H., Daily Totals		$3413.15		$4248.78	$142.21	$177.03

Crew B-32C	Hr.	Daily	Hr.	Daily	Bare Costs	Incl. O&P
1 Labor Foreman (outside)	$44.10	$352.80	$66.25	$530.00	$49.76	$74.55
2 Laborers	42.10	673.60	63.25	1012.00		
3 Equip. Oper. (medium)	56.75	1362.00	84.85	2036.40		
1 Grader, 30,000 Lbs.		1062.00		1168.20		
1 Tandem Roller, 10 Ton		236.75		260.43		
1 Dozer, 200 H.P.		1504.00		1654.40	58.39	64.23
48 L.H., Daily Totals		$5191.15		$6661.43	$108.15	$138.78

Crew B-33A	Hr.	Daily	Hr.	Daily	Bare Costs	Incl. O&P
1 Equip. Oper. (medium)	$56.75	$454.00	$84.85	$678.80	$52.56	$78.68
.5 Laborer	42.10	168.40	63.25	253.00		
.25 Equip. Oper. (medium)	56.75	113.50	84.85	169.70		
1 Scraper, Towed, 7 C.Y.		127.80		140.58		
1.25 Dozers, 300 H.P.		2205.00		2425.50	166.63	183.29
14 L.H., Daily Totals		$3068.70		$3667.58	$219.19	$261.97

Crew B-33B	Hr.	Daily	Hr.	Daily	Bare Costs	Incl. O&P
1 Equip. Oper. (medium)	$56.75	$454.00	$84.85	$678.80	$52.56	$78.68
.5 Laborer	42.10	168.40	63.25	253.00		
.25 Equip. Oper. (medium)	56.75	113.50	84.85	169.70		
1 Scraper, Towed, 10 C.Y.		159.70		175.67		
1.25 Dozers, 300 H.P.		2205.00		2425.50	168.91	185.80
14 L.H., Daily Totals		$3100.60		$3702.67	$221.47	$264.48

Crew B-33C	Hr.	Daily	Hr.	Daily	Bare Costs	Incl. O&P
1 Equip. Oper. (medium)	$56.75	$454.00	$84.85	$678.80	$52.56	$78.68
.5 Laborer	42.10	168.40	63.25	253.00		
.25 Equip. Oper. (medium)	56.75	113.50	84.85	169.70		
1 Scraper, Towed, 15 C.Y.		176.75		194.43		
1.25 Dozers, 300 H.P.		2205.00		2425.50	170.13	187.14
14 L.H., Daily Totals		$3117.65		$3721.43	$222.69	$265.82

Crew B-33D	Hr.	Daily	Hr.	Daily	Bare Costs	Incl. O&P
1 Equip. Oper. (medium)	$56.75	$454.00	$84.85	$678.80	$52.56	$78.68
.5 Laborer	42.10	168.40	63.25	253.00		
.25 Equip. Oper. (medium)	56.75	113.50	84.85	169.70		
1 S.P. Scraper, 14 C.Y.		2395.00		2634.50		
.25 Dozer, 300 H.P.		441.00		485.10	202.57	222.83
14 L.H., Daily Totals		$3571.90		$4221.10	$255.14	$301.51

Crew B-33E	Hr.	Daily	Hr.	Daily	Bare Costs	Incl. O&P
1 Equip. Oper. (medium)	$56.75	$454.00	$84.85	$678.80	$52.56	$78.68
.5 Laborer	42.10	168.40	63.25	253.00		
.25 Equip. Oper. (medium)	56.75	113.50	84.85	169.70		
1 S.P. Scraper, 21 C.Y.		2628.00		2890.80		
.25 Dozer, 300 H.P.		441.00		485.10	219.21	241.14
14 L.H., Daily Totals		$3804.90		$4477.40	$271.78	$319.81

Crew No.	Bare Costs		Incl. Subs O&P		Cost Per Labor-Hour	
Crew B-33F	**Hr.**	**Daily**	**Hr.**	**Daily**	**Bare Costs**	**Incl. O&P**
1 Equip. Oper. (medium)	$56.75	$454.00	$84.85	$678.80	$52.56	$78.68
.5 Laborer	42.10	168.40	63.25	253.00		
.25 Equip. Oper. (medium)	56.75	113.50	84.85	169.70		
1 Elev. Scraper, 11 C.Y.		1133.00		1246.30		
.25 Dozer, 300 H.P.		441.00		485.10	112.43	123.67
14 L.H., Daily Totals		$2309.90		$2832.90	$164.99	$202.35
Crew B-33G	**Hr.**	**Daily**	**Hr.**	**Daily**	**Bare Costs**	**Incl. O&P**
1 Equip. Oper. (medium)	$56.75	$454.00	$84.85	$678.80	$52.56	$78.68
.5 Laborer	42.10	168.40	63.25	253.00		
.25 Equip. Oper. (medium)	56.75	113.50	84.85	169.70		
1 Elev. Scraper, 22 C.Y.		1884.00		2072.40		
.25 Dozer, 300 H.P.		441.00		485.10	166.07	182.68
14 L.H., Daily Totals		$3060.90		$3659.00	$218.64	$261.36
Crew B-33H	**Hr.**	**Daily**	**Hr.**	**Daily**	**Bare Costs**	**Incl. O&P**
.5 Laborer	$42.10	$168.40	$63.25	$253.00	$52.56	$78.68
1 Equipment Operator (med.)	56.75	454.00	84.85	678.80		
.25 Equipment Operator (med.)	56.75	113.50	84.85	169.70		
1 S.P. Scraper, 44 C.Y.		4644.00		5108.40		
.25 Dozer, 410 H.P.		694.25		763.67	381.30	419.43
14 L.H., Daily Totals		$6074.15		$6973.57	$433.87	$498.11
Crew B-33J	**Hr.**	**Daily**	**Hr.**	**Daily**	**Bare Costs**	**Incl. O&P**
1 Equipment Operator (med.)	$56.75	$454.00	$84.85	$678.80	$56.75	$84.85
1 S.P. Scraper, 14 C.Y.		2395.00		2634.50	299.38	329.31
8 L.H., Daily Totals		$2849.00		$3313.30	$356.13	$414.16
Crew B-33K	**Hr.**	**Daily**	**Hr.**	**Daily**	**Bare Costs**	**Incl. O&P**
1 Equipment Operator (med.)	$56.75	$454.00	$84.85	$678.80	$52.56	$78.68
.25 Equipment Operator (med.)	56.75	113.50	84.85	169.70		
.5 Laborer	42.10	168.40	63.25	253.00		
1 S.P. Scraper, 31 C.Y.		3667.00		4033.70		
.25 Dozer, 410 H.P.		694.25		763.67	311.52	342.67
14 L.H., Daily Totals		$5097.15		$5898.88	$364.08	$421.35
Crew B-34A	**Hr.**	**Daily**	**Hr.**	**Daily**	**Bare Costs**	**Incl. O&P**
1 Truck Driver (heavy)	$48.95	$391.60	$73.35	$586.80	$48.95	$73.35
1 Dump Truck, 8 C.Y., 220 H.P.		428.05		470.86	53.51	58.86
8 L.H., Daily Totals		$819.65		$1057.66	$102.46	$132.21
Crew B-34B	**Hr.**	**Daily**	**Hr.**	**Daily**	**Bare Costs**	**Incl. O&P**
1 Truck Driver (heavy)	$48.95	$391.60	$73.35	$586.80	$48.95	$73.35
1 Dump Truck, 12 C.Y., 400 H.P.		572.90		630.19	71.61	78.77
8 L.H., Daily Totals		$964.50		$1216.99	$120.56	$152.12
Crew B-34C	**Hr.**	**Daily**	**Hr.**	**Daily**	**Bare Costs**	**Incl. O&P**
1 Truck Driver (heavy)	$48.95	$391.60	$73.35	$586.80	$48.95	$73.35
1 Truck Tractor, 6x4, 380 H.P.		493.25		542.58		
1 Dump Trailer, 16.5 C.Y.		136.70		150.37	78.74	86.62
8 L.H., Daily Totals		$1021.55		$1279.74	$127.69	$159.97
Crew B-34D	**Hr.**	**Daily**	**Hr.**	**Daily**	**Bare Costs**	**Incl. O&P**
1 Truck Driver (heavy)	$48.95	$391.60	$73.35	$586.80	$48.95	$73.35
1 Truck Tractor, 6x4, 380 H.P.		493.25		542.58		
1 Dump Trailer, 20 C.Y.		151.70		166.87	80.62	88.68
8 L.H., Daily Totals		$1036.55		$1296.24	$129.57	$162.03

Crew No.	Bare Costs		Incl. Subs O&P		Cost Per Labor-Hour	
Crew B-34E	**Hr.**	**Daily**	**Hr.**	**Daily**	**Bare Costs**	**Incl. O&P**
1 Truck Driver (heavy)	$48.95	$391.60	$73.35	$586.80	$48.95	$73.35
1 Dump Truck, Off Hwy., 25 Ton		1379.00		1516.90	172.38	189.61
8 L.H., Daily Totals		$1770.60		$2103.70	$221.32	$262.96
Crew B-34F	**Hr.**	**Daily**	**Hr.**	**Daily**	**Bare Costs**	**Incl. O&P**
1 Truck Driver (heavy)	$48.95	$391.60	$73.35	$586.80	$48.95	$73.35
1 Dump Truck, Off Hwy., 35 Ton		935.20		1028.72	116.90	128.59
8 L.H., Daily Totals		$1326.80		$1615.52	$165.85	$201.94
Crew B-34G	**Hr.**	**Daily**	**Hr.**	**Daily**	**Bare Costs**	**Incl. O&P**
1 Truck Driver (heavy)	$48.95	$391.60	$73.35	$586.80	$48.95	$73.35
1 Dump Truck, Off Hwy., 50 Ton		1771.00		1948.10	221.38	243.51
8 L.H., Daily Totals		$2162.60		$2534.90	$270.32	$316.86
Crew B-34H	**Hr.**	**Daily**	**Hr.**	**Daily**	**Bare Costs**	**Incl. O&P**
1 Truck Driver (heavy)	$48.95	$391.60	$73.35	$586.80	$48.95	$73.35
1 Dump Truck, Off Hwy., 65 Ton		1915.00		2106.50	239.38	263.31
8 L.H., Daily Totals		$2306.60		$2693.30	$288.32	$336.66
Crew B-34I	**Hr.**	**Daily**	**Hr.**	**Daily**	**Bare Costs**	**Incl. O&P**
1 Truck Driver (heavy)	$48.95	$391.60	$73.35	$586.80	$48.95	$73.35
1 Dump Truck, 18 C.Y., 450 H.P.		713.80		785.18	89.22	98.15
8 L.H., Daily Totals		$1105.40		$1371.98	$138.18	$171.50
Crew B-34J	**Hr.**	**Daily**	**Hr.**	**Daily**	**Bare Costs**	**Incl. O&P**
1 Truck Driver (heavy)	$48.95	$391.60	$73.35	$586.80	$48.95	$73.35
1 Dump Truck, Off Hwy., 100 Ton		2736.00		3009.60	342.00	376.20
8 L.H., Daily Totals		$3127.60		$3596.40	$390.95	$449.55
Crew B-34K	**Hr.**	**Daily**	**Hr.**	**Daily**	**Bare Costs**	**Incl. O&P**
1 Truck Driver (heavy)	$48.95	$391.60	$73.35	$586.80	$48.95	$73.35
1 Truck Tractor, 6x4, 450 H.P.		601.75		661.92		
1 Lowbed Trailer, 75 Ton		255.05		280.56	107.10	117.81
8 L.H., Daily Totals		$1248.40		$1529.28	$156.05	$191.16
Crew B-34L	**Hr.**	**Daily**	**Hr.**	**Daily**	**Bare Costs**	**Incl. O&P**
1 Equip. Oper. (light)	$53.00	$424.00	$79.20	$633.60	$53.00	$79.20
1 Flatbed Truck, Gas, 1.5 Ton		196.15		215.76	24.52	26.97
8 L.H., Daily Totals		$620.15		$849.37	$77.52	$106.17
Crew B-34M	**Hr.**	**Daily**	**Hr.**	**Daily**	**Bare Costs**	**Incl. O&P**
1 Equip. Oper. (light)	$53.00	$424.00	$79.20	$633.60	$53.00	$79.20
1 Flatbed Truck, Gas, 3 Ton		820.40		902.44	102.55	112.81
8 L.H., Daily Totals		$1244.40		$1536.04	$155.55	$192.01
Crew B-34N	**Hr.**	**Daily**	**Hr.**	**Daily**	**Bare Costs**	**Incl. O&P**
1 Truck Driver (heavy)	$48.95	$391.60	$73.35	$586.80	$52.85	$79.10
1 Equip. Oper. (medium)	56.75	454.00	84.85	678.80		
1 Truck Tractor, 6x4, 380 H.P.		493.25		542.58		
1 Flatbed Trailer, 40 Ton		186.20		204.82	42.47	46.71
16 L.H., Daily Totals		$1525.05		$2012.99	$95.32	$125.81
Crew B-34P	**Hr.**	**Daily**	**Hr.**	**Daily**	**Bare Costs**	**Incl. O&P**
1 Pipe Fitter	$65.55	$524.40	$98.10	$784.80	$56.52	$84.58
1 Truck Driver (light)	47.25	378.00	70.80	566.40		
1 Equip. Oper. (medium)	56.75	454.00	84.85	678.80		
1 Flatbed Truck, Gas, 3 Ton		820.40		902.44		
1 Backhoe Loader, 48 H.P.		213.75		235.13	43.09	47.40
24 L.H., Daily Totals		$2390.55		$3167.57	$99.61	$131.98

Crew No.	Bare Costs		Incl. Subs O&P		Cost Per Labor-Hour	
Crew B-34Q	**Hr.**	**Daily**	**Hr.**	**Daily**	**Bare Costs**	**Incl. O&P**
1 Pipe Fitter	$65.55	$524.40	$98.10	$784.80	$57.33	$85.80
1 Truck Driver (light)	47.25	378.00	70.80	566.40		
1 Equip. Oper. (crane)	59.20	473.60	88.50	708.00		
1 Flatbed Trailer, 25 Ton		135.55		149.10		
1 Dump Truck, 8 C.Y., 220 H.P.		428.05		470.86		
1 Hyd. Crane, 25 Ton		580.85		638.93	47.69	52.45
24 L.H., Daily Totals		$2520.45		$3318.09	$105.02	$138.25
Crew B-34R	**Hr.**	**Daily**	**Hr.**	**Daily**	**Bare Costs**	**Incl. O&P**
1 Pipe Fitter	$65.55	$524.40	$98.10	$784.80	$57.33	$85.80
1 Truck Driver (light)	47.25	378.00	70.80	566.40		
1 Equip. Oper. (crane)	59.20	473.60	88.50	708.00		
1 Flatbed Trailer, 25 Ton		135.55		149.10		
1 Dump Truck, 8 C.Y., 220 H.P.		428.05		470.86		
1 Hyd. Crane, 25 Ton		580.85		638.93		
1 Hyd. Excavator, 1 C.Y.		795.30		874.83	80.82	88.91
24 L.H., Daily Totals		$3315.75		$4192.93	$138.16	$174.71
Crew B-34S	**Hr.**	**Daily**	**Hr.**	**Daily**	**Bare Costs**	**Incl. O&P**
2 Pipe Fitters	$65.55	$1048.80	$98.10	$1569.60	$59.81	$89.51
1 Truck Driver (heavy)	48.95	391.60	73.35	586.80		
1 Equip. Oper. (crane)	59.20	473.60	88.50	708.00		
1 Flatbed Trailer, 40 Ton		186.20		204.82		
1 Truck Tractor, 6x4, 380 H.P.		493.25		542.58		
1 Hyd. Crane, 80 Ton		1486.00		1634.60		
1 Hyd. Excavator, 2 C.Y.		951.30		1046.43	97.40	107.14
32 L.H., Daily Totals		$5030.75		$6292.82	$157.21	$196.65
Crew B-34T	**Hr.**	**Daily**	**Hr.**	**Daily**	**Bare Costs**	**Incl. O&P**
2 Pipe Fitters	$65.55	$1048.80	$98.10	$1569.60	$59.81	$89.51
1 Truck Driver (heavy)	48.95	391.60	73.35	586.80		
1 Equip. Oper. (crane)	59.20	473.60	88.50	708.00		
1 Flatbed Trailer, 40 Ton		186.20		204.82		
1 Truck Tractor, 6x4, 380 H.P.		493.25		542.58		
1 Hyd. Crane, 80 Ton		1486.00		1634.60	67.67	74.44
32 L.H., Daily Totals		$4079.45		$5246.40	$127.48	$163.95
Crew B-34U	**Hr.**	**Daily**	**Hr.**	**Daily**	**Bare Costs**	**Incl. O&P**
1 Truck Driver (heavy)	$48.95	$391.60	$73.35	$586.80	$50.98	$76.28
1 Equip. Oper. (light)	53.00	424.00	79.20	633.60		
1 Truck Tractor, 220 H.P.		307.10		337.81		
1 Flatbed Trailer, 25 Ton		135.55		149.10	27.67	30.43
16 L.H., Daily Totals		$1258.25		$1707.32	$78.64	$106.71
Crew B-34V	**Hr.**	**Daily**	**Hr.**	**Daily**	**Bare Costs**	**Incl. O&P**
1 Truck Driver (heavy)	$48.95	$391.60	$73.35	$586.80	$53.72	$80.35
1 Equip. Oper. (crane)	59.20	473.60	88.50	708.00		
1 Equip. Oper. (light)	53.00	424.00	79.20	633.60		
1 Truck Tractor, 6x4, 450 H.P.		601.75		661.92		
1 Equipment Trailer, 50 Ton		205.05		225.56		
1 Pickup Truck, 4x4, 3/4 Ton		175.85		193.44	40.94	45.04
24 L.H., Daily Totals		$2271.85		$3009.32	$94.66	$125.39

Crew No.	Bare Costs		Incl. Subs O&P		Cost Per Labor-Hour	
Crew B-34W	**Hr.**	**Daily**	**Hr.**	**Daily**	**Bare Costs**	**Incl. O&P**
5 Truck Drivers (heavy)	$48.95	$1958.00	$73.35	$2934.00	$51.60	$77.27
2 Equip. Opers. (crane)	59.20	947.20	88.50	1416.00		
1 Equip. Oper. (mechanic)	59.15	473.20	88.40	707.20		
1 Laborer	42.10	336.80	63.25	506.00		
4 Truck Tractors, 6x4, 380 H.P.		1973.00		2170.30		
2 Equipment Trailers, 50 Ton		410.10		451.11		
2 Flatbed Trailers, 40 Ton		372.40		409.64		
1 Pickup Truck, 4x4, 3/4 Ton		175.85		193.44		
1 S.P. Crane, 4x4, 20 Ton		582.20		640.42	48.80	53.68
72 L.H., Daily Totals		$7228.75		$9428.10	$100.40	$130.95
Crew B-35	**Hr.**	**Daily**	**Hr.**	**Daily**	**Bare Costs**	**Incl. O&P**
1 Labor Foreman (outside)	$44.10	$352.80	$66.25	$530.00	$52.54	$78.83
1 Skilled Worker	54.85	438.80	82.95	663.60		
1 Welder (plumber)	64.45	515.60	96.45	771.60		
1 Laborer	42.10	336.80	63.25	506.00		
1 Equip. Oper. (crane)	59.20	473.60	88.50	708.00		
1 Equip. Oper. (oiler)	50.55	404.40	75.55	604.40		
1 Welder, Electric, 300 amp		106.40		117.04		
1 Hyd. Excavator, .75 C.Y.		694.25		763.67	16.68	18.35
48 L.H., Daily Totals		$3322.65		$4664.31	$69.22	$97.17
Crew B-35A	**Hr.**	**Daily**	**Hr.**	**Daily**	**Bare Costs**	**Incl. O&P**
1 Labor Foreman (outside)	$44.10	$352.80	$66.25	$530.00	$51.05	$76.60
2 Laborers	42.10	673.60	63.25	1012.00		
1 Skilled Worker	54.85	438.80	82.95	663.60		
1 Welder (plumber)	64.45	515.60	96.45	771.60		
1 Equip. Oper. (crane)	59.20	473.60	88.50	708.00		
1 Equip. Oper. (oiler)	50.55	404.40	75.55	604.40		
1 Welder, Gas Engine, 300 amp		147.00		161.70		
1 Crawler Crane, 75 Ton		2002.00		2202.20	38.38	42.21
56 L.H., Daily Totals		$5007.80		$6653.50	$89.42	$118.81
Crew B-36	**Hr.**	**Daily**	**Hr.**	**Daily**	**Bare Costs**	**Incl. O&P**
1 Labor Foreman (outside)	$44.10	$352.80	$66.25	$530.00	$48.36	$72.49
2 Laborers	42.10	673.60	63.25	1012.00		
2 Equip. Oper. (medium)	56.75	908.00	84.85	1357.60		
1 Dozer, 200 H.P.		1504.00		1654.40		
1 Aggregate Spreader		65.75		72.33		
1 Tandem Roller, 10 Ton		236.75		260.43	45.16	49.68
40 L.H., Daily Totals		$3740.90		$4886.75	$93.52	$122.17
Crew B-36A	**Hr.**	**Daily**	**Hr.**	**Daily**	**Bare Costs**	**Incl. O&P**
1 Labor Foreman (outside)	$44.10	$352.80	$66.25	$530.00	$50.76	$76.02
2 Laborers	42.10	673.60	63.25	1012.00		
4 Equip. Oper. (medium)	56.75	1816.00	84.85	2715.20		
1 Dozer, 200 H.P.		1504.00		1654.40		
1 Aggregate Spreader		65.75		72.33		
1 Tandem Roller, 10 Ton		236.75		260.43		
1 Roller, Pneum. Whl., 12 Ton		345.75		380.32	38.43	42.28
56 L.H., Daily Totals		$4994.65		$6624.68	$89.19	$118.30

Crew No.	Bare Costs		Incl. Subs O&P		Cost Per Labor-Hour	
Crew B-36B	**Hr.**	**Daily**	**Hr.**	**Daily**	**Bare Costs**	**Incl. O&P**
1 Labor Foreman (outside)	$44.10	$352.80	$66.25	$530.00	$50.53	$75.69
2 Laborers	42.10	673.60	63.25	1012.00		
4 Equip. Oper. (medium)	56.75	1816.00	84.85	2715.20		
1 Truck Driver (heavy)	48.95	391.60	73.35	586.80		
1 Grader, 30,000 Lbs.		1062.00		1168.20		
1 F.E. Loader, Crl, 1.5 C.Y.		659.75		725.73		
1 Dozer, 300 H.P.		1764.00		1940.40		
1 Roller, Vibratory, 25 Ton		664.35		730.78		
1 Truck Tractor, 6x4, 450 H.P.		601.75		661.92		
1 Water Tank Trailer, 5000 Gal.		152.25		167.47	76.63	84.29
64 L.H., Daily Totals		$8138.10		$10238.51	$127.16	$159.98
Crew B-36C	**Hr.**	**Daily**	**Hr.**	**Daily**	**Bare Costs**	**Incl. O&P**
1 Labor Foreman (outside)	$44.10	$352.80	$66.25	$530.00	$52.66	$78.83
3 Equip. Oper. (medium)	56.75	1362.00	84.85	2036.40		
1 Truck Driver (heavy)	48.95	391.60	73.35	586.80		
1 Grader, 30,000 Lbs.		1062.00		1168.20		
1 Dozer, 300 H.P.		1764.00		1940.40		
1 Roller, Vibratory, 25 Ton		664.35		730.78		
1 Truck Tractor, 6x4, 450 H.P.		601.75		661.92		
1 Water Tank Trailer, 5000 Gal.		152.25		167.47	106.11	116.72
40 L.H., Daily Totals		$6350.75		$7821.98	$158.77	$195.55
Crew B-36D	**Hr.**	**Daily**	**Hr.**	**Daily**	**Bare Costs**	**Incl. O&P**
1 Labor Foreman (outside)	$44.10	$352.80	$66.25	$530.00	$53.59	$80.20
3 Equip. Oper. (medium)	56.75	1362.00	84.85	2036.40		
1 Grader, 30,000 Lbs.		1062.00		1168.20		
1 Dozer, 300 H.P.		1764.00		1940.40		
1 Roller, Vibratory, 25 Ton		664.35		730.78	109.07	119.98
32 L.H., Daily Totals		$5205.15		$6405.78	$162.66	$200.18
Crew B-37	**Hr.**	**Daily**	**Hr.**	**Daily**	**Bare Costs**	**Incl. O&P**
1 Labor Foreman (outside)	$44.10	$352.80	$66.25	$530.00	$44.25	$66.41
4 Laborers	42.10	1347.20	63.25	2024.00		
1 Equip. Oper. (light)	53.00	424.00	79.20	633.60		
1 Tandem Roller, 5 Ton		255.65		281.21	5.33	5.86
48 L.H., Daily Totals		$2379.65		$3468.82	$49.58	$72.27
Crew B-37A	**Hr.**	**Daily**	**Hr.**	**Daily**	**Bare Costs**	**Incl. O&P**
2 Laborers	$42.10	$673.60	$63.25	$1012.00	$43.82	$65.77
1 Truck Driver (light)	47.25	378.00	70.80	566.40		
1 Flatbed Truck, Gas, 1.5 Ton		196.15		215.76		
1 Tar Kettle, T.M.		156.20		171.82	14.68	16.15
24 L.H., Daily Totals		$1403.95		$1965.98	$58.50	$81.92
Crew B-37B	**Hr.**	**Daily**	**Hr.**	**Daily**	**Bare Costs**	**Incl. O&P**
3 Laborers	$42.10	$1010.40	$63.25	$1518.00	$43.39	$65.14
1 Truck Driver (light)	47.25	378.00	70.80	566.40		
1 Flatbed Truck, Gas, 1.5 Ton		196.15		215.76		
1 Tar Kettle, T.M.		156.20		171.82	11.01	12.11
32 L.H., Daily Totals		$1740.75		$2471.99	$54.40	$77.25
Crew B-37C	**Hr.**	**Daily**	**Hr.**	**Daily**	**Bare Costs**	**Incl. O&P**
2 Laborers	$42.10	$673.60	$63.25	$1012.00	$44.67	$67.03
2 Truck Drivers (light)	47.25	756.00	70.80	1132.80		
2 Flatbed Trucks, Gas, 1.5 Ton		392.30		431.53		
1 Tar Kettle, T.M.		156.20		171.82	17.14	18.85
32 L.H., Daily Totals		$1978.10		$2748.15	$61.82	$85.88

Crew No.	Bare Costs		Incl. Subs O&P		Cost Per Labor-Hour	
Crew B-37D	**Hr.**	**Daily**	**Hr.**	**Daily**	**Bare Costs**	**Incl. O&P**
1 Laborer	$42.10	$336.80	$63.25	$506.00	$44.67	$67.03
1 Truck Driver (light)	47.25	378.00	70.80	566.40		
1 Pickup Truck, 3/4 Ton		110.85		121.94	6.93	7.62
16 L.H., Daily Totals		$825.65		$1194.34	$51.60	$74.65
Crew B-37E	**Hr.**	**Daily**	**Hr.**	**Daily**	**Bare Costs**	**Incl. O&P**
3 Laborers	$42.10	$1010.40	$63.25	$1518.00	$47.22	$70.77
1 Equip. Oper. (light)	53.00	424.00	79.20	633.60		
1 Equip. Oper. (medium)	56.75	454.00	84.85	678.80		
2 Truck Drivers (light)	47.25	756.00	70.80	1132.80		
4 Barrels w/ Flasher		16.40		18.04		
1 Concrete Saw		111.50		122.65		
1 Rotary Hammer Drill		51.70		56.87		
1 Hammer Drill Bit		25.25		27.77		
1 Loader, Skid Steer, 30 H.P.		177.55		195.31		
1 Conc. Hammer Attach.		118.40		130.24		
1 Vibrating Plate, Gas, 18"		31.55		34.70		
2 Flatbed Trucks, Gas, 1.5 Ton		392.30		431.53	16.51	18.16
56 L.H., Daily Totals		$3569.05		$4980.31	$63.73	$88.93
Crew B-37F	**Hr.**	**Daily**	**Hr.**	**Daily**	**Bare Costs**	**Incl. O&P**
3 Laborers	$42.10	$1010.40	$63.25	$1518.00	$43.39	$65.14
1 Truck Driver (light)	47.25	378.00	70.80	566.40		
4 Barrels w/ Flasher		16.40		18.04		
1 Concrete Mixer, 10 C.F.		140.30		154.33		
1 Air Compressor, 60 cfm		152.95		168.25		
1 -50' Air Hose, 3/4"		7.75		8.53		
1 Spade (Chipper)		8.45		9.29		
1 Flatbed Truck, Gas, 1.5 Ton		196.15		215.76	16.31	17.94
32 L.H., Daily Totals		$1910.40		$2658.60	$59.70	$83.08
Crew B-37G	**Hr.**	**Daily**	**Hr.**	**Daily**	**Bare Costs**	**Incl. O&P**
1 Labor Foreman (outside)	$44.10	$352.80	$66.25	$530.00	$44.25	$66.41
4 Laborers	42.10	1347.20	63.25	2024.00		
1 Equip. Oper. (light)	53.00	424.00	79.20	633.60		
1 Berm Machine		271.95		299.14		
1 Tandem Roller, 5 Ton		255.65		281.21	10.99	12.09
48 L.H., Daily Totals		$2651.60		$3767.96	$55.24	$78.50
Crew B-37H	**Hr.**	**Daily**	**Hr.**	**Daily**	**Bare Costs**	**Incl. O&P**
1 Labor Foreman (outside)	$44.10	$352.80	$66.25	$530.00	$44.25	$66.41
4 Laborers	42.10	1347.20	63.25	2024.00		
1 Equip. Oper. (light)	53.00	424.00	79.20	633.60		
1 Tandem Roller, 5 Ton		255.65		281.21		
1 Flatbed Truck, Gas, 1.5 Ton		196.15		215.76		
1 Tar Kettle, T.M.		156.20		171.82	12.67	13.93
48 L.H., Daily Totals		$2732.00		$3856.40	$56.92	$80.34

Crew No.	Bare Costs		Incl. Subs O&P		Cost Per Labor-Hour	
Crew B-37I	**Hr.**	**Daily**	**Hr.**	**Daily**	**Bare Costs**	**Incl. O&P**
3 Laborers	$42.10	$1010.40	$63.25	$1518.00	$47.22	$70.77
1 Equip. Oper. (light)	53.00	424.00	79.20	633.60		
1 Equip. Oper. (medium)	56.75	454.00	84.85	678.80		
2 Truck Drivers (light)	47.25	756.00	70.80	1132.80		
4 Barrels w/ Flasher		16.40		18.04		
1 Concrete Saw		111.50		122.65		
1 Rotary Hammer Drill		51.70		56.87		
1 Hammer Drill Bit		25.25		27.77		
1 Air Compressor, 60 cfm		152.95		168.25		
1 -50' Air Hose, 3/4"		7.75		8.53		
1 Spade (Chipper)		8.45		9.29		
1 Loader, Skid Steer, 30 H.P.		177.55		195.31		
1 Conc. Hammer Attach.		118.40		130.24		
1 Concrete Mixer, 10 C.F.		140.30		154.33		
1 Vibrating Plate, Gas, 18"		31.55		34.70		
2 Flatbed Trucks, Gas, 1.5 Ton		392.30		431.53	22.04	24.24
56 L.H., Daily Totals		$3878.50		$5320.71	$69.26	$95.01
Crew B-37J	**Hr.**	**Daily**	**Hr.**	**Daily**	**Bare Costs**	**Incl. O&P**
1 Labor Foreman (outside)	$44.10	$352.80	$66.25	$530.00	$44.25	$66.41
4 Laborers	42.10	1347.20	63.25	2024.00		
1 Equip. Oper. (light)	53.00	424.00	79.20	633.60		
1 Air Compressor, 60 cfm		152.95		168.25		
1 -50' Air Hose, 3/4"		7.75		8.53		
2 Concrete Mixers, 10 C.F.		280.60		308.66		
2 Flatbed Trucks, Gas, 1.5 Ton		392.30		431.53		
1 Shot Blaster, 20"		206.20		226.82	21.66	23.83
48 L.H., Daily Totals		$3163.80		$4331.38	$65.91	$90.24
Crew B-37K	**Hr.**	**Daily**	**Hr.**	**Daily**	**Bare Costs**	**Incl. O&P**
1 Labor Foreman (outside)	$44.10	$352.80	$66.25	$530.00	$44.25	$66.41
4 Laborers	42.10	1347.20	63.25	2024.00		
1 Equip. Oper. (light)	53.00	424.00	79.20	633.60		
1 Air Compressor, 60 cfm		152.95		168.25		
1 -50' Air Hose, 3/4"		7.75		8.53		
2 Flatbed Trucks, Gas, 1.5 Ton		392.30		431.53		
1 Shot Blaster, 20"		206.20		226.82	15.82	17.40
48 L.H., Daily Totals		$2883.20		$4022.72	$60.07	$83.81
Crew B-38	**Hr.**	**Daily**	**Hr.**	**Daily**	**Bare Costs**	**Incl. O&P**
1 Labor Foreman (outside)	$44.10	$352.80	$66.25	$530.00	$47.61	$71.36
2 Laborers	42.10	673.60	63.25	1012.00		
1 Equip. Oper. (light)	53.00	424.00	79.20	633.60		
1 Equip. Oper. (medium)	56.75	454.00	84.85	678.80		
1 Backhoe Loader, 48 H.P.		213.75		235.13		
1 Hyd. Hammer (1200 lb.)		175.15		192.66		
1 F.E. Loader, W.M., 4 C.Y.		789.35		868.28		
1 Pvmt. Rem. Bucket		63.05		69.36	31.03	34.14
40 L.H., Daily Totals		$3145.70		$4219.83	$78.64	$105.50
Crew B-39	**Hr.**	**Daily**	**Hr.**	**Daily**	**Bare Costs**	**Incl. O&P**
1 Labor Foreman (outside)	$44.10	$352.80	$66.25	$530.00	$44.25	$66.41
4 Laborers	42.10	1347.20	63.25	2024.00		
1 Equip. Oper. (light)	53.00	424.00	79.20	633.60		
1 Air Compressor, 250 cfm		201.50		221.65		
2 Breakers, Pavement, 60 lb.		107.00		117.70		
2 -50' Air Hoses, 1.5"		48.50		53.35	7.44	8.18
48 L.H., Daily Totals		$2481.00		$3580.30	$51.69	$74.59

Crew No.	Bare Costs		Incl. Subs O&P		Cost Per Labor-Hour	
Crew B-40	**Hr.**	**Daily**	**Hr.**	**Daily**	**Bare Costs**	**Incl. O&P**
1 Pile Driver Foreman (outside)	$56.20	$449.60	$87.30	$698.40	$55.24	$84.58
4 Pile Drivers	54.20	1734.40	84.20	2694.40		
2 Equip. Oper. (crane)	59.20	947.20	88.50	1416.00		
1 Equip. Oper. (oiler)	50.55	404.40	75.55	604.40		
1 Crawler Crane, 40 Ton		1206.00		1326.60		
1 Vibratory Hammer & Gen.		2271.00		2498.10	54.33	59.76
64 L.H., Daily Totals		$7012.60		$9237.90	$109.57	$144.34
Crew B-40B	**Hr.**	**Daily**	**Hr.**	**Daily**	**Bare Costs**	**Incl. O&P**
1 Labor Foreman (outside)	$44.10	$352.80	$66.25	$530.00	$46.69	$70.01
3 Laborers	42.10	1010.40	63.25	1518.00		
1 Equip. Oper. (crane)	59.20	473.60	88.50	708.00		
1 Equip. Oper. (oiler)	50.55	404.40	75.55	604.40		
1 Lattice Boom Crane, 40 Ton		2028.00		2230.80	42.25	46.48
48 L.H., Daily Totals		$4269.20		$5591.20	$88.94	$116.48
Crew B-41	**Hr.**	**Daily**	**Hr.**	**Daily**	**Bare Costs**	**Incl. O&P**
1 Labor Foreman (outside)	$44.10	$352.80	$66.25	$530.00	$43.63	$65.50
4 Laborers	42.10	1347.20	63.25	2024.00		
.25 Equip. Oper. (crane)	59.20	118.40	88.50	177.00		
.25 Equip. Oper. (oiler)	50.55	101.10	75.55	151.10		
.25 Crawler Crane, 40 Ton		301.50		331.65	6.85	7.54
44 L.H., Daily Totals		$2221.00		$3213.75	$50.48	$73.04
Crew B-42	**Hr.**	**Daily**	**Hr.**	**Daily**	**Bare Costs**	**Incl. O&P**
1 Labor Foreman (outside)	$44.10	$352.80	$66.25	$530.00	$47.46	$71.67
4 Laborers	42.10	1347.20	63.25	2024.00		
1 Equip. Oper. (crane)	59.20	473.60	88.50	708.00		
1 Equip. Oper. (oiler)	50.55	404.40	75.55	604.40		
1 Welder	57.45	459.60	90.10	720.80		
1 Hyd. Crane, 25 Ton		580.85		638.93		
1 Welder, Gas Engine, 300 amp		147.00		161.70		
1 Horz. Boring Csg. Mch.		314.25		345.68	16.28	17.91
64 L.H., Daily Totals		$4079.70		$5733.51	$63.75	$89.59
Crew B-43	**Hr.**	**Daily**	**Hr.**	**Daily**	**Bare Costs**	**Incl. O&P**
1 Labor Foreman (outside)	$44.10	$352.80	$66.25	$530.00	$46.69	$70.01
3 Laborers	42.10	1010.40	63.25	1518.00		
1 Equip. Oper. (crane)	59.20	473.60	88.50	708.00		
1 Equip. Oper. (oiler)	50.55	404.40	75.55	604.40		
1 Drill Rig, Truck-Mounted		771.10		848.21	16.06	17.67
48 L.H., Daily Totals		$3012.30		$4208.61	$62.76	$87.68
Crew B-44	**Hr.**	**Daily**	**Hr.**	**Daily**	**Bare Costs**	**Incl. O&P**
1 Pile Driver Foreman (outside)	$56.20	$449.60	$87.30	$698.40	$54.19	$83.04
4 Pile Drivers	54.20	1734.40	84.20	2694.40		
2 Equip. Oper. (crane)	59.20	947.20	88.50	1416.00		
1 Laborer	42.10	336.80	63.25	506.00		
1 Crawler Crane, 40 Ton		1206.00		1326.60		
1 Lead, 60' High		209.25		230.18		
1 Hammer, Diesel, 15K ft.-lbs.		617.05		678.76	31.75	34.93
64 L.H., Daily Totals		$5500.30		$7550.33	$85.94	$117.97
Crew B-45	**Hr.**	**Daily**	**Hr.**	**Daily**	**Bare Costs**	**Incl. O&P**
1 Equip. Oper. (medium)	$56.75	$454.00	$84.85	$678.80	$52.85	$79.10
1 Truck Driver (heavy)	48.95	391.60	73.35	586.80		
1 Dist. Tanker, 3000 Gallon		330.15		363.17		
1 Truck Tractor, 6x4, 380 H.P.		493.25		542.58	51.46	56.61
16 L.H., Daily Totals		$1669.00		$2171.34	$104.31	$135.71

Crew No.	Bare Costs		Incl. Subs O&P		Cost Per Labor-Hour	
Crew B-46	**Hr.**	**Daily**	**Hr.**	**Daily**	**Bare Costs**	**Incl. O&P**
1 Pile Driver Foreman (outside)	$56.20	$449.60	$87.30	$698.40	$48.48	$74.24
2 Pile Drivers	54.20	867.20	84.20	1347.20		
3 Laborers	42.10	1010.40	63.25	1518.00		
1 Chain Saw, Gas, 36" Long		41.25		45.38	.86	.95
48 L.H., Daily Totals		$2368.45		$3608.97	$49.34	$75.19
Crew B-47	**Hr.**	**Daily**	**Hr.**	**Daily**	**Bare Costs**	**Incl. O&P**
1 Blast Foreman (outside)	$44.10	$352.80	$66.25	$530.00	$46.40	$69.57
1 Driller	42.10	336.80	63.25	506.00		
1 Equip. Oper. (light)	53.00	424.00	79.20	633.60		
1 Air Track Drill, 4"		1058.00		1163.80		
1 Air Compressor, 600 cfm		421.50		463.65		
2 -50' Air Hoses, 3"		76.70		84.37	64.84	71.33
24 L.H., Daily Totals		$2669.80		$3381.42	$111.24	$140.89
Crew B-47A	**Hr.**	**Daily**	**Hr.**	**Daily**	**Bare Costs**	**Incl. O&P**
1 Drilling Foreman (outside)	$44.10	$352.80	$66.25	$530.00	$51.28	$76.77
1 Equip. Oper. (heavy)	59.20	473.60	88.50	708.00		
1 Equip. Oper. (oiler)	50.55	404.40	75.55	604.40		
1 Air Track Drill, 5"		1127.00		1239.70	46.96	51.65
24 L.H., Daily Totals		$2357.80		$3082.10	$98.24	$128.42
Crew B-47C	**Hr.**	**Daily**	**Hr.**	**Daily**	**Bare Costs**	**Incl. O&P**
1 Laborer	$42.10	$336.80	$63.25	$506.00	$47.55	$71.22
1 Equip. Oper. (light)	53.00	424.00	79.20	633.60		
1 Air Compressor, 750 cfm		538.25		592.08		
2 -50' Air Hoses, 3"		76.70		84.37		
1 Air Track Drill, 4"		1058.00		1163.80	104.56	115.02
16 L.H., Daily Totals		$2433.75		$2979.84	$152.11	$186.24
Crew B-47E	**Hr.**	**Daily**	**Hr.**	**Daily**	**Bare Costs**	**Incl. O&P**
1 Labor Foreman (outside)	$44.10	$352.80	$66.25	$530.00	$42.60	$64.00
3 Laborers	42.10	1010.40	63.25	1518.00		
1 Flatbed Truck, Gas, 3 Ton		820.40		902.44	25.64	28.20
32 L.H., Daily Totals		$2183.60		$2950.44	$68.24	$92.20
Crew B-47G	**Hr.**	**Daily**	**Hr.**	**Daily**	**Bare Costs**	**Incl. O&P**
1 Labor Foreman (outside)	$44.10	$352.80	$66.25	$530.00	$45.33	$67.99
2 Laborers	42.10	673.60	63.25	1012.00		
1 Equip. Oper. (light)	53.00	424.00	79.20	633.60		
1 Air Track Drill, 4"		1058.00		1163.80		
1 Air Compressor, 600 cfm		421.50		463.65		
2 -50' Air Hoses, 3"		76.70		84.37		
1 Gunite Pump Rig		317.90		349.69	58.57	64.42
32 L.H., Daily Totals		$3324.50		$4237.11	$103.89	$132.41
Crew B-47H	**Hr.**	**Daily**	**Hr.**	**Daily**	**Bare Costs**	**Incl. O&P**
1 Skilled Worker Foreman (out)	$56.85	$454.80	$86.00	$688.00	$55.35	$83.71
3 Skilled Workers	54.85	1316.40	82.95	1990.80		
1 Flatbed Truck, Gas, 3 Ton		820.40		902.44	25.64	28.20
32 L.H., Daily Totals		$2591.60		$3581.24	$80.99	$111.91

Crew No.	Bare Costs		Incl. Subs O&P		Cost Per Labor-Hour	
Crew B-48	**Hr.**	**Daily**	**Hr.**	**Daily**	**Bare Costs**	**Incl. O&P**
1 Labor Foreman (outside)	$44.10	$352.80	$66.25	$530.00	$47.59	$71.32
3 Laborers	42.10	1010.40	63.25	1518.00		
1 Equip. Oper. (crane)	59.20	473.60	88.50	708.00		
1 Equip. Oper. (oiler)	50.55	404.40	75.55	604.40		
1 Equip. Oper. (light)	53.00	424.00	79.20	633.60		
1 Centr. Water Pump, 6"		232.90		256.19		
1 -20' Suction Hose, 6"		25.20		27.72		
1 -50' Discharge Hose, 6"		17.90		19.69		
1 Drill Rig, Truck-Mounted		771.10		848.21	18.70	20.57
56 L.H., Daily Totals		$3712.30		$5145.81	$66.29	$91.89
Crew B-49	**Hr.**	**Daily**	**Hr.**	**Daily**	**Bare Costs**	**Incl. O&P**
1 Labor Foreman (outside)	$44.10	$352.80	$66.25	$530.00	$50.12	$75.61
3 Laborers	42.10	1010.40	63.25	1518.00		
2 Equip. Oper. (crane)	59.20	947.20	88.50	1416.00		
2 Equip. Oper. (oilers)	50.55	808.80	75.55	1208.80		
1 Equip. Oper. (light)	53.00	424.00	79.20	633.60		
2 Pile Drivers	54.20	867.20	84.20	1347.20		
1 Hyd. Crane, 25 Ton		580.85		638.93		
1 Centr. Water Pump, 6"		232.90		256.19		
1 -20' Suction Hose, 6"		25.20		27.72		
1 -50' Discharge Hose, 6"		17.90		19.69		
1 Drill Rig, Truck-Mounted		771.10		848.21	18.50	20.35
88 L.H., Daily Totals		$6038.35		$8444.34	$68.62	$95.96
Crew B-50	**Hr.**	**Daily**	**Hr.**	**Daily**	**Bare Costs**	**Incl. O&P**
2 Pile Driver Foremen (outside)	$56.20	$899.20	$87.30	$1396.80	$52.35	$80.15
6 Pile Drivers	54.20	2601.60	84.20	4041.60		
2 Equip. Oper. (crane)	59.20	947.20	88.50	1416.00		
1 Equip. Oper. (oiler)	50.55	404.40	75.55	604.40		
3 Laborers	42.10	1010.40	63.25	1518.00		
1 Crawler Crane, 40 Ton		1206.00		1326.60		
1 Lead, 60' High		209.25		230.18		
1 Hammer, Diesel, 15K ft.-lbs.		617.05		678.76		
1 Air Compressor, 600 cfm		421.50		463.65		
2 -50' Air Hoses, 3"		76.70		84.37		
1 Chain Saw, Gas, 36" Long		41.25		45.38	22.96	25.26
112 L.H., Daily Totals		$8434.55		$11805.73	$75.31	$105.41
Crew B-51	**Hr.**	**Daily**	**Hr.**	**Daily**	**Bare Costs**	**Incl. O&P**
1 Labor Foreman (outside)	$44.10	$352.80	$66.25	$530.00	$43.29	$65.01
4 Laborers	42.10	1347.20	63.25	2024.00		
1 Truck Driver (light)	47.25	378.00	70.80	566.40		
1 Flatbed Truck, Gas, 1.5 Ton		196.15		215.76	4.09	4.50
48 L.H., Daily Totals		$2274.15		$3336.17	$47.38	$69.50
Crew B-52	**Hr.**	**Daily**	**Hr.**	**Daily**	**Bare Costs**	**Incl. O&P**
1 Carpenter Foreman (outside)	$55.15	$441.20	$82.85	$662.80	$48.73	$72.97
1 Carpenter	53.15	425.20	79.85	638.80		
3 Laborers	42.10	1010.40	63.25	1518.00		
1 Cement Finisher	49.95	399.60	73.45	587.60		
.5 Rodman (reinf.)	56.40	225.60	84.90	339.60		
.5 Equip. Oper. (medium)	56.75	227.00	84.85	339.40		
.5 Crawler Loader, 3 C.Y.		584.50		642.95	10.44	11.48
56 L.H., Daily Totals		$3313.50		$4729.15	$59.17	$84.45
Crew B-53	**Hr.**	**Daily**	**Hr.**	**Daily**	**Bare Costs**	**Incl. O&P**
1 Equip. Oper. (light)	$53.00	$424.00	$79.20	$633.60	$53.00	$79.20
1 Trencher, Chain, 12 H.P.		156.95		172.65	19.62	21.58
8 L.H., Daily Totals		$580.95		$806.25	$72.62	$100.78

Crew No.	Bare Costs		Incl. Subs O&P		Cost Per Labor-Hour	
Crew B-54	**Hr.**	**Daily**	**Hr.**	**Daily**	**Bare Costs**	**Incl. O&P**
1 Equip. Oper. (light)	$53.00	$424.00	$79.20	$633.60	$53.00	$79.20
1 Trencher, Chain, 40 H.P.		401.85		442.04	50.23	55.25
8 L.H., Daily Totals		$825.85		$1075.64	$103.23	$134.45
Crew B-54A	**Hr.**	**Daily**	**Hr.**	**Daily**	**Bare Costs**	**Incl. O&P**
.17 Labor Foreman (outside)	$44.10	$59.98	$66.25	$90.10	$54.91	$82.15
1 Equipment Operator (med.)	56.75	454.00	84.85	678.80		
1 Wheel Trencher, 67 H.P.		1126.00		1238.60	120.30	132.33
9.36 L.H., Daily Totals		$1639.98		$2007.50	$175.21	$214.48
Crew B-54B	**Hr.**	**Daily**	**Hr.**	**Daily**	**Bare Costs**	**Incl. O&P**
.25 Labor Foreman (outside)	$44.10	$88.20	$66.25	$132.50	$54.22	$81.13
1 Equipment Operator (med.)	56.75	454.00	84.85	678.80		
1 Wheel Trencher, 150 H.P.		1324.00		1456.40	132.40	145.64
10 L.H., Daily Totals		$1866.20		$2267.70	$186.62	$226.77
Crew B-54C	**Hr.**	**Daily**	**Hr.**	**Daily**	**Bare Costs**	**Incl. O&P**
1 Laborer	$42.10	$336.80	$63.25	$506.00	$49.42	$74.05
1 Equipment Operator (med.)	56.75	454.00	84.85	678.80		
1 Wheel Trencher, 67 H.P.		1126.00		1238.60	70.38	77.41
16 L.H., Daily Totals		$1916.80		$2423.40	$119.80	$151.46
Crew B-54D	**Hr.**	**Daily**	**Hr.**	**Daily**	**Bare Costs**	**Incl. O&P**
1 Laborer	$42.10	$336.80	$63.25	$506.00	$49.42	$74.05
1 Equipment Operator (med.)	56.75	454.00	84.85	678.80		
1 Rock Trencher, 6" Width		429.85		472.83	26.87	29.55
16 L.H., Daily Totals		$1220.65		$1657.64	$76.29	$103.60
Crew B-54E	**Hr.**	**Daily**	**Hr.**	**Daily**	**Bare Costs**	**Incl. O&P**
1 Laborer	$42.10	$336.80	$63.25	$506.00	$49.42	$74.05
1 Equipment Operator (med.)	56.75	454.00	84.85	678.80		
1 Rock Trencher, 18" Width		1005.00		1105.50	62.81	69.09
16 L.H., Daily Totals		$1795.80		$2290.30	$112.24	$143.14
Crew B-55	**Hr.**	**Daily**	**Hr.**	**Daily**	**Bare Costs**	**Incl. O&P**
2 Laborers	$42.10	$673.60	$63.25	$1012.00	$43.82	$65.77
1 Truck Driver (light)	47.25	378.00	70.80	566.40		
1 Truck-Mounted Earth Auger		390.20		429.22		
1 Flatbed Truck, Gas, 3 Ton		820.40		902.44	50.44	55.49
24 L.H., Daily Totals		$2262.20		$2910.06	$94.26	$121.25
Crew B-56	**Hr.**	**Daily**	**Hr.**	**Daily**	**Bare Costs**	**Incl. O&P**
1 Laborer	$42.10	$336.80	$63.25	$506.00	$47.55	$71.22
1 Equip. Oper. (light)	53.00	424.00	79.20	633.60		
1 Air Track Drill, 4"		1058.00		1163.80		
1 Air Compressor, 600 cfm		421.50		463.65		
1 -50' Air Hose, 3"		38.35		42.19	94.87	104.35
16 L.H., Daily Totals		$2278.65		$2809.24	$142.42	$175.58

Crew No.	Bare Costs		Incl. Subs O&P		Cost Per Labor-Hour	
Crew B-57	**Hr.**	**Daily**	**Hr.**	**Daily**	**Bare Costs**	**Incl. O&P**
1 Labor Foreman (outside)	$44.10	$352.80	$66.25	$530.00	$48.51	$72.67
2 Laborers	42.10	673.60	63.25	1012.00		
1 Equip. Oper. (crane)	59.20	473.60	88.50	708.00		
1 Equip. Oper. (light)	53.00	424.00	79.20	633.60		
1 Equip. Oper. (oiler)	50.55	404.40	75.55	604.40		
1 Crawler Crane, 25 Ton		1129.00		1241.90		
1 Clamshell Bucket, 1 C.Y.		68.55		75.41		
1 Centr. Water Pump, 6"		232.90		256.19		
1 -20' Suction Hose, 6"		25.20		27.72		
20 -50' Discharge Hoses, 6"		358.00		393.80	37.78	41.56
48 L.H., Daily Totals		$4142.05		$5483.02	$86.29	$114.23
Crew B-58	**Hr.**	**Daily**	**Hr.**	**Daily**	**Bare Costs**	**Incl. O&P**
2 Laborers	$42.10	$673.60	$63.25	$1012.00	$45.73	$68.57
1 Equip. Oper. (light)	53.00	424.00	79.20	633.60		
1 Backhoe Loader, 48 H.P.		213.75		235.13		
1 Small Helicopter, w/ Pilot		2080.00		2288.00	95.57	105.13
24 L.H., Daily Totals		$3391.35		$4168.73	$141.31	$173.70
Crew B-59	**Hr.**	**Daily**	**Hr.**	**Daily**	**Bare Costs**	**Incl. O&P**
1 Truck Driver (heavy)	$48.95	$391.60	$73.35	$586.80	$48.95	$73.35
1 Truck Tractor, 220 H.P.		307.10		337.81		
1 Water Tank Trailer, 5000 Gal.		152.25		167.47	57.42	63.16
8 L.H., Daily Totals		$850.95		$1092.09	$106.37	$136.51
Crew B-59A	**Hr.**	**Daily**	**Hr.**	**Daily**	**Bare Costs**	**Incl. O&P**
2 Laborers	$42.10	$673.60	$63.25	$1012.00	$44.38	$66.62
1 Truck Driver (heavy)	48.95	391.60	73.35	586.80		
1 Water Tank Trailer, 5000 Gal.		152.25		167.47		
1 Truck Tractor, 220 H.P.		307.10		337.81	19.14	21.05
24 L.H., Daily Totals		$1524.55		$2104.09	$63.52	$87.67
Crew B-60	**Hr.**	**Daily**	**Hr.**	**Daily**	**Bare Costs**	**Incl. O&P**
1 Labor Foreman (outside)	$44.10	$352.80	$66.25	$530.00	$49.15	$73.60
2 Laborers	42.10	673.60	63.25	1012.00		
1 Equip. Oper. (crane)	59.20	473.60	88.50	708.00		
2 Equip. Oper. (light)	53.00	848.00	79.20	1267.20		
1 Equip. Oper. (oiler)	50.55	404.40	75.55	604.40		
1 Crawler Crane, 40 Ton		1206.00		1326.60		
1 Lead, 60' High		209.25		230.18		
1 Hammer, Diesel, 15K ft.-lbs.		617.05		678.76		
1 Backhoe Loader, 48 H.P.		213.75		235.13	40.11	44.12
56 L.H., Daily Totals		$4998.45		$6592.26	$89.26	$117.72
Crew B-61	**Hr.**	**Daily**	**Hr.**	**Daily**	**Bare Costs**	**Incl. O&P**
1 Labor Foreman (outside)	$44.10	$352.80	$66.25	$530.00	$44.68	$67.04
3 Laborers	42.10	1010.40	63.25	1518.00		
1 Equip. Oper. (light)	53.00	424.00	79.20	633.60		
1 Cement Mixer, 2 C.Y.		105.70		116.27		
1 Air Compressor, 160 cfm		202.80		223.08	7.71	8.48
40 L.H., Daily Totals		$2095.70		$3020.95	$52.39	$75.52
Crew B-62	**Hr.**	**Daily**	**Hr.**	**Daily**	**Bare Costs**	**Incl. O&P**
2 Laborers	$42.10	$673.60	$63.25	$1012.00	$45.73	$68.57
1 Equip. Oper. (light)	53.00	424.00	79.20	633.60		
1 Loader, Skid Steer, 30 H.P.		177.55		195.31	7.40	8.14
24 L.H., Daily Totals		$1275.15		$1840.91	$53.13	$76.70

Crew B-62A	Bare Costs Hr.	Bare Costs Daily	Incl. Subs O&P Hr.	Incl. Subs O&P Daily	Bare Costs	Incl. O&P
2 Laborers	$42.10	$673.60	$63.25	$1012.00	$45.73	$68.57
1 Equip. Oper. (light)	53.00	424.00	79.20	633.60		
1 Loader, Skid Steer, 30 H.P.		177.55		195.31		
1 Trencher Attachment		66.20		72.82	10.16	11.17
24 L.H., Daily Totals		$1341.35		$1913.72	$55.89	$79.74

Crew B-63	Hr.	Daily	Hr.	Daily	Bare Costs	Incl. O&P
4 Laborers	$42.10	$1347.20	$63.25	$2024.00	$44.28	$66.44
1 Equip. Oper. (light)	53.00	424.00	79.20	633.60		
1 Loader, Skid Steer, 30 H.P.		177.55		195.31	4.44	4.88
40 L.H., Daily Totals		$1948.75		$2852.91	$48.72	$71.32

Crew B-63B	Hr.	Daily	Hr.	Daily	Bare Costs	Incl. O&P
1 Labor Foreman (inside)	$42.60	$340.80	$64.00	$512.00	$44.95	$67.42
2 Laborers	42.10	673.60	63.25	1012.00		
1 Equip. Oper. (light)	53.00	424.00	79.20	633.60		
1 Loader, Skid Steer, 78 H.P.		400.10		440.11	12.50	13.75
32 L.H., Daily Totals		$1838.50		$2597.71	$57.45	$81.18

Crew B-64	Hr.	Daily	Hr.	Daily	Bare Costs	Incl. O&P
1 Laborer	$42.10	$336.80	$63.25	$506.00	$44.67	$67.03
1 Truck Driver (light)	47.25	378.00	70.80	566.40		
1 Power Mulcher (small)		221.10		243.21		
1 Flatbed Truck, Gas, 1.5 Ton		196.15		215.76	26.08	28.69
16 L.H., Daily Totals		$1132.05		$1531.38	$70.75	$95.71

Crew B-65	Hr.	Daily	Hr.	Daily	Bare Costs	Incl. O&P
1 Laborer	$42.10	$336.80	$63.25	$506.00	$44.67	$67.03
1 Truck Driver (light)	47.25	378.00	70.80	566.40		
1 Power Mulcher (Large)		326.90		359.59		
1 Flatbed Truck, Gas, 1.5 Ton		196.15		215.76	32.69	35.96
16 L.H., Daily Totals		$1237.85		$1647.76	$77.37	$102.98

Crew B-66	Hr.	Daily	Hr.	Daily	Bare Costs	Incl. O&P
1 Equip. Oper. (light)	$53.00	$424.00	$79.20	$633.60	$53.00	$79.20
1 Loader-Backhoe, 40 H.P.		241.40		265.54	30.18	33.19
8 L.H., Daily Totals		$665.40		$899.14	$83.17	$112.39

Crew B-67	Hr.	Daily	Hr.	Daily	Bare Costs	Incl. O&P
1 Millwright	$55.90	$447.20	$81.00	$648.00	$54.45	$80.10
1 Equip. Oper. (light)	53.00	424.00	79.20	633.60		
1 R.T. Forklift, 5,000 Lb., diesel		269.90		296.89	16.87	18.56
16 L.H., Daily Totals		$1141.10		$1578.49	$71.32	$98.66

Crew B-67B	Hr.	Daily	Hr.	Daily	Bare Costs	Incl. O&P
1 Millwright Foreman (inside)	$56.40	$451.20	$81.70	$653.60	$56.15	$81.35
1 Millwright	55.90	447.20	81.00	648.00		
16 L.H., Daily Totals		$898.40		$1301.60	$56.15	$81.35

Crew B-68	Hr.	Daily	Hr.	Daily	Bare Costs	Incl. O&P
2 Millwrights	$55.90	$894.40	$81.00	$1296.00	$54.93	$80.40
1 Equip. Oper. (light)	53.00	424.00	79.20	633.60		
1 R.T. Forklift, 5,000 Lb., diesel		269.90		296.89	11.25	12.37
24 L.H., Daily Totals		$1588.30		$2226.49	$66.18	$92.77

Crew B-68A	Hr.	Daily	Hr.	Daily	Bare Costs	Incl. O&P
1 Millwright Foreman (inside)	$56.40	$451.20	$81.70	$653.60	$56.07	$81.23
2 Millwrights	55.90	894.40	81.00	1296.00		
1 Forklift, Smooth Floor, 8,000 Lb.		255.90		281.49	10.66	11.73
24 L.H., Daily Totals		$1601.50		$2231.09	$66.73	$92.96

Crew B-68B	Hr.	Daily	Hr.	Daily	Bare Costs	Incl. O&P
1 Millwright Foreman (inside)	$56.40	$451.20	$81.70	$653.60	$59.97	$88.47
2 Millwrights	55.90	894.40	81.00	1296.00		
2 Electricians	61.35	981.60	91.35	1461.60		
2 Plumbers	64.45	1031.20	96.45	1543.20		
1 R.T. Forklift, 5,000 Lb., gas		279.90		307.89	5.00	5.50
56 L.H., Daily Totals		$3638.30		$5262.29	$64.97	$93.97

Crew B-68C	Hr.	Daily	Hr.	Daily	Bare Costs	Incl. O&P
1 Millwright Foreman (inside)	$56.40	$451.20	$81.70	$653.60	$59.52	$87.63
1 Millwright	55.90	447.20	81.00	648.00		
1 Electrician	61.35	490.80	91.35	730.80		
1 Plumber	64.45	515.60	96.45	771.60		
1 R.T. Forklift, 5,000 Lb., gas		279.90		307.89	8.75	9.62
32 L.H., Daily Totals		$2184.70		$3111.89	$68.27	$97.25

Crew B-68D	Hr.	Daily	Hr.	Daily	Bare Costs	Incl. O&P
1 Labor Foreman (inside)	$42.60	$340.80	$64.00	$512.00	$45.90	$68.82
1 Laborer	42.10	336.80	63.25	506.00		
1 Equip. Oper. (light)	53.00	424.00	79.20	633.60		
1 R.T. Forklift, 5,000 Lb., gas		279.90		307.89	11.66	12.83
24 L.H., Daily Totals		$1381.50		$1959.49	$57.56	$81.65

Crew B-68E	Hr.	Daily	Hr.	Daily	Bare Costs	Incl. O&P
1 Struc. Steel Foreman (inside)	$58.15	$465.20	$91.20	$729.60	$57.71	$90.53
3 Struc. Steel Workers	57.65	1383.60	90.45	2170.80		
1 Welder	57.45	459.60	90.10	720.80		
1 Forklift, Smooth Floor, 8,000 Lb.		255.90		281.49	6.40	7.04
40 L.H., Daily Totals		$2564.30		$3902.69	$64.11	$97.57

Crew B-68F	Hr.	Daily	Hr.	Daily	Bare Costs	Incl. O&P
1 Skilled Worker Foreman (out)	$56.85	$454.80	$86.00	$688.00	$55.52	$83.97
2 Skilled Workers	54.85	877.60	82.95	1327.20		
1 R.T. Forklift, 5,000 Lb., gas		279.90		307.89	11.66	12.83
24 L.H., Daily Totals		$1612.30		$2323.09	$67.18	$96.80

Crew B-68G	Hr.	Daily	Hr.	Daily	Bare Costs	Incl. O&P
2 Structural Steel Workers	$57.65	$922.40	$90.45	$1447.20	$57.65	$90.45
1 R.T. Forklift, 5,000 Lb., gas		279.90		307.89	17.49	19.24
16 L.H., Daily Totals		$1202.30		$1755.09	$75.14	$109.69

Crew B-69	Hr.	Daily	Hr.	Daily	Bare Costs	Incl. O&P
1 Labor Foreman (outside)	$44.10	$352.80	$66.25	$530.00	$46.69	$70.01
3 Laborers	42.10	1010.40	63.25	1518.00		
1 Equip. Oper. (crane)	59.20	473.60	88.50	708.00		
1 Equip. Oper. (oiler)	50.55	404.40	75.55	604.40		
1 Hyd. Crane, 80 Ton		1486.00		1634.60	30.96	34.05
48 L.H., Daily Totals		$3727.20		$4995.00	$77.65	$104.06

Crew B-69A	Hr.	Daily	Hr.	Daily	Bare Costs	Incl. O&P
1 Labor Foreman (outside)	$44.10	$352.80	$66.25	$530.00	$46.18	$69.05
3 Laborers	42.10	1010.40	63.25	1518.00		
1 Equip. Oper. (medium)	56.75	454.00	84.85	678.80		
1 Concrete Finisher	49.95	399.60	73.45	587.60		
1 Curb/Gutter Paver, 2-Track		1217.00		1338.70	25.35	27.89
48 L.H., Daily Totals		$3433.80		$4653.10	$71.54	$96.94

Crew B-69B	Bare Costs Hr.	Bare Costs Daily	Incl. Subs O&P Hr.	Incl. Subs O&P Daily	Cost Per Labor-Hour Bare Costs	Cost Per Labor-Hour Incl. O&P
1 Labor Foreman (outside)	$44.10	$352.80	$66.25	$530.00	$46.18	$69.05
3 Laborers	42.10	1010.40	63.25	1518.00		
1 Equip. Oper. (medium)	56.75	454.00	84.85	678.80		
1 Cement Finisher	49.95	399.60	73.45	587.60		
1 Curb/Gutter Paver, 4-Track		791.55		870.71	16.49	18.14
48 L.H., Daily Totals		$3008.35		$4185.10	$62.67	$87.19

Crew B-70	Hr.	Daily	Hr.	Daily	Bare Costs	Incl. O&P
1 Labor Foreman (outside)	$44.10	$352.80	$66.25	$530.00	$48.66	$72.94
3 Laborers	42.10	1010.40	63.25	1518.00		
3 Equip. Oper. (medium)	56.75	1362.00	84.85	2036.40		
1 Grader, 30,000 Lbs.		1062.00		1168.20		
1 Ripper, Beam & 1 Shank		90.50		99.55		
1 Road Sweeper, S.P., 8' wide		715.85		787.43		
1 F.E. Loader, W.M., 1.5 C.Y.		423.50		465.85	40.93	45.02
56 L.H., Daily Totals		$5017.05		$6605.44	$89.59	$117.95

Crew B-70A	Hr.	Daily	Hr.	Daily	Bare Costs	Incl. O&P
1 Laborer	$42.10	$336.80	$63.25	$506.00	$53.82	$80.53
4 Equip. Oper. (medium)	56.75	1816.00	84.85	2715.20		
1 Grader, 40,000 Lbs.		1347.00		1481.70		
1 F.E. Loader, W.M., 2.5 C.Y.		702.10		772.31		
1 Dozer, 80 H.P.		401.40		441.54		
1 Roller, Pneum. Whl., 12 Ton		345.75		380.32	69.91	76.90
40 L.H., Daily Totals		$4949.05		$6297.07	$123.73	$157.43

Crew B-71	Hr.	Daily	Hr.	Daily	Bare Costs	Incl. O&P
1 Labor Foreman (outside)	$44.10	$352.80	$66.25	$530.00	$48.66	$72.94
3 Laborers	42.10	1010.40	63.25	1518.00		
3 Equip. Oper. (medium)	56.75	1362.00	84.85	2036.40		
1 Pvmt. Profiler, 750 H.P.		3448.00		3792.80		
1 Road Sweeper, S.P., 8' wide		715.85		787.43		
1 F.E. Loader, W.M., 1.5 C.Y.		423.50		465.85	81.92	90.11
56 L.H., Daily Totals		$7312.55		$9130.49	$130.58	$163.04

Crew B-72	Hr.	Daily	Hr.	Daily	Bare Costs	Incl. O&P
1 Labor Foreman (outside)	$44.10	$352.80	$66.25	$530.00	$49.67	$74.42
3 Laborers	42.10	1010.40	63.25	1518.00		
4 Equip. Oper. (medium)	56.75	1816.00	84.85	2715.20		
1 Pvmt. Profiler, 750 H.P.		3448.00		3792.80		
1 Hammermill, 250 H.P.		848.80		933.68		
1 Windrow Loader		1427.00		1569.70		
1 Mix Paver, 165 H.P.		2149.00		2363.90		
1 Roller, Pneum. Whl., 12 Ton		345.75		380.32	128.41	141.26
64 L.H., Daily Totals		$11397.75		$13803.61	$178.09	$215.68

Crew B-73	Hr.	Daily	Hr.	Daily	Bare Costs	Incl. O&P
1 Labor Foreman (outside)	$44.10	$352.80	$66.25	$530.00	$51.51	$77.13
2 Laborers	42.10	673.60	63.25	1012.00		
5 Equip. Oper. (medium)	56.75	2270.00	84.85	3394.00		
1 Road Mixer, 310 H.P.		1896.00		2085.60		
1 Tandem Roller, 10 Ton		236.75		260.43		
1 Hammermill, 250 H.P.		848.80		933.68		
1 Grader, 30,000 Lbs.		1062.00		1168.20		
.5 F.E. Loader, W.M., 1.5 C.Y.		211.75		232.93		
.5 Truck Tractor, 220 H.P.		153.55		168.91		
.5 Water Tank Trailer, 5000 Gal.		76.13		83.74	70.08	77.09
64 L.H., Daily Totals		$7781.38		$9869.47	$121.58	$154.21

Crew B-74	Hr.	Daily	Hr.	Daily	Bare Costs	Incl. O&P
1 Labor Foreman (outside)	$44.10	$352.80	$66.25	$530.00	$51.39	$76.95
1 Laborer	42.10	336.80	63.25	506.00		
4 Equip. Oper. (medium)	56.75	1816.00	84.85	2715.20		
2 Truck Drivers (heavy)	48.95	783.20	73.35	1173.60		
1 Grader, 30,000 Lbs.		1062.00		1168.20		
1 Ripper, Beam & 1 Shank		90.50		99.55		
2 Stabilizers, 310 H.P.		2778.00		3055.80		
1 Flatbed Truck, Gas, 3 Ton		820.40		902.44		
1 Chem. Spreader, Towed		83.35		91.69		
1 Roller, Vibratory, 25 Ton		664.35		730.78		
1 Water Tank Trailer, 5000 Gal.		152.25		167.47		
1 Truck Tractor, 220 H.P.		307.10		337.81	93.09	102.40
64 L.H., Daily Totals		$9246.75		$11478.55	$144.48	$179.35

Crew B-75	Hr.	Daily	Hr.	Daily	Bare Costs	Incl. O&P
1 Labor Foreman (outside)	$44.10	$352.80	$66.25	$530.00	$51.74	$77.46
1 Laborer	42.10	336.80	63.25	506.00		
4 Equip. Oper. (medium)	56.75	1816.00	84.85	2715.20		
1 Truck Driver (heavy)	48.95	391.60	73.35	586.80		
1 Grader, 30,000 Lbs.		1062.00		1168.20		
1 Ripper, Beam & 1 Shank		90.50		99.55		
2 Stabilizers, 310 H.P.		2778.00		3055.80		
1 Dist. Tanker, 3000 Gallon		330.15		363.17		
1 Truck Tractor, 6x4, 380 H.P.		493.25		542.58		
1 Roller, Vibratory, 25 Ton		664.35		730.78	96.75	106.43
56 L.H., Daily Totals		$8315.45		$10298.08	$148.49	$183.89

Crew B-76	Hr.	Daily	Hr.	Daily	Bare Costs	Incl. O&P
1 Dock Builder Foreman (outside)	$56.20	$449.60	$87.30	$698.40	$55.13	$84.54
5 Dock Builders	54.20	2168.00	84.20	3368.00		
2 Equip. Oper. (crane)	59.20	947.20	88.50	1416.00		
1 Equip. Oper. (oiler)	50.55	404.40	75.55	604.40		
1 Crawler Crane, 50 Ton		1510.00		1661.00		
1 Barge, 400 Ton		858.85		944.74		
1 Hammer, Diesel, 15K ft.-lbs.		617.05		678.76		
1 Lead, 60' High		209.25		230.18		
1 Air Compressor, 600 cfm		421.50		463.65		
2 -50' Air Hoses, 3"		76.70		84.37	51.30	56.43
72 L.H., Daily Totals		$7662.55		$10149.49	$106.42	$140.97

Crew B-76A	Hr.	Daily	Hr.	Daily	Bare Costs	Incl. O&P
1 Labor Foreman (outside)	$44.10	$352.80	$66.25	$530.00	$45.54	$68.32
5 Laborers	42.10	1684.00	63.25	2530.00		
1 Equip. Oper. (crane)	59.20	473.60	88.50	708.00		
1 Equip. Oper. (oiler)	50.55	404.40	75.55	604.40		
1 Crawler Crane, 50 Ton		1510.00		1661.00		
1 Barge, 400 Ton		858.85		944.74	37.01	40.71
64 L.H., Daily Totals		$5283.65		$6978.14	$82.56	$109.03

Crew B-77	Hr.	Daily	Hr.	Daily	Bare Costs	Incl. O&P
1 Labor Foreman (outside)	$44.10	$352.80	$66.25	$530.00	$43.53	$65.36
3 Laborers	42.10	1010.40	63.25	1518.00		
1 Truck Driver (light)	47.25	378.00	70.80	566.40		
1 Crack Cleaner, 25 H.P.		52.45		57.70		
1 Crack Filler, Trailer Mtd.		154.75		170.22		
1 Flatbed Truck, Gas, 3 Ton		820.40		902.44	25.69	28.26
40 L.H., Daily Totals		$2768.80		$3744.76	$69.22	$93.62

Crew No.	Bare Costs		Incl. Subs O&P		Cost Per Labor-Hour	
Crew B-78	**Hr.**	**Daily**	**Hr.**	**Daily**	**Bare Costs**	**Incl. O&P**
1 Labor Foreman (outside)	$44.10	$352.80	$66.25	$530.00	$43.29	$65.01
4 Laborers	42.10	1347.20	63.25	2024.00		
1 Truck Driver (light)	47.25	378.00	70.80	566.40		
1 Paint Striper, S.P., 40 Gallon		127.10		139.81		
1 Flatbed Truck, Gas, 3 Ton		820.40		902.44		
1 Pickup Truck, 3/4 Ton		110.85		121.94	22.05	24.25
48 L.H., Daily Totals		$3136.35		$4284.59	$65.34	$89.26
Crew B-78A	**Hr.**	**Daily**	**Hr.**	**Daily**	**Bare Costs**	**Incl. O&P**
1 Equip. Oper. (light)	$53.00	$424.00	$79.20	$633.60	$53.00	$79.20
1 Line Rem. (Metal Balls) 115 H.P.		916.50		1008.15	114.56	126.02
8 L.H., Daily Totals		$1340.50		$1641.75	$167.56	$205.22
Crew B-78B	**Hr.**	**Daily**	**Hr.**	**Daily**	**Bare Costs**	**Incl. O&P**
2 Laborers	$42.10	$673.60	$63.25	$1012.00	$43.31	$65.02
.25 Equip. Oper. (light)	53.00	106.00	79.20	158.40		
1 Pickup Truck, 3/4 Ton		110.85		121.94		
1 Line Rem.,11 H.P.,Walk Behind		113.45		124.80		
.25 Road Sweeper, S.P., 8' wide		178.96		196.86	22.40	24.64
18 L.H., Daily Totals		$1182.86		$1613.99	$65.71	$89.67
Crew B-78C	**Hr.**	**Daily**	**Hr.**	**Daily**	**Bare Costs**	**Incl. O&P**
1 Labor Foreman (outside)	$44.10	$352.80	$66.25	$530.00	$43.29	$65.01
4 Laborers	42.10	1347.20	63.25	2024.00		
1 Truck Driver (light)	47.25	378.00	70.80	566.40		
1 Paint Striper, T.M., 120 Gal.		592.95		652.25		
1 Flatbed Truck, Gas, 3 Ton		820.40		902.44		
1 Pickup Truck, 3/4 Ton		110.85		121.94	31.75	34.93
48 L.H., Daily Totals		$3602.20		$4797.02	$75.05	$99.94
Crew B-78D	**Hr.**	**Daily**	**Hr.**	**Daily**	**Bare Costs**	**Incl. O&P**
2 Labor Foremen (outside)	$44.10	$705.60	$66.25	$1060.00	$43.02	$64.61
7 Laborers	42.10	2357.60	63.25	3542.00		
1 Truck Driver (light)	47.25	378.00	70.80	566.40		
1 Paint Striper, T.M., 120 Gal.		592.95		652.25		
1 Flatbed Truck, Gas, 3 Ton		820.40		902.44		
3 Pickup Trucks, 3/4 Ton		332.55		365.81		
1 Air Compressor, 60 cfm		152.95		168.25		
1 -50' Air Hose, 3/4"		7.75		8.53		
1 Breaker, Pavement, 60 lb.		53.50		58.85	24.50	26.95
80 L.H., Daily Totals		$5401.30		$7324.51	$67.52	$91.56
Crew B-78E	**Hr.**	**Daily**	**Hr.**	**Daily**	**Bare Costs**	**Incl. O&P**
2 Labor Foremen (outside)	$44.10	$705.60	$66.25	$1060.00	$42.86	$64.38
9 Laborers	42.10	3031.20	63.25	4554.00		
1 Truck Driver (light)	47.25	378.00	70.80	566.40		
1 Paint Striper, T.M., 120 Gal.		592.95		652.25		
1 Flatbed Truck, Gas, 3 Ton		820.40		902.44		
4 Pickup Trucks, 3/4 Ton		443.40		487.74		
2 Air Compressors, 60 cfm		305.90		336.49		
2 -50' Air Hoses, 3/4"		15.50		17.05		
2 Breakers, Pavement, 60 lb.		107.00		117.70	23.80	26.18
96 L.H., Daily Totals		$6399.95		$8694.07	$66.67	$90.56

Crew No.	Bare Costs		Incl. Subs O&P		Cost Per Labor-Hour	
Crew B-78F	**Hr.**	**Daily**	**Hr.**	**Daily**	**Bare Costs**	**Incl. O&P**
2 Labor Foremen (outside)	$44.10	$705.60	$66.25	$1060.00	$42.75	$64.22
11 Laborers	42.10	3704.80	63.25	5566.00		
1 Truck Driver (light)	47.25	378.00	70.80	566.40		
1 Paint Striper, T.M., 120 Gal.		592.95		652.25		
1 Flatbed Truck, Gas, 3 Ton		820.40		902.44		
7 Pickup Trucks, 3/4 Ton		775.95		853.54		
3 Air Compressors, 60 cfm		458.85		504.74		
3 -50' Air Hoses, 3/4"		23.25		25.57		
3 Breakers, Pavement, 60 lb.		160.50		176.55	25.28	27.81
112 L.H., Daily Totals		$7620.30		$10307.49	$68.04	$92.03
Crew B-79	**Hr.**	**Daily**	**Hr.**	**Daily**	**Bare Costs**	**Incl. O&P**
1 Labor Foreman (outside)	$44.10	$352.80	$66.25	$530.00	$43.53	$65.36
3 Laborers	42.10	1010.40	63.25	1518.00		
1 Truck Driver (light)	47.25	378.00	70.80	566.40		
1 Paint Striper, T.M., 120 Gal.		592.95		652.25		
1 Heating Kettle, 115 Gallon		98.65		108.52		
1 Flatbed Truck, Gas, 3 Ton		820.40		902.44		
2 Pickup Trucks, 3/4 Ton		221.70		243.87	43.34	47.68
40 L.H., Daily Totals		$3474.90		$4521.47	$86.87	$113.04
Crew B-79A	**Hr.**	**Daily**	**Hr.**	**Daily**	**Bare Costs**	**Incl. O&P**
1.5 Equip. Oper. (light)	$53.00	$636.00	$79.20	$950.40	$53.00	$79.20
.5 Line Remov. (Grinder) 115 H.P.		475.57		523.13		
1 Line Rem. (Metal Balls) 115 H.P.		916.50		1008.15	116.01	127.61
12 L.H., Daily Totals		$2028.08		$2481.68	$169.01	$206.81
Crew B-79B	**Hr.**	**Daily**	**Hr.**	**Daily**	**Bare Costs**	**Incl. O&P**
1 Laborer	$42.10	$336.80	$63.25	$506.00	$42.10	$63.25
1 Set of Gases		171.55		188.71	21.44	23.59
8 L.H., Daily Totals		$508.35		$694.71	$63.54	$86.84
Crew B-79C	**Hr.**	**Daily**	**Hr.**	**Daily**	**Bare Costs**	**Incl. O&P**
1 Labor Foreman (outside)	$44.10	$352.80	$66.25	$530.00	$43.12	$64.76
5 Laborers	42.10	1684.00	63.25	2530.00		
1 Truck Driver (light)	47.25	378.00	70.80	566.40		
1 Paint Striper, T.M., 120 Gal.		592.95		652.25		
1 Heating Kettle, 115 Gallon		98.65		108.52		
1 Flatbed Truck, Gas, 3 Ton		820.40		902.44		
3 Pickup Trucks, 3/4 Ton		332.55		365.81		
1 Air Compressor, 60 cfm		152.95		168.25		
1 -50' Air Hose, 3/4"		7.75		8.53		
1 Breaker, Pavement, 60 lb.		53.50		58.85	36.76	40.44
56 L.H., Daily Totals		$4473.55		$5891.02	$79.88	$105.20
Crew B-79D	**Hr.**	**Daily**	**Hr.**	**Daily**	**Bare Costs**	**Incl. O&P**
2 Labor Foremen (outside)	$44.10	$705.60	$66.25	$1060.00	$43.24	$64.94
5 Laborers	42.10	1684.00	63.25	2530.00		
1 Truck Driver (light)	47.25	378.00	70.80	566.40		
1 Paint Striper, T.M., 120 Gal.		592.95		652.25		
1 Heating Kettle, 115 Gallon		98.65		108.52		
1 Flatbed Truck, Gas, 3 Ton		820.40		902.44		
4 Pickup Trucks, 3/4 Ton		443.40		487.74		
1 Air Compressor, 60 cfm		152.95		168.25		
1 -50' Air Hose, 3/4"		7.75		8.53		
1 Breaker, Pavement, 60 lb.		53.50		58.85	33.90	37.29
64 L.H., Daily Totals		$4937.20		$6542.96	$77.14	$102.23

Crew No.	Bare Costs		Incl. Subs O&P		Cost Per Labor-Hour	
Crew B-79E	**Hr.**	**Daily**	**Hr.**	**Daily**	**Bare Costs**	**Incl. O&P**
2 Labor Foremen (outside)	$44.10	$705.60	$66.25	$1060.00	$43.02	$64.61
7 Laborers	42.10	2357.60	63.25	3542.00		
1 Truck Driver (light)	47.25	378.00	70.80	566.40		
1 Paint Striper, T.M., 120 Gal.		592.95		652.25		
1 Heating Kettle, 115 Gallon		98.65		108.52		
1 Flatbed Truck, Gas, 3 Ton		820.40		902.44		
5 Pickup Trucks, 3/4 Ton		554.25		609.67		
2 Air Compressors, 60 cfm		305.90		336.49		
2 -50' Air Hoses, 3/4"		15.50		17.05		
2 Breakers, Pavement, 60 lb.		107.00		117.70	31.18	34.30
80 L.H., Daily Totals		$5935.85		$7912.52	$74.20	$98.91

Crew B-80	Hr.	Daily	Hr.	Daily	Bare Costs	Incl. O&P
1 Labor Foreman (outside)	$44.10	$352.80	$66.25	$530.00	$46.61	$69.88
1 Laborer	42.10	336.80	63.25	506.00		
1 Truck Driver (light)	47.25	378.00	70.80	566.40		
1 Equip. Oper. (light)	53.00	424.00	79.20	633.60		
1 Flatbed Truck, Gas, 3 Ton		820.40		902.44		
1 Earth Auger, Truck-Mtd.		200.50		220.55	31.90	35.09
32 L.H., Daily Totals		$2512.50		$3358.99	$78.52	$104.97

Crew B-80A	Hr.	Daily	Hr.	Daily	Bare Costs	Incl. O&P
3 Laborers	$42.10	$1010.40	$63.25	$1518.00	$42.10	$63.25
1 Flatbed Truck, Gas, 3 Ton		820.40		902.44	34.18	37.60
24 L.H., Daily Totals		$1830.80		$2420.44	$76.28	$100.85

Crew B-80B	Hr.	Daily	Hr.	Daily	Bare Costs	Incl. O&P
3 Laborers	$42.10	$1010.40	$63.25	$1518.00	$44.83	$67.24
1 Equip. Oper. (light)	53.00	424.00	79.20	633.60		
1 Crane, Flatbed Mounted, 3 Ton		236.10		259.71	7.38	8.12
32 L.H., Daily Totals		$1670.50		$2411.31	$52.20	$75.35

Crew B-80C	Hr.	Daily	Hr.	Daily	Bare Costs	Incl. O&P
2 Laborers	$42.10	$673.60	$63.25	$1012.00	$43.82	$65.77
1 Truck Driver (light)	47.25	378.00	70.80	566.40		
1 Flatbed Truck, Gas, 1.5 Ton		196.15		215.76		
1 Manual Fence Post Auger, Gas		53.85		59.23	10.42	11.46
24 L.H., Daily Totals		$1301.60		$1853.40	$54.23	$77.22

Crew B-81	Hr.	Daily	Hr.	Daily	Bare Costs	Incl. O&P
1 Laborer	$42.10	$336.80	$63.25	$506.00	$49.27	$73.82
1 Equip. Oper. (medium)	56.75	454.00	84.85	678.80		
1 Truck Driver (heavy)	48.95	391.60	73.35	586.80		
1 Hydromulcher, T.M., 3000 Gal.		275.45		303.00		
1 Truck Tractor, 220 H.P.		307.10		337.81	24.27	26.70
24 L.H., Daily Totals		$1764.95		$2412.41	$73.54	$100.52

Crew B-81A	Hr.	Daily	Hr.	Daily	Bare Costs	Incl. O&P
1 Laborer	$42.10	$336.80	$63.25	$506.00	$44.67	$67.03
1 Truck Driver (light)	47.25	378.00	70.80	566.40		
1 Hydromulcher, T.M., 600 Gal.		116.95		128.65		
1 Flatbed Truck, Gas, 3 Ton		820.40		902.44	58.58	64.44
16 L.H., Daily Totals		$1652.15		$2103.49	$103.26	$131.47

Crew B-82	Hr.	Daily	Hr.	Daily	Bare Costs	Incl. O&P
1 Laborer	$42.10	$336.80	$63.25	$506.00	$47.55	$71.22
1 Equip. Oper. (light)	53.00	424.00	79.20	633.60		
1 Horiz. Borer, 6 H.P.		182.35		200.59	11.40	12.54
16 L.H., Daily Totals		$943.15		$1340.18	$58.95	$83.76

Crew No.	Bare Costs		Incl. Subs O&P		Cost Per Labor-Hour	
Crew B-82A	**Hr.**	**Daily**	**Hr.**	**Daily**	**Bare Costs**	**Incl. O&P**
2 Laborers	$42.10	$673.60	$63.25	$1012.00	$47.55	$71.22
2 Equip. Opers. (light)	53.00	848.00	79.20	1267.20		
2 Dump Truck, 8 C.Y., 220 H.P.		856.10		941.71		
1 Flatbed Trailer, 25 Ton		135.55		149.10		
1 Horiz. Dir. Drill, 20k lb. Thrust		538.65		592.51		
1 Mud Trailer for HDD, 1500 Gal.		309.00		339.90		
1 Pickup Truck, 4x4, 3/4 Ton		175.85		193.44		
1 Flatbed Trailer, 3 Ton		70.30		77.33		
1 Loader, Skid Steer, 78 H.P.		400.10		440.11	77.67	85.44
32 L.H., Daily Totals		$4007.15		$5013.31	$125.22	$156.67

Crew B-82B	Hr.	Daily	Hr.	Daily	Bare Costs	Incl. O&P
2 Laborers	$42.10	$673.60	$63.25	$1012.00	$47.55	$71.22
2 Equip. Opers. (light)	53.00	848.00	79.20	1267.20		
2 Dump Truck, 8 C.Y., 220 H.P.		856.10		941.71		
1 Flatbed Trailer, 25 Ton		135.55		149.10		
1 Horiz. Dir. Drill, 30k lb. Thrust		641.20		705.32		
1 Mud Trailer for HDD, 1500 Gal.		309.00		339.90		
1 Pickup Truck, 4x4, 3/4 Ton		175.85		193.44		
1 Flatbed Trailer, 3 Ton		70.30		77.33		
1 Loader, Skid Steer, 78 H.P.		400.10		440.11	80.88	88.97
32 L.H., Daily Totals		$4109.70		$5126.11	$128.43	$160.19

Crew B-82C	Hr.	Daily	Hr.	Daily	Bare Costs	Incl. O&P
2 Laborers	$42.10	$673.60	$63.25	$1012.00	$47.55	$71.22
2 Equip. Opers. (light)	53.00	848.00	79.20	1267.20		
2 Dump Truck, 8 C.Y., 220 H.P.		856.10		941.71		
1 Flatbed Trailer, 25 Ton		135.55		149.10		
1 Horiz. Dir. Drill, 50k lb. Thrust		815.80		897.38		
1 Mud Trailer for HDD, 1500 Gal.		309.00		339.90		
1 Pickup Truck, 4x4, 3/4 Ton		175.85		193.44		
1 Flatbed Trailer, 3 Ton		70.30		77.33		
1 Loader, Skid Steer, 78 H.P.		400.10		440.11	86.33	94.97
32 L.H., Daily Totals		$4284.30		$5318.17	$133.88	$166.19

Crew B-82D	Hr.	Daily	Hr.	Daily	Bare Costs	Incl. O&P
1 Equip. Oper. (light)	$53.00	$424.00	$79.20	$633.60	$53.00	$79.20
1 Mud Trailer for HDD, 1500 Gal.		309.00		339.90	38.63	42.49
8 L.H., Daily Totals		$733.00		$973.50	$91.63	$121.69

Crew B-83	Hr.	Daily	Hr.	Daily	Bare Costs	Incl. O&P
1 Tugboat Captain	$56.75	$454.00	$84.85	$678.80	$49.42	$74.05
1 Tugboat Hand	42.10	336.80	63.25	506.00		
1 Tugboat, 250 H.P.		717.50		789.25	44.84	49.33
16 L.H., Daily Totals		$1508.30		$1974.05	$94.27	$123.38

Crew B-84	Hr.	Daily	Hr.	Daily	Bare Costs	Incl. O&P
1 Equip. Oper. (medium)	$56.75	$454.00	$84.85	$678.80	$56.75	$84.85
1 Rotary Mower/Tractor		366.75		403.43	45.84	50.43
8 L.H., Daily Totals		$820.75		$1082.22	$102.59	$135.28

Crew B-85	Hr.	Daily	Hr.	Daily	Bare Costs	Incl. O&P
3 Laborers	$42.10	$1010.40	$63.25	$1518.00	$46.40	$69.59
1 Equip. Oper. (medium)	56.75	454.00	84.85	678.80		
1 Truck Driver (heavy)	48.95	391.60	73.35	586.80		
1 Telescoping Boom Lift, to 80'		383.55		421.90		
1 Brush Chipper, 12", 130 H.P.		392.80		432.08		
1 Pruning Saw, Rotary		26.40		29.04	20.07	22.08
40 L.H., Daily Totals		$2658.75		$3666.63	$66.47	$91.67

Crews - Standard

Crew No.	Bare Costs		Incl. Subs O&P		Cost Per Labor-Hour	

Crew B-86	Hr.	Daily	Hr.	Daily	Bare Costs	Incl. O&P
1 Equip. Oper. (medium)	$56.75	$454.00	$84.85	$678.80	$56.75	$84.85
1 Stump Chipper, S.P.		194.45		213.90	24.31	26.74
8 L.H., Daily Totals		$648.45		$892.70	$81.06	$111.59

Crew B-86A	Hr.	Daily	Hr.	Daily	Bare Costs	Incl. O&P
1 Equip. Oper. (medium)	$56.75	$454.00	$84.85	$678.80	$56.75	$84.85
1 Grader, 30,000 Lbs.		1062.00		1168.20	132.75	146.03
8 L.H., Daily Totals		$1516.00		$1847.00	$189.50	$230.88

Crew B-86B	Hr.	Daily	Hr.	Daily	Bare Costs	Incl. O&P
1 Equip. Oper. (medium)	$56.75	$454.00	$84.85	$678.80	$56.75	$84.85
1 Dozer, 200 H.P.		1504.00		1654.40	188.00	206.80
8 L.H., Daily Totals		$1958.00		$2333.20	$244.75	$291.65

Crew B-87	Hr.	Daily	Hr.	Daily	Bare Costs	Incl. O&P
1 Laborer	$42.10	$336.80	$63.25	$506.00	$53.82	$80.53
4 Equip. Oper. (medium)	56.75	1816.00	84.85	2715.20		
2 Feller Bunchers, 100 H.P.		1177.20		1294.92		
1 Log Chipper, 22" Tree		552.00		607.20		
1 Dozer, 105 H.P.		633.85		697.24		
1 Chain Saw, Gas, 36" Long		41.25		45.38	60.11	66.12
40 L.H., Daily Totals		$4557.10		$5865.93	$113.93	$146.65

Crew B-88	Hr.	Daily	Hr.	Daily	Bare Costs	Incl. O&P
1 Laborer	$42.10	$336.80	$63.25	$506.00	$54.66	$81.76
6 Equip. Oper. (medium)	56.75	2724.00	84.85	4072.80		
2 Feller Bunchers, 100 H.P.		1177.20		1294.92		
1 Log Chipper, 22" Tree		552.00		607.20		
2 Log Skidders, 50 H.P.		1808.40		1989.24		
1 Dozer, 105 H.P.		633.85		697.24		
1 Chain Saw, Gas, 36" Long		41.25		45.38	75.23	82.75
56 L.H., Daily Totals		$7273.50		$9212.77	$129.88	$164.51

Crew B-89	Hr.	Daily	Hr.	Daily	Bare Costs	Incl. O&P
1 Equip. Oper. (light)	$53.00	$424.00	$79.20	$633.60	$50.13	$75.00
1 Truck Driver (light)	47.25	378.00	70.80	566.40		
1 Flatbed Truck, Gas, 3 Ton		820.40		902.44		
1 Concrete Saw		111.50		122.65		
1 Water Tank, 65 Gal.		101.90		112.09	64.61	71.07
16 L.H., Daily Totals		$1835.80		$2337.18	$114.74	$146.07

Crew B-89A	Hr.	Daily	Hr.	Daily	Bare Costs	Incl. O&P
1 Skilled Worker	$54.85	$438.80	$82.95	$663.60	$48.48	$73.10
1 Laborer	42.10	336.80	63.25	506.00		
1 Core Drill (Large)		114.30		125.73	7.14	7.86
16 L.H., Daily Totals		$889.90		$1295.33	$55.62	$80.96

Crew B-89B	Hr.	Daily	Hr.	Daily	Bare Costs	Incl. O&P
1 Equip. Oper. (light)	$53.00	$424.00	$79.20	$633.60	$50.13	$75.00
1 Truck Driver (light)	47.25	378.00	70.80	566.40		
1 Wall Saw, Hydraulic, 10 H.P.		85.50		94.05		
1 Generator, Diesel, 100 kW		496.20		545.82		
1 Water Tank, 65 Gal.		101.90		112.09		
1 Flatbed Truck, Gas, 3 Ton		820.40		902.44	94.00	103.40
16 L.H., Daily Totals		$2306.00		$2854.40	$144.13	$178.40

Crew B-89C	Hr.	Daily	Hr.	Daily	Bare Costs	Incl. O&P
1 Cement Finisher	$49.95	$399.60	$73.45	$587.60	$49.95	$73.45
1 Masonry cut-off saw, gas		61.00		67.10	7.63	8.39
8 L.H., Daily Totals		$460.60		$654.70	$57.58	$81.84

Crew No.	Bare Costs		Incl. Subs O&P		Cost Per Labor-Hour	

Crew B-90	Hr.	Daily	Hr.	Daily	Bare Costs	Incl. O&P
1 Labor Foreman (outside)	$44.10	$352.80	$66.25	$530.00	$46.79	$70.14
3 Laborers	42.10	1010.40	63.25	1518.00		
2 Equip. Oper. (light)	53.00	848.00	79.20	1267.20		
2 Truck Drivers (heavy)	48.95	783.20	73.35	1173.60		
1 Road Mixer, 310 H.P.		1896.00		2085.60		
1 Dist. Truck, 2000 Gal.		299.65		329.62	34.31	37.74
64 L.H., Daily Totals		$5190.05		$6904.02	$81.09	$107.88

Crew B-90A	Hr.	Daily	Hr.	Daily	Bare Costs	Incl. O&P
1 Labor Foreman (outside)	$44.10	$352.80	$66.25	$530.00	$50.76	$76.02
2 Laborers	42.10	673.60	63.25	1012.00		
4 Equip. Oper. (medium)	56.75	1816.00	84.85	2715.20		
2 Graders, 30,000 Lbs.		2124.00		2336.40		
1 Tandem Roller, 10 Ton		236.75		260.43		
1 Roller, Pneum. Whl., 12 Ton		345.75		380.32	48.33	53.16
56 L.H., Daily Totals		$5548.90		$7234.35	$99.09	$129.18

Crew B-90B	Hr.	Daily	Hr.	Daily	Bare Costs	Incl. O&P
1 Labor Foreman (outside)	$44.10	$352.80	$66.25	$530.00	$49.76	$74.55
2 Laborers	42.10	673.60	63.25	1012.00		
3 Equip. Oper. (medium)	56.75	1362.00	84.85	2036.40		
1 Roller, Pneum. Whl., 12 Ton		345.75		380.32		
1 Road Mixer, 310 H.P.		1896.00		2085.60	46.70	51.37
48 L.H., Daily Totals		$4630.15		$6044.32	$96.46	$125.92

Crew B-90C	Hr.	Daily	Hr.	Daily	Bare Costs	Incl. O&P
1 Labor Foreman (outside)	$44.10	$352.80	$66.25	$530.00	$48.15	$72.17
4 Laborers	42.10	1347.20	63.25	2024.00		
3 Equip. Oper. (medium)	56.75	1362.00	84.85	2036.40		
3 Truck Drivers (heavy)	48.95	1174.80	73.35	1760.40		
3 Road Mixers, 310 H.P.		5688.00		6256.80	64.64	71.10
88 L.H., Daily Totals		$9924.80		$12607.60	$112.78	$143.27

Crew B-90D	Hr.	Daily	Hr.	Daily	Bare Costs	Incl. O&P
1 Labor Foreman (outside)	$44.10	$352.80	$66.25	$530.00	$47.22	$70.80
6 Laborers	42.10	2020.80	63.25	3036.00		
3 Equip. Oper. (medium)	56.75	1362.00	84.85	2036.40		
3 Truck Drivers (heavy)	48.95	1174.80	73.35	1760.40		
3 Road Mixers, 310 H.P.		5688.00		6256.80	54.69	60.16
104 L.H., Daily Totals		$10598.40		$13619.60	$101.91	$130.96

Crew B-90E	Hr.	Daily	Hr.	Daily	Bare Costs	Incl. O&P
1 Labor Foreman (outside)	$44.10	$352.80	$66.25	$530.00	$47.97	$71.91
4 Laborers	42.10	1347.20	63.25	2024.00		
3 Equip. Oper. (medium)	56.75	1362.00	84.85	2036.40		
1 Truck Driver (heavy)	48.95	391.60	73.35	586.80		
1 Road Mixer, 310 H.P.		1896.00		2085.60	26.33	28.97
72 L.H., Daily Totals		$5349.60		$7262.80	$74.30	$100.87

Crew B-91	Hr.	Daily	Hr.	Daily	Bare Costs	Incl. O&P
1 Labor Foreman (outside)	$44.10	$352.80	$66.25	$530.00	$50.53	$75.69
2 Laborers	42.10	673.60	63.25	1012.00		
4 Equip. Oper. (medium)	56.75	1816.00	84.85	2715.20		
1 Truck Driver (heavy)	48.95	391.60	73.35	586.80		
1 Dist. Tanker, 3000 Gallon		330.15		363.17		
1 Truck Tractor, 6x4, 380 H.P.		493.25		542.58		
1 Aggreg. Spreader, S.P.		848.90		933.79		
1 Roller, Pneum. Whl., 12 Ton		345.75		380.32		
1 Tandem Roller, 10 Ton		236.75		260.43	35.23	38.75
64 L.H., Daily Totals		$5488.80		$7324.28	$85.76	$114.44

Crew No.	Bare Costs		Incl. Subs O&P		Cost Per Labor-Hour	
Crew B-91B	**Hr.**	**Daily**	**Hr.**	**Daily**	**Bare Costs**	**Incl. O&P**
1 Laborer	$42.10	$336.80	$63.25	$506.00	$49.42	$74.05
1 Equipment Oper. (med.)	56.75	454.00	84.85	678.80		
1 Road Sweeper, Vac. Assist.		869.00		955.90	54.31	59.74
16 L.H., Daily Totals		$1659.80		$2140.70	$103.74	$133.79

Crew B-91C	Hr.	Daily	Hr.	Daily	Bare Costs	Incl. O&P
1 Laborer	$42.10	$336.80	$63.25	$506.00	$44.67	$67.03
1 Truck Driver (light)	47.25	378.00	70.80	566.40		
1 Catch Basin Cleaning Truck		536.15		589.76	33.51	36.86
16 L.H., Daily Totals		$1250.95		$1662.17	$78.18	$103.89

Crew B-91D	Hr.	Daily	Hr.	Daily	Bare Costs	Incl. O&P
1 Labor Foreman (outside)	$44.10	$352.80	$66.25	$530.00	$48.94	$73.34
5 Laborers	42.10	1684.00	63.25	2530.00		
5 Equip. Oper. (medium)	56.75	2270.00	84.85	3394.00		
2 Truck Drivers (heavy)	48.95	783.20	73.35	1173.60		
1 Aggreg. Spreader, S.P.		848.90		933.79		
2 Truck Tractors, 6x4, 380 H.P.		986.50		1085.15		
2 Dist. Tankers, 3000 Gallon		660.30		726.33		
2 Pavement Brushes, Towed		174.60		192.06		
2 Rollers Pneum. Whl., 12 Ton		691.50		760.65	32.33	35.56
104 L.H., Daily Totals		$8451.80		$11325.58	$81.27	$108.90

Crew B-92	Hr.	Daily	Hr.	Daily	Bare Costs	Incl. O&P
1 Labor Foreman (outside)	$44.10	$352.80	$66.25	$530.00	$42.60	$64.00
3 Laborers	42.10	1010.40	63.25	1518.00		
1 Crack Cleaner, 25 H.P.		52.45		57.70		
1 Air Compressor, 60 cfm		152.95		168.25		
1 Tar Kettle, T.M.		156.20		171.82		
1 Flatbed Truck, Gas, 3 Ton		820.40		902.44	36.94	40.63
32 L.H., Daily Totals		$2545.20		$3348.20	$79.54	$104.63

Crew B-93	Hr.	Daily	Hr.	Daily	Bare Costs	Incl. O&P
1 Equip. Oper. (medium)	$56.75	$454.00	$84.85	$678.80	$56.75	$84.85
1 Feller Buncher, 100 H.P.		588.60		647.46	73.58	80.93
8 L.H., Daily Totals		$1042.60		$1326.26	$130.32	$165.78

Crew B-94A	Hr.	Daily	Hr.	Daily	Bare Costs	Incl. O&P
1 Laborer	$42.10	$336.80	$63.25	$506.00	$42.10	$63.25
1 Diaphragm Water Pump, 2"		86.80		95.48		
1 -20' Suction Hose, 2"		3.55		3.90		
2 -50' Discharge Hoses, 2"		8.00		8.80	12.29	13.52
8 L.H., Daily Totals		$435.15		$614.18	$54.39	$76.77

Crew B-94B	Hr.	Daily	Hr.	Daily	Bare Costs	Incl. O&P
1 Laborer	$42.10	$336.80	$63.25	$506.00	$42.10	$63.25
1 Diaphragm Water Pump, 4"		105.25		115.78		
1 -20' Suction Hose, 4"		17.05		18.75		
2 -50' Discharge Hoses, 4"		25.30		27.83	18.45	20.30
8 L.H., Daily Totals		$484.40		$668.36	$60.55	$83.55

Crew B-94C	Hr.	Daily	Hr.	Daily	Bare Costs	Incl. O&P
1 Laborer	$42.10	$336.80	$63.25	$506.00	$42.10	$63.25
1 Centrifugal Water Pump, 3"		74.45		81.89		
1 -20' Suction Hose, 3"		8.75		9.63		
2 -50' Discharge Hoses, 3"		9.00		9.90	11.53	12.68
8 L.H., Daily Totals		$429.00		$607.42	$53.63	$75.93

Crew B-94D	Hr.	Daily	Hr.	Daily	Bare Costs	Incl. O&P
1 Laborer	$42.10	$336.80	$63.25	$506.00	$42.10	$63.25
1 Centr. Water Pump, 6"		232.90		256.19		
1 -20' Suction Hose, 6"		25.20		27.72		
2 -50' Discharge Hoses, 6"		35.80		39.38	36.74	40.41
8 L.H., Daily Totals		$630.70		$829.29	$78.84	$103.66

Crew C-1	Hr.	Daily	Hr.	Daily	Bare Costs	Incl. O&P
3 Carpenters	$53.15	$1275.60	$79.85	$1916.40	$50.39	$75.70
1 Laborer	42.10	336.80	63.25	506.00		
32 L.H., Daily Totals		$1612.40		$2422.40	$50.39	$75.70

Crew C-2	Hr.	Daily	Hr.	Daily	Bare Costs	Incl. O&P
1 Carpenter Foreman (outside)	$55.15	$441.20	$82.85	$662.80	$51.64	$77.58
4 Carpenters	53.15	1700.80	79.85	2555.20		
1 Laborer	42.10	336.80	63.25	506.00		
48 L.H., Daily Totals		$2478.80		$3724.00	$51.64	$77.58

Crew C-2A	Hr.	Daily	Hr.	Daily	Bare Costs	Incl. O&P
1 Carpenter Foreman (outside)	$55.15	$441.20	$82.85	$662.80	$51.11	$76.52
3 Carpenters	53.15	1275.60	79.85	1916.40		
1 Cement Finisher	49.95	399.60	73.45	587.60		
1 Laborer	42.10	336.80	63.25	506.00		
48 L.H., Daily Totals		$2453.20		$3672.80	$51.11	$76.52

Crew C-3	Hr.	Daily	Hr.	Daily	Bare Costs	Incl. O&P
1 Rodman Foreman (outside)	$58.40	$467.20	$87.95	$703.60	$52.65	$79.16
4 Rodmen (reinf.)	56.40	1804.80	84.90	2716.80		
1 Equip. Oper. (light)	53.00	424.00	79.20	633.60		
2 Laborers	42.10	673.60	63.25	1012.00		
3 Stressing Equipment		56.70		62.37		
.5 Grouting Equipment		122.08		134.28	2.79	3.07
64 L.H., Daily Totals		$3548.38		$5262.65	$55.44	$82.23

Crew C-4	Hr.	Daily	Hr.	Daily	Bare Costs	Incl. O&P
1 Rodman Foreman (outside)	$58.40	$467.20	$87.95	$703.60	$56.90	$85.66
3 Rodmen (reinf.)	56.40	1353.60	84.90	2037.60		
3 Stressing Equipment		56.70		62.37	1.77	1.95
32 L.H., Daily Totals		$1877.50		$2803.57	$58.67	$87.61

Crew C-4A	Hr.	Daily	Hr.	Daily	Bare Costs	Incl. O&P
2 Rodmen (reinf.)	$56.40	$902.40	$84.90	$1358.40	$56.40	$84.90
4 Stressing Equipment		75.60		83.16	4.72	5.20
16 L.H., Daily Totals		$978.00		$1441.56	$61.13	$90.10

Crew C-5	Hr.	Daily	Hr.	Daily	Bare Costs	Incl. O&P
1 Rodman Foreman (outside)	$58.40	$467.20	$87.95	$703.60	$56.25	$84.51
4 Rodmen (reinf.)	56.40	1804.80	84.90	2716.80		
1 Equip. Oper. (crane)	59.20	473.60	88.50	708.00		
1 Equip. Oper. (oiler)	50.55	404.40	75.55	604.40		
1 Hyd. Crane, 25 Ton		580.85		638.93	10.37	11.41
56 L.H., Daily Totals		$3730.85		$5371.73	$66.62	$95.92

Crew C-6	Hr.	Daily	Hr.	Daily	Bare Costs	Incl. O&P
1 Labor Foreman (outside)	$44.10	$352.80	$66.25	$530.00	$43.74	$65.45
4 Laborers	42.10	1347.20	63.25	2024.00		
1 Cement Finisher	49.95	399.60	73.45	587.60		
2 Gas Engine Vibrators		53.70		59.07	1.12	1.23
48 L.H., Daily Totals		$2153.30		$3200.67	$44.86	$66.68

Crew No.	Bare Costs		Incl. Subs O&P		Cost Per Labor-Hour	
Crew C-6A	**Hr.**	**Daily**	**Hr.**	**Daily**	**Bare Costs**	**Incl. O&P**
2 Cement Finishers	$49.95	$799.20	$73.45	$1175.20	$49.95	$73.45
1 Concrete Vibrator, Elec, 2 HP		47.45		52.20	2.97	3.26
16 L.H., Daily Totals		$846.65		$1227.40	$52.92	$76.71
Crew C-7	**Hr.**	**Daily**	**Hr.**	**Daily**	**Bare Costs**	**Incl. O&P**
1 Labor Foreman (outside)	$44.10	$352.80	$66.25	$530.00	$45.76	$68.48
5 Laborers	42.10	1684.00	63.25	2530.00		
1 Cement Finisher	49.95	399.60	73.45	587.60		
1 Equip. Oper. (medium)	56.75	454.00	84.85	678.80		
1 Equip. Oper. (oiler)	50.55	404.40	75.55	604.40		
2 Gas Engine Vibrators		53.70		59.07		
1 Concrete Bucket, 1 C.Y.		45.80		50.38		
1 Hyd. Crane, 55 Ton		980.20		1078.22	15.00	16.50
72 L.H., Daily Totals		$4374.50		$6118.47	$60.76	$84.98
Crew C-7A	**Hr.**	**Daily**	**Hr.**	**Daily**	**Bare Costs**	**Incl. O&P**
1 Labor Foreman (outside)	$44.10	$352.80	$66.25	$530.00	$44.06	$66.15
5 Laborers	42.10	1684.00	63.25	2530.00		
2 Truck Drivers (heavy)	48.95	783.20	73.35	1173.60		
2 Conc. Transit Mixers		1170.50		1287.55	18.29	20.12
64 L.H., Daily Totals		$3990.50		$5521.15	$62.35	$86.27
Crew C-7B	**Hr.**	**Daily**	**Hr.**	**Daily**	**Bare Costs**	**Incl. O&P**
1 Labor Foreman (outside)	$44.10	$352.80	$66.25	$530.00	$45.54	$68.32
5 Laborers	42.10	1684.00	63.25	2530.00		
1 Equipment Operator, Crane	59.20	473.60	88.50	708.00		
1 Equipment Oiler	50.55	404.40	75.55	604.40		
1 Conc. Bucket, 2 C.Y.		54.70		60.17		
1 Lattice Boom Crane, 165 Ton		2151.00		2366.10	34.46	37.91
64 L.H., Daily Totals		$5120.50		$6798.67	$80.01	$106.23
Crew C-7C	**Hr.**	**Daily**	**Hr.**	**Daily**	**Bare Costs**	**Incl. O&P**
1 Labor Foreman (outside)	$44.10	$352.80	$66.25	$530.00	$46.01	$69.03
5 Laborers	42.10	1684.00	63.25	2530.00		
2 Equipment Operators (med.)	56.75	908.00	84.85	1357.60		
2 F.E. Loaders, W.M., 4 C.Y.		1578.70		1736.57	24.67	27.13
64 L.H., Daily Totals		$4523.50		$6154.17	$70.68	$96.16
Crew C-7D	**Hr.**	**Daily**	**Hr.**	**Daily**	**Bare Costs**	**Incl. O&P**
1 Labor Foreman (outside)	$44.10	$352.80	$66.25	$530.00	$44.48	$66.76
5 Laborers	42.10	1684.00	63.25	2530.00		
1 Equip. Oper. (medium)	56.75	454.00	84.85	678.80		
1 Concrete Conveyer		204.15		224.57	3.65	4.01
56 L.H., Daily Totals		$2694.95		$3963.36	$48.12	$70.77
Crew C-8	**Hr.**	**Daily**	**Hr.**	**Daily**	**Bare Costs**	**Incl. O&P**
1 Labor Foreman (outside)	$44.10	$352.80	$66.25	$530.00	$46.72	$69.68
3 Laborers	42.10	1010.40	63.25	1518.00		
2 Cement Finishers	49.95	799.20	73.45	1175.20		
1 Equip. Oper. (medium)	56.75	454.00	84.85	678.80		
1 Concrete Pump (Small)		410.35		451.38	7.33	8.06
56 L.H., Daily Totals		$3026.75		$4353.39	$54.05	$77.74
Crew C-8A	**Hr.**	**Daily**	**Hr.**	**Daily**	**Bare Costs**	**Incl. O&P**
1 Labor Foreman (outside)	$44.10	$352.80	$66.25	$530.00	$45.05	$67.15
3 Laborers	42.10	1010.40	63.25	1518.00		
2 Cement Finishers	49.95	799.20	73.45	1175.20		
48 L.H., Daily Totals		$2162.40		$3223.20	$45.05	$67.15

Crew No.	Bare Costs		Incl. Subs O&P		Cost Per Labor-Hour	
Crew C-8B	**Hr.**	**Daily**	**Hr.**	**Daily**	**Bare Costs**	**Incl. O&P**
1 Labor Foreman (outside)	$44.10	$352.80	$66.25	$530.00	$45.43	$68.17
3 Laborers	42.10	1010.40	63.25	1518.00		
1 Equip. Oper. (medium)	56.75	454.00	84.85	678.80		
1 Vibrating Power Screed		75.10		82.61		
1 Roller, Vibratory, 25 Ton		664.35		730.78		
1 Dozer, 200 H.P.		1504.00		1654.40	56.09	61.69
40 L.H., Daily Totals		$4060.65		$5194.60	$101.52	$129.86
Crew C-8C	**Hr.**	**Daily**	**Hr.**	**Daily**	**Bare Costs**	**Incl. O&P**
1 Labor Foreman (outside)	$44.10	$352.80	$66.25	$530.00	$46.18	$69.05
3 Laborers	42.10	1010.40	63.25	1518.00		
1 Cement Finisher	49.95	399.60	73.45	587.60		
1 Equip. Oper. (medium)	56.75	454.00	84.85	678.80		
1 Shotcrete Rig, 12 C.Y./hr		245.40		269.94		
1 Air Compressor, 160 cfm		202.80		223.08		
4 -50' Air Hoses, 1"		32.00		35.20		
4 -50' Air Hoses, 2"		115.60		127.16	12.41	13.65
48 L.H., Daily Totals		$2812.60		$3969.78	$58.60	$82.70
Crew C-8D	**Hr.**	**Daily**	**Hr.**	**Daily**	**Bare Costs**	**Incl. O&P**
1 Labor Foreman (outside)	$44.10	$352.80	$66.25	$530.00	$47.29	$70.54
1 Laborer	42.10	336.80	63.25	506.00		
1 Cement Finisher	49.95	399.60	73.45	587.60		
1 Equipment Oper. (light)	53.00	424.00	79.20	633.60		
1 Air Compressor, 250 cfm		201.50		221.65		
2 -50' Air Hoses, 1"		16.00		17.60	6.80	7.48
32 L.H., Daily Totals		$1730.70		$2496.45	$54.08	$78.01
Crew C-8E	**Hr.**	**Daily**	**Hr.**	**Daily**	**Bare Costs**	**Incl. O&P**
1 Labor Foreman (outside)	$44.10	$352.80	$66.25	$530.00	$45.56	$68.11
3 Laborers	42.10	1010.40	63.25	1518.00		
1 Cement Finisher	49.95	399.60	73.45	587.60		
1 Equipment Oper. (light)	53.00	424.00	79.20	633.60		
1 Shotcrete Rig, 35 C.Y./hr		298.00		327.80		
1 Air Compressor, 250 cfm		201.50		221.65		
4 -50' Air Hoses, 1"		32.00		35.20		
4 -50' Air Hoses, 2"		115.60		127.16	13.48	14.83
48 L.H., Daily Totals		$2833.90		$3981.01	$59.04	$82.94
Crew C-9	**Hr.**	**Daily**	**Hr.**	**Daily**	**Bare Costs**	**Incl. O&P**
1 Cement Finisher	$49.95	$399.60	$73.45	$587.60	$46.79	$69.79
2 Laborers	42.10	673.60	63.25	1012.00		
1 Equipment Oper. (light)	53.00	424.00	79.20	633.60		
1 Grout Pump, 50 C.F./hr.		188.45		207.29		
1 Air Compressor, 160 cfm		202.80		223.08		
2 -50' Air Hoses, 1"		16.00		17.60		
2 -50' Air Hoses, 2"		57.80		63.58	14.53	15.99
32 L.H., Daily Totals		$1962.25		$2744.76	$61.32	$85.77
Crew C-10	**Hr.**	**Daily**	**Hr.**	**Daily**	**Bare Costs**	**Incl. O&P**
1 Laborer	$42.10	$336.80	$63.25	$506.00	$47.33	$70.05
2 Cement Finishers	49.95	799.20	73.45	1175.20		
24 L.H., Daily Totals		$1136.00		$1681.20	$47.33	$70.05
Crew C-10B	**Hr.**	**Daily**	**Hr.**	**Daily**	**Bare Costs**	**Incl. O&P**
3 Laborers	$42.10	$1010.40	$63.25	$1518.00	$45.24	$67.33
2 Cement Finishers	49.95	799.20	73.45	1175.20		
1 Concrete Mixer, 10 C.F.		140.30		154.33		
2 Trowels, 48" Walk-Behind		174.00		191.40	7.86	8.64
40 L.H., Daily Totals		$2123.90		$3038.93	$53.10	$75.97

Crews - Standard

Crew No.	Bare Costs		Incl. Subs O&P		Cost Per Labor-Hour	
Crew C-10C	**Hr.**	**Daily**	**Hr.**	**Daily**	**Bare Costs**	**Incl. O&P**
1 Laborer	$42.10	$336.80	$63.25	$506.00	$47.33	$70.05
2 Cement Finishers	49.95	799.20	73.45	1175.20		
1 Trowel, 48" Walk-Behind		87.00		95.70	3.63	3.99
24 L.H., Daily Totals		$1223.00		$1776.90	$50.96	$74.04
Crew C-10D	**Hr.**	**Daily**	**Hr.**	**Daily**	**Bare Costs**	**Incl. O&P**
1 Laborer	$42.10	$336.80	$63.25	$506.00	$47.33	$70.05
2 Cement Finishers	49.95	799.20	73.45	1175.20		
1 Vibrating Power Screed		75.10		82.61		
1 Trowel, 48" Walk-Behind		87.00		95.70	6.75	7.43
24 L.H., Daily Totals		$1298.10		$1859.51	$54.09	$77.48
Crew C-10E	**Hr.**	**Daily**	**Hr.**	**Daily**	**Bare Costs**	**Incl. O&P**
1 Laborer	$42.10	$336.80	$63.25	$506.00	$47.33	$70.05
2 Cement Finishers	49.95	799.20	73.45	1175.20		
1 Vibrating Power Screed		75.10		82.61		
1 Cement Trowel, 96" Ride-On		169.00		185.90	10.17	11.19
24 L.H., Daily Totals		$1380.10		$1949.71	$57.50	$81.24
Crew C-10F	**Hr.**	**Daily**	**Hr.**	**Daily**	**Bare Costs**	**Incl. O&P**
1 Laborer	$42.10	$336.80	$63.25	$506.00	$47.33	$70.05
2 Cement Finishers	49.95	799.20	73.45	1175.20		
1 Telescoping Boom Lift, to 60'		310.90		341.99	12.95	14.25
24 L.H., Daily Totals		$1446.90		$2023.19	$60.29	$84.30
Crew C-11	**Hr.**	**Daily**	**Hr.**	**Daily**	**Bare Costs**	**Incl. O&P**
1 Struc. Steel Foreman (outside)	$59.65	$477.20	$93.55	$748.40	$57.26	$88.92
6 Struc. Steel Workers	57.65	2767.20	90.45	4341.60		
1 Equip. Oper. (crane)	59.20	473.60	88.50	708.00		
1 Equip. Oper. (oiler)	50.55	404.40	75.55	604.40		
1 Lattice Boom Crane, 150 Ton		2269.00		2495.90	31.51	34.67
72 L.H., Daily Totals		$6391.40		$8898.30	$88.77	$123.59
Crew C-12	**Hr.**	**Daily**	**Hr.**	**Daily**	**Bare Costs**	**Incl. O&P**
1 Carpenter Foreman (outside)	$55.15	$441.20	$82.85	$662.80	$52.65	$79.03
3 Carpenters	53.15	1275.60	79.85	1916.40		
1 Laborer	42.10	336.80	63.25	506.00		
1 Equip. Oper. (crane)	59.20	473.60	88.50	708.00		
1 Hyd. Crane, 12 Ton		471.00		518.10	9.81	10.79
48 L.H., Daily Totals		$2998.20		$4311.30	$62.46	$89.82
Crew C-13	**Hr.**	**Daily**	**Hr.**	**Daily**	**Bare Costs**	**Incl. O&P**
1 Struc. Steel Worker	$57.65	$461.20	$90.45	$723.60	$56.08	$86.80
1 Welder	57.45	459.60	90.10	720.80		
1 Carpenter	53.15	425.20	79.85	638.80		
1 Welder, Gas Engine, 300 amp		147.00		161.70	6.13	6.74
24 L.H., Daily Totals		$1493.00		$2244.90	$62.21	$93.54
Crew C-14	**Hr.**	**Daily**	**Hr.**	**Daily**	**Bare Costs**	**Incl. O&P**
1 Carpenter Foreman (outside)	$55.15	$441.20	$82.85	$662.80	$51.36	$76.98
5 Carpenters	53.15	2126.00	79.85	3194.00		
4 Laborers	42.10	1347.20	63.25	2024.00		
4 Rodmen (reinf.)	56.40	1804.80	84.90	2716.80		
2 Cement Finishers	49.95	799.20	73.45	1175.20		
1 Equip. Oper. (crane)	59.20	473.60	88.50	708.00		
1 Equip. Oper. (oiler)	50.55	404.40	75.55	604.40		
1 Hyd. Crane, 80 Ton		1486.00		1634.60	10.32	11.35
144 L.H., Daily Totals		$8882.40		$12719.80	$61.68	$88.33

Crew No.	Bare Costs		Incl. Subs O&P		Cost Per Labor-Hour	
Crew C-14A	**Hr.**	**Daily**	**Hr.**	**Daily**	**Bare Costs**	**Incl. O&P**
1 Carpenter Foreman (outside)	$55.15	$441.20	$82.85	$662.80	$52.88	$79.39
16 Carpenters	53.15	6803.20	79.85	10220.80		
4 Rodmen (reinf.)	56.40	1804.80	84.90	2716.80		
2 Laborers	42.10	673.60	63.25	1012.00		
1 Cement Finisher	49.95	399.60	73.45	587.60		
1 Equip. Oper. (medium)	56.75	454.00	84.85	678.80		
1 Gas Engine Vibrator		26.85		29.54		
1 Concrete Pump (Small)		410.35		451.38	2.19	2.40
200 L.H., Daily Totals		$11013.60		$16359.72	$55.07	$81.80
Crew C-14B	**Hr.**	**Daily**	**Hr.**	**Daily**	**Bare Costs**	**Incl. O&P**
1 Carpenter Foreman (outside)	$55.15	$441.20	$82.85	$662.80	$52.77	$79.17
16 Carpenters	53.15	6803.20	79.85	10220.80		
4 Rodmen (reinf.)	56.40	1804.80	84.90	2716.80		
2 Laborers	42.10	673.60	63.25	1012.00		
2 Cement Finishers	49.95	799.20	73.45	1175.20		
1 Equip. Oper. (medium)	56.75	454.00	84.85	678.80		
1 Gas Engine Vibrator		26.85		29.54		
1 Concrete Pump (Small)		410.35		451.38	2.10	2.31
208 L.H., Daily Totals		$11413.20		$16947.32	$54.87	$81.48
Crew C-14C	**Hr.**	**Daily**	**Hr.**	**Daily**	**Bare Costs**	**Incl. O&P**
1 Carpenter Foreman (outside)	$55.15	$441.20	$82.85	$662.80	$50.37	$75.59
6 Carpenters	53.15	2551.20	79.85	3832.80		
2 Rodmen (reinf.)	56.40	902.40	84.90	1358.40		
4 Laborers	42.10	1347.20	63.25	2024.00		
1 Cement Finisher	49.95	399.60	73.45	587.60		
1 Gas Engine Vibrator		26.85		29.54	.24	.26
112 L.H., Daily Totals		$5668.45		$8495.14	$50.61	$75.85
Crew C-14D	**Hr.**	**Daily**	**Hr.**	**Daily**	**Bare Costs**	**Incl. O&P**
1 Carpenter Foreman (outside)	$55.15	$441.20	$82.85	$662.80	$52.62	$78.99
18 Carpenters	53.15	7653.60	79.85	11498.40		
2 Rodmen (reinf.)	56.40	902.40	84.90	1358.40		
2 Laborers	42.10	673.60	63.25	1012.00		
1 Cement Finisher	49.95	399.60	73.45	587.60		
1 Equip. Oper. (medium)	56.75	454.00	84.85	678.80		
1 Gas Engine Vibrator		26.85		29.54		
1 Concrete Pump (Small)		410.35		451.38	2.19	2.40
200 L.H., Daily Totals		$10961.60		$16278.92	$54.81	$81.39
Crew C-14E	**Hr.**	**Daily**	**Hr.**	**Daily**	**Bare Costs**	**Incl. O&P**
1 Carpenter Foreman (outside)	$55.15	$441.20	$82.85	$662.80	$51.21	$76.85
2 Carpenters	53.15	850.40	79.85	1277.60		
4 Rodmen (reinf.)	56.40	1804.80	84.90	2716.80		
3 Laborers	42.10	1010.40	63.25	1518.00		
1 Cement Finisher	49.95	399.60	73.45	587.60		
1 Gas Engine Vibrator		26.85		29.54	.31	.34
88 L.H., Daily Totals		$4533.25		$6792.34	$51.51	$77.19
Crew C-14F	**Hr.**	**Daily**	**Hr.**	**Daily**	**Bare Costs**	**Incl. O&P**
1 Labor Foreman (outside)	$44.10	$352.80	$66.25	$530.00	$47.56	$70.38
2 Laborers	42.10	673.60	63.25	1012.00		
6 Cement Finishers	49.95	2397.60	73.45	3525.60		
1 Gas Engine Vibrator		26.85		29.54	.37	.41
72 L.H., Daily Totals		$3450.85		$5097.14	$47.93	$70.79

Crew C-14G	Bare Costs Hr.	Bare Costs Daily	Incl. Subs O&P Hr.	Incl. Subs O&P Daily	Cost Per Labor-Hour Bare Costs	Cost Per Labor-Hour Incl. O&P
1 Labor Foreman (outside)	$44.10	$352.80	$66.25	$530.00	$46.87	$69.51
2 Laborers	42.10	673.60	63.25	1012.00		
4 Cement Finishers	49.95	1598.40	73.45	2350.40		
1 Gas Engine Vibrator		26.85		29.54	.48	.53
56 L.H., Daily Totals		$2651.65		$3921.93	$47.35	$70.03

Crew C-14H	Hr.	Daily	Hr.	Daily	Bare Costs	Incl. O&P
1 Carpenter Foreman (outside)	$55.15	$441.20	$82.85	$662.80	$51.65	$77.36
2 Carpenters	53.15	850.40	79.85	1277.60		
1 Rodman (reinf.)	56.40	451.20	84.90	679.20		
1 Laborer	42.10	336.80	63.25	506.00		
1 Cement Finisher	49.95	399.60	73.45	587.60		
1 Gas Engine Vibrator		26.85		29.54	.56	.62
48 L.H., Daily Totals		$2506.05		$3742.74	$52.21	$77.97

Crew C-14L	Hr.	Daily	Hr.	Daily	Bare Costs	Incl. O&P
1 Carpenter Foreman (outside)	$55.15	$441.20	$82.85	$662.80	$49.37	$74.03
6 Carpenters	53.15	2551.20	79.85	3832.80		
4 Laborers	42.10	1347.20	63.25	2024.00		
1 Cement Finisher	49.95	399.60	73.45	587.60		
1 Gas Engine Vibrator		26.85		29.54	.28	.31
96 L.H., Daily Totals		$4766.05		$7136.73	$49.65	$74.34

Crew C-14M	Hr.	Daily	Hr.	Daily	Bare Costs	Incl. O&P
1 Carpenter Foreman (outside)	$55.15	$441.20	$82.85	$662.80	$51.09	$76.53
2 Carpenters	53.15	850.40	79.85	1277.60		
1 Rodman (reinf.)	56.40	451.20	84.90	679.20		
2 Laborers	42.10	673.60	63.25	1012.00		
1 Cement Finisher	49.95	399.60	73.45	587.60		
1 Equip. Oper. (medium)	56.75	454.00	84.85	678.80		
1 Gas Engine Vibrator		26.85		29.54		
1 Concrete Pump (Small)		410.35		451.38	6.83	7.51
64 L.H., Daily Totals		$3707.20		$5378.92	$57.93	$84.05

Crew C-15	Hr.	Daily	Hr.	Daily	Bare Costs	Incl. O&P
1 Carpenter Foreman (outside)	$55.15	$441.20	$82.85	$662.80	$49.34	$73.79
2 Carpenters	53.15	850.40	79.85	1277.60		
3 Laborers	42.10	1010.40	63.25	1518.00		
2 Cement Finishers	49.95	799.20	73.45	1175.20		
1 Rodman (reinf.)	56.40	451.20	84.90	679.20		
72 L.H., Daily Totals		$3552.40		$5312.80	$49.34	$73.79

Crew C-16	Hr.	Daily	Hr.	Daily	Bare Costs	Incl. O&P
1 Labor Foreman (outside)	$44.10	$352.80	$66.25	$530.00	$46.72	$69.68
3 Laborers	42.10	1010.40	63.25	1518.00		
2 Cement Finishers	49.95	799.20	73.45	1175.20		
1 Equip. Oper. (medium)	56.75	454.00	84.85	678.80		
1 Gunite Pump Rig		317.90		349.69		
2 -50' Air Hoses, 3/4"		15.50		17.05		
2 -50' Air Hoses, 2"		57.80		63.58	6.99	7.68
56 L.H., Daily Totals		$3007.60		$4332.32	$53.71	$77.36

Crew C-16A	Hr.	Daily	Hr.	Daily	Bare Costs	Incl. O&P
1 Laborer	$42.10	$336.80	$63.25	$506.00	$49.69	$73.75
2 Cement Finishers	49.95	799.20	73.45	1175.20		
1 Equip. Oper. (medium)	56.75	454.00	84.85	678.80		
1 Gunite Pump Rig		317.90		349.69		
2 -50' Air Hoses, 3/4"		15.50		17.05		
2 -50' Air Hoses, 2"		57.80		63.58		
1 Telescoping Boom Lift, to 60'		310.90		341.99	21.94	24.13
32 L.H., Daily Totals		$2292.10		$3132.31	$71.63	$97.88

Crew C-17	Hr.	Daily	Hr.	Daily	Bare Costs	Incl. O&P
2 Skilled Worker Foremen (out)	$56.85	$909.60	$86.00	$1376.00	$55.25	$83.56
8 Skilled Workers	54.85	3510.40	82.95	5308.80		
80 L.H., Daily Totals		$4420.00		$6684.80	$55.25	$83.56

Crew C-17A	Hr.	Daily	Hr.	Daily	Bare Costs	Incl. O&P
2 Skilled Worker Foremen (out)	$56.85	$909.60	$86.00	$1376.00	$55.30	$83.62
8 Skilled Workers	54.85	3510.40	82.95	5308.80		
.125 Equip. Oper. (crane)	59.20	59.20	88.50	88.50		
.125 Hyd. Crane, 80 Ton		185.75		204.32	2.29	2.52
81 L.H., Daily Totals		$4664.95		$6977.63	$57.59	$86.14

Crew C-17B	Hr.	Daily	Hr.	Daily	Bare Costs	Incl. O&P
2 Skilled Worker Foremen (out)	$56.85	$909.60	$86.00	$1376.00	$55.35	$83.68
8 Skilled Workers	54.85	3510.40	82.95	5308.80		
.25 Equip. Oper. (crane)	59.20	118.40	88.50	177.00		
.25 Hyd. Crane, 80 Ton		371.50		408.65		
.25 Trowel, 48" Walk-Behind		21.75		23.93	4.80	5.28
82 L.H., Daily Totals		$4931.65		$7294.38	$60.14	$88.96

Crew C-17C	Hr.	Daily	Hr.	Daily	Bare Costs	Incl. O&P
2 Skilled Worker Foremen (out)	$56.85	$909.60	$86.00	$1376.00	$55.39	$83.74
8 Skilled Workers	54.85	3510.40	82.95	5308.80		
.375 Equip. Oper. (crane)	59.20	177.60	88.50	265.50		
.375 Hyd. Crane, 80 Ton		557.25		612.98	6.71	7.39
83 L.H., Daily Totals		$5154.85		$7563.27	$62.11	$91.12

Crew C-17D	Hr.	Daily	Hr.	Daily	Bare Costs	Incl. O&P
2 Skilled Worker Foremen (out)	$56.85	$909.60	$86.00	$1376.00	$55.44	$83.80
8 Skilled Workers	54.85	3510.40	82.95	5308.80		
.5 Equip. Oper. (crane)	59.20	236.80	88.50	354.00		
.5 Hyd. Crane, 80 Ton		743.00		817.30	8.85	9.73
84 L.H., Daily Totals		$5399.80		$7856.10	$64.28	$93.53

Crew C-17E	Hr.	Daily	Hr.	Daily	Bare Costs	Incl. O&P
2 Skilled Worker Foremen (out)	$56.85	$909.60	$86.00	$1376.00	$55.25	$83.56
8 Skilled Workers	54.85	3510.40	82.95	5308.80		
1 Hyd. Jack with Rods		35.90		39.49	.45	.49
80 L.H., Daily Totals		$4455.90		$6724.29	$55.70	$84.05

Crew C-18	Hr.	Daily	Hr.	Daily	Bare Costs	Incl. O&P
.125 Labor Foreman (outside)	$44.10	$44.10	$66.25	$66.25	$42.32	$63.58
1 Laborer	42.10	336.80	63.25	506.00		
1 Concrete Cart, 10 C.F.		127.85		140.63	14.21	15.63
9 L.H., Daily Totals		$508.75		$712.88	$56.53	$79.21

Crew C-19	Hr.	Daily	Hr.	Daily	Bare Costs	Incl. O&P
.125 Labor Foreman (outside)	$44.10	$44.10	$66.25	$66.25	$42.32	$63.58
1 Laborer	42.10	336.80	63.25	506.00		
1 Concrete Cart, 18 C.F.		153.45		168.79	17.05	18.75
9 L.H., Daily Totals		$534.35		$741.04	$59.37	$82.34

Crew C-20	Hr.	Daily	Hr.	Daily	Bare Costs	Incl. O&P
1 Labor Foreman (outside)	$44.10	$352.80	$66.25	$530.00	$45.16	$67.60
5 Laborers	42.10	1684.00	63.25	2530.00		
1 Cement Finisher	49.95	399.60	73.45	587.60		
1 Equip. Oper. (medium)	56.75	454.00	84.85	678.80		
2 Gas Engine Vibrators		53.70		59.07		
1 Concrete Pump (Small)		410.35		451.38	7.25	7.98
64 L.H., Daily Totals		$3354.45		$4836.85	$52.41	$75.58

Crew No.	Bare Costs		Incl. Subs O&P		Cost Per Labor-Hour	
Crew C-21	**Hr.**	**Daily**	**Hr.**	**Daily**	**Bare Costs**	**Incl. O&P**
1 Labor Foreman (outside)	$44.10	$352.80	$66.25	$530.00	$45.16	$67.60
5 Laborers	42.10	1684.00	63.25	2530.00		
1 Cement Finisher	49.95	399.60	73.45	587.60		
1 Equip. Oper. (medium)	56.75	454.00	84.85	678.80		
2 Gas Engine Vibrators		53.70		59.07		
1 Concrete Conveyer		204.15		224.57	4.03	4.43
64 L.H., Daily Totals		$3148.25		$4610.03	$49.19	$72.03

Crew C-22	**Hr.**	**Daily**	**Hr.**	**Daily**	**Bare Costs**	**Incl. O&P**
1 Rodman Foreman (outside)	$58.40	$467.20	$87.95	$703.60	$56.71	$85.34
4 Rodmen (reinf.)	56.40	1804.80	84.90	2716.80		
.125 Equip. Oper. (crane)	59.20	59.20	88.50	88.50		
.125 Equip. Oper. (oiler)	50.55	50.55	75.55	75.55		
.125 Hyd. Crane, 25 Ton		72.61		79.87	1.73	1.90
42 L.H., Daily Totals		$2454.36		$3664.32	$58.44	$87.25

Crew C-23	**Hr.**	**Daily**	**Hr.**	**Daily**	**Bare Costs**	**Incl. O&P**
2 Skilled Worker Foremen (out)	$56.85	$909.60	$86.00	$1376.00	$55.26	$83.38
6 Skilled Workers	54.85	2632.80	82.95	3981.60		
1 Equip. Oper. (crane)	59.20	473.60	88.50	708.00		
1 Equip. Oper. (oiler)	50.55	404.40	75.55	604.40		
1 Lattice Boom Crane, 90 Ton		1696.00		1865.60	21.20	23.32
80 L.H., Daily Totals		$6116.40		$8535.60	$76.45	$106.69

Crew C-23A	**Hr.**	**Daily**	**Hr.**	**Daily**	**Bare Costs**	**Incl. O&P**
1 Labor Foreman (outside)	$44.10	$352.80	$66.25	$530.00	$47.61	$71.36
2 Laborers	42.10	673.60	63.25	1012.00		
1 Equip. Oper. (crane)	59.20	473.60	88.50	708.00		
1 Equip. Oper. (oiler)	50.55	404.40	75.55	604.40		
1 Crawler Crane, 100 Ton		2287.00		2515.70		
3 Conc. Buckets, 8 C.Y.		323.25		355.57	65.26	71.78
40 L.H., Daily Totals		$4514.65		$5725.68	$112.87	$143.14

Crew C-24	**Hr.**	**Daily**	**Hr.**	**Daily**	**Bare Costs**	**Incl. O&P**
2 Skilled Worker Foremen (out)	$56.85	$909.60	$86.00	$1376.00	$55.26	$83.38
6 Skilled Workers	54.85	2632.80	82.95	3981.60		
1 Equip. Oper. (crane)	59.20	473.60	88.50	708.00		
1 Equip. Oper. (oiler)	50.55	404.40	75.55	604.40		
1 Lattice Boom Crane, 150 Ton		2269.00		2495.90	28.36	31.20
80 L.H., Daily Totals		$6689.40		$9165.90	$83.62	$114.57

Crew C-25	**Hr.**	**Daily**	**Hr.**	**Daily**	**Bare Costs**	**Incl. O&P**
2 Rodmen (reinf.)	$56.40	$902.40	$84.90	$1358.40	$45.50	$71.00
2 Rodmen Helpers	34.60	553.60	57.10	913.60		
32 L.H., Daily Totals		$1456.00		$2272.00	$45.50	$71.00

Crew C-27	**Hr.**	**Daily**	**Hr.**	**Daily**	**Bare Costs**	**Incl. O&P**
2 Cement Finishers	$49.95	$799.20	$73.45	$1175.20	$49.95	$73.45
1 Concrete Saw		111.50		122.65	6.97	7.67
16 L.H., Daily Totals		$910.70		$1297.85	$56.92	$81.12

Crew C-28	**Hr.**	**Daily**	**Hr.**	**Daily**	**Bare Costs**	**Incl. O&P**
1 Cement Finisher	$49.95	$399.60	$73.45	$587.60	$49.95	$73.45
1 Portable Air Compressor, Gas		38.70		42.57	4.84	5.32
8 L.H., Daily Totals		$438.30		$630.17	$54.79	$78.77

Crew C-29	**Hr.**	**Daily**	**Hr.**	**Daily**	**Bare Costs**	**Incl. O&P**
1 Laborer	$42.10	$336.80	$63.25	$506.00	$42.10	$63.25
1 Pressure Washer		96.95		106.65	12.12	13.33
8 L.H., Daily Totals		$433.75		$612.64	$54.22	$76.58

Crew No.	Bare Costs		Incl. Subs O&P		Cost Per Labor-Hour	
Crew C-30	**Hr.**	**Daily**	**Hr.**	**Daily**	**Bare Costs**	**Incl. O&P**
1 Laborer	$42.10	$336.80	$63.25	$506.00	$42.10	$63.25
1 Concrete Mixer, 10 C.F.		140.30		154.33	17.54	19.29
8 L.H., Daily Totals		$477.10		$660.33	$59.64	$82.54

Crew C-31	**Hr.**	**Daily**	**Hr.**	**Daily**	**Bare Costs**	**Incl. O&P**
1 Cement Finisher	$49.95	$399.60	$73.45	$587.60	$49.95	$73.45
1 Grout Pump		317.90		349.69	39.74	43.71
8 L.H., Daily Totals		$717.50		$937.29	$89.69	$117.16

Crew C-32	**Hr.**	**Daily**	**Hr.**	**Daily**	**Bare Costs**	**Incl. O&P**
1 Cement Finisher	$49.95	$399.60	$73.45	$587.60	$46.02	$68.35
1 Laborer	42.10	336.80	63.25	506.00		
1 Crack Chaser Saw, Gas, 6 H.P.		73.10		80.41		
1 Vacuum Pick-Up System		74.05		81.45	9.20	10.12
16 L.H., Daily Totals		$883.55		$1255.46	$55.22	$78.47

Crew D-1	**Hr.**	**Daily**	**Hr.**	**Daily**	**Bare Costs**	**Incl. O&P**
1 Bricklayer	$52.05	$416.40	$78.90	$631.20	$46.90	$71.10
1 Bricklayer Helper	41.75	334.00	63.30	506.40		
16 L.H., Daily Totals		$750.40		$1137.60	$46.90	$71.10

Crew D-2	**Hr.**	**Daily**	**Hr.**	**Daily**	**Bare Costs**	**Incl. O&P**
3 Bricklayers	$52.05	$1249.20	$78.90	$1893.60	$48.40	$73.31
2 Bricklayer Helpers	41.75	668.00	63.30	1012.80		
.5 Carpenter	53.15	212.60	79.85	319.40		
44 L.H., Daily Totals		$2129.80		$3225.80	$48.40	$73.31

Crew D-3	**Hr.**	**Daily**	**Hr.**	**Daily**	**Bare Costs**	**Incl. O&P**
3 Bricklayers	$52.05	$1249.20	$78.90	$1893.60	$48.18	$73.00
2 Bricklayer Helpers	41.75	668.00	63.30	1012.80		
.25 Carpenter	53.15	106.30	79.85	159.70		
42 L.H., Daily Totals		$2023.50		$3066.10	$48.18	$73.00

Crew D-4	**Hr.**	**Daily**	**Hr.**	**Daily**	**Bare Costs**	**Incl. O&P**
1 Bricklayer	$52.05	$416.40	$78.90	$631.20	$47.14	$71.17
2 Bricklayer Helpers	41.75	668.00	63.30	1012.80		
1 Equip. Oper. (light)	53.00	424.00	79.20	633.60		
1 Grout Pump, 50 C.F./hr.		188.45		207.29	5.89	6.48
32 L.H., Daily Totals		$1696.85		$2484.90	$53.03	$77.65

Crew D-5	**Hr.**	**Daily**	**Hr.**	**Daily**	**Bare Costs**	**Incl. O&P**
1 Bricklayer	52.05	416.40	78.90	631.20	52.05	78.90
8 L.H., Daily Totals		$416.40		$631.20	$52.05	$78.90

Crew D-6	**Hr.**	**Daily**	**Hr.**	**Daily**	**Bare Costs**	**Incl. O&P**
3 Bricklayers	$52.05	$1249.20	$78.90	$1893.60	$47.15	$71.45
3 Bricklayer Helpers	41.75	1002.00	63.30	1519.20		
.25 Carpenter	53.15	106.30	79.85	159.70		
50 L.H., Daily Totals		$2357.50		$3572.50	$47.15	$71.45

Crew D-7	**Hr.**	**Daily**	**Hr.**	**Daily**	**Bare Costs**	**Incl. O&P**
1 Tile Layer	$49.50	$396.00	$72.60	$580.80	$44.15	$64.75
1 Tile Layer Helper	38.80	310.40	56.90	455.20		
16 L.H., Daily Totals		$706.40		$1036.00	$44.15	$64.75

Crew D-8	**Hr.**	**Daily**	**Hr.**	**Daily**	**Bare Costs**	**Incl. O&P**
3 Bricklayers	$52.05	$1249.20	$78.90	$1893.60	$47.93	$72.66
2 Bricklayer Helpers	41.75	668.00	63.30	1012.80		
40 L.H., Daily Totals		$1917.20		$2906.40	$47.93	$72.66

Crew No.	Bare Costs		Incl. Subs O&P		Cost Per Labor-Hour	
Crew D-9	**Hr.**	**Daily**	**Hr.**	**Daily**	**Bare Costs**	**Incl. O&P**
3 Bricklayers	$52.05	$1249.20	$78.90	$1893.60	$46.90	$71.10
3 Bricklayer Helpers	41.75	1002.00	63.30	1519.20		
48 L.H., Daily Totals		$2251.20		$3412.80	$46.90	$71.10
Crew D-10	**Hr.**	**Daily**	**Hr.**	**Daily**	**Bare Costs**	**Incl. O&P**
1 Bricklayer Foreman (outside)	$54.05	$432.40	$81.95	$655.60	$51.76	$78.16
1 Bricklayer	52.05	416.40	78.90	631.20		
1 Bricklayer Helper	41.75	334.00	63.30	506.40		
1 Equip. Oper. (crane)	59.20	473.60	88.50	708.00		
1 S.P. Crane, 4x4, 12 Ton		428.30		471.13	13.38	14.72
32 L.H., Daily Totals		$2084.70		$2972.33	$65.15	$92.89
Crew D-11	**Hr.**	**Daily**	**Hr.**	**Daily**	**Bare Costs**	**Incl. O&P**
1 Bricklayer Foreman (outside)	$54.05	$432.40	$81.95	$655.60	$49.28	$74.72
1 Bricklayer	52.05	416.40	78.90	631.20		
1 Bricklayer Helper	41.75	334.00	63.30	506.40		
24 L.H., Daily Totals		$1182.80		$1793.20	$49.28	$74.72
Crew D-12	**Hr.**	**Daily**	**Hr.**	**Daily**	**Bare Costs**	**Incl. O&P**
1 Bricklayer Foreman (outside)	$54.05	$432.40	$81.95	$655.60	$47.40	$71.86
1 Bricklayer	52.05	416.40	78.90	631.20		
2 Bricklayer Helpers	41.75	668.00	63.30	1012.80		
32 L.H., Daily Totals		$1516.80		$2299.60	$47.40	$71.86
Crew D-13	**Hr.**	**Daily**	**Hr.**	**Daily**	**Bare Costs**	**Incl. O&P**
1 Bricklayer Foreman (outside)	$54.05	$432.40	$81.95	$655.60	$50.33	$75.97
1 Bricklayer	52.05	416.40	78.90	631.20		
2 Bricklayer Helpers	41.75	668.00	63.30	1012.80		
1 Carpenter	53.15	425.20	79.85	638.80		
1 Equip. Oper. (crane)	59.20	473.60	88.50	708.00		
1 S.P. Crane, 4x4, 12 Ton		428.30		471.13	8.92	9.82
48 L.H., Daily Totals		$2843.90		$4117.53	$59.25	$85.78
Crew D-14	**Hr.**	**Daily**	**Hr.**	**Daily**	**Bare Costs**	**Incl. O&P**
3 Bricklayers	$52.05	$1249.20	$78.90	$1893.60	$49.48	$75.00
1 Bricklayer Helper	41.75	334.00	63.30	506.40		
32 L.H., Daily Totals		$1583.20		$2400.00	$49.48	$75.00
Crew E-1	**Hr.**	**Daily**	**Hr.**	**Daily**	**Bare Costs**	**Incl. O&P**
1 Welder Foreman (outside)	$59.45	$475.60	$93.25	$746.00	$56.63	$87.52
1 Welder	57.45	459.60	90.10	720.80		
1 Equip. Oper. (light)	53.00	424.00	79.20	633.60		
1 Welder, Gas Engine, 300 amp		147.00		161.70	6.13	6.74
24 L.H., Daily Totals		$1506.20		$2262.10	$62.76	$94.25
Crew E-2	**Hr.**	**Daily**	**Hr.**	**Daily**	**Bare Costs**	**Incl. O&P**
1 Struc. Steel Foreman (outside)	$59.65	$477.20	$93.55	$748.40	$57.14	$88.49
4 Struc. Steel Workers	57.65	1844.80	90.45	2894.40		
1 Equip. Oper. (crane)	59.20	473.60	88.50	708.00		
1 Equip. Oper. (oiler)	50.55	404.40	75.55	604.40		
1 Lattice Boom Crane, 90 Ton		1696.00		1865.60	30.29	33.31
56 L.H., Daily Totals		$4896.00		$6820.80	$87.43	$121.80
Crew E-3	**Hr.**	**Daily**	**Hr.**	**Daily**	**Bare Costs**	**Incl. O&P**
1 Struc. Steel Foreman (outside)	$59.65	$477.20	$93.55	$748.40	$58.25	$91.37
1 Struc. Steel Worker	57.65	461.20	90.45	723.60		
1 Welder	57.45	459.60	90.10	720.80		
1 Welder, Gas Engine, 300 amp		147.00		161.70	6.13	6.74
24 L.H., Daily Totals		$1545.00		$2354.50	$64.38	$98.10

Crew No.	Bare Costs		Incl. Subs O&P		Cost Per Labor-Hour	
Crew E-3A	**Hr.**	**Daily**	**Hr.**	**Daily**	**Bare Costs**	**Incl. O&P**
1 Struc. Steel Foreman (outside)	$59.65	$477.20	$93.55	$748.40	$58.25	$91.37
1 Struc. Steel Worker	57.65	461.20	90.45	723.60		
1 Welder	57.45	459.60	90.10	720.80		
1 Welder, Gas Engine, 300 amp		147.00		161.70		
1 Telescoping Boom Lift, to 40'		278.90		306.79	17.75	19.52
24 L.H., Daily Totals		$1823.90		$2661.29	$76.00	$110.89
Crew E-4	**Hr.**	**Daily**	**Hr.**	**Daily**	**Bare Costs**	**Incl. O&P**
1 Struc. Steel Foreman (outside)	$59.65	$477.20	$93.55	$748.40	$58.15	$91.22
3 Struc. Steel Workers	57.65	1383.60	90.45	2170.80		
1 Welder, Gas Engine, 300 amp		147.00		161.70	4.59	5.05
32 L.H., Daily Totals		$2007.80		$3080.90	$62.74	$96.28
Crew E-5	**Hr.**	**Daily**	**Hr.**	**Daily**	**Bare Costs**	**Incl. O&P**
2 Struc. Steel Foremen (outside)	$59.65	$954.40	$93.55	$1496.80	$57.48	$89.35
5 Struc. Steel Workers	57.65	2306.00	90.45	3618.00		
1 Equip. Oper. (crane)	59.20	473.60	88.50	708.00		
1 Welder	57.45	459.60	90.10	720.80		
1 Equip. Oper. (oiler)	50.55	404.40	75.55	604.40		
1 Lattice Boom Crane, 90 Ton		1696.00		1865.60		
1 Welder, Gas Engine, 300 amp		147.00		161.70	23.04	25.34
80 L.H., Daily Totals		$6441.00		$9175.30	$80.51	$114.69
Crew E-6	**Hr.**	**Daily**	**Hr.**	**Daily**	**Bare Costs**	**Incl. O&P**
3 Struc. Steel Foremen (outside)	$59.65	$1431.60	$93.55	$2245.20	$57.38	$89.25
9 Struc. Steel Workers	57.65	4150.80	90.45	6512.40		
1 Equip. Oper. (crane)	59.20	473.60	88.50	708.00		
1 Welder	57.45	459.60	90.10	720.80		
1 Equip. Oper. (oiler)	50.55	404.40	75.55	604.40		
1 Equip. Oper. (light)	53.00	424.00	79.20	633.60		
1 Lattice Boom Crane, 90 Ton		1696.00		1865.60		
1 Welder, Gas Engine, 300 amp		147.00		161.70		
1 Air Compressor, 160 cfm		202.80		223.08		
2 Impact Wrenches		104.60		115.06	16.80	18.48
128 L.H., Daily Totals		$9494.40		$13789.84	$74.17	$107.73
Crew E-7	**Hr.**	**Daily**	**Hr.**	**Daily**	**Bare Costs**	**Incl. O&P**
1 Struc. Steel Foreman (outside)	$59.65	$477.20	$93.55	$748.40	$57.44	$89.28
4 Struc. Steel Workers	57.65	1844.80	90.45	2894.40		
1 Equip. Oper. (crane)	59.20	473.60	88.50	708.00		
1 Equip. Oper. (oiler)	50.55	404.40	75.55	604.40		
1 Welder Foreman (outside)	59.45	475.60	93.25	746.00		
2 Welders	57.45	919.20	90.10	1441.60		
1 Lattice Boom Crane, 90 Ton		1696.00		1865.60		
2 Welder, Gas Engine, 300 amp		294.00		323.40	24.88	27.36
80 L.H., Daily Totals		$6584.80		$9331.80	$82.31	$116.65
Crew E-8	**Hr.**	**Daily**	**Hr.**	**Daily**	**Bare Costs**	**Incl. O&P**
1 Struc. Steel Foreman (outside)	$59.65	$477.20	$93.55	$748.40	$57.10	$88.63
4 Struc. Steel Workers	57.65	1844.80	90.45	2894.40		
1 Welder Foreman (outside)	59.45	475.60	93.25	746.00		
4 Welders	57.45	1838.40	90.10	2883.20		
1 Equip. Oper. (crane)	59.20	473.60	88.50	708.00		
1 Equip. Oper. (oiler)	50.55	404.40	75.55	604.40		
1 Equip. Oper. (light)	53.00	424.00	79.20	633.60		
1 Lattice Boom Crane, 90 Ton		1696.00		1865.60		
4 Welder, Gas Engine, 300 amp		588.00		646.80	21.96	24.16
104 L.H., Daily Totals		$8222.00		$11730.40	$79.06	$112.79

Crew No.	Bare Costs		Incl. Subs O&P		Cost Per Labor-Hour	
Crew E-9	**Hr.**	**Daily**	**Hr.**	**Daily**	**Bare Costs**	**Incl. O&P**
2 Struc. Steel Foremen (outside)	$59.65	$954.40	$93.55	$1496.80	$57.31	$89.15
5 Struc. Steel Workers	57.65	2306.00	90.45	3618.00		
1 Welder Foreman (outside)	59.45	475.60	93.25	746.00		
5 Welders	57.45	2298.00	90.10	3604.00		
1 Equip. Oper. (crane)	59.20	473.60	88.50	708.00		
1 Equip. Oper. (oiler)	50.55	404.40	75.55	604.40		
1 Equip. Oper. (light)	53.00	424.00	79.20	633.60		
1 Lattice Boom Crane, 90 Ton		1696.00		1865.60		
5 Welder, Gas Engine, 300 amp		735.00		808.50	18.99	20.89
128 L.H., Daily Totals		$9767.00		$14084.90	$76.30	$110.04

Crew E-10	Hr.	Daily	Hr.	Daily	Bare Costs	Incl. O&P
1 Welder Foreman (outside)	$59.45	$475.60	$93.25	$746.00	$58.45	$91.67
1 Welder	57.45	459.60	90.10	720.80		
1 Welder, Gas Engine, 300 amp		147.00		161.70		
1 Flatbed Truck, Gas, 3 Ton		820.40		902.44	60.46	66.51
16 L.H., Daily Totals		$1902.60		$2530.94	$118.91	$158.18

Crew E-11	Hr.	Daily	Hr.	Daily	Bare Costs	Incl. O&P
2 Painters, Struc. Steel	$45.85	$733.60	$74.45	$1191.20	$46.70	$72.84
1 Building Laborer	42.10	336.80	63.25	506.00		
1 Equip. Oper. (light)	53.00	424.00	79.20	633.60		
1 Air Compressor, 250 cfm		201.50		221.65		
1 Sandblaster, Portable, 3 C.F.		83.80		92.18		
1 Set Sand Blasting Accessories		15.35		16.89	9.40	10.33
32 L.H., Daily Totals		$1795.05		$2661.51	$56.10	$83.17

Crew E-11A	Hr.	Daily	Hr.	Daily	Bare Costs	Incl. O&P
2 Painters, Struc. Steel	$45.85	$733.60	$74.45	$1191.20	$46.70	$72.84
1 Building Laborer	42.10	336.80	63.25	506.00		
1 Equip. Oper. (light)	53.00	424.00	79.20	633.60		
1 Air Compressor, 250 cfm		201.50		221.65		
1 Sandblaster, Portable, 3 C.F.		83.80		92.18		
1 Set Sand Blasting Accessories		15.35		16.89		
1 Telescoping Boom Lift, to 60'		310.90		341.99	19.11	21.02
32 L.H., Daily Totals		$2105.95		$3003.51	$65.81	$93.86

Crew E-11B	Hr.	Daily	Hr.	Daily	Bare Costs	Incl. O&P
2 Painters, Struc. Steel	$45.85	$733.60	$74.45	$1191.20	$44.60	$70.72
1 Building Laborer	42.10	336.80	63.25	506.00		
2 Paint Sprayer, 8 C.F.M.		87.50		96.25		
1 Telescoping Boom Lift, to 60'		310.90		341.99	16.60	18.26
24 L.H., Daily Totals		$1468.80		$2135.44	$61.20	$88.98

Crew E-12	Hr.	Daily	Hr.	Daily	Bare Costs	Incl. O&P
1 Welder Foreman (outside)	$59.45	$475.60	$93.25	$746.00	$56.23	$86.22
1 Equip. Oper. (light)	53.00	424.00	79.20	633.60		
1 Welder, Gas Engine, 300 amp		147.00		161.70	9.19	10.11
16 L.H., Daily Totals		$1046.60		$1541.30	$65.41	$96.33

Crew E-13	Hr.	Daily	Hr.	Daily	Bare Costs	Incl. O&P
1 Welder Foreman (outside)	$59.45	$475.60	$93.25	$746.00	$57.30	$88.57
.5 Equip. Oper. (light)	53.00	212.00	79.20	316.80		
1 Welder, Gas Engine, 300 amp		147.00		161.70	12.25	13.48
12 L.H., Daily Totals		$834.60		$1224.50	$69.55	$102.04

Crew E-14	Hr.	Daily	Hr.	Daily	Bare Costs	Incl. O&P
1 Welder Foreman (outside)	$59.45	$475.60	$93.25	$746.00	$59.45	$93.25
1 Welder, Gas Engine, 300 amp		147.00		161.70	18.38	20.21
8 L.H., Daily Totals		$622.60		$907.70	$77.83	$113.46

Crew E-16	Hr.	Daily	Hr.	Daily	Bare Costs	Incl. O&P
1 Welder Foreman (outside)	$59.45	$475.60	$93.25	$746.00	$58.45	$91.67
1 Welder	57.45	459.60	90.10	720.80		
1 Welder, Gas Engine, 300 amp		147.00		161.70	9.19	10.11
16 L.H., Daily Totals		$1082.20		$1628.50	$67.64	$101.78

Crew E-17	Hr.	Daily	Hr.	Daily	Bare Costs	Incl. O&P
1 Struc. Steel Foreman (outside)	$59.65	$477.20	$93.55	$748.40	$58.65	$92.00
1 Structural Steel Worker	57.65	461.20	90.45	723.60		
16 L.H., Daily Totals		$938.40		$1472.00	$58.65	$92.00

Crew E-18	Hr.	Daily	Hr.	Daily	Bare Costs	Incl. O&P
1 Struc. Steel Foreman (outside)	$59.65	$477.20	$93.55	$748.40	$57.87	$89.95
3 Structural Steel Workers	57.65	1383.60	90.45	2170.80		
1 Equipment Operator (med.)	56.75	454.00	84.85	678.80		
1 Lattice Boom Crane, 20 Ton		1488.00		1636.80	37.20	40.92
40 L.H., Daily Totals		$3802.80		$5234.80	$95.07	$130.87

Crew E-19	Hr.	Daily	Hr.	Daily	Bare Costs	Incl. O&P
1 Struc. Steel Foreman (outside)	$59.65	$477.20	$93.55	$748.40	$56.77	$87.73
1 Structural Steel Worker	57.65	461.20	90.45	723.60		
1 Equip. Oper. (light)	53.00	424.00	79.20	633.60		
1 Lattice Boom Crane, 20 Ton		1488.00		1636.80	62.00	68.20
24 L.H., Daily Totals		$2850.40		$3742.40	$118.77	$155.93

Crew E-20	Hr.	Daily	Hr.	Daily	Bare Costs	Incl. O&P
1 Struc. Steel Foreman (outside)	$59.65	$477.20	$93.55	$748.40	$57.21	$88.73
5 Structural Steel Workers	57.65	2306.00	90.45	3618.00		
1 Equip. Oper. (crane)	59.20	473.60	88.50	708.00		
1 Equip. Oper. (oiler)	50.55	404.40	75.55	604.40		
1 Lattice Boom Crane, 40 Ton		2028.00		2230.80	31.69	34.86
64 L.H., Daily Totals		$5689.20		$7909.60	$88.89	$123.59

Crew E-22	Hr.	Daily	Hr.	Daily	Bare Costs	Incl. O&P
1 Skilled Worker Foreman (out)	$56.85	$454.80	$86.00	$688.00	$55.52	$83.97
2 Skilled Workers	54.85	877.60	82.95	1327.20		
24 L.H., Daily Totals		$1332.40		$2015.20	$55.52	$83.97

Crew E-24	Hr.	Daily	Hr.	Daily	Bare Costs	Incl. O&P
3 Structural Steel Workers	$57.65	$1383.60	$90.45	$2170.80	$57.42	$89.05
1 Equipment Operator (med.)	56.75	454.00	84.85	678.80		
1 Hyd. Crane, 25 Ton		580.85		638.93	18.15	19.97
32 L.H., Daily Totals		$2418.45		$3488.53	$75.58	$109.02

Crew E-25	Hr.	Daily	Hr.	Daily	Bare Costs	Incl. O&P
1 Welder Foreman (outside)	$59.45	$475.60	$93.25	$746.00	$59.45	$93.25
1 Cutting Torch		12.80		14.08	1.60	1.76
8 L.H., Daily Totals		$488.40		$760.08	$61.05	$95.01

Crew E-26	Hr.	Daily	Hr.	Daily	Bare Costs	Incl. O&P
1 Struc. Steel Foreman (outside)	$59.65	$477.20	$93.55	$748.40	$58.91	$91.73
1 Struc. Steel Worker	57.65	461.20	90.45	723.60		
1 Welder	57.45	459.60	90.10	720.80		
.25 Electrician	61.35	122.70	91.35	182.70		
.25 Plumber	64.45	128.90	96.45	192.90		
1 Welder, Gas Engine, 300 amp		147.00		161.70	5.25	5.78
28 L.H., Daily Totals		$1796.60		$2730.10	$64.16	$97.50

Crew No.	Bare Costs		Incl. Subs O&P		Cost Per Labor-Hour	
Crew E-27	**Hr.**	**Daily**	**Hr.**	**Daily**	**Bare Costs**	**Incl. O&P**
1 Struc. Steel Foreman (outside)	$59.65	$477.20	$93.55	$748.40	$57.21	$88.73
5 Struc. Steel Workers	57.65	2306.00	90.45	3618.00		
1 Equip. Oper. (crane)	59.20	473.60	88.50	708.00		
1 Equip. Oper. (oiler)	50.55	404.40	75.55	604.40		
1 Hyd. Crane, 12 Ton		471.00		518.10		
1 Hyd. Crane, 80 Ton		1486.00		1634.60	30.58	33.64
64 L.H., Daily Totals		$5618.20		$7831.50	$87.78	$122.37

Crew F-3	Hr.	Daily	Hr.	Daily	Bare Costs	Incl. O&P
4 Carpenters	$53.15	$1700.80	$79.85	$2555.20	$54.36	$81.58
1 Equip. Oper. (crane)	59.20	473.60	88.50	708.00		
1 Hyd. Crane, 12 Ton		471.00		518.10	11.78	12.95
40 L.H., Daily Totals		$2645.40		$3781.30	$66.14	$94.53

Crew F-4	Hr.	Daily	Hr.	Daily	Bare Costs	Incl. O&P
4 Carpenters	$53.15	$1700.80	$79.85	$2555.20	$53.73	$80.58
1 Equip. Oper. (crane)	59.20	473.60	88.50	708.00		
1 Equip. Oper. (oiler)	50.55	404.40	75.55	604.40		
1 Hyd. Crane, 55 Ton		980.20		1078.22	20.42	22.46
48 L.H., Daily Totals		$3559.00		$4945.82	$74.15	$103.04

Crew F-5	Hr.	Daily	Hr.	Daily	Bare Costs	Incl. O&P
1 Carpenter Foreman (outside)	$55.15	$441.20	$82.85	$662.80	$53.65	$80.60
3 Carpenters	53.15	1275.60	79.85	1916.40		
32 L.H., Daily Totals		$1716.80		$2579.20	$53.65	$80.60

Crew F-6	Hr.	Daily	Hr.	Daily	Bare Costs	Incl. O&P
2 Carpenters	$53.15	$850.40	$79.85	$1277.60	$49.94	$74.94
2 Building Laborers	42.10	673.60	63.25	1012.00		
1 Equip. Oper. (crane)	59.20	473.60	88.50	708.00		
1 Hyd. Crane, 12 Ton		471.00		518.10	11.78	12.95
40 L.H., Daily Totals		$2468.60		$3515.70	$61.72	$87.89

Crew F-7	Hr.	Daily	Hr.	Daily	Bare Costs	Incl. O&P
2 Carpenters	$53.15	$850.40	$79.85	$1277.60	$47.63	$71.55
2 Building Laborers	42.10	673.60	63.25	1012.00		
32 L.H., Daily Totals		$1524.00		$2289.60	$47.63	$71.55

Crew G-1	Hr.	Daily	Hr.	Daily	Bare Costs	Incl. O&P
1 Roofer Foreman (outside)	$48.20	$385.60	$79.55	$636.40	$43.17	$71.25
4 Roofers Composition	46.20	1478.40	76.25	2440.00		
2 Roofer Helpers	34.60	553.60	57.10	913.60		
1 Application Equipment		192.85		212.13		
1 Tar Kettle/Pot		207.85		228.63		
1 Crew Truck		166.25		182.88	10.12	11.14
56 L.H., Daily Totals		$2984.55		$4613.65	$53.30	$82.39

Crew G-2	Hr.	Daily	Hr.	Daily	Bare Costs	Incl. O&P
1 Plasterer	$48.60	$388.80	$72.55	$580.40	$44.37	$66.37
1 Plasterer Helper	42.40	339.20	63.30	506.40		
1 Building Laborer	42.10	336.80	63.25	506.00		
1 Grout Pump, 50 C.F./hr.		188.45		207.29	7.85	8.64
24 L.H., Daily Totals		$1253.25		$1800.10	$52.22	$75.00

Crew G-2A	Hr.	Daily	Hr.	Daily	Bare Costs	Incl. O&P
1 Roofer Composition	$46.20	$369.60	$76.25	$610.00	$40.97	$65.53
1 Roofer Helper	34.60	276.80	57.10	456.80		
1 Building Laborer	42.10	336.80	63.25	506.00		
1 Foam Spray Rig, Trailer-Mtd.		523.85		576.24		
1 Pickup Truck, 3/4 Ton		110.85		121.94	26.45	29.09
24 L.H., Daily Totals		$1617.90		$2270.97	$67.41	$94.62

Crew G-3	Hr.	Daily	Hr.	Daily	Bare Costs	Incl. O&P
2 Sheet Metal Workers	$62.30	$996.80	$94.50	$1512.00	$52.20	$78.88
2 Building Laborers	42.10	673.60	63.25	1012.00		
32 L.H., Daily Totals		$1670.40		$2524.00	$52.20	$78.88

Crew G-4	Hr.	Daily	Hr.	Daily	Bare Costs	Incl. O&P
1 Labor Foreman (outside)	$44.10	$352.80	$66.25	$530.00	$42.77	$64.25
2 Building Laborers	42.10	673.60	63.25	1012.00		
1 Flatbed Truck, Gas, 1.5 Ton		196.15		215.76		
1 Air Compressor, 160 cfm		202.80		223.08	16.62	18.29
24 L.H., Daily Totals		$1425.35		$1980.85	$59.39	$82.54

Crew G-5	Hr.	Daily	Hr.	Daily	Bare Costs	Incl. O&P
1 Roofer Foreman (outside)	$48.20	$385.60	$79.55	$636.40	$41.96	$69.25
2 Roofers Composition	46.20	739.20	76.25	1220.00		
2 Roofer Helpers	34.60	553.60	57.10	913.60		
1 Application Equipment		192.85		212.13	4.82	5.30
40 L.H., Daily Totals		$1871.25		$2982.14	$46.78	$74.55

Crew G-6A	Hr.	Daily	Hr.	Daily	Bare Costs	Incl. O&P
2 Roofers Composition	$46.20	$739.20	$76.25	$1220.00	$46.20	$76.25
1 Small Compressor, Electric		33.80		37.18		
2 Pneumatic Nailers		54.80		60.28	5.54	6.09
16 L.H., Daily Totals		$827.80		$1317.46	$51.74	$82.34

Crew G-7	Hr.	Daily	Hr.	Daily	Bare Costs	Incl. O&P
1 Carpenter	$53.15	$425.20	$79.85	$638.80	$53.15	$79.85
1 Small Compressor, Electric		33.80		37.18		
1 Pneumatic Nailer		27.40		30.14	7.65	8.41
8 L.H., Daily Totals		$486.40		$706.12	$60.80	$88.27

Crew H-1	Hr.	Daily	Hr.	Daily	Bare Costs	Incl. O&P
2 Glaziers	$51.00	$816.00	$76.40	$1222.40	$54.33	$83.42
2 Struc. Steel Workers	57.65	922.40	90.45	1447.20		
32 L.H., Daily Totals		$1738.40		$2669.60	$54.33	$83.42

Crew H-2	Hr.	Daily	Hr.	Daily	Bare Costs	Incl. O&P
2 Glaziers	$51.00	$816.00	$76.40	$1222.40	$48.03	$72.02
1 Building Laborer	42.10	336.80	63.25	506.00		
24 L.H., Daily Totals		$1152.80		$1728.40	$48.03	$72.02

Crew H-3	Hr.	Daily	Hr.	Daily	Bare Costs	Incl. O&P
1 Glazier	$51.00	$408.00	$76.40	$611.20	$45.48	$68.67
1 Helper	39.95	319.60	60.95	487.60		
16 L.H., Daily Totals		$727.60		$1098.80	$45.48	$68.67

Crew H-4	Hr.	Daily	Hr.	Daily	Bare Costs	Incl. O&P
1 Carpenter	$53.15	$425.20	$79.85	$638.80	$49.51	$74.59
1 Carpenter Helper	39.95	319.60	60.95	487.60		
.5 Electrician	61.35	245.40	91.35	365.40		
20 L.H., Daily Totals		$990.20		$1491.80	$49.51	$74.59

Crew No.	Bare Costs		Incl. Subs O&P		Cost Per Labor-Hour	
Crew J-1	**Hr.**	**Daily**	**Hr.**	**Daily**	**Bare Costs**	**Incl. O&P**
3 Plasterers	$48.60	$1166.40	$72.55	$1741.20	$46.12	$68.85
2 Plasterer Helpers	42.40	678.40	63.30	1012.80		
1 Mixing Machine, 6 C.F.		117.35		129.09	2.93	3.23
40 L.H., Daily Totals		$1962.15		$2883.09	$49.05	$72.08
Crew J-2	**Hr.**	**Daily**	**Hr.**	**Daily**	**Bare Costs**	**Incl. O&P**
3 Plasterers	$48.60	$1166.40	$72.55	$1741.20	$47.11	$70.09
2 Plasterer Helpers	42.40	678.40	63.30	1012.80		
1 Lather	52.05	416.40	76.30	610.40		
1 Mixing Machine, 6 C.F.		117.35		129.09	2.44	2.69
48 L.H., Daily Totals		$2378.55		$3493.49	$49.55	$72.78
Crew J-3	**Hr.**	**Daily**	**Hr.**	**Daily**	**Bare Costs**	**Incl. O&P**
1 Terrazzo Worker	$49.40	$395.20	$72.45	$579.60	$44.95	$65.92
1 Terrazzo Helper	40.50	324.00	59.40	475.20		
1 Floor Grinder, 22" Path		104.60		115.06		
1 Terrazzo Mixer		161.15		177.26	16.61	18.27
16 L.H., Daily Totals		$984.95		$1347.13	$61.56	$84.20
Crew J-4	**Hr.**	**Daily**	**Hr.**	**Daily**	**Bare Costs**	**Incl. O&P**
2 Cement Finishers	$49.95	$799.20	$73.45	$1175.20	$47.33	$70.05
1 Laborer	42.10	336.80	63.25	506.00		
1 Floor Grinder, 22" Path		104.60		115.06		
1 Floor Edger, 7" Path		43.95		48.34		
1 Vacuum Pick-Up System		74.05		81.45	9.28	10.20
24 L.H., Daily Totals		$1358.60		$1926.06	$56.61	$80.25
Crew J-4A	**Hr.**	**Daily**	**Hr.**	**Daily**	**Bare Costs**	**Incl. O&P**
2 Cement Finishers	$49.95	$799.20	$73.45	$1175.20	$46.02	$68.35
2 Laborers	42.10	673.60	63.25	1012.00		
1 Floor Grinder, 22" Path		104.60		115.06		
1 Floor Edger, 7" Path		43.95		48.34		
1 Vacuum Pick-Up System		74.05		81.45		
1 Floor Auto Scrubber		178.40		196.24	12.53	13.78
32 L.H., Daily Totals		$1873.80		$2628.30	$58.56	$82.13
Crew J-4B	**Hr.**	**Daily**	**Hr.**	**Daily**	**Bare Costs**	**Incl. O&P**
1 Laborer	$42.10	$336.80	$63.25	$506.00	$42.10	$63.25
1 Floor Auto Scrubber		178.40		196.24	22.30	24.53
8 L.H., Daily Totals		$515.20		$702.24	$64.40	$87.78
Crew J-6	**Hr.**	**Daily**	**Hr.**	**Daily**	**Bare Costs**	**Incl. O&P**
2 Painters	$44.40	$710.40	$66.05	$1056.80	$45.98	$68.64
1 Building Laborer	42.10	336.80	63.25	506.00		
1 Equip. Oper. (light)	53.00	424.00	79.20	633.60		
1 Air Compressor, 250 cfm		201.50		221.65		
1 Sandblaster, Portable, 3 C.F.		83.80		92.18		
1 Set Sand Blasting Accessories		15.35		16.89	9.40	10.33
32 L.H., Daily Totals		$1771.85		$2527.11	$55.37	$78.97
Crew J-7	**Hr.**	**Daily**	**Hr.**	**Daily**	**Bare Costs**	**Incl. O&P**
2 Painters	$44.40	$710.40	$66.05	$1056.80	$44.40	$66.05
1 Floor Belt Sander		50.15		55.16		
1 Floor Sanding Edger		25.15		27.66	4.71	5.18
16 L.H., Daily Totals		$785.70		$1139.63	$49.11	$71.23

Crew No.	Bare Costs		Incl. Subs O&P		Cost Per Labor-Hour	
Crew K-1	**Hr.**	**Daily**	**Hr.**	**Daily**	**Bare Costs**	**Incl. O&P**
1 Carpenter	$53.15	$425.20	$79.85	$638.80	$50.20	$75.33
1 Truck Driver (light)	47.25	378.00	70.80	566.40		
1 Flatbed Truck, Gas, 3 Ton		820.40		902.44	51.27	56.40
16 L.H., Daily Totals		$1623.60		$2107.64	$101.47	$131.73
Crew K-2	**Hr.**	**Daily**	**Hr.**	**Daily**	**Bare Costs**	**Incl. O&P**
1 Struc. Steel Foreman (outside)	$59.65	$477.20	$93.55	$748.40	$54.85	$84.93
1 Struc. Steel Worker	57.65	461.20	90.45	723.60		
1 Truck Driver (light)	47.25	378.00	70.80	566.40		
1 Flatbed Truck, Gas, 3 Ton		820.40		902.44	34.18	37.60
24 L.H., Daily Totals		$2136.80		$2940.84	$89.03	$122.54
Crew L-1	**Hr.**	**Daily**	**Hr.**	**Daily**	**Bare Costs**	**Incl. O&P**
1 Electrician	$61.35	$490.80	$91.35	$730.80	$62.90	$93.90
1 Plumber	64.45	515.60	96.45	771.60		
16 L.H., Daily Totals		$1006.40		$1502.40	$62.90	$93.90
Crew L-2	**Hr.**	**Daily**	**Hr.**	**Daily**	**Bare Costs**	**Incl. O&P**
1 Carpenter	$53.15	$425.20	$79.85	$638.80	$46.55	$70.40
1 Carpenter Helper	39.95	319.60	60.95	487.60		
16 L.H., Daily Totals		$744.80		$1126.40	$46.55	$70.40
Crew L-3	**Hr.**	**Daily**	**Hr.**	**Daily**	**Bare Costs**	**Incl. O&P**
1 Carpenter	$53.15	$425.20	$79.85	$638.80	$57.49	$86.39
.5 Electrician	61.35	245.40	91.35	365.40		
.5 Sheet Metal Worker	62.30	249.20	94.50	378.00		
16 L.H., Daily Totals		$919.80		$1382.20	$57.49	$86.39
Crew L-3A	**Hr.**	**Daily**	**Hr.**	**Daily**	**Bare Costs**	**Incl. O&P**
1 Carpenter Foreman (outside)	$55.15	$441.20	$82.85	$662.80	$57.53	$86.73
.5 Sheet Metal Worker	62.30	249.20	94.50	378.00		
12 L.H., Daily Totals		$690.40		$1040.80	$57.53	$86.73
Crew L-4	**Hr.**	**Daily**	**Hr.**	**Daily**	**Bare Costs**	**Incl. O&P**
2 Skilled Workers	$54.85	$877.60	$82.95	$1327.20	$49.88	$75.62
1 Helper	39.95	319.60	60.95	487.60		
24 L.H., Daily Totals		$1197.20		$1814.80	$49.88	$75.62
Crew L-5	**Hr.**	**Daily**	**Hr.**	**Daily**	**Bare Costs**	**Incl. O&P**
1 Struc. Steel Foreman (outside)	$59.65	$477.20	$93.55	$748.40	$58.16	$90.61
5 Struc. Steel Workers	57.65	2306.00	90.45	3618.00		
1 Equip. Oper. (crane)	59.20	473.60	88.50	708.00		
1 Hyd. Crane, 25 Ton		580.85		638.93	10.37	11.41
56 L.H., Daily Totals		$3837.65		$5713.34	$68.53	$102.02
Crew L-5A	**Hr.**	**Daily**	**Hr.**	**Daily**	**Bare Costs**	**Incl. O&P**
1 Struc. Steel Foreman (outside)	$59.65	$477.20	$93.55	$748.40	$58.54	$90.74
2 Structural Steel Workers	57.65	922.40	90.45	1447.20		
1 Equip. Oper. (crane)	59.20	473.60	88.50	708.00		
1 S.P. Crane, 4x4, 25 Ton		1144.00		1258.40	35.75	39.33
32 L.H., Daily Totals		$3017.20		$4162.00	$94.29	$130.06

Crew No.	Bare Costs		Incl. Subs O&P		Cost Per Labor-Hour	
Crew L-5B	**Hr.**	**Daily**	**Hr.**	**Daily**	**Bare Costs**	**Incl. O&P**
1 Struc. Steel Foreman (outside)	$59.65	$477.20	$93.55	$748.40	$59.83	$90.82
2 Structural Steel Workers	57.65	922.40	90.45	1447.20		
2 Electricians	61.35	981.60	91.35	1461.60		
2 Steamfitters/Pipefitters	65.55	1048.80	98.10	1569.60		
1 Equip. Oper. (crane)	59.20	473.60	88.50	708.00		
1 Equip. Oper. (oiler)	50.55	404.40	75.55	604.40		
1 Hyd. Crane, 80 Ton		1486.00		1634.60	20.64	22.70
72 L.H., Daily Totals		$5794.00		$8173.80	$80.47	$113.53
Crew L-6	**Hr.**	**Daily**	**Hr.**	**Daily**	**Bare Costs**	**Incl. O&P**
1 Plumber	$64.45	$515.60	$96.45	$771.60	$63.42	$94.75
.5 Electrician	61.35	245.40	91.35	365.40		
12 L.H., Daily Totals		$761.00		$1137.00	$63.42	$94.75
Crew L-7	**Hr.**	**Daily**	**Hr.**	**Daily**	**Bare Costs**	**Incl. O&P**
2 Carpenters	$53.15	$850.40	$79.85	$1277.60	$51.16	$76.75
1 Building Laborer	42.10	336.80	63.25	506.00		
.5 Electrician	61.35	245.40	91.35	365.40		
28 L.H., Daily Totals		$1432.60		$2149.00	$51.16	$76.75
Crew L-8	**Hr.**	**Daily**	**Hr.**	**Daily**	**Bare Costs**	**Incl. O&P**
2 Carpenters	$53.15	$850.40	$79.85	$1277.60	$55.41	$83.17
.5 Plumber	64.45	257.80	96.45	385.80		
20 L.H., Daily Totals		$1108.20		$1663.40	$55.41	$83.17
Crew L-9	**Hr.**	**Daily**	**Hr.**	**Daily**	**Bare Costs**	**Incl. O&P**
1 Labor Foreman (inside)	$42.60	$340.80	$64.00	$512.00	$47.81	$72.58
2 Building Laborers	42.10	673.60	63.25	1012.00		
1 Struc. Steel Worker	57.65	461.20	90.45	723.60		
.5 Electrician	61.35	245.40	91.35	365.40		
36 L.H., Daily Totals		$1721.00		$2613.00	$47.81	$72.58
Crew L-10	**Hr.**	**Daily**	**Hr.**	**Daily**	**Bare Costs**	**Incl. O&P**
1 Struc. Steel Foreman (outside)	$59.65	$477.20	$93.55	$748.40	$58.83	$90.83
1 Structural Steel Worker	57.65	461.20	90.45	723.60		
1 Equip. Oper. (crane)	59.20	473.60	88.50	708.00		
1 Hyd. Crane, 12 Ton		471.00		518.10	19.63	21.59
24 L.H., Daily Totals		$1883.00		$2698.10	$78.46	$112.42
Crew L-11	**Hr.**	**Daily**	**Hr.**	**Daily**	**Bare Costs**	**Incl. O&P**
2 Wreckers	$42.10	$673.60	$64.80	$1036.80	$49.10	$74.33
1 Equip. Oper. (crane)	59.20	473.60	88.50	708.00		
1 Equip. Oper. (light)	53.00	424.00	79.20	633.60		
1 Hyd. Excavator, 2.5 C.Y.		1465.00		1611.50		
1 Loader, Skid Steer, 78 H.P.		400.10		440.11	58.28	64.11
32 L.H., Daily Totals		$3436.30		$4430.01	$107.38	$138.44
Crew M-1	**Hr.**	**Daily**	**Hr.**	**Daily**	**Bare Costs**	**Incl. O&P**
3 Elevator Constructors	$85.55	$2053.20	$127.00	$3048.00	$81.28	$120.65
1 Elevator Apprentice	68.45	547.60	101.60	812.80		
5 Hand Tools		50.00		55.00	1.56	1.72
32 L.H., Daily Totals		$2650.80		$3915.80	$82.84	$122.37

Crew No.	Bare Costs		Incl. Subs O&P		Cost Per Labor-Hour	
Crew M-3	**Hr.**	**Daily**	**Hr.**	**Daily**	**Bare Costs**	**Incl. O&P**
1 Electrician Foreman (outside)	$63.35	$506.80	$94.35	$754.80	$64.39	$95.86
1 Common Laborer	42.10	336.80	63.25	506.00		
.25 Equipment Operator (med.)	56.75	113.50	84.85	169.70		
1 Elevator Constructor	85.55	684.40	127.00	1016.00		
1 Elevator Apprentice	68.45	547.60	101.60	812.80		
.25 S.P. Crane, 4x4, 20 Ton		145.55		160.10	4.28	4.71
34 L.H., Daily Totals		$2334.65		$3419.41	$68.67	$100.57
Crew M-4	**Hr.**	**Daily**	**Hr.**	**Daily**	**Bare Costs**	**Incl. O&P**
1 Electrician Foreman (outside)	$63.35	$506.80	$94.35	$754.80	$63.75	$94.94
1 Common Laborer	42.10	336.80	63.25	506.00		
.25 Equipment Operator, Crane	59.20	118.40	88.50	177.00		
.25 Equip. Oper. (oiler)	50.55	101.10	75.55	151.10		
1 Elevator Constructor	85.55	684.40	127.00	1016.00		
1 Elevator Apprentice	68.45	547.60	101.60	812.80		
.25 S.P. Crane, 4x4, 40 Ton		188.55		207.41	5.24	5.76
36 L.H., Daily Totals		$2483.65		$3625.11	$68.99	$100.70
Crew Q-1	**Hr.**	**Daily**	**Hr.**	**Daily**	**Bare Costs**	**Incl. O&P**
1 Plumber	$64.45	$515.60	$96.45	$771.60	$58.00	$86.80
1 Plumber Apprentice	51.55	412.40	77.15	617.20		
16 L.H., Daily Totals		$928.00		$1388.80	$58.00	$86.80
Crew Q-1A	**Hr.**	**Daily**	**Hr.**	**Daily**	**Bare Costs**	**Incl. O&P**
.25 Plumber Foreman (outside)	$66.45	$132.90	$99.45	$198.90	$64.85	$97.05
1 Plumber	64.45	515.60	96.45	771.60		
10 L.H., Daily Totals		$648.50		$970.50	$64.85	$97.05
Crew Q-1C	**Hr.**	**Daily**	**Hr.**	**Daily**	**Bare Costs**	**Incl. O&P**
1 Plumber	$64.45	$515.60	$96.45	$771.60	$57.58	$86.15
1 Plumber Apprentice	51.55	412.40	77.15	617.20		
1 Equip. Oper. (medium)	56.75	454.00	84.85	678.80		
1 Trencher, Chain Type, 8' D		1872.00		2059.20	78.00	85.80
24 L.H., Daily Totals		$3254.00		$4126.80	$135.58	$171.95
Crew Q-2	**Hr.**	**Daily**	**Hr.**	**Daily**	**Bare Costs**	**Incl. O&P**
2 Plumbers	$64.45	$1031.20	$96.45	$1543.20	$60.15	$90.02
1 Plumber Apprentice	51.55	412.40	77.15	617.20		
24 L.H., Daily Totals		$1443.60		$2160.40	$60.15	$90.02
Crew Q-3	**Hr.**	**Daily**	**Hr.**	**Daily**	**Bare Costs**	**Incl. O&P**
1 Plumber Foreman (inside)	$64.95	$519.60	$97.20	$777.60	$61.35	$91.81
2 Plumbers	64.45	1031.20	96.45	1543.20		
1 Plumber Apprentice	51.55	412.40	77.15	617.20		
32 L.H., Daily Totals		$1963.20		$2938.00	$61.35	$91.81
Crew Q-4	**Hr.**	**Daily**	**Hr.**	**Daily**	**Bare Costs**	**Incl. O&P**
1 Plumber Foreman (inside)	$64.95	$519.60	$97.20	$777.60	$61.35	$91.81
1 Plumber	64.45	515.60	96.45	771.60		
1 Welder (plumber)	64.45	515.60	96.45	771.60		
1 Plumber Apprentice	51.55	412.40	77.15	617.20		
1 Welder, Electric, 300 amp		106.40		117.04	3.33	3.66
32 L.H., Daily Totals		$2069.60		$3055.04	$64.67	$95.47
Crew Q-5	**Hr.**	**Daily**	**Hr.**	**Daily**	**Bare Costs**	**Incl. O&P**
1 Steamfitter	$65.55	$524.40	$98.10	$784.80	$59.00	$88.30
1 Steamfitter Apprentice	52.45	419.60	78.50	628.00		
16 L.H., Daily Totals		$944.00		$1412.80	$59.00	$88.30

Crew No.	Bare Costs		Incl. Subs O&P		Cost Per Labor-Hour	
Crew Q-6	**Hr.**	**Daily**	**Hr.**	**Daily**	**Bare Costs**	**Incl. O&P**
2 Steamfitters	$65.55	$1048.80	$98.10	$1569.60	$61.18	$91.57
1 Steamfitter Apprentice	52.45	419.60	78.50	628.00		
24 L.H., Daily Totals		$1468.40		$2197.60	$61.18	$91.57
Crew Q-7	**Hr.**	**Daily**	**Hr.**	**Daily**	**Bare Costs**	**Incl. O&P**
1 Steamfitter Foreman (inside)	$66.05	$528.40	$98.85	$790.80	$62.40	$93.39
2 Steamfitters	65.55	1048.80	98.10	1569.60		
1 Steamfitter Apprentice	52.45	419.60	78.50	628.00		
32 L.H., Daily Totals		$1996.80		$2988.40	$62.40	$93.39
Crew Q-8	**Hr.**	**Daily**	**Hr.**	**Daily**	**Bare Costs**	**Incl. O&P**
1 Steamfitter Foreman (inside)	$66.05	$528.40	$98.85	$790.80	$62.40	$93.39
1 Steamfitter	65.55	524.40	98.10	784.80		
1 Welder (steamfitter)	65.55	524.40	98.10	784.80		
1 Steamfitter Apprentice	52.45	419.60	78.50	628.00		
1 Welder, Electric, 300 amp		106.40		117.04	3.33	3.66
32 L.H., Daily Totals		$2103.20		$3105.44	$65.72	$97.05
Crew Q-9	**Hr.**	**Daily**	**Hr.**	**Daily**	**Bare Costs**	**Incl. O&P**
1 Sheet Metal Worker	$62.30	$498.40	$94.50	$756.00	$56.08	$85.05
1 Sheet Metal Apprentice	49.85	398.80	75.60	604.80		
16 L.H., Daily Totals		$897.20		$1360.80	$56.08	$85.05
Crew Q-10	**Hr.**	**Daily**	**Hr.**	**Daily**	**Bare Costs**	**Incl. O&P**
2 Sheet Metal Workers	$62.30	$996.80	$94.50	$1512.00	$58.15	$88.20
1 Sheet Metal Apprentice	49.85	398.80	75.60	604.80		
24 L.H., Daily Totals		$1395.60		$2116.80	$58.15	$88.20
Crew Q-11	**Hr.**	**Daily**	**Hr.**	**Daily**	**Bare Costs**	**Incl. O&P**
1 Sheet Metal Foreman (inside)	$62.80	$502.40	$95.25	$762.00	$59.31	$89.96
2 Sheet Metal Workers	62.30	996.80	94.50	1512.00		
1 Sheet Metal Apprentice	49.85	398.80	75.60	604.80		
32 L.H., Daily Totals		$1898.00		$2878.80	$59.31	$89.96
Crew Q-12	**Hr.**	**Daily**	**Hr.**	**Daily**	**Bare Costs**	**Incl. O&P**
1 Sprinkler Installer	$63.25	$506.00	$94.80	$758.40	$56.92	$85.33
1 Sprinkler Apprentice	50.60	404.80	75.85	606.80		
16 L.H., Daily Totals		$910.80		$1365.20	$56.92	$85.33
Crew Q-13	**Hr.**	**Daily**	**Hr.**	**Daily**	**Bare Costs**	**Incl. O&P**
1 Sprinkler Foreman (inside)	$63.75	$510.00	$95.55	$764.40	$60.21	$90.25
2 Sprinkler Installers	63.25	1012.00	94.80	1516.80		
1 Sprinkler Apprentice	50.60	404.80	75.85	606.80		
32 L.H., Daily Totals		$1926.80		$2888.00	$60.21	$90.25
Crew Q-14	**Hr.**	**Daily**	**Hr.**	**Daily**	**Bare Costs**	**Incl. O&P**
1 Asbestos Worker	$58.65	$469.20	$90.05	$720.40	$52.77	$81.03
1 Asbestos Apprentice	46.90	375.20	72.00	576.00		
16 L.H., Daily Totals		$844.40		$1296.40	$52.77	$81.03
Crew Q-15	**Hr.**	**Daily**	**Hr.**	**Daily**	**Bare Costs**	**Incl. O&P**
1 Plumber	$64.45	$515.60	$96.45	$771.60	$58.00	$86.80
1 Plumber Apprentice	51.55	412.40	77.15	617.20		
1 Welder, Electric, 300 amp		106.40		117.04	6.65	7.32
16 L.H., Daily Totals		$1034.40		$1505.84	$64.65	$94.11

Crew No.	Bare Costs		Incl. Subs O&P		Cost Per Labor-Hour	
Crew Q-16	**Hr.**	**Daily**	**Hr.**	**Daily**	**Bare Costs**	**Incl. O&P**
2 Plumbers	$64.45	$1031.20	$96.45	$1543.20	$60.15	$90.02
1 Plumber Apprentice	51.55	412.40	77.15	617.20		
1 Welder, Electric, 300 amp		106.40		117.04	4.43	4.88
24 L.H., Daily Totals		$1550.00		$2277.44	$64.58	$94.89
Crew Q-17	**Hr.**	**Daily**	**Hr.**	**Daily**	**Bare Costs**	**Incl. O&P**
1 Steamfitter	$65.55	$524.40	$98.10	$784.80	$59.00	$88.30
1 Steamfitter Apprentice	52.45	419.60	78.50	628.00		
1 Welder, Electric, 300 amp		106.40		117.04	6.65	7.32
16 L.H., Daily Totals		$1050.40		$1529.84	$65.65	$95.61
Crew Q-17A	**Hr.**	**Daily**	**Hr.**	**Daily**	**Bare Costs**	**Incl. O&P**
1 Steamfitter	$65.55	$524.40	$98.10	$784.80	$59.07	$88.37
1 Steamfitter Apprentice	52.45	419.60	78.50	628.00		
1 Equip. Oper. (crane)	59.20	473.60	88.50	708.00		
1 Hyd. Crane, 12 Ton		471.00		518.10		
1 Welder, Electric, 300 amp		106.40		117.04	24.06	26.46
24 L.H., Daily Totals		$1995.00		$2755.94	$83.13	$114.83
Crew Q-18	**Hr.**	**Daily**	**Hr.**	**Daily**	**Bare Costs**	**Incl. O&P**
2 Steamfitters	$65.55	$1048.80	$98.10	$1569.60	$61.18	$91.57
1 Steamfitter Apprentice	52.45	419.60	78.50	628.00		
1 Welder, Electric, 300 amp		106.40		117.04	4.43	4.88
24 L.H., Daily Totals		$1574.80		$2314.64	$65.62	$96.44
Crew Q-19	**Hr.**	**Daily**	**Hr.**	**Daily**	**Bare Costs**	**Incl. O&P**
1 Steamfitter	$65.55	$524.40	$98.10	$784.80	$59.78	$89.32
1 Steamfitter Apprentice	52.45	419.60	78.50	628.00		
1 Electrician	61.35	490.80	91.35	730.80		
24 L.H., Daily Totals		$1434.80		$2143.60	$59.78	$89.32
Crew Q-20	**Hr.**	**Daily**	**Hr.**	**Daily**	**Bare Costs**	**Incl. O&P**
1 Sheet Metal Worker	$62.30	$498.40	$94.50	$756.00	$57.13	$86.31
1 Sheet Metal Apprentice	49.85	398.80	75.60	604.80		
.5 Electrician	61.35	245.40	91.35	365.40		
20 L.H., Daily Totals		$1142.60		$1726.20	$57.13	$86.31
Crew Q-21	**Hr.**	**Daily**	**Hr.**	**Daily**	**Bare Costs**	**Incl. O&P**
2 Steamfitters	$65.55	$1048.80	$98.10	$1569.60	$61.23	$91.51
1 Steamfitter Apprentice	52.45	419.60	78.50	628.00		
1 Electrician	61.35	490.80	91.35	730.80		
32 L.H., Daily Totals		$1959.20		$2928.40	$61.23	$91.51
Crew Q-22	**Hr.**	**Daily**	**Hr.**	**Daily**	**Bare Costs**	**Incl. O&P**
1 Plumber	$64.45	$515.60	$96.45	$771.60	$58.00	$86.80
1 Plumber Apprentice	51.55	412.40	77.15	617.20		
1 Hyd. Crane, 12 Ton		471.00		518.10	29.44	32.38
16 L.H., Daily Totals		$1399.00		$1906.90	$87.44	$119.18
Crew Q-22A	**Hr.**	**Daily**	**Hr.**	**Daily**	**Bare Costs**	**Incl. O&P**
1 Plumber	$64.45	$515.60	$96.45	$771.60	$54.33	$81.34
1 Plumber Apprentice	51.55	412.40	77.15	617.20		
1 Laborer	42.10	336.80	63.25	506.00		
1 Equip. Oper. (crane)	59.20	473.60	88.50	708.00		
1 Hyd. Crane, 12 Ton		471.00		518.10	14.72	16.19
32 L.H., Daily Totals		$2209.40		$3120.90	$69.04	$97.53

Crew No.	Bare Costs		Incl. Subs O&P		Cost Per Labor-Hour	
Crew Q-23	**Hr.**	**Daily**	**Hr.**	**Daily**	**Bare Costs**	**Incl. O&P**
1 Plumber Foreman (outside)	$66.45	$531.60	$99.45	$795.60	$62.55	$93.58
1 Plumber	64.45	515.60	96.45	771.60		
1 Equip. Oper. (medium)	56.75	454.00	84.85	678.80		
1 Lattice Boom Crane, 20 Ton		1488.00		1636.80	62.00	68.20
24 L.H., Daily Totals		$2989.20		$3882.80	$124.55	$161.78
Crew R-1	**Hr.**	**Daily**	**Hr.**	**Daily**	**Bare Costs**	**Incl. O&P**
1 Electrician Foreman	$61.85	$494.80	$92.10	$736.80	$57.35	$85.39
3 Electricians	61.35	1472.40	91.35	2192.40		
2 Electrician Apprentices	49.10	785.60	73.10	1169.60		
48 L.H., Daily Totals		$2752.80		$4098.80	$57.35	$85.39
Crew R-1A	**Hr.**	**Daily**	**Hr.**	**Daily**	**Bare Costs**	**Incl. O&P**
1 Electrician	$61.35	$490.80	$91.35	$730.80	$55.23	$82.22
1 Electrician Apprentice	49.10	392.80	73.10	584.80		
16 L.H., Daily Totals		$883.60		$1315.60	$55.23	$82.22
Crew R-1B	**Hr.**	**Daily**	**Hr.**	**Daily**	**Bare Costs**	**Incl. O&P**
1 Electrician	$61.35	$490.80	$91.35	$730.80	$53.18	$79.18
2 Electrician Apprentices	49.10	785.60	73.10	1169.60		
24 L.H., Daily Totals		$1276.40		$1900.40	$53.18	$79.18
Crew R-1C	**Hr.**	**Daily**	**Hr.**	**Daily**	**Bare Costs**	**Incl. O&P**
2 Electricians	$61.35	$981.60	$91.35	$1461.60	$55.23	$82.22
2 Electrician Apprentices	49.10	785.60	73.10	1169.60		
1 Portable cable puller, 8000 lb.		101.60		111.76	3.17	3.49
32 L.H., Daily Totals		$1868.80		$2742.96	$58.40	$85.72
Crew R-1D	**Hr.**	**Daily**	**Hr.**	**Daily**	**Bare Costs**	**Incl. O&P**
1 Electrician	$61.35	$490.80	$91.35	$730.80	$55.23	$82.23
1 Electrician Apprentice	49.10	392.80	73.10	584.80		
1 Aerial lift		105.00		115.50	6.56	7.22
16 L.H., Daily Totals		$988.60		$1431.10	$61.79	$89.45
Crew R-2	**Hr.**	**Daily**	**Hr.**	**Daily**	**Bare Costs**	**Incl. O&P**
1 Electrician Foreman	$61.85	$494.80	$92.10	$736.80	$57.61	$85.84
3 Electricians	61.35	1472.40	91.35	2192.40		
2 Electrician Apprentices	49.10	785.60	73.10	1169.60		
1 Equip. Oper. (crane)	59.20	473.60	88.50	708.00		
1 S.P. Crane, 4x4, 5 Ton		378.10		415.91	6.75	7.43
56 L.H., Daily Totals		$3604.50		$5222.71	$64.37	$93.26
Crew R-3	**Hr.**	**Daily**	**Hr.**	**Daily**	**Bare Costs**	**Incl. O&P**
1 Electrician Foreman	$61.85	$494.80	$92.10	$736.80	$61.12	$91.08
1 Electrician	61.35	490.80	91.35	730.80		
.5 Equip. Oper. (crane)	59.20	236.80	88.50	354.00		
.5 S.P. Crane, 4x4, 5 Ton		189.05		207.96	9.45	10.40
20 L.H., Daily Totals		$1411.45		$2029.56	$70.57	$101.48
Crew R-4	**Hr.**	**Daily**	**Hr.**	**Daily**	**Bare Costs**	**Incl. O&P**
1 Struc. Steel Foreman (outside)	$59.65	$477.20	$93.55	$748.40	$58.79	$91.25
3 Struc. Steel Workers	57.65	1383.60	90.45	2170.80		
1 Electrician	61.35	490.80	91.35	730.80		
1 Welder, Gas Engine, 300 amp		147.00		161.70	3.67	4.04
40 L.H., Daily Totals		$2498.60		$3811.70	$62.47	$95.29

Crew No.	Bare Costs		Incl. Subs O&P		Cost Per Labor-Hour	
Crew R-5	**Hr.**	**Daily**	**Hr.**	**Daily**	**Bare Costs**	**Incl. O&P**
1 Electrician Foreman	$61.85	$494.80	$92.10	$736.80	$53.61	$80.36
4 Electrician Linemen	61.35	1963.20	91.35	2923.20		
2 Electrician Operators	61.35	981.60	91.35	1461.60		
4 Electrician Groundmen	39.95	1278.40	60.95	1950.40		
1 Crew Truck		166.25		182.88		
1 Flatbed Truck, 20,000 GVW		201.60		221.76		
1 Pickup Truck, 3/4 Ton		110.85		121.94		
.2 Hyd. Crane, 55 Ton		196.04		215.64		
.2 Hyd. Crane, 12 Ton		94.20		103.62		
.2 Earth Auger, Truck-Mtd.		40.10		44.11		
1 Tractor w/Winch		373.50		410.85	13.44	14.78
88 L.H., Daily Totals		$5900.54		$8372.79	$67.05	$95.15
Crew R-6	**Hr.**	**Daily**	**Hr.**	**Daily**	**Bare Costs**	**Incl. O&P**
1 Electrician Foreman	$61.85	$494.80	$92.10	$736.80	$53.61	$80.36
4 Electrician Linemen	61.35	1963.20	91.35	2923.20		
2 Electrician Operators	61.35	981.60	91.35	1461.60		
4 Electrician Groundmen	39.95	1278.40	60.95	1950.40		
1 Crew Truck		166.25		182.88		
1 Flatbed Truck, 20,000 GVW		201.60		221.76		
1 Pickup Truck, 3/4 Ton		110.85		121.94		
.2 Hyd. Crane, 55 Ton		196.04		215.64		
.2 Hyd. Crane, 12 Ton		94.20		103.62		
.2 Earth Auger, Truck-Mtd.		40.10		44.11		
1 Tractor w/Winch		373.50		410.85		
3 Cable Trailers		192.30		211.53		
.5 Tensioning Rig		55.20		60.72		
.5 Cable Pulling Rig		303.57		333.93	19.70	21.67
88 L.H., Daily Totals		$6451.61		$8978.98	$73.31	$102.03
Crew R-7	**Hr.**	**Daily**	**Hr.**	**Daily**	**Bare Costs**	**Incl. O&P**
1 Electrician Foreman	$61.85	$494.80	$92.10	$736.80	$43.60	$66.14
5 Electrician Groundmen	39.95	1598.00	60.95	2438.00		
1 Crew Truck		166.25		182.88	3.46	3.81
48 L.H., Daily Totals		$2259.05		$3357.68	$47.06	$69.95
Crew R-8	**Hr.**	**Daily**	**Hr.**	**Daily**	**Bare Costs**	**Incl. O&P**
1 Electrician Foreman	$61.85	$494.80	$92.10	$736.80	$54.30	$81.34
3 Electrician Linemen	61.35	1472.40	91.35	2192.40		
2 Electrician Groundmen	39.95	639.20	60.95	975.20		
1 Pickup Truck, 3/4 Ton		110.85		121.94		
1 Crew Truck		166.25		182.88	5.77	6.35
48 L.H., Daily Totals		$2883.50		$4209.21	$60.07	$87.69
Crew R-9	**Hr.**	**Daily**	**Hr.**	**Daily**	**Bare Costs**	**Incl. O&P**
1 Electrician Foreman	$61.85	$494.80	$92.10	$736.80	$50.71	$76.24
1 Electrician Lineman	61.35	490.80	91.35	730.80		
2 Electrician Operators	61.35	981.60	91.35	1461.60		
4 Electrician Groundmen	39.95	1278.40	60.95	1950.40		
1 Pickup Truck, 3/4 Ton		110.85		121.94		
1 Crew Truck		166.25		182.88	4.33	4.76
64 L.H., Daily Totals		$3522.70		$5184.41	$55.04	$81.01
Crew R-10	**Hr.**	**Daily**	**Hr.**	**Daily**	**Bare Costs**	**Incl. O&P**
1 Electrician Foreman	$61.85	$494.80	$92.10	$736.80	$57.87	$86.41
4 Electrician Linemen	61.35	1963.20	91.35	2923.20		
1 Electrician Groundman	39.95	319.60	60.95	487.60		
1 Crew Truck		166.25		182.88		
3 Tram Cars		216.15		237.76	7.97	8.76
48 L.H., Daily Totals		$3160.00		$4568.24	$65.83	$95.17

Crew R-11	Bare Costs Hr.	Bare Costs Daily	Incl. Subs O&P Hr.	Incl. Subs O&P Daily	Cost Per Labor-Hour Bare Costs	Cost Per Labor-Hour Incl. O&P
1 Electrician Foreman	$61.85	$494.80	$92.10	$736.80	$58.36	$87.04
4 Electricians	61.35	1963.20	91.35	2923.20		
1 Equip. Oper. (crane)	59.20	473.60	88.50	708.00		
1 Common Laborer	42.10	336.80	63.25	506.00		
1 Crew Truck		166.25		182.88		
1 Hyd. Crane, 12 Ton		471.00		518.10	11.38	12.52
56 L.H., Daily Totals		$3905.65		$5574.98	$69.74	$99.55

Crew R-12	Hr.	Daily	Hr.	Daily	Bare Costs	Incl. O&P
1 Carpenter Foreman (inside)	$53.65	$429.20	$80.60	$644.80	$49.91	$75.30
4 Carpenters	53.15	1700.80	79.85	2555.20		
4 Common Laborers	42.10	1347.20	63.25	2024.00		
1 Equip. Oper. (medium)	56.75	454.00	84.85	678.80		
1 Steel Worker	57.65	461.20	90.45	723.60		
1 Dozer, 200 H.P.		1504.00		1654.40		
1 Pickup Truck, 3/4 Ton		110.85		121.94	18.35	20.19
88 L.H., Daily Totals		$6007.25		$8402.74	$68.26	$95.49

Crew R-13	Hr.	Daily	Hr.	Daily	Bare Costs	Incl. O&P
1 Electrician Foreman	$61.85	$494.80	$92.10	$736.80	$59.29	$88.35
3 Electricians	61.35	1472.40	91.35	2192.40		
.25 Equip. Oper. (crane)	59.20	118.40	88.50	177.00		
1 Equipment Oiler	50.55	404.40	75.55	604.40		
.25 Hydraulic Crane, 33 Ton		236.34		259.97	5.63	6.19
42 L.H., Daily Totals		$2726.34		$3970.57	$64.91	$94.54

Crew R-15	Hr.	Daily	Hr.	Daily	Bare Costs	Incl. O&P
1 Electrician Foreman	$61.85	$494.80	$92.10	$736.80	$60.04	$89.45
4 Electricians	61.35	1963.20	91.35	2923.20		
1 Equipment Oper. (light)	53.00	424.00	79.20	633.60		
1 Telescoping Boom Lift, to 40'		278.90		306.79	5.81	6.39
48 L.H., Daily Totals		$3160.90		$4600.39	$65.85	$95.84

Crew R-15A	Hr.	Daily	Hr.	Daily	Bare Costs	Incl. O&P
1 Electrician Foreman	$61.85	$494.80	$92.10	$736.80	$53.63	$80.08
2 Electricians	61.35	981.60	91.35	1461.60		
2 Common Laborers	42.10	673.60	63.25	1012.00		
1 Equip. Oper. (light)	53.00	424.00	79.20	633.60		
1 Telescoping Boom Lift, to 40'		278.90		306.79	5.81	6.39
48 L.H., Daily Totals		$2852.90		$4150.79	$59.44	$86.47

Crew R-18	Hr.	Daily	Hr.	Daily	Bare Costs	Incl. O&P
.25 Electrician Foreman	$61.85	$123.70	$92.10	$184.20	$53.85	$80.18
1 Electrician	61.35	490.80	91.35	730.80		
2 Electrician Apprentices	49.10	785.60	73.10	1169.60		
26 L.H., Daily Totals		$1400.10		$2084.60	$53.85	$80.18

Crew R-19	Hr.	Daily	Hr.	Daily	Bare Costs	Incl. O&P
.5 Electrician Foreman	$61.85	$247.40	$92.10	$368.40	$61.45	$91.50
2 Electricians	61.35	981.60	91.35	1461.60		
20 L.H., Daily Totals		$1229.00		$1830.00	$61.45	$91.50

Crew R-21	Hr.	Daily	Hr.	Daily	Bare Costs	Incl. O&P
1 Electrician Foreman	$61.85	$494.80	$92.10	$736.80	$61.36	$91.37
3 Electricians	61.35	1472.40	91.35	2192.40		
.1 Equip. Oper. (medium)	56.75	45.40	84.85	67.88		
.1 S.P. Crane, 4x4, 25 Ton		114.40		125.84	3.49	3.84
32.8 L.H., Daily Totals		$2127.00		$3122.92	$64.85	$95.21

Crew R-22	Hr.	Daily	Hr.	Daily	Bare Costs	Incl. O&P
.66 Electrician Foreman	$61.85	$326.57	$92.10	$486.29	$56.16	$83.62
2 Electricians	61.35	981.60	91.35	1461.60		
2 Electrician Apprentices	49.10	785.60	73.10	1169.60		
37.28 L.H., Daily Totals		$2093.77		$3117.49	$56.16	$83.62

Crew R-30	Hr.	Daily	Hr.	Daily	Bare Costs	Incl. O&P
.25 Electrician Foreman (outside)	$63.35	$126.70	$94.35	$188.70	$49.66	$74.29
1 Electrician	61.35	490.80	91.35	730.80		
2 Laborers (Semi-Skilled)	42.10	673.60	63.25	1012.00		
26 L.H., Daily Totals		$1291.10		$1931.50	$49.66	$74.29

Crew R-31	Hr.	Daily	Hr.	Daily	Bare Costs	Incl. O&P
1 Electrician	$61.35	$490.80	$91.35	$730.80	$61.35	$91.35
1 Core Drill, Electric, 2.5 H.P.		62.50		68.75	7.81	8.59
8 L.H., Daily Totals		$553.30		$799.55	$69.16	$99.94

Crew W-41E	Hr.	Daily	Hr.	Daily	Bare Costs	Incl. O&P
.5 Plumber Foreman (outside)	$66.45	$265.80	$99.45	$397.80	$55.91	$83.77
1 Plumber	64.45	515.60	96.45	771.60		
1 Laborer	42.10	336.80	63.25	506.00		
20 L.H., Daily Totals		$1118.20		$1675.40	$55.91	$83.77

Historical Cost Indexes

The table below lists both the RSMeans® historical cost index based on Jan. 1, 1993 = 100 as well as the computed value of an index based on Jan. 1, 2020 costs. Since the Jan. 1, 2020 figure is estimated, space is left to write in the actual index figures as they become available through the quarterly *RSMeans Construction Cost Indexes*.

To compute the actual index based on Jan. 1, 2020 = 100, divide the historical cost index for a particular year by the actual Jan. 1, 2020 construction cost index. Space has been left to advance the index figures as the year progresses.

Year	Historical Cost Index Jan. 1, 1993 = 100		Current Index Based on Jan. 1, 2020 = 100		Year	Historical Cost Index Jan. 1, 1993 = 100	Current Index Based on Jan. 1, 2020 = 100		Year	Historical Cost Index Jan. 1, 1993 = 100	Current Index Based on Jan. 1, 2020 = 100	
	Est.	Actual	Est.	Actual		Actual	Est.	Actual		Actual	Est.	Actual
Oct 2020*					July 2005	151.6	63.4		July 1987	87.7	36.7	
July 2020*					2004	143.7	60.1		1986	84.2	35.2	
Apr 2020*					2003	132.0	55.2		1985	82.6	34.6	
Jan 2020*	239.1		100.0	100.0	2002	128.7	53.8		1984	82.0	34.3	
July 2019		232.2	97.1		2001	125.1	52.3		1983	80.2	33.5	
2018		222.9	93.2		2000	120.9	50.6		1982	76.1	31.8	
2017		213.6	89.3		1999	117.6	49.2		1981	70.0	29.3	
2016		207.3	86.7		1998	115.1	48.1		1980	62.9	26.3	
2015		206.2	86.2		1997	112.8	47.2		1979	57.8	24.2	
2014		204.9	85.7		1996	110.2	46.1		1978	53.5	22.4	
2013		201.2	84.1		1995	107.6	45.0		1977	49.5	20.7	
2012		194.6	81.4		1994	104.4	43.7		1976	46.9	19.6	
2011		191.2	80.0		1993	101.7	42.5		1975	44.8	18.7	
2010		183.5	76.7		1992	99.4	41.6		1974	41.4	17.3	
2009		180.1	75.3		1991	96.8	40.5		1973	37.7	15.8	
2008		180.4	75.4		1990	94.3	39.4		1972	34.8	14.6	
2007		169.4	70.8		1989	92.1	38.5		1971	32.1	13.4	
2006		162.0	67.8		1988	89.9	37.6		1970	28.7	12.0	

Adjustments to Costs

The "Historical Cost Index" can be used to convert national average building costs at a particular time to the approximate building costs for some other time.

Time Adjustment Using the Historical Cost Indexes:

$$\frac{\text{Index for Year A}}{\text{Index for Year B}} \times \text{Cost in Year B} = \text{Cost in Year A}$$

Example:

Estimate and compare construction costs for different years in the same city.

To estimate the national average construction cost of a building in 1970, knowing that it cost $900,000 in 2020:

INDEX in 1970 = 28.7

INDEX in 2020 = 239.1

$$\frac{\text{INDEX 1970}}{\text{INDEX 2020}} \times \text{Cost 2020} = \text{Cost 1970}$$

$$\frac{28.7}{239.1} \times \$900{,}000 = .120 \times \$900{,}000 = \$108{,}000$$

The construction cost of the building in 1970 was $108,000.

Note: The city cost indexes for Canada can be used to convert U.S. national averages to local costs in Canadian dollars.

Example:

To estimate and compare the cost of a building in Toronto, ON in 2020 with the known cost of $600,000 (US$) in New York, NY in 2020:

INDEX Toronto = 115.6

INDEX New York = 137.1

$$\frac{\text{INDEX Toronto}}{\text{INDEX New York}} \times \text{Cost New York} = \text{Cost Toronto}$$

$$\frac{115.6}{137.1} \times \$600{,}000 = .843 \times \$600{,}000 = \$505{,}908$$

The construction cost of the building in Toronto is $505,908 (CN$).

*Historical Cost Index updates and other resources are provided on the following website: **http://info.thegordiangroup.com/RSMeans.html**

How to Use the City Cost Indexes

What you should know before you begin

RSMeans City Cost Indexes (CCI) are an extremely useful tool for when you want to compare costs from city to city and region to region.

This publication contains average construction cost indexes for 731 U.S. and Canadian cities covering over 930 three-digit zip code locations, as listed directly under each city.

Keep in mind that a City Cost Index number is a percentage ratio of a specific city's cost to the national average cost of the same item at a stated time period.

In other words, these index figures represent relative construction factors (or, if you prefer, multipliers) for material and installation costs, as well as the weighted average for Total In Place costs for each CSI MasterFormat division. Installation costs include both labor and equipment rental costs. When estimating equipment rental rates only for a specific location, use 01 54 33 EQUIPMENT RENTAL COSTS in the Reference Section.

The 30 City Average Index is the average of 30 major U.S. cities and serves as a national average.

Index figures for both material and installation are based on the 30 major city average of 100 and represent the cost relationship as of July 1, 2019. The index for each division is computed from representative material and labor quantities for that division. The weighted average for each city is a weighted total of the components listed above it. It does not include relative productivity between trades or cities.

As changes occur in local material prices, labor rates, and equipment rental rates (including fuel costs), the impact of these changes should be accurately measured by the change in the City Cost Index for each particular city (as compared to the 30 city average).

Therefore, if you know (or have estimated) building costs in one city today, you can easily convert those costs to expected building costs in another city.

In addition, by using the Historical Cost Index, you can easily convert national average building costs at a particular time to the approximate building costs for some other time. The City Cost Indexes can then be applied to calculate the costs for a particular city.

Quick calculations

Location Adjustment Using the City Cost Indexes:

$$\frac{\text{Index for City A}}{\text{Index for City B}} \times \text{Cost in City B} = \text{Cost in City A}$$

Time Adjustment for the National Average Using the Historical Cost Index:

$$\frac{\text{Index for Year A}}{\text{Index for Year B}} \times \text{Cost in Year B} = \text{Cost in Year A}$$

Adjustment from the National Average:

$$\frac{\text{Index for City A}}{100} \times \text{National Average Cost} = \text{Cost in City A}$$

Since each of the other RSMeans data sets contains many different items, any *one* item multiplied by the particular city index may give incorrect results. However, the larger the number of items compiled, the closer the results should be to actual costs for that particular city.

The City Cost Indexes for Canadian cities are calculated using Canadian material and equipment prices and labor rates in Canadian dollars. Therefore, indexes for Canadian cities can be used to convert U.S. national average prices to local costs in Canadian dollars.

How to use this section

1. Compare costs from city to city.

In using the RSMeans Indexes, remember that an index number is not a fixed number but a ratio: It's a percentage ratio of a building component's cost at any stated time to the national average cost of that same component at the same time period. Put in the form of an equation:

$$\frac{\text{Specific City Cost}}{\text{National Average Cost}} \times 100 = \text{City Index Number}$$

Therefore, when making cost comparisons between cities, do not subtract one city's index number from the index number of another city and read the result as a percentage difference. Instead, divide one city's index number by that of the other city. The resulting number may then be used as a multiplier to calculate cost differences from city to city.

The formula used to find cost differences between cities for the purpose of comparison is as follows:

$$\frac{\text{City A Index}}{\text{City B Index}} \times \text{City B Cost (Known)} = \text{City A Cost (Unknown)}$$

In addition, you can use RSMeans CCI to calculate and compare costs division by division between cities using the same basic formula. (Just be sure that you're comparing similar divisions.)

2. Compare a specific city's construction costs with the national average.

When you're studying construction location feasibility, it's advisable to compare a prospective project's cost index with an index of the national average cost.

For example, divide the weighted average index of construction costs of a specific city by that of the 30 City Average, which = 100.

$$\frac{\text{City Index}}{100} = \text{\% of National Average}$$

As a result, you get a ratio that indicates the relative cost of construction in that city in comparison with the national average.

3. Convert U.S. national average to actual costs in Canadian City.

$$\frac{\text{Index for Canadian City}}{100} \times \text{National Average Cost} = \text{Cost in Canadian City in \$ CAN}$$

4. Adjust construction cost data based on a national average.

When you use a source of construction cost data which is based on a national average (such as RSMeans cost data), it is necessary to adjust those costs to a specific location.

$$\frac{\text{City Index}}{100} \times \frac{\text{Cost Based on}}{\text{National Average Costs}} = \frac{\text{City Cost}}{\text{(Unknown)}}$$

5. When applying the City Cost Indexes to demolition projects, use the appropriate division installation index. For example, for removal of existing doors and windows, use the Division 8 (Openings) index.

What you might like to know about how we developed the Indexes

The information presented in the CCI is organized according to the Construction Specifications Institute (CSI) MasterFormat 2018 classification system.

To create a reliable index, RSMeans researched the building type most often constructed in the United States and Canada. Because it was concluded that no one type of building completely represented the building construction industry, nine different types of buildings were combined to create a composite model.

The exact material, labor, and equipment quantities are based on detailed analyses of these nine building types, and then each quantity is weighted in proportion to expected usage. These various material items, labor hours, and equipment rental rates are thus combined to form a composite building representing as closely as possible the actual usage of materials, labor, and equipment in the North American building construction industry.

The following structures were chosen to make up that composite model:

1. Factory, 1 story
2. Office, 2–4 stories
3. Store, Retail
4. Town Hall, 2–3 stories
5. High School, 2–3 stories
6. Hospital, 4–8 stories
7. Garage, Parking
8. Apartment, 1–3 stories
9. Hotel/Motel, 2–3 stories

For the purposes of ensuring the timeliness of the data, the components of the index for the composite model have been streamlined. They currently consist of:

- specific quantities of 66 commonly used construction materials;
- specific labor-hours for 21 building construction trades; and
- specific days of equipment rental for 6 types of construction equipment (normally used to install the 66 material items by the 21 trades.) Fuel costs and routine maintenance costs are included in the equipment cost.

Material and equipment price quotations are gathered quarterly from cities in the United States and Canada. These prices and the latest negotiated labor wage rates for 21 different building trades are used to compile the quarterly update of the City Cost Index.

The 30 major U.S. cities used to calculate the national average are:

Atlanta, GA
Baltimore, MD
Boston, MA
Buffalo, NY
Chicago, IL
Cincinnati, OH
Cleveland, OH
Columbus, OH
Dallas, TX
Denver, CO
Detroit, MI
Houston, TX
Indianapolis, IN
Kansas City, MO
Los Angeles, CA
Memphis, TN
Milwaukee, WI
Minneapolis, MN
Nashville, TN
New Orleans, LA
New York, NY
Philadelphia, PA
Phoenix, AZ
Pittsburgh, PA
St. Louis, MO
San Antonio, TX
San Diego, CA
San Francisco, CA
Seattle, WA
Washington, DC

What the CCI does not indicate

The weighted average for each city is a total of the divisional components weighted to reflect typical usage. It does not include the productivity variations between trades or cities.

In addition, the CCI does not take into consideration factors such as the following:

- managerial efficiency
- competitive conditions
- automation
- restrictive union practices
- unique local requirements
- regional variations due to specific building codes

City Cost Indexes

	DIVISION	UNITED STATES			ALABAMA														
		30 CITY			ANNISTON			BIRMINGHAM			BUTLER			DECATUR			DOTHAN		
		AVERAGE			362			350 - 352			369			356			363		
		MAT.	INST.	TOTAL	MAT.	INST.	TOTAL	MAT.	INST.	TOTAL	MAT.	INST.	TOTAL	MAT.	INST.	TOTAL	MAT.	INST.	TOTAL
015433	**CONTRACTOR EQUIPMENT**		100.0	100.0		101.9	101.9		104.7	104.7		99.5	99.5		101.9	101.9		99.5	99.5
0241, 31 - 34	**SITE & INFRASTRUCTURE, DEMOLITION**	100.0	100.0	100.0	90.9	89.5	89.9	91.2	94.5	93.5	105.4	84.4	90.9	84.5	88.1	87.0	102.9	84.8	90.4
0310	Concrete Forming & Accessories	100.0	100.0	100.0	85.6	68.7	71.2	90.6	69.7	72.8	82.3	68.8	70.8	90.9	62.8	66.9	90.3	69.1	72.2
0320	Concrete Reinforcing	100.0	100.0	100.0	88.8	71.2	80.3	94.2	71.3	83.1	93.9	70.5	82.6	88.1	67.7	78.2	93.9	71.3	83.0
0330	Cast-in-Place Concrete	100.0	100.0	100.0	82.3	67.7	76.9	103.8	69.3	90.9	80.3	67.8	75.6	101.9	67.2	89.0	80.3	67.7	75.6
03	**CONCRETE**	100.0	100.0	100.0	89.2	70.4	81.0	92.2	71.4	83.1	90.3	70.4	81.6	92.0	67.0	81.0	89.5	70.7	81.2
04	**MASONRY**	100.0	100.0	100.0	89.2	63.1	73.2	87.1	64.3	73.1	93.7	63.1	74.9	85.2	62.4	71.2	95.1	62.9	75.3
05	**METALS**	100.0	100.0	100.0	101.8	94.8	99.7	100.5	94.2	98.7	100.7	94.8	98.9	102.9	93.1	100.0	100.7	95.4	99.1
06	**WOOD, PLASTICS & COMPOSITES**	100.0	100.0	100.0	82.0	69.9	75.3	91.0	70.0	79.4	77.1	69.9	73.1	95.7	62.1	77.2	88.1	69.9	78.1
07	**THERMAL & MOISTURE PROTECTION**	100.0	100.0	100.0	96.0	63.0	81.6	94.0	67.6	82.5	96.1	65.9	82.9	93.1	64.5	80.7	96.0	65.7	82.8
08	**OPENINGS**	100.0	100.0	100.0	94.7	69.2	88.7	102.6	69.8	94.9	94.7	69.5	88.8	106.9	64.6	97.0	94.7	69.7	88.9
0920	Plaster & Gypsum Board	100.0	100.0	100.0	81.8	69.5	73.5	88.9	69.5	75.9	79.3	69.5	72.7	92.7	61.6	71.7	88.7	69.5	75.8
0950, 0980	Ceilings & Acoustic Treatment	100.0	100.0	100.0	77.1	69.5	72.0	84.8	69.5	74.5	77.1	69.5	72.0	83.2	61.6	68.6	77.1	69.5	72.0
0960	Flooring	100.0	100.0	100.0	82.0	69.1	78.3	99.6	69.1	91.0	86.5	69.1	81.6	89.9	69.1	84.1	91.6	69.1	85.3
0970, 0990	Wall Finishes & Painting/Coating	100.0	100.0	100.0	89.1	66.7	75.9	87.0	66.7	75.0	89.1	45.9	63.5	79.4	61.0	68.5	89.1	80.4	84.0
09	**FINISHES**	100.0	100.0	100.0	80.4	68.8	74.1	91.2	69.2	79.3	83.6	66.6	74.4	85.3	63.5	73.5	86.1	70.3	77.6
COVERS	**DIVS. 10 - 14, 25, 28, 41, 43, 44, 46**	100.0	100.0	100.0	100.0	74.0	94.2	100.0	83.6	96.3	100.0	74.0	94.2	100.0	72.9	94.0	100.0	74.0	94.2
21, 22, 23	**FIRE SUPPRESSION, PLUMBING & HVAC**	100.0	100.0	100.0	101.0	51.8	81.7	100.0	66.3	86.8	98.1	67.0	85.9	100.0	64.5	86.1	98.1	65.0	85.1
26, 27, 3370	**ELECTRICAL, COMMUNICATIONS & UTIL.**	100.0	100.0	100.0	97.7	59.9	79.1	98.4	59.7	79.3	99.8	61.7	81.0	94.5	63.8	79.4	98.6	76.1	87.5
MF2018	**WEIGHTED AVERAGE**	100.0	100.0	100.0	95.9	67.7	84.0	97.4	71.8	86.6	96.1	70.6	85.3	97.1	69.2	85.3	96.2	72.7	86.3

	DIVISION	ALABAMA																	
		EVERGREEN			GADSDEN			HUNTSVILLE			JASPER			MOBILE			MONTGOMERY		
		364			359			357 - 358			355			365 - 366			360 - 361		
		MAT.	INST.	TOTAL	MAT.	INST.	TOTAL	MAT.	INST.	TOTAL	MAT.	INST.	TOTAL	MAT.	INST.	TOTAL	MAT.	INST.	TOTAL
015433	**CONTRACTOR EQUIPMENT**		99.5	99.5		101.9	101.9		101.9	101.9		101.9	101.9		99.5	99.5		102.0	102.0
0241, 31 - 34	**SITE & INFRASTRUCTURE, DEMOLITION**	105.9	84.4	91.0	89.9	89.5	89.6	84.2	89.4	87.8	89.9	89.5	89.6	97.7	85.4	89.2	95.6	89.9	91.7
0310	Concrete Forming & Accessories	79.4	68.6	70.2	83.9	69.1	71.3	90.9	65.9	69.6	88.5	64.7	68.3	89.4	68.4	71.5	92.8	69.4	72.8
0320	Concrete Reinforcing	94.0	70.4	82.6	93.5	71.3	82.8	88.1	76.3	82.4	88.1	71.3	80.0	91.5	70.5	81.3	99.7	71.4	86.0
0330	Cast-in-Place Concrete	80.3	67.6	75.6	101.9	67.8	89.2	99.2	67.8	87.5	113.0	67.8	96.1	84.2	66.9	77.7	81.5	68.7	76.7
03	**CONCRETE**	90.7	70.2	81.7	96.5	70.7	85.1	90.8	70.1	81.7	100.1	68.7	86.3	85.5	69.9	78.7	84.1	71.1	78.4
04	**MASONRY**	93.8	62.9	74.8	83.8	63.1	71.1	86.3	63.3	72.2	81.4	63.1	70.2	92.3	61.4	73.3	89.4	63.2	73.3
05	**METALS**	100.7	94.6	98.9	100.7	95.2	99.1	102.9	96.7	101.0	100.7	95.1	99.0	102.8	94.9	100.5	101.7	94.3	99.5
06	**WOOD, PLASTICS & COMPOSITES**	73.8	69.9	71.6	86.7	69.9	77.4	95.7	65.3	79.0	93.0	64.1	77.1	86.7	69.9	77.4	90.3	70.0	79.1
07	**THERMAL & MOISTURE PROTECTION**	96.0	65.4	82.7	93.3	66.3	81.6	93.0	65.9	81.2	93.3	64.4	80.7	95.6	65.4	82.4	94.2	66.9	82.3
08	**OPENINGS**	94.7	69.3	88.8	103.4	69.7	95.6	106.6	68.5	97.7	103.4	66.6	94.8	97.3	69.5	90.8	96.2	69.8	90.0
0920	Plaster & Gypsum Board	78.3	69.5	72.4	85.1	69.5	74.6	92.7	64.8	73.9	89.5	63.6	72.1	85.6	69.5	74.8	85.9	69.5	74.9
0950, 0980	Ceilings & Acoustic Treatment	77.1	69.5	72.0	79.1	69.5	72.7	85.0	64.8	71.4	79.1	63.6	68.6	83.0	69.5	73.9	84.6	69.5	74.4
0960	Flooring	84.5	69.1	80.1	86.4	69.1	81.5	89.9	69.1	84.1	88.4	69.1	82.9	91.1	69.1	84.9	91.7	69.1	85.3
0970, 0990	Wall Finishes & Painting/Coating	89.1	45.9	63.5	79.4	57.4	66.4	79.4	63.7	70.1	79.4	66.7	71.9	92.3	45.9	64.8	91.1	66.7	76.7
09	**FINISHES**	82.9	66.5	74.1	82.9	67.8	74.7	85.7	65.9	75.0	84.0	65.4	73.9	86.2	66.2	75.4	88.3	68.9	77.8
COVERS	**DIVS. 10 - 14, 25, 28, 41, 43, 44, 46**	100.0	73.9	94.2	100.0	82.9	96.2	100.0	82.5	96.1	100.0	73.4	94.1	100.0	82.5	96.1	100.0	83.3	96.3
21, 22, 23	**FIRE SUPPRESSION, PLUMBING & HVAC**	98.1	59.8	83.0	102.3	62.3	86.6	100.0	66.9	87.0	102.3	65.7	87.9	100.0	61.4	84.8	100.0	64.3	86.0
26, 27, 3370	**ELECTRICAL, COMMUNICATIONS & UTIL.**	97.1	56.8	77.2	94.5	59.7	77.4	95.4	65.9	80.8	94.2	59.9	77.3	100.4	56.8	78.9	101.1	76.1	88.8
MF2018	**WEIGHTED AVERAGE**	95.8	68.3	84.2	97.3	70.2	85.9	97.1	71.7	86.4	97.8	69.8	86.0	96.7	68.7	84.9	96.3	73.1	86.5

	DIVISION	ALABAMA									ALASKA								
		PHENIX CITY			SELMA			TUSCALOOSA			ANCHORAGE			FAIRBANKS			JUNEAU		
		368			367			354			995 - 996			997			998		
		MAT.	INST.	TOTAL	MAT.	INST.	TOTAL	MAT.	INST.	TOTAL	MAT.	INST.	TOTAL	MAT.	INST.	TOTAL	MAT.	INST.	TOTAL
015433	**CONTRACTOR EQUIPMENT**		99.5	99.5		99.5	99.5		101.9	101.9		110.5	110.5		112.9	112.9		110.5	110.5
0241, 31 - 34	**SITE & INFRASTRUCTURE, DEMOLITION**	109.8	85.5	93.0	102.7	85.5	90.8	84.8	89.5	88.0	118.4	121.7	120.7	117.8	124.5	122.5	133.6	121.7	125.4
0310	Concrete Forming & Accessories	85.6	67.6	70.2	83.5	69.1	71.2	90.8	69.2	72.4	120.5	116.2	116.8	124.7	115.0	116.4	125.7	116.2	117.6
0320	Concrete Reinforcing	93.9	66.9	80.9	93.9	71.3	83.0	88.1	71.3	80.0	148.4	120.7	135.0	147.2	120.7	134.4	138.9	120.7	130.1
0330	Cast-in-Place Concrete	80.3	67.9	75.7	80.3	67.8	75.6	103.4	67.9	90.2	105.1	117.5	109.7	109.7	115.6	111.9	116.1	117.5	116.6
03	**CONCRETE**	93.6	69.3	82.9	89.0	70.7	81.0	92.7	70.8	83.1	107.5	116.7	111.6	101.4	115.5	107.6	114.1	116.7	115.2
04	**MASONRY**	93.7	63.1	74.9	97.4	63.1	76.3	85.4	63.1	71.7	171.5	119.4	139.5	181.4	117.8	142.3	162.5	119.4	136.0
05	**METALS**	100.6	93.7	98.6	100.6	95.2	99.0	102.1	95.3	100.1	127.3	104.4	120.5	122.6	105.4	117.5	113.1	104.4	110.6
06	**WOOD, PLASTICS & COMPOSITES**	81.7	67.7	74.0	78.9	69.9	73.9	95.7	69.9	81.5	112.3	113.5	112.9	126.9	112.7	119.1	121.9	113.5	117.3
07	**THERMAL & MOISTURE PROTECTION**	96.5	66.1	83.2	95.9	65.9	82.8	93.1	66.3	81.4	176.9	116.9	150.8	186.2	115.8	155.6	188.4	116.9	157.3
08	**OPENINGS**	94.7	67.4	88.3	94.7	69.7	88.8	106.6	69.7	98.0	130.8	115.8	127.3	132.8	114.7	128.6	131.0	115.8	127.5
0920	Plaster & Gypsum Board	83.0	67.3	72.4	81.1	69.5	73.3	92.7	69.5	77.1	144.1	113.8	123.7	177.5	112.9	134.0	158.2	113.8	128.3
0950, 0980	Ceilings & Acoustic Treatment	77.1	67.3	70.5	77.1	69.5	72.0	85.0	69.5	74.6	131.3	113.8	119.5	123.9	112.9	116.5	136.1	113.8	121.0
0960	Flooring	88.4	69.1	83.0	87.0	69.1	81.9	89.9	69.1	84.1	118.3	119.4	118.6	116.7	119.4	117.4	125.0	119.4	123.5
0970, 0990	Wall Finishes & Painting/Coating	89.1	79.0	83.1	89.1	66.7	75.9	79.4	66.7	71.9	113.1	119.1	116.6	111.6	116.9	114.7	108.7	119.1	114.8
09	**FINISHES**	85.2	68.9	76.4	83.7	68.8	75.7	85.6	68.8	76.6	127.0	117.1	121.7	128.6	116.0	121.8	128.9	117.1	122.5
COVERS	**DIVS. 10 - 14, 25, 28, 41, 43, 44, 46**	100.0	82.7	96.1	100.0	70.4	93.4	100.0	82.9	96.2	100.0	114.1	103.2	100.0	113.6	103.0	100.0	114.1	103.2
21, 22, 23	**FIRE SUPPRESSION, PLUMBING & HVAC**	98.1	64.2	84.8	98.1	65.1	85.1	100.0	67.1	87.1	100.5	107.1	103.1	100.2	107.9	103.2	100.5	107.1	103.1
26, 27, 3370	**ELECTRICAL, COMMUNICATIONS & UTIL.**	99.2	65.4	82.6	98.3	76.1	87.3	95.0	59.7	77.6	117.2	109.4	113.3	124.2	109.4	116.9	110.2	109.4	109.8
MF2018	**WEIGHTED AVERAGE**	96.8	70.8	85.8	95.9	72.5	86.0	97.1	71.4	86.3	118.5	113.1	116.2	118.8	113.0	116.3	116.8	113.1	115.2

	DIVISION	ALASKA			ARIZONA														
		KETCHIKAN			CHAMBERS			FLAGSTAFF			GLOBE			KINGMAN			MESA/TEMPE		
		999			865			860			855			864			852		
		MAT.	INST.	TOTAL	MAT.	INST.	TOTAL	MAT.	INST.	TOTAL	MAT.	INST.	TOTAL	MAT.	INST.	TOTAL	MAT.	INST.	TOTAL
015433	CONTRACTOR EQUIPMENT		112.9	112.9		88.2	88.2		88.2	88.2		89.3	89.3		88.2	88.2		89.3	89.3
0241, 31 - 34	SITE & INFRASTRUCTURE, DEMOLITION	169.1	124.6	138.4	71.5	90.1	84.4	91.2	90.1	90.4	109.0	91.0	96.5	71.4	90.1	84.3	97.8	91.0	93.1
0310	Concrete Forming & Accessories	116.3	116.2	116.2	99.2	70.3	74.6	104.6	67.4	72.8	93.5	70.4	73.8	97.4	66.1	70.7	96.5	70.5	74.3
0320	Concrete Reinforcing	110.6	120.7	115.5	105.5	76.9	91.7	105.4	76.9	91.6	110.5	76.9	94.2	105.7	76.9	91.7	111.2	76.9	94.6
0330	Cast-in-Place Concrete	221.8	117.3	182.9	89.8	67.3	81.4	89.8	67.4	81.5	83.3	67.0	77.2	89.5	67.3	81.2	84.0	67.0	77.7
03	CONCRETE	170.9	116.7	147.1	93.4	70.3	83.3	114.1	69.0	94.3	101.9	70.3	88.0	93.1	68.4	82.3	93.1	70.4	83.1
04	MASONRY	188.6	119.4	146.1	96.0	59.4	73.5	96.1	59.4	73.6	101.7	59.3	75.7	96.0	59.4	73.5	101.9	59.3	75.8
05	METALS	122.8	105.4	117.7	103.0	71.6	93.8	103.5	71.7	94.2	105.8	72.5	96.0	103.7	71.6	94.3	106.2	72.6	96.3
06	WOOD, PLASTICS & COMPOSITES	117.1	113.5	115.1	101.8	73.0	85.9	108.2	68.8	86.5	92.7	73.1	81.9	96.4	67.3	80.4	96.7	73.1	83.7
07	THERMAL & MOISTURE PROTECTION	191.6	116.4	158.9	99.9	69.7	86.8	101.8	69.3	87.7	99.1	68.3	85.7	99.9	69.1	86.5	99.0	68.3	85.6
08	OPENINGS	128.6	115.8	125.6	105.8	69.3	97.3	105.9	70.1	97.6	94.9	69.4	88.9	106.0	67.9	97.1	94.9	69.4	89.0
0920	Plaster & Gypsum Board	161.0	113.8	129.2	92.8	72.5	79.1	96.5	68.2	77.5	88.2	72.5	77.6	84.2	66.6	72.4	91.7	72.5	78.7
0950, 0980	Ceilings & Acoustic Treatment	118.4	113.8	115.3	106.8	72.5	83.6	107.6	68.2	81.1	95.5	72.5	80.0	107.6	66.6	80.0	95.5	72.5	80.0
0960	Flooring	116.4	119.4	117.2	89.7	64.5	82.6	92.0	64.5	84.3	98.6	64.5	89.0	88.4	64.5	81.6	100.4	64.5	90.3
0970, 0990	Wall Finishes & Painting/Coating	111.6	119.1	116.0	87.9	58.6	70.5	87.9	58.6	70.5	90.6	58.6	71.6	87.9	58.6	70.5	90.6	58.6	71.6
09	FINISHES	129.2	117.1	122.7	91.3	68.1	78.8	94.5	65.7	78.9	96.6	68.2	81.2	89.9	64.8	76.3	96.4	68.2	81.2
COVERS	DIVS. 10 - 14, 25, 28, 41, 43, 44, 46	100.0	114.2	103.2	100.0	81.8	95.9	100.0	81.3	95.8	100.0	82.1	96.0	100.0	81.2	95.8	100.0	82.1	96.0
21, 22, 23	FIRE SUPPRESSION, PLUMBING & HVAC	98.1	107.2	101.7	97.7	76.8	89.5	100.3	77.0	91.1	96.6	76.9	88.9	97.7	76.8	89.5	100.1	77.0	91.1
26, 27, 3370	ELECTRICAL, COMMUNICATIONS & UTIL.	124.2	109.4	116.9	102.3	66.5	84.7	101.3	60.7	81.3	97.1	62.9	80.2	102.3	62.9	82.9	94.2	62.9	78.7
MF2018	WEIGHTED AVERAGE	128.4	113.4	122.1	98.3	71.8	87.1	102.4	70.5	88.9	99.5	71.4	87.6	98.3	70.4	86.5	98.7	71.5	87.2

	DIVISION	ARIZONA												ARKANSAS					
		PHOENIX			PRESCOTT			SHOW LOW			TUCSON			BATESVILLE			CAMDEN		
		850,853			863			859			856 - 857			725			717		
		MAT.	INST.	TOTAL	MAT.	INST.	TOTAL	MAT.	INST.	TOTAL	MAT.	INST.	TOTAL	MAT.	INST.	TOTAL	MAT.	INST.	TOTAL
015433	CONTRACTOR EQUIPMENT		92.5	92.5		88.2	88.2		89.3	89.3		89.3	89.3		89.7	89.7		89.7	89.7
0241, 31 - 34	SITE & INFRASTRUCTURE, DEMOLITION	98.3	94.3	95.6	78.4	90.1	86.5	111.4	91.0	97.3	92.9	91.0	91.6	73.0	85.7	81.8	77.2	85.8	83.1
0310	Concrete Forming & Accessories	101.2	70.8	75.3	100.8	70.3	74.8	100.6	70.5	74.9	97.1	67.5	71.8	82.5	61.4	64.5	78.6	61.9	64.4
0320	Concrete Reinforcing	109.1	76.9	93.6	105.4	76.9	91.6	111.2	76.9	94.6	92.3	76.9	84.9	84.4	69.4	77.1	91.0	69.4	80.5
0330	Cast-in-Place Concrete	83.6	67.8	77.7	89.8	67.3	81.4	83.3	67.0	77.3	86.5	67.0	79.2	72.8	74.6	73.5	76.4	74.6	75.7
03	CONCRETE	95.5	70.8	84.6	99.0	70.3	86.4	104.2	70.3	89.3	91.6	69.0	81.7	73.2	68.1	70.9	75.7	68.3	72.4
04	MASONRY	94.8	62.1	74.7	96.1	59.4	73.6	101.8	59.3	75.7	88.9	59.3	70.7	94.1	60.8	73.6	104.7	60.8	77.7
05	METALS	107.8	73.7	97.8	103.5	71.6	94.2	105.6	72.5	95.9	107.0	72.6	96.9	94.7	76.0	89.2	101.1	76.1	93.8
06	WOOD, PLASTICS & COMPOSITES	102.1	72.0	85.5	103.3	73.0	86.6	101.2	73.1	85.7	96.9	69.0	81.6	91.1	63.5	75.9	89.2	64.0	75.4
07	THERMAL & MOISTURE PROTECTION	100.3	69.6	86.9	100.5	69.7	87.1	99.3	68.3	85.8	99.9	67.9	86.0	102.7	61.9	84.9	97.5	61.9	82.0
08	OPENINGS	100.7	71.4	93.9	105.9	69.3	97.4	94.2	71.1	88.8	91.4	70.2	86.5	100.0	61.0	90.9	103.1	61.3	93.4
0920	Plaster & Gypsum Board	99.4	71.2	80.4	93.3	72.5	79.3	93.5	72.5	79.4	96.3	68.2	77.4	73.7	62.9	66.5	84.7	63.5	70.4
0950, 0980	Ceilings & Acoustic Treatment	106.5	71.2	82.7	106.0	72.5	83.4	95.5	72.5	80.0	96.4	68.2	77.4	88.1	62.9	71.1	86.3	63.5	70.9
0960	Flooring	102.4	64.5	91.7	90.6	64.5	83.2	102.2	64.5	91.6	91.8	64.8	84.3	81.8	70.9	78.7	86.1	70.9	81.9
0970, 0990	Wall Finishes & Painting/Coating	95.3	57.2	72.8	87.9	58.6	70.5	90.6	58.6	71.6	91.3	58.6	71.9	92.2	54.5	69.9	88.8	54.5	68.5
09	FINISHES	101.0	68.1	83.2	92.0	68.1	79.1	98.6	68.2	82.2	94.5	65.9	79.0	76.3	62.6	68.9	79.8	62.9	70.7
COVERS	DIVS. 10 - 14, 25, 28, 41, 43, 44, 46	100.0	83.2	96.2	100.0	81.8	95.9	100.0	82.1	96.0	100.0	81.7	95.9	100.0	66.4	92.5	100.0	66.5	92.5
21, 22, 23	FIRE SUPPRESSION, PLUMBING & HVAC	100.0	78.5	91.6	100.3	76.8	91.0	96.6	76.9	88.9	100.1	74.8	90.2	96.5	51.2	78.7	96.4	57.5	81.2
26, 27, 3370	ELECTRICAL, COMMUNICATIONS & UTIL.	99.8	60.7	80.5	101.0	62.9	82.2	94.5	62.9	78.9	96.2	60.6	78.7	95.5	60.5	78.3	95.8	58.1	77.2
MF2018	WEIGHTED AVERAGE	100.6	72.3	88.6	99.9	71.3	87.8	99.7	71.5	87.8	97.7	70.2	86.1	91.6	63.6	79.8	94.0	64.7	81.6

	DIVISION	ARKANSAS																	
		FAYETTEVILLE			FORT SMITH			HARRISON			HOT SPRINGS			JONESBORO			LITTLE ROCK		
		727			729			726			719			724			720 - 722		
		MAT.	INST.	TOTAL	MAT.	INST.	TOTAL	MAT.	INST.	TOTAL	MAT.	INST.	TOTAL	MAT.	INST.	TOTAL	MAT.	INST.	TOTAL
015433	CONTRACTOR EQUIPMENT		89.7	89.7		89.7	89.7		89.7	89.7		89.7	89.7		112.7	112.7		92.4	92.4
0241, 31 - 34	SITE & INFRASTRUCTURE, DEMOLITION	72.5	85.7	81.6	77.7	85.5	83.1	77.7	85.7	83.3	80.2	85.7	84.0	96.7	102.4	100.7	85.3	90.3	88.8
0310	Concrete Forming & Accessories	78.4	61.8	64.3	95.7	61.6	66.7	86.5	61.8	65.4	76.3	61.8	64.0	85.7	61.9	65.4	93.1	62.4	66.9
0320	Concrete Reinforcing	84.4	62.1	73.6	85.4	66.3	76.2	84.0	69.4	76.9	89.3	69.4	79.7	81.5	65.1	73.6	90.4	69.5	80.3
0330	Cast-in-Place Concrete	72.9	74.6	73.5	83.3	75.5	80.4	80.8	74.5	78.5	78.1	74.6	76.8	79.4	75.8	78.0	80.7	76.9	79.3
03	CONCRETE	72.9	67.0	70.3	80.1	68.0	74.8	79.7	68.2	74.6	78.9	68.2	74.2	77.3	69.0	73.7	81.0	69.3	75.9
04	MASONRY	85.0	60.8	70.1	91.6	60.8	72.7	94.3	60.8	73.7	78.6	60.8	67.7	86.5	60.8	70.7	87.7	62.1	72.0
05	METALS	94.7	73.6	88.5	96.9	74.9	90.4	95.8	76.0	90.0	101.1	76.1	93.7	91.3	89.4	90.8	97.2	76.1	91.0
06	WOOD, PLASTICS & COMPOSITES	87.7	64.0	74.7	107.1	64.0	83.4	96.8	64.0	78.7	86.7	64.0	74.2	95.2	64.4	78.3	98.4	64.2	79.5
07	THERMAL & MOISTURE PROTECTION	103.5	61.9	85.4	103.8	61.7	85.5	103.0	61.9	85.1	97.8	61.9	82.2	108.4	61.9	88.2	98.5	63.1	83.1
08	OPENINGS	100.0	59.6	90.6	102.0	60.1	92.3	100.8	61.3	91.6	103.1	61.3	93.3	105.6	60.5	95.1	95.6	60.7	87.5
0920	Plaster & Gypsum Board	73.0	63.5	66.6	79.7	63.5	68.8	78.8	63.5	68.5	83.4	63.5	70.0	87.3	63.5	71.2	92.0	63.5	72.8
0950, 0980	Ceilings & Acoustic Treatment	88.1	63.5	71.5	89.8	63.5	72.0	89.8	63.5	72.0	86.3	63.5	70.9	92.1	63.5	72.8	87.3	63.5	71.2
0960	Flooring	78.9	70.9	76.7	88.3	70.9	83.4	84.1	70.9	80.4	85.1	70.9	81.1	59.3	70.9	62.6	89.2	81.0	86.9
0970, 0990	Wall Finishes & Painting/Coating	92.2	54.5	69.9	92.2	52.1	68.5	92.2	54.5	69.9	88.8	54.5	68.5	81.7	54.5	65.6	92.8	52.4	68.9
09	FINISHES	75.5	62.9	68.7	79.6	62.6	70.4	78.3	62.9	70.0	79.6	62.9	70.6	74.6	63.2	68.4	85.3	64.9	74.3
COVERS	DIVS. 10 - 14, 25, 28, 41, 43, 44, 46	100.0	66.1	92.4	100.0	77.7	95.0	100.0	68.5	93.0	100.0	66.2	92.5	100.0	67.0	92.6	100.0	78.2	95.1
21, 22, 23	FIRE SUPPRESSION, PLUMBING & HVAC	96.5	60.0	82.2	100.0	48.0	79.6	96.5	49.1	77.8	96.4	51.5	78.8	100.4	51.2	81.1	99.9	49.1	79.9
26, 27, 3370	ELECTRICAL, COMMUNICATIONS & UTIL.	90.0	53.9	72.2	93.1	57.0	75.3	94.2	57.5	76.1	97.7	65.2	81.7	99.0	62.3	81.0	100.1	66.7	83.7
MF2018	WEIGHTED AVERAGE	90.5	64.1	79.3	94.1	62.6	80.8	92.9	62.8	80.2	93.5	64.4	81.2	93.8	66.6	82.3	94.5	65.4	82.2

DIVISION		ARKANSAS												CALIFORNIA					
		PINE BLUFF			RUSSELLVILLE			TEXARKANA			WEST MEMPHIS			ALHAMBRA			ANAHEIM		
		716			728			718			723			917 - 918			928		
		MAT.	INST.	TOTAL	MAT.	INST.	TOTAL	MAT.	INST.	TOTAL	MAT.	INST.	TOTAL	MAT.	INST.	TOTAL	MAT.	INST.	TOTAL
015433	**CONTRACTOR EQUIPMENT**		89.7	89.7		89.7	89.7		90.9	90.9		112.7	112.7		95.4	95.4		99.2	99.2
0241, 31 - 34	**SITE & INFRASTRUCTURE, DEMOLITION**	82.3	85.6	84.6	74.2	85.7	82.2	92.9	87.8	89.3	103.1	102.8	102.9	100.8	105.3	103.9	97.7	105.5	103.1
0310	Concrete Forming & Accessories	76.1	61.9	64.0	83.1	61.7	64.8	81.4	61.5	64.5	90.6	62.2	66.4	114.4	139.5	135.8	102.9	139.7	134.3
0320	Concrete Reinforcing	90.9	69.4	80.5	84.9	69.4	77.4	90.5	69.3	80.2	81.5	63.3	72.7	105.0	133.6	118.9	95.2	133.5	113.8
0330	Cast-in-Place Concrete	78.1	74.6	76.8	76.3	74.5	75.6	85.2	74.4	81.2	83.2	75.9	80.5	83.3	127.8	99.9	86.7	131.0	103.2
03	**CONCRETE**	79.7	68.3	74.7	76.0	68.1	72.6	77.8	68.1	73.5	83.8	68.8	77.2	94.2	133.1	111.3	96.2	134.3	112.9
04	**MASONRY**	110.7	60.8	80.0	90.9	60.8	72.4	92.6	60.8	73.0	75.3	60.8	66.4	106.9	139.4	126.9	78.2	138.0	115.0
05	**METALS**	101.9	76.1	94.3	94.7	75.9	89.1	93.8	75.8	88.5	90.4	89.1	90.0	79.9	113.2	89.7	107.7	114.2	109.6
06	**WOOD, PLASTICS & COMPOSITES**	86.4	64.0	74.1	92.3	64.0	76.8	93.4	64.0	77.2	100.7	64.4	80.7	99.3	138.0	120.6	97.5	138.3	119.9
07	**THERMAL & MOISTURE PROTECTION**	97.9	62.0	82.3	103.7	61.9	85.5	98.5	61.9	82.6	108.9	61.9	88.4	104.2	131.8	116.2	108.9	134.9	120.2
08	**OPENINGS**	104.3	60.8	94.1	100.0	61.3	91.0	109.2	61.3	98.0	103.2	60.1	93.1	87.8	137.4	99.4	103.9	137.6	111.8
0920	Plaster & Gypsum Board	83.1	63.5	69.9	73.7	63.5	66.8	85.7	63.5	70.7	89.2	63.5	71.9	94.5	139.3	124.6	112.7	139.3	130.6
0950, 0980	Ceilings & Acoustic Treatment	86.3	63.5	70.9	88.1	63.5	71.5	89.6	63.5	72.0	90.2	63.5	72.2	111.0	139.3	130.1	109.3	139.3	129.5
0960	Flooring	84.9	70.9	80.9	81.3	70.9	78.4	86.9	43.4	74.7	61.6	70.9	64.2	104.8	121.1	109.4	99.2	121.1	105.3
0970, 0990	Wall Finishes & Painting/Coating	88.8	52.1	67.1	92.2	54.5	69.9	88.8	54.5	68.5	81.7	54.5	65.6	103.9	120.3	113.6	92.6	120.3	109.0
09	**FINISHES**	79.6	62.6	70.4	76.4	62.9	69.1	81.7	57.2	68.4	75.9	63.2	69.0	101.8	134.1	119.3	99.4	134.2	118.3
COVERS	**DIVS. 10 - 14, 25, 28, 41, 43, 44, 46**	100.0	77.7	95.0	100.0	66.5	92.5	100.0	66.1	92.5	100.0	67.0	92.6	100.0	119.6	104.4	100.0	120.2	104.5
21, 22, 23	**FIRE SUPPRESSION, PLUMBING & HVAC**	100.0	51.4	80.9	96.5	50.9	78.6	100.0	54.5	82.1	96.8	64.6	84.2	96.5	130.9	110.0	99.9	130.9	112.1
26, 27, 3370	**ELECTRICAL, COMMUNICATIONS & UTIL.**	95.9	57.5	77.0	93.1	57.5	75.6	97.6	57.5	77.9	100.5	65.3	83.1	121.0	129.2	125.0	91.5	113.3	102.2
MF2018	**WEIGHTED AVERAGE**	96.0	63.6	82.3	91.7	63.1	79.6	94.8	63.3	81.5	93.3	69.8	83.3	96.7	128.6	110.2	99.3	126.6	110.9

DIVISION		CALIFORNIA																	
		BAKERSFIELD			BERKELEY			EUREKA			FRESNO			INGLEWOOD			LONG BEACH		
		932 - 933			947			955			936 - 938			903 - 905			906 - 908		
		MAT.	INST.	TOTAL	MAT.	INST.	TOTAL	MAT.	INST.	TOTAL	MAT.	INST.	TOTAL	MAT.	INST.	TOTAL	MAT.	INST.	TOTAL
015433	**CONTRACTOR EQUIPMENT**		98.7	98.7		100.2	100.2		97.1	97.1		97.3	97.3		96.9	96.9		96.9	96.9
0241, 31 - 34	**SITE & INFRASTRUCTURE, DEMOLITION**	96.1	106.8	103.5	109.8	106.0	107.2	109.0	102.7	104.7	98.9	103.2	101.8	87.7	102.0	97.6	94.7	102.0	99.7
0310	Concrete Forming & Accessories	105.8	139.2	134.3	113.0	168.1	160.0	111.2	154.1	147.8	102.9	153.3	145.8	107.3	139.9	135.1	102.0	139.9	134.3
0320	Concrete Reinforcing	99.8	133.4	116.0	90.7	135.2	112.2	103.7	136.6	119.6	83.3	134.2	107.9	99.8	133.7	116.2	99.0	133.7	115.8
0330	Cast-in-Place Concrete	88.1	130.2	103.8	110.4	133.5	119.0	94.1	129.6	107.3	95.5	129.2	108.0	81.5	130.2	99.6	92.7	130.2	106.7
03	**CONCRETE**	91.7	133.7	110.2	105.1	148.2	124.0	106.8	140.8	121.7	95.5	139.9	115.0	90.9	134.2	109.9	100.5	134.2	115.3
04	**MASONRY**	92.1	137.5	120.0	122.7	154.9	142.5	103.3	153.4	134.1	96.9	145.0	126.5	73.1	139.5	113.9	81.7	139.5	117.2
05	**METALS**	102.5	112.5	105.4	109.9	116.5	111.9	107.4	116.1	109.9	102.8	115.2	106.4	88.0	114.8	95.9	87.9	114.8	95.8
06	**WOOD, PLASTICS & COMPOSITES**	96.9	138.3	119.7	104.3	173.4	142.3	111.2	157.3	136.6	101.2	157.3	132.1	99.0	138.4	120.7	92.5	138.4	117.8
07	**THERMAL & MOISTURE PROTECTION**	108.5	124.3	115.4	112.2	155.7	131.2	112.6	151.2	129.4	98.9	132.9	113.7	107.2	133.4	118.6	107.5	133.4	118.8
08	**OPENINGS**	92.6	134.3	102.3	91.0	161.7	107.5	103.2	143.2	112.5	94.7	145.1	106.5	87.6	137.6	99.2	87.5	137.6	99.2
0920	Plaster & Gypsum Board	94.4	139.3	124.6	113.4	175.0	154.8	117.4	158.8	145.2	90.9	158.8	136.6	100.3	139.3	126.5	96.2	139.3	125.2
0950, 0980	Ceilings & Acoustic Treatment	100.7	139.3	126.8	101.2	175.0	151.0	113.3	158.8	144.0	98.1	158.8	139.0	114.0	139.3	131.1	114.0	139.3	131.1
0960	Flooring	102.2	121.1	107.5	124.0	143.3	129.4	103.2	143.3	114.4	100.4	121.5	106.3	114.7	121.1	116.5	111.5	121.1	114.2
0970, 0990	Wall Finishes & Painting/Coating	92.5	108.6	102.0	111.6	171.3	146.9	94.1	138.6	120.4	101.3	120.7	112.8	115.1	120.3	118.2	115.1	120.3	118.2
09	**FINISHES**	96.4	133.0	116.2	110.1	165.4	140.0	104.6	152.0	130.3	94.6	146.3	122.6	108.2	134.3	122.4	107.2	134.3	121.9
COVERS	**DIVS. 10 - 14, 25, 28, 41, 43, 44, 46**	100.0	117.7	103.9	100.0	134.0	107.6	100.0	130.8	106.9	100.0	130.8	106.9	100.0	120.5	104.6	100.0	120.5	104.6
21, 22, 23	**FIRE SUPPRESSION, PLUMBING & HVAC**	100.1	128.8	111.4	96.7	169.6	125.3	96.4	130.6	109.9	100.1	130.1	111.9	96.1	131.0	109.8	96.1	131.0	109.8
26, 27, 3370	**ELECTRICAL, COMMUNICATIONS & UTIL.**	107.6	108.1	107.8	99.9	161.8	130.4	98.4	124.9	111.5	97.0	107.4	102.2	98.5	129.2	113.6	98.3	129.2	113.5
MF2018	**WEIGHTED AVERAGE**	98.9	124.6	109.8	103.0	151.9	123.7	102.5	134.1	115.9	98.5	129.4	111.5	93.8	128.8	108.6	95.4	128.8	109.5

DIVISION		CALIFORNIA																	
		LOS ANGELES			MARYSVILLE			MODESTO			MOJAVE			OAKLAND			OXNARD		
		900 - 902			959			953			935			946			930		
		MAT.	INST.	TOTAL	MAT.	INST.	TOTAL	MAT.	INST.	TOTAL	MAT.	INST.	TOTAL	MAT.	INST.	TOTAL	MAT.	INST.	TOTAL
015433	**CONTRACTOR EQUIPMENT**		103.3	103.3		97.1	97.1		97.1	97.1		97.3	97.3		100.2	100.2		96.3	96.3
0241, 31 - 34	**SITE & INFRASTRUCTURE, DEMOLITION**	93.7	109.2	104.4	105.6	102.6	103.5	100.6	102.6	102.0	91.9	103.4	99.8	115.2	106.0	108.8	99.2	101.6	100.8
0310	Concrete Forming & Accessories	105.4	140.1	135.0	101.6	153.6	145.9	98.1	153.7	145.5	113.6	139.3	135.5	102.3	168.2	158.4	105.5	139.8	134.7
0320	Concrete Reinforcing	100.7	133.7	116.6	103.7	134.1	118.4	107.4	134.1	120.3	100.4	133.4	116.4	92.8	136.9	114.1	98.6	133.5	115.5
0330	Cast-in-Place Concrete	84.1	131.4	101.7	105.2	129.2	114.1	94.2	129.3	107.2	82.6	130.1	100.2	104.8	133.5	115.4	96.1	130.4	108.9
03	**CONCRETE**	98.3	134.7	114.3	107.1	140.0	121.6	98.2	140.1	116.6	87.4	133.8	107.8	106.3	148.5	124.8	95.5	134.1	112.5
04	**MASONRY**	88.6	139.5	119.9	104.2	142.9	128.0	101.5	142.9	126.9	94.9	137.4	121.0	131.5	154.9	145.9	97.9	136.8	121.8
05	**METALS**	94.3	115.0	100.4	106.9	114.7	109.2	103.6	114.8	106.9	100.0	113.4	104.0	105.2	117.2	108.7	98.0	113.9	102.7
06	**WOOD, PLASTICS & COMPOSITES**	105.6	138.6	123.7	98.3	157.3	130.8	93.9	157.3	128.8	100.8	138.4	121.5	92.8	173.4	137.2	95.6	138.4	119.1
07	**THERMAL & MOISTURE PROTECTION**	103.4	133.9	116.7	112.0	138.9	123.7	111.6	139.5	123.7	105.6	123.7	113.5	110.8	155.7	130.4	108.6	133.3	119.3
08	**OPENINGS**	98.1	136.9	107.2	102.5	145.6	112.6	101.3	147.4	112.1	89.8	134.3	100.2	91.1	162.1	107.7	92.4	137.6	102.9
0920	Plaster & Gypsum Board	95.6	139.3	125.0	110.0	158.8	142.8	112.7	158.8	143.7	100.5	139.3	126.6	107.6	175.0	152.9	94.5	139.3	124.7
0950, 0980	Ceilings & Acoustic Treatment	121.4	139.3	133.5	112.5	158.8	143.7	109.3	158.8	142.6	99.4	139.3	126.3	103.8	175.0	151.8	100.8	139.3	126.8
0960	Flooring	112.5	122.2	115.3	98.8	119.5	104.6	99.2	133.0	108.7	101.3	121.1	106.8	117.0	143.3	124.4	93.8	121.1	101.5
0970, 0990	Wall Finishes & Painting/Coating	111.6	120.3	116.7	94.1	134.1	117.8	94.1	138.6	120.4	87.2	109.6	100.5	111.6	171.3	146.9	87.2	114.8	103.6
09	**FINISHES**	109.4	134.7	123.1	101.7	147.4	126.4	101.0	150.2	127.7	95.0	133.2	115.6	108.1	165.4	139.1	92.5	133.7	114.8
COVERS	**DIVS. 10 - 14, 25, 28, 41, 43, 44, 46**	100.0	121.0	104.7	100.0	130.8	106.9	100.0	130.8	106.9	100.0	117.9	104.0	100.0	134.0	107.6	100.0	120.4	104.5
21, 22, 23	**FIRE SUPPRESSION, PLUMBING & HVAC**	99.9	131.0	112.1	96.4	130.0	109.6	99.9	131.7	112.4	96.5	128.8	109.2	100.2	169.6	127.5	100.0	131.0	112.2
26, 27, 3370	**ELECTRICAL, COMMUNICATIONS & UTIL.**	96.9	131.0	113.7	95.1	120.7	107.7	97.5	109.9	103.6	95.5	108.1	101.7	99.1	161.8	130.0	101.4	117.9	109.5
MF2018	**WEIGHTED AVERAGE**	98.5	129.8	111.7	101.7	131.3	114.2	100.6	130.6	113.3	95.4	124.4	107.7	103.4	152.0	123.9	98.0	126.7	110.1

	DIVISION	CALIFORNIA																	
		PALM SPRINGS			PALO ALTO			PASADENA			REDDING			RICHMOND			RIVERSIDE		
		922			943			910 - 912			960			948			925		
		MAT.	INST.	TOTAL	MAT.	INST.	TOTAL	MAT.	INST.	TOTAL	MAT.	INST.	TOTAL	MAT.	INST.	TOTAL	MAT.	INST.	TOTAL
015433	CONTRACTOR EQUIPMENT		98.1	98.1		100.2	100.2		95.4	95.4		97.1	97.1		100.2	100.2		98.1	98.1
0241, 31 - 34	SITE & INFRASTRUCTURE, DEMOLITION	89.3	103.6	99.2	106.1	106.0	106.0	97.5	105.3	102.9	122.2	102.6	108.7	114.3	106.0	108.5	96.2	103.6	101.3
0310	Concrete Forming & Accessories	99.8	136.1	130.7	100.5	168.2	158.2	103.4	139.6	134.2	105.0	153.6	146.5	115.5	167.8	160.1	103.5	139.7	134.4
0320	Concrete Reinforcing	109.0	133.5	120.9	90.7	136.7	113.0	106.0	133.6	119.3	137.8	134.1	136.0	90.7	136.7	112.9	105.9	133.5	119.3
0330	Cast-in-Place Concrete	82.7	131.0	100.6	93.4	133.5	108.4	79.1	127.8	97.2	107.2	129.3	115.4	107.4	133.3	117.1	89.9	131.0	105.2
03	CONCRETE	90.5	132.7	109.0	95.3	148.5	118.7	89.9	133.1	108.9	114.0	140.1	125.4	108.0	148.2	125.7	96.4	134.3	113.1
04	MASONRY	76.5	137.7	114.1	105.1	151.5	133.6	93.3	139.4	121.6	131.9	142.9	138.7	122.5	151.5	140.3	77.5	137.7	114.5
05	METALS	108.3	114.1	110.0	102.7	116.9	106.9	80.0	113.2	89.7	103.9	114.7	107.1	102.8	116.6	106.8	107.7	114.2	109.6
06	WOOD, PLASTICS & COMPOSITES	92.1	133.5	114.9	90.4	173.4	136.1	85.4	138.0	114.4	107.8	157.3	135.0	107.5	173.4	143.8	97.5	138.3	119.9
07	THERMAL & MOISTURE PROTECTION	108.1	134.3	119.5	110.0	155.0	129.6	103.8	131.8	116.0	128.2	138.9	132.9	110.6	154.8	129.8	109.0	134.8	120.2
08	OPENINGS	99.9	134.9	108.1	91.1	160.2	107.2	87.8	137.4	99.4	115.9	145.6	122.8	91.1	160.2	107.2	102.6	137.6	110.8
0920	Plaster & Gypsum Board	107.2	134.4	125.5	106.0	175.0	152.4	88.8	139.3	122.8	112.6	158.8	143.7	114.5	175.0	155.2	111.9	139.3	130.4
0950, 0980	Ceilings & Acoustic Treatment	106.0	134.4	125.1	102.0	175.0	151.2	111.0	139.3	130.1	148.8	158.8	155.5	102.0	175.0	151.2	114.2	139.3	131.1
0960	Flooring	101.5	121.1	107.0	115.6	143.3	123.4	98.5	121.1	104.9	87.8	128.6	99.3	126.5	143.3	131.3	102.9	121.1	108.0
0970, 0990	Wall Finishes & Painting/Coating	91.1	120.3	108.4	111.6	171.3	146.9	103.9	120.3	113.6	105.2	134.1	122.3	111.6	171.3	146.9	91.1	120.3	108.4
09	FINISHES	98.0	131.4	116.1	106.5	165.4	138.4	99.0	134.1	118.0	107.2	149.0	129.8	111.6	165.4	140.7	101.0	134.2	119.0
COVERS	DIVS. 10 - 14, 25, 28, 41, 43, 44, 46	100.0	119.6	104.4	100.0	134.0	107.6	100.0	119.6	104.4	100.0	130.8	106.9	100.0	133.8	107.5	100.0	119.3	104.3
21, 22, 23	FIRE SUPPRESSION, PLUMBING & HVAC	96.4	130.9	109.9	96.7	170.6	125.7	96.5	130.9	110.0	100.3	130.1	112.0	96.7	164.0	123.1	99.9	130.9	112.1
26, 27, 3370	ELECTRICAL, COMMUNICATIONS & UTIL.	94.6	110.8	102.6	99.0	176.6	137.2	117.5	129.2	123.2	101.3	120.8	110.9	99.6	133.2	116.2	91.3	114.4	102.7
MF2018	WEIGHTED AVERAGE	97.4	125.3	109.2	99.2	153.8	122.3	94.8	128.6	109.1	107.6	131.5	117.7	102.4	146.2	120.9	99.3	126.6	110.8

	DIVISION	CALIFORNIA																	
		SACRAMENTO			SALINAS			SAN BERNARDINO			SAN DIEGO			SAN FRANCISCO			SAN JOSE		
		942, 956 - 958			939			923 - 924			919 - 921			940 - 941			951		
		MAT.	INST.	TOTAL	MAT.	INST.	TOTAL	MAT.	INST.	TOTAL	MAT.	INST.	TOTAL	MAT.	INST.	TOTAL	MAT.	INST.	TOTAL
015433	CONTRACTOR EQUIPMENT		99.0	99.0		97.3	97.3		98.1	98.1		102.2	102.2		109.8	109.8		98.7	98.7
0241, 31 - 34	SITE & INFRASTRUCTURE, DEMOLITION	95.6	111.3	106.4	112.3	103.2	106.0	76.1	103.6	95.1	105.7	107.8	107.2	116.4	113.1	114.1	131.6	97.9	108.3
0310	Concrete Forming & Accessories	100.5	156.0	147.8	108.9	156.6	149.6	107.1	139.7	134.9	103.8	129.1	125.4	106.1	168.3	159.1	103.9	167.9	158.5
0320	Concrete Reinforcing	86.0	134.4	109.4	99.1	134.6	116.3	105.9	133.5	119.3	102.6	133.4	117.5	105.6	132.1	118.4	94.7	135.0	114.2
0330	Cast-in-Place Concrete	88.6	130.2	104.1	95.0	129.6	107.8	62.1	131.0	87.8	90.5	124.5	103.2	116.9	133.3	123.0	109.4	132.4	118.0
03	CONCRETE	97.9	141.1	116.9	105.8	141.6	121.5	71.1	134.3	98.9	100.2	127.3	112.1	117.8	148.1	131.1	103.9	148.1	123.3
04	MASONRY	106.1	145.9	130.6	95.1	148.4	127.9	83.6	137.7	116.8	86.2	133.4	115.2	140.9	154.6	149.3	131.5	151.5	143.8
05	METALS	100.8	109.8	103.4	102.7	115.8	106.5	107.7	114.2	109.6	94.8	114.1	100.5	110.7	122.4	114.2	101.8	121.6	107.6
06	WOOD, PLASTICS & COMPOSITES	87.1	160.2	127.3	100.6	160.2	133.4	101.2	138.3	121.6	95.0	126.1	112.1	95.1	173.4	138.2	106.0	173.2	143.0
07	THERMAL & MOISTURE PROTECTION	122.2	142.5	131.0	106.2	145.4	123.3	107.1	134.8	119.2	108.2	121.5	114.0	116.5	155.7	133.5	108.0	155.1	128.5
08	OPENINGS	103.9	149.0	114.4	93.5	154.4	107.7	100.0	137.6	108.7	99.8	127.8	106.3	99.5	159.0	113.3	92.6	161.5	108.7
0920	Plaster & Gypsum Board	103.4	161.5	142.5	95.4	161.8	140.1	113.3	139.3	130.8	94.4	126.7	116.1	105.8	175.0	152.4	108.3	175.0	153.2
0950, 0980	Ceilings & Acoustic Treatment	102.0	161.5	142.1	99.4	161.8	141.5	109.3	139.3	129.5	121.1	126.7	124.8	111.7	175.0	154.4	109.3	175.0	153.6
0960	Flooring	115.8	128.6	119.4	95.9	143.3	109.2	105.0	121.1	109.5	108.9	121.1	112.3	117.5	143.3	124.7	92.4	143.3	106.7
0970, 0990	Wall Finishes & Painting/Coating	107.9	134.1	123.4	88.1	171.3	137.3	91.1	120.3	108.4	101.5	118.0	111.2	109.6	174.8	148.2	94.4	171.3	139.9
09	FINISHES	105.3	150.7	129.9	94.6	157.3	128.5	99.4	134.2	118.3	106.6	126.1	117.2	109.9	165.8	140.1	99.9	165.2	135.3
COVERS	DIVS. 10 - 14, 25, 28, 41, 43, 44, 46	100.0	131.7	107.1	100.0	131.2	107.0	100.0	117.7	103.9	100.0	115.7	103.5	100.0	128.4	106.3	100.0	133.4	107.5
21, 22, 23	FIRE SUPPRESSION, PLUMBING & HVAC	100.1	131.4	112.4	96.5	137.6	112.7	96.4	130.9	110.0	100.0	129.4	111.5	100.1	182.8	132.6	99.9	170.5	127.7
26, 27, 3370	ELECTRICAL, COMMUNICATIONS & UTIL.	94.8	120.8	107.6	96.5	132.1	114.0	94.6	112.6	103.5	103.8	104.3	104.0	99.2	182.8	140.4	100.7	176.6	138.1
MF2018	WEIGHTED AVERAGE	100.8	133.0	114.4	99.1	137.2	115.2	95.0	126.3	108.2	99.7	121.7	109.0	107.4	158.4	129.0	102.6	153.5	124.2

	DIVISION	CALIFORNIA																	
		SAN LUIS OBISPO			SAN MATEO			SAN RAFAEL			SANTA ANA			SANTA BARBARA			SANTA CRUZ		
		934			944			949			926 - 927			931			950		
		MAT.	INST.	TOTAL	MAT.	INST.	TOTAL	MAT.	INST.	TOTAL	MAT.	INST.	TOTAL	MAT.	INST.	TOTAL	MAT.	INST.	TOTAL
015433	CONTRACTOR EQUIPMENT		97.3	97.3		100.2	100.2		99.4	99.4		98.1	98.1		97.3	97.3		98.7	98.7
0241, 31 - 34	SITE & INFRASTRUCTURE, DEMOLITION	104.4	103.4	103.7	112.1	106.0	107.9	107.6	111.2	110.1	87.8	103.6	98.7	99.1	103.4	102.1	131.3	97.7	108.1
0310	Concrete Forming & Accessories	115.3	139.7	136.1	106.3	168.2	159.0	111.5	167.8	159.5	107.4	139.7	134.9	106.1	139.7	134.8	103.9	156.9	149.0
0320	Concrete Reinforcing	100.4	133.4	116.4	90.7	136.9	113.0	91.3	135.1	112.5	109.6	133.5	121.2	98.6	133.4	115.4	117.1	134.6	125.6
0330	Cast-in-Place Concrete	102.1	130.2	112.6	104.0	133.5	115.0	121.1	132.3	125.2	79.3	131.0	98.6	95.8	130.2	108.6	108.7	131.4	117.1
03	CONCRETE	103.3	134.0	116.8	104.6	148.5	123.9	128.2	147.4	136.6	88.0	134.3	108.3	95.4	134.0	112.4	106.4	142.6	122.3
04	MASONRY	96.5	135.9	120.7	122.2	154.6	142.1	99.1	154.6	133.2	73.4	138.0	113.1	95.2	135.9	120.2	135.3	148.6	143.4
05	METALS	100.7	113.8	104.6	102.6	117.1	106.9	103.7	112.4	106.3	107.8	114.2	109.7	98.5	113.8	103.0	109.3	120.2	112.5
06	WOOD, PLASTICS & COMPOSITES	103.0	138.4	122.4	97.9	173.4	139.5	95.0	173.2	138.0	103.0	138.3	122.4	95.6	138.4	119.1	106.0	160.4	135.9
07	THERMAL & MOISTURE PROTECTION	106.4	132.6	117.8	110.4	156.4	130.4	115.0	155.4	132.5	108.4	134.9	119.9	105.8	132.6	117.5	107.6	147.9	125.1
08	OPENINGS	91.7	134.3	101.6	91.1	162.1	107.6	101.5	159.7	115.1	99.2	137.6	108.2	93.2	137.6	103.6	93.9	154.4	108.0
0920	Plaster & Gypsum Board	101.1	139.3	126.8	111.7	175.0	154.3	113.2	175.0	154.8	114.6	139.3	131.2	94.5	139.3	124.7	116.8	161.8	147.1
0950, 0980	Ceilings & Acoustic Treatment	99.4	139.3	126.3	102.0	175.0	151.2	109.3	175.0	153.6	109.3	139.3	129.5	100.8	139.3	126.8	110.9	161.8	145.2
0960	Flooring	102.0	121.1	107.4	119.5	143.3	126.2	132.6	143.3	135.6	105.6	121.1	110.0	95.0	121.1	102.4	96.6	143.3	109.7
0970, 0990	Wall Finishes & Painting/Coating	87.2	114.8	103.6	111.6	171.3	146.9	107.8	165.5	141.9	91.1	118.0	107.0	87.2	114.8	103.6	94.5	171.3	139.9
09	FINISHES	96.3	133.7	116.5	108.9	165.4	139.5	112.0	164.6	140.5	100.9	134.0	118.8	93.0	133.7	115.0	102.6	157.4	132.3
COVERS	DIVS. 10 - 14, 25, 28, 41, 43, 44, 46	100.0	126.5	105.9	100.0	134.0	107.6	100.0	133.3	107.4	100.0	120.2	104.5	100.0	119.5	104.3	100.0	131.5	107.0
21, 22, 23	FIRE SUPPRESSION, PLUMBING & HVAC	96.5	131.0	110.0	96.7	165.7	123.8	96.7	182.8	130.5	96.4	130.9	110.0	100.0	131.0	112.2	99.9	137.9	114.8
26, 27, 3370	ELECTRICAL, COMMUNICATIONS & UTIL.	95.5	110.8	103.1	99.0	168.0	133.0	96.2	124.9	110.3	94.6	111.9	103.1	94.5	113.7	103.9	99.8	132.1	115.7
MF2018	WEIGHTED AVERAGE	98.2	125.8	109.9	101.5	152.0	122.9	104.5	149.1	123.4	97.0	126.3	109.4	97.2	126.1	109.4	104.6	137.5	118.5

DIVISION		CALIFORNIA															COLORADO		
		SANTA ROSA			STOCKTON			SUSANVILLE			VALLEJO			VAN NUYS			ALAMOSA		
		954			952			961			945			913 - 916			811		
		MAT.	INST.	TOTAL	MAT.	INST.	TOTAL	MAT.	INST.	TOTAL	MAT.	INST.	TOTAL	MAT.	INST.	TOTAL	MAT.	INST.	TOTAL
015433	CONTRACTOR EQUIPMENT		97.6	97.6		97.1	97.1		97.1	97.1		99.4	99.4		95.4	95.4		90.2	90.2
0241, 31 - 34	SITE & INFRASTRUCTURE, DEMOLITION	101.4	102.6	102.2	100.3	102.6	101.9	129.5	102.6	110.9	96.3	111.1	106.5	115.7	105.3	108.5	141.3	84.4	102.0
0310	Concrete Forming & Accessories	100.7	167.1	157.3	101.9	155.5	147.6	106.2	153.9	146.8	102.3	166.8	157.3	109.8	139.5	135.2	103.8	66.0	71.5
0320	Concrete Reinforcing	104.6	135.1	119.3	107.4	134.1	120.3	137.8	134.1	136.0	92.5	135.0	113.0	106.0	133.6	119.3	114.0	67.9	91.7
0330	Cast-in-Place Concrete	103.3	130.8	113.5	91.7	129.2	105.7	97.5	129.3	109.3	96.4	131.4	109.5	83.4	127.8	99.9	100.1	73.1	90.1
03	CONCRETE	107.1	146.9	124.6	97.3	140.9	116.5	117.0	140.2	127.2	102.4	146.6	121.8	104.4	133.1	117.0	112.5	69.4	93.6
04	MASONRY	101.8	153.5	133.6	101.4	142.9	126.9	130.4	142.9	138.1	77.4	153.5	124.1	106.9	139.4	126.9	130.1	59.5	86.7
05	METALS	108.1	117.7	110.9	103.9	114.7	107.0	102.9	114.7	106.4	103.7	111.7	106.0	79.2	113.2	89.2	105.2	78.5	97.3
06	WOOD, PLASTICS & COMPOSITES	93.6	173.0	137.3	99.5	159.9	132.8	109.6	157.6	136.0	84.8	173.2	133.4	94.1	138.0	118.3	98.5	67.4	81.4
07	THERMAL & MOISTURE PROTECTION	108.9	154.1	128.6	112.1	138.8	123.7	130.2	138.9	134.0	113.1	154.1	130.9	105.0	131.8	116.6	111.9	67.2	92.5
08	OPENINGS	100.8	161.4	114.9	101.3	147.0	112.0	116.8	145.8	123.6	103.2	159.7	116.3	87.7	137.4	99.3	96.3	68.5	89.8
0920	Plaster & Gypsum Board	109.2	175.0	153.5	112.7	161.5	145.5	113.1	159.1	144.1	107.5	175.0	152.9	92.8	139.3	124.1	80.8	66.5	71.2
0950, 0980	Ceilings & Acoustic Treatment	109.3	175.0	153.6	117.5	161.5	147.1	139.9	159.1	152.9	111.1	175.0	154.2	107.8	139.3	129.1	102.7	66.5	78.3
0960	Flooring	102.1	136.5	111.7	99.2	133.0	108.7	88.3	128.6	99.6	126.0	143.3	130.9	101.9	121.1	107.3	107.8	70.1	97.2
0970, 0990	Wall Finishes & Painting/Coating	91.1	165.5	135.2	94.1	134.1	117.8	105.2	134.1	122.3	108.7	165.5	142.3	103.9	120.3	113.6	101.7	76.9	87.0
09	FINISHES	100.0	163.0	134.1	102.6	151.3	129.0	106.7	149.2	129.7	108.4	164.4	138.7	101.2	134.1	119.0	99.8	67.5	82.3
COVERS	DIVS. 10 - 14, 25, 28, 41, 43, 44, 46	100.0	132.4	107.2	100.0	128.0	106.2	100.0	130.9	106.9	100.0	132.9	107.3	100.0	119.6	104.4	100.0	83.0	96.2
21, 22, 23	FIRE SUPPRESSION, PLUMBING & HVAC	96.4	182.2	130.1	99.9	131.7	112.4	96.8	130.1	109.9	100.2	147.2	118.7	96.5	130.9	110.0	96.5	72.4	87.0
26, 27, 3370	ELECTRICAL, COMMUNICATIONS & UTIL.	94.9	124.9	109.7	97.5	114.3	105.8	101.7	120.8	111.1	92.4	127.5	109.7	117.5	129.2	123.2	96.3	63.2	80.0
MF2018	WEIGHTED AVERAGE	101.2	148.4	121.2	100.7	131.3	113.7	107.3	131.6	117.5	100.2	141.6	117.7	97.8	128.6	110.8	103.6	70.3	89.5

DIVISION		COLORADO																	
		BOULDER			COLORADO SPRINGS			DENVER			DURANGO			FORT COLLINS			FORT MORGAN		
		803			808 - 809			800 - 802			813			805			807		
		MAT.	INST.	TOTAL	MAT.	INST.	TOTAL	MAT.	INST.	TOTAL	MAT.	INST.	TOTAL	MAT.	INST.	TOTAL	MAT.	INST.	TOTAL
015433	CONTRACTOR EQUIPMENT		92.2	92.2		90.2	90.2		99.6	99.6		90.2	90.2		92.2	92.2		92.2	92.2
0241, 31 - 34	SITE & INFRASTRUCTURE, DEMOLITION	97.1	91.4	93.2	99.2	86.7	90.6	104.1	102.5	103.0	134.3	84.4	99.9	109.6	91.2	96.9	99.3	90.9	93.5
0310	Concrete Forming & Accessories	107.0	77.4	81.8	97.3	63.5	68.5	106.1	66.4	72.2	109.9	65.7	72.2	104.5	65.8	71.5	107.6	65.7	71.9
0320	Concrete Reinforcing	108.7	68.1	89.0	107.9	67.9	88.6	107.9	70.1	89.6	114.0	67.9	91.7	108.8	68.0	89.1	108.9	68.0	89.2
0330	Cast-in-Place Concrete	113.8	73.9	98.9	116.8	73.4	100.7	121.8	73.6	103.9	115.1	73.1	99.5	128.5	72.5	107.6	111.6	72.5	97.1
03	CONCRETE	105.2	74.9	91.9	108.7	68.4	91.0	111.4	70.2	93.3	114.2	69.3	94.5	116.3	69.2	95.6	103.7	69.1	88.5
04	MASONRY	101.6	65.4	79.4	103.7	62.3	78.3	104.9	64.1	79.9	117.5	60.6	82.5	118.9	63.1	84.6	116.7	63.1	83.8
05	METALS	94.4	77.9	89.6	97.5	77.7	91.7	99.9	79.0	93.8	105.2	78.4	97.3	95.7	77.7	90.4	94.2	77.8	89.4
06	WOOD, PLASTICS & COMPOSITES	107.1	81.4	93.0	96.6	63.6	78.5	108.4	67.1	85.7	108.4	67.4	85.8	104.6	67.0	83.9	107.1	67.0	85.0
07	THERMAL & MOISTURE PROTECTION	108.7	73.2	93.2	109.4	69.7	92.1	107.1	71.7	91.7	111.9	67.5	92.6	109.0	70.6	92.3	108.6	69.6	91.6
08	OPENINGS	95.4	76.2	90.9	99.5	66.4	91.8	102.5	68.8	94.6	103.1	68.5	95.0	95.3	68.3	89.0	95.3	68.2	89.0
0920	Plaster & Gypsum Board	116.3	81.3	92.8	100.3	62.9	75.2	111.7	66.6	81.4	95.3	66.5	75.9	110.3	66.6	80.9	116.3	66.5	82.8
0950, 0980	Ceilings & Acoustic Treatment	91.9	81.3	84.8	100.0	62.9	75.0	104.4	66.6	78.9	102.7	66.5	78.3	91.9	66.6	74.8	91.9	66.5	74.8
0960	Flooring	111.2	76.7	101.5	102.2	69.0	92.9	108.4	76.7	99.5	113.1	69.0	100.7	107.3	76.7	98.7	111.7	76.7	101.8
0970, 0990	Wall Finishes & Painting/Coating	97.1	76.9	85.1	96.8	76.9	85.0	103.4	76.9	87.7	101.7	76.9	87.0	97.1	76.9	85.1	97.1	76.9	85.1
09	FINISHES	100.6	77.7	88.2	97.4	65.3	80.0	102.5	68.9	84.3	102.4	67.4	83.5	99.2	68.6	82.6	100.7	68.5	83.3
COVERS	DIVS. 10 - 14, 25, 28, 41, 43, 44, 46	100.0	84.5	96.5	100.0	82.2	96.0	100.0	86.3	96.9	100.0	82.9	96.2	100.0	82.1	96.0	100.0	82.1	96.0
21, 22, 23	FIRE SUPPRESSION, PLUMBING & HVAC	96.5	74.8	87.9	100.1	72.2	89.1	99.9	73.8	89.7	96.5	58.7	81.6	100.0	71.7	88.9	96.5	73.5	87.4
26, 27, 3370	ELECTRICAL, COMMUNICATIONS & UTIL.	99.6	80.6	90.3	103.2	73.1	88.4	105.0	80.4	92.9	95.8	50.3	73.4	99.6	80.6	90.2	100.0	78.1	89.2
MF2018	WEIGHTED AVERAGE	98.7	77.0	89.6	101.2	71.5	88.6	103.0	75.4	91.3	103.9	65.6	87.8	102.2	73.5	90.1	99.3	73.5	88.4

DIVISION		COLORADO																	
		GLENWOOD SPRINGS			GOLDEN			GRAND JUNCTION			GREELEY			MONTROSE			PUEBLO		
		816			804			815			806			814			810		
		MAT.	INST.	TOTAL	MAT.	INST.	TOTAL	MAT.	INST.	TOTAL	MAT.	INST.	TOTAL	MAT.	INST.	TOTAL	MAT.	INST.	TOTAL
015433	CONTRACTOR EQUIPMENT		93.3	93.3		92.2	92.2		93.3	93.3		92.2	92.2		91.7	91.7		90.2	90.2
0241, 31 - 34	SITE & INFRASTRUCTURE, DEMOLITION	150.9	91.9	110.2	110.4	91.4	97.3	133.7	92.1	105.0	95.9	91.2	92.7	144.0	87.9	105.3	125.2	84.5	97.1
0310	Concrete Forming & Accessories	100.6	65.5	70.7	99.7	66.7	71.5	108.7	66.6	72.8	102.4	65.8	71.2	100.1	65.5	70.6	106.1	63.7	70.0
0320	Concrete Reinforcing	112.8	68.0	91.2	108.9	68.0	89.2	113.2	67.8	91.2	108.7	68.0	89.0	112.7	67.8	91.0	109.0	67.9	89.1
0330	Cast-in-Place Concrete	100.1	72.1	89.6	111.7	73.9	97.6	110.8	73.5	96.9	107.3	72.5	94.3	100.1	72.6	89.8	99.4	73.7	89.8
03	CONCRETE	118.0	68.9	96.4	114.1	70.0	94.7	110.7	69.9	92.8	100.3	69.1	86.6	108.5	69.0	91.1	100.9	68.7	86.7
04	MASONRY	103.1	63.0	78.5	119.7	65.4	86.3	137.8	61.9	91.2	111.9	62.7	81.7	110.0	59.6	79.0	99.5	60.3	75.4
05	METALS	104.8	78.2	97.0	94.3	77.8	89.5	106.5	78.2	98.2	95.7	77.7	90.4	103.9	78.1	96.3	108.3	78.7	99.6
06	WOOD, PLASTICS & COMPOSITES	93.6	67.1	79.0	98.9	67.0	81.3	106.1	67.2	84.7	101.7	67.0	82.6	94.6	67.2	79.6	101.2	63.9	80.7
07	THERMAL & MOISTURE PROTECTION	111.8	68.2	92.8	109.8	71.6	93.2	110.9	69.3	92.8	108.2	70.4	91.8	112.0	67.2	92.5	110.4	67.8	91.8
08	OPENINGS	102.1	68.3	94.2	95.4	68.2	89.0	102.8	68.4	94.7	95.3	68.3	89.0	103.2	68.4	95.1	98.0	66.6	90.7
0920	Plaster & Gypsum Board	121.2	66.5	84.4	107.5	66.5	79.9	135.4	66.6	89.1	108.7	66.6	80.4	79.9	66.5	70.9	84.7	62.9	70.1
0950, 0980	Ceilings & Acoustic Treatment	101.9	66.5	78.0	91.9	66.5	74.8	101.9	66.6	78.1	91.9	66.6	74.8	102.7	66.5	78.3	110.0	62.9	78.3
0960	Flooring	106.9	76.7	98.4	104.9	76.7	97.0	112.4	70.1	100.5	106.2	73.1	96.9	110.4	73.1	100.0	109.1	73.1	99.0
0970, 0990	Wall Finishes & Painting/Coating	101.7	76.9	87.0	97.1	76.9	85.1	101.7	76.9	87.0	97.1	76.9	85.1	101.7	76.9	87.0	101.7	76.9	87.0
09	FINISHES	105.2	68.7	85.4	98.9	69.1	82.8	106.8	67.9	85.8	97.9	67.9	81.6	100.5	68.0	82.9	100.2	66.3	81.9
COVERS	DIVS. 10 - 14, 25, 28, 41, 43, 44, 46	100.0	82.3	96.1	100.0	82.9	96.2	100.0	83.1	96.2	100.0	82.1	96.0	100.0	82.6	96.1	100.0	82.9	96.2
21, 22, 23	FIRE SUPPRESSION, PLUMBING & HVAC	96.5	58.7	81.6	96.5	74.8	87.9	100.0	74.8	90.1	100.0	71.7	88.9	96.5	58.7	81.6	100.0	72.2	89.1
26, 27, 3370	ELECTRICAL, COMMUNICATIONS & UTIL.	93.4	52.3	73.1	100.0	80.6	90.4	95.5	54.6	75.3	99.6	80.6	90.2	95.5	50.3	73.2	96.4	63.3	80.0
MF2018	WEIGHTED AVERAGE	103.9	66.8	88.2	100.9	74.7	89.8	105.7	70.5	90.8	99.3	73.4	88.4	102.6	65.8	87.1	101.8	70.0	88.3

DIVISION		COLORADO			CONNECTICUT															
		SALIDA			BRIDGEPORT			BRISTOL			HARTFORD			MERIDEN			NEW BRITAIN			
		812			066			060			061			064			060			
		MAT.	INST.	TOTAL	MAT.	INST.	TOTAL	MAT.	INST.	TOTAL	MAT.	INST.	TOTAL	MAT.	INST.	TOTAL	MAT.	INST.	TOTAL	
015433	**CONTRACTOR EQUIPMENT**		91.7	91.7		95.7	95.7		95.7	95.7		99.0	99.0		96.1	96.1		95.7	95.7	
0241, 31 - 34	**SITE & INFRASTRUCTURE, DEMOLITION**	134.0	88.2	102.4	106.1	97.8	100.3	105.1	97.7	100.0	101.6	102.4	102.2	102.8	98.5	99.8	105.3	97.7	100.1	
0310	Concrete Forming & Accessories	108.6	65.9	72.2	103.4	116.7	114.7	103.4	116.6	114.6	103.5	116.8	114.8	103.1	116.6	114.6	103.9	116.6	114.7	
0320	Concrete Reinforcing	112.4	67.9	90.9	116.3	144.5	130.0	116.3	144.5	130.0	111.6	144.6	127.5	116.3	144.5	130.0	116.3	144.5	130.0	
0330	Cast-in-Place Concrete	114.7	72.8	99.1	108.5	127.5	115.6	101.6	127.5	111.2	106.5	128.5	114.7	97.8	127.5	108.8	103.3	127.5	112.3	
03	**CONCRETE**	109.1	69.3	91.6	104.1	124.5	113.0	100.9	124.4	111.2	101.1	124.8	111.5	99.1	124.4	110.2	101.7	124.4	111.7	
04	**MASONRY**	138.6	60.6	90.7	109.4	130.9	122.6	100.8	130.9	119.3	100.5	131.0	119.2	100.4	130.9	119.2	102.5	130.9	120.0	
05	**METALS**	103.6	78.6	96.2	98.1	116.6	103.5	98.1	116.5	103.5	103.1	116.0	106.9	95.4	116.5	101.6	94.5	116.5	100.9	
06	**WOOD, PLASTICS & COMPOSITES**	102.4	67.2	83.0	105.0	114.2	110.1	105.0	114.2	110.1	96.2	114.4	106.2	105.0	114.2	110.1	105.0	114.2	110.1	
07	**THERMAL & MOISTURE PROTECTION**	110.8	67.5	92.0	99.7	123.5	110.0	99.8	120.6	108.9	104.4	121.1	111.7	99.8	120.6	108.9	99.8	120.7	108.9	
08	**OPENINGS**	96.3	68.4	89.8	96.4	120.6	102.0	96.4	120.6	102.0	98.1	120.7	103.4	98.5	120.6	103.7	96.4	120.6	102.0	
0920	Plaster & Gypsum Board	80.2	66.5	71.0	115.7	114.3	114.8	115.7	114.3	114.8	102.5	114.3	110.5	117.3	114.3	115.3	115.7	114.3	114.8	
0950, 0980	Ceilings & Acoustic Treatment	102.7	66.5	78.3	102.0	114.3	110.3	102.0	114.3	110.3	98.8	114.3	109.3	107.5	114.3	112.1	102.0	114.3	110.3	
0960	Flooring	115.9	69.9	103.0	92.6	127.0	102.3	92.6	124.5	101.6	96.9	127.0	105.4	92.6	124.5	101.6	92.6	124.5	101.6	
0970, 0990	Wall Finishes & Painting/Coating	101.7	76.9	87.0	92.0	125.5	111.8	92.0	129.1	113.9	98.1	129.1	116.4	92.0	129.1	113.9	92.0	129.1	113.9	
09	**FINISHES**	101.1	67.4	82.8	94.7	118.9	107.8	94.7	118.8	107.8	96.2	119.4	108.8	96.0	118.8	108.4	94.7	118.8	107.8	
COVERS	**DIVS. 10 - 14, 25, 28, 41, 43, 44, 46**	100.0	82.7	96.1	100.0	113.9	103.1	100.0	113.9	103.1	100.0	114.4	103.2	100.0	114.0	103.1	100.0	113.9	103.1	
21, 22, 23	**FIRE SUPPRESSION, PLUMBING & HVAC**	96.5	73.6	87.5	100.1	118.6	107.4	100.1	118.6	107.4	100.0	118.6	107.3	96.6	118.6	105.2	100.1	118.6	107.4	
26, 27, 3370	**ELECTRICAL, COMMUNICATIONS & UTIL.**	95.7	63.2	79.7	93.0	105.0	98.9	93.0	103.7	98.3	92.7	109.4	100.9	92.9	105.8	99.3	93.0	103.7	98.3	
MF2018	**WEIGHTED AVERAGE**	103.1	70.9	89.5	99.3	117.0	106.8	98.5	116.7	106.2	99.6	118.0	107.3	97.2	117.0	105.6	98.1	116.7	106.0	

DIVISION		CONNECTICUT																	
		NEW HAVEN			NEW LONDON			NORWALK			STAMFORD			WATERBURY			WILLIMANTIC		
		065			063			068			069			067			062		
		MAT.	INST.	TOTAL	MAT.	INST.	TOTAL	MAT.	INST.	TOTAL	MAT.	INST.	TOTAL	MAT.	INST.	TOTAL	MAT.	INST.	TOTAL
015433	**CONTRACTOR EQUIPMENT**		96.1	96.1		96.1	96.1		95.7	95.7		95.7	95.7		95.7	95.7		95.7	95.7
0241, 31 - 34	**SITE & INFRASTRUCTURE, DEMOLITION**	105.3	98.4	100.5	97.2	98.1	97.8	105.8	97.6	100.2	106.5	97.7	100.4	105.8	97.8	100.2	105.8	97.7	100.2
0310	Concrete Forming & Accessories	103.2	116.5	114.6	103.1	116.5	114.5	103.4	116.6	114.7	103.4	116.9	114.9	103.4	116.6	114.7	103.4	116.4	114.5
0320	Concrete Reinforcing	116.3	144.5	130.0	91.2	144.5	117.0	116.3	144.5	130.0	116.3	144.5	130.0	116.3	144.5	130.0	116.3	144.5	130.0
0330	Cast-in-Place Concrete	105.1	125.5	112.7	89.5	125.5	102.9	106.7	126.7	114.2	108.5	126.8	115.3	108.5	127.5	115.6	101.3	125.5	110.3
03	**CONCRETE**	116.3	123.7	119.6	89.2	123.7	104.4	103.3	124.1	112.4	104.1	124.3	113.0	104.1	124.4	113.0	100.8	123.6	110.8
04	**MASONRY**	101.1	130.9	119.4	99.3	130.9	118.7	100.6	130.1	118.7	101.4	130.1	119.0	101.4	130.9	119.5	100.6	130.9	119.3
05	**METALS**	94.7	116.4	101.1	94.4	116.4	100.9	98.1	116.5	103.5	98.1	116.8	103.6	98.1	116.5	103.5	97.9	116.3	103.3
06	**WOOD, PLASTICS & COMPOSITES**	105.0	114.2	110.1	105.0	114.2	110.1	105.0	114.2	110.1	105.0	114.2	110.1	105.0	114.2	110.1	105.0	114.2	110.1
07	**THERMAL & MOISTURE PROTECTION**	99.9	120.8	109.0	99.7	120.3	108.7	99.9	123.2	110.0	99.8	123.2	110.0	99.8	121.1	109.1	100.0	120.3	108.8
08	**OPENINGS**	96.4	120.6	102.0	98.7	120.6	103.8	96.4	120.6	102.0	96.4	120.6	102.0	96.4	120.6	102.0	98.8	120.6	103.9
0920	Plaster & Gypsum Board	115.7	114.3	114.8	115.7	114.3	114.8	115.7	114.3	114.8	115.7	114.3	114.8	115.7	114.3	114.8	115.7	114.3	114.8
0950, 0980	Ceilings & Acoustic Treatment	102.0	114.3	110.3	100.2	114.3	109.7	102.0	114.3	110.3	102.0	114.3	110.3	102.0	114.3	110.3	100.2	114.3	109.7
0960	Flooring	92.6	127.0	102.3	92.6	127.0	102.3	92.6	124.5	101.6	92.6	127.0	102.3	92.6	127.0	102.3	92.6	121.1	100.6
0970, 0990	Wall Finishes & Painting/Coating	92.0	125.5	111.8	92.0	129.1	113.9	92.0	125.5	111.8	92.0	125.5	111.8	92.0	129.1	113.9	92.0	129.1	113.9
09	**FINISHES**	94.7	118.9	107.8	93.9	119.3	107.6	94.7	118.5	107.6	94.8	118.9	107.8	94.6	119.3	107.9	94.5	118.3	107.3
COVERS	**DIVS. 10 - 14, 25, 28, 41, 43, 44, 46**	100.0	114.0	103.1	100.0	114.0	103.1	100.0	113.9	103.1	100.0	114.1	103.2	100.0	114.0	103.1	100.0	114.0	103.1
21, 22, 23	**FIRE SUPPRESSION, PLUMBING & HVAC**	100.1	118.6	107.4	96.6	118.6	105.2	100.1	118.6	107.4	100.1	118.6	107.4	100.1	118.6	107.4	100.1	118.4	107.3
26, 27, 3370	**ELECTRICAL, COMMUNICATIONS & UTIL.**	92.9	106.0	99.4	90.5	107.9	99.1	93.0	104.8	98.8	93.0	150.1	121.1	92.6	107.1	99.7	93.0	108.8	100.8
MF2018	**WEIGHTED AVERAGE**	99.9	117.0	107.1	95.2	117.2	104.5	98.8	116.7	106.4	99.0	123.2	109.2	98.9	117.2	106.6	98.7	117.1	106.5

DIVISION		D.C.			DELAWARE									FLORIDA					
		WASHINGTON			DOVER			NEWARK			WILMINGTON			DAYTONA BEACH			FORT LAUDERDALE		
		200 - 205			199			197			198			321			333		
		MAT.	INST.	TOTAL	MAT.	INST.	TOTAL	MAT.	INST.	TOTAL	MAT.	INST.	TOTAL	MAT.	INST.	TOTAL	MAT.	INST.	TOTAL
015433	**CONTRACTOR EQUIPMENT**		105.3	105.3		118.2	118.2		119.0	119.0		118.4	118.4		99.5	99.5		92.8	92.8
0241, 31 - 34	**SITE & INFRASTRUCTURE, DEMOLITION**	103.0	97.8	99.4	106.0	107.5	107.0	104.5	108.7	107.4	102.2	107.8	106.1	117.6	85.0	95.1	95.0	73.2	80.0
0310	Concrete Forming & Accessories	97.7	73.4	77.0	98.0	99.4	99.2	97.0	99.2	98.9	95.9	99.4	98.9	97.1	61.6	66.8	93.4	59.4	64.4
0320	Concrete Reinforcing	103.5	86.8	95.4	101.4	115.1	108.0	97.2	115.1	105.9	101.0	115.1	107.8	96.0	61.0	79.1	94.6	59.9	77.8
0330	Cast-in-Place Concrete	103.6	79.0	94.5	108.1	107.7	107.9	91.2	106.7	97.0	102.6	107.7	104.5	91.8	66.0	82.2	96.3	62.3	83.6
03	**CONCRETE**	106.0	78.6	94.0	100.4	106.0	102.9	93.4	105.7	98.8	97.7	106.0	101.4	90.5	64.8	79.2	93.4	62.3	79.7
04	**MASONRY**	97.6	87.8	91.6	99.0	98.5	98.7	101.0	98.5	99.5	94.4	98.5	96.9	86.4	60.9	70.7	87.6	54.4	67.2
05	**METALS**	103.9	95.6	101.5	102.9	121.9	108.5	104.7	123.1	110.1	103.1	121.9	108.6	101.9	88.9	98.1	96.9	88.2	94.4
06	**WOOD, PLASTICS & COMPOSITES**	95.4	72.2	82.6	92.3	97.6	95.2	91.0	97.4	94.5	85.7	97.6	92.2	93.2	60.1	75.0	78.3	60.6	68.6
07	**THERMAL & MOISTURE PROTECTION**	104.3	87.1	96.8	105.2	110.4	107.4	109.2	109.8	109.4	105.1	110.4	107.4	101.4	65.4	85.7	106.3	60.4	86.3
08	**OPENINGS**	100.7	74.6	94.6	91.2	107.4	95.0	91.2	107.3	94.9	88.6	107.4	93.0	92.9	59.4	85.1	94.2	59.5	86.1
0920	Plaster & Gypsum Board	104.3	71.4	82.2	96.0	97.4	96.9	96.7	97.4	97.2	97.3	97.4	97.4	92.7	59.5	70.4	107.2	60.0	75.5
0950, 0980	Ceilings & Acoustic Treatment	110.9	71.4	84.3	100.2	97.4	98.3	96.0	97.4	96.9	90.8	97.4	95.2	78.3	59.5	65.6	84.8	60.0	68.1
0960	Flooring	95.4	75.5	89.8	96.9	105.4	99.3	91.3	105.4	95.3	96.0	105.4	98.7	99.5	62.7	89.2	97.1	62.7	87.5
0970, 0990	Wall Finishes & Painting/Coating	103.7	73.1	85.6	92.6	119.1	108.3	88.2	119.1	106.4	95.3	119.1	109.4	103.8	61.4	78.7	96.5	58.3	73.9
09	**FINISHES**	98.5	72.8	84.6	95.9	101.6	99.0	89.7	101.5	96.1	94.2	101.6	98.2	91.5	61.1	75.1	91.3	59.4	74.1
COVERS	**DIVS. 10 - 14, 25, 28, 41, 43, 44, 46**	100.0	94.5	98.8	100.0	103.7	100.8	100.0	103.4	100.8	100.0	103.7	100.8	100.0	81.4	95.9	100.0	80.7	95.7
21, 22, 23	**FIRE SUPPRESSION, PLUMBING & HVAC**	100.0	88.4	95.4	100.0	119.4	107.6	100.3	119.3	107.7	100.1	119.4	107.7	99.9	75.6	90.4	100.0	66.7	86.9
26, 27, 3370	**ELECTRICAL, COMMUNICATIONS & UTIL.**	97.7	100.0	98.8	95.1	109.7	102.3	96.9	109.7	103.2	95.1	109.7	102.3	96.7	58.9	78.1	94.7	67.4	81.3
MF2018	**WEIGHTED AVERAGE**	101.1	87.5	95.4	99.0	109.6	103.5	98.3	109.7	103.1	97.9	109.7	102.9	97.2	69.5	85.5	96.1	66.4	83.6

DIVISION		FLORIDA																	
		FORT MYERS			GAINESVILLE			JACKSONVILLE			LAKELAND			MELBOURNE			MIAMI		
		339, 341			326, 344			320, 322			338			329			330 - 332, 340		
		MAT.	INST.	TOTAL	MAT.	INST.	TOTAL	MAT.	INST.	TOTAL	MAT.	INST.	TOTAL	MAT.	INST.	TOTAL	MAT.	INST.	TOTAL
015433	CONTRACTOR EQUIPMENT		99.5	99.5		99.5	99.5		99.5	99.5		99.5	99.5		99.5	99.5		94.9	94.9
0241, 31 - 34	SITE & INFRASTRUCTURE, DEMOLITION	106.7	85.2	91.9	127.1	84.8	97.9	117.6	85.2	95.2	108.7	85.2	92.5	125.6	85.1	97.7	96.5	78.9	84.3
0310	Concrete Forming & Accessories	89.6	63.8	67.6	92.5	57.1	62.4	96.9	60.2	65.6	86.4	63.8	67.1	93.7	61.7	66.4	99.3	60.2	66.0
0320	Concrete Reinforcing	95.7	76.1	86.2	101.8	61.5	82.3	96.0	61.3	79.2	98.0	76.4	87.6	97.1	65.6	81.9	101.6	60.0	81.5
0330	Cast-in-Place Concrete	100.4	66.0	87.6	105.4	65.7	90.6	92.7	65.8	82.7	102.7	66.1	89.1	110.7	66.0	94.1	93.1	63.9	82.2
03	CONCRETE	94.0	68.3	82.7	101.7	62.7	84.6	90.9	64.2	79.2	95.8	68.5	83.8	102.1	65.6	86.1	91.7	63.2	79.2
04	MASONRY	82.5	58.8	67.9	99.3	60.8	75.7	86.1	58.7	69.3	96.8	61.0	74.8	84.2	60.9	69.9	88.0	55.5	68.0
05	METALS	99.0	93.6	97.4	100.7	88.4	97.1	100.4	88.6	96.9	98.9	94.7	97.7	110.6	90.6	104.7	97.1	87.6	94.3
06	WOOD, PLASTICS & COMPOSITES	75.4	63.2	68.7	87.2	54.6	69.3	93.2	58.6	74.2	71.0	63.2	66.7	88.9	60.1	73.0	92.9	60.7	75.2
07	THERMAL & MOISTURE PROTECTION	106.1	63.2	87.4	101.8	62.5	84.7	101.7	62.4	84.6	106.0	63.8	87.6	101.9	64.1	85.5	106.1	61.4	86.6
08	OPENINGS	95.5	64.6	88.3	92.5	56.5	84.1	92.9	58.8	84.9	95.5	64.7	88.3	92.2	60.5	84.8	96.6	59.5	88.0
0920	Plaster & Gypsum Board	103.3	62.7	76.0	89.6	53.8	65.6	92.7	58.0	69.3	100.2	62.7	75.0	89.6	59.5	69.4	96.2	60.0	71.9
0950, 0980	Ceilings & Acoustic Treatment	79.7	62.7	68.2	72.9	53.8	60.1	78.3	58.0	64.6	79.7	62.7	68.2	77.5	59.5	65.4	85.6	60.0	68.3
0960	Flooring	94.1	77.2	89.4	97.0	62.7	87.4	99.5	62.7	89.2	92.2	62.7	83.9	97.2	62.7	87.5	98.6	62.7	88.5
0970, 0990	Wall Finishes & Painting/Coating	101.4	63.2	78.8	103.8	63.2	79.8	103.8	63.2	79.8	101.4	63.2	78.8	103.8	80.9	90.2	100.2	58.3	75.4
09	FINISHES	91.1	66.1	77.6	90.4	58.1	72.9	91.6	60.4	74.7	90.2	63.1	75.6	91.0	63.2	76.0	90.9	59.8	74.1
COVERS	DIVS. 10 - 14, 25, 28, 41, 43, 44, 46	100.0	80.0	95.5	100.0	80.8	95.7	100.0	78.0	95.1	100.0	80.0	95.5	100.0	81.4	95.9	100.0	81.5	95.9
21, 22, 23	FIRE SUPPRESSION, PLUMBING & HVAC	98.1	59.3	82.9	98.6	63.1	84.7	99.9	63.1	85.5	98.1	59.9	83.1	99.9	74.2	89.8	100.0	63.8	85.8
26, 27, 3370	ELECTRICAL, COMMUNICATIONS & UTIL.	96.6	60.0	78.6	97.0	58.6	78.0	96.4	61.9	79.4	95.0	58.5	77.0	97.7	62.7	80.5	98.4	77.9	88.3
MF2018	WEIGHTED AVERAGE	96.5	67.7	84.3	98.8	65.8	84.8	96.9	66.6	84.1	97.1	67.6	84.6	100.1	70.3	87.5	96.7	68.0	84.6

DIVISION		FLORIDA																	
		ORLANDO			PANAMA CITY			PENSACOLA			SARASOTA			ST. PETERSBURG			TALLAHASSEE		
		327 - 328, 347			324			325			342			337			323		
		MAT.	INST.	TOTAL	MAT.	INST.	TOTAL	MAT.	INST.	TOTAL	MAT.	INST.	TOTAL	MAT.	INST.	TOTAL	MAT.	INST.	TOTAL
015433	CONTRACTOR EQUIPMENT		102.0	102.0		99.5	99.5		99.5	99.5		99.5	99.5		99.5	99.5		102.0	102.0
0241, 31 - 34	SITE & INFRASTRUCTURE, DEMOLITION	115.8	89.6	97.7	131.4	85.2	99.5	131.3	85.0	99.3	119.0	85.2	95.7	110.4	85.0	92.9	110.2	89.8	96.1
0310	Concrete Forming & Accessories	101.8	61.5	67.4	96.2	66.2	70.6	94.2	61.6	66.4	93.7	63.7	68.1	92.8	60.8	65.5	99.0	62.2	67.6
0320	Concrete Reinforcing	104.9	65.6	85.9	100.3	67.4	84.4	102.8	67.4	85.7	96.6	76.2	86.7	98.0	76.0	87.4	98.4	61.5	80.5
0330	Cast-in-Place Concrete	111.6	66.7	94.9	97.4	65.9	85.7	120.4	65.3	99.9	105.1	66.1	90.6	103.8	65.8	89.7	92.3	66.7	82.8
03	CONCRETE	100.8	65.7	85.4	100.0	67.9	85.9	109.7	65.6	90.3	95.3	68.4	83.5	97.3	66.9	83.9	90.0	65.3	79.2
04	MASONRY	94.8	60.9	74.0	90.6	60.3	72.0	109.1	59.9	78.9	88.7	61.0	71.6	132.4	58.7	87.1	85.2	60.3	69.9
05	METALS	96.7	89.1	94.5	101.5	90.9	98.4	102.6	90.8	99.1	102.5	93.9	100.0	99.8	93.5	98.0	101.7	87.7	97.6
06	WOOD, PLASTICS & COMPOSITES	91.7	60.2	74.4	92.0	66.6	78.0	90.3	61.1	74.2	92.5	63.2	76.4	79.5	59.3	68.4	95.6	61.2	76.7
07	THERMAL & MOISTURE PROTECTION	109.2	66.0	90.4	102.0	64.1	85.5	101.9	63.2	85.1	100.1	63.8	84.3	106.2	62.2	87.1	96.7	64.0	82.4
08	OPENINGS	97.8	60.6	89.1	91.0	64.5	84.9	91.0	61.4	84.1	97.8	64.2	90.0	94.3	62.5	86.9	97.9	60.2	89.1
0920	Plaster & Gypsum Board	93.3	59.5	70.6	91.9	66.1	74.6	99.6	60.5	73.3	97.2	62.7	74.0	105.6	58.7	74.0	100.0	60.5	73.4
0950, 0980	Ceilings & Acoustic Treatment	89.7	59.5	69.3	77.5	66.1	69.8	77.5	60.5	66.0	83.9	62.7	69.6	81.6	58.7	66.1	86.2	60.5	68.9
0960	Flooring	95.7	62.7	86.4	99.1	72.5	91.6	95.1	62.7	86.0	103.5	54.9	89.9	96.1	60.8	86.2	97.6	61.7	87.5
0970, 0990	Wall Finishes & Painting/Coating	95.6	60.3	74.7	103.8	63.2	79.8	103.8	63.2	79.8	99.0	63.2	77.8	101.4	63.2	78.8	101.6	63.2	78.9
09	FINISHES	93.2	61.1	75.8	92.6	66.8	78.6	92.1	61.6	75.6	96.0	61.8	77.5	92.6	60.5	75.2	94.4	61.8	76.7
COVERS	DIVS. 10 - 14, 25, 28, 41, 43, 44, 46	100.0	81.7	95.9	100.0	80.6	95.7	100.0	79.7	95.5	100.0	80.0	95.5	100.0	79.6	95.4	100.0	81.5	95.9
21, 22, 23	FIRE SUPPRESSION, PLUMBING & HVAC	100.0	56.4	82.9	99.9	63.2	85.5	99.9	62.7	85.3	99.9	58.5	83.6	100.0	59.9	84.2	100.0	65.5	86.5
26, 27, 3370	ELECTRICAL, COMMUNICATIONS & UTIL.	98.4	63.1	81.0	95.5	58.6	77.3	99.0	50.7	75.2	96.9	58.5	78.0	95.0	60.6	78.0	103.7	58.6	81.4
MF2018	WEIGHTED AVERAGE	99.0	66.6	85.3	98.6	68.3	85.8	101.2	65.8	86.2	98.8	67.0	85.4	99.7	66.8	85.8	98.2	67.7	85.3

DIVISION		FLORIDA						GEORGIA											
		TAMPA			WEST PALM BEACH			ALBANY			ATHENS			ATLANTA			AUGUSTA		
		335 - 336, 346			334, 349			317, 398			306			300 - 303, 399			308 - 309		
		MAT.	INST.	TOTAL	MAT.	INST.	TOTAL	MAT.	INST.	TOTAL	MAT.	INST.	TOTAL	MAT.	INST.	TOTAL	MAT.	INST.	TOTAL
015433	CONTRACTOR EQUIPMENT		99.5	99.5		92.8	92.8		93.7	93.7		92.6	92.6		96.6	96.6		92.6	92.6
0241, 31 - 34	SITE & INFRASTRUCTURE, DEMOLITION	110.9	87.0	94.4	91.6	73.2	78.9	107.3	76.7	86.1	102.8	90.8	94.5	99.7	96.0	97.1	95.9	91.9	93.2
0310	Concrete Forming & Accessories	95.4	64.1	68.7	96.7	59.1	64.7	90.6	66.5	70.1	90.0	43.6	50.4	94.3	72.3	75.5	90.9	73.0	75.6
0320	Concrete Reinforcing	94.6	76.5	85.8	97.3	57.5	78.1	96.5	71.3	84.3	94.5	63.5	79.5	93.9	71.3	83.0	94.9	70.7	83.2
0330	Cast-in-Place Concrete	101.5	66.2	88.3	91.5	62.2	80.6	94.6	69.1	85.1	109.7	69.5	94.7	113.1	71.1	97.5	103.6	70.0	91.1
03	CONCRETE	95.9	68.7	83.9	90.1	61.8	77.7	86.9	70.1	79.5	99.6	57.4	81.1	101.8	72.4	88.9	92.5	72.2	83.6
04	MASONRY	88.3	61.0	71.5	87.1	52.2	65.7	92.5	68.7	77.9	75.0	77.0	76.2	87.8	68.9	76.2	88.1	68.8	76.2
05	METALS	99.0	95.0	97.8	95.9	87.0	93.3	106.2	97.4	103.6	95.3	79.1	90.6	96.2	83.5	92.5	95.0	82.6	91.4
06	WOOD, PLASTICS & COMPOSITES	83.3	63.2	72.2	83.1	60.6	70.7	80.1	66.4	72.6	92.5	35.8	61.3	97.8	74.0	84.7	93.8	75.3	83.6
07	THERMAL & MOISTURE PROTECTION	106.5	63.8	87.9	106.0	60.6	86.2	100.6	68.0	86.4	95.4	68.3	83.6	96.8	73.8	86.8	95.1	71.4	84.8
08	OPENINGS	95.5	64.7	88.3	93.8	59.0	85.7	85.7	68.5	81.7	91.3	50.2	81.7	99.4	73.5	93.3	91.4	74.0	87.3
0920	Plaster & Gypsum Board	108.2	62.7	77.6	111.9	60.0	77.0	101.1	66.0	77.5	93.0	34.4	53.6	95.0	73.6	80.6	94.0	75.0	81.2
0950, 0980	Ceilings & Acoustic Treatment	84.8	62.7	69.9	79.7	60.0	66.4	77.5	66.0	69.8	96.6	34.4	54.7	89.7	73.6	78.8	97.5	75.0	82.4
0960	Flooring	97.1	62.7	87.5	98.9	54.9	86.6	99.4	70.8	91.4	96.2	81.9	92.2	99.1	70.8	91.1	96.4	70.8	89.2
0970, 0990	Wall Finishes & Painting/Coating	101.4	63.2	78.8	96.5	58.3	73.9	88.7	94.0	91.8	92.4	94.0	93.3	96.1	97.3	96.8	92.4	90.2	91.1
09	FINISHES	93.9	63.1	77.3	91.2	57.8	73.2	91.7	69.7	79.8	95.0	53.9	72.7	95.0	74.5	83.9	94.7	74.5	83.7
COVERS	DIVS. 10 - 14, 25, 28, 41, 43, 44, 46	100.0	80.0	95.6	100.0	80.7	95.7	100.0	85.9	96.8	100.0	82.7	96.1	100.0	87.2	97.2	100.0	87.0	97.1
21, 22, 23	FIRE SUPPRESSION, PLUMBING & HVAC	100.0	59.9	84.3	98.1	58.5	82.6	100.0	68.9	87.8	96.6	66.3	84.7	100.0	69.4	88.0	100.1	69.0	87.9
26, 27, 3370	ELECTRICAL, COMMUNICATIONS & UTIL.	94.7	63.1	79.1	95.7	67.4	81.8	95.5	61.9	79.0	97.8	63.9	81.1	97.6	71.5	84.7	98.3	67.3	83.1
MF2018	WEIGHTED AVERAGE	97.6	68.4	85.3	95.1	64.1	82.0	96.5	71.9	86.1	95.6	66.9	83.5	98.2	75.0	88.4	96.0	73.9	86.6

DIVISION		GEORGIA																	
		COLUMBUS			DALTON			GAINESVILLE			MACON			SAVANNAH			STATESBORO		
		318 - 319			307			305			310 - 312			313 - 314			304		
		MAT.	INST.	TOTAL	MAT.	INST.	TOTAL	MAT.	INST.	TOTAL	MAT.	INST.	TOTAL	MAT.	INST.	TOTAL	MAT.	INST.	TOTAL
015433	CONTRACTOR EQUIPMENT		93.7	93.7		107.6	107.6		92.6	92.6		103.2	103.2		96.7	96.7		95.5	95.5
0241, 31 - 34	SITE & INFRASTRUCTURE, DEMOLITION	107.2	76.7	86.2	102.6	95.8	97.9	102.5	90.7	94.3	108.6	91.0	96.4	107.0	82.0	89.7	103.7	78.0	85.9
0310	Concrete Forming & Accessories	90.5	66.6	70.1	83.2	64.5	67.3	93.3	40.7	48.5	90.3	66.6	70.1	93.4	72.0	75.2	78.5	52.2	56.1
0320	Concrete Reinforcing	96.6	71.3	84.4	94.1	60.9	78.0	94.4	63.4	79.4	97.8	71.3	85.0	104.1	70.7	88.0	93.7	71.2	82.8
0330	Cast-in-Place Concrete	94.2	68.8	84.8	106.6	68.0	92.2	115.3	68.7	98.0	93.0	68.3	83.8	98.3	69.5	87.6	109.5	68.6	94.3
03	CONCRETE	86.8	70.0	79.4	98.9	67.0	84.9	101.4	55.8	81.4	86.3	69.8	79.1	88.5	72.5	81.5	99.3	63.4	83.5
04	MASONRY	92.5	68.7	77.9	75.8	75.5	75.6	83.2	76.2	78.9	104.8	68.9	82.7	87.4	68.8	75.9	77.8	77.0	77.3
05	METALS	105.9	97.6	103.4	96.4	94.0	95.7	94.6	78.5	89.9	100.9	97.6	100.0	102.1	96.0	100.3	100.0	98.9	99.7
06	WOOD, PLASTICS & COMPOSITES	80.1	66.4	72.6	75.9	65.2	70.0	96.1	33.2	61.5	85.9	66.5	75.2	87.4	73.9	80.0	69.7	47.2	57.3
07	THERMAL & MOISTURE PROTECTION	100.5	68.9	86.8	97.3	70.1	85.5	95.4	67.7	83.3	99.0	70.6	86.6	98.5	70.4	86.3	95.9	67.1	83.4
08	OPENINGS	85.7	69.3	81.9	92.8	65.8	86.5	91.3	48.8	81.4	85.5	69.3	81.7	95.3	73.2	90.2	93.7	58.5	85.5
0920	Plaster & Gypsum Board	101.1	66.0	77.5	81.9	64.7	70.3	94.9	31.8	52.4	104.7	66.0	78.7	103.3	73.6	83.3	83.6	46.2	58.4
0950, 0980	Ceilings & Acoustic Treatment	77.5	66.0	69.8	110.2	64.7	79.5	96.6	31.8	52.9	72.8	66.0	68.3	87.3	73.6	78.0	105.8	46.2	65.6
0960	Flooring	99.4	70.8	91.4	97.0	81.9	92.8	97.8	81.9	93.3	77.9	70.8	75.9	95.6	70.8	88.6	114.6	81.9	105.4
0970, 0990	Wall Finishes & Painting/Coating	88.7	87.4	87.9	82.8	70.1	75.3	92.4	94.0	93.3	90.7	94.0	92.6	88.0	86.1	86.9	90.8	70.1	78.5
09	FINISHES	91.6	68.9	79.3	104.5	68.2	84.9	95.5	52.1	72.0	80.5	69.7	74.7	92.9	73.2	82.3	107.9	57.9	80.9
COVERS	DIVS. 10 - 14, 25, 28, 41, 43, 44, 46	100.0	85.9	96.9	100.0	85.3	96.7	100.0	35.5	85.6	100.0	86.0	96.9	100.0	87.0	97.1	100.0	83.8	96.4
21, 22, 23	FIRE SUPPRESSION, PLUMBING & HVAC	100.1	66.8	87.0	96.7	60.8	82.6	96.6	65.6	84.4	100.1	68.2	87.6	100.1	64.8	86.3	97.2	67.3	85.5
26, 27, 3370	ELECTRICAL, COMMUNICATIONS & UTIL.	95.7	67.7	81.9	106.9	62.2	84.9	97.8	70.8	84.5	94.2	62.7	78.6	99.1	63.7	81.7	98.3	63.4	81.1
MF2018	WEIGHTED AVERAGE	96.5	72.2	86.2	97.5	71.4	86.5	96.2	65.6	83.2	95.2	73.1	85.8	97.3	72.7	86.9	97.8	69.7	85.9

DIVISION		GEORGIA						HAWAII									IDAHO		
		VALDOSTA			WAYCROSS			HILO			HONOLULU			STATES & POSS., GUAM			BOISE		
		316			315			967			968			969			836 - 837		
		MAT.	INST.	TOTAL	MAT.	INST.	TOTAL	MAT.	INST.	TOTAL	MAT.	INST.	TOTAL	MAT.	INST.	TOTAL	MAT.	INST.	TOTAL
015433	CONTRACTOR EQUIPMENT		93.7	93.7		93.7	93.7		97.8	97.8		99.3	99.3		160.5	160.5		94.6	94.6
0241, 31 - 34	SITE & INFRASTRUCTURE, DEMOLITION	116.9	76.7	89.1	113.7	77.0	88.3	142.5	103.1	115.3	150.0	106.6	120.0	186.3	98.5	125.7	86.8	92.0	90.4
0310	Concrete Forming & Accessories	81.7	41.6	47.5	83.5	64.0	66.8	108.2	123.8	121.5	120.2	123.7	123.2	110.5	51.5	60.2	100.2	81.1	83.9
0320	Concrete Reinforcing	98.7	59.5	79.8	98.7	59.8	79.9	143.4	127.0	135.5	166.8	127.0	147.6	253.7	27.4	144.2	107.6	79.5	94.0
0330	Cast-in-Place Concrete	92.6	68.7	83.7	104.7	68.6	91.2	180.3	123.3	159.1	145.0	123.5	137.0	156.3	95.7	133.7	90.6	96.6	92.8
03	CONCRETE	91.2	56.6	76.0	94.5	66.5	82.2	144.7	123.2	135.3	137.9	123.1	131.4	147.9	63.9	111.0	97.6	86.4	92.7
04	MASONRY	97.8	77.1	85.1	98.7	77.0	85.4	148.6	122.1	132.3	135.5	122.2	127.3	206.7	34.8	101.0	126.3	86.2	101.7
05	METALS	105.3	93.0	101.7	104.4	88.9	99.8	110.8	108.1	110.0	124.1	107.1	119.1	143.1	76.0	123.4	109.6	82.7	101.7
06	WOOD, PLASTICS & COMPOSITES	68.7	33.0	49.0	70.2	63.4	66.5	111.4	124.3	118.5	132.8	124.2	128.1	123.4	53.9	85.1	93.5	79.9	86.0
07	THERMAL & MOISTURE PROTECTION	100.8	66.6	86.0	100.6	69.1	86.9	128.7	119.3	124.6	146.3	120.0	134.9	150.7	58.2	110.4	101.4	86.6	94.9
08	OPENINGS	82.5	47.6	74.4	82.8	61.8	77.9	114.1	123.2	116.2	128.1	123.2	126.9	118.4	43.8	101.0	96.9	75.9	92.0
0920	Plaster & Gypsum Board	93.8	31.6	52.0	93.8	62.9	73.0	115.2	124.8	121.7	160.8	124.8	136.6	236.1	42.5	105.8	92.3	79.4	83.6
0950, 0980	Ceilings & Acoustic Treatment	75.9	31.6	46.0	74.0	62.9	66.5	136.6	124.8	128.6	144.8	124.8	131.3	257.2	42.5	112.4	102.8	79.4	87.0
0960	Flooring	93.0	83.5	90.3	94.3	81.9	90.8	103.8	139.3	113.8	121.3	139.3	126.4	121.2	39.5	98.2	92.9	90.7	92.3
0970, 0990	Wall Finishes & Painting/Coating	88.7	94.0	91.8	88.7	70.1	77.7	99.2	143.8	125.6	110.1	143.8	130.1	105.2	31.5	61.6	92.6	39.5	61.2
09	FINISHES	89.4	52.5	69.4	89.0	67.5	77.4	110.8	129.1	120.7	125.5	129.1	127.4	185.1	48.0	110.9	92.7	78.2	84.9
COVERS	DIVS. 10 - 14, 25, 28, 41, 43, 44, 46	100.0	82.2	96.0	100.0	85.5	96.8	100.0	113.7	103.1	100.0	113.6	103.0	100.0	66.0	92.4	100.0	88.6	97.4
21, 22, 23	FIRE SUPPRESSION, PLUMBING & HVAC	100.1	68.5	87.7	97.9	63.6	84.4	100.3	112.4	105.0	100.4	112.3	105.1	102.6	33.5	75.5	100.1	74.2	89.9
26, 27, 3370	ELECTRICAL, COMMUNICATIONS & UTIL.	94.1	57.8	76.2	97.7	63.4	80.8	108.1	124.6	116.2	109.8	124.5	117.1	158.1	35.9	97.9	97.2	68.2	82.9
MF2018	WEIGHTED AVERAGE	96.7	66.2	83.8	96.8	70.0	85.5	114.8	118.4	116.3	119.1	118.5	118.9	137.1	51.4	100.9	101.0	79.9	92.1

DIVISION		IDAHO															ILLINOIS		
		COEUR D'ALENE			IDAHO FALLS			LEWISTON			POCATELLO			TWIN FALLS			BLOOMINGTON		
		838			834			835			832			833			617		
		MAT.	INST.	TOTAL	MAT.	INST.	TOTAL	MAT.	INST.	TOTAL	MAT.	INST.	TOTAL	MAT.	INST.	TOTAL	MAT.	INST.	TOTAL
015433	CONTRACTOR EQUIPMENT		89.7	89.7		94.6	94.6		89.7	89.7		94.6	94.6		94.6	94.6		101.9	101.9
0241, 31 - 34	SITE & INFRASTRUCTURE, DEMOLITION	85.5	86.5	86.2	85.1	92.1	89.9	92.2	87.2	88.7	88.3	92.0	90.9	95.4	91.8	92.9	94.9	96.0	95.7
0310	Concrete Forming & Accessories	108.7	81.4	85.4	94.1	78.3	80.6	113.9	81.1	85.9	100.4	80.5	83.4	101.6	76.5	80.2	84.0	116.0	111.2
0320	Concrete Reinforcing	115.7	96.2	106.3	109.6	77.9	94.3	115.7	96.2	106.3	108.0	79.4	94.2	110.0	77.8	94.4	93.0	102.4	97.5
0330	Cast-in-Place Concrete	98.0	85.2	93.2	86.3	83.4	85.2	101.8	82.0	94.5	93.1	96.4	94.3	95.6	82.3	90.6	98.0	113.7	103.9
03	CONCRETE	103.9	85.3	95.7	89.6	80.2	85.5	107.5	84.1	97.2	96.7	86.0	92.0	104.6	79.0	93.4	91.9	113.6	101.4
04	MASONRY	128.0	87.3	103.0	121.5	87.2	100.4	128.5	84.7	101.6	124.0	86.2	100.8	126.7	82.8	99.7	110.2	121.7	117.3
05	METALS	103.0	88.1	98.6	118.1	81.6	107.4	102.4	88.2	98.3	118.2	82.0	107.6	118.2	81.4	107.4	95.2	123.1	103.4
06	WOOD, PLASTICS & COMPOSITES	96.7	79.8	87.4	87.2	76.3	81.2	102.6	79.8	90.1	93.5	79.9	86.0	94.6	76.3	84.5	81.7	113.0	99.0
07	THERMAL & MOISTURE PROTECTION	158.9	84.3	126.4	100.4	75.3	89.5	159.2	83.1	126.1	101.0	75.0	89.7	101.8	80.5	92.5	95.4	112.1	102.6
08	OPENINGS	113.8	75.9	105.0	99.8	71.3	93.2	106.7	79.6	100.4	97.6	68.8	90.9	100.6	62.4	91.7	91.3	118.7	97.7
0920	Plaster & Gypsum Board	167.0	79.4	108.1	80.3	75.7	77.2	168.9	79.4	108.7	81.9	79.4	80.2	84.0	75.7	78.4	89.1	113.5	105.5
0950, 0980	Ceilings & Acoustic Treatment	136.3	79.4	98.0	104.3	75.7	85.0	136.3	79.4	98.0	110.0	79.4	89.4	106.8	75.7	85.8	83.9	113.5	103.9
0960	Flooring	130.9	77.0	115.8	92.7	77.0	88.3	134.3	77.0	118.2	96.2	77.0	90.8	97.5	77.0	91.7	86.0	120.8	95.8
0970, 0990	Wall Finishes & Painting/Coating	111.2	73.8	89.1	92.6	39.5	61.2	111.2	71.3	87.6	92.5	39.5	61.1	92.6	39.5	61.2	87.6	134.8	115.5
09	FINISHES	158.3	79.5	115.7	90.9	74.0	81.8	159.7	78.9	116.0	93.7	75.9	84.1	94.5	72.9	82.8	86.1	119.0	103.9
COVERS	DIVS. 10 - 14, 25, 28, 41, 43, 44, 46	100.0	93.0	98.4	100.0	87.4	97.2	100.0	92.6	98.3	100.0	88.5	97.4	100.0	85.9	96.9	100.0	104.5	101.0
21, 22, 23	FIRE SUPPRESSION, PLUMBING & HVAC	99.4	85.5	94.0	101.1	81.8	93.5	100.9	85.1	94.6	100.0	73.6	89.6	100.0	72.2	89.1	96.5	105.4	100.0
26, 27, 3370	ELECTRICAL, COMMUNICATIONS & UTIL.	88.7	81.5	85.2	88.3	70.9	79.7	86.9	78.2	82.6	94.4	66.0	80.4	89.6	70.9	80.4	95.9	90.3	93.1
MF2018	WEIGHTED AVERAGE	108.1	84.4	98.1	100.5	79.9	91.8	108.2	83.5	97.8	102.0	78.4	92.0	103.2	76.8	92.1	95.0	109.5	101.1

	DIVISION	ILLINOIS																	
		CARBONDALE			CENTRALIA			CHAMPAIGN			CHICAGO			DECATUR			EAST ST. LOUIS		
		629			628			618 - 619			606 - 608			625			620 - 622		
		MAT.	INST.	TOTAL	MAT.	INST.	TOTAL	MAT.	INST.	TOTAL	MAT.	INST.	TOTAL	MAT.	INST.	TOTAL	MAT.	INST.	TOTAL
015433	**CONTRACTOR EQUIPMENT**		110.3	110.3		110.3	110.3		102.8	102.8		100.3	100.3		102.8	102.8		110.3	110.3
0241, 31 - 34	**SITE & INFRASTRUCTURE, DEMOLITION**	100.9	96.8	98.1	101.3	98.6	99.5	104.0	97.2	99.3	105.4	104.4	104.7	94.8	97.0	96.3	103.5	97.7	99.5
0310	Concrete Forming & Accessories	90.0	108.9	106.1	91.6	114.2	110.9	90.3	116.7	112.8	97.4	159.7	150.5	92.1	117.9	114.1	87.5	114.2	110.2
0320	Concrete Reinforcing	82.2	103.4	92.5	82.2	104.0	92.8	93.0	101.6	97.2	103.8	153.3	127.7	80.6	99.8	89.9	82.2	104.0	92.7
0330	Cast-in-Place Concrete	91.5	102.7	95.7	92.0	119.6	102.3	113.7	110.8	112.6	126.0	155.6	137.0	99.9	114.6	105.4	93.6	118.4	102.8
03	**CONCRETE**	79.9	107.3	91.9	80.4	115.8	96.0	104.1	112.7	107.9	108.0	156.5	129.3	90.3	114.3	100.8	81.4	115.4	96.3
04	**MASONRY**	75.0	111.5	97.4	75.0	122.9	104.5	133.1	125.4	128.4	105.3	163.4	141.0	70.9	122.8	102.8	75.3	123.0	104.6
05	**METALS**	100.9	130.4	109.6	100.9	133.5	110.5	95.2	120.3	102.6	95.9	146.8	110.8	104.9	120.7	109.5	102.1	133.5	111.3
06	**WOOD, PLASTICS & COMPOSITES**	87.7	105.8	97.7	90.2	111.2	101.7	88.5	114.0	102.5	100.0	158.4	132.1	88.7	116.8	104.2	84.8	111.2	99.3
07	**THERMAL & MOISTURE PROTECTION**	90.6	101.3	95.3	90.7	111.5	99.8	96.2	115.6	104.6	95.2	150.7	119.4	95.9	113.3	103.5	90.7	110.7	99.4
08	**OPENINGS**	87.3	116.3	94.1	87.3	119.2	94.8	91.8	116.9	97.7	100.9	169.3	116.9	98.2	118.2	102.9	87.4	118.4	94.6
0920	Plaster & Gypsum Board	93.7	106.2	102.1	95.0	111.6	106.2	91.0	114.5	106.8	101.3	160.0	140.8	96.5	117.4	110.6	92.2	111.6	105.3
0950, 0980	Ceilings & Acoustic Treatment	80.6	106.2	97.8	80.6	111.6	101.5	83.9	114.5	104.5	100.4	160.0	140.6	87.1	117.4	107.5	80.6	111.6	101.5
0960	Flooring	113.7	116.6	114.5	114.6	116.6	115.1	89.2	116.6	96.9	91.3	163.8	111.7	101.8	117.1	106.1	112.6	116.6	113.7
0970, 0990	Wall Finishes & Painting/Coating	103.1	105.0	104.2	103.1	111.6	108.2	87.6	114.9	103.7	96.7	170.3	140.2	95.5	117.8	108.7	103.1	111.6	108.2
09	**FINISHES**	92.1	110.7	102.1	92.5	114.4	104.4	88.1	116.9	103.7	95.8	162.4	131.8	92.2	118.3	106.3	91.6	114.4	103.9
COVERS	**DIVS. 10 - 14, 25, 28, 41, 43, 44, 46**	100.0	106.9	101.5	100.0	105.8	101.3	100.0	109.2	102.0	100.0	129.3	106.5	100.0	108.6	101.9	100.0	108.3	101.9
21, 22, 23	**FIRE SUPPRESSION, PLUMBING & HVAC**	96.3	106.8	100.4	96.3	97.8	96.9	96.5	107.0	100.6	99.9	137.4	114.6	99.9	98.8	99.5	99.9	99.2	99.6
26, 27, 3370	**ELECTRICAL, COMMUNICATIONS & UTIL.**	91.9	107.8	99.8	93.1	106.3	99.6	99.0	95.4	97.2	97.9	136.2	116.8	95.3	103.7	99.5	92.7	99.7	96.2
MF2018	**WEIGHTED AVERAGE**	92.5	109.6	99.7	92.8	111.2	100.5	98.5	110.5	103.6	100.1	145.6	119.3	96.6	110.1	102.3	93.8	110.4	100.9

	DIVISION	ILLINOIS																	
		EFFINGHAM			GALESBURG			JOLIET			KANKAKEE			LA SALLE			NORTH SUBURBAN		
		624			614			604			609			613			600 - 603		
		MAT.	INST.	TOTAL	MAT.	INST.	TOTAL	MAT.	INST.	TOTAL	MAT.	INST.	TOTAL	MAT.	INST.	TOTAL	MAT.	INST.	TOTAL
015433	**CONTRACTOR EQUIPMENT**		102.8	102.8		101.9	101.9		93.7	93.7		93.7	93.7		101.9	101.9		93.7	93.7
0241, 31 - 34	**SITE & INFRASTRUCTURE, DEMOLITION**	99.3	95.9	97.0	97.3	95.8	96.3	101.2	97.2	98.5	94.7	97.1	96.4	96.7	96.8	96.8	100.5	97.9	98.7
0310	Concrete Forming & Accessories	96.5	114.6	111.9	90.2	117.0	113.0	96.8	156.7	147.8	90.5	145.4	137.3	103.7	125.3	122.1	96.1	158.5	149.3
0320	Concrete Reinforcing	83.3	93.8	88.4	92.5	105.6	98.8	103.8	136.6	119.7	104.6	135.1	119.3	92.7	132.9	112.1	103.8	149.1	125.7
0330	Cast-in-Place Concrete	99.6	109.9	103.4	100.9	109.2	104.0	114.6	143.9	125.5	106.8	135.8	117.6	100.8	121.8	108.6	114.6	152.9	128.9
03	**CONCRETE**	91.0	109.9	99.3	94.8	113.0	102.8	101.0	147.9	121.6	95.0	139.7	114.7	95.6	126.1	109.0	101.0	154.1	124.3
04	**MASONRY**	79.1	115.3	101.3	110.4	122.4	117.8	104.5	156.7	136.6	100.6	147.0	129.1	110.4	129.7	122.3	101.2	158.8	136.6
05	**METALS**	101.9	114.9	105.7	95.2	123.6	103.5	94.1	134.7	106.0	94.1	133.4	105.6	95.3	142.0	109.0	95.1	141.6	108.8
06	**WOOD, PLASTICS & COMPOSITES**	91.0	114.0	103.6	88.3	115.4	103.2	96.6	157.7	130.2	90.0	144.7	120.1	103.8	123.2	114.5	95.4	158.2	130.0
07	**THERMAL & MOISTURE PROTECTION**	95.4	109.1	101.4	95.6	110.0	101.9	99.5	144.2	119.0	98.6	139.5	116.4	95.8	122.5	107.4	99.9	147.6	120.7
08	**OPENINGS**	93.3	114.5	98.2	91.3	115.9	97.0	99.3	161.7	113.9	92.4	156.2	107.3	91.3	133.3	101.1	99.4	167.9	115.4
0920	Plaster & Gypsum Board	96.3	114.5	108.5	91.0	115.9	107.8	93.7	159.5	138.0	92.5	146.1	128.6	98.3	124.0	115.6	97.2	160.0	139.5
0950, 0980	Ceilings & Acoustic Treatment	80.6	114.5	103.4	83.9	115.9	105.5	100.9	159.5	140.4	100.9	146.1	131.4	83.9	124.0	110.9	100.9	160.0	140.8
0960	Flooring	103.0	116.6	106.8	89.1	120.8	98.0	89.4	149.8	106.4	86.3	150.7	104.4	95.7	124.8	103.9	89.9	159.5	109.4
0970, 0990	Wall Finishes & Painting/Coating	95.5	109.3	103.7	87.6	98.5	94.0	89.0	163.0	132.8	89.0	135.7	116.6	87.6	133.0	114.5	91.0	163.9	134.1
09	**FINISHES**	91.3	115.1	104.2	87.4	116.3	103.1	91.6	157.4	127.2	90.0	145.8	120.2	90.4	124.7	109.0	92.2	160.1	129.0
COVERS	**DIVS. 10 - 14, 25, 28, 41, 43, 44, 46**	100.0	104.5	101.0	100.0	104.4	101.0	100.0	122.6	105.0	100.0	120.6	104.6	100.0	105.6	101.2	100.0	124.1	105.4
21, 22, 23	**FIRE SUPPRESSION, PLUMBING & HVAC**	96.4	104.5	99.6	96.5	105.7	100.1	100.0	130.0	111.8	96.5	129.6	109.5	96.5	128.4	109.0	99.9	136.2	114.1
26, 27, 3370	**ELECTRICAL, COMMUNICATIONS & UTIL.**	93.5	107.8	100.5	96.7	86.5	91.7	96.9	133.3	114.8	92.4	137.0	114.4	94.0	137.0	115.2	96.7	135.5	115.9
MF2018	**WEIGHTED AVERAGE**	95.1	109.1	101.0	95.7	108.5	101.1	98.3	138.8	115.4	95.0	134.9	111.9	95.9	127.0	109.0	98.3	142.9	117.1

	DIVISION	ILLINOIS																	
		PEORIA			QUINCY			ROCK ISLAND			ROCKFORD			SOUTH SUBURBAN			SPRINGFIELD		
		615 - 616			623			612			610 - 611			605			626 - 627		
		MAT.	INST.	TOTAL	MAT.	INST.	TOTAL	MAT.	INST.	TOTAL	MAT.	INST.	TOTAL	MAT.	INST.	TOTAL	MAT.	INST.	TOTAL
015433	**CONTRACTOR EQUIPMENT**		101.9	101.9		102.8	102.8		101.9	101.9		101.9	101.9		93.7	93.7		105.1	105.1
0241, 31 - 34	**SITE & INFRASTRUCTURE, DEMOLITION**	97.9	96.0	96.6	98.1	96.4	96.9	95.5	94.8	95.0	97.4	96.8	97.0	100.5	97.8	98.6	99.8	100.5	100.3
0310	Concrete Forming & Accessories	93.0	115.9	112.5	94.3	113.6	110.8	91.6	99.1	98.0	97.2	127.7	123.2	96.1	158.5	149.3	93.2	115.6	112.3
0320	Concrete Reinforcing	90.1	104.7	97.2	83.0	86.2	84.5	92.5	99.0	95.6	85.3	134.1	108.9	103.8	149.0	125.7	83.2	99.8	91.2
0330	Cast-in-Place Concrete	97.9	115.6	104.5	99.8	106.6	102.3	98.7	98.4	98.6	100.2	126.6	110.0	114.6	152.8	128.8	94.9	109.2	100.2
03	**CONCRETE**	92.0	114.6	102.0	90.7	107.1	97.9	92.7	100.0	95.9	92.7	129.0	108.7	101.0	154.0	124.3	88.5	111.2	98.5
04	**MASONRY**	109.7	121.5	116.9	98.9	113.2	107.7	110.2	100.2	104.1	85.6	142.3	120.4	101.2	158.7	136.6	80.7	122.2	106.2
05	**METALS**	97.9	124.2	105.6	101.9	112.6	105.1	95.2	118.5	102.1	97.9	141.5	110.7	95.1	141.4	108.7	102.5	119.1	107.3
06	**WOOD, PLASTICS & COMPOSITES**	96.4	113.2	105.6	88.6	114.0	102.6	90.0	97.5	94.1	96.4	123.6	111.3	95.4	158.2	130.0	90.0	114.1	103.3
07	**THERMAL & MOISTURE PROTECTION**	96.3	112.4	103.3	95.3	106.8	100.3	95.5	98.4	96.8	98.8	129.0	111.9	99.9	147.6	120.7	98.3	114.4	105.3
08	**OPENINGS**	96.6	119.5	102.0	94.0	112.1	98.2	91.3	103.3	94.1	96.6	136.5	105.9	99.4	167.9	115.4	99.2	116.4	103.2
0920	Plaster & Gypsum Board	95.5	113.7	107.7	95.0	114.5	108.1	91.0	97.6	95.5	95.5	124.4	114.9	97.2	160.0	139.5	99.9	114.5	109.7
0950, 0980	Ceilings & Acoustic Treatment	89.6	113.7	105.8	80.6	114.5	103.4	83.9	97.6	93.2	89.6	124.4	113.1	100.9	160.0	140.8	91.3	114.5	107.0
0960	Flooring	92.6	119.5	100.2	101.8	113.6	105.1	90.2	94.6	91.4	92.6	124.9	101.7	89.9	159.5	109.4	104.8	117.1	108.3
0970, 0990	Wall Finishes & Painting/Coating	87.6	134.8	115.5	95.5	111.5	105.0	87.6	95.5	92.3	87.6	142.7	120.2	91.0	163.9	134.1	98.0	111.5	106.0
09	**FINISHES**	90.2	118.7	105.6	90.7	114.2	103.4	87.7	97.6	93.0	90.2	128.5	110.9	92.2	160.1	129.0	98.0	115.9	107.7
COVERS	**DIVS. 10 - 14, 25, 28, 41, 43, 44, 46**	100.0	104.4	101.0	100.0	106.9	101.5	100.0	97.5	99.4	100.0	114.2	103.2	100.0	124.1	105.4	100.0	108.3	101.9
21, 22, 23	**FIRE SUPPRESSION, PLUMBING & HVAC**	100.0	100.2	100.0	96.4	102.9	98.9	96.5	97.5	96.9	100.1	116.1	106.4	99.9	136.2	114.1	99.9	103.7	101.4
26, 27, 3370	**ELECTRICAL, COMMUNICATIONS & UTIL.**	97.7	91.5	94.6	91.4	82.2	86.9	89.3	93.1	91.2	98.0	125.6	111.6	96.7	135.5	115.9	98.1	88.7	93.5
MF2018	**WEIGHTED AVERAGE**	97.6	108.8	102.3	95.7	104.2	99.3	94.6	99.5	96.7	96.6	125.5	108.8	98.3	142.8	117.1	97.5	108.3	102.1

	DIVISION	INDIANA																	
		ANDERSON			BLOOMINGTON			COLUMBUS			EVANSVILLE			FORT WAYNE			GARY		
		460			474			472			476 - 477			467 - 468			463 - 464		
		MAT.	INST.	TOTAL	MAT.	INST.	TOTAL	MAT.	INST.	TOTAL	MAT.	INST.	TOTAL	MAT.	INST.	TOTAL	MAT.	INST.	TOTAL
015433	CONTRACTOR EQUIPMENT		94.5	94.5		81.5	81.5		81.5	81.5		110.3	110.3		94.5	94.5		94.5	94.5
0241, 31 - 34	SITE & INFRASTRUCTURE, DEMOLITION	98.7	89.0	92.0	86.5	87.7	87.3	83.2	87.4	86.1	91.9	115.1	107.9	99.8	88.8	92.2	99.3	92.8	94.8
0310	Concrete Forming & Accessories	94.9	77.4	80.0	100.8	82.1	84.9	95.0	80.2	82.4	94.3	78.6	80.9	92.9	73.7	76.5	95.0	111.1	108.7
0320	Concrete Reinforcing	104.9	83.0	94.3	90.4	86.9	88.7	90.8	86.9	88.9	99.0	81.0	90.3	104.9	77.9	91.9	104.9	115.8	110.2
0330	Cast-in-Place Concrete	106.3	74.9	94.6	101.3	75.7	91.7	100.8	76.5	91.8	96.8	83.2	91.7	113.0	75.2	98.9	111.1	112.2	111.5
03	CONCRETE	95.7	78.1	88.0	99.2	80.3	90.9	98.5	79.6	90.2	99.5	80.8	91.3	98.6	75.7	88.6	97.9	112.1	104.1
04	MASONRY	87.4	74.0	79.2	88.9	74.4	80.0	88.8	74.3	79.9	84.3	77.3	80.0	90.7	71.4	78.8	88.8	110.3	102.0
05	METALS	97.7	88.2	94.9	99.1	75.7	92.2	99.1	74.7	91.9	91.9	83.2	89.3	97.7	86.3	94.3	97.7	107.1	100.5
06	WOOD, PLASTICS & COMPOSITES	93.8	77.6	84.9	109.5	82.9	94.9	104.0	80.5	91.1	90.8	77.7	83.6	93.7	73.8	82.7	91.3	109.1	101.1
07	THERMAL & MOISTURE PROTECTION	109.2	75.0	94.4	95.8	77.9	88.0	95.2	78.7	88.0	100.0	81.7	92.1	108.9	76.8	94.9	107.7	106.2	107.0
08	OPENINGS	93.2	75.8	89.2	97.5	80.0	93.4	93.9	78.6	90.3	91.8	76.0	88.1	93.2	71.2	88.1	93.2	114.6	98.2
0920	Plaster & Gypsum Board	106.4	77.3	86.8	98.6	83.2	88.2	95.4	80.7	85.5	94.2	76.7	82.4	105.8	73.4	84.0	99.5	109.7	106.4
0950, 0980	Ceilings & Acoustic Treatment	89.8	77.3	81.4	77.9	83.2	81.5	77.9	80.7	79.8	81.6	76.7	78.3	89.8	73.4	78.7	89.8	109.7	103.2
0960	Flooring	93.5	75.0	88.3	99.6	83.2	95.0	94.4	83.2	91.2	94.1	71.4	87.7	93.5	71.2	87.2	93.5	110.5	98.3
0970, 0990	Wall Finishes & Painting/Coating	92.8	65.8	76.9	84.4	81.1	82.5	84.4	81.1	82.5	90.3	84.4	86.8	92.8	70.2	79.5	92.8	120.7	109.3
09	FINISHES	91.3	75.7	82.9	90.6	82.4	86.2	88.6	81.0	84.5	89.3	77.7	83.0	91.1	72.9	81.3	90.3	111.8	102.0
COVERS	DIVS. 10 - 14, 25, 28, 41, 43, 44, 46	100.0	87.4	97.2	100.0	88.3	97.4	100.0	88.0	97.3	100.0	91.2	98.0	100.0	87.2	97.1	100.0	107.9	101.8
21, 22, 23	FIRE SUPPRESSION, PLUMBING & HVAC	100.0	77.0	91.0	99.7	79.6	91.8	96.2	79.1	89.5	99.9	78.6	91.5	100.0	72.2	89.1	100.0	106.2	102.4
26, 27, 3370	ELECTRICAL, COMMUNICATIONS & UTIL.	87.7	83.2	85.5	99.9	86.6	93.3	99.1	86.7	93.0	95.7	81.4	88.7	88.4	74.9	81.7	99.5	110.1	104.7
MF2018	WEIGHTED AVERAGE	96.0	79.8	89.1	97.8	81.1	90.7	96.1	80.5	89.5	95.5	82.7	90.1	96.6	76.3	88.0	97.5	108.1	102.0

	DIVISION	INDIANA																	
		INDIANAPOLIS			KOKOMO			LAFAYETTE			LAWRENCEBURG			MUNCIE			NEW ALBANY		
		461 - 462			469			479			470			473			471		
		MAT.	INST.	TOTAL	MAT.	INST.	TOTAL	MAT.	INST.	TOTAL	MAT.	INST.	TOTAL	MAT.	INST.	TOTAL	MAT.	INST.	TOTAL
015433	CONTRACTOR EQUIPMENT		86.1	86.1		94.5	94.5		81.5	81.5		100.6	100.6		92.8	92.8		90.6	90.6
0241, 31 - 34	SITE & INFRASTRUCTURE, DEMOLITION	99.8	91.3	93.9	94.9	89.0	90.8	83.9	87.4	86.3	81.2	102.2	95.7	86.6	87.9	87.5	78.5	89.8	86.3
0310	Concrete Forming & Accessories	100.4	85.2	87.4	98.0	78.2	81.1	92.5	78.6	80.7	91.5	78.4	80.3	92.3	76.9	79.2	90.1	76.9	78.9
0320	Concrete Reinforcing	106.1	87.2	97.0	94.9	87.2	91.1	90.4	82.9	86.8	89.7	79.8	84.9	100.0	82.9	91.7	91.0	82.6	86.9
0330	Cast-in-Place Concrete	101.2	84.8	95.1	105.2	81.8	96.5	101.4	77.6	92.5	94.8	74.9	87.4	106.4	73.9	94.3	97.9	73.1	88.7
03	CONCRETE	99.0	84.8	92.8	92.5	81.6	87.7	98.7	78.7	89.9	91.8	78.0	85.7	97.4	77.6	88.7	97.3	76.8	88.3
04	MASONRY	89.8	78.7	83.0	87.1	77.2	81.0	94.0	74.0	81.7	73.8	74.2	74.1	90.5	74.1	80.4	80.2	71.1	74.6
05	METALS	94.7	76.0	89.2	94.2	89.8	92.9	97.4	73.7	90.5	93.9	86.2	91.7	100.8	88.0	97.0	96.0	82.2	91.9
06	WOOD, PLASTICS & COMPOSITES	100.1	86.0	92.4	96.8	77.4	86.1	101.2	79.2	89.1	89.1	78.3	83.2	102.6	77.1	88.6	91.2	77.6	83.7
07	THERMAL & MOISTURE PROTECTION	98.7	82.0	91.4	108.1	77.5	94.8	95.2	77.2	87.4	100.8	76.9	90.4	98.2	76.4	88.7	87.6	72.3	80.9
08	OPENINGS	103.9	81.9	98.8	88.4	76.9	85.7	92.4	76.7	88.8	93.6	75.0	89.3	91.0	75.6	87.4	91.4	76.3	87.9
0920	Plaster & Gypsum Board	96.8	85.8	89.4	111.0	77.1	88.2	93.3	79.4	83.9	71.7	78.4	76.2	94.2	77.3	82.8	92.0	77.5	82.2
0950, 0980	Ceilings & Acoustic Treatment	92.7	85.8	88.0	90.6	77.1	81.5	73.9	79.4	77.6	84.0	78.4	80.2	77.9	77.3	77.5	81.6	77.5	78.8
0960	Flooring	97.2	83.2	93.2	97.3	88.4	94.8	93.3	79.2	89.3	68.9	83.2	72.9	93.7	75.0	88.5	91.6	54.0	81.1
0970, 0990	Wall Finishes & Painting/Coating	96.5	81.1	87.4	92.8	68.8	78.6	84.4	81.4	82.6	85.2	73.0	78.0	84.4	65.8	73.4	90.3	66.3	76.1
09	FINISHES	94.9	84.6	89.4	93.1	79.1	85.5	87.2	79.2	82.8	78.8	78.9	78.8	87.9	75.4	81.1	88.5	71.4	79.2
COVERS	DIVS. 10 - 14, 25, 28, 41, 43, 44, 46	100.0	93.7	98.6	100.0	88.6	97.5	100.0	86.6	97.0	100.0	88.0	97.3	100.0	86.3	97.0	100.0	87.7	97.3
21, 22, 23	FIRE SUPPRESSION, PLUMBING & HVAC	100.1	79.9	92.2	96.5	79.3	89.7	96.2	76.2	88.3	97.0	75.9	88.7	99.7	76.9	90.7	96.4	78.6	89.4
26, 27, 3370	ELECTRICAL, COMMUNICATIONS & UTIL.	101.9	86.7	94.4	92.0	78.3	85.3	98.6	78.6	88.7	93.6	74.1	84.0	91.4	74.0	82.9	94.3	74.6	84.6
MF2018	WEIGHTED AVERAGE	98.7	83.2	92.1	94.2	81.1	88.6	95.8	78.1	88.3	92.5	79.6	87.0	96.2	78.2	88.6	93.9	77.3	86.9

	DIVISION	INDIANA									IOWA								
		SOUTH BEND			TERRE HAUTE			WASHINGTON			BURLINGTON			CARROLL			CEDAR RAPIDS		
		465 - 466			478			475			526			514			522 - 524		
		MAT.	INST.	TOTAL	MAT.	INST.	TOTAL	MAT.	INST.	TOTAL	MAT.	INST.	TOTAL	MAT.	INST.	TOTAL	MAT.	INST.	TOTAL
015433	CONTRACTOR EQUIPMENT		108.2	108.2		110.3	110.3		110.3	110.3		98.8	98.8		98.8	98.8		95.7	95.7
0241, 31 - 34	SITE & INFRASTRUCTURE, DEMOLITION	97.7	94.2	95.3	93.4	115.4	108.6	93.2	115.7	108.7	98.0	91.8	93.7	87.1	92.6	90.9	99.9	91.4	94.0
0310	Concrete Forming & Accessories	95.1	76.0	78.9	95.0	77.1	79.7	96.0	80.5	82.8	96.5	94.2	94.5	84.1	83.9	83.9	102.5	83.9	86.6
0320	Concrete Reinforcing	104.1	83.6	94.2	99.0	83.0	91.3	91.6	82.6	87.2	94.7	97.8	96.2	95.4	85.8	90.8	95.3	80.2	88.0
0330	Cast-in-Place Concrete	108.2	78.4	97.1	93.7	78.3	88.0	102.0	85.0	95.6	107.3	54.8	87.8	107.3	82.2	97.9	107.6	83.4	98.6
03	CONCRETE	97.7	79.7	89.8	102.5	78.8	92.1	108.3	82.6	97.0	95.8	81.5	89.6	94.7	84.3	90.1	95.9	83.8	90.6
04	MASONRY	91.2	74.0	80.6	92.0	73.3	80.5	84.5	79.6	81.5	99.5	71.9	82.5	101.1	72.4	83.5	105.0	80.0	89.6
05	METALS	101.5	102.4	101.7	92.6	84.4	90.2	87.1	84.7	86.4	87.9	99.1	91.2	87.9	95.4	90.1	90.3	93.1	91.2
06	WOOD, PLASTICS & COMPOSITES	96.6	75.2	84.8	93.0	76.7	84.0	93.2	79.2	85.5	91.6	98.1	95.2	77.8	88.4	83.6	98.9	83.6	90.5
07	THERMAL & MOISTURE PROTECTION	103.2	78.7	92.5	100.1	79.1	91.0	100.0	83.5	92.8	104.4	77.1	92.5	104.7	78.8	93.4	105.5	80.1	94.4
08	OPENINGS	93.1	75.5	89.0	92.3	75.3	88.3	89.3	77.1	86.4	94.5	95.8	94.8	98.8	84.2	95.4	99.3	81.2	95.1
0920	Plaster & Gypsum Board	97.7	74.7	82.2	94.2	75.7	81.8	94.0	78.3	83.4	105.0	98.3	100.5	100.9	88.3	92.4	110.3	83.6	92.3
0950, 0980	Ceilings & Acoustic Treatment	94.2	74.7	81.1	81.6	75.7	77.6	76.7	78.3	77.8	99.0	98.3	98.5	99.0	88.3	91.8	101.4	83.6	89.4
0960	Flooring	90.7	86.2	89.5	94.1	75.7	88.9	95.1	83.2	91.7	93.7	69.1	86.8	88.0	79.5	85.6	107.9	84.3	101.3
0970, 0990	Wall Finishes & Painting/Coating	94.3	82.5	87.3	90.3	79.8	84.1	90.3	84.4	86.8	92.8	84.6	87.9	92.8	84.6	87.9	94.7	70.8	80.6
09	FINISHES	92.5	78.3	84.8	89.3	76.9	82.6	88.9	81.3	84.7	93.9	89.7	91.6	90.1	83.6	86.6	99.3	82.7	90.3
COVERS	DIVS. 10 - 14, 25, 28, 41, 43, 44, 46	100.0	88.7	97.5	100.0	88.7	97.5	100.0	91.8	98.2	100.0	93.4	98.5	100.0	89.1	97.6	100.0	93.4	98.5
21, 22, 23	FIRE SUPPRESSION, PLUMBING & HVAC	99.9	75.0	90.1	99.9	76.5	90.7	96.4	81.0	90.4	96.5	83.8	91.5	96.5	77.8	89.2	100.0	81.2	92.6
26, 27, 3370	ELECTRICAL, COMMUNICATIONS & UTIL.	101.3	83.9	92.7	94.1	83.5	88.8	94.5	83.6	89.2	101.1	72.2	86.9	101.8	76.9	89.5	98.8	80.1	89.5
MF2018	WEIGHTED AVERAGE	98.4	81.8	91.4	96.2	81.7	90.1	94.6	84.8	90.4	95.7	84.3	90.9	95.5	82.4	89.9	98.0	83.7	92.0

DIVISION		IOWA																	
		COUNCIL BLUFFS			CRESTON			DAVENPORT			DECORAH			DES MOINES			DUBUQUE		
		515			508			527 - 528			521			500 - 503,509			520		
		MAT.	INST.	TOTAL	MAT.	INST.	TOTAL	MAT.	INST.	TOTAL	MAT.	INST.	TOTAL	MAT.	INST.	TOTAL	MAT.	INST.	TOTAL
015433	**CONTRACTOR EQUIPMENT**		95.0	95.0		98.8	98.8		98.8	98.8		98.8	98.8		102.4	102.4		94.5	94.5
0241, 31 - 34	**SITE & INFRASTRUCTURE, DEMOLITION**	104.5	88.5	93.4	92.9	93.6	93.4	98.4	94.6	95.8	96.6	91.6	93.2	98.3	99.3	99.0	97.7	88.7	91.5
0310	Concrete Forming & Accessories	83.5	73.6	75.0	79.2	86.7	85.6	102.0	95.9	96.8	94.0	71.8	75.1	95.6	89.5	90.4	84.9	81.5	82.0
0320	Concrete Reinforcing	97.3	79.8	88.8	96.3	85.9	91.3	95.3	99.0	97.1	94.7	85.2	90.1	101.0	101.6	101.3	94.0	79.9	87.2
0330	Cast-in-Place Concrete	111.8	79.1	99.6	114.6	85.7	103.8	103.6	95.2	100.5	104.4	76.0	93.8	97.6	92.1	95.6	105.3	83.0	97.0
03	**CONCRETE**	98.1	77.6	89.1	97.7	86.8	92.9	94.0	96.7	95.2	93.8	76.6	86.3	91.0	92.8	91.8	92.7	82.6	88.3
04	**MASONRY**	106.5	76.2	87.8	99.7	79.9	87.5	102.0	92.2	96.0	120.3	69.2	88.9	87.7	89.0	88.5	105.9	68.8	83.1
05	**METALS**	95.2	92.8	94.5	88.4	96.0	90.6	90.3	105.5	94.8	88.0	94.2	89.9	94.1	96.8	94.9	88.9	92.4	89.9
06	**WOOD, PLASTICS & COMPOSITES**	76.5	72.1	74.1	69.7	88.4	80.0	98.9	95.3	96.9	88.5	70.9	78.8	87.6	88.5	88.1	78.3	82.2	80.4
07	**THERMAL & MOISTURE PROTECTION**	104.8	74.8	91.8	106.2	82.2	95.8	104.9	91.6	99.1	104.6	69.9	89.5	98.5	87.6	93.8	105.1	76.5	92.6
08	**OPENINGS**	98.4	76.8	93.3	108.6	85.9	103.3	99.3	96.7	98.7	97.4	77.9	92.9	100.8	89.6	98.2	98.4	82.4	94.6
0920	Plaster & Gypsum Board	100.9	71.8	81.3	95.8	88.3	90.8	110.3	95.4	100.2	103.7	70.3	81.2	92.8	88.3	89.8	100.9	82.1	88.3
0950, 0980	Ceilings & Acoustic Treatment	99.0	71.8	80.6	90.2	88.3	88.9	101.4	95.4	97.3	99.0	70.3	79.6	93.4	88.3	90.0	99.0	82.1	87.6
0960	Flooring	87.0	84.3	86.2	81.3	69.1	77.8	96.1	89.4	94.2	93.3	69.1	86.5	95.2	94.5	95.0	99.1	69.1	90.7
0970, 0990	Wall Finishes & Painting/Coating	89.6	59.8	72.0	83.6	84.6	84.2	92.8	90.5	91.4	92.8	84.6	87.9	91.6	84.6	87.5	93.9	78.4	84.8
09	**FINISHES**	91.0	73.7	81.7	85.1	83.4	84.2	95.7	94.1	94.8	93.5	72.4	82.1	93.6	90.0	91.6	94.8	78.9	86.2
COVERS	**DIVS. 10 - 14, 25, 28, 41, 43, 44, 46**	100.0	91.4	98.1	100.0	91.6	98.1	100.0	96.9	99.3	100.0	88.6	97.5	100.0	95.6	99.0	100.0	92.3	98.3
21, 22, 23	**FIRE SUPPRESSION, PLUMBING & HVAC**	100.0	72.8	89.4	96.3	79.8	89.8	100.0	93.9	97.6	96.5	74.5	87.9	99.8	86.8	94.7	100.0	75.6	90.4
26, 27, 3370	**ELECTRICAL, COMMUNICATIONS & UTIL.**	103.9	82.7	93.4	93.5	76.9	85.3	96.9	87.6	92.3	98.8	47.2	73.4	105.1	84.3	94.8	102.5	76.9	89.9
MF2018	**WEIGHTED AVERAGE**	98.9	79.2	90.6	95.7	84.2	90.8	97.1	94.5	96.0	96.4	73.7	86.8	97.2	90.2	94.3	97.2	79.9	89.9

DIVISION		IOWA																	
		FORT DODGE			MASON CITY			OTTUMWA			SHENANDOAH			SIBLEY			SIOUX CITY		
		505			504			525			516			512			510 - 511		
		MAT.	INST.	TOTAL	MAT.	INST.	TOTAL	MAT.	INST.	TOTAL	MAT.	INST.	TOTAL	MAT.	INST.	TOTAL	MAT.	INST.	TOTAL
015433	**CONTRACTOR EQUIPMENT**		98.8	98.8		98.8	98.8		94.5	94.5		95.0	95.0		98.8	98.8		98.8	98.8
0241, 31 - 34	**SITE & INFRASTRUCTURE, DEMOLITION**	101.4	90.5	93.9	101.5	91.5	94.6	98.0	86.7	90.2	102.6	88.6	92.9	107.7	91.6	96.6	109.6	93.3	98.3
0310	Concrete Forming & Accessories	79.8	78.0	78.3	83.6	71.4	73.2	91.9	87.1	87.8	85.1	77.2	78.3	85.7	37.8	44.9	102.5	75.5	79.4
0320	Concrete Reinforcing	96.3	85.2	90.9	96.2	85.2	90.9	94.7	98.0	96.3	97.3	85.8	91.7	97.3	85.1	91.4	95.3	100.8	98.0
0330	Cast-in-Place Concrete	107.6	41.9	83.1	107.6	71.2	94.0	108.0	63.9	91.6	108.1	82.6	98.6	105.9	55.5	87.2	106.6	89.3	100.2
03	**CONCRETE**	93.1	67.6	81.9	93.3	74.8	85.2	95.4	81.5	89.3	95.6	81.5	89.4	94.7	54.1	76.8	95.3	85.6	91.0
04	**MASONRY**	98.6	51.9	69.9	111.6	68.0	84.8	102.3	55.2	73.3	106.0	74.5	86.6	124.6	52.2	80.1	99.2	70.0	81.2
05	**METALS**	88.5	94.0	90.1	88.5	94.2	90.2	87.9	99.7	91.3	94.2	95.6	94.6	88.1	93.3	89.7	90.3	101.3	93.6
06	**WOOD, PLASTICS & COMPOSITES**	70.1	88.4	80.2	74.0	70.9	72.3	85.7	97.9	92.4	78.3	78.5	78.4	79.1	34.2	54.4	98.9	74.4	85.4
07	**THERMAL & MOISTURE PROTECTION**	105.5	65.2	88.0	105.0	70.5	90.0	105.3	70.3	90.1	104.0	73.3	90.7	104.3	55.2	82.9	104.9	73.4	91.2
08	**OPENINGS**	102.5	75.9	96.3	94.7	77.9	90.8	98.8	92.8	97.4	90.2	76.4	87.0	95.6	46.0	84.0	99.3	81.6	95.1
0920	Plaster & Gypsum Board	95.8	88.3	90.8	95.8	70.3	78.6	101.5	98.3	99.3	100.9	78.3	85.7	100.9	32.5	54.9	110.3	73.8	85.8
0950, 0980	Ceilings & Acoustic Treatment	90.2	88.3	88.9	90.2	70.3	76.8	99.0	98.3	98.5	99.0	78.3	85.0	99.0	32.5	54.2	101.4	73.8	82.8
0960	Flooring	82.6	69.1	78.8	84.6	69.1	80.2	102.2	69.1	92.9	87.7	73.9	83.8	88.9	69.1	83.3	96.1	73.3	89.7
0970, 0990	Wall Finishes & Painting/Coating	83.6	81.3	82.2	83.6	84.6	84.2	93.9	84.6	88.4	89.6	84.6	86.6	92.8	83.6	87.3	92.8	69.1	78.8
09	**FINISHES**	87.0	78.0	82.1	87.6	72.1	79.2	95.9	84.8	89.9	91.1	77.1	83.5	93.3	46.4	67.9	97.2	73.8	84.5
COVERS	**DIVS. 10 - 14, 25, 28, 41, 43, 44, 46**	100.0	84.8	96.6	100.0	88.2	97.4	100.0	87.3	97.2	100.0	88.2	97.4	100.0	78.9	95.3	100.0	91.4	98.1
21, 22, 23	**FIRE SUPPRESSION, PLUMBING & HVAC**	96.3	71.4	86.6	96.3	77.7	89.0	96.5	73.1	87.3	96.5	83.9	91.6	96.5	71.5	86.7	100.0	81.2	92.6
26, 27, 3370	**ELECTRICAL, COMMUNICATIONS & UTIL.**	99.8	69.4	84.8	98.9	47.2	73.4	100.9	71.0	86.2	98.8	78.4	88.7	98.8	47.2	73.4	98.8	74.0	86.6
MF2018	**WEIGHTED AVERAGE**	95.5	73.7	86.3	95.3	74.0	86.3	96.3	78.7	88.9	96.1	81.9	90.1	96.8	62.4	82.3	97.7	81.5	90.9

DIVISION		IOWA						KANSAS											
		SPENCER			WATERLOO			BELLEVILLE			COLBY			DODGE CITY			EMPORIA		
		513			506 - 507			669			677			678			668		
		MAT.	INST.	TOTAL	MAT.	INST.	TOTAL	MAT.	INST.	TOTAL	MAT.	INST.	TOTAL	MAT.	INST.	TOTAL	MAT.	INST.	TOTAL
015433	**CONTRACTOR EQUIPMENT**		98.8	98.8		98.8	98.8		103.7	103.7		103.7	103.7		103.7	103.7		101.9	101.9
0241, 31 - 34	**SITE & INFRASTRUCTURE, DEMOLITION**	107.8	90.4	95.7	106.9	92.7	97.1	111.3	92.0	98.0	106.7	92.7	97.0	109.4	91.7	97.2	103.1	89.7	93.8
0310	Concrete Forming & Accessories	91.8	37.5	45.5	94.4	68.7	72.5	94.8	54.3	60.3	97.5	61.2	66.6	91.1	61.1	65.5	86.2	70.0	72.4
0320	Concrete Reinforcing	97.3	85.1	91.4	96.9	80.3	88.9	98.9	102.6	100.7	97.1	102.6	99.8	94.7	102.3	98.4	97.5	102.8	100.1
0330	Cast-in-Place Concrete	105.9	65.5	90.9	115.2	84.0	103.6	121.0	83.5	107.1	119.5	87.2	107.5	121.6	86.9	108.7	117.0	87.3	106.0
03	**CONCRETE**	95.0	57.4	78.5	99.0	77.2	89.4	110.5	74.3	94.6	108.3	78.7	95.3	109.7	78.5	96.0	103.1	82.8	94.2
04	**MASONRY**	124.6	52.2	80.1	99.4	75.1	84.5	86.4	58.5	69.3	100.5	64.6	78.4	110.7	59.6	79.3	91.8	66.0	76.0
05	**METALS**	88.1	93.0	89.5	90.7	93.0	91.4	95.6	99.2	96.7	92.8	100.0	94.9	94.2	98.6	95.5	95.3	100.2	96.8
06	**WOOD, PLASTICS & COMPOSITES**	85.5	34.2	57.3	87.2	64.4	74.6	93.3	51.2	70.1	96.9	57.4	75.2	88.9	57.4	71.5	84.3	68.8	75.8
07	**THERMAL & MOISTURE PROTECTION**	105.3	55.8	83.8	105.3	77.2	93.0	91.4	62.1	78.7	97.8	65.4	83.7	97.8	66.6	84.2	89.5	75.4	83.3
08	**OPENINGS**	106.7	46.0	92.5	95.1	72.9	90.0	94.1	62.5	86.7	97.9	65.9	90.4	97.8	65.9	90.4	92.1	74.4	88.0
0920	Plaster & Gypsum Board	101.5	32.5	55.1	103.8	63.6	76.7	88.7	50.0	62.6	96.8	56.3	69.6	90.5	56.3	67.5	85.5	68.1	73.8
0950, 0980	Ceilings & Acoustic Treatment	99.0	32.5	54.2	91.8	63.6	72.8	76.5	50.0	58.6	77.3	56.3	63.2	77.3	56.3	63.2	76.5	68.1	70.8
0960	Flooring	91.6	69.1	85.3	89.6	78.9	86.6	92.9	68.4	86.0	87.6	68.4	82.2	83.8	68.4	79.4	88.3	68.4	82.7
0970, 0990	Wall Finishes & Painting/Coating	92.8	57.2	71.7	83.6	81.3	82.2	89.7	57.9	70.9	96.1	57.9	73.5	96.1	57.9	73.5	89.7	57.9	70.9
09	**FINISHES**	94.2	42.3	66.1	90.7	70.9	80.0	86.1	55.7	69.6	84.7	60.8	71.8	82.8	60.8	70.9	83.3	67.6	74.8
COVERS	**DIVS. 10 - 14, 25, 28, 41, 43, 44, 46**	100.0	78.9	95.3	100.0	91.0	98.0	100.0	83.3	96.3	100.0	86.1	96.9	100.0	86.0	96.9	100.0	85.9	96.9
21, 22, 23	**FIRE SUPPRESSION, PLUMBING & HVAC**	96.5	71.5	86.7	99.9	80.1	92.1	96.4	70.3	86.1	96.5	71.1	86.5	100.0	71.1	88.7	96.4	73.5	87.4
26, 27, 3370	**ELECTRICAL, COMMUNICATIONS & UTIL.**	100.4	47.2	74.2	95.4	61.6	78.7	104.1	62.7	83.7	95.5	67.9	81.9	92.9	67.9	80.6	101.4	66.9	84.4
MF2018	**WEIGHTED AVERAGE**	98.2	62.3	83.0	96.8	77.4	88.6	97.9	70.7	86.4	97.3	74.0	87.4	98.6	73.3	87.9	96.1	76.5	87.8

	DIVISION	KANSAS																	
		FORT SCOTT			HAYS			HUTCHINSON			INDEPENDENCE			KANSAS CITY			LIBERAL		
		667			676			675			673			660 - 662			679		
		MAT.	INST.	TOTAL	MAT.	INST.	TOTAL	MAT.	INST.	TOTAL	MAT.	INST.	TOTAL	MAT.	INST.	TOTAL	MAT.	INST.	TOTAL
015433	CONTRACTOR EQUIPMENT		102.8	102.8		103.7	103.7		103.7	103.7		103.7	103.7		100.4	100.4		103.7	103.7
0241, 31 - 34	SITE & INFRASTRUCTURE, DEMOLITION	99.9	89.7	92.8	111.4	92.2	98.2	90.6	92.7	92.0	110.2	92.7	98.1	94.3	90.2	91.5	111.3	92.3	98.2
0310	Concrete Forming & Accessories	102.8	82.9	85.9	95.2	58.9	64.3	86.0	56.6	61.0	105.8	67.8	73.4	99.4	98.4	98.6	91.5	58.8	63.7
0320	Concrete Reinforcing	96.9	100.8	98.8	94.7	102.6	98.5	94.7	102.6	98.5	94.1	100.8	97.3	94.0	105.3	99.4	96.1	102.3	99.1
0330	Cast-in-Place Concrete	108.5	82.4	98.8	94.2	83.8	90.3	87.3	86.9	87.2	122.1	87.1	109.1	92.9	97.4	94.6	94.2	83.5	90.3
03	CONCRETE	98.4	86.6	93.2	99.8	76.5	89.6	83.0	76.5	80.2	110.8	81.3	97.8	90.7	99.7	94.6	101.7	76.3	90.5
04	MASONRY	92.7	56.1	70.2	109.5	58.6	78.2	100.2	64.5	78.3	97.8	64.5	77.3	93.5	98.1	96.3	108.2	53.6	74.6
05	METALS	95.3	99.0	96.4	92.3	99.8	94.5	92.1	99.0	94.1	92.0	98.8	94.0	103.1	106.2	104.0	92.6	98.7	94.4
06	WOOD, PLASTICS & COMPOSITES	103.8	90.5	96.5	93.9	57.4	73.8	84.0	51.4	66.0	107.3	65.9	84.5	99.6	98.5	99.0	89.6	57.4	71.9
07	THERMAL & MOISTURE PROTECTION	90.5	72.8	82.8	98.1	62.8	82.8	96.6	64.3	82.5	97.8	74.9	87.9	90.3	99.3	94.2	98.3	60.9	82.0
08	OPENINGS	92.1	85.9	90.6	97.8	65.9	90.4	97.8	62.6	89.6	95.9	70.2	89.9	93.2	97.4	94.2	97.9	65.9	90.4
0920	Plaster & Gypsum Board	91.2	90.4	90.7	94.3	56.3	68.7	89.2	50.2	63.0	105.0	65.1	78.1	84.8	98.6	94.1	91.4	56.3	67.8
0950, 0980	Ceilings & Acoustic Treatment	76.5	90.4	85.9	77.3	56.3	63.2	77.3	50.2	59.0	77.3	65.1	69.0	76.5	98.6	91.4	77.3	56.3	63.2
0960	Flooring	102.6	66.9	92.5	86.4	68.4	81.3	80.9	68.4	77.3	92.0	66.9	84.9	82.4	95.3	86.0	84.1	68.4	79.6
0970, 0990	Wall Finishes & Painting/Coating	91.3	77.5	83.1	96.1	57.9	73.5	96.1	57.9	73.5	96.1	57.9	73.5	97.1	102.7	100.4	96.1	57.9	73.5
09	FINISHES	88.5	80.7	84.3	84.5	59.3	70.9	80.3	57.3	67.9	87.3	66.1	75.8	83.6	98.3	91.5	83.7	59.3	70.5
COVERS	DIVS. 10 - 14, 25, 28, 41, 43, 44, 46	100.0	86.8	97.1	100.0	84.0	96.4	100.0	85.4	96.7	100.0	87.0	97.1	100.0	94.5	98.8	100.0	84.0	96.4
21, 22, 23	FIRE SUPPRESSION, PLUMBING & HVAC	96.4	67.4	85.0	96.5	67.4	85.1	96.5	71.1	86.5	96.5	70.4	86.3	99.9	99.1	99.6	96.5	69.3	85.8
26, 27, 3370	ELECTRICAL, COMMUNICATIONS & UTIL.	100.6	67.9	84.5	94.6	67.9	81.4	90.5	61.0	75.9	92.3	72.8	82.7	106.2	98.1	102.2	92.9	67.9	80.6
MF2018	WEIGHTED AVERAGE	96.0	77.1	88.0	96.5	72.0	86.1	92.5	71.9	83.8	97.2	76.0	88.2	97.3	98.6	97.8	96.5	71.7	86.0

	DIVISION	KANSAS									KENTUCKY								
		SALINA			TOPEKA			WICHITA			ASHLAND			BOWLING GREEN			CAMPTON		
		674			664 - 666			670 - 672			411 - 412			421 - 422			413 - 414		
		MAT.	INST.	TOTAL	MAT.	INST.	TOTAL	MAT.	INST.	TOTAL	MAT.	INST.	TOTAL	MAT.	INST.	TOTAL	MAT.	INST.	TOTAL
015433	CONTRACTOR EQUIPMENT		103.7	103.7		105.0	105.0		106.7	106.7		97.4	97.4		90.6	90.6		96.8	96.8
0241, 31 - 34	SITE & INFRASTRUCTURE, DEMOLITION	99.8	92.2	94.6	98.4	94.4	95.6	96.8	98.5	97.9	114.0	78.8	89.7	78.7	89.4	86.1	87.5	90.7	89.7
0310	Concrete Forming & Accessories	87.8	61.4	65.3	96.7	66.1	70.7	92.9	57.0	62.3	87.2	89.6	89.3	86.5	78.8	79.9	89.3	80.9	82.2
0320	Concrete Reinforcing	94.1	100.5	97.2	93.5	103.8	98.5	92.4	100.4	96.3	92.5	94.2	93.3	89.7	76.8	83.5	90.6	93.0	91.7
0330	Cast-in-Place Concrete	105.6	83.8	97.5	97.6	86.4	93.4	98.7	79.5	91.5	89.1	94.6	91.1	88.4	69.7	81.4	98.6	68.1	87.2
03	CONCRETE	96.8	77.3	88.2	92.3	80.8	87.3	92.4	73.7	84.2	93.1	93.3	93.2	91.6	75.7	84.6	95.0	78.8	87.9
04	MASONRY	125.4	54.9	82.1	87.7	64.3	73.3	97.5	54.1	70.8	90.7	92.0	91.5	92.7	69.2	78.3	89.7	54.7	68.2
05	METALS	94.0	98.9	95.5	99.5	100.0	99.6	96.2	97.5	96.6	95.1	107.7	98.8	96.7	82.8	92.6	96.0	89.8	94.2
06	WOOD, PLASTICS & COMPOSITES	85.4	60.7	71.8	99.9	65.2	80.8	98.7	54.4	74.3	73.4	87.7	81.3	85.7	80.6	82.9	84.2	88.3	86.5
07	THERMAL & MOISTURE PROTECTION	97.2	62.1	82.0	94.0	74.7	85.6	96.2	60.8	80.8	91.2	88.3	90.0	87.6	77.9	83.4	100.2	67.8	86.1
08	OPENINGS	97.8	67.3	90.7	103.5	74.1	96.7	101.3	63.8	92.5	90.6	89.0	90.2	91.4	77.1	88.1	92.6	87.4	91.4
0920	Plaster & Gypsum Board	89.2	59.8	69.4	99.5	64.2	75.7	95.7	53.1	67.0	59.3	87.6	78.4	87.6	80.6	82.9	87.6	87.6	87.6
0950, 0980	Ceilings & Acoustic Treatment	77.3	59.8	65.5	87.2	64.2	71.7	88.0	53.1	64.5	77.0	87.6	84.2	81.6	80.6	80.9	81.6	87.6	85.7
0960	Flooring	82.2	68.4	78.3	98.2	68.4	89.8	94.3	68.4	87.0	73.8	81.3	75.9	89.5	61.9	81.8	91.6	64.3	83.9
0970, 0990	Wall Finishes & Painting/Coating	96.1	57.9	73.5	92.8	67.7	78.0	94.9	57.9	73.0	91.6	90.4	90.9	90.3	67.6	76.9	90.3	53.9	68.8
09	FINISHES	81.5	61.3	70.6	93.9	65.7	78.6	90.3	57.7	72.7	76.1	88.1	82.6	87.1	74.6	80.3	87.9	75.9	81.4
COVERS	DIVS. 10 - 14, 25, 28, 41, 43, 44, 46	100.0	85.3	96.7	100.0	80.3	95.6	100.0	85.1	96.7	100.0	88.2	97.4	100.0	84.4	96.5	100.0	47.8	88.4
21, 22, 23	FIRE SUPPRESSION, PLUMBING & HVAC	100.0	70.5	88.5	100.0	73.2	89.5	99.8	71.0	88.5	96.2	85.0	91.8	99.9	77.0	90.9	96.4	76.4	88.6
26, 27, 3370	ELECTRICAL, COMMUNICATIONS & UTIL.	92.7	73.8	83.4	105.1	71.9	88.7	95.9	73.8	85.0	91.9	88.3	90.1	94.6	74.7	84.8	92.1	88.2	90.2
MF2018	WEIGHTED AVERAGE	97.2	73.5	87.2	98.6	76.5	89.3	97.0	72.6	86.7	93.2	89.7	91.7	94.6	77.2	87.3	94.4	78.0	87.5

	DIVISION	KENTUCKY																	
		CORBIN			COVINGTON			ELIZABETHTOWN			FRANKFORT			HAZARD			HENDERSON		
		407 - 409			410			427			406			417 - 418			424		
		MAT.	INST.	TOTAL	MAT.	INST.	TOTAL	MAT.	INST.	TOTAL	MAT.	INST.	TOTAL	MAT.	INST.	TOTAL	MAT.	INST.	TOTAL
015433	CONTRACTOR EQUIPMENT		96.8	96.8		100.6	100.6		90.6	90.6		100.1	100.1		96.8	96.8		110.3	110.3
0241, 31 - 34	SITE & INFRASTRUCTURE, DEMOLITION	91.6	91.2	91.3	82.7	101.9	95.9	72.9	89.3	84.3	90.0	96.8	94.7	85.3	91.8	89.7	81.7	114.6	104.5
0310	Concrete Forming & Accessories	85.0	76.0	77.3	84.9	70.8	72.8	81.3	72.1	73.5	102.0	78.5	82.0	85.8	81.8	82.4	92.3	77.1	79.3
0320	Concrete Reinforcing	89.6	92.3	90.9	89.3	78.6	84.1	90.2	83.7	87.0	94.0	84.0	89.2	91.0	92.5	91.7	89.8	84.4	87.2
0330	Cast-in-Place Concrete	91.0	72.5	84.1	94.3	79.6	88.9	79.9	67.6	75.3	91.1	77.0	85.8	94.8	69.6	85.4	78.1	84.5	80.5
03	CONCRETE	84.2	78.0	81.5	93.4	76.1	85.8	83.5	73.1	79.0	87.1	79.3	83.6	91.8	79.7	86.5	88.7	81.3	85.5
04	MASONRY	81.9	60.0	68.4	104.0	72.0	84.3	77.4	61.8	67.8	78.6	74.2	75.9	88.6	56.8	69.0	96.0	78.7	85.4
05	METALS	90.9	89.3	90.5	93.9	88.1	92.2	95.9	85.3	92.7	93.1	86.4	91.1	96.0	89.8	94.2	86.8	87.0	86.9
06	WOOD, PLASTICS & COMPOSITES	72.1	77.6	75.1	81.8	68.3	74.3	80.6	73.9	76.9	101.5	77.7	88.4	80.9	88.3	85.0	88.4	75.8	81.5
07	THERMAL & MOISTURE PROTECTION	103.8	69.5	88.9	101.1	72.1	88.5	87.0	68.5	79.0	101.8	75.7	90.5	100.1	69.5	86.8	99.3	82.8	92.1
08	OPENINGS	87.3	70.5	83.4	94.5	71.8	89.2	91.4	73.0	87.1	98.9	78.4	94.1	93.0	87.2	91.6	89.6	78.5	87.0
0920	Plaster & Gypsum Board	88.4	76.6	80.5	68.2	68.0	68.1	86.3	73.7	77.8	91.6	76.6	81.5	86.3	87.6	87.2	90.9	74.8	80.0
0950, 0980	Ceilings & Acoustic Treatment	74.7	76.6	76.0	84.0	68.0	73.2	81.6	73.7	76.2	90.5	76.6	81.1	81.6	87.6	85.7	76.7	74.8	75.4
0960	Flooring	88.4	64.3	81.6	66.3	83.5	71.1	86.9	73.8	83.2	98.9	78.6	93.2	89.8	64.3	82.6	93.2	76.0	88.4
0970, 0990	Wall Finishes & Painting/Coating	87.2	62.0	72.3	85.2	71.1	76.9	90.3	68.1	77.2	94.2	91.7	92.8	90.3	53.9	68.8	90.3	86.7	88.2
09	FINISHES	83.0	72.5	77.3	77.8	72.7	75.0	85.8	71.6	78.1	93.2	80.2	86.2	87.0	76.4	81.3	87.1	77.9	82.1
COVERS	DIVS. 10 - 14, 25, 28, 41, 43, 44, 46	100.0	89.8	97.7	100.0	85.9	96.8	100.0	70.0	93.3	100.0	58.5	90.8	100.0	48.5	88.5	100.0	55.8	90.1
21, 22, 23	FIRE SUPPRESSION, PLUMBING & HVAC	96.5	74.2	87.7	97.0	76.0	88.8	96.6	77.9	89.3	99.9	81.0	92.5	96.4	77.5	89.0	96.6	79.3	89.8
26, 27, 3370	ELECTRICAL, COMMUNICATIONS & UTIL.	90.0	88.2	89.1	95.6	72.4	84.2	91.9	77.9	85.0	100.6	77.9	89.4	92.1	88.2	90.2	94.0	77.9	86.1
MF2018	WEIGHTED AVERAGE	90.8	78.0	85.4	94.3	77.8	87.3	91.3	75.6	84.7	95.4	80.3	89.0	93.8	78.7	87.4	92.2	81.9	87.8

DIVISION		KENTUCKY																	
		LEXINGTON 403 - 405			LOUISVILLE 400 - 402			OWENSBORO 423			PADUCAH 420			PIKEVILLE 415 - 416			SOMERSET 425 - 426		
		MAT.	INST.	TOTAL	MAT.	INST.	TOTAL	MAT.	INST.	TOTAL	MAT.	INST.	TOTAL	MAT.	INST.	TOTAL	MAT.	INST.	TOTAL
015433	CONTRACTOR EQUIPMENT		96.8	96.8		93.7	93.7		110.3	110.3		110.3	110.3		97.4	97.4		96.8	96.8
0241, 31 - 34	SITE & INFRASTRUCTURE, DEMOLITION	93.8	92.8	93.1	87.8	94.4	92.3	91.8	115.3	108.0	84.4	114.8	105.4	125.2	78.0	92.6	77.9	91.2	87.1
0310	Concrete Forming & Accessories	96.5	73.7	77.1	93.5	79.1	81.2	90.8	76.8	78.9	88.8	81.2	82.3	96.0	85.6	87.2	86.8	76.6	78.1
0320	Concrete Reinforcing	98.4	84.0	91.4	94.8	84.2	89.6	89.8	77.3	83.8	90.4	82.5	86.6	93.0	94.1	93.5	90.2	92.5	91.3
0330	Cast-in-Place Concrete	93.1	84.4	89.9	89.3	70.7	82.4	91.1	83.9	88.4	83.2	81.1	82.4	97.9	89.4	94.8	78.1	87.9	81.7
03	CONCRETE	86.9	79.7	83.8	85.8	77.4	82.1	100.7	79.8	91.5	93.3	81.7	88.2	106.8	89.7	99.3	78.6	83.7	80.8
04	MASONRY	80.6	71.1	74.8	78.7	72.0	74.6	88.8	78.6	82.5	91.5	79.2	83.9	88.5	83.2	85.3	83.3	63.2	71.0
05	METALS	93.3	86.7	91.3	95.5	85.5	92.5	88.3	84.7	87.3	85.3	87.5	86.0	95.0	107.5	98.7	95.9	90.0	94.2
06	WOOD, PLASTICS & COMPOSITES	87.3	71.6	78.6	88.8	80.8	84.4	86.3	75.8	80.5	84.0	81.4	82.6	83.3	87.7	85.7	81.5	77.6	79.3
07	THERMAL & MOISTURE PROTECTION	104.1	73.9	90.9	101.0	75.7	90.0	100.1	76.6	89.9	99.4	82.6	92.1	92.1	80.5	87.0	99.3	71.5	87.2
08	OPENINGS	87.5	74.6	84.5	88.8	76.8	86.0	89.6	76.7	86.6	89.0	81.1	87.1	91.2	85.7	89.9	92.1	77.1	88.6
0920	Plaster & Gypsum Board	97.7	70.4	79.4	93.7	80.6	84.9	89.3	74.8	79.5	88.6	80.6	83.2	63.1	87.6	79.6	86.3	76.6	79.8
0950, 0980	Ceilings & Acoustic Treatment	77.9	70.4	72.9	82.8	80.6	81.3	76.7	74.8	75.4	76.7	80.6	79.3	77.0	87.6	84.2	81.6	76.6	78.2
0960	Flooring	93.4	64.3	85.2	95.0	62.9	86.0	92.6	61.9	84.0	91.5	76.0	87.2	78.1	64.3	74.2	90.1	64.3	82.8
0970, 0990	Wall Finishes & Painting/Coating	87.2	79.7	82.8	93.9	68.1	78.7	90.3	86.7	88.2	90.3	73.1	80.2	91.6	91.7	91.7	90.3	68.1	77.2
09	FINISHES	86.4	72.0	78.6	90.8	74.9	82.2	87.3	74.7	80.5	86.5	79.4	82.7	78.7	82.6	80.8	86.4	73.3	79.3
COVERS	DIVS. 10 - 14, 25, 28, 41, 43, 44, 46	100.0	92.1	98.2	100.0	90.5	97.9	100.0	93.0	98.4	100.0	91.1	98.0	100.0	48.2	88.4	100.0	89.2	97.6
21, 22, 23	FIRE SUPPRESSION, PLUMBING & HVAC	100.0	76.4	90.8	99.9	80.0	92.1	99.9	77.1	91.0	96.6	78.2	89.4	96.2	82.3	90.7	96.6	74.0	87.7
26, 27, 3370	ELECTRICAL, COMMUNICATIONS & UTIL.	92.5	74.7	83.7	96.4	77.9	87.3	94.1	73.7	84.0	96.3	77.3	86.9	94.8	88.3	91.6	92.5	88.2	90.4
MF2018	WEIGHTED AVERAGE	93.1	78.1	86.7	93.9	79.6	87.8	94.7	81.0	88.9	92.4	83.2	88.6	95.8	85.4	91.4	91.6	79.5	86.5

DIVISION		LOUISIANA																	
		ALEXANDRIA 713 - 714			BATON ROUGE 707 - 708			HAMMOND 704			LAFAYETTE 705			LAKE CHARLES 706			MONROE 712		
		MAT.	INST.	TOTAL	MAT.	INST.	TOTAL	MAT.	INST.	TOTAL	MAT.	INST.	TOTAL	MAT.	INST.	TOTAL	MAT.	INST.	TOTAL
015433	CONTRACTOR EQUIPMENT		90.9	90.9		91.9	91.9		89.3	89.3		89.3	89.3		88.8	88.8		90.9	90.9
0241, 31 - 34	SITE & INFRASTRUCTURE, DEMOLITION	98.7	87.7	91.1	102.1	91.2	94.6	101.2	85.7	90.5	102.5	87.9	92.4	103.2	87.1	92.0	98.7	87.6	91.1
0310	Concrete Forming & Accessories	77.8	60.0	62.6	99.0	71.8	75.8	79.0	54.9	58.5	95.6	67.8	71.9	96.4	67.8	72.0	77.4	59.4	62.1
0320	Concrete Reinforcing	92.2	54.3	73.9	93.3	54.3	74.5	92.1	54.3	73.8	93.4	54.3	74.5	93.4	54.3	74.5	91.2	54.3	73.3
0330	Cast-in-Place Concrete	89.0	66.2	80.5	88.8	73.1	82.9	87.6	62.6	78.3	87.1	66.1	79.3	91.7	66.1	82.2	89.0	64.1	79.7
03	CONCRETE	82.8	62.0	73.6	85.5	69.8	78.6	85.8	58.3	73.7	86.8	65.4	77.4	89.0	65.4	78.6	82.6	61.0	73.1
04	MASONRY	109.3	64.3	81.6	91.7	66.7	76.3	95.3	60.0	73.6	95.3	66.6	77.7	94.7	66.6	77.5	103.9	63.1	78.8
05	METALS	92.6	70.9	86.2	96.9	74.0	90.1	88.4	69.3	82.8	87.7	69.6	82.4	87.7	69.6	82.4	92.6	70.8	86.2
06	WOOD, PLASTICS & COMPOSITES	89.2	58.8	72.5	103.0	73.2	86.6	83.0	54.3	67.2	103.7	68.0	84.1	101.9	68.0	83.3	88.6	58.8	72.2
07	THERMAL & MOISTURE PROTECTION	99.0	65.6	84.5	97.7	69.2	85.3	96.7	62.9	82.0	97.3	67.4	84.3	97.1	67.2	84.1	99.0	64.8	84.1
08	OPENINGS	110.8	58.6	98.6	97.9	70.4	91.5	96.2	54.9	86.6	99.8	62.2	91.0	99.8	62.2	91.0	110.7	57.1	98.2
0920	Plaster & Gypsum Board	83.6	58.1	66.4	101.0	72.7	82.0	100.0	53.4	68.7	108.6	67.6	81.0	108.6	67.6	81.0	83.3	58.1	66.3
0950, 0980	Ceilings & Acoustic Treatment	87.1	58.1	67.6	92.9	72.7	79.3	96.9	53.4	67.6	94.4	67.6	76.3	95.3	67.6	76.6	87.1	58.1	67.6
0960	Flooring	85.1	68.2	80.4	93.0	68.2	86.0	89.9	68.2	83.8	98.2	68.2	89.8	98.2	68.2	89.8	84.7	68.2	80.1
0970, 0990	Wall Finishes & Painting/Coating	88.8	59.7	71.6	94.7	59.7	74.0	97.7	59.9	75.3	97.7	59.7	75.2	97.7	59.7	75.2	88.8	59.7	71.6
09	FINISHES	80.9	61.2	70.2	92.2	70.3	80.3	90.4	57.5	72.6	93.5	67.2	79.3	93.7	67.2	79.4	80.7	60.9	70.0
COVERS	DIVS. 10 - 14, 25, 28, 41, 43, 44, 46	100.0	79.1	95.3	100.0	84.2	96.5	100.0	80.7	95.7	100.0	83.3	96.3	100.0	83.3	96.3	100.0	78.9	95.3
21, 22, 23	FIRE SUPPRESSION, PLUMBING & HVAC	100.0	63.2	85.6	99.9	64.1	85.9	96.7	61.0	82.7	100.2	64.2	86.0	100.2	64.5	86.2	100.0	61.9	85.0
26, 27, 3370	ELECTRICAL, COMMUNICATIONS & UTIL.	94.3	61.4	78.1	101.7	58.0	80.2	97.0	69.6	83.5	98.1	63.8	81.2	97.7	65.9	82.0	95.9	57.3	76.9
MF2018	WEIGHTED AVERAGE	95.9	65.6	83.1	96.6	69.3	85.1	93.7	64.4	81.3	95.4	67.9	83.8	95.6	68.2	84.1	95.8	64.4	82.5

DIVISION		LOUISIANA									MAINE								
		NEW ORLEANS 700 - 701			SHREVEPORT 710 - 711			THIBODAUX 703			AUGUSTA 043			BANGOR 044			BATH 045		
		MAT.	INST.	TOTAL	MAT.	INST.	TOTAL	MAT.	INST.	TOTAL	MAT.	INST.	TOTAL	MAT.	INST.	TOTAL	MAT.	INST.	TOTAL
015433	CONTRACTOR EQUIPMENT		88.6	88.6		93.9	93.9		89.3	89.3		98.4	98.4		95.7	95.7		95.7	95.7
0241, 31 - 34	SITE & INFRASTRUCTURE, DEMOLITION	104.3	93.9	97.1	101.1	92.3	95.0	103.6	87.6	92.6	88.2	98.0	95.0	90.6	95.5	94.0	88.3	94.2	92.4
0310	Concrete Forming & Accessories	95.7	68.9	72.8	92.9	60.3	65.1	90.2	64.5	68.2	98.6	78.6	81.5	93.0	76.5	79.0	89.1	78.4	80.0
0320	Concrete Reinforcing	92.9	54.4	74.3	92.5	54.3	74.0	92.1	54.3	73.8	102.6	81.1	92.2	94.1	81.1	87.8	93.1	81.1	87.3
0330	Cast-in-Place Concrete	90.7	69.7	82.9	92.0	65.2	82.0	94.1	64.3	83.0	89.8	115.3	99.3	70.9	114.1	87.0	70.9	114.3	87.1
03	CONCRETE	94.4	66.6	82.2	86.2	61.7	75.4	90.6	63.2	78.5	95.5	92.3	94.1	88.5	91.0	89.6	88.6	91.9	90.1
04	MASONRY	99.8	62.6	77.0	94.5	63.4	75.4	119.9	62.8	84.8	101.0	94.1	96.8	115.8	93.7	102.2	122.5	94.1	105.0
05	METALS	99.1	62.6	88.4	96.7	70.2	88.9	88.4	69.5	82.8	105.4	92.3	101.6	93.7	93.0	93.5	92.1	92.9	92.3
06	WOOD, PLASTICS & COMPOSITES	100.3	71.1	84.2	98.0	59.7	76.9	90.5	65.5	76.8	93.5	75.6	83.6	90.7	73.0	80.9	85.1	75.4	79.8
07	THERMAL & MOISTURE PROTECTION	95.0	68.0	83.3	97.4	65.6	83.6	96.9	65.2	83.1	111.0	100.6	106.5	109.0	100.0	105.1	108.9	100.1	105.1
08	OPENINGS	98.8	64.2	90.7	104.7	56.6	93.5	100.7	56.9	90.5	102.7	78.5	97.1	96.4	77.1	91.9	96.4	78.4	92.2
0920	Plaster & Gypsum Board	96.5	70.5	79.0	91.5	58.8	69.5	101.6	65.0	77.0	109.7	74.5	86.0	116.8	71.9	86.6	112.0	74.5	86.8
0950, 0980	Ceilings & Acoustic Treatment	97.5	70.5	79.3	92.8	58.8	69.9	96.9	65.0	75.4	107.1	74.5	85.1	89.6	71.9	77.7	88.6	74.5	79.1
0960	Flooring	103.6	68.2	93.7	91.2	68.2	84.8	95.8	68.2	88.0	89.9	110.6	95.7	82.5	110.6	90.4	80.9	110.6	89.2
0970, 0990	Wall Finishes & Painting/Coating	103.6	60.6	78.2	87.7	59.7	71.1	98.9	59.9	75.8	96.0	88.9	91.8	90.6	88.9	89.6	90.6	88.9	89.6
09	FINISHES	96.6	68.2	81.2	87.7	61.5	73.5	92.6	64.8	77.6	95.8	84.7	89.8	89.7	83.0	86.1	88.3	84.6	86.3
COVERS	DIVS. 10 - 14, 25, 28, 41, 43, 44, 46	100.0	86.5	97.0	100.0	82.6	96.1	100.0	81.8	95.9	100.0	98.0	99.6	100.0	99.6	99.9	100.0	100.9	100.2
21, 22, 23	FIRE SUPPRESSION, PLUMBING & HVAC	100.1	62.5	85.4	99.9	62.8	85.3	96.7	62.5	83.2	100.0	74.3	89.9	100.3	74.1	90.0	96.7	74.3	87.9
26, 27, 3370	ELECTRICAL, COMMUNICATIONS & UTIL.	102.1	70.6	86.6	101.8	65.5	83.9	95.7	69.6	82.8	101.6	76.8	89.3	100.0	69.4	84.9	98.2	76.8	87.6
MF2018	WEIGHTED AVERAGE	99.0	68.5	86.1	97.1	66.3	84.1	96.0	67.0	83.8	100.4	85.6	94.2	97.1	83.9	91.5	95.9	85.4	91.4

	DIVISION	MAINE																	
		HOULTON			KITTERY			LEWISTON			MACHIAS			PORTLAND			ROCKLAND		
		047			039			042			046			040 - 041			048		
		MAT.	INST.	TOTAL	MAT.	INST.	TOTAL	MAT.	INST.	TOTAL	MAT.	INST.	TOTAL	MAT.	INST.	TOTAL	MAT.	INST.	TOTAL
015433	CONTRACTOR EQUIPMENT		95.7	95.7		95.7	95.7		95.7	95.7		95.7	95.7		99.2	99.2		95.7	95.7
0241, 31 - 34	SITE & INFRASTRUCTURE, DEMOLITION	90.2	94.2	93.0	79.5	94.3	89.7	88.2	95.5	93.3	89.5	94.2	92.8	87.4	100.6	96.5	86.0	94.2	91.7
0310	Concrete Forming & Accessories	96.7	78.4	81.1	88.8	78.7	80.2	98.2	76.6	79.8	93.8	78.4	80.7	99.8	76.8	80.2	94.9	78.4	80.9
0320	Concrete Reinforcing	94.1	81.1	87.8	89.7	81.1	85.5	115.6	81.1	98.9	94.1	81.1	87.8	106.5	81.1	94.2	94.1	81.1	87.8
0330	Cast-in-Place Concrete	70.9	113.2	86.7	70.7	114.4	87.0	72.4	114.1	87.9	70.9	114.3	87.1	85.8	115.2	96.8	72.4	114.3	88.0
03	CONCRETE	89.5	91.5	90.4	82.1	92.1	86.5	88.1	91.0	89.4	89.0	91.9	90.3	91.5	91.4	91.5	86.0	91.9	88.6
04	MASONRY	97.9	94.1	95.6	106.5	94.1	98.9	98.5	93.7	95.5	97.9	94.1	95.6	104.7	93.7	98.0	92.5	94.1	93.5
05	METALS	92.3	92.8	92.4	86.8	93.1	88.6	97.1	93.0	95.9	92.3	92.8	92.5	103.0	92.5	99.9	92.2	92.9	92.4
06	WOOD, PLASTICS & COMPOSITES	94.5	75.4	84.0	88.2	75.4	81.2	96.3	73.0	83.5	91.5	75.4	82.7	95.8	73.1	83.3	92.4	75.4	83.1
07	THERMAL & MOISTURE PROTECTION	109.1	100.1	105.1	107.8	100.1	104.4	108.7	100.0	104.9	109.0	100.1	105.1	113.6	100.6	107.9	108.6	100.1	104.9
08	OPENINGS	96.5	78.4	92.3	96.1	82.0	92.8	99.3	77.1	94.1	96.5	78.4	92.3	96.8	77.2	92.2	96.4	78.4	92.2
0920	Plaster & Gypsum Board	119.0	74.5	89.0	106.1	74.5	84.8	121.8	71.9	88.2	117.4	74.5	88.5	108.0	71.9	83.7	117.4	74.5	88.5
0950, 0980	Ceilings & Acoustic Treatment	88.6	74.5	79.1	101.8	74.5	83.4	99.3	71.9	80.8	88.6	74.5	79.1	107.4	71.9	83.5	88.6	74.5	79.1
0960	Flooring	83.7	110.6	91.2	86.6	110.6	93.4	85.2	110.6	92.3	82.9	110.6	90.7	89.6	110.6	95.5	83.3	110.6	91.0
0970, 0990	Wall Finishes & Painting/Coating	90.6	88.9	89.6	81.8	102.1	93.8	90.6	88.9	89.6	90.6	88.9	89.6	94.6	88.9	91.2	90.6	88.9	89.6
09	FINISHES	90.2	84.6	87.2	91.6	86.0	88.6	92.6	83.0	87.4	89.7	84.6	86.9	96.2	83.2	89.1	89.4	84.6	86.8
COVERS	DIVS. 10 - 14, 25, 28, 41, 43, 44, 46	100.0	97.6	99.5	100.0	100.9	100.2	100.0	99.7	99.9	100.0	97.6	99.5	100.0	100.1	100.0	100.0	100.9	100.2
21, 22, 23	FIRE SUPPRESSION, PLUMBING & HVAC	96.7	74.3	87.9	96.7	80.4	90.3	100.3	74.1	90.0	96.7	74.3	87.9	100.0	74.1	89.9	96.7	74.3	87.9
26, 27, 3370	ELECTRICAL, COMMUNICATIONS & UTIL.	101.9	76.8	89.5	90.3	76.8	83.6	101.9	76.7	89.5	101.9	76.8	89.5	105.0	76.7	91.0	101.8	76.8	89.5
MF2018	WEIGHTED AVERAGE	95.6	85.2	91.2	92.6	87.0	90.2	97.5	84.9	92.2	95.4	85.3	91.1	99.6	85.4	93.6	94.6	85.4	90.7

	DIVISION	MAINE			MARYLAND														
		WATERVILLE			ANNAPOLIS			BALTIMORE			COLLEGE PARK			CUMBERLAND			EASTON		
		049			214			210 - 212			207 - 208			215			216		
		MAT.	INST.	TOTAL	MAT.	INST.	TOTAL	MAT.	INST.	TOTAL	MAT.	INST.	TOTAL	MAT.	INST.	TOTAL	MAT.	INST.	TOTAL
015433	CONTRACTOR EQUIPMENT		95.7	95.7		104.5	104.5		104.1	104.1		107.8	107.8		102.4	102.4		102.4	102.4
0241, 31 - 34	SITE & INFRASTRUCTURE, DEMOLITION	90.0	94.2	92.9	100.0	93.5	95.5	100.1	97.7	98.4	100.1	93.6	95.6	92.0	90.3	90.8	98.8	87.7	91.1
0310	Concrete Forming & Accessories	88.6	78.4	79.9	101.0	78.1	81.5	98.4	76.2	79.5	85.1	76.2	77.5	92.3	81.9	83.5	90.4	73.6	76.1
0320	Concrete Reinforcing	94.1	81.1	87.8	108.6	94.9	102.0	101.2	87.3	94.5	103.5	94.0	98.9	93.7	87.4	90.6	92.9	87.1	90.1
0330	Cast-in-Place Concrete	70.9	114.3	87.1	114.1	79.4	101.2	113.8	77.1	100.1	109.4	76.4	97.2	98.0	82.9	92.4	108.8	64.8	92.4
03	CONCRETE	90.2	91.9	90.9	101.4	82.7	93.2	108.2	79.4	95.5	105.2	80.9	94.5	91.4	84.5	88.4	99.7	74.2	88.5
04	MASONRY	109.8	94.1	100.2	94.8	76.6	83.6	104.7	73.1	85.2	111.1	74.6	88.7	99.3	85.4	90.8	113.5	58.6	79.7
05	METALS	92.3	92.8	92.4	105.5	105.2	105.4	103.8	96.6	101.7	90.2	108.3	95.5	100.7	104.2	101.8	101.0	99.7	100.6
06	WOOD, PLASTICS & COMPOSITES	84.6	75.4	79.5	97.1	77.2	86.1	102.5	76.7	88.3	77.3	75.7	76.4	86.7	81.0	83.6	84.5	80.8	82.5
07	THERMAL & MOISTURE PROTECTION	109.1	100.1	105.2	103.5	83.4	94.8	102.7	80.7	93.2	105.3	81.7	95.0	100.9	81.9	92.6	101.0	73.3	89.0
08	OPENINGS	96.5	78.4	92.3	102.3	83.0	97.8	100.5	80.7	95.9	92.7	82.4	90.3	97.6	84.6	94.6	96.0	83.0	92.9
0920	Plaster & Gypsum Board	112.0	74.5	86.8	94.4	76.8	82.5	100.3	76.2	84.0	95.3	75.2	81.8	99.9	80.8	87.0	99.9	80.6	86.9
0950, 0980	Ceilings & Acoustic Treatment	88.6	74.5	79.1	89.1	76.8	80.8	100.5	76.2	84.1	115.1	75.2	88.2	99.0	80.8	86.7	99.0	80.6	86.6
0960	Flooring	80.6	110.6	89.0	90.7	79.5	87.6	92.8	75.7	88.0	87.9	78.4	85.2	85.8	93.1	87.9	85.0	75.7	82.4
0970, 0990	Wall Finishes & Painting/Coating	90.6	88.9	89.6	99.2	72.7	83.5	101.3	75.3	85.9	104.8	72.5	85.7	98.2	84.7	90.2	98.2	72.5	83.0
09	FINISHES	88.4	84.6	86.4	88.5	77.3	82.4	97.0	75.7	85.5	95.6	75.6	84.8	93.3	84.2	88.4	93.5	74.4	83.2
COVERS	DIVS. 10 - 14, 25, 28, 41, 43, 44, 46	100.0	97.5	99.4	100.0	90.3	97.8	100.0	86.8	97.1	100.0	86.7	97.0	100.0	91.3	98.1	100.0	82.9	96.2
21, 22, 23	FIRE SUPPRESSION, PLUMBING & HVAC	96.7	74.3	87.9	100.0	87.0	94.9	100.1	82.0	93.0	96.5	87.4	93.0	96.4	72.0	86.8	96.4	72.8	87.1
26, 27, 3370	ELECTRICAL, COMMUNICATIONS & UTIL.	101.9	76.8	89.5	101.6	88.9	95.3	99.6	86.9	93.3	97.3	101.9	99.5	98.2	80.8	89.6	97.7	62.3	80.3
MF2018	WEIGHTED AVERAGE	96.0	85.2	91.4	100.4	86.3	94.4	101.8	83.2	93.9	97.3	87.7	93.2	96.9	83.8	91.4	98.6	74.8	88.5

	DIVISION	MARYLAND															MASSACHUSETTS		
		ELKTON			HAGERSTOWN			SALISBURY			SILVER SPRING			WALDORF			BOSTON		
		219			217			218			209			206			020 - 022, 024		
		MAT.	INST.	TOTAL	MAT.	INST.	TOTAL	MAT.	INST.	TOTAL	MAT.	INST.	TOTAL	MAT.	INST.	TOTAL	MAT.	INST.	TOTAL
015433	CONTRACTOR EQUIPMENT		102.4	102.4		102.4	102.4		102.4	102.4		99.7	99.7		99.7	99.7		103.0	103.0
0241, 31 - 34	SITE & INFRASTRUCTURE, DEMOLITION	86.3	89.0	88.1	90.6	90.4	90.4	98.8	87.6	91.1	88.9	85.7	86.7	95.1	85.7	88.6	92.7	102.1	99.2
0310	Concrete Forming & Accessories	96.2	92.3	92.9	91.5	78.8	80.7	104.5	50.2	58.2	93.0	75.5	78.1	99.7	75.4	79.0	105.1	139.7	134.6
0320	Concrete Reinforcing	92.9	117.3	104.7	93.7	87.4	90.6	92.9	65.2	79.5	102.3	93.9	98.3	103.0	93.9	98.6	117.7	156.9	136.7
0330	Cast-in-Place Concrete	88.1	72.2	82.2	93.3	82.9	89.5	108.8	62.9	91.7	112.1	76.8	98.9	125.6	76.7	107.4	99.2	141.8	115.1
03	CONCRETE	84.0	90.6	86.9	87.7	83.1	85.7	100.5	59.2	82.4	102.8	80.5	93.0	113.3	80.5	98.9	105.4	142.7	121.8
04	MASONRY	98.6	67.0	79.2	105.1	85.4	93.0	112.9	55.2	77.4	110.6	75.0	88.7	94.7	75.0	82.6	112.0	146.3	133.1
05	METALS	101.0	114.0	104.8	100.9	104.3	101.9	101.0	90.9	98.1	94.8	104.3	97.6	94.8	104.3	97.5	102.8	136.1	112.6
06	WOOD, PLASTICS & COMPOSITES	91.9	101.3	97.0	85.8	76.7	80.8	102.6	51.4	74.4	84.0	75.0	79.0	90.8	75.0	82.1	103.8	140.2	123.8
07	THERMAL & MOISTURE PROTECTION	100.5	79.4	91.3	101.2	83.0	93.3	101.3	68.5	87.0	108.4	86.5	98.9	109.0	86.5	99.2	107.2	137.1	120.2
08	OPENINGS	96.0	102.1	97.4	95.9	81.2	92.5	96.2	60.8	87.9	84.9	82.1	84.2	85.5	81.5	84.5	99.9	147.3	111.0
0920	Plaster & Gypsum Board	102.8	101.7	102.0	99.9	76.4	84.1	109.4	50.4	69.7	101.9	75.2	83.9	104.8	75.2	84.8	109.0	141.0	130.6
0950, 0980	Ceilings & Acoustic Treatment	99.0	101.7	100.8	101.7	76.4	84.6	99.0	50.4	66.2	123.2	75.2	90.8	123.2	75.2	90.8	100.5	141.0	127.8
0960	Flooring	87.2	75.7	84.0	85.4	93.1	87.6	90.7	75.7	86.5	93.8	78.4	89.5	97.2	78.4	92.0	90.5	163.5	111.0
0970, 0990	Wall Finishes & Painting/Coating	98.2	72.5	83.0	98.2	72.5	83.0	98.2	72.5	83.0	112.7	72.5	88.9	112.7	72.5	88.9	94.3	160.4	133.4
09	FINISHES	93.7	88.5	90.9	93.5	80.4	86.4	96.4	56.2	74.6	95.8	75.0	84.5	97.5	75.1	85.4	98.7	147.0	124.8
COVERS	DIVS. 10 - 14, 25, 28, 41, 43, 44, 46	100.0	56.7	90.4	100.0	90.8	98.0	100.0	76.5	94.8	100.0	85.2	96.7	100.0	83.1	96.2	100.0	119.1	104.3
21, 22, 23	FIRE SUPPRESSION, PLUMBING & HVAC	96.4	78.7	89.5	99.9	83.1	93.3	96.4	70.7	86.3	96.5	87.9	93.1	96.5	87.8	93.1	100.1	127.4	110.8
26, 27, 3370	ELECTRICAL, COMMUNICATIONS & UTIL.	99.4	86.9	93.2	98.0	80.8	89.5	96.6	60.1	78.6	94.6	101.9	98.2	92.2	101.9	97.0	102.6	129.5	115.9
MF2018	WEIGHTED AVERAGE	95.9	86.2	91.8	97.4	85.3	92.3	99.0	66.9	85.4	96.5	86.7	92.4	97.3	86.6	92.8	101.9	134.0	115.5

DIVISION		MASSACHUSETTS																	
		BROCKTON			BUZZARDS BAY			FALL RIVER			FITCHBURG			FRAMINGHAM			GREENFIELD		
		023			025			027			014			017			013		
		MAT.	INST.	TOTAL	MAT.	INST.	TOTAL	MAT.	INST.	TOTAL	MAT.	INST.	TOTAL	MAT.	INST.	TOTAL	MAT.	INST.	TOTAL
015433	**CONTRACTOR EQUIPMENT**		98.1	98.1		98.1	98.1		98.9	98.9		95.7	95.7		97.4	97.4		95.7	95.7
0241, 31 - 34	**SITE & INFRASTRUCTURE, DEMOLITION**	91.1	98.0	95.8	81.2	98.1	92.9	90.3	98.0	95.6	82.7	97.8	93.1	79.5	97.8	92.1	86.4	97.3	93.9
0310	Concrete Forming & Accessories	100.7	121.1	118.1	98.5	120.6	117.4	100.7	120.8	117.8	94.0	118.5	114.9	101.4	121.3	118.4	92.4	117.7	114.0
0320	Concrete Reinforcing	109.7	142.7	125.7	88.0	122.3	104.6	109.7	122.3	115.8	87.9	142.5	114.3	87.9	142.8	114.4	91.5	126.5	108.4
0330	Cast-in-Place Concrete	88.1	136.6	106.2	73.2	136.4	96.7	85.2	136.9	104.4	78.4	136.5	100.0	78.4	136.7	100.1	80.5	122.2	96.0
03	**CONCRETE**	95.1	129.8	110.3	80.7	126.0	100.6	93.8	126.2	108.0	79.5	128.3	101.0	82.3	130.0	103.2	83.0	120.3	99.4
04	**MASONRY**	105.4	135.1	123.6	97.9	135.1	120.8	105.9	135.1	123.8	98.2	131.9	118.9	104.7	136.1	124.0	102.7	119.1	112.8
05	**METALS**	98.8	125.6	106.7	93.6	117.0	100.5	98.8	117.3	104.3	95.3	122.1	103.2	95.3	125.8	104.3	97.8	113.8	102.5
06	**WOOD, PLASTICS & COMPOSITES**	98.1	119.7	109.9	94.7	119.7	108.4	98.1	119.9	110.1	94.5	116.5	106.6	101.0	119.4	111.2	92.5	119.7	107.4
07	**THERMAL & MOISTURE PROTECTION**	103.2	128.4	114.2	102.0	125.7	112.3	103.0	124.9	112.5	103.2	122.1	111.4	103.3	128.3	114.2	103.3	109.9	106.2
08	**OPENINGS**	97.4	128.4	104.6	93.4	118.9	99.3	97.4	118.3	102.2	99.2	126.7	105.6	89.8	128.3	98.7	99.4	120.3	104.3
0920	Plaster & Gypsum Board	92.3	119.9	110.9	87.3	119.9	109.3	92.3	119.9	110.9	111.3	116.7	114.9	114.4	119.9	118.1	112.6	119.9	117.5
0950, 0980	Ceilings & Acoustic Treatment	104.3	119.9	114.9	86.2	119.9	108.9	104.3	119.9	114.9	93.9	116.7	109.3	93.9	119.9	111.5	103.7	119.9	114.6
0960	Flooring	85.0	162.0	106.6	82.6	162.0	104.9	83.8	162.0	105.8	86.7	162.0	107.8	88.2	162.0	108.9	85.9	135.0	99.7
0970, 0990	Wall Finishes & Painting/Coating	85.0	139.2	117.1	85.0	139.2	117.1	85.0	139.2	117.1	86.2	139.2	117.6	87.1	139.2	117.9	86.2	109.3	99.9
09	**FINISHES**	90.8	130.8	112.5	85.2	130.8	109.9	90.5	131.0	112.4	89.5	128.9	110.9	90.2	130.6	112.1	91.7	120.9	107.5
COVERS	**DIVS. 10 - 14, 25, 28, 41, 43, 44, 46**	100.0	109.7	102.2	100.0	109.7	102.2	100.0	110.2	102.3	100.0	104.2	100.9	100.0	109.3	102.1	100.0	102.9	100.7
21, 22, 23	**FIRE SUPPRESSION, PLUMBING & HVAC**	100.4	103.9	101.7	96.8	103.5	99.4	100.4	103.5	101.6	97.0	104.5	99.9	97.0	121.1	106.4	97.0	98.5	97.6
26, 27, 3370	**ELECTRICAL, COMMUNICATIONS & UTIL.**	101.3	97.7	99.5	98.5	100.1	99.3	101.2	97.7	99.4	100.6	104.1	102.3	97.4	124.4	110.7	100.5	96.1	98.4
MF2018	**WEIGHTED AVERAGE**	98.5	116.7	106.2	93.2	115.2	102.5	98.3	114.9	105.3	94.6	116.2	103.7	94.0	124.2	106.8	95.9	109.0	101.4

DIVISION		MASSACHUSETTS																	
		HYANNIS			LAWRENCE			LOWELL			NEW BEDFORD			PITTSFIELD			SPRINGFIELD		
		026			019			018			027			012			010 - 011		
		MAT.	INST.	TOTAL	MAT.	INST.	TOTAL	MAT.	INST.	TOTAL	MAT.	INST.	TOTAL	MAT.	INST.	TOTAL	MAT.	INST.	TOTAL
015433	**CONTRACTOR EQUIPMENT**		98.1	98.1		98.1	98.1		95.7	95.7		98.9	98.9		95.7	95.7		95.7	95.7
0241, 31 - 34	**SITE & INFRASTRUCTURE, DEMOLITION**	87.4	98.1	94.8	91.6	98.0	96.0	90.6	98.1	95.8	88.7	98.0	95.1	91.6	96.7	95.2	91.1	97.1	95.2
0310	Concrete Forming & Accessories	92.7	120.6	116.5	102.5	121.6	118.8	99.0	122.6	119.1	100.7	120.8	117.9	99.0	102.9	102.3	99.3	117.7	115.0
0320	Concrete Reinforcing	88.0	122.3	104.6	108.8	148.3	127.9	109.7	148.1	128.2	109.7	122.4	115.8	90.8	114.1	102.1	109.7	126.5	117.8
0330	Cast-in-Place Concrete	80.4	136.4	101.3	90.7	136.8	107.9	82.4	138.5	103.3	75.1	136.9	98.1	90.0	113.4	98.7	85.8	122.2	99.4
03	**CONCRETE**	86.7	126.0	104.0	96.2	131.1	111.5	88.0	131.9	107.3	89.1	126.3	105.4	89.0	108.3	97.5	89.5	120.2	103.0
04	**MASONRY**	104.3	135.1	123.2	110.1	135.9	125.9	97.3	134.9	120.4	104.2	135.7	123.6	98.0	111.5	106.3	97.6	119.1	110.8
05	**METALS**	95.1	117.0	101.5	98.2	128.4	107.1	98.2	125.0	106.1	98.8	117.4	104.3	98.0	108.2	101.0	100.9	113.7	104.6
06	**WOOD, PLASTICS & COMPOSITES**	88.0	119.7	105.4	101.5	119.7	111.5	100.8	119.7	111.2	98.1	119.9	110.1	100.8	102.5	101.8	100.8	119.7	111.2
07	**THERMAL & MOISTURE PROTECTION**	102.5	125.7	112.6	104.0	128.6	114.7	103.7	128.0	114.3	102.9	125.1	112.6	103.8	104.3	104.0	103.7	109.9	106.4
08	**OPENINGS**	93.9	118.9	99.8	93.4	129.9	101.9	100.4	129.9	107.3	97.4	123.0	103.3	100.4	107.6	102.1	100.4	120.3	105.1
0920	Plaster & Gypsum Board	83.2	119.9	107.9	117.3	119.9	119.1	117.3	119.9	119.1	92.3	119.9	110.9	117.3	102.3	107.2	117.3	119.9	119.1
0950, 0980	Ceilings & Acoustic Treatment	96.2	119.9	112.2	105.5	119.9	115.2	105.5	119.9	115.2	104.3	119.9	114.9	105.5	102.3	103.4	105.5	119.9	115.2
0960	Flooring	80.0	162.0	103.0	88.8	162.0	109.4	88.8	162.0	109.4	83.8	162.0	105.8	89.1	129.8	100.6	88.2	135.0	101.3
0970, 0990	Wall Finishes & Painting/Coating	85.0	139.2	117.1	86.3	139.2	117.6	86.2	139.2	117.6	85.0	139.2	117.1	86.2	109.3	99.9	87.0	109.3	100.2
09	**FINISHES**	86.3	130.8	110.4	93.8	130.8	113.8	93.7	131.6	114.2	90.4	131.0	112.3	93.8	108.6	101.8	93.6	120.9	108.4
COVERS	**DIVS. 10 - 14, 25, 28, 41, 43, 44, 46**	100.0	109.7	102.2	100.0	109.8	102.2	100.0	110.8	102.4	100.0	110.2	102.3	100.0	99.5	99.9	100.0	102.9	100.7
21, 22, 23	**FIRE SUPPRESSION, PLUMBING & HVAC**	100.4	104.2	101.9	100.1	119.7	107.8	100.1	121.2	108.4	100.4	103.4	101.5	100.1	94.2	97.8	100.1	98.5	99.5
26, 27, 3370	**ELECTRICAL, COMMUNICATIONS & UTIL.**	98.9	97.7	98.3	99.7	122.9	111.1	100.1	119.3	109.6	102.1	97.7	99.9	100.1	96.1	98.2	100.1	93.0	96.6
MF2018	**WEIGHTED AVERAGE**	95.7	115.0	103.8	98.4	124.2	109.3	97.5	123.8	108.6	97.7	115.2	105.1	97.7	102.6	99.7	98.2	108.5	102.5

DIVISION		MASSACHUSETTS			MICHIGAN														
		WORCESTER			ANN ARBOR			BATTLE CREEK			BAY CITY			DEARBORN			DETROIT		
		015 - 016			481			490			487			481			482		
		MAT.	INST.	TOTAL	MAT.	INST.	TOTAL	MAT.	INST.	TOTAL	MAT.	INST.	TOTAL	MAT.	INST.	TOTAL	MAT.	INST.	TOTAL
015433	**CONTRACTOR EQUIPMENT**		95.7	95.7		109.1	109.1		96.1	96.1		109.1	109.1		109.1	109.1		97.8	97.8
0241, 31 - 34	**SITE & INFRASTRUCTURE, DEMOLITION**	91.0	97.9	95.7	81.0	92.2	88.7	93.6	81.7	85.3	72.5	91.7	85.8	80.8	92.3	88.7	98.2	101.2	100.2
0310	Concrete Forming & Accessories	99.6	118.5	115.7	96.3	104.1	103.0	96.0	77.3	80.0	96.4	81.5	83.7	96.2	104.7	103.5	100.4	105.3	104.6
0320	Concrete Reinforcing	109.7	155.5	131.9	101.2	103.5	102.3	98.5	80.9	90.0	101.2	102.7	101.9	101.2	103.6	102.3	101.4	104.8	103.0
0330	Cast-in-Place Concrete	85.3	136.5	104.4	88.3	96.4	91.3	86.1	91.0	87.9	84.5	83.5	84.1	86.4	97.2	90.4	104.5	101.3	103.3
03	**CONCRETE**	89.3	130.6	107.5	89.8	102.2	95.2	85.7	82.6	84.3	88.0	87.3	87.7	88.9	102.8	95.0	103.1	103.2	103.2
04	**MASONRY**	97.2	131.9	118.5	99.6	96.4	97.6	99.1	78.4	86.4	99.2	77.7	86.0	99.5	98.2	98.7	102.6	99.8	100.9
05	**METALS**	100.9	127.4	108.7	102.7	115.7	106.5	103.7	83.7	97.9	103.3	114.0	106.4	102.8	115.8	106.6	104.3	94.3	101.3
06	**WOOD, PLASTICS & COMPOSITES**	101.2	116.5	109.6	89.2	106.6	98.8	88.4	75.6	81.3	89.2	81.9	85.2	89.2	106.6	98.8	97.1	106.7	102.4
07	**THERMAL & MOISTURE PROTECTION**	103.8	122.1	111.8	105.3	98.2	102.2	96.1	79.0	88.7	102.7	81.0	93.2	103.6	100.7	102.3	102.6	102.9	102.7
08	**OPENINGS**	100.4	130.2	107.4	94.2	100.8	95.7	86.5	74.1	83.6	94.2	83.5	91.7	94.2	101.1	95.8	97.2	101.9	98.3
0920	Plaster & Gypsum Board	117.3	116.7	116.9	107.4	106.5	106.8	91.7	71.7	78.2	107.4	81.1	89.7	107.4	106.5	106.8	104.1	106.5	105.7
0950, 0980	Ceilings & Acoustic Treatment	105.5	116.7	113.0	84.8	106.5	99.4	78.7	71.7	74.0	85.7	81.1	82.6	84.8	106.5	99.4	96.9	106.5	103.4
0960	Flooring	88.8	154.9	107.4	91.2	107.2	95.7	91.5	70.4	85.6	91.2	77.0	87.2	90.6	101.6	93.6	95.7	106.7	98.8
0970, 0990	Wall Finishes & Painting/Coating	86.2	139.2	117.6	82.7	95.5	90.3	82.3	76.3	78.7	82.7	79.1	80.6	82.7	93.8	89.3	93.1	100.1	97.3
09	**FINISHES**	93.7	127.5	112.0	90.8	104.3	98.1	84.7	75.2	79.6	90.6	79.9	84.8	90.6	103.5	97.6	97.7	105.4	101.9
COVERS	**DIVS. 10 - 14, 25, 28, 41, 43, 44, 46**	100.0	104.2	100.9	100.0	93.7	98.6	100.0	93.7	98.6	100.0	88.0	97.3	100.0	94.1	98.7	100.0	102.4	100.5
21, 22, 23	**FIRE SUPPRESSION, PLUMBING & HVAC**	100.1	104.7	101.9	100.0	91.3	96.6	100.1	82.6	93.2	100.0	78.6	91.6	100.0	100.0	100.0	99.9	102.7	101.0
26, 27, 3370	**ELECTRICAL, COMMUNICATIONS & UTIL.**	100.1	104.1	102.1	97.3	103.0	100.1	96.6	77.7	87.3	96.4	82.5	89.5	97.3	96.1	96.7	99.9	103.5	101.6
MF2018	**WEIGHTED AVERAGE**	98.1	117.0	106.1	97.1	99.8	98.2	95.5	80.4	89.1	96.6	85.3	91.8	96.9	100.9	98.6	100.7	102.0	101.3

DIVISION		MICHIGAN																	
		FLINT			GAYLORD			GRAND RAPIDS			IRON MOUNTAIN			JACKSON			KALAMAZOO		
		484 - 485			497			493, 495			498 - 499			492			491		
		MAT.	INST.	TOTAL	MAT.	INST.	TOTAL	MAT.	INST.	TOTAL	MAT.	INST.	TOTAL	MAT.	INST.	TOTAL	MAT.	INST.	TOTAL
015433	CONTRACTOR EQUIPMENT		109.1	109.1		103.6	103.6		98.3	98.3		91.1	91.1		103.6	103.6		96.1	96.1
0241, 31 - 34	SITE & INFRASTRUCTURE, DEMOLITION	70.2	91.8	85.1	87.4	79.6	82.0	92.9	85.9	88.1	96.4	88.5	90.9	111.0	81.2	90.4	93.9	81.6	85.4
0310	Concrete Forming & Accessories	99.1	81.8	84.4	95.0	73.9	77.0	95.6	75.7	78.6	86.6	78.6	79.8	91.6	82.3	83.7	96.0	77.1	79.9
0320	Concrete Reinforcing	101.2	103.1	102.1	91.9	93.0	92.4	103.4	80.7	92.4	91.6	85.0	88.4	89.3	103.1	96.0	98.5	76.2	87.8
0330	Cast-in-Place Concrete	88.9	85.8	87.8	85.8	78.4	83.1	90.4	89.0	89.9	101.7	67.6	89.0	85.7	89.8	87.2	87.8	91.0	89.0
03	CONCRETE	90.2	88.4	89.4	82.6	80.5	81.6	89.7	81.1	85.9	90.9	76.6	84.6	77.8	89.8	83.1	88.9	81.7	85.7
04	MASONRY	99.7	85.9	91.2	109.7	72.7	87.0	92.0	74.1	81.0	95.4	79.7	85.7	89.2	86.9	87.8	97.6	78.4	85.8
05	METALS	102.8	114.6	106.3	105.5	110.8	107.0	100.8	83.2	95.6	104.8	91.1	100.8	105.7	112.7	107.8	103.7	81.9	97.3
06	WOOD, PLASTICS & COMPOSITES	92.6	80.7	86.1	81.9	73.4	77.2	92.2	73.9	82.2	78.2	78.9	78.6	80.8	79.8	80.2	88.4	75.6	81.3
07	THERMAL & MOISTURE PROTECTION	103.0	84.5	95.0	94.7	74.7	86.0	98.0	71.8	86.6	98.4	74.8	88.1	94.0	88.3	91.5	96.1	79.0	88.7
08	OPENINGS	94.2	82.6	91.5	86.0	78.7	84.3	101.8	73.8	95.2	92.5	69.8	87.2	85.2	85.5	85.3	86.5	73.5	83.4
0920	Plaster & Gypsum Board	109.0	79.8	89.4	91.3	71.8	78.2	99.8	70.2	79.8	51.3	79.0	69.9	90.1	78.3	82.1	91.7	71.7	78.2
0950, 0980	Ceilings & Acoustic Treatment	84.8	79.8	81.5	77.1	71.8	73.5	91.1	70.2	77.0	76.0	79.0	78.0	77.1	78.3	77.9	78.7	71.7	74.0
0960	Flooring	91.2	86.0	89.7	84.9	85.6	85.1	95.7	76.1	90.2	102.0	91.0	98.9	83.4	78.7	82.1	91.5	70.4	85.6
0970, 0990	Wall Finishes & Painting/Coating	82.7	78.3	80.1	78.7	81.6	80.4	89.3	76.9	82.0	94.8	69.6	79.9	78.7	93.8	87.7	82.3	76.3	78.7
09	FINISHES	90.1	81.6	85.5	84.1	76.2	79.8	91.4	75.2	82.6	84.6	80.0	82.1	85.2	81.9	83.4	84.7	75.2	79.6
COVERS	DIVS. 10 - 14, 25, 28, 41, 43, 44, 46	100.0	88.8	97.5	100.0	77.9	95.1	100.0	93.0	98.4	100.0	85.9	96.8	100.0	91.6	98.1	100.0	93.7	98.6
21, 22, 23	FIRE SUPPRESSION, PLUMBING & HVAC	100.0	83.3	93.4	96.8	79.0	89.8	100.0	79.6	92.0	96.7	83.9	91.7	96.8	85.7	92.4	100.1	78.2	91.5
26, 27, 3370	ELECTRICAL, COMMUNICATIONS & UTIL.	97.3	89.1	93.2	94.5	76.3	85.5	102.2	82.9	92.7	100.8	80.7	90.9	98.3	103.0	100.6	96.4	74.1	85.4
MF2018	WEIGHTED AVERAGE	96.8	88.6	93.3	94.5	80.6	88.6	97.9	79.9	90.3	96.5	81.7	90.2	94.0	90.7	92.6	95.8	78.6	88.6

DIVISION		MICHIGAN															MINNESOTA		
		LANSING			MUSKEGON			ROYAL OAK			SAGINAW			TRAVERSE CITY			BEMIDJI		
		488 - 489			494			480, 483			486			496			566		
		MAT.	INST.	TOTAL	MAT.	INST.	TOTAL	MAT.	INST.	TOTAL	MAT.	INST.	TOTAL	MAT.	INST.	TOTAL	MAT.	INST.	TOTAL
015433	CONTRACTOR EQUIPMENT		111.0	111.0		96.1	96.1		89.8	89.8		109.1	109.1		91.1	91.1		97.7	97.7
0241, 31 - 34	SITE & INFRASTRUCTURE, DEMOLITION	90.7	96.5	94.7	91.5	81.7	84.8	84.8	92.6	90.2	73.5	91.7	86.1	81.8	87.5	85.7	95.1	95.1	95.1
0310	Concrete Forming & Accessories	93.8	74.9	77.7	96.3	79.4	81.9	92.3	105.2	103.2	96.3	79.7	82.1	86.6	72.4	74.5	88.0	82.8	83.6
0320	Concrete Reinforcing	104.8	102.9	103.8	99.3	80.7	90.3	91.7	104.8	98.0	101.2	102.7	101.9	93.1	80.2	86.9	97.4	101.8	99.5
0330	Cast-in-Place Concrete	97.9	85.1	93.1	85.8	90.5	87.5	77.3	97.5	84.8	87.3	83.4	85.8	79.3	76.7	78.3	101.8	104.5	102.8
03	CONCRETE	91.4	84.9	88.5	84.2	83.3	83.8	77.1	101.6	87.9	89.3	86.5	88.1	74.8	76.1	75.4	91.0	95.0	92.8
04	MASONRY	94.5	84.6	88.4	96.3	76.2	83.9	93.2	99.5	97.1	101.2	77.7	86.7	93.5	74.8	82.0	98.9	102.4	101.0
05	METALS	101.6	112.7	104.9	101.5	83.1	96.1	106.3	92.2	102.2	102.8	113.8	106.0	104.8	88.9	100.1	91.2	117.9	99.0
06	WOOD, PLASTICS & COMPOSITES	87.0	71.2	78.3	85.4	78.0	81.3	85.0	106.6	96.9	85.9	79.7	82.5	78.2	71.9	74.8	68.8	77.0	73.3
07	THERMAL & MOISTURE PROTECTION	102.0	83.8	94.1	95.0	73.5	85.7	101.2	100.9	101.1	103.9	80.9	93.9	97.3	74.2	87.3	106.3	90.7	99.5
08	OPENINGS	102.7	77.2	96.8	85.8	76.5	83.6	94.0	101.3	95.7	92.4	82.2	90.0	92.5	65.4	86.2	99.7	101.3	100.1
0920	Plaster & Gypsum Board	100.1	70.2	79.9	71.8	74.2	73.4	104.8	106.5	106.0	107.4	78.8	88.2	51.3	71.8	65.1	104.3	76.8	85.8
0950, 0980	Ceilings & Acoustic Treatment	83.9	70.2	74.6	78.7	74.2	75.6	84.1	106.5	99.2	84.8	78.8	80.8	76.0	71.8	73.2	128.8	76.8	93.7
0960	Flooring	97.2	78.7	92.0	90.2	72.3	85.1	88.2	106.7	93.4	91.2	77.0	87.2	102.0	85.6	97.4	90.7	91.4	90.9
0970, 0990	Wall Finishes & Painting/Coating	91.1	76.9	82.7	80.5	79.7	80.0	84.2	93.8	89.9	82.7	79.1	80.6	94.8	39.4	62.1	89.6	105.8	99.2
09	FINISHES	90.2	74.6	81.7	81.1	77.1	78.9	89.7	104.3	97.6	90.5	78.6	84.0	83.5	71.0	76.7	98.2	86.0	91.6
COVERS	DIVS. 10 - 14, 25, 28, 41, 43, 44, 46	100.0	88.8	97.5	100.0	94.5	98.8	100.0	100.0	100.0	100.0	87.7	97.3	100.0	84.4	96.5	100.0	95.6	99.0
21, 22, 23	FIRE SUPPRESSION, PLUMBING & HVAC	99.9	83.7	93.5	99.9	83.4	93.4	96.7	101.4	98.5	100.0	78.3	91.5	96.7	78.9	89.7	96.7	84.8	92.0
26, 27, 3370	ELECTRICAL, COMMUNICATIONS & UTIL.	99.1	87.6	93.4	96.8	73.0	85.1	99.4	103.1	101.3	95.4	84.5	90.0	96.2	76.2	86.4	105.0	100.7	102.8
MF2018	WEIGHTED AVERAGE	98.0	86.8	93.3	94.3	80.0	88.3	95.1	100.3	97.3	96.5	85.1	91.7	93.4	77.7	86.8	96.7	95.4	96.1

DIVISION		MINNESOTA																	
		BRAINERD			DETROIT LAKES			DULUTH			MANKATO			MINNEAPOLIS			ROCHESTER		
		564			565			556 - 558			560			553 - 555			559		
		MAT.	INST.	TOTAL	MAT.	INST.	TOTAL	MAT.	INST.	TOTAL	MAT.	INST.	TOTAL	MAT.	INST.	TOTAL	MAT.	INST.	TOTAL
015433	CONTRACTOR EQUIPMENT		100.2	100.2		97.7	97.7		102.9	102.9		100.2	100.2		108.4	108.4		100.7	100.7
0241, 31 - 34	SITE & INFRASTRUCTURE, DEMOLITION	96.0	99.9	98.7	93.3	95.4	94.7	98.2	101.7	100.6	92.6	99.4	97.3	95.0	108.8	104.5	96.0	97.6	97.1
0310	Concrete Forming & Accessories	88.7	83.6	84.4	84.9	82.8	83.1	98.4	94.6	95.2	96.9	96.6	96.6	99.3	114.0	111.8	99.2	94.8	95.4
0320	Concrete Reinforcing	96.2	102.2	99.1	97.4	101.8	99.6	102.6	102.0	102.3	96.0	109.2	102.4	96.1	109.4	102.5	98.9	108.8	103.7
0330	Cast-in-Place Concrete	110.7	109.1	110.1	98.9	107.5	102.1	94.3	97.7	95.6	101.9	99.5	101.0	95.6	115.3	102.9	94.8	95.3	95.0
03	CONCRETE	94.8	97.1	95.8	88.6	96.1	91.9	95.3	98.1	96.5	90.5	100.8	95.0	97.6	114.3	104.9	91.4	98.6	94.6
04	MASONRY	122.5	116.3	118.7	122.6	109.8	114.8	95.3	105.7	101.7	111.3	110.8	111.0	114.5	119.5	117.6	102.9	103.5	103.3
05	METALS	92.2	118.3	99.9	91.1	117.5	98.9	100.2	119.1	105.8	92.1	121.6	100.7	98.7	124.5	106.3	99.3	123.1	106.3
06	WOOD, PLASTICS & COMPOSITES	84.9	74.2	79.1	65.8	74.4	70.6	93.9	91.9	92.8	94.5	95.4	95.0	96.1	110.8	104.2	95.2	92.9	93.9
07	THERMAL & MOISTURE PROTECTION	104.7	102.4	103.7	106.1	99.4	103.2	97.9	102.2	99.8	105.1	93.4	100.1	101.2	115.9	107.6	105.3	90.0	98.6
08	OPENINGS	86.4	99.8	89.5	99.6	99.9	99.7	103.3	104.6	103.6	90.9	112.2	95.9	99.3	121.5	104.5	98.6	112.5	101.8
0920	Plaster & Gypsum Board	91.5	74.1	79.8	103.7	74.1	83.8	94.5	92.1	92.9	95.6	95.9	95.8	101.1	111.2	107.9	103.2	93.2	96.5
0950, 0980	Ceilings & Acoustic Treatment	58.8	74.1	69.1	128.8	74.1	91.9	89.2	92.1	91.1	58.8	95.9	83.8	103.1	111.2	108.6	93.9	93.2	93.4
0960	Flooring	89.6	85.6	88.4	89.5	85.6	88.4	93.2	123.9	101.8	91.4	81.8	88.7	100.6	114.6	104.5	92.3	81.8	89.3
0970, 0990	Wall Finishes & Painting/Coating	84.0	105.8	96.9	89.6	83.4	85.9	86.0	104.4	96.9	95.0	103.7	100.2	101.3	123.8	114.6	85.3	98.2	92.9
09	FINISHES	82.1	85.2	83.8	97.6	82.5	89.4	89.4	100.2	95.3	83.7	94.9	89.7	98.9	115.0	107.6	90.0	93.1	91.7
COVERS	DIVS. 10 - 14, 25, 28, 41, 43, 44, 46	100.0	97.3	99.4	100.0	97.1	99.4	100.0	94.6	98.8	100.0	97.5	99.4	100.0	105.8	101.3	100.0	101.8	100.4
21, 22, 23	FIRE SUPPRESSION, PLUMBING & HVAC	96.0	88.6	93.1	96.7	87.5	93.1	99.8	94.5	97.7	96.0	84.6	91.5	100.0	110.6	104.2	100.0	91.6	96.7
26, 27, 3370	ELECTRICAL, COMMUNICATIONS & UTIL.	102.7	101.7	102.2	104.8	71.1	88.2	100.3	101.7	101.0	109.1	93.1	101.2	107.2	110.3	108.7	103.0	93.1	98.1
MF2018	WEIGHTED AVERAGE	95.6	98.6	96.8	97.4	92.4	95.3	98.6	101.3	99.7	95.7	98.5	96.9	100.7	114.1	106.3	98.3	98.7	98.5

	DIVISION	MINNESOTA															MISSISSIPPI		
		SAINT PAUL			ST. CLOUD			THIEF RIVER FALLS			WILLMAR			WINDOM			BILOXI		
		550 - 551			563			567			562			561			395		
		MAT.	INST.	TOTAL	MAT.	INST.	TOTAL	MAT.	INST.	TOTAL	MAT.	INST.	TOTAL	MAT.	INST.	TOTAL	MAT.	INST.	TOTAL
015433	CONTRACTOR EQUIPMENT		102.9	102.9		100.2	100.2		97.7	97.7		100.2	100.2		100.2	100.2		100.1	100.1
0241, 31 - 34	SITE & INFRASTRUCTURE, DEMOLITION	96.5	102.5	100.7	91.4	100.8	97.9	94.0	94.9	94.7	90.6	99.7	96.9	84.5	98.9	94.4	105.2	85.9	91.9
0310	Concrete Forming & Accessories	98.6	117.0	114.3	86.0	113.5	109.5	88.8	81.9	82.9	85.8	86.8	86.7	90.1	82.2	83.4	92.8	64.1	68.4
0320	Concrete Reinforcing	105.4	109.9	107.6	96.2	109.4	102.6	97.8	101.6	99.6	95.8	109.2	102.3	95.8	108.1	101.8	93.4	48.8	71.8
0330	Cast-in-Place Concrete	99.9	116.3	106.0	97.7	115.0	104.1	100.9	81.3	93.6	99.2	79.5	91.9	85.8	83.3	84.8	114.2	66.3	96.4
03	CONCRETE	95.3	116.1	104.4	86.5	114.0	98.5	89.8	86.5	88.3	86.3	89.5	87.7	77.6	88.5	82.4	94.4	64.0	81.0
04	MASONRY	105.3	127.4	118.9	107.3	121.8	116.2	98.9	102.3	101.0	112.1	114.5	113.6	122.0	87.9	101.0	90.4	62.8	73.4
05	METALS	100.1	125.0	107.4	92.9	123.1	101.8	91.3	116.7	98.7	92.0	121.5	100.7	91.9	119.5	100.0	90.3	84.5	88.6
06	WOOD, PLASTICS & COMPOSITES	95.0	114.2	105.6	82.1	110.3	97.6	70.0	77.0	73.8	81.8	79.9	80.8	86.3	79.9	82.8	93.1	64.5	77.3
07	THERMAL & MOISTURE PROTECTION	102.0	120.5	110.1	104.9	111.7	107.9	107.1	88.1	98.8	104.7	101.0	103.1	104.6	83.0	95.2	100.0	63.1	83.9
08	OPENINGS	98.2	124.3	104.3	91.4	122.1	98.5	99.7	101.3	100.1	88.4	102.8	91.7	91.9	102.8	94.4	96.3	55.0	86.7
0920	Plaster & Gypsum Board	94.5	115.0	108.3	91.5	111.2	104.7	104.3	76.8	85.8	91.5	79.9	83.7	91.5	79.9	83.7	103.8	64.0	77.0
0950, 0980	Ceilings & Acoustic Treatment	91.6	115.0	107.4	58.8	111.2	94.1	128.8	76.8	93.7	58.8	79.9	73.0	58.8	79.9	73.0	86.4	64.0	71.3
0960	Flooring	92.4	122.8	101.0	86.4	120.9	96.1	90.4	85.6	89.0	87.8	85.6	87.2	90.3	85.6	89.0	94.0	65.0	85.8
0970, 0990	Wall Finishes & Painting/Coating	92.1	129.1	114.0	95.0	123.8	112.0	89.6	83.4	85.9	89.6	83.4	85.9	89.6	103.7	98.0	86.0	46.9	62.8
09	FINISHES	91.1	119.1	106.3	81.5	115.9	100.1	98.1	82.6	89.7	81.6	85.3	83.6	81.8	84.8	83.4	89.4	62.8	75.0
COVERS	DIVS. 10 - 14, 25, 28, 41, 43, 44, 46	100.0	108.7	101.9	100.0	103.2	100.7	100.0	95.5	99.0	100.0	97.3	99.4	100.0	93.7	98.6	100.0	71.9	93.7
21, 22, 23	FIRE SUPPRESSION, PLUMBING & HVAC	99.9	116.7	106.5	99.5	110.5	103.8	96.7	84.4	91.9	96.0	101.6	98.2	96.0	81.0	90.1	100.0	55.3	82.4
26, 27, 3370	ELECTRICAL, COMMUNICATIONS & UTIL.	102.9	115.1	108.9	102.7	115.1	108.8	102.1	71.0	86.8	102.7	83.7	93.4	109.1	93.1	101.2	101.0	53.7	77.7
MF2018	WEIGHTED AVERAGE	99.0	117.6	106.9	95.1	114.2	103.2	96.3	89.3	93.3	94.0	98.0	95.7	94.3	91.2	93.0	96.2	63.9	82.6

	DIVISION	MISSISSIPPI																	
		CLARKSDALE			COLUMBUS			GREENVILLE			GREENWOOD			JACKSON			LAUREL		
		386			397			387			389			390 - 392			394		
		MAT.	INST.	TOTAL	MAT.	INST.	TOTAL	MAT.	INST.	TOTAL	MAT.	INST.	TOTAL	MAT.	INST.	TOTAL	MAT.	INST.	TOTAL
015433	CONTRACTOR EQUIPMENT		100.1	100.1		100.1	100.1		100.1	100.1		100.1	100.1		102.6	102.6		100.1	100.1
0241, 31 - 34	SITE & INFRASTRUCTURE, DEMOLITION	101.4	84.4	89.7	103.9	85.2	91.0	107.5	86.1	92.7	104.5	84.1	90.4	101.1	90.6	93.9	109.6	84.5	92.2
0310	Concrete Forming & Accessories	83.8	43.9	49.8	81.9	46.3	51.6	80.7	62.7	65.4	92.5	44.3	51.4	92.0	64.6	68.6	82.0	59.5	62.8
0320	Concrete Reinforcing	104.2	66.5	85.9	99.9	66.6	83.8	104.7	66.7	86.3	104.2	66.5	86.0	104.0	51.7	78.7	100.6	32.4	67.6
0330	Cast-in-Place Concrete	104.1	58.8	87.2	116.5	61.1	95.9	107.2	65.7	91.8	112.0	58.9	92.2	99.6	66.5	87.3	113.9	60.1	93.9
03	CONCRETE	93.8	55.1	76.8	95.8	57.1	78.8	99.3	66.2	84.8	100.1	55.4	80.4	89.2	64.7	78.4	97.9	56.9	79.9
04	MASONRY	89.1	50.8	65.6	115.4	53.6	77.4	132.2	61.4	88.7	89.8	50.7	65.8	95.7	61.5	74.7	111.5	49.4	73.4
05	METALS	92.5	87.9	91.1	87.4	90.4	88.3	93.6	91.2	92.8	92.5	87.8	91.1	96.9	84.6	93.3	87.5	76.0	84.1
06	WOOD, PLASTICS & COMPOSITES	79.4	44.4	60.1	78.8	45.2	60.3	76.2	62.8	68.8	91.8	44.4	65.7	96.3	65.5	79.4	79.9	64.5	71.4
07	THERMAL & MOISTURE PROTECTION	97.7	52.0	77.8	100.0	55.4	80.6	98.2	62.1	82.5	98.1	54.2	79.0	98.3	62.9	82.9	100.2	56.5	81.2
08	OPENINGS	95.1	48.2	84.1	95.9	48.6	84.9	94.8	58.1	86.2	95.1	48.2	84.1	99.5	56.0	89.4	93.2	51.8	83.5
0920	Plaster & Gypsum Board	89.9	43.3	58.6	94.3	44.2	60.6	89.6	62.3	71.2	100.6	43.3	62.1	90.7	64.9	73.3	94.3	64.0	73.9
0950, 0980	Ceilings & Acoustic Treatment	82.3	43.3	56.0	81.5	44.2	56.3	85.0	62.3	69.7	82.3	43.3	56.0	89.7	64.9	73.0	81.5	64.0	69.7
0960	Flooring	97.8	45.0	83.0	87.9	65.0	81.4	96.2	65.0	87.4	103.5	65.0	92.7	92.1	65.0	84.5	86.6	65.0	80.5
0970, 0990	Wall Finishes & Painting/Coating	94.6	46.9	66.4	86.0	46.9	62.8	94.6	56.7	72.1	94.6	46.9	66.4	89.9	56.7	70.3	86.0	46.9	62.8
09	FINISHES	90.2	43.9	65.1	85.4	49.1	65.8	90.8	62.5	75.5	93.7	48.0	68.9	89.9	64.1	75.9	85.6	60.5	72.0
COVERS	DIVS. 10 - 14, 25, 28, 41, 43, 44, 46	100.0	48.0	88.4	100.0	49.0	88.6	100.0	71.3	93.6	100.0	48.0	88.4	100.0	71.9	93.7	100.0	34.2	85.3
21, 22, 23	FIRE SUPPRESSION, PLUMBING & HVAC	98.4	51.3	79.9	98.0	53.1	80.4	100.0	57.4	83.3	98.4	51.7	80.1	100.0	59.1	83.9	98.1	47.0	78.0
26, 27, 3370	ELECTRICAL, COMMUNICATIONS & UTIL.	96.8	41.6	69.6	98.6	55.1	77.2	96.8	55.9	76.6	96.8	39.0	68.3	102.7	55.9	79.6	100.1	57.9	79.3
MF2018	WEIGHTED AVERAGE	95.3	55.2	78.4	95.9	59.1	80.4	98.7	65.5	84.7	96.6	55.5	79.2	97.3	65.6	83.9	96.1	57.8	79.9

	DIVISION	MISSISSIPPI									MISSOURI								
		MCCOMB			MERIDIAN			TUPELO			BOWLING GREEN			CAPE GIRARDEAU			CHILLICOTHE		
		396			393			388			633			637			646		
		MAT.	INST.	TOTAL	MAT.	INST.	TOTAL	MAT.	INST.	TOTAL	MAT.	INST.	TOTAL	MAT.	INST.	TOTAL	MAT.	INST.	TOTAL
015433	CONTRACTOR EQUIPMENT		100.1	100.1		100.1	100.1		100.1	100.1		107.1	107.1		107.1	107.1		101.5	101.5
0241, 31 - 34	SITE & INFRASTRUCTURE, DEMOLITION	96.3	84.3	88.0	100.7	86.1	90.7	98.9	84.3	88.8	87.6	90.0	89.3	89.3	89.9	89.7	101.2	88.4	92.3
0310	Concrete Forming & Accessories	81.9	45.5	50.9	79.4	62.8	65.2	81.2	45.9	51.1	96.0	94.0	94.3	88.9	82.0	83.0	86.2	95.7	94.3
0320	Concrete Reinforcing	101.2	34.3	68.8	99.9	51.6	76.6	101.9	66.5	84.8	93.4	96.5	94.9	94.6	79.6	87.3	100.4	102.6	101.5
0330	Cast-in-Place Concrete	101.1	58.4	85.2	108.2	66.0	92.5	104.1	67.9	90.6	90.7	95.5	92.5	89.8	86.6	88.6	91.7	86.0	89.6
03	CONCRETE	85.9	50.2	70.2	90.1	63.8	78.6	93.4	59.2	78.4	91.2	96.4	93.5	90.4	85.0	88.0	95.0	94.3	94.7
04	MASONRY	116.7	50.3	75.9	90.0	62.1	72.9	121.7	53.3	79.6	109.6	97.9	102.4	105.9	80.5	90.3	99.2	92.2	94.9
05	METALS	87.6	75.6	84.1	88.5	85.6	87.7	92.4	87.9	91.1	91.8	117.7	99.4	92.9	109.4	97.8	85.1	110.6	92.5
06	WOOD, PLASTICS & COMPOSITES	78.8	46.8	61.2	76.2	62.8	68.8	76.7	45.2	59.4	96.7	94.4	95.4	89.2	80.0	84.1	94.0	97.0	95.7
07	THERMAL & MOISTURE PROTECTION	99.4	54.2	79.8	99.6	62.4	83.4	97.6	53.9	78.6	96.4	99.2	97.6	95.9	85.3	91.3	92.7	92.7	92.7
08	OPENINGS	96.0	41.3	83.2	95.6	54.8	86.1	95.0	51.4	84.9	98.7	98.2	98.5	98.6	76.1	93.4	86.0	96.2	88.3
0920	Plaster & Gypsum Board	94.3	45.8	61.7	94.3	62.3	72.8	89.6	44.2	59.0	101.7	94.6	96.9	101.1	79.8	86.8	98.2	96.9	97.3
0950, 0980	Ceilings & Acoustic Treatment	81.5	45.8	57.4	82.4	62.3	68.8	82.3	44.2	56.6	88.9	94.6	92.8	88.9	79.8	82.8	87.4	96.9	93.8
0960	Flooring	87.9	65.0	81.4	86.5	65.0	80.5	96.5	65.0	87.6	95.7	96.9	96.0	92.4	84.5	90.2	94.9	99.7	96.2
0970, 0990	Wall Finishes & Painting/Coating	86.0	46.9	62.8	86.0	46.9	62.8	94.6	45.2	65.4	95.4	106.3	101.8	95.4	67.1	78.6	94.3	103.2	99.6
09	FINISHES	84.8	49.2	65.5	84.8	61.6	72.3	89.7	48.9	67.6	98.2	95.7	96.8	97.0	79.8	87.7	97.1	97.6	97.4
COVERS	DIVS. 10 - 14, 25, 28, 41, 43, 44, 46	100.0	50.9	89.1	100.0	71.5	93.6	100.0	49.0	88.6	100.0	81.4	95.8	100.0	94.3	98.7	100.0	81.3	95.8
21, 22, 23	FIRE SUPPRESSION, PLUMBING & HVAC	98.0	50.7	79.5	100.0	57.6	83.3	98.5	52.7	80.5	96.5	99.5	97.7	100.0	98.7	99.5	96.6	99.9	97.9
26, 27, 3370	ELECTRICAL, COMMUNICATIONS & UTIL.	97.3	56.0	76.9	100.1	55.6	78.2	96.5	55.0	76.1	99.2	76.8	88.2	99.2	98.5	98.9	94.9	75.8	85.5
MF2018	WEIGHTED AVERAGE	94.4	55.8	78.0	94.6	64.5	81.9	96.6	59.0	80.7	96.4	95.5	96.0	97.0	91.2	94.5	93.7	93.8	93.7

DIVISION		MISSOURI																	
		COLUMBIA			FLAT RIVER			HANNIBAL			HARRISONVILLE			JEFFERSON CITY			JOPLIN		
		652			636			634			647			650 - 651			648		
		MAT.	INST.	TOTAL	MAT.	INST.	TOTAL	MAT.	INST.	TOTAL	MAT.	INST.	TOTAL	MAT.	INST.	TOTAL	MAT.	INST.	TOTAL
015433	CONTRACTOR EQUIPMENT		110.3	110.3		107.1	107.1		107.1	107.1		101.5	101.5		113.1	113.1		104.8	104.8
0241, 31 - 34	SITE & INFRASTRUCTURE, DEMOLITION	94.2	93.8	93.9	90.2	89.7	89.8	85.4	89.8	88.4	92.9	89.5	90.6	94.4	98.7	97.4	102.0	92.6	95.5
0310	Concrete Forming & Accessories	84.4	81.0	81.5	102.4	88.3	90.4	94.3	82.5	84.3	83.4	99.0	96.7	96.3	78.6	81.2	97.4	74.2	77.7
0320	Concrete Reinforcing	85.6	93.3	89.3	94.6	102.9	98.6	92.9	96.5	94.6	100.0	110.9	105.3	92.4	93.3	92.9	103.7	84.3	94.3
0330	Cast-in-Place Concrete	88.6	84.8	87.2	93.7	90.6	92.5	85.9	93.5	88.7	93.9	99.7	96.0	93.9	84.1	90.2	99.3	75.5	90.4
03	CONCRETE	80.8	86.3	83.2	94.1	93.3	93.8	87.6	90.5	88.9	91.1	102.0	95.9	87.5	84.8	86.3	93.9	77.7	86.7
04	MASONRY	136.9	86.4	105.9	106.5	76.4	88.0	101.6	94.6	97.3	94.0	99.4	97.3	96.0	86.5	90.1	92.4	80.2	84.9
05	METALS	96.3	115.4	101.9	91.7	119.2	99.8	91.8	117.2	99.3	85.4	115.1	94.1	95.8	114.0	101.1	87.9	98.8	91.1
06	WOOD, PLASTICS & COMPOSITES	80.3	78.6	79.3	105.2	89.6	96.6	94.9	81.1	87.3	90.4	98.7	95.0	97.1	75.6	85.2	106.1	73.5	88.2
07	THERMAL & MOISTURE PROTECTION	90.6	86.1	88.7	96.6	91.0	94.2	96.2	93.4	95.0	91.8	100.8	95.7	97.5	86.1	92.5	91.9	81.2	87.3
08	OPENINGS	94.4	81.7	91.4	98.6	97.4	98.4	98.7	83.7	95.2	85.7	103.2	89.8	94.5	80.0	91.1	86.9	76.3	84.5
0920	Plaster & Gypsum Board	84.7	78.2	80.3	107.7	89.7	95.6	101.4	81.0	87.7	94.1	98.6	97.1	96.2	74.9	81.9	105.4	72.7	83.4
0950, 0980	Ceilings & Acoustic Treatment	88.5	78.2	81.5	88.9	89.7	89.4	88.9	81.0	83.6	87.4	98.6	95.0	93.3	74.9	80.9	88.3	72.7	77.8
0960	Flooring	90.2	98.6	92.5	98.9	84.5	94.9	95.0	96.9	95.5	90.3	100.8	93.2	97.6	71.5	90.3	120.1	72.2	106.6
0970, 0990	Wall Finishes & Painting/Coating	93.1	81.9	86.4	95.4	71.7	81.4	95.4	94.1	94.6	98.6	107.8	104.0	91.2	81.9	85.7	93.9	75.9	83.2
09	FINISHES	84.0	83.5	83.7	100.0	85.1	92.0	97.7	85.3	91.0	94.8	100.2	97.7	91.5	77.1	83.7	103.8	74.4	87.9
COVERS	DIVS. 10 - 14, 25, 28, 41, 43, 44, 46	100.0	93.6	98.6	100.0	94.0	98.7	100.0	78.8	95.3	100.0	82.7	96.2	100.0	93.6	98.6	100.0	79.5	95.4
21, 22, 23	FIRE SUPPRESSION, PLUMBING & HVAC	100.0	96.9	98.8	96.5	96.8	96.6	96.5	97.7	97.0	96.6	100.8	98.2	100.0	96.9	98.8	100.1	71.1	88.7
26, 27, 3370	ELECTRICAL, COMMUNICATIONS & UTIL.	97.8	81.4	89.7	103.8	98.5	101.2	98.0	76.8	87.5	101.5	101.2	101.4	103.2	81.4	92.4	92.9	66.1	79.7
MF2018	WEIGHTED AVERAGE	96.1	90.7	93.8	97.3	94.2	96.0	95.3	91.6	93.7	93.2	100.8	96.4	96.4	89.9	93.7	95.1	77.7	87.7

DIVISION		MISSOURI																	
		KANSAS CITY			KIRKSVILLE			POPLAR BLUFF			ROLLA			SEDALIA			SIKESTON		
		640 - 641			635			639			654 - 655			653			638		
		MAT.	INST.	TOTAL	MAT.	INST.	TOTAL	MAT.	INST.	TOTAL	MAT.	INST.	TOTAL	MAT.	INST.	TOTAL	MAT.	INST.	TOTAL
015433	CONTRACTOR EQUIPMENT		104.4	104.4		97.4	97.4		99.6	99.6		110.3	110.3		100.4	100.4		99.6	99.6
0241, 31 - 34	SITE & INFRASTRUCTURE, DEMOLITION	94.6	98.6	97.3	88.8	84.9	86.1	76.1	88.5	84.7	92.9	94.1	93.8	92.1	89.3	90.2	79.6	89.1	86.2
0310	Concrete Forming & Accessories	97.2	103.5	102.6	86.9	78.4	79.6	87.4	79.3	80.5	91.6	94.0	93.6	89.6	78.2	79.9	88.2	79.4	80.7
0320	Concrete Reinforcing	98.5	111.1	104.6	93.8	83.3	88.7	96.7	76.0	86.7	86.0	93.4	89.6	84.8	110.3	97.1	96.0	76.0	86.4
0330	Cast-in-Place Concrete	97.3	102.6	99.3	93.6	82.2	89.4	72.0	83.5	76.3	90.6	94.7	92.1	94.7	81.3	89.7	77.0	83.5	79.4
03	CONCRETE	93.8	104.9	98.7	106.9	81.8	95.9	81.5	81.4	81.5	82.5	95.6	88.2	95.7	86.1	91.5	85.2	81.5	83.6
04	MASONRY	99.6	103.3	101.9	113.2	84.8	95.8	104.5	74.1	85.8	110.8	85.3	95.2	117.1	81.6	95.3	104.2	74.1	85.7
05	METALS	94.9	112.8	100.2	91.5	101.0	94.3	92.1	97.8	93.7	95.7	115.9	101.6	94.5	112.8	99.9	92.4	97.9	94.0
06	WOOD, PLASTICS & COMPOSITES	100.1	103.8	102.2	82.2	76.5	79.1	81.3	80.0	80.6	88.0	95.6	92.2	81.6	76.7	78.9	82.9	80.0	81.3
07	THERMAL & MOISTURE PROTECTION	91.8	105.2	97.7	102.2	90.8	97.3	100.5	83.0	92.9	90.9	92.9	91.8	96.1	87.8	92.5	100.7	82.1	92.6
08	OPENINGS	96.5	106.0	98.7	104.1	78.8	98.2	105.0	75.2	98.1	94.4	91.1	93.6	99.4	87.0	96.5	105.0	75.2	98.1
0920	Plaster & Gypsum Board	100.9	104.0	103.0	96.3	76.2	82.8	96.6	79.8	85.3	86.9	95.6	92.8	80.5	76.2	77.6	98.5	79.8	85.9
0950, 0980	Ceilings & Acoustic Treatment	90.5	104.0	99.6	87.3	76.2	79.8	88.9	79.8	82.8	88.5	95.6	93.3	88.5	76.2	80.2	88.9	79.8	82.8
0960	Flooring	95.9	101.8	97.6	73.8	96.4	80.2	87.3	84.5	86.5	93.8	96.4	94.5	72.5	72.9	72.6	87.8	84.5	86.9
0970, 0990	Wall Finishes & Painting/Coating	97.2	107.8	103.5	91.3	78.3	83.6	90.6	67.1	76.7	93.1	90.2	91.4	93.1	103.2	99.1	90.6	67.1	76.7
09	FINISHES	98.1	103.6	101.1	97.4	81.1	88.6	96.9	78.4	86.9	85.4	94.0	90.1	82.9	79.1	80.8	97.5	78.7	87.3
COVERS	DIVS. 10 - 14, 25, 28, 41, 43, 44, 46	100.0	98.4	99.7	100.0	78.3	95.2	100.0	92.2	98.3	100.0	96.7	99.3	100.0	87.9	97.3	100.0	92.2	98.3
21, 22, 23	FIRE SUPPRESSION, PLUMBING & HVAC	100.0	103.4	101.4	96.6	97.1	96.8	96.6	94.9	95.9	96.5	98.6	97.3	96.4	94.9	95.8	96.6	95.0	95.9
26, 27, 3370	ELECTRICAL, COMMUNICATIONS & UTIL.	103.3	101.2	102.2	98.2	76.8	87.6	98.4	98.4	98.4	96.4	79.4	88.0	97.2	101.2	99.2	97.5	98.4	97.9
MF2018	WEIGHTED AVERAGE	97.9	103.8	100.4	98.9	86.5	93.7	95.1	87.7	92.0	94.1	94.3	94.2	96.4	91.5	94.3	95.7	87.8	92.3

DIVISION		MISSOURI									MONTANA								
		SPRINGFIELD			ST. JOSEPH			ST. LOUIS			BILLINGS			BUTTE			GREAT FALLS		
		656 - 658			644 - 645			630 - 631			590 - 591			597			594		
		MAT.	INST.	TOTAL	MAT.	INST.	TOTAL	MAT.	INST.	TOTAL	MAT.	INST.	TOTAL	MAT.	INST.	TOTAL	MAT.	INST.	TOTAL
015433	CONTRACTOR EQUIPMENT		102.8	102.8		101.5	101.5		108.6	108.6		97.9	97.9		97.7	97.7		97.7	97.7
0241, 31 - 34	SITE & INFRASTRUCTURE, DEMOLITION	94.4	91.4	92.4	96.7	87.4	90.3	94.4	97.8	96.7	91.8	93.5	93.0	97.3	93.3	94.5	101.0	93.3	95.7
0310	Concrete Forming & Accessories	97.7	75.4	78.7	96.3	89.1	90.2	99.5	102.5	102.1	99.3	68.3	72.9	86.5	68.2	70.9	99.2	68.2	72.8
0320	Concrete Reinforcing	82.5	92.9	87.5	97.3	110.4	103.6	86.5	105.2	95.5	98.0	81.8	90.2	106.3	81.7	94.4	98.0	81.8	90.2
0330	Cast-in-Place Concrete	96.3	73.7	87.9	92.2	95.5	93.4	101.3	102.5	101.7	118.7	72.7	101.6	130.6	72.5	109.0	138.1	72.5	113.7
03	CONCRETE	92.1	79.0	86.3	90.3	96.0	92.8	97.1	104.0	100.1	98.3	73.1	87.2	101.9	73.0	89.2	106.9	73.0	92.0
04	MASONRY	89.1	79.9	83.5	95.3	88.7	91.2	91.1	107.1	100.9	125.8	83.8	100.0	120.9	82.7	97.4	125.4	83.8	99.8
05	METALS	100.7	101.7	101.0	91.5	113.8	98.0	97.4	119.0	103.8	107.0	89.6	101.9	101.2	89.4	97.7	104.4	89.5	100.0
06	WOOD, PLASTICS & COMPOSITES	89.4	75.0	81.5	106.1	88.7	96.5	97.6	101.0	99.5	90.1	64.6	76.1	76.8	64.6	70.1	91.4	64.6	76.6
07	THERMAL & MOISTURE PROTECTION	94.6	75.5	86.3	92.2	88.7	90.7	94.3	105.5	99.2	108.3	73.7	93.2	107.8	73.3	92.8	108.5	73.7	93.4
08	OPENINGS	101.7	84.3	97.6	89.9	96.7	91.5	100.3	105.8	101.6	98.6	65.9	91.0	96.9	65.9	89.7	99.8	65.9	91.9
0920	Plaster & Gypsum Board	87.8	74.4	78.8	106.9	88.3	94.4	106.5	101.5	103.1	118.3	64.0	81.8	118.8	64.0	81.9	128.7	64.0	85.2
0950, 0980	Ceilings & Acoustic Treatment	88.5	74.4	79.0	94.8	88.3	90.4	90.5	101.5	97.9	96.6	64.0	74.6	104.1	64.0	77.0	105.7	64.0	77.6
0960	Flooring	92.8	71.5	86.8	99.4	99.0	99.3	98.0	99.1	98.3	88.1	80.8	86.0	85.4	90.9	87.0	91.9	80.8	88.8
0970, 0990	Wall Finishes & Painting/Coating	87.6	99.4	94.6	94.3	107.8	102.3	96.4	106.3	102.2	93.8	91.0	92.2	92.3	67.8	77.8	92.3	91.0	91.5
09	FINISHES	87.6	77.6	82.2	100.3	92.8	96.3	100.5	101.9	101.3	92.5	72.3	81.6	93.0	71.8	81.5	96.8	72.3	83.5
COVERS	DIVS. 10 - 14, 25, 28, 41, 43, 44, 46	100.0	90.5	97.9	100.0	94.0	98.7	100.0	102.8	100.6	100.0	91.6	98.1	100.0	91.6	98.1	100.0	91.6	98.1
21, 22, 23	FIRE SUPPRESSION, PLUMBING & HVAC	100.0	69.4	88.0	100.1	87.5	95.2	100.0	105.5	102.2	100.0	76.4	90.7	100.0	71.2	88.7	100.0	71.2	88.7
26, 27, 3370	ELECTRICAL, COMMUNICATIONS & UTIL.	101.3	68.8	85.3	101.5	75.8	88.9	101.8	98.5	100.2	102.3	75.2	88.9	109.2	71.3	90.5	101.7	70.8	86.5
MF2018	WEIGHTED AVERAGE	97.6	79.0	89.7	96.1	90.9	93.9	98.8	104.5	101.2	101.6	78.4	91.8	101.5	76.5	90.9	102.9	76.6	91.8

DIVISION		MONTANA																	
		HAVRE			HELENA			KALISPELL			MILES CITY			MISSOULA			WOLF POINT		
		595			596			599			593			598			592		
		MAT.	INST.	TOTAL	MAT.	INST.	TOTAL	MAT.	INST.	TOTAL	MAT.	INST.	TOTAL	MAT.	INST.	TOTAL	MAT.	INST.	TOTAL
015433	CONTRACTOR EQUIPMENT		97.7	97.7		99.9	99.9		97.7	97.7		97.7	97.7		97.7	97.7		97.7	97.7
0241, 31 - 34	SITE & INFRASTRUCTURE, DEMOLITION	104.2	93.1	96.6	91.1	96.8	95.0	87.5	93.3	91.5	93.5	93.2	93.3	80.5	93.3	89.3	110.2	93.2	98.4
0310	Concrete Forming & Accessories	79.6	67.2	69.0	100.2	68.3	73.0	89.7	68.3	71.4	97.7	67.3	71.8	89.7	68.4	71.5	90.4	67.3	70.7
0320	Concrete Reinforcing	107.1	78.9	93.5	113.0	81.8	97.9	109.1	85.8	97.8	106.7	78.9	93.3	108.1	85.8	97.3	108.2	78.2	93.7
0330	Cast-in-Place Concrete	140.7	71.2	114.9	102.1	73.3	91.4	113.4	72.5	98.2	124.2	71.3	104.5	96.2	72.6	87.4	139.1	70.2	113.5
03	CONCRETE	109.8	71.6	93.0	94.7	73.3	85.3	91.7	73.7	83.8	98.8	71.7	86.9	80.5	73.8	77.6	113.2	71.2	94.8
04	MASONRY	121.9	80.6	96.5	114.2	82.7	94.8	119.6	82.7	96.9	127.6	80.6	98.7	144.5	82.7	106.5	128.9	80.6	99.2
05	METALS	97.3	88.2	94.6	103.2	88.5	98.9	97.2	90.8	95.3	96.5	88.4	94.1	97.7	91.0	95.7	96.6	88.1	94.1
06	WOOD, PLASTICS & COMPOSITES	68.1	64.6	66.2	93.8	64.7	77.8	80.2	64.6	71.6	88.4	64.6	75.3	80.2	64.6	71.6	79.5	64.6	71.3
07	THERMAL & MOISTURE PROTECTION	108.3	66.2	90.0	103.1	73.9	90.4	107.4	73.6	92.7	107.8	67.7	90.3	106.9	75.6	93.3	108.9	67.5	90.9
08	OPENINGS	97.4	65.2	89.9	96.3	65.9	89.2	97.4	66.8	90.3	96.9	65.2	89.5	96.9	66.8	89.9	96.9	65.1	89.5
0920	Plaster & Gypsum Board	114.1	64.0	80.4	113.5	64.0	80.2	118.8	64.0	81.9	127.8	64.0	84.9	118.8	64.0	81.9	121.8	64.0	82.9
0950, 0980	Ceilings & Acoustic Treatment	104.1	64.0	77.0	107.9	64.0	78.3	104.1	64.0	77.0	101.6	64.0	76.3	104.1	64.0	77.0	101.6	64.0	76.3
0960	Flooring	83.0	90.9	85.3	95.9	90.9	94.5	87.1	90.9	88.2	91.8	90.9	91.6	87.1	90.9	88.2	88.4	90.9	89.1
0970, 0990	Wall Finishes & Painting/Coating	92.3	67.8	77.8	97.6	67.8	79.9	92.3	91.0	91.5	92.3	67.8	77.8	92.3	91.0	91.5	92.3	67.8	77.8
09	FINISHES	92.3	71.3	80.9	100.1	71.8	84.8	92.9	74.3	82.8	95.4	71.3	82.3	92.3	74.3	82.6	95.0	71.3	82.1
COVERS	DIVS. 10 - 14, 25, 28, 41, 43, 44, 46	100.0	88.8	97.5	100.0	91.9	98.2	100.0	89.5	97.7	100.0	88.9	97.5	100.0	92.4	98.3	100.0	88.9	97.5
21, 22, 23	FIRE SUPPRESSION, PLUMBING & HVAC	96.5	68.9	85.6	100.0	71.2	88.7	96.5	69.4	85.8	96.5	75.2	88.1	100.0	71.2	88.7	96.5	75.2	88.1
26, 27, 3370	ELECTRICAL, COMMUNICATIONS & UTIL.	101.7	69.2	85.7	109.0	70.8	90.2	106.0	68.1	87.3	101.7	74.4	88.3	107.0	70.2	88.9	101.7	74.4	88.3
MF2018	WEIGHTED AVERAGE	100.4	74.8	89.6	100.9	76.7	90.7	98.1	76.2	88.9	99.2	77.0	89.8	98.6	77.1	89.5	101.4	76.9	91.1

DIVISION		NEBRASKA																	
		ALLIANCE			COLUMBUS			GRAND ISLAND			HASTINGS			LINCOLN			MCCOOK		
		693			686			688			689			683 - 685			690		
		MAT.	INST.	TOTAL	MAT.	INST.	TOTAL	MAT.	INST.	TOTAL	MAT.	INST.	TOTAL	MAT.	INST.	TOTAL	MAT.	INST.	TOTAL
015433	CONTRACTOR EQUIPMENT		95.3	95.3		101.9	101.9		101.9	101.9		101.9	101.9		105.0	105.0		101.9	101.9
0241, 31 - 34	SITE & INFRASTRUCTURE, DEMOLITION	98.2	96.2	96.8	101.8	90.6	94.1	106.8	90.7	95.7	105.4	90.6	95.2	94.9	95.7	95.4	101.5	90.6	94.0
0310	Concrete Forming & Accessories	87.0	55.3	60.0	96.5	75.0	78.2	96.1	68.8	72.9	99.3	72.5	76.5	94.4	75.5	78.3	92.3	55.9	61.2
0320	Concrete Reinforcing	107.4	87.4	97.7	100.4	85.9	93.4	99.8	76.1	88.4	99.8	76.1	88.4	97.6	76.4	87.3	100.1	76.3	88.6
0330	Cast-in-Place Concrete	108.1	81.8	98.3	107.3	81.8	97.8	113.6	77.7	100.2	113.6	74.0	98.9	87.2	83.1	85.6	116.9	73.9	100.9
03	CONCRETE	114.3	70.9	95.2	100.2	80.4	91.5	104.9	74.5	91.5	105.1	74.9	91.8	87.5	79.4	83.9	103.7	67.4	87.7
04	MASONRY	106.0	77.4	88.4	111.1	77.4	90.4	104.3	73.8	85.6	112.8	74.0	88.9	93.6	78.2	84.1	101.9	77.4	86.8
05	METALS	99.1	85.2	95.0	91.4	98.0	93.3	93.2	93.7	93.3	94.0	93.7	93.9	95.0	93.3	94.5	93.9	93.8	93.9
06	WOOD, PLASTICS & COMPOSITES	82.6	48.7	64.0	94.3	74.9	83.6	93.4	66.3	78.5	97.2	72.1	83.4	99.0	75.0	85.8	90.7	49.7	68.1
07	THERMAL & MOISTURE PROTECTION	100.7	75.8	89.9	101.6	80.4	92.4	101.8	78.3	91.5	101.8	77.6	91.3	100.1	81.1	91.8	96.0	76.1	87.3
08	OPENINGS	90.8	58.1	83.2	91.3	73.7	87.2	91.3	66.8	85.6	91.3	69.8	86.3	105.4	68.8	96.9	91.2	55.8	83.0
0920	Plaster & Gypsum Board	79.8	47.5	58.1	93.4	74.3	80.6	92.5	65.5	74.3	94.4	71.4	78.9	102.7	74.3	83.6	91.0	48.4	62.4
0950, 0980	Ceilings & Acoustic Treatment	88.6	47.5	60.9	83.4	74.3	77.3	83.4	65.5	71.3	83.4	71.4	75.3	98.3	74.3	82.1	83.9	48.4	60.0
0960	Flooring	91.4	87.3	90.2	84.0	87.3	84.9	83.7	78.9	82.4	85.0	81.3	84.0	96.3	86.6	93.6	90.0	87.3	89.2
0970, 0990	Wall Finishes & Painting/Coating	155.8	51.7	94.2	74.9	60.1	66.1	74.9	62.5	67.6	74.9	60.1	66.1	94.4	79.3	85.5	87.6	45.0	62.4
09	FINISHES	90.8	59.2	73.7	84.9	75.5	79.8	84.9	69.3	76.5	85.6	72.7	78.6	95.6	77.9	86.0	88.4	59.1	72.5
COVERS	DIVS. 10 - 14, 25, 28, 41, 43, 44, 46	100.0	83.9	96.4	100.0	83.5	96.3	100.0	89.7	97.7	100.0	83.1	96.2	100.0	91.0	98.0	100.0	84.1	96.5
21, 22, 23	FIRE SUPPRESSION, PLUMBING & HVAC	96.7	75.0	88.2	96.6	75.3	88.2	100.1	79.5	92.0	96.6	74.6	87.9	100.0	79.6	92.0	96.5	75.2	88.1
26, 27, 3370	ELECTRICAL, COMMUNICATIONS & UTIL.	93.6	65.6	79.8	97.3	81.4	89.5	95.9	65.7	81.0	95.3	79.4	87.5	109.2	65.7	87.8	95.6	65.7	80.9
MF2018	WEIGHTED AVERAGE	98.7	73.3	87.9	96.0	80.7	89.6	97.4	76.8	88.7	97.1	78.2	89.1	98.3	79.7	90.5	96.3	73.1	86.5

DIVISION		NEBRASKA												NEVADA					
		NORFOLK			NORTH PLATTE			OMAHA			VALENTINE			CARSON CITY			ELKO		
		687			691			680 - 681			692			897			898		
		MAT.	INST.	TOTAL	MAT.	INST.	TOTAL	MAT.	INST.	TOTAL	MAT.	INST.	TOTAL	MAT.	INST.	TOTAL	MAT.	INST.	TOTAL
015433	CONTRACTOR EQUIPMENT		91.7	91.7		101.9	101.9		94.2	94.2		95.0	95.0		97.9	97.9		94.6	94.6
0241, 31 - 34	SITE & INFRASTRUCTURE, DEMOLITION	84.1	89.7	88.0	102.9	90.4	94.3	89.1	93.9	92.4	86.3	95.2	92.4	83.3	97.4	93.0	67.8	91.3	84.0
0310	Concrete Forming & Accessories	83.2	74.1	75.5	94.8	75.2	78.1	93.1	76.7	79.1	83.5	53.1	57.6	106.1	76.3	80.7	111.4	96.3	98.6
0320	Concrete Reinforcing	100.5	66.5	84.1	99.5	76.2	88.3	100.8	76.5	89.0	100.1	66.3	83.7	115.8	116.1	116.0	124.5	114.5	119.7
0330	Cast-in-Place Concrete	107.9	71.3	94.3	116.9	60.6	96.0	91.4	78.6	86.7	103.1	54.0	84.8	98.0	82.2	92.1	95.0	72.4	86.6
03	CONCRETE	98.8	72.4	87.2	103.8	71.4	89.6	89.8	77.8	84.5	101.2	56.9	81.7	101.0	85.3	94.1	98.5	90.8	95.1
04	MASONRY	117.1	77.3	92.7	90.4	76.9	82.1	93.0	80.0	85.0	101.4	76.8	86.3	118.2	68.8	87.8	125.0	67.7	89.8
05	METALS	94.8	80.4	90.6	93.1	93.2	93.2	95.0	83.9	91.7	105.1	79.4	97.5	111.4	94.1	106.3	115.0	93.6	108.7
06	WOOD, PLASTICS & COMPOSITES	78.4	74.4	76.2	92.7	76.9	84.0	93.4	76.6	84.2	77.5	47.4	61.0	93.6	74.4	83.0	104.7	102.3	103.4
07	THERMAL & MOISTURE PROTECTION	101.2	78.3	91.2	95.9	78.4	88.3	96.7	80.8	89.8	96.5	73.9	86.7	118.1	79.1	101.2	114.2	73.6	96.5
08	OPENINGS	92.7	68.7	87.1	90.6	72.1	86.3	99.8	76.8	94.4	92.7	53.4	83.6	101.1	77.6	95.6	102.5	92.6	100.2
0920	Plaster & Gypsum Board	93.2	74.3	80.5	91.0	76.4	81.2	102.9	76.4	85.1	92.5	46.6	61.6	100.8	73.7	82.6	106.9	102.5	103.9
0950, 0980	Ceilings & Acoustic Treatment	97.6	74.3	81.9	83.9	76.4	78.8	101.0	76.4	84.4	100.4	46.6	64.1	96.0	73.7	80.9	94.5	102.5	99.9
0960	Flooring	104.1	87.3	99.4	91.1	78.9	87.7	100.6	87.3	96.9	116.8	87.3	108.5	99.5	67.1	90.4	103.0	67.1	92.9
0970, 0990	Wall Finishes & Painting/Coating	132.5	60.1	89.7	87.6	58.9	70.6	104.2	61.0	78.6	155.3	61.0	99.5	95.8	79.3	86.1	94.2	79.3	85.4
09	FINISHES	101.8	75.2	87.4	88.7	74.8	81.2	100.4	76.9	87.7	109.1	59.3	82.1	96.5	74.0	84.3	94.0	90.4	92.0
COVERS	DIVS. 10 - 14, 25, 28, 41, 43, 44, 46	100.0	85.1	96.7	100.0	61.1	91.3	100.0	90.2	97.8	100.0	56.1	90.2	100.0	100.2	100.0	100.0	91.0	98.0
21, 22, 23	FIRE SUPPRESSION, PLUMBING & HVAC	96.3	74.3	87.7	100.0	73.7	89.6	100.0	80.3	92.3	96.1	73.5	87.2	100.0	77.0	91.0	98.3	76.9	89.9
26, 27, 3370	ELECTRICAL, COMMUNICATIONS & UTIL.	96.2	81.4	88.9	93.9	65.7	80.0	103.9	83.7	94.0	91.4	65.7	78.7	102.2	88.3	95.3	98.8	88.3	93.6
MF2018	WEIGHTED AVERAGE	97.4	77.5	89.0	96.3	75.6	87.6	97.6	81.5	90.8	98.6	69.2	86.2	102.9	82.5	94.3	102.3	85.2	95.0

DIVISION		NEVADA									NEW HAMPSHIRE								
		ELY			LAS VEGAS			RENO			CHARLESTON			CLAREMONT			CONCORD		
		893			889 - 891			894 - 895			036			037			032 - 033		
		MAT.	INST.	TOTAL	MAT.	INST.	TOTAL	MAT.	INST.	TOTAL	MAT.	INST.	TOTAL	MAT.	INST.	TOTAL	MAT.	INST.	TOTAL
015433	CONTRACTOR EQUIPMENT		94.6	94.6		94.6	94.6		94.6	94.6		95.7	95.7		95.7	95.7		98.4	98.4
0241, 31 - 34	SITE & INFRASTRUCTURE, DEMOLITION	73.4	92.5	86.6	76.6	95.5	89.7	73.4	92.6	86.7	80.6	96.1	91.3	74.6	96.1	89.5	88.7	101.2	97.3
0310	Concrete Forming & Accessories	104.3	101.6	102.0	105.2	105.6	105.5	100.6	76.3	79.9	86.6	83.1	83.6	92.1	83.2	84.5	97.2	94.8	95.1
0320	Concrete Reinforcing	123.1	114.7	119.0	113.5	125.9	119.5	116.4	124.2	120.1	89.7	88.6	89.1	89.7	88.6	89.1	100.9	88.7	95.0
0330	Cast-in-Place Concrete	102.0	95.2	99.4	98.7	108.2	102.2	107.9	81.1	97.9	84.7	116.1	96.4	77.8	116.1	92.1	100.9	117.8	107.2
03	CONCRETE	107.5	101.2	104.7	102.4	109.5	105.5	106.7	86.3	97.8	90.4	95.8	92.7	82.8	95.8	88.5	96.8	101.6	98.9
04	MASONRY	130.5	74.8	96.3	117.4	106.2	110.5	124.2	68.7	90.1	89.8	99.5	95.8	89.9	99.5	95.8	97.8	101.4	100.0
05	METALS	115.0	96.5	109.5	124.1	103.9	118.2	116.7	97.4	111.1	94.2	92.1	93.5	94.2	92.1	93.6	101.3	91.8	98.5
06	WOOD, PLASTICS & COMPOSITES	94.8	104.4	100.1	92.4	102.3	97.8	88.4	74.3	80.6	86.6	79.4	82.6	92.4	79.4	85.3	93.3	93.6	93.5
07	THERMAL & MOISTURE PROTECTION	114.7	92.2	104.9	129.4	101.9	117.4	114.2	78.3	98.6	107.3	106.5	107.0	107.1	106.5	106.8	111.3	109.5	110.5
08	OPENINGS	102.4	93.8	100.4	101.4	109.6	103.3	100.2	79.2	95.3	96.1	80.6	92.5	97.2	80.6	93.3	97.0	88.5	95.1
0920	Plaster & Gypsum Board	101.2	104.6	103.5	93.9	102.5	99.7	89.0	73.7	78.7	106.1	78.5	87.5	107.0	78.5	87.9	108.7	93.1	98.2
0950, 0980	Ceilings & Acoustic Treatment	94.5	104.6	101.3	101.0	102.5	102.0	97.8	73.7	81.5	101.8	78.5	86.1	101.8	78.5	86.1	106.1	93.1	97.4
0960	Flooring	100.5	67.1	91.1	92.0	104.1	95.4	97.3	67.1	88.8	85.1	113.1	92.9	87.6	113.1	94.7	96.4	113.1	101.1
0970, 0990	Wall Finishes & Painting/Coating	94.2	112.0	104.7	96.8	119.4	110.2	94.2	79.3	85.4	81.8	90.0	86.6	81.8	90.0	86.6	95.9	90.0	92.4
09	FINISHES	92.9	97.0	95.1	90.8	106.4	99.3	90.8	73.9	81.7	90.2	88.5	89.3	90.6	88.5	89.5	95.9	97.4	96.7
COVERS	DIVS. 10 - 14, 25, 28, 41, 43, 44, 46	100.0	70.7	93.5	100.0	105.1	101.1	100.0	99.8	100.0	100.0	88.5	97.4	100.0	88.5	97.4	100.0	109.5	102.1
21, 22, 23	FIRE SUPPRESSION, PLUMBING & HVAC	98.3	96.2	97.4	100.1	103.0	101.3	100.0	77.1	91.0	96.7	82.2	91.0	96.7	82.2	91.0	100.0	87.2	95.0
26, 27, 3370	ELECTRICAL, COMMUNICATIONS & UTIL.	99.0	93.0	96.0	103.5	107.8	105.6	99.4	88.3	93.9	91.4	75.2	83.4	91.4	75.2	83.4	90.2	75.2	82.8
MF2018	WEIGHTED AVERAGE	103.7	93.2	99.2	105.0	105.1	105.0	103.6	82.6	94.7	94.1	88.4	91.7	93.1	88.4	91.1	98.0	93.2	96.0

DIVISION		NEW HAMPSHIRE															NEW JERSEY		
		KEENE			LITTLETON			MANCHESTER			NASHUA			PORTSMOUTH			ATLANTIC CITY		
		034			035			031			030			038			082, 084		
		MAT.	INST.	TOTAL	MAT.	INST.	TOTAL	MAT.	INST.	TOTAL	MAT.	INST.	TOTAL	MAT.	INST.	TOTAL	MAT.	INST.	TOTAL
015433	CONTRACTOR EQUIPMENT		95.7	95.7		95.7	95.7		99.0	99.0		95.7	95.7		95.7	95.7		93.3	93.3
0241, 31 - 34	SITE & INFRASTRUCTURE, DEMOLITION	88.0	96.2	93.6	74.7	95.2	88.8	87.8	101.3	97.1	89.5	96.3	94.2	83.3	97.0	92.7	88.2	99.7	96.1
0310	Concrete Forming & Accessories	90.7	83.4	84.5	102.4	77.3	81.0	97.5	95.1	95.5	98.5	94.9	95.4	87.9	94.1	93.2	114.8	142.0	138.0
0320	Concrete Reinforcing	89.7	88.6	89.2	90.4	88.6	89.5	108.5	88.7	98.9	112.0	88.7	100.7	89.7	88.7	89.2	81.1	138.0	108.6
0330	Cast-in-Place Concrete	85.1	116.2	96.7	76.3	107.5	87.9	99.2	118.3	106.3	80.5	117.4	94.3	76.3	117.2	91.5	78.5	137.1	100.3
03	CONCRETE	89.9	95.9	92.6	82.2	90.1	85.7	97.0	101.9	99.2	90.0	101.6	95.1	82.4	101.2	90.7	87.7	138.1	109.8
04	MASONRY	92.5	99.5	96.8	100.9	84.1	90.6	94.4	101.4	98.7	94.4	101.4	98.7	90.2	99.8	96.1	104.1	140.0	126.1
05	METALS	94.8	92.5	94.2	94.9	92.2	94.1	102.0	92.2	99.1	100.5	92.7	98.2	96.4	94.3	95.8	98.8	115.0	103.6
06	WOOD, PLASTICS & COMPOSITES	90.9	79.4	84.6	101.9	79.4	89.5	93.6	93.7	93.7	100.2	93.6	96.6	87.8	93.6	91.0	122.8	143.7	134.3
07	THERMAL & MOISTURE PROTECTION	107.9	106.4	107.2	107.2	99.5	103.8	112.5	109.5	111.2	108.3	108.9	108.6	107.8	108.1	107.9	102.1	134.0	116.0
08	OPENINGS	94.8	84.1	92.3	98.1	80.6	94.0	95.7	93.3	95.2	99.0	91.3	97.2	99.6	81.9	95.5	96.4	139.2	106.4
0920	Plaster & Gypsum Board	106.4	78.5	87.6	120.6	78.5	92.3	99.7	93.1	95.3	115.6	93.1	100.5	106.1	93.1	97.4	114.9	144.6	134.9
0950, 0980	Ceilings & Acoustic Treatment	101.8	78.5	86.1	101.8	78.5	86.1	101.2	93.1	95.8	114.1	93.1	100.0	102.7	93.1	96.3	96.4	144.6	128.9
0960	Flooring	87.2	113.1	94.5	97.9	113.1	102.2	93.8	113.1	99.2	91.0	113.1	97.2	85.2	113.1	93.0	97.5	162.5	115.8
0970, 0990	Wall Finishes & Painting/Coating	81.8	103.4	94.6	81.8	90.0	86.6	95.3	119.0	109.3	81.8	102.2	93.9	81.8	102.2	93.9	81.9	141.8	117.4
09	FINISHES	92.1	89.9	90.9	95.5	84.6	89.6	94.8	100.6	97.9	96.8	98.7	97.8	91.1	98.2	95.0	93.5	147.2	122.6
COVERS	DIVS. 10 - 14, 25, 28, 41, 43, 44, 46	100.0	94.7	98.8	100.0	95.6	99.0	100.0	109.7	102.2	100.0	109.4	102.1	100.0	108.7	101.9	100.0	114.9	103.3
21, 22, 23	FIRE SUPPRESSION, PLUMBING & HVAC	96.7	82.3	91.0	96.7	74.1	87.8	100.1	87.3	95.1	100.2	87.3	95.1	100.2	86.3	94.7	99.7	133.1	112.8
26, 27, 3370	ELECTRICAL, COMMUNICATIONS & UTIL.	91.4	75.2	83.4	92.2	49.1	71.0	93.9	79.3	86.7	93.2	79.3	86.4	91.7	75.3	83.6	94.2	143.4	118.5
MF2018	WEIGHTED AVERAGE	94.5	89.0	92.2	94.4	80.2	88.4	98.2	94.5	96.6	97.5	93.7	95.9	94.9	92.4	93.8	96.8	133.2	112.2

DIVISION		NEW JERSEY																	
		CAMDEN			DOVER			ELIZABETH			HACKENSACK			JERSEY CITY			LONG BRANCH		
		081			078			072			076			073			077		
		MAT.	INST.	TOTAL	MAT.	INST.	TOTAL	MAT.	INST.	TOTAL	MAT.	INST.	TOTAL	MAT.	INST.	TOTAL	MAT.	INST.	TOTAL
015433	CONTRACTOR EQUIPMENT		93.3	93.3		95.7	95.7		95.7	95.7		95.7	95.7		93.3	93.3		92.9	92.9
0241, 31 - 34	SITE & INFRASTRUCTURE, DEMOLITION	89.3	99.3	96.2	103.8	101.4	102.2	108.9	101.4	103.7	104.8	101.4	102.5	95.2	101.4	99.5	100.0	100.5	100.4
0310	Concrete Forming & Accessories	105.0	137.3	132.5	96.4	148.3	140.6	108.0	148.5	142.5	96.4	148.3	140.7	100.6	148.4	141.3	101.1	139.0	133.4
0320	Concrete Reinforcing	106.5	128.5	117.1	75.3	156.5	114.6	75.3	156.5	114.6	75.3	156.5	114.6	97.7	156.5	126.2	75.3	156.1	114.4
0330	Cast-in-Place Concrete	76.1	133.8	97.6	81.3	131.5	100.0	69.8	140.3	96.1	79.4	140.3	102.1	63.4	131.6	88.8	70.5	133.8	94.0
03	CONCRETE	88.2	133.2	108.0	86.1	142.3	110.8	83.3	145.5	110.6	84.5	145.4	111.2	81.0	142.2	107.9	85.1	138.6	108.6
04	MASONRY	94.2	135.0	119.3	91.2	145.6	124.7	107.2	145.6	130.8	95.3	145.6	126.2	85.3	145.6	122.4	99.6	135.5	121.6
05	METALS	104.7	111.0	106.5	96.5	125.1	104.9	98.0	125.3	106.0	96.5	125.1	104.9	102.4	122.2	108.2	96.6	121.0	103.7
06	WOOD, PLASTICS & COMPOSITES	110.1	137.9	125.4	94.9	148.3	124.3	109.7	148.3	131.0	94.9	148.3	124.3	95.8	148.3	124.7	96.9	139.3	120.2
07	THERMAL & MOISTURE PROTECTION	102.0	133.0	115.5	103.7	142.8	120.7	104.0	144.1	121.4	103.5	136.7	117.9	103.2	142.8	120.4	103.4	129.0	114.5
08	OPENINGS	98.4	132.0	106.3	102.0	145.1	112.1	100.4	145.1	110.8	99.8	145.1	110.4	98.5	145.1	109.4	94.6	138.5	104.8
0920	Plaster & Gypsum Board	110.9	138.6	129.6	111.5	149.4	137.0	119.0	149.4	139.5	111.5	149.4	137.0	114.6	149.4	138.0	113.7	140.2	131.5
0950, 0980	Ceilings & Acoustic Treatment	106.3	138.6	128.1	92.3	149.4	130.8	94.1	149.4	131.4	92.3	149.4	130.8	102.2	149.4	134.0	92.3	140.2	124.6
0960	Flooring	93.7	162.5	113.0	83.7	188.8	113.2	88.7	188.8	116.8	83.7	188.8	113.2	84.7	188.8	114.0	84.9	176.9	110.8
0970, 0990	Wall Finishes & Painting/Coating	81.9	141.8	117.4	83.8	148.6	122.1	83.8	148.6	122.1	83.8	148.6	122.1	83.9	148.6	122.2	83.9	141.8	118.2
09	FINISHES	93.7	143.7	120.8	90.0	156.2	125.8	93.4	156.2	127.4	89.9	156.2	125.8	92.4	156.6	127.1	91.0	147.2	121.4
COVERS	DIVS. 10 - 14, 25, 28, 41, 43, 44, 46	100.0	114.3	103.2	100.0	131.7	107.1	100.0	131.7	107.1	100.0	131.7	107.1	100.0	131.7	107.1	100.0	114.5	103.2
21, 22, 23	FIRE SUPPRESSION, PLUMBING & HVAC	100.0	126.3	110.3	99.7	135.7	113.8	100.0	137.1	114.6	99.7	135.7	113.8	100.0	135.7	114.0	99.7	131.1	112.1
26, 27, 3370	ELECTRICAL, COMMUNICATIONS & UTIL.	98.4	132.1	115.0	96.0	138.5	116.9	96.6	138.5	117.2	96.0	142.3	118.8	100.5	142.3	121.1	95.7	129.9	112.6
MF2018	WEIGHTED AVERAGE	98.0	127.8	110.6	96.4	137.5	113.7	97.5	138.3	114.7	96.1	138.3	114.0	96.5	137.8	114.0	95.8	130.9	110.7

City Cost Indexes

	DIVISION	NEW JERSEY																	
		NEW BRUNSWICK			NEWARK			PATERSON			POINT PLEASANT			SUMMIT			TRENTON		
		088 - 089			070 - 071			074 - 075			087			079			085 - 086		
		MAT.	INST.	TOTAL	MAT.	INST.	TOTAL	MAT.	INST.	TOTAL	MAT.	INST.	TOTAL	MAT.	INST.	TOTAL	MAT.	INST.	TOTAL
015433	CONTRACTOR EQUIPMENT		92.9	92.9		98.3	98.3		95.7	95.7		92.9	92.9		95.7	95.7		97.0	97.0
0241, 31 - 34	SITE & INFRASTRUCTURE, DEMOLITION	101.1	100.9	100.9	110.0	106.2	107.4	106.6	101.4	103.0	102.6	100.5	101.2	106.3	101.4	102.9	88.6	105.4	100.2
0310	Concrete Forming & Accessories	108.5	148.0	142.2	100.7	148.7	141.6	98.3	148.3	140.9	102.4	139.1	133.6	99.0	148.5	141.2	103.5	138.8	133.6
0320	Concrete Reinforcing	82.1	156.1	117.9	96.2	156.6	125.4	97.7	156.5	126.2	82.1	156.1	117.9	75.3	156.5	114.6	107.7	116.2	111.8
0330	Cast-in-Place Concrete	97.0	136.9	111.8	89.8	140.9	108.8	80.8	140.2	102.9	97.0	135.8	111.5	67.5	140.3	94.6	95.6	134.4	110.0
03	CONCRETE	104.2	143.8	121.6	93.1	145.7	116.2	88.8	145.3	113.7	103.9	139.4	119.5	80.6	145.5	109.1	97.4	132.1	112.6
04	MASONRY	102.5	140.6	125.9	94.9	145.7	126.1	91.7	145.6	124.9	91.4	135.5	118.5	93.8	145.6	125.6	95.4	135.5	120.0
05	METALS	98.9	121.4	105.5	104.2	124.7	110.3	97.5	125.1	105.6	98.9	121.0	105.4	96.5	125.3	104.9	104.3	108.8	105.6
06	WOOD, PLASTICS & COMPOSITES	115.7	148.3	133.6	97.8	148.4	125.6	97.5	148.3	125.5	107.3	139.3	124.9	98.7	148.3	126.0	101.6	139.4	122.4
07	THERMAL & MOISTURE PROTECTION	102.5	138.2	118.0	105.2	144.7	122.4	103.8	136.7	118.1	102.5	131.3	115.0	104.1	144.1	121.5	102.9	135.2	117.0
08	OPENINGS	91.6	145.1	104.1	100.5	145.1	110.9	105.0	145.1	114.3	93.4	141.0	104.5	106.1	145.1	115.2	96.9	130.0	104.6
0920	Plaster & Gypsum Board	113.0	149.4	137.5	106.2	149.4	135.3	114.6	149.4	138.0	107.9	140.2	129.6	113.7	149.4	137.7	101.6	140.2	127.5
0950, 0980	Ceilings & Acoustic Treatment	96.4	149.4	132.2	103.0	149.4	134.3	102.2	149.4	134.0	96.4	140.2	125.9	92.3	149.4	130.8	101.2	140.2	127.5
0960	Flooring	95.1	188.8	121.4	96.7	188.8	122.6	84.7	188.8	114.0	92.5	162.5	112.2	84.9	188.8	114.1	100.1	176.9	121.7
0970, 0990	Wall Finishes & Painting/Coating	81.9	148.6	121.4	94.1	148.6	126.3	83.8	148.6	122.1	81.9	141.8	117.4	83.8	148.6	122.1	94.9	141.8	122.7
09	FINISHES	93.8	156.2	127.6	97.2	156.7	129.4	92.5	156.2	127.0	92.3	144.7	120.7	91.0	156.2	126.3	97.0	147.3	124.2
COVERS	DIVS. 10 - 14, 25, 28, 41, 43, 44, 46	100.0	131.5	107.0	100.0	131.8	107.1	100.0	131.7	107.1	100.0	111.6	102.6	100.0	131.7	107.1	100.0	114.7	103.3
21, 22, 23	FIRE SUPPRESSION, PLUMBING & HVAC	99.7	133.8	113.1	100.0	139.2	115.4	100.0	135.5	114.0	99.7	131.1	112.0	99.8	137.1	114.4	100.1	130.8	112.1
26, 27, 3370	ELECTRICAL, COMMUNICATIONS & UTIL.	94.8	133.7	114.0	105.0	142.3	123.4	100.5	138.5	119.2	94.2	129.9	111.8	96.6	138.5	117.2	102.6	128.7	115.4
MF2018	WEIGHTED AVERAGE	98.7	135.6	114.3	100.3	139.7	117.0	98.0	137.7	114.8	98.1	130.8	111.9	96.4	138.3	114.1	99.7	128.9	112.0

	DIVISION	NEW JERSEY			NEW MEXICO														
		VINELAND			ALBUQUERQUE			CARRIZOZO			CLOVIS			FARMINGTON			GALLUP		
		080, 083			870 - 872			883			881			874			873		
		MAT.	INST.	TOTAL	MAT.	INST.	TOTAL	MAT.	INST.	TOTAL	MAT.	INST.	TOTAL	MAT.	INST.	TOTAL	MAT.	INST.	TOTAL
015433	CONTRACTOR EQUIPMENT		93.3	93.3		107.5	107.5		107.5	107.5		107.5	107.5		107.5	107.5		107.5	107.5
0241, 31 - 34	SITE & INFRASTRUCTURE, DEMOLITION	92.5	99.4	97.3	91.1	97.5	95.5	109.7	97.5	101.3	96.6	97.5	97.2	97.3	97.5	97.5	107.2	97.5	100.5
0310	Concrete Forming & Accessories	99.2	137.6	131.9	101.0	66.2	71.3	98.6	66.2	71.0	98.6	66.1	70.9	101.1	66.2	71.4	101.1	66.2	71.4
0320	Concrete Reinforcing	81.1	131.2	105.3	99.9	71.5	86.1	116.0	71.5	94.5	117.3	71.4	95.1	109.2	71.5	90.9	104.5	71.5	88.5
0330	Cast-in-Place Concrete	84.6	133.9	103.0	90.5	70.1	82.9	94.8	70.1	85.6	94.7	70.1	85.5	91.3	70.1	83.4	86.0	70.1	80.1
03	CONCRETE	92.2	133.8	110.5	92.9	69.8	82.7	116.7	69.8	96.1	104.3	69.7	89.1	96.4	69.8	84.7	103.3	69.8	88.6
04	MASONRY	92.6	135.5	118.9	106.0	60.5	78.1	108.5	60.5	79.0	108.5	60.5	79.0	113.8	60.5	81.1	101.6	60.5	76.4
05	METALS	98.8	112.6	102.8	108.1	90.6	102.9	106.6	90.6	101.9	106.3	90.5	101.6	105.7	90.6	101.3	104.8	90.6	100.7
06	WOOD, PLASTICS & COMPOSITES	103.7	137.9	122.5	102.1	67.2	82.9	93.5	67.2	79.0	93.5	67.2	79.0	102.2	67.2	82.9	102.2	67.2	82.9
07	THERMAL & MOISTURE PROTECTION	101.9	132.2	115.1	101.0	72.8	88.7	106.3	72.8	91.7	105.0	72.8	91.0	101.3	72.8	88.9	102.5	72.8	89.6
08	OPENINGS	92.9	134.9	102.7	98.5	67.0	91.1	96.5	67.0	89.6	96.6	67.0	89.7	100.8	67.0	92.9	100.8	67.0	93.0
0920	Plaster & Gypsum Board	106.4	138.6	128.1	113.0	66.1	81.5	80.2	66.1	70.7	80.2	66.1	70.7	99.4	66.1	77.0	99.4	66.1	77.0
0950, 0980	Ceilings & Acoustic Treatment	96.4	138.6	124.9	99.8	66.1	77.1	102.7	66.1	78.1	102.7	66.1	78.1	98.6	66.1	76.7	98.6	66.1	76.7
0960	Flooring	91.7	162.5	111.6	88.9	66.8	82.7	97.2	66.8	88.7	97.2	66.8	88.7	90.4	66.8	83.8	90.4	66.8	83.8
0970, 0990	Wall Finishes & Painting/Coating	81.9	141.8	117.4	97.8	52.6	71.1	92.6	52.6	68.9	92.6	52.6	68.9	92.1	52.6	68.7	92.1	52.6	68.7
09	FINISHES	91.1	143.9	119.6	92.3	64.7	77.4	94.2	64.7	78.3	92.9	64.7	77.7	90.7	64.7	76.6	92.1	64.7	77.3
COVERS	DIVS. 10 - 14, 25, 28, 41, 43, 44, 46	100.0	114.5	103.2	100.0	85.1	96.7	100.0	85.1	96.7	100.0	85.1	96.7	100.0	85.1	96.7	100.0	85.1	96.7
21, 22, 23	FIRE SUPPRESSION, PLUMBING & HVAC	99.7	126.6	110.3	100.3	69.0	88.0	97.9	69.0	86.5	97.9	68.6	86.4	100.2	69.0	88.0	98.1	69.0	86.6
26, 27, 3370	ELECTRICAL, COMMUNICATIONS & UTIL.	94.2	143.3	118.5	87.9	69.5	78.8	90.1	69.5	80.0	87.9	69.5	78.8	85.9	69.5	77.8	85.3	69.5	77.5
MF2018	WEIGHTED AVERAGE	96.3	129.8	110.5	98.6	72.5	87.6	101.6	72.5	89.3	99.3	72.4	87.9	99.0	72.5	87.8	99.0	72.5	87.8

	DIVISION	NEW MEXICO																	
		LAS CRUCES			LAS VEGAS			ROSWELL			SANTA FE			SOCORRO			TRUTH/CONSEQUENCES		
		880			877			882			875			878			879		
		MAT.	INST.	TOTAL	MAT.	INST.	TOTAL	MAT.	INST.	TOTAL	MAT.	INST.	TOTAL	MAT.	INST.	TOTAL	MAT.	INST.	TOTAL
015433	CONTRACTOR EQUIPMENT		83.5	83.5		107.5	107.5		107.5	107.5		110.4	110.4		107.5	107.5		83.5	83.5
0241, 31 - 34	SITE & INFRASTRUCTURE, DEMOLITION	97.8	77.5	83.8	96.2	97.5	97.1	99.0	97.5	98.0	99.1	102.8	101.6	92.8	97.5	96.1	112.5	77.5	88.3
0310	Concrete Forming & Accessories	95.4	65.2	69.7	101.1	66.2	71.4	98.6	66.2	71.0	100.0	66.3	71.2	101.1	66.2	71.4	98.8	65.1	70.1
0320	Concrete Reinforcing	113.1	71.3	92.9	106.3	71.5	89.5	117.3	71.5	95.1	99.3	71.5	85.9	108.4	71.5	90.5	101.7	71.3	87.0
0330	Cast-in-Place Concrete	89.5	63.1	79.7	88.8	70.1	81.9	94.7	70.1	85.6	94.3	70.9	85.6	87.0	70.1	80.7	95.4	63.1	83.4
03	CONCRETE	82.1	66.5	75.3	94.0	69.8	83.4	105.3	69.8	89.7	92.9	70.0	82.9	92.9	69.8	82.8	86.1	66.4	77.4
04	MASONRY	104.0	60.2	77.1	101.9	60.5	76.5	119.5	60.5	83.3	95.0	60.6	73.9	101.8	60.5	76.4	98.7	60.2	75.0
05	METALS	105.0	83.0	98.5	104.5	90.6	100.4	107.5	90.6	102.6	101.6	89.6	98.1	104.8	90.6	100.7	104.4	83.0	98.1
06	WOOD, PLASTICS & COMPOSITES	82.9	66.2	73.7	102.2	67.2	82.9	93.5	67.2	79.0	99.0	67.2	81.5	102.2	67.2	82.9	93.5	66.2	78.5
07	THERMAL & MOISTURE PROTECTION	92.2	68.2	81.7	100.8	72.8	88.6	105.2	72.8	91.1	103.5	73.7	90.5	100.8	72.8	88.6	90.1	68.2	80.5
08	OPENINGS	92.0	66.4	86.1	97.3	67.0	90.3	96.4	67.0	89.6	98.8	67.0	91.4	97.2	67.0	90.1	90.6	66.4	85.0
0920	Plaster & Gypsum Board	78.5	66.1	70.2	99.4	66.1	77.0	80.2	66.1	70.7	114.1	66.1	81.8	99.4	66.1	77.0	100.6	66.1	77.4
0950, 0980	Ceilings & Acoustic Treatment	87.3	66.1	73.0	98.6	66.1	76.7	102.7	66.1	78.1	95.2	66.1	75.6	98.6	66.1	76.7	86.2	66.1	72.7
0960	Flooring	128.1	66.8	110.9	90.4	66.8	83.8	97.2	66.8	88.7	100.9	66.8	91.3	90.4	66.8	83.8	119.2	66.8	104.5
0970, 0990	Wall Finishes & Painting/Coating	81.8	52.6	64.5	92.1	52.6	68.7	92.6	52.6	68.9	99.8	52.6	71.9	92.1	52.6	68.7	84.5	52.6	65.6
09	FINISHES	103.0	64.0	81.9	90.5	64.7	76.6	93.0	64.7	77.7	98.2	64.8	80.1	90.4	64.7	76.5	102.4	64.0	81.6
COVERS	DIVS. 10 - 14, 25, 28, 41, 43, 44, 46	100.0	82.7	96.1	100.0	85.1	96.7	100.0	85.1	96.7	100.0	85.2	96.7	100.0	85.1	96.7	100.0	82.6	96.1
21, 22, 23	FIRE SUPPRESSION, PLUMBING & HVAC	100.4	68.7	88.0	98.1	69.0	86.6	100.0	69.0	87.8	100.3	69.0	88.0	98.1	69.0	86.6	98.0	68.7	86.5
26, 27, 3370	ELECTRICAL, COMMUNICATIONS & UTIL.	90.1	83.8	87.0	87.4	69.5	78.6	89.2	69.5	79.5	100.3	69.5	85.2	85.7	69.5	77.7	89.5	69.5	79.6
MF2018	WEIGHTED AVERAGE	96.9	71.4	86.1	97.2	72.5	86.8	100.8	72.5	88.9	99.0	72.9	88.0	96.8	72.5	86.6	96.6	69.4	85.1

	DIVISION	NEW MEXICO			NEW YORK														
		TUCUMCARI			ALBANY			BINGHAMTON			BRONX			BROOKLYN			BUFFALO		
		884			120 - 122			137 - 139			104			112			140 - 142		
		MAT.	INST.	TOTAL	MAT.	INST.	TOTAL	MAT.	INST.	TOTAL	MAT.	INST.	TOTAL	MAT.	INST.	TOTAL	MAT.	INST.	TOTAL
015433	CONTRACTOR EQUIPMENT		107.5	107.5		115.6	115.6		117.6	117.6		104.4	104.4		109.4	109.4		100.5	100.5
0241, 31 - 34	SITE & INFRASTRUCTURE, DEMOLITION	96.2	97.5	97.1	80.0	105.1	97.4	94.3	89.3	90.8	97.3	109.7	105.9	118.1	120.6	119.8	97.3	102.3	100.7
0310	Concrete Forming & Accessories	98.6	66.1	70.9	97.3	107.3	105.8	99.4	93.4	94.3	95.7	189.1	175.3	105.1	189.1	176.7	101.3	117.1	114.7
0320	Concrete Reinforcing	115.0	71.4	93.9	101.3	115.2	108.0	95.8	106.3	100.9	94.8	182.0	136.9	97.2	242.5	167.5	98.9	116.4	107.3
0330	Cast-in-Place Concrete	94.7	70.1	85.5	81.0	116.4	94.2	106.3	106.9	106.5	85.6	172.0	117.8	108.3	170.5	131.5	109.1	123.5	114.4
03	CONCRETE	103.6	69.7	88.7	87.7	112.8	98.7	94.3	102.6	97.9	88.0	181.1	128.9	106.4	189.9	143.1	104.9	118.7	111.0
04	MASONRY	119.9	60.5	83.4	88.0	117.9	106.4	104.3	104.9	104.7	88.8	188.9	150.3	114.8	188.8	160.3	112.6	122.2	118.5
05	METALS	106.3	90.5	101.6	102.2	125.0	108.9	95.0	133.6	106.3	86.4	174.6	112.3	103.1	175.0	124.2	97.0	107.5	100.1
06	WOOD, PLASTICS & COMPOSITES	93.5	67.2	79.0	94.6	104.0	99.8	101.7	90.2	95.4	97.2	188.6	147.5	103.7	188.3	150.3	98.6	115.8	108.1
07	THERMAL & MOISTURE PROTECTION	105.0	72.8	91.0	105.4	110.4	107.6	109.2	94.8	102.9	102.5	168.8	131.3	109.3	168.2	134.9	102.5	111.7	106.5
08	OPENINGS	96.4	67.0	89.5	96.4	103.8	98.1	90.4	93.7	91.2	92.3	197.7	116.9	87.8	197.5	113.4	99.8	111.4	102.5
0920	Plaster & Gypsum Board	80.2	66.1	70.7	97.8	103.9	101.9	109.2	89.7	96.1	102.8	191.0	162.1	103.3	191.0	162.3	105.8	116.1	112.7
0950, 0980	Ceilings & Acoustic Treatment	102.7	66.1	78.1	97.1	103.9	101.7	96.6	89.7	92.0	89.1	191.0	157.8	91.8	191.0	158.7	104.2	116.1	112.2
0960	Flooring	97.2	66.8	88.7	90.9	110.3	96.4	103.1	102.4	102.9	99.9	182.5	123.1	110.2	182.5	130.5	96.3	116.6	102.0
0970, 0990	Wall Finishes & Painting/Coating	92.6	52.6	68.9	95.9	104.4	100.9	89.3	106.3	99.4	106.2	167.4	142.4	116.0	167.4	146.4	96.9	117.8	109.3
09	FINISHES	92.8	64.7	77.6	90.6	107.1	99.5	93.6	95.9	94.8	98.0	186.2	145.7	107.1	186.0	149.8	102.0	117.6	110.5
COVERS	DIVS. 10 - 14, 25, 28, 41, 43, 44, 46	100.0	85.1	96.7	100.0	103.8	100.9	100.0	96.6	99.2	100.0	141.5	109.3	100.0	140.8	109.1	100.0	106.5	101.4
21, 22, 23	FIRE SUPPRESSION, PLUMBING & HVAC	97.9	68.6	86.4	100.1	110.3	104.1	100.6	97.0	99.2	100.3	177.9	130.7	99.8	177.9	130.5	100.0	102.4	100.9
26, 27, 3370	ELECTRICAL, COMMUNICATIONS & UTIL.	90.1	69.5	80.0	100.0	108.4	104.1	98.7	101.0	99.8	90.6	184.5	136.9	98.5	184.5	140.9	99.5	105.3	102.4
MF2018	WEIGHTED AVERAGE	99.9	72.4	88.3	96.8	111.2	102.8	97.3	101.5	99.1	93.8	175.3	128.3	101.9	177.3	133.8	100.8	110.3	104.8

	DIVISION	NEW YORK																	
		ELMIRA			FAR ROCKAWAY			FLUSHING			GLENS FALLS			HICKSVILLE			JAMAICA		
		148 - 149			116			113			128			115, 117, 118			114		
		MAT.	INST.	TOTAL	MAT.	INST.	TOTAL	MAT.	INST.	TOTAL	MAT.	INST.	TOTAL	MAT.	INST.	TOTAL	MAT.	INST.	TOTAL
015433	CONTRACTOR EQUIPMENT		120.0	120.0		109.4	109.4		109.4	109.4		112.6	112.6		109.4	109.4		109.4	109.4
0241, 31 - 34	SITE & INFRASTRUCTURE, DEMOLITION	97.4	89.6	92.0	121.2	120.6	120.8	121.3	120.6	120.8	71.4	99.6	90.9	111.1	119.1	116.6	115.5	120.6	119.0
0310	Concrete Forming & Accessories	85.0	96.3	94.6	92.2	189.1	174.8	95.9	189.1	175.3	81.5	99.9	97.2	88.8	157.2	147.1	95.9	189.1	175.3
0320	Concrete Reinforcing	98.6	105.3	101.8	97.2	242.5	167.5	98.9	242.5	168.4	96.9	112.7	104.5	97.2	174.4	134.6	97.2	242.5	167.5
0330	Cast-in-Place Concrete	98.8	106.4	101.6	117.2	170.5	137.1	117.2	170.5	137.1	78.1	111.6	90.6	99.5	161.5	122.6	108.3	170.5	131.5
03	CONCRETE	90.9	103.7	96.5	112.8	189.9	146.7	113.2	189.9	146.9	81.2	107.4	92.8	98.2	161.8	126.2	105.8	189.9	142.8
04	MASONRY	101.4	106.7	104.6	118.6	188.8	161.8	112.8	188.8	159.5	93.4	112.6	105.2	109.0	174.4	149.2	117.2	188.8	161.2
05	METALS	95.5	135.8	107.3	103.2	175.0	124.3	103.2	175.0	124.3	95.5	124.5	104.0	104.7	172.4	124.6	103.2	175.0	124.3
06	WOOD, PLASTICS & COMPOSITES	84.1	94.4	89.8	87.1	188.3	142.8	91.7	188.3	144.9	82.7	96.1	90.1	83.7	154.0	122.4	91.7	188.3	144.9
07	THERMAL & MOISTURE PROTECTION	108.9	95.2	102.9	109.2	168.3	134.9	109.3	168.3	134.9	98.6	107.1	102.3	108.9	157.8	130.2	109.1	168.3	134.8
08	OPENINGS	96.7	95.7	96.4	86.6	197.5	112.5	86.6	197.5	112.5	90.0	99.1	92.1	87.0	178.6	108.3	86.6	197.5	112.5
0920	Plaster & Gypsum Board	99.5	94.2	96.0	91.7	191.0	158.5	94.2	191.0	159.3	91.0	96.0	94.4	91.4	155.6	134.6	94.2	191.0	159.3
0950, 0980	Ceilings & Acoustic Treatment	106.0	94.2	98.1	81.2	191.0	155.2	81.2	191.0	155.2	86.4	96.0	92.9	80.3	155.6	131.1	81.2	191.0	155.2
0960	Flooring	87.6	102.4	91.8	105.0	182.5	126.8	106.5	182.5	127.9	82.3	107.7	89.5	104.0	179.7	125.2	106.5	182.5	127.9
0970, 0990	Wall Finishes & Painting/Coating	97.2	93.3	94.9	116.0	167.4	146.4	116.0	167.4	146.4	89.3	104.4	98.2	116.0	167.4	146.4	116.0	167.4	146.4
09	FINISHES	94.2	96.9	95.7	102.4	186.0	147.6	103.2	186.0	148.0	84.0	101.1	93.2	101.0	161.6	133.8	102.7	186.0	147.8
COVERS	DIVS. 10 - 14, 25, 28, 41, 43, 44, 46	100.0	97.6	99.5	100.0	140.8	109.1	100.0	140.8	109.1	100.0	96.9	99.3	100.0	132.1	107.1	100.0	140.8	109.1
21, 22, 23	FIRE SUPPRESSION, PLUMBING & HVAC	96.6	95.5	96.2	96.2	177.9	128.3	96.2	177.9	128.3	96.7	109.3	101.6	99.8	161.0	123.8	96.2	177.9	128.3
26, 27, 3370	ELECTRICAL, COMMUNICATIONS & UTIL.	96.4	104.1	100.2	105.3	184.5	144.4	105.3	184.5	144.4	94.3	105.0	99.6	98.0	141.0	119.2	97.0	184.5	140.1
MF2018	WEIGHTED AVERAGE	96.3	102.5	98.9	102.2	177.3	133.9	102.1	177.3	133.9	92.0	107.3	98.5	99.9	157.1	124.1	100.3	177.3	132.8

	DIVISION	NEW YORK																	
		JAMESTOWN			KINGSTON			LONG ISLAND CITY			MONTICELLO			MOUNT VERNON			NEW ROCHELLE		
		147			124			111			127			105			108		
		MAT.	INST.	TOTAL	MAT.	INST.	TOTAL	MAT.	INST.	TOTAL	MAT.	INST.	TOTAL	MAT.	INST.	TOTAL	MAT.	INST.	TOTAL
015433	CONTRACTOR EQUIPMENT		90.6	90.6		109.4	109.4		109.4	109.4		109.4	109.4		104.4	104.4		104.4	104.4
0241, 31 - 34	SITE & INFRASTRUCTURE, DEMOLITION	98.8	89.9	92.6	139.6	115.9	123.2	119.2	120.6	120.2	134.9	115.8	121.7	102.7	105.4	104.5	102.2	105.3	104.3
0310	Concrete Forming & Accessories	85.1	90.1	89.3	82.7	130.7	123.7	100.0	189.1	176.0	89.3	130.8	124.7	86.7	138.8	131.1	100.5	135.3	130.1
0320	Concrete Reinforcing	98.9	111.2	104.8	97.3	160.4	127.8	97.2	242.5	167.5	96.5	160.4	127.4	93.5	180.8	135.7	93.6	180.6	135.7
0330	Cast-in-Place Concrete	102.4	105.2	103.4	105.5	146.3	120.7	111.8	170.5	133.7	98.7	146.3	116.5	95.6	150.0	115.8	95.6	149.7	115.7
03	CONCRETE	93.9	99.0	96.2	102.4	140.8	119.3	109.0	189.9	144.5	97.3	140.8	116.4	97.2	150.4	120.6	96.4	148.7	119.4
04	MASONRY	110.2	104.6	106.8	108.1	155.4	137.2	111.7	188.8	159.1	100.8	155.4	134.3	94.6	156.4	132.6	94.6	156.4	132.6
05	METALS	93.0	101.9	95.6	104.0	135.9	113.4	103.1	175.0	124.2	104.0	136.0	113.4	86.2	169.8	110.7	86.4	169.0	110.7
06	WOOD, PLASTICS & COMPOSITES	82.9	86.2	84.7	83.9	124.0	106.0	97.8	188.3	147.6	90.4	124.0	108.9	87.6	132.4	112.3	104.5	128.7	117.8
07	THERMAL & MOISTURE PROTECTION	108.5	95.4	102.8	121.5	145.0	131.7	109.2	168.3	134.9	121.1	145.0	131.5	103.4	147.3	122.5	103.5	144.7	121.4
08	OPENINGS	96.5	93.0	95.7	92.1	142.5	103.9	86.6	197.5	112.5	88.0	142.4	100.7	92.3	168.4	110.1	92.4	166.3	109.6
0920	Plaster & Gypsum Board	89.2	85.8	86.9	91.0	124.9	113.8	99.3	191.0	161.0	91.7	124.8	114.0	98.3	133.2	121.8	111.0	129.3	123.3
0950, 0980	Ceilings & Acoustic Treatment	102.7	85.8	91.3	77.5	124.9	109.4	81.2	191.0	155.2	77.5	124.8	109.4	87.5	133.2	118.3	87.5	129.3	115.7
0960	Flooring	90.3	102.4	93.7	98.6	160.5	116.0	108.2	182.5	129.1	101.0	160.5	117.8	91.2	179.7	116.1	99.2	164.0	117.4
0970, 0990	Wall Finishes & Painting/Coating	98.8	100.6	99.9	116.6	130.9	125.0	116.0	167.4	146.4	116.6	123.4	120.6	104.6	167.4	141.8	104.6	167.4	141.8
09	FINISHES	93.2	92.5	92.9	97.9	135.4	118.2	104.1	186.0	148.4	98.3	134.6	117.9	95.0	148.0	123.7	98.9	142.6	122.5
COVERS	DIVS. 10 - 14, 25, 28, 41, 43, 44, 46	100.0	96.8	99.3	100.0	121.2	104.7	100.0	140.8	109.1	100.0	121.3	104.7	100.0	129.2	106.5	100.0	113.0	102.9
21, 22, 23	FIRE SUPPRESSION, PLUMBING & HVAC	96.5	93.6	95.4	96.6	134.6	111.5	99.8	177.9	130.5	96.6	138.9	113.2	96.8	145.9	116.1	96.8	145.6	116.0
26, 27, 3370	ELECTRICAL, COMMUNICATIONS & UTIL.	95.4	94.9	95.1	95.7	116.3	105.9	97.5	184.5	140.4	95.7	116.3	105.9	88.8	170.6	129.1	88.8	141.7	114.9
MF2018	WEIGHTED AVERAGE	96.4	96.0	96.3	100.6	133.8	114.6	101.6	177.3	133.6	99.1	134.6	114.1	94.1	150.6	118.0	94.4	144.8	115.7

DIVISION		NEW YORK																	
		NEW YORK			NIAGARA FALLS			PLATTSBURGH			POUGHKEEPSIE			QUEENS			RIVERHEAD		
		100 - 102			143			129			125 - 126			110			119		
		MAT.	INST.	TOTAL	MAT.	INST.	TOTAL	MAT.	INST.	TOTAL	MAT.	INST.	TOTAL	MAT.	INST.	TOTAL	MAT.	INST.	TOTAL
015433	CONTRACTOR EQUIPMENT		104.6	104.6		90.6	90.6		94.5	94.5		109.4	109.4		109.4	109.4		109.4	109.4
0241, 31 - 34	SITE & INFRASTRUCTURE, DEMOLITION	106.6	111.9	110.3	100.9	90.8	94.0	109.3	96.7	100.6	135.9	114.8	121.3	114.3	120.6	118.7	112.1	118.6	116.6
0310	Concrete Forming & Accessories	104.3	188.9	176.4	85.0	113.1	108.9	86.8	91.4	90.7	82.7	167.6	155.1	89.0	189.1	174.3	93.2	156.2	146.9
0320	Concrete Reinforcing	100.3	178.3	138.0	97.5	112.0	104.5	101.4	112.0	106.5	97.3	160.0	127.6	98.9	242.5	168.4	99.1	215.6	155.5
0330	Cast-in-Place Concrete	98.9	172.2	126.2	106.0	124.1	112.7	95.5	100.6	97.4	102.2	135.4	114.5	102.9	170.5	128.1	101.2	159.7	123.0
03	CONCRETE	101.6	180.3	136.2	96.2	116.1	105.0	94.3	98.0	95.9	99.7	153.6	123.4	101.5	189.9	140.3	98.7	166.5	128.5
04	MASONRY	102.1	188.8	155.4	117.5	125.1	122.2	88.3	97.4	93.9	100.6	139.5	124.5	106.0	188.8	156.9	114.3	173.3	150.6
05	METALS	97.5	171.0	119.1	95.6	102.4	97.6	99.8	99.5	99.7	104.0	135.6	113.3	103.1	175.0	124.2	105.1	160.6	121.4
06	WOOD, PLASTICS & COMPOSITES	102.0	188.6	149.7	82.8	108.2	96.8	88.8	88.2	88.5	83.9	179.3	136.4	83.8	188.3	141.3	88.7	154.0	124.6
07	THERMAL & MOISTURE PROTECTION	105.2	170.0	133.4	108.6	110.0	109.2	116.0	97.2	107.8	121.4	144.1	131.3	108.8	168.3	134.7	110.0	154.8	129.5
08	OPENINGS	96.0	197.5	119.7	96.5	104.7	98.4	97.4	94.7	96.8	92.1	170.5	110.4	86.6	197.5	112.5	87.0	172.1	106.8
0920	Plaster & Gypsum Board	108.6	191.0	164.0	89.2	108.4	102.1	108.2	87.4	94.2	91.0	181.7	152.0	91.4	191.0	158.4	92.5	155.6	135.0
0950, 0980	Ceilings & Acoustic Treatment	104.5	191.0	162.8	102.7	108.4	106.6	107.3	87.4	93.9	77.5	181.7	147.7	81.2	191.0	155.2	81.1	155.6	131.4
0960	Flooring	101.3	182.5	124.1	90.3	112.1	96.4	104.2	105.1	104.4	98.6	158.6	115.5	104.0	182.5	126.0	105.0	166.9	122.4
0970, 0990	Wall Finishes & Painting/Coating	102.7	167.4	141.0	98.8	111.5	106.3	111.2	95.1	101.7	116.6	123.4	120.6	116.0	167.4	146.4	116.0	167.4	146.4
09	FINISHES	102.9	186.2	148.0	93.3	112.9	103.9	95.0	93.5	94.2	97.7	164.2	133.7	101.3	186.0	147.1	101.5	158.6	132.4
COVERS	DIVS. 10 - 14, 25, 28, 41, 43, 44, 46	100.0	141.5	109.3	100.0	102.7	100.6	100.0	92.7	98.4	100.0	116.5	103.7	100.0	140.8	109.1	100.0	120.1	104.5
21, 22, 23	FIRE SUPPRESSION, PLUMBING & HVAC	100.2	178.3	130.9	96.5	105.1	99.9	96.6	100.3	98.1	96.6	120.3	105.9	99.8	177.9	130.5	100.0	155.9	121.9
26, 27, 3370	ELECTRICAL, COMMUNICATIONS & UTIL.	96.8	184.5	140.0	94.1	98.9	96.4	92.1	91.7	91.9	95.7	121.2	108.3	98.0	184.5	140.6	99.5	134.9	116.9
MF2018	WEIGHTED AVERAGE	99.7	175.2	131.6	97.4	107.5	101.6	97.0	96.6	96.8	99.8	136.8	115.4	100.0	177.3	132.6	100.6	153.5	122.9

DIVISION		NEW YORK																	
		ROCHESTER			SCHENECTADY			STATEN ISLAND			SUFFERN			SYRACUSE			UTICA		
		144 - 146			123			103			109			130 - 132			133 - 135		
		MAT.	INST.	TOTAL	MAT.	INST.	TOTAL	MAT.	INST.	TOTAL	MAT.	INST.	TOTAL	MAT.	INST.	TOTAL	MAT.	INST.	TOTAL
015433	CONTRACTOR EQUIPMENT		117.6	117.6		112.6	112.6		104.4	104.4		104.4	104.4		112.6	112.6		112.6	112.6
0241, 31 - 34	SITE & INFRASTRUCTURE, DEMOLITION	88.8	105.1	100.0	81.4	100.1	94.3	107.0	109.7	108.9	99.4	103.7	102.4	93.0	98.5	96.8	71.9	98.1	90.0
0310	Concrete Forming & Accessories	105.8	100.3	101.2	96.4	107.1	105.5	86.1	189.3	174.1	94.4	140.8	133.9	98.5	89.1	90.5	99.5	86.3	88.2
0320	Concrete Reinforcing	101.4	105.0	103.1	95.9	115.1	105.2	94.8	216.0	153.4	93.6	152.0	121.9	96.8	106.1	101.3	96.8	100.3	98.5
0330	Cast-in-Place Concrete	97.8	103.3	99.9	94.4	115.4	102.2	95.6	172.1	124.1	92.7	138.9	109.9	98.7	102.3	100.0	90.3	101.0	94.3
03	CONCRETE	99.5	103.3	101.2	94.9	112.4	102.6	99.1	186.4	137.5	93.8	141.3	114.7	97.6	98.1	97.8	95.8	95.4	95.6
04	MASONRY	93.6	103.4	99.7	90.7	117.9	107.4	100.9	188.9	154.9	94.5	142.2	123.8	97.0	101.0	99.5	88.4	99.4	95.2
05	METALS	104.6	119.3	108.9	99.7	126.1	107.4	84.6	174.8	111.1	84.6	131.9	98.5	98.6	119.8	104.9	96.7	117.5	102.9
06	WOOD, PLASTICS & COMPOSITES	108.4	100.2	103.9	100.1	103.8	102.1	86.2	188.6	142.6	96.9	142.5	122.0	98.2	86.0	91.5	98.2	82.7	89.7
07	THERMAL & MOISTURE PROTECTION	116.4	99.8	109.2	100.1	109.8	104.3	102.9	168.8	131.5	103.4	140.9	119.7	104.0	94.1	99.7	92.7	94.0	93.3
08	OPENINGS	100.7	99.3	100.3	95.3	103.7	97.3	92.3	197.7	116.9	92.4	150.7	106.0	92.4	89.5	91.8	95.2	86.3	93.1
0920	Plaster & Gypsum Board	103.8	100.1	101.3	100.3	103.9	102.7	98.4	191.0	160.7	102.1	143.5	130.0	96.8	85.6	89.3	96.8	82.2	87.0
0950, 0980	Ceilings & Acoustic Treatment	100.4	100.1	100.2	92.8	103.9	100.3	89.1	191.0	157.8	87.5	143.5	125.2	96.6	85.6	89.2	96.6	82.2	86.9
0960	Flooring	90.4	105.0	94.5	89.6	110.3	95.4	95.1	182.5	119.6	95.0	171.6	116.5	92.0	92.3	92.1	90.0	92.4	90.7
0970, 0990	Wall Finishes & Painting/Coating	97.7	98.4	98.1	89.3	104.4	98.2	106.2	167.4	142.4	104.6	126.9	117.8	91.9	98.7	95.9	85.6	98.7	93.4
09	FINISHES	95.7	101.4	98.8	89.2	107.0	98.9	96.7	186.2	145.2	96.3	145.8	123.1	92.4	90.0	91.1	90.8	87.9	89.2
COVERS	DIVS. 10 - 14, 25, 28, 41, 43, 44, 46	100.0	98.9	99.7	100.0	103.5	100.8	100.0	141.5	109.3	100.0	114.7	103.3	100.0	95.1	98.9	100.0	89.8	97.7
21, 22, 23	FIRE SUPPRESSION, PLUMBING & HVAC	99.9	88.1	95.3	100.2	106.2	102.6	100.3	177.9	130.8	96.8	123.7	107.4	100.3	94.4	98.0	100.3	92.5	97.2
26, 27, 3370	ELECTRICAL, COMMUNICATIONS & UTIL.	99.5	89.9	94.8	98.7	108.4	103.5	90.6	184.5	136.9	94.8	114.6	104.6	98.6	101.0	99.8	96.6	101.0	98.8
MF2018	WEIGHTED AVERAGE	100.3	99.2	99.8	97.0	109.9	102.4	95.6	176.1	129.6	94.1	130.2	109.4	97.8	98.3	98.0	96.0	96.6	96.2

DIVISION		NEW YORK									NORTH CAROLINA								
		WATERTOWN			WHITE PLAINS			YONKERS			ASHEVILLE			CHARLOTTE			DURHAM		
		136			106			107			287 - 288			281 - 282			277		
		MAT.	INST.	TOTAL	MAT.	INST.	TOTAL	MAT.	INST.	TOTAL	MAT.	INST.	TOTAL	MAT.	INST.	TOTAL	MAT.	INST.	TOTAL
015433	CONTRACTOR EQUIPMENT		112.6	112.6		104.4	104.4		104.4	104.4		99.0	99.0		100.4	100.4		104.1	104.1
0241, 31 - 34	SITE & INFRASTRUCTURE, DEMOLITION	79.5	98.6	92.7	97.1	105.4	102.8	104.5	105.4	105.1	96.4	77.8	83.5	98.6	81.9	87.1	98.5	86.3	90.1
0310	Concrete Forming & Accessories	85.3	93.0	91.9	99.2	149.3	141.9	99.4	149.1	141.8	90.5	61.4	65.7	96.6	61.5	66.6	96.2	61.4	66.5
0320	Concrete Reinforcing	97.5	106.1	101.6	93.6	180.8	135.8	97.3	180.8	137.7	95.4	66.0	81.2	99.6	66.8	83.7	104.4	66.8	86.2
0330	Cast-in-Place Concrete	105.1	104.7	105.0	84.9	150.0	109.1	95.0	150.0	115.4	105.9	71.5	93.1	108.0	71.6	94.4	106.3	71.2	93.2
03	CONCRETE	108.8	100.7	105.3	87.6	155.2	117.3	96.6	155.1	122.3	91.1	67.4	80.7	91.2	67.5	80.8	95.2	67.4	83.0
04	MASONRY	89.7	104.8	99.0	93.9	156.4	132.3	97.1	156.4	133.6	84.9	64.4	72.3	88.6	64.4	73.8	84.3	64.4	72.0
05	METALS	96.8	119.8	103.6	86.1	169.8	110.7	94.1	169.9	116.4	101.2	89.9	97.9	102.1	88.9	98.2	117.9	90.2	109.7
06	WOOD, PLASTICS & COMPOSITES	80.3	89.9	85.6	102.4	146.7	126.8	102.3	146.3	126.5	87.6	59.0	71.9	89.9	59.0	72.9	90.6	59.0	73.2
07	THERMAL & MOISTURE PROTECTION	93.0	97.2	94.8	103.2	148.7	123.0	103.5	149.3	123.4	99.1	63.8	83.8	93.7	64.5	81.0	102.8	63.8	85.8
08	OPENINGS	95.2	93.8	94.8	92.4	176.2	112.0	96.0	176.5	114.8	89.0	59.8	82.2	99.0	60.0	89.9	95.6	60.0	87.3
0920	Plaster & Gypsum Board	87.6	89.6	89.0	105.3	147.8	133.9	108.7	147.4	134.7	102.2	57.8	72.3	98.1	57.8	71.0	89.9	57.8	68.3
0950, 0980	Ceilings & Acoustic Treatment	96.6	89.6	91.9	87.5	147.8	128.2	103.7	147.4	133.2	79.4	57.8	64.9	83.5	57.8	66.2	82.8	57.8	66.0
0960	Flooring	83.1	92.4	85.7	97.5	179.7	120.6	96.9	182.5	121.0	94.9	66.8	87.0	95.3	66.8	87.3	102.0	66.8	92.1
0970, 0990	Wall Finishes & Painting/Coating	85.6	93.0	90.0	104.6	167.4	141.8	104.6	167.4	141.8	102.5	57.1	75.7	95.9	57.1	72.9	104.4	57.1	76.4
09	FINISHES	88.2	92.4	90.5	97.1	156.4	129.2	101.0	156.7	131.1	88.8	61.6	74.1	88.3	61.6	73.9	90.4	61.6	74.8
COVERS	DIVS. 10 - 14, 25, 28, 41, 43, 44, 46	100.0	96.4	99.2	100.0	127.5	106.1	100.0	130.7	106.8	100.0	82.5	96.1	100.0	82.5	96.1	100.0	82.5	96.1
21, 22, 23	FIRE SUPPRESSION, PLUMBING & HVAC	100.3	87.8	95.4	100.4	145.9	118.3	100.4	146.0	118.3	100.4	61.3	85.0	99.9	63.1	85.5	100.5	61.3	85.1
26, 27, 3370	ELECTRICAL, COMMUNICATIONS & UTIL.	98.6	90.2	94.4	88.8	170.6	129.1	94.9	170.6	132.2	100.8	57.8	79.6	100.0	60.3	80.4	96.5	57.4	77.2
MF2018	WEIGHTED AVERAGE	97.7	96.8	97.4	93.8	152.8	118.8	97.9	153.0	121.2	96.4	66.6	83.8	97.4	67.6	84.8	100.1	67.2	86.2

	DIVISION	NORTH CAROLINA																	
		ELIZABETH CITY			FAYETTEVILLE			GASTONIA			GREENSBORO			HICKORY			KINSTON		
		279			283			280			270, 272 - 274			286			285		
		MAT.	INST.	TOTAL	MAT.	INST.	TOTAL	MAT.	INST.	TOTAL	MAT.	INST.	TOTAL	MAT.	INST.	TOTAL	MAT.	INST.	TOTAL
015433	CONTRACTOR EQUIPMENT		108.0	108.0		104.1	104.1		99.0	99.0		104.1	104.1		104.1	104.1		104.1	104.1
0241, 31 - 34	SITE & INFRASTRUCTURE, DEMOLITION	102.7	87.9	92.4	95.6	86.1	89.0	96.1	78.1	83.7	98.3	86.4	90.1	95.1	85.1	88.2	93.9	84.9	87.7
0310	Concrete Forming & Accessories	82.6	63.6	66.4	90.1	60.1	64.5	96.4	61.4	66.6	96.0	61.3	66.5	87.1	61.2	65.0	83.6	59.9	63.4
0320	Concrete Reinforcing	102.3	69.8	86.6	99.1	66.8	83.4	95.8	66.8	81.7	103.2	66.8	85.6	95.4	66.7	81.5	94.9	66.7	81.3
0330	Cast-in-Place Concrete	106.5	71.7	93.5	111.1	69.3	95.5	103.6	71.2	91.5	105.5	71.2	92.7	105.9	71.2	93.0	102.2	69.2	90.0
03	CONCRETE	95.4	69.1	83.8	93.1	66.2	81.3	89.8	67.5	80.0	94.7	67.4	82.7	90.9	67.3	80.5	88.1	66.1	78.4
04	MASONRY	95.4	61.0	74.2	88.3	61.0	71.5	89.4	64.4	74.0	81.8	64.4	71.1	74.6	64.4	68.3	81.2	61.0	68.8
05	METALS	103.5	92.3	100.2	122.2	90.2	112.8	101.8	90.2	98.4	110.0	90.2	104.1	101.3	90.0	98.0	100.0	90.0	97.1
06	WOOD, PLASTICS & COMPOSITES	75.1	63.4	68.7	86.5	59.0	71.4	95.4	59.0	75.4	90.2	59.0	73.1	82.4	59.0	69.6	79.3	59.0	68.2
07	THERMAL & MOISTURE PROTECTION	102.0	62.5	84.8	98.6	62.3	82.8	99.3	63.8	83.9	102.6	63.8	85.7	99.5	63.8	84.0	99.3	62.3	83.2
08	OPENINGS	92.9	63.1	86.0	89.1	60.0	82.3	92.3	60.0	84.8	95.6	60.0	87.3	89.0	60.0	82.3	89.1	60.0	82.3
0920	Plaster & Gypsum Board	84.4	61.7	69.1	105.8	57.8	73.5	107.9	57.8	74.2	91.1	57.8	68.7	102.2	57.8	72.3	102.2	57.8	72.3
0950, 0980	Ceilings & Acoustic Treatment	82.8	61.7	68.5	81.9	57.8	65.7	83.5	57.8	66.2	82.8	57.8	66.0	79.4	57.8	64.9	83.5	57.8	66.2
0960	Flooring	94.1	66.8	86.4	95.1	66.8	87.1	98.0	66.8	89.2	102.0	66.8	92.1	94.8	66.8	86.9	92.2	66.8	85.0
0970, 0990	Wall Finishes & Painting/Coating	104.4	57.1	76.4	102.5	57.1	75.7	102.5	57.1	75.7	104.4	57.1	76.4	102.5	57.1	75.7	102.5	57.1	75.7
09	FINISHES	87.7	63.4	74.6	89.7	60.8	74.0	91.1	61.6	75.2	90.6	61.6	74.9	88.9	61.6	74.1	88.7	60.8	73.6
COVERS	DIVS. 10 - 14, 25, 28, 41, 43, 44, 46	100.0	85.4	96.8	100.0	81.4	95.8	100.0	82.5	96.1	100.0	79.8	95.5	100.0	82.5	96.1	100.0	81.3	95.8
21, 22, 23	FIRE SUPPRESSION, PLUMBING & HVAC	96.9	58.6	81.9	100.2	59.5	84.2	100.4	60.1	84.6	100.4	61.3	85.0	96.9	60.2	82.5	96.9	58.3	81.7
26, 27, 3370	ELECTRICAL, COMMUNICATIONS & UTIL.	96.2	64.6	80.6	100.6	57.4	79.3	100.3	60.4	80.6	95.7	57.8	77.0	98.6	60.4	79.8	98.4	55.8	77.4
MF2018	WEIGHTED AVERAGE	96.9	68.3	84.8	100.2	66.1	85.8	97.0	66.8	84.2	98.5	67.2	85.3	94.7	67.3	83.1	94.4	65.5	82.2

	DIVISION	NORTH CAROLINA															NORTH DAKOTA		
		MURPHY			RALEIGH			ROCKY MOUNT			WILMINGTON			WINSTON-SALEM			BISMARCK		
		289			275 - 276			278			284			271			585		
		MAT.	INST.	TOTAL	MAT.	INST.	TOTAL	MAT.	INST.	TOTAL	MAT.	INST.	TOTAL	MAT.	INST.	TOTAL	MAT.	INST.	TOTAL
015433	CONTRACTOR EQUIPMENT		99.0	99.0		106.5	106.5		104.1	104.1		99.0	99.0		104.1	104.1		99.9	99.9
0241, 31 - 34	SITE & INFRASTRUCTURE, DEMOLITION	97.4	76.3	82.8	98.8	90.8	93.3	100.7	86.3	90.8	97.4	77.6	83.7	98.7	86.4	90.2	99.8	97.8	98.4
0310	Concrete Forming & Accessories	97.0	59.8	65.3	96.9	61.0	66.3	88.5	60.9	64.9	91.8	60.1	64.8	97.9	61.4	66.8	110.6	76.6	81.6
0320	Concrete Reinforcing	94.9	64.3	80.1	105.4	66.8	86.7	102.3	66.8	85.1	96.1	66.8	81.9	103.2	66.8	85.6	92.6	98.3	95.3
0330	Cast-in-Place Concrete	109.6	69.2	94.6	109.3	71.3	95.1	104.2	70.5	91.6	105.5	69.3	92.0	108.1	71.2	94.4	106.3	86.6	98.9
03	CONCRETE	93.9	65.6	81.4	93.9	67.2	82.2	95.9	66.9	83.2	91.1	66.2	80.1	96.0	67.4	83.5	93.3	84.4	89.4
04	MASONRY	77.3	61.0	67.3	80.5	63.1	69.8	74.4	63.1	67.4	75.1	61.0	66.4	82.0	64.4	71.2	105.3	83.4	91.8
05	METALS	99.0	89.1	96.1	102.2	89.1	98.3	102.6	90.2	99.0	100.7	90.2	97.6	107.0	90.2	102.1	95.6	94.2	95.2
06	WOOD, PLASTICS & COMPOSITES	96.1	58.9	75.6	90.9	59.2	73.4	82.0	59.0	69.4	89.2	59.0	72.6	90.2	59.0	73.1	106.6	72.7	88.0
07	THERMAL & MOISTURE PROTECTION	99.3	62.3	83.2	97.2	63.8	82.7	102.5	63.2	85.4	99.1	62.3	83.1	102.6	63.8	85.7	109.3	86.0	99.2
08	OPENINGS	89.0	59.4	82.1	97.6	60.1	88.8	92.2	60.0	84.7	89.1	60.0	82.3	95.6	60.0	87.3	103.7	81.9	98.6
0920	Plaster & Gypsum Board	106.9	57.7	73.8	86.5	57.8	67.2	85.8	57.8	67.0	103.7	57.8	72.8	91.1	57.8	68.7	102.4	72.2	82.1
0950, 0980	Ceilings & Acoustic Treatment	79.4	57.7	64.8	83.5	57.8	66.2	80.4	57.8	65.2	81.9	57.8	65.7	82.8	57.8	66.0	108.5	72.2	84.1
0960	Flooring	98.3	66.8	89.4	96.8	66.8	88.4	97.6	66.8	88.9	95.5	66.8	87.5	102.0	66.8	92.1	86.9	53.7	77.6
0970, 0990	Wall Finishes & Painting/Coating	102.5	57.1	75.7	97.8	57.1	73.7	104.4	57.1	76.4	102.5	57.1	75.7	104.4	57.1	76.4	90.9	59.8	72.5
09	FINISHES	90.6	60.7	74.4	90.0	61.4	74.5	88.6	61.3	73.8	89.6	60.8	74.0	90.6	61.6	74.9	96.1	71.1	82.5
COVERS	DIVS. 10 - 14, 25, 28, 41, 43, 44, 46	100.0	81.3	95.8	100.0	82.3	96.1	100.0	82.1	96.0	100.0	81.4	95.8	100.0	82.5	96.1	100.0	94.8	98.8
21, 22, 23	FIRE SUPPRESSION, PLUMBING & HVAC	96.9	58.3	81.7	100.0	60.6	84.5	96.9	59.5	82.2	100.4	59.5	84.3	100.4	61.3	85.0	99.9	77.0	90.9
26, 27, 3370	ELECTRICAL, COMMUNICATIONS & UTIL.	101.7	55.8	79.1	99.1	56.3	78.0	98.0	57.4	78.0	101.1	55.8	78.8	95.7	57.8	77.0	98.7	73.4	86.3
MF2018	WEIGHTED AVERAGE	95.5	64.7	82.5	97.4	67.0	84.5	96.0	66.6	83.6	96.0	65.2	83.0	98.2	67.3	85.2	98.9	81.6	91.6

	DIVISION	NORTH DAKOTA																	
		DEVILS LAKE			DICKINSON			FARGO			GRAND FORKS			JAMESTOWN			MINOT		
		583			586			580 - 581			582			584			587		
		MAT.	INST.	TOTAL	MAT.	INST.	TOTAL	MAT.	INST.	TOTAL	MAT.	INST.	TOTAL	MAT.	INST.	TOTAL	MAT.	INST.	TOTAL
015433	CONTRACTOR EQUIPMENT		97.7	97.7		97.7	97.7		99.9	99.9		97.7	97.7		97.7	97.7		97.7	97.7
0241, 31 - 34	SITE & INFRASTRUCTURE, DEMOLITION	107.5	93.7	98.0	115.7	93.7	100.5	101.3	97.6	98.8	111.5	93.7	99.2	106.5	93.7	97.7	108.9	94.1	98.7
0310	Concrete Forming & Accessories	107.0	71.6	76.8	95.5	71.5	75.1	99.4	72.1	76.1	99.7	71.6	75.7	97.1	71.6	75.4	95.1	71.8	75.2
0320	Concrete Reinforcing	94.8	98.0	96.4	95.7	98.2	96.9	95.7	98.7	97.2	93.4	98.0	95.6	95.4	98.7	97.0	96.6	98.1	97.4
0330	Cast-in-Place Concrete	126.2	81.6	109.6	114.1	81.5	102.0	102.5	85.6	96.2	114.1	81.5	102.0	124.6	81.6	108.6	114.1	81.8	102.1
03	CONCRETE	104.4	80.4	93.8	103.5	80.4	93.3	98.3	82.1	91.2	100.9	80.3	91.8	102.9	80.5	93.1	99.4	80.6	91.1
04	MASONRY	118.0	79.2	94.2	120.1	81.4	96.4	102.7	90.0	94.9	112.1	79.2	91.9	130.9	89.9	105.7	110.1	80.7	92.1
05	METALS	94.7	94.1	94.5	94.6	94.0	94.4	100.2	94.8	98.6	94.6	93.8	94.4	94.6	94.9	94.7	94.9	94.8	94.9
06	WOOD, PLASTICS & COMPOSITES	99.0	67.5	81.7	85.6	67.5	75.7	95.9	67.6	80.4	90.4	67.5	77.8	87.7	67.5	76.6	85.3	67.5	75.5
07	THERMAL & MOISTURE PROTECTION	107.1	82.3	96.3	107.7	83.7	97.2	104.1	87.4	96.8	107.3	83.0	96.7	106.9	85.7	97.7	107.0	83.9	97.0
08	OPENINGS	100.9	79.0	95.8	100.9	79.0	95.8	101.0	79.1	95.9	99.5	79.0	94.7	100.9	79.0	95.8	99.7	79.0	94.8
0920	Plaster & Gypsum Board	118.2	67.0	83.8	109.1	67.0	80.8	101.1	67.0	78.2	110.3	67.0	81.2	110.0	67.0	81.1	109.1	67.0	80.8
0950, 0980	Ceilings & Acoustic Treatment	105.6	67.0	79.6	105.6	67.0	79.6	95.4	67.0	76.3	105.6	67.0	79.6	105.6	67.0	79.6	105.6	67.0	79.6
0960	Flooring	93.8	53.7	82.5	87.4	53.7	77.9	101.2	53.7	87.9	89.2	53.7	79.2	88.0	53.7	78.4	87.1	53.7	77.7
0970, 0990	Wall Finishes & Painting/Coating	86.6	56.4	68.7	86.6	56.4	68.7	93.1	67.4	77.9	86.6	66.0	74.4	86.6	56.4	68.7	86.6	57.5	69.4
09	FINISHES	95.2	66.8	79.8	93.0	66.8	78.8	96.2	68.1	81.0	93.2	67.8	79.5	92.3	66.8	78.5	92.1	66.9	78.5
COVERS	DIVS. 10 - 14, 25, 28, 41, 43, 44, 46	100.0	87.2	97.2	100.0	87.2	97.2	100.0	93.6	98.6	100.0	87.2	97.2	100.0	87.2	97.2	100.0	93.3	98.5
21, 22, 23	FIRE SUPPRESSION, PLUMBING & HVAC	96.6	78.3	89.4	96.6	73.3	87.4	100.0	74.1	89.8	100.1	72.3	89.2	96.6	72.3	87.0	100.1	72.1	89.1
26, 27, 3370	ELECTRICAL, COMMUNICATIONS & UTIL.	94.3	68.3	81.5	101.9	67.9	85.1	98.5	69.4	84.2	97.3	68.3	83.0	94.3	68.3	81.5	100.1	72.6	86.6
MF2018	WEIGHTED AVERAGE	99.2	78.8	90.6	99.9	77.9	90.6	99.7	80.3	91.5	99.4	77.6	90.2	99.2	78.8	90.6	99.3	78.6	90.5

City Cost Indexes

	DIVISION	NORTH DAKOTA			OHIO														
		WILLISTON			AKRON			ATHENS			CANTON			CHILLICOTHE			CINCINNATI		
		588			442 - 443			457			446 - 447			456			451 - 452		
		MAT.	INST.	TOTAL	MAT.	INST.	TOTAL	MAT.	INST.	TOTAL	MAT.	INST.	TOTAL	MAT.	INST.	TOTAL	MAT.	INST.	TOTAL
015433	CONTRACTOR EQUIPMENT		97.7	97.7		88.7	88.7		84.9	84.9		88.7	88.7		95.5	95.5		95.9	95.9
0241, 31 - 34	SITE & INFRASTRUCTURE, DEMOLITION	109.2	91.5	97.0	96.2	94.1	94.8	107.3	85.3	92.1	96.3	93.9	94.7	93.8	94.9	94.6	90.0	98.2	95.7
0310	Concrete Forming & Accessories	101.7	71.3	75.8	102.8	82.8	85.8	95.3	78.9	81.4	102.8	74.3	78.5	98.0	81.5	84.0	101.5	80.2	83.3
0320	Concrete Reinforcing	97.5	98.1	97.8	94.7	91.0	92.9	86.9	89.2	88.0	94.7	75.0	85.2	84.0	88.8	86.3	89.1	77.8	83.6
0330	Cast-in-Place Concrete	114.1	81.4	101.9	102.6	89.0	97.6	111.2	95.8	105.5	103.6	87.2	97.5	100.9	92.4	97.8	96.4	77.5	89.3
03	CONCRETE	100.8	80.2	91.7	99.8	85.9	93.7	101.0	86.3	94.6	100.3	78.7	90.8	95.6	86.9	91.8	94.7	79.0	87.8
04	MASONRY	105.1	80.7	90.1	90.9	89.4	90.0	73.6	97.2	88.1	91.5	80.7	84.9	80.2	90.5	86.5	83.0	79.8	81.0
05	METALS	94.8	93.6	94.4	97.5	80.2	92.4	97.1	80.5	92.3	97.5	73.7	90.5	89.4	89.8	89.5	91.5	81.7	88.6
06	WOOD, PLASTICS & COMPOSITES	92.1	67.5	78.6	105.9	81.1	92.2	87.4	74.0	80.0	106.2	71.8	87.3	100.3	78.0	88.0	102.3	80.1	90.1
07	THERMAL & MOISTURE PROTECTION	107.3	83.5	96.9	104.9	91.1	98.9	102.9	91.5	98.0	106.0	87.2	97.8	104.7	88.4	97.6	102.7	80.9	93.2
08	OPENINGS	101.0	79.0	95.8	107.5	82.7	101.7	96.2	75.8	91.5	101.3	69.8	93.9	88.1	78.0	85.7	98.3	76.4	93.2
0920	Plaster & Gypsum Board	110.3	67.0	81.2	98.3	80.5	86.3	93.1	73.2	79.7	99.2	71.0	80.2	96.6	77.8	83.9	96.1	79.9	85.2
0950, 0980	Ceilings & Acoustic Treatment	105.6	67.0	79.6	90.6	80.5	83.8	105.9	73.2	83.8	90.6	71.0	77.4	98.7	77.8	84.6	93.6	79.9	84.3
0960	Flooring	90.1	53.7	79.9	91.7	83.3	89.3	120.7	75.3	108.0	91.8	74.1	86.8	97.5	75.3	91.3	98.7	78.5	93.0
0970, 0990	Wall Finishes & Painting/Coating	86.6	56.4	68.7	98.3	90.9	93.9	100.1	89.1	93.6	98.3	73.9	83.9	97.4	89.1	92.5	97.0	71.7	82.0
09	FINISHES	93.4	66.8	79.0	95.0	83.6	88.8	99.4	78.7	88.2	95.2	73.4	83.4	96.5	80.4	87.8	95.9	79.1	86.8
COVERS	DIVS. 10 - 14, 25, 28, 41, 43, 44, 46	100.0	87.2	97.1	100.0	90.3	97.8	100.0	87.9	97.3	100.0	88.4	97.4	100.0	86.8	97.1	100.0	88.4	97.4
21, 22, 23	FIRE SUPPRESSION, PLUMBING & HVAC	96.6	73.2	87.4	100.0	87.8	95.2	96.5	81.2	90.5	100.0	78.7	91.6	97.0	92.4	95.2	99.9	76.9	90.9
26, 27, 3370	ELECTRICAL, COMMUNICATIONS & UTIL.	97.7	68.7	83.4	98.6	81.8	90.4	97.9	89.7	93.9	97.9	85.0	91.5	97.2	89.7	93.5	96.3	74.2	85.4
MF2018	WEIGHTED AVERAGE	98.4	77.7	89.6	99.4	86.0	93.8	97.1	84.8	91.9	98.9	79.9	90.9	94.3	88.5	91.8	96.1	80.0	89.3

	DIVISION	OHIO																	
		CLEVELAND			COLUMBUS			DAYTON			HAMILTON			LIMA			LORAIN		
		441			430 - 432			453 - 454			450			458			440		
		MAT.	INST.	TOTAL	MAT.	INST.	TOTAL	MAT.	INST.	TOTAL	MAT.	INST.	TOTAL	MAT.	INST.	TOTAL	MAT.	INST.	TOTAL
015433	CONTRACTOR EQUIPMENT		92.0	92.0		92.3	92.3		88.9	88.9		95.5	95.5		88.3	88.3		88.7	88.7
0241, 31 - 34	SITE & INFRASTRUCTURE, DEMOLITION	94.9	96.2	95.8	101.8	92.3	95.3	90.0	94.2	92.9	90.0	94.5	93.1	101.1	84.9	89.9	95.5	94.3	94.7
0310	Concrete Forming & Accessories	102.3	90.2	92.0	98.7	78.1	81.1	99.9	77.2	80.6	99.9	77.4	80.7	95.3	76.3	79.1	102.8	74.6	78.7
0320	Concrete Reinforcing	95.2	91.4	93.4	99.9	78.7	89.7	89.1	79.1	84.2	89.1	77.2	83.3	86.9	79.3	83.2	94.7	91.3	93.1
0330	Cast-in-Place Concrete	99.5	97.2	98.6	102.1	80.9	94.2	86.5	80.6	84.3	92.6	80.9	88.3	102.1	89.5	97.4	97.7	91.4	95.4
03	CONCRETE	100.1	92.4	96.7	98.6	79.2	90.0	87.0	78.6	83.3	89.9	79.1	85.1	93.9	81.5	88.5	97.5	83.1	91.2
04	MASONRY	96.9	97.9	97.5	91.4	85.2	87.6	79.1	76.4	77.4	79.5	81.2	80.5	98.8	78.1	86.1	87.8	93.1	91.0
05	METALS	99.1	83.3	94.5	98.1	79.3	92.5	90.9	76.8	86.7	90.9	85.6	89.4	97.2	80.1	92.2	98.1	81.6	93.3
06	WOOD, PLASTICS & COMPOSITES	99.8	88.5	93.6	98.5	77.7	87.0	104.2	77.1	89.3	103.0	77.1	88.7	87.3	75.1	80.6	105.9	70.1	86.2
07	THERMAL & MOISTURE PROTECTION	102.0	96.9	99.8	93.6	83.8	89.3	108.8	79.5	96.1	104.8	79.9	94.0	102.4	83.7	94.3	105.9	91.0	99.4
08	OPENINGS	100.3	86.4	97.0	98.3	74.0	92.7	95.0	74.6	90.3	92.8	74.8	88.6	96.3	73.7	91.0	101.3	76.6	95.5
0920	Plaster & Gypsum Board	97.9	88.1	91.3	94.1	77.0	82.6	98.1	76.9	83.9	98.1	76.9	83.9	93.1	74.3	80.5	98.3	69.2	78.7
0950, 0980	Ceilings & Acoustic Treatment	86.3	88.1	87.5	93.5	77.0	82.4	99.6	76.9	84.3	98.7	76.9	84.0	105.9	74.3	84.6	90.6	69.2	76.2
0960	Flooring	93.1	89.8	92.2	95.1	75.3	89.5	101.2	72.4	93.1	98.5	78.5	92.9	119.8	77.2	107.8	91.8	89.8	91.2
0970, 0990	Wall Finishes & Painting/Coating	101.2	90.9	95.1	99.5	79.6	87.7	97.4	70.7	81.6	97.4	71.3	82.0	100.2	76.5	86.2	98.3	90.9	93.9
09	FINISHES	95.3	90.1	92.5	95.0	77.5	85.5	97.7	75.4	85.6	96.7	76.9	86.0	98.7	76.0	86.4	95.0	78.0	85.8
COVERS	DIVS. 10 - 14, 25, 28, 41, 43, 44, 46	100.0	96.4	99.2	100.0	88.2	97.4	100.0	85.1	96.7	100.0	85.3	96.7	100.0	84.4	96.5	100.0	90.9	98.0
21, 22, 23	FIRE SUPPRESSION, PLUMBING & HVAC	100.0	89.9	96.0	100.0	84.1	93.7	100.7	81.2	93.0	100.5	74.9	90.5	96.5	91.7	94.6	100.0	88.4	95.4
26, 27, 3370	ELECTRICAL, COMMUNICATIONS & UTIL.	98.3	92.7	95.5	99.9	80.2	90.2	94.9	75.7	85.4	95.2	75.5	85.5	98.2	75.7	87.1	98.0	78.2	88.2
MF2018	WEIGHTED AVERAGE	99.1	91.6	95.9	98.4	82.0	91.4	94.8	79.2	88.2	94.8	79.5	88.3	97.2	81.7	90.7	98.4	84.7	92.6

	DIVISION	OHIO																	
		MANSFIELD			MARION			SPRINGFIELD			STEUBENVILLE			TOLEDO			YOUNGSTOWN		
		448 - 449			433			455			439			434 - 436			444 - 445		
		MAT.	INST.	TOTAL	MAT.	INST.	TOTAL	MAT.	INST.	TOTAL	MAT.	INST.	TOTAL	MAT.	INST.	TOTAL	MAT.	INST.	TOTAL
015433	CONTRACTOR EQUIPMENT		88.7	88.7		88.7	88.7		88.9	88.9		92.4	92.4		92.4	92.4		88.7	88.7
0241, 31 - 34	SITE & INFRASTRUCTURE, DEMOLITION	91.7	94.0	93.3	95.7	90.6	92.2	90.3	94.2	93.0	141.5	98.2	111.6	99.8	90.9	93.6	96.1	93.9	94.6
0310	Concrete Forming & Accessories	92.1	73.5	76.3	96.6	79.4	81.9	99.9	77.1	80.5	97.6	79.0	81.7	100.3	84.9	87.1	102.8	76.6	80.5
0320	Concrete Reinforcing	86.2	79.0	82.7	92.1	79.0	85.8	89.1	79.1	84.2	89.7	94.9	92.2	99.9	82.9	91.7	94.7	85.2	90.1
0330	Cast-in-Place Concrete	95.1	87.8	92.3	88.5	88.1	88.3	88.8	80.3	85.6	96.1	89.3	93.6	96.9	90.2	94.4	101.7	87.2	96.3
03	CONCRETE	92.1	79.3	86.4	85.4	82.2	84.0	88.1	78.5	83.9	89.4	84.9	87.4	93.2	86.4	90.2	99.4	81.5	91.5
04	MASONRY	89.9	88.9	89.3	92.7	89.9	90.9	79.3	75.9	77.2	80.7	91.1	87.1	98.9	91.8	94.5	91.1	86.6	88.4
05	METALS	98.4	76.0	91.8	97.1	78.9	91.7	90.9	76.8	86.7	93.6	81.8	90.1	97.8	85.3	94.2	97.5	77.9	91.8
06	WOOD, PLASTICS & COMPOSITES	92.8	70.1	80.3	92.8	77.5	84.4	105.6	77.1	89.9	88.1	76.4	81.6	96.9	83.9	89.8	105.9	74.6	88.7
07	THERMAL & MOISTURE PROTECTION	104.1	88.4	97.3	90.5	88.8	89.8	108.7	79.2	95.9	102.0	87.4	95.7	91.0	91.2	91.1	106.1	88.2	98.3
08	OPENINGS	102.3	71.0	95.0	91.2	75.1	87.5	93.2	74.6	88.9	91.3	78.4	88.3	93.8	81.3	90.9	101.3	75.2	95.2
0920	Plaster & Gypsum Board	90.9	69.2	76.3	95.3	77.0	83.0	98.1	76.9	83.9	93.5	75.4	81.3	97.2	83.6	88.0	98.3	73.9	81.9
0950, 0980	Ceilings & Acoustic Treatment	91.4	69.2	76.4	99.5	77.0	84.3	99.6	76.9	84.3	96.7	75.4	82.3	99.5	83.6	88.8	90.6	73.9	79.3
0960	Flooring	86.6	92.7	88.3	94.4	92.7	93.9	101.2	72.4	93.1	122.5	92.7	114.1	94.8	94.4	94.7	91.8	90.0	91.3
0970, 0990	Wall Finishes & Painting/Coating	98.3	77.8	86.2	104.0	77.8	88.5	97.4	70.7	81.6	117.8	87.1	99.6	104.0	87.0	94.0	98.3	80.0	87.5
09	FINISHES	92.4	76.6	83.8	96.3	81.2	88.1	97.7	75.3	85.6	114.5	81.6	96.7	97.0	86.6	91.4	95.1	78.8	86.3
COVERS	DIVS. 10 - 14, 25, 28, 41, 43, 44, 46	100.0	88.3	97.4	100.0	85.9	96.8	100.0	84.9	96.6	100.0	86.2	96.9	100.0	91.3	98.1	100.0	88.5	97.4
21, 22, 23	FIRE SUPPRESSION, PLUMBING & HVAC	96.5	87.5	92.9	96.5	93.1	95.1	100.7	80.9	92.9	96.9	91.3	94.7	100.0	93.6	97.5	100.0	83.9	93.7
26, 27, 3370	ELECTRICAL, COMMUNICATIONS & UTIL.	95.7	90.8	93.3	94.0	90.8	92.4	94.9	80.2	87.7	88.7	106.4	97.4	100.0	103.4	101.7	98.0	74.3	86.3
MF2018	WEIGHTED AVERAGE	96.4	84.2	91.3	94.3	86.7	91.1	94.8	79.7	88.4	96.1	90.0	93.5	97.6	91.2	94.9	98.7	81.9	91.6

	DIVISION	OHIO			OKLAHOMA															
		ZANESVILLE			ARDMORE			CLINTON			DURANT			ENID			GUYMON		
		437 - 438			734			736			747			737			739		
		MAT.	INST.	TOTAL	MAT.	INST.	TOTAL	MAT.	INST.	TOTAL	MAT.	INST.	TOTAL	MAT.	INST.	TOTAL	MAT.	INST.	TOTAL
015433	CONTRACTOR EQUIPMENT		88.7	88.7		80.7	80.7		79.9	79.9		79.9	79.9		79.9	79.9		79.9	79.9
0241, 31 - 34	SITE & INFRASTRUCTURE, DEMOLITION	98.9	90.5	93.1	96.9	92.6	94.0	98.1	91.2	93.3	93.4	88.6	90.1	100.0	91.2	93.9	102.4	90.5	94.2
0310	Concrete Forming & Accessories	93.7	78.3	80.6	88.0	55.4	60.2	86.8	55.4	60.1	82.9	55.0	59.1	89.9	55.6	60.7	92.6	55.1	60.7
0320	Concrete Reinforcing	91.6	92.6	92.1	79.5	67.7	73.8	80.0	67.7	74.0	88.6	63.6	76.5	79.5	67.7	73.8	80.0	63.3	71.9
0330	Cast-in-Place Concrete	93.2	86.5	90.7	96.6	70.5	86.9	93.4	70.5	84.9	88.6	70.3	81.8	93.4	70.6	84.9	93.4	70.0	84.7
03	CONCRETE	89.2	83.5	86.7	86.7	62.8	76.2	86.2	62.8	75.9	84.1	61.8	74.3	86.6	62.9	76.2	89.4	61.7	77.2
04	MASONRY	90.7	86.6	88.2	95.2	57.0	71.8	119.3	57.0	81.0	88.6	62.1	72.3	101.3	57.0	74.1	97.5	55.4	71.7
05	METALS	98.5	84.2	94.3	97.3	62.2	87.0	97.4	62.2	87.0	92.7	60.6	83.2	98.8	62.4	88.1	97.8	59.3	86.5
06	WOOD, PLASTICS & COMPOSITES	88.2	77.5	82.3	97.4	54.4	73.7	96.5	54.4	73.4	87.8	54.4	69.5	99.7	54.4	74.8	102.9	54.4	76.2
07	THERMAL & MOISTURE PROTECTION	90.6	84.5	88.0	100.7	64.8	85.1	100.8	64.8	85.1	96.4	64.6	82.6	100.9	64.8	85.2	101.3	62.2	84.3
08	OPENINGS	91.2	79.0	88.4	103.7	55.0	92.3	103.7	55.0	92.3	96.0	54.0	86.2	104.9	55.9	93.5	103.8	54.0	92.2
0920	Plaster & Gypsum Board	91.2	77.0	81.7	90.5	53.7	65.8	90.2	53.7	65.7	78.5	53.7	61.8	91.1	53.7	66.0	91.1	53.7	66.0
0950, 0980	Ceilings & Acoustic Treatment	99.5	77.0	84.3	89.1	53.7	65.3	89.1	53.7	65.3	83.0	53.7	63.3	89.1	53.7	65.3	89.1	53.7	65.3
0960	Flooring	92.6	75.3	87.7	82.7	57.3	75.6	81.7	57.3	74.9	89.7	51.7	79.0	83.3	72.9	80.4	84.8	57.3	77.1
0970, 0990	Wall Finishes & Painting/Coating	104.0	89.1	95.2	86.4	44.8	61.8	86.4	44.8	61.8	92.8	44.8	64.4	86.4	44.8	61.8	86.4	42.2	60.3
09	FINISHES	95.4	78.2	86.1	82.7	53.5	66.9	82.5	53.5	66.8	83.2	52.4	66.5	83.2	56.7	68.8	84.0	54.5	68.0
COVERS	DIVS. 10 - 14, 25, 28, 41, 43, 44, 46	100.0	83.3	96.3	100.0	79.4	95.4	100.0	79.4	95.4	100.0	79.3	95.4	100.0	79.4	95.4	100.0	79.4	95.4
21, 22, 23	FIRE SUPPRESSION, PLUMBING & HVAC	96.5	89.8	93.9	96.5	67.5	85.1	96.5	67.5	85.1	96.6	67.4	85.2	100.0	67.5	87.2	96.5	63.6	83.6
26, 27, 3370	ELECTRICAL, COMMUNICATIONS & UTIL.	94.1	89.7	92.0	95.0	71.0	83.2	95.9	71.0	83.7	97.3	69.2	83.4	95.9	71.0	83.7	97.6	65.6	81.8
MF2018	WEIGHTED AVERAGE	94.9	85.7	91.0	95.2	65.6	82.7	96.4	65.5	83.3	93.1	65.1	81.3	97.0	66.0	83.9	96.3	63.3	82.4

	DIVISION	OKLAHOMA																	
		LAWTON			MCALESTER			MIAMI			MUSKOGEE			OKLAHOMA CITY			PONCA CITY		
		735			745			743			744			730 - 731			746		
		MAT.	INST.	TOTAL	MAT.	INST.	TOTAL	MAT.	INST.	TOTAL	MAT.	INST.	TOTAL	MAT.	INST.	TOTAL	MAT.	INST.	TOTAL
015433	CONTRACTOR EQUIPMENT		80.7	80.7		79.9	79.9		90.9	90.9		90.9	90.9		85.6	85.6		79.9	79.9
0241, 31 - 34	SITE & INFRASTRUCTURE, DEMOLITION	96.2	92.7	93.7	86.8	90.5	89.4	88.1	88.1	88.1	88.7	87.9	88.1	94.2	98.0	96.8	93.9	90.9	91.8
0310	Concrete Forming & Accessories	92.5	55.6	61.1	81.1	41.4	47.3	93.2	55.2	60.8	97.3	54.8	61.1	92.8	63.8	68.0	89.0	55.4	60.3
0320	Concrete Reinforcing	79.7	67.7	73.9	88.3	63.3	76.2	86.9	67.7	77.6	87.7	62.8	75.7	87.9	67.8	78.2	87.7	67.7	78.0
0330	Cast-in-Place Concrete	90.3	70.6	83.0	77.7	70.0	74.8	81.4	71.4	77.7	82.4	71.1	78.2	90.3	73.4	84.0	91.0	70.5	83.4
03	CONCRETE	83.0	62.9	74.2	75.3	55.4	66.6	79.7	64.0	72.8	81.2	62.8	73.1	86.1	67.6	78.0	86.1	62.7	75.8
04	MASONRY	97.4	57.0	72.6	106.1	56.9	75.9	91.3	57.1	70.3	108.1	49.2	71.9	99.8	57.4	73.8	84.1	57.0	67.5
05	METALS	102.5	62.3	90.7	92.6	59.4	82.9	92.6	77.0	88.0	94.1	73.9	88.2	95.7	63.7	86.3	92.6	62.1	83.6
06	WOOD, PLASTICS & COMPOSITES	102.0	54.4	75.8	85.5	36.2	58.4	100.0	54.6	75.0	104.3	54.6	76.9	92.5	65.1	77.4	95.6	54.4	72.9
07	THERMAL & MOISTURE PROTECTION	100.7	64.8	85.1	96.0	61.2	80.9	96.5	64.1	82.4	96.8	61.1	81.2	92.1	66.9	81.1	96.6	65.0	82.9
08	OPENINGS	106.6	55.9	94.8	96.0	43.9	83.9	96.0	55.0	86.5	97.2	54.0	87.1	100.0	61.8	91.1	96.0	55.0	86.4
0920	Plaster & Gypsum Board	92.8	53.7	66.5	77.6	35.0	48.9	84.2	53.7	63.7	86.4	53.7	64.4	94.7	64.6	74.4	82.9	53.7	63.3
0950, 0980	Ceilings & Acoustic Treatment	96.4	53.7	67.6	83.0	35.0	50.6	83.0	53.7	63.3	92.7	53.7	66.4	87.4	64.6	72.0	83.0	53.7	63.3
0960	Flooring	85.0	72.9	81.6	88.6	57.3	79.8	95.7	51.7	83.3	98.2	33.2	79.9	86.4	72.9	82.6	92.8	57.3	82.8
0970, 0990	Wall Finishes & Painting/Coating	86.4	44.8	61.8	92.8	42.2	62.9	92.8	42.2	62.9	92.8	42.2	62.9	89.5	44.8	63.1	92.8	44.8	64.4
09	FINISHES	84.9	56.7	69.6	82.2	42.4	60.7	85.1	52.2	67.3	88.1	48.8	66.9	86.3	63.1	73.7	84.9	54.4	68.4
COVERS	DIVS. 10 - 14, 25, 28, 41, 43, 44, 46	100.0	79.4	95.4	100.0	77.4	95.0	100.0	76.3	94.7	100.0	76.3	94.7	100.0	80.9	95.8	100.0	79.4	95.4
21, 22, 23	FIRE SUPPRESSION, PLUMBING & HVAC	100.0	67.5	87.2	96.6	63.5	83.6	96.6	63.6	83.7	100.1	61.3	84.9	99.9	67.8	87.3	96.6	63.6	83.7
26, 27, 3370	ELECTRICAL, COMMUNICATIONS & UTIL.	97.6	69.2	83.6	95.8	67.0	81.6	97.1	67.0	82.3	95.4	67.2	81.5	103.4	71.1	87.5	95.4	65.6	80.7
MF2018	WEIGHTED AVERAGE	97.3	65.8	84.0	92.4	60.5	78.9	92.7	65.1	81.1	95.0	62.9	81.4	96.4	68.7	84.7	93.1	64.0	80.8

	DIVISION	OKLAHOMA												OREGON					
		POTEAU			SHAWNEE			TULSA			WOODWARD			BEND			EUGENE		
		749			748			740 - 741			738			977			974		
		MAT.	INST.	TOTAL	MAT.	INST.	TOTAL	MAT.	INST.	TOTAL	MAT.	INST.	TOTAL	MAT.	INST.	TOTAL	MAT.	INST.	TOTAL
015433	CONTRACTOR EQUIPMENT		89.7	89.7		79.9	79.9		90.9	90.9		79.9	79.9		97.3	97.3		97.3	97.3
0241, 31 - 34	SITE & INFRASTRUCTURE, DEMOLITION	74.8	83.8	81.0	96.9	90.9	92.7	94.8	87.2	89.5	98.5	91.2	93.5	105.9	98.5	100.8	96.6	98.5	97.9
0310	Concrete Forming & Accessories	87.0	54.9	59.7	82.8	55.3	59.4	97.4	56.0	62.1	86.9	43.7	50.1	107.5	96.6	98.2	104.2	96.6	97.7
0320	Concrete Reinforcing	88.7	67.7	78.5	87.7	65.4	76.9	88.0	67.7	78.2	79.5	67.7	73.8	91.0	112.8	101.5	94.9	112.8	103.5
0330	Cast-in-Place Concrete	81.4	71.3	77.7	93.9	70.5	85.2	89.7	73.2	83.6	93.4	70.5	84.9	117.6	100.6	111.3	113.9	100.6	108.9
03	CONCRETE	81.9	63.8	73.9	87.7	62.3	76.6	86.2	65.0	76.9	86.5	57.5	73.7	110.5	100.3	106.0	101.3	100.3	100.9
04	MASONRY	91.6	57.1	70.4	107.5	57.0	76.5	92.2	57.1	70.6	90.8	57.0	70.0	105.1	100.6	102.3	102.1	100.6	101.1
05	METALS	92.6	76.7	87.9	92.5	61.3	83.3	97.0	77.1	91.2	97.4	62.1	87.0	108.0	96.0	104.5	108.7	95.9	105.0
06	WOOD, PLASTICS & COMPOSITES	92.4	54.6	71.6	87.7	54.4	69.4	103.5	55.6	77.2	96.6	38.7	64.8	99.6	96.1	97.7	95.4	96.1	95.8
07	THERMAL & MOISTURE PROTECTION	96.6	64.1	82.5	96.6	63.9	82.4	96.8	64.3	82.7	100.9	63.3	84.5	118.3	100.2	110.4	117.4	102.7	111.0
08	OPENINGS	96.0	55.0	86.5	96.0	54.4	86.3	98.9	56.2	88.9	103.7	46.3	90.3	96.9	100.0	97.6	97.2	100.0	97.8
0920	Plaster & Gypsum Board	81.3	53.7	62.8	78.5	53.7	61.8	86.4	54.8	65.1	90.2	37.6	54.8	122.7	95.9	104.6	120.8	95.9	104.0
0950, 0980	Ceilings & Acoustic Treatment	83.0	53.7	63.3	83.0	53.7	63.3	92.7	54.8	67.1	89.1	37.6	54.4	85.7	95.9	92.6	86.6	95.9	92.9
0960	Flooring	92.2	51.7	80.8	89.7	57.3	80.6	97.0	61.6	87.1	81.7	54.6	74.1	105.8	105.7	105.7	104.2	105.7	104.6
0970, 0990	Wall Finishes & Painting/Coating	92.8	44.8	64.4	92.8	42.2	62.9	92.8	51.3	68.2	86.4	44.8	61.8	98.6	77.3	86.0	98.6	69.6	81.4
09	FINISHES	82.9	52.5	66.4	83.5	53.2	67.1	88.0	55.8	70.5	82.6	43.7	61.5	100.1	95.9	97.8	98.6	95.0	96.7
COVERS	DIVS. 10 - 14, 25, 28, 41, 43, 44, 46	100.0	79.6	95.5	100.0	79.4	95.4	100.0	76.4	94.7	100.0	77.7	95.0	100.0	102.0	100.4	100.0	102.0	100.4
21, 22, 23	FIRE SUPPRESSION, PLUMBING & HVAC	96.6	63.6	83.7	96.6	67.5	85.2	100.1	63.7	85.8	96.5	67.5	85.1	96.6	105.2	100.0	100.1	98.6	99.5
26, 27, 3370	ELECTRICAL, COMMUNICATIONS & UTIL.	95.5	67.0	81.5	97.4	71.0	84.4	97.3	67.0	82.4	97.4	71.0	84.4	102.0	99.8	100.9	100.7	99.8	100.2
MF2018	WEIGHTED AVERAGE	92.3	64.9	80.7	94.5	65.2	82.2	95.9	65.8	83.2	95.3	62.9	81.6	102.6	100.2	101.6	101.7	98.7	100.5

	DIVISION	OREGON																	
		KLAMATH FALLS			MEDFORD			PENDLETON			PORTLAND			SALEM			VALE		
		976			975			978			970 - 972			973			979		
		MAT.	INST.	TOTAL	MAT.	INST.	TOTAL	MAT.	INST.	TOTAL	MAT.	INST.	TOTAL	MAT.	INST.	TOTAL	MAT.	INST.	TOTAL
015433	CONTRACTOR EQUIPMENT		97.3	97.3		97.3	97.3		94.9	94.9		97.3	97.3		99.9	99.9		94.9	94.9
0241, 31 - 34	SITE & INFRASTRUCTURE, DEMOLITION	109.8	98.4	102.0	104.0	98.4	100.2	103.0	91.9	95.4	99.1	98.5	98.7	92.6	102.1	99.2	90.6	91.8	91.5
0310	Concrete Forming & Accessories	100.7	96.3	97.0	99.7	96.3	96.8	101.2	96.6	97.3	105.3	96.8	98.1	105.8	96.8	98.1	107.5	95.3	97.1
0320	Concrete Reinforcing	91.0	112.7	101.5	92.6	112.7	102.3	90.3	112.8	101.2	95.6	112.8	103.9	102.7	112.8	107.6	88.1	112.5	99.9
0330	Cast-in-Place Concrete	117.6	97.0	110.0	117.6	100.4	111.2	118.4	97.5	110.7	117.0	100.6	110.9	107.6	101.7	105.4	93.2	97.8	94.9
03	CONCRETE	113.5	98.9	107.1	107.8	100.1	104.4	93.7	99.3	96.2	103.0	100.4	101.8	98.7	100.7	99.6	77.8	98.8	87.0
04	MASONRY	119.1	100.6	107.7	99.0	100.6	100.0	109.9	100.6	104.2	104.0	100.6	101.9	108.2	100.6	103.5	107.7	100.6	103.4
05	METALS	108.0	95.6	104.4	108.3	95.6	104.6	116.3	96.4	110.4	110.1	96.1	106.0	117.1	95.4	110.8	116.2	95.1	110.0
06	WOOD, PLASTICS & COMPOSITES	90.4	96.1	93.5	89.3	96.1	93.0	92.3	96.2	94.5	96.4	96.1	96.3	90.3	96.3	93.6	100.9	96.2	98.3
07	THERMAL & MOISTURE PROTECTION	118.5	96.8	109.1	118.1	96.1	108.5	110.7	95.4	104.1	117.3	100.2	109.9	114.1	100.9	108.3	110.1	91.4	102.0
08	OPENINGS	96.9	100.0	97.6	99.6	100.0	99.7	93.2	100.0	94.8	95.1	100.0	96.3	102.6	100.0	102.0	93.2	89.0	92.2
0920	Plaster & Gypsum Board	117.1	95.9	102.8	116.5	95.9	102.6	102.4	95.9	98.0	120.3	95.9	103.9	115.4	95.9	102.3	109.0	95.9	100.2
0950, 0980	Ceilings & Acoustic Treatment	93.0	95.9	95.0	98.5	95.9	96.7	61.9	95.9	84.8	88.5	95.9	93.5	96.0	95.9	95.9	61.9	95.9	84.8
0960	Flooring	102.8	105.7	103.6	102.2	105.7	103.2	70.8	105.7	80.6	101.9	105.7	102.9	107.4	105.7	106.9	72.9	105.7	82.1
0970, 0990	Wall Finishes & Painting/Coating	98.6	65.2	78.8	98.6	65.2	78.8	88.3	77.3	81.8	98.4	77.3	85.9	97.3	75.2	84.2	88.3	77.3	81.8
09	FINISHES	100.3	94.6	97.2	100.4	94.6	97.3	70.4	96.0	84.2	98.2	95.9	97.0	99.7	95.8	97.6	71.0	96.0	84.5
COVERS	DIVS. 10 - 14, 25, 28, 41, 43, 44, 46	100.0	101.9	100.4	100.0	101.9	100.4	100.0	96.5	99.2	100.0	102.1	100.5	100.0	102.4	100.5	100.0	102.2	100.5
21, 22, 23	FIRE SUPPRESSION, PLUMBING & HVAC	96.6	105.2	100.0	100.1	105.2	102.1	98.7	112.0	104.0	100.1	111.5	104.6	100.1	106.9	102.8	98.7	71.1	87.9
26, 27, 3370	ELECTRICAL, COMMUNICATIONS & UTIL.	100.7	82.6	91.8	104.1	82.6	93.5	92.8	96.8	94.8	100.9	108.7	104.7	109.3	99.8	104.6	92.8	68.1	80.6
MF2018	WEIGHTED AVERAGE	103.5	97.3	100.9	103.3	97.4	100.8	98.6	100.3	99.3	102.1	102.8	102.4	104.4	100.8	102.9	96.2	87.0	92.3

	DIVISION	PENNSYLVANIA																	
		ALLENTOWN			ALTOONA			BEDFORD			BRADFORD			BUTLER			CHAMBERSBURG		
		181			166			155			167			160			172		
		MAT.	INST.	TOTAL	MAT.	INST.	TOTAL	MAT.	INST.	TOTAL	MAT.	INST.	TOTAL	MAT.	INST.	TOTAL	MAT.	INST.	TOTAL
015433	CONTRACTOR EQUIPMENT		112.6	112.6		112.6	112.6		110.7	110.7		112.6	112.6		112.6	112.6		111.8	111.8
0241, 31 - 34	SITE & INFRASTRUCTURE, DEMOLITION	91.7	97.2	95.5	95.1	97.1	96.5	103.6	94.8	97.6	90.5	96.0	94.3	86.3	97.8	94.2	86.3	95.3	92.5
0310	Concrete Forming & Accessories	97.9	108.2	106.7	83.7	82.6	82.8	82.0	81.2	81.3	85.8	96.5	94.9	85.1	94.1	92.8	88.0	76.8	78.5
0320	Concrete Reinforcing	96.8	113.3	104.8	93.8	105.7	99.6	93.0	105.7	99.1	95.8	105.8	100.6	94.4	119.3	106.5	94.5	112.2	103.1
0330	Cast-in-Place Concrete	89.4	99.9	93.3	99.7	86.9	94.9	109.1	85.9	100.5	95.2	90.6	93.5	88.0	96.2	91.0	93.0	93.8	93.3
03	CONCRETE	92.3	107.0	98.8	87.5	89.5	88.4	102.2	88.5	96.2	94.2	97.0	95.5	79.5	100.3	88.7	97.1	90.5	94.2
04	MASONRY	92.1	93.1	92.7	95.7	83.4	88.1	107.1	80.2	90.6	93.1	82.0	86.3	97.6	93.0	94.8	93.8	79.1	84.8
05	METALS	98.9	120.5	105.3	92.9	114.8	99.3	102.2	113.4	105.5	96.7	113.2	101.6	92.6	121.7	101.2	98.3	118.8	104.3
06	WOOD, PLASTICS & COMPOSITES	97.6	111.5	105.3	76.7	81.8	79.5	79.4	81.7	80.7	82.7	101.2	92.9	78.1	94.1	86.9	84.3	75.6	79.5
07	THERMAL & MOISTURE PROTECTION	104.0	107.8	105.7	102.7	90.0	97.2	98.3	87.4	93.6	103.8	88.5	97.2	102.2	94.4	98.8	96.1	83.1	90.4
08	OPENINGS	92.4	107.3	95.9	86.3	85.6	86.1	93.4	85.6	91.6	92.4	95.3	93.1	86.2	99.3	89.3	89.2	81.4	87.4
0920	Plaster & Gypsum Board	94.9	111.8	106.3	86.3	81.2	82.9	97.5	81.2	86.6	87.1	101.2	96.6	86.3	93.9	91.4	108.9	74.9	86.0
0950, 0980	Ceilings & Acoustic Treatment	88.5	111.8	104.2	90.9	81.2	84.4	101.1	81.2	87.7	91.0	101.2	97.9	91.8	93.9	93.2	96.0	74.9	81.8
0960	Flooring	92.0	95.1	92.8	85.3	98.9	89.2	95.0	103.7	97.5	86.2	103.7	91.1	86.3	105.3	91.6	90.5	78.0	87.0
0970, 0990	Wall Finishes & Painting/Coating	91.9	101.8	97.8	87.2	106.2	98.4	97.6	105.7	102.4	91.9	105.7	100.0	87.2	106.2	98.4	92.0	100.7	97.1
09	FINISHES	90.6	105.8	98.8	88.2	87.5	87.8	100.6	87.4	93.5	88.4	99.4	94.3	88.1	97.2	93.0	90.8	78.8	84.3
COVERS	DIVS. 10 - 14, 25, 28, 41, 43, 44, 46	100.0	99.6	99.9	100.0	94.5	98.8	100.0	93.2	98.5	100.0	96.1	99.1	100.0	97.3	99.4	100.0	92.7	98.4
21, 22, 23	FIRE SUPPRESSION, PLUMBING & HVAC	100.3	114.5	105.9	99.8	83.6	93.4	96.6	84.4	91.8	96.8	90.6	94.4	96.3	95.4	96.0	96.7	89.7	94.0
26, 27, 3370	ELECTRICAL, COMMUNICATIONS & UTIL.	98.0	95.8	96.9	88.7	107.7	98.0	94.2	107.7	100.9	92.1	107.7	99.8	89.2	107.7	98.3	92.4	85.8	89.2
MF2018	WEIGHTED AVERAGE	96.7	105.8	100.6	93.3	92.8	93.1	98.8	92.1	96.0	94.9	97.1	95.8	91.3	100.6	95.2	95.1	89.4	92.7

	DIVISION	PENNSYLVANIA																	
		DOYLESTOWN			DUBOIS			ERIE			GREENSBURG			HARRISBURG			HAZLETON		
		189			158			164 - 165			156			170 - 171			182		
		MAT.	INST.	TOTAL	MAT.	INST.	TOTAL	MAT.	INST.	TOTAL	MAT.	INST.	TOTAL	MAT.	INST.	TOTAL	MAT.	INST.	TOTAL
015433	CONTRACTOR EQUIPMENT		91.5	91.5		110.7	110.7		112.6	112.6		110.7	110.7		114.7	114.7		112.6	112.6
0241, 31 - 34	SITE & INFRASTRUCTURE, DEMOLITION	104.8	85.9	91.7	108.4	95.2	99.3	92.1	97.4	95.7	99.6	96.7	97.6	86.6	100.3	96.1	85.0	97.1	93.3
0310	Concrete Forming & Accessories	82.9	126.8	120.3	81.6	83.7	83.4	97.2	85.8	87.5	87.8	88.6	88.5	100.5	85.7	87.9	80.6	88.8	87.6
0320	Concrete Reinforcing	93.6	150.9	121.3	92.3	119.4	105.4	95.8	107.3	101.4	92.3	119.0	105.2	103.4	111.0	107.1	94.0	113.1	103.2
0330	Cast-in-Place Concrete	84.5	129.7	101.3	105.2	94.1	101.1	98.0	88.7	94.5	101.3	95.7	99.2	91.7	96.9	93.6	84.5	95.6	88.6
03	CONCRETE	88.0	131.2	107.0	104.4	94.7	100.1	86.4	91.9	88.8	97.2	97.5	97.4	91.9	95.3	93.4	85.0	96.7	90.2
04	MASONRY	95.4	134.9	119.7	107.8	93.4	98.9	85.0	86.8	86.1	117.5	89.1	100.0	88.8	83.9	85.8	104.7	89.5	95.4
05	METALS	96.5	123.9	104.5	102.2	118.8	107.1	93.1	115.6	99.7	102.1	119.7	107.3	105.1	117.4	108.7	98.7	119.7	104.8
06	WOOD, PLASTICS & COMPOSITES	78.5	125.8	104.5	78.3	81.7	80.2	94.0	84.4	88.7	85.3	87.1	86.3	101.8	85.3	92.7	77.1	87.0	82.6
07	THERMAL & MOISTURE PROTECTION	101.1	131.7	114.4	98.7	92.9	96.1	103.2	88.6	96.8	98.2	92.2	95.6	98.9	100.4	99.6	103.3	100.7	102.2
08	OPENINGS	94.8	133.2	103.7	93.4	88.7	92.3	86.4	88.0	86.8	93.4	95.5	93.9	100.9	86.5	97.5	93.0	93.3	93.0
0920	Plaster & Gypsum Board	85.4	126.5	113.0	96.3	81.2	86.2	94.9	83.9	87.5	98.3	86.8	90.5	113.8	84.7	94.2	85.8	86.6	86.4
0950, 0980	Ceilings & Acoustic Treatment	87.7	126.5	113.8	101.1	81.2	87.7	88.5	83.9	85.4	100.3	86.8	91.2	104.0	84.7	91.0	89.4	86.6	87.5
0960	Flooring	76.2	139.3	93.9	94.8	103.7	97.3	92.2	98.9	94.1	98.4	78.0	92.7	94.7	90.4	93.5	83.5	89.3	85.1
0970, 0990	Wall Finishes & Painting/Coating	91.5	143.7	122.4	97.6	106.2	102.7	98.2	92.3	94.7	97.6	106.2	102.7	95.8	84.8	89.3	91.9	104.1	99.1
09	FINISHES	81.9	130.5	108.2	100.9	88.8	94.4	91.6	88.4	89.9	101.3	88.1	94.1	96.0	85.8	90.5	86.6	90.2	88.6
COVERS	DIVS. 10 - 14, 25, 28, 41, 43, 44, 46	100.0	113.0	102.9	100.0	95.2	98.9	100.0	96.0	99.1	100.0	96.4	99.2	100.0	95.4	99.0	100.0	95.5	99.0
21, 22, 23	FIRE SUPPRESSION, PLUMBING & HVAC	96.3	132.6	110.6	96.6	87.5	93.0	99.8	93.6	97.4	96.6	87.3	92.9	100.1	91.7	96.8	96.8	95.5	96.3
26, 27, 3370	ELECTRICAL, COMMUNICATIONS & UTIL.	91.6	132.4	111.7	94.8	107.7	101.1	90.3	93.1	91.7	94.8	107.7	101.2	99.7	85.8	92.9	92.9	90.0	91.5
MF2018	WEIGHTED AVERAGE	94.0	127.1	108.0	99.3	96.0	97.9	93.2	93.9	93.5	98.7	96.3	97.7	98.7	93.0	96.3	94.4	95.9	95.0

City Cost Indexes

DIVISION		PENNSYLVANIA																	
		INDIANA 157			JOHNSTOWN 159			KITTANNING 162			LANCASTER 175 - 176			LEHIGH VALLEY 180			MONTROSE 188		
		MAT.	INST.	TOTAL	MAT.	INST.	TOTAL	MAT.	INST.	TOTAL	MAT.	INST.	TOTAL	MAT.	INST.	TOTAL	MAT.	INST.	TOTAL
015433	CONTRACTOR EQUIPMENT		110.7	110.7		110.7	110.7		112.6	112.6		111.8	111.8		112.6	112.6		112.6	112.6
0241, 31 - 34	SITE & INFRASTRUCTURE, DEMOLITION	97.8	95.8	96.4	104.1	96.0	98.5	88.8	97.4	94.7	78.4	95.6	90.3	88.9	97.5	94.8	87.5	97.2	94.2
0310	Concrete Forming & Accessories	82.6	92.9	91.4	81.6	82.5	82.4	85.1	89.7	89.0	90.0	86.4	86.9	91.7	110.3	107.6	81.5	89.3	88.2
0320	Concrete Reinforcing	91.7	119.4	105.1	93.0	118.9	105.5	94.4	119.2	106.4	94.1	111.0	102.3	94.0	108.0	100.8	98.4	116.2	107.0
0330	Cast-in-Place Concrete	99.3	94.4	97.5	110.1	86.7	101.4	91.5	94.7	92.7	79.1	96.7	85.7	91.5	99.2	94.3	89.7	93.0	90.9
03	CONCRETE	94.7	99.1	96.6	103.2	91.6	98.1	82.1	97.7	88.9	85.2	95.6	89.8	91.4	106.9	98.2	90.1	96.6	93.0
04	MASONRY	103.8	93.5	97.5	104.7	83.7	91.8	100.1	88.8	93.1	99.6	86.7	91.7	92.1	96.5	94.8	92.0	91.5	91.7
05	METALS	102.3	119.9	107.4	102.2	117.9	106.8	92.7	120.6	100.9	98.3	118.7	104.3	98.6	119.0	104.6	96.8	120.4	103.7
06	WOOD, PLASTICS & COMPOSITES	80.1	94.0	87.7	78.3	81.7	80.2	78.1	89.3	84.3	87.4	85.1	86.2	89.1	112.6	102.0	77.8	87.0	82.9
07	THERMAL & MOISTURE PROTECTION	98.1	94.2	96.4	98.4	88.5	94.1	102.3	92.4	98.0	95.4	101.0	97.8	103.7	111.9	107.3	103.3	91.0	97.9
08	OPENINGS	93.4	95.4	93.9	93.4	88.7	92.3	86.3	96.7	88.7	89.2	86.4	88.5	92.9	106.6	96.1	89.8	91.9	90.3
0920	Plaster & Gypsum Board	97.8	93.9	95.2	96.1	81.2	86.1	86.3	89.0	88.1	111.8	84.7	93.6	88.0	112.9	104.7	86.2	86.6	86.5
0950, 0980	Ceilings & Acoustic Treatment	101.1	93.9	96.2	100.3	81.2	87.5	91.8	89.0	89.9	96.0	84.7	88.4	89.4	112.9	105.2	91.0	86.6	88.1
0960	Flooring	95.6	103.7	97.8	94.8	78.0	90.1	86.3	103.7	91.2	91.5	95.7	92.7	88.9	94.1	90.4	84.1	103.7	89.6
0970, 0990	Wall Finishes & Painting/Coating	97.6	106.2	102.7	97.6	106.2	102.7	87.2	106.2	98.4	92.0	86.6	88.8	91.9	101.0	97.3	91.9	104.1	99.1
09	FINISHES	100.4	96.1	98.1	100.3	83.9	91.4	88.3	93.4	91.0	90.9	87.3	88.9	88.8	106.5	98.4	87.4	92.7	90.3
COVERS	DIVS. 10 - 14, 25, 28, 41, 43, 44, 46	100.0	96.5	99.2	100.0	94.4	98.8	100.0	96.2	99.1	100.0	95.7	99.0	100.0	102.5	100.6	100.0	96.2	99.2
21, 22, 23	FIRE SUPPRESSION, PLUMBING & HVAC	96.6	85.4	92.2	96.6	78.9	89.6	96.3	90.7	94.1	96.7	92.7	95.1	96.8	117.6	104.9	96.8	94.8	96.0
26, 27, 3370	ELECTRICAL, COMMUNICATIONS & UTIL.	94.8	107.7	101.2	94.8	107.7	101.1	88.7	107.7	98.1	93.7	93.3	93.5	92.9	129.4	110.9	92.1	95.2	93.6
MF2018	WEIGHTED AVERAGE	97.6	97.7	97.7	98.9	91.9	95.9	91.8	97.9	94.4	93.8	94.5	94.1	94.9	111.6	102.0	93.8	96.8	95.1

DIVISION		PENNSYLVANIA																	
		NEW CASTLE 161			NORRISTOWN 194			OIL CITY 163			PHILADELPHIA 190 - 191			PITTSBURGH 150 - 152			POTTSVILLE 179		
		MAT.	INST.	TOTAL	MAT.	INST.	TOTAL	MAT.	INST.	TOTAL	MAT.	INST.	TOTAL	MAT.	INST.	TOTAL	MAT.	INST.	TOTAL
015433	CONTRACTOR EQUIPMENT		112.6	112.6		97.5	97.5		112.6	112.6		99.8	99.8		99.7	99.7		111.8	111.8
0241, 31 - 34	SITE & INFRASTRUCTURE, DEMOLITION	86.7	97.7	94.3	97.3	95.1	95.8	85.2	96.3	92.9	99.8	101.1	100.7	103.7	96.1	98.5	81.3	95.8	91.3
0310	Concrete Forming & Accessories	85.1	93.6	92.3	83.7	125.3	119.1	85.1	92.7	91.6	99.8	142.4	136.2	97.5	97.5	97.5	82.0	87.7	86.8
0320	Concrete Reinforcing	93.3	100.2	96.7	95.9	150.8	122.5	94.4	94.9	94.7	100.4	143.1	121.0	93.5	122.9	107.7	93.4	114.1	103.4
0330	Cast-in-Place Concrete	88.8	94.3	90.8	86.1	127.4	101.5	86.3	94.4	89.3	89.8	135.6	106.8	108.6	101.2	105.8	84.3	96.4	88.8
03	CONCRETE	79.9	96.2	87.0	88.0	129.7	106.3	78.3	95.0	85.6	99.4	139.1	116.8	103.8	103.1	103.5	88.7	96.7	92.2
04	MASONRY	97.3	92.6	94.4	108.3	131.2	122.4	96.5	88.7	91.7	100.2	139.7	124.5	100.4	101.4	101.0	93.3	87.6	89.8
05	METALS	92.7	114.3	99.0	100.3	124.0	107.3	92.7	113.2	98.7	103.8	125.3	110.1	103.7	107.4	104.8	98.6	120.4	105.0
06	WOOD, PLASTICS & COMPOSITES	78.1	94.1	86.9	76.2	125.7	103.4	78.1	94.1	86.9	99.1	143.5	123.6	99.6	96.7	98.0	77.2	85.1	81.6
07	THERMAL & MOISTURE PROTECTION	102.2	92.0	97.8	108.7	130.1	118.0	102.1	90.7	97.1	103.4	139.7	119.2	98.6	99.3	98.9	95.6	99.6	97.3
08	OPENINGS	86.3	91.4	87.5	86.5	133.1	97.4	86.3	94.9	88.3	97.6	144.9	108.6	96.8	102.4	98.1	89.2	93.0	90.1
0920	Plaster & Gypsum Board	86.3	93.9	91.4	84.5	126.5	112.7	86.3	93.9	91.4	103.0	144.6	131.0	97.5	96.4	96.7	106.4	84.7	91.8
0950, 0980	Ceilings & Acoustic Treatment	91.8	93.9	93.2	91.1	126.5	114.9	91.8	93.9	93.2	104.2	144.6	131.5	95.2	96.4	96.0	96.0	84.7	88.4
0960	Flooring	86.3	105.3	91.6	86.7	139.3	101.5	86.3	103.7	91.2	97.8	156.2	114.2	103.1	107.3	104.3	87.5	103.7	92.1
0970, 0990	Wall Finishes & Painting/Coating	87.2	106.2	98.4	89.4	143.7	121.5	87.2	106.2	98.4	97.2	152.6	130.0	100.7	111.3	107.0	92.0	104.1	99.1
09	FINISHES	88.2	96.9	92.9	85.3	129.5	109.2	88.0	96.1	92.4	98.3	146.5	124.4	102.4	100.3	101.2	89.2	91.3	90.3
COVERS	DIVS. 10 - 14, 25, 28, 41, 43, 44, 46	100.0	97.3	99.4	100.0	109.1	102.0	100.0	96.7	99.3	100.0	119.2	104.3	100.0	103.0	100.7	100.0	97.5	99.4
21, 22, 23	FIRE SUPPRESSION, PLUMBING & HVAC	96.3	96.8	96.5	96.6	130.6	110.0	96.3	94.2	95.5	100.1	141.2	116.3	100.0	99.8	99.9	96.7	95.8	96.3
26, 27, 3370	ELECTRICAL, COMMUNICATIONS & UTIL.	89.2	98.2	93.7	91.9	142.0	116.6	90.9	107.7	99.2	98.8	159.0	128.4	97.1	111.3	104.1	92.0	91.3	91.6
MF2018	WEIGHTED AVERAGE	91.4	97.8	94.1	94.7	128.0	108.8	91.3	97.8	94.0	100.2	138.7	116.5	100.7	102.7	101.5	93.7	96.1	94.7

DIVISION		PENNSYLVANIA																	
		READING 195 - 196			SCRANTON 184 - 185			STATE COLLEGE 168			STROUDSBURG 183			SUNBURY 178			UNIONTOWN 154		
		MAT.	INST.	TOTAL	MAT.	INST.	TOTAL	MAT.	INST.	TOTAL	MAT.	INST.	TOTAL	MAT.	INST.	TOTAL	MAT.	INST.	TOTAL
015433	CONTRACTOR EQUIPMENT		119.0	119.0		112.6	112.6		111.8	111.8		112.6	112.6		112.6	112.6		110.7	110.7
0241, 31 - 34	SITE & INFRASTRUCTURE, DEMOLITION	101.8	107.7	105.9	92.2	97.2	95.7	82.7	95.7	91.7	86.6	97.3	94.0	92.6	96.7	95.4	98.3	96.5	97.1
0310	Concrete Forming & Accessories	98.8	86.2	88.0	98.0	86.6	88.3	84.3	82.8	83.0	86.5	90.5	89.9	93.1	84.8	86.0	76.5	93.9	91.4
0320	Concrete Reinforcing	97.2	148.6	122.1	96.8	116.3	106.2	95.1	105.8	100.3	97.1	116.4	106.4	96.0	112.3	103.9	92.3	119.5	105.5
0330	Cast-in-Place Concrete	77.2	97.0	84.6	93.4	93.2	93.3	90.0	87.0	88.9	88.0	94.5	90.4	92.2	94.9	93.2	99.3	95.6	97.9
03	CONCRETE	87.1	102.1	93.7	94.1	95.4	94.7	94.6	89.7	92.5	88.8	97.7	92.7	92.0	94.5	93.1	94.4	100.0	96.9
04	MASONRY	97.5	91.5	93.8	92.5	93.8	93.3	97.8	83.9	89.3	90.3	96.3	94.0	93.6	81.0	85.8	119.3	95.4	104.6
05	METALS	100.6	133.0	110.1	100.9	120.5	106.6	96.6	115.1	102.0	98.7	120.9	105.2	98.3	119.4	104.5	102.0	120.2	107.3
06	WOOD, PLASTICS & COMPOSITES	93.7	82.4	87.5	97.6	83.3	89.8	84.5	81.8	83.0	83.5	87.0	85.5	85.4	85.1	85.2	72.8	94.0	84.4
07	THERMAL & MOISTURE PROTECTION	109.0	103.3	106.6	103.9	91.3	98.4	102.9	97.6	100.6	103.5	91.3	98.2	96.8	92.7	95.0	98.0	95.1	96.7
08	OPENINGS	90.9	99.4	92.9	92.4	89.8	91.8	89.6	85.6	88.7	93.0	94.5	93.3	89.3	89.9	89.4	93.3	99.3	94.7
0920	Plaster & Gypsum Board	94.1	81.9	85.9	96.8	82.8	87.4	87.8	81.2	83.4	86.7	86.6	86.6	105.9	84.7	91.6	94.2	93.9	94.0
0950, 0980	Ceilings & Acoustic Treatment	83.7	81.9	82.5	96.6	82.8	87.3	88.5	81.2	83.6	87.7	86.6	87.0	93.6	84.7	87.6	100.3	93.9	96.0
0960	Flooring	91.3	97.9	93.2	92.0	92.9	92.2	89.4	97.9	91.8	86.8	94.1	88.8	88.3	103.7	92.6	92.0	103.7	95.3
0970, 0990	Wall Finishes & Painting/Coating	88.2	101.8	96.3	91.9	108.3	101.6	91.9	106.2	100.4	91.9	104.1	99.1	92.0	104.1	99.1	97.6	110.0	104.9
09	FINISHES	86.8	88.4	87.7	92.4	89.3	90.7	88.0	87.3	87.6	87.5	91.8	89.8	90.2	89.6	89.9	98.7	97.0	97.8
COVERS	DIVS. 10 - 14, 25, 28, 41, 43, 44, 46	100.0	95.8	99.1	100.0	94.6	98.8	100.0	94.6	98.8	100.0	97.2	99.4	100.0	93.1	98.5	100.0	97.2	99.4
21, 22, 23	FIRE SUPPRESSION, PLUMBING & HVAC	100.3	106.8	102.8	100.3	96.0	98.6	96.8	91.4	94.7	96.8	97.9	97.2	96.7	87.4	93.0	96.6	91.7	94.6
26, 27, 3370	ELECTRICAL, COMMUNICATIONS & UTIL.	97.8	91.3	94.6	98.0	95.2	96.6	91.3	107.7	99.4	92.9	141.8	117.0	92.4	91.9	92.1	91.8	107.7	99.7
MF2018	WEIGHTED AVERAGE	96.5	101.5	98.6	97.4	96.5	97.1	94.5	94.6	94.5	94.3	104.7	98.7	94.6	92.7	93.8	97.8	99.8	98.6

DIVISION		PENNSYLVANIA																	
		WASHINGTON			WELLSBORO			WESTCHESTER			WILKES-BARRE			WILLIAMSPORT			YORK		
		153			169			193			186 - 187			177			173 - 174		
		MAT.	INST.	TOTAL	MAT.	INST.	TOTAL	MAT.	INST.	TOTAL	MAT.	INST.	TOTAL	MAT.	INST.	TOTAL	MAT.	INST.	TOTAL
015433	CONTRACTOR EQUIPMENT		110.7	110.7		112.6	112.6		97.5	97.5		112.6	112.6		112.6	112.6		111.8	111.8
0241, 31 - 34	SITE & INFRASTRUCTURE, DEMOLITION	98.3	96.8	97.3	94.0	96.6	95.8	103.4	96.1	98.4	84.5	97.2	93.3	83.7	96.8	92.8	82.4	95.6	91.5
0310	Concrete Forming & Accessories	82.8	94.0	92.4	85.2	84.2	84.3	90.0	126.6	121.2	89.1	85.3	85.8	90.1	85.6	86.3	85.1	86.3	86.1
0320	Concrete Reinforcing	92.3	119.5	105.5	95.1	116.1	105.3	95.0	150.8	122.0	95.8	116.3	105.7	95.3	110.9	102.8	96.0	111.0	103.3
0330	Cast-in-Place Concrete	99.3	95.9	98.0	94.4	88.2	92.1	95.4	129.5	108.1	84.5	93.0	87.6	78.0	90.0	82.5	84.7	96.7	89.2
03	CONCRETE	94.8	100.1	97.2	96.9	92.5	95.0	95.8	131.0	111.3	85.8	94.7	89.7	80.2	92.9	85.8	89.8	95.6	92.4
04	MASONRY	103.6	95.5	98.6	98.7	81.4	88.1	102.8	134.9	122.5	105.0	93.5	97.9	85.9	86.1	86.0	94.5	86.7	89.7
05	METALS	101.9	120.6	107.4	96.6	119.7	103.4	100.3	124.0	107.3	96.8	120.4	103.7	98.3	118.1	104.1	100.0	118.6	105.5
06	WOOD, PLASTICS & COMPOSITES	80.2	94.0	87.8	82.1	84.4	83.4	82.8	125.7	106.4	85.9	81.7	83.6	81.9	85.1	83.7	80.7	85.1	83.1
07	THERMAL & MOISTURE PROTECTION	98.1	95.1	96.8	104.1	86.0	96.2	109.1	127.0	116.9	103.3	90.9	97.9	96.1	88.3	92.7	95.6	101.0	98.0
08	OPENINGS	93.3	99.3	94.7	92.4	90.0	91.8	86.5	133.1	97.4	89.8	88.9	89.6	89.3	89.6	89.4	89.2	86.4	88.5
0920	Plaster & Gypsum Board	97.7	93.9	95.1	86.5	83.9	84.8	85.2	126.5	112.9	87.4	81.2	83.2	106.4	84.7	91.8	106.7	84.7	91.9
0950, 0980	Ceilings & Acoustic Treatment	100.3	93.9	96.0	88.5	83.9	85.4	91.1	126.5	114.9	91.0	81.2	84.4	96.0	84.7	88.4	95.1	84.7	88.1
0960	Flooring	95.7	103.7	97.9	85.8	103.7	90.9	89.6	139.3	103.6	87.7	92.9	89.1	87.2	86.8	87.1	88.9	95.7	90.8
0970, 0990	Wall Finishes & Painting/Coating	97.6	110.0	104.9	91.9	104.1	99.1	89.4	143.7	121.5	91.9	101.8	97.8	92.0	101.8	97.8	92.0	84.8	87.7
09	FINISHES	100.3	97.0	98.5	88.1	89.6	88.9	86.7	130.5	110.4	88.4	87.6	88.0	89.7	86.7	88.0	89.5	87.1	88.2
COVERS	DIVS. 10 - 14, 25, 28, 41, 43, 44, 46	100.0	97.2	99.4	100.0	96.2	99.2	100.0	115.3	103.4	100.0	94.3	98.7	100.0	95.5	99.0	100.0	95.6	99.0
21, 22, 23	FIRE SUPPRESSION, PLUMBING & HVAC	96.6	93.2	95.2	96.8	87.4	93.1	96.6	132.6	110.7	96.8	96.6	96.7	96.7	92.2	94.9	100.2	92.7	97.3
26, 27, 3370	ELECTRICAL, COMMUNICATIONS & UTIL.	94.2	107.7	100.9	92.1	95.2	93.6	91.7	132.4	111.8	92.9	90.0	91.5	92.8	83.4	88.2	93.7	82.9	88.4
MF2018	WEIGHTED AVERAGE	97.5	100.2	98.6	95.5	92.9	94.4	95.8	127.9	109.3	94.0	95.5	94.7	92.5	92.3	92.4	95.2	93.0	94.3

DIVISION		PUERTO RICO			RHODE ISLAND						SOUTH CAROLINA								
		SAN JUAN			NEWPORT			PROVIDENCE			AIKEN			BEAUFORT			CHARLESTON		
		009			028			029			298			299			294		
		MAT.	INST.	TOTAL	MAT.	INST.	TOTAL	MAT.	INST.	TOTAL	MAT.	INST.	TOTAL	MAT.	INST.	TOTAL	MAT.	INST.	TOTAL
015433	CONTRACTOR EQUIPMENT		87.3	87.3		97.9	97.9		100.1	100.1		103.7	103.7		103.7	103.7		103.7	103.7
0241, 31 - 34	SITE & INFRASTRUCTURE, DEMOLITION	81.4	87.9	85.9	86.3	98.0	94.4	89.4	102.2	98.3	128.3	85.2	98.5	123.4	84.7	96.7	108.2	85.2	92.3
0310	Concrete Forming & Accessories	105.9	21.6	34.0	100.7	124.7	121.1	99.9	124.7	121.0	94.0	64.6	68.9	93.1	38.9	46.9	92.3	64.9	69.0
0320	Concrete Reinforcing	114.9	18.7	68.4	109.7	127.1	118.1	103.7	127.1	115.0	94.3	60.6	78.0	93.4	66.7	80.5	93.2	67.3	80.7
0330	Cast-in-Place Concrete	70.9	32.7	56.6	71.7	118.5	89.1	87.5	119.0	99.2	92.9	66.9	83.2	92.9	66.8	83.2	109.1	67.2	93.5
03	CONCRETE	89.8	26.1	61.8	87.5	122.4	102.9	91.8	122.5	105.3	101.0	66.4	85.8	98.4	55.7	79.7	94.9	67.8	83.0
04	MASONRY	110.0	24.2	57.3	99.2	123.1	113.8	103.4	123.1	115.5	79.9	64.8	70.6	94.9	64.8	76.4	96.4	67.5	78.6
05	METALS	116.8	40.6	94.4	98.8	116.7	104.1	103.2	115.9	106.9	102.6	89.1	98.7	102.7	90.7	99.2	104.6	92.1	100.9
06	WOOD, PLASTICS & COMPOSITES	145.4	20.5	76.6	98.0	124.0	112.3	98.4	124.0	112.5	89.7	66.8	77.1	88.4	32.5	57.6	87.2	66.8	76.0
07	THERMAL & MOISTURE PROTECTION	134.1	27.7	87.8	102.6	123.0	111.5	108.3	123.5	114.9	98.1	64.2	83.4	97.8	56.2	79.7	96.8	66.1	83.5
08	OPENINGS	96.8	19.1	78.7	97.4	124.4	103.7	99.0	124.4	104.9	94.0	62.5	86.7	94.1	44.9	82.6	97.6	65.1	90.0
0920	Plaster & Gypsum Board	76.4	25.2	48.0	91.4	124.3	113.6	104.9	124.3	118.0	83.4	65.8	71.6	87.0	30.5	49.0	88.1	65.8	73.1
0950, 0980	Ceilings & Acoustic Treatment	158.8	19.3	64.7	90.3	124.3	113.2	104.1	124.3	117.7	78.8	65.8	70.1	82.1	30.5	47.3	82.1	65.8	71.1
0960	Flooring	117.8	29.7	93.1	83.8	127.3	96.0	86.4	127.3	97.9	97.5	89.8	95.3	99.0	75.8	92.5	98.6	81.9	93.9
0970, 0990	Wall Finishes & Painting/Coating	90.2	24.6	51.4	85.0	118.9	105.0	90.8	118.9	107.4	95.2	66.0	77.9	95.2	56.9	72.5	95.2	70.0	80.3
09	FINISHES	106.0	23.2	61.2	87.7	125.1	107.9	93.7	125.1	110.7	89.0	69.7	78.5	90.1	46.0	66.2	88.2	68.9	77.8
COVERS	DIVS. 10 - 14, 25, 28, 41, 43, 44, 46	100.0	26.6	83.6	100.0	110.2	102.3	100.0	110.3	102.3	100.0	70.3	93.4	100.0	80.1	95.6	100.0	70.3	93.4
21, 22, 23	FIRE SUPPRESSION, PLUMBING & HVAC	88.5	18.8	61.1	100.4	114.8	106.0	100.2	114.8	105.9	96.9	55.2	80.5	96.9	55.6	80.7	100.4	58.4	83.9
26, 27, 3370	ELECTRICAL, COMMUNICATIONS & UTIL.	84.4	25.0	55.2	102.1	95.5	98.9	102.3	95.5	99.0	96.2	61.9	79.3	99.7	30.9	65.8	98.1	58.9	78.8
MF2018	WEIGHTED AVERAGE	98.5	29.8	69.4	97.0	114.7	104.5	99.3	115.0	105.9	97.6	67.1	84.7	98.3	57.5	81.1	98.7	68.2	85.8

DIVISION		SOUTH CAROLINA															SOUTH DAKOTA		
		COLUMBIA			FLORENCE			GREENVILLE			ROCK HILL			SPARTANBURG			ABERDEEN		
		290 - 292			295			296			297			293			574		
		MAT.	INST.	TOTAL	MAT.	INST.	TOTAL	MAT.	INST.	TOTAL	MAT.	INST.	TOTAL	MAT.	INST.	TOTAL	MAT.	INST.	TOTAL
015433	CONTRACTOR EQUIPMENT		105.6	105.6		103.7	103.7		103.7	103.7		103.7	103.7		103.7	103.7		97.7	97.7
0241, 31 - 34	SITE & INFRASTRUCTURE, DEMOLITION	105.9	89.5	94.6	117.6	84.9	95.1	112.6	85.4	93.8	110.0	84.3	92.2	112.4	85.4	93.8	99.2	93.4	95.2
0310	Concrete Forming & Accessories	93.2	64.9	69.1	81.1	64.8	67.2	91.8	64.9	68.8	90.1	64.0	67.9	94.4	64.9	69.2	98.9	75.7	79.1
0320	Concrete Reinforcing	98.3	67.3	83.3	92.8	67.2	80.4	92.8	64.4	79.1	93.5	65.8	80.1	92.8	67.3	80.4	101.0	71.4	86.7
0330	Cast-in-Place Concrete	110.3	67.5	94.4	92.9	67.0	83.2	92.9	67.2	83.3	92.8	66.4	83.0	92.9	67.2	83.3	109.5	78.8	98.1
03	CONCRETE	93.2	67.8	82.1	93.3	67.6	82.0	92.0	67.3	81.2	90.1	66.8	79.9	92.2	67.8	81.5	98.7	76.7	89.1
04	MASONRY	89.9	67.5	76.1	79.9	67.4	72.2	77.9	67.5	71.5	101.6	64.0	78.5	79.9	67.5	72.3	109.4	73.3	87.2
05	METALS	101.6	91.0	98.5	103.5	91.4	99.9	103.5	91.7	100.0	102.7	90.5	99.1	103.5	92.1	100.1	93.8	82.1	90.4
06	WOOD, PLASTICS & COMPOSITES	88.6	66.9	76.7	74.8	66.8	70.4	86.8	66.8	75.8	85.4	66.8	75.2	90.6	66.8	77.5	97.6	74.2	84.7
07	THERMAL & MOISTURE PROTECTION	93.3	66.3	81.5	97.2	66.1	83.7	97.1	66.1	83.6	96.9	59.2	80.5	97.1	66.1	83.6	103.1	77.2	91.8
08	OPENINGS	99.7	65.1	91.7	94.1	65.1	87.3	94.0	64.8	87.2	94.1	63.6	87.0	94.1	65.1	87.3	97.2	62.2	89.0
0920	Plaster & Gypsum Board	88.7	65.8	73.3	78.2	65.8	69.9	82.1	65.8	71.2	81.8	65.8	71.1	84.7	65.8	72.0	110.1	73.9	85.7
0950, 0980	Ceilings & Acoustic Treatment	86.2	65.8	72.5	79.6	65.8	70.3	78.8	65.8	70.1	78.8	65.8	70.1	78.8	65.8	70.1	98.8	73.9	82.0
0960	Flooring	90.7	80.4	87.8	89.9	80.4	87.2	96.3	80.4	91.8	95.2	75.8	89.7	97.6	80.4	92.8	91.1	43.8	77.8
0970, 0990	Wall Finishes & Painting/Coating	93.9	70.0	79.7	95.2	70.0	80.3	95.2	70.0	80.3	95.2	66.0	77.9	95.2	70.0	80.3	89.8	33.4	56.4
09	FINISHES	86.5	68.7	76.8	85.0	68.7	76.2	86.9	68.7	77.0	86.3	67.1	75.9	87.6	68.7	77.4	92.4	65.9	78.1
COVERS	DIVS. 10 - 14, 25, 28, 41, 43, 44, 46	100.0	70.5	93.4	100.0	70.3	93.4	100.0	70.3	93.4	100.0	70.0	93.3	100.0	70.3	93.4	100.0	84.8	96.6
21, 22, 23	FIRE SUPPRESSION, PLUMBING & HVAC	100.0	58.3	83.6	100.4	58.2	83.9	100.4	58.2	83.9	96.9	54.8	80.4	100.4	58.2	83.9	100.0	54.2	82.0
26, 27, 3370	ELECTRICAL, COMMUNICATIONS & UTIL.	98.5	62.8	80.9	96.1	62.8	79.7	98.2	59.8	79.3	98.2	59.9	79.3	98.2	59.8	79.3	101.1	65.3	83.4
MF2018	WEIGHTED AVERAGE	97.6	68.9	85.5	96.9	68.6	84.9	97.0	68.2	84.8	96.7	66.4	83.9	97.2	68.3	85.0	98.6	70.1	86.5

	DIVISION	SOUTH DAKOTA																	
		MITCHELL			MOBRIDGE			PIERRE			RAPID CITY			SIOUX FALLS			WATERTOWN		
		573			576			575			577			570 - 571			572		
		MAT.	INST.	TOTAL	MAT.	INST.	TOTAL	MAT.	INST.	TOTAL	MAT.	INST.	TOTAL	MAT.	INST.	TOTAL	MAT.	INST.	TOTAL
015433	**CONTRACTOR EQUIPMENT**		97.7	97.7		97.7	97.7		99.9	99.9		97.7	97.7		100.9	100.9		97.7	97.7
0241, 31 - 34	**SITE & INFRASTRUCTURE, DEMOLITION**	95.8	93.2	94.0	95.7	93.2	94.0	97.2	96.3	96.6	97.5	92.8	94.3	92.5	98.7	96.8	95.6	93.2	94.0
0310	Concrete Forming & Accessories	97.9	44.2	52.1	88.2	44.6	51.0	99.7	45.6	53.6	106.9	56.6	64.0	103.3	78.1	81.8	84.6	74.5	76.0
0320	Concrete Reinforcing	100.4	70.5	86.0	103.0	70.5	87.3	97.8	100.2	99.0	94.4	100.4	97.3	106.6	100.6	103.7	97.6	70.5	84.5
0330	Cast-in-Place Concrete	106.4	52.3	86.3	106.4	77.2	95.6	103.1	77.8	93.7	105.6	77.2	95.0	88.3	78.8	84.7	106.4	77.2	95.5
03	**CONCRETE**	96.5	53.1	77.4	96.2	62.0	81.2	91.8	67.4	81.1	95.9	72.3	85.5	88.6	82.7	86.0	95.3	75.5	86.6
04	**MASONRY**	97.8	75.5	84.1	106.5	72.8	85.8	106.5	72.6	85.7	106.0	74.0	86.3	91.1	75.6	81.5	132.0	73.7	96.2
05	**METALS**	92.9	81.6	89.6	92.9	82.0	89.7	96.3	89.5	94.3	95.5	91.2	94.2	95.3	92.6	94.5	92.9	81.9	89.6
06	**WOOD, PLASTICS & COMPOSITES**	96.6	33.7	62.0	84.4	33.4	56.4	104.2	34.0	65.6	102.2	48.0	72.3	101.6	76.5	87.8	80.5	74.2	77.0
07	**THERMAL & MOISTURE PROTECTION**	102.8	72.9	89.8	102.9	75.1	90.8	104.1	72.7	90.4	103.6	78.5	92.7	103.2	83.3	94.6	102.6	76.7	91.3
08	**OPENINGS**	96.3	39.6	83.1	98.9	39.0	84.9	99.8	58.4	90.2	101.0	66.1	92.9	104.2	81.8	99.0	96.3	61.5	88.2
0920	Plaster & Gypsum Board	109.0	32.2	57.3	102.2	32.0	54.9	108.1	32.5	57.2	109.8	46.9	67.5	103.6	76.1	85.1	100.5	73.9	82.6
0950, 0980	Ceilings & Acoustic Treatment	95.5	32.2	52.8	98.8	32.0	53.7	98.6	32.5	54.0	100.4	46.9	64.3	97.7	76.1	83.1	95.5	73.9	80.9
0960	Flooring	90.7	43.8	77.6	86.5	46.4	75.3	98.2	31.2	79.4	90.5	72.7	85.5	93.9	75.0	88.6	85.2	43.8	73.6
0970, 0990	Wall Finishes & Painting/Coating	89.8	36.6	58.3	89.8	37.5	58.8	97.1	95.6	96.2	89.8	95.6	93.2	100.3	95.6	97.5	89.8	33.4	56.4
09	**FINISHES**	91.4	41.1	64.2	89.9	41.6	63.7	98.4	45.2	69.6	92.5	61.8	75.9	96.4	79.1	87.1	88.6	64.7	75.7
COVERS	**DIVS. 10 - 14, 25, 28, 41, 43, 44, 46**	100.0	82.0	96.0	100.0	81.9	96.0	100.0	82.3	96.1	100.0	83.5	96.3	100.0	87.0	97.1	100.0	85.8	96.8
21, 22, 23	**FIRE SUPPRESSION, PLUMBING & HVAC**	96.5	49.0	77.8	96.5	69.0	85.7	100.0	76.8	90.9	100.0	76.8	90.9	99.9	70.2	88.2	96.5	52.1	79.0
26, 27, 3370	**ELECTRICAL, COMMUNICATIONS & UTIL.**	99.4	63.4	81.7	101.1	39.8	70.9	103.9	47.7	76.2	97.6	47.7	73.0	102.6	63.4	83.3	98.6	63.4	81.2
MF2018	**WEIGHTED AVERAGE**	96.3	60.9	81.3	96.9	62.9	82.5	99.0	68.5	86.1	98.3	72.0	87.2	97.7	78.4	89.5	97.3	69.0	85.4

	DIVISION	TENNESSEE																	
		CHATTANOOGA			COLUMBIA			COOKEVILLE			JACKSON			JOHNSON CITY			KNOXVILLE		
		373 - 374			384			385			383			376			377 - 379		
		MAT.	INST.	TOTAL	MAT.	INST.	TOTAL	MAT.	INST.	TOTAL	MAT.	INST.	TOTAL	MAT.	INST.	TOTAL	MAT.	INST.	TOTAL
015433	**CONTRACTOR EQUIPMENT**		104.7	104.7		99.5	99.5		99.5	99.5		105.5	105.5		98.8	98.8		98.8	98.8
0241, 31 - 34	**SITE & INFRASTRUCTURE, DEMOLITION**	106.6	93.8	97.8	91.5	84.6	86.7	97.3	81.6	86.5	100.9	94.0	96.2	113.6	81.1	91.2	92.6	83.8	86.5
0310	Concrete Forming & Accessories	95.8	57.3	63.0	80.9	62.4	65.1	81.1	31.9	39.2	87.5	41.0	47.8	82.7	57.5	61.2	94.3	61.7	66.5
0320	Concrete Reinforcing	101.6	65.1	83.9	91.4	65.0	78.6	91.4	64.6	78.4	91.3	67.2	79.7	102.2	61.9	82.7	101.6	61.8	82.3
0330	Cast-in-Place Concrete	97.5	64.1	85.1	92.8	63.8	82.0	105.2	56.7	87.2	102.7	66.0	89.0	78.4	58.0	70.8	91.6	64.0	81.3
03	**CONCRETE**	93.1	62.8	79.8	92.4	65.1	80.4	102.4	48.7	78.8	93.2	56.5	77.0	102.7	60.2	84.0	90.7	64.2	79.0
04	**MASONRY**	98.6	56.3	72.6	111.4	55.5	77.0	106.7	39.4	65.3	111.7	44.4	70.4	111.0	43.2	69.3	76.3	50.7	60.6
05	**METALS**	93.0	88.1	91.6	93.5	88.9	92.1	93.5	87.6	91.8	96.0	89.5	94.1	90.3	86.4	89.2	93.4	86.5	91.4
06	**WOOD, PLASTICS & COMPOSITES**	106.5	56.3	78.9	67.8	63.4	65.4	68.0	29.5	46.8	82.9	39.7	59.1	78.7	62.6	69.8	93.1	62.6	76.4
07	**THERMAL & MOISTURE PROTECTION**	99.2	61.7	82.9	95.6	62.9	81.4	96.1	52.6	77.2	98.0	55.5	79.5	95.3	55.6	78.0	92.8	60.8	78.9
08	**OPENINGS**	102.0	57.5	91.6	91.0	53.3	82.2	91.1	35.5	78.1	97.7	44.3	85.2	98.3	59.5	89.3	95.7	55.2	86.2
0920	Plaster & Gypsum Board	80.0	55.6	63.6	87.4	62.9	70.9	87.4	28.0	47.4	90.0	38.5	55.3	101.0	62.1	74.8	109.1	62.1	77.5
0950, 0980	Ceilings & Acoustic Treatment	97.9	55.6	69.4	75.8	62.9	67.1	75.8	28.0	43.6	83.9	38.5	53.3	93.5	62.1	72.3	94.3	62.1	72.6
0960	Flooring	98.3	55.2	86.2	81.7	53.9	73.9	81.7	50.0	72.8	81.0	54.2	73.5	93.3	43.9	79.4	98.4	49.7	84.7
0970, 0990	Wall Finishes & Painting/Coating	98.0	59.5	75.2	85.0	56.5	68.1	85.0	56.5	68.1	87.0	56.5	68.9	94.9	58.7	73.5	94.9	57.6	72.8
09	**FINISHES**	94.4	56.6	73.9	86.2	60.1	72.1	86.8	36.2	59.4	86.1	43.4	63.0	98.5	55.3	75.1	91.4	58.9	73.8
COVERS	**DIVS. 10 - 14, 25, 28, 41, 43, 44, 46**	100.0	67.8	92.8	100.0	68.0	92.9	100.0	35.9	85.7	100.0	62.0	91.5	100.0	75.8	94.6	100.0	79.3	95.4
21, 22, 23	**FIRE SUPPRESSION, PLUMBING & HVAC**	100.1	59.3	84.1	97.8	73.6	88.3	97.8	67.1	85.8	100.1	61.4	84.9	100.0	55.5	82.5	100.0	60.5	84.5
26, 27, 3370	**ELECTRICAL, COMMUNICATIONS & UTIL.**	100.8	82.9	92.0	94.0	50.6	72.6	95.5	61.3	78.6	100.5	65.7	83.3	91.5	40.8	66.5	97.0	54.9	76.2
MF2018	**WEIGHTED AVERAGE**	98.0	68.1	85.3	94.8	66.5	82.8	96.2	56.9	79.6	97.6	61.4	82.3	98.1	58.6	81.4	94.8	63.7	81.6

	DIVISION	TENNESSEE									TEXAS								
		MCKENZIE			MEMPHIS			NASHVILLE			ABILENE			AMARILLO			AUSTIN		
		382			375, 380 - 381			370 - 372			795 - 796			790 - 791			786 - 787		
		MAT.	INST.	TOTAL	MAT.	INST.	TOTAL	MAT.	INST.	TOTAL	MAT.	INST.	TOTAL	MAT.	INST.	TOTAL	MAT.	INST.	TOTAL
015433	**CONTRACTOR EQUIPMENT**		99.5	99.5		102.0	102.0		105.5	105.5		90.9	90.9		93.9	93.9		92.5	92.5
0241, 31 - 34	**SITE & INFRASTRUCTURE, DEMOLITION**	96.9	81.8	86.5	91.1	95.4	94.0	102.8	98.1	99.6	91.9	87.4	88.8	91.3	91.3	91.3	95.5	90.1	91.7
0310	Concrete Forming & Accessories	88.4	34.1	42.1	94.6	64.7	69.1	95.9	68.1	72.2	94.0	59.2	64.3	95.6	51.7	58.2	95.3	53.0	59.3
0320	Concrete Reinforcing	91.5	65.5	78.9	110.9	66.0	89.2	107.1	66.8	87.6	88.5	51.7	70.7	95.3	49.6	73.2	86.8	47.2	67.6
0330	Cast-in-Place Concrete	102.9	57.9	86.2	96.5	76.7	89.1	88.2	69.0	81.1	84.6	64.9	77.3	83.2	65.7	76.7	94.0	65.9	83.5
03	**CONCRETE**	101.0	50.2	78.7	96.5	70.2	84.9	92.9	69.4	82.6	84.5	60.7	74.1	87.1	57.2	74.0	89.4	57.4	75.4
04	**MASONRY**	110.5	45.1	70.3	92.6	56.3	70.3	87.7	55.9	68.2	99.9	60.6	75.8	97.6	60.2	74.6	91.7	60.7	72.7
05	**METALS**	93.5	88.0	91.9	87.2	82.5	85.8	101.6	85.1	96.7	107.1	70.4	96.3	101.9	68.7	92.1	100.7	67.4	90.9
06	**WOOD, PLASTICS & COMPOSITES**	76.6	31.7	51.9	98.9	65.4	80.4	100.1	70.1	83.6	99.2	60.8	78.1	102.6	51.0	74.2	92.9	52.6	70.7
07	**THERMAL & MOISTURE PROTECTION**	96.1	53.6	77.6	93.0	66.4	81.4	96.9	65.0	83.0	98.3	63.4	83.1	100.1	62.3	83.6	95.1	63.8	81.5
08	**OPENINGS**	91.1	37.2	78.5	99.5	61.3	90.6	101.0	67.5	93.2	101.8	56.7	91.2	103.2	50.8	91.0	101.1	49.6	89.1
0920	Plaster & Gypsum Board	90.9	30.3	50.1	90.9	64.5	73.1	95.9	69.3	78.0	88.2	60.1	69.3	100.7	50.0	66.6	88.9	51.6	63.8
0950, 0980	Ceilings & Acoustic Treatment	75.8	30.3	45.1	90.8	64.5	73.0	94.1	69.3	77.4	99.2	60.1	72.9	100.7	50.0	66.5	93.3	51.6	65.2
0960	Flooring	84.5	53.9	75.9	97.8	55.2	85.9	97.1	57.5	86.0	93.2	67.6	86.0	93.7	63.8	85.3	94.1	63.8	85.6
0970, 0990	Wall Finishes & Painting/Coating	85.0	44.2	60.8	93.2	59.3	73.1	100.7	68.6	81.7	92.8	51.7	68.5	91.0	51.7	67.7	95.7	44.2	65.2
09	**FINISHES**	88.0	37.5	60.7	95.9	61.9	77.5	97.8	66.1	80.6	86.1	60.1	72.0	92.6	53.5	71.4	91.1	53.6	70.8
COVERS	**DIVS. 10 - 14, 25, 28, 41, 43, 44, 46**	100.0	24.0	83.1	100.0	80.5	95.7	100.0	81.1	95.8	100.0	78.9	95.3	100.0	72.7	93.9	100.0	76.5	94.8
21, 22, 23	**FIRE SUPPRESSION, PLUMBING & HVAC**	97.8	58.6	82.4	100.1	70.3	88.4	99.9	75.1	90.2	100.1	52.1	81.3	99.9	51.1	80.7	99.9	58.7	83.8
26, 27, 3370	**ELECTRICAL, COMMUNICATIONS & UTIL.**	95.3	55.4	75.6	102.9	63.6	83.5	93.5	63.2	78.6	97.9	54.0	76.3	100.9	58.9	80.2	97.5	57.2	77.6
MF2018	**WEIGHTED AVERAGE**	96.3	55.1	78.9	96.6	69.8	85.3	98.0	71.8	86.9	97.8	61.4	82.4	98.2	60.0	82.1	97.2	61.4	82.1

	DIVISION	TEXAS																	
		BEAUMONT			BROWNWOOD			BRYAN			CHILDRESS			CORPUS CHRISTI			DALLAS		
		776 - 777			768			778			792			783 - 784			752 - 753		
		MAT.	INST.	TOTAL	MAT.	INST.	TOTAL	MAT.	INST.	TOTAL	MAT.	INST.	TOTAL	MAT.	INST.	TOTAL	MAT.	INST.	TOTAL
015433	CONTRACTOR EQUIPMENT		95.3	95.3		90.9	90.9		95.3	95.3		90.9	90.9		100.6	100.6		106.9	106.9
0241, 31 - 34	SITE & INFRASTRUCTURE, DEMOLITION	90.6	91.8	91.4	101.3	87.3	91.7	81.5	91.9	88.7	101.8	85.8	90.8	144.2	82.8	101.8	106.0	97.6	100.2
0310	Concrete Forming & Accessories	102.3	54.6	61.7	97.6	58.7	64.4	82.1	58.2	61.7	92.6	58.6	63.6	98.4	52.2	59.0	95.0	62.3	67.1
0320	Concrete Reinforcing	85.7	62.0	74.2	81.9	51.1	67.0	87.9	51.2	70.2	88.7	51.0	70.5	79.8	50.9	65.8	91.4	52.3	72.5
0330	Cast-in-Place Concrete	90.7	65.2	81.2	96.8	58.6	82.6	72.6	59.6	67.8	86.7	58.6	76.3	114.1	67.1	96.6	92.6	71.4	84.7
03	CONCRETE	91.0	60.7	77.7	90.3	58.2	76.2	75.8	58.5	68.2	93.4	58.2	77.9	95.2	59.1	79.4	93.3	65.3	81.0
04	MASONRY	102.9	61.7	77.6	122.8	60.6	84.6	143.3	60.6	92.5	103.5	60.2	76.9	81.4	60.7	68.7	99.5	60.2	75.3
05	METALS	99.4	76.2	92.6	102.6	69.8	92.9	99.5	72.1	91.4	104.4	69.6	94.2	97.3	84.4	93.5	101.7	82.6	96.1
06	WOOD, PLASTICS & COMPOSITES	113.6	54.1	80.9	105.8	60.8	81.0	78.6	59.3	68.0	98.6	60.8	77.8	115.6	51.5	80.3	100.8	63.6	80.3
07	THERMAL & MOISTURE PROTECTION	92.8	64.3	80.4	92.0	61.9	78.9	85.2	63.8	75.9	99.0	60.4	82.2	97.7	62.2	82.2	87.3	67.2	78.6
08	OPENINGS	95.0	55.0	85.6	100.4	56.6	90.2	95.7	56.0	86.4	96.1	56.6	86.9	101.3	50.0	89.4	96.5	59.5	87.9
0920	Plaster & Gypsum Board	101.0	53.2	68.9	82.6	60.1	67.5	88.2	58.7	68.3	87.6	60.1	69.1	96.5	50.5	65.5	92.7	62.5	72.4
0950, 0980	Ceilings & Acoustic Treatment	105.3	53.2	70.2	86.6	60.1	68.7	96.7	58.7	71.1	96.8	60.1	72.1	98.8	50.5	66.2	102.4	62.5	75.5
0960	Flooring	116.9	74.6	105.0	75.7	54.6	69.8	86.7	67.0	81.1	91.6	54.6	81.2	107.4	67.7	96.3	94.2	69.4	87.2
0970, 0990	Wall Finishes & Painting/Coating	94.9	53.9	70.6	92.8	51.7	68.4	92.2	57.4	71.6	92.8	51.7	68.5	110.1	44.2	71.1	96.1	54.8	71.6
09	FINISHES	98.0	57.8	76.2	77.6	57.4	66.7	85.4	59.4	71.3	86.2	57.4	70.6	101.9	53.8	75.9	97.9	62.7	78.8
COVERS	DIVS. 10 - 14, 25, 28, 41, 43, 44, 46	100.0	77.8	95.0	100.0	73.4	94.1	100.0	77.0	94.9	100.0	73.4	94.1	100.0	77.8	95.0	100.0	82.5	96.1
21, 22, 23	FIRE SUPPRESSION, PLUMBING & HVAC	100.0	63.0	85.5	96.5	49.5	78.0	96.5	62.4	83.1	96.6	51.7	79.0	100.0	48.7	79.9	99.9	62.3	85.1
26, 27, 3370	ELECTRICAL, COMMUNICATIONS & UTIL.	100.1	63.4	82.0	98.6	46.1	72.7	97.8	63.4	80.8	97.9	55.8	77.1	94.0	63.4	78.9	95.2	63.2	79.4
MF2018	WEIGHTED AVERAGE	97.9	65.5	84.2	97.4	58.7	81.1	95.2	64.9	82.4	97.6	60.3	81.8	98.9	61.4	83.1	98.2	68.0	85.4

	DIVISION	TEXAS																	
		DEL RIO			DENTON			EASTLAND			EL PASO			FORT WORTH			GALVESTON		
		788			762			764			798 - 799, 885			760 - 761			775		
		MAT.	INST.	TOTAL	MAT.	INST.	TOTAL	MAT.	INST.	TOTAL	MAT.	INST.	TOTAL	MAT.	INST.	TOTAL	MAT.	INST.	TOTAL
015433	CONTRACTOR EQUIPMENT		89.9	89.9		100.2	100.2		90.9	90.9		93.9	93.9		93.9	93.9		107.9	107.9
0241, 31 - 34	SITE & INFRASTRUCTURE, DEMOLITION	122.2	85.5	96.9	101.4	79.7	86.4	104.0	85.5	91.2	89.7	92.7	91.8	97.4	92.7	94.2	108.4	90.6	96.1
0310	Concrete Forming & Accessories	95.1	51.0	57.5	105.0	58.8	65.7	98.4	58.6	64.4	95.6	56.7	62.5	98.2	59.5	65.2	90.9	55.8	61.0
0320	Concrete Reinforcing	80.4	47.6	64.5	83.2	51.1	67.7	82.1	51.1	67.1	94.2	51.1	73.4	88.1	50.9	70.1	87.5	59.7	74.0
0330	Cast-in-Place Concrete	122.9	58.4	98.9	74.9	59.7	69.2	102.3	58.5	86.0	73.9	68.0	71.7	89.3	65.9	80.6	96.4	60.6	83.1
03	CONCRETE	115.8	54.0	88.6	70.6	59.7	65.8	94.8	58.1	78.7	82.7	60.5	72.9	86.1	61.0	75.1	92.1	60.3	78.1
04	MASONRY	95.4	60.5	74.0	131.0	59.3	86.9	92.7	60.6	73.0	91.6	62.7	73.8	92.5	59.3	72.1	99.0	60.7	75.5
05	METALS	97.2	67.2	88.4	102.1	84.9	97.1	102.4	69.6	92.7	102.1	69.4	92.5	104.3	69.5	94.1	101.2	91.4	98.3
06	WOOD, PLASTICS & COMPOSITES	96.1	50.4	71.0	117.8	60.9	86.5	113.0	60.8	84.3	90.2	57.1	72.0	99.6	60.9	78.3	95.3	55.9	73.6
07	THERMAL & MOISTURE PROTECTION	94.7	62.0	80.5	90.0	62.4	78.0	92.4	61.9	79.1	91.9	64.4	80.0	89.7	64.4	78.7	84.6	64.4	75.8
08	OPENINGS	96.6	47.8	85.2	118.7	56.2	104.2	70.6	56.6	67.4	96.3	52.8	86.2	101.2	56.6	90.8	99.7	56.0	89.5
0920	Plaster & Gypsum Board	93.3	49.5	63.8	87.3	60.1	69.0	82.6	60.1	67.5	97.3	56.2	69.6	89.8	60.1	69.8	94.8	55.0	68.0
0950, 0980	Ceilings & Acoustic Treatment	95.3	49.5	64.4	90.7	60.1	70.1	86.6	60.1	68.7	94.2	56.2	68.6	92.7	60.1	70.7	100.8	55.0	69.9
0960	Flooring	91.9	54.6	81.4	71.9	54.6	67.1	95.9	54.6	84.3	97.3	69.7	89.5	94.5	64.5	86.0	102.4	67.0	92.5
0970, 0990	Wall Finishes & Painting/Coating	97.8	42.6	65.1	103.3	48.9	71.1	94.2	51.7	69.0	98.6	49.6	69.6	101.2	51.8	72.0	104.6	53.7	74.5
09	FINISHES	94.5	50.3	70.6	76.3	57.2	66.0	84.1	57.4	69.6	92.3	58.1	73.8	89.6	59.5	73.3	95.4	56.9	74.6
COVERS	DIVS. 10 - 14, 25, 28, 41, 43, 44, 46	100.0	73.6	94.1	100.0	79.2	95.4	100.0	71.7	93.7	100.0	76.6	94.8	100.0	79.2	95.4	100.0	78.3	95.2
21, 22, 23	FIRE SUPPRESSION, PLUMBING & HVAC	96.5	59.4	81.9	96.5	55.8	80.5	96.5	49.5	78.0	99.8	65.3	86.3	99.9	55.4	82.4	96.5	62.5	83.2
26, 27, 3370	ELECTRICAL, COMMUNICATIONS & UTIL.	95.9	59.8	78.1	101.0	58.7	80.1	98.5	58.6	78.8	97.5	50.5	74.3	100.2	58.6	79.7	99.6	63.5	81.8
MF2018	WEIGHTED AVERAGE	99.7	60.4	83.1	97.3	62.9	82.7	94.2	60.3	79.8	95.9	63.6	82.3	97.6	62.9	82.9	97.7	66.5	84.5

	DIVISION	TEXAS																	
		GIDDINGS			GREENVILLE			HOUSTON			HUNTSVILLE			LAREDO			LONGVIEW		
		789			754			770 - 772			773			780			756		
		MAT.	INST.	TOTAL	MAT.	INST.	TOTAL	MAT.	INST.	TOTAL	MAT.	INST.	TOTAL	MAT.	INST.	TOTAL	MAT.	INST.	TOTAL
015433	CONTRACTOR EQUIPMENT		89.9	89.9		100.4	100.4		103.3	103.3		95.3	95.3		89.9	89.9		91.8	91.8
0241, 31 - 34	SITE & INFRASTRUCTURE, DEMOLITION	106.8	85.8	92.3	96.4	82.4	86.7	107.7	95.5	99.3	96.8	91.6	93.2	100.2	86.2	90.5	93.7	88.6	90.2
0310	Concrete Forming & Accessories	92.6	51.1	57.3	86.9	57.1	61.5	96.0	58.9	64.4	89.2	54.1	59.2	95.1	52.1	58.5	82.9	58.8	62.3
0320	Concrete Reinforcing	80.9	48.4	65.2	91.9	51.1	72.2	87.3	52.2	70.3	88.2	51.1	70.3	80.4	50.9	66.1	90.9	50.7	71.5
0330	Cast-in-Place Concrete	104.1	58.7	87.2	92.1	59.6	80.0	89.6	68.9	81.9	99.9	59.5	84.9	87.8	65.0	79.3	107.1	58.5	89.0
03	CONCRETE	92.6	54.3	75.8	86.7	58.7	74.4	91.4	62.5	78.7	98.7	56.6	80.2	87.3	57.4	74.2	102.7	58.1	83.1
04	MASONRY	102.2	60.6	76.7	158.6	59.3	97.5	98.0	63.1	76.6	141.3	60.6	91.7	88.9	60.6	71.5	154.2	59.2	95.8
05	METALS	96.7	68.2	88.4	99.2	83.2	94.5	103.9	77.3	96.1	99.4	71.8	91.3	99.6	69.4	90.7	91.8	68.4	84.9
06	WOOD, PLASTICS & COMPOSITES	95.3	50.4	70.6	87.2	58.4	71.4	100.6	59.2	77.8	88.1	54.1	69.4	96.1	51.4	71.5	81.2	60.8	70.0
07	THERMAL & MOISTURE PROTECTION	95.0	62.7	81.0	87.3	61.5	76.1	86.3	68.5	78.5	86.5	63.2	76.4	93.8	63.0	80.4	89.1	60.9	76.8
08	OPENINGS	95.8	48.7	84.8	92.2	55.3	83.6	106.6	56.4	94.9	95.7	52.5	85.6	96.7	49.8	85.8	83.4	56.2	77.0
0920	Plaster & Gypsum Board	92.4	49.5	63.5	82.8	57.5	65.8	94.0	58.0	69.8	92.3	53.2	66.0	94.1	50.5	64.7	81.7	60.1	67.2
0950, 0980	Ceilings & Acoustic Treatment	95.3	49.5	64.4	98.1	57.5	70.7	97.9	58.0	71.0	96.7	53.2	67.4	98.6	50.5	66.1	93.2	60.1	70.9
0960	Flooring	91.7	54.6	81.2	87.7	54.6	78.4	103.5	71.2	94.4	91.3	54.6	81.0	91.8	63.8	83.9	93.7	54.6	82.7
0970, 0990	Wall Finishes & Painting/Coating	97.8	44.2	66.1	90.4	51.7	67.5	102.1	58.6	76.3	92.2	57.4	71.6	97.8	44.2	66.1	82.1	48.9	62.4
09	FINISHES	93.1	50.5	70.0	92.3	56.1	72.7	102.7	60.8	80.1	88.3	54.1	69.8	93.4	52.9	71.5	98.1	57.2	76.0
COVERS	DIVS. 10 - 14, 25, 28, 41, 43, 44, 46	100.0	70.3	93.4	100.0	76.5	94.8	100.0	83.1	96.2	100.0	69.8	93.3	100.0	75.7	94.6	100.0	79.0	95.3
21, 22, 23	FIRE SUPPRESSION, PLUMBING & HVAC	96.5	61.4	82.7	96.5	58.0	81.3	100.0	65.3	86.3	96.5	62.4	83.1	100.0	58.9	83.8	96.4	57.5	81.1
26, 27, 3370	ELECTRICAL, COMMUNICATIONS & UTIL.	93.1	55.6	74.6	92.1	58.7	75.6	101.8	67.8	85.1	97.8	59.1	78.7	96.0	57.1	76.8	92.6	50.9	72.0
MF2018	WEIGHTED AVERAGE	96.1	60.4	81.0	97.4	62.9	82.8	100.4	68.2	86.8	98.7	62.8	83.6	96.4	61.1	81.5	97.5	61.0	82.1

		TEXAS																	
	DIVISION	LUBBOCK			LUFKIN			MCALLEN			MCKINNEY			MIDLAND			ODESSA		
		793 - 794			759			785			750			797			797		
		MAT.	INST.	TOTAL	MAT.	INST.	TOTAL	MAT.	INST.	TOTAL	MAT.	INST.	TOTAL	MAT.	INST.	TOTAL	MAT.	INST.	TOTAL
015433	CONTRACTOR EQUIPMENT		102.9	102.9		91.8	91.8		100.7	100.7		100.4	100.4		102.9	102.9		90.9	90.9
0241, 31 - 34	SITE & INFRASTRUCTURE, DEMOLITION	115.0	86.7	95.5	88.6	90.0	89.5	149.0	82.9	103.3	92.8	82.4	85.6	117.7	85.8	95.6	92.1	88.1	89.3
0310	Concrete Forming & Accessories	92.9	53.0	58.9	85.9	54.1	58.8	99.1	51.2	58.3	86.0	57.1	61.4	96.4	59.1	64.6	93.9	59.2	64.3
0320	Concrete Reinforcing	89.7	51.8	71.4	92.6	66.7	80.1	80.0	50.9	65.9	91.9	51.1	72.2	90.7	51.7	71.8	88.5	51.6	70.7
0330	Cast-in-Place Concrete	84.7	68.2	78.6	95.8	58.3	81.8	124.0	59.7	100.1	86.3	59.6	76.4	90.0	65.9	81.1	84.6	64.9	77.3
03	CONCRETE	83.1	60.1	73.0	95.1	58.6	79.0	102.8	56.2	82.3	82.1	58.8	71.9	87.1	62.1	76.1	84.5	60.7	74.1
04	MASONRY	99.2	61.0	75.7	117.3	60.5	82.4	95.1	60.7	74.0	170.8	59.3	102.3	116.2	60.2	81.8	99.9	60.2	75.5
05	METALS	110.9	85.8	103.5	99.4	73.3	91.7	97.1	84.2	93.3	99.2	83.3	94.5	109.2	85.4	102.2	106.4	70.4	95.8
06	WOOD, PLASTICS & COMPOSITES	98.6	52.1	73.0	88.6	54.2	69.6	114.3	50.5	79.2	86.1	58.4	70.9	103.0	60.9	79.8	99.2	60.8	78.1
07	THERMAL & MOISTURE PROTECTION	88.2	63.5	77.5	88.8	60.0	76.3	97.6	62.9	82.5	87.1	61.5	76.0	88.5	63.3	77.5	98.3	62.7	82.8
08	OPENINGS	108.6	52.4	95.5	63.3	56.6	61.7	100.4	48.6	88.3	92.2	55.3	83.6	110.0	56.7	97.6	101.8	56.7	91.2
0920	Plaster & Gypsum Board	88.4	51.1	63.3	80.6	53.2	62.2	97.5	49.5	65.2	82.8	57.5	65.8	90.0	60.1	69.9	88.2	60.1	69.3
0950, 0980	Ceilings & Acoustic Treatment	100.0	51.1	67.0	88.3	53.2	64.7	98.6	49.5	65.5	98.1	57.5	70.7	97.6	60.1	72.3	99.2	60.1	72.9
0960	Flooring	87.5	69.0	82.3	126.2	54.6	106.1	106.9	72.8	97.3	87.2	54.6	78.1	88.6	63.8	81.7	93.2	63.8	84.9
0970, 0990	Wall Finishes & Painting/Coating	103.4	53.7	74.0	82.1	51.7	64.1	110.1	42.6	70.2	90.4	51.7	67.5	103.4	51.7	72.8	92.8	51.7	68.5
09	FINISHES	88.5	55.5	70.6	106.5	53.9	78.0	102.4	54.1	76.3	91.9	56.1	72.5	89.0	59.4	72.9	86.1	59.3	71.6
COVERS	DIVS. 10 - 14, 25, 28, 41, 43, 44, 46	100.0	78.3	95.2	100.0	70.1	93.3	100.0	75.8	94.6	100.0	79.1	95.3	100.0	73.8	94.1	100.0	73.5	94.1
21, 22, 23	FIRE SUPPRESSION, PLUMBING & HVAC	99.7	53.1	81.4	96.4	58.1	81.4	96.5	48.7	77.7	96.5	60.2	82.2	96.1	47.4	77.0	100.1	52.5	81.4
26, 27, 3370	ELECTRICAL, COMMUNICATIONS & UTIL.	96.7	60.1	78.6	93.9	61.4	77.8	93.8	31.6	63.1	92.2	58.7	75.6	96.7	58.9	78.1	98.0	58.9	78.7
MF2018	WEIGHTED AVERAGE	99.2	62.9	83.9	94.7	62.6	81.1	99.7	56.5	81.4	97.2	63.5	82.9	99.7	62.2	83.8	97.7	61.8	82.6

		TEXAS																	
	DIVISION	PALESTINE			SAN ANGELO			SAN ANTONIO			TEMPLE			TEXARKANA			TYLER		
		758			769			781 - 782			765			755			757		
		MAT.	INST.	TOTAL	MAT.	INST.	TOTAL	MAT.	INST.	TOTAL	MAT.	INST.	TOTAL	MAT.	INST.	TOTAL	MAT.	INST.	TOTAL
015433	CONTRACTOR EQUIPMENT		91.8	91.8		90.9	90.9		94.1	94.1		90.9	90.9		91.8	91.8		91.8	91.8
0241, 31 - 34	SITE & INFRASTRUCTURE, DEMOLITION	94.2	88.8	90.5	97.8	87.7	90.8	101.5	93.7	96.1	86.7	87.6	87.3	83.4	91.1	88.7	92.9	88.7	90.0
0310	Concrete Forming & Accessories	77.5	58.8	61.5	97.9	51.4	58.3	94.7	52.6	58.8	101.2	51.2	58.5	93.0	59.4	64.3	87.6	58.9	63.1
0320	Concrete Reinforcing	90.2	51.0	71.3	81.8	51.6	67.2	89.6	47.7	69.3	82.0	50.8	66.9	90.1	50.8	71.1	90.9	50.8	71.5
0330	Cast-in-Place Concrete	87.5	58.5	76.7	91.4	64.9	81.5	85.7	68.3	79.2	74.9	58.6	68.8	88.2	64.9	79.5	105.2	58.6	87.8
03	CONCRETE	97.1	58.1	80.0	86.1	57.2	73.4	88.0	58.0	74.8	73.4	54.7	65.2	88.0	60.6	75.9	102.3	58.2	82.9
04	MASONRY	111.9	59.2	79.5	119.4	60.6	83.3	89.7	60.8	71.9	130.1	60.6	87.4	175.3	59.2	103.9	164.4	59.2	99.7
05	METALS	99.1	68.5	90.1	102.7	70.3	93.2	102.0	65.4	91.3	102.4	69.2	92.7	91.7	69.0	85.0	99.0	68.5	90.1
06	WOOD, PLASTICS & COMPOSITES	79.3	60.8	69.1	106.1	50.4	75.5	98.1	51.7	72.5	115.7	50.4	79.8	93.7	60.8	75.6	90.3	60.8	74.1
07	THERMAL & MOISTURE PROTECTION	89.4	61.8	77.4	91.8	62.6	79.1	86.8	65.6	77.5	91.3	61.7	78.4	88.6	63.0	77.5	89.2	61.8	77.3
08	OPENINGS	63.3	56.6	61.7	100.5	51.0	88.9	102.1	49.5	89.8	67.3	50.8	63.4	83.3	56.5	77.1	63.2	56.5	61.7
0920	Plaster & Gypsum Board	79.0	60.1	66.3	82.6	49.5	60.3	93.3	50.5	64.5	82.6	49.5	60.3	86.8	60.1	68.8	80.6	60.1	66.8
0950, 0980	Ceilings & Acoustic Treatment	88.3	60.1	69.3	86.6	49.5	61.6	95.2	50.5	65.1	86.6	49.5	61.6	93.2	60.1	70.9	88.3	60.1	69.3
0960	Flooring	116.4	54.6	99.0	75.8	54.6	69.9	101.2	67.7	91.8	97.5	54.6	85.4	102.7	63.8	91.7	128.7	54.6	107.8
0970, 0990	Wall Finishes & Painting/Coating	82.1	51.7	64.1	92.8	51.7	68.4	99.4	44.2	66.7	94.2	44.2	64.6	82.1	51.7	64.1	82.1	51.7	64.1
09	FINISHES	104.0	57.5	78.8	77.4	51.3	63.3	104.2	54.0	77.0	83.3	50.5	65.5	100.9	59.3	78.4	107.5	57.5	80.4
COVERS	DIVS. 10 - 14, 25, 28, 41, 43, 44, 46	100.0	73.5	94.1	100.0	76.0	94.7	100.0	78.0	95.1	100.0	72.3	93.8	100.0	79.1	95.3	100.0	79.0	95.3
21, 22, 23	FIRE SUPPRESSION, PLUMBING & HVAC	96.4	56.7	80.8	96.5	52.2	79.1	100.0	59.8	84.2	96.5	53.2	79.5	96.4	55.5	80.3	96.4	58.2	81.4
26, 27, 3370	ELECTRICAL, COMMUNICATIONS & UTIL.	90.1	48.4	69.5	102.8	54.6	79.1	97.5	59.8	78.9	99.8	52.0	76.2	93.7	54.5	74.4	92.6	53.7	73.4
MF2018	WEIGHTED AVERAGE	94.2	60.4	79.9	97.1	59.4	81.2	98.3	62.3	83.1	92.5	58.5	78.2	96.7	62.0	82.1	97.8	61.6	82.5

		TEXAS															UTAH		
	DIVISION	VICTORIA			WACO			WAXAHACHIE			WHARTON			WICHITA FALLS			LOGAN		
		779			766 - 767			751			774			763			843		
		MAT.	INST.	TOTAL	MAT.	INST.	TOTAL	MAT.	INST.	TOTAL	MAT.	INST.	TOTAL	MAT.	INST.	TOTAL	MAT.	INST.	TOTAL
015433	CONTRACTOR EQUIPMENT		106.6	106.6		90.9	90.9		100.4	100.4		107.9	107.9		90.9	90.9		93.9	93.9
0241, 31 - 34	SITE & INFRASTRUCTURE, DEMOLITION	113.4	88.0	95.8	95.1	88.0	90.2	94.7	82.5	86.3	118.7	90.0	98.9	95.8	88.1	90.5	99.8	89.6	92.7
0310	Concrete Forming & Accessories	90.6	51.3	57.1	99.8	59.2	65.2	86.0	59.0	63.0	85.7	53.1	57.9	99.8	59.2	65.2	103.5	67.2	72.6
0320	Concrete Reinforcing	84.0	50.9	68.0	81.6	47.2	65.0	91.9	51.1	72.2	87.4	50.9	69.7	81.6	51.7	67.1	108.0	84.4	96.6
0330	Cast-in-Place Concrete	108.3	60.5	90.5	81.0	64.9	75.0	91.1	59.8	79.5	111.3	59.5	92.0	86.5	65.0	78.5	86.4	74.3	81.9
03	CONCRETE	99.7	56.7	80.8	79.6	60.0	71.0	85.8	59.7	74.3	103.8	57.1	83.3	82.2	60.8	72.8	105.4	73.2	91.2
04	MASONRY	116.9	60.7	82.4	90.9	60.6	72.3	159.2	59.3	97.8	100.3	60.6	75.9	91.4	60.1	72.2	109.5	60.5	79.4
05	METALS	99.4	87.8	96.0	104.8	68.4	94.1	99.2	83.6	94.6	101.2	87.4	97.1	104.7	70.4	94.6	109.5	82.5	101.5
06	WOOD, PLASTICS & COMPOSITES	98.4	50.5	72.1	113.9	60.8	84.7	86.1	61.0	72.3	88.4	52.7	68.8	113.9	60.8	84.7	85.9	67.1	75.5
07	THERMAL & MOISTURE PROTECTION	88.0	61.1	76.3	92.2	63.7	79.8	87.2	61.6	76.1	85.0	63.7	75.7	92.2	62.9	79.5	103.0	68.8	88.1
08	OPENINGS	99.5	51.1	88.2	78.4	55.6	73.1	92.2	56.7	83.9	99.7	52.3	88.6	78.4	56.7	73.4	92.2	66.1	86.1
0920	Plaster & Gypsum Board	90.9	49.5	63.0	82.7	60.1	67.5	83.2	60.1	67.7	90.4	51.7	64.4	82.7	60.1	67.5	80.6	66.2	70.9
0950, 0980	Ceilings & Acoustic Treatment	101.6	49.5	66.5	87.4	60.1	69.0	99.7	60.1	73.0	100.8	51.7	67.7	87.4	60.1	69.0	104.3	66.2	78.7
0960	Flooring	100.7	54.6	87.7	96.8	63.8	87.5	87.2	54.6	78.1	99.5	66.6	90.2	97.4	71.9	90.2	97.2	60.5	86.9
0970, 0990	Wall Finishes & Painting/Coating	104.8	57.4	76.8	94.2	51.7	69.0	90.4	51.7	67.5	104.6	55.7	75.7	96.1	51.7	69.8	92.6	60.7	73.7
09	FINISHES	92.4	52.0	70.6	83.7	59.3	70.5	92.5	57.6	73.6	94.7	55.2	73.3	84.1	60.9	71.5	93.9	64.9	78.2
COVERS	DIVS. 10 - 14, 25, 28, 41, 43, 44, 46	100.0	69.7	93.2	100.0	78.9	95.3	100.0	79.4	95.4	100.0	70.0	93.3	100.0	73.5	94.1	100.0	84.4	96.5
21, 22, 23	FIRE SUPPRESSION, PLUMBING & HVAC	96.5	62.3	83.1	100.0	58.7	83.8	96.5	58.0	81.4	96.5	61.9	82.9	100.0	52.1	81.2	100.0	69.2	87.9
26, 27, 3370	ELECTRICAL, COMMUNICATIONS & UTIL.	104.9	54.0	79.8	103.0	54.1	78.9	92.1	58.7	75.6	103.9	59.2	81.9	105.0	56.0	80.8	95.2	71.3	83.4
MF2018	WEIGHTED AVERAGE	99.7	62.9	84.2	94.4	62.4	80.9	97.2	63.5	83.0	99.8	64.3	84.8	95.0	61.6	80.9	100.9	71.7	88.5

City Cost Indexes

	DIVISION	UTAH												VERMONT					
		OGDEN			PRICE			PROVO			SALT LAKE CITY			BELLOWS FALLS			BENNINGTON		
		842, 844			845			846 - 847			840 - 841			051			052		
		MAT.	INST.	TOTAL	MAT.	INST.	TOTAL	MAT.	INST.	TOTAL	MAT.	INST.	TOTAL	MAT.	INST.	TOTAL	MAT.	INST.	TOTAL
015433	CONTRACTOR EQUIPMENT		93.9	93.9		92.9	92.9		92.9	92.9		93.8	93.8		95.7	95.7		95.7	95.7
0241, 31 - 34	SITE & INFRASTRUCTURE, DEMOLITION	87.3	89.6	88.9	97.1	87.4	90.4	95.9	88.0	90.5	87.0	89.5	88.7	86.8	96.1	93.3	86.2	96.1	93.1
0310	Concrete Forming & Accessories	103.5	67.2	72.6	106.0	64.2	70.4	105.3	67.2	72.9	105.8	67.2	72.9	96.5	101.7	100.9	94.0	101.6	100.4
0320	Concrete Reinforcing	107.6	84.4	96.4	115.8	84.0	100.4	116.8	84.4	101.1	110.0	84.4	97.6	85.1	86.3	85.7	85.1	86.3	85.6
0330	Cast-in-Place Concrete	87.7	74.3	82.7	86.5	71.6	80.9	86.5	74.3	81.9	95.8	74.3	87.8	89.6	115.9	99.4	89.6	115.8	99.4
03	CONCRETE	94.4	73.2	85.1	106.6	70.8	90.8	104.9	73.2	91.0	114.2	73.2	96.2	88.7	103.7	95.3	88.5	103.6	95.2
04	MASONRY	104.0	60.5	77.3	115.1	65.1	84.4	115.3	60.5	81.6	118.1	60.5	82.7	96.3	86.2	90.1	105.1	86.2	93.5
05	METALS	110.0	82.5	101.9	106.5	82.5	99.5	107.5	82.5	100.1	113.7	82.5	104.5	95.2	91.1	94.0	95.1	90.9	93.9
06	WOOD, PLASTICS & COMPOSITES	85.9	67.1	75.5	89.2	64.0	75.3	87.5	67.1	76.3	87.8	67.1	76.4	102.8	106.8	105.0	100.1	106.8	103.8
07	THERMAL & MOISTURE PROTECTION	101.8	68.8	87.4	104.8	68.3	88.9	104.9	68.8	89.2	110.0	68.8	92.1	100.9	89.1	95.8	100.9	89.1	95.8
08	OPENINGS	92.2	66.1	86.1	96.1	76.3	91.5	96.1	66.1	89.1	94.1	66.1	87.5	100.1	97.5	99.5	100.1	97.5	99.5
0920	Plaster & Gypsum Board	80.6	66.2	70.9	83.4	63.1	69.8	81.2	66.2	71.1	90.9	66.2	74.3	108.9	106.7	107.4	107.9	106.7	107.1
0950, 0980	Ceilings & Acoustic Treatment	104.3	66.2	78.7	104.3	63.1	76.5	104.3	66.2	78.7	97.1	66.2	76.3	97.2	106.7	103.6	97.2	106.7	103.6
0960	Flooring	95.2	60.5	85.4	98.4	56.1	86.5	98.2	60.5	87.6	99.0	60.5	88.1	92.0	98.9	94.0	91.2	98.9	93.4
0970, 0990	Wall Finishes & Painting/Coating	92.6	60.7	73.7	92.6	60.7	73.7	92.6	60.7	73.7	95.7	60.7	75.0	86.4	101.5	95.3	86.4	101.5	95.3
09	FINISHES	92.0	64.9	77.3	95.0	61.7	77.0	94.5	64.9	78.5	93.7	64.9	78.1	90.6	101.8	96.7	90.3	101.8	96.5
COVERS	DIVS. 10 - 14, 25, 28, 41, 43, 44, 46	100.0	84.4	96.5	100.0	81.6	95.9	100.0	84.4	96.5	100.0	84.4	96.5	100.0	91.8	98.2	100.0	91.7	98.2
21, 22, 23	FIRE SUPPRESSION, PLUMBING & HVAC	100.0	69.2	87.9	98.1	63.2	84.4	100.0	69.2	87.9	100.1	69.2	88.0	96.7	87.6	93.1	96.7	87.6	93.1
26, 27, 3370	ELECTRICAL, COMMUNICATIONS & UTIL.	95.5	71.3	83.6	100.2	68.8	84.7	95.7	68.6	82.3	98.1	71.3	84.9	108.3	79.5	94.1	108.3	54.5	81.8
MF2018	WEIGHTED AVERAGE	98.8	71.7	87.4	101.4	70.0	88.1	101.2	71.2	88.5	103.5	71.7	90.1	96.7	92.1	94.8	97.0	88.6	93.4

	DIVISION	VERMONT																	
		BRATTLEBORO			BURLINGTON			GUILDHALL			MONTPELIER			RUTLAND			ST. JOHNSBURY		
		053			054			059			056			057			058		
		MAT.	INST.	TOTAL	MAT.	INST.	TOTAL	MAT.	INST.	TOTAL	MAT.	INST.	TOTAL	MAT.	INST.	TOTAL	MAT.	INST.	TOTAL
015433	CONTRACTOR EQUIPMENT		95.7	95.7		99.0	99.0		95.7	95.7		99.0	99.0		95.7	95.7		95.7	95.7
0241, 31 - 34	SITE & INFRASTRUCTURE, DEMOLITION	87.8	96.1	93.5	92.6	101.0	98.4	85.8	91.9	90.0	89.7	100.8	97.3	90.4	96.1	94.3	85.9	95.2	92.3
0310	Concrete Forming & Accessories	96.7	101.7	101.0	99.4	80.2	83.0	94.6	95.1	95.1	97.3	101.3	100.7	97.0	80.0	82.5	93.5	95.6	95.2
0320	Concrete Reinforcing	84.1	86.3	85.2	107.3	86.2	97.1	85.8	86.1	86.0	101.6	86.2	94.2	106.4	86.1	96.6	84.1	86.2	85.1
0330	Cast-in-Place Concrete	92.5	115.9	101.2	109.6	116.7	112.2	86.9	107.2	94.4	104.5	116.7	109.0	87.8	115.6	98.2	86.9	107.3	94.5
03	CONCRETE	90.7	103.7	96.4	101.7	94.2	98.4	86.2	97.7	91.3	98.5	103.7	100.8	91.4	93.8	92.5	85.9	97.9	91.2
04	MASONRY	104.6	86.2	93.3	103.9	86.4	93.1	104.8	71.5	84.4	101.4	86.4	92.2	87.2	86.3	86.7	131.1	71.5	94.5
05	METALS	95.1	91.1	93.9	102.8	89.8	99.0	95.2	90.0	93.7	101.2	89.9	97.9	101.1	90.3	97.9	95.2	90.4	93.8
06	WOOD, PLASTICS & COMPOSITES	103.1	106.8	105.1	99.2	78.5	87.8	99.0	106.8	103.3	96.0	106.9	102.0	103.2	78.3	89.5	94.1	106.8	101.0
07	THERMAL & MOISTURE PROTECTION	101.0	89.1	95.8	107.4	86.6	98.3	100.7	82.4	92.8	109.4	89.7	100.8	101.1	86.1	94.6	100.6	82.4	92.7
08	OPENINGS	100.1	97.8	99.6	102.6	78.4	96.9	100.1	94.0	98.7	103.1	94.1	101.0	103.2	78.3	97.4	100.1	94.0	98.7
0920	Plaster & Gypsum Board	108.9	106.7	107.4	109.9	77.4	88.1	114.5	106.7	109.2	110.4	106.7	107.9	109.4	77.4	87.9	115.8	106.7	109.6
0950, 0980	Ceilings & Acoustic Treatment	97.2	106.7	103.6	103.9	77.4	86.0	97.2	106.7	103.6	104.9	106.7	106.1	103.2	77.4	85.8	97.2	106.7	103.6
0960	Flooring	92.2	98.9	94.1	101.8	98.9	101.0	95.0	98.9	96.1	98.5	98.9	98.6	92.0	98.9	94.0	98.3	98.9	98.5
0970, 0990	Wall Finishes & Painting/Coating	86.4	101.5	95.3	98.0	88.3	92.3	86.4	88.3	87.5	97.8	88.3	92.2	86.4	88.3	87.5	86.4	88.3	87.5
09	FINISHES	90.8	101.8	96.8	99.1	83.8	90.8	92.2	96.7	94.6	97.9	100.6	99.4	92.0	83.7	87.5	93.3	96.7	95.2
COVERS	DIVS. 10 - 14, 25, 28, 41, 43, 44, 46	100.0	91.8	98.2	100.0	89.1	97.6	100.0	86.7	97.0	100.0	92.2	98.3	100.0	88.6	97.5	100.0	86.8	97.0
21, 22, 23	FIRE SUPPRESSION, PLUMBING & HVAC	96.7	87.6	93.1	100.0	68.3	87.6	96.7	60.5	82.5	96.5	68.3	85.4	100.2	68.2	87.7	96.7	60.5	82.5
26, 27, 3370	ELECTRICAL, COMMUNICATIONS & UTIL.	108.3	79.5	94.1	107.6	53.6	81.0	108.3	54.5	81.8	108.1	53.6	81.3	108.4	53.6	81.4	108.3	54.5	81.8
MF2018	WEIGHTED AVERAGE	97.4	92.1	95.2	101.9	79.8	92.5	96.8	79.0	89.3	100.2	84.4	93.5	98.9	79.4	90.7	98.1	79.3	90.2

	DIVISION	VERMONT			VIRGINIA														
		WHITE RIVER JCT.			ALEXANDRIA			ARLINGTON			BRISTOL			CHARLOTTESVILLE			CULPEPER		
		050			223			222			242			229			227		
		MAT.	INST.	TOTAL	MAT.	INST.	TOTAL	MAT.	INST.	TOTAL	MAT.	INST.	TOTAL	MAT.	INST.	TOTAL	MAT.	INST.	TOTAL
015433	CONTRACTOR EQUIPMENT		95.7	95.7		105.3	105.3		104.1	104.1		104.1	104.1		108.1	108.1		104.1	104.1
0241, 31 - 34	SITE & INFRASTRUCTURE, DEMOLITION	90.3	95.2	93.7	113.7	89.0	96.7	123.9	87.0	98.5	108.2	86.0	92.8	112.7	88.3	95.9	111.4	86.5	94.2
0310	Concrete Forming & Accessories	91.4	96.1	95.4	91.5	71.9	74.8	90.5	71.6	74.4	86.6	64.1	67.5	85.1	61.7	65.2	82.4	71.0	72.7
0320	Concrete Reinforcing	85.1	86.2	85.6	86.6	85.0	85.9	97.8	85.1	91.6	97.7	83.7	91.0	97.1	73.0	85.4	97.8	84.9	91.6
0330	Cast-in-Place Concrete	92.5	108.1	98.3	108.1	77.7	96.8	105.2	77.2	94.8	104.8	44.5	82.3	109.0	75.5	96.5	107.7	74.4	95.3
03	CONCRETE	92.6	98.4	95.2	99.9	77.7	90.1	104.5	77.4	92.6	100.8	62.4	83.9	101.3	70.2	87.7	98.4	76.0	88.6
04	MASONRY	117.2	72.9	90.0	89.8	73.7	79.9	102.3	72.6	84.0	92.3	45.4	63.5	116.9	55.3	79.0	105.1	70.8	84.1
05	METALS	95.2	90.4	93.8	106.0	100.8	104.5	104.6	101.8	103.8	103.4	97.8	101.8	103.7	95.8	101.4	103.7	99.6	102.5
06	WOOD, PLASTICS & COMPOSITES	97.1	106.8	102.4	91.0	70.1	79.5	87.5	70.1	78.0	80.2	69.8	74.5	78.7	59.8	68.3	77.4	70.1	73.4
07	THERMAL & MOISTURE PROTECTION	101.1	83.0	93.2	102.5	81.1	93.2	104.6	80.5	94.1	103.9	58.8	84.3	103.5	69.7	88.8	103.7	78.9	92.9
08	OPENINGS	100.1	94.0	98.7	95.4	74.3	90.5	93.7	74.3	89.2	96.4	67.9	89.7	94.8	65.7	88.0	95.1	73.8	90.1
0920	Plaster & Gypsum Board	106.6	106.7	106.7	98.0	69.2	78.6	94.8	69.2	77.6	91.1	68.9	76.2	91.1	57.9	68.8	91.3	69.2	76.4
0950, 0980	Ceilings & Acoustic Treatment	97.2	106.7	103.6	92.4	69.2	76.8	90.8	69.2	76.2	90.0	68.9	75.8	90.0	57.9	68.4	90.8	69.2	76.2
0960	Flooring	90.2	98.9	92.6	97.1	77.4	91.5	95.6	75.7	90.0	92.3	55.7	82.0	90.8	55.7	80.9	90.8	74.4	86.2
0970, 0990	Wall Finishes & Painting/Coating	86.4	88.3	87.5	115.8	76.0	92.3	115.8	73.1	90.6	102.1	55.3	74.4	102.1	58.6	76.3	115.8	73.1	90.6
09	FINISHES	90.1	97.1	93.9	93.8	72.6	82.3	93.8	71.6	81.8	90.6	62.3	75.3	90.1	59.4	73.5	90.8	71.2	80.2
COVERS	DIVS. 10 - 14, 25, 28, 41, 43, 44, 46	100.0	87.2	97.1	100.0	86.6	97.0	100.0	84.4	96.5	100.0	75.1	94.5	100.0	83.2	96.2	100.0	83.6	96.3
21, 22, 23	FIRE SUPPRESSION, PLUMBING & HVAC	96.7	61.2	82.8	100.4	86.2	94.8	100.4	85.6	94.6	96.8	46.0	76.9	96.8	69.1	86.0	96.8	85.3	92.3
26, 27, 3370	ELECTRICAL, COMMUNICATIONS & UTIL.	108.3	54.3	81.7	96.6	100.2	98.4	94.4	100.2	97.2	96.2	36.3	66.7	96.2	70.4	83.5	98.5	96.9	97.7
MF2018	WEIGHTED AVERAGE	98.2	79.8	90.4	99.6	84.8	93.3	100.4	84.2	93.6	98.2	59.3	81.8	99.4	71.0	87.4	98.7	83.0	92.1

	DIVISION	VIRGINIA																	
		FAIRFAX			FARMVILLE			FREDERICKSBURG			GRUNDY			HARRISONBURG			LYNCHBURG		
		220 - 221			239			224 - 225			246			228			245		
		MAT.	INST.	TOTAL	MAT.	INST.	TOTAL	MAT.	INST.	TOTAL	MAT.	INST.	TOTAL	MAT.	INST.	TOTAL	MAT.	INST.	TOTAL
015433	**CONTRACTOR EQUIPMENT**		104.1	104.1		108.1	108.1		104.1	104.1		104.1	104.1		104.1	104.1		104.1	104.1
0241, 31 - 34	**SITE & INFRASTRUCTURE, DEMOLITION**	122.4	86.7	97.8	108.5	87.3	93.9	110.9	86.4	93.9	106.0	85.3	91.7	119.3	85.0	95.6	106.8	86.7	92.9
0310	Concrete Forming & Accessories	85.1	63.8	67.0	98.9	60.3	66.0	85.1	65.4	68.3	89.6	38.2	45.8	81.4	56.6	60.2	86.6	70.5	72.9
0320	Concrete Reinforcing	97.8	90.3	94.2	92.9	67.9	80.8	98.5	80.1	89.6	96.4	44.1	71.1	97.8	57.3	78.2	97.1	84.5	91.0
0330	Cast-in-Place Concrete	105.2	78.5	95.2	106.8	83.9	98.3	106.8	74.2	94.7	104.8	50.9	84.7	105.2	52.6	85.6	104.8	74.3	93.4
03	**CONCRETE**	104.2	75.2	91.5	101.8	71.5	88.5	98.1	72.7	86.9	99.3	45.8	75.8	102.0	57.1	82.3	99.2	75.8	88.9
04	**MASONRY**	102.2	73.6	84.6	101.8	50.0	69.9	104.0	66.7	81.1	94.9	55.4	70.6	101.4	57.9	74.6	109.3	63.9	81.4
05	**METALS**	103.8	102.5	103.4	101.3	90.8	98.2	103.8	98.0	102.1	103.4	75.2	95.1	103.7	87.2	98.9	103.6	99.8	102.5
06	**WOOD, PLASTICS & COMPOSITES**	80.2	59.2	68.6	93.0	59.8	74.7	80.2	64.6	71.6	83.1	30.6	54.2	76.3	59.2	66.9	80.2	71.5	75.4
07	**THERMAL & MOISTURE PROTECTION**	104.4	79.9	93.7	105.2	69.4	89.6	103.7	76.3	91.8	103.9	57.3	83.6	104.1	69.6	89.1	103.7	70.4	89.2
08	**OPENINGS**	93.7	69.6	88.0	94.3	54.9	85.1	94.8	68.5	88.7	96.4	30.1	80.9	95.0	57.0	86.2	95.0	68.8	88.9
0920	Plaster & Gypsum Board	91.3	57.9	68.9	103.1	57.9	72.7	91.3	63.6	72.7	91.1	28.5	49.0	91.1	57.9	68.8	91.1	70.7	77.4
0950, 0980	Ceilings & Acoustic Treatment	90.8	57.9	68.6	86.5	57.9	67.2	90.8	63.6	72.4	90.0	28.5	48.6	90.0	57.9	68.4	90.0	70.7	77.0
0960	Flooring	92.5	76.7	88.1	95.6	50.5	82.9	92.5	72.4	86.9	93.5	28.2	75.2	90.6	74.4	86.1	92.3	65.4	84.7
0970, 0990	Wall Finishes & Painting/Coating	115.8	73.6	90.8	99.6	57.2	74.5	115.8	57.0	81.0	102.1	31.2	60.2	115.8	48.3	75.9	102.1	55.3	74.4
09	**FINISHES**	92.4	65.7	77.9	90.6	57.7	72.8	91.3	65.0	77.1	90.7	33.9	60.0	91.3	58.9	73.7	90.4	67.8	78.2
COVERS	**DIVS. 10 - 14, 25, 28, 41, 43, 44, 46**	100.0	84.2	96.5	100.0	77.1	94.9	100.0	78.3	95.2	100.0	73.6	94.1	100.0	62.8	91.7	100.0	78.8	95.3
21, 22, 23	**FIRE SUPPRESSION, PLUMBING & HVAC**	96.8	88.1	93.4	96.9	52.0	79.3	96.8	79.9	90.2	96.8	65.7	84.6	96.8	65.4	84.5	96.8	69.5	86.1
26, 27, 3370	**ELECTRICAL, COMMUNICATIONS & UTIL.**	97.3	101.1	99.1	91.2	68.3	79.9	94.5	91.8	93.2	96.2	68.3	82.5	96.4	86.4	91.5	97.2	67.8	82.7
MF2018	**WEIGHTED AVERAGE**	99.5	83.6	92.8	97.9	65.3	84.1	98.3	78.7	90.0	98.1	58.8	81.5	99.0	68.8	86.3	98.7	73.8	88.2

	DIVISION	VIRGINIA																	
		NEWPORT NEWS			NORFOLK			PETERSBURG			PORTSMOUTH			PULASKI			RICHMOND		
		236			233 - 235			238			237			243			230 - 232		
		MAT.	INST.	TOTAL	MAT.	INST.	TOTAL	MAT.	INST.	TOTAL	MAT.	INST.	TOTAL	MAT.	INST.	TOTAL	MAT.	INST.	TOTAL
015433	**CONTRACTOR EQUIPMENT**		108.2	108.2		110.2	110.2		108.1	108.1		108.0	108.0		104.1	104.1		109.6	109.6
0241, 31 - 34	**SITE & INFRASTRUCTURE, DEMOLITION**	107.6	88.4	94.3	106.9	93.8	97.9	111.1	88.4	95.4	106.2	88.2	93.8	105.4	85.7	91.8	103.4	92.8	96.1
0310	Concrete Forming & Accessories	98.1	61.1	66.6	103.6	61.2	67.5	91.8	61.8	66.2	87.8	61.3	65.2	89.6	40.4	47.7	94.2	61.9	66.7
0320	Concrete Reinforcing	92.8	66.2	79.9	101.2	66.2	84.3	92.4	73.6	83.3	92.4	66.2	79.7	96.4	83.8	90.3	102.1	73.6	88.3
0330	Cast-in-Place Concrete	103.8	64.5	89.2	111.0	76.1	98.0	110.2	76.6	97.7	102.8	65.6	89.0	104.8	82.4	96.5	95.3	76.6	88.4
03	**CONCRETE**	98.9	65.0	84.0	102.1	69.0	87.6	104.5	70.8	89.7	97.7	65.5	83.6	99.3	64.7	84.1	94.5	70.7	84.0
04	**MASONRY**	97.3	55.0	71.3	97.1	55.0	71.3	110.2	64.6	82.2	103.0	55.6	73.9	87.9	54.3	67.3	93.3	64.7	75.7
05	**METALS**	103.8	93.1	100.6	105.3	91.8	101.3	101.4	96.7	100.0	102.7	93.1	99.9	103.5	96.0	101.3	105.9	95.3	102.8
06	**WOOD, PLASTICS & COMPOSITES**	91.8	59.8	74.2	99.3	59.7	77.5	83.2	59.5	70.2	79.6	59.8	68.7	83.1	32.0	55.0	93.7	59.7	75.0
07	**THERMAL & MOISTURE PROTECTION**	105.1	67.7	88.9	104.4	70.0	89.5	105.1	72.3	90.9	105.1	68.0	89.0	103.9	66.6	87.6	101.0	72.9	88.8
08	**OPENINGS**	94.6	60.1	86.6	95.2	60.1	87.0	94.0	65.6	87.4	94.7	56.0	85.7	96.4	42.9	83.9	101.7	65.7	93.3
0920	Plaster & Gypsum Board	104.3	57.9	73.1	98.5	57.9	71.2	97.8	57.7	70.8	98.6	57.9	71.2	91.1	30.0	50.0	99.5	57.9	71.5
0950, 0980	Ceilings & Acoustic Treatment	91.4	57.9	68.8	88.8	57.9	68.0	88.2	57.7	67.6	91.4	57.9	68.8	90.0	30.0	49.6	89.5	57.9	68.2
0960	Flooring	95.6	51.1	83.1	96.2	51.1	83.5	91.7	71.7	86.1	88.9	51.7	78.4	93.5	55.7	82.9	94.0	71.7	87.8
0970, 0990	Wall Finishes & Painting/Coating	99.6	59.7	76.0	101.2	59.7	76.6	99.6	58.6	75.3	99.6	59.7	76.0	102.1	48.3	70.3	95.4	58.6	73.6
09	**FINISHES**	91.5	58.4	73.6	90.5	58.3	73.1	89.3	62.5	74.8	88.8	58.6	72.5	90.7	40.9	63.8	91.9	62.5	76.0
COVERS	**DIVS. 10 - 14, 25, 28, 41, 43, 44, 46**	100.0	83.4	96.3	100.0	83.2	96.3	100.0	83.1	96.2	100.0	83.6	96.3	100.0	73.2	94.0	100.0	83.0	96.2
21, 22, 23	**FIRE SUPPRESSION, PLUMBING & HVAC**	100.4	66.8	87.2	100.1	66.8	87.0	96.9	69.1	86.0	100.4	67.1	87.4	96.8	66.0	84.7	100.0	69.1	87.9
26, 27, 3370	**ELECTRICAL, COMMUNICATIONS & UTIL.**	93.6	64.2	79.1	98.0	60.7	79.6	93.8	70.5	82.3	92.1	60.7	76.7	96.2	83.4	89.9	96.8	70.4	83.8
MF2018	**WEIGHTED AVERAGE**	98.8	68.3	85.9	99.9	68.7	86.7	98.7	72.6	87.7	98.3	67.8	85.4	97.8	67.1	84.8	99.2	72.8	88.1

	DIVISION	VIRGINIA									WASHINGTON								
		ROANOKE			STAUNTON			WINCHESTER			CLARKSTON			EVERETT			OLYMPIA		
		240 - 241			244			226			994			982			985		
		MAT.	INST.	TOTAL	MAT.	INST.	TOTAL	MAT.	INST.	TOTAL	MAT.	INST.	TOTAL	MAT.	INST.	TOTAL	MAT.	INST.	TOTAL
015433	**CONTRACTOR EQUIPMENT**		104.1	104.1		108.1	108.1		104.1	104.1		90.2	90.2		99.3	99.3		101.6	101.6
0241, 31 - 34	**SITE & INFRASTRUCTURE, DEMOLITION**	106.0	86.8	92.7	109.3	87.3	94.1	118.1	86.4	96.2	98.1	86.1	89.8	92.1	105.6	101.4	92.8	108.2	103.5
0310	Concrete Forming & Accessories	95.7	71.2	74.8	89.3	61.4	65.5	83.6	66.4	68.9	107.0	64.2	70.5	111.6	104.6	105.7	101.6	104.3	103.9
0320	Concrete Reinforcing	97.5	84.6	91.2	97.1	83.8	90.7	97.1	80.2	88.9	108.4	95.9	102.3	109.2	114.5	111.8	119.5	114.4	117.0
0330	Cast-in-Place Concrete	119.0	86.2	106.8	109.0	85.5	100.2	105.2	63.9	89.8	82.3	82.9	82.5	102.6	109.4	105.1	95.3	109.8	100.7
03	**CONCRETE**	103.5	80.2	93.3	100.7	75.2	89.5	101.5	69.5	87.4	87.6	76.6	82.8	95.1	107.4	100.5	93.7	107.2	99.7
04	**MASONRY**	96.3	65.9	77.6	105.0	55.3	74.5	98.7	69.8	80.9	97.8	91.7	94.0	121.0	101.8	109.2	115.3	101.7	107.0
05	**METALS**	105.8	100.2	104.2	103.6	96.0	101.4	103.8	97.9	102.1	92.8	87.3	91.2	116.8	98.4	111.4	117.8	96.1	111.4
06	**WOOD, PLASTICS & COMPOSITES**	91.9	71.5	80.7	83.1	59.8	70.3	78.7	64.8	71.1	104.4	58.5	79.2	107.5	104.7	105.9	97.5	104.8	101.5
07	**THERMAL & MOISTURE PROTECTION**	103.6	76.0	91.6	103.4	67.5	87.8	104.2	75.9	91.9	159.7	82.9	126.3	112.4	105.9	109.6	108.3	103.0	106.0
08	**OPENINGS**	95.4	68.8	89.2	95.0	58.5	86.5	96.5	68.3	89.9	116.4	65.0	104.4	106.0	107.3	106.3	109.6	104.8	108.5
0920	Plaster & Gypsum Board	98.0	70.7	79.6	91.1	57.9	68.8	91.3	63.8	72.8	154.4	57.3	89.1	112.7	104.9	107.5	106.5	104.9	105.4
0950, 0980	Ceilings & Acoustic Treatment	92.4	70.7	77.8	90.0	57.9	68.4	90.8	63.8	72.6	103.8	57.3	72.4	101.2	104.9	103.7	98.7	104.9	102.9
0960	Flooring	97.1	67.3	88.7	93.1	33.1	76.2	92.0	74.4	87.1	85.4	77.5	83.2	109.9	96.0	106.0	97.6	96.0	97.1
0970, 0990	Wall Finishes & Painting/Coating	102.1	55.3	74.4	102.1	30.7	59.8	115.8	77.6	93.2	84.4	72.1	77.2	97.6	95.2	96.2	99.6	93.1	95.8
09	**FINISHES**	92.7	68.5	79.6	90.6	52.6	70.0	91.8	68.3	79.1	108.2	65.8	85.2	105.4	101.6	103.4	94.9	101.5	98.5
COVERS	**DIVS. 10 - 14, 25, 28, 41, 43, 44, 46**	100.0	79.2	95.4	100.0	78.2	95.1	100.0	82.7	96.1	100.0	89.3	97.6	100.0	103.7	100.8	100.0	99.9	100.0
21, 22, 23	**FIRE SUPPRESSION, PLUMBING & HVAC**	100.4	67.9	87.6	96.8	59.3	82.1	96.8	81.4	90.8	96.7	80.6	90.4	100.0	105.2	102.0	99.9	100.6	100.2
26, 27, 3370	**ELECTRICAL, COMMUNICATIONS & UTIL.**	96.2	60.8	78.8	95.2	68.4	82.0	94.9	91.8	93.4	89.4	95.6	92.5	107.0	104.1	105.6	105.4	97.7	101.6
MF2018	**WEIGHTED AVERAGE**	100.1	73.6	88.9	98.6	67.9	85.6	98.9	79.5	90.7	99.1	81.9	91.8	105.1	104.0	104.6	104.0	101.8	103.1

City Cost Indexes

DIVISION		WASHINGTON																	
		RICHLAND			SEATTLE			SPOKANE			TACOMA			VANCOUVER			WENATCHEE		
		993			980 - 981, 987			990 - 992			983 - 984			986			988		
		MAT.	INST.	TOTAL	MAT.	INST.	TOTAL	MAT.	INST.	TOTAL	MAT.	INST.	TOTAL	MAT.	INST.	TOTAL	MAT.	INST.	TOTAL
015433	CONTRACTOR EQUIPMENT		90.2	90.2		102.4	102.4		90.2	90.2		99.3	99.3		95.8	95.8		99.3	99.3
0241, 31 - 34	SITE & INFRASTRUCTURE, DEMOLITION	100.3	86.3	90.6	97.8	108.9	105.5	99.7	86.3	90.4	95.2	105.1	102.1	105.2	93.1	96.8	104.0	102.8	103.2
0310	Concrete Forming & Accessories	107.1	80.0	84.0	107.4	109.5	109.2	112.0	79.5	84.3	102.3	104.1	103.9	102.7	95.7	96.7	104.0	76.5	80.5
0320	Concrete Reinforcing	103.9	95.4	99.8	109.2	117.3	113.1	104.6	95.3	100.1	107.8	114.3	110.9	108.8	114.8	111.7	108.8	95.9	102.5
0330	Cast-in-Place Concrete	82.5	84.4	83.2	105.9	113.9	108.9	85.7	84.2	85.2	105.5	108.8	106.7	117.8	99.9	111.2	107.7	90.8	101.4
03	CONCRETE	87.2	84.2	85.9	102.8	111.8	106.7	89.2	83.9	86.8	97.0	106.9	101.4	107.2	100.2	104.1	105.3	84.9	96.3
04	MASONRY	98.8	83.7	89.5	123.5	107.2	113.5	99.4	83.7	89.7	117.6	100.2	106.9	118.8	105.5	110.6	120.9	94.1	104.4
05	METALS	93.2	87.1	91.4	116.1	101.9	111.9	95.4	86.6	92.8	118.7	96.5	112.2	116.0	99.3	111.1	116.0	87.3	107.6
06	WOOD, PLASTICS & COMPOSITES	104.6	77.9	89.9	103.4	108.7	106.3	113.5	77.9	93.9	97.1	104.7	101.2	89.5	94.9	92.5	98.7	73.3	84.8
07	THERMAL & MOISTURE PROTECTION	161.1	84.4	127.7	111.6	110.8	111.3	157.4	84.7	125.8	112.1	102.1	107.8	111.8	101.6	107.4	111.6	88.3	101.5
08	OPENINGS	114.4	74.7	105.2	106.1	110.7	107.2	115.0	74.7	105.6	106.7	104.7	106.2	102.8	101.5	102.5	106.2	73.0	98.5
0920	Plaster & Gypsum Board	154.4	77.2	102.5	111.2	109.1	109.8	144.2	77.2	99.1	110.6	104.9	106.8	109.0	95.2	99.7	113.9	72.7	86.2
0950, 0980	Ceilings & Acoustic Treatment	109.2	77.2	87.6	106.6	109.1	108.3	105.6	77.2	86.4	104.5	104.9	104.8	102.4	95.2	97.5	97.8	72.7	80.9
0960	Flooring	85.7	77.5	83.4	109.5	103.4	107.8	84.8	77.5	82.7	102.2	96.0	100.5	107.4	107.1	107.3	105.3	77.5	97.5
0970, 0990	Wall Finishes & Painting/Coating	84.4	75.1	78.9	107.1	95.8	100.4	84.6	76.4	79.8	97.6	93.1	95.0	100.3	79.9	88.2	97.6	70.9	81.8
09	FINISHES	109.5	78.1	92.5	108.9	106.4	107.6	107.3	78.2	91.6	103.6	101.4	102.4	101.3	95.6	98.3	104.7	74.9	88.5
COVERS	DIVS. 10 - 14, 25, 28, 41, 43, 44, 46	100.0	92.2	98.3	100.0	105.2	101.2	100.0	92.2	98.3	100.0	99.6	99.9	100.0	101.5	100.3	100.0	91.2	98.0
21, 22, 23	FIRE SUPPRESSION, PLUMBING & HVAC	100.3	108.6	103.6	99.9	123.0	109.0	100.3	83.7	93.8	100.0	100.6	100.2	100.0	110.7	104.2	96.5	92.0	94.7
26, 27, 3370	ELECTRICAL, COMMUNICATIONS & UTIL.	87.2	92.8	90.0	106.2	118.9	112.5	85.4	77.1	81.3	106.8	97.7	102.3	113.1	107.3	110.2	107.7	94.5	101.2
MF2018	WEIGHTED AVERAGE	99.8	90.0	95.6	106.3	112.5	108.9	100.1	82.4	92.6	105.4	101.3	103.7	106.6	102.8	105.0	105.6	88.7	98.5

DIVISION		WASHINGTON			WEST VIRGINIA														
		YAKIMA			BECKLEY			BLUEFIELD			BUCKHANNON			CHARLESTON			CLARKSBURG		
		989			258 - 259			247 - 248			262			250 - 253			263 - 264		
		MAT.	INST.	TOTAL	MAT.	INST.	TOTAL	MAT.	INST.	TOTAL	MAT.	INST.	TOTAL	MAT.	INST.	TOTAL	MAT.	INST.	TOTAL
015433	CONTRACTOR EQUIPMENT		99.3	99.3		104.1	104.1		104.1	104.1		104.1	104.1		106.8	106.8		104.1	104.1
0241, 31 - 34	SITE & INFRASTRUCTURE, DEMOLITION	97.6	103.8	101.9	100.8	87.4	91.6	99.9	87.4	91.3	106.0	87.4	93.1	100.4	93.0	95.3	106.8	87.4	93.4
0310	Concrete Forming & Accessories	102.9	99.2	99.8	85.0	87.1	86.8	86.6	87.1	87.0	86.1	83.7	84.0	96.0	88.6	89.7	83.6	83.8	83.7
0320	Concrete Reinforcing	108.3	95.4	102.1	98.0	87.7	93.1	96.0	82.5	89.5	96.7	82.4	89.8	106.1	87.8	97.2	96.7	95.3	96.0
0330	Cast-in-Place Concrete	112.7	83.0	101.7	99.2	92.5	96.7	102.5	92.5	98.8	102.1	92.0	98.4	95.9	96.6	96.2	111.9	87.6	102.9
03	CONCRETE	101.8	92.4	97.7	91.8	90.2	91.1	95.1	89.3	92.6	98.2	87.5	93.5	90.7	92.2	91.4	102.2	88.2	96.1
04	MASONRY	110.3	81.0	92.3	91.0	89.0	89.8	89.5	89.0	89.2	100.6	85.8	91.5	88.4	90.7	89.8	104.1	85.8	92.9
05	METALS	116.7	87.3	108.0	98.2	103.3	99.7	103.6	101.5	103.0	103.8	101.6	103.2	96.8	102.5	98.5	103.8	106.0	104.5
06	WOOD, PLASTICS & COMPOSITES	97.4	104.7	101.4	80.2	87.4	84.2	82.1	87.4	85.0	81.4	83.0	82.3	89.3	88.2	88.7	78.0	83.0	80.8
07	THERMAL & MOISTURE PROTECTION	112.3	86.1	100.9	105.0	86.5	96.9	103.6	86.5	96.1	104.0	86.6	96.4	101.2	88.1	95.5	103.8	86.3	96.2
08	OPENINGS	106.2	100.5	104.9	94.5	83.0	91.8	96.9	81.8	93.4	96.9	79.4	92.8	94.9	83.4	92.2	96.9	82.1	93.4
0920	Plaster & Gypsum Board	110.4	104.9	106.7	93.5	87.0	89.1	90.2	87.0	88.0	90.8	82.5	85.2	98.7	87.6	91.2	88.6	82.5	84.5
0950, 0980	Ceilings & Acoustic Treatment	99.6	104.9	103.2	78.1	87.0	84.1	87.5	87.0	87.2	90.0	82.5	84.9	89.6	87.6	88.2	90.0	82.5	84.9
0960	Flooring	103.2	77.5	96.0	91.0	98.9	93.2	90.1	98.9	92.6	89.8	95.2	91.3	97.1	98.9	97.6	88.8	95.2	90.6
0970, 0990	Wall Finishes & Painting/Coating	97.6	76.4	85.1	89.3	92.1	91.0	102.1	89.8	94.8	102.1	88.0	93.7	88.6	91.9	90.5	102.1	88.0	93.7
09	FINISHES	103.1	93.3	97.8	86.3	90.2	88.4	88.7	89.9	89.4	89.7	86.1	87.8	92.8	91.0	91.9	89.0	86.1	87.4
COVERS	DIVS. 10 - 14, 25, 28, 41, 43, 44, 46	100.0	96.1	99.1	100.0	90.3	97.8	100.0	90.3	97.8	100.0	90.6	97.9	100.0	92.5	98.3	100.0	90.6	97.9
21, 22, 23	FIRE SUPPRESSION, PLUMBING & HVAC	100.0	107.4	102.9	97.1	80.3	90.5	96.8	84.0	91.8	96.8	89.7	94.0	100.1	81.4	92.8	96.8	89.8	94.1
26, 27, 3370	ELECTRICAL, COMMUNICATIONS & UTIL.	109.9	92.9	101.5	92.9	84.1	88.6	95.2	84.1	89.8	96.6	89.3	93.0	98.0	87.1	92.6	96.6	89.3	93.0
MF2018	WEIGHTED AVERAGE	105.6	95.4	101.3	95.2	87.7	92.0	97.0	88.1	93.2	98.3	88.8	94.3	96.5	89.4	93.5	98.9	89.5	94.9

DIVISION		WEST VIRGINIA																	
		GASSAWAY			HUNTINGTON			LEWISBURG			MARTINSBURG			MORGANTOWN			PARKERSBURG		
		266			255 - 257			249			254			265			261		
		MAT.	INST.	TOTAL	MAT.	INST.	TOTAL	MAT.	INST.	TOTAL	MAT.	INST.	TOTAL	MAT.	INST.	TOTAL	MAT.	INST.	TOTAL
015433	CONTRACTOR EQUIPMENT		104.1	104.1		104.1	104.1		104.1	104.1		104.1	104.1		104.1	104.1		104.1	104.1
0241, 31 - 34	SITE & INFRASTRUCTURE, DEMOLITION	103.5	87.4	92.4	105.4	88.5	93.7	115.5	87.4	96.1	104.7	88.0	93.2	100.9	88.2	92.1	109.3	88.4	94.9
0310	Concrete Forming & Accessories	85.5	86.9	86.7	96.0	89.3	90.3	83.9	86.6	86.2	85.0	79.5	80.3	83.9	83.8	83.9	88.0	86.0	86.3
0320	Concrete Reinforcing	96.7	91.6	94.2	99.5	93.1	96.4	96.7	82.4	89.8	98.0	95.1	96.6	96.7	95.3	96.0	96.0	82.3	89.4
0330	Cast-in-Place Concrete	106.9	94.0	102.1	108.2	94.0	102.9	102.5	92.3	98.7	104.0	88.3	98.1	102.1	92.0	98.4	104.4	89.0	98.7
03	CONCRETE	98.6	91.3	95.4	96.8	92.6	95.0	105.1	89.0	98.0	95.3	86.5	91.4	94.9	89.8	92.7	100.3	87.5	94.7
04	MASONRY	105.0	87.9	94.5	91.7	93.4	92.7	94.3	89.0	91.1	93.1	83.9	87.5	122.6	85.8	100.0	80.1	85.2	83.2
05	METALS	103.8	104.8	104.1	100.6	105.7	102.1	103.7	101.0	102.9	98.6	104.4	100.3	103.9	106.0	104.5	104.5	101.4	103.6
06	WOOD, PLASTICS & COMPOSITES	80.4	87.4	84.3	90.8	88.7	89.6	78.3	87.4	83.3	80.2	78.6	79.3	78.3	83.0	80.9	81.5	85.6	83.8
07	THERMAL & MOISTURE PROTECTION	103.6	87.9	96.8	105.2	88.4	97.9	104.6	86.4	96.7	105.3	80.8	94.6	103.7	86.6	96.2	103.9	86.7	96.4
08	OPENINGS	95.2	83.9	92.6	93.7	84.8	91.7	96.9	81.8	93.4	96.2	73.8	91.0	98.0	82.1	94.3	95.8	80.4	92.2
0920	Plaster & Gypsum Board	89.9	87.0	87.9	99.5	88.3	91.9	88.6	87.0	87.5	93.5	77.9	83.0	88.6	82.5	84.5	91.1	85.1	87.1
0950, 0980	Ceilings & Acoustic Treatment	90.0	87.0	88.0	78.1	88.3	85.0	90.0	87.0	88.0	78.1	77.9	78.0	90.0	82.5	84.9	90.0	85.1	86.7
0960	Flooring	89.6	98.9	92.2	98.4	103.3	99.7	88.9	98.9	91.7	91.0	98.9	93.2	88.9	95.2	90.7	92.6	91.9	92.4
0970, 0990	Wall Finishes & Painting/Coating	102.1	91.9	96.0	89.3	89.8	89.6	102.1	64.5	79.8	89.3	83.7	86.0	102.1	88.0	93.7	102.1	83.7	91.2
09	FINISHES	89.2	89.9	89.6	89.3	92.0	90.8	90.1	87.2	88.5	86.6	83.0	84.6	88.6	86.1	87.3	90.6	86.8	88.6
COVERS	DIVS. 10 - 14, 25, 28, 41, 43, 44, 46	100.0	91.1	98.0	100.0	92.3	98.3	100.0	65.4	92.3	100.0	81.4	95.8	100.0	83.5	96.3	100.0	91.3	98.1
21, 22, 23	FIRE SUPPRESSION, PLUMBING & HVAC	96.8	83.4	91.6	100.6	90.6	96.7	96.8	80.2	90.3	97.1	80.9	90.8	96.8	89.8	94.1	100.3	88.3	95.6
26, 27, 3370	ELECTRICAL, COMMUNICATIONS & UTIL.	96.6	87.1	91.9	96.6	89.6	93.1	92.8	84.1	88.5	98.6	76.9	87.9	96.7	89.3	93.1	96.7	88.3	92.5
MF2018	WEIGHTED AVERAGE	98.3	89.0	94.3	97.8	92.2	95.4	98.7	86.1	93.4	96.7	84.1	91.4	98.8	89.6	94.9	98.6	88.6	94.4

	DIVISION	WEST VIRGINIA									WISCONSIN								
		PETERSBURG			ROMNEY			WHEELING			BELOIT			EAU CLAIRE			GREEN BAY		
		268			267			260			535			547			541 - 543		
		MAT.	INST.	TOTAL	MAT.	INST.	TOTAL	MAT.	INST.	TOTAL	MAT.	INST.	TOTAL	MAT.	INST.	TOTAL	MAT.	INST.	TOTAL
015433	CONTRACTOR EQUIPMENT		104.1	104.1		104.1	104.1		104.1	104.1		99.0	99.0		99.9	99.9		97.7	97.7
0241, 31 - 34	SITE & INFRASTRUCTURE, DEMOLITION	100.2	88.3	92.0	103.0	88.3	92.9	109.9	88.2	94.9	97.1	101.7	100.3	96.0	99.0	98.1	99.5	95.3	96.6
0310	Concrete Forming & Accessories	87.0	86.4	86.5	83.3	86.8	86.3	89.7	83.8	84.7	99.8	96.2	96.7	98.5	94.7	95.3	108.2	97.5	99.1
0320	Concrete Reinforcing	96.0	91.4	93.8	96.7	95.4	96.1	95.5	95.3	95.4	95.0	137.1	115.4	94.2	114.2	103.9	92.3	109.4	100.5
0330	Cast-in-Place Concrete	102.1	91.8	98.3	106.9	83.5	98.2	104.4	92.0	99.8	105.8	100.2	103.7	99.9	94.8	98.0	103.3	99.9	102.1
03	CONCRETE	95.0	90.2	92.9	98.4	88.2	93.9	100.3	89.8	95.7	96.7	104.7	100.2	93.1	98.4	95.4	96.2	100.6	98.2
04	MASONRY	96.1	85.8	89.8	94.7	87.9	90.5	103.4	85.7	92.5	99.3	100.3	99.9	91.4	96.1	94.3	122.7	97.8	107.4
05	METALS	103.9	104.1	104.0	104.0	105.9	104.5	104.7	106.1	105.1	98.6	112.0	102.5	93.8	105.8	97.3	96.4	104.2	98.7
06	WOOD, PLASTICS & COMPOSITES	82.3	87.4	85.1	77.5	87.4	82.9	83.1	83.0	83.1	96.9	94.1	95.4	100.6	94.2	97.1	107.8	97.2	101.9
07	THERMAL & MOISTURE PROTECTION	103.7	82.5	94.5	103.8	83.8	95.1	104.1	86.7	96.6	105.4	93.4	100.2	104.0	94.6	99.9	106.1	97.4	102.3
08	OPENINGS	98.1	83.6	94.7	98.0	84.5	94.8	96.6	82.1	93.2	98.1	108.3	100.5	103.1	99.3	102.2	99.1	101.2	99.6
0920	Plaster & Gypsum Board	90.8	87.0	88.2	88.3	87.0	87.4	91.1	82.5	85.3	91.6	94.5	93.5	105.5	94.5	98.1	103.3	97.5	99.4
0950, 0980	Ceilings & Acoustic Treatment	90.0	87.0	88.0	90.0	87.0	88.0	90.0	82.5	84.9	84.4	94.5	91.2	90.9	94.5	93.3	83.6	97.5	93.0
0960	Flooring	90.6	95.2	91.9	88.7	98.9	91.6	93.5	95.2	94.0	94.8	118.5	101.4	80.5	106.5	87.8	96.0	114.8	101.3
0970, 0990	Wall Finishes & Painting/Coating	102.1	88.0	93.7	102.1	88.0	93.7	102.1	88.0	93.7	95.0	103.6	100.1	83.3	76.4	79.2	92.0	80.0	84.9
09	FINISHES	89.4	88.7	89.0	88.7	89.5	89.1	90.9	86.1	88.3	92.3	100.7	96.9	87.3	94.9	91.4	91.4	99.1	95.6
COVERS	DIVS. 10 - 14, 25, 28, 41, 43, 44, 46	100.0	55.0	90.0	100.0	91.0	98.0	100.0	84.4	96.5	100.0	98.6	99.7	100.0	94.1	98.7	100.0	95.7	99.0
21, 22, 23	FIRE SUPPRESSION, PLUMBING & HVAC	96.8	83.7	91.7	96.8	83.6	91.6	100.4	89.8	96.2	99.8	95.5	98.1	100.0	88.2	95.4	100.2	83.4	93.6
26, 27, 3370	ELECTRICAL, COMMUNICATIONS & UTIL.	99.7	76.9	88.4	99.0	76.9	88.1	94.1	89.3	91.7	100.6	84.3	92.5	104.3	84.4	94.5	99.0	81.2	90.2
MF2018	WEIGHTED AVERAGE	98.0	85.9	92.9	98.3	87.2	93.6	99.6	89.6	95.4	98.6	98.9	98.7	97.5	94.1	96.1	99.4	93.4	96.9

	DIVISION	WISCONSIN																	
		KENOSHA			LA CROSSE			LANCASTER			MADISON			MILWAUKEE			NEW RICHMOND		
		531			546			538			537			530, 532			540		
		MAT.	INST.	TOTAL	MAT.	INST.	TOTAL	MAT.	INST.	TOTAL	MAT.	INST.	TOTAL	MAT.	INST.	TOTAL	MAT.	INST.	TOTAL
015433	CONTRACTOR EQUIPMENT		97.1	97.1		99.9	99.9		99.0	99.0		101.4	101.4		90.2	90.2		100.2	100.2
0241, 31 - 34	SITE & INFRASTRUCTURE, DEMOLITION	103.0	98.4	99.9	89.8	99.0	96.1	96.0	101.1	99.5	94.7	105.3	102.0	93.6	96.0	95.2	93.3	99.6	97.7
0310	Concrete Forming & Accessories	108.7	106.1	106.5	85.2	94.5	93.2	99.1	95.2	95.7	103.8	95.8	97.0	102.9	113.9	112.3	93.6	92.5	92.6
0320	Concrete Reinforcing	94.8	110.5	102.4	93.9	106.2	99.8	96.3	106.2	101.1	97.8	106.4	101.9	99.9	115.1	107.2	91.3	114.1	102.3
0330	Cast-in-Place Concrete	115.2	102.7	110.5	89.8	96.5	92.3	105.2	99.7	103.2	101.4	100.3	101.0	93.9	112.1	100.7	104.0	102.7	103.5
03	CONCRETE	101.5	105.5	103.3	84.6	97.5	90.3	96.4	98.8	97.4	96.3	99.2	97.6	94.5	112.5	102.4	90.8	100.1	94.9
04	MASONRY	97.1	108.0	103.8	90.5	96.1	93.9	99.4	100.1	99.9	97.4	99.1	98.5	102.3	117.2	111.4	117.9	98.4	105.9
05	METALS	99.6	103.3	100.7	93.7	102.8	96.4	96.0	100.3	97.3	102.8	100.4	102.1	97.5	97.5	97.5	94.1	105.4	97.4
06	WOOD, PLASTICS & COMPOSITES	103.9	105.7	104.9	85.2	94.2	90.2	96.1	94.1	95.0	99.4	94.3	96.6	101.2	112.3	107.3	89.5	90.2	89.9
07	THERMAL & MOISTURE PROTECTION	105.5	104.0	104.9	103.4	93.5	99.1	105.2	93.7	100.2	106.3	100.5	103.7	104.8	112.7	108.2	104.9	100.2	102.9
08	OPENINGS	92.5	105.4	95.5	103.1	92.0	100.5	94.0	91.9	93.5	101.6	98.9	100.9	103.0	111.5	105.0	88.6	90.3	89.0
0920	Plaster & Gypsum Board	80.2	106.4	97.8	100.9	94.5	96.6	90.9	94.5	93.3	101.7	94.5	96.8	96.4	112.6	107.3	91.7	90.5	90.9
0950, 0980	Ceilings & Acoustic Treatment	84.4	106.4	99.2	90.0	94.5	93.0	81.1	94.5	90.1	86.5	94.5	91.9	93.1	112.6	106.3	57.1	90.5	79.6
0960	Flooring	112.8	114.8	113.4	74.7	111.6	85.1	94.4	111.0	99.0	94.3	111.0	99.0	101.7	118.6	106.4	89.7	114.5	96.7
0970, 0990	Wall Finishes & Painting/Coating	106.4	118.7	113.7	83.3	81.7	82.4	95.0	97.7	96.6	94.9	103.6	100.0	104.5	124.3	116.3	95.0	81.7	87.1
09	FINISHES	97.1	109.4	103.7	84.4	96.5	90.9	91.4	98.4	95.2	93.7	99.3	96.7	99.5	116.1	108.5	82.5	95.2	89.4
COVERS	DIVS. 10 - 14, 25, 28, 41, 43, 44, 46	100.0	97.3	99.4	100.0	92.8	98.4	100.0	84.2	96.5	100.0	96.8	99.3	100.0	105.2	101.2	100.0	95.1	98.9
21, 22, 23	FIRE SUPPRESSION, PLUMBING & HVAC	100.0	95.9	98.4	100.0	88.1	95.3	96.3	89.2	93.5	99.8	95.3	98.0	99.8	107.0	102.6	96.0	89.0	93.2
26, 27, 3370	ELECTRICAL, COMMUNICATIONS & UTIL.	100.9	97.1	99.0	104.6	84.4	94.7	100.3	84.9	92.7	101.5	94.0	97.8	100.2	102.4	101.3	102.4	84.9	93.7
MF2018	WEIGHTED AVERAGE	99.4	102.0	100.5	95.9	93.5	94.9	96.7	94.3	95.7	99.7	98.1	99.0	99.2	107.9	102.9	95.4	94.6	95.0

	DIVISION	WISCONSIN																	
		OSHKOSH			PORTAGE			RACINE			RHINELANDER			SUPERIOR			WAUSAU		
		549			539			534			545			548			544		
		MAT.	INST.	TOTAL	MAT.	INST.	TOTAL	MAT.	INST.	TOTAL	MAT.	INST.	TOTAL	MAT.	INST.	TOTAL	MAT.	INST.	TOTAL
015433	CONTRACTOR EQUIPMENT		97.7	97.7		99.0	99.0		99.0	99.0		97.7	97.7		100.2	100.2		97.7	97.7
0241, 31 - 34	SITE & INFRASTRUCTURE, DEMOLITION	91.0	95.3	94.0	86.9	100.9	96.6	96.9	102.1	100.5	103.2	95.3	97.7	90.1	99.3	96.5	86.9	95.8	93.0
0310	Concrete Forming & Accessories	90.7	94.7	94.1	91.3	95.7	95.1	100.1	106.1	105.2	88.3	94.6	93.7	91.6	88.6	89.0	90.1	97.4	96.3
0320	Concrete Reinforcing	92.4	109.3	100.6	96.4	106.4	101.2	95.0	110.5	102.5	92.6	106.4	99.3	91.3	110.2	100.4	92.6	106.2	99.2
0330	Cast-in-Place Concrete	95.8	98.0	96.6	90.3	98.3	93.3	103.8	102.5	103.3	108.6	97.9	104.6	98.0	98.3	98.1	89.3	92.4	90.5
03	CONCRETE	86.2	98.6	91.7	84.2	98.7	90.6	95.8	105.4	100.0	98.0	98.1	98.1	85.5	96.2	90.2	81.4	97.4	88.4
04	MASONRY	104.7	97.8	100.4	98.2	100.2	99.4	99.3	108.0	104.6	121.3	97.6	106.8	117.1	100.4	106.8	104.2	97.6	100.1
05	METALS	94.3	103.5	97.0	96.7	104.3	98.9	100.2	103.3	101.1	94.2	102.5	96.6	95.1	104.6	97.9	94.0	102.9	96.6
06	WOOD, PLASTICS & COMPOSITES	86.7	94.2	90.9	85.8	94.1	90.4	97.2	105.7	101.9	84.2	94.2	89.7	87.9	85.8	86.7	86.2	97.2	92.2
07	THERMAL & MOISTURE PROTECTION	105.0	83.3	95.6	104.5	100.1	102.6	105.5	102.7	104.3	106.0	83.1	96.0	104.5	95.9	100.8	104.8	95.1	100.6
08	OPENINGS	95.6	96.2	95.7	94.1	99.7	95.4	98.1	105.4	99.8	95.6	91.9	94.7	88.0	91.9	88.9	95.8	97.1	96.1
0920	Plaster & Gypsum Board	90.7	94.5	93.2	83.8	94.5	91.0	91.6	106.4	101.5	90.7	94.5	93.2	92.1	86.0	88.0	90.7	97.5	95.3
0950, 0980	Ceilings & Acoustic Treatment	83.6	94.5	90.9	83.6	94.5	90.9	84.4	106.4	99.2	83.6	94.5	90.9	58.8	86.0	77.1	83.6	97.5	93.0
0960	Flooring	87.8	114.8	95.4	90.5	114.5	97.3	94.8	114.8	100.4	87.2	114.5	94.9	90.8	123.1	99.9	87.7	114.5	95.2
0970, 0990	Wall Finishes & Painting/Coating	89.6	104.3	98.3	95.0	97.7	96.6	95.0	115.9	107.4	89.6	80.0	83.9	84.0	105.8	96.9	89.6	83.2	85.8
09	FINISHES	86.5	100.0	93.8	89.2	99.4	94.7	92.3	109.1	101.4	87.4	97.3	92.7	82.1	96.4	89.8	86.2	99.4	93.3
COVERS	DIVS. 10 - 14, 25, 28, 41, 43, 44, 46	100.0	85.6	96.8	100.0	85.6	96.8	100.0	97.3	99.4	100.0	85.9	96.8	100.0	93.9	98.6	100.0	95.7	99.0
21, 22, 23	FIRE SUPPRESSION, PLUMBING & HVAC	96.7	83.1	91.3	96.3	95.8	96.1	99.8	96.0	98.3	96.7	88.6	93.5	96.0	91.1	94.1	96.7	88.8	93.6
26, 27, 3370	ELECTRICAL, COMMUNICATIONS & UTIL.	103.0	79.1	91.2	104.0	94.0	99.1	100.4	98.0	99.2	102.3	78.1	90.4	107.2	98.5	102.9	104.1	78.1	91.3
MF2018	WEIGHTED AVERAGE	95.4	91.9	93.9	95.2	98.0	96.4	98.7	102.3	100.2	97.9	92.2	95.5	95.1	96.5	95.7	94.7	93.4	94.2

City Cost Indexes

	DIVISION	WYOMING																	
		CASPER			CHEYENNE			NEWCASTLE			RAWLINS			RIVERTON			ROCK SPRINGS		
		826			820			827			823			825			829 - 831		
		MAT.	INST.	TOTAL	MAT.	INST.	TOTAL	MAT.	INST.	TOTAL	MAT.	INST.	TOTAL	MAT.	INST.	TOTAL	MAT.	INST.	TOTAL
015433	CONTRACTOR EQUIPMENT		97.3	97.3		94.6	94.6		94.6	94.6		94.6	94.6		94.6	94.6		94.6	94.6
0241, 31 - 34	SITE & INFRASTRUCTURE, DEMOLITION	97.8	94.7	95.7	91.6	90.1	90.6	83.6	89.3	87.6	97.5	89.3	91.9	91.1	89.3	89.8	87.4	89.5	88.9
0310	Concrete Forming & Accessories	101.3	53.6	60.6	102.5	64.6	70.2	93.7	70.5	74.0	97.6	70.8	74.7	92.7	60.3	65.0	99.7	64.7	69.9
0320	Concrete Reinforcing	111.1	83.4	97.7	106.4	83.4	95.3	114.5	83.2	99.3	114.2	83.2	99.2	115.2	83.2	99.7	115.2	81.6	99.0
0330	Cast-in-Place Concrete	103.8	77.8	94.1	97.9	77.4	90.3	99.0	72.8	89.2	99.0	72.9	89.3	99.0	72.8	89.2	99.0	75.0	90.0
03	CONCRETE	103.3	67.9	87.7	100.9	72.8	88.6	101.1	73.8	89.1	116.0	74.0	97.5	110.1	69.2	92.1	101.6	71.7	88.4
04	MASONRY	102.0	64.6	79.0	105.3	65.2	80.7	102.2	59.3	75.8	102.2	59.3	75.8	102.2	59.3	75.8	164.5	61.8	101.4
05	METALS	102.8	80.1	96.1	105.3	80.8	98.1	101.4	79.9	95.1	101.5	80.1	95.2	101.6	79.9	95.2	102.3	79.5	95.6
06	WOOD, PLASTICS & COMPOSITES	95.2	47.8	69.1	91.8	62.6	75.7	82.4	72.7	77.1	86.2	72.7	78.8	81.4	58.7	68.9	91.3	62.6	75.5
07	THERMAL & MOISTURE PROTECTION	112.7	65.9	92.3	107.2	68.0	90.2	108.8	63.9	89.3	110.3	74.9	94.9	109.7	67.7	91.4	108.9	69.5	91.8
08	OPENINGS	109.3	59.4	97.7	107.2	67.7	98.0	111.4	72.3	102.3	111.1	72.3	102.0	111.3	64.6	100.4	111.8	66.5	101.2
0920	Plaster & Gypsum Board	101.7	46.2	64.4	90.6	61.7	71.2	87.6	72.0	77.1	87.9	72.0	77.2	87.6	57.7	67.5	100.2	61.7	74.3
0950, 0980	Ceilings & Acoustic Treatment	105.3	46.2	65.5	98.5	61.7	73.7	100.3	72.0	81.2	100.3	72.0	81.2	100.3	57.7	71.6	100.3	61.7	74.3
0960	Flooring	102.2	73.8	94.2	100.3	73.8	92.8	94.2	44.5	80.2	97.0	44.5	82.3	93.6	60.3	84.2	99.3	59.9	88.3
0970, 0990	Wall Finishes & Painting/Coating	95.7	58.7	73.8	97.9	58.7	74.7	94.6	60.5	74.4	94.6	60.5	74.4	94.6	60.5	74.4	94.6	60.5	74.4
09	FINISHES	99.6	56.3	76.2	96.3	65.1	79.4	91.4	64.6	76.9	93.6	64.6	77.9	92.0	59.5	74.4	94.6	62.7	77.3
COVERS	DIVS. 10 - 14, 25, 28, 41, 43, 44, 46	100.0	92.5	98.3	100.0	87.4	97.2	100.0	91.8	98.2	100.0	91.8	98.2	100.0	77.3	94.9	100.0	86.9	97.1
21, 22, 23	FIRE SUPPRESSION, PLUMBING & HVAC	100.0	70.5	88.4	100.1	73.3	89.6	98.2	71.8	87.8	98.2	71.9	87.8	98.2	71.8	87.8	100.0	73.9	89.8
26, 27, 3370	ELECTRICAL, COMMUNICATIONS & UTIL.	94.1	60.3	77.5	95.2	69.0	82.3	94.0	62.3	78.4	94.0	62.3	78.4	94.0	59.1	76.8	92.5	66.0	79.4
MF2018	WEIGHTED AVERAGE	101.5	68.9	87.7	101.1	72.7	89.1	99.6	71.1	87.5	102.0	71.5	89.1	101.0	68.6	87.3	103.4	71.4	89.9

	DIVISION	WYOMING												CANADA					
		SHERIDAN			WHEATLAND			WORLAND			YELLOWSTONE NAT'L PA			BARRIE, ONTARIO			BATHURST, NEW BRUNSWICK		
		828			822			824			821								
		MAT.	INST.	TOTAL	MAT.	INST.	TOTAL	MAT.	INST.	TOTAL	MAT.	INST.	TOTAL	MAT.	INST.	TOTAL	MAT.	INST.	TOTAL
015433	CONTRACTOR EQUIPMENT		94.6	94.6		94.6	94.6		94.6	94.6		94.6	94.6		100.1	100.1		99.7	99.7
0241, 31 - 34	SITE & INFRASTRUCTURE, DEMOLITION	91.5	90.1	90.5	88.1	89.3	88.9	85.5	89.3	88.1	85.6	89.3	88.1	117.0	94.7	101.6	105.1	90.6	95.1
0310	Concrete Forming & Accessories	100.5	61.6	67.4	95.4	48.4	55.3	95.5	61.1	66.1	95.5	61.1	66.2	125.6	82.3	88.7	106.1	57.8	64.9
0320	Concrete Reinforcing	115.2	83.2	99.7	114.5	82.6	99.1	115.2	83.2	99.7	117.2	82.3	100.3	175.2	86.9	132.5	140.5	57.5	100.4
0330	Cast-in-Place Concrete	102.3	77.4	93.1	103.3	72.7	91.9	99.0	72.7	89.2	99.0	72.8	89.2	157.7	82.1	129.6	114.0	56.2	92.5
03	CONCRETE	109.9	71.4	93.0	106.3	63.7	87.6	101.3	69.5	87.3	101.6	69.4	87.4	138.9	83.5	114.5	110.3	58.3	87.4
04	MASONRY	102.5	62.0	77.6	102.6	51.2	71.0	102.2	59.3	75.8	102.2	59.3	75.8	165.6	89.3	118.7	159.8	57.4	96.9
05	METALS	105.1	80.5	97.8	101.4	79.1	94.8	101.6	79.8	95.2	102.2	79.1	95.4	111.1	91.5	105.3	114.1	73.6	102.2
06	WOOD, PLASTICS & COMPOSITES	93.3	58.7	74.3	84.2	43.0	61.5	84.3	60.0	70.9	84.3	60.0	70.9	116.5	80.8	96.9	97.5	57.9	75.7
07	THERMAL & MOISTURE PROTECTION	110.1	66.7	91.2	109.1	58.1	86.9	108.9	64.8	89.7	108.3	64.8	89.4	115.3	86.1	102.6	111.8	58.4	88.6
08	OPENINGS	112.0	64.6	101.0	110.0	55.9	97.4	111.6	65.3	100.8	104.5	64.8	95.2	90.5	80.9	88.3	84.0	51.2	76.3
0920	Plaster & Gypsum Board	111.2	57.7	75.2	87.6	41.5	56.6	87.6	58.9	68.3	87.8	58.9	68.4	153.5	80.3	104.3	124.2	56.7	78.8
0950, 0980	Ceilings & Acoustic Treatment	103.9	57.7	72.7	100.3	41.5	60.6	100.3	58.9	72.4	101.1	58.9	72.7	94.3	80.3	84.9	116.8	56.7	76.3
0960	Flooring	97.9	60.3	87.4	95.6	43.8	81.0	95.6	44.5	81.3	95.6	44.5	81.3	119.3	87.3	110.3	99.6	41.5	83.3
0970, 0990	Wall Finishes & Painting/Coating	96.8	58.7	74.2	94.6	58.4	73.2	94.6	60.5	74.4	94.6	60.5	74.4	105.6	83.8	92.7	111.3	47.9	73.8
09	FINISHES	99.2	60.4	78.2	92.1	46.7	67.5	91.8	57.1	73.0	92.0	57.2	73.2	111.3	83.5	96.2	107.7	53.8	78.5
COVERS	DIVS. 10 - 14, 25, 28, 41, 43, 44, 46	100.0	93.4	98.5	100.0	81.4	95.9	100.0	76.1	94.7	100.0	76.2	94.7	139.2	65.0	122.7	131.1	57.9	114.8
21, 22, 23	FIRE SUPPRESSION, PLUMBING & HVAC	98.2	73.3	88.4	98.2	71.8	87.8	98.2	71.8	87.8	98.2	71.8	87.8	103.7	93.5	99.7	103.8	64.8	88.5
26, 27, 3370	ELECTRICAL, COMMUNICATIONS & UTIL.	96.1	59.1	77.9	94.0	76.6	85.4	94.0	76.6	85.4	93.1	86.1	89.7	117.0	82.9	100.2	113.5	55.9	85.1
MF2018	WEIGHTED AVERAGE	102.5	70.2	88.8	100.3	67.1	86.2	99.7	70.7	87.5	99.0	71.9	87.6	116.5	87.1	104.1	110.7	62.3	90.2

	DIVISION	CANADA																	
		BRANDON, MANITOBA			BRANTFORD, ONTARIO			BRIDGEWATER, NOVA SCOTIA			CALGARY, ALBERTA			CAP-DE-LA-MADELEINE, QUEBEC			CHARLESBOURG, QUEBEC		
		MAT.	INST.	TOTAL	MAT.	INST.	TOTAL	MAT.	INST.	TOTAL	MAT.	INST.	TOTAL	MAT.	INST.	TOTAL	MAT.	INST.	TOTAL
015433	CONTRACTOR EQUIPMENT		101.5	101.5		99.7	99.7		99.4	99.4		127.1	127.1		100.2	100.2		100.2	100.2
0241, 31 - 34	SITE & INFRASTRUCTURE, DEMOLITION	126.4	92.9	103.3	116.9	94.8	101.6	101.1	92.3	95.0	125.5	118.4	120.6	96.9	93.7	94.7	96.9	93.7	94.7
0310	Concrete Forming & Accessories	145.4	66.1	77.8	127.2	89.1	94.7	99.7	67.9	72.6	125.0	94.9	99.3	132.9	79.2	87.1	132.9	79.2	87.1
0320	Concrete Reinforcing	170.2	54.5	114.3	163.7	85.6	125.9	139.3	47.6	95.0	136.4	82.0	110.1	139.3	72.4	107.0	139.3	72.4	107.0
0330	Cast-in-Place Concrete	109.4	70.2	94.8	130.7	101.8	119.9	134.3	66.9	109.2	136.4	104.3	124.4	105.5	87.5	98.8	105.5	87.5	98.8
03	CONCRETE	118.1	66.6	95.5	121.8	93.2	109.3	120.1	65.1	95.9	126.3	96.8	113.3	107.5	81.5	96.1	107.5	81.5	96.1
04	MASONRY	212.8	60.7	119.3	166.1	93.4	121.4	161.8	65.6	102.6	212.4	89.8	137.1	162.4	76.3	109.5	162.4	76.3	109.5
05	METALS	127.2	78.2	112.8	112.3	92.0	106.3	111.6	76.0	101.1	131.0	102.4	122.6	110.7	85.0	103.1	110.7	85.0	103.1
06	WOOD, PLASTICS & COMPOSITES	150.5	66.8	104.4	120.3	87.7	102.4	89.6	67.4	77.4	102.6	94.4	98.1	131.6	79.0	102.7	131.6	79.0	102.7
07	THERMAL & MOISTURE PROTECTION	128.4	68.7	102.4	121.6	91.4	108.5	116.1	67.6	95.0	132.0	98.0	117.2	114.8	84.1	101.5	114.8	84.1	101.5
08	OPENINGS	100.3	59.9	90.8	87.7	86.6	87.5	82.4	61.3	77.5	82.3	83.7	82.6	89.1	72.3	85.2	89.1	72.3	85.2
0920	Plaster & Gypsum Board	115.4	65.6	81.9	115.8	87.4	96.7	124.3	66.5	85.4	126.4	93.3	104.1	147.6	78.4	101.0	147.6	78.4	101.0
0950, 0980	Ceilings & Acoustic Treatment	123.8	65.6	84.6	103.5	87.4	92.7	103.5	66.5	78.6	149.2	93.3	111.5	103.5	78.4	86.6	103.5	78.4	86.6
0960	Flooring	131.2	61.9	111.7	113.6	87.3	106.2	95.7	58.9	85.4	118.1	84.8	108.7	113.6	86.0	105.8	113.6	86.0	105.8
0970, 0990	Wall Finishes & Painting/Coating	117.9	54.1	80.1	110.3	92.0	99.5	110.3	59.3	80.1	113.7	106.4	109.4	110.3	83.6	94.5	110.3	83.6	94.5
09	FINISHES	122.9	64.3	91.2	108.7	89.3	98.2	103.8	65.5	83.1	123.6	94.8	108.0	112.2	81.2	95.4	112.2	81.2	95.4
COVERS	DIVS. 10 - 14, 25, 28, 41, 43, 44, 46	131.1	60.1	115.3	131.1	66.9	116.8	131.1	60.8	115.5	131.1	92.7	122.6	131.1	75.6	118.8	131.1	75.6	118.8
21, 22, 23	FIRE SUPPRESSION, PLUMBING & HVAC	104.0	78.9	94.1	103.8	96.3	100.9	103.8	79.5	94.3	105.1	90.2	99.2	104.2	84.2	96.4	104.2	84.2	96.4
26, 27, 3370	ELECTRICAL, COMMUNICATIONS & UTIL.	113.9	63.1	88.8	111.5	82.5	97.2	115.8	58.7	87.6	107.1	93.9	100.6	110.2	65.2	88.0	110.2	65.2	88.0
MF2018	WEIGHTED AVERAGE	120.6	70.5	99.4	113.1	90.7	103.6	111.3	70.3	94.0	119.7	95.6	109.5	110.6	80.0	97.7	110.6	80.0	97.7

	DIVISION	CANADA																	
		CHARLOTTETOWN, PRINCE EDWARD ISLAND			CHICOUTIMI, QUEBEC			CORNER BROOK, NEWFOUNDLAND			CORNWALL, ONTARIO			DALHOUSIE, NEW BRUNSWICK			DARTMOUTH, NOVA SCOTIA		
		MAT.	INST.	TOTAL	MAT.	INST.	TOTAL	MAT.	INST.	TOTAL	MAT.	INST.	TOTAL	MAT.	INST.	TOTAL	MAT.	INST.	TOTAL
015433	CONTRACTOR EQUIPMENT		116.8	116.8		100.3	100.3		100.0	100.0		99.7	99.7		99.7	99.7		98.9	98.9
0241, 31 - 34	SITE & INFRASTRUCTURE, DEMOLITION	134.6	101.5	111.7	102.0	93.6	96.2	130.4	90.5	102.8	115.1	94.3	100.7	101.0	90.6	93.8	117.9	92.0	100.0
0310	Concrete Forming & Accessories	123.7	53.1	63.5	134.9	89.8	96.4	121.9	75.6	82.4	125.0	82.4	88.7	105.9	58.0	65.1	111.8	67.8	74.3
0320	Concrete Reinforcing	156.7	46.6	103.5	106.0	94.5	100.4	156.0	48.8	104.2	163.7	85.3	125.8	142.5	57.6	101.4	163.2	47.6	107.3
0330	Cast-in-Place Concrete	152.1	57.6	116.9	107.4	94.7	102.7	127.3	63.6	103.6	117.5	92.8	108.3	110.5	56.3	90.3	122.9	66.8	102.0
03	CONCRETE	136.8	55.4	101.0	101.6	92.4	97.6	149.1	67.5	113.3	115.7	87.0	103.1	113.1	58.4	89.0	131.9	65.0	102.5
04	MASONRY	187.9	54.9	106.2	161.9	89.8	117.6	208.6	74.3	126.1	165.0	85.6	116.2	163.3	57.4	98.3	222.6	65.6	126.1
05	METALS	136.4	79.6	119.8	113.9	91.9	107.4	127.7	75.2	112.2	112.1	90.7	105.8	105.6	73.7	96.2	128.1	75.7	112.7
06	WOOD, PLASTICS & COMPOSITES	104.3	52.5	75.8	131.8	90.3	108.9	128.1	81.4	102.4	118.5	81.6	98.2	95.8	57.9	74.9	115.9	67.4	89.2
07	THERMAL & MOISTURE PROTECTION	137.8	57.4	102.8	112.3	96.4	105.4	133.0	65.9	103.8	121.4	86.2	106.1	121.0	58.4	93.8	130.1	67.6	102.9
08	OPENINGS	85.7	45.8	76.4	87.8	76.3	85.2	106.3	66.6	97.0	89.1	80.6	87.1	85.5	51.2	77.5	91.3	61.3	84.3
0920	Plaster & Gypsum Board	126.3	50.5	75.3	143.9	89.9	107.6	151.2	80.9	103.9	174.9	81.1	111.8	129.5	56.7	80.5	146.6	66.5	92.7
0950, 0980	Ceilings & Acoustic Treatment	128.9	50.5	76.1	116.0	89.9	98.4	124.6	80.9	95.1	107.6	81.1	89.8	106.6	56.7	72.9	131.1	66.5	87.5
0960	Flooring	115.1	55.3	98.3	115.7	86.0	107.3	112.8	49.5	95.0	113.6	86.0	105.8	101.9	63.4	91.1	107.6	58.9	93.9
0970, 0990	Wall Finishes & Painting/Coating	117.3	39.9	71.5	111.3	104.1	107.1	117.8	56.5	81.5	110.3	85.7	95.7	113.8	47.9	74.8	117.8	59.3	83.2
09	FINISHES	119.8	52.3	83.3	114.6	91.6	102.1	122.6	70.2	94.2	117.3	83.6	99.0	108.4	58.2	81.2	120.3	65.5	90.7
COVERS	DIVS. 10 - 14, 25, 28, 41, 43, 44, 46	131.1	58.4	114.9	131.1	81.5	120.1	131.1	60.6	115.4	131.1	64.6	116.3	131.1	57.8	114.8	131.1	60.8	115.5
21, 22, 23	FIRE SUPPRESSION, PLUMBING & HVAC	104.1	59.0	86.4	103.8	81.0	94.9	104.0	65.9	89.0	104.2	94.3	100.3	103.9	64.8	88.5	104.0	79.5	94.4
26, 27, 3370	ELECTRICAL, COMMUNICATIONS & UTIL.	110.0	47.5	79.2	109.9	84.1	97.2	110.9	51.5	81.6	111.6	83.4	97.7	113.9	52.8	83.8	115.0	58.7	87.2
MF2018	WEIGHTED AVERAGE	121.3	60.2	95.5	110.4	87.6	100.7	124.7	68.4	100.9	113.1	87.3	102.2	110.2	62.5	90.0	121.5	70.2	99.8

	DIVISION	CANADA																	
		EDMONTON, ALBERTA			FORT MCMURRAY, ALBERTA			FREDERICTON, NEW BRUNSWICK			GATINEAU, QUEBEC			GRANBY, QUEBEC			HALIFAX, NOVA SCOTIA		
		MAT.	INST.	TOTAL	MAT.	INST.	TOTAL	MAT.	INST.	TOTAL	MAT.	INST.	TOTAL	MAT.	INST.	TOTAL	MAT.	INST.	TOTAL
015433	CONTRACTOR EQUIPMENT		129.1	129.1		101.9	101.9		113.9	113.9		100.2	100.2		100.2	100.2		115.6	115.6
0241, 31 - 34	SITE & INFRASTRUCTURE, DEMOLITION	125.1	121.8	122.8	121.7	95.4	103.5	116.0	100.8	105.5	96.7	93.7	94.6	97.2	93.7	94.8	102.8	103.6	103.4
0310	Concrete Forming & Accessories	128.3	94.9	99.8	125.0	87.5	93.1	126.9	58.6	68.7	132.9	79.1	87.0	132.9	79.0	87.0	124.1	87.7	93.0
0320	Concrete Reinforcing	136.0	82.0	109.9	151.5	81.9	117.8	141.0	57.8	100.8	147.4	72.4	111.1	147.4	72.4	111.1	156.3	77.9	118.3
0330	Cast-in-Place Concrete	145.8	104.3	130.4	173.5	98.2	145.5	112.7	57.5	92.1	104.0	87.5	97.9	107.5	87.4	100.0	99.4	85.1	94.0
03	CONCRETE	130.8	96.8	115.9	139.8	90.6	118.2	116.9	59.7	91.8	107.9	81.4	96.2	109.5	81.4	97.1	111.9	86.1	100.6
04	MASONRY	212.7	89.8	137.2	206.1	86.3	132.5	189.1	58.8	109.0	162.3	76.3	109.5	162.6	76.3	109.6	185.4	86.4	124.6
05	METALS	134.6	102.4	125.1	139.8	90.9	125.5	136.6	83.3	121.0	110.7	84.8	103.1	110.9	84.8	103.2	136.3	99.2	125.4
06	WOOD, PLASTICS & COMPOSITES	101.6	94.4	97.6	114.5	86.7	99.2	106.5	58.2	79.9	131.6	79.0	102.7	131.6	79.0	102.7	103.7	87.8	95.0
07	THERMAL & MOISTURE PROTECTION	143.5	98.0	123.7	129.8	92.6	113.6	137.4	59.4	103.5	114.8	84.1	101.5	114.8	82.6	100.8	137.4	88.7	116.2
08	OPENINGS	80.9	83.7	81.5	89.1	79.5	86.9	88.0	50.3	79.2	89.1	67.9	84.2	89.1	67.9	84.2	88.9	79.2	86.7
0920	Plaster & Gypsum Board	129.8	93.3	105.2	116.6	86.1	96.1	130.3	56.7	80.8	114.4	78.4	90.2	114.4	78.4	90.2	117.8	87.1	97.1
0950, 0980	Ceilings & Acoustic Treatment	144.5	93.3	110.0	112.5	86.1	94.7	137.2	56.7	82.9	103.5	78.4	86.6	103.5	78.4	86.6	139.8	87.1	104.2
0960	Flooring	121.8	84.8	111.4	113.6	84.8	105.5	115.6	66.2	101.7	113.6	86.0	105.8	113.6	86.0	105.8	109.5	81.5	101.6
0970, 0990	Wall Finishes & Painting/Coating	115.6	107.4	110.8	110.4	88.3	97.3	118.8	61.1	84.6	110.3	83.6	94.5	110.3	83.6	94.5	118.4	90.9	102.1
09	FINISHES	123.6	94.9	108.0	111.8	87.4	98.6	125.1	60.4	90.1	107.8	81.2	93.4	107.8	81.2	93.4	117.7	87.5	101.3
COVERS	DIVS. 10 - 14, 25, 28, 41, 43, 44, 46	131.1	93.6	122.8	131.1	90.1	122.0	131.1	58.5	115.0	131.1	75.6	118.8	131.1	75.6	118.8	131.1	68.9	117.3
21, 22, 23	FIRE SUPPRESSION, PLUMBING & HVAC	104.9	90.4	99.2	104.2	93.2	99.9	104.2	73.7	92.2	104.2	84.2	96.3	103.8	84.2	96.1	104.9	82.3	96.0
26, 27, 3370	ELECTRICAL, COMMUNICATIONS & UTIL.	115.0	93.9	104.6	105.4	77.5	91.7	112.2	69.7	91.2	110.2	65.2	88.0	110.9	65.2	88.4	113.3	90.6	102.1
MF2018	WEIGHTED AVERAGE	121.8	95.9	110.9	121.9	88.4	107.7	119.3	69.0	98.0	110.3	79.8	97.4	110.5	79.7	97.5	117.9	88.0	105.2

	DIVISION	CANADA																	
		HAMILTON, ONTARIO			HULL, QUEBEC			JOLIETTE, QUEBEC			KAMLOOPS, BRITISH COLUMBIA			KINGSTON, ONTARIO			KITCHENER, ONTARIO		
		MAT.	INST.	TOTAL	MAT.	INST.	TOTAL	MAT.	INST.	TOTAL	MAT.	INST.	TOTAL	MAT.	INST.	TOTAL	MAT.	INST.	TOTAL
015433	CONTRACTOR EQUIPMENT		115.5	115.5		100.2	100.2		100.2	100.2		103.3	103.3		101.9	101.9		101.6	101.6
0241, 31 - 34	SITE & INFRASTRUCTURE, DEMOLITION	105.8	107.0	106.6	96.7	93.7	94.6	97.3	93.7	94.8	120.1	97.1	104.3	115.1	98.0	103.3	94.2	99.8	98.0
0310	Concrete Forming & Accessories	130.0	94.7	99.9	132.9	79.1	87.0	132.9	79.2	87.1	124.3	82.3	88.5	125.1	82.5	88.7	117.3	88.0	92.3
0320	Concrete Reinforcing	136.9	103.3	120.7	147.4	72.4	111.1	139.3	72.4	107.0	109.3	75.8	93.1	163.7	85.3	125.8	96.4	103.1	99.6
0330	Cast-in-Place Concrete	107.6	102.1	105.6	104.0	87.5	97.9	108.4	87.5	100.6	93.8	91.8	93.1	117.5	92.8	108.3	111.7	97.2	106.3
03	CONCRETE	110.2	99.3	105.4	107.9	81.4	96.2	108.8	81.5	96.8	117.2	85.0	103.0	117.5	87.0	104.1	97.1	94.1	95.8
04	MASONRY	173.2	100.5	128.5	162.3	76.3	109.5	162.7	76.3	109.6	169.4	84.0	116.9	171.9	85.7	118.9	142.2	98.6	115.4
05	METALS	134.7	106.4	126.4	110.9	84.8	103.2	110.9	85.0	103.3	112.8	87.1	105.3	113.5	90.7	106.8	124.1	98.0	116.5
06	WOOD, PLASTICS & COMPOSITES	111.4	94.0	101.8	131.6	79.0	102.7	131.6	79.0	102.7	102.2	80.7	90.3	118.5	81.7	98.3	110.3	85.9	96.9
07	THERMAL & MOISTURE PROTECTION	134.0	101.5	119.8	114.8	84.1	101.5	114.8	84.1	101.5	131.2	81.7	109.7	121.4	87.3	106.6	116.4	98.4	108.5
08	OPENINGS	85.3	93.2	87.1	89.1	67.9	84.2	89.1	72.3	85.2	86.1	78.5	84.3	89.1	80.3	87.1	80.0	87.1	81.7
0920	Plaster & Gypsum Board	131.0	93.1	105.5	114.4	78.4	90.2	147.6	78.4	101.0	101.1	79.8	86.8	177.6	81.2	112.8	107.2	85.6	92.6
0950, 0980	Ceilings & Acoustic Treatment	139.9	93.1	108.4	103.5	78.4	86.6	103.5	78.4	86.6	103.5	79.8	87.5	119.0	81.2	93.5	107.9	85.6	92.9
0960	Flooring	115.3	95.0	109.6	113.6	86.0	105.8	113.6	86.0	105.8	112.5	49.2	94.7	113.6	86.0	105.8	101.3	95.1	99.6
0970, 0990	Wall Finishes & Painting/Coating	113.0	104.1	107.8	110.3	83.6	94.5	110.3	83.6	94.5	110.3	76.4	90.3	110.3	79.2	91.9	104.4	93.9	98.2
09	FINISHES	121.0	95.6	107.3	107.8	81.2	93.4	112.2	81.2	95.4	108.3	75.8	90.7	119.9	82.9	99.9	102.1	89.4	95.2
COVERS	DIVS. 10 - 14, 25, 28, 41, 43, 44, 46	131.1	92.5	122.5	131.1	75.6	118.8	131.1	75.6	118.8	131.1	84.4	120.7	131.1	64.6	116.3	131.1	89.6	121.9
21, 22, 23	FIRE SUPPRESSION, PLUMBING & HVAC	105.2	92.1	100.0	103.8	84.2	96.1	103.8	84.2	96.1	103.8	87.5	97.4	104.2	94.4	100.4	104.2	91.3	99.1
26, 27, 3370	ELECTRICAL, COMMUNICATIONS & UTIL.	108.5	103.6	106.1	112.3	65.2	89.1	110.9	65.2	88.4	114.3	74.2	94.5	111.6	82.1	97.1	109.6	99.8	104.8
MF2018	WEIGHTED AVERAGE	116.3	98.8	108.9	110.5	79.8	97.5	110.8	80.0	97.8	113.0	83.5	100.5	114.1	87.4	102.8	108.6	94.6	102.7

DIVISION		CANADA																	
		LAVAL, QUEBEC			LETHBRIDGE, ALBERTA			LLOYDMINSTER, ALBERTA			LONDON, ONTARIO			MEDICINE HAT, ALBERTA			MONCTON, NEW BRUNSWICK		
		MAT.	INST.	TOTAL	MAT.	INST.	TOTAL	MAT.	INST.	TOTAL	MAT.	INST.	TOTAL	MAT.	INST.	TOTAL	MAT.	INST.	TOTAL
015433	CONTRACTOR EQUIPMENT		100.2	100.2		101.9	101.9		101.9	101.9		116.7	116.7		101.9	101.9		99.7	99.7
0241, 31 - 34	SITE & INFRASTRUCTURE, DEMOLITION	97.2	93.9	94.9	114.5	96.0	101.7	114.5	95.4	101.3	103.5	106.8	105.8	113.2	95.5	101.0	104.5	92.5	96.2
0310	Concrete Forming & Accessories	133.2	81.1	88.8	126.4	87.6	93.3	124.6	78.3	85.1	129.1	89.1	95.0	126.4	78.2	85.3	106.1	66.1	72.0
0320	Concrete Reinforcing	147.4	74.4	112.1	151.5	81.9	117.8	151.5	81.8	117.8	126.7	102.1	114.8	151.5	81.8	117.8	140.5	69.8	106.3
0330	Cast-in-Place Concrete	107.5	89.4	100.7	130.1	98.2	118.2	120.7	94.6	111.0	120.2	100.8	113.0	120.7	94.6	111.0	109.8	68.0	94.3
03	CONCRETE	109.5	83.4	98.0	120.0	90.6	107.1	115.5	85.2	102.2	114.6	96.2	106.5	115.7	85.2	102.3	108.4	68.5	90.9
04	MASONRY	162.6	78.4	110.8	180.1	86.3	122.4	161.6	80.1	111.6	183.8	99.5	132.0	161.6	80.1	111.6	159.4	74.6	107.3
05	METALS	110.7	85.9	103.5	134.1	91.0	121.5	112.8	90.8	106.4	134.7	106.8	126.5	113.0	90.8	106.5	114.1	86.8	106.1
06	WOOD, PLASTICS & COMPOSITES	131.7	81.1	103.9	118.0	86.7	100.8	114.5	77.4	94.1	114.8	86.8	99.4	118.0	77.4	95.7	97.5	65.0	79.6
07	THERMAL & MOISTURE PROTECTION	115.4	86.0	102.6	127.1	92.6	112.1	123.5	88.2	108.1	129.9	99.0	116.5	130.5	88.2	112.1	116.6	72.4	97.4
08	OPENINGS	89.1	69.8	84.6	89.1	79.5	86.9	89.1	74.4	85.7	81.6	87.9	83.1	89.1	74.4	85.7	84.0	61.9	78.8
0920	Plaster & Gypsum Board	114.7	80.5	91.7	106.4	86.1	92.7	102.1	76.6	84.9	134.4	85.8	101.7	104.3	76.6	85.6	124.2	64.1	83.7
0950, 0980	Ceilings & Acoustic Treatment	103.5	80.5	88.0	112.5	86.1	94.7	103.5	76.6	85.3	138.5	85.8	102.9	103.5	76.6	85.3	116.8	64.1	81.2
0960	Flooring	113.6	88.4	106.5	113.6	84.8	105.5	113.6	84.8	105.5	109.9	95.1	105.7	113.6	84.8	105.5	99.6	65.6	90.1
0970, 0990	Wall Finishes & Painting/Coating	110.3	85.9	95.8	110.2	96.4	102.1	110.4	75.1	89.5	113.5	100.5	105.8	110.2	75.1	89.4	111.3	83.6	94.9
09	FINISHES	107.8	83.3	94.5	109.6	88.3	98.0	107.4	78.9	92.0	121.5	91.0	105.0	107.5	78.9	92.0	107.7	67.5	86.0
COVERS	DIVS. 10 - 14, 25, 28, 41, 43, 44, 46	131.1	77.5	119.2	131.1	89.2	121.8	131.1	87.1	121.3	131.1	91.6	122.3	131.1	86.1	121.1	131.1	61.0	115.5
21, 22, 23	FIRE SUPPRESSION, PLUMBING & HVAC	104.3	86.5	97.3	104.1	90.1	98.6	104.2	90.0	98.6	105.1	89.2	98.8	103.8	86.8	97.2	103.8	73.7	92.0
26, 27, 3370	ELECTRICAL, COMMUNICATIONS & UTIL.	110.9	67.0	89.3	107.1	77.5	92.5	104.6	77.5	91.2	103.5	102.0	102.8	104.6	77.5	91.2	117.3	88.5	103.1
MF2018	WEIGHTED AVERAGE	110.7	81.6	98.4	117.0	87.9	104.7	111.5	84.7	100.2	116.3	96.5	108.0	111.7	84.0	100.0	110.9	76.0	96.2

DIVISION		CANADA																	
		MONTREAL, QUEBEC			MOOSE JAW, SASKATCHEWAN			NEW GLASGOW, NOVA SCOTIA			NEWCASTLE, NEW BRUNSWICK			NORTH BAY, ONTARIO			OSHAWA, ONTARIO		
		MAT.	INST.	TOTAL	MAT.	INST.	TOTAL	MAT.	INST.	TOTAL	MAT.	INST.	TOTAL	MAT.	INST.	TOTAL	MAT.	INST.	TOTAL
015433	CONTRACTOR EQUIPMENT		117.7	117.7		98.3	98.3		98.9	98.9		99.7	99.7		99.3	99.3		101.6	101.6
0241, 31 - 34	SITE & INFRASTRUCTURE, DEMOLITION	113.0	105.5	107.8	114.4	90.0	97.5	111.9	92.0	98.1	105.1	90.6	95.1	126.8	93.6	103.9	104.6	99.6	101.2
0310	Concrete Forming & Accessories	130.9	90.4	96.4	109.6	54.6	62.7	111.8	67.8	74.3	106.1	58.0	65.1	146.2	80.0	89.8	122.8	91.3	96.0
0320	Concrete Reinforcing	126.8	94.6	111.3	106.9	61.4	84.9	156.0	47.6	103.6	140.5	57.6	100.4	184.7	84.8	136.4	152.7	103.7	129.0
0330	Cast-in-Place Concrete	124.4	97.4	114.3	117.1	63.7	97.2	122.9	66.8	102.0	114.0	56.3	92.6	118.4	80.1	104.1	129.2	106.4	120.7
03	CONCRETE	116.7	94.3	106.8	101.4	60.0	83.2	131.0	65.0	102.0	110.3	58.4	87.5	132.7	81.4	110.2	118.3	98.8	109.8
04	MASONRY	178.8	89.8	124.1	160.3	56.4	96.4	208.2	65.6	120.5	159.8	57.4	96.9	214.4	81.9	133.0	145.1	101.9	118.6
05	METALS	144.3	102.2	131.9	109.6	75.0	99.4	125.6	75.7	111.0	114.1	73.8	102.3	126.5	90.2	115.8	114.2	98.6	109.6
06	WOOD, PLASTICS & COMPOSITES	111.3	90.8	100.0	99.4	53.2	74.0	115.9	67.4	89.2	97.5	57.9	75.7	153.5	80.2	113.2	117.6	89.4	102.1
07	THERMAL & MOISTURE PROTECTION	125.9	97.2	113.4	114.1	60.1	90.6	130.1	67.6	102.9	116.6	58.4	91.3	136.4	82.7	113.0	117.3	104.1	111.6
08	OPENINGS	87.4	78.7	85.4	85.3	51.4	77.4	91.3	61.3	84.3	84.0	51.2	76.3	98.6	78.2	93.8	85.0	90.2	86.2
0920	Plaster & Gypsum Board	126.2	89.9	101.8	98.2	51.9	67.1	145.0	66.5	92.2	124.2	56.7	78.8	140.6	79.8	99.7	110.7	89.2	96.2
0950, 0980	Ceilings & Acoustic Treatment	146.1	89.9	108.2	103.5	51.9	68.7	123.8	66.5	85.2	116.8	56.7	76.3	123.8	79.8	94.1	103.9	89.2	94.0
0960	Flooring	112.4	90.5	106.2	103.6	53.7	89.6	107.6	58.9	93.9	99.6	63.4	89.5	131.2	86.0	118.5	104.2	97.6	102.4
0970, 0990	Wall Finishes & Painting/Coating	115.0	104.1	108.5	110.3	61.0	81.1	117.8	59.3	83.2	111.3	47.9	73.8	117.8	85.0	98.4	104.4	108.1	106.6
09	FINISHES	121.6	92.8	106.0	103.7	54.8	77.2	118.7	65.5	89.9	107.7	58.2	80.9	125.9	81.9	102.1	103.2	93.7	98.1
COVERS	DIVS. 10 - 14, 25, 28, 41, 43, 44, 46	131.1	82.8	120.4	131.1	57.5	114.7	131.1	60.8	115.5	131.1	57.9	114.8	131.1	63.4	116.0	131.1	90.4	122.0
21, 22, 23	FIRE SUPPRESSION, PLUMBING & HVAC	105.1	81.2	95.7	104.2	70.3	90.9	104.0	79.5	94.4	103.8	64.8	88.5	104.0	92.4	99.4	104.2	91.9	99.3
26, 27, 3370	ELECTRICAL, COMMUNICATIONS & UTIL.	107.9	84.1	96.2	113.4	56.3	85.3	111.3	58.7	85.4	112.9	55.9	84.8	112.6	83.3	98.1	110.6	103.8	107.3
MF2018	WEIGHTED AVERAGE	119.1	90.1	106.9	109.0	63.9	89.9	119.6	70.2	98.7	110.7	62.9	90.5	122.5	85.2	106.8	110.8	97.2	105.0

DIVISION		CANADA																	
		OTTAWA, ONTARIO			OWEN SOUND, ONTARIO			PETERBOROUGH, ONTARIO			PORTAGE LA PRAIRIE, MANITOBA			PRINCE ALBERT, SASKATCHEWAN			PRINCE GEORGE, BRITISH COLUMBIA		
		MAT.	INST.	TOTAL	MAT.	INST.	TOTAL	MAT.	INST.	TOTAL	MAT.	INST.	TOTAL	MAT.	INST.	TOTAL	MAT.	INST.	TOTAL
015433	CONTRACTOR EQUIPMENT		117.8	117.8		100.1	100.1		99.7	99.7		101.9	101.9		98.3	98.3		103.3	103.3
0241, 31 - 34	SITE & INFRASTRUCTURE, DEMOLITION	107.0	107.2	107.1	117.0	94.6	101.5	116.9	94.2	101.3	115.4	93.2	100.1	110.2	90.1	96.3	123.4	97.1	105.3
0310	Concrete Forming & Accessories	124.8	88.4	93.7	125.6	78.7	85.6	127.2	81.0	87.8	126.6	65.7	74.7	109.6	54.4	62.5	115.1	77.5	83.0
0320	Concrete Reinforcing	137.0	102.0	120.1	175.2	86.9	132.5	163.7	85.3	125.8	151.5	54.5	104.6	111.7	61.4	87.3	109.3	75.8	93.1
0330	Cast-in-Place Concrete	122.3	103.0	115.1	157.7	76.6	127.5	130.7	81.8	112.5	120.7	69.8	101.8	106.2	63.6	90.3	117.6	91.8	108.0
03	CONCRETE	116.6	96.7	107.9	138.9	79.9	113.0	121.8	82.5	104.6	109.5	66.2	90.5	96.9	59.9	80.7	127.5	82.8	107.9
04	MASONRY	173.2	99.4	127.8	165.6	87.1	117.3	166.1	88.0	118.1	164.8	59.7	100.2	159.5	56.4	96.1	171.4	84.0	117.7
05	METALS	135.5	108.9	127.7	111.1	91.3	105.3	112.3	90.8	106.0	113.0	78.4	102.8	109.7	74.8	99.4	112.8	87.2	105.3
06	WOOD, PLASTICS & COMPOSITES	108.4	85.7	95.9	116.5	77.3	94.9	120.3	79.0	97.6	118.0	66.8	89.8	99.4	53.2	74.0	102.2	74.1	86.7
07	THERMAL & MOISTURE PROTECTION	137.3	99.7	121.0	115.3	83.5	101.5	121.6	88.0	107.0	114.6	68.2	94.4	114.0	59.0	90.1	124.5	81.0	105.6
08	OPENINGS	90.3	87.5	89.7	90.5	77.6	87.5	87.7	79.9	85.9	89.1	59.9	82.3	84.3	51.4	76.6	86.1	74.9	83.5
0920	Plaster & Gypsum Board	129.1	84.6	99.1	153.5	76.7	101.8	115.8	78.5	90.7	104.0	65.6	78.2	98.2	51.9	67.1	101.1	73.0	82.2
0950, 0980	Ceilings & Acoustic Treatment	142.3	84.6	103.4	94.3	76.7	82.4	103.5	78.5	86.7	103.5	65.6	78.0	103.5	51.9	68.7	103.5	73.0	83.0
0960	Flooring	108.3	90.6	103.4	119.3	87.3	110.3	113.6	86.0	105.8	113.6	61.9	99.1	103.6	53.7	89.6	109.0	67.4	97.3
0970, 0990	Wall Finishes & Painting/Coating	113.5	95.9	103.1	105.6	83.8	92.7	110.3	87.1	96.6	110.4	54.1	77.1	110.3	52.0	75.8	110.3	76.4	90.3
09	FINISHES	122.3	89.0	104.3	111.3	80.8	94.8	108.7	82.6	94.5	107.4	64.1	84.0	103.7	53.8	76.7	107.2	75.0	89.8
COVERS	DIVS. 10 - 14, 25, 28, 41, 43, 44, 46	131.1	89.5	121.9	139.2	63.9	122.4	131.1	64.8	116.3	131.1	59.8	115.2	131.1	57.5	114.7	131.1	83.7	120.6
21, 22, 23	FIRE SUPPRESSION, PLUMBING & HVAC	105.1	90.7	99.4	103.7	92.4	99.3	103.8	95.8	100.7	103.8	78.4	93.8	104.2	63.4	88.2	103.8	87.5	97.4
26, 27, 3370	ELECTRICAL, COMMUNICATIONS & UTIL.	104.4	101.9	103.2	118.5	82.1	100.6	111.5	82.9	97.4	113.0	54.4	84.2	113.4	56.3	85.3	111.5	74.2	93.1
MF2018	WEIGHTED AVERAGE	117.5	96.8	108.7	116.7	85.4	103.4	113.1	87.1	102.1	111.6	69.1	93.6	108.2	62.2	88.8	114.0	82.8	100.8

DIVISION		CANADA																	
		QUEBEC CITY, QUEBEC			RED DEER, ALBERTA			REGINA, SASKATCHEWAN			RIMOUSKI, QUEBEC			ROUYN-NORANDA, QUEBEC			SAINT HYACINTHE, QUEBEC		
		MAT.	INST.	TOTAL	MAT.	INST.	TOTAL	MAT.	INST.	TOTAL	MAT.	INST.	TOTAL	MAT.	INST.	TOTAL	MAT.	INST.	TOTAL
015433	**CONTRACTOR EQUIPMENT**		117.4	117.4		101.9	101.9		128.3	128.3		100.2	100.2		100.2	100.2		100.2	100.2
0241, 31 - 34	**SITE & INFRASTRUCTURE, DEMOLITION**	114.1	104.7	107.6	113.2	95.5	101.0	126.4	119.6	121.7	97.0	93.6	94.6	96.7	93.7	94.6	97.2	93.7	94.8
0310	Concrete Forming & Accessories	127.4	90.5	95.9	139.7	78.2	87.3	131.0	91.5	97.4	132.9	89.8	96.2	132.9	79.1	87.0	132.9	79.1	87.0
0320	Concrete Reinforcing	124.2	94.6	109.9	151.5	81.8	117.8	123.8	88.8	106.9	105.1	94.5	100.0	147.4	72.4	111.1	147.4	72.4	111.1
0330	Cast-in-Place Concrete	131.6	97.0	118.7	120.7	94.6	111.0	153.8	97.7	132.9	109.4	94.7	103.9	104.0	87.5	97.9	107.5	87.5	100.0
03	**CONCRETE**	119.4	94.3	108.4	116.5	85.2	102.7	131.3	94.0	114.9	105.0	92.4	99.5	107.9	81.4	96.2	109.5	81.4	97.1
04	**MASONRY**	173.6	89.8	122.1	161.6	80.1	111.6	205.2	86.7	132.4	162.1	89.8	117.6	162.3	76.3	109.5	162.6	76.3	109.6
05	**METALS**	138.0	103.3	127.8	113.0	90.8	106.5	139.1	101.4	128.0	110.4	91.8	104.9	110.9	84.8	103.2	110.9	84.8	103.2
06	**WOOD, PLASTICS & COMPOSITES**	114.2	90.7	101.3	118.0	77.4	95.7	108.0	92.2	99.3	131.6	90.3	108.9	131.6	79.0	102.7	131.6	79.0	102.7
07	**THERMAL & MOISTURE PROTECTION**	125.1	97.2	112.9	141.4	88.2	118.3	143.8	86.6	118.9	114.8	96.4	106.8	114.8	84.1	101.5	115.2	84.1	101.7
08	**OPENINGS**	88.4	86.0	87.8	89.1	74.4	85.7	87.9	79.8	86.1	88.7	76.3	85.8	89.1	67.9	84.2	89.1	67.9	84.2
0920	Plaster & Gypsum Board	128.5	89.9	102.6	104.3	76.6	85.6	140.8	91.2	107.4	147.4	89.9	108.7	114.2	78.4	90.1	114.2	78.4	90.1
0950, 0980	Ceilings & Acoustic Treatment	131.7	89.9	103.5	103.5	76.6	85.3	154.3	91.2	111.7	102.7	89.9	94.1	102.7	78.4	86.3	102.7	78.4	86.3
0960	Flooring	108.9	90.5	103.8	116.0	84.8	107.2	123.6	95.1	115.6	114.8	86.0	106.7	113.6	86.0	105.8	113.6	86.0	105.8
0970, 0990	Wall Finishes & Painting/Coating	120.8	104.1	111.0	110.2	75.1	89.4	118.0	89.0	100.9	113.4	104.1	107.9	110.3	83.6	94.5	110.3	83.6	94.5
09	**FINISHES**	117.4	92.7	104.0	108.2	78.9	92.4	130.9	92.9	110.3	112.6	91.6	101.2	107.6	81.2	93.3	107.6	81.2	93.3
COVERS	**DIVS. 10 - 14, 25, 28, 41, 43, 44, 46**	131.1	82.5	120.3	131.1	86.1	121.1	131.1	70.9	117.7	131.1	81.5	120.1	131.1	75.6	118.8	131.1	75.6	118.8
21, 22, 23	**FIRE SUPPRESSION, PLUMBING & HVAC**	105.1	81.2	95.7	103.8	86.8	97.2	104.8	86.9	97.8	103.8	81.0	94.9	103.8	84.2	96.1	100.5	84.2	94.1
26, 27, 3370	**ELECTRICAL, COMMUNICATIONS & UTIL.**	114.7	84.1	99.6	104.6	77.5	91.2	110.8	91.5	101.3	110.9	84.1	97.7	110.9	65.2	88.4	111.5	65.2	88.7
MF2018	**WEIGHTED AVERAGE**	118.7	90.4	106.7	112.1	84.0	100.2	123.1	92.4	110.2	110.2	87.6	100.6	110.3	79.8	97.4	109.8	79.8	97.1

DIVISION		CANADA																	
		SAINT JOHN, NEW BRUNSWICK			SARNIA, ONTARIO			SASKATOON, SASKATCHEWAN			SAULT STE MARIE, ONTARIO			SHERBROOKE, QUEBEC			SOREL, QUEBEC		
		MAT.	INST.	TOTAL	MAT.	INST.	TOTAL	MAT.	INST.	TOTAL	MAT.	INST.	TOTAL	MAT.	INST.	TOTAL	MAT.	INST.	TOTAL
015433	**CONTRACTOR EQUIPMENT**		99.7	99.7		99.7	99.7		98.4	98.4		99.7	99.7		100.2	100.2		100.2	100.2
0241, 31 - 34	**SITE & INFRASTRUCTURE, DEMOLITION**	105.2	92.5	96.4	115.5	94.3	100.9	112.7	92.8	98.9	105.6	93.9	97.5	97.2	93.7	94.8	97.3	93.7	94.8
0310	Concrete Forming & Accessories	126.2	63.4	72.7	125.9	87.6	93.3	109.7	90.7	93.5	115.1	86.2	90.5	132.9	79.1	87.0	132.9	79.2	87.1
0320	Concrete Reinforcing	140.5	69.8	106.3	116.5	86.7	102.1	114.2	88.7	101.9	105.1	85.4	95.6	147.4	72.4	111.1	139.3	72.4	107.0
0330	Cast-in-Place Concrete	112.4	68.0	95.9	120.6	94.2	110.8	113.5	93.1	105.9	108.3	80.5	98.0	107.5	87.5	100.0	108.4	87.5	100.6
03	**CONCRETE**	110.9	67.3	91.7	111.1	90.0	101.8	101.3	91.2	96.9	97.4	84.7	91.8	109.5	81.4	97.1	108.8	81.5	96.8
04	**MASONRY**	180.8	74.4	115.4	177.6	90.5	124.1	172.1	86.6	119.5	162.9	90.1	118.2	162.6	76.3	109.6	162.7	76.3	109.6
05	**METALS**	114.0	86.8	106.0	112.3	91.2	106.1	106.5	88.6	101.2	111.5	95.2	106.7	110.7	84.8	103.1	110.9	85.0	103.3
06	**WOOD, PLASTICS & COMPOSITES**	120.7	61.4	88.0	119.4	86.7	101.4	96.0	91.3	93.4	106.5	87.9	96.3	131.6	79.0	102.7	131.6	79.0	102.7
07	**THERMAL & MOISTURE PROTECTION**	116.9	72.2	97.5	121.7	91.5	108.6	117.0	84.7	103.0	120.4	87.3	106.0	114.8	84.1	101.5	114.8	84.1	101.5
08	**OPENINGS**	83.9	58.9	78.0	90.5	83.5	88.8	85.3	79.4	83.9	82.4	86.1	83.2	89.1	67.9	84.2	89.1	72.3	85.2
0920	Plaster & Gypsum Board	137.9	60.3	85.7	143.2	86.4	105.0	115.6	91.2	99.1	107.1	87.6	94.0	114.2	78.4	90.1	147.4	78.4	101.0
0950, 0980	Ceilings & Acoustic Treatment	121.7	60.3	80.3	109.2	86.4	93.9	124.5	91.2	102.0	103.5	87.6	92.8	102.7	78.4	86.3	102.7	78.4	86.3
0960	Flooring	111.4	65.6	98.5	113.6	94.1	108.1	106.5	95.1	103.3	106.5	92.1	102.4	113.6	86.0	105.8	113.6	86.0	105.8
0970, 0990	Wall Finishes & Painting/Coating	111.3	83.6	94.9	110.3	98.5	103.3	113.8	89.0	99.2	110.3	91.3	99.1	110.3	83.6	94.5	110.3	83.6	94.5
09	**FINISHES**	114.0	65.4	87.7	113.5	90.2	100.9	112.1	92.3	101.4	104.7	88.2	95.8	107.6	81.2	93.3	112.0	81.2	95.3
COVERS	**DIVS. 10 - 14, 25, 28, 41, 43, 44, 46**	131.1	61.0	115.5	131.1	66.1	116.6	131.1	69.3	117.4	131.1	87.9	121.5	131.1	75.6	118.8	131.1	75.6	118.8
21, 22, 23	**FIRE SUPPRESSION, PLUMBING & HVAC**	103.8	75.1	92.6	103.8	101.5	102.9	104.4	86.7	97.5	103.8	90.5	98.6	104.2	84.2	96.3	103.8	84.2	96.1
26, 27, 3370	**ELECTRICAL, COMMUNICATIONS & UTIL.**	119.9	88.5	104.4	114.7	85.0	100.0	114.6	91.4	103.2	113.2	83.3	98.5	110.9	65.2	88.4	110.9	65.2	88.4
MF2018	**WEIGHTED AVERAGE**	113.2	75.7	97.3	113.2	91.3	103.9	109.9	88.5	100.9	108.6	88.7	100.2	110.6	79.8	97.6	110.8	80.0	97.8

DIVISION		CANADA																	
		ST. CATHARINES, ONTARIO			ST JEROME, QUEBEC			ST. JOHN'S, NEWFOUNDLAND			SUDBURY, ONTARIO			SUMMERSIDE, PRINCE EDWARD ISLAND			SYDNEY, NOVA SCOTIA		
		MAT.	INST.	TOTAL	MAT.	INST.	TOTAL	MAT.	INST.	TOTAL	MAT.	INST.	TOTAL	MAT.	INST.	TOTAL	MAT.	INST.	TOTAL
015433	**CONTRACTOR EQUIPMENT**		99.5	99.5		100.2	100.2		123.4	123.4		99.5	99.5		98.8	98.8		98.9	98.9
0241, 31 - 34	**SITE & INFRASTRUCTURE, DEMOLITION**	94.2	96.1	95.6	96.7	93.7	94.6	115.4	111.8	112.9	94.4	95.7	95.3	121.8	89.4	99.5	108.0	92.0	96.9
0310	Concrete Forming & Accessories	115.3	93.8	97.0	132.9	79.1	87.0	125.3	85.2	91.1	111.0	88.9	92.1	112.0	52.5	61.3	111.8	67.8	74.3
0320	Concrete Reinforcing	97.2	103.2	100.1	147.4	72.4	111.1	172.1	82.4	128.7	98.0	101.4	99.7	154.0	46.5	102.0	156.0	47.6	103.6
0330	Cast-in-Place Concrete	106.7	98.7	103.7	104.0	87.5	97.9	137.1	98.5	122.8	107.6	95.3	103.0	116.7	54.7	93.6	94.8	66.8	84.4
03	**CONCRETE**	94.8	97.2	95.8	107.9	81.4	96.2	132.7	90.3	114.1	95.0	93.5	94.3	139.1	53.4	101.5	118.1	65.0	94.8
04	**MASONRY**	141.8	100.8	116.6	162.3	76.3	109.5	200.1	89.9	132.4	141.9	94.9	113.0	207.4	54.8	113.7	205.8	65.6	119.6
05	**METALS**	114.2	97.9	109.4	110.9	84.8	103.2	137.9	99.1	126.5	113.6	97.1	108.8	125.6	69.2	109.1	125.6	75.7	111.0
06	**WOOD, PLASTICS & COMPOSITES**	107.8	93.2	99.8	131.6	79.0	102.7	106.9	83.4	94.0	103.5	87.9	94.9	116.4	51.9	80.9	115.9	67.4	89.2
07	**THERMAL & MOISTURE PROTECTION**	116.4	102.0	110.1	114.8	84.1	101.5	140.2	94.7	120.4	115.7	96.3	107.3	129.4	57.3	98.0	130.1	67.6	102.9
08	**OPENINGS**	79.5	91.4	82.3	89.1	67.9	84.2	85.4	74.3	82.8	80.2	86.7	81.7	102.5	45.4	89.2	91.3	61.3	84.3
0920	Plaster & Gypsum Board	100.8	93.1	95.6	114.2	78.4	90.1	150.0	82.1	104.3	100.0	87.6	91.7	145.9	50.5	81.8	145.0	66.5	92.2
0950, 0980	Ceilings & Acoustic Treatment	103.9	93.1	96.6	102.7	78.4	86.3	143.7	82.1	102.2	98.2	87.6	91.0	123.8	50.5	74.4	123.8	66.5	85.2
0960	Flooring	100.2	91.9	97.8	113.6	86.0	105.8	116.8	52.2	98.7	98.3	92.1	96.6	107.6	55.3	92.9	107.6	58.9	93.9
0970, 0990	Wall Finishes & Painting/Coating	104.4	104.1	104.2	110.3	83.6	94.5	120.8	98.6	107.7	104.4	95.3	99.0	117.8	39.9	71.7	117.8	59.3	83.2
09	**FINISHES**	100.1	94.6	97.1	107.6	81.2	93.3	129.2	80.4	102.8	98.3	89.9	93.7	119.8	51.9	83.1	118.7	65.5	89.9
COVERS	**DIVS. 10 - 14, 25, 28, 41, 43, 44, 46**	131.1	69.7	117.4	131.1	75.6	118.8	131.1	68.6	117.2	131.1	89.6	121.9	131.1	57.0	114.6	131.1	60.8	115.5
21, 22, 23	**FIRE SUPPRESSION, PLUMBING & HVAC**	104.2	91.1	99.0	103.8	84.2	96.1	104.8	83.0	96.2	103.6	89.4	98.1	104.0	58.9	86.2	104.0	79.5	94.4
26, 27, 3370	**ELECTRICAL, COMMUNICATIONS & UTIL.**	111.3	101.8	106.6	111.6	65.2	88.7	115.6	81.4	98.7	109.5	102.1	105.9	110.2	47.4	79.2	111.3	58.7	85.4
MF2018	**WEIGHTED AVERAGE**	106.6	95.6	101.9	110.4	79.8	97.4	122.6	87.4	107.7	106.1	93.6	100.8	121.9	57.9	94.9	117.8	70.2	97.7

DIVISION		CANADA																	
		THUNDER BAY, ONTARIO			TIMMINS, ONTARIO			TORONTO, ONTARIO			TROIS RIVIERES, QUEBEC			TRURO, NOVA SCOTIA			VANCOUVER, BRITISH COLUMBIA		
		MAT.	INST.	TOTAL	MAT.	INST.	TOTAL	MAT.	INST.	TOTAL	MAT.	INST.	TOTAL	MAT.	INST.	TOTAL	MAT.	INST.	TOTAL
015433	CONTRACTOR EQUIPMENT		99.5	99.5		99.7	99.7		117.5	117.5		99.7	99.7		99.4	99.4		135.3	135.3
0241, 31 - 34	SITE & INFRASTRUCTURE, DEMOLITION	98.8	96.0	96.9	116.9	93.9	101.0	108.4	107.7	107.9	108.2	93.4	97.9	101.3	92.3	95.1	117.0	122.8	121.0
0310	Concrete Forming & Accessories	122.8	92.5	97.0	127.2	80.0	87.0	128.7	102.5	106.3	153.7	79.1	90.1	99.7	67.9	72.6	131.0	91.6	97.4
0320	Concrete Reinforcing	86.9	102.6	94.5	163.7	84.8	125.6	136.9	103.9	120.9	156.0	72.4	115.6	139.3	47.6	95.0	132.5	95.2	114.5
0330	Cast-in-Place Concrete	117.5	98.0	110.2	130.7	80.2	111.9	108.4	113.7	110.4	98.2	87.4	94.2	135.7	66.9	110.1	123.5	90.6	111.3
03	CONCRETE	101.9	96.2	99.4	121.8	81.5	104.1	110.5	107.1	109.0	120.3	81.4	103.2	120.8	65.1	96.3	119.7	93.0	108.0
04	MASONRY	142.4	100.5	116.7	166.1	81.9	114.4	173.3	108.7	133.6	209.8	76.3	127.8	161.9	65.6	102.7	165.3	85.6	116.3
05	METALS	114.2	97.0	109.1	112.3	90.4	105.9	135.3	109.4	127.7	124.9	84.7	113.1	111.6	76.0	101.1	133.5	109.8	126.6
06	WOOD, PLASTICS & COMPOSITES	117.6	91.3	103.1	120.3	80.3	98.3	114.8	100.5	107.0	168.2	79.0	119.1	89.6	67.4	77.4	110.3	91.0	99.7
07	THERMAL & MOISTURE PROTECTION	116.6	99.3	109.1	121.6	82.7	104.7	135.6	109.4	124.2	128.6	84.1	109.2	116.1	67.6	95.0	141.7	88.6	118.6
08	OPENINGS	78.9	89.8	81.4	87.7	78.2	85.5	83.9	98.9	87.4	100.3	72.3	93.7	82.4	61.3	77.5	84.5	88.7	85.5
0920	Plaster & Gypsum Board	128.0	91.2	103.2	115.8	79.8	91.6	129.1	99.9	109.5	174.3	78.4	109.8	124.3	66.5	85.4	127.2	89.7	102.0
0950, 0980	Ceilings & Acoustic Treatment	98.2	91.2	93.4	103.5	79.8	87.5	139.9	99.9	112.9	122.2	78.4	92.6	103.5	66.5	78.6	151.0	89.7	109.7
0960	Flooring	104.2	98.5	102.6	113.6	86.0	105.8	109.1	100.8	106.8	131.2	86.0	118.5	95.7	58.9	85.4	124.7	88.5	114.6
0970, 0990	Wall Finishes & Painting/Coating	104.4	96.2	99.5	110.3	85.0	95.3	109.8	108.1	108.8	117.8	83.6	97.6	110.3	59.3	80.1	113.7	100.3	105.8
09	FINISHES	104.1	93.9	98.6	108.7	81.9	94.2	117.8	102.8	109.7	129.3	81.1	103.2	103.8	65.5	83.1	128.0	92.1	108.6
COVERS	DIVS. 10 - 14, 25, 28, 41, 43, 44, 46	131.1	69.7	117.4	131.1	63.4	116.0	131.1	95.4	123.2	131.1	75.6	118.8	131.1	60.8	115.5	131.1	90.8	122.1
21, 22, 23	FIRE SUPPRESSION, PLUMBING & HVAC	104.2	91.2	99.1	103.8	92.4	99.3	105.1	99.6	103.0	104.0	84.2	96.2	103.8	79.5	94.3	105.2	87.4	98.2
26, 27, 3370	ELECTRICAL, COMMUNICATIONS & UTIL.	109.6	101.2	105.4	113.2	83.3	98.5	106.8	103.8	105.3	111.6	65.2	88.8	110.5	58.7	85.0	108.7	78.6	93.9
MF2018	WEIGHTED AVERAGE	107.8	95.0	102.4	113.3	85.3	101.4	116.0	104.2	111.0	120.2	79.9	103.2	110.8	70.3	93.7	117.9	92.4	107.2

DIVISION		CANADA																	
		VICTORIA, BRITISH COLUMBIA			WHITEHORSE, YUKON			WINDSOR, ONTARIO			WINNIPEG, MANITOBA			YARMOUTH, NOVA SCOTIA			YELLOWKNIFE, NWT		
		MAT.	INST.	TOTAL	MAT.	INST.	TOTAL	MAT.	INST.	TOTAL	MAT.	INST.	TOTAL	MAT.	INST.	TOTAL	MAT.	INST.	TOTAL
015433	CONTRACTOR EQUIPMENT		106.3	106.3		132.8	132.8		99.5	99.5		125.5	125.5		98.9	98.9		130.7	130.7
0241, 31 - 34	SITE & INFRASTRUCTURE, DEMOLITION	124.3	101.3	108.4	135.7	119.6	124.6	90.8	96.0	94.4	113.7	115.3	114.8	111.7	92.0	98.1	146.3	123.6	130.6
0310	Concrete Forming & Accessories	114.9	88.9	92.7	135.0	56.7	68.2	122.8	90.6	95.3	130.4	65.4	75.0	111.8	67.8	74.3	137.3	75.8	84.8
0320	Concrete Reinforcing	111.7	95.0	103.6	167.7	63.3	117.2	95.2	101.9	98.5	127.3	61.0	95.3	156.0	47.6	103.6	138.6	65.8	103.4
0330	Cast-in-Place Concrete	117.6	87.0	106.2	152.4	71.2	122.2	109.3	99.7	105.7	143.6	72.0	116.9	121.6	66.8	101.2	174.9	87.6	142.4
03	CONCRETE	129.4	89.5	111.9	144.0	64.8	109.1	96.2	95.9	96.0	127.0	68.4	101.3	130.4	65.0	101.7	150.7	79.5	119.4
04	MASONRY	175.0	85.5	120.0	246.5	58.1	130.7	141.9	100.1	116.3	194.6	65.3	115.2	208.1	65.6	120.5	234.9	68.9	132.9
05	METALS	107.7	92.2	103.2	145.1	90.5	129.1	114.2	97.3	109.2	141.6	88.0	125.8	125.6	75.7	111.0	140.5	92.7	126.4
06	WOOD, PLASTICS & COMPOSITES	100.0	88.3	93.6	121.3	55.3	85.0	117.6	89.0	101.9	107.7	66.0	84.8	115.9	67.4	89.2	130.1	77.1	100.9
07	THERMAL & MOISTURE PROTECTION	126.9	86.3	109.2	145.8	63.3	109.9	116.4	99.1	108.9	136.1	70.3	107.5	130.1	67.6	102.9	143.7	78.8	115.5
08	OPENINGS	86.8	82.9	85.9	99.5	53.0	88.6	78.7	88.9	81.1	87.5	59.7	81.0	91.3	61.3	84.3	93.9	65.5	87.3
0920	Plaster & Gypsum Board	109.1	87.7	94.7	178.3	53.0	94.0	111.4	88.8	96.2	133.0	64.2	86.7	145.0	66.5	92.2	192.2	75.5	113.7
0950, 0980	Ceilings & Acoustic Treatment	105.8	87.7	93.6	169.3	53.0	90.9	98.2	88.8	91.8	145.3	64.2	90.6	123.8	66.5	85.2	169.0	75.5	105.9
0960	Flooring	112.0	67.4	99.5	122.4	55.2	103.5	104.2	95.9	101.9	115.9	68.5	102.6	107.6	58.9	93.9	121.9	82.9	111.0
0970, 0990	Wall Finishes & Painting/Coating	113.8	100.3	105.8	120.7	53.1	80.7	104.4	97.1	100.1	113.9	53.5	78.1	117.8	59.3	83.2	123.3	75.9	95.2
09	FINISHES	110.7	86.8	97.8	143.7	55.8	96.1	101.6	92.1	96.5	126.9	65.0	93.4	118.7	65.5	89.9	147.5	76.6	109.1
COVERS	DIVS. 10 - 14, 25, 28, 41, 43, 44, 46	131.1	65.8	116.6	131.1	60.6	115.4	131.1	69.2	117.3	131.1	63.8	116.1	131.1	60.8	115.5	131.1	63.9	116.1
21, 22, 23	FIRE SUPPRESSION, PLUMBING & HVAC	103.9	88.7	97.9	104.3	71.6	91.5	104.2	90.7	98.9	105.2	64.6	89.2	104.0	79.5	94.4	104.8	88.5	98.4
26, 27, 3370	ELECTRICAL, COMMUNICATIONS & UTIL.	112.8	82.1	97.7	130.5	57.1	94.3	114.5	100.4	107.6	113.1	63.4	88.6	111.3	58.7	85.4	125.8	77.4	102.0
MF2018	WEIGHTED AVERAGE	114.1	87.6	102.9	132.2	69.2	105.6	107.1	94.4	101.8	122.0	71.2	100.5	119.5	70.2	98.7	131.4	83.2	111.0

Location Factors - Commercial

Costs shown in RSMeans cost data publications are based on national averages for materials and installation. To adjust these costs to a specific location, simply multiply the base cost by the factor and divide by 100 for that city. The data is arranged alphabetically by state and postal zip code numbers. For a city not listed, use the factor for a nearby city with similar economic characteristics.

STATE/ZIP	CITY	MAT.	INST.	TOTAL
ALABAMA				
350-352	Birmingham	97.4	71.8	86.6
354	Tuscaloosa	97.1	71.4	86.3
355	Jasper	97.8	69.8	86.0
356	Decatur	97.1	69.2	85.3
357-358	Huntsville	97.1	71.7	86.4
359	Gadsden	97.3	70.2	85.9
360-361	Montgomery	96.3	73.1	86.5
362	Anniston	95.9	67.7	84.0
363	Dothan	96.2	72.7	86.3
364	Evergreen	95.8	68.3	84.2
365-366	Mobile	96.7	68.7	84.9
367	Selma	95.9	72.5	86.0
368	Phenix City	96.8	70.8	85.8
369	Butler	96.1	70.6	85.3
ALASKA				
995-996	Anchorage	118.5	113.1	116.2
997	Fairbanks	118.8	113.0	116.3
998	Juneau	116.8	113.1	115.2
999	Ketchikan	128.4	113.4	122.1
ARIZONA				
850,853	Phoenix	100.6	72.3	88.6
851,852	Mesa/Tempe	98.7	71.5	87.2
855	Globe	99.5	71.4	87.6
856-857	Tucson	97.7	70.2	86.1
859	Show Low	99.7	71.5	87.8
860	Flagstaff	102.4	70.5	88.9
863	Prescott	99.9	71.3	87.8
864	Kingman	98.3	70.4	86.5
865	Chambers	98.3	71.8	87.1
ARKANSAS				
716	Pine Bluff	96.0	63.6	82.3
717	Camden	94.0	64.7	81.6
718	Texarkana	94.8	63.3	81.5
719	Hot Springs	93.5	64.4	81.2
720-722	Little Rock	94.5	65.4	82.2
723	West Memphis	93.3	69.8	83.3
724	Jonesboro	93.8	66.6	82.3
725	Batesville	91.6	63.6	79.8
726	Harrison	92.9	62.8	80.2
727	Fayetteville	90.5	64.1	79.3
728	Russellville	91.7	63.1	79.6
729	Fort Smith	94.1	62.6	80.8
CALIFORNIA				
900-902	Los Angeles	98.5	129.8	111.7
903-905	Inglewood	93.8	128.8	108.6
906-908	Long Beach	95.4	128.8	109.5
910-912	Pasadena	94.8	128.6	109.1
913-916	Van Nuys	97.8	128.6	110.8
917-918	Alhambra	96.7	128.6	110.2
919-921	San Diego	99.7	121.7	109.0
922	Palm Springs	97.4	125.3	109.2
923-924	San Bernardino	95.0	126.3	108.2
925	Riverside	99.3	126.6	110.8
926-927	Santa Ana	97.0	126.3	109.4
928	Anaheim	99.3	126.6	110.9
930	Oxnard	98.0	126.7	110.1
931	Santa Barbara	97.2	126.1	109.4
932-933	Bakersfield	98.9	124.6	109.8
934	San Luis Obispo	98.2	125.8	109.9
935	Mojave	95.4	124.4	107.7
936-938	Fresno	98.5	129.4	111.5
939	Salinas	99.1	137.2	115.2
940-941	San Francisco	107.4	158.4	129.0
942,956-958	Sacramento	100.8	133.0	114.4
943	Palo Alto	99.2	153.8	122.3
944	San Mateo	101.5	152.0	122.9
945	Vallejo	100.2	141.6	117.7
946	Oakland	103.4	152.0	123.9
947	Berkeley	103.0	151.9	123.7
948	Richmond	102.4	146.2	120.9
949	San Rafael	104.5	149.1	123.4
950	Santa Cruz	104.6	137.5	118.5

STATE/ZIP	CITY	MAT.	INST.	TOTAL
CALIFORNIA (CONT'D)				
951	San Jose	102.6	153.5	124.2
952	Stockton	100.7	131.3	113.7
953	Modesto	100.6	130.6	113.3
954	Santa Rosa	101.2	148.4	121.2
955	Eureka	102.5	134.1	115.9
959	Marysville	101.7	131.3	114.2
960	Redding	107.6	131.5	117.7
961	Susanville	107.3	131.6	117.5
COLORADO				
800-802	Denver	103.0	75.4	91.3
803	Boulder	98.7	77.0	89.6
804	Golden	100.9	74.7	89.8
805	Fort Collins	102.2	73.5	90.1
806	Greeley	99.3	73.4	88.4
807	Fort Morgan	99.3	73.5	88.4
808-809	Colorado Springs	101.2	71.5	88.6
810	Pueblo	101.8	70.0	88.3
811	Alamosa	103.6	70.3	89.5
812	Salida	103.1	70.9	89.5
813	Durango	103.9	65.6	87.8
814	Montrose	102.6	65.8	87.1
815	Grand Junction	105.7	70.5	90.8
816	Glenwood Springs	103.9	66.8	88.2
CONNECTICUT				
060	New Britain	98.1	116.7	106.0
061	Hartford	99.6	118.0	107.3
062	Willimantic	98.7	117.1	106.5
063	New London	95.2	117.2	104.5
064	Meriden	97.2	117.0	105.6
065	New Haven	99.9	117.0	107.1
066	Bridgeport	99.3	117.0	106.8
067	Waterbury	98.9	117.2	106.6
068	Norwalk	98.8	116.7	106.4
069	Stamford	99.0	123.2	109.2
D.C.				
200-205	Washington	101.1	87.5	95.4
DELAWARE				
197	Newark	98.3	109.7	103.1
198	Wilmington	97.9	109.7	102.9
199	Dover	99.0	109.6	103.5
FLORIDA				
320,322	Jacksonville	96.9	66.6	84.1
321	Daytona Beach	97.2	69.5	85.5
323	Tallahassee	98.2	67.7	85.3
324	Panama City	98.6	68.3	85.8
325	Pensacola	101.2	65.8	86.2
326,344	Gainesville	98.8	65.8	84.8
327-328,347	Orlando	99.0	66.6	85.3
329	Melbourne	100.1	70.3	87.5
330-332,340	Miami	96.7	68.0	84.6
333	Fort Lauderdale	96.1	66.4	83.6
334,349	West Palm Beach	95.1	64.1	82.0
335-336,346	Tampa	97.6	68.4	85.3
337	St. Petersburg	99.7	66.8	85.8
338	Lakeland	97.1	67.6	84.6
339,341	Fort Myers	96.5	67.7	84.3
342	Sarasota	98.8	67.0	85.4
GEORGIA				
300-303,399	Atlanta	98.2	75.0	88.4
304	Statesboro	97.8	69.7	85.9
305	Gainesville	96.2	65.6	83.2
306	Athens	95.6	66.9	83.5
307	Dalton	97.5	71.4	86.5
308-309	Augusta	96.0	73.9	86.6
310-312	Macon	95.2	73.1	85.8
313-314	Savannah	97.3	72.7	86.9
315	Waycross	96.8	70.0	85.5
316	Valdosta	96.7	66.2	83.8
317,398	Albany	96.5	71.9	86.1
318-319	Columbus	96.5	72.2	86.2

STATE/ZIP	CITY	MAT.	INST.	TOTAL
HAWAII				
967	Hilo	114.8	118.4	116.3
968	Honolulu	119.1	118.5	118.9
STATES & POSS.				
969	Guam	137.1	51.4	100.9
IDAHO				
832	Pocatello	102.0	78.4	92.0
833	Twin Falls	103.2	76.8	92.1
834	Idaho Falls	100.5	79.9	91.8
835	Lewiston	108.2	83.5	97.8
836-837	Boise	101.0	79.9	92.1
838	Coeur d'Alene	108.1	84.4	98.1
ILLINOIS				
600-603	North Suburban	98.3	142.9	117.1
604	Joliet	98.3	138.8	115.4
605	South Suburban	98.3	142.8	117.1
606-608	Chicago	100.1	145.6	119.3
609	Kankakee	95.0	134.9	111.9
610-611	Rockford	96.6	125.5	108.8
612	Rock Island	94.6	99.5	96.7
613	La Salle	95.9	127.0	109.0
614	Galesburg	95.7	108.5	101.1
615-616	Peoria	97.6	108.8	102.3
617	Bloomington	95.0	109.5	101.1
618-619	Champaign	98.5	110.5	103.6
620-622	East St. Louis	93.8	110.4	100.9
623	Quincy	95.7	104.2	99.3
624	Effingham	95.1	109.1	101.0
625	Decatur	96.6	110.1	102.3
626-627	Springfield	97.5	108.3	102.1
628	Centralia	92.8	111.2	100.5
629	Carbondale	92.5	109.6	99.7
INDIANA				
460	Anderson	96.0	79.8	89.1
461-462	Indianapolis	98.7	83.2	92.1
463-464	Gary	97.5	108.1	102.0
465-466	South Bend	98.4	81.8	91.4
467-468	Fort Wayne	96.6	76.3	88.0
469	Kokomo	94.2	81.1	88.6
470	Lawrenceburg	92.5	79.6	87.0
471	New Albany	93.9	77.3	86.9
472	Columbus	96.1	80.5	89.5
473	Muncie	96.2	78.2	88.6
474	Bloomington	97.8	81.1	90.7
475	Washington	94.6	84.8	90.4
476-477	Evansville	95.5	82.7	90.1
478	Terre Haute	96.2	81.7	90.1
479	Lafayette	95.8	78.1	88.3
IOWA				
500-503,509	Des Moines	97.2	90.2	94.3
504	Mason City	95.3	74.0	86.3
505	Fort Dodge	95.5	73.7	86.3
506-507	Waterloo	96.8	77.4	88.6
508	Creston	95.7	84.2	90.8
510-511	Sioux City	97.7	81.5	90.9
512	Sibley	96.8	62.4	82.3
513	Spencer	98.2	62.3	83.0
514	Carroll	95.5	82.4	89.9
515	Council Bluffs	98.9	79.2	90.6
516	Shenandoah	96.1	81.9	90.1
520	Dubuque	97.2	79.9	89.9
521	Decorah	96.4	73.7	86.8
522-524	Cedar Rapids	98.0	83.7	92.0
525	Ottumwa	96.3	78.7	88.9
526	Burlington	95.7	84.3	90.9
527-528	Davenport	97.1	94.5	96.0
KANSAS				
660-662	Kansas City	97.3	98.6	97.8
664-666	Topeka	98.6	76.5	89.3
667	Fort Scott	96.0	77.1	88.0
668	Emporia	96.1	76.5	87.8
669	Belleville	97.9	70.7	86.4
670-672	Wichita	97.0	72.6	86.7
673	Independence	97.2	76.0	88.2
674	Salina	97.2	73.5	87.2
675	Hutchinson	92.5	71.9	83.8
676	Hays	96.5	72.0	86.1
677	Colby	97.3	74.0	87.4

STATE/ZIP	CITY	MAT.	INST.	TOTAL
KANSAS (CONT'D)				
678	Dodge City	98.6	73.3	87.9
679	Liberal	96.5	71.7	86.0
KENTUCKY				
400-402	Louisville	93.9	79.6	87.8
403-405	Lexington	93.1	78.1	86.7
406	Frankfort	95.4	80.3	89.0
407-409	Corbin	90.8	78.0	85.4
410	Covington	94.3	77.8	87.3
411-412	Ashland	93.2	89.7	91.7
413-414	Campton	94.4	78.0	87.5
415-416	Pikeville	95.8	85.4	91.4
417-418	Hazard	93.8	78.7	87.4
420	Paducah	92.4	83.2	88.6
421-422	Bowling Green	94.6	77.2	87.3
423	Owensboro	94.7	81.0	88.9
424	Henderson	92.2	81.9	87.8
425-426	Somerset	91.6	79.5	86.5
427	Elizabethtown	91.3	75.6	84.7
LOUISIANA				
700-701	New Orleans	99.0	68.5	86.1
703	Thibodaux	96.0	67.0	83.8
704	Hammond	93.7	64.4	81.3
705	Lafayette	95.4	67.9	83.8
706	Lake Charles	95.6	68.2	84.1
707-708	Baton Rouge	96.6	69.3	85.1
710-711	Shreveport	97.1	66.3	84.1
712	Monroe	95.8	64.4	82.5
713-714	Alexandria	95.9	65.6	83.1
MAINE				
039	Kittery	92.6	87.0	90.2
040-041	Portland	99.6	85.4	93.6
042	Lewiston	97.5	84.9	92.2
043	Augusta	100.4	85.6	94.2
044	Bangor	97.1	83.9	91.5
045	Bath	95.9	85.4	91.4
046	Machias	95.4	85.3	91.1
047	Houlton	95.6	85.2	91.2
048	Rockland	94.6	85.4	90.7
049	Waterville	96.0	85.2	91.4
MARYLAND				
206	Waldorf	97.3	86.6	92.8
207-208	College Park	97.3	87.7	93.2
209	Silver Spring	96.5	86.7	92.4
210-212	Baltimore	101.8	83.2	93.9
214	Annapolis	100.4	86.3	94.4
215	Cumberland	96.9	83.8	91.4
216	Easton	98.6	74.8	88.5
217	Hagerstown	97.4	85.3	92.3
218	Salisbury	99.0	66.9	85.4
219	Elkton	95.9	86.2	91.8
MASSACHUSETTS				
010-011	Springfield	98.2	108.5	102.5
012	Pittsfield	97.7	102.6	99.7
013	Greenfield	95.9	109.0	101.4
014	Fitchburg	94.6	116.2	103.7
015-016	Worcester	98.1	117.0	106.1
017	Framingham	94.0	124.2	106.8
018	Lowell	97.5	123.8	108.6
019	Lawrence	98.4	124.2	109.3
020-022, 024	Boston	101.9	134.0	115.5
023	Brockton	98.5	116.7	106.2
025	Buzzards Bay	93.2	115.2	102.5
026	Hyannis	95.7	115.0	103.8
027	New Bedford	97.7	115.2	105.1
MICHIGAN				
480,483	Royal Oak	95.1	100.3	97.3
481	Ann Arbor	97.1	99.8	98.2
482	Detroit	100.7	102.0	101.3
484-485	Flint	96.8	88.6	93.3
486	Saginaw	96.5	85.1	91.7
487	Bay City	96.6	85.3	91.8
488-489	Lansing	98.0	86.8	93.3
490	Battle Creek	95.5	80.4	89.1
491	Kalamazoo	95.8	78.6	88.6
492	Jackson	94.0	90.7	92.6
493,495	Grand Rapids	97.9	79.9	90.3
494	Muskegon	94.3	80.0	88.3

Location Factors - Commercial

STATE/ZIP	CITY	MAT.	INST.	TOTAL
MICHIGAN (CONT'D)				
496	Traverse City	93.4	77.7	86.8
497	Gaylord	94.5	80.6	88.6
498-499	Iron Mountain	96.5	81.7	90.2
MINNESOTA				
550-551	Saint Paul	99.0	117.6	106.9
553-555	Minneapolis	100.7	114.1	106.3
556-558	Duluth	98.6	101.3	99.7
559	Rochester	98.3	98.7	98.5
560	Mankato	95.7	98.5	96.9
561	Windom	94.3	91.2	93.0
562	Willmar	94.0	98.0	95.7
563	St. Cloud	95.1	114.2	103.2
564	Brainerd	95.6	98.6	96.8
565	Detroit Lakes	97.4	92.4	95.3
566	Bemidji	96.7	95.4	96.1
567	Thief River Falls	96.3	89.3	93.3
MISSISSIPPI				
386	Clarksdale	95.3	55.2	78.4
387	Greenville	98.7	65.5	84.7
388	Tupelo	96.6	59.0	80.7
389	Greenwood	96.6	55.5	79.2
390-392	Jackson	97.3	65.6	83.9
393	Meridian	94.6	64.5	81.9
394	Laurel	96.1	57.8	79.9
395	Biloxi	96.2	63.9	82.6
396	McComb	94.4	55.8	78.0
397	Columbus	95.9	59.1	80.4
MISSOURI				
630-631	St. Louis	98.8	104.5	101.2
633	Bowling Green	96.4	95.5	96.0
634	Hannibal	95.3	91.6	93.7
635	Kirksville	98.9	86.5	93.7
636	Flat River	97.3	94.2	96.0
637	Cape Girardeau	97.0	91.2	94.5
638	Sikeston	95.7	87.8	92.3
639	Poplar Bluff	95.1	87.7	92.0
640-641	Kansas City	97.9	103.8	100.4
644-645	St. Joseph	96.1	90.9	93.9
646	Chillicothe	93.7	93.8	93.7
647	Harrisonville	93.2	100.8	96.4
648	Joplin	95.1	77.7	87.7
650-651	Jefferson City	96.4	89.9	93.7
652	Columbia	96.1	90.7	93.8
653	Sedalia	96.4	91.5	94.3
654-655	Rolla	94.1	94.3	94.2
656-658	Springfield	97.6	79.0	89.7
MONTANA				
590-591	Billings	101.6	78.4	91.8
592	Wolf Point	101.4	76.9	91.1
593	Miles City	99.2	77.0	89.8
594	Great Falls	102.9	76.6	91.8
595	Havre	100.4	74.8	89.6
596	Helena	100.9	76.7	90.7
597	Butte	101.5	76.5	90.9
598	Missoula	98.6	77.1	89.5
599	Kalispell	98.1	76.2	88.9
NEBRASKA				
680-681	Omaha	97.6	81.5	90.8
683-685	Lincoln	98.3	79.7	90.5
686	Columbus	96.0	80.7	89.6
687	Norfolk	97.4	77.5	89.0
688	Grand Island	97.4	76.8	88.7
689	Hastings	97.1	78.2	89.1
690	McCook	96.3	73.1	86.5
691	North Platte	96.3	75.6	87.6
692	Valentine	98.6	69.2	86.2
693	Alliance	98.7	73.3	87.9
NEVADA				
889-891	Las Vegas	105.0	105.1	105.0
893	Ely	103.7	93.2	99.2
894-895	Reno	103.6	82.6	94.7
897	Carson City	102.9	82.5	94.3
898	Elko	102.3	85.2	95.0
NEW HAMPSHIRE				
030	Nashua	97.5	93.7	95.9
031	Manchester	98.2	94.5	96.6

STATE/ZIP	CITY	MAT.	INST.	TOTAL
NEW HAMPSHIRE (CONT'D)				
032-033	Concord	98.0	93.2	96.0
034	Keene	94.5	89.0	92.2
035	Littleton	94.4	80.2	88.4
036	Charleston	94.1	88.4	91.7
037	Claremont	93.1	88.4	91.1
038	Portsmouth	94.9	92.4	93.8
NEW JERSEY				
070-071	Newark	100.3	139.7	117.0
072	Elizabeth	97.5	138.3	114.7
073	Jersey City	96.5	137.8	114.0
074-075	Paterson	98.0	137.7	114.8
076	Hackensack	96.1	138.3	114.0
077	Long Branch	95.8	130.9	110.7
078	Dover	96.4	137.5	113.7
079	Summit	96.4	138.3	114.1
080,083	Vineland	96.3	129.8	110.5
081	Camden	98.0	127.8	110.6
082,084	Atlantic City	96.8	133.2	112.2
085-086	Trenton	99.7	128.9	112.0
087	Point Pleasant	98.1	130.8	111.9
088-089	New Brunswick	98.7	135.6	114.3
NEW MEXICO				
870-872	Albuquerque	98.6	72.5	87.6
873	Gallup	99.0	72.5	87.8
874	Farmington	99.0	72.5	87.8
875	Santa Fe	99.0	72.9	88.0
877	Las Vegas	97.2	72.5	86.8
878	Socorro	96.8	72.5	86.6
879	Truth/Consequences	96.6	69.4	85.1
880	Las Cruces	96.9	71.4	86.1
881	Clovis	99.3	72.4	87.9
882	Roswell	100.8	72.5	88.9
883	Carrizozo	101.6	72.5	89.3
884	Tucumcari	99.9	72.4	88.3
NEW YORK				
100-102	New York	99.7	175.2	131.6
103	Staten Island	95.6	176.1	129.6
104	Bronx	93.8	175.3	128.3
105	Mount Vernon	94.1	150.6	118.0
106	White Plains	93.8	152.8	118.8
107	Yonkers	97.9	153.0	121.2
108	New Rochelle	94.4	144.8	115.7
109	Suffern	94.1	130.2	109.4
110	Queens	100.0	177.3	132.6
111	Long Island City	101.6	177.3	133.6
112	Brooklyn	101.9	177.3	133.8
113	Flushing	102.1	177.3	133.9
114	Jamaica	100.3	177.3	132.8
115,117,118	Hicksville	99.9	157.1	124.1
116	Far Rockaway	102.2	177.3	133.9
119	Riverhead	100.6	153.5	122.9
120-122	Albany	96.8	111.2	102.8
123	Schenectady	97.0	109.9	102.4
124	Kingston	100.6	133.8	114.6
125-126	Poughkeepsie	99.8	136.8	115.4
127	Monticello	99.1	134.6	114.1
128	Glens Falls	92.0	107.3	98.5
129	Plattsburgh	97.0	96.6	96.8
130-132	Syracuse	97.8	98.3	98.0
133-135	Utica	96.0	96.6	96.2
136	Watertown	97.7	96.8	97.4
137-139	Binghamton	97.3	101.5	99.1
140-142	Buffalo	100.8	110.3	104.8
143	Niagara Falls	97.4	107.5	101.6
144-146	Rochester	100.3	99.2	99.8
147	Jamestown	96.4	96.0	96.3
148-149	Elmira	96.3	102.5	98.9
NORTH CAROLINA				
270,272-274	Greensboro	98.5	67.2	85.3
271	Winston-Salem	98.2	67.3	85.2
275-276	Raleigh	97.4	67.0	84.5
277	Durham	100.1	67.2	86.2
278	Rocky Mount	96.0	66.6	83.6
279	Elizabeth City	96.9	68.3	84.8
280	Gastonia	97.0	66.8	84.2
281-282	Charlotte	97.4	67.6	84.8
283	Fayetteville	100.2	66.1	85.8
284	Wilmington	96.0	65.2	83.0
285	Kinston	94.4	65.5	82.2

Location Factors - Commercial

STATE/ZIP	CITY	MAT.	INST.	TOTAL
NORTH CAROLINA (CONT'D)				
286	Hickory	94.7	67.3	83.1
287-288	Asheville	96.4	66.6	83.8
289	Murphy	95.5	64.7	82.5
NORTH DAKOTA				
580-581	Fargo	99.7	80.3	91.5
582	Grand Forks	99.4	77.6	90.2
583	Devils Lake	99.2	78.8	90.6
584	Jamestown	99.2	78.8	90.6
585	Bismarck	98.9	81.6	91.6
586	Dickinson	99.9	77.9	90.6
587	Minot	99.3	78.6	90.5
588	Williston	98.4	77.7	89.6
OHIO				
430-432	Columbus	98.4	82.0	91.4
433	Marion	94.3	86.7	91.1
434-436	Toledo	97.6	91.2	94.9
437-438	Zanesville	94.9	85.7	91.0
439	Steubenville	96.1	90.0	93.5
440	Lorain	98.4	84.7	92.6
441	Cleveland	99.1	91.6	95.9
442-443	Akron	99.4	86.0	93.8
444-445	Youngstown	98.7	81.9	91.6
446-447	Canton	98.9	79.9	90.9
448-449	Mansfield	96.4	84.2	91.3
450	Hamilton	94.8	79.5	88.3
451-452	Cincinnati	96.1	80.0	89.3
453-454	Dayton	94.8	79.2	88.2
455	Springfield	94.8	79.7	88.4
456	Chillicothe	94.3	88.5	91.8
457	Athens	97.1	84.8	91.9
458	Lima	97.2	81.7	90.7
OKLAHOMA				
730-731	Oklahoma City	96.4	68.7	84.7
734	Ardmore	95.2	65.6	82.7
735	Lawton	97.3	65.8	84.0
736	Clinton	96.4	65.5	83.3
737	Enid	97.0	66.0	83.9
738	Woodward	95.3	62.9	81.6
739	Guymon	96.3	63.3	82.4
740-741	Tulsa	95.9	65.8	83.2
743	Miami	92.7	65.1	81.1
744	Muskogee	95.0	62.9	81.4
745	McAlester	92.4	60.5	78.9
746	Ponca City	93.1	64.0	80.8
747	Durant	93.1	65.1	81.3
748	Shawnee	94.5	65.2	82.2
749	Poteau	92.3	64.9	80.7
OREGON				
970-972	Portland	102.1	102.8	102.4
973	Salem	104.4	100.8	102.9
974	Eugene	101.7	98.7	100.5
975	Medford	103.3	97.4	100.8
976	Klamath Falls	103.5	97.3	100.9
977	Bend	102.6	100.2	101.6
978	Pendleton	98.6	100.3	99.3
979	Vale	96.2	87.0	92.3
PENNSYLVANIA				
150-152	Pittsburgh	100.7	102.7	101.5
153	Washington	97.5	100.2	98.6
154	Uniontown	97.8	99.8	98.6
155	Bedford	98.8	92.1	96.0
156	Greensburg	98.7	96.3	97.7
157	Indiana	97.6	97.7	97.7
158	Dubois	99.3	96.0	97.9
159	Johnstown	98.9	91.9	95.9
160	Butler	91.3	100.6	95.2
161	New Castle	91.4	97.8	94.1
162	Kittanning	91.8	97.9	94.4
163	Oil City	91.3	97.8	94.0
164-165	Erie	93.2	93.9	93.5
166	Altoona	93.3	92.8	93.1
167	Bradford	94.9	97.1	95.8
168	State College	94.5	94.6	94.5
169	Wellsboro	95.5	92.9	94.4
170-171	Harrisburg	98.7	93.0	96.3
172	Chambersburg	95.1	89.4	92.7
173-174	York	95.2	93.0	94.3
175-176	Lancaster	93.8	94.5	94.1

STATE/ZIP	CITY	MAT.	INST.	TOTAL
PENNSYLVANIA (CONT'D)				
177	Williamsport	92.5	92.3	92.4
178	Sunbury	94.6	92.7	93.8
179	Pottsville	93.7	96.1	94.7
180	Lehigh Valley	94.9	111.6	102.0
181	Allentown	96.7	105.8	100.6
182	Hazleton	94.4	95.9	95.0
183	Stroudsburg	94.3	104.7	98.7
184-185	Scranton	97.4	96.5	97.1
186-187	Wilkes-Barre	94.0	95.5	94.7
188	Montrose	93.8	96.8	95.1
189	Doylestown	94.0	127.1	108.0
190-191	Philadelphia	100.2	138.7	116.5
193	Westchester	95.8	127.9	109.3
194	Norristown	94.7	128.0	108.8
195-196	Reading	96.5	101.5	98.6
PUERTO RICO				
009	San Juan	98.5	29.8	69.4
RHODE ISLAND				
028	Newport	97.0	114.7	104.5
029	Providence	99.3	115.0	105.9
SOUTH CAROLINA				
290-292	Columbia	97.6	68.9	85.5
293	Spartanburg	97.2	68.3	85.0
294	Charleston	98.7	68.2	85.8
295	Florence	96.9	68.6	84.9
296	Greenville	97.0	68.2	84.8
297	Rock Hill	96.7	66.4	83.9
298	Aiken	97.6	67.1	84.7
299	Beaufort	98.3	57.5	81.1
SOUTH DAKOTA				
570-571	Sioux Falls	97.7	78.4	89.5
572	Watertown	97.3	69.0	85.4
573	Mitchell	96.3	60.9	81.3
574	Aberdeen	98.6	70.1	86.5
575	Pierre	99.0	68.5	86.1
576	Mobridge	96.9	62.9	82.5
577	Rapid City	98.3	72.0	87.2
TENNESSEE				
370-372	Nashville	98.0	71.8	86.9
373-374	Chattanooga	98.0	68.1	85.3
375,380-381	Memphis	96.6	69.8	85.3
376	Johnson City	98.1	58.6	81.4
377-379	Knoxville	94.8	63.7	81.6
382	McKenzie	96.3	55.1	78.9
383	Jackson	97.6	61.4	82.3
384	Columbia	94.8	66.5	82.8
385	Cookeville	96.2	56.9	79.6
TEXAS				
750	McKinney	97.2	63.5	82.9
751	Waxahachie	97.2	63.5	83.0
752-753	Dallas	98.2	68.0	85.4
754	Greenville	97.4	62.9	82.8
755	Texarkana	96.7	62.0	82.1
756	Longview	97.5	61.0	82.1
757	Tyler	97.8	61.6	82.5
758	Palestine	94.2	60.4	79.9
759	Lufkin	94.7	62.6	81.1
760-761	Fort Worth	97.6	62.9	82.9
762	Denton	97.3	62.9	82.7
763	Wichita Falls	95.0	61.6	80.9
764	Eastland	94.2	60.3	79.8
765	Temple	92.5	58.5	78.2
766-767	Waco	94.4	62.4	80.9
768	Brownwood	97.4	58.7	81.1
769	San Angelo	97.1	59.4	81.2
770-772	Houston	100.4	68.2	86.8
773	Huntsville	98.7	62.8	83.6
774	Wharton	99.8	64.3	84.8
775	Galveston	97.7	66.5	84.5
776-777	Beaumont	97.9	65.5	84.2
778	Bryan	95.2	64.9	82.4
779	Victoria	99.7	62.9	84.2
780	Laredo	96.4	61.1	81.5
781-782	San Antonio	98.3	62.3	83.1
783-784	Corpus Christi	98.9	61.4	83.1
785	McAllen	99.7	56.5	81.4
786-787	Austin	97.2	61.4	82.1

Location Factors - Commercial

STATE/ZIP	CITY	MAT.	INST.	TOTAL
TEXAS (CONT'D)				
788	Del Rio	99.7	60.4	83.1
789	Giddings	96.1	60.4	81.0
790-791	Amarillo	98.2	60.0	82.1
792	Childress	97.6	60.3	81.8
793-794	Lubbock	99.2	62.9	83.9
795-796	Abilene	97.8	61.4	82.4
797	Midland	99.7	62.2	83.8
798-799,885	El Paso	95.9	63.6	82.3
UTAH				
840-841	Salt Lake City	103.5	71.7	90.1
842,844	Ogden	98.8	71.7	87.4
843	Logan	100.9	71.7	88.5
845	Price	101.4	70.0	88.1
846-847	Provo	101.2	71.2	88.5
VERMONT				
050	White River Jct.	98.2	79.8	90.4
051	Bellows Falls	96.7	92.1	94.8
052	Bennington	97.0	88.6	93.4
053	Brattleboro	97.4	92.1	95.2
054	Burlington	101.9	79.8	92.5
056	Montpelier	100.2	84.4	93.5
057	Rutland	98.9	79.4	90.7
058	St. Johnsbury	98.1	79.3	90.2
059	Guildhall	96.8	79.0	89.3
VIRGINIA				
220-221	Fairfax	99.5	83.6	92.8
222	Arlington	100.4	84.2	93.6
223	Alexandria	99.6	84.8	93.3
224-225	Fredericksburg	98.3	78.7	90.0
226	Winchester	98.9	79.5	90.7
227	Culpeper	98.7	83.0	92.1
228	Harrisonburg	99.0	68.8	86.3
229	Charlottesville	99.4	71.0	87.4
230-232	Richmond	99.2	72.8	88.1
233-235	Norfolk	99.9	68.7	86.7
236	Newport News	98.8	68.3	85.9
237	Portsmouth	98.3	67.8	85.4
238	Petersburg	98.7	72.6	87.7
239	Farmville	97.9	65.3	84.1
240-241	Roanoke	100.1	73.6	88.9
242	Bristol	98.2	59.3	81.8
243	Pulaski	97.8	67.1	84.8
244	Staunton	98.6	67.9	85.6
245	Lynchburg	98.7	73.8	88.2
246	Grundy	98.1	58.8	81.5
WASHINGTON				
980-981,987	Seattle	106.3	112.5	108.9
982	Everett	105.1	104.0	104.6
983-984	Tacoma	105.4	101.3	103.7
985	Olympia	104.0	101.8	103.1
986	Vancouver	106.6	102.8	105.0
988	Wenatchee	105.6	88.7	98.5
989	Yakima	105.6	95.4	101.3
990-992	Spokane	100.1	82.4	92.6
993	Richland	99.8	90.0	95.6
994	Clarkston	99.1	81.9	91.8
WEST VIRGINIA				
247-248	Bluefield	97.0	88.1	93.2
249	Lewisburg	98.7	86.1	93.4
250-253	Charleston	96.5	89.4	93.5
254	Martinsburg	96.7	84.1	91.4
255-257	Huntington	97.8	92.2	95.4
258-259	Beckley	95.2	87.7	92.0
260	Wheeling	99.6	89.6	95.4
261	Parkersburg	98.6	88.6	94.4
262	Buckhannon	98.3	88.8	94.3
263-264	Clarksburg	98.9	89.5	94.9
265	Morgantown	98.8	89.6	94.9
266	Gassaway	98.3	89.0	94.3
267	Romney	98.3	87.2	93.6
268	Petersburg	98.0	85.9	92.9
WISCONSIN				
530,532	Milwaukee	99.2	107.9	102.9
531	Kenosha	99.4	102.0	100.5
534	Racine	98.7	102.3	100.2
535	Beloit	98.6	98.9	98.7
537	Madison	99.7	98.1	99.0

STATE/ZIP	CITY	MAT.	INST.	TOTAL
WISCONSIN (CONT'D)				
538	Lancaster	96.7	94.3	95.7
539	Portage	95.2	98.0	96.4
540	New Richmond	95.4	94.6	95.0
541-543	Green Bay	99.4	93.4	96.9
544	Wausau	94.7	93.4	94.2
545	Rhinelander	97.9	92.2	95.5
546	La Crosse	95.9	93.5	94.9
547	Eau Claire	97.5	94.1	96.1
548	Superior	95.1	96.5	95.7
549	Oshkosh	95.4	91.9	93.9
WYOMING				
820	Cheyenne	101.1	72.7	89.1
821	Yellowstone Nat'l Park	99.0	71.9	87.6
822	Wheatland	100.3	67.1	86.2
823	Rawlins	102.0	71.5	89.1
824	Worland	99.7	70.7	87.5
825	Riverton	101.0	68.6	87.3
826	Casper	101.5	68.9	87.7
827	Newcastle	99.6	71.1	87.5
828	Sheridan	102.5	70.2	88.8
829-831	Rock Springs	103.4	71.4	89.9
CANADIAN FACTORS (reflect Canadian currency)				
ALBERTA				
	Calgary	119.7	95.6	109.5
	Edmonton	121.8	95.9	110.9
	Fort McMurray	121.9	88.4	107.7
	Lethbridge	117.0	87.9	104.7
	Lloydminster	111.5	84.7	100.2
	Medicine Hat	111.7	84.0	100.0
	Red Deer	112.1	84.0	100.2
BRITISH COLUMBIA				
	Kamloops	113.0	83.5	100.5
	Prince George	114.0	82.8	100.8
	Vancouver	117.9	92.4	107.2
	Victoria	114.1	87.6	102.9
MANITOBA				
	Brandon	120.6	70.5	99.4
	Portage la Prairie	111.6	69.1	93.6
	Winnipeg	122.0	71.2	100.5
NEW BRUNSWICK				
	Bathurst	110.7	62.3	90.2
	Dalhousie	110.2	62.5	90.0
	Fredericton	119.3	69.0	98.0
	Moncton	110.9	76.0	96.2
	Newcastle	110.7	62.9	90.5
	Saint John	113.2	75.7	97.3
NEWFOUNDLAND				
	Corner Brook	124.7	68.4	100.9
	St. John's	122.6	87.4	107.7
NORTHWEST TERRITORIES				
	Yellowknife	131.4	83.2	111.0
NOVA SCOTIA				
	Bridgewater	111.3	70.3	94.0
	Dartmouth	121.5	70.2	99.8
	Halifax	117.9	88.0	105.2
	New Glasgow	119.6	70.2	98.7
	Sydney	117.8	70.2	97.7
	Truro	110.8	70.3	93.7
	Yarmouth	119.5	70.2	98.7
ONTARIO				
	Barrie	116.5	87.1	104.1
	Brantford	113.1	90.7	103.6
	Cornwall	113.1	87.3	102.2
	Hamilton	116.3	98.8	108.9
	Kingston	114.1	87.4	102.8
	Kitchener	108.6	94.6	102.7
	London	116.3	96.5	108.0
	North Bay	122.5	85.2	106.8
	Oshawa	110.8	97.2	105.0
	Ottawa	117.5	96.8	108.7
	Owen Sound	116.7	85.4	103.4
	Peterborough	113.1	87.1	102.1
	Sarnia	113.2	91.3	103.9

Location Factors - Commercial

STATE/ZIP	CITY	MAT.	INST.	TOTAL
ONTARIO (CONT'D)				
	Sault Ste. Marie	108.6	88.7	100.2
	St. Catharines	106.6	95.6	101.9
	Sudbury	106.1	93.6	100.8
	Thunder Bay	107.8	95.0	102.4
	Timmins	113.3	85.3	101.4
	Toronto	116.0	104.2	111.0
	Windsor	107.1	94.4	101.8
PRINCE EDWARD ISLAND				
	Charlottetown	121.3	60.2	95.5
	Summerside	121.9	57.9	94.9
QUEBEC				
	Cap-de-la-Madeleine	110.6	80.0	97.7
	Charlesbourg	110.6	80.0	97.7
	Chicoutimi	110.4	87.6	100.7
	Gatineau	110.3	79.8	97.4
	Granby	110.5	79.7	97.5
	Hull	110.5	79.8	97.5
	Joliette	110.8	80.0	97.8
	Laval	110.7	81.6	98.4
	Montreal	119.1	90.1	106.9
	Quebec City	118.7	90.4	106.7
	Rimouski	110.2	87.6	100.6
	Rouyn-Noranda	110.3	79.8	97.4
	Saint-Hyacinthe	109.8	79.8	97.1
	Sherbrooke	110.6	79.8	97.6
	Sorel	110.8	80.0	97.8
	Saint-Jerome	110.4	79.8	97.4
	Trois-Rivieres	120.2	79.9	103.2
SASKATCHEWAN				
	Moose Jaw	109.0	63.9	89.9
	Prince Albert	108.2	62.2	88.8
	Regina	123.1	92.4	110.2
	Saskatoon	109.9	88.5	100.9
YUKON				
	Whitehorse	132.2	69.2	105.6

R011105-05 Tips for Accurate Estimating

1. Use pre-printed or columnar forms for orderly sequence of dimensions and locations and for recording telephone quotations.
2. Use only the front side of each paper or form except for certain pre-printed summary forms.
3. Be consistent in listing dimensions: For example, length x width x height. This helps in rechecking to ensure that, the total length of partitions is appropriate for the building area.
4. Use printed (rather than measured) dimensions where given.
5. Add up multiple printed dimensions for a single entry where possible.
6. Measure all other dimensions carefully.
7. Use each set of dimensions to calculate multiple related quantities.
8. Convert foot and inch measurements to decimal feet when listing. Memorize decimal equivalents to .01 parts of a foot (1/8″ equals approximately .01″).
9. Do not "round off" quantities until the final summary.
10. Mark drawings with different colors as items are taken off.
11. Keep similar items together, different items separate.
12. Identify location and drawing numbers to aid in future checking for completeness.
13. Measure or list everything on the drawings or mentioned in the specifications.
14. It may be necessary to list items not called for to make the job complete.
15. Be alert for: Notes on plans such as N.T.S. (not to scale); changes in scale throughout the drawings; reduced size drawings; discrepancies between the specifications and the drawings.
16. Develop a consistent pattern of performing an estimate. For example:
    a. Start the quantity takeoff at the lower floor and move to the next higher floor.
    b. Proceed from the main section of the building to the wings.
    c. Proceed from south to north or vice versa, clockwise or counterclockwise.
    d. Take off floor plan quantities first, elevations next, then detail drawings.
17. List all gross dimensions that can be either used again for different quantities, or used as a rough check of other quantities for verification (exterior perimeter, gross floor area, individual floor areas, etc.).
18. Utilize design symmetry or repetition (repetitive floors, repetitive wings, symmetrical design around a center line, similar room layouts, etc.). Note: Extreme caution is needed here so as not to omit or duplicate an area.
19. Do not convert units until the final total is obtained. For instance, when estimating concrete work, keep all units to the nearest cubic foot, then summarize and convert to cubic yards.
20. When figuring alternatives, it is best to total all items involved in the basic system, then total all items involved in the alternates. Therefore you work with positive numbers in all cases. When adds and deducts are used, it is often confusing whether to add or subtract a portion of an item; especially on a complicated or involved alternate.

R011105-10 Unit Gross Area Requirements

The figures in the table below indicate typical ranges in square feet as a function of the "occupant" unit. This table is best used in the preliminary design stages to help determine the probable size requirement for the total project.

Building Type	Unit	Gross Area in S.F. 1/4	Median	3/4
Apartments	Unit	660	860	1,100
Auditorium & Play Theaters	Seat	18	25	38
Bowling Alleys	Lane		940	
Churches & Synagogues	Seat	20	28	39
Dormitories	Bed	200	230	275
Fraternity & Sorority Houses	Bed	220	315	370
Garages, Parking	Car	325	355	385
Hospitals	Bed	685	850	1,075
Hotels	Rental Unit	475	600	710
Housing for the elderly	Unit	515	635	755
Housing, Public	Unit	700	875	1,030
Ice Skating Rinks	Total	27,000	30,000	36,000
Motels	Rental Unit	360	465	620
Nursing Homes	Bed	290	350	450
Restaurants	Seat	23	29	39
Schools, Elementary	Pupil	65	77	90
Junior High & Middle	↓	85	110	129
Senior High	↓	102	130	145
Vocational	↓	110	135	195
Shooting Ranges	Point		450	
Theaters & Movies	Seat		15	

R011105-20 Floor Area Ratios

Table below lists commonly used gross to net area and net to gross area ratios expressed in % for various building types.

Building Type	Gross to Net Ratio	Net to Gross Ratio	Building Type	Gross to Net Ratio	Net to Gross Ratio
Apartment	156	64	School Buildings (campus type)		
Bank	140	72	Administrative	150	67
Church	142	70	Auditorium	142	70
Courthouse	162	61	Biology	161	62
Department Store	123	81	Chemistry	170	59
Garage	118	85	Classroom	152	66
Hospital	183	55	Dining Hall	138	72
Hotel	158	63	Dormitory	154	65
Laboratory	171	58	Engineering	164	61
Library	132	76	Fraternity	160	63
Office	135	75	Gymnasium	142	70
Restaurant	141	70	Science	167	60
Warehouse	108	93	Service	120	83
			Student Union	172	59

The gross area of a building is the total floor area based on outside dimensions.

The net area of a building is the usable floor area for the function intended and excludes such items as stairways, corridors and mechanical rooms. In the case of a commercial building, it might be considered as the "leasable area."

R011105-30 Occupancy Determinations

Function of Space	SF/Person Required
Accessory storage areas, mechanical equipment rooms	300
Agriculture Building	300
Aircraft Hangars	500
Airport Terminal	
Baggage claim	20
Baggage handling	300
Concourse	100
Waiting areas	15
Assembly	
Gaming floors (keno, slots, etc.)	11
Exhibit Gallery and Museum	30
Assembly w/ fixed seats	load determined by seat number
Assembly w/o fixed seats	
Concentrated (chairs only-not fixed)	7
Standing space	5
Unconcentrated (tables and chairs)	15
Bowling centers, allow 5 persons for each lane including 15 feet of runway, and for additional areas	7
Business areas	100
Courtrooms-other than fixed seating areas	40
Day care	35
Dormitories	50
Educational	
Classroom areas	20
Shops and other vocational room areas	50
Exercise rooms	50
Fabrication and Manufacturing areas where hazardous materials are used	200
Industrial areas	100
Institutional areas	
Inpatient treatment areas	240
Outpatient areas	100
Sleeping areas	120
Kitchens commercial	200
Library	
Reading rooms	50
Stack area	100
Mercantile	
Areas on other floors	60
Basement and grade floor areas	30
Storage, stock, shipping areas	300
Parking garages	200
Residential	200
Skating rinks, swimming pools	
Rink and pool	50
Decks	15
Stages and platforms	15
Warehouses	500

R011105-40 Weather Data and Design Conditions

City	Latitude (1)		Winter Temperatures (1)			Winter Degree Days (2)	Summer (Design Dry Bulb) Temperatures and Relative Humidity		
	0	1′	Med. of Annual Extremes	99%	97½%		1%	2½%	5%
UNITED STATES									
Albuquerque, NM	35	0	5.1	12	16	4,400	96/61	94/61	92/61
Atlanta, GA	33	4	11.9	17	22	3,000	94/74	92/74	90/73
Baltimore, MD	39	2	7	14	17	4,600	94/75	91/75	89/74
Birmingham, AL	33	3	13	17	21	2,600	96/74	94/75	92/74
Bismarck, ND	46	5	-32	-23	-19	8,800	95/68	91/68	88/67
Boise, ID	43	3	1	3	10	5,800	96/65	94/64	91/64
Boston, MA	42	2	-1	6	9	5,600	91/73	88/71	85/70
Burlington, VT	44	3	-17	-12	-7	8,200	88/72	85/70	82/69
Charleston, WV	38	2	3	7	11	4,400	92/74	90/73	87/72
Charlotte, NC	35	1	13	18	22	3,200	95/74	93/74	91/74
Casper, WY	42	5	-21	-11	-5	7,400	92/58	90/57	87/57
Chicago, IL	41	5	-8	-3	2	6,600	94/75	91/74	88/73
Cincinnati, OH	39	1	0	1	6	4,400	92/73	90/72	88/72
Cleveland, OH	41	2	-3	1	5	6,400	91/73	88/72	86/71
Columbia, SC	34	0	16	20	24	2,400	97/76	95/75	93/75
Dallas, TX	32	5	14	18	22	2,400	102/75	100/75	97/75
Denver, CO	39	5	-10	-5	1	6,200	93/59	91/59	89/59
Des Moines, IA	41	3	-14	-10	-5	6,600	94/75	91/74	88/73
Detroit, MI	42	2	-3	3	6	6,200	91/73	88/72	86/71
Great Falls, MT	47	3	-25	-21	-15	7,800	91/60	88/60	85/59
Hartford, CT	41	5	-4	3	7	6,200	91/74	88/73	85/72
Houston, TX	29	5	24	28	33	1,400	97/77	95/77	93/77
Indianapolis, IN	39	4	-7	-2	2	5,600	92/74	90/74	87/73
Jackson, MS	32	2	16	21	25	2,200	97/76	95/76	93/76
Kansas City, MO	39	1	-4	2	6	4,800	99/75	96/74	93/74
Las Vegas, NV	36	1	18	25	28	2,800	108/66	106/65	104/65
Lexington, KY	38	0	-1	3	8	4,600	93/73	91/73	88/72
Little Rock, AR	34	4	11	15	20	3,200	99/76	96/77	94/77
Los Angeles, CA	34	0	36	41	43	2,000	93/70	89/70	86/69
Memphis, TN	35	0	10	13	18	3,200	98/77	95/76	93/76
Miami, FL	25	5	39	44	47	200	91/77	90/77	89/77
Milwaukee, WI	43	0	-11	-8	-4	7,600	90/74	87/73	84/71
Minneapolis, MN	44	5	-22	-16	-12	8,400	92/75	89/73	86/71
New Orleans, LA	30	0	28	29	33	1,400	93/78	92/77	90/77
New York, NY	40	5	6	11	15	5,000	92/74	89/73	87/72
Norfolk, VA	36	5	15	20	22	3,400	93/77	91/76	89/76
Oklahoma City, OK	35	2	4	9	13	3,200	100/74	97/74	95/73
Omaha, NE	41	2	-13	-8	-3	6,600	94/76	91/75	88/74
Philadelphia, PA	39	5	6	10	14	4,400	93/75	90/74	87/72
Phoenix, AZ	33	3	27	31	34	1,800	109/71	107/71	105/71
Pittsburgh, PA	40	3	-1	3	7	6,000	91/72	88/71	86/70
Portland, ME	43	4	-10	-6	-1	7,600	87/72	84/71	81/69
Portland, OR	45	4	18	17	23	4,600	89/68	85/67	81/65
Portsmouth, NH	43	1	-8	-2	2	7,200	89/73	85/71	83/70
Providence, RI	41	4	-1	5	9	6,000	89/73	86/72	83/70
Rochester, NY	43	1	-5	1	5	6,800	91/73	88/71	85/70
Salt Lake City, UT	40	5	0	3	8	6,000	97/62	95/62	92/61
San Francisco, CA	37	5	36	38	40	3,000	74/63	71/62	69/61
Seattle, WA	47	4	22	22	27	5,200	85/68	82/66	78/65
Sioux Falls, SD	43	4	-21	-15	-11	7,800	94/73	91/72	88/71
St. Louis, MO	38	4	-3	3	8	5,000	98/75	94/75	91/75
Tampa, FL	28	0	32	36	40	680	92/77	91/77	90/76
Trenton, NJ	40	1	4	11	14	5,000	91/75	88/74	85/73
Washington, DC	38	5	7	14	17	4,200	93/75	91/74	89/74
Wichita, KS	37	4	-3	3	7	4,600	101/72	98/73	96/73
Wilmington, DE	39	4	5	10	14	5,000	92/74	89/74	87/73
ALASKA									
Anchorage	61	1	-29	-23	-18	10,800	71/59	68/58	66/56
Fairbanks	64	5	-59	-51	-47	14,280	82/62	78/60	75/59
CANADA									
Edmonton, Alta.	53	3	-30	-29	-25	11,000	85/66	82/65	79/63
Halifax, N.S.	44	4	-4	1	5	8,000	79/66	76/65	74/64
Montreal, Que.	45	3	-20	-16	-10	9,000	88/73	85/72	83/71
Saskatoon, Sask.	52	1	-35	-35	-31	11,000	89/68	86/66	83/65
St. John's, N.F.	47	4	1	3	7	8,600	77/66	75/65	73/64
Saint John, N.B.	45	2	-15	-12	-8	8,200	80/67	77/65	75/64
Toronto, Ont.	43	4	-10	-5	-1	7,000	90/73	87/72	85/71
Vancouver, B.C.	49	1	13	15	19	6,000	79/67	77/66	74/65
Winnipeg, Man.	49	5	-31	-30	-27	10,800	89/73	86/71	84/70

(1) Handbook of Fundamentals, ASHRAE, Inc., NY 1989
(2) Local Climatological Annual Survey, USDC Env. Science Services Administration, Asheville, NC

R011105-50 Metric Conversion Factors

Description: This table is primarily for converting customary U.S. units in the left hand column to SI metric units in the right hand column. In addition, conversion factors for some commonly encountered Canadian and non-SI metric units are included.

	If You Know		Multiply By		To Find
Length	Inches	x	25.4[a]	=	Millimeters
	Feet	x	0.3048[a]	=	Meters
	Yards	x	0.9144[a]	=	Meters
	Miles (statute)	x	1.609	=	Kilometers
Area	Square inches	x	645.2	=	Square millimeters
	Square feet	x	0.0929	=	Square meters
	Square yards	x	0.8361	=	Square meters
Volume (Capacity)	Cubic inches	x	16,387	=	Cubic millimeters
	Cubic feet	x	0.02832	=	Cubic meters
	Cubic yards	x	0.7646	=	Cubic meters
	Gallons (U.S. liquids)[b]	x	0.003785	=	Cubic meters[c]
	Gallons (Canadian liquid)[b]	x	0.004546	=	Cubic meters[c]
	Ounces (U.S. liquid)[b]	x	29.57	=	Milliliters[c, d]
	Quarts (U.S. liquid)[b]	x	0.9464	=	Liters[c, d]
	Gallons (U.S. liquid)[b]	x	3.785	=	Liters[c, d]
Force	Kilograms force[d]	x	9.807	=	Newtons
	Pounds force	x	4.448	=	Newtons
	Pounds force	x	0.4536	=	Kilograms force[d]
	Kips	x	4448	=	Newtons
	Kips	x	453.6	=	Kilograms force[d]
Pressure, Stress, Strength (Force per unit area)	Kilograms force per square centimeter[d]	x	0.09807	=	Megapascals
	Pounds force per square inch (psi)	x	0.006895	=	Megapascals
	Kips per square inch	x	6.895	=	Megapascals
	Pounds force per square inch (psi)	x	0.07031	=	Kilograms force per square centimeter[d]
	Pounds force per square foot	x	47.88	=	Pascals
	Pounds force per square foot	x	4.882	=	Kilograms force per square meter[d]
Flow	Cubic feet per minute	x	0.4719	=	Liters per second
	Gallons per minute	x	0.0631	=	Liters per second
	Gallons per hour	x	1.05	=	Milliliters per second
Bending Moment Or Torque	Inch-pounds force	x	0.01152	=	Meter-kilograms force[d]
	Inch-pounds force	x	0.1130	=	Newton-meters
	Foot-pounds force	x	0.1383	=	Meter-kilograms force[d]
	Foot-pounds force	x	1.356	=	Newton-meters
	Meter-kilograms force[d]	x	9.807	=	Newton-meters
Mass	Ounces (avoirdupois)	x	28.35	=	Grams
	Pounds (avoirdupois)	x	0.4536	=	Kilograms
	Tons (metric)	x	1000	=	Kilograms
	Tons, short (2000 pounds)	x	907.2	=	Kilograms
	Tons, short (2000 pounds)	x	0.9072	=	Megagrams[e]
Mass per Unit Volume	Pounds mass per cubic foot	x	16.02	=	Kilograms per cubic meter
	Pounds mass per cubic yard	x	0.5933	=	Kilograms per cubic meter
	Pounds mass per gallon (U.S. liquid)[b]	x	119.8	=	Kilograms per cubic meter
	Pounds mass per gallon (Canadian liquid)[b]	x	99.78	=	Kilograms per cubic meter
Temperature	Degrees Fahrenheit		(F-32)/1.8	=	Degrees Celsius
	Degrees Fahrenheit		(F+459.67)/1.8	=	Degrees Kelvin
	Degrees Celsius		C+273.15	=	Degrees Kelvin

[a]The factor given is exact
[b]One U.S. gallon = 0.8327 Canadian gallon
[c]1 liter = 1000 milliliters = 1000 cubic centimeters
1 cubic decimeter = 0.001 cubic meter
[d]Metric but not SI unit
[e]Called "tonne" in England and "metric ton" in other metric countries

R011105-60 Weights and Measures

Measures of Length
1 Mile = 1760 Yards = 5280 Feet
1 Yard = 3 Feet = 36 inches
1 Foot = 12 Inches
1 Mil = 0.001 Inch
1 Fathom = 2 Yards = 6 Feet
1 Rod = 5.5 Yards = 16.5 Feet
1 Hand = 4 Inches
1 Span = 9 Inches
1 Micro-inch = One Millionth Inch or 0.000001 Inch
1 Micron = One Millionth Meter + 0.00003937 Inch

Surveyor's Measure
1 Mile = 8 Furlongs = 80 Chains
1 Furlong = 10 Chains = 220 Yards
1 Chain = 4 Rods = 22 Yards = 66 Feet = 100 Links
1 Link = 7.92 Inches

Square Measure
1 Square Mile = 640 Acres = 6400 Square Chains
1 Acre = 10 Square Chains = 4840 Square Yards = 43,560 Sq. Ft.
1 Square Chain = 16 Square Rods = 484 Square Yards = 4356 Sq. Ft.
1 Square Rod = 30.25 Square Yards = 272.25 Square Feet = 625 Square Lines
1 Square Yard = 9 Square Feet
1 Square Foot = 144 Square Inches
An Acre equals a Square 208.7 Feet per Side

Cubic Measure
1 Cubic Yard = 27 Cubic Feet
1 Cubic Foot = 1728 Cubic Inches
1 Cord of Wood = 4 x 4 x 8 Feet = 128 Cubic Feet
1 Perch of Masonry = 16½ x 1½ x 1 Foot = 24.75 Cubic Feet

Avoirdupois or Commercial Weight
1 Gross or Long Ton = 2240 Pounds
1 Net or Short Ton = 2000 Pounds
1 Pound = 16 Ounces = 7000 Grains
1 Ounce = 16 Drachms = 437.5 Grains
1 Stone = 14 Pounds

Power
1 British Thermal Unit per Hour = 0.2931 Watts
1 Ton (Refrigeration) = 3.517 Kilowatts
1 Horsepower (Boiler) = 9.81 Kilowatts
1 Horsepower (550 ft-lb/s) = 0.746 Kilowatts

Shipping Measure
For Measuring Internal Capacity of a Vessel:
1 Register Ton = 100 Cubic Feet

For Measurement of Cargo:
Approximately 40 Cubic Feet of Merchandise is considered a Shipping Ton, unless that bulk would weigh more than 2000 Pounds, in which case Freight Charge may be based upon weight.

40 Cubic Feet = 32.143 U.S. Bushels = 31.16 Imp. Bushels

Liquid Measure
1 Imperial Gallon = 1.2009 U.S. Gallon = 277.42 Cu. In.
1 Cubic Foot = 7.48 U.S. Gallons

R011110-30 Engineering Fees

Typical **Structural Engineering Fees** based on type of construction and total project size. These fees are included in Architectural Fees.

Type of Construction	Total Project Size (in thousands of dollars)			
	$500	$500-$1,000	$1,000-$5,000	Over $5000
Industrial buildings, factories & warehouses	Technical payroll times 2.0 to 2.5	1.60%	1.25%	1.00%
Hotels, apartments, offices, dormitories, hospitals, public buildings, food stores	↓	2.00%	1.70%	1.20%
Museums, banks, churches and cathedrals	↓	2.00%	1.75%	1.25%
Thin shells, prestressed concrete, earthquake resistive	↓	2.00%	1.75%	1.50%
Parking ramps, auditoriums, stadiums, convention halls, hangars & boiler houses	↓	2.50%	2.00%	1.75%
Special buildings, major alterations, underpinning & future expansion	↓	Add to above 0.5%	Add to above 0.5%	Add to above 0.5%

For complex reinforced concrete or unusually complicated structures, add 20% to 50%.

Typical **Mechanical and Electrical Engineering Fees** are based on the size of the subcontract. The fee structure for both is shown below. These fees are included in Architectural Fees.

	Subcontract Size							
Type of Construction	$25,000	$50,000	$100,000	$225,000	$350,000	$500,000	$750,000	$1,000,000
Simple structures	6.4%	5.7%	4.8%	4.5%	4.4%	4.3%	4.2%	4.1%
Intermediate structures	8.0	7.3	6.5	5.6	5.1	5.0	4.9	4.8
Complex structures	10.1	9.0	9.0	8.0	7.5	7.5	7.0	7.0

For renovations, add 15% to 25% to applicable fee.

R012153-10 Repair and Remodeling

Cost figures are based on new construction utilizing the most cost-effective combination of labor, equipment and material with the work scheduled in proper sequence to allow the various trades to accomplish their work in an efficient manner.

The costs for repair and remodeling work must be modified due to the following factors that may be present in any given repair and remodeling project.

1. Equipment usage curtailment due to the physical limitations of the project, with only hand-operated equipment being used.
2. Increased requirement for shoring and bracing to hold up the building while structural changes are being made and to allow for temporary storage of construction materials on above-grade floors.
3. Material handling becomes more costly due to having to move within the confines of an enclosed building. For multi-story construction, low capacity elevators and stairwells may be the only access to the upper floors.
4. Large amount of cutting and patching and attempting to match the existing construction is required. It is often more economical to remove entire walls rather than create many new door and window openings. This sort of trade-off has to be carefully analyzed.
5. Cost of protection of completed work is increased since the usual sequence of construction usually cannot be accomplished.
6. Economies of scale usually associated with new construction may not be present. If small quantities of components must be custom fabricated due to job requirements, unit costs will naturally increase. Also, if only small work areas are available at a given time, job scheduling between trades becomes difficult and subcontractor quotations may reflect the excessive start-up and shut-down phases of the job.
7. Work may have to be done on other than normal shifts and may have to be done around an existing production facility which has to stay in production during the course of the repair and remodeling.
8. Dust and noise protection of adjoining non-construction areas can involve substantial special protection and alter usual construction methods.
9. Job may be delayed due to unexpected conditions discovered during demolition or removal. These delays ultimately increase construction costs.
10. Piping and ductwork runs may not be as simple as for new construction. Wiring may have to be snaked through walls and floors.
11. Matching "existing construction" may be impossible because materials may no longer be manufactured. Substitutions may be expensive.
12. Weather protection of existing structure requires additional temporary structures to protect building at openings.
13. On small projects, because of local conditions, it may be necessary to pay a tradesman for a minimum of four hours for a task that is completed in one hour.

All of the above areas can contribute to increased costs for a repair and remodeling project. Each of the above factors should be considered in the planning, bidding and construction stage in order to minimize the increased costs associated with repair and remodeling jobs.

R012153-60 Security Factors

Contractors entering, working in, and exiting secure facilities often lose productive time during a normal workday. The recommended allowances in this section are intended to provide for the loss of productivity by increasing labor costs. Note that different costs are associated with searches upon entry only and searches upon entry and exit. Time spent in a queue is unpredictable and not part of these allowances. Contractors should plan ahead for this situation.

Security checkpoints are designed to reflect the level of security required to gain access or egress. An extreme example is when contractors, along with any materials, tools, equipment, and vehicles, must be physically searched and have all materials, tools, equipment, and vehicles inventoried and documented prior to both entry and exit.

Physical searches without going through the documentation process represent the next level and take up less time.

Electronic searches—passing through a detector or x-ray machine with no documentation of materials, tools, equipment, and vehicles—take less time than physical searches.

Visual searches of materials, tools, equipment, and vehicles represent the next level of security.

Finally, access by means of an ID card or displayed sticker takes the least amount of time.

Another consideration is if the searches described above are performed each and every day, or if they are performed only on the first day with access granted by ID card or displayed sticker for the remainder of the project. The figures for this situation have been calculated to represent the initial check-in as described and subsequent entry by ID card or displayed sticker for up to 20 days on site. For the situation described above, where the time period is beyond 20 days, the impact on labor cost is negligible.

There are situations where tradespeople must be accompanied by an escort and observed during the work day. The loss of freedom of movement will slow down productivity for the tradesperson. Costs for the observer have not been included. Those costs are normally born by the owner.

General Requirements | R0129 Payment Procedures

R012909-80 Sales Tax by State

State sales tax on materials is tabulated below (5 states have no sales tax). Many states allow local jurisdictions, such as a county or city, to levy additional sales tax.

Some projects may be sales tax exempt, particularly those constructed with public funds.

State	Tax (%)	State	Tax (%)	State	Tax (%)	State	Tax (%)
Alabama	4	Illinois	6.25	Montana	0	Rhode Island	7
Alaska	0	Indiana	7	Nebraska	5.5	South Carolina	6
Arizona	5.6	Iowa	6	Nevada	6.85	South Dakota	4.5
Arkansas	6.5	Kansas	6.5	New Hampshire	0	Tennessee	7
California	7.25	Kentucky	6	New Jersey	6.625	Texas	6.25
Colorado	2.9	Louisiana	4.45	New Mexico	5.125	Utah	4.85
Connecticut	6.35	Maine	5.5	New York	4	Vermont	6
Delaware	0	Maryland	6	North Carolina	4.75	Virginia	4.3
District of Columbia	6	Massachusetts	6.25	North Dakota	5	Washington	6.5
Florida	6	Michigan	6	Ohio	5.75	West Virginia	6
Georgia	4	Minnesota	6.875	Oklahoma	4.5	Wisconsin	5
Hawaii	4	Mississippi	7	Oregon	0	Wyoming	4
Idaho	6	Missouri	4.225	Pennsylvania	6	Average	5.06 %

Sales Tax by Province (Canada)

GST - a value-added tax, which the government imposes on most goods and services provided in or imported into Canada. PST - a retail sales tax, which five of the provinces impose on the prices of most goods and some services. QST - a value-added tax, similar to the federal GST, which Quebec imposes. HST - Three provinces have combined their retail sales taxes with the federal GST into one harmonized tax.

Province	PST (%)	QST(%)	GST(%)	HST (%)
Alberta	0	0	5	0
British Columbia	7	0	5	0
Manitoba	7	0	5	0
New Brunswick	0	0	0	15
Newfoundland	0	0	0	15
Northwest Territories	0	0	5	0
Nova Scotia	0	0	0	15
Ontario	0	0	0	13
Prince Edward Island	0	0	0	15
Quebec	9.975	0	5	14.975
Saskatchewan	6	0	5	0
Yukon	0	0	5	0

R012909-85 Unemployment Taxes and Social Security Taxes

State unemployment tax rates vary not only from state to state, but also with the experience rating of the contractor. The federal unemployment tax rate is 6.0% of the first $7,000 of wages. This is reduced by a credit of up to 5.4% for timely payment to the state. The minimum federal unemployment tax is 0.6% after all credits.

Social security (FICA) for 2020 is estimated at time of publication to be 7.65% of wages up to $137,100.

R012909-86 Unemployment Tax by State

Information is from the U.S. Department of Labor, state unemployment tax rates.

State	Tax (%)	State	Tax (%)	State	Tax (%)	State	Tax (%)
Alabama	6.80	Illinois	6.93	Montana	6.30	Rhode Island	9.49
Alaska	5.4	Indiana	7.4	Nebraska	5.4	South Carolina	5.46
Arizona	12.76	Iowa	7.5	Nevada	5.4	South Dakota	9.35
Arkansas	14.3	Kansas	7.6	New Hampshire	7.5	Tennessee	10.0
California	6.2	Kentucky	9.3	New Jersey	5.8	Texas	6.5
Colorado	8.15	Louisiana	6.2	New Mexico	5.4	Utah	7.1
Connecticut	6.8	Maine	5.46	New York	9.1	Vermont	7.7
Delaware	8.20	Maryland	7.50	North Carolina	5.76	Virginia	6.21
District of Columbia	7	Massachusetts	12.65	North Dakota	10.74	Washington	7.73
Florida	5.4	Michigan	10.3	Ohio	9	West Virginia	8.5
Georgia	5.4	Minnesota	9.1	Oklahoma	5.5	Wisconsin	12.0
Hawaii	5.6	Mississippi	5.6	Oregon	5.4	Wyoming	8.8
Idaho	5.4	Missouri	8.37	Pennsylvania	11.03	Median	7.40%

R012909-90 Overtime

One way to improve the completion date of a project or eliminate negative float from a schedule is to compress activity duration times. This can be achieved by increasing the crew size or working overtime with the proposed crew.

To determine the costs of working overtime to compress activity duration times, consider the following examples. Below is an overtime efficiency and cost chart based on a five, six, or seven day week with an eight through twelve hour day. Payroll percentage increases for time and one half and double times are shown for the various working days.

Days per Week	Hours per Day	Production Efficiency					Payroll Cost Factors	
		1st Week	2nd Week	3rd Week	4th Week	Average 4 Weeks	@ 1-1/2 Times	@ 2 Times
5	8	100%	100%	100%	100%	100%	1.000	1.000
	9	100	100	95	90	96	1.056	1.111
	10	100	95	90	85	93	1.100	1.200
	11	95	90	75	65	81	1.136	1.273
	12	90	85	70	60	76	1.167	1.333
6	8	100	100	95	90	96	1.083	1.167
	9	100	95	90	85	93	1.130	1.259
	10	95	90	85	80	88	1.167	1.333
	11	95	85	70	65	79	1.197	1.394
	12	90	80	65	60	74	1.222	1.444
7	8	100	95	85	75	89	1.143	1.286
	9	95	90	80	70	84	1.183	1.365
	10	90	85	75	65	79	1.214	1.429
	11	85	80	65	60	73	1.240	1.481
	12	85	75	60	55	69	1.262	1.524

R013113-40 Builder's Risk Insurance

Builder's risk insurance is insurance on a building during construction. Premiums are paid by the owner or the contractor. Blasting, collapse and underground insurance would raise total insurance costs.

R013113-50 General Contractor's Overhead

There are two distinct types of overhead on a construction project: Project overhead and main office overhead. Project overhead includes those costs at a construction site not directly associated with the installation of construction materials. Examples of project overhead costs include the following:

1. Superintendent
2. Construction office and storage trailers
3. Temporary sanitary facilities
4. Temporary utilities
5. Security fencing
6. Photographs
7. Cleanup
8. Performance and payment bonds

The above project overhead items are also referred to as general requirements and therefore are estimated in Division 1. Division 1 is the first division listed in the CSI MasterFormat but it is usually the last division estimated. The sum of the costs in Divisions 1 through 49 is referred to as the sum of the direct costs.

All construction projects also include indirect costs. The primary components of indirect costs are the contractor's main office overhead and profit. The amount of the main office overhead expense varies depending on the following:

1. Owner's compensation
2. Project managers' and estimators' wages
3. Clerical support wages
4. Office rent and utilities
5. Corporate legal and accounting costs
6. Advertising
7. Automobile expenses
8. Association dues
9. Travel and entertainment expenses

These costs are usually calculated as a percentage of annual sales volume. This percentage can range from 35% for a small contractor doing less than $500,000 to 5% for a large contractor with sales in excess of $100 million.

R013113-55 Installing Contractor's Overhead

Installing contractors (subcontractors) also incur costs for general requirements and main office overhead.

Included within the total incl. overhead and profit costs is a percent mark-up for overhead that includes:

1. Compensation and benefits for office staff and project managers
2. Office rent, utilities, business equipment, and maintenance
3. Corporate legal and accounting costs
4. Advertising
5. Vehicle expenses (for office staff and project managers)
6. Association dues
7. Travel, entertainment
8. Insurance
9. Small tools and equipment

R013113-60 Workers' Compensation Insurance Rates by Trade

The table below tabulates the national averages for workers' compensation insurance rates by trade and type of building. The average "Insurance Rate" is multiplied by the "% of Building Cost" for each trade. This produces the "Workers' Compensation" cost by % of total labor cost, to be added for each trade by building type to determine the weighted average workers' compensation rate for the building types analyzed.

Trade	Insurance Rate (% Labor Cost)		Workers' Compensation	% of Building Cost			Workers' Compensation		
	Range	Average		Office Bldgs.	Schools & Apts.	Mfg.	Office Bldgs.	Schools & Apts.	Mfg.
Excavation, Grading, etc.	2.3 % to 16.9%	9.6%		4.8%	4.9%	4.5%	0.46%	0.47%	0.43%
Piles & Foundations	3.3 to 28.0	15.7		7.1	5.2	8.7	1.11	0.82	1.37
Concrete	3.1 to 25.8	14.4		5.0	14.8	3.7	0.72	2.13	0.53
Masonry	3.3 to 52.1	27.7		6.9	7.5	1.9	1.91	2.08	0.53
Structural Steel	4.4 to 31.8	18.1		10.7	3.9	17.6	1.94	0.71	3.19
Miscellaneous & Ornamental Metals	2.9 to 21.4	12.2		2.8	4.0	3.6	0.34	0.49	0.44
Carpentry & Millwork	3.4 to 29.1	16.2		3.7	4.0	0.5	0.60	0.65	0.08
Metal or Composition Siding	4.8 to 124.6	64.7		2.3	0.3	4.3	1.49	0.19	2.78
Roofing	4.8 to 110.4	57.6		2.3	2.6	3.1	1.32	1.50	1.79
Doors & Hardware	3.1 to 29.1	16.1		0.9	1.4	0.4	0.14	0.23	0.06
Sash & Glazing	3.9 to 21.6	12.8		3.5	4.0	1.0	0.45	0.51	0.13
Lath & Plaster	2.7 to 30.1	16.4		3.3	6.9	0.8	0.54	1.13	0.13
Tile, Marble & Floors	2.0 to 19.0	10.5		2.6	3.0	0.5	0.27	0.32	0.05
Acoustical Ceilings	1.7 to 29.7	15.7		2.4	0.2	0.3	0.38	0.03	0.05
Painting	3.3 to 37.4	20.3		1.5	1.6	1.6	0.30	0.32	0.32
Interior Partitions	3.4 to 29.1	16.2		3.9	4.3	4.4	0.63	0.70	0.71
Miscellaneous Items	1.9 to 97.7	10.3		5.2	3.7	9.7	0.54	0.38	1.00
Elevators	1.3 to 9.0	5.1		2.1	1.1	2.2	0.11	0.06	0.11
Sprinklers	1.8 to 14.6	8.2		0.5	—	2.0	0.04	—	0.16
Plumbing	1.4 to 13.3	7.4		4.9	7.2	5.2	0.36	0.53	0.38
Heat., Vent., Air Conditioning	2.8 to 15.8	9.3		13.5	11.0	12.9	1.26	1.02	1.20
Electrical	1.7 to 10.7	6.2		10.1	8.4	11.1	0.63	0.52	0.69
Total	1.3 % to 124.6%	—		100.0%	100.0%	100.0%	15.54%	14.79%	16.13%
		Overall Weighted Average	15.49%						

Workers' Compensation Insurance Rates by States

The table below lists the weighted average Workers' Compensation base rate for each state with a factor comparing this with the national average of 9.9%.

State	Weighted Average	Factor	State	Weighted Average	Factor	State	Weighted Average	Factor
Alabama	12.4%	126	Kentucky	10.5%	108	North Dakota	6.4%	65
Alaska	8.6	88	Louisiana	17.5	178	Ohio	5.5	56
Arizona	7.7	79	Maine	8.0	81	Oklahoma	7.4	76
Arkansas	5.1	52	Maryland	9.7	99	Oregon	7.5	76
California	19.6	200	Massachusetts	8.8	90	Pennsylvania	17.2	175
Colorado	6.2	63	Michigan	6.3	64	Rhode Island	10.0	102
Connecticut	12.4	126	Minnesota	13.4	137	South Carolina	16.2	165
Delaware	11.8	120	Mississippi	9.2	94	South Dakota	7.6	77
District of Columbia	7.7	78	Missouri	12.0	122	Tennessee	6.0	61
Florida	8.4	86	Montana	7.1	73	Texas	4.9	50
Georgia	30.0	307	Nebraska	11.7	120	Utah	5.7	58
Hawaii	8.4	86	Nevada	7.6	77	Vermont	8.9	91
Idaho	7.8	79	New Hampshire	8.8	90	Virginia	6.7	69
Illinois	17.5	179	New Jersey	14.2	145	Washington	7.3	75
Indiana	3.1	32	New Mexico	13.0	132	West Virginia	4.2	42
Iowa	9.9	101	New York	15.6	159	Wisconsin	11.2	115
Kansas	5.6	57	North Carolina	11.6	119	Wyoming	4.8	49
			Weighted Average for U.S. is	9.9% of payroll = 100%				

The weighted average skilled worker rate for 35 trades is 9.8%. For bidding purposes, apply the full value of Workers' Compensation directly to total labor costs, or if labor is 38%, materials 42% and overhead and profit 20% of total cost, carry 38/80 x 9.8% = 4.66% of cost (before overhead and profit) into overhead. Rates vary not only from state to state but also with the experience rating of the contractor.

Rates are the most current available at the time of publication.

R013113-80 Performance Bond

This table shows the cost of a Performance Bond for a construction job scheduled to be completed in 12 months. Add 1% of the premium cost per month for jobs requiring more than 12 months to complete. The rates are "standard" rates offered to contractors that the bonding company considers financially sound and capable of doing the work. Preferred rates are offered by some bonding companies based upon financial strength of the contractor. Actual rates vary from contractor to contractor and from bonding company to bonding company. Contractors should prequalify through a bonding agency before submitting a bid on a contract that requires a bond.

Contract Amount	Building Construction Class B Projects	Highways & Bridges Class A New Construction	Highways & Bridges Class A-1 Highway Resurfacing
First $ 100,000 bid	$25.00 per M	$15.00 per M	$9.40 per M
Next 400,000 bid	$ 2,500 plus $15.00 per M	$ 1,500 plus $10.00 per M	$ 940 plus $7.20 per M
Next 2,000,000 bid	8,500 plus 10.00 per M	5,500 plus 7.00 per M	3,820 plus 5.00 per M
Next 2,500,000 bid	28,500 plus 7.50 per M	19,500 plus 5.50 per M	15,820 plus 4.50 per M
Next 2,500,000 bid	47,250 plus 7.00 per M	33,250 plus 5.00 per M	28,320 plus 4.50 per M
Over 7,500,000 bid	64,750 plus 6.00 per M	45,750 plus 4.50 per M	39,570 plus 4.00 per M

R015423-10 Steel Tubular Scaffolding

On new construction, tubular scaffolding is efficient up to 60′ high or five stories. Above this it is usually better to use a hung scaffolding if construction permits. Swing scaffolding operations may interfere with tenants. In this case, the tubular is more practical at all heights.

In repairing or cleaning the front of an existing building the cost of tubular scaffolding per S.F. of building front increases as the height increases above the first tier. The first tier cost is relatively high due to leveling and alignment.

The minimum efficient crew for erecting and dismantling is three workers. They can set up and remove 18 frame sections per day up to 5 stories high. For 6 to 12 stories high, a crew of four is most efficient. Use two or more on top and two on the bottom for handing up or hoisting. They can also set up and remove 18 frame sections per day. At 7' horizontal spacing, this will run about 800 S.F. per day of erecting and dismantling. Time for placing and removing planks must be added to the above. A crew of three can place and remove 72 planks per day up to 5 stories. For over 5 stories, a crew of four can place and remove 80 planks per day.

The table below shows the number of pieces required to erect tubular steel scaffolding for 1000 S.F. of building frontage. This area is made up of a scaffolding system that is 12 frames (11 bays) long by 2 frames high.

For jobs under twenty-five frames, add 50% to rental cost. Rental rates will be lower for jobs over three months duration. Large quantities for long periods can reduce rental rates by 20%.

Description of Component	Number of Pieces for 1000 S.F. of Building Front	Unit
5′ Wide Standard Frame, 6′-4″ High	24	Ea.
Leveling Jack & Plate	24	↓
Cross Brace	44	↓
Side Arm Bracket, 21″	12	↓
Guardrail Post	12	↓
Guardrail, 7′ section	22	↓
Stairway Section	2	↓
Stairway Starter Bar	1	↓
Stairway Inside Handrail	2	↓
Stairway Outside Handrail	2	↓
Walk-Thru Frame Guardrail	2	↓

Scaffolding is often used as falsework over 15′ high during construction of cast-in-place concrete beams and slabs. Two foot wide scaffolding is generally used for heavy beam construction. The span between frames depends upon the load to be carried with a maximum span of 5′.

Heavy duty shoring frames with a capacity of 10,000#/leg can be spaced up to 10′ O.C. depending upon form support design and loading.

Scaffolding used as horizontal shoring requires less than half the material required with conventional shoring.

On new construction, erection is done by carpenters.

Rolling towers supporting horizontal shores can reduce labor and speed the job. For maintenance work, catwalks with spans up to 70′ can be supported by the rolling towers.

R015433-10 Contractor Equipment

Rental Rates shown elsewhere in the data set pertain to late model high quality machines in excellent working condition, rented from equipment dealers. Rental rates from contractors may be substantially lower than the rental rates from equipment dealers depending upon economic conditions; for older, less productive machines, reduce rates by a maximum of 15%. Any overtime must be added to the base rates. For shift work, rates are lower. Usual rule of thumb is 150% of one shift rate for two shifts; 200% for three shifts.

For periods of less than one week, operated equipment is usually more economical to rent than renting bare equipment and hiring an operator.

Costs to move equipment to a job site (mobilization) or from a job site (demobilization) are not included in rental rates, nor in any Equipment costs on any Unit Price line items or crew listings. These costs can be found elsewhere. If a piece of equipment is already at a job site, it is not appropriate to utilize mob/demob costs in an estimate again.

Rental rates vary throughout the country with larger cities generally having lower rates. Lease plans for new equipment are available for periods in excess of six months with a percentage of payments applying toward purchase.

Rental rates can also be treated as reimbursement costs for contractor-owned equipment. Owned equipment costs include depreciation, loan payments, interest, taxes, insurance, storage, and major repairs.

Monthly rental rates vary from 2% to 5% of the cost of the equipment depending on the anticipated life of the equipment and its wearing parts. Weekly rates are about 1/3 the monthly rates and daily rental rates are about 1/3 the weekly rates.

The hourly operating costs for each piece of equipment include costs to the user such as fuel, oil, lubrication, normal expendables for the equipment, and a percentage of the mechanic's wages chargeable to maintenance. The hourly operating costs listed do not include the operator's wages.

The daily cost for equipment used in the standard crews is figured by dividing the weekly rate by five, then adding eight times the hourly operating cost to give the total daily equipment cost, not including the operator. This figure is in the right hand column of the Equipment listings under Equipment Cost/Day.

Pile Driving rates shown for the pile hammer and extractor do not include leads, cranes, boilers or compressors. Vibratory pile driving requires an added field specialist during set-up and pile driving operation for the electric model. The hydraulic model requires a field specialist for set-up only. Up to 125 reuses of sheet piling are possible using vibratory drivers. For normal conditions, crane capacity for hammer type and size is as follows.

Crane Capacity	Hammer Type and Size		
	Air or Steam	Diesel	Vibratory
25 ton	to 8,750 ft.-lb.		70 H.P.
40 ton	15,000 ft.-lb.	to 32,000 ft.-lb.	170 H.P.
60 ton	25,000 ft.-lb.		300 H.P.
100 ton		112,000 ft.-lb.	

Cranes should be specified for the job by size, building and site characteristics, availability, performance characteristics, and duration of time required.

Backhoes & Shovels rent for about the same as equivalent size cranes but maintenance and operating expenses are higher. The crane operator's rate must be adjusted for high boom heights. Average adjustments: for 150′ boom add 2% per hour; over 185′, add 4% per hour; over 210′, add 6% per hour; over 250′, add 8% per hour and over 295′, add 12% per hour.

Tower Cranes of the climbing or static type have jibs from 50′ to 200′ and capacities at maximum reach range from 4,000 to 14,000 pounds. Lifting capacities increase up to maximum load as the hook radius decreases.

Typical rental rates, based on purchase price, are about 2% to 3% per month.

Erection and dismantling run between 500 and 2000 labor hours. Climbing operation takes 10 labor hours per 20′ climb. Crane dead time is about 5 hours per 40′ climb. If crane is bolted to side of the building add cost of ties and extra mast sections. Climbing cranes have from 80′ to 180′ of mast while static cranes have 80′ to 800′ of mast.

Truck Cranes can be converted to tower cranes by using tower attachments. Mast heights over 400′ have been used.

A single 100′ high material **Hoist and Tower** can be erected and dismantled in about 400 labor hours; a double 100′ high hoist and tower in about 600 labor hours. Erection times for additional heights are 3 and 4 labor hours per vertical foot respectively up to 150′, and 4 to 5 labor hours per vertical foot over 150′ high. A 40′ high portable Buck hoist takes about 160 labor hours to erect and dismantle. Additional heights take 2 labor hours per vertical foot to 80′ and 3 labor hours per vertical foot for the next 100′. Most material hoists do not meet local code requirements for carrying personnel.

A 150′ high **Personnel Hoist** requires about 500 to 800 labor hours to erect and dismantle. Budget erection time at 5 labor hours per vertical foot for all trades. Local code requirements or labor scarcity requiring overtime can add up to 50% to any of the above erection costs.

Earthmoving Equipment: The selection of earthmoving equipment depends upon the type and quantity of material, moisture content, haul distance, haul road, time available, and equipment available. Short haul cut and fill operations may require dozers only, while another operation may require excavators, a fleet of trucks, and spreading and compaction equipment. Stockpiled material and granular material are easily excavated with front end loaders. Scrapers are most economically used with hauls between 300′ and 1-1/2 miles if adequate haul roads can be maintained. Shovels are often used for blasted rock and any material where a vertical face of 8′ or more can be excavated. Special conditions may dictate the use of draglines, clamshells, or backhoes. Spreading and compaction equipment must be matched to the soil characteristics, the compaction required and the rate the fill is being supplied.

R015433-15 Heavy Lifting

Hydraulic Climbing Jacks

The use of hydraulic heavy lift systems is an alternative to conventional type crane equipment. The lifting, lowering, pushing, or pulling mechanism is a hydraulic climbing jack moving on a square steel jackrod from 1-5/8″ to 4″ square, or a steel cable. The jackrod or cable can be vertical or horizontal, stationary or movable, depending on the individual application. When the jackrod is stationary, the climbing jack will climb the rod and push or pull the load along with itself. When the climbing jack is stationary, the jackrod is movable with the load attached to the end and the climbing jack will lift or lower the jackrod with the attached load. The heavy lift system is normally operated by a single control lever located at the hydraulic pump.

The system is flexible in that one or more climbing jacks can be applied wherever a load support point is required, and the rate of lift synchronized.

Economic benefits have been demonstrated on projects such as: erection of ground assembled roofs and floors, complete bridge spans, girders and trusses, towers, chimney liners and steel vessels, storage tanks, and heavy machinery. Other uses are raising and lowering offshore work platforms, caissons, tunnel sections and pipelines.

R015436-50 Mobilization

Costs to move rented construction equipment to a job site from an equipment dealer's or contractor's yard (mobilization) or off the job site (demobilization) are not included in the rental or operating rates, nor in the equipment cost on a unit price line or in a crew listing. These costs can be found consolidated in the Mobilization section of the data and elsewhere in particular site work sections. If a piece of equipment is already on the job site, it is not appropriate to include mob/demob costs in a new estimate that requires use of that equipment. The following table identifies approximate sizes of rented construction equipment that would be hauled on a towed trailer. Because this listing is not all-encompassing, the user can infer as to what size trailer might be required for a piece of equipment not listed.

3-ton Trailer	20-ton Trailer	40-ton Trailer	50-ton Trailer
20 H.P. Excavator	110 H.P. Excavator	200 H.P. Excavator	270 H.P. Excavator
50 H.P. Skid Steer	165 H.P. Dozer	300 H.P. Dozer	Small Crawler Crane
35 H.P. Roller	150 H.P. Roller	400 H.P. Scraper	500 H.P. Scraper
40 H.P. Trencher	Backhoe	450 H.P. Art. Dump Truck	500 H.P. Art. Dump Truck

R024119-10 Demolition Defined

Whole Building Demolition - Demolition of the whole building with no concern for any particular building element, component, or material type being demolished. This type of demolition is accomplished with large pieces of construction equipment that break up the structure, load it into trucks and haul it to a disposal site, but disposal or dump fees are not included. Demolition of below-grade foundation elements, such as footings, foundation walls, grade beams, slabs on grade, etc., is not included. Certain mechanical equipment containing flammable liquids or ozone-depleting refrigerants, electric lighting elements, communication equipment components, and other building elements may contain hazardous waste, and must be removed, either selectively or carefully, as hazardous waste before the building can be demolished.

Foundation Demolition - Demolition of below-grade foundation footings, foundation walls, grade beams, and slabs on grade. This type of demolition is accomplished by hand or pneumatic hand tools, and does not include saw cutting, or handling, loading, hauling, or disposal of the debris.

Gutting - Removal of building interior finishes and electrical/mechanical systems down to the load-bearing and sub-floor elements of the rough building frame, with no concern for any particular building element, component, or material type being demolished. This type of demolition is accomplished by hand or pneumatic hand tools, and includes loading into trucks, but not hauling, disposal or dump fees, scaffolding, or shoring. Certain mechanical equipment containing flammable liquids or ozone-depleting refrigerants, electric lighting elements, communication equipment components, and other building elements may contain hazardous waste, and must be removed, either selectively or carefully, as hazardous waste, before the building is gutted.

Selective Demolition - Demolition of a selected building element, component, or finish, with some concern for surrounding or adjacent elements, components, or finishes (see the first Subdivision (s) at the beginning of appropriate Divisions). This type of demolition is accomplished by hand or pneumatic hand tools, and does not include handling, loading, storing, hauling, or disposal of the debris, scaffolding, or shoring. "Gutting" methods may be used in order to save time, but damage that is caused to surrounding or adjacent elements, components, or finishes may have to be repaired at a later time.

Careful Removal - Removal of a piece of service equipment, building element or component, or material type, with great concern for both the removed item and surrounding or adjacent elements, components or finishes. The purpose of careful removal may be to protect the removed item for later re-use, preserve a higher salvage value of the removed item, or replace an item while taking care to protect surrounding or adjacent elements, components, connections, or finishes from cosmetic and/or structural damage. An approximation of the time required to perform this type of removal is 1/3 to 1/2 the time it would take to install a new item of like kind (see Reference Number R220105-10). This type of removal is accomplished by hand or pneumatic hand tools, and does not include loading, hauling, or storing the removed item, scaffolding, shoring, or lifting equipment.

Cutout Demolition - Demolition of a small quantity of floor, wall, roof, or other assembly, with concern for the appearance and structural integrity of the surrounding materials. This type of demolition is accomplished by hand or pneumatic hand tools, and does not include saw cutting, handling, loading, hauling, or disposal of debris, scaffolding, or shoring.

Rubbish Handling - Work activities that involve handling, loading or hauling of debris. Generally, the cost of rubbish handling must be added to the cost of all types of demolition, with the exception of whole building demolition.

Minor Site Demolition - Demolition of site elements outside the footprint of a building. This type of demolition is accomplished by hand or pneumatic hand tools, or with larger pieces of construction equipment, and may include loading a removed item onto a truck (check the Crew for equipment used). It does not include saw cutting, hauling or disposal of debris, and, sometimes, handling or loading.

R024119-20 Dumpsters

Dumpster rental costs on construction sites are presented in two ways.

The cost per week rental includes the delivery of the dumpster; its pulling or emptying once per week, and its final removal. The assumption is made that the dumpster contractor could choose to empty a dumpster by simply bringing in an empty unit and removing the full one. These costs also include the disposal of the materials in the dumpster.

The Alternate Pricing can be used when actual planned conditions are not approximated by the weekly numbers. For example, these lines can be used when a dumpster is needed for 4 weeks and will need to be emptied 2 or 3 times per week. Conversely the Alternate Pricing lines can be used when a dumpster will be rented for several weeks or months but needs to be emptied only a few times over this period.

Existing Conditions | R0265 Underground Storage Tank Removal

R026510-20 Underground Storage Tank Removal

Underground Storage Tank Removal can be divided into two categories: Non-Leaking and Leaking. Prior to removing an underground storage tank, tests should be made, with the proper authorities present, to determine whether a tank has been leaking or the surrounding soil has been contaminated.

To safely remove Liquid Underground Storage Tanks:

1. Excavate to the top of the tank.
2. Disconnect all piping.
3. Open all tank vents and access ports.
4. Remove all liquids and/or sludge.
5. Purge the tank with an inert gas.
6. Provide access to the inside of the tank and clean out the interior using proper personal protective equipment (PPE).
7. Excavate soil surrounding the tank using proper PPE for on-site personnel.
8. Pull and properly dispose of the tank.
9. Clean up the site of all contaminated material.
10. Install new tanks or close the excavation.

R028213-20 Asbestos Removal Process

Asbestos removal is accomplished by a specialty contractor who understands the federal and state regulations regarding the handling and disposal of the material. The process of asbestos removal is divided into many individual steps. An accurate estimate can be calculated only after all the steps have been priced.

The steps are generally as follows:

1. Obtain an asbestos abatement plan from an industrial hygienist.
2. Monitor the air quality in and around the removal area and along the path of travel between the removal area and transport area. This establishes the background contamination.
3. Construct a two part decontamination chamber at entrance to removal area.
4. Install a HEPA filter to create a negative pressure in the removal area.
5. Install wall, floor and ceiling protection as required by the plan, usually 2 layers of fireproof 6 mil polyethylene.
6. Industrial hygienist visually inspects work area to verify compliance with plan.
7. Provide temporary supports for conduit and piping affected by the removal process.
8. Proceed with asbestos removal and bagging process. Monitor air quality as described in Step #2. Discontinue operations when contaminate levels exceed applicable standards.
9. Document the legal disposal of materials in accordance with EPA standards.
10. Thoroughly clean removal area including all ledges, crevices and surfaces.
11. Post abatement inspection by industrial hygienist to verify plan compliance.
12. Provide a certificate from a licensed industrial hygienist attesting that contaminate levels are within acceptable standards before returning area to regular use.

Concrete | R0330 Cast-In-Place Concrete

R033053-60 Maximum Depth of Frost Penetration in Inches

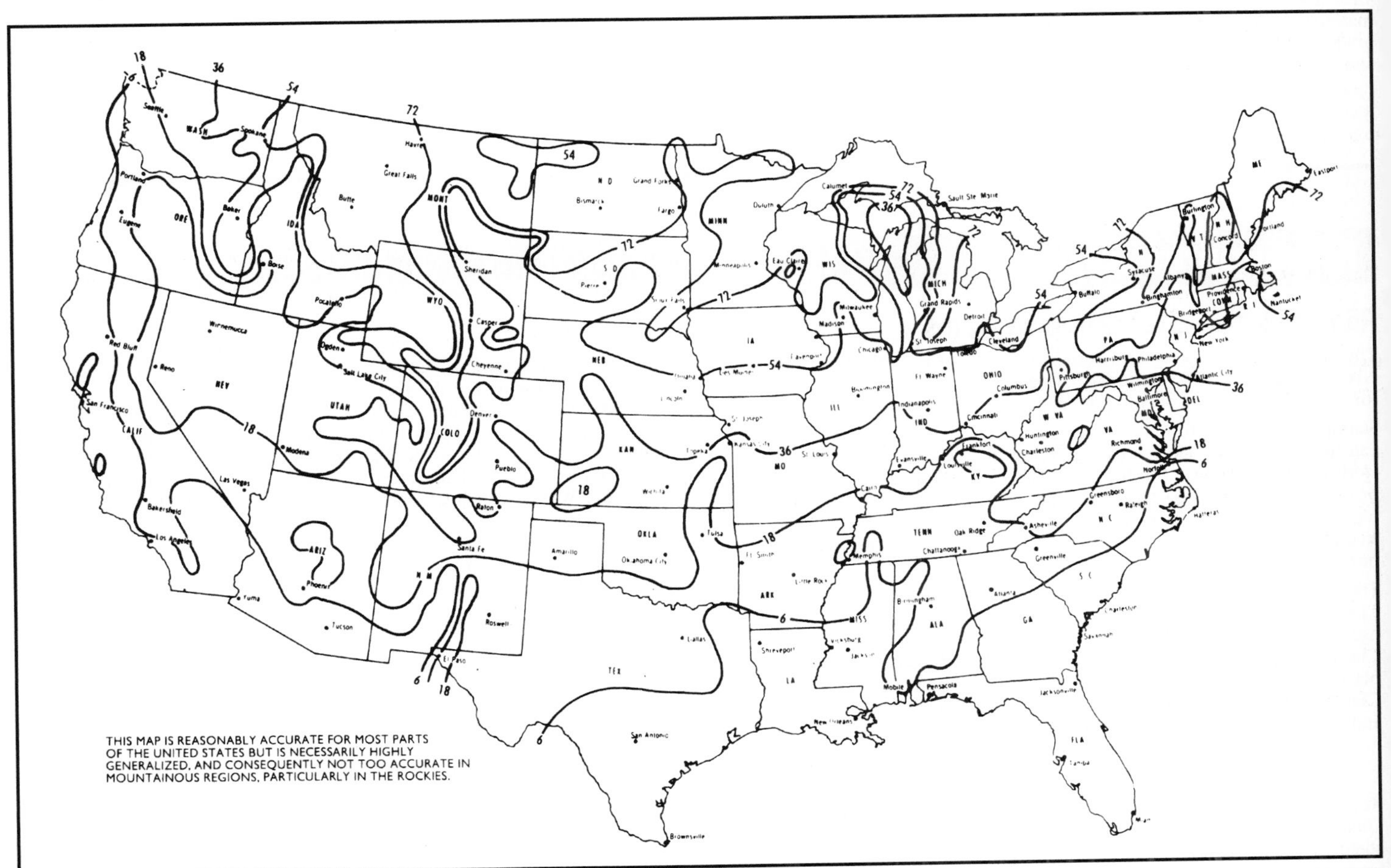

Metals — R0505 Common Work Results for Metals

R050521-20 Welded Structural Steel

Usual weight reductions with welded design run 10% to 20% compared with bolted or riveted connections. This amounts to about the same total cost compared with bolted structures since field welding is more expensive than bolts. For normal spans of 18′ to 24′ figure 6 to 7 connections per ton.

Trusses — For welded trusses add 4% to weight of main members for connections. Up to 15% less steel can be expected in a welded truss compared to one that is shop bolted. Cost of erection is the same whether shop bolted or welded.

General — Typical electrodes for structural steel welding are E6010, E6011, E60T and E70T. Typical buildings vary between 2# to 8# of weld rod per ton of steel. Buildings utilizing continuous design require about three times as much welding as conventional welded structures. In estimating field erection by welding, it is best to use the average linear feet of weld per ton to arrive at the welding cost per ton. The type, size and position of the weld will have a direct bearing on the cost per linear foot. A typical field welder will deposit 1.8# to 2# of weld rod per hour manually. Using semiautomatic methods can increase production by as much as 50% to 75%.

Special Construction — R1311 Swimming Pools

R131113-20 Swimming Pools

Pool prices given per square foot of surface area include pool structure, filter and chlorination equipment, pumps, related piping, ladders/steps, maintenance kit, skimmer and vacuum system. Decks and electrical service to equipment are not included.

Residential in-ground pool construction can be divided into two categories: vinyl lined and gunite. Vinyl lined pool walls are constructed of different materials including wood, concrete, plastic or metal. The bottom is often graded with sand over which the vinyl liner is installed. Vermiculite or soil cement bottoms may be substituted for an added cost.

Gunite pool construction is used both in residential and municipal installations. These structures are steel reinforced for strength and finished with a white cement limestone plaster.

Municipal pools will have a higher cost because plumbing codes require more expensive materials, chlorination equipment and higher filtration rates.

Municipal pools greater than 1,800 S.F. require gutter systems to control waves. This gutter may be formed into the concrete wall. Often a vinyl/stainless steel gutter or gutter/wall system is specified, which will raise the pool cost.

Competition pools usually require tile bottoms and sides with contrasting lane striping, which will also raise the pool cost.

Fire Suppression — R2112 Fire-Suppression Standpipes

R211226-10 Standpipe Systems

The basis for standpipe system design is National Fire Protection Association NFPA 14. However, the authority having jurisdiction should be consulted for special conditions, local requirements, and approval.

Standpipe systems, properly designed and maintained, are an effective and valuable time saving aid for extinguishing fires, especially in the upper stories of tall buildings, the interior of large commercial or industrial malls, or other areas where construction features or access make the laying of temporary hose lines time consuming and/or hazardous. Standpipes are frequently installed with automatic sprinkler systems for maximum protection.

There are three general classes of service for standpipe systems:
Class I A system with 2-1/2″ (65 mm) hose connections for Fire Department use and those trained in handling heavy fire streams.
Class II A system with either 1-1/2″ (40 mm) hose stations for use by trained personnel or a hose connection for the fire department's use.
Class III A system with 1-1/2″ (40 mm) hose stations for trained personnel and 2-1/2″ (65 mm) hose connections for fire department use.

Standpipe systems are also classified by the way water is supplied to the system. The following are the types of systems:
Automatic Dry Standpipe A system containing air or nitrogen under pressure, with a water supply capable of meeting water demands at all times. Opening of a valve allows the air/nitrogen to dissipate, releasing the pressurized dry valve's clapper to open and fill the system with water.
Automatic Wet Standpipe A system containing water with a water supply capable of meeting system demands at all times. Open a valve or connection and water will immediately flow.
Combination Standpipe A system that supplies both sprinkler systems and fire department/hose connections. Manual operation of approved remote control devices located at each hose station.
Manual Dry Standpipe A system not containing water. Water is manually pumped into the system by means of a fire department connection attached to a water supply.
Semiautomatic Dry Standpipe A system with an attached water supply held back at a valve which requires remote activation to open and allow flow of water into the system.
Wet Standpipe A system containing water at all times.

R211226-20 NFPA 14 Basic Standpipe Design

Class	Design-Use	Pipe Size Minimums	Water Supply Minimums
Class I	2 1/2″ hose connection on each floor All areas within 150′ of an exit in every exit stairway Fire Department Trained Personnel	Height to 100′, 4″ dia. Heights above 100′, 6″ dia. (275′ max. except with pressure regulators 400′ max.)	For each standpipe riser 500 GPM flow For common supply pipe allow 500 GPM for first standpipe plus 250 GPM for each additional standpipe (2500 GPM max. total) 30 min. duration 65 PSI at 500 GPM
Class II	1 1/2″ hose connection with hose on each floor All areas within 130′ of hose connection measured along path of hose travel Occupant personnel	Height to 50′, 2″ dia. Height above 50′, 2 1/2″ dia.	For each standpipe riser 100 GPM flow For multiple riser common supply pipe 100 GPM 300 min. duration, 65 PSI at 100 GPM
Class III	Both of above. Class I valved connections will meet Class III with additional 2 1/2″ by 1 1/2″ adapter and 1 1/2″ hose.	Same as Class I	Same as Class I

*Note: Where 2 or more standpipes are installed in the same building or section of building they shall be interconnected at the bottom.

Combined Systems

Combined systems are systems where the risers supply both automatic sprinklers and 2-1/2″ hose connection outlets for fire department use. In such a system the sprinkler spacing pattern shall be in accordance with NFPA 13 while the risers and supply piping will be sized in accordance with NFPA 14. When the building is completely sprinklered the risers may be sized by hydraulic calculation.The minimum size riser for buildings not completely sprinklered is 6″.
The minimum water supply of a completely sprinklered, light hazard, high-rise occupancy building will be 500 GPM while the supply required for other types of completely sprinklered high-rise buildings is 1000 GPM.

General System Requirements

1. Approved valves will be provided at the riser for controlling branch lines to hose outlets.
2. A hose valve will be provided at each outlet for attachment of hose.
3. Where pressure at any standpipe outlet exceeds 100 PSI a pressure reducer must be installed to limit the pressure to 100 PSI. Note that the pressure head due to gravity in 100′ of riser is 43.4 PSI. This must be overcome by city pressure, fire pumps, or gravity tanks to provide adequate pressure at the top of the riser.
4. Each hose valve on a wet system having linen hose shall have an automatic drip connection to prevent valve leakage from entering the hose.
5. Each riser will have a valve to isolate it from the rest of the system.
6. One or more fire department connections as an auxiliary supply shall be provided for each Class I or Class III standpipe system. In buildings having two or more zones, a connection will be provided for each zone.
7. There will be no shutoff valve in the fire department connection, but a check valve will be located in the line before it joins the system.
8. All hose connections street side will be identified on a cast plate or fitting as to purpose.

R211313-10 Sprinkler Systems (Automatic)

Sprinkler systems may be classified by type as follows:

1. **Wet Pipe System.** A system employing automatic sprinklers attached to a piping system containing water and connected to a water supply so that water discharges immediately from sprinklers opened by a fire.
2. **Dry Pipe System.** A system employing automatic sprinklers attached to a piping system containing air under pressure, the release of which as from the opening of sprinklers permits the water pressure to open a valve known as a "dry pipe valve". The water then flows into the piping system and out the opened sprinklers.
3. **Pre-Action System.** A system employing automatic sprinklers attached to a piping system containing air that may or may not be under pressure, with a supplemental heat responsive system of generally more sensitive characteristics than the automatic sprinklers themselves, installed in the same areas as the sprinklers; actuation of the heat responsive system, as from a fire, opens a valve which permits water to flow into the sprinkler piping system and to be discharged from any sprinklers which may be open.
4. **Deluge System.** A system employing open sprinklers attached to a piping system connected to a water supply through a valve which is opened by the operation of a heat responsive system installed in the same areas as the sprinklers. When this valve opens, water flows into the piping system and discharges from all sprinklers attached thereto.
5. **Combined Dry Pipe and Pre-Action Sprinkler System.** A system employing automatic sprinklers attached to a piping system containing air under pressure with a supplemental heat responsive system of generally more sensitive characteristics than the automatic sprinklers themselves, installed in the same areas as the sprinklers; operation of the heat responsive system, as from a fire, actuates tripping devices which open dry pipe valves simultaneously and without loss of air pressure in the system. Operation of the heat responsive system also opens approved air exhaust valves at the end of the feed main which facilitates the filling of the system with water which usually precedes the opening of sprinklers. The heat responsive system also serves as an automatic fire alarm system.
6. **Limited Water Supply System.** A system employing automatic sprinklers and conforming to these standards but supplied by a pressure tank of limited capacity.
7. **Chemical Systems.** Systems using halon, carbon dioxide, dry chemical or high expansion foam as selected for special requirements. Agent may extinguish flames by chemically inhibiting flame propagation, suffocate flames by excluding oxygen, interrupting chemical action of oxygen uniting with fuel or sealing and cooling the combustion center.
8. **Firecycle System.** Firecycle is a fixed fire protection sprinkler system utilizing water as its extinguishing agent. It is a time delayed, recycling, preaction type which automatically shuts the water off when heat is reduced below the detector operating temperature and turns the water back on when that temperature is exceeded. The system senses a fire condition through a closed circuit electrical detector system which controls water flow to the fire automatically. Batteries supply up to 90 hour emergency power supply for system operation. The piping system is dry (until water is required) and is monitored with pressurized air. Should any leak in the system piping occur, an alarm will sound, but water will not enter the system until heat is sensed by a firecycle detector.

Area coverage sprinkler systems may be laid out and fed from the supply in any one of several patterns as shown below. It is desirable, if possible, to utilize a central feed and achieve a shorter flow path from the riser to the furthest sprinkler. This permits use of the smallest sizes of pipe possible with resulting savings.

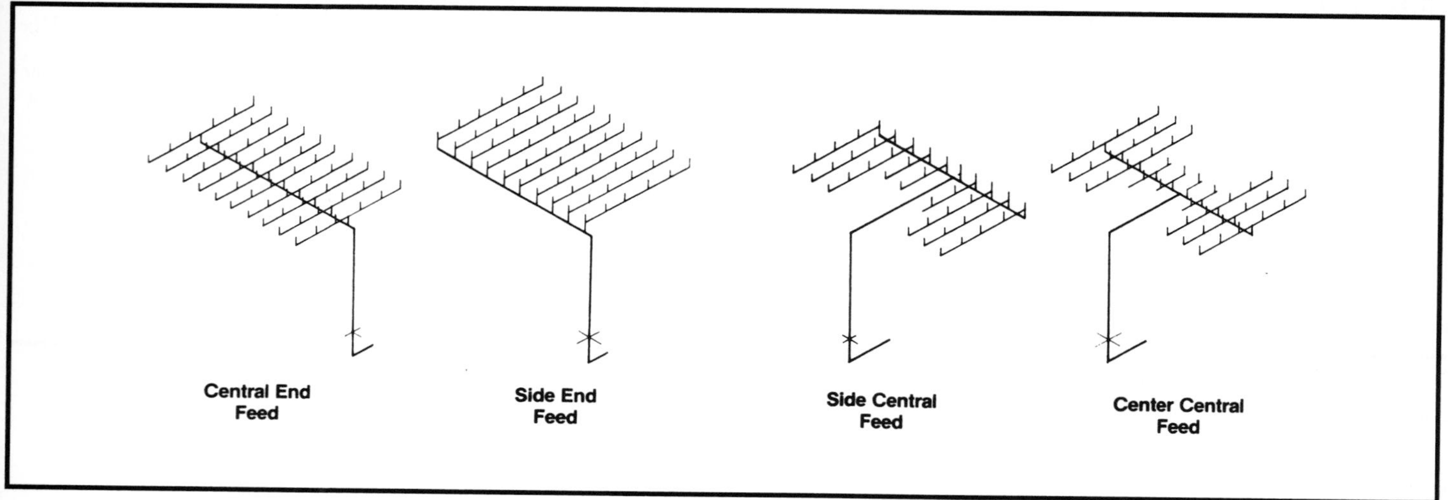

R211313-20 System Classification

System Classification

Rules for installation of sprinkler systems vary depending on the classification of occupancy falling into one of three categories as follows:

Light Hazard Occupancy

The protection area allotted per sprinkler should not exceed 225 S.F., with the maximum distance between lines and sprinklers on lines being 15′. The sprinklers do not need to be staggered. Branch lines should not exceed eight sprinklers on either side of a cross main. Each large area requiring more than 100 sprinklers and without a sub-dividing partition should be supplied by feed mains or risers sized for ordinary hazard occupancy.
Maximum system area = 52,000 S.F.

Included in this group are:

Churches
Clubs
Educational
Hospitals
Institutional
Libraries
(except large stack rooms)
Museums
Nursing Homes
Offices
Residential
Restaurants
Theaters and Auditoriums
(except stages and prosceniums)
Unused Attics

Ordinary Hazard Occupancy

The protection area allotted per sprinkler shall not exceed 130 S.F. of noncombustible ceiling and 130 S.F. of combustible ceiling. The maximum allowable distance between sprinkler lines and sprinklers on line is 15′. Sprinklers shall be staggered if the distance between heads exceeds 12′. Branch lines should not exceed eight sprinklers on either side of a cross main.
Maximum system area = 52,000 S.F.

Included in this group are:

Group 1
Automotive Parking and Showrooms
Bakeries
Beverage manufacturing
Canneries
Dairy Products Manufacturing/Processing
Electronic Plans
Glass and Glass Products Manufacturing
Laundries
Restaurant Service Areas

Group 2
Cereal Mills
Chemical Plants—Ordinary
Confectionery Products
Distilleries
Dry Cleaners
Feed Mills
Horse Stables
Leather Goods Manufacturing
Libraries—Large Stack Room Areas
Machine Shops
Metal Working
Mercantile
Paper and Pulp Mills
Paper Process Plants
Piers and Wharves
Post Offices
Printing and Publishing
Repair Garages
Stages
Textile Manufacturing
Tire Manufacturing
Tobacco Products Manufacturing
Wood Machining
Wood Product Assembly

Extra Hazard Occupancy

The protection area allotted per sprinkler shall not exceed 100 S.F. of noncombustible ceiling and 100 S.F. of combustible ceiling. The maximum allowable distance between lines and between sprinklers on lines is 12′. Sprinklers on alternate lines shall be staggered if the distance between sprinklers on lines exceeds 8′. Branch lines should not exceed six sprinklers on either side of a cross main.
Maximum system area:
Design by pipe schedule = 25,000 S.F.
Design by hydraulic calculation = 40,000 S.F.

Included in this group are:

Group 1
Aircraft hangars
Combustible Hydraulic Fluid Use Area
Die Casting
Metal Extruding
Plywood/Particle Board Manufacturing
Printing (inks with flash points < 100 degrees F
Rubber Reclaiming, Compounding, Drying, Milling, Vulcanizing
Saw Mills
Textile Picking, Opening, Blending, Garnetting, Carding, Combing of Cotton, Synthetics, Wood Shoddy, or Burlap
Upholstering with Plastic Foams

Group 2
Asphalt Saturating
Flammable Liquids Spraying
Flow Coating
Manufactured/Modular Home Building Assemblies (where finished enclosure is present and has combustible interiors)
Open Oil Quenching
Plastics Processing
Solvent Cleaning
Varnish and Paint Dipping

R211313-40 Adjustment for Sprinkler/Standpipe Installations

Quality/Complexity Multiplier (For all installations)

Economy installation, add	0 to 5%
Good quality, medium complexity, add	5 to 15%
Above average quality and complexity, add	15 to 25%

R220102-20 Labor Adjustment Factors

Labor Adjustment Factors are provided for Divisions 21, 22, and 23 to assist the mechanical estimator account for the various complexities and special conditions of any particular project. While a single percentage has been entered on each line of Division 22 01 02.20, it should be understood that these are just suggested midpoints of ranges of values commonly used by mechanical estimators. They may be increased or decreased depending on the severity of the special conditions.

The group for "existing occupied buildings" has been the subject of requests for explanation. Actually there are two stages to this group: buildings that are existing and "finished" but unoccupied, and those that also are occupied. Buildings that are "finished" may result in higher labor costs due to the workers having to be more careful not to damage finished walls, ceilings, floors, etc. and may necessitate special protective coverings and barriers. Also corridor bends and doorways may not accommodate long pieces of pipe or larger pieces of equipment. Work above an already hung ceiling can be very time consuming. The addition of occupants may force the work to be done on premium time (nights and/or weekends), eliminate the possible use of some preferred tools such as pneumatic drivers, powder charged drivers etc. The estimator should evaluate the access to the work area and just how the work is going to be accomplished to arrive at an increase in labor costs over "normal" new construction productivity.

R220105-10 Demolition (Selective vs. Removal for Replacement)

Demolition can be divided into two basic categories.

One type of demolition involves the removal of material with no concern for its replacement. The labor-hours to estimate this work are found under "Selective Demolition" in the Fire Protection, Plumbing and HVAC Divisions. It is selective in that individual items or all the material installed as a system or trade grouping such as plumbing or heating systems are removed. This may be accomplished by the easiest way possible, such as sawing, torch cutting, or sledge hammering as well as simple unbolting.

The second type of demolition is the removal of some items for repair or replacement. This removal may involve careful draining, opening of unions, disconnecting and tagging electrical connections, capping pipes/ducts to prevent entry of debris or leakage of the material contained as well as transporting the item away from its in-place location to a truck/dumpster. An approximation of the time required to accomplish this type of demolition is to use half of the time indicated as necessary to install a new unit. For example: installation of a new pump might be listed as requiring 6 labor-hours so if we had to estimate the removal of the old pump we would allow an additional 3 hours for a total of 9 hours. That is, the complete replacement of a defective pump with a new pump would be estimated to take 9 labor-hours.

R220523-80 Valve Materials

VALVE MATERIALS

Bronze:
Bronze is one of the oldest materials used to make valves. It is most commonly used in hot and cold water systems and other non-corrosive services. It is often used as a seating surface in larger iron body valves to ensure tight closure.

Carbon Steel:
Carbon steel is a high strength material. Therefore, valves made from this metal are used in higher pressure services, such as steam lines up to 600 psi at 850°F. Many steel valves are available with butt-weld ends for economy and are generally used in high pressure steam service as well as other higher pressure non-corrosive services.

Forged Steel:
Valves from tough carbon steel are used in service up to 2000 psi and temperatures up to 1000°F in Gate, Globe and Check valves.

Iron:
Valves are normally used in medium to large pipe lines to control non-corrosive fluid and gases, where pressures do not exceed 250 psi at 450° or 500 psi cold water, oil or gas.

Stainless Steel:
Developed steel alloys can be used in over 90% corrosive services.

Plastic PVC:
This is used in a great variety of valves generally in high corrosive service with lower temperatures and pressures.

VALVE SERVICE PRESSURES

Pressure ratings on valves provide an indication of the safe operating pressure for a valve at some elevated temperature. This temperature is dependent upon the materials used and the fabrication of the valve. When specific data is not available, a good "rule-of-thumb" to follow is the temperature of saturated steam on the primary rating indicated on the valve body. Example: The valve has the number 150S printed on the side indicating 150 psi and hence, a maximum operating temperature of 367°F (temperature of saturated steam and 150 psi).

DEFINITIONS

1. "WOG" – Water, oil, gas (cold working pressures).
2. "SWP" – Steam working pressure.
3. 100% area (full port) – means the area through the valve is equal to or greater than the area of standard pipe.
4. "Standard Opening" – means that the area through the valve is less than the area of standard pipe and therefore these valves should be used only where restriction of flow is unimportant.
5. "Round Port" – means the valve has a full round opening through the plug and body, of the same size and area as standard pipe.
6. "Rectangular Port" – valves have rectangular shaped ports through the plug body. The area of the port is either equal to 100% of the area of standard pipe, or restricted (standard opening). In either case it is clearly marked.
7. "ANSI" – American National Standards Institute.

R220523-90 Valve Selection Considerations

INTRODUCTION: In any piping application, valve performance is critical. Valves should be selected to give the best performance at the lowest cost.

The following is a list of performance characteristics generally expected of valves.

1. Stopping flow or starting it.
2. Throttling flow (Modulation).
3. Flow direction changing.
4. Checking backflow (Permitting flow in only one direction).
5. Relieving or regulating pressure.

In order to properly select the right valve, some facts must be determined.

A. What liquid or gas will flow through the valve?
B. Does the fluid contain suspended particles?
C. Does the fluid remain in liquid form at all times?
D. Which metals does fluid corrode?
E. What are the pressure and temperature limits? (As temperature and pressure rise, so will the price of the valve.)
F. Is there constant line pressure?
G. Is the valve merely an on-off valve?
H. Will checking of backflow be required?
I. Will the valve operate frequently or infrequently?

Valves are classified by design type into such classifications as Gate, Globe, Angle, Check, Ball, Butterfly and Plug. They are also classified by end connection, stem, pressure restrictions and material such as bronze, cast iron, etc. Each valve has a specific use. A quality valve used correctly will provide a lifetime of trouble- free service, but a high quality valve installed in the wrong service may require frequent attention.

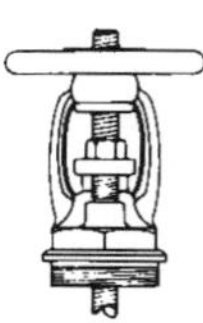

STEM TYPES

(OS & Y)—Rising Stem-Outside Screw and Yoke

Offers a visual indication of whether the valve is open or closed. Recommended where high temperatures, corrosives, and solids in the line might cause damage to inside-valve stem threads. The stem threads are engaged by the yoke bushing so the stem rises through the hand wheel as it is turned.

(R.S.)—Rising Stem-Inside Screw

Adequate clearance for operation must be provided because both the hand wheel and the stem rise.
The valve wedge position is indicated by the position of the stem and hand wheel.

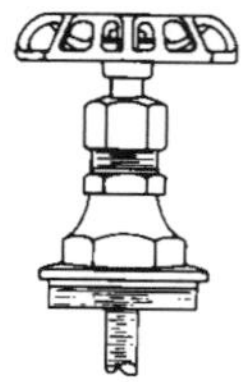

(N.R.S.)—Non-Rising Stem-Inside Screw

A minimum clearance is required for operating this type of valve. Excessive wear or damage to stem threads inside the valve may be caused by heat, corrosion, and solids. Because the hand wheel and stem do not rise, wedge position cannot be visually determined.

VALVE TYPES

Gate Valves

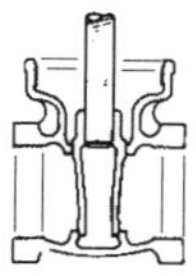

Provide full flow, minute pressure drop, minimum turbulence and minimum fluid trapped in the line.
They are normally used where operation is infrequent.

Globe Valves

Globe valves are designed for throttling and/or frequent operation with positive shut-off. Particular attention must be paid to the several types of seating materials available to avoid unnecessary wear. The seats must be compatible with the fluid in service and may be composition or metal. The configuration of the Globe valve opening causes turbulence which results in increased resistance. Most bronze Globe valves are rising stem-inside screw, but they are also available on O.S. & Y.

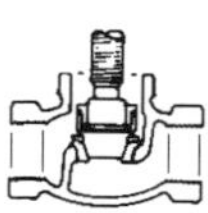

Angle Valves

The fundamental difference between the Angle valve and the Globe valve is the fluid flow through the Angle valve. It makes a 90° turn and offers less resistance to flow than the Globe valve while replacing an elbow. An Angle valve thus reduces the number of joints and installation time.

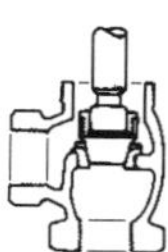

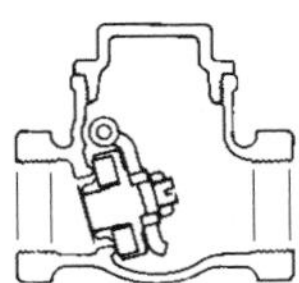

Check Valves

Check valves are designed to prevent backflow by automatically seating when the direction of fluid is reversed.
Swing Check valves are generally installed with Gate-valves, as they provide comparable full flow. Usually recommended for lines where flow velocities are low and should not be used on lines with pulsating flow. Recommended for horizontal installation, or in vertical lines only where flow is upward.

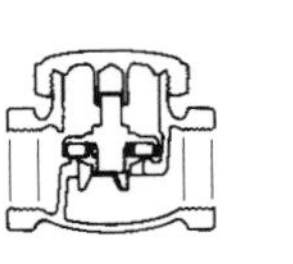
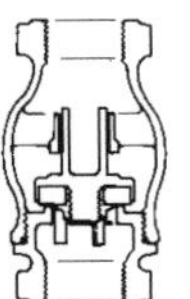

Lift Check Valves

These are commonly used with Globe and Angle valves since they have similar diaphragm seating arrangements and are recommended for preventing backflow of steam, air, gas and water, and on vapor lines with high flow velocities. For horizontal lines, horizontal lift checks should be used and vertical lift checks for vertical lines.

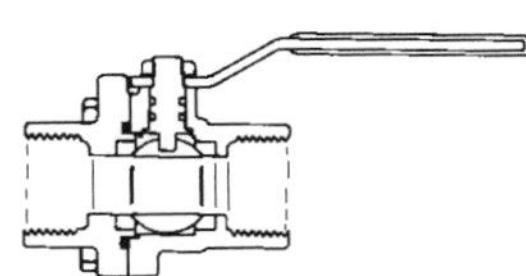

Ball Valves

Ball valves are light and easily installed, yet because of modern elastomeric seats, provide tight closure. Flow is controlled by rotating up to 90° a drilled ball which fits tightly against resilient seals. This ball seats with flow in either direction, and valve handle indicates the degree of opening. Recommended for frequent operation readily adaptable to automation, ideal for installation where space is limited.

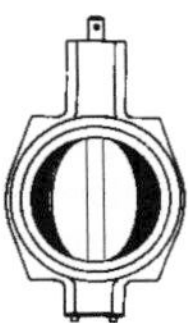

Butterfly Valves

Butterfly valves provide bubble-tight closure with excellent throttling characteristics. They can be used for full-open, closed and for throttling applications.

The Butterfly valve consists of a disc within the valve body which is controlled by a shaft. In its closed position, the valve disc seals against a resilient seat. The disc position throughout the full 90° rotation is visually indicated by the position of the operator.

A Butterfly valve is only a fraction of the weight of a Gate valve and requires no gaskets between flanges in most cases. Recommended for frequent operation and adaptable to automation where space is limited.

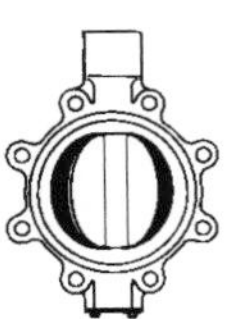

Wafer and Lug type bodies when installed between two pipe flanges, can be easily removed from the line. The pressure of the bolted flanges holds the valve in place.
Locating lugs makes installation easier.

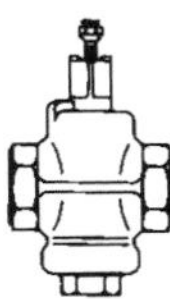

Plug Valves

Lubricated plug valves, because of the wide range of service to which they are adapted, may be classified as all purpose valves. They can be safely used at all pressure and vacuums, and at all temperatures up to the limits of available lubricants. They are the most satisfactory valves for the handling of gritty suspensions and many other destructive, erosive, corrosive and chemical solutions.

R221113-40 Plumbing Approximations for Quick Estimating

Water Control

Water Meter; Backflow Preventer, Shock Absorbers; Vacuum Breakers; Mixer. .... 10 to 15% of Fixtures

Pipe And Fittings .... 30 to 60% of Fixtures

Note: Lower percentage for compact buildings or larger buildings with plumbing in one area.
Larger percentage for large buildings with plumbing spread out.
In extreme cases pipe may be more than 100% of fixtures.
Percentages **do not** include special purpose or process piping.

Plumbing Labor

1 & 2 Story Residential .... Rough-in Labor = 80% of Materials
Apartment Buildings .... Rough-in Labor = 90 to 100% of Materials
Labor for handling and placing fixtures is approximately 25 to 30% of fixtures

Quality/Complexity Multiplier (for all installations)

Economy installation, add .... 0 to 5%
Good quality, medium complexity, add .... 5 to 15%
Above average quality and complexity, add .... 15 to 25%

R221113-50 Pipe Material Considerations

1. Malleable fittings should be used for gas service.
2. Malleable fittings are used where there are stresses/strains due to expansion and vibration.
3. Cast fittings may be broken as an aid to disassembling heating lines frozen by long use, temperature and minerals.
4. A cast iron pipe is extensively used for underground and submerged service.
5. Type M (light wall) copper tubing is available in hard temper only and is used for nonpressure and less severe applications than K and L.
6. Type L (medium wall) copper tubing, available hard or soft for interior service.
7. Type K (heavy wall) copper tubing, available in hard or soft temper for use where conditions are severe. For underground and interior service.
8. Hard drawn tubing requires fewer hangers or supports but should not be bent. Silver brazed fittings are recommended, but soft solder is normally used.
9. Type DMV (very light wall) copper tubing designed for drainage, waste and vent plus other non-critical pressure services.

Domestic/Imported Pipe and Fittings Costs

The prices shown in this publication for steel/cast iron pipe and steel, cast iron, and malleable iron fittings are based on domestic production sold at the normal trade discounts. The above listed items of foreign manufacture may be available at prices 1/3 to 1/2 of those shown. Some imported items after minor machining or finishing operations are being sold as domestic to further complicate the system.

Caution: Most pipe prices in this data set also include a coupling and pipe hangers which for the larger sizes can add significantly to the per foot cost and should be taken into account when comparing "book cost" with the quoted supplier's cost.

R221113-70 Piping to 10′ High

When taking off pipe, it is important to identify the different material types and joining procedures, as well as distances between supports and components required for proper support.

During the takeoff, measure through all fittings. Do not subtract the lengths of the fittings, valves, or strainers, etc. This added length plus the final rounding of the totals will compensate for nipples and waste.

When rounding off totals always increase the actual amount to correspond with manufacturer's shipping lengths.

A. Both red brass and yellow brass pipe are normally furnished in 12′ lengths, plain end. The Unit Price section includes in the linear foot costs two field threads and one coupling per 10′ length. A carbon steel clevis type hanger assembly every 10′ is also prorated into the linear foot costs, including both material and labor.

B. Cast iron soil pipe is furnished in either 5′ or 10′ lengths. For pricing purposes, the Unit Price section features 10′ lengths with a joint and a carbon steel clevis hanger assembly every 5′ prorated into the per foot costs of both material and labor.

Three methods of joining are considered: lead and oakum poured joints or push-on gasket type joints for the bell and spigot pipe and a joint clamp for the no-hub soil pipe. The labor and material costs for each of these individual joining procedures are also prorated into the linear costs per foot.

C. Copper tubing covers types K, L, M, and DWV which are furnished in 20′ lengths. Means pricing data is based on a tubing cut each length and a coupling and two soft soldered joints every 10′. A carbon steel, clevis type hanger assembly every 10′ is also prorated into the per foot costs. The prices for refrigeration tubing are for materials only. Labor for full lengths may be based on the type L labor but short cut measures in tight areas can increase the installation labor-hours from 20 to 40%.

D. Corrosion-resistant piping does not lend itself to one particular standard of hanging or support assembly due to its diversity of application and placement. The several varieties of corrosion-resistant piping do not include any material or labor costs for hanger assemblies (See the Unit Price section for appropriate selection).

E. Glass pipe is furnished in standard lengths either 5′ or 10′ long, beaded on one end. Special orders for diverse lengths beaded on both ends are also available. For pricing purposes, R.S. Means features 10′ lengths with a coupling and a carbon steel band hanger assembly every 10′ prorated into the per foot linear costs.

Glass pipe is also available with conical ends and standard lengths ranging from 6″ through 3′ in 6″ increments, then up to 10′ in 12″ increments. Special lengths can be customized for particular installation requirements.

For pricing purposes, Means has based the labor and material pricing on 10′ lengths. Included in these costs per linear foot are the prorated costs for a flanged assembly every 10′ consisting of two flanges, a gasket, two insertable seals, and the required number of bolts and nuts. A carbon steel band hanger assembly based on 10′ center lines has also been prorated into the costs per foot for labor and materials.

F. Plastic pipe of several compositions and joining methods are considered. Fiberglass reinforced pipe (FRP) is priced, based on 10′ lengths (20′ lengths are also available), with coupling and epoxy joints every 10′. FRP is furnished in both "General Service" and "High Strength." A carbon steel clevis hanger assembly, 3 for every 10′, is built into the prorated labor and material costs on a per foot basis.

The PVC and CPVC pipe schedules 40, 80 and 120, plus SDR ratings are all based on 20′ lengths with a coupling installed every 10′, as well as a carbon steel clevis hanger assembly every 3′. The PVC and ABS type DWV piping is based on 10′ lengths with solvent weld couplings every 10′, and with carbon steel clevis hanger assemblies, 3 for every 10′. The rest of the plastic piping in this section is based on flexible 100′ coils and does not include any coupling or supports.

This section ends with PVC drain and sewer piping based on 10′ lengths with bell and spigot ends and 0-ring type, push-on joints.

G. Stainless steel piping includes both weld end and threaded piping, both in the type 304 and 316 specification and in the following schedules, 5, 10, 40, 80, and 160. Although this piping is usually furnished in 20′ lengths, this cost grouping has a joint (either heli-arc butt-welded or threads and coupling) every 10′. A carbon steel clevis type hanger assembly is also included at 10′ intervals and prorated into the linear foot costs.

H. Carbon steel pipe includes both black and galvanized. This section encompasses schedules 40 (standard) and 80 (extra heavy).

Several common methods of joining steel pipe — such as thread and coupled, butt welded, and flanged (150 lb. weld neck flanges) are also included.

For estimating purposes, it is assumed that the piping is purchased in 20′ lengths and that a compatible joint is made up every 10′. These joints are prorated into the labor and material costs per linear foot. The following hanger and support assemblies every 10′ are also included: carbon steel clevis for the T & C pipe, and single rod roll type for both the welded and flanged piping. All of these hangers are oversized to accommodate pipe insulation 3/4″ thick through 5″ pipe size and 1-1/2″ thick from 6″ through 12″ pipe size.

I. Grooved joint steel pipe is priced both black and galvanized, in schedules 10, 40, and 80, furnished in 20′ lengths. This section describes two joining methods: cut groove and roll groove. The schedule 10 piping is roll-grooved, while the heavier schedules are cut-grooved. The labor and material costs are prorated into per linear foot prices, including a coupled joint every 10′, as well as a carbon steel clevis hanger assembly.

Notes:

The pipe hanger assemblies mentioned in the preceding paragraphs include the described hanger; appropriately sized steel, box-type insert and nut; plus 18″ of threaded hanger rod.

C clamps are used when the pipe is to be supported from steel shapes rather than anchored in the slab. C clamps are slightly less costly than inserts. However, to save time in estimating, it is advisable to use the given line number cost, rather than substituting a C clamp for the insert.

Add to piping labor for elevated installation:

10′ to 14.5′ high	10%	30′ to 34.5′ high	40%
15′ to 19.5′ high	20%	35′ to 39.5′ high	50%
20′ to 24.5′ high	25%	40′ and higher	55%
25′ to 29.5′ high	35%		

When using the percentage adds for elevated piping installations as shown above, bear in mind that the given heights are for the pipe supports, even though the insert, anchor, or clamp may be several feet higher than the pipe itself.

An allowance has been included in the piping installation time for testing and minor tightening of leaking joints, fittings, stuffing boxes, packing glands, etc. For extraordinary test requirements such as x-rays, prolonged pressure or demonstration tests, a percentage of the piping labor, based on the estimator's experience, must be added to the labor total. A testing service specializing in weld x-rays should be consulted for pricing if it is an estimate requirement. Equipment installation time includes start-up with associated adjustments.

R221316-10 Drainage Fixture Units for Fixtures and Groups

Fixture Type	Drainage Fixture Unit Value as Load Factors	Minimum Size of Trap (inches)
Automatic Clothes Washer, commercial (Note A)	3.0	2.0
Automatic Clothes Washer, residential	2.0	2.0
Full Bathroom Group w/bathtub or shower stall (1.6 gpf WC)	5.0	-
Full Bathroom Group w/bathtub or shower stall (WC flushing greater than 1.6 gpf) (Note D)	6.0	-
Bathtub (with or without overhead shower)	2.0	1.5
Bidet	1.0	1.25
Drinking Fountain	0.5	1025
Dishwasher, domestic	2.0	1.5
Emergency floor drains	0.0	2
Floor drains	2.0	2
Kitchen sinks, domestic	2.0	1.5
Lavatory	1.0	1.25
Shower (5.7 gpm or less flow rate)	2.0	1.5
Shower (5.7 gpm - 12.3 gpm flow rate)	3.0	2
Shower (12.3 gpm - 25.8 gpm flow rate)	5.0	3
Shower (25.8 gpm to 55.6 gpm flow rate)	6.0	4
Service Sink	2.0	1.5
Urinal	4.0	Note C
Urinal (1.0 gpf or less)	2.0	Note C
Urinal (nonwater supplied)	0.5	Note C
Water closet, flushometer tank, public or private	4.0	Note C
Water closet, private (1.6 gpf)	3.0	Note C
Water closet, private (flushing greater than 1.6 gpf)	4.0	Note C
Water closet, public (1.6 gpf)	4.0	Note C
Water closet, public (flushing greater than 1.6 gpf)	6.0	Note C

Notes:
A. A showerhead over a bathtub or whirlpool tub attachment does not increase the drainage fixture unit value.
B. Trap size shall be consistent with the fixture outlet size.
C. For fixtures added to a bathroom group, add the DFU value to those additional fixtures to the bathroom group fixture count.

Excerpted from the 2012 *International Plumbing Code*, Copyright 2011. Washington, D.C.: International Code Council. Reproduced with permission. www.ICCSAFE.org

R221316-15 Drainage Fixture Units for Fixture Drains or Traps

Fixture Drain or Trap Size (inches)	Drainage Fixture Unit Value
1.25	1.0
1.5	2.0
2.0	3.0
2.5	4.0
3.0	5.0
4.0	6.0

Notes:
Drainage Fixture Unit values designate the relative load weight or different kinds of fixtures that shall be emplored in estimating the total load carried by a soil or waste pipe.

Excerpted from the 2012 *International Plumbing Code*, Copyright 2011. Washington, D.C.: International Code Council. Reproduced with permission. www.ICCSAFE.org

Plumbing — R2213 Facility Sanitary Sewerage

R221316-20 Allowable Fixture Units (d.f.u.) for Branches and Stacks

Pipe Diam.	Horiz. Branch (not incl. drains)	Stack Size for 3 Stories or 3 Levels	Stack size for Over 3 levels	Maximum for 1 Story building Stack
1-1/2"	3	4	8	2
2"	6	10	24	6
2-1/2"	12	20	42	9
3"	20*	48*	72*	20*
4"	160	240	500	90
5"	360	540	1100	200
6"	620	960	1900	350
8"	1400	2200	3600	600
10"	2500	3800	5600	1000
12"	3900	6000	8400	1500
15"	7000			

*Not more than two water closets or bathroom groups within each branch interval nor more than six water closets or bathroom groups on the stack.

Stacks sized for the total may be reduced as load decreases at each story to a minimum diameter of 1/2 the maximum diameter.

Plumbing — R2240 Plumbing Fixtures

R224000-10 Water Consumption Rates

Fixture Type	Water Supply Fixture Unit Value		
	Hot Water	Cold Water	Combined
Bathtub	1.0	1.0	1.4
Clothes Washer	1.0	1.0	1.4
Dishwasher	1.4	0.0	1.4
Kitchen Sink	1.0	1.0	1.4
Laundry Tub	1.0	1.0	1.4
Lavatory	0.5	0.5	0.7
Shower Stall	1.0	1.0	1.4
Water Closet (tank type)	0.0	2.2	2.2
Full Bath Group w/bathtub or shower stall	1.5	2.7	3.6
Half Bath Group w/W.C. and Lavatory	0.5	2.5	2.6
Kitchen Group w/Dishwasher and Sink	1.9	1.0	2.5
Laundry Group w/Clothes Washer and Laundry Tub	1.8	1.8	2.5
Hose bibb (sillcock)	0.0	2.5	2.5

Notes:
Typically, WSFU = 1GPM

Supply loads in the building water-distribution system shall be determined by total load on the pipe being sized, in terms of water supply fixture units (WSFU) and gallons per minute (GPM) flow rates. For fixtures not listed, choose a WSFU value of a fixture with similar flow characteristics. Water Fixture Supply Units determined the required water supply to fixtures and their service systems. Fixture units are equal to (1) cubic foot of water drained in a 1-1/4" pipe per minute. It is not a flow rate unit but a design factor.

R224000-20 Fixture Demands in Gallons per Fixture per Hour

Table below is based on 140°F final temperature except for dishwashers in public places (*) where 180°F water is mandatory.

Supply Systems for Flush Tanks			Supply Systems for Flushometer Valves		
Load	Demand		Load	Demand	
WSFU	GPM	CU. FT.	WSFU	GPM	CU. FT.
1.0	3.0	0.041040	-	-	-
2.0	5.0	0.068400	-	-	-
3.0	6.5	0.868920	-	-	-
4.0	8.0	1.069440	-	-	-
5.0	9.4	1.256592	5.0	15	2.0052
6.0	10.7	1.430376	6.0	17.4	2.326032
7.0	11.8	1.577424	7.0	19.8	2.646364
8.0	12.8	1.711104	8.0	22.2	2.967696
9.0	13.7	1.831416	9.0	24.6	3.288528
10.0	14.6	1.951728	10.0	27	3.60936
11.0	15.4	2.058672	11.0	27.8	3.716304
12.0	16.0	2.138880	12.0	28.6	3.823248
13.0	16.5	2.205720	13.0	29.4	3.930192
14.0	17.0	2.272560	14.0	30.2	4.037136
15.0	17.5	2.339400	15.0	31	4.14408
16.0	18.0	2.906240	16.0	31.8	4.241024
17.0	18.4	2.459712	17.0	32.6	4.357968
18.0	18.8	2.513184	18.0	33.4	4.464912
19.0	19.2	2.566656	19.0	34.2	4.571856
20.0	19.6	2.620218	20.0	35.0	4.678800
25.0	21.5	2.874120	25.0	38.0	5.079840
30.0	23.3	3.114744	30.0	42.0	5.611356
35.0	24.9	3.328632	35.0	44.0	5.881920
40.0	26.3	3.515784	40.0	46.0	6.149280
45.0	27.7	3.702936	45.0	48.0	6.416640
50.0	29.1	3.890088	50.0	50.0	6.684000

Notes:
When designing a plumbing system that utilizes fixtures other than, or in addition to, water closets, use the data provided in the Supply Systems for Flush Tanks section of the above table.

To obtain the probable maximum demand, multiply the total demands for the fixtures (gal./fixture/hour) by the demand factor. The heater should have a heating capacity in gallons per hour equal to this maximum. The storage tank should have a capacity in gallons equal to the probable maximum demand multiplied by the storage capacity factor.

R224000-30 Minimum Plumbing Fixture Requirements

Classification	Occupancy	Description	Water Closet		Lavatories		Bathtubs/Showers	Drinking Fountains	Other
			Male	Female	Male	Female			
Assembly	A-1	Theaters and other buildings for the performing arts and motion pictures	1:125	1:65	1:200			1:500	1 Service Sink
	A-2	Nightclubs, bars, taverns dance halls	1:40		1:75			1:500	1 Service Sink
		Restaurants, banquet halls, food courts	1:75		1:200			1:500	1 Service Sink
	A-3	Auditorium w/o permanent seating, art galleries, exhibition halls, museums, lecture halls, libraries, arcades & gymnasiums	1:125	1:65	1:200			1:500	1 Service Sink
		Passenger terminals and transportation facilities	1:500		1:750			1:1000	1 Service Sink
		Places of worship and other religious services	1:150	1:75	1:200			1:1000	1 Service Sink
	A-4	Indoor sporting events and activities, coliseums, arenas, skating rinks, pools, and tennis courts	1:75 for the first 1500, then 1:120 for the remainder	1:40 for the first 1520, then 1:60 for the remainder	1:200	1:150		1:1000	1 Service Sink
	A-5	Outdoor sporting events and activities, stadiums, amusement parks, bleachers, grandstands	1:75 for the first 1500, then 1:120 for the remainder	1:40 for the first 1520, then 1:60 for the remainder	1:200	1:150		1:1000	1 Service Sink
Business	B	Buildings for the transaction of business, professional services, other services involving merchandise, office buildings, banks, light industrial	1:25 for the first 50, then 1:50 for the remainder		1:40 for the first 80, then 1:80 for the remainder			1:100	1 Service Sink
Educational	E	Educational Facilities	1:50		1:50			1:100	1 Service Sink
Factory and industrial	F-1 and F-2	Structures in which occupants are engaged in work fabricating, assembly or processing of products or materials	1:100		1:100		See *International Plumbing Code*	1:400	1 Service Sink
Institutional	I-1	Residential Care	1:10		1:10		1:8	1:100	1 Service Sink
	I-2	Hospitals, ambulatory nursing home care recipient	1 per room		1 per room		1:15	1:100	1 Service Sink
		Employees, other than residential care	1:25		1:35			1:100	
		Visitors, other than residential care	1:75		1:100			1:500	
	I-3	Prisons	1 per cell		1 per cell		1:15	1:100	1 Service Sink
		Reformatories, detention and correction centers	1:15		1:15		1:15	1:100	1 Service Sink
		Employees	1:25		1:35			1:100	
	I-4	Adult and child day care	1:15		1:15		1	1:100	1 Service Sink
Mercantile	M	Retail stores, service stations, shops, salesrooms, markets and shopping centers	1:500		1:750			1:1000	1 Service Sink
Residential	R-1	Hotels, Motels, boarding houses (transient)	1 per sleeping unit		1 per sleeping unit		1 per sleeping unit		1 Service Sink
	R-2	Dormitories, fraternities, sororities and boarding houses (not transient)	1:10		1:10		1:8	1:100	1 Service Sink
		Apartment House	1 per dwelling unit		1 per dwelling unit		1 per dwelling unit		1 Kitchen sink per dwelling; 1 clothes washer connection per 20 dwellings
	R-3	1 and 2 Family Dwellings	1 per dwelling unit		1:10		1 per dwelling unit		1 Kitchen sink per dwelling; 1 clothes washer connection per dwelling
	R-3	Congregate living facilities w/<16 people	1:10		1:10		1:8	1:100	1 Service Sink
	R-4	Congregate living facilities w/<16 people	1:10		1:10		1:8	1:100	1 Service Sink
Storage	S-1 and S-2	Structures for the storage of good, warehouses, storehouses and freight depots, low and moderate hazard	1:100		1:100		See *International Plumbing Code*	1;1000	1 Service Sink

Table 2902.1

Heating, Ventilating & A.C. R2305 Common Work Results for HVAC

R230500-10 Subcontractors

On the unit cost pages of the R.S. Means Cost Data data set, the last column is entitled "Total Incl. O&P". This is normally the cost of the installing contractor. In the HVAC Division, this is the cost of the mechanical contractor. If the particular work being estimated is to be performed by a sub to the mechanical contractor, the mechanical's profit and handling charge (usually 10%) is added to the total of the last column.

Heating, Ventilating & A.C. R2356 Solar Energy Heating Equipment

R235616-60 Solar Heating (Space and Hot Water)

Collectors should face as close to due South as possible, but variations of up to 20 degrees on either side of true South are acceptable. Local climate and collector type may influence the choice between east or west deviations. Obviously they should be located so they are not shaded from the sun's rays. Incline collectors at a slope of latitude minus 5 degrees for domestic hot water and latitude plus 15 degrees for space heating.

Flat plate collectors consist of a number of components as follows: Insulation to reduce heat loss through the bottom and sides of the collector. The enclosure which contains all the components in this assembly is usually weatherproof and prevents dust, wind and water from coming in contact with the absorber plate. The cover plate usually consists of one or more layers of a variety of glass or plastic and reduces the reradiation by creating an air space which traps the heat between the cover and the absorber plates.

The absorber plate must have a good thermal bond with the fluid passages. The absorber plate is usually metallic and treated with a surface coating which improves absorptivity. Black or dark paints or selective coatings are used for this purpose, and the design of this passage and plate combination helps determine a solar system's effectiveness.

Heat transfer fluid passage tubes are attached above and below or integral with an absorber plate for the purpose of transferring thermal energy from the absorber plate to a heat transfer medium. The heat exchanger is a device for transferring thermal energy from one fluid to another.

Piping and storage tanks should be well insulated to minimize heat losses.

Size domestic water heating storage tanks to hold 20 gallons of water per user, minimum, plus 10 gallons per dishwasher or washing machine. For domestic water heating an optimum collector size is approximately 3/4 square foot of area per gallon of water storage. For space heating of residences and small commercial applications the collector is commonly sized between 30% and 50% of the internal floor area. For space heating of large commercial applications, collector areas less than 30% of the internal floor area can still provide significant heat reductions.

A supplementary heat source is recommended for Northern states for December through February.

The solar energy transmission per square foot of collector surface varies greatly with the material used. Initial cost, heat transmittance and useful life are obviously interrelated.

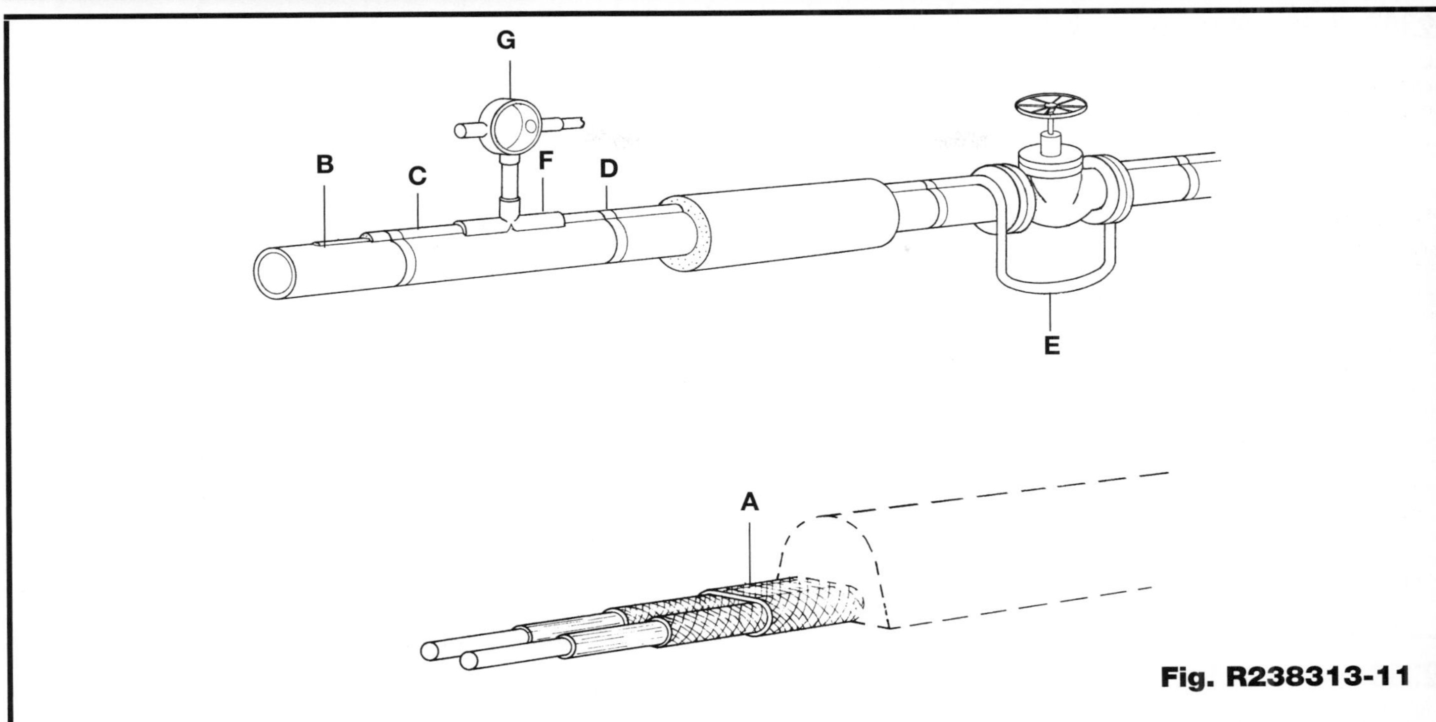

Fig. R238313-11

R238313-10 Heat Trace Systems

Before you can determine the cost of a HEAT TRACE installation, the method of attachment must be established. There are (4) common methods:

1. Cable is simply attached to the pipe with polyester tape every 12′.
2. Cable is attached with a continuous cover of 2″ wide aluminum tape.
3. Cable is attached with factory extruded heat transfer cement and covered with metallic raceway with clips every 10′.
4. Cable is attached between layers of pipe insulation using either clips or polyester tape.

In all of the above methods each component of the system must be priced individually.

Example: Components for method 3 must include:

A. Heat trace cable by voltage and watts per linear foot.
B. Heat transfer cement, 1 gallon per 60 linear feet of cover.
C. Metallic raceway by size and type.
D. Raceway clips by size of pipe.

When taking off linear foot lengths of cable add the following for each valve in the system. (E)

SCREWED OR WELDED VALVE:

Size		Length
1/2″	=	6″
3/4″	=	9″
1″	=	1′ -0″
1-1/2″	=	1′ -6″
2″	=	2′
2-1/2″	=	2′ -6″
3″	=	2′ -6″
4″	=	4′ -0″
6″	=	7′ -0″
8″	=	9′ -6″
10″	=	12′ -6″
12″	=	15′ -0″
14″	=	18′ -0″
16″	=	21′ -6″
18″	=	25′ -6″
20″	=	28′ -6″
24″	=	34′ -0″
30″	=	40′ -0″

FLANGED VALVE:

Size		Length
1/2″	=	1′ -0″
3/4″	=	1′ -6″
1″	=	2′ -0″
1-1/2″	=	2′ -6″
2″	=	2′ -6″
2-1/2″	=	3′ -0″
3″	=	3′ -6″
4″	=	4′ -0″
6″	=	8′ -0″
8″	=	11′ -0″
10″	=	14′ -0″
12″	=	16′ -6″
14″	=	19′ -6″
16″	=	23′ -0″
18″	=	27′ -0″
20″	=	30′ -0″
24″	=	36′ -0″
30″	=	42′ -0″

BUTTERFLY VALVES:

Size		Length
1/2″	=	0′
3/4″	=	0′
1″	=	1′ -0″
1-1/2″	=	1′ -6″
2″	=	2′ -0″
2-1/2″	=	2′ -6″
3″	=	2′ -6″
4″	=	3′ -0″
6″	=	3′ -6″
8″	=	4′ -0″
10″	=	4′ -0″
12″	=	5′ -0″
14″	=	5′ -6″
16″	=	6′ -0″
18″	=	6′ -6″
20″	=	7′ -0″
24″	=	8′ -0″
30″	=	10′ -0″

R238313-10 Heat Trace Systems (cont.)

Add the following quantities of heat transfer cement to linear foot totals for each valve:

Nominal Valve Size	Gallons of Cement per Valve
1/2″	0.14
3/4″	0.21
1″	0.29
1-1/2″	0.36
2″	0.43
2-1/2″	0.70
3″	0.71
4″	1.00
6″	1.43
8″	1.48
10″	1.50
12″	1.60
14″	1.75
16″	2.00
18″	2.25
20″	2.50
24″	3.00
30″	3.75

The following must be added to the list of components to accurately price HEAT TRACE systems:

1. Expediter fitting and clamp fasteners (F)
2. Junction box and nipple connected to expediter fitting (G)
3. Field installed terminal blocks within junction box
4. Ground lugs
5. Piping from power source to expediter fitting
6. Controls
7. Thermostats
8. Branch wiring
9. Cable splices
10. End of cable terminations
11. Branch piping fittings and boxes

Deduct the following percentages from labor if cable lengths in the same area exceed:

150′ to 250′	10%	351′ to 500′	20%
251′ to 350′	15%	Over 500′	25%

Add the following percentages to labor for elevated installations:

15′ to 20′ high	10%	31′ to 35′ high	40%
21′ to 25′ high	20%	36′ to 40′ high	50%
26′ to 30′ high	30%	Over 40′ high	60%

R238313-20 Spiral-Wrapped Heat Trace Cable (Pitch Table)

In order to increase the amount of heat, occasionally heat trace cable is wrapped in a spiral fashion around a pipe; increasing the number of feet of heater cable per linear foot of pipe.

Engineers first determine the heat loss per foot of pipe (based on the insulating material, its thickness, and the temperature differential across it). A ratio is then calculated by the formula:

$$\text{Feet of Heat Trace per Foot of Pipe} = \frac{\text{Watts/Foot of Heat Loss}}{\text{Watts/Foot of the Cable}}$$

The linear distance between wraps (pitch) is then taken from a chart or table. Generally, the pitch is listed on a drawing leaving the estimator to calculate the total length of heat tape required. An approximation may be taken from this table.

	Feet of Heat Trace Per Foot of Pipe															
	Nominal Pipe Size in Inches															
Pitch In Inches	1	1¼	1½	2	2½	3	4	6	8	10	12	14	16	18	20	24
3.5	1.80															
4	1.65															
5	1.46	1.60	1.80													
6	1.34	1.45	1.55	1.75												
7	1.25	1.35	1.43	1.57	1.75											
8	1.20	1.28	1.34	1.45	1.60	1.80										
9	1.16	1.23	1.28	1.37	1.51	1.68										
10	1.13	1.19	1.24	1.32	1.44	1.57	1.82									
15	1.06	1.08	1.10	1.15	1.21	1.29	1.42	1.78								
20	1.04	1.05	1.06	1.08	1.13	1.17	1.25	1.49	1.73							
25		1.04	1.04	1.06	1.08	1.11	1.17	1.33	1.51	1.72						
30				1.04	1.05	1.07	1.12	1.24	1.37	1.54	1.70	1.80				
35					1.06	1.06	1.09	1.17	1.28	1.42	1.54	1.64	1.78			
40						1.05	1.07	1.14	1.22	1.33	1.44	1.52	1.64	1.75		
50							1.05	1.09	1.15	1.22	1.29	1.35	1.44	1.53	1.64	1.83
60								1.06	1.11	1.16	1.21	1.25	1.31	1.39	1.46	1.62
70								1.05	1.08	1.12	1.17	1.19	1.24	1.30	1.35	1.47
80									1.06	1.09	1.13	1.15	1.19	1.24	1.28	1.38
90									1.04	1.06	1.10	1.13	1.16	1.19	1.23	1.32
100										1.05	1.08	1.10	1.13	1.15	1.19	1.23

Note: Common practice would normally limit the lower end of the table to 5% of additional heat and above 80% an engineer would likely opt for two (2) parallel cables.

R260519-90 Wire

Wire quantities are taken off by either measuring each cable run or by extending the conduit and raceway quantities times the number of conductors in the raceway. Ten percent should be added for waste and tie-ins. Keep in mind that the unit of measure of wire is C.L.F. not L.F. as in raceways so the formula would read:

$$\frac{\text{(L.F. Raceway x No. of Conductors) x 1.10}}{100} = \text{C.L.F.}$$

Price per C.L.F. of wire includes:
1. Setting up wire coils or spools on racks
2. Attaching wire to pull in means
3. Measuring and cutting wire
4. Pulling wire into a raceway
5. Identifying and tagging

Price does not include:
1. Connections to breakers, panelboards, or equipment
2. Splices

Job Conditions: Productivity is based on new construction to a height of 15′ using rolling staging in an unobstructed area. Material staging is assumed to be within 100′ of work being performed.

Economy of Scale: If more than three wires at a time are being pulled, deduct the following percentages from the labor of that grouping:

4-5 wires	25%
6-10 wires	30%
11-15 wires	35%
over 15	40%

If a wire pull is less than 100′ in length and is interrupted several times by boxes, lighting outlets, etc., it may be necessary to add the following lengths to each wire being pulled:

Junction box to junction box	2 L.F.
Lighting panel to junction box	6 L.F.
Distribution panel to sub panel	8 L.F.
Switchboard to distribution panel	12 L.F.
Switchboard to motor control center	20 L.F.
Switchboard to cable tray	40 L.F.

Measure of Drops and Riser: It is important when taking off wire quantities to include the wire for drops to electrical equipment. If heights of electrical equipment are not clearly stated, use the following guide:

	Bottom A.F.F.	Top A.F.F.	Inside Cabinet
Safety switch to 100A	5′	6′	2′
Safety switch 400 to 600A	4′	6′	3′
100A panel 12 to 30 circuit	4′	6′	3′
42 circuit panel	3′	6′	4′
Switch box	3′	3′6″	1′
Switchgear	0′	8′	8′
Motor control centers	0′	8′	8′
Transformers - wall mount	4′	8′	2′
Transformers - floor mount	0′	12′	4′

R312316-45 Excavating Equipment

The table below lists theoretical hourly production in C.Y./hr. bank measure for some typical excavation equipment. Figures assume 50 minute hours, 83% job efficiency, 100% operator efficiency, 90° swing and properly sized hauling units, which must be modified for adverse digging and loading conditions. Actual production costs in the front of the data set average about 50% of the theoretical values listed here.

Equipment	Soil Type	B.C.Y. Weight	% Swell	1 C.Y.	1-1/2 C.Y.	2 C.Y.	2-1/2 C.Y.	3 C.Y.	3-1/2 C.Y.	4 C.Y.
Hydraulic Excavator "Backhoe" 15′ Deep Cut	Moist loam, sandy clay	3400 lb.	40%	165	195	200	275	330	385	440
	Sand and gravel	3100	18	140	170	225	240	285	330	380
	Common earth	2800	30	150	180	230	250	300	350	400
	Clay, hard, dense	3000	33	120	140	190	200	240	260	320
Power Shovel Optimum Cut (Ft.)	Moist loam, sandy clay	3400	40	170 (6.0)	245 (7.0)	295 (7.8)	335 (8.4)	385 (8.8)	435 (9.1)	475 (9.4)
	Sand and gravel	3100	18	165 (6.0)	225 (7.0)	275 (7.8)	325 (8.4)	375 (8.8)	420 (9.1)	460 (9.4)
	Common earth	2800	30	145 (7.8)	200 (9.2)	250 (10.2)	295 (11.2)	335 (12.1)	375 (13.0)	425 (13.8)
	Clay, hard, dense	3000	33	120 (9.0)	175 (10.7)	220 (12.2)	255 (13.3)	300 (14.2)	335 (15.1)	375 (16.0)
Drag Line Optimum Cut (Ft.)	Moist loam, sandy clay	3400	40	130 (6.6)	180 (7.4)	220 (8.0)	250 (8.5)	290 (9.0)	325 (9.5)	385 (10.0)
	Sand and gravel	3100	18	130 (6.6)	175 (7.4)	210 (8.0)	245 (8.5)	280 (9.0)	315 (9.5)	375 (10.0)
	Common earth	2800	30	110 (8.0)	160 (9.0)	190 (9.9)	220 (10.5)	250 (11.0)	280 (11.5)	310 (12.0)
	Clay, hard, dense	3000	33	90 (9.3)	130 (10.7)	160 (11.8)	190 (12.3)	225 (12.8)	250 (13.3)	280 (12.0)
				Wheel Loaders				Track Loaders		
				3 C.Y.	4 C.Y.	6 C.Y.	8 C.Y.	2-1/4 C.Y.	3 C.Y.	4 C.Y.
Loading Tractors	Moist loam, sandy clay	3400	40	260	340	510	690	135	180	250
	Sand and gravel	3100	18	245	320	480	650	130	170	235
	Common earth	2800	30	230	300	460	620	120	155	220
	Clay, hard, dense	3000	33	200	270	415	560	110	145	200
	Rock, well-blasted	4000	50	180	245	380	520	100	130	180

R312319-90 Wellpoints

A single stage wellpoint system is usually limited to dewatering an average 15′ depth below normal ground water level. Multi-stage systems are employed for greater depth with the pumping equipment installed only at the lowest header level. Ejectors with unlimited lift capacity can be economical when two or more stages of wellpoints can be replaced or when horizontal clearance is restricted, such as in deep trenches or tunneling projects, and where low water flows are expected. Wellpoints are usually spaced on 2-1/2′ to 10′ centers along a header pipe. Wellpoint spacing, header size, and pump size are all determined by the expected flow as dictated by soil conditions.

In almost all soils encountered in wellpoint dewatering, the wellpoints may be jetted into place. Cemented soils and stiff clays may require sand wicks about 12″ in diameter around each wellpoint to increase efficiency and eliminate weeping into the excavation. These sand wicks require 1/2 to 3 C.Y. of washed filter sand and are installed by using a 12″ diameter steel casing and hole puncher jetted into the ground 2′ deeper than the wellpoint. Rock may require predrilled holes.

Labor required for the complete installation and removal of a single stage wellpoint system is in the range of 3/4 to 2 labor-hours per linear foot of header, depending upon jetting conditions, wellpoint spacing, etc.

Continuous pumping is necessary except in some free draining soil where temporary flooding is permissible (as in trenches which are backfilled after each day's work). Good practice requires provision of a stand-by pump during the continuous pumping operation.

Systems for continuous trenching below the water table should be installed three to four times the length of expected daily progress to ensure uninterrupted digging, and header pipe size should not be changed during the job.

For pervious free draining soils, deep wells in place of wellpoints may be economical because of lower installation and maintenance costs. Daily production ranges between two to three wells per day, for 25′ to 40′ depths, to one well per day for depths over 50′.

Detailed analysis and estimating for any dewatering problem is available at no cost from wellpoint manufacturers. Major firms will quote "sufficient equipment" quotes or their affiliates will offer lump sum proposals to cover complete dewatering responsibility.

Description for 200′ System with 8″ Header		Quantities
Equipment & Material	Wellpoints 25′ long, 2″ diameter @ 5′ O.C.	40 Each
	Header pipe, 8″ diameter	200 L.F.
	Discharge pipe, 8″ diameter	100 L.F.
	8″ valves	3 Each
	Combination jetting & wellpoint pump (standby)	1 Each
	Wellpoint pump, 8″ diameter	1 Each
	Transportation to and from site	1 Day
	Fuel for 30 days x 60 gal./day	1800 Gallons
	Lubricants for 30 days x 16 lbs./day	480 Lbs.
	Sand for points	40 C.Y.
Labor	Technician to supervise installation	1 Week
	Labor for installation and removal of system	300 Labor-hours
	4 Operators straight time 40 hrs./wk. for 4.33 wks.	693 Hrs.
	4 Operators overtime 2 hrs./wk. for 4.33 wks.	35 Hrs.

R314116-40 Wood Sheet Piling

Wood sheet piling may be used for depths to 20′ where there is no ground water. If moderate ground water is encountered Tongue & Groove sheeting will help to keep it out. When considerable ground water is present, steel sheeting must be used.

For estimating purposes on trench excavation, sizes are as follows:

Depth	Sheeting	Wales	Braces	B.F. per S.F.
To 8′	3 x 12′s	6 x 8′s, 2 line	6 x 8′s, @ 10′	4.0 @ 8′
8′ x 12′	3 x 12′s	10 x 10′s, 2 line	10 x 10′s, @ 9′	5.0 average
12′ to 20′	3 x 12′s	12 x 12′s, 3 line	12 x 12′s, @ 8′	7.0 average

Sheeting to be toed in at least 2′ depending upon soil conditions. A five person crew with an air compressor and sheeting driver can drive and brace 440 SF/day at 8′ deep, 360 SF/day at 12′ deep, and 320 SF/day at 16′ deep. For normal soils, piling can be pulled in 1/3 the time to install. Pulling difficulty increases with the time in the ground. Production can be increased by high pressure jetting.

R314116-45 Steel Sheet Piling

Limiting weights are 22 to 38#/S.F. of wall surface with 27#/S.F. average for usual types and sizes. (Weights of piles themselves are from 30.7#/L.F. to 57#/L.F. but they are 15″ to 21″ wide.) Lightweight sections 12″ to 28″ wide from 3 ga. to 12 ga. thick are also available for shallow excavations. Piles may be driven two at a time with an impact or vibratory hammer (use vibratory to pull) hung from a crane without leads. A reasonable estimate of the life of steel sheet piling is 10 uses with up to 125 uses possible if a vibratory hammer is used. Used piling costs from 50% to 80% of new piling depending on location and market conditions. Sheet piling and H piles can be rented for about 30% of the delivered mill price for the first month and 5% per month thereafter. Allow 1 labor-hour per pile for cleaning and trimming after driving. These costs increase with depth and hydrostatic head. Vibratory drivers are faster in wet granular soils and are excellent for pile extraction. Pulling difficulty increases with the time in the ground and may cost more than driving. It is often economical to abandon the sheet piling, especially if it can be used as the outer wall form. Allow about 1/3 additional length or more for toeing into ground. Add bracing, waler and strut costs. Waler costs can equal the cost per ton of sheeting.

Utilities R3311 Water Utility Distribution Piping

R331113-80 Piping Designations

There are several systems currently in use to describe pipe and fittings. The following paragraphs will help to identify and clarify classifications of piping systems used for water distribution.

Piping may be classified by schedule. Piping schedules include 5S, 10S, 10, 20, 30, Standard, 40, 60, Extra Strong, 80, 100, 120, 140, 160 and Double Extra Strong. These schedules are dependent upon the pipe wall thickness. The wall thickness of a particular schedule may vary with pipe size.

Ductile iron pipe for water distribution is classified by Pressure Classes such as Class 150, 200, 250, 300 and 350. These classes are actually the rated water working pressure of the pipe in pounds per square inch (psi). The pipe in these pressure classes is designed to withstand the rated water working pressure plus a surge allowance of 100 psi.

The American Water Works Association (AWWA) provides standards for various types of **plastic pipe.** C-900 is the specification for polyvinyl chloride (PVC) piping used for water distribution in sizes ranging from 4″ through 12″. C-901 is the specification for polyethylene (PE) pressure pipe, tubing and fittings used for water distribution in sizes ranging from 1/2″ through 3″. C-905 is the specification for PVC piping sizes 14″ and greater.

PVC pressure-rated pipe is identified using the standard dimensional ratio (SDR) method. This method is defined by the American Society for Testing and Materials (ASTM) Standard D 2241. This pipe is available in SDR numbers 64, 41, 32.5, 26, 21, 17, and 13.5. Pipe with an SDR of 64 will have the thinnest wall while pipe with an SDR of 13.5 will have the thickest wall. When the pressure rating (PR) of a pipe is given in psi, it is based on a line supplying water at 73 degrees F.

The National Sanitation Foundation (NSF) seal of approval is applied to products that can be used with potable water. These products have been tested to ANSI/NSF Standard 14.

Valves and strainers are classified by American National Standards Institute (ANSI) Classes. These Classes are 125, 150, 200, 250, 300, 400, 600, 900, 1500 and 2500. Within each class there is an operating pressure range dependent upon temperature. Design parameters should be compared to the appropriate material dependent, pressure-temperature rating chart for accurate valve selection.

Change Orders

Change Order Considerations

A change order is a written document usually prepared by the design professional and signed by the owner, the architect/engineer, and the contractor. A change order states the agreement of the parties to: an addition, deletion, or revision in the work; an adjustment in the contract sum, if any; or an adjustment in the contract time, if any. Change orders, or "extras", in the construction process occur after execution of the construction contract and impact architects/engineers, contractors, and owners.

Change orders that are properly recognized and managed can ensure orderly, professional, and profitable progress for everyone involved in the project. There are many causes for change orders and change order requests. In all cases, change orders or change order requests should be addressed promptly and in a precise and prescribed manner. The following paragraphs include information regarding change order pricing and procedures.

The Causes of Change Orders

Reasons for issuing change orders include:

- Unforeseen field conditions that require a change in the work
- Correction of design discrepancies, errors, or omissions in the contract documents
- Owner-requested changes, either by design criteria, scope of work, or project objectives
- Completion date changes for reasons unrelated to the construction process
- Changes in building code interpretations, or other public authority requirements that require a change in the work
- Changes in availability of existing or new materials and products

Procedures

Properly written contract documents must include the correct change order procedures for all parties—owners, design professionals, and contractors—to follow in order to avoid costly delays and litigation.

Being "in the right" is not always a sufficient or acceptable defense. The contract provisions requiring notification and documentation must be adhered to within a defined or reasonable time frame.

The appropriate method of handling change orders is by a written proposal and acceptance by all parties involved. Prior to starting work on a project, all parties should identify their authorized agents who may sign and accept change orders, as well as any limits placed on their authority.

Time may be a critical factor when the need for a change arises. For such cases, the contractor might be directed to proceed on a "time and materials" basis, rather than wait for all paperwork to be processed—a delay that could impede progress. In this situation, the contractor must still follow the prescribed change order procedures including, but not limited to, notification and documentation.

Lack of documentation can be very costly, especially if legal judgments are to be made, and if certain field personnel are no longer available. For time and material change orders, the contractor should keep accurate daily records of all labor and material allocated to the change.

Owners or awarding authorities who do considerable and continual building construction (such as the federal government) realize the inevitability of change orders for numerous reasons, both predictable and unpredictable. As a result, the federal government, the American Institute of Architects (AIA), the Engineers Joint Contract Documents Committee (EJCDC), and other contractor, legal, and technical organizations have developed standards and procedures to be followed by all parties to achieve contract continuance and timely completion, while being financially fair to all concerned.

Pricing Change Orders

When pricing change orders, regardless of their cause, the most significant factor is when the change occurs. The need for a change may be perceived in the field or requested by the architect/engineer *before* any of the actual installation has begun, or may evolve or appear *during* construction when the item of work in question is partially installed. In the latter cases, the original sequence of construction is disrupted, along with all contiguous and supporting systems. Change orders cause the greatest impact when they occur *after* the installation has been completed and must be uncovered, or even replaced. Post-completion changes may be caused by necessary design changes, product failure, or changes in the owner's requirements that are not discovered until the building or the systems begin to function.

Specified procedures of notification and record keeping must be adhered to and enforced regardless of the stage of construction: *before, during,* or *after* installation. Some bidding documents anticipate change orders by requiring that unit prices including overhead and profit percentages—for additional as well as deductible changes—be listed. Generally these unit prices do not fully take into account the ripple effect, or impact on other trades, and should be used for general guidance only.

When pricing change orders, it is important to classify the time frame in which the change occurs. There are two basic time frames for change orders: *pre-installation change orders,* which occur before the start of construction, and *post-installation change orders,* which involve reworking after the original installation. Change orders that occur between these stages may be priced according to the extent of work completed using a combination of techniques developed for pricing *pre-* and *post-installation* changes.

Factors To Consider When Pricing Change Orders

As an estimator begins to prepare a change order, the following questions should be reviewed to determine their impact on the final price.

General

- *Is the change order work* pre-installation *or* post-installation?

  Change order work costs vary according to how much of the installation has been completed. Once workers have the project scoped in their minds, even though they have not started, it can be difficult to refocus. Consequently they may spend more than the normal amount of time understanding the change. Also, modifications to work in place, such as trimming or refitting, usually take more time than was initially estimated. The greater the amount of work in place, the more reluctant workers are to change it. Psychologically they may resent the change and as a result the rework takes longer than normal. Post-installation change order estimates must include demolition of existing work as required to accomplish the change. If the work is performed at a later time, additional obstacles, such as building finishes, may be present which must be protected. Regardless of whether the change occurs pre-installation or post-installation, attempt to isolate the identifiable factors and price them separately. For example, add shipping costs that may be required pre-installation or any demolition required post-installation. Then analyze the potential impact on productivity of psychological and/or learning curve factors and adjust the output rates accordingly. One approach is to break down the typical workday into segments and quantify the impact on each segment.

Change Order Installation Efficiency

The labor-hours expressed (for new construction) are based on average installation time, using an efficiency level. For change order situations, adjustments to this efficiency level should reflect the daily labor-hour allocation for that particular occurrence.

- *Will the change substantially delay the original completion date?*

  A significant change in the project may cause the original completion date to be extended. The extended schedule may subject the contractor to new wage rates dictated by relevant labor contracts. Project supervision and other project overhead must also be extended beyond the original completion date. The schedule extension may also put installation into a new weather season. For example, underground piping scheduled for October installation was delayed until January. As a result, frost penetrated the trench area, thereby changing the degree of difficulty of the task. Changes and delays may have a ripple effect throughout the project. This effect must be analyzed and negotiated with the owner.

- *What is the net effect of a deduct change order?*

  In most cases, change orders resulting in a deduction or credit reflect only bare costs. The contractor may retain the overhead and profit based on the original bid.

Materials

- *Will you have to pay more or less for the new material, required by the change order, than you paid for the original purchase?*

  The same material prices or discounts will usually apply to materials purchased for change orders as new construction. In some instances, however, the contractor may forfeit the advantages of competitive pricing for change orders. Consider the following example:

  A contractor purchased over $20,000 worth of fan coil units for an installation and obtained the maximum discount. Some time later it was determined the project required an additional matching unit. The contractor has to purchase this unit from the original supplier to ensure a match. The supplier at this time may not discount the unit because of the small quantity, and he is no longer in a competitive situation. The impact of quantity on purchase can add between 0% and 25% to material prices and/or subcontractor quotes.

- *If materials have been ordered or delivered to the job site, will they be subject to a cancellation charge or restocking fee?*

  Check with the supplier to determine if ordered materials are subject to a cancellation charge. Delivered materials not used as a result of a change order may be subject to a restocking fee if returned to the supplier. Common restocking charges run between 20% and 40%. Also, delivery charges to return the goods to the supplier must be added.

Labor

- *How efficient is the existing crew at the actual installation?*

  Is the same crew that performed the initial work going to do the change order? Possibly the change consists of the installation of a unit identical to one already installed; therefore, the change should take less time. Be sure to consider this potential productivity increase and modify the productivity rates accordingly.

- *If the crew size is increased, what impact will that have on supervision requirements?*

  Under most bargaining agreements or management practices, there is a point at which a working foreman is replaced by a nonworking foreman. This replacement increases project overhead by adding a nonproductive worker. If additional workers are added to accelerate the project or to perform changes while maintaining the schedule, be sure to add additional supervision time if warranted. Calculate the hours involved and the additional cost directly if possible.

- *What are the other impacts of increased crew size?*

  The larger the crew, the greater the potential for productivity to decrease. Some of the factors that cause this productivity loss are: overcrowding (producing restrictive conditions in the working space) and possibly a shortage of any special tools and equipment required. Such factors affect not only the crew working on the elements directly involved in the change order, but other crews whose movements may also be hampered. As the crew increases, check its basic composition for changes by the addition or deletion of apprentices or nonworking foreman, and quantify the potential effects of equipment shortages or other logistical factors.

- *As new crews, unfamiliar with the project, are brought onto the site, how long will it take them to become oriented to the project requirements?*

  The orientation time for a new crew to become 100% effective varies with the site and type of project. Orientation is easiest at a new construction site and most difficult at existing, very restrictive renovation sites. The type of work also affects orientation time. When all elements of the work are exposed, such as concrete or masonry work, orientation is decreased. When the work is concealed or less visible, such as existing electrical systems, orientation takes longer. Usually orientation can be accomplished in one day or less. Costs for added orientation should be itemized and added to the total estimated cost.

- *How much actual production can be gained by working overtime?*

  Short term overtime can be used effectively to accomplish more work in a day. However, as overtime is scheduled to run beyond several weeks, studies have shown marked decreases in output. The following chart shows the effect of long term overtime on worker efficiency. If the anticipated change requires extended overtime to keep the job on schedule, these factors can be used as a guide to predict the impact on time and cost. Add project overhead, particularly supervision, that may also be incurred.

Days per Week	Hours per Day	Production Efficiency					Payroll Cost Factors	
		1st Week	2nd Week	3rd Week	4th Week	Average 4 Weeks	@ 1-1/2 Times	@ 2 Times
5	8	100%	100%	100%	100%	100%	100%	100%
	9	100	100	95	90	96.25	105.6	111.1
	10	100	95	90	85	92.50	110.0	120.0
	11	95	90	75	65	81.25	113.6	127.3
	12	90	85	70	60	76.25	116.7	133.3
6	8	100	100	95	90	96.25	108.3	116.7
	9	100	95	90	85	92.50	113.0	125.9
	10	95	90	85	80	87.50	116.7	133.3
	11	95	85	70	65	78.75	119.7	139.4
	12	90	80	65	60	73.75	122.2	144.4
7	8	100	95	85	75	88.75	114.3	128.6
	9	95	90	80	70	83.75	118.3	136.5
	10	90	85	75	65	78.75	121.4	142.9
	11	85	80	65	60	72.50	124.0	148.1
	12	85	75	60	55	68.75	126.2	152.4

Effects of Overtime

Caution: Under many labor agreements, Sundays and holidays are paid at a higher premium than the normal overtime rate.

The use of long-term overtime is counterproductive on almost any construction job; that is, the longer the period of overtime, the lower the actual production rate. Numerous studies have been conducted, and while they have resulted in slightly different numbers, all reach the same conclusion. The figure above tabulates the effects of overtime work on efficiency.

As illustrated, there can be a difference between the *actual* payroll cost per hour and the *effective* cost per hour for overtime work. This is due to the reduced production efficiency with the increase in weekly hours beyond 40. This difference between actual and effective cost results from overtime work over a prolonged period. Short-term overtime work does not result in as great a reduction in efficiency and, in such cases, effective cost may not vary significantly from the actual payroll cost. As the total hours per week are increased on a regular basis, more time is lost due to fatigue, lowered morale, and an increased accident rate.

As an example, assume a project where workers are working 6 days a week, 10 hours per day. From the figure above (based on productivity studies), the average effective productive hours over a 4-week period are:

$$0.875 \times 60 = 52.5$$

Depending upon the locale and day of week, overtime hours may be paid at time and a half or double time. For time and a half, the overall (average) *actual* payroll cost (including regular and overtime hours) is determined as follows:

$$\frac{40 \text{ reg. hrs.} + (20 \text{ overtime hrs. x } 1.5)}{60 \text{ hrs.}} = 1.167$$

Based on 60 hours, the payroll cost per hour will be 116.7% of the normal rate at 40 hours per week. However, because the effective production (efficiency) for 60 hours is reduced to the equivalent of 52.5 hours, the effective cost of overtime is calculated as follows:

For time and a half:

$$\frac{40 \text{ reg. hrs.} + (20 \text{ overtime hrs. x } 1.5)}{52.5 \text{ hrs.}} = 1.33$$

The installed cost will be 133% of the normal rate (for labor).

Thus, when figuring overtime, the actual cost per unit of work will be higher than the apparent overtime payroll dollar increase, due to the reduced productivity of the longer work week. These efficiency calculations are true only for those cost factors determined by hours worked. Costs that are applied weekly or monthly, such as equipment rentals, will not be similarly affected.

Equipment

- *What equipment is required to complete the change order?*

  Change orders may require extending the rental period of equipment already on the job site, or the addition of special equipment brought in to accomplish the change work. In either case, the additional rental charges and operator labor charges must be added.

Summary

The preceding considerations and others you deem appropriate should be analyzed and applied to a change order estimate. The impact of each should be quantified and listed on the estimate to form an audit trail.

Change orders that are properly identified, documented, and managed help to ensure the orderly, professional, and profitable progress of the work. They also minimize potential claims or disputes at the end of the project.

Back by customer demand!

You asked and we listened. For customer convenience and estimating ease, we have made the 2020 Project Costs available for download at **RSMeans.com/2020books**. You will also find sample estimates, an RSMeans data overview video, and a book registration form to receive quarterly data updates throughout 2020.

Estimating Tips

- The cost figures available in the download were derived from hundreds of projects contained in the RSMeans database of completed construction projects. They include the contractor's overhead and profit. The figures have been adjusted to January of the current year.
- These projects were located throughout the U.S. and reflect a tremendous variation in square foot (S.F.) costs. This is due to differences, not only in labor and material costs, but also in individual owners' requirements. For instance, a bank in a large city would have different features than one in a rural area. This is true of all the different types of buildings analyzed. Therefore, caution should be exercised when using these Project Costs. For example, for courthouses, costs in the database are local courthouse costs and will not apply to the larger, more elaborate federal courthouses.
- None of the figures "go with" any others. All individual cost items were computed and tabulated separately. Thus, the sum of the median figures for plumbing, HVAC, and electrical will not normally total up to the total mechanical and electrical costs arrived at by separate analysis and tabulation of the projects.
- Each building was analyzed as to total and component costs and percentages. The figures were arranged in ascending order with the results tabulated as shown. The 1/4 column shows that 25% of the projects had lower costs and 75% had higher. The 3/4 column shows that 75% of the projects had lower costs and 25% had higher. The median column shows that 50% of the projects had lower costs and 50% had higher.
- Project Costs are useful in the conceptual stage when no details are available. As soon as details become available in the project design, the square foot approach should be discontinued and the project should be priced as to its particular components. When more precision is required, or for estimating the replacement cost of specific buildings, the current edition of *Square Foot Costs with RSMeans data* should be used.
- In using the figures in this section, it is recommended that the median column be used for preliminary figures if no additional information is available. The median figures, when multiplied by the total city construction cost index figures (see City Cost Indexes) and then multiplied by the project size modifier at the end of this section, should present a fairly accurate base figure, which would then have to be adjusted in view of the estimator's experience, local economic conditions, code requirements, and the owner's particular requirements. There is no need to factor in the percentage figures, as these should remain constant from city to city.
- The editors of this data would greatly appreciate receiving cost figures on one or more of your recent projects, which would then be included in the averages for next year. All cost figures received will be kept confidential, except that they will be averaged with other similar projects to arrive at square foot cost figures for next year.

See the website above for details and the discount available for submitting one or more of your projects.

50 17 00 | Project Costs

			UNIT	UNIT COSTS			% OF TOTAL			
				1/4	MEDIAN	3/4	1/4	MEDIAN	3/4	
01	0000	**Auto Sales with Repair**	S.F.							01
	0100	Architectural		106	119	128	58%	64%	67%	
	0200	Plumbing		8.90	9.30	12.40	4.84%	5.20%	6.80%	
	0300	Mechanical		11.90	15.95	17.60	6.40%	8.70%	10.15%	
	0400	Electrical		18.30	22.50	28.50	9.05%	11.70%	15.90%	
	0500	Total Project Costs	↓	178	186	191				
02	0000	**Banking Institutions**	S.F.							02
	0100	Architectural		160	197	239	59%	65%	69%	
	0200	Plumbing		6.45	9	12.50	2.12%	3.39%	4.19%	
	0300	Mechanical		12.80	17.70	21	4.41%	5.10%	10.75%	
	0400	Electrical		31.50	38	58	10.45%	13.05%	15.90%	
	0500	Total Project Costs	↓	266	299	370				
03	0000	**Court House**	S.F.							03
	0100	Architectural		84.50	166	166	54.50%	58.50%	58.50%	
	0200	Plumbing		3.19	3.19	3.19	2.07%	2.07%	2.07%	
	0300	Mechanical		19.95	19.95	19.95	12.95%	12.95%	12.95%	
	0400	Electrical		25.50	25.50	25.50	16.60%	16.60%	16.60%	
	0500	Total Project Costs	↓	154	284	284				
04	0000	**Data Centers**	S.F.							04
	0100	Architectural		191	191	191	68%	68%	68%	
	0200	Plumbing		10.45	10.45	10.45	3.71%	3.71%	3.71%	
	0300	Mechanical		26.50	26.50	26.50	9.45%	9.45%	9.45%	
	0400	Electrical		25.50	25.50	25.50	9%	9%	9%	
	0500	Total Project Costs	↓	281	281	281				
05	0000	**Detention Centers**	S.F.							05
	0100	Architectural		177	187	198	52%	53%	60.50%	
	0200	Plumbing		18.65	22.50	27.50	5.15%	7.10%	7.25%	
	0300	Mechanical		23.50	34	40.50	7.55%	9.50%	13.80%	
	0400	Electrical		39	46	60	10.90%	14.85%	17.95%	
	0500	Total Project Costs	↓	299	315	370				
06	0000	**Fire Stations**	S.F.							06
	0100	Architectural		100	129	187	49%	54.50%	61.50%	
	0200	Plumbing		10.75	13.95	16.95	4.67%	5.60%	6.30%	
	0300	Mechanical		14.95	21	29.50	6.10%	8.45%	10.20%	
	0400	Electrical		23	29.50	35.50	10.75%	12.75%	14.95%	
	0500	Total Project Costs	↓	210	238	320				
07	0000	**Gymnasium**	S.F.							07
	0100	Architectural		88.50	116	116	57%	64.50%	64.50%	
	0200	Plumbing		2.17	7.10	7.10	1.58%	3.48%	3.48%	
	0300	Mechanical		3.32	30	30	2.42%	14.65%	14.65%	
	0400	Electrical		10.90	21	21	7.95%	10.35%	10.35%	
	0500	Total Project Costs	↓	137	204	204				
08	0000	**Hospitals**	S.F.							08
	0100	Architectural		107	176	191	43%	47.50%	48%	
	0200	Plumbing		7.85	15.05	32.50	6%	7.45%	7.65%	
	0300	Mechanical		52	58.50	76.50	14.20%	17.95%	23.50%	
	0400	Electrical		23.50	47.50	61.50	10.95%	13.75%	16.85%	
	0500	Total Project Costs	↓	250	375	405				
09	0000	**Industrial Buildings**	S.F.							09
	0100	Architectural		45	71.50	232	46%	54%	56.50%	
	0200	Plumbing		1.74	6.55	13.25	2%	3.06%	6.30%	
	0300	Mechanical		4.81	9.15	43.50	4.77%	5.55%	14.80%	
	0400	Electrical		7.35	8.40	70	7.85%	13.55%	16.20%	
	0500	Total Project Costs	↓	80	104	430				
10	0000	**Medical Clinics & Offices**	S.F.							10
	0100	Architectural		89.50	121	161	48.50%	55.50%	62.50%	
	0200	Plumbing		8.90	13.15	21	4.47%	6.60%	8.65%	
	0300	Mechanical		13.95	22.50	44	7.80%	10.95%	16.10%	
	0400	Electrical		20	27	37.50	9.70%	11.65%	14.05%	
	0500	Total Project Costs	↓	166	217	292				

	50 17 00 | Project Costs		UNIT	UNIT COSTS 1/4	UNIT COSTS MEDIAN	UNIT COSTS 3/4	% OF TOTAL 1/4	% OF TOTAL MEDIAN	% OF TOTAL 3/4
11	0000	**Mixed Use**	S.F.						
	0100	Architectural		91.50	129	211	48.50%	57%	62.50%
	0200	Plumbing		6.15	9.35	11.80	3.23%	3.44%	4.18%
	0300	Mechanical		16.40	24.50	47.50	6.10%	13.75%	18.85%
	0400	Electrical		16	26.50	41.50	8.30%	11.40%	14%
	0500	Total Project Costs		189	216	345			
12	0000	**Multi-Family Housing**	S.F.						
	0100	Architectural		76	114	170	56.50%	62%	66.50%
	0200	Plumbing		6.55	11.55	15.05	5.30%	6.85%	8%
	0300	Mechanical		7	9.45	27	4.92%	6.90%	10.40%
	0400	Electrical		9.90	15.45	19.85	6.20%	7.90%	10.25%
	0500	Total Project Costs		111	208	271			
13	0000	**Nursing Home & Assisted Living**	S.F.						
	0100	Architectural		71.50	94	118	51.50%	55.50%	63.50%
	0200	Plumbing		7.75	11.60	12.80	6.25%	7.40%	8.80%
	0300	Mechanical		6.35	9.35	18.30	4.04%	6.70%	9.55%
	0400	Electrical		10.50	16.55	23	7%	10.75%	13.10%
	0500	Total Project Costs		122	160	192			
14	0000	**Office Buildings**	S.F.						
	0100	Architectural		94.50	129	177	56%	61%	69%
	0200	Plumbing		5.10	8	15.05	2.65%	3.56%	5.85%
	0300	Mechanical		11	17	26	5.60%	8.20%	11.10%
	0400	Electrical		12.75	21.50	34	7.75%	10%	12.70%
	0500	Total Project Costs		161	200	282			
15	0000	**Parking Garage**	S.F.						
	0100	Architectural		31.50	38.50	40.50	70%	79%	88%
	0200	Plumbing		1.04	1.09	2.04	2.05%	2.70%	2.83%
	0300	Mechanical		.81	1.24	4.72	2.11%	3.62%	3.81%
	0400	Electrical		2.78	3.05	6.35	5.30%	6.35%	7.95%
	0500	Total Project Costs		38.50	47	51			
16	0000	**Parking Garage/Mixed Use**	S.F.						
	0100	Architectural		103	112	114	61%	62%	65.50%
	0200	Plumbing		3.30	4.32	6.60	2.47%	2.72%	3.66%
	0300	Mechanical		14.10	15.85	23	7.80%	13.10%	13.60%
	0400	Electrical		14.75	21	22	8.20%	12.65%	18.15%
	0500	Total Project Costs		168	175	181			
17	0000	**Police Stations**	S.F.						
	0100	Architectural		116	130	164	49%	56.50%	61%
	0200	Plumbing		15.30	18.40	18.50	5.05%	5.55%	9.05%
	0300	Mechanical		34.50	48.50	50.50	13%	14.55%	16.55%
	0400	Electrical		26	28.50	30.50	9.15%	12.10%	14%
	0500	Total Project Costs		217	267	305			
18	0000	**Police/Fire**	S.F.						
	0100	Architectural		113	113	345	55.50%	66%	68%
	0200	Plumbing		9.05	9.35	34.50	5.45%	5.50%	5.55%
	0300	Mechanical		13.85	22	79	8.35%	12.70%	12.80%
	0400	Electrical		15.75	20	90.50	9.50%	11.75%	14.55%
	0500	Total Project Costs		166	171	625			
19	0000	**Public Assembly Buildings**	S.F.						
	0100	Architectural		115	159	238	57.50%	61.50%	66%
	0200	Plumbing		6.10	8.90	13.40	2.60%	3.36%	4.79%
	0300	Mechanical		12.85	23	35.50	6.55%	8.95%	12.45%
	0400	Electrical		19	26	41	8.60%	10.75%	13%
	0500	Total Project Costs		185	253	375			
20	0000	**Recreational**	S.F.						
	0100	Architectural		106	168	236	53.50%	60%	66%
	0200	Plumbing		7.80	14.75	21.50	3.08%	5%	6.85%
	0300	Mechanical		13.15	20	31.50	5.15%	6.95%	11.70%
	0400	Electrical		15.20	26.50	39.50	7.35%	8.95%	10.75%
	0500	Total Project Costs		168	281	445			

50 17 00 | Project Costs

			UNIT	UNIT COSTS			% OF TOTAL			
				1/4	MEDIAN	3/4	1/4	MEDIAN	3/4	
21	0000	**Restaurants**	S.F.							21
	0100	Architectural		126	193	248	59%	60%	63.50%	
	0200	Plumbing		13.80	31.50	40	7.35%	7.75%	8.95%	
	0300	Mechanical		14.85	17.55	37	6.50%	8.15%	11.15%	
	0400	Electrical		14.80	24	48.50	7.10%	10.30%	11.60%	
	0500	Total Project Costs	↓	208	305	415				
22	0000	**Retail**	S.F.							22
	0100	Architectural		56	86	127	60%	62%	64.50%	
	0200	Plumbing		7	9.25	12.10	5.05%	6.70%	9%	
	0300	Mechanical		6.55	9.25	17.05	5.70%	6.20%	10.20%	
	0400	Electrical		10.45	18.95	31.50	8.05%	11.25%	12.45%	
	0500	Total Project Costs	↓	85.50	114	186				
23	0000	**Schools**	S.F.							23
	0100	Architectural		97	125	164	52.50%	56%	61%	
	0200	Plumbing		7.75	10.60	16.05	3.85%	4.82%	7.25%	
	0300	Mechanical		18.90	26.50	38.50	9.50%	12.35%	15.10%	
	0400	Electrical		17.85	25	32.50	9.45%	11.45%	13.30%	
	0500	Total Project Costs	↓	169	227	300				
24	0000	**University, College & Private School Classroom & Admin Buildings**	S.F.							24
	0100	Architectural		124	153	192	50.50%	55%	59.50%	
	0200	Plumbing		7.05	10.95	15.45	2.74%	4.30%	6.35%	
	0300	Mechanical		26.50	38.50	46	10.10%	12.15%	14.70%	
	0400	Electrical		19.95	28	34	7.65%	9.50%	11.55%	
	0500	Total Project Costs	↓	205	284	375				
25	0000	**University, College & Private School Dormitories**	S.F.							25
	0100	Architectural		81	142	151	54.50%	65%	68.50%	
	0200	Plumbing		10.70	15.10	22.50	6.45%	7.30%	9.15%	
	0300	Mechanical		4.79	20.50	32.50	4.13%	9%	12.05%	
	0400	Electrical		5.70	19.75	30	4.75%	7.35%	12.30%	
	0500	Total Project Costs	↓	119	227	268				
26	0000	**University, College & Private School Science, Eng. & Lab Buildings**	S.F.							26
	0100	Architectural		143	164	193	50.50%	56.50%	58%	
	0200	Plumbing		9.60	15.15	26.50	3.29%	3.95%	8.40%	
	0300	Mechanical		43.50	68.50	70	12.75%	19.40%	23.50%	
	0400	Electrical		29	33.50	38.50	9%	11.55%	13.15%	
	0500	Total Project Costs	↓	290	315	365				
27	0000	**University, College & Private School Student Union Buildings**	S.F.							27
	0100	Architectural		110	289	289	54.50%	54.50%	59.50%	
	0200	Plumbing		16.65	16.65	25	3.13%	4.27%	11.45%	
	0300	Mechanical		31.50	51	51	9.60%	9.60%	14.55%	
	0400	Electrical		27.50	48	48	9.05%	12.80%	13.15%	
	0500	Total Project Costs	↓	217	530	530				
28	0000	**Warehouses**	S.F.							28
	0100	Architectural		47	73.50	174	61.50%	67.50%	72%	
	0200	Plumbing		2.45	5.25	10.10	2.82%	3.72%	5%	
	0300	Mechanical		2.91	16.55	26	4.56%	8.20%	10.70%	
	0400	Electrical		5.30	19.85	33	7.50%	10.10%	18.30%	
	0500	Total Project Costs	↓	70.50	125	243				

Square Foot Project Size Modifier

One factor that affects the S.F. cost of a particular building is the size. In general, for buildings built to the same specifications in the same locality, the larger building will have the lower S.F. cost. This is due mainly to the decreasing contribution of the exterior walls plus the economy of scale usually achievable in larger buildings. The Area Conversion Scale shown below will give a factor to convert costs for the typical size building to an adjusted cost for the particular project.

The Square Foot Base Size lists the median costs, most typical project size in our accumulated data, and the range in size of the projects.

The Size Factor for your project is determined by dividing your project area in S.F. by the typical project size for the particular Building Type. With this factor, enter the Area Conversion Scale at the appropriate Size Factor and determine the appropriate Cost Multiplier for your building size.

Example: Determine the cost per S.F. for a 152,600 S.F. Multi-family housing.

$$\frac{\text{Proposed building area} = 152{,}600 \text{ S.F.}}{\text{Typical size from below} = 49{,}900 \text{ S.F.}} = 2.00$$

Enter Area Conversion Scale at 2.0, intersect curve, read horizontally the appropriate cost multiplier of .94. Size adjusted cost becomes .94 x $208.00 = $195.52 based on national average costs.

Note: For Size Factors less than .50, the Cost Multiplier is 1.1
For Size Factors greater than 3.5, the Cost Multiplier is .90

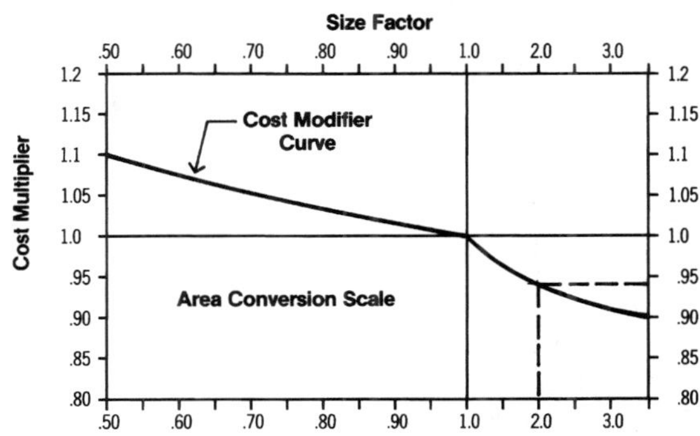

System	Median Cost (Total Project Costs)	Typical Size Gross S.F. (Median of Projects)	Typical Range (Low – High) (Projects)
Auto Sales with Repair	$186.00	24,900	4,700 – 29,300
Banking Institutions	299.00	9,300	3,300 – 38,100
Court House	284.00	47,600	24,700 – 70,500
Data Centers	281.00	14,400	14,400 – 14,400
Detention Centers	315.00	37,800	12,300 – 183,300
Fire Stations	238.00	13,100	6,300 – 49,600
Gymnasium	204.00	52,400	22,800 – 82,000
Hospitals	375.00	87,100	22,400 – 410,300
Industrial Buildings	104.00	22,100	5,100 – 200,600
Medical Clinics & Offices	217.00	20,900	2,300 – 327,000
Mixed Use	216.00	28,500	7,200 – 188,900
Multi-Family Housing	208.00	49,900	2,500 – 1,161,500
Nursing Home & Assisted Living	160.00	38,200	1,500 – 242,600
Office Buildings	200.00	20,500	1,100 – 930,000
Parking Garage	47.00	151,800	99,900 – 287,000
Parking Garage/Mixed Use	175.00	254,200	5,300 – 318,000
Police Stations	267.00	28,500	15,400 – 88,600
Police/Fire	171.00	44,300	8,600 – 50,300
Public Assembly Buildings	253.00	21,000	2,200 – 235,300
Recreational	281.00	28,800	1,000 – 223,800
Restaurants	305.00	6,000	4,000 – 42,000
Retail	114.00	28,700	5,200 – 84,300
Schools	227.00	70,600	1,300 – 410,800
University, College & Private School Classroom & Admin Buildings	284.00	48,300	9,400 – 196,200
University, College & Private School Dormitories	227.00	28,900	1,500 – 126,900
University, College & Private School Science, Eng. & Lab Buildings	315.00	39,800	5,300 – 117,600
University, College & Private School Student Union Buildings	530.00	48,700	42,100 – 50,000
Warehouses	125.00	10,400	600 – 303,800

Abbreviations

A	Area Square Feet; Ampere
AAFES	Army and Air Force Exchange Service
ABS	Acrylonitrile Butadiene Stryrene; Asbestos Bonded Steel
A.C., AC	Alternating Current; Air-Conditioning; Asbestos Cement; Plywood Grade A & C
ACI	American Concrete Institute
ACR	Air Conditioning Refrigeration
ADA	Americans with Disabilities Act
AD	Plywood, Grade A & D
Addit.	Additional
Adh.	Adhesive
Adj.	Adjustable
af	Audio-frequency
AFFF	Aqueous Film Forming Foam
AFUE	Annual Fuel Utilization Efficiency
AGA	American Gas Association
Agg.	Aggregate
A.H., Ah	Ampere Hours
A hr.	Ampere-hour
A.H.U., AHU	Air Handling Unit
A.I.A.	American Institute of Architects
AIC	Ampere Interrupting Capacity
Allow.	Allowance
alt., alt	Alternate
Alum.	Aluminum
a.m.	Ante Meridiem
Amp.	Ampere
Anod.	Anodized
ANSI	American National Standards Institute
APA	American Plywood Association
Approx.	Approximate
Apt.	Apartment
Asb.	Asbestos
A.S.B.C.	American Standard Building Code
Asbe.	Asbestos Worker
ASCE	American Society of Civil Engineers
A.S.H.R.A.E.	American Society of Heating, Refrig. & AC Engineers
ASME	American Society of Mechanical Engineers
ASTM	American Society for Testing and Materials
Attchmt.	Attachment
Avg., Ave.	Average
AWG	American Wire Gauge
AWWA	American Water Works Assoc.
Bbl.	Barrel
B&B, BB	Grade B and Better; Balled & Burlapped
B&S	Bell and Spigot
B.&W.	Black and White
b.c.c.	Body-centered Cubic
B.C.Y.	Bank Cubic Yards
BE	Bevel End
B.F.	Board Feet
Bg. cem.	Bag of Cement
BHP	Boiler Horsepower; Brake Horsepower
B.I.	Black Iron
bidir.	bidirectional
Bit., Bitum.	Bituminous
Bit., Conc.	Bituminous Concrete
Bk.	Backed
Bkrs.	Breakers
Bldg., bldg	Building
Blk.	Block
Bm.	Beam
Boil.	Boilermaker
bpm	Blows per Minute
BR	Bedroom
Brg., brng.	Bearing
Brhe.	Bricklayer Helper
Bric.	Bricklayer
Brk., brk	Brick
brkt	Bracket
Brs.	Brass
Brz.	Bronze
Bsn.	Basin
Btr.	Better
BTU	British Thermal Unit
BTUH	BTU per Hour
Bu.	Bushels
BUR	Built-up Roofing
BX	Interlocked Armored Cable
°C	Degree Centigrade
c	Conductivity, Copper Sweat
C	Hundred; Centigrade
C/C	Center to Center, Cedar on Cedar
C-C	Center to Center
Cab	Cabinet
Cair.	Air Tool Laborer
Cal.	Caliper
Calc	Calculated
Cap.	Capacity
Carp.	Carpenter
C.B.	Circuit Breaker
C.C.A.	Chromate Copper Arsenate
C.C.F.	Hundred Cubic Feet
cd	Candela
cd/sf	Candela per Square Foot
CD	Grade of Plywood Face & Back
CDX	Plywood, Grade C & D, exterior glue
Cefi.	Cement Finisher
Cem.	Cement
CF	Hundred Feet
C.F.	Cubic Feet
CFM	Cubic Feet per Minute
CFRP	Carbon Fiber Reinforced Plastic
c.g.	Center of Gravity
CHW	Chilled Water; Commercial Hot Water
C.I., CI	Cast Iron
C.I.P., CIP	Cast in Place
Circ.	Circuit
C.L.	Carload Lot
CL	Chain Link
Clab.	Common Laborer
Clam	Common Maintenance Laborer
C.L.F.	Hundred Linear Feet
CLF	Current Limiting Fuse
CLP	Cross Linked Polyethylene
cm	Centimeter
CMP	Corr. Metal Pipe
CMU	Concrete Masonry Unit
CN	Change Notice
Col.	Column
CO_2	Carbon Dioxide
Comb.	Combination
comm.	Commercial, Communication
Compr.	Compressor
Conc.	Concrete
Cont., cont	Continuous; Continued, Container
Corkbd.	Cork Board
Corr.	Corrugated
Cos	Cosine
Cot	Cotangent
Cov.	Cover
C/P	Cedar on Paneling
CPA	Control Point Adjustment
Cplg.	Coupling
CPM	Critical Path Method
CPVC	Chlorinated Polyvinyl Chloride
C.Pr.	Hundred Pair
CRC	Cold Rolled Channel
Creos.	Creosote
Crpt.	Carpet & Linoleum Layer
CRT	Cathode-ray Tube
CS	Carbon Steel, Constant Shear Bar Joist
Csc	Cosecant
C.S.F.	Hundred Square Feet
CSI	Construction Specifications Institute
CT	Current Transformer
CTS	Copper Tube Size
Cu	Copper, Cubic
Cu. Ft.	Cubic Foot
cw	Continuous Wave
C.W.	Cool White; Cold Water
Cwt.	100 Pounds
C.W.X.	Cool White Deluxe
C.Y.	Cubic Yard (27 cubic feet)
C.Y./Hr.	Cubic Yard per Hour
Cyl.	Cylinder
d	Penny (nail size)
D	Deep; Depth; Discharge
Dis., Disch.	Discharge
Db	Decibel
Dbl.	Double
DC	Direct Current
DDC	Direct Digital Control
Demob.	Demobilization
d.f.t.	Dry Film Thickness
d.f.u.	Drainage Fixture Units
D.H.	Double Hung
DHW	Domestic Hot Water
DI	Ductile Iron
Diag.	Diagonal
Diam., Dia	Diameter
Distrib.	Distribution
Div.	Division
Dk.	Deck
D.L.	Dead Load; Diesel
DLH	Deep Long Span Bar Joist
dlx	Deluxe
Do.	Ditto
DOP	Dioctyl Phthalate Penetration Test (Air Filters)
Dp., dp	Depth
D.P.S.T.	Double Pole, Single Throw
Dr.	Drive
DR	Dimension Ratio
Drink.	Drinking
D.S.	Double Strength
D.S.A.	Double Strength A Grade
D.S.B.	Double Strength B Grade
Dty.	Duty
DWV	Drain Waste Vent
DX	Deluxe White, Direct Expansion
dyn	Dyne
e	Eccentricity
E	Equipment Only; East; Emissivity
Ea.	Each
EB	Encased Burial
Econ.	Economy
E.C.Y	Embankment Cubic Yards
EDP	Electronic Data Processing
EIFS	Exterior Insulation Finish System
E.D.R.	Equiv. Direct Radiation
Eq.	Equation
EL	Elevation
Elec.	Electrician; Electrical
Elev.	Elevator; Elevating
EMT	Electrical Metallic Conduit; Thin Wall Conduit
Eng.	Engine, Engineered
EPDM	Ethylene Propylene Diene Monomer
EPS	Expanded Polystyrene
Eqhv.	Equip. Oper., Heavy
Eqlt.	Equip. Oper., Light
Eqmd.	Equip. Oper., Medium
Eqmm.	Equip. Oper., Master Mechanic
Eqol.	Equip. Oper., Oilers
Equip.	Equipment
ERW	Electric Resistance Welded

Abbreviations

Abbreviation	Meaning
E.S.	Energy Saver
Est.	Estimated
esu	Electrostatic Units
E.W.	Each Way
EWT	Entering Water Temperature
Excav.	Excavation
excl	Excluding
Exp., exp	Expansion, Exposure
Ext., ext	Exterior; Extension
Extru.	Extrusion
f.	Fiber Stress
F	Fahrenheit; Female; Fill
Fab., fab	Fabricated; Fabric
FBGS	Fiberglass
F.C.	Footcandles
f.c.c.	Face-centered Cubic
f'c.	Compressive Stress in Concrete; Extreme Compressive Stress
F.E.	Front End
FEP	Fluorinated Ethylene Propylene (Teflon)
F.G.	Flat Grain
F.H.A.	Federal Housing Administration
Fig.	Figure
Fin.	Finished
FIPS	Female Iron Pipe Size
Fixt.	Fixture
FJP	Finger jointed and primed
Fl. Oz.	Fluid Ounces
Flr.	Floor
Flrs.	Floors
FM	Frequency Modulation; Factory Mutual
Fmg.	Framing
FM/UL	Factory Mutual/Underwriters Labs
Fdn.	Foundation
FNPT	Female National Pipe Thread
Fori.	Foreman, Inside
Foro.	Foreman, Outside
Fount.	Fountain
fpm	Feet per Minute
FPT	Female Pipe Thread
Fr	Frame
F.R.	Fire Rating
FRK	Foil Reinforced Kraft
FSK	Foil/Scrim/Kraft
FRP	Fiberglass Reinforced Plastic
FS	Forged Steel
FSC	Cast Body; Cast Switch Box
Ft., ft	Foot; Feet
Ftng.	Fitting
Ftg.	Footing
Ft lb.	Foot Pound
Furn.	Furniture
FVNR	Full Voltage Non-Reversing
FVR	Full Voltage Reversing
FXM	Female by Male
Fy.	Minimum Yield Stress of Steel
g	Gram
G	Gauss
Ga.	Gauge
Gal., gal.	Gallon
Galv., galv	Galvanized
GC/MS	Gas Chromatograph/Mass Spectrometer
Gen.	General
GFI	Ground Fault Interrupter
GFRC	Glass Fiber Reinforced Concrete
Glaz.	Glazier
GPD	Gallons per Day
gpf	Gallon per Flush
GPH	Gallons per Hour
gpm, GPM	Gallons per Minute
GR	Grade
Gran.	Granular
Grnd.	Ground
GVW	Gross Vehicle Weight
GWB	Gypsum Wall Board
H	High Henry
HC	High Capacity
H.D., HD	Heavy Duty; High Density
H.D.O.	High Density Overlaid
HDPE	High Density Polyethylene Plastic
Hdr.	Header
Hdwe.	Hardware
H.I.D., HID	High Intensity Discharge
Help.	Helper Average
HEPA	High Efficiency Particulate Air Filter
Hg	Mercury
HIC	High Interrupting Capacity
HM	Hollow Metal
HMWPE	High Molecular Weight Polyethylene
HO	High Output
Horiz.	Horizontal
H.P., HP	Horsepower; High Pressure
H.P.F.	High Power Factor
Hr.	Hour
Hrs./Day	Hours per Day
HSC	High Short Circuit
Ht.	Height
Htg.	Heating
Htrs.	Heaters
HVAC	Heating, Ventilation & Air-Conditioning
Hvy.	Heavy
HW	Hot Water
Hyd.; Hydr.	Hydraulic
Hz	Hertz (cycles)
I.	Moment of Inertia
IBC	International Building Code
I.C.	Interrupting Capacity
ID	Inside Diameter
I.D.	Inside Dimension; Identification
I.F.	Inside Frosted
I.M.C.	Intermediate Metal Conduit
In.	Inch
Incan.	Incandescent
Incl.	Included; Including
Int.	Interior
Inst.	Installation
Insul., insul	Insulation/Insulated
I.P.	Iron Pipe
I.P.S., IPS	Iron Pipe Size
IPT	Iron Pipe Threaded
I.W.	Indirect Waste
J	Joule
J.I.C.	Joint Industrial Council
K	Thousand; Thousand Pounds; Heavy Wall Copper Tubing, Kelvin
K.A.H.	Thousand Amp. Hours
kcmil	Thousand Circular Mils
KD	Knock Down
K.D.A.T.	Kiln Dried After Treatment
kg	Kilogram
kG	Kilogauss
kgf	Kilogram Force
kHz	Kilohertz
Kip	1000 Pounds
KJ	Kilojoule
K.L.	Effective Length Factor
K.L.F.	Kips per Linear Foot
Km	Kilometer
KO	Knock Out
K.S.F.	Kips per Square Foot
K.S.I.	Kips per Square Inch
kV	Kilovolt
kVA	Kilovolt Ampere
kVAR	Kilovar (Reactance)
KW	Kilowatt
KWh	Kilowatt-hour
L	Labor Only; Length; Long; Medium Wall Copper Tubing
Lab.	Labor
lat	Latitude
Lath.	Lather
Lav.	Lavatory
lb.; #	Pound
L.B., LB	Load Bearing; L Conduit Body
L. & E.	Labor & Equipment
lb./hr.	Pounds per Hour
lb./L.F.	Pounds per Linear Foot
lbf/sq.in.	Pound-force per Square Inch
L.C.L.	Less than Carload Lot
L.C.Y.	Loose Cubic Yard
Ld.	Load
LE	Lead Equivalent
LED	Light Emitting Diode
L.F.	Linear Foot
L.F. Hdr	Linear Feet of Header
L.F. Nose	Linear Foot of Stair Nosing
L.F. Rsr	Linear Foot of Stair Riser
Lg.	Long; Length; Large
L & H	Light and Heat
LH	Long Span Bar Joist
L.H.	Labor Hours
L.L., LL	Live Load
L.L.D.	Lamp Lumen Depreciation
lm	Lumen
lm/sf	Lumen per Square Foot
lm/W	Lumen per Watt
LOA	Length Over All
log	Logarithm
L-O-L	Lateralolet
long.	Longitude
L.P., LP	Liquefied Petroleum; Low Pressure
L.P.F.	Low Power Factor
LR	Long Radius
L.S.	Lump Sum
Lt.	Light
Lt. Ga.	Light Gauge
L.T.L.	Less than Truckload Lot
Lt. Wt.	Lightweight
L.V.	Low Voltage
M	Thousand; Material; Male; Light Wall Copper Tubing
M^2CA	Meters Squared Contact Area
m/hr.; M.H.	Man-hour
mA	Milliampere
Mach.	Machine
Mag. Str.	Magnetic Starter
Maint.	Maintenance
Marb.	Marble Setter
Mat; Mat'l.	Material
Max.	Maximum
MBF	Thousand Board Feet
MBH	Thousand BTU's per hr.
MC	Metal Clad Cable
MCC	Motor Control Center
M.C.F.	Thousand Cubic Feet
MCFM	Thousand Cubic Feet per Minute
M.C.M.	Thousand Circular Mils
MCP	Motor Circuit Protector
MD	Medium Duty
MDF	Medium-density fibreboard
M.D.O.	Medium Density Overlaid
Med.	Medium
MF	Thousand Feet
M.F.B.M.	Thousand Feet Board Measure
Mfg.	Manufacturing
Mfrs.	Manufacturers
mg	Milligram
MGD	Million Gallons per Day
MGPH	Million Gallons per Hour
MH, M.H.	Manhole; Metal Halide; Man-Hour
MHz	Megahertz
Mi.	Mile
MI	Malleable Iron; Mineral Insulated
MIPS	Male Iron Pipe Size
mj	Mechanical Joint
m	Meter
mm	Millimeter
Mill.	Millwright
Min., min.	Minimum, Minute

Abbreviations

Abbreviation	Meaning
Misc.	Miscellaneous
ml	Milliliter, Mainline
M.L.F.	Thousand Linear Feet
Mo.	Month
Mobil.	Mobilization
Mog.	Mogul Base
MPH	Miles per Hour
MPT	Male Pipe Thread
MRGWB	Moisture Resistant Gypsum Wallboard
MRT	Mile Round Trip
ms	Millisecond
M.S.F.	Thousand Square Feet
Mstz.	Mosaic & Terrazzo Worker
M.S.Y.	Thousand Square Yards
Mtd., mtd., mtd	Mounted
Mthe.	Mosaic & Terrazzo Helper
Mtng.	Mounting
Mult.	Multi; Multiply
MUTCD	Manual on Uniform Traffic Control Devices
M.V.A.	Million Volt Amperes
M.V.A.R.	Million Volt Amperes Reactance
MV	Megavolt
MW	Megawatt
MXM	Male by Male
MYD	Thousand Yards
N	Natural; North
nA	Nanoampere
NA	Not Available; Not Applicable
N.B.C.	National Building Code
NC	Normally Closed
NEMA	National Electrical Manufacturers Assoc.
NEHB	Bolted Circuit Breaker to 600V.
NFPA	National Fire Protection Association
NLB	Non-Load-Bearing
NM	Non-Metallic Cable
nm	Nanometer
No.	Number
NO	Normally Open
N.O.C.	Not Otherwise Classified
Nose.	Nosing
NPT	National Pipe Thread
NQOD	Combination Plug-on/Bolt on Circuit Breaker to 240V.
N.R.C., NRC	Noise Reduction Coefficient/ Nuclear Regulator Commission
N.R.S.	Non Rising Stem
ns	Nanosecond
NTP	Notice to Proceed
nW	Nanowatt
OB	Opposing Blade
OC	On Center
OD	Outside Diameter
O.D.	Outside Dimension
ODS	Overhead Distribution System
O.G.	Ogee
O.H.	Overhead
O&P	Overhead and Profit
Oper.	Operator
Opng.	Opening
Orna.	Ornamental
OSB	Oriented Strand Board
OS&Y	Outside Screw and Yoke
OSHA	Occupational Safety and Health Act
Ovhd.	Overhead
OWG	Oil, Water or Gas
Oz.	Ounce
P.	Pole; Applied Load; Projection
p.	Page
Pape.	Paperhanger
P.A.P.R.	Powered Air Purifying Respirator
PAR	Parabolic Reflector
P.B., PB	Push Button
Pc., Pcs.	Piece, Pieces
P.C.	Portland Cement; Power Connector
P.C.F.	Pounds per Cubic Foot
PCM	Phase Contrast Microscopy
PDCA	Painting and Decorating Contractors of America
P.E., PE	Professional Engineer; Porcelain Enamel; Polyethylene; Plain End
P.E.C.I.	Porcelain Enamel on Cast Iron
Perf.	Perforated
PEX	Cross Linked Polyethylene
Ph.	Phase
P.I.	Pressure Injected
Pile.	Pile Driver
Pkg.	Package
Pl.	Plate
Plah.	Plasterer Helper
Plas.	Plasterer
plf	Pounds Per Linear Foot
Pluh.	Plumber Helper
Plum.	Plumber
Ply.	Plywood
p.m.	Post Meridiem
Pntd.	Painted
Pord.	Painter, Ordinary
pp	Pages
PP, PPL	Polypropylene
P.P.M.	Parts per Million
Pr.	Pair
P.E.S.B.	Pre-engineered Steel Building
Prefab.	Prefabricated
Prefin.	Prefinished
Prop.	Propelled
PSF, psf	Pounds per Square Foot
PSI, psi	Pounds per Square Inch
PSIG	Pounds per Square Inch Gauge
PSP	Plastic Sewer Pipe
Pspr.	Painter, Spray
Psst.	Painter, Structural Steel
P.T.	Potential Transformer
P. & T.	Pressure & Temperature
Ptd.	Painted
Ptns.	Partitions
Pu	Ultimate Load
PVC	Polyvinyl Chloride
Pvmt.	Pavement
PRV	Pressure Relief Valve
Pwr.	Power
Q	Quantity Heat Flow
Qt.	Quart
Quan., Qty.	Quantity
Q.C.	Quick Coupling
r	Radius of Gyration
R	Resistance
R.C.P.	Reinforced Concrete Pipe
Rect.	Rectangle
recpt.	Receptacle
Reg.	Regular
Reinf.	Reinforced
Req'd.	Required
Res.	Resistant
Resi.	Residential
RF	Radio Frequency
RFID	Radio-frequency Identification
Rgh.	Rough
RGS	Rigid Galvanized Steel
RHW	Rubber, Heat & Water Resistant; Residential Hot Water
rms	Root Mean Square
Rnd.	Round
Rodm.	Rodman
Rofc.	Roofer, Composition
Rofp.	Roofer, Precast
Rohe.	Roofer Helpers (Composition)
Rots.	Roofer, Tile & Slate
R.O.W.	Right of Way
RPM	Revolutions per Minute
R.S.	Rapid Start
Rsr	Riser
RT	Round Trip
S.	Suction; Single Entrance; South
SBS	Styrene Butadiere Styrene
SC	Screw Cover
SCFM	Standard Cubic Feet per Minute
Scaf.	Scaffold
Sch., Sched.	Schedule
S.C.R.	Modular Brick
S.D.	Sound Deadening
SDR	Standard Dimension Ratio
S.E.	Surfaced Edge
Sel.	Select
SER, SEU	Service Entrance Cable
S.F.	Square Foot
S.F.C.A.	Square Foot Contact Area
S.F. Flr.	Square Foot of Floor
S.F.G.	Square Foot of Ground
S.F. Hor.	Square Foot Horizontal
SFR	Square Feet of Radiation
S.F. Shlf.	Square Foot of Shelf
S4S	Surface 4 Sides
Shee.	Sheet Metal Worker
Sin.	Sine
Skwk.	Skilled Worker
SL	Saran Lined
S.L.	Slimline
Sldr.	Solder
SLH	Super Long Span Bar Joist
S.N.	Solid Neutral
SO	Stranded with oil resistant inside insulation
S-O-L	Socketolet
sp	Standpipe
S.P.	Static Pressure; Single Pole; Self-Propelled
Spri.	Sprinkler Installer
spwg	Static Pressure Water Gauge
S.P.D.T.	Single Pole, Double Throw
SPF	Spruce Pine Fir; Sprayed Polyurethane Foam
S.P.S.T.	Single Pole, Single Throw
SPT	Standard Pipe Thread
Sq.	Square; 100 Square Feet
Sq. Hd.	Square Head
Sq. In.	Square Inch
S.S.	Single Strength; Stainless Steel
S.S.B.	Single Strength B Grade
sst, ss	Stainless Steel
Sswk.	Structural Steel Worker
Sswl.	Structural Steel Welder
St.; Stl.	Steel
STC	Sound Transmission Coefficient
Std.	Standard
Stg.	Staging
STK	Select Tight Knot
STP	Standard Temperature & Pressure
Stpi.	Steamfitter, Pipefitter
Str.	Strength; Starter; Straight
Strd.	Stranded
Struct.	Structural
Sty.	Story
Subj.	Subject
Subs.	Subcontractors
Surf.	Surface
Sw.	Switch
Swbd.	Switchboard
S.Y.	Square Yard
Syn.	Synthetic
S.Y.P.	Southern Yellow Pine
Sys.	System
t.	Thickness
T	Temperature; Ton
Tan	Tangent
T.C.	Terra Cotta
T & C	Threaded and Coupled
T.D.	Temperature Difference
TDD	Telecommunications Device for the Deaf
T.E.M.	Transmission Electron Microscopy
temp	Temperature, Tempered, Temporary
TFFN	Nylon Jacketed Wire

Abbreviations

Abbreviation	Meaning
TFE	Tetrafluoroethylene (Teflon)
T. & G.	Tongue & Groove; Tar & Gravel
Th., Thk.	Thick
Thn.	Thin
Thrded	Threaded
Tilf.	Tile Layer, Floor
Tilh.	Tile Layer, Helper
THHN	Nylon Jacketed Wire
THW.	Insulated Strand Wire
THWN	Nylon Jacketed Wire
T.L., TL	Truckload
T.M.	Track Mounted
Tot.	Total
T-O-L	Threadolet
tmpd	Tempered
TPO	Thermoplastic Polyolefin
T.S.	Trigger Start
Tr.	Trade
Transf.	Transformer
Trhv.	Truck Driver, Heavy
Trlr	Trailer
Trlt.	Truck Driver, Light
TTY	Teletypewriter
TV	Television
T.W.	Thermoplastic Water Resistant Wire
UCI	Uniform Construction Index
UF	Underground Feeder
UGND	Underground Feeder
UHF	Ultra High Frequency
U.I.	United Inch
U.L., UL	Underwriters Laboratory
Uld.	Unloading
Unfin.	Unfinished
UPS	Uninterruptible Power Supply
URD	Underground Residential Distribution
US	United States
USGBC	U.S. Green Building Council
USP	United States Primed
UTMCD	Uniform Traffic Manual For Control Devices
UTP	Unshielded Twisted Pair
V	Volt
VA	Volt Amperes
VAT	Vinyl Asbestos Tile
V.C.T.	Vinyl Composition Tile
VAV	Variable Air Volume
VC	Veneer Core
VDC	Volts Direct Current
Vent.	Ventilation
Vert.	Vertical
V.F.	Vinyl Faced
V.G.	Vertical Grain
VHF	Very High Frequency
VHO	Very High Output
Vib.	Vibrating
VLF	Vertical Linear Foot
VOC	Volatile Organic Compound
Vol.	Volume
VRP	Vinyl Reinforced Polyester
W	Wire; Watt; Wide; West
w/	With
W.C., WC	Water Column; Water Closet
W.F.	Wide Flange
W.G.	Water Gauge
Wldg.	Welding
W. Mile	Wire Mile
W-O-L	Weldolet
W.R.	Water Resistant
Wrck.	Wrecker
WSFU	Water Supply Fixture Unit
W.S.P.	Water, Steam, Petroleum
WT., Wt.	Weight
WWF	Welded Wire Fabric
XFER	Transfer
XFMR	Transformer
XHD	Extra Heavy Duty
XHHW	Cross-Linked Polyethylene Wire
XLPE	Insulation
XLP	Cross-linked Polyethylene
Xport	Transport
Y	Wye
yd	Yard
yr	Year
Δ	Delta
%	Percent
~	Approximately
Ø	Phase; diameter
@	At
#	Pound; Number
<	Less Than
>	Greater Than
Z	Zone

A

B

D

Index

E

F

Index

G

H

N

O

P

Q

R

S

T

U

V

Notes

Notes

Notes

Notes

Notes

Division Notes

	CREW	DAILY OUTPUT	LABOR-HOURS	UNIT	BARE COSTS MAT.	BARE COSTS LABOR	BARE COSTS EQUIP.	BARE COSTS TOTAL	TOTAL INCL O&P

Division Notes

I	CREW	DAILY OUTPUT	LABOR-HOURS	UNIT	BARE COSTS MAT.	BARE COSTS LABOR	BARE COSTS EQUIP.	BARE COSTS TOTAL	TOTAL INCL O&P

Division Notes

		CREW	DAILY OUTPUT	LABOR-HOURS	UNIT	BARE COSTS MAT.	LABOR	EQUIP.	TOTAL	TOTAL INCL O&P

Division Notes

		CREW	DAILY OUTPUT	LABOR-HOURS	UNIT	BARE COSTS MAT.	BARE COSTS LABOR	BARE COSTS EQUIP.	BARE COSTS TOTAL	TOTAL INCL O&P

Other Data & Services

A tradition of excellence in construction cost information and services since 1942

Cost Data Selection Guide

The following table provides definitive information on the content of each cost data publication. The number of lines of data provided in each unit price or assemblies division, as well as the number of crews, is listed for each data set. The presence of other elements such as reference tables, square foot models, equipment rental costs, historical cost indexes, and city cost indexes, is also indicated. You can use the table to help select the RSMeans data set that has the quantity and type of information you most need in your work.

Unit Cost Divisions	Building Construction	Mechanical	Electrical	Commercial Renovation	Square Foot	Site Work Landsc.	Green Building	Interior	Concrete Masonry	Open Shop	Heavy Construction	Light Commercial	Facilities Construction	Plumbing	Residential
1	609	444	465	564	0	533	198	365	495	608	550	310	1078	450	217
2	754	278	87	710	0	970	181	397	219	753	737	479	1197	285	274
3	1745	341	232	1265	0	1537	1043	355	2274	1745	1930	538	2028	317	445
4	960	22	0	920	0	724	180	613	1158	928	614	532	1175	0	446
5	1890	158	155	1094	0	853	1788	1107	729	1890	1026	980	1907	204	746
6	2462	18	18	2121	0	110	589	1544	281	2458	123	2151	2135	22	2671
7	1593	215	128	1633	0	580	761	532	523	1590	26	1326	1693	227	1046
8	2140	80	3	2733	0	255	1138	1813	105	2142	0	2328	2966	0	1552
9	2125	86	45	1943	0	313	464	2216	424	2062	15	1779	2379	54	1544
10	1088	17	10	684	0	232	32	898	136	1088	34	588	1179	237	224
11	1096	199	166	540	0	135	56	924	29	1063	0	230	1116	162	108
12	539	0	2	297	0	219	147	1546	14	506	0	272	1565	23	216
13	740	149	157	252	0	365	124	250	77	716	266	109	756	115	103
14	273	36	0	223	0	0	0	257	0	273	0	12	293	16	6
21	127	0	41	37	0	0	0	293	0	127	0	121	665	685	259
22	1165	7543	160	1226	0	2010	1061	849	20	1154	2109	875	7505	9400	719
23	1170	6906	546	940	0	157	865	775	38	1153	98	887	5143	1919	486
25	0	0	14	14	0	0	0	0	0	0	0	0	0	0	0
26	1513	491	10465	1293	0	860	646	1159	55	1439	649	1361	10246	399	636
27	95	0	448	102	0	0	0	71	0	95	39	67	389	0	56
28	143	79	223	124	0	0	28	97	0	127	0	70	209	57	41
31	1511	733	610	807	0	3263	286	7	1216	1456	3280	607	1568	660	616
32	896	49	8	937	0	4523	408	417	361	867	1941	486	1800	140	533
33	1255	1088	565	260	0	3078	33	0	241	532	3213	135	1726	2101	161
34	107	0	47	4	0	190	0	0	31	62	221	0	136	0	0
35	18	0	0	0	0	327	0	0	0	18	442	0	84	0	0
41	63	0	0	34	0	8	0	22	0	62	31	0	69	14	0
44	75	79	0	0	0	0	0	0	0	0	0	0	75	75	0
46	23	16	0	0	0	274	261	0	0	23	264	0	33	33	0
48	8	0	36	2	0	0	21	0	0	8	15	8	21	0	8
Totals	26183	19027	14631	20759	0	21516	10310	16507	8426	24945	17623	16251	51136	17595	13113

Assem Div	Building Construction	Mechanical	Elec-trical	Commercial Renovation	Square Foot	Site Work Landscape	Assemblies	Green Building	Interior	Concrete Masonry	Heavy Construction	Light Commercial	Facilities Construction	Plumbing	Asm Div	Residential
A		15	0	188	164	577	598	0	0	536	571	154	24	0	1	378
B		0	0	848	2554	0	5661	56	329	1976	368	2094	174	0	2	211
C		0	0	647	954	0	1334	0	1641	146	0	844	251	0	3	591
D		1057	941	712	1858	72	2538	330	824	0	0	1345	1104	1088	4	851
E		0	0	85	261	0	301	0	5	0	0	258	5	0	5	391
F		0	0	0	114	0	143	0	0	0	0	114	0	0	6	357
G		527	447	318	312	3378	792	0	0	535	1349	205	293	677	7	307
															8	760
															9	80
															10	0
															11	0
															12	0
Totals		1599	1388	2798	6217	4027	11367	386	2799	3193	2288	5014	1851	1765		3926

Reference Section	Building Construction Costs	Mechanical	Electrical	Commercial Renovation	Square Foot	Site Work Landscape	Assem.	Green Building	Interior	Concrete Masonry	Open Shop	Heavy Construction	Light Commercial	Facilities Construction	Plumbing	Resi.
Reference Tables	yes	yes	yes	yes	no	yes	yes	yes	yes	yes	yes	yes	yes	yes	yes	yes
Models					111			25					50			28
Crews	582	582	582	561		582		582	582	582	560	582	560	561	582	560
Equipment Rental Costs	yes	yes	yes	yes		yes		yes	yes	yes	yes	yes	yes	yes	yes	yes
Historical Cost Indexes	yes	yes	yes	yes	yes	yes	yes	yes	yes	yes	yes	yes	yes	yes	yes	no
City Cost Indexes	yes	yes	yes	yes	yes	yes	yes	yes	yes	yes	yes	yes	yes	yes	yes	yes

2020 Seminar Schedule

877.620.6245

Note: call for exact dates, locations, and details as some cities are subject to change.

Location	Dates	Location	Dates
Seattle, WA	January and August	San Francisco, CA	June
Dallas/Ft. Worth, TX	January	Bethesda, MD	June
Austin, TX	February	Dallas, TX	September
Jacksonville, FL	February	Raleigh, NC	October
Anchorage, AK	March and September	Baltimore, MD	November
Las Vegas, NV	March	Orlando, FL	November
Washington, D.C.	April and September	San Diego, CA	December
Charleston, SC	April	San Antonio, TX	December
Toronto	May		
Denver, CO	May		

Gordian also offers a suite of online RSMeans data self-paced offerings.
Check our website at RSMeans.com/products/training.aspx for more information.

Facilities Construction Estimating

In this two-day course, professionals working in facilities management can get help with their daily challenges to establish budgets for all phases of a project.

Some of what you'll learn:

- Determining the full scope of a project
- Understanding of Means data and what is included in prices
- Identifying appropriate factors to be included in your estimate
- Creative solutions to estimating issues
- Organizing estimates for presentation and discussion
- Special estimating techniques for repair/remodel and maintenance projects
- Appropriate use of contingency, city cost indexes, and reference notes
- Techniques to get to the correct estimate quickly

Who should attend: facility managers, engineers, contractors, facility tradespeople, planners, and project managers.

Mechanical & Electrical Estimating

This two-day course teaches attendees how to prepare more accurate and complete mechanical/electrical estimates, avoid the pitfalls of omission and double-counting, and understand the composition and rationale within the RSMeans mechanical/electrical database.

Some of what you'll learn:

- The unique way mechanical and electrical systems are interrelated
- M&E estimates—conceptual, planning, budgeting, and bidding stages
- Order of magnitude, square foot, assemblies, and unit price estimating
- Comparative cost analysis of equipment and design alternatives

Who should attend: architects, engineers, facilities managers, mechanical and electrical contractors, and others who need a highly reliable method for developing, understanding, and evaluating mechanical and electrical contracts.

Construction Cost Estimating: Concepts and Practice

This one or two day introductory course to improve estimating skills and effectiveness starts with the details of interpreting bid documents and ends with the summary of the estimate and bid submission.

Some of what you'll learn:

- Using the plans and specifications to create estimates
- The takeoff process—deriving all tasks with correct quantities
- Developing pricing using various sources; how subcontractor pricing fits in
- Summarizing the estimate to arrive at the final number
- Formulas for area and cubic measure, adding waste and adjusting productivity to specific projects
- Evaluating subcontractors' proposals and prices
- Adding insurance and bonds
- Understanding how labor costs are calculated
- Submitting bids and proposals

Who should attend: project managers, architects, engineers, owners' representatives, contractors, and anyone who's responsible for budgeting or estimating construction projects.

Training for our Online Estimating Solution

Construction estimating is vital to the decision-making process at each state of every project. Our online solution works the way you do. It's systematic, flexible and intuitive. In this one-day class you will see how you can estimate any phase of any project faster and better.

Some of what you'll learn:

- Customizing our online estimating solution
- Making the most of RSMeans "Circle Reference" numbers
- How to integrate your cost data
- Generating reports, exporting estimates to MS Excel, sharing, collaborating and more

Also offered as a self-paced or on-site training program!

Training for our CD Estimating Solution

This one-day course helps users become more familiar with the functionality of the CD. Each menu, icon, screen, and function found in the program is explained in depth. Time is devoted to hands-on estimating exercises.

Some of what you'll learn:

- Searching the database using all navigation methods
- Exporting RSMeans data to your preferred spreadsheet format
- Viewing crews, assembly components, and much more
- Automatically regionalizing the database

This training session requires you to bring a laptop computer to class.

When you register for this course you will receive an outline for your laptop requirements.

Also offered as a self-paced or on-site training program!

Site Work Estimating with RSMeans data

This one-day program focuses directly on site work costs. Accurately scoping, quantifying, and pricing site preparation, underground utility work, and improvements to exterior site elements are often the most difficult estimating tasks on any project. Some of what you'll learn:

- Evaluation of site work and understanding site scope including: site clearing, grading, excavation, disposal and trucking of materials, backfill and compaction, underground utilities, paving, sidewalks, and seeding & planting.
- Unit price site work estimates—Correct use of RSMeans site work cost data to develop a cost estimate.
- Using and modifying assemblies—Save valuable time when estimating site work activities using custom assemblies.

Who should attend: Engineers, contractors, estimators, project managers, owner's representatives, and others who are concerned with the proper preparation and/or evaluation of site work estimates.

Please bring a laptop with ability to access the internet.

Facilities Estimating Using the CD

This two-day class combines hands-on skill-building with best estimating practices and real-life problems. You will learn key concepts, tips, pointers, and guidelines to save time and avoid cost oversights and errors.

Some of what you'll learn:

- Estimating process concepts
- Customizing and adapting RSMeans cost data
- Establishing scope of work to account for all known variables
- Budget estimating: when, why, and how
- Site visits: what to look for and what you can't afford to overlook
- How to estimate repair and remodeling variables

This training session requires you to bring a laptop computer to class.

Who should attend: facility managers, architects, engineers, contractors, facility tradespeople, planners, project managers, and anyone involved with JOC, SABRE, or IDIQ.

Registration Information

Register early to save up to $100!!!

Register 45+ days before the date of a class and save $50 off each class. This savings cannot be combined with any other promotion or discounting of the regular price of classes!

How to register

By Phone

Register by phone at 877.620.6245

Online

Register online at RSMeans.com/products/seminars.aspx

Note: Purchase Orders or Credits Cards are required to register.

Two-day seminar registration fee - $1,200*.

One-Day Construction Cost Estimating or Building Systems and the Construction Process - $765*.

Government pricing

All federal government employees save off the regular seminar price. Other promotional discounts cannot be combined with the government discount. Call 781.422.5115 for government pricing.

CANCELLATION POLICY:

If you are unable to attend a seminar, substitutions may be made at any time before the session starts by notifying the seminar registrar at 781.422.5115 or your sales representative.

If you cancel twenty-one (21) days or more prior to the seminar, there will be no penalty and your registration fees will be refunded. These cancellations must be received by the seminar registrar or your sales representative and will be confirmed to be eligible for cancellation.

If you cancel fewer than twenty-one (21) days prior to the seminar, you will forfeit the registration fee.

In the unfortunate event of an RSMeans cancellation, RSMeans will work with you to reschedule your attendance in the same seminar at a later date or will fully refund your registration fee. RSMeans cannot be responsible for any non-refundable travel expenses incurred by you or another as a result of your registration, attendance at, or cancellation of an RSMeans seminar.

Any on-demand training modules are not eligible for cancellation, substitution, transfer, return or refund.

AACE approved courses

Many seminars described and offered here have been approved for 14 hours (1.4 recertification credits) of credit by the AACE International Certification Board toward meeting the continuing education requirements for recertification as a Certified Cost Engineer/Certified Cost Consultant.

AIA Continuing Education

We are registered with the AIA Continuing Education System (AIA/CES) and are committed to developing quality learning activities in accordance with the CES criteria. Many seminars meet the AIA/CES criteria for Quality Level 2. AIA members may receive 14 learning units (LUs) for each two-day RSMeans course.

Daily course schedule

The first day of each seminar session begins at 8:30 a.m. and ends at 4:30 p.m. The second day begins at 8:00 a.m. and ends at 4:00 p.m. Participants are urged to bring a hand-held calculator since many actual problems will be worked out in each session.

Continental breakfast

Your registration includes the cost of a continental breakfast and a morning and afternoon refreshment break. These informal segments allow you to discuss topics of mutual interest with other seminar attendees. (You are free to make your own lunch and dinner arrangements.)

Hotel/transportation arrangements

We arrange to hold a block of rooms at most host hotels. To take advantage of special group rates when making your reservation, be sure to mention that you are attending the RSMeans Institute data seminar. You are, of course, free to stay at the lodging place of your choice. (Hotel reservations and transportation arrangements should be made directly by seminar attendees.)

Important

Class sizes are limited, so please register as soon as possible.

***Note: Pricing subject to change.**

Assessing Scope of Work for Facilities Construction Estimating

This two-day practical training program addresses the vital importance of understanding the scope of projects in order to produce accurate cost estimates for facilities repair and remodeling.

Some of what you'll learn:
- Discussions of site visits, plans/specs, record drawings of facilities, and site-specific lists
- Review of CSI divisions, including means, methods, materials, and the challenges of scoping each topic
- Exercises in scope identification and scope writing for accurate estimating of projects
- Hands-on exercises that require scope, take-off, and pricing

Who should attend: corporate and government estimators, planners, facility managers, and others who need to produce accurate project estimates.

Maintenance & Repair Estimating for Facilities

This two-day course teaches attendees how to plan, budget, and estimate the cost of ongoing and preventive maintenance and repair for existing buildings and grounds.

Some of what you'll learn:
- The most financially favorable maintenance, repair, and replacement scheduling and estimating
- Auditing and value engineering facilities
- Preventive planning and facilities upgrading
- Determining both in-house and contract-out service costs
- Annual, asset-protecting M&R plan

Who should attend: facility managers, maintenance supervisors, buildings and grounds superintendents, plant managers, planners, estimators, and others involved in facilities planning and budgeting.

Practical Project Management for Construction Professionals

In this two-day course, acquire the essential knowledge and develop the skills to effectively and efficiently execute the day-to-day responsibilities of the construction project manager.

Some of what you'll learn:
- General conditions of the construction contract
- Contract modifications: change orders and construction change directives
- Negotiations with subcontractors and vendors
- Effective writing: notification and communications
- Dispute resolution: claims and liens

Who should attend: architects, engineers, owners' representatives, and project managers.

Life Cycle Cost Estimating for Facilities Asset Managers

Life Cycle Cost Estimating will take the attendee through choosing the correct RSMeans database to use and then correctly applying RSMeans data to their specific life cycle application. Conceptual estimating through RSMeans' new building models, conceptual estimating of major existing building projects through RSMeans' renovation models, pricing specific renovation elements, estimating repair, replacement and preventive maintenance costs today and forward up to 30 years will be covered.

Some of what you'll learn:
- Cost implications of managing assets
- Planning projects and initial & life cycle costs
- How to use RSMeans data online

Who should attend: facilities owners and managers and anyone involved in the financial side of the decision making process in the planning, design, procurement, and operation of facilities real assets.

Please bring a laptop with ability to access the internet.

Building Systems and the Construction Process

This one-day course was written to assist novices and those outside the industry in obtaining a solid understanding of the construction process - from both a building systems and construction administration approach.

Some of what you'll learn:
- Various systems used and how components come together to create a building
- Start with foundation and end with the physical systems of the structure such as HVAC and Electrical
- Focus on the process from start of design through project closeout

This training session requires you to bring a laptop computer to class.

Who should attend: building professionals or novices to help make the crossover to the construction industry; suited for anyone responsible for providing high level oversight on construction projects.